Handbook of Astrobiology

CRC Press
Taylor & Francis Group
6000 Broken Sound Parkway NW, Suite 300
Boca Raton, FL 33487-2742

First issued in paperback 2020

© 2019 by Taylor & Francis Group, LLC
CRC Press is an imprint of Taylor & Francis Group, an Informa business

No claim to original U.S. Government works

ISBN-13: 978-1-138-06512-3 (hbk)
ISBN-13: 978-0-367-78048-7 (pbk)

Library of Congress Cataloging-in-Publication Data

Names: Kolb, Vera M., editor.
Title: Handbook of astrobiology / edited by Vera M. Kolb.
Description: Boca Raton, Florida : CRC Press, [2019] | Includes
bibliographical references.
Identifiers: LCCN 2018021898 (print) | LCCN 2018047932 (ebook) | ISBN
9781315159966 (eBook General) | ISBN 9781351661119 (eBook Adobe Reader) |
ISBN 9781351661102 (eBook ePub) | ISBN 9781351661096 (eBook Mobipocket) |
ISBN 9781138065123 (hardback : alk. paper)
Subjects: LCSH: Exobiology--Handbooks, manuals, etc.
Classification: LCC QH326 (ebook) | LCC QH326 .H36 2019 (print) | DDC
576.8/39--dc23
LC record available at https://lccn.loc.gov/2018021898

Visit the Taylor & Francis Web site at
http://www.taylorandfrancis.com

and the CRC Press Web site at
http://www.crcpress.com

Contents

SECTION I   Astrobiology: Definition, Scope, and Education

SECTION II   Definition and Nature of Life

SECTION VI RNA and RNA World: Complexity of Life's Origins

SECTION VII   Origin of Life: Early Compartmentalization—Coacervates and Protocells

SECTION VIII   Origin of Life and Its Diversification. Universal Tree of Life. Early Primitive Life on Earth. Fossils of Ancient Microorganisms. Biomarkers and Detection of Life

SECTION IX   Life under Extreme Conditions—Microbes in Space

SECTION X   Habitability: Characteristics of Habitable Planets

SECTION XI   Intelligent Life in Space: History, Philosophy, and SETI (Search for Extraterrestrial Intelligence)

SECTION XII   Exoplanets, Exploration of Solar System, the Search for Extraterrestrial Life in Our Solar System, and Planetary Protection

Preface

Handbook of Astrobiology is designed for astrobiology practitioners to enable them to review the major developments in the field and to learn about the newest discoveries. Since astrobiology is an interdisciplinary field, which draws on various other disciplines such as astronomy, physics, chemistry, biology, geology, planetary science, and philosophy, it is difficult for practitioners to be adequately informed about the new findings in the field, especially if these are not directly related to one's specific area of investigation. Further, rapid developments in astrobiology result in an ever-increasing number of publications, which makes it difficult to stay current in the field. However, a broad background and up-to-date knowledge of the field are important resources for astrobiologists, so that they can place their own findings in a broader context and integrate them with other findings in the field.

This handbook is designed to help astrobiologists reach the above goals. It is the first handbook of astrobiology and will hopefully fill a void in such literature in this broad field.

This handbook comprises 54 chapters divided into 12 sections. The chapters are written by experts in their respective fields. The chapters review the important topics in astrobiology and provide the newest developments.

As the editor, I had the privilege of shaping the content of the handbook by recruiting colleagues whose work I long respected and admired. Further, I have left my imprint on the handbook by following principles taught to me from 1992–1994 at the NASA Specialized Center of Research and Training (NSCORT) in exobiology, in San Diego, by Leslie Orgel and Stanley Miller. I was able to update and develop these principles by regularly attending and eagerly participating in various astrobiology conferences. Based on these experiences, I have built the sections of the handbook such that the content of the chapters intertwines and creates an exciting web of astrobiological knowledge that spans from the past to the present and reaches into the future.

Below I describe briefly the contents of the 54 chapters and 12 sections of the handbook to further illustrate the scope available to the reader.

In *Section 1*, which contains five chapters, the definition and scope of astrobiology are covered, including NASA strategy and the European roadmap, as well as various initiatives in astrobiology education.

Section 2 focuses on the definition and nature of life. Its four chapters cover multiple perspectives of definitions of life; a generalized and universalized definition of life applicable to extraterrestrial environments; synthetic life; and communication as a characteristic of life.

Section 3 covers the origin of life, its history, philosophical aspects, and major developments. It is comprised of three chapters, which analyze the evolutionary hypothesis for emergence of life as opposed to chance or design; a treatise on

Charles Darwin and the plurality of worlds; and the history of the ideas on the origin of life, from Buffon to Oparin.

In *Section 4*, chemical origins of life are discussed, including topics such as chemicals in the universe and their delivery on the early Earth, as well as the geology and atmosphere on the early Earth. This section consists of five chapters, which cover interstellar molecules and their prebiotic potential; formation and delivery of organic materials within the solar system and on early Earth; organic molecules in meteorites and their astrobiological significance; ancient life and crust and mantle evolution; and the atmosphere on the early Earth and its evolution and impact on life.

Section 5 comprises nine chapters which are devoted to the chemical origin of life and prebiotic chemistry. These chapters cover a wide range of interrelated topics, such as prebiotic chemistry that led to life; prebiotic chemical pathways to RNA and the importance of its compartmentalization; the hydrothermal impact crater lakes and origin of life; prebiotic chemistry in hydrothermal vent systems; prebiotic reactions in water, "on water," in supercritical water, as well as "solventless," and in the solid state; the origin of and amplification of chirality, and leading to biological homochirality; phosphorus in prebiotic chemistry; phosphorylation on the early Earth; and silicon and life.

Section 6 addresses RNA, the RNA world, and the complexity of life's origins. It consists of six chapters, which cover the role of RNA and ribozymes in the development of life; three ways of making an RNA sequence and the steps from chemistry to the RNA world; coevolution of RNA and peptides; role of ions in RNA folding and function; the origin of life as an evolutionary process; and a physicochemical view of the complexity of life's origins.

In *Section 7*, the origin of life in conjunction with an early compartmentalization and formation of coacervates and protocells is discussed. The two chapters in this section cover Oparin's coacervates and protocell evolution.

Section 8 consists of seven chapters which cover origin of life and its diversification; the universal tree of life; early primitive life on Earth; fossils of ancient microorganisms; and biomarkers and the detection of life. Specific chapters include topics such as the progenote; LUCA; the root of the cellular tree of life; horizontal gene transfer in microbial evolution; viruses in the origin of life and its subsequent diversification; the work of Carl Woese; biomarkers and their Raman spectral signatures; fossilization of bacteria and the implication for the search for early life forms; and biosignatures in astrobiology missions to Mars.

Section 9 addresses life under extreme conditions, including those in space. Its three chapters cover extremophiles and their natural niches on Earth, microbes in space, and the role of viruses in adaptation of life to extreme environments.

Section 10 deals with habitability and characterization of habitable planets. These topics and the evolution of habitability are discussed in a single chapter in this section.

Section 11 includes four chapters on putative intelligent life in the universe; the history and philosophy of this subject; and SETI (Search for Extraterrestrial Intelligence). Specific topics include philosophical approaches to the origin, evolution, and distribution of intelligent life in space; implications of the Drake equation and the Fermi paradox on the search for intelligent life elsewhere; SETI and its goals and accomplishments; and humanistic implications of discovering life beyond Earth.

The final section of the handbook, *Section 12*, has five chapters which focus on exoplanets, exploration of the solar system, the search for extraterrestrial life, and planetary protection. Specifically, topics include methods for detection of exoplanets and their habitability potential; small bodies in the solar system and their chemical and physical conditions; icy moons and their habitability; searching for extraterrestrial life on Mars and in our solar system in general; and the history, goals and objectives of the planetary protection program.

Colour versions of figures can be downloaded from the CRC Press website https://www.crcpress.com/9781138065123. The password to access the figures is: 9781138065123

This brief overview of the topics covered in the handbook illustrates the comprehensive approach to astrobiology which I desired and hopefully have achieved.

Thanks are expressed to the authors of the chapters, whose contributions made this handbook possible, and to the editorial and production staff of the publisher, notably Rebecca Davies and Georgia Harrison, who facilitated the process of producing the handbook in the record time, thus ensuring that its content stays current.

I dedicate this handbook to the memory of my father, Dr. Martin A. Kolb, and my mother, Dobrila Kolb, who encouraged my early love for science and supported me in my path towards becoming a scientist. This dedication extends to my loved ones, Vladimir, Mirjana, and Natasha Kolb.

Finally, special acknowledgment is due to my teachers of astrobiology, who generously shared their knowledge with me, satisfied my thirst for knowledge, inspired my dreaming about this exciting field, and carried me, so to speak, on their wings. There were many, too many to list here, but the very special ones are Leslie Orgel, Stanley Miller, Gustaf Arrhenius, and, most recently, Benton Clark.

Vera M. Kolb
Kenosha, Wisconsin
June 15, 2018

Editor

Vera M. Kolb earned a BS in chemical engineering and an MS in organic chemistry from Belgrade University, Beograd, Serbia, followed by a PhD in organic chemistry at Southern Illinois University at Carbondale, Illinois. She was a chemistry professor at the University of Wisconsin–Parkside, Kenosha, Wisconsin, from 1985 to 2016, and is now a professor Emeritus. During her first sabbatical leave (1992–1994), she received training in astrobiology (then termed exobiology) at the NASA Specialized Center of Research and Training (NSCORT) in San Diego, where she has worked with Leslie Orgel at the Salk Institute, and Stanley Miller, at the University of California, San Diego, California. She has worked in the field of astrobiology ever since. In 1992 she received the University of Wisconsin–Parkside Award for Excellence in Research and Creative Activity. During her second sabbatical (2002–2003) she studied sugar organo-silicates and their astrobiological importance with Joseph Lambert, at Northwestern University, Evanston, Illinois. Dr. Kolb was inducted in the Southeastern Wisconsin Educators' Hall of Fame in 2002. She is a recipient of numerous research and higher education grants and awards from the Wisconsin Space Grant Consortium/NASA, among others. At this time, she has published over 150 articles, including patents and books, in organic chemistry, medicinal chemistry, and astrobiology. She has edited a book, *Astrobiology, An Evolutionary Approach*, for CRC Press, which was published in 2014. Her second book, *Green Organic Chemistry and its Interdisciplinary Applications*, was published in 2016, also by CRC Press. Since both astrobiology and green chemistry study organic reactions in water, as in the primordial soup for astrobiology and as a benign solvent in green chemistry, the relationship between these two fields speaks of their true interdisciplinary character.

Contributors

Jessica C. Bowman
School of Chemistry and Biochemistry
Georgia Institute of Technology
Atlanta, Georgia

Antoni Camprubí
Departamento de Procesos Litosféricos
Instituto de Geología
Universidad Nacional Autónoma de México
Mexico City, Mexico

Sankar Chatterjee
Geosciences
Museum of Texas Tech University
Lubbock, Texas

Aditya Chopra
Planetary Science Institute
Research School of Astronomy & Astrophysics
Research School of Earth Sciences
Australian National University
Canberra, Australia

Benton C. Clark
Space Science Institute
Boulder, Colorado

Vanessa Colás
Departamento de Procesos Litosféricos
Instituto de Geología
Universidad Nacional Autónoma de México
Mexico City, Mexico

María Colín-García
Departamento de Dinámica Terrestre Superficial
Instituto de Geología
Universidad Nacional Autónoma de México
Mexico City, Mexico

Catharine A. Conley
New Mexico Institute of Mining and Technology
Socorro, New Mexico

Colleen N. Cook
Online and Continuing Education
University of Illinois at Urbana-Champaign
Urbana, Illinois
and
Innovative Education
University of South Florida
Tampa, Florida

Joseph L. Cross
Carl R. Woese Institute for Genomic Biology
University of Illinois at Urbana-Champaign
Urbana, Illinois

Punam Dalai
Department of Polymer Science
University of Akron
Akron, Ohio

David J. Des Marais
Exobiology Branch
NASA—Ames Research Center
Moffett Field, California

Lizanne DeStefano
Center for Education Integrating Science
Mathematics and Computing
Georgia Institute of Technology
Atlanta, Georgia

Steven J. Dick
Scientific Advisory Board
SETI Institute
Board of Directors, METI International

David Dunér
Division of History of Ideas and Sciences
Division of Cognitive Semiotics
Lund University
Lund, Sweden

Howell G.M. Edwards
Emeritus Professor of Molecular Spectroscopy
School of Chemistry and Biosciences
Faculty of Life Sciences
University of Bradford
Bradford, UK

Courtney C. Fenlon
Carl R. Woese Institute for Genomic Biology
University of Illinois at Urbana-Champaign
Urbana, Illinois

Kaitlyn E. Fouke
Denison University
Granville, Ohio

Bruce W. Fouke
Carl R. Woese Institute for Genomic Biology
University of Illinois at Urbana-Champaign
Urbana, Illinois
and
Department of Geology
University of Illinois at Urbana-Champaign
Urbana, Illinois
and
Department of Microbiology
University of Illinois at Urbana-Champaign
Urbana, Illinois
and
Roy J. Carver Biotechnology Center
University of Illinois at Urbana-Champaign
Urbana, Illinois

Kyle W. Fouke
Geology and Environmental Geosciences
Bucknell University
Lewisburg, Pennsylvania

Glenn A. Fried
Carl R. Woese Institute for Genomic Biology
University of Illinois at Urbana-Champaign
Urbana, Illinois

Iris Fry
Department of Humanities and Arts
Technion—Israel Institute of Technology
Haifa, Israel

Daniel G. Gibson
DNA Technology
Synthetic Genomics, Inc.
La Jolla, California

Johann Peter Gogarten
Department of Molecular and Cell Biology
University of Connecticut
Mansfield-Storrs, Connecticut

Adrienne M. Gulley
Carl R. Woese Institute for Genomic Biology
University of Illinois at Urbana-Champaign
Urbana, Illinois

Eric J. Hayden
Department of Biological Science
Boise State University
Boise, Idaho

Keyron Hickman-Lewis
CNRS-Centre de Biophysique Moléculaire
Orléans, France

Paul G. Higgs
Origins Institute and Department of Physics and Astronomy
McMaster University
Hamilton, Ontario, Canada

Clyde A. Hutchison III
Distinguished Professor
Synthetic Biology and Bioenergy Group
J. Craig Venter Institute
La Jolla, California

Clark M. Johnson
Department of Geoscience
University of Wisconsin
Madison, Wisconsin

Shelby L. Jones
Gardiner High School
Gardiner, Montana

James F. Kasting
Department of Geosciences
Penn State University
University Park, Pennsylvania

Tsuneomi Kawasaki
Department of Applied Chemistry
Tokyo University of Science
Tokyo, Japan

Susan B. Kelly
College of Education
University of Illinois at Urbana-Champaign
Urbana, Illinois

Vera M. Kolb
Emerita Professor of Chemistry
University of Wisconsin-Parkside
Kenosha, Wisconsin

Eric J. Korpela
Berkeley SETI Research Center
University of California
Berkeley, California

Martin J. Van Kranendonk
Professor of Geology
Head, School of Biological, Earth and Environmental
  Sciences
Director, Australian Centre for Astrobiology
Director, Big Questions Institute
University of New South Wales Sydney
Kensington, NSW

Sun Kwok
Department of Earth, Ocean and Atmospheric Sciences
University of British Columbia
Vancouver, Canada

Jennifer Lago
School of Geoscience
University of South Florida
Tampa, Florida

Joseph B. Lambert
Research Professor of Chemistry, Trinity University
Clare Hamilton Hall Professor of Chemistry Emeritus
Northwestern University
Department of Chemistry
Trinity University
San Antonio, Texas

Carolyn Lang
School of Geoscience
University of South Florida
Tampa, Florida

Niles Lehman
Department of Chemistry
Portland State University
Portland, Oregon

Charles H. Lineweaver
Planetary Science Institute
Research School of Astronomy & Astrophysics
Research School of Earth Sciences
Australian National University
Canberra, Australia

Erin Louer
Carl R. Woese Institute for Genomic Biology
University of Illinois at Urbana-Champaign
Urbana, Illinois

Claudia C. Lutz
Carl R. Woese Institute for Genomic Biology
University of Illinois at Urbana-Champaign
Urbana, Illinois

Jacalyn Wittmer Malinowsky
Geological Sciences
State University of New York Geneseo
Geneseo, New York

Zita Martins
Centro de Química-Física Molecular-Institute of
  Nanoscience and Nanotechnology (CQFM-IN)
and
Institute for Bioengineering and Biosciences (IBB)
Departamento de Engenharia Química, Instituto Superior
  Técnico (IST)
Universidade de Lisboa
Lisbon, Portugal

Arimasa Matsumoto
Department of Chemistry, Biology and Environmental
  Science
Nara Women's University
Nara, Japan

Sarah R.N. McIntyre
Research School of Astronomy & Astrophysics
Research School of Earth Sciences
Australian National University
Canberra, Australia

Bruce F. Michelson
Department of English
University of Illinois at Urbana-Champaign
Urbana, Illinois

Tom Murphy
Tom Murphy Photography
Livingston, Montana

Alicia Negrón-Mendoza
Departamento de Química de Radiaciones y Radioquimica
Instituto de Ciencias Nucleares
Universidad Nacional Autónoma de México
Mexico City, Mexico

Aharon Oren
Department of Plant and Environmental Sciences
The Alexander Silberman Institute of Life Sciences
The Hebrew University of Jerusalem
Jerusalem, Israel

Fernando Ortega-Gutiérrez
Departamento de Procesos Litosféricos
Instituto de Geología
Universidad Nacional Autónoma de México
Mexico City, Mexico

R. Thane Papke
Department of Molecular and Cell Biology
University of Connecticut
Mansfield-Storrs, Connecticut

Matthew A. Pasek
School of Geoscience
University of South Florida
Tampa, Florida

Juli Pereró
Department of Biochemistry and Molecular Biology
University of Valencia
Institute for Integrative Systems Biology I2SysBio
University of Valencia-CSIC
Paterna, Spain

Anton S. Petrov
School of Chemistry and Biochemistry
Georgia Institute of Technology
Atlanta, Georgia

Nikos Prantzos
Institut d'Astrophysique de Paris
Paris, France

Marilyn J. Roossinck
Professor, Plant Pathology and Environmental Microbiology
   and Biology
Center for Infectious Disease Dynamics
Pennsylvania State University
University Park, Pennsylvania

Ken Rice
Institute for Astronomy
University of Edinburgh
The Royal Observatory
Blackford Hill, Edinburgh

Michael Ruse
Program in the History and Philosophy of Science
Florida State University
Tallahassee, Florida

Frank Ryan
Emeritus Consultant Physician
The Sheffield Teaching Hospitals
Honorary Senior Lecturer
The Medical School
University of Sheffield
Sheffield, England

Nita Sahai
Department of Polymer Science
Department of Geosciences
Integrated Bioscience Program
University of Akron
Akron, Ohio

Daniella Scalice
NASA Astrobiology Program
NASA Ames Research Center
Moffett Field, CA

Alan W. Schwartz
Radboud University Nijmegen
Nijmegen, Netherlands

Mayandi Sivaguru
Carl R. Woese Institute for Genomic Biology
University of Illinois at Urbana-Champaign
Urbana, Illinois

Elizabeth M. Smith
Urbana High School
Urbana, Montana

Hamilton O. Smith
Synthetic Biology and Bioenergy Group
J. Craig Venter Institute
La Jolla, California

Kenso Soai
Department of Applied Chemistry
Tokyo University of Science
Tokyo, Japan

Killivalavan Solai
Center for Innovation in Teaching and
   Learning
University of Illinois at Urbana-Champaign
Urbana, Illinois

Jan J. Spitzer
R&D Manager (retired)
Mallard Creek Polymers, Inc.
Charlotte, North Carolina

Peter Strazewski
Institut de Chimie et Biochimie Moléculaires et
    Supramoléculaires
Université Claude Bernard Lyon 1
Villeurbanne, France

Kenichiro Sugitani
Department of Earth and Environmental Sciences
Graduate School of Environmental Studies
Nagoya University
Nagoya, Japan

Stéphane Tirard
François Viète Center for Epistemology and History of
    Science and Technology
University of Nantes
Nantes, France

Peter J. Unrau
Department of Molecular Biology and Biochemistry
Simon Fraser University
Burnaby, British Columbia, Canada

Steven D. Vance
Jet Propulsion Laboratory
California Institute of Technology
Pasadena, California

Nicholas P. Vasi
Carl R. Woese Institute for Genomic Biology
University of Illinois at Urbana-Champaign
Urbana, Illinois

Kasthuri Venkateswaran
Senior Research Scientist
Biotechnology and Planetary Protection Group
Jet Propulsion Laboratory
California Institute of Technology
Pasadena, California

J. Craig Venter
J. Craig Venter Institute
La Jolla, California

Saúl Villafañe-Barajas
Posgrado en Ciencias de la Tierra
Instituto de Geología
Universidad Nacional Autónoma de México
Mexico City, Mexico

Luis P. Villarreal
University of California
Irvine, California

Frances Westall
CNRS-Centre de Biophysique Moléculaire
Orléans, France

Loren Dean Williams
School of Chemistry and Biochemistry
Georgia Institute of Technology
Atlanta, Georgia

Guenther Witzany
Telos-Philosophische Praxis
Buermoos, Austria

Hikaru Yabuta
Department of Earth and Planetary Systems Science
Hiroshima University
Hiroshima, Japan

Lucy M. Ziurys
Department of Astronomy
Department of Chemistry & Biochemistry
University of Arizona
Tucson, Arizona

Section I

Astrobiology

Definition, Scope, and Education

1.1 Astrobiology
Definition, Scope, and a Brief Overview

Vera M. Kolb

CONTENTS

1.1.1 INTRODUCTION

In this chapter, the definition, scope, and a brief overview of astrobiology are given. Classification of astrobiology as a field of study, its specific goals, its multidisciplinary and interdisciplinary nature, and selected key developments are described. Astrobiology includes numerous research themes, which are expanding as the field evolves. Thus, it is not possible to cover all the developments in a single chapter, even briefly. Instead, an overview of selected astrobiology developments is presented, which is only a sampler of what is offered in this handbook.

1.1.2 ASTROBIOLOGY: DEFINITION, SCOPE, AND A BRIEF HISTORY

Astrobiology is defined as the study of the origin, evolution, distribution, and future of life in the universe (NASA's definition; Des Marais et al. 2008). Astrobiology seeks to answer fundamental questions about the beginning and evolution of life on Earth, possible existence of extraterrestrial life, and the future of life on Earth and beyond. Astrobiology scope is delineated in the NASA's astrobiology roadmap (Des Marais et al. 2008). Specific goals include understanding the emergence of life on Earth, determining how the early life on Earth interacted and evolved with its changing environment, understanding the evolutionary mechanisms and environmental limits of life, exploring habitable environments in our solar system and searching for life, understanding the nature and distribution of habitable environments in the universe, and recognizing signatures of life on the early Earth and on other worlds. The NASA's astrobiology map is being updated as the astrobiology field advances. A similar roadmap has been developed for astrobiology research in Europe (Horneck et al. 2015, 2016). The topic of these roadmaps is further discussed and updated in Section 1 of this handbook.

Astrobiology is a young science that acquired its name only in 1995 (named by Wes Huntress from NASA; Catling 2013). Astrobiology evolved from its predecessor, exobiology, which is the study of the origin of life and of possible life outside Earth (Dick and Strick 2004; Dick 2007). The term exobiology was coined in 1960 (by Joshua Lederberg).

The difference between the two fields is that astrobiology is broader and notably includes the evolution of life on Earth. The exobiology era coincided with the space missions, notably the Viking mission on Mars in 1976, which followed a series of reconnaissance missions in the 1960s and early 1970s. The Viking mission searched for the microbial life on Mars (DiGregorio 1997; Jones 2004), but it did not confirm its existence. The early experiments that used radio astronomy to search for extraterrestrial intelligence by the Search for Extraterrestrial Intelligence (SETI) program (Tarter 2007) did not result in positive findings. In 1996, the analysis of the Martian meteorite ALH84001, which was found in Antarctica, indicated possible presence of the fossils of bacterial life (Goldsmith 1997; Jones 2004), but the results were not convincing enough. Disappointments regarding a lack of positive results in the search for extraterrestrial life (Dick and Strick 2004; Dick 2007, 2012) continue to this date. The only life we know of is the life on Earth. Search for extraterrestrial life and extraterrestrial intelligence is further discussed in Sections 10 through 12 of this handbook.

While astrobiology is a new science, it comes from a very old and long history of ideas, which go all the way back to the antiquity. The history of one of the core ideas of astrobiology, a possibility of extraterrestrial life, is described in the books: "Plurality of worlds: The origins of the extraterrestrial life debate from Democritus to Kant" (Dick 1984), "The extraterrestrial life debate, 1750–1900" (Crowe 1999), "Medieval cosmology: Theories of infinity, place, time, void, and the plurality of worlds" (Duhem 1987), and other sources. A comprehensive coverage of the ideas about extraterrestrial intelligent life is covered in the books: "Extraterrestrials: Science and alien intelligence" (Regis 1987), "Extraterrestrials: Where are they?" (Zuckerman and Hart 1995), "Beyond contact: A guide to SETI and communicating with alien civilizations" (McConnell 2001), and "Civilizations beyond Earth: Extraterrestrial life and society" (Vakoch and Harrison 2013), among other sources.

The future of life on Earth and its potential uniqueness are discussed in the books "The life and death of planet Earth: How the new science of astrobiology charts the ultimate fate of our world" (Ward and Brownlee 2002), and "Rare Earth: Why complex life is uncommon in the universe" (Ward and Brownlee 2004). Search for exoplanets, some of which may be habitable, is a subject of recent studies (Kasting 2010; Summers and Trefil 2017). This subject is further covered in Section 12 of this handbook.

The major hypothesis about the origin of life on Earth, which later became accepted as a key foundation of both exobiology and astrobiology, was proposed independently by A.I. Oparin in 1924 and J. B. S. Haldane in 1929 (Oparin 1965, 1968, 1994; Deamer and Fleischaker 1994; Haldane 1994). This hypothesis states that the origin of life on Earth can be understood based only on the laws of chemistry and physics. Life arose in Earth's distant past by chemical reactions and physical processes under the specific conditions on the early Earth and over a long period of time. Numerous rapid developments based on the Oparin–Haldane proposal followed

(Miller and Orgel 1974; Mason 1991; Brack 2000; Zubay 2000; Chela-Flores 2001, 2011; Fenchel 2002; Lurquin 2003; Gilmour and Sephton 2004; Luisi 2006; Sullivan and Baross 2007; Sullivan and Carney 2007; Kolb 2014a; Longstaff 2015). New developments on chemical origins of life and prebiotic chemistry are covered in Section 5 of this handbook.

1.1.3 CLASSIFICATION OF ASTROBIOLOGY AS A FIELD OF STUDY

Astrobiology is commonly referred to as a field of study rather than a discipline. Dick (2012) examined the question if astrobiology is a separate discipline and if it should be considered as such. This question is still open to discussion. Astrobiology as a discipline would include origins of life (interstellar molecules, complex organics, development of laboratory experiments, life in extreme environments, genomics/phylogenetic relationships), planetary systems (theoretical studies, atmospheric biosignatures, extrasolar planets, circumstellar disks), planetary science (rock biosignatures, biogeochemistry, geochemistry exploration of Mars, Europa, moons, etc.), and SETI (Dick 2012). A common approach is that astrobiology is not a discipline itself but draws upon other disciplines, subdisciplines, and specialized areas of research. Physics, chemistry, biology, geology, and astronomy were the initial disciplinary pillars of astrobiology. Rapid developments in astrobiology and astrobiology-relevant disciplines led to an expanded list, which included cosmology, atmospheric science, oceanography, evolutionary science, paleontology, planetary science, biochemistry, molecular biology, microbiology, and ecology. Astrobiology also seeks insights from history of science and philosophy (Fry 2000). The latter approach is also covered in Section 3 of this handbook.

An all-encompassing goal for astrobiology is to understand cosmic evolution. NASA embraced the idea of cosmic evolution, comprising the following evolutionary sequence: Big Bang, galaxies, stars, biogenic elements, planets, chemical evolution, origin of life, pre-Cambrian biology, complex life, intelligent life, cultural evolution, civilization, science and technology, and the study of life in the universe (Dick 2007). This formidable goal adds many more disciplines and subdisciplines of various fields of study for astrobiology to rely upon. It becomes clearer why it would be difficult to classify astrobiology as a single discipline.

NASA classified astrobiology as a field of study that is multidisciplinary in its content and interdisciplinary in its execution. Astrobiology requires a close coordination of not only various disciplines but also programs, including space missions (Race 2007).

The distinction between multidisciplinary and interdisciplinary studies is described in the book "Interdisciplinary research, process and theory" by Repko (2012). Both types of studies are used to investigate complex problems and issues, which are beyond the ability of any single discipline. These studies are thus well suited for astrobiology. Repko addresses the "multi," "inter," and "disciplinary" parts of these terms. A brief account is given here. In a multidisciplinary approach, a problem is studied from the perspective of several disciplines at one time. This approach tends to be dominated by the method and theory preferred by the main discipline of the investigator. This prevents effective integration of the insights from all the disciplines that were consulted. In contrast, the interdisciplinary approach successfully integrates disciplinary insights to construct a more comprehensive understanding of the problem. While employing a research process that subsumes the methods of relevant disciplines, interdisciplinary research does not privilege any disciplinary method or theory. Thus, the interdisciplinary approach is more powerful but not always easy to achieve (Repko 2012).

Although the modern astrobiology field is commonly perceived as mostly based on the developments in the natural sciences and technology, contribution to astrobiology from other disciplines needs to be considered equally. This is demonstrated in a review paper by Dick (2012) titled "Critical issues in the history, philosophy, and sociology of astrobiology".

In its study of some complex problems, such as the future of life and the humanistic implications of discovery of extraterrestrial life (Dick and Lupisella 2009; Dick 2015), astrobiology may use a transdisciplinary approach. This approach is useful in studies of complex and mega problems, which are at once between the disciplines, across different disciplines, and beyond all disciplines (Repko 2012).

Interdisciplinary and transdisciplinary nature of astrobiology is especially needed for the study of cosmic evolution. Progress in understanding of the individual steps of this evolution has been made based mostly on the multidisciplinary approach. However, a full integration of the insights from the individual disciplines that are involved in these evolutionary steps has not yet been achieved. This is notably the case for the cosmic evolutionary step of the origin of life on Earth.

Astrobiology may also be considered an evolutionary science. The multi-/interdisciplinary nature of astrobiology evolves to include more disciplines and subdisciplines and to achieve a better integration of the disciplinary insights. Astrobiology research tools, which are typically initially borrowed from its contributing disciplines, become more advanced and specialized for astrobiology use, which allows for new research programs to emerge. Technological advances enable space missions that search for extraterrestrial life and habitable places within our solar system. Based on these and many more new developments, critical issues, goals, and objectives of astrobiology evolve accordingly. An evolutionary approach to astrobiology is presented in a recent book (Kolb 2014a).

In the next section, selected key developments in astrobiology are overviewed.

1.1.4   A BRIEF OVERVIEW OF SELECTED KEY DEVELOPMENTS IN ASTROBIOLOGY

As shown in the previous section, astrobiology research themes are numerous and constantly expanding. There are "astro" and "bio" themes, and the organization of their coverage presents

a challenge. A choice taken here is to focus on the themes that are most closely related to the central focus of astrobiology as a study of the origin and evolution of life and on a possibility of life elsewhere.

1.1.4.1   UNDERSTANDING THE CONCEPT OF LIFE

Among key problems that astrobiology is yet to solve is to define life (Popa 2014). Many such definitions have been proposed, but none is universally accepted (Popa 2014). While each definition addresses some features of life, none appears inclusive of all the features. Even the most inclusive definitions may only describe what life does but may not state what life is (Catling 2013). We may not know enough about the theory of life to define it properly (Cleland and Chyba 2007). Our ability to define life is also dependent on our understanding of the origin of life. Currently, the NASA's operational definition of life that is used most often is: "Life is a self-sustained chemical system capable of undergoing Darwinian evolution" (Popa 2014). This definition, however, is not very useful in search for extraterrestrial life, since it would require observation of evolution (Catling 2013). What is desired is a definition of life that will also be applicable to life in general and thus to the putative extraterrestrial life. A universal definition of life in which living beings are considered autonomous systems with open-ended evolutionary capacities has been proposed (Ruiz-Mirazo et al. 2004). A generalized and universalized definition of life has been proposed for the identification of extraterrestrial and artificial life forms (Clark 2004). An examination of the proposal that prebiotic evolution that leads to life is Darwinian continues, with no definitive conclusion (Perry and Kolb 2004). One area of progress is a better understanding of viruses, which are in the twilight zone of life (Villarreal 2004). The question if viruses are alive is also discussed from the philosophical point of view of dialetheism, which allows for viruses to be considered both alive and not alive (Kolb 2010a). The problem of defining life is further discussed in Section 2 of this handbook.

1.1.4.2   CHEMICALS IN THE UNIVERSE AND THEIR DELIVERY TO THE EARLY EARTH

The formation of chemicals in the universe and their delivery to the early Earth (Shaw 2006; Longstaff 2015; Kwok 2016, 2017) were necessary for life to evolve. Many interstellar molecules are prebiotically relevant. Complex organic molecules were formed in space and delivered to the solar system and early Earth. The delivery occurred mostly by the comets (e.g., Chyba et al. 1990) and meteorites. The origin of terrestrial water, which is critical to life, is from comets but also from other sources (O'Brien et al. 2018). Meteorites of the carbonaceous chondrite type (McSween 1999) delivered complex chemical compounds, many of which were important for the emerging life. They include amino acids and nucleobases (Sephton 2004; Schmitt-Kopplin et al. 2010, 2014; Burton et al. 2012). This is further discussed in Section 4 of the handbook.

1.1.4.3 PREBIOTIC CHEMISTRY AND THE RIBONUCLEIC ACID WORLD

In 1953, Stanley Miller published the results of a seminal experiment that showed that the amino acids, which are the building blocks of proteins, can be synthesized in the laboratory under the reaction conditions that simulate those on the early Earth (Miller 1953; Mesler and Cleaves 2016). Miller built a glass apparatus consisting of a series of connected flasks and tubing to simulate the primitive Earth's environment. One flask was filled with water, representing the early ocean. Water could be heated to simulate evaporation. Another flask simulated the early atmosphere. It contained gases methane, hydrogen, ammonia, and water vapor, which reflected the belief at that time that the early Earth's atmospheric conditions were reducing. The source of energy in the experiments was electrical spark generated by Tesla coil. Various amino acids were formed, including glycine and alanine. Miller's experiment showed that organic compounds that are important for life can indeed be formed under prebiotic conditions.

Many more prebiotic experiments by different investigators followed. Much progress has been made by the early 1970s, as shown in the book "The origins of life on earth" (Miller and Orgel 1974). Prebiotic syntheses of amino acids, urea, fatty acids, porphyrins, vitamins, purines, pyrimidines, sugars, and nucleosides have been attempted, and some gave excellent yields.

Further progress in prebiotic syntheses has been achieved in the next three decades (Brack 2000; Bilgen 2004; Orgel 2004; Herdewijn and Kisakürek 2008; Pereró 2012). Some of these syntheses facilitated new proposals on the origins of life. Selected examples are presented here. Miller performed syntheses of amino acids under mildly reducing or non-reducing conditions, which reflected an updated view on the atmospheric conditions on the early Earth. The results were less favorable than those in the reducing atmosphere, but, in combination with other prebiotic factors, these results were judged adequate (Miller 2000). Wächtershäuser proposed that the original source of prebiotic organic material under deep-sea vents conditions may have been a reduction of carbon dioxide with hydrogen sulfide over ferrous sulfide as the reducing agent. He believed that life originated in an iron–sulfur world (Wächtershäuser 2000, 2008; Orgel 2004). Based on the clues from the present-day biology, De Duve (2000) proposed an early thioester world, in which thioesters had a role in primitive (proto) metabolism. Progress in non-enzymatic autocatalysis and self-replication, notably by von Kiedrowski and coauthors, has been made (Burmeister 2000).

The proposal of ribonucleic acid (RNA) world emerged as a leading hypothesis for the origins of life (Joyce and Orgel 1993; Orgel 2004). This hypothesis postulates that the RNA is the original self-replicator that has a capability to carry the genetic information and act as a catalyst. Within this hypothesis, various aspects of RNA were studied. Examples include non-enzymatic polymerization of RNA monomers to RNA, with montmorillonite clay as a catalyst (Ferris 2000), and the enzymatic ribozyme (RNA that acts as an enzyme) catalysis

that might have facilitated polynucleotide replication (James and Ellington 2000; Orgel 2004).

Key discoveries in prebiotic chemistry and the origin of the RNA world up to the early 2000s were reviewed by Orgel (2004). This review includes work on template-directed synthesis of nucleotides by Orgel and his coworkers. Eschenmoser's work is also highlighted. With his coworkers, he synthesized and studied close chemical alternatives to RNA and DNA in a systematic manner to gain an insight into the selection of the naturally occurring forms. They synthesized RNA and DNA analogs in which the sugar component, ribose for RNA and deoxyribose for deoxyribonucleic acid (DNA), were substituted with a different sugar type or a different sugar form (in both RNA and DNA, the sugar form is a five-membered ring). For example, they synthesized homo-DNA, in which the five-membered ring was expanded to a six-membered ring, an analog of glucose. They also prepared pyranosyl-RNA, based on a six-membered sugar form rather than the five-membered form found in RNA. In addition, they varied a sugar type and prepared threose nucleic acid (TNA), which is based on the sugar threose rather than ribose. Eschenmoser put together all the experimental results to determine "chemical etiology" of nucleic acids (Eschenmoser 1999). The latter explains how changes in chemical structures influence hydrogen bonding, folding, and other properties of nucleic acids. For example, while homo-DNA can make a stable double helix via hydrogen bonds, the type of its helix is incompatible with the RNA helix. This prevents formation of a double helix between a strand of homo-DNA and a strand of RNA, which makes homo-DNA unsuitable for the transcription process. This etiology of nucleic acids provided an insight into the selection of the present-day type and form of nucleic acids (Eschenmoser 1999). Benner expanded the genetic alphabet from standard AGCT/U (adenine, guanine, cytosine, thymine/uracil) with synthetic unnatural nucleobases. He also synthesized sulfone-linked RNAs (Benner 2004). The sulfone-based RNA backbone is not charged, as opposed to the natural phosphate backbone. Like Eschenmoser, Benner used the experimental results to gain insight into the structure and function of nucleic acids by considering their hydrogen bonding and folding properties and the reasons why some natural structures are superior to the analogous unnatural ones. As one example, sulfone-linked RNAs are unsuitable, since they fold and aggregate in water. This shows the importance of charged backbones in nucleic acids, since charges prevent undesired folding and aggregation in water. Charged backbones are thus selected in nature.

Prebiotic synthesis of biological compounds (Cleaves 2013, 2014) as well as the work on biochemical pathways as evidence for prebiotic synthesis (McDonald 2014) have advanced. The progress in prebiotic synthesis continued with a new, innovative prebiotic synthesis of nucleotides (Powner et al. 2011; Islam and Powner 2017; Sutherland 2017), which has previously been only marginally successful. Significant new work on the RNA world and its connection to the origins of life has been recently reported (e.g., Robertson and Joyce 2012; Higgs and Lehman 2015; Pressman et al. 2015; Mathis et al.

2017). The advances in prebiotic chemical pathways to RNA and the properties of RNA that enable its self-replication and enzymatic activity are further covered in Sections 5 and 6 of this handbook.

1.1.4.4 Prebiotic Reactions in Hydrothermal Vents

Prebiotic reactions under the conditions of hydrothermal vents gained importance, since such vents on Earth are rich with life. These vents provide both chemicals and energy sources, making them feasible as a niche for the emergence of life. Thus, other bodies in the solar system that possess hydrothermal vents could also be habitable. Chemistry under the conditions of the hydrothermal vents has a substantial prebiotic potential (Holm and Andersson 2000; Orgel 2004; Padgett 2012; Sojo et al. 2016). As one example, components of the citric acid cycle, which are important for development of protometabolism, have been synthesized under these conditions (Orgel 2004). Prebiotic chemistry in hydrothermal vents is further discussed and updated in Section 5 of this handbook.

1.1.4.5 Prebiotic Reactions in Water and in the Solid State

Progress has been made in prebiotic organic reactions in water as a reaction medium, a model for the prebiotic soup. While most organic compounds are not water-soluble, prebiotic reactions are supposed to happen in water, which is the ubiquitous solvent on Earth and is also the medium for life. Recent findings show how prebiotic reactions of water-insoluble organic compounds can occur in water, by a special mechanism called "on-water" reactions. According to this mechanism, insoluble organic materials are driven toward each other in water by the hydrophobic interactions. The proximity between the molecules then facilitates the reaction. Furthermore, selectivity and specificity of the reactions are often improved (Kolb 2014b). This type of reaction is an exception from the usual formation of intractable mixtures in the prebiotic experiments (Benner et al. 2012; Benner 2014).

Many classical organic reactions, which are a bread and butter of prebiotic chemistry, are possible in superheated water and thus under hydrothermal conditions. They have been studied in context of green chemistry (environmentally friendly) (e.g., Kolb 2016). This is covered in more detail in Section 5 of this handbook.

Prebiotic chemistry in water–ice matrix occurs readily (Menor-Salván and Marín-Yaseli 2012). Organic materials and salts concentrate in the liquid phase of the crystalline ice matrix and form eutectic solutions. The concentration effect generally increases the reaction rates. Thus, prebiotic chemistry in the ice world is feasible.

Prebiotic reactions can also occur quite successfully in the solid state. Surprisingly, the reactions are often very fast and specific. This has a direct application to the chemistry on asteroids and meteorites (Kolb 2010b, 2012). These reactions are also covered in Section 5 of this handbook.

1.1.4.6 Phosphorus in Prebiotic Chemistry

Life on Earth requires phosphorus in DNA/RNA and in the chemical energy sources. Thus, the study of the role of phosphorus in prebiotic chemistry is important. A convincing scenario for the incorporation of phosphorus from geologically plausible sources into chemical systems on the early Earth was not initially conceived, leading to the so-called "phosphorus problem." Progress has been made on this front, as a result of several approaches (Fernández-García et al. 2017; Pasek et al. 2017), as well as the discovery of a meteoritic source of reactive phosphorus, which may have been delivered to the early Earth (Maciá 2005; Pasek and Lauretta 2005; Schwartz 2006; Pasek 2014). This subject is further discussed in Section 5 of this handbook.

1.1.4.7 Prebiotic Evolution of Homochirality

The origin of biological homochirality is one of the central problems in understanding chemistry of life. Chiral molecules (or other objects) are not superimposable with their mirror images (such as the left hand and the right hand). Each of the pair of such molecules is termed an enantiomer. Homochirality occurs when all the constituent units are molecules of the same chiral form (enantiomer).

Many key biochemical compounds are chiral. Chiral properties provide an extra dimension in the specificity of molecular recognition, such as between the enzymes and their substrates. Homochirality is common in biology. Thus, nature uses almost exclusively L-amino acids and D-sugars, thus only one out of the two possible forms. A question is posed about the origin of homochirality, if it occurred prebiotically or during the biotic evolution. Prebiotic evolution of homochirality is envisaged to occur in three steps: mirror-symmetry breaking, chiral amplification, and chiral transmission. The mirror-symmetry breaking occurs under the influence of circularly polarized light or other physical causes. It results in a minute enantiomeric imbalance. A slight enantiomeric excess observed in the meteoritic amino acids indicates an asymmetric influence on organic chemical syntheses under abiotic conditions (Cronin and Pizarrello 1997; Pizzarello 2008). The second step, chiral amplification, leads to enantiomeric enrichment. The third step, chiral transmission, enables the transfer of chirality between molecules. Studies on the origin of a single chirality of amino acids and sugars indicate that a partial enantioenrichment in prebiotic world may have been sufficient to trigger chemical and physical processes to enable further amplification of the enantiomeric excess (Hein and Blackmond 2012). Enantioselective automultiplication of chiral molecules can occur via asymmetric catalysis, in which a chiral compound acts as a chiral catalyst for its own production (Soai et al. 2000). Reviews of the prebiotic origins of homochirality are available (Coveney et al. 2012; Ruiz-Mirazo et al. 2014). Origin and chemical evolution of biological homochirality are further discussed in Section 5 of this handbook.

1.1.4.8 Approaches to Reconstructing the Origin of Life

The emergence of life is a historical process, which scientists try to reconstruct. Different approaches exist to this problem. The "metabolism first" approach envisions that the process starts with the development of a primitive metabolism. Others believe that compartmentalization is the essential first step (Deamer 2000, 2007). The "genetics first" proponents believe that the formation of self-replicating polymers, such as RNA, is critical. Peretó discusses problems and controversies of these approaches (Peretó 2005). Most believe that the origin of life is linked to the RNA world, in which RNA has both genetic and enzymatic properties and is capable of self-replication.

The research approaches that are employed for reconstruction of the origin of life are bottom-to-top and top-to-bottom approaches. In the former, prebiotic chemistry is studied experimentally under simulated prebiotic conditions, starting from the simple compounds and moving toward molecular complexity that is characteristic of life (Lahav 1999). This research pathway benefited greatly from new developments in prebiotic synthetic methods and an increased knowledge of the environmental conditions on the early Earth under which these syntheses occurred. Lahav considered different ways of bottom-up reconstruction of early stages of chemical evolution. The reconstruction could be without or with specific biogeochemical constraints. The latter included minerals that acted as catalysts or were involved in energy production and transfer, among other roles (Lahav 1999). The bottom-to-top approach needs to be looked at in conjunction with the conditions on the Early Earth (Burton et al. 2012; Scharf et al. 2015; Domagal-Goldman et al. 2016). Also, the chemistry on meteorites, which are the prebiotic laboratory, needs to be considered.

In the top-to-bottom approach, a reconstruction of the origin of life is attempted, going from its most primitive life forms to the putative RNA world (Lahav 1999; Sephton 2004) and the pre-RNA world (Hazen 2005). Development of new research tools, notably those in molecular biology, facilitated this approach. These tools are responsible for rapid recent advances in the understanding of RNA and the RNA world (Ruiz-Mirazo et al. 2014). Progress has been made in modeling the RNA world by the RNA viral quasispecies (Padgett 2012; Eigen 2013). The newest research on this topic is covered in Sections 5 and 6 of this handbook. It includes various aspects of prebiotic pathways to RNA, coevolution of RNA and peptides, and the role of RNA and ribozymes in the development of life, among others. This is the most significant and most rapidly developing area of research into the origins of life, which thus justifies its extensive coverage in this handbook.

1.1.4.9 Abiotic-to-Biotic Transition

Both bottom-to-top and top-to-bottom approaches to the origins of life lead to a putative transition zone between life and non-life. The nature of this zone is not elucidated, but it is assumed to contain a complex mixture of organic compounds. The mechanism by which life emerged from the transition zone is not known. Progress has been made in the understanding of the nature of life as an organized complexity. Life represents order, which imposes thermodynamic requirements between the life entity (typically a compartmentalized system) and its environment. However, both thermodynamic and kinetic factors constitute driving force for the emergence of life (Pross 2003, 2012). Emergence of order out of chaos has been invoked as a way for an organized system to emerge from chaotic mixtures of components (Kauffman 1993, 1995, 2000). Properties attributed to chemical systems that are transitioning to life are complex chemical behaviors, such as self-organization and self-assembly; formation of auto-catalytic networks and protometabolic cycles; and establishment of self-replicating systems (Padgett 2012; Peretó 2012; Ruiz-Mirazo et al. 2014). Possible scenarios of protobiological events prior to the emergence of the last universal common ancestor (LUCA) have been proposed (Mann 2013). The idea of an alternative living world that existed prior to the LUCA of the organisms, as we know them, is introduced. Peretó discusses proposed stages in the origin of life, which comprise prebiotic, protobiological, and biological stages. Each stage consists of the additional sub-stages, such as the pre-RNA world (Peretó 2005). Joyce (2012) also discusses pathways to life and considers the pre-RNA life, which was based on different types of RNA. Krishnamurthy (2017) took a chemical approach to the transition from prebiotic chemistry to protobiology and analyzed how a diverse pool of prebiotic building blocks could lead to a self-assembling system that is capable of chemical evolution. His analysis encompassed different classes of biomolecules that built the thioester world, protein world, metabolism world, lipid world, and RNA world.

Theoretical and computational approaches have been used to examine if there is an algorithm for abiotic-to-biotic transition (Walker 2014). The nature of biological information, informational limits of evolution, Shannon information content of biopolymers, molecular quasispecies, and replicator equations were addressed. While no definitive conclusion was reached, it appears that the answer to how life emerged from non-life will probably come from some combination of chemistry and information-based formalism. Since life is an organized complexity, studies of complexity and in relation to life are relevant to the understanding of the nature of transition zone to life and to the emergence of life (Waldrop 1992; Holland 1995; Mitchell 2009; Lineweaver et al. 2013). The topics of transition from prebiotic toward biotic world and life are updated and further discussed in Section 6 of this handbook.

1.1.4.10 Early Compartmentalization and Development of Protocells

Since his original proposal in 1924, Oparin made experimental progress in study of coacervates. Coacervates are feasible as prebiotic compartmentalized systems and can serve as prebiotic chemical reactors. In addition, they are capable of undergoing a primitive "reproduction" by splitting into smaller units (Oparin 1968).

It is believed that an early compartmentalization and development of protocells were necessary for the development of life. Compartmentalization could be achieved by forming prebiotic coacervates, micelles, vesicles, and primitive membranes (Deamer 2000; Luisi 2006; Frankel et al. 2014; Kolb 2015). Such compartmentalization enabled encapsulation of macromolecules. These topics are further covered and updated in Section 7 of this handbook.

1.1.4.11 Astrobiology and Evolution of Life

One of the astrobiology goals is to understand the evolution of life, its mechanisms, and evolutionary transitions, starting from the early primitive life on Earth (Fenchel 2002; Agutter and Wheatley 2007). Examples of specific goals are the determination of the evolutionary mechanisms (molecular, genetic, and biochemical) that control and limit evolution, metabolic diversity, and acclimatization of life (Des Marais et al. 2008). Aspects of evolution, such as the role of viruses in the origin and evolution of life (Ryan 2002; Villarreal 2005, 2014; Jalasvuori and Bamford 2014), development of the tree of life (Hug et al. 2016), and horizontal gene transfer (Fenchel 2002; Soucy et al. 2015), are all used to ascertain the origin of life, by the top-to-bottom approach. The topic of astrobiology and evolution is updated and discussed in Section 8 of this handbook.

1.1.4.12 Biosignatures and Detection of Life

Biosignatures are used to detect life, past or present, on Earth or elsewhere. The list of the life features that are used as biosignatures is long (Westall and Cockell 2016). It includes metabolic gases; spectral signatures of biology, such as the red edge of vegetation in earthlight, reflected from the Moon; biomolecules and/or their degradation products; homochiral organic molecules; isotopic fractionation of bioelements, such as C, S, N, and others; biominerals formed by microbial metabolism, such as carbonate and phosphate; morphological features produced by microorganisms and their activities; bacterial fossils; and many more. Discovery of Earth's early bacterial fossils helped pin down the beginning of the biological evolution (Schopf 1999, 2006; Brasier et al. 2015). The issue of the putative biosignatures on Mars is examined, which will be useful for the future mission to Mars (Westall et al. 2015). These topics are updated and discussed in Sections 8 and 12 of this handbook.

1.1.4.13 Life Under Extreme Conditions

One of the rapidly developing areas in astrobiology is the study of life under extreme conditions, including that in space. Extremophiles are microbes that live under extreme conditions on Earth, which are generally hostile to most other life. Extreme conditions include extreme heat or cold, high concentration of salts, and high pressure. Many extremophiles are polyextremophiles. They are adapted to multiple forms of stress (Seckbach et al. 2013). These microbes are good models for the survival under extraterrestrial extreme conditions. Progress in space

missions allowed for the study of such microbes in space. They were found to be able to survive in space under a specific limited exposure (Horneck and Moeller 2014). Adaptation of life to extreme environment includes viruses. These topics are updated and discussed in Section 9 of this handbook.

1.1.4.14 Habitability Within the Solar System and Beyond

Another rapidly developing area of astrobiology is a study of the habitability within the solar system (and beyond). Much of the modern focus of the "astro" part from astrobiology is on the habitability and its limits and on the characteristics of habitable planets (Kasting 2010). The habitability study is facilitated by the advances in planetary sciences and the results of many space missions, notably those on Mars. Planetary science studies planets and processes that formed them. Studies of the Earth, moons, and planets in our solar system provide information about their environments (Beatty et al. 1999; Bennett et al. 2003); these can then be checked against those on Earth, which are hospitable to life. Knowledge about the environments on the Earth's Moon, Mercury, Venus, Mars, Europa, Ganymede, Callisto, Titan, Triton, Pluto, Charon, comets, asteroids, and meteorites is increasing rapidly. As its counterpart, the knowledge about the life on Earth under the extreme condition is also increasing. A combination of these two can guide studies of habitability. For example, Europa is now considered habitable (Jones 2004). The topic of habitability is discussed in Section 10 of this handbook.

Another modern "astro" focus of astrobiology is on the search for planets outside our solar system (exoplanets) (Summers and Trefil 2017). Such search has resulted in a discovery of over 3500 planets so far, and the number is rapidly increasing. Some of these planets may be habitable and may hopefully harbor life. Exoplanets are covered in Section 12 of this handbook.

1.1.4.15 Some Rapidly Advancing Astrobiology Areas

Rapid advances and a huge output of new data in some areas of astrobiology make regular updates mandatory. Examples of such areas are studies of habitable planets and other bodies in the solar systems, search for microbial life within the solar system, analog studies (studying places on Earth with environmental conditions that are extreme for life, which can then serve as laboratories to simulate extraterrestrial environments), developments of instruments for astrobiology research and exploration, missions to Mars and other missions, search for exoplanets, and the planetary protection program. Some of these examples are briefly addressed below.

The planetary protection program develops protocols for the prevention of contamination of Earth with the putative extraterrestrial microbes that may be carried back on the space vehicles or samples that are returned to Earth and for the prevention of contamination of space bodies with microbes from Earth delivered by spaceships, landers, and astronauts (Rummel 2007; Meltzer 2010).

Instruments for the assessment of habitability and search for life on Mars (Conrad 2007) and for the observation and spectroscopic analysis of planets and satellites in the solar system and exoplanets are rapidly developing. Some instruments are designed for a specific mission. For example, Mars Exploration Rover (MER) mission employed panoramic imagers (for visual characterization of environment), microscopic imager (for providing detailed images of rock textures), miniature thermal emission spectrometer (for the identification of minerals), and other instruments to study the surface of Mars (Conrad 2007).

The coverage and updates on these topics are provided in Section 12 of this handbook.

1.1.5 SUMMARY

This chapter provides an overview of astrobiology, which includes its definition, scope, a brief history, and classification as a field of study. In addition, it briefly presents selected topics in astrobiology. Within the narrative of this chapter, contents of various sections of this handbook are announced. Selected key references are cited.

DEDICATION

I dedicate this chapter to the memory of two great scientists, Leslie E. Orgel and Stanley L. Miller, with whom I have worked from 1992 through 1994.

REFERENCES

Agutter, P. S., and D. N. Wheatley. 2007. *About Life: Concepts in Modern Biology*. Dordrecht, the Netherlands: Springer.

Beatty, J. K., C. C. Petersen, and A. Chaikin, Eds. 1999. *The New Solar System*. 4th ed. Cambridge, UK: Sky Publishers & Cambridge University Press.

Benner, S. A. 2004. Understanding nucleic acids using synthetic chemistry. *Acc. Chem. Res.* 37: 784–797.

Benner, S. A. 2014. Paradoxes in the origin of life. *Orig. Life Evol. Biosph.* 44: 339–343.

Benner, S. A., H.-J. Kim, M. A. Carrigan. 2012. Asphalt, water, and the prebiotic synthesis of ribose, ribonucleosides, and RNA. *Acct. Chem. Res.* 45: 2025–2034.

Bennett, J., S. Shostak, and B. Jakosky. 2003. *Life in the Universe*. San Francisco, CA: Addison Wesley.

Bilgen, T. 2004. Metabolic evolution and the origin of life. In *Functional Metabolism: Regulation and Adaptation*, Ed. K. B. Storey, pp. 557–582. Hoboken, NJ: John Wiley & Sons.

Brack, A., Ed. 2000. *The Molecular Origins of Life: Assembling Pieces of the Puzzle*. Cambridge, UK: Cambridge University Press.

Brasier, M. D., J. Antcliffe, H. Saunders, and D. Wacey. 2015. Changing the picture of Earth's earliest fossils (3.5–1.9 Ga) with new approaches and new discoveries. *Proc. Natl. Acad. Sci. U.S.A.* 112: 4959–4864.

Burmeister, J. 2000. Self-replication and autocatalysis. In *The Molecular Origins of Life: Assembling Pieces of the Puzzle*, Ed. A. Brack, pp. 295–311. Cambridge, UK: Cambridge University Press.

Burton, A. S., J. D. Stern, J. E. Elsila, D. P. Glavlin, and J. P. Dworkin. 2012. Understanding prebiotic chemistry thorough the analysis of extraterrestrial amino acids and nucleobases in meteorites. *Chem. Soc. Rev.* 41: 5459–5472.

Catling, D. C. 2013. *Astrobiology: A Very Short Introduction*, 5–9. Oxford, UK: Oxford University Press.

Chela-Flores, J. 2001. *The New Science of Astrobiology: From Genesis of the Living Cell to Evolution of Intelligent Behaviour in the Universe*. Dordrecht, the Netherlands: Kluwer Academic Publishers.

Chela-Flores, J. 2011. *The Science of Astrobiology: A Personal View on Learning to Read the Book of Life*. Dordrecht, the Netherlands: Springer.

Chyba, C. F., P. J. Thomas, L. Brookshaw, and C. Sagan. 1990. Cometary delivery of organic molecules to the Early Earth. *Science* 249: 336–373.

Clark, B. C. 2004. From Mars and machines, to water and worker bees: Application of a GU DoL to identification of extraterrestrial and artificial life forms. In *Workshop and Tutorial Proceedings, Ninth International Conference on the Simulation and Synthesis of Living Systems (Alife IX)*, Eds. M. Bedau, P. Husbands, T. Hutton, S. Kumar, and H. Suzuki. 96–102. Boston, MA, September 12.

Cleaves, H. J. 2013. Prebiotic chemistry: Geochemical context and reaction screening. *Life* 3: 331–345.

Cleaves, H. J. 2014. Prebiotic synthesis of biochemical compounds: An overview. In *Astrobiology: An Evolutionary Approach*, Ed. V. M. Kolb, 83–117. Boca Raton, FL: CRC Press/Taylor & Francis Group.

Cleland, C. E., and C. F. Chyba. 2007. Does 'life' have a definition? In *Planets and Life: The Emerging Science of Astrobiology*, Eds. W. T. Sullivan III, and J. A. Baross, pp. 119–131. Cambridge, UK: Cambridge University Press.

Conrad, P. G. 2007. Instruments and strategies for detecting extraterrestrial life. In *Planets and Life: The Emerging Science of Astrobiology*, Eds. W. T. Sullivan III, and J. A. Baross, pp. 473–482. Cambridge, UK: Cambridge University Press.

Coveney, P. V., J. B. Swadling, J. A. D. Wattis, and H. C. Greenwell. 2012. Theory, modelling and simulation in origins of life studies. *Chem. Soc. Rev.* 41: 5430–5446.

Cronin, J. P., and S. Pizarrello. 1997. Enantiomeric excesses in meteoritic amino acids. *Science* 275: 951–955.

Crowe, M. J. 1999. *The Extraterrestrial Life Debate*, 1750–1900. Mineola, TX: Dover.

De Duve, C. 2000. Clues from present-day biology: The thioester world. In *The Molecular Origins of Life: Assembling Pieces of the Puzzle*, Ed. A. Brack, pp. 219–236. Cambridge, UK: Cambridge University Press.

Deamer, D. W. 2000. Membrane compartments in prebiotic evolution. In *The Molecular Origins of Life: Assembling Pieces of the Puzzle*, Ed. A. Brack, pp. 189–205. Cambridge, UK: Cambridge University Press.

Deamer, D. W. 2007. The origin of cellular life. In *Planets and Life: The Emerging Science of Astrobiology*, Eds. W. T. Sullivan III, and J. A. Baross, pp. 187–209. Cambridge, UK: Cambridge University Press.

Deamer, D. W., and G. P. Fleischaker, Eds. 1994. *Origins of Life: The Central Concepts*. Boston, MA: Jones & Bartlett.

Des Marais, D. J., J. A. Nuth III, L. J. Allamandola et al. 2008. The NASA astrobiology roadmap. *Astrobiology* 8: 715–730.

Dick, S. J. 1984. *Plurality of Worlds: The Origins of the Extraterrestrial Debate from Democritus to Kant.* Cambridge, UK: Cambridge University Press.

Dick. S. J. 2007. From exobiology to astrobiology. In *Planets and Life: The Emerging Science of Astrobiology*, Eds. W. T. Sullivan III, and J. A. Baross, pp. 46–65. Cambridge, UK: Cambridge University Press.

Dick, S. J. 2012. Critical issues in the history, philosophy, and sociology of astrobiology. *Astrobiology* 12: 906–927.

Dick, S. J. 2015. *The Impact of Discovering Life Beyond Earth.* Cambridge, UK: Cambridge University Press.

Dick, S. J., and Lupisella, M. L., Eds. 2009. *Cosmos & Culture: Cultural Evolution in a Cosmic Context.* NASA SP-2009-4802.

Dick, S. J., and J. E. Strick. 2004. *The Living Universe: NASA and the Development of Astrobiology.* New Brunswick, NJ: Rutgers University Press.

DiGregorio, B. E., G. V. Levin, and P. A. Straat. 1997. *Mars: The Living Planet.* Berkeley, CA: Frog Ltd.

Domagal-Goldman, S. D., K. E. Wright, K. Adamala et al. 2016. The astrobiology primer v2.0. *Astrobiology* 16: 561–653.

Duhem, P. 1987. *Medieval Cosmology: Theories of Infinity, Place, Time, Void, and the Plurality of Worlds.* Transl. Ed. R. Ariew. Chicago, IL: The University of Chicago Press.

Eigen, M. 2013. *From Strange Simplicity to Complex Familiarity: A Treatise on Matter, Information, Life and Thought.* Oxford, UK: Oxford University Press.

Eschenmoser, A. 1999. Chemical etiology of nucleic acid structure. *Science* 284: 2118–2124.

Fenchel, T. 2002. *The Origin and Early Evolution of Life.* Oxford, UK: Oxford University Press.

Fernández-García, C., A. J. Coggins, M. A. Powner. 2017. A chemist's perspective on the role of phosphorus at the origins of life. *Life* 7(3): 31. doi:10.3390/life7030031.

Ferris, J. P. 2000. Catalyzed RNA synthesis for the RNA world. In *The Molecular Origins of Life: Assembling Pieces of the Puzzle*, Ed. A. Brack, pp. 255–268. Cambridge, UK: Cambridge University Press.

Frankel, E. A., D. C. Dewey, and C. D. Keating. 2014. Encapsulation of organic materials in protocells. In *Astrobiology: An Evolutionary Approach*, Ed. V. M. Kolb, pp. 217–255. Boca Raton, FL: CRC Press/Taylor & Francis Group.

Fry, I. 2000. *The Emergence of Life on Earth: A Historical and Scientific Overview.* New Brunswick, NJ: Rutgers University Press.

Gilmour, I., and M. A. Sephton, Ed. 2004. *An Introduction to Astrobiology.* Cambridge, UK: Cambridge University Press.

Goldsmith, D. 1997. *The Hunt for Life on Mars.* New York: Dutton.

Haldane, J. B. S. 1994. The origin of life. In *Origins of Life: The Central Concepts*, Eds. D. W. Deamer, and G. P. Fleischaker, pp. 31–71. Boston, MA: Jones & Bartlett.

Hazen, R. M. 2005. *Genesis: The Scientific Quest for Life's Origin.* Washington, DC: Joseph Henry Press.

Hein, J. E., and D. G. Blackmond. 2012. On the origin of single chirality of amino acids and sugars in biogenesis. *Acc. Chem. Res.* 45: 2045–2054.

Herdewijn, P., and M. V. Kisakürek, Eds. 2008. *Origin of Life: Chemical Approach.* Zürich, Switzerland: VHCA & Weinheim, Wiley-VCH.

Higgs, P. G., and N. Lehman. 2015. The RNA World: Molecular cooperation at the origins of life. *Nat. Rev. Genet.* 16: 7–17.

Holland, J. H. 1995. *Hidden Order: How Adaptation Builds Complexity.* Reading, MA: Helix Books.

Holm, N. G., and E. M. Andersson. 2000. Hydrothermal systems. In *The Molecular Origins of Life: Assembling Pieces of the Puzzle*, Ed. A. Brack, pp. 86–99. Cambridge, UK: Cambridge University Press.

Horneck, G., and R. Moeller. 2014. Microorganisms in space. In *Astrobiology: An Evolutionary Approach*, Ed. V. M. Kolb, pp. 283–299. Boca Raton, FL: CRC Press/Taylor & Francis Group.

Horneck, G., P. Rettberg, N. Walter, and F. Gomez. 2015. European landscape in astrobiology, results of the AstRoMap consultation. *Acta Astronaut.* 110: 145–154.

Horneck, G., N. Walter, F. Westall et al. 2016. AstRoMap European astrobiology roadmap. *Astrobiology* 16: 201–243.

Hug, L. A., J. Baker, B. J., K. Anantharaman et al. 2016. A new view of tree of life. *Nat. Microbiol.* 1: 1–6.

Islam, S., and M. W. Powner. 2017. Prebiotic systems chemistry: Complexity overcoming clutter. *Chemistry* 2: 470–501. doi:10.1016/j.chempr.2017.03.001.

Jalasvuori, M., and J. K. H. Bamford. 2014. Evolutionary approach to viruses in astrobiology. In *Astrobiology: An Evolutionary Approach*, Ed. V. M. Kolb, 413–420. Boca Raton, FL: CRC Press/Taylor & Francis Group.

James, K. D., and A. Ellington. 2000. Catalysis in the RNA world. In *The Molecular Origins of Life: Assembling Pieces of the Puzzle*, Ed. A. Brack, pp. 269–294. Cambridge, UK: Cambridge University Press.

Jones, B. W. 2004. *Life in the Solar System and Beyond.* Heidelberg, Germany: Springer-Verlag & Chichester, Praxis Publishing.

Joyce, G. F. 2012. Bit by bit: The Darwinian basis of life. *PLoS Biol.* 10(5): e1001323. doi:10.1371/journal.pbio.1001323.

Joyce, G. F., and L. E. Orgel. 1993. Prospects for understanding of RNA world. In *The RNA World: The Nature of Modern RNA Suggests a Prebiotic RNA World*, Eds. R. F. Gesteland, and J. F. Atkins. Cold Spring Harbor, NY: Cold Spring Harbor University Press.

Kasting, J. 2010. *How to Find a Habitable Planet.* Princeton, NJ: Princeton University Press.

Kauffman, S. A. 1993. *The Origins of Order: Self-Organization and Selection in Evolution.* Oxford, UK: Oxford University Press.

Kauffman, S. A. 1995. *At Home in the Universe: The Search for the Laws of Self-Organization and Complexity.* Oxford, UK: Oxford University Press.

Kauffman, S. A. 2000. *Investigations.* Oxford, UK: Oxford University Press.

Kolb, V. M. 2010a. On the applicability of dialetheism and philosophy of identity to the definition of life. *Int. J. Astrobiol.* 9: 131–136.

Kolb, V. M. 2010b. On the applicability of the green chemistry principles to sustainability of organic matter on asteroids. *Sustainability* 2: 1624–1631. doi:10.3390/su2061624.

Kolb, V. M. 2012. On the applicability of solventless and solid-state reactions to the meteoritic chemistry. *Int. J. Astrobiol.* 11: 43–50. doi:10.1017/S1473550411000310.

Kolb, V. M., Ed. 2014a. *Astrobiology: An Evolutionary Approach.* Boca Raton, FL: CRC Press/Taylor & Francis Group.

Kolb, V. M. 2014b. Prebiotic chemistry: In water and in the solid state. In *Astrobiology: An Evolutionary Approach*, Ed. V. M. Kolb, pp. 199–216. Boca Raton, FL: CRC Press/Taylor & Francis Group.

Kolb, V. M. 2015. Oparin's coacervates as an important milestone in chemical evolution. In *Instruments, Methods, and Missions for Astrobiology XVII (Proceedings of SPIE)*, Eds. R. B. Hoover, G. V. Levin, A. Y. Rozanov, and N. C. Wickramasinghe. Vol. 960604, Article 9606. doi:10.1117/12.2180604.

Kolb, V. M. 2016. *Green Organic Chemistry and its Interdisciplinary Approaches*, pp. 53–63. Boca Raton, FL: CRC Press/Taylor & Francis Group.

Krishnamurthy, R. 2017. Giving rise to life: Transition from prebiotic chemistry to protobiology. *Acc. Chem. Res.* 50: 455–459.

Kwok, S. 2016. Complex organics in space from solar system to distant galaxies. *Astron. Astrophys. Rev.* 24: 8. doi:10.1007/s00159-016-0093-y.

Kwok, S. 2017. Abiotic synthesis of complex organics in the universe. *Nat. Astron.* 1: 642–643.

Lahav, N. 1999. *Biogenesis: Theories of Life's Origin*. Oxford, UK: Oxford University Press.

Lineweaver, C. H., P. C. W. Davies, and M. Ruse, Eds. 2013. *Complexity and the Arrow of Time*. Cambridge, UK: Cambridge University Press.

Longstaff, A. 2015. *Astrobiology: An Introduction*. Boca Raton, FL: CRC Press/Taylor & Francis Group.

Luisi, P. L. 2006. *The Emergence of Life: From Chemical Origins to Synthetic Biology*. Cambridge, UK: Cambridge University Press.

Lurquin, P. F. 2003. *The Origins of Life and the Universe*. New York: Columbia University Press.

Maciá, E. 2005. The role of phosphorus in chemical evolution. *Chem. Soc. Rev.* 34: 691–701.

Mann, S. 2013. The origins of life: Old problems, new chemistries. *Angew. Chem. Int. Ed.* 52: 155–162.

Mason, S. F. 1991. *Chemical Evolution: Origins of the Elements, Molecules and Living Systems*. Oxford, UK: Clarendon Press.

Mathis, C., S. N, Ramprasad, S. I. Walker, and N. Lehman. 2017. Prebiotic RNA network formation: A taxonomy of molecular cooperation. *Life* 7:38. doi:10.3390/life7040038.

McConnell, B. 2001. *Beyond Contact: A Guide to SETI and Communicating with Alien Civilizations*. Sebastopol, CA: O'Reilly & Associates.

McDonald, G. D. 2014. Biochemical pathways as evidence for prebiotic syntheses. In *Astrobiology: An Evolutionary Approach*, Ed. V. M. Kolb, pp. 119–147. Boca Raton, FL: CRC Press/Taylor & Francis Group.

McSween Jr., H. Y. 1999. *Meteorites and their Parent Planets*. 2nd ed. Cambridge, UK: Cambridge University Press.

Meltzer, M. 2010. *When Biospheres Collide: A History of NASA's Planetary Protection Program*. NASA SP-2011-4243.

Menor-Salván, C., and M. R. Marín-Yaseli. 2012. Prebiotic chemistry in eutectic solutions at the water-ice matrix. *Chem. Soc. Rev.* 41: 5404–5415.

Mesler, B., and H. H. Cleaves II. 2016. *A Brief History of Creation: Science and the Search for the Origin of Life*, 172–185. New York: Norton.

Miller, S. L. 1953. A production of amino acids under possible primitive earth conditions. *Science* 117: 528–529.

Miller, S. L. 2000. The endogenous synthesis of organic compounds. In *The Molecular Origins of Life: Assembling Pieces of the Puzzle*, Ed. A. Brack, pp. 59–85. Cambridge, UK: Cambridge University Press.

Miller, S. L., and L. E. Orgel. 1974. *The Origins of Life on the Earth*. Englewood Cliffs, NJ: Prentice Hall.

Mitchell, M. 2009. *Complexity: A Guided Tour*. Oxford, UK: Oxford University Press.

O'Brien, D. P., A. Izidoro, S. A. Jacobson, S. N. Raymond, and D. C. Rubie. 2018. The delivery of water during terrestrial planet formation. *Space Sci. Rev.* 214(1). doi:10.1007/s11214-018-0475-8. Article 47.

Oparin, A. I. 1965. *Origin of Life*. Transl. S. Morgulis. 2nd ed. New York: Dover Publications.

Oparin, A. I. 1968. *Genesis and Evolutionary Development of Life*. Transl. E. Maass. New York: Academic Press.

Oparin, A. I. 1994. The origin of life. Transl. A. Synge. In *Origins of Life: The Central Concepts*, Eds. D. W. Deamer, and G. P. Fleischaker, pp. 31–71. Boston, MA: Jones & Bartlett.

Orgel, L. E. 2004. Prebiotic chemistry and the origin of the RNA world. *Crit. Rev. Biochem. Mol. Biol.* 9: 99–123.

Padgett, J. F. 2012. Autocatalysis in chemistry and the origin of life. In *The Emergence and Organizations and Markets*, Eds. J. F. Padgett, and W. W. Powell, 33–69. Princeton, NJ: Princeton University Press.

Pasek, M. 2014. Role of phosphorus in prebiotic chemistry. In *Astrobiology: An Evolutionary Approach*, Ed. V. M. Kolb, pp. 257–270. Boca Raton, FL: CRC Press/Taylor & Francis Group.

Pasek, M. A., M. Gull, and B. Herschy. 2017. Phosphorylation on the early earth. *Chem. Geol.* 475: 149–170. doi:10.1016/j.chemgeo.2017.11.008.

Pasek, M. A., and D. S. Lauretta. 2005. Aqueous corrosion of phosphide minerals from iron meteorites: A highly reactive source of prebiotic phosphorus on the surface of the early Earth. *Astrobiology* 5: 515–535.

Pereto, J. 2005. Controversies on the origin of life. *Int. Microbiol.* 8: 23–31.

Pereto, J. 2012. Out of fuzzy chemistry: From prebiotic chemistry to metabolic networks. *Chem. Soc. Rev.* 41: 5394–5403.

Perry, R. S., and V. M. Kolb. 2004. On the applicability of Darwinian principles to chemical evolution that led to life. *Int. J. Astrobiol.* 3: 45–53.

Pizzarello, S. 2008. The chemistry that preceded life's origin: A study guide from meteorites. In *Origin of Life: Chemical Approach*, Eds. P. Herdewijn, and M. V. Kisakürek, pp. 231–243. Zürich, Switzerland: VHCA & Weinheim, Wiley-VCH.

Popa, R. 2014. Elusive definition of life: A survey of main ideas. In *Astrobiology: An Evolutionary Approach*, Ed. V. M. Kolb, pp. 325–348. Boca Raton, FL: CRC Press/Taylor & Francis Group.

Powner, M. W., J. D. Sutherland, and J. W. Szostak. 2011. The origins of nucleotides. *SYNLETT* 14: 1956–1964.

Pressman, A., C. Blanco, and I. A. Chen. 2015. The RNA world as a model system to study the origin of life. *Curr. Biol.* 25: R953–R963.

Pross, A. 2003. The driving force for life's emergence: Kinetic and thermodynamic considerations. *J. Theor. Biol.* 220: 393–406.

Pross, A. 2012. *What is Life? How Chemistry Becomes Biology*. Oxford, UK: Oxford University Press.

Race, M. S. 2007. Societal and ethical concerns. In *Planets and Life: The Emerging Science of Astrobiology*, Eds. W. T. Sullivan III, and J. A. Baross, pp. 483–497. Cambridge, UK: Cambridge University Press.

Regis Jr., E., Ed. 1987. *Extraterrestrials: Science and Alien Intelligence*. Cambridge, UK: Cambridge University Press.

Repko, A. F. 2012. *Interdisciplinary Research: Process and Theory*. 2nd ed. pp. 20–22, 73, 94–96. Thousand Oaks, CA: Sage.

Robertson, M. P., and G. F. Joyce. 2012. The origins of the RNA world. *Cold Spring Harb. Perspect. Biol.* 4: a003608.

Ruiz-Mirazo, K., C. Briones, and A. de la Escosura. 2014. Prebiotic systems chemistry: New perspectives for the origins of life. *Chem. Rev.* 114: 285–366.

Ruiz-Mirazo, K., J. Pereto, and A. Moreno. 2004. A universal definition of life: Autonomy and open-ended evolution. *Orig. Life Evol. Biosph.* 34: 323–346.

Rummel, J. D. 2007. Planetary protection: Microbial tourism and sample return. In *Planets and Life: The Emerging Science of Astrobiology*, Eds. W. T. Sullivan III, and J. A. Baross, pp. 498–512. Cambridge, UK: Cambridge University Press.

Ryan, F. 2002. *Darwin's Blind Spot: Evolution beyond Natural Selection*. New York: Houghton Mifflin.

Scharf, C., N. Virgo, H. J. Cleaves II et al. 2015. A strategy for origins of life research. *Astrobiology* 15: 1031–1042.

Schmitt-Kopplin, P., Z. Gabelica, R. D. Gougeon et al. 2010. High molecular diversity of extraterrestrial organic matter in Murchison meteorite revealed 40 years after its fall. *Proc. Natl. Acad. Sci. U.S.A.* 107: 2763–2768.

Schmitt-Kopplin, P., M. Harir, B. Kanawati et al. 2014. Analysis of extraterrestrial organic matter in Murchison meteorite: A progress report. In *Astrobiology: An Evolutionary Approach*, Ed. V. M. Kolb, pp. 63–82. Boca Raton, FL: CRC Press/ Taylor & Francis Group.

Schopf, J. W. 1999. *Cradle of Life: The Discovery of Earth's Earliest Fossils*. Princeton, NJ: Princeton University Press.

Schopf, W. J. Ed. 2002. *Life's Origin: The Beginning of Biological Evolution*. Berkeley, CA: University of California Press.

Schwartz, A. W. 2006. Phosphorus in prebiotic chemistry. *Philos. Trans. R. Soc. Lond. B Biol. Sci.* 361: 1743–1749.

Seckbach, J., A. Oren, and H. Stan-Lotter, Eds. 2013. *Polyextremophiles: Life Under Multiple Forms of Stress*. Dordrecht, the Netherlands: Springer.

Sephton, M. A. 2004. In *An Introduction to Astrobiology*, Eds. I. Gilmour, and M. A. Sephton, pp. 1–41. Cambridge, UK: Cambridge University Press.

Shaw, A. M. 2006. *Astrochemistry: From Astronomy to Astrobiology*. Chichester, UK: John Wiley & Sons.

Soai, K., T. Shibata, and I. Sato. 2000. Enantioselective automultiplication of chiral molecules by asymmetric autocatalysis. *Acc. Chem. Res.* 33: 382–390.

Sojo, V., B. Herschy, A. Whicher, E. Camprubi, and N. Lane. 2016. The origin of life in alkaline hydrothermal vents. *Astrobiology* 16: 181–197.

Soucy, S. M., J. Huang, and J. P. Gogarten. 2015. Horizontal gene transfer: Building the web of life. *Nat. Rev. Genet.* 16: 472–482.

Sullivan III, W. T., and J. A. Barross, Eds. 2007. *Planets and Life: The Emerging Science of Astrobiology*. Cambridge, UK: Cambridge University Press.

Sullivan III, W. T., and D. Carney. 2007. History of astrobiological ideas. In *Planets and Life: The Emerging Science of Astrobiology*, Eds. W. T. Sullivan III, and J. A. Baross, pp. 9–45. Cambridge, UK: Cambridge University Press.

Summers, M., and J. Trefil. 2017. *Exoplanets: Diamond Worlds, Super Earths, Pulsar Planets, and the New Search for Life beyond Our Solar System*. Washington, DC: Smithsonian Books.

Sutherland, J. D. 2017. Studies on the origin of life – The end of the beginning. *Nat. Rev. Chem.* 1 (2017). doi:10.1038/s41470-0762. Article number 0012.

Tarter, J. C. 2007. Searching for extraterrestrial intelligence. In *Planets and Life: The Emerging Science of Astrobiology*, Eds. W. T. Sullivan III, and J. A. Baross, pp. 513–526. Cambridge, UK: Cambridge University Press.

Vakoch, D. A., and A. A. Harrison, Eds. 2013. *Civilizations beyond Earth: Extraterrestrial Life and Society*. New York: Berghahn Books.

Villarreal, L. P. 2004. Are viruses alive? *Sci. Am.* (December): 291(6): 100–105.

Villarreal, L. P. 2005. *Viruses and the Evolution of Life*. Washington, DC: ASM Press.

Villarreal. L. P. 2014. Virolution can help us understand the origin of life. In *Astrobiology: An Evolutionary Approach*, Ed. V. M. Kolb, pp. 421–440. Boca Raton, FL: CRC Press/Taylor & Francis Group.

Wächtershäuser, G. 2000. Origin of life in an iron-sulfur world. In *The Molecular Origins of Life: Assembling Pieces of the Puzzle*, Ed. A. Brack, pp. 206–218. Cambridge, UK: Cambridge University Press.

Wächtershäuser, G. 2008. On the chemistry and evolution of the pioneer organism. In *Origin of Life: Chemical Approach*, Eds. P. Herdewijn, and M. V. Kisakürek, pp. 61–79. Zürich, Switzerland: VHCA & Weinheim, Wiley-VCH.

Waldrop, M. M. 1992. *Complexity: The Emerging Science at the Edge of Order and Chaos*. New York: Touchstone.

Walker, S. I. 2014. Transition from abiotic to biotic: Is there an algorithm for it? In *Astrobiology: An Evolutionary Approach*, Ed. V. M. Kolb, pp. 371–397. Boca Raton, FL: CRC Press/Taylor & Francis Group.

Ward, P. D., and D. Brownlee. 2002. *The Life and Death of Planet Earth: How the New Science of Astrobiology Charts the Ultimate Fate of Our World*. New York: Henry Holt.

Ward, P. D., and D. Brownlee. 2004. *Rare Earth: Why Complex Life is Uncommon in the Universe*. New York: Copernicus Books.

Westall, F., and C. S. Cockell. 2016. Biosignatures for astrobiology. *Orig. Life Evol. Biosph.* 46: 105–106.

Westall, F., F. Foucher, N. Bost et al. 2015. Biosignatures on Mars: What, where, and how? Implications for the search for Martian life. *Astrobiology* 15: 998–1029.

Zubay, G. 2000. *Origins of Life on the Earth and the Cosmos*. 2nd ed. San Diego, CA: Harcourt Academic Press.

Zuckerman, B., and M. H. Hart, Eds. 1995. *Extraterrestrials: Where are They?* 2nd ed. Cambridge, UK: Cambridge University Press.

1.2 Astrobiology Goals
NASA Strategy and European Roadmap

David J. Des Marais

CONTENTS

1.2.1 INTRODUCTION

As stated in Chapter 1.1, astrobiology seeks to understand the origins, evolution, distribution, and future of life in the universe. Because this compelling charter statement is also quite ambitious, a framework is needed to identify key concepts and advocate priorities. An astrobiology roadmap or strategy provides this framework.

Figure 1.2.1 illustrates one example of an area of investigation that is centrally important to astrobiology and also requires a combination of expertise in multiple disciplines.

This figure illustrates that four complementary approaches are being pursued in order to understand the origins of life on Earth and beyond. One can envision a timeline that led up to, included, and followed an origin of life event. Two approaches investigate forward in time and address the processes and

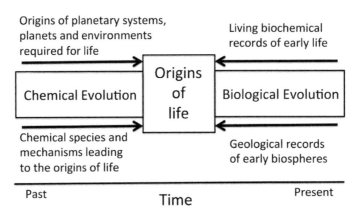

FIGURE 1.2.1 Four disciplinary approaches to understanding the origins of life.

materials that set the stage for life. The fields of astronomy, astrophysics, and planetary science investigate the cosmic origins and distribution of key elements and compounds, the formation of planetary systems, and the maintenance of planetary environments conducive to life. The fields of chemistry, geochemistry, and molecular biology address the environmental conditions and processes leading to an origin of life. Two additional approaches investigate back in time from the present to discern the history of life. Geologists investigate ancient rock records to characterize habitable environments and to identify any "signatures of life" or "biosignatures." Molecular biologists investigate the biochemistry of extant biota in order to trace evolutionary events back in time to infer the earliest attributes of living systems. Astrobiologists strive to identify effective strategies for each of these approaches.

The broad scope of astrobiology also necessitates advances in both technology and access (i.e., missions) to environments beyond Earth. Accordingly, space agencies have supported research in astrobiology that, in turn, informs technology development and missions. A roadmap should highlight research, technologies, and missions that are not just relevant to astrobiology but are also either currently possible or achievable in a realistic near-term time frame. Accordingly, the goals and objectives of the NASA Astrobiology Strategy and the European Roadmap are necessarily also influenced by the opportunities and constraints imposed by the government agencies and programs from which astrobiology research and exploration derive support.

1.2.2 RESEARCH TOPICS AND THEMES

NASA's 2015 Astrobiology Strategy (abbreviated as "NAS" herein; Hays et al., 2015), the European Space Agency (ESA's) 2016 Astrobiology Roadmap (or "EAR"; Horneck et al., 2016), and NASA's 2008 Astrobiology Roadmap (or "NAR"; Des Marais et al., 2008) articulated Themes (NAS), Research Topics (EAR), and Goals (NAR), respectively. Each of these documents delineated a sequence of investigations that addresses the progression of developments leading to the origins of life and its subsequent evolution. The six sections below summarize this sequence, starting with the cosmic origins of prebiotic compounds and planetary systems, then progressing through

the development of habitable planetary environments and biospheres, and ultimately to more complex life. Each section cites the particular Theme, Research Topic, and Goal that address the topic of that section. For example, Section 1.2.3 draws its content from NAS Theme 1, EAR Research Topic 2, and NAR Goal 3.

The intent of this chapter is to provide a concise yet moderately detailed guide to the NAS and EAR. These documents are substantial both in length and in detail—in published form, the NAS and EAR are on 215 and 43 pages, respectively. These documents introduce each of their themes or research topics, summarize recent progress on each, articulate the research objectives, and finally outline approaches toward achieving those objectives. This chapter summarizes principally the research objectives recommended by the NAS and EAR. The NAR is not summarized in detail, as it is an earlier document and its content is substantively captured by the NAS and EAR.

1.2.3 NONBIOLOGICAL ORIGINS OF ORGANIC COMPOUNDS IN SPACE AND IN PLANETARY ENVIRONMENTS (NAS THEME 1, EAR RESEARCH TOPIC 2, AND NAR GOAL 3)

In order to understand how life can begin, it is essential to know what organic compounds might have been available and how they interacted within planetary environments. Any chemical syntheses occurring in the solid crust, hydrosphere, and atmosphere are likely to be key sources of relevant compounds. Interesting organic compounds also occur in interstellar clouds, which are birthplaces of planetary systems. Laboratory simulations of interstellar ices have demonstrated that key prebiotic molecules can be synthesized under conditions occurring in interstellar clouds, and such materials can contribute to nascent planetary systems. Many compounds relevant to biological processes occur in meteorites, interplanetary dust particles, and comets. Substantial inventories of organic matter probably were delivered to Earth during late accretion, providing feedstock for chemical evolution and early life. So, it is important to establish sources of prebiotic organic compounds and to understand their histories in terms of the processes that occur on a newly formed planet. Investigations should include pan-spectral astronomical observations, sample return missions, laboratory studies of extraterrestrial materials, and realistic laboratory simulations of inaccessible cosmic environments.

1.2.3.1 NAS Theme 1: Identifying Abiotic Sources of Organic Compounds

Sources, activities, and fates of organic compounds on the prebiotic Earth. This topic addressed the relationships between monomers utilized by living organisms and prebiotically relevant molecules in a range of environments. Relevant aspects include identifying plausible sources of compounds consisting of the elements C, H, N, O, P, and S, as well as determining the properties, formation pathways, and parent environments of molecules that were utilized by the first life forms.

Role of the environment in the production of organic molecules. This topic addressed various conceivable nonbiological sources of organic matter in planetary environments. This includes the primary mechanisms for organic synthesis within planetary environments, the delivery of organics from space, the relative importance of these two sources, and the roles played by different energy sources in determining the nature of the organic matter produced.

Role of the environment on the stability and accumulation of organic molecules. The survival and modification of small organic molecules in various environments are also important. Key aspects of this include the role of processes in environments that host organic synthesis and the survivability of organics from various sources during delivery through a range of atmospheric compositions. It is also important to understand how enantiomeric excesses arose at the monomer level and what processes were important for determining biological chirality.

Constraints that the rock record places on the environments and abiotic reactions of the early Earth. Earth's earliest environments probably played crucial roles in shaping the course of prebiotic evolution. Therefore, the nature of Earth's early atmosphere, oceans, and crust should be determined. The bombardment history of Earth is probably an important aspect of this. Accordingly, the search for additional rocks of the Archean and older ages should continue in order to characterize early environments.

1.2.3.2 EAR Research Topic 2: Origins of Organic Compounds in Space

Understand the diversity and the complexity of abiotic organics. The EAR emphasizes the need to study the mechanisms for the formation of organics and their evolution under space conditions. It advocates research to understand the role of catalysts associated with the formation of abiotic organic matter in a variety of environments.

Understand better the molecular evolution of abiotic organics present in solar system objects. Research should address the combined role of physical agents such as thermal variations, high-energy particles, photons, and solar wind irradiation. This effort should include organics on the early Earth.

Understand the role of spontaneous inorganic (organic) self-organization processes in molecular evolution. Investigations should identify and characterize self-organized inorganic and organic systems and determine their potential roles in the prebiotic synthesis of biomolecules.

1.2.4 ORIGIN AND EVOLUTION OF PLANETARY SYSTEMS AND HABITABLE ENVIRONMENTS (NAS THEME 6, EAR RESEARCH TOPICS 1 AND 4, AND NAR GOALS 1 AND 2)

A planet or planetary satellite is habitable if it sustains life that originates there or if it sustains life that is carried to it. The goal is to understand the most fundamental environmental requirements for an environment to be habitable. As currently envisioned, an environment must provide chemical building blocks (nutrients), extended regions of liquid water, energy sources to sustain metabolism, and conditions that favor the assembly and maintenance of complex organic molecules. A major long-range goal is to characterize habitable environments beyond the solar system, independent of the presence of life, or to recognize habitability by detecting the presence of life. Terrestrial planets and large satellites might form in a state where they are likely to become habitable. Alternatively, habitable environments might emerge only after a sequence of less probable events. How frequently do habitable environments arise on solid planets, including large satellites? What are the specific indicators of habitability and habitation, and how might such indicators differ between planets having different attributes (e.g., mass, distance from the star, history, and relative abundances of volatile compounds)?

1.2.4.1 NAS Theme 6: Constructing Habitable Worlds

Fundamental ingredients and processes of a planet that can define a habitable environment. These attributes include the following: 1) size of the planet; 2) inventory, distribution, and role of critical elements and minerals within the planet's interior: radiogenic nuclides, water, and the elements C, H, N, O, P, and S; 3) core formation (e.g., separation of Fe/H, Mg/Si, C/O, magnetic field, and siderophile element distributions); 4) geophysical evolution of the interior (e.g., heat budget, rate and manner of mantle-surface exchange, redistribution of bio-relevant compounds, and energy sources); 5) energy from radiogenic, accretionary, tidal, differentiation, and chemical processes; 6) crustal processes and their evolution (e.g., volcanism, plate tectonics, erosion, sedimentation, geochemical reactions and cycles, and atmospheric effects); and 7) plausible energy sources for prebiotic processes.

Exogenic factors in the formation of a habitable planet. These include the following: 1) initial inventory and distribution of volatiles and the mass and oxidation state of the protoplanetary disk; 2) disk processes (e.g., accretion, migration, and stellar activity); 3) properties of the host star; 4) time when the stellar environment permits the formation and preservation of stable environments (ocean, atmosphere, and shielding from solar radiation); 5) abundances of elements and molecules in the protoplanetary disk (e.g., Mg/Si ratio, C/O ratio, Fe/H ratio, and radiogenic isotopes); 6) dynamical environment of planets that are conducive to habitability (system architecture, obliquity, tidal and resonant effects, spin-orbit coupling, satellite systems, impactors, primary and late accretion, bombardment(s), and volatile delivery); 7) processing of water and organic molecules from interstellar clouds through disks to planet formation; 8) chemical evolution of protoplanetary disk interiors, including planet formation (traced by applying spectroscopic methods and identifying molecules); and 9) accretion of terrestrial planets (investigated by applying exoplanet census data to constrain accretion models).

Insights from Earth about general properties of habitability (and what aspects might be missing). Earth's rock record provides some clues about the conditions and processes present when life emerged (e.g., conditions and composition of oceans, atmosphere, and near-subsurface; redox state; abundance and diversity of life across various niches; and tectonic processes). The availability of bio-relevant compounds is influenced by processes associated with the (bio) geochemical cycles of elements that are key for life. These include processes that affected the composition of the atmosphere over time and the roles played by an intrinsic magnetic field in maintaining habitable conditions.

Processes on other types of planets that could create habitable niches. Relevant factors include the following: 1) how processes mimic those on Earth; 2) effects of very large or very sparse atmospheres; 3) the nature of hydrospheres on icy, water-rich, or water-poor planets; 4) processes of heat flow and mantle-crust exchange of materials; 5) effects of alternative styles of rock-atmosphere-ocean cycles; 6) potential reservoirs and processing of prebiotic, bio-relevant, or biological materials; 7) formation

of volatile-rich/potentially habitable moons orbiting giant planets; and 8) how to measure habitability indicators and quantify the relative uncertainties in the observations.

Changes in habitability through time. These include the following parameters: 1) volatile inventory in the aftermath of planet formation and formation-like impact events; 2) evolution and variation of the host star; 3) how long a favorable environment must persist to foster the emergence of life; 4) climate stability, internal heat budget, geochemical inventories and cycles, and tectonic regimes; 5) effects of climate stability, heat flows, chemical inventory, and tectonic regimes on the persistence of habitable environments; 6) effects of orbital dynamics over time (e.g., spin-orbit coupling, stabilization from satellites, changing semi-major axis, and eccentricity); and 7) movement of the host star through the galaxy.

1.2.4.2 EAR Research Topic 1: Origin and Evolution of Planetary Systems

The EAR identified objectives that are very similar to those advocated by the NAS.

Assess the elemental and chemical picture of protoplanetary stellar discs. 1) Understand the metallicity of stars; 2) improve chemical models of protoplanetary disc formation and evolution; 3) improve our understanding of the evolution of circumstellar discs, in relation to their host stars; 4) determine the chemical history of key molecules (such as water and oxygen) in the evolution from molecular clouds to star-planet(s) system; and 5) link chemical processes to disc hydrodynamics and structure.

Understand our solar system better: planet formation, dynamical evolution, and water/organics delivery to Earth and to the other planets/satellites. 1) Understand the transition from planetesimals to planets and satellites (end to end); 2) understand the dynamical evolution of the "young" solar system; 3) improve models of conditions for the survival and/or generation of essential molecules during impacts; 4) identify dynamical processes that can redistribute essential material throughout a system; 5) better understand the effects of post-formation bombardment episodes on Earth and other planetary bodies generally assumed to have been important for the development of life; 6) define the timeline of the formation of the solar system and water/organic delivery on Earth; and 7) interpret the temporal link between solar system evolution and the rise of life on Earth.

Understand better the diversity of exoplanetary systems and the development of habitable environments. 1) Understand the dynamical mechanisms that lead to the observed diversity of exoplanetary architecture, and assess how they affect habitability; 2) identify biomarkers and promising methods of detection; and 3) determine how the study of exoplanets can help to fill the gaps in our understanding of the formation of our own solar system.

1.2.4.3 EAR RESEARCH TOPIC 4: LIFE AND HABITABILITY

The following aspects of this topic address the attributes and exploration of habitable environments: 1) Expand our knowledge of the diversity, adaptability, and boundary conditions of life on Earth. The quest to identify the ultimate boundaries of habitable environments starts with determining the ultimate environmental limits of life as we know it. This knowledge serves as a starting point for defining the fundamental attributes of habitable environments and how to identify them beyond Earth; and 2) assess the habitability of extraterrestrial environments. Mars, icy moons of Jupiter and Saturn, and exoplanets are key exploration targets.

1.2.5 ORIGINS OF LIFE (NAS THEME 2, EAR RESEARCH TOPIC 3, AND NAR GOAL 3)

The origin(s) of life on Earth likely represents one path among many, along which life might emerge. This view forms the foundation for pursuing observations and missions to search for extant or extinct life elsewhere in the universe. Astrobiologists must strive to understand the universal principles that underlie not only the origins of life on Earth but also its possible origins elsewhere. A starting point for identifying these principles is to determine the raw materials of life that can arise from chemical evolution in space and on planets (Section 1.2.3). Then, the mechanisms should be identified by which organic compounds are assembled into more complex molecular structures and how the functions of these structures are coordinated in order to form complex evolving systems that are such a key aspect of life. These systems must be able to perform and coordinate the capture of energy and nutrients from the environment, catalyze the chemical reactions that maintain and grow these systems, and synthesize copies of key biomolecules. Evidence both from biomolecular and fossil records and from diverse extant microorganisms should be explored for clues that can identify the fundamental properties of the living state.

1.2.5.1 NAS THEME 2: SYNTHESIS AND FUNCTION OF MACROMOLECULES IN THE ORIGIN OF LIFE

The following questions address paths that can lead to our DNA/RNA/protein-dominated world.

What is the chemistry of macromolecular formation reactions? The nature and sources of any chemical coupling agents, along with the sources of energy and pathways of energy utilization for enabling these agents, should be identified. It is important to understand how their stability and activity can be maintained in aqueous environments, despite their probable chemical instability in water. The mechanisms whereby high effective concentrations of monomers can facilitate macromolecular formation should be identified. The potential role of noncovalent molecular assemblages in the formation of covalently bonded macromolecules merits attention.

How do information transmission and chemical evolution occur? Plausible chemical mechanisms for information transmission and mutation should be identified. This includes identifying molecules that can be replicated with errors, where the errors themselves are replicable. Perhaps, nontemplated production of informational macromolecules is possible. The physical parameters of pre-ribonucleic acids (RNAs) and preproteins that influence kinetics and the accuracy of replication should be investigated.

What are the chemical alternatives? How and why do they occur? If plausible alternative backbones and side chains exist for functional and informational macromolecules, their properties should be identified. Perhaps, homochirality arose either during monomer synthesis or during polymerization/assembly. The degree of homochirality that is necessary or optimal should be identified.

What is the role of the environment? Conceivably, the mechanisms of polymerization in modern biology (primarily condensation polymerization and aldol reactions) offer clues about the environment in which the polymers evolved. For example, perhaps, oligomerization was driven by cyclic conditions (e.g., water activity, temperature, tidal, and convection). Perhaps, nonaqueous solvents and environments (formamide, hydrocarbons, deep eutectic solvents, micelles, etc.) played key roles in condensation polymerization. Understanding how nonstandard solvents and environments (formamide, hydrocarbons, deep eutectic solvents, micelles, etc.) affect the activity of functional macromolecules might offer key insights.

Macromolecular function: how did physicochemical effects develop over time? The following key aspects merit particular attention: 1) roles that the solubility, solvation, and hydrophobicity of polymers play in the origin of function, 2) types of rudimentary functions that might promote polymer persistence, assembly, and growth, 3) folding properties of macromolecules that are sufficient for function to arise, 4) molecular properties (such as specific binding and catalysis) that are optimal for function, 5) other chemical linkages that exhibit the simultaneous thermodynamic instability and kinetic stability that characterize phosphate diesters.

What are the advanced steps of macromolecular function? The functions that are most likely to emerge early should be identified. Also relevant is the range of chemical functions, besides the ones that contemporary biology proteins and/or RNA could accomplish. Regarding "alternative biochemistries," the biological and protobiological functions of non-natural polymers should be explored.

What led to macromolecular complexity? Key aspects include the following: 1) how polypeptides and polynucleotides began to cooperate; 2) how the genetic code arose; and 3) the forces that selected for regulatory processes in the prebiotic world.

1.2.5.2 EAR RESEARCH TOPIC 3: ROCK-WATER-CARBON INTERACTIONS, ORGANIC SYNTHESIS ON EARTH, AND STEPS TO LIFE

The EAR emphasized understanding key aspects of the cosmic, geological, and geochemical processes that shaped prebiotic planetary environments. The following factors should be better characterized: 1) dynamic redox interactions of rock, water, and carbon in their geological context on planets and moons; 2) transition metals as electron sources and catalysts in geo-organic chemistry; 3) carbon reduction in modern serpentinizing hydrothermal vents; 4) hydrothermal modification of carbon delivered to Earth from space; and 5) role of molecular self-organization, higher-order organization, and cellular organization in the origin of life.

1.2.6 THE LIVING RECORD OF EVOLUTIONARY MECHANISMS AND THE ENVIRONMENTAL LIMITS OF LIFE (NAS THEME 3, EAR RESEARCH TOPIC 4, AND NAR GOAL 5)

Life, as we know it today, is an outcome of interactions between genetic opportunities, metabolic capabilities, and environmental changes. The metabolism and evolution of microorganisms have substantially influenced Earth's habitable environments for most of their existence. Geological, climatologic, and microbial processes have interacted across geological timescales to change the physical-chemical environments on Earth that influenced the path of biological evolution. For example, fundamental environmental changes at the global scale were triggered by the production of molecular oxygen by cyanobacterial photosynthesis and by the colonization of Earth's surface by metazoan life. These changes, in turn, created novel evolutionary opportunities for life, leading ultimately to the rise of extant plants and animals. The "co-evolution" of organisms and their environment has been an intrinsic feature of our biosphere.

1.2.6.1 NAS THEME 3: EARLY LIFE AND INCREASING COMPLEXITY

Theoretical considerations regarding the origin and dynamics of evolutionary processes in living systems. Theoretical work has highlighted the following perspectives that merit attention: 1) "Environmental conditions and genetic context of a genotype can heavily influence its phenotype but their interactions are unclear and unpredictable but absolutely crucial for understanding evolution and the emergence of innovations"; 2) biology can innovate in myriad ways, but we do not yet understand how to manipulate or analyze this crucial capability; 3) interacting networks (e.g., molecular, genetic, epistasis, metabolism, and ecosystems) are central to life, but we know relatively little about their interactions; 4) theoreticians have extensively addressed evolutionary landscapes, but comparatively, little experimental work has been done because of technical difficulties; 5) how mechanisms of selection in evolution change as well as interact across increasing levels of biological complexity (e.g., biomolecules, individuals, communities, and ecosystems); 6) physical, chemical, and developmental constraints on selection in the evolution of developmental complexity (e.g., in multicellular life forms); 7) the extent to which "life [is] a generic outcome of the laws of physics, an inevitable planetary phenomenon if certain conditions of disequilibrium are met."

Fundamental innovations in earliest life. This part of NAS Theme 3 addressed aspects of life's origins that were also identified by Theme 2. For example, it is important to determine how the essential traits of life arise from the geochemical environment, which factors are considered essential for the most basic living systems, and what were the history and order of biological innovations associated with the emergence of life. The following paragraphs summarize topics that were not addressed by Theme 2.

Genomic, metabolic, and ecological attributes of life at the root of the evolutionary tree. This part explores several implications that arise from the observation that present-day organisms share several fundamental commonalities. This implies that life, as we know it, evolved from a "Last Universal Common Ancestor," abbreviated as "LUCA." These commonalities include cellularity and many of the genes, proteins, and biological functions shared by all modern lineages. Most likely, LUCA consisted of many different organisms or populations that readily exchanged their cellular attributes. Understanding such populations might help us understand fundamental aspects of life elsewhere in the universe. The following topics address the evolutionary history and ecological context of LUCA: 1) the amount of time and evolution that occurred between the origin of life and LUCA; 2) the possibility that genomic, proteomic, or metabolic attributes of LUCA might help to elucidate earlier stages of evolution; 3) whether LUCA represents a single individual, species, or population of species; 4) the roles played by horizontal gene transfer in the development of LUCA; and 5) the selection factors that led from LUCA to develop discrete lineages and their evolution.

The LUCA concept also helps to highlight the following objectives: 1) identify the gene families that had evolved by the time of LUCA and the order in which these earliest functions emerged; 2) determine the metabolic networks associated with LUCA and how they might help to constrain the early geochemical environment; 3) determine the nature of early cellular membranes; 4) identify the genetic material of LUCA; and 5) determine the timing of LUCA with respect to the geological and geochemical records.

Dynamics of the subsequent evolution of life. Biological innovations have been mediated by changes in both their biological and physical environments; thus, it is useful to address both intrinsic and extrinsic drivers of evolutionary dynamics. Key intrinsic drivers that merit attention include the following: 1) the evolution of complexity (e.g., advent of chemotrophy, autotrophy, intracellular structure, regulation of internal conditions, and multicellularity); 2) energy acquisition in living systems; and 3) in particular, the origin and early evolution of phototrophy, especially oxygenic photosynthesis, which profoundly affected subsequent evolution and the global environment. Key extrinsic drivers influencing evolutionary dynamics include the following: 1) environmental gradients (temperature, moisture, redox potential, light spectra and intensity, nutrients, etc.) and how those gradients have developed through time; and 2) effects of environmental perturbations on extinctions, biodiversity, and innovations. Of course, many evolutionary events (e.g., the Cambrian "explosion") probably reflect the combined effects of both intrinsic and extrinsic factors.

Common attributes of living systems on Earth. Key research questions include the following: 1) understanding the extent to which common attributes of life on Earth arise from common ancestry (homology) or indicate limited solutions to common problems (convergence); and 2) identifying those attributes common to all life on Earth that are likely to arise elsewhere in our solar system and beyond.

1.2.6.2   EAR RESEARCH TOPIC 4: LIFE AND HABITABILITY

This topic addressed the following key factors that were also identified by NAS Theme 3 (summarized earlier): 1) expand our knowledge of the diversity and adaptability of life on Earth; 2) explore the diversity of life on Earth; 3) explore biological interactions and systems ecology; and 4) expand our understanding of the general principles of life and habitability.

1.2.7   GEOLOGICAL RECORD OF THE EVOLUTION OF OUR BIOSPHERE (NAS THEME 4, EAR RESEARCH TOPIC 4, AND NAR GOAL 4)

Environmental changes have influenced biological evolution, which, in turn, has profoundly altered local and global environments over time. Life interacts with its environment on all spatial scales. Insights into the early evolution of life in a global context can emerge both by documenting the effects of microbial metabolism in the form of proxy geochemical records in ancient rocks and by interpreting the preserved biosignatures of microorganisms. Major transitions in the physical state of Earth that have influenced its biosphere included changes in volcanism, plate tectonics and continents, impacts, and climatic transitions and extremes such as global-scale glaciations. Major extinctions, which were triggered by geological and biological processes and by cosmic processes in some cases, also profoundly affected the evolution of the biosphere. An improved understanding of these interactions will enhance our efforts to detect and characterize any habitable environments and life beyond Earth.

1.2.7.1   NAS THEME 4: CO-EVOLUTION OF LIFE AND THE PHYSICAL ENVIRONMENT

Earth history can indicate how climates, atmospheric compositions, interiors, and biospheres can co-evolve. Interactions between these components can be explored by addressing the following aspects: 1) how interactions between life and its environment affect Earth's ability to retain habitability; 2) how processes in Earth's interior and their surface manifestations have shaped habitable environments and how life has affected these physical processes; 3) how climate and atmospheric processes have shaped habitability and how life has affected these physical relationships; and 4) how studies of Earth's deep history as well as numerical modeling can inform our understanding of future environmental changes.

Characterizing interactions between life and its local environment could inform our understanding of biological and geochemical co-evolutionary dynamics. This topic should be investigated as follows: 1) document how geochemical variations in space and time have affected biological diversity; 2) understand how ecosystems have been structured through time; 3) determine how changing patterns of life through time have impacted the environment; and 4) devise strategies for measuring the co-impacts of life and the environment.

Determining how our current ignorance about microbial life on Earth hinders our understanding of the limits of life. Objectives are as follows: 1) reduce biases associated with current methodologies and database limitations; 2) identify, to a greater extent, the physiological characteristics of uncultured organisms; 3) link newly discovered genomes to the functional roles the microbes play in a community and/or environment; and 4) elucidate catabolic strategies that might be employed by organisms in largely unexplored environments.

1.2.7.2 EAR RESEARCH TOPIC 4: LIFE AND HABITABILITY

This research topic endorses the objectives identified by NAS Theme 4.

Expand our knowledge of the diversity, adaptability, and boundary conditions of life on Earth. Document more thoroughly the diversity of life on Earth in order "to improve our knowledge of the origin, diversity, and limits of life and the habitability of different environments on Earth and elsewhere." Explore biological interactions and systems ecology.

Expand our understanding of the general principles of life and habitability. Characterizing interactions within ecosystems in both extreme and nonextreme environments is essential for understanding the evolution and survival of life as a planetary phenomenon.

1.2.8 SIGNATURES OF HABITABLE ENVIRONMENTS AND LIFE IN THE SOLAR SYSTEM AND BEYOND (NAS THEME 5, EAR RESEARCH TOPIC 5, AND NAR GOAL 7)

The premise that habitable environments and biosignatures will be recognizable within their planetary contexts is a fundamental principle of astrobiology. Localities that exhibit evidence of past or present habitable environments are promising targets to search for biosignatures. A *biosignature* is an object, substance, and/or pattern whose origin specifically requires a biological agent. The usefulness of a biosignature is determined not only by the probability that life created it but also by the improbability that nonbiological processes produced it. For example, a biosignature can be a complex organic molecule and/or structure whose formation is unachievable in the absence of life. A *potential biosignature* is a feature that is consistent with biological processes and that compels investigators to examine it and its local environment more thoroughly in order to determine more conclusively the presence or absence of life. The EAR uses the term *bioindicator* when referring to a potential biosignature.

1.2.8.1 NAS THEME 5: IDENTIFYING, EXPLORING, AND CHARACTERIZING ENVIRONMENTS FOR HABITABILITY AND BIOSIGNATURES

Characterize habitability at different spatial scales. A variety of observable features of planets and planetary systems can help to assess habitability. Examples of such features include the following: 1) the type and metallicity of the parent star(s); 2) the architecture of a planetary system; 3) planet size, mass, and composition; 4) the composition and structure of the atmosphere; 5) the nature, extent, and duration of water activity at regional and local scales and in the near-subsurface; and 6) evidence of temporal changes in surface and near-surface environments, including understanding how they form and evolve. A key objective is to develop instrumentation to characterize such features. Such efforts will be guided by a better understanding of the range of parameters that influence habitability. An enhanced ability to quantify the relative merits of planetary systems, individual planets, and potential landing sites will enhance our search for evidence of life elsewhere.

Enhance the utility of biosignatures to search for life in the solar system and beyond. Biosignatures include the following features that, ideally, are formed exclusively by biological processes: 1) Organic compounds having diagnostic molecular structures (e.g., porphyrins, steranes, and hopanoids); 2) other chemicals and their abundance patterns (e.g., consequences of metabolism, disequilibria, and biological redox reactions); 3) molecular stable isotopic abundance patterns (e.g., enzymatic isotopic discrimination during redox processes); 4) minerals (e.g., those whose stability is created by biological processes); 5) microscopic structures and textures (e.g., cells and biofilms); 6) macroscopic physical structures and textures (e.g., stromatolites); 7) atmospheric gases (e.g., molecular oxygen in certain global environments); 8) surface spectral reflectance features (e.g., arising from forest canopies); 9) temporal variability (e.g., chemical or spectral changes caused by active metabolism); and 10) technosignatures (e.g., radio signals created by technology). The following objectives seek to enhance the utility of biosignatures: 1) Characterize better the processes influencing biosignature formation, preservation, and destruction; 2) determine how the properties of habitable environments can affect the attributes of biosignatures and their detectability (i.e., identify any contextual information needed to enhance the confidence in the interpretation of biosignatures); 3) better define the fundamental characteristics of life (even as we do not know it) that may translate into biosignatures; 4) identify any potential abiotic mimics of biosignatures; and 5) identify new types of biosignatures and how they can be detected.

Identify habitable environments and search for life within the solar system. Some habitable environments on Earth offer opportunities to study extraterrestrial habitable environments by analogy. Extreme environments on Earth can reveal how environmental stresses influence the persistence and activity of life and can help us understand the potential for preserving biosignatures in such environments.

Mars and Earth might have hosted relatively similar environments, particularly in the distant past; therefore, Mars has become a prime exploration target for astrobiology. Efforts should continue to determine where, when, and for how long habitable conditions were maintained on the surface of early Mars and whether habitable environments exist on Mars today. This requires an improved understanding of the major processes that degrade or preserve evidence of habitable environments and biosignatures. These efforts enhance efforts to seek signs of life.

The icy worlds in our solar system present some of the clearest evidence beyond Earth for liquid water on a global scale. Future space missions should characterize the processes related to subsurface oceans. This includes determining the key energy sources that affect these water reservoirs and how they evolved over time. Observations should seek to document the roles that interior and/or atmospheric chemical processes play in the formation of prebiotic compounds on the icy worlds.

Recent discoveries of diverse planetary systems indicate that the search for habitable planets and any evidence life in deep space holds enormous potential. Astrobiologists can augment efforts by astronomers to observe and characterize potentially habitable exoplanets. They can help to identify the diversity of detectable biosignatures that might be expected for habitable exoplanets. This effort includes, for example, recognizing atmospheric biosignatures and distinguishing them from potential false positives.

1.2.8.2   EAR Research Topic 5: Biosignatures as Facilitating Life Detection

Distinguish life from nonlife. As with NAS 5, earlier, this objective seeks to identify and evaluate the various categories of biosignatures for their capacity to distinguish life from nonlife. Our knowledge of the environmental conditions conducive to the origin and persistence of life is required to characterize the context ("background noise") against which we might detect the presence of any biosignatures. Measurements should be developed to achieve these objectives more effectively by remote observations and by landed spacecraft.

Follow the energy: Identify energy sources, redox couples, and photoreactions. This objective focuses on energy, both as a necessary resource in a habitable environment and as an approach to life detection. It addresses how life interacts with photochemical reactions and redox reactions and how biological activity might be identified as a component in such interactions. For example, biosignatures in

atmospheres might arise as redox pairs in a variety of environmental conditions on other planets.

Follow the data: Evaluate the potential for life in different planetary environments (from microscale to planets). This objective focuses on super-Earths—planets ranging in size up to 10 Earth masses—and asks how attributes such as plate tectonics, climate, photochemistry, and mineralogy might affect the development of life and the detection of any biosignatures in their atmospheres. The objective identifies super-Earths orbiting cooler stars (M dwarfs and K dwarfs) as key targets of exploration.

Follow biosignatures with time: Reach a better understanding of the evolution and preservation of biosignature assemblages with time. The various categories of biosignatures differ regarding the potential for their preservation under a range of environmental and geological conditions. Important examples include the following: 1) Remote detection of atmospheric biosignatures depends upon their survival against photochemical destruction, etc.; 2) preservation of biosignatures in the rock record depends upon their preservation against destruction by oxidation, radiation, temperature, and pressure; and 3) attributes of planetary environments probably influence the nature of life that might develop there and the categories of biosignatures that life creates.

1.2.9   ASTROBIOLOGY AT NASA

Funding agencies and their associated scientific communities have shaped the scope, operating principles, and current priorities of astrobiology research and exploration. Sections 1.2.9 and 1.2.10 address this topic for the astrobiology communities in the U.S. and Europe, respectively.

NASA is the lead agency for supporting astrobiology-related research and flight missions in the U.S. The agency works closely with other organizations in the U.S. Government; it also collaborates with other nations that support planetary exploration and astronomy.

The NAS advocates the roles played by astrobiology in spaceflight missions; it also articulates several operating principles of astrobiology.

1.2.9.1   Spaceflight Missions

Mars. During its history, the climate of Mars has been more similar to that of Earth than has the climate of any other planet in our solar system. This similarity and the proximity of Mars make it a principal focus for astrobiology. Characterizing any habitable environments and life on Mars has become a key goal of NASA's Mars Exploration Program (MEPAG, 2015). For example, the two Mars Exploration Rovers, the Curiosity rover, and the MAVEN orbiter have gathered evidence to determine past and present environmental conditions. NASA's "Mars 2020" rover will extend these efforts and also carefully select samples that will ultimately be returned to Earth to search for evidence of past life. NASA-sponsored astrobiology research supports these missions by quantifying the qualities of environments that enhance their habitability, by elucidating the variety and attributes of biosignatures that make them definitive signs of life and by understanding the environments and processes that favor preservation of biosignatures.

Icy worlds in our solar system. Discoveries by recent missions of subsurface liquid water on Jupiter's moons and icy plumes and organic matter on the Saturn's moons indicate that these bodies might harbor habitable environments. Astrobiology has become a key driver for future missions to these bodies. As with Mars, astrobiology research can help to characterize any evidence for habitable environments and biosignatures while taking into account the factors that could affect their preservation. Astrobiologists are participating in the planned mission to Europa and are formulating mission concepts for missions to the moons of Saturn.

Exoplanets. Exoplanets orbit stars other than our own. The discovery that exoplanets are common and diverse indicates that Earth-like habitable worlds might be observed directly. This would enable a far more comprehensive search for distant life. But characterizing the climates of newly discovered promising planets remains far more challenging than their discovery. Astrobiologists must collaborate with astronomers to develop instrumentation and numerical models to characterize the geological processes and atmospheres of these distant worlds.

1.2.9.2 Astrobiology Principles

In addition to articulating key Themes and Goals, the NAS and NAR explicitly stated the following key attributes of astrobiology that distinguish it from more traditional scientific disciplines. The EAR articulates a vision that is fully consistent with these attributes.

Interdisciplinary research and exploration. "Astrobiology is multidisciplinary in its content and interdisciplinary in its execution. Its success depends critically upon the close coordination of diverse scientific disciplines and programs, including space missions."

Stewardship. "Astrobiology encourages planetary stewardship through an emphasis on protection against forward and back biological contamination and recognition of ethical issues associated with exploration."

Societal engagement. "Astrobiology recognizes a broad societal interest in its endeavors, especially in areas such as achieving a deeper understanding of life, searching for extraterrestrial biospheres, assessing the societal implications of discovering other examples of life, and envisioning the future of life on Earth and in space."

Inspire future generations. "The intrinsic public interest in astrobiology offers a crucial opportunity to educate and inspire the next generation of scientists, technologists, and informed citizens; thus, a strong emphasis upon education and public outreach is essential."

1.2.9.3 Humanities and Social Sciences

The above principles illustrate how astrobiology extends beyond the natural sciences. The humanities and social sciences can also make meaningful contributions to the field. The NAS provides the following examples of such contributions: 1) Epistemology might help to identify comparative standards of evidence in astrobiology-related fields and to address the challenge of envisioning a meaningful definition of life. 2) The social sciences can help to understand the basis of the public's interest in, and support for, astrobiology. 3) The legal profession can help to address the costs and benefits of planetary protection protocols as well as the legal implications of major discoveries in astrobiology. 4) Professional educators are exploring strategies whereby astrobiology can enhance skills in creative and critical thinking, cross-disciplinary awareness, and science, technology, engineering, and mathematics (STEM), from K-12 through postgraduate educational levels.

1.2.10 ASTROBIOLOGY IN EUROPE

The EAR advocates principles and mission objectives similar to those summarized earlier for the NAS. But, of course, there are qualitative differences between Europe and ESA versus the U.S. and NASA. It is essential that research and exploration efforts are coordinated not only across scientific disciplines but also across international boundaries.

1.2.10.1 Biotechnology

Astrobiological investigations of extremophiles and the origins and evolution of life on Earth are relevant to research in medicine, biotechnology, and industry. Methodologies developed in these other fields can, in turn, be applied to astrobiology.

1.2.10.2 A Pan-European Platform for Astrobiology Research

Space science in Europe has traditionally been strong and has maintained parity with space-faring countries outside of Europe. However, particular areas need improvement. The following objectives of a proposed "pan-European astrobiology platform" are quoted directly from the EAR:

- "Encourage and capitalize on multidisciplinary research and projects that include diverse space sciences and/or Earth science disciplines;

- Allow a rational, possibly prioritized, access to relevant research infrastructures and equipment, possibly through development of specific networks;
- Allow the implementation of (and mobilize around) a prioritized research plan, taking into consideration the strengths and expertise available in different countries and allowing them to be further improved and shared;
- Provide a platform for the management, access, and archiving of relevant scientific data;
- Facilitate the design and development of new techniques and technologies;
- Provide stable funding and support allowing scientific objectives to be addressed in the medium to long term;
- Catalyze and facilitate undergraduate and graduate education and help with the development of PhD programs and postdoctoral opportunities;
- Facilitate exchange of scientific and technical staff;
- Represent a European anchor for international relations; and
- Represent a focal point for coherent public outreach and engagement."

The pan-European astrobiology platform also should embody the following attributes:

- "Be science-driven with a strong representation of the community in its governance.
- Evolve organically, starting with pilot initiatives such as a network of institutes and laboratories, a framework for PhD students/postdoc exchange, a framework for the use of infrastructures and access to field sites dedicated to collaborative projects, and eventually grow towards a more institutionalized structure, using the experience gained through pilots.
- Be flexible and able to adapt to the European and international landscape as well as evolving priorities and breakthrough discoveries.
- Be decentralized as much as possible to reap the benefits of European diversity without involving a heavy administrative burden and overhead costs.

- Be, as much as possible, in a position to provide end-to-end funding with a common (set of) scheme(s) to avoid the complication of various funding procedures and principles."

1.2.11 ASTROBIOLOGY ROADMAPS AND STRATEGIES ARE LIVING DOCUMENTS

What is a 'living document?' Evolution is an essential attribute of life; it must also be an essential aspect of astrobiology roadmaps and strategic plans. U.S. President Eisenhower quoted the following words of military wisdom: "Plans are worthless, but planning is everything. The very definition of 'emergency' is that it is unexpected, therefore it is not going to happen the way you are planning." The same perspective applies to research and exploration—we cannot predict scientific discoveries and other relevant developments. Yet, insightful roadmaps and strategic plans can enhance the likelihood of discoveries that will, in turn, drive the ongoing evolution of our science and our strategies and roadmaps.

REFERENCES

Des Marais, D.J., J.A. Nuth III, L.J. Allamandola, A.P. Boss, J.D. Farmer, T.M. Hoehler, B.M. Jakosky, V.S. Meadows, A. Pohorille, and A.M. Spormann. 2008. The NASA astrobiology roadmap. *Astrobiology* 8: 715–730.

Hays, L., L. Achenbach, J. Bailey, R. Barnes, J. Baross, C. Bertka, P. Boston et al. 2015. *Astrobiology Strategy*. NASA. https://nai.nasa.gov/media/medialibrary/2015/10/NASA_Astrobiology_Strategy_2015_151008.pdf.

Horneck, G., N. Walter, F. Westall, J.L. Grenfell, W.F. Martin, F. Gomez, S. Leuko et al. 2016. AstRoMap European astrobiology roadmap. *Astrobiology* 16 (3): 201–243. doi:10.1089/ast.2015.1441.

MEPAG, 2015. *Mars Science Goals, Objectives, Investigations, and Priorities: 2015 Version.* https://mepag.jpl.nasa.gov/reports/MEPAG%20Goals_Document_2015_v18_FINAL.pdf.

1.3 Online, Classroom and Wilderness Teaching Environments
Reaching Astrobiology Learners of All Ages Around the World

Bruce W. Fouke, Kyle W. Fouke, Tom Murphy, Colleen N. Cook,
Bruce F. Michelson, Joseph L. Cross, Lizanne DeStefano,
Glenn A. Fried, Killivalavan Solai, Jacalyn Wittmer Malinowsky,
Susan B. Kelly, Claudia C. Lutz, Erin Louer, Courtney C. Fenlon,
Nicholas P. Vasi, Adrienne M. Gulley, and Mayandi Sivaguru

CONTENTS

1.3.1   Introduction 27
1.3.2   Hybrid Textbook 28
1.3.3   Open Educational Resources and Global Outreach 30
     1.3.3.1   SciFlix 30
     1.3.3.2   Emergence of Life Massive Open Online Course 30
1.3.4   Higher Education 31
     1.3.4.1   GEOL 111 Emergence of Life 31
     1.3.4.2   Illinois Campus Honors Program 31
1.3.5   Youth Outreach 32
     1.3.5.1   Research Outreach Carbonate Hot-Spring Simulation 34
     1.3.5.2   Scholar-Athlete Sports in Space Science Camps 35
     1.3.5.3   University of Illinois 4-H Academy 37
     1.3.5.4   Outreach to Underrepresented Communities 37
1.3.6   Teaching Teachers 38
     1.3.6.1   Student-Teacher-Scientist Partnership 38
1.3.7   Teaching in the Wild 38
1.3.8   Public Policy and International Engagement 39
     1.3.8.1   Chicago Field Museum of Natural History 39
     1.3.8.2   Arctic Change Course in Stockholm, Sweden, and Svalbard, Norway 40
1.3.9   Lifelong Learning 41
     1.3.9.1   Illinois Osher Lifelong Learning Institute and Yellowstone Forever Institute 41
1.3.10   Summary 42
Acknowledgments 42
References 43

1.3.1 INTRODUCTION

The clearest way into the Universe is through a forest wilderness.

—John Muir, 1890 (Wolfe 1966)

The search for life throughout the cosmos originates from the most basic of human desires: to understand the universe within and around us. That quest lies at the heart of the imagination and spirit of the NASA Astrobiology Institute (NAI). A highly interdisciplinary organization dedicated to interstellar science, the NAI seeks to answer three fundamental questions (Des Marais et al. 2008):

- How does life begin and evolve?
- Does life exist elsewhere in the universe?
- What is the future of life on Earth and beyond?

Founded in 1997, the NAI is designed to produce transformative research that: (1) infuses astrobiology science into ongoing and future NASA missions; (2) nurtures a coherent and interactive astrobiology research community; and (3) establishes dynamic, long-lasting Education and Public Outreach (EPO) programs that connect the public with the latest astrobiology scientific discoveries (Hays et al. 2015; National Research Council 2018; NASA Astrobiology Institute 2018).

A recent NAI program, *Towards a Universal Biology: Constraints from Early and Continuing Evolutionary Dynamics of Life on Earth,* was awarded in 2012 to a consortium led by the University of Illinois Urbana-Champaign that included Baylor University College of Medicine and the University of California, Davis (*Illinois-Baylor-Davis NAI*). The program was housed in, and orchestrated from, the Carl R. Woese Institute for Genomic Biology (IGB) on the University of Illinois Urbana-Champaign campus. The Carl R. Woese IGB opened the 186,000 sq. ft. open floor plan in Spring 2007, with the explicit intent of serving as a dynamic platform for cross-disciplinary genome-enabled research and education. The Illinois-Baylor-Davis NAI initiative has addressed multiple goals of the NASA Astrobiology Roadmap (Des Marais et al. 2008), including *Biochemical Adaptation to Extreme Environments* and *Biosignatures to be sought in Solar System Materials.* Major evolutionary transitions through geological time that occur in living matter, act to constrain the diversity of life, and govern the way in which energy and information are utilized by life have been studied. A unifying theme in these projects derives from NASA-sponsored research of the late Professor Carl R. Woese at the University of Illinois. As the first scientist to identify the Archaea, Woese established a three-domain Tree of Life, providing our first insights into the emergence of the Last Universal Common Ancestor (LUCA; Woese and Fox 1977; Woese et al. 1990). Of special importance to NASA missions, this Tree of Life is a fundamental framework for deciphering how life has evolved on Earth. The Tree of Life informs the search for extraterrestrial life forms, as well as their classification, as interplanetary explorations unfold.

The Illinois-Baylor-Davis NAI team has implemented an ambitious EPO effort for the rapid distribution of astrobiology research results. At the outset, the team identified and addressed learning challenges that impede understanding in scientific disciplines and targeted specific diverse demographic audiences by using multifaceted methods of content delivery to reach learners of all ages from around the world. These initiatives have been dedicated to the inclusion of new and challenging interdisciplinary scientific and cultural astrobiology content from active NAI research. Simply put, the goal has been to bring this work to learners around the world on their own terms. With that purpose in mind, dynamic delivery approaches were chosen for effectiveness in a wide variety of global learning environments. The resultant Illinois-Baylor-Davis NAI EPO program, which has already achieved considerable scope and scale, provides valuable insight for developing future astrobiology outreach initiatives that emphasize lifelong learning and global engagement.

This review highlights the accomplishments of the Illinois-Baylor-Davis NAI EPO, which has engaged more than 100,000 students of all ages from around the world via integrated online, classroom, and wilderness learning opportunities. These efforts have included the development and ongoing offering of: (1) a *Hybrid Textbook*, integrating the basics of astrobiology science with nature photography in Yellowstone National Park, with 6,000 hardcopies and more than 10,000 eBook versions distributed worldwide at no cost; (2) an *Open Educational Resources and Global Outreach*, providing a series of short two-minute videos called *SciFlix* and a free Massive Open Online Course (MOOC) on the Coursera platform that has enrolled 50,000 students internationally; (3) *Higher Education*: learning experiences for thousands of students on the University of Illinois campus, including an accredited 3-credit-hour course in the university's Campus Honors Program, with a capstone field experience in Yellowstone National Park; an accredited 3-credit-hour classroom and laboratory Campus Honors Program course that has already served more than 500 students of the University of Illinois; and special outreach to underrepresented communities on the University of Illinois campus; (4) a *Youth Outreach* program that has served an additional thousands of students annually, with the help of classroom and laboratory lectures, laboratories, workshops, and camps, which included a Research Outreach Carbonate Hot-Spring Simulation (ROCHSS) laboratory teaching apparatus, Scholar-Athlete "Sports in Space" Summer Camps, and a 4-H Academy Summer Camp; (5) *Teaching Teachers*: workshops and field trips for a Student-Teacher-Scientist Partnership (STSP) program; (6) *Teaching in the Wild*: a series of outdoor astrobiology lectures and nature hikes taught in Yellowstone National Park in collaboration with the National Park Service (NPS), as well as Emergence of Life displays for exhibit at the Field Museum of Natural History in Chicago, Illinois; (7) *Public Policy and International Engagement*: a 2012 course with 38 undergraduate students that studied arctic climate change in Stockholm, Sweden, and Svalbard, Norway; and (8) *Lifelong Learning*: a blended course through the Osher Lifelong Learning Institute (OLLI) on the University of Illinois campus that to date has engaged more than 300 learners 50 years of age or older, in addition to lifelong learner wilderness courses with the Yellowstone Forever Institute (YFI).

1.3.2 HYBRID TEXTBOOK

A hybrid textbook, *The Art of Yellowstone Science: Mammoth Hot Springs as a Window on the Universe,* has been written for use in both formal and informal educational settings by science and non-science secondary and higher education students, lifelong learners, and visitors of all ages to Yellowstone National Park. Available in both hardcopy and eBook formats, *The Art of Yellowstone Science* (Fouke and Murphy 2016; Figure 1.3.1), is unique in its focus on the geological and biological history of the region; moreover, as a textbook, it is outstanding for the quality and abundance of its nature photography and instructional illustrations. A supplementary

FIGURE 1.3.1 (Top left) Cover of the *The Art of Yellowstone Science* (From Fouke, B.W. and Murphy, T., *The Art of Yellowstone Science: Mammoth Hot Springs as a Window on the Universe*, Crystal Creek Press, 2016.) written to complement and support the Illinois-Baylor-Davis NAI EPO activities. (Top right) Field photograph by of a frosted bison when the air temperature was −37°C. (Bottom) Travertine terraces at Mammoth Hot Springs. Photographs by Tom Murphy.

interactive website, with additional teaching materials, lesson plans, and interactive exercises, is openly available on the University of Illinois servers (Fouke 2016). More than 6,000 hard copies and 10,000 eBooks copies have been distributed at no cost to middle schools and high schools around the world. Hard copies have been provided to science teachers and libraries of hundreds of collaborating secondary schools, colleges, and universities across the nation and internationally. A lecture series, underway at museums, public, libraries, civic organizations, and universities nationally and internationally, has reached thousands of attendees.

The Art of Yellowstone Science showcases how fundamental understandings of the Earth's biosphere are rapidly changing, propelled by analyses of DNA, RNA, proteins, and other life molecules in a variety of natural and manmade environments. Microorganisms, essential to the healthy functioning and evolution of all living things, are proving critical to solving global challenges in everything from environmental sustainability, human medicine, food security, and energy generation to the search for life elsewhere in the universe. These insights

are presented to showcase the elegant interconnectedness of life and the contextual history of an ever-changing Earth. This narrative requires an integration of many fields of study in the sciences and humanities, and *The Art of Yellowstone Science* demonstrates how both visual art and science originate from the same human desire to understand the world in which we live. Photographic art is melded with natural sciences to emphasize the importance of observation and the need for a willingness to embrace the unexpected. Because biological evolution is the ultimate expression of integrated life and Earth processes, Mammoth Hot Springs is studied as a window on the universe.

Co-authors Bruce Fouke and Tom Murphy first met at Mammoth Hot Springs in the winter of 2008, when NPS rangers requested a photographic record of the fieldwork being done there, for use in educational displays at the then-new Old Faithful Visitor Center. From this encounter began an extraordinary collaboration founded on a passion for the wilderness. *The Art of Yellowstone Science* contains 260 field and laboratory photographs that were shot over the course of 60 years. Fouke contributes his 30 years of integrated scientific research and teaching that focuses on hot springs, coral reefs, subsurface hydrocarbons, Roman aqueducts, human kidney stones, and astrobiology, bringing thousands of students into the field around the world. Murphy contributes three decades of nature photography and wilderness insights. In residence near Yellowstone since 1975, Murphy established Wilderness Photography Expeditions in 1986 and has built an internationally renowned photography seminar company, teaching natural history photography in Yellowstone and other locales around the globe.

The Art of Yellowstone Science is composed of seven chapters with the following content:

Chapter 1 *Wilderness Laboratory*: It explains why Mammoth is such an extraordinary site for artistic inspiration and scientific inquiry.

Chapter 2 *Perspectives*: It considers the basic approaches required to complete astrobiology science in complex natural environments.

Chapter 3 *Dynamics*: It delves into the details of water-microbe-mineral interactions at Mammoth Hot Springs in Yellowstone National Park.

Chapter 4 *Emergence of Life*: It explains the work of Carl Woese, his development of the three-domain Tree of Life, and its scientific implications.

Chapter 5 *Deep Time*: It reviews major epochs that have shaped life and Earth interactions.

Chapter 6 *Human History*: It reviews the westward expansion of the United States and events leading to the founding and scientific analyses of Yellowstone National Park.

Chapter 7 *Worldwide Laboratory*: It explains how results from research on the natural laboratory of Mammoth Hot Springs provide a window for understanding natural processes around the world and throughout the universe.

The Art of Yellowstone Science relates how Mammoth Hot Springs has long served as a natural laboratory for astrobiology research that investigates how microbes form fossilized "biomarkers" in ancient calcium carbonate ($CaCO_3$) limestone (*travertine*) rock deposits. Similar to conditions found on Earth hundreds of millions of years ago, the water temperature, chemistry, and flow at Mammoth still support heat-loving (thermophilic) microbes that evolved in the earliest periods of life on Earth. Requiring a combination of field and laboratory experimentation, this research has expanded into practical applications: coral reef preservation, oil and gas exploration, the functioning of the Roman aqueducts, and human kidney stone disease. The overarching goal is to understand the unifying physical, chemical, and biological themes of biomineralization in a natural system like Mammoth on every level, from the single microbial cell to the entire ecosystem, and how these components and dimensions interact. After decades of investigation, Mammoth is beginning to reveal how it came into existence, how its vitality and scope can best be conserved, and how its secrets contribute to the well-being of humanity. *The Art of Yellowstone Science* emphasizes that the time is now for curious, inquisitive, caring, and good-hearted people to come together from all fields and sectors of science, art, and society to preserve nature and wilderness. Ordinary citizens everywhere, enticed by direct experience with natural settings or moved by photographs and paintings of the world around us, must join the quest to better understand Earth. For our survival on this water-rich planet, we must know much more about how physical, chemical, and biological components of natural systems have evolved to work together and to incorporate that knowledge into our everyday life.

1.3.3 OPEN EDUCATIONAL RESOURCES AND GLOBAL OUTREACH

1.3.3.1 SciFlix

A critical need exists to reduce the time lag between scientific discovery in university, government, and industry research centers and to disseminate that knowledge to formal and informal learning environments around the world. Illinois-Baylor-Davis NAI EPO program activities have been designed to reduce this delay. One effective approach has been the creation of a series of *SciFlix*, free 2-minute videos developed in collaboration with the Illinois Center for Innovation in Teaching and Learning (CITL) group. The *SciFlix* series targets secondary and higher education teachers and their students, as well as lifelong learners, to deliver clear, concise summaries of key advances in astrobiology science. For ease of access, these videos have been made available on YouTube. Because each *SciFlix* video and its supporting curriculum materials made available online have been designed to augment state and federal K-12 curriculum requirements in the natural sciences, decisions about the use of *SciFlix* reside with teachers, who can include these short-duration videos into lesson plans as they see fit. These resources are also further supported with free 1-day workshops to encourage teacher adoption. The *SciFlix* series also integrates with other federally funded educational programs

such as the Global Learning and Observations to Benefit the Environment (GLOBE) program, Project WILD (conserving wildlife through education), the Worldwide Water Education (WET) program, and other initiatives commissioned by the National Academy of Science (NAS), the National Research Council (NRC), the National Science Teachers Association (NSTA), and a variety of non-governmental agencies.

1.3.3.2 EMERGENCE OF LIFE MASSIVE OPEN ONLINE COURSE

An *Emergence of Life* Massive Open Online Course (*EoL MOOC*) was developed, filmed, and produced over a 2-year period in collaboration with the staff of the Illinois CITL and has been offered on the Coursera platform since 2015 (Fouke 2015). Summarizing and updating the research of Carl R. Woese that established the three-domain Tree of Life, this no-tuition course delves into the implications of the Tree of Life for NASA astrobiology missions while exploring evolutionary history. Interviews with Woese, in which he explains how he approached and accomplished his scientific discoveries (Figure 1.3.2), are complemented with interviews of several of his renowned contemporaries and protégés in the fields of astrobiology, molecular microbiology, and evolutionary biology.

The target audience of the EoL MOOC are learners of all ages and incentives from around the world. More than 50,000 participants representing 160 countries have enrolled to date, of which approximately 25% finished the entire course.

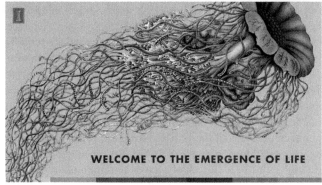

FIGURE 1.3.2 (Top) Interview with Professor Carl R. Woese in the *Emergence of Life* Massive Open Online Course. (Bottom) Website home page for the GEOL 111 *Emergence of Life* course. Filming and production by Colleen Cook and the CITL staff.

Participants include middle and high school students, college and graduate students, postgraduate students, professionals, lifelong learners, and teachers, who receive continuing professional development units (CPDUs) for completing the course. There are two examples of the project's global impact: in 2015, Fouke was completing research SCUBA on a coral reef off the coast of Na Trang, Vietnam. When he returned to shore at the end of a dive, he encountered a group of local high school students, two of whom had completed the EoL MOOC. Greeting Fouke by name, they asked him detailed questions about the Tree of Life as he emerged from the water. In another instance: via the EoL MOOC website, a single mother of five children, living in Dehradun, India, has engaged in regular weekly discussions about the Tree of Life with a retired petroleum geologist living in Tampa, Florida.

Though the EoL MOOC was initially offered with fixed start and end dates, after 1 year, it was transitioned into a constantly available "On Demand" format. Comprehensive teaching assessments were completed by the University of Illinois CITL staff. The 8-week EoL MOOC is composed of 80 video lectures, each approximately 4 to 7 minutes in duration, which cover the following topics:

Week 1: *Geological Time, the Nature of Science and the Early Earth*
Week 2: *The Tree of Life and Early Earth Environments*
Week 3: *Fossilization and Precambrian Life-Earth Interaction*
Week 4: *Paleozoic Life after the Advent of Skeletons*
Week 5: *Paleozoic Plants, Reptiles and the Transition to Land*
Week 6: *Mesozoic Reign of Dinosaurs and the Development of Flight*
Week 7: *Cenozoic Mammals and Global Environmental Change*
Week 8: *Astrobiology and the Search for Life in the Cosmos*

1.3.4 HIGHER EDUCATION

1.3.4.1 GEOL 111 EMERGENCE OF LIFE

An 8-week version of the EoL MOOC, Geology 111 *Emergence of Life* (GEOL 111 EoL), with significantly expanded content and requirements, was adopted in 2015 as a formal full-tuition 3-hour course in the University of Illinois Department of Geology (*GEOL*) through the College of Liberal Arts and Sciences. The target audience is formally enrolled science and non-science undergraduate students from disciplines across the University of Illinois campus. The course is offered in each spring semester and has an enrollment of 250 students. In 2016, GEOL 111 EoL became the first online course to be accredited as a General Education fulfillment. GEOL 111 EoL significantly expands beyond the composition and structure of the EoL MOOC and is intended to appeal to students from a wide variety of disciplines. This includes those taking GEOL 111 EoL for general education credit as well as those taking

it as majors in the natural sciences, who take the course to broaden their background in earth system science and evolution. The course also makes full use of multiple asynchronous and synchronous methods of communication with the instructor and graduate student teaching assistants. In addition, the contributions of women and minorities who have worked on the origin and evolution of life are included and emphasized (e.g., the critically important yet overlooked accomplishments in x-ray diffraction analyses of DNA by Rosalind Franklin in the laboratory of Crick and Watson). GEOL 111 EoL also includes extended video recordings of instructor presentations, guest lectures, and virtual field trips to the coral reefs of Curaçao in the Caribbean, the hot springs of Yellowstone National Park, the dinosaur fossils of the Field Museum in Chicago, and an animated trip to Mars. In addition to learning from the instructor and teaching assistants through regular discussion forums, students are expected to learn from each other through peer-reviewed essays and reflections.

1.3.4.2 ILLINOIS CAMPUS HONORS PROGRAM

A formally accredited astrobiology and geobiology course has been offered for each year of the Illinois-Baylor-Davis NAI EPO program. *Yellowstone Biocomplexity*, a 3-hour credit seminar for advanced undergraduate students of all majors in the Illinois Campus Honors Program, has filled to capacity every offering, achieving a total enrollment to date of almost 100. The centerpiece of the course is to study about feedback interactions among thermophilic (heat-loving) microbes, extreme water conditions, and rapid mineral precipitation in terrestrial and marine hot springs around the world. Studies of the modern-day ecosystems are evaluated as a means to better interpret the ancient fossil record of microbes on Earth while aiding the search for life on other planets. The semester-long Illinois campus-based portion of each semester includes an extensive essay-based midterm examination and an oral presentation-based final examination. The seminar culminates with fieldwork on these astrobiology dynamics and interactions at the hot springs at Mammoth. Students work in groups of four to develop an experiment to be conducted from the boardwalks at Mammoth, using infrared thermometers and cell phone cameras. Resulting data is used to test meaningful astrobiology-relevant hypotheses, which the students then orally present and defend 1 week after their return to campus. These students have completed their work even under challenging field conditions (low temperatures and heavy snow), requiring them at times to conduct their research on skis and snowshoes.

Mammoth Hot Springs is a unique federally protected natural laboratory for these types of courses, which are offered under strictly controlled NPS educational permits. Mammoth contains a high diversity of microbial communities and travertine deposits that are systematically distributed along the outflow drainage channels of each hot-spring system (Fouke 2011). Students focus on tracking ways in which the environment influences and controls microbial life and, in turn, ways in which microbial life rises to influence and alter the environment. Students are also exposed to system-scale dynamics and learn how linked carbonate-microbe systems grow, retreat, and change across large spatial and temporal scales. Gaining first-hand experience with four basic parameters within the Mammoth geo-ecosystem, students have learned to quantify, contextually integrate, track, and predict key mineral-water-microbe interactions. These include (1) spring water temperature; (2) spring water pH; (3) hot-spring water flow rate and dynamics; and (4) all basic contextual observations of the substrates flooring the drain outflow systems (i.e., travertine and microbial mat color, shape, size, growth rates, and distance along the drainage system, away from the vent). Concepts behind these basic modeling approaches to parameters are applicable to other sciences and can be translated to a broad range of grade levels. Several students have continued their Mammoth project as directed research in future semesters, contributed as co-authors to publications, and eventually joined the Fouke laboratory to complete their graduate degrees.

1.3.5 YOUTH OUTREACH

The current focus in the United States on standardized testing in secondary schools has produced unintended consequences observable in undergraduate populations of 4-year universities across the country. Professors and researchers in the sciences are concerned about a decline of critical thinking skills and curiosity, both of which are essential to scientific inquiry. At the same time, public secondary schools face constricted budgets, preventing teachers from enriching their curricula beyond state and national requirements. To address these

issues directly, the Illinois-Baylor-Davis NAI EPO has developed a series of in-person classroom, laboratory, and field lectures, as well as summer workshops and camps, which have been presented to thousands of secondary education-level students. These teaching efforts, focused on the core astrobiology principles included in the EoL online courses and textbook, have been presented in secondary schools in the United States (Illinois, Wyoming, Idaho, Montana, California, and Washington, D.C.), Europe (Sweden, The Netherlands, and Italy), and Asia (Vietnam).

In this outreach, emphasis falls on five cornerstone illustrations from *The Art of Yellowstone Science* textbook (Fouke and Murphy 2016). These have proven effective in teaching cutting-edge astrobiology scientific principles and approaches that nurture curiosity and interest in students. The science content of these illustrations is also aligned with the Next Generation Science Standards (NGSS) (2018) in a variety of subject areas, including biology, chemistry, physics, and geology. These connections and free downloadable teaching materials are available online. The five illustrations receiving special emphasis include (1) *Tree of Life* (Figure 1.3.3), integrating geological time, specific geological events, the Tree of Life, and the origin of life in deep seafloor spreading centers; (2) *Tree of Life Comparisons* (Figure 1.3.4), illustrating the difference between the Woese three-domain Tree of Life and other constructs, now outmoded, that still appear in textbooks; (3) *Powers of 10* (Figure 1.3.5), an organizational principle relating examples from the Yellowstone ecosystem to immense hierarchical dimensions of space and time evaluated in astrobiology scientific studies; (4) *Scientific Inquiry* (Figure 1.3.6), graphically representing the

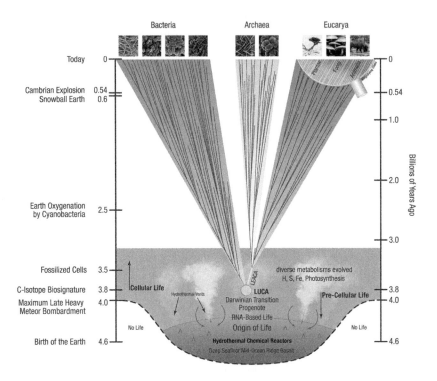

FIGURE 1.3.3 The Tree of Life in a framework of geological time and geobiological events. Content by Bruce Fouke and graphic design by Killivalavan Solai. (From Fouke, B.W. and Murphy, T., *The Art of Yellowstone Science: Mammoth Hot Springs as a Window on the Universe*, Crystal Creek Press, 2016.)

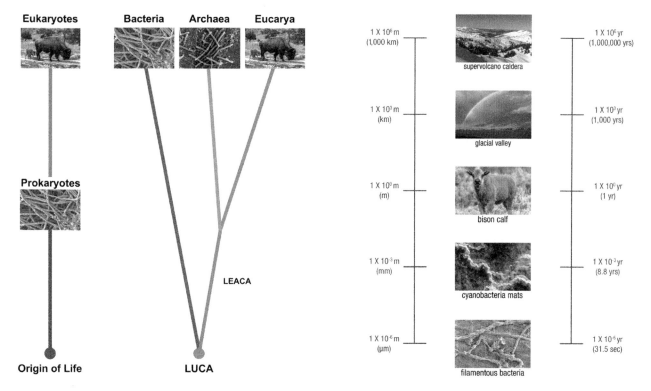

FIGURE 1.3.4 (Left) Diagram illustrating the single-trunk Prokaryote-Eukaryote Tree of Life. (Right) The three-domain Tree of Life constructed from molecular phylogeny. Content by Bruce Fouke and graphic design by Killivalavan Solai. (From Fouke, B.W. and Murphy, T., *The Art of Yellowstone Science: Mammoth Hot Springs as a Window on the Universe*, Crystal Creek Press, 2016.)

FIGURE 1.3.5 Powers of Ten space and time composition of Mammoth Hot Springs—from microbes to bison to glaciers and the supervolcano. Content by Bruce Fouke and graphic design by Killivalavan Solai. (From Fouke, B.W. and Murphy, T., *The Art of Yellowstone Science: Mammoth Hot Springs as a Window on the Universe*, Crystal Creek Press, 2016.)

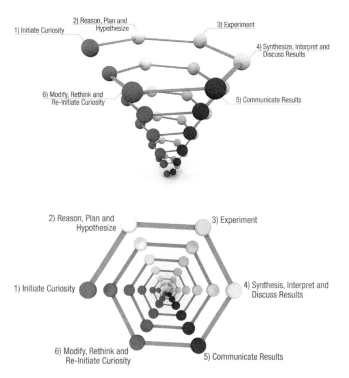

FIGURE 1.3.6 Scientific inquiry, a systematic approach that generates new questions and spirals ever more closely toward the truth and knowledge of nature. (Top) Side view of the Scientific Inquiry spiral. (Bottom) Top view of the Scientific Inquiry spiral. Content by Bruce Fouke and graphic design by Killivalavan Solai. (From Fouke, B.W. and Murphy, T., *The Art of Yellowstone Science: Mammoth Hot Springs as a Window on the Universe*, Crystal Creek Press, 2016.)

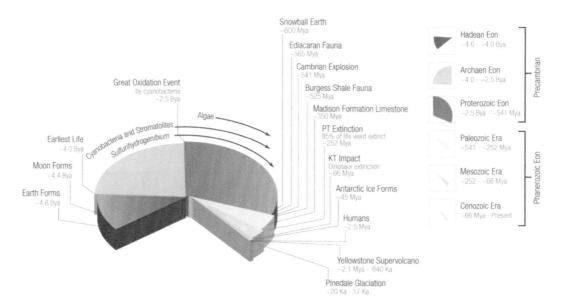

FIGURE 1.3.7 Overview of the geological history of Earth since it formed 4.6 billion years ago. Content by Bruce Fouke and graphic design by Killivalavan Solai. (From Fouke, B.W. and Murphy, T., *The Art of Yellowstone Science: Mammoth Hot Springs as a Window on the Universe*, Crystal Creek Press, 2016.)

Scientific Method as an endless spiraling process of inquiry and discovery; and (5) *Deep Geological Time* (Figure 1.3.7), coalescing the deep geological time into a handful of select ages, time intervals, and historical events.

1.3.5.1 RESEARCH OUTREACH CARBONATE HOT-SPRING SIMULATION

An experimental laboratory apparatus, constructed at the University of Illinois IGB, permits real-time, in-person, and online teaching of hot-spring astrobiology principles to secondary and higher education students. Named the *Research Outreach Carbonate Hot-Spring Simulation* (ROCHSS), this experiment simulates Mammoth Hot Springs in Yellowstone National Park by continuously heating and transporting 1,200 L of mineral- and CO_2-charged water along a 1-m-long precipitation runway (Figure 1.3.8). This apparatus allows students to manipulate and track a basic suite of physical, chemical, and biological parameters (i.e., temperature, flow rate, microbial composition, and water chemistry), using remote-controlled probes and video cameras. Students are therefore able to test directly astrobiology-relevant hypotheses pertinent to understanding the formation of biosignatures in the rock record on Earth, with likely relevance to other planets. The ROCHSS is designed to (1) expose 6–12 students and educators to university-level interdisciplinary scientific research applicable to a variety of middle and high school subject areas; (2) allow 6–12 students and educators to directly interact with university faculty, staff, and students; and (3) develop a sustainable experimental platform that can be used for both teaching and research.

The ROCHSS water begins in a pressurized mixing tank (Figure 1.3.8). Here, carbon dioxide (CO_2), $CaCO_3$, sodium chloride (NaCl), and other minerals and salts are dissolved

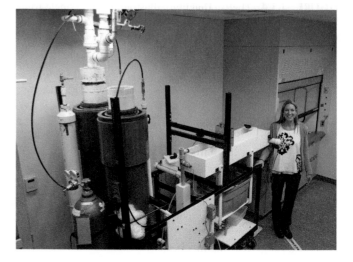

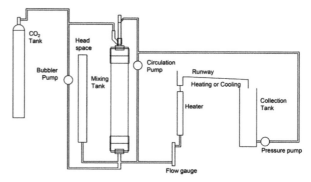

FIGURE 1.3.8 The ROCHSS laboratory teaching experiment. (Top) The ROCHSS instrument under construction at the University of Illinois Urbana-Champaign. (Bottom) Schematic of the ROCHSS experiment. Design and construction by Glenn Fried (shown working in top photograph).

to make a solution very similar to the water that erupts from the ground at Mammoth Hot Springs (Fouke 2011). At this point, the experiment either runs under default settings or is inoculated with living microbial cultures from our University of Illinois IGB laboratories. Next, the water travels through a heater, where the temperature can reach 70°C and pH can reach 6. It then flows along a runway (Figure 1.3.8), where $CaCO_3$ in the form of travertine is deposited and microbial mats grow simultaneously. As the hot water further cools and continues to degas CO_2, microbial mats upregulate proteins that serve to catalyze the travertine growth rate (Fouke 2011). Water flowing out from the downstream end of the travertine deposition runway is collected, pumped back into the mixing tank, and recycled with a circulating pump, while CO_2 is continuously bubbled through the recharge tanks (Figure 1.3.8).

The ROCHSS project was designed to support a broad range of teachers and students in learning basic hot-spring astrobiology principles, either in person at the IGB or online by accessing the experiment virtually. This virtual capability has dramatically expanded the reach of this University of Illinois campus project to national and international learning communities and can be easily coordinated into the EoL online courses. The online presence of the ROCHSS experiments requires that all sensors and cameras be controlled remotely with secure connections to a website that automatically organizes experimental data into an accessible database. This process includes streaming data from sensors that measure pH, temperature, ion concentration, CO_2, and flow. Cameras also feed streaming video and still images directly to the website, allowing for real-time monitoring and tracking of the experiment. At the conclusion of each trial, these data are "packaged" as downloadable files with associated time-lapse videos. Prior experimental data sets are also stored and accessible via the online portal for future reference and experimental comparisons. The versatility of this innovative platform will permit the ROCHSS to be utilized for a wide range of scientific topics, principles, and over a prolonged period of time.

Beyond accessing the experiment, educators and students can also access teaching materials, examples of experimental approaches, and laboratory exercises created using the five cornerstone illustrations from *The Art of Yellowstone Science*. These materials are linked to the astrobiology concepts presented in the EoL online course and are aligned to both state and federal NGSS (2018). Lessons cover up-to-date concepts on the Tree of Life, scientific inquiry, units of measurement, Yellowstone's ecosystems and history, and a classroom outreach project. By connecting these foundational illustrations and the concepts they represent to the NGSS Physical Science, Life Science, Earth and Space Science, and Engineering Design Standards, the site supports diverse teachers in making connections to a variety of existing secondary education courses, including biology, chemistry, physics, and environmental science.

In addition to providing teaching materials, students have a chance to take an active role in the experimental design process using ROCHSS and to submit proposals for future experiments. Because these proposals need to include the specific experimental parameters to be measured, a hypothesis to be tested, a statement about the expected outcomes, and the astrobiology relevance and significance of the overall experiment, they provide important introductory experience in thinking like a scientist. When a proposal receives approval, IGB staff set the requested experimental parameters and run the experiment, monitored in real time by the students. The ROCHSS tests can run from a few days to several weeks in duration, given the millimeters-per-day growth rates of the travertine observed at Mammoth (Fouke 2011). Throughout these experiments, emphasis is placed on the concept that microorganisms are essential to the healthy functioning and evolution of all living things and that microbes are critical to solving global challenges in environmental sustainability, human medicine, food security, energy generation, and space exploration.

The IGB ROCHSS team will also host three in-person and online professional development days each year for middle and high school teachers. The goals of these workshops are (1) to support teachers in creating additional lessons aligned with the NGSS; and (2) to connect middle school and high school teachers with scientists, professors, graduate students, and the IGB staff. During these workshops, educators will have the opportunity to ask pertinent astrobiology questions, develop realistic and meaningful hypotheses to use in their classrooms, collaborate with other educators, and earn CPDUs. This professional development, identified as important by middle school, high school, and university staff, is vital to the ROCHSS project, as it provides a rare venue for middle school and high school teachers to discuss vertical alignment of standards and content, connect with university faculty, and build a sense of community in the ROCHSS program. Eventually, a group of teachers and students will also be included in a field trip to experience Mammoth and Yellowstone (Houseal et al. 2010).

1.3.5.2 SCHOLAR-ATHLETE SPORTS IN SPACE SCIENCE CAMPS

The Illinois-Baylor-Davis NAI EPO will partner with the Illinois IGB and the Illinois Division of Intercollegiate Athletics (DIA) in Summer 2018 to offer three 4-day Scholar-Athlete "Sports in Space" Science Camps for secondary education-level students. A total of 180 campers will participate from the Illinois DIA Men's and Women's Tennis, Men's and Women's Gymnastics, and Men's Golf camps. If this experiment is successful, future camps could include thousands of campers from other Illinois DIA Division I sports, possibly including basketball, football, baseball, volleyball, and soccer. During each workshop, students will be exposed to cutting-edge astrobiology science in the integrated fields of genomic biology, medicine, geobiology, astrobiology, nutrition, kinesiology, and information technology through hands-on learning experiences. These experiences will challenge students to consider (1) the role of the human microbiome and its possible implications on health and athletic performance; and (2) key concepts of space, including sports in space. Throughout the duration of these camps, each scholar-athlete will be asked to apply these concepts to their own athletic performance. Their camp experiences will aim to (1) increase campers' awareness of Science, Technology, Engineering, and Mathematics (STEM) fields and research and increase their desire to pursue a STEM major in college; (2) have campers engage with interdisciplinary science underway at the University of Illinois and to build teamwork and laboratory inquiry skills in a variety of genomic-driven astrobiology scientific disciplines; and (3) provide campers with tools to evaluate their own health and performance long after the camp has concluded (Figure 1.3.9).

Building directly upon astrobiology insights contextualized by the Tree of Life, these scholar-athletes will learn about recent medical studies that estimate that as much as 90% of the total number of cells comprising the human body are single-celled microorganisms (*microbes*). These microbial communities (Bacteria, Archaea, and Eucarya) create a *human microbiome* that interacts with the mammalian host (Eucarya). They will also learn how the microbiome concept has fundamentally impacted all aspects of human medicine and will reshape how sports medicine is practiced in the future. From food and nutrient assimilation to muscle strength and fitness, the human microbiome is a fundamentally important concept for athletes to understand and incorporate into their training to maximize their health, strength, and athletic performance. The University of Illinois Department of Kinesiology is conducting research specifically on the influence of the human microbiome on sports performance (e.g., Allen et al. 2017). As a result, Kinesiology

faculty and graduate students will assist our IGB team in presenting at these camps. During the Scholar-Athlete "Sports in Space" Science Camps, students will plate and grow microbial communities collected from their bodies. They will also record and analyze basic body measurements, including temperature, weight, CO_2 exchange, hydration, food nutrition, and athletic achievement, to gauge and evaluate performance and efficiency of their bodies over the duration of each camp. In turn, these factors will be related to fundamental principles of Yellowstone astrobiology and experimentation, using the ROCHSS laboratory. For instance, CO_2, temperature, nutrients, and extent of dehydration are the factors that influence microbial communities in both Mammoth Hot Springs (extent of travertine deposition) and the human body (sports performance). These linkages will be carefully explained to further emphasize the global interconnectedness of life and Earth processes.

Each camp includes 2 contact teaching hours each day, for a total of 4 consecutive days, as follows:

Day 1 *Nutrition*: Scholar-athletes are introduced to nutrition awareness and are asked to evaluate their eating habits based on their intake of water, sugar, protein-to-carbohydrate ratio, and their consumption of processed versus natural foods.

Day 2 *Fuel and Hydration*: Scholar-athletes learn about the role of the human microbiome in nutrition and sports; run experiments on their own bodies, including testing their substrate utilization with a certified sports nutritionist; and learn how to optimize their nutritional and water intake.

Day 3 *Microbiomes, ROCHSS and Genomic Sciences*: Scholar-athletes will evaluate the human microbiome

FIGURE 1.3.9 Scholar-Athlete outreach on the University of Illinois Urbana-Champaign campus. Urbana High School scholar-athletes in a lecture (left) and using microscopes (right) at the Carl R. Woese Institute for Genomic Biology at the University of Illinois Urbana-Champaign.

within the context of the Tree of Life, run real-time experiments on the ROCHSS laboratory device that utilize the same parameters that they have measured on their bodies, and have in-depth discussions about the role of their microbiome on sports performance.

Day 4 *Exercise and Sports in Space*: Scholar-athletes will evaluate human activity, health, and performance at low or zero gravity and calculate basic parameters such changes in the gravitational force that influence the way in which their sport will need to be played under extraterrestrial environmental conditions; campers will then predict how their chosen sport would need to be modified to be played on a spacecraft or other planets.

1.3.5.3 University of Illinois 4-H Academy

Closely following the content and approach presented in the EoL MOOC, a 4-day residential course for 30 secondary school students was offered to expose secondary education students to astrobiology sciences and work with university researchers through the University of Illinois 4-H Academy. Entitled *Yellowstone: A Window on How the Universe Works* (Figure 1.3.10), this course offered and explored fundamental astrobiology concepts and approaches, including (1) the structure and implications of the Tree of Life; (2) how Scientific Inquiry is conducted; and (3) key events in the Deep Geological history of the universe. In the process, students took a "virtual" trip to Yellowstone through high-definition videos and interactive laboratories. Students also scrutinized the dynamics of a flowing fountain on the Illinois campus (Figure 1.3.10), integrating their powers of observation with strategic photographic skills. Students were also provided with an overview of biotechnologies being used in the life and natural sciences at the Illinois Roy J. Carver Biotechnology Center. In addition, students also visited Illinois IGB laboratories in which cutting-edge genomics, proteomics, bioinformatics, and other analyses are underway.

1.3.5.4 Outreach to Underrepresented Communities

The University of Illinois Graduate College is committed to fostering an inclusive community, and Fouke serves on its coordinating Equity and Access Commission. The Illinois-Baylor-Davis NAI EPO program has been actively engaged with making astrobiology learning experiences a part of four major outreach programs within the Graduate College. The *Community of Scholars Program* is a campus-wide initiative that invites admitted students from these populations to visit and tour the campus, meet peers, and get to know faculty. The *Grad Mentoring @Illinois Program* is a mentoring network dedicated to increased recruitment, retention, and successful academic and career outcomes for underrepresented graduate students. The *Summer Pre-Doctoral Institute* (SPI) provides incoming graduate students from these constituencies with an 8-week orientation to graduate study before they arrive at Illinois for the fall semester. The *Summer Research*

FIGURE 1.3.10 4-H Summer Academy residential camp presented on the University of Illinois campus in June 2016. (Top) Advertisement for the camp. (Bottom) Teaching assistant demonstrates techniques for measuring hot-spring water discharge in Yellowstone.

Opportunities Program (SROP) provides minority undergraduate students with an opportunity to explore careers in research, providing each student with an experience to strengthen his or her knowledge base, skills, and understanding of graduate work. The Graduate College has also worked with the Illinois-Baylor-Davis NAI EPO team to build connections with minority-enrollment institutions in Chicago, including Northeastern Illinois University and Chicago State University. The Illinois NAI E/PO program has also coordinated with the Graduate College office of the *Educational and Equity Programs* (EEP). The NAI faculty also worked work with the *American Indian Studies* (AIS) Institute to work with Native American populations in the Midwest. Working relationships have also been forged with the Illinois *Center for African Studies* and the *Center for Latin America and Caribbean Studies* to provide astrobiology learning experiences to students from those ethnic and racial backgrounds.

The University of Illinois campus is also a national leader in support and outreach to persons with disabilities. Formal coordination has been established between the Illinois-Baylor-Davis NAI EPO program and the Illinois *Division of Disability Resources and Educational Services* (DRES) to integrate these students. The mission of DRES is to ensure that qualified individuals with disabilities are afforded an equal opportunity to participate in and benefit from the programs, services, and activities of the university through the provision of effective auxiliary aids and services, the establishment of innovative educational services, and the pursuit of interdisciplinary disability research.

1.3.6 TEACHING TEACHERS

1.3.6.1 STUDENT-TEACHER-SCIENTIST PARTNERSHIP

An STSP program was established in collaboration with the Yellowstone NPS, in which a series of 1-day workshops were held to teach teachers for several years at the Illinois IGB and at Mammoth Hot Springs. Each of these workshops had an enrollment of 20 secondary education teachers, who received CPDUs for participating. Emphasis was placed on the astrobiology scientific themes presented in the EoL online courses and the textbook, with these resources utilized throughout the workshops. After returning to their home classrooms, teachers presented these concepts as modules of 1 or 2 weeks in working with their students. Periodically throughout the ensuing academic year, the Illinois-Baylor-Davis NAI team of researchers held live question-and-answer sessions with each classroom on Google Hangout. An online supplemental curriculum was developed specifically for this project and made available online from Illinois servers, which included (1) real-time engagement with active astrobiology research in Yellowstone; (2) discussion of hypothesis-driven research by using an iterative systems model that focuses on field research best practices; (3) focus on systems-level, astrobiology-relevant water-microbe-mineral interactions at Mammoth, including the geological "plumbing" under the system, carbonate cycling, and microbial ecosystems within the hot springs; (4) familiarity with astrobiology principles and protocols intended to protect Earth from extraterrestrial inoculation; and (5) sharing of findings with student colleagues and partners.

1.3.7 TEACHING IN THE WILD

Illinois-Baylor-Davis NAI EPO activities have embraced Yellowstone National Park as a dynamic natural classroom and laboratory within which to offer astrobiology learning experiences that are contextualized by the grandeur of wilderness. In such a unique context, students can be immersed in the complexity, scale, and relevance of natural processes. There is no substitute for astrobiology exploration and teaching in the wild, where scientific encounter with the primordial origins of our humanity encourages students to contemplate urgent challenges to the survival of species, including our own (Fouke 2014).

Thus far in the course of the Illinois-Baylor-Davis NAI EPO program, more than 50 lectures have been offered to visitors and ranger groups throughout Yellowstone National Park, in collaboration with the NPS and the Youth Conservation Corps (YCC). Locations for these talks have included Mammoth Hot Springs, Old Faithful, Fishing Bridge, the Norris Geyser Basin, and the Lamar Valley. In addition to the previously described astrobiology themes of the EoL online courses and textbook, these talks focus on teaching visitors to view Yellowstone hot springs as a natural laboratory in which to develop new scientific understandings of global processes (Figure 1.3.11).

Though Mammoth Hot Springs and coral reef ecosystems may seem like wildly different and unrelated environments, closer examination indicates that these awe-inspiring places exhibit a host of striking similarities and scientific parallels. The spring water emerging from Mammoth Hot Springs in northern Yellowstone is derived from rain and snowmelt runoff in the Gallatin Mountains, which flow down along faults into the rock subsurface. Heated by the Yellowstone supervolcano to approximately 100°C, the water chemically dissolves deeply buried 350-million-year-old Mississippian-age marine limestone and evaporates deposits called the Madison Formation. The water then flows back up to the surface to emerge from vents at a temperature of 73°C (Fouke 2011), in a cycle requiring hundreds to thousands of years. During this hydrologic journey, the Mammoth water evolves into a salty chemical fluid that is remarkably similar to seawater. Furthermore, much of the travertine that precipitates to form the classic millimeter- to centimeter- to meter-scale terraced steps of Mammoth is composed of a mineralogical form of $CaCO_3$ called aragonite. This is the same mineral that corals use to precipitate and grow their skeletons in warm shallow tropical seas. Furthermore, several of the microbes identified in the 73°C–25°C hot-spring drainage patterns at Mammoth are similar, and sometimes closely related, to microbes inhabiting coral tissues, coral mucus, and seawater (Fouke 2011).

Results of our field-based controlled experiments at Mammoth are therefore being used to predict how corals will respond to future global warming and associated increases in sea surface temperature. Heat-loving (thermophilic) microbes living at 62°C–71°C at Mammoth can respond to rapid shifts in water flow rate and temperature by changing the rate at which travertine rock is deposited on the floor of the spring outflow drainage channels. Biogeochemical analyses further suggest that the microbes do this by producing different types of membrane-bound proteins under changing water temperatures and flow conditions (Fouke 2011). These proteins, in turn, change the level and distribution of cell surface energy that controls the rate at which ions in the spring combine to form a solid travertine mineral precipitate. These dynamics established at Mammoth are now being used to establish new interpretations of how density banding in the aragonite skeleton of tropical corals (similar to tree rings) reflects coral response to the changing sea surface temperature (Fouke and Murphy 2016).

FIGURE 1.3.11 Yellowstone wilderness learning experiences at Mammoth, including partnering with National Park Service rangers (Top Left), University of Illinois Campus Honors Program students (Top Right), and Student-Teacher-Scientist Partnership high school students.

1.3.8 PUBLIC POLICY AND INTERNATIONAL ENGAGEMENT

An ongoing series of lectures and displays has been developed for general audiences; this series presents astrobiology science themes highlighted in the EoL online courses and *The Art of Yellowstone Science* textbook. More than 20,000 people have participated across the United States, including (1) at Yellowstone, in collaboration with the NPS and YFI; (2) at the Field Museum in Chicago, Illinois; (3) at the Mayo Clinic in Rochester, Minnesota; (4) at libraries in small towns and large cities; (5) at organization gatherings such as 4-H and the Rotary Club; (6) at various public and private colleges and universities; and (7) at the European Union and other embassies in Washington, D.C.

1.3.8.1 Chicago Field Museum of Natural History

Illinois-Baylor-Davis NAI EoL displays were constructed on the Illinois campus, transported to Chicago, and shown as part of a 3-day Wilderness of Illinois IGB *World of*

Genomics event presented in Stanley Hall, the largest venue at the Field Museum of Natural History. Situated between a complete *Tyrannosaurus rex* named Sue and the warring bull elephants, these NAI exhibits received more than 15,000 visitors, according to museum's estimates. In addition to NAI, partners for the event included Carl Zeiss Microscopes, the Yellowstone NPS, and the YFI (the Park's official education and fundraising non-profit partner). Featuring four Carl Zeiss petrographic microscopes and manned by 15 members of our NAI research group, the display included (1) the Emergence of Life on Earth and potentially on other planets; (2) microbial biomarkers in hot springs and potentially on other planets; and (3) the revolutionary NASA-supported work on molecular phylogeny, pioneered by Carl R. Woese. To further maximize the outreach impact of these exhibits, more than 500 students were bused to the field museum from schools in the Chicago Public School System, including Fenger Academy, Schurz High School, Gwendolyn Brooks College Prep, and Prosser Career Academy. More than 80% were students from underrepresented populations (Figure 1.3.12).

FIGURE 1.3.12 (Top) *Emergence of Life* and *Universal Tree of Life* interactive displays in the Stanley Hall of the Field Museum in Chicago, Illinois. (Bottom) Lectures being given to 500 Chicago Public School students at the field museum.

1.3.8.2 ARCTIC CHANGE COURSE IN STOCKHOLM, SWEDEN, AND SVALBARD, NORWAY

NASA runs multiple ongoing satellite-based missions that monitor Earth and that directly integrate with NAI planetary research goals (NASA Missions 2018). Upon receiving the NAI funding in 2012, an opportunity immediately emerged to develop a new broadly interdisciplinary course targeting university undergraduates from around the world. The course, as part of the Illinois-Baylor-Davis NAI EPO, fully integrated astrobiology science with Earth-based NASA satellite missions. This was made possible via a formal exchange agreement between the University of Illinois Urbana-Champaign and the KTH Royal Institute of Technology in Stockholm, Sweden (formally called the Illinois-Sweden Program for Educational and Research Exchange *INSPIRE*). Entitled *Arctic Change*, this 8-week formally accredited university undergraduate course within the INSPIRE program had a total enrollment of 17 students from the University of Illinois and 18 students from KTH. *Arctic Change* included 7 weeks of instruction in Stockholm and 1 week of instruction in Longyearbyen on the remote glaciated Svalbard (Spitsbergen) island archipelago, north of the Arctic Circle. *Arctic Change* offered a unique international integration of classroom, laboratory, and field experiences that linked Nordic humanities with environmental, life, and space sciences. The course contextualized these concepts with the Tree of Life and basic astrobiology sciences, while tracking these connections through deep geological time. This progressive breadth of cross-disciplinary education in a single field-based course is unparalleled in other universities around the world.

Arctic Change emphasized biocomplex relationships that arise from dynamic interactions between the physical, chemical, biological, and social components of the Svalbard ecosystem. Cornerstone parameters such sea surface temperature, air temperature, glacier thickness and retreat, polar bear

FIGURE 1.3.13 *Arctic Change* field course in Svalbard, Norway, that included students from the University of Illinois Urbana-Champaign and the KTH Royal University in Stockholm, Sweden.

and reindeer ecology, the emergence of infectious disease, changing human behavior, and cultural history were evaluated across multiple spatial (microns to thousands of kilometers) and temporal (nanoseconds to eons) scales. These were then directly compared to astrobiology-relevant life phenomena, such as NAI studies of the Antarctic Dry Valleys (e.g., Gilichinsky et al. 2007), that are fundamental to the search for life on other planets. Emergent phenomena and feedbacks through time were used as a framework to better understand the development and logistics of Nordic society, language, and culture in its own unique environmental setting.

The *Arctic Change* course emphasized the powerful influence of the ecology and evolution of microorganisms when discussing natural resources and land use by human inhabitants in the Arctic (e.g., food, fuel, water purification, and erosion control). This was evaluated with cutting-edge molecular sequencing and OMICS science tools, which are also central to the role of new biotechnical initiatives throughout the Nordic countries. Discussions and exercises were consistently linked back to the Tree of Life. To reinforce these principles, students were brought to tour the KTH-Karolinska SciLifeLab in Stockholm prior to departing for Svalbard. *Arctic Change* students were from internationally diverse cultural and academic backgrounds. Student participants were majoring in everything from oceanography, geology,

hydrology, microbiology, ecology, physics, chemistry, and the biosciences, as well as contemporary and historical Danish, Norwegian, Swedish, and Svalbard cultures, languages, and societies (Figure 1.3.13).

1.3.9 LIFELONG LEARNING

1.3.9.1 ILLINOIS OSHER LIFELONG LEARNING INSTITUTE AND YELLOWSTONE FOREVER INSTITUTE

In order to strengthen and more widely disseminate astrobiology science to lifelong learners, the Illinois-Baylor-Davis NAI EPO program presented a series of new hybrid courses on the *Emergence of Life* and *The Art of Yellowstone Science*. These were offered through the University of Illinois OLLI and the YFI to participants who were 50 years of age and older. Future emphasize needs to be continually placed on learners in order to provide inspiring educational opportunities for this rapidly growing cohort of accomplished and experienced people around the world. The Illinois-Baylor-Davis NAI EPO courses at the OLLI were offered in the spring semester for 8 weeks, with one class meeting each week. The linked objectives of astrobiology, Yellowstone science, and art naturally lend themselves perfectly to the OLLI classroom presentations and seminars. As a result, the maximum enrollment

FIGURE 1.3.14 Lifelong learners participating in Yellowstone Forever Institute course at Mammoth Hot Springs in Yellowstone National Park.

of 103 students per course was quickly reached, with long waiting lists each time the course was offered. Mirrored courses were also presented multiple times through the YFI at Mammoth Hot Springs (Figure 1.3.14).

1.3.10 SUMMARY

An NAI's program entitled *Towards a Universal Biology: Constraints from Early and Continuing Evolutionary Dynamics of Life on Earth* was awarded in 2012 to a consortium led by the University of Illinois Urbana-Champaign and included the Baylor University College of Medicine and the University of California, Davis. A key initiative in this work has been to further advance the work of Professor Carl R. Woese, whose pioneering research was the first to identify the Archaea, establish a three-domain Tree of Life, and provide our first glimpse of the evolutionary events that led to the emergence of the Last Universal Common Ancestor. The three-domain Tree of Life also provided the fundamental framework required to search for, identify, compare, and interpret extraterrestrial life forms that might be discovered in the foreseeable future. A progressive EPO effort that directly translated NAI team's ongoing astrobiology research

discoveries to online, classroom, and wilderness learning opportunities was completed. Mammoth Hot Springs in Yellowstone National Park was fully utilized as a dynamic classroom and natural laboratory for astrobiology research-inspired teaching opportunities, which have included development and presentation of (1) *Hybrid Textbook*; (2) *Open Education Resources and Global Outreach*; (3) *Higher Education*; (4) *Youth Outreach*; (5) *Teaching Teachers*; (6) *Teaching in the Wild*; (7) *Public Policy and International Engagement*; and (8) *Lifelong Learning*.

ACKNOWLEDGMENTS

Financial support was provided by the National Aeronautics and Space Administration (NASA) through the NAI's EPO initiative under Cooperative Agreement No. NNA13AA91A, issued by the Science Mission Directorate. Partial support was also provided by the National Science Foundation Biocomplexity in the Environment Coupled Biogeochemical Cycles Program (EAR 0221743), the National Science Foundation Geosciences Postdoctoral Research Fellowship Program (EAR 0000501), the Petroleum Research Fund of the American Chemical Society Starter Grant Program (34549-G2), and the University of Illinois Urbana-Champaign Critical Research Initiative. In addition, several organizations partnered to make these EPO activities a reality, including the Illinois College of Liberal Arts and Sciences (LAS), the Illinois Carl R. Woese IGB, the National Science Foundation (NSF), the NPS, the YFI, the Mayo Clinic, the Carl Zeiss Microscopy, and several middle and high schools across North America and Europe. Conclusions in this study are those of the authors, and they do not necessarily reflect those of these funding agencies and partners.

The NAI EPO field activities were completed under a series of YNP NPS education permits and research permit number 3060R. We are indebted to the many Yellowstone NPS Rangers who made this NAI EPO program possible, including Bob Fuhrmann, Ann Rodman, Christie Hendrix, Brian Suderman, Hank Heasler, Stacey Sigler, and Sarah Haas. We thank you for your pivotal role in supporting and allowing integrated research and teaching activities in Yellowstone and for fulfilling your absolutely vital roles of stewardship, protection, oversight, and partnership.

The ongoing guidance and support of Daniella Scalice at the NAI and of Hilarie Davis were crucially important in advancing all stages of this NAI EPO program. Sincere thanks also for the immense efforts and diligence of the entire Illinois CITL staff in creating the online courses, and special thanks to Jason Mock, Robert Dignan, and Liam Moran Gerard. Invaluable and significant contributions toward this work were also provided by all members of the Fouke laboratory research group at Illinois. We also acknowledge and appreciate the permission from parents and teachers to present the photographs of high school students who participated in our wilderness courses.

REFERENCES

Allen, J.M., Mailing, L.J., Niemiro, G.M., Moore, R., Cook, M.D., White, B.A., Hoslcher, H.D., and Woods, J.A., 2017. Exercise alters gut microbiota and function in lean and obese humans. *Medicine & Science in Sports & Exercise*, 48(9), 1688–1698.

Des Marais, D.J., Nuth III, J.A., Allamondola, L.J., Boss, A.P., Farmer, J.D., Hoehler, T.M., Jakosky, B.M. et al. 2008. The NASA astrobiology roadmap. *Astrobiology*, 8(4), 715–730.

Fouke, B.W., 2011. Hot-spring Systems Geobiology: Abiotic and biotic influences on travertine formation at Mammoth Hot Springs, Yellowstone National Park, USA. *Sedimentology*, 58(1), 170–219.

Fouke, B.W., 2014. A sense of the Earth. In Eds. Winkelmes, M.A., and Burton, A. *An Illinois Sampler*, Champaign, IL: University of Illinois Press, pp. 1–7.

Fouke, B.W., 2015. *Emergence of Life*, Coursera Website, https://www.coursera.org/learn/emergence-of-life.

Fouke, B.W., 2016. *The Art of Yellowstone Science*, University of Illinois Urbana-Champaign Website, http://artofyellowstonescience.igb.illinois.edu.

Fouke, B.W., and Murphy, T., 2016. *The Art of Yellowstone Science: Mammoth Hot Springs as a Window on the Universe*, Crystal Creek Press, 300 p.

Gilichinsky, D.A., Wilson, G.S., Friedmann, E.I., Mckay, C.P., Sletten, R.S., Rivkina, E.M., Vishnivetskaya, T.A. et al. 2007. Microbial populations in Antarctic permafrost: Biodiversity, state, age, and implication for astrobiology. *Astrobiology*, 7(2), 275–311.

Hays, L., Ed., 2015. *NASA Astrobiology Strategy*. NASA Headquarters NASA/SP-2015-3710, 236 p.

Houseal, A., Fouke, B.W., Sanford, R., and Furhmann, R., 2010. Mammoth hot springs: Where change is constant. *Yellowstone Science*, 18(3), 15–22.

NASA Astrobiology Institute, 2018. *Education and Outreach.* https://nai.nasa.gov/education-and-outreach/products-and-resources/.

NASA Missions, 2018. *Earth Missions List.* https://www.nasa.gov/content/earth-missions-list.

National Research Council, 2018. *Assessment of the NASA Astrobiology Institute*, The National Academies Press, Washington, DC, 80 p.

Next Generation Science Standards, 2018. https://www.nextgenscience.org.

Woese, C.R., and Fox, G.E., 1977. Phyogenetic structure of the prokaryotic domain: The primary kingdoms. *Proceedings of the National Academy of Sciences of the United States of America*, 74(11), 5088–5090.

Woese, C.R., Kandler, O., and Wheelis, M.L., 1990. Towards a natural system of organisms: Proposal for the domains Archaea, Bacteria, and Eucarya. *Proceedings of the National Academy of Sciences of the United States of America*, 87, 4576–4579.

Wolfe, L.M., 1966. *John of the Mountains: The Unpublished Journals of John Muir*. University of Wisconsin Press, Madison, WI, 459 p.

1.4 Astrobiology as a Medium of Science Education

Sun Kwok

CONTENTS

1.4.1 ASTROBIOLOGY AS AN INTERDISCIPLINARY APPROACH OF TEACHING SCIENCE

Astrobiology is the study of the origin, evolution, and distribution of life in the universe. Through the study of how life developed on Earth, we extrapolate this knowledge to extraterrestrial objects and speculate on how life could have developed under different conditions. Using remote observations with telescopes and direct physical experimentation with space probes, we are actively searching for evidence of life beyond Earth.

Living organisms on Earth are made of cells, and their constituents can be broken down into carbohydrates, lipids, proteins, and nucleic acids. These biomolecules are made of chemical elements, origins of which are found in stars. The present life forms on Earth are the result of billion years of biological evolution, which occurred through periods of geological transformation of Earth. The emergence of life altered the composition of the Earth's atmosphere and the biological evolution on Earth was interrupted by external events such as cometary and asteroid impacts, resulting in major extinctions. Earth, being one of the four terrestrial planets, was formed out of solid debris of the primordial solar nebula, which is a remnant of interstellar clouds and stellar ejecta. The birth-to-death life cycles of stars are part of the galactic evolution and cosmic evolution of the universe.

Astrobiology is an interdisciplinary subject that draws from research in astronomy, biology, biochemistry, chemistry, geology, microbiology, physics, and planetary science. It also touches upon the disciplines of history, philosophy, and sociology. Education in astrobiology therefore helps students develop the awareness that all sciences are related. Astrobiology studies the origin of life on Earth and the search for extraterrestrial life (biology); both are governed by universal physical laws (chemistry) but could develop under different rules of biochemistry. The evolution of life on Earth occurred in a changing physical environment (geology), which has been influenced by external events in the solar system (planetary science). Earth has changed greatly over time as the result of interactions between the crust, oceans, and atmosphere (earth system science and global change). Life on Earth must adjust constantly to adapt to these changes in order to survive (ecology). In a larger context, Earth is just a rocky planet that has billions of counterparts all over the Galaxy (astronomy). The possible existence of extraterrestrial life and the potential consequence of contact with alien civilizations have major social implications (sociology and philosophy).

Study of astrobiology therefore gives students a broader view of science than traditional science courses and offers them a grand perspective of the frontier of science.

1.4.2 ASTROBIOLOGY AS A VEHICLE TO BROADEN THE MIND

While many students have the perception that science is dull and boring, astrobiology discusses topics that are of great interest to students and the general public. One example is extraterrestrial life. Most people have a simplistic view of extraterrestrial life, assuming it to be similar to ours and to the inhabited extraterrestrial worlds having geological structures, climates, and daylight cycles similar to those of the Earth. This naïve view is commonly depicted in popular science fiction novels and movies such as *Star Wars* and *Star Trek*. Astrobiology offers an opportunity to look deeper into these issues. Our science explores and shows that there are many possible evolutionary paths to life, and extraterrestrial life may have totally different appearances from terrestrial living organisms. Exoplanets that harbor such life may have diverse atmospheric and surface conditions, as well as different daily and seasonal arrangements from Earth.

An unidentified flying object (UFO) is another very popular topic. Many assume that aliens visit us in mechanical flying objects with shining visible lights. Through the study of biological and cosmic evolutions, students in an astrobiology class will learn that any visiting alien civilization will be much more advanced than us and that their methods of visit are likely to be well beyond our present technological knowledge.

The above two examples of extraterrestrial life and UFO help illustrate the common pitfalls in popular thinking. Scientists are conservative in their methods but bold in their imagination. Others are careless in their methods but conservative in their interpretations of observed events.

The study of astrobiology also forces us to seek deeper and more fundamental meaning of seemingly obvious concepts, such as the definition of life. Although the distinction between life and non-life seems instructively self-evident, a precise scientific definition that can be broadly applied to extraterrestrial life is not simple (see Chapter 1.1).

The study of alien worlds may have impacts on the understanding of our own Earth system. Our present awareness of global warming originated with the study of the greenhouse effect on the atmosphere of Venus. Will observations of other planets and planetary satellites help us learn more about past and future climate changes?

There are obvious religious implications on the question of extraterrestrial life. Does the existence of extraterrestrial life, in particular intelligent life, cause any conflict with existing religious doctrines? If such life exists, are humans really the chosen people created by God?

Astrobiology can also have a linkage to the past. If advanced intelligent civilizations are indeed common in the Galaxy, would they have already visited us? Is there any anthropological evidence for past visits by alien civilizations? It is possible that alien civilizations did visit the Earth, without us knowing it (Shklovskii and Sagan 1966)?

The exploration of other worlds by humans also raises ethical, political, and economic issues. Is it within our rights to visit, and possibly contaminate, other worlds? Should we be allowed to exploit the resources in the Moon and other planetary bodies? Will our visits to asteroids, planets, and planetary satellites disrupt the environment of these objects to the extent that it is irreversible? If we send returning probes or humans to other planets, we could be bringing back alien life forms that may pose danger to life on Earth. Colonization of Mars may require us bringing Earth-based plants and other species to the planet; what are the potential consequences? What kinds of safeguards or precautions we should exercise before engaging in these activities (Chon-Torres 2017)? Given the current high public interest in human inhabitation of the Moon and Mars, we need to have a thorough discussion on these issues before embarking on such endeavors.

The study of astrobiology can also serve as a vehicle for the discussion of possible development of human society in the future. Over the last decade, we have seen tremendous progress in artificial intelligence. Will a time come when there will be no distinction between natural and artificial intelligence? Will our world be taken over by the intelligent machines? What will be the role of humans in such a society?

On a longer timescale, should humans be engaged in planetary engineering? It is conceivable that we could alter the atmosphere of Venus to make the planet habitable. Chemical or biological means can be introduced to convert carbon dioxide in the Venus atmosphere to oxygen, so that the surface temperature can be lowered and oxygen can be available for human respiration. Is planetary engineering inherently different from the way in which we have transformed planet Earth in the last several centuries?

The above examples on religion, ethics, social change, etc. can serve as excellent forums to bring in students in humanities, social sciences, engineering, and science together. A course in astrobiology will challenge students' minds and open their imagination to wider perspectives.

1.4.3 ASTROBIOLOGY AND OUR PLACE IN THE UNIVERSE

Humans have asked the question: "How important are we?" since the beginning of our existence. The search for answers to "where is our place in the universe" began as early as humans developed the intellectual abilities to think, and this quest is still ongoing. Five-thousand years ago, our ancestors already began careful observations of the heavens and recognized that celestial objects have regular patterns of behavior and are highly predictable. Extensive ancient monuments were constructed to mark the repeatability of celestial events such as the extreme north and south positions of sunrise and moonrise (Selin 2000).

The search for the relationship between heaven and Earth and the place of humans in it was the greatest driver for the development of intellectual thought. Through the use of simple observing instruments, our ancestors were able to achieve a highly sophisticated degree of theoretical understanding of the universe. Over 2000 years ago, ancient thinkers already knew that Earth was round and a free-floating object in space, not attached to anything. They also knew that the Sun's motion through the stellar constellation along the path of the ecliptic was responsible for the seasons. By adopting a model that puts the spherical Earth at the center of a giant celestial sphere upon which all stars reside, ancient astronomers were able to accurately predict the positions and times of sunrise and sunset at all locations on Earth. Ancient astronomers knew that the Moon shines by reflected sunlight and the phases of the Moon are the result of geometric relationship between the Sun, the Moon, and the Earth. Through observations and mathematical analysis, Greek astronomers were able to estimate the physical sizes of the Earth and the Moon and the distance to the Moon to great degrees of accuracy (Kwok 2017).

The integration of Ptolemy cosmology, Aristotle natural philosophy, and Christian doctrines gave rise to a model that humans are located between heaven (which is located outside of the celestial sphere) and hell (which is located in the middle of the Earth). This system of belief was challenged only after the proposal of the heliocentric model of the universe by Copernicus (Kuhn 1957, Koestler 1959). The realization that Earth is just one of the planets and the possibility of existence of other worlds similar to our own, as raised by Giordano Bruno, greatly weakened the foundation of the Christian faith and the authority of the Church.

Recently, we are able to go beyond the model of the physical structure of the universe and incorporate the scientific study of life into the question of search for our place in the universe. The origin and evolution of life are considered in the context of the physical universe. We now believe that Earth is a tiny speck of dust in the vicinity of an ordinary star, one of about 100 billion stars in our Milky Way Galaxy, which itself is just one of 100 billion galaxies in the universe. We have learned that planets are common in our Galaxy (Lissauer et al. 2014, Kaltenegger 2017) and organic compounds are widely present in the universe (Kwok 2011). The fact that the building blocks of life are ubiquitously present suggests that life could also be ubiquitously present.

The historical evolution of our beliefs on this subject and the speculation of how things could develop in the future represent a fascinating topic of class discussion. Changing interpretations of the structure of the universe over human history have overturned established social structures and how we perceive ourselves in the context of the Earth and the universe. The discovery of extraterrestrial life will have an equal, if not greater, impact on society.

May be, the most significant implications on the question of "our place in the universe" and the search of extraterrestrial life are changing our perception of ourselves: not as citizens of certain nations, member of certain races, or adherents of certain regions, but as common inhabitants of planet Earth. The education value of this concept goes far beyond what any humanities or social science course can deliver.

1.4.4 CURRENT STATE OF SCIENCE EDUCATION

In most universities today, science is taught by discipline. Students enroll in programs or majors of a specific discipline such as physics, chemistry, or biology. Students often perceive these subjects as segregated and not related to each other. In research-intensive universities, emphasis is placed on preparing students for graduate work, earning a PhD degree, and becoming the next generation of researchers.

The emphasis on specialty training is particularly strong in Russia, Asia (China, Japan, and Korea) and certain countries in Europe. Students and young researchers are encouraged to dig deeper into increasingly narrower subject areas. In order to give students a broader education, many North American universities have out-of-major requirements to supplement the major courses. For example, it is common to have distribution requirements for undergraduate science students to take courses in humanities and social sciences, such as psychology, economics, and history and for non-science majors to take science courses such as physics and biology. However, these courses are often introductory courses in these disciplines and are not designed specifically to be relevant to the out-of-major students.

An alternative approach to university education is to adopt a whole-person training philosophy, where the emphasis is to develop students as intellectual beings capable of analyzing the increasing complex problems of the modern world. The goal is not so much to impart factual knowledge of a particular discipline but to train students to be aware, to observe, to analyze, and to produce solutions to unfamiliar problems.

1.4.5 A NEW APPROACH OF TEACHING SCIENCE

Although the knowledge content of astrobiology is interesting in its own right, I am keener to use the teaching of astrobiology as a tool to develop students' intellectual abilities. In our education system, science courses often focus on the description of current models. Courses in biology and chemistry are commonly laden with facts, and courses in mathematics and physics are filled with abstract derivations. It is not to say that these are not useful, but these may have missed an important element of science education. Most science courses inform students on our current interpretations on how things are, with little attention paid to how scientists came to the present points of view. In fact, science is about the process of rational thinking and creativity. What we consider to be the truth is constantly evolving and has certainly changed greatly over the history of humankind. The essence of science is not so much about the current view of our world but how we changed from one set of views to another. Our goal of science education should not be about the outcome but about the process (Preface, Kwok 2017). We should try to help students develop the ability to observe, deduce patterns, and hypothesize on the origin of these patterns. Understanding the process of scientific thinking benefits students in all areas of study, not just science students.

This approach of teaching was used by me to teach a common core course at the University of Hong Kong between 2010 and 2018 to about 100 students per year drawn from all faculties of the university. The content of the course is designed to inform the students that our world views are the result of observations, pattern recognition, logical deductions, predictions, and further experimentation. This process is applicable not only to science but also to our everyday lives. By going through the historical examples discussed in the course, students learn to be suspicious of authority, be less willing to conform to the current way of thinking, and be bolder to explore new ideas. These potential effects on students have far greater impact on their lives than any knowledge content that science courses may impart.

During my tenure as president of the Astrobiology Commission of the International Astronomical Union, I tried to promote this approach of science education and use astrobiology as a tool to develop new ways of science teaching, in particular in the developing countries. As traditions are strong, change is not easy. However, the rapid development of the field of astrobiology offers a better opportunity as a vehicle for education reform than other more established or entrenched disciplines of science. Being at the wide-open frontier of interdisciplinary science, astrobiology can make a stronger case, so that the next generation of scientists be educated differently.

ACKNOWLEDGMENTS

My interest in astrobiology began when I first read the book "Intelligent life in the Universe" by Shklovskii and Sagan in 1973. Much of the content of the book is still relevant today.

REFERENCES

Chon-Torres, O.A. 2017, Astrobioethics, *Int. J. Astrobiol.*, 17:51–56.

Kaltenegger, L. 2017, How to characterize habitable worlds and signs of life, *Annu. Rev. Astron. Astrophys.*, 55(1):433–485.

Koestler, A. 1959, *The Sleepwalkers*, Hutchinson, London, UK.

Kuhn, T.S. 1957, *The Copernican Revolution*, Harvard University Press, Cambridge, MA.

Kwok, S. 2011, *Organic Matter in the Universe*, Wiley, Hoboken, NJ.

Kwok, S. 2017, *Our Place in the Universe*, Springer, Berlin, Germany.

Lissauer, J.J., Dawson, R.I., and Tremaine, S. 2014, Advances in exoplanet science from Kepler, *Nature*, 513:336–344.

Selin, H. 2000, *Astronomy across Cultures*, Kluwer Academic, Dordrecht, the Netherlands.

Shklovskii, I.S., and Sagan, C. 1966, *Intelligent Life in the Universe*, Holden-Day, San Francisco, CA.

1.5 Astrobiology-as-Origins-Story
Education and Inspiration across Cultures

Daniella Scalice

CONTENTS

1.5.1 ASTROBIOLOGY'S VALUE TO EDUCATION IN MAINSTREAM US CULTURE

Astrobiology explores our origins like no other part of NASA does, and the potential future discovery of extraterrestrial life will mark a major revolution in all of science, much like the Copernican revolution of the sixteenth century, in which we learned that we are not at the center of the universe. In this sense, astrobiology may be the most important and impactful investment that NASA has made. It is imperative that we mobilize astrobiology content across cultures and contemplative traditions toward making meaning of the search for and discovery of life elsewhere in the universe and furthering authentic dialogue around our origins.

Astrobiology is a multidisciplinary, diverse, and expansive endeavor, involving many seemingly disparate investigations. If astronomy, biology, chemistry, geology, paleontology, physics, and planetary science can be thought of as individual threads, then as the intersections of and spaces between these disciplines are explored by astrobiologists, the threads are brought together and woven into an interdisciplinary tapestry depicting a story of our origins, our place in the universe, and our search to find life elsewhere.

This interdisciplinary nature of astrobiology has special relevance in education. Astrobiology provides an umbrella, a larger context in which many fundamental science concepts such as gravity and volcanism can be taught. In alignment with teaching standards developed by states and professional education organizations in the US, such as the Next Generation Science Standards (National Academies Press 2013), educators are required to teach these concepts, introduced at the appropriate time with respect to cognitive ability and conveyed at increasing levels of depth, as students progress through the grade levels. While discipline-based concepts are typically taught in isolation, at both the K-12 and college levels, it has been shown that learning outcomes are enhanced when concepts are tied together in a larger, interdisciplinary context, such as the search for life elsewhere in the universe (Jones 2010, Howes et al. 2013, Hye 2017).

The huge questions guiding astrobiology (Are we alone? Is there life elsewhere?) cannot be addressed by one scientific discipline alone. Each discipline is a contributor of its knowledge and methodologies, and astrobiologists must communicate across discipline boundaries and access each other's expertise in order to successfully address these vast and ancient questions (Hays 2015). This collaborative essence is also a valuable tool in education, as many of today's best instructional techniques employ problem-based and/or team-based approaches (Dochy et al. 2003, Laal and Ghodsi 2011).

The astrobiology community, as primarily funded by NASA, has capitalized on the unique value that astrobiology brings to education. Numerous learning materials such as curricula, hands-on activities, games, web interactives, online courses, films, and books, as well as programs for youth, teachers, and the general public, in places such as classrooms, after-school settings, and museums have been developed and implemented over the past 20 years (for more, see https://astrobiology.nasa.gov/education/). But there is much more value to be had.

1.5.2   ASTROBIOLOGY-AS-ORIGINS-STORY

To weave that tapestry story of our place in the universe, as described in the other chapters of this book, astrobiology starts at the first moments of our universe. Astrobiology traces our origins from the Big Bang as the source of hydrogen— the simplest element. Over time, nuclear fusion in the cores of the first stars turned that hydrogen into many of the rest of the elements, especially the ones that make up all life on Earth, including our very selves. Astrobiology follows those elements into planets and moons, studying the process of planetary system formation, in which the star and all the bodies that orbit it form together, from the same source material and at the same time.

Astrobiology investigates the relationship between life and its planet (or moon). It champions the idea of a planet as a "container" that facilitated the emergence of life as a transition from geochemistry to biochemistry, supplying the required raw materials and energies. And where did those raw materials come from? The cores of stars that lived out their lives long ago.

Astrobiology goes on to explore the dance between planet and life—the co-evolution wherein planet influences life's evolutionary trajectory; then life innovates, adapts, and influences back; and on and on. It even looks into the relationship of all life at the molecular level, seeking to define the Last Universal Common Ancestor, at once appreciating the diversity—yet common ancestry—of all life, and the shared provenance of all things in the universe.

When we see the threads coming together in this way, it's easy to recognize astrobiology as an origins' story. Today's scientific community is at the heart of a modern, secular culture that has advanced to the point of using its core methodologies to articulate an origins' story. This is good and right. It is a natural, cultural phenomenon to seek to understand humanity's place in the larger world, in the cosmos. Astrobiology is not the first origins' story and perhaps won't be the last, but it should claim its place in the great cannon of cosmologies and origin stories of humanity—what Joseph Campbell has called "mankind's one great story" (Campbell 1991).

1.5.3   NEW DIMENSIONS OF VALUE TO EDUCATION—DUAL LEARNING

In terms of education then, seeing astrobiology as a cultural expression opens up whole new ways in which learning can be facilitated, beyond using it as an interdisciplinary context in which standardized, discipline-based concepts and skills can be more effectively taught.

Because most cultures have highly developed and intricate cosmologies and ways of knowing (aka methodologies to generate knowledge), astrobiology-as-origins-story (and its underlying way of knowing, the scientific method) can be utilized as a means of connecting to learners and bringing together people and ideas. In such connecting, the goal should never be to take something away from someone, to antagonize a pre-existing system of knowledge, or to disabuse any culture

or individual of knowledge they generated and hold dear, nor should it be to verify or validate cultural "myths" and beliefs. Such a perspective must also acknowledge and respect that each culture or contemplative tradition will have unique origins' stories and content. Assumptions about similarity of material or ways of working across cultures must be checked.

In this approach of meeting learners where they are with respect to their origins' stories and offering astrobiology-as-origins-story as another expression of "mankind's one great story," there is no value judgment woven into the paradigm, and therefore, no one body of knowledge can be proclaimed as better or truer. Instead, they co-exist in what is perhaps a new level of interdisciplinarity—a meta level—for astrobiology.

Rather than put origins' stories at odds, it is possible to create—and my colleagues and I have co-created—an environment wherein two bodies of knowledge and ways of knowing are equal—a "dual-learning" environment that facilitates the learning of both (see Section 1.5.5). When both bodies of knowledge are presented as equal in a non-threatening, dominance-free paradigm, learners are motivated to gain knowledge about both, and neither is diminished. Learners are supported to examine the relationship between the two from their own perspectives and decide for themselves if there are resonances or irreconcilable differences. Either conclusion is valid, as, again, the point is not for one to dominate or be revealed as absolute truth.

Many Science, Technology, Engineering, and Mathematics (STEM) programs and the pursuit of STEM educational pathways and careers pressurize youth to leave their communities. In many cases, opportunities for American Indian/Alaska Native (AI/AN) youth such as taking part in an internship, pursuing advanced degrees, and accepting job offers require them to move away, off tribal lands. Those communities consequently experience a "brain-drain" phenomenon (Kapur and McHale 2005), while some cultural priorities and the desire of many young Native people are often to keep all tribal members, especially the best and brightest, at home and involved in tribal advancement and the preservation and perpetuation of cultural knowledge and language.

This mandatory exodus, even if the intention is for the student to return and be involved in tribal advancement, harkens back to the early "management" strategies of non-Native government organizations in the US, which instituted boarding schools, wherein Native youth were forcibly separated from their families to receive an education and were forbidden to speak their languages (Adams 1995). Thus, the modern STEM paradigm, which pulls Native youth out of their communities, reinforces, perhaps unawares, the attitudes and practices of colonization.

Today, Native youth at least have the choice to leave to pursue STEM (and other career paths), versus in the past, when they were forcibly put into the boarding schools, but why should Native youth have to choose at all between a STEM identity and a cultural identity? Why does it have to be an either/or situation? How can we move toward yes/and?

Experiencing a dual-learning environment can foster the simultaneous development of both cultural identity and

STEM identity and ensure content learning gains of both STEM and cultural concepts (Demmert and Towner 2003). It can also support Native youth to consider STEM approachable and non-threatening. But much remains to be changed and evolved toward decolonizing and deinstitutionalizing STEM, so that Native youth who do pursue it don't have to sacrifice their cultural identity in the process.

1.5.4 NASA AND THE NAVAJO NATION PARTNERSHIP

It was with this perspective of astrobiology-as-origins-story that I met the opportunity, when it came along in 2005, to work with the Navajo Nation. I was invited to participate in a seminal Round Table between NASA's Office of Education and the Navajo Nation, designed to engender dialogue around the question, Is the Navajo Nation interested in partnering with NASA on education initiatives? This kind of "asking first" approach demonstrates inherent respect for the sovereignty of the Navajo Nation and Diné culture, as well as lack of assumption that everyone wants what NASA/science has to offer.

The Round Table convened educators, tribal government officials, community leaders, medicine people, and other stakeholders from across the Navajo Nation. Through dialogue, it was ascertained that educational initiatives with NASA were indeed desired. In line with their emerging tribal legislation at the time, the Sovereignty in Education Act, and through a reexamination of Diné Fundamental Law, the value expressed was that scientific and cultural knowledge should be taught together, as equals. This value emerged as a core philosophy, inspiring the term "dual-learning environment," and began guiding what would become a long-term, multifaceted partnership, now in its 14th year. It was also made clear that NASA was not being invited into partnership to "fix" or "solve" anything but rather to represent science completely and accurately in a dual-learning paradigm.

1.5.5 CO-CREATION AND DUAL LEARNING IN ACTION

My Diné colleagues and I formed the NASA and the Navajo Nation project team, standing solidly upon the foundation laid down by the Round Table (Allen 2016). Numerous meetings and ceremonies led to the co-creation of the first educator guide, "Sǫʼ Baa Haneʼ – Story of the Stars." Both cultural and scientific knowledge and stories relating to stars were explored, and the scientific content was chosen for its ability to align with the cultural content. The concept of stellar evolution, with a focus on the process of star formation and nuclear fusion as the source of heavier elements, was selected to be part of the hands-on activities in the booklet. The general theme of cycle/recycle is present throughout and is explored in more detail in the second educator guide.

The "Cycles in the Cards" activity, in particular, embodies the dual-learning environment and offers learners a potential connection between Diné origins' stories and those of science,

as told through stellar evolution. In this activity, cards are used (in sets of four) to represent all the parts and phases of creation and are laid down on a table in the order told in a Diné origins' story called "Dahadíníisą́ / The Beginning." As the story unfolds, each of the four cards in each set is placed into a different section of the table, corresponding to the four cardinal directions: east, south, west, and north. Finally, all the cards are placed, and piles form in each area. Once this is completed, four cards representing four phases of a star's life cycle are each placed, one onto each pile, aligning to what is known about each direction.

A card with an image of a star-forming region such as contained in many nebulae is placed in the east direction; a card with an artist conception of a circumstellar disk, representing the early stages after a star is formed, is placed in the south direction; a card with an image of our Sun, representing the main sequence of a star's "life" is placed in the west direction; and a card with an image of a supernova remnant is placed in the north direction. In this way, the life cycle of a star and the process by which elements are created and made available to the next generation of stars and planetary systems map directly to the processes and knowledge inherent in the Diné origins' story. Indeed, if the scientific method can be seen as having four phases: inquiry/hypothesis, experimentation/observation, data collection/analysis, and conclusions/next questions, then these map into the pattern as well.

As the partnership grew and years passed, another educator guide was co-created, and numerous workshops for Diné teachers and summer camps for Diné youth were co-implemented. The team also held a "Cultural Immersion Experience" for managers of academic and federal education programs, designed to serve Native youth, with the goal of potentially altering programmatic designs that don't reflect a full appreciation of the needs of those whom they are intended to serve.

The second educator guide, "Tłʼéhonaaʼéí Nihemá Navajo Moon," deepens the dual-learning philosophy. In it, the scientific concepts of planetary system formation, the electromagnetic spectrum, and spectroscopy find resonance with the Diné concept of K'é, a system of relationship and kinship, and with the knowledge embedded in the Diné cradleboard.

In the teacher workshops and student summer camps, the dual-learning environment is anchored by the presence of both a scientist and a medicine man. In the camps, field sites that are at once geologically significant and culturally sacred are visited, and both experts share knowledge and stories on the trails. Reflecting upon the experience at Canyon de Chelly National Monument, one student remarked, "Both the rocks themselves and the petroglyphs tell a story." Of Chaco Culture National Historic Park, where the scientist teaches about the Western Inland Seaway that was once there and the students learn about the former human occupants of the area, another student reflected, "At Chaco, the people may be gone, but the spirits are still there…and the ancient sea is gone, but the rocks record its presence."

Current efforts are focused on the development of a new hands-on activity, wherein the aspect of the Diné cosmology that involves a progression through four worlds and a scientific timeline from the Big Bang to the present day are brought together. The team is also exploring new field sites and developing new curricula for future summer camps.

In all these ways, the dual-learning environment is thus co-created, and learners are free to find and/or generate connections between the two bodies of knowledge and the two ways of knowing. Our team is presently in the process of publishing data collected from these programs, which assess the extent to which both cultural and scientific concepts are learned, as well as the extent to which learners have made connections between the two.

1.5.6 MORALS OF THE STORY?

I have learned from the indigenous knowledge and scholarship to which I've been exposed that the universe is relational, meaning that the fundamental context in which all things exist is one of community, interconnectedness, and relationship (Thayer-Bacon 2002, Cajete 2010). In my view, the origins story, as told through astrobiology, and the epistemology through which it has been developed share this understanding of relationality. And I'm not the only one. As Carl Sagan has said, "The cosmos is also within us. We are made of starstuff. We are a way for the cosmos to know itself."

The methodologies of science may be reductionist, but the output, especially in the case of the current scientific thinking about our cosmic origins and the emergence, evolution, and diversity of life on Earth, is systemic, holistic, and relational in nature. What could be more relational than Newton's Third Law of Motion, which states that every action has an equal and opposite reaction, or than the main principle of the First Law of Thermodynamics that matter and energy are neither created nor destroyed, only transformed? If astrobiology-as-origins-story is the output, can we move beyond to identify outcomes or morals of the story?

There are clear themes or concepts resident in astrobiology-as-origins-story that I think of as teachings or lessons from the science. The knowledge that the atoms that make up our Earth and us were made in a star that lived and died long ago intimately connects us not only to the universe and its cyclical processes but also inextricably and immutably to the solar system, the Earth, and to one another. As Neil de Grasse Tyson says, "We are all connected…to each other—biologically, to the Earth—chemically, to the rest of the Universe—atomically." We can also look at the process of stellar nuclear fusion, the process by which most atoms heavier than hydrogen were made, and see creative potential in the harshest environment imaginable, the hyper-pressured core of a star.

Planetary system formation as a phenomenon concurrent with star formation shows us that all the material in the star, planets, moons, asteroids, comets, etc., in our (and any given) solar system formed from a common source (nebulae, the giant molecular clouds) and according to the same processes. Herein lies more evidence of connectivity and relationship. If we find life elsewhere in our solar system, for example, on Mars or on Enceladus, in this way, we'll be related to it, too.

As humans, we think of the conditions in which early life may have emerged as "extreme," such as the high temperatures of hot springs or vents, the crushing pressures and lack of sunlight on the seafloor, an atmosphere without free oxygen, and bombardments at the surface. We continue to investigate such extreme environments here on the modern Earth and study how life innovated and adapted in order to spread and survive and thrive in these environments.

We look at the co-evolutionary relationship between life and Earth and see times when the environmental conditions presented life with challenges, and in turn, life innovated new biochemical strategies and influenced the environment. This is perhaps most perfectly illustrated by the microbial invention of oxygenic photosynthesis ~2.5 billion years ago, through which oxygen was put into the atmosphere for the first time, likely paving the way for multicellular life and the Cambrian explosion, and the asteroid impact at the K/T boundary ~65 million years ago not only wiped out the dinosaurs but also opened up environmental conditions that ultimately gave rise to the time of mammals. And it continues today, as we humans and our industries emit carbon dioxide and other greenhouse gases into the atmosphere, causing environmental conditions to change again. How will life on Earth respond?

In looking at the origin and evolution of life in this way, the themes of life's resiliency, tenacity, and ability to innovate new strategies—under environmental pressure—to make a living and grow are clear. Again, we can see the theme of creative potential in times of disruption, adverse conditions, and even catastrophe. And the idea of cause/effect and the relational nature of the universe are on full display.

This knowledge is begging for us to act accordingly. Neil de Grasse Tyson goes on to say, "Whatever you do, say, or even think influences others, and they influence you." Interconnectedness and relationality can show us the way to respect and stewardship. If we impact or harm one part of the system, there will be reverberations throughout the system, and we will eventually feel the repercussions ourselves. This has implications in how we approach the discovery of life elsewhere—will we see it as other and unrelated or embrace it as family and strive to protect and preserve it? There are implications too in how we behave toward our planet and, most importantly, how we treat one another.

1.5.7 ASTROBIOLOGY FOR THE INCARCERATED

Within the perspective of astrobiology-as-origins-story representing a relational universe, it stands to reason that our actions have consequences, for which we must take responsibility. There is perhaps no place more exemplary of consequences than a prison or jail. There is also perhaps no place more rich with transformative potential. What shifts might we see if astrobiology-as-origins-story and the themes, lessons, and teachings therein are shared as educational programs for incarcerated populations?

Founded in 2016, Astrobiology for the Incarcerated (AfI) is a partnership between NASA's Astrobiology Program, the Initiative to Bring Science Programs to the Incarcerated (INSPIRE) at the University of Utah, and the Sustainability in Prisons Project (SPP) at the Evergreen State College in Olympia, WA. The partnership leverages the prior work, experience, and expertise of INSPIRE and SPP, as well as much larger studies of exposure to educational programs behind bars, which show a large reduction in the likelihood of returning and an equally large increase in post-release employment (Davis et al. 2013). The AfI team and I bring astrobiology lectures to incarcerated adults and hands-on astrobiology activities to incarcerated youth across the US. We will also be deploying astrobiology imagery via short-format films into special housing units where live programming isn't allowed.

The goals are that the story, teachings, and images will broker connections for inmates to the cycles of the cosmos, our common origins, and the overall idea of a relational universe and that those connections will result in a greater capacity for positive relationship. When, in 1969, NASA provided us our first glimpse of Earth from space (in the image that became known as "Earthrise"), a reframing of our perspective catalyzed a movement of environmental stewardship (McGuinness 2009). In a similar vein, astrobiology-as-origins-story can help inmates contemplate their place in the world and inspire stewardship of self and relationships. The AfI team is currently in the process of collecting data from a pilot of our activities to assess the impact of our efforts on the inmates in terms of knowledge gains and attitudinal shifts.

1.5.8 CONCLUSIONS

The questions of astrobiology—the pursuit of knowledge about our origins, life on Earth, and the possibility of life elsewhere in the universe—are not exclusive to science. We must tread lightly, as we bring the scientific stories of our place in the universe and the teachings therein into "first contact" with those of other cultures and contemplative traditions. We must be relational and approach with humility and respect. Using astrobiology-as-origins-story and the teachings within it in educational environments should be undertaken with even more care; science should not be wielded as a sword to cut down or diminish other bodies of knowledge or epistemologies. Rather, we can create a safe space, promote their co-existence, and let them speak and listen to each other. We can be present to that unfolding and hear what is being said. As have I, you may find that these bodies of knowledge and ways of knowing have more in common than not. The etymology of the very word "universe" is telling us, if you listen closely enough, that no matter what ways of knowing we use, we are all speaking "one verse," telling "one story," and singing "one song."

REFERENCES

Adams, D.W. 1995. *Education for Extinction: American Indians and the Boarding School Experience, 1875–1928.* Lawrence, KS: University of Kansas Press.

Allen, J. 2016. Weaving Diné Knowledge with NASA Science for Community Education. *Indian Country Today.* Available at: https://indiancountrymedianetwork.com/education/native-education/weaving-din-knowledge-with-nasa-science-for-community-education/.

Cajete, G.A. 2010. Contemporary Indigenous Education: A Nature-Centered American Indian Philosophy for a 21st Century World. *Futures* 42: 1126–1132.

Campbell, J. 1991. *The Power of Myth.* New York: Anchor.

Davis, L., Bozick, R., Steele, J., Saunders, J., and Miles, J. 2013. *Evaluating the Effectiveness of Correctional Education: A Meta-Analysis of Programs at Provide Education to Incarcerated Adults.* Washington DC: Bureau of Justice Administration, Rand Corporation.

Demmert, W.G., and Towner, J. C. 2003. A Review of the Research Literature on the Influences of Culturally Based Education on the Academic Performance of Native American Students. Final Paper. Northwest Regional Educational Lab. Available at: https://files.eric.ed.gov/fulltext/ED474128.pdf.

Dochy, F., Segers, M., Van den Bossche, P., and Gijbels, D. 2003. Effects of Problem-Based Learning: A Meta-Analysis. *Learning and Instruction* 13: 533–568.

Hays, L.E. (Ed.). 2015. NASA Astrobiology Strategy 2015. Available at: http://astrobiology.nasa.gov/uploads/filer_public/01/28/01283266-e401-4dcb-8e05-3918b21edb79/nasa_astrobiology_strategy_2015_151008.pdf.

Howes, A., Kaneva, D., Swanson, D., and Williams, J. 2013. Re-Envisioning STEM Education: Curriculum, Assessment and Integrated, Interdisciplinary Studies. A Report for the Royal Society's Vision for Science and Mathematics. Available at: https://royalsociety.org/~/media/education/policy/vision/reports/ev-2-vision-research-report-20140624.pdf.

Hye, S.Y. 2017. Why Teach Science with an Interdisciplinary Approach: History, Trends, and Conceptual Frameworks. *Journal of Education and Learning* 6: 66–77.

Jones, C. 2010. Interdisciplinary Approach - Advantages, Disadvantages, and the Future Benefits of Interdisciplinary Studies. *ESSAI* 7: Article 26. Available at: https://dc.cod.edu/essai/vol7/iss1/26.

Kapur, D., and McHale, J. 2005. *Give Us Your Best and Brightest: The Global Hunt for Talent and its Impact on the Developing World.* Washington, DC: Brookings Institution Press.

Laal, M., and Ghodsi S.M. 2011. Benefits of Collaborative Learning. *Procedia – Social and Behavioral Sciences* 3: 486–490.

McGuinness, M. 2009. Science Wednesday: Earthrise – The Picture That Inspired the Environmental Movement. Environmental Protection Agency Blog. Available at: https://blog.epa.gov/blog/2009/07/science-wednesday-earthrise/.

National Academies Press. 2013. *Next Generation Science Standards: For States, By States.* Washington, DC: The National Academies Press.

Thayer-Bacon, B.J. 2002. Native American Philosophies as Examples of W/holistic Relational (e)pistemologies. Educational Resources Information Center, US Department of Education. Available at: https://files.eric.ed.gov/fulltext/ED474264.pdf.

Section II

Definition and Nature of Life

2.1 Defining Life
Multiple Perspectives

Vera M. Kolb

CONTENTS

2.1.1 INTRODUCTION

This chapter presents various perspectives on defining life. First, an overview of the most common approaches to defining life is given, accompanied by a selection of definitions of life that illustrate these approaches. Some definitions of life also address the origin of life. Advantages and disadvantages of minimal versus expanded definitions of life are evaluated. Finally, borderline cases of life (such as viruses) are addressed.

2.1.2 BACKGROUND ON DEFINITIONS OF LIFE

One of the key goals of astrobiology is to define life, which is important not only for understanding the origin and evolution of life on Earth but also for the search for extraterrestrial life. Many definitions of life have been proposed. They are given in a chronological collection of 95 definitions from 1855 to 2002 (Popa 2004), 77 life definitions from 1871 to 2002 (Popa 2014), and 78 definitions from 1999 to 2000 (Pályi et al. 2002). Although a large number of definitions of life have been proposed, none is universally accepted (Popa 2014). While each definition addresses some key features of life, none appears inclusive of all the features. Minimalistic definition of life was also sought (Trifonov 2011), as well as the universal definition of life, which would be applicable to extraterrestrial life also (Ruiz-Mirazo et al. 2004). Many inclusive definitions describe what life does but do not state what life is. Our ability to define life is also dependent on our understanding of the origin of life, which is a subject to controversies (Pereto 2005). A comprehensive definition of life should also address its origins. A philosophical problem exists that we may not know enough about the theory of life to define it (Cleland and Chyba 2002, 2007). Currently, the NASA's definition of life is used most often. It is: "Life is a self-sustained chemical system capable of undergoing Darwinian evolution" (Joyce 1994, 2002; cited in Popa 2004, 2014). A comprehensive definition of life should also address the borderline cases, such as viruses. The latter are in the twilight zone of life, according to Villarreal (2004).

The task of defining life has also been undertaken by the philosophers, alone or in cooperation with scientists (Cleland and Chyba 2002, 2007; Gayon 2010; Malaterre 2010; Tirard 2010; Tirard et al. 2010; Weber 2015). Some scientists applied philosophical principles in defining life and the origin of life (Kolb 2005, 2007, 2010, 2012, 2016; Perry and Kolb 2004). These examples testify to the interdisciplinary nature of astrobiology, which now solidly includes philosophy as a discipline (see Chapter 1.1 for the discussion on the interdisciplinarity of astrobiology).

Definitions of life often reflect the original disciplinary training of the authors or their current investigational area of expertise and their understanding of the origins of life. Examples include chemically oriented authors (Benner et al. 2004; Benner 2010). Specific questions that were examined include types of polymers that support Darwinian evolution (Benner 2010), a repeating dipole as a universal structural feature for the catalysis in water, universal structural features of metabolites in water, and the charge requirement for folding of the genetic materials with repeating charges, among others (Benner et al. 2004). These examples provide us with chemical descriptors for life.

To define life, one must understand it first, at least at some level. The ultimate experimental way to understand life is to make it in the laboratory. Work on the synthetic life has been quite successful (Venter 2007, 2013; Ball 2016). While we still do not have synthetic life produced *ab initio*, much progress has been made, such as the synthesis of a minimal bacterial genome (Hutchinson et al. 2016). Philosophical aspects of the synthetic biology are recently reviewed, from chemical and ethical points of view (Schummer 2016). New molecular biology tools, such as the Clustered Regularly Interspaced Short Palindromic Repeats (CRISPR) gene-editing method, provide additional ways to tinker with the building blocks of life (Callaway 2016).

In the next section, different ways to define life are addressed.

2.1.3 DIFFERENT WAYS OF DEFINING LIFE AND FACTORS INVOLVED

Defining life is important for astrobiology, specifically for understanding biology and the origins of life on Earth, in searching for the putative extraterrestrial life, and for placing the artificial life in a context of the biological life.

In general, different types of definitions exist (Churchill 1986, 1990; Gupta 2015). Examples include lexical, stipulative, and operational definitions. A lexical definition explains the meaning of a word as it is currently used and given in the dictionaries. A stipulative definition assigns a meaning to a word, with the objective of clarifying arguments or discussion in a specific context. A stipulative definition represents an explicit decision to use a word in a given way. Since stipulative definitions represent proposals to use a word in a certain way, they can be judged as useful or not, clear or not, but not as true or false. Operational definitions are often used in the experimental sciences. An operational definition states the meaning of a term by connecting its proper use to some observable condition (Churchill 1986, 1990). Scientific definitions are typically stipulative. One example is the NASA's definition of life: "Life is a self-sustained chemical system capable of undergoing Darwinian evolution" (Popa 2004, 2014; Gayon 2010).

Some definitions of life are expanded dictionary definitions of life, in which the additional description is added for the use by specialists. Some other definitions aspire to capture only the perceived essential characteristics of life, which are then specified. Most valuable definitions attempt to include the origins of life. Many definitions are the mixtures of these types.

Defining life is often influenced by the three historical conceptions of life. They are: life as animation (Aristotle), life as mechanism (Descartes), and life as organization (Kant) (Gayon 2010). Some contemporary definitions of life can be easily traced to these three conceptions.

Philosophical visions about life, such as holism (generalism), reductionism (minimalism), vitalism and spiritualism, and mechanistic reductionism and dialectical materialism, also influence the process of defining life (Popa 2010, 2014). The holism regards life as a collective property. It considers

its function and purpose. The reductionism aspires to explain life by the fundamental and minimalistic mechanisms and forces. The vitalism and spiritualism view life as the result of some esoteric force or supernatural will, which are beyond the power of observation and understanding. Contemporary science rejects vitalism and spiritualism. However, the latter views may still be present in the contemporary thinking about life, but in a non-transparent way. Élan vital may be substituted by another concept that is difficult to define or decipher but sounds less vitalistic ("spontaneous emergence of organizational complexity," as one example). Mechanistic reductionism and dialectical materialism consider life as a fully anticipated phenomenon, which is the result of the laws of nature (Popa 2010). These philosophical views may be revealed in the definitions of life.

Life is often defined by a list of properties that are common to all life. Such list typically includes chemical components, thermodynamic properties, metabolism, compartmentation via a semi-permeable membrane, reproduction and replication, and a combination of such properties (Gayon 2010).

Autopoiesis is a concept proposed and developed by Maturana and Varela as the sole requirement for life (Maturana and Varela 1980). Autopoiesis (from Greek roots, meaning self-making) refers to life's continuous production of itself. Autopoiesis involves metabolism, which is believed to be essential for life (Margulis and Sagan 1995). An autopoietic unit is defined as an open system that is capable of sustaining itself because of an inner network of reactions that regenerate the system's components (Maturana and Varela 1980; Luisi 2006, 2016). Maturana stated: "When you regard a living system you always find a network of processes or molecules that interact in such a way as to produce the very network that produced them and that determine its boundary. Such a network I call autopoietic." Luisi lists three criteria for autopoietic system: (1) boundary of its own making; (2) self-maintenance; and (3) self-generation via a network or reactions that are generated by the system itself. If these criteria are fulfilled, the system is alive (Luisi 2016).

There are individual life forms and entities that are considered alive but cannot reproduce, such as sterile organisms. Also, self-reproduction of individuals may not be possible if the type of reproduction must include a partner, such as in the sexual reproduction. Likewise, individual humans at the life stages during which they are not fertile, such as babies and very old people, are not capable of reproducing. The requirement for reproduction is clearer when put in the context of species, rather than an individual. Various stages on the path of the life cycle of the individuals and borderline cases of life can all be covered by the term "life forms," even though these forms may not exhibit all typical characteristics for life (Liesch and Kolb 2007; Kolb and Liesch 2008).

In his article titled "Why is the definition of life so elusive? Epistemological considerations" Tsokolov analyzes the problems of defining life (Tsokolov 2009). The problems include attempts to define life with terms that may not be clearly defined. Examples include the terms "complexity" and "self-organization." Defining life via living and with terms that seem intuitively understandable but are not sufficiently defined (e.g., complexity) may create confusion. Many definitions of life are descriptive, since they use a list of various characteristics of life. Such lists often include the uptake of nutrients and energy from the environment, metabolism, growth, reproduction, adaptation, heredity, autocatalytic reactions, chirality, compartmentation, disequilibrium, decrease in entropy, feedback loops, homeostasis, and genetic codes. Such descriptions are not ideal, since they may include terms that are not necessary or sufficient for life and that may not be characteristic of all living systems. The author also considers defining "minimal" life. For example, a minimal living cell has certain properties. If all the properties are not present, the system is not yet living. If more properties are present, the system is not minimal. The author shows the "minimal life" on a scale for continuous evolution from inanimate to living. However, the scale is arbitrary, since the borders in the continuity are arbitrary. The author also considers the concept of "arbitrarily alive" for various borderline cases of life.

Another important characteristic of life is its apparent purposefulness, which is called teleonomy (Pross 2012). It refers to the general behavior of living beings, in which they act in their own behalf and follow their "agenda," such as feeding and reproducing. Teleonomy is evident at all levels of life, from the single-cell to multicellular levels. As one example, bacteria "swim" toward a high-glucose region, in search for food. Pross points out the similarity between the teleonomic character of life and life's élan vital, which is a discredited concept. This is the case even when élan vital is replaced with seemingly scientific terms, such as "emergent properties of complex systems." Pross (2003, 2012) places teleonomy on a scientifically transparent basis in his proposal of physicochemical characterization of teleonomy via dynamic kinetic stability (DKS).

Life cycles of even the most primitive life are complicated. They include at various stages life forms that by themselves may not have properties of fully developed life. Kolb and Liesch (2008) included these forms in their definition of life: "Life is a chemical phenomenon which occurs in space and time as a succession of life forms which combined have a potential to metabolize, reproduce, interact with the environment, including other life forms, and are the subject of natural selection." However, a question remains about the identity of individual life forms and how they relate to the fully developed life. This is best handled via the philosophy of identity, which is described in the next subsection.

2.1.3.1 PHILOSOPHICAL PROBLEM OF IDENTITY AND ITS APPLICATION TO THE DEFINITION OF LIFE

The identity problem is an old philosophical problem, which has been applied to defining life (Kolb 2010). To start with, "The Ship of Theseus" paradox, described by Plutarch in 75 ACE, raises the question of whether a ship that has all its parts replaced, plank by plank, remains the same ship. This paradox can also be applied to the living organisms. For example, one can raise questions if an organism stays the

same despite its cell turnover and if a cell that has its constituent molecules replaced in a catabolic/metabolic cycle remains the same. Different ways to handle this problem have been proposed (Kolb 2010). The identity or sameness is covered here. Identity is what makes an entity definable, recognizable, and distinguishable from other entities. When one looks at a living entity at this particular moment, in its present space, and compares it with what it was in the past when it was different (such as during its developmental life stages), and possibly also in a different space, one faces a problem in defining its identity. Either one can say that its identity remains the same, although the living entity was qualitatively different in the past, or that it is different. The latter option means that the living entities would have to have an infinite number of identities, over a continuum of time and the associated space. This would not be useful. One way out of this problem is to use the so-called numerical identity, by which an individual living entity is numerically the same, namely "one"; however, at any time and point in time/space, it may exist as a different life form. Identity of living entities during a spatiotemporal succession of its life forms can be handled by considering the identity over time (Gallois 2003, 2016). Gallois proposed that things can be identical at one time but distinct at another time. This would represent "occasional identities." Gallois also addressed the identity issues in the case of a dividing amoeba, the subject that was discussed and further developed by Kolb (2010). The consideration of the identity problem has resulted in a new definition of life: "Life of an organism is the sum of its life forms over a period of time. We set the integral of time from the birth of the organism to its death" (Kolb 2010).

In the next section, selected examples of definitions of life are provided, together with their focus and strengths.

2.1.4 REPRESENTATIVE EXAMPLES OF TYPES OF DEFINITIONS OF LIFE AND THEIR FOCUS AND STRENGTHS

In this section, selected examples of definitions of life are listed. Each definition is followed by a short statement about its focus and strength. Definitions are listed in no special order. Note that the bibliographical source for a number of these definitions is Popa (2004).

Life 1a: the quality that distinguishes a vital and functional being from a dead body; b: a principle or force that is considered to underlie the distinctive quality of animate beings (compare vitalism); c: an organismic state characterized by capacity for metabolism, growth, reaction to stimuli, and reproduction ... 5 a: the period from birth to death (Merriam-Webster's Collegiate® Dictionary, Tenth Edition, 1993). This is an example of a lexical definition of life.

Life as a system capable of (1) self-organization, (2) self-replication, (3) evolution through mutation, (4) Metabolism, and (5) concentrative encapsulation (Arrhenius, 2002; source Popa 2004). Strength:

It gives a list of essential properties of life and is comprehensive.

Life is the process of existence of open non-equilibrium complete systems, which are composed of carbon-based polymers and are able to self-reproduce and evolve on basis of template synthesis of their polymer components (Altstein, 2002; source Popa 2004). Strength: It is specific about thermodynamic and chemical requirements.

Life may be described as "a flow of energy, matter and information" (Baltscheffsky, 1997; source Popa 2004). Strength: It includes flow of information.

Life is an expected, collectively self-organized property of catalytic polymers (Kauffmann, 1993; source Popa 2004). Strength: It includes catalysis and self-organization; life is expected.

Life is a self-sustained chemical system capable of undergoing Darwinian evolution (NASA's working definition of life; Joyce 1994, 2002; source Popa 2004). Strength: It includes Darwinian evolution and self-sustainability as criteria for life.

Life is defined as "a material system that can acquire, store, process, and use information to organize its activities" (Dyson, 2000; source Popa 2004). Strength: It includes information.

Any living system must comprise four distinct functions: (1) increase of complexity; (2) directing the trends of increased complexity; (3) preserving complexity; and (4) recruiting and extracting the free energy needed to drive the three preceding motions (Anbar, 2002; source Popa 2004). Strength: A focus on complexity.

Any system capable of replication and mutation is alive (Oparin, 1961; source Popa 2004). Strength: It gives key properties of life.

Life is self-reproduction with variations (Trifonov 2011). Strength: This is a minimal definition of life. It is like Oparin's definition (Oparin 1961; source Popa 2004).

Life is *metabolizing material informational system* with the *ability* of *self-reproduction* with *changes* (*evolution*), which requires *energy* and *suitable environment* (Trifonov 2011). This is a composite definition of nine groups of defining terms that were used in 123 definitions of life. Strength: It is comprehensive.

Life is a chemical system capable of transferring its molecular information independently (self-reproduction) and also capable of making some accidental errors to allow the system to evolve (evolution) (Brack, 2002; source Popa 2004). Strength: Chemical accidental errors enable evolution.

Life is synonymous with the possession of genetic properties, that is, the capacities for self-replication and mutation (Horowitz, 2002; source Popa 2004). Strength: A focus on key properties of life.

Life is what the scientific establishment (probably after some healthy disagreement) will accept as life (Friedman, 2002, paraphrasing Theodosius Dobzhansky; source Popa 2004). Strength: It reveals a drawback of stipulatory definitions, which can change.

Life is a new quality brought upon an organic chemical system by a dialectic change resulting from an increase in quantity of complexity of the system. This new quality is characterized by the ability of temporal self-maintenance and self-preservation (Kolb, 2002; source Popa 2004). Strength: It includes the origin of life as the result of abiotic-to-biotic transition via a dialectic change, and as such is expected.

Life is a chemical phenomenon that occurs in space and time as a succession of life forms that combined have a potential to metabolize, reproduce, interact with the environment, including other life forms, and are the subject of natural selection (Kolb and Liesch 2008). Strength: It includes life forms; comprehensive.

We propose that life of an organism is the sum of its life forms over a period of time. We set the integral of time from the birth of the organism to its death (Kolb 2010). Strength: It includes life forms of an organism.

It's alive if it can die (Lauterbur, 2002; source Popa 2004). Strength: It includes death as a criterion for life.

No physiology is held to be scientific if it does not consider death an essential factor of life. Life means dying (Engels ca. 1880; source Popa 2004). Strength: It includes death in defining life.

Life (L) is a total sum (Σ) of all acts of communication (C) executed by a sender-receiver at all its levels of compartmental organization. Thus $L = \Sigma C$ (De Loof 2015). Strength: It includes communication as a criterion for life.

Life is synonymous with the possession of genetic properties. Any system with the capacity to mutate freely and to reproduce its mutation must almost inevitably evolve in directions that will ensure its preservation. Given sufficient time, the system will acquire the complexity, variety, and purposefulness that we recognize as alive (Horowitz, 1986; source Popa 2004). Strength: It gives the essential properties of life, and it includes purposefulness.

The characteristics that distinguish most living things from non-living things include a precise kind of organization; a variety of chemical reactions that we term metabolism; the ability to maintain an appropriate internal environment, even when the external environment changes (a process referred to as homeostasis); and movement, responsiveness, growth, reproduction, and adaptation to environmental change (Vilee et al., 1989; source Popa 2004). Strength: It gives a detailed list of the essential properties; it includes homeostasis.

Any definition of life that is useful must be measurable. We must define life in terms that can be turned into measurables and then turn these into a strategy that can be used to search for life. So, what are these? (1) structures; (2) chemistry; (3) replication with fidelity; and (4) evolution (Nealson, 2002; source Popa 2004). Strength: A focus on measurables to define life.

Life is a metabolic network within a boundary (Maturana and Varela, 1980; reformulated by Luisi, 1993; source Popa 2004). All that is living must be based on autopoiesis, and if a system is discovered to be autopoietic, that system is defined as living; that is, it must correspond to the definition of minimal life (Maturana and Varela 1980). Strength: It gives autopoiesis as a criterion for life.

A living system is a system capable of self-production and self-maintenance through a regenerative network of processes that takes place within a boundary of its own making and regenerates itself through cognitive or adaptive interactions with the medium (Damiano and Luisi 2010; this is a reformulation of the original Maturana and Varela definition of living; Maturana and Varela 1980). Strength: Focus on autopoiesis.

Life is a historical process "as the mode of existence of ribosome encoding organisms (cells) and capsid encoding organisms (viruses) and their ancestors" (Forterre 2010). Strength: This definition includes viruses.

The selection of the definitions of life reflects their diversity and different approaches. They justify Popa's statement: "We may never agree on a definition of life which will remain forever subject of a personal perspective. The measure of one's scientific maturity may actually be his/her latest definition of life and the acceptance that it cannot be ultimate" (Popa 2010).

2.1.5 MINIMAL VERSUS EXPANDED DEFINITIONS: PROS AND CONS

In his paper titled "Vocabulary of definitions of life suggests a definition," Trifonov analyzed the vocabulary of 123 definitions of life (Trifonov 2011). The analysis revealed nine groups of defining terms. A composite definition of these is: "*Life* is *metabolizing material informational system* with *ability* of *self-reproduction* with *changes* (*evolution*), which requires *energy* and *suitable environment*." Out of these nine groups, (self) reproduction and evolution (variation) appear to be a minimal set for a concise and inclusive definition: "Life is self-reproduction with variations." This definition is quite similar to Oparin's definition: "Any system capable of replication and mutation is alive" (Oparin 1961; source Popa 2004).

Trifonov's paper elicited lively responses. Some were critical of his approach, some attempted to supplement the minimal definition of life, and some presented their own views on the matter. Examples of the responses to Trifonov's paper are given, with selected arguments: (1) complexity could be used as a basis for defining life (Jagers op Akkerhuis 2012); (2) whether or not viruses are alive, they are clearly on a continuum between living and non-living; the existence of such a continuum complicates the search for an all-purpose definition of life (Hansma 2012); (3) life is a process, not a system (Di Mauro 2012); (4) all known biological replicators exhibit digital character, since they are polymers consisting of multiple types of monomers; digital properties are necessary for life (Koonin 2012); (5) the definition includes self-replication, which is an activity, but does not specify any quality that the agent needs to satisfy; thus, life simulations can be classified as actual forms of life (Macagno 2012); (6) vocabulary analysis does not adequately address complex words, which may not be often used, such as "cryptic information" and "energy dissipative systems" (Popa 2012); and (7) a disappointing feature of the minimal definition of life is its failure to connect with the problem of the origin or emergence of life (Tang 2012).

As pointed out by Tang, good definitions of life should connect with the emergence of life (Tang 2012). Higgs has addressed this subject in his paper titled "Chemical evolution and the evolutionary definition of life" (Higgs 2017). He suggests that a good place to put the conceptual boundary between non-life and life is between the chemical evolution and the biological evolution.

A question may be posed if a minimal definition is better than a more expanded one. The quest for a minimal definition reflects a desire to identify the essential requirements for life, which would be both necessary and sufficient for it. However, the minimal definition works by considering the lowest common denominator for life. Thus, when considering all life from bacteria to humans, highly developed intelligence of humans as a criterion for life is not included, since bacteria do not have it (Kolb 2007). This impoverishes our thinking and the overall insight of what it means to be alive at every level of life. Minimal definitions are not usually suitable to include borderline cases for life. More elaborate definitions, which include a lengthy list of the life's characteristics, may be more useful in describing different types of life and perhaps even the gray areas of life. A disadvantage may be that at least some of these characteristics are non-essential.

In the next subsection, another philosophical approach to the definition of life, the application of Aristotelian principles, is presented.

2.1.5.1 APPLICATION OF ARISTOTELIAN PRINCIPLES TO THE DEFINITION OF LIFE

This discussion is based on the paper "On the applicability of the Aristotelian principles to the definition of life" (Kolb 2007). Here, it is specifically geared toward the examples of the definitions of life from this chapter and the arguments' *pro* and *con* for the minimalistic definition of life. Aristotle considered a living thing by reference to a list of characteristic "life functions," such as self-nutrition, growth, decay, reproduction, appetite, sensation or perception, self-motion, and thinking. In "De Anima," Aristotle states: "Some natural bodies are alive and some are not—by 'life' I mean self-nourishment, growth, and decay." Even today, life is often defined via a list of its characteristics, as seen in the examples given in Section 2.2.4. In his "Categories," Aristotle developed a theory of classification of existing things. Aristotle placed an individual living organism higher in the classification than its species or genera. The reason for his belief is that the individual substance does not lose its qualities as it becomes part of a species and genera, but the converse is not true. Thus, manhood, which would be a generic description of the properties of all men in the species, is not contained in the primary substance, which is an individual man. General is not present in specific, and the abstract is not present in the real. This has a direct relevance to the question if minimalistic definitions of life, which may be abstract, are better than the descriptions of the real-life characteristics. Aristotle's arguments are still valid today.

2.1.6 BORDERLINE CASES OF LIFE: ARE VIRUSES ALIVE? ANSWERS PROVIDED VIA DIALETHEISM AND FUZZY LOGIC

Some biological forms or entities do not exhibit critical features of life, such as self-reproduction and metabolism. Such examples include viruses, which can reproduce only with the help of their host, and they do not carry out metabolism themselves. Because of these deficiencies, and their relative simplicity, viruses are generally not considered to be alive by a large majority of the definitions of life. Some scientists state explicitly that viruses cannot be considered alive, since they are not autopoietic (Margulis and Sagan 1995; Luisi 2016). However, Forterre disagrees and sees viruses as biological entities (Forterre 2010). In his article "Defining life: The virus viewpoint," Forterre suggests the following definition of life, which includes viruses: life is a historical process "as the mode of existence of ribosome encoding organisms (cells) and capsid encoding organisms (viruses) and their ancestors" (Forterre 2010). Viruses do replicate, albeit with the assistance of the host, they evolve and have evolutionary relationship with the host (Villarreal 2005), are ancient, and may have a link to the origins of life (Villarreal 2014), among other properties that are typically considered biological. The discovery of a giant virus that falls ill through infection by another virus (a "virophage") has been used to suggest that viruses are alive, since they can get sick (Pearson 2008). The question if viruses are alive or not is still not resolved to everybody's satisfaction. Villarreal saw viruses in a twilight zone of life (Villarreal 2004). Villarreal's classic paper on the subject so far has not been surpassed. However, progress has been made in understanding viruses in general and if they are alive. For example, Forterre notices that viruses are no more confused with their virions and that they are viewed more often as complex living entities (Forterre 2010).

When considering that most of the definitions of life do not include viruses as alive, one should also reassess the process of making definitions in general. Most definitions of life are stipulative. Most of the experts in the field decide that the definition of life should capture the properties of most of life. Since viruses do not share the key features of most of life, such as self-reproduction and metabolism, they are not considered alive. Such a decision is practical, since viruses are indeed borderline cases. Stipulative definitions in general exclude cases in gray areas. Likewise, in the lexical definitions and in other definitions in which a list of various properties of life is given, only the most common properties are given. These definitions tend to cover the lowest common denominator in all life, from lowly bacteria to highly intelligent life. Bacteria do not exhibit high intelligence, so the latter property is typically not included in the definitions of life (Kolb 2007). In case of viruses, one goes even lower on the complexity scale, as compared with bacteria.

Another important factor in answering the question if viruses are alive or not is that definitions of life are often built based on the distinction between living and non-living. Aristotelian classical logic, which allows for only two answers, true and false, can deal only with the alive and non-alive options. However, some believe that there is no sharp boundary between these two options, but, instead, there are degrees and modes of "lifeness" (the state or quality of having a life) (Gayon 2010). Thus, in addition to "alive" and "non-alive," a category of "more-or-less-alive" is introduced (Bruylants et al. 2010; Gayon 2010; Malaterre 2010). This requires a departure from the Aristotelian logic. Examples include application of fuzzy logic (Cintula et al. 2017), according to which there would be degrees of being alive (alive = 1; non-alive = 0; partially alive = higher than 0 up to 1) (Bruylants et al. 2010; Malaterre 2010), which can be applied to viruses, and dialetheism, which we address next.

Dialetheism (Priest 1995, 2006; Priest and Berto 2017) does not follow the Aristotelian logic. Priest examines the limits and boundaries of the mind, thought, concepts, expressions, descriptions, conceptions, and knowing. These limits and boundaries cannot normally be crossed; yet, they may be crossed under the circumstances of transitions. Transcendence beyond these limits may create contradictions. Priest believes that some contradictions may be true and gives a simple example of a true contradiction. He considers a person leaving the room through an open door to go outside. At some point in this transition, as the person exits the room by passing through the open door, the person will be both inside and outside the room. It is important to note that this is true only for the transition situation. Thus, both the statement and its negations could be true. This belief is termed dialetheism (from Greek a two-way truth) (Priest 1995, 2006). Dialetheism would allow for both alive and non-alive options to apply to viruses (Kolb 2010). Viruses are considered non-alive based on the criterion that they cannot self-reproduce. This is true for the virion form of viruses. However, when virions penetrate their hosts, their reproduction becomes possible via the host. Thus, in their hosts, viruses become alive, if we accept assisted reproduction as fulfilling the reproduction criteria for life. Thus, it appears that viruses can be both non-alive and alive, which is a contradiction in the Aristotelian logic. However, if we accept dialetheism, we can propose that viruses can be both non-alive and alive, as they are transitioning from the virion phase to the reproductive phase inside their hosts.

However, the non-Aristotelian logic is necessary only if one sticks to the presently available definitions of life, which exclude viruses. If additional definitions of life are formulated to include viruses, and if they become accepted, the approach of multivalued logic may become obsolete.

Finally, definitions evolve, in a predictable way. For example, stipulative definitions, which represent proposals to use a word in a certain way, can be expanded to include new findings. As more people accept the expanded stipulative definitions, they may become more widely used and, in some cases, can even become lexical (Churchill 1990). If definitions of life are formulated to include the most current knowledge of life and its borderline cases, they will become generally accepted. This is not the case at this time.

2.1.7 CONCLUSIONS

Different methods of defining life are reviewed. Examples of definitions of life are given, and their strengths are described. Minimalistic definitions of life are important, since they presumably capture the essence of life. More elaborate and descriptive definitions are also valuable. They are more suitable for describing diverse life entities than the minimalistic definitions. Descriptive definitions do not try to exclude life properties that may be deemed as non-essential, which may be wise. Problem in classification of the borderline cases of life, such as viruses, is briefly reviewed. A part of the problem resides with the definitions themselves, which address most life but not its gray areas. This problem may be surpassed if one departs from the Aristotelian logic, which would allow only two states, alive and non-alive, and uses the non-Aristotelian logic, such as dialetheism and fuzzy logic. Dialetheism allows for the states of being both alive and non-alive during the transitions, and the fuzzy logic assigns the state of life on a scale of 1 (alive) to 0 (non-alive) and the cases in between (less than 1 and higher than 0). Definitions may evolve to address new knowledge about life and its borderline cases. At some point, such inclusive definitions of life will be universally accepted, but this is presently not the case.

ACKNOWLEDGMENTS

Special thanks are due to Ben Clark for his detailed review of this chapter and many helpful comments. Donald A. Cress and Joseph Pearson have provided useful input about different definitions, as formulated in logic.

REFERENCES

Ball, P. 2016. Man made: A history of synthetic life. *Distillations* 2: 14–23.

Benner, S. A. 2010. Defining life. *Astrobiology* 10: 1021–1030.

Benner, S. A., A. Ricardo, and M. A. Carrigan. 2004. Is there a common chemical model for life in the universe? *Curr. Opin. Chem. Biol.* 8: 672–689.

Bruylants, G., K. Bartik, and J. Reisse. 2010. Is it useful to have a clear-cut definition of life? On the use of fuzzy logic in prebiotic chemistry. *Orig. Life Evol. Biosphere* 40: 137–143.

Callaway, E. 2016. Race to design life heats up. Craig Venter's minimal-cell triumph comes as the CRISPR gene-editing method provides alternative ways to tinker with life's building blocks. *Nature* 531: 557–558.

Churchill, R. P. 1986. *Becoming Logical, An Introduction to Logic*, pp. 106–121. New York: St. Martin's Press.

Churchill, R. P. 1990. *Logic, An Introduction*, 2nd ed., pp. 122–140. New York: St. Martin's Press.

Cintula, P., C. G. Fermüller, and C. Noguera. 2017. Fuzzy logic. The Stanford Encyclopedia of Philosophy (Fall 2017 Edition), E. N. Zalta (Ed.). https://plato.stanford.edu/archives/fall2017/entries/logic-fuzzy/.

Cleland, C. E., and C. F. Chyba. 2002. Defining life. *Orig. Life Evol. Biosphere* 32: 387–393.

Cleland, C. E., and C. F. Chyba. 2007. Does 'life' have a definition? In *Planets and Life: The Emerging Science of Astrobiology*, W. T. Sullivan III, and J. A. Baross (Eds.), pp. 119–131. Cambridge, UK: Cambridge University Press.

Damiano, L., and P L. Luisi. 2010. Towards an autopoietic redefinition of life. *Orig. Life Evol. Biosphere* 40: 145–149.

De Loof, A. 2015. How to deduce and teach the logical and unambiguous answer, namely L = ΣC, to "What is life?" Using the principles of communications? *Commun. Integr. Biol.* 8(5): e1059977. doi:10.1080/19420889.2015.1059977.

Di Mauro, E. 2012. Trifonov's meta-definition of life. *J. Biomol. Struct. Dyn.* 29: 601–602.

Forterre, P. 2010. Defining life: The virus viewpoint. *Orig. Life Evol. Biosphere* 40: 151–160.

Gallois, A. 2003. *Occasions of Identity, The Metaphysics of Persistence, Change, and Sameness*, pp. 1–10, 16–19, 25–28, 58–59, 64, 70–77, 101–110, 255–264. Oxford, UK: Oxford University Press.

Gallois, A. 2016. Identity over time. The Stanford Encyclopedia of Philosophy (Winter 2016 Edition), E. N. Zalta (Ed.). https://plato.stanford.edu/archives/win2016/entries/identity-time/.

Gayon, J. 2010. Defining life: Synthesis and conclusions. *Orig. Life. Evol. Biosphere* 40: 231–244.

Gupta, A. 2015. Definitions. The Stanford Encyclopedia of Philosophy (Summer 2015 Edition), E. N. Zalta (Ed.). https://plato.stanford.edu/archives/sum2015/entries/definitions/.

Hansma, H. G. 2012. Life = self-reproduction with variations? *J. Biomol. Stuct. Dyn.* 29: 621–622.

Higgs, P. G. 2017. Chemical evolution and the evolutionary definition of life. *J. Mol. Evol.* 84: 225–235.

Hutchinson III, C. A., P-Y. Chuang, V. N. Noskov et al. 2016. Design and synthesis of a minimal bacterial genome. *Science* 351: 1414.

Jagers op Akkerhuis, G. A. J. M. 2012.The role of logic and insight in the search for a definition of life. *J. Biomol. Struct. Dyn.* 39: 619–620.

Kolb, V. M. 2005. On the applicability of the principle of the quantity-to-quality transition to chemical evolution that led to life. *Int. J. Astrobiol.* 4: 227–232.

Kolb, V. M. 2007. On the applicability of the Aristotelian principles to the definition of life. *Int. J. Astrobiol.* 6: 51–57.

Kolb, V. M. 2010. On the applicability of dialetheism and philosophy of identity to the definition of life. *Int. J. Astrobiol.* 9: 131–136.

Kolb, V. M. 2012. On the laws for the emergence of life from the abiotic matter. *Proc. SPIE Instruments, Methods, and Missions for Astrobiology XV* 8521: 852109. doi:10.1117/12.924817.

Kolb, V. M. 2016. Origins of life: Chemical and philosophical approaches. *Evol. Biol.* 43: 506–515.

Kolb, V. M., and P. J. Liesch. 2008. Abiotic, biotic and in-between. *Proc. SPIE* 7097: 70970A. doi:10.1117/12.792668.

Koonin, E. V. 2012. Defining life: An exercise in semantics or a route to biological insights? *J. Biomol.Struct. Dyn.* 29: 603–605.

Liesch, P. J., and V. M. Kolb. 2007. Living strategies of unusual life forms on Earth and the relevance to astrobiology. *Proc. SPIE* 6694: 66941F. doi:10.1117/12.731346.

Luisi, P. L. 2006. *The Emergence of Life: From Chemical Origins to Synthetic Biology*, pp. 17–37. Cambridge, UK: Cambridge University Press.

Luisi, P. L. 2016. *The Emergence of Life: From Chemical Origins to Synthetic Biology*, 2nd ed., pp. 119–156, 247–261. Cambridge, UK: Cambridge University Press.

Macagno, F. 2012. Classifying the properties of life. *J. Biomol. Struct. Dyn.* 29: 627–629.

Malaterre, C. 2010. On what it is to fly can tell us something about what it is to live. *Orig. Life. Evol. Biosphere* 40: 169–177.

Margulis, L., and D. Sagan. 1995. *What is life?* pp. 23–24. New York: Simon & Schuster.

Maturana, H. R., and F. J. Varela. 1980. Autopoiesis and cognition: The realization of the living. In *Boston Studies in the Philosophy of Science*, Vol. 42, R. S. Cohen, and M. W. Wartofsky (Eds.). Dodrecht, the Netherlands: Springer.

Pályi, G., Zucchi, C., and Caglioti, L. (Eds.). 2002. *Fundamentals of Life*, pp. 15–55. Paris, France: Elsevier.

Pearson, H. 2008. 'Virophage' suggests viruses are alive. *Nature* 454: 677.

Peretó, J. 2005. Controversies on the origin of life. *Int. Microbiol.* 8: 23–31.

Perry, R. S., and V. M. Kolb. 2004. On the applicability of Darwinian principles to chemical evolution that led to life. *Int. J. Astrobiol.* 3: 45–53.

Popa, R. 2004. *Between Necessity and Probability: Searching for the Definition and Origin of Life*, pp. 197–205. New York: Springer. This book has biographical information for the citations of the definitions of life.

Popa, R. 2010. Necessity, futility and the possibility of defining life are all embedded in its origin as a punctuated-gradualism. *Orig. Life Evol. Biosphere* 40: 183–190.

Popa, R. 2012. Merits and caveats of using a vocabulary approach to define life. *J. Biomol. Struct. Dyn.* 29: 607–608.

Popa, R. 2014. Elusive definition of life: A survey of main ideas. In *Astrobiology: An evolutionary approach*, V. M. Kolb (Ed.), pp. 325–348. Boca Raton, FL: CRC Press/Taylor & Francis.

Priest, B. 2006. What's so bad about contradictions? In *The Law of Non-Contradiction, New Philosophical Essays*, G. Priest, J. C. Beall, and B. Armour-Garb (Eds.), pp. 23–38. Oxford, UK: Oxford University Press.

Priest, G. 1995. *Beyond the Limits of Thought*, pp. 3–8, 56–70, 104–105. Cambridge, UK: Cambridge University Press.

Priest, G., and F. Berto 2017. Dialetheism. The Stanford Encyclopedia of Philosophy (Spring 2017 Edition), E. N. Zalta (Ed.). https://plato.stanford.edu/archives/spr2017/entries/dialetheism/.

Pross, A. 2003. The driving force for life's emergence: Kinetic and thermodynamic considerations. *J. Theor. Biol.* 220: 393–406

Pross, A. 2012. *What is Life? How Chemistry Becomes Biology.* Oxford, UK: Oxford University Press.

Ruiz-Mirazo, K., J. Peretó, and A. Moreno. 2004. A universal definition of life: Autonomy and open-ended evolution. *Orig. Life Evol. Biosphere* 34: 323–346.

Schummer, J. 2016. Are you playing God? Synthetic biology and the chemical ambition to create artificial life. *HYLE – Int. J. Phil. Chem.* 22: 149–172.

Tang, B. L. 2012. A minimal or concise set of definition of life is not useful. *J. Biomol. Struct. Dyn.* 29: 613–614.

Tirard, S. 2010. Origin of life and definition of life, from Buffon to Oparin. *Orig. Life Evol. Biosphere* 40: 215–220.

Tirard, S., M. Morange, and A. Lazcano. 2010. The definition of life: A brief history of an elusive scientific endeavour. *Astrobiology* 10: 1003–1009.

Trifonov, E. N. 2011. Vocabulary of definitions of life suggests a definition. *J. Biomol. Struct. Dyn.* 29: 259–266.

Tsokolov, S. A. 2009. Why is the definition of life so elusive? Epistemological considerations. *Astrobiology* 9: 401–412.

Venter, J. C. 2007. *A Life Decoded. My Genome: My Life.* New York: Penguin Books.

Venter, J. C. 2013. *Life at the Speed of Light: From the Double Helix to the Dawn of Digital Life.* New York: Penguin Books.

Villarreal, L. P. 2004. Are viruses alive? *Sci. Amer.* 291(6): 100–105.

Villarreal, L. P. 2005. *Viruses and the Evolution of Life.* Washington, DC: ASM Press.

Villarreal. L. P. 2014. Virolution can help us understand the origin of life. In *Astrobiology: An Evolutionary Approach*, V. M. Kolb (Ed.), pp. 421–440. Boca Raton, FL: CRC Press/Taylor & Francis.

Weber, B. 2015. Life. The Stanford Encyclopedia of Philosophy (Spring 2015 Edition), E. N. Zatta (Ed.). https://plato.stanford.edu/entries/life/;https://plato.stanford.edu/archives/spr2015/entries/life/.

2.2 A Generalized and Universalized Definition of Life Applicable to Extraterrestrial Environments

Benton C. Clark

CONTENTS

2.2.1 INTRODUCTION: WHY A DEFINITION OF LIFE?

The need for, and a proposed version of, an operational definition that intends to capture the essence of life in its most fundamental terms was presented at the ALIFE IX Conference in Boston (Clark 2004). This chapter draws heavily on the workshop-tutorial materials of that conference. However, those materials were not included in the published conference proceedings, and for that reason, the survey of previous attempts of defining life by Trifonov and his sources did not include this viewpoint (Trifonov 2011, 2012). What follows entails updates of the concept and its relevance to the work of Trifonov and those who have provided published comments on his work. In addition, this chapter addresses the reticence by some scientists and philosophers to accept the premise that a definition of life (DoL) is possible (Cleland and Chyba 2002, 2007) or useful (Szostak 2012; Tessera 2012).

Definitions of life now appear profusely in the philosophical, biological, and now even the computer literature. Biologists often independently construct lists of favorite characteristics of life as they best know it. For decades, a typical biologist's approach has been to narrow down favorite attributes to a half dozen or so to advocate the keystones of a living system (Oparin 1965; Mazia 1966; deDuve 1991; Hickman et al. 2001; Koshland 2002). One problem has been that there is no general consensus—only three or four of these key attributes appear on all lists. A greater problem is that there are at least 104 attributes of life in the literature, many of which are non-essential, such as irritability.

Most DoLs have not been sufficiently generalized and universalized (GU) for application to alien life or artificial constructs of new forms of life (generalized, in the sense of being applicable to all currently known forms of life and non-life; universalized, by encompassing conceivable new or novel forms of life). Some DoLs seem simpler, yet are not universal. For example, the so-called NASA DoL, "Life is a self-sustained chemical system capable of undergoing Darwinian evolution," (Joyce 1994) is precise and succinct but prescribes that the search for life, for example, on Mars, be focused on evidence for chemical functionality and an apparent requirement for independence ("self-sustained"). Furthermore, the assumption that only Darwinian evolution can be employed by an alien form of life is both fundamentally unprovable and difficult to observe.

2.2.2 NEED FOR A DEFINITION OF LIFE

The need for a DoL is sometimes questioned. "We'll know it when we see it." The Viking missions to Mars in 1976 adopted specific criteria but obtained inconclusive results on possible metabolic-like activities of chemically stimulated samples of Martian soil, leading to a controversy that persists to this day as to whether these missions did or did not detect life.

Billion-dollar class space missions to Mars, Europa, Titan, Enceladus, Ceres, and other planetary objects of astrobiological interest are aimed toward the ultimate objective of detecting life or evidence of past life or prebiotic activity. At these costs, it is incumbent upon astrobiologists to provide measurements whose results can be traced to well-defined fundamental objectives. Certainly, a more-accepted definition of what constitutes life as compared with non-life, including guidance on criteria that may be used to uniquely infer the presence of life activities, is needed for the fiduciary accountability of budgeting public funds for space exploration, both within and beyond our solar system.

A DoL is also of profound importance to understanding nature's transition from inanimate milieu to the first organism. The Origin scenario remains unknown but cannot be well investigated without some concept of what is being sought.

Finally, for the sake of intellectual clarity and pedagogical responsibility to the student, it is time for biologists writing the textbooks of the future to come up with something better than: "Life is cells; cells are compartments containing proteins, nucleic acids, and other biochemicals; cells divide; read further to learn the difference between mitosis and meiosis."

2.2.3 ELUSIVE THEORY OF LIFE

A proposal that defining life be postponed until "a theory of life" comes into being (Cleland and Chyba 2002, 2007) takes as its analogy the historical delay in recognizing that the chemical composition of water is H_2O. This line of reasoning can be challenged on several grounds. Water is a specific substance, whereas it is clear that life is a phenomenon, with many different manifestations in bulk chemical makeup.

The statement that water is H_2O, and vice versa, embodies an entire hierarchy of hidden sub-definitions. We must know that H_2O is a *molecule*; that molecules are made of *atoms*; that *quantum mechanical* (QM) rules govern the energetics and stability of *chemical bonds*, which in this case are *covalent*; that the bonding is related to QM configuration *orbitals* of *electrons* surrounding a charged *nucleus*; etc., etc., etc. Italicized terms in the preceding sentence require lengthy definitions, some of which will lead to additional new terms recognized only by specialists. Even QM theory has not yet been successfully applied to quantitatively predict all the physicochemical properties of water.

The thought has been that "H_2O" explains the substance "water." It does not. In its form most valuable to life, water is a liquid. But it also may exist as one of several ices (amorphous or specific crystalline habits), as gaseous vapor, as a supercritical fluid, or even as a plasma. Much more serious

is that even in the ultrapure state, the liquid substance is not solely H_2O molecules. Rather, its pH is 7.0, which means that small but definite amounts of H^+ and OH^- are intimately associated with it. These "non-H_2O entities" cannot be removed. Are they water or not? The chemist would say they certainly are not—they are hydrogen and hydroxyl ions—and would further complicate the story with species such as H_3O^+ and processes such as hydrogen bonding. Owing to the ever-present galactic cosmic ray flux plus emissions of additional ionizing radiation from long-lived natural radioisotopes in soil and rock, the physicist would then point out other species that are present in this water substance, including peroxides and the solvated electron. Do we really understand all that is there to know about this substance? No, but this does not mean that we cannot define what it is for now, perhaps taking extra care to distinguish between water, the molecule (H_2O); water, the liquid (aqua); the various forms of ice; etc. Yet, every human on planet Earth is intimately familiar with water substance, its powerful effects as the dominant fluid of nature, as well as its extraordinary values to humanity, including its absolute necessity for sustenance (typically by drinking).

We may not yet understand the origins of life, but we also do not yet understand the origin of our oceans of water on Earth—whether it is mainly due to endogenous sources such as lava and magmatic degassing of rock in the Earth's mantle or to exogenous delivery of cometary ices and hydrous asteroids during terminal accretion of the planet.

Another example of definition difficulties is "the flu." We all are well aware of the influenza, regardless of whether we have ever directly experienced it. As a potential scourge to humanity, it is second to none. It drove to its end the First World War, but by exacting a tragic toll of 20 million human lives. Any attempt at a strict definition of "influenza" immediately encounters two critical difficulties. A broad range of outward attributes (i.e., symptoms of the disease: fever, body aches, loss of appetite or intestinal distress, and general weakness and malaise) render diagnosis difficult, especially in the face of numerous other similar but distinctly different infectious diseases (e.g., the common cold). The manifestations of influenza are thus inexact. In addition, the core cause of the disease, the influenza viral genome, comes in many different RNA sequences. As a zoonose, the next dangerous flu epidemic is likely to involve a totally new invention, a genotype never before spread on planet Earth. We cannot specify its precise composition. However, we can specify most of its sequence and many of its properties in advance. And we understand the general phenomenon in terms of how it operates, how it comes about, and how to classify it (an orthomyxovirus). The lack of a formula as precise as H_2O or lack of an unambiguous symptoms fingerprint (attributes list) does not put off a whole suite of on-going activities, ranging from investigations of its functions, studies of its etiology, and a continual search to detect new forms as soon as they appear for preparations for prevention of infectious disease in humans. Likewise, bacteria, protista, and metazoa are far more resistant to definition by their composition than simple viruses.

2.2.4 DATA BASE FOR GENERALIZATION

The full inventory of species in the biosphere constitutes a broad and diverse Data Base of Life. It is sometimes stated, however, that we cannot understand the nature of life because "we have only one example of it." This refers, of course, to the surprising discoveries in similarities in the biochemical fundamentals of all known living organisms. A restricted set of nucleotides and amino acids, and especially their polymers, permeate our terrestrial biosphere. The mechanics of our genetic underpinnings seem to be nearly the same in all organisms, great or small, ancient or modern.

This point of view neglects, however, the astounding diversity of phenotypes that spring forth from these biochemical and genetic foundations. Seahorses and thoroughbred race horses—one a fish and the other a mammal—have little in common except the shape of their heads and the nature of their cells. Conversely, bacteria and archaea have morphologies and sizes that fall within a very restricted universe of modalities. Similar-appearing organisms can differ dramatically in terms of which energy sources they utilize (solar photons, or a wide range of chemical redox couples) and ecological interactions (independent vs. various degrees of symbioses: parasitism, commensalism, and mutualism). Life in its breadth is far more astonishing than what the science fiction writers for *Star Trek* and *Star Wars* have so far been able to imagine. Diversity in shape, size, and functionality, just in our single biosphere, provides countless examples of different implementations of life.

A wide variety of phenotypes of organisms is recognized as alive, in contrast to certain natural but unrelated phenomena having similarities. A few key examples are provided in Table 2.2.1.

In the first column are numerous organisms and ecological units that are undeniably "alive," both in common usage and scientific acceptance. In the second column, "not alive," a set of opposite cases, is presented, some of which have been previously considered "alive" by at least some cultures, but all of which, under the illumination of scientific understanding, are phenomena or objects that are not categorized as being in the realm of the living. The third column is a set of cases about

which there has been uncertainty or debate but which can in fact be categorized by this DoL. It is the first two columns that were mainly utilized to construct this GU DoL.

Reproduction cannot be the sole criterion for life. Crystals reproduce and so do fires and even lightning—in all three cases, accelerated by auto-catalysis. These phenomena also transduce energy ("metabolize") and grow in size. In contrast to the beliefs of the ancients, we now clearly understand the science, albeit imperfectly, behind all three phenomena. They are not life. To overcome the nettlesome problem that these are not alive, we must invoke some other property. In this DoL, the choice is to recognize that all known life forms contain an instruction set that is tied to their reproductive uniqueness and metabolic activities. These instructions are advantageously kept separated from structure and the numerous required and optional functionalities (von Neumann 1966; Poundstone 1984; Benner 2010); however, in principle, they need not be physically distinct—e.g., one-biopolymer cell (Benner 1999) and the RNA World (Pressman et al. 2015).

Reproduction also is not the essence of all living organisms. The classic example is the mule, an offspring of donkey (male) and horse (female), which itself is sterile because of a miss-match in ploidy between the two parental species. In many eusocial insect communities, the workers are non-reproductive (ants, bees, and wasps), yet they are the organisms that are essential for the propagation of the colony.

2.2.5 GENERALIZED AND UNIVERSALIZED DEFINITION OF LIFE

A simple definition of lightning is "an electrical discharge in air, arising naturally." Yet, the phenomenon of lightning is anything but simple. Cumulonimbus cloud formation; aerosol physics; electrostatics of charging; Maxwell's equations of electromagnetism; ionization potentials; coronal and arc discharges; streamers; bifurcation ratios; excited states of atoms; photon production and propagation; etc., all come into play.

Similarly, simplified DoLs are "system with evolvable self-replication" or "reproduction with heritance." Yet, life

---

TABLE 2.2.1
Cases in the Data Base (Columns Not Meant to Always Correlate)

Alive	Not Alive	Questionable
Homo sapiens	Hair, fingernails, urine	Comatose human, Frankenstein monster
Amoeba	Fire, volcanoes, lightning	Organs, bone, erythrocyte
Drosophila melanogaster	Rivers, clouds, earthquakes	Virus, viroid, mitochondrion
Bacillus subtilis	Crystals, minerals	Crystal defect pattern
Lichen	Weathering rinds	Spore, cyst, seed
Birds and bees	Nest, honeycomb	Prions, organelles
Butterflies and trees	Cocoon silk, cellulose	DNA, srTn, exons, introns
Endoliths	Chondrite	Martian meteorites
Rainforest	Mahogany table	Mother Earth
Microbial mats	Fossil stromatolites	Computer virus, evolutionary algorithms
Coral reef	Coral jewelry	Replicating robots

on Earth is gloriously complex and proficient in its ability to survive, to transform, and to spread. As previously described (Clark 2001a, 2002), the fundamentals of life can be generalized and then universalized by stripping away dependencies on the specificities of terrestrial organisms, such as physical configurations (morphologies), scales of size or time, biochemical modalities, and even chemistry itself.

The DoL adopted here is in two parts because of recognition of the need to distinguish specialist organisms, which may have no *direct* role in reproduction, from the essential reproductive unit. To highlight this distinction, we combine the two individual words of "life" and "form" into a single, all-representative word encompassing the minimum reproductive set of organisms, the Lifeform (Lf). We formulate separate definitions for Life, Lifeform, and Organism. Accordingly, **Life** is any phenomenon of reproduction and activity that has the following characteristics:

> **A Lifeform is a single *organism*, or a collection of specialist organisms, whose ability to reproduce is enabled by a set of indispensable yet modifiable *instructions* embedded in the Lifeform.**

Although this statement is simple and succinct, it alone is insufficient as a DoL because of the question of what, exactly, is an *organism*. It furthermore leaves unanswered whether there can be non-reproductive organisms. For example, in the case of humans, a very large fraction of the population consists of non-reproductive persons—including pre-puberty children, non-sexually active adults, males and females with age-related cessation of fertility, and persons born sterile or who become infertile, etc.—who are all nonetheless "alive" and enormously contributive to the success of the Lf. Thus, the following is essential to complete this DoL:

> **An Organism is any *physical entity* produced by a Lifeform that can, in a *suitable environment*, affect the *flow and/or conversion of energy* to perform active functions *guided by a subset of the Lifeform's instruction set.***

By definition, an organism is "alive." Some organisms are not reproductive, and some Lfs need more than one type of organism (as in sexual reproduction) to produce a single new organism. The act of reproduction is typically non-trivial. It can require the active functions of accumulation and sequestration of raw materials, their processing into useful materials, an elaborate construction process, and the mechanics of separating the new organism from the parent organism(s). Many organisms on this planet today are multi-tasking bundles of prodigious and manifold functionalities.

The organism is guided by instructions, but not all instructions are necessarily embedded in each individual organism. Forms of communication, such as mating rituals and the worker bee's waggle dance, are in-effect instructions via which an organism's activities are beneficial to the Lf. The ubiquitous sensing and response systems built into most organisms will automatically influence their activities, based upon environmental conditions and stimuli.

Biological reproduction is the process of copying the instructions and using them in the production of a new organism. The newly created instructions may be, for example, a faithful (cloned) or mutated copy of previous instructions or created by combining copies of instructions from two or more organisms (e.g., sexual reproduction and horizontal gene transfer).

The Lf encompasses the fundamental unit of biological reproduction.

Our definition emphasizes that environment is critical. Although most organisms are highly efficient under certain conditions, they may enter into a cryptobiotic or dormant state for ultimate survival and will succumb (die) if their environment causes damage, such that they cannot be revived. It is not just physicochemical aspects of the natural environment that are critical but also ecological features—the abundance and types of other organisms that compete directly or indirectly or that provide nutrients or a favorable environment.

It may be difficult to detect whether an entity is alive if it is simply dormant. Likewise, it may also be difficult to determine if an organism that was formerly alive is now dead rather than dormant. The condition of death may be considered as an irreversible transition to a state where an organism is no longer able to control energy to perform useful functions. The only way to ascertain this state may be, in certain cases, by providing it with an optimum, suitable environment.

The reason instructions are so crucial is that crystals, as one example, can suffer dislocations and other defects that propagate because of growth via templating. They contain no instructions per se. These differences are important for distinguishing between the substances that can grow by templating and those that can grow by spontaneous assembly but without instructional content. However, in view of self-replicating transposons (srTns) and viruses, the main distinction must be that the structural templates (genomes) encode critical instructions, which, once decoded, manifest separately from the template itself.

Compared with other definitions, certain specific departures of this DoL from others are notable. The phrase "in a suitable environment" not only makes explicit that which is not always implicit but also emphasizes the importance of organisms that can only live given the existence of other, different organisms or completely different Lfs.

2.2.5.1 THIS DEFINITION IS GENERALIZED

Extreme parasites such as viruses are sometimes considered not to be alive because of a false premise that most organisms are generally independent. The vast majority actually depend upon predation, parasitization, mutualism, or commensalism. Like well-structured computer programs, these hierarchies can be nested several layers deep—a man keeps a dog, the dog harbors worms, the worms contain bacteria, and the bacteria are infested by plasmids and phage. All are alive. Most astoundingly, the non-coding DNA that infiltrates so many genomes has been found to contain "mobile elements"—proliferating short sequences of DNA, some of which encode the keys to their own reproduction. It is therefore the conclusion of this GU DoL that srTns and the like are alive, representing simply another, deeper level in the broad scheme of life.

Parasites vastly outnumber primary organisms, and in that sense, they can be considered even *more* successful. Only a few classes of microbes (e.g., certain anaerobic chemo- and photo-lithoautotrophs) are fully independent of other organisms or their products. For example, the term "free living cell" is often applied to cells that may not be parasitic on a living organism but nonetheless are highly dependent on the availability of nutrients which are in the environment only because of the activities of other organisms. Thus, the synthesis of a minimal bacterial genome by knocking out all genes not required for growth has reduced the number of "essential" genes to less than 500, but the laboratory growth medium used in such investigations is rich in small-molecule metabolites, including vitamins and glucose (Hutchison et al. 2016). Truly independent life grows successfully on substances existing in the geosphere, hydrosphere, and/or atmosphere that were never metabolized by other organisms.

In the notable effort by Trifonov in 2011 to cull a minimalist DoL from 123 definitions by other workers, including the extensive compilations and analysis by Popa in 2004 and more recently in 2015, he at first summarizes properties close to those herein in a definition-1 of life, including the phrases "informational systems," "self-reproduction," "metabolizing … energy," and the often neglected "suitable environment" (Popa 2004, 2014; Trifonov 2011; Kolb 2018). However, Trifonov then eliminates several of these attributes to postulate a streamlined definition-2, that life is "any system capable of replication and mutation."

However, any such definition-2 lacks identification of key requirements to the reader, as also pointed out in comments by others (Egel 2012). By not identifying the need for some input of energy, it neglects the fact that the reproductions performed by life would otherwise violate the inevitable increases in entropy (Schrödinger 1944), as specified in the Second Law of Thermodynamics. Invoking mutation without specifying the transmission and replication of a content of "information," where the mutatable elements (genes) are stored, neglects the breakthrough discoveries of DNA and RNA as the informational molecules at the heart of life on Earth. And eliminating the important condition of a "suitable environment" would seem to render "not alive" status to all dormant organisms, non-reproductive organisms, and highly parasitic organisms. Thus, although we agree with Trifonov (2011, 2012) that the more streamlined version-2 definition is the most fundamental core essence of life, it is not preferred because of the lack of recognition of the key components cited earlier.

An even greater departure from previous work is the distinction between organisms and what is defined here as the Lf, which highlights that for some forms of life, a collection of different organisms is needed to create a reproductive unit, for example, the male/female pair in those species locked into obligatory bisexual reproduction.

A class that is most instructive includes the example of the worker honey bee. Only the queen and drone honey bee can accomplish reproduction, while the vast number of bees in the hive community are female "workers," whose reproductive organs are non-functional. Nonetheless, the worker bee is very much alive by virtue of any number of traits, including vigorous metabolism, the ability to sting, purposeful unnatural movement (counter-current flight), and even intelligence. Here, we term such an entity an organism, but not an Lf. Instead, the Lifeform (as one word) is the assemblage of queen-drone-worker honeybee.

In the case of obligatory parasitism, the host is part of the Lf, willing or not. Thus, an organism can contribute to more than one Lf in certain cases.

A final detail is the difference between organisms and their non-living products. Considering the bird, its nest, and an egg, the first and third are alive, but the second is not. What is special about the nest? Passivity, but especially the lack of instructions. The nest was built, most obviously, according to a plan. This plan, a subset of instructions, is embedded in the bird, not in the nest. As a result, the nest is evidence of an organism as a biomarker but is not an organism itself. We can call such products *Tools* of the organism, to discriminate against non-useful products such as excreted wastes. Tools may be passive (shell, fur, and ant stick) or active (venom and triggered spines). The erythrocyte (red blood "cell") is not only non-reproductive but also contains no genome or other instructions, and in this DoL, it is not considered alive, because even though it performs a metabolic function, it has no genome and none of the instruction set. Likewise, even though tractors or self-driving automobiles can control energy to perform functions, they are not alive, because they do not contain even a portion of the genome (instruction set) of the humans, who created them.

2.2.5.2 This Definition is Universalized

In order to universalize this definition, the key statements are carefully divorced from the plethora of important capabilities that so many of the most successful organisms on Earth have developed, yet that might not be essential in a less-competitive world. Even if a fundamentally different form of life once existed on Earth, it surely was subsumed (and/or consumed) by the prolific, versatile, and prodigious organisms of our current biosphere. Alternative chemistries for life have been proposed, and various forms of artificial life are possible, in which the relative importance of the roles of chemical and physical phenomena is reversed. Based upon terrestrial life, we recognize a need for organics and liquid water, yet our planet is atypically well endowed with these ingredients compared with many other bodies in this and assuredly other solar systems. In Table 2.2.2, a number of attributes that may *not* be essential to all forms of life in the universe are listed.

In order for any DoL to be fundamental, the one-hundred or so attributes of life must be derivable from it. For this GU DoL, some attributes are intrinsic (e.g., reproduction implies growth and organization, and growth implies capture of matter and energy). Most other attributes have been shown to be the plausible outcomes of evolution. Although a DoL that

TABLE 2.2.2

Some Attributes a Universalized DoL Does
***Not* Explicitly Require**

- Scale: Neither on size, nor on speed or time of existence
- H_2O or liquid water
- Organic chemistry or even chemistry
- Conventional nutrients or common energy sources
- Homeostasis or self-repair
- Mobility or motility
- Responsivity, sensors
- Hierarchical structure, intrinsic complexity
- Independence of other Lfs
- Intelligence
- Other *vital* factors: Life imperative, volition, purpose, soul, Gaia, or "The Force"

ignores evolution is almost unthinkable (although some do), it is nonetheless the case here that evolution is a capability, not an essence, because the replication of instructions is unlikely to be error-free. The acquisition of valuable attributes and the invention of new features and strategies by prolonged evolution include everything from motility to senescence, from apoptosis to phototrophy, and from homeostasis to morphogenesis.

2.2.6 FORMALISM OF THIS DEFINITION OF LIFE

If a DoL cannot be considered definitive without stringent specification, an approach would be to express it mathematically. The definition for water stated as "H_2O" is actually an equation:

$$\text{water molecule} = 2 \text{ hydrogen atoms} + 1 \text{ oxygen atom}$$

or,

$$H_2O = 2*H + 1*O$$

Life is a process. As Poundstone has discussed, it is a recursive process (Poundstone 1984). Similarly, this DoL can be expressed mathematically, as a recursive relationship. With separation of dependent and independent variables:

$$I_{n+1} = r\ (I_n, O_n, E)$$

$$O_{n+1} = R\ (I_{n+1}, O_n, E)$$

where O_n is the parent organism with embedded instruction set I_n; the O_{n+1} and I_{n+1} are the progeny organism and instructions, respectively; E represents environmental factors; and r and R designate appropriate functions of these variables. These relationships are meant to capture the process that the instruction set to produce the progeny is from a copy of the instruction set of the parent, which is then operated on by the parent organism to produce the progeny organism. In biology, copying the genome (instruction set, I) is often called

replication (r), while making copies at the organism level is often termed reproduction (R).

Generalizing to the case where multiple organisms, j_{max}, are required to constitute the reproductive unit,

$$I_{n+1} = r\ (I_{n,j}, O_{n,j}, E)$$

$$O_{n+1} = R\ (I_{n+1,j}, O_{n,j}, E)$$

where the r and R functionalities include assimilation of all contributory organisms and instructions ($j = 1$ to j_{max}).

The Lf can similarly be expressed mathematically as the set of O and I:

$$L_f = \left\{ O_j,\ I_j \right\}$$

realizing that the I_j is embedded in the O_j and that some organisms contain minimal instructions, whereas others (e.g., viroids and srTn) are naked instructions.

2.2.7 APPLICATIONS OF THIS DEFINITION OF LIFE

This DoL can be applied to several questions, issues, and opportunities related to various forms of life.

2.2.7.1 ALIEN LIFE

When astronomers and planetary scientists search for H_2O in other places of our universe, they observe for such characteristics as the emission or absorption spectra of H, OH, and H_2O (respectively Lyman alpha, in the far ultraviolet [UV] wavelengths; bond stretch at near-to-mid infrared [IR] wavelengths; and rotational bands at microwave frequencies). These are special attributes of the water substance, as are phenomena such as expansion upon freezing, high specific heat, high latent heats for phase transformations, sharp melting points, and specific crystalline phases when solid.

Likewise, searches for extraterrestrial life will generally settle on certain attributes that are most amenable to measurement and reasonably indicative of living functions. The Viking life-detection instrument results remain inconclusive, but in the main because of a highly restricted set of criteria based upon chemical reactions, which, although indicative of metabolism in tested Earth soils, are now thought by many to reflect the presence of thermally labile but highly reactive inorganic chemical components in a Martian soil whose composition is exotic by terrestrial standards. In fact, nearly all attempts to identify life by one or a few attributes can be shown to have pitfalls, as delineated elsewhere (Clark 2001b, 2003, 2018). The important lesson to recognize is the difficulty of establishing that something is alive under remote or robotic *in situ* experiments. Searches for life must accordingly include a robust number of tests for multiple attributes or else for the core characteristic of reproduction with instructions.

2.2.7.2 ORIGIN OF LIFE

A recurring quandary in the hypotheses and experimental investigations into the progression of abiotic evolution to the first *bona fide* Lf is how to judge whether an entity that is on the path to become "alive" actually has reached that state. Different camps of scientists studying prebiotic evolution embrace concepts of "metabolism-first," "replication-first," "compartmentalization-first," and various combinations of these. Others promote the idea that it is not even possible to define what life is and hence avoid making judgments (Cleland and Chyba 2002, 2007). It has also been pointed out that such definitions are difficult to employ in studies of the pathways to the origin of life (Szostak 2012; Tessera 2012).

However, a useful analogy is to compare "alive" and "not alive" to the common concepts of "daytime" and "nighttime." Although we clearly recognize the difference between day and night, there are the intermediate states of "dawn" and "dusk," when the Sun is neither all the way present nor totally absent. These intermediate states are somewhat difficult to define rigorously, without being arbitrary, yet we do not have difficulty in recognizing both day and night. Thus, the allusion to "we will know it when we see it" is not so arbitrary, as sometimes seems, in that systems being investigated could be deemed "alive," "not alive," or "incipient."

Another example of intermediate states may be drawn from the common concepts of solid, liquid, and gas. When asked, most educated persons would judge an ice cube as being in the solid phase. Yet, we know that if such a solid is placed into a sealed chamber that is kept cold while being evacuated and then closed off from the pump with a valve, a sensitive pressure gauge will measure a gradual increase in pressure as the ice sublimates and the chamber is filled with a vapor, whose equilibrium concentration is dictated by the temperature of the ice. Thus, what is considered to be in the solid state is also partly in the gaseous state. Furthermore, if a light is shone on the ice, or a small amount of a suitable salt is sprinkled onto its surface, the system now manifests all three phases—solid, liquid film, and gas vapor. Yet, there are many measurement methods that can determine which regions are best described as being in which phase. Most difficult to define, however, is the thin-film liquid phase, which is intermediate between the other two phases. Perhaps, the expedient and most correct path to follow is to explicitly describe the phases for which a determination is not conclusively apparent. Likewise, as an abiotic chemical system evolves from non-living to creating living entities, the various stages could be described without recourse to pre-determined nomenclature which could be inadequate, misleading, and/or superfluous.

The motivation often seems to have been to be able to state with precision whether life has been created at some certain point, or not, in the laboratory or by prebiotic nature itself. Certainly, when some reproductive entity finally became firmly established on Earth, it was a tipping point in that the evidence suggests an unbroken chain for several billion years in which the biosphere evolved and flourished. However, in analogy with the human reproductive cycle, is the critical tipping point ovulation, or ejaculation, or fertilization, or embryogenesis, or fetal development, or birth of baby, or puberty, and so forth? Clearly, these are all "critical events" to borrow a phrase from space exploration. In a successful space mission to Mars, there are multiple critical events ranging from rocket launch, spacecraft separation, navigation to Mars, atmospheric entry, and landing to deployment of instruments and sample-processing equipment. Virtually, all critical events are essential for the overall success of the mission.

Perhaps, the most constructive approach to assessing the origin of life is to continue focusing on the various scenarios and series of critical events that must occur until achievement of the final, fully capable reproductive system. This system must include not only the physical form necessary but also the functional capabilities and the information package that enable it to evolve as it spreads and thus always produce some organisms that can survive one or more of the plethora of insults that the exogenous environments will hurl at them.

2.2.7.3 COMPUTERS AND NETWORKS

"Soft Life" is taken to be that generated by software. Once the final software program step is written, debugged, and properly executed in a code whose overall functionality includes reproduction enabled by instructions and active control of a flow of energy, the criteria for life have been met. Of course, such soft entities do need a suitable environment, which entails a computer, server, or mobile electronic device. In addition, in order for the progeny programs to spread, there needs to be inter-computer connections, preferably a large network (e.g., the Internet). As has been advocated by some, software programs that operate in computers in effect create physical entities, since they occupy physical locations in memory elements and affect the flow, for example, of read-write energy. What is often called "virtual life" is not ethereal, after all. Even as they multiply in memory via a program simulating population growth, or as a "computer virus" (CV) propagated over a network, various physical phenomena occur (electromagnetic waves, electrical currents, and switched states of magnetic domains), creating new physical entities independent of their parent. Computer viruses are alive, not as parasites of organic bodies but as the inanimate tools of *technosapiens*, the "suitable environment" of interconnected general-purpose computers. True, they readily are migrated from one physical assemblage to another, as the computer manages its storage space, but at any instant, they exist in at least one copy in a physically specifiable form. This DoL is invariant with timescale. Teleportation may not be possible for metazoans, but it is commonplace for computer organisms.

Evolutionary algorithms are not reproductively proliferative, in general, and hence are not alive by this DoL. They do accomplish evolution in the sense of adaptation by feedback, and they often are designed to mimic biologically known enhanced methods for speeding evolution, such as chromosome crossovers and gene duplications. Self-expanding program codes are also well known and relatively simple to generate but generally represent growth rather than reproduction.

2.2.7.4 Creating Chemical Life in the Laboratory

Laboratory efforts at Test Tube Life may formally satisfy this DoL and also the NASA DoL by way of explicitly demonstrating Darwinian evolution. Already more than three decades ago, the Spiegelman team achieved accelerated evolution in the test tube to create new variants of the RNA virus $Q\beta$ (Spiegelman et al. 1965; Mills et al. 1967). Much of the current work on this general topic, for example, by several groups (Rasmussen et al. 2004, 2016; Joyce 2007; Powner et al. 2011; Robertson and Joyce 2012; Pressman et al. 2015; Islam and Powner 2017; Sutherland 2017; Szostak 2017), focus on achieving results consistent with concepts of the RNA-World model, using success criteria similar or identical to the NASA DoL. Correspondingly, the use of informational macromolecules and suitable translation chemistries satisfies the first part of the GU DoL given previously. By using molecular evolution for achieving enhanced efficiencies in catalysis of one or more functions of the system and by providing substrates from which energy can be extracted, these experiments also satisfy the second part of this definition. However, the "suitable environment" is an artificial one, not a natural one, and the ultimate challenge will be to develop an assemblage of molecules that conceivably have resulted from a natural environment that at one time (or more) occurred on the early Earth. Although originally highly contrived, these approaches are showing considerable progress.

Rebuilding of functional viruses and genomes for cells by using purely *in vitro* systems is an example of synthetic biology investigations that are actively creating new versions of life.

2.2.7.5 Citizen Science

This DoL is more easily discussed with the lay public than many, if not most, definitions posited previously. Some DoLs are actually obtuse even to the science community. Although it is not a requirement imposed on scientists, in general, to use popular terminology, it is worth acknowledging that research into the origins of life, although fascinating, is seldom at an intrinsically high priority for funding through public support.

In this definition, an analogy to a soccer team could be made—the "Lifeform" is the team itself, and the players are the "organisms." Other organisms in this Lf are the coaches and support personnel. The opposing team is another instance of the same type of Lf. The referees constitute a totally different Lf but are parasitic on the soccer teams. There are also artifacts/tools, such as the soccer ball, the playing field, turf, line markings, and goal. A set of formal rules constitutes instructions that are essential to defining and guiding the game of soccer. These instructions are subject to revision, that is, evolution. Of course, we don't see any of these Lfs actively reproducing, but we know in fact that, somehow, they do, because there appears to be more and more soccer teams all the time. Likewise, it is rare to observe many organisms actively reproducing in nature, because those acts are rare compared with daily life. Yet, they somehow always occur, or the Lf would not persist.

A DoL should also pass the reasonableness test. Thus, to say that a mule is not alive because it cannot self-reproduce seems to fly in the face of common intuition. Furthermore, little children are eminently alive, yet not capable of reproduction.

And to say that a virus is not alive, despite the fact that the professionals who conduct the scientific study of viruses are biologists, biochemists, geneticists, and medical and pharmacology researchers, seems likewise to violate the common-sense logic of any thoughtful external observer.

2.2.7.6 Machine Life

Finally, we must most seriously contemplate machine life and its potential ramifications. This GU DoL is fully consistent with the concept of a "mesorg," a *m*echanized, *e*lectronics-based, *s*oftware-driven *org*anism. To be classified as being "alive," such a machine must include enabling instructions sufficient for its own reproduction. Robots, regardless of type—industrial, recreational, or financial—are distinctly not alive, because none of the instructions for their manufacture are embedded within them. The production of robots is not recursive. Indeed, the concept of "six-sigma" manufacturing is to strive for exact duplication of a specific product, fully faithful to a standard design. On the other hand, as a result of product development engineering, a new model that is only a variant of a previous robot (or any artifact) may be put into production.

In this sense, products evolve at the hands of humans. As such, they remain under strict control by their creators. Furthermore, if the designs, procedures, and software, carefully locked away somewhere in a safe (or server), were somehow destroyed, there would be no possibility of faithful reproduction of the same product, except possibly by reverse engineering. A widespread Lf, on the other hand, confers massively parallel redundancy, with distributed organisms capable of propagating the basic design.

A mesorg need not, by virtue of "a suitable environment," fabricate all of its own components from scratch—it is not necessary to construct electronic devices by extracting silicon from silica sand, to invent new software languages and subprograms, or to reinvent wheels, gears, and latches. Hardware designs and software can be borrowed and then modified until the transition occurs, such that reproduction occurs in a manner that is controlled, at least in part, by instructions embedded in each mesorg itself. This departure from the "complete description" specification in the von Neumann/Poundstone views of life is an important extension (von Neumann 1966; Poundstone 1984); however, it in no way is in disagreement with their penetrating insights overall.

With a mere modicum of intelligence, somewhere between a honeybee community and an assembly worker in a human factory, a mesorg could strike out on its own. Darwinian- and Lamarckian-type evolutions could occur, including the potential for speciation. From a minimalist mesorg, initially highly dependent upon *technosapiens*, such evolution as well as pirating could evolve measures of progressive independence, just as humankind has achieved an amazing degree of success in taming its own environment. Although such an Lf might always be dependent on the technologies of humankind, it would nonetheless be alive and could exhibit many of the attributes of other Lfs and organisms.

2.2.8 ETHICS OF CREATING NEW LIFEFORMS

The chain of sequential propagation with every progeny slightly or significantly different from its parent(s) creates ever-expanding lines of independent diversity. Changes that are beneficial to Lf XYZ may be competitive or otherwise threatening to one or more other organisms, including its own parents. Thus, we reach the attribute most usually not explicit, although many of us are aware of it implicitly. That attribute is conspicuously self-centric with respect to our own Lf. Variously called hazardous, dangerous, or deleterious, that attribute is certainly not the essence of life. But it is an attribute of several natural Lfs, the predators and parasites of humankind, as well as what we do to our supporting species (crops and domestic animals).

All artificial Lfs, whether wet, soft, or hard, have the potential to do damage to the biosphere and therefore require sober control. Genetically engineered organisms and machines that could become alive should be a concern to *Homo sapiens*, just as are those many forms of life that are already problematical (from viruses to microbes to mosquitoes to dangerous animals) and that we either protect against, domesticate, or have exterminated.

ACKNOWLEDGMENTS

Any novel formulations above are the responsibility of the author, but an enormous body of work by numerous researchers underlies the basis for the discussions, clarifications, and examples herein. The encouragement, thoughtful review, and editorial assistance by V. M. Kolb are gratefully acknowledged.

REFERENCES

Benner, S. A. 1999. How small can a microorganism be? In *Proceedings of the Workshop on Size Limits of Very Small Microorganisms*, pp. 126–135. Washington, DC: National Research Council, National Academy Press.

Benner, S. A. 2010. Defining life. *Astrobiology* 10: 1021–1030.

Clark, B. C. 2001a. Mesorg: The advent of inorganic life. *1st NASA Astrobiology Institute Science Conference*, Carnegie Institution of Washington, DC, April 11–12, 2001.

Clark, B. C. 2001b. Astrobiology's central dilemma: How can we detect life if we cannot even define it? *AAS/Division for Planetary Sciences (DPS) Annual Meeting, (Abs) Session 60.10*, New Orleans, LA, November 30, 2001.

Clark, B. C. 2002. Separating attributes from essentials in a generalized and universalized definition of life. *Plenary Presentation, Astrobiology Science Conference (AbSciCon-02)*, NASA/Ames Research Center, April 9, 2002. (Abs) Astrobiol. J.

Clark, B. C. 2003. Identifying life: Pitfalls and conundrums. NASA Astrobiology Institute, Arizona State University, Tempe, AZ, February 9–11, 2003. (Abs) Astrobiol. J.

Clark, B. C. 2004. From Mars and machines, to water and worker bees: Application of a GU DoL to identification of extraterrestrial and artificial life forms. In *Workshop and Tutorial Materials, Ninth International Conference on the Simulation and Synthesis of Living Systems (ALIFE IX)*, M. Bedau, P. Husbands, T. Hutton, S. Kumar, and H. Suzuki (Eds.), pp. 96–102. Boston, MA, September 12, 2004.

Clark, B. C. 2019. Searching for Extraterrestrial Life in Our Solar System, *Handbook of Astrobiology*, Chapter 12.4, pp. 801–818. Boca Raton, FL: CRC Press.

Cleland, C. E., and C. F. Chyba. 2002. Defining life. *Orig. Life Evol. Biosphere* 32: 387–393.

Cleland, C. E., and C. F. Chyba. 2007. Does 'life' have a definition? In *Planets and Life: The Emerging Science of Astrobiology*, W. T. Sullivan III, and J. A. Baross (Eds.), pp. 119–131. Cambridge, UK: Cambridge University Press.

deDuve, C. 1991. *Blueprint for a Cell: The Nature and Origin of Life*. Burlington, NC: Neil Patterson Publishers.

Egel, R. 2012. On the misgivings of anthropomorphic consensus polling in defining the complexity of life. *J. Biomol. Struct. Dyn.* 29(4): 615–616.

Hickman Jr., C. P., L. S. Roberts, and A. Larson. 2001. *Integrated Principles of Zoology*, 11th ed. Boston, MA: McGraw-Hill.

Hutchison III, C. A., R. Y. Chuang, V. N. Noskov et al. 2016. Design and synthesis of a minimal bacterial genome. *Science* 351: 1414. doi:10.1126/science.aad6253.

Islam, S., and M. Powner. 2017. Prebiotic systems chemistry: Complexity overcoming clutter. *Chem* 2: 470–501. doi:10.1016/j.chempr.2017.03.001.

Joyce, G. F. 1994. Forward. In *Origins of Life: The Central Concepts*, D. W. Deamer and G. R. Fleischaker (Eds.), p. xi. Boston, MA: Jones & Bartlett.

Joyce, G. F. 2007. Forty years of in vitro evolution. *Angew. Chem. Int. Ed.* 46: 6420–6436.

Kolb, V. M. 2019. Defining life: Multiple perspectives, *Handbook of Astrobiology*, Chapter 2.1, pp. 57–64. Boca Raton, FL: CRC Press.

Koshland Jr., D. E. 2002. The seven pillars of life. *Science* 295: 2215–2216.

Mazia, D. 1966. What is life? In *Biology and Exploration of Mars*, C. S. Pittendrigh, W. Vishnjac, and J. P. T. Pearman (Eds.). Washington DC: NAS, NRC Publ. 1296.

Mills, D. R., R. L. Peterson, and S. Spiegelman. 1967. An extracellular Darwinian experiment with a self-duplicating nucleic acid molecule. *Proc. Natl. Acad. Sci.* 58: 217–224.

Oparin, A. I. 1965. *Origin of Life*, Translated by S. Morgulis, 2nd ed. New York: Dover Publications.

Popa, R. 2004. *Between Necessity and Probability: Searching for the Definition and Origin of Life*, pp. 197–205. New York: Springer.

Popa, R. 2014. Elusive definition of life: A survey of main ideas. In *Astrobiology: An Evolutionary Approach*, V. M. Kolb (Ed.), pp. 325–348. Boca Raton, FL: CRC Press/Taylor & Francis Group.

Poundstone, W. 1984. *The Recursive Universe: Cosmic Complexity and the Limits of Scientific Knowledge*. New York: William Morrow.

Powner, M. W., J. D. Sutherland, and J. W. Szostak. 2011. The origins of nucleotides. *Synlett* 14: 1956–1964.

Pressman, A., C. Blanco, and I. A. Chen. 2015. The RNA world as a model system to study the origin of life. *Curr. Biol.* 25: R953–R963. doi:10.1016/j.cub.2015.06.016.

Rasmussen, S., L. Chen, D. Deamer, D. Krakauer, N. Packard, P. Stadler, and M. Bedau. 2004. Transitions from nonliving and living matter. *Science* 303: 963–965.

Rasmussen S., A. Constantinescu, and C. Svaneborg. 2016. Generating minimal living systems from non-living materials and increasing their evolutionary abilities. *Philos. Trans. R. Soc B* 371: 20150440. doi:10.1098/rstb.2015.0440.

Robertson, M., and G. F. Joyce. 2012. The origins of the RNA world. *Cold Spring Harb. Perspect. Biol.* 2(4): a003608.

Schrödinger, E. 1944. *What is Life?* Cambridge, UK: Cambridge University Press; reprinted 1967.

Spiegelman, S., I. Haruna, I. B. Holland, G. Beaudreau, and D. R. Mills. 1965. The synthesis of a self-propagating and infectious nucleic acid with a purified enzyme. *Proc. Natl. Acad. Sci.* 54: 919–927.

Sutherland, J. D. 2017. Studies on the origin of life – the end of the beginning. *Nat. Rev. Chem.* 1: 0012. doi:10.1038/s41470-0762.

Szostak, J. W. 2012. Attempts to define life do not help to understand the origin of life. *J. Biomol. Struct. Dyn.* 29: 599–600.

Szostak, J. W. 2017. The narrow road to the deep past: In search of the chemistry of the origin of life. *Angew. Chem. Intl. Ed.* 56: 11037–11043.

Tessera, M. 2012. Is a n+1 definition of life useful? *J. Biomol. Struct. Dyn.* 29: 635–636.

Trifonov, E. N. 2011. Vocabulary of definitions of life suggests a definition. *J. Biomol. Struct. Dyn.* 29: 259–266.

Trifonov, E. N. 2012. Definition of Life: Navigation through uncertainties. *J. Biomol. Struct. Dyn.* 29: 67–650.

von Neumann, J. 1966. *Theory of Self-Reproducing Automata.* Edited and completed by A. W. Burks. Urbana, IL: University of Illinois Press.

2.3 Synthetic Cells and Minimal Life

*Daniel G. Gibson, Clyde A. Hutchison III,
Hamilton O. Smith, and J. Craig Venter*

CONTENTS

2.3.1 INTRODUCTION

This chapter presents the current state of the art in creating microorganisms with synthetic genomes. Cloning of bacterial genomes in yeast as centromeric plasmids and transplantation of genomes from one bacterial species to another are described. Methods of DNA synthesis and genome assembly, as well as progress on automation of DNA synthesis, are reviewed. The classification of genes as non-essential, quasi-essential, or essential and identification of those required for minimal life are discussed. Synthesis of a minimal bacterial genome and its installation into a compatible cytoplasm to produce a minimal synthetic cell are described. Progress with synthetic yeast chromosomes is reviewed. Finally, a digital-to-biological converter (DBC) that uses DNA sequence information stored in a computer to direct the automated production of biological materials such as DNA, RNA, proteins, and bacteriophages is described.

2.3.2 DEFINITION AND BRIEF HISTORY OF SYNTHETIC CELLS

Cells are the fundamental units of life. They reproduce and carry out multiple chemical processes by virtue of the genetic information carried in their genomes. Recent advances in DNA synthesis now make it possible to synthesize entire genomes of some single-cell microorganisms. A synthetic cell is one that is controlled by such a chemically synthesized genome. It will generally contain various features that do not occur naturally, such as changes in codon usage, modular design, and addition or removal of genes. Cells whose genomes are copied from natural DNA templates by using DNA polymerases or cells with genetic modifications introduced by genetic engineering, even if some parts of the genome were made synthetically, are not included.

Although a synthetic cell could be of any type, it is currently feasible to synthesize only the genomes of single-cell microorganisms, specifically bacteria and yeast. Generally, the properties of bacteria are completely determined by information carried in their DNA genomes. Thus, whether a synthetic cell is created by installing a synthetic genome into an existing recipient cell and replacement of the recipient genome, or alternatively, was somehow created entirely from non-living cell parts plus a synthetic genome, the result would be the same (Figure 2.3.1). Examination of the resulting cell after several generations of growth would not reveal the path by which it was created. This holds strictly for bacteria and archaea but may not precisely hold for eukaryotes, since mitochondria also contain essential information, and epigenetic information such as DNA methylation patterns may not be easily reproduced in the synthetic genome.

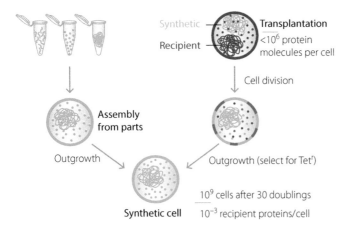

FIGURE 2.3.1 The final synthetic cell is the same, regardless of the path by which it was created. The path on the left involves assembling the synthetic cell from non-living parts, including the synthetic genome. The non-living parts might be those specified by the synthetic genome or by the recipient cell or any other set of parts compatible with assembly of a living cell programmed by the synthetic (blue) genome. In the path on the right, a synthetic genome (blue) is transformed into a recipient cell (red), followed by genome segregation and outgrowth of daughter cells containing only the synthetic genome. After many doublings, the original parts of the recipient cell have been replaced by those specified by the new (blue) genome. The original parts of the recipient cell (red) are diluted following loss of the recipient (red) genome. A mycoplasma cell contains fewer than a million protein molecules. Thirty doublings following the loss of the recipient genome will produce ~10^9 daughter cells. So, after 30 doublings in the absence of the recipient genome, a typical cell will contain no protein molecules at all from the original recipient cell.

Synthetic cells are currently made from genes that are found in existing natural organisms. No one yet has the knowledge to create genes *de novo* specifying proteins or RNA with new functions; however, this may change at some point in the future. The design of synthetic cells is thus entirely dependent on knowledge of the genetic repertoire of existing organisms. The first complete gene complement of a living cell, that of *Haemophilus influenzae* Rd, was determined in 1995 (Fleischmann et al. 1995). Since then, thousands of organisms have been sequenced, and the repertoire of genes and genomes in GenBank is enormous.

Sequencing technology has evolved rapidly, especially with the development of highly parallel, chip-based next-generation sequencing (NGS) (Attia and Saeed 2016). However, DNA synthesis capability lags far behind. In 1970, Khorana and coworkers were the first to develop the chemistry for synthesizing DNA and apply it to the synthesis of the gene for yeast alanine transfer RNA (tRNA) (Agarwal et al. 1970), but it was not until 2002 that a full genome, that for the 7.5 kb polio virus, was synthesized as a DNA molecule that could be transcribed to make the RNA genome of the virus particle (Cello et al. 2002; Wimmer 2006). In 2003, the bacteriophage phiX174 genome (5.3 kb) was synthesized by

whole-genome assembly from 256 oligonucleotides (Smith et al. 2003). Only in the past dozen years has DNA synthesis advanced to the point that small bacterial genomes can be made in the laboratory by chemical synthesis (Gibson et al. 2008a, 2010a; Hutchison et al. 2016).

The smallest known bacterial genomes belong to the genus *Mycoplasma*. The smallest among these is *Mycoplasma genitalium*, a wall-less bacterium that inhabits the human urogenital tract and can be grown independently in the laboratory. Its genome was synthesized in 2008 and was propagated as a yeast centromeric plasmid (Gibson et al. 2008a). In 2010, the *Mycoplasma mycoides capri* genome was synthesized and activated by transplanting into the closely related *Mycoplasma capricolum* to yield a synthetic cell, syn1.0, that differed only slightly from the wild-type cell (Gibson et al. 2010a). The syn1.0 genome was subsequently used to create the first nearly minimal genome, as described in Section 2.3.6 (Hutchison et al. 2016).

2.3.3 PROPAGATION OF GENOMES IN YEAST (*SACCHAROMYCES CEREVISIAE*)

Sections of a bacterial genome can be synthesized in the test tube, but since DNA is fragile and shears easily when handled in solution, it is impractical to assemble complete genomes in solution. The discovery that complete bacterial genomes can be propagated as centromeric plasmids in yeast cells has solved this and other problems in the generation of synthetic genomes (Gibson et al. 2008a; Benders et al. 2010). Propagation of bacterial genomes in yeast has become invaluable because:

1. A complete synthetic genome can be assembled by recombination in yeast from overlapping synthetic genome segments that share terminal DNA sequences.
2. Propagation in yeast permits cloning of a synthetic genome by selecting single yeast colonies. This is important because the production of synthetic DNA is an error-prone process and cloning is essential to ensure a pure source of the desired sequence.
3. Growth of yeast carrying a synthetic genome provides a source of a cloned synthetic genome for further experiments, including sequence verification.

To clone a bacterial genome (or segments thereof) in yeast, it is generally necessary to add several genetic elements. These may be combined into a tri-shuttle vector (Benders et al. 2010) that includes:

1. A yeast selectable marker. The *HIS3* gene has usually been used, and the genome has been introduced into a histidine auxotroph of yeast, lacking a functional *HIS3* gene. Growth on medium lacking histidine allows the selection of yeast cells stably carrying the bacterial genome.

2. A yeast centromere. *CEN3* or *CEN6* may be used. This allows the bacterial genome to duplicate and segregate as a chromosome in the yeast nucleus.

3. A yeast origin of DNA replication or autonomously replicating sequence (ARS).

4. Sequences required for replication and selection of the cloned bacterial DNA in *Escherichia coli*, for example, the pUC19 origin and the *TetM* gene. The *TetM* gene also serves as a selectable marker to allow isolation of bacterial cells into which the vector has integrated.

5. Such vectors also sometimes carry a transposase gene to facilitate random integration into the genome.

A synthetic genome can be assembled by transforming yeast with a set of overlapping synthetic DNA fragments that span the entire genome (Figure 2.3.2c). The ends of fragments adjacent on the genome overlap, typically by 40–200 bp. One of the fragments contains the yeast vector (described earlier). These fragments assemble by genetic recombination between their overlapping terminal sequences. A synthetic *M. genitalium* genome was assembled from as many as 25 overlapping fragments in this way (Gibson et al. 2008b), but typically synthetic mycoplasma genomes have been assembled from 8–10 segments.

It is also possible to clone complete genomes, isolated from bacterial cells, into yeast (Figure 2.3.2a and b). A yeast vector carrying a transposase gene can be introduced into the bacterial cell by polyethylene glycol (PEG)-mediated transformation or by electroporation, depending on the type of cell (Figure 2.3.2a). Following transformation, cells with the integrated vector are selected by plating in the presence of tetracycline. Individual colonies are picked, and the integration site can be determined by sequencing from the vector into the adjacent genomic sequence. Because colonies with the integrated vector are selected, one can be sure that the vector is not disrupting any gene or other sequence that is necessary for viability of the cell. The genome can be released from the cell by lysis of the bacteria in agarose plugs, to prevent shearing of the genome during lysis. Then, the genome can be released from the agarose plug by melting and treating with β-agarase. Next, the genome is introduced, by PEG-mediated transformation, into a yeast strain carrying an auxotrophic marker in the *HIS3* gene. Selection of yeast carrying the bacterial genome is accomplished by selecting yeast colonies on a -His plate (Benders et al. 2010).

As an alternative to random integration of a yeast vector, the vector can be designed to recombine with the bacterial genome at a specific location (Figure 2.3.2b). If the genome is cleaved at the designed vector integration site, then recombination is facilitated when the isolated cleaved genome is co-transformed with the yeast vector.

Genomes can be transferred directly from bacteria to yeast under conditions that promote cell fusion (Karas et al. 2014). In some situations, this is a useful alternative to transplantation of an isolated DNA genome.

The completeness of a bacterial genome carried in yeast can be verified by using multiplex polymerase chain reaction

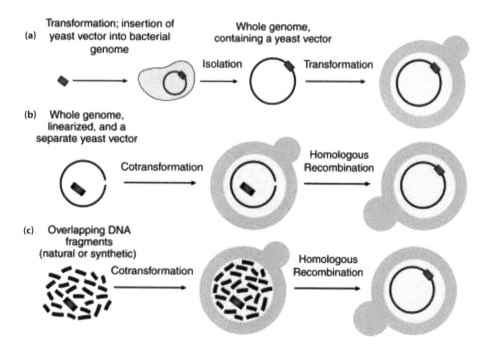

FIGURE 2.3.2 Three methods for cloning mycoplasma genomes in yeast. In order to be propagated by yeast, upon transformation, the bacterial genome must contain several yeast sequences (vector, black bar). (a) These can be incorporated into the bacterial genome by transformation of the bacterium. (b) Alternatively, they can be inserted by co-transformation of yeast with the vector and the bacterial genome. In this case, the two must share overlapping sequences, so that yeast can combine them by homologous recombination. (c) The bacterial genome can also be cloned by assembling multiple overlapping fragments.

(PCR) to detect junctions between adjacent overlapping segments. The mycoplasma genome is inherently quite stable when propagated in yeast. However, to avoid deletions in the mycoplasma genome, unnecessary propagation of the genome in yeast should be avoided.

The cloning of bacterial genomes into yeast has been accomplished for a variety of bacteria, including several *Mycoplasma* species (0.6–1.1 Mb) (Benders et al. 2010), *Acholeplasma laidlawii* (1.5 Mb) (Karas et al. 2012), *Prochlorococcus marinus* (1.6 Mb) (Tagwerker et al. 2012), and *H. influenzae* (1.8 Mb) (Karas et al. 2013). For genomes with a higher G + C content, it has been necessary to insert ARS elements into genomic sequences to permit cloning of large segments (Noskov et al. 2012).

2.3.4 TRANSPLANTATION OF GENOMES FROM ONE SPECIES TO ANOTHER

A bacterial genome can be propagated in yeast and gently extracted, but it is of little value unless it can be activated in a suitable recipient cell. The process of genome transplantation has been developed at the J. Craig Venter Institute (JCVI) to achieve activation of complete synthetic genomes. So far, transplantation has only been reported in several species of the genus *Mycoplasma* (Lartique et al. 2007; Labroussaa et al. 2016) and in the closely related mollicute *Mesoplasma florum* (Baby et al. 2018). Other bacterial genomes could be engineered in yeast cells and then transferred piecemeal into bacterial cells by using genetic recombination.

In genome transplantation, a donor genome is prepared by gently lysing cells embedded in agarose plugs. The cells can be mycoplasma or yeast cells, carrying either a natural or a synthetic genome. The genome to be transplanted carries a selectable genetic marker such as a drug-resistance gene (typically *TetM*) to allow selection of cells carrying a transplanted genome. Cellular proteins are digested in the agarose plug, and then, the plug is digested with β-agarase. Recipient cells for the transplantation are usually a different mycoplasma species than the donor genome and are tetracycline-sensitive. The recipient cells are grown under standardized conditions, washed, and resuspended in calcium chloride. The donor genome preparation is mixed with the recipient cells under transformation conditions by using PEG. Growth medium is added, and the cells are incubated at 37°C for several hours. The cells are then washed and plated on solid medium containing tetracycline, under selective conditions.

Under selection for the tetracycline-resistance marker (*TetM*) on the donor genome, the recipient cell genome is lost following repeated rounds of cell division. All cellular components are replaced by those coded by the donor genome, as shown in Figure 2.3.1 (right-hand pathway). Tetracycline-resistant colonies carry only the transplanted genome, which can be verified by genome sequencing. Typically, the donor genome is either *M. mycoides* or a synthetic genome based on *M. mycoides* genes and the recipient is *M. capricolum*. These genomes are easily distinguishable at the sequence level.

The published synthetic genomes, such as JCVI-syn1.0, JCVI-syn2.0, and JCVI-syn3.0, have all been verified by complete genome sequencing (Gibson et al. 2010a; Hutchison et al. 2016).

2.3.5 CHEMICAL SYNTHESIS OF DNA

Synthetic biology is genetic engineering that uses chemical synthesis of DNA as the major method for constructing genetic parts. Breakthroughs in low-cost, accurate, and rapid DNA synthesis are having wide-reaching implications for the growing synthetic biology community. Cost, accuracy, and speed have been major inhibitory factors to the development of the field and to widespread applications. Synthetic biologists are synthesizing and expressing genetic elements to provide a sustainable means for producing desirable products such as new and improved therapeutic drugs, vaccines, biosensors, biofuels, biochemicals, food ingredients, and cosmetics. With chemical synthesis of genes, pathways, and genomes becoming faster, more precise, and only pennies per base pair, the technology is becoming more prevalent. Researchers are picking promoters, genes, terminators, and regulatory sequences from vast genomic and metagenomic databases and beginning to experiment with various pathway constructions on an empirical basis. This, in turn, is significantly accelerating progress, reducing development time, and increasing the probability of success in developing valuable products derived from synthetic DNA.

DNA can be synthesized in the laboratory, either enzymatically or chemically, but there is an important difference. Enzymatic synthesis requires template DNA, whereas chemical synthesis does not. This lack of necessity for a template allows tremendous flexibility in the design of new sequences. For example, codon usage of a gene can readily be changed, or genes from any source can be added, deleted, or rearranged in a genome in any fashion compatible with life.

Chemical synthesis of DNA always starts with oligonucleotide building blocks of a few dozen nucleotides in length that are made by chemical joining of single nucleotides. All the DNA in a synthetic genome is derived originally from chemically synthesized oligonucleotides. Duplex DNA molecules of several hundred base pairs in length are made by joining oligonucleotides together. Subsequently, duplex DNA molecules are joined to form larger and larger DNAs in steps that employ enzymatic synthesis; however, the precursor DNA is from synthetic oligonucleotides. The steps in producing genome-sized DNA molecules are described in Sections 2.3.5.1 through 2.3.5.5 and Figure 2.3.5.

2.3.5.1 OLIGONUCLEOTIDE SYNTHESIS

Oligonucleotides are the building blocks of synthetic biology. They are the starting material, from which genes, pathways, and even genomes are synthesized, and these, in turn, enable products, such as medicines, biofuels, and biochemicals, to be developed. Khorana developed chemistry for oligonucleotide synthesis in the 1960s (Khorana 1968, 1972; Caruthers et al. 1972). His original methods were greatly improved during subsequent decades (Caruthers 2013). Currently, DNA oligonucleotides are most frequently made inexpensively by computer-controlled solid-phase automated synthesizers. To make a desired oligonucleotide, phosphoramidite 2′-deoxynucleoside dA, dC, dG, and T building blocks are sequentially coupled to the growing oligonucleotide chain in the order specified. The product is then released from solid support, deprotected, and collected. Oligonucleotides always contain a low frequency of errors, because coupling steps are not 100% efficient and accurate, and thus, achieving error-free synthetic oligonucleotides becomes the most significant factor as the oligonucleotides are assembled into larger DNA molecules.

Oligo providers, such as Integrated DNA Technologies (IDT), offer individual tubes of oligos, which contain a unique sequence, at a price of ~$0.10/base. The cost remains high owing to inefficient reagent use during the chemical synthesis process. Oligos are generally produced at the nanomole scale, which translates to consumption of only 0.01% of each oligo with oligo assembly methods. Microarray chip technologies offer a low-cost source of oligonucleotides and have the potential to significantly decrease the overall synthesis costs by more than an order of magnitude (Kosuri et al. 2010). However, these oligos are of very low concentration and in a single, complex mixture containing thousands of unique oligonucleotides, which make them difficult and unreliable to process and consequently reduce turnaround time.

Although oligonucleotide synthesis is drastically better than what it was 40 years ago, this process continues to produce a fraction of unintended DNA sequences, which will be more prevalent in longer synthetic DNA fragments. Excluding microchip DNA technologies, which traditionally have significantly higher error rates, the synthesis error rate for the assembly of standard oligonucleotides is generally about one error per 1000 bp. Thus, it should be expected that if oligonucleotide errors are not weeded out early on in the building process by, for example, cloning and sequencing or with an error correction procedure, most, if not all, DNA fragments above 10 kb will contain errors. Cloning and sequencing procedures that identify error-free DNA products or subassemblies can be very effective but reduce turnaround time and increase DNA synthesis costs.

2.3.5.2 Assembly of Oligonucleotides to Make Pieces of Duplex DNA

Recent methods, including our own, have reduced the bottlenecks for the assembly of oligonucleotides into genes and genes into pathways and genomes. Overlapping oligonucleotides (oligos), typically 20–50 nucleotides in length, are generally assembled into larger duplex DNA molecules by either of two methods. If adjacent oligos perfectly overlap each other without gaps, they can be ligated together to form assemblies of varying size, up to a few hundred base pairs in length (Smith et al. 2003). If the oligos overlap with single-stranded gaps, then polymerase cycle assembly (PCA) is used. This is essentially set up as a PCR, except that there is no amplification, since all of the oligos are at approximately the same concentration, including the terminal 5′-oligos. If the terminal two 5′-oligos are in excess over those in the interior of the assembled product, then both assembly and some amplification can take place.

Synthetic Genomics, Inc. (SGI's) Archetype™ software is configured to allow for a DNA sequence to be instantly converted into a synthesis paradigm, starting from the overlapping oligonucleotide sequences that need to be synthesized. In general, oligonucleotides are designed to contain (1) 30-bp overlaps between adjacent 60-base single-stranded (ss)DNA oligos; (2) 40-bp overlaps between adjacent dsDNA fragments; (3) universal primer-binding domains for PCR amplification of the assembly intermediates; and (4) restriction sites to release the primer-binding domains following PCR amplification.

Adjacent oligos are combined ~50 at a time to build ~1.5 kb DNA fragments. Following oligo assembly and PCR amplification, error-containing DNAs are selectively removed by an endonuclease (identified in Archetype™ by SGI's scientists) to reduce error rates caused by imperfect oligonucleotide synthesis. This "error correction" process starts by denaturing and annealing the PCR products. Because the majority of the DNA molecules in the population contain the correct DNA sequence at every position, heteroduplex DNA will be formed at sites containing a substitution, deletion, or insertion.

These regions are recognized and cleaved by the endonuclease. Because intact molecules amplify more efficiently than nuclease-digested DNA, PCR can be used to enrich for error-free synthetic fragments. It has been found that this error correction procedure can reduce error rates by 1–2 orders of magnitude per round. For example, one round of error correction can reduce the average number of errors per 30 kb construct from about 30 to 3 or 0.3. If necessary, multiple rounds of error correction will be used. These methods allow one to avoid using a biological intermediate, such as *E. coli*, to filter out error-containing DNA, which slows down the DNA construction process.

2.3.5.3 Enzymatic Assembly of Pieces of Duplex DNA: From Small to Large

Following error correction and PCR, the synthetic DNA fragments are separated from their primer-binding domains by restriction enzyme digestion to expose overlapping sequence with adjacent fragments. Several in vitro enzymatic reactions capable of assembling genome-size molecules from multiple overlapping DNA fragments have been previously described (Gibson et al. 2009; Gibson 2011). The simplest of these recombination methods can be carried out as a single isothermal reaction. The assembly reaction mixture in this system (commonly referred to as Gibson Assembly) contains a 5′ exonuclease, a DNA polymerase, and a DNA ligase working in

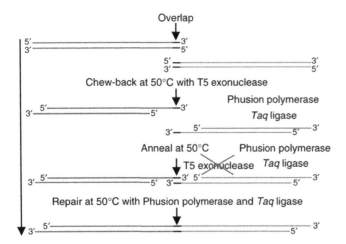

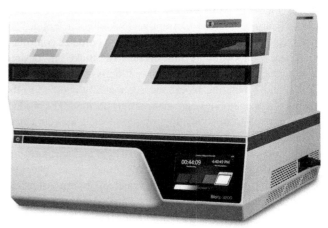

FIGURE 2.3.3 Two adjacent DNA fragments sharing terminal sequence overlaps (black) are joined into a covalently sealed molecule in a one-step isothermal reaction. T5 exonuclease removes nucleotides from the 5′ ends of double-stranded DNA molecules, complementary single-stranded DNA overhangs anneal, Phusion DNA polymerase fills the gaps, and *Taq* DNA ligase seals the nicks. T5 exonuclease is heat-labile and is inactivated during the 50°C incubation.

FIGURE 2.3.4 The BioXp DNA printer. The desired gene sequence is submitted to sgidna.com. Custom reagents are received, the BioXp is loaded, and the run is begun. Oligonucleotides are assembled, and errors are corrected. After the run is completed, the duplex DNA tiles are collected.

harmony to accomplish the seamless joining of multiple DNA fragments during a brief 50°C reaction (Figure 2.3.3).

Other in vitro methods that are both seamless and sequence-independent include circular polymerase extension cloning (CPEC), In-Fusion (Clontech®), uracil-specific excision reagent (USER), and sequence- and ligation-independent cloning (SLIC) (Hughes and Ellington 2017). Although these methods have been useful in the assembly of DNA up to the size of small genetic pathways, they have not been used in the construction of larger DNA molecules such as whole genomes. The Gibson Assembly system does not rely on thermocycling to bring overlapping parts together, and the assembled sub-fragments are covalently joined with DNA ligase, reducing the likelihood of fragmentation of large DNA molecules. In addition, ligated DNA assembly products are more efficiently moved into host organisms such as *E. coli*. Gibson Assembly was used to rapidly synthesize the entire 16,520-bp mouse mitochondrial genome from 600 overlapping 60-base oligonucleotides (Gibson et al. 2010b). It was also used in combination with yeast assembly to synthesize syn1.0 and syn3.0.

2.3.5.4 Automation of DNA Synthesis

A variety of DNA synthesis and assembly methodologies that lead from oligonucleotides up to whole chromosomes have been previously developed. Since first reporting on the Gibson Assembly method, several improvements have been made, including (1) increased robustness of the reactions by redesigning how oligonucleotides assemble; (2)

increased synthesis speed by assembling more oligos at once; and (3) increased accuracy by introducing an endonuclease-based, error-correction step. By reducing error rates, speed is increased, in turn, as cloning and sequencing requirements are reduced or eliminated. By improving the robustness of these procedures, it was found that the processes described earlier were highly amenable to automation. With this in mind, a "DNA printer" was developed, called the BioXp (Figure 2.3.4); it combines these workflows to build highly accurate synthetic genes. The synthetic DNA fragments that are produced on the instrument can be stitched together by Gibson Assembly to build larger DNA constructs.

2.3.5.5 Assembly of a Synthetic Bacterial Genome

The procedures described previously can be combined to generate complete synthetic bacterial genomes, as shown in Figure 2.3.5. A typical workflow may be as follows: (1) Gibson Assembly of 1.5 kb BioXp products, five at a time, into a vector, and then transformation into *E. coli*; (2) identification of error-free 7 kb "cassettes" on an Illumina MiSeq DNA sequencer; (3) yeast assembly of as many as 20 cassettes to generate sub-genomic assemblies; (4) rolling circle amplification (RCA) from positively screened yeast clones to generate microgram quantities of DNA in vitro; and (5) whole-genome assembly in yeast from the RCA-generated material.

This whole-genome synthesis workflow was used to create the syn3.0 minimal genome and can currently be carried out in less than 3 weeks, which is about two orders of magnitude faster than the first reported synthesis of a bacterial genome by our group in 2008.

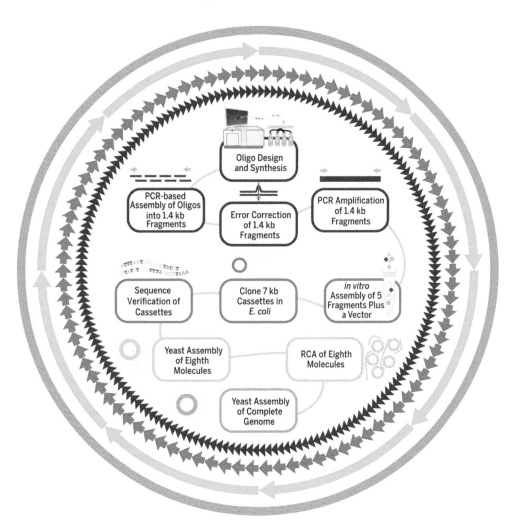

FIGURE 2.3.5 Strategy for whole-genome synthesis. Overlapping oligonucleotides are designed, chemically synthesized, and assembled into 1.4 kb fragments (red). Following error correction and PCR amplification, five fragments are assembled into 7 kb cassettes (blue). Cassettes are sequence verified and then assembled in yeast to generate eighth molecules (green). The eight molecules are amplified and assembled in yeast to generate the complete genome (orange).

2.3.6 MINIMAL SYNTHETIC BACTERIAL CELLS

Scientists have long sought to understand the minimal genetic requirements for life (Morowitz 1984; Fraser et al. 1995; Mushegian and Koonin 1996). There is no unique solution to the set of genes required for a minimal cell. The set will be determined by cell type, cell architecture, and the growth environment. For example, *E. coli* can be minimized for growth in either a rich or a minimal medium. Fewer genes will be required in the rich medium. Eukaryotic cells have complex inner structures and organelles requiring more genes than bacteria with relatively simple architectures.

In attempting to define a minimal set of genetic functions for life, it makes sense to start with a natural cell that already has a small genome. Mycoplasmas have the simplest architecture and smallest genomes among the known bacteria and are considered ideal targets for minimization. They have undergone reductive evolution in the restrictive environments of

animal hosts and have lost many genes that would be required in harsher surroundings. The 580 kb genome sequence of *M. genitalium* was determined by Fraser et al. (1995). It has the smallest genome of any known natural cell that can be independently grown in the laboratory. With only 485 protein-coding and 43 RNA-coding genes, the question immediately arose as to how many of these were essential. In 1996, Mushegian and Koonin compared the gene contents of the first two sequenced bacteria, *M. genitalium* and *H. influenzae*, with the idea that if there was a core set of genetic functions necessary for life, each of the genomes would contain that set. They found 256 orthologous genes in common and suggested that as the minimal set (Mushegian and Koonin 1996). To test this experimentally, Hutchison et al. (1999) introduced global transposon mutagenesis for identifying non-essential genes in *M. genitalium* and the closely related *Mycoplasma pneumoniae* and estimated that there were between 265 and 351 essential genes in *M. genitalium*. In a follow-up study, individual *M. genitalium* clones with single-gene transposon knockouts were isolated. The total number of different viable-gene knockouts approached 101 asymptotically (Glass et al. 2006). In the 1999 study, the authors proposed that one way to determine the minimal requirements for life would be to "create and test a cassette-based artificial chromosome." However, DNA synthesis was not sufficiently developed at that time to pursue this "bottom up" approach.

By 2008, DNA synthesis was sufficiently advanced to completely synthesize the 580 kb *M. genitalium* genome and propagate it as a yeast centromeric plasmid (Gibson et al. 2008a), but attempts to activate the genome by installation in a recipient cell were unsuccessful. Concurrently, the somewhat-larger natural 1.1 Mb genome of *M. mycoides capri* was successfully installed into a related mycoplasma strain, *M. capricolum*, by transplantation (Lartique et al. 2007).

The stage was thus set to create the first synthetic cell. Synthesis of the *M. mycoides* genome was carried out in stages as described previously and shown in Figure 2.3.5. The synthetic genome with slight modifications from the natural genome was assembled and propagated in yeast. The genome was then recovered from yeast and transplanted into *M. capricolum* to produce the synthetic cell syn1.0 (Gibson et al. 2010a). Although possessing a considerably larger genome than *M. genitalium*, syn1.0 became the target for minimization. The genome could be propagated and genetically manipulated in yeast, and the genome was readily recoverable from yeast and transplantable to obtain cells at various stages of minimization.

2.3.6.1 CLASSIFICATION OF GENES AS ESSENTIAL, QUASI-ESSENTIAL, OR NON-ESSENTIAL

How does one design a minimal genome? The key is to be able to identify non-essential genes that can be removed without effecting viability. Genes can be classified into three categories based on whether inactivation results in cell death, impaired growth, or no discernable defect (Smith et al. 1996; Badarinarayana et al. 2001; Lluch-Senar et al. 2015; Hutchison et al. 2016).

A gene is essential if when inactivated, the cell cannot be indefinitely propagated. The cell may continue to divide for a time, but once the essential gene product is sufficiently diluted among the progeny, no more growth occurs. In this case, the gene specifies an essential function that is not supplied by any other gene in the cell. On the other hand, if this essential function is supplied by each of two different genes, either one of these two genes is classed as non-essential. They form a synthetic lethal pair (Dobzhansky 1949). Only when one of the two is inactivated does the function reveal itself as essential. A highly redundant genome will appear to have few essential genes, even though it has many essential functions.

A gene is non-essential if it can be inactivated without affecting the viability or growth rate of the cell. It can specify either a non-essential genetic function or an essential genetic function. In the latter case, as discussed previously, it is dispensable because another gene in the cell supplies the same essential function.

A third class of genes is called quasi-essential. When inactivated, these genes result in impaired growth. The degree of growth impairment can vary from modest to severe. Inactivation of a quasi-essential gene in a cell results in its gradual loss from a culture when grown competitively with cells having mutations in non-essential genes.

Genes are usually inactivated by deletion or by disruption with a transposon. Global transposon mutagenesis by using the Tn5 transposome (a protein-DNA complex of the transposase and the transposon) carrying a selectable antibiotic marker is widely used (Hutchison et al. 2016). In a single transformation experiment, a population of thousands of individual cells with single Tn5 insertions can be produced. These are plated on selective solid medium to obtain the population of viable cells with Tn5 inserts. This initial population (P0) is then passaged several times in liquid medium. During passage, slower growing cells with Tn5 insertions in quasi-essential genes gradually decrease in number relative to the normally growing cells with Tn5 inserts in non-essential genes. After the fourth passage (P4), usually amounting to about 40–50 generations of growth, the culture will mostly contain cells with Tn5 insertions in non-essential genes. Quasi-essential genes contain relatively fewer Tn5 hits in P4 than in P0. Locations of the Tn5 insertions in the genome sequence are typically determined by sequencing the junctions between Tn5 and the genome sequence (Hutchison et al. 1999, 2016).

In Figure 2.3.6, a section of Tn5 insertion map illustrating the three categories of genes is shown for syn1.0. There are 432 non-essential, 229 quasi-essential, and 240 essential genes in syn1.0 by this assay. The quasi-essential genes are further divided into three sub-groups (ie, i, and in), based on the degree of growth impairment.

2.3.6.2 STEPWISE SYNTHESIS OF A MINIMAL CELL

Without prior knowledge of the final minimal content of essential and quasi-essential genes, minimization is best carried out in a series of design, build, and test (DBT) cycles, as illustrated in Figure 2.3.7 (Hutchison et al. 2016). The syn1.0 genome was divided into eight approximately equal pieces, such that each could be designed, built, and tested for viability in the presence of the other seven wild-type pieces. An initial DBT cycle was based on available information from the literature, from deletion analysis, and from incomplete transposon mutagenesis information. Synthesis of each of the eight pieces yielded only one feebly viable segment. A second DBT cycle utilized information on quasi-essential genes and successfully yielded all eight viable pieces. However, when the eight pieces were joined together, the genome was not viable on transplantation. Only certain combinations of the designed pieces in combination with the other pieces taken from syn1.0 were viable. The reason for this became apparent when it was discovered that certain essential functions were carried redundantly by two different genes, each on a separate piece of the genome. These pairs of synthetic lethal genes (Dobzhansky 1949) had been

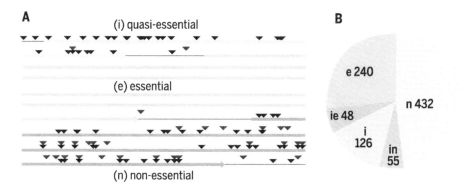

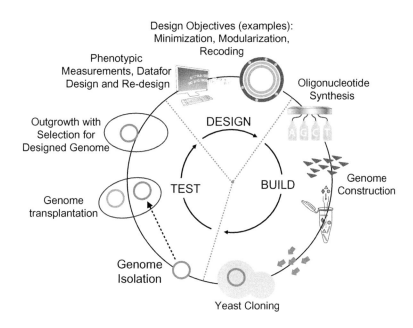

FIGURE 2.3.6 (a) Examples of the three gene classifications based on Tn5 mutagenesis data. The gene *MMSYN1_0128* (dark blue arrow starting at the right end of the top line) has P0 Tn5 inserts (open bars) and is a quasi-essential "i" gene. The next gene (*MMSYN1_0129*) has no inserts and is an essential "e" gene. The last gene (*MMSYN1_0130*) has both P0 (open bars) and P4 (red bars) inserts and is a non-essential "n" gene. (b) Number of syn1.0 genes in each Tn5 mutagenesis classification group. "i" genes are divided into three categories, "ie," "i," and "in," depending on the degree of growth impairment. The "n" genes and "in" genes are candidates for deletion in reduced-genome designs.

FIGURE 2.3.7 The JCVI design-build-test (DBT) cycle for bacterial genomes. The design objective is genome minimization. Starting from syn1.0, a reduced genome is designed by removing non-essential genes, as judged by global Tn5 gene disruption. Each of eight reduced segments is tested in the context of a 7/8 syn1.0 genome and in combination with other reduced segments. At each cycle, gene essentiality is re-evaluated by Tn5 mutagenesis of the smallest viable assembly of reduced and syn1.0 segments that give robust growth.

considered non-essential, since one member of the pair could be removed without effect. However, if both were removed, the cell lost the essential function and died. Restoration of one member of each synthetic lethal pair allowed all eight pieces to be combined into a viable cell. This cell, syn2.0, was examined by Tn5 mutagenesis, and an additional 42 genes were identified as apparently non-essential. These were removed in a third DBT cycle to yield syn3.0, a close approximation to a minimal cell (Hutchison et al. 2016). The syn3.0 genome is 531 kb in size and has 438 protein-coding and 35 RNA-coding genes (Figure 2.3.8).

Comparison of syn3.0 with the starting syn1.0 cell revealed several differences. Its doubling rate in culture was about 3 hours compared with 1 hour for syn1.0. Colonies were similar in appearance on solid media but were about half the diameter. In static liquid culture, syn1.0 remained in suspension, whereas Syn3.0 formed a mat in the bottom of the culture tube. The most striking changes were observed microscopically. Syn1.0 appeared as uniform spheres with a diameter of about 400 nm. Syn3.0, in contrast, was strikingly pleiomorphic in appearance, with long filaments, large vesicles, and a spectrum of sphere sizes, both smaller and larger than syn1.0 (Figure 2.3.9). Apparently, some genes affecting cell division and morphology were deleted in syn3.0, without compromising viability.

How close is syn3.0 to a minimal cell? Comparison of the non-essential gene content for the four mycoplasmas *M. genitalium* (580 kb) (Glass et al. 2006), *M. pneumoniae* (816 kb)

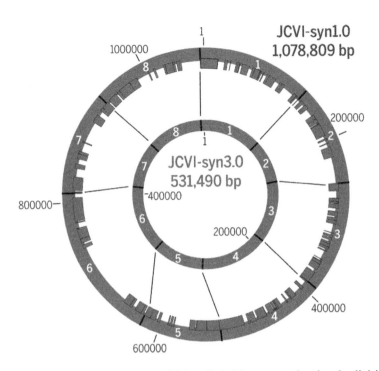

FIGURE 2.3.8 Comparison of the syn1.0 (large circle) and syn3.0 (small circle) genomes showing the division of each into eight segments. The bars inside the large circle show regions that were retained in syn3.0.

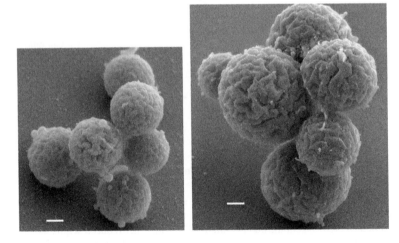

FIGURE 2.3.9 Electron micrograph of a cluster of syn1.0 cells showing uniform spheres (left), and a cluster of syn3.0 cells showing spheres of varying sizes (right). Scale bar is 200 nm.

(Lluch-Senar et al. 2015), *Mycoplasma pulmonis* (963 kb) (French et al. 2008), and *M. mycoides* JCVI-syn1.0 (1080 kb) (Gibson et al. 2010a) shows that the number of non-essential genes varies directly with genome size (Figure 2.3.10). When non-essential genes are plotted against genome size, extrapolation to non-essential genes = 0 yields a minimal genome size of about 423 kb. Plotting non-essential genes versus total genes yields an extrapolated value of 412 genes when non-essential genes = 0. Based on these extrapolated values, syn3.0 is not yet minimal. However, as a working minimal cell, it is important to retain a reasonable growth rate. The Syn3.0 genome is already substantially smaller than that of *M. genitalium*, yet its doubling time at 3 hours is five times

faster. Removal of more genes might inevitably result in slower growth.

2.3.6.3 Genetic Requirements of a Minimal Cell

The syn3.0 genome is 531 kb in size and has 438 protein-coding and 35 RNA-coding genes compared with the 1079 kb syn1.0 genome with 862 protein-coding and 39 RNA-coding genes. Table 2.3.1 lists six major gene functional categories as well as the sub-categories under each major category. Genes involved in expression of genetic information such as transcription and translation are mostly retained; only 22 genes were deleted, whereas 195 genes were kept in syn3.0. Genes involved in

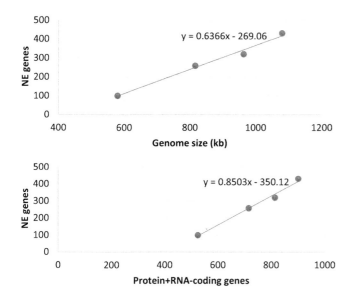

FIGURE 2.3.10 Non-essential (NE) protein-coding genes were determined by transposon mutagenesis in *M. genitalium* (580 kb), *M. pneumoniae* (816 kb), *M. pulmonis* (963 kb), and *M. mycoides* JCVI-syn1.0 (1080 kb). Plotting NE genes versus genome size yields an extrapolated value of 423 kb for the minimal genome size when NE = 0. Plotting NE genes versus total genes yields an extrapolated value of 412 genes when NE = 0.

preservation of genome information such as DNA replication and cell division were also largely retained; 34 genes were kept in syn3.0, and 13 genes were deleted. Among the genes involved in cell membrane structure and transport, 92 were retained and 77 were deleted from syn3.0. Only 73 genes involved in metabolism and energy production were retained, and 109 genes were deleted. A total of 73 genes specified mobile elements and DNA restriction systems, and these were all completely dispensable. Finally, among genes that could not be assigned to functional categories, 79 were retained and 134 were deleted.

All sequenced organisms currently contain a substantial fraction of genes with unknown or imprecisely known functions. In syn3.0, 65 genes are of unknown function (although seven genes have functional assignments as lipoproteins or membrane proteins) and 84 have generic functional assignments. Among the generics, the majority have been assigned to functional categories and belong to families such as hydrolases, ATPases, ATP transporters, etc., but their precise substrates are not known. In Table 2.3.1, only 79 genes have not been assigned to a functional category in syn3.0. Fifty-eight of these are of unknown function, and 21 are generic. A major goal for the future is to assign precise functions to those genes with unknown or generic functions. Only then will a complete understanding of life be possible (Alberts 2011).

2.3.7 SYNTHETIC YEAST CELLS

An international consortium of scientists from the USA, China, the UK, Singapore, and Australia has undertaken DNA synthesis of the yeast *Saccharomyces cerevisiae* genome (Annaluru et al. 2015; Richardson et al. 2017). *Saccharomyces*

TABLE 2.3.1

A List of All the Syn1.0 Genes Either Kept or Deleted in Syn3.0

Functional Category	Keep	Delete
Expression of Genome Information	**195**	**22**
Ribosome biogenesis	14	1
Transcription	9	0
RNA metabolism	7	0
Protein folding	3	0
Translation	89	2
RNA (ribosomal RNAs [rRNAs], tRNAs, small RNAs)	35	4
rRNA modification	12	3
tRNA modification	17	2
Regulation	9	10
Preservation of Genome Information	**34**	**13**
DNA topology	5	0
Chromosome segregation	3	0
DNA metabolism	3	0
DNA replication	16	2
DNA repair	6	8
Cell division	1	3
Cell Membrane	**92**	**77**
Protein export	10	0
Lipid salvage and biogenesis	21	4
Cofactor transport and salvage	21	4
Efflux	7	3
Membrane transport	31	32
Transport and catabolism of non-glucose carbon sources	2	34
Cytosolic Metabolism	**73**	**109**
Glucose transport and glycolysis	15	0
Nucleotide salvage	19	8
Metabolic processes	10	10
Redox homeostasis	4	4
Proteolysis	10	11
Lipoprotein	15	72
Acylglycerol breakdown	0	4
Mobile Elements and DNA Restriction	**0**	**73**
Unassigned	**79**	**134**
Total	**473**	**428**

cerevisiae has been extensively characterized genetically. It has 16 linear chromosomes, approximately 12 megabases of non-redundant sequence, and 6000 genes (Engel et al. 2014). Complete synthesis of the genome is within reach of current DNA synthesis capability and is currently well underway.

Yeast researchers have agreed on certain design features for the synthetic chromosomes. Only conservative changes will be made to ensure that the synthetic yeast will have the same fitness and normal growth of the wild-type cells. No destabilizing elements will be included to ensure genome stability and avoid rearrangements. The synthetic yeast will also have genetic flexibility to facilitate future studies. A complete design of the yeast genome incorporating these and other features was recently announced (Richardson et al. 2017).

Chromosome III was the first to be synthesized and installed (Annaluru et al. 2014). Design changes included TAG/TAA stop codon replacements; deletion of introns, mobile elements, and silent mating loci; and deletion of subtelomeric regions. In addition, 98 loxPsym sites were inserted to enable genome scrambling. The designer chromosome was synthesized in 12 "minichunks" with alternating selectable markers. Native chromosome III was replaced by 11 successive minichunk transformation rounds by using the high recombination frequency of yeast. Chromosome III size was reduced from 316,617 to 272,871 bp, without detectable loss of fitness of the synIII yeast strain.

So far, all of the yeast chromosomes have been designed and six (II, III, V, VI, X, and XII) have been synthesized (Richardson et al. 2017; Wang et al. 2018). When all the chromosomes have been synthesized and installed to create a synthetic yeast cell carrying the Sc2.0 genome, the focus will be on producing a yeast cell with a minimal genome, a truly daunting task. With around 5000 estimated non-essential genes mixed in and around at least another 1000 essential genes, how many can be removed without substantially compromising viability? One approach will be to introduce Cre recombinase into the cell to promote "scrambling" of genes and segments flanked by loxPsym sites (Shen et al. 2016). Survivors would be sequenced to discover genes and segments that can be deleted. Only yeast with small deletions are likely to be viable. This "top down" approach seems unlikely to proceed quickly, as the number of possible combinations is huge. However, much is likely to be learned. With the rapid progress in automated DNA synthesis, a bottom-up approach might ultimately prove to be the winner.

2.3.8 A DIGITAL-TO-BIOLOGICAL CONVERTER

In his book "Life at the Speed of Light," J. Craig Venter speculates on the possibility of biological teleportation (Venter 2013), the sending of DNA sequence information for a life form to a distant planet, and then recreating the living organism by means of a machine located at the destination. As a step toward this goal, his company, SGI, has recently developed an automated machine called a digital-to-biological converter (DBC) that can receive digital sequence information stored on a computer and use it to direct the automated production of DNA, RNA, proteins, or virus particles (Boles et al. 2017). The efforts to create synthetic cells have led to the development of ways to write DNA faster, more accurately, and more reliably. Because of the robustness of these technologies, the process could be automated to move workflows out of scientists' hands and onto a machine. In 2013, the first DNA printer, the BioXp, was built (Figure 2.3.4).

The next step was to develop a prototypical DBC, an essential first step to bringing on-demand and distributed manufacturing in biology to fruition. Unlike the BioXp, which starts with premanufactured small pieces of DNA (oligonucleotides), the DBC instrument was designed to receive digitized DNA code and convert it into whatever biological materials the code specified, which could be DNA, RNA, proteins, viruses, or even living cells.

To build the DBC, scientists worked together with software and instrumentation engineers to collapse multiple laboratory workflows into a single box. This included software algorithms to predict what DNA to build; chemistry to link the G, A, T, and C building blocks into short stretches of DNA; enzymatic assembly reactions to stitch those DNA fragments into much longer ones; and biology to convert DNA into other biological entities such as proteins (Boles et al. 2017).

A schematic diagram of the prototype is shown in Figure 2.3.11. It made therapeutic drugs and vaccines that, when tested, were as good as those from traditional methods. Laboratory workflows and processes that once took weeks or months were now carried out in only 1–2 days, without any human intervention, and were simply activated by the receipt of an email that could be sent from anywhere in the world.

Less than 200 years ago, it could take several weeks to get a one-way written message to someone. This all began to change in the 1840s with the invention of the fax machine—the first device that could transmit an image over a wire. The DBC can be compared to the fax machine, but rather than receiving images and documents, the DBC receives the instructions for printing biological materials.

Consider how fax machines have evolved. The prototype of the 1840s is unrecognizable compared with the fax machines of today. In the 1980s, most people still didn't know what a fax machine was, and it was difficult to grasp the concept of reproducing an image instantly on the other side of the world. But, nowadays, smartphones can do everything that fax machines can do, and this rapid exchange of digital information is taken for granted.

Digital-to-biological converters will be useful for distributed manufacturing of medicines, starting from DNA sequence. A day can now be imagined when it is routine for people to have a DBC connected to their home computer or smartphone as a means to download their prescriptions such as insulin or antibody therapies or for farmers to instantly download the most updated therapies for infested vegetables, fruits, and livestock.

It is easy to imagine DBCs located in strategic places around the world for rapid response to disease outbreaks such as the flu. For example, the Centers for Disease Control and Prevention (CDC) in Atlanta, Georgia, could prevent a flu outbreak by emailing a DNA sequence for a vaccine against it to DBCs in developing countries on the other side of the world, where a vaccine is manufactured on the front lines. This flu vaccine could even be tailored to match the specific strain circulating in that area. Sending DNA for instructing vaccines around the world in a digital file, rather than stockpiling those same vaccines and shipping them out, promises to save thousands of lives. With the cost and turnaround times to read DNA dropping exponentially, one can imagine the DBC being central to precision medicine, where every hospital in the world has a DBC to manufacture treatments tailored to a patient's DNA, at the patient's bedside.

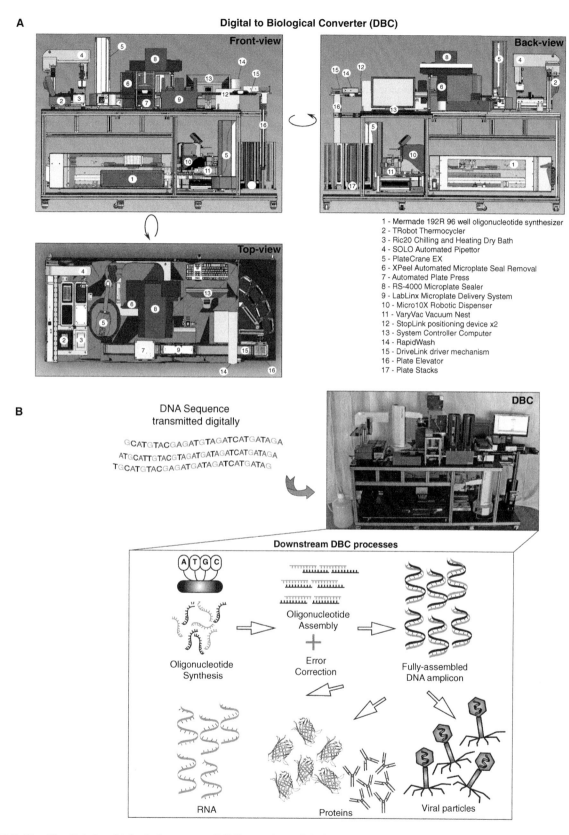

A

Digital to Biological Converter (DBC)

1 - Mermade 192R 96 well oligonucleotide synthesizer
2 - TRobot Thermocycler
3 - Ric20 Chilling and Heating Dry Bath
4 - SOLO Automated Pipettor
5 - PlateCrane EX
6 - XPeel Automated Microplate Seal Removal
7 - Automated Plate Press
8 - RS-4000 Microplate Sealer
9 - LabLinx Microplate Delivery System
10 - Micro10X Robotic Dispenser
11 - VaryVac Vacuum Nest
12 - StopLink positioning device x2
13 - System Controller Computer
14 - RapidWash
15 - DriveLink driver mechanism
16 - Plate Elevator
17 - Plate Stacks

B

DNA Sequence
transmitted digitally

GCATGTACGAGATGTAGATCATGATAGA
ATGCATTGTACGTAGATGATAGATCATGATAGA
TGCATGTACGAGATGATAGATCATGATAG

FIGURE 2.3.11 The digital-to-biological converter (DBC) constitutes fully integrated instruments and underlying software that are coordinated to perform a series of tasks without human intervention. Before a run, the DBC is preloaded with reagents required to make a designated biopolymer from digitally transmitted DNA sequence. The DBC process commences by generating overlapping oligonucleotide sequences that are subsequently chemically synthesized. These oligonucleotides are then assembled and processed to build a high-quality synthetic DNA amplicon. This DNA amplicon may be further converted into RNA, proteins, or viral particles by user-designated instructions. Examples of products generated on the DBC include influenza virus genes, antibody polypeptides, an RNA vaccine, and φX174 viral particles.

The applications go as far as the imagination will go. One can imagine DBCs being placed on other planets. Scientists on Earth could send the digital instructions to the DBC for a new medicine to treat an infectious disease or to make the planet more habitable for humans. This could include digital instructions to assemble synthetic organisms that make oxygen, food, fuel, and building materials. With digital information traveling at the speed of light, it would take only minutes for those instructions to be sent from Earth to Mars, compared with months to physically deliver those samples on a spacecraft.

For now, one must be satisfied beaming new medicines across the globe, fully automated and on-demand to save lives from emerging deadly infectious diseases and to create personalized cancer medicines for those who do not have time to wait.

REFERENCES

Agarwal, K. L., H. Buchi, M. H. Caruthers et al. 1970. Total synthesis of the gene for an alanine transfer ribonucleic acid from yeast. *Nature* 227: 27–34.

Alberts, B. 2011. A grand challenge in biology. *Science* 333: 1200.

Annaluru, N., H. Muller, L. A. Mitchell et al. 2014. Total synthesis of a functional designer eukaryotic chromosome. *Science* 344: 55–58.

Annaluru, N., S. Ramalingam, S. Chandrasegaran. 2015. Rewriting the blueprint of life by synthetic genomics and genome engineering. *Genome Biol* 16: 125.

Attia, T. H., M. A. Saeed. 2016. Next generation sequencing technologies: A short review. *Next Gener Seq Appl* S1: 006. doi:10.4172/2469-9853.S1-00.

Baby, V., J.-C. Lachance, J. Gagnon et al. 2018. Inferring the minimal genome of *Mesoplasma florum* by comparative genomics and transposon mutagenesis. *Mol Biol Physiol.* doi:10.1128/mSystems.00198-17.

Badarinarayana, V., P. W. Estep 3rd, J. Shendure et al. 2001. Selection analyses of insertional mutants using subgenic-resolution arrays. *Nat Biotechnol* 19: 1060–1065.

Benders, G. A., V. N. Noskov, E. A. Denisova et al. 2010. Cloning whole bacterial genomes in yeast. *Nucleic Acids Res* 38: 2558–2569.

Boles, K. S., K. Kannan, J. Gill et al. 2017. Digital-to-biological converter for on-demand production of biologics. *Nat Biotechnol* 35: 672–675.

Cello, J., A. V. Paul, E. Wimmer. 2002. Chemical synthesis of poliovirus cDNA: Generation of infectious virus in the absence of natural template. *Science* 297: 1016–1018.

Caruthers, M. H. 2013. The chemical synthesis of DNA/RNA: Our gift to science. *J Biol Chem* 288: 1420–1427.

Caruthers, M. H., K. Kleppe, J. H. Van de Sande et al. 1972. CXV. Total synthesis of the structural gene for an alanine transfer RNA from yeast. Enzymic joining to form the total DNA duplex. *J Mol Biol* 72: 475–492.

Dobzhansky, T. 1949. Genetics of natural populations; recombination and variability in populations of Drosophila pseudoobscura. *Genetics* 31: 269–290.

Engel, S. R., F. S. Dietrich, D. G. Fisk. 2014. The reference genome sequence of Saccharomyces cerevisiae: Then and now. *G3 (Bethesda)* 4: 389–398.

Fleischmann, R. D., M. D. Adams, O. White et al. 1995. Whole-genome random sequencing and assembly of *Haemophilus influenzae* Rd. *Science* 269: 496–512.

Fraser, C. M., J. D. Gocayne, O. White et al. 1995. The minimal gene complement of *Mycoplasma genitalium*. *Science* 270: 397–403.

French, C. T., P. Lao, A. E. Loraine et al. 2008. Large-scale transposon mutagenesis of *Mycoplasma pulmonis*. *Mol Microbiol* 69: 67–76.

Gibson, D. G. 2011. Enzymatic assembly of overlapping DNA fragments. *Methods Enzymol* 498: 349–361.

Gibson, D. G., G. A. Benders, C. Andrews-Pfannkoch et al. 2008a. Complete chemical synthesis, assembly, and cloning of a *Mycoplasma genitalium* genome. *Science* 319: 1215–1220.

Gibson, D. G., G. A. Benders, K. C. Axelrod et al. 2008b. One-step assembly in yeast of 25 overlapping DNA fragments to form a complete synthetic *Mycoplasma genitalium* genome. *Proc Natl Acad Sci U S A* 105: 20404–20409.

Gibson, D. G., J. L. Glass, C. Lartique et al. 2010a. Creation of a bacterial cell controlled by a chemically synthesized genome. *Science* 329: 52–56.

Gibson D. G., H. O. Smith, C. A. Hutchison, J. C. Venter, and C. Merryman. 2010b. Chemical synthesis of the mouse mitochondrial genome. *Nat Methods* 7: 901–903.

Gibson, D. G., L. Young, R.-Y. Chuang et al. 2009. Enzymatic assembly of DNA molecules up to several hundred kilobases. *Nat Methods* 6: 343–345.

Glass, J. I., N. Assad-Garcia, N. Alperovich et al. 2006. Essential genes of a minimal bacterium. *Proc Natl Acad Sci U S A* 103: 425–430.

Hutchison, C. A. 3rd, P.-Y. Chuang, V. N. Noskov et al. 2016. Design and synthesis of a minimal bacterial genome. *Science* 351: 1414, aad6253.

Hutchison, C. A., S. N. Peterson, S. R. Gill et al. 1999. Global transposon mutagenesis and a minimal Mycoplasma genome. *Science* 286: 2165–2169.

Karas, B. J., J. Jablonovic, E. Irvine et al. 2014. Transferring whole genomes from bacteria to yeast spheroplasts using entire bacterial cells to reduce DNA shearing. *Nat Protoc* 9: 743–750.

Karas, B. J., J. Jablanovic, L. Sun et al. 2013. Direct transfer of whole genomes from bacteria to yeast. *Nat Methods* 10: 410–412.

Karas, B. J., C. Tagwerker, I. T. Yonemoto et al. 2012. Cloning the *Acholeplasma laidlawii* PG-8A genome in *Saccharomyces cerevisiae* as a yeast centromeric plasmid. *ACS Synth Biol* 1: 22–28.

Khorana, H. G. 1968. Synthesis in the study of nucleic acids. The Fourth Jubilee Lecture. *Biochem J* 109: 709–725.

Khorana, H. G. 1972. Nucleic acid synthesis in the study of the genetic code. In *Nobel Lectures, Physiology or Medicine 1963–1970*. Amsterdam, the Netherlands: Elsevier.

Kosuri, S., N. Eroshenko, E. LeProust et al. 2010. A scalable gene synthesis platform using high-fidelity DNA microchips. *Nat Biotechnol* 28: 1295–1299.

Hughes R. A., A. D. Ellington. 2017. Synthetic DNA synthesis and assembly: Putting the synthetic in synthetic biology. *Cold Spring Harb Perspect Biol.* doi:10.1101/cshperspect.a023812.

Labroussaa, F., A. Lebaudy, V. Baby et al. 2016. Impact of donor-recipient phylogenetic distance on bacterial genome transplantation. *Nucleic Acids Res* 44: 8501–8511.

Lartique, C., J. I. Glass, N. Alperovich et al. 2007. Genome transplantation in bacteria: Changing one species to another. *Science* 317: 632–638.

Lluch-Senar, M., J. Delgado, W. H. Chen et al. 2015. Defining a minimal cell: Essentiality of small ORFs and ncRNAs in a genome-reduced bacterium. *Mol Syst Biol* 11: 780.

Morowitz, H. 1984. The completeness of molecular biology. *Isr J Med Sci* 20: 750–753.

Mushegian, A. R., E. V. Koonin 1996. A minimal gene set for cellular life derived by comparison of complete bacterial genomes. *Proc Natl Acad Sci U S A* 93: 10268–10273.

Noskov, V. N., B. J. Karas, L. Young et al. 2012. Assembly of large, high G+C bacterial DNA fragments in yeast. *ACS Synth Biol* 1: 267–273.

Richardson, S. M., L. A. Mitchell, G. Stracquadanio et al. 2017. Design of a synthetic yeast genome. *Science* 355: 1040–1044.

Shen, Y., G. Stracquadanio, Y. Wang et al. 2016. SCRaMbLE generates designed combinatorial stochastic diversity in synthetic chromosomes. *Genome Res* 26: 36–49.

Smith, H. O., C. A. Hutchison, C. Pfannkoch, J. C. Venter. 2003. Generating a synthetic genome by whole genome assembly: PhiX174 bacteriophage from synthetic oligonucleotides. *Proc Natl Acad Sci U S A* 100: 15440–15445.

Smith, V., K. N. Chou, D. Lashkari et al. 1996. Functional analysis of the genes of yeast chromosome V by genetic footprinting. *Science* 274: 2069–2074.

Tagwerker, C., C. L. Dupont, B. J. Karas et al. 2012. Sequence analysis of a complete 1.66 Mb Prochlorococcus marinus MED4 genome cloned in yeast. *Nucleic Acids Res* 40: 10375–10383.

Venter, J. C. 2013. *Life at the Speed of Light: From the Double Helix to the Dawn of Digital Life.* New York: Penguin Books.

Wang, L., S. Jiang, C. Chen et al. 2018. Synthetic genomics: From DNA synthesis to genome design. *Angew Chem Int Ed Engl* 57: 1748–1756.

Wimmer, E. 2006. The test-tube synthesis of poliovirus: The simple synthesis of a virus has far reaching societal implications. *EMBO Reports* 7: S3–S9.

2.4 Communication as the Main Characteristic of Life

Guenther Witzany

CONTENTS

2.4.1 INTRODUCTION

With an expertise in the philosophies of science and language, I turned my focus to the philosophy of biology and life sciences in the mid-1980s. I found it interesting that the results of the philosophy of science debates did not reach any biological discipline, although philosophy of science is the essential discipline for the foundations and justifications of scientific theory building and methodology. At the center of these more than half-a-century lasting debates was the question how to define a scientific sentence in contrast to metaphysical ones. "Scientific" meant empirically based, experimentally testable, and theoretically formalizable, because mathematics was assumed to be the only exact science that could depict material (physical and chemical) reality (Whitehead and Russel 1910–1913; Goedel 1931; Carnap 1939).

The crucial question remained how to construct sentences that are scientific and not metaphysical, that is, the question how to define language, the ultimate prerequisite of any human utterance. The debate lasted from 1920 to 1980, with its most prominent proponents such as early Wittgenstein, Carnap, Hilbert, Russel, Whitehead, Goedel, Shannon, Weaver, Turing, von Neumann, late Wittgenstein, Austin, Searle, Chomsky, and Habermas (Witzany 2010). The results of these debates are part of the history of science:

- The concept of a coherent axiomatic system with error-free logical sentences is impossible, in principle.
- The concept of an exact scientific language was a pipe dream.
- Natural languages do not speak themselves; there are always real-life individuals in populations that generate and use such languages to coordinate and organize the real-life world, and the usage of natural languages is, therefore, part of communication, that is, a kind of social interaction.
- Living agents that use natural languages are principally able to generate new sign sequences that cannot be predicted or deduced from former ones and for which no algorithm is available, in principle.

- The meaning (semantics) of the signs and sign sequences in natural languages depends on the real-life context (pragmatics) within which signs are used and not on its syntax.

As a consequence, formalizable, that is, mathematical, theories of (context-free) languages are not an appropriate tool for explaining the essential features of natural languages used in communication processes. They cannot identify the context dependence of meaning and its deep grammar that helps to transport different and even contradictory meanings to the superficial grammar, both represented by identical sign sequences. Additionally, describing the social character of real-life organisms is not within their expertise, because social interacting organisms don't behave like formalizable abiotic elements, and, for the inherent feature of generating new sequences, new behavior, and new interactional motifs, no algorithm is available in principle (Witzany 1995, 2010, 2011a). As we will see at the end of this contribution, this must have serious consequences in the explanatory model on how to describe the sign-mediated interactions in non-cellular early RNA-world and emergence of the genetic code, genetic information, and its regulation.

2.4.2 EMPIRICAL FACTS ABOUT COMMUNICATION

2.4.2.1 COMMUNICATION IS SOCIAL INTERACTION

Communication designates social interaction. Social interacting living agents need some tools so that interaction may lead to coordination and organization of common behavior to reach goals. In contrast to physical-chemical interactions on an abiotic planet, communicative interactions on biotic planets are mediated by signs. Such signs must be uttered by bodily expressed movements, phonetics, audiovisuality, tactility (e.g., vibrational), or odor (semiochemical). This means that sign-mediated utterances may also be transported as body movement patterns (e.g., series of gestures).

The use of signs in communicative interactions must be learned somehow (Morris 1946). This means that the use must follow an acquired competence. The competence of living agents is inherited or learned or both. In any case, it needs some social interaction experience to trigger this competence into an actual available behavior. In concrete real-life social experiences, living agents learn to designate a message for non-self agents in a real-life context. In social experiences, living agents learn to use more than one sign and combine multiple different signs into sign sequences to communicate complex content. Social experiences are the essential background to learning contextual, combinatorial, and content-based rules on how to use signs in social interactions (Witzany 2000).

2.4.2.2 COMMUNICATION IS SIGN-MEDIATED ACCORDING TO RULES

Communication is a kind of social interaction. It is an empirically proven fact that communication could not be invented by a singular living agent for one time only once. Communication needs signs to communicate context-relevant content. Such signs can be combined in a line-up to sign sequences. The use of signs needs some competence to combine sign sequences correctly. This competence of rule-following depends on social interaction experience (context). Communication is the essential interaction to common understanding of content (meaning). Social interaction experience is embedded into a cultural background history.

2.4.2.3 MEANING DEPENDS ON CONTEXT

The meaning of a sign sequence or a "word" or any other sign used to communicate is a social function (Mead 1934). The behavioral reaction to a communicative utterance emerges semantic meaning. Not to forget that natural communication depends on the context. Also, meaning-generating social events are context-dependent. For example, the humble dance in bees gives direction and energy costs in food gathering. Exactly the same dance figures are used during the hive search, where context transports the completely different meaning of an appropriate hive (Witzany 2000). Another example: The sentence "Shooting of the hunters" in human communication may transmit the message that hunters shoot and, conversely, that hunters are shot. Without context, contradictory semantics of sign sequences cannot be identified. Syntactic analyses alone are insufficient to identify semantic content (Witzany and Baluška 2012b).

2.4.2.4 COMMUNICATION INITIATES DE NOVO GENERATION

Communication as rule-governed sign-mediated interactions are different from interactions in a purely physical-chemical world without any biotic agents. In such interactions, signs and rules of sign use as well as sign-using living agents are not existent. In contrast, communicating living agents share a limited repertoire of signs that are used according to a limited number of rules that must be followed to generate correct sign sequences to designate context-dependent content. Most interesting is the fact that such rules—although rather conservative—may be changed in extreme cases or if adaptation is necessary. Rule-following by living agents is rather flexible in contrast to natural laws to which living agents abide strictly. This means that communication is the essential tool to generate new signs, sign sequences, new rules for sign use, and generation of new content according to unexpected contextual circumstances (Witzany 2015). Communicating living agents are able, in principle, to generate new communicative patterns for better or innovative adaptation to a new and unforeseeable situation.

2.4.2.5 EXPERTISE ON LANGUAGE AND COMMUNICATION

Curiously, many biologists are not very familiar with the current definitions of "language" and "communication," although they are commonly used in molecular biology to speak about "genetic code," "code without commas" (Crick et al. 1957), "nucleic acid language," "recognition sequences," "transcription," "translation," "amino acid language," "immune responses," "intercellular communication," etc. The above-listed theoretical and empirically tested results on "language" and "communication" have been ignored until today. If we speak now about (1) the three categories of signs (index, icon, and symbol); (2) the three complementary non-reducible levels of semiotic rules' syntax, pragmatics, and semantics; and (3) communication as rule-governed sign-mediated interactions, it can easily be seen that all these categories are nearly unknown in biology, especially in molecular biology, cell biology, genetics, bioinformatics, and related disciplines (Schroedinger 1944; Brenner 2012; Eigen 2013).

All the previously summarized characteristics of communication processes are results of empirical investigations (Habermas 1984, 1987; Austin 1975; Searle 1976; Tomasello 2008). These are no hypotheses or theoretical constructions but empirically proven facts. Some derive from investigations on cognition differences between apes and humans, some from investigations of communicative interactions and speech acts, and some from investigations of language constructions and applications. There is no doubt that language, communication, and social interacting living agents are within the expertise of empirical social sciences.

In the last 100 years, there has been an abundance of theories and concepts on language and communication by natural sciences, such as mathematics, physics, chemistry, information theory, cybernetic systems theory and its derivatives, "mathematical theory of language," "cognitive revolution," and "artificial intelligence" (Wiener 1948; Shannon and Waever 1949; Turing 1950; Chomsky 1964, 1965; von Neumann 1966; Nowak and Krakauer 1999). With the results of pragmatic action theory, all of them have been essentially falsified, because none of them could coherently explain the steps from a single biotic agent to a commonly shared understanding as a prerequisite for common coordination and organizational behavior or "how to make the move from a state of private consciousness to a state of mutual agreement and cooperation" (McCarthy 1984).

In this respect, this contribution will demonstrate that communication is the main characteristic not only of humans but also of life generally, based on a common methodological assumption of natural sciences and social sciences by presenting empirical data that can be proved by experiments and observations.

2.4.3 BIOCOMMUNICATION IN ALL DOMAINS OF LIFE

If communication is the main characteristic of life, it must be possible to identify communicative actions throughout all domains of life. Until the middle of the last century, language and communication were thought be special tools of only humans. Meanwhile, we know an abundance of examples of non-human languages and communication processes.

Therefore, the description of communication processes must be valid in principle in all organisms, from the simplest akaryote up to the humans. Whereas we identified main characteristics of communication, its (1) social character, (2) dependence on signs according, (3) the three kinds of rules (combinatorial, context-specific, and content-coherent), we must draw our attention from the decades-long narrative suggested by information theory and systems theory, that is, the sender-receiver narrative (coding-decoding), which was wrong in several aspects, as outlined elsewhere (Witzany 1995). Communication is not only an information-transfer process but more than this interaction, mediated mainly by utterances represented by signs. If we speak, or any other organism uses signs to signal something, we do something (Austin 1975; Searle 1976; Habermas 1994).

Interactions between living organisms that are based on signs (signals, icons, and symbols) used according to combinatorial, context-dependent, and content-coherent rules are the very fundamental techniques to coordinate any common behavior and organize division of labor. Although the lifeworld of organisms is rather restricted to their species-specific habitats, symbiotic partners, geologically determined niche constructions, the signals they share, and the behavioral motifs they like to use, we can find communication in all organisms of all domains of life (Witzany 2011c, 2012a, 2014a, 2017a; Witzany and Baluška 2012a; Witzany and Nowacki 2016) (Figure 2.4.1).

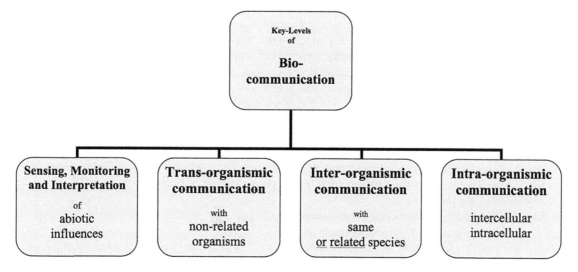

FIGURE 2.4.1 The biocommunication approach identified four levels in which cellular organisms are involved since the start of their life until death.

But biocommunication is not restricted to a species-specific lifeworld. Additionally, we must think of cell-cell communication processes in parallel within the organismal bodies, between tissues and organs, to coordinate interactions of the whole body. Last but not least, from the organisms that communicate within their body and between the same or related members of their species, we can also find communication processes between non-related organisms, as in attack and defense behaviors or more than this in an abundance of symbiotic and symbiogenetic interactions (Witzany 2006).

2.4.3.1 HUMANS AND THEIR ANCESTORS

To learn a natural language means to communicate basic everyday needs with community members. This is how we learn what a word means. "The meaning of a word is its use" (Wittgenstein 1953), or Ch. S. Peirce: to identify "…meaning, we have, therefore, simply to determine what habits it produces" (Peirce 1923). We can understand words and their sequences, because we have learned a practice of interaction, which includes learning from the community as to which words are combined with which customs or interactional patterns.

Although great apes understand many aspects of social interactions in their lifeworlds, including causal and intentional relationships, there is a crucial difference from human understanding: In contrast even with human infants, great apes cannot participate in shared intentionality or co-operative communication (Tomasello 2008; Bohn et al. 2016; Engelmann et al. 2017).

Another crucial difference between humans and great apes is that humans designate situations and entities for other people in a language-like manner. These other persons then try to understand why active agents wants to share information and why they want them to know this could be relevant to them. Besides the ability to participate in shared intentionality, this requires a variety of complex and recursive logical conclusions about the intentions of others.

The empirical result is that although several animal species can represent situations and other entities cognitively in an abstract way, only humans can generate actively distinct perspectives of the same situation. Additionally, only humans draw social recursive conclusions that are self-reflexive also with regard to the intentional states of others. Also, several animal species evaluate and feed back their actions in respect of their intentional goals, but only humans evaluate their behavior with respect to the normative perspectives of others or their group identity.

The individual intentionality changed into co-operation by means of a common intentionality, which made new forms of cognitive representation, that is, perspectivistic and symbolic, possible. The conclusions changed into social recursive ones, which means they focused not on individual perspectives but on the perspective of all group members.

This new form of co-operation within common intentionality emerged from common goals (not individual ones) and commonly shared attention. These are the ingredients of social coordination, which is different from individual coordination with each other. Group hunting in great apes is characterized by the fact that each individual ape tries to catch the prey. This means the group behavior remains in the "ego-status" for every individual. But the so-called co-operative turn in humans means that the group members now act in a group perspective without "ego-status." The decisions that are taken focus on group benefit, not individual benefit. This was the result of experiments with 3-year-old children, in contrast to great apes, which remain in the "ego-status," although they act together in groups. The children co-operated in several experimental setups, and it was shown that the common goal is so important that individual children who reached their goals early did not stop their actions until all the others had reached their goals. Similar behavior could not be observed in great apes (Tomasello 2008; Bohn et al. 2016; Engelmann et al. 2017).

Additionally, humans communicate with others about content that is not actually present. This means language must serve as symbolic (conventionally derived) representation of information that is not actual but abstract. This is a crucial difference from their ancestors (Bohn et al. 2015).

The speech acts in humans differ fundamentally from their evolutionary ancestors. We can differentiate superficial grammar and deep grammar in human speech acts, or, in the words of John Searle, the founder of speech act theory, we can find locutionary acts done by utterances (designated by their grammatic-semantic sequence); illocutionary acts (what we want to do with what we say), which cannot be identified through analyses of the words that are said but depend on mutual expectations, intentions, emotions, roles, etc.; and the performative speech acts that are characterized by what is done by the one who is addressee of what is said. Especially the illocutionary speech act is nothing that can be identified easily in great apes. Also, in humans, there is no algorithm available to identify such crucial determinants of what is expected and intended. So, natural human communication in everyday life and colloquial speech are nothing that can be computed or predicted easily.

Natural signs used by humans are not only auditory and visual (gestures) but also body movements and tactile. Beyond conscious sign generation, there are inherent unconscious messages transported, as in all other animals by semiochemicals, that is, pheromone and other hormone-like odors, although odor receptors and related tissues are largely lost during retroviral infection events in the African primate populations prior to the chimpanzee-human split (Villarreal 2009). This caused a strong selective pressure to evolve alternative communication tools to perpetuate social life and social order without pre-dominant odor determinants. Additionally, we must not forget that 95% of the cells that a human body assembles are persistent settlers, in most cases, symbionts, as demonstrated on the skin or in the human gut, without which humans cannot survive. Natural signs produced by humans to the symbionts depend on cell-cell communication in tissues and organs.

2.4.3.2 BEES

To take an example for communication as the main characteristic of animal life, we can take any species from this kingdom. More appropriate would be animal species that obviously share a coherent swarm behavior, as represented by some fish, insect, or bird species. I will take bees for example, because, for successful investigation of bee language and communication, Karl von Frisch received the Nobel Prize in 1973. If we speak about language and communication, we usually think of humans that talk to each other and communicate to organize common goals and to coordinate common behavior (von Frisch 1971).

But since Karl von Frisch received Nobel Prize for detection and investigation of bee languages and dialects, broader research communities and experts in bee breeding noticed that even non-human social animals might communicate to reach complex behavioral patterns. Since Karl von Frisch's work, it has been evident that the highly complex social behavior of bee swarms is organized and coordinated by sign-mediated interactions, that is, communication. If communication processes are disturbed, they may have fatal consequences for bee colonies, especially if we think of special bees' behavior such as foraging, their search for a new hive, and the semantics of bee dances. As in every other natural language, the same sign sequences may have different meanings in different contexts. This means that bees with a limited repertoire of signs can transport different messages, which trigger different response behaviors, with far-reaching consequences (Witzany 1995).

As in every other natural language, bee languages also differ in habitat-dependent dialects. Small groups may generate specific signs according to niche adaptations, which are not present in other habitats, or similar signs may transport different meanings. For example, when mixing Austrian and Italian bees from certain regions, von Frisch exemplified that a special bee movement pattern (dance) that represents the distance to a nutrition field meant 100 meters for Austrian bees but 500 meters for Italian bees, which led to struggle between both groups, because one group did not find what was designated.

Most interestingly, the language of honey bees in colder hemispheres is the only known non-human language that uses body movements that represent symbolic signs.

Natural signs in bee languages are moving patterns, "dances" of various kinds, and motifs that transport all differentiated messages being relevant for social bee life to the other swarm-members. Besides this, hormonal (semiochemical), tactile, vibrational, and even visual signs are used.

2.4.3.3 OTHER ANIMALS

Similar to the general categories of biocommunication, communication processes within the animal kingdom have been investigated in chimpanzees, elephants, wolfs, dogs, rodents, mouse, rats, spiders, ants, termites, crows, parrots, birds, salamanders, chelonians, cetaceae, fish, cephalopods such as octopus, corals, nematodes (Witzany 2014a), and many others not listed here. Although the detailed investigations according to the conceptual categories listed previously are missing, they can be expected to be undertaken within next decades.

Natural signs in nearly all animals are generated to coordinate reproduction cycles and social membership roles, as well as in mating, kinship welfare, attack, defense, etc., by semiochemicals, that is, hormones, in combination with auditory, visual, and tactile signs that are rather species-specific.

2.4.4 BIOCOMMUNICATION OF PLANTS

Plants have traditionally been viewed as growth automatons. Now, we recognize that the coordination of growth and development in plants is possible only by using signs rather than pure mechanics. Understanding the use of signs in communication processes requires a differentiated perspective (Baluška et al. 2006). Chemical molecules are used as signs. They function as signals, messenger substances, information carriers, and memory medium in solid, liquid, or gaseous form (Witzany and Baluška 2012a).

Plants are sessile organisms that actively compete for environmental resources, both above and below the ground (Bais et al. 2004). They assess their surroundings, estimate how much energy they need for particular goals, and then realize the optimum variant. They take measures to control certain environmental resources. They perceive themselves and can distinguish between self and non-self. This capability allows them to protect their territory. They process and evaluate information and then modify their behavior accordingly.

To understand these competences, we will notice that this is possible due to parallel communication processes within the plant (intraorganismic), between same and different plant species (interorganismic), and between plants and non-plants (transorganismic). Intraorganismic communication involves sign-mediated interactions in cells (intracellular) and between cells (intercellular).

Intercellular communication processes are crucial in coordinating growth and development and shape and dynamics. Such communication must function on both the local level and between widely separated plant parts. This allows plants to react in a differentiated manner to its current developmental status and physiological influences. Chemical communication is either vesicular trafficking or cell-cell communication via the plasmodesmata. Moreover, numerous signal molecules are produced in or controlled by the cell walls. Physical communication takes place through airborne, electrical, hydraulic, and mechanical signs (Blande and Glinwood 2016).

Natural signs that enable the communication processes between tissues and cells in plants are incredibly complex and encompass nucleic acids, oligonucleotides, proteins and peptides, minerals, oxidative signals, gases, mechanical signals, electrical signals, fatty acids, oligosaccharides, growth factors, several amino acids, various secondary metabolite products (up until now, 100,000 different metabolites are identified), and simple sugars (Witzany and Baluška 2012a).

2.4.5 BIOCOMMUNICATION OF FUNGI

The evolutionary forerunners to animals and plants are fungi, which are represented by single as well as multicellular species. Currently, it is estimated that there are at least 1.5 million fungal species, out of which about 300,000 are described in the scientific literature. It is estimated that fungi account for at least one-fourth of the global biomass. As with plants, fungi are sessile organisms that can live for extremely long periods or extend over large areas: one example (*Armillaria gallica*) has been found, which covers as much as 15 ha and has an age of approximately 1500 years (Casselman 2007). Another example (*Armillaria ostoyae*) covers 900 ha and has an estimated age of 2400 years (Casselman 2007). Higher fungi are modular hyphal organisms in that they reproduce by clonation or are parasexual. They establish interlocking networks. The symbiotic relation to plants, especially in the rhizosphere, is indispensable, because without fungal degradants, plants cannot uptake soluble nutrients (Witzany and Baluška 2012a).

Natural signs in fungi are restricted to semiochemicals that are generated and interpreted to coordinate further reactions. To exemplify such signaling process, we can look at the following examples: (i) Mitogen-activated protein kinase (MAPK) signaling is involved in cell integrity, cell wall construction, pheromones/mating, and osmoregulation (Dohlman and Slessareva 2006; Yu et al. 2008). (ii) Cyclic adenosine monophosphate/protein kinase A (cAMP/PKA) system is involved in fungal development and virulence. (iii) RAS protein is involved in cross-talk between signaling cascades. (iv) Calcium-calmodulin-calcineurin are involved in cell survival under oxidative stress, high temperature, and membrane/cell wall perturbation. (v) Rapamycin is involved in the control of cell growth and proliferation (Fernandes et al. 2005). (vi) Aromatic alcohols such as tryptophol and phenethyl alcohol are used as quorum-sensing molecules. (vii) A variety of volatiles (alcohols, esters, ketones, acids, and lipids) and non-volatile inhibitory compounds (Farnesol and H_2O_2) (Leeder et al. 2011). To date, 400 different secondary metabolites have been documented. These are known to contain mycotoxins and are used both for defensive and aggressive behaviors.

2.4.6 BIOCOMMUNICATION OF SINGLE-CELLED EUKARYOTES

Unicellular eukaryotes represent an own kingdom of life. The crucial difference to their evolutionary forerunners, that is, akaryotes, is their rather complex cellular structure, which includes several parts being absent in akaryotes, such as the nucleus, which contains the genome. The Serial Endosymbiotic Theory successfully explained this evolutionary event (Margulis 2004; Witzany 2006). They evolved signaling to modulate fundamental activities such as acquisition of nutrients and reproduction. They can be found in abundance in aqueous and soil environments; they assess their surroundings, estimate how much energy they need for particular goals, and then realize the optimum variant. They take measures to control certain environmental resources (Witzany and Nowacki 2016). They perceive themselves and can distinguish between self and non-self. They process and evaluate information and then modify their behavior accordingly. In order to generate an appropriate response behavior, protozoa must be able to not only sense but also interpret and memorize important indices from the abiotic environment and adapt to them accordingly. This is decisive in coordinating growth and development, mating, shape, and dynamics.

Parasitic protozoa cause some of the most severe infectious human diseases worldwide.

Natural signs in unicellular eukaryotes are semiochemicals, such as hormones and secondary metabolites that serve as signs within interactions of various motifs, such as reproduction, mating, feeding, attack, and defense. In this respect, it is important to interconnect semiochemicals with the concrete interactional motif (Luporini et al. 1995, 2006). Identical signals may be used in several interaction motifs with quite different meanings, and this indicates the context dependence of signals. Quite common are hormone receptors. Signaling also includes cAMP in different life-cycle stages, calcium-mediated adenylyl cyclase (AC) regulation, as well as cyclic nucleotide (guanylyl cyclase) signaling involvement in exflagellation, phosphodiesterases (PDEs), and cyclic-nucleotide-activated protein kinases are essentially conserved in protozoa. Ca^{2+}-signaling pathways and intracellular Ca^{2+} channels are present in nearly all unicellular eukaryotes.

2.4.7 BIOCOMMUNICATION OF AKARYOTES (BACTERIA AND ARCHAEA)

Bacteria and non-related akaryotic archaea have been assumed to be the most primitive organisms and consequently have been investigated as single-cell individuals, determined by mechanistic input-output reactions. Now, this picture has changed. Today, we know that bacteria and archaea are part of a community that interacts in a highly sophisticated manner (Crespi 2001). The production and the exchange of messenger molecules enable unicellular organisms to coordinate their behavior like a multicellular organism (Shapiro 1998; Bassler 1999; Schauder and Bassler 2001; Ben Jacob et al. 2004).

The coordinated community, for example, of oral bacteria in humans, relies on intra- and interspecies communication. This community encompasses ca. 500 different species, some of which co-operate while others compete (Kolenbrander et al. 2002). The complexity of potential interactions in the oral cavity and the number of possibilities reach unimaginable proportions if we assume that each of the 500 bacteria species can regulate its genes in response to host-produced molecules and interact with all other bacteria species (Kolenbrander et al. 2002).

The medium of every bacterial and archaeal coordination is communication, that is, sign-mediated interaction (Dunn and Handelsmann 2002). A wide range of chemical molecules serves as signs through which bacterial and archaeal communities coordinate to reach a "quorum," which is the starting point for decision-making: one of many different behavioral patterns will thereby be organized, such as biofilm organization, bioluminescence, virulence, and sporulation (Sharma et al. 2003). Quorum sensing includes not only chemotaxis but also interpretation, which means that the incoming signs are measured against the background memory of the species colony in their real-life world (Kaiser and Losick 1993; Losick and Kaiser 1997). Interpretation before decision-making, coordination, and organization, such as fruiting body formation and co-operative hierarchical organization, is context-dependent (Witzany 2017a).

Natural signs in bacterial and archaea communications are semiochemicals involved in producing, releasing, detecting, and responding to small hormone-like molecules termed autoinducers. Acylated homoserine lactone (AHL) and peptides represent the two major classes of known bacterial cell-cell signaling molecules. AHLs are the products of LuxI-type autoinducer synthases, short peptides that often contain chemical modifications, phosphorylation cascades that ultimately impinge on DNA-binding transcription factors responsible for regulation of target genes (Fuqua et al. 1996). In general, bacteria keep their AHL and peptide quorum-sensing conversations private, by each species of bacteria producing and detecting a unique AHL (AHLs differ in their acyl side-chain moieties), peptide, or combination thereof. Archaea share some molecular biological features with Eukarya that are not found in any bacteria, such as ATP production, protein secretion, cell division and vesicles formation, and protein modification pathways (Woese and Fox 1977; Woese et al. 1990; Forterre 1997). Archaea therefore seem to share a strong evolutionary relationship with the Eukarya (Garrett and Klenk 2007; Garrett et al. 2011).

2.4.8 BIOCOMMUNICATION OF VIRUSES

With drawing our attention now to viruses and their relatives, we seem to move from cellular life to sub-cellular elements. Current knowledge about the virosphere and their roles in evolution ("virolution") indicates interactions of RNA viruses, DNA viruses, viral swarms, and viral and RNA-based sub-viral networks that co-operate and coordinate (regulate) within cellular genomes either as replication-relevant co-players or suppression-relevant silencers (Ryan 2009; Stedman 2013, 2015; Seligmann and Raoult 2016). Some represent infection-derived modular tools of non-coding RNAs, which have built consortia of complementary agents that function together (Tycowski et al. 2015; Heaton and Cullen 2017; Inuoue et al. 2017).

Viruses have long been accepted only as disease-causing, epidemic phenomena, with lytic and therefore extremely dangerous consequences for infected organisms. However, new research has corrected this picture (Villarreal 2009; Witzany 2012b, Berliner et al. 2018). Viruses are part of the living world, in most cases integrated in the cytoplasm or the nucleoplasm of cells, without harming the host. Viruses are on their way to representing the best examples of symbiotic relationships, because there is no living being since the start of life that has not been colonized by them, in most often cases in the form of multiple colonizations (Villarreal and Witzany 2010, 2015, 2018).

Today, we can identify several key players that coordinate and organize the genetic content compositions of host organisms (Koonin 2016). They include endogenous viruses and defectives, transposons, retrotransposons, long terminal repeats, non-long terminal repeats, long interspersed nuclear elements, short interspersed nuclear elements, group I introns, group II introns, phages, and plasmids (Weiner 2006; Jurka et al. 2007; Slotkin and Martienssen 2007; Lambowitz and Zimmerly 2011; McNeil et al. 2016; Belfort 2017). These are

just some examples that use genomic DNA as their preferred living habitat. This means that DNA is not a solely stable genetic storage medium that serves as an evolutionary protocol but is also a species-specific ecological niche for viral RNAs.

Persistent viral lifestyle is the most dominant biological lifestyle on this planet (Villarreal 2005). From this perspective, the total number of cellular organisms looks like small islands in an ocean of the global virosphere. Viruses and virus-derived parts represent the most abundant genetic information on the planet, overrepresenting cellular genetic information 10 times. If we only consider prokaryotic life, we have a number of prokaryote viruses (phages) of 10^{31}, which means if we line up the length of their virions, we get 40 million light-years (Rohwer et al. 2014). Importantly, a key feature of this viral lifestyle is that only few need to remain as functional agents, such as mammalian endogenous retroviruses needed for the syncytia, which regulate mammalian pregnancy (Perot et al. 2012), not to forget the role of persistent viruses in any kind of host immunity functions.

In most cases parts of infectious agents remain as defectives such as LTRs, non-LTRs, SINEs, LINEs, and Alu's. As defectives they are exapted for cellular needs such as regulation tools in all steps and fine-tuned substeps of cellular functions, such as transcription, translation, epigenetics, repair, and immunity (Slotkin and Martienssen 2007; Lambowitz and Zimmerly 2011; Witzany 2011b; Chalopin et al. 2012; Conley and Jordan 2012; Roossinck 2015).

The abundance of such genetic agents has been identified during the last 40 years as obligate inhabitants of all genomes, whether prokaryotic or eukaryotic. They infect, insert, and delete; some cut and paste; others copy and paste; and both spread within the genome. They change host genetic identities by insertion, recombination, or epigenetic regulation or re-regulation of genetic content, and co-evolve with the host to interact in a modular manner (Geuking et al. 2009; Shapiro 2011). Together with non-coding RNAs, they shape both genome architecture and regulation. In this respect, they are the agents of change, not only over evolutionary time but also in real time as domesticated agents.

Viruses can co-operate, that is, they interact to build groups that invade host genomes and even compete as a group for limited resources such as host genomes (Villarreal and Witzany 2013a, 2013b). This leads to an extraordinary effective result and a key behavioral motif that is able to integrate a persistent lifestyle into cellular host organisms, the "addiction" modules: former competing viral groups are counterbalancing each other, together with the host immune system (Villarreal 2012a). Although rather stable under certain circumstances, this addiction balance can also get out of balance, which means the competing viral features may become virulent again. But when stable, we can find such counter-regulating paired genes of the addiction modules, as in the restriction/modulation (RM) systems as well as in the toxin/antitoxin (TA) systems (Harms et al. 2018). Insertion/deletion functions represent similar modules as do the RM systems. This "infectious" colonization by new addiction modules is a main process in generating new sequence space without error replication and therefore in the evolution, conservation, and plasticity of host genetic identities. In this perspective, the genetic identities of cellular life throughout all domains of life are edited by their genetic parasites.

More recently, it has been found that phages communicate to decide whether to initiate lysogeny or not. Some phages produce communication peptides that are released. In subsequent infections, progeny phages measure the concentration of this peptide and lysogenize if the concentration is high. Interestingly, different phages encode different versions of the communication peptide, representing a phage-specific peptide communication code for lysogeny decisions (Erez et al. 2017). Virus-to-virus interactional motifs generally range from conflict to co-operation in various forms dependent on situational context (Díaz-Muñoz et al. 2017).

Consequently, the biocommunication approach has to integrate coherent natural genome editing of the genetic code by viruses (Witzany 2009). In this perspective, the genetic code of living organisms did not result out of chance mutations (error-replication events) that are biologically selected. The error-replication narrative has problems to explain the sudden emergence of new species, new phenotypic traits, and genome innovations as a sudden single event.

2.4.9 BIOCOMMUNICATION OF RNA STEM-LOOP GROUPS

To go one step deeper to the roots of life, we meet the viroids, short strands of circular, single-stranded RNA virus without a protein coat (Flores et al. 2012, 2014; Diener 2016). Here, we are only at the RNA strand level, which clearly shows infective and host-manipulating properties. This fits into the RNA world of RNA stem-loop groups, RNA group identities, and selection relevant biotic behavior (Robertson and Joyce 2012; Lehman 2015).

Recently, it has been found that single stem loops interact in a purely physical-chemical mode, without selective forces (Hayden and Lehman 2006; Smit et al. 2006). But if these single RNA stem loops build groups, they transcend the purely physical-chemical interaction pattern and emerge biological selection, biological identities of self/non-self identification and preclusion, and immune functions, dynamically varying (adapting) their membership roles (Marraffini and Sontheimer 2010; Vaidya et al. 2013; Krupovic and Koonin 2016).

A single alteration in a base-pairing RNA stem that leads to a new bulge may dynamically alter not just this single stem-loop but also the whole group identity of which this stem loop is part (Villarreal and Witzany 2013a, 2013b). Simple self-ligating RNA stem loops can build much larger groups of RNA stem loops that serve to increase complexity (Briones et al. 2009; Gwiazda et al. 2012). This may lead to ribozymatic consortia, which later on build success stories, such as (1) the merger of the two subunits of transfer RNAs, (2) RNA-dependent RNA polymerases for replication of RNA through RNA, and (3) the subunits of ribosomal RNAs originally evolved for different reasons than the later exapted and conserved modes (Dick and Schamel 1995; Sun

and Caetano-Anolles 2008; Fujishima and Kanai 2014; Kanai 2015).

The RNA stem-loop groups not only generate and constitute nucleotide sequences that serve as information storage media but also primarily interact (Nicholson and White 2014). The generation of RNA stem-loop consortia results in real entities, not just genetic syntax, and they are active in contributing to the identity of such groups and rejecting agents that do not fit in this identity. In several motifs, the interactions depend on the context in which an interaction takes place (Doudna et al. 1989; Przybilski and Hammann 2007; Popovic et al. 2015). This may be evolutionary conserved in DNA, as demonstrated in the co-operative behavior of the two ribosomal subunits. The activity is crucial for the result of co-operation (Doudna and Rath 2002).

Significantly, mixtures of RNA fragments that self-ligate into self-replicating ribozymes spontaneously form co-operative networks (Robertson and Joyce 2014). For example, three-member networks show highly co-operative growth dynamics. When such co-operative networks compete directly against selfish autocatalytic cycles, the former grow faster, indicating the ability of RNA populations to evolve greater complexity through co-operation (Vaidya 2012; Vaidya et al. 2012). In this respect, co-operation clearly outcompetes selfishness. Therefore, the primacy of selfish gene hypothesis is outdated on the RNA level.

RNA groups are able to act as de novo producers of nucleic acid sequences, identify sequence-specific target sites, coherently integrate such sequences into pre-existing ones (without destruction of former content arrangements), recombine according to adaptational needs, and mark sequence sites to vary meaning epigentically or identify sequences to be marked for excision or deletion (Bushman 2003; Martinez et al. 2017). In all these processes, the genetic identity of the genetic parasite and/or the host genome may vary, with far-reaching consequences in terms of the function, co-operation, and coordination of various regulatory networks. Natural genome editing is therefore far from being a random-like process as a result of error replication (mutations).

Non-coding RNAs interact with DNA, RNA, and proteins and play important roles in nuclear organization, transcription, posttranscriptional, and epigenetic processes (Zuckerkandl and Cavalli 2007; Mattick 2009; Mercer and Mattick 2013). Non-coding RNAs are transcribed in both the sense and anti-sense directions and may be expressed in a cell type, subcellular compartment, developmental stage, or an environmental stimuli-specific, that is, context-dependent, manner (Zinad et al. 2017). Specific RNA polymerases overlap in transcriptional contents, which means that each nucleotide can participate in varying transcriptional content arrangements according to varying contexts (Mattick 2010, 2011).

Non-coding RNAs can be regulated in a varying manner, coordinated or independently, autonomously or functionally interrelated, and can regulate individual genes as well as large genetic networks; they can precisely control the spatiotemporal deployment of genes that are executing neuronal processes with extreme cell specificity (Bartel 2004; Matera et al. 2007). Various classes of non-coding RNAs target each other for post-transcriptional regulation via alternative splicing, polyadenylation, 5′ capping, non-templated modifications, and RNA editing. RNA editing especially can transmit environmental information to the epigenome and therefore enable neuronal plasticity with learning and memory (Kandel 2001; Qureshi and Mehler, 2012). Additionally, non-coding RNAs (ncRNAs) can undergo nuclear-cytoplasmic, nuclear-mitochondrial, and axodendritic trafficking via ribonucleoprotein complexes that promote the spatiotemporal distribution and function of various combinations of ncRNAs, messenger RNAs (mRNAs), and RNA-binding proteins (Clark et al. 2013; Mercer and Mattick 2013).

If we think of biocommunication as sign-mediated interaction, the question arises what are the signs used in RNA stem-loop interactions? RNAs themselves represent signs as their four bases of the RNA alphabet. In these interacting agents the binding sites are the bases themselves that serve as identity indices. As identity indices they are relevant for interactional motifs such as integration or preclusion (self and non-self differentiation).

2.4.10 COMMUNICATION AT THE ORIGIN OF LIFE

How did prebiotic chemistry start life? From the biocommunication approach, there is a variety of RNA agents that share the competence to act on nucleic acid sequences (genetic code) by several techniques, such as de novo generation, ligation/degradation, insertion/deletion, silencing, amplification, epigenetic markings, editing, splicing, and kissing. This is in tune with the general assumption that no natural code codes itself but needs a consortium of competent agents that act on this code (Witzany 2011b, 2012, 2014b; Witzany and Baluška 2012b). Now, we can ask how these agent consortia evolve and how they interact, co-operate, or compete, or even both, at different times, depending on the varying real-life contexts.

2.4.10.1 FROM "DEAD" TO "LIVING" BY SOCIAL INTERACTING RNAS

It has been suggested that a "dead" state controlled by prebiotic chemistry and a "living" state controlled by autocatalytic replication should be differentiated (Higgs and Lehman 2014).

Recent empirical data and experiments demonstrated that single RNA stem loops react only in a physical-chemical way according to natural laws, without any biotic feature assembling a "dead" state clearly. If we have a group of RNA-set loops that interact and initiate group building, biological selection emerges, which resembles a "living" state (Vaidya 2012).

As demonstrated experimentally co-operative RNA stem loops are more successful in biological selection processes than selfish agents.

From the biocommunication approach, the autocatalytic set narrative of Stuart Kauffman looks rather coherent, because it

represents a kind of social interaction. Unfortunately, the autocatalytic set concept reduces behavioral motifs to an explanatory model of systems theory, that is, the interacting RNA stem loops as a "system" that functions according to mathematical (formalizable) features. As we have seen in the introduction, behavioral motifs of interacting living agents are not within the expertise of mathematical theories or their derivatives, for several serious reasons (Witzany 1995, 2011a).

Also, the in-between definition of abiotic and biotic features such as "molecular co-operation" does not function as an appropriate explanatory tool, because if we look at biotic co-operation, we deal with living agents that co-operate. RNA "entities" consist of molecules, but molecules in a prebiotic world cannot co-operate like agents representing living organisms that co-operate based on communicative interactions, which is absent in abiotic matter. Molecules in a prebiotic world only react strictly according to physical/chemical laws. No biological selection is present here.

The reaction between single-stranded RNAs and their fold-back capability to form double strands (stems), including single-stranded regions (loops), that are free to interact in cis (kissing loop to increase complex functional motifs) and trans (integration or warding of infectious non-self agents) are based on a physical/chemical feature (complementary base pairing). Although their biotic interactions depend on complementary base pairing, they overrule physical/chemical laws by their biotic behavior such as biological selection, group building, group identity, and self/non-self differentiation, all of which are absent in abiotic planets.

In RNA groups that interact on identity motifs such as integration or warding off and self and/or non-self RNAs, the RNAs represent living agents underlying biological selection. This phenomenon is clearly absent in abiotic planets. The primary goal is the interaction becoming part of co-operative networks that may also be competing and restricting (Popovic et al. 2015), as documented in the predominant ligase function, in contrast to the nuclease function (Hayden and Lehman 2006; Díaz Arenas and Lehman 2010). This means the turn to life is also a behavioral pattern between becoming part of an identity of RNA groups that reject those that do not fit into this identity. The unique aspect of such RNA stem loops is that they may act both as templates for replication and as catalysts, if they are spatially separated.

2.4.10.2 THE ORIGIN OF RNA GROUP IDENTITY AND THE ORIGIN AS A GENETIC CODE

Humans consist of molecules. Yes, we need an abundance of oxygen atoms (that we call air) to transport vocal sound and talk to each other to organize everyday life. But this does not help us understand human language and communication. Detailed investigations and measurements of the molecular movements of oxygen atoms when humans talk to each other do not tell us anything about intentions being expressed by several different behavioral motifs.

If the interactions of individual RNA stem loops lead to groups of RNA consortia, we can look at the crucial event of them becoming more than physics and chemistry, which determine abiotic planets exclusively. The group interactions of RNA stem loops make the genotype evolve into semiotic biology, which means the emergence of natural code rules (syntax, pragmatics, and semantics). The characters get their natural language-like feature *only through the social interactions* of RNAs. Their emergence as semiotic signs and as characters for a language/code-like property, which represents information about RNA-group identities, crucially depends on the social interactions. Although statistically measurable, this semiotic feature cannot be substantiated by quantifiable analyses, because it is a socially interacting feature (Witzany 2016).

All such interactions depend on the ability to identify appropriate interaction partners and therefore differentiate between self and non-self, which means differentiating identity-sharing agents against proponents, which represent non-identity and must be fought, precluded, and warded off (Marraffini and Sontheimer 2010). Others may represent co-operation partners; some are actually interaction partners. Others remain as silenced enemies, which could destroy identity. Importantly, later on, in another context, they may be useful tools for co-opting or ex-apting processes. At different times, formerly fought non-identity proponents may be useful tools for generating new partnership (Villarreal 2011, 2012a, 2012b, 2015, 2016).

Several DNA- and RNA-degrading processes are not investigated under the assumption that they serve as useful resources for re-use and identity-building modules (Vaidya et al. 2013; Villarreal 2015, 2016). Additionally, important and key features of this agent-based co-operativity are the fast-changing identities, which lead to the question of how a former enemy becomes a new co-operation partner. What does this mean for former co-operative partners, which may turn into enemies, in fast-changing functional partnerships and the concurrent changes in organization and syntax order? These interaction processes start from the beginning of RNA life (Hayden and Lehman 2006; Mattick 2011; Yarus 2011; Higgs and Wu 2012; Higgs and Lehman 2015), that is, life without unicellular organism. But group identity and co-operativity

of an RNA collective also require opposite functions for the genesis of life (social behavior of agents). This is an essential part of the "gangen hypothesis," which describes RNA groups emerging and acting like gangs (Villarreal 2015) in the context of the virosphere and the world of cell-based organisms.

The genetic code with its typical language-like features, that is, characters assembled according to syntactic, pragmatic, and semantic (combinatorial, context-dependent, and content-coherent) rules, takes the stage of a real, natural language, with the interactional group building of various RNA stem loops. The interactional and group-building co-operativity of the RNA stem loops constitutes the genetic code as a real natural code, not its physico-chemical key characters alone.

The crucial difference of the biocommunication perspective to that of molecular biology is that it is not the molecular aspect solely, but the agent-groups' aspect. The RNA stem loops are part of a group identity, which co-operates or precludes and, by compartmentalization, gets the same first stages of conservation. Until DNA was invented—reverse transcriptase is the most appropriate candidate for this—conservation would have been a rare event, instead of being a constantly evolutionary innovation process of maximum productivity of genetic novelties (Witzany 2008; Moelling et al. 2017). Too much productivity in former concepts has been termed "error catastrophe." Pure physico-chemical variations based on replication are correctly termed error. In a social RNA perspective as the biocommunication approach, this is not an error but productivity of novelties. The paradigmatic differences between the concepts of molecular biology and the biocommunication approach are shown in Table 2.4.1.

2.4.10.3 REPETITIVE SEQUENCE SYNTAX IS ESSENTIAL IN RNA COMMUNICATION

The repetitive syntax comprises the main characteristics of nucleotide sequences that represent interactions of RNA stem loops (Shapiro and Sternberg 2005). Stem-loop group building is and was based on repeat syntax and the complementary binding rules of the four bases of the RNA alphabet. Also, the

---

TABLE 2.4.1
Different Paradigms Investigating and Defining Life

Key Terms	Molecular Biology	Biocommunication
"Dead"	Pre-biotic chemical reactions	No sign-mediated interactions
"Living"	Replication/biological selection (molecular reactions)	Sign-mediated interactions (social events)
Determinants	Natural laws	Semiotic rules
RNA-ensembles	Molecular co-operation	Identity-groups integrate or preclude non-self agents
Viruses	Escaped selfish parasites	Essential agents of all life
Biological selection	Fittest type	Fittest consortium
Genetic code	Genetic material	Genetic text (according syntax, pragmatics, semantics)
Communication	Information transfer (via coding/decoding mechanisms)	Social interactions (of agent-groups mediated by signs according semiotic rules)

The molecular biological paradigm explains life by the physical-chemical and information theoretical (mathematical theory of language) properties of living organisms and life processes; the biocommunication approach explains life as a social event of cellular and sub-cellular agents that communicate, that is, based on interactions that are sign-mediated according to three levels of semiotic rules.

modifications on RNA sequences—after being transcribed out of DNA storage medium, such as loop kissing, RNA editing, splicing, and others—depend on repetitive syntax (Witzany 2017b). The non-repetitive syntax—main characteristics of protein-coding sequences—represents another code, an evolutionarily later-derived code for DNA storage of protein blueprints, not relevant for social RNA interactions such as group building, recognition (identifying), attack, and defense against genetic parasites (which rely also on repetitive sequence order). The main reason is that repetitive syntax represents a recognition pattern completely different from non-repetitive ones (Jurka et al. 2007). If an invading RNA species meets a host genome, it will sense and use repetitive sequences, which are much easier for insertion than non-repetitive ones. This is why genetic parasites insert more into non-coding repetitive syntax structures than in protein-coding non-repetitive syntax structures. This means that insertions with relevant evolutionary drive are inserted into regulation-relevant genome space, not in protein coding regions.

2.4.10.4 DNA Sequence Syntax with Variable Meanings

Consequently, the biocommunication approach has to integrate coherent natural genome editing of the genetic code of living organisms. In this perspective, the genetic code of living organisms cannot be the result of chance mutations (error-replication events) that are biologically selected. The error-replication narrative has serious problems to explain the sudden emergence of new species, new phenotypic traits, and genome innovations as a sudden single event.

The genetic syntax, that is, the nucleotide arrangement of genetic information, is not unequivocal. This means that a given genetic sequence does not offer its final meaning. There are several processual steps in which the same sequence syntax may get several different meanings (functions). In natural languages and codes, this is a usual procedure and—additionally—it saves energy costs: A given sign sequence may represent various meanings/functions according its contextual use. It is not necessary to generate a new sequence for every function. This indicates the primacy of pragmatics in determining the sequence structure, not its syntax (Doudna et al. 1989; Cech 2012; Witzany 2014b). Several context-dependent natural modifications of the meaning (function) of a given DNA sequence are currently known and investigated. Let's have a look at the various techniques to modify meaning (function) of given genetic sequences.

The currently known natural processes of generating variable natural meaning functions out of a given DNA sequence are the context-dependent markings in epigenetics, which also play a key role to understand and coherently explain memory and learning as competencies of living organisms in all domains of life (Witzany 2018). Additionally, we may think of complex regulatory control via RNA editing, alternative splicing, the various roles of tRNA-derived fragments, ribosomal frameshifting, the interaction motif of kissing loops, bypassing, translation, the role of competing endogenous RNAs, and diversity-generating retroelements, not to forget pseudoknotting (Maizels and Weiner 1994; Keam and Hutvagner 2015; Arkhipova 2017; Yablonovitch et al. 2017).

Most interestingly, the base pairing in pseudoknots tends to be strictly context-sensitive, and base pairing overlaps in sequence positions. Additionally, the emergence of single-nucleotide bulge loops can hardly be predicted. This currently places limitations on algorithm-based prediction models, such as dynamic programming, and on stochastic context-free grammars (Lyngsø and Pedersen 2004). It indicates the language/code nature of nucleic acid language, which represents the possibility of coherent de novo generation and context-dependent alterations for a diversity of different meanings (functions) relating to the same syntax structures.

All these natural techniques fulfilled by RNA consortia must be mentioned if we think about artificially manipulating genes of living organisms. Without these historically evolved interwoven interaction networks, the natural competence of gene conservation cannot be understood coherently. Genes and their regulatory networks are not at all molecular bricks that can be combined and divided in a lego-like manner. If synthetic biology tries to generate living organisms artificially, it should be sure to guarantee the whole network coherence. Otherwise, low-level manipulation will be the result and may cause far-reaching ethical problems, which are not within the expertise of natural sciences in general.

2.4.11 CONCLUSIONS

If we look at life in contrast to abiotic matter, we will identify cellular organisms that coordinate and organize their life actively by communicative actions, which means that they use signs in various forms to interact. The use of signs is essential. The sign use is governed by three levels of rules: how to combine signs to sign sequences for more complex information (syntax); the context dependence, which determines meaning of the used sign sequence (pragmatics); and content coherence, in which signs are used to designate something (semantics). All communication processes in the cellular world share these features.

If we go a step deeper, communication forms change dramatically. In the RNA world, the evolutionary produced sign sequences, which we would term a kind of "writing down" what happened (as evolutionary protocol), are a side result of constant interacting RNAs, stabilizing their functional forms by proteins. The communication in the RNA world is a different one to that of the cellular world, another essential level. Communicating RNAs share common motifs to reach their goals, described earlier. The signs in this form of communication are the interacting agents themselves, as they represent a natural code alphabet of four nucleotides, which are combined according to semiotic rules coherent with the physical chemical structures. The interacting RNAs resemble both a syntax like alphabetic order, dominated by repeat sequences, and context-dependent interaction that represent (and are the result of) biological selection, which is absent in abiotic matter. The RNAs serve as editors of RNA sequence only by their group interactions, which lead to group identities. Competing RNA group identities initiate biological selection. Several smaller RNA groups may be combined to form bigger elements such as tRNA and even ribosome and other ribonucleoproteins (RNPs) with essential functions in all life, as they serve as crucial tools for gene regulation in transcription, translation, repair, immunity, inheritance, etc. in any known cellular organism.

The very vivid and highly active meeting place of these two levels of communication, (1) cellular sign-mediated interactions and (2) RNA stem-loop group interactions, are (3) competent viruses or, better, the virospheres, which use both communication systems, perfectly adapted to the whole life on this planet. They are the driving force of life based on (a) RNA world communication and (b) cellular communication, which can be edited by viruses in an abundance of ways. As masters of the technique of natural genome editing in all the cellular world, they do not need a cell-independent replication. Interestingly, they do not damage cellular life during persistent integration but serve as the key for all forms of immunity for their host organisms constantly adapting to the high productivity of viruses. They serve as key, based on the RNA world communication, which they use actively for their tremendous repertoire of techniques to invade, manipulate, and regulate cellular life forms. Together, these three agent-based lifeworlds, RNA world, virosphere, and cellular organisms, constitute life as we know it. We may come closer to understand the secret of life now, if we do not take life mechanistically again.

REFERENCES

Arkhipova, I.R. 2017. Using bioinformatic and phylogenetic approaches to classify transposable elements and understand their complex evolutionary histories. *Mob. DNA.* 8: 19.

Austin, J.L. 1975. *How to Do Things with Words*. London, UK: Harvard University Press.

Bais, H.P., Park, S.W., Weir, T.L. et al. 2004. How plants communicate using the underground information superhighway. *Trends Plant Sci.* 9: 26–32.

Baluška, F., Mancuso, S., and D. Volkmann (Eds.). 2006. *Communication in Plants*. Heidelberg, Germany: Springer.

Bartel, D.P. 2004. MicroRNAs: Genomics, biogenesis, mechanism, and function. *Cell* 116: 281–297.

Bassler, B.L. 1999. How bacteria talk to each other: Regulation of gene expression by quorum sensing. *Curr. Opin. Microbiol.* 2: 582–587.

Belfort, M. 2017. Mobile self-splicing introns and inteins as environmental sensors. *Curr. Opin. Microbiol.* 38: 51–58.

Ben Jacob, E., Becker, I., Shapira, Y., and H. Levine. 2004. Bacterial linguistic communication and social intelligence. *Trends Microbiol.* 12: 366–372.

Berliner, A., Mochizuki, T., and K.M. Stedman. 2018. Astrovirology: Viruses at large in the universe. *Astrobiology* 18: 207–223.

Blande, J.D., and R. Glinwood (Eds.). 2016. *Deciphering Chemical Language of Plant Communication*. Cham, Switzerland: Springer.

Bohn, M., Call, J., and M. Tomasello. 2015. Communication about absent entities in great apes and human infants. *Cognition* 145: 63–72.

Bohn, M., Call, J., and M. Tomasello. 2016. The role of past interactions in great apes' communication about absent entities. *J. Comp. Psychol.* 130: 351–357.

Brenner, S. 2012. Turing centenary: Life's code script. *Nature* 482: 461.

Briones, C., Stich, M., and S.C. Manrubia. 2009. The dawn of the RNA world: Toward functional complexity through ligation of random RNA oligomers. *RNA* 15: 743–749.

Bushman, F.D. 2003. Targeting survival: Integration site selection by retroviruses and LTR-retrotransposons. *Cell* 115: 135–138.

Carnap, R. 1939. *Foundations of Logic and Mathematics*. Chicago, IL: University of Chicago Press.

Casselman, A. 2007. Strange but true: The largest organism on earth is a fungus. *Scientific American*, October 4.

Cech, T.R. 2012. The RNA worlds in context. *Cold Spring Harb. Perspect. Biol.* 4(7): a006742.

Chalopin, D., Tomaszkiewicz, M., Galians, D., and J.N. Volff. 2012. LTR retroelement-derived protein-coding genes and vertebrate evolution. In *Viruses: Essential Agents of Life*, G. Witzany (Ed.), pp. 269–282. Dordrecht, the Netherlands: Springer.

Chomsky, N. 1964. *Current Issues in Linguistic Theory*. London, UK: The Hague, Mouton.

Chomsky, N. 1965. *Aspects of the Theory of Syntax*. Cambridge, MA: MIT Press.

Clark, M.B., Choudhary, A., Smith, M.A. et al. 2013. The dark matter rises: The expanding world of regulatory RNAs. *Essays Biochem.* 54: 1–16.

Conley, A.B., and I.K. Jordan. 2012. Endogenous retroviruses and the epigenome. In *Viruses: Essential Agents of Life*, G. Witzany (Ed.), pp. 309–323. Dordrecht, the Netherlands: Springer.

Crespi, B.J. 2001. The evolution of social behavior in microorganisms. *Trends Ecol. Evol.* 16: 178–183.

Crick, F.H., Griffith, J.S., and L.E. Orgel. 1957. Codes without commas. *Proc. Natl. Acad. Sci. U S A* 43: 416–421.

Díaz Arenas, C., and N. Lehman. 2010. Quasispecies-like behavior observed in catalytic RNA populations evolving in a test tube. *BMC Evol. Biol.* 10: 80.

Díaz-Muñoz, S.L., Sanjuán, R., and S. West. 2017. Sociovirology: Conflict, cooperation, and communication among viruses. *Cell Host Microbe.* 22: 437–441.

Dick, T.P., and W.A. Schamel. 1995. Molecular evolution of transfer RNA from two precursor hairpins: implications for the origin of protein synthesis. *J. Mol. Evol.* 41: 1–9.

Diener, T.O. 2016. Viroids: "living fossils" of primordial RNAs? *Bio Direct.* 11: 15.

Dohlman, H.G., and J.E. Slessareva. 2006. Pheromone signaling pathways in yeast. *Sci. Signal.* 364: cm6.

Doudna, J.A., Cormack, B.P., and J.W. Szostak. 1989. RNA structure, not sequence, determines the 5′splice-site specificity of a group I intron. *Proc. Natl. Acad. Sci. U S A* 86: 7402–7406.

Doudna, J.A., and V.L. Rath. 2002. Structure and function of the eukaryotic ribosome: The next frontier. *Cell* 109: 153–156.

Dunn, A.K., and J. Handelsman. 2002. Toward an understanding of microbial communities through analysis of communication networks. *Antonie van Leeuwenhoek* 81: 565–574.

Eigen, M. 2013. *From Strange Simplicity to Complex Familiarity a Treatise on Matter, Information, Life, and Thought*. Oxford, UK: Oxford University Press.

Engelmann, J.M., Clift, J.B., Herrmann, E., and M. Tomasello. 2017. Social disappointment explains chimpanzees' behaviour in the inequity aversion task. *Proc. Biol. Sci.* 284(1861): 20171502.

Erez, Z., Steinberger-Levy, I., Shamir, M. et al. 2017. Communication between viruses guides lysis-lysogeny decisions. *Nature* 541: 488–493.

Fernandes, L., Araujo, M.A.M., Amaral, A. et al. 2005. Cell signaling pathways in Paracoccidioides brasiliensis – inferred from comparisons with other fungi. *Genet. Mol. Res.* 4: 216–231.

Flores, R., Gago-Zachert, S., Serra, P. et al. 2014. Viroids: Survivors from the RNA world? *Annu. Rev. Microbiol.* 68: 395–414.

Flores, R., Serra, P., Minoia, S. et al. 2012. Viroids: From genotype to phenotype just relying on RNA sequence and structural motifs. *Front. Microbiol.* 3: 217.

Forterre, P. 1997. Archaea: What can we learn from their sequences? *Curr. Opin. Genet. Dev.* 7: 764–770.

Fujishima, K., and A. Kanai. 2014. tRNA gene diversity in the three domains of life. *Front. Genet.* 5: 142.

Fuqua, C., Winans, S.C., and E.P. Greenberg. 1996. Census and consensus in bacterial ecosystems: The LuxR-LuxI family of quorum sensing transcriptional regulators. *Ann. Rev. Microbiol.* 50: 727–751.

Garrett, R., and H.P. Klenk. 2007. *Archaea: Evolution, Physiology, and Molecular Biology*. Maiden, MA: Blackwell Publishing.

Garrett, R.A., Vestergaard, G., and S.A. Shah. 2011. Archaeal CRISPR-based immune systems: Exchangeable functional modules. *Trends Microbiol.* 19: 549–556.

Geuking, M.B., Weber, J., Dewannieux, M. et al. 2009. Recombination of retrotransposon and exogenous RNA virus results in nonretroviral cDNA integration. *Science* 323: 393–396.

Goedel, K. 1931. Über formal unentscheidbare Sätze der Principia Mathematicaund verwandter Systeme. *Monatshefte fur Mathematik und Physik* 38: 173–198.

Gwiazda, S., Salomon, K., Appel, B., and S. Muller. 2012. RNA self-ligation: From oligonucleotides to full length ribozymes. *Biochimie* 94: 1457–1463.

Habermas, J. 1984. *The Theory of Communicative Action. Reason and the Rationalization of Society*, Vol. 1. Boston, MA: Beacon Press.

Habermas, J. 1987. *The Theory of Communicative Action. Lifeworld and System: A Critique of Functionalist Reason*, Vol. 2. Boston, MA: Beacon Press.

Habermas, J. 1994. Actions, speech acts, linguistically mediated interactions and the lifeworld. In *Philosophical Problems Today*, G. Floistad (Ed.), Vol. 1, pp. 45–74. Amsterdam, the Netherlands: Kluwer.

Harms, A., Brodersen, D.E., Mitarai, N., and K. Gerdes. 2018. Toxins, targets, and triggers: An overview of toxin-antitoxin biology. *Mol. Cell.* 70: 768–784.

Hayden, E.J., and N. Lehman. 2006. Self-assembly of a group I intron from inactiveoligonucleotide fragments. *Chem. Biol.* 13: 909–918.

Heaton, N.S., and B.R. Cullen. 2017. Viruses hijack a long non-coding RNA. *Nature.* 552: 184–185.

Higgs, P.G., and N. Lehman. 2015. The RNA World: Molecular cooperation at the origins of life. *Nat. Rev. Genet.* 16: 7–17.

Higgs, P.G., and M. Wu. 2012. The importance of stochastic transitions for the origin of life. *Orig. Life Evol. Biosphere* 42: 453–457.

Inoue, Y., Saga, T., Aikawa, T. et al. 2017. Complete fusion of a transposon and herpesvirus created the Teratorn mobile element in medaka fish. *Nat. Commun.* 8(1): 551.

Jurka, J., Kapitanov, V.V., Kohany, O., and M.V. Jurka. 2007. Repetitive sequences in complex genomes: Structure and evolution. *Annu. Rev. Genomics Hum. Genet.* 8: 241–259.

Kanai, A. 2015. Disrupted tRNA Genes and tRNA Fragments: A Perspective on tRNA Gene Evolution. *Life (Basel)* 5: 321–331.

Kandel, E.R. 2001. The molecular biology of memory storage. A dialogue between genes and synapses. *Science* 294: 1030–1038.

Kaiser, D., and R. Losick. 1993. How and why bacteria talk to each other. *Cell* 73: 873–885.

Keam, S.P., and G. Hutvagner. 2015. tRNA-derived fragments (tRFs): Emerging new roles for an ancient RNA in the regulation of gene expression. *Life (Basel).* 5: 1638–1651.

Kolenbrander, P.E., Andersen, R.N., Blehert, D.S. et al. 2002. Communication among oral bacteria. *Microbiol. Mol. Biol. Rev.* 66: 486–505.

Koonin, E.V. 2016. Viruses and mobile elements as drivers of evolutionary transitions. *Philos. Trans. R. Soc. Lond. B Biol. Sci.* 371(1701): 20150442.

Krupovic, M., and E.V. Koonin. 2016. Self-synthesizing transposons: Unexpected key players in the evolution of viruses and defense systems. *Curr. Opin. Microbiol.* 31: 25–33.

Lambowitz, A.M., and S. Zimmerly. 2011. Group II introns: Mobile ribozymes that invade DNA. *Cold Spring Harb. Perspect. Biol.* 3(8): a003616.

Leeder, A.C., Palma-Guerrero, J., and N.L. Glass. 2011. The social network: Deciphering fungal language. *Nat. Rev. Microbiol.* 9: 440–451.

Lehman, N. 2015. The RNA world: 4,000,000,050 years old. *Life (Basel)* 5: 1583–1586.

Losick, R., and D. Kaiser. 1997. Why and how bacteria communicate. *Sci. Am.* 2: 52–58.

Luporini, P., Vallesi, A., Alimenti, C., and C. Ortenzi. 2006. The cell type-specific signal proteins (pheromones) of protozoan ciliates. *Curr. Pharm. Des.* 12: 3015–3024.

Luporini, P., Vallesi, A., Miceli, C., and R.A. Bradshaw. 1995. Chemical signaling in ciliates. *J. Eukaryot. Microbiol.* 42: 208–212.

Lyngsø, R.B., and C.N. Pedersen. 2004. RNA pseudoknot prediction in energy-based models. *J. Comput. Biol.* 7: 409–427.

Maizels, N., and A.M. Weiner. 1994. Phylogeny from function: Evidence from the molecular fossil record that tRNA originated in replication, not translation. *Proc. Natl. Acad. Sci. U S A* 91: 6729–6734.

Margulis, L. 2004. Serial endosymbiotic theory (SET) and composite individuality. Transition from bacterial to eukaryotic genomes. *Microbiol. Today* 31: 173–174.

Marraffini, L.A., and E.J. Sontheimer. 2010. Self versus non-self discrimination during CRIPR RNA-directed immunity. *Nature* 463: 568–571.

Martinez, G., Choudury, S.G., and R.K. Slotkin. 2017. tRNA-derived small RNAs target transposable element transcripts. *Nucleic Acids Res.* 45: 5142–5152.

Matera, A.G., Terns, M., and M.P. Terns. 2009. Non-coding RNAs: Lessons from the small nuclear and small nucleolar RNAs. *Nat. Rev. Mol. Cell Biol.* 8: 209–220.

Mattick, J.S. 2010. RNA as a substrate for epigenome-environment interactions: RNA guidance of epigenetic processes and the expansion of RNA editing in animals underpins development, phenotypic plasticity, learning and cognition. *Bioessays* 32: 548–552.

Mattick, J.S. 2011. The double life of RNA. *Biochimie* 93(11): viii–ix.

McCarthy, T. 1984. Translator's introduction. In *The Theory of Communicative Action*, p. ix. Boston, MA: Beacon Press.

McNeil, B.A., Semper, C., and S. Zimmerly. 2016. Group II introns: Versatile ribozymes and retroelements. *WIREs RNA* 7: 341–355.

Mead, G.H. 1934. *Mind, Self, and Society*. Chicago, IL: The University of Chicago Press.

Mercer, T.R., and J.S. Mattick. 2013. Structure and function of long non-coding RNAs in epigenetic regulation. *Nat. Struct. Mol. Biol.* 20: 300–307.

Moelling, K., Broecker, F., Russo, G., and S. Sunagawa. 2017. RNase H as gene modifier, driver of evolution and antiviral defense. *Front. Microbiol.* 8: 1745.

Morris, C. 1946. *Signs, Language, and Behavior.* New York: Prentice Hall.

Nicholson, B.L., and K.A. White. 2014. Functional long-range RNA–RNA interactions inpositive-strand RNA viruses. *Nat. Rev. Microbiol.* 12: 493–504.

Nowak, M.A., and D.C. Krakauer. 1999. The evolution of language. *Proc. Acad. Sci. U S A* 96: 8023–8033.

Peirce, C.S. 1923. How to make our ideas clear. In *Chance, Love, and Logic: Philosophical Essays*, R.C. Morris (Ed.), pp. 41–42. New York: Harcourt, Brace and World.

Perot, P., Bolze, P.A., and F. Mallet. 2012. From viruses to genes: Syncytins. In *Viruses: Essential Agents of Life*, G. Witzany (Ed.), pp. 325–361. Dordrecht, the Netherlands: Springer.

Popovic, M., Fliss, P.S., and M.A. Ditzler. 2015. In vitro evolution of distinct self-cleavingribozymes in diverse environments. *Nucleic Acids Res.* 43: 7070–7082.

Przybilski, D.M., and C. Hammann. 2007. The tolerance to exchanges of theWatson–Crick base pair in the hammerhead ribozyme core is determined by surrounding elements. *RNA* 13: 1625–1630.

Qureshi, I.A., and M.F. Mehler. 2012. Emerging roles of non-coding RNAs in brainevolution, development, plasticity and disease. *Nat. Rev. Neurosci.* 13: 528–541.

Robertson, M.P., and G.F. Joyce. 2012. The origins of the RNA world. *Cold Spring Harb. Perspect. Biol.* 4(5): a003608.

Robertson, M.P., and G.F. Joyce. 2014. Highly efficient self-replicating RNA enzymes. *Chem. Biol.* 21: 238–245.

Rohwer, F., Youle, M., Maughan, H., and N. Hisakawa. 2014. *Life in Our Phage World.* San Diego, CA: Wholon.

Roossinck, M.J. 2015. Metagenomics of plant and fungal viruses reveals an abundance of persistent lifestyles. *Front. Microbiol.* 5: 1–3.

Ryan, F. 2009. *Virolution.* London, UK: Harper Collins.

Schauder, S., and B.L. Bassler. 2001. The languages of bacteria. *Genes Dev.* 15: 1468–1480.

Schroedinger, E. 1944. *What is Life? The Physical Aspect of the Living Cell.* London, UK: Cambridge University Press.

Searle, J.R. 1976. *Speech Acts. An Essay in the Philosophy of Language.* Cambridge, UK: Cambridge University Press

Seligmann, H., and D. Raoult. 2016. Unifying view of stem-loop hairpin RNA as origin of current and ancient parasitic and non-parasitic RNAs, including in giant viruses. *Curr. Opin. Microbiol.* 31: 1–8.

Shannon, C.E., and W. Weaver. 1949. *The Mathematical Theory of Communication.* Urbana, IL: University of Illinois Press.

Shapiro, J.A. 2011. *Evolution: A View from the 21st Century.* New York: Financial Times Prentice Hall.

Shapiro, J.A., and R.V. Sternberg. 2005. Why repetitive DNA is essential to genome function. *Biol. Rev.* 80: 1–24.

Sharma, A., Sahgal, M., and B.N. Johri. 2003. Microbial communication in the rhizosphere: Operation of quorum sensing. *Curr. Sci.* 85: 1164–1172.

Slotkin, R.K., and R. Martienssen. 2007. Transposable elements and the epigenetic regulation of the genome. *Nat. Rev. Genet.* 8: 272–285.

Smit, S., Yarus, M., and R. Knight. 2006. Natural selection is not required to explain universal compositional patterns in rRNA secondary structure categories. *RNA* 12: 1–14.

Stedman, K. 2013. Mechanisms for RNA capture by ssDNA viruses: Grand theft RNA. *J. Mol. Evol.* 76: 359–364.

Stedman, K.M. 2015. Deep recombination: RNA and ssDNA virus genes in DNA virus and host genomes. *Annu. Rev. Virol.* 2: 203–217.

Sun, F.J., and G. Caetano-Anollés. 2008. Transfer RNA and the origins of diversified life. *PLoS One* 3(7): e2799.

Tomasello, M. 2008. *Origins of Human Communication.* Cambridge, MA: MIT Press.

Turing, A. 1950. Computing machinery and intelligence. *Mind* 59: 433–460.

Tycowski, K.T., Guo, Y.E., Lee, N. et al. 2015. Viral noncoding RNAs: More surprises. *Genes Dev.* 29: 567–584.

Vaidya, N. 2012. Spontaneous cooperative assembly of replicative catalytic RNA systems. Dissertations and Theses. Paper 934. doi:10.15760/etd.934.

Vaidya, N., Manapat, M.L., Chen, I.A. et al. 2012. Spontaneous network formation among cooperative RNA replicators. *Nature* 491: 72–77.

Vaidya, N., Walker, S.I., and N. Lehman. 2013. Recycling of informational units leads to selection of replicators in a prebiotic soup. *Chem. Biol.* 20: 241–252.

Villarreal, L.P. 2005. *Viruses and the Evolution of Life.* Washington, DC: ASM Press.

Villarreal, L.P. 2009. *Origin of Group Identity: Viruses, Addiction and Cooperation.* New York: Springer.

Villarreal, L.P. 2011. Viral ancestors of antiviral systems. *Viruses* 3: 1933–1958.

Villarreal, L.P. 2012a. The addiction module as social force. In *Viruses: Essential Agents of Life*, G. Witzany (Ed.), pp. 107–145. Dordrecht, the Netherlands: Springer.

Villarreal, L.P. 2012b. *Viruses and Host Evolution: Virus-Mediated Self Identity.* Austin, TX: Landes Bioscience and Springer Science+Business Media.

Villarreal, L.P. 2015. Force for ancient and recent life: Viral and stem-loop RNA consortia promote life. *Ann. N. Y. Acad. Sci.* 1341: 25–34.

Villarreal, L.P. 2016. Viruses and the placenta: the essential virus first view. *APMIS* 124: 20–30.

Villarreal, L.P., and G. Witzany. 2010. Viruses are essential agents within the roots and stem of the tree of life. *J. Theor. Biol.* 262: 698–710.

Villarreal, L.P., and G. Witzany. 2013a. The DNA habitat and its RNA inhabitants: At the dawn of RNA sociology. *Genomics Insights* 6: 1–12.

Villarreal, L.P., and G. Witzany. 2013b. Rethinking quasispecies theory: From fittest type to cooperative consortia. *World J. Biol. Chem.* 4: 70–79.

Villarreal, L.P., and G. Witzany. 2015. When competing viruses unify: Evolution, conservation, and plasticity of genetic identities. *J. Mol. Evol.* 80: 305–318.

Villarreal L.P., and G. Witzany. 2018. Editorial: Genome invading RNA-networks. *Front. Microbiol.* 9: 581.

von Frisch, K.V. 1971. *Bees: Their Vision, Chemical Senses and Language.* Ithaca, NY: Cornell University Press.

von Neumann, J. 1966. *Theory of Self-Reproducing Automata.* Urbana, IL: University of Illinois Press.

Weiner, A.M. 2006. SINEs and LINEs: Troublemakers, saboteurs, benefactors, ancestors. In *The RNA World*, R.F. Gesteland, T.R. Cech, and J.F. Atkins (Eds.), 3rd ed., pp. 507–534. New York: Cold Spring Harbor Laboratory Press.

Whitehead, A.N., and B. Russell. 1910/1912/1913. *Principia Mathematica*. Cambridge, UK: Cambridge University Press.

Wiener, N. 1948. *Cybernetics, or Control and Communication in the Animal and the Machine*. New York: Wiley.

Witzany, G. 1995. From the "logic of the molecular syntax" to molecular pragmatism. Explanatory deficits in Manfred Eigen's concept of language and communication. *Evol. Cogn.* 1: 148–168.

Witzany, G. 2000. *Life: The Communicative Structure*. Norderstedt, Germany: LoB.

Witzany, G. 2006. Serial endosymbiotic theory (SET): The biosemiotic update. *Acta Biotheor.* 54: 103–117.

Witzany, G. 2008. The viral origins of telomeres and telomerases and their important role in eukaryogenesis and genome maintenance. *Biosemiotics* 1: 191–206.

Witzany, G. (Ed.). 2009. *Natural Genetic Engineering and Natural Genome Editing*. New York: Annals of the New York Academy of Sciences.

Witzany, G. 2010. *Biocommunication and Natural Genome Editing*. Dordrecht, the Netherlands: Springer.

Witzany, G. 2011a. Can mathematics explain the evolution of human language? *Commun. Integr. Biol.* 4: 1–5.

Witzany, G. 2011b. The agents of natural genome editing. *J. Mol. Cell Biol.* 3: 181–189.

Witzany, G. (Ed.). 2011c. *Biocommunication in Soil Microorganisms*. Heidelberg, Germany: Springer.

Witzany, G. (Ed.). 2012a. *Biocommunication of Fungi*. Dordrecht, Germany: Springer.

Witzany, G. (Ed.) 2012b. *Viruses: Essential Agents of Life*. Dordrecht, Germany: Springer.

Witzany, G. (Ed.). 2014a. *Biocommunication of Animals*. Dordrecht, Germany: Springer.

Witzany, G. 2014b. Language and communication as universal requirements for life. In *Astrobiology: An Evolutionary Approach*, V. Kolb (Ed.), pp. 349–369, Boca Raton, FL: CRC Press.

Witzany, G. 2015. Life is physics and chemistry and communication. *Ann. N.Y. Acad. Sci.* 1341: 1–9.

Witzany, G. 2017a. *Biocommunication of Archaea*. Dordrecht, the Netherlands: Springer.

Witzany, G. 2017b. Two genetic codes: Repetitive syntax for active non-coding RNAs; non-repetitive syntax for the DNA archives. *Commun. Integr. Biol.* 10(2): e1297352.

Witzany, G. 2018. *Evolution of Genetic Information without Error-Replication*. World Scientific (forthcoming).

Witzany, G., and F. Baluška (Eds.). 2012a. *Biocommunication of Plants*. Heidelberg, Germany: Springer.

Witzany, G., and F. Baluška. 2012b. Life's code script does not code itself. The machine metaphor for living organisms is outdated. *EMBO Rep.* 13: 1054–1056.

Witzany, G., and M. Nowacki. 2016. *Biocommunication of Ciliates*. Dordrecht, the Netherlands: Springer.

Wittgenstein, L. 1953. *Philosophical Investigations*. Oxford, UK: Basil Blackwell.

Woese, C.R., and G.E. Fox. 1977. Phylogenetic structure of the prokaryotic domain: The primary kingdoms. *Proc. Natl. Acad. Sci. U S A* 74: 5088–5090.

Woese, C.R., Kandler, O., and M.L. Wheelis. 1990. Towards a natural system of organisms: Proposal for the domains Archaea, Bacteria, and Eukarya. *Proc. Natl. Acad. Sci. U S A* 87: 4576–4579.

Yablonovitch, A.L., Deng, P., Jacobson, D., and J.B. Li. 2017. The evolution and adaptation of A-to-I RNA editing. *PLoS Genet.* 13(11): e1007064.

Yarus, M. 2011. *Life from an RNA World: The Ancestor Within*. Cambridge, MA: Harvard University Press.

Yu, R.C., Pesce, C.G., Colman-Lerner, A. et al. 2008. Negative feedback that improves information transmission in yeast signalling. *Nature* 456: 755–761.

Zinad, H.S., Natasya, I., and A. Werner. 2017. Natural antisense transcripts at the interface between host genome and mobile genetic elements. *Front. Microbiol.* 8: 2292.

Zuckerkandl, E., and G. Cavalli. 2007. Combinatorial epigenetics, 'junk DNA', and the evolution of complex organisms. *Gene* 390: 232–242.

Section III

Origin of Life

History, Philosophical Aspects, and Major Developments

3.1 Philosophical Aspects of the Origin-of-Life Problem
Neither by Chance Nor by Design

Iris Fry

CONTENTS

3.1.1 INTRODUCTION

The scientific study of the origin of life is based on the philosophical postulate that living systems emerged on the primordial Earth by natural means. This fundamental presupposition is implemented in the search for physical and chemical processes that might have brought about the emergence of life under prebiotic conditions. The naturalistic conception of the origin of life is by now part of the general scientific worldview that developed historically beginning in the 16th and 17th centuries. This worldview was stimulated first by the rise of the physical sciences and was joined during the nineteenth century by the development of modern biology. Specific cultural, social, and political changes constituted the framework within which this worldview grew. One of the major tenets that came to characterize this intellectual and cultural transformation was the rejection of supernatural purposes as causal explanations of natural phenomena. These scientific and philosophical developments reached their culmination when Darwin's theory of evolution, published in mid-nineteenth century, came into fruition in the first half of the twentieth century. It was the "neo-Darwinian synthesis," taking place in the 1940s, that established natural selection as the major evolutionary mechanism and rejected all purposive explanations of evolution previously entertained (Mayr 1991:135, 138; McMullin 1998:704; Provine 1988:58–62).

The scientific field investigating the emergence of life still faces enormous challenges and is plagued by various dividing issues. There is as yet no agreed-upon theory accounting for the origin of life, and no suggested scenario has yet been simulated in full in the laboratory. However, it is due to the robustness of the scientific-evolutionary worldview that the presupposition that life arose by natural means is no longer just one option among others but constitutes the basis for research of the origin of life (see, Searle 1994:85–86, 90–91). At the same time, the epistemological status of the naturalistic philosophical presupposition should be noted. This presupposition makes a general claim about the natural and the supernatural that transcends our possible experience. There is no empirical proof or disproof of the belief in the sufficiency of natural mechanisms to explain the emergence of life. Yet, as an inherent part of the scientific-evolutionary worldview, it is not a dogmatic belief. Unlike belief in the divine, the naturalistic conception developed in interaction with wide-ranged empirical data, grew stronger historically based on such data, and, furthermore, is continuously giving rise to testable empirical hypotheses that advance research. As part of a general scientific framework that proved itself fruitful in accounting for natural phenomena, the conception that life emerged naturally thus reached the status of a fully justified "given" (Fry 2012).

In fact, the commitment of origin-of-life researchers goes further than the naturalistic postulate. The realization that the emergence of life itself was an evolutionary process is also shared by most researchers in the field (Pascal et al. 2013:1–3). It is well established today that the "end point" of the emergence process, the cells constituting the Last Universal Common Ancestor at the root of the evolutionary tree, were already highly complex, organized systems. Within a naturalistic framework, the only feasible process that could have produced such a complex organization could not have been an all-or-none chance emergence but a gradual succession of increasingly complex stages (see, Peretó 2012:5394). What I suggest to call, *the evolutionary hypothesis of the emergence of life* maintains that physical and chemical prebiotic processes under specific geochemical conditions gave rise to a chemical infrastructure that could have gradually evolved toward primitive living systems. What could have been the nature of such evolutionary processes prior to the establishment of living systems? On the face of it, the very possibility of evolution prior to life is paradoxical. Didn't natural selection depend on the existence of a population of genetic replicating and mutating systems catalyzed by encoded enzymes?

It was the rise of molecular biology and experiments conducted in the 1970s and 1980s that revealed that natural selection and evolution could have taken place not only in populations of living cells but also in populations of RNA molecules in the test tube (Eigen et al. 1981:82–85; Orgel 1973; Safhill et al. 1970). Though these experiments were carried out under conditions far removed from prebiotic conditions, they drove the search for the most primitive chemical systems compatible with such conditions and with the basic evolutionary requirements of reproduction, division, and selection. Concomitantly, the discovery of the interdependence of genetic material and protein enzymes in extant cells and the specter of the chicken-and-egg problem became principal stimuli for research in the origin-of-life field.

Various prebiotic evolvable systems, some of them very different from the replicating genetic systems familiar to us from extant organisms, are suggested and studied in the field. The nature of these systems, the details of their prebiotic synthesis, and the various factors involved in their evolution are still issues contested among investigators. However, based on data from numerous experiments and on insights gained from theoretical models, study in the field is guided by several shared hypotheses: Notwithstanding the role of random events and randomly changing geochemical conditions, the gradual construction of evolvable structures was highly constrained and channeled by the characteristics of prebiotic chemistry and the nature of geochemical locals on ancient Earth (see, among many, Copley et al. 2007; de Duve 2005:21–24; Ferris 2006). Inorganic and organic catalysts, physical and chemical selection events, and the interaction between physico-chemical selection and primitive forms of natural selection led to the emergence of Darwinian evolution "in a series of stages, step by step" (Szostak 2012a:600).

The objective of this chapter is to shed light on the philosophical aspects of the origin-of-life problem. This aim will be tackled by examining the two main ideas, opposed to the evolutionary hypothesis of the origin of life, supernatural design, and chance. The position advanced by proponents of intelligent design (ID), among the most vocal current creationists, will be one of my targets. The various aspects of the ID movement, including its strategy and tactics in fighting science in general and biological evolution in particular, have been well documented and explored (Forrest 2007; Forrest and Gross 2004; Pennock 2000). I will concentrate here mainly on the contentions made by IDers on the origin of life. As part of their attack, they repudiate the notion of prebiotic natural selection as "a contradiction in terms" (Meyer 2004:244–248). However, this is just another pretext to deny the very possibility of a natural origin of life (Behe 1996; Bradley and Thaxton 1994; Meyer 2009). I will call attention to the most fundamental anti-scientific tenet of ID creationism—the belief in supernatural purpose and intention as a causal agent in the natural world (Johnson 1993:115).

Remarkably, the denial of the evolutionary conception of the origin of life makes for curious bed fellows: Richard Dawkins, the renowned evolutionary biologist, also rejects the relevance of evolution to the origin of life, contending that natural selection could have proceeded only once life had originated (Dawkins 2006:137, 140). Dawkins, a defender of evolutionary naturalism, obviously rejects the creationist viewpoint. However, instead of divine teleology, he suggests that the origin of life could have been an extremely improbable single-step chance event (Dawkins 1986:140, 2006:135). My discussion will point to the deep division between Dawkins's position and the conception and practice of origin-of-life investigators. Most significantly, Dawkins's chance thesis conflicts with the various scenarios that suggest viable gradual alternatives leading to life. Although only two examples of such scenarios will be described briefly in this philosophically oriented chapter, detailed examination of major scenarios in the field that serve as case studies of the evolutionary hypothesis of the origin of life can be found in Chapter 6.5 of this volume.

The philosophical dimension of Dawkins's position will be examined in juxtaposition to the ideas of ID creationists. The question of chance versus design will be further illuminated by considering the views of Thomas Nagel, an eminent New York University philosopher, who under the influence of ID creationists mounted an attack against the possibility of a natural origin of life. My conclusion will be that despite Dawkins's intentions, his chance position does not provide an answer to the challenge of design but rather weakens the scientific case. It is the hypothesis of the evolutionary emergence of life that obviates both the chance and the design solutions.

3.1.2 A HISTORICAL INTERLUDE: WHAT DID DARWIN THINK?

The evolutionary hypothesis of the origin of life, guiding investigations in the field, obviously assumes a continuous transition between the emergence of life and biological evolution. Proponents of ID, from their creationist perspective, attack the origin of life and biological evolution as a unified target, claiming that both could not have happened without divine intervention (see, Johnson 1993:103). One of their major arguments against biological evolution is based on their characterization of complex organized biological systems as un-evolvable "irreducible complex systems." As will be shown later, this is also their argument against a natural origin of life (Behe 1996:187–205). It is remarkable that several philosophers and historians of biology, as part of their repudiation of creationism, insist on considering the origin of life and biological evolution as two separate categories. This position is often justified by citing Darwin's views on the origin of life. In his witness testimony against the ideas of "creation science" in the 1981 McLean v. Arkansas trial, philosopher Michael Ruse stated that "The ultimate origins of life and the subsequent evolution of life were separated by Darwin (he never mentions the former in the *Origin*). They have been kept separate ever since" (Ruse 1996:27). Ruse (Ruse 2001:65, see, Ruse 2006:53) changed his view in more recent books, describing the origin of life as an "extension of Darwinism, for scientists are now starting to think in terms of the ways in which selection might have taken over the process." Robert Pennock, philosopher and historian of the ID movement, criticized prominent IDer Michael Behe for including in the definition of evolution also the emergence of life by natural means and especially for wrongly attributing this perception of evolution also to Darwin (Pennock 2000:160–161). To refute Behe, Pennock referred to Darwin's quote in the *Origin* that "it hardly concerns us how life itself originated" (Darwin 1988 [1872]:151). So, what did Darwin think on the origin of life? It might seem peculiar that Darwin (1859) hardly discussed the subject of the origin of life in his major opus *The Origin of Species* (1859). As attested especially in Darwin's post *Origin* letters, there were two major reasons why he avoided the subject in the *Origin*. First, Darwin perceptively realized and repeatedly stated that the origin-of-life question was beyond the powers of the science of his day (see, among many, Darwin 1863, 1959 [1882]). Second, realizing the inflammatory potential of the origin-of-life subject in the theological context of his day, Darwin didn't wish to alienate his public even more. So much so that at the end of the second edition of the *Origin*, departing from the general tenor of the work, Darwin used the biblical term of the creation of life (Darwin 1860:484). However, in a letter

to his close friend Joseph Hooker, Darwin later expressed his regret that he had given in to public opinion, emphasizing that all he meant to say was that the origin-of-life subject is "some wholly unknown process" (Darwin 1863).

Despite his apparent "silence" in the *Origin*, Darwin was preoccupied with the question of the original appearance of life, and quite a few of his comments on the subject reflected his feeling that the origin of life was a natural process (see, Peretó et al. 2009:404). Darwin's most famous allusion to the subject appeared in 1871 in another private letter to Hooker, where he raised a hypothesis that he considered extremely daring ("But if – and oh! What a big if!"). He imagined the synthesis in "a warm little pond" on the ancient Earth of "a protein compound…ready to undergo still more complex changes" from inorganic compounds under the influence of various sources of energy (Darwin 1871). Furthermore, in a most revealing letter in 1882, the last year of his life, Darwin explicitly stated that though "no evidence worth anything has as yet, in my opinion, been advanced in favor of a living being, being developed from inorganic matter, yet I cannot avoid believing the possibility of this will be proved some day in accordance with the law of continuity" (Darwin 1882). In light of Darwin's warm-little-pond musings in his letter to Hooker in 1871, "the law of continuity" of 1882 might have suggested a natural mechanism of emergence.

Clearly, Darwin was not silent on the question of the origin of life. His (mostly private) consideration of a natural origin of life and his realization of the incompetence of the science of his day to face the enormous complexity of the problem should not surprise us. Relying on Darwin's separation of the origin of life and its evolution is thus not a convincing argument against creationists. Moreover, anachronism in the history of science is a bad policy, or "poor history," as put by historian of evolution David Kohn (1989:221). Kohn speaks of Darwin's secularization of biology by changing the biological meaning of teleology (Kohn 1989:220–221, 229), but at the same time, he shows convincingly that Darwin's context was still theological and that Darwin was "a theologically transitional figure" (Kohn 1989:232). Even Darwin cannot be read out of the cultural and social context of his time.

3.1.3 FROM OPARIN AND HALDANE VIA THE CHICKEN-AND-EGG PUZZLE TO GENETIC AND METABOLIC INFRASTRUCTURES

Based on the far-reaching naturalistic implications of Darwin's theory, the major Darwinians in the last decades of the nineteenth century believed that a natural origin of life was an essential element of a consistent general evolutionary worldview (Haeckel 1902:256–258; Kamminga 1988). However, by the beginning of the twentieth century, new cytological studies revealed the heterogeneity of the cell content and the role of the nucleus. The rise of biochemistry led to the discovery and isolation of enzymes (Fry 2000:58–59). This new awareness of the complexity of the simplest cells convinced many scientists of the impassable gap between the living and the inanimate (Oparin 1967:203). In addition, in the beginning of the twentieth century, the mechanism of natural selection came under heavy attack, especially as a result of the uncertainties surrounding the causes and nature of mutations (Bowler 1989:246; Kellog 1907:1–7). This also led to an impasse in the scientific perception of the origin of life. Despite these obstacles and the avoidance of the problem by most of the scientific community (Muller 1966:494), several scientists, notably the biochemist Alexander Oparin in the Soviet Union and the geneticist and biochemist J. B. S. Haldane in Britain, independently published in the 1920s pioneering theories based on the evolutionary conception of the origin of life (Haldane 1967 [1929]; Oparin 1967 [1924]).

The empirical aspects of Oparin's and Haldane's scenarios were inspired by the new disciplines of biochemistry and genetics. They also relied on new data from the study of geochemistry and astronomy in their description of the prebiotic environment. Oparin, the biochemist, contended that cell-like structures that engaged in primitive metabolism could have been synthesized on the primordial Earth, whereas Haldane, the geneticist and biochemist, suggested the prebiotic synthesis of gene-like or virus-like molecules. These structures, as they both hypothesized, could have undergone evolutionary processes, leading to "protoorganisms" and later to life (see, Fry 2006). By assuming that such prebiotic structures could have evolved, Oparin and Haldane unknowingly foreshadowed theoretical ideas that were made possible and empirical data that became available only after the rise of molecular biology (Eigen et al. 1981:82–85; Orgel 1973; Safhill et al. 1970). Their theories, especially Oparin's, contributed to a scientific breakthrough in the 1950s. Oparin's hypothesis concerning the prebiotic synthesis of organic compounds was successfully implemented in the laboratory for the first time in the early 1950s by Stanley Miller (1953). Miller's experiments, and others that followed, and the rise of molecular biology in the 1950s and 1960s led to the establishment of a new experimental field devoted to the origin of life (Fry 2006).

One of the crucial discoveries of molecular biology that transformed the thinking on the origin-of-life problem was the functional interdependence of nucleic acids and proteins. It was realized that nucleic acids cannot be synthesized and cannot function without protein enzymes. Proteins, in order to be formed, have to be translated from nucleic acids. How could such a "chicken-and-egg" circle first emerge? Which came first, proteins or nucleic acids? From the early 1960s, origin-of-life scientists explored the two options of nucleic-acid-first (e.g., Orgel 1994) and proteins-first (e.g., Fox and Dose 1977), in order to solve the puzzle and establish a basis for an evolutionary process. The discovery in the early 1980s of RNA molecules in extant cells that function as enzymes, given the name "ribozymes," seemed to have overcome the conundrum. The idea of RNA as both chicken and egg dramatically changed the scene. It led to the hypothesis, now backed by strong evolutionary data (Benner et al. 2012:4; Cech 2012:2), that an "RNA world," in which ribozymes functioned both as genetic material and as enzymes, existed at one stage or another during the emergence of life (Gilbert 1986). However, despite numerous experiments throughout the years aimed to reconstruct the emergence of such an RNA world,

enormous challenges still remain, especially associated with the prebiotic synthesis of nucleotides and of a replicating RNA ribozyme (Joyce and Orgel 2006:23–24).

Currently, genetically oriented researchers (proponents of the gene-first notion) are divided into supporters of "RNA-first" and "RNA-later" scenarios. The latter doubt the feasibility of the spontaneous prebiotic synthesis of RNA and examine alternative genetic molecules as more probable candidates (Orgel 2003). Metabolism-first theorists reject gene-first models on the grounds that the prebiotic emergence of a replicating genetic polymer was extremely improbable (de Duve 1991:112–113; Shapiro 2007). Instead, they suggest various scenarios in which metabolic autocatalytic cycles could have emerged spontaneously to function as an evolvable infrastructure. Such metabolic infrastructures are supposed to have engaged in an evolutionary process entirely different from the familiar genetic-based process but still based on reproduction, variation, and selection (Lancet and Shenhav 2009:239, 243; Segré et al. 2001:137–139; Smith et al. 2009:445–446; Vasas et al. 2012; Wächtershäuser 1992:89, 111–112, 2007:594–595). Yet, many practitioners in the field doubt the viability of these scenarios. It is the potential of such metabolic cycles to diversify in order to undergo Darwinian evolution that is mainly questioned (Eigen 1971:498–503; Maynard Smith and Szathmáry 1985:68, 71). Unlike the essentially unlimited diversity of replicating polymers, in terms of base sequence, the ability of metabolic cycles to diversify in order to provide competing metabolic variations is very limited (Eschenmoser 2007:12828; Peretó 2012:5398; Vasas et al. 2010).

Some researchers who consider the RNA world as crucial in the emergence of life promote the hypothesis that nucleotides and their assembled sequences could have emerged only with the help of a previously growing protometabolism (de Duve 1991:133–143, 2005:73–75; Martin and Russell 2007:1893). Preparatory metabolic theories postpone the onset of natural selection to the rise of RNA. The emergence of prebiotic metabolic networks is assumed to have been physically and chemically determined to a large extent by specific geochemical conditions, by inorganic and organic catalysts, and by physical and chemical selection processes (de Duve 2005:23–24; Russell et al. 2014). Another noteworthy scenario suggests that replication and metabolism could have co-evolved in a specific geochemical site and could have interactively led to the emergence of the RNA world (Copley et al. 2007). This mechanism entails a multi-stage process, during which mutual catalysis among small molecules, such as amino acids and nucleotides, led to the gradual synthesis of longer and more efficient catalysts, to the rise of protometabolism, and to the emergence of RNA as the dominant macromolecule because of its unique combination of catalytic and template-replication capabilities (Copley et al. 2007:430–431). This scenario is based on physical and chemical selection processes and the interaction between such processes and emerging forms of natural selection. Significantly, it is emphasized that the proposed processes toward the RNA world were not chance-like but rather highly channeled. As the mutual catalysis in protometabolic reaction networks led to more effective catalysts, the "emergence of RNA as the dominant macromolecule became inevitable" (Copley et al. 2007:440, 431, 432). Thus, the various theories currently entertained by researchers in the field, including those examined in Chapter 6.5 and not mentioned here for lack of space, presuppose different versions of the evolutionary thesis of the emergence of life.

3.1.4   "A PIECE OF STATISTICAL LUCK" OR AN EVOLUTIONARY PROCESS?

In his seminal book, *The Blind Watchmaker* (1986), Richard Dawkins discussed at length the question of the emergence of life, which he identified with the rise of a "minimal machinery of replication and replicator power" (Dawkins 1986:141). He returned to this question, though in less detail, in his later books, maintaining his original view that the origin of life was "a single-step chance event," entirely different from the process of biological evolution based on cumulative selection (Dawkins 1986:139–141, 2006:139–140). According to Dawkins (1986:140), his chance-event claim is an inevitable conclusion of "what seems to be a paradox" that lies at the heart of the origin of life. This paradox is no other than a version of the "chicken-and-egg" dilemma, already apparent at this first step: Natural selection, the mechanism responsible for biological evolution, depends on the existence of replicating molecules and a catalytic machinery enabling this replication. However, such a complex system could not have emerged without evolution by natural selection. This paradox led Dawkins (1986:140) to conclude that there is no other way to account for the source of a self-replicating molecule but to postulate "a single-step chance event in the origin of cumulative selection itself." Dawkins (1986:158) admitted that before he gave the matter a deeper thought, his feeling was that "it would take a miracle to make randomly jostling atoms join together into a self-replicating molecule." However, he realized that we are allowed to include the whole universe in order to answer the question "how improbable, how miraculous, a single event…we are allowed to postulate" in order to account for the presence of life on Earth (Dawkins 1986:141). In an argument he titled later the "Planetary version of the Anthropic Principle" (Dawkins 2006:134), and based on the estimated number of planets in the universe, Dawkins (1986:145) concluded that we should not be surprised at all by a chance-like origin of life on Earth. Since there are around a billion billion planets in the universe, a model that predicts that "life will arise on *one* planet in a billion billion [will] give us a good and entirely satisfying explanation for the presence of life here" (Dawkins 2006:137–138). Adding to his calculation the estimate that life could have emerged on Earth during about a billion years (based on findings of the first "fossil organisms" from around 3.5 billion years ago), we have about a billion billion billion planet years "to play with." This amount of time can explain nicely how "random

thermal jostling of atoms and molecules resulted in a self-replicating molecule" (Dawkins 1986:144–145).

Obviously, Dawkins's conclusion leads the science of the origin of life to a dead-end. No hypothesis aiming to reconstruct a possible emergence process in the laboratory can account in scientific terms for a "random jostling of atoms" of such improbabilities. Several leading researchers in the field have described the lucky-accident claim as "sterile" or as "stifling research" (Cairns-Smith 1986:7; de Duve 2005:74; Orgel 1973:192). The ideas of the eminent molecular biologist Jacques Monod on the origin of life and its probability come to mind. In his book *Chance and Necessity*, published in 1970, when origin-of-life research was taking its first steps, Monod expressed his conviction that the origin of the linked system of nucleic acids and proteins (the chicken-and-egg vicious circle) poses a "veritable enigma." In his view, it is entirely possible that the origin of life occurred only once, "which would mean that its *a priori* probability was 'virtually zero'" (Monod 1974:136). Monod was fully aware of the unfortunate bearing of his conclusion on the scientific study of the origin of life (Ibid.). Dawkins, unlike Monod, seems strangely to ignore the significance of his one-in-billion-billion probability notion for the possibility of finding a specific mechanism of emergence. His claim is that we should not be afraid of such enormous numbers because an event or events of such staggeringly improbability are still feasible and calculable in a universe in which these numbers are based on astronomical facts (Dawkins 1986:143–144).

It is quite surprising that despite what seems like a consistent position on the origin of life, Dawkins occasionally vacillated between very different estimates of the probability of the event. In his various books, he discussed several theories that attempted to offer plausible explanations for the origin of life such as the primeval soup theory (Dawkins 1978:16) and Cairns-Smith's clay model (Dawkins 1986:148–157). Describing the subject of the origin of life as "a flourishing, if speculative subject for research," he astonishingly commented at one point that "I shall not be surprised if, within the next few years, chemists report that they have successfully midwifed a new origin of life in the laboratory" (Dawkins 2006:137). In his examination of the RNA-world theory, Dawkins (2009:421) asserted that he found this theory "plausible, and I think it quite likely that chemists will, within the next few decades, simulate in the laboratory a full reconstruction of the events that launched natural selection on its momentous way four billion years ago." The picture gets more confusing when Dawkins considers the subject of extraterrestrial life. He takes very seriously the famous question raised by the distinguished physicist Enrico Fermi in 1950, "Where is everybody"? Why haven't we detected any signs, even by radio signals, of the existence of life elsewhere in the universe? (Dawkins 2009:421). Assuming unequivocally that once life existed, the evolution of intelligence was probable (Dawkins 1986:146), Dawkins saw the failure to detect signs of extraterrestrial intelligence as an indication of the "paucity of life in the universe" (Dawkins 1986:165). Dawkins's guess was that "life in the universe is very rare" and we should expect that "no plausible theory of the origin of life exists" (Dawkins 1986:145, 2009:422). His conclusion was, thus, that the various origin-of-life theories that he examined err "on the side of being too plausible" (Dawkins 1986:165).

How can one account for Dawkins's fluctuating views on the likelihood of origin-of-life theories? Do these fluctuations reflect the still-prevailing uncertainty and controversy over the questions whether the emergence of life on Earth was "easy" or "difficult" and whether life, intelligent or not, is common or rare in the universe? In my view, the problem lies in Dawkins's focus of interest, which is neither on the scientific questions that occupy researchers in the field, nor on the specifics of the "pet theories" that scientists suggest (Dawkins 1986:147). On his own admittance, the theories that he discussed were chosen "arbitrarily...any one of the modern theories would have served the same purpose" (Dawkins 1986:139). Rather, Dawkins's main objective is to show that chance on a cosmological scale, replacing the inapplicable alternative of natural selection, is an adequate naturalistic answer (Dawkins 1986:141, 147, 2006:139). This and other philosophical implications of Dawkins's chance solution will be discussed presently. It is important first to point to the gap separating current research of the origin of life and Dawkins's perception of the subject.

3.1.5 ORIGIN-OF-LIFE RESEARCH VERSUS DAWKINS'S LUCKY-ACCIDENT SOLUTION

Disregarding the option of an evolutionary hypothesis, Dawkins, as we saw, considers "chance" a valid account for the emergence of life. Based on calculations very different from Dawkins's, active researchers reject the "statistical luck" solution (see, e.g., de Duve 1991:137; Eigen 1992:10–11, 12–13, 22; Miyakawa and Ferris 2003:8202; Pascal et al. 2013:2). A few examples will suffice to demonstrate why they would find Dawkins's "jostling of atoms" equivalent to a miracle. Eminent biochemist and origin-of-life theorist, Christian de Duve, calculated that the probability of a small protein enzyme, for example, human cytochrome C (104 amino acids), coming together by chance, assuming that all possible combinations of 104 amino acids have the same probability, is 20^{-104}, or about 10^{-135}. De Duve concluded that "there simply was not enough time, space, and matter on the prebiotic Earth, even in the entire universe, to try more than an infinitesimal fraction of the possible amino acid combinations" (de Duve 1991:137–138, 137, note 4). In another example, considering the probability of the synthesis of a specific RNA sequence made of 100 nucleotides, needed for an active ribozyme, out of all possible sequences, it is concluded that "synthesizing all of the [sequences] in one molecular unit over 1 billion years would lead to the synthesis of a mass of nucleic acid representing several tens that of the Earth per day" (Pascal et al. 2013:2). Another distinguished researcher, physical chemist Manfred Eigen, calculated the number of alternative arrangements of the four nucleotides of a single gene, with only a thousand symbols. He reached the staggering number of 4^{1000}, or about 10^{602}. The volume of the whole universe

(calculated on the basis of astronomical data) and the entire material of the cosmos lag far behind the combined volume and mass of this number of variants (Eigen 1992:10).

In distinction, Dawkins did not refer specifically to what, according to his own view, will be needed chemically in order to overcome the "paradox" of the origin of life. He did not calculate the probabilities of the chance assembly of a genetic polymer and a protein replicating enzyme or of two ribozyme molecules, one acting as an RNA template and the other as a replicase ribozyme. The examples cited earlier, few of the many found in the literature, by referring to such specific requirements, demonstrate why investigators regard the contention that the origin of life was a chance event as a "scientifically respectable version of the special miracle" (Cairns-Smith 1986:7).

In addition to the strong doubts cast upon Dawkins's position as a result of specific calculations, the rejection of the "statistical luck" option by origin-of-life researchers rests mainly on their suggested scenarios.

3.1.5.1 SINGLE-STEP SELECTION, CUMULATIVE-SELECTION, AND EVOLUTIONARY SCENARIOS OF THE ORIGIN OF LIFE

The conflict between the evolutionary hypothesis of the origin of life and Dawkins's view may be further clarified by Dawkins's distinction between "single-step selection" and "cumulative selection" (Dawkins 1986:45–49). A non-random arrangement of pebbles on a beach by the action of waves and a sieve separating objects non-randomly according to size are among Dawkins's examples of single-step selection events. Single-step selection applies to physical and chemical events, in which entities are selected on the basis of physical and chemical features such as size and stability. Most importantly, each single-step selection event starts afresh. In cumulative selection, on the other hand, reproduction or "memory" is involved, and hence, the selected product is retained in the next generation and serves as the starting point for the next selection event. Dawkins asserts that for reasons of probability, the innumerous cases of complex organization manifested by living organisms could not have been the result of a concatenation of chance-like events but only of the gradual process of the accumulation of selected small advantages dependent on the mechanism of natural selection (Dawkins 1986:43–74; see, Cairns-Smith 1986:45–49). As we saw, according to Dawkins, the case of the origin of life is different. Whereas single-step selection events, that is, "lucky chance," could never have been enough to explain biological evolution, "the origin of life was (or could have been) a unique event which had to happen only once" (Dawkins 1986:163, 2006:139). Given enough time and space, in fact, given the whole universe and its billion billion planets, a complex replicating molecular system at the origin of cumulative selection could have been the result of a single-step highly improbable chance event (Dawkins 1986:140).

Thus, the realms of the emergence of life and of biological evolution are separated by Dawkins and are kept distinct.

Based on this dichotomous view, the origin of Darwinian evolution is presented as an all-or-none transformation. In distinction, scenarios suggested by origin-of-life researchers describe various means by which what might be described by Dawkins as single-step selection events could have contributed to the step-wise growth of complexity and the gradual emergence of Darwinian evolution. Major scenarios in the field are examined in Chapter 6.5 of this volume, and their evolutionary characteristics are pointed out. Two examples out of these scenarios will be concisely presented here in order to demonstrate the difference between Dawkins's position and the scientific study of the origin of life: First, the empirical study in the framework of an RNA-first position, conducted in the laboratory of Jack Szostak, and second, the theoretical protometabolic scenario suggested by Christian de Duve. Another scenario, not discussed here, of the emergence of life in hydrothermal fields, suggested by Deamer, Damer, and colleagues (Damer and Deamer 2015), is a clear example of the role of physico-chemical selection in the evolutionary emergence of life (see Chapter 6.5, Section 6.5.9).

3.1.5.2 The Evolutionary Emergence of RNA-Containing Protocell

Experimental work in progress during more than two decades in the laboratory of leading origin-of-life researcher Jack Szostak was aimed to study the origin of cellular life. It explored models of cellular compartments containing replicating RNAs, supposedly analogous to protocells that emerged during the origin of life. Physico-chemical features and processes that were crucial to the emergence of replicating vesicles, replicating RNAs, and their interaction were studied in these experiments. Among them are the chemical composition of membranes, mineral catalysis of the assembly and growth of vesicles, encapsulation of RNA adsorbed to minerals, chemical conditions affecting the stability of both RNA and vesicles, vesicle growth and division affected by internal osmotic pressure and by other mechanisms, and the impact of internal osmotic pressure on competition among vesicles. All these processes, and many more, are hypothesized to have contributed to the emergence of Darwinian evolution during the origin of life (Chen et al. 2004:1746; Szostak 2012a).

Not able to review here, in detail, Szostak's work in progress, I will consider only two major instances demonstrating how the evolutionary hypothesis shaped Szostak's scenario: First, the change in attitude of Szostak and other RNA-world researchers toward the need for an early protocell, and second, the decision of the Szostak group to consider anew, after years of failed trials, a non-enzymatic alternative to RNA replication by a ribozyme replicase, originally thought to be the basis of the RNA-world hypothesis.

3.1.5.2.1 RNA-World Research and Compartments

The concept of some sort of "primitive cell" engulfing emerging metabolism characterized early metabolism-first theories, notably Oparin's. Continuing the Oparin tradition in later years, the division of protocells, representing reproduction, was included by "metabolists" in their evolutionary scenarios (e.g., Lancet and Shenhav 2009; Morowitz 1992:153, 1999:44). In distinction, "geneticists" considered proto-cellular membranes as possible barriers, preventing the easy access of organic molecules from the "soup," and preferred to postpone their emergence (see, Eigen et al. 1981:91–92; Orgel 2004:117–118). Reflecting on the development of his own professional interest in protocells, Szostak commented that "geneticists" traditionally "neglected the cellular nature of life," and work within the RNA-word program continued this tradition. For example, most "template copying experiments have been…performed in solution," not within a membrane-engulfed compartment (Service 2013:1034; Szostak 2012b:1, 2). However, the need for spatial localization is now commonly accepted by RNA-world researchers, not only because it allows the concentration of the ingredients of life but especially also because it enables the emergence of Darwinian evolution (Monnard 2007:389). It is realized that a population of RNA replicating molecules free in solution could not have evolved into more efficient or accurate replicators. "By keeping molecules that are closely related together, advantageous mutations can lead to preferential replication" (Szostak et al. 2001:387). Within a compartment, better replicators will be able to replicate each other more efficiently and will have an advantage over other compartments (Service 2013:1034; Szostak et al. 2001:387). Indeed, this change in perception proved itself fruitful. As shown in more detail in Chapter 6.5, the Szostak group demonstrated growth, division, and competition among vesicles containing RNA replicating molecules as a result of various physico-chemical mechanisms (Budin and Szostak 2011:5249; Chen et al. 2004; Hanczyc et al. 2003; Szostak 2011:2897; Zhu and Szostak 2009:5706–5709).

3.1.5.2.2 Alternative Route to RNA Replicase

Following the numerous difficulties encountered by researchers in their attempts to reconstruct a ribozyme replicase in the laboratory, Szostak and colleagues wondered whether RNA-catalyzed RNA replication could have emerged through a series of simpler steps, "driven purely by physical and chemical processes" (Schrum et al. 2010:8; Szostak 2012b:1). They thus decided to try to replicate RNA chemically, that is, without enzymes, in the "simplest form of genetic replication" (Blain and Szostak 2014:11.2). This reaction was studied in the 1970s and early 1980s in the Orgel laboratory (prior to the discovery of ribozymes). However, full cycles of chemical replication of RNA could not be achieved without the participation of a protein enzyme (Orgel 1994:60, 2004:111–112). Among the many obstacles preventing full replication was the fact that the complementary strand produced in the non-enzymatic copying process always contained a random mixture of "non-natural" 2'-5' and "natural" 3'-5' phosphodiester linkages, unlike the homogenous 3'-5' linkages in contemporary RNA. After the discovery of ribozymes, it was commonly assumed that a random mixture of 2'-5' and 3'-5' linkages will prevent the spatial folding of the RNA chain needed for ribozyme activity, thus preventing an origin of life based on RNA catalysis (Engelhart et al. 2013:390). Another major problem encountered by Orgel and others in the studies of non-enzymatic RNA replication was the difficulty to separate the strands after copying the template, because of the thermal stability of the duplex.

However, the experiments conducted by the Szostak group revealed that the heterogenous chains that were copied non-enzymatically from an RNA template and that contained low to moderate percentage of 2'-5' linkages not only allowed easier thermal separation of the duplex but were also capable of folding into spatial structures that showed catalytic properties (Engelhart et al. 2013). The authors thus suggested that the random incorporation of "wrong," "non-natural" 2'-5' linkages "far from being problematic" might have, in fact, contributed to the emergence of RNA as the first genetic polymer of life (Engelhart et al. 2013:391). Furthermore, when non-enzymatic RNA replication became the leading hypothesis in studying the origin of cellular life, an alternative model of an evolutionary pathway was suggested (Adamala and Szostak 2013). The early model of a protocell was assumed by Szostak and others to enable Darwinian evolution based on an RNA replicase, self-replicating within a vesicle. Now, chemical replication of RNA implied that the first evolved RNA catalysts, rather than improving replication, might have functioned as metabolic ribozymes (Szostak 2012b:1, 12). Experiments conducted by the Szostak group focused on a metabolic function that "conferred an advantage on its host cell" (Szostak 2012b:12). More specifically, a model was constructed, in which a metabolic ribozyme catalyzed the synthesis of a possible membrane component that led to an accelerated vesicle growth and division (Adamala and Szostak 2013:500).

To recapitulate, the system under study in Szostak's genetic scenario consists of an RNA polymer engulfed by a vesicle. As hypothesized and tested experimentally, both elements could not have been synthesized prebiotically without what Dawkins might have called "single-step selection events." These were part and parcel of the processes carried out by inorganic and organic catalysts, affected by physico-chemical selection under specific geochemical conditions. Such was the case also with the processes of vesicle growth and division and the processes of competition among vesicles. Even in the non-enzymatic chemical replication of RNA, the random distribution of 2'-5' linkages results from pure physico-chemical factors. None of these events and processes is endowed with "memory" and in itself cannot guarantee an accumulation effect. However, all these "single-step selection" events resulted in the development of a coordinated genetic system, which began to undergo primitive forms of natural selection. Remarkably, Szostak's experiments attempt to follow not just the preparation for but also the step-wise development of Darwinian evolution.

3.1.5.3 DE DUVE'S THEORY OF PROTOMETABOLISM AND THE RISE OF THE RNA WORLD

3.1.5.3.1 The Multimer Hypothesis

The eminent biochemist, Christian de Duve, regarded the RNA world as a crucial, though not the earliest, stage in the way to life. De Duve contended that the synthesis of

nucleotides, their assembled RNA sequences, and the RNA world could have emerged only with the help of a previously growing protometabolism, whose function depended on various kinds of prebiotic catalysts (de Duve 1991:133–143, 2003:562, 567–569, 2005:75). In most origin-of-life scenarios, the focus was usually on the role of minerals, especially metal ions and clays, as prebiotic catalysts. Inspired by the work of Sidney Fox on "proteinoids" (Fox and Dose 1977), de Duve raised the hypothesis that heterogenous, peptide-like molecules that he called "multimers," made of amino acids and occasionally also hydroxy acids, were the major catalysts sustaining primordial metabolism (de Duve 2003:567–568, 2005:22). It should be noted that experiments attesting to the catalytic capabilities of small organic molecules, especially short peptides, even single amino acids, in carrying out prebiotic syntheses are increasingly being conducted in various laboratories (Adamala et al. 2014; Wieczorek et al. 2013). According to de Duve, a "thioester world" was a critical chemical stage enabling the synthesis of multimers and the emergence of a primordial metabolism. In de Duve's scenario, sulfur compounds called thioesters could have been formed from the condensation of amino acids (and hydroxy acids) with thiols (alcohol-type compounds containing sulfur) that were all prebiotically available. This condensation required energy and was favored under acidic and hot conditions (de Duve 1991:116). De Duve (2005:64) indeed supported a volcanic setting at hydrothermal vents for the origin of life. Once formed, thioesters could have assembled to form multimers with the help of the energy derived from the thioester bond. De Duve (1991:113–116) hypothesized that among the multimers were probably a number of crude catalysts that could have catalyzed the synthesis of various organic compounds, including additional thioesters and multimers. Thus, like other metabolic theories, de Duve's theory is based on an autocatalytic process that drove the growth of protometabolism. In addition to catalyzing their own formation, multimers could have also catalyzed the formation of other protoenzymes. The increasing catalytic capabilities of the growing protometabolic network could have eventually led to the synthesis of nucleotides and oligonucleotides and to the rise of the RNA world (de Duve 1991:115).

3.1.5.3.2 What Kind of Evolutionary Hypothesis?

What are the evolutionary aspects of de Duve's theory? De Duve's protometabolic theory postpones the onset of natural selection to the rise of RNA replication. In this, it differs from scenarios that propose a metabolic evolvable infrastructure such as an autocatalytic metabolic cycle engaged in reproduction, variation, and selection (Lancet and Shenhav 2009:239, 243; Smith et al. 2009:445–446; Vasas et al. 2012; Wächtershäuser 1992:89, 111–112, Wächtershäuser 2007:594–595). In not suggesting the step-wise emergence of the process of natural selection itself, de Duve's theory also differs from gene-first theories, for example, Szostak's theory, that portray the emergence of a replicating RNA, within a membrane-engulfed compartment, as part of the gradual emergence of Darwinian evolution (Szostak 2012a). Indeed, de Duve acknowledged that his multimer hypothesis was criticized for not including a mechanism of replication and selection. This is true, he believed, of all processes that preceded the emergence of the first replicative machinery (de Duve 2005:23).

Thus, how did de Duve envisage the emergence and growth of protometabolism up to the emergence of replicating RNA? How did he account for the selection of active catalysts without a mechanism such as natural selection? (de Duve 2003:568, 2005:23)? First, it should be noted that he was utterly opposed to the chance thesis (1991:212–213). He did not refer specifically to Dawkins but argued with and rejected Jacques Monod's similar notion of the "fantastic luck hypothesis," which he viewed as implausible and as stifling research, just like the ID idea (de Duve 2005:73–74). De Duve contended that without natural selection, his suggested scenario depended on two elements, chemistry and environment: First, deterministic, reproducible chemical events resulted in an active set of catalysts as part of the multimer mixture, arising spontaneously under the prevailing conditions. Second, these conditions remained stable long enough to ensure an uninterrupted supply of catalysts, until they could be dispensed with (de Duve 2005:23–24). De Duve's "chemical determinism," his alternative to natural selection prior to the emergence of RNA, will be made clear by pointing to his concept of randomness and the way it was implemented in his multimer hypothesis (de Duve 1991:140–143). De Duve insisted that "randomness" in the context of the origin of life was not the commonly assumed "statistical randomness." "Random events or processes" involved in the synthesis of multimers were not statistically random or chance-like. Owing to stringent physical and chemical constraints, not all possible multimer sequences were equally probable. De Duve pointed to the "significant differences in bond energies and, especially in activation energies…between different associations…[that affected] the probabilities of their formation" (de Duve 1991:141). Also, differences among the multimers in solubility, stability, and ability to fold affected their survival and catalytic potential. Owing to physico-chemical selection channeling the synthesis and function of the catalytic multimers, de Duve claimed that only very rare peptides out of the theoretically possible ones could have been synthesized and only very rare peptides out of those produced could have survived. Thus, in this process, the end products were fully determined by the initial conditions (de Duve 1991:137–145, 213).

De Duve also sought to deal with the general problem of the multistage growth of complex biological molecules during the emergence of life (de Duve 1991:141). Without a "continuous renewal mechanism," such as replication, to counter spontaneous decay phenomena (see, Luisi 2015:915), how could intermediates survive and reach effective concentrations (de Duve 1991:141)? De Duve's proposed mechanism to overcome this problem was the selection and stabilization of multimeric crude catalysts by binding to substrates (de Duve 1991:143–145). He noted that this mechanism, unlike "regular" physico-chemical selection, "does not just rely on favorable coincidence. It is truly selective, in that it contains a built-in

feedback loop that actually makes it save what is useful" (de Duve 1991:143). At the same time, unlike natural selection, this mechanism "can operate with random chemistry of production," selecting by differential survival, in distinction to the complex machinery of replication, selecting by differential replication (Ibid.).

We have noticed, when discussing Szostak's scenario, how "single-step-selection events," to use Dawkins's parlance were important in the emergence of an evolvable genetic infrastructure. It is also demonstrated in de Duve's theory that Dawkins's clear-cut distinction between single-step selection and cumulative selection does not apply to the origin of life. The synthesis and selection of multimers serve as another example of the involvement of "single-step-selection events" in the construction of a metabolic infrastructure leading to the rise of RNA. De Duve's proposed mechanism for the protection of physically and chemically selected multimers by the binding to substrates is a clear rebuttal of Dawkins's distinction (de Duve 1991:140–144).

3.1.6 CHANCE OR DESIGN?

3.1.6.1 THE PHILOSOPHICAL IMPLICATIONS OF DAWKINS'S CHANCE HYPOTHESIS

A clarifying comment is called for from the very beginning of this discussion. In the following examination, I suggest that Dawkins's views on the origin of life involve a philosophical element that contributed to his improbable-chance conclusion. However, there is no doubt that Dawkins would deny any philosophical interpretation of his views, which he considers purely scientific (Dawkins 2006:48, 50–51,109; see Fry 2012:670–671). It could be added in passing that Dawkins is not the only scientist to adopt a similar negative attitude, either explicitly or implicitly, toward the involvement of philosophy in science (see, Fry 2012). Thus, the attribution of a philosophical dimension to Dawkins's ideas is all mine.

Dawkins's blind-watchmaker thesis, presented forcefully in his 1986 book and since then in numerous books and public appearances throughout his career, was aimed to reject the "myth that Darwinism is a theory of 'chance'" and to assert that biological evolution through natural selection is the antithesis of chance (Dawkins 1986:xv, 49). Biological complexity could have come about only through cumulative selection of small advantageous changes. However, Dawkins pointed out that the same also seems to be true for the machinery of replication needed at the origin of life, the very beginning of the evolutionary process (Dawkins 1986:141). This realization, according to Dawkins, leads either to an infinite regress or to a scientific conundrum. His solution to overcome this dilemma was the one-in-a-billion-billion chance hypothesis. As we saw, his suggestion that, given enough space and time, "random thermal jostling of atoms and molecules resulted in a self-replicating molecule" (Dawkins 1986:144–145) is commonly rejected by origin-of-life researchers, notwithstanding their various specific commitments. During the first decades of the twentieth century, before the molecular intricacies of the living cell were known and when the notion that life emerged during vast geological times was still popular, scientists were much more cavalier about attributing various stages of the emergence process to chance (Fry 2000:71, 74–76, 102, 193). This was still occasionally the case during the "optimistic years" of the 1960s and 1970s, especially in gene-first theories (Fry 2000:193–194), and the "lucky accident" notion was still considered, for example, by Monod in 1970 and by Francis Crick and several other scientists in the 1980s (Fry 1995). A minority position at its outset, it is now considered completely obsolete on both theoretical and empirical grounds (Fry 2000:102, 2009).

How can Dawkins's unique position be explained? I have noted earlier that Dawkins's focus of interest in his discussion of the origin of life is not on actual scientific questions that preoccupy investigators in their efforts toward a solution (Dawkins 1986:147). Rather, when describing the "paradox at the heart of the origin of life" (Dawkins 1986:140–141), Dawkins expressed his worry that this obstacle might be conceived as a threat to the understanding of life without a designer, that is, to his "blind watchmaker" thesis (Dawkins 1986:141, 147). The "gap theologians," he pointed out, indeed insist on the need for a designer to cross the huge gap of the origin of life (Dawkins 2006:134–135). I believe that this challenge to the blind-watchmaker thesis and Dawkins's wish to fend off this threat are at the heart of Dawkins's position on the origin of life. Presenting his argument, he insists that chance on a cosmological scale is an adequate naturalistic answer and that "this statistical argument completely demolishes any suggestion that we should postulate design to fill the gap" (Dawkins 2006:139). The philosophical question to be considered is whether Dawkins's chance thesis indeed constitutes a convincing rebuttal of the design challenge. In order to comment on the question of chance versus design, the views expressed by IDers on the origin of life and the highly relevant ideas of philosopher Thomas Nagel will be examined.

3.1.6.2 Intelligent Design Proponents and Thomas Nagel on the Origin of Life

3.1.6.2.1 "Irreducibly Complex Systems"

In *Darwin's Black Box* (1996), Michael Behe, a professor of biochemistry and a leading promoter of ID views, offered to the ID movement an invaluable gift—an argument for design in molecular language. Behe's terms of discussion have been in use by proponents of the movement ever since, upheld as representing a "scientific" argument for ID (Dembski 1999:147–148; see, Nagel 2012a:10). Behe (1996:3–6) argued that the discoveries of biochemistry, by opening the black box of the cell and revealing the existence of complex molecular machines, created an unsurmountable challenge for the Darwinian theory. Each cell, according to Behe, contains intricate systems, whose function depends on the interaction of their various components. If a single component in such a system is missing, the system ceases to function. Obviously,

he claimed, such systems could not have evolved gradually. Any earlier system that is missing a part is, by definition, non-functional and thus could not have been favored by natural selection (Behe 1996:39). Such systems also could not have emerged by chance. The probability that "irreducibly complex systems" could have arisen "as an integrated unit in one fell swoop" is nil. Denying the possibility of an alternative natural mechanism that could have accomplished such a feat and relying on an analogy with the power of intelligent agents, that is, humans, to design such intricate machines, Behe (1996:187–205, 219) concluded that these systems had to result from the action of an intelligent designer. Space will not allow me to dwell on the various mechanisms that could have led to the natural evolution of functional systems of interdependent parts, brought forth in the literature by evolutionary biologists and cell biologists (see, among many, Coyne 2006; Gishlick 2004; Miller 1999:129–161; Musgrave 2004).

The same "irreducible-complexity" logic guided Behe in his account of the origin of life. In his "simplest possible design scenario," he suggested that "nearly four billion years ago the designer made the first cell already containing all of the irreducible complex biochemical systems…(One can postulate that the designs for systems that were to be used later, such as blood clotting, were present but not "turned on") (Behe 1996:227–228, 231). Behe indicated that the defining feature of intelligent design is its purposeful nature. Biochemical systems, either turned on or off, "were designed not by the laws of nature…rather, they were planned. The designer knew what the systems would look like when they were completed, then took steps to bring the systems about" (Behe 1996:193). In the same vein, ID proponents, the scientists Walter Bradley and Charles Thaxton, suggested that the information for the first nucleic acids on Earth came from some intelligence and that life originated from "a who rather than from a what" (Bradley and Thaxton 1994:209). The opposition between Dawkins and IDers is clear. The contention that integrated biochemical systems could not have arisen naturally by a "one-fell-swoop" event is IDers' justification for an intervention of a purposeful intelligent designer. A naturalistic origin of life via a "single-step chance event" is Dawkins's anti-designer argument. Significantly, however, both deny the possibility of a gradual, evolutionary emergence of life.

3.1.6.2.2 Thomas Nagel on the Origin of Life

Thomas Nagel, the esteemed New York University professor, has developed, since the 1970s, his criticism of science for its materialist reductionism and its in-principle failure to solve the mind-body problem. In recent years and especially in his book, *Mind and Cosmos: Why the Materialist Neo-Darwinian Conception on Nature is Almost Certainly False* (2012), Nagel extended his criticism, claiming that science has also failed in accounting for the origin of life and its evolution. He found it "highly implausible that life is a result of a sequence of physical accidents combined with natural selection" (Nagel 2012a:6). Nagel (2012a:10) mentioned that his expanded criticism of science has been stimulated by "the attack on Darwinism mounted…from a religious perspective by the defenders of intelligent design." He pointed out that though he is not "drawn to the explanation by the actions of a designer" on the grounds of his own atheistic position (Nagel 2012a:12n, 2012b:63), he favors the critical "negative part of intelligent design position" (Nagel 2012a:10–11). Nagel was particularly impressed with a book published by Stephen Meyer, *Signature in the Cell: DNA and the Evidence for Intelligent Design* (2009). Nagel (2009) even suggested to nominate Meyer's book to the Times Literary Supplement list of top books of the year. In his recommendation, Nagel (2009) called upon atheists and anti-ID theists to be "instructed by [Meyer's] careful presentation" of the origin-of-life problem and of the present state of research on "non-purposive chemical explanations of the origin of life." Following IDers, Nagel (2012a:5, 6–9) came to find the materialistic explanation of biological evolution "hard to believe." As for the scientific accounts of the emergence of life from "dead matter," he felt positive that it is a complete failure, because the only possible mechanism of natural selection is ruled out in this case (Nagel 2012a:9–10).

3.1.6.2.2.1 Stephen Meyer's "Thorough Search"

In his *Signature* book and other publications, Stephen Meyer, a senior fellow at the Discovery Institute, the major organization behind the ID movement, defined the "mystery of the origin of life as the DNA enigma," that is, the origin of biological information "needed to build the first living organisms" (Meyer 2009:13). In order to show that ID is the best and "only causally adequate explanation" of the origin of biological information (Meyer 2009:171), Meyer conducted a "thorough search" of theories and scenarios suggested by origin-of-life investigators. It will be impossible to indicate all the distortions in Meyer's review. I believe the following example will suffice. Meyer shared with his readers a moment of discovery, when "the day of the breakthrough came" in the mid-1990s while he was looking at slides of the double helix. There and then Meyer discovered that "there are no [direct] chemical bonds *between the bases* along the longitudinal axis in the center of the helix. Yet, it is precisely along this axis of the DNA molecule that the genetic information is stored" (Meyer 2009:241, emphasis in the text). Meyer then indicated the "devastating implications for self-organization models" that attempt to explain the DNA enigma on the basis of "bonding affinities" (Meyer 2009:244), while regrettably no direct bonding exists between the bases in DNA. This is all figment of Meyer's imagination. No origin-of-life scientist claimed or could have made such a claim. In fact, it is embarrassing to discover that while Meyer completed a Ph.D. dissertation on origin-of-life studies in 1990 and while he was devoting himself from the late 1980s to "developing the case for intelligent design upon the discovery of the …digital code stored in the DNA molecule" (Meyer 2009:4–5), he was all along ignorant of the basic structure of DNA, with which every undergraduate student of biology must be familiar. It also might be of interest to readers of this chapter that following Meyer's critique of the improbability involved in any claim for prebiotic natural selection, he added that "nevertheless,

during the 1980s, Richard Dawkins…[in 1986] attempted to resuscitate prebiotic natural selection as an explanation for the origin of biological information" (Meyer 2004:246–247)!

Yet, despite Meyer's unreliable account of the current state of origin-of-life research, Nagel was fully convinced by his demonstration that "nonpurposive chemical explanations" cannot provide an answer to the origin of life. Nagel noted that another source shaping his views on the origin of life was "an important paper" by Roger White, an MIT philosopher and linguist, claiming that attempts by scientists to account for the origin of life in terms of "nonpurposive principles of physics and chemistry" are "based on a confusion" (Nagel 2012a:ix; 89).

3.1.6.2.2.2 White: "Biased Toward the Marvelous"

In his paper, "Does origins of life research rest on a mistake?" (2007), White states his conviction that the insistence of origin-of-life researchers on finding a natural explanation for the origin of life reveals no less than a confused psyche. According to White, researchers try as hard as they can to find convincing mechanisms based on "non-intentional biasing" (White 2007:462). However, the crucial stage in the origin of life, the appearance of self-replicating molecules, is not more likely to happen as a result of non-intentional bias compared to chance. "Translating" his abstract text into prebiotic chemical terms, White contends that physical and chemical constraints that could have enabled physico-chemical selection (e.g., differential chemical affinities between diverse building blocks; differential thermodynamic stability of molecular structures; and catalysis by, e.g., metal ions, specific minerals, and short peptides) could not have made the outcome of self-replicating molecules more probable. The reason, as stated by White is that "what makes certain molecular configurations stand out from the multitude of possibilities seems to be that they are capable of developing into something which strikes us as *rather marvelous*, namely a world of living creatures. But there is no conceivable reason that blind forces of nature or physical attributes should be *biased toward the marvelous*" (White 2007:467, emphasis added). White's reasoning is clearly teleological and anti-evolutionary. Indeed, "blind forces of nature" are not "biased toward the marvelous," nor are they preordained to produce life. Yet, it just so happened because of "physical attributes" that certain configurations of molecules in specific geophysical conditions on the primordial Earth about four billion years ago interacted in certain ways, obeying the "blind forces of nature," and resulted in specific chemical outcomes that could have evolved into living systems. It certainly could have been otherwise, here on Earth. It probably was otherwise on myriad of other planets.

Rejecting "non-intentional biasing" as contributing to an explanation of the origin of life, we are left, according to White, with the alternatives of chance and "intentional biasing," that is, "an agent." White's psychological diagnosis as to why most origin-of-life scientists will not embrace chance, despite the fact that a natural non-intentional biasing does not work, suggests that scientists cannot resist the "intuition that people really have" upon contemplating the molecular mechanisms from which life developed (White 2007:475). It is a "gut reaction to the data" that such molecular mechanisms "were designed by an agent" (White 2007:474). White goes as far as claiming that "it is just the reaction that everyone has" [!]. However, acting upon such an intuition is considered illegitimate scientifically. According to White, this explains the insistence on looking for a physico-chemical non-purposive mechanism that promises no advantage over chance (White 2007:475). White's anti-naturalistic and anti-evolutionary perspective strengthened Nagel's conviction that physical and chemical means were insufficient to guarantee the choice of the desired "marvelous" molecules out of all the possible ones (Nagel 2012a:89–91).

3.1.6.3   Dawkins's Chance Position: Scientifically Unnecessary and Philosophically Damaging

Convinced by the arguments of White and IDers, Nagel came to deny the possibility of a naturalistic origin of life. However, denying also explanations based on chance and on intelligent design (Nagel 2012a:91), he raised the alternative of a "natural teleology," assuring that "the universe is rationally governed...also through principles that imply that things happen because they are on a path that leads toward certain outcomes – notably the existence of living, and ultimately of conscious, organisms" (Nagel 2012a:67). The question how natural could such a "natural teleology" be and the possibility of an internal contradiction in a cosmic teleology, without a "rational agent" or a God that makes the choices of specific physical parameters (Plantinga 2012:6; White 2007:466), though relevant to our discussion, cannot be elaborated upon here. Yet, coming back to Dawkins's position on the origin of life, I suggest that by assuming to raise a solution distinct from a natural solution but also from design, Nagel might contribute another perspective from which to evaluate Dawkins's chance-versus-design contention.

Interestingly, unlike Stephen Meyer and Roger White, who erroneously attributed to Dawkins, respectively, the "resuscitation of prebiotic natural selection" and the rejection of a chance account of the origin of life (Meyer 2004:246–247; White 2007:459, 474), Nagel (2010a:19–26) confronted Dawkins on his actual chance attitude to the origin of life. First, Nagel noted that Dawkins acknowledged the problem, or "paradox," inherent in the attempt to provide a natural explanation for the origin of life, that is, the need for and, at the same time, the impossibility of applying evolution prior to life (Nagel 2010a:23, 24). Nagel and Dawkins thus agree in rejecting any involvement of an evolutionary process in the origin of life. However, Nagel saw Dawkins's chance solution as no more than "pure hand waving" and "a desperate device to avoid the demand for a real explanation" (Nagel 2010a:24, 25). As to Dawkins's one-in-a-billion-billion statistics, Nagel criticized his calculation for being too probable: The genetic material is so staggeringly complex that "no one has a theory that would support anything remotely near such a high probability as one in a billion billion" (Nagel 2010a:24).

Referring to the argument from design, Nagel commented that it was prompted by the "overwhelming improbability" of life coming into existence by chance, "simply by the purposeless laws of physics" (Nagel 2010a:24). Claiming that science has no alternative to chance, Nagel asserted that *the problem of the origin of life is the reason* "*why the argument from design is still alive*" (Nagel 2010a:25, emphasis added). Though Nagel is mistaken and science does have an evolutionary alternative to chance, his evaluation of Dawkins's attempt to account for the origin of life by a single chance event accurately addresses Dawkins's problem. Here, in a nutshell, in Nagel's words, is the reason why Dawkins's position fails in confronting the argument from design.

Dawkins's contention that chance on a cosmological scale is a valid explanation of the origin of life is not based on a serious examination of scenarios offered by researchers in the field. His position thus draws attention away from such evolutionary scenarios and strengthens the IDers' cause of bringing back the argument from design, this time based on "the DNA enigma." By implying that chance and design are the only explanatory options of the emergence of life, Dawkins unwillingly encourages both religious and "natural" teleological views.

3.1.7   CONCLUSION

This chapter called attention to the active role of philosophical presuppositions in the scientific study of the origin of life. This is generally true of science but most notable when science tackles a problem whose philosophical dimension challenges deep-seated traditional beliefs. As pointed out here, the scientific worldview that grew historically in the last few hundred years became strongly established when the theory of evolution and the rejection of "big teleology" were fully accepted by the scientific community in the first half of the twentieth century. It is this scientific-evolutionary worldview, based on the interaction between the theoretical and empirical achievements of science and the philosophical tenets of naturalism, that underlies the scientific study of the emergence of life. The attack by proponents of ID on this study, under the guise of theoretical and empirical critique (see Behe 1996; Meyer 2009), is in fact a philosophical-theological rejection of the naturalistic tenets, a continuation in molecular garb of William Paley's argument from design.

Though the pioneering theories of Oparin and Haldane raised the possibility of an evolutionary origin of life, only following the discoveries of molecular biology; the demonstration of *in vitro* molecular evolution and, more recently, the study of ribozymes, of various organic and inorganic compartments, and of inorganic and organic catalytic molecules involved in metabolism; and the discoveries of geochemical sites that might have been conductive to the emergence of life could the evolutionary hypothesis of the emergence of life become a fruitful research tool. Maintaining that constrained prebiotic processes could have given rise to an evolvable infrastructure and that Darwinian evolution itself emerged gradually through physical and chemical processes, the evolutionary hypothesis is also not neutral philosophically. By guiding the various scenarios suggested in the field, it implements the philosophical postulate of the natural emergence of life in active research.

Dawkins's contention that the origin of life was the result of a highly improbable single chance event is a principled denial of this implementation. Though a most prominent defender of evolutionary naturalism, Dawkins's clear-cut separation between the origin of life and biological evolution conflicts with the continuity of the physical, chemical, and biological underlying origin-of-life research. As commented previously, Dawkins seems to regard the "event" of the origin of life as a problem that luckily had to happen only once, opening the way to the working of the blind watchmaker. Interestingly, Dawkins suggested that "Darwin probably (and in my view rightly) saw the origin of primitive life as a relatively easy problem compared with the one he solved" (Dawkins 2004:574). This interpretation of Darwin's attitude to the origin of life seems to fit Dawkins's inclination to put the origin-of-life "event" in parentheses.

Another case of, so to speak, "marginalizing" the origin-of-life problem, but from a different perspective, is encountered in the position of various scientists and philosophers of science. It is a reaction to the fact that researchers have not yet reached an agreement as to the path by which life emerged on Earth, nor have they managed so far to simulate this process in the laboratory. Consequently, the scientific status of origin-of-life research, compared with biological evolution, is often perceived as too weak to stand a chance against creationists' attacks (de Duve 2002:288; Miller 1999:276–277; Ruse 2001:65; Scott 1996:515–516). In my view, judging the science investigating the origin of life exclusively on the basis of its current empirical achievements misses the point. Clearly, there are huge objective hurdles hindering the solution of the origin-of-life problem. Most significantly, traces of early-Earth's geochemistry were eradicated by geological and biological processes of over 4 billion years (Pizzarello and Shock 2010:1). The uncertainty about the specific site where life could have emerged, and its physical and chemical conditions will be difficult to dispel (Eschenmoser 2007:12822). There are enormous difficulties in searching for remnants of analogous processes on other planets (Benner et al. 2012), and the interdisciplinary nature of the problem requires active engagement of diverse scientists (Eschenmoser 2007:12821). However, disregarding the interaction between the theoretical, empirical, and philosophical aspects, demonstrated, in particular, in the evolutionary hypothesis of the origin of life, ignores the major basis for the most promising hypotheses raised in the field.

Regretting that researchers still have not provided facts "to demonstrate an entirely material origin of life" (Easterbrook 1997:893) wrongly assumes that there are "facts" to demonstrate or refute an entirely natural emergence of life. The philosophical naturalistic postulate cannot be proved or disproved empirically. However, it is robustly established by now as a result of its historically developed interdependence with empirical data. It is neither doubted by researchers nor put to the test afresh in their experiments, despite the fact that the problem of the origin of life was not yet solved. Moreover, despite the lack of solution, the scientific investigation of the origin of life, far from being a "problem" or the "soft underbelly of evolutionary biology" (Scott 1996:515–516), is central to the evolutionary worldview. It is not by chance that Nagel, from his philosophical anti-evolutionary perspective, contended that "the profoundly nonteleological character" of the failed evolutionary naturalism of our day is especially expressed in its attempt to provide a natural explanation to the origin of life (Nagel 2010b:15).

Indeed, it will be appropriate to conclude this chapter by reiterating the "nonteleological" or rather anti-teleological character of the evolutionary hypothesis of the origin of life. This hypothesis is incompatible not only with the idea that life was created by a designer for a purpose but also with Nagel's "natural-teleology" that postulates the existence of teleological principles, ensuring the bound-to-happen emergence and evolution of living and eventually conscious organisms (Nagel 2012a:67). Assuming such teleological principles, the emergence of life is claimed to have resulted from a "path toward a certain outcome," which is predetermined and guaranteed (Nagel 2012a:93). In distinction, the scenarios raised by investigators in the field, referred to shortly in this chapter and in more detail in Chapter 6.5, though evaluating prebiotic physical and chemical constraining factors as "preparatory" for the emergence of life, do not imply purpose, plan, or, in general, teleological tendencies. As put by Christian de Duve, "chemistry has no prescience." It did not produce nucleotides "for the purpose" of synthesizing RNA. RNA itself was not "destined to be the first bearer of information in nascent life" (de Duve 2005:74–75; see, Luisi 2015:908, 909). Only in retrospect can we judge these outcomes as relevant biologically. Only from the perspective of extant life, and assuming the continuity between prebiotic chemistry and biochemistry, can we try to discern by theoretical considerations and simulation experiments what were the relevant constraints and what were the various catalyzed reactions "preparatory" for the emergence of life. Under different physical and chemical circumstances, other results, less "marvelous" (White 2007:467), could have been produced, as might have been the case on other planets in the solar system and beyond.

The major objective of this chapter was to analyze the philosophical aspects of the origin-of-life problem in light of the challenges faced by the evolutionary hypothesis, the chance thesis, and supernatural and natural teleology. As part of this analysis, Dawkins's contention that his highly-improbable-chance thesis provides an answer to the notion of design and IDers' and Nagel's claims to counter chance by divine or "natural" telos were also examined. We can conclude that there is no real dichotomy between chance and telos. Chance and telos are the two horns of a faulty dilemma. The true conflict is between chance and purpose on the one hand and an evolutionary emergence of life on the other hand.

REFERENCES

Adamala, K., and J. W. Szostak. 2013. Competition between model protocells driven by an encapsulated catalyst. *Nat. Chem.* 5:495–501.

Adamala, K., F. Anella, R. Wieczorek, P. Stano, C. Chiarabelli, and P. L. Luisi. 2014. Open questions in origin of life: Experimental studies on the origin of nucleic acids or proteins with specific and functional sequences by a chemical synthetic biology approach. *Comput. Struct. Biotech. J.* 9(14):e201402004.

Behe, M. J. 1996. *Darwin's Black Box*. New York: The Free Press.

Benner, S. A., H. J. Kim, and Z. Yang. 2012. Setting the stage: The history, chemistry, and geobiology behind RNA. *Cold Spring Harb. Perspect. Biol.* 4:a003541.

Blain, J. C., and J. W. Szostak. 2014. Progress toward synthetic cells. *Annu. Rev. Biochem.* 83:11.1–11.26.

Bowler, P. J. 1989. *Evolution: The History of an Idea*. Berkeley, CA: University of California Press.

Bradley, W., and C. Thaxton. 1994. Information and the origin of life. In *The Creation Hypothesis*, Ed. J. P. Moreland, pp. 173–210. Downers Grove, IL: InterVarsity Press.

Budin, R., and J. W. Szostak. 2011. Physical effects underlying the transition from primitive to modern cell membranes. *PNAS* 108(13):5249–5254.

Cairns-Smith, A. G. 1986. *Seven Clues to the Origin of Life*. Cambridge, UK: Cambridge University Press.

Cech, T. R. 2012. The RNA worlds in context. *Cold Spring Harb. Prspect. Biol.* 4:a006742.

Chen, I. A., R. W. Roberts, and J. W. Szostak. 2004. The emergence of competition between model protocells. *Science* 305:1474–1476.

Copley, S. D., E. Smith, and H. J. Morowitz. 2007. The origin of the RNA world: Co-evolution of genes and metabolism. *Bioorg. Chem.* 35:430–443.

Coyne, J. 2006. Intelligent design: The faith that dare not speak its name. In *Intelligent Thought*, Ed. J. Brockman, pp. 3–23. New York: Vintage Books.

Damer, B., and D. Deamer. 2015. Coupled phases and combinatorial selection in fluctuating hydrothermal pools: A scenario to guide experimental approaches to the origin of cellular life. *Life* 5(1):872–887. doi:10.3390/life5010872.

Darwin, C. 1859. *On the Origin of Species by Means of Natural Selection, or the Preservation of Favored Races in the Struggle for Life*. London, UK: John Murray.

Darwin, C. 1860. *On the Origin of Species by Means of Natural Selection, or the Preservation of Favored Races in the Struggle for Life*. 2nd ed. London, UK: John Murray.

Darwin, C. 1863. Letter to J. D. Hooker, 29 March. In *The Correspondence of Charles Darwin*, Vol. 11, pp. 277–278. Cambridge, UK: Cambridge University Press.

Darwin, C. 1871. Letter to J. D. Hooker, February 1st. In *The Life and Letters of Charles Darwin, Including an Autobiographical Chapter*, Ed. F. Darwin. 1887. Vol. 3, pp. 168–169. London, UK: John Murray.

Darwin, C. 1988 [1872]. *Origin of Species*. 6th ed. New York: New York University Press.

Darwin, C. 1882. Letter to D. Mackintosh, February 28. Letter 13711, DAR. 146:335, Cambridge, UK: Cambridge University Library.

Darwin, C. 1959 [1882]. Letter of Darwin to G. C. Wallich, 28 March. In *Some Unpublished Letters of Charles Darwin*, Ed. G. De Beer, Vol. 14, pp. 12–66. Notes and Records of the Royal Society of London.

Dawkins, R. 1978. *The Selfish Gene*. Oxford, UK: Oxford University Press.

Dawkins, R. 1986. *The Blind Watchmaker*. London, UK: Penguin Books.

Dawkins, R. 2004. *The Ancestor's Tale*. London, UK: Weidenfeld and Nicolson.

Dawkins, R. 2006. *The God Delusion*. Boston, MA: Houghton Mifflin.

Dawkins, R. 2009. *The Greatest Show on Earth*. London, UK: Bantam Press.

De Duve, C. 1991. *Blueprint for a Cell*. Burlington, NC: Neil Patterson.

De Duve, C. 2002. *Life Evolving*. Oxford, UK: Oxford University Press.

De Duve, C. 2003. A research proposal on the origin of life. *Orig. Life Evol. Biosph.* 33:559–574.

De Duve, C. 2005. *Singularities*. Cambridge, UK: Cambridge University Press.

Dembski, W. A. 1999. *Intelligent Design*. Downers Grove, IL: InterVarsity Press.

Easterbrook, G. 1997. Science and god: A warming trend? *Science* 277:890–893.

Eigen, M. 1971. Self-organization of matter and the evolution of biological macromolecules. *Naturwissenschaften* 58:465–523.

Eigen, M. 1992. *Steps Towards Life*. Oxford, UK: Oxford University Press.

Eigen, M., W. Gardiner, P. Schuster, and R. Winkler-Oswatitsch. 1981. The origin of genetic information. *Sci. Am.* 244(4):78–118.

Engelhart, A. E., M. W. Powner, and J. W. Szostak. 2013. Functional RNAs exhibit tolerance for non-heritable 2'–5' versus 3'–5' backbone heterogeneity. *Nat. Chem.* 5:390–394.

Eschenmoser, A. 2007. The search for the chemistry of life's origin. *Tetrahedron* 63:12821–12844.

Ferris, J. P. 2006. Montmorillonite-catalyzed formation of RNA oligomers: The possible role of catalysis in the origins of life. *Philos. Trans. R. Soc. Lond. B Biol. Sci.* 361:1777–1786.

Forrest, B. 2007. Understanding the intelligent design creationist movement: Its true nature and goals. In *A Position Paper from the Center for Inquiry, Office of Public Policy*. Washington, DC: Center for Inquiry, Inc.

Forrest, B., and P. R. Gross. 2004. *Creationism's Trojan Horse: The Wedge of Intelligent Design*. Oxford, UK: Oxford University Press.

Fox, S. W., and K. Dose. 1977. *Molecular Evolution and the Origin of Life*. New York: Marcel Dekker.

Fry, I. 1995. Are the different hypotheses on the emergence of life as different as they seem? *Biol. Philos.* 10:389–417.

Fry, I. 2000. *The Emergence of Life on Earth. A Historical and Scientific Overview*. New Brunswick, NJ: Rutgers University Press.

Fry, I. 2006. The origins of research into the origin of life. *Endeavour* 30(1):25–29.

Fry, I. 2009. Philosophical aspects of the origin-of-life problem: The emergence of life and the nature of science. In *Exploring the Origin, Extent, and Future of Life: Philosophical, Ethical, and Theological Perspectives*, Ed. C. M. Bertka, pp. 61–79. Cambridge, UK: Cambridge University Press.

Fry, I. 2012. Is science metaphysically neutral? Studies in his. *Philos. Biol. Biomed. Sci.* 43:665–673.

Gilbert, W. 1986. The RNA world. *Nature* 319:618.

Gishlick, A. D. 2004. Evolutionary path to irreducible systems: The avian flight apparatus. In *Why Intelligent Design Fails?* Eds. M. Young, and T. Edis, pp. 58–71. New Brunswick, NJ: Rutgers University Press.

Haeckel, E. 1902. *The Riddle of the Universe*. J. McCabe, trans. New York: Harper & Brothers.

Haldane, J. B. S. 1967. The origin of life. In *The Origin of Life*, Ed. J. D. Bernal, pp. 242–249. Appendix II. London, UK: Weidenfeld & Nicolson. (Originally published in 1929 in Rationalist Annual, 3–10).

Hanczyc, M. M., S. M. Fujikawa, and J. W. Szostak. 2003. Experimental models of primitive cellular compartments: Encapsulation, growth, and division. *Science* 302:618–621.

Johnson, P. E. 1993. *Darwin on Trial*. Downers Grove, IL: InterVarsity Press.

Joyce, G. F., and L. E. Orgel. 2006. Progress toward understanding the origin of the RNA world. In *The RNA World*. 3rd ed., Eds. R. F. Gesteland, T. R. Cech, and J. F. Atkins, pp. 23–56. Cold Spring Harbor, NY: Cold Spring Harbor Laboratory.

Kamminga, H. 1988. Historical perspective: The problem of the origin of life in the context of developments in biology. *Orig. Life Evol. Biosph.* 18:1–11.

Kellog, V. L. 1907. *Darwinism Today*. New York: Henry Holt.

Kohn, D. 1989. Darwin's ambiguity: The secularization of biological meaning. *Br. J. Hist. Sci.* 22:215–239.

Koonin, E. V. 2007. An RNA-making reactor for the origin of life. *PNAS* 104(22):9105–9106.

Lancet, D., and B. Shenhav. 2009. Compositional lipid protocells: Reproduction without polynucleotides. In *Protocells: Bridging Nonliving and Living Matter*, Eds. S. Rasmussen, M. A. Bedau, L. Chen, et al., pp. 233–252. Cambridge, MA: The MIT Press.

Luisi, L. L. 2015. Chemistry constraints on the origin of life. *Isr. J. Chem.* 55:906–918.

Martin, W., and M. J. Russell. 2007. On the origin of biochemistry at an alkaline hydrothermal vent. *Philos. Trans. R. Soc. Lond. B Biol. Sci.* 362:1887–1925.

Maynard Smith, J., and E. Szathmáry. 1985. *The Major Transitions in Evolution*. Oxford, UK: W. H. Freeman.

Mayr, E. 1991. *One Long Argument*. Cambridge, MA: Harvard University Press.

McMullin, E. 1998. Evolution and special creation. In *The Philosophy of Biology*, Eds. D. L. Hull, and M. Ruse, pp. 698–733. Oxford, UK: Oxford University Press.

Meyer, S. C. 2004. DNA and the origin of life: information, specification, and explanation. In *Darwinism, Design, and Public Education*, Eds. J. A. Campbell, and S. C. Meyer, pp. 223–285. East Lansing, MI: Michigan State University Press.

Meyer, S. C. 2009. Signature in the cell: DNA and the evidence for intelligent design. New York: HarperONe.

Miller, K. R. 1999. *Finding Darwin's God*. New York: Cliff Street Books.

Miller, S. L. 1953. A production of amino acids under possible primitive Earth conditions. *Science* 117:528–529.

Miyakawa, S., and J. P. Ferris. 2003. Sequence- and regioselectivity in the montmorillonite-catalyzed synthesis of RNA. *J. Am. Chem. Soc.* 125:8202–8208.

Monnard, P-A. 2007. Question 5: Does the RNA-world still retain its appeal after 40 years of research? *Orig. Life Evol. Biosph.* 37:387–390.

Monod, J. 1974. *Chance and Necessity*. Glasgow, Scotland: Collins, Fontana Books.

Morowitz, H. J. 1992. *Beginnings of Cellular Life*. New Haven, CT: Yale University Press.

Morowitz, H. J. 1999. The theory of biochemical organization, metabolic pathways, and evolution. *Complexity* 4:39–53.

Muller, H. J. 1966. The gene material as the initiator and the organizing basis of life. *Am. Nat.* 100(915):493–517.

Musgrave, J. 2004. The evolution of the bacterial flagellum. In *Why Intelligent Design Fails?* Eds. M. Young, and T. Edis, pp. 72–84. Brunswick, NJ: Rutgers University Press.

Nagel, T. 2009. Thomas Nagel and Stephen Meyer's signature in the cell. *Times Lit. Suppl.* (TLS) (November 27, 2009). https://www.the-tls.co.uk/articles/public/thomas-nagel-and-stephen-c-meyers-signature-in-the-cell/ (accessed December 24, 2017).

Nagel, T. 2010a. Dawkins and atheism. In *Secular Philosophy and the religious Temperament*, Ed. T. Nagel, pp. 19–26. Oxford, UK: Oxford University Press. (Originally published in The New Republic, October 23, 2006).

Nagel, T. 2010b. Secular philosophy and the religious temperament. In *Secular Philosophy and the Religious Temperament*, Ed. T. Nagel, pp. 3–17. Oxford, UK: Oxford University Press.

Nagel, T. 2012a. *Mind & Cosmos: Why the Materialist Neo-Darwinian Conception of Nature is Almost Certainly False*. Oxford, UK: Oxford University Press.

Nagel, T. 2012b. A philosopher defends religion. N.Y. Rev. Books Nov. 8:62–63.

Oparin, A. I. 1967. The origin of life. A. Synge, trans. In *The Origin of Life*, Ed. J. D. Bernal, pp. 199–234. Appendix I. London, UK: Weidenfeld & Nicolson. (Originally published in Moscow in 1924).

Orgel, L. E. 1973. *The Origins of Life*. New York: Wiley.

Orgel, L. E. 1994. The origin of life on earth. *Sci. Am.* 271(4):53–61.

Orgel, L. E. 2003. Some consequences of the RNA world hypothesis. *Orig. Life Evol. Biosph.* 33:211–218.

Orgel, L. E. 2004. Prebiotic chemistry and the origin of the RNA world. *Crit. Rev. Biochem. Mol. Biol.* 39:99–123.

Pascal, R., A. Pross, and J. D. Sutherland. 2013. Towards an evolutionary theory of the origin of life based on kinetics and thermodynamics. *Open Biol.* 3:130156. doi:10.1098/rsob.130156.

Pennock, R. T. 2000. *The Tower of Babel*. Cambridge, MA: The MIT Press.

Peretó, J. 2012. Out of fuzzy chemistry: From prebiotic chemistry to metabolic networks. *Chem. Soc. Rev.* 41:5394–5403.

Peretó, J., J.L. Bada, and A. Lazcano. 2009. Charles Darwin and the origin of life. *Orig. Life Evol. Biosph.* 39:395–406.

Pizzarello, S., and E. Shock. 2010. The organic composition of carbonaceous meteorites: The evolutionary story ahead of biochemistry. *Cold Spring Harb. Perspect. Biol.* 2(3):a002105. doi:10.1101/cshperspect.a002105.

Plantinga, A. 2012. Why Darwinist materialism is wrong. *New Repub.* https://newrepublic.com/article/110189/why-darwinist-materialism-wrong (November 16) (Accessed, December 24, 2017).

Provine, W. 1988. Progress in evolution and the meaning of life. In *Evolutionary Progress*, Ed. M. Nitecki, pp. 49–74. Chicago, IL: The University of Chicago Press.

Ruse, M. 1996. Prologue: a philosopher day in court. In But is it science? Ed. M. Ruse, pp. 13–35. Amherst, NY: Prometheus Books.

Ruse, M. 2001. *Can a Darwinian be a Christian?* Cambridge, UK: Cambridge University Press.

Ruse, M. 2006. *Darwinism and its Discontents*. Cambridge, UK: Cambridge University Press.

Russell, M. J., L. M. Barge, R. Bhartia et al. 2014. The drive to life on wet and icy worlds. *Astrobiology* 14(4):308–343.

Safhill, R., H. Schneider-Bernloehr, L. E. Orgel, and S. Spiegelman. 1970. *In vitro* selection of bacteriophage Qβ variants resistant to ethidium bromide. *J. Mol. Biol.* 51:531.

Schrum, J. P., T. F. Zhu, and J. W. Szostak. 2010. The origins of cellular life. *Cold Spring Harb. Perspect. Biol.* doi:10.1101/cshperspect.a002212.

Scott, E. C. 1996. Creationism, ideology and science. *Ann. N. Y. Acad. Sci.* 775:505–522.

Searle, J. R. 1994. *The Rediscovery of the Mind*. Cambridge, MA: The MIT Press.

Segré, D., D. Ben Eli, D. Deamer, and D. Lancet. 2001. The lipid world. *Orig. Life Evol. Biosph.* 31:119–145.

Service, R. F. 2013. The life force. *Science* 342:1032–1034.

Shapiro, R. 2007. A simpler origin of life. *Sci. Am.* 296(6):47–53.

Smith, E., H. J. Morowitz, and S. D. Copley. 2009. Core metabolism as a self-organizing system. In *Protocells: Bridging Nonliving and Living Matter*, Ed. S. Rasmussen, M. A. Bedau, L. Chen et al., pp. 433–460. Cambridge, MA: The MIT Press.

Szostak, J. W. 2011. An optimal degree of physical and chemical heterogeneity for the origin of life? *Philos. Trans. R. Soc. Lond. B Biol. Sci.* 366:2894–2901.

Szostak, J. W. 2012a. Attempts to define life do not help to understand the origin of life. *J. Biomol. Struct. Dyn.* 29:599–600.

Szostak, J. W. 2012b. The eightfold path to non-enzymatic RNA replication. *J. Syst. Chem.* 3:2. http://www.systchem.com/contents/3/1/2.

Szostak, J. W., D. P. Bartel, and P. L. Luisi. 2001. Synthesizing life. *Nature* 409:387–390.

Vasas, V., E. Szathmáry, and M. Santos. 2010. Lack of evolvability in self-sustaining autocatalytic networks constraints metabolism-first scenarios for the origin of life. *PNAS* 107(4):1470–1475.

Vasas, V., C. Fernando, M. Santos, S. Kauffman, and E. Szathmáry. 2012. Evolution before genes. *Biol. Direct* 7:1. doi:10.1186/1745-6150-7-1.

Wächtershäuser, G. 1992. Groundwork for an evolutionary biochemistry: The iron-sulfur world. *Prog. Biophys. Mol. Biol.* 58:85–201.

Wächtershäuser, G. 2007. On the chemistry and evolution of the pioneer organism. *Chem. Biodiv.* 4:584–602.

White, R. 2007. Does origin-of-life research rest on a mistake? *Nous* 41(3):453–477.

Wieczorek, R., M. Dçrr, A. Chotera, P. L. Luisi, and P.-A. Monnard. 2013. Formation of RNA phosphodiester bond by histidine-containing dipeptides. *Chembiochem* 14:217–223.

Zhu, T. F., and J. W. Szostak. 2009. Coupled growth and division of model protocell membranes. *J. Am. Chem. Soc.* 131:5705–5713.

3.2 Charles Darwin and the Plurality of Worlds

Are We Alone?

Michael Ruse

CONTENTS

No sooner had Mr. Darwin peered through the microscope on one of the finest specimens when he started up from his seat and exclaimed:

"Almighty God! what a wonderful discovery! Wonderful!"

And after a pause of silent reflection he added:

"Now reaches life down!" (Rachel 1881)

3.2.1 THE PLURALITY OF WORLDS

Charles Darwin's *On the Origin of Species by Natural Selection, or the Preservation of Favoured Races in the Struggle for Life* was published in 1859. It came at the end of a decade when the possibility of extraterrestrials – beings on other planets, especially intelligent beings – had been discussed very extensively, in both Britain and America. In 1853, publishing anonymously, William Whewell – tidologist, textbook writer, historian, and philosopher of science and moral theorist – had produced *On the Plurality of Worlds*, a work in which he argued that our world alone – Planet Earth – has sentient beings. The rest of space is empty and lifeless (Whewell [1853] 2001).

Everyone knew the identity of the author, but Whewell then was Master of Trinity College, Cambridge, and was shortly to become the tutor of the Prince of Wales, the future Edward the Seventh (Todhunter 1876). He thought it would be unseemly to be too publicly identified with a controversy on this topic, and in assuming that there would be a controversy, he was right, although whether many thought the less of him for it is a moot point. He certainly stirred up a lively debate, and those who did think the less of him undoubtedly did so already. These included Sir David Brewster, general Scottish man of science and biographer of Isaac Newton. He had already taken umbrage at what he considered Whewell's rather cavalier attitude to Scottish culture, a topic on which he tended to take umbrage. His response to Whewell (under his own name!) had the glorious title *More Worlds than One, the Creed of the Philosopher and the Hope of the Christian*. "Philosopher" here meant scientist, as in "natural philosopher" rather than "moral philosopher," which was more Whewell's field (Brewster 1854).

What Brewster's title tells us is that this controversy was as much theological as it was scientific or (what we would call) philosophical. Whewell had not put pen to paper because he was worried about origin of life questions or that sort of thing. Did all life forms have the same pattern kind of question? Whewell was no evolutionist and – for all that, when a young man, Darwin had been mentored by Whewell and so naturally later was to send his old teacher a copy of the *Origin* – remained a non-believer on this subject. So, particularly, Whewell was not asking whether in development organisms always took the same path. What worried him was revealed theology – religion through faith – and the possibility that there might be intelligent organisms other than those down here on Earth. If there were, did they too have their Adams and Eves? Did their first people succumb to temptation and fall? Are these beings tainted therefore by original sin? And, worst of all, did this mean that Jesus had to come down on their planets and die over and over again on the Cross? Even as we are now, is Jesus somewhere in the universe being nailed up?

The problem Whewell faced was that if the universe other than our planet is empty, then why on earth did God create it.

We need the Sun and perhaps the Moon, but not much else. This goes flatly against the main premise of natural theology – religion through reason – that the world shows such evidence of design that it points unhesitatingly to a Designer, namely God (Ruse 2017b). It is true that usually the big evidence of design is to be found in organisms – teeth for chewing, eyes for seeing, flowers for pollination – but Whewell, of all people, 20 years earlier had contributed to a series on natural theology, the *Bridgewater Treatises*, and his contribution had been about astronomy. He had become a public figure based on his lively arguments about the ubiquity of design, in all things! Now, he had to backtrack or at least find some way of wriggling out of the dilemma (Whewell 1833).

This Whewell did, and it really took some gall, by arguing that straightforward design is not so ubiquitous anyway. He drew on work by anatomists, particularly his good friend at the Royal College of Surgeons, Richard Owen (Owen 1848, 1849), who was showing that isomorphisms – what Owen was to call "homologies" – are widespread in the animal kingdom but have no obvious functional value. Particularly popular were the analogies between the bone structures of the forelimb of the human, the horse, the bat, the bird, the mole, and the porpoise – different functions, often very different functions, but the same bones in the same order. Of course, the paradox is that Whewell was jumping from the teleological fat and into the evolutionary fire. Homologies are the best of all possible evidence for evolution. The aged Immanuel Kant worried about them in the *Third Critique* (Kant [1790] 1928), and Darwin was to make much of them in the *Origin*. Probably, Owen had already been converted to evolution on the basis of them. He sensibly kept quiet, because he did not want to upset his non-evolutionary sponsors like Whewell.

To defend his revealed theology, Whewell was turning his natural theology – he claimed that homologies are evidence of a Creator, who likes symmetry and order (Shades of Plato's *Phaedo*) – into good support for evolution. Brewster spotted this and, in response, peopled the whole universe. He even had intelligent beings living on the Sun, which is ridiculous, but which points, as does Whewell's thinking, to an important point about science. In his *The Structure of Scientific Revolutions* (1962), Thomas Kuhn argues that one of the major reasons for a change of paradigms, a revolution, is that the old paradigm starts to break down (Kuhn 1962). It reveals inner contradictions and the like. Whewell and Brewster illustrate this perfectly. In their entanglements, caused by their desire to maintain the old, anti-evolutionary harmony between revealed and natural theology, they put the whole system under great strain. Let no one deny the honors to Darwin, but the ground for the new was being prepared by the crumbling of the old.

3.2.2 DARWINIAN THEORY

The *Origin of Species* is deceptive. For a work of science, it is written in a highly user-friendly fashion. This is not Newton's *Principia*. But it is a very carefully constructed piece of work, one that shows a sophisticated appreciation of the dicta of the leading methodologists of the day, the physicist John F. W. Herschel and the polymath William Whewell (Ruse 1975, 1979). Darwin's problem, obviously, was to convince the reader of events that happened in the past, as hidden from us today as are the waves that supposedly lie behind visible light, as it bends its way through prisms and splits into many colors when coxed in the right way. A pertinent analogy, because just when Darwin was becoming an evolutionist and seeking a cause, in the 1830s, the wave theory of light (of the Dutchman Christian Huygens) was just then conquering the particle theory, due to Newton and popular for over a hundred years.

The quest was for what Newton had called a "true cause," a *vera causa*. Herschel (1830), somewhat of an empiricist, said that what you need is an analogy, from what you know to what you don't know but presume to exist or to be true. In the case of light, we have physical waves as in water – the waves you make in the bathtub – and most obviously sound, and Herschel argued that this prepares the way to accept light waves. Whewell (1840), somewhat of a rationalist, said that no direct evidence was needed for a true cause. We work rather as a detective looking for clues that point to the culprit. If all the clues point the same way, then we can be pretty sure that we know who did it. Whewell called this a "consilience of inductions."

Darwin covered his options, offering evidence in both modes that his cause, natural selection, was a *vera causa*. He opened the *Origin* by arguing that the work of animal and plant breeders in making new forms is analogous to what goes on in nature.

That most skilful breeder habitually speak of an animal's organization as something quite plastic, which they can model almost as they please. If I had space, I could quote numerous passages to this effect from highly competent authorities. Youatt, who was probably better acquainted with the works of agriculturalists than almost any other individual and who was himself a very good judge of an animal, speaks of the principle of selection as "that which enables the agriculturist, not only to modify the character of his flock, but to change it altogether. It is the magician's wand, by means of which he may summon into life whatever form and mould he pleases." Lord Somerville, speaking of what breeders have done for sheep, says "It would seem as if they had chalked out upon a wall a form perfect in itself, and then had given it existence." Most skillful breeder, Sir John Sebright, used to say, with respect to pigeons, that "he would produce any given feather in three years, but it would take him six years to obtain head and beak." In Saxony, the importance of the principle of selection in regard to merino sheep is so fully recognized that men follow it as a trade: the sheep are placed on a table and are studied, like a picture by a connoisseur; this is done three times at intervals of months, and each time, the sheep are marked and classed, so that the very best may ultimately be selected for breeding (Darwin 1859, p. 31).

Darwin made much of the fact that pigeons have such variety, yet clearly, all come from common stock.

Then, having postulated that new variations are constantly appearing in populations – not uncaused but undirected – he was ready for his key inferences. These were put in (quasi- or proto-) deductive form. As Herschel and Whewell pointed out non-stop, this was the form of the gravitational theory of Newton. First was the Malthusian element to a struggle for existence.

A struggle for existence inevitably follows from the high rate at which all organic beings tend to increase. Every being, which, during its natural lifetime, produces several eggs or seeds, must suffer destruction during some period of its life and during some season or occasional year; otherwise, on the principle of geometrical increase, its numbers would quickly become so inordinately great that no country could support the product. Hence, as more individuals are produced than can possibly survive, there must, in every case, be a struggle for existence, either one individual with another of the same species or with the individuals of distinct species or with the physical conditions of life (Darwin 1859, pp. 63–64).

Then, second, to natural selection.

Let it be borne in mind how infinitely complex and close-fitting are the mutual relations of all organic beings to each other and to their physical conditions of life. Can it, then, be thought improbable, seeing that variations useful to man have undoubtedly occurred, that other variations useful in some way to each being in the great and complex battle of life should sometimes occur in the course of thousands of generations? If such do occur, can we doubt (remembering that many more individuals are born than can possibly survive) that individuals having any advantage, however slight, over others would have the best chance of surviving and of procreating their kind? On the other hand, we may feel sure that any variation in the least degree injurious would be rigidly destroyed. I call this preservation of favorable variations and the rejection of injurious variations as Natural Selection (pp. 80–81).

All-important is the fact that this cause – natural selection – points to adaptation, to design-like features. The eye is created as it is in order to see, the flower to attract pollinators, the fangs of the snake to kill, and the instincts of the nest-building bird to promote and continue life. It is not just change, but change of a particular cause.

Under nature, the slightest difference of structure or constitution may well turn the nicely balanced scale in the struggle for life and so be preserved. How fleeting are the wishes and efforts of man! how short his time! And, consequently, how poor will his products be, compared with those accumulated by nature during whole geological periods? Can we wonder, then, that nature's productions should be far "truer" in character than man's productions and that they should be infinitely better adapted to the most complex conditions of life and should plainly bear the stamp of far higher workmanship? (pp. 83–84).

Having shown that this should all lead to the venerable tree of life, Darwin was ready for the third and final part of the *Origin*, although in length well over half. Darwin showed the

causal power of natural selection by going in a Whewellian consilience right through the life sciences: social behavior, the fossil record (paleontology), geographical distributions (biogeography), anatomy and morphology, systematics, and embryology. Why is there a progressive fossil record? Evolution through natural selection. Why do organisms fit the Linnaean system? Because it reflects the tree of life. Picking up on homology, why are there isomorphisms between the bones of different species? Evolution through natural selection. *Déjà vu*, all over again. Why are there such similarities between the organisms of very different adult forms? Because selection only works on the adults. This all done, Darwin was ready for his final famous passage.

It is interesting to contemplate an entangled bank, clothed with many plants of many kinds, with birds singing on the bushes, with various insects flitting about, and with worms crawling through the damp earth, and to reflect that these elaborately constructed forms, so different from each other and dependent on each other in so complex a manner, have all been produced by laws acting around us… Thus, from the war of nature, from famine and death, the most exalted object that we are capable of conceiving, namely, the production of the higher animals, directly follows. There is grandeur in this view of life, with its several powers, having been originally breathed into a few forms or into one, and that, whilst this planet has gone cycling on according to the fixed law of gravity, from so simple a beginning endless forms most beautiful and most wonderful have been, and are being, evolved (pp. 489–490).

3.2.3 RECEPTION

For completeness, I will talk about the reception of Darwin's ideas, even though I don't intend to spend much time on the topic. The received position is that, despite some well-known opposition – notoriously the Bishop of Oxford, Soapy Sam Wilberforce (son of William Wilberforce of slave trade abolition fame) – evolution as such was accepted quickly in both Britain and North America, the evangelicals of the American South – using the bible read literally as consolation for the defeat in the Civil War – being the notable exception. (Anything to separate them from the Yankees.) However, natural selection was far less successful and only came into its own in the 1930s, when Mendelian genetics had been developed and could be shown to be the complement of natural selection, not its rival.

This story is only partly true (Richards and Ruse 2016; Ruse 2017a). In the world of fast-breeding organisms, insects, there was much profitable theorizing and empirical study using selection. Particularly successful was the work showing how selection was behind industrial melanism, where insects (predated on by birds) develop adaptive coloring that tracks the coloring of the trees on which they perch – as the trees get darker, thanks to industrial soot, the insects (butterflies and moths) get darker too. One enthusiast even wrote to Darwin about this.

My dear Sir,

The belief that I am about to relate something which may be of interest to you, must be my excuse for troubling you with a letter.

Perhaps among the whole of the British Lepidoptera, no species varies more, according to the locality in which it is found, than does that Geometer, Gnophos obscurata. They are almost black on the New Forest peat; grey on limestone; almost white on the chalk near Lewes; and brown on clay, and on the red soil of Herefordshire.

Do these variations point to the "survival of the fittest"? I think so. It was, therefore, with some surprise that I took specimens as dark as any of those in the New Forest on a chalk slope; and I have pondered for a solution. Can this be it?

It is a curious fact, in connexion with these dark specimens, that for the last quarter of a century the chalk slope, on which they occur, has been swept by volumes of black smoke from some lime-kilns situated at the bottom: the herbage, although growing luxuriantly, is blackened by it.

I am told, too, that the very light specimens are now much less common at Lewes than formerly, and that, for some few years, lime-kilns have been in use there.

These are the facts I desire to bring to your notice.

I am, Dear Sir, Yours very faithfully,
A. B. Farn

Letter from Albert Brydges Farn on November 18, 1878 (Darwin Correspondence Project, 11747).

The other place fascinatingly where natural selection made a splash – and a secondary mechanism that Darwin introduced of competition for mates, sexual selection, even more – was in the realm of poetry and fiction (Ruse 2017a). After all, what is this most often but stories of elaborate mating dances, which may or may not be successful? This poem is by a young woman, in the 1880s. It is entitled "Natural Selection" although truly, it is more about sexual selection.

I HAD found out a gift for my fair,
I had found where the cave men were laid:
Skulls, femur and pelvis were there,
And spears that of silex they made.
But he ne'er could be true, she averred,
Who would dig up an ancestor's grave—
And I loved her the more when I heard
Such foolish regard for the cave.
My shelves they are furnished with stones,
All sorted and labelled with care;
And a splendid collection of bones,
Each one of them ancient and rare;
One would think she might like to retire

To my study—she calls it a "hole"!
Not a fossil I heard her admire
But I begged it, or borrowed, or stole.
But there comes an idealess lad,
With a strut and a stare and a smirk;
And I watch, scientific, though sad,
The Law of Selection at work.
Of Science he had not a trace,
He seeks not the How and the Why,
But he sings with an amateur's grace,
And he dances much better than I.
And we know the more dandified males
By dance and by song win their wives—
'Tis a law that with *avis* prevails,
And ever in *Homo* survives.
Shall I rage as they whirl in the valse?
Shall I sneer as they carol and coo?
Ah no! for since Chloe is false
I'm certain that Darwin is true.
(Naden 1999, pp. 207–208)

All a good joke, but serious for all that. (I note that this is by a woman. In the first hundred years of professional evolutionary biology after Darwin, I do not think I have ever encountered a female author. In the realm of creative writing, at least half of the authors writing on evolution and using ideas of selection are females – George Eliot, Emily Dickinson, Mrs. Humphrey Ward, Edith Wharton, Edna St. Vincent Millay – the list is nigh endless.)

3.2.4 WHAT DARWIN LEFT OUT AND WHY

So, go back to Darwin and the *Origin* and focus on our interests. Two things are to be noted, again not by chance but by very careful design. They are both related to the fact that Darwin was a very skilled methodologist and equally to the fact that Darwin appreciated that sometimes saying nothing is a better strategy than plunging in half-baked, as it were. First, humans. Darwin knew that, as soon as he published, his theory would be considered in the light of possible human evolution. And so, it proved. At once, it was known as the monkey theory or the gorilla theory, and people started writing on this. In 1863, Darwin's "bulldog," Thomas Henry Huxley, published *Man's Place in Nature*, locating us firmly in the animal world (Huxley 1863). Darwin himself was essentially silent, but he never had any doubts that we are part of the animal world. The first evidence, in late 1838, that Darwin had hit on natural selection is a comment in a private notebook applying natural selection to humans, and not just to our bodies but also to our mental abilities. In his *Autobiography*, written toward the end of his life, Darwin reaffirmed this belief. "I am inclined to agree with Francis Galton [Darwin's half cousin] in believing that education and environment produce only a small effect on the mind of any one, and that most of our qualities are innate" (Darwin 1958, p. 43). From the first to the last, Darwin was (in today's terms) a human sociobiologist or evolutionary psychologist. In the *Origin*, however, Darwin wanted to get his main theory into the public domain. So, not to appear cowardly, he simply said "light will be thrown on man and his history" (p. 488).

If he had been left to his own devices, Darwin may never have written on humans – it was not to be. Darwin was spurred into publishing the *Origin* by the arrival in 1858 of an essay by a young naturalist, Alfred Russel Wallace – an essay that contained just about the same ideas that Darwin had been hugging privately for 20 years (Wallace 1858). In the 1860s, Wallace turned to spiritualism, arguing that only supernatural forces could explain human evolution (Wallace 1870). Horrified, Darwin again put pen to paper, and in 1871, he published *The Descent of Man and Selection in Relation to Sex*. Much of the early part of the book is predictable, as Darwin argued that his theory applies to us. What makes the *Descent* rather odd is that most of it is not about humans at all! It is a very detailed and lengthy discussion of the second mechanism of "sexual selection," where the competition is within a species for mates – as the peacock having a ludicrously large and flamboyant tail to attract the peahen. The reason for the imbalance is simple. Wallace argued that certain human features, such as hairlessness, could not have been produced by natural selection and hence demanded spirit forces. Darwin agreed that natural selection could not do the job but argued that sexual selection could! Hairlessness was all a matter of people choosing those mates they found most attractive, all of which led to some very Victorian sentiments.

Man is more courageous, pugnacious, and energetic than woman and has a more inventive genius. His brain is absolutely larger, but whether relatively to the larger size of his body, in comparison with that of woman, has not, I believe, been fully ascertained.

Continuing:

Male and female children resemble each other closely, like the young of so many other animals in which the adult sexes differ; they likewise resemble the mature female much more closely than the mature male. The female, however, ultimately assumes certain distinctive characters and, in the formation of her skull, is said to be intermediate between the child and the man (Darwin 1871, pp. 2, 317).

Darwin was a great revolutionary. He was no rebel against the mores and norms of Victorian society!

The second topic that Darwin eschewed in the *Origin* was the origin of life problem. It was not customary to avoid the topic in evolutionary discussions. Darwin's grandfather Erasmus Darwin (1801) had been an evolutionist at the end of the eighteenth century, and he had talked of it, and so too did the great French evolutionist Jean Baptiste de Lamarck. In his *Philosophie Zoologique* of 1809, the year of Charles Darwin's birth, he speculated that electricity (lightening) on ponds of chemically saturated water would start off life. Then, in 1844, the publisher Robert Chambers, writing anonymously, penned *The Vestiges of the Natural History of Creation* (Chambers, 1844). He argued for evolution, thus incurring the wrath of the respectable scientific community, including Whewell (1845). This, no doubt, confirmed to Darwin the wisdom of keeping quiet about his speculations. There was much on the origin of life, spontaneous generation, involving mites on electrical

batteries, and such things. Bringing the scorn of the professional scientists, who knew only too well how dirty fingers can contaminate things. If this were not enough, by the end of the 1850s, in France, Louis Pasteur was showing how most, if not all, claims about natural origins of life were unsubstantiated. (Contrary to popular opinion, Pasteur believed in natural origins – just that they had not yet been proven.)

Darwin wisely stayed away from the topic, speaking only of "this view of life, with its several powers, having been originally breathed into a few forms or into one." Take that as you will. He made a smart move, because it was not a big charge brought against his theory. This didn't mean that Darwin had no ideas on the topic. Later, to his friend Joseph Hooker, he speculated in a very Lamarckian fashion.

It is often said that all the conditions for the first production of a living organism are now present, which could ever have been present.—But if (& oh what a big if) we could conceive in some warm little pond with all sorts of ammonia & phosphoric salts,—light, heat, electricity &c present, that a protein compound was chemically formed, ready to undergo still more complex changes, at the present day such matter w^d be instantly devoured, or absorbed, which would not have been the case before living creatures were formed.— (Letter from Darwin to Hooker, February 1, 1871, Darwin Correspondence, DCP-LETT-7471)

I don't think Darwin had any objections to life elsewhere in the universe. After all, why would he? It would be grist for his mill. But judging from his writings, public and private, not to mention his correspondence, he seems to have stayed away from the subject. This makes sense. We know from a private reading list that, in 1854, Darwin read Whewell's *Plurality of Worlds*. But, apart from the natural desire to get distance from Whewell, given the latter's strident anti-evolutionism, Darwin would have seen how greatly the whole topic is plunged into theological controversy, natural and revealed religion. Apart from his lack of personal interest in such topics – he protested truly that although circumstances forced him to think much about religion, it was not really a topic that excited him (unlike, say, Thomas Henry Huxley) – he would see how fatal it would be to the presentation of his theory if it were draped in sheets of religious controversy. A better metaphor would be strangled by the coils of the aggressive religious anaconda. As a kind of epilogue to Darwin's life's work, the quotation in the story at the beginning of this essay shows that almost at the end of his life (Darwin died in the spring of 1882), he was shown evidence that convinced him that probably this Earth of ours was seeded by life from elsewhere. He reacted strongly and favorably to this information, although to be honest, Darwin was rather given to reacting strongly and favorably to ideas from people whom he liked.

3.2.5 ALIENS OF OUTER SPACE

Turn now from Darwin and broaden the discussion. Even if Darwin didn't have a lot to say, others did and embraced with enthusiasm aspects of Darwin's thinking, using them in fact and fiction – almost always fiction, even when presented as fact – to speculate on life elsewhere. One thinks, for instance, of the most famous case of all, H. G. Wells' story *The War of the Worlds*, with those pesky Martians coming in to do a job on us – at least, until they got wiped out by our diseases. I discern something of a common pattern in fiction about extraterrestrials. There will be evolution or development of mobile, talking animals. Take the *Star Wars* movies. You get all kinds of non-humans, such as the Wookie Chewbacca, Jabba the Hutt, and my all-time favorite, the truly dreadful little Kowakian monkey-lizard, Salacious B. Crumb. Anyone who has had Cairn Terriers knows the type and the reason why, for all that they are absolutely appalling in behavior, there is something about them that makes them deeply loved. Often, alien beings are really intelligent, like Chewbacca and Jabba. They can be much brighter than us. Yoda, the teacher of the Jedi, is an example.

In such fiction, humans also have a way of making an appearance. Sometimes, there is no real question about their authenticity. Are they real humans? The humans in *Planet of the Apes* are all in or descendants from our species today. Technically, whether all such human-like beings are truly in our species is a happily discussed question. With respect to *Star Wars*, some say sternly that, given that *Star Wars* takes place "a long, long time ago" in a "galaxy far, far away," they can't be human. However, common sense and the droids think otherwise. In *The Empire Strikes Back*, C-3PO says that Luke is "quite clever, you know ... for a human being." Since C-3PO is fluent in 7 million languages, one presumes that he knows.

The question at issue is whether, in the world of Charles Darwin, these various life forms are even scientifically possible. First up, obviously, is whether on all and any planets or planet substitutes, natural selection is going to be operating and the main cause of any kind of change. Richard Dawkins (1983) has discussed this very problem. He starts with the ubiquitous design-like nature of organisms and argues that any cause must be able to produce such a nature. The only two competitors here are Darwinian selection – the blacksmith's arms are strong because, in early times, strong laborers had more kids than weak laborers – and Lamarckian inheritance of acquired characteristics – the blacksmith's arms are strong because his dad worked hard at the forge and built up strong muscles, which he then passed on. We know that Lamarckism is false, so that just leaves selection. All other competitors for the job simply leave the design problem unsolved. To suppose saltations – jumps from one form to another – is to drop design. Jumps, what we today would call macromutations, almost always mess things up. Being albino, for instance, can be a terrible life choice, because you then have absolutely no adaptive camouflage. You stand out, and the predators can spot you at once. Dawkins concludes that where you get life, you get selection. The ancestors of Salacious Crumb really did outreproduce their (no doubt, warm and friendly, and boring) rivals.

Would selection produce consciousness and eventually intelligence? Could selection produce humans? This seems to imply that some kind of force is operative for progress, driving organisms up from (as was said) the monad to the man. Practically, all evolutionists, at least up to Charles Darwin,

assumed without question that evolution is progressive. Thus, Erasmus Darwin, his grandfather:

> Organic Life beneath the shoreless waves
> Was born and nurs'd in Ocean's pearly caves;
> First forms minute, unseen by spheric glass,
> Move on the mud, or pierce the watery mass;
> These, as successive generations bloom,
> New powers acquire, and larger limbs assume;
> Whence countless groups of vegetation spring,
> And breathing realms of fin, and feet, and wing.
> Thus the tall Oak, the giant of the wood,
> Which bears Britannia's thunders on the flood;
> The Whale, unmeasured monster of the main,
> The lordly Lion, monarch of the plain,
> The Eagle soaring in the realms of air,
> Whose eye undazzled drinks the solar glare,
> Imperious man, who rules the bestial crowd,
> Of language, reason, and reflection proud,
> With brow erect who scorns this earthy sod,
> And styles himself the image of his God;
> Arose from rudiments of form and sense,
> An embryo point, or microscopic ens!
> (Darwin 1803, pp. 1, 11, 295–314)

Notions of biological progress, running up from the blob to the human, make the very backbone (to use an apt metaphor) of this vision, shown as Darwin explicitly tied his biology into his philosophy. The idea of organic progressive evolution "is analogous to the improving excellence observable in every part of the creation; such as the progressive increase of the wisdom and happiness of its inhabitants" (Darwin 1794–1796, pp. 2, 247–242).

At the time of the *Origin*, the man who in respects was an even-better-known English evolutionist than Charles Darwin was the wildly enthusiastic Herbert Spencer.

> Now, we propose in the first place to show that this law of organic progress is the law of all progress. Whether it be in the development of the Earth, in the development of Life upon its surface, in the development of Society, of Government, of Manufactures, of Commerce, of Language, of Literature, of Science, or of Art, this same evolution of the simple into the complex, through successive differentiations, holds throughout. From the earliest traceable cosmical changes down to the latest results of civilization, we shall find that the transformation of the homogeneous into the heterogeneous is that in which Progress essentially consists (Spencer 1857, pp. 2–3).

Always a bit iffy about selection – for all that he discovered a decade before the *Origin* was published – Spencer simply assumed that evolution progresses upward. In this, he was like Lamarck and Erasmus Darwin. One suspects that in Spencer's case, he was strongly influenced by the German Romantics – especially Friedrich Schelling – who saw history progressively, for Hegel particularly to the German state (Ruse 2013).

Of course, the English knew that it could not be so, but they were receptive to ideas of world spirits or forces acting for the good. Thus, William Wordsworth in *Tintern Abbey*:

And I have felt
A presence that disturbs me with the joy
Of elevated thoughts; a sense sublime
Of something far more deeply interfused,
Whose dwelling is the light of setting suns,
And the round ocean, and the living air,
And the blue sky, and in the mind of man,
A motion and a spirit, that impels
All thinking things, all objects of all thought,
And rolls through all things.
(Wordsworth, "Tintern Abbey," 1798)

3.2.6 CAN NATURAL SELECTION BE A FORCE FOR PROGRESS?

Now for the sixty-four-thousand-dollar question, or questions: Is natural selection a force that promotes progress – monad to man – or, at least, is it a force that can, in the right circumstances, make for progress? The answer is "not obviously." On the one hand, the new variations coming into populations, the building blocks of evolution – what today we would call "mutations" – are random. Not random in the sense of uncaused – no one from Darwin on thought that, and most put it down to things like disturbances in reproduction and such things – but random in the sense of not occurring according to need. The climate turns dark, as in industrial pollution. There is no guarantee that the next variation will make you dark. As like as not, it might make you pink. (Today's evolutionists speak to this issue, arguing, in the tradition of eminent twentieth-century evolutionist Theodosius Dobzhansky (1951), that selection can maintain variation in a population. If a new selective need occurs, you are not waiting on new variations. Already, you have a library at your disposal – with luck, a mutation for darkness or perhaps for making yourself very unpleasant to taste, so birds will avoid young.) On the other hand, natural selection is opportunistic. It doesn't care about advance. It cares about surviving and reproducing. Paleontologist Jack Sepkoski puts the point colorfully. "I see intelligence as just one of a variety of adaptations among tetrapods for survival. Running fast in a herd while being as dumb as shit, I think, is a very good adaptation for survival" (Ruse 1996, p. 486). Cow power rules supreme!

This said, straight off after the *Origin* appeared, Darwinians (starting with Darwin) went with the tide. No self-respecting Victorian was going to take seriously the idea that we humans – we Europeans – are still down at the warthog level or below. Darwinians built in the thoroughly value-impregnated belief that evolution – evolution through natural selection – leads progressively up to humankind. This was Darwin himself, 20 years before the *Origin*, just before he discovered natural selection.

We see gradation to mans mind in Vertebrate Kindgdom in more instincts in rodents than in other animals & again in mans mind, in different races. being unequally developed. ? is not Elephant intellectually developed amongst Pachydermata like man amongst Monkeys or dogs in Carnivora.—Man in his arrogance thinks himself a great work worthy the interposition of a deity, more humble & I believe truer to consider him created from animals (Barrett et al., 1987, C 196).

There is progress to humans, if not quite as specially as many presume. Notice how Darwin put things in an evolutionary context and described his position not as superior to another evolutionary position, Lamarck, for instance, but to the Christian position. This is a man who thought in a theological mode. After the discovery of selection, there is still progress albeit with an increasing recognition that selection changes the rules of the game. You could not now get a teleological upward direction à la Spencer, even if you wanted it.

The enormous <u>number</u> of animals in the world depends of their varied structure & complexity. —hence as the forms became complicated, they opened fresh means of adding to their complexity. —but yet there is no necessary tendency in the simple animals to become complicated although all perhaps will have done so from the new relations caused by the advancing complexity of others. —It may be said, why should there not be at any time as many species tending to dis-developement (some probably always have done so, as the simplest fish), my answer is because, if we begin with the simplest forms & suppose them to have changed, their very changes ton tend to give rise to others (E 95-7).

The *Origin* does not discuss humans in detail, but it still makes clear Darwin's belief in progress: "The inhabitants of each successive period in the world's history have beaten their predecessors in the race for life, and are, in so far, higher in the scale of nature; and this may account for that vague yet ill-defined sentiment, felt by many palæontologists, that organisation on the whole has progressed" (Darwin 1859, p. 345). The closing passage to the book repeats that sentiment. The question is how selection brings this about. Through competition, obviously, but precisely how? By the third edition of the *Origin*, 1861, Darwin was confident that he had the answer. He invoked what today's Darwinians call "arms races" – lines compete against each other, and one line gets better and better. Eventually, this comparative improvement is translated into some form of absolute improvement, *Homo sapiens*.

If we look at the differentiation and specialization of the several organs of each being, when adult (and this will include the advancement of the brain for intellectual purposes), as the best standard of highness of organization, natural selection clearly leads toward highness, for all physiologists admit that the specialization of organs, inasmuch as they perform in this state their functions better, is an advantage to each being, and hence, the accumulation of variations tending toward specialization is within the scope of natural selection (Darwin 1861, p. 134).

Natural selection cannot guarantee anything, but everything is probably going to be just fine. The great success of capitalism and British industry leads the way.

3.2.7   BUT CAN SELECTION PRODUCE ALIEN HUMANS? NO!

It seems that natural selection can and does produce intelligent beings all the way up to humans. I confess that even if this can happen, I would think selection would more likely produce humanoids – beings like humans but not necessarily identical to us. There might be at least as many Wookies in the universe as there are humans (or human-identical beings). Perhaps, Jabba the Hutt and his conspecifics outnumber humans ten to one! There is incidentally an interesting point of theology here. Humans are made in the image of God. Whatever else in the Judeo-Christian story, humans are special. We are not warthogs. It seems to me that humans are necessary to the story (Ruse 2001, 2015). They had to appear. But would a Wookie be adequate? Could Jesus come to Earth as a Wookie? One rather doubts he could come as a conspecific of Jabba the Hutt, although we would certainly satisfy the qualification of original sin. I suppose that if you insist that humans must be humans – perhaps you would allow a greenish tint to the skin – then if you are into multiverses, given enough time anything will happen, and so you will eventually get humans. Perhaps, an infinite number of times, although this then does give rise to the worry of Whewell that perhaps Jesus will be overworked. It does also give rise to the thought of billions and billions of organisms that didn't quite make it. Multiverses filled with human-like failures. Brave New Worlds without the alphas. Is this truly possible given an all-powerful, all-loving God? I guess, since He let the Holocaust occur, it might well be.

Darwin thought that selection could produce humans. How often this would occur is unanswered and perhaps unanswerable, but once it could and did occur. Not all Darwinians in the next hundred years or so shared this optimism. Alfred Russel Wallace is perhaps an atypical exception, because he thought that humans are produced by spirit forces. He wrote about the possibility of life elsewhere, arguing adamantly that selection could not do the job.

> Now, the numerous conditions and sub-conditions essential to the development of the higher organic life, which must all exist simultaneously and which must all have continued to exist for enormous periods of time, are improbable in various degrees, as clearly shown by the fact that the great majority of them are not present in the degree required in any other planet of the solar system. The actual degree of improbability of each of these conditions cannot be determined except vaguely, but anyone who will carefully consider and weigh my very imperfect exposition of them will, I think, admit that they are usually very considerable, the chances against each of them being in some cases, perhaps, ten to one and in others being ten thousand to one or even more. But if we take the whole of the simultaneous conditions requisite as only fifty in number, and the chances against each occurring simultaneously with the rest to be only ten to one, an estimate that seems to me absurdly low because many of them are quantitative within narrow limits, then the chances against the simultaneous occurrence of the whole fifty would be a million raised to the eighth power (1,000,000⁸) or a

million multiplied by a million eight times successively to 1. These figures are suggested merely to give some indication to the general reader of the way in which the chances against any event happening more than once mount up to unimaginable numbers when the event is a highly complex one (Wallace 1903, p. 336).

This set a pattern. Darwinian evolutionists for the next century were not at all opposed to life elsewhere in the universe – as for Darwin, the origin of life was not their thing – but they didn't want it evolving into humans or even often intelligent beings. The paleontologist George Gaylord Simpson, writing in 1964 (5 years after the centenary of the *Origin*), spoke for many.

> The assumption, so freely made by astronomers, physicists, and some bio-chemists, that once life gets started anywhere, humanoids will eventually and inevitably appear is plainly false. The chance of duplicating man on any other planet is the same as the chance that the planet and its organisms have had a history identical in all essentials with that of the Earth through some billions of years. Let us grant the unsubstantiated claim of millions or billions of possible planetary abodes of life; the chances of such historical duplication are still vanishingly small.

Continuing:

> I cannot share the euphoria current among so many, even among certain biologists (some of them now ex-biologists converted to exobiologists). The reasons for my pessimism are given here only in barest suggestion. They will not, I know, convince all or indeed many. There are too many emotional factors and, to put it bluntly, selfish interests opposed to these conclusions. In fact, I myself would like to be proved wrong, but a rational view of the evidence seems now to make the following conclusions logically inescapable:
>
> 1. There are certainly no humanoids elsewhere in our solar system.
> 2. There is probably no extraterrestrial life in our solar system, but the possibility is not wholly excluded as regards Mars.
> 3. There probably are forms of life on other planetary systems somewhere in the universe, but if so, it is unlikely that we can learn anything, whatever about them, even as to the bare fact of their real existence.
> 4. It is extremely improbable that such forms of life include humanoids and apparently as near impossible as does not matter that we could ever communicate with them in a meaningful and useful way if they did exist (Simpson 1964, p. 774).

Others felt the same way. Stephen Jay Gould (1981), notoriously, was against biological progress. He hated the idea, because he felt that it implied some peoples are better than others, and as a Jew, he was very mindful of how that particular belief had played out earlier in the twentieth century. In his bestseller, *Wonderful Life* (Gould 1989), he wrote of the "... staggeringly improbable series of events, sensible enough in retrospect and subject to rigorous explanation, but utterly unpredictable and quite unrepeatable... Wind back the tape of life to the early days of the Burgess Shale;

starting point, and the chance becomes vanishingly small that anything like human intelligence would grace the replay" (p. 14).

3.2.8 BUT CAN SELECTION PRODUCE ALIEN HUMANS? YES!

Yet, there are those who have endorsed thoughts of progress, including progress up to humans. Paradoxically, elsewhere, Gould (1985) counts himself in this number. "When we use 'evolutionary theory' to deny categorically the possibility of extraterrestrial intelligence, we commit the classic fallacy of substituting specifics (individual repeatability of humanoids) for classes (the probability that evolution elsewhere might produce a creature in the general class of intelligent beings.) I can present a good argument from "evolutionary theory" against the repetition of anything like a human body elsewhere; I cannot extend it to the general proposition that intelligence in some form might pervade the universe."

More generally, we find a range. Very much in the Spencerian tradition of being ardent for progress and simply unbothered by the causes is today's most distinguished evolutionary biologist, the eminent ant specialist, and sociobiologist, Edward O. Wilson. "The overall average across the history of life has moved from the simple and few to the more complex and numerous. During the past billion years, animals as a whole evolved upward in body size, feeding and defensive techniques, brain and behavioral complexity, social organization, and precision of environmental control — in each case farther from the nonliving state than their simpler antecedents did" (Wilson 1992, p. 187). The enthusiasm is for absolute progress, leading to humankind. In his great book, *Sociobiology: The New Synthesis*, Wilson offers a tale of social evolution from colonial invertebrates, through the Hymenoptera, onto the vertebrates and primates, and now humans. "Man has intensified these vertebrate traits while adding unique qualities of his own. In so doing he has achieved an extraordinary degree of cooperation with little or no sacrifice of personal survival and reproduction." The big question is not whether progress occurred, but why it occurred. "Exactly how he alone has been able to cross to this fourth pinnacle, reversing the downward trend of social evolution in general, is the culminating mystery of all biology" (Wilson 1975, p. 382). Parenthetically, one might add that Wilson is an ardent Spencerian. His intellectual grandfather at Harvard, the supervisor of his supervisor, was W. H. Wheeler, a fellow ant specialists and Spencer groupie. In Wilson's lab, there was a picture of Spencer on the wall, hung more prominently than that of Darwin, but neither as prominent as the picture of Wilson receiving the National Medal of Science from then-president Jimmy Carter!

Most, however, try to offer Darwinian reasons for the upward progress. Thomas Henry Huxley (Huxley [1893] 2009), although a keen supporter of Darwin, as a morphologist and paleontologist was never that keen on selection and, as one living toward the end of the nineteenth century, when things were going wrong (like overcrowded cities and tensions between the great powers), was also a bit dubious about progress. His grandson, Julian Huxley, felt no such doubts. If humans are the end point, the apotheosis of evolution, how did Huxley think we progressed up to us? Arms races! In the spirit of Darwin, it was Julian Huxley who developed the idea, before the Great War, making the case in a little book, *The Individual in the Animal Kingdom* (Huxley 1912). Then, after the War, restating the case in *Animal Biology*, a textbook published jointly with Haldane Deeply influenced by the progressionist thinking of the French philosopher Henri Bergson, stressing the competition between evolving lines leading to improvement, obviously inspired by the pre-War competition between the British and German navies, Huxley gave a graphic description of an arms race couched in terms of naval, military technology (Huxley and Haldane 1927). "The leaden plum-puddings were not unfairly matched against the wooden walls of Nelson's day." Now, however, obviously having in mind, the then-huge competition between Britain and Germany, "though our guns can hurl a third of a ton of sharp-nosed steel with dynamite entrails for a dozen miles, yet they are confronted with twelve-inch armor of backed and hardened steel, water-tight compartments, and targets moving thirty miles an hour. Each advance in attack has brought forth, as if by magic, a corresponding advance in defence." Likewise, in nature, "If one species happens to vary in the direction of greater independence, the inter-related equilibrium is upset, and cannot be restored until a number of competing species have either given way to the increased pressure and become extinct, or else have answered pressure with pressure, and kept the first species in its place by themselves too discovering means of adding to their independence" (Huxley 1912, pp. 115–116). Eventually, "it comes to pass that the continuous change which is passing that through the organic world appears as a succession of phases of equilibrium, each one on a higher average plane of independence than the one before, and each inevitably calling up and giving place to one still higher."

Today, Richard Dawkins stands in this tradition. "Directionalist common sense surely wins on the very long time scale: once there was only blue-green slime and now there are sharp-eyed metazoa" (Dawkins and Krebs 1979, p. 508). He too finds the key in arms races. Perhaps, expectedly, as one who embraced computer technology early and enthusiastically, Dawkins notes that, more and more, today's arms races rely on computer technology rather than brute power, and – in the animal world – he finds this translated into ever-bigger and more efficient brains. No need to hold your breath about who has won. Dawkins invokes a notion known as an animal's EQ, standing for "encephalization quotient" (Jerison 1973). This is a kind of cross-species measure of intelligence quotient (IQ) that takes into account the amount of brain power needed simply to get an organism to function (whales require much bigger brains than shrews, because they need more computing power to get their bigger bodies to function) and that then scales according to the surplus left over. Dawkins (1986) writes, "The fact that humans have an EQ of 7 and hippos an EQ of 0.3 may not literally mean that humans are 23 times as clever as hippos! But the EQ as measured is probably telling us *something* about how much 'computing power' an animal probably has in its head, over and above the irreducible amount of computing power needed for the routine running of its large or small body" (p. 189).

As always, it is the analogy with human Progress that is the key.

> Computer evolution in human technology is enormously rapid and unmistakably progressive. It comes about through at least partly a kind of hardware/software coevolution. Advances in hardware are in step with advances in software. There is also software/software coevolution. Not only did the advances in software made possible improvements in short-term computational efficiency – although they certainly do that – they also make possible further advances in the evolution of the software. So, the first point is just the sheer adaptedness of the advances of software for efficient computing. The second point is the progressive thing. The advances of software, open the door – again, I wouldn't mind using the word "floodgates" in some instances – or open the floodgates, to further advances in software (Conference talk, Melbu, Norway, 1989; printed in Ruse 1996, p. 469).

A rather different but still supposedly Darwinian approach to the question involves the notion of channeling or external direction. Using an argument that seems to have swayed Gould, the paleontologist Simon Conway Morris, as a Christian, is very keen to argue for the inevitability of the appearance of humans. He argues that only certain areas of what we might call "morphological space" are welcoming to life forms (the center of the Sun would not be, for instance) and that this constrains the course of evolution (Conway Morris 2003). Again and again, organisms take the same route into a pre-existing niche. The saber-toothed, tiger-like organisms are a nice example, where the North American placental mammals (real cats) were matched right down the line by South American marsupials (thylacosmilids). There existed a niche for organisms that were predators, with cat-like abilities and shearing/stabbing-like weapons. Darwinian selection found more than one way to enter it – from the placental side and from the marsupial side. It was not a question of beating out others but of finding pathways that others had not found.

Conway Morris argues that, given the ubiquity of convergence, we must allow that the historical course of nature is not random but strongly selection-constrained along certain pathways and to certain destinations. Most particularly, some kind of intelligent being was bound to emerge. After all, our own very existence shows that a kind of cultural adaptive niche exists – a niche that prizes intelligence and social abilities. "If brains can get big independently and provide a neural machine capable of handling a highly complex environment, then perhaps there are other parallels, other convergences that drive some groups towards complexity." Continuing: "We may be unique, but paradoxically those properties that define our uniqueness can still be inherent in the evolutionary process. In other words, if we humans had not evolved then something more-or-less identical would have emerged sooner or later" (Conway Morris 2003, p. 196).

Finally, let us mention a non-Darwinian solution to the problem, one acknowledged to be more in a Spencerian vein. Duke University colleagues, paleontologist Daniel McShea and philosopher Robert Brandon (2010), promote what they proudly call "Biology's First Law." Named the "zero-force evolutionary law" or ZFEL, its general formulation runs: "In any evolutionary system in which there is variation and heredity, there is a tendency for diversity and complexity to increase, one that is always present but that may be opposed or augmented by natural selection, other forces, or constraints acting on diversity or complexity" (p. 4). It is something apparently with the status in evolutionary biology of Newton's First Law of Motion – a kind of background condition of stability, even though somewhat paradoxically, their law suggests perpetual motion.

Although the authors are fairly (let us say) generic on their understandings of complexity and diversity – number of parts and number of kinds – the claims made are grandiose if familiar. Given the natural tendency of life to complexity – parts tend to be added on – this generates new organic variations and hence types, and so, one gets a version of (what the theoretical biologist Stuart Kauffman (1993) has called) "order for free." As is usual in these discussions, it is not always obvious whether the claim is that adaptation is created in this way or if adaptation is now irrelevant. Probably, more the former: "we raise the possibility that complex adaptive structures arise spontaneously in organisms with excess part types. One could call this self-organization. But it is more accurately described as the consequence of the explosion of combinatorial possibilities that naturally accompanies the interaction of a large diversity of arbitrary part types" (p. 124).

Expectedly, all of these approaches have attracted criticism. Not every Darwinian biologist is that enthused by arms races. The fossil evidence, for instance, does not show unambiguously that prey and predators have become ever faster. And even if arms races are ubiquitous, it does not follow that intelligence will always emerge. Having high intelligence means having large brains and having large brains means having ready access to large chunks of protein, the bodies of other animals. There were no clever vegans in the Pleistocene. Being clever, however, isn't everything. Sometimes – as cows and horses demonstrate – it is just easier to get your food in other ways, especially if you are living on grassy savannahs. Remember the cautionary words of Jack Sepkoski.

Even if it exists, why should we or anyone else necessarily or even probably enter the culture niche? Life is full of missed opportunities. Maybe Gould is right, and most times, evolution would have gone other ways and avoided culture entirely. Warthogs rule supreme. Julian Huxley always argued that now that humans occupy the culture niche, no other animal is going to be able to enter (Huxley 1942). Perhaps, other animals – dinosaurs – would have prevented our animals – mammals and then primates – from making their way to the door. In any case, many wonder if it is right to think that niches are just waiting out there, ready to be conquered and entered. Do organisms not create niches as much as find them? There was hardly a niche for head lice, for instance, until vertebrates like us humans came along. Should we expect that there was a niche for culture, just waiting there, such as dry land or the open air? Perhaps, there are other niches not yet invented. We cannot imagine something other than consciousness, but remember the wise warning of J. B. S. Haldane (1927). "Now my own suspicion is that the Universe is not only queerer than we suppose, but queerer than we *can* suppose" (p. 286). For all their talk about analogy, Christians tend to think that their God can get up to some clever tricks, way beyond their ken. Perhaps, these are not all supernatural abilities but simply abilities that were omitted from our evolution. Perhaps, far from being the best, we are a short sidepath and very limited in the true scheme of things. No more than in the case of arms races do we get much guarantee of either human emergence or a sense that we are in some way superior and, for this reason, we won.

Zero-force evolutionary law (ZFEL)? One can only say that if you believe in this, then you are ready for the tooth fairy. Ideas like this seem to appeal to and only to those who spend their times in front of computers. If they spent less time running simulated scenarios and more time looking at the real world, they would think otherwise. Blind law leads to blind results, and supposing that laws are not really blind is to suppose like Wallace that there is some kind of intelligence behind it all. These days, popular among philosophers is an Aristotelian force, an Unmoved Mover, to which everything strives teleologically. I just don't see the evidence for it, I just don't. In my world, Murphy's Law rules supreme. If it can go wrong, it will go wrong. I am not sure that either arms races or niches did do the job, but I can see how they could and how natural selection was vital.

3.2.9 CONCLUSION

That brings us to the end of the discussion. As one looks at it – the relevance of Darwinism to problems of astrobiology – one might at first wonder if there is much to say on the topic, especially given that for Darwin (except perhaps at the very end), it seems to have been such a non-issue. A non-issue not through ignorance – he had read Whewell – but through non-importance or relevance. However, as I hope I have shown, if one persists, there are both some interesting findings historically, especially about Darwin's strategy of getting his ideas presented and accepted and why alien beings were excluded, and some interesting findings philosophically, about Darwinian evolution and the concept of progress and how it impinges on the question of extraterrestrials generally and humanoids specifically. In a sense, it is the best of all topics, because it is clear that there is a lot more work to be done. That is for another place and another time. Perhaps – *pace* Simpson – our intelligent friends across the galaxy can lend a hand.

REFERENCES

Barrett, P. H., P. J. Gautrey, S. Herbert, D. Kohn, and S. Smith, editors. 1987. *Charles Darwin's Notebooks, 1836–1844*. Ithaca, NY: Cornell University Press.

Brewster, D. 1854. *More Worlds than One: The Creed of the Philosopher and the Hope of the Christian*. London, UK: Camden Hotten.

Chambers, R. 1844. *Vestiges of the Natural History of Creation*. London, UK: Churchill.

Conway Morris, S. 2003. *Life's Solution: Inevitable Humans in a Lonely Universe*. Cambridge, UK: Cambridge University Press.

Darwin, C. 1859. *On the Origin of Species by Means of Natural Selection, or the Preservation of Favoured Races in the Struggle for Life*. London, UK: John Murray.

Darwin, C. 1861. *On the Origin of Species*. 3rd ed. London, UK: John Murray.

Darwin, C. 1871. *The Descent of Man, and Selection in Relation to Sex*. London, UK: John Murray.

Darwin, C. 1958. *The Autobiography of Charles Darwin, 1809–1882*. Edited by N. Barlow. London, UK: Collins.

Darwin, E. 1794–1796. *Zoonomia; or, The Laws of Organic Life*. London, UK: Joseph Johnson.

Darwin, E. 1801. *Zoonomia; or, The Laws of Organic Life*. 3rd ed. London, UK: J. Johnson.

Darwin, E. 1803. *The Temple of Nature*. London, UK: Joseph Johnson.

Dawkins, R. 1983. Universal Darwinism. *Evolution from Molecules to Men*. D. S. Bendall (Ed.), pp. 403–425. Cambridge, UK: Cambridge University Press.

Dawkins, R. 1986. *The Blind Watchmaker*. New York: Norton.

Dawkins, R., and J. R. Krebs. 1979. Arms races between and within species. *Proceedings of the Royal Society of London, B* 205: 489–511.

Dobzhansky, T. 1951. *Genetics and the Origin of Species*. 3rd ed. New York: Columbia University Press.

Gould, S. J. 1981. *The Mismeasure of Man*. New York: Norton.

Gould, S. J. 1985. *The Flamingo's Smile: Reflections in Natural History*. New York: Norton.

Gould, S. J. 1989. *Wonderful Life: The Burgess Shale and the Nature of History*. New York: W. W. Norton Co.

Haldane, J. B. S. 1927. *Possible Worlds and Other Essays*. London, UK: Chatto & Windus.

Herschel, J. F. W. 1830. *Preliminary Discourse on the Study of Natural Philosophy*. London, UK: Longman, Rees, Orme, Brown, Green, and Longman.

Huxley, J. S. 1912. *The Individual in the Animal Kingdom*. Cambridge, UK: Cambridge University Press.

Huxley, J. S. 1942. *Evolution: The Modern Synthesis*. London, UK: Allen & Unwin.

Huxley, J. S., and J. B. S. Haldane. 1927. *Animal Biology*. Oxford, UK: Oxford University Press.

Huxley, T. H. 1863. *Evidence as to Man's Place in Nature*. London, UK: Williams and Norgate.

Huxley, T. H. [1893] 2009. *Evolution and Ethics, edited with an Introduction by Michael Ruse*. Princeton, NJ: Princeton University Press.

Jerison, H. 1973. *Evolution of the Brain and Intelligence*. New York: Academic Press.

Kant, I. [1790] 1928. *The Critique of Teleological Judgement*. Translator J. C. Meredith. Oxford, UK: Oxford University Press.

Kauffman, S. A. 1993. *The Origins of Order: Self-Organization and Selection in Evolution*. Oxford, UK: Oxford University Press.

Kuhn, T. 1962. *The Structure of Scientific Revolutions*. Chicago, IL: University of Chicago Press.

Lamarck, J. B. 1809. *Philosophie Zoologique*. Paris, France: Dentu.

McShea, D. W., and R. Brandon. 2010. *Biology's First Law: The Tendency for Diversity and Complexity to Increase in Evolutionary Systems*. Chicago, IL: University of Chicago Press.

Naden, C. 1999. *Poetical Works of Constance Naden*. Kernville, CA: High Sierra Books.

Owen, R. 1848. *On the Archetype and Homologies of the Vertebrate Skeleton*. London, UK: Voorst.

Owen, R. 1849. *On the Nature of Limbs*. London, UK: Voorst.

Rachel, G. W. 1881. Fossil organisms in meteorites. *Science* 2(50): 275–277.

Richards, R. J., and M. Ruse. 2016. *Debating Darwin*. Chicago, IL: University of Chicago Press.

Ruse, M. 1975. Darwin's debt to philosophy: an examination of the influence of the philosophical ideas of John F.W. Herschel and William Whewell on the development of Charles Darwin's theory of evolution. *Studies in History and Philosophy of Science* 6: 159–181.

Ruse, M. 1979. *The Darwinian Revolution: Science Red in Tooth and Claw*. Chicago, IL: University of Chicago Press.

Ruse, M. 1996. *Monad to Man: The Concept of Progress in Evolutionary Biology*. Cambridge, MA: Harvard University Press.

Ruse, M. 2001. *Can a Darwinian be a Christian? The Relationship between Science and Religion*. Cambridge, UK: Cambridge University Press.

Ruse, M. 2013. *The Gaia Hypothesis: Science on a Pagan Planet*. Chicago, IL: University of Chicago Press.

Ruse, M. 2015. *Atheism: What Everyone Needs to Know*. Oxford, UK: Oxford University Press.

Ruse, M. 2017a. *Darwinism as Religion: What Literature Tells Us About Evolution*. Oxford, UK: Oxford University Press.

Ruse, M. 2017b. *On Purpose*. Princeton, NJ: Princeton University Press.

Simpson, G. G. 1964. The nonprevalence of humanoids. *Science* 143(3608): 769–775.

Spencer, H. 1857. Progress: Its law and cause. *Westminster Review* LXVII: 244–267.

Todhunter, I. 1876. *William Whewell, DD. Master of Trinity College Cambridge: An Account of his Writings with Selections from his Literary and Scientific Correspondence*. London, UK: Macmillan.

Wallace, A. R. 1858. On the tendency of varieties to depart indefinitely from the original type. *Journal of the Proceedings of the Linnean Society, Zoology* 3: 53–62.

Wallace, A. R. 1870. *Contributions to the Theory of Natural Selection: A Series of Essays*. London, UK: Macmillan.

Wallace, A. R. 1903. *Man's Place in the Universe*. London, UK: Chapman & Hall.

Whewell, W. 1833. *Astronomy and General Physics (Bridgewater Treatise, 3)*. London, UK: Pickering.

Whewell, W. 1840. *The Philosophy of the Inductive Sciences*. London, UK: Parker.

Whewell, W. 1845. *Indications of the Creator*. London, UK: Parker.

Whewell, W. [1853] 2001. *Of the Plurality of Worlds. A Facsimile of the First Edition of 1853: Plus Previously Unpublished Material Excised by the Author Just Before the Book Went to Press; and Whewell's Dialogue Rebutting His Critics, Reprinted from the Second Edition*. Chicago, IL: University of Chicago Press.

Wilson, E. O. 1975. *Sociobiology: The New Synthesis*. Cambridge, MA: Harvard University Press.

Wilson, E. O. 1992. *The Diversity of Life*. Cambridge, MA: Harvard University Press.

3.3 An Early History from Buffon to Oparin

Stéphane Tirard

CONTENTS

3.3.1 INTRODUCTION

The issue of the origin of life as a scientific matter dates back to the eighteenth century. For a very long period, the beginning of life on Earth was only a part of cosmogonies and religious creationist dogmas. With the first scientific revolution, the place of Earth in the universe, and therefore the place of humans in the universe, was completely reconsidered in scientific terms. The Copernican Revolution, which placed the Sun at the center of the solar system, was the first such reconsideration. Moreover, theories, such as Descartes', provoked a new thought on the history of the formation of Earth. However, the issue of the origin of life was neglected for a long time because of the fixist conceptions of the living beings. Since the eighteenth century, the primordial beginnings of life on Earth became a central issue in life sciences. This chapter focuses on this emergence, from the Enlightenment, notably represented by Buffon, to the interwar period, focusing on Oparin and Haldane. It shows how it became intrinsic to the general issue of the theories of the evolution of living beings.

3.3.2 THE FIRST DEBATES ON SPONTANEOUS GENERATION

Spontaneous generation was a conception of the emergence of life in current nature. According to these early views, living organisms occurred from inert matter and particularly from decayed matter. Spontaneous generation originated in Antiquity, its history being marked by three main experimental phases between the seventeenth and nineteenth centuries.

In 1668, Francesco Redi (1626–1697), a naturalist and a physician from Pisa (Italy), published *Esperienze intorne all generazione degli insetti*, a book about his experiments on the spontaneous generation of flies. First, the Italian naturalist exposed samples of meat in two groups of jars. The jars from the first group were left open, in order for the meat to come in direct contact with the air from the room. Those from the second group were covered with caps. Redi observed that fly larvae appeared in the first group of jars, while nothing appeared in the second group. Based on this finding, he concluded that the maggots came from the exterior, from the eggs laid by flies. Some critics argued against this conclusion, according to which the cap prevented air from entering the jars and, in the absence of air, the phenomenon of spontaneous generation could not occur. He thus repeated his experiments and replaced the caps with gauze. The result was the same. The jars closed with gauze never contained maggots, while the others did. Therefore, Redi was able to claim that the spontaneous generation of flies did not exist. This was the first experimental approach of spontaneous generation. There were naturalists who continued to claim its existence.

During the eighteenth century, spontaneous generation was the reason for a scientific debate between Joseph Needham (1713–1781) and Lazarro Spallanzani (1729–1799) (Spallanzani 1769). The former argued for the spontaneous generation of animalcules, which the latter rejected. They discussed their respective experiments. Needham, who was determined to observe spontaneous generation, criticized Spallanzani's method, arguing that heating the mixtures destroyed the vegetative force.

This debate was closely linked to another fierce one regarding generation. Needham was indeed convinced that generation could be explained through the epigenetic phenomenon—namely, the complete formation of the embryo after fecundation from an unorganized egg—while Spallanzani claimed that eggs contained a preformed organism prior to fecundation.

This debate fascinated naturalists. For instance, the French naturalist Buffon, on the one hand, supported spontaneous generation and used it in his theories. On the other hand, Charles Bonnet (1720–1793) supported Spallanzani. Enlightenment philosophers were also interested in this issue. French philosophers also debated on it. Voltaire, for one, strongly criticized spontaneous generation, while Denis Diderot (1713–1784) supported it in *Le rêve de D'Alembert* (*The D'Alembert's Dream*) as well as in his *Elements de physiologie* (*Elements of physiology*), in which he claimed, "Nature has made only a very small number of beings that she has infinitely varied, perhaps from a single one, by combining, mixing and dissolving from which all the others have been formed" (Diderot 1994, p. 1261).

3.3.3 BUFFON: SPONTANEOUS GENERATIONS AND THE HISTORY OF EARTH

Georges Louis Leclerc, Comte de Buffon (1707–1788) was the most important French naturalist of this century. He developed a set of concepts regarding living beings and implied them in his writings on the history of Earth. Spontaneous generation was one such concept.

Buffon occupied an influent position for a long time, as Intendant of the King's garden as well as part of the King's natural history cabinet. Moreover, between 1749 and 1789, he published, together with his collaborators, a huge encyclopedia about nature, *Histoire naturelle*, comprising 22 volumes and 7 supplements.

The volumes focused mainly on living beings, comprising very famous monographs of species as well as important conceptual perspectives. Buffon's central concept regarded the constitution of living matter. He claimed that all the living beings, that is, plants and animals, are constituted with organic molecules, which, according to Buffon, are microscopic, alive and indestructible. Organic molecules circulate from a living being to another by alimentation, when animals eat animals and animals eat plants. Then, when living beings die, the organic molecules return to the soil and can be absorbed by plant roots, thus beginning a new cycle. Buffon also formulated a concept of species. In the chapter dedicated to the donkey (Buffon 1753, vol 4), he claimed that each species is constituted by a set of living beings characterized by the fact that they can reproduce among themselves and have a fertile offspring. Moreover, the organic molecules take the imprint of the "inner mould" characteristic to each individual and ensure the perpetuation of species. According to Buffon, organic

molecules are transmitted in the semen during reproduction, the characteristic of the inner mould thus being transmitted. It is also of great significance that Buffon was a fixist naturalist and all his concepts allowed him to justify that each species, related to its own original prototype, does not evolve into another species, even if it can vary through the ages.

Buffon's theory of Earth evolved continuously throughout his career, and it was particularly revealed in two volumes of his *Histoire naturelle*. The first, in 1749, was dedicated to a *Theorie de la Terre*, in which he presented a cyclic history of Earth. On the other hand, in 1778, in the seventh supplement, entitled *Les époques de la nature* (*The epochs of Nature*), he proposed a non-cyclic history of Earth, claiming a succession of irreversible transformations due to the irreversible cooling of the planet. Buffon calculated that the age of the Earth, in this linear history, was 75,000 years (Roger 1988). Several aspects of *The Epochs* particularly regarded the origin of life. The first epoch was that of the formation of planets. The second epoch regarded the solidification on Earth. After 2936 years of cooling, Earth was solid and its temperature was compatible with the presence of living beings. These emerged during the third epoch. According to Buffon, life is a physical property of matter; therefore, a big quantity of organic molecules was available from this very early period. During the fifth epoch, large species, such as elephants, were produced by spontaneous generations in the northern regions of the planet. Man emerged only during the seventh epoch (Buffon 1778).

To sum up, Buffon claimed that the origin of species depends on spontaneous generation. The first species emerged as soon as Earth was sufficiently cooled. The first species to emerge were the bigger ones and were progressively followed by the smaller ones.

3.3.4 JEAN-BAPTISTE LAMARCK: SPONTANEOUS GENERATIONS AND THE PERPETUAL BEGINNING OF LINEAGES

Jean-Baptiste Monet de Lamarck (1744–1829) presented a developed theory of the transformation of species in 1802. He explained this process through the influence of environment and the habits of living beings. The principles of his theory were operable at all the levels of complexity of the chain of living beings, including the simplest of the animalcules, which, according to him, emerged by spontaneous generation.

Lamarck began his naturalist career as a botanist, and, in 1779, he published his *Flore de France*. At that time, he was deeply interested in chemistry. However, he never carried out any practical work in this field, but developed his theoretical ideas in several books. During the last decades of the eighteenth century, he struggled to impose his ideas that never took into account the results of Lavoisier's chemical revolution. In 1893, Lamarck became professor of zoology of animals without vertebrae at the Museum of Natural History in Paris. Lamarck was the inventor of the word invertebrate. In 1801, Lamarck (1801) published the first book on this topic: *Le système des animaux sans vertèbres* (*System of animals without vertebrae*).

The following year, in his *Hydrogéologie ou recherches sur l'influence qu'ont les eaux sur la surface du globe terrestre...* (*Hydrogeology or research on the influence that water has on the surface of Earth...*), he suggested an analogy between the role of the fluids on Earth's surface and the fluids that would shape the interior of organisms. In *Hydrogéologie*, as in his previous books, he did not hesitate to claim one of his main chemical concepts, namely that living bodies are the "primary cause of the existence of all compounds" (Lamarck 1797, p. 238). He added that "all compounds we observe on our globe, owe their origins, directly or indirectly, to the organic faculties of living beings" (Lamarck 1797, p. 386). The logical consequence of this idea is that the primordial beginning of life and spontaneous generation are impossible.

Later that year, Lamarck (1802) definitively abandoned his fixist conceptions and presented the first complete version of his theory about the transformation of living beings. Surprisingly, and in complete contradiction with his previous point of view, he used the notion of spontaneous generation for the first time. He presented his theory in a book, *Recherches sur l'organisation des corps vivans*, comprising two parts: the first is the lecture that he read at the Museum d'Histoire Naturelle in 1802, and the second, longer part, is an explanation of his new theory. The book begins with the description of the chain of living beings, from the most complex to the simplest. Lamarck said that, in his description, he followed the reverse order of the order followed by the nature. Taking group by group, he finally arrived at the animalcules, the last and simplest group of polyps, claiming that some of them were probably products of nature, by way of spontaneous generations. Therefore, Lamarck described a process of "animalization" of matter and explained how the fluids of the environment transform the inert gelatinous matter. He distinguished between two types of fluids: "subtle fluids," that is, "fluids which cannot be contained," heat and electricity, and fluids "which can be contained," water and gases. The former fluids give the simplest organization to gelatinous matter and are the cause of the vital organism, which is a sort of an excitation of the matter in which molecules are kept apart. The latter, with their permanent movements through matter, also shape it. At this stage, the gelatinous matter acquires contractility, an ability compared by Lamarck to a property of animalcules. Moreover, fluids passing through the gelatinous matter leave some particles, which the naturalist compared to nutrition. The next stage of shaping is the formation of a hole corresponding to the intestine of polyps. Therefore, according to Lamarck, this process initiates the chain of living that he previously described. Lamarck's process of animalization resumed all the principles of his theory. He explained the transformations of the more complex animals with the effects of habits, the actions of inner fluids, and the interactions with environment.

It is very important to stress upon the fact that Lamarck claimed that spontaneous generation is a permanent process in today's nature. Consequently, chains of living beings are permanently initiated, and the diversity of the groups that we currently observe in nature correspond to as many stages of the transformation of different chains (Figure 3.3.1).

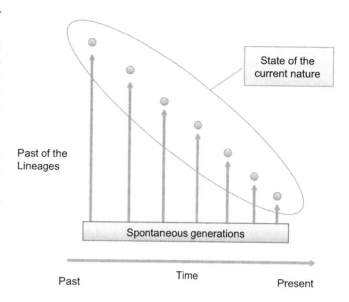

FIGURE 3.3.1 Spontaneous generations and lineages according to Lamarck.

Moreover, there are two crucial questions to be asked regarding the origins of life: did Lamarck believe in the primordial spontaneous generation on the primitive lifeless Earth? In Lamarck's writings, there are very rare passages about a sort of an absolute beginning of living beings. In *Recherches sur l'organisation des corps vivans*, he said, about the simplest forms of life, that nature began and is still in the process of beginning its production by this extremity. Further, in the same text, he wrote that life has not always been as we see it today. He probably referred to a period in the past, when only very simple living beings were present on Earth. Moreover, it is interesting to notice that he wrote that animal life began in the sea. Therefore, did Lamarck consider a primordial beginning? If so, the permanent spontaneous generation would be the permanent repetition of the primordial one (Tirard 2006).

Nevertheless, Lamarck's ideas did not dominate the life sciences of the first part of the nineteenth century, and, during the following decades, whenever spontaneous generation was mentioned, his transformist conception was usually not often mentioned.

3.3.5 LOUIS PASTEUR: THE REJECTION OF SPONTANEOUS GENERATION

In the middle of the nineteenth century, a very important debate took place in France between partisans of the spontaneous generation, led by Félix Pouchet (1800–1872), and opponents, led by Louis Pasteur (1822–1895). At the end of the 1850s, Pouchet, a biologist at the museum of Rouen, sent several successive notes to the Académie des sciences de Paris, and in 1859, he published a book: *Hétérogénie ou Traité de la génération spontanée* (*The Heterogeny or Spontaneous Generation Treatise*) (Pouchet 1859). In 1862, in light of the discussion provoked by these texts, the academy decided to award him a prize for the work carried out on the issue of spontaneous generation: "Try, through well done experiments, to throw a new light on the question of spontaneous generations." His main opponent was Pasteur, a physicist who focused on fundamental issues about life. Before turning to spontaneous generation, Pasteur focused on fermentation and, in his capacity as a crystallographer, discovered the dissymmetry of chemical molecules on living beings, a very interesting characteristic, which would become an important issue in the study of the origins of life.

Therefore, the award launched constituted the beginning of a long opposition between two groups supporting Pouchet and Pasteur, respectively. The two scientists opposed their own experimental results, which were based on heating glass balloons containing organic mixtures. Their results and conclusions were in complete contradiction. Their balloons containing alterable solutions produced opposite results. Pouchet observed spontaneous generation systematically, which was not the case for Pasteur, in whose balloons microorganisms did not occur spontaneously. Pasteur's conclusion was that spontaneous generation could not occur in such conditions. He showed that if the boiling mixture was left unclosed in the balloon, no germs could appear. However, they became present as soon as the balloon was opened. He concluded that the germs that were present in the air also got inside the balloon.

In 1864, in a famous lecture held at the University of Paris, Pasteur presented his conclusions and referred to a drop of water coming from one of his balloons, waiting for spontaneous generation:

> "But it's mute! It has been mute for several years since these experiences began. Ah! It is because I have removed from her, and that I am still removing at this moment, the only thing that man has not been given to produce, I have removed from her the germs that float in the air, I have removed from her life, because life is the germ and the germ is life. The doctrine of spontaneous generation will never rise from the mortal blow that this simple experience brings it." (Latour 1989, p. 427)

In France, Pasteur's conclusions led to the complete abandonment of spontaneous generation. In other countries, however, the debate continued for a few more years, notably in England with Henry Charlton Bastian (1837–1915). This professor of pathological anatomy at the University College of London argued for the existence of spontaneous generation throughout his career. He notably supported heterogeny, that is, the possibility that every organic matter coming from a living being can produce another living being eventually from another species.

To conclude, spontaneous generation was the subject of experiments or scientific debates several times in history: during the seventeenth century with Redi, during the eighteenth century with Spallanzani and Needham, and during the nineteenth century with notably Pasteur, Pouchet, and Bastian. It appears that despite the relevance of certain pieces of evidence, it was difficult for certain scientists to abandon this theory. However, the history of spontaneous generation cannot be limited to the history of a progressive denying of an ancient error. Indeed, the issue of spontaneous generation was also present in general thought on life and on its history, as it was demonstrated in Buffon's and Lamarck's works.

3.3.6 DARWIN'S CONCEPTION OF A UNIQUE ORIGIN OF LIFE

The British naturalist Charles Darwin (1809–1882) published his book, *The Origin of Species*, in 1859, after more than 20 years of preparation. He proposed to explain the descent of living beings with modifications, thanks to the concepts of variation and natural selection. Darwin's theory, which explained the history of species on Earth, could logically include the beginning of this history, that is, the first living beings. However, two main remarks can be made regarding Darwin's ideas on the origin of life.

The first regards a very small part of the origins of life in *The Origins of Species*. In this book, or in any of his other books, he did not develop the issue of the beginning of life on Earth, mainly because there was no evidence of this phenomenon (particularly no paleontological evidence). There were only a few lines in the conclusion of his book where he linked his general explanation of the evolution of living beings to the issue of the beginning of life:

> "It is interesting to contemplate an entangled bank, clothed with many plants of many kinds, with birds singing on the bushes, with various insects flitting about, and with worms crawling through the damp earth, and to reflect that these elaborately constructed forms, so different from each other in so complex a manner, have all been produced by laws acting around us. These laws, taken in the largest sense, being Growth by Reproduction; inheritance which is almost implied by reproduction; Variability from the indirect and direct action of the external conditions of life, and from use and disuse, a Ratio of Increase so high as to lead to a Struggle for Life, and as a consequence to Natural Selection, entailing Divergence of Character and the Extinction of less-improved forms. Thus, from the war of nature, from famine and death, the most exalted object which we are capable of conceiving, namely, the production of the higher animals, directly follows. There is a grandeur in this view of life, with its several powers, having been originally breathed into a few forms or into one; and that, whilst this planet has gone cycling on according to the fixed law of gravity, from so simple a beginning endless forms most beautiful and most wonderful have been, and are being, evolved." (Darwin 1859, pp. 489–490)

Therefore, it is clear what Darwin thought about the origin of life and about the simple early organisms. He claimed that all the living beings could come from a unique and simple ancestor, an idea that was in accordance with the permanent and divergent modification of species, which characterized the Darwinian evolutionary process. However, he never published anything more about the origin of the first living beings.

It is important to notice that, as a concession to a conservative environment, he introduced a small but important change in the following editions of his famous book, in the last sentence, with a mention to the creator (we underline):

> "There is a grandeur in this view of life, with its several powers, having been originally breathed by the Creator into a few forms or into one; and that, whilst this planet has gone cycling on according to the fixed law of gravity, from so simple a beginning endless forms most beautiful and most wonderful have been, and are being, evolved." (Darwin 1860, p. 490)

The second remark regards a private letter (1871), in which Darwin explained to his friend, the botanist Joseph Dalton Hooker (1817–1911), his thoughts on the origin of life on Earth. They discussed about spontaneous generations, and Darwin wrote this very interesting paragraph:

> "It is often said that all the conditions for the first production of a living organism are now present, which could have been present. But if (and oh what a big if) we could conceive in some warm little pond with all sort of ammonia and phosphoric salts, - light, heat, electricity &c. present, that a protein compound was chemically formed, ready to undergo still more complex changes, at the present day such matter would be instantly devoured, or absorbed, which would not have been the case before living creatures were formed." (Quoted in Calvin 1969, p. 4)

This quote is important for at least four reasons. First, it is the only time that Darwin has ever claimed a process for the beginning of life. Second, the process of the complexification of matter is clearly described, step by step, from mineral to organic matter. Third, he demonstrated the impossibility of spontaneous generation. This reason has a great significance: it would be life itself that would prevent life to occur again. Fourth, this impossibility of the repetition of the process of the origin of life is characteristic of a historical process, each step depending on the previous steps, without any possibility of reversibility.

Therefore, in this quote, Darwin put the issue of the origin of life in modern terms. Several other biologists would do the same in the following decades.

3.3.7 THE BEGINNING OF EVOLUTIONARY ABIOGENESIS

During the second part of the nineteenth century, with the emergence of evolutionary theories, some scientists made proposals regarding a progressive evolution from inert matter to living matter (Kamminga 1991). These claims were in accordance with the new concepts that became central in biology and in the issue of the origins of life. The first of them was the cellular theory, which was formulated in two distinct time periods. In 1838, Matthias Schleiden (1804–1881) claimed that plants were composed of cells that have a nucleus. According to him, cells appeared due to the matter of a pre-existing cell. One year later, Theodor Schwann (1810–1882) generalized this claim to include animals and claimed that the matter constituting a new cell could be an intercellular liquid. During the 1850s, Robert Remak (1815–1865) and Virchow (1821–1902), independently of each other, stated that each cell came from the division of a pre-existing cell.

The composition of cells, and particularly of their protoplasm, was also a central issue. In 1868, Thomas Huxley (1825–1895), a British biologist close to Darwin, claimed that albuminous bodies were "The Physical Basis of Life," as he wrote a lecture on this issue in 1868. Moreover, he was opposed to the idea of spontaneous generations (Huxley 1892).

The importance of the chemical nature of living matter was particularly illustrated in the debate about a substance named *Bathybius haeckeli* after its discovery in 1857 on the floor of the North Atlantic Ocean by the Britannic ship Cyclops. For almost 20 years, there had been a discussion on the nature of this substance, that some scientists, such as the German biologist Ernst Haeckel (1834–1919), considered to be a primitive form of life. Finally, in 1876, the chemist John Young Buchanan (1849–1925) subjected it to analysis and determined that it was calcium sulfate and not living matter.

Haeckel, one of the main German zoologists of the second part of the nineteenth century, was always very interested in the issue of the chemical evolution of matter at the origin of life on Earth. He claimed that albuminous protoplasm could be at the origin of *Monera*, which, according to him, could be the most primitive living structure (Haeckel 1879, 1897).

During the same period, the British philosopher Herbert Spencer (1820–1903) developed a huge view of the world by using all sciences, including psychology and sociology, which he published in several books throughout the second part of the nineteenth century. Spencer's thought on evolution was closer to Lamarck's than to Darwin's. He did not agree with natural selection and claimed that transformations of organisms were led by environmental conditions. When it comes to the origin of life, Spencer was strongly against spontaneous generation and, what is more, discussed the notion of the first organism (Tirard 2011). Spencer argued for a very progressive process of evolution and claimed that it was impossible to define the absolute limits between the stages of the evolutionary process. Therefore, according to him, there was no first organism. However, he was interested in this process and, in the 1898 issue of the *Principles of Biology*, presented the steps of the genesis of the characteristics of organic matter. He supported the notion of "protyle" suggested by Crookes, which would be the fundamental compound of organic matter. Its formation, like the formation of all the compounds present in nature, would be a stage of the cosmic evolution. He insisted on the fact that there was a chemical evolution before the evolution of life (Spencer 1898).

Therefore, all these authors, in different ways, were convinced by the possibility of a progressive evolution of matter at the origin of life. Using a word proposed by Huxley, this family of conceptions corresponded to evolutionary abiogenesis.

The turn of the century was also marked by chemical experiments in the field of the organic synthesis. Berthelot's work was particularly emblematic. Toward the turn of the century, D. Berthelot and H. Gaudechon (1910), O. Baudisch (1913) and E. Baly et al. (1922), for example, obtained amino acids by exposing simple mineral or organic mixtures to various energy sources. Similarly, sugars had been commonly obtained since the second half of the nineteenth century.

The post-war authors interested in the origins of life sometimes cited these results. They reported these results to show that syntheses were possible.

3.3.8 PANSPERMIA

During the second part of the nineteenth century, some scientists supported theories of panspermia regarding the origin of life on Earth. These theories claimed that life on Earth came from germs arriving from space. Indeed, fundamentally, in panspermia theories, the presence of life is eternal in the universe and germs containing life are elsewhere in space. Several propositions had been made in the past, and the scientists who claimed this possibility during this period were notably H.E. Richter in 1865 and William Thomson, Lord Kelvin (1824–1907), in 1871 (Thomson 1872). In an address as president of the British Association for the Advancement of Science, the latter criticized Darwin's theory and argued that the age of Earth, which he calculated himself, was not sufficient for the duration of the life evolution process predicted by the famous naturalist. According to Thomson, the only way to conceive the formation of living beings on Earth was by the importation of germs of life from space. He therefore argued that they were imported on Earth by meteorites. Thomson's view opened a period in which panspermia had some supporters, such as H. von Helmholtz (1821–1894) (Kamminga 1991). At the beginning of the twentieth century, Svante Arrhenius (1859–1927) advanced the most elaborated theory of panspermia, and he described how very tiny germs could travel from a planet to another (Arrhenius 1910).

Finally, panspermia was contradicted in 1910 by the French biologist Paul Becquerel, who tested the resistance of organisms (seeds, bacteria, and so on) experimentally in extreme conditions, simulating those of space (Tirard 2013). He showed that they could not resist being exposed to ultraviolet (UV) rays. Becquerel's work quieted down the last panspermia supporters (Becquerel 1910).

3.3.9 1920s–1940s: THE PERIOD OF THE GREAT HYPOTHESES

During the interwar period, developed hypotheses, constructed as complete scenarios of the origins of life on Earth, were formulated. Two names are particularly noticeable. The first is that of Alexander Ivanovitch Oparin (1894–1980), a young Russian scientist specialized in plant physiology. In 1923, in Moscow, he delivered a lecture on the origins of life, and 1 year later, he published a booklet on this issue (Oparin 1924). Published in Russian, this book was only locally diffused; however, it constituted the foundation of his thought on the origins of life, which he developed throughout his life. In this book, Oparin integrated the origins of life into a general presentation of the evolution of Earth. This allowed him to describe the primitive conditions under which the process of the appearance of life could take place. Oparin described how chemical reactions could lead in the primitive atmosphere containing carbon dioxide, and then in the primitive ocean,

from carbon dioxide to organic substances. Finally, according to him, this organic matter could constitute some bits of gel. Therefore, for the first time, a complete scenario of an evolutionary process regarding the origins of life was claimed.

A few years later, in 1929, John Burdon Sanderson Haldane (1892–1964), a biochemist and geneticist, wrote, independently from Oparin, a paper in the Rationalist Annual on the issue of the origin of life (Haldane 1929). Like Oparin's booklet, this text was a broad approach to the process of the origin of life, as well as a general thought on life and its limits. Haldane was mostly interested in the nature of viruses. He stated, especially regarding the origin of life, that the primitive atmosphere contained carbon dioxide and that organic molecules could be synthesized under the effect of the Sun's energy in the "Hot dilute soup" constituted by the primitive sea. He wrote, "when ultraviolet light acts on a mixture of water, carbon dioxide, and ammonia, a vast variety of organic substances are made, including sugars and apparently some of the materials from which proteins are built." According to him, on primitive Earth, "the whole sea was a vast chemical laboratory." The result could be the formation of "half-living molecules," the first step in the process toward life. On several points, Haldane's scenario was pretty close to Oparin's. They both first tried to establish the primitive conditions of the atmosphere and then imagined organic syntheses due to the molecules present in the atmosphere and diluted in the primitive sea (Tirard 2017).

After publishing this paper, Haldane returned to the issue of the origin of life occasionally; however, it was not his specialty. He was best known as a very famous geneticist. In comparison, Oparin focused on this issue throughout his life. In 1936, he published an influent book, translated into English in 1938. It largely developed the ideas of his 1924 booklet (Oparin 1924). Oparin presented the evolution of Earth for the purpose of better understanding the conditions in which the chemical process of the chemical evolution could occur and lead to life. He changed his mind about carbon dioxide in the primitive atmosphere and claimed that it was absent. He then analyzed the chemical reactions that might occur in the environment. This book is based on many references, one of them drawing on (H. G. Bungenberg de Jong 1883–1977), a Belgian biochemist who described microscopic structures in colloidal solutions, which he named coacervates (Bungenberg de Jong 1932). Oparin used them to design models of primitive cellular structures. When it comes to the primitive atmosphere, Oparin's book from 1936 introduced a modification of the atmosphere: from this moment, he claimed that it did not contain carbon dioxide and that it was reducing. This book had an important influence in the middle of the century and was often quoted by specialists in the origins of life (Oparin 1938).

Oparin's and Haldane's proposals marked the interwar period. Their most important and common contribution is represented by the fact that they included the origin of life in the general process of the evolution of Earth and therefore in the primitive conditions in which the evolution of matter occurred.

Just after World War II (WWII), the physician John Desmond Bernal (1901–1971), who was deeply involved in the X-ray diffraction method and interested in biological molecules, held a lecture entitled "The basis of life," which he published in a paper in 1949 as well as in a book in 1951 (Bernal 1951). He made a synthesis of the issue of the origins of life, mainly referring to Oparin, and conducted a revision of some aspects of the scenario. Bernal added his personal views and made an interesting claim. He noticed that molecules were able to disperse in the liquid environment. Moreover, he argued that chemical reactions had to be catalyzed. Therefore, he claimed that catalysis could be possible if the reactions happened at the clayed bottom of aquatic zones, such as estuaries.

3.3.10 CONCLUSION

During the nineteenth century, the issue of the origins of life on Earth became a part of the issue of the evolution of living beings. From the second half of this century, after spontaneous generation was abandoned and the Darwinian theory of evolution was formulated, many biologists admitted that life resulted from a progressive evolution of matter that could be named evolutionary abiogenesis. A second important period was the interwar one, during which more developed scenarios were constructed, notably by Oparin and Haldane. The third remarkable period in the history of the issue of the origins of life is that of its experimental exploration in the second half of the twentieth century. This is the subject of the next chapters.

REFERENCES

Arrhenius S. 1910. *L'évolution des mondes.* Paris, France: Ch. Béranger.

Baly E. C. C., I. Heilbron, and D. P. Hudson. 1922. Photocatalysis Part. II: The photosynthesis of nitrogen compound from nitrate and carbone dioxide. *J. Chem. Soc.* 121: 1078.

Baudisch O. 1913. Uber nitrat-und nitrit-assimilation. *Z. Angew. Chem.* 26: 612.

Becquerel P. 1910. L'action abiotique de l'ultraviolet et l'hypothèse de l'origine cosmique de la vie. *C. R. Ac. Sc. T.* 151: 86–88.

Bernal J. D. 1951. *The Physical Basis of Life.* London, UK: Routledge and Kegan Paul.

Berthelot D., and H. Gaudechon. 1910. Synthèse photochimique des hydrates de carbone aux dépens du gaz carbonique et de l'eau en l'absence de chlorophylle. *C. R. Ac. Sc.* 150, 1169, 1327, 1690.

Buffon. 1778. Les époques de la nature, *Histoire naturelle générale et particulière*, supplément, tome cinquième. Paris, France: Imprimerie Royale.

Buffon. 1753, *Histoire naturelle générale et particulière*, Paris, France: Imprimerie Royale, Vol 4.

Bungenberg de Jong H. G. 1932. Die Koacervation und ihre Bedeutung für die Biologie. *Protoplasma* 15: 110–176.

Calvin M. 1969. *Chemical Evolution: Molecules Evolution towards the Origin of Living Systems on the Earth and Elsewhere.* Oxford, UK: Clarendon.

Darwin C. 1859. *The Origin of Species by Means of Natural Selection or the Preservation of Favoured Races in the Struggle for Life.* London, UK: John Murray.

Darwin C. 1860. *The Origin of Species by Means of Natural Selection or the Preservation of Favoured Races in the Struggle for Life.* London, UK: John Murray.

Diderot D. 1994. Éléments de physiologie, *Oeuvres, Tome 1, Philosophie.* Paris, France: Robert Laffont. 1253–1317.

Haeckel E. 1897. Le monisme lien entre la religion et la science, *Profession de foi d'un naturaliste.* Paris, France: Schleicher Frères.

Haeckel E. 1879. *Histoire de la création des êtres organisés d'après les lois naturelles, - tr. de l'allemand par Ch. Letourneau et revu sur la septième édition allemande.* Paris, France: Librairie Schleicher Frères.

Haldane J. B. S. 1929. The origin of life. *Rationalist Annu.* 148: 3–10.

Huxley T. 1892. « Les bases physiques de la vie » et « Biogenèse et abiogenèse », *Les problèmes de la biologie.* Paris, France: Baillière.

Kamminga H. 1991. The origin of life on Earth: Theory, history and method. *Uroboros* 1 (1): 95–110.

Lamarck J. B. 1797. *Mémoires de physique et d'histoire naturelle...* Paris, France: chez l'auteur.

Lamarck J. B. 1801. *Système des Animaux sans vertèbres....* Paris, France: Deterville.

Lamarck J. B. 1802. *Recherches sur l'organisation des corps vivants....* Paris, France: Maillard.

Latour B. 1989. Pasteur et Pouchet: hétérogénèse de l'histoire des sciences. M. Serres (Ed.), *Éléments d'histoire des sciences.* Paris, France: Bordas, pp. 423–445.

Oparin A. I. 1924. *Proiskhozhdenie zhizny* (The origin of life, Ann Synge Trans.). J. D. Bernal (Ed.), *The Origin of Life.* London, UK: Weidenfeld & Nicholson.

Oparin A. I. 1938. *The Origin of Life.* New York: Macmillan Publishers.

Pouchet F. 1859. *Hétérogénie ou Traité de la génération spontanée.* Paris, France: Baillère.

Roger J. 1988. Buffon les Epoques de la nature Edition critique. Paris: Edition du Muséum. lxv.

Spallanzani L. 1769. *Nouvelles recherches sur les découvertes microscopiques et la génération des corps organisés.* Paris, France: Lacombe. Published in French, translated by Abbé Regley, Needham's notes and comments.

Spencer H. 1898. *The Principles of Biology.* London, UK: Williams and Norgate.

Thomson W. 1872. Address of Sir William Thomson (President). *British Association for the Advancement of Science, Edinburgh, Report – 1871.* London, UK: John Murray.

Tirard S. 2006. Génération spontanée. P. Corsi, J. Gayon, G. Gohau and S. Tirard (Eds.), *Lamarck: Philosophe de la nature.* Paris: Presses Universitaires de France, pp. 65–104.

Tirard S. 2011. Spencer et les origines de la vie. La double induction comme méthode. D. Becquemont et D. Ottavi (Eds.), *Penser Spencer.* Paris: Presses Universitaire de Vincennes, pp. 81–95.

Tirard S. 2013. The Debate over Panspermia: The case of the French Botanists and plant physiologists at the beginning of the Twentieth Century D. Düner (Ed.), *The History and Philosophy of Astrobiology: Perspectives on the Human Mind and Extraterrestrial Life.* Cambridge, UK: Scholars Publishing, pp. 213–222.

Tirard S. 2017. J.B.S. Haldane and the Origin of Life. *Journal of Genetics* 96 (5): 735–739.

Section IV

Chemical Origins of Life

Chemicals in the Universe and Their Delivery on the Early
Earth; Geology and Atmosphere on the Early Earth

4.1 Interstellar Molecules and Their Prebiotic Potential

Lucy M. Ziurys

CONTENTS

4.1.1 INTERSTELLAR MOLECULES: AN OVERVIEW

The past 45 years of the study of interstellar molecules has clearly shown that molecular content of the universe is far more extensive than previously thought. The observation of abundant species in many harsh environments such as around dying stars, in planetary nebulae (PNe), and in diffuse clouds has demonstrated that molecular material in interstellar gas is remarkably robust. The recent identifications of interesting organics such as CH_3NCO (Halfen et al. 2015), CH_3CONH_2 (Hollis et al. 2006), and $CHOCH_2OH$ (Halfen et al. 2006) suggest that even more complicated prebiotic molecules may be present in interstellar space. Certainly, the discovery of fullerene species such as C_{60} and C_{70} (Cami et al. 2010) is solid evidence that compounds with large numbers of carbons can be formed in astrophysical environments. A picture is emerging of interstellar space as a vast synthetic factory with as-yet unknown limits in complexity, which must, at some unspecified level, affect the origin of prebiotic and possibly biotic compounds.

The chemical richness of interstellar space has only recently been realized, with the development of radioastronomy in the 1960s, which enabled the measurement of gas-phase molecular rotational spectra at high resolution. Prior to that time, a few diatomic molecules such as CH, CH^+, and CN were known to exist in diffuse gas, identified by their electronic transitions observed at optical wavelengths (e.g., McKellar 1940). The first molecules detected in the radio were at centimeter wavelengths, starting with the OH radical (Weinreb et al. 1963) and followed by H_2O, NH_3, and H_2CO (e.g., Snyder et al. 1969). Perhaps, the most significant discovery, however, was that of carbon monoxide, CO, in 1970 (Wilson et al. 1970), which was identified in the millimeter regime with the fledging 36-ft telescope on Kitt Peak, AZ.

The observation of CO was in fact a key factor in initiating the field of molecular astrophysics, as this wavelength region is best suited for molecular measurements. Subsequent molecule detections quickly followed, using millimeter telescopes, including that of HCN, HNC, HCO^+ (known as "Xogen"), H_2S, CS, SiO, CH_3CH, and $HNCO$ (e.g., Snyder and Buhl 1971; Wilson et al. 1971). The field of astrochemistry had begun.

Today, more than 170 individual chemical compounds have been identified in the interstellar medium (ISM) via their gas-phase spectra. The current list of confirmed interstellar molecules is presented in Table 4.1.1, sorted by the number of atoms. As shown in the table, a wide variety of chemical compounds are found in interstellar space, ranging from simple diatomic molecules, such as N_2, CS, and CH, to more complicated organic species, such as methyl formate, $HCOOCH_3$, and dimethyl ether, $(CH_3)_2O$, with fullerenes being the most complex. Many of the known interstellar molecules are species commonly found on Earth, such as water, methane, acetylene, and ethanol, but about 50% are free radicals and molecular ions, including CCN, C_4H, H_3O^+, C_8H^-, and

TABLE 4.1.1
Known Interstellar Molecules[a]

2		3		4	5	6	7	8	9	10	11	12	>12
H_2	HF	C_3	H_2O	$c\text{-}C_3H$	C_5	C_5H	C_6H	CH_3C_3N	CH_3C_4H	CH_3C_5N	HC_9N	$c\text{-}C_6H_6$	$c\text{-}C_6H_5CN$
CH	SH	C_2H	H_2S	$l\text{-}C_3H$	C_4H	$l\text{-}H_2C_4$	CH_2CHCN	$HCOOCH_3$	CH_3CH_2CN	$(CH_3)_2CO$	CH_3C_6H	$i\text{-}C_3H_7CN$	C_{60}
CH^+	SH^+	C_2O	SO_2	C_3N	C_4Si	C_2H_4	CH_3C_2H	CH_3COOH	$(CH_3)_2O$	$(CH_2OH)_2$	C_2H_5OCHO	$n\text{-}C_3H_7CN$	C_{70}
CN	PO	C_2S	N_2H^+	C_3O	$l\text{-}C_3H_2$	CH_3CN	HC_5N	C_7H	CH_3CH_2OH	CH_3CH_2CHO	CH_3OCOCH_3		C_{60}^+
CO	AlO	CH_2	HNO	C_3S	$c\text{-}C_3H_2$	CH_3NC	CH_3CHO	C_6H_2	HC_7N				
CO^+	TiO	HCN	NH_2	C_2H_2	H_2CCN	CH_3OH	CH_3NH_2	CH_2OHCHO	C_8H				
CP	ArH^+	HCO	H_3^+	$HCCN$	CH_4	CH_3SH	$c\text{-}C_2H_4O$	$l\text{-}HC_6H$	CH_3CONH_2				
SiC	SiS	HCO^+	N_2O	$HCNH^+$	HC_3N	HC_3NH^+	H_2CCHOH	CH_2CHCHO?	C_8H^-				
CS	OH^+	HCS^+	H_2O^+	$HNCO$	HC_2NC	HC_2CHO	C_6H^-	CH_2CHCN	C_3H_6				
CF^+	HCl^+	HOC^+	TiO_2	$HNCS$	$HCOOH$	NH_2CHO	CH_3NCO	H_2NCH_2CN	HC_7O				
CN^-	NS^+	HNC	Si_2C	$HOCO^+$	H_2CNH	C_5N	HC_5O	CH_3CHNH					
C_2	N_2	$MgCN$	H_2Cl^+	H_2CO	H_2CCO	$l\text{-}HC_4H$		CH_3SiH_3					
OH	O_2	$MgNC$		H_2CN	H_2NCN	$l\text{-}HC_4N$							
NS		$NaCN$		H_2CS	HNC_3	$c\text{-}H_2C_3O$							
HCl		OCS		$c\text{-}SiC_3$	H_2COH^+	SiH_3CN							
SO		$c\text{-}SiC_2$		CH_3	C_4H^-	C_5N^-							
SO^+		CO_2		C_3N^-	$HCOCN$	$HNCHCN$							
SiO		$SiCN$		$HCNO$	$HNCNH$								
NO		$AlNC$		$HOCN$	CH_3O								
$NaCl$		$SiNC$		$HSCN$	CH_3Cl								
KCl		HCP		C_3H^+	SiH_4								
$AlCl$		CCP		$HMgNC$									
AlF		KCN		PH_3									
PN		$FeCN$		NH_3									
SiN		C_2N		H_3O^+									
NH		$AlOH$		$HCCO$									

[a] By number of atoms.

H_2COH^+. These species are very short-lived under terrestrial conditions and often function as intermediates in chemical reactions.

The gas-phase molecules are usually mixed with dust grains, which typically are 0.5–10 microns in size (e.g., Jones 2001). The composition of the grains is harder to deduce. Some are composed of aliphatic and aromatic carbonaceous materials, while others are silicates. Many grains develop ice mantles typically consisting of water, CO, and methanol. It is also interesting that most interstellar molecules contain carbon, and there is also a significant number with silicon. Carbon comes in the form of long acetylenic chains (HC_3N, HC_5N, C_7H, C_8H, etc.) and simple rings (c-C_3H_2 and CH_2OCH_2). Those containing metals such as magnesium, sodium, and iron are rare, but do exist, and phosphorus-containing compounds currently number six. This rather strange mix of chemical compounds attests to the exotic, but at the same time mundane, nature of interstellar chemistry.

Part of the reason for the presence of unusual compounds in the ISM is that the physical conditions are quite "non-terrestrial." Typical temperatures in molecular gas fall in the range 10–100 K, while densities are extremely low: 10^2–10^7 particles/cc. At 50 K, 10^7 particles/cc (or 10^7 cm^{-3}) corresponds to 10^{-13} atm. Interstellar species therefore mostly exist in a cold vacuum. On the other hand, these conditions can vary. Toward regions of star formation, molecular gas can reach T > 1000 K, and densities can increase by orders of magnitude. Bombardment by ultraviolet (UV) radiation or X-rays, both from stars, can substantially alter the physical conditions and energize nearby molecular material, impacting chemical processes (e.g., Gerin et al. 2010).

4.1.1.1 SETTING THE STAGE: THE INTERSTELLAR MOLECULAR LIFE CYCLE

Not all known interstellar species exist in the same regions of space, which complicates our interpretation of the chemistry. Molecular material in interstellar space is cycled through various phases, as shown in Figure 4.1.1, which repeat many times in the lifetime of the galaxy, thought to be ~13.5 billion years.

This "Molecular Life Cycle" can be thought to begin in stars. Most stars, namely those with masses between 1 and 10 M_{sun}, will deplete hydrogen in their cores, leaving the so-called "Main Sequence," will begin to burn hydrogen in a shell around a helium core, and then ignite helium in the core, converting it to carbon in the "triple alpha" reaction. These processes, which define the Red Giant phase, have two important consequences for these stars: (1) very convective atmospheres can develop, which mix H-burning intermediates such as ^{13}C to the stellar surface, and (2) substantial mass loss occurs. As the helium in the core becomes depleted, the stars develop a layered structure around a now-carbon core consisting of a H-burning shell and a He-burning shell, entering the Asymptotic Giant Branch (AGB) (e.g., Herwig 2005). The AGB is an important stage in molecule formation and in astrobiology for two reasons. First, mass loss significantly increases to 10^{-6}–$10^{-4} M_{sun}$/year, creating an outflowing stellar envelope, which quickly cools and allows the formation of both gas-phase molecules and dust (e.g., Glassgold 1996). Second, the very convective envelope can bring the carbon formed in the He-burning shell to the stellar surface, converting material with O > C to that with C > O. Recent calculations suggest that as much as 80% of the carbon arose from

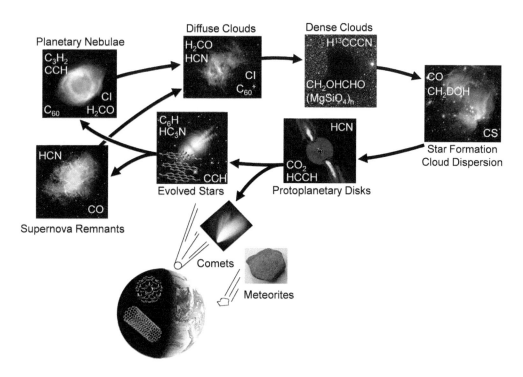

FIGURE 4.1.1 The interstellar life cycle showing how molecular material is passed from one class of objects to another: from stars to diffuse clouds, onto dense clouds, and eventually to protostellar disks and planets.

AGB stars, at least in the solar neighborhood (Mattsson 2010). The basis of organic chemistry may trace its roots back to these unique objects.

Mass loss continues on the AGB until the star has blown off almost all its initial mass and nucleosynthesis ceases, thus entering the PN stage, the next step in the cycle. The core of the progenitor star evolves over the next 10^4 years into a white dwarf (Kwok 2000), which becomes a copious emitter of UV radiation. The remnant circumstellar shell continues to flow away from the star, bringing fresh material into the ISM. In fact, mass loss from these types of stars is responsible for about 85% of interstellar matter (Dorschner and Henning 1995). Recent observations have clearly shown that the gas-phase molecules from the prior evolved star/circumstellar envelope stage survive into the late stages of the PN phase and thus are ejected, along with circumstellar dust grains, into the diffuse ISM. Here, they coalesce into what are called diffuse clouds. These clouds are defined by extremely low densities (~10–100 cm^{-3}) and warm temperatures (T ~ 30–100 K), with partial attenuation of the ambient stellar radiation field (e.g., Snow and McCall 2006). These objects will collapse or accrete into denser clouds over a period of 10^6–10^7 years (e.g., Inoue and Inutsuka 2012), evolving with their molecular material. Dense clouds, defined by gas densities near 10^4–10^6 cm^{-3}, span a wide range of masses, from 10^3–10^6 M$_{sun}$ (so-called "giant molecular clouds" [GMCs]) to a few M$_{sun}$ (e.g., Bok globules). They are colder than their diffuse counterparts, with T ~ 10–75 K, and are sufficiently self-shielding, so that they are typically rich in molecular material and are referred to as "molecular clouds." Dense clouds contain a large fraction of the mass in our galaxy (e.g., Ferrière 2001), as much as half of the total gaseous matter in the inner 8–9 kiloparsecs (kpc).

Dense clouds will collapse into stars, incorporating molecular material into protostellar disks. The disks may evolve into solar systems. Some of the primordial molecular matter will be processed and some preserved and condensed into comets, asteroids, and meteorites. Some of this molecular "debris" will continue to survive in a large "bubble" of icy material surrounding a given planetary system; for the solar system, this bubble is the Oort cloud. Other pristine interstellar molecular material will end up on planet surfaces through bombardment by solar system bodies, including comets and meteorites. The central stars of the solar systems will eventually exhaust the supply of hydrogen in their cores, begin to burn helium, and move onto the giant branches, likely engulfing surrounding planets and reinitiating the molecular cycle.

Very massive stars will not pass through the giant branches but rather continue to burn elements beyond helium and then become supernovae (SNe). Supernova explosions also return matter to the diffuse ISM, but only about 10% of the total material. The ejecta for SNe are initially quite hot (T > 10,000 K). At about 6000 K, the material is all atomic in composition (Clayton and Meyer 2018), but as the gas cools, molecules, perhaps even long carbon chains, likely form, which in turn lead to grain production. It is accepted that certain types of grains originate in SNe, but the contribution of gas-phase molecules by SNe to the diffuse ISM in not known. Supernovae

(Type II) are also thought to be important sources of ^{12}C in the ISM, although, perhaps, not as significant as AGB stars (Mattsson 2010).

Because gas-phase and solid-state molecular materials are passed from one type of object to another, a complete view of interstellar and pre-biotic chemistry must consider the entire cycle. Studies of pre-solar grains have shown that material found in meteorites can be traced back to both AGB stars and SNe (e.g., Lodders and Amari 2005). As will be discussed, the different types of sources harbor distinct physical conditions and therefore spawn varied chemical processes, which influence the next step in the cycle.

4.1.2 IDENTIFICATION OF INTERSTELLAR MOLECULES

4.1.2.1 A Brief Overview of Molecular Spectroscopy

Molecules are typically identified in interstellar sources by measuring their gas-phase spectra with telescopes. Molecular spectra can be very complicated, because chemical compounds consist of multiple atoms. Therefore, in addition to excitation of electrons within a molecule, which creates electronic transitions, the nuclei can rotate and undergo vibrational motions. Pure rotational spectra occur in the microwave, millimeter, and sub-millimeter regimes (~10 GHz–1 THz) and involve transitions between the lowest energy levels (E$_{rot}$ ≤ 100 cm^{-1}). Because interstellar gas is typically cold, population of rotational levels is heavily favored, which readily occurs by collisions with H$_2$. As a consequence, about 90% of all known interstellar molecules have been detected via their pure rotational transitions. Rotational spectra can also exhibit additional splittings, known as fine and hyperfine structures, arising from unpaired electrons and nuclei with net spin, which can aid in molecule identification. Vibrational transitions, in contrast, occur in the infrared (IR) region. They are sufficiently high in energy (E$_{vib}$ ~ 100–5000 cm^{-1}) and usually require an external field for excitation, as from a young or old star or in a shock wave (e.g., Goto et al. 2003). Molecules will rotate as they vibrate, and therefore, vibrational transitions consist of rotational structure, typically labeled as P, Q, and R branches, which in some cases may be resolved. Electronic transitions, which involve promotion of an electron to a higher-lying molecular orbital, are the highest-energy transitions (E$_{elec}$ > 10,000 cm^{-1}). They occur typically in the optical and UV regions of the electromagnetic spectrum (400–700 nm). In the ISM, they usually require background stellar radiation for the excitation mechanism and are almost always seen only in diatomic molecules. Electronic transitions have both rotational (P, Q, and R branches) and vibrational structures, often resolved in interstellar spectra. (For further detail, see Herzberg 1971; Bernath 2005).

Spectra can also be obtained for molecular material in the solid state (e.g., see Kwok 2000). However, there is no free rotation in solids, and vibrational motion occurs within a crystal lattice or amorphous material. Therefore, solid-state spectra, which are typically measured in the IR, do not have the specificity found in the gas phase. They are typically used to identify types of bonds rather than discreet molecules (aromatic C–C bends, SiO stretch, etc.).

4.1.2.2 SPECTROSCOPIC LIMITATIONS OF TELESCOPES

Spectroscopic measurements in the radio- and millimeter-wave regions have definite advantages over those in the IR and visible regions. First, there is, for the most part, less obscuration by the Earth's atmosphere at these longer wavelengths. Space-borne telescopes such as the upcoming James Webb Space Telescope (JWST) can ameliorate this problem, but these instruments are very expensive and have short lifetimes. Furthermore, radio- and millimeter-wave telescopes, which are diffraction-limited antennas, typically employ heterodyne mixing techniques to obtain spectral-line data, coupled with a multiplexing spectrometer as the "backend." This combination of instrumentation results in high spectral resolution capability, typically 1 part in 10^6–10^8. Therefore, the measurement of rotational spectra of interstellar molecules, which are present in cold, quiescent gas, can generate a true "fingerprint" for unambiguous identification, as illustrated in Figure 4.1.2.

Here, a spectrum of the $N = 2 \rightarrow 1$ rotational transition of the CN radical is shown, observed toward the molecular cloud G19.6. This molecule has one unpaired electron, and the nitrogen atom has a nuclear spin, both of which interact to generate a unique fine structure/hyperfine pattern, indicated under the spectrum in blue. The spectrum clearly demonstrates the presence of CN in this source. For molecules without fine and/or hyperfine structure, multiple rotational lines can be observed for unequivocal identification. At radio and millimeter wavelengths, resolution is typically limited by the spectral linewidth generated by the velocity motions within a given astronomical source.

In the case of optical and IR spectroscopy, different instrumentation is used. The telescopes in these cases are essentially "light buckets," and molecular spectra are created by use of a grating or prism. Typical resolution of modern spectrographs is 1 part in 10^3–10^4, with the very best being 1 part in 10^5. Therefore, for IR and optical spectroscopic measurements, resolution is determined by the instrumentation, not the astronomical source.

4.1.3 BASIC CHEMISTRY OF INTERSTELLAR MOLECULES

Given the cold and highly rarified conditions in the ISM, and the extremely long timescales, it is very difficult to accurately simulate interstellar chemical processes in the laboratory. Formulation of interstellar chemistry has relied heavily on observations and theory. However, identification of interstellar molecules requires precise knowledge of their spectra, and therefore, experimental laboratory spectroscopy has played a major role in our understanding of the chemistry in space.

4.1.3.1 IMPORTANT GAS-PHASE REACTIONS

Because of the very low densities, chemical reactions in molecular gas are typically restricted to two-body processes. There is usually no third body or container wall to stabilize a reaction complex. Furthermore, with temperatures near 10 K, reactions cannot have activation barriers; otherwise, they cannot proceed. A general class of reactions fulfilling these criteria (two-body, no barriers) are ion-molecule processes of the general form (e.g., Herbst and Klemperer 1976):

$$A + B^+ \rightarrow C + D^+ \tag{4.1.1}$$

These types of reactions (about 50% have no barriers) proceed very quickly under interstellar conditions, typically at the collisional or Langevin rate of k ~ 10^{-9} cm^3 s^{-1}. They are readily initiated by the cosmic-ray ionization of H_2, leading to H_2^+. This species then reacts with another H_2, producing H_3^+:

$$H_2^+ + H_2 \rightarrow H_3^+ + H \tag{4.1.2}$$

H_3^+ is a key ion in interstellar chemistry and readily donates its proton to other neutral molecules to form a series of ions, HCO^+, N_2H^+, H_3O^+, $HCNH^+$, and so forth. Several of these ions are in fact prominent interstellar molecules and have provided good evidence for the ion-molecule scheme. Not all ion-molecule reactions are barrierless, however. An important example involves carbon. The process $C^+ + H_2$ is endothermic, and therefore, incorporating carbon into molecular form requires other processes.

Radiative association is another important class of reactions. Here, two smaller gas-phase species collide to form a larger molecule while emitting a photon, that is:

$$A + B^+ \rightarrow AB^+ + h\nu \tag{4.1.3}$$

Reactions of this type can also occur between two small neutral molecules, but such processes typically have activation

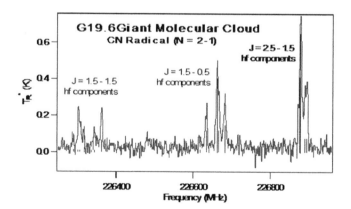

FIGURE 4.1.2 An example of a pure rotational spectrum of the $N = 2 \rightarrow 1$ transition of the CN molecule near 226 GHz, showing all the respective fine and hyperfine components, observed toward the dense cloud G19.6. The red trace is the actual observed spectrum; the blue lines under the data show the predicted laboratory frequencies and relative intensities. The match is virtually exact.

barriers, except when radicals and atoms are involved (e.g., Gerlich and Horning 1992). Those with ion-neutral reactants are not nearly as fast as classic ion-molecule reactions, typically having rates of k ~ 10^{-17}–10^{-12} cm^3 s^{-1}. Nonetheless, radiative association can become quite important if faster routes do not exist, such as the case with C$^+$. The process C$^+$ + H$_2$ → CH$_2^+$ + $h\nu$ is believed, in fact, to be the major route that incorporates carbon into molecules.

Reactions between two neutral molecules can also be significant in interstellar chemistry, particularly if one or both of the reactants are radicals:

$$A + B \rightarrow C + D \qquad (4.1.4)$$

Neutral-neutral rates can be as high as 10^{-11}–10^{-10} cm^3 s^{-1} and thus can compete in certain cases with ion-molecule reactions. The formation of N$_2$O, for example, is thought to occur from the neutral-neutral route NO + NH → N$_2$O + H (Halfen et al. 2001).

Another set of reactions imperative to interstellar molecules is dissociative recombination. In this case, a molecular ion combines with an electron, producing neutral fragments:

$$A^+ + e^- \rightarrow B + C \qquad (4.1.5)$$

These types of processes are essential for the production of neutrals after species formation by ion-molecule pathways and proceed relatively quickly, with k ~ 10^{-7} s^{-1}. A good example of such a reaction is the recombination of HCNH$^+$, which then forms HCN and HNC. However, product branching ratios are often unknown and may favor smaller fragments (e.g., Geppert et al. 2006).

These four reaction types form large gas-phase networks that are predicted to lead to the majority of interstellar molecules, creating a chemistry that is kinetically controlled. There are instances, however, where three-body collisions do occur in molecular gas, and thermodynamics is the chemical basis for molecule production. "LTE chemistry" occurs in the inner regions of circumstellar envelopes, for example. It also may take place near imbedded protostars.

4.1.3.2 Surface Processes

Dust grains are mixed with gas-phase molecules; therefore, surface chemistry can also occur. There is more uncertainty in the role of reactions on grain surfaces, as solid-state spectroscopy has its limitations. Nonetheless, it is recognized that H$_2$, the most abundant interstellar molecule, must form on grain surfaces (e.g., Herbst and Millar 2008). All two-body gas-phase processes leading to this species are simply too inefficient, while hydrogen atoms can readily migrate over a surface. Furthermore, energy is released when two H atoms combine to form H$_2$ on the surface, allowing efficient ejection from the grain.

For molecules larger than H$_2$, the exact role of surface chemistry is less defined. It is clear that grains can acquire molecular mantles composed of water, methanol, and other species, as deduced from IR astronomical measurements

(e.g., Jones and Nuth 2011). In fact, it is remarkable that more molecular material is not frozen onto grain surfaces, particular at 10 K. Mantle formation does not imply chemical processing. On the other hand, there are larger "organic"-type molecules, such as CH$_3$CH$_2$CN and CH$_2$CHCN, that, according to chemical modeling, cannot be produced efficiently in the gas phase. It is postulated that hydrogenation of "backbone" species on grain surfaces creates these molecules (e.g., Hickson et al. 2016). The precursor molecules, in this case C$_3$N or HC$_3$N, are created in the gas phase; they freeze onto grains and acquire hydrogen atoms, which are labile on the surface. Gas-phase chemistry and surface chemistry are thus closely linked. Defining their relative contributions remains one of the frontiers of interstellar chemistry.

4.1.3.3 Influence of Star Formation and Destruction

The classic structure of ion-molecule chemistry can be substantially altered by stars. First, when these objects form as protostars, or die as SNe, they can induce shock waves from mass ejection, which can propagate into nearby molecular material. These shocks compress, accelerate, and abruptly heat the affected gas to temperatures as high as a few thousand degrees, often drastically altering its chemical composition, depending on the type of shock (e.g., Bachiller and Gutierrez 1997). The main result is that endothermic reactions can occur, preferentially producing species such as H$_2$O, SiO, and CH$^+$ (e.g., Ziurys et al. 1989; Viti et al. 2011). Very fast shocks (v > 50 km/s) can also break up dust grains, releasing molecular mantles and/or refractory material and also modifying the local chemical composition.

Young hot stars can also impact the chemistry of adjacent clouds through their strong emission of UV radiation, creating a "photon-dominated region" (PDR). Here, the high UV flux will fragment and ionize molecules, creating a high abundance of ions and radicals, such as CCH and CN (e.g., Cuadrado et al. 2015). Photon-dominated regions host a unique chemical environment, as studies of the Orion Bar has shown (e.g., van der Wiel et al. 2009).

The gravitational collapse of young stars, which are usually imbedded in molecular clouds, can result in the elevation of the density and temperature of the surrounding gas, producing a "hot core" or "corino" (e.g., Cesaroni et al. 2011). Temperatures can typically achieve T ~ 200–500 K, with densities increasing to n ~ 10^7–10^8 cm^{-3}; under such conditions, reactions with energy barriers can occur, once again altering the molecular composition. Observations suggest that abundances of certain sulfur-bearing molecules are enhanced, along with CH$_3$CN and HCOOCH$_3$ (e.g., Xu and Wang 2013). Products of surface chemistry may also be released with elevated temperatures.

4.1.3.4 Chemical Modeling and Reaction Networks

For comparison with observations, and to test chemical theory, gas-phase molecular abundances are computationally predicted from large reaction networks, using the "rate equation"

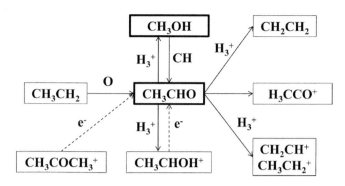

FIGURE 4.1.3 A simplified chemical network showing the reaction pathways leading to the production and destruction of CH_3CHO in dense clouds, including ion-molecule reactions and dissociative electron recombination.

method (e.g., Stantcheva and Herbst 2004). Here, hundreds, often thousands, of chemical reactions are coupled together, and the abundance of a given molecule is balanced between its production and destruction pathways, as a function of time. Often, the calculations are allowed to achieve a steady state. A key factor in this modeling is the accurate assessment of reaction rates and product branching ratios; unfortunately, these quantities are often unknown, perhaps as many as 50% in a given model. Another complication is unusual temperature dependence in many reactions; some rates will actually increase as the temperature drops to ~10 K (e.g., see Gerlich and Horning 1992). Large databases have been created to tabulate reaction rates, based on experimental values, theory, or reasonable assumptions, such as the University of Manchester Institute of Science and Technology (UMIST) database (UMIST RATE2012/astrochemistry.net) and KIDA. However, additional experimental data is clearly needed.

Often, insight can be gleamed about a given molecule's abundance by looking at a reaction sub-network, as shown in Figure 4.1.3. In this figure, the major possible formation and destruction routes are shown for CH_3CHO, applicable to molecular clouds. The main pathways leading to this species are the neutral-neutral reactions involving $CH_3CH_2 + O$ and $CH_3OH + CH$, as well as dissociative recombination of $CH_3COCH_3^+$ and CH_3CHOH^+. The molecule is destroyed by reacting with H_3^+, which leads to various products, including CH_2CH_2 and H_3CCO^+.

4.1.4 MOLECULES IN CIRCUMSTELLAR ENVIRONMENTS

Circumstellar envelopes are a major site for molecule formation in interstellar space. They are most prominent in AGB stars, as well as supergiants (precursors to SNe), and are created by mass loss, as previously mentioned. Mass loss occurs because of depletion of hydrogen in the stellar core, which causes it to contract and release energy, resulting in expansion of the surrounding hydrogen sphere. This process creates a red giant or supergiant star, depending on the mass; red giants evolve into AGB stars, which have significantly higher

mass loss (see Herwig 2013 and references therein). Once the mass loss begins, dust grains can form in the winds that emerge from the stellar photosphere. Radiation pressure on these grains can further enhance mass loss, along with shocks caused by stellar pulsations.

4.1.4.1 BASIC CHEMICAL SCENARIO OF CIRCUMSTELLAR ENVELOPES

The molecular envelope forms from very hot gas (T ~ 2000–4000 K) that leaves the photosphere and creates a large shell around the star, typically extending out to $r \sim 1000\ R_*$ (R_* = stellar radius and r = radial distance from the star). The gas cools as it flows outward, and molecules form, along with dust, as mentioned. There are large temperature and density gradients across the envelope ($n \sim r^{-2}$ and $T \sim r^{-0.5\ -\ 0.7}$), which foster varied chemistries. The material adjacent to the photosphere is hot (T ~ 1000 K) and sufficiently dense ($n \sim 10^{10}\ cm^{-3}$), such that three body reactions can occur. Chemical processes therefore occur under conditions of Local Thermodynamic Equilibrium (LTE), which means that the most thermodynamically stable species are formed. Molecules such as HCCH, HCN, SiO, H_2O, and CH_4 are in high abundance in this zone and form a group of "parent" compounds (e.g., Glassgold 1996). Dust grains also condense out of the gas under LTE conditions in this region; condensation starts within a few stellar radii and continues until about 15–20 R_*. Prominent minerals predicted to form in the shells include fosterite (Mg_2SiO_4), corundum (Al_2O_3), and silicon carbide, depending on the C/O ratio (Lodders and Fegley 1999). As dust forms and radiation pressure comes into play, the gas from the star is accelerated to its terminal velocity by ~20 R_*. Such velocities are typically ~10–20 km/s for AGB stars and somewhat higher for supergiants (e.g., Tenenbaum et al. 2010). The material therefore is transported to the outer envelope sufficiently quickly (~1000 years), so that LTE abundances become "frozen" in the intermediate shell, as densities and temperatures drop (e.g., McCabe et al. 1979). In the outer envelope, the temperature and densities have decreased considerably (T ~ 25 K and $n \sim 10^4\ cm^{-3}$), such that UV photons from the diffuse stellar background can penetrate the gas/dust mixture, initiating photochemistry. In this third zone, free radicals, such as CN and CCH, are readily created, which then initiate other radical-neutral or radical-radical reactions (e.g., Anderson and Ziurys 2014). Negative ions are also created through radiative attachment, $A + e \rightarrow A- + h\nu$ (e.g., Thaddeus et al. 2008). A diagram of a cross-section of a circumstellar shell is shown in Figure 4.1.4, displaying the three distinct chemical zones.

4.1.4.2 CHEMICAL CONSEQUENCES OF CARBON ENRICHMENT

The ISM is oxygen-rich, with O > C by about a factor of 1.5. Therefore, stars are typically oxygen-rich, which is reflected in the molecular composition of their envelopes (Tenenbaum et al. 2010). However, as AGB stars advance in their evolution, they

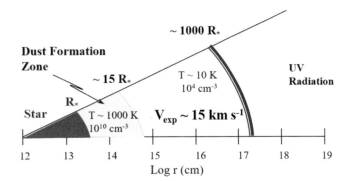

FIGURE 4.1.4 Idealized picture of a cross-section of an AGB circumstellar envelope, plotted as a function of radial distance (log scale, in cm) from the center of the star. The star is shown in red, and the dust acceleration zone is shown in cream. Representative temperatures and densities are shown for the dust zone, in which LTE chemistry occurs, and the outer envelope, where photochemistry is dominant. R_* designates stellar radius.

TABLE 4.1.2
Representative Circumstellar Molecules

Carbon-Rich			Oxygen-Rich	
SiO	CO	SiC	H_2O	SiO
SiS	AlF	AlCl	SO_2	SiS
CS	SiN	KCl	H_2S	CS
CN	HCP	CH_3CN		CN
HCN	SiC_2	CCP		HCN
HNC	CCH	SiC	NH_3	HNC
NaCl	NaCN	AlCl	SO	NaCl
PN	$l\text{-}C_3H$	KCl	AlO	PN
HCO^+	$c\text{-}C_3H$	HC_3N	AlOH	HCO^+
PH_3	C_3N	$c\text{-}C_3H_2$	TiO	NS
CH_2NH	H_2CO	C_4H	TiO_2	PO
CP	H_2CS	CH_3CCH	CO	
C_4H^-	HC_5N	CCN		
C_5N	HC_7N	C_8H		

become thermally unstable, as the H- and He-burning shells thin out and thermal pulses develop. The pulses cause the highly convective hydrogen shell to reach into the helium-burning shell and dredge up the carbon-rich material created there by the triple-α process. This mechanism, so-called "Third Dredge-Up," converts an initially oxygen-rich star to a carbon-rich one, such that C > O in the stellar envelope, by roughly a factor of 1.5 (Mowlavi 1999). It is not clear that all AGB stars undergo Third Dredge-Up, as the envelopes of many such objects remain oxygen-rich. Those stars that do become carbon-rich, however, develop a very unique chemistry in their envelopes – one of the few sets of objects in the galaxy with C > O.

The C/O ratio has a major influence on molecular composition of a circumstellar envelope. First, carbon-rich shells of AGB stars appear to have a more diverse chemical composition than their oxygen-rich analogs, with more complex molecules, numbering up to 11 atoms. (Note that C_{60} is observed in ejecta of post-AGB stars). They also contain a larger number of distinct chemical compounds, roughly by a factor of 3–4, with many exotic species. Long, acetylenic chain-type species have been observed in carbon-rich shells, with the formulas HC_nN, C_nH, and C_nN, where $n = 1,2,3, ...$, as well as more common compounds such as HCCH, CH_3CN, and CH_3CCH. These objects are also sources of rare metal (in the chemist's sense) and phosphorus-bearing molecules, including FeCN, NaCl, AlNC, CCP, and HCP (e.g., Ziurys et al. 2016). They additionally are the most prominent hosts of silicon-containing compounds (SiS, SiO, SiC_2, SiN, SiC_3, SiC, etc.). In contrast, molecules with up to only four atoms have thus far been observed in oxygen-rich envelopes, and the compounds observed are typically simple, such as HCN, CS, SO_2, and H_2O. However, these sources do contain gas-phase refractory oxides, including AlO, TiO, AlOH, and TiO_2 (e.g., Tenenbaum and Ziurys 2010; Tenenbaum et al. 2010; Kamiński et al. 2013).

An illustration of the chemical differences between carbon- and oxygen-rich envelopes is given in Table 4.1.2. Here, a representative sample of molecules is presented for the carbon-rich envelope of the AGB star IRC+10216 and the oxygen-rich shell of the supergiant VY Canis Majoris (VY CMa). These objects are chosen because they are the best studied stars in terms of molecular observations.

The dust grain composition also reflects the C/O ratio. As mentioned, grains are thought to form under LTE conditions. Indeed, some grain types are found that are predicted by LTE models (Mg_2SiO_4 and Al_2O_3), but others, such as FeS and FeSi, are not (Lodders and Fegley 1999). In oxygen-rich stars, amorphous silicates are found, as identified by their IR bands at 9.7 and 18 microns (SiO stretching and Si-O-O bending modes). These grains are thought to condense directly out of gaseous SiO, a common circumstellar molecule. Crystalline silicates are also found, but are less abundant, with typical formulas $Mg_{2-2x}Fe_{2x}SiO_4$ (olivine) and $Mg_{1-x}Fe_xSiO_3$ (pyroxene), as well as refractory oxides such as corundum, Al_2O_3, and spinel, $MgAl_2O_4$. In contrast, carbon-rich environments favor the formation of silicon carbide (SiC), identified by the 11.3-μm feature (Si-C stretch), and carbonaceous grains, which might be amorphous, graphitic, and/or organic aromatic/aliphatic compounds (Kwok 2004).

Circumstellar envelopes with C > O are perhaps the richest sources of carbon-carbon bonds. They concentrate this element into gas-phase molecular or solid-state form and prevent it from being locked into CO. For example, in IRC+10216, the carbon chain species alone account for about 10% of the available carbon, based on cosmic abundances. The unusual environment of these envelopes apparently preserves the carbon enrichment that results from Third Dredge-Up and therefore makes them central to astrobiology. They pass this enrichment onto PNe and then subsequently into interstellar clouds. These sources are also unique in that they contain the majority of the known phosphorus- and metal-bearing compounds found in the ISM, with C-P, C-Fe, and C-Mg bonds. It is most interesting from the biological aspect that the most common carrier of metallic elements in the ISM in the gas phase involves bonds to the CN moiety (FeCN, MgCN, KCN,

NaCN, and AlNC). It should also be noted that oxygen-rich envelopes of both supergiant and AGB stars typically contain PO, the only molecule thus far securely identified in interstellar gas with a phosphorus-oxygen bond (Tenenbaum et al. 2007; Ziurys et al. 2018).

4.1.5 THE SURPRISINGLY MOLECULAR CONTENT OF PLANETARY NEBULAE

The AGB stars evolve into PNe. The former circumstellar envelope flows away from the dying star, now a white dwarf, creating a nebula seen in so many optical images. The nebula itself contains most of the original stellar material, as a white dwarf typically has a mass of ~0.5 M_{sun}. The amount of matter in the nebula is thus substantial, between approximately 0.5 M_{sun} and 7.5 M_{sun}. A major question is the nature of this material, which is the primary source of matter for the general ISM.

4.1.5.1 OBSERVATIONS PROVE THEORY INCORRECT

For many years, it was thought that virtually all of the molecular material from the former AGB shell would be photodissociated and returned to the atomic state in PNe (e.g., Redman et al. 2003; Kimura et al. 2012). White dwarfs are very strong emitters of UV radiation, with fields several orders of magnitude higher than that found in the ambient ISM (Kwok 2000). The bright images of PNe observed in the optical region, resulting from atomic emission lines of [O III], [OII], CII, Ne II, He II, [N II], and Hα, support the idea of widespread molecular destruction, as shown in Figure 4.1.5. However,

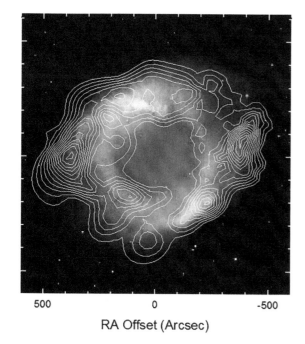

FIGURE 4.1.5 Contour map showing the distribution of HCO+ in the Helix nebula, as traced by the J = 1 → 0 rotational transition, superimposed over the optical image of the object (reconstructed from Zeigler et al. 2013). The data show that HCO+ is widespread throughout the nebula, even where bright, ionized atomic gas also exists.

the ionized, rarified gas in PNe can only account for about ~0.1–0.5 M_{sun} of the nebular material, creating a "missing mass" problem (Kimura et al. 2012).

Recent observations, however, have shown that many PNe are rich in molecular gas and contain large abundances of polyatomic molecules (e.g., Zack and Ziurys 2013; Edwards et al. 2014; Schmidt and Ziurys 2016, 2017a, 2017b). Both C_{60} and C_{70} have been identified in several PNe as well (e.g., Cami et al. 2010; García-Hernández et al. 2012), providing additional evidence for a molecular environment. The list of known molecules in PNe has certainly expanded in the past 5 years, as well as the number of molecule-rich nebulae. The current list of known molecules in PNe, besides fullerenes, includes HCN, HNC, CN, CS, H_2CO, HCO^+, N_2H^+, SO, SO_2, SiO, c-C_3H_2, and CCH. Moreover, mapping observations of H_2 and CO show that the molecular material appears to be present in dense, dusty clumps, apparently imbedded in an ionized medium (e.g., O'Dell et al. 2007). The gas-phase molecules, mixed with dust grains and probably macromolecules, are protected in the clumps, which are self-shielding. As a consequence, not only do gas-phase molecules survive throughout the lifetime of the nebulae, but their measured abundances are fairly constant with nebular age (e.g., Edwards et al. 2014; Schmidt and Ziurys 2017a, 2017b). For example, one of the oldest known PN, the "Helix," has high abundances of CCH, H_2CO, c-C_3H_2, HCN, HCO^+, and HNC, with values comparable to those in both middle-aged nebulae, such as M2-48 and M3-55, and the youngest ones, such as K4-47, K3-17, and K3-45 (Schmidt and Ziurys 2017b). Furthermore, the widespread molecular distribution in the Helix is coincident with the atomic gas and traces the whole of the nebula, as shown in Figure 4.1.5.

Here, the distribution of HCO^+ in the Helix nebula is shown, as traced by its fundamental rotational transition. The contours show the integrated intensity of the transition, in units of 0.033 K km/s (see Zeigler et al. 2013 for details), superimposed over the optical image of the nebula. Current chemical models predict that molecular abundances should drop by orders of magnitude as the nebulae evolve and be nonexistent in very old objects such as the Helix (e.g., Redman et al. 2003). Such predictions are simply not correct, and new modeling is needed.

4.1.5.2 ORIGIN OF THE MOLECULAR MATERIAL IN PLANETARY NEBULAE

The molecules in the PNe may, in part, be residual species from the remnant AGB envelope. Abundances of species such as CCH and HCN in PNe are certainly lower than those in AGB envelopes by factors of 5 and 100–1000, respectively (Schmidt and Ziurys 2016, 2017b). Other molecules, such as HCO^+, on the other hand, are more abundant in PNe. There are obviously chemical processes occurring in the nebulae that are altering AGB abundances, including photo destruction. New synthetic routes, may take place in the early PN or protoplanetary phase (see Schmidt and Ziurys 2106, 2017a). The observations suggest that molecular abundances are altered in the protoplanetary phase and then are "frozen," as the nebulae evolve. There is also a possible contribution from the destruction of dust grains or large molecules such as C_{60}, which may be countering molecule photodissociation. Independent of the processes, the molecular content in at least some PNe is unexpectedly high. As a consequence, PNe ejecta must contain substantial amounts of molecular material, mixed with dust.

4.1.5.3 Further Impact of Carbon Enrichment in the Asymptotic Giant Branch Stage

Not all PNe harbor the same molecular species. Like their precursors, circumstellar envelopes, the C/O ratio is reflected in the molecular content of the nebulae. There are also estimates of this ratio from measurements of atomic transitions, but these are rare (see Schmidt and Ziurys 2017a). It is clear that both carbon- and oxygen-rich PNe exist, along with some where C ~ O. Carbon-rich nebulae are identified by the presence of CCH, HCN, c-C_3H_2, and, in the most extreme cases, HC_3N, as seen in K4-47, probably the best example of this type (Schmidt et al. in preparation). The Helix nebula also appears to be a carbon-rich type, as it contains widespread CCN and c-C_3H_2. Oxygen-rich PNe, on the other hand, do not contain CCH but harbor oxides such as SiO, SO, and SO_2, as observed in M2-48 (Edwards and Ziurys 2014). The Red Spider, NGC 6537, contains both CCH and SO, and appears to be an intermediate case. Additional surveys of molecules in PNe are needed to further quantify such classifications, but the distinctions are clearly present.

Carbon enrichment from AGB stars is therefore passed on to PNe and preserved in gas-phase molecules in the form of C–C, C–H, C–N, and C–O bonds, as well as in macromolecules such as fullerenes. This material is then ejected into the diffuse ISM, bringing along the enrichment. Therefore, the expectation is that molecules in diffuse clouds contain more carbon that might be expected in an environment where O > C. Observations of species such as C_2, C_3, CCH, and c-C_3H_2 toward diffuse clouds support this idea, as well as the recent identification of C_{60}^+ (Liszt et al. 2006; Snow and McCall 2006; Walker et al. 2015).

4.1.6 MOLECULES IN DIFFUSE CLOUDS: ANOTHER SURPRISE

Diffuse clouds are another class of objects with unusual molecular content. These clouds are formed from stellar ejecta (principally AGB stars) and are characterized by low densities (~10–100 cm^{-3}) and warm temperatures (T ~ 30–100 K). It is thought that the ambient stellar UV radiation field is somewhat attenuated, such that the H_2 content is >0.1 of the total available hydrogen. Also, the gas-phase carbon is predicted to be predominantly in the form of C^+ (e.g., Snow and McCall 2006).

4.1.6.1 Unexpected Polyatomic Molecules in Diffuse Gas

It has been universally thought that the chemistry in diffuse clouds would be limited to principally diatomic species, as verified by sophisticated chemical models (e.g., van Dishoeck and Black 1986). First, the densities are so low that it is difficult to build molecules through the usual gas-phase reactions in the ion-molecule scheme. Second, there is not sufficient self-shielding to protect the molecules that are created from the destructive effects of UV photons. It has been therefore very surprising to find that a host of polyatomic molecules exists in diffuse clouds. Many of these species remained "hidden" for decades, because the densities are so low in diffuse clouds that collisional excitation, the principal route to rotational lines and therefore molecule detection, is inefficient. Clever measurements of absorption spectra against background quasars resulted in the identification of numerous polyatomic compounds, including HCN, HNC, CCH, HCO^+, H_2CO, SO_2, N_2H^+, and c-C_3H_2, along with the diatomics CN, CS, SiO, SO, and CO (e.g., Liszt et al. 2006). Interestingly, these molecules are almost an identical set to what has been observed in PNe and strengthen the chemical connection. The species in diffuse clouds also have abundances (f ~ 10^{-9}–10^{-8}, relative to H + H_2) that are about a factor of 10–100 less than that in PNe (e.g., Ziurys et al. 2016), consistent with the gradual dispersion of the dense clumps of gas and dust ejected from the nebulae. Calculations suggest that a clump with n ~ 3×10^5 cm^{-3} will dissipate into diffuse gas in about 10^7 years (e.g., Tenenbaum et al. 2009).

4.1.6.2 Large Macromolecules and Intermediate Species

Diffuse clouds also are sources of the Diffuse Interstellar Bands (DIBs), which are observed in absorption in the visible region against background stars. The DIBs have been known for decades but have remained largely unidentified. A few are now attributable to C_{60}^+ (e.g., Walker et al. 2015); the rest remain a mystery but are likely arising from some large complex compounds. A series of IR emission features known as the Unidentified Infrared Bands (UIBs) are also seen in diffuse gas, as well as in PNe and proto-PN (e.g., Snow and McCall 2006, Chapter 4.2 of this book). Based on the wavelengths of the bands (3.3, 6.2, 7.7, 8.6, and 11.3 microns), the UIBs have been attributed to aromatic C–H and C–C stretching and bending modes. The identification of the UIBs is a subject of debate but is likely due to some large organic macromolecules with multiple functional groups (e.g., Kwok and Zhang 2011). More details on the UIBs can be found in Chapter 4.2 of this book.

The UIBs, the DIBs, and C_{60}^+, as well as other smaller species seen in the millimeter region, all indicate the existence of a far more complex chemistry within diffuse clouds than ever conceived. This chemistry appears to be based on carbon, suggesting that the previous carbon enrichment is, to some degree, preserved. There is also a large jump in chemical

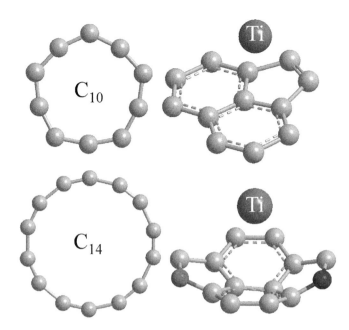

FIGURE 4.1.6 Possible carbon chains and "buckybowls" that might be present in the interstellar medium and represent intermediate complexity between smaller molecules and C_{60}.

complexity from the three- to five-atom molecules to C_{60}^+ and beyond, so intermediate species likely exist. Perhaps, compounds such as "buckybowls" and carbon fragments with a metal or other heteroatom are also present. Examples are shown in Figure 4.1.6. There is clearly more molecular material to be discovered.

4.1.7 COLLAPSE TO DENSE CLOUDS AND THE RISE OF "ORGANIC" CHEMISTRY

Gravitational collapse and/or accretion of diffuse clouds leads to the formation of dense clouds on timescales of about 10 million years (e.g., Inoue and Inutsuka 2012 and references

therein). These clouds are categorized by their high gas densities ($n \sim 10^4$–10^6 cm^{-3}), relative to the general ISM, and by their colder temperatures (T ~ 10–75 K). Hydrogen is in the form of H_2. Their elevated densities enable ion-molecule reactions to proceed efficiently, and "molecular clouds" have rich chemical compositions.

4.1.7.1 GIANT MOLECULAR CLOUDS AND LOWER-MASS OBJECTS

From the chemical viewpoint, molecular clouds can be classified into two types. There are the so-called "giant molecular clouds" (GMCs). These objects contain between 10^3 M$_{sun}$ and 10^6 M$_{sun}$, are 10–100 pc in size (1 pc = 3.26 light-years or 3.086 × 10^{13} km), and usually have T ≥ 50 K. They are not gravitationally stable and collapse to within 10^6–10^7 year to create stars and solar systems (e.g., Lequeux 2013). They often contain protostellar cores, which emit in the IR region and heat the surrounding gas to T ~ 100–300 K. Densities are higher in these cores as well, on the order of 10^7–10^8 cm^{-3}. These clouds usually exhibit spectra rich in molecular lines, partly because of their warmer temperatures, which accelerate the chemistry, evaporate grain mantles, and populate more of the molecular rotational manifolds. An example of such a typical GMC spectrum is displayed in Figure 4.1.7. Here, a portion of the 3-mm region from the Orion Molecular Cloud 1 is shown near 110 GHz, with molecular identifications indicated on the data. The spectral lines are virtually continuous across the spectrum.

There are also less massive, dense clouds, with $M \sim 1 - 10^3$ M$_{sun}$, the most compact of which are Bok globules. These globules, such as Barnard 68, have masses between 1 M$_{sun}$ and 50 M$_{sun}$ and are no larger than ~0.3 pc in size. Typically, lower-mass stars form in these types of clouds, which tend to be more quiescent than GMCs. The very coldest (T ≤ 10 K), dense clouds of this type, such as TMC-1, are sometimes referred to as "dark clouds."

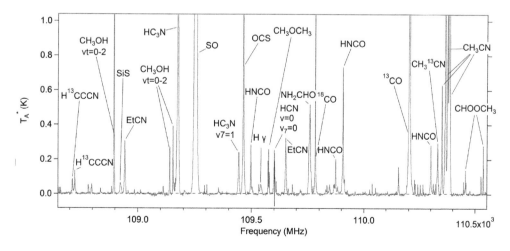

FIGURE 4.1.7 A typical spectrum exhibited by dense molecular clouds with star formation, in this case Orion Molecular Cloud 1, observed at 110 GHz with the Arizona Radio Observatory 12-m telescope. The molecular carriers of many of the spectral features are labeled, but some fractions are unidentified. A variety of organic-type compounds are present in this spectrum, including CH_3OH, EtCN, and $HCOOCH_3$. (From Bernal et al., in preparation.)

4.1.7.2 Chemical Processes and Trends to Higher Complexity

Ion-molecule and associated reactions characterize dense clouds, along with surface processes. The less massive clouds contain the simpler species, such as HCO^+, N_2H^+, CS, HCN, HC_3N, CN, CCH, c-C_3H_2, and H_2CO (e.g., Kaifu et al. 2004). Typical abundances, relative to H_2, fall in the range $f \sim 10^{-10}$–10^{-8} for common polyatomic species. Also, a few of these clouds, as viewed by their molecular composition, are apparently carbon-rich, with C > O, such as TMC-1. This source contains many of the acetylenic chains and other carbon-bearing molecules observed in carbon-rich AGB envelopes, including HC_3N, HC_5N, HC_7N, and HC_9N, as well as HCCNC, HNCC, C_4H, C_6H, and C_3N (e.g., Kaifu et al. 2004). Clouds like TMC-1 suggest that the chemical environment was inherited from a carbon-rich star.

Giant molecular clouds, in contrast, foster a chemistry that appears to further integrate oxygen. These clouds produce compounds of the more classic "organic" type: alcohols (CH_3OH and CH_3CH_2OH), carboxylic acids (HCOOH and CH_3COOH), ketones (CH_3COCH_3), aldehydes (CH_3CHO), ethers (CH_3OCH_3), esters (CH_3OCOH), pseudo-peptides (CH_3CONH_2), and even sugars ($CHOCH_2OH$). They also are the only source class besides AGB stars that are currently known to contain phosphorus-bearing species. However, the phosphorus inventory includes only PN and PO (e.g., Ziurys 1987; Rivilla et al. 2016).

Representative examples of GMCs are SgrB2(N), Orion-KL, and W51 (e.g., Belloche et al. 2016; Halfen et al. 2017). Typical abundances of these species are 10^{-10}–10^{-7}, an overall increase relative to the less massive clouds. The preponderance of C-O bonds may be a result of the vaporization of icy grain mantles by young protostars or simply because in the general ISM, O > C and material has been mixed. The organic molecules are certainly small by chemists' standards, but the true extent of the chemical complexity in GMCs is not known. Because the spectra are so complex (c.f. Figure 4.1.7), identification of new molecules becomes increasingly difficult due to line confusion. Furthermore, laboratory spectral studies of larger molecules also face mounting challenges owing to the presence of multiple conformers near the same energy, complications of internal rotation, and floppy molecular structures, as seen in CH_3NCO (e.g., Halfen et al. 2015). Surface processes also play an important role in GMCs, because elevated temperatures from imbedded protostars liberate compounds formed on grain surfaces. Shocks, elevated gas temperatures, and PDRs are also prominent in molecular synthesis in GMCs, as discussed.

Dust grains in these clouds are thought to consist of amorphous silicates, hydrocarbon material, and graphite, some which appear to have mantles of water, CO, CO_2, CH_3OH, and other species (e.g., Jones 2001). Evidence for these grain types arises principally from observations of IR vibrational bands.

4.1.7.3 Isotope Enrichment and Chemical Fractionation

Molecules in dense clouds are known to be enriched in the heavier isotopes because of chemical fractionation, in particular in deuterium, ^{13}C, and ^{15}N. For the most part, such enhancements occur in very cold gas (T ~ 10 K), where differences in zero-point energies between competing isotopes become significant. Such energy differences are most extreme for deuterium. For example, the reaction of $H_3^+ + DH \rightarrow H_2D^+ + H_2$ is exothermic by 230 K (e.g., Millar 2005), effectively funneling the deuterium into the products. Furthermore, the cosmic deuterium and hydrogen (D/H) ratio of $\sim 10^{-5}$ is fixed by the "Big Bang," unlike $^{12}C/^{13}C$ and $^{14}N/^{15}N$ ratios, which are highly influenced by stellar nucleosynthesis. Ratios as high as D/H ~ 0.001–0.1 have been observed in molecules in dense clouds, with multiply-deuterated species such as CD_3OH and D_2CO, where the enrichment is even higher. These high ratios are often found in cold clouds, particularly in their dense cores, but sometimes also in "hot cores." The enhancement in the latter case is thought to be a result of previous "fossil" low-temperature chemistry, followed by grain condensation and then subsequent evaporation (e.g., Neill et al. 2013).

There is also some evidence of enrichment of molecules in ^{13}C and ^{15}N, but the zero-point energy differences are not nearly as great. For example, the exchange reaction $^{15}N^+ + C^{14}N$, leading to $C^{15}N$, is exothermic by only ~23 K (see Adande et al. 2012). Therefore, high enrichments of the heavier isotopes for carbon and nitrogen are not observed. There is some evidence of chemical fractionation in cold clouds (e.g., Taniguchi et al. 2016), but as the gas warms, such fractionation is erased (e.g., Halfen et al. 2017).

4.1.8 THE NEXT STAGE: PROTOPLANETARY DISKS

The gravitational collapse of dense clouds leads to star formation. As motions in the cloud become synchronized, conservation of angular momentum results in the creation of a disk around the young star (e.g., Williams and Cieza 2011). Mass falls from the vertical direction to form the disk and is compressed by the gas pressure. Disks can extend out to 1000 AU and have a flared structure. Over a period of ~3–10 million years, these disks will evolve from having a gas/dust ratio near 100 to <1, as planetesimals and eventually planets form.

4.1.8.1 Disks: A Complex Physical and Hence Chemical Structure

Protostellar (or protoplanetary) disks are postulated to contain both vertical and radial temperature and density gradients, with vastly different radiation fields at various disk locations. Disks are therefore quite complicated, and a wide range of chemical processes occur, including photochemistry, ion-molecule reactions, neutral-neutral pathways, and surface reactions (e.g., Henning and Semenov 2013). In the radial direction, disks consist of a warm inner zone, the site of planet formation, typically <20 AU and a cold outer region. In the inner disk, T ~ 150–5000 K, such that grain mantles are vaporized, and densities approach 10^{12} cm^{-3}. Here, the chemical processes are near LTE, and three-body collisions and neutral-neutral reactions play an important role, producing molecules such as H_2O, CO, and HCN. There is a balance between molecule production and thermal dissociation, as well as destruction by radiation from the young star. In the outer zone, the key chemical drivers are radiation and cosmic rays, which promote ion-molecule reactions. "Freeze-out" becomes important as temperatures decrease, pulling molecules from the gas phase and increasing the density of solid particles. At about 150 K, water condenses onto grain surface, defining the so-called "snow line," the radial distance from the protostar at which ice grains form, typically a few AU. CO and other molecules also have "snow lines," but much further out (≥20 AU). Recently, it has been speculated that a "centrifugal barrier" also exists in the outer disk, namely the radial distance at which the gravitational pull of the central star is balanced by the centrifugal force, such that infall no longer occurs (e.g., Sakai 2015). This barrier is thought to define a change in chemical conditions, possibly indicating the transition from unprocessed to processed material.

In the perpendicular direction, the outer disk can be divided into three regions. There is the dense midplane, typically obscured from observations, where planets/asteroids can form and molecules are frozen out onto grains. Here, it is thought that T < 20 K, such that most molecules condense onto grains and surface hydrogenation plays a dominant chemical role. Then, there is the disk surface layer, which is bombarded by UV and, in certain cases, X-ray photons. Photodissociation, ion-molecule reactions, and radical-radical processes are important in this PDR-like material. In between the two layers, protected from UV and X-ray radiations, is the intermediate zone, which is warm (T ~ 30–100 K) and is thought to have the richest molecular content. Common interstellar molecules such as CO, CS, HCO$^+$, and HCN are present there.

4.1.8.2 Molecular Content and Abundances in the Disks

Observational work of molecules in protoplanetary disks is in its infancy. Nonetheless, a variety of species have been identified thus far, and more are likely to occur with Atacama Large Millimeter Array (ALMA) and other new facilities. Infrared spectroscopy has revealed the presence of CO_2, H_2O, CO, OH, HCN, and HCCH (e.g., Williams and Cieza 2011), which are observed principally through ro-vibrational transitions. These higher-excitation transitions are thought to trace the hotter gas in the inner disk, within a few AU of the star. Radio- and millimeter-wave molecular measurements typically sample the more extended material, namely the outer disk >30 AU. Detections include H_2CO, CS, C_2H, c-C_3H_2, HCN, HNC, CN, DCN, CCH, HCO$^+$, and DCO$^+$ (e.g., Guilloteau et al. 2016). The deuterated molecules likely arise in regions that are cold and favor ion-molecule reactions (e.g., Bergin 2000). Abundances, relative to H_2, span the range $f \sim 10^{-5}$ for CO to $f \sim 10^{-11}$–10^{-9} for other species (see Henning and Semenov 2013). Thus, overall abundances are less than that in dense molecular clouds. Also, the organic content is not nearly as varied as in dense clouds, with the apparent absence of molecules such as CH_3CH_2OH and $HCOOCH_3$. However, the failure to detect the larger organic species may be a function of telescope sensitivity. It is quite possible that further ALMA studies will reveal that disks have a similar organic complexity to dense clouds. Dust grains appear to be chiefly silicate in composition (amorphous and crystalline), along with carbonaceous types. They have also been found to contain ice and CO_2 mantles.

4.1.9 INTO THE SOLAR SYSTEM: CONNECTING INTERSTELLAR MOLECULAR MATERIAL TO PLANETS

The rich chemistry in protoplanetary disks can significantly influence the composition of planets and their atmospheres, as well as that of asteroids and comets (Aikawa and Herbst 1999; Henning and Semenov 2013). A major uncertainty in the chemistry of protostellar disks is the extent of the processing of molecular material inherited from the primordial dense cloud. Given the physical conditions in disks, with the presence of high temperatures, influx of UV and X-ray photons and turbulent motions, molecular material will be processed, at least in the inner disk; there is also evidence of such processing from meteoritic studies (e.g., Pizzarello and Shock 2010). The degree of processing in the outer disk and the radial distance at which the material remains primordial are not currently known. It can be explored to some extent by studying solar system bodies, namely comets, meteorites, and asteroids.

4.1.9.1 Molecular Composition of Comets

Comets are thought to be among the most primitive bodies in the solar system. They formed from the accretion of icy grains in the outer solar nebula. Many potential comets were incorporated into the giant planets, but others survived, some past the orbit of Neptune, while others were ejected into the Oort cloud (e.g., Cochran et al. 2015). The Oort cloud contains the pristine remains of the original molecular cloud, which spawned the solar system and extends as far out as 100,000 AU.

The chemical composition of comets has been studied by spectroscopic observations, chiefly in the IR and radio/millimeter regions, of gas-phase molecules evaporated from the icy comet nucleus (Biver et al. 2014; Marboeuf and Schmitt 2014). A wide range of chemical compounds have been identified, including simpler compounds such as CO, OH, CS, HCN, CH_4, NH_3, H_2S, and HCCH (e.g., Bockelêe-Morvan et al. 2000), as well the more complex organic species, including H_2CO, CH_3OH, CH_3CN, C_2H_2, C_2H_6, CH_3CHO, HC_3N, $HCOOH$, $HNCO$, $(CH_2OH)_2$, and NH_2CHO (e.g., Biver et al. 2014). A few more organic molecules have been identified in mass spectroscopic data of surface dust, vaporized from the nucleus, from the Rosetta mission; these molecules include CH_3CONH_2, $CHOCH_2OH$, and CH_3NCO (Goesmann et al. 2015). This work prompted the identification of CH_3NCO in interstellar gas. The solid-state component of comets, derived from spectroscopic and scattering measurements, suggests a mixture of amorphous carbon, carbonates, and aromatic hydrocarbons, as well as ice and silicates (Lisse et al. 2006). Studies by Rosetta indicate the presence of non-volatile organic material similar to the "Insoluble Organic Material" found in meteorites, as well as amorphous and crystalline silicates (e.g., Capaccioni et al. 2015).

The chemical composition of comets thus reflects for the most part that found in dense molecular clouds. With the exception of CO, the abundances of compounds found in comets are remarkably similar to those in dense clouds within an order of magnitude (e.g., Mumma and Charnley 2011). There is also a lack of deuterium, ^{13}C, or ^{15}N enrichment in cometary molecules, suggesting that these compounds formed in warmer gas, likely in the GMC that formed the solar system. The evidence indicates that, for the most part, the gas-phase chemical compounds simply condensed out of the nascent cloud and froze onto grains surfaces, which subsequently aggregated to create larger bodies. Any processing that may have occurred is better described as "recycling," because the basic organic content of the interstellar cloud has been preserved.

4.1.9.2 A Link to Meteorites: Organics in Carbonaceous Chondrites

Certain types of meteorites, namely carbonaceous chondrites, are known to contain a wide variety of soluble organic compounds, as well as insoluble, kerogen-like material called "Insoluble Organic Matter" (IOM) (e.g., Pizzarello et al. 2006; Alexander et al. 2007). Many carbonaceous chondrites undergo aqueous processing in the solar system, which is reflected in their organic content, while others appear to be quite pristine. The most primitive, unaltered meteorites contain a wide variety of organic compounds, including α-type amino acids (glycine, isovaline, alanine, and α-aminoisobutyric acid), aldehydes, ketones, hydroxyamino acids, and tertiary amines (e.g., Martins et al. 2007; Elsila et al. 2012). The latter two compound types are very reactive species and likely signify pristine material (Pizzarello and Holmes 2009). Insoluble Organic Matter is macromolecular in nature, with a typical elemental composition $C^{100}H^{71}N^3O^{12}S^2$. It represents the bulk of the organic matter in these types of meteorites, as much as 2% by weight. In addition, there has been one report of phosphorus compounds in meteorites; phosphonic acids, R-POOH, have been thought be have been extracted from the Murchison meteorite (Cooper et al. 1992).

The soluble organic compounds found in unaltered meteorites, and processed ones as well, on average contain many more atoms and are certainly more complex than those found in GMCs. However, simpler species common in dense clouds, such as H_2CO, CH_3CHO, NH_2CH_3, and even NH_3, have been extracted from primitive meteoritic material. These compounds are quite volatile at room temperature and so are likely present in some derivative material in the meteorites. Insoluble Organic Matter is composed of both aromatic and aliphatic carbons. The aromatic material appears to consist of one to a few benzene-like rings, while the aliphatic sections have highly branched short chains (Alexander et al. 2007). Oxygen-containing functional groups are also found in the IOM, mostly carbonyl and carboxylic types. All considered, the same types of functional groups (–COOH, C=O, –NH$_2$, –CH$_3$, etc.) are found in both interstellar and meteoritic molecules, with an emphasis on carbon in the elemental composition, suggesting a connection.

Analysis of isotopic ratios can provide clues to the origin of the meteoritic organics, in particular D/H, $^{12}C/^{13}C$, and $^{14}N/^{15}N$. Studies of both the soluble matter and the IOM have shown that they are often heavily enriched in deuterium, often in ^{15}N, and have some minor enhancement in ^{13}C (e.g., Pizzarello 2007). For example, in 2-aminoisobutyric acid from the primitive meteorite GRA 95229, a high deuterium enrichment of D/H ~ 0.0012 (or δD = +7200‰) has been found (Pizzarello and Holmes 2009). In certain "hotspots" within IOM samples, values of D/H ~ 0.003 and $^{14}N/^{15}N$ ~ 65 have been measured, where $^{14}N/^{15}N_{standard}$ ~ 272 (e.g., Busemann et al. 2006). These enrichments suggest an origin in a very cold environment, where isotope effects can occur. As discussed, molecules in dense clouds show incredibly high deuterium enrichments, which are subsequently preserved on grains and are therefore found in the much warmer environments of hot cores. Such extreme D enhancements have not yet been found in meteoritic compounds, suggesting that there was partial dilution in the protostellar disk but that some enrichment remained, as dust grains aggregated into meteorites. Not all compounds extracted from meteorites show anomalous D/H ratios, either (e.g., see Pizzarello and Holmes 2009). Enrichments in ^{13}C and ^{15}N are more difficult to interpret, as effects in these isotopes are not as extreme, and these anomalies could also be caused by nucleosynthesis. Nonetheless, the fact that any strong D enrichment is seen in meteoritic organic material is suggestive that this matter had at least in part an interstellar origin.

4.1.9.3 Delivery to Planetary Surfaces

It has long been recognized that the Earth and presumably Earth-like planets in other solar systems are bombarded by extraterrestrial bodies, in particular comets, meteorites, and interplanetary dust particles (IDPs) (e.g., Flynn et al. 2003). Furthermore, early Earth was thought to be quite hot and therefore lost most of its volatiles, which contained the bulk of the available carbon. Therefore, a significant fraction of the present surface inventory of organic material was likely brought back to Earth through such bombardment (e.g., Dauphas and Marty 2002; Brack 2014). This matter can be traced back to interstellar molecules, which, despite possible alteration in the protostellar disk, provided the basic organic material that was deposited on solar system bodies and then delivered to the surface of Earth. The carbon enrichment from AGB stars that was preserved in interstellar matter, including macromolecules such as fullerenes, thus may have provided the basis of terrestrial organic chemistry.

4.1.10 INTERSTELLAR MOLECULES AND LIFE

After decades of observations, it has become evident that organic molecular material is present throughout the galaxy in large quantities and survives under harsh conditions. Furthermore, the molecular material is not formed in isolation but is cycled from one phase in the ISM to another. This cycling enables carbon enrichment from the AGB stars to be preserved at some level and seed the organic chemistry of diffuse and then dense clouds. Although there is processing in the protostellar disk, the basic interstellar organic content remains and is condensed onto surfaces that aggregate to create solar system bodies. Some pristine interstellar organic material is still preserved, as indicated by isotopic enrichment found in meteoritic compounds, which are both discrete molecules and polymers; the latter may be of importance in the formation of cell membranes. In addition to organic molecules, phosphorus-bearing molecules are found in circumstellar and interstellar gas, indicating that this element is still active chemically and not all condensed into inert crystalline forms such as apatite. Mixed into this organic soup are metals. Some relevant metals such as iron, magnesium, and potassium have been found in gas-phase molecules, notably bound to the CN moiety. Metallic grains are also present in carbonaceous chondrites, perhaps providing catalytic sites for more complex chemistry needed to produce larger biological molecules.

Do interstellar molecules play role in prebiotic chemistry? Given their widespread abundance and survival on delivery to planet surfaces, the answer is "very likely." The necessary ingredients are all present in interstellar material, including phosphorus, metals, and a broad organic content. However, the degree of chemical complexity contributed by interstellar molecules is still unknown, and many pieces of the puzzle remain missing.

ACKNOWLEDGMENTS

This work was in part supported by NSF grant AST-1515568 and NASA Agreement No. NNX15AD94G issued through the Science Mission Directorate interdivisional initiative Nexus for Exoplanet System Science (NExSS).

REFERENCES

Adande, G.R., and L.M. Ziurys. 2012. Millimeter-wave observations of CN and HNC and their ^{15}N isotopologues: A new evaluation of the ^{14}N/^{15}N ratio across the galaxy. *Astrophys. J.* 744: 194 (15 pp).

Aikawa, Y., and E. Herbst. 1999. Molecular evolution in protoplanetary disks. Two-dimensional distributions and column densities of gaseous molecules. *Astro. Astrophys.* 351: 233–246.

Alexander, C.M.O'D., M. Fogel, H. Yabuta, and G.D. Cody. 2007. The origin and evolution of chondrites recorded in the elemental and isotopic compositions of their macromolecular organic matter. *Geochim. Cosmochim. Acta.* 71: 4380–4403.

Anderson, J.K., and L.M. Ziurys. 2014. Detection of CCN ($X^2\Pi_r$) in IRC+10216: Constraining carbon-chain chemistry. *Astrophys. J.* 795: L1 (6 pp).

Bachiller, R. and M.P. Gutierrez. 1997. Shock chemistry in the young bipolar outflow L1157. *Astrophys. J. (Lett.).* 487: L93–L96.

Belloche, A., H.S.P. Müller, K.M. Menten et al. 2016. Exploring molecular complexity with ALMA (EMoCA): Deuterated complex organic molecules in Sagittarius B2(N2). *Astron. Astrophys.* 587: A91 (66 pp).

Bergin, E.A. 2000. Chemical models of collapsing envelopes. *IAU Symp.* 197: 51 (10 pp).

Bernath, P.F. 2005. *Spectra and Atoms and Molecules.* (Oxford, UK: Oxford University Press).

Biver, N., D. Bockelée-Morvan, V. Debout et al. 2014. Complex organic molecules in comets C/2012 F6 (Lemmon) and C/2013 R1 (Lovejoy): Detection of ethylene glycol and formamide. *Astron. Astrophys.* 566: L5 (5 pp).

Bockelée-Morvan, D., D.C. Lis, J. Wink et al. 2000. New molecules found in comet C/1995 O1 (Hale-Bopp): Investigating the link between cometary and interstellar material. *Astron. Astrophys.* 353: 1101–1114.

Brack, A. 2014. *Extraterrestrial Delivery of Organic Compounds. Encyclopedia of Astrobiology.* (Berlin, Germany: Springer-Verlag).

Busemann, H., A.F. Young, C.M.O'D Alexander, P. Hoppe, S. Mukhopadhyay, and L.R. Nittler, 2006. Interstellar chemistry recorded in organic matter from primitive meteorites. *Science* 312: 727–730.

Cami, J., J. Bernard-Salas, E. Peeters, and S. Malek. 2010. Detection of C_{60} and C_{70} in a young planetary nebula. *Science* 329: 1180–1182.

Capaccioni, F., G. Filacchione, S. Erard et al. 2015. The nucleus and coma of 67P/Churyumov-Gerasimenko: Highlights of the Rosetta-VIRTIS results. *EGU General Assembly Conference Abstracts* 17: 12375.

Cesaroni, R., M.T. Beltrán, Q. Zhang, H. Beuther, and C. Fallscheer. 2011. Dissecting a hot molecular core: The case of G31.41 + 0.3. *Astron. Astrophys.* 533: A73 (14 pp).

Clayton, D.D., and B.S. Meyer. 2018. Graphite grain-size spectrum and molecules from core-collapse supernovae. *Geochem. Cosmochim. Acta.* 221: 47–59.

Cochran, A. L., A. Levasseur-Regourd, M. Cordiner et al. 2015. The composition of comets. *Spa. Sci. Rev.* 197: 9–46.

Cooper, G.W., W.M. Onwo, and J.R. Cronin. 1992. Alkyl phosphonic acids and sulfonic acids in the Murchison meteorite. *Geochim. Cosmochim. Acta.* 56: 4109–4115.

Cuadrado, S., J.R. Goicoechea, P. Pilleri, J. Cernicharo, A. Fuente, and C. Joblin. 2015. The chemistry and spatial distribution of small hydrocarbons in UV-irradiated molecular clouds: The Orion Bar PDR. *Astron. Astrophys.* 575: A82 (15 pp).

Dauphas, N., and B. Marty. 2002. Inference on the nature and the mass of Earth's late veneer from noble metals and gases. *J. Geophys. Res.* 107: E12 (7 pp).

Dorschner, J., and T. Henning. 1995. Dust metamorphosis in the galaxy. *Astron. Astrophys. Rev.* 6: 271–333.

Edwards, J.L., E.G. Cox, and L.M. Ziurys. 2014. Millimeter observations of CS, HCO+, and CO toward five planetary nebulae: Following molecular abundance with nebular age. *Astrophys. J.* 791: 79 (15 pp).

Edwards, J.L., and L.M. Ziurys. 2014. Sulfur and silicon-bearing molecules in planetary nebulae: The case of M2-48. *Astrophys. J. Lett.* 794: L27 (6 pp).

Elsila, J.E., S.B. Charnley, A.S. Burton, D.P. Glavin, and J.P. Dworkin. 2012. Compound specific carbon, nitrogen, and hydrogen isotope rations for amino acids in CM and CR meteorotes and their use in evaluating potential formation pathways. *Meteorit. Planet. Sci.* 47: 1517–1536.

Ferrière, K.M. 2001. The interstellar environment of our galaxy. *Rev. Mod. Phys.* 73: 1031–1066.

Flynn, G.J., L.P. Keller, M. Feser, S. Wirick, and C. Jacobsen. 2003. Origin of organic matter in the solar system: Evidence from the interplanetary dust particles. *Geochim. Cosmochim. Acta.* 67: 4791–4806.

García-Hernández, D.A., E. Villaver, P. García-Lario. 2012. Infrared study of fullerene planetary nebulae. *Astrophys. J.* 760: 107 (16 pp).

Geppert, W.D., M. Hamberg, R.D. Thomas et al. 2006. Dissociative recombination of protonated methanol. *Faraday Discuss.* 133: 177–190.

Gerin, M., M. de Luca, J. Black et al. 2010. Interstellar OH+, H_2O+ and H_3O+ along the sight-line to G10.6−0.4. *Astron. Astrophys.* 518: L110 (5 pp).

Gerlich, D. and S. Horning. 1992. Experimental investigations of radiative association processes as related to interstellar chemistry. *Chem. Rev.* 92: 1509–1539.

Glassgold, A.E. 1996. Circumstellar photochemistry. *Ann. Rev. Astron. Astrophys.* 34: 241–278.

Goesmann, F., H. Rosenbauer, J.H. Bredehoeft et al. 2015. Organic compounds on comet 67P/Churyumov–Gerasimenko revealed by COSAC mass spectrometry. *Science* 349: aab0689 (3 pp).

Goto, M., T. Usuda, N. Takato et al. 2003. Carbon isotope ratio in $^{12}CO/^{13}CO$ toward local molecular clouds with near-infrared high-resolution spectroscopy of vibrational transition bands. *Astrophys. J.* 598: 1038–1047.

Guilloteau, S., L. Reboussin, A. Dutrey et al. 2016. Chemistry in disks. X. The molecular content of protoplanetary disks in Taurus. *Astron. Astrophys.* 592: A124 (29 pp).

Halfen, D.T., A.J. Apponi, and L.M. Ziurys. 2001. Evaluating the N/O chemical network: The distribution of N_2O and NO in the Sagittarius B2 complex. *Astrophys. J.* 561: 244–253.

Halfen, D.T., V.V. Ilyushin, and L.M. Ziurys. 2015. Interstellar detection of methyl isocyanate CH_3NCO in Sgr B2(N): A link from molecular clouds to comets. *Astrophys. J. Lett.* 812: L5 (8 pp).

Halfen, D.T., A.J. Apponi, N.J. Woolf, R. Polt, and L.M. Ziurys. 2006. A systematic study of Glycolaldehyde in Sgr B2(N) at 2 and 3 millimeters: Criteria for detecting large interstellar molecules. *Astrophys. J.* 639: 237–245.

Halfen, D.T., N.J. Woolf, and L.M. Ziurys. 2017. The $^{12}C/^{13}C$ ratio in Sgr B2 (N): Constraints for galactic chemical evolution and isotopic chemistry. *Astrophys. J.* 845: 845, 158 (11 pp).

Henning, Th., and D. Semenov. 2013. Chemistry in protoplanetary disks. *Chem. Rev.* 113: 9016–9042.

Herbst, E. and W. Klemperer. 1976. The formation of interstellar molecules. *Phys. Today* 29: 32–39.

Herbst, E. and T.J. Millar. 2008. The chemistry of cold interstellar cores, In *Low Temperatures and Cold Molecules.* Ed. I.W.M. Smith (London, UK: Imperial College Press), p. 1.

Herwig, F. 2005. Evolution of asymptotic giant branch stars. *Ann. Rev. Astron. Astrophys.* 43: 435–479.

Herwig, F. 2013. *Evolution of Solar and Intermediate-Mass Stars. Planets, Stars and Stellar Systems.* 4 (Dordrecht, the Netherlands: Springer), p. 397.

Herzberg, G. 1971. *Spectra and Structure of Simple Free Radicals.* (New York: Dover).

Hickson, K.M., V. Wakelam, and J. Loison. 2016. Methylacetylene (CH_3CCH) and propene (C_3H_6) formation in cold dense clouds: A case of dust grain chemistry. *Mol. Astrophys.* 3: 1–9.

Hollis, J.M., F.J. Lovas, A.J. Remijan, P.R. Jewell, V.V. Ilyushin, and I. Kleiner. 2006. Detection of acetamide (CH_3CONH_2): The largest interstellar molecule with a peptide Bond. *Astrophys. J.* 693: L25–L28.

Inoue, T., and S. Inutsuka. 2012. Formation of turbulent and magnetized molecular clouds via accretion flows of H I clouds. *Astrophys. J.* 759: 35 (14 pp).

Jones, A.P. 2001. Interstellar and circumstellar grain formation and survival. *Philos. Trans. R. Soc. London, Ser. A* 359: 1961–1972.

Jones, A.P., and J.A. Nuth III. 2011. Dust destruction in the ISM: A re-evaluation of dust lifetimes. *Astron. Astrophys.* 530: A44 (12 pp).

Kaifu, N., M. Ohishi, K. Kawaguchi et al. 2004. A 8.8–50GHz complete spectral line survey toward TMC-1 I. Survey data. *PASJ* 56: 69–173.

Kamiński, T., C.A. Gottlieb, K. Menten et al. 2013. Pure rotational spectra of TiO and TiO_2 in VY Canis Majoris. *Astron. Astrophys.* 551: A113 (13 pp).

Kimura, R.K., R. Gruenwald, and I. Aleman. 2012. Molecular chemistry and the missing mass problem in planetary nebulae. *Astron. Astrophys.* 541: A112 (11 pp).

Kwok, S. 2000. *The Origin and Evolution of Planetary Nebulae.* (Cambridge, UK: Cambridge University Press).

Kwok, S. 2004. The synthesis of organic and inorganic compounds in evolved stars. *Nature* 430: 985–991.

Kwok, S., and Y. Zhang. 2011. Mixed aromatic–aliphatic organic nanoparticles as carriers of unidentified infrared emission features. *Nature* 479: 80–83.

Lequeux, J. (2013). *Birth, Evolution and Death of Stars.* (Singapore: World Scientific).

Lisse, C.M., J. Van Cleve, A.C. Adams et al. 2006. Spitzer spectral observations of the deep impact ejecta. *Science* 313: 635–640.

Liszt, H.S., R. Lucas, R., and J. Pety. 2006. Comparative chemistry of diffuse clouds. V. Ammonia and formaldehyde. *Astron. Astrophys.* 448: 253–259.

Lodders, K., and B. Fegley, Jr. 1999. Condensation chemistry of circumstellar grains. Giant branch stars. *IAU Symp.* 191: 279.

Lodders, K., and S. Amari. 2005. Presolar grains from meteorites: Remnants from the early times of the solar system. *Chem. Erde Geochem.* 65: 93–166.

McCabe, E.M., R. Connon Smith, and R.E.S. Clegg. 1979. Molecular abundances in IRC+10216. *Nature* 218: 263–266.

McKellar, A. 1940. Evidence for the molecular origin of some hitherto unidentified interstellar lines. *PASP* 52: 187–192.

Marboeuf, U., and B. Schmitt. 2014. How to link relative abundances of gas species in coma of comets to their chemical composition? *Icarus* 242: 225–248.

Martins, Z., C.M.O'D. Alexander, G.E. Orzechowska, M.L. Fogel, and P. Ehrenfreund. 2007. Indigenous amino acids in primitive CR meteorites. *Meteorit. Planet. Sci.* 42: 2125–2136.

Mattsson, L. 2010. The origin of carbon: Low-mass stars and an evolving, initially top-heavy IMF? *Astron. Astrophys.* 515: A68 (12 pp).

Millar, T.J. 2005. Deuterium in interstellar space: Deuterium in interstellar clouds. *Astron. Geophys.* 46: 2.29–2.32.

Mowlavi, N. 1999. On the third dredge-up phenomenon in asymptotic giant branch stars. *Astron. Astrophys.* 344: 617–631.

Mumma, M.J., and S.B. Charnley. 2011. The chemical composition of comets—emerging taxonomies and natal heritage. *Ann. Rev. Astron. Astrophys.* 49: 471–524.

Neill, J.L., S. Wang, E.A. Bergin, et al. 2013. The abundance of H_2O and HDO in Orion Kl from Herschel/HIFI. *Astrophys. J.* 770: 142 (18 pp).

O'Dell, C.R., W.J. Henney, and G. Ferland. 2007. Determination of the physical conditions of the knots in the Helix Nebula from Optical and Infrared Observations. *Astronom. J.* 133: 2343–2356.

Pizzarello, S., G.W. Cooper, and G.J. Flynn. 2006. The nature and distribution of the organic material in carbonaceous chondrites and interplanetary dust particles. *Meteorites and the Early SolarSystem II.* D.S. Lauretta and H.Y. McSween (Eds.) (Tucson, AZ: University of Arizona Press).

Pizzarello, S. 2007. Question 2: Why astrobiology? *Orig. Life Evol. Biosph.* 37: 341–344.

Pizzarello, S., and W. Holmes. 2009. Nitrogen-containing compounds in two CR 2 Meteorites: ^{15}N composition, molecular distribution, and precursor molecules, *Geochem. Cosmochim. Acta.* 73: 2150–2162.

Pizzarello, S., and E. Shock 2010. The organic composition of carbonaceous meteorites: The evolutionary story ahead of biochemistry. *Cold Spring Harb. Perspect. Biol.* 2: a0021051 (19 pp).

Rivilla, V.M, F. Fontani, M.T. Beltrán et al. 2016. The first detections of the key prebiotic molecule PO in star-forming regions. *Astrophys. J.* 826: 161 (8 pp).

Redman, M.P., S. Viti, P. Cau, and D.A. Williams. 2003. Chemistry and clumpiness in planetary nebulae. *MNRAS* 345: 1291–1296.

Sakai, N. 2015. Protostellar disk formation traced by chemistry. *ASP Conf. Ser.* 499: 199–204.

Schmidt, D.R., and L.M. Ziurys. 2016. Hidden molecules in planetary nebulae: New detections of HCN and HCO+ from a multi-object survey. *Astrophys. J.* 817: 175 (17 pp).

Schmidt, D.R., and L.M. Ziurys. 2017a. New detections of HNC in planetary nebulae: Evolution of the [HCN]/[HNC] ratio. *Astrophys. J.* 835: 79 (12 pp).

Schmidt, D.R., and L.M. Ziurys. 2017b. New identifications of the CCH radical in planetary nebulae: A connection to C_{60}? *Astrophys. J.* 850: 123 (10 pp).

Snow, T.P., and B.J. McCall. 2006. Diffuse atomic and molecular clouds. *Annu. Rev. Astron. Astrophys.* 44: 367–414.

Stantcheva, T., and E. Herbst. 2004. Models of gas-grain chemistry in interstellar cloud cores with a stochastic approach to surface chemistry. *Astron. Astrophys.* 432: 241–251.

Snyder, L.E., D. Buhl, B. Zuckerman, and P. Palmer. 1969. Microwave detection of interstellar formaldehyde. *Phys. Rev. Lett.* 22: 679–681.

Snyder, L.E., and D. Buhl. 1971. Observations of radio emission from interstellar hydrogen cyanide. *Astrophys. J. Lett.* 163: L47–L53.

Taniguchi, K., H. Ozeki, Hiroyuk, M. Saito et al. 2016. Implication of formation mechanisms of HC_5N in TMC-1 as studied by ^{13}C isotopic fractionation. *Astrophys. J.* 817: 147 (7 pp).

Tenenbaum, E.D., N.J. Woolf, and L.M. Ziurys. 2007. Identification of phosphorus monoxide ($X^2\Pi_r$) in VY Canis Majoris: Detection of the first P–O bond in space. *Astrophys. J. Lett.* 666: L29–L32.

Tenenbaum, E.D., S.N. Milam, N.J. Woolf, and L.M. Ziurys. 2009. Molecular survival in evolved planetary nebulae: Detection of H_2CO, c-C_3H_2, and C_2H in the helix. *Astrophys. J. Lett.* 704: L108–L112.

Tenenbaum, E.D., J.L. Dodd, S.N. Milam, N.J. Woolf, and L.M. Ziurys. 2010. Comparative spectra of oxygen-rich vs. carbon-rich circumstellar shells: VY Canis Majoris and IRC+10216 at 215–285 GHz. *Astrophys. J.* 720: L102–L107.

Tenenbaum, E.D., and L.M. Ziurys. 2010. Exotic metal molecules in oxygen-rich envelopes: Detection of AlOH ($X^1\Sigma+$) in VY Canis Majoris. *Astrophys. J. Lett.* 712: L93–L96.

Thaddeus, P., C.A. Gottlieb, H. Gupta et al. 2008. Laboratory and astronomical detection of the negative molecular ion C_3N^-. *Astrophys. J.* 677: 1132–1139.

van der Wiel, M.H.D., F.F.S. van der Tak, S. Ossenkopf, et al. 2009. Chemical stratification in the Orion Bar: JCMT spectral legacy survey observations. *Astron. Astrophys.* 498: 161–165.

van Dishoeck, E., and J.H. Black. 1986. Comprehensive models of diffuse interstellar clouds—Physical conditions and molecular abundances. *Astrophys. J.* 62: 109–145.

Viti, S., I. Jimenez-Serra, J.A. Yates et al. 2011. L1157-B1: Water and ammonia as diagnostics of shock temperature. *Astrophys. J. Lett.* 740: L3 (5 pp).

Walker, G.A.H., D. Bohlender, J.P. Maier, and E.K. Campbell. 2015. Identification of more interstellar $C_{60}+$ bands. *Astrophys. J. Lett.* 812: L8 (5 pp).

Weinreb, S., A.H. Barrett, M.L. Meeks, and J.C. Henry. 1963. Radio observations of OH in the interstellar medium. *Nature* 200: 829–831.

Williams, J.P., and Cieza, 2011. Protoplanetary disks and their evolution. *Annu. Rev. Astron. Astrophys.* 49: 67–117.

Wilson, R.W., K.B. Jefferts, and A.A. Penzias. 1970. Carbon monoxide in the Orion Nebula. *Astrophys. J. Lett.* 161: L43–L44.

Wilson, R.W., A.A. Penzias, K.B. Jefferts, M. Kutner, and P. Thaddeus. 1971. Discovery of interstellar silicon monoxide. *Astrophys. J. Lett.* 167: L97–L100.

Xu, J. and J. Wang. 2013. Kinematics and chemistry of the hot core in G20.08–0.14N. *MNRAS* 432: 2385–2396.

Zack, L.N., and L.M. Ziurys. 2013. Chemical complexity in the helix nebula: Multi-line observations of H_2CO, HCO+, and CO. *Astrophys. J.* 756: 112 (14 pp).

Zeigler, N.R., L.N. Zack, N.J. Woolf, and L.M. Ziurys. 2013. The helix nebula viewed in HCO+: Large scale mapping of the $J = 1 \rightarrow 0$ transition. *Astrophys. J.* 778: 16 (9 pp).

Ziurys, L.M. 1987. Detection of interstellar PN: The first phosphorus-bearing species observed in molecular clouds. *Astrophys. J. Lett.* 321: L81–L84.

Ziurys, L.M., P. Friberg, and W.M. Irvine. 1989. Interstellar SiO as a tracer of high temperature chemistry. *Astrophys. J.* 343: 201–207.

Ziurys, L.M., D.T. Halfen, W.D. Geppert, and Y. Aikawa. 2016. Following the interstellar history of carbon: From the interiors of stars to the surfaces of planets. *Astrobiology* 16: 997–1012.

Ziurys, L.M., D.R. Schmidt, and J.J. Bernal. 2018. New circumstellar sources of PO and PN: The increasing role of phosphorus chemistry in oxygen-rich stars. *Astrophys. J.* 856: 169 (11 pp).

4.2 Formation and Delivery of Complex Organic Molecules to the Solar System and Early Earth

Sun Kwok

CONTENTS

4.2.1 INTRODUCTION

Life as we know is based on chemical building blocks of organic molecules. Although organic molecules extracted from living organisms were once believed to possess a special component called "vitality," we no longer believe "vitality" is a real physical entity, as many organic molecules extracted from living organisms can also be artificially synthesized in the laboratory. Instead, organic molecules are just molecules with the elements carbon (C) and hydrogen (H) as major constituents while also containing other elements such as nitrogen (N), oxygen (O), sulfur (S), and phosphorus (P). From a chemical point of view, there is no intrinsic difference between organic and other molecules.

Since the 1950s, we have learned that all chemical elements (except H and some helium, He) in our bodies originated from nucleosynthesis in the interior of stars (Burbidge et al. 1957). This is the first established link between stars and life on Earth. Advances in infrared and millimeter-wave observing capabilities in the 1960s led to the discovery of molecules in space and to our realization that stars can synthesize molecules and minerals during their late stages of evolution (Ziurys et al. 2016). The most surprising aspect of these discoveries was the detection of complex organics in the circumstellar envelopes of evolved stars. Infrared spectroscopic observations have shown that common, ordinary stars can synthesize complex organics during their last stage of evolution in the planetary nebulae phase, before their nuclear fuel supplies are exhausted in the following white dwarf phase. This organic synthesis takes place under an extremely low ($<10^6$ cm^{-3}) density environment and occurs over very short (10^4 year) timescales. These organics are ejected into the interstellar medium and quite probably have spread across the Milky Way Galaxy (Kwok 2004).

In this chapter, we will summarize the observational evidence of stellar organic synthesis, discuss the chemical structure of these organics, explore the possible relationship between stellar organics and those found in the solar system, and speculate on the delivery of stellar organics to Earth and its implications on the origin of life on Earth.

4.2.2 LATE STAGES OF STELLAR EVOLUTION

Stars like our Sun maintain their luminosities by generating energy through fusing H into He in the core. After the supply of H is exhausted in the core, H is converted into He in a shell surrounding the inert He core. The outer envelope of the star expands, and the luminosity increases to several hundred times the current solar luminosity ($L_\odot$). This stage is called the red giant phase. Further contraction of the core leads to the ignition of He in a shell, and He is converted into carbon (C) through the triple-α reaction. The envelope of the star expands to a size more than one astronomical unit (AU, the distance between the Earth and the Sun) and the luminosity climbs to more than 3000 $L_\odot$. As the result of size expansion, the surface temperature of the star drops to 3000 K, with its emerging radiation shifting from the visible to the infrared, peaking at a wavelength of ~1 μm. This is referred to in astronomical nomenclature as the asymptotic giant branch (AGB) phase.

A strong stellar wind develops during the late AGB phase, ejecting mass in the stellar envelope into interstellar space. For stars with initial mass under eight times the mass of the Sun, this mass loss process can remove most of the envelope mass before the ignition of C and avoids the fate of becoming a supernova. This path of evolution is followed by over 95% of all stars in our Milky Way Galaxy.

After the complete depletion of the H envelope by mass loss, a faster wind develops and compresses and accelerates the previously ejected circumstellar material into a high-density shell. As the core is gradually exposed by its diminishing envelope, the star increases its temperature from 3000 K to over 100,000 K (Figure 4.2.1). The increasing ultraviolet (UV) radiation output from the star photoionizes the circumstellar gas,

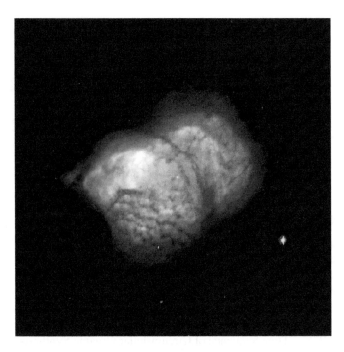

FIGURE 4.2.2 A composite color image of the planetary nebula NGC 7027. The color of the nebula is due to emission lines from ions of H, O, and N. Unidentified infrared emission (UIE) bands were first discovered in the spectrum of this nebula.

and the resulting strong atomic emission lines create a bright optical nebula called planetary nebula (Figure 4.2.2). Planetary nebula is a short-lived phase of stellar evolution, lasting only about 20,000 years, before it disperses into the interstellar medium. The hot central star (core of the progenitor AGB star) gradually burns out its H fuel and becomes a white dwarf. A detailed description of the late stages of evolution leading to the formation of planetary nebulae is given in Kwok (2000).

4.2.3 CIRCUMSTELLAR SYNTHESIS OF MOLECULES AND MINERALS

Carbon atoms synthesized in the core of AGB stars are dredged up to the stellar surface through the process of convection. In the low-temperature atmosphere of AGB stars, simple molecules such as C_2, C_3, and CN can form. As the envelope is ejected by the stellar wind, further chemical reactions have been observed to take place in the circumstellar envelope, leading to a rich variety of chemical species. As the gas expands and cools, atoms combine to form molecules and gas condenses directly into solid phase to form micron-size solid grains. Solid particles in the stellar wind can be identified by their lattice vibrational modes in the infrared through infrared spectroscopy. The rotational transitions of gas-phase molecular species can be detected by millimeter- and sub-millimeter-wave spectroscopy. As of 2018, approximately 80 molecular species have been detected in the circumstellar environment of evolved stars. The detected molecular species include inorganics (CO, SiO, SiS, NH_3, and AlCl, etc.), organics (C_2H_2, CH_4, H_2CO, CH_3CN, etc.), radicals (CN, C_2H, C_3, HCO^+, etc.), chains (HCN, HC_3N, HC_5N, etc.), and rings (C_3H_2) (Ziurys 2006; Ziurys et al. 2016).

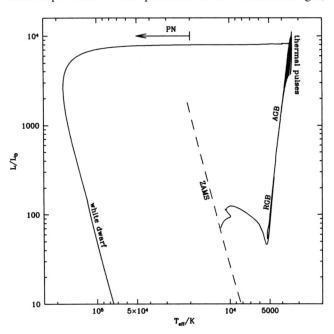

FIGURE 4.2.1 The evolutionary track of a 3-solar-mass star on a plot of luminosity (vertical axis) versus temperature (horizontal axis) from the zero age main sequence (ZAMS), through the red giant branch (RGB) and asymptotic giant branch (AGB) to planetary nebulae (PN) and ending as a white dwarf (Figure from T. Blöcker). The beginning of the planetary nebulae phase is indicated by the end of an arrow, when the star reaches a temperature of ~20,000 K, with sufficient output of UV photons to photoionize the surrounding nebula. The evolutionary stage between the end of the AGB and the beginning of planetary nebulae phases is called proto-planetary nebulae phase (Kwok 1993). During this phase, the circumstellar nebulae have no atomic line emission and are only illuminated by scattered starlight.

Micron-size solid-state particles (commonly referred to as dust in the astronomical literature) can be directly detected by their thermal radiation in the infrared. The first circumstellar solids discovered were amorphous silicates, which were identified by their Si–O stretching and the Si–O–Si bending modes at 9.7 and 18 μm, respectively (Woolf and Ney 1969). These features are detected in over 4,000 O-rich AGB stars by the Low-Resolution Spectrometer (LRS) on board of the *Infrared Astronomical Satellite* (*IRAS*) all-sky survey (Kwok et al. 1997). These features can be in emission or in self-absorption, depending on the amount of mass in the circumstellar envelope (Figure 4.2.3).

A variety of refractory oxides (corundum α-Al_2O_3, spinel $MgAl_2O_4$, and rutile TiO_2) are also detected in the envelopes of AGB stars (Posch et al. 1999, 2002). Crystalline silicates such as pyroxenes and olivines have sharper features and are also detected in the circumstellar environment (Jäger et al. 1998). The wide detection of minerals in stars by infrared techniques led to the emergence of the field of astromineralogy (Henning 2009).

In C-rich stars, where the atmospheric C abundance exceeds that of O, all O atoms are tied up in CO and the surplus C atoms form C-based molecules and solids. The most common C-based solid is silicon carbide (SiC), for which 11.3 μm emission feature is observed in over 700 C-rich stars in the *IRAS LRS* survey (Kwok et al. 1997). For highly evolved C-rich stars, the dust component is believed to be dominated by amorphous carbon, which emits a strong featureless continuum in the infrared (Volk et al. 2000). The circumstellar dust can completely obscure the optical surface (photosphere) of the star and convert all its energy output to the infrared. These highly evolved stars have no optical counterparts and can be detected only in the infrared. They are referred to as extreme carbon stars. It is also during this very late AGB evolution stage that the molecule acetylene (C_2H_2) is seen (Figure 4.2.4).

In the terrestrial environment, the formation of molecules and solids is the result of three- or multi-body collisions. The densities of the stellar winds, however, are much too low (<10^6 H atoms per c.c.) for such processes. From observations of novae, we learn that silicate dust condenses as soon as the radiation temperature drops below the condensation temperature, regardless of the density conditions. The emergence of a dust spectrum from a pure-gas spectrum in novae can take place over a time span as short as a few days (Ney and Hatfield 1978). From mapping of molecular distribution in the circumstellar envelopes of AGB stars by millimeter-wave interferometric techniques, we can set limits to the formation time of circumstellar molecules as hundreds of years, based on the sizes of the molecular emitting regions and expansion velocities of the envelopes. These observations suggest that

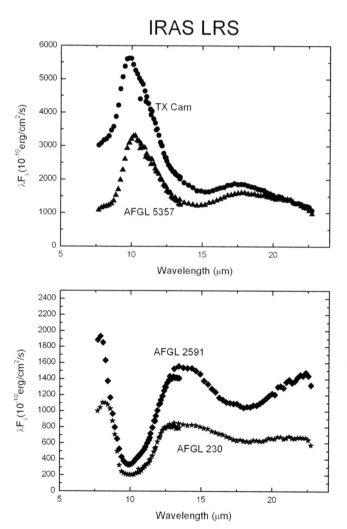

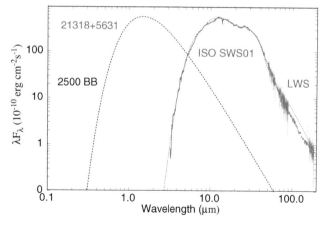

FIGURE 4.2.3 Infrared spectra of amorphous silicates. The 9.7 and 18 μm features of silicates can be seen both in emission (top panel) and in self-absorption (bottom panel). Data obtained from the Low-Resolution Spectrometer observations from the *Infrared Astronomical Satellite*.

FIGURE 4.2.4 The extreme carbon star IRAS 21318 + 5631 is an example of a star so obscured by its own ejected circumstellar dust envelope that the central star becomes undetectable in the optical region. Its infrared spectrum (solid line) is completely due to dust emission and has a color temperature of 300 K. The dotted line represents the theoretical fit to the spectrum, based on a one-dimensional (1-D) radiation transfer model, with a hidden 2500 K central star (dashed line) as the energy source. The absorption feature near the peak of the spectrum is the 13.7 μm band of acetylene.

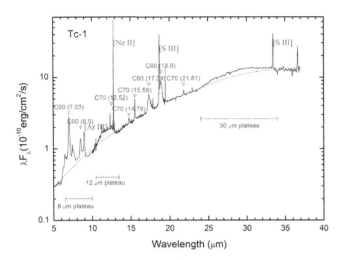

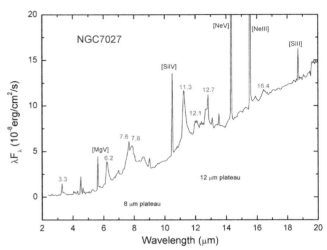

FIGURE 4.2.5 The *Spitzer* IRS spectrum of planetary nebula Tc-1 shows C_{60} and C_{70} emission bands as well as broad emission plateau features at 8, 12, and 30 μm. The narrow lines are atomic lines. No UIE bands are seen in this object.

FIGURE 4.2.6 *Infrared Space Observatory* (*ISO*) spectrum of the planetary nebula NGC 7027 showing the UIE bands (labeled in red, with the wavelength of peak emission in units of microns). Broad emission plateaus around 8 and 12 μm as well as a strong underlying continuum can be seen. The narrow lines (in blue) are atomic lines.

molecule and solid syntheses in the circumstellar environment can be extremely efficient, in spite of the low-density environment.

Since the initial discovery of fullerenes in the laboratory (Kroto et al. 1985), there have been strong interests in searching for this molecule in space. C_{60} is now unambiguously detected in planetary nebulae (Cami et al. 2010; García-Hernández et al. 2010), in reflection nebulae (Sellgren et al. 2010), and in proto-planetary nebulae (Zhang and Kwok 2011, 2013). These cage-like molecules are detected through either the C–C or the C–H stretching vibrational modes in the infrared. With 60 or more C atoms, fullerenes are the heaviest molecular species detected in the envelopes of stars. The paths of synthesis of C_{60} have been suggested to be either top-down (as breakdown products of complex organics) or bottom-up (built up from small C-based molecules) (Bernard-Salas et al. 2012; García-Hernández et al. 2012). As broad plateau emission features always accompany C_{60} features (Figure 4.2.5), it is possible that C_{60} synthesis is related to the unidentified infrared emission (UIE) phenomenon (see Section 4.2.4).

In a circumstellar environment, H can easily attach to the fullerene molecule and become hydrogenated fullerene (or fullerane, $C_{60}H_m$, $m = 1$–60). The theoretical vibrational spectra of fulleranes can be calculated and compared with observed astronomical spectra, resulting in the suggestion that fullerance may be common in the circumstellar environment (Zhang and Kwok 2013; Zhang et al. 2017).

4.2.4 UNIDENTIFIED INFRARED EMISSION BANDS

Due to the low-density and high-UV-radiation background of the interstellar medium, complex organic molecules and solids were not believed to be able to exist in space. The detection of the UIE features in the spectra of planetary nebulae, now attributed to organics, came as a complete surprise. A family

of broad infrared bands at 3.3, 6.2, 7.7, 8.6, and 11.3 μm was first found in the spectrum of the planetary nebula NGC 7027 (Russell et al. 1977) (Figures 4.2.2 and 4.2.6). Shortly after discovery, it was recognized that the UIE bands probably arise from the vibrational modes of organic compounds (Knacke 1977; Duley and Williams 1979). More specifically, the UIE features are suggested to be originating from stretching and bending vibrational modes of aromatic compounds (Duley and Williams 1981).

Also present in spectra are emission features around 3.4 μm, which arise from symmetric and anti-symmetric C–H stretching modes of methyl and methylene groups (Puetter et al. 1979; Geballe et al. 1992). The bending modes of these groups also manifest themselves at 6.9 and 7.3 μm (De Muizon et al. 1990; Chiar et al. 2000). In addition, there are weaker unidentified emission features at 15.8, 16.4, 17.4, 17.8, and 18.9 μm, which probably arise from C-skeleton vibrational modes.

The emission bands themselves are often accompanied by strong, broad emission plateaus features at 6–9, 10–15, and 15–20 μm. The first two plateau features have been identified as superpositions of in-plane and out-of-plane bending modes emitted by a mixture of aliphatic side groups attached to aromatic rings (Kwok et al. 2001).

The UIE bands are extremely prominent in the spectra of C-rich planetary nebulae (Figure 4.2.6). They are not seen in AGB stars but emerge during the proto-planetary nebula phase. Their existence suggests that organic compounds can be efficiently synthesized in the circumstellar environment over very short (10^3 year) timescales.

The UIE and the plateau emission features lie on top of a strong continuum that extends from the near-infrared to millimeter wavelengths. This continuum emission must be due to thermal emission from micron-size solid grains, similar to those observed in AGB stars (Figure 4.2.4). In planetary nebulae or

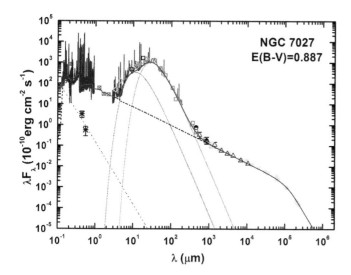

FIGURE 4.2.7 The spectral energy distribution of the planetary nebula NGC 7027 from wavelength 0.1 μm (ultraviolet) to 10 cm (radio). The continuum emission of the object is shown in a red line. The continuum shortward of wavelengths of 3 μm is due to bound-free gas emission. Between wavelengths 1 μm and 1 mm, the continuum is due to solid-state particle (dust) emission. At wavelengths longward of 1 mm, free-free gas emission dominates. We can see that most of the energy output is emitted by a dust component that is approximated by a sum of two blackbodies (dotted lines) of 246 K and 97 K. The narrow lines above the continuum are atomic lines and UIE bands.

active galaxies, most of the output energy of the object is emitted through this continuum component (Figure 4.2.7). Even in the diffuse interstellar medium, the strengths of the UIE features are strongly correlated with the dust continuum, suggesting that the heating source for the UIE carriers and the dust continuum must be the same (Kahanpää et al. 2003).

4.2.5 CARRIER OF THE UNIDENTIFIED INFRARED EMISSION BANDS

Noting that the astronomical UIE spectra resemble the spectra of automobile exhaust soot particles, Allamandola et al. (1985) suggested that the UIE carriers are simple polycyclic aromatic hydrocarbon (PAH) molecules. It is now widely believed in the astronomical community that the UIE features are due to infrared fluorescence of far-UV-pumped PAH molecules, each containing ~50 C atoms (Tielens 2008). These small gas-phase PAH molecules can be excited by a single incoming UV photon stochastically to a high temperature over a short time interval, therefore accounting for the strengths of the UIE features under low-temperature interstellar conditions (Sellgren 1984). By engaging a mixture of hundreds of PAH molecules of different sizes, structures (compact, linear, and branched), and ionization states, and broadened by artificial broad intrinsic line profiles, the PAH model can fit the astronomical UIE spectra.

The fitting of the astronomical UIE features by PAH molecules is not straightforward. Although PAH molecules were known to have C–H stretching modes around 3.3 μm, the actual wavelengths of the C–H stretching is shortward of the 3.3 μm UIE band (Sakata et al. 1990; Kwok and Zhang

2013). Polycyclic aromatic hydrocarbon molecules generally have out-of-plane bending modes around 10–14 μm, but to fit the observed UIE 11.3 μm feature requires exotic mix of PAH molecules (Sadjadi et al. 2015b). The assignments of the 6.2, 7.7, and 8.6 μm UIE features to PAH molecules are even more difficult. It was suggested that the 6.2 μm feature is due to C–C stretching and the 8.6 μm feature is due to C–H in-plane bending modes of PAH molecules (Léger and Puget 1984). The 7.7 μm UIE feature has no obvious counterpart in PAH spectra and has been suggested to be due to blending of several C–C stretching modes (Allamandola et al. 1989). Since C–C stretching modes in PAH molecules are generally very weak, it was later proposed that the 6.2 and 7.7 μm bands are due to ionized PAH molecules, which tend to show stronger C–C stretching modes (Hudgins and Allamandola 1999; van Diedenhoven et al. 2004). However, experimental results show that the C–C vibrational modes of PAH ions generally occur at wavelengths longer than 6.2 μm (Bauschlicher 2002; Hudgins et al. 2005). By increasing the size of the molecules, the peak of the band can shift to shorter wavelengths but never as short as 6.2 μm (Hudgins et al. 2005). In order to account for this discrepancy, it was suggested that some of the C atoms in specific positions in the ring be replaced by nitrogen (N), and this modified PAH model is referred to as the PANH model.

In addition, the PAH hypothesis suffers from the following problems: (1) PAH molecules have well-defined sharp features, but the UIE features are broad; (2) PAHs are primarily excited by UV, with little absorption in the visible, but UIE features are seen in proto-planetary nebulae and reflection nebulae, objects with very little UV background radiation, and the shapes and peak wavelengths of UIE features are independent of temperature of the exciting stars (Uchida et al. 2000); (3) the strong and narrow PAH gas-phase features in the UV are not seen in interstellar extinction curves to very low upper limits (Clayton et al. 2003; Gredel et al. 2011; Salama et al. 2011); (4) no specific PAH molecule has been detected, in spite of the fact that the vibrational and rotational frequencies of PAH molecules are well known; (5) there are great difficulties in reconciling the band positions or relative intensities of laboratory PAH spectra with astronomical UIE spectra (Wagner et al. 2000); (6) the large number of free parameters in the PAH model fitting suggests that such fittings are not very meaningful (Zhang and Kwok 2015).

In response to these criticisms, the PAH model has been revised to incorporate ionization states and large sizes to increase the absorption cross-sections in the visible; introduce dehydrogenation, superhydrogenation and minor aliphatic side groups to explain the aliphatic features; and appeal to a large mixture of different PAH molecules to explain the lack of detection of individual PAH molecules. Since known PAH molecules have problems reproducing the wavelengths of the UIE bands, a large mixture of diverse PAH molecules is needed to fit the observed astronomical spectra. Heteroatoms such as N and O are also introduced to explain the 6.2 and 11.3 μm features. The PAH hypothesis therefore has moved away from the chemical definition of PAH molecules to a hybrid to save the hypothesis.

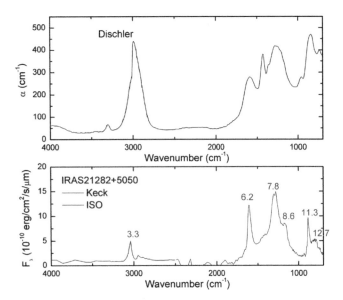

FIGURE 4.2.8 Laboratory infrared spectra (inverted from absorption) of hydrogenated amorphous carbon (top, from Dischler et al. 1983) compared with the astronomical spectrum of the planetary nebula IRAS 21282 + 5050 (bottom panel). The UIE bands are labeled by their wavelengths in micrometers.

In order to identify possible carriers of the UIE bands, it would be useful to see what carbonaceous products can naturally exist in the interstellar medium. By introducing H into graphite and diamond (both crystalline forms of pure carbon), a variety of amorphous C–H alloys can be created (Robertson and O'Reilly 1987; Jones et al. 1990; Jones 2012a, 2012b, 2012c; Jones et al. 2013). Different geometric structures with long and short ranges can be created by varying the aromatic to aliphatic and C to H ratios. The infrared spectra of these amorphous carbonaceous materials (Dischler et al. 1983) show resemblance to the astronomical UIE bands seen in planetary nebulae and proto-planetary nebulae (Figure 4.2.8). Since these amorphous carbonaceous solids have absorption bands in the visible, they can be easily excited by visible light from stars.

By the early 1980s, it was known that carbon clusters can be produced by laser vaporization of graphite, followed by supersonic expansion into an inert gas. The employment of this technique has led to the discovery of fullerene, a new form of carbon (Kroto et al. 1985). Since then, various techniques based on laser pyrolysis of gas-phase hydrocarbons, followed by condensation, have been used to create laboratory counterparts of cosmic organic dust (Jäger et al. 2009). These include the quenching of plasma of 4-torr methane (Sakata et al. 1987), hydrocarbon flame or arc discharge in a neutral of hydrogenated atmosphere (Colangeli et al. 2003; Mennella et al. 2003), laser ablation of graphite in a hydrogen atmosphere (Scott and Duley 1996; Mennella et al. 1999), infrared laser pyrolysis of gas-phase hydrocarbon molecules (Herlin et al. 1998), and photolysis of methane at low temperatures (Dartois et al. 2004).

Amorphous carbonaceous solids are known to be naturally produced through the process of combustion. Soot is formed by igniting a mixture of gas-phase hydrocarbons with oxygen,

resulting in amorphous structures consisting of islands of aromatic rings linked by aliphatic chains (Pino et al. 2008).

In addition to soot, there are also other natural decayed products of living organisms such as coal and kerogen that have similar amorphous mixed aromatic/aliphatic properties (Painter et al. 1981). The infrared spectra of soot (Keifer et al. 1981), coal (Guillois et al. 1996; Ibarra et al. 1996; Papoular 2001), and petroleum and asphaltenes (Cataldo et al. 2002, 2013), all show spectral features similar to the astronomical UIE bands.

As the results of these laboratory developments, alternate models for the UIE phenomenon have been proposed. These include hydrogenated amorphous carbon (HAC, Duley 1993); soot and carbon nanoparticles (Hu and Duley 2008); quenched carbonaceous composite particles (QCCs, Sakata et al. 1987); kerogen and coal (Papoular et al. 1989); petroleum fractions (Cataldo et al. 2002), and mixed aromatic/aliphatic organic nanoparticles (MAONs, Kwok and Zhang 2011, 2013). In the coal, petroleum, and MAON models, other elements such as O, S, and N are also incorporated into the hydrocarbon compounds. In a natural environment such as circumstellar envelopes of evolved stars, where a mix of cosmic elements is present, it is expected that any organic product of synthesis will contain other elements beyond C and H. A schematic illustration of part of a MAON structure is shown in Figure 4.2.9.

The advantages of such amorphous models are that the vibrational bands are naturally broad and do not need to be artificially broadened, as in the case of PAH molecules. The disadvantages are that although some experimental spectra of these materials have been obtained in the laboratory, the exact vibrational modes of the features are not known.

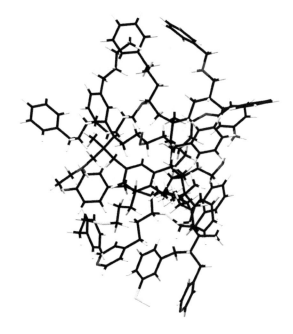

FIGURE 4.2.9 A three-dimensional (3-D) illustration of a possible partial structure of a MAON particle. Carbon atoms are represented in black, hydrogen in light gray, sulfur in yellow, oxygen in red, and nitrogen in blue. There are 169 C, 225 H, 7 N, 4 O, and 3 S atoms in this example.

Theoretical calculations are needed to analyze the vibrational modes of HACs, QCCs, and MAONs (Sadjadi et al. 2015a). If these molecules are too large to be transiently heated by single-photon excitation, alternative excitation mechanism such as chemical excitation needs to be considered (Duley and Williams 1988, 2011).

4.2.6 OTHER UNIDENTIFIED CIRCUMSTELLAR SPECTRAL PHENOMENA

There are a number of unidentified circumstellar spectral phenomena whose carriers are likely to be organic compounds. The 30 μm feature was discovered in C-rich AGB stars and planetary nebulae (Forrest et al. 1981), and the 21 μm feature was discovered in proto-planetary nebulae (Kwok et al. 1989). High-resolution *Infrared Space Observatory (ISO)* and *Spitzer* observations have found that all observed 21 μm features have the same intrinsic profile and peak wavelength (20.1 μm) (Volk et al. 1999; Hrivnak et al. 2009). An example of the 21 and 30 μm features is shown in Figure 4.2.10. Strong 8 and 12 μm plateau features are seen, suggesting that there is a link between the 21 and 30 μm features and the UIE phenomenon.

These two unidentified emission features can carry a large fraction of the total energy output of the central stars – up to 8% for the 21 μm and 20% for the 30 μm features (Hrivnak et al. 2000). This implies that carrier must be made of common elements. The fact that the features are seen only in C-rich (based on photospheric absorption spectrum) objects suggests that the carrier is carbonaceous.

The extended red emission (ERE) is a broadband optical emission seen in reflection nebulae and planetary nebulae. It is likely to be the result of photoluminescence from semiconductor solid particles (Witt et al. 1998). Some of the over

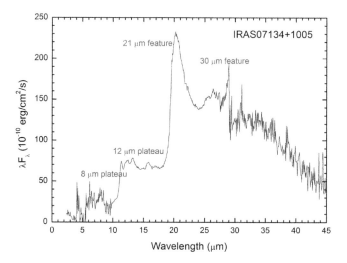

FIGURE 4.2.10 Combined *ISO* SWS and LWS and *Spitzer* IRS spectra of the proto-planetary nebula IRAS07134 + 1005, showing the 21 and 30 unidentified emission features. No atomic line is seen in the infrared spectrum of this proto-planetary nebula. Also seen in the spectra are broad plateau emission features around 8 and 12 μm. All these features sit on top of a strong continuum.

500 unidentified diffuse interstellar bands (DIBs, Sarre 2006) observed in absorption in the diffuse interstellar medium are also seen in circumstellar envelopes (Van Winckel et al. 2002), suggesting that some of the carriers of DIBs are also synthesized in circumstellar envelopes.

4.2.7 TIME SEQUENCE OF CIRCUMSTELLAR ORGANIC SYNTHESIS

The detection of simple molecules in early AGB stars to complex organics in planetary nebulae suggests that the synthesis of organic dust is not a breakdown of biological compounds but a bottom-up synthesis from simple molecules. The formation of the linear molecule acetylene in highly evolved AGB stars (Volk et al. 2000) is the first step for the formation of benzene (Cernicharo et al. 2001), the simplest aromatic molecule, in proto-planetary nebulae. The emergence of aromatic and aliphatic bands in the transition from AGB to planetary nebulae suggests that small rings group into larger aromatic islands, attached with aliphatic side groups. Since circumstellar chemical synthesis must occur over timesscales much shorter than the dynamical timescale of expansion of the nebula and evolution timescales of the central star, we know that circumstellar organic synthesis occurs at 10^3–10^4 year timescales.

4.2.8 RELATIONSHIP BETWEEN STELLAR, INTERSTELLAR, AND SOLAR SYSTEM ORGANICS

Although it was commonly believed that solar system objects were made of minerals, metals, and ices, we now know that meteorites, asteroids, comets, planetary satellites, and interplanetary dust particles contain organic materials. Since the early discoveries of hydrocarbons and amino acids in meteorites (Nagy et al. 1961; Kvenvolden et al. 1970), we now know that meteorites contain rich assortment of prebiotic compounds, which are entirely of abiotic origin. In the soluble component of carbonaceous chondrites, almost all biologically relevant organic compounds can be found (Schmitt-Kopplin et al. 2010).

The insoluble organic matter (IOM) of carbonaceous chondrites is composed of highly substituted single-ring aromatics, substituted furan/pyran functional groups, highly branched oxygenated aliphatics, and carbonyl groups (Cody et al. 2011). The 3.4 μm feature found in proto-planetary nebulae, in planetary nebulae, and in the diffuse interstellar medium is also detected in meteorites (Cronin and Pizzarello1990; Ehrenfreund et al. 1991), interplanetary dust particles (Flynn et al. 2003), comets (Keller et al. 2006), and the haze of Titan (Kim et al. 2011) and Saturn (Kim et al. 2012). The relative elemental abundance in the IOM ranges from $C_{100}H_{70}N_3O_{12}S_2$ for Murchison meteorite to $C_{100}H_{46}N_{10}O_{15}S_{4.5}$ for the Tagish Lake meteorite (Pizzarello and Shock 2017). The detection of anomalous isotopic ratios in elements in IOM suggests that at least part of the IOM is of presolar origin.

4.2.9 DELIVERY OF STELLAR ORGANICS TO THE SOLAR SYSTEM

Since most (~95%) stars in our galaxy go through the AGB and planetary nebulae phases of evolution, large amounts of mineral and organic solids are produced by ordinary stars over the last 10 billion years in our Milky Way Galaxy. These products are ejected into the interstellar medium and spread throughout the galaxy, and evolved stars therefore represent a major source of molecules and solids in the galaxy. Direct evidence for stellar grains to have traveled across the galaxy to arrive at the solar system and to Earth can be found in the form of presolar grains in meteorites. Presolar grains are grains of diamonds, silicon carbide, corundum, and spinel that have isotopic abundances not typical of the solar system but consistent with origin in AGB stars (Zinner, 1998; Davis 2011). Since macroscopic organics such as MAON, kerogen, and IOM are extremely sturdy, they are therefore likely to be able to transverse across interstellar space, even under high-UV-radiation background and dynamical shock conditions. Stardust therefore plays an important role in the chemical enrichment of the galaxy and possibly also of the primordial solar system (Kwok 2004; Ziurys et al. 2016).

Possible links between solar system organics and stellar organics can be tested by isotopic ratios. The isotopic signatures of IOM in the Murchison meteorite are consistent with at least in part being presolar in origin (Pizzarello and Shock 2017). Further work is needed to determine the origin of solar system organics, whether these organics were made in the early solar system or were delivered from stellar sources.

4.2.10 ORGANIC RESERVOIR ON EARTH

The Earth has a large reservoir of organics, almost all of which are biological in origin. The total amount of C in the living biosphere (Earth's crust, ocean, and atmosphere) is about 1000 Gigatons (GT). In addition, there are about 4000 GT of fossil fuels in the form of petroleum, coal, and natural gases. The majority of which – 15,000,000 GT – are in the form of kerogen, a macromolecular compound found in sedimentary rocks (Falkowski et al. 2000). Although kerogen and fossil fuels are all remnants of past life, there is a small amount of hydrocarbons found in hydrothermal vents that is of abiological origin (Sherwood Lollar et al. 2002).

The possibility that there may be primordial organics in Earth was suggested by Gold (1999). At that time of Gold's writing, gas-phase organic molecules had already been known to be widely present in interstellar space. Gold suggested that the early Earth might have trapped inside a large reservoir of methane, which could serve as a precursor of fossil fuels, which later flowed from the interior to the surface. This theory was never taken seriously by geologists, because interstellar methane gas would have great difficulty surviving the high temperature and shock during the coalescence process of Earth formation.

At a more general level, although abiotic theories of the origin of oil were seriously discussed in the Soviet Union, they were never popular in the West, in part because of the well-established links between petroleum and life in the form of nickel- and vanadium-porphyrin complexes in petroleum, iron-porphyrin and hemoglobin in animals, and magnesium chlorophyll in plants. However, in addition to biological formation of fossil fuels, there exists the additional possibility of primordial organics deep inside the Earth. If the primordial hydrocarbon was in the form of a macromolecular compound, its chances of survival during Earth's formation would have been much higher than that in the case for methane. If such primordial macro-organics are embedded deep inside the Earth, they would be difficult for us to discover with present technology. If such primordial organics does exist and can be retrieved, the economic impact could be immense.

4.2.11 EFFECTS OF STAR DUST ON THE ORIGIN OF LIFE ON EARTH

Since the Oparin–Haldane hypothesis (Haldane 1929; Oparin 1938) and the Miller–Urey experiment (Miller 1953), it has been commonly assumed that life on Earth originated from chemical reactions of simple ingredients in a suitable environment (primordial soup). The theory of endogenous synthesis as the basis of origin of life has been the dominant theory in the past 50 years. However, with the observation of stellar synthesis of organics and the wide presence of complex organics among solar system objects, the exogenous delivery hypothesis has been gaining acceptance. The most widely discussed agents for exogenous delivery are comets, asteroids, interplanetary dust particles, and micrometeorites. The influx rate at present for micrometeorites is estimated to be ~30,000 tons year^{-1}, but the rates were much higher during the early 500 million years of the Earth's history during the period of heavy early bombardment. Stellar MAON-like materials embedded in comets and asteroids have a good chance of survival upon impact.

We cannot rule out the possibility that the planetismals that aggregated to form the primordial Earth may have contained remnants of macromolecular star dust. These organics could well survive the heat and shock conditions during the Earth-formation process. If this is the case, then there could exist primordial organics deep inside the interior of the Earth (Kwok 2017).

Under suitable temperature and pressure conditions, for example, those in hydrothermal vents, organic components such as hydrocarbons; dicarboxylic acids; N-, O-, and S-containing aromatic compounds; as well as ammonia can be released from macromolecular compounds (Yabuta et al. 2007; Pizzarello et al. 2011). These prebiotic materials could form the ingredients of the first steps to life in the early Earth. Whether life on Earth originated on the surface in water, or deep inside the Earth, is an interesting question for further exploration.

4.2.12 CONCLUSIONS

In the early twentieth century, the discipline of astrophysics was born out of development of atomic and nuclear physics. The discipline of astrochemistry, or the study of molecules and minerals in space, began only 50 years ago as the result of advances in observing capabilities in the infrared and millimeter-wave regions of the electromagnetic spectrum. The discipline of astrobiology, trying to bridge the gap between molecules and life, is still in its infancy. The unexpected discovery of common presence of organics in the solar system and in the interstellar medium gives us the optimism and hope for the future of this new discipline.

From the analysis of organic matter in meteorites and other solar system objects, we now know that abiotic synthesis can create a wide range of organic compounds far beyond those found on Earth. This hints that the degrees of complexity and diversity of prebiotic organics are much larger than previously thought (Meringer and Cleaves 2017). It is therefore not surprising that there still remains a number of unexplained astronomical spectral phenomena, as the carrier of these phenomena may be organic compounds that we are unfamiliar with in the terrestrial environment.

The circumstellar envelopes of evolved stars provide the only laboratory where such abiotic organic synthesis is directly observed to take place. Based on spectroscopic observations of evolved stars in different stages of evolution, we learn that organic molecules and solids can be synthesized in near-vacuum conditions over very short timescales. This suggests that our understanding of chemical reactions based on terrestrial conditions is inadequate to explain chemistry at work in the universe.

From the fact that UIE features are seen in galaxies, with redshifts as high as 2, we know that organic synthesis took place soon after the nucleosynthesis of the element carbon, as early as 10 billion years ago (Kwok 2011). Since our Sun and our solar system came into existence only 4.6 billion years ago, life could have emerged a very long time ago elsewhere in our galaxy and beyond. Since terrestrial biochemistry only represents a very small branch of the possible biochemical pathways in the universe, life as we know it can be totally different from life evolved from similar prebiotic ingredients but under different conditions and pathways. If some of these extraterrestrial lives had developed into intelligent life, the level of extraterrestrial intelligence is far beyond our imagination. Our present study of the abiotic synthesis of complex organics in the circumstellar environment therefore represents a small but important step in our understanding of intelligent life in the universe.

ACKNOWLEDGMENTS

SK thanks Yong Zhang, Chih-Hao Hsia, and SeyedAbdolreza Sadjadi for helpful discussions.

REFERENCES

Allamandola, L. J., A. G. G. M. Tielens, and J. R. Barker. 1985. Polycyclic aromatic hydrocarbons and the unidentified infrared emission bands – auto exhaust along the Milky Way. *Astrophys. J.* 290:L25–L28.

Allamandola, L. J., A. G. G. M. Tielens, and J. R. Barker. 1989. Interstellar polycyclic aromatic hydrocarbons: The infrared emission bands, the excitation/emission mechanism and the astrophysical implications. *Astrophys. J. Suppl. Ser.* 71:733–775.

Bauschlicher, C. W., Jr. 2002. The infrared spectra of $C_{96}H_{24}$, $C_{96}H^{+}_{24}$, and $C_{96}H^{+}_{25}$. *Astrophys. J.* 564:782–786.

Bernard-Salas, J., J. Cami, E. Peeters et al. 2012. On the excitation and formation of circumstellar fullerenes. *Astrophys. J.* 757:41.

Burbidge, E. M., G. R. Burbidge, W. A. Fowler, and F. Hoyle. 1957. Synthesis of the elements in stars. *Rev. Mod. Phys.* 29:547–650.

Cami, J., J. Bernard-Salas, E. Peeters, and S. E. Malek. 2010. Detection of C_{60} and C_{70} in a young planetary nebula. *Science* 329:1180–1182.

Cataldo, F., D. A. García-Hernández, and A. Manchado. 2013. Far- and mid-infrared spectroscopy of complex organic matter of astrochemical interest: Coal, heavy petroleum fractions and asphaltenes. *Mon. Not. R. Astron. Soc.* 429:3025–3039.

Cataldo, F., Y. Keheyan, and D. Heymann. 2002. A new model for the interpretation of the unidentified infrared bands (UIBS) of the diffuse interstellar medium and of the protoplanetary nebulae. *Int. J. Astrobiol.* 1:79–86.

Cernicharo, J., A. M. Heras, A. G. G. M. Tielens et al. 2001. Infrared Space Observatory's discovery of C_4H_2, C_6H_2, and benzene in CRL 618. *Astrophys. J.* 546(2):L123–L126.

Chiar, J. E., A. G. G. M. Tielens, D. C. B. Whittet et al. 2000. The composition and distribution of dust along the line of sight toward the Galactic Center. *Astrophys. J.* 537:749–762.

Clayton, G. C., K. D. Gordon, F. Salama et al. 2003. The role of polycyclic aromatic hydrocarbons in ultraviolet extinction. I. probing small molecular polycyclic aromatic hydrocarbons. *Astrophys. J.* 592:947–952.

Cody, G. D., E. Heying, C. M. O. Alexander et al. 2011. Establishing a molecular relationship between chondritic and cometary organic solids. *Proc. Natl. Acad. Sci. U.S.A.* 108(48):19171–19176.

Colangeli, L., Th. Henning, J. R. Brucato et al. 2003. The role of laboratory experiments in the characterisation of silicon-based cosmic material. *Astron. Astrophys. Rev.* 11(2–3):97–152.

Cronin, J. R. and S. Pizzarello. 1990. Aliphatic hydrocarbons of the Murchison meteorite. *Geochim. Cosmochim. Acta* 54:2859–2868.

Dartois, E., G. M. Muñoz Caro, D. Deboffle, and L. d'Hendecourt. 2004. Diffuse interstellar medium organic polymers: Photo-production of the 3.4, 6.85 and 7.25 μm features. *Astron. Astrophys.* 423:L33–L36.

Davis, A. M. 2011. Cosmochemistry special feature: Stardust in meteorites. *Proc. Natl. Acad. Sci. U.S.A.* 108:19142–19146.

De Muizon, M. J., L. B. d'Hendecourt, and T. R. Geballe. 1990. Polycyclic aromatic hydrocarbons in the near-infrared spectra of 24 IRAS sources. *Astron. Astrophys.* 227(2):526–541.

Dischler, B., A. Bubenzer, and P. Koidl. 1983. Bonding in hydrogenated hard carbon studied by optical spectroscopy. *Solid State Commun.* 48(2):105–108.

Duley, W. W. 1993. Infrared spectra of interstellar carbon solids. In *Astronomical Infrared Spectroscopy: Future Observational Directions*, ed. S. Kwok, ASP Conference Series, p. 241.

Duley, W. W. and D. A. Williams. 1979. Are there organic grains in the interstellar medium? *Nature* 277:40–41.

Duley, W. W. and D. A. Williams. 1981. The infrared spectrum of interstellar dust – Surface functional groups on carbon. *Mon. Not. R. Astron. Soc.* 196:269–274.

Duley, W. W. and D. A. Williams. 1988. Excess infrared emission from large interstellar carbon grains. *Mon. Not. R. Astron. Soc.* 231:969–975.

Duley, W. W. and D. A. Williams. 2011. Excitation of the aromatic infrared emission bands: Chemical energy in hydrogenated Amorphous Carbon Particles? *Astrophys. J. Lett.* 737:L44.

Ehrenfreund, P., F. Robert, L. d'Hendecourt, and F. Behar. 1991. Comparison of interstellar and meteoritic organic matter at 3.4 microns. *Astron. Astrophys.* 252:712–717.

.

Falkowski, P., R. J. Scholes, E. Boyle et al. 2000. The global carbon cycle: A test of our knowledge of Earth as a system. *Science* 290:291–296.

Flynn, G. J., L. P. Keller, M. Feser, S. Wirick, and C. Jacobsen. 2003. The origin of organic matter in the solar system: Evidence from the interplanetary dust particles. *Geochim. Cosmochim. Acta* 67:4791–4806.

Forrest, W. J., J. R. Houck, and J. F. McCarthy. 1981. A far-infrared emission feature in carbon-rich stars and planetary nebulae. *Astrophys. J.* 248:195–200.

García-Hernández, D. A., A. Manchado, P. Garcia-Lario et al. 2010. Formation of fullerenes in H-containing planetary nebulae. *Astrophys. J.* 724:L39–L43.

García-Hernández, D. A., E. Villaver, P. Garcia-Lario et al. 2012. Infrared study of fullerene planetary nebulae. *Astrophys. J.* 760:107.

Geballe, T. R., A. G. G. M. Tielens, S. Kwok, and B. J. Hrivnak. 1992. Unusual 3 micron emission features in three proto-planetary nebulae. *Astrophys. J.* 387:L89–L91.

Gold, T. 1999. *The Deep Hot Biosphere*. Copernicus Books, New York.

Gredel, R., Y. Carpentier, G. Rouillé, M. Steglich, F. Huisken, and Th. Henning. 2011. Abundances of PAHs in the ISM: Confronting observations with experimental results. *Astron. Astrophys.* 530:26.

Guillois, O., I. Nenner, R. Papoular, and C. Reynaud. 1996. Coal models for the infrared emission spectra of proto-planetary nebulae. *Astrophys. J.* 464:810–817.

Haldane, J. B. S. 1929. The origin of life. *Rationalist Annu.* 148:3–10.

Henning, T. 2009. Astromineralogy. Berlin, Germany: Springer.

Herlin, N., I. Bohn, C. Reynaud, M. Cauchetier, A. Galvez, and J.-N. Rouzaud. 1998. Nanoparticles produced by laser pyrolysis of hydrocarbons: Analogy with carbon cosmic dust. *Astron. Astrophys.* 330:1127–1135.

Hrivnak, B. J., K. Volk, and S. Kwok. 2000. 2–45 micron infrared spectroscopy of carbon-rich proto-planetary nebulae. *Astrophys. J.* 535:275–292.

Hrivnak, B. J., K. Volk, and S. Kwok. 2009. A Spitzer study of 21 and 30 μm emission in several galactic carbon-rich proto-planetary nebulae. *Astrophys. J.* 694:1147–1160.

Hu, A. and W. W. Duley. 2008. Spectra of carbon nanoparticles: Laboratory simulation of the aromatic CH emission feature at 3.29 μm. *Astrophys J.* 677:L153–L156.

Hudgins, D. M. and L. J. Allamandola. 1999. The spacing of the interstellar 6.2 and 7.7 micron emission features as an indicator of polycyclic aromatic hydrocarbon size. *Astrophys. J. Lett.* 513:L69–L73.

Hudgins, D. M., C. W. Bauschlicher, and L. J. Allamandola. 2005. Variations in the peak position of the 6.2 μm interstellar emission feature: A tracer of N in the interstellar polycyclic aromatic hydrocarbon population. *Astrophys. J.* 632:316–332.

Ibarra, J., E. Muñoz, and R. Moliner. 1996. FTIR study of the evolution of coal structure during the coalification process. *Org. Geochem.* 24(6):725–735.

Jäger, C., F. J. Molster, J. Dorschner, T. Henning, H. Mutschke, and L. Waters. 1998. Steps toward interstellar silicate mineralogy – IV. The crystalline revolution. *Astron. Astrophys.* 339(3):904–916.

Jäger, C., F. Huisken, H. Mutschke, I. L. Jansa, and T. Henning. 2009. Formation of polycyclic aromatic hydrocarbons and carbonaceous solids in gas-phase condensation experiments. *Astrophys. J.* 696(1):706–712.

Jones, A. P. 2012a. Variations on a theme – The evolution of hydrocarbon solids. I. Compositional and spectral modelling – The eRCN and DG models. *Astron. Astrophys.* 540:1.

Jones, A. P. 2012b. Variations on a theme – The evolution of hydrocarbon solids. II. Optical property modelling – the optEC(s) model. *Astron. Astrophys.* 540:2.

Jones, A. P. 2012c. Variations on a theme – The evolution of hydrocarbon solids (corrigendum). III. Size-dependent properties – the optEC(s)(a) model. *Astron. Astrophys.* 545:3.

Jones, A. P., L. Fanciullo, M. Köhler, L. Verstraete, V. Guillet, M. Bocchio, and N. Ysard. 2013. The evolution of amorphous hydrocarbons in the ISM: Dust modelling from a new vantage point. *Astron. Astrophys.* 558:62.

Jones, A. P., W. W. Duley, and D. A. Williams. 1990. The structure and evolution of hydrogenated amorphous carbon grains and mantles in the interstellar medium. *QJRAS* 31:567–582.

Kahanpää, J., K. Mattila, K. Lehtinen, C. Leinert, and D. Lemke. 2003. Unidentified infrared bands in the interstellar medium across the Galaxy. *Astron. Astrophys.* 405:999–1012.

Keifer, J. R., M. Novicky, M. S. Akhter, A. R. Chughtai, and D. M. Smith. 1981. The nature and reactivity of the elemental carbon (soot) surface as revealed by Fourier transform infrared (FTIR) Spectroscopy. *1981 International Conference on Fourier Transform Infrared Spectroscopy, (SPIE)*, p. 5.

Keller, L. P., S. Bajt, G. A. Baratta et al. 2006. Infrared spectroscopy of Comet 81P/Wild 2 samples returned by Stardust. *Science* 314:1728–1731.

Kim, S. J., A. Jung, C. K. Sim et al. 2011. Retrieval and tentative identification of the 3 μm spectral feature in Titan's haze. *Planet. Space Sci.* 59:699–704.

Kim, S. J., C. K. Sim, D. W. Lee, R. Courtin, J. I. Moses, Y. C. Minh. 2012. The three-micron spectral feature of the Saturnian haze: Implications for the haze composition and formation process. *Planet. Space Sci.* 65:122–129.

Knacke, R. F. 1977. Carbonaceous compounds in interstellar dust. *Nature* 269:132–134.

Kroto, H. W., J. R. Heath, S. C. Obrien, R. F. Curl, and R. E. Smalley. 1985. C_{60}: Buckminsterfullerene. *Nature* 318:162–163.

Kvenvolden, K., J. Lawless, K. Pering et al. 1970. Evidence for extra-terrestrial amino-acids and hydrocarbons in the murchison meteorite. *Nature* 228:923–926.

Kwok, S. 1993. Proto-planetary nebulae. *Annu. Rev. Astron. Astrophys.* 31:63–92.

Kwok, S. 2000. *Origin and Evolution of Planetary Nebulae*. New York: Cambridge University Press.

Kwok, S. 2004. The synthesis of organic and inorganic compounds in evolved stars. *Nature* 430:985–991.

Kwok, S. 2011. *Organic Matter in the Universe*. Hoboken, NJ: John Wiley & Sons.

Kwok, S. 2017. Abiotic synthesis of complex organics in the Universe. *Nat. Astron.* 1(10):642.

Kwok, S. and Y. Zhang. 2013. Unidentified infrared emission bands: PAHs or MAONs? *Astrophys. J.* 771:5

Kwok, S. and Y. Zhan. 2011. Mixed aromatic-aliphatic organic nanoparticles as carriers of unidentified infrared emission features. *Nature* 479:80–83.

Kwok, S., K. M. Volk, and B. J. Hrivnak. 1989. A 21 micron emission feature in four proto-planetary nebulae. *Astrophys. J.* 345:L51–L54.

Kwok, S., K. Volk, and P. Bernath. 2001. On the origin of infrared plateau features in proto-planetary nebulae. *Astrophys. J.* 554:L87–L90.

Kwok, S., K. Volk, and W. P. Bidelman. 1997. Classification and Identification of IRAS sources with low-resolution spectra. *Astrophys. J. Suppl. Ser.* 112:557–584.

Léger, A. and J. L. Puget. 1984. Identification of the 'unidentified' IR emission features of interstellar dust? *Astron. Astrophys.* 137:L5–L8.

Mennella, V., G. A. Baratta, A. Esposito, G. Ferini, and Y. J. Pendleton. 2003. The effects of ion irradiation on the evolution of the carrier of the 3.4 micron interstellar absorption band. *Astrophys. J.* 587:727–738.

Mennella, V., J. R. Brucato, L. Colangeli, and P. Palumbo. 1999. Activation of the 3.4 micron band in carbon grains by exposure to atomic hydrogen. *Astrophys. J.* 524:L71–L74.

Meringer, M. and H. J. Cleaves. 2017. Exploring astrobiology using "in silico" molecular structure generation. *Philos. Trans. R. Soc. Lond., Ser. A: Math. Phys. Eng. Sci.* 375(2109):20160344.

Miller, S. L. 1953. A production of amino acids under possible primitive earth conditions. *Science* 117:528–529.

Nagy, B., D. J. Hennessy, and W. G. Meinschein. 1961. Mass spectroscopic analysis of the Orgueil meteorite: Evidence for biogenic hydrocarbons. *Ann. N.Y. Acad. Sci.* 93:27–35.

Ney, E. P. and B. F. Hatfield. 1978. The isothermal dust condensation of Nova Vulpeculae 1976. *Astrophys. J.* 219:L111–L115.

Oparin, A. I. 1938. *The Origin of Life.* MacMillan, New York.

Painter, P. C., R. W. Snyder, M. Starsinic et al. 1981. Concerning the application of FT-IR to the study of coal: A critical assessment of band assignments and the application of spectral analysis programs. *Appl. Spectrosc.* 35(5):475–485.

Papoular, R. 2001. The use of kerogen data in understanding the properties and evolution of interstellar carbonaceous dust. *Astron. Astrophys.* 378:597–607.

Papoular, R., J. Conrad, M. Giuliano, J. Kister, and G. Mille. 1989. A coal model for the carriers of the unidentified IR bands. *Astron. Astrophys.* 217:204–208.

Pino, T., E. Dartois, A.-T. Cao et al. 2008. The 6.2 μm band position in laboratory and astrophysical spectra: A tracer of the aliphatic to aromatic evolution of interstellar carbonaceous dust. *Astron. Astrophys.* 490:665–672.

Pizzarello, S. and E. Shock. 2017. Carbonaceous chondrite meteorites: The chronicle of a potential evolutionary path between stars and life. *Origins Life Evol. Biosphere* 47(3):249–260.

Pizzarello, S., L. B. Williams, J. Lehman, G. P. Holland, and J. L. Yarger. 2011. Abundant ammonia in primitive asteroids and the case for a possible exobiology. *Proc. Natl. Acad. Sci. U.S.A.* 108:4303–4306.

Posch, T., F. Kerschbaum, H. Mutschke, J. Dorschner, and C. Jäger. 2002. On the origin of the 19.5 μm feature. Identifying circumstellar Mg–Fe-oxides. *Astron. Astrophys.* 393:L7–L10.

Posch, T., F. Kerschbaum, H. Mutschke et al. 1999, On the origin of the 13 μm feature. A study of ISO-SWS spectra of oxygen-rich AGB stars. *Astron. Astrophys.* 352:609–618.

Puetter, R. C., R. W. Russell, S. P. Willner, and B. T. Soifer. 1979. Spectrophotometry of compact H II regions from 4 to 8 microns. *Astrophys. J.* 228:118–122.

Robertson, J. and E. P. O'Reilly 1987. Electronic and atomic structure of amorphous carbon. *Phys. Rev. B* 35(6):2946–2957.

Russell, R. W., B. T. Soifer, B. T., and S. P. Willner. 1977. The 4 to 8 micron spectrum of NGC 7027. *Astrophys. J.* 217:L149–L153.

Sadjadi, S., Y. Zhang, and S. Kwok. 2015a. A theoretical study on the vibrational spectra of polycyclic aromatic hydrocarbon molecules with aliphatic sidegroups. *Astrophys. J.* 801:34.

Sadjadi, S., Y. Zhang, and S. Kwok. 2015b. On the origin of the 11.3 micron unidentified infrared emission feature. *Astrophys. J.* 807:95.

Sakata, A., S. Wada, T. Onaka, and A. T. Tokunaga. 1987. Infrared spectrum of quenched carbonaceous composite (QCC). II – A new identification of the 7.7 and 8.6 micron unidentified infrared emission bands. *Astrophys. J.* 320:L63–L67.

Sakata, A., S. Wada, T. Onaka, and A. T. Tokunaga. 1990. Quenched carbonaceous composite. III – Comparison to the 3.29 micron interstellar emission feature. *Astrophys. J.* 353:543–548.

Salama, F., G. A. Galazutdinov, J. Krełowski et al. 2011. Polycyclic aromatic hydrocarbons and the diffuse interstellar bands: A survey. *Astrophys. J.* 728:154.

Sarre, P. J. 2006. The diffuse interstellar bands: A major problem in astronomical spectroscopy. *J. Mol. Spectrosc.* 238:1–10.

Schmitt-Kopplin, P., Z. Gabelica, R. D. Gougeon et al. 2010. High molecular diversity of extraterrestrial organic matter in Murchison meteorite revealed 40 years after its fall. *Proc. Natl. Acad. Sci. U.S.A.* 107:2763–2768.

Scott, A. and W. W. Duley. 1996. The decomposition of hydrogenated amorphous carbon: A connection with polycyclic aromatic hydrocarbon molecules. *Astrophys. J.* 472:L123–L125.

Sellgren, K. 1984. The near-infrared continuum emission of visual reflection nebulae. *Astrophys. J.* 277:623–633.

Sellgren, K., M. W. Werner, J. G. Ingalls et al. 2010. C_{60} in reflection nebulae. *Astrophys. J.* 722:L54–L57.

Sherwood Lollar, B., T. D. Westgate, J. A. Ward, G. F. Slater, and G. Lacrampe-Couloume. 2002. Abiogenic formation of alkanes in the Earth's crust as a minor source for global hydrocarbon reservoirs. *Nature* 416, 522–524.

Tielens, A. G. G. M. 2008. Interstellar polycyclic aromatic hydrocarbon molecules. *Annu. Rev. Astron. Astrophys.* 46:289–337.

Uchida, K. I., K. Sellgren, M. W. Werner, and M. L. Houdashelt. 2000. Infrared space observatory mid-infrared spectra of reflection nebulae. *Astrophys. J.* 530(2):817–833.

van Diedenhoven, B., E. Peeters, C. Van Kerckhoven et al. 2004. The profiles of the 3–12 micron polycyclic aromatic hydrocarbon features. *Astrophys. J.* 611:928–939.

Van Winckel, H., M. Cohen, and T. R. Gull. 2002. The ERE of the red rectangle revisited. *Astron. Astrophys.* 390: 147–154.

Volk, K., G.-Z. Xiong, and S. Kwok. 2000. Infrared space observatory spectroscopy of extreme carbon stars. *Astrophys. J.* 530: 408–417.

Volk, K., S. Kwok, and B. J. Hrivnak. 1999. High-resolution infrared space observatory spectroscopy of the unidentified 21 micron feature. *Astrophys. J.* 516:L99–L102.

Wagner, D. R., H. Kim, and R. J. Saykally. 2000. Peripherally hydrogenated neutral polycyclic aromatic hydrocarbons as carriers of the 3 micron interstellar infrared emission complex: Results from single-photon infrared emission spectroscopy. *Astrophys. J.* 545:854–860.

Witt, A. N., K. D. Gordon, and D. G. Furton. 1998. Silicon nanoparticles: Source of extended red emission? *Astrophys. J.* 501:L111–L115.

Woolf, N. J. and E. P. Ney. 1969. Circumstellar infrared emission from cool stars. *Astrophys. J.* 155:L181–L184.

Yabuta, H., L. B. Williams, G. D. Cody, C. M. O. D. Alexander, and S. Pizzarello. 2007. The insoluble carbonaceous material of CM chondrites: A possible source of discrete organic compounds under hydrothermal conditions. *Meteorit. Planet. Sci.* 42:37–48.

Zhang, Y. and S. Kwok. 2011. Detection of C_{60} in the protoplanetary nebula IRAS 01005 + 7910. *Astrophys. J.* 730:126.

Zhang, Y. and S. Kwok. 2013. On the detections of C_{60} and derivatives in circumstellar environments. *Earth, Planets Space* 65:1069–1081.

Zhang, Y. and S. Kwok. 2015. On the viability of the PAH model as an explanation of the unidentified infrared emission features. *Astrophys. J.* 798:37.

Zhang, Y., S. Sadjadi, C.-H. Hsia, and S. Kwok. 2017. Search for hydrogenated C_{60} (fulleranes) in circumstellar envelopes. *Astrophys. J.* 845: 76.

Zinner, E. 1998. Stellar nucleosynthesis and the isotopic composition of presolar grains from primitive meteorites. *Annu. Rev. Earth Planet. Sci.* 26:147–188.

Ziurys, L. M. 2006. Interstellar chemistry special feature: The chemistry in circumstellar envelopes of evolved stars: Following the origin of the elements to the origin of life. *Proc. Natl. Acad. Sci. U.S.A.* 103:12274–12279.

Ziurys, L. M., D. T. Halfen, W. Geppert, and Y. Aikawa. 2016. Following the interstellar history of carbon: From the interiors of stars to the surfaces of planets. *Astrobiology* 16:997–1012.

4.3 Organic Molecules in Meteorites and Their Astrobiological Significance

Zita Martins

CONTENTS

4.3.1 INTRODUCTION

Life emerged on the early Earth ~3.8 to ~3.5 billion years ago (Schidlowski 1988; Rosing 1999; Furnes et al. 2004; Rosing and Frei 2004; Westall et al. 2006, 2011a, 2011b; Sugitani et al. 2009; Javaux et al. 2010; Wacey et al. 2010). The building blocks necessary for the first living organisms to self-assemble and for life to emerge may have been provided by two main sources: endogenous and exogenous (Cleaves and Lazcano 2009). The atmosphere of the primitive Earth, before the rise of the oxygen ~2.5 billion years ago, is thought to have been relatively non-reducing, that is, mainly composed of molecular nitrogen, carbon dioxide and water vapour (Kasting et al. 1993; Olson 2006; Zahnle et al. 2010). In these conditions, only limited amounts of organic compounds may have been produced (Schlesinger and Miller 1983a, 1983b;

Plankensteiner et al. 2004; Cleaves et al. 2008). Submarine hydrothermal systems may have also contributed to the synthesis of organic molecules on the early Earth. However, there is little consensus on the real likelihood of this (Bernhardt et al. 1984; White 1984; Miller and Bada 1988; Bada et al. 1995; Imai et al. 1999; McCollom et al. 1999; Ito et al. 2006; Aubrey et al. 2009; Kopetzki and Antonietti 2011).

The primitive Earth may have also received the necessary organic molecules via exogenous delivery, as it was bombarded by comets, asteroids and their fragments (i.e., meteorites, micrometeorites and interplanetary dust particles [IDPs]) between 4.56 and 3.8 billion years ago (Schidlowski 1988; Chyba and Sagan 1992; Schopf 1993). Comets are known to contain several extra-terrestrial molecules (Crovisier and Bockelée-Morvan 1999; Bockelée-Morvan et al. 2004;

Crovisier et al. 2009; Mumma and Charnley 2011). Comet Wild-2 contains glycine (Elsila et al. 2009), while comet 67P/Churyumov–Gerasimenko has carbon-rich species (e.g., alcohols, carbonyls, amines, nitriles, amides and isocyanates), the polymer polyoxymethylene (Goesmann et al. 2015; Wright et al. 2015), phosphorus and glycine (Altwegg et al. 2016). Although, to date, no other amino acids besides glycine have been detected in comets, amino acid precursors have been observed in comets (Bockelée-Morvan et al. 2000; Mumma et al. 2003; Crovisier et al. 2009). Furthermore, *ab initio* molecular dynamics simulations showed that shock waves passed into ice mixtures, representative of comets, could theoretically yield amino acids (Goldman et al. 2010). This was confirmed by hypervelocity impact shock experiments simulating comet impacts, which proved that the impact of comets produces α-amino acids (Martins et al. 2013) and that glycine oligomerisation up to trimers is possible (Sugahara and Mimura 2014). Micrometeorites and IDPs are a set of extra-terrestrial particles known to contain organic molecules, including ketones, aliphatic and polycyclic aromatic hydrocarbons (PAHs) and their alkylated derivatives. Amino acids were also identified at low abundances, or their detection was tentatively. In any case, it is not certain whether any of these molecules are indigenous to the micrometeorites and IDPs or terrestrial contamination (Clemett et al. 1993, 1998; Brinton et al. 1998; Flynn et al. 2003, 2004; Glavin et al. 2004; Keller et al. 2004; Matrajt et al. 2004, 2005). Ultra-carbonaceous Antarctic Micrometeorites (UCAMMs) are also a set of extra-terrestrial particles that contain up to ~80% of carbonaceous material, which is one of the highest organic matter contents detected in an extra-terrestrial body (Duprat et al. 2010; Dartois et al. 2013). They contain unusually high nitrogen- and deuterium-rich organic matter (Dartois et al. 2013). Meteorites have also been extensively analysed, and they contain extra-terrestrial organic matter, including organic compounds that presently have a role in terrestrial biochemistry.

4.3.2 METEORITES

Meteorites are extra-terrestrial objects that survive the passage through the Earth's atmosphere and impact the Earth's surface. Most are chondrites (more than 86%), which are primitive samples that have not been melted and are named after the chondrules that almost all of them contain (i.e., silicate millimetre-sized spherules, formed by melting or partial melting before accretion into the meteorite parent bodies). The remaining meteorites (less than 14%) have experienced melting and differentiation (McSween and Huss 2015). Chondrites are divided into ordinary (O), enstatite (E), carbonaceous (C), Rumuruti (R) and Kakangari (K) chondrites. The O, E and C chondrites are all further subdivided, according to their bulk chemical composition (Bischoff et al. 2011; Krot et al. 2014; McSween and Huss 2015). Carbonaceous chondrites contain a large variety of groups named after their type specimen (CI, CM, CK, CO, CR and CV), two unusual groups that have been affected by impact processes

(CH and CB) as well as grouplets and ungrouped carbonaceous chondrites (Choe et al. 2010 and references therein). Chondrites are also classified and grouped into petrographic types, according to the intensity of thermal metamorphism or aqueous alteration that occurred on the meteorite parent body and range from types 1 to 6 (McSween 1979; Browning et al. 1996). A petrologic type from 3 to 1 indicates increasing aqueous alteration, due to the melting of ices in the meteorite parent bodies, which occurred most likely due to the heating from the decay of short-lived radionuclides (e.g., ^{26}Al) (Grimm and McSween 1989). A petrologic type from 3 to 6 indicates increasing thermal metamorphism, that is, is the adjustment of the minerals (with the formation of anhydrous secondary phases, in most cases) in response to the increased temperatures in the meteorite parent body. Chondrites may be further subdivided into petrologic subtypes (Sears et al. 1980; Browning et al. 1996; Rubin et al. 2007; Alexander et al. 2013). Thermal metamorphism and aqueous alteration seem to influence the chemical content of chondrites (Browning et al. 1996; Palmer and Lauretta 2011; Vinogradoff et al. 2017).

4.3.3 CHEMICAL COMPOSITION AND ORGANIC MATTER IN CARBONACEOUS CHONDRITES

The chemical compositions of carbonaceous chondrites match the chemistry of the Sun more closely than of any other class of chondrites, with their bulk chemical compositions being very similar to that of the solar photosphere, except for the gaseous elements (e.g., H and He) (Lodders 2003). Therefore, carbonaceous chondrites are the most primitive meteorites, having preserved the history of the early solar system. They are usually composed of chondrules, calcium-aluminium-rich inclusions (CAIs) (most are up to 1-centimetre irregular-shaped white inclusions, composed mainly of oxides and silicates of calcium, aluminium, magnesium and titanium) and a fine-grained matrix of phyllosilicates, oxides, sulphides and carbonates (King et al. 2017). Carbonaceous chondrites, as the name indicates, have relatively high carbon content (~3.5 wt.%) (Alexander et al. 2017), which is present in different forms, including silicon carbide, graphite, diamonds, carbonate and organic matter (Smith and Kaplan 1970; Lewis et al. 1987; Fredriksson and Kerridge 1988; Grady et al. 1988; Amari et al. 1990, 1993; Anders and Zinner 1993; Zinner et al. 1995; Hoppe et al. 1996; Benedix et al. 2003). More than 70% of the organic matter is composed of solvent-insoluble organic matter (IOM) (Gardinier et al. 2000; Cody et al. 2002; Cody and Alexander 2005), which can be isolated from the meteorite by using hydrochloric acid (to remove carbonates) and hydrofluoric acid (to remove silicates) (Robert and Epstein 1982). Alternatively, caesium fluoride may be used at room temperature (Cody et al. 2002). The remaining less than 30% is a mixture of solvent-soluble organic compounds, which may be extracted and isolated by using organic solvents of different polarities.

4.3.4 INSOLUBLE ORGANIC MATTER IN CARBONACEOUS CHONDRITES

The insoluble organic material in carbonaceous chondrites contains carbon, hydrogen, nitrogen, oxygen and sulphur. These elements vary within carbonaceous chondrites, with the bulk compositions of their IOMs (normalised to 100 carbons) in the range of $C_{100}H_{75-79}N_{3-4}O_{11-17}S_{1-3}$ for the most primitive chondrites, $C_{100}H_{70}N_3O_{12}S_2$ for Murchison (a CM2) and $C_{100}H_{70-80}N_{3-4}O_{15-20}S_{1-4}$ for CR chondrites (Pizzarello et al. 2006; Alexander et al. 2007, 2017). The IOM contains graphitic flakes, spheres, tubes (Nakamura et al. 2002; Garvie and Busek 2004) and macromolecular organic compounds. The determination of the functional groups of the macromolecular organic compounds may be done by using thermal or chemical degradation methods, which are used to break the IOM down into smaller components. Chemical degradation studies (e.g., oxidation reactions with $Na_2Cr_2O_7$, CuO or RuO_4) have shown that the IOM is composed of aromatic compounds, bridged by short, highly branched aliphatic linkages that contain ether or ester functional groups (Hayatsu et al. 1977, 1980; Remusat et al. 2005a; Huang et al. 2007).

Pyrolysis has also been used to identify the main components of the IOM in carbonaceous chondrites, and two fractions have been identified: one labile (i.e., liberated by pyrolysis and enriched in deuterium D, ^{13}C and ^{15}N) and one refractory (i.e., more resistant to pyrolysis and depleted in ^{13}C and ^{15}N) (Sephton et al. 2003, 2004). Aromatic molecules with 1–3 rings and short aliphatic side chains have been identified as the dominant fraction in IOM pyrolysates (Komiya et al. 1993; Komiya and Shimoyama 1996; Remusat et al. 2005b; Wang et al. 2005). On the other hand, Okumura and Mimura used gradual and stepwise pyrolysis and found that the linkage and substituent portion of the IOM consisted of saturated and unsaturated aliphatic hydrocarbons up to C_8, substituted and non-substituted aromatic hydrocarbons with up to six rings, sulpho and thiol groups and carboxyl and hydroxyl groups (Okumura and Mimura 2011). Remusat et al. (2005b) performed pyrolysis analysis to reveal ester and ether functional groups as linkages between the aromatic units, which is in agreement with chemical degradation studies. In addition to conventional pyrolysis, hydrous pyrolysis (i.e., high temperature in the presence of pressurised water) and hydropyrolysis (i.e., high-pressure H_2) have been used. The free radical that is formed during pyrolysis when an IOM bond is cleaved will react with the available hydrogen from the water or the H_2, respectively, preserving the structural features of the pyrolysates (Sephton et al. 1998, 2000, 2004, 2005, 2015; Sephton and Gilmour 2000, 2001; Yabuta et al. 2007). Hydrous pyrolysis analysis of the IOM of the carbonaceous meteorite Murray (CM2) revealed the presence of C_3-C_{17} alkyl dicarboxylic acids; N- and O-containing hydroaromatic and N-, O- and S-containing aromatic compounds; a series of PAHs of up to five rings, together with non-condensed aromatic species such as substituted benzenes, biphenyl and terphenyls, as well as their substituted homologs; and hydrated PAHs (Yabuta et al. 2007). On the other hand, hydropyrolysis analysis of the IOM

of the carbonaceous meteorite Murchison released one- to six-ring aromatic hydrocarbons and their methyl-substituted homologues (Sephton et al. 2004, 2005, 2015).

Non-destructive techniques have also been used to analyse the IOM: nuclear magnetic resonance (NMR) (Cronin et al. 1987; Gardinier 2000; Cody et al. 2002, 2011; Cody and Alexander 2005), Raman spectroscopy (Quirico et al. 2003; Bonal et al. 2006, 2007, 2016), high-resolution transmission electron microscopy (HRTEM) (Garvie and Buseck 2004; Derenne et al. 2005), Fourier-transform infrared spectroscopy (FTIR) (Pendleton et al. 1994; Kebukawa et al. 2011; Orthous-Daunay et al. 2013), X-ray absorption near-edge spectroscopy (XANES) (Cody et al. 2008; Orthous-Daunay 2010; Bose et al. 2014; Le Guillou and Brearley 2014; Le Guillou et al. 2014) and electron paramagnetic resonance (EPR) (Binet et al. 2002, 2004a, 2004b; Gourier et al. 2008). Nuclear magnetic resonance analysis has shown that the IOM contains small, highly substituted aromatic units and that the aliphatic component is short and highly branched (Gardinier et al. 2000; Cody et al. 2002; Cody and Alexander 2005). Furthermore, two-dimensional solid-state H and ^{13}C NMR analysis showed that the IOM of Murchison is composed primarily of highly substituted single-ring aromatics, substituted furan/pyran moieties, highly branched oxygenated aliphatics and carbonyl groups (Cody et al. 2011).

Carbonaceous chondrites contain small regions of the IOM that are extremely isotopically enriched relative to the surrounding matter (the so called "hotspots"), with δD values of up to ~20,000‰ and $\delta^{15}N$ values of 2000–3200‰ (Busemann et al. 2006; Floss et al. 2014). D and ^{15}N enrichments of the bulk and of hotspots suggest that the IOM was formed in the interstellar medium (ISM) and/or in the outer solar system, being subsequently modified by nebular and parent body processes (Robert and Epstein 1982; Yang and Epstein 1983; Remusat et al. 2006; Alexander et al. 2007). Indeed, it has been suggested that there is a connection between the origin of chondritic IOM and comets (Alexander et al. 2017). This is supported by the elemental compositions of the comet Halley CHON particles being ~$C_{100}H_{80}N_4O_{20}S_2$ (Kissel and Krueger 1987) as well as by the refractory carbon in comet 67P/Churyumov–Gerasimenko that is similar to chondritic IOM (Fray et al. 2016).

4.3.5 SOLUBLE ORGANIC COMPOUNDS IN CARBONACEOUS CHONDRITES

The soluble organic matter (SOM) in the CM2 chondrite Murchison has a bulk composition of ~$C_{100}H_{155}O_{20}N_3S_3$, with a high molecular diversity of tens of thousands of different molecular compositions (Schmitt-Kopplin et al. 2010). Carbonaceous chondrites contain an extensive list of indigenous soluble organic compounds: carboxylic acids, amino acids, diamino acids, dipeptides, diketopiperazines, sulphonic and phosphonic acids, purines, pyrimidines, sugars and sugar-related compounds, hydrocarbons, alcohols, amines, amides, aldehydes and ketones. Four approaches may be used to determine if these organic compounds are indigenous to the carbonaceous chondrites (i.e., that they are not terrestrial contamination) (Martins et al. 2007a; Martins and Sephton 2009): (1) detection of molecules that are unusual in the terrestrial environment; (2) comparison of the absolute abundances of organic molecules in the meteorites with the abundances found in the fall site; (3) determination of enantiomeric ratios for chiral molecules and (4) measurement of the compound-specific carbon, hydrogen or nitrogen isotopic compositions. Stable isotope compositions are expressed in δ (‰) values, according to the following formula:

$$\delta\left(\permille\right) = \frac{\left(R_{sample} - R_{standard}\right)}{R_{standard}} \times 1000$$

where R stands for D/^{1}H for hydrogen, ^{13}C/^{12}C for carbon and ^{15}N/^{14}N for nitrogen. The standards used are the Vienna standard mean ocean water (VSMOW) for hydrogen, Vienna Pee Dee Belemnite (VPDB) for carbon and atmospheric nitrogen for nitrogen. The soluble organic compounds indigenous to carbonaceous meteorites are enriched in D, ^{13}C and ^{15}N (e.g., Yuen et al. 1984; Krishnamurthy et al. 1992; Cronin et al. 1993; Cooper et al. 1997). D and ^{15}N enrichments of the SOM suggest low-temperature formation of precursors in the ISM (e.g., gas-phase ion-molecule reactions), in which chemical fractionation is efficient (Tielens 1983; Yang and Epstein 1983; Millar et al. 1989; Terzieva and Herbst 2000; Sandford et al. 2001; Robert 2003; Aléon and Robert 2004).

4.3.5.1 MONOCARBOXYLIC ACIDS

Carboxylic acids are the most abundant class of soluble compounds in carbonaceous chondrites. Monocarboxylic acids were first reported in carbonaceous meteorites by Yuen and Kvenvolden (1973), followed by studies from Lawless and Yuen (1979) and Yuen et al. (1984). The CM2 Murchison and Murray were analysed, and they contained a limited number of straight- and branched-chain monocarboxylic acids with two to eight carbon atoms and abundances ranging from about 10 to 60 parts per million (ppm) for each individual compound (Yuen and Kvenvolden 1973). Shimoyama and co-authors analysed the monocarboxylic acid content of a few Antarctic carbonaceous chondrites. Yamato (Y)-791198 (CM2) contained aliphatic and aromatic monocarboxylic acids. The aliphatic monocarboxylic acids included saturated straight and branched-chain structures, ranging from 2 to 12 carbon atoms, with most structural isomers of C_4, C_5 and C_6 identified. The aromatic monocarboxylic acids included benzoic acid, methylbenzoic acid and phenylacetic acid (Shimoyama et al. 1986). Y-74662 (CM2) had several monocarboxylic acids, including aliphatic and aromatic structures. The aliphatic carboxylic acids contained saturated straight- and branched-chain structures in the range of C_2 to C_{12}. The four aromatic carboxylic acids were benzoic acid and o-, m- and p-methylbenzoic acids (Shimoyama et al. 1989). Y-793321 (CM2) was also analysed, and it contained no

monocarboxylic acids (Shimoyama et al. 1989). Naraoka et al. (1999) reported more than 30 monocarboxylic acids in Asuka (A)-881458 (CM2), including aliphatic and aromatic acids with various structural isomers. The straight-chain aliphatic monocarboxylic acids present in A-881458 had 2-12 carbon atoms, while the branched-chain structures had 4-9 carbon atoms. Aromatic monocarboxylic acids included benzoic acid and all three isomers of toluic acid. Two other CM2 chondrites (A-881280 and A-881334) were depleted in monocarboxylic acids, and Naraoka et al. (1999) suggested that these two chondrites may have been subjected to aqueous alteration or metamorphism on their meteorite parent bodies. More recently, Huang et al. (2005) identified more than 50 monocarboxylic acids on Murchison, using solid-phase microextraction (SPME). Martins et al. (2006) detected aromatic monocarboxylic carboxylic acids in the Murchison meteorite, including benzoic acid and all three isomers of methylbenzoic acid and hydroxybenzoic acid, while Orgueil (CI1) contained only benzoic acid and very small amounts of methylbenzoic acids and methylhydroxybenzoic acids. Aponte et al. (2011) analysed the monocarboxylic acid content of six carbonaceous chondrites with a range of classifications: Murchison (CM2), Elephant Moraine (EET) 87770 (CR2), Allan Hills (ALH) 83034 (CM1), ALH 83033 (CM2), Meteorite Hills (MET) 00430 (CV3) and Wisconsin Range (WIS) 91600 (CM2). These authors found an overall abundance of branched monocarboxylic acids over straight-chain ones and that EET 87770 and Murchison contained high concentrations and a complete suite of monocarboxylic acids. The remaining four meteorites (ALH 84034, ALH 84033, MET 00430 and WIS 91600) were depleted of high-molecular-weight branched monocarboxylic acids relative to their straight-chain counterparts (Aponte et al. 2011). Ivuna (CI1) was also analysed for these compounds, and it contained monocarboxylic acids with chain lengths from 1 to 10 carbon atoms, with few and less branched molecular species when compared with the Murchison meteorite (Monroe and Pizzarello 2011).

4.3.5.1.1 Stable Isotope Measurements of Monocarboxylic Acids

The bulk δ^{13}C and δD values of the monocarboxylic acids' fraction in the Murchison meteorite ranged from −3‰ to +6.7‰ and from +377‰ to +697‰, respectively (Epstein et al. 1987; Krishnamurthy et al. 1992). Stable isotope values of individual monocarboxylic acids were also obtained. Yuen et al. (1984) obtained the first δ^{13}C values for six monocarboxylic acids (acetic acid, propionic acid, isobutyric acid, butyric acid, isovaleric acid and valeric acid), ranging from +4.5‰ to +22.7‰, while Huang et al. (2005) obtained δ^{13}C values for 48 individual monocarboxylic acids in the Murchison meteorite, ranging from −31.8‰ to +32.5‰. These authors also obtained δD values for the monocarboxylic acids in Murchison, which ranged from −76.5‰ to + 2024.0‰ (Huang et al. 2005). They noted that, in general, straight-chain monocarboxylic acids had a relationship between increasing chain length and enrichment in lighter isotopes, supporting a kinetically controlled carbon-addition reaction involving radicals, as has been suggested by Yuen et al. (1984). Huang et al. observed that the branched monocarboxylic acids were significantly more enriched in the heavier isotopes of carbon and deuterium than the straight-chain ones, which suggested different synthetic pathways for the two sets of monocarboxylic acids (Huang et al. 2005). More recently, Aponte et al. (2011) determined the δD and $δ^{13}C$ values for straight- and branched-chain monocarboxylic acids for six carbonaceous chondrites: Murchison, EET 87770, ALH 83034, ALH 83033, MET 00430 and WIS 91600. The δD values ranged from −76.6 to +1778.7‰ (Murchison), +302.5 to +1632.2‰ (EET 87770), −212.2 to +432.4‰ (ALH 83034), −65.1 to +646.2‰ (ALH 83033), −156 to +539.1‰ (MET 00430), −37.8 to +787.6‰ (WIS 91600), while the $δ^{13}C$ values ranged from −25.9 to +11.3‰ (Murchison), −70.0 to +7.5‰ (EET 87770), −64.0 to +12.2‰ (ALH 83034), −79.3 to +0.4‰ (ALH 83033), −80.8 to +5.5‰ (MET 00430), −36.8 to +8.6‰ (WIS 91600). Aponte et al. noted that Murchison and EET 87770 had the highest deuterium enrichments in their monocarboxylic acids; this was consistent with their relatively pristine nature. These authors also noted that branched monocarboxylic acids were more D-enriched than their straight-chain counterparts and that generally δD values decreased with increasing number of carbons (Aponte et al. 2011). Regarding the $δ^{13}C$ for the monocarboxylic acids, the values were mostly in the range for terrestrial matter (Sephton and Botta 2005). Aponte et al. (2011) suggested two possible causes: (1) contribution of monocarboxylic acids from terrestrial sources (in which case, only the carbon isotopic compositions of the meteoritic monocarboxylic acids were lowered, and not the hydrogen isotopic compositions), or (2) the isotopic ratios reflect the original values (which would indicate different precursor compositions and synthetic pathways). The free monocarboxylic acids of the Lonewolf Nunataks (LON) 94101 (CM2) meteorite were also analysed, and they had δD values ranging from +381‰ to +1720‰ (Aponte et al. 2014a).

Overall, the carboxylic acids present in carbonaceous meteorites are enriched in deuterium and ^{13}C; this is consistent with an interstellar origin or formation from interstellar precursors. Ethanoic acid and formic acid have been detected in the ISM (Zuckerman et al. 1971; Winnewisser et al. 1975; Turner 1989; Wootten et al. 1992; Mehringer et al. 1997). Data from Huang et al. (2005) and Aponte et al. (2011) support the suggestion by Cronin and Chang (1993) that random gas-phase reactions involving radicals and ions were responsible for the production of meteoritic monocarboxylic acids, prior to parent body accretion and processing. Alternatively, carboxylic acids may be formed by processing on the meteorite parent body by two methods: (1) Via hydrolysis of precursor molecules: Carboxylic and dicarboxylic acids can be formed via hydrolysis of carboxamides (Cooper and Cronin 1995). Also, α-hydroxycarboxylic acids could be formed via the Strecker-cyanohydrin reaction (Peltzer and Bada 1978; Peltzer et al. 1984; Cronin et al. 1993), which is supported by the presence of α-amino acids and iminodicarboxylic acids in carbonaceous meteorites (Lerner and Cooper 2005). However, the δD

and $δ^{13}C$ values of α-hydroxycarboxylic acids were significantly lower than the ones for α-amino acids (Epstein et al. 1987; Pizzarello et al. 1991), which would suggest different precursors for these two classes of compounds. (2) By the oxidation of hydrocarbons from the IOM or the oxidation of free hydrocarbons by oxidised fluids or minerals, via alcohol and ketone intermediaries (Martins et al. 2006).

4.3.5.2 DICARBOXYLIC ACIDS AND HYDROXYCARBOXYLIC ACIDS

Branched- and straight-chained aliphatic dicarboxylic acids were first detected in the Murchison meteorite (Lawless and Zeitman 1974). Further studies confirmed this detection and that the aliphatic dicarboxylic acids had up to nine carbon atoms (Martins et al. 2006; Peltzer et al. 1984; Shimoyama and Shigematsu 1994). The Murchison meteorite also had aromatic dicarboxylic acids, including methylphthalic acid and all three isomers of phthalic acid (Martins et al. 2006). Dicarboxylic acids were found in Ivuna, with linear alkyl-chain lengths of 4 to 14 carbon atoms and with several branched species up to 8 carbon atoms (Monroe and Pizzarello 2011). Dicarboxylic acids present in Sutter's Mill (CM2) were almost exclusively linear species of 3 to 14 carbon atoms, with concentrations typically less than 10 nmole/g of meteorite (Pizzarello and Garvie 2014).

Peltzer and Bada first detected seven α-hydroxycarboxylic acids in the Murchison meteorite, at the same concentration as hydrolysed amino acids (14.6 ppm in total) (Peltzer and Bada 1978). This number was later expanded by Cronin et al. (1993), who found almost 60 α-hydroxymonocarboxylic acids, with carbon chains through C_8. Cronin et al. also found more than 50 α-hydroxydicarboxylic acids in Murchison, with carbon chains up to C_8 or C_9. These authors showed that the hydroxycarboxylic acids and the hydroxydicarboxylic acids had a carbon-chain structure that corresponded to α-amino acids (α-aminocarboxylic acids and α-aminodicarboxylic acids, respectively), which suggested that they all share a common Strecker-cyanohydrin reaction pathway during aqueous alteration on the meteorite parent body (Peltzer and Bada 1978; Peltzer et al. 1984; Lerner et al. 1993). Hydrocarboxylic acids were also investigated in the Graves Nunataks (GRA) 95229 (CR2) and LaPaz Icefield (LAP) 02342 (CR2) meteorites (Pizzarello et al. 2008, 2010). Results showed that 2-H-2-OH acids have distributions close to those of amino acids, supporting the hypothesis that these two groups of compounds were formed from common aldehyde precursors. Monroe and Pizzarello detected several hydroxycarboxylic acids in Ivuna, which were the most abundant group of compounds in this meteorite (Monroe and Pizzarello 2011).

4.3.5.2.1 Stable Isotope Measurements of Dicarboxylic Acids and Hydroxycarboxylic Acids

The bulk $δ^{13}C$ and δD values of the combined hydroxydicarboxylic and dicarboxylic acids fractions in Murchison were −6‰ and +357‰, respectively, while the bulk $δ^{13}C$ and δD values of the hydroxycarboxylic acid fraction in the same meteorite were +4‰ and +573‰, respectively (Cronin et al. 1993). Dicarboxylic acids from Murchison had δ¹³C values ranging from +19.1‰ to +28.1‰ and δD values ranging from +389‰ to +1550.7‰ (Pizzarello et al. 2001; Pizzarello and Huang 2002). ¹³C and D isotopic enrichments were detected for several of the hydroxycarboxylic acids in the GRA 95229 (CR2) and LAP 02342 (CR2) meteorites, and the highest δD value (+3450‰) was displayed by GRA 95229 2-OH-2-methylbutyric acid (Pizzarello et al. 2010).

4.3.5.3 Pyridine Carboxylic Acids

Pyridine carboxylic acids were detected in eight CM2 carbonaceous chondrites: ALH 85013, Dominion Range (DOM) 03183, DOM 08003, EET 96016, LAP 02333, LAP 02336, Lewis Cliff (LEW) 85311 and WIS 91600. Pyridine monocarboxylic acids negatively correlated with the degree of aqueous alteration of CM2-type meteorite parent bodies, suggesting that aqueous alteration may have destroyed some of these compounds (Smith et al. 2014).

4.3.5.3.1 Stable Isotope Measurements of Pyridine Carboxylic Acids

Pizzarello et al. (2004) have determined δ¹³C values for nicotinic acid and methylnicotinic acid (+20.3‰ ± 1.7‰ and +20.3‰ ± 1.2‰, respectively) in Murchison. δD values were also determined for nicotinic acid in Murchison (+129‰) and for nicotinic methyl homologue in Murray (+621‰) Pizzarello and Huang (2005).

4.3.5.4 Amino Acids

Kvenvolden et al. (1970) analysed the amino acid content of samples of the Murchison meteorite that had fallen in Australia the year before. These authors detected almost equal amounts of the enantiomers of D- and L-valine, proline, alanine and glutamic acid, which suggested an extra-terrestrial origin for these compounds. Furthermore, the presence of the non-proteinogenic amino acids 2-methylalanine and sarcosine was also indicative of a possible abiogenic synthesis. Samples of the Murchison meteorite were re-analysed in the next few years by different groups. Oró et al. (1971) found similar results as the ones published by Kvendolden et al. in 1970. Further studies by Kvenvolden et al. (1971) found a total of 18 amino acids in Murchison, including 12 non-proteinogenic amino acids. Lawless found at least 35 amino acids (i.e., 17 amino acids in addition to the 18 already detected in the Murchison meteorite by Kvenvolden et al. (1971)), which included a wide variety of linear and cyclic difunctional and polyfunctional amino acids (Lawless 1973). Since then, Murchison has been extensively analysed for amino acids (e.g., Glavin et al. 2006, 2010; Martins and Sephton 2009), and more recently, a new family of extra-terrestrial hydroxy amino acids has been detected in this meteorite (Koga and Naraoka 2017). Analyses by several groups have determined a total amino acid abundance of around 60 ppm and identified more than 80 different amino acids, with carbon number from C_2 through C_8, complete structural diversity and a decrease in concentration with increasing carbon number within homologous series. Overall, they can be divided into two structural types (i.e., monoamino alkanoic acids and monoamino dialkanoic acids), with branched-chain amino acid isomers more abundant than straight-chain ones and with structural α configuration more abundant than γ configuration, which in turn is more abundant than the β configuration (e.g., Cronin et al. 1988; Cronin and Chang 1993; Martins and Sephton 2009).

Other carbonaceous chondrites have also been extensively analysed for amino acids (e.g., Martins and Sephton 2009). The CM2 chondrites have total amino acid abundances that vary, ranging from 71 ppm for Y-791198, (Shimoyama et al. 1985; Shimoyama and Ogasawara 2002) to nearly depleted for Y-79331 and Belgica (B-) 7904 (Shimoyama and Harada 1984). Other CM2 chondrites include Murray (Cronin and Moore 1971; Botta et al. 2002), Y-74662 (Shimoyama et al. 1979a, 1979b), LEW 90500 (Botta and Bada 2002; Glavin et al. 2006), Essebi, Mighei, Nogoya (Cronin and Moore 1976; Botta et al. 2002), Allan Hills (ALH) 77306 (Cronin et al. 1979; Holzer and Oró 1979; Kotra et al. 1979), LON 94102 (Glavin et al. 2010), ALH 83100 (Glavin et al. 2006), Sutter's Mill (Burton et al. 2014) and Paris (Martins et al. 2015). CM1 chondrites (ALH 88045, Meteorite Hills (MET) 01070, LAP 0227 and Scott Glacier (SCO) 06043) are almost depleted of amino acids (Botta et al. 2007; Glavin et al. 2010). Aqueous alteration in the parent body of CM chondrites seems to contribute, at least partially, to the amino acid content and distribution, for example, the relative abundance of β-alanine increases with increasing aqueous alteration in the parent body of CM chondrites (Glavin et al. 2006; Martins et al. 2007a, 2015; Elsila et al. 2016). Increasing aqueous alteration may lead to the destruction of α-amino acids (previously formed by the Strecker-cyanohydrin synthesis; Peltzer and Bada 1978; Peltzer et al. 1984; Lerner et al. 1993) and the synthesis of β- and γ-amino acids in the parent body of carbonaceous chondrites (Cooper and Cronin 1995; Cronin and Chang 1993). Indeed, the CI1 chondrites Orgueil and Ivuna have high abundances of β-alanine and γ-aminobutyric acid (γ-ABA) (Lawless et al. 1972a; Ehrenfreund et al. 2001; Burton et al. 2014). Other two CI1 carbonaceous chondrites, Y-86029 and Y-980115, were analysed by liquid chromatography with fluorescence detection and time-of-flight mass spectrometry and contained low levels of L-amino acids, suggesting a terrestrial origin. Their lack of amino acids was explained by the fact that these two meteorites showed evidence of parent body heating, in combination with aqueous alteration (Burton et al. 2014).

The amino acid content of several CR chondrites has been determined. The CR2 EET 92042 and GRA 95229 were analysed for the first time for amino acids, and they had total concentrations of 180 ppm and 249 ppm, respectively, with the α-amino acids glycine, isovaline, α-AIB and alanine as the most abundant amino acids (Martins et al. 2007a). Similar amino acid abundances and distributions were found for the CR2 Miller Range (MIL) 090657 (Burton et al. 2016). Linear α-amino acids were the most abundant amino acids in the CR2 LAP 02342 (Pizzarello and Holmes 2009), while glycine and α-AIB were the most abundant for Queen Alexandra Range (QUE) 99177, which had total amino acid abundances of 81 ppm (Glavin et al. 2010). Shiṣr 033, a CR2 chondrite from the Omani desert, had a small fraction of indigenous amino acids, as indicated by the presence of α-AIB, but overall, its amino acid content was similar to the one from the meteorite fall site soil (Martins et al. 2007b). Glycine, γ-ABA, and L-glutamic acid were the most abundant amino acids in the CR2 Renazzo (Botta et al. 2002). Finally, the CR1 Grosvenor Mountains (GRO) 95577 had very low amino acid content (i.e., total amino acid abundance of 900 parts per billion [ppb]), which included α-AIB, isovaline, γ-ABA and α-AIB (Martins et al. 2007a).

The CV3 Allende and Mokoia were found to be depleted in amino acids (Cronin and Moore 1971, 1976). Several years later, Burton et al. analysed several type 3 CV and CO carbonaceous chondrites: ALH 84028 (CV3), EET 96026 (CV3), LAP 02206 (CV3), GRA 06101 (CV3), Larkman Nunataks (LAR) 06317 (CV3), Allan Hills A (ALHA) 77307 (CO3), MIL 05013 (CO3) and DOM 08006 (CO3). They found total amino acid concentrations ranging from 300 ppb to 3200 ppb, with small, straight-chain, amine terminal (n-ω-amino) amino acids as the dominant amino acids detected (Burton et al. 2012). Similar results were found by Chan et al. (2012), who analysed the CO3 carbonaceous chondrites Colony and Ornans. They found high relative abundances (to glycine) of β-alanine and γ-ABA, suggesting that these amino acids may be formed from gas-grain reactions after the meteorite parent body cooled to much lower temperatures or during the cooling process (Chan et al. 2012). Burton et al. (2015) also found a similar amino acid distribution for R and CK chondrites. LAR 04318 (CK4), ALH 85002 (CK4), EET 92002 (CK5) and LAR 06872 (CK6) were depleted in amino acids, and NWA 5956 (CK3) experienced significant terrestrial amino acid contamination. PCA 82500 (CK4/5), LEW 87009 (CK6), LAP 03834 (R3) and LAP 031135 (R4) contained low levels of the straight-chain, amino-terminal (n-ω-amino) acids β-alanine and γ-ABA (Burton et al. 2015).

While most non-proteinogenic amino acids present in carbonaceous chondrites are racemic (i.e., equal amounts of D- and L-enantiomers), this is not true for all, and L-enantiomeric excesses have been reported for a few non-proteinogenic amino acids in Murchison, Murray and Orgueil (Martins and Sephton 2009). Cronin and Pizzarello found L-enantiomer excess for α-methylisoleucine and α-methylalloisoleucine (7.0% and 9.1%, respectively), as well as for isovaline and α-methylnorvaline in the Murchison meteorite (Cronin and Pizzarello 1997). Further analysis found L-enantiomer excesses (1.0%–9.2%) in six α-methyl-α-amino alkanoic acids (2-amino-2,3-dimethylpentanoic acid (both diastereomers), isovaline, α-methyl norvaline, α-methyl valine, and α-methyl norleucine) from Murchison and Murray by gas chromatography-mass spectroscopy (GC-MS) (Pizzarello and Cronin 2000). L-enantiomeric excesses of isovaline were found to range from 0% to 15.2%, using GC-MS, with a high variation within different stones of the same meteorites and within the same stone (Pizzarello

et al. 2003). In order to eliminate any experimental bias for these observations (i.e., interference from other C_5 amino acid isomers), Glavin and Dworkin used liquid chromatography-fluorescence detection (LC-FLD)/Time-of-Flight mass spectrometry (LC/ToFMS) and found L-enantiomeric excesses for isovaline of 18.5 ± 2.6% in Murchison and of 15.2 ± 4.0% in Orgueil (Glavin and Dworkin 2009). They found no L-enantiomeric excesses for isovaline in the unaltered Antarctic CR meteorites EET 92042 and QUE 99177, which suggested that amplification of a small initial L-enantiomeric isovaline excess occurred during aqueous alteration on the meteorite parent bodies. Similar conclusion was obtained by Martins et al. (2015), who analysed Paris, one of the least aqueously altered CM chondrites, and they found that isovaline was racemic (L-enantiomer excess = 0.35% ± 0.5%), supporting the hypothesis that aqueous alteration is responsible for the high L-enantiomer excess of isovaline observed in the most aqueously altered carbonaceous chondrites. Multidimensional GC, which is an improved method to accurately quantify amino acid enantiomers, was used by Myrgorodska et al. (2016), who found an L-enantiomeric excess of 4.61% ± 0.83% for isovaline in Murchison.

4.3.5.4.1 Stable Isotope Measurements of Amino Acids

Stable isotope analyses of individual amino acids in meteorites show that, with a few exceptions due to terrestrial contamination, they are enriched in D, ^{13}C and ^{15}N and have values in agreement with other meteoritic polar organic compounds (Sephton and Botta 2005). Engel and Macko (1997) determined $\delta^{15}N$ values for individual amino acids in Murchison, ranging from +37‰ to +184‰. Pizzarello et al. (2003) found individual D- and L-isovaline $\delta^{13}C$ values in Murchison of around +18‰, while the $\delta^{13}C$ values of more than 30 individual amino acids in the same meteorite ranged from +4.9‰ to +52.8‰ (Pizzarello et al. 2004). These authors observed that the α-methyl-α-amino acids were more enriched in ^{13}C than the corresponding α-H-α-amino acids and that the ^{13}C content declined with increasing chain length for the α-amino acids, which was in agreement with the trends observed for carboxylic acids and alkanes (Pizzarello et al. 2004). Ehrenfreund et al. (2001) determined the $\delta^{13}C$ values for β-alanine and glycine in Orgueil to be +18‰ and +22‰, respectively. $\delta^{13}C$ values were also determined for amino acids in EET 92042 and GRA 95229, and these ranged from +31.6‰ to +50.5‰ (Martins et al. 2007a). Overall, amino acids present in carbonaceous chondrites are outside the carbon isotopic range of terrestrial amino acids (−70.5‰ to +11.25‰) (Scott et al. 2006).

The δD values of more than 40 amino acids present in Murchison and Murray ranged from +181‰ to +3419‰ and from +60‰ to +3604‰, respectively, with higher D values for amino acids with branched alkyl chain (Pizzarello and Huang 2005). The enrichment in D (as well as in ^{15}N) for amino acids in carbonaceous chondrites is in agreement with chemical fractionation of the amino acid precursors in the ISM (Terzieva and Herbst 2000; Sandford et al. 2001; Aléon and Robert 2004).

4.3.5.5 Diamino Acids

Diamino acids were detected in the Murchison meteorite by
GC-MS analyses (Meierhenrich et al. 2004). These included
DL-2,3-diaminopropanoic acid, DL-2,4-diaminobutanoic acid,
4,4′-diaminoisopentanoic acid, 3,3′-diaminoisobutanoic acid
and 2,3-diaminobutanoic acid, and the chiral diamino acids
were racemic. Meierhenrich et al. (2004) pointed out the poly-
condensation reactions of diamino acids into early peptide
nucleic acid material as one feasible pathway for the prebiotic
evolution of DNA and RNA.

4.3.5.6 Dipeptides and Diketopiperazines

The Y-791198 and Murchison meteorites were analysed for
dipeptides and diketopiperazines (Shimoyama and Ogasawara
2002). Y-791198 contained 11 pmol/g and 18 pmol/g of gly-
cylglycine (Gly-Gly) and cyclo(Gly-Gly), respectively, while
Murchison contained 4 pmol/g and 23 pmol/g, respectively,
of the same compounds. No other dipeptide and diketopiperazine were detected in this study. Shimoyama and Ogasawara
suggested that the glycine dimers were formed by condensation of glycine monomers (Shimoyama and Ogasawara 2002).

4.3.5.7 Sulphonic and Phosphonic Acids

Alkyl phosphonic acids and alkyl sulphonic acids with carbon atoms up to C_4 were detected in the Murchison meteorite
(Cooper et al. 1992). Five of the eight possible alkyl phosphonic acids and seven of the eight possible alkyl sulphonic
acids have been identified, with abundances decreasing with
increasing carbon number. Concentrations range downward
from approximately 380 nmol/gram in the alkyl sulphonic
acid series and from 9 nmol/gram in the alkyl phosphonic
acid series (Cooper et al. 1992).

4.3.5.7.1 Stable Isotope Measurements of Sulphonic Acids

Cooper et al. (1997) observed that the $\delta^{13}C$ values of sulphonic
acids in the Murchison meteorite decreased with increasing carbon number, with $\delta^{13}C$ values of +29.8‰, +9.1‰, –0.4‰ and
–0.9‰ for the methyl, ethyl, isopropyl, and n-propyl sulphonic
acids, respectively. δD values were also obtained in the same
study: +483‰, +787‰, +536‰, and +852‰ for the methyl,
ethyl, isopropyl, and n-propyl sulphonic acids, respectively.

4.3.5.8 Purines and Pyrimidines

Purines and pyrimidines have been detected in several carbonaceous meteorites (Hayatsu 1964; Hayatsu et al. 1968,
1975; Folsome et al. 1971, 1973; Lawless et al. 1972b; Van
der Velden and Schwartz, 1977; Stoks and Schwartz 1979,
1981; Shimoyama et al. 1990; Martins et al. 2008; Callahan
et al. 2011). Purines and triazines were first isolated from the
Orgueil meteorite and identified by paper chromatography by
Hayatsu (1964). These results were confirmed by Hayatsu co-
authors, who also detected guanylurea in the same meteorite

(Hayatsu et al. 1968). Folsome et al. (1971) extended the search
to the Murchison meteorite and found 4-hydroxypyrimidine
and two of its methyl-substituted isomers. These authors later
extended their analysis to the Murchison, Murray and Orgueil
meteorites and detected 4-hydroxypyrimidine, N,N,C-alkyl-
ketohexahydropyrimidines and a heterogenous class of (speculative) pyrimidines (Lawless et al. 1972b; Folsome et al.
1973), but no purines, triazines or guanylurea, which contrasted with the results from Hayatsu and co-authors (Hayatsu
1964; Hayatsu et al. 1968). In order to address this discrepancy in the results, Hayatsu et al. (1975) analysed Murchison
by using mild extraction conditions (water or formic acid)
and detection by mass spectrometry as well as paper chromatography and thin-layer chromatography. They only detected
aliphatic amines and C_2-C_6 alkylpyridines. However, when
they used drastic extraction conditions (hot, 3-6 M HCl or
CF_3COOH), adenine, guanine, melamine, cyanuric acid, guanylurea and urea were detected, which suggested that these
molecules were released from the macromolecular material.
Furthermore, Hayatsu and co-authors suggested that the dis-
crepancies were due to inadequate extraction conditions by
Folsome and co-authors (Hayatsu et al. 1975). A couple of
years later, Van der Velden and Schwartz (1977) again analysed the Murchison meteorite, this time using dual-column,
ion-exclusion chromatography and ultraviolet (UV) spectroscopy. They detected for the first time xanthine in a formic
acid extract. Guanine and hypoxanthine were also tentatively
detected. Also, after silylation of a water extract, hydroxypyrimidines (4-hydroxypyrimidine, 4-hydroxy-2-methylpyrim-
idine and 4-hydroxy-6-methylpyrimidine) were identified,
which suggested that these compounds previously detected
in the Murchison meteorite were the result of contaminants present in the silylation reagent (Van der Velden and
Schwartz 1977). Following this study, Stoks and Schwartz
(1979) reinvestigated the presence of pyrimidines in extracts
from Murchison, Murray and Orgueil by using fractionation
techniques (i.e., activated charcoal, which separated the
nucleobases from other organic compounds) and ion-exclusion chromatography with UV spectroscopy. They reported
for the first time the presence of uracil in water and formic
acid extracts of all three meteorites. Two years later, the same
meteorites were analysed by Stoks and Schwartz (1981) by
using GC, cation- and anion-exclusion liquid chromatography-mass spectrometry. Xanthine, adenine, hypoxanthine and
guanine were detected in in formic acid extracts of all three
meteorites. Hydroxypyrimidines and s-triazines were not
detected in any of the extracts, and Stoks and Schwartz (1981)
suggested that these molecules, previously detected by other
authors, were formed during the experimental procedure.
The search for purines and pyrimidines was then extended
to other meteorites by other groups. Shimoyama et al. (1990)
detected guanine and possible xanthine and hypoxanthine in
Y-74662 and Y-791198 in formic acid extracts by using high-
performance liquid chromatograph (HPLC) with UV spectroscopy. Y-793321 and B-7904 were also analysed, but no
nucleobases were detected. Callahan et al. (2011) searched
for nucleobases and nucleobase analogues in formic acid

extracts of several meteorites by liquid chromatography-mass spectrometry, including Orgueil, MET 01070, Scott Glacier (SCO) 06043 (CM1), ALH 83100, LEW 90500, LON 94102, Murchison, GRO 95577, EET 92042, GRA 95229 and QUE 99177 (CR3). Most of these meteorites contained adenine, while Murchison, LEW 90500 and LON 94102 had the most abundant concentrations of purines when compared with the other meteorites. Guanine, hypoxanthine, xanthine, adenine, purine and 2,6-diaminopurine were also detected in some of these meteorites (Callahan et al. 2011).

4.3.5.8.1 Stable Isotope Measurements of Purines and Pyrimidines

The variations in the content of purines and pyrimidines in fragments of the same meteorite indicated that the detected purines and pyrimidines may had a terrestrial origin, either from direct contamination (since the meteorite fell to the Earth until it was analysed) and/or formed from impurities in the reagents used during the experimental procedure. This hypothesis remained opened until the beginning of the twenty-first century. Martins et al. (2008) performed for the first time compound-specific carbon isotope measurements of these compounds by using gas chromatography-combustion-isotope ratio mass spectrometry (GC-C-IRMS), which is the ultimate way to determine the origin of purines and pyrimidines. These authors optimised the extraction and purification methods and detected purines and pyrimidines in formic acid extracts of the Murchison meteorite and of a soil collected in the proximity of the meteorite fall site. Carbon isotope ratios were obtained for uracil and xanthine in Murchison ($\delta^{13}C = +44.5‰$ and $+37.7‰$, respectively). On the other hand, uracil and thymine in the meteorite fall site had $\delta^{13}C$ values of $-10.6‰$ and $-15.9‰$, respectively, and xanthine was not detected. These results clearly showed that purines and pyrimidines detected in the Murchison meteorite were enriched in ^{13}C, that their $\delta^{13}C$ values fell outside of the terrestrial range and that they were indigenous to the meteorite.

4.3.5.9 Sugars and Sugar-Related Compounds

Degens and Bajor (1962) detected sugars (70 ppm) in a sample of the Murchison meteorite by using paper chromatography. Sugars were also detected in eight carbonaceous chondrites (Orgueil, Cold Bokkeveld [CM2], Mighei [CM2], Murray, Mokoia [CV3], Felix [CO3.3], Lancé [CO3.5] and Warrenton [CO3.7]) in small quantities (5-26 ppm), with mannose and glucose as the most abundant (Kaplan et al. 1963). Several sugar-related compounds (sugar alcohols, sugar mono-acids, sugar di-acids and deoxysugar acids) and the simplest sugar molecule dihydroxyacetone were later detected in the Murchison and Murray meteorites as their trimethylsilyl (TMS) and/or tertiary butyl-dimethylsilyl (TBDMS) derivatives by using GC-MS (Cooper et al. 2001). Their indigeneity was indicated by the decrease of abundances with increase of carbon number within a class of compounds, almost complete structural diversity and stable isotopic values ($\delta^{13}C$ $-5.89‰$ and δD $+119‰$) of a combined fraction of neutral polyols (Cooper et al. 2001).

4.3.5.10 Hydrocarbons

The presence of *n*-alkanes in meteorites was first reported in analyses of Orgueil in 1961, and it was associated to biogenic activity (Nagy et al. 1961). Following this, Meinschein and co-authors analysed benzene extracts of the same meteorite and supported this conclusion by suggesting that the presence of the *n*-alkanes was evidence for biological activity in the meteorite parent body (Meinschein 1963a, 1963b; Meinschein et al. 1963). This was received with scepticism by other scientists, who did not agree with these conclusions (Anders 1962). Oró et al. (1966) analysed Orgueil, Murray, Mokoia, Alais (CI1), Ivuna, Al Rais (CR2-anomalous), Cold Bokkeveld, Mighei, Nogoya, Santa Cruz (CM2), Grosnaja (CV3), Kaba (CV3), Lancé (CO3.5), Ornans (CO3.4), Karoonda (CK4) and Warrenton (CO3.7) by using GC-MS. It was observed that all these meteorites contained paraffinic hydrocarbons; alkanes from C_{15} to about C_{25} were found in Orgueil, Murray and Mokoia; there was no predominance of odd over even carbon-number hydrocarbons, except in the C_{22}-C_{27} normal alkane range of Orgueil; pristane and phytane, which are known to be degradation products of chlorophyll, were found in all meteorites, except Ornans; and the distribution of paraffinic hydrocarbons in some of these meteorites was found similar to that of microfossil-bearing pre-Cambrian rocks. Therefore, Oró et al. (1966) questioned whether the detected *n*-alkanes were indigenous to the meteorite or indeed terrestrial contamination. Support for the terrestrial contamination origin for the *n*-alkanes detected in meteorites came from further studies from Oró and Nooner (1967) and Nooner and Oró (1967). Their results showed that aliphatic hydrocarbons from several carbonaceous chondrites had the same chromatographic patterns as isoprenoids and other aliphatic hydrocarbons present in ancient sediments, crude oil, terrestrial graphite and other terrestrial samples. In addition, the external parts of the meteorites contained more aliphatic hydrocarbons than the inside, which supported a terrestrial contamination source (Nooner and Oró 1967; Oró and Nooner 1967). On the other hand, Studier et al. (1968, 1972) supported the hypothesis that the *n*-alkanes were indigenous to meteorites and were formed by abiotic processes. They found that the aliphatic hydrocarbons in the Murchison meteorite above C_8 included *n*-alkanes, mono- and dimethylalkanes and alkenes, while below C_8, the *n*-alkanes were absent. Isoprenoids from C_{17} to C_{20} occurred in a surface rinse and appeared to be terrestrial contaminants (Studier et al. 1972). Similar distribution was found in the Orgueil and Murray meteorites: aliphatic hydrocarbons above C_{10} included paraffins and 2-methyl-, 3-methyl- and other slightly branched paraffins, while below C_{10}, the aliphatic hydrocarbons were not present (Studier et al. 1968). These authors considered the aliphatic hydrocarbon distribution found in the carbonaceous meteorites similar to the one synthesised from CO and D_2 in the presence of iron meteorite powder. According to them, this type of catalytic reactions (i.e., Fischer-Tropsch reaction) could have taken place in the solar nebula, converting CO to less volatile carbon compounds and further condensing in the inner solar system.

However, the Fischer-Tropsch hypothesis was not supported by experimental data, as the necessary catalysts were formed much later (i.e., by aqueous activity in the meteorite parent body, and not on the solar nebula) (Kerridge et al. 1979). The final word on the origin of the aliphatic hydrocarbons came years later from stable isotope measurements, which clearly indicated a terrestrial contamination origin.

Aromatic hydrocarbons were detected in the Murchison meteorite (Oró et al. 1971; Pering and Ponnamperuma 1971; Studier et al. 1972; Basile et al. 1984; Deamer and Pashley 1989). Basile et al. (1984) detected more than 30 aromatic hydrocarbons in solvent extracts of the Murchison meteorite, including several alkylated aromatic hydrocarbons. Pyrene and fluoranthene were also found in solvent extracts of Murchison (Deamer and Pashley 1989). Other carbonaceous chondrites were analysed for aromatic hydrocarbons, including Cold Bokkeveld (Commins and Harington 1966; Sephton et al. 2001), Orgueil (Commins and Harington 1966; Olson et al. 1967; Sephton et al. 2001), Murray, Santa Cruz, Boriskino (CM2), Mokoia (Olson et al. 1967), Y-791198 (Naraoka et al. 1988), Y-74662 and Y-793321 (Shimoyama et al. 1989) and Paris (Martins et al. 2015). Orgueil had relatively low-molecular-weight aromatic hydrocarbons, while Murray, Santa Cruz, Boriskino and Mokoia had higher-molecular-weight ones (Olson et al. 1967). Y-791198 contained up to 44 aromatic hydrocarbons and related compounds in the range of naphthalene and pyrene, including almost all structural isomers of methyl- and dimethyl-naphthalene (Naraoka et al. 1988). Y-74662 included around 50 aromatic hydrocarbons with various structural isomers: the smallest was naphthalene, while the largest was methylfuoranthene and/or methylpyrene. No aromatic hydrocarbon was detected in Y-793321 (Shimoyama et al. 1989). Supercritical fluid extraction was used to extract hydrocarbons from Orgueil and Cold Bokkeveld, which revealed a complex mixture of free non-polar organic molecules, including tetrahydronaphthalenes and aromatic hydrocarbons (Sephton et al. 2001). More recently, the Paris meteorite was analysed for both aliphatic and aromatic hydrocarbons. It contained n-alkanes ranging from C_{16} to C_{25} and 3- to 5-ring nonalkylated PAHs (Martins et al. 2015). The distribution of alkylated PAHs in CM2 chondrites seems to be related to the degree of aqueous alteration on its parent body (Martins et al. 2015).

4.3.5.10.1 Stable Isotope Measurements of Hydrocarbons

Solvent-soluble hydrocarbons in the Murchison meteorite were analysed after separation by silica gel chromatography into aliphatic, aromatic and polar hydrocarbon fractions. The aliphatic hydrocarbon fraction had δD values that ranged from +103‰ to +280‰ and $\delta^{13}C$ values that ranged from –13‰ to –5‰. Pyrene, fluoranthene, phenanthrene and acenaphthene were the most abundant components in the aromatic fraction, which had δD values ranging from +244‰ to +468‰ and $\delta^{13}C$ values ranging from –6‰ to –5‰. Aromatic ketones, nitrogen and sulphur heterocycles were identified in the polar fraction, with δD values of +751‰ to +947‰ and $\delta^{13}C$ values of +5‰ to +6‰ (Krishnamurthy et al. 1992). The carbon isotopic compositions of individual n-alkanes in several meteorites (Orgueil, Cold Bokkeveld, Murchison, Vigarano (CV3), Allende and Ornans) were determined years later. $\delta^{13}C$ values for the individual n-alkanes ranged from –25.3‰ to –38.7‰, which means that they had a similar range as the n-alkanes in petroleum, clearly indicating a terrestrial origin for these compounds present in meteorites (Sephton et al. 2001).

Yuen et al. (1984) determined for the first time the $\delta^{13}C$ value (–28.7‰) of benzene in Murchison. Years later, Gilmour and Pillinger determined the $\delta^{13}C$ values of individual PAHs in the Murchison meteorite (ranging from –22.3‰ to –5.9‰) (Gilmour and Pillinger 1994). They observed that more complex PAHs became progressively enriched in ^{12}C, which is consistent with a mechanism involving the synthesis of higher-molecular-weight compounds from lower homologues (Gilmour and Pillinger 1994). Those values were later confirmed by Sephton et al. (1998), who obtained $\delta^{13}C$ values for aromatic hydrocarbons in Murchison, ranging from –28.8‰ to –5.8‰. Naraoka et al. (2000) analysed the A-881458 meteorite, which contained more than 70 individual PAHs. Their carbon isotopic compositions ranged from –26‰ to +8‰ (Naraoka et al. 2000). More recently, Huang et al. (2015) reported the δD and $\delta^{13}C$ values for solvent-soluble aromatic compounds (PAHs and heteropolycyclic aromatic compounds [HACs]) in Murchison and LON 94101 (CM2). The $\delta^{13}C$ values for Murchison ranged from –5‰ to –23‰, while for LON 94101, these values ranged from –1‰ to –18‰. The average δD values for PAHs and HACs in Murchison were +260‰ and +312‰ and in LON 94101 were +436‰ and +903‰ (Huang et al. 2015).

4.3.5.11 ALCOHOLS, ALDEHYDES AND KETONES

Alcohols and carbonyl compounds have been identified in carbonaceous chondrites. Jungclaus et al. (1976a) detected these compounds in the Murchison meteorite, including methanol, ethanol, 2-propanol, butyl alcohols, acetaldehyde, propionaldehyde, formaldehyde, acetone, 2-butanone, 2-pentanone and 3-pentanone. Aldehydes and ketones have been detected in Murchison (Jungclaus et al. 1976a; Pizzarello and Holmes 2009), Ivuna (Monroe and Pizzarello 2011), LAP 02342 and GRA 95229 (Pizzarello and Holmes 2009). Pizzarello and Holmes (2009) confirmed the results of Jungclaus et al. (1976a) for aldehydes and ketones in the Murchison meteorite and extended the list of these compounds, determining the total abundance of aldehydes and ketones to be 200 nmol/g. Ivuna had high abundances of formaldehyde and acetone (Monroe and Pizzarello 2011), while LAP 02342 and GRA 95229 contained both aldehydes and ketones (total abundances of 534 nmol/g and 3168 nmol/g for LAP 02342 and GRA 95229, respectively). Acetone was the most abundant carbonyl-containing compound in these meteorites, but the longer-chain-length ketones were less abundant than the corresponding molecular weight aldehydes (Pizzarello and Holmes 2009).

4.3.5.12 Amines and Amides

Jungclaus et al. (1976b) reported for the first time aliphatic amines in water extracts of the Murchison meteorite. Pizzarello et al. (1994) confirmed this detection, showing that Murchison contained a mixture of both primary and secondary isomers through C_5 at a total concentration of ≥ 100 nmol/g. Ammonia and tertiary amines were detected in LAP 02342 and GRA 95229, in which ammonia was the most abundant, followed by sec-butyl, butyl and isopropyl amines (Pizzarello and Holmes 2009). Aponte et al. (2014b) extracted aliphatic monoamines from the Murchison meteorite by using a novel GC method suited for the separation of aliphatic primary and secondary amines, including the enantiomeric separation of chiral amines. Results showed a complete suite of structural isomers, with a larger concentration of methylamine and ethylamine and decreasing amine concentrations with increasing carbon number. The same method was again applied by Aponte et al. (2015) to the Orgueil meteorite. Twelve amines were detected, with concentrations ranging from 1.1 nmol/g to 332 nmol/g. The enantiomeric composition for the chiral monoamines (R)- and (S)-sec-butylamine in Orgueil was also measured and was found to be racemic within experimental error (Aponte et al. 2015). Further analyses of aliphatic monoamines in hot acid-water extracts included the carbonaceous chondrites LAP 02342 (CR2), GRA 95229 (CR2), LON 94101 (CM2), LEW 90500 (CM2) and ALH 83100 (CM1/2) (Aponte et al. 2016). Results showed that (1) these CR2 chondrites contained higher concentrations of monoamines when compared with the CM2 chondrites analysed in the same study; (2) the concentration of monoamines decreased with increasing carbon number; (3) isopropylamine was the most abundant monoamine in these CR2 chondrites, while methylamine was the most abundant amine in these CM2 and CM1/2 chondrites; and (4) sec-butylamine, 3-methyl-2-butylamine and sec-pentylamine had racemic compositions (Aponte et al. 2016).

Linear and cyclic aliphatic amides were detected in the Murchison meteorite by Cooper and Cronin (1995), including monocarboxylic acid amides, dicarboxylic acid monoamides, hydroxy acid amides, lactams, carboxy lactams, lactims, N-acetyl amino acids and substituted hydantoins. Numerous isomers and homologues through at least C_8 were observed, except for the N-acetyl amino acids and hydantoins. These authors noted that the detected cyclic amides, given their potential for hydrogen-bonded pair formation, might be considered candidate bases for a primitive sequence coding system (Cooper and Cronin 1995).

4.3.5.12.1 Stable Isotope Measurements of Amines

The $\delta^{13}C$ values of aliphatic monoamines from the Murchison meteorite were measured and ranged from +21‰ to +129‰, showing a decrease in ^{13}C with increasing carbon number. Aponte et al. (2014b) suggested that this relationship may be consistent with a chain-elongation mechanism under kinetic control, similar to the previously proposed mechanism for meteoritic amino acids. The compound-specific stable carbon isotopic ratios ($\delta^{13}C$) of five monoamines detected in Orgueil were also measured and had $\delta^{13}C$ values ranging from –20‰ to +59‰ (Aponte et al. 2015). Aponte et al. (2016) also analysed the LAP 02342 (CR2), GRA 95229 (CR2), LON 94101 (CM2), LEW 90500 (CM2) and ALH 83100 (CM1/2) meteorites. They found that the $\delta^{13}C$ values of monoamines in these CR2 chondrite did not correlate with the number of carbon atoms. On the other hand, in these CM2 and CM1/2 chondrites, the ^{13}C enrichment decreases with increasing monoamine carbon number. All amines in these CR2, CM2 and CM1/2 chondrites were enriched in ^{13}C, which supported their formation in cold interstellar environments (Aponte et al. 2016). Pizzarello and Yarnes (2016) determined $\delta^{13}C$ values of sec-butylamine in two CR meteorites (GRA 95229 and LAP 02342) and confirmed their indigeneity ($\delta^{13}C \sim +10‰$).

4.3.6 SUMMARY

This chapter provides an overview of the IOM and the solvent-soluble organic molecules present in carbonaceous meteorites. Their sources and formation are discussed. The delivery of these organic molecules to the early Earth and the contribution to the origin of life are briefly mentioned.

REFERENCES

Aléon, J., and F. Robert. 2004. Interstellar chemistry recorded by nitrogen isotopes in Solar System organic matter. *Icarus* 167: 424–430.

Alexander, C. M. O'D., M. Fogel, H. Yabuta et al. 2007. The origin and evolution of chondrites recorded in the elemental and isotopic compositions of their macromolecular organic matter. *Geochim. Cosmochim. Acta* 71: 4380–4403.

Alexander, C. M. O.'D., K. T. Howard, R. Bowden et al. 2013. The classification of CM and CR chondrites using bulk H, C and N abundances and isotopic compositions. *Geochim. Cosmochim. Acta* 123: 244–260.

Alexander, C. M. O'D., G. D. Cody, B. T. De Gregorio, L. R. Nittler, J. Davidson, et al. 2017. The nature, origin and modification of insoluble organic matter in chondrites, the major source of Earth's C and N. *Chemie der Erde … Geochemistry* 77: 227–256.

Altwegg, K., H. Balsiger, A. Bar-Nun et al. 2016. Prebiotic chemicals - amino acid and phosphorus - in the coma of comet 67P/ Churyumov-Gerasimenko. *Sci. Adv.* 2: e1600285-e1600285.

Amari, S., A. Anders, A. Virag et al. 1990. Interstellar graphite in meteorites. *Nature* 345: 238–240.

Amari, S., P. Hoppe, E. Zinner et al. 1993. The isotopic compositions and stellar sources of meteoritic graphite grains. *Nature* 365: 806–809.

Anders, E. 1962. Meteoritic hydrocarbons and extraterrestrial life. *Ann. N.Y. Acad. Sci.* 93: 651–657.

Anders, E., and E. Zinner. 1993. Interstellar grains in primitive meteorites - Diamond, silicon carbide, and graphite. *Meteoritics* 28: 490–514.

Aponte, J. C., M. R. Alexandre, Y. Wang et al. 2011. Effects of secondary alteration on the composition of free and IOM-derived monocarboxylic acids in carbonaceous chondrites. *Geochim. Cosmochim. Acta* 75: 2309–2323.

Aponte, J. C., R. Tarozo, M. R. Alexandre et al. 2014a. Chirality of meteoritic free and IOM-derived monocarboxylic acids and implications for prebiotic organic synthesis. *Geochim. Cosmochim. Acta* 131: 1–12.

Aponte, J. C., J. P. Dworkin, and J. E. Elsila. 2014b. Assessing the origins of aliphatic amines in the Murchison meteorite from their compound-specific carbon isotopic ratios and enantiomeric composition. *Geochim. Cosmochim. Acta* 141: 331–345.

Aponte, J. C., J. P. Dworkin, and J. E. Elsila. 2015. Indigenous aliphatic amines in the aqueously altered Orgueil meteorite. *Meteorit. Planet. Sci.* 50: 1733–1749.

Aponte, J. C., H. L. McLain, J. P. Dworkin et al. 2016. Aliphatic amines in Antarctic CR2, CM2, and CM1/2 carbonaceous chondrites. *Geochim. Cosmochim. Acta* 189: 296–311.

Aubrey, A. D., H. J. Cleaves, and J. L. Bada. 2009. The role of submarine hydrothermal systems in the synthesis of amino acids. *Orig. Life Evol. Biosph.* 39: 91–108.

Bada, J. L., S. L. Miller, and M. Zhao. 1995. The stability of amino acids at submarine hydrothermal vent temperatures. *Orig. Life Evol. Biosph.* 25: 111–118.

Basile, B. P., B. S. Middleditch, and J. Oró. 1984. Polycyclic aromatic hydrocarbons in the Murchison meteorite. *Org. Geochem.* 5: 211–216.

Benedix, G. K., L. A. Leshin, J. Farquhar et al. 2003. Carbonates in CM2 chondrites: Constraints on alteration conditions from oxygen isotopic compositions and petrographic observations. *Geochim. Cosmochim. Acta* 67: 1577–1588.

Bernhardt, G., H. D. Ludemann, R. Jaenicke et al. 1984. Biomolecules are unstable under "black smoker" conditions. *Naturwissenschaften* 71: 583–586.

Binet, L., D. Gourier, S. Derenne et al. 2002. Heterogeneous distribution of paramagnetic radicals in insoluble organic matter from the Orgueil and Murchison meteorites. *Geochim. Cosmochim. Acta* 66: 4177–4186.

Binet, L., D. Gourier, S. et al. 2004a. Diradicaloids in the insoluble organic matter from the Tagish Lake meteorite: Comparison with the Orgueil and Murchison meteorites. *Meteorit. Planet. Sci.* 39: 1649–1654.

Binet, L., D. Gourier, S. Derenne et al. 2004b. Occurence of abundant diradicaloid moieties in the insoluble organic matter from the Orgueil and Murchison meteorites: A fingerprint of its extraterrestrial origin?. *Geochim. Cosmochim. Acta* 68: 881–891.

Bischoff, A., N. Vogel, and J. Roszjar. 2011. The Rumuruti chondrite group. *Chem. Erde Geochem.* 71: 101–133.

Bockelée-Morvan, D., J. Crovisier, M. J. Mumma et al. 2004. The composition of cometary volatiles. In *Comets II*, Eds. M. Festou, H. U. Keller, H. A. Weaver, pp. 391–423. University of Arizona Press, Tucson, AZ.

Bockelée-Morvan, D., D. C. Lis, J. E. Wink et al. 2000. New molecules found in comet C/1995 O1 (Hale-Bopp). Investigating the link between cometary and interstellar material. *Astron. Astrophys.* 353: 1101–1114.

Bonal, L., M. Bourot-Denise, E. Quirico et al. 2007. Organic matter and metamorphic history of CO chondrites. *Geochim. Cosmochim. Acta* 71: 1605–1623.

Bonal, L., E. Quirico, M. Bourot-Denise et al. 2006. Determination of the petrologic type of CV3 chondrites by Raman spectroscopy of included organic matter. *Geochim. Cosmochim. Acta* 70: 1849–1863.

Bonal, L., E. Quirico, L. Flandinet et al. 2016. Thermal history of type 3 chondrites from the Antarctic meteorite collection determined by Raman spectroscopy of their polyaromatic carbonaceous matter. *Geochim. Cosmochim. Acta* 189: 312–337.

Bose, M., T. J. Zega, and P. Williams. 2014. Assessment of alteration processes on circumstellar and interstellar grains in Queen Alexandra Range 97416. *Earth Planet. Sci. Lett.* 399: 128–138.

Botta, O., and J. L. Bada. 2002. Extraterrestrial organic compounds in meteorites. *Surv. Geophys.* 23: 411–467.

Botta, O., D. P. Glavin, G. Kminek et al. 2002. Relative amino acid concentrations as a signature for parent body processes of carbonaceous chondrites. *Orig. Life Evol. Biosph.* 32: 143–163.

Botta, O., Z. Martins, and P. Ehrenfreund. 2007. Amino acids in Antarctic CM1 meteorites and their relationship to other carbonaceous chondrites. *Meteorit. Planet. Sci.* 42: 81–92.

Brinton, K. L. F., C. Engrand, D. P. Glavin et al. 1998. A search for extraterrestrial amino acids in carbonaceous Antarctic micrometeorites. *Orig. Life Evol. Biosph.* 28: 413–424.

Browning, L. B., H. Y. McSween, M. E. Zolensky. 1996. Correlated alteration effects in CM carbonaceous chondrites. *Geochim. Cosmochim. Acta* 60: 2621–2633.

Burton, A. S., T. Cao, K. Nakamura-Messenger et al. 2016. Organic analysis in the Miller Range 090657 CR2 chondrite: Part 2 amino acid analyses. *47th Lunar and Planetary Science Conference, LPI Contribution No. 1903*, p. 2987.

Burton, A. S., J. E. Elsila, M. P. Callahan et al. 2012. A propensity for n-ω-amino acids in thermally altered Antarctic meteorites. *Meteorit. Planet. Sci.* 47: 374–386.

Burton, A. S., D. P. Glavin, J. E. Elsila et al. 2014. The amino acid composition of the Sutter's Mill CM2 carbonaceous chondrite. *Meteorit. Planet. Sci.* 49: 2074–2086.

Burton, A. S., H. McLain, D. P. Glavin et al. 2015. Amino acid analyses of R and CK chondrites. *Meteorit. Planet. Sci.* 50: 470–482.

Busemann, H., A. F. Young, C. M. O'D. Alexander et al. 2006. Interstellar chemistry recorded in organic matter from primitive meteorites. *Science* 312: 727–730.

Callahan, M. P., K. E. Smith, H. J. Cleaves et al. 2011. Carbonaceous meteorites contain a wide range of extraterrestrial nucleobases. *Proc. Natl. Acad. Sci. U.S.A.* 108: 13995–13998.

Chan, H.-S., Z. Martins, and M. A. Sephton. 2012. Amino acid analyses of type 3 chondrites Colony, Ornans, Chainpur, and Bishunpur. *Meteorit. Planet. Sci.* 47: 1502–1516.

Choe, W. H., H. Huber, A. E. Rubin et al. 2010. Compositions and taxonomy of 15 unusual carbonaceous chondrites. *Meteorit. Planet. Sci.* 45: 531–554.

Chyba, C., and C. Sagan. 1992. Endogenous production, exogenous delivery and impact-shock synthesis of organic molecules: An inventory for the origins of life. *Nature* 355: 125–132.

Cleaves, H. J., J. H. Chalmers, A. Lazcano et al. 2008 A reassessment of prebiotic organic synthesis in neutral planetary atmospheres. *Orig. Life Evol. Biosph.* 38: 105–115.

Cleaves, H. J., and A. Lazcano. 2009. The origin of biomolecules. In *Chemical Evolution II: From Origins of Life to Modern Society*, Eds. L. Zaikowski, J. M. Friedrich, S. R. Seidel, pp. 17–43. Oxford University Press, New York, NY.

Clemett, S. J., X. D. F. Chillier, S. Gillette et al. 1998. Observation of indigenous polycyclic aromatic hydrocarbons in 'giant' carbonaceous Antarctic micrometeorites. *Orig. Life Evol. Biosph.* 28: 425–448.

Clemett, S. J., C. R. Maechling, R. N. Zare et al. 1993. Identification of complex aromatic molecules in individual interplanetary dust particles. *Science* 262: 721–725.

Cody, G. D., and C. M. O. Alexander. 2005. NMR studies of chemical structural variation of insoluble organic matter from different carbonaceous chondrites groups. *Geochim. Cosmochim. Acta* 69: 1085–1097.

Cody, G. D., C. M. O. Alexander, and F. Tera. 2002. Solid-state (1H and 13C) nuclear magnetic resonance spectroscopy of insoluble organic residue in the Murchison meteorite: A self-consistent quantitative analysis. *Geochim. Cosmochim. Acta* 66: 1851–1865.

Cody, G. D., C. M. O. D. Alexander, H. Yabuta et al. 2008. Organic thermometry for chondritic parent bodies. *Earth Planet. Sci. Lett.* 272: 446–455.

Cody, G. D., E. Heying, C. M. O. Alexander et al. 2011. Establishing a molecular relationship between chondritic and cometary organic solids. *Proc. Natl. Acad. Sci. U.S.A.* 108: 19171–19176.

Cooper, G. W., and J. R. Cronin. 1995. Linear and cyclic aliphatic carboxamides of the Murchison meteorite: Hydrolyzable derivatives of amino acids and other carboxylic acids. *Geochim. Cosmochim. Acta* 59: 1003–1015.

Cooper, G., N. Kimmich, W. Belisle et al. 2001. Carbonaceous meteorites as a source of sugar-related organic compounds for the early Earth. *Nature* 414: 879–883.

Cooper, G. W., W. M. Onwo, and J. R. Cronin. 1992. Alkyl phosphonic acids and sulfonic acids in the Murchison meteorite. *Geochim. Cosmochim. Acta* 56: 4109–4115.

Cooper, G. W., M. H. Thiemens, T. L. Jackson et al. 1997. Sulfur and hydrogen isotope anomalies in meteorite sulfonic acids. *Science* 277: 1072–1074.

Commins, B. T., and J. S. Harington. 1966. Polycyclic aromatic hydrocarbons in carbonaceous meteorites. *Nature* 212: 273–274.

Cronin, J. R., and C. B. Moore. 1971. Amino acid analyses of the Murchison, Murray, and Allende carbonaceous chondrites. *Science* 172: 1327–1329.

Cronin, J. R., and C. B. Moore. 1976. Amino acids of the Nogoya and Mokoia carbonaceous chondrites. *Geochim. Cosmochim. Acta* 40: 853–857.

Cronin, J. R., and S. Chang. 1993. Organic matter in meteorites: Molecular and isotopic analyses of the Murchison meteorites. In *The Chemistry of Life's Origin*. Eds J. M. Greenberg, C. X. Mendoza-Gomez, and V. Pirronello, pp. 209–258. Kluwer Academic Publishers, the Netherlands.

Cronin, J. R., and S. Pizzarello. 1997. Enantiomeric excesses in meteoritic amino acids. *Science* 275: 951–955.

Cronin, J. R., S. Pizzarello, and C. B. Moore. 1979. Amino acids in an Antarctic carbonaceous chondrite. *Science* 206: 335–337.

Cronin, J. R., S. Pizzarello, and D. P. Cruikshank. 1988. Organic matter in carbonaceous chondrites, planetary satellites, asteroids and comets. In *Meteorites and the Early Solar System*. Eds. J. F. Kerridhe, M. S. Matthews, pp. 819–857. University of Arizona Press, Tucson, AZ.

Cronin, J. R., S. Pizzarello, S. Epstein et al. 1993. Molecular and isotopic analyses of the hydroxy acids, dicarboxylic acids, and hydroxydicarboxylic acids of the Murchison meteorite. *Geochim. Cosmochim. Acta* 57: 4745–4752.

Cronin, J. R., S. Pizzarello, and J. S. Frye. 1987. 13C NMR spectroscopy of the insoluble carbon of carbonaceous chondrites. *Geochim. Cosmochim. Acta* 51: 299–303.

Crovisier, J., and D. Bockelée-Morvan 1999. Remote observations of the composition of cometary volatiles. *Space Sci. Rev.* 90: 19–32.

Crovisier, J., N. Biver, D. Bockelee-Morvan et al. 2009. The chemical diversity of comets. *Earth Moon Planets* 105: 267–272.

Dartois, E., C. Engrand, R. Brunetto et al. 2013. UltraCarbonaceous Antarctic micrometeorites, probing the Solar System beyond the nitrogen snow-line. *Icarus* 224: 243–252.

Deamer, D. W., and R. M. Pashley. 1989. Amphiphilic components of the Murchison carbonaceous chondrite: Surface properties and membrane formation. *Orig. Life Evol. Biosph.* 19: 21–28.

Degens, E. T., and M. Bajor. 1962. Amino acids and sugars in the Bruderheim and Murray Meteorite. *Naturwissenschaften* 49: 605–606.

Derenne, S., J.-N. Rouzaud, C. Clinard et al. 2005. Size discontinuity between interstellar and chondritic aromatic structures: A high-resolution transmission electron microscopy study. *Geochim. Cosmochim. Acta* 69: 3911–3917.

Duprat, J., E. Dobrica, C. Engrand et al. 2010. Extreme deuterium excesses in ultracarbonaceous micrometeorites from Central Antarctic snow. *Science* 328: 742–745.

Ehrenfreund, P., D. P. Glavin, O. Botta et al. 2001. Extraterrestrial amino acids in Orgueil and Ivuna: Tracing the parent body of CI type carbonaceous chondrites. *Proc. Natl. Acad. Sci. U.S.A.* 98: 2138–2141.

Elsila, J. E., J. C. Aponte, D. G. Blackmond et al. 2016. Meteoritic amino acids: Diversity in compositions reflects parent body histories. *ACS Cent. Sci.* 2: 370–379.

Elsila, J. E., D. P. Glavin, and J. P. Dworkin. 2009. Cometary glycine detected in samples returned by Stardust. *Meteorit. Planet. Sci.* 44: 1323–1330.

Engel, M. H., and S. A. Macko. 1997. Isotopic evidence for extraterrestrial non-racemic amino acids in the Murchison meteorite. *Nature* 389: 265–268.

Epstein, S., R. V. Krishnamurthy, J. R. Cronin et al. 1987. Unusual stable isotope ratios in amino acid and carboxylic acid extracts from the Murchison meteorite. *Nature* 326: 477–479.

Floss, C., C. Le Guillou, and A. Brearley. 2014. Coordinated NanoSIMS and FIB-TEM analyses of organic matter and associated matrix materials in CR3 chondrites. *Geochim. Cosmochim. Acta* 139: 1–25.

Flynn, G. J., L. P. Keller, M. Feser et al. 2003. The origin of organic matter in the solar system: Evidence from the interplanetary dust particles. *Geochim. Cosmochim. Acta* 67: 4791–4806.

Flynn, G. J., L. P. Keller, C. Jacobsen et al. 2004. An assessment of the amount and types of organic matter contributed to the Earth by interplanetary dust. *Adv. Space Res.* 33: 57–66.

Folsome, C. E., J. Lawless, M. Romiez et al. 1971. Heterocyclic compounds indigenous to the Murchison meteorite. *Nature* 232: 108–109.

Folsome, C. E., J. Lawless, M. Romiez et al. 1973. Heterocyclic compounds recovered from carbonaceous chondrite. *Geochim. Cosmochim. Acta* 37: 455–465.

Fray, N., A. Bardyn, H. Cottin et al. 2016. High-molecular-weight organic matter in the particles of comet 67P/Churyumov-Gerasimenko. *Nature* 538: 72–74.

Fredriksson, K., and J. F. Kerridge. 1988. Carbonates and sulfates in CI chondrites - Formation by aqueous activity on the parent body. *Meteoritics* 23: 35–44.

Furnes, H., N. R. Banerjee, K. Muehlenbachs et al. 2004. Early life recorded in archean pillow lavas. *Science* 304: 578–581.

Gardinier, A., S. Derenne, F. Robert et al. 2000. Solid state CP/MAS 13C NMR of the insoluble organic matter of the Orgueil and Murchison meteorites: Quantitative study. *Earth Planet. Sci. Lett.* 184: 9–21.

Garvie, L. A. J., and P. R. Buseck. 2004. Nanosized carbon-rich grains in carbonaceous chondrite meteorites. *Earth Planet. Sci. Lett.* 224: 431–439.

Gilmour, I., and C. T. Pillinger. 1994. Isotopic compositions of individual polycyclic aromatic hydrocarbons from the Murchison meteorite. *Mon. Not. R. Astron. Soc.* 269: 235–240.

Glavin, D. P., and J. Dworkin. 2009. Enrichment of the amino acid L-isovaline by aqueous alteration on CI and CM meteorite parent bodies. *Proc. Natl. Acad. Sci. U.S.A.* 106: 5487–5492.

Glavin, D. P., J. P. Dworkin, A. Aubrey et al. 2006. Amino acid analyses of Antarctic CM2 meteorites using liquid chromatography-time of flight-mass spectrometry. *Meteorit. Planet. Sci.* 41: 889–902.

Glavin, D. P., G. Matrajt, and J. L. Bada. 2004. Re-examination of amino acids in Antarctic micrometeorites. *Adv. Space Res.* 33: 106–113.

Glavin, D. P., M. P. Callahan, J. P. Dworkin et al. 2010. The effects of parent body processes on amino acids in carbonaceous chondrites. *Meteorit. Planet. Sci.* 45: 1948–1972.

Goesmann, F., H. Rosenbauer, J. H. Bredehöft et al. 2015. Organic compounds on comet 67P/Churyumov-Gerasimenko revealed by COSAC mass spectrometry. *Science* 349: aab0689.

Goldman, N., E. J. Reed, L. E. Fried et al. 2010. Synthesis of glycine-containing complexes in impacts of comets on early Earth. *Nat. Chem.* 2: 949–954.

Gourier, D., F. Robert, O. Delpoux et al. 2008. Extreme deuterium enrichment of organic radicals in the Orgueil meteorite: Revisiting the interstellar interpretation?. *Geochim. Cosmochim. Acta* 72: 1914–1923.

Grady, M. M., I. P. Wright, P. K. Swart et al. 1988. The carbon and oxygen isotopic composition of meteoritic carbonates. *Geochim. Cosmochim. Acta* 52: 2855–2866.

Grimm, R. E., and H. Y. McSween. 1989. Water and the thermal evolution of carbonaceous chondrite parent bodies. *Icarus* 82: 244–280.

Hayatsu, R. 1964. Orgueil meteorite: Organic nitrogen contents. *Science* 146: 1291–1293.

Hayatsu, R., E. Anders, M. H. Studier et al. 1975. Purines and triazines in the Murchison meteorite. *Geochim. Cosmochim. Acta* 39: 471–488.

Hayatsu, R., M. H. Studier, A. Oda et al. 1968. Origin of organic matter in early solar system-II. Nitrogen compounds. *Geochim. Cosmochim. Acta* 32: 175–190.

Hayatsu, R., S. Matsuoka, E. Anders et al. 1977. Origin of organic matter in the early solar system. VII - The organic polymer in carbonaceous chondrites. *Geochim. Cosmochim. Acta* 41: 1325–1339.

Hayatsu, R., R. E. Winans, R. G. Scott et al. 1980. Phenolic ethers in the organic polymer of the Murchison meteorite. *Science* 207: 1202–1204.

Holzer, G., and J. Oró. 1979. The organic composition of the Allan Hills carbonaceous chondrite 77306 as determined by pyrolysis-gas chromatography-mass spectrometry and other methods. *J. Mol. Evol.* 13: 265–270.

Hoppe, P., R. Strebel, P. Eberhardt et al. 1996. Type II supernova matter in a silicon carbide grain from the Murchison meteorite. *Science* 272: 1314–1316.

Huang, Y., M. R. Alexandre, and Y. Wang. 2007. Structure and isotopic ratios of aliphatic side chains in the insoluble organic matter of the Murchison carbonaceous chondrite. *Earth Planet. Sci. Lett.* 259: 517–525.

Huang, Y., J. C. Aponte, J. Zhao et al. 2015. Hydrogen and carbon isotopic ratios of polycyclic aromatic compounds in two CM2 carbonaceous chondrites and implications for prebiotic organic synthesis. *Earth Planet. Sci. Lett.* 426: 101–108.

Huang, Y., Y. Wang, M. R. Alexandre et al. 2005. Molecular and compound-specific isotopic characterization of monocarboxylic acids in carbonaceous meteorites. *Geochim. Cosmochim. Acta* 69: 1073–1084.

Imai, E., H. Honda, H. Hatori et al. 1999. Autocatalytic synthesis of oligoglycine in a simulated submarine hydrothermal system. *Orig. Life Evol. Biosph.* 29: 249–259.

Ito, M., L. P. Gupta, H. Masuda et al. 2006. Thermal stability of amino acids in seafloor sediment in aqueous solution at high temperature. *Org. Geochem.* 37: 177–188.

Javaux, E. J., C. P. Marshall, and A. Bekker. 2010. Organic-walled microfossils in 3.2-billion-year-old shallow-marine siliciclastic deposits. *Nature* 463: 934–938.

Jungclaus, G. A., G. U. Yuen, C. B. Moore et al. 1976a. Evidence for the presence of low molecular weight alcohols and carbonyl compounds in the Murchison meteorite. *Meteoritics* 11: 231–237.

Jungclaus, G., J. R. Cronin, C. B. Moore et al. 1976b. Aliphatic amines in the Murchison meteorite. *Nature* 261: 126–128.

Kaplan, I. R., E. T. Degens, and J. H. Reuter. 1963. Organic compounds in stony meteorites. *Geochim. Cosmochim. Acta* 27: 805–808.

Kasting, J. F., D. H. Eggler, and S. P. Raeburn. 1993. Mantle redox evolution and the oxidation state of the Archean atmosphere. *J. Geol.* 101: 245–257.

Kebukawa, Y., C. M. O'D. Alexander, and G. D. Cody. 2011. Compositional diversity in insoluble organic matter in type 1, 2 and 3 chondrites as detected by infrared spectroscopy. *Geochim. Cosmochim. Acta* 75: 3530–3541.

Keller, L. P., S. Messenger, G. J. Flynn et al. 2004. The nature of molecular cloud material in interplanetary dust. *Geochim. Cosmochim. Acta* 68: 2577–2589.

Kerridge, J. F., A. L. Mackay, and W. V. Boynton. 1979. Magnetite in CI carbonaceous meteorites: Origin by aqueous activity on a planetesimal surface. *Science* 205: 395–397.

King A. J., P. F. Schofield, S. S. Russell. 2017. Type 1 aqueous alteration in CM carbonaceous chondrites: Implications for the evolution of water-rich asteroids. *Meteorit. Planet. Sci.* 52: 1197–1215.

Kissel, J., and F. R. Krueger. 1987. The organic component in dust from comet Halley as measured by the PUMA mass spectrometer on board Vega 1. *Nature* 326: 755–760.

Koga, T., and H. Naraoka. 2017. A new family of extraterrestrial amino acids in the Murchison meteorite. *Sci Rep.* 7: 636–643.

Komiya, M., and A. Shimoyama. 1996. Organic compounds from insoluble organic matter isolated from the Murchison carbonaceous chondrite by heating experiments. *Bull. Chem. Soc. Jpn.* 69: 53–58.

Komiya, M., A. Shimoyama, and K. Harada. 1993. Examination of organic compounds from insoluble organic matter isolated from some Antarctic carbonaceous chondrites by heating experiments. *Geochim. Cosmochim. Acta* 57: 907–914.

Kopetzki, D., and M. Antonietti. 2011. Hydrothermal formose reaction. *New J. Chem.* 35: 1787–1794.

Kotra, R. K., A. Shimoyama, C. Ponnamperuma et al. 1979. Amino acids in a carbonaceous chondrite from Antarctica. *J. Mol. Evol.* 13: 179–184.

Krishnamurthy, R. V., S. Epstein, J. R. Cronin et al. 1992. Isotopic and molecular analyses of hydrocarbons and monocarboxylic acids of the Murchison meteorite. *Geochim. Cosmochim. Acta* 56: 4045–4058.

Krot, A. N., K. Keil, E. R. D. Scott et al. 2014. Classification of meteorites and their genetic relationships. In *Meteorites and Cosmochemical Processes, Volume 1 of Treatise on Geochemistry*, Ed. A. M. Davis, pp. 1–63. Oxford: Elsevier.

Kvenvolden, K., J. Lawless, K. Pering et al. 1970. Evidence for extraterrestrial amino acids and hydrocarbons in the Murchison meteorite. *Nature* 228: 923–926.

Kvenvolden, K. A., J. G. Lawless, and C. Ponnamperuma. 1971. Nonprotein amino acids in the Murchison meteorite. *Proc. Natl. Acad. Sci. U.S.A.* 68: 486–490.

Lawless, J. G. 1973. Amino acids in the Murchison meteorite. *Geochim. Cosmochim. Acta* 37: 2207–2212.

Lawless, J. G., and G. U. Yuen. 1979. Quantification of monocarboxylic acids in the Murchison carbonaceous meteorite. *Nature* 282: 396–398.

Lawless, J. G., and B. Zeitman. 1974. Dicarboxylic acids in the Murchison meteorite. *Nature* 251: 40–42.

Lawless, J. G., C. E. Folsome, and K. A. Kvenvolden. 1972b. Organic matter in meteorites. *Sci. Am.* 226: 38–46.

Lawless, J. G., K. A. Kvenvolden, E. Peterson, 1972a. Physical sciences: Evidence for amino acids of extraterrestrial origin in the Orgueil meteorite. *Nature* 236: 66–67.

Le Guillou, C., S. Bernard, A. Brearley et al. 2014. Evolution of organic matter in Orgueil, Murchison and Renazzo during parent body aqueous alteration: In situ investigations. *Geochim. Cosmochim. Acta* 131: 368–392.

Le Guillou, C., and A. Brearley. 2014. Relationships between organics, water and early stages of aqueous alteration in the pristine CR3.0 chondrite MET 00426. *Geochim. Cosmochim. Acta* 131: 344–367.

Lerner, N. R., and G. W. Cooper. 2005. Iminodicarboxylic acids in the Murchison meteorite: Evidence of Strecker reactions. *Geochim. Cosmochim. Acta* 69: 2901–2906.

Lerner, N. R., E. Peterson, and S. Chang. 1993. The Strecker synthesis as a source of amino acids in carbonaceous chondrites - Deuterium retention during synthesis. *Geochim. Cosmochim. Acta* 57: 4713–4723.

Lewis, R. S., T. Ming, J. F. Wacker et al. 1987. Interstellar diamonds in meteorites. *Nature* 326: 160–162.

Lodders, K. 2003. Solar system abundances and condensation temperatures of the elements. *Astrophys. J.* 591: 1220–1247.

Martins, Z., and M. A. Sephton. 2009. Extraterrestrial amino acids. In *Amino Acids, Peptides and Proteins in Organic Chemistry*, Ed. A. B. Hughes, pp. 3–42. Wiley. -VCH Verlag GmbH & Co. KGaA, Weinheim, Germany

Martins, Z., C. M. O'D. Alexander, G. E. Orzechowska et al. 2007a. Indigenous amino acids in primitive CR meteorites. *Meteorit. Planet. Sci.* 42: 2125–2136.

Martins, Z., J. S. Watson, M. A. Sephton et al. 2006. Free dicarboxylic and aromatic acids in the carbonaceous chondrites Murchison and Orgueil. *Meteorit. Planet. Sci.* 41: 1073–1080.

Martins, Z., B. A. Hofmann, E. Gnos et al. 2007b. Amino acid composition, petrology, geochemistry, ^{14}C terrestrial age and oxygen isotopes of the Shisr 033 CR chondrite. *Meteorit. Planet. Sci.* 42: 1581–1595.

Martins, Z., O. Botta, M. L. Fogel et al. 2008. Extraterrestrial nucleobases in the Murchison meteorite. *Earth Planet. Sci. Lett.* 270: 130–136.

Martins, Z., M. C. Price, N. Goldman et al. 2013. Shock synthesis of amino acids from impacting cometary and icy planet surface analogues. *Nat. Geosci.* 6: 1045–1049.

Martins, Z., P. Modica, Z. Zanda et al. 2015. The amino acid and hydrocarbon contents of the Paris meteorite: Insights into the most primitive CM chondrite. *Meteorit. Planet. Sci.* 50: 926–943.

Matrajt, G., S. Pizzarello, S. Taylor et al. 2004. Concentration and variability of the AIB amino acid in polar micrometeorites: Implications for the exogenous delivery of amino acids to the primitive Earth. *Meteorit. Planet. Sci.* 39: 1849–1858.

Matrajt, G., G. M. Muñoz Caro, E. Dartois et al. 2005. FTIR analysis of the organics in IDPs: Comparison with the IR spectra of the diffuse interstellar medium. *Astron. Astrophys.* 433: 979–995.

McCollom, T. M., G. Ritter, and B. R. T. Simoneit. 1999. Lipid synthesis under hydrothermal conditions by Fischer-Tropsch-type reactions. *Orig. Life Evol. Biosph.* 29: 153–166.

McSween, H. Y. 1979. Alteration in CM carbonaceous chondrites inferred from modal and chemical variations in matrix. *Geochim. Cosmochim. Acta* 43: 1761–1765, 1767–1770.

McSween, H. Y. Jr., and G. R. Huss. 2015. In Cosmochemistry. Cambridge University Press.

Mehringer, D. M., L. E. Snyder, and Y. T. Miao. 1997. Detection and confirmation of interstellar acetic acid. *Astrophys. J.* 480: L71–L74.

Meierhenrich, U. J., G. M. Muñoz Caro, J. H. Bredehöft et al. 2004. Identification of diamino acids in the Murchison meteorite. *Proc. Natl. Acad. Sci. U.S.A.* 101: 9182–9186.

Meinschein, W. G. 1963a. Benzene extracts of the Orgueil meteorite. *Nature* 197: 833–836.

Meinschein, W. G. 1963b. Hydrocarbons in terrestrial samples and the Orgueil meteorite. *Space Sci. Rev.* 2: 653–679.

Meinschein, W. G., B. Nagy, and D. J. Hennessy. 1963. Evidence in meteorites of former life: The organic compounds in carbonaceous chondrites are similar to those found in marine sediments. *Ann. N.Y. Acad. Sci.* 108: 553–579.

Miller, S. L., and J. L. Bada. 1988. Submarine hot springs and the origin of life. *Nature* 334: 609–611.

Millar, T. J., A. Bennett, and E. Herbst. 1989. Deuterium fractionation in dense interstellar clouds. *Astrophys. J.* 340: 906–920.

Monroe, A. A., and S. Pizzarello. 2011. The soluble organic compounds of the Bells meteorite: Not a unique or unusual composition. *Geochim. Cosmochim. Acta* 75: 7585–7595.

Mumma, M. J., and S. B. Charnley. 2011. The chemical composition of comets - Emerging taxonomies and natal heritage. *Annu. Rev. Astron. Astrophys.* 49: 471–524.

Mumma, M. J., M. A. Disanti, N. Dello Russo et al. 2003. Remote infrared observations of parent volatiles in comets: A window on the early solar system. *Adv. Space Res.* 31: 2563–2575.

Myrgorodska, I., C. Meinert, Z. Martins et al. 2016. Quantitative enantioseparation of amino acids by comprehensive two-dimensional gas chromatography applied to non-terrestrial samples. *J. Chromatogr. A* 1433: 131–136.

Nagy, B., W. G. Meinschein, and D. J. Hennessy. 1961. Mass spectroscopic analysis of the Orgueil meteorite: Evidence for biogenic hydrocarbons. *Ann. N.Y. Acad. Sci.* 93: 25–35.

Nakamura, K., M. E. Zolensky, S. Tomita et al. 2002. Hollow organic globules in the Tagish Lake meteorite as possible products of primitive organic reactions. *Int. J. Astrobiol.* 1: 179–189.

Naraoka, H., A. Shimoyama, and K. Harada. 1999. Molecular distribution of monocarboxylic acids in Asuka chondrites from Antarctica. *Orig. Life Evol. Biosph.* 29: 187–201.

Naraoka, H., A. Shimoyama, and K. Harada. 2000. Isotopic evidence from an Antarctica carbonaceous chondrite for two reaction pathways of extraterrestrial PAH formation. *Earth Planet. Sci. Lett.* 184: 1–7.

Naraoka, H., A. Shimoyama, M. Komiya et al. 1988. Hydrocarbons in the Yamato-791198 carbonaceous chondrite from Antarctica. *Chem. Lett.* 17: 831–834.

Nooner, D. W., and J. Oró. 1967. Organic compounds in meteorites - I. Aliphatic hydrocarbons. *Geochim. Cosmochim. Acta* 31: 1359–1394.

Okumura, F., and K. Mimura. 2011. Gradual and stepwise pyrolyses of insoluble organic matter from the Murchison meteorite revealing chemical structure and isotopic distribution. *Geochim. Cosmochim. Acta* 75: 7063–7080.

Olson, J. M. 2006. Photosynthesis in the Archean era. *Photosynth. Res.* 88: 109–117.

Olson, R. J., J. Oró, and A. Zlatkis. 1967. Organic compounds in meteorites – II. Aromatic hydrocarbons. *Geochim. Cosmochim. Acta* 31: 1935–1948.

Oró, J., and D. W. Nooner. 1967. Aliphatic hydrocarbons in meteorites. *Nature* 213: 1085–1087.

Oró, J., D. W. Nooner, A. Zlatkis et al. 1966. Paraffinic hydrocarbons in the Orgueil, Murray, Mokoia, and other meteorites. *Life Sci. Space Res.* 4: 63–100.

Oró, J., J. Gibert, H. Lichtenstein et al. 1971. Amino acids, aliphatic and aromatic hydrocarbons in the Murchison meteorite. *Nature* 230: 105–106.

Orthous-Daunay, F.-R., E. Quirico, P. Beck et al. 2013. Mid-infrared study of the molecular structure variability of insoluble organic matter from primitive chondrites. *Icarus* 223: 534–543.

Orthous-Daunay, F. R., E. Quirico, L. Lemelle et al. 2010. Speciation of sulfur in the insoluble organic matter from carbonaceous chondrites by XANES spectroscopy. *Earth Planet. Sci. Lett.* 300: 321–328.

Palmer, E. E., and D. S. Lauretta. 2011. Aqueous alteration of kamacite in CM chondrites. *Meteorit. Planet. Sci.* 46: 1587–1607.

Peltzer, E. T., and J. L. Bada. 1978. a-Hydroxycarboxylic acids in the Murchison meteorite. *Nature* 272: 443–444.

Peltzer, E. T., J. L. Bada, G. Schlesinger et al. 1984. The chemical conditions on the parent body of the Murchison meteorite: Some conclusions based on amino, hydroxy, and dicarboxylic acids. *Adv. Space Res.* 4: 69–74.

Pendleton, Y. J. 1994. Near-infrared absorption spectroscopy of interstellar hydrocarbon grains. *Astrophys. J.* 437: 683–696.

Pering, K. L., and C. Ponnamperuma. 1971. Aromatic hydrocarbons in the Murchison meteorite. *Science* 173: 237–239.

Pizzarello, S., and J. R. Cronin. 2000. Non-racemic amino acids in the Murchison and Murray meteorites. *Geochim. Cosmochim. Acta* 64: 329–338.

Pizzarello, S., and Y. Huang. 2002. Molecular and isotopic analyses of Tagish Lake alkyl dicarboxylic acids. *Meteorit. Planet. Sci.* 37: 687–696.

Pizzarello, S., and Y. Huang. 2005. The deuterium enrichment of individual amino acids in carbonaceous meteorites: A case for the presolar distribution of biomolecule precursors. *Geochim. Cosmochim. Acta* 69: 599–605.

Pizzarello, S., and W. Holmes 2009. Nitrogen-containing compounds in two CR2 meteorites: ^{15}N composition, molecular distribution and precursor molecules. *Geochim. Cosmochim. Acta* 73: 2150–2162.

Pizzarello, S., and L. A. J. Garvie. 2014. Sutter's Mill dicarboxylic acids as possible tracers of parent-body alteration processes. *Meteorit. Planet. Sci.* 49: 2087–2094.

Pizzarello, S., and C. T. Yarnes. 2016. Enantiomeric excesses of chiral amines in ammonia-rich carbonaceous meteorites. *Earth Planet. Sci. Lett.* 443: 176–184.

Pizzarello, S., R. V. Krishnamurthy, S. Epstein et al. 1991. Isotopic analyses of amino acids from the Murchison meteorite. *Geochim. Cosmochim. Acta* 55: 905–910.

Pizzarello, S., X. Feng, S. Epstein et al. 1994. Isotopic analyses of nitrogenous compounds from the Murchison meteorite: Ammonia, amines, amino acids, and polar hydrocarbons. *Geochim. Cosmochim. Acta* 58: 5579–5587.

Pizzarello, S. G., W. Cooper, and G. J. Flynn. 2006. The nature and distribution of the organic material in carbonaceous chondrites and interplanetary dust particles. In *Meteorites and the Early Solar System II*, Eds. D. S. Lauretta, H. Y. McSween Jr., pp. 625–651. University of Arizona Press, Tucson, AZ.

Pizzarello, S., Y. Huang, and M. R. Alexandre. 2008. Molecular asymmetry in extraterrestrial chemistry: Insights from a pristine meteorite. *Proc. Natl. Acad. Sci. U.S.A.* 105: 3700–3704.

Pizzarello, S., Y. Huang, L. Becker et al. 2001. The organic content of the Tagish Lake meteorite. *Science* 293: 2236–2239.

Pizzarello, S., Y. Huang, and M. Fuller. 2004. The carbon isotopic distribution of Murchison amino acids. *Geochim. Cosmochim. Acta* 68: 4963–4969.

Pizzarello, S., Y. Wang, and G. M. Chaban. 2010. A comparative study of the hydroxy acids from the Murchison, GRA 95229 and LAP 02342 meteorites. *Geochim. Cosmochim. Acta* 74: 6206–6217.

Pizzarello, S., M. Zolensky, and K. A. Turk. 2003. Nonracemic isovaline in the Murchison meteorite: Chiral distribution and mineral association. *Geochim. Cosmochim. Acta* 67: 1589–1595.

Plankensteiner, K., H. Reiner, B. Schranz et al. 2004. Prebiotic formation of amino acids in a neutral atmosphere by electric discharge. *Angew. Chem. Int. Ed. Engl.* 43: 1886–1888.

Quirico, E., P. I. Raynal, and M. Bourot-Denise. 2003. Metamorphic grade of organic matter in six unequilibrated ordinary chondrites. *Meteorit. Planet. Sci.* 38: 795–811.

Remusat, L., S. Derenne, and F. Robert. 2005a. New insight on aliphatic linkages in the macromolecular organic fraction of Orgueil and Murchison meteorites through ruthenium tetroxide oxidation. *Geochim. Cosmochim. Acta* 69: 4377–4386.

Remusat, L., S. Derenne, F. Robert et al. 2005b. New pyrolytic and spectroscopic data on Orgueil and Murchison insoluble organic matter: A different origin than soluble? *Geochim. Cosmochim. Acta* 69: 3919–3932.

Remusat, L., F. Palhol, F. Robert et al. 2006. Enrichment of deuterium in insoluble organic matter from primitive meteorites: A solar system origin?. *Earth Planet. Sci. Lett.* 243: 15–25.

Robert, F. 2003. The D/H ratio in chondrites. *Space Sci. Rev.* 106: 87–101.

Robert, F., and S. Epstein. 1982. The concentration and isotopic composition of hydrogen, carbon and nitrogen in carbonaceous meteorites. *Geochim. Cosmochim. Acta* 46: 81–95.

Rosing, M. T. 1999. ^{13}C-Depleted carbon microparticles in >3700-Ma sea-floor sedimentary rocks from west Greenland. *Science* 283: 674–676.

Rosing, M. T., and R. Frei. 2004. U-rich Archaean sea-floor sediments from Greenland - indications of >3700 Ma oxygenic photosynthesis. *Earth Planet. Sci. Lett.* 217: 237–244.

Sandford, S. A., M. P. Bernstein, and J. P. Dworkin. 2001. Assessment of the interstellar processes leading to deuterium enrichment in meteoritic organics. *Meteorit. Planet. Sci.* 36: 1117–1133.

Schidlowski, M. 1988. A 3,800-million-year isotopic record of life from carbon in sedimentary rocks. *Nature* 333: 313–318.

Schlesinger, G., and S. L. Miller. 1983a. Prebiotic synthesis in atmospheres containing CH_4, CO and CO_2. II. Hydrogen cyanide, formaldehyde and ammonia. *J. Mol. Evol.* 19: 383–390.

Schlesinger, G., S. L. Miller. 1983b. Prebiotic synthesis in atmospheres containing CH_4, CO, and CO_2. I. Amino acids. *J. Mol. Evol.* 19: 376–382.

Schmitt-Kopplin, P., Z. Gabelica, R. D. Gougeon et al. 2010. High molecular diversity of extraterrestrial organic matter in Murchison meteorite revealed 40 years after its fall. *Proc. Natl. Acad. Sci. U.S.A.* 107: 2763–2768.

Schopf, J. W. 1993. Microfossils of the Early Archean: New evidence of the antiquity of life. *Science* 260: 640–646.

Scott, J. H., D. M. O'Brien, D. Emerson et al. 2006. An examination of the carbon isotope effects associated with amino acid biosynthesis. *Astrobiology* 6: 867–880.

Sears, D. W., J. N. Grossman, C. L. Melcher et al. 1980. Measuring metamorphic history of unequilibrated ordinary chondrites. *Nature* 287: 791–795.

Sephton, M. A., and O. Botta. 2005. Recognizing life in the Solar System: Guidance from meteoritic organic matter. *Int. J. Astrobio.* 4: 269–276.

Sephton, M. A., and I. Gilmour. 2000. Aromatic moieties in meteorites: Relics of interstellar grain processes? *Astrophys. J.* 540: 588–591.

Sephton, M. A., and I. Gilmour. 2001. Pyrolysis-gas chromatography-isotope ratio mass spectrometry of macromolecular material in meteorites. *Planet. Space Sci.* 49: 465–471.

Sephton, M. A., C. T. Pillinger, I. Gilmour. 1998. $\delta^{13}C$ of free and macromolecular aromatic structures in the Murchison meteorite. *Geochim. Cosmochim. Acta* 62: 1821–1828.

Sephton, M. A., C. T. Pillinger, and I. Gilmour. 2000. Aromatic moieties in meteoritic macromolecular materials: Analyses by hydrous pyrolysis and $\delta13C$ of individual compounds. *Geochim. Cosmochim. Acta* 64: 321–328.

Sephton, M. A., C. T. Pillinger, and I. Gilmour. 2001. Normal alkanes in meteorites: Molecular $\delta^{13}C$ values indicate an origin by terrestrial contamination. *Precambrian Res.* 106: 47–58.

Sephton, M. A., A. B. Verchovsky, P. A. Bland et al. 2003. Investigating the variations in carbon and nitrogen isotopes in carbonaceous chondrites. *Geochim. Cosmochim. Acta* 67: 2093–2108.

Sephton, M. A., G. D. Love, W. Meredith et al. 2005. Hydropyrolysis: A new technique for the analysis of macromolecular material in meteorites. *Planet. Space Sci.* 53: 1280–1286.

Sephton, M. A., G. D. Love, J. S. Watson et al. 2004. Hydropyrolysis of insoluble carbonaceous matter in the Murchison meteorite: New insights into its macromolecular structure. *Geochim. Cosmochim. Acta* 68: 1385–1393.

Sephton, M. A., J. A. Watson, W. Meredith et al. 2015. Multiple cosmic sources for meteorite macromolecules? *Astrobiology* 15: 779–786.

Shimoyama, A., and K. Harada. 1984. Amino acid depleted carbonaceous chondrites (C2) from Antarctica. *Geochem. J.* 18: 281–286.

Shimoyama, A., and R. Ogasawara. 2002. Dipeptides and diketopiperazines in the Yamato-791198 and Murchison carbonaceous chondrites. *Orig. Life Evol. Biosph.* 32: 165–179.

Shimoyama, A., and R. Shigematsu. 1994. Dicarboxylic acids in the. Murchison and Yamato-791198 carbonaceous chondrites. *Chem. Lett.* 3: 523–526.

Shimoyama, A., C. Ponnamperuma, and K. Yanai. 1979a. Amino acids in the Yamato carbonaceous chondrite from Antarctica. *Nature* 282: 394–396.

Shimoyama, A., C. Ponnamperuma, and K. Yanai. 1979b. Amino acids in the Yamato-74662 meteorite, an Antarctic carbonaceous chondrite. *Nat. Inst. Polar Res.* 15: 196–205.

Shimoyama, A., K. Harada, and K. Yanai. 1985. Amino acids from the Yamato-791198 carbonaceous chondrite from Antarctica. *Chem. Lett.* 8: 1183–1186.

Shimoyama, A., S. Hagishita, and K. Harada. 1990. Search for nucleic acid bases in carbonaceous chondrites from Antarctica. *Geochem. J.* 24: 343–348.

Shimoyama, A., H. Naraoka, H. Yamamoto et al. 1986. Carboxylic acids in the Yamato-791198 carbonaceous chondrite from Antarctica. *Chem. Lett.* 15: 1561–1564.

Shimoyama, A., H. Naraoka, M. Komiya et al. 1989. Analyses of carboxylic acids and hydrocarbons in Antarctic carbonaceous chondrites, Yamato-74662 and Yamato-793321. *Geochem. J.* 23: 181.

Smith, J. W., and I. R. Kaplan. 1970. Endogenous carbon in carbonaceous meteorites. *Science* 167: 1367–1370.

Smith, K. E., M. P. Callahan, P. A. Gerakines et al. 2014. Investigation of pyridine carboxylic acids in CM2 carbonaceous chondrites: Potential precursor molecules for ancient coenzymes. *Geochim. Cosmochim. Acta* 136: 1–12.

Stoks, P. G., and A. W. Schwartz. 1979. Uracil in carbonaceous meteorites. *Nature* 282: 709–710.

Stoks, P. G., and A. W. Schwartz. 1981. Nitrogen-heterocyclic compounds in meteorites: Significance and mechanisms of formation. *Geochim. Cosmochim. Acta* 45: 563–569.

Studier, M. H., R. Hayatsu, and E. Anders. 1968. Origin of organic matter in early solar system -I. Hydrocarbons. *Geochim. Cosmochim. Acta* 32: 151–173.

Studier, M. H., R. Hayatsu, and E. Anders. 1972. Origin of organic matter in early solar system - V. Further studies of meteoritic hydrocarbons and a discussion of their origin. *Geochim. Cosmochim. Acta* 36: 189–215.

Sugahara, H., and K. Mimura. 2014. Glycine oligomerization up to triglycine by shock experiments simulating comet impacts. *Geochem. J.* 48: 51–62.

Sugitani, K., K. Grey, T. Nagaoka et al. 2009. Taxonomy and biogenicity of Archaean spheroidal microfossils (ca. 3.0 Ga) from the Mount Goldsworthy-Mount Grant area in the northeastern Pilbara Craton, Western Australia. *Precambrian Res.* 173: 50–59.

Rubin, A. E., J. M. Trigo-Rodr%guez, H. Huber et al. 2007. Progressive aqueous alteration of CM carbonaceous chondrites. *Geochim. Cosmochim. Acta* 71: 2361–2382.

Terzieva, R., and E. Herbst. 2000. The possibility of nitrogen isotopic fractionation in interstellar clouds. *Mon. Not. R. Astron. Soc.* 317: 563–568.

Tielens, A. G. G. M. 1983. Surface chemistry of deuterated molecules. *Astron. Astrophys.* 119: 177–184.

Turner, B. E. 1989. A molecular line survey of Sagittarius B2 and Orion-KL from 70 to 115 GHz. I - The observational data. *Astrophys. J. Suppl. Ser.* 70: 539–622.

Van der Velden, W., and A. W. Schwartz. 1977. Search for purines and pyrimidines in Murchison meteorite. *Geochim. Cosmochim. Acta* 41: 961–968.

Vinogradoff, V., C. Le Guillou, S. Bernard et al. 2017. Paris vs. Murchison: Impact of hydrothermal alteration on organic matter in CM chondrites. *Geochim. Cosmochim. Acta* 212: 234–252.

Wang, Y., Y. Huang, C. M. O'D. Alexander et al. 2005. Molecular and compound-specific hydrogen isotope analyses of insoluble organic matter from different carbonaceous chondrite groups. *Geochim. Cosmochim. Acta* 69: 3711–3721.

Wacey, D., N. McLoughlin, M. J. Whitehouse et al. 2010. Two coexisting sulfur metabolisms in a ca. 3400 Ma sandstone. *Geology* 38: 1115–1118.

Westall, F., S. T. de Vries, W. Nijman et al. 2006. The 3.466 Ga "Kitty's Gap Chert," an early Archaean microbial ecosystem. Special Paper of the Geological Society of America 405:105–131.

Westall, F., B. Cavalazzi, L. Lemelle et al. 2011a. Implications of in situ calcification for photosynthesis in a ~3.3-Ga-old microbial biofilm from the Barberton greenstone belt, South Africa. *Earth Planet. Sci. Lett.* 310: 468–479.

Westall, F., F. Foucher, B. Cavalazzi et al. 2011b. Early life on Earth and Mars: A case study from ~3.5 Ga-old rocks from the Pilbara, Australia. *Planet. Space Sci.* 59: 1093–1106.

White, R. H. 1984. Hydrolytic stability of biomolecules at high temperature and its implication for life at 250%C. *Nature* 310: 430–432.

Winnewisser, G., and E. Churchwell. 1975. Detection of formic acid in Sagittarius B2 by its 2/11/-2/12/ transition. *Astrophys. J.* 200: L33–L36.

Wootten, A., G. Wlodarczak, J. G. Mangum et al. 1992. Search for acetic acid in interstellar clouds. *Astron. Astrophys.* 257: 740–744.

Wright, I. P., S. Sheridan, S. J. Barber et al. 2015. CHO-bearing organic compounds at the surface of 67P/Churyumov-Gerasimenko revealed by Ptolemy. *Science* 349: aab0673.

Yabuta, H., L. B. Williams, G. D. Cody et al. 2007. The insoluble carbonaceous material of CM chondrites: A possible source of discrete organic compounds under hydrothermal conditions. *Meteorit. Planet. Sci.* 42: 37–48.

Yang, J., and S. Epstein. 1983. Interstellar organic matter in meteorites. *Geochim. Cosmochim. Acta* 47: 2199–2216.

Yuen, G., N. Blair, D. J. Des Marais et al. 1984. Carbon isotope composition of low molecular weight hydrocarbons and monocarboxylic acids from Murchison meteorite. *Nature* 307: 252–254.

Yuen, G. U., and K. A. Kvenvolden. 1973. Monocarboxylic acids in Murray and Murchison carbonaceous meteorites. *Nature* 246: 301–303.

Zahnle, K., L. Schaefer, and B. Fegley. 2010. Earth's earliest atmospheres. *Cold Spring Harb. Perspect. Biol.* 2: a003467.

Zinner, E., S. Amari, B. Wopenka et al. 1995. Interstellar graphite in meteorites: Isotopic compositions and structural properties of single graphite grains from Murchison. *Meteoritics* 30: 209.

Zuckerman, B., J. A. Ball, and C. A. Gottlieb. 1971. Microwave detection of interstellar formic acid. *Astrophys. J.* 163: L41–L45.

4.4 Ancient Life and Plate Tectonics

Clark M. Johnson and Martin J. Van Kranendonk

CONTENTS

4.4.1 INTRODUCTION

Of key interest to astrobiology is the question: was the initiation of plate tectonics a critical component to shaping Earth's early biosphere? In the last decade, there has been dramatic advances in research areas that help address this question. Research in Archean and Paleoproterozoic paleobiology has provided a rich microfossil record (e.g., Knoll et al. 2016; Javaux and Lepot 2018) that may be integrated with environmental proxy data that indicate changes in redox conditions (e.g., Lyons et al. 2014). Geochemical data, as well as insights from molecular phylogeny, highlight important changes in electron donors and acceptors over Earth's history. In addition, our understanding of the geologic history of plate tectonics and of the evolution of the continental crust has advanced greatly, in part due to the rapid growth of U-Pb, O, and Hf isotope and trace element studies of zircons (e.g., Roberts and Spencer 2015; Van Kranendonk and Kirkland 2016; Hawkesworth et al. 2017; Iizuka et al. 2017), which leaves little doubt that plate tectonics has played an important role in influencing Earth's biosphere as we know it today. Tied to this is understanding the history of emergent crust early in Earth's history (e.g., Korenaga et al. 2017), which affects nutrient availability and weathering controls on climate. All of these components contribute to understanding the relations between plate tectonics and evolution of early life.

We focus this chapter on the evolution of the solid Earth, continental crust, plate tectonics, and the availability of key nutrients, such as P, for life. Starting with a review of evidence for plate tectonics on the early Earth, we address the issue of evolution of the continental crust, with specific attention to the composition of the crust and if it was emergent, based on the premise that such crust exerts a primary control on nutrient delivery to the biosphere. We then turn to the water envelope, exploring evolution of seawater as a recorder of weathering inputs to the marine environment. Our focus is largely on the Archean and Paleoproterozoic, with emphasis on primary productivity and oxygenic photosynthesis.

4.4.2 CRUST-MANTLE EVOLUTION AND ORIGIN OF PLATE TECTONICS

Planetary heat loss after completion of accretion and solidification of any magma ocean or mantle may be considered to be achieved via two end-member mechanisms: (1) a "stagnant lid" (e.g., O'Neill and Debaille 2014), where a thick conductive crust is immobile and mantle heat is primarily lost through magmatic resurfacing of that crust (e.g., as occurs on Venus: Herrick 1994; see also Fischer and Gerya 2016), and (2) "plate tectonics," where mantle convection ascends to very shallow levels, allowing extensive mantle melting that is accompanied by surface spreading, development of mid-oceanic ridges, and density-driven subduction after plate cooling (e.g., Korenaga 2013). The mantle potential temperature (T_P) is expected to be variable over time, reflecting the balance of radioactive heat production to heat loss, the latter of which may be impeded by a stagnant lid relative to that under plate tectonics. Whereas classical thermal modeling shows an asymptotic heat loss curve for Earth over time (Labrosse and Jaupart 2007), petrologic data from Archean lavas and other thermal modeling suggests that T_P could have *increased* from the time of Earth formation to the late Archean, after which a decrease in T_P would occur, arising from the possible onset of modern-style plate tectonics (Herzberg et al. 2010; Korenaga 2013). Modern-style plate tectonics is used here to refer to steep and deep subduction of cold oceanic lithosphere, down to the core-mantle boundary. This contrasts with a possible Archean style of plate tectonics, whereby thicker, more buoyant oceanic lithosphere was involved in shallow subduction at destructive plate margins.

Over the rest of Precambrian time, inferred swings in T_P have been ascribed to the supercontinent cycle, which affects mantle temperature through episodic pulses of widespread (and rapid) subduction, which cools the mantle, followed by pulses of mantle heating that accompany supercontinent disaggregation (Van Kranendonk and Kirkland 2016). The implication for life of varying T_P over time may be profound, because percent melting of the mantle scales with T_P, producing both thick crust at high T_P (which may have been emergent) and low-density (Mg-rich) depleted and buoyant lithosphere that may be related to the development, stabilization, and preservation of cratons (Griffin and O'Reilly 2007; Herzberg and Rudnick 2012); both components bear on recent proposals that life may have originated on land that was emergent (e.g., Van Kranendonk et al. 2017), as well as on habitats and nutrient supplies for early life.

The role of water is critical both in facilitating the operation of plate tectonics (e.g., Bercovici 1998) and in the subduction process, where the water flux of slab dehydration into the overlying mantle wedge allows for melting in relatively cold subduction zones (e.g., Gill 1981). Production of magmas in subduction zones is the primary mechanism for generating silicic crust at sufficient volumes to develop increasing continental crust through time. The efficiency of this process depends upon normal faulting (that brings water to depths of 4 km: i.e., Lécuyer 2014) and on hydrothermal alteration of oceanic crust that is supported by an ambient ocean mass. As summarized by Campbell and Taylor (1983): "The Earth is the only planet with abundant granite and continents because it is the only planet with abundant water," to which we should add "and plate tectonics."

The question of when plate tectonics started on Earth has been the subject of much research and debate. The dramatic discovery of detrital zircons of Hadean age from Western Australia that had elevated $\delta^{18}O$ values, indicating the presence of liquid water on earth by ~4.2 Ga (Mojzsis et al. 2001; Wilde et al. 2001), turned the focus to the Hadean. The occurrence of zircons with these isotopic compositions as well as inferred crystallization temperatures have been used as evidence for plate tectonics in the Hadean (e.g., Harrison 2009). Such proposals are controversial, as they are based on what is now largely discredited data (Kemp et al. 2010), and conflict with models of stagnant lid tectonics for the Hadean (O'Neill and Debaille 2014) and evidence of vertical tectonics in some Archean terrains (e.g., Collins et al. 1998; Robin and Bailey 2009). More direct evidence for when an early form of plate tectonics seems to have arisen and when modern-style plate tectonics commenced comes directly from the Archean rock record.

Evidence for Archean plate tectonics was recognized quite early on from Archean high-grade gneiss terrains, based on the following: the presence of large-scale recumbent isoclinal folds associated with crustal thickening (Bridgwater et al. 1974; Myers 1976; Wilks 1988; Hanmer and Greene 2002); voluminous sodic granitoids derived from high-pressure melting of basalt (Martin et al. 2005; Rapp et al. 1991); high-pressure metamorphism (Riciputi et al. 1990; Harley 2003; Tappe et al. 2011); and structures and geochronological evidence consistent with terrane accretion (Nutman et al. 2002; Windley and Garde 2009). Over the past three decades, evidence for Archean plate tectonics has also been found in some granite-greenstone belts in the form of fossil subduction zones, accreted terranes, rift sequences, and subduction-zone magmatism (Card 1990; Calvert et al. 1995; White et al. 2003; Smithies et al. 2005a; Brown 2006; Stachel et al. 2006; Wyman et al. 2006; Van Kranendonk et al. 2007a, 2010).

However, an absence of hallmark characteristics of modern subduction-accretion zones in many granite-greenstone terrains (Hamilton 1998; McCall 2003; Stern 2005), the autochthonous nature of some major greenstone successions (Van Kranendonk et al. 2007b, 2015a), dome-and-keel structural map patterns (Bouhallier et al. 1995; Van Kranendonk et al. 2004; Van Kranendonk 2011), modeling of crustal thermal behavior (Sandiford et al. 2004), and suggestions that mantle roots form through in-situ melting events rather than by subduction stacking (Griffin and O'Reilly 2007) indicate at least some crustal development as volcanic plateaus, in many cases developed on older continental basement (e.g., Blenkinsop et al. 1993; Bleeker et al. 1999; Van Kranendonk et al. 2002, 2007a, 2015b; Barnes et al. 2012; Smithies et al. 2018). Indeed, many studies suggest that some Archean crust contains features that cannot be ascribed to uniformitarian, Phanerozoic-type, plate tectonics, but rather formed as a result of large-scale intracrustal differentiation accompanying periods of mantle plume-related magmatism (Stein and Hofmann 1994; Whalen et al. 2002; Rey et al. 2003; Smithies et al. 2005b; Van Kranendonk et al. 2009). The fact that different processes have been recognized from studies of different pieces of Archean crust indicates that there was no single Archean tectonic process, but rather that – as with modern Earth – Archean continental crust formed through a variety of processes, including plate tectonics and mantle-derived upwelling, and the probable interaction between these two end-member processes (e.g., Van Kranendonk 2010).

Much of the evidence for Archean plate tectonics comes from the Neoarchean, except for the well-documented case of the Isukasia Terrane of the North Atlantic Craton in Western Greenland (Nutman et al. 2002; Polat et al. 2002). An onset of modern-style plate tectonics at 3.2–3.0 Ga was first suggested for the Pilbara Craton of Western Australia, based on geochemical, geochronological, and structural grounds (Smithies et al. 2005a; Van Kranendonk et al. 2007a, 2007b), but this is now supported by a variety of evidence from different terrains and global datasets from around the world (e.g., Shirey and Richardson 2011; Dhuime et al. 2012; Naeraa et al. 2012; Van Kranendonk et al. 2015a; Van Kranendonk and Kirkland 2016). Significantly, this time coincides with a crossover point between the modeled heat production from the Earth and the amount of heat that can be lost through conduction via the oceanic crust (Labrosse and Jaupart 2007).

Temporal variations in the age and chemical and isotopic compositions of the continental crust have been closely connected to discussions on when plate tectonics began on Earth. It has long been known that histograms of U-Pb ages for crustal rocks show distinct peaks, and these have been interpreted to record episodic subduction and/or periodicity in the supercontinent cycle (e.g., O'Neill et al. 2007; Condie and Aster 2010; Condie et al. 2015). With the rapid increase in database size of combined U-Pb, O, trace element, and Hf isotope data from igneous and detrital zircons, periodicity is now clear in terms of age and isotopic compositions, which provide insight into primitive orogenic cycles and crustal recycling (e.g., Dhuime et al. 2012; Van Kranendonk and

Kirkland 2013, 2016; Roberts and Spencer 2015; Gardiner et al. 2016; Hawkesworth et al. 2017; Iizuka et al. 2017). This record, however, has been interpreted quite differently, where some workers argue that periodicities do not reflect changes in crustal growth but mixing of sources and crustal reworking (e.g., Hawkesworth et al. 2017), whereas others interpret periodicities in U-Pb ages and isotopic compositions to be correlated with episodic changes in mantle dynamics and continental amalgamation and dispersal (e.g., Van Kranendonk and Kirkland 2013, 2016; Gardiner et al. 2016; Puetz et al. 2017). Arguments against periodicity in continental crustal growth point to the expected smooth cooling of the mantle, but, in detail, modeling shows that T_P and crustal growth may be episodic due to the interplay between mantle creep, lateral heat flow, mantle cooling, and punctuations in the effect of water on the peridotite solidus (Van Kranendonk and Kirkland 2016; Walzer and Hendel 2017).

In the context of the zircon record, most pertinent to the evolution of the biosphere are O isotope compositions, because they provide evidence for weathering processes and fluid-rock interaction. Such evidence is somewhat indirect in the sense that a high-$\delta^{18}O$ zircon reflects crystallization from a magma that assimilated sedimentary rocks (so-called "S-type" magmas). The preponderance of Archean zircons with high-$\delta^{18}O$ values, however, suggests that the sedimentary reservoir was extensive enough that the probability would be so great that high-$\delta^{18}O$ crust would be intersected by ascending magmas. A dominance of mantle-like $\delta^{18}O$ values for Archean zircons, characteristic of "I-type" magmas, compared with elevated $\delta^{18}O$ values for Proterozoic and Phanerozoic zircons, has been taken to indicate limited weathering and crustal recycling in the Archean, as compared with younger times (e.g., Valley et al. 2005). More recent compilations have generally confirmed this view, where the rise in $\delta^{18}O$ values of igneous and detrital zircons in the transition from the Archean to the Proterozoic is taken to record a marked increase in crustal thickening and reworking during the onset of collisional tectonics and initiation of the supercontinent cycle (Van Kranendonk and Kirkland 2013; Spencer et al. 2014). It must be remembered, however, that high-$\delta^{18}O$ zircons are known back to 4.3 Ga (e.g., Cavosie et al. 2005), and recent work on very old cratons documents high-$\delta^{18}O$ crust of 3.6–3.8 Ga (Bolhar et al. 2017) and 3.2 Ga (Van Kranendonk et al. 2015a) age. An important component to this discussion is recognition that biases exist in the detrital zircon record in terms of magma composition and temperature, which affect zircon saturation in magmas, where it is likely that the proportion of felsic Archean or Hadean crust is markedly underestimated from the zircon record (Keller et al. 2017).

Changes in the composition of the continental crust through time have been assessed through direct studies of igneous and metamorphic rocks, although more commonly through use of sedimentary rocks, based on the premise that they represent samples of large areas of exposed crust at various times in the past (e.g., Taylor and McLennan 1985). Broadly, such studies suggest that Archean crust exposed to weathering was mafic in bulk composition (high MgO and low SiO_2) but that Proterozoic and Phanerozoic crust exposed to weathering became more silicic (low MgO and high SiO_2) over time. Such conclusions have been reinforced through studies of trace elements, mapping these to predicted major element compositional changes for magmatic evolution (e.g., Dhuime et al. 2015; Tang et al. 2016a). These studies have, in turn, concluded that the apparent increase in abundance of silicic continental crust between 3.0 Ga and 2.5 Ga coincided with the initiation of plate tectonics, solidifying the expected connection between silicic magma genesis and subduction processes (Van Kranendonk and Kirkland 2016).

Returning to an isotopic approach, which has the capability of capturing a wider range of information than measured bulk compositions, Greber et al. (2017) reported on a new proxy for tracing crustal composition, stable Ti isotopes. They noted that $^{49}Ti/^{47}Ti$ ratios, expressed as $\delta^{49}Ti$ values, correlate strongly with SiO_2 contents in igneous rocks, reflecting the well-known process of increasing crystallization of Fe-Ti oxides with SiO_2 in magmas. Although it is not yet clear if there may be differences in $\delta^{49}Ti$ values for I- and S-type magmas, which can change saturation of different Fe-Ti oxides, the $\delta^{49}Ti$ values of shales have been invariant since 3.5 Ga, and this is explained by a consistently high abundance of silicic crust since the Paleoarchean (Greber et al. 2017). Drawing again on the connection between abundance of silicic continental crust and plate tectonics, an early initiation of plate tectonics >3.5 Ga is proposed (Greber et al. 2017), which is consistent with evidence from West Greenland but contrasts with the more commonly proposed time of ~3 Ga or younger from other datasets.

We conclude this section with a discussion of the emergence of Archean continental crust, which is dependent on its compositional and thermal structure. If the early Archean continental crust was thin and mafic (e.g., Dhuime et al. 2015), isostatic principles would predict that such crust would be largely submerged. Thermal modeling appears to support this conclusion (Flament et al. 2008); however, it is important to note that there exist examples of exposed land surfaces by 3.5 Ga (Buick et al. 1995; Djokic et al. 2017) and that commonly cited thermal modeling assumes that the crustal geotherm was relatively high (Flament et al. 2008); such an assumption is not supported by petrologic or thermobarometric studies of Archean terrains (Burke and Kidd 1978; Galer and Mezger 1998; Brown 2007; Satkoski et al. 2017), which indicate that Archean crust was relatively thick, on the order of ~40 km or more (Champion and Smithies 2007; Van Kranendonk et al. 2015b), with crustal geotherms similar to those of today. This line of argument reflects the fact that the majority of heat loss from the Earth occurs in the oceans and not through the continents. There is increasing evidence, therefore, that early Archean crust was likely more silicic and emergent than previously thought, a model that can be tested through study of the chemical and isotopic compositions of seawater – this would then affect how we view the nutrient delivery to an early biosphere.

4.4.3 THE ATMOSPHERE-WATER ENVELOPE

Perhaps the canonical change in the surface of the Precambrian Earth is the rise of atmospheric oxygen, the first stage of which has been termed the "Great Oxidation Event" (GOE) and has been proposed to have occurred in the Paleoproterozoic at ~2.3 Ga (e.g., Bekker et al. 2004; Holland 2006). There is a vast and extensive literature on the rise of oxygen in the atmosphere (e.g., Holland 1984; Canfield 2005; Lyons et al. 2014, to name a very small fraction). It is commonly accepted that the primary driver of the rise in atmospheric oxygen was oxygenic photosynthesis (e.g., Farquhar et al. 2010), but much debate exists as to when this metabolism first evolved, as oxygen levels were suppressed by the presence of vast reductive sinks until the GOE. Fischer et al. (2016), for example, caution about relying too heavily on redox proxies in the rock record and emphasize that all known phototrophic groups were derived late in the phyla in which they occur, suggesting a relatively late evolution. An increasing body of data from geochemical redox proxies and field evidence, however, suggests that oxygenic photosynthesis arose much earlier than the GOE (e.g., Lyons et al. 2014; Homann et al. 2015). In this context, the oldest known marine redox gradient, inferred to record a gradient in ocean oxygen contents, dates back to 3.2 Ga (Satkoski et al. 2015), suggesting a >3.2 Ga origin for oxygenic photosynthesis, an age supported by molecular clock studies of cyanobacteria (Schirrmeister et al. 2015). Interestingly, this age corresponds with a rapid evolutionary innovation via genetic expansion in microbial life (David and Alm 2011), as explored more fully in this section.

If we accept an origin for oxygenic photosynthesis far earlier than the GOE, the first question is: what kept atmospheric oxygen levels low for perhaps one billion years? The solution lies in understanding the sources and sinks of oxygen. The obvious source for O_2 is oxygenic photosynthesis, which requires the reactants CO_2 and H_2O, presumably always in excess abundance in the Archean Earth. Much of the discussion of primary productivity in the early Earth has therefore focused on limiting nutrients. Over geologic timescales, P is the limiting nutrient over N (Tyrell 1999), and many models for oxygen evolution in the Precambrian explicitly assess P as the key limiting nutrient (e.g., Planavsky et al. 2010; Jones et al. 2015; Laasko and Schrag 2017; Poulton 2017; Reinhard et al. 2017). Additional sources of oxygen include burial of organic C in sedimentary rocks, a source that becomes quite important if the early biosphere was limited in size by nutrients such as P (e.g., Hayes and Waldbauer 2006; Husson and Peters 2017). In addition, subduction of organic C into the mantle may be a source for oxygen (Duncan and Dasgupta 2017); however, this is not detected in the redox state of the mantle (e.g., Nicklas et al. 2018).

Keeping oxygen "in check" before the GOE requires sinks, and numerous options have been proposed. The interplay between CH_4 and O_2 is important not only in moderating the levels of these two gases in the atmosphere and ocean, but also in terms of climate in the early Earth when solar luminosity was lower (e.g., Olson et al. 2013, 2016; Daines and

Lenton 2016; Ozaki et al. 2017). Reaction between O_2 and aqueous Fe^{2+} in the oceans prior to the GOE may produce banded iron formations and was likely an important sink for oxygen; however, oxidation of Fe^{2+} may also occur by anoxygenic phototrophy (e.g., Konhauser et al. 2017). Oxygen consumption by reduced volcanic gases has also been extensively discussed (e.g., Kasting et al. 1993; Holland 2002; Kump and Barley 2007; Gaillard et al. 2011). Early mafic crust has been proposed as an oxygen sink, where the GOE is interpreted to record a decrease in O_2 sinks through emergence of granitic continental crust near the Archean/Proterozoic boundary (e.g., Kamber 2010; Lee et al. 2016; Smit and Mezger 2017).

Turning to the water envelope (oceans), as far as is known, water provides a first-order control on the presence or absence of a biosphere on a planet. To some degree, water might be considered the ultimate "limiting nutrient" for life. Looking beyond the presence of water, the supply of nutrients, as well as electron donors and acceptors, to the marine realm is considered the major driver of the biosphere. The chemistry of the oceans reflects an array of inputs and sinks, with major influences today via continental weathering and hydrothermal systems (e.g., Holland 1984). The conventional view that the continents in the early Archean were small, mafic in composition, and predominantly (though not exclusively) submerged puts severe constraints on the availability of key components for the early biosphere, leading to the model that the earliest oceans were "mantle buffered" (e.g., Shields 2007). A variation on this theme of a "mantle-dominated" early Archean ocean can be found in proposals of emergent, but mafic/ultramafic, oceanic crust (e.g., Kamber 2010). The previous discussion, however, suggests that we must now reconsider an important role for early, emergent, and silicic continental crust in determining the chemistry of the early Archean oceans.

The radiogenic isotopic composition of Sr of seawater (expressed as $^{87}Sr/^{86}Sr$ ratios) has long been used as a tracer of the relative contributions of element fluxes to the oceans from oceanic and continental crust through time (e.g., Shields 2007). Because the residence time of Sr in the oceans is long (i.e., several million years), relative to mixing times of the oceans (e.g., Holland 1984), seawater $^{87}Sr/^{86}Sr$ ratios are globally homogeneous at any one time but change over m.y. timescales due to changes in the proportion of mantle and crustal components. In essence, seawater $^{87}Sr/^{86}Sr$ ratios serve as a global-averaged recorder of mantle and crustal inputs to the oceans. Most commonly, marine carbonates are used to infer seawater $^{87}Sr/^{86}Sr$ ratios, but other lithologies such as barite are also useful. Unfortunately, the seawater $^{87}Sr/^{86}Sr$ record for the Archean has been plagued with difficulties due to the effects of carbonate alteration and incorporation of radiogenic clays, issues that are particularly problematic for very old rocks (Shields and Veizer 2002). The result has been that many workers gave up on determining the seawater Sr isotope curve for the early Archean.

Recently, however, the Archean Sr isotope seawater curve has been re-investigated using marine barite and carbonate, where the effects of hydrothermal alteration and clay contamination have been assessed using C, O, S, and

Rb isotopic analysis (Satkoski et al. 2016, 2017). This work shows that early Archean seawater had significantly higher $^{87}Sr/^{86}Sr$ ratios than the mantle. Satkoski et al. (2016) estimated a Sr oceanic/continental flux ratio of 2–3 at 3.2 Ga, about 5–7 times lower than that of previous estimates that assumed mantle-like $^{87}Sr/^{86}Sr$ ratios for early Archean seawater (Kamber 2010). Additional work on 2.94, 2.80, and 2.6–2.55 Ga marine carbonates produces similarly low oceanic/continental Sr flux ratios (Kamber and Webb 2001; Satkoski et al. 2017), suggesting that the contribution of evolved continental crust to seawater chemistry has been much higher than previously thought, back to at least 3.2 Ga. Significant weathering of silicic continental crust is consistent with evidence for high-$\delta^{18}O$ crust of 3.6–3.8 Ga age, as determined in zircons (see Section 4.4.2; Bolhar et al. 2017), as well as the chemical alteration indices of 3.2 Ga clastic deposits (Hessler and Lowe 2006).

With the majority of attention on O_2 in the atmosphere, relatively few studies have attempted to quantify O_2 contents in the photic zone of the oceans. Recently, however, multiple isotopic systems (S, Fe, Cr, Mo, and U) now consistently indicate free O_2 in the photic zone, at least episodically, as far back as 3.2 Ga, continuing through the Meso- and Neoarchean, rising during the GOE, then returning to low, but significant levels post-GOE (e.g., Czaja et al. 2012, 2013; Canfield et al. 2013; Li et al. 2013; Planavsky et al. 2014; Scott et al. 2014; Satkoski et al. 2015; Tang et al. 2016b; Yang et al. 2017; Eickmann et al. 2018). These results suggest a more pervasively oxygenated photic zone before the GOE than envisioned by early proposals of "oxygen oases" (Kasting et al. 1992). Modeling shows that an Archean photic zone that has O_2 on the order of 1–10 μM could exist beneath an anoxic atmosphere (Olson et al. 2013). A key component to the consumption of O_2 may have been aerobic respiration and methanotrophy, the latter of which is suggested by the highly negative $\delta^{13}C$ values of Neoarchean organic C (Hayes 1994; Daines and Lenton 2016); however, much of the most negative $\delta^{13}C$ values appear to have developed within lakes (Flannery et al. 2016). If photic zone O_2 sinks were significant, primary productivity in the Archean may have been much higher than previously thought, based on assumed limitations in P (e.g., Laakso and Schrag 2017), consistent with evidence from $\delta^{13}C$ values from carbonates and organic carbon through to the earliest Precambrian (e.g., Hayes et al. 1983). In addition, there is increasing evidence for a diverse biosphere on land in the Neoarchean, with a major role for methanotrophy, which would consume oxygen (e.g., Hayes 1994; Stüeken et al. 2017). Broadly, the Meso- to Neoarchean biosphere now appears to have been remarkably diverse long before the GOE.

A proposed link between Precambrian Earth geodynamics and fluctuating atmospheric oxygen levels suggests that the supercontinent cycle had a major influence on the source of reductive sinks, which varied in step with the supercontinent cycle and controlled (in part) the levels of atmospheric oxygen (Van Kranendonk et al. 2012; Van Kranendonk and Kirkland 2016). In this model, Earth has operated in a conditioned duality between two states. Periods of supercontinent aggregation are accompanied by rapid, global subduction, resulting in increased levels of reducing volcanic gasses, a greenhouse atmosphere, and low levels of atmospheric oxygen. Formed supercontinents are accompanied by mantle cooling, arising from the preceding global subduction of cool oceanic lithosphere and from mantle insulation from core heat by slab graveyards across the core-mantle boundary. This results in cooling of the crust, leading to periodic ice ages – sometimes global and long-lived (Hoffman et al. 1998; Kirschvink et al. 2000) – accompanied by a rise in atmospheric oxygen. But the supercontinents insulate the mantle, and in the second state, conductive heating of the slab graveyards returns the mantle to warm conditions, leading to a renewed burst of volcanic activity, supercontinent breakup, melting of ice ages, and a return to a greenhouse atmosphere. These states cycle over time and help drive biological diversification and evolution.

4.4.4 SUMMARY

We summarize temporal variations in a number of solid and surface Earth parameters from 4 to 2 Ga in Figure 4.4.1. As discussed by Konhauser et al. (2017), there is increasing evidence that the shallow oceans (photic zone) became progressively enriched in O_2 from the Paleo- to Neoarchean, perhaps reaching ~10% of present ocean level (POL) before the GOE. The first evidence for photic zone oxygenation at ~3.2 Ga correlates with an increase in $^{87}Sr/^{86}Sr$ ratios for seawater, which is inferred to record the initiation of plate tectonics, and fits some models for the first rapid increase in emergent land. The relatively high proportion of felsic continental crust between 3.5 Ga and 3 Ga, relative to previous estimates, is consistent with the elevated $^{87}Sr/^{86}Sr$ ratios for seawater during this time. This, in turn, suggests a significant P_2O_5 flux from the continents, even in the Paleo- and Mesoarchean. From 3 Ga to 2 Ga, the P_2O_5 flux from the continents to the oceans increased, in part reflecting the decrease in T_P, which decreases percent mantle melting, which, in turn, further enriches the crust in incompatible elements such as P. Collectively, these changes indicate an increasing size and diversity of the biosphere, which are likely to have been driven by continental growth via plate tectonics that began in the early Archean.

In terms of specific metabolisms during the time period illustrated in Figure 4.4.1, anoxygenic phototrophy is likely to have preceded oxygenic photosynthesis (e.g., Widdel et al. 1993), and evidence for this can be found in Fe isotope compositions of Paleoarchean iron formations (Czaja et al. 2013; Li et al. 2013). Based on Fe and U isotopes in Mesoarchean iron formations, it has been proposed that oxygenic photosynthesis arose by 3.2 Ga (Satkoski et al. 2015). Microbial Fe reduction has been identified as an early metabolism that likely was extensive by the Mesoarchean (Konhauser et al. 2017). An active methane cycle, including methanogenesis and methanotrophy, was also important prior to the GOE and was closely tied to O_2 balances, as discussed earlier. Looking past the GOE, oxygenic photosynthesis was firmly established over anoxygenic phototrophy as the major driver of

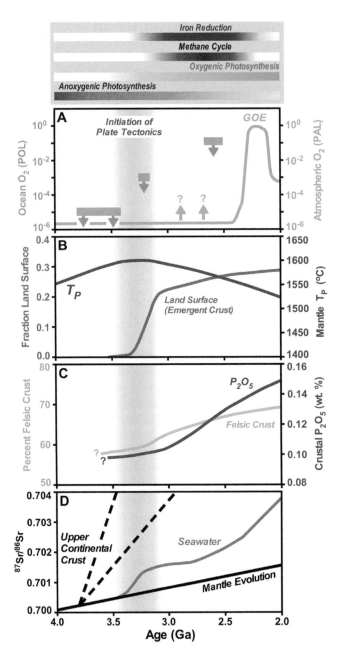

FIGURE 4.4.1 Changes in the solid and surface Earth from 4 Ga to 2 Ga. Upper gray panel shows temporal changes in various microbial pathways. Initiation of plate tectonics inferred to have occurred between 3.5 Ga and 3.2 Ga (see text). (a) Variations in ocean O_2 (relative to present ocean level, POL) and atmospheric O_2 (relative to present atmosphere level, PAL). (Adapted from Konhauser et al., based on previous work [Czaja, A.D. et al., *Geochim. Cosmochim. Acta*, 86, 118–137, 2012; *Earth Planet. Sci. Lett.*, 363, 192–203, 2013; Li, W. et al., *Geochim. Cosmochim. Acta*, 120, 65–79, 2013; Satkoski, A.M. et al., *Earth Planet. Sci. Lett.*, 430, 43–53, 2015; Konhauser, K.O. et al., *Earth Sci. Rev.* 172, 140–177, 2017.) (b) Changes in mantle potential temperature (T_P) and fraction of land surface relative to total Earth surface area. T_P (Korenaga 2013) modified to change transition from stagnant lid to plate tectonics at ~3.3 Ga. Land surface adapted from a fast plate tectonic model, with constant continental crust density (Korenaga et al. 2017), and crustal growth model from Campbell (2003), parameters that favor early emergent crust. (c) Proportions of felsic crust (of total continental crust) and wt. % P_2O_5 for total continental crust, based on modeling Ti isotope variations in shales. (From Greber, N.D. et al., *Science*, 357, 1271–1274, 2017.) (d) $^{87}Sr/^{86}Sr$ evolution curves for mantle and seawater from several sources. (From Shields, G., and Veizer, J., *Geochem. Geophys. Geosyst.*, 3, 2002; Kamber and Webb 2001; Satkoski, A.M. et al., *Earth Planet. Sci. Lett.*, 454, 28–35, 2016; *Geochim. Cosmochim. Acta*, 209, 216–232, 2017.) The highest slope upper continental crustal evolution curve calculated using >3.5 Ga juvenile upper crust (Rb = 71 ppm and Sr = 251 ppm; Condie 1993), evolving from the mantle at 3.8 Ga. The lower slope upper continental crust curve based on extrapolation of modern silicate weathering composition ($^{87}Sr/^{86}Sr$ = 0.7178; Palmer and Edmond 1989), back to the mantle at 3.8 Ga.

primary productivity, methane was effectively scrubbed from the atmosphere, and microbial Fe reduction contracted in its footprint to suboxic environments.

REFERENCES

Barnes, S.J., M.J. Van Kranendonk, and I. Sonntag. 2012. Geochemistry and tectonic setting of basalts from the Eastern Goldfields Superterrane. *Australian Journal of Earth Sciences* 59 (5):707–735.

Bekker, A., H.D. Holland, P.L. Wang, et al. 2004. Dating the rise of atmospheric oxygen. *Nature* 427 (6970):117–120.

Bercovici, D. 1998. Generation of plate tectonics from lithosphere–mantle flow and void–volatile self-lubrication. *Earth and Planetary Science Letters* 154:139–151.

Bleeker, W., J.W. Ketchum, V.A. Jackson, and M.E. Villeneuve. 1999. The Central Slave Basement Complex, Part I: Its structural topology and autochthonous cover. *Canadian Journal of Earth Sciences* 36 (7):1083–1109.

Blenkinsop, T.G., C.M. Fedo, M.J. Bickle, et al. 1993. Ensialic origin for the Ngezi Group, Belingwe greenstone belt, Zimbabwe. *Geology* 21 (12):1135.

Bolhar, R., A. Hofmann, A.I.S. Kemp, M.J. Whitehouse, S. Wind, and B.S. Kamber. 2017. Juvenile crust formation in the Zimbabwe Craton deduced from the O-Hf isotopic record of 3.8–3.1 Ga detrital zircons. *Geochimica et Cosmochimica Acta* 215:432–446.

Bouhallier, H., D. Chardon, and P. Choukroune. 1995. Strain patterns in Archaean dome-and-basin structures: The Dharwar craton (Karnataka South India). *Earth and Planetary Science Letters* 135:57–75.

Bridgwater, D., V.R. McGregor, and J.S. Myers. 1974. A horizontal tectonic regime in the Archean of Greenland and its implications for early crustal thickening. *Precambrian Research* 1:179–197.

Brown, M. 2006. Duality of thermal regimes is the distinctive characteristic of plate tectonics since the Neoarchean. *Geology* 34:961–964.

Brown, M. 2007. Metamorphic conditions in orogenic belts: A record of secular change. *International Geology Review* 49:193–234.

Buick, R., J.R. Thornett, N.J. McNaughton, J.B. Smith, M.E. Barley, and M. Savage. 1995. Record of emergent continental crust ~3.5 billion years ago in the Pilbara Craton of Australia. *Nature* 375:574–577.

Burke, K., and W.S. Kidd. 1978. Were Archean continental geothermal gradients much steeper than those of today? *Nature* 272 (5650):240–241.

Calvert, A.J., E.W. Sawyer, W.J. Davis, and J.N. Ludden. 1995. Archean subduction inferred from seismic images of a mantle suture in the Superior Province. *Nature* 375:670–674.

Campbell, I.H. 2003. Constraints on continental growth models from Nb/U ratios in the 3.5 Ga barberton and other archaean basalt-komatiite suites. *American Journal of Science* 303:319–351.

Campbell, I.H., and S.R. Taylor. 1983. No water, no granites–no oceans, no continents. *Geophysical Research Letters* 10 (11):1061–1064.

Canfield, D.E. 2005. The early history of atmospheric oxygen: Homage to Robert M. Garrels. *Annual Review of Earth and Planetary Sciences* 33 (1):1–36.

Canfield, D.E., L. Ngombi-Pemba, E.U. Hammarlund, et al. 2013. Oxygen dynamics in the aftermath of the Great Oxidation of Earth's atmosphere. *Proceedings of the National Academy of Sciences United States of America* 110 (42):16736–16741.

Card, K.D. 1990. A review of the Superior Province of the Canadian Shield, a product of Archean accretion. *Precambrian Research* 48:99–156.

Cavosie, A.J., J.W. Valley, and S.A. Wilde. 2005. Magmatic δ18O in 4400–3900 Ma detrital zircons: A record of the alteration and recycling of crust in the Early Archean. *Earth and Planetary Science Letters* 235 (3–4):663–681.

Champion, D.C., and R.H. Smithies. 2007. Geochemistry of Paleoarchean rocks of the East Pilbara Terrane, Pilbara Craton, Western Australia: Implications for early Archean crustal growth. In *Earth's Oldest Rocks*, edited by M.J. Van Kranendonk, R.H. Smithies, and V. Bennet. Amsterdam, the Netherlands: Elsevier.

Collins, W.J., M.J. Van Kranendonk, and C. Teyssier. 1998. Partial convective overturn of Archaean crust in the east Pilbara Craton, Western Australia: Driving mechanisms and tectonic implications. *Journal of Structural Geology* 20:1405–1424.

Condie, K., S.A. Pisarevsky, J. Korenaga, and S. Gardoll. 2015. Is the rate of supercontinent assembly changing with time? *Precambrian Research* 259:278–289.

Condie, K.C. 1993. Chemical composition and evolution of the upper continental crust: Contrasting results from surface samples and shales. *Chemical Geology* 104:1–37.

Condie, K.C., and R.C. Aster. 2010. Episodic zircon age spectra of orogenic granitoids: The supercontinent connection and continental growth. *Precambrian Research* 180:227–236.

Czaja, A.D., C.M. Johnson, B.L. Beard, E.E. Roden, W. Li, and S. Moorbath. 2013. Biological Fe oxidation controlled deposition of banded iron formation in the ca. 3770Ma Isua Supracrustal Belt (West Greenland). *Earth and Planetary Science Letters* 363:192–203.

Czaja, A.D., C.M. Johnson, E.E. Roden, et al. 2012. Evidence for free oxygen in the Neoarchean ocean based on coupled iron–molybdenum isotope fractionation. *Geochimica et Cosmochimica Acta* 86:118–137.

Daines, S.J., and T.M. Lenton. 2016. The effect of widespread early aerobic marine ecosystems on methane cycling and the Great Oxidation. *Earth and Planetary Science Letters* 434:42–51.

David, L.A., and Alm. E.J. 2011. Rapid evolutionary innovation during an Archaean genetic expansion. *Nature* 469:93–96.

Dhuime, B., A. Wuestefeld, and C.J. Hawkesworth. 2015. Emergence of modern continental crust about 3 billion years ago. *Nature Geoscience* 8:552–555.

Dhuime, B., C.J. Hawkesworth, P.A. Cawood, and C.D. Storey. 2012. A change in the geodynamics of continental growth 3 billion years ago. *Science* 335:1334–1336.

Djokic, T., M.J. Van Kranendonk, K.A. Campbell, M.R. Walter, and C.R. Ward. 2017. Earliest signs of life on land preserved in c. 3.5 Ga hot spring deposits. *Nature Communications* 8:15263.

Duncan, M.S., and R. Dasgupta. 2017. Rise of Earth's atmospheric oxygen controlled by efficient subduction of organic carbon. *Nature Geoscience* 10 (5):387–392.

Eickmann, B., A. Hofmann, M. Wille, T.H. Bui, B.A. Wing, and R. Schoenberg. 2018. Isotopic evidence for oxygenated Mesoarchaean shallow oceans. *Nature Geoscience* 11 (2):133–138.

Farquhar, J., A.L. Zerkle, and A. Bekker. 2010. Geological constraints on the origin of oxygenic photosynthesis. *Journal of Geodynamics* 107:11–36.

Fischer, R., and T. Gerya. 2016. Early Earth plume-lid tectonics: A high-resolution 3D numerical modelling approach. *Journal of Geodynamics* 100:198–214.

Fischer, W.W., J. Hemp, and J.E. Johnson. 2016. Evolution of oxygenic photosynthesis. *Annual Review of Earth and Planetary Sciences* 44 (1):647–683.

Flament, N., N. Coltice, and P.F. Rey. 2008. A case for late-Archaean continental emergence from thermal evolution models and hypsometry. *Earth and Planetary Science Letters* 275 (3–4):326–336.

Flannery, D.T., A. Allwood, and M.J. Van Kranendonk. 2016. Lacustrine facies dependence of highly 13C-depleted organic matter during the Global Age of Methanotrophy. *Precambrian Research* 285:216–241.

French, K.L., C. Hallmann, J.M. Hope, et al. 2015. Reappraisal of hydrocarbon biomarkers in archean rocks. *Proceedings of the National Academy of Sciences United States of America* 112:5915–5920.

Gaillard, F., B. Scaillet, and N.T. Arndt. 2011. Atmospheric oxygenation caused by a change in volcanic degassing pressure. *Nature* 478 (7368):229–232.

Galer, S.J.G., and K. Mezger. 1998. Metamorphism, denudation and sea level in the Archean and cooling of the Earth. *Precambrian Research* 92 (1998):389–412.

Gardiner, N.J., C.L. Kirkland, and M.J. Van Kranendonk. 2016. More than average: The juvenile Hf signal as a record of global supercycles. *Scientific Reports* 6:38503.

Gill, J.B. 1981. *Orogenic Andesites and Plate Tectonics*. Berlin, Germany: Springer.

Greber, N.D., N. Dauphas, A. Bekker, M.P. Ptacek, I.N. Bindeman, and A. Hoffmann. 2017. Titanium isotopic evidence for felsic crust and plate tectonics 3.5 billion years ago. *Science* 357:1271–1274.

Griffin, W.L., and S. O'Reilly. 2007. Cratonic lithospheric mantle: Is anything subducted? *Episodes* 30:43–53.

Hamilton, W. 1998. Archean magmatism and deformation were not the products of plate tectonics. *Precambrian Research* 91:143–179.

Hanmer, S., and Greene, D.C. 2002. A modern structural regime in the Paleoarchean (~3.64 Ga); Isua Greenstone Belt, southern West Greenland. *Tectonophysics* 346:201–222.

Harley, S.L. 2003. Archean to Pan-African crustal development and assembly of East Antarctica: Metamorphic characteristics and tectonic implications. In *Proterozoic East Gondwana: Supercontinent Assembly and Breakup*, edited by M. Yoshida, and B.F. Windley. London, UK: Geological Society Special Publications.

Harrison, T.M. 2009. The hadean crust: Evidence from >4 Ga zircons. *Annual Review of Earth and Planetary Sciences* 37 (1):479–505.

Hawkesworth, C.J., P.A. Cawood, B. Dhuime, and A.I.S. Kemp. 2017. Earth's continental lithosphere through time. *Annual Review of Earth and Planetary Sciences* 45:169–198.

Hayes, J.M. 1994. Global methanotrophy at the Archean–Proterozoic transition. In *Early Life on Earth. Nobel Symposium*, edited by S. Bengtson. New York: Columbia University Press.

Hayes, J.M., and J.R. Waldbauer. 2006. The carbon cycle and associated redox processes through time. *Philosophical Transactions of the Royal Society B: Biological Sciences* 361 (1470):931–950.

Hayes, J.M., I.R. Kaplan, and K.W. Wedeking. 1983. Precambrian organic geochemistry, preservation of the record. In *The Earth's Earliest Biosphere: Its Origin and Evolution*, edited by J.W. Schopf. Princeton, NJ: Princeton University Press.

Herrick, R.R. 1994. Resurfacing history of venus. *Geology* 22:703–706.

Herzberg, C., and R. Rudnick. 2012. Formation of cratonic lithosphere: An integrated thermal and petrological model. *Lithos* 149:4–15.

Herzberg, C., K. Condie, and J. Korenaga. 2010. Thermal history of the Earth and its petrological expression. *Earth and Planetary Science Letters* 292 (1–2):79–88.

Hessler, A.M., and D.R. Lowe. 2006. Weathering and sediment generation in the Archean: An integrated study of the evolution of siliciclastic sedimentary rocks of the 3.2Ga Moodies Group, Barberton Greenstone Belt, South Africa. *Precambrian Research* 151 (3–4):185–210.

Hoffman, P.F., A.J. Kaufman, G.P. Halverson, and D.P. Schrag. 1998. A neoproterozoic snowball Earth. *Science* 281:1342–1346.

Holland, H.D. 1984. *The Chemical Evolution of the Atmosphere and Oceans*. Princeton, NJ: Princeton University Press.

Holland, H.D. 2002. Volcanic gases, black smokers, and the great oxidation event. *Geochimica et Cosmochimica Acta* 66 (21):3811–3826.

Holland, H.D. 2006. The oxygenation of the atmosphere and oceans. *Philosophical Transactions of the Royal Society B* 361:903–915.

Homann, M., C. Heubeck, A. Airo, and M.M. Tice. 2015. Morphological adaptations of 3.22 Ga-old tufted microbial mats to Archean coastal habitats (Moodies Group, Barberton Greenstone Belt, South Africa). *Precambrian Research* 266:47–64.

Husson, J.M., and S.E. Peters. 2017. Atmospheric oxygenation driven by unsteady growth of the continental sedimentary reservoir. *Earth and Planetary Science Letters* 460:68–75.

Iizuka, T., T. Yamaguchi, K. Itano, Y. Hibiya, and K. Suzuki. 2017. What Hf isotopes in zircon tell us about crust–mantle evolution. *Lithos* 274–275:304–327.

Javaux, E.J., and K. Lepot. 2018. The Paleoproterozoic fossil record: Implications for the evolution of the biosphere during Earth's middle-age. *Earth-Science Reviews* 176:68–86.

Jones, C., S. Nomosatryo, S.A. Crowe, C.J. Bjerrum, and D.E. Canfield. 2015. Iron oxides, divalent cations, silica, and the early earth phosphorus crisis. *Geology* 43 (2):135–138.

Kamber, B.S. 2010. Archean mafic–ultramafic volcanic landmasses and their effect on ocean–atmosphere chemistry. *Chemical Geology* 274 (1–2):19–28.

Kamber, B.S., and G.E. Webb. 2001. The geochemistry of late Archaean microbial carbonate: Implications for ocean chemistry and continental erosion history. *Geochimica Et Cosmochimica Acta* 65 (15):2509–2525.

Kasting, J.F., D.H. Eggler, and S.P. Raeburn. 1993. Mantle redox evolution and the oxidation state of the archean atmosphere. *Journal of Geology* 101:245–257.

Kasting, J.F., J.W. Schopf, and C. Klein. 1992. Models relating to Proterozoic atmospheric and oceanic chemistry. In *The Proterozoic Biosphere, A Multidisciplinary Study*. Cambridge: Cambridge University Press.

Keller, C.B., P. Boehnke, and B. Schoene. 2017. Temporal variation in relative zircon abundance throughout Earth history. *Geochemical Perspectives Letters* 3:179–189.

Kemp, A.I.S., S.A. Wilde, C.J. Hawkesworth, et al. 2010. Hadean crustal evolution revisited: New constraints from Pb–Hf isotope systematics of the Jack Hills zircons. *Earth Planetary Science Letters* 296:45–56.

Kirschvink, J.L., E.J. Gaidos, L.E. Bertani, et al. 2000. Paleoproterozoic snowball Earth: Extreme climatic and geochemical global change and its biological consequences. *Proceedings of the National Academy of Sciences* 97:1400–1405.

Knoll, A.H., K.D. Bergmann, and J.V. Strauss. 2016. Life: The first two billion years. *Philosophical Transactions of the Royal Society B* 371:20150493.

Konhauser, K.O., N.J. Planavsky, D.S. Hardisty, et al. 2017. Iron formations: A global record of Neoarchaean to Palaeoproterozoic environmental history. *Earth-Science Reviews* 172:140–177.

Korenaga, J. 2013. Initiation and evolution of plate tectonics on Earth: Theories and observations. *Annual Review of Earth and Planetary Sciences* 41 (1):117–151.

Korenaga, J., N.J. Planavsky, and D.A.D. Evans. 2017. Global water cycle and the coevolution of the Earth's interior and surface environment. *Philosophical Transactions. Series A, Mathematical, Physical, and Engineering Sciences* 375 (2094). doi:10.1098/rsta.2015.0393.

Kump, L.R., and M.E. Barley. 2007. Increased subaerial volcanism and the rise of atmospheric oxygen 2.5 billion years ago. *Nature* 448 (7157):1033–1036.

Laakso, T.A., and D.P. Schrag. 2017. A theory of atmospheric oxygen. *Geobiology* 15 (3):366–384.

Labrosse, S., and C. Jaupart. 2007. Thermal evolution of the Earth: Secular changes and fluctuations of plate characteristics. *Earth Planetary Science Letters* 260:465–481.

Lécuyer, C. 2014. Water and plate tectonics. In *Water on Earth: Physiochemical and Biological Properties*. London, UK: John Wiley & Sons.

Lee, C.T.A., L.Y. Yeung, N.R. McKenzie, Y. Yokoyama, K. Ozaki, and A. Lenardic. 2016. Two-step rise in atmospheric oxygen linked to the growth of continents. *Nature Geoscience* 9:417–424.

Li, W., A.D. Czaja, M.J. Van Kranendonk, B.L. Beard, E.E. Roden, and C.M. Johnson. 2013. An anoxic, Fe(II)-rich, U-poor ocean 3.46 billion years ago. *Geochimica et Cosmochimica Acta* 120:65–79.

Lyons, T.W., C.T. Reinhard, and N.J. Planavsky. 2014. The rise of oxygen in Earth's early ocean and atmosphere. *Nature* 506 (7488):307–315.

Martin, H., R.H. Smithies, R. Rapp, J.F. Moyen, and D. Champion. 2005. An overview of adakite, tonalite–trondhjemite–granodiorite (TTG), and sanukitoid: Relationships and some implications for crustal evolution. *Lithos* 79:1–24.

McCall, J.G.H. 2003. A critique of the analogy between Archean and Phanerozoic tectonics based on regional mapping of the Mesozoic–Cenozoic plate convergent zone in the Makran, Iran. *Precambrian Research* 127:5–18.

Mojzsis, S.J., T.M. Harrison, and R.T. Pidgeon. 2001. Oxygen-isotope evidence from ancient zircons for liquid water at the Earth's surface 4300 Myr ago. *Nature* 409:178–181.

Myers, J.S. 1976. Granitoid sheets, thrusting and Archean crustal thickening in West Greenland. *Geology* 4:265–268.

Næraa, T., A. Scherstén, M.T. Rosing, et al. 2012. Hafnium isotope evidence for a transition in the dynamics of continental growth 3.2 Gyr ago. *Nature* 485:627–630.

Nicklas, R.W., I.S. Puchtel, and R.D. Ash. 2018. Redox state of the Archean mantle: Evidence from V partitioning in 3.5–2.4 Ga komatiites. *Geochimica et Cosmochimica Acta* 222:447–466.

Nutman, A.P., C.R.L. Friend, and V.C. Bennett. 2002. Evidence for 3650–3600 Ma assembly of the northern end of the Itsaq Gneiss complex, Greenland: Implications for early Archean tectonics. *Tectonics* 21 (1):5–1–5–7.

O'Neill, C., A. Lenardic, L. Moresi, T.H. Torsvik, and C.T.A. Lee. 2007. Episodic Precambrian subduction. *Earth and Planetary Science Letters* 262 (3–4):552–562.

O'Neill, C., and V. Debaille. 2014. The evolution of Hadean-Eoarchaean geodynamics. *Earth Planetary Science Letters* 406:49–58.

Olson, S.L., C.T. Reinhard, and T.W. Lyons. 2016. Limited role for methane in the mid-Proterozoic greenhouse. *Proceedings of the National Academy of Sciences* 113 (41):11447–11452.

Olson, S.L., L.R. Kump, and J.F. Kasting. 2013. Quantifying the areal extent and dissolved oxygen concentrations of Archean oxygen oases. *Chemical Geology* 362:35–43.

Ozaki, K., E. Tajika, P.K. Hong, Y. Nakagawa, and C.T. Reinhard. 2017. Effects of primitive photosynthesis on Earth's early climate system. *Nature Geoscience* 11 (1):55–59.

Palmer, M., and J. Edmond. 1989. The strontium isotope budget of the modern ocean. *Earth and Planetary Science Letters* 92 (1):11–26.

Planavsky, N.J., O.J. Rouxel, A. Bekker, et al. 2010. The evolution of the marine phosphate reservoir. *Nature* 467 (7319):1088–1090.

Planavsky, N.J., C.T. Reinhard, X. Wang, et al. 2014. Low mid-Proterozoic atmospheric oxygen levels and the delayed rise of animals. *Science* 346:635–638.

Polat, A., A.W. Hofmann, and M.T. Rosing. 2002. Boninite-like volcanic rocks in the 3.7–3.8 Ga Isua greenstone belt, West Greenland: Geochemical evidence for intra-oceanic subduction zone processes in the early Earth. *Chemical Geology* 184:231–254.

Poulton, S.W. 2017. Early phosphorus redigested. *Nature Geoscience* 10 (2):75–76.

Puetz, S.J., K.C. Condie, S. Pisarevsky, A. Davaille, C.J. Schwarz, and C.E. Ganade. 2017. Quantifying the evolution of the continental and oceanic crust. *Earth-Science Reviews* 164:63–83.

Rapp, R.P., E.B. Watson, and C.F. Miller. 1991. Partial melting of amphibolite/eclogite and the origin of Archean trondhjemites and tonalites. *Precambrian Research* 51:1–25.

Reinhard, C.T., N.J. Planavsky, B.C. Gill, et al. 2017. Evolution of the global phosphorus cycle. *Nature* 541 (7637):386–389.

Rey, P.F., P. Philippot, and N. Thebaud. 2003. Contribution of mantle plumes, crustal thickening and greenstone blanketing to the 2.75–2.65 Ga global crisis. *Precambrian Research* 127:43–60.

Riciputi, L.R., J.W. Valley, and V.R. McGregor. 1990. Conditions of Archean granulite metamorphism in the Godthab-Fiskenaesset region, southern West Greenland. *Journal of Metamorphic Geology* 8:171–190.

Roberts, N.M.W., and C.J. Spencer. 2015. The zircon archive of continent formation through time. *Geological Society, London, Special Publications* 389 (1):197–225.

Robin, C.M.I., and R.C. Bailey. 2009. Simultaneous generation of Archean crust and subcratonic roots by vertical tectonics. *Geology* 37 (6):523–526.

Sandiford, M., M.J. Van Kranendonk, and S. Bodorkos. 2004. Conductive incubation and the origin of dome-and-keel structure in Archean granite-greenstone terrains: A model based on the eastern Pilbara Craton, Western Australia. *Tectonics* 23 :TC1009 1–19.

Satkoski, A.M., D.R. Lowe, B.L. Beard, M.L. Coleman, and C.M. Johnson. 2016. A high continental weathering flux into Paleoarchean seawater revealed by strontium isotope analysis of 3.26 Ga barite. *Earth and Planetary Science Letters* 454:28–35.

Satkoski, A.M., N.J. Beukes, W. Li, B.L. Beard, and C.M. Johnson. 2015. A redox-stratified ocean 3.2 billion years ago. *Earth and Planetary Science Letters* 430:43–53.

Satkoski, A.M., P. Fralick, B.L. Beard, and C.M. Johnson. 2017. Initiation of modern-style plate tectonics recorded in Mesoarchean marine chemical sediments. *Geochimica et Cosmochimica Acta* 209:216–232.

Schirrmeister, B.E., M. Gugger, and P.C.J. Donoghue. 2015. Cyanobacteria and the great oxidation event: Evidence from genes and fossils. *Paleontology* 58:769–785.

Scott, C., B.A. Wing, A. Bekker, et al. 2014. Pyrite multiple-sulfur isotope evidence for rapid expansion and contraction of the early Paleoproterozoic seawater sulfate reservoir. *Earth and Planetary Science Letters* 389:95–104.

Shields, G., and J. Veizer. 2002. Precambrian marine carbonate isotope database: Version 1.1. *Geochemistry, Geophysics, Geosystems* 3 (6):1–12.

Shields, G.A. 2007. The marine carbonate and chert isotope records and their implications for tectonics, life and climate on the early Earth. In *Earths Oldest Rocks Developments in Precambrian Geology*, edited by M.J. van Kranendonk, H. Smithies, and V.C. Bennett, pp. 971–983, Amsterdam, the Netherlands: Elsevier.

Shirey, S.B., and Richardson, S.H. 2011. Start of the Wilson cycle at 3 Ga shown by diamonds from subcontinental mantle. *Science* 333:434–436.

Smit, M.A., and K. Mezger. 2017. Earth's early O2 cycle suppressed by primitive continents. *Nature Geoscience* 10 (10):788–792.

Smithies, R.H., D.C. Champion, M.J. Van Kranendonk, H.M. Howard, and A.H. Hickman. 2005a. Modern-style subduction processes in the Mesoarchaean: Geochemical evidence from the 3.12 Ga Whundo intraoceanic arc. *Earth and Planetary Science Letters* 231:221–237.

Smithies, R.H., M.J. Van Kranendonk, and D.C. Champion. 2005b. It started with a plume – early Archaean basaltic proto-continental crust. *Earth and Planetary Science Letters* 238:284–297.

Smithies, R.H., T.J. Ivanic, J.R. Lowrey, et al. 2018. Two distinct origins for Archean greenstone belts. *Earth and Planetary Science Letters* 487:106–116.

Spencer, C.J., P.A. Cawood, C.J. Hawkesworth, T.D. Raub, A.R. Prave, and N.M.W. Roberts. 2014. Proterozoic onset of crustal reworking and collisional tectonics: Reappraisal of the zircon oxygen isotope record. *Geology* 42 (5):451–454.

Stachel, T., A. Banas, K. Muehlenbachs, S. Kurszlaukis, and E.C. Walker. 2006. Archean diamonds from Wawa (Canada): Samples from deep cratonic roots predating cratonization of the Superior Province. *Contributions to Mineralogy and Petrology* 151:737–750.

Stein, M., and A.W. Hofmann. 1994. Mantle plumes and episodic crustal growth. *Nature* 372:63–68.

Stern, R. 2005. Evidence from ophiolites, blueschists, and ultrahigh-pressure metamorphic terranes that the modern episode of subduction tectonics began in Neoproterozoic time. *Geology* 33:557–560.

Stueken, E.E., R. Buick, R.E. Anderson, J.A. Baross, N.J. Planavsky, and T.W. Lyons. 2017. Environmental niches and metabolic diversity in Neoarchean lakes. *Geobiology* 15 (6):767–783.

Tang, D., X. Shi, X. Wang, and G. Jiang. 2016b. Extremely low oxygen concentration in mid-Proterozoic shallow seawaters. *Precambrian Research* 276:145–157.

Tang, M., K. Chen, and R.L. Rudnick. 2016a. Archean upper crust transition from mafic to felsic marks the onset of plate tectonics *Science* 351 (6271):372–375.

Tappe, S., K.A. Smart, D.G. Pearson, A. Steenfelt, and Simonetti A. 2011. Craton formation in Late Archean subduction zones revealed by first Greenland eclogites. *Geology* 39:1103–1106.

Taylor, S.R., and S.M. McLennan. 1985. *The Continental Crust: Its Composition and Evolution: An Examination of the Geochemical Record Preserved in Sedimentary Rocks.* Oxford, UK: Blackwell Scientific Publications.

Tyrell, T. 1999. The relative influences of nitrogen and phosphorus on oceanic primary production. *Nature* 400:525–531.

Valley, J.W., J.S. Lackey, A.J. Cavosie, et al. 2005. 4.4 billion years of crustal maturation: Oxygen isotope ratios of magmatic zircon. *Contributions to Mineralogy and Petrology* 150 (6):561–580.

Van Kranendonk, M.J. 2010. Two types of Archean continental crust: Plume and plate tectonics on early Earth. *American Journal of Science* 310:1187–1209.

Van Kranendonk, M.J. 2011. Cool greenstone drips, hot rising domes, and the role of partial convective overturn in Barberton greenstone belt evolution. *Journal of African Earth Sciences* 60:346–352.

Van Kranendonk, M.J., A. Kröner, E. Hegner, and J. Connelly. 2009. Age, lithology and structural evolution of the c. 3.53 Ga Theespruit Formation in the Tjakastad area, southwestern Barberton Greenstone Belt, South Africa, with implications for Archean tectonics. *Chemical Geology* 261:114–138.

Van Kranendonk, M.J., A.H. Hickman, R.H. Smithes, and D.R. Nelson. 2002. Geology and tectonic evolution of the Archean North Pilbara Terrain, Pilbara Craton, Western Australia. *Economic Geology* 97:695–732.

Van Kranendonk, M.J., and C.L. Kirkland. 2016. Conditioned duality of the Earth system: Geochemical tracing of the supercontinent cycle through Earth history. *Earth-Science Reviews* 160:171–187.

Van Kranendonk, M.J., and C.L. Kirkland. 2013. Orogenic climax of Earth: The 1.2–1.1 Ga Grenvillian Superevent. *Geology* 41:735–738.

Van Kranendonk, M.J., C.L. Kirkland, and J. Cliff. 2015b. Oxygen isotopes in Pilbara Craton zircons support a global increase in crustal recycling at 3.2 Ga. *Lithos* 228–229:90–98.

Van Kranendonk, M.J., D. Deamer, and T. Djokic. 2017. Life springs. *Scientific American* (August):28–35.

Van Kranendonk, M.J., R.H. Smithies, A.H Hickman, and D.C. Champion. 2007a. Chapter 4.1 Paleoarchean development of a continental nucleus: The East Pilbara terrane of the Pilbara Craton, Western Australia. *Developments in Precambrian Geology* 15:307–337.

Van Kranendonk, M.J., R.H. Smithies, A.H. Hickman, and D.C. Champion. 2007b. Secular tectonic evolution of Archean continental crust: Interplay between horizontal and vertical processes in the formation of the Pilbara Craton, Australia. *Terra Nova* 19:1–38.

Van Kranendonk, M.J., R.H. Smithies, A.H. Hickman, M.T.D. Wingate, and S. Bodorkos. 2010. Evidence for Mesoarchean (~3.2 Ga) rifting of the Pilbara Craton: The missing link in an early Precambrian Wilson cycle. *Precambrian Research* 177:145–161.

Van Kranendonk, M.J., R.H. Smithies, W.L. Griffin, et al. 2015a. Making it thick: A volcanic plateau model for Paleoarchean continental lithosphere of the Pilbara and Kaapvaal cratons. In *Continent Formation Through Time*, edited by N.M.W. Roberts, M.J. Van Kranendonk, S. Parman, S. Shirey, and P.D. Clift. London, UK: Geological Society London Special Publications.

Van Kranendonk, M.J., W. Altermann, B.L. Beard, et al. 2012. A chronostratigraphic division of the Precambrian: Possibilities and challenges. In *The Geologic Time Scale 2012*, edited by F.M. Gradstein, J.G.Ogg, M.D. Schmitz, and G.J. Ogg. Boston, MA: Elsevier.

Van Kranendonk, M.J., W.J. Collins, A.H. Hickman, and M.J. Pawley. 2004. Critical tests of vertical vs horizontal tectonic models for the Archaean East Pilbara Granite-Greenstone Terrane, Pilbara Craton, Western Australia. *Precambrian Research* 131:173–211.

Walzer, U., and R. Hendel. 2017. Continental crust formation: Numerical modelling of chemical evolution and geological implications. *Lithos* 278–281:215–228.

Whalen, J.B., J.A. Percival, V.J. McNicoll, and F.J. Longstaffe. 2002. A mainly crustal origin for tonalitic granitoid rocks, Superior Province, Canada: Implications for late Archean tectonomagmatic processes. *Journal of Petrology* 43:1551–1570.

White, D.J., G. Musacchio, H.H. Helmstaedt, et al. 2003. Images of a lower-crustal oceanic slab: Direct evidence for tectonic accretion in the Archean western Superior province. *Geology* 31:997–1000.

Widdel, F., S. Schnell, S. Heising, A. Ehrenreich, B. Assmus, and B Schink. 1993. Ferrous iron oxidation by anoxygenic phototrophic bacteria. *Nature* 362 (6423):834–836.

Wilde, S.A., J.W. Valley, W.H. Peck, and C.M. Graham. 2001. Evidence from detrital zircons for the existence of continental crust and oceans on the Earth 4.4 Gyr ago. *Nature* 409:175–178.

Wilks, M.E. 1988. The Himalayas – a modern analogue for Archean crustal evolution. *Earth and Planetary Science Letters* 87:127–136.

Windley, B.F., and A.A. Garde. 2009. Arc-generated blocks with crustal sections in the North Atlantic craton of West Greenland: Crustal growth in the Archean with modern analogue. *Earth-Science Reviews* 93:1–30.

Wyman, D.A., J.A. Ayer, R.V. Conceição, and R.P. Sage. 2006. Mantle processes in an Archean orogen: Evidence from 2.67 Ga diamond-bearing lamprophyres and xenoliths. *Lithos* 89:300–328.

Yang, S., B. Kendall, X. Lu, F. Zhang, and W. Zheng. 2017. Uranium isotope compositions of mid-Proterozoic black shales: Evidence for an episode of increased ocean oxygenation at 1.36 Ga and evaluation of the effect of post-depositional hydrothermal fluid flow. *Precambrian Research* 298:187–201.

4.5 Atmosphere on Early Earth and Its Evolution as It Impacted Life

James F. Kasting

CONTENTS

4.5.1 INTRODUCTION

The nature of Earth's Hadean (>4.0 Ga) atmosphere remains poorly known, as we have no direct samples of it to study and as the geologic record itself is nearly nonexistent during this time period. But we can make inferences about the Hadean atmosphere based partly on theory and partly on geologic evidence from the ensuing Archean Eon (4.0–2.5 Ga). Whether or not this lack of knowledge about conditions during the Hadean presents a problem for origin of life studies depends on when life on Earth originated. The earliest convincing evidence for life comes from stromatolites (laminated structures created by mat-forming organisms), which date back to 3.3–3.5 Ga (Buick et al. 1981; Hofmann et al. 1999; Allwood et al. 2006, 2009). But ^{13}C-depleted organic carbon has been found in sedimentary rocks near Isua, West Greenland (Rosing 1999), and this is interpreted by some authors as being biologically generated. So, life may well have originated back during the time for which we have no direct record.

One might ask what good it does to speculate about early atmospheric composition if we have no way of checking the results. The answer is that some theories of the origin of life (e.g., Oparin 1938; Miller 1953, 1955; Powner et al. 2009; Sutherland 2016) depend critically on this parameter. Other theories in which life originates within submarine hydrothermal vents (e.g., Russell and Hall 1997; Martin et al. 2008) are less sensitive to the nature of the early atmosphere. But even those theories depend indirectly on atmospheric composition, because it helps to define free-energy gradients that are needed to power early metabolisms. In this chapter, I will briefly summarize constraints on early atmospheric composition and climate and then discuss how the atmosphere might have played a role in establishing these critical free-energy gradients. I will also employ the methodology of (Hoehler

2004) to argue that some of these free-energy gradients may have been less useful to the biota than previously advertised (including by this author) because of kinetic barriers imposed by the limited concentrations of required reactants. This, in turn, may have helped guide biological evolution toward more productive methods of carbon fixation, particularly ones involving photosynthesis.

4.5.2 COMPOSITION OF THE EARLY ATMOSPHERE

During the Archean Eon, the geologic record provides some information about both atmospheric composition and climate. Most importantly, it tells us that the atmosphere contained very little free O_2, as evidenced by the survival of detrital reduced minerals such as pyrite (FeS_2) and uraninite (UO_2) and by the preservation of sulfur mass-independent fractionation (MIF) (Cloud 1972; Farquhar et al. 2000; Pavlov and Kasting 2002; Holland 2006; Catling and Kasting 2017, Ch. 10). The mechanism for producing the sulfur MIF is under reinvestigation; it may result from elemental sulfur chain formation rather than from SO_2 photolysis (Babikov 2017; Harman et al. 2018). But the conclusion from (Pavlov and Kasting 2002) remains valid: sulfur must exit the atmosphere in a variety of different compounds in order to produce MIF, and this implies O_2 concentrations $<10^{-5}$ Present Atmospheric Level (PAL).

The lack of free O_2 in the Archean atmosphere, and presumably in the Hadean as well, does *not* imply that the atmosphere was strongly reduced, i.e., rich in hydrogen-containing compounds such as CH_4 and NH_3. A strongly reduced atmosphere was favored for many years by biologists interested in the origin of life (e.g., Oparin 1938) and by geochemists who performed early origin-of-life experiments (Miller 1953, 1955). (Harold

Urey's name is usually attached to Miller's experiments also, as Miller was his graduate student.) Some authors continue to support this idea (Shaw 2008, 2014), but theory suggests that Earth's mantle was oxidized relatively early in its history (Wade and Wood 2005) and that the dominant volcanic gases were H_2O, CO_2, and N_2 (Rubey 1951; Walker 1977; Holland 1984; Kasting 1993; Catling and Kasting 2017, Ch. 7). The early atmosphere may have been formed by impact degassing of incoming planetesimals (Zahnle et al. 2007, 2010), but this does not change the basic conclusion. Strongly reduced atmospheres on rocky planets turn into weakly reduced atmospheres within geologically short time periods as a consequence of photolysis of H_2O, followed by escape of hydrogen to space. Strongly reduced atmospheres can persist on gas giant planets such as Jupiter and Saturn, because hydrogen does not escape and because photolysis of CH_4, NH_3, H_2S, and H_2O is balanced by reformation of these molecules deep within the planet's atmosphere.

The composition of weakly reduced atmospheres can be estimated by using photochemical models. A typical result is shown in Figure 4.5.1.

The atmosphere depicted has a surface pressure of 1 bar, like today's atmosphere. The specification of a 1-bar surface pressure is arbitrary; the real Hadean atmosphere may have had a pressure of several bars or more as a result of release of CO_2 during impacts (Kasting 1990; Zahnle et al. 2007, 2010; Wong et al. 2017), combined with the smaller estimated surface area of continents on which silicate weathering could occur (Flament et al. 2008). But CO_2 should have been removed by seafloor weathering, as well, at rates that are difficult to estimate but that could have been relatively fast (Sleep and Zahnle 2001; Coogan and Gillis 2013; Coogan and Dosso 2015). By the late Archean (2.7 Ga), atmospheric pressure appears to have been relatively low: Som et al. (2012) estimated a surface pressure <2 bar based on fossil raindrop imprints, whereas this same group has

estimated <0.5 bar based on gas vesicles in submarine basalts (Som et al. 2016). It remains to be seen whether these empirical estimates of surface pressure are robust.

The H_2 mixing ratio in the atmosphere depicted in Figure 4.5.1 was determined by balancing the outgassing rates of H_2 and other reduced gases with escape of hydrogen to space, following the methodology outlined in Chapter 8 of (Catling and Kasting 2017). A total H_2 outgassing flux of 1.7×10^{12} mol/year was assumed (ibid., Table 7.3), and the escape rate was set by the diffusion-limited flux (ibid, eqs. 8.6 and 8.7). Global redox balance was ensured by making sure that the net transfer of reducing power from the atmosphere to the ocean was zero.

4.5.3 CLIMATE ON THE EARLY EARTH

If the composition of the Hadean atmosphere is uncertain, then the climate is even more so. In a predominantly CO_2-N_2 atmosphere, the surface temperature depends most strongly on the CO_2 partial pressure, along with solar luminosity. Solar luminosity can be estimated from the formula (Gough 1981):

$$\frac{S}{S_0} = \frac{1}{1 + 0.4\left(t/4.6\right)} \quad (4.5.1)$$

Here, S is the solar flux at Earth's orbit, S_0 is the modern value, ~1365 W/m^2, and t is time before present in Ga. Although dated, Gough's formula agrees well with more recent models (Bahcall et al. 2001). Equation 4.5.1 predicts that the Sun was about 80% of its present luminosity at 2.8 Ga. If one combines that with the 0.2-bar CO_2 atmosphere shown in Figure 4.5.1, this would produce a mean surface temperature of ~0°C (Kasting, 2013). Prior to 2.8 Ga, significantly higher CO_2 partial pressures would have been required to keep Earth's surface from freezing. Even 10 bar of CO_2 would only have produced a mean surface temperature of ~85°C at 4.6 Ga, when the Sun was ~70% of its present brightness (Kasting and Ackerman 1986). So, the Hadean surface temperature could have been either quite warm or quite cold, depending on how fast CO_2 was removed from the atmosphere by seafloor weathering.

Once we get into the Archean, the climate record improves, but interpretations remain controversial. Oxygen isotopes in cherts suggest that the climate was extremely warm, 70 ± 15°C at 3.3 Ga (Knauth and Lowe 2003; Knauth 2005). But this result is in apparent conflict with evidence for glaciation at 3.5 Ga (De Wit and Furnes 2016), 2.9 Ga (Young et al. 1998), and 2.7 Ga (Ojakangas et al. 2014). The glacial evidence is more convincing, as the oxygen isotope record could be influenced by changes in seawater oxygen isotopic composition over time (Jaffres et al. 2007; Veizer and Prokoph 2015). (For counter-arguments, see Gregory and Taylor 1981 and Muehlenbachs 1998.) If we accept that continental glaciers were present at least some of the time during the Archean, then it seems likely that mean surface temperatures were within 10 or 15 degrees of the present value, 15°C.

The Archean geologic record also provides some direct constraints on atmospheric CO_2 by way of paleosols (ancient soils).

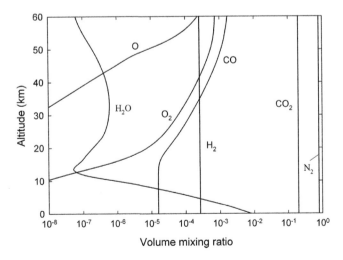

FIGURE 4.5.1 Vertical profiles of major atmospheric species for a typical early Archean model atmosphere. A surface pressure of 1 bar was assumed. Other parameters for the calculation are described in the text. (Reprinted from Kasting, J.F., *Geol. Soc. Am. Spec. Pap.*, 504, 19-28, 2014. With permission of the Geological Society of America.)

Soils are usually not preserved, because they erode away on relatively short timescales. But occasionally, a soil will become covered by a lava flow from a nearby volcano, preventing it from eroding. If this sequence is later buried, it can be preserved in the geologic record. One can then analyze the amount of silicate weathering that occurred within the soil and use this to estimate atmospheric CO_2. This type of analysis has been done for one Archean paleosol (Driese et al. 2011) and for several younger Proterozoic paleosols (Sheldon 2006) (see Figure 4.5.2).

The Driese et al. result suggests 10-50 PAL of CO_2 at 2.7 Ga. Here, 1 PAL is defined at 370 ppmv CO_2, so this prediction corresponds to a CO_2 partial pressure of about 0.004–0.02 bar. This is a factor of 10-50 lower than the 0.2 bar of CO_2 required to keep the Earth from freezing at 2.8 Ga. If correct, this would imply that other greenhouse gases, e.g., CH_4, would have been needed to keep the surface temperature above freezing at this time. That is not impossible. Several years ago, we showed that 1000 ppmv of CH_4 combined with 0.02 bar of CO_2 could have produced a mean surface temperature of ~286 K, not far from today's value of 288 K (Haqq-Misra et al. 2008). Such a CH_4 concentration could plausibly have been produced by methanogens living within an anaerobic marine biosphere (Kharecha et al. 2005). (See further discussion of this topic in Section 4.5.4.1 below.) So, relatively low atmospheric CO_2 concentrations are not necessarily inconsistent with the Archean climate record.

On the other hand, a more recent reanalysis of this same suite of paleosols by (Kanzaki and Murakami 2015) predicts Archean and Proterozoic CO_2 levels 10–20 times higher than those estimated by (Driese et al. 2011) and (Sheldon 2006), both of whom used a similar methodology to analyze their samples. This methodology, developed originally by (Holland and Zbinden 1988), essentially assumes that every CO_2

molecule that enters the soil reacts with a silicate mineral. It thus really produces only a lower bound on atmospheric CO_2. If the efficiency of silicate weathering reactions is less than assumed, then higher pCO_2 values are allowed. So, it seems likely that the Sheldon's and Driese's estimates of pCO_2 are too low and that those of (Kanzaki and Murakami 2015), which themselves have large error bars, may be more accurate. In either case, the problem of countering reduced solar luminosity during the Archean appears to have viable solutions.

4.5.4 REDOX COUPLES FOR ORIGINATING AND SUSTAINING LIFE

As mentioned at the beginning of this chapter, one of the main reasons that we care about the nature of the early atmosphere is that it helped to establish free-energy gradients that could have been critical to originating and then sustaining life. Let us briefly consider how such gradients fit into the broader picture of life's origin.

Theories of the origin of life are sometimes divided into two classes: (1) information (or genes) first, and (2) metabolism first. Researchers pursuing the first class of hypotheses stress that nothing can be considered alive unless it possesses the ability to pass on its characteristics to its offspring. Modern organisms do this by using DNA as the information molecule. But RNA is similarly able to encode information, and because it is single-stranded, it can perform catalysis as well (Cech and Bass 1986). Thus, most researchers agree that once RNA had evolved, with its four nucleobases for storing information, life would have been off to a good start.

Another view of the origin of life holds that metabolism is the key concept. Organisms depend on sources of thermodynamic free energy to drive their metabolisms. Some proponents of this view (e.g., Smith and Morowitz 2016) argue that life *arises* in response to free-energy gradients in the environment. This concept plays a major role in the hydrothermal vent model for life's origin (Russell and Hall 1997; Martin et al. 2008; Lane et al. 2010), as strong free-energy gradients are arguably present in such environments. These gradients involve both redox gradients that drive transfer of electrons and pH gradients that drive transfer of protons. Both types of gradients play important roles in modern biochemistry. In the hydrothermal vent model, pH gradients between alkaline vent fluids and slightly acidic seawater constitute the most likely initial source of free energy for life. To move out of the vents, though, organisms would have needed to take advantage of redox gradients to obtain their free energy. In this chapter, I shall focus on these gradients.

Importantly, redox gradients themselves are constrained in their biological use by two different factors (Hoehler 2004 and references therein). The first is the available thermodynamic free energy. Hoehler terms this the *biological energy quantum* (BEQ). This constraint comes from the fact that the energy needed to synthesize adenosine triphosphate (ATP) from adenosine diphosphate (ADP) is 35.6 kJ/mol. ATP is the universal energy currency of metabolism, and so, all organisms need to be able to make it. Organisms are able to store electrochemical energy internally and then combine

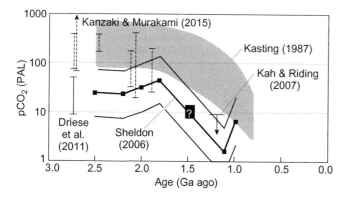

FIGURE 4.5.2 CO_2 estimates from paleosols compared with those from the climate model of (Kasting 1987). (Modified from Sheldon, N.D., *Precambrian Res.*, 147, 148-155, 2006.) Sheldon's estimates (with error bars) are shown by the dark squares connected by the solid black line. The downward arrow at 1.2 Ga is the upper limit on pCO_2 from cyanobacterial sheath calcification (Kah and Riding 2007). The bar in the middle left is the paleosol estimate at 2.7 Ga from (Driese et al. 2011). Bars in the upper left are paleosol estimates from (Kanzaki and Murakami 2015). (Reprinted from Catling, D. and Kasting, J.F., *Atmospheric Evolution on Inhabited and Lifeless Worlds*, Cambridge, UK, Cambridge University Press, 2017. With permission.)

n different quanta together to produce the required total yield to make ATP; however, in practice, n is a small number. Thus, the minimum Gibbs free energy required to power actively growing cultures is about 20 kJ/mol, whereas static or starving cultures can subsist on 12–15 kJ/mol. By analogy with electrical systems, BEQ can be associated with the *voltage* of a system. Indeed, this connection is direct, because voltage differentials, E^0, and free-energy changes, ΔG^0, are related through the Nernst equation, $E^0 = -\Delta G^0/nF$, where n is the number of moles and F is Faraday's constant, 9.6485×10^4 C/mol.

Hoehler also defines a *maintenance energy* (ME) requirement for different microbial ecosystems (Hoehler 2004). This is a kinetic limitation, rather than a thermodynamic one: If critical metabolic substrates (e.g., H_2 for methanogenesis) are not available in sufficient quantity, then the rate of metabolism in an organism/culture will not be adequate to allow that culture to survive or grow. By analogy to electrical systems, ME is like *current*. Maintenance energy is more difficult to calculate than BEQ, because it depends on the size of the organisms and the nature of the metabolic reaction. But it has been estimated for a few anaerobic metabolisms, including methanogenesis, and I will use these values in the remainder of this subsection. Importantly, the relative values of ME for growth, maintenance, and survival are 1000, 1, and 0.001–0.01, respectively, according to Hoehler. The ME for growth of laboratory microbial cultures is surprisingly high, ~127 J/g dry biomass/h, as compared with a typical human energy budget of 12.5 J/g dry biomass/h. But microbial populations in natural environments are able to persist at ME values several orders of magnitude lower, so it may be the maintenance ME threshold that is most relevant for the present analysis.

4.5.4.1 H_2-Based Methanogenesis

Let us now apply these concepts to specific metabolic reactions that have been proposed for early life. One reaction that has been recognized by many authors is H_2-based methanogenesis:

$$CO_2 + 4H_2 \rightarrow CH_4 + 2H_2O(\ell) \qquad (4.5.2)$$

Virtually, all modern methanogens are capable of performing this reaction. Phylogenetic sequencing suggests that H_2-using methanogens evolved sometime during the Archean (House et al. 2003; Battistuzzi et al. 2004), if not earlier. This reaction is strongly exergonic under standard conditions; it has $\Delta G_R^0 = -130.5$ kJ/mol at 298 K. (Thermodynamic data are taken from the National Institute of Standards (NIST) Chemistry Webbook on-line. I use gas phase concentrations for convenience [except for water], even though the reaction is assumed to occur in aqueous solution. This is allowed, because at Henry's Law equilibrium, the free energy of a gas and its dissolved counterpart are the same.)

The actual amount of free energy available to organisms from reaction (4.5.2) depends on the concentrations of the reactants and the products. The free energy of the reaction, ΔG_R, is related to the standard free-energy change by:

$$\Delta G_R = \Delta G_R^0 + RT \ln Q \qquad (4.5.3)$$

Here, R is the universal gas constant, 0.008314 kJ/mol, T is temperature in Kelvin, and Q is the reaction quotient, which for this reaction is given by $Q = pCH_4/(pCO_2 \cdot pH_2^4)$. The activity of liquid water is assumed to be unity.

To calculate the energy available from this reaction, we need to estimate the amount of methane in the early atmosphere. That, in turn, depends on its source. The present CH_4 flux, which is mostly biological, is estimated to be 535 Tg(CH_4)/year (Houghton et al. 1994) or 3.3×10^{13} mol/year. This is accompanied by a much smaller abiotic CH_4 flux produced by serpentinization reactions deep within submarine hydrothermal systems (McDermott et al., 2015). By ratioing to the better-known flux of H_2 produced by these systems, Catling and Kasting (2017, p. 211) estimated an abiotic CH_4 flux of 3×10^{10} mol/year, about 1000 times smaller than the biological flux. Converting to photochemists' units (10^{12} mol/year = 3.74×10^9 molecules cm^{-2}s^{-1}), the abiotic CH_4 flux is ~1×10^8 cm^{-2}s^{-1}. Putting this flux into a photochemical model similar to that used to produce Figure 4.5.1 gives an atmospheric CH_4 mixing ratio of 5×10^{-7}, or 0.5 ppmv (Catling and Kasting 2017, Figure 9.9). I shall use that value for the analysis.

Now, let us calculate the free-energy yield from reaction (4.5.2). For a case that might be representative of the late Archean atmosphere, we can use the values in Figure 4.5.1: $pCO_2 = 0.2$ atm and $pH_2 = 2.6 \times 10^{-4}$ atm. Evaluating at 298 K, we have $Q = 5.5 \times 10^8$, $RT \ln Q \cong 50$ kJ/mol, and hence, $\Delta G_R \cong -80$ kJ/mol. This is almost equal to the ~97 kJ/mol estimated to be available for methanogenesis within modern submarine hydrothermal vent plumes (McCollom 2000). Methanogens thrive within these modern plumes, and so, from a BEQ standpoint, the global Archean surface ocean should have readily been able to sustain methanogens. Implementation of this BEQ limit in a model led Kharecha et al. (2005) to predict that a methanogen-based ecosystem (their Case 1) or a methanogen- and acetogen-based system (their Case 2) could have produced a methane flux comparable to the present-day flux.

From an ME's standpoint, the prospects for widespread methanogenesis in the Archean oceans is less clear. Hoehler (2004) argues that a minimum internal dissolved H_2 concentration for methanogens is ~10^{-8} M. For 0.5-um diameter cells, this requires $[H_2] = 3 \times 10^{-8}$ M externally for maintenance and 1000 times that level for growth. The value of pH_2 from Figure 4.5.1 is 2.6×10^{-4} atm, and the solubility of H_2 is ~7.8×10^{-4} M/atm (Kharecha et al. 2005), so the ambient dissolved H_2 concentration would be 2×10^{-7} M. This exceeds the maintenance ME requirement by a factor of 7 but is more than two orders of magnitude lower than the corresponding growth requirement. In the Case 1 model of Kharecha et al., the methanogens draw down atmospheric H_2 to as low as 40 ppmv (4×10^{-5} atm), which would put ambient $[H_2]$ right at the maintenance ME value. Given that these requirements are not very precise, it is not obvious what this implies about the viability of a global, methanogen-based biosphere. But it may be that the CH_4 production rates estimated by Kharecha et al. for their Case 1 and Case 2 biospheres are somewhat overestimated. By contrast, their Case 3 biosphere, which assumed that primary production was dominated by anaerobic, H_2-utilizing phototrophs, should be largely unaffected by either the BEQ or ME constraint. So, if anoxygenic photosynthesis originated early, as seems likely, then the prediction of high methane levels in the Archean atmosphere should remain valid.

Note that the ME limitation for methanogenesis is much less severe in modern hydrothermal vents. There, the concentration of dissolved H_2 ranges from ~2 mM in hot, axial vent fluids (von Damm 1990, 1995) to ~15 mM in cooler, alkaline off-axis vents (Cannat et al. 2010). Even if mixing with ambient seawater dilutes these concentrations by a factor of 70 (for axial vent fluids) or 500 (for off-axis fluids), $[H_2]$ remains above the growth threshold. Thus, it is easy to see why methanogens thrive in these environments.

Reaction (4.5.2) has also been proposed as a possible driver for the origin of life (Lane et al. 2010). Or, more accurately, these authors have proposed that this reaction allowed organisms to escape their dependence on the pH gradients in the hydrothermal vents and thereby to spread out to the global environment. But this suggestion seems unlikely for at least two reasons: First, reaction (4.5.2) involves a transfer of eight electrons, which may not have been possible until sophisticated biochemical machinery evolved to perform this task. And, second, Nitschke and Russell (2013) argue that neither methanogenesis nor its cousin, acetogenesis, appears to be particularly ancient, based on the observation that the C_1 branch of the Wood-Ljungdahl (WL) pathway is not conserved between acetogens and methanogens. The WL pathway produces acetyl-coenzyme A (CoA) in these and other organisms and is considered primitive because it starts from CO_2 and H_2. By contrast, the part of the WL pathway that involves CO *is* conserved between these two groups (*ibid.*), suggesting that CO metabolism may be quite ancient. But CO is arguably more available in the surface environment than in vents. This leads naturally to the next section of this chapter.

4.5.4.2 The Water-Gas Shift Reaction

A second chemical reaction that could have been used metabolically over much of the early Earth's surface environment is the reaction of CO with H_2O to produce CO_2 and H_2:

$$CO + H_2O\,(\ell) \rightarrow CO_2 + H_2 \qquad (4.5.4)$$

This reaction, which is used industrially in the manufacture of ammonia, hydrocarbons, methanol, and hydrogen, is sometimes referred to as the "water-gas shift" reaction.

I have analyzed this reaction previously (Kasting 2014), so the present discussion can be brief. Under standard conditions, this reaction has a Gibbs free-energy change, ΔG_R^0, of −20.1 kJ/mol.

For a case that might be representative of the late Archean atmosphere, we can use the gas concentrations shown in Figure 4.5.1: $pCO_2 = 0.2$ atm, $pCO = 1.6 \times 10^{-5}$ atm, and $pH_2 = 2.6 \times 10^{-4}$ atm. Evaluating at 298 K, we get $Q = 3.25$ and $\Delta G_R \cong -17$ kJ/mol. While less than the 20 kJ/mol required for active growth, this is still well above BEQ requirement of 12–15 kJ/mol for static ecosystems.

From the standpoint of ME, the reaction is less favorable. The minimum substrate concentration is proportional to $(D\Delta G_R)^{-1}$, where D is the diffusivity of the substrate (CO in this case) and ΔG_R is energy of the reaction (Hoehler 2004). The free energy for reaction (4.5.4) is lower than that for reaction (4.5.2) by a factor of (17/80), and the diffusivity of CO is lower than that of H_2 by a factor of (1.9/5) (Kharecha et al. 2005, Table 1); thus, the minimum substrate concentration is higher by a factor of 12. For H_2, the ME maintenance limit was 3×10^{-8} M, so for CO, the corresponding limit should be ~4.6×10^{-7} M. With a CO solubility of 1×10^{-3} mol/L/atm (*ibid.*), the dissolved CO concentration corresponding to Figure 4.5.1 is only 1.6×10^{-8} M. This is a factor of 30 smaller than the ME limit. This suggests that CO abundances equal to those shown in Figure 4.5.1 could *not* have sustained a global biosphere.

Before giving up on CO, though, let us consider why the composition of the early atmosphere should favor this reaction and why we may have underestimated it in the above analysis. The reason that this free-energy gradient should have existed is three-fold: First, CO would have been produced in copious quantities in the early stratosphere by photolysis of CO_2: $CO_2 + h\nu \rightarrow CO + O$. Second, the direct recombination of CO with O is spin-forbidden and slow; hence, CO would have flowed down into the troposphere, where it could recombine with OH produced by water vapor photolysis: $CO_2 + OH \rightarrow CO_2 + H$. And third, H_2 was always escaping to space; hence, its atmospheric concentration was limited by the balance between escape and outgassing from volcanoes. All of these factors tend to drive reaction (4.5.4) to the right.

With this in mind, let us also remember that the free-energy yield estimated here is appropriate for some time during the late Archean. The amount of free energy available during the Hadean could have been significantly higher, particularly if large amounts of CO were generated by impacts (Kasting 1990). Indeed, the early atmosphere could have been CO-dominated if the CO production rate from impacts exceeded the rate at which H_2O was photolyzed to produce OH (*ibid.*). Furthermore, reaction (4.5.4) involves a transfer of only two electrons and hence should be biochemically simpler than reaction (4.5.2). As mentioned in the previous section, the branch of the WL pathway that involves CO is considered to be evolutionarily ancient (Nitschke and Russell 2013). Others have also identified this reaction as being of possible early biological importance (Huber and Wachtershauser 1997, 1998). So, even if this reaction was not a viable source of energy for an Archean global biosphere, it might well have been important during the earliest stages of biological evolution, particularly if life originated within the surface environment, as opposed to a midocean ridge hydrothermal vent system. The jury is still out on that important question. We may learn the answer to this question from studies of Europa and Enceladus, both of which are uninhabitable at their surfaces but are expected to have hydrothermal vent systems similar to those on Earth.

4.5.4.3 Nitrate as an Energy Source for Powering Metabolism

A third source of thermodynamic free energy on the early Earth would have been reactions between nitrogen oxides produced photochemically within the atmosphere and reductants such as H_2, Fe^{+2}, and CH_4 (Nitschke and Russel 2013; Wong et al. 2017). Nitschke and Russell argue that the reaction of NO with CH_4, a form of *methanotrophy*, may have been one of the earliest bioenergetic pathways.

We can get some idea of the energetics of these reactions by looking at a representative reaction in which nitrate reacts with molecular hydrogen to form N_2:

$$NO_3^- + 4H^+ + 3e^- + H_2 \rightarrow \frac{1}{2}N_2 + 3H_2O \quad (4.5.5)$$

This reaction is not, to my knowledge, performed by any individual organism; however, modern consortia of microorganisms accomplish essentially this same thing. Nitrate would have been produced in the primitive atmosphere by a sequence of reactions initiated by lightning. In the high-temperature core of a lightning bolt, N_2 would have reacted with CO_2 to produce nitric oxide:

$$CO_2 + \frac{1}{2}N_2 \leftrightarrow CO + NO \quad (4.5.6)$$

The rate at which NO was produced can be estimated by scaling to the modern atmosphere, in which N_2 reacts with O_2 to produce NO under these circumstances. The modern NO production rate by lightning, in photochemists' units, is about 6×10^8 $cm^{-2}s^{-1}$ (Wong et al. 2017). These authors estimate prebiotic NO production rates of $7 \times 10^6 – 2 \times 10^9$ $cm^{-2}s^{-1}$ by using a three-dimensional general circulation model coupled to a photochemical model. By comparison, Kasting (1990) estimated an NO production rate of 5×10^8 $cm^{-2}s^{-1}$ for a 0.2-bar CO_2 atmosphere and 2×10^9 $cm^{-2}s^{-1}$ for a 2-bar CO_2 atmosphere. Let us use the lower of these two values to remain consistent with the atmosphere shown in Figure 4.5.1. A rate of 5×10^8 $cm^{-2}s^{-1}$ is equivalent to ~1.3×10^{11} mol/year, or 1.9 Tg(N)/year.

The next step in forming nitrate is to react NO with H to produce nitroxyl, or HNO. HNO then dissolves in rainwater and goes into the ocean, where it disproportionates to form nitrate, nitrite, and N_2O (Mancinelli and McKay 1988). Assume for simplicity that it all forms nitrate. Nitrate will then build up in the ocean, until it encounters a sink that balances production. Wong et al. (2017) assume that the ultimate sink for nitrate on the prebiotic Earth would be cycling through the hot, axial hydrothermal vents at the midocean ridges. (A similar assumption was made by Kasting et al. [1993, Sect. 5*iv*]. We reevaluate this assumption below.) According

to these authors, the rate of seawater circulation through the axial vents is 7.2×10^{12} kg/year. The mass of the oceans is 1.4×10^{21} kg, so the entire ocean circulates through the vents in about 200 Myr. The loss rate of nitrate is equal to the flow rate through the vents, multiplied by the nitrate concentration in seawater. Balancing this loss with the production rate from the previous paragraph yields a dissolved nitrate concentration of ~20 mM. Kasting et al. (1993) assumed a faster flow rate through the ridges, corresponding to a 10-Myr lifetime and predicted about 1 mM of nitrate in the oceans.

Now, let us return to reaction (4.5.5), the reaction of nitrate with H_2. Using electrochemical data (www2.ucdsb.on.ca/tiss/stretton/database/Standard_reduction_potentials), I calculate $\Delta G_R^0 \cong -92.6$ kJ/mol for the reduction of NO_3^- to gas-phase NO. Then, reduction of NO to N_2 yields an additional -323.8 kJ/mol, giving a ΔG_R^0 for reaction (4.5.5) of -416.4 kJ/mol. This is an enormous number. To calculate the reaction quotient for reaction (4.5.5), we need to put everything in units of mol/l. Assuming an ocean pH of 7 (because of the high pCO_2) and using Henry's Law constants of 7.8×10^{-4} mol/L/atm for H_2 and 7×10^{-3} mol/L/atm for N_2, I get $Q = 2.2 \times 10^{10}$, $RT \ln Q = 59$ kJ/mol, and $\Delta G_R \cong -357$ kJ/mol. This is still an enormous number. If correct, this would imply that nitrate was an extremely potent oxidant on the early Earth.

There is a reason, however, to doubt these numbers. Dissolved ferrous iron should have oxidized spontaneously in the presence of solar ultraviolet (UV) radiation on the early Earth (Braterman et al. 1983), releasing H_2. If nitrate (or nitrite) was present, then it seems likely that it would have been used as the oxidant in place of H_2O. As nitrate was being deposited directly into the surface ocean, its actual lifetime may have been only a few days. Assume 30 days to be generous. In that case, the nitrate concentration would be reduced by a factor of 0.1 year/(2×10^8 year) = 5×10^{-10}. So, instead of 20 mM nitrate, one would have 0.01 nM nitrate. Q would increase by a factor of 1/(5×10^{-10}) = 2×10^9, $RT \ln Q$ would increase by 53 kJ/mol, and ΔG_R would be -304 kJ/mol. This number remains enormous. The BEQ constraint for microbial metabolism should easily be satisfied by reaction (4.5.5) or any number of related reactions between nitrogen oxides and available reductants.

But now, consider this reaction from the ME standpoint. Take -320 kJ/mol as an estimate for ΔG_R. This is four times higher than the corresponding value for methanogenesis. But the diffusivity of nitrate, 1.7×10^{-5} cm^2/s (http://bionumbers.hms.harvard.edu), is a factor of three lower than that for H_2, so to first order, the minimum nitrate concentration required for maintenance should be the same as that for H_2, 3×10^{-8} M. This constraint would be easily satisfied if nitrate were as abundant as estimated by (Wong et al. 2017). But if the much lower concentration calculated above is correct, then the average concentration of nitrate in the oceans would be less than the ME requirement by a factor of 3000. Thus, given a short lifetime, nitrate could not have sustained a global biosphere. The question of whether the lifetime of nitrate was short or long could readily be answered by irradiating water containing Fe^{+2} and NO_3^- with UV light, following (Braterman et al. 1983), and seeing whether nitrate is consumed in the process.

This same analysis suggests that nitrogen oxides might *not* have made a good redox partner for CH_4 within hydrothermal vents, as suggested by (Nitschke and Russell 2013). These compounds might have been abundant in the surface ocean, following thunderstorms, but they would not have been around long enough to make it into the deep ocean and circulate through submarine vents. The lifetime of nitrate should be reevaluated as part of pursuing this hypothesis.

4.5.4.4 OTHER POTENTIAL OXIDANTS ON THE EARLY EARTH

Finally, nitrogen oxides would not have been the only oxidants produced within a weakly reduced primitive atmosphere. Sulfate and hydrogen peroxide would have been generated photochemically (Pavlov and Kasting 2002; Segura et al. 2003), and O_2 would have been produced by photochemistry and lightning (Haqq-Misra et al. 2011). Sulfate is a less potent oxidant than nitrate, and I shall simply neglect it here. O_2 is a powerful oxidant, though, so let us briefly consider it. The reaction

$$H_2 + \frac{1}{2} O_2 \rightarrow H_2O(\ell) \qquad (4.5.7)$$

has a standard free energy of -237 kJ/mol. Surface concentrations of O_2, $\sim10^{-13}$ atm, are too low to be seen in Figure 4.5.1. However, as with nitrate, localized O_2 concentrations could have been much higher than this in the immediate aftermath of thunderstorms. And downward transport of O_2 from the stratosphere in the polar night region could have produced a surface O_2 concentration of $\sim3 \times 10^{-6}$ atm (*ibid.*). Using this concentration, along with $pH_2 = 2.6 \times 10^{-4}$ atm from Figure 4.5.1, gives $Q = 1/(pH_2 \cdot pO_2^{0.5}) = 1.3 \times 10^9$, $RT \ln Q = 52$ kJ/mol, and $\Delta G_R \cong -185$ kJ/mol. As with nitrate, this reaction is extremely favorable from a BEQ standpoint.

Now, calculate the ME limit. For a solubility of 2.6×10^{-4} mol/L/atm and $pO_2 = 3 \times 10^{-6}$ atm, the dissolved O_2 concentration is 8×10^{-10} M. The diffusivity of O_2 is about 2.5 times lower than that of H_2, but the free-energy change for reaction (4.5.7) is 2.5 times higher than that for methanogenesis; thus, the critical $[O_2]$ for maintenance should be like that of H_2, 3×10^{-8} M. Thus, the ambient dissolved O_2 concentration is ~40 times below the ME limit. Like nitrate, O_2 would not have been able to sustain a global biosphere. And, like nitrate, O_2 would probably have been consumed within the surface ocean during UV oxidation of ferrous iron, so it would be unlikely to have been present in appreciable concentrations in the deep ocean. Thus, O_2 is also not a probably redox partner for reactions important to the origin or early evolution of life.

4.5.5 CONCLUSIONS

Earth's Archean atmosphere was probably a weakly reduced mixture of CO_2 and N_2, with smaller amounts of H_2, CO, and CH_4. The Hadean atmosphere was similar except that it was richer in CO_2 (and maybe CO) and poorer in CH_4.

Climate during the Archean was probably similar to that of more recent times, with periods of glaciation interspersed with periods of warmth. The faintness of the young Sun was offset by the increased greenhouse effect from CO_2 and CH_4. Climate during the Hadean is essentially unknown. Theory suggests that a multi-bar CO_2 atmosphere produced warm surface temperatures early on but that drawdown of CO_2 by seafloor weathering cooled the Earth to near-modern temperatures by ~3.5 Ga.

Earth's atmosphere was always more oxidized than gases coming out of Earth's interior because of photolysis of H_2O, followed by escape of hydrogen to space. The redox gradients set up in this manner could have played an important role in originating and sustaining life. Of the reactions discussed here, the water-gas shift reaction, in which CO combines with H_2O to produce CO_2 and H_2, shows the most promise of being relevant to the origin of life, particularly if life originated in the surface environment, as opposed to hydrothermal vents. Methanogenesis was probably a somewhat later invention (though still early in biological evolution), and it may have played a major role in producing the methane that helped to keep the Archean climate warm. Although strong oxidants such as nitrate and O_2 were produced by photochemistry and lightning within the Archean and Hadean atmospheres, these species were likely short-lived and are unlikely to have played a major role in either originating or sustaining life. Laboratory experiments to simulate UV radiation impinging on a Fe^{+2}-rich early ocean are needed to determine whether this last proposition is correct.

REFERENCES

Allwood, A. C., J. P. Grotzinger, A. H. Knoll, I. W. Burch, M. S. Anderson, M. L. Coleman, and I. Kanik. 2009. Controls on development and diversity of Early Archean stromatolites. *Proceedings of the National Academy of Sciences of the United States of America* 106 (24): 9548–9555.

Allwood, A. C., M. R. Walter, B. S. Kamber, C. P. Marshall, and I. W. Burch. 2006. Stromatolite reef from the early Archaean era of Australia. *Nature* 441 (7094): 714–718.

Babikov, D. 2017. Recombination reactions as a possible mechanism of mass-independent fractionation of sulfur isotopes in the Archean atmosphere of Earth. *Proceedings of the National Academy of Sciences of the United States of America* 114 (12): 3062–3067.

Bahcall, J. N., M. H. Pinsonneault, and S. Basu. 2001. Solar models: Current epoch and time dependences, neutrinos, and helioseismological properties. *Astrophysical Journal* 555 (2): 990–1012.

Battistuzzi, F. U., A. Feijao, and S. B. Hedges. 2004. A genomic timescale of prokaryote evolution: Insights into the origin of methanogenesis, phototrophy, and the colonization of land. *BMC Evolutionary Biology* 4: 44.

Braterman, P. S., A. G. Cairns-Smith, and R. W. Sloper. 1983. Photooxidation of hydrated Fe^{+2} - Significance for banded iron formations. *Nature* 303: 163–164.

Buick, R., J. S. R. Dunlop, and D. I. Groves. 1981. Stromatolite recognition in ancient rocks - An appraisal of irregularly laminated structures in an early Archean chert-barite unit from North-Pole, Western-Australia. *Alcheringa* 5 (3-4): 161–181.

Cannat, M., F. Fontaine, and J. Escartin. 2010. Serpentinization and associated hydrogen and methane fluxes at slow spreading ridges. In *Diversity of Hydrothermal Systems on Slow Spreading Ocean Ridges*, Edited by P. A. Rona, C. W. Devey, J. Dyment, and B. J. Murton.

Catling, D., and J. F. Kasting. 2017. *Atmospheric Evolution on Inhabited and Lifeless Worlds*. Cambridge, UK: Cambridge University Press.

Cech, T. R., and B. L. Bass. 1986. Biological catalysis by RNA. *Annual Review of Biochemistry* 55: 599–629.

Cloud, P. E. 1972. A working model of the primitive Earth. *American Journal of Science* 272: 537–548.

Coogan, L. A., and S. E. Dosso. 2015. Alteration of ocean crust provides a strong temperature dependent feedback on the geological carbon cycle and is a primary driver of the Sr-isotopic composition of seawater. *Earth and Planetary Science Letters* 415: 38–46.

Coogan, L. A., and K. M. Gillis. 2013. Evidence that low-temperature oceanic hydrothermal systems play an important role in the silicate-carbonate weathering cycle and long-term climate regulation. *Geochemistry, Geophysics, Geosystems* 14 (6): 1771–1786.

de Wit, M. J., and H. Furnes. 2016. 3.5-Ga hydrothermal fields and diamictites in the Barberton Greenstone Belt-Paleoarchean crust in cold environments. *Science Advances* 2 (2): e1500368.

Driese, S. G., M. A. Jirsa, M. Ren, S. L. Brantley, N. D. Sheldon, D. Parker, and M. Schmitz. 2011. Neoarchean paleoweathering of tonalite and metabasalt: Implications for reconstructions of 2.69 Ga early terrestrial ecosystems and paleoatmospheric chemistry. *Precambrian Research* 189: 1–17.

Farquhar, J., H. Bao, and M. Thiemans. 2000. Atmospheric influence of Earth's earliest sulfur cycle. *Science* 289: 756–758.

Flament, N., N. Coltice, and P. F. Rey. 2008. A case for late-Archaean continental emergence from thermal evolution models and hypsometry. *Earth and Planetary Science Letters* 275 (3-4): 326–336.

Gough, D. O. 1981. Solar interior structure and luminosity variations. *Solar Physics* 74: 21–34.

Gregory, R. T., and H. P. Taylor. 1981. An oxygen isotope profile in a section of Cretaceous oceanic crust, Samail Ophiolite, Oman: Evidence for d^{18}O buffering of the oceans by deep (>5 km) seawater-hydrothermal circulation at mid-ocean ridges. *Journal of Geophysical Research* 86: 2737–2755.

Haqq-Misra, J., J. F. Kasting, and S. Lee. 2011. Availability of O_2 and H_2O_2 on pre-photosynthetic Earth. *Astrobiology* 11 (4): 293–302.

Haqq-Misra, J. D., S. D. Domagal-Goldman, P. J. Kasting, and J. F. Kasting. 2008. A revised, hazy methane greenhouse for the early Earth. *Astrobiology* 8: 1127–1137.

Harman, C. E., A. A. Pavlov, D. Babikov, and J. F. Kasting. 2018. Chain formation as a mechanism for mass-Independent fractionation of sulfur Isotopes in the Archean atmosphere. *Earth and Planetary Science Letters* 496: 238–247.

Hoehler, T. M. 2004. Biological energy requirements as quantitative boundary conditions for life in the subsurface. *Geobiology* 2 (4): 205–215.

Hofmann, H. J., K. Grey, A. H. Hickman, and R. I. Thorpe. 1999. Origin of 3.45 Ga coniform stromatolites in Warrawoona Group, Western Australia. *Geological Society of America Bulletin* 111 (8): 1256–1262.

Holland, H. D. 1984. *The Chemical Evolution of the Atmosphere and Oceans*. Princeton, NJ: Princeton University Press.

Holland, Heinrich D. 2006. The oxygenation of the atmosphere and oceans. *Philosophical Transactions of the Royal Society B - Biological Sciences* 361 (1470): 903–915.

Holland, H. D., and E. A. Zbinden. 1988. Paleosols and the evolution of the atmosphere: Part I. In *Physical and Chemical Weathering in Geochemical Cycles*, Edited by A. Lerman and M. Meybeck. Dordrecht, the Netherlands: Reidel.

Houghton, J. T., and et al. 1994. *Climate Change, 1994: Radiative Forcing of Climate Change and an Evaluation of the IPCC IS92 Emission Scenarios*. Cambridge, UK: Cambridge University Press.

House, C. H., B. Runnegar, and S. T. Fitz-Gibbon. 2003. Geobiological analysis using whole genome-based tree building applied to the Bacteria, Archea, and Eukarya. *Geobiology* 1: 15–26.

Huber, C., and G. Wachtershauser. 1997. Activated acetic acid by carbon fixation on (Fe, Ni)S under primordial conditions. *Science* 276 (5310): 245–247.

Huber, C., and G. Wachtershauser. 1998. Peptides by activation of amino acids with CO on (Ni, Fe) surfaces: Implications for the origin of life. *Science* 281: 670–672.

Jaffres, J. B. D., G. A. Shields, and K. Wallmann. 2007. The oxygen isotope evolution of seawater: A critical review of a long-standing controversy and an improved geological water cycle model for the past 3.4 billion years. *Earth-Science Reviews* 83 (1-2): 83–122.

Kah, L. C., and R. Riding. 2007. Mesoproterozoic carbon dioxide levels inferred from calcified cyanobacteria. *Geology* 35 (9): 799–802.

Kanzaki, Y., and T. Murakami. 2015. Estimates of atmospheric CO_2 in the Neoarchean-Paleoproterozoic from paleosols. *Geochimica et Cosmochimica Acta* 159: 190–219.

Kasting, J. F. 1987. Theoretical constraints on oxygen and carbon dioxide concentrations in the Precambrian atmosphere. *Precambrian Research* 34: 205–229.

Kasting, J. F. 1990. Bolide impacts and the oxidation state of carbon in the Earth's early atmosphere. *Origins of Life* 20: 199–231.

Kasting, J. F. 1993. Earth's early atmosphere. *Science* 259: 920–926.

Kasting, J. F. 2013. How was early Earth kept warm? *Science* 339: 44–45.

Kasting, J. F. 2014. Atmospheric composition of Hadean-early Archean Earth: The importance of CO. *Geological Society of America Special Papers* 504: 19–28.

Kasting, J. F., and T. P. Ackerman. 1986. Climatic consequences of very high CO_2 levels in the earth's early atmosphere. *Science* 234: 1383–1385.

Kasting, J. F., D. P. Whitmire, and R. T. Reynolds. 1993. Habitable zones around main sequence stars. *Icarus* 101: 108–128.

Kharecha, P., J. F. Kasting, and J. L. Siefert. 2005. A coupled atmosphere-ecosystem model of the early Archean Earth. *Geobiology* 3: 53–76.

Knauth, L. P. 2005. Temperature and salinity history of the Precambrian ocean: Implications for the course of microbial evolution. *Palaeogeography, Palaeoclimatology, Palaeoecology* 219: 53–69.

Knauth, P., and D. R. Lowe. 2003. High Archean climatic temperature inferred from oxygen isotope geochemistry of cherts in the 3.5 Ga Swaziland Supergroup, South Africa. *GSA Bulletin* 115: 566–580.

Lane, N., J. F. Allen, and W. Martin. 2010. How did LUCA make a living? Chemiosmosis in the origin of life. *Bioessays* 32 (4): 271–280.

Mancinelli, R. L., and C. P. McKay. 1988. The evolution of nitrogen cycling. *Origins of Life* 18: 311–325.

Martin, W., J. Baross, D. Kelley, and M. J. Russell. 2008. Hydrothermal vents and the origin of life. *Nature Reviews Microbiology* 6 (11): 805–814.

McCollom, T. M. 2000. Geochemical constraints on primary productivity in submarine hydrothermal vent plumes. *Deep-Sea Research Part I - Oceanographic Research Papers* 47 (1): 85–101.

McDermott, J. M., J. S. Seewald, C. R. German, and S. P. Sylva. 2015. Pathways for abiotic organic synthesis at submarine hydrothermal fields. *Proceedings of the National Academy of Sciences of the United States of America* 112 (25): 7668–7672.

Miller, S. L. 1953. A production of amino acids under possible primitive Earth conditions. *Science* 117: 528–529.

Miller, S. L. 1955. Production of some organic compounds under possible primitive Earth conditions. *Journal of the American Chemical Society* 77: 2351–2361.

Muehlenbachs, K. 1998. The oxygen isotopic composition of the oceans, sediments, and the seafloor. *Chemical Geology* 145: 263–273.

Nitschke, W., and M. J. Russell. 2013. Beating the acetyl coenzyme A-pathway to the origin of life. *Philosophical Transactions of the Royal Society B - Biological Sciences* 368 (1622): 20120258.

Ojakangas, R. W., R. Srinivasan, V. S. Hegde, S. M. Chandrakant, and S. V. Srikantia. 2014. The Talya Conglomerate: An Archean (similar to 2.7 Ga) Glaciomarine Formation, Western Dharwar Craton, Southern India. *Current Science* 106 (3): 387–396.

Oparin, A. I. 1938. *The Origin of Life*. New York, NY: Macmillan.

Pavlov, A. A., and J. F. Kasting. 2002. Mass-independent fractionation of sulfur isotopes in Archean sediments: Strong evidence for an anoxic Archean atmosphere. *Astrobiology* 2: 27–41.

Powner, M. W., B. Gerland, and J. D. Sutherland. 2009. Synthesis of activated pyrimidine ribonucleotides in prebiotically plausible conditions. *Nature* 459 (7244): 239–242.

Rosing, M. T. 1999. ^{13}C-depleted carbon microparticles in >3700-Ma sea-floor sedimentary rocks from West Greenland. *Science* 283: 674–676.

Rubey, W. W. 1951. Geological history of seawater. An attempt to state the problem. *Geological Society of America Bulletin* 62: 1111–1148.

Russell, M. J., and A. J. Hall. 1997. The emergence of life from iron monosulphide bubbles at a submarine hydrothermal redox and pH front. *Journal of the Geological Society* 154: 377–402.

Segura, A., K. Krelove, J. F. Kasting, D. Sommerlatt, V. Meadows, D. Crisp, M. Cohen, and E. Mlawer. 2003. Ozone concentrations and ultraviolet fluxes on Earth-like planets around other stars. *Astrobiology* 3: 689–708.

Shaw, G. H. 2008. Earth's atmosphere - Hadean to early Proterozoic. *Chemie Der Erde-Geochemistry* 68 (3): 235–264.

Shaw, G. H. 2014. Evidence and arguments for methane and ammonia in Earth's earliest atmosphere and an organic compound-rich early ocean. In *Earth's Early Atmosphere and Surface Environment, Geological Society of American Special Papers* 504: 1–10.

Sheldon, N. D. 2006. Precambrian paleosols and atmospheric CO_2 levels. *Precambrian Research* 147: 148–155.

Sleep, N. H., and K. Zahnle. 2001. Carbon dioxide cycling and implications for climate on ancient Earth. *Journal of Geophysical Research* 106: 1373–1399.

Smith, E., and H. J. Morowitz. 2016. *The Origin and Nature of Life on Earth: The Emergence of the Fourth Geosphere*. Cambridge, UK: Cambridge University Press.

Som, S. M., R. Buick, J. W. Hagadorn, T. S. Blake, J. M. Perreault, J. P. Harnmeijer, and D. C. Catling. 2016. Air pressure 2.7 billion years ago limited to less than half modern levels by gas bubbles in basalt. *Nature Geoscience* 9: 448.

Som, S. M., D. C. Catling, J. P. Harnmeijer, P. M. Polivka, and R. Buick. 2012. Air density 2.7 billion years ago limited to less than twice modern levels by fossil raindrop imprints. *Nature* 484 (7394): 359–362.

Sutherland, J. D. 2016. The origin of life-out of the blue. *Angewandte Chemie-International Edition* 55 (1): 104–121.

Veizer, J., and A. Prokoph. 2015. Temperatures and oxygen isotopic composition of Phanerozoic oceans. *Earth-Science Reviews* 146: 92–104.

Von Damm, K. L. 1990. Seafloor hydrothermal activity: Black smoker chemistry and chimneys. *Annual Review of Earth and Planetary Sciences* 18: 173–204.

Von Damm, K. L. 1995. Controls on the chemistry and temporal variability of seafloor hydrothermal fluids. In *Seafloor Hydrothermal Systems: Physical, Chemical, Biological, and Geological Interactions*, Edited by S. E. Humphris, R. A. Zierenberg, L. S. Mullineaux, and R. E. Thomson. Washington, DC: American Geophysical Union.

Wade, J., and B. J. Wood. 2005. Core formation and the oxidation state of the Earth. *Earth and Planetary Science Letters* 236 (1-2): 78–95.

Walker, J. C. G. 1977. *Evolution of the Atmosphere*. New York: Macmillan.

Wong, M. L., B. D. Charnay, P. Gao, Y. L. Yung, and M. J. Russell. 2017. Nitrogen oxides in early Earth's atmosphere as electron acceptors for life's emergence. *Astrobiology* 17 (10): 975–983.

Young, G. M., V. von Brunn, D. J. C. Gold, and W. E. L. Minter. 1998. Earth's oldest reported glaciation; physical and chemical evidence from the Archean Mozaan Group (~2.9 Ga) of South Africa. *The Journal of Geology* 106: 523–538.

Zahnle, K., N. Arndt, C. Cockell, A. Halliday, E. Nisbet, F. Selsis, and N. H. Sleep. 2007. Emergence of a habitable planet. *Geology and Habitability of Terrestrial Planets* 24: 35–78.

Zahnle, K., L. Schaefer, and B. Fegley. 2010. Earth's earliest atmospheres. *Cold Spring Harbor Perspectives in Biology* 2 (10): a004895.

Section V

Chemical Origin of Life
Prebiotic Chemistry

5.1 Prebiotic Chemistry That Led to Life

Juli Peretó

CONTENTS

5.1.1 INTRODUCTION

Life belongs to the very fabric of the universe.

Christian de Duve (de Duve 1991)

Without doubt, explaining how life appeared on our planet is highly problematic, so daunting that some think it impossible. This fundamental question would fall outside the domain of science if the origin of life were the result of a highly improbable, almost unrepeatable phenomenon. In this case, we would lose all hope of finding an answer. However, the materialistic and evolutionary view adopted by contemporary scientists suggests that the origin of life is a chemical enigma of a historical nature, which, despite its intrinsic contingencies, we can understand and even test experimentally. It is a chemical enigma because the origin of life sprang from the transition from inert chemical matter (*i.e.,* cosmo- and geochemistry) to the most primitive biochemical systems. It is historical since this is an evolutionary transition, *that is,* a long series of successive, contingent stages of increasing complexity, each one genealogically depending on the previous one (Lazcano 2010b), and since this took place on our planet in the past, more than 3,500 million years ago (see this Handbook, Section 8).

These chemical and historical facets of the riddle of life's beginnings were obvious to Charles Darwin (Peretó et al. 2009). He was always reluctant to publicly express his ideas on the origin of life, but in a letter dated 1871 to his friend and colleague, the botanist Joseph Hooker, Darwin described a set of conditions and ingredients necessary to spark evolution: "It is often said that all the conditions for the first production of a living organism are now present, which could ever have been present. But if (and oh what a big if) we could conceive in some warm little pond with all sorts of ammonia and phosphoric salts, –light, heat, electricity, etc. present, that a protein compound was chemically formed, ready to undergo still more complex changes, at the present day such matter would be instantly devoured, or absorbed, which would not have been the case before living creatures were formed." (Letter reproduced in full in Peretó et al. 2009.)

Like Darwin, we accept that the primitive Earth was governed by the same laws of physics and chemistry as they are today. Those chemical events of the past are, thus, amenable to scientific examination and understanding, and therefore, any narration of the transition from inert to living matter must be compatible with all our scientific knowledge of the material world. Eventually, any model we propose must not only be plausible but also empirically verifiable. This does not mean that the answer to the origin-of-life enigma will be a detailed account of what exactly happened somewhere on the primitive Earth. This is something we will never know for certain in full detail, not only because of the complete absence of a chemical record of the first steps of life and the many unknowns about Archaean Earth's physicochemical constraints but, essentially, also because of the historical nature of the problem, including emergent phenomena difficult to reduce to physics and chemistry and requiring other types of explanatory accounts. Rather, we seek a full explanation of how the origin of life *might have happened* on the primitive Earth or, in other words, an experimental demonstration of the outstanding potential of chemistry to generate life, a goal considered a Holy Grail in chemical science (Krishnamurthy 2017).

The prospect of reproducing the earliest steps of life in a laboratory is not new. Over a century ago, several pioneers of synthetic biology, such as Alfonso Herrera, John Burke, and Stéphane Leduc, tried to cross the boundaries dividing inert from living matter (Peretó and Català 2007; Peretó 2016). There is an epistemological continuity between these premature and naïve attempts to synthesize life and the ultimate objective of the scientific quest to reveal the origin of life, framed within an evolutionary and chemical context. This approach is represented by the original proposals of Aleksandr I. Oparin (1924) and John B. S. Haldane (1929), who independently argued that any proposal on the natural emergence of life would be mere speculation, unless we are able to offer empirical support for its plausibility on early Earth.

The classic research on prebiotic chemistry – the chemistry that led to life – initiated by Stanley L. Miller and Harold C. Urey in the United States in the 1950s and inspired by Oparin's heterotrophic theory (Lazcano 2010a), promoted the notion of a primitive Earth rich in organic materials that could be a source of both raw materials and food for the earliest life forms. Thus, the chemical landscape on a young abiotic Earth (generated both from endogenous sources and from extraterrestrial delivery) offered a wide range of compounds, whose accumulation in the oceans shaped the primitive soup. Did life originate from an immense chemical diversity? Or rather, were there prebiotic processes providing a fairly reduced chemical repertoire that anticipated the subset of molecules operating in biochemistry? In this chapter, I will show that although these fundamental questions are not yet resolved, they are greatly clarified by current advances in prebiotic systems chemistry – an approach dealing with complex multicomponent, interacting reactions and processes (Powner and Sutherland 2011; Ruiz-Mirazo et al. 2014).

5.1.2 IS TERRESTRIAL BIOCHEMISTRY ONE AMONG MANY?

Life is the outcome of a complex network of chemical reactions and molecular interactions. Biological complexity is organized around a small set of organic molecules – amino acids, nucleotides, lipid molecules, and a suite of metabolic intermediates – used to build polymers and supramolecular architectures – proteins, nucleic acids, membranes, ribosomes, etc. The whole system is sustained by an almost universal metabolic network (Smith and Morowitz 2016), which, by gathering matter and energy from the environment, drives the idiosyncratic and paradoxical performance of life, namely, autonomous chemical systems able to make imperfect copies of themselves and, hence, evolve (Ruiz-Mirazo et al. 2004). But, how did this peculiar chemical phenomenon begin? We do not know for sure, but let's start by quoting Richard

Dawkins (Dawkins 2009, p. 419): it must have been whatever it took to get natural selection started. Once the extraordinary unfolding of natural selection has begun, the Darwinian evolutionary theory, conveniently updated and extended, explains much of what happened along the history of life. And for natural selection, a population of entities that replicated with imperfections was needed, resulting in a variety of survival and reproductive capacities. Although the specific characteristics of these "infra-biological" or "pre-biological" entities are not yet well determined, there is a broad consensus that natural selection was operative before cells were endowed with DNA as the genetic material, during the era of the so-called RNA world (de Duve 2005b). This situation can occur in primitive cells, in which catalytic RNAs (ribozymes) facilitate processes like the synthesis of monomers for their own replication (self-replicative ribozymes) and the synthesis of membrane components (Szostak et al. 2001). It is fascinating to see how – albeit in restricted circles – the development of molecular biology in the twentieth century ran parallel to the discussion on how evolution could have occurred using RNA before DNA (Lazcano 2010a, 2012).

But before the advent of RNA, was there some other kind of selection process taking place in this immense sea of chemical possibilities, or was RNA the result of chemical determinism? Is terrestrial biochemistry just one among many? Was the use of RNA the result of a frozen accident, a matter of chance? It is reasonable to suppose that a variety of cosmic and geological processes took place on the primitive Earth, contributing to an extraordinarily diverse inventory of organic materials. When we scrutinize interstellar space, or analyze pieces of extraterrestrial matter such as meteorites or comets, we discover the astounding richness of organic cosmochemistry (see this handbook, Section 4). Likewise, when we simulate primitive atmospheric chemistry, as in the classic Miller experiments and its derivatives (see this handbook, Chapter 5.5), all the evidence points toward notable chemical wealth (Cleaves 2015). Processes of chemical selection would operate on this chemodiversity, aided by minerals and by the environmental conditions. We know very little about how the biological subset of amino acids, nitrogen bases, sugars, or lipids, all with a certain isomerism, could have been selected to ignite the earliest cellular systems.

However, the prevalent view of life emerging from a complex and diverse mixture of chemical building blocks was questioned by some evolutionary biochemists and, more recently, by Günther Wächtershäuser (Wächtershäuser 1988), Harold Morowitz (Morowitz 1992; Smith and Morowitz 2016), and Christian de Duve (de Duve 1993, 2011). With diverse emphasis and on different grounds, these scientists suggest that prebiotic chemistry was predisposed toward synthesizing a few components, which would be almost the same as those configuring the contemporary metabolic repertoire. In fact, chemical determinism in the origin of life is a central theme in de Duve's thought and is one of the questions he posed in his posthumous book, which recounts the intellectual journey of this exceptional biologist (de Duve and Vandenhaute, Chapter 9, 2013). In short, de Duve argues that under certain environmental constraints, the outcome of chemical evolution would almost always be the same and that the specific polymers and systems that eventually turned into evolving cell populations would not have randomly emerged by chance from an immense ocean of molecular possibilities (Fry 2019a,b).

Thus, the "cosmic imperative" for life's emergence proposed by de Duve is at the philosophical antipodes of the proposals considering the origin of life as an unlikely chance event (Fry 2000; Lazcano 2017). In his influential book *Chance and Necessity*, molecular biologist Jacques Monod accepted the almost-proven notion of an early Earth rich in organic compounds and the possible emergence of biopolymers, such as proteins and nucleic acids (Monod 1970). Yet, Monod attributed an exceedingly low chance to the emergence of the genetic code and the functional organization of a cell having the capacity to evolve by Darwinian mechanisms. He famously stated that the planet was not pregnant with life and that, as a consequence, ours was the only inhabited planet in the universe. Clearly, Monod was astounded by the advances in molecular biology in his time, showing that the intimacies of universal cell mechanisms were much more complex than expected, thus making the question of their natural emergence still more inextricable (Monod 1970). The historicity of the process of biogenesis would comprise not only contingencies of the type that Monod was conceiving (essentially, stochastic and irreversible events) but also the emergence of intermediate systems that allowed for further evolutionary transitions to take place. Nevertheless, as we will discuss in Section 5.1.4, recent discoveries in prebiotic systems chemistry indicate that the path leading from chemistry to the earliest evolving cells may be simpler than what Monod could ever have imagined (Islam and Powner 2017; Ruiz-Mirazo et al. 2017; Sutherland 2017).

5.1.3 THE THREE PILLARS OF CLASSIC PREBIOTIC CHEMISTRY

During the last six decades, many investigations in classic prebiotic chemistry have been conditioned by the search for reasonable mechanisms underlying the direct and robust synthesis of biomonomers. Miller's spark discharge experiments elegantly showed the ease of amino acid synthesis from mixtures of simple molecules, such as H_2O, NH_3, CH_4, and H_2, purportedly representing the chemical composition of a primitive reducing atmosphere (Miller 1953). It is likely that the amino acids obtained in this kind of setting are a product of the Strecker mechanism, a well-known reaction between ammonia and *in situ* produced cyanide **1** and aldehydes (Miller 1957).

Meanwhile, the traditional approach to investigating the prebiotic synthesis of RNA precursors has researched the conjunction of nucleobases – purines are easily derived from cyanide **1** (Figure 5.1.1) oligomerization (Oró 1960) and pyrimidines can be obtained from cyanoacetylene **6** (Figure 5.1.2), for instance (Robertson and Miller 1995) – and ribose, which is just one of the myriad products of the formose reaction of formaldehyde **2a** (Figure 5.1.1) oligomerization (Breslow 1959). The three main avenues of prebiotic research during the twentieth century (Strecker-type chemistry for amino acid synthesis, cyanide chemistry for nucleobases, and the formose reaction for sugars) have been identified as the three pillars of prebiotic chemistry (Islam and Powner 2017). These so-called pillars underpin scientists' endeavors in their quest to find prebiotically plausible pathways for monomers en route to forming the major biopolymers, *that is*, proteins and nucleic acids. Nevertheless, despite the great intellectual and scientific benefits of this classical approach, there are some caveats (see Figure 5.1.1).

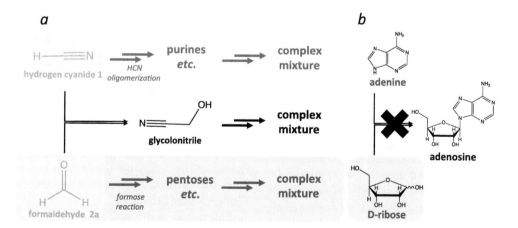

FIGURE 5.1.1 Prebiotic chemistry classical approaches. (a) Hydrogen cyanide **1** oligomerization produces purines among other compounds (blue background) and formaldehyde **2a** produces sugars (among many others, pentoses) through the formose reaction (red background); during the Miller synthesis of amino acids, the formation of glycolonitrile may impede the simultaneous synthesis of purines and pentoses (yellow background). (b) The traditional abiotic pathway to nucleosides starting from preformed nucleobases and pentoses has been proven extraordinarily difficult.

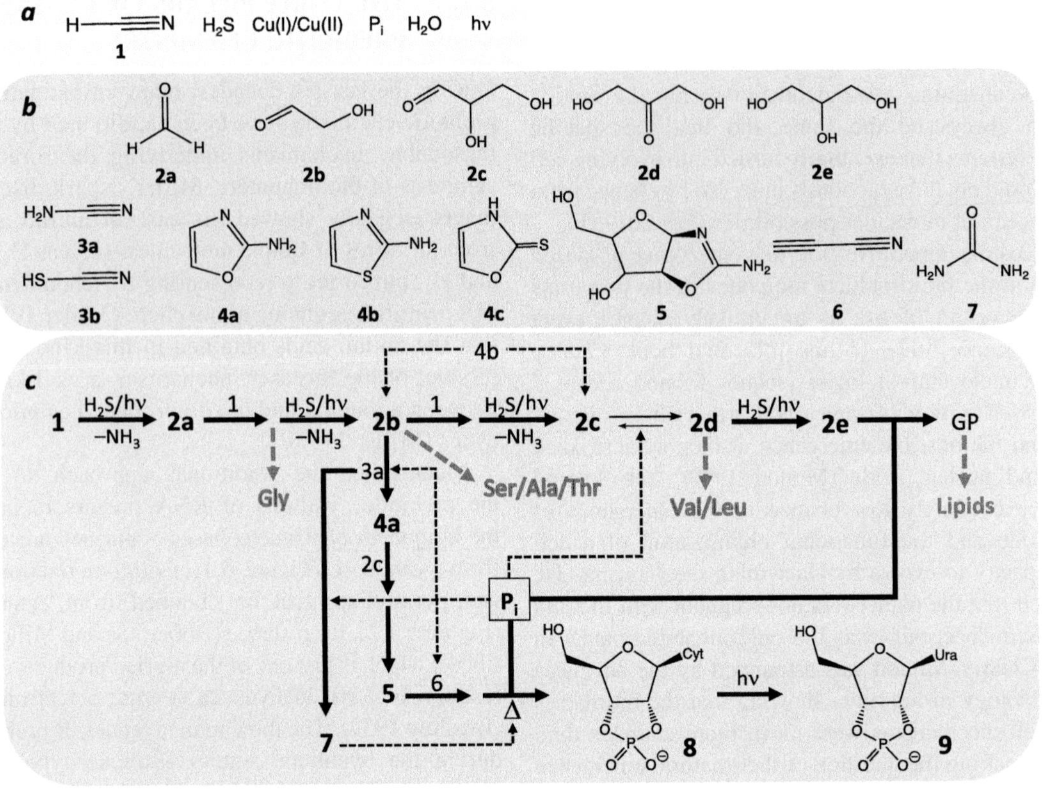

FIGURE 5.1.2 The cyanosulfidic protometabolism. (a) Feedstocks of cyanosulfidic protometabolism: hydrogen cyanide **1**, hydrogen sulfide (H₂S), copper ions, inorganic phosphate (Pᵢ), water, and UV light (hv). (b) Some intermediates in the cyanosulfidic protometabolism: formaldehyde **2a**, glycolaldehyde **2b**, glyceraldehyde **2c**, dihydroxyacetone **2d**, glycerol **2e**, cyanamide **3a**, thiocyanic acid **3b**, 2-aminooxazole **4a**, 2-aminothiazole **4b**, 2-thiooxazole **4c**, ribose aminooxazoline **5**, cyanoacetylene **6**, and urea **7**. (c) The pathway to some proteinogenic amino acids, and glycerolphosphate (GP) as precursor of lipids (gray dashed arrows), as well as cyclic ribonucleotides of cytosine **8** and uracil **9**. Black dashed arrows indicate catalytic activity.

For example, (1) sometimes, the required concentrations of reactants to produce the desired products are too high to be realistic in primitive settings; (2) there is a fundamental incompatibility between the kind of chemistry required for nucleobase and sugar syntheses – the highly favorable and rapid formation of glycolonitrile and its further oligomerization may prevent the independent oligomerization of cyanide **1** and formaldehyde **2a** within the same setting; however, a strict pH control could prevent the appearance of complex polymers (Cleaves et al. 2008) (Figure 5.1.1a); (3) even if we conceive a geochemical scenario, where both cyanide **1** and formaldehyde **2a** oligomerizations proceed, we are confronted with ongoing processes driven by thermodynamic and kinetic forces toward the generation of intractable (and uncontrollable) mixtures of products, sometimes insoluble polymers, and, consequently, the interesting molecules (nucleobases and pentoses) are actually scarce intermediates of these reactions (Figure 5.1.1a); (4) last, but not least, even assuming that pentoses and nucleobases somehow simultaneously accumulate, we know that the reaction between these two moieties required to generate a nucleoside with a suitable isomerism is one of the most elusive processes in the entire history of classic prebiotic chemistry (Figure 5.1.1b).

5.1.4 A PREFERRED ROUTE TO BIOMONOMERS?

Prebiotic systems chemistry has graphically been shown as the prebiotic rocket emerging from a historical phase of more conventional and short-range experimental flights (Lehman 2017). Recently, diverse laboratories have contributed to developing a new paradigm in prebiotic chemistry based on complex combinations that paradoxically give relatively clean products instead of intractable mixtures (Powner and Sutherland 2011; Szostak 2011; Ruiz-Mirazo et al. 2014; Islam and Powner 2017). In short, prebiotic systems chemistry exploits molecular cooperation (Lehman 2017) through the reactivity, catalytic interactions, and specific physical behavior of molecules within complex, multicomponent chemical systems. In this section, I will summarize some notable results of this new approach.

5.1.4.1 The Cyanosulfidic Protometabolism

Classical prebiotic chemistry pointed toward an apparent dead end because of the incompatibility of nitrogenous (nucleobases) and oxygenous (sugars) chemistries; however, the results of a systems approach led by John D. Sutherland and collaborators opened up brand new avenues for the prebiotic synthesis of ribonucleotides, amino acids, and lipid precursors. The work on a proposed cyanosulfidic protometabolism (Patel et al. 2015) is a true *tour de force* in origins-of-life research, since it proposes a reaction for the synthesis of two pyrimidine ribonucleotides, the aminonitrile precursors for 12 proteinogenic amino acids, glycerolphosphate (GP, a precursor of membrane phospholipids), and (remarkably) almost nothing else. The simple starting materials of this small repertoire of organic molecules are hydrogen cyanide (HCN **1**), hydrogen sulfide (H₂S), and inorganic phosphate (H₂PO₄⁻, Pᵢ) in an aqueous solution in the presence of copper ions and ultraviolet (UV) light (Figure 5.1.2). All these ingredients can be considered prebiotically plausible.

The cyanosulfidic scheme incorporates previously successful processes, in particular the synthesis of pyrimidine ribonucleotides, based on a strategy avoiding the direct reaction between ribose and nucleobase to form a *N*-glycosidic bond (Figure 5.1.1b), through the synthesis of 2-aminooxazole **4a**, a key intermediate derived from C2 (glycoladehyde **2b**) and C3 (glyceraldehyde **2c**) sugars (Figure 5.1.2), and containing the elusive C-N bond (asterisk in Figure 5.1.3) in its structure (Powner et al. 2009).

It is remarkable that some components of the system act as catalyzers. For instance, Pᵢ is not only a precursor for ribonucleotide and GP formation but also acts as a buffer and as a general acid-based catalyst in critical steps, such as the synthesis of 2-aminooxazole **4a** from glycolaldehyde **2b** and cyanamide **3a** or the transformation of cyanamide **3a** into urea **7**, a compound that will likewise catalyze the final step in cyclic ribonucleotide of cytosine **8** synthesis (Figure 5.1.2c). This mixture of adducts, products, and

catalyzers beautifully illustrates the prebiotic systems approach.

However, even in the case of such effective abiotic synthesis, certain selective mechanisms and sequential processes are required for the accumulation of the right ribonucleotides and amino acids. Remarkably, 2-aminothiazole **4b**, plausibly present in a cyanosulfidic environment, selectively crystallizes the required C2 and C3 sugars (**2b** and **2c** in Figure 5.1.2c) in the presence of complex mixtures of aldoses and ketoses, acting as a chemical chaperone for ribonucleotide synthesis. Furthermore, 2-aminothiazole **4b** specifically excludes the synthesis of α,α-disubstituted amino acids, favoring instead the synthesis of proteinogenic α-amino acids precursors. Thus, a common physicochemical mechanism, in which 2-aminothiazole **4b** selects the "right" precursors by chemical reaction and precipitation, would explain the prebiotic synthesis of specific biomonomers out of complex chemical mixtures (Islam et al. 2017). It is noteworthy that the role of 2-aminothiazole **4b** in this crystallization-driven selective process was deduced from a systems chemical analysis applied to cysteine (one of the proteinogenic amino acids missing in Sutherland's cyanosulfidic scheme) (Islam et al. 2017; Islam and Powner 2017).

The original cyanosulfidic scenario (Patel et al. 2015) explains the origin of two canonical pyrimidine (cytosine **8** and uracil **9**, Figure 5.1.2) ribonucleotides. Matthew Powner and coworkers have added plausibility to the whole model by including the divergent synthesis of both pyrimidine and purine ribonucleotides (Stairs et al. 2017): 2-aminooxazole **4a** is the common precursor to pyrimidine ribonucleotides, whereas the generationally related 2-thiooxazole **4c** serves as starting material for the synthesis of 8-oxo-adenine **11** and 8-oxohypoxanthine **12** ribonucleotides (Figure 5.1.4).

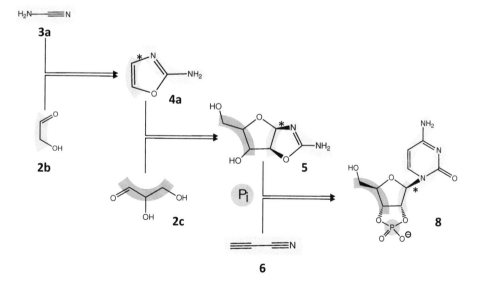

FIGURE 5.1.3 Pyrimidine ribonucleotide synthesis in a cyanosulfidic protometabolism. Asterisk (*) indicates the bond in 2-aminooxazole **4a** that will become the N-glycosidic bond in the final ribonucleotide **8**.

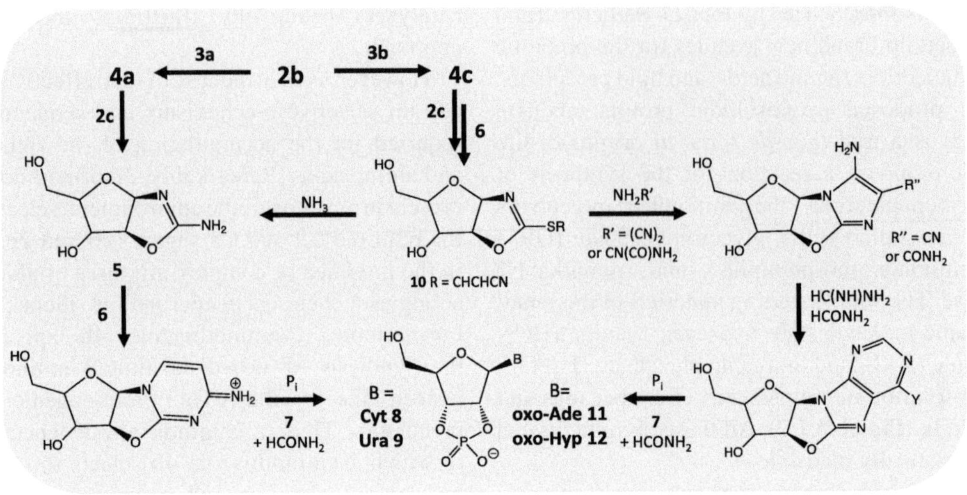

FIGURE 5.1.4 Simultaneous synthesis of purine and pyrimidine ribonucleotides in a cyanosulfidic protometabolism. Simultaneous assembly of cytidine-2′,3′-cyclic phosphate **8**, uridine-2′,3′-cyclic phosphate **9**, 8-oxo-adenosine-2′,3′-cyclic phosphate **11**, and 8-oxo-inosine-2′,3′-cyclic phosphate **12**. Divergent syntheses take place at the level of glycolaldehyde **2b** and thione **10**. For sake of simplicity, stereochemical details in the diverse compounds are omitted.

Thus, researchers have synthesized both purine and pyrimidine ribonucleotides efficiently and simultaneously under plausible prebiotic conditions for the first time. The easy synthesis of ribonucleotides based on oxo-purines would suggest that these monomers could precede the canonical ones (based on adenine and guanine) in a primitive genetic alphabet.

At the moment of preparing this chapter, the synthesis in the cyanosulfidic scenario of one important component in early protocells is missing: the amphiphilic molecules (*e.g.*, fatty acids) that would self-assemble in vesicles, generating the first compartments. Although the prebiotic presence of this kind of molecules is completely acceptable by other means, including abiotic synthesis and extraterrestrial delivery (Cleaves 2015), a continuous protometabolic source of amphiphiles is needed. It is of note that the cyanosulfidic scenario supplies both stereoisomers of GP (Figure 5.1.2c), necessary building blocks for the origin of the two phospholipid classes, namely bacterial and archaeal types of membrane lipids (Peretó et al. 2004). A further challenging issue to be addressed is the emergence of polymers from the monomers generated in the protometabolism and their functional integration with lipid vesicles into a highly organized "multiple polymer world" (Ruiz-Mirazo 2015a).

The cyanosulfidic scenario has been criticized because successful organic syntheses imply the sequential adjustment of specific laboratory conditions. Notwithstanding, all the conditions considered in the experiments, as well as the feedstocks, are prebiotically plausible, and one can imagine that all the constraints could be met over space and time on the primitive planet. In a comprehensive review, Sutherland has explained in detail the general strategy of his systems approach and the fruitful dialog between laboratory experiments and the geochemical constraints (Sutherland 2016). In short, several plausible geochemical settings and processes, including impacts of meteorites and comets or water streams encountering a diversity of minerals, are invoked in a working model substantiated by the experimental results, and then, the refined model heuristically informs and suggests new experiments and simulations. This bidirectional conversation between prebiotic systems chemistry and geochemistry of the Archean Earth will eventually deliver a specific chemical definition of Darwin's "so simple a beginning of life" – as he put it in the last paragraph of (Darwin 1859) – and a refined and realistic version of his metaphoric little pond.

5.1.4.2 Prebiotic Prefiguration of Metabolism

At first sight, we do not see any obvious correlation between the repertoire of classic prebiotic mechanisms (*e.g.*, Strecker synthesis and formose reaction) and the network of chemical transformations operative in extant enzymatic-dependent metabolic pathways. This apparent discontinuity has been compared to a dense cloud (Lazcano and Miller 1999) or an opaque barrier (Orgel 2003) separating prebiotic chemistry from biochemistry. However, looking in more detail for possible parallelisms in specific mechanisms, there are a few cases where enzymatic steps in metabolism resemble abiotic transformations. In this context, the term *chemomimetic biosynthesis* was suggested in reference to the enzymatic mechanism for riboflavin biosynthesis, which resembles the organic process taking place *in vitro* without enzymes (Eschenmoser and Loewenthal 1992). These authors perceived this case as evidence of how an enzyme could co-opt and optimize a previously existing non-enzymatic chemical transformation, suggesting certain chemical determinism and continuity with prebiotic chemistry in the origin of metabolism. But, in fact, the repertoire of possible chemomimetic reactions in metabolism identified so far is disappointingly low (Peretó 2012 and references therein).

This lack of strict mechanistic parallelism, however, does not preclude the operation of a *principle of congruence* between a (non-enzymatic) protometabolism and a modern-like (enzymatic) metabolism, as proposed by de Duve (1993), elaborated in several works and summarized in (de Duve 2005a), with a conservation of a general outline of the mutual relationships between the main metabolic substrates and products. Therefore, it is supposed that a chemical determinism was operative in the first stages of the origin of metabolism (see also the discussions on this topic by Iris Fry, Chapters 3.1 and 6.5 in this handbook). For de Duve, the term *protometabolism* is synonymous with prebiotic chemistry, *that is,* the early chemical processes that preceded and gave rise to the extant metabolism, and this notion has also been adopted by Sutherland and coworkers recently. In this particular sense, protometabolism was the necessary chemical landscape for the emergence of "useful" catalysts in an RNA world.

The development of these initial chemistries into something that resembles more closely a proper metabolic organization possibly involved a series of stages in which oligomerization processes ought to be included, so that functional compounds, such as peptides, oligonucleotides, amphiphiles, and so on, take the stage. Furthermore, pre-Darwinian selection processes (*e.g.,* molecular chaperones) must have started having an influence on the underlying chemistry. In other words, evolutionary dynamics will start dictating some of the changes and innovations taking place in those protobiological systems, according to the result of a complex history of interactions with similar systems and the environment at large. On these lines, it is important to notice that the replacement of a protometabolism by a modern (enzymatic) metabolism required the emergence of catalysts by a selection process based on their *usefulness* under certain conditions. For instance, in an RNA world, only the ribozymes fitting with existing substrates to be transformed into suitable products would be selected for, since an *evolving* catalyst would have a meaningful existence, as such, only if an appropriate substrate is present and ready to be transformed into a product that benefits the system as a whole. The initial suite of substrates would be the product of more primitive catalysts (mineral surfaces; metal ions; small organic molecules, such as amino acids and some cofactor precursors; short polymers, such as non-coded peptides referred to by de Duve as "multimers"; lipid vesicles; etc.) or, put in simple terms, the stable and robust outcome of a prevalent abiotic process, such as the cyanosulfidic scenario. Therefore, one could envision some initial stages in which chemical determinism was more prominent, to be later replaced, or modulated, by other processes in which evolutionary and historical aspects became increasingly influential.

From a bioenergetics perspective, de Duve's protometabolic scenario is driven by the reactivity of the thioester bond, used before the adoption of polyphosphates and nucleotides as energy currencies (de Duve 1991). Possible traces of an ancient metabolism, independent of the use of phosphate but enriched in reactions catalyzed by iron-sulfur clusters or thermodynamically enabled by thioesters, have been described using a systems-biology approach on a biosphere-level (or meta-metabolic) network (Goldford et al. 2017). The formation of iron-sulfur clusters under prebiotic conditions, driven by UV light, has also recently received further experimental support (Bonfio et al. 2017).

5.1.4.3 Focusing the Metabolic Image

The emergence of ribozymes, as subjects of natural selection, added a new dimension to the pure chemical processes, since adaptation allowed catalysts' optimization. Thus, the first enzymatic catalysts under natural selection (either ribozymes in an RNA world or, later on, genetically encoded protein enzymes) would play a key role in the transition from a dirty protometabolism (de Duve's *gemisch*) to a cleaner metabolism. But, how dirty was the abiotic chemical landscape? The results of prebiotic systems chemistry (as discussed in Section 5.1.4.1) point out that certain chemical determinism would result in a more restricted abiotic repertoire than was imagined before. We now examine some empirical examples consistent with a protometabolic scenario.

Glycolaldehyde **2b** and glyceraldehyde **2c** (C2 and C3 sugars, respectively, nucleotide precursors in the cyanosulfidic protometabolism; see Figure 5.1.2) are also related to the structure of some intermediates of glycolysis, one of the most ancient metabolic pathways. Particularly, a robust and efficient synthesis of phosphoenolpyruvate **16** (PEP) from C2 **2b** and C3 **2c** sugars has been described (Coggins and Powner 2017) (Figure 5.1.5). The glycolytic intermediate PEP **16** is not only the metabolite with the highest phosphoryl transfer potential, and thus involved in the coupling between glycolysis and ATP synthesis, but also a key precursor in the metabolism of sugars and aromatic amino acids. Under plausible prebiotic conditions, other glycolytic intermediates, such as 3-phosphoglycerate **14**, 2-phosphoglycerate **15**, and pyruvate **17**, as well as serine precursor phosphoserine, are also obtained, clearly connecting the modern lower glycolysis (involving C3 intermediates) with a putative protometabolic network (Figure 5.1.5).

The oxidative decarboxylation of pyruvate C3 to acetate C2 is also observed, approaching a connection of another sub-network represented by the tricarboxylic acid (TCA) or Krebs cycle, which is through acetyl CoA **18**, or coenzyme A (CoA)-activated acetate (Figure 5.1.5), in extant metabolisms.

Incidentally, if the construction of the glycolytic pathway started from its lower segment, allowing the connection of ribonucleotide and amino acid metabolisms, and the access to other sugars, such as hexoses, was a posterior innovation, we would have a historical explanation for one of the most intriguing stoichiometric arrangements in universal metabolism: for the synthesis of C2 fragments (acetyl CoA **18**, metabolic precursor, for instance, of fatty acids and isoprenoids), C6 sugars must follow the fragmentation in C3 moieties up to pyruvate **17** that will suffer oxidative decarboxylation to acetyl CoA **18** with an

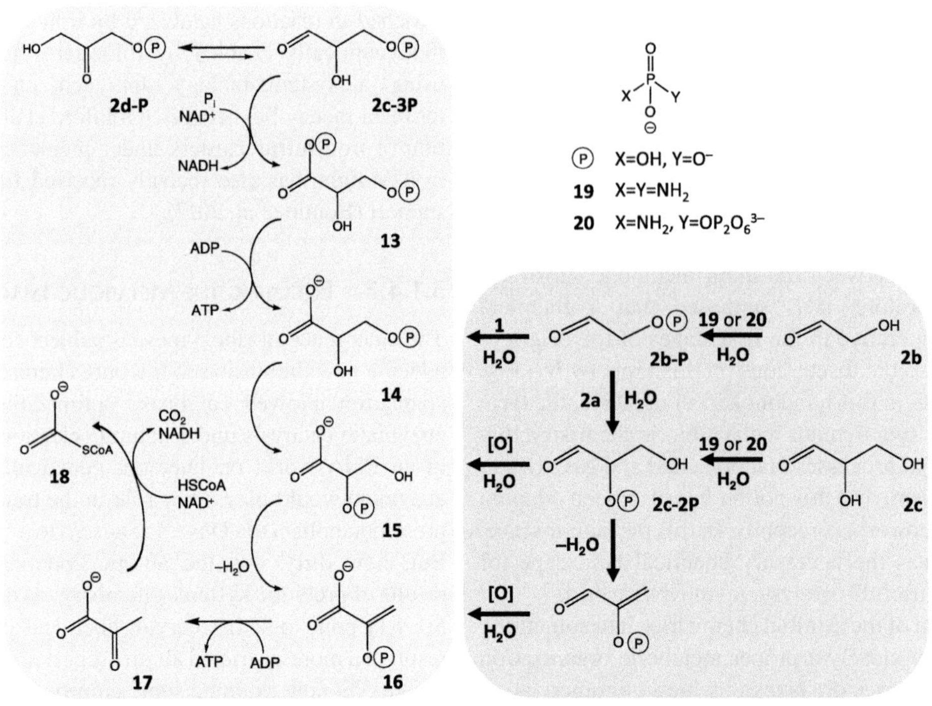

FIGURE 5.1.5 Synthesis of glycolytic intermediates in a cyanosulfidic protometabolism. A short sequence of mild and robust reactions (bold arrows, right panel) starting from glycoladehyde **2b**, glyceraldehyde **2c**, and amidophosphates **19** or **20** produces some glycolytic intermediates such as 3-phosphoglycerate **14**, 2-phosphoglycerate **15**, and phosphoenolpyruvate **16**. Left panel shows the glycolytic pathway of enzymatic transformations from triosephosphates (dihydroxyacetonephosphate **2d-P** and glyceraldehyde-3-phosphate **2c-3P**) to pyruvate **17** and acetyl CoA **18**.

unescapable loss of 33% of carbon as CO_2 (Figure 5.1.5). In contrast, stoichiometric logic indicates that a more efficient process would fragment hexoses C6 into three C2 moieties (acetyl CoA **18**). This arrangement is absent in natural metabolic networks, because evolution had to incorporate C6 into a previous metabolic groundwork based on C3/C2 intermediates; however, this has been obtained in a remarkable example of metabolic engineering by using an artificial combination of enzymatic steps (Bogorad et al. 2013).

The preceding example, taken from synthetic organic chemistry (Coggins and Powner 2017), is complemented by the work of Markus Ralser and colleagues, who have explored the non-enzymatic conversions between the intermediate metabolites of central pathways such as glycolysis, pentose phosphate pathway (PPP) (Keller et al. 2014, 2016) and TCA cycle (Keller et al. 2017). Their strategy involves screening hundreds or thousands of possible conditions, inspired by the composition of Archaean sediments, looking for chemical transformations between intermediates. Albeit in some cases, the prebiotic plausibility of the assayed initial substrate is doubtful and the yields are very low, it is remarkable that they were able to reconstruct the basic topology of the pathways by using very simple catalysts such as Fe(II) ions and phosphate (glycolysis and PPP) or sulfate radicals (TCA cycle) at 70 °C. The aldol condensation of glyceraldehyde-3-phosphate

2c-3P and dihydroxyacetonephosphate **2d-P** (Figure 5.1.5) to generate fructose-1,6-bisphosphate – mimicking a non-enzymatic gluconeogenic reaction – has also been observed in ice (Messner et al. 2017). In addition to these indications of primitiveness of some fundamental chemistry of the central metabolic pathways, Joseph Moran and coworkers have shown that non-enzymatic catalysis by transition metals can promote sets of reactions that could be involved in primitive anabolic processes such as a reverse TCA cycle (Muchowska et al. 2017). All these experiments reinforce the notion of the establishment of networks of non-enzymatic transformations between organic molecules under plausible geochemical conditions. The topology of these networks strikingly resembles that of the metabolic core.

However interesting the research on the prebiotic plausibility of some metabolic intermediates and their non-enzymatic transformations is, we urgently need examples of sustainable non-enzymatic catalytic cycles as a model of primitive stages of metabolic cycles emergence. In that sense and following a bottom-up strategy, Springsteen et al. (2018) have described two linked protometabolic cycles that oxidize glyoxylate **19** to CO_2 and generate some intermediates shared with the metabolic TCA cycle.

The bicycle described by Ram Krishnamurthy and collaborators and shown in Figure 5.1.6a represents a simpler version of the overall metabolic process of carbon oxidation and contains some noteworthy analogies with modern

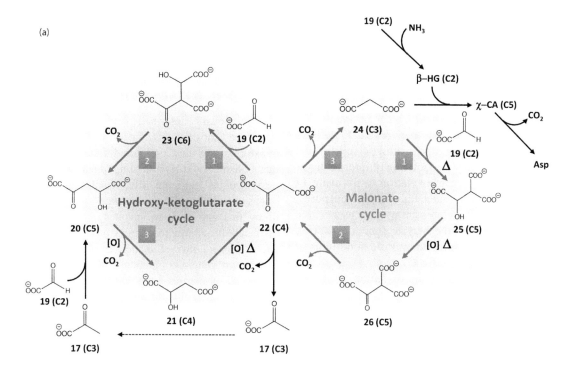

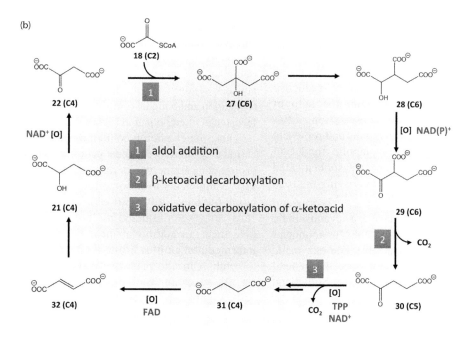

FIGURE 5.1.6 Mimicking TCA cycle, Springsteen-Krishnamurthy's bicycle oxidizes glyoxylate to CO_2. (a) Two concatenated abiotic catalytic cycles with oxaloacetate **22** (OAA) as common intermediate allow the complete oxidation of glyoxylate **19** under mild conditions during several hours. The 4-hydroxy-2-ketoglutarate **20** (HKG) and the malonate **24** (MLN) cycles provide a model of simpler versions of metabolic cycles. Additional intermediates involved in the HKG cycle are malate **21**, OAA **22**, and oxalomalate **23**, whereas in the MLN cycle, in addition to OAA **22** and MLN **24**, 3-carboxy-malate **25** and 3-carboxy-oxaloacetate **26** also participate. The scheme also shows the anaplerotic synthesis of HKG **20** from pyruvate **17** and glyoxylate **19** and the synthesis of aspartate (Asp) from glyoxylate **19**, ammonia, and MLN **24** through the intermediates α-hydroxyglycine (α–HG) and β-carboxyaspartate (β–CA). Note that, in each cycle, the addition of glyoxylate **19** (reaction 1) and the decarboxylation of a β- and an α-ketoacid (reaction 2 and 3, respectively) follow the same order as in the modern TCA cycle, as shown in panel (b). (b) Metabolic TCA cycle allows the oxidation of acetate (activated as acetyl CoA **18**) to CO_2. After condensation of acetate to OAA **22** (reaction 1), two successive decarboxylations follow: the spontaneous decarboxylation of a β-ketoacid, oxalosuccinate **29** (reaction 2), and the decarboxylation of an α-ketoacid, α-ketoglutarate **30** (reaction 3), with the help of cofactors such as thiamine. A series of reactions synthesizes OAA **22** from succinate **31**, allowing the catalytic functioning of the cycle. Other symbols: Δ, 50 °C; [O], hydrogen peroxide; and TPP, thiamine pyrophosphate.

biochemical pathways: (1) the bicycle shares three reaction types with modern TCA cycle, namely, aldol addition and decarboxylation of β- and α-ketoacids, operating in the same order over the intermediates of the cycle but without the involvement of enzymes and cofactors (in fact, it uses H_2O_2 as oxidant, instead of complex organic coenzymes) (compare panel a and b in Figure 5.1.6); (2) an anaplerotic reaction (addition of glyoxylate **19** to pyruvate **17**) allows the sustained functioning of the catalytic cycle and compensates for the spontaneous depletion of intermediates by oxaloacetate **22** (OAA) decarboxylation to pyruvate **17** or the consumption of intermediates for synthetic purposes, since (3) some intermediate skeletons may serve as a source of amino acids. Remarkably, the bicycle functions during several hours in aqueous medium under mild conditions of pH (7.0–8.5) and temperature (≤50 °C).

The modern TCA cycle allows the oxidation of acetate to CO_2 after addition of acetyl CoA **18** to OAA **22** and the successive decarboxylation of a β-ketoacid and an α-ketoacid with the support of sophisticated enzymes and cofactors, *that is*, NAD(P)⁺ and thiamine (Figure 5.1.6b). In contrast, the abiotic bicycle adds glyoxylate **19** to OAA **22** or to malonate **24** (MLN) and decarboxylates a β-ketoacid (oxalomalate **23** or 3-carboxy-oxaloacetate **26**) and an α-ketoacid (4-hydroxy-2-ketoglutarate **20** or OAA **22**) only with the help of H_2O_2 (Figure 5.1.6a). Furthermore, the chemistry involved in these abiotic cycles allows a net carboxylation (pyruvate **17** to OAA **22**), without requiring CO_2/ATP, as in modern metabolism, and a reductive amination (OAA **22** to aspartate), without the help of an external reductant. These notable observations offer simpler alternatives to metabolic mechanisms and represent an encouraging and plausible protometabolic model of reactions operating before the emergence of sophisticated polymeric enzymes and their cofactors and paving the road for more complex metabolic cycles.

In summary, the subset of products of cyanosulfidic protometabolism is congruent with the biomonomer spectrum, whereas geochemically plausible non-enzymatic transformations between metabolites reproduce the general topology of the central metabolism (Ralser 2014; Sutherland 2016), and extant metabolic networks may contain fossils of a phosphate-independent biochemistry (Goldford et al. 2017). It is of note that the chemosynthetic strategies observed in the scenario of the cyanosulfidic protometabolism also are comparable to those of extant pathways of the central metabolism (Sutherland 2017). Furthermore, the discovery of plausible protometabolic sustained catalytic cycles, sharing remarkable resemblances and commonalities with extant metabolic pathways, starts promising research views into the primitive stages of life evolution (Springsteen et al. 2018). Thus, today, we have more empirical reasons to conclude that de Duve's *gemisch* was perhaps less dirty than initially supposed and that, as he put it, protometabolism must have prefigured metabolism. The two are *congruent*; they followed similar pathways (de Duve 2005a, p. 19). Prebiotic systems chemistry is opening a tunnel through Orgel's "opaque barrier."

5.1.5 THE TOTAL SYNTHESIS OF PROTOCELLS

The previously described panorama of a protometabolic landscape mostly dominated by deterministic chemical transformations that prefigured fundamental parts of extant metabolisms is still at a long distance of what we could recognize as minimal life. Evolution through transition stages involving the emergence of intermediate systems filling this gap must be invoked. Those intermediate forms of metabolism have to address problems related not only to energetics and flux of matter, from geochemical feedstocks to the own system components, but also to the emergence of functional properties such as active compartments, kinetic control, and heredity, through "major prebiotic transitions," in which all these organizational and evolutionary aspects must be integrated (Shirt-Ediss et al. 2017). Diverse theoretical efforts have tried to reduce the complexity of life to its simplest components and elementary behaviors (Letelier et al. 2011). A remarkable example is the chemoton model of Gánti (2003), based on the observation that a minimal living cell is the result of the organizational and functional coupling between three autocatalytic subsystems: (1) a self-reproductive compartment (boundary in Figure 5.1.7) (2) a self-replicative genetic polymer (template in Figure 5.1.7), and (3) a self-sustained chemical network synthesizing the necessary components for polymer replication and boundary growth out of available environmental feedstocks (metabolism in Figure 5.1.7) (Peretó 2012).

Thus, the chemoton could represent the minimal organization exhibited by the most primitive entities emerged from protometabolic networks and endowed with all the necessary components to become the subject of Darwinian evolution (de Duve 2005a, 2005b). These theoretical "minimal cells" may be considered as "protocells," provided that we give them an additional meaning of historical (emergent, intermediate) entities filling the temporal gap between protometabolism and primitive cells, similar to extant bacteria. As an example of this kind of protocell (or minimal cell), Figure 5.1.7 presents a sketch of a ribocell, inspired by (Szostak et al. 2001). A more elementary use of the term "protocell" refers to any experimental or computational construct composed of a self-assembled compartment with other chemical (or biochemical) components in order to simulate or reconstruct the possible historical steps of acquisition of complexity (Ruiz-Mirazo 2015b).

Actually, the empirical exploration of the reconstruction of a simple protocell is a remarkable research program in contemporary biology (Szostak et al. 2001), with deep historical and philosophical roots (Peretó 2016), sound conceptual grounds (Letelier et al. 2011; Pascal and Pross 2017; Ruiz-Mirazo et al. 2017; Shirt-Ediss et al. 2017), and good prospects of success (Ruiz-Mirazo et al. 2014;

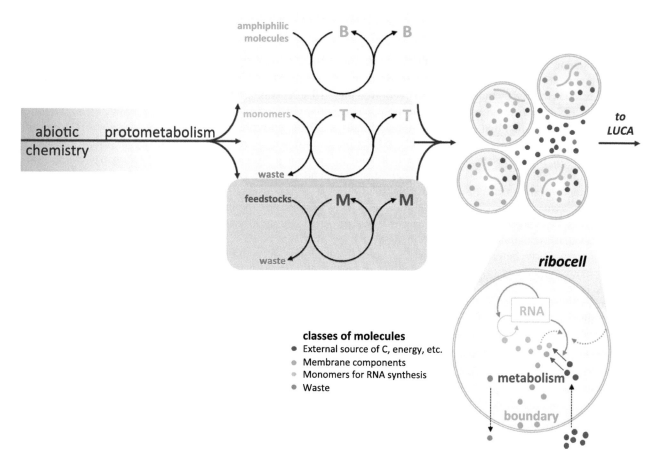

FIGURE 5.1.7 The pathway for total synthesis of a protocell. Bottom-up, constructive approaches in synthetic biology seek to combine autocatalytic cycles like self-reproductive lipid vesicles (boundaries B), self-replicative templates (T), and self-sustained sets of reactions (metabolism M) in protocellular constructs. The scheme shows, as an example, a possible ribocell with an RNA showing ribozymic activities (gray arrows) for its own replication (blue arrow), as well as the synthesis of ribonucleotide monomers for replication and membrane precursors (red arrows). Small molecules or the lipid membrane could also help ribozymes in their catalytic functions (dashed gray arrows).

de la Escosura et al. 2015). We can perceive the synthesis of a living entity from simpler, non-living components as a bottom-up strategy within the emerging discipline of synthetic biology (Porcar and Pereto 2014), also qualified as constructive biology (Kaneko 2006; Ichihashi et al. 2010; Ichihashi and Yomo 2016) or a way of knowing biology by making life (Keller 2009). This strategy has notably progressed in the field of chemistry of nucleic acid replication (*e.g.*, non-enzymatic oligomerization of RNA and ribozyme self-replication) as well as in biophysics of membranes of a simple nature (*e.g.*, fatty acid and phospholipid vesicles) (Blain and Szostak 2014; Szostak 2017). Alas, the experimental implementation of autocatalytic networks of reactions allowing the autonomy of the whole system is still in its infancy (Pereto 2012; Blain and Szostak 2014; Ruiz-Mirazo et al. 2014; Scharf et al. 2015); however, some remarkable advances in this direction have been achieved (Hardy et al. 2015).

The study of ribozyme chemistry and nucleic acid replication in the context of the RNA world model has

attracted much attention among researchers, and there was a time when some considered compartments as latecomers and referred to an acellular RNA world (Pereto 2005). Today, the general perception of the role played by early compartments is very different, and the functional significance of protocell boundaries goes beyond its task as a mere closed space (Walde et al. 2014; Hanczyc and Monnard 2017). As a remarkable example, fatty acids, used as a model of primitive amphiphilic molecules, spontaneously form vesicles that provide both a local environment for the location of hydrophobic substrates and acid-based catalysis promoting peptide synthesis (Murillo-Sánchez et al. 2016).

There are also noteworthy examples of the emergence of new properties in more integrative artificial systems, for instance, when mixing two subsystems such as template replication encapsulated in lipid vesicles or when the supramolecular system is assisted by catalytic interactions with small molecules (Blain and Szostak 2014; Ruiz-Mirazo et al. 2014; Hardy et al. 2015; Ichihashi and Yomo 2016). Although results are promising, there are still many problems to overcome, such as the adequate permeability of the boundary for accessing feedstocks and releasing waste and the apparent incompatibility of physicochemical conditions allowing stability of lipid vesicles and efficient nucleic acid replication simultaneously. In this context, a paradigmatic case is offered by magnesium ions, required for non-enzymatic RNA replication and ribozyme polymerase at a millimolar concentration that is incompatible with fatty acid membrane stability (Szostak 2012, 2017). Chelation chemistry by small molecules (*e.g.,* citrate, albeit with a doubtful prebiotic plausibility) or by short peptides (de Duve's multimers) seems to offer some possible solutions to this critical issue (Szostak 2017). Another way to advance and overcome some of these problems is by using alternative compartmentation: instead of lipid vesicles, some laboratories assay heterogeneous systems, such as water-in-oil emulsions. Using this approach, Tetsuya Yomo and coworkers have described a simple artificial chemical system that exhibits Darwinian evolution (Ichihashi et al. 2013), a notable success of the synthetic approach.

5.1.5.1 A Note of Caution: Origins of Life ≠ Last Universal Common Ancestor

When we refer to protocell research, we are talking about bottom-up strategies, from chemistry up to biology, in reconstructing the earliest steps, from protometabolism to fully fledged biological entities. Now, the reverse strategy is represented by comparative biochemistry of present prokaryotic biodiversity (or top-down approach), which aims to reconstruct the cenancestor or last common ancestor (LCA), also known as Last Universal Common Ancestor (LUCA) (Figures 5.1.7 and 5.1.8).

Comparative genomics and phylogenomic analysis recovers a rather metabolically complex LUCA (Becerra et al. 2007; Weiss et al. 2016). However, top-down approaches are plagued with methodological issues, such as dealing with the effect of horizontal gene transfers and lineage-specific gene losses (Becerra et al. 2007). Even apparently straightforward questions, such as whether LUCA was endowed with an autotrophic metabolism, remain controversial (*cf.* Becerra et al. 2014; Weiss et al. 2016). Notwithstanding the obvious scientific interest of the top-down approach to the phenotypic profile of LUCA, we cannot justify extrapolating the reconstructed metabolism and putative environmental constraints to the origins of life, as presented, for instance by (Weiss et al. 2016). We certainly do not know the time it took from protometabolism and the emergence of protocells to LUCA, but the huge distance between them is evident in terms of biochemical complexity and environmental constraints (Arrhenius et al. 1999; Cornish-Bowden and Cárdenas 2017). Phylogenomic analyses cannot deliver conclusions about a period beyond the invention of the coded macromolecules used in those analyses. In other words, phylogenomic studies based on proteins cannot give direct information on stages before the emergence of the genetic code and the translation machinery. Protometabolism occurred before the earliest stages of the RNA world and, for profound reasons, will remain inaccessible to protein phylogenies. On the other hand, prebiotic systems chemistry, in intimate correlation with geo- and cosmochemistry, is the most appropriate way to project the chemical logic of protometabolism upon extant metabolic virtuosities, thus searching for the transition from chemistry to biology and revealing the very origins of biological evolution.

Finally, we should contemplate the "origin of life" not as a historical singular event but rather as a long process of expansion of increasingly complex (autonomous, evolutionary) chemical systems. Introducing an updated

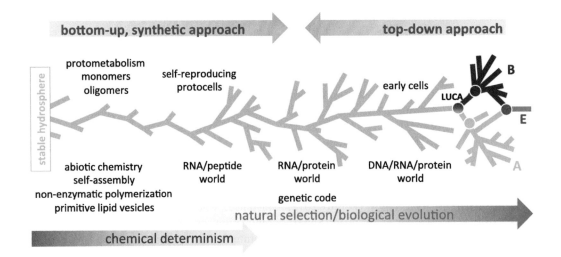

FIGURE 5.1.8 The origins of life and LUCA are not the same. The scheme illustrates some intermediate stages from protometabolism to LUCA through systems of increasing complexity, with chemical determinism playing a major role before the emergence of RNA as genetic material and catalytic player in an RNA world. Two main research approaches, from chemistry up to primitive biological entities (bottom-up) and from extant biodiversity down to early life (top-down), are indicated.

discussion on *"what is life?"* may have a heuristic effect in prebiotic systems chemistry (Ruiz-Mirazo et al. 2004), as it would illuminate our views on the intermediate stages endowed with different degrees of "aliveness" and that populate the trajectory between protometabolism and the first fully live cells – at this moment, the reader should ask him or herself where to locate this stage in Figure 5.1.8 according to his or her own life definition.

5.1.6 SUMMARY AND PROSPECTS

Prebiotic systems chemistry is on the way to superseding the classical controversies surrounding the origin of life based on dichotomist alternatives, such as the early *vs.* late emergence of compartments, metabolism-first *vs.* genetics-first models, and autotrophic *vs.* heterotrophic scenarios (Peretó 2005). A more integrative approach is rapidly evolving in this field.

Experimental results in prebiotic systems chemistry suggest that protometabolism on Archaean Earth generated a small set of biomonomers that prefigured extant biochemistry. There are, nevertheless, extraordinary challenges ahead when trying to implement systems with increasing levels of complexity. For instance, we need to know more about the dynamic coupling of lipid vesicles with heterogeneous chemical mixtures of small molecules and polymers and their experimental evolution into more robust systems, aiming to simulate protocells with Darwinian behavior. The merging of systems chemistry with experimental evolution methodologies is an absolute must if we are to bridge the vast gap between protometabolism and functionally integrated protocells (Ruiz-Mirazo et al. 2017; Shirt-Ediss et al. 2017). In short, solutions hardly accessible by rational design can be found by the own system's dynamics. In this context, merging laboratory, computational, robotics, and microfluidics technologies is compulsory to explore in depth the transition from simple chemical mixtures to systems with complex behaviors (Scharf et al. 2015; Cronin 2017), also opening up the field to chemical models that imitate biological behaviors, albeit assaying with abiological molecular components (Miljanić 2017; Taylor et al. 2017). Certain aspects, however, still require significant and persistent research efforts, which include finding the paths for the emergence of autocatalytic networks of reactions exhibiting dynamics of prebiotic relevance (Semenov et al. 2016; Wong and Huck 2017) or looking for empirical explanations for the origin of biochemical homochirality (Hein et al. 2012), one of the most colossal unknowns in origins-of-life research.

As we will never know exactly how it happened, the scientific study of the origins of life is an open and permanent debate that has maintained, maintains, and will continue to maintain the interest of many generations of scientists. This is particularly true for synthetic chemists, among others, since the field of prebiotic systems chemistry urgently needs the enthusiasm and audacity of young investigators (Wagner and Blackmond 2016; Krishnamurthy and Snieckus 2017; Tor 2017) engaged in one of the major synthetic breakthroughs:

finding a second example of life through the chemical synthesis of an artificial cell (Solé 2016).

ACKNOWLEDGMENTS

The author wishes to thank several colleagues for reading and criticizing earlier versions of the manuscript, in alphabetical order: C. García-Ferris, R. Krishnamurthy, A. Lazcano, A. Moreno, M. W. Powner, K. Ruiz-Mirazo, and J. D. Sutherland. The final version, including any mistake or misunderstanding, is the author's sole responsibility. Proofreading by F. Barraclough is also acknowledged.

REFERENCES

Arrhenius, G., J. L. Bada, G. F. Joyce, A. Lazcano, S. Miller, and L. E. Orgel. 1999. Origin and ancestor: Separate environments. *Science* 283 (5403): 792. http://www.ncbi.nlm.nih.gov/pubmed/10049121.

Becerra, A., L. Delaye, S. Islas, and A. Lazcano. 2007. The very early stages of biological evolution and the nature of the last common ancestor of the three major cell domains. *Annu. Rev. Ecol. Evol.* 38: 361–379.

Becerra, A., M. Rivas, C. García-Ferris, A. Lazcano, and J. Peretó. 2014. A phylogenetic approach to the early evolution of autotrophy: The case of the reverse TCA and the reductive acetyl-CoA pathways. *Int. Microbiol.* 17 (2): 91–97. doi:10.2436/20.1501.01.211.

Blain, J. C., and J. W. Szostak. 2014. Progress toward synthetic cells. *Annu. Rev. Biochem.* 83 (1): 615–640. doi:10.1146/annurev-biochem-080411-124036.

Bogorad, I. W., T. S. Lin, and J. C. Liao. 2013. Synthetic non-oxidative glycolysis enables complete carbon conservation. *Nature* 502 (7473): 693–697.

Bonfio, C., L. Valer, S. Scintilla, S. Shah, D. J. Evans, L. Jin, J. W. Szostak, D. D. Sasselov, J. D. Sutherland, and S. S. Mansy. 2017. UV-light-driven prebiotic synthesis of iron–sulfur clusters. *Nat. Chem.* doi:10.1038/nchem.2817.

Breslow, R. 1959. On the mechanism of the formose reaction. *Tetrahedron Lett.* 1: 22–26.

Cleaves, H. J. 2015. Prebiotic synthesis of biochemical compounds. An overview. In *Astrobiology. An Evolutionary Approach*, edited by V. M. Kolb, pp. 83–117. Boca Raton, FL: CRC Press/Taylor & Francis Group.

Cleaves, H. J., J. H. Chalmers, A. Lazcano, S. L. Miller, and J. L. Bada. 2008. A reassessment of prebiotic organic synthesis in neutral planetary atmospheres. *Orig. Life Evol. Biosph.* 38 (2): 105–115. doi:10.1007/s11084-007-9120-3.

Coggins, A. J., and M. W. Powner. 2017. Prebiotic synthesis of phosphoenol pyruvate by α-phosphorylation-controlled triose glycolysis. *Nat. Chem.* 9: 310–317.

Cornish-Bowden, A., and M. L. Cárdenas. 2017. Life before LUCA. *J. Theor. Biol.* 434: 68–74. doi:10.1016/j.jtbi.2017.05.023.

Cronin, L. 2017. A new genesis for origins research? *Chem* 2: 601–603.

Darwin, C. R. 1859. *On the Origin of Species*. London, UK: John Murray.

Dawkins, R. 2009. *The Greatest Show on Earth. The Evidence for Evolution*. London, UK: Bantam Press.

de Duve, C. 1991. *Blueprint for a Cell. The Nature and Origin of Life*. Burlington, NC: Neil Patterson.

de Duve, C. 1993. Co-chairman's remarks. The RNA world: Before and after? *Gene* 135: 29–31.

de Duve, C. 2005a. *Singularities: Landmarks on the Pathways of Life.* Cambridge, UK: Cambridge University Press.

de Duve, C. 2005b. The onset of selection. *Nature* 433: 581–82. doi:10.1038/433581a.

de Duve, C. 2011. Life as a cosmic imperative? *Philos. Trans. R. Soc. Lond. A Math. Phys. Eng. Sci.* 369: 620–623. doi:10.1098/rsta.2010.0312.

de Duve, C., and J. Vandenhaute. 2013. *Sur la science et au-delà.* Paris, France: Odile Jacob.

de la Escosura, A., C. Briones, and K. Ruiz-Mirazo. 2015. The systems perspective at the crossroads between chemistry and biology. *J. Theor. Biol.* 381: 11–22.

Eschenmoser, A., and E. Loewenthal. 1992. Chemistry of potentially prebiological natural products. *Chem. Soc. Rev.* 21: 1–16.

Fry, I. 2000. *The Emergence of Life on Earth. A Historical and Scientific Overview.* New Brunswick, NJ: Rutgers University Press.

Fry, I. 2019a. Philosophical aspects of the origin-of-life question: Neither by chance not by design. In V. Kolb (Ed.), *Handbook of Astrobiology.* Boca Raton, FL: CRC Press/Taylor & Francis Group. pp. 109–124.

Fry, I. 2019b. The origin of life as an evolutionary process: Representative case studies. In V. Kolb (Ed.), *Handbook of Astrobiology.* Boca Raton, FL: CRC Press/Taylor & Francis Group. pp. 437–462.

Gánti, T. 2003. *The Principles of Life, with Commentary by James Griesemer and Eörs Szathmáry.* Oxford, UK: Oxford University Press.

Goldford, J. E., H. Hartman, T. F. Smith, and D. Segrè. 2017. Remnants of an ancient metabolism without phosphate. *Cell* 168: 1126–1134. doi:10.1016/j.cell.2017.02.001.

Haldane, J. B. S. 1929. The origin of life. *Ration. Annu.* 148: 3–10.

Hanczyc, M. M., and P.-A. Monnard. 2017. Primordial membranes: More than simple container boundaries. *Curr. Open. Chem. Biol.* 40: 78–86. doi:10.1016/j.cbpa.2017.07.009.

Hardy, M. D., J. Yang, J. Selimkhanov, C. M. Cole, L. S. Tsimring, and N. K. Devaraj. 2015. Self-reproducing catalyst drives repeated phospholipid synthesis and membrane growth. *Proc. Nat. Acad. Sci. U.S.A.* 112: 8187–8192. doi:10.1073/pnas.1506704112.

Hein, J. E., D. Gherase, and D. G. Blackmond. 2012. Chemical and physical models for the emergence of biological homochirality. *Top. Curr. Chem.* 333: 83–108. doi:10.1007/128_2012_397.

Ichihashi, N., and T. Yomo. 2016. Constructive approaches for understanding the origin of self-replication and evolution. *Life* 6: 26.

Ichihashi, N., K. Usui, Y. Kazuta, T. Sunami, T. Matsuura, and T. Yomo. 2013. Darwinian evolution in a translation-coupled RNA replication system within a cell-like compartment. *Nat. Comm.* 4: 2494. doi:10.1038/ncomms3494.

Ichihashi, N., T. Matsuura, H. Kita, T. Sunami, H. Suzuki, and T. Yomo. 2010. Constructing partial models of cells. *Cold Spring Harb. Perspect. Biol.* 2: a004945.

Islam, S., and M. W. Powner. 2017. Prebiotic systems chemistry: Complexity overcoming clutter. *Chem.* 2: 470–501.

Islam, S., D. K. Bučar, and M. W. Powner. 2017. Prebiotic selection and assembly of proteinogenic amino acids and natural nucleotides from complex mixtures. *Nat. Chem.* 9: 584–589.

Kaneko, K. 2006. *Life: An Introduction to Complex Systems Biology.* Berlin, Germany: Springer.

Keller, E. F. 2009. Knowing as making, making as knowing: The many lives of synthetic biology. *Biol. Theor.* 4: 333–339.

Keller, M. A., A. V. Turchyn, and M. Ralser. 2014. Non-enzymatic glycolysis and pentose phosphate pathway-like reactions in a plausible archean ocean. *Mol. Syst. Biol.* 10: 725.

Keller, M. A., A. Zylstra, C. Castro, A. V. Turchyn, J. L. Griffin, and M. Ralser. 2016. Conditional iron and pH-dependent activity of a non-enzymatic glycolysis and pentose phosphate pathway. *Sci. Adv.* 2: e1501235. doi:10.1126/sciadv.1501235.

Keller, M. A., D. Kampjut, S. A. Harrison, and M. Ralser. 2017. Sulfate radicals enable a non-enzymatic krebs cycle precursor. *Nat. Ecol. Evol.* 1: 83. doi:10.1038/s41559-017-0083.

Krishnamurthy, R, and V. Snieckus. 2017. Prebiotic organic chemistry and chemical pre-biology: Speaking to the synthetic organic chemists. *Synlett* 28: 27–29.

Krishnamurthy, R. 2017. Giving rise to life: Transition from prebiotic chemistry to protobiology. *Acc. Chem. Res.* 50: 455–459.

Lazcano, A. 2010a. Historical development of origins research. *Cold Spring Harb. Perspect. Biol.* 2: a002089. doi:10.1101/cshperspect.a002089.

Lazcano, A. 2010b. Which way to life? *Orig. Life Evol. Biosph.* 40: 161–167. doi:10.1007/s11084-010-9195-0.

Lazcano, A. 2012. The biochemical roots of the RNA world: From zymonucleic acid to ribozymes. *Hist. Phils. Life Sci.* 34: 407–423.

Lazcano, A. 2017. L'émergence de la vie: quelques notes sur l'origine de l'information biologique. In *Colloquium de Cerisy: Sciences de la vie, sciences de l'information*, edited by T. Gaudin, D. Lacroix, M. C. Maurel, and J. C. Pomerol, 3–14. Paris, France: ISTE Editions.

Lazcano, A., and S. L. Miller. 1999. On the origin of metabolic pathways. *J. Mol. Evol.* 49: 424–431.

Lehman, N. 2017. Reaction: Systematic hope for life's origins. *Chem* 2: 604–605.

Letelier, J. C., M. L. Cárdenas, and A. Cornish-Bowden. 2011. From L'Homme Machine to metabolic closure: Steps towards understanding life. *J. Theor. Biol.* 286: 100–113. doi:10.1016/j.jtbi.2011.06.033.

Messner, C. B., P. C. Driscoll, G. Piedrafita, M. F. L. De Volder, and M. Ralser. 2017. Nonenzymatic gluconeogenesis-like formation of fructose 1,6-bisphosphate in ice. *Proc. Nat. Acad. Sci. U.S.A.* 114: 7403–7407. doi:10.1073/pnas.1702274114.

Miljanić, O Š. 2017. Small-molecule systems chemistry. *Chem* 2: 502–524.

Miller, S. L. 1953. A production of amino acids under possible primitive earth conditions. *Science* 117: 528–529.

Miller, S. L. 1957. The mechanism of synthesis of amino acids by electric discharges. *Biochim. Biophys. Acta* 23: 480–489.

Monod, J. 1970. *Le hasard et la necessité. Essai sur la philosophie naturelle de la biologie moderne.* Paris, France: Éditions du Seuil.

Morowitz, H. J. 1992. *Beginnings of Cellular Life. Metabolism Recapitulates Biogenesis.* New Haven, CT: Yale University Press.

Muchowska, K. B., S. J. Varma, E. Chevallot-Beroux, L. Lethuillier-Karl, G. Li, and J. Moran. 2017. Metals promote sequences of the reverse Krebs cycle. *Nat. Ecol. Evol.* doi:10.1038/s41559-017-0311-7.

Murillo-Sánchez, S., D. Beaufils, J. M. González Mañas, R. Pascal, and K. Ruiz-Mirazo. 2016. Fatty acids' double role in the prebiotic formation of a hydrophobic dipeptide. *Chem. Sci.* 7: 3406–3413.

Oparin, A. I. 1924. *Proiskhozhedenie zhizni.* Moscow, Russia: Mosckovskii Rabochii.

Orgel, L E. 2003. Some consequences of the rna world hypothesis. *Orig. Life Evol. Biosph.* 33: 211–218.

Oró, J. 1960. Synthesis of adenine from ammonium cyanide. *Biochim. Biophys. Res.* 2: 407–412.

Pascal, R., and A. Pross. 2017. A roadmap toward synthetic protolife. *Synlett* 28: 30–35.

Patel, B. H., C. Percivalle, D. J. Ritson, C. D. Duffy, and J. D. Sutherland. 2015. Common origins of RNA, protein and lipid precursors in a cyanosulfidic protometabolism. *Nat. Chem.* 7: 301–307.

Peretó, J. 2005. Controversies on the origin of life. *Int. Microbiol.* 8: 23–31.

Peretó, J. 2012. Out of fuzzy chemistry: From prebiotic chemistry to metabolic networks. *Chem. Soc. Rev.* 41: 5394–5403. doi:10.1039/c2cs35054h.

Peretó, J. 2016. Erasing borders: A brief chronicle of early synthetic biology. *J. Mol. Evol.* 83: 176–183. doi:10.1007/s00239-016-9774-4.

Peretó, J., and J. Català. 2007. The renaissance of synthetic biology. *Biol. Theor.* 2: 128–130. doi:10.1162/biot.2007.2.2.128.

Peretó, J., J. L. Bada, and A. Lazcano. 2009. Charles Darwin and the origin of life. *Orig. Life Evol. Biosph.* 39: 395–406. doi:10.1007/s11084-009-9172-7.

Peretó, J., P. López-García, and D. Moreira. 2004. Ancestral lipid biosynthesis and early membrane evolution. *Trends Biochem. Sci.* 29: 469–977. doi:10.1016/j.tibs.2004.07.002.

Porcar, M., and J. Peretó. 2014. *Synthetic Biology: From iGEM to the Artificial Cell.* Dordrecht, the Netherlands: Springer. doi:10.1007/978-94-017-9382-7.

Powner, M. W., and J. D. Sutherland. 2011. Prebiotic chemistry: A new modus operandi. *Philos. Trans. R. Soc. Lond. B: Biol. Sci.* 366: 2870–7287. doi:10.1098/rstb.2011.0134.

Powner, M. W., B. Gerland, and J. D. Sutherland. 2009. Synthesis of activated pyrimidine ribonucleotides in prebiotically plausible conditions. *Nature* 459: 239–242.

Ralser, M. 2014. The RNA world and the origin of metabolic enzymes. *Biochem. Soc. Trans.* 42: 985–988. doi:10.1042/BST20140132.

Robertson, M. P., and S. L. Miller. 1995. An efficient prebiotic synthesis of cytosine and uracil. *Nature* 375: 772.

Ruiz-Mirazo, K. 2015a. Opening the systemic avenue from chemistry to biology. *Mapping Ignorance.* https://mappingignorance.org/2015/04/27/opening-the-systemic-avenue-from-chemistry-to-biology/.

Ruiz-Mirazo, K. 2015b. Protocell. In *Encyclopedia of Astrobiology*, edited by M. Gargaud, W. M. Irvine, R. Amils, H. J. Cleaves, D. Pinti, J. Cernicharo, D. Rouan, T. Spohn, S. Tirard, and M. Viso, 2nd ed. Berlin, Germany: Springer.

Ruiz-Mirazo, K., C. Briones, and A. de la Escosura. 2014. Prebiotic systems chemistry: New perspectives for the origins of life. *Chem. Rev.* 114: 285–366.

Ruiz-Mirazo, K., C. Briones, and A. de la Escosura. 2017. Chemical roots of biological evolution: The origins of life as a process of development of autonomous functional systems. *Open Biol.* 7: 170050. doi:10.1098/rsob.170050.

Ruiz-Mirazo, K., J. Peretó, and A. Moreno. 2004. A universal definition of life: Autonomy and open-ended evolution. *Orig. Life Evol. Biosph.* 34: 323–346. doi:10.1023/B:ORIG.0000016440.53346.dc.

Scharf, C., N. Virgo, H. J. Cleaves, M. Aono, N. Aubert-Kato, A. Aydinoglu, A. Barahona, et al. 2015. A strategy for origins of life research. *Astrobiology* 15: 1031–1042. doi:10.1089/ast.2015.1113.

Semenov, S. N., L. J. Kraft, A. Ainla, M. Zhao, M. Baghbanzadeh, V. E. Campbell, K. Kang, J. M. Fox, and G. M. Whitesides. 2016. Autocatalytic, bistable, oscillatory networks of biologically relevant organic reactions. *Nature* 537: 656–660. doi:10.1038/nature19776.

Shirt-Ediss, B., S. Murillo-Sánchez, and K. Ruiz-Mirazo. 2017. Framing major prebiotic transitions as stages of protocell development: Three challenges for origins-of-life research. *Beilstein J. Org. Chem.* 13: 1388–1395. doi:10.3762/bjoc.13.135.

Smith, E., and H. J. Morowitz. 2016. *The Origin and Nature of Life on Earth: The Emergence of the Fourth Geosphere.* Cambridge, UK: Cambridge University Press.

Solé, R. 2016. Synthetic transitions: Towards a new synthesis. *Philos. Trans. R. Soc. B: Biol. Sci.* 371: 20150438. doi:10.1098/rstb.2015.0438.

Springsteen, G., J. R. Yerabolu, J. Nelson, C. J. Rhea, R. Krishnamurthy. 2018. Linked cycles of oxidative decarboxylation of glyoxylate as protometabolic analogs of the citric acid cycle. *Nat. Comm.* 9: 91. doi:10.1038/s41467-017-02591-0.

Stairs, S., A. Nikmal, D. K. Bučar, S. L. Zheng, J. W. Szostak, and M. W. Powner. 2017. Divergent prebiotic synthesis of pyrimidine and 8-oxo-purine ribonucleotides. *Nat. Comm.* 8: 15270. doi:10.1038/ncomms15270.

Sutherland, J. D. 2016. The origin of life—Out of the blue. *Angew. Chem. Int. Ed.* 55: 104–121.

Sutherland, J. D. 2017. Studies on the origin of life—The end of the beginning. *Nat. Rev. Chem.* 1: 12.

Szostak, J. W. 2011. An optimal degree of physical and chemical heterogeneity for the origin of life? *Philos. Trans. R. Soc. B: Biol. Sci.* 366: 2894–2901.

Szostak, J. W. 2012. The eightfold path to non-enzymatic RNA replication. *J. Syst. Chem.* 3: 2.

Szostak, J. W. 2017. The narrow road to the deep past: In search of the chemistry of the origin of life. *Angew. Chem. Int. Ed.* 56: 11037–11043. doi:10.1002/anie.201704048.

Szostak, J. W., D. P. Bartel, and L. Luisi. 2001. Synthesizing life. *Nature* 409: 387–390. doi:10.1038/35053176.

Taylor, J. W., S. A. Eghtesadi, L. J. Points, T. Liu, and L. Cronin. 2017. Autonomous model protocell division driven by molecular replication. *Nat. Comm.* 8: 237. doi:10.1038/s41467-017-00177-4.

Tor, Y. 2017. Catalyst: On organic systems chemistry and origins research. *Chem* 2: 448–450.

Wächtershäuser, G. 1988. Before enzymes and templates: Theory of surface metabolism. *Microbiol. Rev.* 52: 452–484.

Wagner, A. J., and D. G. Blackmond. 2016. The future of prebiotic chemistry. *ACS Cent. Sci.* 2: 775–777.

Walde, P., H. Umakoshi, P. Stano, and F. Mavelli. 2014. Emergent properties arising from the assembly of amphiphiles. Artificial vesicle membranes as reaction promoters and regulators. *Chem. Comm.* 50: 10177–10197. doi:10.1039/c4cc02812k.

Weiss, M. C., F. L. Sousa, N. Mrnjavac, S. Neukirchen, M. Roettger, S. Nelson-Sathi, and W. F. Martin. 2016. The physiology and habitat of the last universal common ancestor. *Nat. Microbiol.* 1: 16116.

Wong, A. S. Y., and W. T. S. Huck. 2017. Grip on complexity in chemical reaction networks. *Beilstein J. Org. Chem.* 13: 1486–1497. doi:10.3762/bjoc.13.147.

5.2 Prebiotic Chemical Pathways to RNA and the Importance of Its Compartmentation

Peter Strazewski

CONTENTS

5.2.1 INTRODUCTION

5.2.1.1 HIERARCHICAL REDUCTION IN THE RESEARCH OF ORIGINS

"Hierarchical reduction" (Pross 2012) of problem solving in the logics of the natural sciences means that, in order to fully solve a problem, which always includes the problem of finding the true origin of its processes, we need to navigate outside the field, in a domain that embraces and integrates the field, rather than being simply part of it. For example, our perceived reality has three spatial dimensions, so we can easily visualize dynamic one-dimensional (1D) and two-dimensional (2D) objects, such as vibrating violin strings and tambourine membranes, but we reach our limits when mentally picturing dynamic three-dimensional (3D) objects, such as those that we call atomic or molecular "orbitals," and our "universe." We then prefer to resort to holographic projections from a reduced dimensionality.

Applying this logical rule to "major transitions" in our perceived reality, that is, the large decisive steps in evolution (Maynard Smith and Szathmáry 1995), we can state that the *origins* of the physical cosmos, biotic life, human language, and societal economy are studied best by logicians and mathematicians for physics, including cosmology; by astronomers, geologists, and chemists for biology, including astrobiology and the origin of life; by biologists for linguistics, including etymology and the origin of language; and by psychologists for economy, politics, and wars, including the origin of human trade, respectively (Figure 5.2.1). For example, a neurobiologist working on the origin of human language will *think* like a linguist in terms of expected results but *exploit* "Darwinian" algorithms to generate these results. A psychologist will *study* human behavior in a family community and *translate* its effects to societies that are at (trade) war with one another.

In this set theory of scientific relations, chemistry bears a very special role, since it operates at the border between the inanimate and the animate states of matter (outermost white border in Figure 5.2.1). Robert Pascal and Addy Pross have recently put forward that physics embraces both states of matter, "dead and alive," only when thermodynamic (and mechanic) stability

FIGURE 5.2.1 Euler diagram of the hierarchic relationship between sciences: inner ∈ outer zones. Relations of the inner zones are more complex, and their elements act more autonomously than those of the outer zones. The more inward bound the zone, the larger the distance from, and more complex the relationship to, the fundament of logics. Diffuse gray borders define abstract and very general elements, concepts, and rules. Dark borders delineate inanimate ("inert", "dead") elements observed in nature. White borders demarcate animate ("alive") elements. Dynamic relationships (rules) within gray and dark borders are, in principle, predictable properties. Dynamic relationships within white borders are best described as a combination of properties and not exactly predictable "behavior." First, second, and third white borders mark out elements of increasingly higher autonomy ("intelligence").

is looked upon as one of the two fundamental ways of persisting in time, the other way being the exponential reproduction ("replication") of individual units. A population or colony of similar reproducing units will *persist, despite the relative instability* and eventual degradation of its individual members. The most fundamental notion in physics must therefore be *persistence* rather than stability (Pascal and Pross 2015). Hence, a chemist or geochemist that aims at *understanding* the transition from one state to the other, for example, the origin of life on Earth, will have to *explore* first the "bricks" that ultimately make up a living system. Prebiotic chemistry is a very active science of prominent age; its task is to provide the bricks of life from "natural" and "plausible" chemical reaction conditions. A second step will consist of experimentally realizing abiogenesis from such or similar bricks (cf. Chapter 6.3).

5.2.1.2 SYSTEMIC MEMORY: PRINCIPLES OF COPYING AND ENCODING

The bricks that biotic nature—as we know it on Earth—uses to maintain a systemic memory throughout many generations of reproduction of individual system units, which themselves individually almost fully degrade and ultimately vanish, are composed of nucleic acids. Nucleic acids can harbor and transmit an astonishingly large number of information through more or less faithfully copying long strings termed "polymers." In biotic nature, these strings are composed of four different (but similar) letters, that is, information-theoretical *quaternary digits* ("quits," not Qbits) realized by the nucleotides A, G, C, and U or T. Natural *quits* are pairwise complementary to one another through the Watson-Crick rules (G-C, A-U, or A-T), which give the grounds for faithful template-directed

copying, as during cellular replication (double copying of complementary single-stranded DNA), or complement copying, as in transcription or reverse transcription, of strings of nucleic acids of virtually deliberate length. Upon translation, however, these strings of *quits*, rather than being recognized one by one, as during complement copying, can be read out by anticodons—parts of transfer RNA bound to ribosomes—as a series of *three-letter words* termed "base triplets," that is, information-theoretical unitary blocks of 3-*quit* "quytes" ($3Q$), that are chained up in heterogeneously and almost deliberately long sentences termed "reading frames" (genes). The grammar and syntax (gene regulation, message editing, and epigenetics) used in these sentences are then a matter of *system unit type* and *network organization*, that is, cell (germ line or somatic), organism, species, interaction with other species, ecological niches and traits, and so on.

All known natural (biotic) genomes can be divided into a "coding fraction," made up of the sum of all reading frames, and a "non-coding fraction" composed of all untranslated but nevertheless "permissively transcribed" regulator and not transcribed spacer nucleic acids. Known natural total genome sizes range from 600 kilobases (kb) in *Mycoplasma genitalium* through 2.9–3.4 gigabases (Gb) in chicken, mice, chimpanzees, and humans to 250 Gb in *Psilotum nudum* (skeleton fork fern). Known amounts of natural genes span between 481 in *M. genitalium* (average distance between genes $đ = 1.2$ kb) through 16,736–23,493 in chicken, mice, chimpanzees, and humans ($đ = 141$–177 kb) to about 27,000–50,000 genes, many of them are gene repeats, in plants such as cress, rice, and maize ($đ ≈ 4$–50 kb). Known coding fractions can thus take anything between 2% in *Homo sapiens* to 90% in *M. genitalium*.

According to Claude Shannon (Equation 5.2.1), the

$$\text{Shannon Information} \left(\text{SI}\right) = \text{Log}_b N \qquad (5.2.1)$$

where b denotes digit multiplicity (the number of different digits)—$b = 2$ for *bit* (binary digit), $b = e \approx 2.718\ldots$ for *nit/nat/nepit* (natural or Neperian digit), $b = 3$ for *trit* (trinary or ternary digit), $b = 4$ for *quit* (quaternary digit), etc.—and N denotes the number of digits in a string, for example, the polynucleotide sequence length. The maximal amount of information, viz. the degree of minimal diversity, stored in the coding fraction of the *M. genitalium* genome equals $SI = \text{Log}_4 (0.9 \cdot 6 \cdot 10^5) = \text{Log}_4 10 \cdot \text{Log}_{10} (0.9 \cdot 6 \cdot 10^5) \approx 1.66 \cdot 5.73 \approx 9.5$, whereas the SI stored in 2% of the human genome merely is $SI = \text{Log}_4 (0.02 \cdot 3.2 \cdot 10^9) \approx 7.8$. Total genome sizes result in maximal information contents $SI = 9.6$ for *M. genitalium* and, ignoring gene repeats, $SI = 15.8$ for *H. sapiens*, which indicates that the evolution from one to the other species undeniably prevailed in the non-coding fraction of their genomes, i.e. in their long non-coding RNA (Morillon 2018).

Biotic nucleic acids such as RNA and DNA, whether translated or not, are used in known animate systems as *quit* carriers that are relatively easy to copy through molecular templating, irrespective of whether this copying is assisted by enzymes or not. Complementary or self-complementary readouts, that is, copying and encoding rules as we know them from biotic genome replication and the universal genetic code, are reducible to the hydrogen donor-acceptor patterns that are being exposed from the so-called Watson-Crick face of the natural nucleobases of each nucleotide (Strazewski 1990; Strazewski and Tamm 1990). These patterns are not limited to the natural nucleobases. Other N-heterocycles may furnish different patterns and thus distinct pairing preferences (Krishnamurthy 2012). Therefore, from a purely chemical point of view, molecular template variants such as binary- or ternary-digit memory polymers composed of strings of subsequent *bits* or *trits*, respectively, are well imaginable (Figure 5.2.2). These *bits*

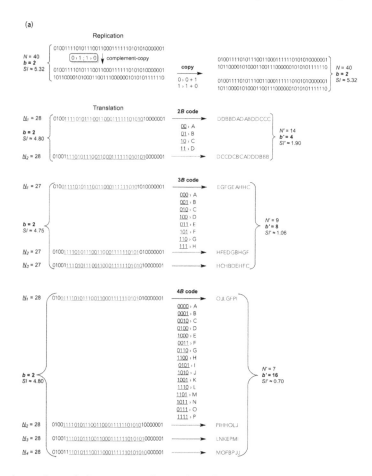

FIGURE 5.2.2 Template copying and translating an exemplary string of total $N = 40$ residues, composed of b digits, storing, according to Equation 5.2.1, a maximum amount of Shannon Information (*SI*). Upper part: Replication. Complement-copying rules (small frames) by virtue of a minimal requirement for self-complementary digits, that is, *bits* {0,1}, *trits* {0,1,2}, and *quits* {0,1,2,3}. Lower part: Translation of consecutive blocks (B = bytes, T = trytes, Q = quytes) of two to four digits (a *bits*, b *trits*, and c *quits*). The reading frame (underligned digits) is translated into products (strings of letters) of a condensed residue number N' and higher digit multiplicity (diversity) $b' = b^{2-4}$, and thus, lower SI'. Frameshifts $\underline{N}_{1-2}$ for two-digit blocks, $\underline{N}_{1-3}$ for three-digit blocks, and $\underline{N}_{1-4}$ for four-digit blocks generate alternative translation products of the same length N', diversity b', and SI' but different sequences. (a) Binary digit (*bit*) strings, replicated and translated from 2–*bit* bytes (2B), 3–*bit* bytes (3B), and 4–*bit* bytes (4B).

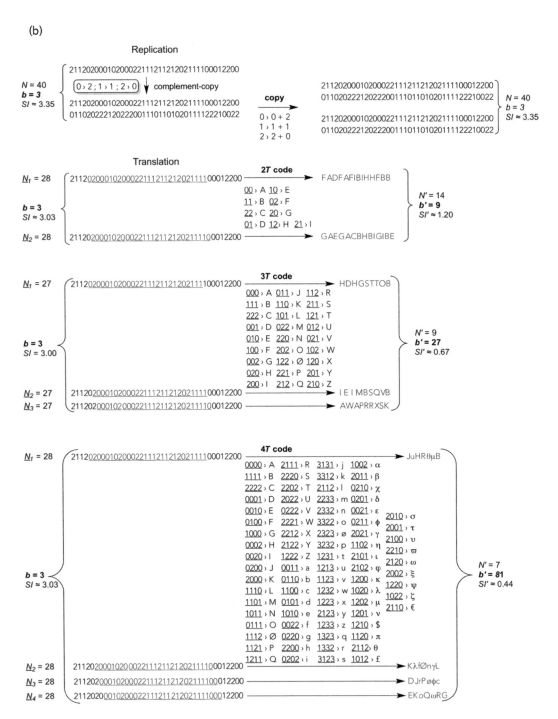

FIGURE 5.2.2 (*Continued*) (b) Ternary digit (*trit*) strings, replicated and translated from 2–*trit trytes* (2*T*), 3–*trit trytes* (3*T*), and 4–*trit trytes* (4*T*). For a *trit* string to bear the same *SI* as that of a *bit* string, it needs to be about 8.65-fold longer (higher *N*) than a *bit* string; for example, $\text{Log}_2 40 \approx \text{Log}_3 346$.

or *trits* could be complementary through different pairing modes. In principle, each *bit* or *trit* could be strictly self-complementary in its pairing property: 0 pairs only with 0, 1 pairs only with 1, and 2 pairs only with 2. Chemically more likely, alternative *bit* genomes could be composed of, for instance, only G and C or only A and U, where one digit (0 or 1) is complementary to the other (Figure 5.2.2a). Alternative *trit* genomes could bear two digits that are

complementary to one another (e.g., 0 pairs with 2) and a third strictly self-complementary digit (1 pairs only with 1) and thus be composed of, say, G, C, and X, the latter being a strictly self-complementary nucleotide (Figure 5.2.2b). In addition, the coding fractions of such low-digit memory polymers could be read out, for example, as 4-*bit bytes* (4*B*) or 3-*trit trytes* (3*T*). Biotic 3-*quit quytes* (3*Q*: natural base triplets) comprise $b' = b^3 = 4^3 = 64$ values, that is, offer 64

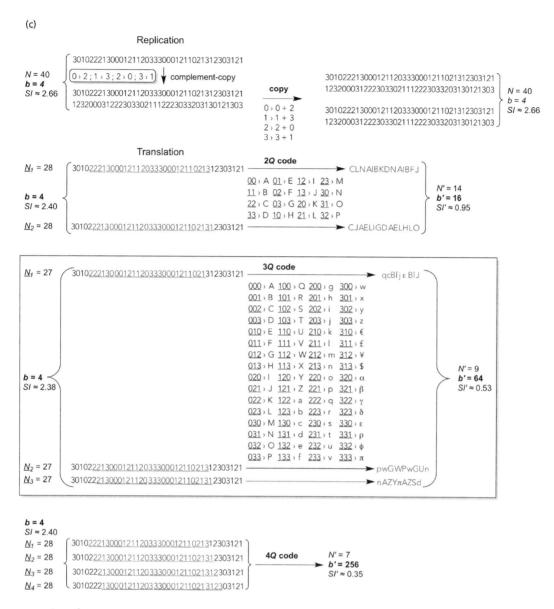

FIGURE 5.2.2 (*Continued*) (c) Quaternary digit (*quit*) strings, replicated and translated from 2–*quit quytes* (2Q), 3–*quit quytes* (3Q), and 4–*trit quytes* (4Q, code and translation products not shown). For a *quit* string to bear the same *SI* as that of a *bit* string, it needs to be 400-fold longer (higher *N*) than a *bit* string; for example, $\text{Log}_2 40 = \text{Log}_4 1600$. Framed: the current natural (biotic) memory systems are *quit* strings being translated from reading frames of consecutive 3Q, utilizing $b' = 64$ codons that are reduced, mainly for fidelity reasons, to $b'_{eff} = 21$ effectively translated digits, viz. 20 different proteinogenic amino acids and a stop signal (i.e., lack of amino acid).

different triplet "codons" (large frame in Figure 5.2.2c). So do 6-*bit bytes* of binary memory polymers (6B: $b' = 2^6 = 64$, not shown), but such long codons would necessitate hexaplet anticodons for translational readout. Shorter 5-*bit bytes* (5B) generate $b' = 32$ different pentaplet codons. Chemically more realistic are 4-*bit bytes* (4B), giving rise to $b' = 16$ different quadruplet codons, and 3-*bit bytes* (3B), giving merely $b' = 8$ different triplet codons (Figure 5.2.2a). In ternary-digit memory polymers, blocks of 3-*trit trytes* (3T) produce $b' = 3^3 = 27$ different triplet codons, whereas 4-*trit trytes* (4T) generate $b' = 3^4 = 81$ different quadruplet codons. The latter set of codons would suffice for an even larger than natural (biotic) diversity of translated digits b', cf. 4T code in Figure 5.2.2b versus 3Q code in Figure 5.2.2c.

These are simply numerical–combinatorial guidelines "degenerate" (redundant) and stop codons exempt. The modern-day ribosomal translation mechanism has established a universal genetic code based on 64 different 3-*quit quytes*, that is, triplet codons that are currently occupied by merely 20 "proteinogenic" amino acids and usually 3 stop codons, unless a biocompatible "expanded alphabet" for triplet codons has been artificially introduced at selected positions by using synthetic nucleotides that offer a distinct pairing selectivity that may differ from the natural Watson-Crick rules (Picirilli et al. 1990; Strazewski and Tamm 1990; Cornish et al. 1995; Liu et al. 2003; Malyshev et al. 2014; Georgiadis et al. 2015; Winiger et al. 2016). Most of the 20 proteinogenic amino acids are coded by a set of faster and slower, thus, more and less erroneously translated, redundant codons being read by more abundant and rarer "isoaccepting" anticodon triplets, respectively, all carrying the same amino acid.

However, these copying and encoding principles (Figure 5.2.2) show that unbranched molecular strings (1D polymers) composed of a limited number of different monomeric complementary residues (monomers), that is, strings that bear a relatively *low multiplicity* of digits (low *b*), are likely to be a *general feature of memory keepers* in any animate system. The lower the digit multiplicity the smaller the diversity of a given string length (*N*), the simpler the template, and less ambiguous to copy and replicate the string on a molecular level (Szathmáry 1992). This generates fewer errors in template-copied memory polymers and thus a higher replication error threshold for a given spreading "quasi-species" (similar genome population), and this eventually imposes a weaker selection pressure on the maximal genome length of any evolved organism (Eigen 1971; Biebricher and Eigen 2005; Kun et al. 2005).

The opposite is true when it comes to *functional translation products*; there, the higher their digit multiplicity (high *b'*), the *higher the diversity* (low *SI'*) of a given resulting string length (lower *N'*). The high diversity is further multiplied by the number of reading frame shifts (*Nb*) that give rise to an encoded set of a completely different choice of translated string sequences (Figure 5.2.2 bottom). This generates translated string polymers that are inherently difficult to copy through direct templating, since the complement rules—analogous to the Watson-Crick base-pairing rules—needed to be as manifold and exclusive as the digits are diverse. On the other hand, the longer the translation blocks (codons, translated words) in the messenger nucleic acids, the more compact the generated diversity of the "secondary" polymer, which allows for more diverse "molecular functions" at a given secondary string length *N'*. In other words, we should expect evolution from alien and very early biota to string polymers of both kinds, low-digit and high-digit variants, where the more diverse latter is encoded by the simpler former.

Alternative nucleic acids composed of fewer letters, *bits* or *trits* rather than *quits*, might be considered in extra-terrestrial biota and/or during early periods of the origin of life on Earth. They might encode a smaller or larger choice of proteinogenic amino acids—or some other molecular equivalent of a functionally more diverse polymer than nucleic acids—by translating from shorter or longer *bytes* or *trytes,* respectively, as mentioned earlier. In principle, alternative nucleic acids could also form triple complements through triple-strand formation or even higher-order supramolecular string associations. The chemical reality, as expressed in pairing/tripling/quadrupling properties of such alternative nucleic acids, would expectedly impose grave consequences on their copying and translation fidelity and thus on the number of genes and maximal genome length, all of which is out of scope of this chapter. In principle, memory strings could also be extended to higher than quaternary digit multiplicities (not to be confused with an "expanded alphabet" of triplet codons) and translated using longer than four-digit blocks (pentaplet, hexaplet, etc. codons). In the reality of macromolecules offered by nature, however, more diverse higher-digit memory polymers are likely to be copied more erroneously, since the monomers

would necessarily be more similar to one another, again limiting the replication error threshold, maximal number of genes, and total genome length. In addition, longer codons than quadruplets are at higher risk of being misread because of spontaneous frameshifting and mispairing, which would produce more erroneously assembled proteins (secondary polymers) and necessitate a more elaborate and costly error correction effort by the system. Yet, alien biota that would provide linear memory polymers that were markedly more rigid than RNA or DNA, thus perhaps less prone to frameshifting and mispairing, should not be ruled out *a priori*, not for chemical reasons. The overall energetic "cost" at the available energy influx needed to generate such polymers, to keep their replication error threshold high, and also to keep the erroneously produced secondary polymers under a liveable limit, are probably much more preventive factors than the huge choice of bricks chemistry can offer.

The prebiotic chemistry on our planet apparently produced prebiotic bricks that could condense under prebiotic reaction conditions into 1D polymers (nucleic acids) that could form double strands, at least locally in certain string zones, through the spontaneous association (hybridization) of pair-wise complementary digits, cf. complement copying in Figure 5.2.2c. The digit multiplicity of the first replicating nucleic acids (*bit, trit, quit,* etc.) is unknown, although there is a consensus on *bit* polymers having preceded modern natural (biotic) nucleic acids that are generally *quit* polymers. For reasons that shall be elucidated in the sections further on, these prebiotic bricks are purine and pyrimidine ribonucleosides that, under appropriate prebiotic reaction conditions being present on this planet some 3.6 Gya, could condense with phosphate and polymerize into RNA and similar RNA-like linear polymers. At least a part of the early nucleic acid single strands could synthesize 3*Q*-translated secondary polymers, viz. polypeptides and proteins very early on, or else, we would hardly expect the genetic code to be universal (Theobald 2010). Apparently, on Earth, RNA proved to be the most successful "primary" 1D (memory) polymer. Not only can its monomer sequence be easily copied and faithfully reproduced, but also more faithful and streamlined information storage carriers can be derived from RNA by its deoxygenation to DNA. Most importantly, RNA, not DNA, can direct and catalyze the linking of amino acids into defined strings of polypeptides, that is, take an essential part in catalyzing the controlled dehydration of amino acids to produce amide bonds, a process termed peptidyl transfer (PT). Strong evidence suggests that uncoded PT preceded coded PT and, thus, that RNA could grow polypeptide chains from amino acids before a recognition system eventually emerged—from RNA, too—that allowed RNA-directed PT to profit from specific codon-anticodon interactions, thus, to translate genetic information (Section 6).

5.2.1.3 COPYING VERSUS COMPARTMENTATION

Long unfolded template polymers need to be soluble and disaggregated for copying. Self-folded polymer zones require the costly effort of denaturation (unfolding). From a prebiotic chemical point of view, RNA is probably ancestral to DNA, although perhaps not exclusively (Section 5.2.2.3). The molecular reason for why, out of all previously discussed possibilities, not base-paired (dehybridized) RNA and DNA single-strands constitute the ideal systemic memory polymer is that they are periodically polyanionic and thus very well water-soluble and reasonably unfolded (because of being hydrophilic throughout) when compared with other polymers, at almost deliberate chain (string) lengths. An inevitable consequence of this physico-chemical property is that RNA and DNA are inherently difficult to compartment in "units" that separate bulk solvent from "self" content. Hans Kuhn (later with Jürg Waser), Sidney Walter Fox with Klaus Dose, and William Hargreaves with David Deamer have realized in the 1970s that semipermeable compartments are prerequisite for "biology to happen," that is, what was called "molecular Darwinism" or "molecular evolution" to start from a prebiotic mixture of chemical substances. Kuhn, Waser, and others assumed that porous mineral (rock or clay) micro-cavities played the earliest compartmenting role for nucleic acids (Kuhn 1972; Kuhn and Waser 1981; Greenwood Hansma 2017). Kuhn postulated that, at an early stage of production of secondary polymers (polypeptides) from mineral-caged primary RNA or DNA (–) strands, that is, the pairing complements of the translated (+) strands, the translation products would rarely be soluble and catalytically functional, or not at all. Such (–)-strand polypeptides would have been made up of "gibberish" amino acid sequences that most likely misfolded and coagulated, ending up as aggregated polypeptide "coacervates" that had encapsulated the RNA or DNA from which they originated. Waste peptides of that sort might have eventually become the translated source of primitive organic compartments. Fox and Dose, on the other hand, proposed that microspheres composed of "proteinoids," viz. untranslated insoluble polypeptide mixtures generated from heating certain amino acids, could form the first "protocellular" vesicular objects defined by semipermeable membranes (Fox and Dose 1972, 1977). Hargreaves and Deamer showed for the first time that phospholipids and "fatty" acids would aggregate under simulated prebiotic hydration conditions and form membranes that would spontaneously close to give micrometer-sized vesicles and thus define an isolated aqueous interior "lumen" that could, in principle, harbor foreign macromolecules (Hargreaves et al. 1977; Hargreaves and Deamer 1978). Membranogenic lamellar-phase bilayered amphiphiles, usually originating from hydrating lipids, are ever since the most convincing candidates for primitive dynamic protocellular compartments.

Unlike clay or volcanic mineral pores, or even (viscous) proteinoid coacervates, lamellar-phase amphiphiles can, depending on temperature and composition, adopt either a disordered liquid-like phase (2D liquid) within the bilayered lamella that make up the membranes, or they aggregate in a more ordered but soft gel phase. Both phases transform into one another non-linearly with temperature, through a phase transition involving an abrupt molecular reorganization on a supramolecular level. Spontaneously separated 2D zones of both phases, liquid disordered and gel, can co-exist in the same membrane. In addition, these membrane-bound amphiphiles are in a more or less rapid dynamic exchange with amphiphiles outside the membranes. Thus, vesicles composed of lamellar-phase amphiphiles such as fatty acids (Hargreaves and Deamer 1978), long-chain *n*-alkanoic ("fatty") alcohols (Apel et al. 2002), and glyceryl single- and double-chain esters are able to grow in size (Bachmann et al. 1990), fuse with one another (Chen et al. 2004), divide into smaller vesicles through different mechanisms (Budin and Szostak 2011; Briers et al. 2012), spontaneously cluster into "colonies" (Carrara et al. 2012), and interact in diverse ways with the outside solution—especially when the membranes are composed of mixtures of chemically different lipids that form mobile heterogeneous-phase zones separated by temporary membrane defects (Baumgart et al. 2003). For example, the import of externally added ribose and nucleoside triphosphates (NTPs) inside vesicles composed of mixtures of single-chain amphiphiles has been achieved that way (Mansy et al. 2008).

Despite being covered with polar (phosphate, carbohydrate, and amino acid) headgroups at both the outside and inside membrane surfaces, each being about 0.7–1.0 nm deep and often carrying net negative or zwitterionic (never net positive) electric charge, all lipid vesicles bear closed hydrophobic zones between the inside and outside polar headgroups. This inner 2.5- to 3.5-nm-thick part of lipid bilayers renders the membranes virtually impermeable for large electrically charged hydrophilic molecules—the higher the electric charge density and size of the solute, the slower the membrane-crossing rate (Deamer 2017). Large charged macromolecules such as RNA and DNA are therefore lipophobic and cannot cross lipid membranes easily, irrespective of the direction, whether from the outside or inside. So, once they find themselves, for whatever reason, encapsulated in a lipid vesicle (Chen et al. 2004; Mansy et al. 2008; Kurihara et al. 2015; Engelhart et al. 2016; Tsuji et al. 2016), they shall not leak out easily, unless the membrane architecture dissolves (sometimes termed "denaturation"). Notwithstanding this intrinsic physico-chemical incompatibility between lipids and nucleic acids, self-evolvable compartments, being able to grow in size and compositional diversity, are still at the heart of every so-called Darwinian protocellular and cellular behavior (Part VII), which operates over long time periods, with competing variants of reproducing (cellular) units that harbor variants of replicating systemic memory polymers inside them (Szathmáry et al. 2005). Hence, once the prebiotic chemical pathways leading from primary chemicals to RNA and DNA are reasonably clarified and experimentally reproduced (Section 5.2.2), future research must focus on plausibly prebiotic compartmentation pathways of RNA and DNA into lipid vesicles (Section 5.2.3).

5.2.2 PREBIOTIC CHEMISTRY OF GENETIC BIOMOLECULES

5.2.2.1 BRICKS OF LIFE

The prebiotic bricks of life, as we know it, are more or less of the composite type and can be roughly divided into four chemical compound classes: (1) α-amino acids; (2) carbohydrate-type polyhydroxyaldehydes, ketones, and polyols; (3) *N*-heterocycles; and (4) long-alk(en)yl-chain fatty acids and alcohols. They should all covalently assemble ("condense") in water, or neat, under prebiotically and geochemically plausible reaction conditions, from repeated units of similar fragments or molecules into macromolecules that bear a resemblance to the known bio-macromolecules, the proteins (oligopeptides), the carbohydrates (oligosaccharides), the nucleic acids (oligo-nucleotides), and the lipid amphiphiles (phospholipids).

Currently and experimentally, prebiotic syntheses giving reasonable yields of α-amino acids (Miller 1953; Ritson and Sutherland 2013; Parker et al. 2014; Coggins and Powner 2017; Islam et al. 2017) and *N*-heterocycles (Oró 1960; Oró and Kimball 1961; Sanchez et al. 1966; Saladino et al. 2001; Al-Azmi et al. 2003; Orgel 2004; Menor-Salván et al. 2009; Saladino et al. 2009; Salván and Marín-Yaseli 2013; Saladino et al. 2015) are mastered best, followed by those of lipids (Hargreaves et al. 1977; Hargreaves and Deamer 1978; Rao et al. 1982, 1987; McCollom et al. 1998; McCollom and Simoneit 1999; Deamer and Dworkin 2005; Rushdi and Simoneit 2006; Simoneit et al. 2007; Fernández-García and Powner 2017; Fiore et al. 2017) and carbohydrates (Butlerow 1861; Kim et al. 2011; Ritson and Sutherland 2012). For too long, and unhappily at the bottom of this heap, were specific prebiotic syntheses of nucleosides and nucleotides. Nucleosides are composed of two precursor classes, one of which, the *N*-heterocycles, is particularly stable and usually readily accessed. The other compound class, prebiotic carbohydrates, is rather problematic from a stability point of view (Kim et al. 2011). The importance of finding convincing prebiotic pathways to synthesize nucleosides and nucleotides cannot be overestimated. One can almost say that the world is waiting for the answer for more than 60 years.

So, which carbon-, oxygen-, and nitrogen-containing small molecules should be considered plausibly prebiotic? Cyanamide $H_2N-C\equiv N$ has been used by many prebiotic chemists, starting in 1964 (Steinman et al. 1964), for its structural simplicity and its manifold reactivity once the conditions are appropriate (Eschenmoser 2004, Fiore and Strazewski 2016). A geochemically plausible provenance of large amounts of cyanamide on primordial Earth may have been similar to that of its industrial production. Elementary carbon was nucleosynthesized in high amounts by the young Sun. This carbon (density of C: $d_{20°C}$ = 1.8–3.5 g/cm³, Mp > 3500°C), being markedly less volatile than dinitrogen, dioxygen, or even water, would accumulate and condense into essentially "coke" at a distance in the solar nebula, where now our planet Earth orbits, together with abundant rock-forming minerals mainly composed of the oxides and hydroxides of sodium, magnesium, aluminum, silicium, potassium, calcium, titanium and iron in smaller amounts of strontium, zirconium, barium, and others. While cooling down, the Earth's crust separated from the bulk iron (Fe: d_{Mp} = 6.98 g/cm³, Mp 1538°C), which eventually gravitated to form the planet's core. Nevertheless, even today, there is carbon hidden as iron carbide Fe_7C_3 in the Earth's core (Fe_3C: $d_{20°C}$ = 4.93 g/cm³, Mp ~3140°C) (Chen et al. 2014). The atmosphere of that time was already rich in dinitrogen (almost 90%), contained perhaps 10% carbon dioxide, much lesser amounts of water vapor and carbon monoxide, and only trace amounts of noble gases, dihydrogen, dioxygen, sulfur dioxide, and ozone (trioxygen).

The majority of the oxygen atoms were immobilized in the minerals or covalently bound to hydrogen. When calcium carbonate was present on the hot planetary surface, it would decarboxylate to give carbon dioxide and calcium oxide, viz. partly molten "quicklime" or "burnt lime" (CaO: $d_{20°C}$ = 3.34 g/cm³, Mp > 2580°C). With passing time, the temperatures fell. Burnt lime reacted above 2000°C with coke to give carbon monoxide and calcium carbide (Equation 5.2.2). Industrial CaC_2 is produced since 1892 from the same starting materials at 2300°C in an electric arc furnace. The procedure was invented by Henri Moissan (Nobel Prize in Chemistry 1906) and by Thomas Willson (*Union Carbide®*). When calcium carbide rose to the planet's surface through convection (CaC_2: $d_{20°C}$ = 2.22 g/cm³, Mp 2160°C, $Bp_{1\ atm}$ 2300°C), it reacted with dinitrogen gas to give (back) coke and calcium cyanamide (Equation 5.2.3). This is an exothermic reaction, known as the Frank-Caro process for the industrial production of "nitrolime." The reaction is conducted above 1000°C, where it is exergonic, $\Delta G_f°(1273\ K)$ = –22.6 ± 1.4 kJ/mol, and the composition of nitrolime equilibrates at around 20 wt% C, 18 wt% CaO, and 62 wt% $CaCN_2$ (Nagakura et al. 1998; Yamanaka et al. 2000). Calcium cyanamide on dry spots of the primordial Earth would stay as a liquid or solid reservoir ($CaCN_2$: $d_{20°C}$ = 2.29 g/cm³, Mp 1340°C). Once the temperatures fell enough for allowing liquid water to arrive and slightly acidify the crust with carbonic acid, calcium cyanamide would transform into limestone (calcium carbonate) and separate from aqueous cyanamide (Equation 5.2.4). Upon periodic night-day evaporation of water, cyanamide could crystallize (Mp 44°C). If during that period unreacted calcium and other carbides were still penetrating the water phase, carbide would immediately eliminate acetylene gas (Equation 5.2.5). The resulting burnt lime would hydrolyze to "slaked lime" (calcium hydroxide) and, upon slow addition of carbonic acid, again transform into limestone, mostly as calcite and aragonite (Equation 5.2.6). Hence, this could have been a huge geochemical planetary prebiotic nitrogen-fixation process, resulting in virtually unlimited amounts of cyanamide. If cyanamide would co-deposit with large amounts of liquid ammonia ($Bp_{1\ atm}$ –33°C, Mp –77°C), solid guanidine feedstocks could form (Mp 50°C). At elevated temperatures, dry cyanamide could be transported by sublimation or would react with itself, especially above 150°C, to give the dimer 2-cyanoguanidine (Mp 209.5°C) and the stable cyclic trimer melamine (Mp 345°C). In the presence of other organic compounds, higher oligomers and polymers could form (Pankratov and Chesnokova 1989). But, given an excess of liquid water, cyanamide would inevitably hydrolyze to urea (Equation 5.2.7) and, over geological timescales, to ammonium cyanate (Equation 5.2.8, Shaw and Bordeaux 1955), unstable carbamic acid (Callahan et al. 2005), and finally evaporate water, ammonia, and carbon dioxide (Equation 5.2.9). Of note, this way of generating cyanamide and urea is independent of exposing carbon monoxide and ammonia ices to galactic cosmic rays, as has been recently proposed (Förstel et al. 2016), or of heating calcium and magnesium cyanides to 400°C and 660°C, respectively (Pincass 1922; Seifer 1962a, 1962b). The overall process consists of the deposition of calcite and aragonite—"chalk" rock, that is, the main non-silicate minerals of today's Earth crust—generated from transforming primordial hot burnt lime and calcium carbide, entropically driven by the oxidative release of elemental solid carbon as carbon monoxide and solid or liquid carbide as acetylene gases and accompanied by the reductive hydrolysis of nitrogen gas to ammonia (Equation 5.2.10).

$$>2000°C \quad CaO + 3C \rightleftarrows CaC_2 + CO \qquad (5.2.2)$$

$$>1000°C \quad CaC_2 + N_2 \rightleftarrows CaCN_2 + C \qquad (5.2.3)$$

$$\leq100°C \quad CaCN_2 + H_2O + CO_2 \rightarrow$$
$$CaCO_3 + H_2N-C\equiv N \qquad (5.2.4)$$

$$\leq100°C \quad CaC_2 + H_2O \rightarrow CaO + HC\equiv CH \qquad (5.2.5)$$

$$\leq100°C \quad CaO + H_2CO_{3,\,aq} \rightleftarrows Ca(OH)_2 + CO_{2,\,aq} \rightarrow$$
$$CaCO_3 + H_2O \qquad (5.2.6)$$

$$\leq100°C \quad H_2N-C\equiv N + H_2O \rightarrow (H_2N)_2C=O \qquad (5.2.7)$$

$$\geq100°C \quad (H_2N)_2C=O \rightarrow NH_3 + HN=C=O \rightleftarrows$$
$$NH_4(N=C=O) \qquad (5.2.8)$$

$$\leq100°C \quad HN=C=O + H_2O \rightleftarrows H_2N-COOH \rightarrow$$
$$NH_3 + CO_2 \qquad (5.2.9)$$

$$CaC_2 + CaO + 2C + CO_2 + 4H_2O + N_2 \rightarrow$$
$$2CaCO_3 + CO + HC\equiv CH + 2NH_3 \qquad (5.2.10)$$

Carbon monoxide (CO) is the isoelectronic oxygen analogue of dinitrogen (N_2), hydrogen cyanide (HCN), and acetylene or ethyne (HCCH). Cyanamide (H_2NCN) is the isoelectronic dinitrogen analogue of carbon dioxide (CO_2) and isocyanic acid (HNCO). Guanidine [$(H_2N)_2CNH$] is the isoelectronic trinitrogen analogue of urea [$(H_2N)_2CO$] and carbonic acid (H_2CO_3). Electrophilic formaldehyde (H_2CO), isocyanic acid, cyanamide, and nucleophilic hydrogen cyanide, which had arrived from the solar nebula (Öberg et al. 2015) and was immobilized in cyanoferrate minerals (Sutherland 2016), as well as carbon monoxide, acetylene, guanidine, and large amounts of urea, were the geochemically metastable carbon feedstock reagents that could generate a great many of more complex organic products further down the prebiotic reaction network (examples in Figure 5.2.3). Most important for the chemistry discussed in this chapter (cf. Equations 5.2.11 and 5.2.12) are the homologation products of hydrogen cyanide $(HCN)_{2-5}$ up to adenine; the homologation products of cyanamide $(H_2NCN)_{2-3}$ up to melamine; cyanogen $N\equiv C-C\equiv N$, acetylene $HC\equiv CH$, and cyanoacetylene $N\equiv C-C\equiv CH$ and their partially oxidized, reduced, or hydrolyzed products such as glyoxal $O=CH-CH=O$, glyoxylic acid $O=CH-COOH$, acetaldehyde $H_3C-CH=O$, cyanoacetaldehyde $N\equiv C-CH_2-CH=O$, and 2-amino-2-cyanoacetamide $N\equiv C-CH(NH_2)-CONH_2$; many different aldolization products, that is, carbohydrates and their partly reduced or oxidized derivatives, including glycerol, ribose, 2-deoxyribose (Oró and Cox 1962), and pyruvic acid (Coggins and Powner 2017); and many small

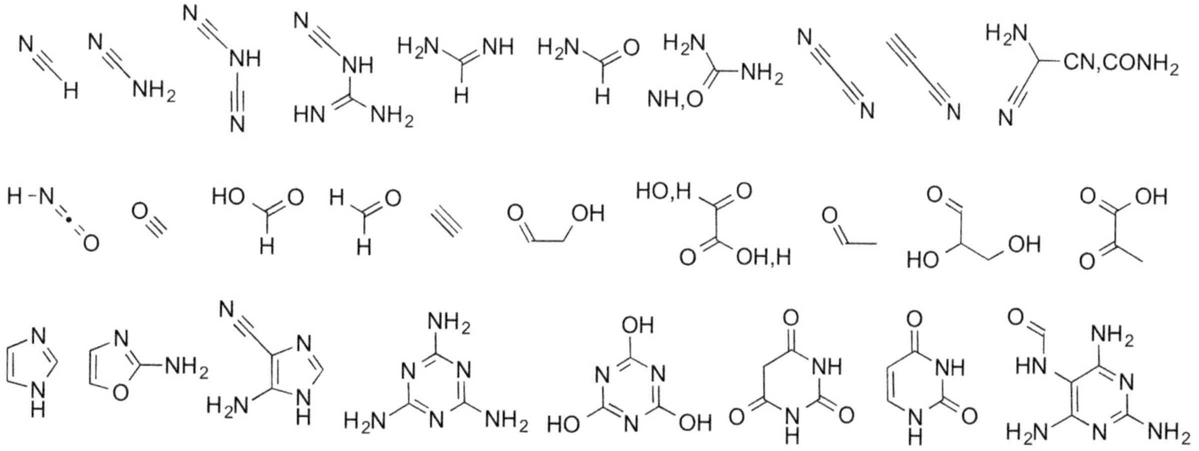

FIGURE 5.2.3 Prebiotic H-, N-, and O-containing C_1-, C_2-, and C_3-feedstock reagents, and some of their most important cyclic products. First row (from left to right): Hydrogen cyanide (hydrocyanic acid, from electric discharge on nitrogen and methane or interstellar clouds); cyanamide (stable tautomer of carbodiimide, from carbide and nitrogen, see text); dicyanamide (deamination product of 2-cyanoguanidine); 2-cyanoguanidine (dicyandiamide) = $(H_2NCN)_2$; formamidine; formamide; RNH = guanidine/RO = urea; cyanogen (dicyan, oxalonitrile); cyanoacetylene (2-propynenitrile, from electric discharge on nitrogen and methane); and RCN = 2-aminomalononitrile = $(HCN)_3$/RCONH₂ = 2-amino-2-cyanoacetamide. Second row (from left to right): Isocyanic acid (unstable deamination product of urea); formic acid; formaldehyde; acetylene (ethyne, from carbide and water); glycolaldehyde; 2x OH = oxalic acid/2x H = glyoxal (hydrolysis product of acetylene); glyceraldehyde; and pyruvic acid. Third row (from left to right): Imidazole (from formaldehyde, glyoxal and ammonia, or from photo-irradiation of cyanoacetylene and ammonia); 2-aminooxazole (from cyanamide and glycolaldehyde); 5-amino-4-cyanoimidazole (5-aminoimidazole-4-carbonitrile) = $(HCN)_4$; melamine (2,4,6-triamino-s-triazine) = $(H_2NCN)_3$; cyanuric acid (s-triazinetriol = major tautomer in the absence of melamine, see also Figure 5.2.4D); barbituric acid (pyrimidine-2,4,6(1H,3H,5H)-trione); uracil (pyrimidine-2,4(1H,3H)-dione, from cyanoacetylene to cyanoacetaldehyde plus urea and water); and 2,4,6-triamino-N5-formylaminopyrimidine (from 2-malononitrile, guanidine and formamide or formic acid).

N-heterocycles, that is, imidazoles, oxazoles, pyridines, pyrimidines, purines, and pterins. The richness of this list of "bricks of life" becomes much higher when other hetero-elements, mainly sulfur and phosphorous, are included (Section 5.2.2.3).

5.2.2.2 NUCLEOSIDES

Much has been written about intuitive pathways leading from very simple organic compounds to the known nucleosides and nucleotides. For the average organic chemist who spends his or her scientific life on synthesizing molecules and, for doing so, works through the most evident "retro-syntheses," that is, ways to cut down the target molecule into simpler starting compounds, the most intuitive means of getting nucleic acids "from scratch" should roughly follow the Equations 5.2.11 through 5.2.15:

$$\text{formaldehyde} + \text{base catalysis} = \text{carbohydrates} \quad (5.2.11)$$

$$\text{cyanide/cyanoacetylene} \left(+ \text{water} \right) = N\text{-heterocycles} \quad (5.2.12)$$

$$\text{carbohydrate} + N\text{-heterocycle} = \text{nucleoside} \quad (5.2.13)$$

$$\text{nucleoside} + \text{phosphate} - \text{water} = \text{nucleotide} \quad (5.2.14)$$

$$x \cdot \text{nucleotide} - \left(x - 1 \right) \cdot \text{water} = \text{nucleic acid} \quad (5.2.15)$$

The first two equations are based on solid experimental knowledge pioneered by Alexander/Aleksander Mikhaylovich/Mikhaïlovitch Butlerov/Butlerow/Boutlerow—Александр Михайлович Бутлеров—for the formose reaction (Equation 5.2.2; Boutlerow/Butlerow 1861) and, respectively (Equation 5.2.3), the works of Juan Oró (purines) and Leslie Orgel (pyrimidines) and their collaborators at the time (Oró 1960; Oró and Kimball 1961; Sanchez et al. 1966). These reactions have been exploited by several generations of chemists for many decades. A prebiotic alternative to the formose reaction has been proposed by Ramanarayanan Krishnamurthy and collaborators, according to which ketoses, rather than aldose-ketose mixtures, are produced from dihydroxyfumaric acid, but the corresponding pentulofuranosyl nucleic acids do not hydridize through base pairing (Sagi et al. 2011, 2012; Stoop et al. 2013). Perhaps, the most authoritative recent essay on the problems that every chemist encounters, when trying to experimentally reconstruct the above pathways, was written by Steven Benner et al. (2012). Their arguments concentrated on the rich reaction network of the formose reaction and how this reaction would favor tetroses and pentoses over hexoses, longer -oses, branched polycarbohydrates, and intractable "tar." By reversibly scavenging furanosidic cis-diols as cyclic borate esters, typical formose run-off side reactions are suppressed in the presence of plausibly prebiotic borate minerals or boric acid (Prieur 2001). To tame the formose reaction is a formidable challenge ever since Boutlerow/Butlerow (1861), who himself needed to stop the reaction by adding water and cooling down the mixture before it turned brown. Using borates to obtain useful yields of tetroses and pentoses, among which were arabinose and ribose, is a formidable achievement (Kim et al. 2011). The same run-off side reactions—based on ever-present aldose-ketose isomerizations and collateral aldolizations of the intermediate ene-1,2-diols—could be suppressed by yet another modification. Albert Eschenmoser and collaborators obtained preferentially pentose-2,4-diphosphates over tetrose-2,4-diphosphates and hexose-2,4,6-triphosphates (72% total pentose, 36% ribose, 19% total hexose, and 4% total tetrose) when carrying out the formose reaction in the presence of 0.08 M glyceraldehyde-2-phosphate and 0.04 M formaldehyde in, albeit prebiotically questionable, 2 N NaOH (Müller et al. 1990). This kind of taming of the formose reaction was not wholly plausible in terms of pH >13, albeit locally imaginable, cf. Equation 5.2.6 at high temperature. It did set a new standard for the chemical analysis of the formose reaction mechanisms and how to interfere with the formation of carbohydrate ene-1,2-diols (Eschenmoser and Loewenthal 1992).

But there are more "demons" lurking on the pathway from formaldehyde and cyanide or cyanoacetylene to the nucleic acids. Let us go straightaway to the most difficult chemical problem, which is—once the ribose problem has been sufficiently solved and all sorts of N-heterocycles being more or less similar to the natural nucleobases have also been obtained—the formation of the nucleoside's N-glycosidic bond. Much of the research has been carried out under simulated primitive Earth conditions. The Eo-Archean Sun 4.0–3.6 Gya was smaller and its radiation was at least 20% weaker than today, albeit with a larger fraction in the broad-band ultraviolet (UV) range, but the solar wind is thought to have blown more intensely on the surface of the early Earth because of the absence of today's atmospheric dioxygen and protective ozone gases. In the early 1960s, Cyril Ponnamperuma and collaborators reported on the prebiotic synthesis of minute amounts of adenosine from irradiating aqueous solutions of adenine and ribose and of deoxyadenosine from irradiating aqueous solutions of adenine and 2-deoxyribose, under a mercury lamp emitting narrow-band UV light of 254-nm wavelength (Ponnamperuma et al. 1963; Ponnamperuma and Kirk 1964). At this wavelength, however, the N-glycosidic bond of guanosine, for instance, is unstable (Crespo-Hernández and Arce 2000). γ-Photons at 0.93-nm and 1.06-nm wavelengths—the 1.33-MeV and 1.17-MeV beams generated from $^{60}\text{Co} \rightarrow {}^{60}\text{Ni}$—were shown to transform deaerated, that is, oxygen-depleted, aqueous solutions of deoxyadenosine to the corresponding 5-N-formylaminopyrimidine (FaPy) nucleoside, through the hydrolytic ring opening of the imidazole part of the adenine residue (Raoul et al. 1995). Pyrimidine and purine derivatives obtained in the same pot, that is, mixtures of uracil, thymine, 5-hydroxymethyluracil, cytosine, N-formylpurine, N,N-diformyladenine, and N-(hydroxyacetyl)- and N-(2,3-dihydroxypropionyl) purines and -adenines, have been reported from heating pure formamide (FA) at 160°C in the presence of TiO_2, a powdered form of the abundant mineral rutile (Saladino et al.

2003). Under the same conditions, purine and pyrimidine 2′-deoxynucleotides were shown to degrade into 8-oxopurines, dihydroxylated pyrimidines, and FaPy derivatives. In the presence of ammonia, one of FaPy's further degradation products is FA. Formamide is thought to have played, in small amounts, crucial catalytic and, in large amounts, dehydrating and solvating roles on prebiotic Earth (Saladino et al. 2009). Raffaele Saladino, Ernesto Di Mauro, and their collaborators have recently experimentally simulated the presumed conditions on an early Earth being submerged in a persistent flux of slow protons supposed to arrive from the solar wind. They obtained complex mixtures of products from the irradiation of pure FA at –30°C with a 170-MeV proton beam (7.29 fm wavelength = 7.29 · 10^{-6} nm), in the presence of various catalytically active meteorite powders. Trimethylsilyl-derivatized crude reaction mixtures were analyzed by GC-MS, revealing the presence of a plethora of polyhydroxy alcohols, aldehydes, carboxylic acids, amino acids, urea, guanidine, nitriles, N-heterocycles, and, most notably, what appeared to be pyrimidine nucleosides and purine nucleosides in nanogram to microgram amounts per milliliter FA (Saladino et al. 2015). As eyebrow-lifting or -raising these recent results might be, still no convincing prebiotic synthesis starting from N-heterocycles and carbohydrates that would lead to significant amounts of reasonably enriched nucleosides or nucleotides has been demonstrated to date.

Benner et al. also proposed high-boiling ($Bp_{1\,atm}$ 210°C) and low-melting (Mp 2°C–3°C) FA as the prebiotic salvage reagent and solvent. Hot FA is thought to promote the regioselective N-ribosylation of N-formylheterocycles, O-phosphorylation of the resulting nucleosides (Schoffstall 1976, Schoffstall et al. 1982, Furukawa et al. 2015), as well as the activation of mononucleotides through O-formylphosphate-mixed anhydrides to give nucleic acids (Benner et al. 2012). The *umpolung* from nucleophilic water to electrophilic FA as prebiotic solvent was evoked. The weak electrophilicity of FA's carbon atom—it's a carboxamide after all—versus the weak nucleophilicity of its oxygen atom—thought to generate O-formidoylphosphate (not O-formylphosphate) in supposedly FA-catalyzed phosphorylation reactions (Xu et al. 2017)—is a matter of ongoing debate and so are its long-term reactivity profile, geochemical abundance, and prebiotic role in general (Hudson et al. 2012; Pietrucci and Saitta 2015; Pino et al. 2015; Bada et al. II 2016; Šponer et al. 2016). Mechanisms involving FA as a catalyst still need experimental support.

Be it as it may, many chemists have believed for long that there should be a way of *preferentially* building the required carbon-nitrogen bond, that is, between the carbohydrate's anomeric carbon atom from the right side—the β-face of the furanose ring—and the required nitrogen atom of both the pyrimidine base (N1) and the purine base (N9). Alas! No, not spontaneously in water, not without protective groups and modern directive synthetic strategy or enzymes. Something out there steadfastly and successfully hampers this chemical challenge. For us chemists it is as if, every time when we try to *throw a dart* at the bull's eye, we find ourselves instead struggling to *catch a javelin* more or less

frontally—it is a nightmare. The trials under plausibly prebiotic reaction conditions but, without more or less harsh irradiation, have started in the early 1970s. Leslie Orgel and his collaborators at the time succeeded in adding cyanamide to a pentose (ribose and arabinose) to make the corresponding 2-aminooxazole that would react with cyanoacetylene to give, after partial hydrolysis, pyrimidine nucleosides in moderate yields and with the wrong stereoselectivity at the anomeric center for ribose (Sanchez and Orgel 1970). The Orgel group has also attempted to obtain purine nucleosides from N-ribosylating purines under neat conditions (Figure 5.2.4**B**). They obtained nucleosides in low yields and disastrous regioselectivities with respect to the purine's nitrogen atoms (Fuller et al. 1972). Orgel's approach of obtaining pyrimidine nucleosides from arabinosyl and ribosyl 2-aminooxazoles turned out to be absolutely pioneering and highly influential (see Section 5.2.2.3), but the low yields and regioselectivity problems could not be overcome until shortly. The only successful regio- and reasonably stereoselective direct ribosylation and 5-phosphoribosylation of a pyrimidine derivative were recently achieved by Nicolas Hud and collaborators (Figure 5.2.4**C**). Pyrimidine-2,4,6(1H,3H,5H)-trione, alias barbituric acid, is an oxidized form of pyrimidine-2,4(1H,3H)-dione, alias uracil (Figure 5.2.3, third row). Thanks to a favorable 6-keto to 5-en-6-ol tautomerization of barbituric acid, good yields of the corresponding C(5)-nucleoside and C(5)-nucleotide were obtained under plausibly prebiotic reaction conditions (Cafferty et al. 2016). Melamine (2,4,6-triamino-s-triazine), the cyclic trimer of cyanamide and base-pairing partner of barbituric acid and cyanuric acid—the latter locked as tautomeric s-triazinetrione when mixed with melamine (Whitesides et al. 1995, Avakyan et al. 2016)—is the other N-heterocycle that could be easily and selectively ribosylated and 5-phosphoribosylated under the same conditions, however, at one of the exocyclic nitrogen atoms (Figure 5.2.4**D**). In spite of this, there is no pyrimidine or purine that would be directly and regioselectively N1- and N9-ribosylated, respectively—for that matter, regioselectively N-glycosylated at any endocyclic nitrogen atom—to give the corresponding nucleoside or nucleotide, respectively, under plausibly prebiotic conditions. The "natural" N-heterocycles (biotic nucleobases) are simply not nucleophilic enough at their endocyclic nitrogen atoms to be efficiently N-glycosylated enzyme-free in water (Figure 5.2.4**A** and **B**).

Thus far, the short answer to the question: "Why did it take so long to obtain nucleosides under plausibly prebiotic conditions?" should be: "Because *separately* joining prebiotic N-heterocycles with prebiotic sugars was too seducing an intuition." It was nourished by several seminal results from the past. They were recently comprehensively discussed by Sutherland (2016). This idea (Equation 5.2.13) was kept alive by our almost insurmountable confidence in what we scientists see in biotic nature. For example, knowing that pyrimidine nucleosides are biosynthesized from the natural N-heterocycle orotate (6-carboxyuracil, Figure 5.2.4**A**, R^1 = O, R^2 = COOH) and 5-phospho-α-D-ribofuranosyl-1-pyrophosphate (PRPP: the chemically activated carbohydrate, Figure 5.2.4**A**,

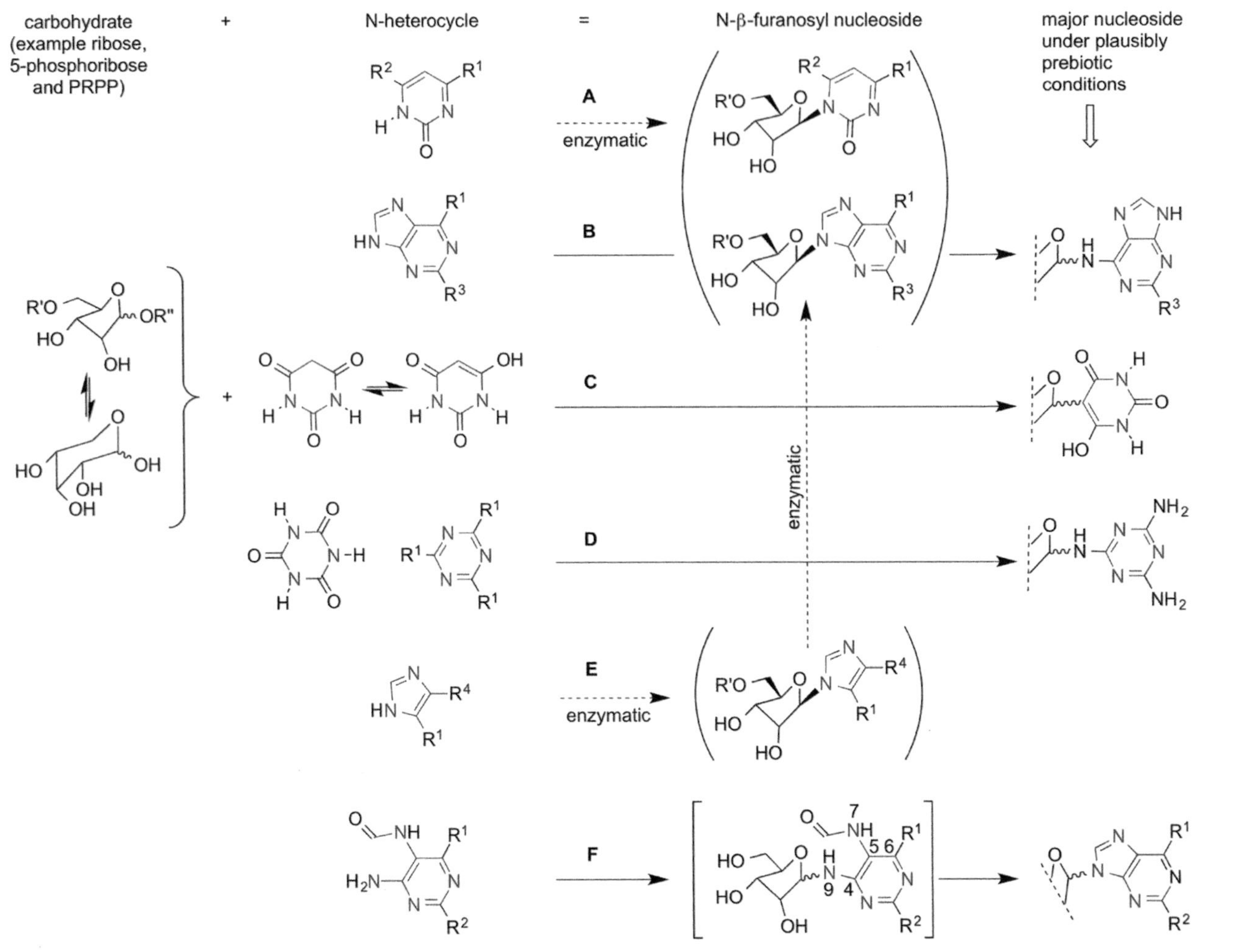

FIGURE 5.2.4 Summary of prebiotic nucleoside syntheses (continuous arrows) according to the concept carbohydrate + *N*-heterocycle = nucleoside, as in Equation 5.2.13 and enzymatic pathways (dotted arrows). First column: Any tetrose, pentose, or hexose as furanose, pyranose, or corresponding carbohydrate phosphate, for example, R′ = H or phosphate, R″ = H or diphosphate. Second column: *N*-heterocycles R^1 = NH$_2$, OH (= rare lactim tautomer shown for depicting convenience, major tautomer = lactam), R^2 = H (uracil, cytosine) or COOH (orotic acid/orotate = 6-carboxyuracil), R^3 = H or NH$_2$ (adenine, guanine, 2,6-diaminopurine), and R^4 = H, CN, or CONH$_2$ (5-aminoimidazole, 5-amino-4-cyanoimidazole/5-aminoimidazole-4-carbonitrile, 5-amino-4-carboxamidoimidazole/5-aminoimidazole-4-carboxamide). Third column: desired (biotic) nucleoside: round parentheses indicate absence or very low yields in prebiotic mixtures. Fourth column: obtained major product(s). **A**: Pyrimidines do not form nucleosides under plausibly prebiotic conditions. Natural pyrimidine nucleotides are biosynthesized from orotate (R^2 = COOH) and PRPP (R′ = phosphate, R″ = diphosphate). **B**: Purines (R^1 as for pyrimidines) do not form (endocyclic) N9-glycosylated nucleosides under plausibly prebiotic conditions. Instead, exocyclic R^1 = NH$_2$ is preferentially glycosylated. **C**: The enol tautomer (right) of barbituric acid (left) is readily (5-phospho)ribosylated to the corresponding C5-nucleos(t)ide. **D**: Cyanuric acid trioxo tautomer (left), forced by base pairing (not shown) with melamine, and cyanuric acid triol tautomer in aqueous solution (R^1 = OH, right). Melamine (R^1 = NH$_2$, right) is readily (5-phospho)ribosylated to the corresponding exocyclic N^2-nucleos(t)ide. Mixtures of products from **C** and **D** form long supramolecular fibers and have been proposed as alternative genetic material (Cafferty et al. 2016). **E**: Imidazoles and carbohydrates do not form nucleosides under plausibly prebiotic conditions, but the natural purine nucleotides are biosynthesized from AIR (R^1 = NH$_2$, R^4 = H) and PRPP (R′ = phosphate, R″ = α-diphosphate). **F**: Purine nucleosides do form from 6-amino-5-formylaminopyrimidines and carbohydrates.

R′ = PO$_3^{2-}$, R″ = α-triphosphate), we tend to assume that a similar but prebiotic pyrimidine compound, for example, cytosine or uracil (Figure 5.2.4A, R^1 = NH$_2$ or O, R^2 = H), can be plausibly phosphoribosylated—but it can't. Purine nucleosides are biosynthesized from PRPP to give first the imidazole *N*-glycoside 5′-monophosphate 5-aminoimidazole β-D-ribonucleotide (AIR). The purine is then completed on the aminoimidazole part by stepwise building up the pyrimidine component of the purine (Figure 5.2.4E, R′ = PO$_3^{2-}$,

R^1 = NH$_2$, R^4 = H). Hence, we imagine that a similar but direct prebiotic "imidazole-first" synthesis should be possible (Figure 5.2.4E, R′ = H, R^1 = NH$_2$, R^4 = CN or CONH$_2$), but it shall not, not directly at least (cf. Section 5.2.2.3), despite the fact that the well-understood HCN homologation pathway to adenine (HCN)$_5$ leads through 5-amino-4-cyanoimidazole (HCN)$_4$ (Figure 5.2.3, third row).

An answer to the "imidazole-first" purine nucleoside question was recently given by Hendrick Zipse, Thomas

Carell, and collaborators (Becker et al. 2016). They proposed a "pyrimidine-first" purine nucleoside synthesis strategy, according to which three different FaPy derivatives would be mixed with solid carbohydrates and heated under plausibly prebiotic neat conditions (100°C, 8 hours). The carbohydrate readily added regioselectively to the exocyclic free amino group next to an endocyclic N-atom of the FaPy producing a Schiff base intermediate (not shown). The subsequent ring closure of the N-glycoside generated intermediate furanosidic (Figure 5.2.4F) and pyranosidic FaPy nucleosides with varying stereoselecitivities at the anomeric center. Upon addition of water and a choice of plausibly prebiotic additives, such as simple amines, basic amino acids, borax, or a carbonate-borate buffer, and heating the mixtures in a sealed tube at 100°C for 1 day to 1 week, the intermediate FaPy nucleosides ring-closed to the purine nucleosides. These researchers obtained, for the first time in considerable yields (up to 60%), prebiotic purine nucleosides with the expectedly absolute N9-regioisomeric selectivity and, especially in the presence of borax or borate, with the "correct" (biotic) β-furanosidic stereoisomeric preference. In addition, several carbohydrate variants, that is, pyranosides versus furanosides and tetrosides versus pentosides, were produced from adding dry FaPy to different carbohydrate precursor mixtures, that is, to glycolaldehyde and glyceraldehyde that were kept together beforehand for 30 minutes at ambient temperature in the presence of slaked lime (calcium hydroxide), and to ammonia, formaldehyde, and glycolaldehyde, after heating the solution without FaPy for 1 hour at 65°C, respectively, thus under prebiotic formose reaction conditions. These protocols necessitated the separation of calcium sulfate by filtration and repeated lyophilization before proceeding with the imidazole ring closure at 100°C.

This "pyrimidine-first" synthesis of purines is rooted in Wilhelm Traube's work, *Der Aufbau der Xanthinbasen aus der Cyanessigsäure. Synthese des Hypoxanthins und Adenins*, where the first synthesis of hypoxanthine and adenine appeared in the literature (Traube 1904). The absolutely regioselective formylation of the 5-amino group of 4,5,6-triamino-2-thiopyrimidine, to become N7 in purine nucleosides, was reproduced several times since (originally in boiling formic acid, later at higher temperatures in FA) (Bendich et al. 1948; Haley and Maitland 1951; Robins et al. 1953; Trinks 1987; Koch 1992), was then proven by a crystal structure of 2,4,6-triamino-5-formylamino-pyrimidine (Koch 1992), and is now elegantly explained through an *ab initio* pK_a calculation and X-ray analysis of 4,5,6-triaminopyrimidinium chloride (Becker et al. 2016). It transpires from two dissertations from the Eschenmoser group that the key prebiotic purine (and pterine) nucleoside precursors could all be FaPy derivatives (Trinks 1987; Koch 1992). In the presence of carbohydrates, 6-amino-5-FaPys readily generate exocyclic N-Schiff bases that cyclize twice to give first FaPy nucleosides and then purine nucleosides. In the absence of carbohydrates, FaPys cyclize to purines that are not glycosylated, but the FaPys can be partly recovered in hot FA (Saladino et al. 2003).

There is no question about the possible abiotic provenance of FaPy precursors that lead to 2,6-diaminopurine nucleoside ($R^1 = R^2 = NH_2$, cf. in Figure 5.2.4F) and natural guanosine, since both FaPy precursors can be obtained under prebiotically plausible conditions from guanidine (Trinks 1987), if not urea (Koch 1992). The pyrimidine ring closes with 2-aminomalononitrile and with 2-amino-2-cyanoacetamide to form the required 2,4,6-triaminopyrimidine and 2,6-diaminopyrimidine-4-one, respectively. The C_1-precursor guanidine forms from cyanamide and ammonia and both three-carbon 2-aminonitriles are intermediate products on the HCN homologation pathway (Figure 5.2.3). The FaPy C_1-precursor for adenosine ($R^1 = NH_2$, $R^2 = H$, Figure 5.2.4F) would be formamidine, viz. the addition product of two molecules of ammonia on dehydrated formic acid. Formamidine, however, rather than ring-closing with 2-aminomalononitrile to form the obligatory 4,6-diaminopyrimidine, apparently prefers to integrate the imidazole ring of 4-aminoimidazole-5-carboxamide under similar reaction conditions (Sanchez et al. 1966; Schwartz and Hornyak-Hamor 1971; Trinks 1987; Koch 1992). So, the only known access to an FaPy precursor of adenosine is (thus far) not prebiotic, since the currently available reaction protocols proceed from malononitrile and thiourea rather than guanidine. Both molecules are plausibly prebiotic, but the reaction works only with the assistance of anhydrous alcoholate (Traube 1904). In Traube's hands, nitrosylation of the resulting crude 4,6-diamino-2-thiopyrimidine with nitrous acid, subsequent reduction of the C5-oxime to 4,5,6-triamino-2-thiopyrimidine with hydrogen sulfide, and regioselective N^5-formylation in hot formic acid allowed for the cyclization of this FaPy to create the adjacent imidazole ring. The resulting 2-thioadenine was oxidized and eliminated in hot 20% sulfuric acid containing 3% H_2O_2 to give adenine and H_2SO_4. This protocol was much later optimized and modified by a final reductive desulfuration over Raney-Ni (Bendich et al. 1948). An alternative reductive desulfuration of 4,5,6-triamino-2-thiopyrimidine with Raney-Ni or Ni_2B (amorphous synthetic "nickel boride") was carried out by Clark et al. (1968).

Hence, there is still work to be done to find the missing links for fully prebiotic adenosine to be generated on the "pyrimidine-first" purine nucleoside pathway. We should anyway refrain from intuitively assuming that there is only one prebiotic way to RNA, as shown in the next section.

5.2.2.3 Nucleotides

In all previously described works, other crucial prebiotic elements were missing: sulfur, thiols, and phosphorous in one of their oxidized forms. In plausibly prebiotic and geochemical environments on the young Earth, phosphate was mostly locked up in very stable and quite insoluble apatite minerals, such as hydroxylapatite $Ca_5(PO_4)_3(OH)$. The role of phosphate for the prebiotic production of nucleosides, ribonucleotides, and other vital organophosphates—either neat in more soluble evaporite minerals or as a eutectic liquid in the presence of water, ammonium formate, and urea—is currently a hot subject in prebiotic chemistry (Burcar et al.

2016; Kim et al. 2016; Karki et al. 2017). In this handbook, Chapters 5.7 and 5.8 are devoted to phosphorous in prebiotic chemistry, so let us concentrate here on the synthesis of nucleotides from simple building blocks and ignore for the moment the questions: how highly soluble phosphate concentrations could be achieved in truly prebiotic mixtures and how they would react with nucleosides when present.

When the old prebiotic pyrimidine synthesis work of Sanchez and Orgel 1970 was taken up by the Sutherland group and now in the presence of high concentrations of inorganic phosphate, a whole new "prebiotic universe" opened up. Said Sutherland in Venice 2005: let's make prebiotically plausible nucleosides and nucleotides in a sequence of reactions, where *N-heterocycles are allowed to form in the presence of inorganic phosphate and sugar precursors*, instead of separately synthesizing prebiotic *N*-heterocycles and prebiotic sugars. Exploiting the homologation of HCN in concert with unsaturated or otherwise highly reactive prebiotic nitriles (Figure 5.2.3) *and* with the homologation of formaldehyde in "one pot," being "geologically linked" to other "pots" (Sutherland 2016), does produce prebiotic carbohydrates, as if under classic Butlerow conditions. But, while the sugars keep their electrophilicity and α-enolisability at almost all stages of H_2CO homologation, as in a classical formose reaction, the compositional outcome of an expanded formose reaction mixture is strongly modified by the presence of nitrogen-rich intermediates. This change of outcome takes its beginning in the reaction between C-electrophilic cyanamide and the sufficiently nucleophilic α-hydroxy group of hydroxyaldehydes and ketones (carbohydrates) when anchimerically assisted by the opposite reactivity next door, viz. the nucleophilic attack of the amino group of cyanamide on the carbonyl group of the aldehyde or ketone, which in the end leads to all sorts of 2-aminooxazoles, for example, from pure glycolaldehyde to 2-aminooxazole *per se* (cf. first reaction in Figure 5.2.5).

At neutral pH, this reaction works particularly well in the presence of phosphate as a general acid-base catalyst,

and excess cyanamide is rapidly converted to urea under these conditions (see side products in the second row of Figure 5.2.5). In turn, 2-aminooxazole adds through its relatively nucleophilic C5-atom to the aldehyde function of aldoses, whereby the adjacent α-hydroxy group attacks imine-like C4, thus readily forming various bicyclic 2-aminooxazoles, technically 3a,5,6,6a-tetrahydrofuro[2,3-*d*]oxazol-2-amine derivatives (cf. second reaction in Figure 5.2.5). From pure glyceraldehyde and 2-aminooxazole, a diastereoisomeric mixture of bicyclic pentose derivatives is generated that way, where the *ribo*- and *arabino*-configurations prevail. It is of note that the natural β-anomeric pyrimidine nucleotides are derived from the corresponding *arabino*-configured bicyclic 2-aminooxazole and that the *ribo*-configured bicyclic 2-aminooxazole crystallizes in enantiomorphic forms as "kryptoracemate" or "conglomerate" and hence offers a spontaneous way of separating the enantiomers. If, for some reason, a scalemic (non-racemic) mixture of *ribo*-configured bicyclic 2-aminooxazole was locally present in solution, that is, one of the enantiomers appeared enriched because of its formation from scalemic (enantioenriched) glyceraldehyde, for example, it could spontaneously crystallize virtually enantiopure and thus persist over much longer time periods than if the compound stayed in solution (Powner et al. 2009; Patel et al. 2015; Sutherland 2016).

Mixtures of carbohydrates are, of course, more typical for a prebiotic formose reaction. A mixture of carbohydrates, including aldoses and ketoses, reacts with cyanamide to give a mixture of 2-aminooxazoles. A common short ketose, dihydroxyacetone, reacts with two cyanamides to form a C4-spiro-*bis*-(2-aminooxazole) compound, whereas with 2-aminooxazole, it forms a bicyclic C5-disubstituted 2-aminooxazole (both shown in the second row in Figure 5.2.5). Formaldehyde and gylcolaldehyde react to the corresponding C5-substituted 2-aminooxazoles, none of which are useful precursors for pentofuranosyl nucleotides, albeit tetrafuranosyl nucleotides and particularly 3',2'-α-threose nucleic acids

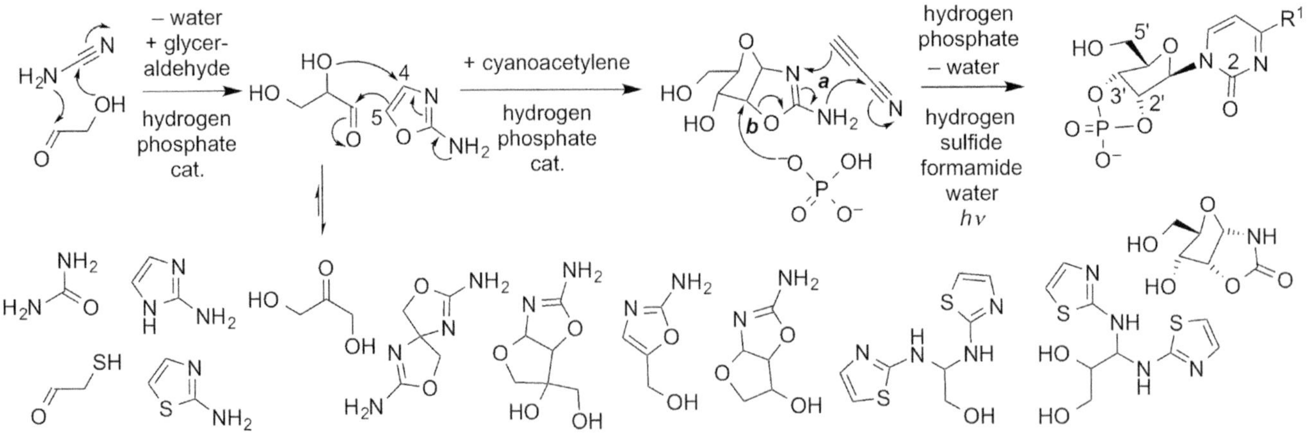

FIGURE 5.2.5 Prebiotic pyrimidine nucleoside 2',3'-cyclic phosphate synthesis. First row: Streamlined summary with pure starting materials. Second row: Identified side products when carbohydrate and other mixtures are used. $R^1 = NH_2$, OH = rare lactim tautomer shown for depicting convenience, major tautomer = lactam. For explanations, see text.

(TNAs) might, in principle, have had an alternative functional role in alien or early biota (Eschenmoser 1999; Schöning et al. 2000; Eschenmoser 2004; Pinheiro et al. 2012; Yu et al. 2012). However, here, the problem is how to form and accumulate (enrich) useful pentofuranosyl nucleotide precursors from mixtures of carbohydrates in the presence of cyanamide. In the presence of prebiotically plausible β-mercaptoacetaldehyde, for instance, the addition to cyanamide results in 2-aminothiazole instead of 2-aminooxazole, whereas, when phosphate and ammonium are present, 2-aminoimidazoles compete with 2-aminooxazoles (Fahrenbach et al. 2017). Interestingly, mixtures of short carbohydrates react differently with 2-aminothiazole. This *N,S*-heterocycle was found to play the role of a "chemical chaperone" in sequestering glycolaldehyde from carbohydrate mixtures through plausibly prebiotic crystallization events (Islam et al. 2017). In those mixtures, in the (temporary) absence of cyanamide, aldehyde functions are preferentially captured as their corresponding di(thiazol-2-yl)aminals (second row in Figure 5.2.5), showing that the exocyclic amino group of 2-aminothiazole is much more nucleophilic than that of 2-aminooxazole, and it outweighs the nucleophilicity of its C5, too (Powner et al. 2012). Out of the two mentioned aminals, the first shown is glycolaldehyde aminal. Both, glyceraldehyde and dihydroxyketone are shown in the first and second rows, respectively. After a prebiotically plausible physical separation of the solids from the liquids, this "prebiotic mother liquor" continues to react over longer time periods to form the homologue glyceraldehyde aminal (shown). In the absence of the former solid, this glyceraldehyde aminal accumulates through spontaneous crystallization as well. The physically separated glycolaldehyde aminal crystals can now be exposed to fresh aqueous cyanamide to form per aminal two molecules of 2-aminothiazole and one 2-aminooxazole in a quantitative yield. Furthermore, when the homologous glyceraldehyde aminal crystals are exposed to this reaction mixture, one aminal is transformed in a high yield into two 2-aminothiazoles and one bicyclic pentose 2-aminooxazole, the same as is obtained from mixing pure glyceraldehyde with hydrogen phosphate and cyanamide. Hence, the C5-atom of 2-aminooxazole preferentially reacts with aldehydes under aqueous close to neutral prebiotic reaction conditions, even in the presence of 2-aminothiazole; the latter transforms aldehydes to di(thiazol-2-yl)aminals only in the absence of 2-aminooxazole. This explains under what circumstances di(thiazol-2-yl)aminals can offer chaperone services to a formose reaction, namely, when 2-aminothiazole exposure, crystallization events, and cyanamide exposure can occur separately from one another. A subsequent solid-liquid separation (crystallization) leaves pure *ribo*-configured bicyclic 2-aminooxazole crystallized from the last liquid. All initially formed tetroses, pentoses, and hexoses do not participate in this sequential reaction-crystallization-reaction-crystallization process; they remain minor leftovers in the liquid that has separated from pure *ribo*-configured bicyclic 2-aminooxazole.

The other decisive reactivity in such mixtures is based on the readiness of 2-aminooxazoles to ring-close with cyanoacetylene to pyrimidines (Ferris et al. 1968). Hydrogen cyanide and cyanoacetylene are the major products after the exposure of a mixture of ammonia and methane to powerful electric discharges (Sanchez et al. 1966). Best in the presence of 1 M phosphate buffer, bicyclic 2-aminooxazoles add cyanoacetylene to form 2,2'-anhydropyrimidines, that is "oxygen-tethered" pyrimidine nucleosides (not shown intermediate after reaction *a*), and subsequently ring-open this tether through a stereoinversion at C2' with (hydrogen) phosphate (reaction *b*) to end up as various pyrimidine nucleoside 2',3'-cyclic phosphates (cf. third reaction in Figure 5.2.5). Of course, if pure bicyclic pentose 2-aminooxazoles are brought in contact with cyanoacetylene and phosphate, they all transform into pentosyl pyrimidine nucleoside 2',3'-cyclic phosphates. If nature starts from pure crystalline, perhaps enantiopure bicyclic *ribo*-configured 2-aminooxazole, and then, the "wrong" anomeric α-configuration results (not shown). Only diastereoisomerically pure bicyclic *arabino*-configured 2-aminooxazole will generate the biotic β-furanosyl pyrimidine ribonucleoside 2',3'-cyclic phosphates, 2',3'-cCMP and 2',3'-cUMP, which can be racemic, scalemic, or enantiopure (Figure 5.2.5). In the absence of phosphate, the conversion from ribo- to arabinofuranosidic tethered derivatives (not shown) requires UV light and long photoanomerization times, cleaves in parts the *N*-glycosidic bond, and also generates high amounts of the bicyclic *ribo*-configured oxazolidin-2-one, that is, hydrolyzed 2-aminooxazole (Figure 5.2.5) and correspondingly low yields of the desired β-furanosyl cytidine. Alternatively, if the starting compound was the minor, less persistent because non-crystalline bicyclic *arabino*-configured 2-aminooxazole, then satisfactory yields of both 2',3'-cCMP and, after photoanomerization, 2',3'-cUMP are generated (Powner et al. 2009). However, in the presence of aqueous FA and hydrosulfide HS⁻, being abundant only in alkaline waters as the conjugate base of plausibly prebiotic hydrogen sulfide (H$_2$S) (pK_a = 7.0), the reaction with bicyclic *ribo*-configured 2-aminooxazole involves the formation of α-configured 2-thiocytidine (α-2-thioC) by opening the tether at C2 rather than through stereoinversion at C2' (not shown). The β-furanosyl anomers of 2-thiocytidine (β-2-thioC), cytidine, 2',3'-cCMP, and 2',3'-cUMP are quite readily generated through the photoanomerization of α-2-thioC in the presence of FA and phosphate, followed by partial hydrolysis at 60°C (Xu et al. 2017). Having alleviated the anomerization problem with the help of hydrosulfide, other problems have been evoked, namely, the poor water solubility of hydrogen sulfide and low abundance of hydrosulfide at neutral pH, as well as the expected high reactivity between cyanoacetylene and hydrosulfide to form the adduct (Z)-3-mercaptoacrylonitrile HS–CH=CH–C≡N. Cyanoacetylene is the key ingredient to transform bicyclic pentose 2-aminooxazoles into 2',3'-cCMP and 2',3'-cUMP. Therefore, the intermediate photoanomerization precursor 2,2'-anhydro-α-cytidine can only be *generated in the absence of hydrosulfide* from pure crystalline, perhaps enantiopure bicyclic *ribo*-configured 2-aminooxazole (reaction *a* in Figure 5.2.5). Once FA and hydrosulfide arrive after that, the ring opening and the smooth photoanomerization can take place. But what happens when cyanoacetylene meets hydrosulfide and bicyclic *ribo*-configured 2-aminooxazole? And what if, instead of hydrosulfide, "hydrated SO$_2$" viz. highly water-soluble bisulfite HSO$_3^-$ (pK_a = 7.2), was more easily available on the early Earth?

Before we dive into subsequent investigations, let us reflect for a moment upon the pyrimidine chemistry thus far. What we learn from the previously described experiments is that short prebiotic carbohydrates act as efficient scavengers of intermediate *N*-heterocycles before these intermediates continue to react downhill to give stable dead-end products such as cytosine and uracil. It turns out that ribo- and arabinofuranosyl scaffolds can hold this promise best. They are ideally configured for the formation of cyclic nucleoside phosphates, thus providing a prebiotic reservoir of, in principle, sufficient free enthalpy for the spontaneous polymerization to RNA, *vide infra* (Figure 5.2.7). The same argument holds the other way around: few nitrogen-containing organic key compounds that result from prebiotic high-energy reaction conditions act as efficient scavengers of intermediate short carbohydrates before these intermediates continue to react downhill to give intractable dead-end products that taste like caramel, sweet as it is, and eventually look like tar. For this mutual reaction interference between carbohydrates and *N*-heterocycles to be efficient and productive, two more elements have been identified: phosphorous and sulfur. The phosphate anion catalyzes important reactions between small acyclic molecules as general acid-base reactions at close to neutral pH, and both thiols and phosphate act as nucleophiles on different carbon atoms of nucleotide key precursors. Phosphate prefers to attack carbohydrate residues, whereas thiols favorably interfere with unsaturated parts of organic compounds. The sunny side of the thus-far described systems chemistry experiments: The farther from chemical equilibrium chemists work, the more complex the chemical mixtures can be, which, despite this complexity, generate clear-cut results, that is, few major products. One way of avoiding thermodynamic overall equilibrium is to work with spatio-temporal separations. Prebiotically plausible sequential reaction, crystallization, and re-solubilization/reaction events help generating the virtually enantiopure pyrimidine nucleoside key precursor from scalemic carbohydrate mixtures. Another important result is the successful experimental bifurcation or divergence of the prebiotic pathway from carbohydrates and organic nitriles to the amino acids and to the nucleotides (Ksander et al. 1987; Eschenmoser 2004; Ritson and Sutherland 2013; Patel et al. 2015; Coggins and Powner 2016; Sutherland 2016; Islam et al. 2017). The downside thus far is that there is apparently no viable *direct* way in sight that would lead from 2-aminooxazoles to the purine nucleotides.

One of several possible "small bricks of life" that contains more than merely nitrogen and oxygen "hetero-elements" has led to a remarkable conclusion for a possible divergent pathway that might lead to pyrimidine and purine nucleotides. The plausibly prebiotic combination of hydrogen cyanide and sulfur results in the quantitative formation of thiocyanic acid HS–C≡N (Bartlett and Skoog 1958), the mono-sulfur analogue of cyanamide. Therefore, thiocyanic acid is regarded as a plausibly

prebiotic compound (Chyba and Sagan 1992; Thaddeus 2006; Ritson and Sutherland 2013). Thiocyanic acid reacts with glycolaldehyde in a different way from that of cyanamide, thus generating crystalline 2-thioxazole, not 2-aminothiazole, in high yields (first reaction in Figure 5.2.6). Unlike the reaction to 2-aminooxazole, there is no need for phosphate to buffer this reaction. Analogously to the bicyclic 2-aminooxazoles, similar pentose preferences are observed for the reaction of 2-thioxazole, or its *NH* tautomer, with glyceraldehyde to form bicyclic oxazolidinone-2-thiones, from which the *arabino*-configured diastereoisomer crystallizes (second reaction in Figure 5.2.6). The aminolysis of *S*-benzylated bicyclic pentose oxazolidinone-2-thiones derived from xylose, ribose, arabinose, and fructose with anthranilic acid (*ortho*-aminobenzoic acid) was pioneered by Patrick Rollin and collaborators, who were the first to synthesize quinazolinedione nucleosides from D-arabinose, albeit not in a plausibly prebiotic context (Girniene et al. 2001, 2004). Matthew Powner and collaborators took up this route and discovered that, while *S*-benzyl or *S*-methyl derivatives were unsuitable for an efficient aminolysis by ammonia, the plausibly prebiotic *S*-cyanovinyl residue could be readily formed and subsequently displaced with ammonia, 2-aminomalononitrile, or 2-amino-2-cyanoacetamide (p$K_{a, ammonium}$ ≈ 9.2, 6.5, and 3.4, respectively). The reactivity of bicyclic oxazolidinone-2-thiones is not the same as that of bicyclic 2-aminooxazoles (compare the second reaction in Figures 5.2.5 and 5.2.6). The superior and softer nucleophilicity of sulfur, when compared with nitrogen, favors conjugate attack on cyanoacetylene, forming an intermediate bicyclic pentose 2-*S*-cyanovinyloxazolidine-2-sulfide adduct (oxazolidine-2-thioacrylnitrile, not shown), thus offering (*Z*)-3-mercaptoacrylonitrile HS–CH=CH–C≡N as an excellent leaving group. This constitutes the bifurcation point on the divergent pathway to pyrimidine nucleotides upon exposure to ammonia (left downward reaction in Figure 5.2.6) and to oxypurine nucleotides upon exposure to the other bricks of life (Figure 5.2.3), viz. 2-aminomalononitrile or 2-amino-2-cyanoacetamide (third reaction in Figure 5.2.6). The crude aminolysis yields, as determined by proton NMR spectroscopy, were roughly 20%–45%, along with lower amounts of pentose oxazolidine-2-thione from the reaction with aqueous ammonia and both pentose oxazolidine-2-thione and pentose oxazolidine-2-one (second row in Figure 5.2.5) roughly 1:1 from displacements by 2-aminomalononitrile or 2-amino-2-cyanoacetamide. The latter two aminolytic products are tricyclic pentosyl "oxygen-tethered" aminocyano- and aminocarboxamidoimidazoles. All *arabino*-configured bicyclic and tricyclic pentose derivatives, that is, the oxazolidine-2-thione, oxazolidine-2-thioacrylnitrile and both cyano- and acetamidoimidazoles, are crystalline. This opens a plausibly prebiotic pathway to an "imidazole-first" synthesis for N9-β-ribofuranosyl-8-oxopurine nucleotides (Stairs et al. 2017).

The *dot on the i* was put by ring-closing the pyrimidine residue of the above tricyclic imidazoles in adding formamidine in FA at 100°C to form tetracyclic pentosyl "oxygen-tethered" 2′,8-anhydropurine nucleosides in up to 60% yield (right downward reaction in Figure 5.2.6). This tether was opened during prebiotic phosphorylation in urea at 140°C, where

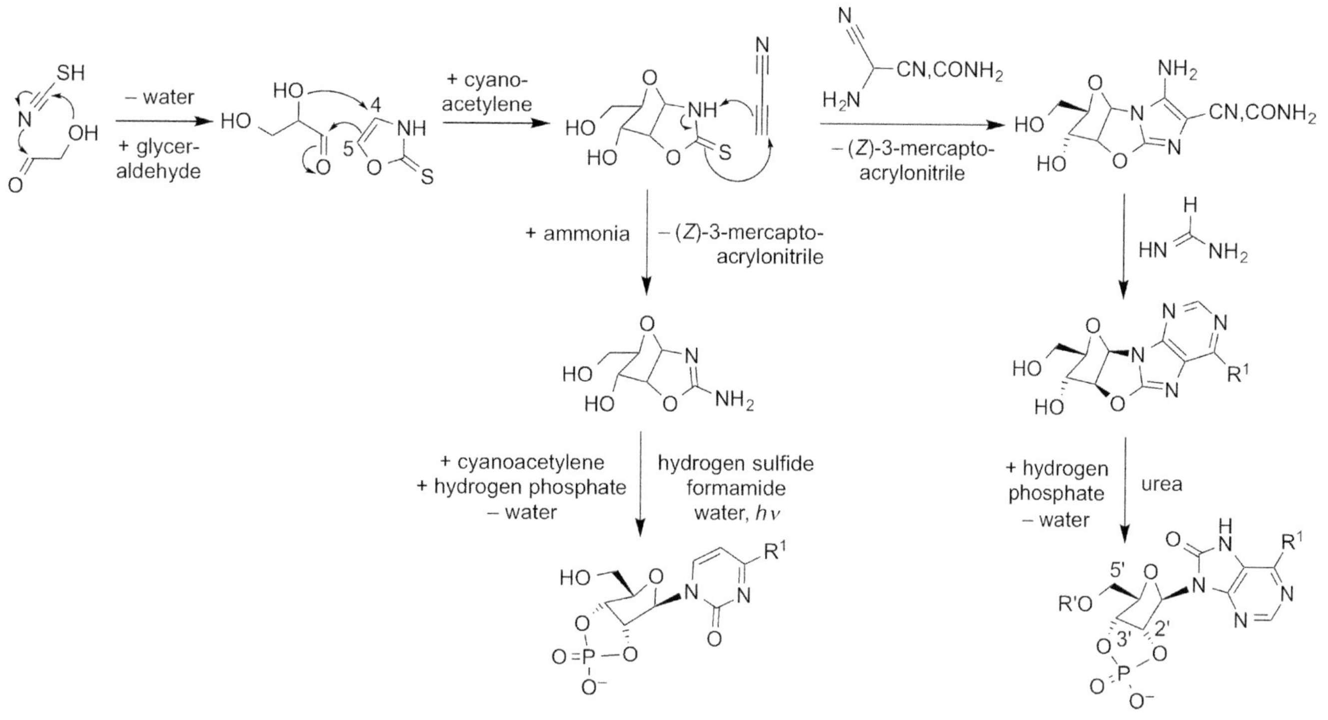

FIGURE 5.2.6 Divergent prebiotic pyrimidine nucleoside and oxopurine nucleoside 2′,3′-cyclic phosphate synthesis. The bent arrows of the first reaction are a strong oversimplification and meant only to illuminate the overall constitutional outcome of a multi-step reaction mechanism. Most probably, the reaction sequence involves the initial attack of sulfur on the aldehyde group to form a hemithioacetal, which gives anchimeric assistance to the nucleophilic attack of the α-hydroxy group on the nitrile carbon atom. A first reasonably stable intermediate is expected to be 2-imino-1,3-oxathiolan-4-ol. This intermediate can reversibly dehydrate to 1,3-oxathiol-2-imine, but the first formed C–S bond can also ring-open again to generate O-(2-oxoethyl) carbamothioate O=CHCH$_2$OC(=S)NH$_2$, which, upon rotation and nitrogen attack on the aldehyde, can ring-close and dehydrate to the aromatic 2-thiooxazole, thence, its *NH* tautomer (shown). R^1 = NH$_2$, OH = rare lactim tautomer shown for depicting convenience, major tautomer = lactam, R′ = H or phosphate. For explanations, see text.

8-oxopurine nucleoside 2′,3′-cyclic phosphates are the major products. In mixtures containing both bicyclic pentosyl oxazolidine-2-thiones and pentosyl 2-aminooxazoles, the former can be quantitatively and chemoselectively re-cyanovinylated at the sulfur in the presence of the latter or with excess cyanoacetylene, thus concomitant synthesis of anhydropyrimidine nucleoside and anhydropurine nucleoside precursors can be achieved with remarkable efficiency. This is where the field presently stands (September 2017).

This newest work is highly complementary to the prebiotic syntheses of pyrimidine nucleotides (Figure 5.2.5), as well as to the "pyrimidine-first" purine nucleoside synthesis that thus far lacks prebiotic access to adenosine (Figure 5.2.4**F**). The complementarity of this "imidazole-first" 8-oxopurine nucleoside synthesis can be tracked down to the different reactivities of formamidine versus guanidine or urea (Figure 5.2.3). Formamidine, the diamide of formic acid HCOOH, furnishes the C2 atom of adenosine and inosine. Guanidine, the triamide of carbonic acid H$_2$CO$_3$, serves to create the same C2 atom in guanosine and 2,6-diaminopurine nucleosides. Formamidine is more nucleophilic than guanidine, albeit its conjugate acid is expected to be less resonance-stabilized, and thus, formamidine is somewhat less basic. Formamidine reacts with 2-aminomalononitrile to the corresponding 5-amino-1*H*-imidazole-4-carbonitrile,

not 4,5,6-triaminopyrimidine (cf. Section 5.2.2.2). But when an imidazole is already present, as in the tricyclic oxygen-tethered pentose imidazoles (Figure 5.2.6), then formamidine reacts at high temperatures, thus ring-closing to the imidazole-annealed pyrimidine, whereas guanidine apparently does not. If it did after all, oxygen-tethered 8-oxoguanosine nucleosides were in prebiotic reach on an "imidazole-first" pathway. It is intriguing to see how this strategy of avoiding the separate synthesis of carbohydrates and *N*-heterocycles (Equation 5.2.13) navigates through a chemical space where the oxidation states of the obtained purine nucleotides do not correspond (yet) to the biotic variants. All biotic pyrimidine nucleotides bear the same oxidation state at the corresponding carbon atoms—in particular, that of C2 is the same as that of urea, guanidine, cyanamide, carbonic acid, and carbon dioxide, or, for that matter, of thiocyanic acid. Hence, many (all biotic) pyrimidines can, in principle, be built up from these bricks, with no need for prebiotic redox reactions. All biotic purine nucleotides bear the same oxidation state at C8, being the same as in formic acid, FA, and formamidine but different from 8-oxopurines. C8 in pentosyl 8-oxopurines is provided from thiocyanic acid, that is, an isoelectronic sulfur-nitrogen equivalent of CO$_2$. In contrast, the oxidation state of C2 varies in biotic purine nucleosides; it is either reduced, as in adenosine and inosine, or oxidized, as in guanosine and 2,6-diaminopurine nucleoside. Therefore, the prebiotic pathway(s) to the biotic purine nucleosides must either involve a prebiotic redox reaction, or else, of course, two chemically different pathways are taken for one and the other, as described in this chapter. Prebiotic redox reactions have been relatively recently worked out for the divergent prebiotic synthesis of carbohydrates, and thus also nucleotides, and α-amino acids from formaldehyde and hydrogen cyanide through a so-called "cyanosulfidic chemistry out of the blue" (Sutherland 2016). Photoreductions from UV light at 254 nm intended to mimic the effect of the early Earth's Sun light (Ranjan and Sasselov 2016), using cuprous cyanide (CuCN) and sodium hydrosulfide (NaHS) or hydrogen sulfide (H$_2$S) (Ritson and Sutherland 2013; Patel et al. 2015; Sutherland 2016; Ritson et al. 2016). Organic nitriles, formed, for example, from prebiotic HCN homologation, are usually quite inert, but, upon photoirradiation in the presence of Cu(I) and HS$^-$, they thiolyze to thioamides (R–CSNH$_2$) that are chemoselectively reduced to the imines that rapidly hydrolyze to the aldehydes. Prebiotic Strecker reactions (Miller 1953; Parker et al. 2014) furnish α-aminonitriles that equilibrate with α-hydroxynitriles ("cyanhydrines"). Cyanhydrines are thus mildly reduced to the corresponding α-hydroxyaldehydes, that is, to short carbohydrates. Given the importance of the sequential addition, exclusive presence, and separate prebiotic access of reactive nitriles, such as cyanamide, cyanoacetylene, and 2-aminomalononitrile, from that of organothiols or -thiones, such as β-mercaptoacetaldehyde and pentose oxazole-2-thiones, this calls for testing plausibly prebiotic photoreduction conditions (Ranjan and Sasselov 2016) on mixtures containing 8-oxopurine nucleosides or, even more so, on mixtures of both tetracyclic pentosyl oxygen-tethered 2′,8-anhydropurine nucleosides and 8-oxopurine nucleosides. Which carbon atom is more easily reduced, tethered, or untethered C2? If the tethered C2, then we shall encounter a stereochemical problem at C2′ of the crystalline *arabino*-configured 2′,8-anhydropurine nucleoside. Besides, are there ways of photooxidizing C8 of the above nucleosides, perhaps through a prebiotic dismutation between C2 and C8? These questions remain elusive. If 8-oxopurine nucleosides won't react to the biotic purine nucleosides, then the polymerization and base-pairing properties of 8-oxopurine nucleotides (OP) need much deeper and more systematic investigation. Perhaps, indeed, 1D polymers containing OP, C, and G (or U) could have been early *trit* (or *quit*) memory polymers.

5.2.2.4 NUCLEIC ACIDS

The repeated and sequential dehydration of nucleotides to generate polymeric chains of nucleic acids (Equation 5.2.15) under plausibly prebiotic conditions is still an open question, although several methods are being discussed and investigated since long. Two general routes differ in the kind of nucleotides with which to begin the polymerization, either from acyclic 5′-phosphates, such as the biotic nucleoside monophosphates (NMPs), or from the 3′,5′-cyclic nucleoside phosphates and 2′,3′-cyclic nucleoside phosphates. From a purely thermodynamic point of view, the former needs strong chemical activation by an *umpolung* reagent that can transform a dianionic foremost nucleophilic phosphate residue into an electrophile, where the phosphorous atom can be attacked in water by a moderately nucleophilic hydroxy group of another NMP. The NMP pathway is the biotic buildup of nucleic acids and promotes step-wise chain growth in the 5′→3′-direction

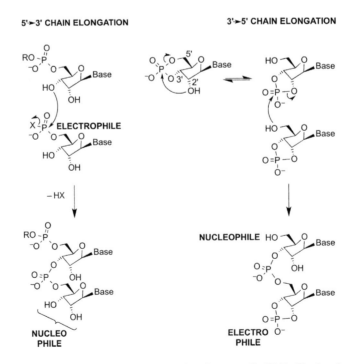

FIGURE 5.2.7 Principal pathways for RNA chain growth. R = negative charge or 5′-pRNA, X = leaving group (nucleofuge). For explanations, see text.

(Figure 5.2.7 left). The latter cyclic phosphates should be able to profit from their strained-ring tension energy and a diminished negative (maximally single) charge for a hydroxy group attack by another nucleotide monomer, without the absolute requirement for an additional activator X. With both an electrophilic site and a nucleophilic site present in each oligo- and polymer molecule, multiple ligations at both the termini and step-wise chain growth in the $3'{\rightarrow}5'$-direction are possible (Figure 5.2.7 right).

While the mere principles of RNA chain growth from the nucleotides seem straightforward and clear-cut, the chemical reality is far from simple (Costanzo et al. 2011, 2012; Burcar et al. 2013). The problems can be divided into three classes, where one is purely chemical. Hydroxy groups are poor nucleophiles and compete with a huge excess of water molecules, which are also poor nucleophiles; they also compete with anionic phosphate groups, which are stronger nucleophiles, especially the dianionic terminal phosphates (Liu et al. 2016). When an inactivated nucleophilic phosphate group attacks an activated electrophilic phosphate group, hydrolytically labile organic "pyrophosphates" (phosphoric acid anhydrides) are generated; these eventually hydrolyze back to the starting materials. Biotic nature uses NTPs as monomers, where X leaves as inorganic pyrophosphate $P_2O_7^{4-}$ (Figure 5.2.7 left). Pyrophosphate is a good leaving group once it is tightly complexed, that is, when it "clamps" one or two divalent metal ions, such as magnesium and manganese, that are best provided enzymatically and ideally positioned by the polymerase (Wu et al. 2017). The required concentration of magnesium ions without enzyme is much too high to be useful for RNA polymerization. Nucleotide polymerizations are endergonic processes, that is, energy-consuming "uphill" reactions, so the reversed reaction, the magnesium-catalyzed RNA strand cleavage, is usually favored over the forward reaction. What other plausibly prebiotic activator might have been taken over by biotic pyrophosphate?

Many different charge-neutral leaving groups have been explored ever since imidazole was proposed as nucleofuge X and phosphorimidazolides as the corresponding P-electrophiles (Weimann et al. 1968). Most often, an electrophilic dehydrating agent, such as cyanamide, cyanoimidazole, or a carbodiimide, would be combined with charge-neutral, somewhat-labile 5'-mononucleotidic acids (better than with 5'-mononucleotides) and a nitrogenous nucleophile as the ultimate leaving group. The longest homopolynucleotides have been obtained from separately (nonprebiotically) prepared adenosine-5'-phosphorimidazolide, when the polymerization was carried out in the presence of montmorillonite clay (Ferris and Ertem 1993; Ferris et al. 1996; Ferris 2002). The chain lengths are much shorter from mixed nucleoside-5'-phosphorimidazolides, despite optimized conditions and the use of acid-treated montmorillonite (Coari et al. 2017). The clay-catalyzed reactions of mixtures of separately or *in situ* prepared nucleoside-5'-phosphorimidazolides showed that the extent of polymerization was greatest in the presence of adenylate. The pyrimidine nucleotides generally were less reactive than the purine nucleotides, and heteropolymers containing both pyrimidines and purines tended to favor those with higher purine content. When activated adenylate was present, it tended to dominate the polymerization products. Polymerizations in the absence of solid catalysts, for example, mixed ribonucleoside-5'-phosphorimidazolides that have been achieved in ice eutectic phases, show a greatly diminished preference for purines (Monnard et al. 2003).

Many alternative non-anionic prebiotic X groups, which have been tested for enzyme-free and ribozyme-free polymerizations, can be found among the biotic bricks in Figure 5.2.3. But the favorite prebiotic "pet nucleofuge" remains imidazole, albeit often as a derivative at the C2 position, that is, an imidazole, where the hydrogen between the endocyclic nitrogen atoms has been replaced by something else, for example, by a C2-alkyl or C2-amino group (Wu and Orgel 1992; Li et al. 2017). It is of note that enzyme-free primer extensions from separately prepared ribonucleoside-5'-phospho-(2-aminoimidazolides), that is, using plausibly prebiotic 2-aminoimidazole as the leaving group (Fahrenbach et al. 2017), show a greatly enhanced reaction rate with templated nucleotidic hydroxy groups when compared with other C2-imidazole derivatives. Another important discovery was the enhanced reactivity of 5'-nucleotides, when activated in the presence of possibly prebiotic N-alkyl derivatives of imidazole (derivatives at the N1-position), owing to the *in situ* activation of NMPs as N1-alkylimidazolium-N3-phosphoramidates, rather than phosphoimidazolides, which allows for more efficient chain extensions with all four NMPs (Röthlingshöfer and Richert 2010; Kervio et al. 2014; Jauker et al. 2015a). These rate enhancements are explained by the positively charged X group, making the activated monomers zwitterionic, rather than monoanionic, at neutral pH, and thus, making them more electrophilic.

For the biomimetic RNA $5'{\rightarrow}3'$-chain extension pathway, 5'-NMPs need to be generated from the nucleosides under plausibly prebiotic phosphorylation conditions. Two different tendencies are reported in the recent literature. On the one hand, prebiotic nucleoside phosphorylations from ammonium dihydrogen phosphate and urea, either neat or in FA and in the absence or presence of triols such as glycerol and cytidine, generate, over several days at 140°C, stable majority amounts of 2',3'-cyclic nucleoside phosphates and varying lower amounts of 2',3'-cyclic 5'-nucleoside diphosphates (Stairs et al. 2017). This confirms that the 2',3'-cyclic nucleoside phosphates are irreversibly formed even at these elevated temperatures and over prolonged time periods, while 5'-phosphates that form first, and much lower amounts of 3',5'-cyclic esters (Schoffstall 1976; Schoffstall et al. 1982; Furukawa et al. 2015), may be reversibly transported to other hydroxy groups. On the other hand, phosphorylations at 65°C and 85°C in urea-ammonium formate-water eutectic liquids containing FA generate about twice the amount of 5'-NMPs with respect to 2',3'-cyclic nucleoside phosphates (Burcar et al. 2016), and geochemical phosphorylations at 85°C in the presence of borate minerals suppress 2',3'-cyclic nucleoside phosphates almost completely, through the reversible formation of 2′,3′-cyclic nucleoside borates that are transformed into 5′-NMPs in 13%–33% yields (Kim et al. 2016). Generally, prebiotic nucleoside phosphorylations are not very efficient; they usually generate 10%–30% nucleotides, and molar yields above 45% are the far-out exception. Low steady-state activated NMP concentrations and/or high competitive inhibition from the unreactive non-activated NMPs and nucleosides thus limit enzyme-free step-wise RNA or DNA growth in the 5′→3′-direction to usually one-digit or maximally low two-digit nucleotide lengths. The literature is, let's say, rich in enzyme-free template-directed nucleotide chain elongation experiments, and to summarize, adequately all results are out of scope of this chapter. Excellent reviews on the topic have been published recently (Fernández-García et al. 2017; Karki et al. 2017). What should be pointed out here is that this problem class of weak chemical reactivity of nucleosides in water, for both phosphorylations and phosphodiester syntheses, may be linked to the second and third problem classes and hopefully then be solved. They are rooted in the difficulty to take over RNA and DNA *de novo* synthesis by template-directed RNA extension, alias primer extension, and hence, to achieve step-wise sequence copying from sufficiently stable complementary base-paired parts. Of course, first, we do need RNA or DNA strings of a certain (unknown) minimal chain length before useful RNA and DNA sequences can be copied and translated, as depicted in Figure 5.2.2. *De novo* chain synthesis and extension could indeed occur at quite high temperatures and would not much depend on the identity of the *N*-heterocycles being attached to the carbohydrates through the *N*-glycosidic bond. However, the temperatures must have dropped very significantly before template-directed RNA extensions could have set in (Monnard et al. 2003, Attwater et al. 2010; Attwater et al. 2013); or else, the solvent viscosity must have been at least locally remarkably higher on the prebiotic Earth, possibly generated by periodic water evaporation (He et al. 2017). Hence, what might have been a stockable monomer reservoir of nonetheless conveniently spring-loaded mononucleotides for *de novo* nucleic acid synthesis—the 2′,3′-cyclic nucleoside phosphates—must have become useless for primer extensions. Yet, the ultimate reason for RNA to be ever present in all known forms of life is their ability to synthesize proteins, a function for which RNA needs to be folded (transfer RNA, ribosomal RNA). In contrast, in order to be copied by template primer extension, unfolded RNA is prerequisite—so how do we join these two loose ends?

The second problem class is the huge numbers of isomers that can be formed once prebiotic monomer polymerization begins. The greatest concerns are all linked to the required efficiency of template-directed nucleotide copying:

1. The effects of different nucleic acid backbones that have been generated from prebiotically available monomers, for example, mixtures of tetrose-, pentose-, and hexose nucleotides; mixtures of furanosyl and pyranosyl nucleotides; and mixtures of deoxyribonucleotides and ribonucleotides in the same polymer.

2. The effects of huge numbers of diastereoisomers that have been generated from racemic or scalemic mixtures of monomers, most notably, D-ribonucleotides and L-ribonucleotides, but also from α- and β-anomeric nucleotides in the same polymer.

3. The ultimately required preponderance of 3′–5′-linkages over 2′–5′-linkages in RNA, once the first and second concerns are resolved.

The literature is rich in proposed solutions, but experimentally, much more needs to be done. There are indeed prebiotic pathways that generate, for example, 3′–5′-linkages rather than 2′–5′-linkages in dinucleotides, or separate enantiomeric monomers through spontaneous enantiomorphous crystallization. But it is an entirely different problem to master astronomically huge numbers of isomers in polymers with concepts that are viable for small molecules. To cut a long and sometimes debatable story short, it seems that all these concerns could be solved by the same physically fundamental notion or law of persistence (Pascal and Pross 2015). The concept is convincing; the sloppy version sounds like: Over long time periods, everything lasts as long as something longer-lasting takes over. If not under non-equilibrium conditions and in a world where animate matter has emerged and evolved at astonishing rates, this statement sounds almost trivial. Still, a polymer of the nucleic acid kind will always be susceptible to ever-grinding hydrolysis, besides photolysis, thermolysis, and radiolysis. What transpires from recent experimental works is that, for nucleic acid polymers to persist longer in a plausibly prebiotic environment, the polymer needs to be homogeneously built. It seems that nucleic acid polymers that are not composed of identical repeat units won't last as long. The polymers that are best protected from hydrolysis are those that fold best upon themselves or hybridize best with their complements. However, for template-directed copying of nucleotide strands, the hybridized, that is, the base-paired, parts need to be separated from one another—the more stable the base pair is, the more energy-costly the strand separation becomes. A delicate balance between base-pairing stability and dehybridization effort was possibly a driving force for maintaining both prebiotic DNA and RNA at early stages of biotic evolution (Rajamani et al. 2010; Leu et al. 2011; Chen and Nowak 2012; Kumar 2016). Polymers that do bear the odd foreign monomer, anomer, isomer, and internucleotide linkage can nevertheless be functional to a certain more or less limited degree, but those that have a totally homogenous backbone will persist over the longest time periods. This can be shown by theory, but to unambiguously demonstrate it by systematic experimentation is one of the most tedious works that needs to be carried out and on by chemists (Joshi et al. 2013; Engelhart et al. 2013; Szabat et al. 2016; Gavette et al. 2016).

The third problem class is one of amplification efficiency. How can a system produce relatively huge amounts of DNA or RNA, all having the virtually identical nucleotide sequence to, at first, one single DNA or RNA molecule? Again, it is out of scope of this chapter to discuss this deeply (see Chapters 5.5 and 5.6 and Part VI of this handbook). It is obviously a dynamical problem: How to copy enzyme-free a nucleotide polymer sequence and thus ultimately generate two identical nucleic acid molecules from one, where now, both mother and daughter molecules are copied again to produce not fewer than four identical molecules, and so forth? This is synonymous to how to obtain an exponential growth of "population" size from a self-copying or else-copied starting molecule. The problem in a real chemical, if not prebiotic, environment is one that is probably best circumscribed by "strand inhibition," although there is much more to it than just this. Let us ignore here huge amounts of highly challenging scientific work over several decades and state that it has become obvious that the only way to master this problem class is to work far from thermodynamic equilibrium. In other words, spatio-temporal separations must gain crucial importance in future experiments. Reaction conditions must be tested where compound mixtures are periodically furnished with fresh chemical activation energy, monomers, dehydration or coupling agents, and additives and where the reaction products should be more or less rapidly and efficiently separated from the starting molecules (Chapter 6.3, Strazewski 2015b). Regarding the chemistry of enzyme-free nucleic acid copying, amplification, and ultimately gene replication, the best solution thus far was pioneered by Günter von Kiedrowski and collaborators (Luther et al. 1998). The core concept was taken up and modified by Clemens Richert and collaborators, albeit for a different reason than exponential growth rates (Deck et al. 2011). Both explored the immobilization of a first polymer molecule on an insoluble surface (microscale beads), made the immobilized hybridized polymer grow in chain length, provoked the detachment of the products, in order to move away the starting molecules from the daughter molecules, and then cycled the procedure several times. This guarantees for efficient and truly exponential growth in the number of copied molecules of considerable nucleotide length and accessible sequence diversity.

And here comes the "chemical devil" back again: Enzyme-free template-directed DNA or RNA copying has been best achieved with chemical tricks being, at first sight at least, incompatible with a prebiotic scenario. The most helpful trick for growing long chains efficiently was to use synthetic monomers bearing significantly stronger nucleophilic groups than the natural carbohydrate-derived hydroxy groups (Mansy et al. 2008). 3'-Amino-3'-deoxynucleotides react much faster with 5'-activated monomers; the hydrolytic competition in the absence of enzymes won't win under such conditions and copied nucleic acid-like polymers result. But "amino sugars" (aminodeoxysaccharides), irrespective of the nitrogen's position in the carbohydrate other than anomeric, are not the usual outcome of an enzyme-free formose reaction in the presence of ammonia or other amines. The biotic nucleic acids bear phosphodiester—not phosphoramidate—linkages. This difficulty becomes even more apparent in experiments where foreign primary amino groups are present in NMP coupling reactions. When amino acids and NMPs are mixed and a dehydrating water-soluble coupling reagent, plausibly prebiotic or not, starts polymerization reactions, one of the easiest detectable product molecules is indeed a

phosphoramidate, one that results from the rapid capture of an amino acid by a 5'-nucleoside phosphate (Jauker et al. 2015b; Griesser et al. 2017). What happens next is moderately rapid peptide chain growth on an NMP scaffold. N-terminal 5'-phosphoramidate-linked peptido-NMPs with impressive peptide lengths are detected. The apparition of peptido-RNA, that is, slow RNA-chain growth, can be detected in strongly diminishing amounts up to five or so nucleotides. It is obvious that, under identical reaction conditions, amino acids condense to peptides much faster than NMPs condense to RNA, irrespective of the $2'/3'–5'$ homogeneity of internucleotide linkages. One way to slow down amino acid-to-peptide couplings, or, for that matter, reactions with any amines, with respect to NMP-phosphodiester couplings could be to lower the pH down to values where amino groups are abundantly protonated and thus their nucleophilicity suppressed, whereas hydroxy groups would still keep their reactivity. But this does not seem to be a robust prebiotic situation, and other problems, such as solubility and aggregation, would have to be traded off in exchange. Again, we hit the problem of minimal interference of the scientist. Nucleic acid chains can be grown, even in a template-directed manner, up to impressive nucleotide lengths, with sufficient coupling rates for all four biotic *quits* and, when appropriately separated from the products through immobilization, probably with exponential amplification rates. However, these longer oligo- or polynucleotides can only form in the absence of amino groups, such as the amines that were needed to build up the nucleotides from simple bricks (cf. first row in Figure 5.2.3), or the amino acids from prebiotic Strecker reactions. Yet, the most important and never-ending activity of biotic RNA, even while the cells are dividing into two, is to synthesize proteins. Amino acids and nucleotides have been a nice pair ever since Methusalem. So, how do these two loose ends meet? Somehow, the comparingly ready-to-couple amino acids must have accumulated under prebiotic and activating (coupling) conditions and grown peptide chains apart from the much slower nucleotides. The need for the compartmentation and isolation of nucleotides and nucleic acids from the environment becomes more and more obvious.

5.2.3 COMPARTMENTATION OF NUCLEIC ACIDS

5.2.3.1 Self-Evolvable Surfaces

The reaction between a given number of molecules in a given volume of liquid can gain a factor of 1000 or more in catalytic efficiency—regardless of product inhibition and the catalytic turnover numbers—when all the molecules are somehow attached to a minimal surface that holds this volume without decreasing their mobility on that surface (2D mobility). When all these surface-bound molecules are lined up on a string instead, on which they can move only back and forth (1D mobility), the effective molarity could gain another factor 1000. The strongest power of catalysis, that is, the acceleration of a reaction rate $v = k \;][[\text{educts}]$, is not necessarily due to the change of the effective reaction rate constant k by increasing $k_{cat} \gg k$ through an appropriate complexation of a rate-limiting transition state on the reaction pathway. Such complexations, of course, are the usual way in which enzymatic reactions occur. There are also other more refined mechanisms, such as lifting the ground-state energy of the catalyst-substrate complex with respect to the uncomplexed specimens and/or catalyzing the rate-limiting elementary step through vibrational preferences and precise atomic orientation in "near-attack-conformation" complexes (Bruice 2002; Schowen 2003; Hur and Bruice 2003; Sadiq and Coveney 2015). But creating "entropy traps" (Westheimer 1962; Page and Jencks 1971), thus, enhancing the entropic component of the "effective molarity" (Kirby 1980; Mandolini 1986; Chi et al. 1995; Cacciapaglia et al. 2004; Hunter and Anderson 2009; Sinclair et al. 2009) of the starting compounds [[educts] through confinement (Lopez-Fontal et al. 2016, 2018), for example, by their more or less specific concentration on a surface, will do a good job, not negligible at all and plausibly prebiotic too (Strazewski 2009). How deep into biotic systems this effect may reach can be seen in the modern-day ribosomal biosynthesis of proteins by aminoacyl-transfer RNAs. In addition to being severely demobilized in cellular zones of locally ultra-high viscosity *in vivo*, these universal amino acid carriers can, only partly and highly reversibly of course, line up on messenger RNAs, viz. molecular 1D strings, and thus have an enhanced effective molarity. Ribosomes line up on the same messenger RNAs and form polysomes. The prokaryotic *in vivo* translation rate, that is, protein synthesis in terms of numbers of peptide bonds formed per second and ribosome that constantly select their "cognate" anticodons (correct for each codon) from a choice of 40 different comparingly large, about 76 nucleotides long slowly diffusing aminoacyl-transfer RNAs, is synchronized with its *in vivo* transcription rate, that is, the chain growth rate of the corresponding messenger RNAs, where the RNA polymerase selects at each template deoxyribonucleotide position the complementary ribonucleotide from a mere choice of four different fast-diffusing NTPs per codon. Most bacteria synthesize that way on the average one peptide bond per synthesis of one codon, or, the other way around, they extend the messenger RNA on the DNA template by about three nucleotides during the time of synthesis of one peptide bond. Both processes, transcription and translation, profit from reducing the dimensionality of at least parts of the starting material, although all cellular single-stranded DNAs (ssDNAs) and much of the messenger RNAs (mRNAs) are floating in the cytoplasm. Polysomes can interact with membrane surfaces and thus concentrate close to the membranes of organelles or the endoplasmic reticulum, where the nascent peptides cross the membrane through the ribosomal nascent peptide tunnel and the ribosome-bound transmembrane proteins and fold into their native conformation on the other side of the membrane. Hence, like for the sepharose or magnetic bead-bound DNA polymers that were immobilized to alleviate strand inhibition (Luther et al. 1998) or inactivated NMP

inhibition (Deck et al. 2011), biotic nature has chosen to synthesize proteins close to and interacting with membrane surfaces. This "choice" could only emerge from self-evolving primitive systems; therefore, it should not astonish if prebiotic nature already has taken advantage of enhancing the effective molarity of nucleotides and explored lipid membranes as self-evolving compartments, not only for isolated polymerization processes (Ross and Deamer 2016) but also to reduce their dimensionality through membrane-bound macromolecular interactions. Today, an estimated 20%–30% of all proteins are membrane proteins.

5.2.3.2 Trapping, Anchoring, and Growth of Nucleic Acids in Vesicles

The question of how RNA could be spontaneously compartmented in lipid vesicles, without designing elaborate chemical scaffolds on purpose, is being approached in three ways. A *rudimentary strategy* has been used by Hanczyc et al. and was deeper explored by David Deamer and collaborators. It consists of beginning with initially premixed lipids and nucleic acids, or their precursor NMPs, which become trapped within vesicles as they form. When pentadecaadenylate A_{15} was adsorbed to montmorillonite clay particles and added to myristoleate micelles made from Z-tetradec-9-enoic acid at pH 8.5, the covered particles were trapped in the aqueous interior of the resulting giant vesicles and A_{15} also became compartmented (Hanczyc et al. 2003). Deamer took a step backward and argued that it should be possible to drive phosphodiester bond formation in the absence of activated substrates by producing conditions in which water can be removed from the reactants. Upon hydration of dry lipid films, these amphiphiles draw water molecules away from bulk water to tightly hydrate the polar headgroups of nascent lipid membranes. In addition, Deamer evoked the lamellarity of dry lipid films (Walde 2006) that would produce large numbers of sheet-type surfaces, within which NMPs, when present, would be co-oriented in a liquid crystalline-like fashion and therefore their effective molecularity would be advantageously enhanced for condensation reactions (Rajamani et al. 2008; Olasagasti et al. 2011; Deamer 2012; Black et al. 2013; Mungi and Rajamani 2015). Indeed, long RNA-like polymers were detected when dry phospholipid films were cycled through repeated hydration-evaporation cycles in the presence of NMPs. In a typical reaction, the reactants were exposed to one to seven cycles of wetting and drying. A stream of carbon dioxide or nitrogen gas was used to dry the samples while they were exposed to varying experimental parameters, including temperature (60–90°C), time (30–120 min), lipid composition, and mole ratio of mononucleotide to lipid. After each drying cycle, the samples were dispersed in 1 mM HCl and allowed to rehydrate for 15–20 min. During this time, lipid vesicles reformed and components underwent mixing, and the mixtures were then exposed to a further dehydration cycle. The starting pH was 6.8, and this decreased to 2.2 at the end of seven cycles. Later, other amphiphiles and nucleotide components were explored, and it was shown, for example, that AMP and UMP performed far better than GMP or CMP for solubility reasons and confirmed that the pH needed to be dangerously acidic for the formation of polymeric chains to be efficient. More detailed later analyses have shown that large parts of the polymers contained, in fact, many "abasic sites," that is, positions where the nucleobases eliminated off from the polymer due to low pH, and that the chains were polynucleosides rather than polynucleotides (ethers not phosphodiesters). This, of course, sets early limits to a prebiotic lamellar dehydration and orientation-driven RNA synthesis. In any case, nucleotide trapping in nascent lipid vesicles may be effective when possible, but it is a quite crude and unselective process. What is missing in this approach is the evolvability of such a chemical system. A priori, there seems nothing particularly evolvable when nucleotides and nucleic acids are being trapped within nascent lipid vesicles, as if by force. The transformation from this first unselective compartmentation and subsequent self-evolvable fundamental "function" of compartmented RNA are not evident at all. Of course, once the DNA and dNTPs or RNA and NTPs are inside with the help of synthetic additives and by directive means, compartmented nucleic acids unfold astonishing properties during growth and division of the phospholipid vesicles (Nourian et al. 2012; Kurihara et al. 2015; Tsuji et al. 2016).

A *discriminatory strategy* is to directly select RNA sequences that bear a significant affinity to lipid membranes, despite the usual lipophobicity of ordinary RNA. Mike Yarus and collaborators impressively demonstrated that it is possible to select in vitro from "random pools" RNA molecules that do interact firmly with phospholipid vesicles (Khvorova et al. 1999; Vlassov et al. 2001; Vlassov and Yarus 2002; Janas and Yarus 2003; Vlassov 2005; Janas et al. 2006, 2012). Particular RNA loop structures and lipid membrane defects are important components for RNA-lipid interactions. The best binding RNAs were finicky hetero-trimers, that is, specifically folded RNA that would only anchor to phospholipid membranes when two distinctly different RNAs complexed with one another in a molar 2:1 ratio. Once anchored to membranes, the RNA aggregates oligomerized further, yielding larger, irregular rope-like structures that preferred the edges of altered lipid patches. These properties could be interpreted in terms of RNA-RNA kissing-loop interactions, and the RNA effects on membranes were explained in terms of an RNA preference for irregular lipid conformations. Structure-dependent RNA binding appeared for rafted liquid-ordered domains. Binding to more highly ordered gel-phase membranes was stronger, but much less RNA secondary structure-dependent. All modes of RNA-membrane association seem to be electrostatic and headgroup-directed and require divalent cations. Such unexpected RNA-lipid membrane interactions are fascinating and could have played a role in prebiotic environments, but, usually, these RNAs are so extremely rare that they appear untraceable in random pools of RNA. It seems that nature could profit from the variety of natural RNA pools for a spontaneous compartmentation but only if such RNA could find a way to cross the membranes and get inside.

A *mediator strategy* proposes an acquired further attraction between RNA and lipids, viz. with the help of a mediator compound bearing intermediate properties that will smoothen the mechanism of compartmentation of largely lipophobic nucleic acids and thus may increase the selectivity for particularly useful RNA, or DNA if present (Bada 2013; Gavette et al. 2016; Kumar 2016), to be compartmented but to a much less restrictive degree than RNA in the absence of the mediator (Vlassov 2005; Strazewski 2007, 2009, 2014, 2015a). For reasons that are described and elaborated in Chapter 6.3, the mediators could be prebiotic lipophilic peptides. Upon addition of chemical activators and amino acids inside lipid vesicles, such peptides are expected to grow in length, variety, and eventually competence of broad catalytic specificities (Jensen 1976; Szilágyi et al. 2012). One such catalysis could be the acceleration of the PT reaction between aminoacyl-RNA and peptidyl-RNA, taking place in a lipid bilayer, that is, being anchored to the surface from the inside of lipid vesicles. The other vital catalysis is a template-dependent RNA polymerization reaction. This approach should allow for the evolution of translation and, from there, to protein-assisted RNA or DNA replication. This hypothesis is testable. First experimental examinations have been carried out and published. We found that a hydrophobic peptide served as an efficient anchoring device for a lipophobic RNA hairpin. Just how well the lipophilicity of this peptide compensated for the lipophobicity of the RNA was an unexpected outcome of this study (Le Chevalier Isaad et al. 2014; Strazewski 2015a).

5.2.4 SUMMARY

In this chapter, the transdisciplinary field of astrobiology focuses on the organic chemist's crucial role in finding answers to questions about the origin of life through scientific experimentation on a molecular level. Organic chemistry has a particular responsibility, because it operates very close to the animated, living state of matter. A characteristic feature of all known life-forms is the ability to maintain a systemic memory in the *gestalt* of nucleic acids throughout successive generations. The principles of replication and translation are analyzed for more general cases than what is known from biology. Another characteristic feature of life is its cellular, compartmented composition. The requirement for the spontaneous compartmentation of nucleic acids is discussed. The most important molecules that are needed for a spontaneous and plausibly prebiotic assembly of nucleosides, nucleotides, and nucleic acids are presented, with a special focus on ribonucleosides and RNA. Experimental results from the past 10 years are discussed in greater detail. The chapter ends with an overview on the kinetic effect of reduced dimensionality and a short update on the spontaneous and plausibly prebiotic interactions of nucleotides and nucleic acids with lipid vesicles.

ACKNOWLEDGMENTS

I thank Michele Fiore (Lyon) for his enthusiasm in finding literature references that were destined for other articles but have found their way into this chapter. I thank my wife Marzena for her patience and smiling attitude, even though we still do not understand how I could invest so many months in writing without being paid for it.

REFERENCES

Al-Azmi, A., A.-Z. A. Elassar, and B. L. Booth. 2003. The chemistry of diaminomaleonitrile and its utility in heterocyclic synthesis. *Tetrahedron* 59: 2749–2763.

Apel, C. L., M. N. Mautner, and D. W. Deamer. 2002. Self-assembled vesicles of monocarboxylic acids and alcohols: Conditions for stability and for encapsulation of biopolymers. *Biochim. Biophys. Acta* 1559: 1–9.

Attwater, J., A. Wochner, V. B. Pinheiro, A. Coulson, and P. Holliger. 2010. Ice as a protocellular medium for RNA replication. *Nat. Commun.* 1: 76.

Attwater, J., A. Wochner, and P. Holliger. 2013. In-ice evolution of RNA polymerase ribozyme activity. *Nat. Chem.* 5: 1011–1018.

Avakyan, N., A. A. Greschner, F. Aldaye, C. J. Serpell, V. Toader, A. Petitjean, and H. F. Sleiman. 2016. Reprogramming the assembly of unmodified DNA with a small molecule. *Nat. Chem.* 8: 368–376.

Bachmann, P. A., P. Walde, and P.-L. Luisi. 1990. Self-replicating reverse micelles and chemical autopoiesis. *J. Am. Chem. Soc.* 112: 8200–8201.

Bada, J. 2013. New insights into prebiotic chemistry from Stanley Miller's spark discharge experiments. *Chem. Soc. Rev.* 42: 2186–2196.

Bada, J. L., J. H. Chalmers, and H. J. Cleaves II. 2016. Is formamide a geochemically plausible prebiotic solvent? *Phys. Chem. Chem. Phys.* 18: 20085–20090.

Bartlett, J. K., and D. A. Skoog. 1958. Colorimetric determination of elemental sulfur in hydrocarbons. *Anal. Chem.* 26: 1008–1011.

Baumgart, T., S. T. Hess, and W. W. Webb. 2003. Imaging coexisting fluid domains in biomembrane models coupling curvature and line tension. *Nature* 425: 821–824.

Becker, S., I. Thoma, A. Deutsch, T. Gehrke, P. Mayer, H. Zipse, and T. Carell. 2016. A high-yielding, strictly regioselective prebiotic purine nucleoside formation pathway. *Science* 352: 833–836.

Bendich, A., J. F. Tinker, and G. B. Brown. 1948. A synthesis of isoguanine labeled with isotopic nitrogen. *J. Am. Chem. Soc.* 70: 3109–3113.

Benner, S. A., H.-J. Kim, and M. A. Carrigan. 2012. Asphalt, water, and the prebiotic synthesis of ribose, ribonucleosides, and RNA. *Acc. Chem. Res.* 45: 2025–2034.

Biebricher, C. K., and M. Eigen. 2005. The error threshold. *Virus Res.* 107: 117–127.

Black, R. A., M. C. Blosser, B. L. Stottrup, R. Tavakley, D. W. Deamer, and S. L. Keller. 2013. Nucleobases bind to and stabilize aggregates of a prebiotic amphiphile, providing a viable mechanism for the emergence of protocells. *Proc. Natl. Acad. Sci. U.S.A.* 110: 13272–13276.

Boutlerow, A. M. 1861. Formation synthétique d'une substance sucrée. *C. R. Acad. Sci.* 53: 145–147.

Briers, Y., P. Walde, M. Schuppler, and M. J. Lössner. 2012. How did bacterial ancestors reproduce? Lessons from L-Form cells and giant lipid vesicles. *Bioessays* 34: 1078–1084.

Bruice, T. C. 2002. A view at the millennium: The efficiency of enzymatic catalysis. *Acc.Chem. Res.* 35: 139–148.

Budin, I., and J. W. Szostak. 2011. Physical effects underlying the transition from primitive to modern cell membranes. *Proc. Natl. Acad. Sci. U.S.A.* 108: 5249–5254.

Burcar, B. T., L. M. Cassidy, E. M. Moriarty, P. C. Joshi, K. M. Coari, and L. B. McGown. 2013. Potential pitfalls in MALDI-TOF MS analysis of abiotically synthesized RNA oligonucleotides. *Origins Life Evol. Biosphere* 43: 247–261.

Burcar, B., M. Pasek, M. Gull, B. J. Cafferty, F. Velasco, N. V. Hud, and C. Menor-Salván. 2016. Darwin's warm little pond: A one-pot reaction for prebiotic phosphorylation and the mobilization of phosphate from minerals in a urea-based solvent. *Angew. Chem. Int. Ed.* 55: 13249–13253.

Butlerow, A. M. 1861. Bildung einer zuckerartigen Substanz durch Synthese (Formation of a sugar-like substance by synthesis). *Justus Liebigs Ann. Chem.* 120: 295–298.

Cacciapaglia, R., S. Di Stefano, and L. Mandolini. 2004. Effective molarities in supramolecular catalysis of two-substrate reactions. *Acc. Chem. Res.* 37: 113–122.

Cafferty, B. J., D. M. Fialho, J. Khanam, R. Krishnamurthy, and N. V. Hud. 2016. Spontaneous formation and base pairing of plausible prebiotic nucleotides in water. *Nat. Commun.* 7: 11328.

Callahan, B. P., Y. Yuan, and R. Wolfenden. 2005. The burden borne by urease. *J. Am. Chem. Soc.* 127: 10828–10829.

Carrara, P., P. Stano, and P.-L. Luisi. 2012. Giant vesicles "colonies": A model for primitive cell communities. *ChemBioChem* 13: 1497–1502.

Chen, I., R. W. Roberts, and J. W. Szostak. 2004. The emergence of competition between model protocells. *Science* 305: 1474–1476.

Chen, I. A., and M. A. Nowak. 2012. From prelife to life: How chemical kinetics become evolutionary dynamics. *Acc. Chem. Res.* 45: 2088–2096.

Chen, B., Z. Li, D. Zhang, J. Liu, M. Y. Hu, J. Zhao, W. Bi, E. E. Alp, Y. Xiao, P. Chow, and J. Li. 2014. Hidden carbon in Earth's inner core revealed by shear softening in dense Fe_7C_3. *Proc. Natl. Acad. Sci. U.S.A.* 111:17755–17758.

Chi, X., A. J. Guerin, R. A. Haycock, C. A. Hunter, and L. D. Sarson. 1995. The thermodynamic of self-assembly. *J. Chem. Soc., Chem. Commun.* 2563–2565.

Chyba, C., and C. Sagan. 1992. Endogenous production, exogenous delivery and impact shock synthesis of organic molecules: An inventory for the origins of life. *Nature* 355: 125–132.

Clark, J., R. K. Grantham, and J. Lydiate.1968. Heterocyclic studies. Part V. Desulphuration of heterocyclic thiols with nickel boride. *J. Chem. Soc. C* 1122–1124.

Coari, K. M., R. C. Martin, K. Jain, and L. B. McGown. 2017. Nucleotide selectivity in abiotic RNA polymerization reactions. *Origins Life Evol. Biosphere* 47: 305–321.

Coggins, A. J., and M. W. Powner. 2017. Prebiotic synthesis of phosphoenol pyruvate by α-phosphorylation-controlled triose glycolysis. *Nat. Chem.* 9: 310–317.

Costanzo, G., S. Pino, G. Botta, R. Saladino, and E. Di Mauro. 2011. May cyclic nucleotides be a source for abiotic RNA synthesis? *Origins Life Evol. Biosphere* 41: 559–562.

Costanzo, G., R. Saladino, G. Botta, A. Giorgi, A. Scipioni, S. Pino, and E. Di Mauro. 2012. Generation of RNA molecules by a base-catalysed click-like reaction. *ChemBioChem* 13: 999–1008.

Cornish, V. W., D. Mendel, and P. G. Schultz. 1995. Probing protein structure and function with an expanded genetic code. 1995. *Angew. Chem. Int. Ed. Engl.* 34: 621–633.

Crespo-Hernández, C. E., and R. Arce. 2000. Mechanism of formation of guanine as one of the major products in the 254 nm photolysis of guanine derivatives: Concentration and pH effects. *Photochem. Photobiol.* 71: 544–550.

Deamer, D. W. 2017. The role of lipid membranes in life's origin. *Life* 7: 5.

Deamer, D. W. 2012. Liquid crystalline nanostructures: Organizing matrices for non-enzymatic nucleic acid polymerization. *Chem. Soc. Rev.* 4: 5375–5379.

Deamer, D. W., and J. P. Dworkin. 2005. Chemistry and physics of primitive membranes. *Top. Curr. Chem.* 259: 1–27.

Deck, C., M. Jauker, and C. Richert. 2011. Efficient enzyme-free copying of all four nucleobases templated by immobilized RNA. *Nat. Chem.* 3: 603–608.

Eigen, M. 1971. Self-organization of matter and the evolution of biological macromolecules. *Naturwissenschaften* 58: 465–523.

Engelhart, A. E., K. P. Adamala, and J. W. Szostak. 2016. A simple physical mechanism enables homeostasis in primitive cells. *Nat. Chem.* 8: 448–453.

Engelhart, A. E., M. W. Powner, and J. W. Szostak. 2013. Functional RNAs exhibit tolerance for non-heritable 2'–5' versus 3'–5' backbone heterogeneity. *Nat. Chem.* 5: 390–394.

Eschenmoser, A. 1999. Chemical etiology of nucleic acid structure. *Science* 284: 2118–2124.

Eschenmoser, A. 2004. The TNA-family of nucleic acid systems: Properties and prospects. *Origins Life Evol. Biosphere* 34: 277–306.

Eschenmoser, A., and E. Loewenthal. 1992. Chemistry of potentially prebiological natural products. *Chem. Soc. Rev.* 21: 1–16.

Fahrenbach, A. C., C. Giurgiu, C. P. Tam, L. Li, Y. Hongo, M. Aono, and J. W. Szostak. 2017. Common and potentially prebiotic origin for precursors of nucleotide synthesis and activation. *J. Am. Chem. Soc.* 139: 8780–8783.

Fernández-García, C., and M. W. Powner. 2017. Selective acylation of nucleosides, nucleotides, and glycerol-3-phosphocholine in water. *Synlett* 28: 78–83.

Fernández-García, C., A. J. Coggins, and M. W. Powner. 2017. A chemist's perspective on the role of phosphorus at the origins of life. *Life* 7: 31.

Ferris, J. P., and G. Ertem. 1993. Montmorillonite catalysis of RNA oligomer formation in aqueous solution. A model for the prebiotic formation of RNA. *J. Am. Chem. Soc.* 115: 12270–12275.

Ferris, J. P., A. R. Hill Jr., R. Liu, and L. E. Orgel. 1996. Synthesis of long prebiotic oligomers on mineral surfaces. *Nature* 381: 59–61.

Ferris, J. P. 2002. Montmorillonite catalysis of 30–50 mer oligonucleotides: Laboratory demonstration of potential steps in the origin of the RNA world. *Origins Life Evol. Biosphere* 32: 311–332.

Ferris, W. D., R. A. Sanchez, and L. E. Orgel. 1968. Studies in prebiotic synthesis. III. Synthesis of pyrimidines form cyanoacetylene and cyanate. *J. Mol. Biol.* 33: 693–704.

Fiore, M., W. Madanamoothoo, A. Berlioz-Barbier, O. Maniti, A. Girard-Egrot, R. Buchet, and P. Strazewski. 2017. Giant vesicles from rehydrated crude mixtures containing unexpected mixtures of amphiphiles formed under plausibly prebiotic conditions. *Org. Biomol. Chem.* 15: 4231–4240.

Fiore, M., and P. Strazewski. 2016. Prebiotic lipidic amphiphiles and condensing agents on the early Earth. *Life* 6: 17.

Förstel, M., P. Maksyutenko, B. M. Jones, B.-J. Sun, A. H. H. Chang, and R. I. Kaiser. 2016. Synthesis of urea in cometary model ices and implications for Comet 67P/Churyumov–Gerasimenko. *Chem. Commun.* 52: 741–744.

Fox, S. W., and K. Dose. 1972. *Molecular Evolution and the Origin of Life*. San Francisco, CA: W. H. Freeman & Co Ltd.

Fox, S. W., and K. Dose. 1977. *Molecular Evolution and the Origin of Life*, revised ed. New York, Basel: Marcel Dekker.

Fuller, W. D., R. A. Sanchez, and L. E. Orgel. 1972. Studies in prebiotic synthesis VI. Synthesis of purine nucleosides. *J. Mol. Biol.* 67: 25–33.

Furukawa, Y., H.-J. Kim, D. Hutter, and S. A. Benner. 2015. Abiotic regioselective phosphorylation of adenosine with borate in formamide. *Astrobiology* 15: 259–267.

Gavette, J. V., M. Stoop, N. V. Hud, and R. Krishnamurthy. 2016. RNA–DNA chimeras in the context of an RNA world transition to an RNA/DNA world. *Angew. Chem. Int. Ed.* 55: 13204–13209.

Georgiadis, M. M., I. Singh, W. F. Kellett, S. Hoshika, S. A. Benner, and N. G. J. Richards. 2015. Structural basis for a six nucleotide genetic alphabet. *J. Am. Chem. Soc.* 137: 6947–6955.

Girniene, J., D. Gueyrard, A. Tatibouët, A. Sackus, and P. Rollin. 2001. Base-modified nucleosides from carbohydrate derived oxazolidinethiones: A five-step process. *Tetrahedron Lett.* 42: 2977–2980.

Girniene, J., G. Apremont, A. Tatibouët, A. Sackus, and P. Rollin. 2004. Small libraries of fused quinazolinone-sugars: Access to quinazolinedione nucleosides. *Tetrahedron* 60: 2609–2619.

Greenwood Hansma, H. 2017. Better than membranes at the origin of life? *Life* 7: 28.

Griesser, H., M. Bechthold, P. Tremmel, E. Kervio, and C. Richert. 2017. Amino acid-specific, ribonucleotide-promoted peptide formation in the absence of enzymes. *Angew. Chem. Int. Ed.* 56: 1224–1228.

Haley, C., and P. Maitland. 1951. Organic reactions in aqueous solution at room temperature. Part I. The influence of pH on condensations involving the linking of carbon to nitrogen and of carbon to carbon. *J. Chem. Soc.* 3155–3174.

Hanczyc, M. M., S. M. Fujikawa, and J. W. Szostak. 2003. Experimental models of primitive cellular compartments: Encapsulation, growth, and division. *Science* 302: 618–622.

Hargreaves, W. R., Mulvihill, S. J., and D. W. Deamer. 1977. Synthesis of phospholipids and membranes in prebiotic conditions. *Nature* 266: 78–80.

Hargreaves, W. R., and D. W. Deamer. 1978. Liposomes from ionic, single-chain amphiphiles. *Biochemistry* 17: 3759–3768.

He, C., I. Gállego, B. Laughlin, M. A. Grover, and N. V. Hud. 2017. A viscous solvent enables information transfer from gene-length nucleic acids in a model prebiotic replication cycle. *Nat. Chem.* 9: 318–324.

Hudson, J. S., J. F. Eberle, R. H. Vachhani, L. C. Rogers, J. H. Wade, R. Krishnamurthy, and G. Springsteen. 2012. A unified mechanism for abiotic adenine and purine synthesis in formamide. *Angew. Chem. Int. Ed.* 51: 5134–5137.

Hunter, C. A., and H. L. Anderson. 2009. What is cooperativity? *Angew. Chem. Int. Ed.* 48: 7488–7499.

Hur, S., and T. C. Bruice. 2003. The near attack conformation approach to the study of the chorismate to prephenate reaction. *Proc. Natl. Acad. Sci. U.S.A.* 100: 12015–12020.

Islam, S., D.-K. Bučar, and M. W. Powner. 2017. Prebiotic selection and assembly of proteinogenic amino acids and natural nucleotides from complex mixtures. *Nat. Chem.* 9: 584–589.

Janas, T, and M. Yarus. 2003. Visualization of membrane RNAs. *RNA* 9: 1353–1361.

Janas, T., T. Janas, and M. Yarus. 2006. Specific RNA binding to ordered phospholipid bilayers. *Nucleic Acids Res.* 34: 2128–2136.

Janas, T., T. Janas, and M. Yarus. 2012. Human tRNA[Sec] associates with HeLa membranes, cell lipid liposomes, and synthetic lipid bilayers. *RNA* 18: 2260–2268.

Jauker, M., H. Griesser, and C. Richert. 2015a. Copying of RNA sequences without pre-activation. *Angew. Chem. Int. Ed.* 54: 14559–14563.

Jauker, M., H. Griesser, and C. Richert. 2015b. Spontaneous formation of RNA strands, peptidyl RNA, and cofactors. *Angew. Chem. Int. Ed.* 54: 14564–14569.

Jensen, R. A. 1976. Enzyme recruitment in evolution of new function. *Annu. Rev. Microbiol.* 30: 409–425.

Joshi, P. C., M. F. Aldersley, and J. P. Ferris. 2013. Progress in demonstrating homochiral selection in prebiotic RNA synthesis. *Adv. Space Res.* 51: 772–779.

Karki, M., C. Gibard, S. Bhowmik, and R. Krishnamurthy. 2017. Nitrogenous derivatives of phosphorus and the origins of life: Plausible prebiotic phosphorylating agents in water. *Life* 7: 32.

Kervio, E., B. Claasen, U. Steiner, and C. Richert. 2014. The strength of the template effect attracting nucleotides to naked DNA. *Nucleic Acids Res.* 42: 7409–7420.

Khvorova, A., Y. G. Kwak, M. Tamkun, I. Majerfeld, and M. Yarus. 1999. RNAs that bind and change the permeability of phospholipid membranes. *Proc. Natl. Acad. Sci. U.S.A.* 96: 10649–10654.

Kim, H.-J., Y. Furukawa, T. Kakegawa, A. Bita, R. Scorei, and S. E. Benner. 2016. Evaporite borate-containing mineral ensembles make phosphate available and regiospecifically phosphorylate ribonucleosides. Borate as a problem solver in prebiotic chemistry. *Angew. Chem. Int. Ed.* 55: 15816–15820.

Kim, H.-J., A. Ricardo, H. I. Illangkoon, M. J. Kim, M. A. Carrigan, F. Frye, and S. A. Benner. Synthesis of carbohydrates in mineral-guided prebiotic cycles. 2011. *J. Am. Chem. Soc.* 133: 9457–9468.

Kirby, A. J. 1980. Effective molarities for intramolecular reactions. *Adv. Phys. Org. Chem.* 17: 183–278.

Koch, K. E. 1992. Über strukturelle Zusammenhänge zwischen Blausäure und natürlichen Purinen und Pteridinen (On structural relations between hydrogen cyanide and natural purines and pteridines). *Thesis*, no. 9968 (dir. A. Eschenmoser), ETH Zürich, ETH e-collection; doi:10.3929/ethz-a-000694124.

Krishnamurthy, R. 2012. Role of pK_a of nucleobases in the origins of chemical evolution. *Acc. Chem. Res.* 45: 2035–2044.

Ksander, G., G. Bold, R. Lattmann, C. Lehmann, T. Früh, Y.-B. Xiang, K. Inomata, H.-P. Buser, J. Schreiber, E. Zass, and A. Eschenmoser. 1987. 105. Chemie der α-Aminonitrile. 1. Mitteilung: Einleitung und Wege zu Uroporphyrinogen-octanitrilen (105. Chemistry of the α-aminonitriles. First communication: Introduction and pathways to uroporphyrogen-octanitriles). *Helv. Chim. Acta* 70: 1115–1172.

Kuhn, H. 1972. Self-Organization of molecular systems and evolution of the genetic apparatus. *Angew. Chem. Int. Ed.* 11: 798–811.

Kuhn, H., and J. Waser. 1981. Molecular self-organization and the origin of life. *Angew. Chem. Int. Ed.* 20: 500–520.

Kumar, V. 2016. Evolution of specific 3'–5'-linkages in RNA in pre-biotic soup: A new hypothesis. *Org. Biomol. Chem.* 14: 10123–10133.

Kun, Á., M. Santos, and E. Szathmáry. 2005. Real ribozymes suggest a relaxed error threshold. *Nat. Genet.* 37: 1008–1011.

Kurihara, K., Y. Okura, M. Matsuo, T. Toyota, K. Suzuki, and T. Sugawara. 2015. A recursive vesicle-based model protocell with a primitive model cell cycle. *Nat. Commun.* 6: 8352.

Le Chevalier Isaad, A., P. Carrara, P. Stano, K. S. Krishnakumar, D. Lafont, A. Zamboulis, R. Buchet, D. Bouchu, F. Albrieux, and P. Strazewski. 2014. A hydrophobic disordered peptide spontaneously anchors a covalently bound RNA hairpin to giant lipidic vesicles. *Org. Biomol. Chem.* 12: 6363–6373.

Leu, K., B. Obermayer, S. Rajamani, U. Gerland, and I. A. Chen. 2011. The prebiotic evolutionary advantage of transferring genetic information from RNA to DNA. *Nucleic Acids Res.* 39: 8135–8147.

Li, L., N. Prywes, C. P. Tam, D. K. O'Flaherty, V. S. Lelyveld, E. C. Izgu, A. Pal, and J. W. Szostak. 2017. Enhanced nonenzymatic RNA copying with 2-aminoimidazole activated nucleotides. *J. Am. Chem. Soc.* 139: 1810–1813.

Liu, H., J. Gao, S. R. Lynch, Y. D. Saito, L. Maynard, and E. T. Kool. 2003. A four-base paired genetic helix with expanded size. *Science* 302: 868–871.

Liu, Z., L. Rigger, J.-C. Rossi, J. D. Sutherland, and R. Pascal. 2016. Mixed anhydride intermediates in the reaction of 5(4H)-oxazolones with phosphate esters and nucleotides. *Chem. Eur. J.* 22: 14940–14949.

Lopez-Fontal, E., L. Milanesi, and S. Tomas. 2016. Multivalence cooperativity leading to "all-or-nothing" assembly: The case of nucleation-growth in supramolecular polymers. *Chem. Sci.* 7: 4468–4475.

Lopez-Fontal, E., A. Grochmal, T. Foran, L. Milanesi, and S. Tomas. 2018. Ship in a bottle: Confinement-promoted self-assembly. *Chem. Sci.* 9: 1760–1768.

Luther, A., R. Brandsch, and G. von Kiedrowski. 1998. Surface-promoted replication and exponential amplification of DNA analogues. *Nature* 396: 245–248.

Malyshev, D. A., K. Dhami, T. Lavergne, T. Chen, N. Dai, J. M. Foster, I. R. Corrêa Jr., and F. E. Romesberg. 2014. A semisynthetic organism with an expanded genetic alphabet. *Nature* 509: 385–388.

Mandolini, L. 1986. Intramolecular reactions of chain molecules. *Adv. Phys. Org. Chem.* 22: 1–111.

Mansy, S. S., J. P. Schrum, M. Krishnamurthy, S. Tobé, D. A. Treco, and J. W. Szostak. 2008. Template-directed synthesis of a genetic polymer in a model protocell. *Nature* 454: 122–125.

Maynard Smith, J., and E. Szathmáry. 1995. *The Major Transitions in Evolution:* Oxford, UK: Oxford University Press.

McCollom, T. M., G. Ritter, and B. R. T. Simoneit. 1998. Lipid synthesis under hydrothermal conditions by Fischer-Tropsch-type reactions. *Origins Life Evol. Biosphere* 29: 153–166.

McCollom, T. M., and B. R. T. Simoneit. 1999. Abiotic formation of hydrocarbons and oxygenated compounds during thermal decomposition of iron oxalate. *Origins Life Evol. Biosphere* 29: 167–186.

Menor-Salván, C., D. M. Ruiz-Bermejo, M. I. Guzmán, S. Osuna-Esteban, and S. Veintemillas-Verdaguer. 2009. Synthesis of pyrimidines and triazines in ice: Implications for the prebiotic chemistry of nucleobases. *Chem. Eur. J.* 15: 4411–4418.

Miller, S. L. 1953. Production of amino acids under possible primitive Earth conditions. *Science* 117: 528–529.

Monnard P.-A., A. Kanavarioti, and D. W. Deamer. 2003. Eutectic phase polymerization of activated ribonucleotide mixtures yields quasi-equimolar incorporation of purine and pyrimidine nucleobases. *J. Am. Chem. Soc.* 125: 13734–13740.

Morillon, A. 2018. Long non-coding RNA. *The Dark Side of the Genome.* London, UK: ISTE Press.

Müller, D., S. Pitsch, A. Kittaka, E. Wagner, C. E. Wintner, and A. Eschenmoser. 1990. 135. Chemie von α-Aminonitrilen. Aldomerisierung von Glycolaldehyd-phosphat zu racemischen Hexose-2,4,6-triphosphaten und (in Gegenwart von Formaldehyd) racemischen Pentose-2,4-diphosphaten: Rac-Allose-2,4,6-triphosphat und rac-Ribose-2,4-diphosphat sind die Reaktionshauptprodukte. *Helv. Chim. Acta* 73: 1410–1468.

Mungi, C. V., and S. Rajamani. 2015. Characterization of RNA-like oligomers from lipid-assisted nonenzymatic synthesis: Implications for origin of informational molecules on Early Earth. *Life* 5: 65–84.

Nagakura, S., H. Izawa, H. Ezawa, S. Iwamuara, H. Saho, and R. Kubo (Eds.). 1998. *Rikagaku-Jiten*, 5th ed. Tokyo, Japan: Iwanami-shoten.

Nourian, Z., W. Roelofsen, and C. Danelon. 2012. Triggered gene expression in fed-vesicle microreactors with a multifunctional membrane. *Angew. Chem. Int. Ed.* 51: 3114–3118.

Öberg, K. I., V. V. Guzmán, K. Furuya, C. Qi, Y. Aikawa, S. M. Andrews, R. Loomis, and D. J. Wilner. 2015. The comet-like composition of a protoplanetary disk as revealed by complex cyanides. *Nature* 520: 198–201.

Olasagasti, F., H. J. Kim, N. Pourmand, and D. W. Deamer. 2011. Non-enzymatic transfer of sequence information under plausible prebiotic conditions. *Biochimie* 93: 556–561.

Orgel, L. 2004. Prebiotic chemistry and the origin of the RNA world. *Crit. Rev. Biochem. Mol. Biol.* 39: 99–123.

Oró, J. 1960. Synthesis of adenine from ammonium cyanide. *Biochem. Biophys. Res. Commun.* 2: 407–412.

Oró, J., and A. C. Cox. 1962. Non-enzymic synthesis of deoxyribose. *Fed. Proc.* 21: 80.

Oró, J, and A. P. Kimball. 1961. Synthesis of purines under possible primitive Earth conditions. I. Adenine from hydrogen cyanide. *Arch. Biochem. Biophys.* 94: 217–227.

Page, M. I., and W. P. Jencks. 1971. Entropic contributions to rate accelerations in enzymic and intramolecular reactions and the chelate effect. *Proc. Natl. Acad. Sci. U.S.A.* 18: 1678–1683.

Pankratov, V. A., and A. E. Chesnokova. 1989. Polycyclotrimerization of cyanamides. *Russ. Chem. Rev.* 58: 879–890.

Parker, E. T., M. Zhou, A. S. Burton, D. P. Glavin, J. P. Dworkin, R. Krishnamurthy, F. M. Fernández, and J. L. Bada. 2014. A plausible simultaneous synthesis of amino acids and simple peptides on the primordial Earth. *Angew. Chem. Int. Ed.* 53: 1–6.

Pascal, R., and A. Pross. 2015. Stability and its manifestation in the chemical and biological worlds. *Chem. Commun.* 51: 16160–16165.

Patel, B. H., C. Percivalle, D. J. Ritson, C. D. Duffy, and J. D. Sutherland. 2015. Common origins of RNA, protein and lipid precursors in a cyanosulfidic protometabolism. *Nat. Chem.* 7: 301–307.

Picirilli, J., T. Krauch, S. E. Moroney, and S. A. Benner. 1990. Enzymatic incorporation of a new base pair into DNA and RNA extends the genetic alphabet. *Nature* 343: 33–37.

Pietrucci, F., and A. M. Saitta. 2015. Formamide reaction network in gas phase and solution via a unified theoretical approach: Toward a reconciliation of different prebiotic scenarios. *Proc. Natl. Acad. Sci. U.S.A.* 112: 15030–15035.

Pinheiro, V. B., A. I. Taylor, C. Cozens, M. Abramov, M. Renders, S. Zhang, J. C. Chaput, J. Wengel, S. Y. Peak-Chew, S. H. McLaughlin, P. Herdewijn, and P. Holliger. 2012. Synthetic genetic polymers capable of heredity and evolution. *Science* 336: 341–4.

Pino, S., J. E. Šponer, G. Costanzo, S. Pino, R. Saladino, and E. Di Mauro. 2015. From formamide to RNA, the path is tenuous but continuous. *Life* 5: 372–384.

Pincass, H. 1922. Die Bildung von calciumcyanamid aus ferrocyancalcium (The formation of calcium cyanamide out of calcium cyanoferrate). *Chemiker-Ztg.* 46: 661.

Ponnamperuma, C., R. Mariner, and C. Sagan. 1963. Formation of adenosine by ultra-violet irradiation of a solution of adenine and ribose. *Nature* 198:1199–1200.

Ponnamperuma, C., and P. Kirk. 1964. Synthesis of deoxyadenosine under simulated primitive Earth conditions. *Nature* 203: 400–401.

Powner, M. W., B. Gérland, and J. D. Sutherland. 2009. Synthesis of activated pyrimidine ribonucleotides in prebiotically plausible conditions. *Nature* 45: 239–242.

Powner, M. W., S. L. Zheng, and J. W. Szostak. 2012. Multicomponent assembly of proposed DNA precursors in water. *J. Am. Chem. Soc.* 134: 13889–13895.

Prieur B. P. 2001. Étude de l'activité prébiotique potentielle de l'acide borique (Studies on the potential prebiotic activity of boronic acid). *C. R. Acad. Sci. Ser. IIC Chem.* 4: 667–670.

Pross, A. 2012. *What is Life? How Chemistry Becomes Biology.* Oxford, UK: Oxford University Press.

Rajamani, S., J. K. Ichida, T. Antal, D. A. Treco, K. Leu, M. A. Nowak, J. W. Szostak, and I. A. Chen. 2010. Effect of stalling after mismatches on the error catastrophe in nonenzymatic nucleic acid replication. *J. Am. Chem. Soc.* 132: 5880–5885.

Rajamani, S., A. Vlassov, S. Benner, A. Coombs, F. Olasagasti, and D. W. Deamer. 2008. Lipid-assisted synthesis of RNA-like polymers from mononucleotides. *Origins Life Evol. Biosphere* 38: 57–74.

Ranjan, S., and D. D. Sasselov. 2016. Influence of the UV environment on the synthesis of prebiotic molecules. *Astrobiology* 16: 68–88.

Rao, M., J. Eichberg, and J. Oró. 1982. Synthesis of phosphatidylcholine under possible primitive Earth conditions. *J. Mol. Evol.* 18: 196–202.

Rao, M., J. Eichberg, and J. Oró. 1987. Synthesis of phosphatidylethanolamine under possible primitive earth conditions. *J. Mol. Evol.* 25: 1–6.

Raoul, S., M. Bardet, and J. Cadet. 1995. γ-Irradiation of 2'-deoxyadenosine in oxygen-free aqueous solutions: Identification and conformational features of formamidopyrimidine nucleoside derivatives. *Chem. Res. Toxicol.* 8: 924–933.

Ritson, D. J., J. Xu, and J. D. Sutherland. 2016. Thiophosphate – A versatile prebiotic reagent? *Synlett* 28: 64–67.

Ritson, D., and J. D. Sutherland. 2012. Prebiotic synthesis of simple sugars by photoredox systems chemistry. *Nat. Chem.* 4: 895–899.

Ritson, D., and J. D. Sutherland. 2013. Synthesis of aldehydic ribonucleotide and amino acid precursors by photoredox chemistry. *Angew. Chem. Int. Ed.* 52: 5845–5847.

Robins, R. K., K. J. Dille, C. H. Willits, and B. E. Christensen. 1953. Purines. II. The synthesis of certain purines and the cyclization of several substituted 4,5-diaminopyrimidines. *J. Am. Chem. Soc.* 75: 263–266.

Ross, D. S., and D. Deamer. 2016. Dry/wet cycling and the thermodynamics and kinetics of prebiotic polymer synthesis. *Life* 6: 28.

Röthlingshöfer, M., and C. Richert. 2010. Chemical primer extension at submillimolar concentrations of deoxynucleotides. *J. Org. Chem.* 75: 3945–3952.

Rushdi, A. I., and B. R. T. Simoneit. 2006. Abiotic condensation synthesis of glyceride lipids and wax esters under simulated hydrothermal conditions. *Origins Life Evol. Biosphere* 36: 93–108.

Sadiq, S. K., and P. V. Coveney. 2015. Computing the role of near attack conformations in an enzyme-catalyzed nucleophilic bimolecular reaction. *J. Chem. Theory Comput.* 11: 316–324.

Sagi, V. N., P. Karri, F. Hu, and R. Krishnamurthy. 2011. Diastereoselective self-condensation of dihydroxyfumaric acid in water: Potential route to sugars. *Angew. Chem. Int. Ed.* 50: 8127–8130.

Sagi, V. N., V. Punna, F. Hu, G. Meher, and R. Krishnamurthy. 2012. Exploratory experiments on the chemistry of the "Glyoxylate Scenario": Formation of ketosugars from dihydroxyfumarate. *J. Am. Chem. Soc.* 134: 3577–3589.

Saladino, R., C. Crestini, G. Constanzo, R. Negri, and E. Di Mauro. 2001. A possible prebiotic synthesis of purine, adenine, cytosine, and 4(3H)-pyrimidinone from formamide: Implications for the origin of life. *Bioorg. Med. Chem.* 9: 1249–1253.

Saladino, R., C. Crestini, F. Ciciriello, S. Pino, G. Constanzo, and E. Di Mauro. 2009. From formamide to RNA: The roles of formamide and water in the evolution of chemical information. *Res. Microbiol.* 160: 441–448.

Saladino, R., E. Carota, G. Botta, M. Kapralov, G. N. Timoshenko, A. Y. Rozanov, E. Krasavin, and E. Di Mauro. 2015. Meteorite-catalyzed syntheses of nucleosides and of other prebiotic compounds from formamide under proton irradiation. *Proc. Natl. Acad.Sci. U.S.A.* 112: E2746–E2755.

Saladino, R., U. Ciambecchini, C. Crestini, G. Constanzo, R. Negri, and E. Di Mauro. 2003. One-pot TiO$_2$-catalyzed synthesis of nucleic bases and acyclonucleosides from formamide: Implications for the origin of life. *ChemBioChem* 4: 514–521.

Salván, C., and M. R. Marín-Yaseli. 2013. A new route for the prebiotic synthesis of nucleobases and hydantoins in water/ice solutions involving the photochemistry of acetylene. *Chem. Eur. J.* 19: 6488–6497.

Sanchez, R. A., and L. E. Orgel. 1970. Studies in prebiotic synthesis V. Synthesis and photoanomerization of pyrimidine nucleosides. *J. Mol. Biol.* 47: 531–43.

Sanchez, R. A., J. P. Ferris, and L. E. Orgel. 1966. Cyanoacetylene in prebiotic synthesis. *Science* 154: 784–785.

Schoffstall, A. M. 1976. Prebiotic phosphorylation of nucleosides in formamide.*Origin. Life* 7: 399–412.

Schoffstall, A. M., R. J. Barto, and D. L. Ramos. 1982. Nucleoside and deoxynucleoside phosphorylation in formamide solutions.*Origin. Life* 12: 143–151.

Schöning, K., P. Scholz, S. Guntha, X. Wu, R. Krishnamurthy, and A. Eschenmoser. 2000. Chemical etiology of nucleic acid structure: The alpha-threofuranosyl-(3'→2') oligonucleotide system. *Science* 290: 1347–1351.

Schowen, R. L. 2003. How an enzyme surmounts the activation energy barrier. *Proc. Natl. Acad. Sci. U.S.A.* 100: 11931–11932.

Schwartz, J., and M. Hornyak-Hamor. 1971. *Hung. Teljes* (Hungarian patent application) 6998, September 27.

Seifer, G. B. 1962a. Thermal decomposition of alkali cyanoferrates(II). *Russ. J. Inorg. Chem.* 7: 640–643.

Seifer, G. B. 1962b. Thermal decomposition of alkaline earth metal and magnesium cyanoferrates(II). *Russ. J. Inorg. Chem.* 7: 1187–1189.

Shaw, W. H. R., and J. J. Bordeaux. 1955. The decomposition of urea in aqueous media. *J.Am. Chem. Soc.* 77: 4729–4733.

Simoneit, B. R. T., A. I. Rushdi, and D. W. Deamer. 2007. Abiotic formation of acylglycerols under simulated hydrothermal conditions and self-assembly properties of such lipid products. *Adv. Space Res.* 40: 1649–1656.

Sinclair, A. J., V. del Amo, and D. Philp. 2009. Structure–reactivity relationships in a recognition mediated [3 + 2] dipolar cycloaddition reaction. *Org. Biomol. Chem.* 7: 3308–3318.

Šponer, J. E., J. Šponer, O. Nováková, V. Brabec, O. Šedo, Z. Zdráhal, G. Costanzo, S. Pino, R. Saladino, and E. Di Mauro. 2016. Emergence of the first catalytic oligonucleotides in a formamide-based origin scenario. *Chem. Eur. J.* 22: 3572–3586.

Stairs, S., A. Nikmal, D.-K. Bučar, S.-L. Zheng, J. W. Szostak, and M. W. Powner. 2017. Divergent prebiotic synthesis of pyrimidine and 8-oxo-purine ribonucleotides. *Nat. Commun.* 8: 15270.

Steinman, G., R. M. Lemmon, and M. Calvin. 1964. Cyanamide: A possible key compound in chemical evolution. *Proc. Natl. Acad. Sci. U.S.A.* 52: 27–30.

Stoop, M., G. Meher, P. Karri, and R. Krishnamurthy. 2013. Chemical etiology of nucleic acid structure. The pentulofuranosyl oligonucleotide systems: (1'→3')-β-L-ribulo, (4'→3')-α-L-xylulo, and (1'→3')-α-L-xylulo nucleic acids. *Chem. Eur. J.* 19: 15336–15345.

Strazewski, P. 1990. The biological equilibrium of base pairs. *J. Mol. Evol.* 30: 116–124.

Strazewski, P., and C. Tamm. 1990. Replication experiments with nucleotide base analogues. *Angew. Chem. Int. Ed. Engl.* 29: 36–57.

Strazewski, P. 2007. Question 6: How did translation occur? *Origins Life Evol. Biosphere.* 37: 399–401.

Strazewski, P. 2009. Adding to Hans Kuhn's thesis on the emergence of the genetic apparatus: Of the Darwinian advantage to be neither too soluble, nor too insoluble, neither too solid, nor completely liquid. *Colloids Surf. B* 74: 419–425.

Strazewski, P. 2014. RNA as major components in chemical evolvable systems. In: *Chemical Biology of Nucleic Acids: Fundamentals and Clinical Applications.* V. A. Erdmann, W. T. Markiewicz, J. Barciszewski (Eds.), Series: RNA Technologies. Springer, chapter 1, pp. 1–24. https://www.springer.com/fr/book/9783642544514.

Strazewski, P. 2015a. Amphiphilic peptidyl-RNA. In: *DNA in Supramolecular Chemistry and Nanotechnology.* E. Stulz, G. H. Clever (Eds,), Wiley-VCH, chapter 4.3, pp. 294–307. doi:10.1002/9781118696880.ch4.3

Strazewski, P. 2015b. Omne vivum ex vivo … omne? How to feed an inanimate evolvable chemical system so as to let it self-evolve into increased complexity and life-like behaviour. *Isr. J. Chem.* 55: 851–864.

Sutherland, J. D. 2016. The origin of life—Out of the blue. *Angew. Chem. Int. Ed.* 55: 104–121.

Szabat, M., D. Gudanis, W. Kotkowiak, Z. Gdaniec, R. Kierzek, and A. Pasternak. 2016. Thermodynamic features of structural motifs formed by β-L-RNA. *PLoS ONE* 11: e0149478.

Szathmáry, E. 1992. What is the optimum size for the genetic alphabet? *Proc. Natl. Acad. Sci. U.S.A.* 98: 2614–2618.

Szathmáry, E., M. Santos, and C. Fernando. 2005. Evolutionary potential and requirements for minimal protocells. Springer: Berlin, Heidelberg. *Top. Curr. Chem.* 259: 167–211.

Szilágyi, A., Á. Kuhn, and E. Szathmáry. 2012. Early evolution of efficient enzymes and genome organization. *Biol. Direct* 7: 38.

Thaddeus, P. 2006. The prebiotic molecules observed in the interstellar gas. *Philos. Trans. R.Soc. B* 361: 1681–1687.

Theobald, D. L. 2010. A formal test of the theory of universal common ancestry. *Nature* 465: 219–222.

Traube, W. 1904. Der Aufbau der Xanthinbasen aus der Cyanessigsäure. Synthese des Hypoxanthins und Adenins (The buildup of the xanthine bases from cyanacetic acid: Synthesis of hypoxanthine and adenine). *Liebigs Ann. Chem.* 331: 64–88.

Trinks, U. P. 1987. Zur Chemie der Aminopyrimidine (On the chemistry of aminopyrimidines). *Thesis*, no. 8368 (dir. A. Eschenmoser), ETH Zürich, ETH e-collection; doi:10.3929/ethz-a-000413538.

Tsuji, G., S. Fujii, T. Sunami, and T. Yomo. 2016. Sustainable proliferation of liposomes compatible with inner RNA replication. 2016. *Proc. Natl. Acad. Sci. U.S.A.* 113: 590–595.

Vlassov, A., A. Khvorova, and M. Yarus. 2001. Binding and disruption of phospholipid bilayers by supramolecular RNA complexes. *Proc. Natl. Acad. Sci. U.S.A.* 98: 7706–7711.

Vlassov, A., and M. Yarus. 2002. Interaction of RNA with phospho-lipid membranes. *Mol. Biol.* (Moscow) 36: 389–393.

Vlassov, A. 2005. How was membrane permeability produced in an RNA world? *Origins Life Evol. Biosphere* 35: 135–149.

Walde, P. 2006. Surfactant assemblies and their various possible roles for the origin(s) of life. *Origins Life Biosphere* 36: 109–150.

Weimann, B. J., R. Lohrmann, L. E. Orgel, H. Schneider-Bernlöhr, and J. E. Sulston. 1968. Template-directed synthesis with ade-nosine-5'-phosphorimidazolide. *Science* 161: 387.

Westheimer, F. H. 1962. Mechanisms related to enzyme cataly-sis. In: *Advances in Enzymology and Related Areas of Microbiology*. 2006. F. F. Nord, John Wiley & Sons, chapter 9, pp. 441–482.

Whitesides, G. M., E. E. Simanek, J. P. Mathias, C. T. Seto, D. N. Chin, M. Mammen, and D. M. Gordon. 1995. Noncovalent synthesis: Using physical-organic chemistry to make aggre-gates. *Acc. Chem. Res.* 28: 37–44.

Winiger, C. B., M.-J. Kim, S. Hoshika, R. W. Shaw, J. D. Moses, M. F. Matsuura, D. L. Gerloff, and S. A. Benner. 2016. Polymerase interactions with wobble mismatches in synthetic genetic systems and their evolutionary implications. *Biochemistry* 55: 3847–3850.

Wu, T., and L. E. Orgel. 1992. Nonenzymatic template-directed syn-thesis on hairpin oligonucleotides. 3. Incorporation of adenos-ine and uridine residues. *J. Am. Chem. Soc.* 114: 7963–7969.

Wu, W.-J., Yang W., and Tsai, M.-D. 2017. How DNA polymerases catalyse replication and repair with contrasting fidelity. *Nat. Rev. Chem.* 1: 0068.

Xu, J., M. Tsanakopoulou, C. J. Magnani, R. Szabla, J. E. Šponer, J. Šponer, R. W. Góra, and J. D. Sutherland. 2017. A prebi-otically plausible synthesis of pyrimidine β-ribonucleosides and their phosphate derivatives involving photoanomerization. *Nat. Chem.* 9: 303–309.

Yamanaka, M., Y. Fujita, A. McLean, and M. Iwase. 2000. A ther-modynamic study on $CaCN_2$. *High Temp. Mater. Process.* 19: 275–279.

Yu, H., S. Zhang, and J. C. Chaput. 2012. Darwinian evolution of an alternative genetic system provides support for TNA as an RNA progenitor. *Nat. Chem.* 4: 183–187.

5.3 The Hydrothermal Impact Crater Lakes

The Crucibles of Life's Origin

Sankar Chatterjee

CONTENTS

5.3.1 INTRODUCTION

Impact cratering was the primal force in the early history of our planet, before the onset of plate tectonics. It has shaped the surface architecture, composition, and rheology of the lithosphere and enhanced the emergence of life. The explorations of the Earth, Moon, Mercury, and Mars allow reasonable inferences about the physical conditions of prebiotic Earth. During the Hadean and early Archean Eons, impacts and collisions were the dominant geodynamic processes that played a

key role in the crustal, tectonic, thermal, magmatic, and environmental evolutions of young Earth. For its first half-billion years, the Hadean Earth was violent and desolate, enduring the punishing rain of extraterrestrial impacts, which scorched the globe and widely reprocessed the crustal surface through mixing and burial by impact-generated melt (Marchi et al. 2014). It was a hot molten globe with an atmosphere filled with water vapor, carbon dioxide, nitrogen, hydrogen sulfide, sulfur dioxide, and other noxious gases. This first stage of Earth's history included the initial accretion, core/mantle differentiation, the development of a magma ocean, and an undifferentiated mafic crust (Hawkesworth et al. 2016).

Following the Hadean, the second stage in Earth's evolution occurred during the Eoarchean Eon in a pre-plate tectonic regime (Tang et al. 2016). It was marked by elevated mantle temperatures compared to the present day, which resulted in a lithosphere weakened by the emplacement of melts. This inhibited subduction and hence plate tectonics, and magmatism was driven by impact-induced mantle upwellings that percolated the lithosphere. The crust was at least 15–20 km thick during this period; remelting could take place by meteorite impact, and the resultant magmas represented the high silica component that characterizes the Archean crust (Grieve et al. 2006; Hawkesworth et al. 2016). Although few grains of detrital zircons from the Jack Hills of western Australia are known to be Hadean (~4.4 Ga) (Harrison et al. 2008; Wilde et al. 2001), the oldest rock formations exposed on the surface of the Earth are exclusively Archean. It is in the Archean that the first earthly ecosystems are found, with their clues to life's earliest days on the planet.

During the Eoarchean (~4 Ga), the pummeling continued, but the Earth became more tranquil, as the frequency of impacts slowed down. Earth cooled, clouds formed, and torrential rain submerged the lowlands to form global oceans, and as the crust began to harden, it was followed by erosion and sediment recycling. Shortly after Earth began to cool, the planet's first outer layer, the lithosphere, was a single, solid, deformable shell, lacking any plate tectonic activity (Stern 2005; Tang et al. 2016). It was a one-plate world. The Archean Eon witnessed the production of early continental crust and global oceans, the emergence of life, and fundamental changes to the atmosphere. Around 4 billion years ago, the primordial oceans covered more than 95% of the Earth's surface surrounding the emerging protocontinents (Mojzsis and Harrison 2000). The focus of this study is to determine the likely site of life's beginnings on young Earth—the crucible of life—and to reconstruct its environment during the Eoarchean to Paleoarchean Eons (~4–3.2 Ga).

The early terrestrial planets and moons in the inner solar system experienced a cataclysm known as the Late Heavy Bombardment (LHB), about ~4.1–3.8 Ga, when a disproportionately large number of asteroids collided and crashed onto these rocky planets. The early Earth was bombarded by projectiles at the same time that large impact basins were forming on the Moon and Mercury. Earth, being a bigger target with a stronger gravitational pull, would have suffered hundreds of blows from terrestrial objects of that size. Although these early craters on Earth are no longer visible, the Earth was actually rocked by 15–20 times more impacts than our Moon! (Kring 2000; Marchi et al. 2014). The impacting objects were mostly asteroids from the asteroid belt located between the orbits of Mars and Jupiter. Comets that reside in a huge cloud at the outer reaches of the solar system appear to have comprised a small fraction of the projectiles. Impact events on the Eoarchean crust may have produced thousands of craters on early Earth, resembling the surfaces of the Moon and Mercury. But unlike our planetary neighbors, the crater basins on early Earth were filled with water and cosmic building blocks, and there developed a complex system of subsurface hydrothermal vents that were the fundamental crucibles that led to the origin of life.

Impacts on a water-rich planet such as Earth or even Mars can generate hydrothermal activity—specifically, underwater areas boiling with heat and spewing chemicals. Over 20,000 hydrothermal crater lakes (with diameters ranging from 5 km to >1000 km) dotted the Eoarchean crust, inadvertently becoming the perfect crucibles for the prebiotic chemistry of early life (Chatterjee 2015, 2016, in press; Cockell 2006; Kring 2000; Marchi et al. 2014; Osiniski et al. 2013). Unlike the Moon and Mercury, physical evidence of Earth's early bombardments has been erased by weathering and plate tectonics over the eons.

Large meteorite impacts trigger widespread catastrophe, killing life on a global scale; this happened during the extinction of the dinosaurs 66 million years ago at the Cretaceous-Paleogene boundary (Alvarez et al. 1980). But contradistinct, comets and asteroids—the very objects impacting the inner planets and moons—contain the building blocks of life (Chyba and Sagan 1992). These impacts delivered the cosmic, organic compounds to Earth and were very likely a boon to early life on the planet: instead of wiping it out or preventing it from originating, they may have helped kick-start life. Like the cosmic dance of *Shiva*, impacts not only deliver death but also foster life. Perhaps, the bulk of the organic molecules necessary for life's origins was brought by meteorite impacts on the young Earth. Similarly, impacts may have created vast underground habitats. Impact-generated crater lakes on the Eoarchean crust may have created novel and protective habitats from meteoritic assaults, where the building blocks of life could be concentrated, integrated, selected, encapsulated, polymerized, and organized into successively more complex and information-rich molecules through symbiosis, leading to the emergence of the first cells. These crater lakes may have provided protective sanctuaries for the emerging thermophiles during the ongoing bombardments (Chatterjee 2015, 2016; Cockell 2006; Cockell and Lee 2002; Osiniski et al. 2013).

The continuous impacts of meteorites, and their cosmic composition, strongly support the conclusions that the building blocks began to accumulate in the oceans and the lakes, where the hydrothermal energy could drive the synthesis of ever more complex organic compounds. This period of LHB overlaps with the evidence of the earliest life on Earth, thus hinting at some causal connection between impacts and the origin of life. It's becoming apparent in recent years that most of the biosphere was brought to the primitive Earth by an intense bombardment of meteorites (Delsemme 1998). This included atmosphere, hydrosphere, and those volatile carbon compounds needed for the emergence of life. Prebiotic molecules, delivered by the meteorites, would have provided the "seed" for the origin of the Earth's first life.

The Eoarchean crust is the most valuable archive for unraveling the mystery of the origin of life. A symbiotic view of the geochemical origin of life has been presented in previous publications (Chatterjee 2015, 2016, in press). In this view, life arose through five hierarchical stages of increasing molecular complexity: cosmic, geologic, chemical, information, and biological. The main thrust of this article is to elaborate the first two stages—cosmic and geologic—determining the role of impacts for biogenesis and exploring the likely site of life's origin on young Earth.

5.3.2 DELIVERY OF THE BUILDING BLOCKS OF LIFE

Biological evolution was preceded by a long phase of geochemical evolution, during which the precursors of life molecules accumulated on young Earth and began to interact. One of the intriguing questions associated with the origin of life is related to the original source of these chemical building blocks: Were they created by chemical reactions of primordial atmospheric and crustal ingredients on the young Earth, or did they, perhaps, hitch a ride on meteorites that germinated our and other planets in our solar system with the same biomolecules? Many terrestrial formation models suggest Earth would have formed extremely poor volatiles and was deficient in the organic compounds necessary for life synthesis (Delsemme 2001). Surprisingly, the chemical makeup of life is much closer to the chemistry of meteorites and stardust than it is to that of our rocky planet during this early history. Comets and carbonaceous asteroids are rich in organic molecules, which are required for the emergence of life on early Earth. Recent research in space exploration and astrobiology provides strong evidence that meteorite impacts may have sparked life on early Earth. Life's biochemical building blocks can form in the freezing, harsh, zero-gravity environment of deep space, bolstering the odds that meteorite strikes triggered life (Bernstein et al. 1999; Chatterjee 2016; Chyba and Sagan 1992; Cockell 2006; Deamer et al. 2002; Kring 2000; Osiniski et al. 2013). Unlike space, the early Earth's high gravity and hot temperature were not conducive to create these building blocks.

One of the most exciting possible sources of these raw materials may have been the asteroids and comets that bombarded Earth during its first billion years. Consequently, while life itself likely arose on Earth, the building blocks of life may well have had an extraterrestrial origin. Perhaps, the important raw material needed to build life came from space and was delivered by meteorites. Many of these complex biomolecules such as lipid membranes, amino acids, nucleotides, phosphorous, and sugars have been detected in meteorites (Bernstein et al. 1999). Moreover, these building blocks of life are abundant in interstellar space. The astrobiological origin of these building blocks occurred in an unusual interstellar, freezing, zero-gravity environment during the explosion of a nearby star (Boss and Keiser 2010). Complex organic molecules, precursors to life, have been detected everywhere in space, in comets, in carbonaceous asteroids, and in interstellar dust. The NASA STARDUST mission detected these building blocks in interstellar dust particles (Materese et al. 2013).

In interstellar space, far from the Sun, temperatures are freezing, leading to abundant ice and other frozen gases such as methane, ammonia, carbon dioxide, carbon monoxide, and even ethyl alcohol. Through the heavy meteorite bombardment, the volatiles on Earth arrived from this distant reservoir in the outer solar system. New experiments simulating conditions in deep space reveal that the complex building blocks could have been created in icy interplanetary dust and then carried to Earth, jump-starting life. Interstellar ice simulation experiments at the NASA Ames Research Center in a cryogenic laboratory at 40 K ($-233°C$) have shown that the ultraviolet (UV) radiation processing of presolar ices leads to more complex organic compounds. All the major building blocks of life, including ribose and related sugars, were synthesized via irradiation of interstellar ice analogs (silicate minerals coated with simple molecules such as H_2O, CO, CO_2, NH_3, and CH_3OH) (Deamer et al. 2002; Meinert et al. 2016; Sandford et al. 2014). Ribose is the central molecular subunit of RNA, and its prebiotic origin in interstellar environments is significant. These experimental findings support the identification of organic molecules in comet samples taken in situ by the Philae Lander part of the cometary Rosetta Mission (Meierhenrich 2015). In another laboratory simulation of interstellar ices, the building blocks of comets, that is, a frozen mixture of water, methanol, ammonia, and carbon monoxide, were subjected to UV radiation (Dworkin et al. 2001). This combination yielded organic compounds that formed bubbles when immersed in water. These bubbles are reminiscent of cell membranes that enclose and concentrate the chemistry of life, separating it from the outside world. Many of the organic compounds formed in these experiments are also present in meteorites, cometary and asteroidal dusts, some of which are necessary to the origin of life. These experiments suggest that the building blocks of life may have begun in a rather unusual freezing environment of interstellar space during the planetary formation and that they were delivered to early Earth in the suitable environments provided by meteorites. In space, when atoms are locked in ice, this bond-breaking process can make molecular fragments recombine into unusually complex structures, which would not be possible if these fragments were free to drift apart. Everywhere in space, complex organic compounds are forming on these ice grains, especially in UV-rich regions around young suns (Bernstein et al. 1999). Much of the organic compounds found on early Earth had an interstellar heritage. Carbon-based compounds are being synthesized everywhere in the Milky Way, including diffuse interstellar clouds, giant molecular clouds, and protoplanetary disks, in which new solar systems are formed. Meteorites and interstellar dusts provide a record of the chemical processes that occurred before life began on Earth; they are like manna from heaven, triggering and accelerating biosynthesis on Earth. Most of the material of biospheres—as our bodies—is stardust, a gift from the meteorites.

There is a paradigm shift in the origin of life research from the Miller-type experiment that demonstrated that several organic compounds could be formed spontaneously by simulating the condition of Earth's early atmosphere (Miller 1953) to cosmic building blocks. Experiments by NASA suggest that the building blocks of life originated in an unusual, extreme environment of cold temperature in an interstellar medium with very low density, which was missing in early Earth. This may explain why the Miller-type experiment in laboratory failed to produce similar diverse arrays of building blocks. Finding these biomolecules in an interstellar gas cloud means that important building blocks of life such as membrane-forming molecules, amino acids, nucleobases, sugars, phosphorous, and other organic compounds—the crucial components of cells—were seeded on the primordial Earth via carbonaceous chondrites, comets, and interstellar dusts (Bernstein et al. 1999; Chyba and Sagan 1992). These prebiotic organic compounds were delivered to early Earth by the impacting meteorites that contained the building blocks of life. Moreover, cosmic dusts or micrometeorites (<2 mm) delivered some of these biomolecules to early Earth. They contain organic compounds—such as polycyclic aromatic hydrocarbons, aliphatic chains, and amino acids—some of which were used by life synthesis. They represent the largest mass flux of extraterrestrial material (30,000 tons per year), raining down on Earth daily, carpeting the land and water surface with fine dust (Motrajit et al. 2006). Though these building blocks from space were also delivered to other planets in the solar system, our Goldilocks planet appears to have been the only habitable environment, ready to jump-start biosynthesis.

Carbonaceous chondrites, comets, and micrometeorites in interstellar clouds are extremely rich in the biogenic compounds, those volatile elements like carbon and nitrogen that are essential to organic life. Carbonaceous chondrites are primitive stony asteroids that contain a rich diversity of organic molecules, including those relevant to biogenesis, such as sugars, carboxylic acids, aldehydes, ketones, nucleobases, amino acids, and membrane-forming lipids (Deamer 2011). These asteroids experienced liquid water for the first 10,000 years of their history, during which time their amino acids were synthesized. Yet, there is no evidence for peptides or nucleic acids in these asteroids, and in general, it appears that prebiotic synthesis in these objects did not proceed beyond monomers. In complete contrast, starting with these cosmic monomers, prebiotic chemistry proceeded much further on Earth, leading first to polymers and then to the first cells. Can these molecules survive the catastrophic impacts during the collision with Earth? Obviously, they did. Carbonaceous chondrites collected across the planet still, to this day, contain those essential building blocks, clearly demonstrating that the hitchhiking organic molecules withstood the extreme pressures and temperatures of fiery crashes onto Earth's surface; they remained intact and survived despite the tremendous shock wave and other violent conditions (Callahan et al. 2011).

Among the most celebrated carbonaceous chondrites is the Murchison meteorite that fell to Earth near Murchison, Victoria, in Australia in 1969. Some of the building blocks detected in the Murchison meteorite are essential components of living cells. These include pyrimidine and purine nucleobases, sugars, and phosphates for making nucleic acids, amino acids, and peptides for synthesizing proteins, membrane-forming compounds such as long-chain monocarboxylic

acids, and carbohydrates (Callahan et al. 2011; Damer and Deamer 2015; Deamer 2011; Pizzarello and Cronin 2000). Carbonaceous chondrites recovered from Antarctica contain numerous indigenous organic compounds, nucleobases, and amino acids (Burton et al. 2013). Carbon-rich asteroids may have been important sources for the organic compounds required for the emergence of life. Impacts may have increased the diversity of complex prebiotic chemical species through shock metamorphism. The carbonaceous asteroids were important contributors to life's building blocks on early Earth (Bernstein et al. 1999; Chyba and Sagan 1992).

Unlike asteroids, comets are cosmic snowballs of frozen gases, rock, and dust. Like asteroids, they are leftovers from the dawn of the solar system around 4.6 billion years ago and consist mostly of ice coated with dark organic matter. The study of the chemical composition of comets provides key information about the raw materials of organic compounds present in interstellar space, which are of prebiotic interest. Comets contain prodigious quantities of organic compounds, including amino acids, adenine, ketones, quinones, carboxylic acids, hydrogen cyanide, polycyclic amino aromatic hydrocarbons, thioformaldehyde, acetaldehyde, sugars, cyanogens, cyanide, methanol, and ethanol (Bernstein et al. 1999). HCN is a key molecule in the prebiotic synthesis of amino acids, nucleobases, and sugars. The numerous complex organic molecules such as ethyl alcohol and sugar have been detected in comet C/2014 Q2 (Lovejoy) (Biver et al. 2015). Recently, the Rosetta mission of the European Space Agency (ESA) craft Philae Lander already has detected water and organic molecules on the surface of 67P/Churyumov–Gerasimenko, a 4-km-wide comet; this may shed new light on the composition of Earth's building blocks (Goesmann et al. 2015). Sixteen organic compounds were identified, divided into six classes of organic molecules (alcohols, carbonyls, amines, nitriles, amides, and isocyanates). Of these, four organic compounds were detected for the first time on a comet (methyl isocyanates, acetone, propionaldehyde, and acetamide). The presence of complex organic compounds in comets and asteroids implies that the early solar system fostered the formation of prebiotic material in noticeable concentrations. There is thus ample evidence that a number of biogenic compounds can form spontaneously in interstellar space and on comets and carbonaceous asteroids. Hence, the origin of life as an extraordinary event is the product of two worlds: interstellar space and Earth. The building blocks of life came from interstellar space via meteorites (exogenous delivery), but first life was synthesized and evolved in the womb of our planet (endogenous production) (Chyba and Sagan 1992).

Meteorites were much more numerous in the LHB period, about ~4.1–3.8 Ga, than they are today, and they were correspondingly more likely to collide with early Earth. These collisions would have seeded our planet with organic compounds from the very beginning. Organic compounds are also found in interplanetary dust particles, which still rain down on Earth. Meteorite impacts played two distinct roles in biogenesis: the delivery of key ingredients of life and the creation of hydrothermal impact crater lakes for prebiotic synthesis (Chatterjee 2015, 2016; Chyba and Sagan 1992; Cockell 2006; Kring 2000; Osinski et al. 2013). These post-impact crater lakes became the incubators for biogenesis and early life.

Large impacts created tremendous shock waves and other violent conditions, causing volcanic activity in a water-rich geological system on early Earth environments and vaporizing oceans as they hit (Sleep et al. 1989; Zahnle et al. 2007). Smaller ones would have survived the impact, bringing with them prebiotic molecules. Comets would have been ideal packages for delivering terrestrial volatiles, including some important precursors for prebiotic synthesis. Because of the icy cushion of comets, the building blocks of life would have remained intact upon impact, while others fused with other chemicals to form more complex organic compounds (Delsemme 2001). There might well have been multiple delivery systems of organic compounds over the millions of years of the Eoarchean—by comets, asteroids, and interstellar dusts. These cosmic ingredients were then mixed with various biomolecules produced in the hydrothermal crater basin to form the primordial soup (Chyba and Sagan 1992; Damer and Deamer 2015). The initial stages in the origin of life required a continuous source of organic molecules, both cosmic and terrestrial, to act as precursors to the biopolymers (Deamer 2011). Today, a constant rain of interplanetary dust particles delivers the same kind of organic compounds that might have contributed to the formation of the primordial soup on early Earth (Delsemme 1998, 2001).

It is fascinating to consider that most of the building blocks that led to the first life on Earth may well have been of extraterrestrial origin. Most likely, such cosmic biomolecules provided the crucial ingredients of life, but the "cooking" was done entirely in the hot crucibles on early Earth. Of these myriads of cosmic ingredients percolating in the prebiotic soup, a few organic compounds would be selected through the hydrothermal vents for making more complex biomolecules and polymers, leading to the assembly of cell components; the rest would be discarded. The picture that emerges suggests that most of the molecular components we see in living cells were available in prebiotic conditions. Slowly, over millions of years, organized systems of contained reactions have gained the ability to grow and may self-replicate. Life began when a system of encapsulated polymers initiated symbiosis and was then able to capture energy and nutrients from the environment and use that energy to grow and reproduce (Chatterjee 2016; Deamer 2011). Life arose rapidly during the early Eoarchean Eon, in the narrow window of some 100 million years.

5.3.3 LIQUID WATER FROM STONE: THE MOLECULE OF LIFE

Liquid water is essential for the kind of delicate chemistry that makes life possible. A bacterial cell is 70% water by weight, which serves as life's solvent. Liquid water is a universal solvent, a mediator of life's chemical reactions for transporting and mixing organic molecules. Water provides the matrix and universal medium for life. Water is a small, highly polar molecule. As a result, it is the ideal solvent for the chemistries of life—dissolving many molecules and transporting them to reaction sites, while preserving their integrity. Ice or water vapor will not accomplish the same function. Ice is fairly common in the universe, found everywhere, from vast interstellar clouds to many planets and moons. In contrast, liquid water is rare in the universe. Thus, the origin of life hinges greatly around the source and availability of water on the early Earth. Water is kept in the liquid form on Earth by the precise combination of atmospheric pressure and internal heat. From the geologic record, it appears that Earth was a watery planet about 4 billion years ago, or even earlier (Mojzsis et al. 2001).

Astronomers know that interstellar water is abundantly available to young planetary systems, and our young Earth collected plenty of it. Despite the abundance of water on Earth, there is a great deal of controversy about the origin of water on our planet: whether water was present at the formation of the planet through the degassing of volcanoes, or whether it arrived later, perhaps by asteroids or comets. The composition of the ocean offers some clues to its origin. Evidence is mounting that the planet's water arrived early during the LHB, or even earlier, aboard meteorites. Delsemme (2001) championed the view that water was brought to primitive Earth by an intense bombardment of comets. New analysis of the isotopic ratio of hydrogen of the carbonaceous chondrites suggests that the water did not originate in the comets in the outer solar system, but in asteroid belts, between Mars and Jupiter. The deuterium-to-hydrogen (D/H) ratio in ocean water, approximately 150 ppm, is similar to the average of carbonaceous chondrites, which suggests that the principal sources of water might be delivered by carbonaceous chondrite-like sources, not from icy comets (Sarafian et al. 2014). Water vapor streaming off the comet contains a higher percent of deuterium than what the water on Earth has (Alexander et al. 2012). Most likely, carbonaceous asteroids were the dominant sources of Earth's volatiles, as well as the preferred vehicle for the arrival of water on early Earth. This view of an asteroidal origin of water has been reinforced by the recent study of the D/H ratio of the comet 67P/Churuymov-Gerasimenko, which negates the cometary origin of water on Earth (Altwegg et al. 2014).

5.3.4 PALEOECOLOGY OF LIFE'S BEGINNINGS: CLUES FROM HYPERTHERMOPHILIC MICROBES

The geologic site of life's beginnings is one of the key tenets to discovering where and how life originated. The habitats of hyperthermophiles (superheat-loving microbes), which are the most primitive living organisms, may shed new light on the oldest ecosystems on our planet—the crucible of life. Discovered in 1977, submarine hydrothermal vents astounded many scientists when it was discovered that the hyperthermophilic bacteria and archaea thrive in these deep, dark, anaerobic, hostile, and volcanic environments. They developed the unusual ability to utilize the chemical nutrients that rise from the hot vent fluids interfacing with cooler seawater as a source of energy (McCollom and Shock 1997). Today, hyperthermophiles are found in geothermally heated subterranean rocks such as the boiling hot springs of the Yellowstone National Park, hydrothermal impact crater lakes, and submarine hydrothermal vents along the mid-ocean ridge. They grow optimally above 80°C and exhibit an upper temperature border of growth up to 113°C. Hyperthermophiles have certain heat-stable enzymes (that are very important in biotechnology) and unusual rigid membranes that are specially geared to working in high temperatures. Based on their growth requirements, hyperthermophiles were probably the most primitive living organisms that could have existed on early Earth about 4 billion years ago (Stetter 2006).

Because of their antiquity, hyperthermophiles may provide clue to the paleoecology and ancestry of life's beginnings. Darwin's doctrine of common descent proclaims that all life on Earth arose from a common ancestor, which is the central pillar of modern evolutionary biology (Darwin 1859). Trace back the separate lines of descent of all organisms that ever lived, and they will converge to a single point of origin—the beginning of life. Nothing concrete could be said about the nature of this ancestor initially, but it was intuitively assumed to be simple, bacteria-like organism.

Molecular phylogeny provides compelling evidence for Darwin's theory of universal common ancestor. Based on the pioneering work of Carl Woese, the conserved molecule such as 16S ribosomal RNA (rRNA) is widely used as a powerful tool for determining microbial relationships (Woese 1987, 1994; Woese et al. 1990); molecular phylogeny suggests that all living organisms cluster into three domains—bacteria, archaea, and the eukarya within the "universal tree." Moreover, these domains are descended from a single ancestral form, the last universal common ancestor (LUCA). This venerable ancestor was a single-cell, possibly a hyperthermophilic, microbe and is estimated to have lived some 4 billion years ago. The deepest and shortest phylogenetic branches are represented by Aquificales and Thermotogales within the bacteria and the Nanoarchaeota, Pyrodictiaceae, and Methanopyraceae within the archaea (Stetter 2006). At present, about 90 species of hyperthermophilic bacteria and archaea are known, which had been isolated from different terrestrial and marine thermal areas of the world. The LUCA is genetically closest to the base of the tree of life. The roots of the 16S rRNA tree in the domains of bacteria and Archaea are all populated by hyperthermophiles, suggesting that the emergence of life may have occurred in hydrothermal systems. If so, examining the environment and geochemistry of hydrothermal systems should provide a glimpse of the oldest ecology of our planet.

Hyperthermophiles are now regarded as the most "slowly-evolving" of all extant life, the first to have diverged from the LUCA. New studies suggest that the LUCA itself was a hyperthermophilic chemoautotroph (Akanuma et al. 2013; Ciccarelli et al. 2006; Gaucher et al. 2003, 2008; Weiss et al. 2016). This view is supported by recent analyses of resurrected proteins that predicted the hyperthermophilic adaptation of the LUCA, favoring the view that hydrothermal systems may have been the crucible of life for early biosphere evolution (Gaucher et al. 2003; Nisbet and Sleep 2001; Pace 1997; Theobald 2010). The LUCA's genes pointed to an organism that lived in deep, hot, oxygen-free, mineral-rich hydrothermal vent environments. The distribution of hyperthermophiles at the roots of the tree suggests that hydrothermal systems may have been the site for life's beginnings. Most evidence suggests that the LUCA stored genetic information by using DNA, built proteins, and used ATP as its currency for energy. The LUCA was a single-cell microbe with a few hundred genes that fed on hydrogen gas and lived in a world devoid of oxygen, bolstering strong suspicions that early life on Earth may have evolved under hyperthermophilic conditions in and around hydrothermal vents.

Current theories on the possible geologic sites for biogenesis range from frozen oceans (Bada and Lazcano 2002) to various hydrothermal systems, including tidal pools and hot springs (Deamer 2011), hydrothermal impact crater lakes (Chatterjee 2016; Cockell 2006; Osiniski et al. 2013), and submarine hydrothermal vents (Baross and Hoffman 1985; Martin and Russell 2007; Wachterhäuser 1993), to name a few. However, hydrothermal vents stand out among the possible environments for life's origin, holding particular promise for understanding the transition from geochemistry to biochemistry.

Several environmental and geochemical features of early Earth favor the supposition that hydrothermal vents were the likely site for life's beginnings—the long-sought incubators. These include the following: (1) the hyperthermophiles, the most primitive organisms known, which presently inhabit such environments, are phylogenetically close to the LUCA (Gaucher et al. 2003, 2008); (2) subterranean hydrothermal vents were likely the only protected environment that would have shielded prebiotic synthesis and early life from large asteroid impacts, which vaporized the ocean's surface and annihilated the ecosystems (Sleep et al. 1989); (3) hydrothermal vents harbor reactive geochemical environments with far-from-equilibrium conditions, being rich in gradients of redox, pH, and temperature, and facilitate prebiotic synthesis (Sousa et al. 2013); (4) the paleoenvironment of early Earth was extremely hot, battered by meteorites that could trigger hydrothermal systems (Marchi et al. 2014); the obvious implication is that the first life evolved in our planet under hyperthermophilic conditions, presumably when the Earth was much hotter than it is at present; (5) mixing of hydrothermal fluids and water at the basin floor combines with slow reaction kinetics for oxidation/reduction reactions, providing a chemosynthetic source for metabolic energy and ATP for hyperthermophiles (Holm 1972); and (6) the earliest fossil evidence supports a hot start to the origin of life. The paleoenvironment of the oldest sedimentary rocks containing evidence of early life (~4 Ga) suggests that hydrothermal settings may have hosted the hyperthermophilic microbes, the oldest life-forms (Dodd et al. 2017; Furnes et al. 2004; Hofmann 2011; Nisbet and Sleep 2001).

Once hyperthermophilic life emerged, it adapted to increasingly cooler environments, as represented in the fossil records, leading to thermophilic anoxygenic photosynthetic bacteria and finally to mesophilic cyanobacteria near the surface of the water in the 10°C–40°C temperature range, where the abundant free energy of sunlight could be directly harvested.

5.3.5 EVOLUTION OF THE EOARCHEAN CONTINENTAL CRUST

Today, the high-standing continental crust covers nearly a third of the Earth's surface. It is buoyant—being less dense than the crust under the surrounding oceans; this explains why most of the continents are not submerged. The origin of the Eoarchean continental crust is one of the key objectives to understanding the provenance of life's origin. There is little consensus with regard to processes by which Archean continental cratons formed, but the impact model can provide insight and a viable mechanism into how and when it was formed.

Proterozoic continental crust grew mainly along subduction zones, when one tectonic plate subducts beneath another into the Earth's mantle and causes magma to rise to the surface. Recent study suggests that plate tectonics began during Mesoarchean through the Neoarchean time, as indicated by a drastic change of crustal composition. The Archean crust evolved from a highly mafic bulk composition before ~3 Ga to a felsic bulk composition by 2.5 Ga. This upper-crust transition from mafic to felsic marks the onset of plate tectonics at ~3 Ga (Stern 2005; Tang et al. 2016). With the onset of Proterozoic Eon (~2.5 Ga), the crust had already assumed much of its present makeup, and modern plate tectonic cycling began. Sizable later additions to the continental crust occurred throughout the Proterozoic.

However, there was no plate tectonic regime during the early Archean time; Earth was a one-plate planet. In the Eoarchean Eon, Earth's first stable continents did not form by subduction or plate tectonics. At that time, the oceanic crust was hotter, thicker, and more buoyant and was unable to be subducted. On the other hand, the profound change in crustal composition of the continental crust 4 billion years ago appears to be linked to a novel kind of tectonic regime. In the absence of plate tectonics, we need an alternative geodynamic model to explain the origin of the Eoarchean continental crust.

The origin of the Earth's fundamental crustal dichotomy of low-density continental and high-density oceanic crust before the onset of plate tectonics remains obscure. However, cratering was still a very significant mechanism of crustal modification during the LHB (Marchi et al. 2014). Impact cratering was the major geodynamic force during the Eoarchean and Paleoarchean times that must have created small clusters of granitic protocontinents by mantle upwelling (Glikson 2010; Grieve et al. 2006). These protocontinents were like small

islands surrounded by vast oceans. It has become widely recognized that impact cratering is a ubiquitous geological process that affects all planetary bodies with solid surfaces (Kring 2000; Melosh 1989). The surface of the Moon attests to the importance of large-scale impacts in its crustal evolution. Studies of these craters indicate the largest and most violent among them were produced during the Eoarchean. Archean cratons consist of crustal granite-greenstone terrains. It is evident that if an LHB occurred on the Moon, the Earth must have been subjected to an impact flux at least as intense as that recorded on the Moon. So far, no unequivocal record of LHB on the early Earth has been found because of recycling and resurfacing of Archean crust by erosion and plate tectonics, but inferences of contemporary impact cratering events from the records of the Moon and Mercury are strong and convincing. Impactors that bombarded the Earth, Moon, and Mercury during the LHB were dominated by asteroids. The oldest known impact structure from the Archean is the Maniitsoq crater (~3 Ga) of the greenstone belt in West Greenland (Garde et al. 2012). Most impact structures are known from the Paleoproterozoic, such as the Suavjärvi crater in Russia (~2.4 Ga), the Vredefort dome (~2 Ga) of South Africa, and the Sudbury basin (~1.8 Ga) of Canada (Cockell and Lee 2002).

Bolide impact events were both larger and more frequent during the LHB than they are today. The terrestrial effect of the LHB lasted through the Archean Eon. The bolide impact hypothesis for the origin of Archean cratons requires a large bolide to pierce the thin Archean lithosphere and cause massive melting in the sublithospheric mantle (Hansen 2015). Grieve et al. (2006) calculated that early large impact events could lead to the formation of a felsic crust on the early Earth to form protocontinents. They found that on Earth, impact melt volumes exceed transient crater diameters greater than 500 km. The bolide, which penetrates the crust, also triggers a mantle response that, in turn, affects the surface, creating a plume. The impact would create massive partial melting of the sublithospheric mantle, at which point the region of excavated and thinned felsic crust shortens and thickens to form protocontinents (Hansen 2007, 2015). Thus, the initial basaltic Hadean crust would have reprocessed to produce pockets of felsic crust by crustal fractionation in the early Archean time through the impacting process.

Eoarchean rocks are known from the oldest granite gneisses, called the greenstone belts, including the Nuvvuagittuq craton of Canada, the Isua craton of Greenland, the Kaapvaal craton in South Africa, and the Pilbara craton in Australia. They are composed of volcanic and hydrothermal sedimentary rocks intruded by granitoids. Archean cratons represent the remnants and relics of the highly eroded giant Archean hydrothermal crater lakes, as the cross-section morphology of the greenstone belts suggests (Chatterjee 2015, 2016). In these basins, the oldest fossils of hyperthermophilic life have been detected in the hydrothermal chert beds in the younger sequences. These four major Eoarchean greenstone belts were the nuclei of the emerging protocontinents and are valuable archives for understanding the paleoecology of the earliest life. Of these, the Nuvvuagittuq craton of Canada and the Isua craton of Greenland were located in the northern hemisphere and were proximate geographically, if not sutured together. Similarly, the Kaapvaal craton of South Africa and the Pilbara craton in Australia were positioned in the southern hemisphere and would be clustered into one supercontinent Vaalbara (Zegers et al. 1998). General similarities and correlation of the greenstone rocks between the Kaapvaal craton and the Pilbara craton by using various constraints such as lithostratigraphy, geochronology, structures, paleomagnetism, impact ejecta layers, and microfossils suggest that they were once part of a larger Vaalbara supercontinent as far back as 3.6 Ga. In Vaalbara, the Pilbara craton is juxtaposed next to the southwestern margin of the present-day Kaapvaal craton.

The greenstone gneisses were a product of the deformation of the original granitic rock, combined with layered intrusions, and provide a rare and useful reference model for early Eoarchean history. They represent the relics of the felsic protocontinents. Sediments in early Archean greenstones are largely cherts and volcanic sediments erupted in shallow water in hydrothermal vent settings (De Wit and Ashwal 1997). The newly formed crust of the greenstone belts of young Earth was presumably saturated with thousands of impact craters resembling the surfaces of the Moon and Mercury. The pockmarked surfaces of the Eoarchean crust were completely erased and resurfaced by weathering and plate tectonics over the eons.

The protocontinents of the greenstone belts in the Nuvvuagittuq craton of Canada, the Isua craton of Greenland, the Kaapvaal craton of South Africa, and the Pilbara craton in western Australia were the target of frequent impacts during the LHB, creating thousands of terrestrial impact crater lakes, the ideal crucibles for life synthesis. Meteorite impacts gave ancient Earth thousands of hydrothermal crater lakes that may have become the incubators for prebiotic chemistry and the emergence of life. Thus, the impacts during the LHB created not only the continental crust but also the innumerable hydrothermal crater lakes, where first biosynthesis began.

5.3.6 CRUCIBLES FOR LIFE'S BEGINNINGS

Today, Earth is teeming with life, but when it formed from the spinning disk, it was a hot, violent, and sterile planet for the first half-billion years. We know from the fossil record that life began nearly 4 billion years ago, but there is still intense debate about where on the Earth life began. During the last three decades, the discovery of hyperthermophiles living at high temperature and under extreme pressure in submarine hydrothermal vents and thriving on a chemical soup rich in hydrogen, carbon dioxide, and sulfur has inspired the imagination and hinted at novel scenarios for life's origin. The theory of deep-sea hydrothermal vents as a possible site for life's beginning was appealing, because simple metabolic reactions emerged near ancient seafloor hot springs, enabling the leap from a non-living to a living world (Sousa et al. 2013; Wachterhäuser 1993). To test this hypothesis, one must first reconstruct the paleoenvironment on the young planet back when life emerged.

During this catastrophic time of the LHB, the giant meteorites that pelted the planet vaporized the ocean surface when they contacted water and created innumerable crater basins when they struck land (Sleep et al. 1989; Zahnle et al. 2007). Since water is so vital for life, we can say that life emerged in a watery environment in our young planet. But the ocean was not conducive to the conditions for biosynthesis. There was only one place for the refuge and nourishment of the building blocks and early life—subsurface hydrothermal vent systems. Such vents are attractive sites for the origin of life, owing to the protection they would have afforded against the bombardment of comets and asteroids (Sleep et al. 1989).

The hot-start hypothesis is supported by several lines of evidence, as discussed earlier. Among various hydrothermal settings for the incubators of life, three possible locations have been proposed: (1) submarine hydrothermal vents, (2) hot springs and tidal pools, and (3) subsurface hydrothermal crater lakes. Hydrothermal systems can develop anywhere on the crust where water coexists with a magmatic heat source. All these hydrothermal sites are geochemically reactive habitats, where microbial life thrives today around the super-heated water that supports diverse chemosynthetic ecosystems (Wiegel and Adams 1998). Variations in the temperature and density of fluids drive the convective circulation in the crust, producing large-scale transfers of energy and material. Hydrothermal systems revealed vast and previously unknown domains of microbial habitats on the Earth. The geothermal vents harbor rich ecosystems, the energy source of which stems from chemicals and volcanism, not from sunlight. In these extreme environments, the microbial and geothermal interactions are tightly interconnected, providing many of the basic constituents for the primordial synthesis of organic molecules and for the evolution of fundamental metabolic processes of some modern prokaryotic chemoautotrophs. The biochemistry of these hyperthermophiles might, in turn, provide clues about the kinds of reaction that initiated the chemistry of life. Hydrothermal systems have prevailed throughout geologic history, and ancient hydrothermal deposits could provide clues to understanding Earth's earliest biosphere. Modern hydrothermal systems support a wide range of microbial habitats and provide good comparisons for paleontological interpretation of ancient hydrothermal systems.

There is a great deal of controversies about the likely site for life's beginnings among three kinds of hydrothermal settings: submarine hydrothermal vents, hot springs and tidal pools, and hydrothermal crater lakes. Here, we evaluate these three hydrothermal sites.

5.3.6.1 SUBMARINE HYDROTHERMAL VENTS

Discovery of hydrothermal vents, and their associated lifeforms, on the deep ocean floor has opened up the possibility that live emerged at the ocean depths. The mid-ocean ridges are pockmarked with hot springs—the hydrothermal vents. Hot chemical-rich water is welling up below from the sea floor and pumping out through cracks of the ridges. These hydrothermal vents are densely populated by strange animals.

The chemicals found in these vents and the energy they could provide could have fueled many of the chemical reactions necessary for the emergence of life. This novel habitat is too dark at the ocean floor for photosynthesis to occur, so organisms survive by chemosynthesis, whereby energy is derived from chemical reactions. The cosmic building blocks, along with spewing gases from the hydrothermal vents, entered into a complex series of far-from-equilibrium chemical reactions in an extremely hot, dark, and highly reducing environment. Submarine hydrothermal vents are generally considered the likely habitat for the origin and early evolution of life (Baross and Hoffman 1985; Martin et al. 2008; Sousa et al. 2013; Wachterhäuser 1993). Like hot springs and geyser on land, hydrothermal vents form in volcanically active areas—often on mid-ocean ridges. To maintain life, the hydrothermal system must be open, accepting energetic input constantly or periodically in a non-equilibrium state.

Two types of submarine hydrothermal vents—acidic black smokers and the alkaline Lost City—have been considered as possible cradles of early life. Black smokers are deep, hot, energy-rich extreme environments of submarine hydrothermal vents at the ocean floor (Baross and Hoffman 1985; Nisbet and Sleep 2001; von Damm 1990; Wachterhäuser 1993). At the mid-ocean ridges where tectonic plates are spreading apart and magma wells up to the surface, intense geothermal heating generates hot springs on the sea floor. Seawater seeps through the cracks, dissolving metals and minerals, as it becomes superheated from the upwelling magma and erupts as a geyser from a hydrothermal vent. During the cycling of seawater along the axis of the vent, geothermal energy is transferred into chemical energy in the form of reduced inorganic compounds. The spewing water of these black smokers, which can reach temperatures of 400°C, contains chemicals and minerals that precipitate on contact with cold seawater, forming a black chimney around the vent. High temperature and acidic (pH 2–3) fluids characterize these black smokers. The vent spewing reduced gases (H_2S, H_2, CH_4, CO_2, and NH_3) dissolved in hydrothermal fluids facilitate chemosynthetic reaction for microbes, a highly efficient microbial utilization of geothermal energy. Here, the hyperthermophiles flourish and form the base of the food chain. These heat-loving microbes support diverse organisms, including fish, clams, giant worm tubes, limpets, and shrimps.

Although black smokers provide perfect conditions for primitive hyperthermophilic life, the most basic problem of high-temperature and low-pH living is molecular stability during the emergence of life. Under both regimes, the bonds that hold together vital cell components, such as nucleic acids, proteins, and membranes, face the danger of being torn apart as quickly as they are produced during prebiotic stage, to prevent further biogenesis. However, once these hyperthermophiles appeared, they developed a high degree of thermostability because of their special high-temperature enzymes and tough membranes (Alberts et al. 2000). This thermal paradox suggests that the crucible of life was less heat-intense than the black smokers. Black smokers are not only very hot and acidic, but they are also short-lived, lasting only a few

decades (Sousa et al. 2013). The origin of life is most fruitfully approached as a long process during which the capacity to evolve facilitates itself in a step-by-step fashion rather than as a series of low-probability events.

An alternative view is that life might have originated in alkaline, off-ridge submarine hydrothermal vents. These giant white carbonate chimneys called the Lost City develop from exothermic reaction of seawater with minerals such as olivine as a result of serpentinization and release white carbonate minerals (Kelley et al. 2005; Martin and Russell 2007; Martin et al. 2008; Sousa et al. 2013). Unlike black smokers that reside directly on the seafloor spreading zone, Lost City is located several kilometers away from the mid-ocean ridge; it is highly alkaline (pH 9–11) and has a relatively cooler temperature (40°C–90°C) than that of the black smokers. The vent fluids of the Lost City system are very different from those of the black smokers. Lost City vents release methane, ammonia, hydrogen, as well as calcium and traces of acetate, molybdenum, and tungsten into the surrounding water; they do not produce significant amounts of carbon dioxide, hydrogen sulfides, or metals, which are major outputs of black smokers. The Lost City system may have been active for 120,000 years (Sousa et al. 2013), longer than the black smokers. The denizens of the Lost City are hyperthermophilic bacteria, methane-cycling archaea, and other micro- and macro-organisms.

The Lost City remains an attractive hypothesis to explain the energy flow across the rock walls before the inception of cell membranes. These vents have resulted in two chemical imbalances with the invention of cellular pumps. First, when the hydrogen-rich alkaline water from the vents met the acidic ocean water, a natural proton gradient or chemiosmotic coupling could have been generated within the pores of rocks, which could have been used as energy source. Second, an electron transfer could have occurred when the hydrogen and methane-rich vent fluid met the carbon-dioxide-rich ocean water, generating an electrical gradient. Combined with the natural proton gradient from the vent, the porous alkaline rocks were the ideal place for metabolism to begin. Once life had harnessed chemical energy of the alkaline vent water, it started making complex molecules such as RNA. Eventually, it created its own membrane and became a true cell and escaped from the porous rock into the open water (Lane and Martin 2012; Martin and Russell 2007).

However, the Lost City hypothesis faces several challenges. First, the Lost City fluid does not contain an abundant supply of catalytic metals (iron, copper, manganese, zinc, nickel, etc.), which are needed for polymerization of biomolecules. Second, the oldest ecosystems on Earth, as preserved in Eoarchean greenstone belts in Canada, Greenland, Australia, and South Africa, do not show any evidence of the Lost City environments. Third, the first cells probably developed in zinc-rich environments; the cytoplasm is rich in potassium, zinc, manganese, and phosphate ions, which are not widespread in the Lost City environments (Mulkidjanian et al. 2012). Fourth, nucleic acids such as RNA are unstable in highly alkaline pH, raising the possibility that RNA must have arisen in neutral or mildly acidic environment. This observation clearly contradicts the hypothesis that RNA evolved in the vicinity of alkaline (pH 9–11) hydrothermal vents (Bernhardt and Tate 2012).

Supports of the submarine vent hypothesis are waning in recent years (Deamer 2011; Hazen 2005). Both submarine hydrothermal vent theories—the black smokers and the Lost City—suffer from the "concentration problem" of organic compounds. The cosmic ingredients would be dispersed and diluted rather than concentrated in the vastness of the Eoarchean global ocean, before they can assemble into the complex molecules of life. A sufficient concentration of reactants is difficult to imagine in the open oceans. A protective barrier of the cradle of life is needed for prebiotic synthesis. One crucial precondition for the origin of life is that comparatively simple biomolecules must have had the opportunities to form more complex molecules by segregation and concentration of chemical compounds. In open oceans, cosmic and terrestrial chemicals could not have mixed, concentrated, selected, or organized into more complex molecules, thus inhibiting prebiotic synthesis. Moreover, at deep sea, wet and dry cycles of condensation reactions mediated by sunlight for polymerization of nucleic acids and proteins would not be available (Deamer 2011).

The presence of the submarine hydrothermal vents as a likely incubator is difficult to explain in one-plate Eoarchean Earth. How did the submarine hydrothermal vents such as the black smokers or the Lost City originate without plate tectonics? Today, they occur along or near the axis of the spreading ridge. But if the plate tectonics did not start before 3 Ga (Tang et al. 2016), there was no spreading ridge in the oceans; we have to seek alternative hydrothermal systems on land, not in ocean.

Mulkidjanian and co-workers, working on the chemical makeup of living cells, have discovered that the chemistry of modern cells provides important clues to the original environment in which life evolved (Mulkidjanian et al. 2012). It turns out that all cells contain a lot of phosphate, potassium, and other metals—but hardly any sodium. In contrast, seawater is rich in sodium but deficient in potassium and phosphates. The composition of the living cell does not match that of the ocean water. On the other hand, the inorganic chemistry of cell protoplasm mirrors the environment of freshwater ponds and lakes. These authors concluded that first life began on land, not in sea. Once life does evolve, then both black smokers and the Lost City provide ready habitats for hyperthermophiles; but these environments are not supportive of prebiotic synthesis.

An alternative theory for the terrestrial geothermal sites of life's origin includes hot springs and tidal pools (Deamer 2011) and hydrothermal crater lakes (Chatterjee 2016; Cockell 2006; Farmer 2000; Osiniski et al. 2013). These hydrothermal sites are conceptually similar to the central idea of Darwin's "warm little pond" that life on Earth originated on land.

5.3.6.2 Darwin's "Warm Little Pond"

Darwin, primarily concerned with the evolution of complex life-forms from simpler ones, said little about the ultimate origin of life. Such ancient origins, he believed, were lost in the mists of time. His evolutionary theory was not about the start of life but about the processes that gave rise to biodiversity—the tree of life (Darwin 1859). Darwin was reluctant to publish his views on life's origin. However, in a private letter in 1871 to his young botanist friend Joseph Hooker, Darwin speculated that life could have sprung from a "warm little pond" rich in nutrients:

> "It is often said that all the conditions for the first production of a living organism are now present, which could ever have been present. But if (and oh what a big if) we could conceive in some warm little pond, with all sorts of ammonia and phosphoric salts, light, heat, electricity, etc., present, that a protein compound was chemically formed ready to undergo still more complex changes, at the present day such matter would be instantly devoured or absorbed, which would not have been the case before living creatures were formed."

Despite Darwin's early musing of the "warm little pond" for life synthesis, the marine-origin theory for life dominated the field for many decades, in part as a reaction to Alexander Oparin's landmark work. Oparin suggested that primordial soup was synthesized as a result of the combination of reducing atmospheric gases and some form of energy (lightning or UV rays), which would concentrate in the oceans to form amino acids and other building blocks and then combine to more complex molecules and finally to the first cells (Oparin 1924). For more than nine decades, the ocean origin was the favorite explanation, which was reinforced by the discovery of the submarine hydrothermal vents in the 1970s as the likely incubators.

Recent studies suggest that life on Earth really did start on land in a "warm little pond" and not in the oceans—just as Darwin speculated more than 140 years ago. Mulkidjanian and co-workers (2012) suggested that modern cells may reflect the chemical condition in which the first cells emerged. In all living cells, the cytoplasm is rich in potassium, zinc, manganese, and phosphate ions, which are not widespread in marine environments, and it has lower amounts of sodium ions than outside. Such conditions are found only where hot hydrothermal fluid brings the ions to the surface—places such as geysers, fumaroles, crater basins, and other geothermal fields. The chemistry of modern cellular fluid mimics the original environment in which life first evolved. Since oceans and cellular fluid are chemically dissimilar, these authors claim that it is unlikely that life evolved there. Contradistinct, the chemical nature of volcanic pools, or "warm little ponds," more closely resembles the cell's cytoplasm composition. For cells to synthesize proteins, their molecular machines need a lot of potassium. Life cannot live without synthesizing proteins, so it must maintain high potassium levels. But in ancient seawater—as well as in our modern ocean—sodium outnumbers potassium 40 to 1. The high K^+/Na^+ ratio and relatively high concentrations of Zn, Mn, and the phosphorous compounds of living cells support the geothermal terrestrial origin of life. Mulkidjanian and his colleagues concluded that terrestrial springs, such as those in Yellowstone Park, are much better environments for the first origin of life. This finding challenges the widespread view that life originated in the submarine hydrothermal vents (Baross and Hoffmann 1985; Martin et al. 2008), positing instead terrestrial sites such as "comet pond" (Clark 1998) or "fluctuating hydrothermal pools," where evaporation and replenishment by variable hot springs and precipitation occur (Damer and Deamer 2015), or hydrothermal crater lakes (Chatterjee 2015, 2016; Cockell 2006; Osiniski et al. 2013).

If all the impacting meteorites were small, the environment of early Earth might have approximated the "warm little pond." However, the evidence preserved in the densely packed craters of the lunar highlands suggests otherwise. The environment of the early Earth appears to be far more extreme and inhospitable during the LHB than the benign picture conjured up by the "warm little pond." Nonetheless, the "warm little pond" has literal relevance in understanding the alternative land-based environment for the origin of life.

Using the cratering history of early Earth, hydrothermal crater lakes would be ubiquitous and numerous on the Eoarchean crust and appear to be the ideal crucibles for life's origin. They share all the necessary chemical components of the cellular fluid, first proposed by Mulkidjanian and co-workers for the terrestrial origin of life (Mulkidjanian et al. 2012). The paleoecology of the earliest fossils favors deep and large hydrothermal crater lakes as the likely crucible, rather than small surface hot springs (Dodd et al. 2017; Hofmann 2011; Knauth and Lowe 2003). Unlike the submarine hydrothermal vents, terrestrial geothermal fields such as crater lakes are conducive to condensation reactions for polymerization of nucleic acids and proteins and enable the involvement of solar light as an additional energy source (Chatterjee 2016; Cockell 2006; Deamer 2011; Osiniski et al. 2013). The mineral substrates at the crater floor, such as clays and pyrites, were excellent catalysts and helped in the polymerization of nucleic acids and proteins, respectively (Chatterjee 2016; Hazen 2005). Moreover, subsurface hydrothermal vents at the crater floor provided an escape network for emerging life from the continual bolide bombardment. Once life originated at hydrothermal crater lakes, it subsequently invaded the oceans.

5.3.6.3 Post-Impact Crater Lakes with Hydrothermal Systems

It has been suggested in recent years that post-impact crater lakes with hydrothermal systems may have provided habitats for the origin and evolution of early life on Earth (Chatterjee 2015, 2016, in press; Cockell 2006; Mulkidjanian et al. 2012; Osiniski et al. 2013). We propose that life originated in a neutral pH milieu of terrestrial, hydrothermal crater lakes, where the building blocks of life, derived from meteorite impacts, began to concentrate. In addition, continuous infall of micrometeorites containing building the blocks of life, as well as iron, manganese, and silicates, formed a blanket around the lifeless surface of newly formed crater lakes and began to interact with biomolecules at the basins. Most micrometeorites were compositionally similar to carbonaceous chondrites (Greshake et al. 1988). Inside the crater basin, the hydrothermal vent provided a continuous stream of chemical compounds and energy, which were mixed with cosmic ingredients by convection current to form sticky, brownish, primordial prebiotic soup. Hydrothermal crater vents provided the selection, concentration, and organization of specific organic molecules into successively more-information-rich biopolymers and finally into the first cells (Chatterjee in press). The transition from chemistry to biology inside the hydrothermal crater lakes occurred around 4 billion years ago.

These crater lakes possess all the advantages of deep-sea hydrothermal vents that have been previously proposed for life synthesis, such as a prolonged circulation of heated water for mixing and concentration of the prebiotic soup, various chemicals and ATP for chemosynthesis, abundant catalytic surfaces of mineral substrates with nanopores and pockets for polymerization, and nutrients for primitive life (Chatterjee 2016; Hazen 2005; Huber and Wachterhäuser 1988; Wacey et al. 2011; Wachterhäuser 1993; Wiegel and Adams 1998). In addition, in these sequestered, hydrothermal crater lakes, cosmic and terrestrial chemicals were mixed, concentrated, and linked together to grow into larger and more complex ones by convective currents; the hydrothermal chemistry was powered by solar, tidal, and chemical energies, including ATP. The convective current in the crater basins caused the water and cosmic ingredients to move constantly, creating a thick primordial soup rich in organic molecules, which ultimately led to the first organisms in a step-by-step process of molecular symbiosis (Chatterjee 2015, 2016, in press). Hydrothermal impact crater lakes were the perfect crucibles for the concentration and cooking of these cosmic and terrestrial ingredients, where life began to brew and synthesize. Because of their abundance in the Eoarchean crust, they probably provided the ideal habitats for the origin and evolution of life on early Earth and possibly on other planets such as Mars (Farmer 2000; Osiniski et al. 2013).

One advantage of the multiple and interconnected networks of crater basins for biogenesis may be the close spatial proximity of different reaction sites to create more complex molecules in a feedback loop (Figure 5.3.1). Crater basins provide additional geochemical and environmental advantages over the earlier proposed submarine vents (Chatterjee 2016):

1. The continental excavation by impacts, forming the crater lakes, exposed a variety of rock types, including sialic, mafic, and ultramafic, which provided a dynamic fluid that mixed with variable temperatures, magmas, and pH conditions (Figure 5.3.1). This allowed a dynamic fluid that favored the prebiotic synthesis in reducing and reactive habitats; this mix might be less available in a purely volcanic and tectonic venue.

2. Sequestered high crater rims, which physically compartmentalized biomolecules, were ideal for concentration and complex reactions; contradistinct, in the global ocean, most of the organic compounds from cosmic and volcanic sources would have been diluted and dispersed, inhibiting prebiotic synthesis.

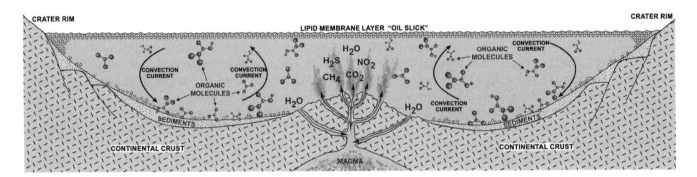

FIGURE 5.3.1 The crucible of life. During the early Archean period (~4 Ga), freshwater crater basins with hydrothermal vent system at their central peak would have offered a protective sanctuary for the origin of life. The boiling water was rich in organic molecules brought by meteorites. On the water's surface, primitive lipid membranes and hydrocarbons float like an oil slick. The minerals along the floor of the basin acted as a catalytic surface for the concentration and polymerization of monomers. The bubbling, biotic soup was thoroughly mixed by convection currents. These same currents also circulated some of the lipid membranes down to the basin floor, where they attached to the porous mineral layers, encapsulating biopolymers such as RNA and amino acids. Heat, gases, and chemical energy such as ATP released from the hydrothermal vent brewed and condensed the prebiotic soup, which began to collect at the mineral substrate, at the bottom of the basin. (After Chatterjee, S., *Phys. Chem. Chem. Phys.*, 18, 20033–20046, 2016.)

3. The chemical content of vent water, which originates in a complex set of reactions between lake water and hot, newly minted impact melt rock, nourishes a chain of living microbial communities, and very likely, it did so during biogenesis. Impact-induced fracturing at the floor of the crater basin increased the surface area available for the concentration of polynucleotides and polypeptides (Cockell 2006).

4. Terrestrial crater basins with low ionic composition are compatible with those of modern cells, thus favoring the likely site for life origin (Mulkidjanian et al. 2012).

5. The hydrothermal reactions were the main source of the metal-rich sediments and nodules that carpet the crater floor. The diverse metal sulfides, clays, and zeolites that were available in crater basins functioned as catalysts for complex sequences of reactions and polymerization of amino acids and nucleotides (Chatterjee 2016; Cockell 2006; Hazen 2005). The "Iron-Sulfur World" inside the hydrothermal vents acted as an early form of metabolism before nucleic acids appeared and promoted a variety of chemical reactions (Wachterhäuser 1993). Mineral surface bonding and pore spaces are seen as functional precursors of cellular enclosure and of bonding enzyme surfaces.

6. Wet and dry cycles of tidal pools, near the surface of the crater, favored condensation reactions of polymers (Damer and Deamer 2015; Deamer 2011).

7. High temperate ranges (~60°C–90°C) and pH values near neutrality (pH 5–8) are ideal for stabilizing the membrane vesicles of emerging hyperthermophilic microbes (Damer and Deamer 2015; Deamer 2011).

8. Additionally, solar and tidal sources of energy were available for the evaporation and concentration of polymers (Mulkidjanian et al. 2012).

9. Shocked rocks from impacts would offer shelter against harmful UV radiation when Earth still lacked a protective ozone layer (Osiniski et al. 2013).

10. Because of the huge amount of energy from impacts, impact melt rocks could keep the vent environments hot and reactive for a long time during biosynthesis (Osiniski et al. 2013).

11. Hydrothermally altered and precipitated rocks would have provided life-long sources of nutrients and habitats, even when hydrothermal activity ceased (Osiniski et al. 2013). All hydrothermal crater lakes are rich in H_2S, a source of nourishment for the emerging thermophiles.

12. Impact craters could provide protected sedimentary basins and hospitable environments for sustaining communities of primitive organisms (Osiniski et al. 2013).

These 12 geochemical and environmental characteristics favor hydrothermal crater lakes as the likely site for life's origin rather than the submarine hydrothermal vents model.

Hydrothermal crater lakes encompass a multiplicity of physical, chemical, and mineralogical gradients that may shape the structure of the microbial communities inhabiting these systems. The environmental complexity of the crater basin was a necessary requirement for the origin of biologic complexity, because a wide range of environmental conditions was required to produce a living cell from organic precursors. The anoxic crater lakes in a pre-plate tectonic Earth are unique in geologic history and are very different from the submarine black smokers or the Lost City. The cold freshwater of the crater lake is heated by hot magma from the central peak and reemerges from the vents, reaching a relatively moderate temperature (40°C–90°C). The fluctuating temperature gradient created a convection current within the lake water, mixing the assorted chemicals, both cosmic and vent-spewing reduced gases (such as H_2S, H_2, CH_4, CO_2, and NH_3). These mixed hydrothermal fluids formed a complex solution of thick, prebiotic soup. The alternate wet/dry cycles of the lake surface facilitated condensation reactions for the polymerization of monomers such as nucleic acids and nucleotides.

The ecosystem of a hydrothermal crater vent includes a gradient of temperature, nutrient abundance, a chemical environment, and pH, even within a single thermal regime.

5.3.6.3.1 Morphology and Distribution of the Impact Crater Lakes

Impact craters are formed when meteorites smash into a planet or a moon with a hard-crustal surface. All the bodies in our solar system have been heavily bombarded by meteorites throughout their early history. The landscapes of the Moon, Mars, and Mercury have beautifully preserved the LHB record because the surfaces of these relatively small planetary bodies have remained unchanged for millions of years, without any plate tectonic activity. Crater formation is mediated by an energy-transfer mechanism, where the kinetic energy of the impactor is transferred into heat, fracturing, displacing, and excavating the target rocks. Impactor velocity ranges from 10 to 70 km/s, with an average of 20 km/s on Earth. Over 150 impact craters have been identified on our planet, in a wide diversity of biomes and target rocks (Grieve 1990).

Hydrothermal impact crater lakes are good indicators for the presence of subsurface water and appear to be the most plausible location for the beginning of life (Chatterjee 2016; Cockell 2006; Osiniski et al. 2013). Like hot springs and geyser on land, hydrothermal vents formed in impact induced crater lakes. Large impacts that excavated huge craters also shattered the central peak across the diameter, creating volcanically driven geothermal vents in concert with ground water. As rains filled up the crater basins, underwater hydrothermal vents developed and crater lakes formed. The reduced gases from hydrothermal vents mixed with the cosmic biomolecules created ideal prebiotic soup for biogenesis.

Impact craters are divided into two groups based on their morphology: simple and complex craters (Grieve 1990). Simple craters are relatively small, up to 3 km in diameter, with uplifted and overturned rim rocks, surrounding a bowl-shaped depression, partially filled by breccia, with a maximum depth at the center. Complex craters and basins are large, generally 4 km or more in diameter, with a distinct central uplift in the form of a peak and/or ring, an annular trough, and a slumped outer rim, with relatively low depth/diameter ratio. Any hypervelocity impact capable of forming a complex crater lake can potentially generate a hydrothermal system (Figure 5.3.2). Small and complex crater lakes, sequestered by crater rims, are ideal sites for the concentration of reactants during prebiotic synthesis.

Terrestrial hydrothermal impact crater lakes have been proposed as the most likely environments for biogenesis (Chatterjee 2015, 2016; Cockell 2006; Osiniski et al. 2013), and, by analogy, on Mars (Farmer 2000). These crater lakes could have also provided a refuge for the earliest thermophillic life during late, giant-impact events. Craters were presumably plentiful on the Eoarchean crust; this period also overlaps with the evidence of the earliest life on Earth. Moreover, modern subaerial hydrothermal lakes are widely colonized by hyperthermophilic, thermophilic, and mesophilic bacteria and archaea.

Every impact onto water-bearing crustal surfaces generates a long-term hydrothermal activity in the resulting craters; this is due to the transfer of kinetic energy from the impactor to the target. Impact events trigger shock pressures and temperatures that can melt substantial volumes of target material. There are three main potential sources of heat for creating impact-generated hydrothermal systems (Osiniski et al. 2013): (1) impact melt rocks and impact melt-bearing breccias; (2) elevated geothermal gradients in central uplifts; and (3) energy deposited in central uplifts due to the passage of the shock wave. Any

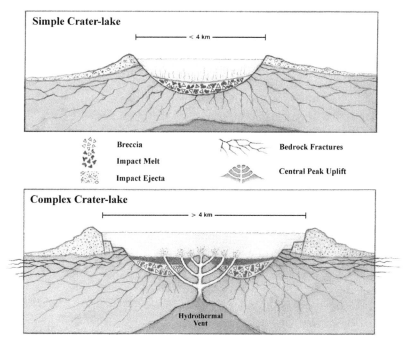

FIGURE 5.3.2 Impact craters are divided into two groups based on morphology: simple and complex. Complex craters (>2–4-km diameter) with central peaks generated a dynamic hydrothermal system, creating the ideal environment for these first crucibles of life.

hypervelocity impact capable of forming a complex crater (>4-km diameter) can potentially generate a hydrothermal system. It appears that small complex hydrothermal crater lake would enhance the concentration of biomolecules than the larger ones and would be an ideal site for life synthesis.

Impact craters are of high interest in planetary exploration because they are considered as possible sites for the evidence of life. Evidence for impact-generated hydrothermal activity is recognized at over 70 of the ~180 craters on Earth, from the ~1.8-km-diameter Lonar Crater structure in India to the ~250 km diameter Sudbury impact structure in Canada. Complex impact craters with central peaks sustain a hydrothermal system for a longer period than the simple, bowl-shaped crater (Figure 5.3.2). In large craters, thermal activity causes convection of ground and meteoric water. There is an extensive zone of fractured rocks at the floor, favorable to the circulation of the prebiotic soup. Moreover, the shock deformation of mineral surfaces stimulates reactions between minerals and active fluids (Cockell 2006; Osiniski et al. 2013). Impact-generated silicate debris in the crater basins would produce clay minerals, and hydrothermal fluid would precipitate pyrites; both minerals played a critical catalytic role in the organic synthesis of prebiotic biopolymers such as nucleic acids and proteins (Chatterjee 2016; Hazen 2005).

In terrestrial setting, the interaction of melt rock with the ground water generates a hydrothermal system (Figure 5.3.2). Eventually, the groundwater and rainwater would fill up the crater to form a lake, but the floor of the crater sustained the hydrothermal system for a long period. The crater rim rises considerably above the lake level and forms an ideal sequestered basin for the concentration of biomolecules. Impacts on a water-rich planet like Earth or even Mars can generate hydrothermal activity, that is, underwater areas boiling with heat (Farmer 2000). The central peak of a complex crater might be fractured due to impact with elevated geothermal gradient, spewing chemicals similar to the axis of the submarine hydrothermal vent, and might be an important source of heat for creating a hydrothermal system. The small complex crater (~5-km diameter) like the Gardnos crater of Norway and Gow crater of Canada would be ideal for prolonged biosynthesis. Near the surface of the crater, evaporative heating and drying organic compounds are proposed as the concentrating mechanism (Cockell 2006; Deamer 2011).

There is no constraint on the length of time required for the origin of life. The duration of impact-generated hydrothermal systems is poorly known. In general, the larger the crater, the longer the duration of hydrothermal activity. For example, in the 4-km-diameter Kärdla crater in Estonia, a hydrothermal system lasted for several thousand years, whereas in the 24-km-diameter Haughton crater in Canada, a hydrothermal system was maintained for more than 10,000 years. Sudbury crater (~250 km) probably retained hydrothermal activity at least for ~2 myr (Abramov and Kring 2004; Osiniski et al. 2013). However, during the Eoarchean time, the crust was relatively thin and the heat flow was higher through the crust than that at present, so these crater lakes possibly retained hydrothermal activity for a longer period. Moreover, the continental crust had greater heat production than it does today, given the greater abundance of heat-producing isotopes in the early part of Earth's history. The gradual cooling of the crater basin was advantageous for biogenesis, creating simple to complex organic compounds at different thermal gradients.

5.3.6.3.2 Microbial Colonization in Hydrothermal Impact Crater Lakes

The inhabitants of current hydrothermal crater lakes represent relict microbial communities that have remained distinct from other surrounding terrestrial organisms for millions of years, retaining their stamp of extreme antiquity. Hydrothermal crater lakes today harbor rich ecosystems and variable energy sources, which stem mainly from vents and impact melt rocks, and are favorable habitats for microbial colonization (Parnell et al. 2004; Reysenbach and Cady 2001). Many of these microbes are holdovers from the Archean ecosystems and have retained their ancestral characteristics. Thus, the hydrothermal crater lakes we have today are a refuge for novel derivatives of ancient forms of young Earth. The high-temperature vents of modern crater basins are perhaps the oldest ecosystem reminiscent of early Earth. Inside the crater basins are reactive gases, dissolved elements, and chemical and thermal gradients that range bottom to top from hyperthermophilic, to thermophilic, to mesophilic on the water surface (Figures 5.3.3 and 5.3.4).

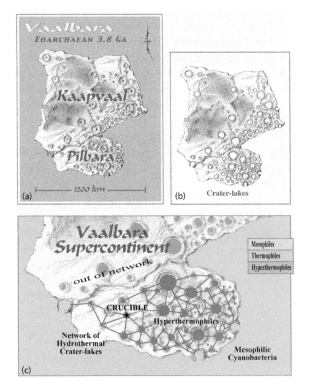

FIGURE 5.3.3 Hydrothermal crater lakes become the first sites for the origin of life: (a) the highly cratered surface of Earth's first supercontinent Vaalbara is pockmarked like the Moon; (b) from rain and aquifer, crater lakes with hydrothermal systems, rich in cosmic ingredients, are formed; (c) An underground network forms within the bedrock fractures, interconnecting the various closely linked crater lakes, exchanging heat and life-building chemical ingredients.

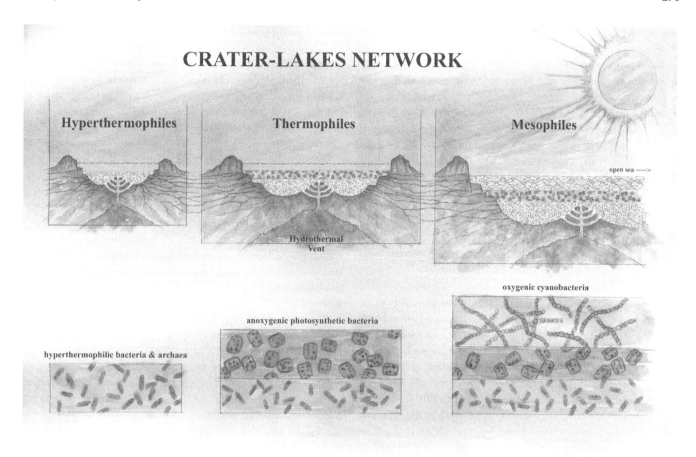

FIGURE 5.3.4 Along the top of the illustration, three cross-sections of crater lakes of different sizes show (1) the underground fissure networks that connect the closely spaced crater basins, and (2) the vertical gradients of microbial communities. Within these percolating crater lakes, three microbial zones form. The first hyperthermophilic zone, along the bottom of the crater lake, is where bacteria and archaea first emerge. Above that, the thermophilic zone forms, and in that layer, anoxygenic photosynthetic bacteria evolve. Within the top mesophilic zone, as the crater lake merges with the ocean, oxygenic cyanobacteria evolve and begin to photosynthetically harness the Sun's energy—Earth's first oxygen is the by-product of photosynthesis. Along the bottom of the illustration, those three microbial zones are enlarged.

The vent microorganisms are self-sufficient. Vent chemicals sustain hyperthermophiles, while photoautotrophs require solar power that comes easily in a planktonic lifestyle near the surface.

There are several examples of impact crater lakes that harbored microbial communities. The Siljan impact crater in Sweden, which formed during the Late Devonian (~378 Ma), is the largest known impact structure in western Europe, with a diameter of 52 km. The structure has preserved remnants of biofilm-forming fossilized microbial communities (Cockell 2006).

The impact that created the Ries crater (~24-km diameter) in Germany formed about 14 Ma; microbial trace fossils occur there as tubular features in impact glasses. The tubules have complex forms—consistent with tunneling behavior of microbes—and contain organic molecules associated with biologic activity. The Ries crater may have generated hydrothermal activity as long as 10,000 years, giving microbes enough time to colonize (Sapers et al. 2015).

The Miocene Haughton impact structure of Devon Island in the Canadian high Arctic (~20-km diameter), with its hydrothermal vent systems, contains the fossil of microbial communities (Parnell et al. 2004). The target rocks excavated at the site included massive gypsum-bearing carbonate rocks of the Ordovician age. Microbial communities such as cyanobacteria were found in the highly porous and shocked gypsum crystals. The community shows two species of cyanobacteria: *Gloeocapsa alpine* and *Nostoc commune*. With more empty spaces, microorganisms can better colonize porous minerals and extract nutrients from them more easily. Several features of the Haughton crater, specifically its freezing environment, serve as a potential Martian analog in the ongoing search for life, thus initiating the NASA's Haughton-Mars Project.

The Lonar crater of India is one of the youngest and best-preserved impact structures on Earth. The 1.8-km-diameter simple crater formed entirely within the Deccan Traps around 50,000 years ago, making it an important analog for small craters on the basaltic surfaces of the Moon and Mars. The impact event that generated the Lonar crater probably tapped groundwater supplies in the underlying basaltic aquifer, producing the fluidized ejecta blanket. As water from this lake began to interact with the hot, porous impact melt deposits, a hydrothermal system developed. Samples from the cores drilled into the Lonar crater flow show that some clay mineral formed at temperatures between 130°C and 200°C in extremely hot, hyperthermophilic environments (Hagerty and Newsom 2003). The Lonar crater is a hypersaline and hyperalkaline soda lake with a diverse microbial community. Environmental constraints have favored a distinctive mesophilic microbial community: methylotrophs, anoxygenic purple sulfur and non-sulfur photosynthetic bacteria, and oxygenic cyanobacteria. The microbial assemblage includes largely *Proteobacteria* (30%), *Actinobacteria* (24%), *Firmicutes* (11%), and *Cyanobacteria* (5%), but other microbes such as *Bacteroidetes* (1.12%), *BD1-5* (0.5%), *Nitrospirae* (0.41%), and *Verrucomicrobia* (0.28%) were detected in minor abundances in the ecosystem (Paul et al. 2016).

5.3.6.3.3 Reconstruction of the Archean Impact Crater Lakes

Although hydrothermal crater lakes seem to be the likely incubators for the origin of life, and the Eoarchean environments during the LHB support this view, no pristine craters from this age have been preserved. However, the greenstone belts containing the oldest fossil record may shed new light on the ecology and environment of these basins. The origins of Eoarchean greenstone belts in Canada, Greenland, Australia, and South Africa (>3.5 Ga) are controversial, often cited as the result of plate tectonics (Condie 1981). However, a recent study suggests that plate tectonics did not start before ~3 Ga, about 500 million years later than the origin of the greenstone basins (Tang et al. 2016). It appears that some megatectonic process other than plate tectonics contributed to the origin of greenstone belts.

Bickel et al. (1994) argued that greenstone belts are not oceanic crust, but continental. In cross-section, the central part of greenstone belts shows an anticlinal structure representing the central peak of an ancient crater, which is flanked on either side by synclinoriums, dominated by ultramafic, basic, and andesitic rocks. The volcano-sedimentary sequences in two synclinoriums containing basalt, chert, and the Algoma-type banded iron formation, all suggest a hydrothermal crater environment (Furnes et al. 2004). Bilateral symmetry of the greenstone belts resembles the geometry of a complex crater. Similarly, the environment of Eoarchean greenstone belts suggests that these basins may be the relics of ancient hydrothermal crater lakes (Chatterjee 2016; Green 1972). Studies of ancient hydrothermal impact crater lakes can provide important constraints for reconstructing the history of hyperthermophilic ecosystems on early Earth.

A felsic hydrothermal crater setting for Earth's oldest fossils, found in the Dresser Formation (~3.5 Ga) of the Pilbara craton, has been proposed (Van Kranendonk 2006). The habitat combines a high-energy input for chemosynthesis. Recently Djokic and co-workers concluded that the fossil-bearing horizons of the Dresser Formation were formed in terrestrial hot springs or crater lakes, not in the ocean (Djokic et al. 2017).

There is a growing body of evidence that the Archean Vaalbara supercontinent, including the Pilbara craton of Australia and the Kaapvaal craton of South Africa, yield the

oldest microfossils that preserve multiple impact-cratering events during 3.5–2.5 Ga, in the form of glass spherules (Byerly et al. 2002; Glikson 2010). This crucial geologic record from the ancient Vaalbara continent offers the possibility for extrapolating earlier impact-cratering events during the LHB, which must have created innumerable crater basins with hydrothermal systems hosting the ideal niches for early life.

Presumably, the Eoarchean crusts of the Nuvvuagittuq craton, the Isua craton, the Pilbara craton, and the Kaapvaal craton, the nuclei of ancient continents, were heavily pockmarked during LHB by post-impact craters of varying sizes, as those on the Moon and Mercury. Unlike the Moon and Mercury, those primordial craters were filled up with water, forming innumerable lakes with hydrothermal systems. These Eoarchean craters—the crucibles of life—were deformed, modified, and even erased by later erosion and plate tectonic activity.

Environmental conditions and habitats at the Eoarchean surface of the Nuvvuagittuq craton, the Isua craton, the Pilbara craton, and the Kaapvaal craton were conducive for the emergence and early evolution of life. These greenstone belts with hydrothermal crater lakes were the ideal location for biosynthesis (Figure 5.3.3). Interstellar particles, micrometeorites, small comets, and chondrites were suitable carriers for the safe delivery of the cosmic biomolecules to these crater lakes. Earth's young atmosphere slowed down these carriers of life's first building blocks such as fine dust, as they lightly settled upon the crater surface. These crater lakes became enriched with cosmic ingredients and were mixed by the convection current of the hydrothermal systems to form a concentrated, prebiotic soup.

Instead of a single crucible, the closely spaced crater basins on the Vaalbara supercontinent were interconnected through an extensive underground network of tunnels and cracks that interlinked closely spaced craters. Cosmic ingredients and temperature gradients could move from one crater to another through these elaborate underground networks, thus increasing the chances for the right crucibles to form life (Figure 5.3.4). These crater lakes were separated and isolated by raised rims on the surface but were interlinked through underground cracks and crevices. These networks connected craters, ranging in size from 5- to 500-km diameter, had a higher probability of forming the ideal crucible systems for the origin of life than a single crater. The bootstrapping network of crater lakes becomes increasingly suited to facilitate the prebiotic synthesis, enhancing the condition for biosynthesis. This network of impact crater lakes invokes a system something like the Mono Lake in California, where a series of lakes from higher to lower elevation have a linked flow of groundwater.

A smaller complex crater, such as the Gow crater of Canada (~5-km diameter), has the advantage of a rapid concentration of building block molecules. Chemical reactions work best in small crucibles when the biomolecules involved are closely crowded together, constantly bumping into one another, and eventually linking to form complex molecules. On the other hand, larger craters (~500-km diameter), with their larger vents, have a higher energy input for chemosynthesis and can retain hydrothermal activity for more than millions of years. Thus, a network of craters, both small and large, has the optimal possibility for processing the life synthesis. Among these hydrothermal crater networks, a single, small crater with a dynamic central peak would be the ideal site for the concentration and synthesis of biomolecules, assisted by the other craters in maintaining heat and a supply of ingredients. This small crater was interconnected with adjacent larger craters by an extensive underground network, exchanging heat and the chemicals necessary for life processing. The reduced gases from the hydrothermal vents mixed with the cosmic biomolecules, creating the thick ideal prebiotic soup for biogenesis. Once life emerged in a small hydrothermal crater, it migrated from crater to crater through the underground system or by overflowing, ultimately spilling into the ocean where mesophilic photosynthesis first developed. These oldest crater-basin ecosystems on Earth are preserved as relics and remnants in the Eoarchean greenstone belts in Canada, Greenland, Australia, and South Africa.

Differentiating Archean lacustrine deposits from their marine equivalents is challenging. The depositional environment of the widespread occurrence of stromatolites in the Dresser Formation (~3.49 Ga) of the Pilbara craton has been interpreted to be a terrestrial, hydrothermal crater lake environment, perhaps a felsic volcanic caldera, but not a shallow marine basin (Van Kranendonk 2006). The volcanic caldera may actually represent the central peak of a hydrothermal impact crater in the Vaalbara supercontinent that contains a 20-km-thick succession of dominantly volcanic rocks of the Pilbara Group associated with hydrothermal chert layers and sedimentary rocks. Similarly, Westall and co-workers claimed that thermophilic, anoxygenic, and photosynthetic bacteria likely built the desiccation-cracked mats of the Paleoarchean (~3.33 Ga) Josefsdal Chert of the Barberton Greenstone Belt, South Africa, located in a nearshore hydrothermal setting that was periodically subaerial (Westall et al. 2015). Perhaps, the Josefsdal Chert was deposited in a hydrothermal crater lake near a coastal region that was periodically flooded. If so, then life should have been able to withstand the transition from periodically exposed settings of lakes to a coastal environment (Figure 5.3.4).

A similar transition from lacustrine to a coastal environment is documented in the Frotescue Group of Australia. During the Late Archean, the Pilbara craton of western Australia was an emergent landmass upon which the Frotescue Group was deposited. The group consists of 6.5 km-thick flood basalts, volcaniclastic and siliclastic sediments, and carbonates that presumably accumulated in a continental setting like the Warrawoona Group. Awramik and co-workers suggested that the ~2.72 Ga old Meentheena Member of the Tumbiana Formation, Frotescue Group, was formed in a giant lake (~680-km diameter) in the Pilbara craton. Both the volcaniclastic and carbonate sediments and stromatolites from the Tumbiana Formation have been interpreted as a hydrothermal lacustrine environment (Awramik and Buchheim

2009; Buick 1992). Thus, the Josefsdal Chert of the Kaapvaal craton of South Africa and the Tumbiana Formation of the Pilbara craton are direct evidences that giant hydrothermal lake systems existed during the Archean time. Carbon isotope data suggest that both bacteria and archaea inhabited the Tumbiana lakes. The abundance and diversity of complex microbial communities with oxygenic cyanobacteria and stromatolites indicate that, by Archean time, cyanobacteria had adapted to the shallow-water lacustrine environment, indicating the antiquity of oxygenic photosynthesis. These environments were dynamic and underwent repeated transgression and regression.

Impact-induced hydrothermal systems are very diverse, with the widest range of pH and temperature gradients possible (Cockell 2006; Kring 2000; Nisbet and Sleep 2001; Osiniski et al. 2013). The dominant mineral assemblages include zeolites, calcite, clays, and pyrite: zeolites and calcite play crucial roles as catalysts in chiral selection of amino acids and sugars, whereas clays and pyrite are important for polymerization of monomers such as nucleotides and amino acids (Hazen 2005). Hydrothermal crater lakes hosted many microenvironments, offering a number of possible niches for prebiotic chemistry. The convection current inside the crater basin mixed hot, concentrated prebiotic soup thoroughly and caused simple chemicals to grow into larger, more complex ones by combinatorial chemistry, with a chaotic mix of energy sources and organic compounds released from vents (Deamer 2011; Martin et al. 2014). Reactive vent molecules such as H_2S, CH_4, NH_3, and H_2S, as well as metabolic energy such as ATP mediate energy flow in the chemical evolution. Infinite combinations must have taken place in the vent environments, and an infinite number must have dissolved, rejected, and vanished. If life assembled from combinations of these already-stable and selected biomolecules, rather than a random combination of raw molecules from scratch, the process would have been much more efficient. Natural monomers such as nucleotides and amino acids were selected from a large pool of cosmic ingredients from the biotic soup for their cooperation and molecular recognition and were polymerized at pores and pockets of mineral substrates to create RNAs and protein enzymes respectively. These biopolymers were encapsulated by lipid membranes to initiate endosymbiosis. The final assembly of the first cell was produced through combinatorial chemistry and natural selection (Chatterjee 2016, in press; Damer and Deamer 2015; Deamer 2011).

5.3.6.3.4 Post-Impact Crater Lakes on Mars

Building on the recent NASA exploration of Martian life, the hydrothermal crater lake of early Earth is the terrestrial equivalent of the habitable environment of the Gale Crater of Mars, which is characterized by neutral pH, low salinity, and variable redox states of both iron and sulfur compounds (Grotzinger et al. 2014). The long-term habitat stability and the wide distribution of the hydrothermal crater basin have major implications for understanding the evolution of microbial life on other planets.

Impact craters on Mars larger than 1 km exist by the hundreds of thousands, and many of them were hydrothermal crater lakes in its early history. Because subsurface fluids and crustal heat sources could have coexisted on Mars, hydrothermal deposits are important targets in NASA's planetary exploration and their ongoing search for extraterrestrial life. Mars has always been situated at the outer edge of our solar system's habitable zone, and the probability of finding life is high. Hydrothermal environments appear to have been widespread on Mars early in the planet's history and are considered high targets in the exploration for a Martian fossil record (Farmer 2000). Martian surface features have been divided into three age groups, the Noachian, Hesperian, and Amazonian on the basis of intersection relations and the numbers of superimposed impact craters (Carr and Head 2010). Four billion years ago during the Noachian Period, when life was arising on Earth, Mars enjoyed a warmer climate, with abundant liquid water on its surface; river valley networks, large crater lakes, and oceans may have been present. Noachian-aged surfaces on Mars are scarred by many large impact craters. Dried-up river beds, impact crater lake deposits, polar ice caps, and minerals that form in the presence of water suggest that Mars had habitable environments for microorganisms. The habitats for microorganisms in many impact craters on Earth could be useful in the search for life on Mars, given the extraordinary number of craters there. Low temperatures and extremely low water availability govern the surface of the planet, but impact craters could provide a sheltered habitat. The ability of impacts to create hydrothermal systems is a major reason why NASA scientists are looking for ancient craters (such as the Gale crater) to search for early Martian life. Although Mars looks desolate today, it may have harbored life in the past. The long-term habitat stability and the wide distribution of hydrothermal crater basins throughout the solar system have major implications for evolution of microbial life on other planets.

5.3.7   ARCHEAN BIOSIGNATURES AND MICROFOSSILS: THE DAWN OF LIFE

Impact-generated hydrothermal systems have prevailed throughout the geological history of Earth, and their ancient deposits provide clues to understanding Earth's earliest biosphere (Parnell et al. 2004; Reysenbach and Cady 2001). The hydrothermal systems of the Eoarchean greenstone belts are promising sites for preserving the earliest evidence of life.

We live on a planet that records its own history, encrypted in the physical and chemical features of sedimentary rocks. Part of this history is paleontological (Knoll 2003). Paleontology is the most important tool for studying the emergence of life in the oldest volcano-sedimentary rocks. Biosignatures of earliest life fall into three categories: (1) carbonaceous remains of microbial cells or chemofossils; (2) bona fide cellular fossils; and (3) microbially influenced sedimentary structures such as stromatolites. Burial is an integral part of fossilization, in part, because walls and envelopes are more likely to survive microbial decay. Moreover, thin layers of sediment would have sealed microorganisms from the overlying water levels, whereas mineral precipitation can fossilize microorganisms in three-dimensional detail.

The search for the earliest life on Earth relies on finding ancient volcano-sedimentary rocks where biosignatures are still preserved. Biosignatures such as chemofossil, microfossil, and stromatolite records in the Archean are very patchy and fragmentary and often controversial. The oldest fossils discovered have been found in the sedimentary deposits of shallow-water hydrothermal habitats in four Archean greenstone belts: the Nuvvuagittuq craton of Canada, the Isua craton of Greenland, the Pilbara craton of Australia, and the Kaapvaal craton of South Africa. The biosignatures from these cratons are described in the subsequent sections. These earliest Archean fossil records are both difficult to find and difficult to decipher because of their microscopic size. However, Early-Mid Archean biosignatures demonstrate that microbial life was abundant and possibly even more diverse than currently believed.

5.3.7.1   CHEMOFOSSILS: INDIRECT EVIDENCE FOR LIFE

Microbes may be recorded by distinctive chemical signatures. The first traces of life on Earth date back to early Archean, around 4 Ga in the form of chemofossils, but the microfossil record only extends to ~3.8 Ga. The best record of early life in sedimentary rocks from the Archean greenstone belts comes from carbon and sulfur isotopes in the form of chemical fingerprints, called chemofossils (Grassineau et al. 2006). Metabolic processes produce distinctive isotopic fractionations of carbon and sulfur isotopes. The residual $\partial^{13}C$ and $\partial^{34}S$ signatures in the sediments, found both in organic matter and in associated minerals, are particularly useful tools to understand the nature and extent of microbial activity. In living organisms, the light isotope of carbon (^{12}C) dominates the heavier isotope of carbon (^{13}C), whereas in inorganic limestones, carbon (^{13}C) is dominant. Another biosignature of early life is the sulfur isotope ratio. Intense biological sulfur cycling characterizes modern microbial mats as well as their ancient counterparts—stromatolites. A growing body of evidence suggests that life existed on Earth around 4 billion years ago in the form of chemofossils. The earliest indications of life on Earth come from isotopic measurements of biogenic carbon (^{12}C), preserved in a 4.1-billion-year-old zircon from Jack Hills, western Australia (Bell et al. 2015).

The next record of biogenic carbon (^{12}C) comes from the early Archean Akilia and Isua supracrustal rocks of Greenland, about 300 million years later than the carbon inclusions in Jack Hills (Fedo and Whitehouse 2002; Mojzsis et al. 1996; Rosing 1999). Unfortunately, these greenstone rocks of Greenland have been subjected to intense metamorphism, to the degree that any microfossils that may have existed have long since been destroyed by heat and pressure. The Greenland graphitic residues are enriched in ^{12}C to an extent that indicates chemical traces of early microbial life (Mojzsis et al. 1996; Rosing 1999). Carbon isotopes of this type, however, cannot tell us which organisms were present in Greenland.

5.3.7.2 MICROFOSSILS AND STROMATOLITES: DIRECT EVIDENCE FOR LIFE

Microfossils are the preserved remains of microbial organisms. In the young Earth, the earliest life found are single-celled prokaryotes, such as bacteria and archaea. They are microscopic, only a few tens of microns in size, and need electron microscope to study. Early microbes are preserved as microfossils, in the forms of spheroids, ellipsoids, or filaments, given the presence of highly durable envelopes or extracellular membranes that resist postmortem decay. Similarly, stromatolites provide important proxy information on the distribution of microbial mat communities; they are laminated sedimentary structures accreted as a result of microbial growth, movement, or metabolism. As such, they are trace fossils of microbial activity and thus give less evidence for life than microfossils.

The record of Archean microfossils is sparse. Of the few authentic fossil assemblages, most are from shallow-water hydrothermal settings, and they are typically associated with laminated, stromatolitic sedimentary rocks. Some of the earliest habitable environments may have been hydrothermal crater lakes. Microfossils from deep-sea hydrothermal systems have not been reported in Precambrian rocks, although hyperthermophilic microbes are ubiquitous in modern seafloor hydrothermal settings (Rasmussen 2000). The microfossils from the Archean volcano-sedimentary rocks of the greenstone belts are discussed in the next section.

5.3.7.3 MICROFOSSILS FROM THE NUVVUAGITTUQ CRATON OF CANADA

The putative earliest microfossils (~3.7-Ga-old) from the Nuvvuagittuq craton on the eastern shore of Hudson Bay of Canada have been discovered in the form of tiny filaments and tubes in hydrothermal setting; they provided a habitat for Earth's first life-forms reminiscent of hyperthermophilic bacteria (Cates and Mojzsis 2007; Dodd et al. 2017). Today, they live around hydrothermal vents and grow as filaments, feeding on iron compounds and creating tube-shaped cavities in the sediment. Similar filaments contain iron compounds in the Nuvvuagittuq rocks and are attached to round clumps that resemble the tiny anchors that bacteria use to hold on to rock surfaces. The fossils were also surrounded by minerals containing phosphorous, an element incorporated into the building blocks of life and released by decaying organisms. The rocks also contain organic carbon (carbon-12) that has been created by bacteria. Although the recognition of ancient microfossils through morphology turns out to be difficult, there are chemical traces of life within Precambrian rocks that very often provide positive identification (Brasier et al. 2002; Schopf et al. 2007; van Zullen et al. 2002).

5.3.7.4 STROMATOLITES FROM THE ISUA CRATON OF GREENLAND

The Isua rocks are known to contain a unique biogenic carbon signature (Mojzsis et al. 1996; Rosing 1999), but it was unclear whether the signature had been created by ancient life-forms or changes caused by heat and pressure. Recently, Nutman and co-workers reported stromatolites of possible bacterial origin from 3.7-billion-year-old Isua Greenstone Belt in Greenland in a shallow marine carbonate setting (Nutman et al. 2016). They were identified by rare earth and yttrium trace elements, rather than a hydrothermal, lacustrine, or estuarine environments. These sedimentary rocks, created by ancient storms, contain evidence of ripple marks. Sequestered in the middle of the sedimentary layers are structures resembling stromatolites—microbially mediated macroscopic structures. The biogenic stromatolitic structures are about 1–4 cm high and have an overall conical shape with low amplitude and internally a finely layered texture. Stromatolitic structures are compositionally distinct from interlayered bedded sedimentary structures, but there are no organic or cellular remains. As the product of cyanobacterial communities, Greenland stromatolites suggest that life had attained a certain degree of sophistication and complexity around 3.7 billion years ago. The chemistry of life favors a terrestrial origin, rather than oceanic. By that time, cyanobacterial life may have invaded shallow seas from the hydrothermal crater basins and used solar energy for photosynthesis. It follows from this find that some considerable amount of time had probably passed with the emergence of cyanobacteria since the origin of life itself. The fossil evidence of cyanobacterial stromatolites from the Isua Greenstone Belt however push the origin of hyperthermophilic life back further, near the start of the Earth's sedimentary record; this corroborates with the genetic molecular clock that placed life's origin in the Hadean Eon (>4 Ga) (Hedges 2002). This find, if confirmed, would make these stromatolitic fossils the oldest direct evidence of microbial life on Earth.

5.3.7.5 THE VAALBARA SUPERCONTINENT OF AUSTRALIA AND SOUTH AFRICA

The Warrawoona Group of the Pilbara craton in western Australia and the Onverwacht Group of the Kaapvaal craton in South Africa host some of the oldest and best-preserved early Archean microfossils in the world from the hydrothermal volcano-sedimentary rocks. These microfossils provide crucial evidence for the habitat and nature of early life on Earth in the ancient Vaalbara continent. The stunning correlation of the Onverwacht Group with the Warrawoona Group suggests that South Africa and western Australia were spatially contiguous in the Earth's oldest supercontinent Vaalbara that formed beginning 3.6 Ga and was broken apart by ~2.8 Ga (De Koch et al. 2009; Zegers et al. 1998). The rock types of Onverwacht and Warrawoona groups consist of several kilometer-thick piles of impact-generated basaltic to komatiitic lava flows interbedded with sediments that were completely silicified to chert by hydrothermal activity. Fossil microorganisms from these chert horizons suggest that single-celled bacterial and archaeal lives in the form of microscopic filaments and spheres were already thriving ~3.5 Ga, leaving a much narrower window than previously thought for life to develop in the Archean world. Those early cells were complex enough to imply that long chains of evolutionary steps preceded the fossil record. The finding of microfossils from the greenstone belts of western Australia and South Africa shrink that window to a mere 400-million-year time span. In the Vaalbara supercontinent, there is evidence for four large bolide impacts that created large craters, deformed the target rocks, and altered the environment. These impacts left their imprint for the next 300 million years between 3.5 Ga and 3.2 Ga in both the Pilbara and Kaapvaal cratons, as documented by four horizons of spherule/ejecta layers when life had diversified (Byerly et al. 2002).

It appears that early life was tougher and more resilient than is usually believed and survived catastrophic bombardments in benthic vent environments. Given the variety of fossil biosignatures likely to survive in hydrothermal deposits, the possibility that ancient microbial life survived in hydrothermal niches has important implications for tracing the earliest forms of life (Furnes et al. 2004; Raysenbach and Cady 2001; Westall et al. 2001). Hydrothermal cherts associated with well-preserved pillow lavas (~3.5 Ga) are a major component of Archean greenstone belts that show promising sites for tracing ancient microbial activity. The presence of microbial remains in hydrothermally silicified crater floor sedimentary rocks would not be surprising, as hydrothermal crater systems in the early Archean have been suggested as favorable habitats for early life (Chatterjee 2015, 2016, in press; Cockell 2006; Osiniski et al. 2013). In Vaalbara supercontinent, both the Pilbara and the Kaapvaal cratons are two pristine Archean greenstone belts (~3.6–2.7 Ga) of continental lithosphere that have yielded the earliest known life on Earth.

5.3.7.6 Microfossils and Stromatolites from the Pilbara Craton of Australia

The Pilbara craton of western Australia is one of the rare geological regions that provide insight into the early evolution of life on Earth (Wacey 2012). The 30-km-thick Pilbara Supergroup contains one of the best and most prolific sections of Archean volcano-sedimentary rocks and is a valuable site for searching early cellular life. The oldest sequence is the Warrawoona Group (~3.5–3.4 Ga), which is interspersed with thin chert horizons and felsic volcanic rocks and is overlain by the Kelly Group. There are three potential horizons within the Warrawoona volcaniclastic rocks that may contain evidence for Earth's earliest life. These formations, from the oldest to the youngest, are the Dresser Formation (~3.49 Ga), the Apex Chert (~3.46 Ga), and the Strelley Pool Formation (~3.43 Ga). Carbonaceous remains and possible remnants of cell walls from the hydrothermal black cherts of the Warrawoona Group suggest that hyperthermophilic bacteria and archaea

were the major contributors to the ancient microbial activity. Moreover, the diverse Warrawoona fossils indicate that the two domains of life, bacteria and archaea, were already split from the LUCA during this time, supporting the view that life originated on Earth at least 600 million years earlier, as indicated by the chemofossils from Jack Hills of western Australia (Bell et al. 2015).

The Dresser Formation (~3.49 Ga) at the North Pole area of the Pilbara craton is a complex volcano-sedimentary unit, long known for the earliest record of stromatolites, but their biogenicity has been debated over three decades. Chert-barite beds of the Dresser Formation are a very promising source of microbial remains due to the occurrence of biogenic carbons within low-stress thermal stress environments. Hyperthermophilic microfossils that lived in methanogenic vent environments have been found in bedded chert and hydrothermal black silica veins in the Dresser Formation (Ueno et al. 2006). Two types of filaments are found in the bedded chert, unbranched and tubular, perhaps representing methanogens or methane-producing hyperthermophilic archaea. Recently, Djokic and co-workers discovered a diverse variety of microbial activity and stromatolites from the Dresser Formation in terrestrial hot spring deposits (Djokic et al. 2017). They posited that the deposits were formed on land, not in the ocean, by identifying the presence of geyserite—a mineral deposit formed from near-boiling-temperature, silica-rich fluids that is only found in terrestrial hot springs and hydrothermal craters. This shows that a diverse variety of life existed in freshwater, as well as on land, very early in Earth's history. While stromatolites are commonly found in warm, shallow marine environments, such as Shark Bay in western Australia, they are also found in many freshwater hot springs around the world.

Noffke and co-workers (2013) reinterpreted the stromatolites from the Dresser Formation. They identified various morphotypes of stromatolites, including domal, nodular, wavy-laminated, conical, and stratiform, that thrived in coastal flat environments. They suggested that the microbial mats that produced these stromatolites could be anoxygenic photosynthetic bacteria. If this interpretation is correct, complex photosynthetic communities appeared fairly quickly through mutation, evolving from hyperthermophiles during the origin and early evolution of life. This discovery suggests that microbial life invented photosynthetic machinery from H_2S earlier than previously thought and changed its habitat from benthic to planktonic.

Perhaps the most famous and controversial evidence for Archean life comes from a chert unit within the Apex Basalt Formation (~3.46 Ga) of the Pilbara craton. The microbial assemblage of the Apex chert includes 11 taxa of filamentous microorganisms, which are allied to cyanobacteria (Schopf 1993, 1999). However, doubts about the biogenicity of the Apex microfossils have been raised (Brasier et al. 2002) and rebutted (Schopf et al. 2007). In my estimation, the Apex microbial fossils are among the oldest evidence of bacterial remains, but a cyanobacterial affinity cannot be decided by the filamentous morphology alone. Given that the filamentous microfossils are encased in hydrothermal silica, barite, and native metals, it seems likely that the Apex microbiota were ancient hyperthermophilic bacteria and archaea, not cyanobacteria (Pinti et al. 2009; Raysenbach and Cady 2001; Schopf et al. 2007).

The Strelley Pool Formation (~3.43) has emerged as one of the best sequences for the study of early life on Earth, where both stromatolites and microfossils have been discovered. The formation is sandwiched between the Warrawoona and the Kelly Groups in a shallow water environment. The stromatolites complex morphology occurred in a silicified carbonate unit which was most likely biologically mediated (Wacey 2012).

Discovery of filamentous microfossils of sulfur-based bacteria from the Sulfur Springs Group (~3.2 Ga) indicates the proliferation of thermophilic bacterial life in hydrothermal vent environments (Rasmussen 2000; Wacey et al. 2011). The fossils are very clearly preserved, showing cell-like structures all of a similar size. The cells are clustered in groups, and associated with pyrite grains. These microfossils possible represent anoxygenic photosynthetic bacteria like our modern sulfur bacteria. Apparently, these primitive anoxygenic bacteria used bacteriochlorophyll pigment to tap sunlight and produced glucose and sulfur as by-products. Pyrite crystals associated with the microfossils are very likely the by-products. The sulfur bacterium has been thought to be an important transition during the evolution of cyanobacteria.

5.3.7.7 Microfossils and Stromatolites from the Kaapvaal Craton of South Africa

Like the Pilbara craton of Australia, the Kaapvaal craton of South Africa is another rare archive for the earliest remains of life, especially the emergence of primitive hyperthermophilic bacteria and archaea from the hydrothermal cherts associated with pillow lavas (Furnes et al. 2004). The Barberton Greenstone Belt of the Kaapvaal craton contains the oldest known volcano-sedimentary sequences that have yielded evidence for Archean life. The 23-km-thick Swaziland Supergroup of the eastern Transvaal is one of the oldest, relatively unmetamorphosed volcanic and sedimentary sequences on Earth. The Swaziland Supergroup is divided into the basal volcanic-sedimentary Onverwacht Group, followed in the middle by the succeeding sedimentary Fig Tree Group, with the Moodie Group at the highest level, all containing tantalizing evidence of early life. Strata of the Onverwacht Group (~3.5 Ga) are composed mainly of volcaniclastic chert alternating with basalt layers, mimicking the sequences of the Warrawoona Group of the Pilbara craton, Australia. Both groups contain four distinct layers of impact spherules and are regarded as promising niches for the Earth's oldest fossils (Byerly et al. 2002).

Paleontological and geochemical studies have provided evidence for the existence, as early as 3.8 Ga, of a rich microbial ecosystem, possibly hyperthermophilic methanogenic archaea and bacteria (Cockell 2006; Dodd et al. 2017; Farmer 2000; Kring 2000; Osinski et al. 2011; Parnell et al. 2004; Sapers et al. 2015; Ueno et al. 2006). In this environment,

microbes colonized in the impact glasses of melt rocks, extracting energy and nutrients from the glass by dissolving it and leaving indirect biomarkers that reveal their former presence. The microbial community is directly preserved as filamentous and spherical structures of carbonaceous matter and microbial mats in cherts (Banerjee et al. 2006; Glikson et al. 2008; Walsh 1992; Westall et al. 2001). Westall and co-workers reviewed the bacterial fossils from the finely laminated Onverwacht cherts (~3.3–3.5 Ga) and recognized small spherical and rod-shaped structures of biogenic origin (Westall et al. 2001). Walsh and co-workers reported long, filamentous microfossils from the Onverwacht Group (Walsh and Lowe 1985). In the 3.5-billion-year-old lavas from the Barberton Greenstone Belt in South Africa, Banerjee and co-workers have discovered mictrotubules apparently bored by hyperthermophiles that resemble modern microbe borings (Banerjee et al. 2006). The microbes etched their way into the lava flows of the seafloor in the Archean Eon. The margins of the tube contain organic light carbon, suggesting that microbial life colonized these underwater volcanic rocks around hydrothermal vents. This biologically mediated corrosion of synthetic glass is a well-known phenomenon in the microbial world (Furnes et al. 2004). The geologic setting of the Onverwacht Group suggests that underwater volcanic rocks were the likely habitats of early hyperthermophilic microbial life in much the same way as modern black smokers or hydrothermal crater lakes.

Tice and Lowe (2004) found anoxygenic photosynthetic microbial mats, from the Buck Reef Chert (~3.4 Ga) of the upper Onverwacht Group, preserved in shallow marine environments, supersaturated with silica. Although these microfossils have been altered by metamorphism, they show filamentous structures, and the organic matter preserved in these rocks appears to be of biological, not hydrothermal, origin. The bacteria that inhabited the Earth at that time were anaerobic photosynthetic bacteria (such as purple and green bacteria) existing in a primitive ecosystem devoid of molecular oxygen.

The overlying Fig Tree Group consists of terrigenous clastic sedimentary units interstratified with volcaniclastic and volcanic rocks. It has yielded stromatolites, kerogens, and spheroidal microfossils from the chert bed (Byerly et al. 1986). Knoll and his mentor documented a well-preserved population of spheroidal carbonaceous microfossils that contains not only isolated individuals and common paired cells but also the intermediate stages of binary fission (Knoll and Barghoorn 1977). Another tantalizing discovery of early Archean life comes from the 3.2-billion-year-old Moodies Group in the form of bacterial cells associated with stromatolites in tidal-flat settings.

Noffke and co-worker described the microbial mats of anoxygenic bacteria from the siliciclastic tidal deposits of the Moodies Group (~3.2 Ga) of the of Kaapvaal craton associated with stromatolites (Noffke et al. 2006). These stromatolites reveal carpet-like, laminated fabric characteristics of microbial mats. This discovery suggests that siliciclastic tidal-flat settings became the habitat of thriving ecosystems of anoxygenic photosynthetic bacteria. The microbial mats of anoxygenic bacteria (such as green and purple bacteria) created the sedimentary structures in the tidal flats of the Moodie Group. Similarly, Javaux and co-worker reported the oldest and largest Archean, organic-walled spheroidal microfossils from the siliclastic deposits of the Moodies Group (Javaux et al. 2010). These large organic-walled microfossils suggest a cyanobacterial affinity.

In summary, life's first dawn on the young Earth began with hyperthermophilic bacteria and archaea and evolved to anoxygenic photosynthesizers and finally to oxygenic cyanobacteria; this phenomenon is fully documented in the fossil record of the Pilbara craton and the Kaapvaal craton of ancient Vaalbara supercontinent. Recreating that Vaalbara ecology, the evolution of microbial life is illustrated in the cross-section of three crater basins that flow from terrestrial to shallow marine setting and are dynamically interconnected by underground networks; this illustrates the emergence of Earth's first microbial life during the Archean Eon (Figure 5.3.4).

5.3.8   EVOLUTION OF THE ARCHEAN BIOSPHERE

A growing body of geochemical, stromatolitic, and morphological evidence suggests that the Archean microbial diversity arose around 3.2 Ga. Contemporary microbial mats provide insight into ancient microbial life and the processes of mineralization that led to the formation of certain sedimentary rocks. During the early Archean time, the land would be barren for visible life. Only in shallow freshwater crater lakes and intertidal marine basins, the accumulation of biofilms and microbial mats thrived, which were highly structured communities of prokaryotes. They were layered communities that continue to develop today as carpets, often many centimeters thick (Figure 5.3.4). The top surface layer, the canopy, primarily consists of oxygen-producing bacteria, the cyanobacteria. The cyanobacteria, together with anoxygenic phototrophs that carry out photosynthesis, sustain a remarkable diverse "understory" of other microorganisms. At the bottom of the crater lakes or deep oceans, at the hydrothermal vents, hyperthermophilic microbes proliferate in hot, deep, dark, and anaerobic environments, where sunlight is not available. The layered microbial habitats usually exhibit pH values close to neutrality (pH 6–8).

5.3.8.1   THE EARLIEST MICROBIAL MATS

Three distinct microbial regimes can be reconstructed from the Archean fossil records in stratified sequences from the bottom to the top: In the earliest stage, microbial mats of coexisting hyperthermophilic bacteria (such as Thermotogales) and archaea (such as Methanococcales) emerged concurrently in the early Archean and independently adapted a benthic, hyperthermophilic lifestyle around the vent of the hydrothermal crater basin (Figure 5.3.5). These hyperthermophiles were chemosynthetic and anaerobic, harnessing energy that was

stored in chemicals such as iron and hydrogen sulfide from the vent. An experimental study suggests evidence for the hyperthermophilicity of ancestral life (Akanuma et al. 2013). Recent phylogenetic analysis suggests that the first enzymes were fully adapted in those hot environment (Nguyen et al. 2017).

In the second stage of microbial evolution, anoxygenic photosynthetic bacteria evolved first in a thermophilic and then in a mesophilic environment. Nisbet and Fowler (1999) suggested that photosynthesis began as an adaptation of thermotaxis, a behavior in which an organism directs its locomotion up or down a gradient of temperature. There was a gradual change of temperature gradient and niche from extremely hot vent sites (hyperthermophilic) to the moderate temperature (thermophilic) near the surface of the crater basin, where the primitive, anoxygenic, photosynthetic bacteria (such as green sulfur bacteria and purple bacteria) began to appear in a hydrogen-sulfur world. These anoxygenic microbes initially tapped infrared thermal radiation from the hydrothermal vents below the photic zone. Geothermal light at otherwise dark and deep environments may have provided a selective advantage for the evolution of photosynthesis from a chemosynthetic microbial ancestor that used light-sensing molecules. Soon, the motile forms spread to the shallow water of the hydrothermal systems, where they shifted from a vent source of infrared radiation to direct solar radiation, using bacteriophyll pigment during photosynthesis. Green sulfur bacteria have a very high tolerance of sulfur; they are exclusively anaerobic and very oxygen-sensitive. In contrast, purple bacteria can tolerate oxygen and occur at a higher level, just below the cyanobacterial mat.

In the third stage of microbial evolution, advanced, oxygenic, photosynthetic bacteria (such as cyanobacteria) emerged at the water surface of the crater lake, adapting to a normal-water-temperature niche (mesophilic). As they began to harness the visible solar energy by using chlorophyll, cyanobacteria began to invade the global oceans and produced oxygen. The evolution of oxygenic photosynthesis in cyanobacteria, using chlorophyll rather than bacteriophyll, was crucial in creating the modern biosphere (Nisbet and Fowler 1999).

These vertically stratified temperature regimes and niche partitions in crater basins encouraged different kinds of microbes to adapt thermotolerance and diversify (Figure 5.3.5). As cyanobacterial life invaded the ocean surfaces from the hot- to the normal-temperature regime, the associated enzymes coevolved to keep life's chemical reactions going and shifted their optimal temperature range gradually to cooler environment (Nguyen et al. 2017). These researchers used a technique called ancestral sequence reconstruction to figure out what the enzyme's genes might have looked like at different points in the last 3 billion years. They edited *Escherichia coli*'s genes to make the bacteria produce those probable ancient enzymes and then looked at how the reincarnated molecules held up under different temperatures.

Both kinds of photosynthetic bacteria, anoxygenic and oxygenic, created distinctive stromatolite horizons in the Pilbara and Kaapvaal sequences that indicate their microbial activity

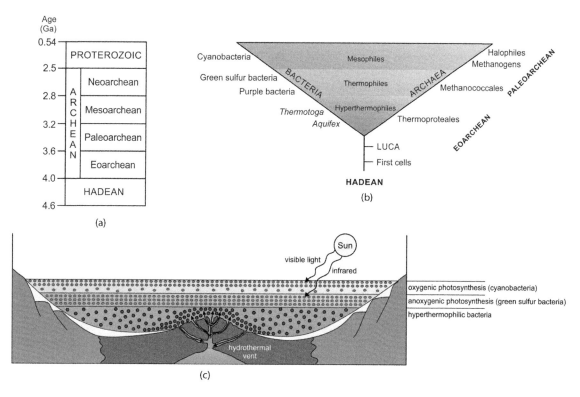

FIGURE 5.3.5 The evolution of the Archean biosphere (a) delineates the geological timescale of the Archean Eon, and (b) illustrates the origin and early evolution of life in an Archean hyperthermophilic, benthic crater lake. Two distinct domains of life, bacteria and archaea, are preserved in the fossil record (~3.5 Ga) of the ancient Vaalbara continent. Both domains show the gradual reduction of thermotolerance over time: from hyperthermophilic to thermophilic to mesophilic. In the first hyperthermophilic habitat, both bacteria and archaea appear. In the second, thermophilic habitat bacteria evolve as anoxygenic photosynthesizers. In the final mesophilic habitat, bacteria congregate at the upper surface of the crater lake; they begin to tap solar energy and evolve into oxygenic cyanobacteria. Over time, these cyanobacteria spread globally through the ocean and begin to produce oxygen. Hyperthermophilic archaea, on the other hand, evolve through two stages: as thermophilic Methanococcales and then as mesophilic Methanogens and Halophiles, the latter thriving in the hypersaline environment of ponds and lakes. (c) A cross-section of a hydrothermal crater lake showing the evolution of photosynthesis. This occurs through three stages of an evolving microbial community: first, on the bottom, hyperthermophilic bacteria emerge; next, in the thermophilic middle stage, the anoxygenic photosynthetic green sulfur bacteria appear; and in the final stage, at the upper mesophilic level, oxygenic photosynthesizing cyanobacteria form and begin the production of oxygen.

and distinctive habitats. Anoxygenic photosynthesis predates the oxygenic photosynthesis in standard phylogenetic models (Nisbet and Sleep 2001). Microbial mats are a unique ecological niche representative of early life on Earth. This is due, in part, to the persistence of fossilized mat counterparts, stromatolites. Rapid nutrient cycling across microgradients, coupled with putative niche differentiation within mat layers, enables diverse metabolic processes to occur in spatial proximity. On a broad scale, these early ecosystems of cyanobacteria on the top layer of the water surface played a major role in the oxygenation of the ocean and atmosphere, paving the way for oxygen-dependent life such as respiratory bacteria.

The hallmark of early bacterial and archaeal adaptation to novel environments, from benthic to planktonic, with decreased thermotolerance, might be linked to physiological differentiation and ecological opportunity, whereby evolved organisms interact with their environments differently than their ancestors. In bacteria, physiological change may be effected by a single-gene acquisition, perhaps mediated by horizontal gene transfer, as well as mutation (Doolittle 2000). Genes acquired by lateral transfer play a major role

in the evolution of microbial organisms that adapted to different habitats. Acquired genes can confer novel functions, allowing effective competition in new environments by virtue of the novel biochemical processes they encode. The early hyperthermophilic bacteria and archaea were entirely benthic and chemosynthetic and depended upon the chemical nutrients and energy from the vent environments. As these bacteria invaded the planktonic lifestyle near the water surface to become photosynthetic, they tapped new source of abundant energy from the Sun and became more mobile and widespread. Ecological differences in the hydrothermal crater lakes in stratified sequences allow for genetic isolation and rapid speciation of the microbial community in the Archean world (Figure 5.3.4).

5.3.8.2 Metabolic Evolution of Archean Microbial Communities

Metabolism mediates the flow of matter and energy through the biosphere. Carbon is so important to living things that organisms are sometimes referred to as "carbon-based life."

The acquisition of cellular carbon from CO_2 is a prerequisite for life, and it marked the transition from the prebiotic to the organic world. Early microbes that synthesized carbons contributed substantially to the evolution of planetary environments through numerous important geochemical processes. Early Archean microbes were essentially autotrophs. Autotrophs are the organisms that can derive some or all of their organic carbon and nutrients from inorganic sources. Autotrophs are capable of manufacturing their own food (such as sugars and carbohydrates). They take energy from the environment in the form of inorganic chemicals or sunlight. In the archaean, we see two major kinds of autotrophs: Chemoautotrophs, such as thermophiles, obtain chemical energy through oxidation in vent environments, and photoautotrophs, such as sulfur bacteria and cyanobacteria, obtain energy from sunlight. The chlorophylls of photoautotrophs capture solar energy and convert it to chemical energy stored in sugars and other organic molecules. The process is called photosynthesis. Autotrophs were the dominant microbes in the early Archean ecosystems and may shed light on the evolution of photosynthesis. Autotrophs are the producers of the food chain; they synthesize their own energy, creating organic compounds, which are consumed by heterotrophs. The Eoarchean ecology was dominated by autotrophs.

Autotrophs govern the isotopic composition of the living world. The isotopic makeup of autotrophs is largely controlled by the mix of isotopes in the gases and simple nutrients they take in from environment. For example, compared with atmospheric CO_2, autotrophs are enriched with organic carbon ^{12}C. As discussed earlier, isotopic composition of chemofossils from the Eoarchean deposits provides reliable clues to the presence of autotrophs in the young Earth. Here, we discuss three kinds of autotrophs that prevailed during the early Archean biosphere: hyperthermophilic microbes, anoxygenic photosynthetic bacteria, and oxygenic photosynthetic bacteria. Much of our understanding of metabolic evolution of Archean microbes is based on our observations of extant life, yet we have no idea either of the degree to which extant life reflects early life or of how similar both might be.

5.3.8.2.1 Chemoautotrophs: Hyperthermophiles

The volcanic and associated sedimentary rocks containing the earliest record of life document an anaerobic Earth (Westall et al. 2015). Anaerobic conditions are mandatory for the emergence of hyperthermophiles. No prebiotic reactions can take place in the presence of oxygen because of the immediate oxidation of the reduced carbon compounds. This means that first cellular life was anaerobic and chemosynthetic; indeed, life continued to be anaerobic and hyperthermophilic until the advent of mutations that led to the innovation of photosynthetic metabolism. The first traces of life on Earth date back to early Archean, between 4.1 Ga and 3.8 Ga, in the form of chemofossils, microfossils, and stromatolites, which provide little clue to the habitat and nature of early life during the LHB. This is the time interval when the first cellular life and the LUCA originated in harsh environments (Woese 1994). Because the photic zone was vulnerable for large asteroid

impacts, life must have originated in sheltered benthic environment of hydrothermal crater lakes. The LUCA was hyperthermophilic, as molecular phylogeny suggests (Gaucher et al. 2003, 2008; Pace 1997). Small hydrothermal crater lakes may have hosted the first living systems on Earth; this is consistent with the hyperthermophilic origin of life (Chatterjee 2015, 2016, in press; Cockell 2006; Osiniski et al. 2013).

The oldest fossil evidence suggests that hyperthermophilic organisms were already present on Earth at least 3.8 billion years ago or even earlier (Dodd et al. 2017). Around 3.5 Ga, life was diversified into two distinct domains, bacteria and archaea, as revealed from the fossil record. The universal tree of life shows that both kinds of hyperthermophilic microbes evolved early, but archaea were more derived than bacteria (Woese 1987). Soon, there was a drastic change in the metabolism of microbes. Instead of thermal energy from the vent, they began to harness new source of energy from the Sun, when anoxygenic photosynthetic bacteria and oxygenic photosynthetic bacteria appeared in quick succession (Figure 5.3.5). These asexual microbes remained the only forms of life for more than 1 billion years, until the appearance of the first eukaryotes. It may have taken nearly half a billion years to evolve from the LUCA, with over 500 genes, to full-blown microbes such as Cyanobacteria, which contain more than 3000 genes (Koonin 2003). Bacteria and archaea shared genetic information by horizontal gene transfer that led to the rapid evolution of these two domains (Doolittle 2000). As the LUCA survived and evolved in the harsh environment of early Archean crater basins, it gave rise to two distinct domains, bacteria and archaea, as revealed from the microfossil record from the Pilbara and Kaapvaal cratons (~3.5 Ga), very early in the history of life.

Cumulative fossil evidences from the Nuvvuagittuq craton, the Isua craton, the Pilbara craton, and the Kaapvaal craton provide an unusual window into the origin and ecology of the Archean life. Surprisingly, three broad groups of fossil microbes, which are known from the Archean fossil record, include hyperthermophilic bacteria and archaea, thermophilic anoxygenic phototrophs, and mesophilic oxygenic phototrophs; they all survive today in hydrothermal settings, which provide critical information about the habitats of early microbial organisms (Figure 5.3.4). For examples, some hyperthermophilic communities in the Yellowstone National Park can grow into thick mats. Within those mats, hyperthermophile species may migrate up or down, depending on the water and air temperatures and other conditions, demonstrating that these communities are dynamic and ever-changing. This migratory behavior of early hyperthermophiles in hydrothermal crater vent environment may give rise to photosynthetic bacteria, as they migrate up to the water surface. Similarly, a diverse group of anoxygenic photosynthetic bacteria thrives in habitats characterized by extreme temperature, pH, or salinity. These anoxygenic phototrophs are optimally adapted to the conditions of their habitats and are ideal models for defining the physiochemical limits of photosynthesis. For example, the carbonate- and sulfide-rich Mammoth Hot Springs of the Yellowstone National Park contain rare examples of laminated microbial mats formed by purple sulfur bacteria, which are precursors to cyanobacteria. They are anoxygenic photosynthesizers, which means that they harness energy for metabolism from light. These bacteria use hydrogen sulfide in chemosynthetic reactions, producing sulfur instead of oxygen as a by-product. Finally, there are many examples today of living cyanobacteria inhabiting geothermal habitats, as in the Archean world. For example, the hydrothermal Lonar Crater Lake in India and Haughton crater lake in Canada show cyanobacterial diversity in pelagic and benthic habitats (Hagerty and Newsom 2003; Parnell et al. 2004; Paul et al. 2016). All these modern examples of living bacteria in geothermal setting could be used to reconstruct the habitats of three distinct communities of early microbes in Archean time (Figure 5.3.5).

The hyperthermophilic microbes are of the size of a typical prokaryotic cell ($\theta \sim 0.5$–$2~\mu m$) and employ a variety of different morphotypes, including rods, thin filaments, disc, and cocci. Once life evolved, groundwater networks dispersed these early microbes, some going to larger hydrothermal crater lakes (Figure 5.3.4). There is evidence that hyperthermophily arose independently in the archaea and bacteria during the Archean Eon because of their similar adaptation at the high-temperature regime (Figure 5.3.5b) (Boussau et al. 2008; Nisbet and Sleep 2001). Hyperthermophilic bacteria (such as Thermotogales) and archaea (such as Methanococcales) appeared concurrently around the vent of the hydrothermal crater basin. The hydrogen-based ecosystem was the early driving energy. Hyperthermophilic bacteria and archaea probably used hydrogen and carbon dioxide, coupled with sulfur, for metabolism, whereas methanogenic archaea produced methane as hydrogen reduced carbon dioxide. Although the recognition of ancient microfossils just by morphology turned out to be very difficult, molecular phylogeny of microbes and their habitats for early life may provide important clue about their affinity. Microbial fossils from the Archean are consistent with hyperthermophily, based on morphological similarities with living prokaryotes. Carbonaceous matter from the hydrothermal black cherts (~3.5 Ga) of the Pilbara and Kaapvaal cratons has yielded transmission electron microscopy images that are suggestive of hyperthermophilic microbial life (Glikson et al. 2008). In these hot, Archean pillow lavas, only hyperthermophilic bacteria and archaea could survive (Furnes et al. 2004). An early evolution of archaea is supported by the discovery of ^{13}C-depleted methane in hydrothermal fluid inclusions in cherts from the Pilbara craton. Ueno et al. (2006) reported evidence for the existence of an ancient population of methane-producing archaea (Methanococcales) from the Dresser Formation (~3.5 Ga) of the Pilbara craton. Archaea produces only methane, suggesting that the split between the two domains of life, bacteria and archaea, must have occurred at least 3.5 Ga.

Life began very early in Earth's history, about 4 billion years ago, and achieved remarkable level of metabolic sophistication before the end of the Archean (~2.5 Ga). It thus appears that the Archean fossils from the Nuvvuagittuq, Isua, Pilbara, and Kaapvaal cratons provide crucial evidence for

three phases of environmental temperatures and niche adaptation over 300 million years: in the first period, hyperthermophilic bacteria and archaea adopted high thermotolerance around the vent of the crater basin, nourished by chemicals from spewing vents (Arndt and Nisbet 2012). In the second period, the thermotolerance decreased with the emergence thermophilic anoxygenic photosynthetic bacteria near the water surface; they used hydrogen sulfide to produce glucose and sulfur as by-products, powered by infrared. In the third phase, cyanobacteria adapted to the normal temperature regime at the water surface to become mesophilic; they used water to build glucose and oxygen as by-products, powered by visible sunlight. It thus appears that chemosynthesis preceded phototrophy (Figure 5.3.5).

5.3.8.2.2 Photoautotrophs: Anoxygenic Photosynthetic Bacteria

The advent of photosynthesis is one of the central events in the development of Archean microbial communities. Nisbet and co-worker suggested that photosynthesis began as an adaptation of thermotaxis (Nisbet and Fowler 1999). We know very little about the origin of photosynthesis, but Archean fossils give some clue. Early hyperthermophiles were entirely chemosynthetic and benthic and lived in proximity to the vents for energy and nutrition. Phototrophy is a process by which organisms trap light energy (photons) and store it as chemical energy in the form of ATP. There is suggestive evidence that photosynthetic organisms were present in the younger sequences of the Kaapvaal and Pilbara cratons (~3.4–3.2 Ga) in the form of stromatolites and microfossils, which are inferred from morphology or geological context (Javaux et al. 2010; Noffke et al. 2006; Westall 2004).

Photoautotrophic bacteria exhibit great diversity with regard to the biochemistry of photosynthesis. There are two kinds of photosynthetic bacteria: primitive (anoxygenic) photosynthetic bacteria (such as sulfur bacteria) and advanced (oxygenic) photosynthetic bacteria (such as cyanobacteria). Phylogenetic analysis supports that anoxygenic photosynthesis developed first. Though closely related, the two forms of photosynthetic bacteria, primitive anoxygenic and advanced oxygenic organisms, differ in their metabolism (Blankenship 2010; Schopf 1999). The primitive photosynthetic bacteria (such as green sulfur bacteria and purple bacteria) absorbed near-infrared rather than visible sunlight and produced sulfur or sulfate compounds rather than oxygen. They differ from oxygenic photosynthesis in the nature of the terminal reactants (e.g., hydrogen sulfide rather than water) and in the by-product generated (e.g., elemental sulfur instead of molecular oxygen). Therefore, water is not used as an electron donor. Anoxygenic phototrophs use molecules such as H_2S, as opposed to H_2O. Their pigments (possibly bacteriochlorophylls) were predecessors to chlorophyll. H_2S is supplied in the vent environment by geothermal activity, and the metabolism of anoxygenic phototrophs produces sulfur as a by-product and builds glucose, the universal cellular fuel. They use infrared instead of visible light for metabolism. The origin and evolution of photosynthesis have long remained enigmatic, owing to a lack of sequence information of photosynthesis genes across the entire photosynthetic domain. Xiong and co-worker obtained new sequence information from the green sulfur bacterium *Chlorobium* and the green nonsulfur bacterium *Chloroflexus* and demonstrated conclusively for the first time that the major lineages of pigment (bacteriophyll) involved in anoxygenic photosynthesis arose before the development of oxygenic photosynthesis (chlorophyll) (Xiong et al. 2000)

The appearance of anoxygenic photosynthesis would have made shallow-level and subaerial hydrothermal systems much more productive. Anoxygenic photosynthetic bacteria in the Archean were still thermophilic enough to drive a primordial sulfur cycle. As hyperthermophiles originated in superheat vent environments, the association of anoxygenic bacteria with hyperthermophiles suggests that anoxygenic bacteria appeared in the surface environment of the crater lakes to trap solar energy while still adapting to the thermophilic lifestyle. They acquired global significance only after sulfate concentrations had considerably increased in the Proterozoic oceans. There are many examples of living anoxygenic phototrophs such as *Chloroflexus*, *Rosieflexus*, *Thermochromatium*, and *Chlorodium*, which are adapted to thermophilic habitat. Hydrothermal vents have the potential for hosting anoxygenic photosynthetic life (such as green sulfur bacteria), using infrared radiation emitted by hot water. Eventually, these anoxygenic photosynthetic bacterial would tap infrared radiation directly from the Sun as they became planktonic. There is little doubt that anoxygenic photosynthesis preceded the oxygenic photosynthesis, as suggested by the Archean fossil record from the Pilbara and the Kaapvaal cratons. The development of anoxygenic and then oxygenic photosynthesis allowed life to escape the hydrothermal setting. By about ~3.2 Ga, most of the principal biochemical pathways that sustain the modern biosphere had evolved.

5.3.8.2.3 Photoautotrophs: Oxygenic Photosynthetic Bacteria

By the late Archean, a more complicated form of photoautotrophy, the oxygenic photosynthesis, evolved, as manifested by cyanobacteria associated with their stromatolites. In cyanobacterial synthesis, the pigment is chlorophyll, hydrogen is always provided by water, and glucose and oxygen are given off as by-product. Solar radiation was overwhelmingly the new source of energy for cyanobacteria to access different chemistries rather than the energy provided by other sources, such as hydrothermal vents, because of the unique characteristics of photochemistry that differentiate it from conventional hyperthermal chemistry. The ecological transition of the Archean life from the benthic thermophilic life to the anoxygenic photosynthetic bacteria in the crater vent environment and finally to planktonic mesophilic cyanobacteria in the shallow marine environment was rather rapid, as documented in the Pilbara and Kaapvaal cratons (Figures 5.3.4 and 5.3.5b, c). However, we are left with the possibility that the photic (sunlit) zone could have been destroyed intermittently throughout the Archean Eon, as revealed by the spherule ejecta layers

between 3.5 and 3.2 Ga (Byerly et al. 2002). Perhaps, after the cessation of large bolide impacts after 3.2 Ga, obligate photosynthetic cyanobacteria could continuously evolve. This is the time when the paleontological record indicates highly evolved photosynthetic systems, transforming the atmosphere, and permitting the evolution of eukaryotes. The Archean fossils from South Africa and western Australia shed new light on the origin and early evolution of photosynthesis from their hyperthermophilic bacterial ancestors.

One of the most important developments in Earth's history is the change from the anaerobic environment of the early Earth to the aerobic and highly oxidizing environment that we have today, with 21% atmospheric oxygen. Although the "great oxidation event" that led to the oxygen bloom occurred on our planet in the early Proterozoic (~2.3 Ga) with the proliferation of cyanobacteria, the chlorophyll-based photosynthesis originated fairly early in the bacterial domain in the Archean (~3.2 Ga). This led to the onset of global oxygenation in the atmosphere-biosphere system in the early history of the Earth. The accumulation of oxygen in our atmosphere was an exceedingly protracted process. It is still enigmatic how the early cyanobacteria developed the ability to oxidize water and produce oxygen as a by-product (Figure 5.3.5). Photosynthesis generates a distinctive chemical signature in the form of a carbon isotopic, composed of the organic material produced. During photosynthesis, ^{12}C (carbon-12 in CO_2) is selectively absorbed by the photosynthetic machinery, compared with ^{13}C (carbon-13); that is, it reacts faster with the enzymatic system and is therefore selectively incorporated into the organic compounds produced. The carbon-12 isotope ratio has been used to trace the early evidence of life in Greenland ~3.8 Ga (Mojzsis et al. 1996).

Phylogenetic studies of 3,983 gene families across the three domains of life onto a geological timeline suggest the rapid evolutionary innovation of microbial life during a brief period in the Archean Eon around 3 Ga, which coincides with a rapid diversification of bacterial lineages and genetic expansion; the microbial gene expansion was mediated by gene duplication and horizontal gene transfer. Genes arising after this expansion show the increasing use of molecular oxygen and redox-sensitive transition metals such as iron and nickel (David and Alm 2011; Doolittle 2000).

5.3.8.3 RADIATION OF ARCHEAN MICROBIAL COMMUNITIES

Life has been built on the evolution and innovation of microbial metabolisms. Microbes have become integral components of the biogeochemical cycles that drive our planet. Life may have originally started on the crater floor of deep and dark hydrothermal vent environments, when oxygen was not yet present in the water or atmosphere, but life would ultimately spread into vast sunlight-exposed ocean surfaces. Early microbial evolution refers to that phase of biological history that began with the emergence of life-forms such as hyperthermophiles, followed by purple and green sulfur bacteria in the anaerobic environment and the explosive diversification of aerobic photosynthesizers such as cyanobacteria. Initially, hyperthermophiles and anoxygenic photosynthesizers clustering around hydrothermal vents may have been great contributors to microbial metabolisms in early Archean. Eventually, oxygenic photosynthesis would evolve and eclipse its anoxygenic cousins, becoming the driver of the global carbon cycle.

The development of oxygenic photosynthesis by cyanobacteria around 3.2 billion years ago, or even earlier, transformed Earth forever. This grand biogeochemical shift set into motion the evolution of subsequent microbial metabolisms and lifestyles. The abundance of life on Earth is almost entirely due to oxygenic photosynthesis. Once life emerged on Earth, it proliferated across the global ocean by mutation through Darwinian evolution. The ability of oxygenic photosynthetic bacteria to capture hydrogen by splitting water was an extraordinary innovation in the microbial community. The virtually unlimited supply of hydrogen in water freed life from its sole dependence on the abiotic chemical sources found in hydrothermal vents. Communities sustained by oxygenic photosynthesis could thrive wherever supplies of sunlight, water, and nutrients were sufficient. Three billion years later, photosynthesis continues to play a vital role in regulating atmospheric levels of oxygen and carbon dioxide. Photosynthetic microbial communities have left a relatively robust fossil record because of their extremely high productivity on stable continental shelves; this contributed to sediments with excellent potential for long-term preservation. Although cyanobacteria emerged during the mid-Archean or even earlier, their microfossil record is robust throughout the Proterozoic.

The oxygen, as the result of photosynthetic activity by cyanobacteria, was so pervasive and so productive that it eventually reached levels sufficient to drive the emergence of respiratory bacteria, eukaryotes, and complex multicellular life in the Proterozoic. The endosymbiotic merging of cyanobacteria with respiratory bacteria drove a complex eukaryote biosphere (Margulis 1981). The origin of eukaryote cells is considered a milestone in the evolution of life, because they comprise all multicellular organisms—plants, fungi, animals, and us. Without eukaryotes, we wouldn't be here to discuss the question of the origin of life. The fossil record doesn't tell much about the time of their origin. The oldest fossil record of eukaryotes goes back 2.7 billion ago, with the discovery of eukaryotic biomarkers in ancient oil. The oldest eukaryotic body fossil is *Grypania spiralis* from the 2.1-billion-year-old Negaunee Iron Formation of Michigan. Molecular evidence suggests that three domains of life—bacteria, archaea, and eukaryotes—appeared in the Archean Eon (Arndt and Nisbet 2012). For eukaryotes to thrive, oxygen had to be present in the atmosphere in higher amounts than what existed on early Earth. A great diversity of microbial and aerobic habitats must have existed in the late Archean time. Eukaryotes are more closely related to archaea than to bacteria, at least in terms of nuclear DNA and genetic machinery (Woese 1987; Woese et al. 1990).

5.3.9 SUMMARY

Earth is approximately 4.6 billion years old. Massive bolide bombardments of the planet took place during the first 500 million years, followed by the LHB (~4.1–3.8 Ga). The young Earth, upon which first life was eventually to appear, was very different than today. It was dominated by violent meteoritic impacts associated with strong volcanic and hydrothermal activities. Comets and carbonaceous asteroids delivered large amount of building blocks of life to the primitive Earth, thus playing a vital role in the origin of life. Extraterrestrial bolide impacts were more numerous and larger during the LHB, which created thousands of impact crater lakes with hydrothermal systems across the protocontinents. But instead of sterilizing the planet, the impacts encouraged microbial life to emerge. In these sequestered crater lakes, the cosmic building blocks of life concentrated and were churned by convective currents, producing more complex organic compounds. The chemicals and energy found in these hydrothermal crater lakes fueled many of the chemical reactions necessary for prebiotic synthesis and the resulting emergence of life. The chemical nature of the hydrothermal crater lakes, with their neutral pH and high K^+/Na^+ ratio, resembles the living cell's cytoplasm more closely than that of the submarine hydrothermal vents. There are striking parallels between the chemistry of the living cell's cytoplasm, present in terrestrial hydrothermal systems, and the core energy metabolic reactions of some modern microbial communities. The hypotheses that life originated in a very hot environment is supported by present-day hyperthermophiles, which are the direct descendants of the LUCA. Here, I argue that Eoarchean hydrothermal crater lakes provided the stable, anaerobic, and hot environments in which hyperthermophilic life could have emerged and become established. These terrestrial hydrothermal crater lakes were geochemically reactive in the far-from-equilibrium setting and contained an assortment of cosmic and terrestrial organic compounds. The hydrothermal vent was powered by volcanic, chemical, and solar energies and created convection cells that drove the prebiotic synthesis into more complex organic molecules, initiating molecular symbiosis and providing physically and chemically diverse and secured habitats for microbial life. The widespread occurrence of hydrothermal crater lakes on the Eoarchean crust has led to the suggestion that such environments may have constituted the crucible of life on Earth. It was on this chaotic and violent Earth, before the onset of plate tectonics, around 4 billion years ago, that first life probably emerged in hydrothermal crater lakes. The atmosphere was anoxic, consisting mostly of CO_2.

Other settings have been proposed for the prebiotic synthesis for life, most notably the submarine hydrothermal vents, but this popular theory has several drawbacks: (1) large global oceans extended from pole to pole during the Eoarchean; in this vast ocean, cosmic and terrestrial biomolecules could not be concentrated into a compact environment because of the vastness of the oceans; organic compounds would be diluted and dissipated, prohibiting the assembly of the complex molecules of life; (2) submarine hydrothermal vents could not be formed at the spreading ridge axis, because plate tectonics did not start at that time, and there was no spreading ridge; (3) the chemistry of modern cellular fluid is drastically different from that of ocean water, thus affirming that life could not have started in the ocean.

The paleoecology Earth's earliest fossil records favor impact-generated hydrothermal systems as the most likely incubators for biosynthesis. Impacts during the LHB played a seminal role in shaping the Eoarchean and Paleoarchean volcano-sedimentary sequences (evident in the greenstone belts in Canada, Greenland, South Africa, and western Australia) and in the emergence of early microbial life. Fossil records suggest that both hyperthermophilic bacteria and archaea appeared simultaneously, followed by the evolution of anoxygenic photosynthetic bacteria and then oxygenic photosynthetic bacteria. The arrival of oxygenic photosynthesis allowed life to escape the hydrothermal setting and invade an utterly new frontier—broad continental shelves. Cyanobacteria contributed to the geological processes by producing vast amounts of carbonate sediments and stromatolitic structures in the shallow seas; they formed oxygen as the by-product, transforming the ocean and the atmosphere around 3.2 billion years ago and triggering an explosive evolution. This development led to the origin of eukaryotes.

ACKNOWLEDGMENTS

I thank Vera M. Kolb for inviting me to contribute this article in *Handbook of Astrobiology* volume and for critically editing the manuscript. I owe an immense debt of gratitude to the many authors, who have helped me with their thoughtful, well-documented, and enlightening expositions in the origin of life research. I thank David Deamer for his insights and Oliver McRae for reading the manuscript and helpful suggestions. I thank Oliver McRae, Volkan Sarigul, and Jeff Martz for illustrations. This work was supported by P. W. Horn Professor grant from Texas Tech University.

REFERENCES

Abramov, O., and Kring, D. A. 2004. Numerical modeling of an impact-induced hydrothermal system at the Sudbury crater. *J. Geophys. Res.* 109. doi:10/1029/2003JE002213.

Akanuma, S., Nakajima, Y., Yokobori, S., Kimura, M., Nemoto, N., Mase, T., Miyazono, K., Tanokura, M., and Yamagishi, A. 2013. Experimental evidence for the thermophilicity of ancestral Life. *Proc. Nat. Acad. Sci. USA.* 110: 11067–11072.

Alberts, S. V., Van de Goosenberg, J. L. C. M., Driessen, A. J. M., and Konings, W. N. 2000. Adaptations of the archaeal cell membrane to heat stress. *Front. Biosci.* 5: 813–820.

Alexander, C. M. O'D., Bowden, R., Fogel, M., Howard, K. T., Herd, C. D. K., Nittler, L. R. 2012. The provenances of asteroids, and their contributions to the volatile inventories of the terrestrial planets. *Science* 337: 721–723.

Altwegg, K., Balsiger, H., Bar-Nun, A. et al. 2014. 67P/Churyumov-Gerasimenko, a Jupiter family comet with a high D/H ratio. *Science.* doi:10.1126/Science.1261952.

Alvarez, L. W., Alvarez, W., Asaro, F., and Michel, M. V. 1980. Extraterrestrial cause of the Cretaceous–Tertiary extinction. *Science* 208: 1095–1108.

Arndt, N. T., Nisbet, E. G. 2012. Processes on the young earth and habitats of early life. *Ann. Rev. Earth Planet. Sci.* 40: 521–549.

Awramik, S. M., and Buchheim, H. P. 2009. A giant, Late Archean lake system: The Meentheena member (Tumbiana Formation; Fortescue Group), Western Australia. *Precambrian Res.* 174: 215–240.

Bada, J. L., and Lazcano, A. 2002. Some like it hot, but not the first biomolecules. *Science* 296: 1982–1983.

Banerjee, N. R., Furnes, H., Muehlenbachs, K., Staudigel, H., and de Wit, M. 2006. Preservation of ~3.4–3.5 Ga microbial biomarkers in pillow lavas and hyaloclastites from the Barberton Greenstone Belt, South Arica. *Earth Planet. Sci. Lett.* 241: 707–722.

Blankenship, R. E. 2010. Early evolution of photosynthesis. *Plant Physiol.* 154: 434–438.

Baross, J. A., and Hoffman, S. E. 1985. Submarine hydrothermal vents and associated gradient environment as sites for the origin and evolution of life. *Orig. Life* 15: 327–345.

Bell, E. A., Boehnke, P., Harrison, T. M., and Mao, W. L. 2015. Potentially biogenic carbon preserved in a 4.1 billion-year-old zircon. *Proc. Nat. Acad. Sci. USA.* 112: 14518–14521.

Bernhardt, H. S., and Tate, W. P. 2012. Primordial soup or vinaigrette: Did the RNA world evolve at acidic pH? *Biol. Dir.* www.biology-direct.com/content/7/1/4.

Bernstein, M. P., Sandford, S. A., and Allamonda, L. J. 1999. Life's fur flung raw material. *Sci. Amer.* 263 (7): 42–49.

Byerly, G. R., Lowe, D. L., and Walsh, M. M. 1986. Stromatolites from the 3,300–3,500-Myr Swaziland Supergroup, Barberton mountain Land, South Africa. *Nature* 319: 489–491.

Byerly, G. R., Lowe, D. L., Wooden, J. L., and Xie, X. 2002. An Archean impact layer from the Pilbara and Kaapvaal Cratons. *Science* 297: 125–1327.

Bickel, M. J., Nisbet, E. G., and Martin, A. 1994. Archean Greenstone belts are not oceanic crust. *J. Geol.* 102: 121–137.

Biver, N., Bocklee-Morvan, D., Moreno, R., Crovisier, J., Colom, P., Lis, D. C., Sandquist, A., Boissier, J., Despois, D., and Milan, S. N. 2015. Ethyl alcohol and sugar in comet C/2014 Q2 (Lovejoy). *Sci. Adv.* 1. e1500863.

Boss, A. P., and Kreiser, S. A. 2010. Who pulled the trigger: A supernova or an asymptotic giant branch star? *Astro. J. Lett.* 717: L1–L5.

Boussau, B., Blanquart, S., Necsulea, A., Lartillot, N., and Gouy, M. 2008. Parallel adaptations to high temperatures in the Archean eon. *Nature* 456: 942–946.

Brasier, M. D., Green, O. R., Jephcoat, A. P., Kleppe, A. K., Van Kranendonk, M. J., Lindsay, J. F., Steele, A., and Grassineau, N. V. 2002. Questioning the evidence of Earth's oldest fossils. *Nature* 416: 76–81.

Buick, R. 1992. The antiquity of oxygenic photosynthesis – Evidence from stromatolites in sulfate-deficient Archean lakes. *Science* 255: 74–77.

Burton, A. S., Elsila, J. E., Hein, J. E., Glavin, D. P., and Dworkin, J. P. 2013. Extraterrestrial amino acids identified in metal-rich CH and CB carbonaceous chondrites from Antarctica. *Meteorit. Planet. Sci.* 48: 390–402.

Callahan, M. P., Smith, K. E., Cleaves, S. H., Ruzicka, J., Stern, J. C., Glavid, D. P., Huse, C. H., and Dworkin, J. P. 2011. Carbonaceous meteorites contain a wide range of nucleobases. *Proc. Nat. Acad. Sci. USA.* 108: 13995–13998.

Carr, M. H., and Head, J. W. H. 2010. Geologic history of Mars. *Earth Planet. Sci. Lett.* 293: 185–203.

Cates, N. L., and Mojzsis, S. J. 2007. Pre-3750 Ma supracrustal rocks from the Nuvvuagittuq supracrustal belt, northern Quebec. *Earth Planet. Sci. Lett.* 255: 9–21.

Chatterjee, S. 2015. The RNA/protein world and the endoprebiotic origin of life. In *Earth, Life, and System*, B. Clarke (Ed.), pp. 39–79. New York: Fordham University Press.

Chatterjee, S. 2016. A symbiotic view of the origin of life at hydrothermal impact crater-lakes. *Phys. Chem. Chem. Phys.* 18: 20033–20046.

Chatterjee, S. (unpublished). The origin of life in hydrothermal impact crater-lakes: A hierarchical model of the prebiotic synthesis. In *Earth and Life II*, J. A. Talent (Ed.). Dordrecht, the Netherlands: Springer.

Chyba, C., and Sagan, C. 1992. Endogenous production, exogenous delivery and impact-shock synthesis of organic molecules: An inventory for the origin of life. *Nature* 355: 125–132.

Ciccarelli, F. D., Doerks, T., von Mering, C., Creevey, C. J., Snel, B., and Bork, P. 2006. Toward automated reconstruction of a highly-resolved tree of life. *Science* 311: 1283–1287.

Clark, B. C. 1998. Origins of life. *Evol. Biol.* 18: 209–238.

Cockell, C. S. 2006. The origin and emergence of life under impact bombardment. *Philos. Trans. R. Soc. Lond. B: Biol. Sci.* 361: 1845–1856.

Cockell, C., and Lee, P. 2002. The biology of impact craters – A review. *Biol. Rev.* 77: 279–310.

Condie, K. C. 1981. *Archean Greenstone Belts*. Amsterdam, the Netherlands: Elsevier.

Damer, B., and Deamer, D. W. 2015. Coupled phases and combinatorial selection in fluctuating hydrothermal pools: A scenario to guide experimental approaches to the origin of cellular life. *Life* 5: 872–887.

Darwin, C. 1859. *On the Origin of Species*. London, UK: John Murray.

David, L. A., and Alm, E. J. 2011. Rapid evolutionary innovation during an Archean genetic expansion. *Nature* 469: 93–96.

De Wit, M. J., and Ashwal, L. D. 1997. *Greenstone Belts*. Gloucestershire, UK: Clarendon Press.

Deamer, D. W. 2011. *First Life*. Berkeley, CA: University of California Press.

Deamer, D. W., Dworkin, J. P., Sandford, S. A., Bernstein, M. P., and Allamandola, L. J. 2002. The first cell membranes. *Astrobiology* 2: 371–381.

De Koch, M. O., Evans, D. A. D., Beukes, N. J. 2009. Validating the existence of Vaalbara in the Neoarchean. *Precambrian Res.* 174: 145–154.

Delsemme, A. H. 1998. *Our Cosmic Origins*. Cambridge, UK: Cambridge University Press.

Delsemme, A. H. 2001. Argument for the cometary origin of the Biosphere. *Sci. Am.* 89(9): 431–442.

Djokic, T., Van Kranendonk, M. J., Campbell, K. A., Walter, M. R., and Ward, C. R. 2017. Earliest signs of life on land preserved in ca, 3.5 Ga hot spring deposits. *Nat. Commun.* doi:10.1038/ncomms15263.

Dodd, M. S., Papineau, D., Grenne, T., Slack, J. F., Rittner, M., Prajno, F., O'Neil, J. O., Little, C. T. S. 2017. Evidence for early life in Earth's oldest hydrothermal vent precipitates. *Nature* 543: 60–65.

Doolittle, W. F. 2000. Uprooting the tree of life. *Sci. Am.* 282(2): 90–95.

Dworkin, J. P., Deamer, D. W., Sandford, S. A., Allamonda, L. J. 2001. Self-assembling amphiphilic molecules: Synthesis in simulated interstellar/precometary ices. *Proc. Nat. Acad. Sci. USA.* 98: 815–819.

Farmer, J. D. (2000) Hydrothermal systems: Doorways to early biosphere evolution. *GSA Today* 10: 1–9.

Fedo, C. M., and Whitehouse, M. J. (2002). Metasomatic origin of quartz-pyroxene rock, Akilia, Greenland, and implication for Earth's life. *Science* 296: 1448–1452.

Furnes, H., Banerjee, N. R., Muehlenbachs, K., Staudigel, H., and de Wit M. 2004. Early life recorded in Archean pillow lavas. *Science* 304: 578–581.

Garde, A. A., McDonald, I., Dyck, B., and Keulen, N. 2012. Searching for giant, ancient impact structures on Earth: The Mesoarchaean Maniitsoq structure, West Greenland. *Earth Planet. Sci. Lett.* 337: 197–210.

Gaucher, E. A., Govindan, S., and Ganesh, O. K. 2008. Paleotemperature trend for Precambrian life inferred from resurrected proteins. *Nature* 451: 704–707.

Gaucher, E. A., Thomson, J. M., Burgan, M. F., and Benner, S. A. 2003. Inferring palaeoenvironment of ancient bacteria on the basis of resurrected proteins. *Nature* 425: 285–288.

Glikson, A. Y. 2008. Field evidence of Eros-scale asteroids and impact forcing of Precambrian geodynamic episodes. *Earth Planet. Sci. Lett.* 267: 558–570.

Glikson, A. Y. 2010. Archean asteroid impacts, banded iron formations and MIF-S anomalies. *Icarus* 207: 39–44.

Glikson, M., Duck, L. J., Golding, S., Hoffman, A., Bolhar, R., Webb, R, Baiano, J., and Sly, L. 2008. Microbial remains in some earliest Earth rocks: Comparison with a potential modern analogue. *Precambrian Res.* 164: 187–200.

Grassineau, N. V., Abell, P., Appel, P. W. U., Lowry, D., and Nisbet, E. G. 2006. Early life signatures in sulfur and carbon isotopes from Isua, Barberton, Wabigon (Sleep Rock) and Balingwe Greenstone belts (3.8 to 2.7 Ga). *Geol. Soc. Am. Mem.* 198: 33–52.

Green, D. H. 1972. Archean Greenstone belts may include terrestrial equivalents of lunar maria. *Earth Planet. Sci. Lett.* 15: 263–270.

Grieve, R. A. F. 1990. Impact cratering on earth. *Scient. Amer.* 261: 66–73.

Grieve, R. A. F., Cintala, M. J., and Therriault, A. M. 2006. Large-scale impacts and the evolution of Earth's crust: The early years. In *Processes of the Early Earth*, W. U. Reimold, and R. L. Gibson (Eds.), pp. 23–31. Boulder, CO: Geological Society of America Special Paper.

Goesmann, F., Rosenbauer, H., Bredhoft, J. H. et al. 2015. CHO-bearing organic compounds on comet 67P/Churyumov–Gerasimenko reveale by COSAC mass spectrometry. *Science* 349. doi:10.1126/science.aab0689.

Greshake, A., Kloek, W., Arndt, P., Maetz, M., Flynn, G. J,. Bajt, S., and Bischoff, A. 1988. Heating experiments stimulating atmospheric entry heating of micrometeorites: Clues to their parent body source. *Meteorit. Planet. Sci.* 33: 267–290.

Grotzinger, J. P., Sumner, D. Y., Kah, L. C. et al. (2014) A habitable fluvio-lacustrine environment at Yellowknife Bay, Gale Crater, Mars. *Science* 343. doi:10.1126/science 1242777.

Hagerty, J. J., and Newsom, H. E. 2003. Hydrothermal alteration at the Lonar Lake impact structure, India: Implications for impact cratering on Mars. *Meteorit. Planet. Sci.* 38: 365–381.

Hawkesworth, C. J., Cawood, P. A., and Dhuime, B. 2016. Tectonics and crustal evolution. *GSA Today* 26: 4–11.

Hansen, V. L. 2007. Subduction on early earth. *Geology* 35: 1059–1062.

Hansen, V. L. 2015. Impact origins of Archean Cratons. *Lithosphere* 7. doi:10.1130/L371.1.

Harrison, T. M., Schmitt, A. K., McCulloch, M. T., Lovera, and O. M. 2008. Early (> 4.5 Ga) formation of terrestrial crust: Lu-Hf, $\partial^{18}O$, and Ti thermometry results for Hadean zircons. *Earth Planet. Sci. Lett.* 268: 476–486.

Hazen, R. M. 2005. *Genesis: The Scientific Quest for Life*. Washington, DC: Joseph Henry Press.

Hedges, S. B. 2002. The origin and evolution of model organisms. *Nat. Rev. Genet.* 3: 838–849.

Hofmann, A. 2011. Archean hydrothermal systems in the Barberton Greenstone Belt and their significance as a habitat for early life. In *Earliest Life on Earth: Habitats, Environments and Methods of Detection*, S. D. Golding and M. Glikson (Eds.), pp. 51–78. New York: Springer Science.

Holm, N. G. 1972. Why are hydrothermal vent systems proposed as possible environments for the origin of life? *Origin Life Evol. Biol.* 22: 5–14.

Huber, C., and Waterhauser, G. 1988. Peptides by activation of amino acids with CO on (Ni,Fe)S: Implications for origin of life. *Science* 281: 670–672.

Javaux, E. J., Marshall, C. P., and Bekker, A. 2010. Organic-walled microfossils in 3.23-billion-year-old shallow-marine siliclastic deposits. *Nature* 463: 934–938.

Kelley, D. S., Karson, J. A., Fruh-Green, G. L. et al. 2005. A serpentine-hosted ecosystem: The Lost City hydrothermal field. *Science* 307: 1428–1434.

Knauth, L. P., and Lowe, D. R. 2003. High Archean climatic temperature inferred from oxygen isotope geochemistry of cherts in the 3.5 Ga Swaziland Supergroup, South Africa. *Geol. Soc. Am. Bull.* 115: 566–580.

Knoll, A. H. 2003. *Life on a Young Planet: The First Three Billion Years of Evolution on Earth*. Princeton, NJ: Princeton University Press.

Knoll, A. H., and Barghoorn, E. S. 1977. Archean microfossils from the early Precambrian of South Africa. *Science* 156: 508–512.

Kring, D. A. 2000. Impact events and their effect on the origin, evolution, and distribution of life. *GSA Today* 10(8): 1–7.

Lane, N., and Martin, W. F. 2012. The origin of membrane bioenergetics. *Cell* 151: 1406–1416.

Marchi, S., Bottke, W. F., Elkins-Tanton, L. T., Bierhaus, M., Wuennemann, K., Morbidelli, A., and Kring, D. A. 2014. Widespread mixing and burial of Earth's Hadean crust by asteroid impacts. *Nature* 511: 578–582.

Margulis, L. 1981. *Symbiosis in Cell Division*. San Francisco, CA: Freeman.

Martin, W., and Russel, M. J. 2007. On the origin of biochemistry at an alkaline hydrothermal vent. *Philos. Trans. R. Soc. Lond. B: Biol. Sci.* 362: 1887–1926.

Martin, W., Baross, J. Kelley, D., Russell, and M. J. 2008. Hydrothermal vents and the origin of life. *Nat. Rev. Microbiol.* 6: 805–814.

Martin, W. F., Sousa, F. L., and Lane, N. 2014. Energy at life's origin. *Science* 344: 1092–1093.

Materese, C. K., Nuevo, M., Bera, P. P., Lee, T. J., and Sanford, S. A. 2013. Thymine and other prebiotic molecules produced from the ultraviolet photo-irradiation of pyrimidine in simple astrophysical ice analogs. *Astrobiology* 13: 948–962. doi:10.1089/ast.2013.1044.

McCollom, T. M., and Shock, E. L. 1997. Geochemical constraints on chemolithoautotrophic metabolism by microorganisms in seafloor hydrothermal systems. *Geochim. Cosmochim. Acta* 61: 4375–439.

Meierhenrich, U. J. 2015. *Comets and Their Origin*. Weinheim, Germany: Wiley-VCH.

Meinert, C., Myrgorodska, I., Mercellus, P. D., Buhse, T., Nahon, L., Hoffmann, S. V., d'Hendercourt, L. L. S., and Meierhenrich, U. J. 2016. Ribose and related sugars from ultraviolet irradiation of interstellar ice analogs. *Science* 352: 208–212.

Melosh, H. J. 1989. Impact cratering: A geologic process. New York: Oxford University Press.

Miller, S. L. 1953. A production of amino acids under possible primitive Earth conditions. *Science* 117: 528–529.

Mojzsis, S. J., Arrhenius, G., McKeegan, K. D., Harrison, T. M., Nutma, A. P., and Friend, C. R. L. 1996. Evidence for life on Earth before 3,800 million years ago. *Nature* 384: 55–59.

Mojzsis, S. J., and Harrison, T. M. 2000. Vestiges of a beginning. Clues to the emergent biosphere recorded in the oldest known sedimentary rocks. *GSA Today* 10: 1–5.

Mojzsis, S. J., Harrison, T. M., and Pidgeon, T. T. 2001. Oxygen-isotope evidence from ancient zircons for liquid water at the Earth's surface 4,300 Myr ago. *Nature* 409: 178–181.

Motrajit, G., Guan, Y., Leshin, L., Taylor, S., Genge, M., Joswiak, D., and Brownlee, D. 2006. Oxygen isotope measurements of bulk unmelted Antarctic micrometeorites. *Geochim. Cosmochim. Acta* 70: 4007–4018.

Mulkidjanian, A. Y., Bychkov, A. B., Diprova, D. V., Galperin, M. Y., and Koonin, E. B. 2012. Origin of first cells at terrestrial, anoxic geothermal fields. *Proc. Nat. Acad. Sci. USA.* 109: E821–E830.

Nisbet, E. G., and Fowler, C. M. R. 1999. Archaean metabolic evolution of microbial mats. *Proc. R. Soc. Lond. B* 266: 2375–2382.

Nisbet, E. G., and Sleep, N. H. 2001. The habitat and nature of early life. *Nature* 409: 1083–1091.

Nguyen, V., Wilson, C., Hoemberger, M., Stiller, J. B., Agavonof, R. V., Kutter, S., English, J., Theobold, D. L., and Kern, D. 2017. Evolutionary drivers of thermoadaptation in enzyme catalysis. *Science* 355: 289–294.

Noffke, N., Daniel, C., Waxey, D., and Hazen, R. M. 2013. Microbially induced sedimentary structures recording an ancient ecosystem in the ca. 3.48 billion-year-old Dresser Formation, Pilbara, western Australia. *Astrobiol.* 13: 1103–1124.

Noffke, N., Eriksson, K. A., Hazen, R. M., and Simpson, E. L. 2006. A new window into Early Archean life: Microbial mats in Earth's oldest siliciclastic tidal deposits (3.2 Ga Moodies Group, South Africa). *Geology* 34: 253–256.

Nutman, A. P, Bennett, V. C., Friend, C. R., Van Kranendrok, M. J., and Chivas, A. R. 2016. Rapid emergence of life shown by discovery of 3,700-million-year-old microbial structures. *Nature* 537: 535–538.

Oparin, A. I. 1924. *Origin of Life*. Moscow, Russia: Moscovskii Rabochii [in Russian].

Osiniski, G. R., Tornabene, L. L., Banerjee, N. R., Cockell, C. S., Flemming, R., Izawa, M. R. M., McCutcheon, J., Parnell, J., Preston, L. J., Pickersgill, A. E., Pontefract, A, Sapers, H. M., and Southam, G. 2013. Impact-generated hydrothermal systems on Earth and Mars. *Icarus* 224: 347–363.

Osiniski, G. R., Tornabene, L. L., and Grieve, R. A. F. 2011. Impact-ejecta emplacement on the terrestrial planets. *Earth. Planet. Sci. Lett.* 310: 167–181.

Pace, N. R. 1997. A molecular view of microbial diversity and the biosphere. *Science* 276: 734–740.

Parnell, J., Lee, P., Cockell, C. S., Osinski, and G. R. 2004. Microbial colonization in impact-generated hydrothermal sulphate deposits, Haughton impact structure, and implications for sulfate on Mars. *Int. J. Astrobiol.* 3: 247–256.

Paul, D., Kumbhare, S. V., Mhatre, S. S., Chowdhury, S. P., Shett, S. A., Marathe, N. P., Bhute, S., and Shouche, Y. S. 2016. Exploration of microbial diversity and community structure of Lonar Lake: The only hypersaline meteorite crater-lake within basalt rock. *Front. Microbiol.* doi:10.3389/fmicb.2015.01553.

Pinti, D. L., Mineau, R., and Clement, V. 2009. Hydrothermal alteration and microfossil artifacts of the 3,465-million-year-old Apex Chert. *Nat. Geosci.* 2: 640–643.

Pizzarello, J. R., and Cronin, J. R. 2000. Non-racemic amino acids in the Murchison and Murray meteorites. *Geochem. Cosmochem. Acta* 64: 329–338.

Rasmussen, B. 2000. Filamentous microfossils in a 3,235-million-year-old volcanogenic massive sulfide deposit. *Nature* 405: 676–679.

Reysenbach, A. L., and Cady, S. L. 2001. Microbiology of ancient and modern hydrothermal systems. *Trends Microbiol.* 9: 79–86.

Rosing, M. T. 1999. ^{13}C-depleted carbon microparticles in >3700-Ma sea-floor sedimentary rocks from West Greenland. *Science* 283: 674–676.

Sandford, S. A., Bera, P. B., Lee, T. J., Materese, K., and Nuevo. M. 2014. Photosynthesis and photo-stability of nucleic acids in prebiotic extraterrestrial environments. *Top. Curr. Chem.* doi:10.1007/128_2013_499.

Sapers, H. M., Osinski, G. R., Banerjee, N. R., Preston, L. J. 2015. Enigmatic tubular features in impact glass. *Geology* 43: 635–638.

Sarafian, A. R., Nielson, S. G., Marschall, H. R., McCubbin, F. M., and Monteleone, B. D. 2014. Early accretion of water in the inner solar system from a carbonaceous chondrite-like source. *Science* 346: 623–626.

Schopf, J. W. 1993. Microfossils of the Early Archean Apex Chert: New evidence for the antiquity of life. *Science* 260: 640–646.

Schopf, J. W. 1999. *Cradle of Life: The Discovery of Earth's Earliest Fossils*. Princeton, NJ: Princeton University Press.

Schopf, J. W., Tripathi, A., Kudryavatev, A. B., Czaja, A. D., and Tripathi, A. B. 2007. Evidence of Archean life: Stromatolites and microfossils. *Precambrian Res.* 158: 141–155.

Sleep, N. H., Zahnle, K. J., Kasting, J. F., and Morowitz, H. J. 1989. Annihilation of ecosystems by large asteroid impacts on the early earth. *Nature* 342: 139–142.

Sousa, F., Thiegart, T., Landan, G., Nelson-Sathi, S., Pereira, I. A. C., Allen, J. F., Lane, N., and Martin, W. F. 2013. Early bioenergetic evolution. *Philos. Trans. R. Soc. Lond. B: Biol. Sci.* 368: 2013988. doi:10.1098/rstb.2013.0088.

Stern, R. J. 2005. Evidence from ophiolites, blueschists, and ultra-high-pressure metamorphic terranes that the modern episode of subduction tectonics began in Neoproterozoic time. *Geology* 33: 5557–5560.

Stetter, K. O. 2006. Hyperthermophiles in the history of life. *Philos. Trans. R. Soc. Lond. B: Biol. Sci.* 361: 1837–1843.

Tang, M., Chen, K., and Rudnick, R. L. 2016. Archean upper crust transition from mafic to felsic marks the onset of plate tectonics. *Science* 351: 373–375.

Theobald, D. L. 2010. A formal test theory of the theory of universal common ancestor. *Nature* 465: 219–222.

Tice, M. M., and Lowe, D. R. 2004. Photosynthetic microbial mats in the 3,146-Myr-old ocean. *Nature* 430: 549–552.

Ueno, Y., Yamada, K., Yoshida, N., Maruyama, S., and Isozakai, Y. 2006. Evidence from fluid inclusions for microbial methanogenesis in the early Archean era. *Nature* 440: 516–519.

Van Kranendonk, M. J. 2006. Volcanic degassing, hydrothermal circulation and the flourishing of early life on Earth: New evidence from the Warrawoona Group, Pilbara Craton, western Australia. *Earth Sci. Rev.* 74: 197–240.

van Zullen, M. A., Lepland, A., and Arrhenius, G. 2002. Reassessing the evidence for the earliest traces of life. *Nature* 418: 627–630.

Von Damm, K. L. 1990. Seafloor hydrothermal activity: Black smoker chemistry and chimneys. *Ann. Rev. Earth Planet. Sci.* 18: 173–204.

Wachterhäuser, G. 1993. The cradle of chemistry of life: On the origin of natural products in a pyrite-pulled chemoautotrophic origin of life. *Pure Appl. Chem.* 65: 1343–1348.

Wacey, D. 2012. Earliest evidence for life on Earth: An Australian perspective. *Aust. J. Earth Sci.* 59: 153–166.

Wacey, D., Kilburn, M. S., Saunders, M., Cliff, J., and Brasier, M. D. 2011. Microfossils of sulfur-metabolizing cells in 3.4-billion-year-old rocks of Western Australia. *Nat. Geosci.* 4: 698–702.

Walsh, M. M. 1992. Microfossils and possible microfossils from the Early Archean Onverwacht Group, Barberton Mountain Land, South Africa. *Precambrian Res.* 54: 271–292.

Walsh, M. M., and Lowe, D. R. 1985. Filamentous microfossils from the 3,5000-Myr-old Onverwacht Group, Barberton Mountain Land, South Africa. *Nature* 314: 530–532.

Weiss, M. C., Sousa, F. L., Mrnjavac, N., Neukirchen, S., Roettger, M., Nelson-Sathi, S., and Martin, W. F. 2016. The physiology and habitat of the last common ancestor. *Nat. Microbiol.* 1: 161116. doi:10.1038/nmicrobiol.2016.116.

Westall, F. 2004. Early life on Earth: The ascent of fossil record. In *Astrobiology: Future Perspective* Ehrenfreund, P. et al. (Eds.) pp. 287–316, Kluwer, Deventer, the Netherlands.

Westall, F., Campbell, K. A., Breheret, J. G., Foucher, F., Gautret, P., Hubert, A., Sorfieul, S., Grassineau, N., and Guido, D. M. 2015. Archean (3.33 Ga) microbe-sediment systems were diverse and flourished in a hydrothermal context. *Geology* 43: 615–618.

Westall, F., de Wit, M. J., Dann, J., van der Gaast, S., Ronde, C. E. J., and Gerneke, D. 2001. Early Archean fossil bacteria and biofilms in hydrothermally-influenced sediments from the Barberton Greenstone Belt, South Africa. *Precambrian Res.* 106: 93–116.

Wilde, S. A., Valley, J. W., Peck, W. H., and Graham, C. M. 2001. Evidence from detrital zircons for the existence of continental crust and oceans on the Earth 4.4 Gyr ago. *Nature* 409: 175–178.

Wiegel, J., and Adams, M. W. W. 1998. *Thermophiles: The Keys to the Molecular Evolution of Life*. London, UK: Taylor & Francis Group.

Woese, C. R. 1987. Bacterial evolution. *Microbiol. Rev.* 51: 221–271.

Woese, C. R. 1994. The universal ancestor. *Proc. Nat. Acad. Sci. USA.* 99: 6864–6859.

Woese, C. R., Kandler, O., and Wheelis, M. L. 1990. Towards a natural system of organisms: Proposals for the domains Archaea, Bacteria, and Eukarya. *Proc. Nat. Acad. Sci. USA.* 87: 4576–4579.

Xiong, J., Fischer, W. M., Inoue, K., Nakahara, M., and Bauer, C. E. 2000. Molecular evidence for the early evolution of photosynthesis. *Science* 289: 1724–1730.

Zahnle, K., Arndt, N., Cockell, C., Halliday, A., Nisbet, E., Selsis, F., and Sleep, N. H. 2007. Emergence of a habitable planet. *Space Sci. Rev.* 129: 35–78.

Zegers, T. E., De Wit, M. J., Dann, J., White, S. H. 1998. Vaalbara, Earth's oldest assembled continent? a combined structural, geochronological, and palaeomagnetic test. *Terra Nova* 10: 250–259.

5.4 Prebiotic Chemistry in Hydrothermal Vent Systems

María Colín-García, Saúl Villafañe-Barajas, Antoni Camprubí,
Fernando Ortega-Gutiérrez, Vanessa Colás, and Alicia Negrón-Mendoza

CONTENTS

5.4.1 INTRODUCTION

In this chapter, the role of hydrothermal systems in prebiotic evolution is exposed. When Oparin and Haldane presented the first scientific explanation about the origin of life in the late 1920 decade, neither the Earth's structure nor its dynamics were understood. Oparin started exposing his ideas about the origin of life at a time when plate tectonics was not an existing theory, and yet, he considered a possible relationship between igneous processes and chemical evolution (Oparin 1936 in Holm 1992). Later, the first experiments performed by Miller and Urey in 1953 highlighted the role of the atmosphere and ultraviolet (UV) radiation in chemical evolution (Miller 1953). Such experiments considered some of the geological variables known at the time, which were only a promising start in comparison with the present wealth of knowledge about the Earth's dynamics and history. However, the arrival of new knowledge led to new paradigms.

Little is known about the first stages of the Earth, since geological evidence has been wiped out by erosion, modified by metamorphism, or concealed by later rocks. The oldest the geological records the poorest its preservation Step by step, geologists have contributed to solve the puzzle and new evidence is still comes to come to better understand those epochs. Even so, the study of the origin of life has matured based in part on the growing geological evidence. One amazing breakthrough that brought with it a completely new wealth of knowledge was the discovery of submarine hydrothermal vents in the 1980 decade. Among other side effects, such discovery promoted many attempts in order to connect these systems with the origin of life on Earth.

Hydrothermal vents are environments that could have been important for chemical evolution since they harbor a great combination of physical, chemical and geological conditions that probably conditioned prebiotic synthesis. In this chapter, the role of such systems in prebiotic chemistry will be discussed. With this aim, we focus on (1) characterizing the link between the origin of life and hydrothermal systems, (2) the variety of geological settings that could have accounted for chemical evolution, by describing the main geological variables that could have contributed to chemical evolution, and (3) presenting a review of the types of experiments that have been performed in order to model prebiotic reactions in hydrothermal systems.

5.4.2 THE LINK BETWEEN THE ORIGIN OF LIFE AND HYDROTHERMAL SYSTEMS

The study of the origin of life is a very complex matter, provided that the emergence of life depended largely on variables and events that cannot be observed in the meantime. In fact, many of them have left little or absolutely no trace in the geological record (Cleaves 2013). Some facts await more evidence for their fully understanding; among them are (1) the nature of the atmosphere, (2) the amount and action of different types of energy, and, most importantly, the available amount of them for organic synthesis, (3) the temperature at the planet's surface, and (4) the pH of the oceans and their salinity, to name some of the most relevant unknowns (Delaye and Lazcano 2005). All these variables certainly conditioned the reaction of molecules on Earth and other bodies in the solar system; of course, they determined the emergence of life and its evolution in our planet. In fact, even if the time, place and mode in the origin of life are not fully understood (Walker 2017), life in this planet came out from the fortunate convergence of physical, chemical and geological processes.

As emphasized by Walker (2017), the research on origin of life started historically as a chemical problem. This was a natural consequence of the development of organic chemistry and biochemistry on the twentieth century that made possible the understanding of the metabolic pathways and the genetic coding of living beings (Schwille 2017). The main attributes of life are known, although life itself is not fully understood. The incomplete knowledge of the rules that underlie the biological processes has not thwarted the research nonetheless. The first experiments were thus designed to synthesize organic molecules known to be relevant for living beings: amino acids, nucleic bases, sugars, and carboxylic acids, among the most important ones.

The geologic time scale on Earth defined the Hadean Eon, as the time running from the origin of the Earth as a planet ~4.65 to 4.0 Ga (International Commission on Stratigraphy 2017), when the oldest vestiges of continental crust had been found. If the planet is as old as 4.6 Ga (Jacobsen 2003) and the oldest true fossils were dated at 3.4 Ga (Wacey et al. 2011a, 2011b), life must have appeared and evolved much earlier than the latter. The inclusion trails in sandstones of the Strelley Pool in Australia (dated at 3.4 Ga) when analyzed by nanoscale secondary ion mass spectrometry (NanoSIMS) indicate that their formation is the result of biological activity (Wacey et al. 2008). Evidence suggests that life emerged in a relatively short geological time in our planet (Bada and Lazcano 2009). If the available geochemical evidence is analyzed, there is a gap of 1 Ga during which the chemical reactions that preceded the life must have occurred. However, there is not much information about the rhythm or rate of chemical evolution (Cleaves 2013).

Lazcano and Miller (1994) suggested that it took a few million years for life to originate from a completely inert world. The precedent hypothetical period before the origin of life has been called *"chemical evolution."* This term includes all the chemical processes that led to the formation of relevant organic compounds, those that structurally, genetically, or metabolically constitute present living beings. This is a fundamental point, provided that life is crucially organized along these three axes: structure, metabolism, and genetics (Ruiz-Mirazo et al. 2014).

5.4.2.1 PREBIOTIC CHEMISTRY

The synthesis of complex molecules must have required simpler molecules as raw material and the action of one or more energy sources to induce reactions. If those conditions were fulfilled then the synthesis of more complex organics would have been possible. An energy source could be considered as any physical or chemical process that triggered chemical prebiotic reactions (Stüeken et al. 2013). The relevance of a source was directly related to its availability, abundance, and its ability to promote reactions in specific environments on prebiotic Earth. In chemical evolution studies, to determine the best suitable location for chemical reactions to occur constitutes a major question. Different environments were subjected to diverse energy sources.

Undoubtedly, the most abundant energy source on primitive Earth by far was cosmic rays; these are charged particles (electrons, neutrons, and atomic nuclei) accelerated by astrophysical sources and that originate beyond the Earth's atmosphere, and they constituted the main source of ionizing radiation (Ferrari and Szuszkiewicz 2009). The advantage of ionizing radiation is their ability to form free radicals that quickly react to form more stable species. This energy source was effective for promoting reactions in almost every environment due to its penetrating power.

For a long time, UV radiation, emitted by the Sun, was considered the most important energy source for the early Earth. It was capable to generate photolytic reactions in the atmosphere; this kind of reactions is still important on the planet and determines the atmospheric chemistry. Another energy source is shock energy originated by the impacts of different objects, such as meteorites and comets, and it was very important during the early stages of the planet, when the planetesimals that did not accrete impacted the inner planets, including Earth (Chyba and Sagan 1992).

Radioactivity due to the decay of radionuclides was also relevant, even if it was undervalued for many years, since it was thought that its main effects were confined to the lithosphere (Miller 1993). However, it is now known that radiogenic atoms were also abundantly produced in the atmosphere, and some of them (*i.e.,* ^{40}K) might have also been present in water bodies in which they could have promoted many chemical reactions (Draganić et al. 1991). Geothermal energy, responsible for processes such as mantle convection, plate tectonics, and hydrothermal circulation (Stüeken et al. 2013), has been traditionally underestimated as well. Geothermal energy generated by volcanoes and hydrothermal springs was capable of promoting the synthesis by pyrolysis of organic compounds; even so, besides its direct contribution as heat, this energy also can result in electrochemical energy (Stüeken et al. 2013).

Nature itself is a continuum and environments are not isolated but connected, which must also have been the case in primitive environments. Accordingly, energy sources must have acted simultaneously to promote chemical reactions. This means that some specific environment must have been under the influence of different energy sources, and that the products that were synthesized by a sole energy source could be modified later by another one.

Some environments have been considered as the most plausible for chemical synthesis on prebiotic Earth. The supply of organics and the presence of energy must be a requisite in them. The atmosphere, the water bodies, the interfaces, and the hydrothermal environments (Cleaves 2013) all meet the requisites of holding organics and energy. They also represent the main plausible scenarios in which chemical reactions could have occurred on early Earth. For instance, the composition of the atmosphere determined the nature of the chemical reactions that could have taken place in it. The atmospheric composition was very important not only for the synthesis, but it also affected the amount of radiation that reached the surface (Cnossen et al. 2007) and determined its density. Consequently, it also determined the erosion that experienced bodies (such as comets, asteroids, and dust) when entering the atmosphere (Chyba and Sagan 1992). As a result, the atmosphere composition also affected the type and flow of organic compounds that reached the surface of early Earth (Cleaves 2013). Another relevant factor was the availability of continental areas, which allowed the concentration of some organics by evaporation. In those environments, the action of solar radiation was also combined with the presence of minerals that could have acted as surface concentrators or catalyst of reactions.

However, depending on the aggregation state, the type and abundance of molecules, and the availability of energy, some places surely were more efficient in promoting synthesis than others. In chemical evolution, to determine the best suitable location for chemical reactions to occur constitutes a major question, and here relies the importance of analyzing contemporary natural environments. Just by understanding the variables present in such places, more realistic experiments can be designed in prebiotic chemistry studies. Hydrothermal vents constitute natural complex environments in which many variables interact and can be characterized in active systems. Some environmental variables in such systems would have contributed to the increase in complexity of molecules on primitive Earth, whereas others would have inhibited it (Cleaves 2013), even in association within a single hydrothermal manifestation. As explained in the next section, several types of hydrothermal systems may meet the physicochemical requirements for hosting and promoting prebiotic reactions.

5.4.2.2 The Relevance of Oceans and Hydrothermal Vents for Prebiotic Chemistry

The primitive hydrosphere must have played a fundamental role in chemical evolution. In fact, the presence of liquid water on the surface for long periods of time is a necessary condition for life to emerge and evolve (Komiya et al. 2008). Some hydrothermal processes would have occurred on primitive oceans, and many organic compounds could have formed, diffused, and eventually reacted to form more complex molecules in oceans. The composition and physicochemical properties of oceans (including, but not restricted to, pH, salinity, temperature, etc.) surely inhibited some reactions and favored others.

Oceans were formed very soon (Kasting 1993a). Direct evidence, such as turbiditic sandstone beds, shales, conglomerates, silica sediments, and banded iron formations (BIFs), indicates the unequivocal presence of water at about 3.8 Ga (Nutman et al. 1984). However, there is evidence for the presence of liquid water during the Hadean time at 4.4 Ga: $\delta^{18}O$ positive anomalies in zircons can be explained by the presence of liquid water (Wilde et al. 2001; Valley et al. 2002; Cavosie et al. 2005). The formation of the oceans began shortly after the planetary accretion period ended. As stated above, isotopic evidence of liquid water and oceans on the surface of the Earth goes back to a little more than 4.4 Ga (Wilde et al. 2001), but what happened between this time and the first geologic (not isotopic) record of sediments deposited in the ~3.8 Ga ancient seas of the Isua Greenstone Belt in southwest Greenland (Mojzsis et al. 1996) is not known and probably will never be known; the oldest continental crust and hence the first possible evidence of plate tectonics processes and deep recycling of materials, which constitute the life engine of the modern Earth, goes back only to the 4.03 Ga metamorphic rocks that constitute the Acasta Gneiss of northern Canada (Bowring and Williams 1999).

Probably, one of the latest catastrophic events that contributed most to the destruction of the oldest continents was the so-called Late Heavy Bombardment (LHB), variously estimated to have occurred between 4.1 billion and 3.9 billion years based on a time scale calibrated for the Moon, where the record was not erased by subsequent geological evolution as happened on the Earth. Models associated to the effects of LHB on the environmental conditions of the epoch are divided between those that (a) consider afortunate cosmic event that may have triggered the first emergence of life because of some massive contribution of biogenic compounds, such as methane and water, to the outer layers of the primitive Earth, as well as by the building of ideal environmental niches where chemical evolution eventually originated life (e.g., Gladman et al. 2005; Cockell 2006), or (b) those that blame such event for having caused the total extinction of former living matter that possibly evolved in the preceding 500 million yearselapsed between the emplacement of the earliest oceans (4.4 Ga) and the peak of the LHB (~3.9 Ga) (Ryder 2003). Continuous bombardments must have generated a highly dense vapor atmosphere; when the temperature changed, the dispersed water (in the form of vapor) condensed to form the oceans (Kasting 1993b).

Geological evidence indicates that both plate tectonics and liquid water on the surface would have been present at 4.0 Ga (Bowring et al. 1990; Bowring and Williams 1999; Iizuka et al. 2006, 2007). There is sound evidence for plate tectonics and the presence of an open sea at 3.8 Ga (Komiya et al. 1999). One direct consequence of plate tectonics was the formation of hydrothermal systems.

Due to the lack of geological evidence, the physicochemical conditions of the primitive oceans (pH, temperature, ionic strength, etc.) are ill defined. However, some inferences can be made from the available wealth of data. The oceanic pH depended strongly, as ever, on the chemical composition of the atmosphere and the lithosphere. When examining sedimentary deposits, Holland (2003) found that the most abundant minerals in Archean oceans were calcite [$CaCO_3$], aragonite [$CaCO_3$], and dolomite [$CaMg(CO_3)_2$], which means that the ocean must have been saturated with salts containing high amounts of calcium and magnesium. If the atmospheric values of pCO_2 are considered, the pH of the Archean oceans must have been close to ≥ 6.5 (Holland 2003). The initial salinity of the oceans probably ranged between 1.5 and 2 times the present values, and it must have remained high during the Archean, due to the absence of continental cratons, which are deemed necessary to sequester large quantities of halite [NaCl] and/or brines (Knauth 2005).

The oceanic temperature was estimated to have ranged between 0°C and 100°C during the Archean, as evidenced by the occurrence of pillow lavas, and cross-bedding or other sedimentary structures (Knauth 2005). It is assumed that the first ocean must have been completely anoxic (just like the atmosphere) and that the oxygen content increased little by little, although in a variable way. The solubility of oxygen in water is governed by temperature and salinity, and these variables have changed drastically through time on Earth (Knauth 2005). Estimates based on the content of cerium and carbonate mineral anomalies indicate that the concentration of oxygen in seawater was very low until at least 1.9 Ga (Komiya et al. 2008). This, with low or null oxygen concentration, allowed easily oxydizable organic molecules to remain in the environment (Parker et al. 2011). In this context, in primitive Earth, it has been estimated that the concentration of many organic compounds dissolved in the ocean would have ranged between 0.003 M and 0.03 M (Miller and Orgel 1974). Despite such low concentrations, some key organic reactions might have formed nonetheless.

5.4.3 GEOLOGICAL SETTINGS FOR HYDROTHERMAL SYSTEMS

After the discovery of the first marine hydrothermal springs (Ballard and Van Andel 1977; Corliss et al. 1979) some researchers proposed those environments as possible niches

for the origin of life (Corliss et al. 1980; Corliss et al. 1981; Baross and Hoffman 1985). Corliss et al. (1981) suggested that hydrothermal systems could be *"ideal sites for abiotic synthesis"* [sic] due to the high availability of geochemical variables (*i.e.,* water, rocks, gases, thermal energy, gradients in chemical concentration, temperature, and pH). The combination of these variables would eventually lead to the formation of "protocells" and the proliferation of the *"first organism in the vicinity of hydrothermal vents"* [sic]. As soon as these ideas were published, several experiments questioned the viability of these environments as plausible niches for the origin of life, due to the high instability of organic molecules at the conditions present in hydrothermal springs, namely, high pressures and temperatures (Bernhardt et al. 1984; White 1984; Miller and Bada 1988). Subsequently, it was proposed that hydrothermal systems could have been involved in the process of chemical evolution, but they were hardly propitious places for the origin of life (Miller and Bada 1988; Bada and Lazcano 2002; Becerra et al. 2007).

Later, the discovery of a different type of hydrothermal system, the "Lost City Hydrothermal Field" (LCHF) (Kelley et al. 2001, 2002; Proskurowski et al. 2008), rekindled such discussion. The LCHP is located on an ultramafic basement with smoother physicochemical conditions than black smokers, in association with lower temperature and alkaline fluids. This discovery reinforced the idea that important geochemical processes would have been likely in ancient settings and that they probably contributed to the synthesis of organics on primitive oceans (Kelley et al. 2001, 2002; Martin et al. 2008; Proskurowski et al. 2008; Konn et al. 2015; McDermott et al. 2015).

To support the idea of hydrothermal systems as "niches for chemical evolution," it is fundamental at first to identify the natural settings where hydrothermal systems may occur. Then, to understand the physicochemical processes that occur, it is necessary to identify the geochemical variables involved on them (Table 5.4.1) and to evaluate those that could be involved in prebiotic chemical reactions. Such steps are necessary to guide systematic experimental studies that, in turn, will determine if such environments served as places for synthesis and complication or, instead, these were the environments in which decomposition predominated. It is necessary to take into account the geochemical variables in laboratory experiments, even if that means to increase the complexity of experiments (Holm and Andersson 2005; Kawamura 2011). It is also relevant to consider that some variables must have contributed to the increase in complexity of molecules on primitive Earth whereas other variables would have limited it (Cleaves 2013). For instance, during the LHB, Earth was exposed to a constant impact of bodies that might have terminated organics. However, already synthesized organics might have lingered on for subsequent reactions "protected" by hydrothermal vents and deep sediments (Holm 1992).

Many geological settings may harbor hydrothermal activity whose likeliness to be associated with chemical evolution needs to be tested. For that matter, in this section the geological, chemical and physical characteristics of selected settings

TABLE 5.4.1

Some Geochemical Variables at Hydrothermal Springs; All Are Relevant for Studies of Chemical Evolution and Origin of Life

Energy Sources	Dissolved Elements	Gradients	Minerals
Geothermal energy	Cl	Temperature	Sulfides
Thermal energy and	Na	Concentration	Sulfates
electrochemical	Ca	Salinity	Oxides
energy	K	pH	Halides
Radiation	Ba	Eh	Carbonates
From radionuclides	Fe	Pressure	Phosphates
embedded on	Mn	Mineralogy	Silicates
minerals or in	Cu		Native elements
radionuclides in	Zn		
water	Pb		
	Co		
	Cd		
	Ni		
	Mg		
	Ag		
	Hg		
	As		
	Sb		

that are widely distributed in time and space are described and evaluated. First of all, though, it is necessary to comprehensively describe some key geological notions.

A hydrothermal system is an environment where hot fluids circulate below the Earth's surface and may (or not) reach the surface, as hot springs or vents. There are two main components of a hydrothermal system: a heat source and a fluid phase. In addition, fluid circulation requires faults, fractures, and permeable lithologies (*e.g.*, Pirajno 2009). The most common heat sources comprise anomalously high geothermal gradients due to a broad panoply of magmatic foci or tectonic setting that implies crustal thinning. Fluid phases in such systems are typically aqueous (hence the hydro- prefix) that comprise vapor, supercritical fluids or water with very variable salinity (brines) or volatile contents, but these can also be CO_2, H_2S, hydrocarbons, and several other species (Roedder 1990), and even molten sulfur. The sources for such fluids are very variable as well: magmatic, sedimentary-diagenetic, meteoric, oceanic, metamorphic, etc. Such sources determine the physical and chemical characteristics of such fluids, which, in turn, determine the ability of fluids to mobilize and carry ions in solution and, therefore, the mineralogy of precipitates that were formed from them in key geological environments.

Therefore, hydrothermal systems can be classified according to their tectonic setting, the characteristics of their emplacement, including rock permeability, the sources for fluids and their physical and chemical major variables (*i.e.*, anions and cations in solution, pH, Eh, temperature, pressure, density, oxygen and sulfur fugacity, etc.), and the resulting mineral associations, among other geological variables. Upon such characteristics and their natural variations,

geological environments are typified, classified, and categorized for the sake of reasonably guiding further research or exploration endeavors. Such is the essential usefulness of the emphasis in branding types of mineral deposits (for fossil instances) and their active manifestations. Our understanding of the processes that govern the formation of (fossil) mineral deposits is boosted by the understanding of their presently active analogues, and *vice versa*. It is not always possible to resort to uniformitarianism to explain any geological process, but when conceptual models are wisely used, they can provide clues to the past or the present, which would not be obtained otherwise. Such is the reason for grouping fossil and present examples of key hydrothermal environments in this chapter (see Table 5.4.2). As for the matter of prebiotic synthesis, a variety of such environments have been invoked as likely sites to harbor the plausible chemical reactions thus involved. These can be classified between submarine and subaerial hydrothermal systems, and similar systems associated with impact cratering. The latter will not be discussed here; see Chapter 5.3 in this book (Chatterjee 2018) and references therein.

Mineral parageneses are a key for the definition of physicochemical parameters that characterize the conditions for mineral precipitation, including timing. Mineralization in submarine magmatic-hydrothermal systems (volcanogenic massive sulfide [VMS]-type systems) is a product of the chemical and thermal exchange among the ocean, the lithosphere, and the magmas emplaced within it and the fluids exsolved from them. Mineralization in submarine sedimentary-exhalative (SEDEX)-type systems is a product of the chemical and thermal exchange among the ocean, the lithosphere, and sedimentary brines. Mineralization in epithermal systems is a product of the chemical and thermal exchange among magmas and magmatic fluids, the lithosphere, and meteoric water, besides other plausible minor contributors. Mineralization in submarine seafloor serpentinization systems is a product of the chemical and thermal exchange among the ocean, the lithosphere, the mantle, and the fluids derived from them all. Mineralization in shallow submarine to subaerial hydrothermal systems may be the result of chemical and thermal exchange among the ocean, the lithosphere, magmas, magmatic fluids, sedimentary brines, and meteoric water. In all these cases, hydrothermal minerals can be found in many different ways, with regard to (1) type of mineralized structure (vein, veinlet, stockwork, lens, stratum, layer, etc.), (2) geometric relationship between mineralized structures and host rocks (stratiform, stratabound, and crosscutting), (3) time relationship between mineralized structures and host rocks (syngenetic and epigenetic), (4) types of crystal aggregates (massive, disseminated, laminated, banded, etc.), and (5) grain size (fine to coarse). Several aspects in each category may coexist in each model for mineral deposition, and most of them may be exposed on the surface during the lifetime of hydrothermal systems due to syndepositional faulting, diapirism, or slope sliding. The latter, no matter how obvious it may be, means that some mineral association on the surface does not necessarily have been deposited there. This broadens the spectrum of mineral species that could have had a role in

TABLE 5.4.2
Comparison between Shallow Hydrothermal Environments (Both Fossil Ore Deposits and Present Examples)

Type of Deposit and Metallic Associations	Tectonic Setting	Association with Volcanism	Main Types of Mineralizing Fluids and Mechanisms of Formation	Range of Temperatures	Other key Characteristics of Fluids and Geological Processes	Depth	Sources for Sulfur	Mineral Variability in Surface Exposures during Hydrothermal Activity[a]	Age Distribution	Fossil Examples	Presently Active Examples	Further Readings
Volcanogenic massive sulfide deposits (VMS); Cu(-Pb-Zn-Ag-Au)	Arc to back-arc settings (Kuroko and Besshi types), and mid-ocean ridges (Cyprus type).	Close time, space, and genetic association between volcanism and hydrothermal activity.	Magmatic, fresh marine, and modified marine water (evolved within the crust). Seafloor venting.	~100°C to >500°C. Important temperature gradients occur from central to peripheral vents.	Euxinic environment; pH very acidic to alkaline; reduced to oxidized fluids; broad variations in salinity (lower than seawater up to >50 wt.% NaCl equiv.); no bubbling unless in atypical shallow locations; sources for fluids are magmatic, fresh, and deeply evolved seawater. Possible associated submarine brine and molten sulfur pools.	Generally ~2 km below sea level; known shallow examples.	Magmatic, also from seawater sulfate.	Sphalerite, galena, pyrite, chalcopyrite, pyrrhotite, wurtzite, bornite, covellite, arsenopyrite, marcassite, tetrahedrite-tennantite, löllingite, acanthite, cubanite, boulangerite, proustite-pyrargyrite, stannite, cassiterite, realgar, orpiment, cinnabar, microcrystalline varieties of silica, hematite, barite, anhydrite, gypsum, Mn and Fe oxides and hydroxides, native sulfur, talc, chrysotile, chlorite, illite, smectite, montmorillonite, kaolinite, nontronite, jarosite, calcite, dolomite, siderite, ankerite, kutnohorite, native sulfur, cerussite. Possible precipitation of silica gels.	Since the Archean (ca. 3.5 Ga); numerous Paleozoic examples.	Pilbara craton (Australia), Noranda, Abitibi, Windy Craggy (Canada), Troodos ophiolite (Cyprus), Hokuroku and the Green Tuff region (Japan), Semail ophiolite (Oman), Iberian Pyrite Belt (Spain, Portugal), Almadén (Spain).	Mid-ocean ridges; Lucky Strike, Lost City, East Pacific Rise at 21°N, Sonne Field.	Franklin (1996), Barrie and Hannington (1999), Allen et al. (2002), Franklin et al. (2005), Hannington et al. (2005), Piercey (2011)
Sedimentary-exhalative deposits (SEDEX) and subaqueous brine pools; Zn-Pb-Ba(-Cu), Fe-Mn	Advanced stages of continental rifting, failed rifts, and passive continental margins.	None, but volcanic rocks can be present in the hosting sedimentary series.	Sedimentary brines, fresh marine, and modified marine water. Seafloor venting and precipitation from seawater combined.	~50°C to <300°C. Important temperature gradients possibly occurred (stratiform deposits are frequently separated from their hydrothermal feeders).	Euxinic environment; pH very acidic to alkaline; reduced to oxidized fluids; broad variations in salinity (lower than seawater up to <50 wt.% NaCl equiv. in brine pools); no bubbling unless in atypical shallow locations; sources for fluids are sedimentary brines fresh and deeply evolved seawater. Possible associated submarine brine pools.	Between a few hundreds of m and >1000 m below sea level.	Seawater sulfate.	Pyrite, galena, sphalerite, microcrystalline varieties of silica, barite, anhydrite, gypsum, pyrrhotite, marcassite, chalcopyrite, arsenopyrite, löllingite, siderite, ankerite, kutnohorite, dolomite, magnesite, calcite, aragonite, hematite, Mn and Fe oxides and hydroxides, Cu-Fe-Ag-Pb sulfosalts, nontronite, talc, jarosite, stevensite, atacamite, caminite, anglesite, montmorillonite, chlorite, apatite, cassiterite, scheelite, celsian, native sulfur, native silver.	Since the late Paleoproterozoic (ca. 1.8 Ga); peaking during the Mesoproterozoic and Paleozoic.	Broken Hill, Mount Isa, McArthur River (Australia), Sullivan (Canada), Matahambre (Cuba), Rammelsberg, Meggen (Germany), Rajpura-Dariba (India), Molango (México), Gamsberg-Aggeneys (South Africa), Red Dog, Anarraaq (USA).	Atlantis II Deep (Red Sea), Salton Sea, lakes in the East Africa Rift System (?).	Russell et al. (1981), Goodfellow et al. (1993), Lydon (1996), Jorge et al. (1997), Hannington et al. (2005), Leach et al. (2005), Lyons et al. (2006), van der Zwan et al. (2015), Schardt (2016), Sangster (2018)
Shallow submarine or sublacustrine to subaerial exhalative deposits; Mn-Fe-Cu-Zn-Co-Ba-Hg-As-Sb-Ag-Pb	Incipient to advanced stages of continental rifting, failed rifts, and passive continental margins; also shallow manifestations in volcanic arcs (similar to VMS deposits except for their depth).	Close time, space and genetic association between volcanism and hydrothermal activity, or none at all due to high heat flux related to crustal thinning.	Fresh marine and modified marine water to magmatic. Seafloor venting to subaerial precipitation.	≤100°C, eventually higher in magma-derived systems.	Oxidized to euxinic environments; pH very acidic to very alkaline (normally moderately acidic to alkaline); reduced to oxidized fluids; in general, salinities lower than seawater; bubbling very common; sources for fluids highly variable upon their geological location.	<~200 m below sea (or lake) level to subaerial.	Seawater sulfate or magmatic.	Opal, calcite, aragonite, barite, celestine, Mn and Fe oxides and hydroxides, microcrystalline varieties of silica, braunite, rhodochrosite, pyrite, marcasite, cinnabar, carlinite, orpiment, realgar, zeolites, illite, smectite, montmorillonite, chlorite, talc, kaolinite, zeolites, celadonite, apatite, native sulfur, native mercury, Pb-Cu-Ag oxychlorides. Precipitation of silica gels.	Since the Mesoproterozoic.	El Boleo-Lucifer and Concepción peninsula (Mexico), Wafangzi (China), Tatra Mts. (Poland), Valle del Azogue (Spain), sedimentary phosphorites (?).	East Africa Rift System (including alkaline lakes), Milos island (Greece), Concepción bay (Mexico), Kueishantao (Taiwan), Kraternaya Bight (Russia), Bay of Plenty (New Zealand), Lihir island (Papua-New Guinea).	Tiercelin et al. (1993), Fan et al. (1999), Tarasov et al. (2005), Canet and Prol-Ledesma (2006), Conly et al. (2011), Papavassiliou et al. (2017)

(Continued)

TABLE 5.4.2 (Continued)

Comparison between Shallow Hydrothermal Environments (Both Fossil Ore Deposits and Present Examples)

Type of Deposit and Metallic Associations	Tectonic Setting	Association with Volcanism	Main Types of Mineralizing Fluids and Mechanisms of Formation	Range of Temperatures	Other key Characteristics of Fluids and Geological Processes	Depth	Sources for Sulfur	Mineral Variability in Surface Exposures during Hydrothermal Activity[a]	Age Distribution	Fossil Examples	Presently Active Examples	Further Readings
Metalliferous black shales (and closely associated phosphorites); Zn-Cu-Pb(-Mo-Au-Ni-PGE-U-V-P-Se), Cu-Co-Zn, Sb-W-Hg, etc.	Intracontinental rift-related sedimentary basins.	Unclear; may be correlated with mantle plumes; may occur without associated magmatism or associated with mafic volcanism.	Sedimentary brines, fresh marine, and modified marine water. Seafloor venting? Precipitation from seawater? (Impact-related?).	~100°C to >300°C.	Speculative or ill-defined; euxinic environment (but some authors claim otherwise); pH acidic to alkaline; reduced to oxidized fluids; sedimentary to magmatic (related to mantle plumes, according to some authors) sources for fluids.	Between a few hundreds of m and >1000 m below sea level.	Seawater sulfate.	Pyrite, vaesite, gersdorffite, jordisite, millerite, polydymite, sphalerite, chalcopyrite, galena, clausthalite, illite, smectite, montmorillonite, kaolinite, apatite and other phosphates (particularly in sedimentary phosphorites), collophane, Mn and Fe oxides and hydroxides, dolomite, calcite, fluorite, barite, anhydrite, gypsum. Precipitation of colloidal phases (e.g., Ni-Mo sulfides and carbides).	Since the middle Paleoproterozoic (ca. 2.1 Ga) or earlier, peaking during the Paleozoic, associated with worldwide anoxic events (?) after the global oceanic oxygenation.	Willyama Supergroup (Australia), Selwyn Basin (Canada), Niutitang Formation (China), Bohemian massif (Czech Rep., Germany, Poland), Outokumpu, Talvivaara (Finland), Franceville Series (Gabon).	Black Sea, Caspian Sea, Ontong Java Plateau.	Pašava (1993), Coffin and Eldholm (1994), Mossman et al. (2005), Laznicka (2006), Johnson et al. (2017)
Seafloor serpentinization; Ni-Co-Cu-Fe, PGE-Au-Ag, Cr(-Ti-V)	Mid-ocean ridges; supra-subduction zones (back-arc and fore-arc settings), and incipient stages of continental rifting.	Close time, space and genetic association between volcanism (usually the spreading centers) and hydrothermal activity.	Magmatic, fresh marine, and modified marine water (evolved within the crust and/or slab-derived fluids). Precipitation from seawater, seafloor venting and ultramafic-mafic host-rock combined.	<450°C to 500°C. Important temperature gradients occur from the seafloor to the oceanic crust inward.	Euxinic environment; pH acidic to alkaline; reduced to oxidized fluids; board variations in salinity (lower than seawater up to <40 wt.% NaCl equiv.); no bubbling; sources for fluids are magmatic (related to the dehydration of subduction slab in forearc settings), fresh and deeply evolved seawater.	Generally >2000 m below sea level; known shallow examples.	Seawater sulfate, also magmatic.	Olivine, pyroxene, amphibole, plagioclase, serpentinite group minerals (antigorite, chrysotile, lizardite), (Fe)-brucite, talc, chlorite, fuchsite, Mg-rich clays (smectite, sepiolite, palygorskite) zeolite, amphibole, aragonite, calcite, dolomite, magnesite, garnet, clinozoisite, prehnite, pumpellyite, lawsonite, barite, spinel, magnetite, hematite, pyrite, calcopyrite, pentaldite, Co-pentlandite, hazdewoodtie, millerite, marcasite, valleite, violarite, Fe-Ni-(Co)-alloys, bornite, covellite, Platinum Group Minerals (PGM), native gold and minor Mg-Fe-hydroxychlorides (iowaite, pyroaurite-sjögernite group).	Since the Phanerozoic (ophiolites) or earlier (ca. 3.5 Ga; Archean greenstone belts).	Greenstone belts: Isua (Greenland), Barberton (South Africa), Jormua Complex (Finland). Ophiolites: Semail (Oman), Troodos (Cyprus), Zambales (Philippines), Mineoka (Japan), Leka (Norway).	Mid-ocean ridges: Mid-Atlantic Ridge at Lost City, Kane Fracture Zone and 15°20N Fracture Zone, Hess Deep, Iberian Abisal Plain; Forearc seamounts: Izu-Ogasawara-Mariana	Frost (1985), Bach et al. (2004, 2006), Früh-Green et al. (2004), Deschamps et al. (2013), Frost et al. (2013)
Epithermal deposits; Au-Ag, Ag-Pb-Zn-Au	Continental and island arcs; transform boundaries (minor setting).	Close time, space and genetic association between volcanism and hydrothermal activity.	Magmatic, fresh meteoric, and modified meteoric water. Epigenetic structures and subaerial precipitation (sinters).	~50°C to ~400°C. Important temperature gradients occur from central to peripheral springs.	Oxidized environment; pH very acidic to mildly alkaline; reduced to oxidized fluids; generally low salinities; bubbling and formation of aerosols common in geothermal fields; sources for fluids are magmatic, fresh to deeply evolved meteoric water.	Between >1500 m and subaerial paleosurfaces.	Magmatic, also sedimentary or metasedimentary.	Silica and carbonate minerals in sinters (mostly opal and calcite), kaolinite, alunite, native sulfur, barite, anhydrite, gypsum, celestine, Mn and Fe oxides and hydroxides, montmorillonite, illite, smectite, chlorite, epidote, calcite, dolomite, zeolites, adularia, quartz, chalcedony, cristobalite, tridymite, pyrite, cinnabar, realgar, orpiment, stibnite, and other minor sulfides and halides. Precipitation of silica gels.	Vastly Cenozoic; known Archean and Paleoproterozoic examples.	Campbell (Canada), Emperor (Fiji), Kelian (Indonesia), Hishikari (Japan), Fresnillo, Pachuca, Tayoltita (México), Yanacocha (Perú), Lepanto, Baguio (Philippines), Comstock Lode, Summitville, Creede (USA).	Campi Flegrei (Italy), Los Azufres, Cerro Prieto (México), Taupo Volcanic Zone, White Island (New Zealand), Yellowstone, Steamboat Springs, The Geysers (USA).	Corbett and Leach (1998), Sillitoe and Hedenquist (2003), Simmons et al. (2005), Camprubí and Albinson (2006, 2007), Hedenquist and Taran (2013), Sillitoe (2015)

Source: Based loosely on Misra, K. C., *Understanding Mineral Deposits*, Kluwer Academic Publishers, Dordrecht, the Netherlands, 845 p. 1999; Jébrak, M., and Marcoux, E., *Géologie des ressources minérales*, Ministère des ressources naturelles et de la faune du Québec, Québec, Canada, 667 p. 2008; Pirajno, F., *Hydrothermal Processes and Mineral Systems*, Springer, East Perth, Australia, 1250 p. 2009.

[a] Besides those in host rocks. The listed minerals would belong to precipitates or alteration assemblages on the surface or near it, but also to deeper areas that could be exposed by faulting or syndepositional slope sliding.

(?) Indicates debatable locations.

chemical evolution, as one must account equally for minerals that were deposited on the surface and those that were formed somewhere else and eventually made available on the surface while hydrothermal activity still is on. The last (but not least) relevant characteristics for the matter would be the mineral species itself and its particular variation in chemical composition. However, in spite of the geological complexity of each family of depositional environments, many mineral families, groups and classes are shared by most of them (Table 5.4.2), namely silica-group minerals (or silica gels), clay minerals, serpentine minerals and other phyllosilicates, sulfides (especially base-metal sulfides, including arsenides, antimonides, etc.), sulfates, oxides (especially Fe and Mn oxides and hydroxides), phosphates, halides, native elements (including alloys), and many miscellaneous silicates (zeolites, feldspars, amphiboles, scapolites, epidote-group minerals, etc.). It is important to note that these are minerals that form as hydrothermal precipitates or replacements (passive and reactive associations, respectively), but also rock-forming minerals in host rocks should be added to the list above. Among the latter, zircon can be considered as a "cosmopolitan" mineral due to its resilience.

5.4.3.1 Subaqueous Hydrothermal Systems

Submarine vents in association with magmas at or near mid-ocean ridges are usually described as the likeliest hydrothermal systems to be associated with the emergence of life on Earth (*i.e.*, Corliss et al. 1981; Nisbet and Sleep 2001). A common case for the formation of submarine hydrothermal vents occurs when seawater migrates through fractures into the crust and reaches the vicinity of a magmatic intrusion. While approaching the magmas, water is heated up by the anomalously high thermal gradient induced by their emplacement. Additionally, magmas release aqueous fluids upon their cooling down. Therefore, hydrothermal fluids in magma-related systems may come from magmatic, marine or meteoric sources. Similarly, sulfur, iron, copper, zinc, and other metals may come from magmas, ocean water or host rock leachates. The dissolved minerals nourish chemosynthetic bacteria that constitute the base of the food chain for a variety of invertebrates, including large tubeworms (*e.g.*, Levin 2009).

The importance of submarine hydrothermal vents for studies related to the origin of life lies in: (1) providing hot water to surficial environments, (2) upwelling fluids interact with seawater, provide agents for chemical imbalance, thus potentially allowing the synthesis of organic compounds, (3) they produce a rapid crystallization of carbonates and silicates at low temperatures; this increases the local potential to preserve microbial organisms as fossils and their chemical signatures, despite later diagenetic or low-grade metamorphic processes (Pope et al. 2006), and (4) even debatable, some authors suggest that the oldest forms of terrestrial life might have been autotrophic-thermophiles (Pope et al. 2006).

Submarine or sublacustrine hydrothermal systems can be divided into those linked to magmatism as both source for heat, water, and other chemical components, and those associated with venting of basinal brines or other fluids. Such environments correspond, respectively, to volcanogenic massive sulfide (or VMS; Figure 5.4.1) and sedimentary-exhalative (or SEDEX; Figure 5.4.2) deposits and their present-day analogues, which are well-established types of ore deposits and depositional environments, both submarine. Another case, similar to the latter, is constituted by hydrothermal systems that are associated with early stages of continental rifting or lingering activity in more tectonically evolved environments (*e.g.*, along the eastern side of the Baja California peninsula; Canet and Prol-Ledesma 2006; Camprubí et al. 2008). Such systems may occur between shallow submarine and subaerial manifestations, including intertidal venting (Figure 5.4.3). Seafloor serpentinization and paleo-hydrothermal submarine systems associated with metalliferous deposits in black shales are also likely to be accounted among keen environments for prebiotic synthesis. Most theoretical and experimental approaches to prebiotic reactions have been carried out considering VMS-like hydrothermal systems, while neglecting the others. The following account does not intend to be exhaustive, but demonstrative of the variability of hydrothermal scenarios that might have had a role in chemical evolution. Some authors even proposed that BIFs could be related to hydrothermal activity (*e.g.*, Ohmoto et al. 2006), which could be at least the case of Algoma-type deposits (Jébrak and Marcoux 2008).

5.4.3.1.1 Volcanogenic Massive Sulfide-Type Systems

The so-called "black smokers" are the most conspicuous submarine hydrothermal manifestations in VMS-type systems (Figure 5.4.1) and are also interpreted to occur in SEDEX-type systems (Figure 5.4.2; see Larter et al. 1981; Boyce et al. 1983; Russell et al. 1989). These are hydrothermal fumaroles with abundant sulfides and sulfates in suspension that upon precipitation form mounds along favorable faults or within submarine calderas. In VMS systems associated with divergent margins, these are usually located close to mid-ocean ridges. In these fumaroles, due to their proximity to magmas, water may attain temperatures above 400°C and low pH (see Table 5.4.2; also, Von Damm et al. 1985; Camprubí et al. 2017). The latter facilitates the leaching of iron and other metals as the water seeps through the country rocks, which is a feature shared with the other systems hereby described. Such fluids come in contact with cold seawater, thus generating a rapid nucleation of sulfides and other minerals and resulting in a turbid suspension resembling a cloud of black smoke. Salinities of sulfide-rich ore-forming fluids are generally up to 20 wt.% NaCl equiv. although extremely high-salinity polysaline brines may eventually reach the sea bottom as well (up to 65 wt.% NaCl and ~44 wt.% KCl, at >500°C; Camprubí et al. 2017). That being the case, it is likely that brine pools can be associated with VMS-type systems. Another type of hydrothermal vents is dubbed "white smokers," which are generally more distant from their heat source than the black ones (Figure 5.4.1). Black and white smokers may coexist in the same submarine hydrothermal field, but they generally represent proximal and distal vents, respectively, to the main

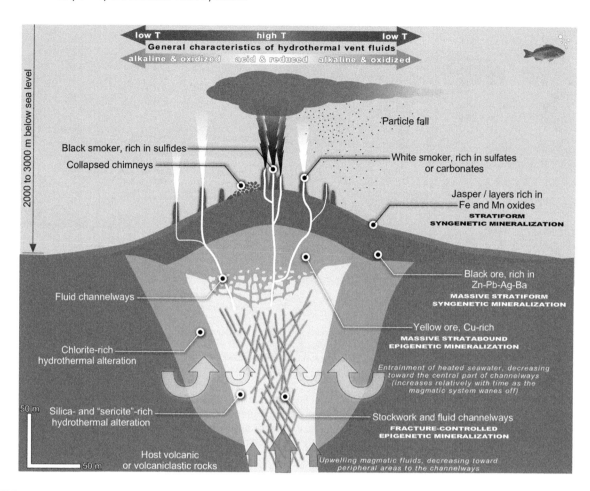

FIGURE 5.4.1 Structural section that combines evidence from active submarine magmatic-hydrothermal vents and from fossil volcanogenic massive sulfide (VMS) deposits, especially in Kuroko-type settings, including all typical styles of mineralization and hydrothermal assemblages. "Sericite," not a valid mineral species name, stands for filosilicates that range between illite and fine-grained muscovite. Key: T = temperature. (Adapted from Colín-García, M. et al., *Bol. Soc. Geol. Mex.*, 68, 599–620, 2016, and based on Lydon, J. W., Volcanogenic massive sulphide deposits, Part 2: Genetic models, in *Ore Deposit Models*, R. G. Roberts and P. A. Sheahan (Eds.), pp. 155–181, Geological Association of Canada, Geoscience Canada, Reprint Series, 3, St. John's, Canada, 1988, and Hannington, M. D. et al., Sea-floor tectonics and submarine hydrothermal systems, in *Economic Geology One Hundredth Anniversary*, Vol. 1905–2005, J. W. Hedenquist, J. F. H. Thompson, and R. J. Goldfarb (Eds.), pp. 111–142, Society of Economic Geologist, 2005.)

upflow zone (Figure 5.4.1). However, white smokers correspond mostly to waning stages of such hydrothermal fields, as magmatic heat sources become progressively more distant from the source (due to magma crystallization) and hydrothermal fluids become dominated by seawater instead of magmatic water (see references in Table 5.4.2 for VMS systems). The temperature in white smokers can be as low as 40°C to 75°C and are alkaline (pH ranges between 9 and 9.8; Kelley et al. 2001). Mineralizing fluids from this type of vents are rich in calcium, and they form dominantly sulfate-rich (*i.e.*, barite and anhydrite) and carbonate deposits. These may form giant chimneys, the largest of which stand almost 60 m above the bottom of the ocean at the LCHF (Kelley et al. 2001). Hydrothermal fluids in this location contain methane, ethane, and propane and organic acids in the form of formate and acetate; these are produced in association with this hydrothermal system (Proskurowski et al. 2008). The global frequency distribution of depths for the occurrence of actualistic examples of VMS deposits (magmatic-hydrothermal seafloor

vents, either black or white smokers) shows a dominant range between 2000 and 3000 m (Figure 3 in Colín-García et al. 2016; Figures 2 and 3 in Hannington et al. 2005), which is consistent with calculated depths of fossil examples (see Table 5.4.2 and references therein). In spite of this, much shallower examples are known both in present-day manifestations as well as in fossil VMS deposits (*e.g.*, Camprubí et al. 2017).

Different mineral associations precipitate during the typical stages of mineralization that characterize the life span of VMS-type systems (Figure 5.4.1). Comprehensive reviews of this subject have been published (*e.g.*, Franklin et al. 1981, 2005; Lydon 1988; Ohmoto 1996; Barrie and Hannington 1999; Hannington et al. 2005). Minerals present in a hydrothermal system or a fossil VMS deposit are deposited passively or reactively. Mineral associations may vary (1) in different mineralized structures, either syngenetic (namely, passive precipitation in chimneys, mounds, and stratiform deposits) or epigenetic (structures that correspond to feeder channels, and replacements of host rocks or pre-existing

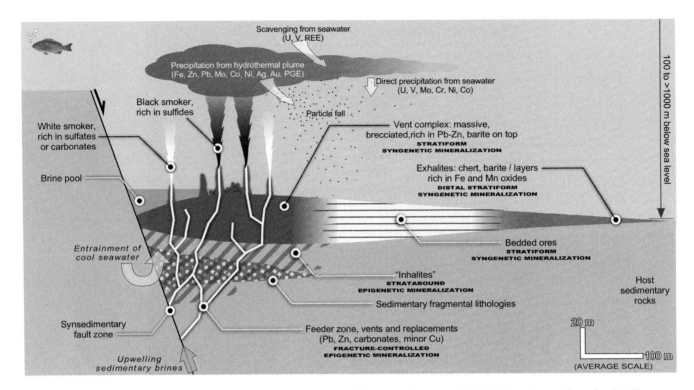

FIGURE 5.4.2 Structural section that combines evidence from active submarine non-magmatic hydrothermal vents, from fossil sedimentary-exhalative (SEDEX) deposits and from an analogy with VMS deposits (see Figure 5.4.1), including all typical styles of mineralization and hydrothermal assemblages. It is important to note that this model corresponds to proximal-to-vent deposits in the model of (Lydon 1996), because feeder zones in most SEDEX deposits, which are commonly more distal to vents than this model exemplifies, have not been found.

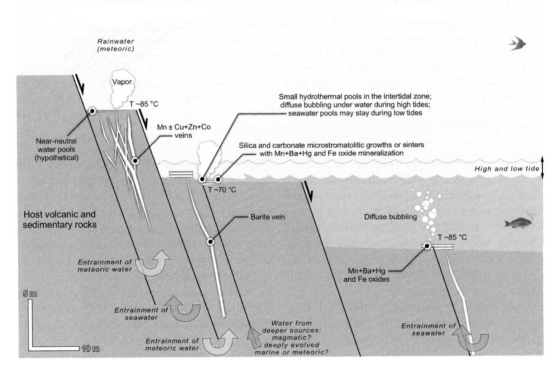

FIGURE 5.4.3 Structural section that combines evidence from active submarine non-magmatic (?) hydrothermal vents and from fossil Mn ± Cu + Zn + Co deposits that formed during early stages of continental rifting, as based in the models for evidence in the eastern Baja California Peninsula in Mexico. The intertidal and submarine hydrothermal manifestations illustrated here would be those around the Concepción Bay area and the ore deposits would be El Boleo, Lucifer, and those in the Concepción peninsula. (Model adapted from Canet, C., and Prol-Ledesma, R. M., *Bol. Soc. Geol. Mex.*, 58, 83–102, 2006; Camprubí, A. et al., *Island Arc*, 17, 6–25, 2008; and references therein.) The (?) here indicates a dubious origin or debatable origin for these vents.

massive sulfide bodies), or *structural zonation*, (2) from proximal to distal associations with respect to venting areas within the same stratigraphic horizon, or *horizontal zonation*, (3) from deep to shallow associations (*i.e.*, stockworks to mounds), or *vertical zonation*, (4) from early and climactic to late stages of mineralization (dominated by sulfides, and sulfates or oxides, respectively), or *temporal zonation*, and (5) in various volcano sedimentary contexts, depending essentially on the composition of volcanic rocks and, ultimately, on the tectonomagmatic context. The most common minerals in ore-bearing associations of VMS deposits (nonmetamorphosed or oxidized) and their modern analogues are summarized in Table 5.4.2. The most common hydrothermal alteration assemblages are chloritic (including Mg-rich ones) and phyllic (dominated by "sericite," mostly illite), and also silicification, deep and shallow talcose alteration, and ferruginous (including Fe oxides, carbonates, and sulfides) alteration. Although not necessarily associated with these systems, molten sulfur lakes may occur in association with submarine volcanism (de Ronde et al. 2015). The precipitation of silica and iron oxide gels in VMS-type systems is also possible (Grenne and Slack 2003).

5.4.3.1.2 Sedimentary-Exhalative-Type Systems

Sedimentary-exhalative systems and those associated with metalliferous black shales provide all geological and physicochemical characteristics that would have favored prebiotic reactions as effectively as VMS systems, such as temperature gradients, euxinic environments, and a wide range of depths of formation (see Table 5.4.2). Sedimentary-exhalative-type systems also provide strong mineralogical and geochemical gradients, and similar pH and Eh gradients to those in VMS-type systems. In addition, the occurrence of submarine brine pools in association with these environments (Figure 5.4.2), such as those interpreted for ore deposits (Boyce et al. 1983) and present-day examples found in the Red Sea (*e.g.*, Schardt 2016), provides strong salinity (and geochemical) contrasts among them, seawater, and more dilute vent fluids.

The problem in the involvement of SEDEX systems with prebiotic reactions resides in the age of the oldest examples of such systems, as no known deposits are older than late Paleoproterozoic (*ca.* 1.8 Ga; Lydon 1996). Metalliferous black shales can be significantly older (middle Paleoproterozoic, *ca.* 2.1 Ga or older; Mossman et al. 2005) than SEDEX or shallow submarine/sublacustrine deposits (see Section 5.4.3.1.3). However, neither type has yet been found to be old enough as to be coeval with prebiotic processes or, least of all, be involved with them. In contrast, Archean VMS deposits are numerous (*ca.* 3.5 Ga; Barrie and Hannington 1999). The striking lack of Archean SEDEX deposits can be associated with the *limiting effect of high reduced iron contents on the activity of reduced sulfur in anoxic oceans* (*sic*, Goodfellow 1992) in which *metals in hydrothermal fluids [...] were dispersed because a lack of reduced sulfur to precipitate them* (*sic*, Misra 1999). Therefore, it is likely that SEDEX-type hydrothermal systems did effectively exist during the Archean, despite being unable to generate

sulfide deposits because reduced sulfur in the oceans would have been previously "sequestered" by iron to precipitate iron sulfides directly from seawater. After the oxygenation of Earth's oceans SEDEX deposits formed, during worldwide anoxic events of the Paleozoic, as might also be the case for Proterozoic deposits (Misra 1999).

Sedimentary-exhalative-type systems (Figure 5.4.2) span different mineralogical styles that are similar to those in VMS-type systems, although their mineral associations are typically more diverse than in the latter. Comprehensive reviews of this subject are published (*e.g.*, Russell et al. 1981; Goodfellow et al. 1993; Hannington et al. 2005; Leach et al. 2005; Lyons et al. 2006; Sangster 2018). Massive and layered sulfide deposits, and exhalites are similar to syngenetic stratiform associations in VMS deposits, whereas "inhalites" are similar to epigenetic stratabound associations in VMS deposits. Exhalites can be constituted by silica-rich, nontronite, montmorillonite, chlorite, carbonate, phosphate, and Fe-Mn oxide-hydroxide layers, and all deposits normally show strong mineralogical banding. Although the SEDEX term is usually reserved to Pb-Zn deposits, their mineralogical diversity allows to classify them into barite, Sb-W-Hg in black shales, Fe-Mn, Sn, and base-metal deposits (Jorge et al. 1997). The latter are subdivided into (type I) deposits with metallic sulfides alone, (type II) deposits with metallic sulfides and exhalites (chert + barite), and (type III) deposits with metallic sulfides, Fe-Mn oxide-hydroxide layers, and exhalites (Sangster and Hillary 1998). Unlike VMS-type systems, feeder zones are rarely found (supra-vent systems; Figure 5.4.2). The most common minerals in ore-bearing associations of SEDEX deposits (non-metamorphosed or oxidized) and their modern analogues are summarized in Table 5.4.2. A particular feature found in association with active SEDEX-type systems is the occurrence of submarine brine pools (*e.g.*, Schardt 2016). Owing to the relatively shallow character that these systems may have, gasohydrothermal activity (bubbling) can be common.

5.4.3.1.3 Shallow Subaqueous Hydrothermal Systems

Hydrothermal systems that are associated with early stages of continental rifting (and later stages as well) share some similarities with SEDEX systems, starting with their specific tectonic setting (early to late stages of continental rifting, respectively). The former, however, stretches between shallow submarine (up to ~200 m deep; neritic zone) or sublacustrine to subaerial hydrothermal to gasohydrothermal systems, including intertidal environments, at temperatures <100°C. The associated aqueous fluids have generally weaker gradients in pH and Eh than in VMS- and SEDEX-type environments, although the genetic variability of these systems, between magmatic-derived and sedimentary-marine-meteoric, implies a large chemical variation within this group of environments. Also, intertidal environments provide an interesting gradient, which is implied by low- and high-tide cycles, with thermochemical gradients between upwelling fluids and seawater that can be pooled during low tides (Figure 5.4.3). In addition, the occurrence of gasohydrothermal fluids in this environment induces generalized diffuse bubbling (*e.g.*, Tarasov et al. 1990; Vázquez-Figueroa et al. 2009; Canet et al. 2010). Diffuse bubbling is normally absent in VMS- and SEDEX-type environments due to high pressures from the water column above them, unless these are atypically shallow. Such gases are rich in CO_2 in most cases, in N_2 and CH_4 in systems that interacted with sedimentary rocks, and in H_2S in association with fumarolic activity (Canet and Prol-Ledesma 2006, and references therein). Little is known about the oldest possible ages of the fossil deposits generated in this environment because of their scarcity in the geological record, but the oldest known example is the Mesoproterozoic Wafangzi Deposit in China (Fan et al. 1999). However, with regard to the most likely geodynamic setting for these systems, which encompasses incipient to advanced stages of continental rifting, the crustal dynamics in the Archean could have furnished plenty of such systems.

Shallow submarine/sublacustrine to subaerial hydrothermal systems are as mineralogically diverse as their possible geological settings. Some of them could be ascribed to VMS- or SEDEX-type environments and their characteristic tectonic settings (see Table 5.4.2 and Canet and Prol-Ledesma 2006). However, a tectonic setting that is seemingly the most characteristic for these systems, which is represented by incipient stages of continental rifting and into advanced stages but confined to the resulting passive margins (Figure 5.4.3). Sedimentary phosphorites may perhaps be the largest deposits in association with such environments (*e.g.*, those formed during the breakup of Pangea in NE Mexico and during the Miocene rifting-off of the Baja California peninsula; see Camprubí [2013]), and a possible hydrothermal origin has been postulated for them (*e.g.*, Slack et al. 2015; Caird et al. 2017), although it is not a necessary condition (Jébrak and Marcoux 2008). Such phosphorites are constituted by amorphous to cryptocrystalline apatite-group phases, generally carbonate-fluorapatite or francolite and rarely hydroxilapatite (Jébrak and Marcoux 2008). Note that phosphorites can be associated with SEDEX deposits and metalliferous black shales as well. However, the most common minerals in shallow settings are silica-group minerals and microcrystalline varieties, carbonates, barite, Fe-Mn oxides and hydroxides and pyrite. Silica minerals and carbonates are most characteristic forming sinters, microstromatolitic growths and veins. Intertidal environments are characteristic to this group of environments (Figure 5.4.3), which means high and low tides twice a day, but also fully subaerial pools are possible in association with Mn $\pm$ Cu+Zn+Co deposits. Both tide variations onto hydrothermal venting and permanent subaerial ponds imply the plausibility of concentration mechanisms. The latter is also valid for epithermal systems (see Section 5.4.3.2). Besides all the possible minerals in the various settings in which such hydrothermal systems occur, the precipitation of silica gels is also possible (Tiercelin et al. 1993). Owing to the shallow character of these systems, gasohydrothermal activity (diffuse bubbling; Figure 5.4.3) is very common.

5.4.3.1.4 Seafloor Serpentinization Systems

The seafloor serpentinization processes consist in the hydration of the oceanic lithosphere, which is mainly formed by mantle peridotites (*i.e.*, dunites, harzburgites, lherzolites, and piroxenites) and lower crustal plutonic rocks. These rocks comprise ~20% or more of the ocean crust and are usually located in the vicinity of spreading ridges, passive margins, arc-subduction environments and could constitute the host rocks for VMS-type systems. The serpentinization process constitute a sequence of exothermic reactions at variable temperatures (below 450°C–500°C), which provide strong mineralogical and geochemical gradients controlled by pH and Eh changes in hydrothermal fluids that were derived from seawater and/or the mantle (*e.g.*, Früh-Green et al. 2004; Alt and Shanks 2006; Klein et al. 2009; Deschamps et al. 2013; Frost et al. 2013; see Table 5.4.2). The process starts with the hydration of olivine to form serpentinite and ferroan brucite, in presence of high pH and reducing fluids, thus promoting the dissolution of sulfides from the peridotitic host rock. The evolution of the system results in the formation of magnetite, from the breakdown of ferroan brucite in presence of SiO_2-rich, acidic, and oxidizing fluids, subsequently releasing significant amounts of H_2 and forming sulfides, oxides, and metal alloys (*e.g.*, Alt and Shanks 1998, 2003; Bach et al. 2004, 2006). A byproduct of serpentinization under highly reducing conditions is the venting of methane- and hydrogen-rich fluids, which can lead to the formation of carbonate mounds and chimneys similar as those described in the LCHF (Kelley et al. 2001). Mixing zones of these reduced hydrothermal fluids with seawater above peridotite outcrops (subseafloor mixing zones, according to Klein et al. 2015) represent niches for chemolithoautotrophic microbial communities associated with the serpentinization of the oceanic mantle (*e.g.*, Alt and Shanks 1998; Kelley et al. 2001; Früh-Green et al. 2004). Fossil examples of seafloor serpentinization systems are found as ancient sections of the Earth's oceanic crust (*i.e.*, ophiolitic sequence defined at the Penrose Conference, 1972) (Juteau and Maury 1997). The temporal frequency of ophiolites ranges between the Archean (*i.e.*, greenstone belts; de Wit 2004) and the Phanerozoic (Table 5.4.2; Misra 1999), thus suggesting that the modern plate tectonics, the formation of oceanic crust, and the seafloor serpentinization systems operated since the early Archean (Furnes et al. 2014).

The mineralogical styles shown in seafloor serpentinization systems are similar to those described in VMS- and SEDEX-type systems, while their mineral assemblages depend largely on the primary mineral association of the altered peridotite rocks. For comprehensive reviews on this subject see (Sakai et al. 1990; Alt and Shanks 1998, 2003; Kelley et al. 2001; Klein and Bach 2009; Frost et al. 2013; Schwarzenbach et al. 2014, 2016; Klein et al. 2015). Sulfide-rich serpentinites, talc-rich serpentinites, carbonate-rich serpentinites (associated with magnesite deposits), listvenites (*i.e.*, CO_2- and SiO_2-rich serpentinites), and rodingites (hydrated gabbros with calcium silicates) are products of the hydrothermal alteration (*i.e.*, carbonation and/or silicification) of serpentinites. These serpentinites are enriched in sulfides, oxides, alloys, and metallic and non-metallic elements (Co, Ni, Fe, Cu, Zn, Mn, Na, Ca, K, and S) relative to peridotitic protoliths (Uçurum 1998, 2000; Hansen et al. 2005; Tsikouras et al. 2006; Nasir et al. 2007; Falk and Kelemen 2015; Hinsken et al. 2017), which are key minerals/elements to promote prebiotic reactions. The types of serpentines listed previously, especially fossil examples associated with magnesite deposits, are currently the subject of numerous studies due to their usefulness as a "natural laboratory" for CO_2 sequestration and storage (Hansen et al. 2005; Kelemen et al. 2011; Beinlich et al. 2012; Power et al. 2013; Falk and Kelemen 2015). The most common minerals in ore-bearing associations of seafloor serpentinization deposits (non-metamorphosed or oxidized) and their modern analogues are summarized in Table 5.4.2.

5.4.3.2 SUBAERIAL HYDROTHERMAL SYSTEMS

Aside from those mentioned above for shallow submarine/sublacustrine to subaerial exhalative deposits in continental rifts, hydrothermal manifestations are abundant in subaerial settings, particularly in association with convergent plate boundaries (continental and island volcanic arcs), but also occur in transform boundaries. For instance, hydrothermal activity is known to occur in the Salton Sea, in association with the San Andreas Fault System but, unlike volcanic arcs, this case is normally placed among modern equivalents to SEDEX deposits and their associated subaqueous brine pools (see Table 5.4.2 and Figure 5.4.2). The most relevant and numerous recent/active hydrothermal fields are found in geothermal and magmatic-hydrothermal contexts, which are normally considered as the modern analogues of low-sulfidation and high-sulfidation epithermal deposits, respectively (which, in both cases, may include intermediate-sulfidation deposits); (Simmons et al. 2005; Camprubí and Albinson 2006, 2007; Sillitoe 2015). The uppermost part of such systems has a tendency to display wide variations in temperature, salinity, volatile content, pH, and redox potential (Figure 5.4.4) and hence the broad range in reactivity between the associated hydrothermal fluids and host rocks. Additionally, such differences in physicochemical variables and reactivity determine broad mineralogical variations on the surface or near it. The above variables are largely controlled by the vertical or lateral nearness of hydrothermal discharge zones to their parental intrusions (see Figure 1 in Sillitoe 2015) and the geological and hydrological characteristics in each area (White and Hedenquist 1990). Besides the broad temperature and salinity gradients that may occur in the actual variety of such environments, the occurrence of deep hypogene low- to intermediate-sulfidation fluids (generally near-neutral and reduced; geothermal context) or high- to intermediate-sulfidation fluids (acidic and oxidized; magmatic-hydrothermal context) determines (1) the possible zonation of alteration assemblages around the fluid conduits, and (2) the mineralogy of the mineral precipitates (if any) that may occur on the surface. In addition to the

"original" physicochemical characteristics of hydrothermal fluids, their chemical characteristics may vary depending on the occurrence of (relatively) near-surface boiling, which may generate H_2SO_4-rich fluids locally in steam-heated grounds (shallow hypogene acidic fluids), independently from the composition of pre-boiling fluids (e.g., Sillitoe 2015). This means that hydrothermal fluids of any kind that undergo boiling may generate acidic fluids upon condensation of boiled-off steam, and the associated alteration assemblages and surficial hydrothermal features.

Acidic fluids from either deep or shallow hypogene sources generate alteration assemblages that result from extremely reactive to relatively mild reactions between fluids and host rocks, from proximal to distal areas to hydrothermal upflow. No surface sinter deposits, either carbonate- or silica-rich, can be expected from highly reactive high-sulfidation type fluids. In this environment, silica is the only residue after extreme acid leaching of every other mineral, or as a late overprint. Common hydrothermal manifestations of the high-sulfidation type are high-temperature solfataras and fumaroles centered on recent volcanic edifices, and hyper-acidic crater lakes. Near-neutral low-sulfidation type fluids, on the contrary, may develop sinter deposits in hot spring environments unless the hydrothermal discharge occurs in high-relief terrains. Common hydrothermal manifestations of the low-sulfidation type are hot springs and geysers. Common manifestations associated with steam-heated grounds are fumaroles, steaming grounds, and mud pots (or mud "volcanoes"). The position of the groundwater table is sensitive to seasonal variations in rainwater availability, climatic and tectonic changes, or several other phenomena (e.g., Sillitoe 2015).

Paleosurface features for the various types of subaerial/sublacustrine magmatic-hydrothermal systems (i.e., epithermal-like systems; Figure 5.4.4) and their mineralogy were summarized in detail by Sillitoe (2015) as (1) steam-heated grounds, with opal/chalcedony, alunite, kaolinite and smectite, (2) groundwater table silicification, with opal/chalcedony, (3) lacustrine amorphous silica sediments, with opal and cristobalite, (4) hydrothermal eruption craters and breccias, with illite and smectite, (5) hot spring sinter, with opal/chalcedony, (6) hot spring travertine, with calcite and aragonite, (7) hydrothermal chert, with opal/chalcedony, and (8) silicified lacustrine sediments, with opal/chalcedony. See their occurrence and nature schematized in Figure 11 by Sillitoe (2015), as replacements, open-space, or on-surface (subaerial or subaqueous) precipitation. Cases 6 to 8 (particularly case 6) occur distally to their hydrothermal upflow zone, which implies that their temperatures are lower than in proximal features and their chemical characteristics are attenuated by interaction with meteoric water. Such features and their particular mineral assemblages may be found topping various hydrothermal alteration assemblages in association with either acidic or near-neutral to alkaline fluids (high-sulfidation and intermediate- to low-sulfidation fluids, respectively), but not necessarily. Alteration assemblages in the uppermost part of these systems are characteristically zoned as follows, from the central portion of hydrothermal upflow outwards into

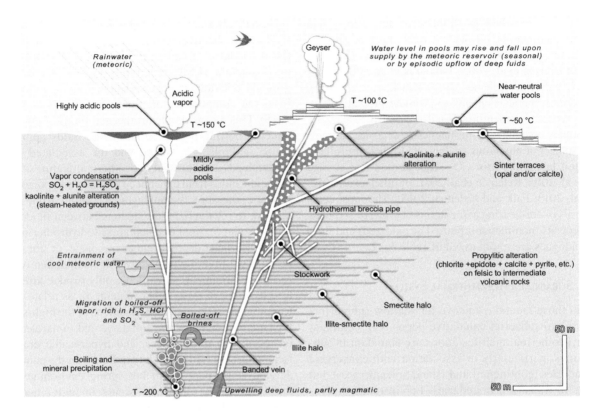

FIGURE 5.4.4 Structural section that combines evidence from active geothermal fields with neutral to alkaline fluids and from the upper-most portion of fossil low-sulfidation epithermal deposits, including all typical styles of mineralization and hydrothermal assemblages. (Loosely based on Buchanan, L. J., Precious metal deposits associated with volcanic environments in the Southwest, in: *Relations of Tectonics to Ore Deposits in the Southern Cordillera*, Vol. 14, W. R. Dickinson and W. D. Payne (Eds.), pp. 237–262, Arizona Geological Society Digests, 1981; Corbett, G. J., and Leach, T. M., *Southwest Pacific Rim Gold-Copper Systems: Structure, Alteration and Mineralisation*, Society of Economic Geologists, Boulder, CO, Special Publications No. 6. 238 p, 1998; Albinson, T. et al., Controls on formation of low-sulfidation epithermal deposits in Mexico: Constraints from fluid inclusion and stable isotope data, in: *New Mines and Discoveries in Mexico and Central America*, T. Albinson and C. E. Nelson (Eds.), pp. 1–32, Society of Economic Geologist Special Publications No. 8, 2001.)

non-altered host rocks (see Stoffregen 1987; Corbett and Leach 1998; Camprubí and Albinson 2006, 2007):

1. High-sulfidation systems: residual quartz (with opal, cristobalite and tridymite), advanced argillic (from silica + alunite to alunite + kaolinite outwards), argillic (from kaolinite + silica to kaolinite + silica + smectite outwards), and illite- or smectite-rich phyllic, from montmorillonite to chlorite-rich propylitic alteration, including zeolites and carbonates (calcite and dolomite), in association with the most alkaline fluids;
2. Intermediate- to low-sulfidation systems: phyllic to propylitic alteration, with the same mineral assemblages as those described for high-sulfidation systems.

Additionally, all the features in the uppermost portions of epithermal deposits and their modern analogues may have anomalously high concentrations of Mn, As, Sb, Hg, Tl, Se, Au, Ag, Ga, and W (Hedenquist et al. 2000; Sillitoe 2015). These anomalies occur in association with minerals such as pyrite, cinnabar, stibnite, orpiment, realgar, native sulfur, livingstonite, corderoite, several amorphous phases, and, exceptionally,

borates (Sillitoe 2015, and references therein). The precipitation of silica gels in epithermal systems is very common. The occurrence in present-day geothermal fields (broadly regarded as the active examples of epithermal deposits) of phenomena such as geysers and phreatomagmatic eruptions add another interesting feature for prebiotic reactions: the formation of aerosols and the occurrence of minute mineral particles alongside them (see subsequent sections).

Other possible hydrothermal environments associated with volcanism that might have harbored similar conditions as epithermal-type systems are Algoma-type BIFs, which were formed during the Archean. Such deposits are rich in magnetite, hematite, siderite, pyrite, pyrrhotite, quartz (as cherts), chlorites, amphiboles, feldspars, biotite, chalcopyrite, and apatite, and the precipitation of colloidal silica would have also occurred (Jébrak and Marcoux 2008).

5.4.4 PREBIOTIC EXPERIMENTS RELATED TO HYDROTHERMAL VENTS

Even if hydrothermal vent systems effectively harbored chemical synthesis, it is necessary to remind that the formation of relevant molecules is only the first step toward the origin of life. The permanence of molecules in these environments is a crucial factor to guarantee the formation of more complex molecules (even polymers) that are necessary for the emergence of living beings. In the forthcoming sections some experiments that aimed to reproduce the environmental conditions in some hydrothermal systems are discussed.

Although hydrothermal systems are the combination of a broad number of geochemical variables, the vast majority of experiments that consider the "conditions of hydrothermal springs" have been initially oriented to test the effect of high temperatures on prebiotic reactions. Thermal energy has been considered as a synonym of energy in hydrothermal systems and other energy sources have been neglected. Therefore, the research in the matter started by testing the stability of organic molecules at high temperatures, which is one of the most straightforward possibilities. Lately, pressure has also been incorporated in simulations as many of the experiments deal with the simulation of deep submarine environments where the pressure is a critical variable. In submarine environments pressure affects the physical state of fluids: water and other molecules (CO_2 and CH_4) behave as supercritical fluids, hence their high efficiency as solvents for organics (Holm 1992).

5.4.4.1 Stability of Organics at High Temperatures and Pressures

The stability of organic molecules exposed to high temperatures is a fundamental issue. The endurance of molecules allows to increase their concentration and the chance for further chemical reactions. The formation of polymers, such as nucleic acids, proteins, and polysaccharides, requires a low decomposition rate of the monomers that constitute them. Some groups of molecules are seemingly good candidates to be the objects in these experiments; the tested molecules include amino acids, sugars, nitrogenous bases, and organic acids, among others (Table 5.4.3). Half-life values ($t_{1/2}$) provide an estimate of the time span molecules may last in the environments under the tested environmental conditions, for an excellent review of half-lives values see Weber (2004) and references therein.

Amino acids are essential molecules for life and their behavior has been largely studied. Amino acids at high temperatures (*i.e.,* 250°C) and pressures (*i.e.,* 250 atm) are very short-lived, as they are stable for only a few hours (Abelson 1954; Bernhardt et al. 1984; White 1984; Miller and Bada 1988; Bada et al. 1995). These results would suggest that hydrothermal springs in such conditions would not fit the requirements for being considered "niches for chemical evolution." However, it is essential to emphasize that most of these experiments do not incorporate many of the physicochemical parameters (*e.g.,* oxidation state, pH, dissolved gas species, presence of inorganic surfaces, etc.). Also, such studies did not recreate the entire natural variability in temperature and pressure and their existing gradients even at the scale of a single vent. Consequently, the stability of the organic molecules could have been largely underestimated (Holm and Andersson 2005).

Other relevant molecules are sugars, which are essential constituents of nucleic acids and the basis for structural polymers in cells. The studies regarding the thermolysis of sugars (Larralde et al. 1995) showed that these compounds are very unstable ($t_{1/2}$ ribose $\approx$ 77 min) at temperatures as low as 100°C. The same behavior is observed for nitrogenous bases. The half-life of adenine is $t_{1/2} \approx 1$ year at 100°C (Levy and Miller 1998), which is a very short availability for further reactions. Similarly, considering the effect of pH and high pressures (*e.g.,* 7.2 MPa), uracil and adenine are even more labile ($t_{1/2} < 1$ h) than in the previous case (White 1984; Balodis et al. 2012). All in all, this only emphasizes the high lability of organic molecules at relatively high temperatures, as they have much lower fusion and decomposition points than inorganic compounds.

5.4.4.2 Temperature Gradients

Even if energy is necessary for promoting chemical reactions, the temperatures of some actual hydrothermal fluids may attain up to 400°C (Von Damm 2013; James et al. 2014), which are excessively high for organic molecules. It is an immense problem, considering that these high temperatures could have existed in some environments during the Archean (Shibuya et al. 2010). However, as explained in sections above, such temperatures are not constant in time and space; high-temperature vents are commonly located in submarine areas at temperatures nearing 0°C, which implies the occurrence of strong thermal gradients. *In situ* studies at present-day hydrothermal vent systems demonstrate the existence of such gradients: just 15 cm around the hydrothermal vents temperature drops at least 30°C from the vent (Chevaldonné et al. 1991; Bates et al. 2010). The spatial variability of the temperature, associated with the turbulent mixing of the fluids in this type of environments, could considerably increase the stability of the organic molecules.

However, temperature is not the only variable that may present broad gradients, such is also the case of pressure, pH, Eh, ionic concentration, mineralogy, etc. Such gradients, besides representing an undeniable fact in natural hydrothermal systems, could also have contributed to increase the stability of organic molecules. Recently, attention has been paid to gradients in the experimental field and some attempts have been made indeed.

A first approximation is to estimate the half-life of molecules at different temperatures. Levy and Miller (1998) demonstrated that at 100°C, nucleobases have short half-lives ($t_{1/2} \approx 1$ year for A and G, $t_{1/2} \approx 12$ years for U, and $t_{1/2} \approx 19$ days for C). However, the half-lives of same molecules at 0°C leapt various orders of magnitude, to the point that all of them presented values in the order of 10^6 years. In other experiments, Larralde et al. (1995) tested the stability of sugars and found that the stability ribose increased from 73 min at 100°C to 44 years at 0°C. Such experiments can only encourage working with temperature gradients, ideally in congruence with those determined in natural examples.

The design and use of flow reactors brought a new perspective in the field, as they permit the generation of thermal

TABLE 5.4.3

Some Relevant Experiments in Prebiotic Chemistry Simulating Physicochemical Conditions of Hydrothermal Vents

Type of Study	Variables Used (Ω)	Description of the Experiments	Organic Molecule Used	Mineral	Main Findings	Reference
Decomposition	T	Solutions of amino acids (AAs) sealed in Pyrex glass	Ala, Arg. HCl, Asp, Cys, Glu, Gly, His. HCl, HyPro, Ile, Leu, Lys. HCl, Met, Phe, Pro, Ser, Thr, Tyr, Val	N.A.	Interconversion of amino acids (i.e., methylamine from glycine; ethylamine from alanine; glycine, alanine, and ethanolamine from serine, etc.). There is an order of relative thermal stability at temperatures between 216°C and 280°C: (1) Asp, Thr, Ser, Arg. HCl; (2) Lys-HCl, His-HCl, Met; (3) Tyr, Gly, Val, Leu, Ile; (4) Ala, Pro, Hyp. Glu.	Vallentyne (1964)
Decomposition	T	Amino acid in water solutions	Glu, PCA (pyroglutamic acid)	N.A.	Kinetic parameters	Povoledo and Vallentyne (1964)
Decomposition	T	Hydrolysis and decomposition of biomolecules in solutions at 250°C (pH = 7, potassium phosphate buffer).	Free amino acids, nucleic bases, peptide bonds, phosphodiester bonds	N.A	The t½ of peptide bonds on alanine oligomers is close to 7 min. A fast breakdown of other oligomers (Ala-Asp and Glu-Ala). Short half-life for proteins (1.08 s for (Ala)$_3$).	White (1984)
Decomposition	T, P	Amino acid decomposition (6 h of incubation at 250°C and 260 bar).	Ala, Arg, Asp, Glu, Gly, His, Ile, Leu, Lys, Met, Phe, Ser, Thr, Tyr, Val	N.A.	Amino acids are drastically affected by high temperature and pressure. Some amino acids are almost quantitatively transformed or decomposed (Asp, Glu, Ser, Thr, Cys, Trp); apolar amino acids as well as His, Lys, Arg and Phe are partially degraded.	Bernhardt et al. (1984)
Decomposition	T, P, pH	Decomposition of solutions of amino acids at 250°C, 265 atm, and pH = 7.	Ala, Leu, Ser, Asp	N. A.	Leu t½ is 15–20 min; Asp (<1 min), and Ser (few minutes) decomposed more rapid than Leu; Ala is more stable than Leu. Gly is produced during the heating experiment.	Miller and Bada (1988)
Decomposition	T, P, M	Amino acid and oligomers at high temperatures, at both high and low pressure.	Gly, Di-Gly, L-Ala, L-Glu	Magnetite	Dipeptide hydrolysis and amino acid decomposition have a first-order rate law. Magnetite accelerates the decomposition.	Qian et al. (1993)
Decomposition	T, P, M	Kinetics of thermal decarboxylation of aqueous solutions of acetic acid and sodium acetate (335°C and 355°C) in contact with surfaces as potential catalysts. Different pressures were used.	Acetic acid and sodium acetate	Quartz, fused quartz, calcite, natural pyrite, titanium oxide, Au, Ca-montmorillonite, Fe- montmorillonite, hematite, synthetic pyrite, and magnetite.	The decarboxylation of acetic acid and acetate catalyzed by the cleavage of the C-C bond, while the acetate molecule is adsorbed onto a surface. Oxidation of acetic acid occur with hematite and defected magnetite.	Bell et al. (1994)
Decomposition	T, P, M, pH	Amino acids, at high temperatures (240°C), buffered solutions (pH = 7).	Ala, Gly + Leu, ethylamine	Quartz-fayalite-magnetite mixture	Amino acids are irreversibly destroyed by heating at 240°C. Equilibrium thermodynamic calculations are not applicable to organics under submarine vent conditions.	Bada et al. (1995)
Decomposition	T	Determination of half-lives of sugars (aldopentoses and aldohexoses). Decomposition at pH 4-8 and, temperatures from 40°C to 120°C.	Ribose	N.A.	Ribose t½ are very short (73 min at pH 7.0 and 100°C and 44 years at pH 7.0 and 0°C). The other sugars also have short half-lives (2-deoxyribose, ribose 5-phosphate, and ribose 2,4-bisphosphate).	Larralde et al. (1995)
Decomposition	T, redox	Amino acids at different temperatures, controlling the oxidation state of the environment.	Ala, bAla, aABA, Asp, Glu, Gly, Leu, Ser, Val	N.A.	The decomposition rate is lower in high hydrogen fugacity environments.	Kohara et al. (1997)

(Continued)

TABLE 5.4.3 (Continued)

Some Relevant Experiments in Prebiotic Chemistry Simulating Physicochemical Conditions of Hydrothermal Vents

Type of Study	Variables Used (Ω)	Description of the Experiments	Organic Molecule Used	Mineral	Main Findings	Reference
Decomposition	T, pH	Half-lives of nucleic bases at 100°C and pH = 7.	Adenine, Guanine, Cytosine, Uracil	N.A.	At 100°C the half-lives for nucleobases are very short. half-For A is 1 yr, G is 0.8 yr, U is 12 yr, and T is 56 yr. C has half-life of 19 days. At 0°C, the half-life of A is 6×10^5 yr, G is 1.3×10^6 yr, U is 3.8×10^8 yr, and T is 20×10^8 yr.	Levy and Miller (1998)
Decomposition	T, P, M, redox	Stability of amino acids, under redox buffered hydrothermal conditions; 200°C and 50 bar in Teflon-coated autoclaves.	Ala, Asp, Leu, Ser	Pyrite-pyrrhotite-magnetite (PPM) to constrain the oxygen fugacity. K-feldspar-muscovite-quartz (KMQ) to control the hydrogen ion activity.	Gly, and Ala were formed, from Ser. Decomposition rates of Leu, Ala and Asp lower in experiments containing the PPM assemblage.	Andersson and Holm (2000)
Decomposition	T, P	Influence of P (22.2 and 40.0 MPa). and T (250°C, 300°C, 374°C, 400°C) in the processes.	Gly	N.A.	Formation of Di-Gly, Tri-Gly (traces), diketopiperazine and a 433 Da product. P and T influence both dimerization and decomposition. Maximum dimers formation at 350°C–375°C 22.2 and 40 Mpa.	Alargov et al. (2002)
Decomposition	T, P	High temperature (200°C–340°C) and high pressure (20 MPa), in a continuous-flow tubular reactor.	Ala, Asp, Leu, Phe, Ser	N.A.	Degradation rates Asp> Ser>Phe>Leu>Ala. Two main reaction paths: *deamination* (Asp) to produce ammonia and organic acids, and *decarboxylation* to produce carbonic acid and amine. Production of glycine and alanine from serine.	Sato et al. (2004)
Decomposition	T, P	Hydrothermal reaction kinetics. Followed by a custom-built spectrophotometric reaction cell. *In situ* observations.	Asp	Reactor type. Non-inert (Ti-6-4/Au reactor) Inert reactor (Au reactor).	The reaction kinetics of Asp is complicated, and highly dependent on experimental conditions (P, T, catalytic surfaces).	Cox and Seward (2007a)
Decomposition	T, P	Hydrothermal reaction kinetic. A custom spectrophotometric reaction cell was constructed. *In situ* observations. Experiments performed at 120°C–165°C and 20 bar.	α-Ala, β-Ala, Gly	N.A.	Under certain hydrothermal conditions, α-Ala, Gly, and β-Ala undergo dimerization and cyclization reaction pathways.	Cox and Seward (2007b)
Decomposition	T, P	Decomposition of the amino acids sub- and supercritical water. The effect of T (250°C–450°C), and residence time (2.5–35 s), P (34 and 24 MPa), and reactant concentration (1.0% and 2.0%, w:v).	Ala, Gly	N.A.	Decarboxylation and amino acid deamination reactions were proposed for both molecules. Ala is decomposed in lactic acid and ethylamine. Gly is decomposed in methylamine	Klingler et al. (2007)
Decomposition	T, M	The effect of iron oxide and sulfide minerals on decomposition reactions of amino acids.	Nva, Ala	Iron oxide and sulfide minerals. Mineral assemblage hematite-magnetite-pyrite (HMP) and pyrite-pyrrhotite-magnetite (PPM).	Nva decomposes by (1) decarboxylation followed by oxidative deamination, and by (2) deamination directly to valeric acid. Ala decomposes in acetic and propionic acids, CO_2 and NH_3. Minerals accelerated decomposition rates. Decomposition is faster in presence HMP than PPM. Surface catalysis and production of dissolved sulfur compounds are probably responsible of the decomposition.	McCollom (2013)

(Continued)

(Continued)

TABLE 5.4.3 (Continued)
Some Relevant Experiments in Prebiotic Chemistry Simulating Physicochemical Conditions of Hydrothermal Vents

Type of Study	Variables Used (Ω)	Description of the Experiments	Organic Molecule Used	Mineral	Main Findings	Reference
Decomposition	T, M	Role of relevant minerals and mineral mixtures in the thermal behavior of an amino acid (200°C–250°C).	Gly	Mineral matrices: montmorillonite, nontronite, kaolinite, salts, artificial sea salt, gypsum, magnesite, picritic basalt, and three samples that simulate the Martian regolith.	Glycine intercalated in some phylosilicates was well protected against thermomelanoid, survived unaltered or been transformed into the cyclic dipeptide (DKP) and linear peptides up to $(Gly)_6$.	Dalai et al. (2017)
Decomposition	T, P, M, redox	Decomposition of aspartate (200°C and 15.5 bars in gold capsules) with and without a mineral product of serpentinization, and at reducing conditions (NH_4Cl and $H_{2(aq)}$).	Aspartate	Brucite	The reaction products vary significantly depending on the reaction conditions. Fluids including just aspartate formed: fumarate, maleate, malate, acetate, and succinate and glycine (both in traces). Under reducing conditions, the main product was succinate and amino acids glycine, α-alanine, and β-alanine.	Estrada et al. (2017)
Stability	T, M_d	The thermal stability of amino acids in seafloor hydrothermal systems is tested.	Ala, Asp, Gaba, Glu, Gly, Leu, Met, Ser.	Carbonaceous ooze. Calcite, with minor amounts of quartz and huntite and traces of illite, smectite and chlorite.	The upper limit temperature for the stable presence was 150°C and 200°C. AAs cannot be synthesized or survive at temperatures higher than 250°C.	Ito et al. (2006)
Stability	T, pH	Reactions of amino acids under subcritical water conditions (220°C–290°C).	Ala, Arg, Asp, Cys, Glu, Gly, His, Ile, Leu, Lys, Met, Phe, Pro, Ser, Thr, Tyr, Val.	N.A.	A decrease in the overall stability in amino acids mixtures. Most of the amino acids decompose at acidic and near-natural pH, stable at basic pH.	Abdelmoez et al. (2007)
Stability	T, pH, M	Evaluation of the thermal stability of amino acids under alkaline hydrothermal conditions (an aqueous solution of NaCl and Na_2CO_3) at elevated temperature (100°C–300°C).	Ala, Arg, Asp, Bala, Gaba, Glu, Gly, His, Ile, Lys, Met, Or, Phe, Pro, Ser, Thr, Tyr, Val	Siliceous ooze	Compared with decomposition at neutral conditions, the decomposition rates are lower under alkaline conditions.	Yamaoka et al. (2007)
Stability	T, P, redox	Stability of adenine under hydrothermal conditions, at 300°C under fugacities of CO_2, N_2, and H_2.	Adenine	Iron	The gases improve the stability of adenine. The concentration of adenine decreased rapidly during the first 24 h of the experiment, then kept decreasing slowly. Adenine was still present in the hydrothermal solution after ~200 h.	Franiatte et al. (2008)
Stability	T, M	Effect of the mineralogical and chemical properties of host sediments on the thermal stability of amino acids.	Ala, Arg, Asp, B-Ala, Gaba, Glu, Gly, His, Ile, Leu, Lys, Met, Phe, Pro, Ser, Thr, Tyr, Val	Siliceous ooze: silica minerals (mostly quartz and minor opaline silica); moderate amounts of calcite and minor amounts of smectite and illite. Montmorillonite Saponite (synthesized).	Amino acids protected from decomposition by amorphous silica and silicate minerals via adsorption and/or binding. The optimal temperature for amino acids was below 150°C. Amino acids are more stable at higher temperatures when associated with silicates.	Ito et al. (2009)

TABLE 5.4.3 (Continued)
Some Relevant Experiments in Prebiotic Chemistry Simulating Physicochemical Conditions of Hydrothermal Vents

Type of Study	Variables Used (Ω)	Description of the Experiments	Organic Molecule Used	Mineral	Main Findings	Reference
Stability	T, P, pH	Ionization constants of nucleic bases at 250°C and 7.2 MPa	Adenine, Uracil	N.A.	Uracil and adenine decomposition occurred by one-step and two-step processes. Phosphate buffer solution enhances the stability of nucleic acid bases.	Balodis et al. (2012)
Stability	T, pH, redox	The effects of temperature (25°C, 150°C, 200°C, and 250°C) pH (6 and 10) and redox state (13 mM aqueous H_2) of hydrothermal fluids. Reaction times from 3 to 36 min.	Glu	N.A.	Glutamic acid at high-temperatures cyclizes and forms pyroglutamate. The formed products (succinate, formate, carbon dioxide, and ammonia) depend on the temperature, pH, and the redox state.	Lee et al. (2014)
Oligomerization	T, M	Amino acids.	Gly, Phe, Tyr	(Ni,Fe)S surfaces.	The formation of oligopeptides was pH dependent. Dipeptide formation (L-Phe, L-Tyr, D,L-Tyr and Gly).	Huber and Wächtershäuser (1998)
Oligomerization	T, P, pH, M	Amino acids in a flow reactor with temperature gradients (T = 250°C, at 24 MPa, pH 2.5)	Gly+Ala	$CuCl_2$	Exponential growth of the products, as a consequence of previous cycles formation, function as templates for the next cycle. At least six different oligopeptides were detected; Ala-Gly, Gly-Ala, Ala-Ala, Gly-Ala-Ala, Ala-Ala-Ala, Ala-Ala-Ala-Ala.	Ogata et al. (2000)
Oligomerization	T, pH, S_f	Oligomerization in a flow reacton (temperature gradient from 110°C to 0°C at pH 3). Dissolved $ZnCl_2$	AMP	N.A.	Synthesis of oligonucleotides in the absence of condensing agents.	Ogasawara et al. (2000)
Oligomeization	T	Amino acid solution heating at different temperatures (200°C–350°C) in a supercritical flow reactor. The stability of some amino acids (ω- and α-amino acids) under hydrothermal conditions was explored.	Gly	N.A.	Oligomers, up to tetra-Gly, formed at 200°C–350°C. No glycine oligopeptides were produced at 400°C. ω-Amino acids and glutamic acid exhibited higher stability than other α-amino acids.	Islam et al. (2003)
Oligomerization	T	Hydrothermal stability of alanine oligopeptides.	(Ala)$_3$, (Ala)$_4$, (Ala)$_5$	N.A.	Small excess of oligopeptides longer than the starting ones. Elongation of (Ala)$_4$ and (Ala)$_5$ was possible in Ala excess. Elongation is competitive with degradation.	Kawamura (2005)
Oligomerization	T	Amino acid solution heating at different T (160°C, 220°C, and 260°C).	Gly	Gold hydrothermal reaction cells.	Peptide synthesis (di-Gly, tri-Gly) is favored in hydrothermal fluids. Rapid recycling of products from cool into near-supercritical fluids will enhance peptide chain elongation.	Lemke et al. (2009)
Oligomerization	T	In a hydrothermal microflow reactor, the synthesis of oligopeptide-like molecules at 250°C–310°C.	Glu, Asp		Synthesis of oligopeptide-like molecules of length up to 20-mers from Glu and Asp.	Kawamura and Shimahashi (2008)
Oligomerization	T, pH	Dimerization rate of glycine, the effects of pH (ranging from 3.1 to 10.9) and temperature (120°C, 140°C, 160°C, and 180°C) were tested.	Gly	N.A.	Dimerization increases at basic pH (7–10). The dimerization rate increases with temperature (150°C). Gly dimerizes most under alkaline pH (~9.8) at about 150°C.	Sakata et al. (2010)

(Continued)

TABLE 5.4.3 (Continued)
Some Relevant Experiments in Prebiotic Chemistry Simulating Physicochemical Conditions of Hydrothermal Vents

Type of Study	Variables Used (Ω)	Description of the Experiments	Organic Molecule Used	Mineral	Main Findings	Reference
Oligomerization	T	The production of phosphodiester bond and formation of mononucleotides capable of base pairing after hydration–dehydration cycles (85°C).	AMP; UMP	N.A.	The cycles of hydration and dehydration drive the synthesis of ester bonds. Oligomers resembling RNA are synthesized. Some of the products have properties suggesting secondary structures, including duplex species stabilized by hydrogen bonds.	DeGuzman et al. (2014)
Oligomerization	T, M, pH	The effect of elemental composition, pH, presence of clay, doping with small organic compounds, ribonucleotide activation on RNA oligomerization.	AMP, Imidazole-activated AMP (ImpA)	Iron-sulfide synthesized chimneys. Montmorillonite.	Nucleotide oligomerization—for both the activated and unactivated nucleotide—can occur in synthetic alkaline hydrothermal chimneys. Generation of oligomers (up to 4 units) with imidazole-activated ribonucleotides.	Burcar et al. (2015)
Synthesis	T, M	Heating of NH_4HCO_3 solution with C_2H_2, H_2 and O_2 (produced in situ).	Acetylene	Calcium carbide. Calcium.	Amino acids (Gly, Ala, Asp, Glu, Pro, Ser, Leu, Ile, Lys, Val, Thr) and amines formed at 200°C–275°C, no formation at <150°C.	Marshall (1994)
Synthesis	T	Formation of lipids through Fischer-Tropsch-type synthesis of aqueous solutions.	Formic acid or oxalic acid	N.A.	Heating at 175°C for 2–3 days, lipid compounds from C_2 to $>C_{35}$ (n-alkanols, n-alkanoic acids, n-alkanes and alkanones).	McCollom et al. (1999)
Synthesis	T, P, M	Reactivity of organic acids and acid anions (325°C, 350 bars) in the presence of the mineral assemblages.	Acetic acid, sodium acetate, valeric acid	(1) Hematite + magnetite + pyrite (HMP). (2) Pyrite + pyrrhotite + magnetite (PPyM) (3) Hematite + magnetite (HM)	Acetic acid and acetate decompose by decarboxylation and oxidation. Reactions are catalyzed by minerals: magnetite promotes decarboxylation; hematite promotes oxidation The oxidation reaction is much faster. Valeric acid decomposed faster than acetic acid under similar conditions.	McCollom and Seewald (2003b)
Synthesis		Decomposition of formic acid and formate and the production of formate from CO_2 reduction (175°C–260°C). Experiments conducted in gold-TiO2 reactors	Formic acid and formate	Hematite, Magnetite. Serpentinized olivine, Ni-Fe alloy	Minerals had no effect on the stability of formic acid or formate. The quantity of formate in hydrothermal fluids could be controlled by an equilibrium with dissolved CO_2 at the oxidation state and pH of the fluid.	McCollom and Seewald (2003a)
Synthesis	T, P, M	Conversion of CO_2 into organic compounds in hydrothermal conditiions (300°C and 30 MPa).	CO_2 and H_2	Cobalt-bearing magnetite.	Formation of CH_4, C_2H_6, and C_3H_8, but also n-C_4H_{10} and n-C_5H_{12}.	Ji et al. (2008)
Synthesis	T, M	N-bearing molecules, to synthesize ammonia, at different T (200°C, 70°C, and 22°C).	Dinitrogen, nitrate	Fe and Ni metal, awaruite ($Ni_{80}Fe_{20}$) and tetrataenite ($Ni_{50}Fe_{50}$), alloys bearing Fe and Ni.	Nitrite and nitrate are converted to ammonium rapidly. The reaction of dinitrogen is slower. Reduction is strongly temperature-dependent. Metals were more reactive than alloys.	Smirnov et al. (2008)
Synthesis	T	Amino acids synthesis in function of temperature, heating time, starting material composition and concentration.	NH_4HCO_2	Fe and Ni.	Amino acids synthesized (Gly, Ala, Asp, Glu, Ser) from simple precursors under submarine hydrothermal systems conditions. Degradation is privileged in such conditions. Synthesis at lower temperatures.	Aubrey et al. (2009)

(Continued)

TABLE 5.4.3 (Continued)
Some Relevant Experiments in Prebiotic Chemistry Simulating Physicochemical Conditions of Hydrothermal Vents

Type of Study	Variables Used (Ω)	Description of the Experiments	Organic Molecule Used	Mineral	Main Findings	Reference
Synthesis/ Precipitation	T, M	Pyrophosphate synthesis in inorganic precipitates simulating hydrothermal chimney structures in thermal and/or ionic gradients.	$FeCl_2 \cdot 4H_2O$, $Na_2S \cdot 9H_2O$, K_2HPO_4, Sodium silicate solution (Na_2O/26.5% SiO_2), Na-acetyl phosphate	Iron mineral films.	Poi was synthesized. Iron-rich membranes with incorporated phosphates were generated.	Barge et al. (2014)
Cycling	T, S	Laboratory simulation of hydrothermal pools under cycles of hydration and dehydration at 85°C in an atmosphere of CO_2 and monovalent salts	Mixtures of AMP and UMP	N. A.	1:1 are cycled in the presence of monovalent salts, a polymerization reaction yields a product with	Da Silva et al. (2015)
Adsorption	T, M	Adsorption–desorption experiments at 80°C for 10 days.	Lys	Na-smectite (>90%), a small amount of cristobalite (<10%) and traces of calcite and quartz.	Thermal treatment originates stronger smectite–lysine binding, by H bonds between NH_3^+ lysine groups and smectite basal O atoms.	Cuadros et al. (2009)
Microsphere formation	T, P	Aqueous solution with amino acids (Gly, Ala, Val and Asp) in glass tubes, heated at 200°C, 250°C, 300°C, and 350°C, at 134 atm, buffered (pH 7.2).	Gly, Ala, Val, Asp	N.A.	Formation of microspheres at temperatures above 250°C. Polar amino acids are needed for the microsphere formation. Microspheres are made of peptide-like polymers.	Yanagawa and Kojima (1985)
Reaction	T	Decarboxylation of an amino acid solution as a function of pH.	α-Ala	N.A.	Arrhenius parameters were determined. The addition of KCl resulted in a reduction of the decarboxylation rate.	Li et al. (2002)
Reactivity	T, M	Pyruvate reactions in presence of transition-metal sulfide minerals, at moderate temperatures (25°C–110°C).	Pyr	Pyrrhotite, troilite, arsenopyrite, pyrite, marcasite, sphalerite, chalcopyrite	Amino acids and fatty acids were formed. Formation of lactate, propionate, and alanine, among others.	Novikov and Copley (2013)
Concentration	T, M	To test if channels within the mineral could act as act as natural Clusius–Dickel thermal diffusion column and increase local amphiphile concentrations.	Oleic acid	Borosilicate microcapillaries.	Microcapillaries act as a thermal diffusion column and concentrated the molecule. Vesicle formation.	Budin et al. (2009)
Mineral precipitation	T, pH	Raman spectroscopy to study ancient hydrothermal iron sulfide formation (growth temperatures from 40°C to 80°C).	Aqueous alkaline solutions containing bisulfide and silicate injected into iron (II) solutions.	N.A.	Formation of mackinawite and greigite iron sulfide phases. Mackinawite was probably the dominant catalyst in ancient pre-biotic chemistry.	White et al. (2015)

Source: Colín-García, M. et al., *Bol. Soc. Geol. Mex.,* 68, 599–620, 2016.

Note: All amino acids are abbreviated according to the IUPAC indications.

Key: ABA = aminobutyric acid, Ala = alanine, AMP = adenosine monophosphate, Arg = arginine, Asp = asparagine, Cys = cysteine, Gaba = gamma-aminobutyric acid, Glu = glutamic acid, Gly = glycine, His = histidine, Hyp = hydroxyproline, Ile = isoleucine, Leu = leucine, Lys = lysine, Met = methionine, Nva = norvaline, Orn = ornithine, PCA = pyroglutamic acid, Phe = phenylalanine, Pro = proline, Pyr = pyruvate, Ser = serine, Thr = threonine, Tyr = tyrosine, Val = valine.

(Ω) Variables used refer to: T = temperature, P = pressure, M = minerals, S = salts, redox estate, and pH.

gradients. The use of those equipment showed that it is not only possible for organic molecules to remain in these environments, but it is also feasible to form more complex molecules, such as oligomers. This was tested for amino acids that formed oligopeptides (Imai et al. 1999; Ogata et al. 2000; Islam et al. 2003); similarly, oligonucleotides were formed from the repeatedly circulation of nucleotides between hot and cold regions, even in the absence of condensation agents (Ogasawara et al. 2000). These experiments have been completed with theoretical approaches that suggest that thermal gradients in mineral pores could favour the accumulation of molecules (Braun and Libchaber 2002; Baaske et al. 2007; Mast et al. 2013; Niether et al. 2016).

5.4.4.3 Redox State and Dissolved Gases

Hydrothermal vents harbor a high diversity of chemical species. Currently, some of these dissolved species in hydrothermal fluids are the fuel for microbial communities thriving in these environments (Martin et al. 2008). It has been proposed that the wide availability of gases and/or dissolved elements (CO_2, H_2S, H_2, CH_4, NH_3, Co, Fe, Mg, SO_4^{2-}, and Mn) (Tivey 2007; Martin et al. 2008) in hydrothermal systems could have been the basis of the first metabolic routes (Wächtershäuser 1990; Russell and Martin 2004; Martin et al. 2008). Certainly, dissolved chemical species in these environments may have promoted or inhibit the reactivity of organic molecules.

More recently, several research groups have delved into the effect of various geochemical variables on the stability of organic molecules. In general, the experiments suggest that the geochemical variables are intrinsically related to the fate of the organic molecules. That implies that in these environments small changes in environmental conditions could have promoted chemical changes. Investigations (Kohara et al. 1997) and (Lee et al. 2014) showed that the decomposition rate of some amino acids (*i.e.,* glycine, glutamic acid, and alanine) was much slower in environments with high hydrogen fugacity. In other experiments, the decomposition of amino acids decreases under buffered conditions, as the pyrite-pyrrhotite-magnetite system acts as a redox buffer (Andersson and Holm 2000). However, the redox condition also affects the decomposition pathways. There are two possible decomposition reactions of amino acids under hydrothermal conditions: *deamination* to produce ammonia and organic acids, and *decarboxylation* to produce carbonic acid and amines (Sato et al. 2004). It has been shown that some amino acids (*i.e.,* glycine and alanine) may preferably undergo dimerization and then cyclization reactions, instead of decomposition, depending on the experimental conditions (mineral surface, temperature, residence time, and redox state; (Cox and Seward 2007b, 2007c). In addition, depending on temperature and pressure, there may be a selection of the decomposition mechanism (Klingler et al. 2007).

Experiments with nitrogenous bases (Franiatte et al. 2008) showed that under an atmosphere containing CO_2, N_2, and H_2, adenine was present after $\approx$ 200 hours of heating at 300°C. On the other hand, the decomposition reactions (*e.g.,* dehydrogenation) may depend on the redox state of the system.

5.4.4.4 pH Gradients

There is a great debate about the role of proton gradients in the origin of life, as the electrochemical gradients across the membranes could have boosted metabolism (Jackson 2016; Lane 2017). It is very feasible that the pH values in hydrothermal vent systems could also be involved in the chemical behavior of organic molecules. On-site measurements showed that hydrothermal fluids at high temperatures (*e.g.,* >350°C) and with acidic pH (close to 5) change their pH toward neutrality when they come in contact with oceanic water (Ding et al. 2005). Similarly, theoretical studies explain that it is possible to have a gradient of up to 6 pH units (that is, six orders of magnitude) at a micrometric scale in hydrothermal vents (Möller et al. 2017). Experimentally, Sakata et al. (2010) showed that dimerization of glycine is most efficient at alkaline pH ($\approx$ 9.8) due to differences in dissociation states of the molecule (*i.e.,* Gly$^{\pm}$ and Gly$^-$ fractions are approximately equal at this pH). Also, the decarboxylation of alanine is three times higher in pH values where the zwitterion predominates, although the presence of dissolved ions (*i.e.,* KCl) reduces it (Li et al. 2002).

5.4.4.5 Experiments That Include Minerals

One of the most important characteristics of hydrothermal vents is their high mineralogical diversity. The broad diversity of minerals in hydrothermal springs of any type could be a crucial parameter in the formation and production of complex organic molecules in prebiotic experiments (Colín-García et al. 2016). The role that minerals could have played in prebiotic reactions is complex but can be envisioned as follows: (1) catalysts of reactions, both decomposition and formation of more complex species, (2) templates where organics can organize and constitute more complex molecules, (3) protective agents, sheltering molecules from decomposition in the media, and (4) concentrating agents—the concentration of organics in contact with solids increases and some organic reactions can be promoted. The specific role that minerals play depends on variables including: chemical composition of minerals and their impurities, solubility of minerals, redox potential in the environment, pH, temperature, ionic force, and, of course, the characteristics of the organic molecule. Minerals affect at different degrees the thermal decomposition of organics; actually, many experiments have been performed in order to specifically understand the effect of minerals on the chemical reactions related to prebiotic chemistry experiments (Table 5.4.3).

In order to evaluate the effect of the mineralogical and chemical properties of host sediments on the thermal stability of amino acids (Ito et al. 2006) reported almost full decomposition of amino acids (90.1% at 200°C and 99.7% at 300°C) in experiments including calcareous sediments and

NaCl solutions. Also, they suggested that the temperature roof for having stable amino acids probably varies between 150°C and 200°C (Ito et al. 2006). Later, the same group demonstrated that at the very same conditions siliceous ooze served as a better protection for amino acid than calcareous sediments (Ito et al. 2009). Yamaoka et al. (2007) also studied the effect of siliceous ooze in the decomposition of amino acids, but under alkaline hydrothermal conditions (by enrichment of Na_2CO_3 and volatile gases). They found that decomposition is inhibited in alkaline conditions and, even more, amino acids remained even after heating for 240 hours at 300°C. Dalai et al. (2017) investigated how minerals and mineral mixtures change the formation of a black water-soluble thermal polymer ("thermomelanoid") of glycine. When the experiment was carried out in the presence of phyllosilicates, these minerals precluded the formation of the polymer; instead, glycine remained unaltered or was even transformed to the cyclic dipeptide diketopiperazine (DKP) or polymerized up to Gly6.

It has also been tested the role of sulfides in prebiotic reactions. Novikov and Copley (2013) demonstrated that pyrite favors the synthesis of amino acids and fatty acids from pyruvate (an important precursor for organic molecules and already synthesized under simulated hydrothermal vent conditions). Qian et al. (1993) showed that magnetite increases the decomposition of amino acids at high pressures and temperatures. In addition, the presence of mineral substrates (*e.g.*, iron oxides and sulfides) accelerates the decomposition rates of some amino acids (norvaline; McCollom 2013).

A series of experiments (McCollom and Seewald 2003a) showed that minerals such as hematite, magnetite, serpentinized olivine, and NiFe alloy had little effect on the stability of some species (formate and formic acid, the simplest organic acid, and acid anion present in natural waters). However, minerals influenced both the pH and concentration in CO_2 of fluids, and, in turn, dissolved CO_2 determined the amount of formate that was present. Acetic acid and acetate (intermediates in metabolism) decompose rapidly in the presence of minerals containing metal ions (*e.g.*, calcium- and iron-bearing montmorillonite, pyrite, hematite, and magnetite) as they act as catalysts in the decarboxylation reaction (Bell et al. 1994).

Stereoselectivity is also promoted by minerals. Fuchida et al. (2017) heated (at 120°C) DL-alanine in presence of olivine and water to investigate the formation of the diastereoisomers of diketopiperazine (DKP). Olivine was an efficient catalyst to form DKP and determined the preferential formation of one of these dipeptides.

Studies that included other types of variables demonstrated that minerals can also have a representative role in the behavior of organic molecules. For example, the presence of dissolved ions may favor the protection of amino acids against thermal decomposition, probably due to the formation of complexes with dissolved ions (*e.g.*, Ca^{2+} and Mg^{2+}) that, in turn, increase their sorption onto mineral surfaces (Dalai et al. 2017). Likewise, thermal treatment from a hydrated phase can favor the concentration of amino acids in clays due to the increase in hydrogen bonds between the tetrahedron

bed and the amino group of the amino acid (Cuadros et al. 2009). Also, the stability of formate and formic acid could be affected more by dissolved CO_2 contents and the pH of the fluid than by the presence of some minerals (*e.g.*, hematite, magnetite, and serpentine; McCollom and Seewald 2003b).

5.4.5 SUMMARY OF RELEVANT GEOLOGICAL AND MINERALOGICAL ASPECTS FOR EXPERIMENTAL STUDIES WITH REGARD TO PREBIOTIC SYNTHESIS

Physicochemical variables such as temperature, pressure, composition of solutes in aqueous brines or supercritical fluids, pH, and Eh are obvious key factors that control prebiotic synthesis. In addition, the role of minerals in prebiotic chemistry is acknowledged nowadays as no lesser factor (see the previous sections; Schoonen et al. 2004; Cleaves et al. 2012). The broad variety of factors that have an influence on prebiotic chemistry come from an even broader variety of natural environments (Figures 5.4.5 through 5.4.8, which stretch far more generously than those visible in deep submarine VMS-type hydrothermal vents, either acidic or alkaline (see Table 5.4.2). In other words, many natural settings interesting as models for prebiotic studies can be found outside VMS-type hydrothermal vents.

Natural systems keep resisting experimental simplifications but also have a great potential for furnishing new possibilities for experimental patterns. It is in such spirit that we may now summarize from the above the following variables and "situations," as it were, in order to direct future experimental endeavors as much close to nature as possible.

1. Physicochemical gradients: temperature gradients, pH gradients, Eh gradients, salinity gradients, ionic force gradients, pCO_2 gradients, and O_2 and S_2 fugacity gradients (virtually all hydrothermal systems in early sections of this paper). All natural hydrothermal systems show strong natural variations (Hazen and Sverjensky 2010) in all these cases with respect to (A) upwelling hydrothermal fluids themselves, and (B) the contrast between them and the environment they may encounter (whether it is a rock, air, seawater, a lake, or a puddle). As consequence, the value distribution of all variables may behave fractally across space and time—for instance, temperature decreases and pH increases from the central to the peripheral vents in VMS and SEDEX mounds, but they also do so in the lifetime of a particular vent and as the whole hydrothermal system wanes. The most obvious differences that arise for a given variable when comparing two or more types of hydrothermal systems have to do with how extreme is the range of variation of a given physicochemical variable, as such general ranges overlap in all the types of systems considered hereby. For example, all ranges of temperatures in Table 5.4.2 are those of

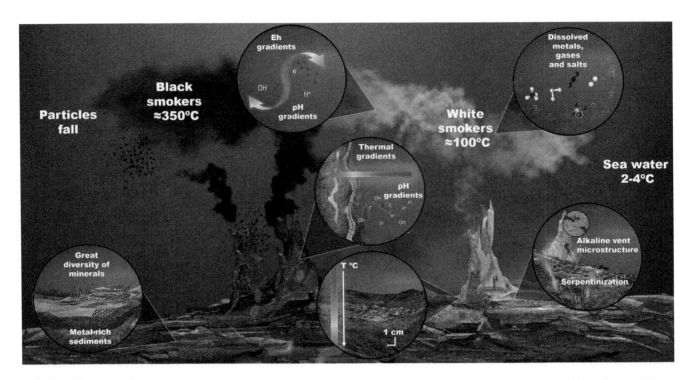

FIGURE 5.4.5 Representation of structures, processes, and gradients of key variables on the seafloor in association with hydrothermal systems of the volcanogenic massive sulfide (VMS) type, as of Figure 5.4.1.

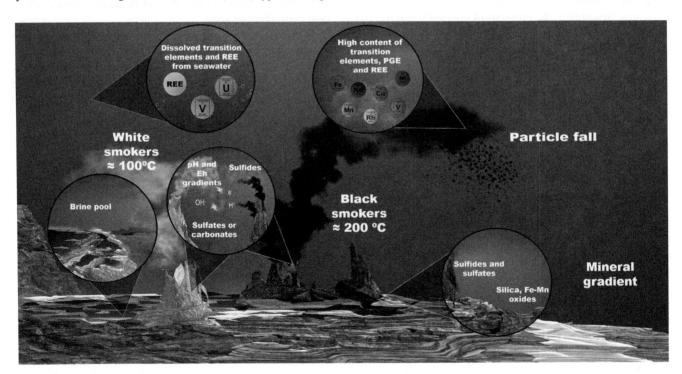

FIGURE 5.4.6 Representation of structures, processes, and gradients of key variables on the seafloor in association with hydrothermal systems of the sedimentary-exhalative (SEDEX) type, as of Figure 5.4.2.

typical stages of mineral deposition (ore associations in particular), but the general variation of temperature stretches down to the environmental temperature for each case (~0°C in deep seawater and ~25°C in continental systems). In addition, all the (paleo-) hydrothermal systems considered in this paper have

a multi-episodic behavior, which means that each hydrothermal pulse has physicochemical characteristics that may differ greatly from those of preceding or later pulses.

2. Mineralogical gradients across a mineral deposit or compositional gradients within the same mineral

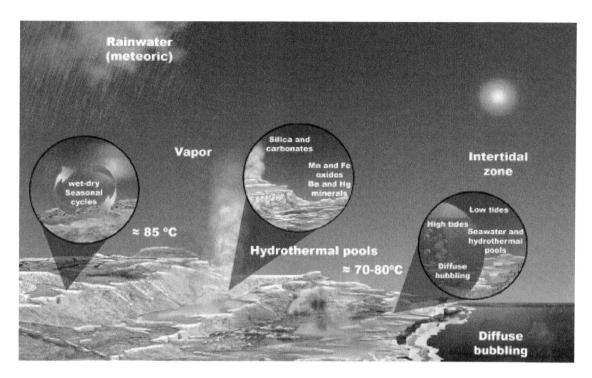

FIGURE 5.4.7   Representation of structures, processes, and gradients of key variables in shallow submarine to subaerial hydrothermal systems that occur during early stages of continental rifting, as of Figure 5.4.3.

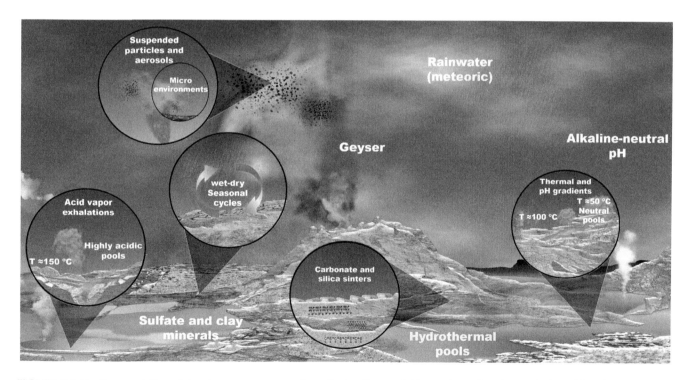

FIGURE 5.4.8   Representation of structures, processes, and gradients of key variables on the surface in association with active geothermal fields and epithermal deposits, as of Figure 5.4.4.

(virtually all systems). Most hydrothermal systems share minerals from about the same mineralogical hierarchical categories (silica, carbonates, sulfates, sulfides, clay minerals, etc.), but the distribution of such minerals may vary among different types of hydrothermal systems and even among different vents of the same system. Strong mineralogical contrast may occur at very variable stretches, from millimeters to meters, even fractally in the same deposit (*e.g.*, within the internal banding of a single mineralized chimney and on the surface from core to rim of venting in a VMS mineralized vent). There is no straightforward explanation for such variation, which is multifactorial in nature. It may even happen that high-temperature fluids cross low-temperature mineral associations and vice versa. Additionally, minerals show natural variations in composition that may vary among different types of hydrothermal systems (*i.e.*, sphalerite may attain very different Fe contents) and even down to crystal scale.

3. Grain-size gradients (virtually all systems). Ever since their formation, crystals may naturally vary between coarse-grained to cryptocrystalline, even into amorphous phases. This is an important aspect for experimental matters because as grain size decreases, the overall specific surface of a mineral aggregate increases. Also, the grain size of a mineral aggregate can be relatively homogeneous or inhomogeneous. Grain size may be highly variable since the very moment in which a mineral aggregate is deposited, but grain sizes can be greatly modified during the life span of a hydrothermal system. Indeed, minerals can be "ground" by fracturing and deformed by diapiric ascent or slope sliding; the grain size of a mineral may increase due to thermal recrystallization (or replacement by another mineral), and gels and amorphous phases may eventually crystallize. However, these processes do not necessarily occur during the life span of a hydrothermal system, and these are issues that need to be addressed case by case.

4. Occurrence of silica and/or Fe-Mn oxide gels, or colloidal phases (VMS, shallow subaqueous systems, epithermal, black shales, SEDEX, Algoma-type BIF). Can the presence of silica and other inorganic species from gels have any role in the oligomerization of organic species? It has been demonstrated that amino acids interact with the surface of silica by different mechanism (Rimola et al. 2013 and reference therein). Moreover, once adsorbed, amino acids react to give DKP cyclic intermediates, which constitute a full research area, the study of the thermal transformations of amino acids catalyzed by silica (Rimola et al. 2013 and reference therein).

5. Occurrence of liquid interfaces with highly contrasting physicochemical characteristics. Such is the case of the interfaces among submarine brine pools, less saline hydrothermal fluids, and even less saline and cooler seawater (VMS and SEDEX); among some of these and hot molten sulfur (VMS and submarine volcanoes); or between two aqueous fluids with contrasting salinities, which will not mix easily on the continental surface (geothermal/epithermal). The experimental work concerning prebiotic synthesis is generally carried out between a liquid phase with dissolved organic molecules and a solid phase, which is the most common situation in natural hydrothermal systems. However, it is yet to be tested whether some of the situations described previously may have a role in prebiotic chemistry.

6. Seasonal gradients for water supply (epithermal and other continental systems) or invasion/retreat of seawater upon low and high tides twice a day (shallow subaqueous systems). These gradients imply the variation in time of water availability either by (A) seasonality of meteoric recharge that may not compensate the evaporation of puddled water, or (B) the occurrence of puddles in intertidal areas filled with upwelling hydrothermal fluids and variable amounts of seawater. The relevance and potential of fluctuations in wet/dry or hydration/dehydration cycles for prebiotic synthesis have been already pointed out by Damer and Deamer (2015).

7. Occurrence of diffuse bubbling (epithermal, shallow subaqueous systems, including exceptionally shallow VMS and SEDEX systems) or aerosols with or without the occurrence of minute mineral particles alongside them in association with geysers, mud pots, or phreatomagmatic eruptions (geothermal/epithermal and other subaerial to shallow submarine/sublacustrine hydrothermal manifestations). It has been demonstrated that in bubbles and aerosols, many chemical reactions occur, and both types of interfaces represent heterogeneous microenvironments. Bubbles become more stable as they adsorb dissolved materials, thus concentrating these; afterward, the stabilized bubbles that contain organics act as nucleation sites for larger bubbles. This mechanism increases the concentration of organics (Lerman 2010) that could react to form more complex molecules.

8. Transition metals, which are available in aqueous solution elsewhere in all the hydrothermal systems described in the previous sections, may be effective catalysts for abiotic reactions that are involved in prebiotic synthesis by means of natural gases at temperatures $\leq 200°C$ (Horita 2001).

5.4.6 FINAL CONSIDERATIONS

Hydrothermal systems in oceanic (submarine) or continental settings (subaerial, sublacustrine) may have played an important role in the synthesis of organic compounds in the early Earth. In order to understand the role of the various geological environments in prebiotic chemistry, a multidisciplinary approach is necessary, which includes chemistry and geology. There are many environments, both ancient and recent, that can be called hydrothermal. Although many experiments have been conducted to simulate the conditions of the most conspicuous environments (mostly VMS-type systems, both white and black smokers), little effort has been devoted to understanding the possible contributions to organic synthesis stemming from other likely hydrothermal environments. In this work, the possible role of other hydrothermal environments is highlighted, provided that their characteristics also provide physicochemical variables, a broad variety of minerals, and a natural variability that could have been relevant in the synthesis of organic molecules from inorganic compounds. In order to contribute to a consensus on whether the participation of hydrothermal environments to chemical evolution was constructive or destructive, it is necessary to design more accurate experiments that combine not only the known physicochemical variables, but also their gradients in space and time. In this sense, recent literature in this matter increasingly aims at reproducing physicochemical conditions that are closer to natural gradients than earlier work. Although multivariable approaches to prebiotic chemistry are hard to manage, they will surely prove to be worthwhile endeavors, especially those that get as close as possible to the natural diversity of hydrothermal systems. In addition, the approach on which this paper relies provides many possibilities for conducting experiments that use heterogeneous interfaces of many kinds that would be firmly based on actual natural hydrothermal systems.

ACKNOWLEDGMENTS

The support of DGAPA-PAPIIT (No. IG100116 and IA203217) is acknowledged.

REFERENCES

Abdelmoez, W., Nakahasi, T., and H. Yoshida. 2007. Amino acid transformation and decomposition in saturated subcritical water conditions. *Ind. Eng. Chem. Res.* 46: 5286–5294.

Abelson, P. H. 1954. Amino acids in fossils. *Science* 119: 576.

Alargov, D., Deguchi, S., Tsujii, K., and K. Horikoshi. 2002. Reaction behaviors of glycine under super- and subcritical water conditions. *Orig. Life Evol. Biosph.* 32: 1–12.

Albinson, T., Norman, D. I., Cole, D., and B. A. Chomiak. 2001. Controls on formation of low-sulfidation epithermal deposits in Mexico: Constraints from fluid inclusion and stable isotope data. In: *New Mines and Discoveries in Mexico and Central America*, T. Albinson and C. E. Nelson (Eds.), *Soc. Econ. Geol. Spec. Publ.* pp. 1–32. No. 8. Boulder, CO.

Allen, R. L., Weihed, P., Blandell, D. et al. 2002. Global comparisons of volcanic-associated massive sulphide districts. *Geol. Soc. Lond. Spec. Publ.* 204: 13–37.

Alt, J. C., and W. C. Shanks. 1998. Sulfur in serpentinized oceanic peridotites: Serpentinization processes and microbial sulfate reduction. *J. Geophys. Res. B: Solid Earth* 103: 9917–9929.

Alt, J. C., and W. C. Shanks. 2003. Serpentinization of abyssal peridotites from the MARK area, Mid-Atlantic Ridge: Sulfur geochemistry and reaction modeling. *Geochim. Cosmochim. Acta* 67: 641–653.

Alt, J. C., and W. C. Shanks. 2006. Stable isotope compositions of serpentinite seamounts in the Mariana forearc: Serpentinization processes, fluid sources and sulfur metasomatism. *Earth Planet. Sci. Lett.* 242: 272–285.

Andersson, E., and N. G. Holm. 2000. The stability of some selected amino acids under attempted redox constrained hydrothermal conditions. *Orig. Life Evol. Biosph.* 30: 9–23.

Aubrey, A. D., Cleaves, H. J., and J. L. Bada. 2009. The role of submarine hydrothermal systems in the synthesis of amino acids. *Orig. Life Evol. Biosph.* 39: 91–108.

Baaske, P., Weinert, F. M., Duhr, S., Lemke, K. H., Russell, M. J., and D. Braun. 2007. Extreme accumulation of nucleotides in simulated hydrothermal pore systems. *PNAS* 104: 9346–9351.

Bach, W., Garrido, C. J., Paulick, H., Harvey, J., and M. Rosner. 2004. Seawater-peridotite interactions: First insights from ODP Leg 209, MAR 15 N. *Geochem. Geophys. Geosyst.* 5: Q09F26.

Bach, W., Paulick, H., Garrido, C. J., Ildefonse, B., Meurer, W. P., and S. E. Humphris. 2006. Unraveling the sequence of serpentinization reactions: Petrography, mineral chemistry, and petrophysics of serpentinites from MAR 15 N (ODP Leg 209, Site 1274). *Geophys. Res. Lett.* 33: L13306.

Bada, J. L., and A. Lazcano. 2002. Some like it hot, but not the first biomolecules. *Science* 296: 1982–1983.

Bada, J. L., and A. Lazcano. 2009. The origin of life. In *Evolution: The First Four Billion Years*, M. Ruse and J. Travis (Eds.), pp. 49–79. Cambridge, MA: Harvard University Press.

Bada, J. L., Miller, S. L., and M. Zhao. 1995. The stability of amino acids at submarine hydrothermal vent temperatures. *Orig. Life Evol. Biosph.* 25: 111–118.

Ballard, R. D., and T. H. Van Andel. 1977. Morphology and tectonics of the inner rift valley at lat 36°50′N on the Mid-Atlantic Ridge. *GSA Bull.* 88: 507–530.

Balodis, E., Madekufamba, M., Trevani, L. N., and P. R. Tremaine. 2012. Ionization constants and thermal stabilities of uracil and adenine under hydrothermal conditions as measured by in situ UV–visible spectroscopy. *Geochim. Cosmochim. Acta.* 93: 182–204.

Barge, L. M., Doloboff, I. J., Russell, M. J., Vander Velde, D., White, L. M., Stucky, G. D., Baum, M. M., Zeytounian, J., Kidd, R., and I. Kanik. 2014. Pyrophosphate synthesis in iron mineral films and membranes simulating prebiotic submarine hydrothermal precipitates. *Geochim. Cosmochim. Acta* 128: 1–12.

Baross, J. A., and S. E. Hoffman. 1985. Submarine hydrothermal vents and associated gradient environments as sites for the origin and evolution of life. *Orig. Life Evol. Biosph.* 15: 327–345.

Barrie, C. T., and M. D. Hannington. 1999. Classification of volcanic-associated massive sulfide deposits based on host-rock deposition. *Rev. Econ. Geol.* 8: 1–11.

Bates, A. E., Lee, R. W., Tunnicliffe, V., and M. D. Lamare. 2010. Deep-sea hydrothermal vent animals seek cool fluids in a highly variable thermal environment. *Nat. Commun.* 1: 14.

Becerra, A., Delaye, L., Islas, S., and A. Lazcano. 2007. The very early stages of biological evolution and the nature of the last common ancestor of the three major cell domains. *Annu. Rev. Ecol. Evol. Syst.* 38: 361–379.

Beinlich, A., Plümper, O., Hövelmann, J., Austrheim, H., and B. Jamtveit. 2012. Massive serpentinite carbonation at Linnajavri, N–Norway. *Terra Nova* 24: 446–455.

Bell, J. L. S., Palmer, D. A., Barnes, H. L., and S. E. Drummond. 1994. Thermal decomposition of acetate: III. Catalysis by mineral surfaces. *Geochim. Cosmochim. Acta* 58: 4155–4177.

Bernhardt, G., Lüdemann, H. D., Jaenicke, R., König, H., and K. O. Stetter. 1984. Biomolecules are unstable under "black smoker" conditions. *Naturwissenschaften* 71: 583–586.

Bowring, S. A., and I. S. Williams. 1999. Priscoan (4.00–4.03 Ga) orthogneisses from northwestern Canada. *Contrib. Mineral. Petrol.* 134: 3–16.

Bowring, S., Housh, T., and C. Isachsen. 1990. *The Acasta Gneisses: Remnant of Earth's Early Crust*, Vol. 1, pp. 319–343. LPI Conference on the Origin of the Earth.

Boyce, A. J., Coleman, M. L., and M. J. Russell. 1983. Formation of fossil hydrothermal chimneys and mounds from Silvermines, Ireland. *Nature* 306: 545–550.

Braun, D., and A. Libchaber. 2002. Trapping of DNA by thermophoretic depletion and convection. *Phys. Rev. Lett.* 89: 188103.

Buchanan, L. J. 1981. Precious metal deposits associated with volcanic environments in the Southwest. In: *Relations of Tectonics to Ore Deposits in the Southern Cordillera*, Vol. 14, W. R. Dickinson and W. D. Payne (Eds.), pp. 237–262. Tucson, AZ: Arizona Geological Society Digests.

Budin, I., Bruckner, R. J., and J. W. Szostak. 2009. Formation of protocell-like vesicles in a thermal diffusion column. *J. Am. Chem. Soc.* 131: 9628–9629.

Burcar, B. T., Barge, L. M., Trail, D., Watson, E. B., Russell, M. J., and L. B. McGown. 2015. RNA oligomerization in laboratory analogues of alkaline hydrothermal vent systems. *Astrobiology* 15: 509–522.

Caird, R. A., Pufahl, P. K., Hiatt, E. E., Abram, M. B., Rocha, A. J. D., and T. K. Kyser. 2017. Ediacaran stromatolites and intertidal phosphorite of the Salitre Formation, Brazil: Phosphogenesis during the Neoproterozoic Oxygenation Event. *Sediment. Geol.* 350: 55–71.

Camprubí, A. 2013. Tectonic and metallogenic history of Mexico. In *Tectonics, Metallogeny, and Discovery: The North American Cordillera and Similar Accretionary Settings*, Vol. 17, M. Colpron, T. Bissig, B. G. Rusk, and J. F. H. Thompson (Eds.), pp. 201–243. Boulder, CO: Society of Economic Geologist Special Publications.

Camprubí, A., and T. Albinson. 2006. Depósitos epitermales en México: Actualización de su conocimiento y reclasificación empírica. *Bol. Soc. Geol. Mex.* 58: 27–81.

Camprubí, A., and T. Albinson. 2007. Epithermal deposits in México – an update of current knowledge, and an empirical reclassification. In *Geology of México: Celebrating the Centenary of the Geological Society of Mexico*, Vol. 422, S. A. Alaniz-Álvarez and A. F. Nieto-Samaniego (Eds.), pp. 377–415. Boulder, CO: Special Paper of the Geological Society of America.

Camprubí, A., Canet, C., Rodríguez-Díaz, A. A., Prol-Ledesma, R. M., Villanueva-Estrada, R. E., Blanco-Florido, D., and A. López-Sánchez. 2008. Geology, mineral deposits and hydrothermal activity in Bahía Concepción, Baja California Sur, Mexico. *Island Arc* 17: 6–25.

Camprubí, A., González-Partida, E., Torró, L., Alfonso, P., Miranda-Gasca, M. A., Martini, M., Canet, C., and F. González-Sánchez. 2017. Mesozoic volcanogenic massive sulfide (VMS) deposits in Mexico. *Ore Geol. Rev.* 81: 1066–1083.

Canet, C., and R. M. Prol-Ledesma. 2006. Procesos de mineralización en manantiales hidrotermales submarinos someros, ejemplos en México. *Bol. Soc. Geol. Mex.* 58: 83–102.

Canet, C., Prol-Ledesma, R. M., Dando, P. R., Vázquez-Figueroa, V., Shumilin, E., Birosta, E., Sánchez, A., Robinson, C. J., Camprubí, A., and E. Tauler. 2010. Discovery of massive gas seepage along the Wagner Fault, northern Gulf of California. *Sediment. Geol.* 228: 292–303.

Cavosie, A., Valley, J., and S. Wilde. 2005. Magmatic δ18O in 4400–3900 Ma detrital zircons: A record of the alteration and recycling of crust in the Early Archean. *Earth Planet. Sci. Lett.* 235: 663–681.

Chatterjee, S. 2018. Hydrothermal impact crater lakes and the origin of life. In *Handbook of Astrobiology*, V. Kolb (Ed.). Boca Raton, FL: Taylor & Francis Group.

Chevaldonné, P., Desbruyères, D., and M. L. Haître. 1991. Time-series of temperature from three deep-sea hydrothermal vent sites. *Deep-Sea Res. A. Oceanogr. Res. Pap.* 38: 1417–1430.

Chyba, C., and C. Sagan. 1992. Endogenous production, exogenous delivery and impact-shock synthesis of organic molecules: An inventory for the origins of life. *Nature* 355: 125–132.

Cleaves, H. 2013. Prebiotic chemistry: Geochemical context and reaction screening. *Life* 3: 331–345.

Cleaves II, H. J., Scott, A. M., Hill, F. C., Leszczynski, J., Sahai, N., and R. Hazen. 2012. Mineral–organic interfacial processes: Potential roles in the origins of life. *Chem. Soc. Rev.* 41: 5502–5525.

Cnossen, I., Sanz-Forcada, J., Favata, F., Witasse, O., Zegers, T., and N. F. Arnold. 2007. Habitat of early life: Solar X-ray and UV radiation at Earth's surface 4–3.5 billion years ago. *J. Geophys. Res. Planets* 112: E02008.

Cockell, C. S. 2006. The origin and emergence of life under impact bombardment. *Philos. Trans. R. Soc. Lond. Ser. B.* 361: 1845–1856.

Coffin, M. F., and O. Eldholm. 1994. Large igneous provinces: Crustal structure, dimensions, and external consequences. *Rev. Geophys.* 32: 1–36.

Colín-García, M., Heredia, A., Cordero, G., Camprubí, A., Ortega-Gutiérrez, F., Negrón-Mendoza, A., Beraldi, H., and S. Ramos-Bernal. 2016. Hydrothermal vents and prebiotic chemistry: A review. *Bol. Soc. Geol. Mex.* 68: 599–620.

Conly, A. G., Scott, S. D., and H. Bellon. 2011. Metalliferous manganese oxide mineralization associated with the Boléo Cu–Co–Zn district, Mexico. *Econ. Geol.* 106: 1173–1196.

Corbett, G. J., and T. M. Leach. 1998. *Southwest Pacific Rim Gold-Copper Systems: Structure, Alteration and Mineralisation.* Boulder, CO: Society of Economic Geologists. Special Publications No. 6. 238 p.

Corliss, J. B., Dymond, J., Gordon, L. I. et al. 1979. Submarine thermal aprirngs on the galápagos rift. *Science* 203: 1073–1083.

Corliss, J. B., Baross, J. A., and Hoffman, S. E. 1980. Submarine hydrothermal systems: A probable site for the origin of life. Corvallis, OR: School of Oceanography, Oregon State University.

Corliss, J. B., Baross, J. A., and S. E. Hoffman. 1981. An hypothesis concerning the relationship between submarine hot springs and the origin of life on Earth. *Oceanol. Acta* 4: 59–69.

Cox, J. S., and T. M. Seward. 2007a. The reaction kinetics of alanine and glycine under hydrothermal conditions. *Geochim. Cosmochim. Acta* 71: 2264–2284.

Cox, J. S., and T. M. Seward. 2007b. The hydrothermal reaction kinetics of aspartic acid. *Geochim. Cosmochim. Acta* 71: 797–820.

Cox, J. S., and T. M. Seward. 2007c. The reaction kinetics of alanine and glycine under hydrothermal conditions. *Geochim. Cosmochim. Acta* 71: 2264–2284.

Cuadros, J., Aldega, L., Vetterlein, J., Drickamer, K., and W. Dubbin. 2009. Reactions of lysine with montmorillonite at 80°C: Implications for optical activity, H+ transfer and lysine–montmorillonite binding. *J. Colloid Interface Sci.* 333: 78–84.

Da Silva, L., Maurel, M. C., and D. Deamer. 2015. Salt-promoted synthesis of RNA-like molecules in simulated hydrothermal conditions. *J. Mol. Evol.* 80: 86–97.

Dalai, P., Pleyer, H. L., Strasdeit, H., and S. Fox. 2017. The influence of mineral matrices on the thermal behavior of glycine. *Orig. Life Evol. Biosph.* 47: 427–452.

Damer, B. and D. Deamer. 2015. Coupled phases and combinatorial selection in fluctuating hydrothermal pools: A scenario to guide experimental approaches to the origin of cellular life. *Life* 5: 872–887.

de Ronde, C. E. J., Chadwick, W. W. Jr., Ditchburn, R. G., Embley, R. W., Tunnicliffe, V., Baker, E. T., Walker, S. L., Ferrini, V. L., and S. M. Merle. 2015. Molten sulfur lakes of intraoceanic arc volcanoes. In *Volcanic Lakes*, D. Rouwet, B. Christenson, F. Tassi, and J. Vandemeulebrouck (Eds.), pp. 261–288. Berlin, Germany: Springer-Verlag.

de Wit, M. J. 2004. *Archean Greenstone Belts Do Contain Fragments of Ophiolites, Developments in Precambrian Geology*, Vol. 13, pp. 599–614. Elsevier.

DeGuzman, V., Vercoutere, W., Shenasa, H., and D. Deamer. 2014. Generation of oligonucleotides under hydrothermal conditions by non-enzymatic polymerization. *J. Mol. Evol.* 78: 251–262.

Delaye, L., and A. Lazcano. 2005. Prebiological evolution and the physics of the origin of life. *Phys. Life Rev.* 2: 47–64.

Deschamps, F., Godard, M., Guillot, S., and K. Hattori. 2013. Geochemistry of subduction zone serpentinites: A review. *Lithos* 178: 96–127.

Ding, K., Seyfried, W. E., Zhang, Z., Tivey, M. K., Von Damm, K. L., and A. M. Bradley. 2005. The in situ pH of hydrothermal fluids at mid-ocean ridges. *Earth Planet. Sci. Lett.* 237: 167–174.

Draganić, I. G., Bjergbakke, E., Draganić, Z. D., and K. Sehested. 1991. Decomposition of ocean waters by potassium-40 radiation 3800 Ma ago as a source of oxygen and oxidizing species. *Precambrian Res.* 52: 337–345.

Estrada, C. F., Mamajanov, I., Hao, J., Sverjensky, D. A., Cody, G. D., and R. M. Hazen. 2017. Aspartate transformation at 200°C with brucite [Mg(OH)$_2$], NH$_3$ and H$_2$: Implications for prebiotic molecules in hydrothermal systems. *Chem. Geol.* 457: 162–172.

Falk, E. S., and P. B. Kelemen. 2015. Geochemistry and petrology of listvenite in the Samail oph iolite, Sultanate of Oman: Complete carbonation of peridotite during ophiolite emplacement. *Geochim. Cosmochim. Acta.* 160: 70–90.

Fan, D., Ye, J., and J. Li. 1999. Geology, mineralogy, and geochemistry of the Middle Proterozoic Wafangzi ferromanganese deposit, Liaoning Province, China. *Ore Geol. Rev.* 15: 31–53.

Ferrari, F., and E. Szuszkiewicz. 2009. Cosmic rays: A review for astrobiologists: *Astrobiology* 9: 413–436.

Franiatte, M., Richard, L., Elie, M., Nguyen-Trung, C., Perfetti, E., and D. E. LaRowe. 2008. Hydrothermal stability of adenine under controlled fugacities of N$_2$, CO$_2$ and H$_2$. *Orig. Life Evol. Biosph.* 38: 139–148.

Franklin, J. M. 1996. Volcanic-associated massive sulphide base metals. In: *Geology of Canadian Mineral Deposit Types*, Vol. 8, O. R. Eckstrand, W. D Sinclair, and R. I. Thorpe (Eds.), pp. 158–183. Ottawa, Canada: Geological Survey of Canada, Geology of Canada.

Franklin, J. M., Gibson, H. L., Jonasson, I. R., and A. G. Galley. 2005. Volcanogenic massive sulfide deposits. In *Economic Geology One Hundredth Anniversary*, Vol. 1905–2005, J. W. Hedenquist, J. F. H. Thompson, and R. J. Goldfarb (Eds.), pp. 523–560. Littleton, CO: Society of Economic Geologist.

Franklin, J. M., Lydon, J. W., and D. M. Sangster. 1981. Volcanic-associated massive sulfide deposits. In *Economic Geology Seventy-Fifth Anniversary*, B. J. Skinner (Ed.), pp. 485–627. Lancaster, PA: Economic Geology Publishing Company.

Frost, B. R. 1985. On the stability of sulfides, oxides, and native metals in serpentinite. *J. Petrol.* 26: 31–63.

Frost, B. R., Evans, K. A., Swapp, S. M., Beard, J. S., and F. E. Mothersole. 2013. The process of serpentinization in dunite from New Caledonia. *Lithos* 178: 24–39.

Früh-Green, G. L., Connolly, J. A., Plas, A., Kelley, D. S., and B. Grobety. 2004. Serpentinization of oceanic peridotites: Implications for geochemical cycles and biological activity. In *The Subseafloor Biosphere at Mid-ocean Ridges*, W. S. D Wilcock, E. F., Delong, D. S. Kelley, J. A. Baross, and S. C. Cary (Eds.), pp. 119–136. Washington, DC: American Geophysical Union.

Fuchida, S., Naraoka, H., and H. Masuda. 2017. Formation of diastereoisomeric piperazine-2,5-dione from DL-Alanine in the Presence of Olivine and Water. *Orig. Life Evol. Biosph.* 47: 83–92.

Furnes, H., de Wit, M., and Y. Dilek. 2014. Precambrian greenstone belts host different ophiolite types. In *Evolution of Archean Crust and Early Life, Modern Approaches in Solid Earth Sciences*, Vol. 7, Y. Dilek and H. Furnes (Eds.), pp. 1–22. Dordrecht, the Netherlands: Springer.

Gladman, B., Dones, L., Levison, H. F., and J. A. Burns. 2005. Impact seeding and reseeding in the inner solar system. *Astrobiology* 5: 483–496.

Goodfellow, W. D. 1992. Chemical evolution of the oceans as discerned from the temporal distribution of sedimentary exhalative (SEDEX) Zn-Pb-Ag deposits. In *International Geological Congress*, Kyoto, Japan, Programs and Abstracts, p. 185.

Goodfellow, W. D., Lydon, J. W., and R. J. W. Turner. 1993. Geology and genesis of stratiform sediment-hosted (SEDEX) zinc-lead-silver sulphide deposits. *Geol. Assoc. Can. Spec. Pap.* 40: 201–252.

Grenne, T., and J. F. Slack. 2003. Paleozoic and Mesozoic silica-rich seawater: Evidence from hematitic chert (jasper) deposits. *Geology* 31: 319–322.

Hannington, M. D., de Ronde, C. E. J., and S. Petersen. 2005. Sea-floor tectonics and submarine hydrothermal systems. In *Economic Geology One Hundredth Anniversary*, Vol. 1905–2005, J. W. Hedenquist, J. F. H. Thompson, and R. J. Goldfarb (Eds.), pp. 111–142. Littleton, CO: Society of Economic Geologists.

Hansen, L. D., Dipple, G. M., Gordon, T. M., and D. A. Kellett. 2005. Carbonated serpentinite (listwanite) at Atlin, British Columbia: A geological analogue to carbon dioxide sequestration. *Can. Mineral.* 43: 225–239.

Hazen, R. M., and D. A Sverjensky. 2010. Mineral surfaces, geochemical complexities, and the origins of life. *Cold Spring Harb. Perspect Biol.* 2: a002162.

Hedenquist, J. W., Arribas, A. Jr., and E. Urien-Gonzalez. 2000. Exploration for epithermal gold deposits. *Rev. Econ. Geol.* 13: 245–277.

Hedenquist, J. W., and Y. A. Taran. 2013. Modeling the formation of advanced argillic lithocaps: Volcanic vapor condensation above porphyry intrusions. *Econ. Geol.* 108: 1523–1540.

Hinsken, T., Bröcker, M., Strauss, H., and F. Bulle. 2017. Geochemical, isotopic and geochronological characterization of listvenite from the Upper Unit on Tinos, Cyclades, Greece. *Lithos* 282: 281–297.

Holland, H. D. 2003. The geologic history of seawater. In *Treatise on Geochemistry*, Vol. 6, K. K. Turekian and H. D. Holland (Eds.), pp. 583–625. Oxford, UK: Elsevier Science.

Holm, N. G. 1992. Why are hydrothermal systems proposed as plausible environments for the origin of life? In *Marine Hydrothermal Systems and the Origin of Life: Report of SCOR Working Group 91*, N. G. Holm (Ed.), pp. 5–14. Dordrecht, the Netherlands: Springer.

Holm, N. G., and E. Andersson. 2005. Hydrothermal simulation experiments as a tool for studies of the origin of life on Earth and other terrestrial planets: A review. *Astrobiology* 5: 444–160.

Horita, J. 2001. Carbon isotope exchange in the system CO$_2$–CH$_4$ at elevated temperatures. *Geochim. Cosmochim. Acta* 65: 1907–1919.

Huber, C., and G. Wächtershäuser. 1998. Peptides by activation of amino acids with CO on (Ni, Fe)S surfaces. *Orig. Life Evol. Biosph.* 281: 670–672.

Iizuka, T., Horie, K., Komiya, T., Maruyama, S., Hirata, T., Hidaka, H., and B. F. Windley. 2006. 4.2 Ga zircon xenocryst in an Acasta gneiss from northwestern Canada: Evidence for early continental crust. *Geology* 34: 245–248.

Iizuka, T., Komiya, T., and S. Maruyama. 2007. Chapter 3.1: The early Archean Acasta Gneiss Complex: Geological, geochronological and isotopic studies and implications for early crustal evolution. In *Developments in Precambrian Geology*, M. J. van Kranendonk, R. H. Smithies, and V. C. Bennett (Eds.), pp. 127–147. Amsterdam, the Netherlands: Elsevier.

Imai, E.-I., Honda, H., Hatori, K., Brack, A., and K. Matsuno. 1999. Elongation of oligopeptides in a simulated submarine hydrothermal system. *Science* 283: 831–833.

Islam, M. N., Kaneko, T., and K. Kobayashi. 2003. Reaction of amino acids in a supercritical water-flow reactor simulating submarine hydrothermal systems. *Bull. Chem. Soc. Jpn.* 76: 1171–1178.

Ito, M., Gupta, L. P., Masuda, H., and H. Kawahata. 2006. Thermal stability of amino acids in seafloor sediment in aqueous solution at high temperature. *Org. Geochem.* 37: 177–188.

Ito, M., Yamaoka, K., Masuda, H., Kawahata, H., and L. P. Gupta. 2009. Thermal stability of amino acids in biogenic sediments and aqueous solutions at seafloor hydrothermal temperaturas. *Geochem. J.* 43: 331–341.

Jackson, J. B. 2016. Natural pH gradients in hydrothermal alkali vents were unlikely to have played a role in the origin of life. *J. Mol. Evol.* 83: 1–11.

Jacobsen, S. B. 2003. How old is planet Earth? *Science* 300: 1513–1514.

James, R. H., Green, D. R. H., Stock, M. J., Alker, B. J., Banerjee, N. R., Cole, C., German, C. R., Huvenne, V. A. I., Powell, A. M., and D. P. Connelly. 2014. Composition of hydrothermal fluids and mineralogy of associated chimney material on the East Scotia Ridge back-arc spreading centre. *Geochim. Cosmochim. Acta* 139: 47–71.

Jébrak, M., and E. Marcoux. 2008. *Géologie des ressources minérales*. Québec, Canada: Ministère des ressources naturelles et de la faune du Québec, 667 p.

Ji, F., Zhou, H., and Q. Yang. 2008. The abiotic formation of hydrocarbons from dissolved co_2 under hydrothermal conditions with cobalt-bearing magnetite. *Orig. Life Evol. Biosph.* 38: 117–125.

Johnson, S. C., Large, R. R., Coveney, R. M., Kelley, K. D., Slack, J. F., Steadman, J. A., Gregory, D. D., Sack, P. J., and S. Meffre. 2017. Secular distribution of highly metalliferous black shales corresponds with peaks in past atmosphere oxygenation. *Miner. Deposita* 52: 791–798.

Jorge, S., Melgarejo, J. C., and M. P. Alfonso. 1997. Sedimentos exhalatixos y sus derivados metamórficos. In *Atlas de Asociaciones Minerales en Lámina Delgada*, J. C. Melgarejo (Ed.), pp. 287–308. Barcelona, Spain: Edicions Universitat de Barcelona – Fundació Folch.

Juteau, T., and R. Maury. 1997. *Géologie de la Croûte Océanique*. Petrologie et Dynamique: Endogénes, 569 p. Dunod, Montrouge (Haus-de-Seine), France.

Kasting, J. F. 1993a. Earth's early atmosphere. *Science* 259: 920–926.

Kasting, J. F. 1993b. Evolution of the Earth's atmosphere and hydrosphere. In *Organic Geochemistry Topics in Geobiology*, Vol. 11. M. H. Engel, and S. A. Macko (Eds.), pp. 611–623. Boston, MA: Springer.

Kawamura, K. 2005. Behaviour of RNA under hydrothermal conditions and the origins of life. *Int. J. Astrobiol.* 3: 301–309.

Kawamura, K. 2011. Development of micro-flow hydrothermal monitoring systems and their applications to the origin of life study on Earth. *Anal. Sci.* 27: 675–675.

Kawamura, K., and M. Shimahashi. 2008. One-step formation of oligopeptide-like molecules from Glu and Asp in hydrothermal environments. *Naturwissenschaften* 95: 449–454.

Kelemen, P. B., Matter, J., Streit, E. E., Rudge, J. F., Curry, W. B., and J. Blusztajn. 2011. Rates and mechanisms of mineral carbonation in peridotite: Natural processes and recipes for enhanced, in situ CO_2 capture and storage. *Annu. Rev. Earth Planet. Sci.* 39: 545–576.

Kelley, D. S., Karson, J. A., Blackman, D. K. et al. 2001. An off-axis hydrothermal vent field near the Mid-Atlantic Ridge at 30°N. *Nature* 412: 145–149.

Kelley, D. S., Baross, J. A., and Delaney, J. R. 2002. Volcanoes, fluids, and life at mid-ocean ridge spreading centers. *Annu. Rev. Earth Planet Sci.* 30: 385–491.

Klein, F., and W. Bach. 2009. Fe–Ni–Co–O–S phase relations in peridotite–seawater interactions. *J. Petr.* 50: 37–59.

Klein, F., Bach, W., Jöns, N., McCollom, T., Moskowitz, B., and T. Berquó. 2009. Iron partitioning and hydrogen generation during serpentinization of abyssal peridotites from 15 N on the Mid-Atlantic Ridge. *Geochim. Cosmochim. Acta* 73: 6868–6893.

Klein, F., Humphris, S. E., Guo, W., Schubotz, F., Schwarzenbach, E. M., and W. D. Orsi. 2015. Fluid mixing and the deep biosphere of a fossil Lost City-type hydrothermal system at the Iberia Margin. *PNAS* 112: 12036–12041.

Klingler, D., Berg, J., and H. Vogel. 2007. Hydrothermal reactions of alanine and glycine in sub- and supercritical water. *J. Supercrit. Fluids* 43: 112–119.

Knauth, L. P. 2005. Temperature and salinity history of the Precambrian ocean: Implications for the course of microbial evolution. *Palaeogeogr. Palaeoclimatol. Palaeoecol.* 219: 53–69.

Kohara, M., Gamo, T., Yanagawa, H., and K. Kobayashi. 1997. Stability of amino acids in simulated hydrothermal vent environments. *Chem. Lett.* 26: 1053–1054.

Komiya, T., Hirata, T., Kitajima, K., Yamamoto, S., Shibuya, T., Sawaki, Y., Ishikawa, T., Shu, D., Li, Y., and J. Han. 2008. Evolution of the composition of seawater through geologic time, and its influence on the evolution of life. *Gondwana Res.* 14: 159–174.

Komiya, T., Maruyama, S., Masuda, T., Nohda, S., Hayashi, M., and K. Okamoto. 1999. Plate tectonics at 3.8–3.7 Ga: Field evidence from the Isua accretionary complex, southern West Greenland. *J. Geol.* 107: 515–554.

Konn, C., Charlou, J. L., Holm, N. G., and O. Mousis. 2015. The production of methane, hydrogen, and organic compounds in ultramafic-hosted hydrothermal vents of the mid-atlantic ridge. *Astrobiology* 15: 381–399.

Lane, N. 2017. Proton gradients at the origin of life. *BioEssays* 39: 1600217.

Larralde, R., Robertson, M. P., and S. L. Miller. 1995. Rates of decomposition of ribose and other sugars: Implications for chemical evolution. *PNAS* 92: 8158–8160.

Larter, R. C. L., Boyce, A. J., and M. J. Russell. 1981. Hydrothermal pyrite chimneys for the Ballynoe baryte deposits, Silvermines, County Tipperary, Ireland. *Miner. Deposita* 16: 309–318.

Lazcano, A., and S. L. Miller. 1994. How long did it take for life to begin and evolve to cyanobacteria? *J. Mol. Evol.* 39: 546–554.

Laznicka, P. 2006. *Giant Metallic Deposits, Future Sources of Industrial Minerals*. Heidelberg, Germany: Springer, 732 p.

Leach, D. L., Sangster, D. F., Kelley, K. D., Large, R. R., Garven, G., Allen, C. R., Gutzmer, J., and S. Walters. 2005. Sediment-hosted lead-zinc deposits: A global perspective. In *Economic*

Geology One Hundredth Anniversary, Vol. 1905–2005, J. W. Hedenquist, J. F. H Thompson, and R. J. Goldfarb (Eds.), pp. 561–607. Littleton, CO: Society of Economic Geologists.

Lee, N., Foustoukos, D. I., Sverjensky, D. A., Cody, G. D., and R. M. Hazen. 2014. The effects of temperature, pH and redox state on the stability of glutamic acid in hydrothermal fluids. *Geochim. Cosmochim. Acta* 135: 66–86.

Lemke, K. H., Rosenbauer, R. J., and D. K. Bird. 2009. Peptide synthesis in early Earth hydrothermal systems. *Astrobiology* 9: 141–146.

Lerman, L. 2010. The primordial bubble: Water, symmetry-breaking, and the origin of life. In *Water and Life: The Unique Properties of Water*, R. M. Lynden-Bell, S. C. Morris, J. D. Barrow, J. L. Finney, and C. Harper (Eds.), pp. 259–290. Boca Raton, FL: CRC Press.

Levin, H. 2009. *The Earth through Time*. Hoboken, NJ: John Wiley & Sons, 624 p.

Levy, M., and S. L. Miller. 1998. The stability of the RNA bases: Implications for the origin of life. *PNAS* 95: 7933–7938.

Li, J., Wang, X., Klein, M. T., and T. B. Brill. 2002. Spectroscopy of hydrothermal reactions, 19: pH and salt dependence of decarboxylation of?-alanine at 280–330°C in an FT-IR spectroscopy flow reactor. *Int. J. Chem. Kinet.* 34: 271–277.

Lydon, J. W. 1988. Volcanogenic massive sulphide deposits, Part 2: Genetic models. In *Ore Deposit Models*, R. G. Roberts and P. A. Sheahan (Eds.), pp. 155–181. St. John's, Canada: Geological Association of Canada, Geoscience Canada, Reprint Series, 3.

Lydon, J. W. 1996. Sedimentary exhalative sulphides (sedex). In *Geology of Canadian Mineral Deposit Types*, Vol. 8, O. R. Eckstrand, W. D.Sinclair, and R. I. Thorpe (Eds.), pp. 130–152. Ottawa-Ontario, Canada: Geological Survey of Canada, Geology of Canada.

Lyons, T. W., Gellatly, A. M., McGoldrick, P. J., and L. C., Kah. 2006. Proterozoic sedimentary exhalative (SEDEX) deposits and links to evolving global ocean chemistry. *Geol. Soc. Am. Mem.* 198: 169–184.

Marshall, W. L. 1994. Hydrothermal synthesis of amino acids. *Geochim. Cosmochim. Acta* 58: 2099–2106.

Martin, W., Baross, J., Kelley, D., and M. J. Russell. 2008. Hydrothermal vents and the origin of life. *Nat. Rev. Microbiol.* 6: 805–814.

Mast, C. B., Schink, S., Gerland, U., and D. Braun. 2013. Escalation of polymerization in a thermal gradient. *PNAS* 110: 8030–8035.

McCollom, T. M. 2013. The influence of minerals on decomposition of the n-alkyl-α-amino acid norvaline under hydrothermal conditions. *Geochim. Cosmochim. Acta* 104: 330–357.

McCollom, T. M., Ritter, G., and Simoneit, B. R. T. 1999. Lipid synthesis under hydrothermal conditions by Fischer-Tropsch-type reactions. *Orig. Life Evol. Biosph.* 29: 153–166.

McCollom, T. M., and J. S., Seewald. 2003a. Experimental constraints on the hydrothermal reactivity of organic acids and acid anions: I. Formic acid and formate. *Geochim. Cosmochim. Acta* 67: 3625–3644.

McCollom, T. M., and J. S. Seewald. 2003b. Experimental study of the hydrothermal reactivity of organic acids and acid anions: II. Acetic acid, acetate, and valeric acid: *Geochim. Cosmochim. Acta* 67: 3645–3664.

McDermott, J. M., Seewald, J. S., German, C. R., and S. P. Sylva. 2015. Pathways for abiotic organic synthesis at submarine hydrothermal fields. *PNAS* 112: 7668–7672.

Miller, S. L. 1953. A production of amino acids under possible primitive Earth conditions. *Science* 117: 528–529.

Miller, S. L. 1993. The prebiotic synthesis of organic compounds on the early Earth. In *Organic Geochemistry: Principles and Applications*, M. H. Engel, and S. A. Macko (Eds.), pp. 625–636. Boston, MA: Springer US.

Miller, S. L., and L. E. Orgel. 1974. *The Origins of Life on the Earth*. Englewood Cliffs, NJ: Prentice-Hall.

Miller, S. L., and J. L. Bada. 1988. Submarine hot springs and the origin of life. *Nature* 334: 609–611.

Misra, K. C. 1999. *Understanding Mineral Deposits*. Dordrecht, the Netherlands: Kluwer Academic Publishers, 845 p.

Mojzsis, S. J., Arrhenius, G., McKeegan, K. D., Harrison, T. M., Nutman, A. P., and C. R. L. Friend. 1996. Evidence for life on Earth before 3,800 million years ago. *Nature* 384: 55–59.

Möller, F. M., Kriegel, F., Kieß, M., Sojo, V., and D. Braun. 2017. Steep pH gradients and directed colloid transport in a microfluidic alkaline hydrothermal pore. *Angew. Chem.* 56: 2340–2344.

Mossman, D. J., Gauthier-Lafaye, F., and S. E. Jackson. 2005. Black shales, organic matter, ore genesis and hydrocarbon generation in the Paleoproterozoic Franceville Series, Gabon. *Precambrian Res.* 137: 253–272.

Nasir, S., Al Sayigh, A. R., Al Harthy, A., Al-Khirbash, S., Al-Jaaidi, O., Musllam, A., Al-Mishwat, A., and S. Al-Bu'saidi. 2007. Mineralogical and geochemical characterization of listwaenite from the Semail Ophiolite, Oman. *Chem Erde.* 67: 213–228.

Niether, D., Afanasenkau, D., Dhont, J. K. G., and S. Wiegand. 2016. Accumulation of formamide in hydrothermal pores to form prebiotic nucleobases. *PNAS* 113: 4272–4277.

Nisbet, E. G., and N. H. Sleep. 2001. The habitat and nature of early life. *Nature* 409: 1083–1091.

Novikov, Y., and S. D. Copley. 2013. Reactivity landscape of pyruvate under simulated hydrothermal vent conditions. *PNAS* 110: 13283–13288.

Nutman, A. P., Allaart, J. H., Bridgwater, D., Dimroth, E., and M. Rosing. 1984. Stratigraphic and geochemical evidence for the depositional environment of the early Archaean Isua supracrustal belt, southern West Greenland. *Precambrian Res.* 25: 365–396.

Ogasawara, H., Yoshida, A., Imai, E., Honda, H., Hatori, K., and K. Matsuno. 2000. Synthesizing oligomers from monomeric nucleotides in simulated hydrothermal environments. *Orig. Life Evol. Biosph.* 30: 519–526.

Ogata, Y., Imai, E., Honda, H., Hatori, K., and K. Matsuno. 2000. Hydrothermal circulation of seawater through hot vents and contribution of interface chemistry to prebiotic synthesis: *Orig. Life Evol. Biosph.* 30: 527–537.

Ohmoto, H. 1996. Formation of volcanogenic massive sulfide deposits: The Kuroko perspective. *Ore Geol. Rev.* 10: 135–177.

Ohmoto, H., Watanabe, Y., Yamaguchi, K. E., Naraoka, H., Haruna, M., Kakegawa, T., Hayashi, K.-I., and Y. Kato. 2006. Chemical and biological evolution of early Earth: Constraints from banded iron formations. In *Evolution of Early Earth's Atmosphere, Hydrosphere, and Biosphere – Constraints from Ore Deposits*, Vol. 198, S. E., Kesler and H. Ohmoto (Eds.), pp. 291–333. Geological Society of America Memoir.

Papavassiliou, K., Voudouris, P., Kanellopoulos, C., Glasby, G., Alfieris, D., and I. Mitsis. 2017. New geochemical and mineralogical constraints on the genesis of the Vani hydrothermal manganese deposit at NW Milos Island, Greece: Comparison with the Aspro Gialoudi deposit and implications for the formation of the Milos manganese mineralization. *Ore Geol. Rev.* 80: 594–611.

Parker, E. T., Cleaves, H. J., Callahan, M. P., Dworkin, J. P., Glavin, D. P., Lazcano, A., and J. L. Bada. 2011. Prebiotic synthesis of methionine and other sulfur-containing organic compounds on the primitive Earth: A contemporary reassessment based on an unpublished 1958 Stanley Miller experiment. *Orig. Life Evol. Biosph.* 41: 201–212.

Pašava, J. 1993. Anoxic sediments—an important environment for PGE: An overview. *Ore Geol. Rev.* 8: 425–445.

Piercey, S. J. 2011. The setting, style, and role of magmatism in the formation of volcanogenic massive sulfide deposits. *Miner. Deposita* 46: 449–471.

Pirajno, F. 2009. *Hydrothermal Processes and Mineral Systems.* East Perth, Australia, Springer, 1250 p.

Pope, K. O., Kieffer, S. W., and D. E. Ames. 2006. Impact melt sheet formation on Mars and its implication for hydrothermal systems and exobiology. *Icarus* 183: 1–9.

Povoledo, D., and J. R. Vallentyne. 1964. Thermal reaction kinetics of the glutamic acid-pyroglutamic acid system in water. *Geochim. Cosmochim. Acta.* 28: 731–734.

Power, I. M., Wilson, S. A., and G. M. Dipple. 2013. Serpentinite carbonation for CO_2 sequestration. *Elements* 9: 115–121.

Proskurowski, G., Lilley, M. D., Seewald, J. S., Früh-Green, G. L., Olson, E. J., Lupton, J. E., Sylva, S. P., and D. S., Kelley. 2008. Abiogenic hydrocarbon production at lost city hydrothermal field. *Science* 319: 604–607.

Qian, Y., Engel, M. H., Macko, S. A., Carpenter, S., and J. W. Deming. 1993. Kinetics of peptide hydrolysis and amino acid decomposition at high temperatura. *Geochim. Cosmochim. Acta* 57: 3281–3293.

Rimola, A., Costa, D., Sodupe, M., Lambert, J.-F., and P. Ugliengo. 2013. Silica surface features and their role in the adsorption of biomolecules: computational modeling and experiments. *Chem. Rev.* 113: 4216–4313.

Roedder, E. 1990. Fluid inclusion analysis—prologue and epilogue. *Geochim. Cosmochim. Acta* 54: 495–508.

Ruiz-Mirazo, K., Briones, C., and A. de la Escosura. 2014. Prebiotic Systems Chemistry: New Perspectives for the Origins of Life. *Chem. Rev.* 114: 285–366.

Russell, M. J., and W. Martin. 2004. The rocky roots of the acetyl-CoA pathway. *Trends Biochem. Sci.* 29: 358–363.

Russell, M. J., Hall, A. J., and D. Turner. 1989. In vitro growth of iron sulphide chimneys: Possible culture chambers for origin-of-life experiments. *Terra Nova* 1: 238–241.

Russell, M. J., Solomon, M., and J. L. Walshe. 1981. The genesis of sediment-hosted, exhalative zinc + lead deposits. *Miner. Deposita* 16: 113–127.

Ryder, G. 2003. Bombardment of the Hadean Earth: Wholesome or Deleterious? *Astrobiology* 3: 3–6.

Sakai, R., Kusakabe, M., Noto, M. and T. Ishii. 1990. Origin of waters responsible for serpentinization of the Izu-Ogasawara-Mariana forearc seamounts in view of hydrogen and oxygen isotope ratios. *Earth Planet. Sci. Lett.* 100: 291–303.

Sakata, K., Kitadai, N., and T. Yokoyama. 2010. Effects of pH and temperature on dimerization rate of glycine: Evaluation of favorable environmental conditions for chemical evolution of life. *Geochim. Cosmochim. Acta* 74: 6841–6851.

Sangster, D. F. 2018. Toward an integrated genetic model for vent-distal SEDEX deposits. *Miner. Deposita* 53: 509–527.

Sangster, D. F., and E. M. Hillary. 1998. SEDEX lead-zinc deposits: Proposed sub-types and their characteristics. *Explor. Min. Geol.* 7: 341–357.

Sato, N., Quitain, A. T., Kang, K., Daimon, H., and K. Fujie. 2004. Reaction kinetics of amino acid decomposition in high-temperature and high-pressure water. *Ind. Eng. Chem. Res.* 43: 3217–3222.

Schardt, C. 2016. Hydrothermal fluid migration and brine pool formation in the Red Sea: Atlantis II Deep. *Miner. Deposita* 51: 89–111.

Schoonen, M., Smirnov, A., and C. Cohn. 2004. A perspective on the role of minerals in prebiotic synthesis. *AMBIO* 33: 539–551.

Schwarzenbach, E. M., Gazel, E., and M. J. Caddick. 2014. Hydrothermal processes in partially serpentinized peridotites from Costa Rica: Evidence from native copper and complex sulfide assemblages. *Contrib. Mineral. Petrol.* 168: 1079.

Schwarzenbach, E. M., Caddick, M. J., Beard, J. S., and R. J. Bodnar. 2016. Serpentinization, element transfer, and the progressive development of zoning in veins: Evidence from a partially serpentinized harzburgite. *Contrib. Mineral. Petrol.* 171: 5.

Schwille, P. 2017. How simple could life be? *Angew. Chem. Int. Ed.* 56: 10998–11002.

Shibuya, T., Komiya, T., Nakamura, K., Takai, K., and S. Maruyama. 2010. Highly alkaline, high-temperature hydrothermal fluids in the early Archean ocean. *Precambrian Res.* 182: 230–238.

Sillitoe, R. H. 2015. Epithermal paleosurfaces. *Miner. Deposita* 50: 767–793.

Sillitoe, R. H., and J. W. Hedenquist. 2003. Linkages between volcanotectonic settings, ore-fluid compositions, and epithermal precious metal deposits. *Soc. Econ. Geol. Spec. Publ.* 10: 314–343.

Simmons, S. F., White, N. C., and D. A. John. 2005. Geological characteristics of epithermal precious and base metal deposits. In *Economic Geology One Hundredth Anniversary*, Vol. 1905–2005, J. W Hedenquist, J. F. H. Thompson, and R. J. Goldfarb (Eds.), pp. 485–522. Littleton, CO: Society of Economic Geologist.

Slack, J. F., Selby, D., and J. A. Dumoulin. 2015. Hydrothermal, biogenic, and seawater components in metalliferous black shales of the brooks range, Alaska: Synsedimentary metal enrichment in a carbonate ramp setting. *Econ. Geol.* 110: 653–675.

Smirnov, A., Hausner, D., Laffers, R., Strongin, D. R., and M. A. Schoonen. 2008. Abiotic ammonium formation in the presence of Ni-Fe metals and alloys and its implications for the Hadean nitrogen cycle. *Geochem. Trans.* 9: 5.

Stoffregen, R. E. 1987. Genesis of acid-sulfate alteration and Au–Cu–Ag mineralization at Summitville, Colorado. *Econ. Geol.* 82: 1575–1591.

Stüeken, E. E., Anderson, R. E., Bowman, J. S., Brazelton, W. J., Colangelo-Lillis, J., Goldman, A. D., Som, S. M., and J. A. Baross. 2013. Did life originate from a global chemical reactor? *Geobiology* 11: 101–126.

Tarasov, V. G., Gebruk, A. V., Mironov, A. N., and L. I. Moskalev. 2005. Deep-sea and shallow-water hydrothermal vent communities: Two different phenomena? *Chem. Geol.* 224: 5–39.

Tarasov, V. G., Propp, M. V., Propp, L. N., Zhirmunsky, A. V., Namsaraev, B. B., Gorlenko, V. M., and D. A. Starynin. 1990. Shallow-water gasohydrothermal vents of Ushishir volcano and the ecosystem of Kraternaya Bight (the Kurile Islands). *Mar. Ecol.*, 11: 1–23.

Tiercelin, J. J., Pflumio, C., Cartec, M., Boulègue, J., Gente, P., Rolet, J., Coussement, C., Stetter, K. O., Huber, R., Buku, S., and W. Mifundu. 1993. Hydrothermal vents in the Lake Tanganyika, East Africa Rift System. *Geology* 21: 499–502.

Tivey, M. K. 2007. Generation of seafloor hydrothermal vent fluids and associated mineral deposits. *Oceanography* 20: 50–65.

Tsikouras, B., Karipi, S., Grammatikopoulos, T. A., and K. Hatzipanagiotou. 2006. Listwaenite evolution in the ophiolite melange of Iti Mountain (continental Central Greece). *Eur. J. Mineral.* 18: 243–255.

Uçurum, A. 1998. Application of the correspondence-type geostatistical analysis on the Co, Ni, As, Ag and Au concentrations of the listwaenites from serpentinites in the Divrigi and Kuluncak Ophiolitic Mélanges. *Turk. J. Earth Sci.* 7: 87–96.

Uçurum, A. 2000. Listwaenites in Turkey: Perspectives on formation and precious metal concentration with reference to occurrences in east-central Anatolia. *Ofioliti* 25: 15–29.

Vallentyne, J. R. 1964. Biogeochemistry of organic matter—II Thermal reaction kinetics and transformation products of amino compounds. *Geochim. Cosmochim. Acta* 28: 157–188.

Valley, J. W., Peck, W. H., King, E. M., and S. A. Wilde. 2002. A cool early Earth. *Geology* 30: 351–354.

van der Zwan, F. M., Devey, C. W., Augustin, N., Almeev, R. R., Bantan, R. A., and A. Basaham. 2015. Hydrothermal activity at the ultraslow- to slow-spreading Red Sea Rift traced by chlorine in basalt. *Chem. Geol.* 405: 63–81.

Vázquez-Figueroa, V., Canet, C., Prol-Ledesma, R. M., Sánchez, A., Dando, P., Camprubí, A., Robinson, C. J., and G. Hiriart Le Bert. 2009. Batimetría y características hidrográficas (Mayo, 2007) en las Cuencas de Consag y Wagner, Norte del Golfo de California, México. *Bol. Soc. Geol. Mex.* 61: 119–127.

Von Damm, K. L. 2013. Controls on the chemistry and temporal variability of seafloor hydrothermal fluids. In *Seafloor Hydrothermal Systems: Physical, Chemical, Biological, and Geological Interactions*, pp. 222–247. Washington, DC: American Geophysical Union.

Von Damm, K. L., Edmond, J. M., Grant, B, Measures, C. I., Walden, B., and R. F. Weiss. 1985. Chemistry of submarine hydrothermal solutions at 21 °N, East Pacific Rise. *Geochim. Cosmochim. Acta* 49: 2197–2220.

Wacey, D., Kilburn, M. R., Saunders, M., Cliff, J., and M. D. Brasier. 2011a. Microfossils of sulphur-metabolizing cells in 3.4-billion-year-old rocks of Western Australia. *Nat. Geosci.* 4: 698–703.

Wacey, D., Mcloughlin N., and M. D. Brasier. 2008. Looking through windows onto the earliest history of life on earth and mars. In *From Fossils to Astrobiology: Records of Life on Earth and Search for Extraterrestrial Biosignatures*, J. Seckbach and M. Walsh (Eds.), pp. 39–68. Dordrecht, the Netherlands: Springer.

Wacey, D., Saunders, M., Cliff, J., and M. D. Brasier. 2011b. Microfossils of sulphur-metabolizing cells in 3.4-billion-year-old rocks of Western Australia. *Nature* 4: 698–702.

Wächtershäuser, G. 1990. The case for the chemoautotrophic origin of life in an iron-sulfur world. *Orig. Life Evol. Biosph.* 20: 173–176.

Walker, S. I. 2017. Origins of life: A problem for physics, a key issues review. *Rep. Prog. Phys.* 80: 092601.

Weber, A. L. 2004. Kinetics of organic transformations under mild aqueous conditions: Implications for the origin of life and its metabolism. *Orig. Life Evol. Biosph.* 34: 473–495.

White, N. C., and Hedenquist, J. W. 1990. Epithermal environments and styles of mineralization: Variations and their causes, and guidelines for exploration. *J. Geochem. Explor.* 36: 445–474.

White, L. M., Bhartia, R., Stucky, G. D., Kanik, I., and M. J. Russell. 2015. Mackinawite and greigite in ancient alkaline hydrothermal chimneys: Identifying potential key catalysts for emergent life. *Earth Planet. Sci. Lett.* 430: 105–114.

White, R. H. 1984. Hydrolytic stability of biomolecules at high temperatures and its implication for life at 250°C. *Nature* 310: 430–432.

Wilde, S. A., Valley, J. W., Peck, W. H., and C. M. Graham. 2001. Evidence from detrital zircons for the existence of continental crust and oceans on the Earth 4.4 Gyr ago. *Nature* 409: 175–178.

Yamaoka, K., Kawahata, H., Gupta, L. P., Ito, M., and H. Masuda. 2007. Thermal stability of amino acids in siliceous ooze under alkaline hydrothermal conditions. *Org. Geochem.* 38: 1897–1909.

Yanagawa, H., and K. Kojima. 1985. Thermophilic microspheres of peptide-like polymers and silicates formed at 250°C. *J. Biochem.* 97: 1521–1524.

5.5 Prebiotic Reactions in Water, "On Water," in Superheated Water, Solventless, and in the Solid State

Vera M. Kolb

CONTENTS

5.5.1 INTRODUCTION

Prebiotic reactions have been studied in different media. These include reactions in the aqueous medium at normal temperature and pressure, such as in the prebiotic soup; reactions in the gas phase, such as in the interstellar space; reactions in the solid state, such as on asteroids and meteorites; and reactions in the aqueous systems under high pressures and temperatures, such as in the hydrothermal vents (e.g., Miller and Orgel 1974; Mason 1991; Holm and Anderson 2000; Hazen et al. 2002; Shaw 2006; Martin et al. 2008; Braakman 2013). In this chapter, prebiotic reactions in such media are discussed, with the exception of the gas phase, which is addressed in Section IV of this handbook.

Water is a commonly assumed prebiotic reaction medium on the early Earth. This assumption is based on other underlaying assumptions. First, since life, as we know it, is based on water, it is then assumed that the pre-biological reactions, to be prebiotically feasible, need to occur in water. Second, even though some prebiotic organic materials may have been formed in the gas or solid phases within the solar system, upon their delivery to the early Earth, they would eventually end up in the prebiotic soup and primordial oceans. As a consequence, prebiotic chemists have focused mostly on the syntheses of the water-soluble organic compounds. The latter are typically small molecules, generally five carbon atoms or less, which have hydrophilic groups, such as hydroxyl, carboxyl, and amino. In addition, charged organic compounds, such as salts of carboxylic acids (carboxylates), of phenols (phenolates), and of amines (amine salts), are also water-soluble, providing

that the hydrophobic portion of these salts is not disproportionally large (e.g., Solomons and Fryhle 2013). Examples of water-soluble organic compounds that are prebiotically relevant include simple sugars, polyols, amino acids, and alcohols. These water-soluble compounds can react in water. Such reactions are classical prebiotic in-water reactions. The choice of the water-soluble chemicals has limited the synthetic options under prebiotic conditions, since most organic compounds are either insoluble in water or are only poorly soluble (Kolb 2014). Also, it has been assumed that non-aqueous solvents, which are commonly used in organic syntheses, such as aliphatic and aromatic hydrocarbons, ethers, and alcohols, were not available on the prebiotic Earth as pools of liquids. They could be found only in relatively small amounts as compared with the large bodies of water. This has also created a problem for the prebiotic chemists, since they could not tap to the wealth of the chemical synthetic pathways that have been developed for the non-aqueous organic chemistry.

In the introductory chapter of this handbook, Chapter 1.1, the multidisciplinary and interdisciplinary nature of astrobiology was discussed. It was shown that astrobiology, as a field of science, clearly benefits from the input of other disciplines. Relatively recently, a new subfield of chemistry has been developed, that of green chemistry, and most relevant for this chapter, green organic chemistry (e.g., Lancaster 2002; Ahluwalia 2012; Zhang and Cue 2012; Kolb 2016). Within this subfield, numerous chemical syntheses that occur under the green (environmentally safe) conditions have been reported (e.g., Ahluwalia 2012). These include organic reactions in water, including superheated water, without solvent ("solvent-less"), and in the solid state, as reviewed in the recent book by Kolb (2016). These new developments have been recently applied to prebiotic chemistry (e.g., Kolb 2010, 2012, 2014, 2015). These are reviewed in this chapter.

5.5.2 ORGANIC REACTIONS "ON WATER" AND THEIR PREBIOTIC POTENTIAL

It has been known for quite some time that many organic compounds that are not soluble in water are still capable of reacting in water. Such organic reactions occur in the heterogeneous system, comprising a suspension or an emulsion of organic compounds in water. Initially, such reactions were viewed as a curiosity and as an exception. However, after a number of examples have been discovered, they became a subject of an extensive study (e.g., Rideout and Breslow 1980; Breslow 1991, 2006). Such heterogeneous reactions were termed "on water" reactions. They are distinct from "in water" reactions, which occur in solution, and are thus homogenous.

The importance of these "on water" reactions grew, as they were shown to be extremely efficient (often giving quantitative yields), quite specific, and accelerated in the presence of water, as compared with the organic solvents. Many such reactions have been studied as green chemistry reactions (e.g., Liu and Wang 2010; Kolb 2016). The discovery of the mechanism of such reactions removed them from the category of unexplained exceptions (e.g., Rideout and Breslow 1980; Breslow 1991, 2006; Lindström 2002; Narayan et al. 2005; Pirrung et al. 2008; Chanda and Fokin 2009; Manna and Kumar 2013).

The mechanism of these "on water" reactions and their acceleration are briefly reviewed here. For a detailed coverage, which is beyond the scope of this chapter, the reader is directed to the literature in the field (e.g., Rideout and Breslow 1980; Breslow 1991, 2006; Otto and Engberts 2000; Lindström 2002, 2007; Klijn and Engberts 2005; Narayan et al. 2005; Sarma et al. 2006; Pirrung et al. 2008; Shapiro and Vigalok 2008; Chanda and Fokin 2009; Manna and Kumar 2013; Kolb 2014, 2016).

The mechanism can be conveniently explained on a classical example of Diels-Alder reaction. This reaction is a cycloaddition reaction between a diene and a dienophile, which results in the formation of carbon-carbon bonds and the six-membered cyclic compound (Solomons and Fryhle 2013). In the past, this reaction was typically performed in organic solvents. However, when Diels-Alder reaction was attempted between two water-insoluble components, anthracene-9-carbinol (a diene) and *N*-ethylmaleimide (a dienophile), the cycloaddition product formed rapidly (Figure 5.5.1) (work by Breslow and co-workers; see Kolb [2016] and references therein for a review) (e.g., Breslow 1991; Kolb 2016). Since various water-insoluble dienes and dienophiles are potential prebiotic compounds, the discovery that Diels-Alder reaction may be performed in water makes it prebiotically feasible.

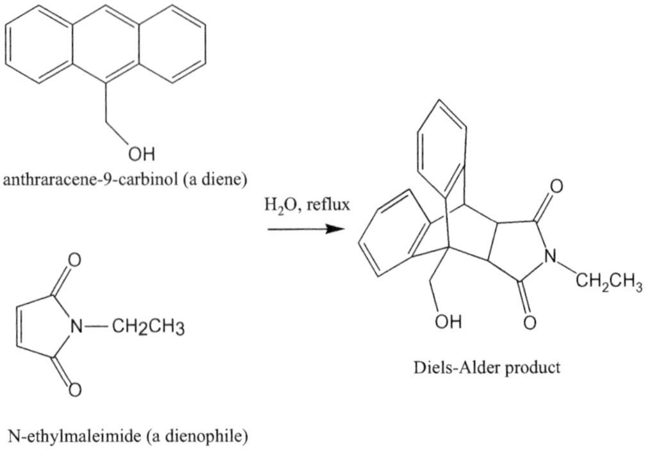

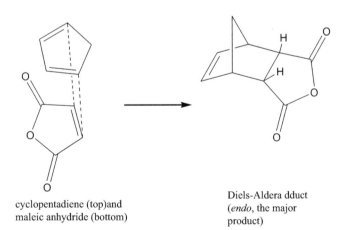

FIGURE 5.5.1 Diels-Alder "on water" reaction between anthracene-9-methanol and N-ethylmaleimide, which yields the Diels-Alder adduct.

This reaction's success in water was explained by a hydrophobic effect in which the water-insoluble reactants are pushed toward each other in an attempt to escape water. A close proximity of the reactants that is thus achieved allows for an efficient molecular orbital overlap between the diene and dienophile. This is shown conveniently on a simpler example of a diene and dienophile (e.g., Solomons and Fryhle 2013) in Figure 5.5.2.

Such an overlap results in a faster and more selective reaction. The operation of the hydrophobic effect is supported by the effect of various salts on the reaction (e.g., Kumar 2001). Some salts cause an increase in the hydrophobic effect and the reaction rates (e.g., LiCl, NaCl, KCl, NaBr, $MgCl_2$, $CaCl_2$, and Na_2SO_4). Some other salts decrease both (e.g., guanidinium chloride, LiI, and ClO_4^-). Additives other than salts may also influence hydrophobic interaction (e.g., urea, ethanol, as anti-hydrophobic) (see Kolb [2016] and the references therein for a review). These findings would have an immediate applicability to the prebiotic soup model, since pure water as

FIGURE 5.5.2 Orbital overlap in Diels-Alder reaction between cyclopentadiene and maleic anhydride, which results in the major product, which is *endo*.

its medium is unrealistic, while the presence of salts, including amino acids and small polar molecules, is expected.

Performing the Diels-Alder reaction "on water," as opposed to the organic solvents, may result in a very large increase of the reaction rate, for example, more than 10,000 times (Otto and Engberts 2000). Although the Diels-Alder reaction in organic solvents shows an intrinsic selectivity toward the *endo* product, water enhances this selectivity further. This may be attributed to the difference in stability of the transition states for the *endo,* as opposed to the *exo* isomer in water. Hydrophobic effects are assumed to favor stabilization of the *endo* transition state, since it is more compact. In contrast, the transition state for the *exo* product is more extended. A concise explanation is that "hydrophobic packing" favors the *endo* product.

Before the discovery of the "on water" reactions, the reactions between water-insoluble starting materials in the aqueous media have not been examined by the prebiotic chemists, even when the starting materials were prebiotically feasible. Many such reactions exist, in addition to the already mentioned Diels-Alder reaction. Examples include 1.3-dipolar cycloadditions, Claisen rearrangement, Passerini and Ugi reactions, nucleophilic substitution reactions, and nucleophilic opening of three-membered rings, among others (Chanda and Fokin 2009). Although numerous "on water" reactions have been reported, they cannot be automatically assumed to be prebiotically relevant. Only a limited number of the "on water" reactions can be classified as potentially prebiotic at this time, by virtue of their starting materials when they are prebiotically feasible. For the cases in which they are not, modification of the starting materials to fit prebiotic criteria is possible in principle but has not been studied in most cases.

In addition to Diels-Alder reaction, which has a broad synthetic utility and a possible prebiotic relevance, some multicomponent "on water" reactions (e.g., Chanda and Fokin 2009; Huang et al. 2012; Kolb 2016) are prebiotically promising (e.g., Kolb 2014). The multicomponent reactions typically occur fast and give a single product of high purity, in an almost quantitative yield (e.g. Pirrung and Sarma 2004). One such example is the Passerini reaction, which is a three-component reaction. Some multicomponent reactions have more components, for example, even seven (Huang et al. 2012). Multicomponent reactions are an ideal type of prebiotic reactions, since they give a clean product, and thus, the common problem of prebiotic chemistry, namely an intractable mixture of products, is avoided. The Passerini reaction is described next.

The ease by which this reaction occurs and the purity of its single product are so outstanding that this reaction is now introduced in the undergraduate organic laboratory (e.g., Hooper and DeBoef 2009; Williamson and Masters 2011) as an example of a green "on water" reaction. This reaction also has a prebiotic potential. An example of the Passerini reaction is shown in Figure 5.5.3.

Figure 5.5.3 shows the structures of three components, an aldehyde (benzaldehyde), an isocyanide (*tert*-butyl isocyanide), and a carboxylic acid (benzoic acid). The categories of these compounds are all prebiotically feasible, since these or similar compounds have either been found on the meteorites, such as

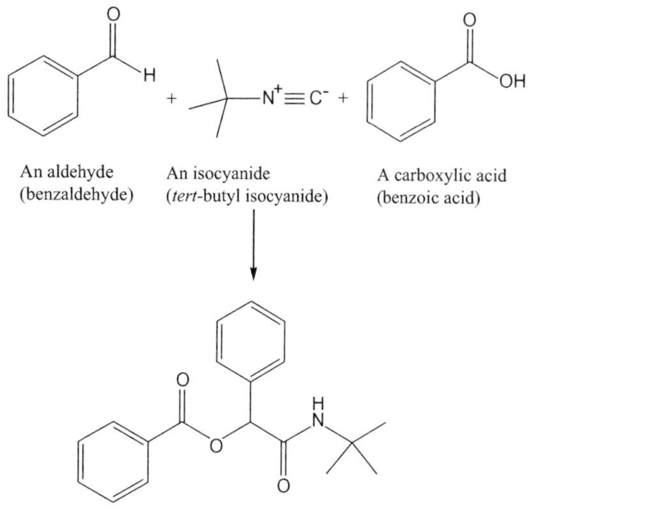

The Passerini reaction product (benzoic acid, *tert*-butylcarbamoyl-phenyl-methyl ester)

FIGURE 5.5.3 The Passerini three-component reaction between an aldehyde (benzaldehyde), an isocyanide (*tert*-butyl isocyanide), and a carboxylic acid (benzoic acid), which results in a single product (*tert*-butylcarbamoyl-phenyl-methyl ester of benzoic acid).

Murchison (e.g., Schmitt-Kopplin et al. 2010, 2014) or could be reasonably easily synthesized by the known simulated prebiotic reactions (e.g., Miller and Orgel 1974; Mason 1991). The product itself, *tert*-butylcarbamoyl-phenyl-methyl ester of benzoic acid, is a solid product, which forms rapidly in water. The reaction is very fast, which is also prebiotically desirable, since side reactions of the components with other compounds that may be present in the prebiotic soup may be avoided. The Passerini reaction can build prebiotic chemical diversity easily, without side products, when the derivatives of the starting materials are employed.

Recently, Sutherland and his co-workers have successfully applied multicomponent reactions to various aspects of prebiotic synthesis of nucleotides and their components, which were previously notoriously difficult. In some applications, they have used the isocyanide chemistry that was modeled by the multicomponent Ugi reaction (related to the Passerini reaction), in which an isocyanide, an aldehyde, and ammonium chloride were reacted (Mullen and Sutherland 2007). Their work continues and is a highly successful application of multicomponent reactions, such as a chemoselective assembly of purine precursors in water (Powner et al. 2010) and potentially prebiotic Passerini-type reactions of phosphates (Sutherland et al. 2008), among others. These contributed to a new modus operandi of prebiotic chemistry (Powner and Sutherland 2011) and an outstanding advancement of the prebiotic chemistry synthetic tools.

5.5.3 ORGANIC REACTIONS IN SUPERHEATED WATER AND THEIR PREBIOTIC SIGNIFICANCE

Various prebiotically significant reactions have been studied under the simulated hydrothermal conditions in the past (Orgel 2004). The impetus for such studies was that the hydrothermal vents are rich in life, despite high pressures and temperatures. As one example, components of the citric acid cycle have been synthesized under these conditions (Orgel 2004). Since hydrothermal vents provide both chemicals and energy sources, they are considered as localities that are hospitable for the emergence of life on Earth and on other bodies in the solar system that may harbor such vents.

The subject of hydrothermal vents and the origin of life has been reviewed (e.g., Holm and Anderson 2000; Hazen et al. 2002; Orgel 2004; Martin et al. 2008; Padgett 2012; Sojo et al. 2016) and is further updated in Section V of this handbook.

Organic chemistry under hydrothermal conditions has been studied extensively (Avola et al. 2013), including feasibility of biochemical reactions and metabolic cycles under these conditions (Braakman 2013). Organic chemical possibilities in superheated water were systematically explored (Savage 1999; Katritzky et al. 2001; Siskin and Katritzky 2001; da Silva 2007; Liotta et al. 2007; Savage and Rebacz 2010; Kruse and Dinjus 2012). Many organic reactions in superheated water not only are prebiotically feasible but also offer clear prebiotic advantages (Kolb 2015). Examples of such reactions are shown in Section 5.5.3.2, after a brief review of the properties of superheated water in Section 5.5.3.1.

5.5.3.1 Properties of Superheated Water

In this subsection, properties of water and how they change when water is superheated are briefly reviewed (Li and Chan 2007; Kolb 2016). Water is capable of making hydrogen-bonded networks, owing to the large electronegativity difference between oxygen and hydrogen. Oxygen has lone electron pairs, and it acts as a hydrogen-bond acceptor, while partially positive hydrogen behaves as a hydrogen-bond donor. The resulting hydrogen-bonded networks are responsible for the relatively high boiling point of water, among other properties. Figure 5.5.4 depicts such a network.

Water exhibits high polarity, which is reflected in its high dielectric constant (78.4 at 25°C). Water dissociates to proton (H^+, which hydrates to H_3O^+) and hydroxide (OH^-) ions. Since concentrations of both acidic protons and basic hydroxide ions are the same, water is neutral. The ionic product of water ($Kw = [H^+][OH^-]$) is very small (10^{-14} at room temperature). This means that the concentrations of both protons and hydroxide ions are very small (10^{-7} M).

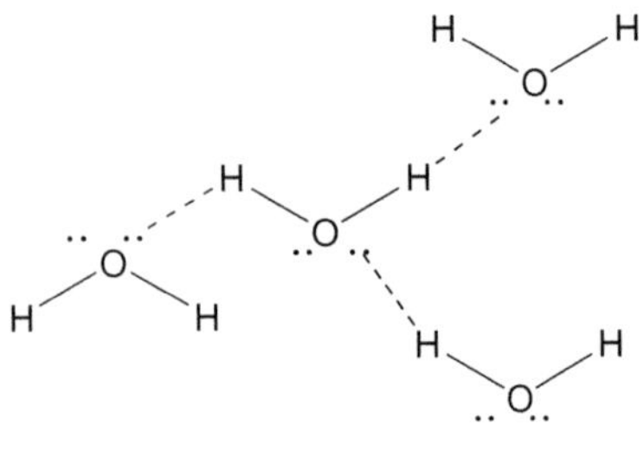

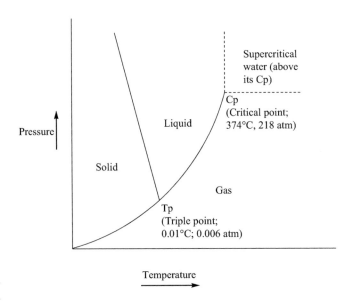

FIGURE 5.5.5 Phase diagram of water (simplified, on a free scale).

These properties of water are dramatically changed when water becomes superheated. A simplified phase diagram for water is shown in Figure 5.5.5. It includes the areas of superheated water.

The points that are labeled on the diagram are T_p (triple point) and C_p (critical point). The T_p is the unique pressure and temperature condition where all three phases (solid, liquid, and gas) coexist (0.01°C; 0.006 atm). At the C_p (374°C; 218 atm), the boundary line between the liquid and gas phases terminates, and these two phases become indistinguishable and form a single phase. The region in the phase diagram between the dashed lines, which start at the C_p, shows the supercritical fluid phase. The supercritical water (SCW) is a single uniform phase at and above the C_p. The high temperatures of superheated water are achieved in conjunction with an increase in pressure. The superheated water region also includes the near-critical water (NCW), at temperatures of 200°C–300°C, and the upper critical solution temperature (UCST), above which the organic substrate and water are miscible in all proportions (Liotta et al. 2007). The term high-temperature water (HTW) is also used, for water above its normal boiling point (100°C; 1 atm) (Savage and Rebacz 2010).

Properties of water, such as density, dielectric constant, and Kw, differ between water at the ambient conditions (25°C; 1 atm) and the HTW and SCW. This is shown in Table 5.5.1.

The values from Table 5.5.1 show that the HTW, at 250°C and 50 atm, has decreased values of density and dielectric constant but has a substantially increased value of Kw, as compared with water at the ambient conditions. For the SCW at 400°C, these properties vary with pressure. For example, the SCW at 250 atm shows a dramatic decrease in Kw as compared with water at the ambient conditions, while at 500 atm, the Kw is substantially increased.

The increased value of Kw of the HTW under specific conditions means that water becomes more acidic (and basic) than the ambient water. This is quite significant for the reactions that require acid (or base) catalysis. While at the ambient

FIGURE 5.5.4 Hydrogen-bonded network of water molecules, which are presented as their Lewis structures.

TABLE 5.5.1

Properties of Superheated Water as Compared with Water at Ambient Conditions

Property	Ambient Water	High-Temperature Water	Supercritical	Water
Temperature (°C)	25	250	400	400
Pressure (atm)	1	50	250	500
Density (kg/m³)	997	800	170	580
Dielectric constant	78.5	27.1	5.9	10.5
Kw	10^{-14}	$10^{-11.2}$	$10^{-19.4}$	$10^{-11.9}$

Source: Savage, P. D., and Rebacz, N. A., Water under extreme conditions for green chemistry, in *Handbook of Green Chemistry, Volume 5,* C.-J. Li (Ed.), pp. 331–361, Wiley-VCH, Weinheim, Germany, 2010; Kolb, V. M., *Green Organic Chemistry and Its Interdisciplinary Applications,* CRC Press/Taylor & Francis, Boca Raton, FL, 2016.

conditions an external catalyst needs to be added, in HTW, the H⁺ (or OH⁻) provided by the heated water may be sufficient. When the water is cooled, the ambient ion concentrations are restored, thus representing a self-neutralizing catalytic medium (Liotta et al. 2007). Such *in situ* catalysis is highly significant for prebiotic chemistry, as it facilitates reactions without an externally added catalyst.

As the temperature is increased from 25°C to 300°C, under these near-critical conditions, water loses ca. 55%–60% of its hydrogen-bonding network (Liotta et al. 2007). In addition to reducing the number of hydrogen bonds per water molecule, the increase in temperature also reduces the persistence of hydrogen bonds, as they form and break at faster rates. This leads to the shorter hydrogen-bond lifetime (Savage and Rebacz 2010).

The increase in temperature is accompanied by a decrease of the dielectric constant. The value of the dielectric constant for the NCW at 300°C is ca. 20, which represents a 75% reduction from the value for the ambient water (Liotta et al. 2007). Significantly, this value for the dielectric constant corresponds most closely to a moderately polar solvent, such as acetone (dielectric constant of 21.4 at 25°C). This enables enhanced solubility of nonpolar organic species in NCW (Liotta et al. 2007). This is of a high significance for prebiotic chemistry. It means that the organic compounds that are not water-soluble will became so in the superheated water.

5.5.3.2 Prebiotic Potential of Organic Reactions in Superheated Water

A large variety of organic reactions have been reported to occur in superheated water (Savage 1999; Katritzky et al. 2001; Siskin and Katritzky 2001; Akiya and Savage 2002; da Silva 2007; Li and Chan 2007; Kruse and Dinjus 2012; Avola et al. 2013). Possibilities also exist for new reaction pathways, since the reaction mechanism sometimes is changed in superheated water. Some such reactions were recently reviewed as potentially prebiotic (Kolb 2015). They include some common reactions that could be used for the synthesis of more complicated molecules. They are shown in Table 5.5.2.

Chemical equations for some of the reactions from Table 5.5.2 are shown in Figure 5.5.6.

The reactions in superheated water that are shown represent just a small number of the reactions that are known to occur successfully in this medium and that have a prebiotic potential. However, the great wealth of the reaction possibilities in superheated water is virtually unexplored by prebiotic chemists. The reactions that are shown here illustrate two important points. The first is that the reactions that have been impossible to perform under commonly simulated prebiotic conditions can now be added to the prebiotic synthetic repertoire. Examples include Friedel-Crafts and other reactions that require acid/base catalysis. For example, traditional chemistry uses Lewis acid catalysts, such as AlCl₃, for the Friedel-Crafts reaction, but such a catalyst

TABLE 5.5.2

Examples of Organic Reactions in Superheated Water That Are Prebiotically Relevant. No Externals Catalysts, Acids, or Bases Are Necessary

Reaction Type	Starting Materials and Products	References
Hydrolysis	Amides (to corresponding carboxylic acid), nitriles (to corresponding amides and carboxylic acids), esters (to corresponding acids and alcohols), and aryl ethers (to phenols).	Kruse and Dinjus (2012)
Rearrangements	Beckmann: oximes rearranged to lactam; Pinacol: pinacol rearranged to pinacolone; Claisen: allyl phenyl ether rearranged to 2-allyl phenol.	Savage and Rebacz (2010), Liotta et al. (2007), Kruse and Dinjus (2013)
Alkylations (Friedel-Crafts type)	Phenol and *t*-butyl alcohol, to give phenol with *t*-butyl group in the ring.	Liotta et al. (2007), Avola et al. (2013)
Condensations (Aldol type, such as Claisen-Schmidt and related reactions)	Benzaldehyde and acetone, benzaldehyde and 2-butanone (to give dehydrated aldol).	Liotta et al. (2007), Savage and Rebacz (2010), Avola et al. (2013)
Additions (Diels-Alder type)	Cyclopentadiene and dimethylmaleate to give the Diels-Alder adduct.	Avola et al. (2013)
Dehydrations	Alcohols to alkenes, various examples.	Avola et al. (2013)

Source: Kolb, V. M., Prebiotic reactions in superheated water, *Proceedings of the SPIE 9606, Instruments, Methods, and Missions for Astrobiology XVII,* 96060D, 2015; *Green Organic Chemistry and Its Interdisciplinary Applications,* CRC Press/Taylor & Francis, Boca Raton, FL, 2016.

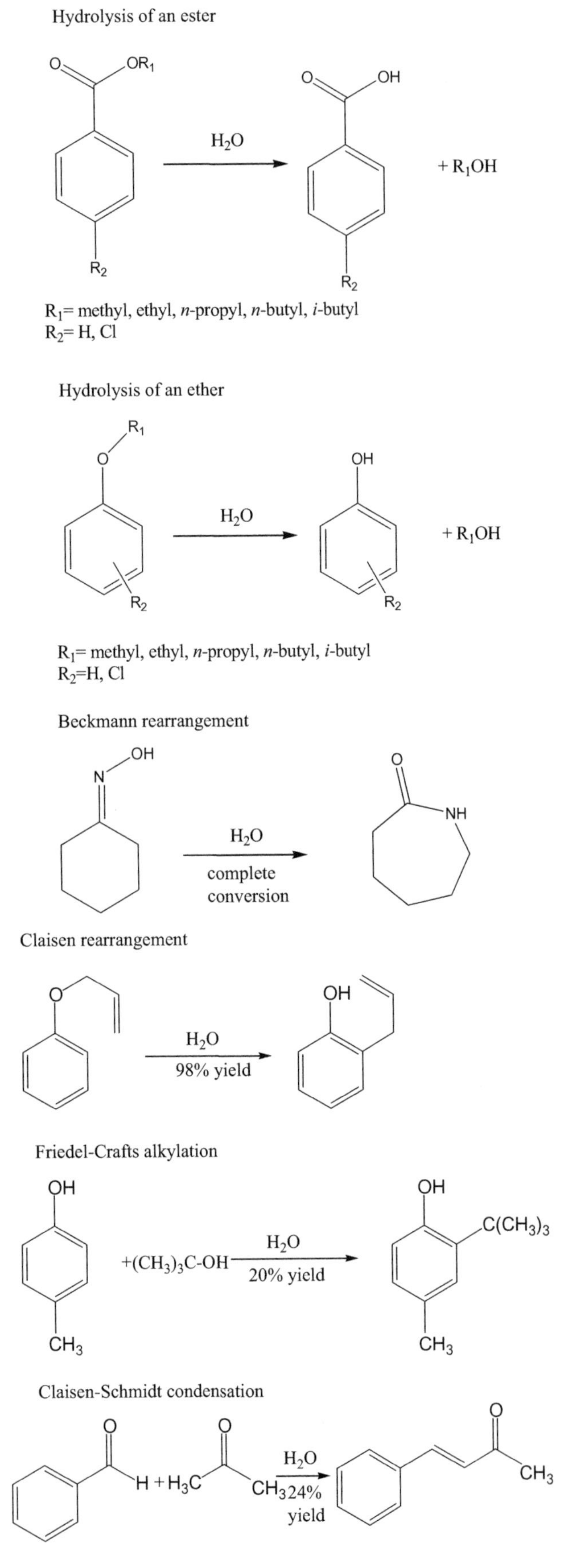
Hydrolysis of an ester
H_2O
$+ R_1OH$
R_1= methyl, ethyl, *n*-propyl, *n*-butyl, *i*-butyl
R_2= H, Cl
Hydrolysis of an ether
H_2O
$+ R_1OH$
R_1= methyl, ethyl, *n*-propyl, *n*-butyl, *i*-butyl
R_2=H, Cl
Beckmann rearrangement
H_2O
complete conversion
Claisen rearrangement
H_2O
98% yield
Friedel-Crafts alkylation
$+(CH_3)_3C\text{-}OH$
H_2O
20% yield
Claisen-Schmidt condensation
H_2O
24% yield

FIGURE 5.5.6 Chemical equations for some of the reactions from Table 5.5.2.

would fall apart in water. The superheated water provides H^+ catalyst intrinsically and self-neutralizes upon cooling. The second is that hydrothermal vents provide a medium that is capable of producing rich organic chemistry, based on a myriad of the reactions that have been successfully performed in this medium. Prebiotic chemists need to get fully acquainted with these reactions and should add them to the prebiotic synthetic toolbox.

5.5.4 SOLVENTLESS AND SOLID-STATE REACTIONS, AND THEIR PREBIOTIC POTENTIAL

5.5.4.1 General

Solventless and solid-state reactions may have been critical for the formation of organic compounds that are found on the meteorites, such as the Murchison meteorite, which is a carbonaceous chondrite. Meteorites are believed to be one of the sources of organic compounds on the early Earth (Shaw 2006; Longstaff 2015; Kwok 2016, 2017). A large number of organic compounds has been identified on the Murchison meteorite and is getting larger with better analytical techniques. For example, Schmitt-Kopplin et al. (2010, 2014) reported tens of thousands of different molecular compositions on the Murchison meteorite, by an application of ultra-high-resolution analytical methods, which combine state-of-the-art mass spectrometry, liquid chromatography, and nuclear magnetic resonance. Classes of organic compounds found on the Murchison meteorite include aliphatic and aromatic hydrocarbons, carboxylic acids, hydroxyl carboxylic acids, dicarboxylic acids, carboxamides, amino acids, aldehydes, ketones, alcohols, amines, purines and pyrimidines, nitrogen heterocycles, sulfonic acids, and phosphonic acids, among many others. Since water on asteroids was available only sporadically, chemical pathways in the absence of water need to be examined. Solid-state and solventless conditions have been proposed for chemical reactions on asteroids and the resulting meteoritic chemistry (Kolb 2010, 2012), as well as for general prebiotic chemistry (Kolb 2014).

Solventless reactions occur without solvent, as the term indicates. This term is popular in the green chemistry literature. Such reactions occur when two or more solid compounds are mixed. Solid compounds act as impurities to each other. The impurities become incorporated into the crystalline lattice of a pure compound and cause a breakup of the regular crystalline pattern, which results in the lowering of the melting point. In some instances, the melting point is lowered to the extent that melting of the mixture occurs. When this happens, chemical reactions may proceed in the melted state, thus eliminating the need for the solvent. Such reactions are known as solventless (e.g., Raston and Scott 2000; Cave et al. 2001; Raston 2004; Kolb 2016). Solid-state reactions occur when the solid compounds are mixed but do not melt. However, a very small amount of melted material may occur at the interface of the solids that are mixed, which is not visually observable, unless the product is colored. This blurs the division between solventless and solid-state reactions (Rothenberg et al. 2001; Kolb 2016). Other variations exist. For example, when a liquid and a solid component come in contact, the suspension may become reactive. Sometimes, a gas-solid reaction is successful. All of these scenarios could be fruitful on asteroids (Kolb 2010, 2012, 2014).

The asteroids are presumably rich in organic compounds, as deduced by the analysis of Murchison and similar carbonaceous chondrite meteorites that originate from them. While some of these compounds may be potentially reactive toward each other, some others may not be. However, the unreactive compounds could act as chemically inert additives to the potentially reactive mixture of compounds and could lower the melting points of such mixtures to cause their melting. This would enable the reactions to occur in the melted state. The temperature estimates on the asteroids are in a range from 25°C to 100°C. This is consistent with the conditions of aqueous alterations and is friendly to the organic reactions and preservation of the organic compounds. The heat to maintain this temperature range could be provided by the radioactive processes or impacts (Kolb 2010, 2012).

5.5.4.2 EXAMPLES OF PREBIOTICALLY RELEVANT SOLVENTLESS AND SOLID-STATE REACTIONS

Major types of organic reactions, such as oxidations; reductions; eliminations; substitutions; condensations; cyclizations; rearrangements; and formation of carbon-carbon, carbon-oxygen, and carbon-nitrogen bonds, can occur as solventless or in the solid state (Toda 1993; Toda 1995; Raston 2000; Tanaka and Toda 2000; Cave et al. 2001; Rothenberg et al. 2001; Raston 2004; Tanaka 2009; Kolb 2016). These reactions often occur rapidly and are quite efficient.

In the reported studies, both terms "solventless" and "solvent-free" are used. In this section, the terms are used as given by the authors of the study, to facilitate the literature searches by the key words.

Experimental procedures for solvent-free reactions (Tanaka 2009) generally involve just mixing of the reactants. Such a procedure is prebiotically feasible. This is also the case for many solid-state reactions (e.g., Toda 1993, 1995; Kaupp 2005). If a catalyst is needed, the reaction needs to be explored with the catalysts that were available on the early Earth. This is possible in principle. For example, a non-prebiotic acidic catalyst could be substituted with a naturally occurring acidic clay.

Selected examples of solventless and solid-state reactions that are of prebiotic interest are given here. More examples are given by Kolb (2012, 2014).

One such reaction is the formation of ethers in the solid state, from the solid mixture of alcohols in the presence of an acid catalyst. Ethers with the appropriate structures could be used as membranes for the primitive life. The precedent is found in some Archaea's membranes, which are ethers, in contrast to the esters, which are found in the lipid membranes of the majority of contemporary life (Kolb 2012 and the references therein). An example of a solid-state reaction of an alcohol to form an ether under the acid catalysis with p-toluenesulfonic acid (TsOH) (Toda et al. 1990) is shown in Figure 5.5.7.

Etherification of alcohols

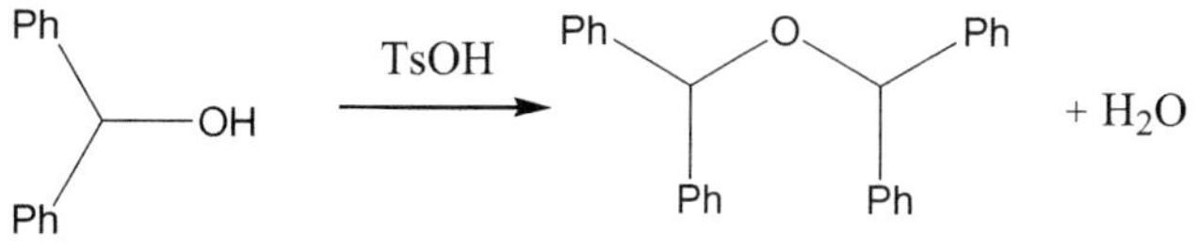

FIGURE 5.5.7 Solvent-free etherification of alcohols.

Alcohols are prebiotically feasible compounds. However, the corresponding ethers are difficult to make prebiotically in the aqueous solution, since water needs to be driven off during the condensation reaction, which is a dehydration. The solid-state etherification reaction provides a viable prebiotic path to the ethers. The TsOH could be substituted by a prebiotic acid catalyst, such an acidic clay.

Another prebiotically significant reaction, a carbon-carbon bond formation by the oxidative coupling of a phenol (a naphthol) to a binol ([1,1'-binaphthalene]-2,2'-diol), occurs in the solid state (Tanaka 2009). It is shown in Figure 5.5.8.

Phenol itself can be synthesized under prebiotic conditions, from sodium bicarbonate ($NaHCO_3$) in water, with iron as a catalyst, under hydrothermal conditions (Tian 2007). The example from Figure 5.5.8 is especially significant. When a phenol (2-naphthol) undergoes the solid-state oxidative coupling with $FeCl_3$, it gives a binol ([1,1'-binaphthalene]-2,2'-diol), which is not planar, owing to the steric hindrance of the two hydroxyl groups. Instead, it is twisted, and thus asymmetric, which makes it chiral (Carey and Sundberg 2007). The binol forms in the reaction as a racemic mixture of two enantiomers. Under the prebiotic conditions, a possible enantiomeric enrichment may be achieved under the influence of the available chiral molecules or minerals (Kolb 2012, 2014).

Aldol-type condensation reactions, in which carbon-carbon bonds are made, play an important role in chemistry and biology (e.g., in the citric-acid cycle). Such reactions are thus of a great prebiotic importance (Kolb 2012, 2014). An example of aldol reaction in which the basic alumina is used as a catalyst, under solvent-free conditions, is shown in Figure 5.5.9.

The reaction is clean, since there is no cross-reactivity and a single product is obtained. The molecules of aldehyde are not reactive between themselves, since they cannot form the

Oxidative coupling of naphthols

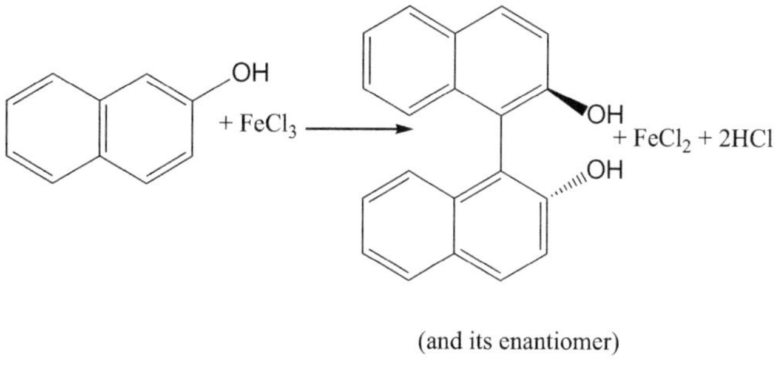

(and its enantiomer)

FIGURE 5.5.8 Solid-state oxidative coupling of a phenol (a naphthol) to a binol.

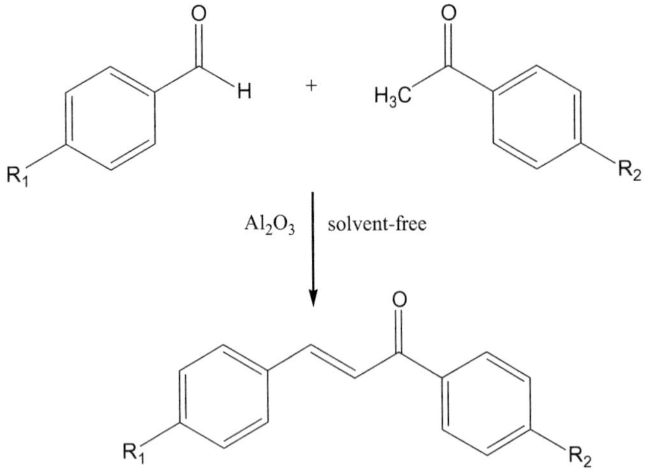

FIGURE 5.5.9   Solvent-free aldol condensation.

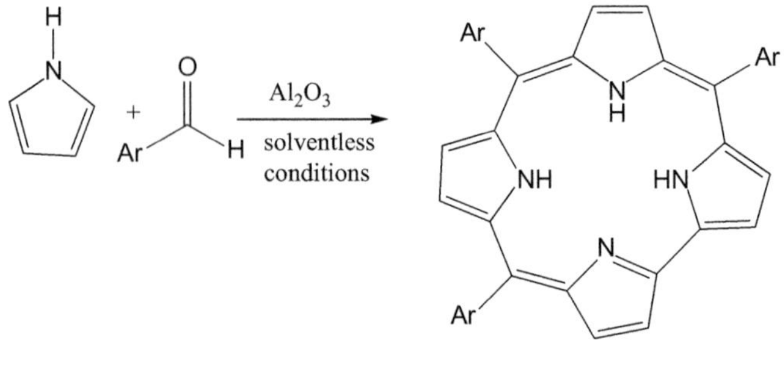

FIGURE 5.5.11   Solventless synthesis of a corrole by condensation of pyrrole and aldehyde.

needed α-carbanion (enolate). The ketone molecules can form the α-carbanion, but the latter does not give the condensation product with other ketone molecules, owing to the unfavorable thermodynamics. Thus, the only reaction occurring is between the ketone carbanion and the aldehyde, to give the shown product. This example also shows that a belief that in prebiotic chemistry "everything reacts with everything else" can be misleading. Single products may be obtained, owing to the intrinsic limitations in the reactivities of the starting materials, as shown in this example (Kolb 2012, 2014).

Esterification is another general reaction of importance to the prebiotic chemistry (Kolb 2012, 2014). When the starting materials for this reaction, an alcohol and a carboxylic acid, react to give an ester, the equilibrium of the reaction does not favor the ester product. Thus, the equilibrium needs to be shifted toward the product. This can be accomplished by removing water or by using a strong acid catalyst with dehydrating properties, such as concentrated H_2SO_4. Such requirements make esterification in the prebiotic soup infeasible. However, the solvent-free esterification, shown in Figure 5.5.10 (Tanaka 2009, p. 288), appears to be feasible (Kolb 2012, 2014).

This reaction requires perchlorates, which are found on Earth, especially under arid conditions, and also on Mars. $MgSO_4$ could act as a recyclable dehydrating agent (Kolb 2012, 2014).

Another important synthesis is that of the corroles, which are porphyrin analogues. They are important prebiotically, since they may serve as primitive enzymes, as they have a metal complexing site in the center of the molecule (Kolb 2012, 2014). A solventless synthesis of a corrole is shown in Figure 5.5.11 (Tanaka 2009).

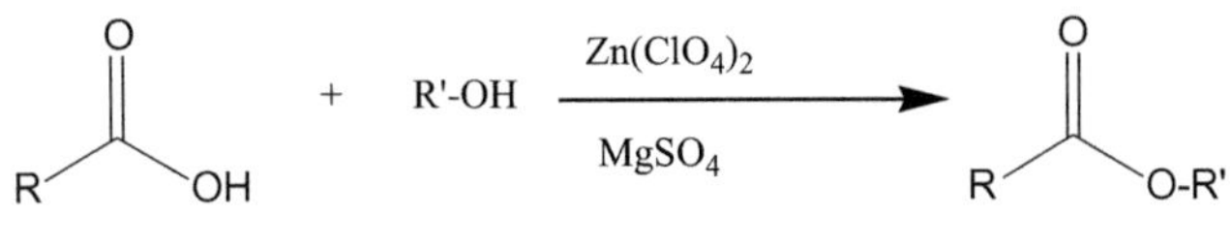

This reaction occurs between pyrrole and an aromatic aldehyde, under solventless conditions, with Al_2O_3 as catalyst, to give a corrole product. A diversity of the structures of corroles can be achieved by employing differently substituted aromatic aldehydes (Kolb 2012, 2014).

As the final example, a synthesis of a macromolecule with a cavity is shown. It occurs readily by a cyclocondensation of resorcinol with an aldehyde under the solvent-free conditions (Tanaka 2009) (Figure 5.5.12).

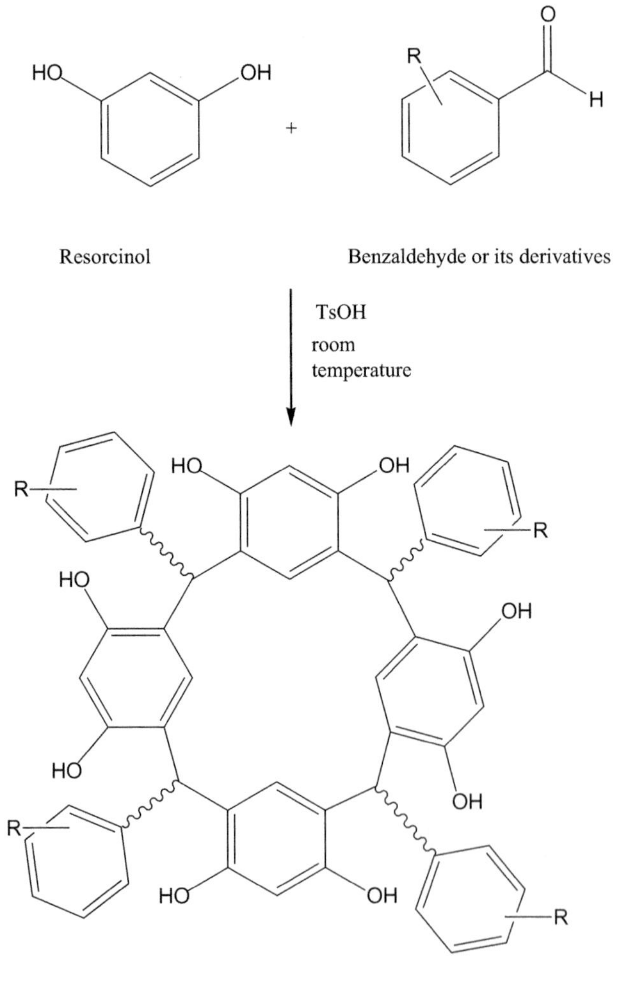

FIGURE 5.5.10   Solvent-free esterification.

FIGURE 5.5.12   Solvent-free cyclocondensation of resorcinol and a benzaldehyde, which gives a macromolecule with a cavity.

This remarkable reaction occurs at room temperature. The large cavity inside the molecule could find some enzymatic role under the prebiotic conditions. The TsOH as a catalyst could be substituted with an acidic clay, which is prebiotically more feasible (Kolb 2012, 2014).

5.5.5 CONCLUSIONS

In this chapter, prebiotically feasible reactions "on water," under hydrothermal conditions, solventless, and solid-solid are reviewed. Such reactions expand the repertoire of prebiotic chemistry. Among "on water" reactions, Diels-Alder and Passerini multicomponent reactions are highlighted. Although the starting materials for "on water" reactions are not water-soluble, the reactions nevertheless occur smoothly, in a selective manner, and give high yields. Reactions in superheated water are also significant for prebiotic chemistry. Properties of superheated water, under certain conditions of temperature and pressure, enable water to dissolve organic compounds, since water changes its polarity and behaves as acetone. In addition, superheated water can also act as an acid/base catalyst, as its Kw changes. Solventless and solid-solid reactions are relevant to the chemistry on asteroids, in addition to the general prebiotic chemistry. Selected examples of the key organic reactions that occur under these conditions are shown. They include esterifications, etherifications, aldol-type condensations, and preparations of the large cyclic molecules. The latter could serve as primitive enzymes.

ACKNOWLEDGMENTS

Thanks are expressed to the Wisconsin Space Grant Consortium/NASA for their continuous support of my experimental work in prebiotic chemistry. Special thanks are due to my research students, who have performed experimental work, especially in the area of the "on water" and solid-state chemistry.

REFERENCES

Ahluwalia, V. K. 2012. *Green Chemistry: Environmentally Benign Reactions*, 2nd ed. Boca Raton, FL: CRC Press/Taylor & Francis.

Akiya, N., and P. E. Savage. 2002. Roles of water for chemical reactions in high-temperature water. *Chem. Rev.* 102: 2725–2750.

Avola, S., M. Guillot, D. da Silva-Perez, S. Pellet-Rostaing, W. Kunz, and F. Goettmann. 2013. Organic chemistry under hydrothermal conditions. *Pure. Appl. Chem.* 85(1): 89–103.

Braakman, R. 2013. Mapping metabolism onto the prebiotic organic chemistry of hydrothermal vents. *Proc. Natl. Acad. Sci. U.S.A.* 110: 13236–13237.

Breslow, R. 1991. Hydrophobic effects on simple organic reactions in water. *Acc. Chem. Res.* 24: 159–164.

Breslow, R. 2006. The hydrophobic effect in reaction mechanism studies and in catalysis by artificial enzymes. *J. Phys. Org. Chem.* 19: 813–822.

Carey, F.A., and R. J. Sundberg. 2007. *Advanced Organic Chemistry. Part B: Reactions and Synthesis*, 5th ed. New York: Kluwer/Plenum Publisher.

Cave, G. W. V., C. L. Raston, and J. L. Scott. 2001. Recent advances in solventless organic reactions: towards benign synthesis with remarkable versatility. *Chem. Commun.* 21: 2159–2169.

Chanda, A., and V. V. Fokin. 2009. Organic synthesis 'on water'. *Chem. Rev.* 109: 725–748.

da Silva, F de C. 2007. Organic reactions in superheated water. *Res. J. Chem. Environ.* 11: 72–73.

Hazen, R. M., N. Boctor, J. A., Brandes et al. 2002. High pressure and the origin of life. *J. Phys. Condens. Matter* 14: 11489–11494.

Holm, N. G., and E. M. Andersson. 2000. Hydrothermal systems. In *The Molecular Origins of Life: Assembling Pieces of the Puzzle*, A. Brack (Ed.), pp. 86–99. Cambridge, UK: Cambridge University Press.

Hooper, M. H., and B. DeBoef. 2009. A green multicomponent reaction for the organic chemistry laboratory: the aqueous Passerini reaction. *J. Chem. Ed.* 86: 1077–1079.

Huang, Y., A. Yazbak, and A. Dömling. 2012. Multicomponent reactions. In *Green Techniques for Organic Synthesis and Medicinal Chemistry*, W. Zhang, and B. W. Cue Jr. (Eds.), pp. 497–522. Chichester, UK: Wiley.

Katritzky, A. R., D. A. Nichols, M. Siskin, R. Murugan, and H. Balasubramanian. 2001. Reactions in high-temperature aqueous media. *Chem. Rev.* 101: 837–892.

Kaupp, G. 2005. Organic solid-state reactions with 100% yield. *Top. Curr. Chem.* 254: 95–183.

Klijn, J. E., and J. B. F. N. Engberts. 2005. Fast reactions 'on water'. *Nature* 435: 746–747.

Kolb, V. M. 2010. On the applicability of the green chemistry principles to sustainability of organic matter on asteroids. *Sustainability* 2: 1624–1631.

Kolb, V. M. 2012. On the applicability of solventless and solid state reactions to the meteoritic chemistry. *Int. J. Astrobiol.* 11: 43–50.

Kolb, V. M. 2014. Prebiotic chemistry in water and in the solid state. In *Astrobiology: An Evolutionary Approach*, V. M. Kolb (Ed.), pp. 199–216. Boca Raton, FL: CRC Press/Taylor &Francis.

Kolb, V. M. 2015. Prebiotic reactions in superheated water. *Proceedings of the SPIE 9606, Instruments, Methods, and Missions for Astrobiology XVII*, 96060D. doi:10.1117/12.2180606.

Kolb, V. M. 2016. *Green Organic Chemistry and Its Interdisciplinary Applications*. Boca Raton, FL: CRC Press/Taylor & Francis.

Kruse, A. and E. Dinjus. 2012. Sub- and supercritical water. In *Water in Organic Synthesis*, S. Kobayashi (Ed.), pp. 749–771. Stuttgart, Germany: Thieme.

Kumar, A. 2001. Salt effects on Diels–Alder reaction kinetics. *Chem. Rev.* 101: 1–19.

Lancaster, M. 2002. *Green Chemistry: An Introductory Text*, pp. 135–154. Cambridge, UK: Royal Society of Chemistry.

Kwok, S. 2017. Abiotic synthesis of complex organics in the universe. *Nat. Astron.* 1: 642–643.

Kwok, S. 2016. Complex organics in space from Solar System to distant galaxies. *Astron. Astrophys. Rev.* 24: 8. doi:10.1007/s00159-016-0093-y.

Li, C.-J., and T-H. Chan. 2007. *Comprehensive Organic Reactions in Aqueous Media*, 2nd ed. Hoboken, NJ: Wiley.

Lindström, U. M. (Ed.). 2007. *Organic Reactions in Water: Principles, Strategies and Applications*. Oxford, UK: Blackwell Publishing.

Lindström, U. M. 2002. Stereoselective organic reactions in water. *Chem. Rev.* 102: 2751–2772.

Liotta, C. L., J. P. Hallett, P. Pollet, and C. A. Eckert. 2007. Reactions in near critical water. In *Organic Reactions in Water, Principles, Strategies and Applications*, U. M. Lindström, U. M. (Eds.), pp. 256–300. Oxford, UK: Blackwell.

Liu, L., and D. Wang. 2010. "On water" for green chemistry. In *Handbook of Green Chemistry. Volume 5: Reactions in Water*, C.-J. Li (Ed.), pp. 207–228. Weinheim, Germany: Wiley-VCH.

Longstaff, A. 2015. *Astrobiology: An Introduction*. Boca Raton, FL: CRC Press/Taylor &Francis.

Manna, A., and A. Kumar. 2013. Why does water accelerate organic reactions under heterogeneous condition? *J. Phys. Chem.* 117: 2446–2454.

Mason, S.F. 1991. *Chemical Evolution, Origins of the Elements, Molecules and Living Systems*. Oxford, UK: Oxford University Press.

Martin, W., J. Baross, D. Kelley, and M. J. Russell. 2008. Hydrothermal vents and the origin of life. *Nat. Rev., Microbiol.*, 6: 805–814.

Miller, S.L., and L. E. Orgel. 1974. *The Origins of Life on Earth*. Englewood Cliffs, NJ: Prentice Hall.

Mullen, L. B., and J. D. Sutherland. 2007. Simultaneous nucleotide activation and synthesis of amino acid amides by a potentially prebiotic multi-component reaction. *Angew. Chem. Int. Ed.* 46: 8063–8066.

Narayan, S., J. Muldoon, M. G. Finn, V. V. Fokin, H. H. Kolb., and K. B. Sharpless. 2005. "On water": unique reactivity of organic compounds in aqueous suspension. *Angew. Chem., Int. Ed.* 44: 3275–3279.

Orgel, L. E. 2004. Prebiotic chemistry and the origin of the RNA world. *Crit. Rev. Biochem. Mol. Biol.* 9: 99–123.

Otto, S., and J. B. F. N. Engberts. 2000. Diels–Alder reactions in water. *Pure Appl. Chem.* 72: 1365–1372.

Padgett, J. F. 2012. Autocatalysis in chemistry and the origin of life. In *The Emergence and Organizations and Markets*, J. F. Padgett, and W. W. Powell (Eds.), pp. 33–69. Princeton, NJ: Princeton University Press.

Pirrung, M. C., and K. D. Sarma. 2004. Multicomponent reactions are accelerated in water. *J. Amer. Chem. Soc.* 126: 444–445.

Pirrung, M. C., K. D. Sarma, and J. Wang. 2008. Hydrophobicity and mixing effects on select heterogeneous, water-accelerated synthetic reactions. *J. Org. Chem.* 73: 8723–8730.

Powner, M. W., and J. D. Sutherland. 2011. Prebiotic chemistry: A new modus operandi. *Phil.Trans. R. Soc. B* 366: 2870–2877.

Powner, M. W., J. D. Sutherland, and J. W. Szostak. 2010. Chemoselective multicomponent one-pot assembly of purine precursors in water. *J. Amer. Chem. Soc.* 132: 16677–16688.

Raston, C. L. 2004. Versatility of 'alternative' reaction media: Solventless organic syntheses. *Chemistry in Australia*, 71(4): 10–13.

Raston, C. L., and J. L. Scott. 2000. Chemoselective, solvent-free aldol condensation reaction. *Green Chem.* 2: 49–52. doi:10.1039/A907688C.

Rideout, D. C., and R. Breslow. 1980. Hydrophobic acceleration of Diels–Alder reaction. *J. Amer.Chem. Soc.* 102: 7816–7817.

Rothenberg, G., A. P. Downie, C. L. Raston, and J. L. Scott. 2001. Understanding solid/solid organic reactions. *J. Amer. Chem. Soc.* 123, 8701–8708.

Sarma, D., S. S. Pawar, S. S. Deshpander, and A. Kumar. 2006. Hydrophobic effects in a simple Diels–Alder reaction in water. *Tetrahedron Lett.* 47: 3957–3958.

Savage, P. E. 1999. Organic reactions in supercritical water. *Chem. Rev.* 99: 603–621.

Savage, P. D., and N. A. Rebacz. 2010. Water under extreme conditions for green chemistry. In *Handbook of Green Chemistry, Volume 5*, C.-J. Li (Ed.), pp. 331–361. Weinheim, Germany: Wiley-VCH.

Shapiro, N., and A. Vigalok. 2008. Highly efficient organic reactions "on water", "in water", and both. *Angew. Chem.* 120: 2891–2894.

Shaw, A. M. 2006. *Astrochemistry: From Astronomy to Astrobiology*, pp. 157–192. New York, NY: Wiley.

Schmitt-Kopplin, P., M. Harir, B. Kanawati, et al. 2014. Analysis of extraterrestrial organic matter in Murchison meteorite: a progress report. In *Astrobiology: An Evolutionary Approach*, V. M. Kolb (Ed.), pp. 63–82. Boca Raton, FL: CRC Press/ Taylor & Francis.

Schmitt-Kopplin, P., Z. Gabelica, R. D., Gougeon et al. 2010. High molecular diversity of extraterrestrial organic matter in Murchison meteorite revealed 40 years after its fall. *Proc. Natl. Acad. Sci. U.S.A*, 107: 2763–2768.

Siskin, M., and A. R. Katritzky. 2001. Reactivity of organic compounds in superheated water: General background. *Chem. Rev.* 101: 825–835.

Sojo, V., B. Herschy, A. Whicher, E. Camprubi, and N. Lane. 2016. The origin of life in alkaline hydrothermal vents. *Astrobiology* 16: 181–197.

Solomons, G. and C. Fryhle. 2013. *Organic Chemistry*, 11th ed. New York: Wiley.

Sutherland, J. D., L. B. Mullen, and F. Buchet. 2008. Potential prebiotic Passerini-type reactions of phosphates. *Synlett* 14: 2161–2163.

Tanaka, K. 2009. *Solvent-Free Organic Synthesis*, 2nd ed. Weinheim, Germany: Wiley-VCH.

Tanaka, K., and F. Toda. 2000. Solvent-free organic syntheses. *Chem. Rev.* 100: 1025–1074.

Tian, G., H. Yuan, Y. Mu, C. He, and S. Feng. 2007. Hydrothermal reactions from sodium hydrogen carbonate to phenol. *Org. Lett.* 9: 2019–2021.

Toda, F. 1993. Solid state organic reactions. *Synlett* 5, 303–312, doi 10.1055/51993-2244.

Toda, F. 1995. Solid state organic chemistry: Efficient reactions, remarkable yields, and stereoselectivity. *Acc. Chem. Res.* 28: 480–486.

Toda, F., H. Takumi, and M. Akehi. 1990. Efficient solid-state reactions of alcohols: Dehydration, rearrangement, and substitution. *J. Chem. Soc. Chem. Commun.* 0, 1270–1271. doi:10.1039/C39900001270.

Williamson, K. L., and K.M. Masters. 2011. *Macroscale and Microscale Organic Experiments*, 6th ed., pp. 699–701. Belmont, CA: Brooks/Cole.

Zhang, W., and B. W. Cue Jr. (Eds.). 2012. *Green Techniques for Organic Synthesis and Medicinal Chemistry*. Chichester, UK: Wiley.

5.6 The Origin and Amplification of Chirality Leading to Biological Homochirality

Kenso Soai, Arimasa Matsumoto, and Tsuneomi Kawasaki

CONTENTS

5.6.1 IMPLICATIONS OF HOMOCHIRALITY AND SELF-REPLICATION FOR THE ORIGIN OF LIFE

Homochirality of biomolecules, such as seen in L-amino acids and D-sugars, is not a mere phenomenon of life but one of the essential features of life. Left and right hands are mirror images and are not superimposable. In the same manner, L- and D-amino acids are mirror images and are not superimposable, that is, they are chiral compounds (Figure 5.6.1). In general, when a carbon atom has four different substituents or atoms, the carbon atom becomes a stereogenic (chiral) center and is termed an asymmetric carbon atom (Eliel and Wilen 1994).

Why is homochirality essential for life? Consider the situation of shaking hands. When two people use their right hands to shake hands, it is normal; however, if one uses the left hand and the other the right, the situation becomes very different. In peptides, for example, L-alanyl-L-alanine and D-alanyl-L-alanine are diastereomers and have different properties, such as melting points. Peptides and proteins are composed of L-amino acids by forming peptide bonds. If D-amino acids are irregularly incorporated, the conformation of the protein changes and the enzymes would not be able to exhibit their functions. DNA is composed of D-deoxyribose and forms a helix. If L-deoxyribose is irregularly incorporated, the normal double helix structure would not be formed and the gene information would not be transferred. Thus, the homochirality of components is essential for living organisms and the origin of life.

Then, how did biomolecules become highly enantioenriched? How was the first chiral organic compound formed? It should be mentioned that usual chemical reactions using achiral reactants and without the intervention of any chiral factor afford a mixture with an equal ratio of D- and L-products, that is, racemates. Therefore, the origin of the chirality of organic compounds has been a subject of considerable attention for 170 years, ever since the discovery of molecular chirality by Pasteur (Gal 2008).

Several theories have been proposed as the origin of chirality of organic compounds (Weissbuch et al. 1984; Kondepudi et al. 1990; Inoue 1992; Bolli et al. 1997; Siegel 1998; Feringa and van Delden 1999; Ribó et al. 2001; Hazen and Sholl 2003; Mislow 2003; Pizzarello and Weber 2004; Raval 2009; Breslow and Cheng 2010; Weissbuch and Lahav 2011; Ernst 2012; Gellman et al. 2013; Gonzalez-Campo and Amabilino 2013; Saito and Hyuga 2013; Olsson et al. 2015; Miyagawa et al. 2017). However, because the enantiomeric excesses (ee) induced by the proposed mechanisms have been very low in

FIGURE 5.6.1   (a) General structure of L- and D-amino acids (R = CH₃: alanine). (b) The general structure of asymmetric carbon, i.e., C (carbon) with four different substituents (a ≠ b ≠ c ≠ d). (c) D-Deoxyribose (natural) and L-deoxyribose (unnatural) in nucleosides.

many cases, the amplification of low ee to high ee is necessary (Girard and Kagan 1998; Green et al. 1999; Zepik et al. 2002; Viedma 2005; Soloshonok et al. 2012; Soai et al. 2014a). In this chapter, we focus on asymmetric autocatalysis and the origin of homochirality. Asymmetric autocatalysis is a reaction in which a chiral product acts as a chiral catalyst for its own production (Figure 5.6.2) (Soai et al. 1995, 2000, 2014b, 2017; Shibata et al. 1999; Avalos et al. 2000; Todd 2002; Sato et al. 2003a; Soai and Kawasaki 2008; Bissette and Fletcher 2013). The process is a catalytic self-replication of a chiral compound. In the asymmetric autocatalysis of pyrimidyl alkanol in the addition of diisopropylzinc (*i*-Pr₂Zn) to pyrimidine-5-carbaldehyde, the ee amplifies from extremely low to >99.5% ee (the Soai reaction). Various chiral factors, such as circularly polarized light (CPL) and chiral inorganic crystals, trigger asymmetric autocatalysis and afford the product with high ee. Moreover, spontaneous absolute asymmetric synthesis without the intervention of any chiral factor has been achieved by the Soai reaction.

Self-replication, that is, automultiplication, at the individual and cellular levels is another essential feature of life. Self-replication at the molecular level has been examined (Sievers and von Kiedrowski 1994; Wintner et al. 1994; Lee et al. 1996; Ashkenasy et al. 2004, 2017). Asymmetric autocatalysis of pyrimidyl alkanol, that is, the Soai reaction, is unique in that the process is a catalytic self-replication of a chiral compound.

As to the theories of the origin of homochirality, mechanisms other than asymmetric autocatalysis, including crystallization with racemization, self-disproportionation by distillation, crystallization with abrasion (Viedma ripening), and amino acids in meteorites, have been reported. These are beyond the scope of this article, and readers are encouraged to consult the corresponding articles and reviews.

5.6.2   ASYMMETRIC AUTOCATALYSIS OF 5-PYRIMIDYL ALKANOL WITH AMPLIFICATION OF ENANTIOMERIC EXCESS: THE SOAI REACTION

Frank (1953) proposed a scheme of asymmetric autocatalysis without mentioning the actual chemical structure of compounds. We found in 1995, for the first time, the asymmetric autocatalysis with amplification of ee as a real chemical reaction in the enantioselective addition of *i*-Pr₂Zn to pyrimidine-5-carbaldehyde (Soai et al. 1995). After further investigation of the substituent effect, that is, by the introduction of a 2-alkynyl group at the 2-position of the pyrimidine ring (Shibata et al. 1999), the isopropylzinc alkoxide of 2-alkynyl-5-pyrimidyl alkanol **1** (Matsumoto et al. 2015a, 2016a) formed *in situ* acted as a practically perfect asymmetric autocatalyst during the course of enantioselective addition of *i*-Pr₂Zn to pyrimidine-5-carbaldehyde **2** (Figure 5.6.3). It has been demonstrated that 5-pyrimidyl alkanol **1** shows remarkable amplification of ee in the addition reaction of *i*-Pr₂Zn to aldehyde **2**, therefore, an initial slight enantiomeric imbalance, as low as approximately 0.00005% ee, was significantly amplified to nearly enantiopure (>99.5% ee) in only three consecutive asymmetric autocatalyses of **1** by the reaction between **2** and *i*-Pr₂Zn (Sato et al. 2003a). During these three consecutive reactions, the initial slight excess of the (*S*)-enantiomer **1** was automultiplied by a factor of *ca.* 630,000, whereas the slightly minor initial (*R*)-**1** was automultiplied by a factor of less than 1,000. Therefore, the significant asymmetric amplification of pyrimidyl alkanol **1** has been realized during its formation, without the intervention of any other chiral auxiliary.

5.6.3   MECHANISTIC INSIGHT OF THE ASYMMETRIC AUTOCATALYSIS OF PYRIMIDYL ALKANOL

The mechanism of this significant asymmetric amplification of pyrimidyl alkanol **1** has not yet been fully revealed. However, many research groups have contributed to understanding and gradually unveiling this unique amplification mechanism. The amplification mechanism is considered an analogy of the positive nonlinear effect in nonautocatalytic asymmetric reactions (Puchot et al. 1986; Kitamura et al. 1989), such as the dimer as a catalyst or a nonreactive reservoir (Figure 5.6.4).

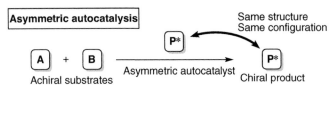

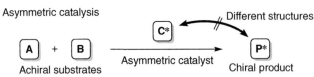

FIGURE 5.6.2   The concept of asymmetric autocatalysis in comparison with conventional asymmetric catalysis.

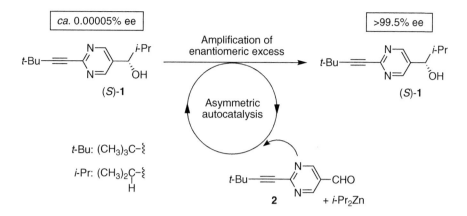

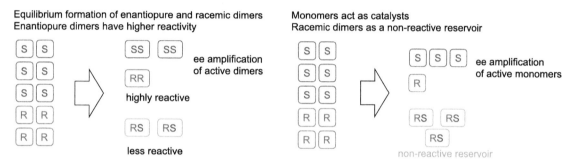

FIGURE 5.6.3 The Soai reaction. Asymmetric autocatalysis with significant amplification of ee from *ca.* 0.00005% ee to nearly enantiopure (>99.5% ee).

Equilibrium formation of enantiopure and racemic dimers
Enantiopure dimers have higher reactivity

ee amplification
of active dimers

highly reactive

less reactive

Monomers act as catalysts
Racemic dimers as a non-reactive reservoir

ee amplification
of active monomers

non-reactive reservoir

FIGURE 5.6.4 Amplification of ee via oligomerization and possible oligomer units.

In the asymmetric autocatalysis of pyrimidyl alkanol, the formation of higher oligomeric structures via Zn–O–Zn–O square (Figure 5.6.5a) and macrocyclic Zn–N coordination (Figure 5.6.5b) has been suggested. In early kinetic studies (Blackmond et al. 2001; Sato et al. 2001a, 2003b) and experimental observation of ee reaction profiles, the dimers appear to act as an actual catalyst rather than a nonreactive reservoir. Nuclear magnetic resonance (NMR) studies suggest an equilibrium between homochiral and racemic square dimers (Gridnev et al. 2004), and density functional theory (DFT) calculation studies suggest no significant stability difference between homochiral and heterochiral dimers (Gridnev and Brown 2004). According to these results and a more detailed kinetics study (Buono and Blackmond 2003), the reaction cycle in which an active catalyst dimer with two aldehydes afforded tetramer resting state and released two new active dimers was proposed by a DFT

calculation (Schiaffino and Ercolani 2008). However, recent studies on the reaction model suggest that the simple dimer catalyst model is not enough to explain the significant amplification of ee (Micheau et al. 2012). Furthermore, the existence of higher oligomer species via N–Zn coordination was also supported by NMR observations (Quaranta et al. 2010; Gehring et al. 2012).

We have recently observed the actual oligomeric structure of the asymmetric autocatalyst by using single-crystal X-ray diffraction analysis (Matsumoto et al. 2015a, 2016a). Crystallization of both enantiopure and racemic pyrimidyl alkanol **1** with an excess amount of *i*-Pr₂Zn afforded the tetrameric structure of the isopropylzinc alkoxide of alkanol **1**. In this tetrameric structure, two zinc alkoxides form a square dimer by the Zn–O–Zn–O structure, and the two dimers are bridged by pyrimidine ring coordination to the alkoxide Zn atom of each other to form a 12-membered macrocyclic

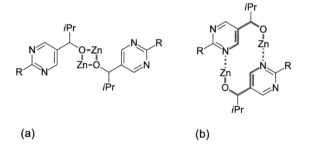

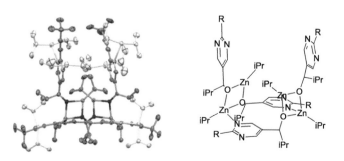

FIGURE 5.6.5 Planar description of dimeric structures of asymmetric autocatalyst with (a) Zn–O–Zn–O square and (b) macrocyclic Zn–N coordination.

FIGURE 5.6.6 Single-crystal X-ray structure of the isopropylzinc alkoxide of the enantiopure asymmetric autocatalyst.

structure (Figure 5.6.6). In the presence of an almost stoichiometric amount of *i*-Pr$_2$Zn, the one-dimensional oligomeric structure was obtained by further formation of the macrocyclic structure of Zn–O–Zn square dimers. These results also support the involvement and importance of higher-order oligomer structures for the significant asymmetric amplification and spontaneous symmetry breaking.

5.6.4 ASYMMETRIC AMPLIFICATION INDUCED BY CHIRAL COMPOUNDS IN CONJUNCTION WITH ASYMMETRIC AUTOCATALYSIS

Asymmetric autocatalysis of 5-pyrimidyl alkanol **1** can significantly enhance the slight enantiomeric imbalance. Therefore, if the initial asymmetry of the autocatalyst **1** could be introduced by chiral compounds other than the autocatalyst **1**, it can be expected to obtain pyrimidyl alkanol **1** with detectable ee by the subsequent significant asymmetric amplification during autocatalytic formation of **1** by the reaction between aldehyde **2** and *i*-Pr$_2$Zn. The absolute configuration of **1** should be controlled by the absolute handedness of the originally used external chiral factor; thus, it has been found that various chiral compounds, even those whose asymmetric induction power is assumed to be weak, can act as chiral initiators of asymmetric autocatalysis (Soai et al. 2014a).

Meanwhile, propylene oxide is known as the first interstellar chiral compound, found in the Sagittarius B2 star-formation region (McGuire et al. 2016). The enantioselective addition of *i*-Pr$_2$Zn to pyrimidine-5-carbaldehyde **2** was examined using (*S*)- and (*R*)-(+)-propylene oxide with low to high ee as a source of chirality (Figure 5.6.7) (Kawasaki et al. 2004). When the addition of *i*-Pr$_2$Zn to pyrimidine-5-carbaldehyde **2** was conducted in the presence of (*R*)-(+)-propylene oxide with 97% ee, (*S*)-5-pyrimidyl alkanol **1** with 96% ee was obtained after the asymmetric autocatalysis with amplification of ee. By contrast, in the presence of (*S*)-(−)-propylene oxide with 97% ee, oppositely configured (*R*)-**1** with 94% ee was formed. Even when propylene oxide with low (2–3%) ee was utilized, the same stereochemical relationship could be observed, that is, the (*R*)-(+) isomer with only 3% ee and (*S*)-(−)-isomer with 2% ee were found to serve as chiral initiators in the asymmetric autocatalysis, producing (*S*)-**1** and (*R*)-**1** with high ee, respectively. Thus, the chiral interstellar chiral

compound propylene oxide, acting as the source of chirality, is correlated with highly enantioenriched 5-pyrimidyl alkanol **1** *via* the asymmetric autocatalysis with amplification of ee. Styrene oxide also is a chiral initiator of asymmetric autocatalysis (Kawasaki et al. 2004).

It should also be mentioned that extraterrestrial amino acids detected in meteorites often exhibit enantioenrichments (Cronin and Pizzarello 1997). This shows that space is a candidate for the place of origin of homochirality. In fact, the Rosetta spacecraft of the European Space Agency (ESA) with a probe was launched to perform the *in situ* analysis of chirality of amino acids on a comet (Myrgorodska et al. 2015).

The existence in space of polycyclic aromatic hydrocarbons (PAHs) has been discussed (Tielens 2005). For example, fullerenes have been detected in meteorites (Becker et al. 1994) and the planetary nebula (Cami et al. 2010). Helicenes are chirally relevant to PAHs with right- and left-handed helical structures, and the asymmetric photosynthesis of [6]helicine has been reported by irradiation with CPL, followed by oxidation (Kagan et al. 1971). We have reported that chiral [6] and [5]helicenes acted as highly efficient chiral inducers in the enantioselective synthesis of a pyrimidyl alkanol **1** by the addition of *i*-Pr$_2$Zn to pyrimidine-5-carbaldehyde **2** in combination with asymmetric autocatalysis (Figure 5.6.8) (Sato et al. 2001b). It should be noted that the asymmetric induction using chiral hydrocarbons without heteroatoms as a catalyst or ligand would be a challenge from the synthetic point of view.

When pyrimidine-5-carbaldehyde **2** was treated with *i*-Pr$_2$Zn in the presence of (*P*)-(+)-[6]helicene, after the amplification of ee by asymmetric autocatalysis, (*S*)-alkanol **1** was formed in 95% ee. On the other hand, (*M*)-(−)-[6]helicene induced the production of (*R*)-**1** with 93% ee. Even if the ee values of (*P*)-(+)- and (*M*)-(−)-[6]helicenes were lowered to *ca*. 0.13% and 0.54% ee, asymmetric autocatalysis afforded (*S*)- and (*R*)-**1** with the same stereochemical relationships, respectively. (*P*)-(+)- and (*M*)-(−)-[5]helicenes also gave (*S*)- and (*R*)-**1**, respectively. Therefore, one of the PAHs, [6] and [5]helicenes could also act as an origin of chirality of asymmetric autocatalysis. 1,1′-Binaphthyl (Sato et al. 2002) and finite single-wall carbon nanotube molecules (Hitosugi et al. 2014) can also be utilized as chiral inducers for asymmetric autocatalysis.

Since the pioneering work by Miller (1953), it has been demonstrated that racemic amino acids could be synthesized

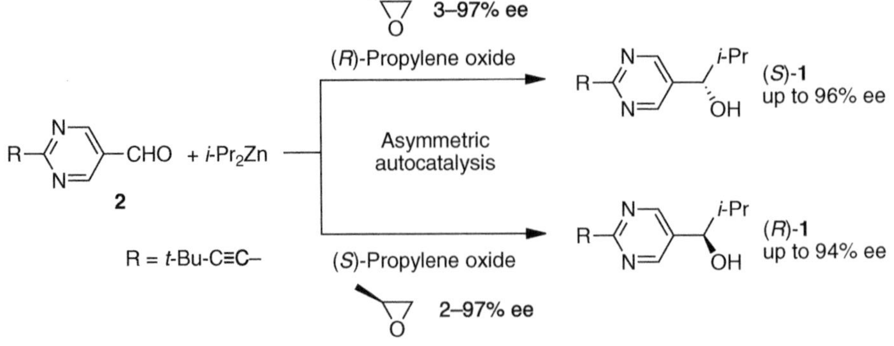

FIGURE 5.6.7 Highly enantioselective synthesis induced by propylene oxide in conjunction with asymmetric autocatalysis.

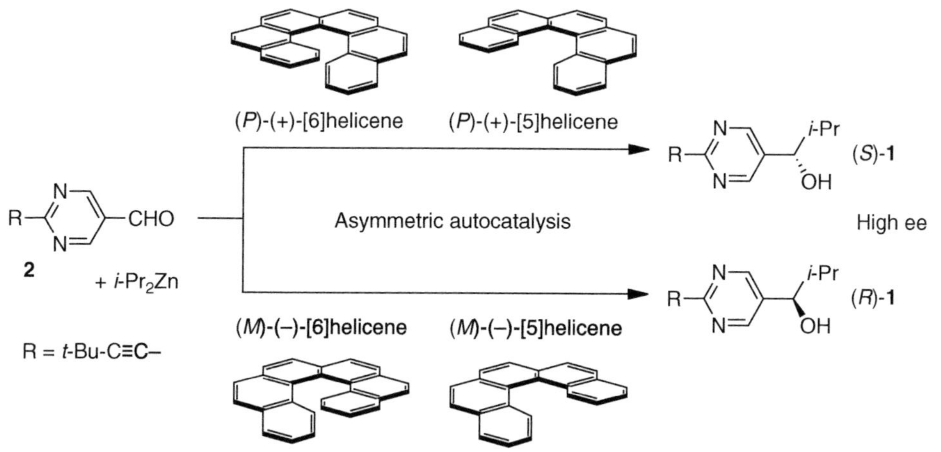

FIGURE 5.6.8 Asymmetric autocatalysis triggered by [6] and [5]helicenes with helical chirality.

under primitive earth conditions with an electric discharge (Miller and Urey 1959). Formation of amino acids and asymmetric induction by irradiation with CPL (Meierhenrich et al. 2005) have also been reported under simulated interstellar conditions (Bernstein et al. 2002; Munoz Caro et al. 2002). Furthermore, enantiomeric enrichment of amino acids was detected in a meteorite (Cronin and Pizzarello 1997; Engel and Macko 1997). Thus, in combination with a suitable amplification pathway, that is, asymmetric autocatalysis, highly enantioselective synthesis induced by the prebiotic amino acids becomes an important approach toward understanding the origin of biological homochirality, such as in L-amino acids.

We have demonstrated the asymmetric autocatalysis induced by enantioenriched proteinogenic amino acids (Shibata et al. 1998; Sato et al. 2007). When asymmetric autocatalysis was initiated in the presence of L-alanine with high ee, the production of (S)-5-pyrimidyl alkanol **1** with 92% ee was observed, whereas D-alanine gave (R)-alkanol **1** with 90% ee (Figure 5.6.9). Further investigations were conducted using alanine with low ee. When L-alanine with ca. 10% ee was used as the chiral initiator, the (S)-alkanol **1** with 94% ee was obtained. Even when the ee of L-alanine was as low as ca. 1% and ca. 0.1% ee, (S)-**1** was produced with high ee. On the other hand, asymmetric autocatalysis in the presence of D-alanine with low ee (ca. 10%, 1%, and 0.1% ee) gave oppositely configured (R)-**2**. Therefore, even if the enantiomeric

imbalance of alanine is as low as ca. 0.1% ee, it can work as the origin of chirality in asymmetric autocatalysis to give a large amount of highly enantioenriched organic compound **1**.

Serine has been considered one of the prebiotic molecules (Ring et al. 1972), and a chiral effect in the formation of the aggregate has been reported (Nanita and Cooks 2006). It has also been reported that the crystallization of racemic serine from aqueous sulfuric acid gives chiral crystals of DL-diserinium sulfate monohydrate that contain an equimolar amount of D- and L-serine with (P)- or (M)-crystal chirality. We have demonstrated that the chiral crystal formed from racemic serine can induce initial ee toward the autocatalyst **1** by the reaction between **2** and i-Pr₂Zn to initiate the asymmetric autocatalytic amplification of ee (Figure 5.6.10) (Kawasaki et al. 2011a). Thus, the enantioselective addition of i-Pr₂Zn to **2** was conducted in the presence of enantiomorphs of racemic serine as a heterogeneous chiral source, because dissolution of the crystal of serine causes the disappearance of the chirality. As a result, the (P)-crystal of rac-serine induced the production and propagation of (S)-pyrimidyl alkanol **1** with 94% ee by asymmetric autocatalysis with amplification of ee. On the other hand, the reaction between aldehyde **2** and i-Pr₂Zn in the presence of (M)-crystal afforded (R)-**1** with 95% ee. The ee value was achieved to be greater than 99.5% ee by an additional round of asymmetric autocatalysis. Therefore, highly enantioselective synthesis was achieved utilizing the crystal chirality, including racemic serine.

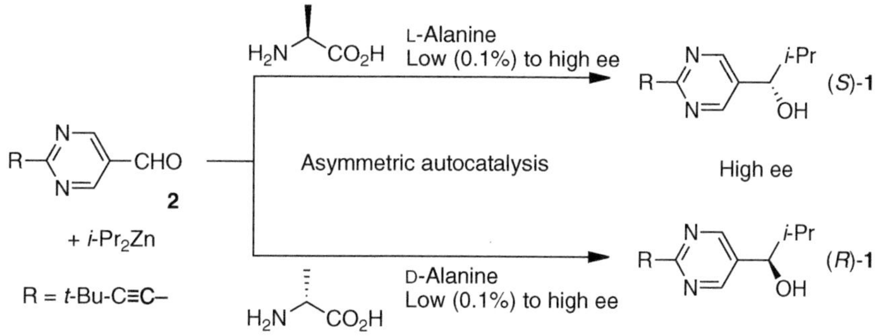

FIGURE 5.6.9 L- and D-Alanine-induced asymmetric autocatalysis to give (S)- and (R)-5-pyrimidyl alkanols **1** with high enantiomeric purity.

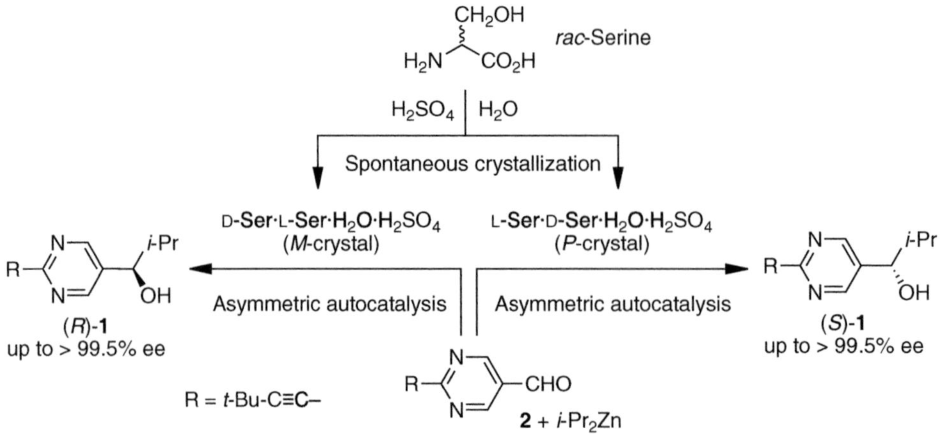

FIGURE 5.6.10   Asymmetric autocatalysis initiated with a chiral crystal of racemic serine.

5.6.5   ENANTIOSELECTIVE SYNTHESIS INDUCED BY CIRCULARLY POLARIZED LIGHT

The occurrence of strong CPL in the infrared region is observed in the star-formation region of the Orion nebula (Bailey et al. 1998). As investigated previously, left (*l*)- and right (*r*)-handed CPL is an external chiral physical force to induce ee in chiral compounds (Kagan et al. 1971; Balavoine et al. 1974; Bonner and Rubenstein 1987; Suarez and Schuster 1995). Thus, CPL has been proposed as a possible candidate for the origin of chirality. However, the degree of ee induced by irradiation with CPL is usually too small to be associated with overwhelming enantioenrichment in the compounds found in nature because of the very small anisotropy factors of organic compounds. The irradiation by CPL of racemic organic compounds such as leucine induces enantioenrichment with only *ca.* 2% ee (Inoue 1992). The correlation between these low ee and the homochirality of bioorganic compounds has not been clear. Therefore, we examined the direct irradiation by CPL of the racemic asymmetric autocatalyst **1** (Figure 5.6.11). Direct irradiation of *rac-***1** by *l*-CPL and the subsequent asymmetric autocatalysis afford highly enantioenriched (*S*)-alkanol **1** with >99.5% ee. By contrast, irradiation with *r*-CPL gives (*R*)-**1** with >99.5%

ee. The overall reaction provides the first direct correlation of the handedness of CPL with that of an organic compound with significantly high ee.

5.6.6   ASYMMETRIC AUTOCATALYSIS TRIGGERED BY ISOTOPICALLY CHIRAL COMPOUNDS AND ISOTOPOMERS OF METEORITIC AMINO ACIDS

Many apparently achiral organic molecules may be chiral by considering the randomly labeled isotopes in the enantiotopic moiety. However, it is extremely difficult to recognize the chirality arising from the isotopic substitution because the chirality originates from the small difference between the numbers of neutrons in the atomic nuclei. We have reported asymmetric autocatalysis triggered by the chiral isotopomers arising from carbon isotope ($^{12}C/^{13}C$) substitution (Figure 5.6.12) (Kawasaki et al. 2009a). The carbon isotope chirality induces enantioselectivity in asymmetric autocatalysis; therefore, enantiomerically amplified alkanol **1** can be obtained with chirality corresponding to that of the carbon isotope substitution. This is the first example of such a chiral effect, that is, asymmetric induction by chiral compounds arising from the substitution of carbon isotopes ($^{12}C/^{13}C$).

On the other hand, the simplest achiral α-amino acid, glycine, was detected in the coma of a comet (Sandford et al. 2006; Altwegg et al. 2016) and was also synthesized under prebiotic conditions (Miller 1953). Moreover, in addition to achiral glycine and α-methylalanine (Pizzarello and Huang 2005), L-enriched chiral α-methyl-substituted amino acids were identified in meteorites as deuterium-enriched forms (Engel and Macko 1997). We focused on the hydrogen isotope chirality in the apparently achiral meteoritic amino acids. When one of the hydrogen atoms of the methylene group of glycine or one methyl group of α-methylalanine is enantiomerically deuterated, these compounds become chiral because of the deuterium substitution. Thus, asymmetric

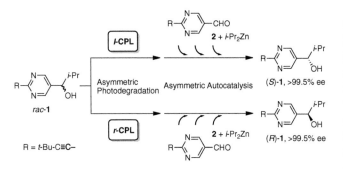

FIGURE 5.6.11   Asymmetric autocatalysis initiated by direct irradiation with CPL.

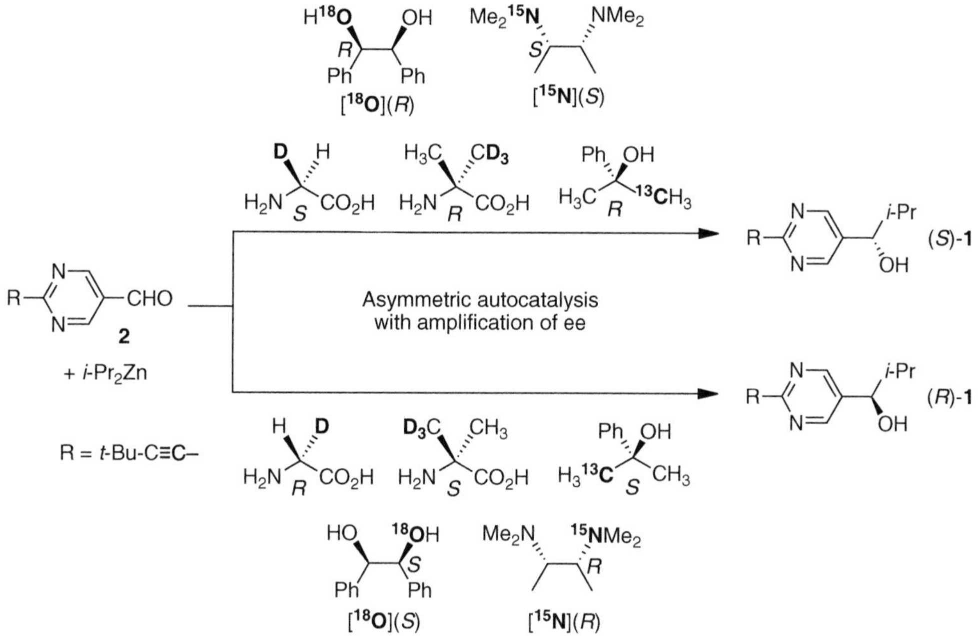

FIGURE 5.6.12 Asymmetric autocatalysis triggered by the chiral isotopomers with hydrogen (D/H), nitrogen (^{15}N/^{14}N), carbon (^{13}C/^{12}C), and oxygen (^{18}O/^{16}O) isotope substitution.

autocatalysis triggered by chiral meteoritic amino acids would become one approach toward the extraterrestrial origin of biological homochirality.

We have asymmetrically synthesized the isotopomers of glycine-α-*d* and α-methyl-d_3-alanine, which were used as chiral initiators in asymmetric autocatalysis (Figure 5.6.12) (Kawasaki et al. 2009b). When *i*-Pr$_2$Zn was reacted with aldehyde **2** in the presence of (*S*)-glycine-α-*d*, (*S*)-**1** with high ee was obtained. In the presence of (*R*)-glycine-α-*d*, on the other hand, (*R*)-**1** was formed after the asymmetric autocatalysis. Further investigations show that α-methyl-d_3-alanine also acts as a chiral source in asymmetric autocatalysis with amplification of ee. We have reported that the partially deuterated α-methylalanine derivatives could also work as highly efficient chiral triggers for asymmetric autocatalysis (Kawasaki et al. 2011b).

Nitrogen isotope (^{15}N) enrichment was also reported in the meteoritic organic compounds (Engel and Macko 1997). We have demonstrated that the chiral nitrogen isotopomer of a diamine can also act as a chiral trigger for asymmetric autocatalysis to afford highly enantioenriched alkanol **1** with the corresponding absolute configurations to that of N^2,N^2,N^3,N^3-tetramethyl-2,3-butanediamine containing nitrogen isotope (^{14}N/^{15}N) chirality (Figure 5.6.12) (Matsumoto et al. 2016b). The addition of *i*-Pr$_2$Zn to pyrimidine-5-carbaldehyde **2** in the presence of [^{15}N](*S*)-diamine afforded (*S*)-alkanol **1**. By contrast, (*R*)-alkanol **1** was obtained from the reaction initiated with [^{15}N](*R*)-diamine. The ee was amplified to >99.5% ee by a further asymmetric autocatalytic reaction by using the obtained pyrimidyl alkanol **1**. In addition, asymmetric autocatalysis

has enormous power to discriminate and amplify the chirality of the isotopomers with ^{18}O/^{16}O substitutions (Kawasaki et al. 2011c; Matsumoto et al. 2013).

As described, the present results should increase the implications of chiral isotopomers in the study of the origin of biological chirality.

5.6.7 ASYMMETRIC AUTOCATALYSIS WITH AMPLIFICATION OF ENANTIOMERIC EXCESSES INITIATED WITH A CHIRAL CRYSTAL OF ACHIRAL NUCLEOBASES CYTOSINE AND ADENINE

Nucleobases are essentially achiral flat molecules and are considered to have existed on the prebiotic earth before the RNA world (Gilbert 1986). For example, cytosine could be efficiently synthesized from interstellar molecules, that is, hydrogen cyanide and cyanoacetylene (Robertson and Miller 1995; Shapiro 1999). The isolation of adenine from meteorites (Callahana et al. 2011) and prebiotic synthesis, that is, pentamerization of hydrogen cyanide under prebiotic conditions (Oro 1960; Schwartz et al. 1982), has been reported. Therefore, the investigation of enantioselective synthesis based on crystal chirality arising from an achiral nucleobase is an important experimental approach to understanding the origins of both biological homochirality and life.

Cytosine crystallizes spontaneously from methanol into a chiral form (space group: $P2_12_12_1$) that exhibits either plus or minus Cotton effects in the solid-state circular dichroism (CD) spectra

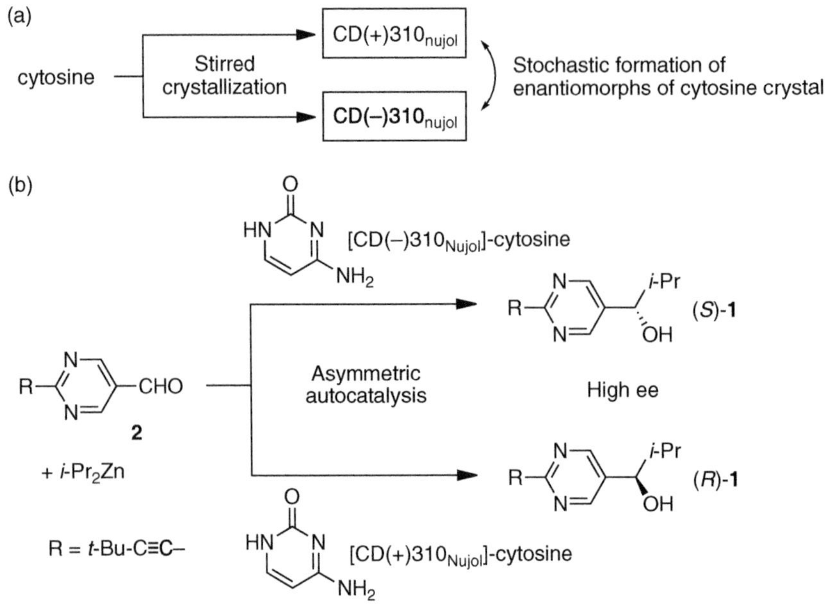

FIGURE 5.6.13 (a) Stirred crystallization of cytosine. (b) Highly enantioselective asymmetric autocatalysis induced by a spontaneously generated chiral crystal of achiral nucleobase cytosine.

at *ca.* 310 nm in Nujol. The enantiomorphs of cytosine can be obtained spontaneously using stirred crystallization (Figure 5.6.13a) (Kawasaki et al. 2008a). Chiral crystals of cytosine have been used as chiral initiators for asymmetric autocatalysis. In the presence of [CD(−)310$_{Nujol}$]-cytosine crystals, the reaction between aldehyde **2** and *i*-Pr$_2$Zn gives enantioenriched (*S*)-**1** in conjunction with asymmetric autocatalysis (Figure 5.6.13b). By contrast, [CD(+)310$_{Nujol}$]-cytosine crystals trigger the production of (*R*)-**1**. Furthermore, we have demonstrated the enantiospecific formation of chiral cytosine crystals by the dehydration of crystal water of achiral cytosine monohydrate under thermal and vacuum conditions (Kawasaki et al. 2010a; Mineki et al. 2013).

Crystallization of adenine from aqueous nitric acid produces adeninium dinitrate (adenine·2HNO$_3$) in enantiomorphic form (space group: $P2_12_12_1$). The chirality of adenine·2HNO$_3$ can be discriminated using solid-state CD analysis with Nujol (Mineki et al. 2012). Thus, one crystal exhibited a positive cotton effect at 250 nm ([CD(+)250$_{Nujol}$]-adenine·2HNO$_3$), while the other enantiomorph had a negative Cotton effect at the same wavelength ([CD(−)250$_{Nujol}$]-adenine·2HNO$_3$).

The enantiomorphs of adenine were subjected to the enantioselective addition of *i*-Pr$_2$Zn to pyrimidine-5-carbaldehyde **2** as a source of chirality (Figure 5.6.14). The reaction of **2** with *i*-Pr$_2$Zn in the presence of finely powdered crystalline [CD(−)250$_{Nujol}$]-adeninium dinitrate resulted in the formation of (*S*)-5-pyrimidyl alkanol **1** with >99.5 ee after the asymmetric amplification. By contrast, with enantiomorphous [CD(+)250$_{Nujol}$]-adeninium dinitrate as the chiral initiator, the opposite (*R*)-**1** was obtained with >99.5% ee. The absolute configuration of the produced alkanol **1** with high ee obtained was controlled efficiently by the crystal chirality of the adenine salt in conjunction with asymmetric autocatalysis. Therefore, the crystal chirality of the prebiotic nucleobases cytosine and adenine can be responsible for the synthesis of nearly enantiopure organic compounds in conjunction with asymmetric autocatalysis.

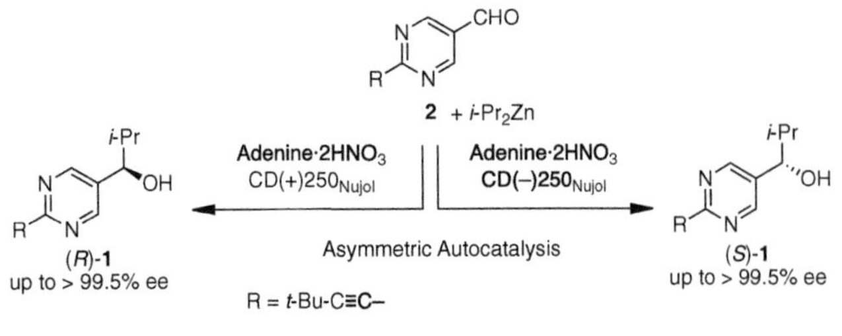

5.6.8 ASYMMETRIC AUTOCATALYSIS WITH AMPLIFICATION OF ENANTIOMERIC EXCESSES TRIGGERED BY CHIRAL CRYSTALS COMPOSED OF ACHIRAL ORGANIC COMPOUNDS

In addition to cytosine and adenine, chiral crystallization of achiral compounds is not such a rare phenomenon (Matsuura and Koshima 2005; Pidcock 2005). Various chiral crystals of achiral compounds, such as the co-crystal of tryptamine and *p*-chlorobenzoic acid (Kawasaki et al. 2005), hippuric acid (Kawasaki et al. 2006a), benzil (Kawasaki et al. 2008b), tetraphenyl ethylene (Kawasaki et al. 2010b), benzene triester (Kawasaki et al. 2013), and ethylenediamine sulfate (Matsumoto et al. 2015b), act as chiral initiators in asymmetric autocatalysis (Table 5.6.1). These results clearly suggest that although the chiral induction of asymmetric autocatalysis by the chiral crystal is not specific to some crystal, the crystal chirality of achiral compounds has efficiently worked as a chiral trigger of homochirality in conjunction with asymmetric autocatalysis.

FIGURE 5.6.14 Asymmetric autocatalysis induced by a chiral crystal of adeninium dinitrate.

TABLE 5.6.1

Asymmetric Autocatalysis Triggered by Chiral Crystals of Achiral Organic Compounds

R = t-Bu-C≡C–

Enantiomorphous Crystals as Chiral Triggers

Entry	Structure	Space Group	Helicity (Axis)	Solid-State CD	Configuration of Product Pyrimidyl Alkanol 1	References
1		$P2_12_12_1$	M (c)	$CD(-)225_{Nujol}$	S	Kawasaki et al. (2005)
			P (c)	$CD(+)225_{Nujol}$	R	
2		$P2_12_12_1$	–	$CD(+)260_{Nujol}$	S	Kawasaki et al. (2006a)
			–	$CD(-)260_{Nujol}$	R	
3		$P3_121$ or $P3_221$	–	$CD(-)400_{KBr}$	S	Kawasaki et al. (2008b)
			–	$CD(+)400_{KBr}$	R	
4		$P2_1$	–	$CD(+)270_{Nujol}$	S	Kawasaki et al. (2010b)
			–	$CD(-)270_{Nujol}$	R	
5		$P6_1$	P	$CD(+)224_{KBr}$	S	Kawasaki et al. (2013)
		$P6_5$	M	$CD(-)224_{KBr}$	R	
6		$P4_12_12$	P	–	S	Matsumoto et al. (2015b)
		$P4_32_12$	M	–	R	

5.6.9 ASYMMETRIC INDUCTION BY THE CHIRAL SURFACE OF MINERALS

Chiral minerals provide effective environments for the discrimination of chiral molecules. Their possible roles in the origin of biological homochirality have been discussed previously (Hazen 2004). In the earth's crust, there is a wide variety of chiral minerals that could serve as accessible chiral surfaces in the prebiotic evolution of chiral organic molecules. However, only a very small asymmetric induction has been reported in the asymmetric adsorption of chiral compounds on quartz (Bonner et al. 1974). We have reported asymmetric autocatalysis triggered by d- and l-quartz. When pyrimidine-5-carbaldehyde **2** was treated with i-Pr$_2$Zn in the presence of d- and l-quartz powder, (S)- and (R)-**1** with a high enantioenrichment were obtained in conjunction with asymmetric autocatalysis (Figure 5.6.15) (Soai et al. 1999). In addition, chiral ionic crystals of sodium chlorate (Sato et al. 2000) and sodium bromate (Sato et al. 2004) also act as chiral triggers of asymmetric autocatalysis of 5-pyrimidyl alkanol **1**.

Cinnabar is a chiral mineral of mercury(II) sulfide. Recently, we have demonstrated that enantiomorphic P and M crystals of cinnabar act as chiral triggers for asymmetric autocatalysis to afford (R)- and (S)-enantiomers of **1**, respectively (Figure 5.6.15) (Shindo et al. 2013). When pyrimidine-5-carbaldehyde **2** was treated with i-Pr$_2$Zn in the presence of cinnabar, which had right-handed P-helicity, (R)-pyrimidyl alkanol **1** was obtained in 88% ee and 87% yield. By contrast, in the presence of the (M)-HgS, the opposite enantiomer of (S)-**1** was isolated in 92% ee and 91% yield. Almost enantiopure (S)- and (R)-**1** with >99.5% ee were obtained by applying consecutive asymmetric autocatalysis.

Meanwhile, the chiral enantiotopic surface of achiral minerals has also been considered an origin of chirality. Therefore, increasing attention has been focused on two-dimensional surface chirality. Hazen reported the enantiomer-selective adsorption of racemic amino acids on the enantiotopic faces of the achiral mineral calcite (CaCO$_3$) (Hazen et al. 2001). Certain metal surfaces, such as artificially prepared Cu(643),

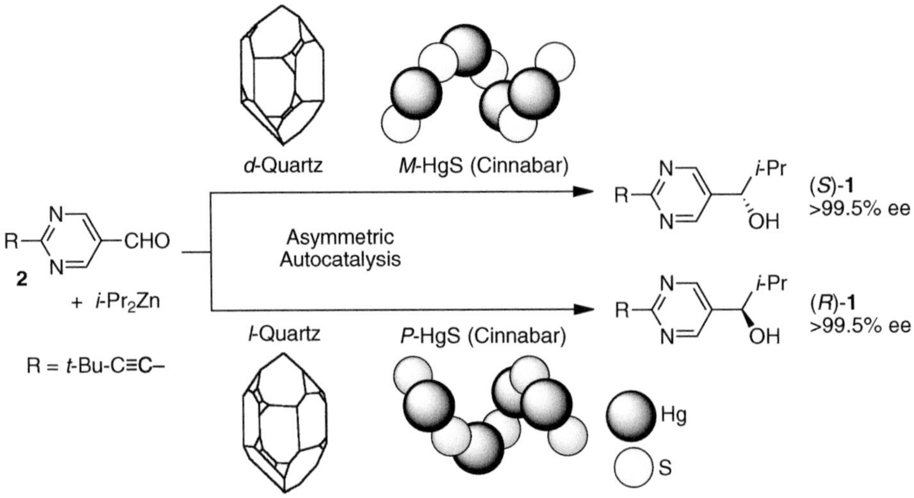

FIGURE 5.6.15 Enantioselective synthesis of 5-pyrimidyl alkanol **1** on the chiral surface of the chiral minerals quartz and cinnabar in conjunction with asymmetric amplification by asymmetric autocatalysis.

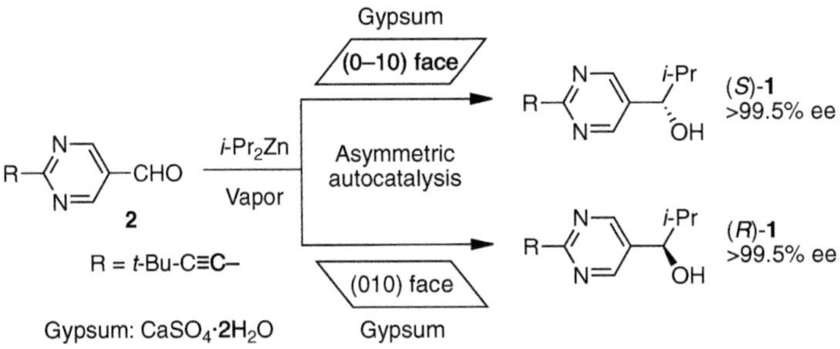

FIGURE 5.6.16 Enantioselective synthesis on the enantiotopic face of achiral gypsum in combination with asymmetric autocatalysis.

become enantiotopic. Enantiomer-selective desorption and decomposition have been reported on these enantiotopic metal surfaces (Gellman et al. 2013).

We have demonstrated that the chiral (010) and (0–10) surfaces of achiral mineral gypsum ($CaSO_4 \cdot 2H_2O$) could act as chiral environments to induce asymmetry in the asymmetric autocatalysis (Figure 5.6.16) (Matsumoto et al. 2017). On the (0–10) surface of gypsum, the reaction of adsorbed aldehyde **2** with $i\text{-Pr}_2Zn$ vapor afforded alkanol **2** in the (S)-configuration. On the other hand, (R)-alkanol **1** was obtained through the reaction on the opposite (010) face. The ee values of pyrimidyl alkanol **1** could increase to >99.5% ee during further asymmetric autocatalysis with significant amplification of ee. It is considered that the slightly imbalanced Re and Si face orientations of aldehyde **2** on the enantiotopic (010) and (0–10) faces of gypsum are the origins of asymmetric induction.

5.6.10 SPONTANEOUS ABSOLUTE ASYMMETRIC SYNTHESIS: AMPLIFICATION OF THE FLUCTUATION OF ENANTIOMERIC EXCESSES BY ASYMMETRIC AUTOCATALYSIS

The reaction of achiral substrates forming enantioenriched products without the intervention of any chiral factor has been defined as spontaneous absolute asymmetric synthesis

(Mislow 2003). Spontaneous absolute asymmetric synthesis has been proposed as one of the origins of homochirality. However, it is widely accepted that common reactions without using a chiral factor always give a 1:1 mixture of two enantiomers, that is, a racemate. Based on statistics, the numbers of the two enantiomers fluctuate. As described in the preceding section, the Soai reaction can amplify extremely low ee to >99.5% ee. We have reported asymmetric autocatalysis of (S)- and (R)-alkanols during the reaction of pyrimidine-5-carbaldehyde and $i\text{-Pr}_2Zn$ without adding any chiral substance (Soai et al. 1996) in the mixed solvent including diethyl ether (Soai et al. 2003) and in the presence of achiral silica gel (Kawasaki et al. 2006b) (Figure 5.6.17). Further investigations on spontaneous absolute asymmetric synthesis revealed that reactions in the presence of achiral amines spontaneously afforded pyrimidyl alkanol **1** with either (S)- or (R)-configurations (Suzuki et al. 2010). An approximately stochastic distribution for the formation of either (S)- or (R)-5-pyrimidyl alkanol **1** was observed, for example, formation of S 19 times and R 18 times for the 37 experiments (Soai et al. 2003). The fluctuation of ee produced spontaneously and stochastically in the initial stage of the addition of $i\text{-Pr}_2Zn$ to achiral pyrimidine-5-carbaldehyde **2** forming racemic alkanol **1** can be amplified to a detectable value during the asymmetric autocatalysis. This stochastic behavior in the formation of (S)- and (R)-**1** fulfills one of the conditions necessary for spontaneous absolute asymmetric synthesis.

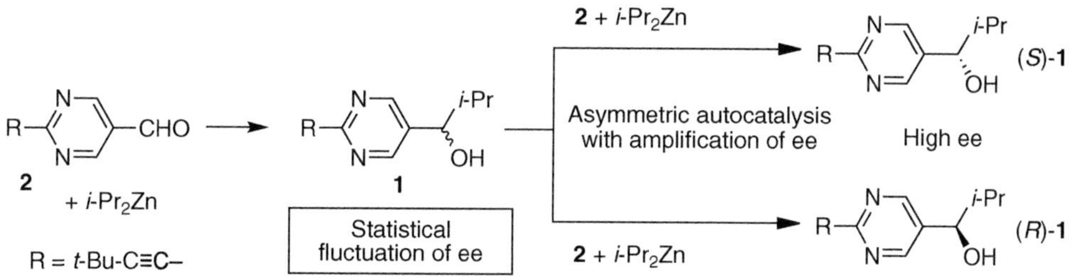

FIGURE 5.6.17 Spontaneous absolute asymmetric synthesis: amplification of the statistical fluctuation of ee in the initially forming racemic asymmetric autocatalyst.

5.6.11 ASYMMETRIC AUTOCATALYSIS AS A CHIRALITY SENSOR OF METEORITE SAMPLES

As mentioned above, asymmetric autocatalysis is a highly sensitive reaction to discriminate the chirality of a variety of materials, including chiral compounds such as isotopomers and chiral surfaces of minerals. In addition, without the addition of chiral materials, asymmetric autocatalysis affords (S)- and (R)-**1** with high ee in almost the same probability by the amplification of stochastic enantiomeric imbalances at the beginning of the reaction.

Thus, we have used asymmetric autocatalysis to determine the chirality inside a meteorite, that is, the autocatalytic reaction was repeatedly performed in the presence of meteorite samples (Kawasaki et al. 2006c). As a result, we could determine that chirality remains in Murchison and Allende powders after extraction with water and solvents in addition to the Murray kerogen-like insoluble organic materials (IOM) after demineralization. On the other hand, the Murray IOM samples after hydrothermolytic treatment (IOM-H) gave both (S)- and (R)-pyrimidyl alkanol **1** in equal probability, that is, stochastic results were obtained, which indicates the absence of chiral elements in the IOM-H sample. Chirality is not found in Murchison powders from which all organic components had been removed by O_2 plasma at low temperature. These results provide a powerful approach to detect the extraterrestrial origin of chirality.

5.6.12 SUMMARY

As described, we have shown that 5-pyrimidyl alkanol is a highly enantioselective asymmetric autocatalyst in the reaction between $i\text{-Pr}_2\text{Zn}$ and pyrimidine-5-carbaldehyde, with significantly large amplification of ee from extremely low (*ca.* 0.00005% ee) to a nearly enantiopure state (>99.5% ee) without the assistance of any other chiral auxiliary. Chiral compounds such as interstellar chiral propylene oxide and chiral isotopomers of meteoritic amino acids can act as chiral initiators of asymmetric autocatalysis to afford enantioenriched alkanol with a handedness corresponding to that of chiral initiators. The spontaneous absolute asymmetric synthesis was realized in the formation of enantioenriched pyrimidyl alkanol; therefore, stochastic imbalance of the initially forming racemate could also be asymmetrically amplified. Furthermore, various sources of chirality, such as CPL, chiral crystals of achiral nucleobases, and enantiotopic surfaces of achiral mineral gypsum, have been directly correlated with nearly enantiopure pyrimidyl alkanol using asymmetric autocatalysis. The asymmetric autocatalysis of 5-pyrimidyl alkanol, that is, the Soai reaction (Pályi et al. 2012), is a unique reaction leading to overwhelming enantioenrichments, as observed in biological compounds such as L-amino acids and D-sugars. According to the new definition of life proposed by Carroll, the Soai reaction is the simplest form of life, that is, a self-amplifying autocatalytic reaction (Carroll 2009).

REFERENCES

Altwegg, K., H. Balsiger, A. Bar-Nun et al. 2016. Prebiotic chemicals-amino acid and phosphorus—In the coma of comet 67P/Churyumov-Gerasimenko. *Sci. Adv.* 2: e1600285.

Ashkenasy, G., R. Jagasia, M. Yadav et al. 2004. Design of a directed molecular network. *Proc. Natl. Acad. Sci. U.S.A.* 101: 10872–10877.

Ashkenasy, G., T.M. Hermans, S. Otto et al. 2017. Systems chemistry. *Chem. Soc. Rev.* 46: 2543–2554.

Avalos, M., R. Babiano, P. Cintas et al. 2000. Chiral autocatalysis: Where stereochemistry meets the origin of life. *Chem. Commun.* 887–892.

Bailey, J., A. Chrystosmou, J.H. Hough et al. 1998. Circular polarization in star-formation regions: Implications for biomolecular homochirality. *Science* 281: 672–674.

Balavoine, G., A. Moradpour, and H.B. Kagan. 1974. Preparation of chiral compounds with high optical purity by irradiation with circularly polarized light, a model reaction for the prebiotic generation of optical activity. *J. Am. Chem. Soc.* 96: 5152–5158.

Becker, L., J.L. Bada, R.E. Winans et al. 1994. Fullerenes in Allende meteorite. *Nature* 372: 507.

Bernstein, M.P., J.P. Dworkin, S.A. Sandford et al. 2002. Racemic amino acids from the ultraviolet photolysis of interstellar ice analogues. *Nature* 416: 401–403.

Bissette, A.J., and S.P. Fletcher. 2013. Mechanisms of autocatalysis. *Angew. Chem. Int. Ed.* 52: 12800–12826.

Blackmond, D.G., C.R. McMillan, S. Ramdeehul et al. 2001. Origins of asymmetric amplification in autocatalytic alkylzinc additions. *J. Am. Chem. Soc.* 123: 10103–10104.

Bolli, M., R. Micura, and A. Eschenmoser. 1997. Pyranosyl-RNA: Chiroselective self-assembly of base sequences by ligative oligomerization of tetranucleotide-2',3'-cyclophosphates (with a commentary concerning the origin of biomolecular homochirality). *Chem. Biol.* 4: 309–320.

Bonner, W.A., and E. Rubenstein. 1987. Supernovae, neutron stars and biomolecular chirality. *BioSystems* 20: 99–111.

Bonner, W.A., P.R. Kavasmaneck, F.S. Martin et al. 1974. Asymmetric adsorption of alanine by quartz. *Science* 186: 143–144.

Breslow, R., and Z.-L. Cheng. 2010. L-Amino acids catalyze the formation of an excess of D-glyceraldehyde, and thus of other D sugars, under credible prebiotic conditions. *Proc. Natl. Acad. Sci. U.S.A.* 107: 5723–5725.

Buono, F.G., and D.G. Blackmond. 2003. Kinetic evidence for a tetrameric transition state in the asymmetric autocatalytic alkylation of pyrimidyl aldehydes. *J. Am. Chem. Soc.* 125: 8978–8979.

Callahana, M.P., K.E. Smith, H.J. Cleaves II et al. 2011. Carbonaceous meteorites contain a wide range of extraterrestrial nucleobases. *Proc. Natl. Acad. Sci. U.S.A.* 108: 13995–13998.

Cami, J., J.B. Salas, E. Peeters et al. 2010. Detection of C_{60} and C_{70} in a young planetary nebula. *Science* 329: 1180–1182.

Carroll, J.D. 2009. A new definition of life. *Chirality* 21: 354–358.

Cronin, J.R., and S. Pizzarello. 1997. Enantiomeric excesses in meteoritic amino acids. *Science* 275: 951–955.

Eliel, E.L., and S.H. Wilen. 1994. *Stereochemistry of Organic Compounds*. New York: John Wiley & Sons.

Engel, M.H., and S.A. Macko. 1997. Isotopic evidence for extraterrestrial non-racemic amino acids in the Murchison meteorite. *Nature* 389: 265–268.

Ernst, K.-H. 2012. Molecular chirality at surfaces. *Phys. Status Solid. B* 249: 2057–2088.

Feringa, B.L., and R.A. van Delden. 1999. Absolute asymmetric synthesis: The origin, control, and amplification of chirality. *Angew. Chem. Int. Ed.* 38: 3418–3438.

Frank, F.C. 1953. On spontaneous asymmetric synthesis. *Biochim. Biophys. Acta* 11: 459–463.

Gal, J. 2008. When Did Louis Pasteur present his memoir on the discovery of molecular chirality to the academie des sciences? Analysis of a discrepancy. *Chirality* 20: 1072–1084.

Gehring, T., M. Quaranta, B. Odell et al. 2012. Observation of a transient intermediate in Soai's asymmetric autocatalysis: Insights from ^{1}H NMR turnover in real time. *Angew. Chem. Int. Ed.* 51: 9539–9542.

Gellman, A.J., Y. Huang, X. Feng et al. 2013. Superenantioselective chiral surface explosions. *J. Am. Chem. Soc.* 135: 19208–19214.

Gilbert, W. 1986. Origin of life: The RNA world. *Nature* 319: 618.

Girard, C., and H.B. Kagan. 1998. Nonlinear effects in asymmetric synthesis and stereoselective reactions: Ten years of investigation. *Angew. Chem. Int. Ed.* 37: 2923–2959.

Gonzalez-Campo, A., and D.B. Amabilino. 2013. Biomolecules at interfaces-chiral, naturally. *Top. Curr. Chem.* 333: 109–156.

Green, M.M., J.-W. Park, T. Sato et al. 1999. The macromolecular route to chiral amplification. *Angew. Chem. Int. Ed.* 38: 3138–3154.

Gridnev, I.D., J.M. Serafimov, and J.M. Brown. 2004. Solution structure and reagent binding of the zinc alkoxide catalyst in the Soai asymmetric autocatalytic reaction. *Angew. Chem. Int. Ed.* 43: 4884–4887.

Gridnev, I.D., and J.M. Brown. 2004. Asymmetric autocatalysis: Novel structures, novel mechanism? *Proc. Natl. Acad. Sci. U.S.A.* 101: 5727–5731.

Hazen, R.M., and D.S. Sholl. 2003. Chiral selection on inorganic crystalline surfaces. *Nat. Mater.* 2: 367–374.

Hazen, R.M. 2004. Chiral crystal faces of common rock-forming minerals. In *Progress in Biological Chirality*, G. Pályi, and C. Zucchi (Eds.), chap. 9, p. 137. Oxford, UK: Elsevier.

Hazen, R.M., T.R. Filley, and G.A. Goodfriend. 2001. Selective adsorption of L- and D-amino acids on calcite: Implications for biochemical homochirality. *Proc. Natl. Acad. Sci. U.S.A.* 98: 5487–5490.

Hitosugi, S., A. Matsumoto, Y. Kaimori et al. 2014. Asymmetric autocatalysis initiated by finite single-wall carbon nanotube molecules with helical chirality. *Org. Lett.* 16: 645–647.

Inoue, Y. 1992. Asymmetric photochemical reactions in solution. *Chem. Rev.* 92: 741–770.

Kagan, H., A. Moradpour, J.F. Nicoud et al. 1971. Photochemistry with circularly polarized light. The synthesis of optically active hexahelicene. *J. Am. Chem. Soc.* 93: 2353–2354.

Kawasaki, T., M. Shimizu, K. Suzuki et al. 2004. Enantioselective synthesis induced by chiral epoxides in conjunction with asymmetric autocatalysis. *Tetrahedron: Asymmetry* 15: 3699–3701.

Kawasaki, T., T. Sasagawa, K. Shiozawa et al. 2011a. Enantioselective synthesis induced by chiral crystal composed of DL-serine in conjunction with asymmetric autocatalysis. *Org. Lett.* 13: 2361–2363.

Kawasaki, T., Y. Matsumura, T. Tsutsumi et al. 2009a. Asymmetric autocatalysis triggered by carbon isotope (${}^{13}C/{}^{12}C$) chirality. *Science* 324: 492–495.

Kawasaki, T., M. Shimizu, D. Nishiyama et al. 2009b. Asymmetric autocatalysis induced by meteoritic amino acids with hydrogen isotope chirality. *Chem. Commun.* 4396–4398.

Kawasaki, T., H. Ozawa, M. Ito et al. 2011b. Enantioselective synthesis induced by compounds with chirality arising from partially deuterated methyl groups in conjunction with asymmetric autocatalysis. *Chem. Lett.* 40: 320–321.

Kawasaki, T., Y. Okano, E. Suzuki et al. 2011c. Asymmetric autocatalysis: Triggered by chiral isotopomer arising from oxygen isotope substitution. *Angew. Chem. Int. Ed.* 50: 8131–8133.

Kawasaki, T., K. Suzuki, Y. Hakoda et al. 2008a. Achiral nucleobase cytosine acts as an origin of homochirality of biomolecules in conjunction with asymmetric autocatalysis. *Angew. Chem. Int. Ed.* 47: 496–499.

Kawasaki, T., Y. Hakoda, H. Mineki et al. 2010a. Generation of absolute controlled crystal chirality by the removal of crystal water from achiral crystal of nucleobase cytosine. *J. Am. Chem. Soc.* 132: 2874–2875.

Kawasaki, T., K. Jo, H. Igarashi et al. 2005. Asymmetric amplification using chiral cocrystals formed from achiral organic molecules by asymmetric autocatalysis. *Angew. Chem. Int. Ed.* 44: 2774–2777.

Kawasaki, T., K. Suzuki, K. Hatase et al. 2006a. Enantioselective synthesis mediated by chiral crystal of achiral hippuric acid in conjunction with asymmetric autocatalysis. *Chem. Commun.* 1869–1871.

Kawasaki, T., Y. Harada, K. Suzuki et al. 2008b. Enantioselective synthesis utilizing enantiomorphous organic crystal of achiral benzils as a source of chirality in asymmetric autocatalysis. *Org. Lett.* 10: 4085–4088.

Kawasaki, T., M. Nakaoda, N. Kaito et al. 2010b. Asymmetric autocatalysis induced by chiral crystals of achiral tetraphenylethylenes. *Orig. Life Evol. Biosph.* 40: 65–78.

Kawasaki, T., M. Uchida, Y. Kaimori et al. 2013. Enantioselective synthesis induced by the helical molecular arrangement in the chiral crystal of achiral tris(2-hydroxyethyl) 1,3,5-benzenetricarboxylate in conjunction with asymmetric autocatalysis. *Chem. Lett.* 42: 711–713.

Kawasaki, T., K. Suzuki, M. Shimizu et al. 2006b. Spontaneous absolute asymmetric synthesis in the presence of achiral silica gel in conjunction with asymmetric autocatalysis. *Chirality* 18: 479–482.

Kawasaki, T., K. Hatase, Y. Fujii et al. 2006c. The distribution of chiral asymmetry in meteorites: An investigation using asymmetric autocatalytic chiral sensors. *Geochim. Cosmochim. Acta* 70: 5395–5402.

Kitamura, M., S. Okada, S. Suga et al. 1989. Enantioselective addition of dialkylzincs to aldehydes promoted by chiral amino alcohols. Mechanism and nonlinear effect. *J. Am. Chem. Soc.* 111: 4028–4036.

Kondepudi, D.K., R.J. Kaufman, and N. Singh. 1990. Chiral symmetry breaking in sodium chlorate crystallization. *Science* 250: 975–976.

Lee, D.H., J.R. Granja, J.A. Martinez et al. 1996. A self-replicating peptide. *Nature* 382: 525–528.

Matsumoto, A., T. Abe, A. Hara et al. 2015a. Crystal structure of isopropylzinc alkoxide of pyrimidyl alkanol: Mechanistic insights for asymmetric autocatalysis with amplification of enantiomeric excess. *Angew. Chem. Int. Ed.* 54: 15218–15221.

Matsumoto, A., S. Fujiwara, T. Abe et al. 2016a. Elucidation of the structures of asymmetric autocatalyst based on x-ray crystallography. *Bull. Chem. Soc. Jpn.* 89: 1170–1177.

Matsumoto, A., H. Ozaki, S. Harada et al. 2016b. Asymmetric induction by nitrogen $^{14}N/^{15}N$ isotopomer in conjunction with asymmetric autocatalysis. *Angew. Chem. Int. Ed.* 55: 15246–15249.

Matsumoto, A., S. Oji, S. Takano et al. 2013. Asymmetric autocatalysis triggered by oxygen isotopically chiral glycerin. *Org. Biomol. Chem.* 11: 2928–2931.

Matsumoto, A., T. Ide, Y. Kaimori et al. 2015b. Asymmetric autocatalysis triggered by chiral crystal of achiral ethylenediamine sulfate. *Chem. Lett.* 44: 688–690.

Matsumoto, A., Y. Kaimori, M. Uchida et al. 2017. Achiral inorganic gypsum acts as an origin of chirality through its enantiotopic surface in conjunction with asymmetric autocatalysis. *Angew. Chem. Int. Ed.* 56: 545–548.

Matsuura, T., and H. Koshima. 2005. Introduction to chiral crystallization of achiral organic compounds spontaneous generation of chirality. *J. Photochem. Photobio. C* 6: 7–24.

McGuire, B., P.B. Carroll, R.A. Loomis et al. 2016. Discovery of the interstellar chiral molecule propylene oxide (CH₃CHCH₂O). *Science* 352: 1449–1452.

Meierhenrich, U.J., L. Nahon, C. Alcarez et al. 2005. Asymmetric vacuum UV photolysis of the amino acids leucine in solid state. *Angew. Chem. Int. Ed.* 44: 5630–5636.

Micheau, J.-C., C. Coudret, J.-M. Cruz et al. 2012. Amplification of enantiomeric excess, mirror-image symmetry breaking and kinetic proofreading in Soai reaction models with different oligomeric orders. *Phys. Chem. Chem. Phys.* 14: 13239–13248.

Miller, S.L. 1953. A production of amino acids under possible primitive earth conditions. *Science* 117: 528–529.

Miller, S.L., and H.C. Urey. 1959. Organic compound synthesis on the primitive earth. *Science* 130: 245–251.

Mineki, H., Y. Kaimori, T. Kawasaki et al. 2013. Enantiodivergent formation of a chiral cytosine crystal by removal of crystal water from an achiral monohydrate crystal under reduced pressure. *Tetrahedron: Asymmetry* 24: 1365–1367.

Mineki, H., T. Hanasaki, A. Matsumoto et al. 2012. Asymmetric autocatalysis initiated by achiral nucleic acid base adenine: Implications on the origin of homochirality of biomolecules. *Chem. Commun.* 48: 10538–10540.

Mislow, K. 2003. Absolute asymmetric synthesis: A commentary. *Collect. Czech. Chem. Commun.* 68: 849–864.

Miyagawa, S., K. Yoshimura, Y. Yamazaki et al. 2017. Asymmetric Strecker reaction arising from the molecular orientation of an achiral imine at the single-crystal face: Enantioenriched L- and D-amino acids. *Angew. Chem. Int. Ed.* 56: 1055–1058.

Munoz Caro, G.M., U.J. Meierhenrich, W.A. Schutte et al. 2002. Amino acids from ultraviolet irradiation of interstellar ice analogues. *Nature* 416: 403–406.

Myrgorodska, I., C. Meinert, Z. Martins et al. 2015. Molecular chirality in meteorites and interstellar ices, and the chirality experiment on board the ESA cometary rosetta mission. *Angew. Chem. Int. Ed.* 54: 1402–1412.

Nanita, S.C., and R.G. Cooks. 2006. Serine octamers: Cluster formation, reactions, and implications for biomolecule homochirality. *Angew. Chem. Int. Ed.* 45: 554–569.

Olsson, S., P.M. Björemark, T. Kokoli et al. 2015. Absolute asymmetric synthesis: Protected substrate oxidation. *Chem. Eur. J.* 21: 5211–5219.

Oro, J. 1960. Synthesis of adenine from ammonium cyanide. *Biochem. Biophys. Res. Commun.* 2: 407–412.

Pályi, G., C. Zicchi, and L. Caglioti (Eds.). 2012. *The Soai Reaction and Related Topic.* Modena, Italy: Academia Nationale di Scienze Lettere e Arti Modena.

Pidcock, E. 2005. Achiral molecules in non-centrosymmetric space groups. *Chem. Commun.* 3457–3459.

Pizzarello, S., and A.L. Weber. 2004. Prebiotic amino acids as asymmetric catalysts. *Science* 303: 1151.

Pizzarello, S., and Y. Huang. 2005. The deuterium enrichment of individual amino acids in carbonaceous meteorites: A case for the presolar distribution of biomolecule precursors. *Geochim. Cosmochim. Acta* 69: 599–605.

Puchot, C., O. Samuel, E. Dunach et al. 1986. Nonlinear effects in asymmetric synthesis. Examples in asymmetric oxidations and aldolization reactions. *J. Am. Chem. Soc.* 108: 2353–2357.

Quaranta, M., T. Gehring, B. Odell et al. 2010. Unusual inverse temperature dependence on reaction rate in the asymmetric autocatalytic alkylation of pyrimidyl aldehydes. *J. Am. Chem. Soc.* 132: 15104–15107.

Raval, R. 2009. Chiral expression from molecular assemblies at metal surfaces: Insights from surface science techniques. *Chem. Soc. Rev.* 38: 707–721.

Ribó, J.M., J. Crusats, F. Sagués et al. 2001. Chiral sign induction by vortices during the formation of mesophases in stirred solutions. *Science* 292: 2063–2066.

Ring, D., Y. Wolman, N. Friedmann et al. 1972. Prebiotic synthesis of hydrophobic and protein amino acids. *Proc. Natl. Acad. Sci. U.S.A.* 69: 765–768.

Robertson M.P., and S.L. Miller. 1995. An efficient prebiotic synthesis of cytosine and uracil. *Nature* 375: 772–774.

Saito, Y., and H. Hyuga. 2013. Homochirality: Symmetry breaking in systems driven far from equilibrium. *Rev. Mod. Phys.* 85: 603–621.

Sandford, S.A., J. Aléon, C.M.O'D. Alexander et al. 2006. Organics captured from comet 81P/Wild 2 by the stardust spacecraft. *Science* 314:1720–1724.

Sato, I., H. Urabe, S. Ishiguro et al. 2003a. Amplification of chirality from extremely low to greater than 99.5% ee by asymmetric autocatalysis. *Angew. Chem. Int. Ed.* 42: 315–317.

Sato, I., D. Omiya, K. Tsukiyama et al. 2001a. Evidence of asymmetric autocatalysis in the enantioselective addition of diisopropylzinc to pyrimidine-5-carbaldehyde using chiral pyrimidyl alkanol. *Tetrahedron: Asymmetry* 12: 1965–1969.

Sato, I., D. Omiya, H. Igarashi et al. 2003b. Relationship between the time, yield, and enantiomeric excess of asymmetric autocatalysis of chiral 2-alkynyl-5-pyrimidyl alkanol with amplification of enantiomeric excess. *Tetrahedron: Asymmetry* 14: 975–979.

Sato, I., R. Yamashima, K. Kadowaki et al. 2001b. Asymmetric induction by helical hydrocarbons. *Angew. Chem., Int. Ed.* 40: 1096–1098.

Sato, I., S. Osanai, K. Kadowaki et al. 2002. Asymmetric autocatalysis of pyrimidyl alkanol induced by optically active 1,1'-binaphthyl, an atropisomeric hydrocarbon, generated from spontaneous resolution on crystallization. *Chem. Lett.* 31: 168–169.

Sato, I., Y. Ohgo, H. Igarashi et al. 2007. Determination of absolute configurations of amino acids by asymmetric autocatalysis of 2-alkynylpyrimidyl alkanol as a chiral sensor. *J. Organomet. Chem.* 692: 1783–1787.

Sato, I., K. Kadowaki, and K. Soai. 2000. Asymmetric synthesis of an organic compound with high enantiomeric excess induced by inorganic ionic sodium chlorate. *Angew. Chem. Int. Ed.* 39: 1510–1512.

Sato, I., K. Kadowaki, Y. Ohgo et al. 2004. Highly enantioselective asymmetric autocatalysis induced by chiral ionic crystals of sodium chlorate and sodium bromate. *J. Mol. Cat. A: Chem.* 216: 209–214.

Schiaffino, L., and G. Ercolani. 2008. Unraveling the mechanism of the Soai asymmetric autocatalytic reaction by first-principles calculations: Induction and amplification of chirality by self-assembly of hexamolecular complexes. *Angew. Chem. Int. Ed.* 47: 6832–6835.

Schwartz, A.W., H. Joosten, and A.B. Voet. 1982. Prebiotic adenine synthesis via HCN oligomerization in ice. *Biosystems* 15: 191–193.

Shapiro, R. 1999. Prebiotic cytosine synthesis: A critical analysis and implications for the origin of life. *Proc. Natl. Acad. Sci. U.S.A.* 96: 4396–4401.

Shibata, T., S. Yonekubo, and K. Soai. 1999. Practically perfect asymmetric autocatalysis using 2-alkynyl-5-pyrimidylalkanol. *Angew. Chem. Int. Ed.* 38: 659–661.

Shibata, T., J. Yamamoto, N. Matsumoto et al. 1998. Amplification of a slight enantiomeric imbalance in molecules based on asymmetric autocatalysis. The first correlation between high enantiomeric enrichment in a chiral molecule and circularly polarized light. *J. Am. Chem. Soc.* 120: 12157–12158.

Shindo, H., Y. Shirota, K. Niki et al. 2013. Asymmetric autocatalysis induced by cinnabar: Observation of the enantioselective adsorption of a 5-pyrimidyl alkanol on the crystal surface. *Angew. Chem. Int. Ed.* 52: 9135–9138.

Siegel, J.S. 1998. Homochiral imperative of molecular evolution. *Chirality* 10: 24–27.

Sievers, D., and G. von Kiedrowski. 1994. Self-replication of complementary nucleotide-based oligomers. *Nature* 369: 221.

Soai, K., T. Kawasaki, and A. Matsumoto. 2014a. Asymmetric autocatalysis of pyrimidyl alkanol and its application to the study on the origin of homochirality. *Acc. Chem. Res.* 47: 3643–3654.

Soai, K., T. Shibata, H. Morioka et al. 1995. Asymmetric autocatalysis and amplification of enantiomeric excess of a chiral molecule. *Nature* 378: 767–768.

Soai, K., T. Kawasaki, and A. Matsumoto. 2014b. The origins of homochirality examined by using asymmetric autocatalysis. *Chem. Rec.* 14: 70–83.

Soai, K., A. Matsumoto, and T. Kawasaki. 2017. Asymmetric autocatalysis and the origins of homochirality of organic compounds. An overview. In *Advances in Asymmetric Autocatalysis and Related Topic*, G. Pályi, R. Kurdi, and C. Zucchi (Eds.), Chap. 1, pp. 1–30. Cambridge, MA: Elsevier.

Soai, K., T. Shibata, and I. Sato. 2000. Enantioselective automultiplication of chiral molecules by asymmetric autocatalysis. *Acc. Chem. Res.* 33: 382–390.

Soai, K., and T. Kawasaki. 2008. Asymmetric autocatalysis with amplification of chirality. *Top. Curr. Chem.* 284: 1–33.

Soai, K., S. Osanai, K. Kadowaki et al. 1999. *d*- and *l*-Quartz-promoted highly enantioselective synthesis of a chiral compound. *J. Am. Chem. Soc.* 121: 11235–11236.

Soai, K., T. Shibata, and Y. Kowata. 1996. Production of optically active pyrimidylalkyl alcohol by spontaneous asymmetric synthesis. Jpn. Kokai Tokkyo Koho JP 19960121140 19960418, 1996. An abstract is readily available as JPH09268179 from the European Patent Office. http://worldwide.espacenet.com.

Soai, K., I. Sato, T. Shibata et al. 2003. Asymmetric synthesis of pyrimidyl alkanol without adding chiral substances by the addition of diisopropylzinc to pyrimidine-5-carbaldehyde in conjunction with asymmetric autocatalysis. *Tetrahedron: Asymmetry* 14: 185–188.

Soloshonok, V.A., C. Roussel, O. Kitagawa et al. 2012. Self-disproportionation of enantiomers via achiral chromatography: A warning and an extra dimension in optical purifications. *Chem. Soc. Rev.* 41: 4180–4188.

Suarez, M., and G.B. Schuster. 1995. Photoresolution of an axially chiral bicyclo[3.3.0]octan-3-one: Phototriggers for a liquid-crystal-based optical switch. *J. Am. Chem. Soc.* 117: 6732–6738.

Suzuki, K., K. Hatase, D. Nishiyama et al. 2010. Spontaneous absolute asymmetric synthesis promoted by achiral amines in conjunction with asymmetric autocatalysis. *J. Syst. Chem.* 1: 5.

Tielens, A.G.G.M. 2005. *The Physics and Chemistry of the Interstellar Medium*. Cambridge, UK: Cambridge University Press.

Todd, M.H. 2002. Asymmetric autocatalysis: Product recruitment for the increase in the chiral environment (PRICE). *Chem. Soc. Rev.* 31: 211–222.

Viedma, C. 2005. Chiral symmetry breaking during crystallization: Complete chiral purity induced by nonlinear autocatalysis and recycling. *Phys. Rev. Lett.* 94: 065504.

Weissbuch, I., L. Addadi, Z. Berkovitch-Yellin et al. 1984. Spontaneous generation and amplification of optical activity in α-amino acids by enantioselective occlusion into centrosymmetric crystals of glycine. *Nature* 310: 161–164.

Weissbuch, I., and M. Lahav. 2011. Crystalline architectures as templates of relevance to the origins of homochirality. *Chem. Rev.* 111: 3236–3267.

Wintner, E.A., M.M. Conn, and J.J. Rebek. 1994. Studies in molecular replication. *Acc. Chem. Res.* 27: 198–203.

Zepik, H., E. Shavit, M. Tang et al. 2002. Chiral amplification of oligopeptides in two-dimensional crystalline self-assemblies on water. *Science* 295: 1266–1269.

5.7 Phosphorus in Prebiotic Chemistry— An Update and a Note on Plausibility

Alan W. Schwartz

CONTENTS

5.7.1 BACKGROUND

I last reviewed the subject matter of this chapter in 2006 (Schwartz 2006). That paper included a historical survey of previous literature, in which the topic of phosphorus and its place in biology and prebiotic chemistry was addressed, which of course is still appropriate and recommended as background. Much attention was placed in that review on what was referred to as the "phosphorus problem." It is useful to summarize that issue briefly by considering the possible geological sources of phosphorus, which must be the starting point for any discussion of prebiotic chemistry involving the element (see Table 5.7.1).

With the possible exception of schreibersite (a special case to be discussed later), all of the calcium-containing minerals listed in the table share a common property. The low equilibrium concentrations of phosphate, which might be liberated in aqueous solution (and especially in the presence of the many other possible terrestrial sources of Ca^{2+} or other alkaline earth metals), lead directly to a potential problem for prebiotic chemistry, first recognized by Addison Gulick (1955). In 2006, I described a number of possible solutions for this problem.

These included the possible volatilization of P_4O_{10} from apatite at high temperatures to produce condensed phosphate (Yamagata et al. 1991), the corrosion (hydrolysis reactions) of meteoritic schreibersite to produce phosphite and other products (Pasek and Lauretta 2005), and the consequences of reduction of phosphate to phosphite by lightning strikes, either in volcanic dust clouds or to the ground (De Graaf and Schwartz 2000). The conclusion reached then is summarized here:

"A variety of pathways now are available as possible solutions for what this author used to refer to as the 'phosphate problem'.

The 'problem' seems to have evolved gradually into the less perplexing one of having to decide which of several possible mechanisms are most likely to have operated on the primitive Earth." (Schwartz 2006). The purpose of the present note is to update the previous review and add some more recent data. First, however, it seems necessary to summarize some *older* data. Perhaps because of misplaced modesty, combined with enthusiasm for newer ideas, an earlier hypothesis of mine (which in retrospect still seems quite attractive) was not covered in the previous review and will be discussed briefly here.

5.7.2 PHOSPHORYLATION WITH APATITE: UREA, AMMONIUM CHLORIDE, AND AMMONIUM OXALATE

It was demonstrated (Lohrmann and Orgel 1971) that mixtures of urea and ammonium chloride were effective in solubilizing and activating hydroxylapatite as a source of phosphate for the phosphorylation of nucleosides. To introduce a personal note here, I should comment that I met Leslie Orgel for the first time at the Third International Conference on the Origin of Life, at Pont-à-Mousson, France, which was held in 1970. I had presented a talk entitled "Phosphate: Solubilization and Activation on the Primitive Earth" (Schwartz 1971), which was a theoretical study of the possible role of simple complexing agents that strongly bind calcium (Schwartz and Deuss 1971), of which the prime example is oxalic acid ($H_2C_2O_4$). During the break, Orgel (to whom I was introduced by Sid Fox) remarked that he and Lohrmann thought they had "solved the problem." The paper to which he was referring was soon published (Lohrmann and Orgel 1971); it reported that mixtures of urea, ammonium chloride, and various inorganic phosphates, including hydroxylapatite,

TABLE 5.7.1
Major Phosphate Minerals and Their Occurrences

Fluorapatite	$Ca_5(PO_4)_3F$	Common Accessory Mineral in Terrestrial Igneous Deposits (see Figure 5.7.1)
Hydroxyapatite	$Ca_5(PO_4)_3(OH)$	Terrestrial sedimentary deposits such as phosphorites
Francolite	$Ca_5(PO_4)_{2.5}(CO_3)_{0.5}F$	Terrestrial sedimentary deposits such as phosphorites
Schreibersite	$(Fe, Ni)_3P$	Meteorites
Whitlockite	$Ca_9(Mg, Fe)(PO_4)_6PO_3OH$	Meteorites
Chlorapatite	$Ca_5(PO_4)_3Cl$	Meteorites

could phosphorylate nucleosides, and thus, these conditions clearly solubilized the phosphate. Urea is indeed an indisputably plausible prebiotic molecule, but so is ammonium oxalate, since both urea and ammonium oxalate are products of the oligomerization of HCN (Ferris et al. 1978). Additionally, prebiotic oxalate could have been formed by a number of mechanisms. It had been established that oxalic acid can be generated photochemically from CO_2 and formic acid (Getoff 1962), and later, it was reported that oxalic acid was formed by irradiation of bicarbonate solution (Draganic et al. 1991). (For review of related literature, see Chittenden and Schwartz 1981).

Later experiments indeed demonstrated that phosphorylation of nucleosides by fluorapatite (Figure 5.7.1) occurred as a result of the evaporation of dilute solutions containing ammonium oxalate and that inclusion in the reaction mixture of the products of cyanogen hydrolysis was even more effective than oxalate alone (Schwartz et al. 1975). Cyanogen (C_2N_2) is a convenient stand-in for HCN, as it is available in cylinders and produces a similar set of hydrolysis products after reaction in water (particularly oxalate and urea).

FIGURE 5.7.1 Crystals of fluorapatite from Durango, Mexico. (Courtesy of author.)

5.7.3 PHOSPHORYLATION WITH STRUVITE

It was later argued (Handschuh and Orgel 1973) that, in the presence of concentrations of ammonium ion greater than 0.01 M, struvite ($MgNH_4PO_4·6H_2O$) could precipitate rather than apatite, which could lead to the possibility of loss of ammonia upon heating to generate condensed phosphate and/or phosphorylate nucleosides. This proposal runs into several difficulties. The difficulty of maintaining an atmosphere on the primitive earth containing significant amounts of ammonia has been pointed out (Abelson 1966; Ferris and Nicodem 1972; Kuhn and Atreya 1979; Levine et al. 1982), so it is far from clear whether struvite would actually ever be precipitated on the prebiotic earth. (Also see the discussion in Section 5.7.7 on the plausibility of prebiotic chemical pathways and their interrelationships). The relevance of struvite has also been questioned more recently. It has been argued, on the basis of theoretical calculations, that struvite would not have been abundant on the earth (Gull and Pasek 2013). The same paper reports the results of some model phosphorylations (of glycerol and of choline chloride), which are somewhat unclear in their applicability, since they compare struvite to monetite (an acidic calcium phosphate, $CaHPO_4$), rather than apatite, and are conducted in the *presence of water*. The most interesting property of struvite, if formed, would be its activity in phosphorylation reactions under *drying conditions* (Handschuh et al. 1973). Such conditions are, by the way, crucial to the effectiveness of ammonium salts of phosphoric acid in phosphorylation reactions; heating leads to loss of ammonia and generates an acidic phase.

5.7.4 PHOSPHORYLATION STARTING WITH SCHREIBERSITE

The meteoritic phosphide mineral schreibersite was proposed as a possible starting point for prebiotic phosphorus chemistry (Pasek and Lauretta 2005). The hydrolysis reactions of this mineral ("corrosion" in the terminology of Pasek and Lauretta) lead to the formation of a number of interesting reduced phosphorus oxyacids, including phosphite. It has been established that phosphite is not only more soluble than phosphate as an alkaline earth salt (Schwartz 2006) but also more reactive (De Graaf and Schwartz 2005). More recently, evidence has been presented that carbonate samples from the Archean limestone deposits from Coonterunah, Australia, show the presence of phosphite, as determined by extraction and ^{31}P NMR studies (Pasek et al. 2013). This appears to be the first direct evidence supporting the hypothesis that meteoritic schreibersite might have had a role in prebiotic chemistry and is a very exciting result.

On the other hand, another report (Britvin et al. 2015) makes a similar claim but is essentially irrelevant. The abstract states: "The results of the present study *could* [emphasis added] provide a new insight on the terrestrial origin of natural phosphides—the most likely source of reactive prebiotic phosphorus at the times of the early Earth."

Unfortunately, the deposit studied (the Hatrurim Formation near the Dead Sea in Israel and Jordan) is a recent formation, and the authors discuss a possible metamorphic event that occurred in the formation between 2 million and 4 million years ago. It is clear that this formation might serve as a *model* for a prebiotic event (which may be what the authors intended to say), but the inference is an analogy rather than evidence for a historical process. In the past several years, however, a number of other publications have appeared that considerably strengthen the case for schreibersite in prebiotic phosphorus chemistry.

5.7.5 PHOSPHORYLATION IN DEEP EUTECTIC SOLVENT SYSTEMS

Deep eutectic solvent (DES) systems have been studied as a medium for phosphorylation (Gull et al. 2014; La Cruz et al. 2016), with application to reactions on the surfaces of synthetic schreibersite samples, and have been proposed to be prebiologically plausible:

> "Deep eutectic solvents arise spontaneously by evaporation of water, and hence these may be formed merely by the drying down of solutions on minerals." (La Cruz et al. 2016).

For example, evaporation of water solutions containing urea and choline chloride are said to lead to the formation of a 2:1 urea and choline chloride eutectic, in which phosphorylation studies have been conducted. It would be convincing if experiments had actually demonstrated the plausibility of DES systems, but while these papers suggest that this has been established, a little investigation does not support this inference. For example:

> "We believe that these reactions may be prebiotically relevant. The DES system we employed here has shown significance for other biochemical reactions and from the prebiotic perspective, as eutectic melts of urea and choline can form spontaneously by drying these mixtures (Austin and Waddell 1999; Miller 1953; Miller and Schlesinger 1993)…" (Gull et al. 2014).

Some of these citations are rather strange. Austin and Waddell merely report the synthesis of "choline like molecules" such as ethanolamine and N-methylethanolamine. Work by Miller (1953) does not appear to be relevant here. The inference in the above citation is therefore rather carelessly drawn. I do not necessarily question the possible importance of DES systems—such systems might possibly arise through the evaporation of aqueous solutions containing these two organic compounds—but this simple mechanism has not yet been demonstrated. Published data are available for the urea–choline chloride phase diagram (Morrison et al. 2009), but it is not clear what effect water would have on the system, nor whether a 2:1 urea–choline chloride phase would be formed in this way, since a urea–water eutectic is also possible (Durickovic et al. 2009); the combination would lead to a more complex case. Actual experiments need to be conducted.

5.7.6 PHOSPHORYLATION IN UREA–AMMONIUM FORMATE–WATER

A better case has been made for the urea–ammonium formate–water (UAFW) eutectic system (Burcar et al. 2016). Although these results are intriguing, here too, the argument is not clear-cut. In an elaborate series of experiments in which (among other factors studied) the eutectic system was partially reformed by addition of water and re-evaporation, hydroxyapatite, struvite, and several other phosphates were used to phosphorylate nucleosides. However, questions remain. For example, what would the results have been if other starting conditions had been used? In all the experiments described, a 1:2 starting ratio of urea to ammonium formate was used (the eutectic composition). The suggested plausibility of this system really depends on uncertainties, such as this, being clarified. It would be valuable to determine the full phase diagram for the three-component UAFW system.

5.7.7 PLAUSIBILITY AND INTERRELATIONSHIPS OF KEY PREBIOTIC CHEMICAL PATHWAYS—AMMONIA AND THE CENTRAL IMPORTANCE OF HYDROGEN CYANIDE

As summarized earlier, the solubilization and activation of terrestrial phosphate minerals seems to require, or at least is greatly stimulated by, the participation of products of hydrogen cyanide oligomerization. Stanley Miller (1953), in his iconic work on amino acid synthesis from reduced gas mixtures, was the first to identify HCN as an important precursor in the process (Miller 1957). John Oró went a considerable step further by demonstrating that adenine was a minor product of the extended oligomerization of HCN (Oró 1960).[1]

In an extremely thorough study by Robert Sanchez, James Ferris, and Leslie Orgel (1967), the necessary conditions to permit the oligomerization of HCN were described. Hydrogen cyanide in water will oligomerize in mildly alkaline solution if present at a concentration substantially greater than 0.01 M. At concentrations greater than 0.1 M, polymerization reactions are more important than hydrolysis, and reaction to produce tetramer and higher oligomers is rapid. A plausible mechanism for concentrating HCN solutions was already known to be freezing, which produces a highly concentrated solution (74.5 M!) of HCN in the form of a eutectic phase (Coates and Hartshorne 1931). This remarkable effect (shown in modified form as Figure 5.7.2) may have had a crucial influence in prebiotic chemistry (Sanchez et al. 1966). Perhaps, it is reasonable to suggest the possibility that such low-temperature conditions may have been *required* to permit the concentration of HCN in a eutectic phase and to have permitted key synthetic processes to occur. A related problem is that of ammonia availability for prebiotic chemistry. While synthesis of HCN (as well as C_2N_2) has been shown to occur in less highly reduced and therefore more plausible

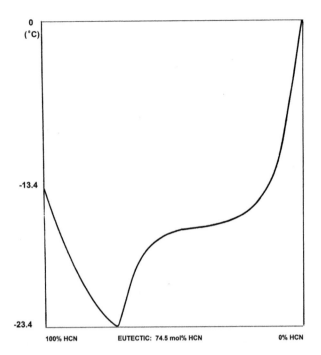

FIGURE 5.7.2 The HCN–water eutectic. This melting point diagram is originally from Coates and Hartshorne and has been redrawn, modified, and simplified to emphasize the portion of the diagram above the eutectic point. (For the original diagram, see: Coates, J.E. and Hartshorne, N.H., *J. Chem. Soc.*, 657–665, 1931.)

methane-nitrogen atmospheres (Toupance et al. 1975), the problem of ammonia stability complicates the question.

The susceptibility of ammonia to photolytic destruction is potentially an extremely serious problem for prebiotic chemistry, which has probably not been appreciated widely enough (Ferris and Nicodem 1972; Kuhn and Atreya 1979; Levine et al. 1982). Henderson-Sellers and Schwartz pointed out that a possible prebiotic source of ammonia (photochemical generation on natural minerals), which was not considered by Ferris and Nicodem, might have contributed to a steady-state concentration of ammonia (ammonium ions) sufficient to have permitted prebiotic chemistry to occur (Henderson-Sellers and Schwartz 1980). The cycling of HCN through the hydrosphere (via hydrolysis) must have constituted another such source. In fact, it seems obvious that such cycling was indispensable to the origin of life, an idea that has been expressed before: "Alternatively, we may suppose that cyanide never reached a concentration approaching 10^{-2} M but that nitrogen was constantly recycled through the oceans in the form of cyanide" (Sanchez et al. 1967).

The lesson seems to be clear; prebiotic chemical cycles were not only finely tuned but intricately *interdependent*.

ENDNOTE

[1] Incidentally, after spending some years trying to sort out the complexities of the pathways of HCN oligomerization (Voet and Schwartz 1983), I am less enthusiastic than Leslie Orgel was, when he remarked that the formation of a complex structure such as adenine by such a

simple reaction was a *remarkable* result (Orgel 2004). It certainly was an *important* result but should now be viewed against a background of the enormous complexity, which this reaction system produces. I am reminded of a remark attributed to Harold Urey, who—when asked what he expected from Miller's planned experiment—answered "Beilstein" (Wills and Bada 2000). Beilstein is the classical compendium of all known organic molecules and at the time would have amounted to more than 500 volumes.

REFERENCES

Abelson, P. H. 1966. Chemical events on the primitive Earth. *Proc. Natl. Acad. Sci. U.S.A.* 55: 1365–1372.

Austin, S. M., and T. G. Waddell. 1999. Prebiotic synthesis of Vitamin B6-type compounds. *Origins Life Evol. Biosphere* 29: 287–296.

Britvin, S. N., M. N. Murashko, Y. Vapnik, Y. S. Polekhovsky, and S. V. Krivovichev. 2015. Earth's phosphides in Levant and insights into the source of Archean prebiotic phosphorus. *Sci. Rep.* 5: 8355–8360.

Burcar, B., M. Pasek, M. Gull et al. 2016. Darwin's warm little pond: A one-pot reaction for prebiotic phosphorylation and the mobilization of phosphate from minerals in a urea-based solvent. *Angew. Chem. Int. Ed.* 55: 12249–13253.

Chittenden, G. J. F., and A. W. Schwartz. 1981. Prebiotic photosynthetic reactions. *BioSystems* 14: 75–32.

Coates, J. E., and N. H. Hartshorne. 1931. Studies on hydrogen cyanide Part III. The freezing points of hydrogen cyanide- water mixtures. *J. Chem. Soc.* 657–665.

Draganic, Z. D., S. I. Vujošević, A. Negrón-Mendoza et al. 1991. Radiolysis of aqueous solutions of ammonium bicarbonate over a large dose range. *Int. J. Rad. Appl. Instrum. C* 38: 317–321. doi:10.1016/1359–0197(91)90100-G.

Durickovic, I., L. Thiébaud, P. Bourson, T. Kauffmann, and M. Marchettia. 2009. Spectroscopic characterization of urea aqueous solutions: Experimental phase diagram of the urea–water binary system. *Int. J. Pharm.* 378: 136–139.

Ferris, J. P, and D. E. Nicodem. 1972. Ammonia photolysis and the role of ammonia in chemical revolution. *Nature* 238: 268–269.

Ferris, J. P., P. C. Joshi, E. H. Edelson, and J. G. Lawless. 1978. HCN: A plausible source of purines, pyrimidines and amino acids on the primitive Earth. *J Mol Evol* 11: 293–311.

Getoff, N. 1962. Effect of ferrous ions on the formation of oxalic acid from aqueous carbon dioxide by ultraviolet irradiation. *Naturforschung* 18B: 169–170.

De Graaf, R. M., and A. W. Schwartz. 2000. Reduction and activation of phosphate on the primitive Earth. *Origins Life Evol. Biosphere* 30: 405–410.

De Graaf, R. M., and A. W. Schwartz. 2005. Thermal synthesis of nucleoside H-phosphonates under mild conditions. *Origins Life Evol. Biosphere* 35: 1–10.

Gulick, A. 1955. Phosphorus as a factor in the origin of life. *Am. Sci.* 43: 479–489.

Gull, M., M. Zhou, F. M. Fernández, and M. A. Pasek. 2014. Prebiotic phosphate ester syntheses in a deep eutectic solvent. *J. Mol. Evol.* 78: 109–117.

Gull, M., and M. A. Pasek. 2013. Is Struvite a prebiotic mineral? *Life* 3(2), 321–330. doi:10.3390/life3020321.

Handschuh, G. J., R. Lohrmann, and L. E. Orgel. 1973. The effect of Mg^{2+} and Ca^{2+} on urea-catalyzed phosphorylation reactions. *J. Mol. Evol.* 2: 251–262.

Handschuh, G. J., and L. E. Orgel. 1973. Struvite and prebiotic phosphorylation. *Science* 149: 483–484.

Henderson-Sellers, A., and A. W. Schwartz. 1980. Chemical evolution and ammonia in the early Earth's atmosphere. *Nature* 287: 526–528.

Kuhn, W. R., and S. K. Atreya. 1979. Ammonia photolysis and the greenhouse effect in the primordial atmosphere of the earth. *Icarus* 37: 207–213

La Cruz, N. L., D. Qasim, H. Abbott-Lyon et al. 2016. The evolution of the surface of the mineral schreibersite in prebiotic chemistry. *Phys. Chem. Chem. Phys.* 18: 20160–20167.

Levine, J. S., and T. R. Augustsson. 1982. The prebiological paleoatmosphere: Stability and composition. *Origins Life Evol. Biosphere* 12: 245–259.

Lohrmann, R., and L. E. Orgel. 1971. Urea-inorganic phosphate mixtures as prebiotic phosphorylating agents. *Science* 171: 490–494.

Miller, S. L. 1953. A Production of amino acids under possible primitive earth conditions. *Science* 117: 528–529.

Miller, S. L. 1957. The mechanism of synthesis of amino acids by electric discharges. *Biochim. Biophys. Acta* 23: 480–489.

Miller, S. L., and G. Schlesinger. 1993. Prebiotic syntheses of vitamin coenzymes: I. Cysteamine and 2-mercaptoethanesulfonic acid (Coenzyme M). *J. Mol. Evol.* 36: 302–307.

Morrison, H. G., C. C. Sun, and S. Neervannan, S. 2009. Characterization of thermal behavior of deep eutectic solvents and their potential as drug solubilization vehicles. *Int. J. Pharm* 378: 136–139.

Orgel, L. E. 2004. Prebiotic Adenine Revisited: Eutectics and Photochemistry. *Origins Life Evol. Biosphere* 34: 361–369.

Oró, J. 1960. Synthesis of adenine from ammonium cyanide. *Biochem. Biophys. Res. Commun.* 2: 407–412.

Pasek, M. A., and D. S. Lauretta. 2005. Aqueous corrosion of phosphide minerals from iron meteorites: A highly reactive source of prebiotic phosphorus on the surface of the early Earth. *Astrobiology* 5: 515–535. doi:10.1089/ast.2005.5.515.

Pasek, M. A., J. P. Harnmeijer, R. Buick, M. Gull, and Z. Atlasa. 2013. Evidence for reactive reduced phosphorus species in the early Archean ocean. *Proc. Natl. Acad. Sci. U.S.A.* 110: 10089–10094.

Schwartz, A. W. 2006. Phosphorus in prebiotic chemistry. *Phil. Trans. R. Soc. B* 361: 1743–1749.

Schwartz, A. W. 1971. Phosphate: Solubilization and activation on the primitive earth. In *Chemical Evolution and the Origin of Life*, R. Buvet and C. Ponnamperuma (Eds.), pp. 207–215. Amsterdam, the Netherlands: North-Holland.

Schwartz, A. W., and H. Duess. 1971. Concentrative processes and the origin of biological phosphates. In *Theory and Experiment in Exobiology*. vol. 1, A. W. Schwartz (Ed.), pp. 73–81. Groningen, the Netherlands: Wolters Noordhoff.

Schwartz, A. W., M. van der Veen, T. Bisseling, and J. J. F. Chittenden. 1975. Prebiotic nucleotide synthesis-demonstration of a geologically plausible pathway. *Origins Life Evol. Biosphere* 6: 163–168.

Sanchez, R., J. Ferris, J., and L. E. Orgel. 1966. Conditions for purine synthesis: Did prebiotic synthesis occur at low temperatures? *Science* 153: 72–73.

Sanchez, R., J. Ferris, J., and L. E. Orgel. 1967. Studies in prebiotic synthesis II. Synthesis of purine precursors and amino acids from aqueous hydrogen cyanide. *J. Mol. Biol.* 80: 223–253.

Toupance, G., F. Raulin, and R. Buvet. 1975. Formation of prebiological compounds in models of the primitive Earth's atmosphere. I: CH_4-NH_3 and CH_4-N_2 atmospheres. *Origins Life Evol. Biosphere* 6: 83–90.

Voet, A. B., and A. W. Schwartz. 1983. Prebiotic adenine synthesis from HCN- Evidence for a newly discovered major pathway. *Bioinorg. Chem.* 12: 8–17.

Wills, C., and J. Bada. 2000. *The Spark of Life: Darwin and the Primeval Soup*. Oxford, UK: Oxford University Press.

Yamagata, Y., H. Watanabe, M. Saitoh, and T. Namba. 1991. Volcanic production of polyphosphates and its relevance to prebiotic chemistry. *Nature* 352: 516–519.

5.8 Phosphorylation on the Early Earth
The Role of Phosphorus in Biochemistry and Its Bioavailability

Carolyn Lang, Jennifer Lago, and Matthew A. Pasek

CONTENTS

5.8.1 INTRODUCTION

The element phosphorus (P) is critical to life as we know it. On the modern Earth, phosphorus is often the limiting reagent in an ecosystem (Karl and Björkman 2014), and a phosphorus-limited system will continue to reproduce and grow until exhausting the available phosphorus (Benitez-Nelson 2000). On the other extreme, if there is an abundance of phosphorus in an area, the organisms will continue to reproduce and consume other available nutrients. Often, this abundance leads to algal blooms that can exhaust the oxygen and cause eutrophication (Smayda 2008). To maintain a healthy ecosystem, a source for phosphorus must be available, and the source must not overwhelm the ecosystem with an overabundance of phosphorus. The unique chemical characteristics of phosphorus, including its structure and geochemical abundance, make it well suited as a key nutrient. Given the strong limitation of available phosphorus in biology today, it seems plausible that phosphorus may have been a critical element for early organisms and the chemistry for the origin of life.

We note here a need for a distinction between "phosphorus", "phosphorous", "orthophosphate", and "phosphate". The element is termed "phosphorus". Other than when referring to the acid, H_3PO_3, "phosphorous" is typically a misspelling, except where it archaically refers adjectivally to bonding structures. We typically call C-C bonds "carbon bonds" instead of "carbonous bonds". "Orthophosphate" typically refers to inorganic phosphate, often free in solution or bonded to minerals. "Phosphate" is a specific bonding structure of phosphorus, where phosphate forms four single bonds and one double bond, supported by varying resonance structures, to four oxygen atoms.

The structure of phosphate, and of similar compounds bearing P, O, and H, is shown in Figure 5.8.1. In phosphate, the phosphorus has an oxidation state of +5. Phosphite, hypo-

phosphate, and hypophosphite are distinct phosphorus speciations where the phosphorus employs a different oxidation state (+3, +4, and +1 respectively) and the species are chemically distinct.

Living organisms depend on phosphorus as an essential macronutrient. For example, in the human body, phosphorus participates in nearly all biochemical functions and is crucial for bones and teeth, for the formation of high-energy compounds, as a buffer to maintain homeostasis, and in the synthesis of genetic material, phospholipids, and phosphoproteins. Approximately 80%–85% of the phosphate in the human body is in the form of orthophosphate as the mineral component of bone and teeth (Wang et al. 1992). The remaining phosphate is present in a variety of vital inorganic and organic compounds found in both intracellular and extracellular tissues. These compounds are involved in metabolism, structure, and genetic reproduction (Dorozhkin 2014). Prokaryotic organisms are similarly enriched in phosphorus with respect to dry weight (Lange and Heijnen 2001), despite lacking a calcium phosphate endoskeleton.

Nucleic acids are biochemical macromolecules that primarily store, synthesize, and transfer genetic material. Ribonucleic acid (RNA) and deoxyribonucleic acid (DNA) are two nucleic acids, having backbones of sugars alternating with phosphates, used in all known life. These are orthophosphate diesters that are formed when an alcohol (including sugars) is linked to a phosphate by two bridging oxygen atoms (Figure 5.8.2). The bonds of these diesters are relatively low in energy and are ionized, giving the backbone of these important structures stability in aqueous environments while hindering their diffusion through the phospholipid cellular membrane (Nelson et al. 2008).

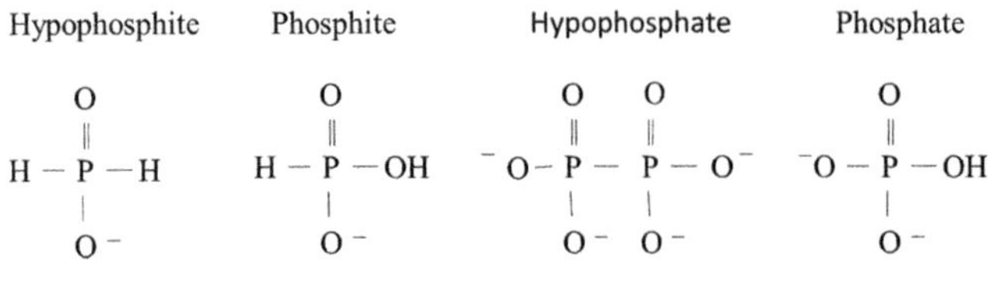

Hypophosphite	Phosphite	Hypophosphate	Phosphate

FIGURE 5.8.1 Phosphorus molecules consisting of H-P-O arrangements.

High-energy biochemical molecules store and deliver energy to chemical reactions needed by organisms. Many of the high-energy molecules are organic polyphosphates which transfer energy within the intermediaries of metabolism by breaking one of the phospho-anhydride or polyphosphate bonds by hydrolysis (reaction with water) yields energy and drives metabolism. The typical example of a metabolic phosphorylated molecule is adenosine triphosphate (ATP, Figure 5.8.2). It contains a high-energy anhydride bond between each phosphate group. The products formed when the terminal phosphate bond is broken are adenosine diphosphate (ADP) and a phosphate group. The ADP and the phosphate group are stabilized by the electrostatic repulsion of the positively charged phosphates and the negatively charged oxygen, and by the ionization and resonance of the products, and because the products have greater entropy than the reactants (Nelson et al. 2008). The phosphate group is then reattached to the ADP by phosphorylation, a condensation reaction, to reform the ATP.

One of phosphorus's properties that makes it indispensable to modern biochemical reactions is its ability to form two bonds while remaining ionized (Westheimer 1987). This ionization provides distinct advantages to specific biomolecules. For example, the molecule retains a charge even when bridging two organics, such as sugars in the primary nucleic acids RNA and DNA. The resulting ionization prevents diffusion of the nucleic acid out of the cell membrane, itself composed of a bilayer of phospholipids. This compartmentalization by phospholipids helps to give local stability to RNA and DNA, which store central genetic material. The charge also prevents hydrolysis of the bonds by water or hydroxide, giving the nucleic acids additional stability in either basic or neutral solutions. This resulting stability gives the individual bonds half-lives estimated as millions of years and results in nucleic acids (which are millions of nucleotides long) requiring only yearly repair (Benner 2004). The ionization also enhances dissolution of phosphorylated organics in water, as the negative charge on the phosphorylated organic increases the interactions with the polar solvent—water—and increases the solubility of the phosphorylated organics (Pasek and Kee 2011). As many biochemical molecules rely on water for transportation, increased solubility increases the number of available biomolecules for catalysis (Pasek et al. 2017).

Organophosphates lack stereochemical handedness, conferring a structural trait that makes phosphorus indispensable for biology; phosphate is achiral. In RNA, the two oxygen share the charge via resonance with no resulting stereoisomers. As life is selective when it comes to chirality, this achirality is especially pertinent. Therefore, it is difficult to replace organophosphates as many of the suggested replacement molecules are chiral or induce a handedness when linking the nucleotides (Westheimer 1987).

To summarize, phosphorus is a key biologic element due to several of its unique properties. It plays a role in biological structure as phosphate biominerals (bone) and phospholipid cell membranes and is an important structural element in nucleic acids, the key genetic material of all life as we know it. It is important to metabolism as ATP, also known as the energy currency of life. The question is, "In which of these roles is phosphorus so critical that there is not much evidence of alternatives near the onset of life?" The next section seeks to address this question.

5.8.2 TOP DOWN AND BOTTOM UP APPROACHES FOR IDENTIFYING THE CRITICAL PHOSPHATE BIOMOLECULES

Historically, scientists have approached questions surrounding the origins of life in two ways, termed "top down" and "bottom up." The "top down" approach falls principally in the field of biology. Biological researchers hope to gain insight into the fundamental features of life—what it did and where it lived—by using their knowledge of modern organisms—their biochemical pathways, environmental habitats, and core genetic material—to address what life might have looked like at the very start of modern biology. Such an approach has been immensely successful and has provided the key data that led to the RNA world insight (Woese 1967; Orgel 1968; Pace and Marsh 1985; Gilbert 1986).

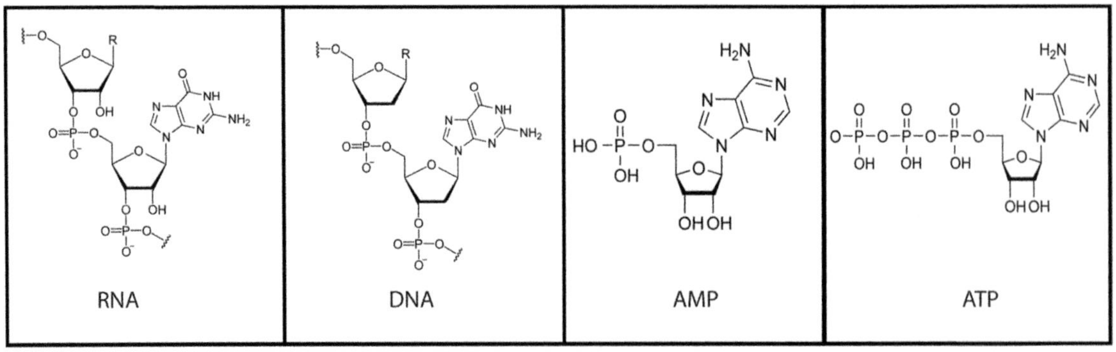

FIGURE 5.8.2 Examples of key phosphorylated biomolecules.

The "bottom up" approach is typically in the realm of chemists, including biochemists, geochemists, and cosmochemists who seek to define the simple chemical reactions that may have led to the components of modern life. Such information may include prebiotic reactions that lead to sugars, the competition and selection of polymers, and the environmental conditions leading to organic reactions of interest (Oró 1961; Ferris et al. 1968; Mizuno and Weiss 1974; Krishnamurthy et al. 2000; Ricardo et al. 2004). As such, researchers take knowledge of the world/solar system, before the onset of life, and try to find a plausible route to making a simple, plausible biochemistry.

Typically, how and when phosphates first became part of life has been explored using the bottom up approach. Researchers use a combination of geochemistry and biochemistry, with the goal of forming organophosphates molecules. However, a "top down" approach may be beneficial because it allows us to elucidate the fundamental chemical services of phosphorus and phosphate, to review the various roles of phosphorus in biology, and to seek places where phosphate has possible substitutes in order to constrain the places where nothing but phosphate will work. These phosphate-essential biomolecules can then be the targets of the bottom up approach for chemical synthesis.

Constraining the biomolecules in the top-down approach involves answering four questions. A "no" to all of the questions points to a molecule with a deeper, fundamental requirement for phosphorous in life: (1) Is the biomolecule used by only a portion of modern organisms? (2) Are there replacements for the phosphorous in modern biochemistry that work well enough? (3) Are there any indications that there were alternatives to the phosphorous in early life? (4) Are there routes that might be imagined on early Earth that lead to the replacement of phosphorous compounds by less stable but easier-to-build compounds?

First, we investigate phosphate as a biomineral. The phosphate mineral apatite makes up the internal supporting structure of vertebrate organisms, as bones and teeth, and is produced by some brachiopods and protists as a biomineral (Omelon et al. 2013) and by one group of plants as a deterrent against herbivores (Ensikat et al. 2016). In addition to apatite, there are many other phosphate biominerals, including struvite ($NH_4MgPO_4 \cdot 6H_2O$), brushite ($CaHPO_4 \cdot 2H_2O$), and vivianite ($Fe_3(PO_4)_2 \cdot 8H_2O$), most of which are produced as waste products by microbes.

The use of apatite as a biomineral is limited to a few branches of Eukarya and other organisms produce many other biominerals. For example, various mollusks, echinoderms, and arthropods produce calcite and aragonite. Also, iron oxides and sulfides, carbonates, and silica are used extensively to support cell structure and are produced by organisms as a waste mineral. Other researchers identify bioapatite as a young (~550 million years ago) biogenic mineral with an origin dependent on the chordate life cycle. We can conclude biomineralization is not a phosphorus-dependent function.

Second, we investigate are the phospholipids. Phospholipids make up the cell membranes of nearly all modern life. Phospholipids are composed of either two long-chain fatty acids (carboxylic acids) or two ethers linked to a glycerol molecule. On one end of the glycerol molecule is a phosphate, which itself may be a diester, linking to an ethanolamine, choline, or similar molecule.

In contrast to biominerals where phosphates are one of many, organophosphates are the major variety of cell membrane lipid, thus fulfilling the first criteria for our top-down approach. However, with respect to phosphorous, biology explores variations in lipid chemistry more than it does metabolism and nucleic acids. Many lipids do not involve phosphate at all, including the sphingolipids, the sulfolipids, and the aminolipids. Both the sphingolipids and aminolipids are used to supplement phospholipid chemistry, but the sulfolipids can completely replace them. This replacement tends to occur most commonly in those environments where phosphate is limited (Van Mooy et al. 2009). The synthesis of phospholipids requires dissolved phosphate, glycerol, and fatty acids to condense, and, though the membranes formed are robust, simple fatty acids also form membranes. To this end, lipids appear to not require phosphate as part of cell membranes. However, if given the "choice," most biochemical systems do appear to preferentially make phospholipids, pointing to a greater utility in biochemistry.

Phospholipids are beneficial to cells, and replacement of phosphate is done only out of necessity. The preference for phospholipids likely is a result of their stability and of the stability of the membranes they form. However, prebiotic chemists have proposed that if less stability can be tolerated in a tradeoff for ease of formation, then simple fatty acids were likely precursors of phospholipids. Thus, our criterion 4 suggests that phospholipids were post-origin phosphorylated biomolecules. There are advantages of using phospholipids, but those advantages were likely evolutionarily selected.

Third, we investigated the metabolites. The most critical of these in biochemistry today is the molecule ATP. It is the dominant energy currency for all life as we know it: storing energy during catabolism and releasing energy during anabolism. It is the most widespread nucleotide in this role, although other nucleotide triphosphates (guanosine triphosphate, uridine triphosphate, and cytosine triphosphate) also have minor roles in metabolism, primarily in sugar transfer and protein synthesis. In contrast, ATP is used in nearly all biochemical reactions requiring an energy input.

This use is universal, hence fulfilling our criterion 1. All organisms require phosphate—as ATP—in their metabolic processes. Some metabolic processes require GTP, UTP, or CTP instead, but these are much less common than ATP-requiring processes. Aside from ATP, each of these nucleotide triphosphates tends to serve a specific metabolic role (e.g., glycosylation), and not one is as widespread as ATP.

In life today, ATP is generated primarily by a proton gradient and a molecular machine, the enzyme ATP synthase. The proton gradient is formed through metabolic processes as organisms dump protons outside of their cell membranes (or organelles within cell membranes that do the same). Then, these protons are slowly let back in and, in the process, drive a mechanical enzyme that sticks phosphate on ADP to make ATP.

The reverse process is called a proton pump and trades chemical energy for pushing protons (H+) out of a cell. This process is generally powered by ATP (breaking down to form ADP); however, there are a suite of enzymes that break a much simpler molecule—pyrophosphate—down to phosphate. These enzymes are not widespread, and the use of an alternative molecule to ATP—that of pyrophosphate—is limited to this main role thus far (Baltscheffsky 1967; Baltscheffsky and Baltscheffsky 1995).

The assembly of two phosphates forms pyrophosphate. Continuing to build up phosphates yields a biopolymer termed polyphosphate. Polyphosphates have an energy potential similar to that of ATP but are composed exclusively of phosphate. Polyphosphates are ubiquitous in cells, where they serve as phosphate and energy reservoirs. Organisms, when both energy and phosphate are in abundance, turn ATP into polyphosphate and then transfer the polyphosphate into small cellular sacks called vacuoles. From these vacuoles, polyphosphate can then be extracted and the energy can be recovered by transferring a phosphate to ADP to make ATP when either energy or phosphate is required to maintain cellular function. Many scientists have argued that polyphosphate is ancient and possibly a precursor to ATP in metabolic processes (Kornberg 1995; Achbergerova and Nahalka 2011). The main issue with polyphosphates is that they are highly ionized and sometimes less soluble than the nucleotide triphosphates. In modern life, they do not participate in metabolism without the ATP intermediate.

Our second criterion is that there be no replacements that work well enough. Indeed, modern biochemistry points to a uniqueness of phosphorus in biochemical metabolism. Nothing really substitutes for phosphorus in biochemical processes. Some molecular changes are possible, but even so, the dominant biomolecule driving metabolic reactions is ATP.

Our third criterion deserves close analysis. Recently, work by Goldford et al. (2017) pointed out that there exists a suite of reactions that they identified as being close to a "core" metabolism that does not rely on phosphate for the key steps. They postulated that this core was ancient and involved, in lieu of phosphates, molecules termed thioesters. Such a scenario had indeed been envisioned by others interested in prebiotic chemistry (De Duve 1991; De Duve and Miller 1991), who argued that sulfur could form high-energy bonds capable of metabolism for the origin of life. To this end, the replacement of phosphate in metabolism may have existed, and there is some evidence that alternatives to phosphorylated biomolecules can perform the key roles of life (though not conclusively so). Several plausible prebiotic alternatives have been outlined, suggesting that our top-down perspective of the requirement of phosphate in metabolism falls somewhere around our criteria 3 and 4. Phosphate presently is required throughout all biological metabolism, but there may have been ancient ways around it.

The final major role of phosphorus in biomolecules is in nucleic acids. The two nucleic acids used by all life today are DNA and RNA, which differ in their sugar backbones by an oxygen (DNA has one less O on each sugar), and in their nucleobases (uracil in RNA is replaced by thymidine in DNA). The phosphorus backbone comparatively remains unchanged in all nucleic acids. The DNA and RNA easily meet the first and second criteria in our top-down view of prebiotic chemistry: phosphorus is critical to all nucleic acids.

There is a slight exception to stating that *phosphate* is critical to all nucleic acids. In recent studies, Wang et al. (2007) demonstrated that, in lieu of phosphate, DNA in some bacteria consisted of a thiophosphate (SPO_3^{3-}) linker. This change was proposed to be due to an evolved defense mechanism that modifies existing DNA to make it less susceptible to attack by nuclease enzymes. Other alternatives to phosphate, such as arsenate (Wolfe-Simon et al. 2011), have not been definitively proven and, for reasons of chemical stability, are highly unlikely.

Biochemicals reactions are all contingent on DNA and RNA, and, currently, there is no core biochemical pathway that doesn't rely on these molecules, unlike metabolism and its potential non-phosphorus reactions (criterion 3). In this respect, a requirement for phosphorus in nucleic acids seems to be even stronger than the requirement for phosphorus in metabolic molecules. However, Hud et al. (2013) outlined an alternative view of nucleic acids and took a reductionist approach to nucleic acids, which they proposed to be composed of three parts: an ionized linker, trifunctional connector, and a recognition unit. These correspond to phosphate, ribose/deoxyribose, and nucleobases, respectively, in the context of RNA and DNA. Hud et al. (2013) postulated that each of these characteristic molecules could have been replaced at some point in early life, and hence, modern day nucleic acids are not representative of ancient nucleic acids. They provide the molecule glyoxylate as a potential substitute for phosphate in nucleic acids as an example of changing out the ionized linker (Bean et al. 2006).

In summary (Figure 5.8.3), a top-down approach highlights two sets of phosphorylated molecules that merit focus as goal molecules for a bottom-up approach: high-energy polyphosphates such as ATP or the simpler polyphosphates,

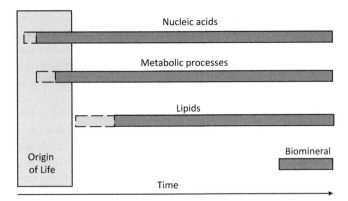

FIGURE 5.8.3 A top-down view of phosphorus in biology. The likely addition point of each major phosphorus-bearing molecule is shown, based on our criteria for inclusion. The "origin of life" block may be either a singular event or a continuum. Therefore, molecules shown starting within this box were potentially present at the origin of life (but not necessarily so, since the singular event may have been prior to the synthesis of these molecules).

and phosphorylated nucleic acids. Phospholipids are unlikely to have been critical in prebiotic chemistry, comparatively, and it is almost certain that phosphate biominerals were not part of the origin of life. Next, we investigate routes to forming nucleic acids and similar compounds.

5.8.3 THE PHOSPHORYLATION REACTION

A major goal of prebiotic chemistry is to find plausible, simple routes to forming the molecules that are presumed to have been important in the origin of life on the Earth. It is likely, though not proven, that organophosphates were in the suite of chemicals necessary for the origin of life. In the prior section, we identified the polyphosphates and nucleotides (or sugar/alcohol phosphates) as potential target molecules for prebiotic chemistry. To this end, finding a plausible route to forming these molecules merits closer review.

The phosphorylation reaction is simplified broadly as:

$$R^1OH + R^2PO_3^{2-} = R^1\text{-}OPO_3^{2-} + R^2H \text{ (Reaction 1)}$$

In this reaction, an organic compound, R^1OH, reacts with a phosphate, $R^2PO_3^{2-}$, to yield an organophosphate (R^1–OPO_3^{2-}), releasing the attached functional group of the reagent phosphate. This reaction has been left ambiguous, so that it encompasses most sample reactions that have been discussed in the literature (Keefe and Miller 1995; Pasek and Kee 2011; Pasek et al. 2017).

A variety of routes to perform this reaction have been proposed, all of which vary the conditions of this reaction to increase the yield of organophosphate product.

5.8.4 OVERCOMING THE ISSUES OF PHOSPHORYLATION

One major hurdle in the origin of life study is a problem termed "the water problem." Unquestionably, all life as we know it now requires water. Water is also ubiquitous on the Earth, and water is assumed to have been readily available on the early Earth, due to its ubiquity in the solar system and throughout the universe. In a world where water would be abundant, a challenge arises: many reactions in prebiotic chemistry require a dehydration step. For instance, the formation of nucleotides requires a nucleobase bonding to a sugar that also bonds to a phosphate. Each of these steps releases

water. Furthermore, the polymerization of nucleotides to form nucleic acids occurs via a condensation reaction, as does the assembly of amino acids into polypeptides. Releasing water as a product is unfavorable in an aqueous solvent. This chemistry leads to a conundrum in the study of prebiotic chemistry: how to overcome the so-called "water problem." To overcome this hurdle, researchers have employed different methods, including the use of condensing agents, high temperatures, reduced phosphorus minerals, and water-free (or low-water-activity) solvents. While researchers have had varying successes using different methods, not all of these routes would have been plausible on the early Earth.

One significant route to promoting phosphorylation reactions is in the use of condensing agents. Condensing agents have been utilized extensively in prebiotic chemistry, and their main action is by making R^1 in Reaction 1 a good leaving group when reacting with OH. Some of the condensing agents extensively studied have been cyanide-containing molecules and condensed phosphates. Dehydration reactions have been demonstrated in aqueous environments aided by the addition of condensing agents. Condensing agents, when added to the milieu, allow for a dehydration reaction to occur in a water solvent. One of the simplest condensing agents, cyanamide, has been demonstrated to aid in the phosphorylation of uridine (Lohrmann and Orgel 1968). Cyanogen has proven successful in aiding the phosphorylation of ribose (Halmann et al. 1969) by using the condensed phosphate trimetaphosphate (TMP). Urea has also been shown to aid in phosphorylation of uridine (Lohrmann and Orgel 1971; Osterberg et al. 1973).

A second route to promoting the phosphorylation reaction is through the addition of heat. In order to drive a condensation reaction in a water solvent, some have utilized heating nucleotides in the presence of phosphates to drive the reaction (Ponnamperuma and Mack 1965). Although this process afforded appreciable yields of nucleosides, the plausibility of the temperature used (160°C) is questionable on the surface of the early Earth and would lead to decomposition of compounds if they are left at these temperatures. Any researcher conducting a thermal experiment must take into account the temperature needed to overcome the activation barrier of the reaction and the temperature limit at which degradation of the products occurs too fast to be prebiotically plausible.

In lieu of activating the organic compound in the phosphorylation reaction ("energizing" R^1), phosphorylation can be promoted by activating the phosphate ion ("energizing" R^2),

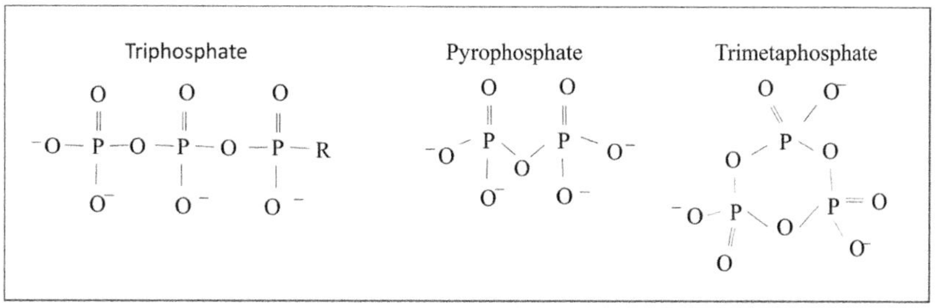

FIGURE 5.8.4 Examples of condensed phosphates that have been used as phosphorylating agents.

through the formation of a condensed phosphate (Figure 5.8.4). A condensed phosphate is one in which the phosphate anion is attached to a second phosphate anion (or more) via the oxygen. Condensed phosphates have been shown to phosphorylate nucleosides with appreciable yields (Chung et al. 1971; Krishnamurthy et al. 1999; Cheng et al. 2002). Trimetaphosphate and cyclic TMP have been utilized extensively by prebiotic chemists as both a condensing agent and a source for phosphorus. Krishnamurthy et al. (1999) demonstrated that TMP, when reacted with ammonia, yields amidotriphosphate, which then was able to phosphorylate glycoaldehyde under mild conditions in appreciable yields. In 2017, the Krishnamurthy group went on to demonstrate that diamidophosphate, derived from TMP, is able to phosphorylate nucleosides, fatty acids, sugars, and amino acids (Karki et al. 2017; Gibard et al. 2018). The results from this study showed not just phosphorylation of a single compound but also oligomerization of nucleotides to nucleic acids.

Although reactive phosphates and polyphosphates are able to provide chemists with a "one stop" route to phosphorylation, the availability of such phosphates outside of a laboratory is questionable. Keefe and Miller (1995) noted that the availability of such reactive phosphates on the early Earth is unlikely or limited to volcanic fumaroles (Yamagata et al. 1991). Hazen (2013), who has provided an extensive study of the mineralogy of the Hadean, has limited phosphate minerals to a few orthophosphate minerals (such as apatite, merrillite, and whitlockite) due to the absence of biology and a young Earth. Therefore, activated phosphate minerals that would have been geologically plausible on the early Earth would only have been found in a few niche settings (such as volcanic fumaroles and meteorites).

A specific route to activating phosphates is in the use of reduced-oxidation-state phosphorus compounds (Pasek and Lauretta 2005; Bryant and Kee 2006; Gull et al. 2015). Reduced-oxidation-state phosphorus species (bearing a formal charge on phosphorus less than the typical $+5$ oxidation state) and condensed phosphates have demonstrated the ability to phosphorylate organics. Gulick (1955) was the first to propose that phosphite minerals (oxidation state $+3$) may have been available in the Hadean Earth. He based his theory on a reducing atmosphere; however, the idea was abandoned by geochemists working at the time (Miller and Urey 1959). Pasek and Lauretta (2005) demonstrated that reduced-oxidation-state phosphorus compounds were common to meteorites and reacted to form a variety of intriguing phosphorus compounds. Specifically, the phosphide mineral schreibersite, $(Fe, Ni)_3P$, is found in meteorites and may be a source of reactive phosphorus on the early Earth following the Late Heavy Bombardment (Pasek et al. 2013). Pasek and his coworkers demonstrated the formation of condensed reactive phosphates from the reaction of the mineral schreibersite with water. In addition, when these phosphide minerals are added to solutions of organic solutes, phosphorylation occurs at low yields (Gull et al. 2015), likely through a metaphosphate intermediate (Pasek et al. 2017).

The use of alternative solvents in condensation reactions has been studied for decades. This presents our final route to

phosphorylation, which is to minimize the activity of water (the product R^2-H in our Reaction 1). Schoffstall (1976) first published a paper on the use of several solvents in place of water, demonstrating that only formamide had the capability of phosphorylating adenosine. This study used phosphate sources with questionable availability in the Hadean: KH_2PO_4, polyphosphates, TMP, and ammonium phosphates. No phosphorylation was detected with the calcium phosphates. Saladino along with his coworkers also used formamide as an alternative solvent (Costanzo et al. 2007; Saladino et al. 2009, 2015). This research group found that formamide condenses to yield nucleobases and yields phosphorylated nucleotides. They have a caveat: at some point in time, life must have transitioned to a water-based solvent. The rebuttal to this work comes from Bada et al. (2016), arguing that a source of pure formamide on the early Earth was unlikely. Although formamide as an alternative solvent provides promising results, a prebiotic geological environment in which formamide could form and accumulate is problematic. Also, pure formamide is a hygroscopic compound; in water, formamide will rapidly undergo hydrolysis to form formic acid and ammonia.

Other water-free alternative solvents previously studied are the deep eutectic solvents of urea and choline chloride. Gull et al. (2014) showed that adding soluble phosphates and the nucleobases of adenine or uridine to this eutectic mixture results in phosphorylated products. While this method gives promising results in overcoming the challenges of a water solvent, the prebiotic relevance of choline chloride is questionable. More recent work in the use of alternative solvents has been explored by Burcar et al. (2016), using a low water-activity solvent of urea, ammonium formate, and water. Burcar and coworkers found that under mild heating, the three-component solvent mixture forms formamide in the solution and was able to demonstrate phosphorylation of nucleosides within this milieu. The compounds used—urea, ammonium formate, and water—all would have been available on the early Earth under a variety of environmental settings.

The above workarounds to the "water problem" demonstrate that the production of organophosphates in water is problematic, for much the same reason as condensation reactions as a whole are problematic. In this respect, the "phosphate problem" is a subset of the "water problem." Water is difficult to form as a product of a chemical reaction when it is also the solvent. However, a second issue with phosphates also merits comments: the poor solubility of phosphate under many geochemical conditions and its general geochemical rarity. Thus, the "phosphate problem" bears the aspects of the "water problem" but includes an additional major caveat.

Phosphorus is a minor element in most rocks. On the Earth, 95% of the total abundance of phosphorus is concentrated in the core, giving it a phosphorus concentration of 3000 ppm. Of the remaining sources of phosphorus on Earth, the biosphere contains 6600 ppm phosphorus, while the mantle and crust are relatively depleted in phosphorus, with 90 ppm and 650 ppm, respectively (Pasek et al. 2017). While there are 16 known allotropes of elemental phosphorus (Greenwood and Hershaw 1984), neither any of the allotropes are stable nor do any occur naturally on Earth. Virtually all natural phosphorus on Earth's surface occurs as a phosphate, with the phosphorus oxidation state of +5. Of the lithospheric (crust + mantle) phosphorus, most of it is in one of the several hundred phosphate minerals—with most of these in the apatite family—and all of the commercially important minerals are in the apatite family. The exceptions to +5 phosphorus are rare and are found either in meteorites or in areas where a high-energy event reduced phosphate minerals or where highly reducing/high-temperature environments have reduced phosphate minerals. The main phosphate group of minerals is the apatite minerals, which can also be classified as orthophosphate minerals, as they bear a phosphate group (PO_4^{3-}). The orthophosphate minerals are mostly bound in rocks and soils, and the availability of phosphorus to an ecosystem is regulated by the dissolution of phosphate minerals available in nearby rocks and soil and any available phosphorus from decaying organic matter.

The phosphorus on Earth is cycled through several primary cycles of varying lengths from weeks to millions of years. There are two rapid biological cycles: one small stable cycle on land and one in water. These fast cycles are intertwined with each other and with the much longer inorganic cycle (Benitez-Nelson 2000). In this respect, the phosphorus cycle differs significantly from the other major biogenic element cycles (C, N, H, O, and S), which all bear a significant volatile process.

5.8.5   WHAT ENVIRONMENTS WORK FOR PHOSPHORYLATION?

Taking the previously described caveats on phosphorus and its reactions, we can posit that there are two main issues associated with phosphorylation: its low reactivity and its low solutional availability. A successful prebiotic chemistry that incorporates solutions to both would be ideal. These concerns highlight that the "phosphate problem" is the "water problem," coupled with the low solubility and cosmic abundance of phosphorus.

The cosmic rarity of phosphorus is not as obvious from its general geochemical behavior. Phosphorus sinks to the metallic cores of stony planets and asteroids, and it is also concentrated in the thin lithosphere of stony planets (where life likely arose). In contrast, it is mostly absent from the massive middle part of a planet, termed the mantle. The concentration of phosphorus on planetary surfaces is further abetted by the delivery of meteorites to a planet during bombardment (Pasek and Lauretta 2008). Thus, phosphorus is typically present on planetary surfaces and was presumably present on most rocky terrestrial planets. It is unlikely that the issue of abundance is terribly problematic for stellar systems like our own.

We highlighted two primary target molecules as goal molecules for prebiotic phosphorylation: nucleic acids and polyphosphates. Nucleic acids are formed from a combination of $HCN/HCONH_2$ chemistry for the nucleobases (Oró 1961; Orgel 2004) and from CH_2O chemistry for the ribose (Breslow 1959). The precursors to these compounds do not mix well (Danger et al. 2014), but routes to forming the products from other reactants have been identified (Powner

et al. 2009). Nonetheless, with respect to phosphate, both the nucleic acids and the polyphosphates must form through loss of water at some point in their reaction.

Our target reactions for formation of the organophosphates of interest is one that allowed organics to condense with phosphate. Perhaps, this reaction occurred in the presence of water, or perhaps, it occurred in a solvent where water was a constituent. We propose that the two most likely routes are those that use activated phosphorus sources or those that use non- or semi-aqueous solvents.

Activated phosphate sources must be placed in the context of general geochemical availability. The various geochemically available activated phosphates are rare, with only a few known polyphosphate minerals (see Pasek et al. 2017 for a review). There are no known TMP minerals, and the putative source of polyphosphates in fumaroles—P_4O_{10}—has never been directly observed (Yamagata et al. 1991). However, recently, a few activated phosphorus sources have been identified: the mineral schreibersite, the amidophosphates, and oxidation products of the compounds phosphite and hypophosphite. The mineral schreibersite is common to meteorites, and its oxidation and its ability to phosphorylate are a consequence of its high potential energy under aqueous conditions. The amidophosphates are robust reactants in phosphorylation (Gibard et al. 2018), but an obvious geologic source is yet to be identified. If one were to be found, these might prove to be the best route to making organophosphates. The oxidation products of phosphite and hypophosphite are polyphosphates (Pasek et al. 2008), which are known phosphorylating agents. Recently, a robust route to the formation of phosphite on the early Earth was identified: reaction of phosphate with iron(II) (Herschy et al. 2018).

Non-aqueous or semi-aqueous solvents are a second plausible route to phosphorylation. More generally, environments with low water activity have been shown to definitively produce phosphorylated biomolecules. The problem with non-aqueous solvents is their geochemical production, isolation, and concentration. In contrast, the semi-aqueous solvents may be preferable, given the ubiquity of water in the universe. The chemistry shown by Burcar et al. (2016) is also promising, as their route to phosphorylating organic molecules is coupled to the transformation of the common mineral apatite into the more soluble but rarer mineral struvite. Figure 9.5.5 displays a compilation of the environments conducive to reducing and/or condensing phosphorus species (Figure 5.8.5).

5.8.6   SUMMARY

The element phosphorus appears to have been crucial, due to several key properties that have been outlined in prior work, to the development of life on the Earth. (Westheimer 1987; Pasek et al. 2017). Some of the strongest evidences of a primordial phosphorus biochemistry are demonstrated by the fact that phosphate-based metabolism and nucleic acids are common to all known life. While phosphate-based membranes exhibit enhanced fitness, compared with the fitness of simpler fatty-acid membranes, and may have been present early on as well, they are less of a requirement for life than nucleic acids and nucleotide phosphates.

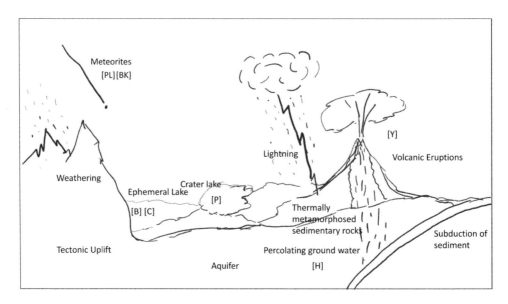

FIGURE 5.8.5 A schematic of the environments of phosphorylation. The key environments include phosphate-reacting environments, such as volcanic fumaroles, as identified by Yamagata et al. (1991), denoted with [Y], ephemeral lakes with non-aqueous solvents such as formamide (Costanzo et al. 2007, denoted with [C], and semi-aqueous solvents including urea-ammonium formate-water mixtures (Burcar et al. 2016, denoted [B]). In addition, the oxidation of reduced phosphorus compounds produces polyphosphates capable of forming organophosphates and likely occurred in lakes, especially crater lakes (denoted [P] for Pasek et al. 2008). Such a reaction requires reduced phosphorus sources such as meteorites (Pasek and Lauretta 2005 and Bryant and Kee 2006, denoted [PL] and [BK], respectively), lightning, and redox reactions occurring within sedimentary rocks (Herschy et al. 2018, denoted [H]).

To this end, the molecules likely needed by prebiotic chemistry were simple alcohol phosphates and polyphosphates akin to nucleic acids and nucleotide triphosphates present in life today. The formation of these molecules has been demonstrated in reactions that include high-chemical-energy condensing agents and polyphosphates formed by drying and the oxidation of reduced-oxidation-state phosphorus compounds, as well as in solutions that bear significant non-aqueous components or those completely water-free. Plausible geochemistry suggests that polyphosphates and semi-aqueous solvents may be the best routes to forming organophosphates. If so, the challenge of organophosphate production may be approaching a solution.

REFERENCES

Achbergerova, L., and J. Nahalka. 2011. Polyphosphate – An ancient energy source and active metabolic regulator. *Microb. Cell Fact.* 10: 63.

Bada, J.L., J.H. Chalmers, and H.J. Cleaves II. 2016. Is formamide a geochemically plausible prebiotic solvent? *Phys. Chem. Chem. Phys.* 18: 20085–20090.

Baltscheffsky, M. 1967. Inorganic pyrophosphate and ATP as energy donors in chromatophores from *Rhodospirillum rubrum*. *Nature* 216: 241–243.

Baltscheffsky, M., and H. Baltscheffsky. 1995. Alternative photophosphorylation, inorganic pyrophosphate synthase and inorganic pyrophosphate. *Photosynth. Res.* 46: 87–91.

Bean, H.D., F. Anet, I.R. Gould, and N. Hud. 2006. Glyoxylate as a backbone linkage for a prebiotic ancestor of RNA. *Orig. Life Evol. Biosph.* 36: 39–63.

Benitez-Nelson, C.R. 2000. The biogeochemical cycling of phosphorus in marine systems. *Earth-Sci. Rev.* 51: 109–135.

Benner, S.A. 2004. Understanding nucleic acids using synthetic chemistry. *Acc. Chem. Res.* 37: 784–797.

Breslow, R. 1959. On the mechanism of the formose reaction. *Tetrahedron Lett.* 21: 22–26.

Bryant, D.E., and T.P. Kee. 2006. Direct evidence for the availability of reactive, water soluble phosphorus on the early Earth. H-Phosphinic acid from the Nantan meteorite. *Chem. Commun.* 22: 2344–2346.

Burcar, B., M. Pasek, M. Gull et al. 2016. Darwin's warm little pond: A one-pot reaction for prebiotic phosphorylation and the mobilization of phosphate from minerals in a urea-based solvent. *Angew. Chem. Int. Ed. Engl.* 55: 13249–13253.

Cheng, C., C. Fan, R. Wan et al. 2002. Phosphorylation of adenosine with trimetaphosphate under simulated prebiotic conditions. *Orig. Life Evol. Biosph.* 32: 219–224.

Chung, N.M., R. Lohrmann, L.E. Orgel, and J. Rabinowitz. 1971. The mechanism of the trimetaphosphate-induced peptide synthesis. *Tetrahedron* 27: 1205–1210.

Costanzo, G., R. Saladino, C. Crestini et al. 2007. Nucleoside phosphorylation by phosphate minerals. *J. Biol. Chem.* 282: 16729–16735.

Danger, G., A. Rimola, N. Abou Mrad et al. 2014. Formation of hydroxyacetonitrile (HOCH$_2$CN) and polyoxymethylene (POM) – Derivatives in comets from formaldehyde (CH$_2$O) and hydrogen cyanide (HCN) activated by water. *Phys. Chem. Chem. Phys.* 16: 3360–3370.

De Duve, C. 1991. *Blueprint for a Cell: The Nature and Origin of Life*. Carolina Biological Supply Company. Burlington, NC: Neil Patterson Publishers.

De Duve, C., and S.L. Miller. 1991. Two-dimensional life? *PNAS* 88: 10014–10017.

Dorozhkin, S.V. 2014. Calcium orthophosphate coatings on magnesium and its biodegradable alloys. *Acta Biomater.* 10: 2919–2934.

Ensikat, H., T. Geisler, and M. Weigend. 2016. A first report of hydroxylated apatite as structural biomineral in Loasaceae – plants' teeth against herbivores. *Sci. Rep.* 6: 26073.

Ferris, J.P., R.A. Sanchez, and L.E. Orgel. 1968. Studies in prebiotic synthesis. III. Synthesis of pyrimidines from cyanoacetylene and cyanate. *J. Mol. Biol.* 33: 693–704.

Gibard, C., S. Bhowmik, M. Karki et al. 2018. Phosphorylation, oligomerization and self-assembly in water under potential prebiotic conditions. *Nat. Chem.* 10: 212.

Gilbert, W. 1986. The RNA world. *Nature* 319: 618.

Goldford, J.E., H. Hartman, T.F. Smith, and D. Segrè. 2017. Remnants of an ancient metabolism without phosphate. *Cell* 168: 1126–1134.

Greenwood, N.N., and A. Earnshaw. 1984. *Chemistry of the Elements*. Elmsford, NY: Pergamon Press.

Gulick, A. 1955. Phosphorus as a factor in the origin of life. *Am. Sci.* 43: 479–489.

Gull, M., M. Zhou, F.M. Fernandez, and M.A. Pasek. 2014. Prebiotic phosphate ester syntheses in a deep eutectic solvent. *J. Mol. Evol.* 78: 109–117.

Gull, M., M.A. Mojica, F.M. Fernández et al. 2015. Nucleoside phosphorylation by the mineral schreibersite. *Sci. Rep.* 5: 17198.

Halmann, M., R.A. Sanchez, and L.E. Orgel. 1969. Phosphorylation of D-ribose in aqueous solution. *J. Org. Chem.* 34: 3702–3703.

Hazen, R.M. 2013. Paleomineralogy of the Hadean Eon: A preliminary species list. *Am. J. Sci.* 313: 807–843.

Herschy, B., S.J. Chang, R. Blake et al. 2018. Archean phosphorus liberation induced by iron redox geochemistry. *Nat. Commun.* 9: 1346.

Hud, N.V., B.J. Cafferty, R. Krishnamurthy, and L.D. Williams. 2013. The origin of RNA and "my grandfather's axe". *Chem. Biol.* 20: 466–474.

Karki, M., C. Gibard, S. Bhowmik, and R. Krishnamurthy. 2017. Nitrogenous derivatives of phosphorus and the origins of life: Plausible prebiotic phosphorylating agents in water. *Life* 7: 32.

Karl, D.M., and K.M. Björkman. 2014. Dynamics of dissolved organic phosphorus. In *Biogeochemistry of Marine Dissolved Organic Matter* (2nd ed.), pp. 233–334. Amsterdam, the Netherlands: Academic Press.

Keefe, A.D., and S.L. Miller. 1995. Are polyphosphate or phosphate esters prebiotic reagents? *J. Mol. Evol.* 41: 693–702.

Kornberg, A. 1995. Inorganic polyphosphate: Toward making a forgotten polymer unforgettable. *J. Bacteriol.* 177: 491–496.

Krishnamurthy, R., G. Arrhenius, and A. Eschenmoser. 1999. Formation of glycolaldehyde phosphate from glycoaldehyde in aqueous solution. *Orig. Life Evol. Biosph.* 29: 333–354.

Krishnamurthy, R., S. Guntha, and A. Eschenmoser. 2000. Regioselective α-phosphorylation of aldoses in aqueous solution. *Angew. Chem. Int. Ed. Engl.* 39: 2281–2285.

Lange, H.C., and J.J. Heijnen. 2001. Statistical reconciliation of the elemental and molecular biomass composition of *Saccharomyces cerevisiae*. *Biotechnol. Bioeng.* 75: 334–344.

Lohrmann, R., and L.E. Orgel. 1968. Prebiotic synthesis: Phosphorylation in aqueous solution. *Science* 101: 64–66.

Lohrmann, R., and L.E. Orgel. 1971. Urea-inorganic phosphate mixtures as prebiotic phosphorylating agents. *Science* 171: 490–494.

Miller, S.L., and H.C. Urey. 1959. Organic compound synthesis on the primitive earth. *Science* 130: 245–251.

Mizuno, T., and A.H. Weiss, A.H. 1974. Synthesis and utilization of formose sugars. *Adv. Carbohyd. Chem. Biochem.* 29: 173–227.

Nelson, D.L., A.L. Lehninger, and M.M. Cox. 2008. *Lehninger Principles of Biochemistry*. New York: W.H. Freeman.

Omelon, S., M. Ariganello, E. Bonucci et al. 2013. A review of phosphate mineral nucleation in biology and geobiology. *Calcif. Tissue Int.* 93: 382–396.

Orgel, L.E. 1968. Evolution of the genetic apparatus. *J. Mol. Biol.* 38: 381–393.

Orgel, L.E. 2004. Prebiotic chemistry and the origin of the RNA world. *Crit. Rev. Biochem. Mol. Biol.* 9: 99–123.

Oró, J. 1961. Mechanism of synthesis of adenine from hydrogen cyanide under plausible primitive earth conditions. *Nature* 191: 1193–1194.

Osterberg, R., L.E. Orgel, and R. Lohrmann 1973. Further studies of urea-catalyzed phosphorylation reactions. *J. Mol. Evol.* 2: 231–234.

Pace, N.R., and T.L. Marsh. 1985. RNA catalysis and the origin of life. *Orig. Life Evol. Biosph.* 16: 97–116.

Pasek, M.A., T.P. Kee, D.E. Bryant et al. 2008. Production of potentially prebiotic condensed phosphates by phosphorus redox chemistry. *Angew. Chem. Int. Ed. Engl.* 120: 8036–8038.

Pasek, M.A., and D.S. Lauretta. 2005. Aqueous corrosion of phosphide minerals from iron meteorites: A highly reactive source of prebiotic phosphorus on the surface of the early Earth. *Astrobiology* 5: 515–535.

Pasek, M.A., and D.S. Lauretta. 2008. Extraterrestrial flux of potentially prebiotic C, N, and P to the early Earth. *Orig. Life Evol. Biosph.* 38: 5–21.

Pasek, M.A., and T.P. Kee. 2011. On the origin of phosphorylated biomolecules. In *Origins of Life: The Primal Self-Organization*, pp. 57–84. Springer, Berlin, Germany.

Pasek, M.A., J.P. Harnmeijer, R. Buick et al. 2013. Evidence for reactive reduced phosphorus species in the early Archean ocean. *PNAS* 110: 10089–10094.

Pasek, M.A., M. Gull, and B. Herschy. 2017. Phosphorylation on the early earth. *Chem. Geol.* 475: 149–170.

Ponnamperuma, C., and R. Mack. 1965. Nucleotide synthesis under possible primitive earth conditions. *Science* 148: 1221–1223.

Powner, M.W., B. Gerland, and J.D. Sutherland. 2009. Synthesis of activated pyrimidine ribonucleotides in prebiotically plausible conditions. *Nature* 459: 239.

Ricardo, A., M.A. Carrigan, A.N. Olcott, and S.A. Benner. 2004. Borate minerals stabilize ribose. *Science* 303: 196.

Saladino, R., C. Crestini, F. Ciciriello et al. 2009. From formamide to RNA: The roles of formamide and water in the evolution of chemical information. *Res. Microbiol.* 160: 441–448.

Saladino, R., E. Carota, G. Botta et al. 2015. Meteorite-catalyzed syntheses of nucleosides and of other prebiotic compounds from formamide under proton irradiation. *PNAS* 112: E2746–E2755.

Schoffstall, A.M. 1976. Prebiotic phosphorylation of nucleosides in formamide. *Orig. Life* 7: 399–412.

Smayda, T.J. 2008. Complexity in the eutrophication–harmful algal bloom relationship, with comment on the importance of grazing. *Harmful Algae* 8: 140–151.

Van Mooy, B.A.S., H.F. Fredricks, B.E. Pedler et al. 2009. Phytoplankton in the ocean use non-phosphorus lipids in response to phosphorus scarcity. *Nature* 458: 69.

Wang, L., S. Chen, T. Xu et al. 2007. Phosphorothioation of DNA in bacteria by *dnd* genes. *Nat. Chem. Biol.* 3: 709.

Wang, Z-M., R.N. Pierson Jr, and S.B. Heymsfield. 1992. The five-level model: A new approach to organizing body-composition research. *Am. J. Clin. Nut.* 56: 19–28.

Westheimer, F.H. 1987. Why nature chose phosphates. *Science* 235: 1173–1178.

Woese, C. 1967. *The Genetic Code*. New York: Harper & Row, 200 pp.

Wolfe-Simon, F., J.S. Blum, T.R. Kulp et al. 2011. A bacterium that can grow by using arsenic instead of phosphorus. *Science* 332: 1163–1166.

Yamagata, Y., H. Watanabe, M. Saitoh, and T. Namba. 1991. Volcanic production of polyphosphates and its relevance to prebiotic evolution. *Nature* 352: 516.

5.9 Silicon and Life

Joseph B. Lambert

CONTENTS

5.9.1 INTRODUCTION

Silicon (Si) is the second most abundant element in the Earth's crust (ca. 27%) after oxygen, the third must abundant element in the Earth as a whole (ca. 15%) after iron and oxygen, and the sixth most abundant element in the Milky Way Galaxy (ca. 0.065%), in which hydrogen and helium are the most abundant (Wikipedia 2018a). With an atomic number of 14, it is located just below carbon in the Periodic Table in the 14th column (formerly Group 4). Carbon and silicon have similar s- and p-electron structures, whereby the outer electrons comprise two s electrons and two p electrons. In addition, silicon has available empty d orbitals. The four outer electrons in both cases are capable of hybridization to form four equivalent sp^3 orbitals, enabling the formation of four bonds with a tetrahedral geometry, as in the structures in Figures 5.9.1 and 5.9.2. The ligands on carbon and silicon are illustrated as hydrogen, but many other elements can serve as well, such as carbon, silicon, oxygen, nitrogen, sulfur, and the halogens.

With four bonding opportunities and a plethora of potential bonding partners, both carbon and silicon can produce a wide variety of simple and complex compounds. In particular, they may bond repeatedly with like-functionalized atoms to form linear chains of molecules. Figure 5.9.3 illustrates polymethylene (more commonly, polyethylene), a linear polymer composed of methylene units, $-(CH_2)_n-$, and Figure 5.9.4 illustrates poly(dimethylsiloxane), a linear polymer composed of dimethylsiloxane units, $-(SiMe_2O)_n-$. The structures are illustrated with three units of the monomer building block, which extends further in either direction. The functionalities were chosen because they represent possibly the commonest and simplest such polymeric materials. Nonetheless, the analogous carbon polymer for siloxanes exists as polyoxymethylenes [or polyacetals, such as $-(CH_2O)_n-$], as does the analogous silicon version of polyethylenes as polysilanes, such as $-(SiMe_2)_n-$. The structures may be augmented by replacing one or more side ligands (H in Figure 5.9.3 or Me in Figure 5.9.4) with additional chains of the same monomer to yield branched polymers. Alternatively, the ends of the polymers may be brought around and bonded to each other

to form molecular rings, as in Figure 5.9.5 for six methylene units (cyclohexane) or in Figure 5.9.6 for three dimethylsiloxane units (hexamethyltrisiloxane).

With just these three structural motifs (chains, branching, and rings), together with replacement of the ligands with other organic groups (ethyl, hexyl, phenyl, vinyl, and so on) and atoms (oxygen for ethers, nitrogen for amines, and so on), the potential exists for building millions of different organic (carbon-based) or organosilane (silicon-based) molecules.

Since carbon is the core element for the molecules of life that exist on Earth, it seems logical that silicon should be able to serve as the core element for a parallel biochemistry for what would be called silicon-based life. This idea goes back at least to the late 19th century, when the science fiction writer H. G. Wells wrote "One is startled towards fantastic imaginings by such a suggestion: visions of silicon-aluminium organisms—why not silicon-aluminium men at once?—wandering through an atmosphere of gaseous sulphur, let us say, by the shores of a sea of liquid iron some thousand degrees or so above the temperature of a blast furnace" (Wells 1894) The geneticist Haldane (1927) speculated more specifically that silicon-based life-forms would be constructed of silicates. A number of recent writers have considered the possibilities of life based on silicon and generally found the concept unlikely (Dessy 1998; Rampelotto 2010; Schulze-Makuch 2013; Darling 2014; Jacob 2016; Wikipedia 2018b). Certain characteristics of silicon chemistry are distinct from carbon chemistry, rendering silicon an unlikely or even impossible core element for life.

5.9.2 POLYSILANES AND CARBOSILANES

Polysilanes are the silicon analogues of polymethylenes (Figure 5.9.7), and polycarbosilanes contain a mixture of carbon and silicon in the main and side chains. Until about 1960, only short-chain silanes were known, but a flowering of polysilane chemistry soon thereafter resulted in the development of a robust chemistry of even long-chain polysilanes, along with branched and multiply branched (dendritic) polysilanes (West 1989; Lambert et al. 2003). Some authors have touted polysilanes as a possible basis for life, since they are

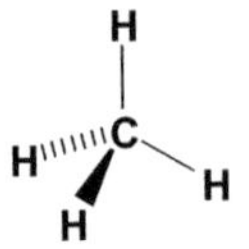

FIGURE 5.9.1

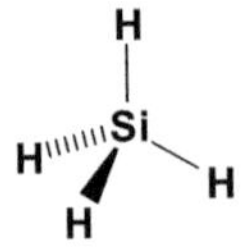

FIGURE 5.9.2

——(CH₂CH₂CH₂)——

FIGURE 5.9.3

——(SiMe₂OSiMe₂OSiMe₂O)——

FIGURE 5.9.4

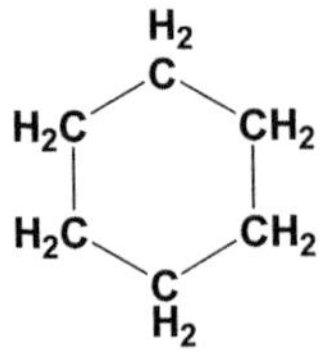

FIGURE 5.9.5

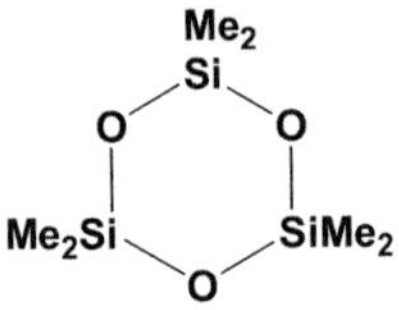

FIGURE 5.9.6

—— (SiMe₂SiMe₂SiMe₂)——

FIGURE 5.9.7

analogous to long-chain hydrocarbons (paraffins) in terms of branching and complexity and their d orbitals permit some electronic properties reminiscent of unsaturation (double bonds). Nonetheless, a number of characteristics militate against polysilanes as a basis for life.

In general, bonds between two silicon atoms and between silicon and carbon (in polysilanes and polycarbosilanes, respectively) are thermodynamically and kinetically less stable than carbon-carbon bonds. The simplest silane ligand is the hydrogen atom, so that, for example in Figure 5.9.7, Si—Me ligands are replaced by Si—H. The Si–H bond is somewhat weaker than the C—H bond, for example, 95 kcal mol⁻¹ in Me₃SiH and Et₃SiH, 92 kcal mol⁻¹ in SiH₄, and 84 kcal mol⁻¹ in (Me₃Si)₃SiH (Fleming et al. 1997), compared with a typical C—H bond dissociation energy of 104 kcal mol⁻¹ in methane (CH₄) (Blanksby and Ellison 2003). Moreover, the Si—H bond is far more reactive than the C—H bond, as it is sensitive

to hydrolysis, oxidation by numerous reagents, and reduction by free radical reagents (Brook 2000a). In general, the C—Si bond (89.4 kcal mol⁻¹ in Me₄Si) is weaker than the C—C bond (90.4 kcal mol⁻¹ in CH₃CH₃), and the Si—Si bond is still weaker (80.5 kcal mol⁻¹ in Me₃SiSiMe₃ and 74 kcal mol⁻¹ in H₃SiSiH₃) (Corey 1989). The Si—C bond reacts with oxidizing reagents (C—Si → C—O), nucleophiles (C—Si → C—C), and electrophiles (C—Si → C—H, C—halogen, C—C) (Brook 2000b). Disilanes and other molecules containing the Si—Si bond are sensitive to electrophilic cleavage and to photolysis (West 1989). Thus, organosilanes of all types are less stable and more reactive than hydrocarbons and would be less practical as a platform for a biochemistry.

Possibly, even more devastating for a potential silicon lifeform is a serious structural deficit. Silicon engages in only weak multiple bonds. In contrast to the robust chemistry of alkenes and aromatics in organic chemistry, analogous structures did not even exist in organosilicon chemistry until the last decades of the 20th century, and the molecules proved to be unstable and highly reactive. The first stable, crystalline molecules containing the C=Si and Si=Si bonds were not reported until 1981 (Raabe and Michl 1985). The poor stability of unsaturated organosilicon compounds results primarily from poor overlap of the p orbitals, so that the resulting π bonds are consequently weaker. In the C=C double bond and other unsaturated organic compounds, the single (σ) bond is augmented by a second bond involving overlap between two parallel p bonds to form the π bond, as in Figure 5.9.8. There are two causes for the weakness of π bonds in C=Si systems. First, the C—Si bond is significantly longer than the C—C bond because of the larger atomic radius of silicon. Second, whereas the p orbitals on carbon are from the second row (2p), those on silicon are from the third row (3p). Thus, there is a misfit between parallel 2p and 3p orbitals, as in Figure 5.9.9. Both factors allow less effective p-p overlap that constitutes π bonding. Although the fit theoretically is appropriate in Si=Si bonds, the bond length between the Si atoms is larger (C—C 1.54 Å, C—Si 1.90 Å, and Si—Si 2.34 Å) (Corey 1989). Over greater distances, p orbitals overlap less and form weaker π bonds.

The absence of available unsaturation in organosilane chemistry is highlighted by consideration of the three principal classes of molecules in our (carbon-based) biochemistry. (1) Proteins and their amino acid constituents provide a structural basis for life, catalysis of the reactions of life,

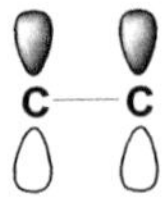

FIGURE 5.9.8

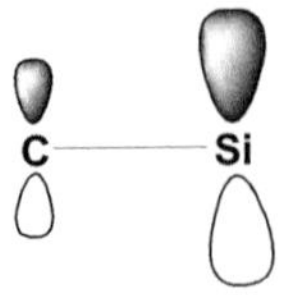

FIGURE 5.9.9

transportation of molecules within cells, various aspects of reproduction, and many other activities. Proteins comprise linear and branched chains of amino acids (Figure 5.9.10) that are connected by the peptide linkage (Figure 5.9.11), in which peptide resonance is manifested by the two illustrated resonance forms. Amino acids and both resonance forms of the peptide unit contain unsaturation, so that silicon analogues are prohibited. (2) Carbohydrates comprise monomeric sugars that can catenate to form very large molecules. The building block is the polyhydroxy Figure 5.9.12, in which the number n varies, 4 for hexoses and 3 for pentoses. The end functionalities normally react to generate a cyclic form, which is illustrated by β-glucose (Figure 5.9.13). Carbohydrates can serve as structural components (cellulose and chitin), energy storage (glycogen and starch), energy transfer (ATP), the backbone of polynucleotide genetic material, and in many other functions. The aldehyde unit in Figure 5.9.12 is a key structural feature, which cannot exist in silicon analogues. (3) Nucleic acids are the building blocks of DNA and RNA (the critical molecules in reproduction), as well as important components of energy transfer (ATP) and other biochemical functions. Nucleic acids contain three components: a pentose sugar, a phosphate side chain, and a nitrogen heterocyclic side chain. Figure 5.9.14 illustrates a nucleic acid with thymine as the heterocycle and ribose as the sugar (thymidylic acid). Polynucleic acids such as DNA and RNA can be constructed by reaction of the hydroxy group of the —CH_2OH unit with a free hydroxy group on the phosphate unit. The nucleotide repeating unit contains not

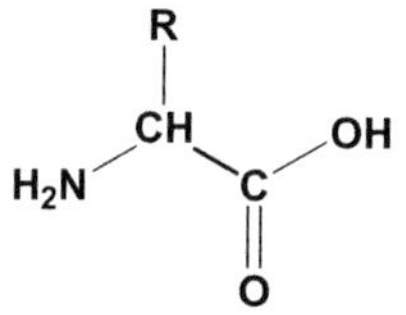

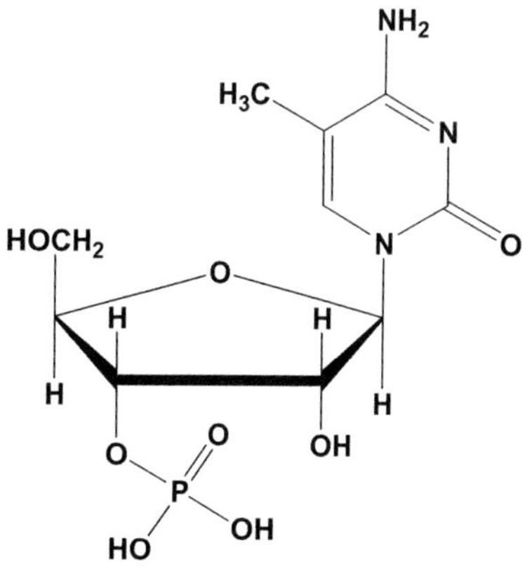

FIGURE 5.9.14

only a carbohydrate, in which the aldehyde function is necessary, but also the heterocycle, which contains unsaturation. Thus, silicon analogues are prohibited again.

These three core functionalities of biochemistry contain unsaturation and cannot admit a silicon parallel. Similarly, most vitamins are excluded from having silicon analogues. The remnant biochemistry that involves only saturated molecules would comprise paraffins (waxes), fats (but not fatty acids), and most steroids (but their earthly biosynthesis from unsaturated molecules such as squalene obviates their silicon analogues). It is extremely doubtful that the biochemistry required to attain life could be constructed from molecules lacking unsaturation.

5.9.3 POLYSILICATES

One thing silicon has going for it is strong bonds to electronegative elements such as oxygen, nitrogen, and fluorine. The Si—F bond at 135 kcal mol^{-1} is one of the strongest known, and the Si—N bond at 100 kcal mol^{-1} is about 20 kcal mol^{-1} stronger than the Si—C bond. The Si—O bond at 128 kcal mol^{-1} is nearly 40 kcal mol^{-1} stronger than the C—O bond (Corey 1989). These properties derive primarily from the highly electropositive nature of silicon, so that bonds with highly electronegative elements have a strong, stabilizing polar component ($^{\delta+}$Si—O$^{\delta-}$). When silicon is fully oxygenated, SiO_4, not only are the bonds extremely stable but they can also propagate fourfold from Si and twofold from O to form large, three-dimensional networks, known as silicates or polysilicates. When the Si—O bonds are diluted with Si—C bonds, the systems are still quite stable. The class of compounds called silsesquioxanes has the core silicon unit of SiO_3R, and silicones (polysiloxanes) have the core unit of SiO_2R_2, as in Figure 5.9.4. The repeating unit and overall molecular formula of silicones is R_2SiO, which is analogous to the formula of ketones, $R_2C=O$, so Kipping coined the term *silicones* by analogy.

The monomeric form of the silicon analogue of carbonyl compounds has the structure $R_2Si=O$, a molecular component that has defied synthesis despite its deceptive simplicity. Thus, no silicon analogues of aldehydes, ketones, amides, esters, or

FIGURE 5.9.10

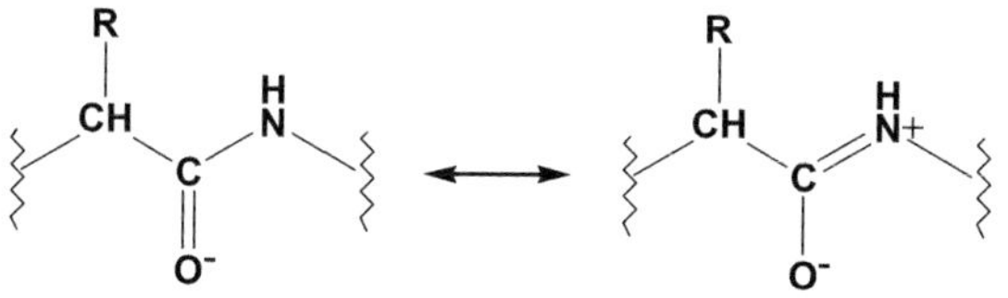

FIGURE 5.9.11

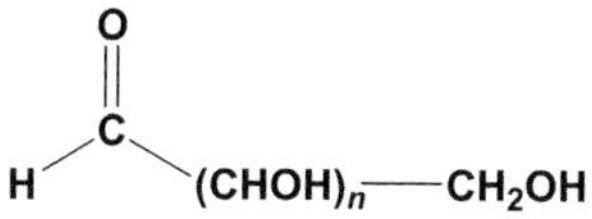

FIGURE 5.9.12

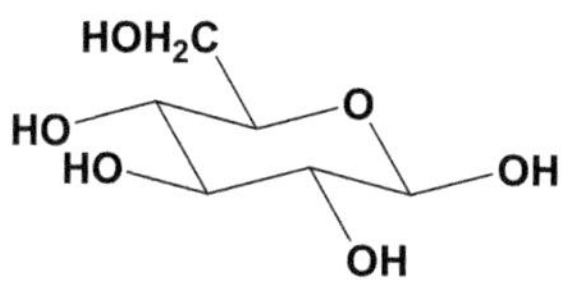

FIGURE 5.9.13

carboxylic acids can exist, all molecules central to our bio-chemistry. The silicon analogue of carbon dioxide, O=Si=O, also cannot exist, so that respiration, as employed by both plants and animals, would be impossible. Silicon dioxide is unstable with respect to polymerization, which leads to three-dimensional silicate matrices. The wide abundance of silicon and oxygen on Earth and other rocky planets results in large amounts of silicate minerals as a sort of thermodynamic end product. The overall molecular formula of pure silicates is SiO_2, so that sand, for example, is often said to have that struc-ture, in essence the polymerized analogue of CO_2. Moreover, silicate minerals are inert to reaction with water, whereas the equilibrium between CO_2 and H_2O is fundamental to respira-tion in most life-forms. Thus, the strong Si—O bond militates against the existence of common silicon structures in either the gas or the liquid phase. In a highly reductive atmosphere such as H_2, the silane functionality, as in SiH_4, can exist but, as the analogue of methane CH_4 in the carbon universe, offers little ability to carry out the necessary functions of life that have already been enumerated.

The question of chirality, or handedness, is central to life, because it allows the development of highly specific reactions. These terms refer to the nonsuperimposability of mirror images. When molecules have that property, the mirror-image forms, or enantiomers, are distinct molecules that can have highly specific properties. The most common structure that generates nonsuper-imposability is the tetrahedral unit with four different ligands, CWXYZ or SiWXYZ (Figure 5.9.15). The chirality of both amino acids (Figure 5.9.10) and carbohydrates (Figure 5.9.13) is key to their highly specific functions. Although silicon allows the same sort of specificity in theory, in practice, chiral silanes are extremely rare, as they apparently lack the functionality that permits separation of the enantiomers.

Silicates, however, have another property that may permit chiral environments. Even if the molecules themselves are unlikely to form such materials, there is evidence that sili-cate crystalline lattices themselves may exhibit chirality, so that their surfaces might serve as catalysts for the formation of chiral carbon compounds (Washington 2000; Hazen and Sholl 2003). Here, silicon is not providing the chemical build-ing blocks for life but may be fulfilling an important role in the formation of carbon-based biomolecules.

Silicon also provides an important structural role in some of Earth's organisms. Many of the organismal structural ele-ments are carbon-based (organic). Thus, the polycarbohydrate cellulose is the main component of the cell walls of green plants and algae; the protein collagen is the main component of animal connective tissue such as bone and tendons; the pro-tein keratin comprises hair, horns, claws, hooves, and nails;

and the protein cartilage is the main component of the exo-skeleton of arthropods and insects and the scales of fish. Many structural elements also are inorganic, including calcium car-bonate in sea shells, pearls, snails, and eggs and the phos-phate mineral hydroxyapatite in bones and teeth. In addition to these carbon- and phosphorus-based inorganic structural elements, silicates are structurally important for a number of organisms. The shells of diatoms and radiolarians and the spicules of sponges are composed of silica. Silica occurs as a structural element of many plants in the form known as phy-toliths, for example, in many grasses, conifers, horsetails, and rice husks. The wide presence of silica in organisms implies that there is a biochemistry for its uptake from inorganic sili-cate in the soil or water.

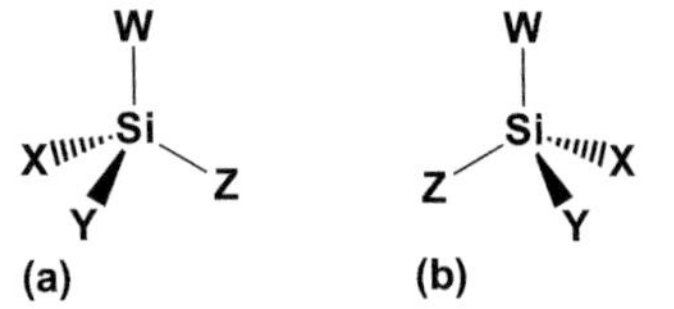

5.9.4   SILICATE AS A PREBIOTIC MEDIATOR

The development of life required the existence and avail-ability of the biomolecules such as carbohydrates, proteins, and nucleic acids. Miller (1953) reported experiments based on simple reagents and mild conditions that produced essen-tial amino acids. Miller and Urey (1959) followed up with methods to produce other types of biomolecules. Two groups described how the formose reaction can generate carbohy-drates from small aldehydes (Gabel and Ponnamperuma 1967; Reid and Orgel 1967). Unfortunately, this reaction is carried out normally under highly basic conditions that decompose the carbohydrates as soon as they are formed. Lambert et al. (2010) found that sodium silicate improves the yields of carbohydrates in the formose reaction. As the sugars are formed, they react with silicate to form sugar sili-cates, which are stable to the basic conditions. The sugars not only are sequestered in a stable form but also are selected according to structure and stereochemistry. The high pH of the silicate-mediated formose reaction might be a drawback, because there are few environments with such characteris-tics. Lambert and Gurusamy-Thangavelu (2015) also have demonstrated that natural aluminosilicate clays can simulate the effects of strong base at nearly neutral pH, whereby the clays under base-free conditions produce carbohydrates that are not formed without the clay, particular at temperatures above 80°C.

5.9.5   SUMMARY

Silicon is a poor platform for the molecular basis of life, which requires structure, metabolism, growth, adaptation, response to stimuli, and reproduction. Bonds involving silicon are relatively weak compared with carbon, except in relation to highly electronegative elements. Highest structural vari-ability is favored with higher valences, as supplied by tetrava-lent elements. Concatenation of silicon with other tetravalent elements, including carbon and other silicon atoms, produces weak and highly reactive Si—Si and Si—C bonds, in con-trast with the stability of C—C bonds. Moreover, silicon pro-duces exceedingly weak double bonds (Si=C, Si=Si, Si=O), unknown in nature on Earth, which form the heart of almost

FIGURE 5.9.15

all biomolecules, including carbohydrates, proteins, nucleic acids, and fatty acids. The silicon analogue of carbon dioxide does not exist, as it is unstable with respect to polymerization to polysilicates, which in general lack functionality on which life might be based.

Although organisms that could be characterized as being silicon-based cannot exist, silicon can play important biological roles within a carbon-based biology. The stability and strength of silicate solids make them suitability to serve as biological structures, as in phytoliths and rice husks. The three-dimensionality of aluminosilicate clays can create chirality, which is useful to improve reaction specificity. Silicates and aluminosilicates can sequester carbohydrate molecules, as formed in the formose reaction, which otherwise are unstable and subject to decomposition. Thus, silicon has biological roles to play but cannot be the central element of life.

REFERENCES

Blanksby, S. J., and G. B. Ellison. 2003. Bond dissociation energies of organic molecules. *Acc. Chem. Res.* 36, 255–263.

Brook, M. A. 2000a. *Silicon in Organic, Organometallic, and Polymer Chemistry.* New York: John Wiley & Sons, Chapter 7.

Brook, M. A. 2000b. *Silicon in Organic, Organometallic, and Polymer Chemistry.* New York: John Wiley & Sons, Chapter 16.

Corey, J. Y. 1989. Historical overview and comparison of silicon with carbon. In *The Chemistry of Organic Silicon Compounds,* S. Patai and Z. Rappaport (Eds.), Vol. 2, p. 6. Chichester, UK: John Wiley & Sons.

Darling, D. 2014. Silicon-based life. *Encyclopedia of Science.* http://www.daviddarling.info/encyclopedia/S/siliconlife.html.

Dessy, R. 1998. Could silicon be the basis for alien life forms, just as carbon is on Earth? *Sci. Ame.* https://www.scientificamerican.com/article/could-silicon-be-the-basi/.

Fleming, I., A. Barbero, and D. Walter. 1997. Stereochemical control in organic synthesis using silicon-containing compounds. *Chem. Rev.* 97, 2063–2192.

Gabel, N. W., and C. Ponnamperuma. 1967. Model for origin of monosaccharides. *Nature* 216, 453–455.

Haldane, J. B. S. 1927. The last judgement. In *Possible Worlds and Other Essays,* pp. 287–312. London, UK: Chatto & Windus.

Hazen, R. M., and D. S. Sholl. 2003. Chiral selection on inorganic crystalline surfaces. *Nat. Mat.* 2, 367–374.

Jacob, D. T. 2016. There is no silicon-based life in the solar system. *Silicon* 8, 175–176.

Schulze-Makuch, D. 2013. Is Silicon-based life possible? *Air & Space Magazine,* October 24.

Lambert, J. B., J. C. Pflug, H. Wu, and X. Liu. 2003. Dendritic polysilanes. *J. Organomet. Chem.* 685, 113–121.

Lambert, J. B., S. A. Gurusamy-Thangavelu, and K. Ma. 2010. The silicate-mediated formose reaction: Bottom-up synthesis of sugar silicates. *Science* 327, 984–986.

Lambert, J. B., and S. A. Gurusamy-Thangavelu. 2015. Roles of silicon in life on Earth and elsewhere. In *Astrobiology. An Evolutionary Approach,* V. M. Kolb (Ed.), Chapter 7, pp. 149–161. Boca Raton, FL: CRC Press/Taylor & Francis Group.

Miller, S. L. 1953. Production of amino acids under possible primitive Earth conditions. *Science* 117, 528–529.

Miller, S. L., and H. C. Urey. 1959. Organic compound synthesis on the primitive Earth. *Science* 130, 245–251.

Raabe, G., and J. Michl. 1985. Multiple bonding to silicon. *Chem. Rev.* 85, 419–509.

Rampelotto, R. 2010. The search for life on other planets: Sulfur-based, silicon-based, ammonia-based life. *J. Cosmol.* 5, 818–827.

Reid, C., and L. E. Orgel. 1967. Model for origin of monosaccharides: Synthesis of sugars in potentially prebiotic conditions. *Nature* 216, 455–457.

Washington, J. 2000. The possible role of volcanic aquifers in prebiologic genesis of organic compounds and RNA. *Orig. Life Evol. Biosph.* 30, 53–79.

Wells, H. G. 1894. Another basis for life. *Saturday Review,* December 22, p. 676.

West, R. 1989. Polysilanes. In *The Chemistry of Organic Silicon Compounds,* S. Patai and Z. Rappaport (Eds.), Vol. 2, pp. 1207–1240. Chichester, UK: John Wiley & Sons.

Wikipedia. 2018a. *Abundance of the Chemical Elements.* https://en.wikipedia.org/wiki/Abundance_of_the_chemical_elements (accessed February 8, 2018).

Wikipedia. 2018b. *Hypothetical Types of Biochemistry.* https://en.wikipedia.org/wiki/Hypothetical_types_of_biochemistry (accessed February 8, 2018).

Section VI

RNA and RNA World

Complexity of Life's Origins

6.1 Transitions
RNA and Ribozymes in the Development of Life

Eric J. Hayden, Niles Lehman, and Peter J. Unrau

CONTENTS

6.1.1 INTRODUCTION

In this chapter, we will discuss the role of chemical transitions from one pre-living state to another. We will focus on the popular paradigm of the RNA World as a model system to understand the stages that primordial chemical systems went through to become simple, and then more complex, life. In particular, we will survey the evolutionary pressures that populations of RNA went through to become more life-like.

6.1.2 RNA AND RIBOZYMES

RNA, or ribonucleic acid, which refers most often to the 3'-5'-linked single polyanionic chains of β-D-ribonucleotides, is a water-soluble macromolecule that plays a central role in contemporary metabolism. It is typically, but not always, synthesized in a transcription reaction by the protein enzyme RNA polymerase from a complementary DNA template. This molecule turns out to hold many clues for the origins of life (OoL) on the Earth and elsewhere in the universe.

6.1.2.1 THE RNA WORLD HYPOTHESIS

The RNA World hypothesis was introduced in Section 1 of this handbook, and recent advances in prebiotic chemical pathways to RNA and the features of RNA that enable its reproduction and enzymatic activity are covered in Sections 5 and 6 of this handbook. This hypothesis postulates that RNA, or something very chemical similar, was the first informational polymer to acquire the characteristics most commonly attributed to life, namely evolution and concerted metabolism. While the RNA World takes many forms, and the simplest view—that an RNA molecule emerged spontaneously from a pre-biotic milieu, with the ability to self-replicate—has been challenged on many fronts (*e.g.*, Shapiro 1987; Lehman 2008; Branscomb and Russell 2013; Hud et al. 2013), the hypothesis undeniably serves as a scaffold to begin to make and test hypotheses about how life could originate from a molecular perspective. The RNA World's popularity in this regard derives from four very powerful observations: (1) that ATP is not only the energy currency of the cell but also a fragment of RNA; (2) that protein enzymes often depend on ribonucleotide-like cofactors for activity (White 1976); (3) that the core of the ribosome where polypeptide synthesis occurs is composed entirely of RNA (Nissan et al. 2000); and (4) that RNA can be a catalytic molecule and hence can simultaneously possess both a genotype and a phenotype (Kruger et al. 1982; Guerrier-Takada et al. 1983; Gilbert 1986). It is the last of these four points on which we will elaborate in this chapter, with an eye toward understanding how evolutionary transitions from one state to another are key phenomena to describe and understand.

6.1.2.2 CATALYTIC RNA

One of the greatest surprises in biology in the latter quarter of the twentieth century was that RNA could be a catalyst. Prior to the discoveries of Cech and Altman that RNA could speed up the rates of a chemical reaction, RNA was thought by many to serve an intermediary role between DNA and proteins. Starting with the discoveries of self-splicing introns (Kruger et al. 1982) and with the catalytic ability of the RNA portion of RNase P (Guerrier-Takada et al. 1983), we have now discovered 13 distinct classes of catalytic RNAs throughout biology. The search is underway continually for additional catalytic RNAs, termed ribozymes, and by the time this handbook is published, there may, in fact, be more than 13 known classes. The majority of these classes of ribozymes catalyze a single type of bond conversion, phosphor-transfer reactions (including bond cleavage). However, the ribosome speeds up the rate of peptide bond formation by many orders of magnitude over the uncatalyzed rate by exploiting an anhydrous environment in its RNA core. Thus, even in a biological setting, RNA has evolved to catalytically make and break substrate chemical bonds. In the laboratory, the repertoire of RNA catalysis has been extended to myriad chemical chemistries, including the Diels-Alder reaction, aminoacylation, and porphyrin metalation, just to name a few (Chen et al. 2007). Moreover, hundreds of ligand-specific RNAs have been selected as binding molecules; these are termed aptamers. These observations suggest that RNA has a latent ability to serve as a broad-spectrum catalyst, even if the majority of biological catalysis was eventually subsumed by protein enzymes. A major question remains: How this and other prebiotic transitions were accomplished by geochemical and evolutionary mechanisms?

6.1.3 TRANSITION FROM A CHEMICAL WORLD TO A BIOLOGICAL WORLD

Life on the Earth formed some 4.0–4.4 billion years ago (see Chapter 4.6) and was initially, by definition, a purely chemical process. In the sections that follow, we will discuss some of the key transitions from chemistry to biology—however defined—that must have taken place.

6.1.3.1 PREBIOTIC SCAFFOLDS AND THE EMERGENCE OF COMPLEXITY BY A PREBIOTIC PROCESS

The formation of relevant chemicals in the universe and their delivery to the early Earth were necessary events during the formation of the Earth and during the Hadean Eon (4.6–4.0 Gya). The tiny subset of chemical systems that could have led to life must have been both non-equilibrium systems and those capable of possessing some form of evolutionary character. The former requirement has been discussed at length (Pross 2012; Branscomb and Russell 2013). The latter, though, is often equated to Darwinian processes that are apparent in cellular life but not so clearly defined in chemical collections. Wu and Higgs (2009) have made inroads in reconciling this dichotomy by delineated "dead" and "living" states for nascent life. Similarly, Nowak and Ohtsuki (2008) describe "pre-life" by a system of monomers that can polymerize to produce an informational and evolvable polymer. In this chapter, we amalgamate both of these ideas and others into two simple diagrams that indicate a sharp transition from the many unsuccessful attempts to make this transition to the single successful example that we know of with extant terran life. First, we trace the rate of energy consumption (power density) of the living or pre-living system as a function of time it takes for the system to reproduce itself (Figure 6.1.1). A sharp transition at the OoL would occur when the system is able to capture local energy and successfully direct that into reproduction, rather than random dissipation as heat. One of the more remarkable facts about biology is that the power density of a bacteria is approximately equal to that of a jet turbine engine. Stated simply, the free-energy flux across a bacterial surface in exponential growth corresponds to a power density within the cell of about 0.2 MW/m^3 (Lane and Martin 2010). When information can capture energy, life can begin.

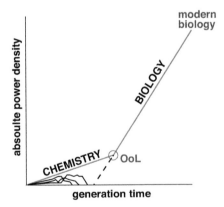

FIGURE 6.1.1 The chemistry-to-biology transition. The green lines represent the ultimately successful lineage leading to extant life. The black lines represent failed lineages that did not make the transition: chemical systems that did not acquire evolvable informational capacity. The dotted line represents an extrapolation of a "biological system" down to zero cells.

6.1.3.2 Populating Sequence Space with Functional RNAs

A second way to visualize the chemistry-to-biology transition is to focus on complexity. Using the RNA World as a model, the embodiment of evolvable information in polymeric form was a key component of the transition discussed earlier. We have already posited that this transition was coincident with a parallel changeover from an analog characterization of the chemical state to a digital one (Baum and Lehman 2017). In the analog stage, the chemical nature of the backbone was the dominant feature; selection would have taken place between polyanions and polycations (Benner 2016); between ribose and its alternatives such as threose, arabanose, and even peptides; and between linear chains and branched chains (Eschenmoser 1999). At this stage, information storage would have been poor, as the information content (*i.e.*, the reduction in ambiguity compared to random conformations) would have been minimal. With the inclusion of side chains with specific hydrogen bonding patterns, a later, digital stage of pre-living systems could arise (Benner 2016; Baum and Lehman 2017). As side-chain complementarity (*e.g.*, A with U and G with C) became more defined with a selection process occurring among possible side chains (Krishnamurthy 2015), then the informational capacity of the polymer could sharply improve, leading to a more rapid cycle of selection. A relationship between polymer length and complexity would have iteratively fed back on one another (Figure 6.1.2). Eventually, some type of template-directed RNA replication catalyst came into being. This chemical innovation would have been the canonical point at which biology, with exponential replication and Darwinian-type selection, could fully explode. However, there was likely a previous period during which the broad-scale shuffling of blocks of information, via recombination, was a dominant means of creating diversity prior to the emergence of genomes capable of exponential growth with mutation and selection *per se* (Lehman and Unrau 2005; Lehman 2008; Pesce et al. 2016).

Once some mechanism for template-directed polymerization came into existence, then Darwinian evolution could firmly take hold. For such evolution, a system of inheritable point mutations, with intervening selection and amplification, is necessary (cf. Joyce 1989). For polymerization, either an RNA replicase (Johnston et al. 2001; Cheng and Unrau 2010) or a protein RNA polymerase would be necessary. Note that the most effective RNA replicase ribozymes today are laboratory-derived evolutionary derivatives of a particular class of ligase ribozyme (Zaher and Unrau 2007; Mutchler et al. 2015; Samanta and Joyce 2017) (Figure 6.1.3). With any such polymerase, errors would accrue during the copying of any RNA, and those errors would generate genetic diversity on which natural selection could act. This stage of life, however, could be either encapsulated as cells (or protocells) or still be free in a solution. Theory predicts, from many points of view, that replication would be far more efficient—and be able to circumvent

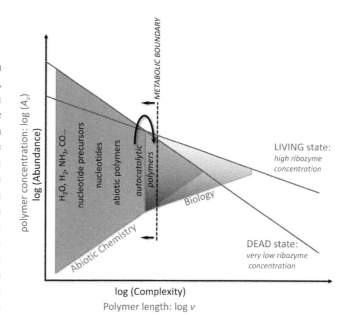

FIGURE 6.1.2 Metabolic need drives the crystallization of complex information-rich polymers at the transition from prebiotic chemistry to biology. The black vertical line/text defines a metabolic boundary where ribozymes of high complexity (light grey triangle) consume abiotically available compounds (dark grey triangle), driving the emergence of a complex biological metabolism that consumes ever-simpler compounds. Over evolutionary time, the metabolic boundary moves to the simplest compounds, as reflected by modern metabolism. In green, two views of the line between non-life and life are overlain, in which the living and the dead states display distinct relationships between polymer abundance and length. (From Wu, M. and Higgs, P.G., *Astrobiology*, 11, 895–906, 2011.)

Eigen's error threshold and escape parasitic extinction if encapsulated, and thus, it is likely that life would have been cellular at this point. This was recently demonstrated experimentally by Matsumura et al. (2016). In any event, a key realization is that multi-level selection would now be operating.

Both individual RNA sequences and the populational context in which they exist and "segregate" as alternative alleles would exert influence over which sequences would possess a fitness advantage and hence be reproduced disproportionately greater than others. RNAs at this stage of evolution would squarely fall within the quasispecies concept (Eigen and Schuster 1978). Here, individual genotypes would flounder or flourish depending on the distribution and frequencies of closely related sequences. An entire cloud of similar sequences would be moving through the space of all possible sequences (see Sections 6.1.4 and 6.1.5), with interrelated fates. This has been shown empirically for RNA viruses (Domingo et al. 2006), for ligase ribozymes evolving *in vitro* (Díaz Arenas and Lehman 2010), and for aptamers (Jiménez et al. 2013). These clouds of similar sequences set the stage for the mechanisms by which networks of RNAs interact, to be considered in the following sections.

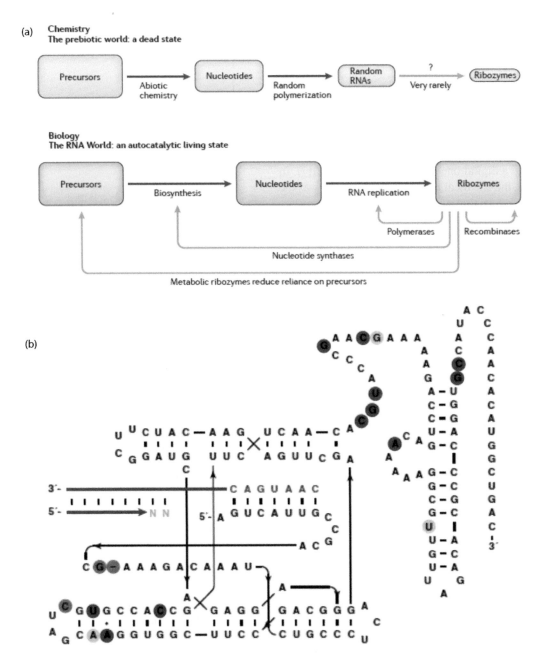

FIGURE 6.1.3 (a) The transition to biological catalysis and metabolism. The abiotic synthesis of high-diversity random polymers rarely leads to high-functioning ribozyme polymers. Once ribozymes are capable of sustaining evolution (shown here as recombinases and polymerases), metabolic processes capable of using prebiotic precursors become essential. (From Higgs, P.G., and Lehman, N., *Nat. Rev. Genet.*, 16, 7–17, 2015.) (b) The most efficient known (laboratory-derived) RNA polymerase ribozyme is derived from the class I ligase ribozyme. This ribozyme can catalyze the polymerization of dozens of nucleotides onto the 3′ end of a primer (shown in pink). Nucleotides in circles are those that were mutated relative to a less efficient version of the ribozyme. (From Horning, D.P. and Joyce, G.F., *Proc. Natl. Acad. Sci. U.S.A.*, 113, 9786–9791, 2016.)

6.1.4 RNA WORLDS: THE POSSIBLE AND THE PROBABLE

6.1.4.1 IN VITRO SELECTIONS—THE POSSIBLE RNA WORLD

Ribozymes filled in a missing link to life's origins, because they provided both the heritable properties of DNA and the catalytic properties of proteins. RNA, it was proposed, could greatly simplify origins, because only a single type of macromolecule needed to rise from the prebiotic soup (Crick 1968; Orgel 1968). While the discovery of ribozymes was quite serendipitous, it quickly became intentional. Researchers began devising creative strategies to pull RNA sequences with specific catalytic properties out of large random pools of RNA oligonucleotides. The initial goal, it seemed, was simply to see the extent of chemical reactions that could be catalyzed by RNA molecules.

Some of these research goals had biotechnology motivations, but many *in vitro* selections were aimed at reactions that seemed critical for RNA-based life, such as the chemistry required for RNA replication and metabolic reactions (*e.g.*, Figure 6.1.3b). This body of work has demonstrated that RNA can catalyze a plethora of chemical reactions, that even short RNA sequences can provide considerable catalysis, and that RNA sequences with catalytic capacity are not exceedingly rare within random pools of RNA sequences (Chen et al. 2007; Joyce 2007). These *in vitro* selected ribozymes have resoundingly demonstrated that complex chemistry is possible with RNA alone.

6.1.4.2 THE PROBABLE RNA WORLD: DISTRIBUTION AND CONNECTION OF FUNCTIONS

To truly evaluate the RNA World scenario, we need to ask not only *What is possible*? but also *What is probable*? The probability of finding a given RNA function in a random pool of sequences remains difficult to determine and depends on several factors such as the complexity of the structure, the stringency of the selection, and the length and complexity of the random pool (Sabeti et al. 1997; Legiewicz et al. 2005). Nevertheless, the data that does exist suggest that all given ribozymes exist in a very low frequency in a random pool, with sufficient length diversity. Obviously, a ribozyme of a particular length cannot exist in a pool that only contains RNA shorter than this length, although some combinations of small functional motifs can be present in only a zeptomole of material (Knight and Yarus 2003). Nevertheless, the emergence and coalescence of many different ribozymes simultaneously from a prebiotic soup could be just as unlikely as the simultaneous emergence of primitive ribosomes and DNA polymerases, thus negating the perceived advantage of RNA as a simpler route to life. An alternative scenario is that crude RNA reproduction emerged first, in some form, and then evolution took over. Evolutionary innovations, defined as new forms and functions that help survival or replication, can then occur through the forces of evolution, namely replication with mutation, drift, gene flow between populations, and selection. Under this scenario, it becomes critical to understand how new RNA functions emerge through the process of evolution. This rest of this section will review the research that has contributed to our understanding of this topic and will describe future research directions that are still needed to evaluate the role of evolutionary innovations of RNA in life's origins.

6.1.4.3 SEQUENCE SPACE AND FITNESS LANDSCAPES

The concept of sequence space (SS) is useful for discussing evolutionary innovations. For our discussion here, we will define sequence space (SS) as the space of all possible RNA nucleotide sequences. Each unique sequence of nucleotides occupies a discrete point in this space. While the space is of high dimension and is discrete because there can be no partial mutations, every point in SS has a set of nearest neighbors that allows a sequence to move through sequence space in a pseudo-continuous fashion. For RNA of length $n = 10$, every point in SS has $3n = 30$ neighbors, because each position can be changed to three different nucleotides via point mutation. When insertions and deletions are considered, a total of $5n + 4 = 54$ nearest neighbors exist for this particular example. The size and dimensionality of SS increases exponentially as longer RNA molecules are considered. For even modest-sized RNA molecules, such as length $n = 100$ nucleotides, there exist so many unique sequences ($4^{100} = 1.6 \times 10^{60}$) that they could not be synthesized, even with all the carbon on the Earth as the mass of such diversity if realized would be 9×10^{37} kg, exceeding the mass of the Earth by some 13 orders of magnitude. This is important because it means that life has not sampled all possible genotypes, and no experiment in a lab or in a computer could hope to. Instead, we should aim to understand the properties of sequence spaces that support life and promote innovations and how these properties change through different evolutionary scenarios.

6.1.4.4 COMPUTATIONAL WORK: THEORY OF GENOTYPE NETWORKS

To objectively evaluate the challenge of evolutionary innovations under RNA World scenarios, we must attempt to understand how functions are distributed and connected within RNA genotype space. Considerable computational experiments have contributed greatly to our understanding on this topic. For computational experiments, genotype is defined by the nucleotide sequence, and phenotype is typically defined as an optimal secondary structure, which is predicted from the most energetically favorable combinations of base-pair interactions (termed the minimum-free-energy [MFE] structure). Collectively, these experiments have shown that there are vastly more genotypes than there are phenotypes, because many sequences fold into the same MFE structure. Put another way, each individual RNA structure can tolerate many consecutive changes in nucleotide sequence, without changing the most probable structure (Schuster et al. 1994). RNA molecules possess *mutational robustness*. There exist webs of RNA genotypes that are well connected by individual mutational changes and all form the same genotype. These webs have been termed *genotype networks* (Figure 6.1.4).

The robustness of RNA structures to mutations, facilitated by large genotype networks, suggests that the specific sequence required for a function is flexible. Evolving populations can move throughout sequence space along their genotype network, even while under selection to maintain a structure or function (Sumedha et al. 2007). However, as evolving populations move within their own genotype network, they may encounter different phenotypes, enabling evolutionary innovations (Figure 6.1.4). This has led to the idea that seemingly "neutral" evolution that maintains a molecular phenotype can nevertheless pave the way for new adaptations (Wagner 2008).

A few experimental examples have tested the prediction that genotype networks can promote innovations. For example, the genotype networks of the self-cleaving Hepatitis Delta Virus (HDV) ribozyme and the class III ligase ribozyme were shown to come in close proximity in at least one point in sequence

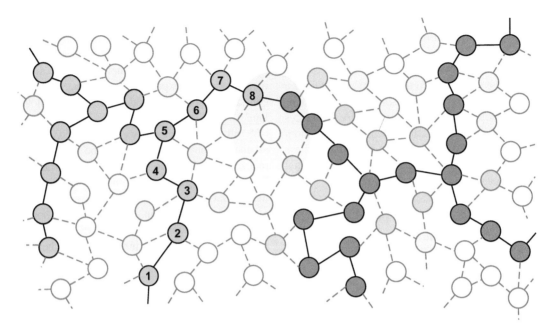

FIGURE 6.1.4   An evolutionary innovation in sequence space (SS). Circles represent different genotypes, and lines connect genotypes that differ by a single mutation. Circles with the same color have the same phenotype and belong to the same genotype network. Fitness is represented by the intensity of the color. Solid black lines indicate nearly neutral mutational steps, while dashed gray lines indicate deleterious mutational steps. Numbered circles follow a path of eight nearly neutral mutations that maintain the blue phenotype but allow the discovery of the purple phenotype. The yellow region highlights where the two genotype networks are proximal and a mutation can enable an evolutionary innovation, defined as the emergence of a new phenotype.

space (Schultes and Bartel 2000) and are described in more detail in Section 6.1.5. Related work studied the sequence space between two RNA aptamers that bind either ATP or GTP. Only a few mutations were required to switch specificity (Huang and Szostak 2003). Finally, a study of the genotype network of the *Azoarcus* group I ribozyme (Figure 6.1.5) demonstrated that a population that had spread out into its genotype network adapted faster to the challenge presented by a modified substrate than did a population that started from a smaller region of genotype space (Hayden et al. 2011). Taken together, all these experiments support the ability of genotype networks to promote RNA innovation within SS.

6.1.4.5   Experimental Fitness Landscapes

One important parameter that is missing from our discussion of genotype networks so far is ribozyme fitness. Quantitatively, once a selective pressure (P) is defined experimentally (or biologically), then in principle, the fitness (W_P) can be defined as the probability of a ribozyme surviving this condition. As it is rare that a ribozyme needs to survive only one selective pressure, one can consider either the elemental fitness of a particular selective pressure or the total fitness, which would include the probability that, in a particular environment, the ribozyme is replicated ($W_{Replication}$). To the extent that selective pressures are independent, these individual fitness values are multiplicative in nature ($W_{total} = W_{P1}...W_{Pn}W_{Replication}$). For any given ribozyme, the effect of a mutation for a specified selective

pressure P_1 can therefore fall on a continuous spectrum, from beneficial to approximately neutral to deleterious. As the effect of this mutation under a second selective pressure P_2 need not correlate with the phenotype induced by P_1, it is entirely possible that a mutation might be beneficial or neutral in one instance and detrimental in another (Gibson and Dworkin 2004). Despite this complexity, for a given overall selective pressure, a fitness landscape can be defined when each genotype is assigned a fitness value.

Due to the underlying complexity of sequence space, fitness landscapes are very high-dimensional. However, a common metaphor is to imagine them as three-dimensional landscapes where the height on a two-dimensional landscape is determined by fitness. While this image fails to capture the very high dimensionality of actual SS landscapes, it provides a simple way visualize movement in SS. More realistically, in this picture, a set of flat topped "mesas" exists, where a set of genotypes all have approximately the same fitness and are bounded by genotypes with lesser fitness. This is the expectation of the quasispecies phenomenon. Increasing selective pressure on such a metaphorical landscape results in populations climbing "up-hill," where the flat population of equivalent genotypes found in a weaker selective pressure are steadily differentiated from each other into a complicated set of "optimal" peaks.

Assigning fitness to a genotype requires an appropriate experimental assay. The utilization of next-generation sequencing to analyze RNA function has greatly expanded the number of genotypes that can be assigned a fitness, increasing the size of fitness landscapes (Pitt and Ferre-D'Amare 2010;

(a)   (b)

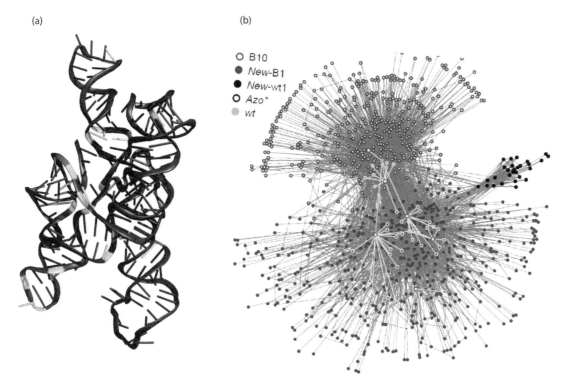

FIGURE 6.1.5 The *Azoarcus* ribozyme and its mutational landscape. (a) Nucleotide positions with non-zero mutation accumulation were mapped onto the crystal structure of the *Azoarcus* ribozyme, by replacing the b-factor column with P value from regression analysis, with red being lowest P values and blue being higher than $P = 0.05$. (b) Networks produced from selected sequences. Each node represents genotypes determined by clustering several sequences at a desired similarity cutoff (*i.e.*, 97%). Edges connect nodes if within two nodes there are sequences that differ by less than a specific edit distance. The green node represents the starting wild-type sequence. (From Hayden, E.J. et al., *Nature*, 474, 92–95, 2011.)

Kobori and Yokobayashi 2016; Xulvi-Brunet et al. 2016). In fact, in one case, all sequences of length 25 nucleotides were assayed for their ability to bind a small molecule, GTP (Jiménez et al. 2013). The results not only largely confirm the presence of vast genotype networks, but also show that the space is divided into clusters of sequences with very high fitness that are connected by networks of lower fitness. The effect of such landscape properties on evolutionary innovations and whether such properties extend to other functions remain unknown.

Mutational robustness should predict the size and connection of genotype networks, even when they cannot be exhaustively measured. Computational work on a few structures, and experiments on the *Azoarcus* ribozyme have measured robustness and epistasis. The former work has shown that natural RNA structures are more robust to mutations than 99% of random structures (Jörg et al. 2008), and had also showed that *in vitro* selected ribozymes structures are less robust to mutations than natural structures (Meyers et al. 2004; Hayden et al. 2012, 2015). In concert, laboratory experiments on ribozymes have shown that mutational robustness can be measured and can be altered by *in vitro* selection. For example, selection for more active variants of the *Azoarcus* ribozyme resulted in the unintentional reduction in mutational robustness (Hayden et al. 2012).

6.1.5 NEUTRAL NETWORKS

The digital information encoded and decoded *via* Watson-Crick (WC) pairing helps to ensure that RNA folds in a highly hierarchical fashion. The primary sequence of an RNA immediately after transcription or denaturation has the potential to rapidly find and stitch together self-complementary regions of sequence into stable helical secondary structure elements that then slowly relax into a final tertiary shape. It is this final fold that is typically associated with a functional RNA. Such a folding hierarchy exists, from the smallest stem-loop structures (Crick 1958) to the ribosome (Davis and Williamson 2017), the largest functional RNA known. Even random-sequence RNAs are highly likely to be structured as a result of the intrinsic self-recognition properties of RNA (Schultes et al. 2005).

The existence of RNA secondary structures along the folding pathway from primary sequence to final functional RNA implies that functional RNAs are highly degenerate in their primary sequences. Equivalent secondary structures for any specific RNA are easily found as in the absence of any other constraint, because in principle, any particular WC pair has three equivalent digital matches (*i.e.*, a C-G pair can be G-C, A-U, or U-A at the digital level of base recognition). This simple fact implies that RNA secondary structure motifs are highly degenerate. This degeneracy rises still further for RNA folds that involve multiple helical elements as typically such helices can be arbitrarily spaced from each other while still having the potential to form a desired secondary structure (Sabeti et al. 1997). The requirement that these otherwise-equivalent secondary structures correctly form a tertiary fold places a constraint on primary sequence regions that must stay invariant in order to adopt a functional fold. The self-cleaving hammerhead ribozyme, for example, consists of three arbitrary RNA helices that converge on a common central core defined by 12 critical residues required for RNA cleavage. Conservatively, assuming a minimal primary sequence of 43 nt for this ribozyme, there are nevertheless some 10^{12} nearly equivalent secondary structures (Figure 6.1.6) that are all predicted to fold and potentially be functional self-cleaving hammerhead ribozymes (Sabeti et al. 1997). Consistent with this high degeneracy, selections for self-cleaving ribozymes have overwhelmingly returned hammerhead ribozymes (Salehi-Ashtiani and Szostak 2001). Likewise, many important biological RNAs are also highly degenerate. The 6S regulatory RNA, for example, which resembles an open DNA promoter complex, has 30 absolutely conserved nucleotide residues in sequences that span the gamma proteobacteria (Barrick et al. 2005; Shephard et al. 2010), with these residues being correctly positioned by at least 28 bp of arbitrary stem-loop structure. This yields a helical degeneracy for this RNA of at least 10^{17} sequences. Such high degeneracies are typical for RNA folds but can be hard to visualize. To set the scale for

such large degeneracies, picture a hockey arena filled to the brim with sand; this would correspond to ~10^{15} sand grains.

Not only are RNA secondary structures highly degenerate, but these degenerate sequences are also typically connected to each other through sequence space, as previously discussed. Defining the distance between two sequences as the number of mutations required to move in SS from one to the other allows the introduction of the concept of a *neutral network*. RNA secondary structural degeneracy largely results from the fact that the secondary structure—and to a large extent the tertiary structure—of a RNA helical element is independent of its sequence. As a consequence, no member of a population of molecules sharing a common secondary structure can be further than two mutations away from another degenerate member. Thus, a helix containing a C-G pair is substantially equivalent to one containing a A-U pair at the same position. This mutational distance can be less when wobble pairs are considered, as a C-G pair is two mutations away from an A-U pair at that position but only one mutation away from a U-G wobble pair. Likewise, when helical structures can tolerate helical bulges caused by the insertion or deletion of a single nucleotide within a helix, mutational distances can be pseudo-continuous. Thus, the huge reservoir of degeneracy that exists for any functional RNA is not scattered randomly throughout sequence space but can be thought of as a neutral network of nearest neighbors that are capable of stretching a surprisingly large distance through sequence space, like a cosmic filament.

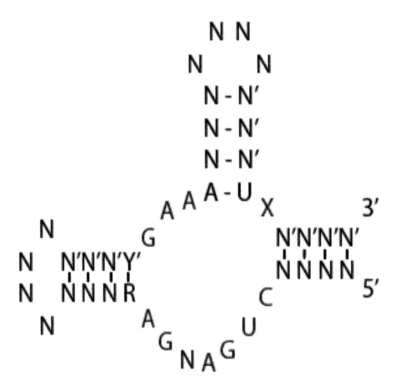

FIGURE 6.1.6 The hammerhead ribozyme has a small catalytic core connected to three arbitrary helices, giving the motif huge sequence degeneracy, shown here with two helices closed by 4-nt loop structures. R: any purine, Y: any pyrimidine, X: A, C, or U. N: arbitrary base. Prime indicates a pairing requirement.

When fitness is once again added into this picture, the concept of a neutral network makes more precise the previously discussed and more limiting three-dimensional picture of a "mesa." This implies that, in addition to a high degeneracy of sequences, there is also a high degeneracy of mutational paths that link together these degenerate sequences into a set of trajectories that natural evolution can move along by the conservative strategy of mutating, as little as one or at times two nucleotides at a time. Remarkably, experimental realizations of such a beautiful idea have been obtained.

6.1.5.1 Boundaries Between Distinct Neutral Networks

How might two distinct folds of the same sequence length occupy sequence space? A more accurate mental picture resembles the mixing of spaghetti and fettuccine into a bowl, no strand of spaghetti is that far from a fettuccine strand. While this does not capture the high dimensionality of the problem, neutral networks with one function are commonly neighbors with a network of another. This was demonstrated clearly by Schultes and Bartel (2000), when they moved 80 nt through sequence space, one point mutation at time, keeping the function of the HDV self-cleaving ribozyme (Figure 6.1.7a) for some 40 distinct point mutations, before moving abruptly to the fold of a new ribozyme. This class II ligase ribozyme (Figure 6.1.7a) was then moved a further 40 nt, all the time keeping within this fold's neutral network (Schultes and Bartel 2000). Interestingly, one single intersection sequence had the properties, albeit weakly, of both the HDV and the ligase, suggesting that this sequence was balanced energetically between either fold and was therefore optimal for neither

(Figure 6.1.5b). In the following three sections, we will discuss other types of intersections between distinct neutral networks that have occurred with surprising frequency in the experimental literature.

6.1.5.2 Point-Mutation-Induced Divergence in Network Function: Conserved Structural Motifs

The notion that mutational networks connect motifs of similar function is a powerful one that appears to be actively used, at least as evidenced by modern biology. A surprising finding in the riboswitch RNA gene-regulation fields was the discovery of adenine- and guanine-sensing riboswitches that share an entirely common secondary and tertiary structure fold. These two riboswitches are distinguished by a single U (found in the adenine riboswitch core) to C (found in the guanine riboswitch) mutation (Mandal and Breaker 2004). Thus, in sequence space, the neutral networks for these two networks are always one nucleotide distant from each other, with discovery of either by evolution, guaranteeing with near certainty the immediate evolutionary discovery of the other via a single point mutation. While painting a picture of parallel sheets of adenine and guanine riboswitch networks stretching through sequence space, this example is not alone. The *glmS* riboswitch, which requires glucosamine-6-phosphate as a small coenzyme to promote self-cleavage can, *via* the clever *in vitro* selection of only three point mutations, adopt an identical fold that now achieves self-cleavage in the complete absence of coenzyme (Lau and Ferré-D'Amaré 2013). Together, these two examples suggest that structurally related but functionally distinct ribozymes must, as a consequence of

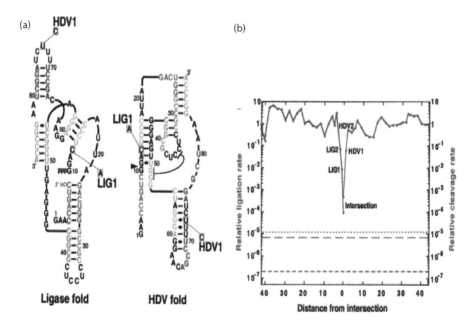

FIGURE 6.1.7   A neutral network between two ribozymes. (a) A ligase ribozyme and the HDV ribozyme were endpoints of a study to find a path through sequence space from one RNA to the other. (b) The neutral network that resulted; there is even an intersection sequence that retains (weakly) the catalytic activities of both ribozymes. (From Schultes, E.A. and Bartel, D.P., *Science*, 289, 448–452, 2000.)

their shared sequence motifs, have neutral networks that are extensively proximal to each other and may in fact be quite common in evolution.

Before embracing this picture fully, it is worth noting that experimental attempts to test this idea with more complex ribozyme functions have not always been productive. Using a highly degenerate library of sequences having the proposed secondary structure of a pyrimidine nucleotide synthase ribozyme, *in vitro* selection did not result in purine nucleotide synthase ribozymes having the same secondary structure as the starting ribozyme fold (Lau et al. 2004). Here, it may be important that the chemistry of pyrimidine nucleotide synthesis is thought to be considerably more challenging than that of purine nucleotide synthesis, implying that the discovery of considerably simpler motifs was favored in this case over the discovery of motifs having an identical secondary structure scaffold.

6.1.5.3 Catalytic Promiscuity: One Sequence, One Fold, Two Functions

While we have discussed intersection sequences that can fold into two structures, each with a distinct function (Section 6.1.5.1), and sequences that can rapidly adopt a new function while maintaining the same fold (Section 6.1.5.2), a surprisingly large number of ribozymes appear to have a single fold, with many distinct promiscuous functions. Uncontrolled promiscuity is clearly dangerous for complex metabolic processes where chemical reactions need to be precisely controlled, but early in evolution, such sequences might have played a useful role in helping to discover new catalytic function from old. Evolution would then have had the responsibility of moving such sequences away from sites of promiscuity toward that of unique function.

Perhaps, the best characterized of these bifunctional RNAs consists of the phosphodiester-forming and -cleaving ribozymes. A large number, but not all (*e.g.*, RNase P mediates cleavage by the attack of water), self-cleaving ribozymes generate a cyclic phosphate upon cleavage. The thermodynamic difference in energy between a phosphodiester linkage and a cyclic phosphate is close to zero, allowing RNA cleaving motives to either strongly favor the forward reaction (*e.g.*, the HDV ribozyme) or tolerate cleavage and re-ligation (*e.g.*, the hammerhead ribozyme). This balance of cleavage and religation properties is very important for the rolling-circle replication of the viral RNAs from where these ribozymes originate. These viruses generate linear transcripts from a circular genome that must be cleaved, but, at least, a small fraction of these cleaved products must re-ligate to allow replication of the opposite viral strand (Soll et al. 2001). This form of promiscuity, which exploits the near-equivalent energetics of products and reactants, is not perhaps unexpected and, in fact, is the basis for spontaneous RNA-RNA recombination that could have generated diversity prior to the advent of more complex chemistry (Lutay et al. 2007; Lehman 2008).

More complicated forms of ribozyme promiscuity have interesting potential to diversify the chemistry of an early RNA World ecosystem. The ribosome, which has evolved for over 3.5 billion years to promote peptide-bond formation, is known

under laboratory conditions to promote a range of chemistries that relies on spatial proximity of the reactants (Petrov et al. 2015). Purine nucleotide synthesis ribozymes selected for the ability to promote phosphoribosyl pyrophosphate (PRPP)-dependent nucleotide synthesis were found to also promote a distinct Schiff-base type of chemistry (Lau and Unrau 2009). RNA polymerase ribozymes that have been evolved in the laboratory to copy RNA templates using NTPs (Horning and Joyce 2016) were recently demonstrated to have the surprising ability to use dNTPs in addition (Samanta and Joyce 2017). This type of unexpected promiscuity could have surprising implications for the evolution of DNA genomes, even early in the evolution of an RNA World. In short, RNA sequence space can be traversed by evolving populations that take advantage of a wide variety of adaptive mechanisms, including neutral evolution, epistasis, exaptations, and parallel evolution.

6.1.5.4 Coevolution of Neutral Network Systems

Neutral networks in this chapter have been defined by selective pressures that are defined by external boundary conditions. Evolving multicomponent RNA systems are much more complex than this and respond to selective pressure from both the external environments and their coevolving partners. In this picture, a network of functionally distinct RNAs is a set of RNA sequences coevolving within SS simultaneously in response to externally defined and constantly changing selective pressures. For any one functional RNA sequence, then a complex local environment is defined by both external selection and local selective pressures that at least partially must be defined by the shape and function of the specific RNA fold and by its coevolving sequences (Figure 6.1.8). Biologically, this is best exemplified by the ribosomes, where the small and large subunits have coevolved in lock step over billions of years while maintaining the common function of protein synthesis, so as to systematically change their external environment from an RNA World to a protein-dominated one. In this framework, sequence space, RNA shape space, and RNA function are interconnected via evolution. Some typical biological examples of this connection are given in Figure 6.1.9.

Experimental evidence suggests that RNA can respond both positively and negatively to changes in external selective pressure. The pathways to higher-fitness genotypes of the group I *Azoarcus* self-splicing intron were studied under three distinct selective pressures (Hayden and Wagner 2012). These selective pressures changed the fitness of individual genotypes such that some mutational pathways were only beneficial in some environments. Interestingly, RNase P, which cleaves pre-tRNA transcripts, is an RNA-protein complex in *Escherichia coli*. In the absence of the C5 protein, this ribozyme is still functional but no longer recognizes distinct pre-tRNAs homogeneously, highlighting how the presence of a peptide sequence in the environment can alter the function of a ribozyme (Sun et al. 2016). Perhaps, an RNA able to interact with the RNaseP RNA could be selected artificially in the future to recover this homogeneity in the complete absence of

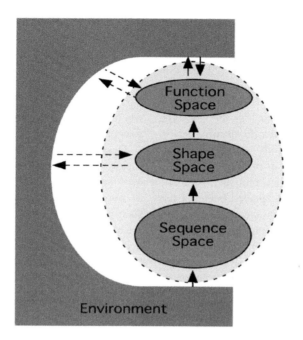

FIGURE 6.1.8 A system-based approach for the analysis of functional RNA evolution. An RNA system (dotted oval) contains RNA sequences (as points in sequence space) that have definable shapes and functions (phenotypes) once the RNA system is placed into a specific environment (solid horseshoe). Interactions with this environment (paired arrows) define the selective pressure that drives the evolution of the RNA system. In this example, the environment is responsible for RNA replication (by, e.g., a protein polymerase) and thus controls the sequence diversity available to be sampled (bottom vertical arrow). Complex environments can be difficult to characterize and may generate a multicomponent selective pressure (dotted paired arrows). In contrast, *in vitro* selection can often be used to apply a particular pressure (solid paired arrows), making the relationship between selection pressure and RNA functionality simpler to understand. (From Wang, Q. and Unrau, P.J., Evolving an understanding of RNA function by *in vitro* approaches, in *The Chemical Biology of Nucleic Acids*, G. Mayer (Ed.), pp. 355–376, John Wiley & Sons, Chichester, UK, 2010.)

Mapping from sequence space to shape space, to function space

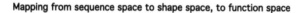
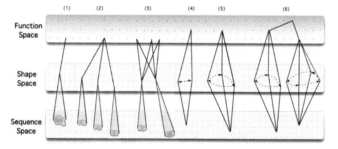

FIGURE 6.1.9 Mapping between sequence, shape space, and function space. Some of the common mapping relationships from sequence to shape to function space are shown: (1) One-to-one-to-one relationship: a set of sequences can fold into a common structure (one point in shape space) that performs a unique function (one point in function space), the hammerhead ribozyme being a good example. (2) Many-to-many-to-one relationship: many different sequence motifs (not directly connected in sequence space) can fold into distinct structures that each perform the same function. The functions of RNA ligation and cleavage are typically satisfied in this way. (3) Many-to-one-to-many relationship: a group of similar sequences can fold into a structure that performs many different functions that are in common with a distinct second fold. For example, two capping ribozymes have distinct secondary structures and yet perform the same set of chemical reactions. When the thermodynamic equilibrium of shapes is also considered as in (4), (5), (6), a point in sequence space often samples more than one point in shape space, each representing one of the substrates of the RNA system. The equilibrium between those states depending on environmental conditions. (4) Two states representing a conformational change before and after a binding event, such as occurs commonly with aptamers and some allosteric ribozymes. (5) Three structures representing three states of a ribozyme, before, during, and after the transition state for a particular reaction. (6) Multi-component RNA systems, such as the ribosome, have subunits that each move through a set of states in order to complete a catalytic cycle. Here, the large and small subunits of the ribosome provide discrete contributions (peptide bond formation and mRNA triplet decoding, respectively) to the overall function of peptide bond formation. (From Wang, Q. and Unrau, P.J., Evolving an understanding of RNA function by *in vitro* approaches, in *The Chemical Biology of Nucleic Acids*, G. Mayer (Ed.), pp. 355–376, John Wiley & Sons, Chichester, UK, 2010.)

protein factors and to enable a model system for RNA-RNA coevolution.

All the observations presented above reinforce the notion that a specific functional RNA can respond to external selection by moving through sequence space (Figure 6.1.8) while preserving its original function (Figure. 6.1.9; type 5). But how might highly cooperative systems such as the ribosome initially evolve? Artificially created networks of three RNAs have displayed a level of cooperation so as to ensure their own replication, even in the presence of molecular parasites that might have been expected to overwhelm such altruistic behavior (Vaidya et al. 2012). This unexpected behavior has a number of ramifications for the spontaneous emergences of metabolic type networks of RNA interactions early in the evolution of life. Moreover, this work strongly suggests that even in the absence of compartmentalization, RNA networks could have existed and presumably gained the evolutionary

traction required to form fully functional metabolic networks once encapsulated. While such multicomponent RNA systems have not been subjected to evolution, the presence of extensive hydrogen bonding between each of the three RNAs is shown in Figure 6.1.10, indicating that coevolution of such systems would sample an interesting three-way neutral network, where mutations in one network would require coevolution in a second and possibly a third.

Despite the evidence that RNAs are capable of cooperation, such systems are vulnerable to external changes in environment. Most specifically, maintaining a high effective concentration of polymers and the smaller substrates required to make such polymers could easily be imagined to be disrupted by, for example, simple dilution. Compartmentalization has long been recognized as an effective solution to this important problem (cf. Szathmáry and Demeter 1987; Bartel and Unrau 1999) and is particularly effective against parasitic

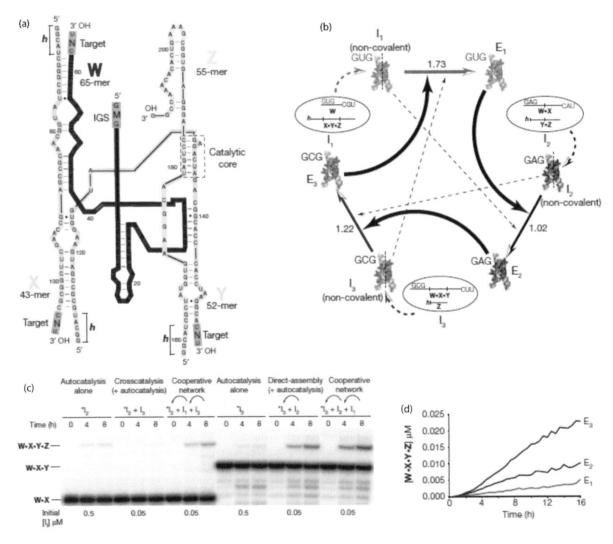

FIGURE 6.1.10 Cooperative covalent assembly of recombinase ribozymes. (a) The *Azoarcus* RNA, broken into four fragments that can spontaneously and covalently reassemble when mixed together. (b) The cooperative action of three separate subsystems. In each, the ribozyme has been broken into two pieces but cannot self-assemble due to mutational mismatches. However, when all three subsystems are combined, the collective set can assemble and reproduce. (c) Gel electrophoretogram showing this obligatory cooperation. (d) Each subsystem, when together, assembles at characteristic rates. (From Vaidya, N. et al., *Nature*, 491, 72–77, 2012.)

sequences, as demonstrated by Matsumura et al. (2016). In this framework, compartmentalization serves as a fundamental environmental selective pressure, without which modern life could not be contemplated. Imagine a protocell containing an RNA replicase and a non-functional extra RNA molecule that is neutral; its presence neither hurts nor helps the replication of the protocell. The extra molecule is free to mutate, and because it has no sequence requirements, it does not contribute to the information capacity of the system. Now, imagine that a selection pressure for GTP sequestration or membrane creation arises from a drop in the concentration of GTP in the environment that becomes rate limiting for replication. This drop could come from a lack of production (chemical environment) and/or because other protocells are depleting the local GTP (genetic environment). One of the several randomly evolving extra molecules happens to be an aptamer for GTP and can sequester GTP, increasing its local concentration inside the protocell. This provides the protocell with a growth advantage, and now, the sequence is under selection. In this scenario, a novel function (GTP binding) now provides the system with not only greater functionality but also greater information capacity. This simple example illustrates several points: (1) molecules spatially confined can become cooperative for survival; (2) environmental change is needed to promote innovations; and (3) innovations expand the information capacity of a system. It remains to be seen if such systems can be manifested in the laboratory (Figure 6.1.11).

6.1.5.5 FUTURE RESEARCH—SEASCAPES— GENOME BY ENVIRONMENT INTERACTIONS/ EVOLUTIONARY EQUILIBRIUM

Experimentally, much remains to be learned about the coevolution of RNAs within sequence space. While several data sets contribute to our understanding, intentional and systematic laboratory studies are lacking. Here, we speculate briefly into potentially interesting directions that are yet to be explored experimentally.

A coevolving system of RNAs appears highly likely to place evolutionary pressure on specific parts of the coevolving neutral networks in response to changes in external selective pressures: How, then, does such a system pick it up a notch when a specific neutral network is under strong selective pressure for enhanced function? For example, it is well known that the hammerhead ribozyme, even when selected for optimal self-cleavage, is not a tremendously fast ribozyme. How would an evolving network of RNAs in all likelihood replace the hammerhead with a faster ribozyme (such as the HDV) when required? Future experiments could explore connections between the neutral networks of two or more self-cleaving ribozymes that catalyze the same reaction but do so with different supporting secondary and tertiary folds.

As previously discussed, sequence space is rugged when selective pressure is high enough to distinguish even between nearest neighbors. If selective pressure is low, then a substantial

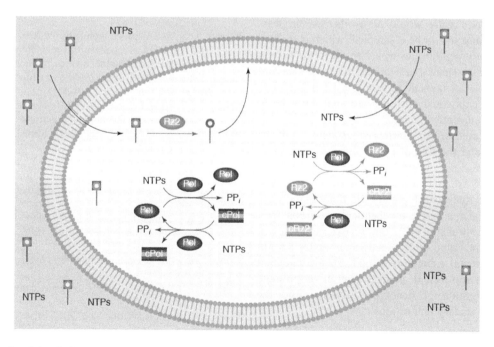

FIGURE 6.1.11 A minimal ribo-organism could eventually be constructed in the laboratory, provided that suitable ribozymes and membrane components can be identified. Although not an autotroph, and likely to be different from the first life in the RNA World, such an organism would be capable of sustained evolution. A polymerase ribozyme (Pol) uses externally supplied nucleotide triphosphates (NTPs) to replicate itself and its reverse complement (cPol) in an autocatalytic cycle (purple arrows). The polymerase ribozyme also makes copies of a second ribozyme and its complement (Rz2 and cRz2, respectively; green arrows). This ribozyme promotes a reaction that facilitates growth and eventual division of the membrane-like compartment. In this example, the second ribozyme matures a membrane precursor (square heads) into an active membrane component (round heads). Insertion of such a compound into the membrane is imagined to lead to its growth and eventual bifurcation, leading to a form of cellular replication. (From Bartel, D.P. and Unrau, P.J., *Trends Biochem. Sci.*, 15, M9–M13, 1999.)

fraction of the neutral network for the motif can be occupied, provided that the evolutionary system can populate the network. Such a sustained population can be imagined to be in a near-equilibrium state, with mutations moving sequences around the neutral network. At the edges of the network, two things can occur: (1) the RNAs are of no functional relevance and become rapidly extinct or more interesting; (2) they can interact with functionally relevant RNA motifs. In the instance that this motif is a faster RNA ribozyme fold than the neutral network previously occupied, this network will also be in a near-equilibrium state with the slower neutral network. While the second neutral network is intrinsically capable of faster catalysis at this stage, such activity is never realized due to the absence of selective pressure. As faster activities require a more complex RNA fold, the more functional RNAs will tend to be longer in sequence. Since longer RNAs are at a slight replicative disadvantage in low selection pressure conditions, they will only tend to be found at the extreme boundary of its intersection with the slower neutral network.

When selective pressure is raised (in this chapter most easily imagined as being driven by the demand for higher metabolic capacity envisioned in Figure 6.1.2), the near-equilibrium state between the two neighboring neutral networks will be violated. At this point, the slower network cannot sustain metabolism and is replaced by the faster ribozyme activity. At this transitional point, the faster neutral network will become heavily populated, driving the apparently sudden emergence of a new RNA fold in the RNA population. Relaxing selection pressure would allow the slower, less-information-rich ribozyme to be resampled, which if tracked again to evolutionary equilibrium would result in the system depopulating the faster neutral network. While such ideas are thought experiments for now, there is no reason in principle why such experiments could not be performed with the *in vitro* technologies currently in hand.

Another important set of future experiments will be needed to understand the role of environmental changes to genotype networks and evolutionary innovations. While several experiments are now expanding the size of genotype space that is being analyzed, typically the fitness landscape is only studied under a single biochemical condition (pH, temperature, ionic strength, etc.). Future experiments will require studying the same fitness landscapes under several different environmental conditions to understand the extent to which these environmental changes restrict or promote evolutionary exploration of genotype networks and the emergence of evolutionary innovations.

6.1.6 SUMMARY

This chapter provides a review of the role of RNA in the OoL, with special attention to how function can evolve once a diversity of sequences comes into existence. RNA can be a catalytic molecule, and the RNA World hypothesis suggests that, at some point, populations of RNA guided their own reproduction by using this catalytic prowess. We do not yet understand fully which catalytic functions came into being in

what order during the history of the development of life on the Earth. However, we are beginning to define the powerful evolutionary mechanisms that could have guided RNA populations through a network of interactions, as they searched sequence space for ever more potent means of survival and reproduction.

DEDICATION

We dedicate this chapter to our immediate ancestors in the scientific tree of investigators, namely Thomas Jukes, Gerald Joyce, Andreas Wagner, and David Bartel. Without their inspiration and guidance, we would not have made the advances that we did, which we hope to pass on to the next cohort of origins researchers.

GLOSSARY

Mutational Robustness – *the capacity of a phenotype to remain unchanged upon mutation*
Genotype Network – *the collection of all genotypes with the same phenotype*
Neutral Network – *the collection of all genotypes with the same secondary structure that have the same phenotype*

REFERENCES

Benner, S.A. 2016. Unusual hydrogen bonding patterns and the role of the backbone in nucleic acid information transfer. *ACS Cent. Sci.* 2: 882–884.

Barrick, J., N. Sudarsan, Z. Weinberg, W. Ruzzo, and R.R. Breaker. 2005. 6S RNA is a widespread regulator of eubacterial RNA polymerase that resembles an open promoter. *RNA* 11: 774–784.

Bartel, D.P., and P.J. Unrau. 1999. Constructing an RNA world. *Trends Biochem. Sci.* 15: M9–M13.

Baum, D.A., and N. Lehman. 2017. Life's late digital revolution and why it matters for the study of the origins of life. *Life* 7: 34.

Branscomb, E., and M.J. Russell. 2013. Turnstiles and bifurcators: The disequilibrium converting engines that put metabolism on the road. *Biochim. Biophys. Acta.* 1827: 62–78.

Chen, X., N. Li, and A.D. Ellington. 2007. Ribozyme catalysis of metabolism in the RNA world. *Chem. Biodivers.* 4: 633–655.

Cheng, L.K.L., and P.J. Unrau. 2010. Closing the circle: Replicating RNA with RNA. *Cold Spring Harb. Perspect. Biol.* 2: a002204.

Crick, F.H. 1958. On protein synthesis. *Symp. Soc. Exp. Biol.* 12: 138–163.

Crick, F.H. 1968. The origin of the genetic code. *J. Mol. Biol.* 38: 367–379.

Davis, J.H., and J.R. Williamson. 2017. Structure and dynamics of bacterial ribosome biogenesis. *Philos. Trans. R. Soc. Lond., Ser. B* 372: 20160181.

Díaz Arenas C., and N. Lehman. 2010. Quasispecies-like behavior observed in catalytic RNA populations evolving in a test tube. *BMC Evol. Biol.* 10: 80.

Domingo, E., V. Martin, C. Perales, A. Grande-Pérez, J. García-Arrianza, and A. Arias. 2006. Virus as quasispecies: Biological implications. *Curr. Top. Microbiol. Immunol.* 299: 51–82.

Eigen, M., and P. Schuster. 1978. The hypercycle: A principle of natural self-organization. Part B: The abstract hypercycle. *Naturwissenschaften* 65: 7–41.

Eschenmoser, A. 1999. Chemical etiology of nucleic acid structure. *Science* 284: 2118–2124.

Gibson G., and I. Dworkin. 2004. Uncovering cryptic genetic variation. *Nat. Rev. Genet.* 5: 681–690.

Gilbert, W. 1986. The RNA world. *Nature* 319: 618.

Guerrier-Takada, C., K. Gardiner, T. Marsh, N. Pace, and S. Altman. 1983. The RNA moiety of ribonuclease P is thecatalytic subunit of the enzyme. *Cell* 35: 849–857.

Hayden, E.J., D.P. Bendixsen, and A. Wagner. 2015. Intramolecular phenotypic capacitance in a modular RNA molecule. *Proc. Natl. Acad. Sci. U.S.A.* 112: 12444–12449.

Hayden, E.J., E. Ferrada, and A. Wagner. 2011. Cryptic genetic variation promotes rapid evolutionary adaptation in an RNA enzyme. *Nature* 474: 92–95.

Hayden, E.J., and A. Wagner. 2012. Environmental change exposes beneficial epistatic interactions in a catalytic RNA. *Proc. R. Soc. B Biol. Sci.* 279: 3418–3425.

Hayden, E.J., C. Weikert, and A. Wagner. 2012. Directional selection causes decanalization in a group I ribozyme. *PLoS ONE* 7: e45351.

Higgs, P.G., and N. Lehman. 2015. The RNA world: Molecular cooperation at the origins of life. *Nat. Rev. Genet.* 16: 7–17.

Horning, D.P., and G.F. Joyce. 2016. Amplification of RNA by an RNA polymerase ribozyme. *Proc. Natl. Acad. Sci. U.S.A.* 113: 9786–9791.

Hud, N.V., B.J. Cafferty, R. Krishnamurthy, and L.D. Williams. 2013. The origin of RNA and "my grandfather's axe". *Chem. Biol.* 20: 466–474.

Huang, Z., and J.W. Szostak. 2003. Evolution of aptamers with a new specificity and new secondary structures from an ATP aptamer. *RNA* 9: 1456–1463.

Jiménez, J.I., R. Xulvi-Brunet, G.W. Campbell, R. Turk-MacLeod, and I.A. Chen. 2013. Comprehensive experimental fitness landscape and evolutionary network for small RNA. *Proc. Natl. Acad. Sci. U.S.A* 110: 14984–14989.

Johnston, W.K., P.J. Unrau, M.S. Lawrence, M.E. Glasner, and D.P. Bartel. 2001. RNA-catalyzed RNA polymerization: Accurate and general RNA-templated primer extension. *Science* 292: 1319–1325.

Jörg, T., O.C. Martin, and A. Wagner. 2008. Neutral network sizes of biological RNA molecules can be computed and are not atypically small. *BMC Bioinformatics* 9: 464.

Joyce, G.F. 1989. Amplification, mutation and selection of catalytic RNA. *Gene* 82: 83–87.

Joyce, G.F. 2007. Forty years of in vitro evolution. *Angew. Chem. Int. Ed.* 46: 6420–6436

Knight, R., and M. Yarus. 2003. Finding specific RNA motifs: Function in a zeptomole world? *RNA* 9: 218–230.

Kobori, S., and Y. Yokobayashi. 2016. High-throughput mutational analysis of a twister ribozyme. *Angew. Chem. Int. Ed. Engl.* 55: 10354–10357.

Krishnamurthy, R. 2015. On the emergence of RNA. *Isr. J. Chem.* 55: 837–850.

Kruger, K., P.J. Grabowski, A.J. Zaug, J. Sands, D.E. Gottchling, and T.R. Cech. 1982. Self-splicing RNA: Autoexcision and autocyclization of the ribosomal RNA intervening sequence of *Tetrahymena*. *Cell* 31: 147–157.

Lane, N., and W. Martin. 2010. The energetics of genome complexity. *Nature* 467: 929–934.

Lau, M.W.L., K.E.C. Cadieux, and P.J. Unrau. 2004. Isolation of fast purine nucleotide synthase ribozymes. *J. Am. Chem. Soc.* 126: 15686–15693.

Lau, M.W.L., and A.R. Ferré-D'Amaré. 2013. An in vitro evolved glmS ribozyme has the wild-type fold but loses coenzyme dependence. *Nat. Chem. Biol.* 9: 805–810.

Lau, M.W.L., and P.J. Unrau. 2009. A promiscuous ribozyme promotes nucleotide synthesis in addition to ribose chemistry. *Chem. Biol.* 16: 815–825.

Legiewicz, M., C. Lozupone, R. Knight, and M. Yarus. 2005. Size, constant sequences, and optimal selection. *RNA* 11: 1701–1709.

Lehman, N. 2008. A recombination-based model for the origin and early evolution of genetic information. *Chem. Biodivers.* 5: 1707–1717.

Lehman, N., and P.J. Unrau. 2005. Recombination during in vitro evolution. *J. Mol. Evol.* 61: 245–252.

Lutay, A.V., M.A. Zenkova, and V.V. Vlassov. 2007. Nonenzymatic recombination of RNA: Possible mechanism for the formation of novel sequences. *Chem. Biodivers.* 4: 762–767.

Mandal, M., and R.R. Breaker. 2004. Adenine riboswitches and gene activation by disruption of a transcription terminator. *Nat. Struct. Mol. Biol.* 11: 29–35.

Matsumura, S., Á. Kun, M. Ryckelynck, F. Coldren, A. Szilágyi, F. Jossinet, C. Rick, P. Nghe, E. Szathmáry, and A.D. Griffiths. 2016. Transient compartmentalization of RNA replicators prevents extinction due to parasites. *Science* 354: 1293–1296.

Meyers, L.A., J.F. Lee, M. Cowperthwaite, and A.D. Ellington. 2004. The robustness of naturally and artificially selected nucleic acid secondary structures. *J. Mol. Evol.* 58: 681–691.

Mutchler, H., A. Wochner, and P. Holliger. 2015. Freeze-thaw cycles as drivers of complex ribozyme assembly. *Nat. Chem.* 7: 502–508.

Nissan, P., J. Hansen, N. Ban, P.B. Moore, and T.A. Steitz. 2000. The structural basis of ribosome activity in peptide bond synthesis. *Science* 289: 920–930.

Nowak, M., and H. Ohtsuka. 2008. Prevolutionary dynamics and the origin of evolution. *Proc. Natl. Acad. Sci. U.S.A.* 105: 14924–14927.

Orgel, L.E. 1968. Evolution of the genetic apparatus. *J. Mol. Biol.* 38: 381–393.

Pesce, D., N. Lehman, and A.G.M. de Visser. 2016. Sex in a test tube: Benefits of in vitro recombination. *Phil. Trans. Royal Soc.* B 371: 20150529.

Petrov A.S. et al. 2015. History of the ribosome and the origin of translation. *Proc. Natl. Acad. Sci. U.S.A.* 112: 15396–15401.

Pitt, J.N., and A.R. Ferre-D'Amare. 2010. Rapid construction of empirical RNA fitness landscapes. *Science* 330: 376–379.

Pross, A. 2012. *What is Life? How Chemistry Becomes Biology.* Oxford, UK: Oxford University Press.

Sabeti, P.C., P.J. Unrau, and D.P. Bartel. 1997. Accessing rare activities from random RNA sequences: The importance of the length of molecules in the starting pool. *Chem. Biol.* 4: 767–774.

Salehi-Ashtiani, K., and J.W. Szostak. 2001. In vitro evolution suggests multiple origins for the hammerhead ribozyme. *Nature* 414: 82–84.

Samanta, B., and G.F. Joyce. 2017. A reverse transcriptase ribozyme. *eLife* 6: e31153.

Schultes, E.A., and D.P. Bartel. 2000. One sequence, two ribozymes: Implications for the emergence of new ribozyme folds. *Science* 289: 448–452.

Schultes, E.A., A. Spasic, U. Mohanty, and D.P. Bartel. 2005. Compact and ordered collapse of randomly generated RNA sequences. *Nat. Struct. Mol. Biol.* 12: 1130–1136.

Schuster, P., W. Fontana, P.F. Stadler, and I.L. Hofacker. 1994. From sequences to shapes and back: A case study in RNA secondary structures. *Proc. Biol. Sci.* 255: 279–284.

Shapiro, R. 1987. *Origins: A Skeptic's Guide to the Creation of Life on the Earth.* New York: Bantam New Age.

Shephard, L., N. Dobson, and P.J. Unrau. 2010. Binding and release of the 6S transcriptional control RNA. *RNA* 16: 885–892.

Soll, D., S. Nishimura, and P.B. Moore (Eds.). 2001. *RNA*. Oxford, UK: Pergamon Press.

Sumedha, S., O.C. Martin, and A. Wagner. 2007. New structural variation in evolutionary searches of RNA neutral networks. *Biosystems* 90: 475–485.

Sun L., F.E. Cambell, N.H. Zahler, and M.E. Harris. 2006. Evidence that substrate-specific effects of C5 protein lead to uniformity in binding and catalysis by RNase P. *EMBO J.* 25: 3998–4007.

Szathmáry, E., and L. Demeter. 1987. Group selection of early replicators and the origin of life. *J. Theor. Biol.* 128: 463–486.

Vaidya, N., M.L. Manapat, I.A. Chen, R. Xulvi-Brunet, E.J. Hayden, and N. Lehman. 2012. Spontaneous network formation among cooperative RNA replicators. *Nature* 491: 72–77.

Wagner, A. 2008. Neutralism and selectionism: A network-based reconciliation. *Nat. Rev. Genet.* 9: 965–974.

Wang, Q., and P.J. Unrau. 2010. Evolving an understanding of RNA function by *in vitro* approaches. pp. 355–376. In *The Chemical Biology of Nucleic Acids* (Ed. G. Mayer). Chichester, UK: John Wiley & Sons.

White, H.B. 1976. Coenzymes as fossils of an earlier metabolic state. *J. Mol. Evol.* 7: 101–104.

Wu, M., and P.G. Higgs. 2009. Origin of self-replicating biopolymers: Autocatalytic feedback can jump-start the RNA world. *J. Mol. Evol.* 69: 541–554.

Wu, M., and P.G. Higgs. 2011. Comparison of the roles of momoner synthesis, polymerization, and recombination in the origin of autocatalytic sets of biopolymers. *Astrobiology* 11: 895–906.

Xulvi-Brunet, R., G.W. Campbell, S. Rajamani, J.I. Jiménez, and I.A. Chen. 2016. Computational analysis of fitness landscapes and evolutionary networks from *in vitro* evolution experiments. *Methods* 106: 86–96.

Zaher, H.S., and P.J. Unrau. 2007. Selection of an improved RNA polymerase ribozyme with superior extension and fidelity. *RNA* 13: 1017–1026.

6.2 Three Ways to Make an RNA Sequence

Steps from Chemistry to the RNA World

Paul G. Higgs

CONTENTS

6.2.1 INTRODUCTION AND OBJECTIVES

It is widely believed that life on Earth passed through an RNA World stage in which RNA sequences acted as both genes and catalysts (Gilbert 1986; Joyce 2002; Higgs and Lehman 2015). The universality of ribosomal protein synthesis in current cellular life and the essential roles of rRNA, tRNA, and mRNA in this process make it clear that some form of RNA replication existed very early in evolutionary history, prior to the origin of genetically encoded proteins and the translation mechanism. The simplest version of the RNA World hypothesis for the origin of life therefore proposes that life emerged due to the formation of catalytic RNAs (ribozymes) that were able to self-replicate (Cheng and Unrau 2010). This is supported by the large number of ribozymes with diverse functions that have been synthesized in the lab and studied experimentally. In particular, autocatalytic RNA systems of ligases (Lincoln and Joyce 2009) and recombinases (Hayden et al. 2008) have been demonstrated, in which specific sequences catalyze formation of the same sequences. However, it is usually envisaged that the RNA World would have required an RNA polymerase ribozyme that used another strand as a template to synthesize a complementary strand. If the polymerase could operate with any general template sequence, then it would be able to synthesize further copies of itself, as well as replicate any other ribozymes with different functions. Hence, it would form the basis of a genetic system that could evolve and diversify. There has been considerable progress in developing polymerase ribozymes in the laboratory (Johnston et al. 2001; Lawrence and Bartel 2005; Zaher and Unrau 2007; Wochner et al. 2011;

Attwater et al. 2013). The best current polymerases are able to replicate sequences of their own length but require specific template sequences and therefore cannot yet self-replicate.

Supporters of the RNA World see this as evidence that a self-replicating RNA polymerase could have arisen on the early Earth, given the large amount of time available. Nevertheless, many alternative scenarios are also possible. In this chapter, I discuss possible scenarios for the origin of life from the point of view of mathematical modeling and evolutionary theory. Of course, models are always much simpler than reality and always leave out many details, but defining a simple model forces us to focus on the key aspects of a problem and to properly define the questions we are asking. Results from a simulation or analytical solution of a model tell us clearly what happens if we make the assumptions on which the model is based. This helps to avoid confusion from speculative verbal arguments.

The theoretical models discussed here describe replication of strands of a biopolymer that is built up from a small alphabet of monomers that have complementary pairing. It is assumed that a strand can act as a template for synthesis of its complementary strand, and that sequence information is passed on in this process. It is assumed that all or most sequences have the ability to be a template. Additionally, at least some of the sequences have the ability to fold to a specific structure that is a functional catalyst. It may be that functional sequences are very rare in the space of possible sequences, so that it might take a very long time for non-living chemistry to find a functional sequence. This also means that maintaining accurate sequence replication is important, because errors in replication will often generate non-functional sequences. I will call the polymer RNA, but similar polymers with different backbone chemistry and/or different base pairs might also function in a similar way, so these models do not rule out alternative nucleic acids existing prior to or alongside RNA. However, we *are* excluding the possibility that the principal polymers were proteins, because complementary pairing of proteins is not known to occur and because the evidence from current biology is that proteins are synthesized by ribosomes, whose catalytic part is made of RNA.

Since I am operating under the premise that life *did* originate on Earth and that it wasn't a miracle, then the first functional sequences must have been synthesized by non-living chemistry. This means that there must have been an environment in which biopolymer sequences could be synthesized chemically that were sufficiently long to have catalytic function. The most obvious function that would be required in the RNA World would be a polymerase ribozyme that catalyzed RNA replication. However, there are also scenarios in which other kinds of catalysts could have come first. One of the main aims of this chapter is to consider *which* kinds of catalysts came first.

6.2.2   THE ORIGIN OF LIFE AS A SPATIALLY LOCALIZED STOCHASTIC EVENT

There are three qualitatively different ways to make an RNA strand, as in Figure 6.2.1. The spontaneous or *s* reaction is spontaneous chemical synthesis from monomers by polymerization, without a template strand. The replication or *r* reaction is template-directed synthesis, (i.e. synthesis of a strand that is complementary to a template strand, without the presence of a ribozyme or other biological catalyst). The catalysis or *k* reaction is template-directed synthesis that is catalyzed by a polymerase ribozyme. All three of these cases may proceed one monomer at a time or by formation of short oligomers that then join to form longer ones. Experimental support for these three reactions will be discussed in Section 6.2.7. The models discussed in this chapter assume that all these three reactions can occur to some extent.

The simplest possible model for RNA replication incorporating the *s*, *r*, and *k* reactions (Wu and Higgs 2012) can be defined by a single equation for the rate of change of the concentration, ϕ, of a single type of replicating strand.

$$\frac{d\phi}{dt} = (s + r\phi + k\phi^2)(1 - \phi) - v\phi \qquad (6.2.1)$$

The distinction of these three reactions is that the *s* reaction occurs at a rate independent of the current strand concentration, the *r* reaction occurs at a rate proportional to ϕ (the template concentration), and the *k* reaction occurs at a rate proportional to ϕ^2 (the product of the template and catalyst concentrations). Replication is limited by resources (*i.e.*, finite numbers of monomers), which is included via the (1 − ϕ) term. The concentration scale is defined such that

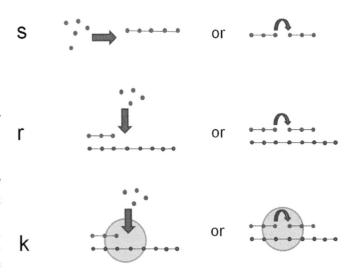

FIGURE 6.2.1   Three categories of reaction that form RNA strands. The *s* reaction (meaning "spontaneous") is the formation of a strand by spontaneous polymerization of monomers or by linking oligomers. The *r* reaction (meaning "replication") is the non-enzymatic template-directed synthesis of a complementary strand, either by addition of monomers or by ligation of two oligomers that are bound to the same template. The *k* reaction (meaning "catalysis") is the ribozyme-catalyzed template-directed synthesis of a complementary strand. The ribozyme is a separate strand from the template and the substrate oligomers and monomers.

the maximum possible concentration when all resources are used is $\phi = 1$. The final $-v\phi$ term represents breakdown of the strands or loss of the strands from the system.

For appropriate values of the reaction rate parameters, there are two possible stable states for ϕ, which we have termed living and dead. In the dead state, ϕ is very small. The spontaneous synthesis term is balanced by the breakdown rate; hence, $\phi \approx s/v$. In≈ the living state, the catalytic replication term, $k\phi^2$, is large, and replication becomes resource limited; hence, ϕ is close to 1. We have assumed that the strand is a self-replicating ribozyme with a high k. This means that it can sustain itself by the k reaction if it manages to achieve a high-enough concentration to get started. If k is large enough, there is a living state solution of Equation 6.2.1, even if s and r are zero. This means that the ribozyme sustains itself by its own replication, which is why we call it "*living.*" On the other hand, in the dead state, the existence of the strand depends on the very small rate of chemical synthesis, s, and the strand is not self-sustaining.

We also found living and non-living states in more detailed models for RNA polymerization and replication that we studied previously (Wu and Higgs 2009, 2011). The model described by Equation 6.2.1 was introduced because it captures this idea in a very simple way. In this picture, the origin of life is a transition from the non-living to the living stable state. It requires a rare stochastic event for this to happen. In a well-mixed solution with very large numbers of molecules, this is extremely unlikely, but in a spatially distributed model, where molecules interact with local neighbors on a surface, this becomes much more likely. A small patch of replicators can arise in one location, which can then spread across the surface. We have compared the waiting times for replication to become established in the spatial and well-mixed version of this model (Wu and Higgs 2012).

For the discussion in this chapter, we will use a slightly more detailed model for RNA replicators (Shay et al. 2015; Kim and Higgs 2016; Tupper and Higgs 2017), in which there are three types of strand: a polymerase P, the complementary strand to the polymerase C, and a non-functional strand X. The figures in this chapter use a version of this model with a two-dimensional lattice, in which each site may be either empty or occupied by only one type of strand (Kim and Higgs 2016). The model proceeds in short time steps in which stochastic events occur on each site with the following rates. Non-functional strands appear at a small rate s. All strands replicate at a rate r. Strands that are next to a polymerase are copied by the polymerase at rate k. All strands die at rate v. When replication occurs, new strands are placed on vacant lattice sites next to the template. If replication is accurate, a P template produces a C, and a C template produces a P. If an error occurs in replication, an X is produced instead of a P or C. The probability that an error occurs somewhere in the whole template is M. Replicating an X always produces another X (i.e. beneficial mutations that create functional strands are not included).

In this model, chemistry creates a mixture of random strands. If s is low, the density of the random strands is low. If r is also low, a strand will usually be destroyed faster than it produces new strands by the r reaction. Within this mixture, after some very long time, we suppose that an unusual sequence is created by chemistry that functions as a polymerase. Such a polymerase can spread if its catalytic rate k is large enough and if its error rate M is small enough. Figure 6.2.2 illustrates the spread of a patch of polymerases and complements that has arisen from the introduction of a single polymerase. This patch will continue to grow and fill the whole surface, even if we set s and r to zero from this point on.

How likely this scenario is to occur in the real world depends on the distribution of lengths of strands that can be synthesized chemically; on how long a sequence has to be,

in order to have significant catalytic ability; and on how rare catalytic sequences are, among random sequences. We do not know these things with any certainty. Some people have a gut reaction that polymerase ribozymes are just "too complicated" to have arisen from random chemistry. However, it should be remembered that rare events are not equivalent to miracles. If an event occurs only once on a whole planet in a billion years, then it is quite likely to have occurred once on a planet of the Earth's age (4.5 billion years). There are now known to be many potentially habitable planets in our own neighborhood of the galaxy (Petigura et al. 2013); hence, with numbers such as those above, life would be frequent in our galaxy. We will return to some of these questions later, after considering another possibility that can arise in a chemical system with s, r, and k reactions.

6.2.3 CHEMICAL EVOLUTION IS ENABLED BY THE r REACTION

We already emphasized that a replicator with high k can sustain itself, even if s and r are zero. However, a replicator with high r can also sustain itself if s and k are zero. In Equation 6.2.1, it is easy to see that if s and k are zero, there is a stationary solution with non-zero replicator concentration if $r > v$. In a spatial model, the minimum replication rate necessary for survival, r_{min}, is greater than v, because the replicators grow in clusters and cannot grow into sites that are already occupied. For a square lattice in which growth occurs into the eight surrounding lattice sites, r_{min} is approximately $1.4v$. The numerical constant does not matter very much, however. For any model of this type, there is always an r_{min}, such that replicators can survive if $r > r_{min}$. This says that the rate of the templating reaction has to be fast enough that a typical strand generates at least one surviving complementary strand before it is hydrolyzed or lost from the system.

It might also seem unlikely that this scenario could occur in the real world. We are used to the idea that organisms have very large genomes that are very accurately replicated because highly evolved polymerase proteins catalyze the replication. However, on the prebiotic Earth, we don't need to replicate whole genomes. If oligomers of just a few tens of nucleotides could be replicated by the r reaction, this would have huge importance for understanding the origin of life.

Figure 6.2.3 illustrates replicating strands with r above and below r_{min}. New random sequences are generated by the s reaction at a low rate. Each of these is colored a different shade of green to emphasize that these are all different. The complementary (plus and minus) strands of each sequence are given the same color. Replication produces patches of related sequences of the same color. If $r < r_{min}$, these patches consist of just a few strands. If $r > r_{min}$, the patches fill the whole surface and the different strands compete for resources/space.

Figure 6.2.3 is an example of what I will call "chemical evolution." I will use this term in a precise way to indicate a stage on the path to the origin of life that is different from "just chemistry" and different from fully fledged biological

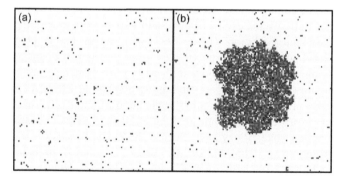

FIGURE 6.2.2 Model of the origin and spread of RNA polymerases (Kim and Higgs 2016). Red sites are polymerases, and orange sites are complementary strands. Non-functional strands are black. (a) The non-living state has a low density of non-functional strands. A single polymerase has been added in the center of this figure. (b) A living patch of polymerases and complements has spread from a single polymerase. Parameters $s = 0.01$, $r = 0.4$, $k = 25$, and $v = 1$.

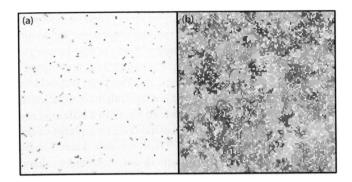

FIGURE 6.2.3 Model of chemical evolution of non-functional RNA templates. (a) $r = 0.4$. Sequences cannot spread. (b) $r = 5$. Patches of different sequences (illustrated by different shades of green) spread across the whole surface. Empty sites are gray. $s = 0.01$ and $v = 1$ in both figures.

evolution (Higgs 2017). Evolution via natural selection is often considered to be a defining feature of life. If sustainable replication occurs, whether by the r or the k reaction, then evolution is possible. A variety of sequences will arise that replicate at different rates, and replication will pass on the properties of the parent sequence. If there is no replication of either kind, then it is still possible to have selection to a limited extent, but there is no evolution. For example, suppose that spontaneous chemistry generates a range of random oligomers and that these oligomers have different mean life times because they differ in physical properties such as melting temperature of secondary structures, strength of binding to surfaces, and susceptibility to hydrolysis. These oligomers will end up with different concentrations in the mixture. In other words, selection can happen to a certain extent, but when there is no replication present, the differences in concentrations remain small and proportional to the difference in the physical properties. However, if the oligomers can replicate (by the r reaction), then differences in concentrations build up and the steady-state concentrations vary over a range that is much greater than the range of differences of the physical properties, as I have previously shown (Higgs 2017). This has also been emphasized by Chen and Nowak (2012), who use the term "pre-life" to describe the case of selection without replication, where small differences in rate constants give rise to small differences in concentrations.

The oligomer system described earlier shows evolution by natural selection when replication is present. Nevertheless, it is still much simpler than the kind of evolution that most biologists are used to thinking about. For this reason, I will distinguish between what I will call "chemical evolution" and "biological evolution." The first distinguishing feature of chemical evolution is that it deals with molecules that are simple enough to be generated by spontaneous synthesis, as well as by replication. In biological evolution (i.e., the evolution of genes and proteins in modern organisms), we are dealing with sequences that are too long to be created from scratch; hence,

chemical synthesis is not relevant. The only way to make another copy is by replicating an existing sequence. Variant sequences arise by mutations, i.e. by errors in replicating existing sequences. On the other hand, if we consider short oligomers, as envisaged in the previous paragraph, then these are still simple enough to be made chemically from scratch, as well as by replication of existing oligomers. Thus, random chemical synthesis is a significant source of new sequence diversity, which in some cases could be larger than the diversity generated by errors in replication.

The second distinguishing feature of chemical evolution is that selection acts mostly on physical properties of the molecule rather than encoded properties. We are used to thinking of biological evolution, where selection acts on the encoded properties of gene sequences. For example, a particular gene sequence makes a particular protein enzyme that makes a color pigment that makes a flower more attractive to the bee. However, at the chemical evolution level, selection acts directly on the properties of the molecule. In the oligomer example, properties such as surface binding and susceptibility to hydrolysis are relevant. These are properties possessed by all short oligomers. In contrast, having an encoded function is a rare property of long information-carrying molecules. Most DNA sequences do not encode a protein with a useful function.

In the RNA World, the separation of genotype and phenotype is not as clear as it is in modern organisms, because replicating RNAs have to act as both a template and a catalyst. However, if we have reached the stage of functional polymerase ribozymes (as in Section 6.2.2), then we have reached what I consider to be biological evolution. The ribozyme would be quite long (maybe a hundred or more nucleotides), and it could not be synthesized again by random polymerization. It would be relatively sensitive to mutations, with many deleterious mutations that would destroy its function and a small number of beneficial mutations that might improve its catalytic ability. If we added the possibility of beneficial mutations to the polymerase system in Figure 6.2.2, then better ribozymes would evolve.

The common evolutionary definition of life states that "life is a self-sustained chemical system capable of undergoing Darwinian evolution" (Joyce 1994). This means that an RNA World sustained by a polymerase ribozyme would count as living. On the other hand, if there is no polymerase, as in Figure 6.2.3, then we are envisaging a situation where the r reaction is sufficient to allow replication of short oligomers but insufficient to replicate long sequences with encoded function (like ribozymes). Such a system still satisfies the evolutionary definition of life because it has Darwinian evolution with heredity and natural selection. However, I would prefer not to call it living. I have advocated (Higgs 2017) a stricter evolutionary definition of life, where the requirement is for biological evolution and not just chemical evolution. Biological evolution is defined (in a non-circular way) by the requirement that it involves sequences that are too long for spontaneous chemical synthesis to be relevant and are long enough to have encoded function. Additionally, biological evolution is complex enough to be open ended and not repeatable, whereas chemical evolution occurs in a relatively small sequence space and is likely to result in the same outcome if we start again from the same initial conditions.

Whether a system of oligomers replicating by the r reaction counts as alive is, of course, simply a matter of definition. Here, I want to turn away from the definition question and consider the significance of chemical evolution for the origin of life. The essential point so far is that, if short sequences can replicate without the necessity of a biological catalyst, this makes a lot of difference to scenarios for the RNA World and the origin of life, as we shall now consider.

6.2.4 SCENARIOS FOR THE ORIGIN OF LIFE AT LOW r AND HIGH r

The way we envisage the origin of life depends a lot on how effective was the r reaction. Table 6.2.1 summarizes important differences between cases where r is above and below r_{min}. First, if $r < r_{min}$, a single short sequence formed spontaneously will tend to be destroyed faster than it replicates. Therefore, there is no heredity and evolution. There may be a diverse mixture of sequences, but the common components of this mixture will stay the same over time, because short oligomers will be continually reformed. However, we are presuming that functional ribozymes must be quite long and that they are rare in sequence space. In the low-r scenario, it is necessary for relatively long sequences to be formed chemically. As a round number, let us suppose that there is a polymerase ribozyme of length 100 nucleotides. If we (optimistically) presume that chemistry generates an appreciable total concentration of 100-mers, there are still 4^{100} possible 100-mers, so there will never be more than a handful of copies of any one sequence. Of course, there are many sequences that fold to a given secondary structure, as we know from computational studies of RNA folding (Szilagyi et al. 2014) and also from the diversity of biological molecules such as tRNAs (Jühling et al. 2009), whose sequences differ widely, despite having the same structure and function in different organisms. Thus, we are searching for *any* sequence that has a ribozyme function,

not for one specific RNA out of 4^{100}. Even accounting for this, however, it is likely that functional sequences will be rare in the mixture generated by chemistry. This means that the origin of life is stochastic: we may have to wait for some unpredictably long time before any functional sequence is generated at all.

For replication to begin with a polymerase ribozyme, there needs to be a minimum of one polymerase and one complement (or two polymerases). Therefore, it is necessary for r to be non-zero, even in the low-r scenario. This is what the simulation shows in Figure 6.2.2. If a single P or C arises by the s reaction, there is an appreciable probability that the complement (C or P) is created by the r reaction and that the P begins to copy the C before either of them is destroyed. Once we get beyond a handful of sequences of both strands, the spread of the polymerase is deterministic because not all copies of the polymerase will be destroyed at the same time. Although not every single polymerase that arises will generate a stable spreading cluster, we have shown (Shay et al. 2015) that there is a relatively high probability of the cluster spreading, assuming that the k rate of the polymerase is high. Hence, the major stochastic element of this scenario is waiting for the polymerase with high k, rather than the difficulty of the polymerase spreading once it arises.

It should also be clear that, in the low-r scenario, the first ribozyme must be a polymerase—that is, it must have a k rate. Many other beneficial functions of ribozymes can be envisaged, for example, nucleotide synthetases; or ribozymes that make sugars, nucleobases, and lipids; or any other ribozyme with a function that contributes to the formation or survival of RNAs in its vicinity. However, if the ribozyme has some beneficial property that does not directly create another copy of itself, and if it cannot replicate by the r reaction, then the beneficial property will disappear after the finite life time of the original molecule.

A variation on this argument is possible if r is only slightly below r_{min}. Suppose a ribozyme arises with a beneficial function, such as a nucleotide synthetase. It causes an increase in concentration of nucleotides in its vicinity, which increases the rate of the r reaction. If the original r is not too low, then the increase might cause it to rise above r_{min}, in which case a cluster of sequences containing the nucleotide synthetase

TABLE 6.2.1

Comparison of Scenarios for the Origin of Life, with Low and High Rates of Template-Directed Replication r

Low-r Scenario ($r < r_{min}$)	High-r Scenario ($r > r_{min}$)
No significant chemical evolution before the origin of ribozymes. Biological evolution is enabled by the polymerase.	Significant chemical evolution before the origin of ribozymes.
The origin of life is stochastic. Wait for a rare sequence and then wait for the establishment of a growing patch.	Chemical evolution of short strands is deterministic. Still need to wait for a rare sequence to initiate catalysis.
The first ribozyme must be a polymerase.	The first ribozyme could be a polymerase or nucleotide synthase or any beneficial function. Maybe there was no polymerase at all.
Once the k reaction is established, s and r are no longer required. Life can move outside the initial environment.	Constrained to an environment where $r > r_{min}$ unless a polymerase subsequently evolves.

can spread. In this variant scenario, the nucleotide synthetase is affecting its local environment. It is important that the nucleotides it generates remain close by, so that it directly benefits itself and copies of itself on neighboring sites. In this case, it can spread in a cluster in the same way as a polymerase that performs the k reaction. The variant case is also similar to the original low-r case, because it requires waiting in the non-living state for a long time until the beneficial function arises in the mixture of random sequences.

Let us now contrast the high-r case with the low-r case. If $r > r_{min}$, sequences do not need to have a function in order to spread. Any sequence that is a template can replicate via the r reaction. Chemical evolution is possible, as described in Section 6.2.3. Since sequences are likely to differ in their ability to be templates, then we expect short strands to evolve with sequences that are efficient templates. This differs from the low-r scenario in that the concentration of strands will be high even before ribozyme function has evolved, and there will be chemical evolution of non-random properties of the sequences. So far, this scenario is not a stochastic one. We would expect the same simple physical properties (templating ability, stability against hydrolysis, *etc.*) to evolve in a repeatable way.

Thus, if $r > r_{min}$, we reach a state with simple replicating strands with no function other than to be templates. This system would not count as living, but if a functional ribozyme arises and spreads within this state, then we switch from chemical evolution to biological evolution, and the system becomes living, by my definition (Higgs 2017). In the chemically evolving system, there will be clusters with multiple copies of each sequence. If a ribozyme does arise by random chemistry, there will usually be many copies of the functional strand and its complement. Hence, the spread of a beneficial ribozyme, when it arises, should also be fairly deterministic. Nevertheless, this does not get around the fact that long sequences may be rare in the mixture and that the fraction of functional sequences of a given length may be very small, which is the same issue, whether r is high or low. Hence, there may still be an unpredictably long waiting time until the first catalyst arises.

Another important point is that it is easier for ribozymes to evolve with functions other than polymerases if r is high than it is if r is low. A template strand can spread and reach multiple copies, even if it has no function. If it does have a beneficial function of any type, it is relatively easy for this to feed back to the replication rate of this strand. If a nucleotide synthetase arises, then any slight increase in the r rate that this causes will be beneficial. It does not have to tip the balance from below to above r_{min}, since the chemical environment is already giving us an r greater than r_{min}. The beneficial effects of the ribozyme need not be so local, as in the previous case, because there will be relatively large clusters of each sequence, even before the function arises. A computational model that simulates this case is the metabolic replicator model (Könnyű et al. 2008; Könnyű and Czárán 2013; Czárán et al. 2015). A metabolic replicator is any ribozyme that increases the local rate of replication by the r reaction. It has been shown that systems with multiple types of metabolic replicators that contribute in different ways to increasing the replication rate can evolve together in a surface-based model. Szostak (2012) has reviewed the possibility that RNA replication could be sustained by the r reaction and has also pointed out that it is possible for metabolic ribozymes to arise before polymerases in this case. He envisages RNA replication inside protocells and therefore suggests that a ribozyme catalyzing lipid synthesis would confer a benefit on the protocells that contain it.

If r is sufficient for only short oligomers to replicate, then it is still possible for chemical evolution of the short oligomers to occur. However, if replication of metabolic ribozymes is to be sustained by the r reaction, then the r reaction must function efficiently for sequences that are long enough to be ribozymes. These other ribozymes are potentially as long and as sensitive to mutational error as a polymerase would be. Thus, even if r is high, we cannot avoid the necessity to accurately replicate long sequences.

In the low-r case, the polymerase can spread under its own k reaction. Therefore, it can spread into environments where there is no s or r reaction at all. The example in Figure 6.2.4a shows a polymerase cluster spreading under the k reaction in the case where s and r are zero. In contrast, in the high-r case, replication is provided "for free" by the chemistry of the r reaction. Therefore, other metabolic functions can evolve without a polymerase, but the system will be constrained to the environment in which $r > r_{min}$, unless it subsequently evolves a polymerase. We have discussed the possibility that other ribozymes could evolve first in the high-r case. But it could also be the case that a polymerase evolves first in the high-r case too. Figure 6.2.4b shows a simulation in which a single polymerase is added into a region of non-functional strands replicating via the r reaction in the high-r case. A cluster of polymerases and complements is able to spread and displace the non-functional strands.

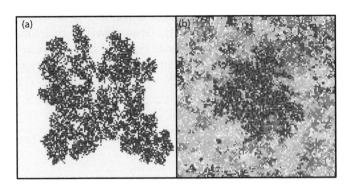

FIGURE 6.2.4 (a) Spread of a polymerase in an environment where s and r are zero. The polymerase must survive against non-functional parasites (black) that arise by deleterious mutation. In this example, $M = 12\%$. (b) Spread of a polymerase into an environment with $s = 0.01$ and $r = 5$. The polymerase must outcompete non-functional strands (green) created by chemistry, which are common in this case.

6.2.5   COOPERATION IN RNA SYSTEMS

An essential point for evolution in the RNA World is that a polymerase ribozyme can copy another template strand but cannot copy itself. The survival of the polymerases therefore requires cooperation between a group of molecules that replicate one another. Non-functional strands act as parasites of polymerases because polymerases spend their time replicating the non-functional strands, instead of replicating other polymerases, and because non-functional strands consume resources (monomers and space *etc.*) that could otherwise be used by the polymerases. There are two distinct ways in which non-functional strands arise, illustrated in Figure 6.2.4. In the low-*r* case, the density of random strands created by *s* and *r* reactions is low. The principal source of non-functional strands is deleterious mutations. In Figure 6.2.4a, *s* and *r* are zero, so the only non-functional strands are deleterious mutations (shown in black), which are created with a mutation probability $M = 12\%$ on replication of either a polymerase or complement. In the high-*r* case in Figure 6.2.4b, the density of random strands made by *s* and *r* reactions is high (shown in green). This example has $M = 0$, so the only non-functional strands are made by chemical synthesis and not by mutation (although mutation could easily be added to this simulation as well). Recall that whether the main source of diversity came from chemical synthesis or mutation was a key factor used in Sections 6.2.3 and 6.2.4 to distinguish chemical and biological evolution. In both parts of Figure 6.2.4, the polymerases are functioning as cooperators in the evolutionary sense. Each polymerase benefits its neighbors and does not directly benefit itself. We now consider the factors that allow the evolution of cooperative polymerases.

The simplest description of a chemical reaction system is "well-mixed," *that is,* molecules encounter one another randomly in proportion to their concentration. For the case of polymerases, complements, and parasites that we are discussing, a set of time-dependent differential equations for the concentrations in the well-mixed case can be easily written down (Shay et al. 2015; Kim and Higgs 2016). However, this approach leads to the conclusion that, unless the mutation rate is exactly zero, parasite strands always drive the polymerases to extinction. If *s* and *r* are zero, then replication of the parasites depends on the polymerases; hence, the parasites also become extinct if the polymerases disappear. This result depends on the assumption that the polymerase replicates parasite and functional strands at an equal rate but that mutation always favors the parasites. The latter point is true because it is easy for errors in replication of functional strands to create non-functional parasites but very unlikely that a beneficial mutation in a random strand will create a functional polymerase. Thus, for any $M > 0$, the parasites always accumulate and destroy the system. Any real molecular replication process would never be perfectly accurate, but even if we do set *M* to zero in the model, a polymerase is only neutral with respect to other non-functional strands, because the benefit it provides to the other strands is equal to the benefit that it provides to itself. Hence, a polymerase cannot spread by natural selection in a well-mixed system.

In order for a polymerase to survive and spread, it needs to encounter other polymerases more frequently than it would in a well-mixed system. Spatial structure is one way in which this can occur. In the two-dimensional surface models shown in Figures 6.2.2–6.2.4, a newly synthesized strand is placed on a neighboring site to the template strand. This means that there is clustering of polymerases and templates. Polymerases are more likely to be next to other polymerases and complements and less likely to be next to parasites. Hence, clustering gives an advantage to polymerases that can outweigh the mutational disadvantage. In spatial models, a polymerase can survive as long as *M* is less than some critical error rate M_{max}.

The term error threshold is usually used to denote the maximum error rate at which a functional replicator can survive (Eigen et al. 1988). The original theory describes independent replicators (equivalent to the *r* reaction in our terminology). In this case, the advantage of the functional sequence is that it has a higher *r* rate. In the RNA polymerase models that we are discussing here, replication is via the *k* reaction and the advantage of the polymerase comes from spatial clustering, not from faster replication. These two cases are qualitatively similar, so we will also use the error threshold term. We have discussed the details of error thresholds in polymerase models elsewhere (Shay et al. 2015; Kim and Higgs 2016; Tupper and Higgs 2017). Figure 6.2.4a illustrates a case where $M < M_{max}$. The polymerases can spread, although there is always a finite concentration of parasites mixed in, because parasites are continually being created by mutation. Figure 6.2.4b illustrates the case where $M = 0$. Parasites are present due to the *r* reaction. The polymerase spreads because of the advantage of spatial clustering, whereas it would be only neutral in the well-mixed case.

There are many other two-dimensional computational models that predict that cooperative replicators can survive as a result of clustering and spatial pattern formation (Boerlijst and Hogeweg 1991; McCaskill et al. 2001; Szabo et al. 2002; Takeuchi and Hogeweg 2009, 2012). It should be realized, however, that there is nothing special about two dimensions. A three-dimensional lattice model would behave in a similar way (although would be more difficult to visualize). What is important is that spatial clustering only builds up if diffusion is slow. In previous papers, we have allowed diffusion of strands into neighboring sites. If diffusion becomes faster than the birth and death of strands, then the system becomes well mixed and the parasites destroy the system (Shay et al. 2015; Kim and Higgs 2016). It seems reasonable to envisage two-dimensional systems where strands are bound to surfaces and where diffusion along the surface might be slow. A mineral surface might actually catalyze the RNA polymerization process, as is known in experiments with Montmorillonite clay (Huang and Ferris 2006; Joshi et al. 2009). Systems of RNA between stacked lipid bilayers are close to two-dimensional systems (Rajamani et al. 2008; Mungi and Rajamani 2015). It might also be possible to envisage three-dimensional systems where diffusion would be slow, such as a network of pores in a rock (Koonin and Martin 2005). However, in an open-water system in three dimensions, diffusion would lead to mixing on a scale faster than replication.

Other aspects of the parasite problem are that shorter sequences will replicate faster and there will be a tendency for short parasites to arise that replicate much faster than full-length functional polymerases. This is reminiscent of the experiments in which Qβ RNA is replicated in vitro by a polymerase protein (Mills et al. 1967), where short RNAs arise that are rapidly copied by the protein but have lost the information on the original viral genome. We have simulated a case in which termination of replication can occur prior to the end of the sequence (Tupper and Higgs 2017). This creates parasites of all lengths shorter than the polymerase. The polymerase is still able to survive in a spatial model, and we observe interesting traveling wave patterns in which parasites chase polymerases but do not manage to destroy them. The error threshold for termination errors (which create short parasites) is much lower than for point mutations (which create parasites of equal length to the polymerases).

One way in which a polymerase might lessen the impact of parasites is to use a tag (Wu et al. 2017), which is a short sequence at the beginning of a strand, recognized by the polymerase. Only sequences that begin with the tag sequence are replicated. The complement of the tag sequence has to be at the end of the strand, so that the complementary strand also has the tag at the beginning. This mechanism makes almost no difference to parasites generated by point mutations, because the tags will inevitably be much shorter than the functional part of the ribozyme and any parasite that has a mutation in the main part of the sequence will still have the tags. However, for short parasites created by termination errors, the end tag would be missing and the complementary strand of the parasite would not be recognized by the polymerase. Thus, the tag mechanism would go a long way toward stabilizing the polymerase against short parasites that have arisen via termination errors.

An alternative way in which polymerases can survive against parasites is via compartmentalization in protocells (lipid vesicles). In protocell models, strands within the same compartment are well mixed, but they do not interact with strands in other compartments. Protocells divide randomly when they reach a specified maximum size. The time taken to reach this size depends on the composition. Hence, compartments with few parasitic strands will grow and divide faster than compartments that are overrun with parasites. Group selection at the protocell level benefits polymerases and can outweigh individual selection at the molecular level, which would benefit parasites. Once again, a variety of computational models shows that compartmentalization allows the survival of cooperative replicators (Szathmáry et al. 2005; Takeuchi and Hogeweg 2009; Zintzaras et al. 2010; Bianconi et al. 2013). This effect has also been realized in experimental systems involving vesicles containing RNAs that are replicated by a protein polymerase (Matsumura et al. 2016).

The evolution of cooperation is a very broad field in evolutionary biology, because it spans many different levels, from molecules to cells, to multicellular organisms, to social groups. The major transitions in evolution are important stages in the history of life on Earth in which entities of higher complexity have arisen via cooperation of entities at a lower level (Szathmáry and Maynard Smith 1995). A system of five mechanisms for the evolution of cooperation has been proposed (Nowak 2006), and RNA polymerase systems exhibit several of these mechanisms.

In addition to the cooperation between groups of polymerases discussed so far, RNAs can also exhibit cooperation in two other senses (Higgs and Lehman 2015). First, we can envisage the formation of an autocatalytic set of mutually dependent strands, where no single strand can replicate itself but the set as a whole can replicate. Experimental systems of this nature have been realized with RNAs (Lincoln and Joyce 2009; Vaidya et al. 2012; Mathis et al. 2017). These examples are close to the general theoretical models of autocatalytic sets (Hordijk and Steel 2013). Second, strands that assemble into a larger complex via secondary structure formation are cooperative if the complex has a function but none of the component strands has a function on its own, for example, recombinase ribozymes assembled from smaller strands (Hayden et al. 2008).

6.2.6 EVOLUTIONARY QUESTIONS

In this section, we will consider how RNA replicators might evolve toward modern organisms. It should be realized that, although modern organisms contain complex genetic networks, these do not resemble the networks of reactions in autocatalytic sets. In autocatalytic set models, each type of strand catalyzes a specific reaction that is required for replication of the set. Each strand in these models is part of the replication process, even though none of them can replicate independently. In contrast, in the genetic systems of modern organisms, there is one system for DNA replication that can replicate the whole genome. Only a very small number of genes are directly connected to replication. The majority of genes have functions related to metabolism or structure of the organism. These genes contribute to fitness, but they are not part of the replication machinery. If we want to understand the pathway from RNA replicators to modern organisms, then we are led to consider a system of multiple ribozymes having different functions, only one of which is a polymerase. If an RNA polymerase that works with general template sequences has already arisen, then it is possible for ribozymes with different functions to work in cooperation with the polymerase, because all these other ribozymes would be copied by the same polymerase.

Figure 6.2.5 shows a hypothetical schematic network of chemical reactions that would be required to generate a supply of nucleotides from small molecules that are abundant in the environment. There would presumably be many reaction steps (thick straight arrows) involving organic molecules that are precursors to nucleotide synthesis. In order for the RNA World to get going at all, all these steps must have occurred chemically to some extent in the environment in which life evolved. Thin curly arrows indicate ribozymes that catalyze some or all of these steps. A polymerase ribozyme would catalyze the last of these steps: synthesis of new strands from nucleotides. If any of the steps was catalyzed by ribozymes, it would lead to faster replication by one means or another.

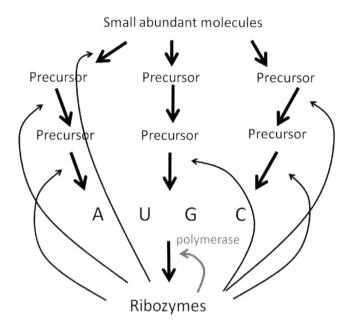

Small abundant molecules

FIGURE 6.2.5 Can we envisage a system of many different cooperating ribozymes? Only one of these ribozymes needs be a polymerase. The others would be metabolic ribozymes that control all the steps of nucleic acid synthesis.

There is a long-standing debate about whether life arose via metabolism or replication first. The evolutionary definition of life requires Darwinian evolution, which means that the origin of life is more or less synonymous with the origin of sequence replication. This is especially true if we use the criterion of biological evolution that I have proposed (Higgs 2017), which requires that there must be molecules sufficiently long that they can only be synthesized by copying previous molecules and not by spontaneous synthesis. Under this definition, proposals of compositional evolution of small molecules that might occur without sequence information (Markovitch and Lancet 2012) would not qualify as life. However, the real scientific question here is not whether a small molecule reaction network should be defined as living but what was the nature of the small molecule reaction network that generated RNAs. It is clear that many steps are required to build up nucleotides. It has often been proposed that this can arise only if there is a small molecule metabolism preceding replication (Shapiro 2000, 2006), but this still has little experimental support (Anet 2004). Although some kind of reaction network must have existed, the reaction network that created nucleotides need not have been autocatalytic or evolving—it simply needed to make lots of nucleotides in a repeatable way, until sequences that initiated the autocatalysis were eventually formed. In my view, therefore, autocatalysis and life begin at the point where replication of autocatalytic sequences is possible and metabolism begins at the point where a significant number of reaction steps in the precursor network become catalyzed by ribozymes.

In Section 6.2.4, we argued that it is most likely that the polymerase was first, according to the low-r picture, whereas according to the high-r picture, a ribozyme that catalyzed either polymerization or any of the metabolic steps could have been selected first, if its effects were sufficiently localized to benefit itself. If a network such as Figure 6.2.5 is to evolve, it must be possible to build it up by stepwise addition of new ribozymes. We have started to do this by studying the conditions under which a nucleotide synthetase can be added to a replicating system controlled by a polymerase (Kim and Higgs 2016). The two ribozymes are on separate strands, and each one is subject to deleterious mutations that create non-functional parasites. It is clear that the functions of these two ribozymes are complementary, as the nucleotide synthetase produces monomers that speed up the polymerization rate and the polymerase copies the synthetase. If there is a supply of monomers by chemical synthesis, then the polymerase can survive on its own (as in all the examples in the figures of this chapter). If there is no chemical supply of monomers, then the only monomers present are those created by the synthetase. In this case, each ribozyme is completely dependent on the other. We have shown (Kim and Higgs 2016) that there are parameter ranges where the two independent cooperating ribozymes can coexist, including cases where there is no chemical supply of monomers. This demonstrates that building up more complex ribozyme systems in a surface model is possible in principle. It also shows that the two-ribozyme system could move into environments where the single-polymerase system could not go alone. Other computational models (Ma et al. 2007, 2010; Ma and Hu 2012) have also considered replicating RNA systems controlled by a variety of different kinds of ribozyme functioning singly or in association and have emphasized the important role that nucleotide synthetases could have played.

However, there are considerable difficulties associated with getting ribozymes to work together, even when they have complementary functions (Kim and Higgs 2016). First, we have shown that the nucleotide synthetase can coexist with the polymerase only if its length is within a range that is comparable to the length of the polymerase. If the synthetase is too long, it replicates too slowly and dies out, leaving the polymerase working alone. If the synthetase is too short, it replicates too fast, which causes the extinction of both ribozymes. Another important factor is the diffusion rate of the monomers produced by the synthetase. If monomers diffuse rapidly, they benefit non-functional synthetases as much as functional ones. In this case, the functional synthetases are destroyed by parasites. If monomers diffuse too slowly, they do not reach the sites where the polymerase can use them. We have not yet tried to simulate a system with more than two independent ribozymes, but we expect that it will become increasingly difficult to find regions of coexistence as more ribozymes are added. Other studies of metabolic replicators (Könnyű et al. 2008; Könnyű and Czárán 2013; Czárán et al. 2015) have found coexistence of multiple ribozymes in surface-based models. However, these models do not have a polymerase (replication is r not k), so there is no requirement for metabolic ribozymes to occupy neighboring sites to the polymerase, as there is in our model of polymerases plus synthetases.

A quick glance at real biology gives us two important clues as to how life solved the problem of building up genetic systems with multiple functions: cells and chromosomes. We are currently investigating protocell models in order to compare these systems with surface models. Our work in progress suggests that it is considerably easier to get independent replicators to coexist in compartments than on surfaces. It also appears that error thresholds are higher in protocell models than in equivalent surface models. Protocell models also solve many of the problems related to diffusion of small molecules such as monomers and metabolites, as there is free mixing within the compartment, but useful molecules produced within one compartment do not benefit other compartments. The existence of membranes in the protocell brings additional problems, however. There needs to be a way to synthesize the lipids of the membranes (unless these are already present in large amounts from prebiotic chemistry), and there needs to be a way to transport small polar molecules across membranes, when required. The other important factor seen in biology is linkage of genes onto chromosomes. If we link two genes with complementary functions together, there is an advantage that they are always found together when needed together, but there is a disadvantage that a longer strand replicates more slowly than a shorter one. Mathematical models have previously shown (Maynard Smith and Szathmary 1993) that linkage can be favored in protocell models. We also expect that linkage can be favored in surface models. There are a lot of remaining questions about how chromosomes with multiple genes can be built up and whether this requires a compartmentalized protocell system to be already in operation.

The usual view of the RNA World is that it sowed the seeds of its own downfall, *that is,* ribozymes were doing just fine until ribosomes evolved and protein synthesis began, at which point proteins took over the catalytic roles and nucleic acids retained the genetic role. However, from the relics of the RNA World remaining in modern organisms (Jeffares et al. 1998), there is little evidence about how wide a range of metabolic functions might have been possessed by the RNA World. When we consider an RNA-based organism with a metabolic network such as Figure 6.2.5, we are speculating without any direct evidence from modern biology that these ribozymes existed. If we begin with a comparative evolutionary study of extant organisms, then we are immediately struck by the fact that there is one very obvious ribozyme—the ribosome— shared by all forms of cellular life. Ribosomal RNA genes, together with a few protein-coding genes closely associated with translation, are almost the only genes that are homologous across all domains of life. It is therefore clear that the ribosome appeared very early in evolution. I want to finish this section by considering the hypothesis that the ribosome could have evolved without needing to pass through a stage with RNA-catalyzed RNA replication.

Fox and co-workers studying the evolution of the ribosome have noted that both the peptidyl transferase center (the core part of the ribosome that catalyzes peptide bond formation) and tRNAs are less than 100 nucleotides in length and that

"their importance supports the notion that the translation machinery was originally a discovery of the RNA World" (Fox 2010). They have proposed that a protoribosome, having the ability to synthesize peptides, may have been one of the first ribozymes and that "the peptides produced by the protoribosome may have acquired the ability to replicate RNA before RNA-based RNA replicases could evolve in the RNA World. Alternatively, an RNA-peptide complex may have been the first to accomplish the task. In either case, the transition from the RNA World to an RNA-Protein World may have occurred much more rapidly and easily than is traditionally thought" (Fox et al. 2012). This hypothesis acknowledges that RNA was important early on and that some kinds of ribozymes existed (in particular the protoribosome), but it argues that there was never an RNA polymerase composed solely of RNA. This view has been elaborated recently by (Bowman et al. 2015), who propose that "There simply were not any sophisticated enzymes (ribozymes or other catalytic polymers) predating the emergence of the ribosome. In this scenario, the ancestral ribosome arose when building blocks for RNA, or proto-RNA, were provided by abiotic processes."

When evaluating this hypothesis from the perspective of a theoretical biologist, my first question is to ask whether I could get a computational model to work in this way. I have not actually done this, but I am fairly sure that I could. The model would be rather similar to the high-r scenario, where the nucleotide synthetase evolves first. We would need to propose a ribozyme (the protoribosome) that synthesizes peptides. The replication of the ribozyme would have to be by the r reaction, because there is no polymerase. The peptides produced by the ribozyme would have to benefit the ribozyme somehow. With a nucleotide synthetase, it is clear that the increased concentration of monomers makes the replication go faster. With peptides, it is less clear what the benefit is. We could propose that peptides bind to RNA strands in their neighborhood and protect them from degradation—making their life time longer. Or, we could propose that peptides bind to double-stranded pieces of RNA in a non-specific way, making the rate of polymerization faster (this would be a primitive kind of peptide polymerase). Either of these things could be put into a model, and I am pretty sure that we could demonstrate survival and spread of the protoribosome within the model.

However, even though this scenario seems to pass the evolutionary logic test, it still appears unlikely. It should be remembered that this will only work if the r reaction is fast enough and accurate enough to replicate strands long enough to be ribozymes and that we do not have clear evidence that random peptides do benefit RNA strands in either of the ways proposed above. It is important that the mechanism has to work with random peptides and not with encoded proteins. While it is clear that specific encoded protein sequences could bind and stabilize the protoribosome and that specific proteins could act as RNA polymerases, it seems highly unlikely that a mixture of random peptides—all different from one another—could do these things. If we hypothesize that the protoribosome was actually decoding mRNAs, then we dramatically increase the complexity of the protoribosome. It would have to have parts that associate with tRNAs and mRNAs and would have to move processively along the mRNA. There would have to be a set of tRNAs and a way of charging them with specific amino acids. All these different RNA strands would have to be replicated without a polymerase. Any argument that requires the whole translation system and genetic code to come into being simultaneously seems untenable. So, any argument that puts the peptidyl transferase center before (or instead of) an RNA polymerase would have to establish that random peptides give a direct benefit to the ribozyme.

In conclusion, let us remember how simple it would be if there were a single ribozyme that could increase the rate of RNA polymerization in some way. The first polymerase does not need to be processive. Given that the r reaction must be happening, there will be sites where short oligomers are bound to a longer template. Maybe the first polymerase simply binds to regions of double strands and stabilizes them for long enough to enable subsequent oligomers to join to the end of the strand. Maybe it recognizes places where the ends of two short oligomers are side by side on the same template, and it stabilizes this configuration until the bond is formed. These would be relatively simple mechanisms by which the first ribozyme with a k-type function might operate. It should also be remembered that the r reaction enables the evolution of specific RNA sequences that are selected for function, whereas the scenarios in which peptides act as polymerases (as in the above quote from Fox et al. 2012) would have to work with random peptides.

6.2.7 GETTING BACK TO REAL CHEMISTRY

Although a lot can be concluded from simulations of simple models, for these results to be meaningful, it is necessary to establish that real molecules can do the things assumed in the simulations. We wish to emphasize that there is experimental support for all three of the reactions we have called s, r, and k.

Spontaneous RNA synthesis is difficult in aqueous solution because hydrolysis tends to proceed faster than polymerization. The balance can be tipped by using monomers activated with triphosphates or imidazoles, and there has been significant success with clay-catalyzed polymerization of activated nucleotides (Huang and Ferris 2006; Joshi et al. 2009; Burcar et al. 2015). There has been considerable recent interest in the idea that the origin of life might be in hydrothermal pools on land rather than in the deep sea (Damer and Deamer 2015; Pearce et al. 2017). Under such conditions, wetting and drying cycles can drive polymerization (DeGuzman et al. 2014; Mamajanov et al. 2014; Da Silva et al. 2015; Forsythe et al. 2015). While dry conditions favor polymerization via condensation reactions, molecules become immobile under completely dry conditions, in which case the reactions cannot occur. In the wet phase of the cycle, hydrolysis is faster than polymerization, but diffusion brings new molecules into proximity, which can then react in the next dry phase. Simulations show that this mechanism gives longer polymers than could be formed in either the wet or the dry phase alone (Higgs 2016). X-ray scattering analysis of nucleotides trapped between lipid bilayers also helps to illustrate the influence of

the lipid layers on RNA polymerization during wetting and drying cycles (Toppozini et al. 2013; Himbert et al. 2016).

Non-enzymatic template-directed replication has also been observed in the laboratory (Deck et al. 2011; Olasagasti et al. 2011). An important issue here is the relative rates of incorporation of correctly matching Watson-Crick pairs, GU wobble pairs, and mismatch errors. These rates have also been measured experimentally (Rajamani et al. 2010; Leu et al. 2011, 2013). Another important factor associated with non-enzymatic replication is that it will tend to give rise to stable double helices. Repeated replication therefore requires a mechanism to separate the strands, and it seems likely that temperature cycling could have played a role (Szostak 2012), and possibly also wet-dry cycling, as in the experiments with spontaneous polymerization. Cycling conditions of either kind keep the reactions out of equilibrium, which is probably important in creating long strands.

We have also studied the kinetics of templating reactions (Tupper et al. 2017) and proposed that templating can solve three types of problems associated with the RNA World by the same mechanism: the origin of homochirality, the selection of a limited set of ribonucleotides, and the regioselectivity of the backbone. The common feature of these problems is that uniform strands are better templates than non-uniform ones. Firstly, in the chirality problem, strands composed of only one of the two enantiomers are good templates, but mixed chirality strands are not. When the template-directed synthesis rate is high, the racemic mixture becomes unstable and chiral symmetry breaking occurs, resulting in a system dominated by one enantiomer or the other. Secondly, templating leads to the selection of a small alphabet of monomers with a particular backbone chemistry, for example, four kinds of nucleotides built from ribose sugars, rather than some random mixture of other sugars that would be expected in prebiotic chemical synthesis. Thirdly, templating favors strands with uniform bond structure (such as the regular $3'-5'$ bonds, rather than the mixtures of $2'-5'$ and $3'-5'$ bonds). We have shown that these latter two cases are closely analogous to the problem of chiral symmetry breaking (Tupper et al. 2017).

Ribozyme-catalyzed replication (k reaction) has also been studied experimentally, with gradual progress in the development of polymerases by in vitro evolution (Johnston et al. 2001; Lawrence and Bartel 2005; Zaher and Unrau 2007; Wochner et al. 2011; Attwater et al. 2013). Beyond polymerases, the list of functions that can be catalyzed by nucleic acids continues to expand (Breaker and Joyce 2014). The current situations with regard to the s, r, and k reactions seem to be rather similar: there is experimental support for all three, but there is not a cast-iron demonstration of any of them, and important problems remain. For the s reaction—can we show that wet-dry cycling can be done under conditions that do not lead to depurination of nucleotides? For the r reaction—can we achieve accurate replication of long strands on a time scale faster than hydrolysis of the strand? For the k reaction—can we find a polymerase that will replicate a template strand with its own sequence? Although the evidence is incomplete, the RNA World hypothesis has much greater experimental and theoretical support than any of the alternatives. Therefore, personally, I am sticking to an "orthodox" RNA World point of view. That means it really was RNA first, not some alternative polymer; there really was a polymerase ribozyme; and the essential step for the origin of life was the origin of ribozyme-catalyzed RNA replication. Let us hope that concrete experimental information will lead either to more firm support of this view or to a well-supported alternative hypothesis in the not-too-distant future.

REFERENCES

Anet, F.A.L. 2004. The place of metabolism in the origin of life. *Curr. Opin. Chem. Biol.* 8: 654–659.

Attwater, J., A. Wochner, and P. Holliger. 2013. In-ice evolution of RNA polymerase ribozyme activity. *Nat. Chem.* 5: 1011–1018.

Bianconi, G., K. Zhao, I.A. Chen, and M.A. Nowak. 2013. Selection for replicases in protocells. *PLoS Comput. Biol.* 9(5): e1003051.

Boerlijst, M.C., and P. Hogeweg. 1991. Spiral wave structure in prebiotic evolution: Hypercycles stable against parasites. *Physica D* 48: 17–28.

Bowman, J.C., Hud, N.V., and L.D. Williams. 2015. The ribosome challenge to the RNA World. *J. Mol. Evol.* 80: 143–161.

Breaker, R.R., and G.F. Joyce. 2014. The expanding view of RNA and DNA function. *Chem. Biol.* 21: 1059–1065.

Burcar, B.T., M. Jawed, H. Shah, and L.B. McGown. 2015. In situ imidazole activation of ribonucleotides for abiotic RNA oligomerization reactions. *Orig. Life Evol. Biosph.* 45: 31–40.

Chen, I.A., and M.A. Nowak. 2012. From prelife to life: How chemical kinetics become evolutionary dynamics. *Acc. Chem. Res.* 45: 2088–2096.

Cheng, L.K.L., and P.J. Unrau. 2010. Closing the circle: Replicating RNA with RNA. *Cold Spring Harb. Perspect. Biol.* doi:10.1101/cshperspect.a002204.

Czárán, T., B. Könnyű, and E. Szathmáry. 2015. Metabolically coupled replicator systems: Overview of an RNA-world model concept of prebiotic evolution on mineral surfaces. *J. Theor. Biol.* 381: 39–54.

Damer, B., and D. Deamer. 2015. Coupled phases and combinatorial selection in fluctuating hydrothermal pools: A scenario to guide experimental approaches to the origin of cellular life. *Life* 5: 872–887.

Deck, C., M. Jauker, and C. Richert. 2011. Efficient enzyme-free copying of all four nucleobases templated by immobilized RNA. *Nat. Chem.* 3: 603–608.

DeGuzman, V., W. Vercoutere, H. Shenasa, and D. Deamer. 2014. Generation of oligonucleotides under hydrothermal conditions by non-enzymatic polymerization. *J. Mol. Evol.* 78: 251–262.

Da Silva, L., M.C. Maurel, and D. Deamer. 2015. Salt-promoted synthesis of RNA-like molecules in simulated hydrothermal conditions. *J. Mol. Evol.* 80: 86–97.

Eigen, M., J. McCaskill, and P. Schuster. 1988. Molecular quasispecies. *J. Phys. Chem.* 92: 6881–6891.

Forsythe, J.G., S.S. Yu, I. Mamajanov, M.A. Grover, R. Krishnamurthy, F.M. Fernandez, and N.V. Hud. 2015. Ester-mediated amide bond formation driven by wet-dry cycles: A possible path to polypeptides on the prebiotic earth. *Angew. Chem. Ind. Ed.* 54: 9871–9875.

Fox, G.E. 2010. Origin and evolution of the ribosome. *Cold Spring Harb. Perspect. Biol.* doi: 10.1101/cshperspect.a003483

Fox, G.E., Q. Tran, and A. Yonath. 2012. An exit cavity was crucial to the polymerase activity of the early ribosome. *Astrobiology* 12: 57–60.

Gilbert, W. 1986. The RNA world. *Nature* 319: 618.

Hayden, E.J., G. von Kiedrowski, and N. Lehman. 2008. Systems chemistry on ribozyme self-construction: Evidence for anabolic autocatalysis in a recombination network. *Angew. Chem. Int. Ed.* 47: 8424–8428.

Higgs, P.G., and N. Lehman. 2015. The RNA World: Molecular cooperation at the origins of life. *Nat. Rev. Genet.* 16: 7–17.

Higgs, P.G. 2016. The effects of limited diffusion and wet-dry cycling on reversible polymerization reactions: Implications for prebiotic synthesis of nucleic acids. *Life* 6: 24.

Higgs, P.G. 2017. Chemical evolution and the evolutionary definition of life. *J. Mol. Evol.* 84: 225–235.

Himbert, S., M. Chapman, D.W. Deamer, and M.C. Rheinstadter. 2016. Organization of nucleotides in different environments and the formation of pre-polymers. *Sci. Rep.* 6: 31285.

Hordijk, W., and M. Steel. 2013. A formal model of autocatalytic sets emerging in an RNA replicator system. *J. Syst. Chem.* 4: 3.

Huang, W., and J.P. Ferris. 2006. One-step, regioselective synthesis of up to 50-mers of RNA oligomers by montmorillonite catalysis. *J. Am. Chem. Soc.* 128: 8914–8919.

Jeffares, D.C., A.M. Poole, and D. Penny. 1998. Relics from the RNA World. *J. Mol. Evol.* 46: 18–36.

Johnston, W.K., P.J. Unrau, M.S. Lawrence, M.E. Glasner, and D.P. Bartel. 2001. RNA-catalyzed RNA polymerization: Accurate and general RNA-templated primer extension. *Science* 292: 1319–1325.

Joshi, P.C., M.F. Aldersley, J.F. Delano, and J.P. Ferris. 2009. Mechanism of montmorillonite catalysis in the formation of RNA oligomers. *J. Am. Chem. Soc.* 131: 13369–13374.

Joyce, G.F. 1994. Foreword. In D. Deamer and G. Fleischaker (Eds.), *Origins of Life: The Central Concepts*, Jones & Bartlett, Boston, MA, pp. xi–xii.

Joyce, G.F. 2002. The antiquity of RNA-based evolution. *Nature* 418: 214–221.

Jühling, F., M. Mörl, R.K. Hartmann, M. Sprinzl, P.F. Stadler, and J. Pütz. 2009. tRNAdb 2009: Compilation of tRNA sequences and tRNA genes. *Nucleic Acids Res.* 37: D159–D162.

Kim, Y.E., and P.G. Higgs. 2016. Co-operation between polymerases and nucleotide synthetases in the RNA world. *PLoS Comput. Biol.* 12(11): e1005161.

Könnyű, B., T. Czárán, and E. Szathmáry. 2008. Prebiotic replicase evolution in a surface-bound metabolic system: Parasites as a source of adaptive evolution. *BMC Evol. Biol.* 8: 267.

Könnyű, B., and Czárán, T. 2013. Spatial aspects of prebiotic replicator coexistence and community stability in a surface-bound RNA world model. *BMC Evol. Biol.* 13: 204.

Koonin, E.V., and W. Martin. 2005. On the origin of genomes and cells within inorganic compartments. *Trends Genet.* 21: 647–654.

Lawrence, M.S., and Bartel, D.P. 2005. New ligase-derived RNA polymerase ribozymes. *RNA* 11(8): 1173–1180.

Leu, K., B. Obermayer, S. Rajamani, U. Gerland, I.A. Chen. 2011. The prebiotic evolutionary advantage of transferring information from RNA to DNA. *Nucleic Acids Res.* 39: 8135–8147.

Leu, K., E. Kervio, B. Obermayer, R. Turk-MacLeod, C. Yuan, J.M. Luevano, E. Chen, U. Gerland, C. Richert, and I.A. Chen. 2013. Cascade of reduced speed and accuracy after errors in enzyme-free copying of nucleic acid sequences. *J. Am. Chem. Soc.* 135: 354–366.

Lincoln, T.A., and G.F. Joyce. 2009. Self-sustained replication of an RNA enzyme. *Science* 323: 1229–1232.

Ma, W.T., C.W. Yu, W.T. Zhang, and J.M. Hu. 2007. Nucleotide synthetase ribozymes may have emerged first in the RNA world. *RNA* 13: 2012–2019.

Ma, W.T., C.W. Yu, W.T. Zhang, and J.M. Hu. 2010. A simple template-dependent ligase ribozyme as the RNA replicase emerging first in the RNA World. *Astrobiology* 10: 437–447.

Ma, W.T., and J.M. Hu. 2012. Computer simulation on the cooperation of functional molecules during the early stages of evolution. *PLoS One* 7: e35454.

Mamajanov, I., P.J. MacDonald, J. Ying et al. 2014. Ester formation and hydrolysis during wet-dry cycles: Generation of far-from-equilibrium polymers in a model prebiotic reaction. *Macromolecules* 47: 1334–1343.

Markovitch, O., and D. Lancet. 2012. Excess mutual catalysis is required for effective evolvability. *Artif. Life* 18: 243–266.

Mathis, C., S.N. Ramprasad, S.I. Walker, and N. Lehman. 2017. Prebiotic RNA network formation: A taxonomy of molecular cooperation. *Life* 7: 38.

Matsumura, S., A. Kun, M. Ryckelynck, F. Coldren, A. Szilagyi, F. Jossinet, C. Rick, P. Nghe, E. Szathmary, and A.D. Griffiths. 2016. Transient compartmentalization of RNA replicators prevents extinction due to parasites. *Science* 354: 1293–1296.

Maynard Smith, J., and E. Szathmary. 1993. The origin of chromosomes I. Selection for linkage. *J. Theor. Biol.* 164: 437–446.

McCaskill, J.S., R.M. Füchslin, and S. Altmeyer. 2001. The stochastic evolution of catalysts in spatially resolved molecular systems. *Biol. Chem.* 382: 1343–1363.

Mills, D. R., R.L. Peterson, and S. Spiegelman. 1967. An extracellular Darwinian experiment with a self-duplicating nucleic acid molecule. *Proc. Nat. Acad. Sci. U.S.A.* 58: 217–224.

Mungi, C.V., and S. Rajamani. 2015. Characterization of RNA-like polymers from lipid-assisted nonenzymatic synthesis: Implication for origin of informational molecules on the early Earth. *Life* 5: 65–84.

Nowak, M.A. 2006. Five rules for the evolution of cooperation. *Science* 314: 1560–1563.

Olasagasti, F., H.J. Kim, M. Pourmand, and D.W. Deamer. 2011. Non-enzymatic transfer of sequence information under plausible prebiotic conditions. *Biochimie* 93: 556–561.

Pearce, B.K.D., R.E. Pudritz, D.A. Semenov, and T.K. Henning. 2017. Origin of the RNA world: The fate of nucleobases in warm little ponds. *Proc. Natl. Acad. Sci. U.S.A.* 114: 11327–11332.

Petigura, E.A., A.W. Howard, and G.W. Marcy. 2013. Prevalence of Earth-size planets orbiting Sun-like stars. *Proc. Natl. Acad. Sci. U.S.A.* 110: 19273–19278.

Rajamani, S., A. Vlasov, S. Benner, A. Coombs, F. Olasagasti, and D. Deamer. 2008. Lipid-assisted synthesis of RNA-like polymers from mononucleotides. *Orig. Life Evol. Biosph.* 38: 57–74.

Rajamani, S., J.K. Ichida, T. Antal, D.A. Treco, K. Leu, M.A. Nowak, J.W. Szostak, and I.A. Chen. 2010. Effect of stalling after mismatches on the error catastrophe in nonenzymatic nucleic acid replication. *J. Am. Chem. Soc.* 132: 5880–5885.

Shapiro, R. 2000. A replicator was not involved in the origin of life. *IUBMB Life* 49: 173–176.

Shapiro, R. 2006. Small molecule interactions were central to the origin of life. *Q. Rev. Biol.* 81: 105–125.

Shay, J.A., C. Huynh, and P.G. Higgs. 2015. The origin and spread of a cooperative replicase in a prebiotic chemical system. *J. Theor. Biol.* 364: 249–259.

Szabo, P., I. Scheuring, T. Czaran, and E. Szathmary. 2002. In silico simulations reveal that replicators with limited dispersal evolve towards higher efficiency and fidelity. *Nature* 420: 340–343.

Szathmáry, E., and J. Maynard Smith. 1995. *The Major Transitions in Evolution*. Oxford, UK: Oxford University Press.

Szathmáry, E., M. Santos, and C. Fernando. 2005. Evolutionary potential and requirements for minimal protocells. *Top. Curr. Chem.* 259: 167–211.

Szilagyi, A., A. Kun, and E. Szathmary. 2014. Local neutral networks help maintain inaccurately replicating ribozymes. *PLoS One* 9(10): e109987.

Szostak, J.W. 2012. The eightfold path to non-enzymatic RNA replication. *J. Syst. Chem.* 3: 2.

Takeuchi, N., and P. Hogeweg. 2009. Multilevel selection in models of prebiotic evolution II: A direct comparison of compartmentalization and spatial self-organization. *PLoS Comput. Biol.* 5(10): e1000542.

Takeuchi, N., and P. Hogeweg. 2012. Evolutionary dynamics of RNA-like replicator systems: A bioinformatic approach to the origin of life. *Phys. Life Rev.* 9: 219–263.

Toppozini, L., H. Dies, D.W. Deamer, and M.C. Rheinstadter. 2013. Adenosine monophosphase forms ordered arrays in multilamellar lipid matrices: Insights into assembly of nucleic acid for primitive life. *PLoS One* 8(5): e62810.

Tupper, A.S., K. Shi, and P.G. Higgs. 2017. The role of templating in the emergence of RNA from the prebiotic chemical mixture. *Life* 7: 41.

Tupper, A.S., and P.G. Higgs. 2017. Error thresholds for RNA replication in the presence of both point mutations and premature termination errors. *J. Thoer. Biol.* 428: 34–42.

Vaidya, N., M.L. Manapat, I.A. Chen, R. Xulvi-Brunet, E.J. Hayden, and N. Lehman. 2012. Spontaneous network formation among cooperative RNA replicators. *Nature* 491: 72–77.

Wochner, A., J. Attwater, A. Coulson, and P. Holliger. 2011. Ribozyme-catalyzed transcription of an active ribozyme. *Science* 332: 209–212.

Wu, M., and P.G. Higgs. 2009. Origin of self-replicating biopolymers: Autocatalytic feedback can jump-start the RNA world. *J. Mol. Evol.* 69: 541–554.

Wu, M., and P.G. Higgs. 2011. Comparison of the roles of momoner synthesis, polymerization, and recombination in the origin of autocatalytic sets of biopolymers. *Astrobiology* 11: 895–906.

Wu, M., and P.G. Higgs. 2012. The origin of life is a spatially localized stochastic transition. *Biol. Direct* 7: 42.

Wu, S., C. Yu, W. Zhang, S. Yin, Y. Chen, Y. Feng, and W. Ma. 2017. Tag mechanism as a strategy for the RNA replicase to resist parasites in the RNA World. *PLoS One* 12(3): e0172702.

Zaher, H.S., and P.J. Unrau. 2007. Selection of an improved RNA polymerase ribozyme with superior extension and fidelity. *RNA* 13: 1017–1026.

Zintzaras, E., M. Santos, and E. Szathmáry. 2010. Selfishness versus functional cooperation in a stochastic protocell model. *J. Theor. Biol.* 267: 605–613.

6.3 Coevolution of RNA and Peptides

Peter Strazewski

CONTENTS

6.3.1 INTRODUCTION

6.3.1.1 SYSTEMS CHEMISTRY OF BIOMOLECULES

One of the most obvious relationships between biomolecular classes in living cells is the mutual interaction between nucleic acids and proteins. Of course, there is a plethora of interaction networks on the small molecule-metabolic level, and a mind-boggling spatio-temporal organization of all molecules and ions on a supra-supramolecular level of any living cell. These include "crowds" of membrane-affine macromolecules that are located in the vicinity of, or anchored to a surface of, the cell or some cellular organelle, all of which creates ion-conducting transport channels and three-dimensional (3D) spaces for the enrichment of crucial, chemically more or less labile metabolites (Spitzer and Poolman 2009; Spitzer 2011, see Chapter 6.6). However, all "nodes" of these metabolic transformation networks are kinetically strongly favored over other, in consequence, kinetically suppressed chemical reaction networks by proteins. A number of very slow chemical transformations are actually made possible by proteins, but many reactions are not nil without enzymes, only much slower, thence confronted with a much fiercer competition of unproductive "side reactions."

Nucleic acids are synthesized by proteins where the polymerization process is driven by nucleoside triphosphate (NTP) alcoholysis, that is, when pyrophosphate anions in triphosphoric anhydrides of NTPs are replaced by 3'-hydroxy groups of mono-, oligo-, or polynucleotides. In order to prevent the reversed reaction, the system makes sure that pyrophosphate is permanently hydrolyzed to phosphate. Proteins are synthesized by nucleic acids, and their production is *indirectly* dynamically controlled through RNA-dependent gene regulation and by catalytically active RNA. By far, the most abundant *direct* RNA catalysis is the ribosomal biosynthesis of proteins, being at its core a carboxyester aminolysis termed "peptidyl transfer" (PT) reaction (cf. Sections 5.2.1.3 and 5.2.3.2). The modern-day biosynthesis of proteins depends on the universal genetic code, also termed "primary code." The "secondary code" is much less straightforward but prerequisite for the primary code to be operative, since it warrants for the attachment of all proteinogenic amino acids to their "cognate" (appropriate) transfer RNAs. This process is catalyzed by 20 different amino acid-specific proteins, aminoacyl RNA synthetases (ARS), that all recognize and bind their set of "isoaccepting" transfer RNAs, but also require, bind, and hydrolyze a high-energy metabolite, adenosine 5'-triphosphate (ATP). Ribosomes require another high-energy metabolite, guanosine triphosphate (GTP), in order to proceed reasonably error-free in the "wanted" (functional) direction. The ATP is produced by the membrane protein ATP synthase and is driven by a proton concentration difference being maintained across the membrane. Other NTPs are produced from many different enzymatically catalyzed reactions, one of which is a phosphate exchange with ATP. Hence, the mutual control between nucleic acids and proteins through the genetic code is connected to all other actors of the living cell through high-energy metabolites that need to be constantly "replenished" (re-phosphorylated). Many small metabolites, irrespective of whether they are phosphorylated or not, are composed of a distinct and well-controlled number of reduced carbon atoms—reduced with respect to the oxidation number +IV of carbon dioxide being the thermodynamically most stable form of carbon in the presence of other elements.

The chemical reaction networks realized by all small molecules, that is, metabolites of higher or lower chemical energy, guarantee for a stepwise controlled management of the cell's

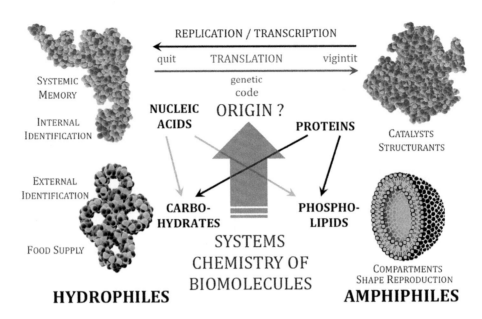

FIGURE 6.3.1 Four principal macromolecular compound classes, their interrelationships, and their main functions in living systems. Left: Hydrophiles. Right: Amphiphiles. Arrows indicate control by (begin) over synthesis of (end). Quit = quaternary digit (polymer); vigintit = 20 digit (polymer), cf. Section 5.2.1.3. All processes are coupled to metabolic cycles (not shown).

carbon atoms and energy household. These reaction networks help the cells to remain reasonably robust against changes and, to varying degrees, independent of the surroundings, including other cells and the non-cellular environment. The fundaments of both the cellular energy household and the intercellular communication are composed of smaller and larger carbohydrates, respectively. For example, the transformation of critical, and easily "digestible," monosaccharides is catalyzed by the enzymes of the Calvin cycle. On the macromolecular level, oligosaccharides in the form of glycoproteins and glycolipids serve as extracellular antennae by which cells are more or less roughly identified from the outside. Inside a cell, a very special kind of carbohydrate congener, the nucleic acids, serves to precisely identify any cell from within. The "splendid isolation" of the cellular inside from the environmental outside of any cell is provided by a semipermeable membrane composed of a phospholipid bilayer spherically closed into a vesicular shape; this *per se* quite flexible boundary is stabilized by cell wall carbohydrates ("glycans") and membrane proteins. The supramolecular aggregation property of relatively small and multicompositional phospholipids furnishes a direct physico-chemical means of spontaneous shape replication through a process termed "growth and division."

This extremely reductionist compositional and systemic view of the organization of any living cell holds true, without significant amendments, ever since Francis Crick and his comrades of the time—late 1950s—began to formulate and elaborate on the main message of the Central Dogma (Cobb 2017). A dogma of any variety and language, be it the "Central Dogma," "Dogme 95" (von Trier et al. 1995), or "Dogma 13" (Strazewski 2014), is by definition firmly linked to infallibility, being in turn, and also by self-definition, reserved for non-scientists and non-artists. Therefore, this specific wording from the mouth, feather, or paintbrush of a scientist or artist

needs, again by definition, to be taken with a dash of humor. Nonetheless, the "basic assumption" (Crick 1970) on the flow direction of information in a living cell persists like Einstein's Relativity Theory—with and without humor. The questions here are, how can such a systemic network of biomolecules, as a matter of fact, of any kind of molecules (Cronin and Walker 2016) arise spontaneously, that is, originate from "dead" matter? And how can any genetic code, universal or not, get established spontaneously by itself (Figure 6.3.1)?

6.3.1.2 METAPHORS VERSUS REAL THINGS

The above questions set us back to a definition problem, namely what do we mean by saying "dead or alive"? Erwin Schrödinger's dead *and* alive cat was originally meant to be of sub-atomic size, but twenty-first-century experimental physics and chemistry teach us that the "wave character," that is, the property of constructive and destructive interference, thus superposition of different states, can be clearly observed on the molecular and macromolecular size scales of objects. In principle, all nature should be seen as a "superposition of multiple states" being very delicate and easy to disturb—they readily "decohere" upon the slightest disturbance (Haroche and Wineland 2012). Physics does not give us other than probabilistic grounds, for which of all possible decisions is taken by nature when a superposition of multiple states decoheres. It is pure chance that decides about the fate of things coming into being, that is, becoming observable and "real."

Let us use the above metaphor of the superposition of a dead and alive cat. From a human observer's point of view, nature is gaining momentum, far from probabilistic, as time goes by, from inanimate (inert) to animate (alive) and from abiotic to biotic, provided the conditions such as temperature, pressure, energy source, and fluctuations are appropriate. Therefore,

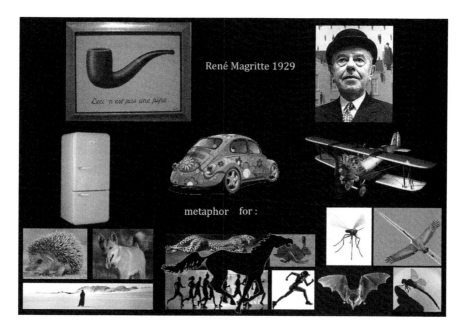

FIGURE 6.3.2 The difference between models of the reality and reality itself. *Ceci n'est pas une pipe* (Magritte 1929) translates into: "This is not a pipe," meaning "you cannot smoke with this pipe." The fact that the photographic image of René Magritte (upper right) cannot and did not paint this picture (upper left © ADAGP, Paris and DACS, London 2018) should be neglected for the sake of the argument (see text). Creating models of the reality (machines) is not the same as creating the reality. (Source: Reproduced with the permission of Artimage DACS © ADAGP, Paris and DACS, London 2018.)

I should like to illustrate here how I usually introduce to my students the difference between dead and alive (Figure 6.3.2). It is based on the vision of René Magritte, a Belgian painter of the early twentieth century (1898–1967), who unequivocally illustrated that we should not confuse a metaphor with the reality.

The metaphor that I am using to get to my point is that, if inanimate matter is a purposefully designed metaphor, then animate matter is self-emergent reality. For instance, a fridge, a mechanically reasonably solid and hopefully long-lasting electric machine, is an *inanimate model* (a metaphor) for any warm-blooded animal—like a desert hedgehog, a dingo, or a person that lives in a hot desert—specialized in transforming chemical energy into exported heat, which is closely linked to a "function," that is, the purpose of keeping an interior comparingly cool and thermostated, when the outside is ferociously hot. Animals are long-lasting, too, albeit thermodynamically sometimes less stable than many man-made machines. The difference between the "model" and the "real thing" is that you need somebody from the outside to reproduce a fridge, while animals reproduce themselves, without any external designer. Reproduction of itself is the way of how thermodynamically less stable but alive "fridges" stay long-lasting. Individually, we may be short-lasting, but we self-reproduce and our *genre* is long-lasting. The same concept and difference can be applied to other functions and machines, such as motor cars being metaphors for fast running-animals and airplanes for flying animals and so forth. One necessity of this fundamental dichotomy is that, probably (unproven but intuitive), self-reproducing matter cannot be, at the same time, thermodynamically very stable, or else, it could not be dynamically self-evolvable and thus could not adequately and rapidly adapt to changing surroundings. When sometimes, challenged by the odd student, I must respond to the question: "What about self-repairing robots?"—note that to "re-pair" could mean to "re-make one (whole) out of two (parts)"—I usually answer: "Show me a robot that self-reproduces by somehow giving birth to at least one similar but slightly different and also self-reproducible baby robot—hence, a robot that makes at least two out of one—and I'll be happy to call her alive."

The main reason for the above, perhaps bizarre, introduction into the concepts of life is the important fact that we humans, scientists, and *chemists* just *adore metaphors*. We are not listening to René Magritte and are definitely in danger of confusing models with the reality.

6.3.1.3 Synthetic Cells versus Abiogenesis

Of course, there is a very good reason for why many scientists prefer to produce and study models. It is much easier to understand the functioning of a good model than to decipher, for example, all the mechanisms taking place in a living creature or even unravel a few mechanisms being part of a much more complex chemical system. Sometimes, it is clearly impossible to do otherwise. A cosmologist or experimental particle physicist won't desire—perhaps only in his or her wildest dreams—to create a real, not *in silico*, "synthetic universe." Given our fervor for an early "inflationary," spontaneous and sudden cosmic expansion of the total space volume over at least 26 orders of magnitude, it may be too dangerous to do so at this stage of research. There are physicists that would like to produce, in a particle accelerator, small experimental mini-black holes, which should be very

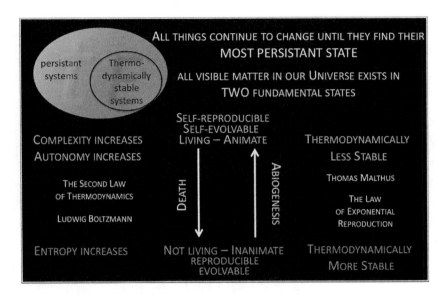

FIGURE 6.3.3 Two fundamental states of visible matter, their interrelationships, and their driving forces in the physical reality. Upper left: There are more ways of persisting over a long period of time than being thermodynamically stable (Pascal and Pross 2015). Absolute qualities, such as "long," "short," "high," "low," and so forth, are avoided and replaced by relative descriptions, such as "increase," "decrease," "more," and "less."

short-living, if possible at all. Usually, cosmologists and particle physicists hardly ever confuse a model with the reality; too different are the looks of both.

Things look different on a less energetic scale of experimentation. Living individuals die relatively easily. From what we know about biotic nature on Earth, we can be quite certain that the same holds true for any extraterrestrial life within our universe. As mentioned earlier, the persistence of animate matter over long periods of time is based on the strategy to self-reproduce, provided the surrounding conditions permit productive (fertile) self-reproduction. This strategy is not based on thermodynamic stability, quite on the contrary (Figure 6.3.3).

Evidently, a piece of cloth, wood, coal, granite, iron, or lead is less persistent than a slowly radiating black hole; so is a barrel of petrol when compared with the same amount of water. Yet, all are thermodynamically quite stable, can persist due to their stability over long periods of time under reasonably isolated conditions, and are, despite their very different provenance, definitely inanimate. The "lifetime" of inanimate matter should be compared with that of a living species, not with an individual.

Two open questions arise from this holistic view of physics, one is the difficulty of quantification of complexity, autonomy, and the precise lifetime of a species when compared with that of thermodynamic stability and entropy of inanimate matter. The main reason for this difficulty is the lack of any reference system for animate matter, owing to the permanent change of both organisms and surrounding environment.

The other open question is the obvious disparity between the ways of transformation. Although the mathematical rules that drive both ways (arrows in Figure 6.3.3) are known since long (Malthus 1798; Boltzmann 1877), experimentally, only one way, the transformation from the animate to the inanimate states, death, has been and is being explored and studied many times. The other direction, abiogenesis,

still awaits experimental proof. My personal conviction on the importance of truly knowing both ways of transformation is the ultimate reason for writing this book chapter. I think that abiogenesis reconstructed by humans would not only be the greatest-ever achievement in the history of the natural sciences. Also, it is a necessity for modern society to be able to make a great leap forward in its development and a responsibility for ourselves and for all non-humans, including inanimate matter (see also Section III of this handbook). We humans need to show it to ourselves by experiment that life can emerge from inanimate matter. Maybe this is not really the place for personal opinions of the non-scientific kind, but, since it may be the only occasion for me to write down this opinion, let me state here that, in a society where experimental, that is, synthetic abiogenesis will be known for let's say two generations, our grand-grand children won't be astonished anymore about how actually quite easy it is to animate from the truly inanimate and how extremely delicate such an emergent synthetic life is—given the enormous extant biotic competition and the lack of eons of self-evolution for the establishment of a new biosphere. Concerning exobiology and astrobiology, creating, on purpose, life from the inanimate might be an option to survive in some form elsewhere. Most importantly for us, in the not-so-far future, when a majority of human society knows about this possibility and can now and again remake life in laboratories—outside, I believe, our biotic life is too strong to let any new life persist—the major reason for religious wars will become naturally extinct and definitely obsolete. Other ways of solving problems will impose themselves; this archetype will become history, and religion will gain the importance and character of taste: extremely important in everyday life ("I like marmalade," "I hate oysters") but no reason at all for violent behavior against others ("You must hate oysters or I'll kill you"). This should change the world quite a bit.

I am not a philosopher, but it seems quite evident that at least the lucky part of society that has rejoiced over two or more generations the absence of wars on its territories is, by virtue of a highly and globally developed technology, at the verge of a "big leap," not unlike the ancient Greek society more than 2400 years ago, when the first European philosophers (Parmenides, Aristotle, Theophrastus, Diogenes, Socrates, and Plato) wrote about the reality; not unlike during the Renaissance, when the need of careful and analytical dissection of corpses emerged from pure curiosity (Leonardo da Vinci), became first allowed and then established (Andreas Vesalius), and gave the grounds for the development of modern medicine; and not unlike during the era of Enlightenment, when human curiosity and the confidence in oneself had been encouraged against current opinion for a long-enough time (Francis Bacon, René Descartes, John Locke, and Baruch Spinoza), so that the construction of machines could gain significant momentum, which paved the way to modern technology. These three phases can be roughly circumscribed and extremely reduced to "observation" (meaning mainly theoretical analysis), "analysis" (meaning experimental analysis), and "synthesis," respectively, where each activity is prerequisite for, and permanently accompanies, the next. Of course, there are anxieties; of course, we make mistakes. One mistake that we should work out a little bit more is related to our natural desire to control the design as much as possible, while not being able to control the outcome in a way that would correspond to the effort put into the design. Highly complex, especially animate systems are alas! highly non-predictable as well.

My temper was sometimes violent, and my passions vehement; but by some law in my temperature they were turned not towards childish pursuits but to an eager desire to learn, and not to learn all things indiscriminately. I confess that neither the structure of languages, nor the code of governments, nor the politics of various states possessed attractions for me. It was the secrets of heaven and earth that I desired to learn; and whether it was the outward substance of things or the inner spirit of nature and the mysterious soul of man that occupied me, still my inquiries were directed to the metaphysical, or in it highest sense, the physical secrets of the world... (Wollstonecraft Godwin Shelley 1818, Chapter 2)

He held up the curtain of the bed; and his eyes, if eyes they may be called, were fixed on me. His jaws opened, and he muttered some inarticulate sounds, while a grin wrinkled his cheeks. He might have spoken, but I did not hear; one hand was stretched out, seemingly to detain me, but I escaped and rushed downstairs. I took refuge in the courtyard belonging to the house which I inhabited, where I remained during the rest of the night, walking up and down in the greatest agitation, listening attentively, catching and fearing each sound as if it were to announce the approach of the demoniacal corpse to which I had so miserably given life. (Wollstonecraft Godwin Shelley 1818, Chapter 5).

To end this introduction, and to round up what was unwound in seeming digressions, let us take a brief look from a distance at the research strategies that synthetic biology, systems biology, and systems chemistry have chosen in the past 15 years.

Much has been undertaken by following a top-down approach in building more and more complex *models* of cells that are capable of producing designed components through designed genetic-control circuits. Such artificial cells are either very close to natural living cells or stripped off their non-essential genes to survive as so-called minimal cells, but they are not living for many generations. The top-down approach of synthetic biology is considered as designed assembly of artificial cells from components of once-animate origins.

A pioneering vision for the creation from bottom-up of living synthetic cells was published by Szostak, Bartel, and Luisi in 2001. The authors proceeded from designing a protocell, providing it with a synthetic RNA replicase and furnishing the RNA with the ability of catalyzing the synthesis of vesicle membrane components:

The first challenge on the path to a synthetic life form is to imagine a collection of molecules that is simple enough to form by self-assembly, yet sufficiently complex to take on the essential properties of a living organism... We believe that within this framework structures can be found that are both indisputably alive and yet simple enough to be amenable to total synthesis. We note that solutions found in the laboratory need not be chemically similar or even directly relevant to the actual molecular assemblies that led to the origin of life on Earth [...] The first major synthetic intermediates are an RNA replicase and a self-replicating vesicle [...] Such simple protocells would be nearly, but not quite, alive. When fed small-molecule precursors for membrane and RNA synthesis, they would grow and divide, and improved replicases would evolve. However, a vesicle carrying an improved replicase would itself not have improved capacity for survival or reproduction. For this to happen, an RNA-coded activity is needed that imparts an advantage in survival, growth or replication for the membrane component. A simple example would be a ribozyme that synthesizes amphipathic lipids and so enables the membrane to grow. (Szostak et al. 2001)

This bold plan seems to be rooted in creating an autopoietic system (Maturana and Varela 1980) with implementations of the concept of the chemoton (Gánti 1971, 2003) and notwithstanding the hypothesis of an "RNA World." More than 15 years of experimental research after this idea have taught us to be critical and to constantly re-think the "best" initial conditions of such an endeavor (Blain and Szostak 2014; Buddingh and van Hest 2017). In particular, it seems now that even this inventive bottom-up approach probably bears too much human design and is reaching too far into some desired future that might prove different from what we would expect at a first sight.

It is still true, of course, that, in order to be conceived as "alive" in a conventional sense and consensus, protocells have to contain some kind of readable information-carrying "memory polymer" (Section 5.2.1.3), parts of which, at least, need to be reproduced, that is, their sequence of digits "replicated" and spread over the dividing cells. However, there is no need for self-replication, as beautiful and "simple" such a property appears. It is perfectly acceptable, when the memory polymer is replicated by some other molecule(s), even by some other molecular class of compounds. What counts is the mere replication rate k and spreading of a set of "sequence of digits" upon every cell division, resulting in an exponentially growing ($p = 1$) population density x with time t (Equation 6.3.1).

$$dx/dt = k \cdot x^p \qquad (6.3.1)$$

Exponential growth rates are mandatory for a so-called Darwinian evolution to set in, according to which different reproducing populations compete with each other when sharing the same ecological niche, thus competing at least partly for the same resources. Under initial conditions, when the resources can be considered unlimited and the degradation, decay, or individual "death" rates are negligible, a faster replicating population can completely outcompete and make extinct a slower-replicating population only if the growth rates of reproducing populations are exponential or higher than exponential $1 \leq p \leq 2$ (Equation 6.3.1), that is, hyperbolic with time (Lotka 1921; Volterra 1938; Eigen and Schuster 1979; Szathmáry 1991; Szathmáry and Maynard Smith 1997; von Kiedrowski and Szathmáry 2012; Szilágyi et al. 2013, 2017). Under subexponential initial (resource-rich) growth conditions $0 < p < 1$ (termed parabolic), a "survival of everyone" regime is sustained and no Darwinian selection is possible. The selection rules in these three growth order regimes are independent of the stoichiometry of growth (1 parent gives 1, 2, 3 or more children) and include changes in the fitness landscape (fertility, health). Exponential growth order means that, overall and on the average, every new generation produces the same number of offspring as the previous generation per parent. Hyperbolic growth, such as the human race is supporting right now, is when every generation gives off, on the average, more fertile progeny than the previous generation per parent. Parabolic growth means, of course, that the brood becomes on the average fewer and fewer per parent. Enzyme-free template-directed (e.g., RNA) replication follows parabolic growth dynamics, owing to the fact that newly grown strands "scavenge" their complementary single strands through stable base pairing, thus inhibiting somewhat their templating ability. The extinction of a sub-population within competing parabolic replicator systems can only be caused by a below-critical abundance (concentration or density), sequence-selective decay rates, or a separation into different ecological niches, for example, a complete spatial separation of the competitors. In that case, the absence of a competitor in a given area is of course not due to its extinction. Irrespective of whether systemic memory carriers, such as RNA, are self-replicating or being replicated by someone else, an inanimate mixture of a great many of different compounds and different compartments must first somehow segregate, "demix," and diminish its *chemical diversity*, or else, a living state of matter will never be achieved. Conversely, once early replicators, that is, chemical systems that are driven by the doubling of memory polymers, have become operative, the coexistence of populations of different replicators is prerequisite for the maintenance of a sufficient amount of *inheritable diversity* of, at first, highly mutable (unstable) digital information, or

else, the survival of such replicators under early biotic environmental conditions can hardly be sustained over long periods of time and many generations (Szilágyi et al. 2017). One of the current difficulties in modeling evolutionary population dynamics is to properly outline the scope of "selectability" of replicators, the emergence of Darwinian selection through the extinction of competing sub-populations of coexisting replicators while maintaining the survivors stable, for example, stable against parasites, yet still evolvable over space and time in the sense that the survivors may integrate more different replicating (memory) polymers, without making the whole system collapse. It turns out from the research of the past decade that spatially explicit systems of cooperating replicators that are irrevocably coupled to ("fed by") metabolic reaction networks are incomparingly more robust than replicators devoid of metabolism.

However, a memory polymer that cannot carry out or give rise to a function that is of use for the survival of the whole system, is not much of a memory polymer. This is the reason why the RNA World hypothesis was so attractive over a long period of time. Experiments on the "ribozyme" activity of many RNA molecules have shown a plethora of different catalytic promises of this compound class. When we look closer at these fantastic experimental achievements, including ribozymic RNA copying, aminoacylation of RNA, and limited (without turnover) aminoacyl transfer, some of us still cannot help scratching our heads. Synthetic catalytic RNA still seems to be best at catalyzing the cleavage of RNA "in *cis*" (intramolecularly, self-cleaving) and second-best, when appropriately designed, at catalyzing cleavage "in *trans*." Intermolecular RNA strand cleavage works only upon sufficient base-pair-stabilized hybridization between the ribozyme and the substrate RNA. All other ribozymic catalyses, such as the reversed ligation reaction or the above-mentioned (trans)aminoacylation ribozymes, are comparingly less well performing. The typical catalytic rates of RNA self-cleavage that can be achieved *in vitro* with synthetic ribozymes are worrying, since the rate constants at 25°C under optimized conditions, that is, high divalent salt concentrations, including magnesium, manganese, and calcium salts, are in the range of one cleavage event per hour to one event per week, per molecule of ribozyme.

The chemical nature of RNA and similar nucleic acids—the facts that they are periodically polyanionic and very well water soluble at virtually any nucleotide length and that they do fold upon themselves quite readily without the need of a complementary strand, as in double-stranded DNA, is advocating against the RNA World, meaning against a self-contained functioning and copying of nucleic acids, without the need of peptides or proteins. Nucleic acids are good at hybridizing with nucleic acids. Nucleic acids are bad at "denaturing," that is, unfolding nucleic acids. Hybridizing and folding through base pairing is the opposite of denaturing and unfolding. But, to be amenable to both intermolecular "cross" catalysis and RNA strand copying, unfolding and exposing its nucleotide sequence is prerequisite. No unfolding, no readout—"no" may be replaced with "slow." In extant biotic systems, all nucleic acid denaturing work is executed by proteins. Most of these proteins are "helicase" or "unwindase" enzymes of some sort. However, not only highly evolved and well ordered (folded) enzymes are known to unfold nucleic acids. An intrinsically disordered protein assembling in organelle-like liquid droplets has been identified as being able to efficiently stabilize single-stranded DNA and RNA (Nott et al. 2016). There is only one exception to this unique dehybridizing property of proteins. The binding of messenger RNA and exposure of two codons for peptidyl- and aminoacyl-tRNA anticodon readout is carried out by the "decoding center" located at the "neck" of the small ribosomal subunit termed 30S subunit in bacterial and 40S subunit in eukaryotic ribosomes (Demeshkina et al. 2012). The decisive molecular contacts between the mRNA codons and the 30S/40S subunit are mediated by ribosomal RNA, 16S rRNA in bacteria and 18S rRNA in eukarya. However, when mRNAs are partly self-folded, in spite of the degeneracy of the genetic code, which allows for a certain choice of nucleotides that all code for the same protein, then these folded parts may constitute a problem for the ribosome—ribosomal "stalling" during translation is a commonly observed phenomenon. It is of note that the main catalytic function of RNA is executed by folded tRNA and folded rRNA (see also Chapter 6.4), whereby the covalent chemistry of peptide bond formation is carried out at the single-stranded 2′,3′-terminus of tRNA. It is also of note that the ribosome depends on many ribosomal proteins; 52 bacterial and 82 eukaryotic different ribosomal proteins enhance its structural integrity, and even more proteins are needed for its proper functioning. Not only are all ARS enzymes proteins, so are the "elongation factors" that make sure that hydrolytically labile aminoacyl-tRNAs are safely transported from their ARS to the ribosome and are reliably selected for the cognate codon. The ribosome will only produce proteins in a processive manner when it always moves "translocates" forward and never backward, which would otherwise catalyze the cleavage of the nascent peptide bonds, instead of their synthesis. This translocation is guaranteed by a protein (an elongation factor) that mimics the shape of a tRNA being bound to another elongation factor. The correct final disconnection of the nascent protein from the last tRNA can only occur thanks to protein enzymes, "termination factors." Most of these translational protein enzymes are binding and hydrolyzing GTP as a thermodynamic driving force. The attachment of whole ribosomes to the lipid membranes again is mediated by proteins, through which the nascent protein moves to the other side of the membrane of the corresponding organelle, the "rough" part of the endoplasmic reticulum (ER). There, the nascent proteins are most often captured and isolated from their immediate environment by protein chaperones called "chaperonins," until they slowly but safely fold into their "native" conformation being usually required for their function, and so, they won't be degraded too early by protein-degrading proteins. This cellular work and mutual dependence between RNA and proteins seem so extremely interwoven and intimate that it is really hard to imagine how one compound class could ever do without the other (Carter 2015). Any purist vision of a "protein-only" or "RNA-only" or, for that matter, "anything-only" world looks like an act of desperation for order rather than a theory based on experimental scientific work.

In addition, we should take into account a realistic prebiotic situation, on the early Earth or elsewhere, where amino acids are always chemically more stable and more persistent than nucleotides. It is very likely that, at an early stage of molecular complexification, both prebiotic synthetic pathways shared the same chemical transformation routes, namely, from the first C-C-bond-forming reactions to the point where α-aminonitriles diverge to α-amino acids and α-hydroxyaldehydes, that is, small carbohydrates (cf. Section 5.2.2). If this is the case, a pure or even less pure RNA World seems even more improbable. Instead of stubbornly clinching to this sharp "RNA World blade" that is swimming over deep waters, we should take much more care of how any spontaneous RNA-catalyzed synthesis of peptides from the amino acids can emerge, in order to get hold of useful primordial functions provided by such first peptides. We should try to convince ourselves by experiment that it might indeed be possible to let first self-evolve a kind of primitive RNA-driven protein or peptide production before the advent of RNA replication. Given the initial absence of any systemic memory that would require RNA replication, a protocellular approximate (error-prone) reproduction would indeed ensue the provision of the same RNA molecules over long time periods and many protocellular generations over and over again and of the same amino acids and other small molecules as well. But no dogma forbids this highly "heterotrophic" scenario devoid, at first, of any systemic memory. Compartmented protometabolic reaction cycles thus gain importance, not because they provide the first building blocks—being the essence of an "autotrophic" origin of protocells—and *not yet* for the dynamic stabilization of cooperating parabolic replicators (*vide supra*), but for the upkeep of a sufficient level of compartment chemical energy, in order to maintain good growth conditions, for instance, for compartmented peptides from internal aminoacyl-RNA. The upside of this "no memory-scenario" is that, once RNA-driven peptides would fold into structures capable of unfolding RNA and generating hydrophobic "pockets" around short parts of single-stranded RNA, then an RNA-dependent RNA polymerase activity would be within close reach, if chemical energy from, for example, NTPs, could be coupled into this process. When such a mutual protein-RNA cross-catalysis—peptides from RNA catalysis, RNA from protein catalysis—emerges, then eventually, a huge acceleration of population growth of compartments that would host these activities is expected to literally shoot off exponentially from the throng of no-memory compartments.

Finally, it is true that a point must be reached where the membrane components should be synthesized by the system itself (Gánti 1971, 2003; Szostak et al. 2001), in order to become reasonably independent of external lipid sources. However, to join glyceryl monophosphate with fatty acids, or phosphate-containing polar headgroups with diacylglycerols, or any such dehydration reaction that would generate a strong amphiphile, is by far easier to catalyze in water than to actually polymerize methylene groups (CH_2) into long hydrophobic chains, which is at the heart of any membrano-genic amphiphile. A minimal length of the hydrophobic zone in an amphiphile, for its stable assembly into bilayer membranes that could close into vesicles, is at least 9 or 10 carbon atoms long. In a biotic system, this C–C bond-forming oligo-merization process is carried out and cycled, for a fatty acid chain length of C_{24}, up to two dozens of times consecutively. The fatty acid synthetase machinery is huge in size (2.6 MDa, 260×230 Å, 100% protein), slightly larger than the bacterial ribosome (2.5 MDa, 250 Å diameter hemisphere, 65 weight % rRNA and 35 weight % ribosomal proteins) and is highly elab-orated, cooperative, and genetically streamlined to make sure that a controlled "homeostatic" mixture of different phospho-lipids and glycolipids is provided from within a cell to feed into its membranes.

There is thus far no chemically reasonable enzyme-free process known to work under conditions that would synthesize membranogenic hydrophobic carbon chains *and* support the presence of RNA. Fischer-Tropsch-type processes, which gen-erate in water carbon chain lengths of the required size, neces-sitate temperatures between 150°C and 250°C (McCollom et al. 1998). Of course, once the carbon chains are "heterotrophi-cally" fed from the outside, a protein or even ribozyme might be able to attach through condensation, that is, dehydration, one or two alkanols or alkanoic acids to glycerol or glycerol phos-phate. But the extant degree of autonomy, that is, independence of macromolecular food from the outside, can, for chemical reasons, only be a late "Darwinian" achievement, as far as our current knowledge stands now. Unless, of course, some bright chemist will come up with a mild fatty-acid-like enzyme-free synthesis from acetate C_2 and CO_2 to give malonate C_3, that reacts with acetate to produce back CO_2 and acetoacetate C_4, that gets reduced to β-hydroxybuturate, then dehydrated and reduced to give butyrate, and so forth: $C_4 + C_3 - C_1 = C_6$; $C_6 + C_3 - C_1 = C_8$; $C_8 + C_3 - C_1 = C_{10}$ etc., perhaps in the presence of peptide-stabilized iron-sulfur clusters (Bonfio et al. 2017), yet still in water and, most importantly, without hydro-phobic aggregation *en route* that would stop the whole process before it formed long enough chains that could assemble into membranes. They say that miracles happen, but I would not bet on it (Fry et al. 2017), but, of course, never say never.

To conclude on this introduction, I tried to show that the times are ripe for chemists to investigate the transformation of matter from the inanimate to an animate state by labora-tory experiment. Such experiments should not be confused with the production of artificial or synthetic cells. While the latter efforts are extremely important for us to understand the detailed mechanisms that take place in living organ-isms, it is highly unlikely that we can generate through this "high-design" approach actually living and self-evolving cells *de novo* that could survive over many generations in an appropriate environment being well isolated from extant biotic life, on some other habitable planet, for instance, or in the laboratory. For reasons that are beyond mere scientific curiosity, which include the belief in full human responsi-bility over all domains of our perceived reality, we should convince ourselves that it is indeed possible to create life

truly *ex inanimo*. For this endeavor to be successful, we cannot keep patching up and energizing "dead but not yet degraded" formerly living material, or else, we will not overcome the "Frankenstein threshold," below which no reproductive self-evolution can emerge. We need to start from conditions that are less complex and preclude far-less-designed properties that would supposedly provide "func-tions." However, it is true that we cannot start yet another historical origin of life beginning either with metal oxides, graphite, sulfur, phosphorous, hydrogen, and nitrogen gases at temperatures above 2000°C, or even with carbon monox-ide, formaldehyde, acetylene, ammonia, hydrogen cyanide, sulfide, phosphate, and water between 1000°C and 100°C (cf. Section 5.2.2.1). We simply do not have this kind of timescale at our disposal.

A major issue is, therefore, to figure out by experiment what kinds of initial functions, that is, a minimal set of functions, are needed to make and keep a complex chemical mixture alive. In my opinion, we should not assume *a priori* that *replication* belongs to the minimal set of functions, even less so the self-con-sistent synthesis of cell membranes. Both functions are indeed absolutely required for an autopoietic, self-reproducing state of matter, but they still could be *emergent rather than prerequi-site*. So, what are the functions that can be considered to belong to the minimal set of a self-reproducing system? This question can be boiled down to the minimal autonomy needed for a sustainable complex chemical compartmented mixture devoid of any systemic memory. What kind of complexity of matter needs to be inanimately provided long, constantly, or regularly enough for a systemic memory to emerge? Let us consider a working hypothesis that states that, when bits and pieces of oligonucleotides and oligopeptides can be kept compartmented under conditions that allow them to grow in chain length, thus sequential diversity, then, after some time shorter than the aver-age life time of a researcher, a translational code will emerge between the two classes of compounds that will keep particu-larly useful peptide and RNA sequences more abundant than the background abundance carried over from dividing mother cells to daughter cells. This, in turn, will keep some of the compartments that harbor such "selected" RNA and peptides more and more persistent and abundant, when compared with the other compartments that do not. There is growing experi-mental evidence that peptides and nucleic acids—those that do interact with lipids—can render membranes more resistant and robust against mechanical and osmotic stress. Lipophilic pep-tides, which may form spontaneously without any instruction or catalysis by RNA, are quite likely to firmly interact with lipidic vesicles, whereas the lipophobicity of many kinds of nucleic acids, most prominently RNA, precludes any significant inter-action with lipid membranes (cf. Section 5.2.3.2), unless consid-erable human intervention and design are involved (Kurokawa et al. 2017). Therefore, one primordial function prerequisite for not replicating but *quasi* self-reproductive protocells might be the selection of those peptides, out of the astronomical number of abiotic oligopeptide sequences, that are able to import RNA into the interior of lipidic vesicles and perhaps anchor them to the inner surface of the membrane. Since vesicles that harbor imported RNA and lipophilic peptides are not yet alive and do not yet harbor replicating RNA, the terms "function" and "selection" are a bit of a stretch. But the experimental abiogenesis is a bit of a stretch, too, between inanimate chemistry and animate biology.

6.3.2 EXPERIMENTAL ABIOGENESIS

Most important in this experimental setup is the fact that we start with a mixture of chemically different compounds, rather than beginning with merely one kind of compound, for example RNA, and only one kind of amphiphile, or perhaps a uniform and rigid physical barrier, where the compound can evolve in micro- or nano-compartments. (Strazewski 2015, p. 853)

The experimental conditions need to keep the macromolecules, RNA and peptides, under hydrolytic stress and, at the same time, the compartments under mechanical and/or osmotic stress, for example. This could be achieved when giant lipid vesicles (GVs) are kept on a membrane filter that can be dipped into an aqueous phase containing any kind of molecules and into which any kind of solid or aqueous liquid solution can be added from above (Figure 6.3.4). In addition, the population of GVs should be submitted to changing temperatures and atmospheric conditions and therefore be placed in a convenient incubator.

6.3.2.1 DYNAMIC COMPARTMENTS

Above everything else, always provide the system with many possible variants—variants in chemical composition, property, reactivity, but also in shape and size—so it can choose whatever is best in a given situation or phase of development, "best" being whatever leads to longer persistence of the vesicles over time. With respect to the extent of replicative property at the beginning, I think that we can start with virtually none whatsoever. We should not provide a system with ribozymes that already know how to copy certain RNA sequences once

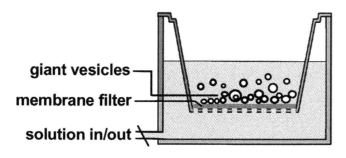

FIGURE 6.3.4 Fueling and feeding giant vesicles. Schematic drawing that should allow for (a) the growth and division of giant lipid vesicles when fed with lipid mixtures, and (b) growth of biomolecules that are encapsulated, when fed with the corresponding monomers and fueled with large amounts of dehydrating agents that can pass through the membrane. Stresses imposed on the system can be any leakage of molecules from the inside, osmotic pressures, shearing forces due to filtration, pH difference between the inside and outside of the vesicles, high salt concentrations, and so forth, in the absence or presence of oxygen and/or carbon dioxide gases at varying temperatures and humidity.

or twice, and try to evolve replicators from there. It suffices to start with the most simple physico-chemical replicative property, the one of spontaneously reproducible shape. Lipid-like amphiphiles do organise themselves into compartments, and other supramolecular architectures, spontaneously. When the compartments are fed with more lipids, they reproduce more of those shapes already being present. Grow and divide is at the same time an intrinsic physico-chemical property of lipidic supramolecular architectures (vesicular, hexagonal, cubic, etc.) and an ubiquitous cellular behaviour in the living domain." (Strazewski 2015, p. 853)

6.3.2.2 FUEL

A chemical energy gradient of the maximally atom-sparing kind is needed for keeping the system environment locally out of thermodynamic equilibrium. Through compartmentation we seek to economise on all macromolecules, viz. avoid any "organised waste" of large elaborate compounds including the added amphiphiles that make up the compartments of the system. We prefer to recycle them. But we shall not compromise on the delivered amounts and waste of small-molecular-weight-high-energy compounds, the fuel that keeps the system running. Such fuel should show a prebiotically plausible identity, in other words, should be composed of geologically and geochemically feasible minerals and compounds. They should not be too reactive, not too destructive when brought into contact with the macromolecules and compartments of the system. Basically, they should be "anhydrides" of some sort (not to be understood uniquely as organic chemical anhydride functional groups) that mostly hydrolyse into small diffusible waste products. Only mostly, because the residual fraction of fuel compounds, the one that will not react directly with water but rather eliminate water molecules from organic compounds, shall be the motor of the self-evolving process, a motor of quite low energy conversion efficiency at first. But this motor, too, should be able to self-evolve. (Strazewski 2015, p. 853)

6.3.2.3 FOOD

"We shall feed the compartments with more than lipids and chemical energy: with amino acids, peptides, nucleosides, nucleotides and" carbohydrates, the *ad hoc* assumption being *"that all these compound classes were available from natural prebiotic reservoirs and could locally accumulate at the same time and space for certain time periods once water was present."* (Strazewski 2015, p. 854)

In view of the stunning success of the first enzyme-free protometabolic reactions in water, catalyzed by iron, zinc, and other transition metal salts (Muchowska et al. 2017), compound mixtures belonging to the Krebs cycle, that is, acetate, pyruvate, oxaloacetate, malate, fumarate, succinate, α-ketoglutarate, oxalosuccinate, isocitrate, aconitate, and citrate, could also be periodically fed. We should keep in mind that too high transition metal concentrations may be detrimental to the GVs but that a limited amount of this kind of "chemical stress" might permit to generate more robust populations. One way or another, protometabolic networks need to be coupled into the system (Bonfio et al. 2017).

6.3.2.4 RNA-Directed Peptide Synthesis in the Lipid Bilayer

At this point, usually, or let's say quite often, some of my colleagues begin to have serious scientific doubts in the feasibility of such experiments—not to say, they are probably rolling their eyes behind my back. Especially in grant applications, anonymous reviewers are asking for reasons why RNA-directed peptide synthesis should emerge from the experiment? For an obvious purpose, I try to satisfy such longings, although I think that the main reason for myself is not explicable on a molecular level. Any chemical rationale should not evidently contradict any solid knowledge on molecular chemistry and, of course, physics. I imagine that if Harold Urey and Stanley Miller had been asked, in advance of their experiments, why they think that α-amino acids would form from methane and ammonia when submitted to spark discharges, they would, maybe, not argue with their real main reason—in the sense of motivation—but rather come up with the possibility of transiently formed aldehydes and ammonium cyanide undergoing the Strecker reaction. Who knows.

It is of course an open question how exactly RNA can start producing peptides from chemically activated α-amino acids without instruction from a proto-messenger RNA and without catalysis by a proto-ribosome. One of my own imaginations on how on a molecular level this could come about was described in considerable detail in Strazewski (2015). That "mechanism" replaces some of the function of a future proto-ribosome with a lipid bilayer and therefore places the primordial PT reaction two-dimensionally bounded and reversibly immobilized to the inner part of vesicle membranes. The immersion of the reactive atoms, that is, the nucleophilic α-amino group of an aminoacyl-RNA and the α-carboxyester function of a close-by peptidyl-RNA, into the hydrophobic phase of a lipid membrane should strongly enhance the reactivity of the locally desolvated reactive sites and lead to spontaneous and sequential aminolyses of ribonucleotidyl ester bonds to give peptide bonds. Since this mechanism is biomimetic but purely putative, I would like to direct the interested reader to this article and concentrate here on what arguments and thoughts have found their way into this hypothesis since 2015, albeit all of it still stands on purely hypothetical grounds and absolutely needs experimental testing.

Hence, despite early successes in the spontaneous aminoacylation of nucleoside-5′-phosphates (Biron et al. 2005), the last few years have shown that the C-terminal esterification of α-amino acids with ribonucleotides does not lead to peptide bonds that are in any way directed by RNA. The chemically activated amino acids do get esterified with nucleotides but re-hydrolyze too rapidly and, during their chemical activation, form peptides independently of the nucleotides (Liu et al. 2014). On the other hand, plausibly prebiotic reaction conditions have been found, where chemically activated α-amino acids react with nucleoside-5′-phosphates to form N-terminally linked nucleotidyl amino acid phosphoramidates (Jauker et al. 2015; Griesser et al. 2017). What's more, such nucleotidyl amino acid phosphoramidates grow peptide chains and, comparingly slowly, short RNA chains on this first "composite monomer." Experimentally, thus far, the access to N-terminally linked peptido-RNA is more readily achieved than to the more biomimetic C-terminally linked peptidyl-RNA. The current efforts concentrate now on finding reaction conditions that permit the rearrangement from peptido-RNA to peptidyl-RNA. Second, the evolution of a RNA-driven PT reaction at protocellular membranes might well be accompanied by the growth of unbound RNA chains within the vesicles' lumen that could step by step take over the hosting role of the membrane in promoting the PT reaction. Thus, a proto-ribosomal RNA activity could emerge and considerably accelerate the membrane-associated PT reactions. However, these are all hypotheses, not yet confirmed by experiment. The literature is full of unproven hypotheses, and the above hypothesis is only one more. This is why I wish to get back to my main argument that experimental abiogenesis with minimal initial design and open-ended expected results must be the way to obtain the answers to our questions on the origin of life: We need to create life in order to really understand it, as Richard Feynman wrote on his blackboard shortly before his death: "What I cannot create, I do not understand."

"Let us not forget that inside the GVs the system continues to produce longer peptides and longer RNAs" at ever increasing "crowded" concentrations, *"and that both compound families bear useful functionalities in their populations. In such dynamic-flow off-equilibrium experiments anything that proves useful for the GV hosts, for example peptide pores that facilitate the import of useful and export of harmful compounds, becomes noticeable as an enhanced persistence of the corresponding GVs. It all depends on the food delivered by the environment. This at first total dependence of the GVs on sustainable food (heterotrophism in a biotic world) contrasts true autopoiesis. The lack of robustness is potentially lethal. The permanent danger diminishes sharply with the appearance of a 'systemic memory.' The passing-on of inherited information can only be functional and robust for the whole system after a genetic code has established. Proto-ribosomes, proto-ARS and proto-RNA polymerases can then better self-evolve; also proto-fatty acid and nucleoside triphosphate synthetases. Eventually, true protocells, living cells are expected to appear"* ex inanimo. [...]

The farther we get carried away with deriving in a chemically logical sequence what might happen, the more divergent territory we are occupying. Nonetheless, divergence is what we are ultimately seeking for. I am not particularly fond of skating on thin predictive ice. Instead of hypothesizing on the order of foreseeable events, on whether a ribosome (RNA-dependent amino acid polymerase) has to be ribozymic, a RNA polymerase or ARS must be proteic, what comes first, what second and why our genetic code is as it is, I much prefer to see what actually happens, and how far I can get, and how different the results will be when we restart the same way with a new batch, or after we restart racemically instead of enantiomerically pure, and so forth. (Strazewski 2015, p. 857)

6.3.3 SUMMARY

The importance of being able to create truly self-evolving cells *ex inanimo* cannot be underestimated, for our knowledge on the natural sciences and, should the necessity arise, for being able to safely seed extraterrestrial life, in particular farther away than on Mars. Current research on artificial and synthetic cells concentrates more on designed approaches than on actually animating inanimate complex chemical systems. A bottom-up and more pro-active approach would be to test a general experimental setup, as proposed here. Its aim is to work in the laboratory toward self-evolvable chemical systems that, according to what we know about life's chemical processes, should eventually become alive. The experiment is based on reasonably complex molecular compound classes that are known from extant biotic life: amino acids, peptides, nucleotides, lipids, and small organic molecules. The most challenging realistic goals are the observation of the emergence of one or several "genetic" codes and of "genetic" replication.

ACKNOWLEDGMENTS

I thank the *Volkswagen Foundation* for the funding of the research program "Molecular Life" within their call *Life?—A Fresh Scientific Approach to the Basic Principles of Life*. I hope that closer funding agencies will start opening their funds to experimental abiogenesis proposals in the nearest possible future.

REFERENCES

Biron, J.-P., A. L. Parkes, R. Pascal, and J. D. Sutherland. 2005. Expeditious, potentially primordial, aminoacylation of nucleotides. *Angew. Chem. Int. Ed.* 44: 6731–6734.

Blain, J. C., and J. W. Szostak. 2014. Progress toward synthetic cells. *Annu. Rev. Biochem.* 83: 615–640.

Boltzmann, L. E. 1877. Über die Beziehung zwischen dem zweiten Hauptsatz der mechanischen Wärmetheorie und der Wahrscheinlichkeitsrechnung respektive den Sätzen über das Wärmegleichgewicht. *Sitzungsber. d. k. Akad. der Wissenschaften zu Wien II* 76: 428.

Bonfio, C., L. Valer, S. Scintilla, S. Sha, D. J. Evans, L. Jin, J. W. Szostak, D. D. Sasselov, J. D. Sutherland, and S. S. Mansy. 2017. UV-light-driven prebiotic synthesis of iron-sulfur clusters. *Nat. Chem.* 9: 1229–1234.

Buddingh, B. C., and J. C. M. van Hest. 2017. Artificial cells: Synthetic compartments with life-like functionality and adaptivity. *Acc. Chem. Res.* 50: 769–777.

Carter, Jr., C. W. 2015. What RNA world? Why a peptide/RNA partnership merits renewed experimental attention. *Life* 5: 294–320.

Cobb, M. 2017. 60 years ago, Francis Crick changed the logic of biology. *PLoS Biol.* 15(9): e2003243.

Crick, F. H. C. 1970. Central dogma of molecular biology. *Nature* 227: 561–563.

Cronin, L., and S. I. Walker. 2016. Beyond prebiotic chemistry: What dynamic network properties allow the emergence of life? *Science* 352: 1174–1175.

Demeshkina, N., L. Jenner, E. Westhof, M. Yusupov, and G. Yusupova. 2012. A new understanding of the decoding principle on the ribosome. *Nature* 484: 256–259.

Eigen, M., and P. Schuster. 1979. *The Hypercycle – A Principle of Matural Self-Organization*. Springer, Berlin, Germany.

Fry, S., B. McIntyre, T. Hodgson, J. Finnemore, and R. Hill. 2017. *The Hippopotamus*. J. Jencks (dir.), movie, 89 minutes. The Electric Shadow Company, London, UK.

Gánti, T. 1971 (original version in Hungarian). *The Principles of Life*. Oxford University Press, Oxford, UK (2003).

Gánti, T. 2003. *Chemoton Theory: Theory of Living Systems*. Kluwer Academic/Plenum Publishers, New York.

Griesser, H., M. Bechthold, P. Tremmel, E. Kervio, and C. Richert. 2017. Amino acid-specific, ribonucleotide-promoted peptide formation in the absence of enzymes. *Angew. Chem. Int. Ed.* 56: 1224–1228.

Haroche, S., and D. J. Wineland. 2012. The Nobel Prize in Physics for ground-breaking experimental methods that enable measuring and manipulation of individual quantum systems.

Jauker, M., H. Griesser, and C. Richert. 2015. Spontaneous formation of RNA strands, peptidyl RNA, and cofactors. *Angew. Chem. Int. Ed.* 54: 14564–14569.

Kurokawa, C., K. Fujiwara, M. Morita et al. 2017. DNA cytoskeleton for stabilizing artificial cells. *Proc. Natl. Acad. Sci. U.S.A.* 114: 7228–7233.

Liu, Z., D. Beaufils, J.-C. Rossi, and R. Pascal. 2014. Evolutionary importance of the intramolecular pathways of hydrolysis of phosphate ester mixed anhydrides with amino acids and peptides. *Sci. Rep.* 4: 7440.

Lotka, A. J. 1921. Note on the economic conversion factors of energy. *Proc. Natl. Acad. Sci. U.S.A.* 7: 192–197.

McCollom, T. M., G. Ritter, and B. R. T. Simoneit. 1998. Lipid synthesis under hydrothermal conditions by Fischer-Tropsch-type reactions. *Orig. Life Evol. Biosph.* 29: 153–166.

Magritte, R. 1929. Canvas 59×65 cm, oil painting. Los Angeles County Museum of Art.

Malthus, T. R. 1798–1826. *An Essay on the Principle of Population*. 1st–6th ed. J. Murray, London, UK.

Maturana, H., and F. Varela. 1980. *Autopoiesis and Cognition. The Realization of the Living*. D. Reidel Publishing Company, Dordrecht, the Netherlands.

Muchowska, K. B., S. J. Varma, E. Chevallot-Beroux, L. Lethuillier-Karl, G. Li, and J. Moran. 2017. Metals promote sequences of the reverse Krebs cycle. *Nat. Ecol. Evol.* 1: 1716–1721.

Nott, T. J., T. D. Craggs, and A. J. Baldwin. 2016. Membraneless organelles can melt nucleic acid duplexes and act as biomolecular filters. *Nat. Chem.* 8: 569–575.

Pascal, R., and A. Pross. 2015. Stability and its manifestation in the chemical and biological worlds. *Chem. Commun.* 51: 16160–16165.

Spitzer, J. 2011. From water and ions to crowded biomacromolecules: *In vivo* structuring of a prokaryotic cell. *Microbiol. Mol. Biol. Rev.* 75: 491–506.

Spitzer, J., and B. Poolman. 2009. The role of biomacromolecular crowding, ionic strength, and physicochemical gradients in the complexities of life's emergence. *Microbiol. Mol. Biol. Rev.* 73: 371–388.

Strazewski, P. 2014. RNA as major components in chemical evolvable systems. In: *Chemical Biology of Nucleic Acids: Fundamentals and Clinical Applications*, V. A. Erdmann, W. T. Markiewicz, J. Barciszewski (Eds.), Series: RNA

Technologies. V. A. Erdmann, J. Barciszewski. Springer, Chapter 1, pp. 1–24. Springer, Berlin, Germany.

Strazewski, P. 2015. Omne vivum ex vivo … omne? How to feed an inanimate evolvable chemical system so as to let it self-evolve into increased complexity and life-like behaviour. *Isr. J. Chem.* 55: 851–864.

Szathmáry, E. 1991. Simple growth laws and selection consequences. *Trends Ecol. Evol.* 6: 366–370.

Szathmáry, E., and J. Maynard Smith. 1997. From replicators to reproducers: The first major transitions leading to life. *J. Theor. Biol.* 187: 555–571.

Szilágyi, A., I. Zachar, and E. Szathmáry. 2013. Gause's principle and the effect of resource partitioning on the dynamical coexistence of replicating templates. *PLoS One* 9(8): e1003193.

Szilágyi, A., I. Zachar, I. Scheuring, Á. Kun, B. Könnyű, and T. Czárán. 2017. Ecology and evolution in the RNA world dynamics and stability of prebiotic replicator systems. *Life* 7: 48.

Szostak, J. W., D. P. Bartel, and P. L. Luisi. 2001. Synthesizing life. *Nature* 409: 387–390.

Volterra, V. 1938. Population growth, equilibria, and extinction under specified breeding conditions: A development and extension of the theory of the logistic curve. *Hum. Biol.* 10: 1–11.

von Kiedrowski, G., and E. Szathmáry. 2012. *The Monetary Growth Order.* arXiv:1204.6590v1 [q-fin. GN].

von Trier, L., T. Vinterberg, K. Levring, S. Kragh-Jacobsen, and J.-M. Barr. 1995. *Dogme 95.* Denmark.

Wollstonecraft Godwin Shelley, M. 1818. Victor Frankenstein. In: *Frankenstein, or the Modern Prometeus.* Lackington, Hughes, Harding, Mavor & Jones, London, UK.

6.4 Role of Cations in RNA Folding and Function

Jessica C. Bowman, Anton S. Petrov, and Loren Dean Williams

CONTENTS

6.4.1 CATIONS IN WATER: RELEVANCE TO ORIGIN OF LIFE AND EVOLUTION ON EARTH

Inorganic metal cations are abundant in cells and are essential to life on Earth. Metal cations are vital for a variety of biological processes. Fundamental biological processes such as electron transfer, photosynthesis, and respiration require these ions (Bertini et al. 2007). Cations stabilize transition states in the replication of DNA, transcription of DNA to RNA, and ligation and cleavage of the nucleic acids by both proteins and ribozymes (Lykke-Andersen and Christiansen 1998; Steitz 1999; Doherty and Dafforn 2000; Lee et al. 2000; Yin and Steitz 2004; Ellenberger and Tomkinson 2008). Protein sequences encoded in messenger RNA are translated by amino-acid-bearing tRNAs at the ribosome, neither of which are transcribed, folded, assembled, or functional in the absence of inorganic cations. Few natural biological manipulations of the nucleic acid phosphodiester backbone proceed *in vitro* without divalent cations, and none are known to proceed in the absence of monovalent cations. Divalent cations stabilize three-dimensional structures of proteins, catalyze redox reactions, stabilize cell membranes, and transmit electrical current in the nervous system. Here, we focus on the interactions of metal ions with nucleic acids and explain their role in the translation system.

The earliest available data suggests that translation, one of the most central processes of life, originated with dependence on inorganic cations. The catalytic center of the ribosomal large subunit (LSU), responsible for peptide bond formation, is stabilized by Mg^{2+}. Within 20 Å of this center, about 20% of the phosphate oxygens of the RNA phosphodiester backbone interact directly with Mg^{2+} (Klein et al. 2004; Hsiao et al. 2009). Mg^{2+} bridges RNA nucleotides from disparate regions of the RNA sequence to form a structure that is inaccessible with monovalent cations alone (Hsiao and Williams 2009; Lenz et al. 2017). These Mg^{2+}-RNA bridges are highly conserved in position and structure through all three major evolutionary branches of life, suggesting that they preceded, or at least coincided with, the last universal common ancestor (LUCA).

The availability of biologically useful inorganic ions is dependent on the existence of certain minerals and on their solubility in water. The composition of the atmosphere affects cation solubility, especially partial pressures of $O_{2(g)}$ and $CO_{2(g)}$. An influx of $O_{2(g)}$ to an aqueous system diminishes the solubility of redox-sensitive ions such as Cu^{2+}, Fe^{2+}, and Mn^{2+}. Similarly, $CO_{2(g)}$ decreases pH, mobilizing ions such as Ca^{2+} and Na^+ from soils and precipitants. Fe^{2+} has been proposed and demonstrated to be a functional early Earth analogue for Mg^{2+} in nucleic-acid-processing steps, due to its similarity in size and charge density (Athavale et al. 2012; Okafor et al. 2017). Yet, in the pH range that sustains most life, $Fe^{2+}_{(aq)}$ precipitates as $Fe(OH)_{2(s)}$ (Morgan and Lahav 2007) or $Fe(III)_{(s)}$ species (Derry 2015) in a complex mechanism and at a rate that is dependent on the concentration of dissolved $O_{2(g)}$. Together with well-documented substitutions of Cu^{2+}, Zn^{2+}, and Mn^{2+} for Fe^{2+} in catalytic proteins (Torrents et al. 2002;

Wolfe-Simon et al. 2006; Anjem et al. 2009; Cotruvo and Stubbe 2011; Martin and Imlay 2011; Ushizaka et al. 2011; Aguirre and Culotta 2012; Harel et al. 2014), these observations suggest that changes in Earth's atmosphere over evolution may have altered the ions of key ion-dependent biological processes.

Water's interactions with ions are as important to biology as the ions themselves. Cations observed at the atomic level in three-dimensional molecular structures are almost always at least partially hydrated. Water molecules are coordinated, oriented, polarized, and acidified by ions in a manner that depends on the properties of the ion—mainly size and charge density (Hribar et al. 2002; Collins et al. 2007). Hydrated cations mitigate the negative charge on nucleic acids by long-range electrostatic and shorter-range hydrogen-bonding interactions with backbone phosphates. Partial dehydration allows direct coordination of the cation by phosphates (up to four for Mg^{2+}) of the nucleic acid backbone, while remaining cation-coordinated water molecules are available to participate in hydrogen bonding. Cation hydration mitigates the thermodynamic favorability of direct coordination by phosphodiesters by requiring unfavorable dehydration. Together, these interactions provide access to a range of complex and finely tunable conformational states of the RNA. The functional roles of cations in extant biology are interdependent with the unique properties of water.

6.4.2 HYDRATED CATIONS

6.4.2.1 COORDINATION BY WATER

Water molecules are oriented and polarized by cations, forming acidic metal hydrates $(Me(H_2O)_x)^{n+}$ with elevated potential for participation in hydrogen bonding and function in enzymatic activity and proton donation. The extent of attraction between a cation and water molecules is given by the cation's hydration enthalpy. A large, negative hydration enthalpy indicates a stronger attraction than a small, negative one. The hydration enthalpies of cations vary in magnitude by cation identity, roughly based on size and charge. Hydration enthalpies of Zn^{2+} and Mg^{2+}, for example, are large in magnitude (~ -450 kcal/mol) compared with those of Na^{2+} and K^+ (~ -100 kcal/mol) (Rashin and Honig 1985).

Biologically relevant metal hydrates include those of Ca^{2+}, Fe^{2+}, K^+, Mg^{2+}, Mn^{2+}, Na^+, and Zn^{2+}. These cations are coordinated by multiple water molecules in their hydrate form. The inherent polarity of each water molecule is increased by coordination with a cation, increasing the acidity of the water hydrogens. Experimentally derived pK_as (Table 6.4.1) indicate a relative hydrate acidity of $Zn^{2+} > Fe^{2+} > Mn^{2+} > Mg^{2+} > Ca^{2+} > Na^+ > K^+$ (Baes and Mesmer 1976; Stumm and Morgan 1996; Mähler and Persson 2012; Jackson et al. 2015), where the most acidic hydrate $[Zn(H_2O)_6]^{2+}$ has, by definition, the largest relative abundance of potentially reactive oxyanion species in water near physiological pH (~ 7.4). All of these hydrates are more acidic than bulk water (pK_a 15.7) (Baes and Mesmer 1976). The standard free energy for ionization of an uncoordinated water molecule ($H_2O \rightarrow OH^- + H^+$) is large and positive (382.8 kcal/mol); coordination in a hexahydrate of Mg^{2+}, Mn^{2+}, or Zn^{2+} reduces the energy of ionization by more than half (Bock et al. 1999).

Cations constrain water molecules to specific geometries. Though the number of water ligands for a hydrated cation is not fixed, especially when considering the dynamics of water molecules in solution, crystallographic structures reveal dominant coordination numbers (Table 6.4.1). Coordination by water ligands compresses metal ions: hydration decreases the effective ionic radius of the cation progressively with each water molecule added, while metal-oxygen bond distances increase (Shannon 1976; Bock et al. 1999). The effective ionic radius of a hexahydrated cation is also related to spin state and cation identity. Comparison of ionic radii, assuming hexahydrate coordination and low spin, yields relative atomic sizes $Fe^{2+} < Mn^{2+} < Mg^{2+} \cong Zn^{2+} < Ca^{2+} \cong Na^+ < K^+$ (Shannon 1976). The radius of a hexahydrated Fe^{2+} ion is less than half the radius of a hexahydrated K^+ ion. With the exception of Zn^{2+}, the radii of cation hexahydrates mentioned here correlate inversely with hydrate acidity. The availability of d-orbital electrons in the transition metals (Fe^{2+}, Mn^{2+}, and Zn^{2+}) is known to affect hydrate pKa (Roychowdhury-Saha and Burke 2006).

TABLE 6.4.1
Properties of a Subset of Biologically Relevant Hydrated Cations

Cation	Effective Ionic Radii (Å) in a Hexahydrate (Shannon 1976)	Common Hydrate Coordination (Baes and Mesmer 1976; Jackson et al. 2015)	Experimentally Derived pK_a (Baes and Mesmer 1976; Jackson et al. 2015)
Fe^{2+}	0.61[a], 0.78[b]	$[Fe(H_2O)_6]^{2+}$	9.3–9.5
Mn^{2+}	0.67[a], 0.83[b]	$[Mn(H_2O)_6]^{2+}$	10.6–11.0
Mg^{2+}	0.72	$[Mg(H_2O)_6]^{2+}$	11.2–11.4
Zn^{2+}	0.74	$[Zn(H_2O)_6]^{2+}$	9.0
Ca^{2+}	1.00	$[Ca(H_2O)_8]^{2+}$	12.7–12.9
Na^+	1.02	$[Na(H_2O)_{6-8}]^+$	14.2
K^+	1.38	$[K(H_2O)_{6-8}]^+$	14.5

[a] Low spin (estimate);
[b] High spin.

Water molecule dynamics are suppressed by cations. The lifetime of a water molecule in the first coordination shell of a cation is generally longest for ions with the greatest charge density (e.g., highly charged with small ionic radius) (Diebler et al. 1969). It may be expected, based on ionic radii alone, that the relative exchange rate for the cations of Table 6.4.1 is $Fe^{2+} < Mn^{2+} < Mg^{2+} \cong Zn^{2+} < Ca^{2+} \cong Na^+ < K^+$. Exchange rates have been estimated experimentally as $Mg^{2+} < Fe^{2+} \cong Mn^{2+} < Zn^{2+} < Ca^{2+} \cong Na^+ \cong K^+$ (Diebler et al. 1969), where rates within one order of magnitude are considered within the range of uncertainty. The rate of exchange of water from the first coordination shell of K^+ (on the order of ~10^9 waters per second) occurs four orders of magnitude faster than from the first shell of Mg^{2+} (Diebler et al. 1969).

Ions affect water enthalpically and entropically. In pure water, intramolecular interactions are dominated by hydrogen bonds. Hydrogens and electron lone pairs on each water participate in hydrogen-bonding interactions with surrounding water molecules. The high electronegativity of the oxygen makes each water molecule a dipole, with partial negative and positive charges. The partially negative oxygen of a water dipole interacts electrostatically with adjacent cations. Conversely, the partially positive charge of the dipole (bearing two hydrogens) interacts electrostatically with anions. Ions organize first-shell water molecules, affecting surrounding water molecules. The geometric perturbation of waters around an ion in aqueous solution reflects the outcome of competition between the ion and surrounding water molecules (or between electrostatic and hydrogen-bonding interactions) for the ion-constrained water. Theoretical studies suggest that the outcome of this competition strongly favors electrostatic interactions when a cation is small and densely charged, which disrupts canonical water-water hydrogen bonds (Hribar et al. 2002). Large, less-charge-dense cations have a greater tendency to share their most proximal

water molecules with other water molecules and even promote canonical hydrogen bonding among surrounding waters (Hribar et al. 2002). The combined result is a difference in the distance of propagation of ion-induced organization of water. One report suggests that small, charge-dense cations reorganize up to two layers of water molecules around the ion (<5 Å), whereas large, less-charge-dense cations perturb only one layer (Collins et al. 2007). Differences in the organization of water around ions may explain, if only in part, the lack of interchangeability of ions in some biochemical processes.

6.4.2.2 DEHYDRATION BY PHOSPHATE OXYANIONS

Biology employs partially hydrated cations that constrain ligands other than water in the first coordination shell. Ligand binding to hexahydrated Mg^{2+}, for example, displaces a water molecule in the reaction (Mayaan et al. 2004):

$$L^{-q} + \left[Mg\left(H_2O\right)_6 \right]^{2+} \rightarrow \left[Mg\left(H_2O\right)_5(L) \right]^{2-q} + H_2O \quad (6.4.1)$$

Single substitution of a water molecule for an alternate first-shell ligand type alters the structure of the Mg^{2+} hydrate (Mayaan et al. 2004). If the ligand (L) is the oxyanion OH^-, for example, the interatomic distance between Mg^{2+} and the oxygen atom of L is slightly less than the distance between Mg^{2+} and each of its remaining five waters oxygens in $[Mg(H_2O)_5(OH)]^+$. In addition, the interatomic distance between Mg^{2+} and the remaining five water oxygens increases slightly relative to $[Mg(H_2O)_6]^{2+}$. That is, the oxyanion ligand OH^- is held more tightly by Mg^{2+} than the neutral waters in $[Mg(H_2O)_5(OH)]^+$, and the neutral waters are less constrained (and possibly less acidic) than those in hexahydrated Mg^{2+} (Figure 6.4.1).

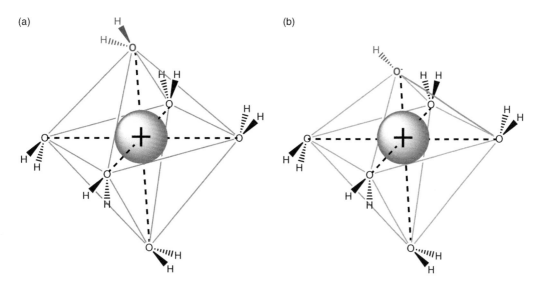

FIGURE 6.4.1 Small oxyanion ligands of cations are held more tightly than neutral water. A conceptual model of the difference in interatomic distances between a charge-dense cation and a water ligand or oxyanion ligand. (a) A hexahydrated cation with octahedral coordination geometry and six roughly equal cation-water oxygen interatomic distances. (b) The interatomic distance between the cation and an oxyanion ligand, in this case OH^-, is less than remaining cation-water distances. The magenta sphere is a generic charge-dense cation. Neutral water ligand and oxyanion ligand are shown in red font. Remaining cation-water distances are increased slightly in panel B, relative to panel A. The change in bond length is exaggerated for clarity.

Phosphates and phosphate esters are common first-shell ligands of cations in biology. Like water, they are oriented, polarized, and constrained by cations. Most biological phosphate is covalently bound in adenosine mono-, di-, or tri- phosphate (AMP, ADP, or ATP); deoxyribonucleic acid (DNA); or ribonucleic acid (RNA) (Figure 6.4.2), and these molecules are well-documented partners of cations, especially Mg^{2+}. The transition states in condensation of the nucleic acid phosphodiester backbone by polymerases, ligases, and ribozymes, for example, are directly stabilized by divalent cations (Lykke-Andersen and Christiansen 1998; Steitz 1999; Doherty and Dafforn 2000; Lee et al. 2000; Yin and Steitz 2004; Ellenberger and Tomkinson 2008; Butcher 2011; Johnson-Buck et al. 2011).

Computation suggests that Mg^{2+} binds phosphate or phosphate esters more tightly than water molecules. In work by

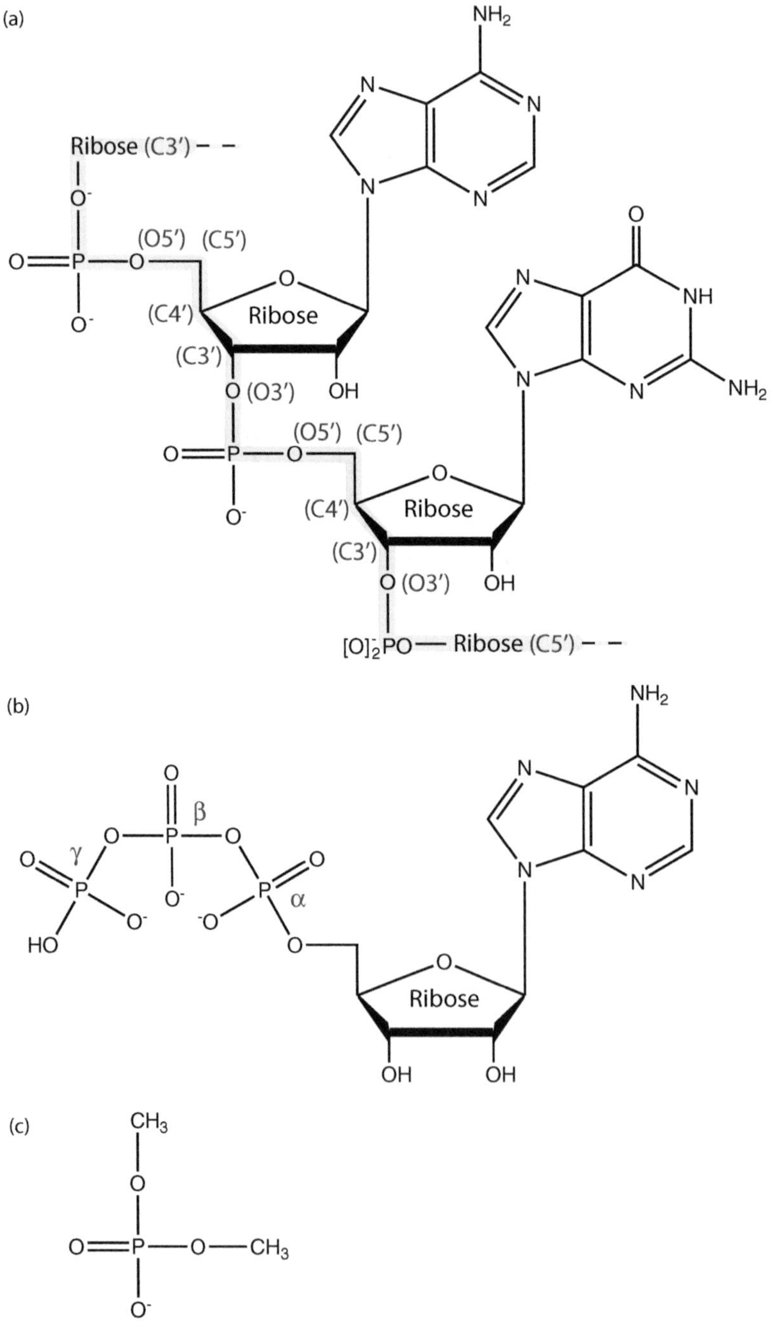

FIGURE 6.4.2 Common biological ligands of cations. (a) Two monomers of RNA connected by the phosphodiester backbone (yellow highlight), with conventional ribose atom designations in parentheses. (b) Adenosine triphosphate (ATP) with conventional phosphate designations. (c) Dimethyl phosphate. Dimethyl phosphate, as discussed, is a molecular model representing the phosphodiester connecting unit of the nucleic acid backbone. Hydrogens on the ribose sugars of panels (a) and (b) have been omitted for clarity.

Mayaan et al. (2004), substitution of a single water molecule in Mg^{2+} hexahydrate for a biologically derived ligand resulted in ligand-dependent changes in Mg^{2+}-ligand interatomic distances and remaining cation-water interatomic distances, consistent with Figure 6.4.1. Mg^{2+} coordinated all evaluated small oxygen- and phosphate-bearing ligands with shorter cation-ligand oxygen interatomic distances than cation-water oxygen interatomic distances in single-substitution reactions (Mayaan et al. 2004). These ligands included free phosphate species hydrogen phosphate (HPO_4^{2-}) and dihydrogen phosphate ($H_2PO_4^-$), which dominate at physiological pH of 7.4, and dimethyl phosphate. Since the electronic structure of dihydrogen phosphate is somewhat different than phosphodiester, the dimethyl-phosphate anion ($CH_3-O-PO_2-O-CH_3$)$^-$ (Figure 6.4.2) is often used as a model system representing the phosphodiester connection between the base-bearing sugars of nucleic acids. In dimethyl phosphate, one methyl group is an analogue for the $3'$ ribose carbon and the other an analogue for the $5'$ carbon of the adjacent ribose sugar. Reported relative coordination distances for Mg^{2+} are OH- < dimethyl phosphate < free phosphate species < water (Mayaan et al. 2004). In other words, Mg^{2+} binds to dimethyl phosphate more tightly than to free phosphate or water.

The geometry of the nucleic acid backbone is cation-dependent. Work by Panteva et al. aiming, in part, to computationally replicate experimentally derived thermodynamic parameters for the interactions of Mg^{2+}, Mn^{2+}, or Zn^{2+} with RNA, derived parameters for interactions of these cations with the nucleic acid analogue dimethyl phosphate. Their results suggest that contact and minimum energy distances between cation and phosphate follow a trend in which dimethyl phosphate oxyanion-Mg^{2+} and -Zn^{2+} distances are approximately equal and less than that of Mn^{2+} (Panteva et al. 2015), suggesting that dimethyl phosphate binds more tightly to Mg^{2+} and Zn^{2+} than to Mn^{2+}. Fujimoto et al. (1994) reported cation-dependent variations in the nucleic acid hydrodynamic radius, in which Na^+ slightly decreased while divalent cations Mg^{2+} and Mn^{2+} substantially increased the hydrodynamic radius of DNA. These observations suggest cation-specific structural changes in nucleic acids, perhaps phosphodiester bond stabilization and/or changes in hydration.

Mg^{2+} polarizes first-shell phosphate ligands, shortening phosphodiester bonds and leaving phosphorus more susceptible to nucleophilic attack. Upon substitution of a Mg^{2+}-constrained first-shell water with dimethyl phosphate in computational analyses, Mayaan et al. (2004) observed corresponding changes in the phosphate ligand. While the interatomic bond distance between phosphorus and the phosphate oxygen interacting with Mg^{2+} increased, bond distances between the phosphorus atom and phosphate oxygens covalently bound to the methyl groups decreased, suggesting that Mg^{2+} polarizes phosphate, drawing negative charge away from phosphate oxygens not directly coordinated by the cation. The magnitude of decrease in the length of the phosphodiester bonds of dimethyl phosphate has been calculated to be cation-specific, where the change induced by Ni^{2+} > Mg^{2+} > Na^+ (Zhang et al. 2015). Application of this observation to

nucleic acids suggests that first-shell coordination of a backbone phosphate with cations may increase the reactivity of phosphorus while rigidifying the backbone by shortening the phosphodiester bonds in a cation-dependent manner. The expected change in bond lengths based on these calculations are within the range of precision of most nucleic acid structures but are believed to be corroborated by a few very-high-resolution structures.

6.4.3 CATIONS AS RNA LIGANDS

6.4.3.1 Conceptual Frameworks for Binding

Cations stabilize RNA through a continuum of associations, not just first-shell interactions. Williams and co-workers classified cation interactions with RNA as chelated, condensed, glassy, or diffuse, based on six criteria: relative population, number of first-shell RNA ligands, extent of mobility, dimensionality of mobility, contribution to RNA thermodynamic stability, and contribution to non-canonical RNA structure (Bowman et al. 2012).

The most populous ions in solutions of RNA are diffuse ions, which are fully hydrated without first-shell or first-shell water interactions with RNA and are no more restricted in mobility than bulk water. Diffuse cations are much more abundant than partially hydrated cations that directly coordinate RNA. The phosphodiester backbone of RNA is negatively charged and therefore self-repellant; diffuse cations electrostatically neutralize the RNA backbone. They interact with RNA by numerous weak, long-range electrostatic interactions (Manning 1969), contributing to the overall thermodynamic stability of nucleic acids and their complexes. These ions are undetectable in crystallographic structures in part due to their positional disorder. Diffuse ions do not stabilize specific non-canonical RNA structures.

Condensed ions (Manning 1969) have slightly attenuated mobility due to electrostatic or hydrogen-bonding interactions of one or more ion-coordinated waters with RNA or an RNA ligand. Their relative population and rate and dimensionality of diffusion are less than that of diffuse ions but are greater than glassy ions. Condensed ions neutralize charge that is more localized to the RNA than diffuse ions. Their population is sensitive to a local base pairing of nucleic acids and affects the geometry and conformational dynamics of the nucleic acid grooves (Lavery and Pullman 1981). Condensed ions can be visualized as a shell or coating on the surface of a three-dimensional RNA, extending beyond the van der Waals radius of the RNA.

Glassy cations have at least one first-shell RNA ligand and restricted mobility. They are typically found in the interior of the RNA, encased within the external three-dimensional surface of the native structure, such that the dimensionality of diffusion is reduced. The mobility of glassy cations is less than that of condensed cations. Glassy cations may contribute to stabilization of specific RNA structural states that are inaccessible in their absence. Divalent ions, especially Mg^{2+}, play a special role in the stability of RNA, due to their ability to mediate strong tertiary interactions. Charge-dense Mg^{2+} is glassy, with one first-shell RNA ligand, while less-charge-dense cations such as Na^+ and K^+ require four to five first-shell RNA ligands to similarly restrict diffusion.

Chelated cations, bearing more than one first-shell RNA ligand in the case of Mg^{2+}, are the least frequent ions in RNA structures but often have the most significance for the structure-function relationship. Chelated cations are typically at the interior of the three-dimensional surface of the native RNA. Their rate of diffusion is dependent on the number of first-shell RNA ligands. These cations are the least mobile of the four classes and, being rigidly bound by the RNA, are difficult to remove *in vitro* without disturbing the RNA structure.

The mode of Mg^{2+}-RNA interaction is influenced by local RNA structure. Experimental work suggests that sites of Mg^{2+} chelation in RNAs are not random. Linear RNAs are less likely than folded RNAs to form first-shell interactions with Mg^{2+}, suggesting secondary structure and even sequence dependence (Porschke 1979). However, Mg^{2+} forms first-shell interactions less readily with double-stranded RNA than with compact single-stranded RNA (Kankia 2003, 2004). In many cases, site-specific Mg^{2+} binding and RNA conformation are interdependent: the conformational space in which an RNA achieves its native structure, including chelation of Mg^{2+}, is not available without Mg^{2+}, even with high concentration of monovalent cations (Draper et al. 2005; Grilley et al. 2006).

6.4.3.2 A QUANTUM MECHANICAL PERSPECTIVE ON FIRST-SHELL RNA-CATION INTERACTIONS

First-shell RNA-Mg^{2+} complexes occur so frequently in ribosome and ribozyme x-ray structures that they appear to be fundamental units of RNA folding (Petrov et al. 2011). These complexes often appear in a singular and defined geometry, referred to here as the bidentate chelation complex (or bidentate clamp, Figure 6.4.3). Bidentate clamps are characterized by a 10-membered ring system (Mg^{2+}–O$^-$–P–O5′–C5′–C4′–C3′–O3′–P–O$^-$), which includes a single Mg^{2+} chelated by phosphates of adjacent nucleotides (Hsiao et al. 2008; Hsiao and Williams 2009). Tri- and tetra-dentate RNA-Mg^{2+} chelates have been noted (Hsiao et al. 2008).

Petrov et al. (2011) characterized the energetics of bidentate RNA-Mg^{2+} clamps computationally using density functional theory (DFT). This work explains a special role of magnesium ions in the formation of tertiary interactions with RNA due to magnesium's enhanced binding ability with phosphodiester linkages and nucleobases. The Mg^{2+}-RNA interactions are amplified quantum effects of polarization and charge transfer due to high charge to radius ratio, which is the highest of all alkali and alkali-earth cations, making magnesium a unique ion in this group.

Bidentate chelation refers to cation association with two anions of the same ligand (in this case, the oxyanions of RNA phosphates) in the first coordination shell of the cation. The defined geometry and high stability of RNA-Mg^{2+} bidentate clamps have been attributed to the tightly packed octahedral structure of the Mg^{2+} first coordination shell (Brown 1992;

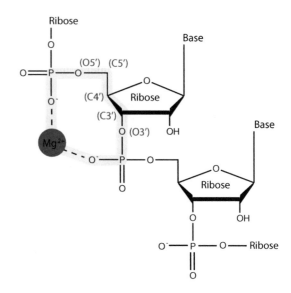

FIGURE 6.4.3 The bidentate clamp characterized by a 10-membered ring system (Mg^{2+}–O$^-$–P–O5′–C5′–C4′–C3′–O3′–P–O$^-$), which includes a single Mg^{2+} chelated by phosphates of adjacent nucleotides. (From Petrov et al. 2011; Hsiao, C. et al., Complexes of nucleic acids with group I and II cations, in *Nucleic Acid Metal Ion Interactions*, edited by N. Hud, pp. 1–35, The Royal Society of Chemistry, London, UK, 2008; Hsiao, C. and Williams, L.D., *Nucleic Acids Res.*, 37, 3134–3142, 2009.)

Bock et al. 1999) and to the resulting non-canonical RNA backbone confirmation (Klein et al. 2004; Hsiao et al. 2008; Hsiao and Williams 2009). Using DFT, Petrov et al. explored the reaction:

$$RNA^{2-} + \left[Mg\left(H_2O\right)_6 \right]^{2+} \rightarrow \left[RNA\text{-}Mg\left(H_2O\right)_4 \right]_{complex} + 2H_2O$$

$$(6.4.2)$$

In this reaction, two waters of hexahydrated magnesium are displaced as the cation is chelated by phosphate oxyanions on consecutive RNA nucleotides. Petrov et al. modeled the reaction by computationally optimizing the RNA structure in the absence and presence of $[Mg(H_2O)_6]^{2+}$, followed by optimization of the bidentate RNA-$Mg(H_2O)_4$ complex. Optimization was repeated with $[Na(H_2O)_6]^+$ or $[Ca(H_2O)_6]^{2+}$ in place of the Mg^{2+} hexahydrate, with the objective of understanding the uniqueness of the role of Mg^{2+} in complex stability.

RNA forms a more stable bidentate chelation complex with Mg^{2+} than with Na^+ or Ca^{2+} (Petrov et al. 2011). The RNA-$Mg(H_2O)_4$ complex is ~10 kcal/mol more stable than the RNA-$Ca(H_2O)_4$ complex. The $[RNA-Na(H_2O)_4]^-$ complex approaches neutral interaction energy and is unstable. The ionic radius of hexahydrated $Mg^{2+} < Ca^{2+} < Na^+$, therefore Mg^{2+} forces first-shell oxyanion RNA ligands closer together than Ca^{2+} or Na^+. The close proximity of the RNA oxyanions coordinated by Mg^{2+} incurs an electrostatic penalty of ~10 kcal/mol from the opposing negative charges on the two oxygens; this is mitigated by other favorable energy components.

The octahedral coordination geometry of hexahydrate metals is distorted by RNA phosphate oxyanion ligands.

Distances from metal to oxygen are smaller for RNA phosphates than for water in Mg^{2+} and Ca^{2+} RNA chelation complexes. The metal-to-RNA phosphate oxyanion interatomic distances in the Na^+ RNA chelation complex are polymorphic. Optimization with Na^+ yields a monodentate structure with only one RNA phosphate oxyanion ligand in the first coordination shell of Na^+.

Energies of binding are substantial for small, charge-dense Mg^{2+} complexes but are less so for Ca^{2+} due to its larger size and for Na^+ due to its smaller charge. Decomposition of the interaction energy of bidentate clamps reveals that, for all cations modeled, electrostatic components make the greatest contributions to complex stability, followed by polarization (of the RNA phosphates) and charge transfer. There is a ~20 kcal/mol difference in the electrostatic contribution to bidentate clamp stability by Mg^{2+} compared with Ca^{2+}; and a ~220 kcal/mol difference by Na^+ compared with Mg^{2+}. The polarization component is less than half of the electrostatic contribution for all cations. Charge transfer components include a net reduction of positive charge on each cation (Mg^{2+}: 0.2 e^- > Ca^{2+}: 0.1 e^- > Na^+: 0.1 e^-) (Petrov et al. 2011).

Magnesium's high charge density brings phosphate oxyanions of adjacent RNA nucleotides into closer proximity than Ca^{2+} or Na^+, inducing non-canonical RNA backbone conformations found especially in catalytic RNAs. Complex formation may be driven, in part, by magnesium's preference for RNA's negatively charged and polarizable phosphate oxyanions over water. Though electrostatic energy components dominate stability, polarization and charge transfer components are not negligible. The small size of Mg^{2+} creates a symmetrical distortion of the metal's octahedral coordination geometry, in which phosphate oxyanions are bound more tightly than water oxygens.

6.4.4 STRUCTURAL ROLES IN THE RIBOSOME

6.4.4.1 A Mg^{2+} Scaffold for the Ribosome Peptidyl Transferase Center

All proteins of living organisms are synthesized by the ribosome, a highly conserved and ancient RNA-protein enzyme. The ribosome is the site of execution of the genetic code. With the help of aminoacyl tRNA synthetases and transfer RNAs, messenger RNA is decoded at the small subunit (SSU), while peptide bonds are made in the ribosomal LSU at the catalytic center, the peptidyl transferase center (PTC). The bacterial LSU contains two ribosomal RNA (rRNA) polymers, the 23S rRNA and the 5S rRNA, as well as numerous ribosomal proteins. The LSU rRNA of *Thermus thermophilus* is made up of more than 2,900 nucleotides, subdivided into seven domains (Domains 0, I, II, III, IV, V, and VI; Figure 6.4.4), based on the native structure common to all ribosomes (Petrov et al. 2014). Bokov, Steitz, Williams, and others observed that the occurrence of Mg^{2+} in the LSU is greatest in Domain V near the PTC (Hansen et al. 2001; Klein et al. 2004; Bokov and Steinberg 2009; Hsiao et al. 2009).

Hsiao and Williams (2009) reported a framework of highly conserved Mg^{2+} microclusters in the LSU. These microclusters include the 10-membered Mg^{2+}-RNA ring structure (Figure 6.4.3), modeled using DFT by Petrov et al. (2011), but in microclusters, the Mg^{2+} ion is one of a pair of Mg^{2+} ions chelated by a common bridging phosphate. The bridging phosphate, which has the form $Mg^{2+}_a-O^--P-O^--Mg^{2+}_b$, is part of the 10-membered chelation ring ($Mg^{2+}_a-O^--P-O5'-C5'-C4'-C3'-O3'-P-O^--$) (Hsiao and Williams 2009). As in Petrov et al. (2011), the phosphates of the chelation ring are contributed by adjacent nucleotides, but in a microcluster, at least one additional phosphate oxyanion interacts with at least one of the Mg^{2+} ions.

Mg^{2+}-RNA microclusters support long-range intra- and inter-domain interactions relevant to the global structure of the ribosome. Hsiao and Williams (2009) found three ancient Mg^{2+}-RNA microclusters supporting the PTC and a fourth located near the LSU exit tunnel (Figure 6.4.4). These microclusters are conserved among archaea (represented by *Haloarcula marismortui*), bacteria (represented by *T. thermophilus*), eukarya, and mitochondria. Two of the ancient microclusters make intra-domain nucleotide connections, while the other two connect very remote rRNA nucleotides of multiple domains of the LSU. References to specific nucleotides are given as *T. thermophilus* residues in the *Escherichia coli* numbering convention.

Microcluster D2 connects Domain V of the LSU, near the PTC, to Domain II of the LSU (Figure 6.4.5). Here, the 10-membered ring is made of up of the phosphate groups and ribose atoms of A783 and A784 in Domain II. One Mg^{2+} of this microcluster interconnects these nucleotides to the phosphate of A2589 in Domain V near the PTC. The other Mg^{2+} links the phosphate of A784 to the phosphate of G2588. Five phosphate oxygens of four nucleotides representing two disparate locations in the rRNA sequence are bridged by these two Mg^{2+}.

Microcluster D4 connects Domain IV of the LSU to Domain II (Figure 6.4.6). There are two 10-membered rings in this microcluster. One ring is made up of the phosphate groups and ribose atoms of C1782 and A1783; the other is made up of the phosphate groups and ribose atoms of A1783 and A1784. All of these nucleotides are in LSU Domain IV. The phosphate of A1780, a non-consecutive nucleotide of Domain IV, chelates one of these Mg^{2+}. One Mg^{2+} of this microcluster is also chelated by the phosphate of U740 in Domain II.

Microclusters D1 and D3 make long-range intra-domain connections of the rRNA (Figure 6.4.4). Microcluster D1 links RNA nucleotides of Domain V that are not adjacent in the linear sequence or secondary structure; microcluster D3 links non-adjacent nucleotides of Domain III. In microcluster D1, the 10-membered ring is made up of the phosphate groups and ribose atoms of C2498 and C2499. The phosphate of A2448 (50 nucleotides away in the primary sequence) bridges both of the microcluster magnesiums. In microcluster D3, located near the exit tunnel, the 10-membered ring is made of up of the phosphate groups and ribose atoms of

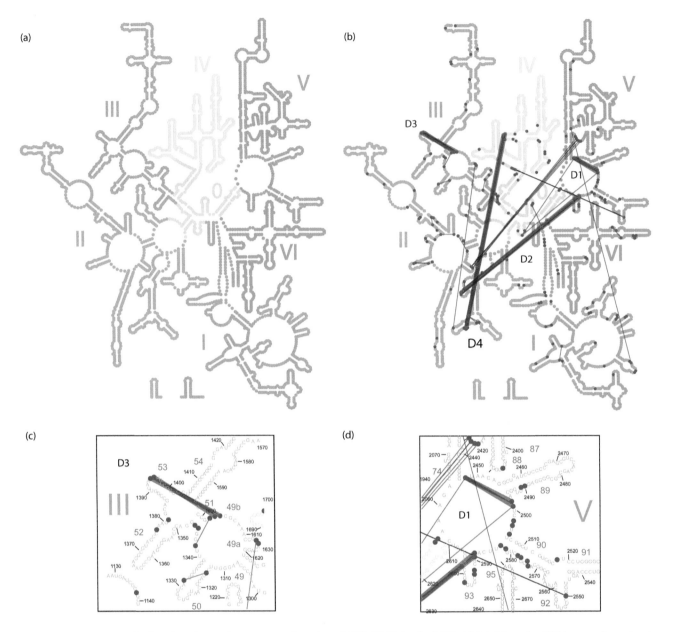

FIGURE 6.4.4 Ancient Mg^{2+} microclusters interconnect domains of 23S rRNA in the LSU. (a) Divisions in rRNA secondary structure Domains 0, I, II, III, IV, V, and VI are represented by changes in nucleotide color. (b) Microclusters D1, D2, D3, and D4 are shown as thick gray lines connecting nucleotide phosphates sharing a common first-shell Mg^{2+}. Other first-shell Mg^{2+} interactions are shown as thin gray or black lines (some have been omitted for clarity). Nucleotides are colored spheres. Mg^{2+} ions are magenta spheres. (c) Mg^{2+} microcluster D3 of Domain III. (d) Mg^{2+} microcluster D1 of Domain V. Secondary structure data generated with RiboVision (Bernier et al. 2014); microcluster data from Hsiao. (From Hsiao, C. and Williams, L.D., *Nucleic Acids Res.*, 37, 3134–3142, 2009.)

A1603, along with the phosphate group of C1604. Phosphates of U1394 and U1395 chelate the Mg^{2+} of this microcluster.

Mg^{2+} microclusters cause bases to unstack, priming them for short- and long-range rRNA interactions. Microcluster D2 unstacks base A784 of Domain II, which forms a pocket for ribosomal protein L2, and base A2587 of Domain V, which intercalates with unstacked bases C1782 and A1783 from microcluster D4 of Domain IV (Figure 6.4.5). Microcluster D4 induces a large array of unstacked bases in Domain IV, some of which intercalate with A2587 (unstacked in D2) of Domain V (Figure 6.4.6). Microcluster D1 unstacks G2447 from A2448 and A2497 from C2498 in Domain V.

Microcluster D3 unstacks base U1602 from A1603 and base A1393 from U1394 in Domain III.

The Mg^{2+}-RNA microcluster appears to be an ancient motif, with roles in rRNA folding and function, though it does not participate directly in peptide bond formation. The microclusters and rRNA elements linked by microclusters are conserved over evolution among archaea, bacteria, eukarya, and mitochondrial rRNA. Mg^{2+}-RNA microclusters have been identified in other rRNAs, including the bacterial SSU rRNA (Wimberly et al. 2000), and the P4-P6 domain of the *Tetrahymena thermophila* Group I intron ribozyme (Cate et al. 1997).

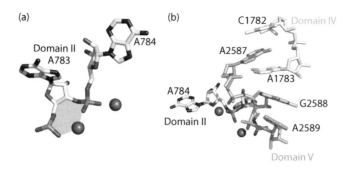

FIGURE 6.4.5 Inter-domain Mg²⁺-RNA microcluster D2 of 23S rRNA. (a) The D2 microcluster is a 10-membered ring (composed of a single Mg²⁺, the phosphate and ribose backbone atoms of nucleotide A783, and the phosphate of A784) and a second Mg²⁺ ion bridged by the common phosphate of A784. The area of the 10-membered ring is shown gray. RNA carbons are light gray, oxygens red, phosphorus orange, and nitrogen blue. Magnesium ions are magenta spheres. D2 microcluster nucleotides A783 and A784 are in Domain II of the 23S rRNA secondary structure. (b) The D2 microcluster mediates short- and long-range interactions in the 23S rRNA. The two Mg²⁺ of D2 are chelated by nucleotide phosphates of Domain II (atoms colored as in panel [a]) and Domain V rRNA (pink). Microcluster formation induces a non-canonical RNA backbone conformation, unstacking A2587 of Domain V, which intercalates with C1782 and A1783 of Domain IV (yellow). Structure of *T. thermophilus* [PDB 2J01 (obsolete). Superseded by 4V51]; microcluster data from Hsiao. (From Hsiao, C. and Williams, L.D., *Nucleic Acids Res.*, 37, 3134–3142, 2009.)

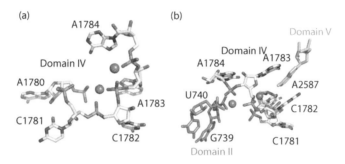

FIGURE 6.4.6 Inter-domain Mg²⁺-RNA microcluster D4 of 23S rRNA. (a) The D4 microcluster has two consecutive 10-membered rings composed of two Mg²⁺ and the phosphate and ribose backbone atoms of nucleotides C1782 and A1783, and A1783 and A1784. The phosphate of A1780 also chelates one of the Mg²⁺. (b) The D4 microcluster mediates short- and long-range interactions in the 23S rRNA. One Mg²⁺ of D4 is chelated by the phosphate of U740 of Domain II (blue). C1782 and A1783 intercalate with A2587 of Domain V (pink). Structure of *T. thermophilus* [PDB 2J01 (obsolete). Superseded by 4V51]; microcluster data from Hsiao. (From Hsiao, C. and Williams, L.D., *Nucleic Acids Res.*, 37, 3134–3142, 2009.)

6.4.4.2 RIBOSOMAL RNA FOLDING IN THE ABSENCE OF PROTEINS

The 23S rRNA of the ribosome LSU is a single polymer, thousands of nucleotides in length. Folding of the native ribosome to its catalytic state *in vivo* occurs with cations, including Mg²⁺; many ribosomal proteins; and post-transcriptional modifications.

It has been shown that the 23S rRNA, without ribosomal proteins or post-transcriptional modifications, can fold to form RNA-RNA interactions resembling those of the native ribosome (Lenz et al. 2017). Lenz et al. used a chemical footprinting technique called Selective 2′ Hydroxyl Acylation Analyzed by Primer Extension (SHAPE) (Mortimer and Weeks 2007) to assess the reactivity of each ribose 2′-OH in the 23S rRNA sequence to an electrophile. SHAPE reactivity can be viewed as a measure of a nucleotide's mobility or tendency to adopt a reactive conformation, meaning that the least-reactive nucleotides are the most likely participants in canonical hydrogen-bonding interactions.

Lenz et al. (2017) found that Na⁺ alone supports formation of native-like rRNA secondary structure in the 23S rRNA. In other words, regions of rRNA such as helices, which are base-paired in the x-ray crystallographic structure, show minimal reactivity by SHAPE, whereas bases in or near the bulges and loops of rRNA, which should be more mobile and/or unpaired, are reactive.

Structural collapse of 23S rRNA to form the tertiary interactions of the native ribosome LSU requires Mg²⁺ (Lenz et al. 2017). Addition of Mg²⁺ alters the pattern of reactivity, observed with Na⁺ alone. Some bases or regions of bases at or near bulges and loops become more, and others less, reactive in the presence of Mg²⁺ and Na⁺ than in the presence of Na⁺ alone. The reactivities in helical regions, unreactive in the presence of Na⁺ alone, are largely unaltered in the presence of Mg²⁺. However, non-helical regions showed changes in reactivity upon addition of Mg²⁺. Inspection of regions of crystallographic structures of native ribosomes showing altered reactivity by SHAPE suggests that these regions are largely associated with or are proximal to (1) known RNA-RNA tertiary interactions or (2) known RNA-protein interactions. Regions of altered reactivity are not limited to nucleotides expected to form first-shell tertiary interactions with Mg²⁺, as observed by Hsiao and Williams (2009).

Magnesium-induced tertiary interactions include formation of an intricate inter-domain rRNA interaction network. Nucleotides within secondary structure domains come into close proximity in three-dimensional space on folding and formation of the native 23S rRNA structure, interacting, in part, via base-pairing, base stacking, and phosphate-Mg^{2+} interactions. Of the 21 possible combinations of domain-to-domain interactions, the native 23S rRNA (as represented by bacterium *T. thermophilus*) has 14 interactions. Lenz et al. (2017) found that 12 of the 14 domain-domain interactions are detectable as Mg^{2+}-induced by SHAPE in the absence of ribosomal proteins. Here, one domain-domain interaction may be represented by a series of nucleotide-nucleotide or nucleotide-Mg^{2+} interactions.

Mg^{2+} appears necessary to position rRNA nucleotides for native interactions with ribosomal proteins and form a catalytic structure. Folding of the LSU 23S rRNA in the presence of Mg^{2+} induces a native-like structure in which the rRNA has collapsed to form much of the essential inter-domain architecture.

6.4.4.3 Ribosomal RNA Folding in the Presence of Proteins

Divalent cation-rRNA backbone interactions induce conformational changes that accommodate other ligands. Magnesium-microcluster D2 forms a pocket in the ribosome for ribosomal protein L2 (now known as uL2; Ban et al. 2014). Hsiao et al. found that Mg^{2+}-microcluster D2 connects Domain V of the 23S rRNA, near the PTC, to Domain II of the 23S rRNA (Hsiao and Williams 2009). The 10-membered ring of this microcluster is made of up A783 and A784 in Domain II. One Mg^{2+} forms a phosphate-mediated tertiary interaction with A2589 in Domain V near the PTC. The other Mg^{2+} forms a phosphate-mediated tertiary interaction with G2588 of Domain V. The D2 microcluster unstacks base A784 of Domain II, forming a pocket for ribosomal protein uL2 (Figure 6.4.5).

Ribosomal protein uL2 is unique in that its interaction with 23S rRNA appears to be more extensively mediated by Mg^{2+} than the interactions of other LSU proteins. The two Mg^{2+} ions of the D2 microcluster have (in sum) 12 first-shell ligands: five RNA phosphate oxyanions and seven waters (Petrov et al. 2012). Protein uL2 interacts directly with four of these waters through two highly conserved amino acids, alanine (or valine) and asparagine, interposed by a conserved methionine. The conserved uL2 sequence is most frequently alanine-methionine-asparagine (AMN) (Petrov et al. 2012). In *T. thermophilus,* these are residues Ala225, Met226, and Asn227 (Figure 6.4.7).

The observed interaction at the D2 microcluster is protein-water-Mg^{2+}-RNA, where the protein-water components interact via hydrogen bonds between a protein carbonyl and an acidic Mg^{2+}-bound water. In addition to directly coordinating the phosphates of A783, G784, G2588 and A2589 of the D2 microcluster, the Mg^{2+} ions, through their water ligands, interact with lone pairs of electrons on the backbone carbonyl of Ala225 and the side-chain carbonyl of Asn227 (Figure 6.4.7). Petrov et al. (2012) dissected and examined the stability of

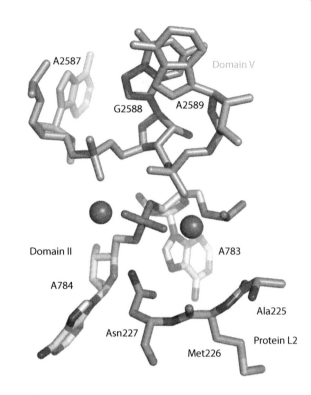

FIGURE 6.4.7 The microcluster D2-ribosomal protein L2 interaction, an RNA-Mg^{2+}-water-protein interaction. Phosphates of A783, A784, G2588, and A2589 coordinate two Mg^{2+} ions and form microcluster D2. These magnesium ions are also chelated by the backbone carbonyl of Ala225 and the side-chain carbonyl of Asn227. Structure of *T. thermophilus* [PDB 2J01 (obsolete). Superseded by 4V51] adapted from Petrov et al. (From Petrov, A.S. et al., *J. Phys. Chem. B*, 116, 8113–8120, 2012.)

each component of the D2 microcluster protein-water-Mg^{2+}-RNA interaction by using quantum methods.

In Mg^{2+} microcluster D2, the Mg^{2+} ions stabilize amino acid-water interactions, while the RNA ligands of the Mg^{2+} attenuate them slightly. Though Ala225 and Asn227 each have favorable and roughly equal interaction energies with water alone, interaction energies with water bound to Mg^{2+} in a hexahydrate (e.g., $[Mg(H_2O)_6]^{2+}$) are four-fold more favorable. The charge transfer component of the interaction energies implies that the charge on the amino acid carbonyl is transferred to the Mg^{2+} by polarization of the mediating water. To understand the effect of the RNA phosphates on the stability of the complex, two dimethyl phosphate molecules (Figure 6.4.2) were used to simulate chelation of $[Mg(H_2O)_4]^{2+}$ by the RNA bidentate clamp. Chelation of the metal by the RNA analog decreases the overall stability of the amino acid-water-Mg^{2+}-RNA complex by roughly one third, though interaction energies remained favorable. In other words, chelation of the Mg^{2+} by the phosphates of the RNA analog decreases polarization of the remaining water ligands of the Mg^{2+} and thus the acidity of the protein-mediating water, slightly attenuating the strength of the water-protein interaction. Even so, the amino acid-water-Mg^{2+}-RNA complex remains more than two-fold more stable than the amino acid-water complex, and the charge transfer component of the interaction energy remains strong enough to infer that some charge from the amino acid carbonyl remains with the Mg^{2+}.

The D2 microcluster stabilizes a segment of protein uL2 through orientation and polarization of water molecules. The AMN segment of protein uL2 interacts with the D2 microcluster through six hydrogen bonds. Four of these are with Mg^{2+}-polarized water molecules. The backbone carbonyl oxyanion of Ala225 hydrogen bonds (H-bonds) with two Mg^{2+}-bound waters of the same D2 Mg^{2+} and one 2'OH of rRNA A782. The side-chain carbonyl of Asn227 H-bonds with one water of each D2 Mg^{2+}, while the side-chain amine of Asn227 shares a hydrogen with a phosphate oxyanion of A784. To determine the role of the Mg^{2+} in supporting this interaction, Petrov et al. (2012) used DFT to examine the stability of the hydrated L2 segment, AMN-water, with and without the Mg^{2+} and RNA of the D2 microcluster. The AMN-water-Mg^{2+}-RNA complex is four-fold more stable than AMN-water, even though AMN does not interact directly with the Mg^{2+}. In the case of uL2, neither Mg^{2+} nor RNA alone stabilizes the interaction with ribosomal protein; it is rather the orientation and polarization of the water molecules by Mg^{2+} that enhance stability.

6.4.5 RIBOZYME FUNCTION

6.4.5.1 RIBOZYMES WITH MG^{2+}, FE^{2+}, AND OTHER CATIONS

Cations in the LSU play, to current knowledge, strictly structural roles in their support of the peptidyl transfer reaction. Although Mg^{2+} is necessary for folding and function of the LSU, there is not a direct role for any Mg^{2+} ion in the mechanism of peptidyl transferase (Noller et al. 1992). This is not the case for all catalytic RNAs.

The transition states of many ribozymes (see also Chapters 6.4 and 6.6) are directly stabilized by divalent metals (Butcher 2011; Johnson-Buck et al. 2011). In the activity of the *in vitro* evolved L1 ligase ribozyme (Robertson and Scott 2007), for example, the 5' α-phosphate of a nucleotide triphosphate (Figure 6.4.2) is polarized by a Mg^{2+} ion, priming the α-phosphorus for nucleophilic attack by an acidic 3'-OH of the adjacent nucleotide. Nucleophilic attack of the 3'-OH on the 5' α-phosphate completes the phosphodiester bond necessary for backbone ligation, generating a pyrophosphate leaving group (Robertson and Scott 2007). In the hammerhead ribozyme, an acidic metal-coordinated water at the active site is believed to donate a hydrogen, stabilizing the oxyanion leaving group during cleavage (Scott et al. 1995; Murray et al. 1998; Scott 2007). Unlike the L1 ligase ribozyme, which requires the divalent metal for activity, the hammerhead ribozyme retains some activity in the absence of divalent metals in high (e.g., >1.0 M) Na^+ concentration (Nakano et al. 2014) or low Na^+ concentration (100 mM) at freezing temperature (Lie and Wartell 2015).

The hammerhead ribozyme, in particular, has a promiscuous relationship with cations. Computed barriers in the free-energy landscapes of "metal migration" within the active site of the self-cleaving hammerhead ribozyme are generally

larger for Mg^{2+} and Zn^{2+} than for Mn^{2+} (Panteva et al. 2015). This observation is consistent with experimental work showing that at pH 7, in the presence of 0.1 M NaCl, the catalytic rate of self-cleavage by the hammerhead is greater with Mn^{2+} than with Mg^{2+} (Hunsicker and DeRose 2000).

Compaction and formation of secondary structure in the Group I intron P4-P6 domain ribozyme of *T. thermophila* are supported by Na^+ alone, but divalent cations are needed to induce tertiary structure (Cate et al. 1997; Deras et al. 2000). Athavale et al. (2012) used SHAPE to probe the P4-P6 domain in the presence of Na^+ and observed low SHAPE reactivity, indicative of canonical base pairing, in regions observed in the crystallographic structure to be helical, while loop regions (unpaired in the structure) tended to be much more reactive. Addition of mM concentrations of Mg^{2+} caused distinct changes in SHAPE reactivity, suggestive of specific magnesium binding and formation of tertiary RNA structure (Figure 6.4.8).

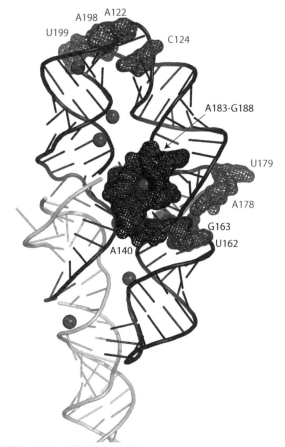

FIGURE 6.4.8 SHAPE reactivity of the Group I intron P4-P6 domain ribozyme of *T. thermophila* is altered in the presence of Mg^{2+} or Fe^{2+}. Sites of pronounced increases in SHAPE reactivity in the presence of Mg^{2+} or Fe^{2+} and Na^+ (compared with Na^+ alone) are red; decreases are blue. RNA for which there is moderate to no change in SHAPE reactivity with Mg^{2+} or Fe^{2+} is black. RNA for which SHAPE data are not available is shown in gray. Mg^{2+} ions are magenta spheres. SHAPE data are adapted from Athavale et al. (2012), displayed on PDB 1GID (Cate et al. 1996). The structure was crystallized in the presence of Co^{2+} (not shown), which was not present in SHAPE reactions by Athavale et al. (From Athavale, S.S. et al., *PLoS One*, 7, e38024, 2012.)

Most of the residues having an increase in SHAPE reactivity in the presence of Mg^{2+} (compared with Na^+ alone) bear a C2′-endo sugar pucker in structure PDB 1GID (Cate et al. 1996). This geometry of the ribose, typical of DNA, but rare and restricted to non-helical regions of RNA structures (Auffinger and Westhof 1997), was previously demonstrated to have enhanced reactivity by SHAPE (Cate et al. 1996; Vicens et al. 2007). The region with a decrease in SHAPE reactivity in the presence Mg^{2+} is consistent with packing of phosphates by Mg^{2+} in a core inaccessible to solvent (Cate et al. 1997). Mg^{2+} microclusters are present in this region of the P4-P6 RNA (Cate et al. 1997; Hsiao and Williams 2009).

$Fe^{2+}_{(aq)}$ oxidizes and precipitates rapidly in the presence of atmospheric oxygen, but under anoxic conditions, it induces folding of P4-P6 domain RNA to a near-native state. Athavale et al. (2012) demonstrated that the pattern of SHAPE reactivity in P4-P6 domain RNA in the presence of Fe^{2+} under anoxic conditions closely resembles that in the presence of Mg^{2+}. Regions of RNA with pronounced increases in SHAPE reactivity upon the addition of Mg^{2+} (compared with Na^+ alone) also increase in reactivity upon addition of Fe^{2+}. Parallel decreases in SHAPE reactivity were also observed in other regions of the RNA, suggesting that, in anoxia, Fe^{2+} replaces Mg^{2+}, without perturbation of the native structure of the P4-P6 ribozyme.

Catalysis by the L1 ligase or hammerhead ribozyme occurs at an enhanced rate in the presence of Fe^{2+}. Athavale et al. (2012) compared the catalytic rate of the L1 ligase ribozyme in Na^+, Na^+/Mg^{2+}, and Na^+/Fe^{2+}. As expected, the L1 ligase was inactive in Na^+ alone. Micromolar concentrations of Mg^{2+} or Fe^{2+} induced catalytic activity. Surprisingly, the rate of Fe^{2+}-induced catalysis was 25 times that of Mg^{2+}-induced catalysis. A similar affect was observed for the hammerhead ribozyme: catalysis in the presence of Fe^{2+} initiated at a rate three times higher than in the presence of Mg^{2+}. The interchangeability of Mg^{2+} and Fe^{2+} in RNA folding and function is supported by theoretical computational analyses that revealed only subtle differences in the coordination geometries of these RNA-divalent metal (RNA-M^{2+}) complexes. Furthermore, the calculations demonstrated that stability of RNA-M^{2+} complexes is enhanced when Mg^{2+} is replaced with Fe^{2+} (Petrov et al. 2011; Athavale et al. 2012), possibly due to the availability of d-orbitals in Fe^{2+}.

6.5.6 CONCLUSIONS

Fundamental biochemical processes of life as we know it depend on inorganic cations. Inorganic cations in association with RNA are always hydrated to some extent, and the importance of water in their interactions cannot be overstated. The unique properties of each cation—principally size and charge—alter the mobility and physicochemical properties of first-shell waters, which in turn affect surrounding water and non-water molecules. The energetic penalties associated with cation dehydration are a part of the net energy of reaction for every first-shell interaction with a biological ligand.

Inorganic cations have a special relationship with the phosphodiester backbone of nucleic acids. Backbone phosphates are anionic and highly polarizable, coordinating some cations with more compact geometry than others. The intricacies of a cation's individual properties can be viewed as magnified by interactions with phosphate esters, due to phosphate's enhanced polarizability relative to water. Magnesium stabilizes RNA through a continuum of associations, from neutralization of bulk negative charge in solution by mobile hexahydrates to compact first-shell coordination of up to four RNA backbone phosphates, while the range of associations of Na^+ and other large, monovalent cations is less diverse (Bowman et al. 2012).

The role of inorganic cations in protein translation provides an essential perspective on their importance in biology and relevance to the origin of life. Multiple high-resolution crystal structures of the ribosome reveal that the core structure of the LSU is essentially unchanged throughout extant life, including the locations of cations and their binding sites' geometries (Hsiao and Williams 2009). When compared with other functional RNAs (ribozymes), the rRNA of the LSU is immense in length and complexity. An understanding of the principles underlying folding of such a large RNA to the universal conformation of peptide bond synthesis informs our understanding of LUCA and the constraints under which it might have evolved.

The LSU provides every indication that it has been dependent on Mg^{2+} or Mg^{2+}-like divalent cations for folding and function since, and likely before, LUCA. Reoccurring motifs of rRNA in first-shell interactions with Mg^{2+} help to organize the three-dimensional structure of the LSU, though Mg^{2+} does not participate in catalysis directly. A recent study suggests that Mg^{2+} may also play an indirect role in stabilization of pre- and post-peptidyl transfer states (Polikanov et al. 2014). For perspective, Mg^{2+} ions observed in a crystal structure of the ribosome are shown in Figure 6.4.9.

These motifs include the bidentate RNA-Mg^{2+} clamps and Mg^{2+} microclusters (Hsiao and Williams 2009; Petrov et al. 2011). In an RNA-Mg^{2+} clamp, the phosphates of two consecutive nucleotides chelate a single Mg^{2+}, altering the geometry of the metal hydrate and in turn the local properties of the RNA. Computation suggests that Na^+ is incapable of stabilizing this type of RNA geometry. In a Mg^{2+} microcluster, one phosphate of an RNA-Mg^{2+} clamp serves as a bridge to a second Mg^{2+} ion, which, in three-dimensional structures, coordinates and restrains the phosphate of nucleotide(s) that are hundreds to more than a thousand nucleotides away in the primary sequence. Mg^{2+} microclusters induce non-canonical geometries to RNA bases, priming them for interaction with nucleotides from other domains and ribosomal proteins. Twelve of the 14 interdomain interactions in the LSU 23S rRNA of *T. thermophilus* are Mg^{2+}-induced (Lenz et al. 2017).

The Mg^{2+} dependencies of the LSU extend to interactions of first-shell waters with rRNA and ribosomal proteins. The first-shell waters of Mg^{2+} also stabilize the LSU. In one case, waters coordinating the Mg^{2+} of the D2 microcluster participate in hydrogen-bonding interactions with side-chain and backbone carbonyls of two highly conserved residues

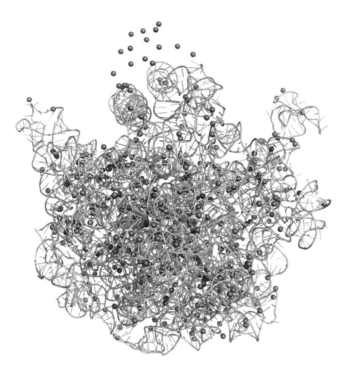

FIGURE 6.4.9 Prevalence of Mg²⁺ (magenta spheres) in the 23S rRNA crystal structure. Structure of *T. thermophilus* [PDB 2J01 (obsolete). Superseded by 4V51].

of ribosomal protein uL2 (Petrov et al. 2012). Mg²⁺ substantially increases the favorability and stability of the uL2-water interactions through orientation and polarization of the water molecules.

The ribosome is our telescope to origins and evolution of life (Gutell et al. 1994). Evolution appears to have perpetuated inorganic ion-dependent mechanisms that are essential and somewhat resilient to ion substitution. Three of the four ribozymes discussed, the Group I Intron P4-P6 domain, the hammerhead, and the ribosome LSU, are natural ribozymes. The P4-P6 domain and hammerhead ribozymes accommodate a broader range of cations in function and folding than the *in vitro* selected L1 ligase (Hunsicker and DeRose 2000; Doudna and Cech 2002; Travers et al. 2007; Athavale et al. 2012), yet the most central ribozyme of life, the ribosome, has just recently been shown to catalyze peptidyl transferase with a cation other than Mg²⁺ (Bray et al. submitted). Fe²⁺, which has been largely insoluble and toxic to life in the absence of protein chaperones since the Great Oxidation Event, appears to be a capable and, in some cases, an even superior substitute for Mg²⁺ in RNA folding and function.

Cation substitutions over evolution extend beyond nucleic acids. An almost-singular mechanism with transition states stabilized by divalent cations underpins a variety of nucleic-acid-processing proteins (Lykke-Andersen and Christiansen 1998; Steitz 1999; Doherty and Dafforn 2000; Lee et al. 2000; Yin and Steitz 2004; Ellenberger and Tomkinson 2008). As a group, but with species-specific

exceptions, these enzymes generally tolerate partial to complete substitutions of Mg²⁺ for Mn²⁺ and occasionally other cations. One analysis ties the early evolution of eukaryotes to an oxic Earth and prokaryotes to an anoxic one, based on trends in metal substitutions in metallo-proteins (Dupont et al. 2006). Another result of this work is that the overall abundance of metallo-proteins was found to scale with proteome size, suggesting that biology's dependence on inorganic cations is not vestigial.

ACKNOWLEDGMENTS

We thank Roger M. Wartell for critical review of an early version of the manuscript and Afrah Ghauri and Chieri Ito for careful editorial review. This work was supported by the National Aeronautics and Space Administration (NNX16AJ29G and NNX16AJ28G) and the National Science Foundation (1724274).

REFERENCES

Aguirre, J. D., and V. C. Culotta. 2012. Battles with iron: Manganese in oxidative stress protection. *Journal of Biological Chemistry* 287 (17): 13541–13548. doi:10.1074/jbc.R111.312181.

Anjem, A., S. Varghese, and J. A. Imlay. 2009. Manganese import is a key element of the OxyR response to hydrogen peroxide in *Escherichia coli*. *Molecular Microbiology* 72 (4): 844–858. doi:10.1111/j.1365-2958.2009.06699.x.

Athavale, S. S., A. S. Petrov, C. Hsiao et al. 2012. RNA folding and catalysis mediated by iron (II). *PLoS One* 7 (5): e38024. doi:10.1371/journal.pone.0038024.

Auffinger, P., and E. Westhof. 1997. Rules governing the orientation of the 2′-hydroxyl group in RNA11.Edited by I. Tinoco. *Journal of Molecular Biology* 274 (1): 54–63. doi:10.1006/jmbi.1997.1370.

Baes, C. F., and R. E. Mesmer. 1976. *Hydrolysis of Cations*. New York: Wiley.

Ban, N., R. Beckmann, J.H.D. Cate et al. 2014. A new system for naming ribosomal proteins. *Current Opinion in Structural Biology* 24: 165–169. doi:10.1016/j.sbi.2014.01.002.

Bernier, C., A. S. Petrov, C. Waterbury et al. 2014. RiboVision: Visualization and analysis of ribosomes. *Faraday Discuss* 169 (1): 195–207. doi:10.1039/C3FD00126A.

Bertini, I., H. B. Gray, E. I. Stiefel, and J. S. Valentine. 2007. *Biological Inorganic Chemistry, Structure and Reactivity*. Sausalito, CA: University Science Books.

Bock, C. W., A. K. Katz, G. D. Markham, and J. P. Glusker. 1999. Manganese as a replacement for magnesium and zinc: Functional comparison of the divalent ions. *Journal of the American Chemical Society* 121 (32): 7360–7372.

Bokov, K., and S. V. Steinberg. 2009. A hierarchical model for evolution of 23S ribosomal RNA. *Nature* 457 (7232): 977–980.

Bowman, J. C., T. K. Lenz, N. V. Hud, and L. D. Williams. 2012. Cations in charge: Magnesium ions in RNA folding and catalysis. *Current Opinion in Structural Biology* 22: 262–272. doi:10.1016/j.sbi.2012.04.006.

Bray, M., T. K. Lenz, J. C. Bowman, A. S. Petrov, A. R. Reddi, N. V. Hud, L. D. Williams, and J. B. Glass. Submitted. Ferrous iron folds rRNA and mediates translation.

Brown, I. D. 1992. Chemical and steric constraints in inorganic solids. *Acta Crystallographica Section B: Structural Science* 48: 553–572.

Butcher, S. E. 2011. The spliceosome and its metal ions. *Metal Ions in Life Sciences* 9: 235–251.

Cate, J. H., A. R. Gooding, E. Podell, K. Zhou, B. L. Golden, C. E. Kundrot, T. R. Cech, and J. A. Doudna. 1996. Crystal structure of a group I ribozyme domain: Principles of RNA packing. *Science* 273 (5282): 1678–1685.

Cate, J. H., R. L. Hanna, and J. A. Doudna. 1997. A magnesium ion core at the heart of a ribozyme domain. *Nature Structural Biology* 4 (7): 553–558.

Collins, K. D., G. W. Neilson, and J. E. Enderby. 2007. Ions in water: Characterizing the forces that control chemical processes and biological structure. *Biophysical Chemistry* 128(2–3): 95–104.

Cotruvo, J. A., and J. Stubbe. 2011. Class I ribonucleotide reductases: Metallocofactor assembly and repair *in vitro* and *in vivo*. *Annual Review of Biochemistry* 80: 733–767. doi:10.1146/annurev-biochem-061408-095817.

Deras, M. L., M. Brenowitz, C. Y. Ralston, M. R. Chance, and S. A. Woodson. 2000. Folding mechanism of the Tetrahymena ribozyme P4-P6 domain. *Biochemistry* 39 (36): 10975–10985.

Derry, L. A. 2015. Causes and consequences of mid-Proterozoic anoxia. *Geophysical Research Letters* 42 (20): 8538–8546.

Diebler, H., M. Eigen, G. Ilgenfritz, G. Maass, and R. Winkler. 1969. Kinetics and mechanism of reactions of main group metal ions with biological carriers [Review]. *Pure and Applied Chemistry* 20 (1): 93–116. doi:10.1351/pac196920010093.

Doherty, A. J., and T. R. Dafforn. 2000. Nick recognition by DNA ligases. *Journal of Molecular Biology* 296 (1): 43–56.

Doudna, J. A., and T. R. Cech. 2002. The chemical repertoire of natural ribozymes. *Nature* 418 (6894): 222–228.

Draper, D. E., D. Grilley, and A. M. Soto. 2005. Ions and RNA folding. *Annual Review of Biophysics and Biomolecular Structure* 34: 221–243.

Dupont, C. L, S. Yang, B. Palenik, and P. E. Bourne. 2006. Modern proteomes contain putative imprints of ancient shifts in trace metal geochemistry. *Proceedings of the National Academy of Sciences of the United States of America* 103 (47): 17822–17827.

Ellenberger, T., and A. E. Tomkinson. 2008. Eukaryotic DNA ligases: Structural and functional insights. *Annual Review of Biochemistry* 77: 313.

Fujimoto, B. S., J. M. Miller, N. S. Ribeiro, and J. M. Schurr. 1994. Effects of different cations on the hydrodynamic radius of DNA. *Biophysical Journal* 67 (1): 304–308. doi:10.1016/S0006-3495(94)80481-3.

Grilley, D., A. M. Soto, and D. E. Draper. 2006. Mg^{2+}–RNA interaction free energies and their relationship to the folding of RNA tertiary structures. *Proceedings of the National Academy of Sciences of the United States of America* 103 (38): 14003.

Gutell, R. R., N. Larsen, and C. R. Woese. 1994. Lessons from an evolving rRNA: 16S and 23S rRNA structures from a comparative perspective. *Microbiological Reviews* 58 (1): 10–26.

Hansen, J. L., T. M. Schmeing, D. J. Klein, J. A. Ippolito, N. Ban, P. Nissen, B. Freeborn, P. B. Moore, and T. A. Steitz. 2001. Progress toward an understanding of the structure and enzymatic mechanism of the large ribosomal subunit. *Cold Spring Harbor Symposia on Quantitative Biology* 66: 33–42.

Harel, A., Y. Bromberg, P. G. Falkowski, and D. Bhattacharya. 2014. Evolutionary history of redox metal-binding domains across the tree of life. *Proceedings of the National Academy of Sciences of the United States of America* 111 (19): 7042–7047. doi:10.1073/pnas.1403676111.

Hribar, B., N. T. Southall, V. Vlachy, and K. A. Dill. 2002. How ions affect the structure of water. *Journal of the American Chemical Society* 124 (41): 12302–12311. doi:10.1021/ja026014h.

Hsiao, C., S. Mohan, B. K. Kalahar, and L. D. Williams. 2009. Peeling the onion: Ribosomes are ancient molecular fossils. *Molecular Biology and Evolution* 26 (11): 2415–2425. doi:10.1093/molbev/msp163.

Hsiao, C., M. Tannenbaum, H. VanDeusen, E. Hershkovitz, G. Perng, A. Tannenbaum, and L. D. Williams. 2008. Complexes of nucleic acids with group I and II cations. In *Nucleic Acid Metal Ion Interactions*, edited by N. Hud, pp. 1–35. London, UK: The Royal Society of Chemistry.

Hsiao, C., and L. D. Williams. 2009. A recurrent magnesium-binding motif provides a framework for the ribosomal peptidyl transferase center. *Nucleic Acids Research* 37 (10): 3134–3142. doi:10.1093/nar/gkp119.

Hunsicker, L. M., and V. J. DeRose. 2000. Activities and relative affinities of divalent metals in unmodified and phosphorothioate-substituted hammerhead ribozymes. *Journal of Inorganic Biochemistry* 80 (3): 271–281. doi:10.1016/S0162-0134(00)00079-9.

Jackson, V. E., A. R. Felmy, and D. A. Dixon. 2015. Prediction of the pKa's of aqueous metal ion +2 complexes. *The Journal of Physical Chemistry A* 119 (12): 2926–2939. doi:10.1021/jp5118272.

Johnson-Buck, A. E., S. E. McDowell, and N. G. Walter. 2011. Metal ions: Supporting actors in the playbook of small ribozymes. *Metal Ions in Life Sciences* 9: 175–196.

Kankia, B. I. 2003. Binding of Mg^{2+} to single-stranded polynucleotides: Hydration and optical studies. *Biophysical Chemistry* 104 (3): 643–654.

Kankia, B. I. 2004. Inner-sphere complexes of divalent cations with single-stranded poly(rA) and poly(rU). *Biopolymers* 74 (3): 232–239. doi:10.1002/bip.20082.

Klein, D. J., P. B. Moore, and T. A. Steitz. 2004. The contribution of metal ions to the structural stability of the large ribosomal subunit. *RNA* 10 (9): 1366–1379.

Lavery, R., and B. Pullman. 1981. The molecular electrostatic potential, steric accessibility and hydration of Dickerson's B-DNA dodecamer d(CpGpCpGpApApTpTpCpGpCpG). *Nucleic Acids Research* 9 (15): 3765–3777.

Lee, J. Y., C. Chang, H. K. Song, J. Moon, J. K. Yang, H.-K. Kim, S.-T. Kwon, and S. W. Suh. 2000. Crystal structure of NAD^{+}-dependent DNA ligase: Modular architecture and functional implications. *The EMBO Journal* 19 (5): 1119–1129.

Lenz, T. K., A. M. Norris, N. V. Hud, and L. D. Williams. 2017. Protein-free ribosomal RNA folds to a near-native state in the presence of Mg^{2+}. *RSC Advances*. doi:10.1039/c7ra08696b.

Lie, L., and R. M. Wartell. 2015. Ligation of RNA oligomers by the Schistosoma mansoni hammerhead ribozyme in frozen solution RNA. 82 (2–3), 81–92.

Lykke-Andersen, J., and J. Christiansen. 1998. The C-terminal carboxy group of T7 RNA polymerase ensures efficient magnesium ion-dependent catalysis. *Nucleic Acids Research* 26 (24): 5630–5635. doi:10.1093/nar/26.24.5630.

Mähler, J., and I. Persson. 2012. A study of the hydration of the alkali metal ions in aqueous solution. *Inorganic Chemistry* 51 (1): 425–438. doi:10.1021/ic2018693.

Manning, G. S. 1969. Limiting laws and counterion condensation in polyelectrolyte solutions I. Colligative properties. *The Journal of Chemical Physics* 51 (3): 924–933. doi:10.1063/1.1672157.

Martin, J. E., and J. A. Imlay. 2011. The alternative aerobic ribonucleotide reductase of *Escherichia coli*, NrdEF, is a manganese-dependent enzyme that enables cell replication during periods of iron starvation. *Molecular Microbiology* 80 (2): 319–334. doi:10.1111/j.1365-2958.2011.07593.x.

Mayaan, E., K. Range, and D. M. York. 2004. Structure and binding of Mg(II) ions and di-metal bridge complexes with biological phosphates and phosphoranes. *JBIC Journal of Biological Inorganic Chemistry* 9 (8): 1034–1035. doi:10.1007/s00775-004-0608-2.

Morgan, B., and O. Lahav. 2007. The effect of pH on the kinetics of spontaneous Fe(II) oxidation by O_2 in aqueous solution—basic principles and a simple heuristic description. *Chemosphere* 68 (11): 2080–2084. doi:10.1016/j.chemosphere.2007.02.015.

Mortimer, S. A., and K. M. Weeks. 2007. A fast-acting reagent for accurate analysis of RNA secondary and tertiary structure by SHAPE Chemistry. *Journal of the American Chemical Society* 129 (14): 4144–4145.

Murray, J. B., D. P. Terwey, L. Maloney, A. Karpeisky, N. Usman, L. Beigelman, and W. G. Scott. 1998. The structural basis of hammerhead ribozyme self-cleavage. *Cell* 92 (5): 665–673. doi:10.1016/S0092-8674(00)81134-4.

Nakano, S.-i., Y. Kitagawa, D. Miyoshi, and N. Sugimoto. 2014. Hammerhead ribozyme activity and oligonucleotide duplex stability in mixed solutions of water and organic compounds. *FEBS Open Bio* 4: 643–650. doi:10.1016/j.fob.2014.06.009.

Noller, H. F., V. Hoffarth, and L. Zimniak. 1992. Unusual resistance of peptidyl transferase to protein extraction procedures. *Science* 256 (5062): 1416–1419. doi:10.1126/science.1604315.

Okafor, C. D., K. A. Lanier, A. S. Petrov, S. S. Athavale, J. C. Bowman, N. V. Hud, and L. D. Williams. 2017. Iron mediates catalysis of nucleic acid processing enzymes: Support for Fe(II) as a cofactor before the Great Oxidation Event. *Nucleic Acids Research* 45 (7): 3634–3642. doi:10.1093/nar/gkx171.

Panteva, M. T., G. M. Giambaşu, and D. M. York. 2015. Force field for Mg^{2+}, Mn^{2+}, Zn^{2+}, and Cd^{2+} ions that have balanced interactions with nucleic acids. *The Journal of Physical Chemistry B* 119 (50): 15460–15470. doi:10.1021/acs.jpcb.5b10423.

Petrov, A. S., C. R. Bernier, B. Gulen, C. C. Waterbury, E. Hershkovitz, C. Hsiao, S. C. Harvey, N. V. Hud, G. E. Fox, R. M. Wartell, and L. D. Williams. 2014. Secondary structures of rRNAs from all three domains of life. *PLoS One* 9 (2): e88222. doi:10.1371/journal.pone.0088222.

Petrov, A. S., C. R. Bernier, C. L. Hsiao et al. 2012. RNA-magnesium-protein interactions in large ribosomal subunit. *Journal of Physical Chemistry B* 116 (28): 8113–8120. doi:10.1021/jp304723w.

Petrov, A. S., J. C. Bowman, S. C. Harvey, and L. D. Williams. 2011. Bidentate RNA-magnesium clamps: On the origin of the special role of magnesium in RNA folding. *RNA* 17 (2): 291–297. doi:10.1261/rna.2390311.

Polikanov, Y. S., T. A. Steitz, and C. A. Innis. 2014. A proton wire to couple aminoacyl-tRNA accommodation and peptide bond formation on the ribosome. *Nature Structural and Molecular Biology* 27 (9): 787–793. doi:10.1038/nsmb.2871.

Porschke, D. 1979. The mode of Mg^{++} binding to oligonucleotides. Inner sphere complexes as markers for recognition? *Nucleic Acids Research* 6 (3): 883–898.

Rashin, A. A., and B. Honig. 1985. Reevaluation of the Born model of ion hydration. *Journal of Physical Chemistry* 89 (26): 5588–5593.

Robertson, M. P., and W. G. Scott. 2007. The structural basis of ribozyme-catalyzed RNA assembly. *Science* 315 (5818): 1549–1553. doi:10.1126/science.1136231.

Roychowdhury-Saha, M., and D. H. Burke. 2006. Extraordinary rates of transition metal ion-mediated ribozyme catalysis. *RNA* 12 (10): 1846–1852. doi:10.1261/rna.128906.

Scott, W. G., J. T. Finch, and A. Klug. 1995. The crystal structure of an all-RNA hammerhead ribozyme: A proposed mechanism for RNA catalytic cleavage. *Cell* 81 (7): 991–1002. doi:S0092-8674(05)80004-2.

Scott, W. G. 2007. Ribozymes. *Current Opinion in Structural Biology* 17 (3): 280–286.

Shannon, R. 1976. Revised effective ionic radii and systematic studies of interatomic distances in halides and chalcogenides. *Acta Crystallographica Section A* 32 (5): 751–767. doi:10.1107/S0567739476001551.

Steitz, T. A. 1999. DNA polymerases: Structural diversity and common mechanisms. *Journal of Biological Chemistry* 274 (25): 17395–17398. doi:10.1074/jbc.274.25.17395.

Stumm, W., and J. J. Morgan. 1996. *Aquatic Chemistry: Chemical Equilibria and Rates in Natural Waters*, 3rd ed. New York: John Wiley & Sons.

Torrents, E., P. Aloy, I. Gibert, and F. Rodriguez-Trelles. 2002. Ribonucleotide reductases: Divergent evolution of an ancient enzyme. *Journal of Molecular Evolution* 55 (2): 138–152. doi:10.1007/s00239-002-2311-7.

Travers, K. J., N. Boyd, and D. Herschlag. 2007. Low specificity of metal ion binding in the metal ion core of a folded RNA. *RNA* 13 (8): 1205–1213. doi:10.1261/rna.566007.

Ushizaka, S., K. Kuma, and K. Suzuki. 2011. Effects of Mn and Fe on growth of a coastal marine diatom Talassiosira weissflogii in the presence of precipitated Fe(III) hydroxide and EDTA-Fe(III) complex. *Fisheries Science* 77 (3): 411–424.

Vicens, Q., A. R. Gooding, A. Laederach, and T. R. Cech. 2007. Local RNA structural changes induced by crystallization are revealed by SHAPE. *RNA* 13 (4): 536–548.

Wimberly, B. T., D. E. Brodersen, W. M. Clemons, Jr., R. J. Morgan-Warren, A. P. Carter, C. Vonrhein, T. Hartsch, and V. Ramakrishnan. 2000. Structure of the 30S ribosomal subunit. *Nature* 407 (6802): 327–339.

Wolfe-Simon, F., V. Starovoytov, J. R. Reinfelder, O. Schofield, and P. G. Falkowski. 2006. Localization and role of manganese superoxide dismutase in a marine diatom. *Plant Physiology* 142 (4): 1701–1709. doi:10.1104/pp.106.088963.

Yin, Y. W., and T. A. Steitz. 2004. The structural mechanism of translocation and helicase activity in T7 RNA polymerase. *Cell* 116 (3): 393–404. doi:10.1016/S0092-8674(04)00120-5.

Zhang, C., C. Lu, Q. Wang, J. W. Ponder, and P. Ren. 2015. Polarizable multipole-based force field for dimethyl and trimethyl phosphate. *Journal of Chemical Theory and Computation* 11 (11): 5326–5339. doi:10.1021/acs.jctc.5b00562.

6.5 The Origin of Life as an Evolutionary Process
Representative Case Studies

Iris Fry

CONTENTS

6.5.1 INTRODUCTION

The question of the origin of life is among the most difficult problems faced by science today. Evaluating what might have been the chemistry of life's emergence around 4 billion years ago involves inherent uncertainties. Thus, the scientific goal is to reconstruct a possible scenario, under assumed prebiotic environmental conditions, that could have led to the first living systems. Optimistically, as Earth science, astronomy, and other relevant fields of science gain more knowledge about conditions on the primordial Earth, the gap between the possible and the actual will become narrower. Various scenarios delineating possible mechanisms of emergence are being raised by researchers, but none of them has as yet gained the consensus of the origin-of-life community and none has as yet been simulated in full in the laboratory. Investigators are nevertheless convinced that the transition from chemical compounds on the primordial Earth to the first living systems *was an evolutionary process* (Peretó 2012; Szostak 2012a; Pascal et al. 2013). As was asserted by the late Leslie Orgel, among the most prominent researchers in the field, "[A]ny "living" system must come into existence either as a consequence of a long evolutionary process or a miracle" (Orgel 1973, p. 192). According to what I suggest to call *the evolutionary hypothesis of the origin of life,* the living systems at the origin of biological evolution did not arise through a staggeringly improbable chance event. Rather, they resulted from a long series of stages growing in complexity. Clearly, an evolutionary process prior to the existence of life was different from the process of biological evolution. According to the evolutionary hypothesis suggested in this chapter, prebiotic evolution involved highly constrained physical and chemical processes that led to the rise of chemical infrastructures capable first of "a poor man's natural selection" (Vasas et al. 2012, p. 1) and gradually of full-blown natural selection.

The nature of the chemical structures that could have undergone evolution during the emergence of life and the details of their prebiotic synthesis are still debated. However, theoretical models and experiments conducted in various laboratories point to various physical and chemical factors in specific environmental conditions that could have changed the odds of prebiotically synthesized molecules and molecular systems away from statistical randomness and thus constrained and channeled the emergence process (De Duve 1991, p. 141; Copley et al. 2007, pp. 432–433). The crucial role of prebiotic inorganic and organic catalysts and of physico-chemical selection events in allowing the emergence of evolvable infrastructures and their gradual evolution is noted in various scenarios in the field. Growing understanding in the last few decades of the chemistry and functions of RNA, the study of primitive cellular-like structures and of prebiotic catalysts and their possible role in early metabolism, and the discovery of sites on Earth that might have been conducive to the rise of life, all combined to make the evolutionary hypothesis of the origin of life a robust research framework.

My aim in this paper is to examine major theoretical and empirical lines of investigation based on the heuristic, fruitful value of the evolutionary conception. Various scenarios attack the enormous challenges still facing the field from different perspectives. I will also indicate how they differ in their evaluation of possible evolutionary processes leading to the emergence of life. Chapter 3.1 of this volume examines views rejecting the possibility of evolution prior to the existence of life, most notably, those of the renowned evolutionary biologist Richard Dawkins (1986, pp. 139–145, 2006, pp. 134–141). Dawkins's view that life arose by a highly improbable single-step chance event is not only contested by numerous probability calculations offered by researchers in the field, but also most significantly by the more probable alternatives suggested by the case studies examined here.

6.5.2 A SHORT HISTORICAL BACKGROUND: EARLY EVOLUTIONARY THEORIES OF THE ORIGIN OF LIFE

Several scientists, notably the biochemist Alexander Oparin in the Soviet Union and the geneticist and biochemist J. B. S. Haldane in England, independently published pioneering theories on the origin of life in the 1920s. Their theories (Oparin

1953 [1936], 1967 [1924]; Haldane 1967 [1929]) were based on new developments in biochemistry and genetics, on the chemistry of colloids, and on new data from geochemistry and astronomy. Their heterotrophic hypotheses upheld that organic compounds were synthesized from inorganic molecules in a primordial reducing atmosphere under the impact of various energy sources. They both contended that these organic building blocks that dissolved in the ancient ocean, described by Haldane as "a hot dilute soup" (Haldane 1967, p. 246), were transformed into structures that could have undergone evolutionary processes, leading to "protoorganisms" and later to life (see Fry 2000, pp. 65–78). A lesser-known contribution to the study of the origin of life was made already in the 1910s, in a few papers published by the American physicist and psychophysiologist Leonard Thompson Troland (see Muller 1966, pp. 495–498).

As a biochemist, Oparin saw metabolism, supported by interacting enzymatically catalyzed reactions, as the defining characteristic of life. He hypothesized that organic polymers synthesized in the "soup" self-organized into cell-like structures that selectively absorbed organic material from the solution, grew in size, and divided. Oparin saw these structures, the so-called "coacervates," as endowed with metabolism-enabling organization that could have been transmitted from "parents" to "offspring" during these divisions. Based on such "heredity," assured via reproduction through division, and variations among the coacervates, Oparin (1953, pp. 193–195) envisaged primitive processes of natural selection and the evolution of metabolism. Haldane was first and foremost inspired by the rise of genetics and the discovery of viruses and saw self-reproduction as the defining feature of life. He hypothesized that large self-reproducing organic molecules synthesized in the soup (which functioned as a "vast chemical laboratory") were analogous to genes and constituted the first "living or half-living things" (Haldane 1967, p. 247). Leonard Troland's abstract analysis focused on the logic of the emergence of life and represented in a most straightforward way what later came to be called the "replication-first" approach. Moreover, unknowingly, he predicted the future discovery of the ribozymes. He suggested the appearance in the ancient ocean of a "genetic enzyme" capable of catalyzing both its own formation and the synthesis of an envelope around itself. The reproduction of such "enzymes" was also the basis for variation (mutation) and led to natural selection and evolution. Troland acknowledged the high improbability of the spontaneous synthesis of such a "genetic enzyme" but insisted that all that was needed was the appearance of a single copy of such entity, the probability of which was also aided by the vast stretches of time available for the process (Troland 1914, pp. 102–104, 105, 110–112; Fry 2006, p. 27).

Oparin's biochemical conception characterized life as a multi-molecular, multi-functional metabolic system, whereas Haldane's and Troland's genetic theories identified early life with molecular self-reproduction. These two concepts later became the two main lines of research in the origin-of-life field, and the conflict between their supporters grew in intensity, especially with the establishment of molecular biology. These two "camps," the "metabolism-first" and the "gene-first," clearly suggested very different prebiotic evolvable structures.

6.5.2.1 Evolution in the Test Tube

The discoveries of the structure of DNA by Watson and Crick in the early 1950s and of the mechanisms by which nucleic acids function as the hereditary material transformed the study of evolution and consequently also the study of the emergence of life. Several researchers raised the question whether natural selection and evolution might occur not only in a population of organisms but also in a population of molecules. Beginning in the 1960s and 1970s, experiments carried out in various laboratories demonstrated Darwinian evolution "in the test tube" (*in vitro*) among genetic molecules capable of replicating and mutating. The American molecular biologist Sol Spiegelman and his group isolated a single strand of viral RNA and its replicating enzyme and, upon addition of the four types of nucleotides, showed RNA replication outside the cell (Mills et al. 1967). They also examined the impact of changing the "environmental conditions" in the reaction mixture. For example, when replication was carried under high temperatures or in the presence of an RNA-degrading enzyme, mutants that could form spatial structures more resistant to high temperatures or to a degrading enzyme were selected (Safhill et al. 1970). Extended experiments exploring molecular evolution *in vitro* and its application to the study of the origin of life were also conducted in the laboratory of the German physical chemist Manfred Eigen at the Max Planck Institute in Göttingen (Eigen et al. 1981, pp. 82–85).

The general conclusion from these and later experiments was that any group of entities could evolve by natural selection, provided it conformed to the specific conditions of reproduction, variations, inheritance, relative advantage conferred by some of these variations, and competition (Orgel 1973, pp. 145–159; Cairns-Smith 1986, pp. 2–3). It was nevertheless realized that a realistic model for the origin of life on the ancient Earth cannot involve a sophisticated replicating enzyme like the one isolated by Spiegelman and Eigen from highly evolved cells. The quest was thus for much more primitive systems, more fitting to primordial conditions.

6.5.2.2 The Chicken and Egg Conundrum—A Dead-End or Maybe Not?

Researchers searching for such systems were facing a challenge raised by a major discovery of molecular biology—the interdependence of nucleic acids and proteins in terms of their synthesis and function. How could such a "chicken and egg" circle first emerge? Which came first, proteins or nucleic acids? From the early 1960s, origin-of-life scientists explored the two options of nucleic-acid-only (or nucleic-acid-first) and proteins-first, in order to "break the vicious circle" and establish a basis for an evolutionary process. These two options came to be described as the gene-first versus metabolism-first lines of research.

Leslie Orgel and his group at the Salk Institute in California conducted, for many years, experiments aiming to explore the possibility of natural selection and evolution in a system without a replicating enzyme. Although early experiments showed promise, further results were disappointing. For several empirical reasons, a full cycle of non-enzymatic replication of oligonucleotides could not be achieved (Orgel 1994). Parallel to Orgel's nucleic-acid-only experiments, an evolutionary protein-first, metabolic scenario was formulated by Sidney Fox. He demonstrated, under supposedly prebiotic conditions, the synthesis of amino acids and peptide-like oligomers, or "proteinoids," manifesting weak catalytic activity. He also achieved the synthesis of microspheres out of these proteinoids that were engaged in some metabolic activity (Fox and Dose 1977; see, De Duve 2003, pp. 567–568). Fox, similar to several later proponents of metabolic theories, claimed that these cell-like systems could have served as infrastructure for evolution during the emergence of life through Oparin-like alternative processes of inheritance and variation without nucleic acids (see, Fry 2011, pp. 8–9).

Awareness of the implications of the chicken-and-egg problem for attaining a scientific understanding of the emergence of life resulted in diverse reactions in the 1960s and 1970s. A highly pessimistic attitude was adopted by the eminent French biologist Jacques Monod. In his book *Chance and Necessity* published in 1970, Monod described the chicken-and-egg problem as unsolvable and a "veritable enigma" (Monod 1974 [1970], p. 135) and acknowledged that this conclusion precluded a scientific solution to the origin-of-life problem (1974, pp. 135–136). The puzzle of the chicken-and-egg problem motivated a very different reaction, expressed in three seminal publications by Carl Woese (1967), Francis Crick (1968), and Leslie Orgel (1968). All three were committed to evolutionary thinking as means to search for a scientific solution to the origin-of-life problem, including the origin of the genetic code and the mechanism of translation. Crick and Orgel in their papers and Woese in his book examined molecules and mechanisms in extant cells from an evolutionary perspective and raised an unconventional idea that was formulated almost 20 years later as the RNA-world hypothesis (Woese 1967, pp. 179–195). Based on several factors, especially RNA's tertiary structure, each of them suggested that RNA molecules could have also functioned on the ancient Earth as enzymes (see Crick 1968, pp. 371–372; Orgel 1968, p. 387). And yet, as long as there was no experimental indication suggesting otherwise, the common belief was that only proteins could have functioned as catalysts and nucleic acids acted only as replicators (see Orgel 1968, p. 387). Thus, the chicken-and-egg puzzle continued to cast a huge shadow on efforts to understand the origin of life (see also, Crick 1981, p. 72; Jacob 1982, pp. 305–306; Mayr 1982, pp. 583–584).

6.5.3 THE RNA WORLD: HIGH HOPES, ENORMOUS DIFFICULTIES

In the early 1980s, RNA molecules functioning as enzymes, given the name "ribozymes," were unexpectedly discovered in extant cells (Kruger et al. 1982; Guerrier-Takada et al. 1983). In 1986, Walter Gilbert coined the term the "RNA World," referring to a hypothetical ancient chemical "world" in which RNA molecules were both chicken and egg, both catalysts and replicators, catalyzing their own replication and later also the synthesis of proteins and DNA (Gilbert 1986). The first discovered ribozymes were found to catalyze the cutting and joining of RNA segments. Since then, various ribozymes were also isolated in *in vitro* evolution experiments. Experimental efforts in several laboratories revealed that "the ribosome uses RNA catalysis to perform the key activity of protein synthesis in all extant organisms, so it must have done so in the Last Universal Common Ancestor (LUCA)" (Cech 2012, p. 2). This conclusion is among the strongest indications that an RNA world did exist during the emergence of life (see Benner et al. 2012, p. 4).

It was first optimistically assumed that prebiotic chemistry enabled the synthesis of RNA nucleotides; their assembly into RNA sequences; and the presence in this RNA pool of various ribozymes, especially a self-replicating ribozyme, an RNA replicase. If such was the case, the RNA world would have been indeed "the molecular biologist's dream," serving as basis for early evolution through natural selection (Joyce and Orgel 2006, pp. 23–24). This optimistic picture was called into question following several decades of failed experiments to show that the RNA world could have been a "prebiotic product," that is, the first genetic "world" to emerge on Earth by physico-chemical prebiotic means (Anastasi et al. 2007, p. 721). Though some of the originally anticipated goals, especially the synthesis of oligonucleotides, were achieved to some extent (see Ferris 2005; Monnard 2007; Pino et al. 2008), a failure to demonstrate prebiotic synthesis of activated nucleotides still poses an enormous obstacle. Also, so far, no ribozyme replicase has been found in extant cells or has been evolved *in vitro*. The late Orgel, a veteran proponent of the view that life could have begun only on the basis of the emergence of a replicating system, that is, *the gene-first view*, came to adopt the *RNA-later position*. He and his colleague Joyce described the reality of the *RNA-first option* as "the prebiotic chemist's nightmare" (Joyce and Orgel 2006, pp. 23–24). Assuming that RNA could not have been a prebiotic product, an earlier, pre-RNA "world" would have required, first, an easier prebiotic synthesis of its components. It should have nevertheless been based on some sort of genetic system; should have involved various catalysts, including a "replicase"; and eventually should have evolved into the RNA world (see Orgel 2003, p. 213). Various pre-RNA candidates were considered (e.g., Nielsen 1993, 2009; Eschenmoser 1999, 2007a, pp. 12823–12827). Some of these candidates do form stable double helices and cross-pair with RNA strands; however, they do not appear to be synthesized prebiotically more readily than RNA (Orgel 2004, pp. 114–115; Anastasi et al. 2007, pp. 733–736).

Insisting on the evolutionary nature of the emergence of life, but viewing organic chemistry in general and nucleotides in particular as prohibitively "high-tech" for prebiotic chemistry, the Scottish physical chemist and crystallographer Graham Cairns-Smith suggested a radical "out of the box" idea. Already in the 1970s, he envisaged a "low-tech" genetic system completely unrelated to RNA, made of clay minerals that were and are among the most ubiquitous materials

on Earth. Constituting a sort of scaffolding, such minerals could have, according to Cairns-Smith, replicated, mutated, and evolved to catalyze the synthesis of organic molecules (Cairns-Smith 1986, 2008). Cairns-Smith's "scaffolding" conception was a key theoretical and philosophical contribution to the evolutionary conception of the origin of life. Yet, his specific clay scenario was criticized by a number of researchers (Hazen 2005, pp. 160–164; Morowitz 1992, p. 27; Orgel 1994, p. 61, 1998, p. 493; Yarus 2011, p. 3). Most practitioners in the field are convinced of the continuity between primitive life and extant life in terms of organic chemistry and biochemistry (e.g., Morowitz 1992, pp. 90–91; Orgel 2003, p. 213, 2004, p. 100; De Duve 2005a, pp. 17–21; Martin and Russell 2007, p. 1922).

Convinced of the need to find an evolutionary path and to overcome the RNA-nucleotides-prebiotic hurdle, Orgel (2003) also raised an idea of a pre-RNA "organism." Taking into account the central role of amino acids in biochemistry, combined with the easy synthesis of amino acids and peptides under prebiotic conditions, Orgel (2003, p. 214) suggested that "it is worth exploring the possibility that a self-replicating peptide was a precursor of RNA because it seems to be the only major, biochemistry-based, game in town." Orgel was inspired by the work of Ulf Diederichsen and colleagues in organic chemistry, who produced self-replicating polymers, in which the monomers were amino acids whose β hydrogen was replaced by a nitrogen base. Specifically, the polymers' backbone was made of alanines and the substituting bases were guanine and cytosine (Diederichsen 1996, 1997). Orgel pointed to the major achievement of Diederichsen's work that overcame the structural obstacle posed by forming a pairing structure of two homochiral peptide chains. By synthesizing chains in which L- and D-alanines alternated, a very stable hydrogen-bond interaction between the bases guanine and cytosine was made possible (Orgel 2003, p. 215, 2004, pp. 115–116).

Orgel (2004, p. 116), though indicating that the monomers of alanyl nucleic acid (ANA) play no role in biochemistry, nevertheless found Diederichsen's ideas attractive in the context of prebiotic chemistry. Not only was the use of the two amino acid enantiomers advantageous in a prebiotic racemic environment, but Orgel (2003, p. 217) also saw the relevance of this work in suggesting "a generic pairing structure." Thus, in his motivation to find a prebiotically plausible genetic material that would provide a basis for replication and evolution, Orgel also suggested to explore pairing structures made of amino acids with side chains whose specific interactions are based on charge or size. He wondered whether the stability of nucleic acid double helix, provided by the non-specific stacking interactions between bases, could have been achieved without the bases on the surface of a mineral (Orgel 2003, p. 217, 2004, p. 116). Several research groups have demonstrated self-replication of peptides, for example, through the autocatalytic ligation of two shorter peptides (Lee et al. 1996, 1997; Issac and Chmeilewski 2002). Orgel (2008, pp. 11–12) noted in his last 2008 paper that this is an interesting direction to explore. However, he questioned the prebiotic relevance of these and previous experiments on the self-replication of peptides. More recent work demonstrated the ability of short peptides that self-assemble into β-sheet structures to self-replicate (Bourbo et al. 2011; Brack 2014).

It is important to note that despite the enormous challenges still faced by the RNA-first and RNA-later approaches, they are among the most favored lines of research in the field, owing to the evolutionary explanation they provide to the emergence of life (see Schwartz 2013, p. 784).

6.5.3.1 RNA-First, Despite It All

6.5.3.1.1 Possible Synthesis of Ribonucleotides

Supporters of the RNA-first option acknowledge the fact that almost half a century of effort by researchers did not lead to a prebiotically plausible generation and oligomerization of activated nucleotides. They claim, however, that a large number of potential chemical routes toward these goals from prebiotic feedstock molecules were not explored yet (Anastasi et al. 2007, pp. 721–722). Furthermore, based on the fact that, so far, convincing pre-RNA candidates were not found (Anastasi et al. 2007, pp. 733–737) and on the clear advantage of the "real thing," nucleotides and oligonucleotides (James and Ellington 1995, pp. 519–520; see Eschenmoser 2007b, p. 559), it is considered premature to conclude that the RNA-first option is not viable. Expanding on several previous experimental leads (see, Orgel 2004, pp. 105–106), recent work by John Sutherland and colleagues has overcome some of the most difficult problems in a possible prebiotic synthesis of nucleotides (Powner et al. 2009; Sutherland 2010). This line of research questions the long-held "dogma" in prebiotic chemistry that RNA nucleotides were originally assembled from their components, nucleobase, ribose, and phosphate. It is suggested now to replace this dogma by a novel approach, by which free ribose and the free bases are bypassed in the synthetic procedure and are produced from a common precursor. The synthesis of activated pyrimidine ribonucleotides was successfully carried out according to this approach (Powner et al. 2009). Significantly, the synthesis followed the methodology of "systems chemistry," in which "reactants from different stages of a pathway are allowed to interact" (Szostak 2009, p. 171). In subsequent work, fragments of ribose were synthesized using similar systems chemistry, involving simple prebiotic oxygenous and nitrogenous components (Ritson and Sutherland 2012). So far, however, the synthesis of activated purine ribonucleotides was not achieved (but see Powner et al. 2010). Criticism was raised against "an excessive Deux ex machina" type of prebiotic chemistry assumed in the synthesis of activated pyrimidine ribonucleotides (Benner et al. 2012, p. 7). It was also pointed out that, though overcoming the problem of ribose synthesis, the sequence of reactions suggested by the Sutherland group raised other questions (Schwartz 2013, pp. 784–785).

As will be seen later, several metabolically inclined scenarios suggest that organic compounds in general and nucleobases and nucleotides in particular had a good chance of being synthesized and oligomerized on the surface of minerals and in interaction with other organic substrates under the conditions at hydrothermal vents (see, e.g., Wächtershäuser 1992; Copley et al. 2007; Martin and Russell 2007). The delivery of nucleobases to Earth, together with other organic compounds, within carbonaceous chondrite meteorites was also proposed (Martin et al. 2008b, but see a more cautious evaluation of the exogeneous-organics claim, Pizzarello and Shock 2010; Damer and Deamer 2015, p. 875).

It might be of interest that following the Sutherland line of study on ribose, mentioned earlier, the Sutherland group engaged in experiments aimed to show that precursors of ribonucleotides, amino acids, and lipids can be commonly derived from hydrogen cyanide and hydrogen sulfide under certain prebiotic conditions. They also suggested a geochemical scenario that could have enabled such a common origin (Patel et al. 2015; Sutherland 2016).

6.5.3.1.2 Experimental Synthesis of Oligonucleotides

Assuming the discovery of a plausible prebiotic mechanism for the synthesis of ribonucleotides, the next stage in an RNA-world scenario must have been the polymerization of these monomers that somehow underwent activation to produce RNA sequences. The work of James Ferris and colleagues during the last decades has yielded a growing body of knowledge on the synthesis of short-, medium-, and longer-size RNA oligomers on the surface of minerals, in particular some montmorillonite clays (Huang and Ferris 2006; Joshi et al. 2009; Brack 2017). Another well-researched method of concentrating and polymerizing activated ribonucleotides uses ice eutectic phases prepared from dilute solutions of monomers and metal ions, Mg^{+2} and Pb^{+2}, which act as catalysts. In such an environment, "two phases co-exist and form the eutectic-phase system: a solid (the ice crystals made of pure water) and a liquid phase containing most solutes" (Monnard and Ziock 2008, p. 1524). In the experiments performed by Monnard and colleagues, this particular environment allowed for the concentration of the activated nucleotide monomers out of the aqueous medium, and their condensation catalyzed by the metal ions. When this mixture is frozen and maintained for days at $-18°C$, that is, below its freezing point, the monomers are concentrated as eutectics in the ice matrix (Monnard et al. 2003; Monnard and Ziock 2008). Polymerization in this case probably occurs not by adsorption to ice surfaces but in cavities between ice crystals, with monomers and catalysts being highly concentrated. Such experiments are viewed as possible simulation of ice deposits on the primordial Earth that facilitated the synthesis of short- and medium-size RNA or RNA-like sequences (Monnard et al. 2003, pp. 13735, 13738, 13739).

It was reported recently that condensation of activated RNA monomers into short RNA chains was achieved via the catalysis by the dipeptide seryl-histidine (ser-his) in water-ice eutectic phase. As noted by the authors: "Because peptides are much more likely products of spontaneous condensation than nucleotide chains, their potential as catalysts for the formation of RNA is interesting from the origin-of-life perspective" (Wieczorek et al. 2013). The dipeptide ser-his was chosen as a catalyst of the condensation of activated RNA monomers, based on its record as the smallest known peptide catalyzing the dissociation of peptide and phosphodiester bonds and as a catalyst of the condensation of amino acids. Since "all chemical reactions are in principle reversible," the possibility of the peptide catalyzing also the condensation of phosphodiester bonds, so far not checked, was put to the test (Wieczorek et al. 2013).

Lipids were also demonstrated to catalyze polymerization of RNA under conditions of wetting and drying cycles (Rajamani et al. 2008). Using a different technique, it was found by Di Mauro and his group that spontaneous ligation of short RNA segments (10–24 monomers) in water at acidic pH and moderate temperatures led to the formation of longer RNA sequences. Under these conditions, short segments capable of forming double-stranded structures tended to ligate at their ends. This mechanism overcomes the problems associated with polymerization in water by monomer addition, which is slow and results in an equilibrium favoring degradation over synthesis (Pino et al. 2008).

Against the contention that the origin of life was a "highly improbable chance event" (see Dawkins 1986, pp. 139–141), it should be indicated that the RNA products of the above-mentioned polymerization and elongation procedures under prebiotic conditions are not statistically random and depend, among other factors, on the monomers involved (see, Miyakawa and Ferris 2003; Monnard 2007, p. 389; Pino et al. 2008). In the framework of their studies of the synthesis of RNA on montmorillonite clays, it was found by Ferris and colleagues that both the sequences, for example, purines versus pyrimidines, and the regioselectivity, 3′-5′ versus 2′-5′ phosphodiester bonds, followed certain "rules for RNA elongation on montmorillonite" (Miyakawa and Ferris 2003, p. 8204; see also, Brack 2017, pp. 219–220). Most significantly, longer oligomers can form only under catalysis because "the catalyst limits the number of reaction pathways and there are sufficient monomers to make the longer oligomers" (Miyakawa and Ferris 2003, p. 8206). Based on these data, it seems that the prebiotic synthesis of RNA (and in extension, of other polymers that might have been involved in the origin of life) was not statistically random (Miyakawa and Ferris 2003, p. 8202).

6.5.3.1.3 Small-Size Ribozymes?

Recent evaluations of the catalytic functions performed by RNAs in extant organisms in what is referred to as the "contemporary RNA world," point to "very small RNAs" that can act as ribozymes. It was hypothesized that some of these ribozymes could have functioned similarly in a primordial RNA world (Cech 2012, p. 3). Studies of ribosomal RNA and the functions involved in the mechanism of translation have led to similar conclusions. Harry Noller, a leading contributor to these discoveries, pointed out that only three nucleotides of 16S rRNA constitute the ribosomal structure active in sensing whether Watson-Crick pairing between bases are made. Noller commented that "it is not difficult to imagine assembling such a mechanism from small, rudimentary RNAs of the kind that have been suggested to have populated the

RNA world" (Noller 2012, p. 5). The group of Michael Yarus, another prominent contributor to the field, has isolated a tiny ribozyme, five nucleotides in length, whose active center consists of three nucleotides that catalyze the aminoacylation-of-tRNA reaction performed today by the protein enzymes aminoacyl-tRNA synthetases. This ribozyme also supported peptidyl-RNA synthesis (Turk et al. 2010). The researchers claimed that these results support the hypothesis that "minuscule RNA enzymes participated in early forms of translation" (Turk et al. 2010, p. 4585).

Clearly, the possibility that small RNAs could have accelerated various reactions on the primordial Earth raises the chance that the RNA world could have been a "prebiotic product" (see probability calculations in Turk et al. 2010, p. 4588). This is even more probably the case if it is taken into account that the assembly of these short RNAs might have involved catalysis by minerals, metal ions, and short peptides. Yarus noted that the most intriguing possibility raised by these results is that a ribozyme catalyzing phosphodiester transfer "…may exist somewhere near this size. *This would make the polymerase/replicase needed to initiate Darwinian evolution of RNAs, the founding event of the RNA world, much more likely*" (Turk et al. 2010, p. 4588, emphasis added). Unlike Yarus's optimistic evaluation, other researchers doubt whether the structure of such a small ribozyme could have been compatible with the complex functions required of an RNA replicase (Schrum et al. 2010, p. 7).

6.5.3.1.4 Alternative Route to RNA Replicase?

Discussing the many difficulties involved in the *in vitro* evolution of an RNA replicase, Jack Szostak and his colleagues considered seriously the option that it might have been easier and thus more likely for life to begin if "RNA-catalyzed RNA replication could have emerged gradually in a series of simpler steps" (Schrum et al. 2010, p. 8; see also Szostak 2012b, pp. 12–13). They, thus, embarked recently on attempts to replicate RNA chemically, that is, without enzymes, in the "simplest form of genetic replication" (Blain and Szostak 2014, p. 11.3).

This reaction was studied for many years in the Orgel (1994, 2004) laboratory (see, Section 6.5.2.2) and was proved most inefficient. When Orgel began his studies of non-enzymatic replication, he was interested to find out whether replication without a protein enzyme was at all possible. Ribozymes have not yet been discovered. Szostak's interest today, however, is in the possibility of replication without a polymerizing/replicating ribozyme. Among the many difficulties considered as hindering a full replication of RNA sequences was the fact that the complementary strand produced in the non-enzymatic copying process always contained a random mixture of non-heritable "non-natural" 2′-5′ and "natural" 3′-5′ phosphodiester linkages, unlike the homogenous 3′-5′ linkages in contemporary RNA. After the discovery of ribozymes, it was commonly assumed that a random mixture of 2′-5′ and 3′-5′ linkages will prevent the spatial folding of the RNA chain needed for ribozyme activity, thus preventing an origin of life based on RNA catalysis (Engelhart et al. 2013, p. 390). And yet, facing the many problems associated with the emergence of an RNA replicase, revisiting chemical template replication of RNA presented itself as a plausible path (Blain and Szostak 2014, p. 11.13). (As will be seen shortly, this was especially the case in the context of the current research of prebiotic protocells).

In a series of experiments by Szostak and colleagues, it was found that "RNAs containing remarkably high proportion of randomly distributed 2′-5′ linkages retain the ability to fold, recognize ligands and catalyze reactions" (Engelhart et al. 2013, p. 391). More specifically, the self-cleavage activity of the hammerhead ribozyme, synthesized to contain various percentages of 2′-5′ linkages, retained most of its activity at the prebiotically plausible 10% level, though the rate of reaction was slowed. Though specific point substitutions of 2′-5′ linkages in the catalytic core of the ribozyme led to substantial or total decrease of activity, still with a low to moderate level of randomly dispersed 2′-5′ linkages, only a fraction in a pool of functional RNAs would be seriously inactivated (Engelhart et al. 2013, p. 392).

Furthermore, heterogeneous mixtures of linkages seemed to overcome another major problem that was encountered by Orgel and others in the studies of non-enzymatic RNA replication. Full replication of an RNA strand, that is, repeated cycles of template copying, depends on the post-copying separation of strands. Due to the thermal stability of long RNA duplexes, strand separation was hard to achieve. It has been known for quite some time that the presence of 2′-5′ linkages destabilizes the RNA duplex structure, lowering its "melting" (i.e., separating) temperature. The recent results of the Szostak group indicate that the presence of low to moderate 2′-5′ linkages has multiple effects: it effectively lowers the melting temperature of long RNA duplexes while being, at the same time, compatible with non-enzymatic RNA copying and ribozyme activity (Engelhart et al. 2013, p. 393). Thus, contrary to a long-held view, 2′-5′ linkages in RNA *far from being problematic, are, in fact, an essential feature that allowed RNA to emerge as the first genetic polymer of life*" (Engelhart et al. 2013, p. 391, emphasis added).

6.5.4 ORIGIN OF LIFE: COMPARTMENTS AND EVOLUTION

The concept of some sort of "primitive cell" engulfing emerging metabolism characterized metabolism-first theories from the first half of the twentieth century. The most known cell-like structures were suggested by Oparin (1953, pp. 163–195) and Fox (1984). Already in these early theories, such "primitive cells" were supposed to be involved in evolutionary scenarios (see Sections 6.5.2 and 6.5.2.2). In contrast, the realization of the need for a protocell came late to gene-first proponents. Early-twentieth-century "geneticists" supported the notion of a "naked gene" (Troland 1914, 1917; Haldane 1967, p. 274). Dealing with early membranes posed a serious problem for gene-first scenarios that traditionally envisioned early cells as heterotrophic and thus considered membranes as possible barriers (see Orgel 2004, pp. 117–118). In distinction, many metabolism-first theorists embraced an autotrophic

conception (see, among many, Morowitz 1992; Morowitz et al. 2000). Continuing previous traditions, with the rise of the RNA-world concept, discussions first "neglected the cellular nature of life" (see, Pohorille and Wilson 1995; Szostak 2012b, pp. 1, 2).

Beginning in the 1970s, several pioneering groups conducted experiments designed to examine the prebiotic relevance of phospholipids and membranes (see Hargreaves et al. 1977; Morowitz et al. 1988). Further studies demonstrated that macromolecules can be encapsulated by vesicles under prebiotic conditions (Deamer and Barchfeld 1982; Oberholzer et al. 1995). During these years and later, the vital question of the origin of amphiphiles, membrane-forming long-chain molecules containing both polar and non-polar groups on the same molecule, was explored. In a seminal study, it was found that amphiphilic molecules are present in carbonaceous meteorites, for example, the Murchison meteorite, and can self-assemble into membrane structures (Deamer and Pashley 1989). The synthesis of amphiphiles, as well as other biogenic building blocks, is assumed to have also occurred by geochemical processes on Earth (see Damer and Deamer 2015, p. 875; Deamer 2017).

In the last few decades, the question of the origin of cellular life is raised by "geneticists" and "metabolists" alike (Ruiz-Mirazo et al. 2014). This question has become a major aspect of the emergence of life on Earth (Rasmussen et al. 2009) and, for some, especially metabolists, its defining aspect (Morowitz 1992, p. 9; see, Luisi 2015, p. 913). There is a consensus now among RNA-world researchers that RNA or pre-RNA macromolecules had to be isolated and confined by some sort of membrane (or other delimiting structures), "both to concentrate the ingredients of life and to promote a Darwinian process" (Monnard 2007, p. 389; Service 2013, p. 1034). It is realized that a population of RNA replicating molecules free in solution could not have evolved into more efficient or accurate replicators. "By keeping molecules that are closely related together, advantageous mutations can lead to preferential replication." Within a compartment, better replicators will be able to replicate each other more efficiently and will have an advantage over other compartments (Szostak et al. 2001, p. 387; Service 2013, p. 1034).

6.5.5 A CASE STUDY: RNA-ORIENTED PROTOCELL RESEARCH GUIDED BY AN EVOLUTIONARY PERSPECTIVE

Work in progress reported in a series of papers by the Szostak laboratory during close to two decades explored mechanisms of encapsulation, growth, and division in experimental models of cellular compartments supposedly analogous to primitive protocells. These experiments were explicitly guided by the need to achieve a setting enabling Darwinian evolution in a protocell population. The work was originally led by a theory proposing that the origin of cellular life depended on the following elements: a membrane engulfing a protocell; a replicating genetic molecule within a protocell; the compatibility between these two elements; and their growing interdependence (Szostak et al. 2001). Not ignoring the serious criticism addressed at this experimental study (see, Orgel 2003, pp. 213, 217) and the obstacles it faces, acknowledged by the Szostak group itself (see, Schrum et al. 2010, pp. 12–13), the theoretical and empirical aspects of these studies can serve as an instructive case demonstrating the heuristic value of the evolutionary perspective in the study of the emergence of life (see, e.g., Budin and Szostak 2011; Szostak 2011, p. 2895).

The work examined here was based on a heterotrophic conception, assuming the passage of organic building blocks from the prebiotic environment into the protocell. Permeability of vesicles to activated nucleotides is thus crucial for the compatibility of RNA and membrane required for an integrated, viable protocell. Relying on the ubiquity of phospholipids in biological membranes and their self-assembly in aqueous environment to form stable vesicles, phospholipids were considered "obvious candidates for prebiotic membrane components" (Hargreaves et al. 1977, p. 78). However, since phospholipid membranes are known to bar the entrance of polar and charged molecules, for example, activated nucleotides, the Szostak group tested the permeability of primitive membranes made of fatty acids (Mansy et al. 2008). They found that nucleotides that contained highly polar groups, for example, ATP, could not cross the fatty-acids membrane, but the less polar imidazole-activated nucleotides were permeable. If indeed this obstacle could have been overcome prebiotically, the authors contended that "extremely simple heterotrophic protocells could have emerged within a prebiotic environment rich in complex nutrients" (Mansy et al. 2008, p. 4).

Referring, in one of their first papers, to the catalytic function of montmorillonite clay in polymerizing activated ribonucleotides, the Szostak group reported that montmorillonite also accelerated vesicle assembly and growth when fatty acids were supplied as tiny lipid single-layered micelles to a mixture of fatty-acid vesicles (Hanczyc et al. 2003). Moreover, it was also found that RNA adsorbed to montmorillonite particles can be encapsulated within vesicles. These results pointed to the possibility of an assembly of protocell-like structures. However, a serious problem interfering with a plausible prebiotic fatty-acid vesicle in which RNA could have replicated is the incompatible effect of Mg^{+2} or other divalent cations on RNA on the one hand and on membranes on the other: RNA replication, both enzymatic and non-enzymatic, requires high concentrations of magnesium ions, which tend to destroy fatty-acid vesicles. Szostak and coworkers found, however, that while vesicles made of certain fatty acids were unstable in the presence of divalent cations, the addition of the glycerol ester of these acids stabilized the membrane (Chen et al. 2005). This combined composition also allowed the rapid diffusion of Mg^{+2} ions into the vesicle and their equilibration across the membrane. Furthermore, the presence of Mg^{+2} ions increased the permeability of the membrane to nucleotides, such as uridine monophosphate. This could have been advantageous to an encapsulated polymerizing RNA using the increased supply of nucleotides, while the polymeric product would be retained within the vesicle (Chen et al. 2005, pp. 13216, 13220). The downside of the glycerol ester-containing

stable membrane was the decrease of its ability to grow at the expense of both empty vesicles and added micelles, dynamic behaviors previously manifested by fatty-acid membranes (Chen and Szostak 2004).

A line of studies was devoted to processes of vesicle growth and division driven by various environmental physical and chemical processes. For example, some experiments explored the development of osmotic pressure on the inside of the membrane, exerted by encapsulated RNA (Chen and Szostak 2004). It was shown that internal osmotic pressure and the resulting extension of the membrane caused by sucrose, short oligonucleotides, and tRNA led to the uptake of additional membrane components from other vesicles in the near environment that were subjected to a lower osmotic pressure. This process was described by the authors as competition among vesicles for a limited resource (Chen and Szostak 2004, Supporting Online Text; see also, Budin and Szostak 2011, p. 5249). Szostak and colleagues pointed to the possibility that a simple physical mechanism, such as the exertion of osmotic pressure that led to competitive vesicle growth, could have mediated "a coordinated interaction between genome and compartment boundary" (Chen and Szostak 2004, p. 1474). Furthermore, a replicating RNA inside a vesicle and the resulting membrane growth due to increase of osmotic pressure could have *played an important role in the emergence of Darwinian evolution during the origin of cellular life*: "A faster replicase genotype would thus produce the higher-level phenotype of faster cellular growth" (Chen and Szostak 2004, p. 1746). It should be noted that, at this stage of the work, ribozyme replicase was still considered crucial for the evolution of protocells. Focus on non-enzymatic chemical replication of RNA (see Section 6.5.3.1.4) came later.

Difficulties arose when a plausible mechanism for the division of osmotically swollen vesicles was explored (Adamala and Szostak 2013, p. 495). A prebiotically feasible division process could not be demonstrated with the small unilamellar vesicles (around 100 nm in diameter) used in the osmotic pressure experiments. However, when experimenting with larger multilamellar vesicles (several microns in diameter), a division mechanism based on processes of growth and transformation into thread-like vesicles was demonstrated (Zhu and Szostak 2009; Szostak 2011, pp. 2896–2897). Applying mild shear forces by gently agitating the suspension of the transformed thread-like vesicles led to their division into multiple smaller spherical daughter vesicles. It is suggested that a prebiotic process that might have triggered such division could have been wind-driven waves on the surface of a pond (Szostak 2011, p. 2897). Cycles of growth and division were shown to repeat themselves (Zhu and Szostak 2009, pp. 5706–5709).

6.5.5.1 An Alternative Evolutionary Pathway: Possible Role for Metabolic Ribozymes

When non-enzymatic RNA replication became the leading hypothesis in studying the origin of cellular life, an alternative evolutionary pathway to the osmotic-pressure-driven process suggested itself. In this context, the first ribozyme to have evolved need not have improved replication but could have "carried out a metabolic function...that conferred an advantage on its host cell" (Szostak 2012b, p. 12). Attention was drawn, among other activities, to a metabolic ribozyme that could have catalyzed the synthesis of a possible membrane component, for example, phospholipids (2012b, p. 12). It was demonstrated that vesicles with mixed membranes of fatty acids and phospholipids grew faster, were transformed into thread-like vesicles, and thus favored division (Budin and Szostak 2011, p. 5250; Szostak 2011, p. 2898).

A hydrophobic dipeptide, AcPheLeuNH$_2$, synthesized inside a vesicle, was shown to exert a similar effect on membrane growth as did phospholipids (Adamala and Szostak 2013). This hydrophobic peptide was synthesized by the dipeptide ser-his, known to catalyze peptide bonds between amino acids (see, Wieczorek et al. 2013). The produced hydrophobic dipeptide then localized to the fatty-acid membrane and in a similar way to phospholipids drove vesicle growth at the expense of pure fatty-acid vesicles and micelles in its environment. Moreover, it was shown that such competitive growth also caused vesicles containing the hydrophobic dipeptide to be transformed to a filamentous form and to undergo division (Adamala and Szostak 2013, pp. 498, 499). The authors thus raised the possibility that metabolic ribozymes could have synthesized such hydrophobic peptides, leading to accelerated protocell growth and thereby conferring a strong selective advantage (Adamala and Szostak 2013, p. 500).

Realizing, however, that the synthesis of hydrophobic peptides by ribozymes might have been hampered by the polarity of nucleic acids, the Szostak group further suggested a way that might have circumvented this difficulty. A ribozyme could have catalyzed instead the synthesis of peptides such as ser-his, leading in this indirect way to the synthesis of a functional hydrophobic end product. "The synthesis of intermediate catalytic peptides [e.g. ser-his] could be an effective strategy for the synthesis of membrane-modifying products" (Adamala and Szostak 2013, p. 500).

6.5.5.2 Primordial Chemical RNA Replication "Too Slow and Inefficient"?

Following their results, Szostak and colleagues claimed that a plausible prebiotic scenario for the evolution of primitive protocells capable of growth and division seems to suggest itself. However, nucleic acid replication within replicating vesicles is still elusive. On the one hand, based on the work discussed previously, non-enzymatic mode of replication might have led to the evolution of replicating RNA molecules and functional ribozymes. Yet, this possibility clashes with the realization that primordial chemical non-enzymatic RNA replication was probably too slow and inefficient to produce RNAs that were long enough to function as ribozymes. Relying on previous reports that shorter oligonucleotides can be assembled into non-covalent active complex, the Szostak laboratory studied the assembly of three model systems, one aptamer and two ribozymes, from their component oligonucleotides (Adamala

et al. 2015). It was found that subsets of short oligonucleotides first assembled into inactive complexes, forming a "primer-template" configuration. In the next step, these complexes were converted to fully functional, stable complexes by non-enzymatic template-directed primer extension (Adamala et al. 2015, pp. 484–487).

The authors also suggested that non-enzymatic primer extension may have allowed for the coexistence of short, unstructured oligonucleotides that could have functioned as templates and longer sequences that form functional complexes that could have been good catalysts (Adamala et al. 2015, pp. 487–488). Previous findings of the Szostak group on the random incorporation of 2′-5′ linkages during non-enzymatic RNA copying (Engelhart et al. 2013) already alluded to the conflict between optimal templates that should be completely unfolded and good ribozymes that must have a stable folded structure. Since different copies of the same sequence will contain 2′-5′ linkages in different locations, some copies would fold more readily and function as active ribozymes, whereas others would remain unfolded and serve as better templates for replication (Engelhart et al. 2013, p. 393). A question remains about possible synergism between the effects of 2′-5′ linkages and those of non-enzymatic primer extension to produce distinct sets of template and catalytic oligonucleotides in primitive cells (Adamala et al. 2015, p. 488).

In summary, the list of remaining problems on the way to a possible gene-first evolutionary emergence of cellular life, enumerated and acknowledged by Szostak and coworkers, is daunting (see, e.g., Schrum et al. 2010, pp. 12–13; Szostak 2012b, pp. 12–13; Adamala et al. 2015, p. 483). It is nevertheless clear from the work examined here that the knowledge gained so far resulted from applying the evolutionary hypothesis of the origin of life to detailed mechanisms and processes. It appears that further possible progress will depend on an extension of the same program (e.g., Schrum et al. 2010, pp. 12–13).

6.5.6 A CASE STUDY: METABOLISM-FIRST AND THE POSSIBILITY OF EVOLUTION

Many of the evolutionary mechanisms suggested currently by "metabolists" continue the Oparin tradition. Oparin's "metabolic" ideas in the 1920s and 1930s predated the rise of molecular biology. However, even later, he persisted in rejecting the genetic conception and instead of the "evolution of the molecule," that is, DNA and RNA, he emphasized the "self-reproduction of the entire system" (Oparin 1965, p. 96). Based on data accumulated during the last decades on prebiotic chemistry and the hurdles faced by the prebiotic synthesis of RNA and RNA-like molecules, supporters of metabolism-first theories today contend that the gene-first conception is extremely implausible (De Duve 1991, pp. 112–113; Shapiro 2007; Peretó 2012, p. 5396; Luisi 2015, pp. 906, 917). Though relying on more sophisticated chemistry and computer simulation techniques, metabolism-first theories fundamentally share the Oparin principles based on the growth and inaccurate division of protocellular units, competition for resources among these units, and selection of the metabolically more active and hence faster-growing assemblies (Fox 1984; Kauffman 1986, 1993; Segré et al. 2001).

In a paradigmatic paper delineating a theory of biochemical organization, metabolic pathways, and evolution, Harold Morowitz emphasized the "continuity between biogenesis and evolution" (Morowitz 1999, p. 39). He analyzed the emergence and evolution of metabolic cycles within membrane-enclosed vesicles in terms of natural selection and Darwinian competition, stressing that these processes did not necessarily require genetic macromolecules. According to the metabolic conception, information embodied in genetic polymers was a late development, a consequence rather than a prerequisite for prebiotic evolution (Morowitz 1992, p. 154). Most current metabolic scenarios assume the spontaneous emergence of autocatalytic metabolic cycles within a protocell (Morowitz 1992, 1999; Kauffman 1993, pp. 309, 323; Lancet and Shenhav 2009), within inorganic compartments in the vicinity of hydrothermal vents (Copley et al. 2007; Smith et al. 2009), or on the surface of nickel and iron sulfide minerals (Wächtershäuser 1988, 1992). Russell and Martin's theory of the development of metabolic reaction chains in inorganic enclosures within hydrothermal pores (Martin and Russell 2007) will be discussed in Section 6.5.7.

In distinction to "genetic" theories, "metabolic" models adopt an altogether different, "compositional" concept of information and inheritance (Segré et al. 2001, p. 137). Unlike the autocatalysis of a single genetic polymer, autocatalysis of a whole cycle is achieved by mutual catalysis of its chemical components when a certain level of chemical complexity of the system is reached (Morowitz 1992, pp. 153–154). A "reproduction" of the components of an autocatalytic cycle is suggested in the case of the reverse citric acid cycle, considered to be the most ancient metabolic network. The reduction of carbon provided by CO_2 and completed by the cycle leads to doubling of its constituents (Morowitz 1999, pp. 46–47). Chance differences arising in a group of such networks, if persisting through the process of "replication" and division, could have provided variant networks participating in a Darwinian-type competition for common energy and nutrients resources (Morowitz 1992, p. 153, 1999, p. 44).

A metabolic mode of reproduction and variation was also hypothesized by Günter Wächtershäuser in his theory of a "pyrite-pulled-chemo-autotrophic-origin." Wächtershäuser's autotrophic theory was among the first theories to locate the emergence of life in the vicinity of submarine hydrothermal vents. In few of his earlier works (Wächtershäuser 1988, 1990, 1992), Wächtershäuser suggested the emergence of the reverse citric acid cycle attached to and catalyzed by the mineral pyrite, FeS_2, in the environment of hydrothermal vents. This "surface metabolist," comprising a core pyrite crystal and an attached layer of synthesized and interacting organic molecules, demonstrated growth-related division (Wächtershäuser 1992, pp. 104–107). Wächtershäuser proposed that variations could have resulted when rare products branched out of autocatalytic cycles and intervened catalytically in the working of the cycles and also in their own production (Wächtershäuser

1992, pp. 111–112). In more recent papers, Wächtershäuser no longer referred to the reverse citric acid cycle. He spoke about a "pioneer organism" in which "metabolic reproduction" resulted from an autocatalytic feedback mechanism (Wächtershäuser 2007). Metallo-catalysts synthesized small organic molecules, which in their turn enhanced the competence of metallo-catalysts to produce more organic molecules. When new organic molecules were produced in this way, not only reproduction but also metabolic innovation and growth in metabolic complexity were achieved (Huber et al. 2012).

A metabolic theory formulated by Kauffman (1986, 1993) postulated the emergence of self-reproducing metabolic systems consisting of interacting catalytic polymers, mainly peptides, within compartments. Based on mathematical analysis of a set of interacting catalytic peptides, Kauffman contended that when certain parameters, for example, length of polymers, number of possible polymers, and number of reactions within the set, are met, a stage is reached when all polymers catalyze the production of all the other polymers and the whole set becomes collectively autocatalytic (Kauffman 1993, pp. 309, 322). Kauffman also suggested several mechanisms that could account for selection among mutant sets, especially based on an Oparin-like division of the containing compartments (Kauffman 1993, pp. 331–332).

More recently, several authors, including Kauffman and Szathmáry, suggested a new version of the previous mathematical model of autocatalytic sets made of catalytic polymers (Vasas et al. 2012). The authors contended that under certain conditions of chemical evolution, rare reactions can lead to the rise of "viable autocatalytic cores" that allow the system to become evolvable. Unlike the original model, this version delineates a separation of the chemical system under scrutiny to "cores" and "periphery." In a core, "all species catalyze the production of all other species including themselves," and can thus be compared to a "genotype." The periphery that consists of non-autocatalytic molecular species that are catalyzed by the core is described as analogous to a "phenotype" (Vasas et al. 2012, p. 2). It is further noted that while the mechanism suggested previously by Kauffman (1986, 1993) did not fulfill the conditions for evolution, molecular systems characterized by the newly suggested structure can become evolvable. In distinction to Kauffman's previous model, which described a one-core system, "independent viable autocatalytic cores embedded in a large molecular network" allow for the possibility of competition and selection (Vasas et al. 2012, p. 5). Indeed, when many cores become enclosed in a compartment, competition between them ensues. Between-compartment competition can arise "due to the phenotypic effects of cores and their periphery at the compartment level." This model thus presumes to explain how "a poor man's natural selection could have operated prior to genetic templates" (Vasas et al. 2012, p. 1).

6.5.6.1 ARE METABOLIC CYCLES PLAUSIBLE?

As indicated in Section 6.5.6, the main motivation to formulate metabolism-first theories is their authors' conviction that the prebiotic emergence of a replicating genetic polymer was extremely improbable. Critics of this assessment retort that the emergence of evolvable metabolic cycles is itself implausible (Anet 2004; Pross 2004; Eschenmoser 2007a, p. 12838; Peretó 2012, p. 5398). Metabolist scenarios face the double challenge of, first, establishing that an autocatalytic cycle can plausibly be organized under prebiotic conditions, and, second, of demonstrating evolvability of a group of such cycles. Whether these questions can be answered experimentally is questioned even by some metabolists (Lancet and Shenhav 2009, pp. 245–247; Pohorille 2009, p. 569). Quite a few researchers do not regard computer models and simulations, predominant in metabolic theories, as substitute to experimental results (Anet 2004, p. 656; Pross 2004, p. 312; Eschenmoser 2007a, p. 12838; Orgel 2008, pp. 10–11). Most metabolists, however, do consider such computer simulations as valid demonstrations (Smith et al. 2009, p. 445).

Genetically inclined researchers raised various points of criticism against metabolism-first models, focusing especially on the potential of metabolic autocatalytic systems, lacking a genetic template, to undergo Darwinian evolution (Eigen 1971, pp. 498–503; Maynard Smith and Szathmáry 1985, pp. 68, 71; Vasas et al. 2010). It is worthwhile to follow more closely Orgel's analysis of the plausibility of metabolic cycles. A veteran gene-first proponent, Orgel's position within the genetic camp was unique. Along with Crick and Woese, he independently suggested the RNA-world idea already in the 1960s (see Section 6.5.2.2). For years, he was a leading researcher of this theory and at the same time among its most severe critics (see Section 6.5.3). Being a convinced "geneticist," he nevertheless knew firsthand that this conviction was "with good reason been challenged repeatedly" (Orgel 2000, p. 12503). Realizing the enormous challenge of solving the problem of the origin of life, he didn't shy from considering the possibility that the genetic conception is not the right track after all. Referring to the theories of Kauffman (1986), Wächtershäuser (1988), and De Duve (1991), he wondered whether perhaps, "systems of high complexity can develop without any need for a genome in the usual sense" (Joyce and Orgel 2006, p. 21). Orgel acknowledged that demonstrating the existence of a complex, non-enzymatic metabolic cycle would be a major step toward solving the origin-of-life problem and demonstrating that a family of such cycles could evolve "would transform the subject" (Orgel 2008, p. 12).

Thus, in several of his papers (e.g., Orgel 2000), and especially in his last one, published posthumously (2008), Orgel carried out a thorough investigation of metabolic theories. As indicated earlier (in Section 6.5.6), the reverse (reductive) citric acid cycle was considered by Morowitz and colleagues and by Wächtershäuser in his earlier work (1988, 1990, 1992), as the core of early biochemistry and a possible metabolic infrastructure for evolution. Before considering the possibility of the evolvability of such an autocatalytic metabolic cycle, Orgel suggested that, first, the chemical plausibility of the prebiotic self-organization of such a cycle should be examined. Though several metabolic reactions that could have formed part of core metabolic chains were demonstrated (Cody et al. 2000; see also, Huber et al. 2003; Huber and Wächtershuäser 2006),

Orgel indicated that unlike genetic scenarios, based on a substantial body of experimental work, metabolic scenarios lack similar empirical basis and their evaluation has to rely on considerations of chemical plausibility (Orgel 2008, pp. 11–12; Peretó 2012, p. 5398). Orgel's starting point was the fact that reactions involved in autocatalytic metabolic cycles in extant biochemistry depend on catalysis by specific enzymes. Orgel asked whether on the primordial Earth, prebiotic available catalysts such as mineral could have replaced enzymes.

Analyzing in detail the different reactions involved in the reverse citric acid cycle, the extent of specificity required by their catalysts, and the possible disruption of the cycle by side reactions, Orgel concluded that though a skilled chemist could have constructed such catalysts, it is questionable whether they could have been found on the ancient Earth. Whereas "it is not completely impossible" that specific catalysts for the various reactions could be found, "the chance of a full set of such catalysts occurring at single locality…in the absence of catalysts for disruptive side reactions seems remote in the extreme" (Orgel 2008, p. 8). This critique was addressed at the scenarios of both Wächtershäuser (1988) and Smith and Morowitz (2004).

Orgel also analyzed in detail Kauffman's autocatalytic peptide-cycle model. The various pitfalls found by Orgel in this model will not be reviewed here (Orgel 2008, pp. 10–12; see, also, Lifson 1997; Anet 2004). Suffice it to say that indicating that Kauffman assigned to peptides properties that they do not have, Orgel concluded that Kauffman took for granted that it is possible to "write down on paper a closed peptide cycle and a set of catalyzed ligations leading from monomeric amino acids to the peptides of the cycles" and to expect "that [such] cycle would self-organize spontaneously and come to dominate the chemistry of a reaction system" (Orgel 2008, p. 10). (It should be added that Orgel, who died in 2007, could not have reviewed Kauffman's new version of his model [Vasas et al. 2012]. For the response to Orgel's 2008 critique of Kauffman's original model, see, Vasas et al. 2012, p. 5). Orgel did mention the experimental demonstration by Ghadiri and colleagues of peptide cycles of the type suggested by Kauffman; and he did refer to the possibility of the self-replication of peptides (Lee et al. 1996; Ashkenasy et al. 2004). However, he pointed out that the "carefully designed peptides" used in these experiments were not relevant to the claims made by Kauffman of a spontaneous self-organization of polymerizing amino acids in a prebiotic setting (Orgel 2008, p. 11; see Luisi 2015, p. 915).

Based on his detailed analysis of various metabolic theories (Morowitz's, Wächtershäuser's, and Kauffman's), Orgel concluded that the most serious challenge to these theories—the lack of specificity of non-enzymatic catalysts—casts a heavy doubt on the possibility of a prebiotic self-organization of non-enzymatic cycles (Orgel 2008, p. 12). According to Orgel, the lack of chemical plausibility is even more evident in claims for the evolvability of such cycles (Ibid.). Evolvability cannot arise unless a cycle can become more complex. Orgel mentioned a few ways by which this could have happened: a development of another autocatalytic cycle from a constituent of the core cycle or the rise of a side reaction generating a catalyst for a reaction of the core cycle (Orgel 2008, p. 9). It should be noted that these mechanisms were indeed suggested by metabolists in their arguments for evolvability of metabolic cycles (Morowitz 1992, pp. 153–154; Wächtershäuser 1992, p. 89, 111–112, 2007, pp. 594–595; Shenhav et al. 2003, pp. 22–23; Pohorille 2009, p. 579; Smith et al. 2009, pp. 445–446). However, Orgel commented that such claims should be substantiated experimentally and not by faith alone. Generally speaking, he insisted that metabolic theories cannot be justified by the inadequacy of genetic theories and must be subjected to as severe a criticism, as has "rightly been applied to genetic theories" (Orgel 2008, p. 12). It should be restated that Orgel was a severe critic not only of metabolism-first theories. His last paper ended with a warning: "Solutions offered by supporters of geneticist or metabolist scenarios that are dependent on 'if pigs could fly' hypothetical chemistry are unlikely to help" (Orgel 2008, p. 12).

Yet, Orgel still pointed to the possible relevance of metabolic reactions to the evolutionary hypothesis not associated with the self-organization of complex metabolic cycles but as a source of organic molecules. He referred to a few simple cycles experimentally supported (Schwartz and Goverde 1982; Weber 2007) and considered the possibility that more simple cycles could be discovered (Orgel 2008, p. 12). He saw the relevance of these cycles and of various metabolic chain reactions to the origin of life in their potential to synthesize organic molecules needed to build a primitive genetic system (Ibid.). In this context, Orgel mentioned the successful active research on prebiotic syntheses catalyzed by metal sulfides under hydrothermal conditions (Cody et al. 2000; Wächtershäuser 2007). However, in considering the role of metabolism in preparing the way for a genetic system, Orgel pointed to the critical gap separating the chemistry of metabolism and that of replication: prebiotic syntheses tend to form complex mixtures of products, while polymer replication schemes require reasonably pure monomers. "No solution to the origin of life problem will be possible until [this] gap is closed…" (Orgel 2008, p. 12).

In light of Orgel's crucial alert to this gap, it is worthwhile to call attention to the role of physico-chemical constraints in preparing the ground for the emergence of an evolvable infrastructure. An illustration of this role is provided by the work of the origin-of-life chemist Alan Schwartz, who pointed to several attempts to search for mechanisms of separation and selectivity, in order to get individual sugars, for example, ribose, out of mixtures of closely related products (Schwartz 2013, pp. 785–786). Schwartz noted, however, that the resulting products of such mechanisms were still mixtures, rather than individual sugars, and a process of complex separation was still missing (Schwartz 2013, p. 756). It is in this context that Schwartz discussed the possible role of physico-chemical processes in various geochemical sites on the ancient Earth that might have helped in attaining purer reactants out of heterogenous products of prebiotic syntheses (Schwartz 2013, p. 787; see Schwartz and Henderson-Sellers 1983).

Schwartz called attention to the phenomenon called geochromatography that results in the selection and concentration of potential reactants. This phenomenon was discovered in the early 1990s by Wing and Bada (1991), who studied the carbonaceous chondrite Ivuna. The researchers explained the specific repertoire of organic compounds on this meteorite as the result of "geochromatography," a process driven by the flow of water on the parent body of the meteorite and the selective transport to the meteorite of some but not all organic components "of an originally more complex mixture of products" (Schwartz 2013, pp. 786–787). In distinction to chromatography carried in the laboratory, geochromatography occurring under natural conditions, on the parent bodies of meteorites and quite possibly on the primordial Earth, is more complex and multidimensional and thus might have involved highly selective processes. Schwartz suggested that such processes might have "contributed to the probability of the production of relatively homogenous samples of organic reactants" (Schwartz 2013, p. 787). He envisaged a number of separation and purification steps that would have made a self-assembly of a polymer or of a structure much more probable compared with a reaction out of an extremely complex mixture. (For a somewhat-different attitude to the question whether complex mixtures obtained in prebiotic reactions should be separated into pure and homogenous building blocks for the assembly of nucleotides, membranes, protocells, etc., see Szostak 2011).

6.5.6.2 A Possible Role for Organocatalysis

The possible role of small organic molecules such as amino acids, short peptides, nucleotides, and cofactors in catalyzing synthetic prebiotic reactions has been suggested recently (Copley et al. 2007; Adamala et al. 2014). Smith et al. (2009) contended that such organic catalysts could have constituted an intermediate stage between catalysis by mineral surfaces and by macromolecules, that is, RNA and proteins, and could have functioned in chemical reaction networks (Smith et al. 2009, pp. 445, 444). These ideas were greatly inspired by the growing field of organocatalysis that focuses on catalytic asymmetric syntheses mediated by small organic molecules (Barbas 2008, p. 42). Experimental work is now widening the scope of organocatalysis beyond the study of the emergence of homochiral biological building blocks (Eschenmoser 2011, p. 12462) to a wider array of synthetic reactions that might have taken place during the emergence of life (Wieczorek et al. 2017). For example, the dipeptide ser-his was found to catalyze the condensation of amino acids (Luisi 2015, p. 911). This dipeptide also catalyzes the synthesis of RNA phosphodiester bond, forming short oligonucleotides in water-eutectic phase (Wieczorek et al. 2013), and amino acids and oligopeptides were shown to catalyze the condensation of amino acids and nucleotides (Adamala et al. 2014). It is noteworthy that the amino acid proline was found to catalyze a wide variety of substrates, a generality highly unusual with enzymes (Movassaghi and Jacobsen 2002; Morowitz et al. 2005). Researchers in the organocatalysis field comment that, for a long time, chemists could not imagine that a small molecule like proline could match the reactivity of enzymes (Movassaghi and Jacobsen 2002, p. 1905; Barbas 2008, p. 43).

It was mentioned previously that Orgel's major criticism of the likelihood of prebiotic self-organization of autocatalytic cycles was based on the non-specificity of prebiotic molecules and mineral catalysts. Was Orgel's skepticism vis-à-vis metabolic autocatalytic cycles also a reflection of this attitude? Still, with no experimental demonstration, as yet, of the viability of prebiotic autocatalytic cycles, it is not clear whether the catalytic prowess of small organic molecules could have contributed not only to various prebiotic reactions but also to the self-organization of cycles.

6.5.6.3 ESCHENMOSER ON THE "GENETIC POTENTIAL" OF METABOLIC CYCLES

It is of interest to comment shortly on the position on metabolic cycles of the renowned origin-of-life chemist, Albert Eschenmoser. Similar to Orgel, he was doubtful as to the reality of such cycles because of "problems of the efficiency and selectivity of organocatalysts" (Eschenmoser 2007b, p. 570). Yet, unlike Orgel, he chose to refrain from a definitive judgment as to the plausibility of metabolic cycles, noting the "renaissance" brought about by the field of organocatalysis and the results and insights gained from this field (Eschenmoser 2011, pp. 12461–12462). Among Eschenmoser's motivations in discussing metabolic scenarios was his wish to find a common ground between the metabolic and genetic viewpoints and to evaluate the "genetic," that is, evolvable aspects of metabolic theories (Eschenmoser 2007a, 2007b). Eschenmoser (2011, p. 12461) called attention to the "genetic potential" in Wächtershäuser's and Morowitz's theories, both emphasizing the "replicator" aspect of their hypothetical autocatalytic cycle. However, even assuming the existence of such cycles and their reproduction as a whole, this would have been a necessary condition but not a sufficient one. For evolution, such autocatalytic cycles would have to be able to diversify in order to provide competing metabolic variations. Like Orgel (see, 6.5.6.1), it was on this point that Eschenmoser (2007a, p. 12828, 2011, p. 12461) found metabolic scenarios seriously lacking. He examined the possibility, raised by Wächtershäuser (e.g., Wächtershäuser 1992), of the contingent emergence of catalytic loops feeding into existing autocatalytic cycles and thus creating a new cycle (Eschenmoser 2007b, pp. 555–559). Nevertheless, he still reaffirmed the outstanding capabilities of informational polymers compared with metabolic cycles (see Pereto 2012, p. 5398; Vasas et al. 2010). Like Orgel, he thus emphasized the more probable role of such cycles in producing organic materials for the synthesis of genetic polymers (Eschenmoser 2007b, p. 559).

In summary, the question whether metabolism-first scenarios can suggest valid mechanisms for the evolutionary emergence of life is still open. According to eminent origin-of-life researchers, such as Orgel and Eschenmoser, who gave a serious consideration to these theories, such evolutionary mechanisms are highly doubtful. Nevertheless, it is clear that the search for plausible metabolic modes of reproduction, variation, and selection in theories across the metabolic board (e.g., Kauffman and Szatmáry's, Lancet's, Morowitz's, Wächthershäuser's) played and still plays an important role in guiding metabolism-first research.

6.5.7 A CASE STUDY: PROTOMETABOLISM AS PREPARATION FOR RNA AND ITS EVOLUTION

6.5.7.1 ALKALINE HYDROTHERMAL VENTS AND THE ORIGIN OF LIFE

Following the discovery in the late 1970s of submarine vents and of flourishing extant organisms in and around the "Black-Smokers" chimneys in these sites (Corliss et al. 1979), Michael Russell and colleagues were among the first to raise the possibility of an autotrophic emergence of life in these sites (Baross and Hoffman 1985; Russell et al. 1988; Kaschke et al. 1994). Such an environment might have offered emerging life energy and nutrients, as well as numerous advantages, compared with the traditional soup scenario. Among those merits were independence from the primordial atmosphere, protection from extraterrestrial impacts, availability of inorganic catalysts and isolating compartment-like structures made of catalytic minerals comprising the chimneys and mounds, and the critical role of membrane-spanning electrochemical gradients (Russell and Hall 1997). From early on and up to Russell's most recent works, his conception of life's origin at hydrothermal vents was of a "drive to life," based on the disequilibria between reducing agents (H_2 and CH_4) and electron acceptors (among others, CO_2 and Fe^{+3}) and between an acidic external ocean and alkaline hydrothermal fluid (Russell et al. 2014).

An early criticism of the submarine hypothesis was the claim that organic molecules could not have been synthesized and kept intact at the very high temperatures of these sites that might exceed 350°C (Miller and Bada 1988). However, hydrothermal vents with colder temperatures were discovered in 2000, farther from oceanic spreading centers (Kelly et al. 2001, 2005). Following the discovery of such a warm, alkaline system of vents, named the Lost City hydrothermal field (LCHF), bearing H_2-rich water in the range of 40°C–90°C, and the hypothesis that such vents could have also existed billions of years ago (Martin and Russell 2007, pp. 1887–1888; Martin et al. 2008a, p. 812), such geochemical sites were suggested as optimal for the origin of biochemistry and the first living systems (Martin and Russell 2007; Russell et al. 2014, pp. 313–314). It was hypothesized that the microporous internal structure of the vents restrained reaction products from diffusing into the ocean. It was further suggested that such sufficient concentration of organic compounds might have led to the rise of an RNA world (Martin and Russell 2007, p. 1890).

The discovery of hydrothermal vents, and especially the Lost City off-ridge-vent system, led to intensive study of the geochemical process of serpentinization, a sequence of geochemical reactions that take place when sea water circulates through cracks in the oceanic crust. The term serpentinization derives from "serpentine," one of a group of minerals, produced as a result of this process. The chemistry of the vents, both of the hot, acidic, "black smokers" and of the cooler alkaline systems, such as Lost City, is dominated by these geochemical reactions. Sea water, rich in CO_2, upon interacting with the abundant Fe^{+2} in the crust (especially in the form of the mineral olivine made of magnesium and iron silicate) generates vast amounts of H_2, while Fe^{+2} is oxidized to Fe^{+3}. The released H_2 reduces CO_2 to methane and other reduced carbon compounds (Martin et al. 2008a, p. 809). The processes of serpentinization are also responsible for the alkaline nature of the hydrothermal fluids that feed the mounds due to the dissolution of calcium, magnesium, and other metals by the circulating seawater in the crust that produces soluble hydroxides (Mielke et al. 2010, pp. 799–800).

Not only attesting to the enormous complexity of the study of the origin of life but also reflecting the complexity of the chemistry of hydrothermal vents, still vastly underexplored (Martin et al. 2014, p. 1093), the theory developed by Russell, Martin, and colleagues underwent many changes and represents a theory in progress (Russell et al. 2014; see Mazur 2014). The scenario suggested in the earlier papers (Russell et al. 1994; Russell and Hall 1997) and clearly formulated by Martin and Russell (2007) described the process of the emergence of life in compartment structures made of metal sulfides in an alkaline vent at a redox-pH-and-temperature gradient between sulfide-rich hydrothermal fluid and iron (II)-containing water at the primordial ocean floor. It was hypothesized that the fossilized chimneys and mounds at the ocean floor that contained FeS, NiS, and other metal sulfides and were formed by the process of serpentinization might have served as primordial catalytic inorganic enclosures (Martin and Russell 2007, p. 1890). This hypothesis was strengthened by breakthrough experiments by several groups demonstrating that FeS and NiS catalyze the synthesis of key organic compounds (see Heinen and Lowers 1996; Huber and Wächtershäuser 1997).

Russell originally supported Wächtershäuser's early hypothesis of the reverse citric acid cycle as the metabolic cycle providing organic synthesis with energy and reduced carbon compounds (Russell and Hall 1997). Later, Russell and Martin came to consider a primitive form of the Wood-Ljungdahl pathway, a linear acetyl-CoA pathway of CO_2 fixation, to be a better candidate for the first biochemical pathway to arise within a hydrothermal mound (see, Russell and Martin 2004, p. 363). Whether the reverse citric acid cycle or the acetyl-CoA pathways are considered to be the most antique biochemical pathway of CO_2 reduction in early life, the idea common to many practitioners in the origin-of-life field is the continuity between primitive and extant life in terms of organic chemistry and biochemistry (e.g., Morowitz 1992, pp. 90–91; Orgel 2003, p. 213, 2004, p. 100; De Duve 2005a, pp. 17–21; Copley et al. 2007, p. 433; Martin and Russell 2007, p. 1922). In the acetyl-CoA linear pathway, operating in various eubacteria and archaebacteria, CO_2 is the electron acceptor and H_2 is the donor. Thioester is the product, readily hydrolyzed to produce acetate. Importantly, this mechanism is exergonic, and the reactions in this pathway today are catalyzed by FeS and (Fe, Ni)S centers in proteins.

Exploring the physical and chemical dynamics of extant alkaline hydrothermal vents, for example, the temperature range in such systems, and especially the chemistry at such sites led to a reevaluation of the possible conditions in a Hadean hydrothermal system compared with those in extant sites. Most importantly, the mineral composition of today's enclosed structures and the precipitate mounds at Lost City lack sulfides and are made mainly of carbonates (Mielke et al. 2010). This fact, as acknowledged by Russell and colleagues, "is a potential threat to the alkaline hydrothermal hypothesis" (Mielke et al. 2010, p. 801), clashing with a fundamental premise of the original theory that metal sulfides as catalytic constituents of inorganic membranes are central to the synthesis of organic compounds and the emergence of life. A possible answer to this challenge was the suggestion that the geochemical conditions at the vent-ocean interface were markedly different in the Hadean ocean compared with today's oxic conditions. Back in the past, "the Hadean ocean was replete with Fe (II), and therefore FeS chimneys would have been abundant at that time" (Martin et al. 2008a, p. 809).

Trying to dispel at least some of the uncertainties and to reproduce the relevant conditions in the ancient ocean, a hydrothermal flow reactor, mimicking an assumed ancient natural reactor, was built at the Jet Propulsion Laboratory at the California Institute of Technology (Mielke et al. 2010, 2011). The first experimental test was aimed to investigate whether hydrogen sulfide and silica anions could have been dissolved from iron sulfides and basaltic rock wool in a representation of a Hadean ocean crust through the passage of the hydrothermal fluid (Mielke et al. 2010, p. 801, 2011, p. 934). Such anions were expected to precipitate as chimneys and catalytic compartments made of FeS, containing also nickel, cobalt, and other metals, through interaction with a Fe^{+2}-containing solution representing the Hadean ocean (Mielke et al. 2010, p. 801). The authors reported that "the reactor operated to plan," sulfide and silica were dissolved in the solution, which upon transfer to the ferrous "oceanic" solution produced chimneys and compartments dosed with iron sulfide (Ibid.). However, analyzing their results in detail (especially, following the rise and fall of peaks in sulfide concentrations released in the experiment), Russell and colleagues were led to reconsider the contention in their original model that the first compartments of life consisted purely of sulfides. They still contended that there would be enough sulfide to ensure a component of catalytic iron (and nickel-cobalt) sulfide within the walls of compartments, which will ensure "a growing network of reactions requisite for the onset of metabolism." However, compartments would be mainly composed of carbonates, silicates, ferrous hydroxide, and other materials (Mielke et al. 2010, p. 808). Furthermore, they came to the conclusion that deposits of sulfides at alkaline vents at the Hadean era depended on processes in the oceanic crust beneath "and perhaps proximity to a 400°C hydrothermal spring discharging high concentration…of hydrogen sulfide" (Mielke et al. 2010, p. 801, 2011, p. 946). They emphasized, that "the generation of iron sulfide/hydroxy-rich chimneys would have been a rather rare, but inevitable, event" (Mielke et al. 2011, p. 946).

The Russell group is convinced that the emergence of biochemistry from geochemistry could not have taken place without "the potential across iron-sulfide bearing membranes." It was electrochemical and thermal gradients, created across such membranes, that enabled the synthesis and storage of organic molecules and the development of proto-metabolism that led to the emergence of life (Mielke et al. 2011, p. 946). The rarity of iron sulfide membranes revealed in their experiments could have undermined this conviction. However, it is suggested that this threat could have been overcome by the dynamics of hydrothermal vents on the ancient Earth. Russell's assumption of inevitability stemmed from the estimates that during the Hadean era, approximated at hundreds of millions of years "between the condensation of the first ocean and the Archean with its signs of life" (Mielke et al. 2011, p. 944), there were hundreds to thousands hydrothermal alkaline vents, each one producing billions of assorted compartments over their lifetimes of tens of thousands of years (Mielke et al. 2011, pp. 946, 944–945; see also Russell et al. 2014, p. 328).

The authors plan to use the reactor to attempt the synthesis of methane and other reduced carbon compound (Mielke et al. 2010, pp. 800, 808). Following a similar line of research, a more recent electrochemical reactor to simulate conditions in alkaline hydrothermal vents was built by Herschy et al. (2014). They tested simulated vent structures, containing catalytic Fe(Ni)S mineral, precipitated from a mixture of alkaline solution containing, among other components, Na_2S (representing vent fluid) with an acidic solution of $FeCl_2$, $NiCl_2$, and sodium bicarbonate (representing the acidic ocean). Low yields of formic acid and very low yields of formaldehyde, expected products of the reduction of CO_2, were detected in the experiments. The authors acknowledged that the formation of formaldehyde "is variable and inconsistent between runs and these methods are still being optimized." Nevertheless, they considered the results a proof of principle (Herschy et al. 2014, p. 221) of the potential of alkaline vents to reduce CO_2 with H_2 (and to form simple organic molecules) "using natural proton gradients across thin, semi-conducting inorganic barriers (Herschy et al. 2014, p. 219).

6.5.7.2 The Emergence and Evolution of the RNA World in Alkaline Hydrothermal Vents

Up to this point, our discussion of the theory of the emergence of life in alkaline hydrothermal vents revolved around the theory's main guiding question: what was the source of the reduced organic compounds on which the origin of life depended (Martin et al. 2008a, p. 805)? Rejecting the alternatives of the organic soup and of Wächtershäuser's surface metabolism, the theory of Russell, Martin, and colleagues focused on the spontaneous synthesis of organic compounds at the vents as an inevitable outcome of geochemical processes (Russell and Martin 2004, p. 362). Yet, the synthesis and concentration of organic compounds and the emergence of protometabolic reactions within the enclosed structures at the vents were conceived as just the first stage that might have led to the rise of a self-replicating system at the same site.

Russell, Martin, and colleagues considered the RNA world as an essential element in the origin of life that provided a basis for Darwinian selection and the evolutionary emergence of life (Martin and Russell 2007, p. 1893). However, they rejected the notion that the RNA world could have originated *de novo* on the ancient Earth, pointing not only to the need for prior protometabolism but also to the relics left by such protometabolism in extant biochemistry (Martin and Russell 2007, pp. 1901, 1918; Martin et al. 2008a, p. 811). Their model explored the possibility that a primitive form of the acetyl-CoA pathway of CO_2 fixation could have produced the reduced carbon and nitrogen constituents of purines and pyrimidines (Martin and Russell 2007, pp. 1893, 1908).

In several experiments, Russell and colleagues simulated an alkaline-hydrothermal-vent system in order to deal with the "intermediate stage," once nucleotides existed, referring to "the transition from a solution of small organic molecules to a population of RNAs" (Koonin 2007, p. 9106). These experiments provided data relevant to the emergence of the RNA world (Baaske et al. 2007; Burcar et al. 2015). Baaske and colleagues have demonstrated in a simulated pore system the extreme accumulation of nucleotides and polynucleotides (e.g., DNA of up to 1000 base pairs) in plugged pores of various sizes. Such concentration and accumulation are driven solely by the strong thermal gradients present in hydrothermal vents simulated in the experiment and the structure of the pores. Resulting from an interaction between fluid shuttling along the pore by thermal convection and molecules drifting across the pore by thermodiffusion, "millimeter-sized pores accumulate even single nucleotides more than 10^8-fold into micrometer-sized regions" (Baaske et al. 2007, p. 9346). In addition to providing a concentration mechanism for prebiotically produced small molecules and polymers, these results also raised the possibility that such compartments under the hydrothermal vents conditions "could be veritable 'reactors' for RNA synthesis" (Koonin 2007, p. 9105). High concentrations of nucleotides would push the equilibrium toward polymerization (Koonin 2007, p. 9105; see, Martin and Russell 2007, p. 1890). Another experiment demonstrated nucleotide oligomerization within synthetic hydrothermal iron-sulfide chimneys (Burcar et al. 2015). Notably, dimers of inactivated AMP and up to tetramers of activated AMP were synthesized. No oligomerization was detected with any other nucleotide. It was found that the addition of montmorillonite clay to the chimney-forming system improved oligomerization of inactivated AMP, an observation not reported before in aqueous solution (Burcar et al. 2015, pp. 517–519). Control experiments indicated the importance of the chimney structure itself to the process (Burcar et al. 2015, p. 515).

6.5.7.3 Selection and Evolution within and among Compartments

Assuming that a population of self-replicating RNA molecules became available out of protometabolic processes, Martin and Russell noted "the dramatic transition in the nature of the chemistry at the vent…[when] natural selection sets in" (Martin and Russell 2007, p. 1910). The stage in which populations of versatile RNA molecules in different compartments evolved independently (Koonin 2007, pp. 9105–9106) was speculated to have been ensued by processes involving a wider system of compartments. Porosity could have enabled movements of organic molecules of different sizes among compartments, and new compartments were continuously being formed by mineral deposits at the ocean interface (Koonin and Martin 2005, p. 649). These attributes of the vent system, evidenced in active current sites and assumed to exist in ancient vents, could have allowed for "organic molecules [invading] new territories, thereby generating discrete units on which selection could act" (Ibid.). Outlining further evolutionary stages, it was indicated "how Darwinian selection and evolution of genetic organization" culminated in the LUCA, still confined in an inorganic hydrothermal enclosure (Martin and Russell 2003; Koonin and Martin 2005).

6.5.7.4 Protometabolism and the Evolutionary Hypothesis of the Emergence of Life

An important difference between the Russell scenario and the scenarios examined here previously, genetic and metabolic alike, should be noted. These scenarios suggested that evolution through natural selection evolved gradually and that various primitive modes of selection took part in the emergence of a full-blown Darwinian system. The theory of Russell and colleagues, on the other hand, focuses on the earlier stages of the emergence of life as an inevitable outcome of deterministic geological and geochemical processes at alkaline hydrothermal vents (Russell et al. 2014, p. 308), postponing the introduction of selection and evolution to the rise of RNA.

6.5.7.4.1 De Duve's and Russell's Protometabolic Scenarios: A Comparison

In a similar fashion to Russell and colleagues, the eminent biochemist Christian de Duve regarded the RNA world as a crucial, though not the earliest, stage on the way to life. And in a similar fashion to the Russell theory, de Duve's protometabolic theory postpones the onset of natural selection to the rise of RNA. According to de Duve, nucleotides and their assembled sequences could have emerged only with the help of a previously growing protometabolism (de Duve 1991, pp. 133–143, 2005a, pp. 156–157, 2005b). A critical stage of de Duve's scenario was the prebiotic synthesis of peptide-like catalytic "multimers" with the help of thioester bond energy. De Duve suggested that thioesters were formed from the endergonic condensation of amino acids and other carboxylic acids with thiols (alcohol-type compounds containing sulfur), favored under acidic and hot conditions at hydrothermal vents (de Duve 1991, pp. 113–116). He thus supported such a volcanic setting on the ocean floor for the origin of life (de Duve 2005a, p. 64). De Duve argued that it was probable that among the synthesized multimers were a number of crude catalysts that could have catalyzed the synthesis of various organic compounds, including additional thioesters and multimers (de Duve 1991, pp. 113–116). Thus, like other metabolic

theories, de Duve's theory is based on an autocatalytic process that drove the growth of protometabolism.

It is crucial to note that de Duve insisted that, in the context of the origin of life, "random event or process" did not mean "statistically random" or chance-like but rather "not directed by genetic information" (de Duve 1991, p. 141). Unlike a chance-like or statistically random situation, in de Duve's scenario, not all possible multimer sequences were equally probable due to stringent physical and chemical constraints. Differences among the multimers in solubility, stability, or ability to fold affected their survival. Some of the physically and chemically selected multimers were further protected by being bound to substrates. This mechanism interactively selected both catalysts and substrates that constituted the emerging protometabolism (de Duve 1991, pp. 140–144). Due to physico-chemical selection channeling the synthesis and function of the catalytic multimers, de Duve claimed that only very rare peptides out of the theoretically possible ones could have been synthesized and only very rare peptides out of those produced could have survived. Thus, in this process, the end products were fully determined by the initial conditions (de Duve 1991, pp. 137–145, 213, 2005a, p. 24).

De Duve also sought to deal with the general problem of the multistage growth of complex biological molecules during the emergence of life. Without a "continuous renewal mechanism," such as replication, to encounter spontaneous decay phenomena (see Luisi 2015, p. 915), how could intermediates survive and reach effective concentrations (de Duve 1991, p. 141)? His proposed mechanism to overcome this problem was the selection and stabilization of multimeric crude catalysts by binding to substrates (de Duve 1991, pp. 143–145). He noted that this mechanism, unlike "regular" physical and chemical selection, "does not just rely on favorable coincidence. It is truly selective, in that it contains a built-in feedback loop that actually makes it save what is useful" (de Duve 1991, p. 143). At the same time, unlike natural selection, this mechanism "can operate with random chemistry of production," selecting by differential survival, in distinction to the complex machinery of replication, selecting by differential replication (Ibid.). In summary, de Duve contended that "chemical determinism had to suffice" in the protometabolic stage, during which the chemical infrastructure for natural selection was gradually being built. He also assumed that a continuous supply of catalysts could have been ensured as long as geochemical conditions remained stable (De Duve 2005a, pp. 23–24). Thus, in de Duve's scenario, the process of the emergence of life could be viewed as divided into two by the rise of RNA and the onset of natural selection. Prior to this stage, the emergence and growth of protometabolism depended on deterministic and reproducible chemical processes made possible by the stability of geochemical conditions.

A similar combination of de Duve's two elements, physico-chemical determinism and stability of conditions, was also crucial in the first stage of Russell and Martin scenario, as it ensured prolonged, developing metabolic activity leading to the RNA world (Martin and Russell 2007, p. 1901; see, Lane 2012, p. 36). With a strong deterministic commitment, the Russell theory views life, "like other self-organizing systems in the universe, as an inevitable outcome of particular disequilibria" (Russell et al. 2014, p. 308). According to this theory, "metabolism was forced to emerge in such fine compartments" to resolve such disequilibria (Russell et al. 2014, p. 314). It was the "attempts" to reach equilibrium between the hydrothermal fluid and the Hadean ocean, "frustrated" by the spontaneous precipitation of the membrane between them, that provided the energy needed for chemical syntheses of organic compounds and for protometabolic reactions (Russell and Hall 1997, p. 378). As to the stability of conditions, hydrothermal reactors "sustainable over geological time-scales provided the continuity of conditions conducive to organic-chemical reactions" (Martin and Russell 2003, p. 65, 2007, p. 1901). The authors emphasized that the building blocks of biochemistry had to have "a continuous source of reduced carbon and energy and would have to remain concentrated at their site of sustained synthesis over extended times" (Russell and Martin 2004, p. 362). Addressing, like de Duve, the question of effective concentrations of intermediates in order to enable the growth of organic complexity, Martin and Russell referred first to the prevention of diffusion by the inorganic compartments. The possibility of a product nevertheless diffusing and thus discontinuing a reaction chain is countered by the argument of "an uninterrupted geochemical supply of the basic starting compounds" (Martin and Russell 2007, p. 1907). This could have led to the channeling of products into more complex organic compounds and possibly to the RNA world (Martin and Russell 2007, pp. 1890, 1907).

6.5.8 A CASE STUDY: COEVOLUTION OF GENES AND METABOLISM LEADING TO THE RNA WORLD

According to de Duve and Russell and colleagues, RNA and the RNA world could have emerged only with the help of a previously growing protometabolism. A proposal that replication and metabolism could have co-evolved in a hydrothermal-vent scenario and could have interactively led to the emergence of the RNA world was raised by Copley et al. (2007). The mechanism suggested by the authors was supposed to overcome the enormous challenge of the emergence of RNA from small molecules available on the primordial Earth. Their hypothesis entails a multi-stage process during which mutual catalysis among small molecules, such as amino acids and nucleotides, led to the rise of proto-metabolism, to the gradual synthesis of longer and more efficient catalysts, and to the emergence of RNA as the dominant macromolecule due to its unique combination of catalytic and template-replication capabilities (Copley et al. 2007, pp. 430–431). Copley and colleagues favored hydrothermal vents as the site for the origin of life, owing to the characteristics of these locations: compartmentalized structures lined with catalytic metal sulfides; small source molecules (e.g., CO_2, H_2, H_2S, and NH_3) vented into the porous structures; and, as attested by experiments (Cody et al. 2000), the possibility of catalyzed production of organic molecules, such as pyruvate, under high temperatures. They

further suggested that such synthesized molecules could have diffused through the walls of the porous structures to cooler regions of the vents, closer to the ocean water, allowing for the synthesis of more fragile molecules (Copley et al. 2007, p. 432).

The authors posited that to launch their multi-stage process toward the RNA world, the "core of the core" of necessary reactions must have included a pathway for the synthesis of organic compounds from CO_2 and H_2, such as the reverse citric acid cycle (Smith and Morowitz 2004) or the acetyl-CoA (Wood-Ljungdahl) pathway of CO_2 fixation (see Russell and Martin 2004; Martin and Russell 2007, p. 1887) and pathways for the synthesis of ribose, purines, pyrimidines, simple amino acids, and cofactors. Furthermore, it was contended by Copley and colleagues that the synthesis of nucleotides could have been achieved by "a surprisingly small" metabolic module that "was simpler than might be supposed." All the components in this module can be synthesized from pyruvate (which, as said, can be made under hydrothermal conditions), with the addition of a few other compounds, such as NH_3 and CO_2 (Copley et al. 2007, p. 433).

The earliest *monomer stage* of the long process was suggested to include α-keto acids such as pyruvate, simple amino acids, NADH, ribose, purines, and pyrimidines. The synthesis of these compounds was assumed to have been catalyzed at the vents by mineral surfaces and available small molecules. Mineral surfaces could have helped to concentrate and retain reactants; to polarize functional groups that facilitated attacks by other charged groups; and to accelerate, even if modestly, rates of reactions (Copley et al. 2007, pp. 434–436). The significance to the emergence of life of the ability of small organic molecules to catalyze reactions, explored by the growing field of organocatalysis (see, 6.5.6.2), was noted by Eschenmoser and by Smith and colleagues (Smith et al. 2009; Eschenmoser 2011, pp. 12461–12462). Acknowledging the limited systematic knowledge of the catalytic capabilities of small molecules, Copley and colleagues nevertheless pointed to their possible role at establishing a monomer network (see Smith et al. 2009, pp. 444). They also suggested several chemical means that could have overcome the weak and non-specific binding interactions between small molecules, enhancing the probability of productive encounters. They mentioned, among these chemical means, the covalent attachment of substrates to catalytic molecules (Copley et al. 2007, p. 436) and predicted that the monomer stage led to increased concentrations of activated monomers (Copley et al. 2007, pp. 434–438).

The hypothesis guiding the proposal of the following stages was that mutual catalysis of molecules within the networks and selection of communities with increased concentrations of activated monomers led to ever-larger and more-effective catalysts. Such a selection led, first, to the formation of *multimers* made of amino acids, nucleotides, and cofactors (note difference from de Duve's "multimers") that combined randomly and also contained alternative sugars and bases and alternative backbone linkages. New options for catalysis at this stage became available, possibly catalysis of reactions by dinucleotides or dipeptides, such as ser-his, and by the cofactor NADH (Copley et al. 2007, p. 438). At the next stage, a community of catalytic monomers and multimers containing a higher concentration of activated nucleotides allowed the synthesis of random, 3–10 nucleotide-long oligonucleotides, defined by the authors as *micro-RNAs*. These RNAs were more efficient catalysts owing to their increased size. Additionally, direct base pairing between complementary oligomers resulted in replication of catalytic components of the network. Notably, other non-replicable catalysts in the system, for example, amino acids, peptides, and cofactors, depended on the replication of RNA catalysts that promoted their synthesis (Copley et al. 2007, pp. 439–440). In such a way, the unique advantage of RNA within the community of catalysts and the interdependence between replication and catalysis were established.

The authors suggested that, at this stage, RNA replication proceeded by template-directed ligation. They considered as possible an early emergence of RNA replicase that catalyzed such ligation and thus overcame the slow rate of non-enzymatic ligation. However, in distinction to the common characterization of RNA replicase as "self-replicating," Copley and colleagues emphasized the evolutionary logic underlying the emergence of this replicase, based on its ability to enhance the function of the entire system. Its selective advantage was the replication of "all the catalytic RNA molecules required to maintain the protometabolic network" (Copley et al. 2007, p. 440). Referring also to the possible selection for homochirality at this stage, the conclusion was that this phase of micro-RNAs "would have been the stage at which the die was cast, and emergence of RNA as the dominant macromolecule became inevitable" (Ibid.). The emergence of 11–40 nucleotides-long RNA molecules by the same selection process would have led to a *mini-RNA* stage. Forming secondary structures, these oligomers could function as ribozymes, such as those found to catalyze aminoacyl-RNA and peptidyl-RNA synthesis (see, Illangasekare and Yarus 1999). RNA replication by template-directed ligation was eventually replaced by a primer-extension mechanism. At the following *macro-RNA* stage, large RNA molecules could have formed tertiary structures such as the ribosome. At this stage, all the metabolic reactions sustaining the RNA world could have been catalyzed by ribozymes. However, RNA catalysts interacted with catalysts such as amino acids, metals, and organic cofactors (Copley et al. 2007, pp. 440–441).

Copley, Smith, and Morowitz asserted that metabolism and replication were intertwined "from the very beginning," a conclusion obviating the famous chicken-and-egg problem. They again called attention to their claim that at all stages "replication was a function of a community of molecules that collectively generated monomers necessary to create lager molecules, rather than a property of a particular self-replicating RNA molecule" (Copley et al. 2007, pp. 441–442). This view of "replication," contrary to the "genetic" focus on an information-carrying sequence, brings to mind conceptions of the collective self-replication of metabolic cycles described previously (Morowitz 1992, pp. 153–154; Segré et al. 2001, p. 137; Lancet and Shenhav 2009). However, in distinction to traditional metabolic scenarios, Copley and colleagues' model of "co-evolution" incorporated both metabolic and genetic conceptions, without, nevertheless, giving up their commitment to the primacy of metabolism.

Another hypothesis raised by Copley, Smith, and Morowitz crucial to their evolutionary view of the origin of life suggested that in addition to the obvious function of prebiotic catalysts in accelerating reaction rates, not less important was their role in channeling flux of materials through specific pathways, thus contributing to the emergence of sparse metabolic networks (Copley et al. 2007, pp. 432–433). They pointed out that though chemical diversity in a prebiotic environment seemed to be advantageous for the emergence of life, a sparse network of reactions in such an environment might have offered strong merits. "Given a limited amount of carbon, a sparse network allows the accumulation of individual components at higher concentrations than could be achieved if the carbon were distributed among an enormous number of compounds. The increased concentrations of specific compounds result in proportional increase in rates of second-order reactions… Further…there is less potential for unproductive side reactions" (Copley et al. 2010, pp. 3345–3346). It is significant that the hydrothermal-vent setting favored by the authors for the origin of life is highly compatible with the "sparseness advantage": in distinction to ponds and lagoons that could have been supplied by an abundance of organic constituents from space but lacked abundance of catalysts, at hydrothermal vents, the opposite situation of "sparse set of inputs but an abundance of catalytic minerals" held (Copley et al. 2010, pp. 3347–3348).

More recently, Copley and Novikov attempted to explore experimentally the sparseness hypothesis by following the "catalytic landscape" of pyruvate under simulated hydrothermal-vent conditions. The study focused on the products that could form from pyruvate in cooler regions of the vent that were "more conductive to biogenic molecules" (Novikov and Copley 2013, pp. 13283–13284). They developed a novel pressurized reactor that could simulate vent chemistry, while exploring changing parameters and various catalytic minerals (Novikov and Copley 2013). Copley and Novikov identified in their experiments several products formed of pyruvate in a simplified system containing a varied range of transition metal sulfides, CO_2, H_2, H_2S, and, in some cases, NH_4Cl, at 25°C or 110°C, within the pressurized reactor. A particular interest of the investigators, the carboxylation of pyruvate to oxaloacetate, an intermediate in the tricarboxylic acid cycle, was not detected in the experiments. On the other hand, the reduction of pyruvate to lactate by minerals, not demonstrated previously under hydrothermal conditions, was achieved. Thiolated products were easily formed, and the reductive amination of pyruvate to alanine was also observed with different minerals, with the addition of NH_4Cl, at strikingly different levels (Novikov and Copley 2013, pp. 13284–13287). In control experiments with no minerals, the majority of pyruvate was recovered intact. The products depended not only on the presence or absence of additional substrates, for example, ammonia, but also on the nature of the catalytic minerals and the temperature (Novikov and Copley 2013, pp. 13284–13285). Most importantly, the researchers found that several reaction pathways in which specific metal sulfides and small molecules catalyzed the synthesis of products from pyruvate were preferentially favored, "allowing accumulation of higher concentrations of few components rather than low concentrations of many components" (Novikov and Copley 2013, p. 13283). Thus, dependent on the availability of mineral catalysts and the interaction between catalysts and co-reactants, a reaction network favoring non-random specific pathways was beginning to be shaped (Copley et al. 2007, p. 432; Novikov and Copley 2013, p. 13288).

6.5.8.1 What Kind of Evolution?

The description of the model by Copley, Smith, and Morowitz is replete with terms usually applied in the context of a Darwinian process: co-evolution, evolutionary logic, selection, favored communities, etc. Yet, this model does not propose a prebiotic buildup of either a genetic or a metabolic infrastructure capable of engaging in a primitive mode of replication/reproduction/division, mutation, and selection. Rather, the many-staged process claimed to have gradually led to the emergence of the RNA world was based in each of its stages on physical and chemical constraints that channeled the products toward a certain direction. These channeling effects were achieved via the catalytic abilities of minerals and the catalysis of reactions by, among others, amino acids, nucleotides; and growingly by di-amino acids, di-nucleotides, peptides, and short RNA sequences. All these catalyzed reactions depended on the physical and chemical properties of these organic catalysts and their interactions with their substrates. The availability of inorganic and organic catalysts and specific reactants at the vent site, together with environmental conditions such as temperature and pH, contributed collectively to a process of evolution governed by physico-chemical selection that favored communities with better chances of sustaining their constituents. Such selected communities were able to synthesize more effective catalysts, including replicating RNAs. The channeled process toward the RNA world, according to the proposed mechanism, was not a chance-like process (Copley et al. 2007, pp. 431–432). As the mutual catalysis in protometabolic reaction networks led to more effective catalysts, the "emergence of RNA as the dominant macromolecule became inevitable" (Copley et al. 2007, p. 440).

6.5.9 A CASE STUDY: "SYSTEM-FIRST" EVOLUTIONARY SCENARIO IN A "HYDROTHERMAL FIELD" SETTING

An origin-of-life scenario that explicitly focuses on processes of physical and chemical selection in molecular systems was suggested by researchers David Deamer and Bruce Damer (Damer and Deamer 2015). Considering the properties of hydrothermal sites as conducive to the emergence of life, the authors favor "hydrothermal fields," associated with small ponds on volcanic land masses emerging from the ocean, rather than submarine hydrothermal vents, as a more suitable

environmental candidate. Deamer and colleagues believe that for cellular life to begin, a fluid-phase membrane made of amphiphilic compounds was more advantageous than a mineral-made compartment. Most importantly, the fluctuating environment of hydration/dehydration cycles characterizing hydrothermal fields is the basis for their scenario. (For a more detailed comparison between alkaline hydrothermal vents and hydrothermal fields, see Deamer and Georgiou 2015, p. 1093).

Their scenario takes place in a hydrothermal field on a volcanic island rising out of the Earth's first ocean, resembling sites like Kamchatka, Hawaii, Iceland, and Yellowstone today (Damer 2016). Extant hydrothermal fields are characterized by cycles of hydration and dehydration in clay-lined freshwater pools undergoing evaporation and replenishment by hot springs and rainfall (Damer and Deamer 2015, pp. 873–874). According to Damer and Deamer, it is highly likely that various organic molecules such as amino acids, nucleobases, and membrane-forming amphiphiles have been synthesized by geochemical processes associated with volcanism, or delivered to Earth by comets and meteorites and were available at such sites (Martin et al. 2008b, pp. 135–136; Damer and Deamer 2015, p. 875; Deamer 2017). As a guide to possible synthetic reactions on the ancient Earth, Damer and Deamer rely on organic compounds synthesized by non-biological processes in carbonaceous meteorites in the early solar system (see Pizzarello et al. 2013). Such organic compounds, from either endogenic or exogenic source, are considered to have been dissolved by precipitation and accumulated in hydrothermal pools.

The authors, first, suggest how hydration-dehydration cycles drove the assembly of vesicles with encapsulated polymers. During the dry phase, dehydrated amphiphiles self-assembled into multilamellar structures on mineral surfaces at the pool edges and functioned as scaffolding that captured and concentrated organic monomers between amphiphilic layers (Deamer 2012; Toppozini et al. 2013). Condensation reactions between the monomers then led to the creation of polymers that resembled peptides, oligonucleotides, and possible complexes of the two (Damer and Deamer 2015, p. 881). Numerous laboratory simulations have demonstrated the polymerization of monomers, for example, mononucleotides, under dehydration-phase conditions (Rajamani et al. 2008; Deamer 2012). The mechanism of condensation involves the concentration of monomers up to the point when water activity becomes sufficiently low. "Water molecules become leaving groups and ester bonds are synthesized. Activation energy is provided by the elevated temperatures of the hydrothermal site" (Damer and Deamer 2015, p. 876). The polymerization of amino acids into oligopeptides under similar conditions was also reported (Forsythe et al. 2015).

At the hydrated phase of the scenario, lipid vesicles were produced when water interacted with the amphiphilic multilamellar matrix and polymers synthesized at the dry phase were captured in the vesicles (see Deamer and Barchfeld 1982). Damer and Deamer have calculated that trillions of vesicles of a diameter of typical bacteria could have been produced from a few milligrams of amphiphiles. The produced vesicles or "protocells," some empty and some containing polymers, were each unique in its composition, both in respect to their membranes and to their contained polymers. With each vesicle representing an "experiment," this resulted in "a natural version of combinatorial chemistry in which a few vesicles survive while most are dispersed" (Deamer 2014, p. 637; Damer and Deamer 2015, p. 876). Selection of the vesicles that began at the hydrated phase was first determined by membrane stability, effected, for example, by the random presence of a polymer that stabilized the membrane. Specific polymers could have interacted with specific amphiphiles by electrostatic forces, hydrogen bonding, and hydrophobic interactions. In the next dehydration phase, the surviving protocells fused again with the multilamellar matrix and the stabilizing polymers were distributed to the next generation of protocells. The authors also noted that vesicles composed of strongly associated amphiphiles and polymers would have persisted even during the dehydration and fusion phase (Damer and Deamer 2015, p. 878).

The most important element in the hydrothermal-field scenario is what Damer and Deamer describe as "coupling" of the contents of the protocells between the hydrated and dehydrated phases, which repeat themselves indefinitely. "The system is continuously experimenting, tending toward combinations of polymers that are more stable than other combinations" (Damer and Deamer 2015, pp. 879–880). Furthermore, it is hypothesized that, in addition to such processes of physical selection, given a continuous "coupled phase cycles" over extended time span, rare combinations of functional polymers will emerge. Among them, polymers, using their own sequences as templates, will catalyze growth by polymerization. The rare protocells encapsulating such polymers will take over in a process of chemical selection between molecular systems. This evolutionary process will give rise to molecular systems that "…could plausibly possess the necessary structures and functions for the transition to cellular life" (Damer and Deamer 2015, pp. 872–873, 881).

6.5.9.1 Functional Properties of Proposed Polymers

Damer and Deamer acknowledged that the hypothesis of the emergence of functioning systems is so far speculative, in distinction to the physical selection parts of the scenario based on multiple observations and experiments (noted previously). However, they suggested predictions that could be tested to check the validity of their overall model. First, they proposed a list of the functional properties of polymers that could have emerged gradually, participating in the continuous evolution of protocells toward living systems (Damer and Deamer 2015, pp. 881–882).

1. S-polymers—bind to and stabilize a membrane-bound protocell.
2. P-polymers—form pores in the bilayer membrane that will allow access of potential nutrients to the protocell.

3. M-polymers—catalyze reactions of a primitive metabolism among molecules that entered the protocell, providing energy and products for polymerization reactions.
4. R-polymers—undergo a primitive version of replication, using their own sequence as a template. C-polymers, a subset of this group, catalyze their own replication.
5. F-polymers—provide feedback control for various processes.
6. D-polymers—initiate and control the division of a protocell, following the duplication of its contents. The authors note that this function requires the joint action of all the other functional polymers. Most importantly, when protocells can divide in solution, the scaffolding of the dehydrated phase will not be needed. At this point, "the transition to life would occur" (Damer and Deamer 2015, p. 882).

It is predicted that when various functional polymers accumulate in a protocell, it is possible that interactions between polymers may produce novel emergent functions. For example, interactions between oligopeptides and oligonucleotides could produce a primitive version of coding that guides catalyzed polymerization and replication (Ibid.).

6.5.9.2 From Physical and Chemical Selection to Natural Selection

The Deamer and Damer scenario described here aims to account for the gradual emergence of complex functional protocellular systems from organic compounds on the ancient Earth, without relying, for most of the process, on the Darwinian mechanism of replication, division, and natural selection. Damer and Deamer suggest that it was the capturing and concentrating of organic monomers and their condensation into polymers by the mineral-amphiphilic scaffolding during the anhydrous phase that obviated the need for "the complex and high-risk process of replication and division in the bulk phase" (Damer and Deamer 2015, pp. 879, 882). Only when the long, continuous process of repeating cycles and coupled phases culminated in the emergence of functioning molecular systems supporting both metabolism and replication, controlled protocell division in solution took over (2015, p. 882). At this stage, natural selection proper could have taken center stage, instead of the previously exclusive physico-chemical selection processes. It should be emphasized that Damer's and Deamer's focus all along is on the emergence and evolution of molecular systems. Their theory is thus a "system first" rather than "gene-first" or "metabolism-first" theory (Deamer 2014, p. 637; Damer and Deamer 2015, pp. 873, 881).

From an evolutionary perspective, it is noteworthy that features commonly associated with the Darwinian mechanism, such as the emergence of novel adaptive properties in response to selective hurdles, are hypothesized to arise as a result of physico-chemical processes and to be physically and chemically selected. Thus, for example, membrane-stabilizing polymers and polymers forming pores in the membrane are synthesized by chance in the dry phase and are encapsulated by vesicles in the hydrated phase. However, these polymer types are selected as adaptations to imposed hurdles of vesicle survival and of the need for transport of nutrients across membrane barrier. Not only polymers ensuring stability and porosity of protocells but also combinations of more complex polymers featuring intricate functions, such as catalysis of template replication, are hypothesized to participate in an evolutionary process, in which physical and chemical selection plays a major role (Damer and Deamer 2015, pp. 879–880).

6.5.10 CONCLUSION

This chapter was aimed to demonstrate that origin-of-life researchers commonly evaluate the emergence of life as an evolutionary process. It was also intended, through the discussion of a few representative case studies, to show what could have been the characteristics of a prebiotic evolutionary process. No consensual theory is yet available to account for the origin of life on Earth, nor is there an experimental confirmation of such theory. However, the various scenarios discussed here demonstrate the crucial role played by the evolutionary hypothesis of the origin of life in pursuit of a solution. As a reminder, what I suggest to call "the evolutionary hypothesis of the origin of life" maintains that the transition from a lifeless environment to the first living systems involved a long series of stages growing in complexity. Prebiotic evolution constituted in highly constrained and channeled physical and chemical processes that led to the emergence of infrastructural entities capable of gradually engaging in Darwinian evolution.

The notion that the emergence of RNA and an RNA world was a necessary stage at one point or another on the way to life is supported by the scenarios considered in this chapter. The many challenges involved in the prebiotic synthesis of RNA led to the formulation of metabolism-first and proto-metabolism models as alternatives to the gene-first conception. Yet, the "great explanatory power the [RNA world] hypothesis provides for understanding how Darwinian evolution might have begun" is acknowledged across the board (Schwartz 2013, p. 784). Similarly, the idea that compartmentalization was necessary for the evolutionary emergence of life is accepted by "geneticists" and "metabolists" alike (Szostak et al. 2001, p. 387; Monnard 2007, pp. 387–390, 389; Luisi 2015, pp. 906–918, 913). On the other hand, the question of the prebiotic geochemical setting in which life could have emerged and the various implications of such a setting (e.g., nature of compartments and source of organic molecules) is still a bone of contention (e.g., Copley et al. 2010; Damer and Deamer 2015).

This CONCLUSION, however, should first and foremost comment on the evolutionary aspects of the scenarios discussed in this chapter, noting their differences and especially their unifying framework. *Gene-first and metabolism-first*

models, despite their fundamental divisions, hypothesize that evolution through natural selection emerged gradually and that primitive modes of natural selection took part in the emergence of a full-blown Darwinian system. Experiments performed by the *Szostak group* on lipid vesicles containing a genetic polymer showed primitive forms of division, competition, and selection of such vesicles. Data from other laboratories on the role of physical and chemical factors, for example, involvement of mineral and organic catalysis in the prebiotic synthesis of an RNA polymer and an engulfing vesicle (e.g., Deamer and Barchfeld 1982; Miyawaka and Ferris 2003; Huang and Ferris 2006; Monnard 2007, p. 389; Joshi et al. 2009; Brack 2017), were extended through Szostak's experiments also to the behavior of growing and dividing vesicles (Hanczyc et al. 2003). These processes were determined by, among others, the chemical nature of the membrane, affecting its physical behavior; internal osmotic pressure within the vesicle and the resulting extension of the membrane; exertion of mild shear forces, for example, wind-driven waves; and, notably, catalysis by clay minerals and small organic molecules, such as ser-his (Chen and Szostak 2004; Szostak 2011). Experiments by this group also established the possibility of a non-enzymatic, chemical replication of RNA, and revealed that RNA copies produced by such chemical replication showed catalytic activities (Szostak 2011; Adamala and Szostak 2013; Engelhart et al. 2013). This experimental work substantiated Szostak's conviction that "*Darwinian evolution itself emerged in a series of stages, step by step*" (Szostak 2012a).

Hypotheses concerning primitive forms of reproduction, division, and selection already in early stages of the emergence of life are also promoted by several *metabolism-first models* (Morowitz 1992; Segré et al. 2001; Vasas et al. 2012). However, in distinction to Szostak's gene-first scenario, "metabolic" models discussed here adopt an altogether different, "compositional" concept of information and inheritance (Segré et al. 2001, p. 137) and apply the gradual emergence of the Darwinian framework to autocatalytic metabolic cycles within a protocell (Morowitz 1992, 1999; Kauffman 1993, pp. 309, 323; Lancet and Shenhav 2009), within inorganic compartments in the vicinity of hydrothermal vents (Smith et al. 2009), or on the surface of nickel and iron sulfide minerals (Wächtershäuser 1988, 1992). Metabolism-first scenarios allot a major role to physico-chemical constraints and physico-chemical selection. Wächtershäuser's "surface-metabolist" scenario provides a clear example in his hypothesis of chemical selection processes on a pyrite surface that were due to the differential propensity for detachment of the various organic molecules formed on the pyrite surface (Wächtershäuser 1992, p. 96). In a later version of his scenario, Wächtershäuser spoke about a "pioneer organism," in which "metabolic reproduction" resulted from an autocatalytic feedback mechanism (Wächtershäuser 2007). Metallo-catalysts synthesized small organic molecules, which in their turn enhanced the competence of metallo-catalysts to produce more organic molecules (Huber et al. 2012). Interestingly, Wächtershäuser referred to the "conversion from an early, direct, chemically deterministic mechanism of evolution by ligand feedback to...a genetic mechanism of evolution by sequence variation [as] truly an evolution of the mechanism of evolution. It surely must have occurred in multiple phases..." (Wächtershäuser 2007, p. 597).

The approach to the involvement of natural selection in the emergence of life is different in the *proto-metabolism case studies* of Russell and de Duve, which postpone the onset of natural selection to the emergence of RNA. Both scenarios emphasize the necessary preparatory role of metabolic synthetic reactions in producing organic molecules, including the building blocks of RNA, and in setting the stage for natural selection (De Duve 1991, pp. 133–143, 2005a, pp. 156–157, 2005b; Martin and Russell 2007, pp. 1890, 1893, 1910). De Duve contends that during the stage in which the chemical infrastructure for natural selection was gradually being built, protometabolic reactions growing in complexity depended on deterministic chemical processes, whose continuity was made possible by the stability of geochemical conditions (de Duve 2005a, p. 23–24). Importantly, he called attention to the fact that the products of protometabolic reactions were not statistically random, owing to stringent physical and chemical constraints. (De Duve 1991, pp. 137–145, 213; 2005a, p. 24). Thus, in de Duve's protometabolic scenario, the onset of natural selection depended on a prior evolutionary process involving physical and chemical selection (1991, pp. 143–145).

According to the Russell and Martin scenario, metabolism grew out of the geochemistry at alkaline hydrothermal vents, and a necessary condition for this emergence was the electrochemical and thermal gradients across membranes in the porous structures containing iron sulfides (Russell et al. 2011, p. 946, 2014, p. 308). Theoretical and experimental work (Baaske et al. 2007; Burcar et al. 2015) also raised the possibility that compartments under the hydrothermal vents conditions "could be veritable 'reactors' for RNA synthesis" (Koonin 2007, p. 9105). Thus, this proto-metabolic model involves an early phase, referred to by Russell and colleagues as "the kind of thermodynamically driven evolutionary process" (Russell et al. 2014, p. 326), in which the highly constrained geochemical conditions at the vent determined the emergence and growth of metabolism, and a later stage, in which natural selection within and among compartments took over. The overall process is described by the researches as "an evolutionary transition from geochemical to biochemical processes" (Martin et al. 2008a, p. 805).

The role of physical and chemical selection is notable in the scenario of *Copley and colleagues,* in which genes and metabolism are suggested to have *co-evolved toward an RNA world.* The emergence of protometabolic reaction networks, described by the researchers, is based predominantly on the catalytic abilities of minerals and small organic molecules (Copley et al. 2007). Mutual catalysis of molecules within the networks and selection of communities with increased concentrations of activated monomers are suggested to have led to ever-larger and more-effective catalysts. A crucial element of the co-evolution model is the hypothesis that nucleotides could have been synthesized in the vents by "a surprisingly small"

metabolic module (Copley et al. 2007, p. 433) and could have been part of the emergence of protometabolism from the very start. Each of the many stages depicted by the authors depended on physico-chemical constraints that channeled the products in a non-random direction, as demonstrated by their experiments on the "catalytic landscape" of pyruvate in hydrothermal vents (Novikov and Copley 2013, p. 13283). It is important to note that the overall evolutionary process detailed by the investigators is governed by physico-chemical selection that favored communities with better chances of increasing the levels of monomeric building blocks (Copley et al. 2007, p. 431). Such selected communities were able to synthesize longer and more effective catalysts, including replicating RNAs, thus resulting gradually in a transition to evolution through natural selection.

From the above concluding comments, it is apparent that the case studies examined in this chapter share the hypothesis that the emergence of Darwinian evolution depended to a large extent on physico-chemical constraints, especially on catalysis by minerals and small organic molecules and on physico-chemical selection processes. This unifying theme is brought to its most explicit expression in the *"system-first" scenario in a hydrothermal-field setting,* proposed by Damer and Deamer. These investigators focus specifically on processes of physical selection of more stable polymers and on chemical selection of rare combinations of functional polymers, taking place in the fluctuating environment of hydration/dehydration cycles characterizing hydrothermal-field sites.

It is important to note that the evolutionary hypothesis of the origin of life provides a decisive rebuttal to the argument against the possibility of prebiotic evolution, as formulated by the prominent evolutionary biologist Richard Dawkins. In his argument, discussed in Chapter 3.1 of this volume, Dawkins contends that the first living system, that is, a replicating apparatus at the root of the evolutionary tree, had to emerge by a highly improbable, single-step chance event (Dawkins 1986, p. 140). According to Dawkins, physical and chemical selection events, lacking the element of heredity or "memory," are transient and cannot, in principle, lead to an accumulation of selected small advantages necessary for an evolutionary process (Dawkins 1986, pp. 43–74). The scenarios described in this chapter "beg to differ." Based on theoretical considerations and empirical data, they raise the possibility of stepwise processes that offer an alternative to the notion that the origin of life was the result of a single improbable chance event. The hypothesis raised by Damer and Deamer is particularly illuminating as an answer to Dawkins's contention. According to their scenario, in the "coupling" of the contents of protocells between the hydrated and dehydrated phases, which repeat themselves indefinitely, "reactions do not occur just once and then approach equilibrium, but instead reaction products accumulate and can undergo increasingly complex interactions" (Deamer and Georgiou 2015, p. 1092). Not less important, their scenario is based on the statistics of large numbers. Near-infinite number of combinations of randomly synthesized polymers encapsulated in protocells, each constituting an "experiment," can arise in the physico-chemical setting of a hydrothermal pool and, being physically and chemically selected, may gradually lead to the onset of natural selection.

We still don't know whether any of the scenarios examined here will eventually be able to provide an answer to the question of the emergence of life on Earth. It seems clear, however, that the evolutionary hypothesis of the origin of life is a fruitful hypothesis, leading researchers toward new avenues in the search for a solution.

REFERENCES

Adamala, K., and J. W. Szostak. 2013. Competition between model protocells driven by an encapsulated catalyst. *Nat. Chem.* 5:495–501.

Adamala, K., F. Anella, R. Wieczorek, P. Stano, C. Chiarabelli, and P. L. Luisi. 2014. Open questions in origin of life: Experimental studies on the origin of nucleic acids or proteins with specific and functional sequences by a chemical synthetic biology approach. *Comput. Struct. Biotechnol. J.* 9(14):e201402004.

Adamala, K., A. E. Engelhart, and J. W. Szostak. 2015. Generation of functional RNAs from inactive oligonucleotide complexes by non-enzymatic primer extension. *J. Am. Chem.* 137:483–489.

Anastasi, C., F. F. Buchet, M. A. Crow et al. 2007. RNA: Prebiotic product, or biotic invention? *Chem. Biodivers.* 4:721–739.

Anet, F. 2004. The place of metabolism in the origin of life. *Curr. Opin. Chem. Biol.* 8:654–659.

Ashkenasy, G., R. Jagasia, M. Yadav, and M. R. Ghadiri. 2004. Design of a directed molecular network. *PNAS* 101:10872–10877.

Baaske, P., F. M. Weinert, S. Duhr, K. H. Lemke, M. J. Russell, and D. Braun. 2007. Extrem accumulation of nucleotides in simulated hydrothermal pore systems. *PNAS* 104(22):9346–9351.

Barbas, C. F. III. 2008. Organocatalysis lost: Modern chemistry, ancient chemistry, and an unseen biosynthetic apparatus. *Angew. Chem. Int. Ed.* 47:42–47.

Benner, S. A., Kim, H. J., and Z. Yang. 2012. Setting the stage: The history, chemistry, and geobiology behind RNA. *Cold Spring Harb. Perspect. Biol.* 4:a003541.

Blain, J. C., and J. W. Szostak. 2014. Progress toward synthetic cells. *Annu. Rev. Biochem.* 83:11.1–11.26.

Brack, A. 2014. Where do we go from here? Astrobiology editorial board opinions. *Astrobiology* 14(8):638–639. doi:10.1089/ast.2014.1405.

Brack, A. 2017. Farewell to a friend. *Orig. Life Evol. Biosph.* 47:219–222.

Bourbo, V., M. Matmor, E. Shtelman, B. Rubinov, N. Ashkenasy, and G. Ashkenasy. 2011. Self-assembly and self-replication of short amphiphilic β sheet peptides. *Orig. Life Evol. Biosph.* 41:563–567. doi:10.1007/s 11084-011-9257-y.

Budin, R., and J. W. Szostak. 2011. Physical effects underlying the transition from primitive to modern cell membranes. *PNAS* 108(13):5249–5254.

Corliss, J. B., J. Dymond, L. I. Gordon et al. 1979. Submarine thermal springs on the Galápagos rift. *Science* 203:1073–1083.

Crick, F. H. C. 1968. The origin of the genetic code. *J. Mol. Biol.* 38:367–379.

Crick, F. 1981. *Life Itself.* New York: Simon & Schuster.

Baross, J. A., and S. E. Hoffman. 1985. Submarine hydrothermal vents and associated gradient environments as sites for the origin and evolution of life. *Orig. Life Evol. Biosph.* 15:327–345.

Burcar, B. T., L. M. Barge, D. Trail, E. B. Watson, M. J. Russell, and L. B. McGown. 2015. RNA oligomerization in laboratory analogues of alkaline hydrothermal vent systems. *Astrobiology* 15(7):509–522.

Cairns-Smith, A. G. 1986. *Seven Clues to the Origin of Life.* Cambridge, UK: Cambridge University Press.

Cairns-Smith, A. G. 2008. Chemistry and the missing era of evolution. *Chem. Eur. J.* 14:3830–3839.

Cech, T. R. 2012. The RNA worlds in Context. *Cold Spring Harb. Prspect. Biol.* 4:a006742.

Chen, I. A., and J. W. Szostak. 2004. A kinetic study of the growth of fatty acid vesicles. *Biophys. J.* 87(2):988–998.

Chen, I. A., K. Salehi-Ashtiani, and J. W. Szostak. 2005. RNA catalysis in model protocell vesicles. *J. Am. Chem. Soc.* 127(38):13213–13221.

Cody, G. D., N. Z. Boctor, T. R. Filley et al. 2000. Primordial carbonylated iron-sulfur compounds and the synthesis of pyruvate. *Science* 289:1337–1340.

Copley, S. D., E. Smith, and H. J. Morowitz. 2007. The origin of the RNA world: Co-evolution of genes and metabolism. *Bioorganic Chem.* 35:430–443.

Copley, S. D., E. Smith, and H. J. Morowitz. 2010. How life began: The emergence of sparse metabolic networks. *J. Cosmol.* 10:3345–3361.

Damer, B. 2016. A field trip to the Archaean in search of Darwin's warm little pond. *Life* 6(2):21. doi:10.3390/life6020021.

Damer, B., and D. Deamer. 2015. Coupled phases and combinatorial selection in fluctuating hydrothermal pools: A scenario to guide experimental approaches to the origin of cellular life. *Life* 5(1):872–887. doi:10.3390/life5010872.

Dawkins, R. 1986. *The Blind Watchmaker.* London, UK: Penguin Books.

Dawkins, R. 2006. *The God Delusion.* Boston, MA: Houghton Mifflin.

Deamer, D. W. 2012. Liquid crystalline nanostructures: Organizing matrices for non-enzymatic nuleic acid polymerization. *Chem. Soc. Rev.* 41:5375–5379.

Deamer, D. 2014. Where do we go from here? Astrobiology editorial board opinions. *Astrobiology* 14(8):637. doi:10.1089/ast.2014.1405.

Deamer, D. W. 2017. The role of lipid membranes in life's origin. *Life* 7(1):5. doi:10.3390/life7010005.

Deamer, D. W., and G. L. Barchfeld. 1982. Encapsulation of macromolecules by lipid vesicles under simulated prebiotic conditions. *J. Mol. Evol.* 18:203–201.

Deamer, D. W., and R. M. Pashley. 1989. Amphiphilic components of carbonaceous meteorites. *Orig. Life Evol. Biosph.* 19:21–38.

Deamer, D. W., and C. D. Georgiou. 2015. Hydrothermal conditions and the origin of cellular life. *Astrobiology* 15(12):1091–1095.

De Duve, C. 1991. *Blueprint for a Cell.* Burlington, NC: Neil Patterson.

De Duve, C. 2003. A research proposal on the origin of life. *Orig. Life Evol. Biosph.* 33:559–574.

De Duve, C. 2005a. *Singularities.* Cambridge, UK: Cambridge University Press.

De Duve, C. 2005b. The onset of selection. *Nature* 433:581–582.

Diederichsen, U. 1996. Pairing properties of alanyl peptide nucleic acids containing an amino acid backbone with alternating configuration. *Angew. Chem. Int. Ed. Engl.* 35:445–448.

Diederichsen, U. 1997. Alanyl PNA: Evidence for linear band structures based on guanine-cytosine base pairs. *Angew. Chem. Int. Ed. Engl.* 36(17):1886–1889.

Eigen, M. 1971. Self-organization of matter and the evolution of biological macromolecules. *Naturwissenschaften* 58:465–523.

Eigen, M., W. Gardiner, P. Schuster, and R. Winkler-Oswatitsch. 1981. The origin of genetic information. *Sci. Am.* 244(4):78–118.

Engelhart, A. E., M. W. Powner, and J. W. Szostak. 2013. Functional RNAs exhibit tolerance for non-heritable 2′–5′ versus 3′–5′ backbone heterogeneity. *Nat. Chem.* 5:390–394.

Eschenmoser, A. 1999. Chemical etiology of nucleic acid structure. *Science* 284:2118–2124.

Eschenmoser, A. 2007a. The search for the chemistry of life's origin. *Tetrahedron* 63:12821–12844.

Eschenmoser, A. 2007b. On a hypothetical generational relationship between HCN and constituents of the Reductive Citric Acid Cycle. *Chem. Biodivers.* 4:554–573.

Eschenmoser, A. 2011. Etiology of potentially primordial biomolecular structures: From vitamin B_{12} to the nucleic acids and an inquiry into the chemistry of life's origin: A retrospective. *Angew. Chem. Int. Ed.* 50(52):12412–12472.

Ferris, J. P. 2005. Catalysis and prebiotic synthesis. *Rev. Mineral. Geochem.* 59:187–210.

Fry, I. 2000. *The Emergence of Life on Earth. A Historical and Scientific Overview.* New Brunswick, NJ: Rutgers University Press.

Fry, I. 2006. The origins of research into the origin of life. *Endeavour* 30(1):25–29.

Fry, I. 2011. The role of natural selection in the origin of life. *Orig. Life Evol. Biosph.* 41:3–16.

Forsythe, J. G., S. S. Yu, I. Mamajanov et al. 2015. Ester-mediated amide bond formation driven by wet-dry cycles: A possible path to polypeptides on the prebiotic Earth. *Angew. Chem. Int. Ed. Engl.* 54:9871–9875.

Fox, S. W. 1984. Proteinoid experiments and evolutionary theory. In *Beyond Neo-Darwinism*, M. W. Ho, and P. T. Saunders (Eds.), pp. 15–60. New York: Academic Press.

Fox, S. W., and K. Dose. 1977. *Molecular Evolution and the Origin of Life*. New York: Marcel Dekker.

Gilbert, W. 1986. The RNA world. *Nature* 319:618.

Guerrier-Takada, C., K. Gardiner, T. Marsh, N. Pace, and S. Altman. 1983. The RNA moiety of ribonuclease P is the catalytic subunit of the enzyme. *Cell* 35:849–857.

Haldane, J. B. S. 1967. The origin of life. In *The Origin of Life*, J. D. Bernal (Ed.), pp. 242–249. Appendix II. London, UK: Weidenfeld & Nicolson. (Originally Published in 1929 in Rationalist Annual, 3–10).

Hanczyc, M. M., S. M. Fujikawa, and J. W. Szostak. 2003. Experimental models of primitive cellular compartments: Encapsulation, growth, and division. *Science* 302:618–621.

Hargreaves, W. R., S. J. Mullvihill, and D. W. Deamer. 1977. Synthesis of phospholipids and membranes in prebiotic conditions. *Nature* 266:78–80.

Hazen, R. M. 2005. *Genesis*. Washington, DC: Joseph Henry Press.

Heinen, W., and A. M. Lauwers. 1996. Organic sulfur compounds resulting from the interaction of iron sulfide, hydrogen sulfide and carbon dioxide in an anaerobic aqueous environment. *Orig. Life Evol. Biosph.* 26:131–150.

Herschy, B., A. Whicker, E. Camprubi et al. 2014. An origin-of-life reactor to simulate alkaline hydrothermal vents. *J. Mol. Evol.* 79:213–227.

Huang, W., and J. P. Ferris. 2006. One-step, regioselective synthesis of up to 50-mers of RNA oligomers by montomorillonite catalysis. *J. Am. Chem. Soc.* 128:8914–8919.

Huber, C., and G. Wächtershäuser. 1997. Activated acetic acid by carbon fixation on [Fe, Ni]S under primordial conditions. *Science* 276:245–247.

Huber, C., W. Eisenreich, S. Hecht, and G. Wächtershäuer. 2003. A possible primordial peptide cycle. *Science* 301:938–940.

Huber, C., and G. Wächtershäuser. 2006. α-Hydroxy and α-amino acids under possible hadean, volcanic origin-of-life conditions. *Science* 314:630–632.

Huber, C., F. Kraus, M. Hanzlik, W. Eisenreich, and G. Wächtershäuser. 2012. Elements of metabolic evolution. *Chem. Eur. J.* 18:2063–2080.

Illangasekare, M., and M. Yarus. 1999. A tiny RNA that catalyzes both aminoacyl-RNA and peptidyl-RNA synthesis. *RNA* 5:1482–1489.

Issac, R., and J. Chmielewski. 2002. Approaching exponential growth with a self-replicating peptide. *J. Am. Chem. Soc.* 124:6808–6809.

Jacob, F. 1982. *The Logic of Life*, trans. B. E. Spillman. New York: Pantheon Books.

James, K. D., and A. D. Ellington. 1995. A search for missing links between self-replicating nucleic acids and the RNA world. *Orig. Life Evol. Biosph.* 25:515–530.

Joshi, P. C., M. F. Aldersley, J. W. Delano, and J. P. Ferris. 2009. Mechanism of montmorillonite catalysis in the formation of RNA oligomers. *J. Am. Chem. Soc.* 131:13369–13374.

Joyce, G. F., and L. E. Orgel. 2006. Progress toward understanding the origin of the RNA world. In *The RNA World*, 3rd ed., R. F. Gesteland, T. R. Cech, and J. F. Atkins (Eds.), pp. 23–56. Cold Spring Harbor, NY: Cold Spring Harbor Laboratory.

Kaschke, M., M. J. Russell, and W. J. Cole. 1994. [FeS/FeS$_2$]. A redox system for the origin of life. *Orig. Life Evol. Biosph.* 24:291–307.

Kauffman, S. A. 1986. Autocatalytic sets of proteins. *J. Theor. Biol.* 119:1–24.

Kauffman, S. A. 1993. *The Origins of Order: Self-Organization and Selection in Evolution*. New York: Oxford University Press.

Kelly, D. S., J. A. Karson, D. K. Blackman et al. 2001. An off-axis hydrothermal vent field near the Mid-Atlantic Ridge at 30°N. *Nature* 412:145–149.

Kelly, D. S., J. A. Karson, G. L. Früh-Green et al. 2005. A serpentinite-hosted ecosystem: The Lost City hydrothermal field. *Science* 307:1428–1434.

Koonin, E. V. 2007. An RNA-making reactor for the origin of life. *PNAS* 104(22):9105–9106.

Koonin, E. V., and W. Martin. 2005. On the origin of genomes and cells within inorganic compartments. *Trends Gen.* 21(12):647–653.

Kruger, K., P. J. Grabowski, A. J. Zaug, J. Sands, D. E. Gottschling, and T. R. Cech. 1982. Self-splicing RNA: Auto excision and autocyclization of the ribosomal RNA intervening sequence of *Tetrahymena.Cell* 31:147–157.

Lancet, D., and B. Shenhav. 2009. Compositional lipid protocells: Reproduction without polynucleotides. In *Protocells: Bridging Nonliving and Living Matter*, S. Rasmussen, M. A. Bedau, L. Chen et al. (Eds.), pp. 233–252. Cambridge, MA: The MIT Press.

Lane, N. 2012. Life: Inevitable or fluke? *New Sci.* 14(2870):33–37.

Lee, D. H., J. R. Cranja, J. A. Martinez, K. Severin, and M. R. Ghadiri. 1996. A self-replicating peptide. *Nature* 382:525–528.

Lee, D. H., K. Severin, Y. Yokobayashi, and M. R. Ghadiri. 1997. Emergence of symbiosis in peptide self-replication through a hypercyclic network. *Nature* 390:591–594.

Lifson, S. 1997. On the crucial stages in the origin of animate matter. *J. Mol. Evol.* 44:1–8.

Luisi, P. L. 2015. Chemistry constraints on the origin of life. *Isr. J. Chem.* 55:906–918.

Mansy, S. S., and J. W. Szostak. 2008. Themostability of model protocell membranes. *PNAS* 105(36):13351–13355.

Martin, W., and M. J. Russell. 2003. On the origins of cells: A hypothesis for the evolutionary transitions from abiotic geochemistry to chemoautotrophic prokaryotes, and from prokaryotes to nucleated cells. *Phils. Trans. R. Soc. Lond. B: Biol. Sci.* 358:59–83.

Martin, W., and M. J. Russell. 2007. On the origin of biochemistry at an alkaline hydrothermal vent. *Philos. Trans. R. Soc. Lond. B: Biol. Sci.* 362:1887–1925.

Martin, W., J. Baross, D. Kelly, and M. J. Russell. 2008a. Hydrothermal vents and the origin of life. *Nat. Rev. Microbiol.* 6:805–814.

Martin, W. F., F. L. Sousa, and N. Lane. 2014. Energy at life's origin. *Science* 344(6188):1092–1093.

Martin, Z., O. Bota, M. L. Fogel et al. 2008b. Extraterrestrial nucleobases in the Murchison meteorite. *Earth Planet. Sci. Lett.* 270:130–136.

Mayr, E. 1982. *The Growth of Biological Thought.* Cambridge MA: Harvard University Press.

Maynard Smith, J., and E. Szathmáry. 1985. *The Major Transitions in Evolution.* Oxford, UK: W. H. Freeman.

Mazur, S. 2014. *'Oomph' and the Origin of Life at Hydrothermal Vents.* https://www.huffingtonpost.com/suzan-mazur/oomph-origin-of-life-at-h_b_5646294.html (accessed January 4, 2018).

Mielke, R. E., M. J. Russell, P. R. Wilson et al. 2010. Design, fabrication, and test of a hydrothermal reactor for origin-of-life experiments. *Astrobiology* 10(8):799–810.

Mielke, R. E., K. J. Robinson, L. M. White et al. 2011. Iron-sulfide-bearing chimneys as potential catalytic energy traps at life's emergence. *Astrobiology* 11(10):933–950.

Miller, S. L., and J. L. Bada. 1988. Submarine hot springs and the origin of life. *Nature* 334:155–176.

Mills, D. R., R. L. Peterson, and S. Spiegelman. 1967. An extracellular Darwinian experiment with a self-duplicating nucleic acid molecule. *PNAS* 58:217.

Miyakawa, S., and J. P. Ferris. 2003. Sequence-and regioselectivity in the montmorillonite-catalyzed synthesis of RNA. *J. Am. Chem. Soc.* 125:8202–8208.

Monnard, P. A. 2007. Question 5: Does the RNA-world still retain its appeal after 40 years of research? *Orig. Life Evol. Biosph.* 37:387–390.

Monnard, P. A., A. Kanavarioti, and D. W. Deamer. 2003. Eutectic phase polymerization of activated ribonucleotide mixtures yields quasi-equimolar incorporation of purine and pyrimidine nucleobases. *J. Am. Chem. Soc.* 125:13734–13740.

Monnard, P. A., and H. Ziock. 2008. Eutectic phase in water-ice: A self-assembled environment conductive to metal-catalyzed nonenzymatic RNA polymerization. *Chem. Biodivers* 5:1521–1539.

Monod, J. 1974. *Chance and Necessity.* Glasgow, UK: Collins, Fontana Books.

Morowitz. H. J. 1999. The theory of biochemical organization, metabolic pathways, and evolution. *Complexity* 4:39–53.

Morowitz, H. J. 1992. *Beginnings of Cellular Life.* New Haven, CT: Yale University Press.

Morowitz, H. J., B. Heinz, and D. W. Deamer. 1988. The chemical logic of a minimal protocell. *Orig. Life Evol. Biosph.* 18:281–287.

Morowitz H. J., J. D. Kostelnik, J. Yang, and G. D. Cody. 2000. The origin of intermediary metabolism. *PNAS* 97(14):7704–7708.

Morowitz, H., V. Srinivasan, S. Copley, and E. Smith. 2005. The simplest enzyme revisited. *Complexity* 10(5):12–13. doi:10.1002/cplx.200087.

Movassaghi, M., and E. N. Jacobsen. 2002. The simplest "enzyme." *Science* 298:1904–1905.

Muller, H. J. 1966. The gene material as the initiator and the organizing basis of life. *Am. Nat.* 100(915):493–517.

Nielsen, P. E. 1993. Peptide nucleic acid (PNA): A model structure for the primordial genetic material. *Orig. Life Evol. Biosph.* 23:323–327.

Nielsen, P. E. 2009. Peptide nucleic acid as prebiotic and abiotic genetic material. In *Protocells: Bridging Nonliving and Living Matter*, S. Rasmussen, M. A. Bedau, L. Chen et al. (Eds.), pp. 336–346. Cambridge, MA: The MIT Press.

Noller, H. F. 2012. Evolution of protein synthesis from an RNA world. *Cold Spring Harb. Perspect. Biol.* 4:a003681.

Novikov, Y., and S. D. Copley. 2013. Reactivity landscape of pyruvate under simulated hydrothermal vent conditions. *PNAS* 110(33):13283–13288.

Oberholzer, T., R. Wick, P. L. Luisi, and C. K. Biebricher. 1995. Enzymatic RNA replication in self-reproducing vesicles: An approach to a minimal cell. *Biochem. Biophys. Res. Commun.* 207:250–257.

Oparin, A. I. 1953. *Origin of Life*, trans. S. Margulis. New York: Dover Publications. (Originally Published in Moscow in 1936).

Oparin, A. I. 1965. History of the subject matter of the conference. In *The Origin of Prebiological Systems and of Their Matrices*, S. W. Fox (Ed.). New York: Academic Press.

Oparin, A. I. 1967. The origin of life. A. Synge, trans. In *The Origin of Life*, J. D. Bernal (Ed.), pp. 199–234. Appendix I. London, UK: Weidenfeld and Nicolson (Originally Published in Moscow in 1924).

Orgel, L. E. 1968. Evolution of the genetic apparatus. *J. Mol. Biol.* 38:381–393.

Orgel, L. E. 1973. *The Origins of Life.* New York: Wiley.

Orgel, L. E. 1994. The origin of life on earth. *Sci. Am.* 271(4):53–61.

Orgel, L. E. 2000. Self-organizing biochemical cycles. *PNAS* 97:12503–12507.

Orgel, L. E. 2003. Some consequences of the RNA world hypothesis. *Orig. Life Evol. Biosph.* 33:211–218.

Orgel, L. E. 2004. Prebiotic chemistry and the origin of the RNA world. *Crit. Rev. Biochem. Mol. Biol.* 39:99–123.

Orgel, L. E. 2008. The implausibility of metabolic cycles on the prebiotic earth. *PLoS Biol.* 6(1):e18:5–13.

Pascal, R., A. Pross, and J. D. Sutherland. 2013. Towards an evolutionary theory of the origin of life based on kinetics and thermodynamics. *Open Biol.* 3:130156. doi:10.1098/rsob.130156.

Patel, B. H., C. Percivalle, D. J. Ritson, C. D. Duffy, and J. D. Sutherland. 2015. Common origins of RNA, protein and lipid precursors in a cyanosulfidic protometabolism. *Nat. Chem.* 7:301–307.

Pereto, J. 2012. Out of fuzzy chemistry: From prebiotic chemistry to metabolic networks. *Chem. Soc. Rev.* 41:5394–5403.

Pino, S., F. Ciciriello, G. Costanzo, and E. Di Mauro. 2008. Nonenzymatic RNA ligation in water. *J. Biol. Chem.* 283(52):36494–36503.

Pizzarello, S., and E. Shock. 2010. The organic composition of carbonaceous meteorites: The evolutionary story ahead of biochemistry. *Cold Spring Harb. Perspect. Biol.* 2(3):a002105.

Pizzarello, S., S. K. Davidowski, G. P. Holland, and L. B. Williams. 2013. Processing of meteoritic organic materials as a possible analog of early molecular evolution in planetary environments. *PNAS* 110(39):15614–15619.

Pohorille, A. 2009. Early ancestors of existing cells. In *Protocells: Bridging Nonliving and Living Matter*, S. Rasmussen, M. A. Bedau, L. Chen et al. (Eds.), pp. 563–581. Cambridge, MA: The MIT Press.

Pohorille, A., and M. A. Wilson. 1995. Molecular dynamics of simple membrane-water interfaces: Structure and functions in the beginnings of cellular life. *Orig. Life Evol. Biosph.* 21:21–46.

Powner, M. W., B. Garland, and J. D. Sutherland. 2009. Synthesis of activated pyrimidine ribonucleotides in prebiotically plausible conditions. *Nature* 459:239–242.

Powner, M. W., J. D. Sutherland, and J. W. Szostak. 2010. Chemoselective multicomponent one-pot assembly of purine precursors in water. *J. Am. Chem. Soc.* 132(46):16677–16688.

Pross, A. 2004. Causation and the origin of life. Metabolism or replication first? *Orig. Life Evol. Biosph.* 34:307–321.

Rajamani, A., S. Vlassov, A. Benner, R. Coombs, F. Olasagasti, and D. Deamer. 2008. Lipid assisted synthesis of RNA-like polymers from mononucleotides. *Orig. Life Evol. Biosph.* 38:57–74.

Rasmussen, S., M. A. Bedau, L. Chen et al. (Eds.). 2009. *Protocells: Bridging Nonliving and Living Matter*. Cambridge, MA: The MIT Press.

Ritson, D., and J. D. Sutherland. 2012. Prebiotic synthesis of simple sugars by photoredox systems chemistry. *Nat. Chem.* 4:895–899.

Ruiz-Mirazo, K., C. Briones, and A. de la Escosura. 2014. Systems chemistry: New perspectives for the origin of life. *Chem. Rev.* 114(1):285–366.

Russell, M. J., and A. J. Hall. 1997. The emergence of life from iron monosuphide bubbles at a submarine hydrothermal redox and pH front. *J. Geol. Soc. London* 154:377–402.

Russell, M. J., and W. Martin. 2004. The rocky roots of the acetyl-CoA pathway. *Trends Biochem. Sci.* 29:358–363.

Russell, M. J., A. J. Hall, A. G. Cairns-Smith, and P. S. Braterman. 1988. Submarine hot springs and the origin of life. *Nature* 336:117.

Russell, M. J., R. M. Daniel, A. J., Hall, and J. Sherringham. 1994. A hydrothermally precipitated catalytic iron sulphide membrane as a first step toward life. *J. Mol. Evol.* 39:231–243.

Russell, M. J., L. M. Barge, R. Bhartia et al. 2014. The drive to life on wet and icy worlds. *Astrobiology* 14(4):308–343.

Safhill, R., H. Schneider-Bernloehr, L. E. Orgel, and S. Spiegelman. 1970. *In vitro* selection of bacteriophage Qβ variants resistant to ethidium bromide. *J. Mol. Biol.* 51:531.

Schrum, J. P., T. F. Zhu, and J. W. Szostak. 2010. The origins of cellular life. *Cold Spring Harb. Perspect. Biol.* doi:10.1101/cshperspect.a002212.

Schwartz, A. W., and A. Henderson-Sellers. 1983. Glaciers, volcanic islands and the origin of life. *Precambrian Res.* 22:167–174.

Schwartz, A. W., and M. Goverde. 1982. Accelaration of HCN oligomerization by formaldehyde and related compounds: Implications for prebiotic syntheses. *J. Mol. Evol.* 18:351–353.

Schwartz, A. W. 2013. Evaluating the plausibility of prebiotic multistage syntheses. *Astrobiology* 13(8):784–789.

Segré, D., D. Ben Eli, D. Deamer, and D. Lancet. 2001. The lipid world. *Orig. Life Evol. Biosph.* 31:119–145.

Service, R. F. 2013. The life force. *Science* 342:1032–1034.

Smith, E., and H. J. Morowitz. 2004. Universality in intermediary metabolism. *PNAS* 101:13168–13173.

Shenhav, B., D. Segré, and D. Lancet. 2003. Mesobiotic emergence: Molecular and ensemble complexity in early evolution. *Adv. Complex Syst.* 6:15–35.

Smith, E., H. J. Morowitz, and S. D. Copley. 2009. Core metabolism as a self-organizing system. In *Protocells: Bridging Nonliving and Living Matter*, S. Rasmussen, M. A. Bedau, L. Chen et al. (Eds.), pp. 433–460. Cambridge, MA: The MIT Press.

Sutherland, J. D. 2010. Ribonucleotides. *Cold Spring Harb. Perspect. Biol.* 2. doi:10.1101/cshperspect.a005439.

Sutherland, J. D. 2016. The origin of life–Out of the blue. *Angew. Chem. Int. Ed.* 55:104–121.

Szostak, J. W. 2009. System chemistry on early Earth. *Nature* 459:171–172.

Szostak, J. W. 2011. An optimal degree of physical and chemical heterogeneity for the origin of life? *Philos. Trans. R. Soc. Lond. B: Biol. Sci.* 366:2894–2901.

Szostak, J. W. 2012a. Attempts to define life do not help to understand the origin of life. *J. Biomol. Struct. Dyn.* 29:599–600.

Szostak, J. W. 2012b. The eightfold path to non-enzymatic RNA replication. *J. Syst. Chem.* 3:2. http://www.systchem.com/contents/3/1/2.

Szostak, J. W., D. P. Bartel, and P. L. Luisi. 2001. Synthesizing life. *Nature* 409:387–390.

Toppozini, L., H. Dies, D. W. Deamer, and M. C. Rheinstädter. 2013. Adenosine monophosphate forms ordered arrays in multiammelar lipid matrices: Insights into assembly of nucleic acid for primitive life. *PLoS ONE* 8. doi:10.1371/journal.pone.0062810.

Troland, L. T. 1914. The chemical origin and regulation of life. *Monist* 24:92:133.

Troland, L. T. 1917. Biological enigmas and the theory of enzyme action. *Am. Nat.* 51(606):321–350.

Turk, R. M., N. V. Chumachenko, and M. Yarus. 2010. Multiple translational products from a five-nucleotide ribozyme. *PNAS* 107(10):4585–4589.

Vasas, V., E. Szathmáry, and M. Santos. 2010. Lack of evolvability in self-sustaining autocatalytic networks constraints metabolism-first scenarios for the origin of life. *PNAS* 107(4):1470–1475.

Vasas, V., C. Fernando, M. Santos, S. Kauffman, and E. Szathmáry. 2012. Evolution before genes. *Biol. Direct* 7:1. doi:10.1186/1745-6150-7-1.

Wächtershäuser, G. 1988. Before enzymes and templates: Theory of surface metabolism. *Microbiol. Rev.* 52:452–484.

Wächtershäuser, G. 1990. Evolution of the first metabolic cycles. *PNAS* 87(1):200–204. doi:10.1073/pnas.87.1.200.

Wächtershäuser, G. 1992. Groundwork for an evolutionary biochemistry: The iron-sulfur world. *Prog. Biophys. Mol. Biol.* 58:85–201.

Wächtershäuser, G. 2007. On the chemistry and evolution of the pioneer organism. *Chem. Biodivers* 4:584–602.

Weber, A. L. 2007. The sugar model: Autocatalytic activity of the triose-ammonia reaction. *Orig. Life Evol. Biosph.* 37:105–111.

Wieczorek, R., M. Dçrr, A. Chotera, P. L. Luisi, and P-A, Monnard. 2013. Formation of RNA phosphodiester bond by histidine-containing dipeptides. *Chembiochem* 14:217–223.

Wieczorek, R., K. Adamala, T. Gasperi, F. Polticelli, and P. Stano. 2017. Small and random peptides: An unexplored reservoir of potentially functional primitive organocatalysts. The case of seryl-histidine. *Life* 7(2):19. doi:103390/life7020019.

Wing, M. R., and J. L. Bada. 1991. Geochromatography on the parent body of the carbonaceous chondrite Ivuna. *Geochim. Cosmochim. Acta* 55:2937–2942.

Woese, C. R. 1967. *The Genetic Code*. New York: Harper & Row.

Yarus, M. 2011. Getting past the RNA world: The initial Darwinian ancestor. *Cold Spring Harb. Perspect. Biol.* 3:a003590.

Zhu, T. F., and J. W. Szostak. 2009. Coupled growth and division of model protocell membranes. *J. Am. Chem. Soc.* 131:5705–5713.

6.6 The Complexity of Life's Origins
A Physicochemical View

Jan J. Spitzer

CONTENTS

6.6.1   INTRODUCTION

Alice laughed: "There's no use trying," she said; "one can't believe impossible things"

(Carroll 1999)

So enigmatic are the complexities of living states of matter that we tilt at windmills to gain an understanding: we arm ourselves with quixotic ideas, hopeful of winning the battle. In this review, I lay out a *physicochemical* line of attack: it constrains speculations; it clarifies assumptions; and it brings into focus the molecular mystery of "becoming alive and evolving." It delineates complex yet realistic processes of life's origins—a "jigsaw puzzle" yet to be fully worked out (Spitzer 2017).

I organize this review as follows. In Section 6.6.2, I re-consider traditional predicaments of origins research—the questions of life's natural "designers," Schrödinger's unknown biophysical laws, and the questions related to "defining life." These re-appraisals suggest that life's emergence can be viewed as a "jigsaw puzzle" based on existing physicochemical laws and cyclic prebiotic processes driven by diurnal and tidal energies. The updated version of the puzzle is summarized in Section 6.6.3. Then I take a top-down view in Section 6.6.4, and discuss the physicochemical complexity of today's bacterial cells (Spitzer 2011) in order to exploit the assumption that first Archaean organisms were similar to today's bacterial cells. Next, in Section 6.6.5, I discuss experimental ideas based on bottom-up (chemical) and top-down (biochemical) processes energized by cyclic temperature and water activity gradients (Spitzer 2013b, 2014); their cyclic character caused proto-nucleic acids to evolve as unwinding and rewinding double helices—replayed today in polymerase chain reactions (PCRs). In Section 6.6.6, I suggest that the same cyclic energies brought about repeated fusions of first cells with environmental nucleic acids and with each other—a natural mechanism of cellular evolution analogous to genetic engineering.

I conclude that Earth's diurnal and tidal cycles drove phase-separations and prebiotic chemical reactions towards cellular life in the microspaces (compartments) of consolidating geochemical matrices (Spitzer and Poolman 2009). The "compartments first" paradigm (Morowitz 1992) is *unconditionally* required to constrain diffusional decay to disorder mandated by the second law of thermodynamics. Orgel (2004) expressed this requirement intuitively as "molecules that stay together, evolve together." Further, the cyclic "triple" chemical coevolution of proto-proteins and proto-nucleic acids *with and within* the micro-compartments must proceed under high macromolecular crowding at relatively high ionic strengths. Under these conditions, the excluded volume effect of biomacromolecular crowding becomes commensurate with hydration and screened electrostatic forces at a distance of about one nano-meter. Only at such close distances can cyclic chemical evolution progress to today's molecular recognition (e.g., substrate-enzyme binding), to the complementarity of hydrogen bonding of nucleobases (genetics), to protein folding and the assembly of biomacromolecular "machines" (bio-catalysis), and to the repeatable rebuilding of cellular "architecture" of dividing cells (Harold 2005).

6.6.2 CELLULAR COMPLEXITY CREATES PREDICAMENTS

Origins research has been a very broad and complex effort involving many scientific disciplines, as evidenced in this handbook and in many books and compendia (e.g., Deamer and Fleischaker 1994; Lahav 1999; Brack 2000; Bedau and Cleland 2010). Such wide transdisciplinary research has inevitably engendered many diverging viewpoints, including those addressing "complexity."

I use the term "complexity" in an informal meaning of "being very complicated" (Spitzer 2013a; Whitesides and Ismagilov 1999). In the physicochemical sense, the complexity refers to concentrated (crowded) mixtures of many kinds of interacting molecules undergoing cyclic phase separations and chemical reactions in permeable containers. "Molecules" are taken broadly, including ions, macromolecules, and polyelectrolytes; "phase separation" has the normal physicochemical meaning of the appearance of liquid and solid phases. However, in the presence of macromolecules and surfactant (amphiphilic) molecules, phase separations of "semi-solid" phases at low water activities become tantalizingly complicated; complex materials then appear with diverse morphologies of submicron domains—crystalline, amorphous, and liquid, including interpenetrating networks (Spitzer 2011).

In the cytological sense, the complexity refers to the processes of the cell cycle. I borrow the *eukaryotic* term "cell cycle" (which has well-defined stages) to describe also the growth and division of prokaryotic cells as representatives of the "first" life. In this case, the cell cycle is a complex physicochemical process of the nucleoid being replicated while attached to the growing cell envelope, with subsequent nucleoid segregation and cell envelope re-organization into two cells.

Because physicochemical complexities of life's emergence have not been explored in much detail, it is constructive to re-appraise some traditional views about the nature and origins of life.

6.6.2.1 THE "BLIND WATCHMAKERS"

How Darwinian evolution fits into molecular mechanisms of life's emergence has not yet been adequately explained. That Darwinian evolution is "blind," even at the microbial level, is uncontroversial: genetic variations occur *before* natural selection—before the "blind watchmaker" can act on them (Dawkins 1987)—cf. Section 8 of this handbook. That genetic variations arise from errors in replications of nucleic acids (nucleoids and chromosomes) is also uncontroversial, cf. Section 6.6.6.

But how does planetary prebiotic *chemical evolution* operate? How does the "clutter" or "gunk" of prebiotic molecular inventory chemically react, phase-separate, and evolve into "proto-cells" and then into single-cell ancient organisms? After all, the major product of Stanley Miller experiments is "goo" of many kinds of chemical compounds; after all, organic synthetic reactions often produce complex ("intractable") mixtures of chemicals (Miller 1953; Shapiro 1986, 2007; Schwartz 2007; Robertson and Joyce 2012; Scherer et al. 2016). Because Stanley Miller experiments generate a multitude of carbon compounds in less than a week, the chemical complexification of Earth was a *very fast* evolutionary process, with many parallel chemical reactions in different geochemical locales (Figure 6.6.1).

Experimental evidence thus suggests that prebiotic Earth was a complex dynamic reservoir of molecules: heterogeneous, phase-separating mixtures in molecularly homogeneous multicomponent electrolyte solutions, undergoing chemical reactions driven by diurnal and tidal energies (Earth-Moon rotations). From astrochemical observations, we also know that complex mixtures of many kinds of molecules are being created and occur throughout the cosmos, cf. Section 4.

Regardless of the chemical details of prebiotic molecular inventory, physical chemistry and planetary sciences do provide a plausible "principle" of life's emergence: living states of matter arose through (natural) planetary energies acting *repeatedly* on prebiotic chemical matter (Spitzer and Poolman 2009; Spitzer et al. 2015; Spitzer 2017). Such cyclic—stoking—energies create non-random dynamic "patterns," the first necessity for a life to arise (Whitesides and Boncheva 2002; Hazen 2005). Cyclic energies can drive phase separations, that is, submicron colloidal nucleations

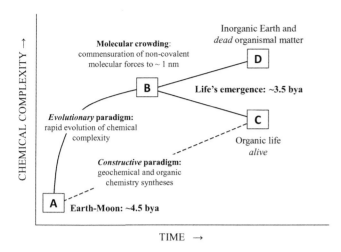

FIGURE 6.6.1 Evolution of Earth's physicochemical complexity and the emergence of life. After the formation of the Earth-Moon planetary system [A] the physicochemical complexity of prebiotic processes grew very fast, until a "self-purifying" process (phase-separating in localized micro-spaces) [B] began to evolve toward cellular life, currently simplified to only five nucleobases, two derivatives of ribose, phosphates, 20 amino acids, two-tail hydrophobic surfactants, and multi-ionic aqueous electrolyte of relatively high ionic strength. After the appearance of life [C], the chemical complexity of Earth [C + D] grew fast also through dead organismal matter [D].

toward micro-scale agglomerations and crystallizations; they can drive the reverse phenomena of re-dispersions and dissolutions; they can drive chemical reactions back and forth, for example, condensation/hydrolysis reactions; and today, cyclic temperatures drive amplifications of DNA segments in cell-free PCR protocols.

Thus, in general, cycling temperatures and the gradients of water activity and chemical and electrochemical potentials play an essential part in the emergence of life. As a corollary, life is extremely unlikely to emerge *spontaneously* (in the sense of chemical thermodynamics and kinetics) from a "prebiotic soup" of molecules, representing a one-of-a-kind, highly improbable event of kinetically arrested accumulation of high free energy in localized molecules of just the right kind, which self-reacted and self-organized, creating first "molecular replicators" or first "simple proto-cells." Though statistically imaginable and described as a "near miracle," this view has now been abandoned. It is noteworthy that it was entertained by influential scientists, such as Monod and Crick (Fry 2000).

6.6.2.2 Schrödinger's Laws of Life

The above clarification that cellular life emerged in a "blind" cycling process through chemical reactions and phase separations is closely related to cellular metabolism—the energies that cells require for reproduction. This is one issue that Schrödinger addressed in his book "What is life?," originally published in 1944 (Schrödinger 2012); the other topic was the physicochemical nature of the gene.

On metabolism, Schrödinger (incorrectly) suggested that an organism "feeds on negative entropy" and that it extracts "order" from the environment. He assumed (writing for a general audience) that it was not worthwhile to make a rigorous distinction between free energy, the combination of the first and second laws of chemical thermodynamics, and entropy, the second law (Perutz 1991), being focused on the statistical nature of the second law and of quantum mechanics. Since then, however (e.g., Hinshelwood 1946), it has been uncontroversial that biological cells do obey the first and the second laws of thermodynamics, albeit through mechanistically complex couplings of thermodynamically spontaneous reactions with non-spontaneous reactions, summarized and arranged in biochemical metabolic cycles. Many such cycles (glycolytic, tricarboxylic, biosynthetic pathways, etc.) are described in textbooks (Zubay 1998; Schaechter et al. 2005; Kim and Gadd 2008; Alberts et al. 2010), including the creation and dissipation of free energy stored in the membrane potential (Mitchell 1979; Harold 1986; Nicholls and Ferguson 2002). Fundamentally then, biological cells operate as *open thermodynamic systems*; the cells exchange molecules and energies with the environment—they cannot exist without it; they are not autonomous—they are dynamically *united with the environment* from which they emerged; and their genome must fit the environment—or they die.

On the question of the gene, Schrödinger brought forth the idea of molecularly coded genetic information. He described a gene imaginatively as a large and extremely stable, isomeric molecule, the isomers of which create aperiodic crystals of the "hereditary substance" (chromosomal DNA) that can be mutated by quanta of radiation. Though such a model explains radiation-induced mutations, it does not explain how the "hereditary substance" gets replicated; Schrödinger therefore suggested that other physical laws (yet-to-be-discovered) are required.

Schrödinger's prediction of new laws of biophysics, not directly related to biochemistry, has been pursued through abstract physical and philosophical concepts, for example, processes of self-assembly arising from energy flows far from equilibrium, criticality phenomena, autopoiesis, teleonomy, dynamic kinetic stability, and theories of metabolic autocatalytic cycles and of hyper-cycles of replicating molecular quasi-species (Eigen and Schuster 1979; Kauffman 1993; Bak and Paczuski 1995; Solé and Goodwin 2000; Luisi 2003; Pross 2005; Morowitz and Smith 2007; Pascal and Pross 2015). These theoretical concepts have remained remote from the replicating and segregating nucleoids in growing and dividing cells.

It is noteworthy that Pasteur's "microbiological law"—all life only from life—is valid also in thermodynamic and quantum mechanical sense: a single bacterial cell may not self-assemble spontaneously from the "dead" (complex) mixture of its natural biomolecules and "start living." As Crick observed already in 1966, it might be more important to find out how to re-assemble a living bacterial cell from its natural components than to deploy organic chemistry to synthesize all its biomacromolecules (Crick 1966). Or, in Harold's language, the question is how to bring about the dynamical molecular "architecture that distinguishes a living cell from a soup of the chemicals of which it is composed" (Harold 2005). Remarkably, the Crick-Harold challenge remains open.

6.6.2.3 Defining Life

The conceptual issues with life's "blind designers" and Schrödinger's unknown biophysical laws are enmeshed in efforts to define life—in the question "what is life?" This topic is reviewed in Section 2 of this handbook; the jigsaw puzzle in the next section gives a physicochemical view. The puzzle makes the meanings of biological terms, for example, "being alive and evolving," relatively clear and positioned for further elucidations. Thus, the jigsaw puzzle can be viewed as a "theory that provides meaning" (Benner 2010) or at least as a "theory that is beginning to provide meaning."

More generally, however, *physicochemical* properties of molecules and related laws (Atkins 1978, 2003, 2007, 2011) provide the footing—a firm foundation, from which to consider any definition of life, even though the value of such definitions is uncertain (Luisi 1998; Cleland and Chyba 2002; Lazcano 2008; Benner 2010; Cleaves 2010; Szostak 2012; Trifonov 2012). Depending on the temperature range, molecules exhibit a wide range of molecular motions (a prerequisite for chemical reactions), including translational diffusion, without which life cannot exist. Molecular diffusion was discovered qualitatively already in the early nineteenth century by Robert Brown; much later, Einstein and Perrin quantified Brownian motion, cementing the view that all molecules are real, albeit very small objects (Newburgh et al. 2006). As such, they are subject *only* to physicochemical laws, principally chemical thermodynamics and quantum chemistry. Therefore, definitions of life that impute *biological* characteristics, for example, Darwinian evolution, to chemical systems become tautological. Neither bacterial cells *in vivo*, nor the molecules comprising them have any inherent capability for Darwinian evolution—cells *cannot help but* exhibit Darwinian evolution because of molecular (and supra-macromolecular) errors in the transmission of genetic information, cf. Section 6.6.6. Similarly, no physicochemical properties of molecules indicate that molecules can *spontaneously* self-organize in "cycles" and self-replicate (Orgel 2000). They can do so only in reproducing cells or in designed cell-free systems, which eventually reach equilibrium, cf. Shapiro's golfer's analogy in Section 6.6.3.1.

6.6.3 THE PHYSICOCHEMICAL JIGSAW PUZZLE OF LIFE'S ORIGINS

The metaphor of a "jigsaw puzzle" (Brack 2000; Higgs and Lehman 2015), in a physicochemical adaptation, has proved useful in tying together wide-ranging themes of origins research (Spitzer 2017). Such themes are the prebiotic Earth environments, prebiotic chemistry, self-assembly processes, energetics, and bio-informational polymers (Deamer and Fleischaker 1994). The "roadmaps" of NASA and of the European Space Agency suggest similar topics as discussed in Section 1 of this handbook. Since "life's origins" is a very broad research area (Smith and Szathmáry 1999; Knoll 2003; Lane 2003; Mesler and Cleaves 2016), the physicochemical jigsaw puzzle

is intentionally limited and focused only on the question of how Hadean chemistry evolved into Archaean microbiology.

6.6.3.1 The Evolutionary Nature of the Puzzle

A physicochemical jigsaw puzzle avoids any particular reaction schemes, such as those of organic synthetic chemistry, geochemistry, or atmospheric photochemistry. Any such reaction schemes must include a natural mechanism of how the desired components get separated from the overall chemical inventory of the prebiotic Earth. The three-phase "line" boundary of atmosphere, hydrosphere, and lithosphere seems the most natural geochemical location for a life to emerge: at these "beach and rocky shores," prebiotic molecules of wide-ranging water solubility can accumulate, chemically react, and phase-separate ("self-purify") under the reactive conditions of cyclic solar energies and hydration-dehydration cycles of tidal seawater. That cyclic (rhythmic) or fluctuating environments played a role in life's origins has been suggested before, though the concept has not been developed from a physicochemical standpoint (Lahav 1999; Deamer 2011). Thus, Earth acts as a huge chemical reactor, both on chemical engineering "geo-scales" and on molecular scales of cyclic chemical reactions and phase separations (Spitzer and Poolman 2009; Stüeken et al. 2013).

It remains an experimental fact that organic syntheses involving bonding of carbon atoms, including Stanley Miller-type experiments, yield many organic compounds, often in complex "tarry" mixtures, depending on the initial mix of starting materials and the chosen reaction conditions (Shapiro 1986). In this sense, multi-step syntheses of organic chemistry (of natural products) cannot mimic unconstrained processes of prebiotic environments. Shapiro described the issue this way: "*The analogy that comes to mind is that of a golfer, who having played a golf ball through an 18-hole course, then assumed that the ball could also play itself around the course in his absence*" (Shapiro 2007). Though synthetic organic chemistry has demonstrated a number of plausible prebiotic reaction schemes, this approach has now run its course. It has been described as the "end of the beginning" of origins research (Sutherland 2017). The reason is, as Shapiro implies, that synthetic strategies of organic chemistry discount the diffusional effects of the second law—they require "Maxwell's demons" on prebiotic Earth to separate (purify) molecules for pre-conceived chemical reactions. However, Shapiro's golfer's analogy concerns equally any other *constructive* (non-evolutionary) paradigms (Figure 6.6.1), such as the designs and constructions of artificial life or of simple or minimal proto-cells. Their primary goal is to "design, synthesize and engineer life," rather than to illuminate life's complex organismal nature and its *natural* evolutionary emergence from abiotic chemistry (Woese and Goldenfeld 2009).

The physicochemical jigsaw puzzle of life's emergence is updated and expanded here into two parts, shown in Figure 6.6.2a and 6.6.2b.

These figures split the original puzzle into the bottom-up chemical part and the top-down microbiological part. The Earth-Moon rotations provide two cyclic driving energies: solar radiation and seawater tides, designated as [A] in Figures 6.6.1

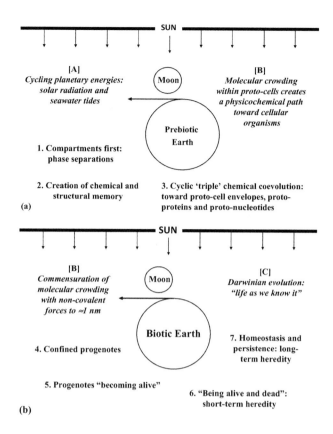

FIGURE 6.6.2   (a) The bottom-up chemical part of the jigsaw puzzle.
[A] Cyclic planetary energies are the "creators" of life. Jigsaw puzzle pieces #1: Life originated in a "compartments first" paradigm, required by the nature a biological cell as an open thermodynamic system. Such compartments came about through phase separations and consolidations of geochemical sediments. #2: The lagging responses of the compartments to the driving planetary energies create chemical and structural memory, a prerequisite for a cellular emergence of life. #3: The chemistries of proto-nucleic acids, proto-proteins, and proto-cell envelopes coevolved. [B] Molecular crowding in proto-cells signifies the possibility of a physicochemically defined path toward cellular life.
(b) The top-down microbiological part of the jigsaw puzzle.
Molecular crowding in proto-cells [B] signifies the evolution of interacting proto-macromolecular surfaces at about 1 nm separations. Jigsaw puzzle piece #4: *Confined progenotes* represent "triple" coevolution of proto-metabolism, proto-genetics, and proto-cell envelopes toward today's biochemistry; onset of the evolution of homo-chirality and of the cytoplasmic excess of potassium; Woese's concept of progenotes (Woese 1998; Koonin 2014) is extended by specifying permeable confinements to satisfy the requirement of an open thermodynamic system. #5: *"Progenotes becoming alive"*; proto-metabolism, proto-transcription, and proto-translation have begun to get stabilized under cyclic gradients. #6: *"Being alive and dead"*; proto-metabolism, proto-transcription, proto-translation, and proto-replication begin to show short-term heredity over a few generations under cycling gradients in still evolving unstable proto-cell-envelopes; onset of heterotrophy; the age of LUCAs with fixed genetic code; the rooting of the Tree of Life (Doolittle 1999). #7: *Homeostasis and persistence*; proto-ribosomes, proto-nucleoid, and proto-cell envelope functionally unify and stabilize and partly decouple from environmental cyclic energies; establishment of persistent cell cycles less dependent on cyclic environmental gradients, that is, energized also by chemical free energy of nutrients. [C] *"Prokaryotic life as we know it."* Cellular Darwinian evolution via molecular errors in template replications of DNA and via fusions of cells with environmental nucleic acids and with themselves (natural genetic engineering); prokaryotic evolution with plasmids and viruses, and the start of evolution of proto-eukaryotic cellular structures with multiple membranes and nucleoids, relentlessly driven by cyclic environmental conditions.

and 6.6.2a. The prebiotic part is largely unchanged from the first version of the puzzle (Spitzer 2017), but it does emphasize the fundamental *physicochemical* condition for a life to emerge [B]: *only* when macromolecular crowding becomes commensurate with hydrogen bonding and screened electrostatic forces at about 1 nm, the prebiotic chemical evolution can proceed toward confined progenotes and first cellular organisms (Figures 6.6.1 and 6.6.2).

The microbiological part (Figure 6.6.2b) is substantially expanded to describe the emergence of cellular life. This part of the puzzle is quite tentative, dealing with biological issues such as "becoming alive" and "being persistent"; the suggested

experiments in Section 6.6.5.2 will throw more light on these very complicated supra-macromolecular issues. The end point [C] signifies the emergence of single-cell organisms, or "life as we know it"; the creation of genetic variations; and their natural selection via interactions with the environment (Darwinian evolution).

6.6.3.2   PROVISIONAL OBSERVATIONS AND CONCLUSIONS

The phenomena and conditions described in the updated puzzle in Figure 6.6.2 show that life's emergence was a highly complex planetary process. Nevertheless, even at this early stage of assembling and solving the puzzle, we can make some provisional conclusions about the processes, physicochemical conditions, and principles through which first prokaryotic-like organisms came about:

1. The emergence of life is a *cyclic evolutionary process of reacting and phase-separating molecules*, the results of which are contingent on Earth-like prebiotic chemical compositions and cyclic energies (solar radiation and seawater tides).
2. Cycling planetary energies are the "blind designers" of prebiotic chemical evolution, while natural selection is the "blind watchmaker" of organismal Darwinian evolution.
3. Chemical prebiotic evolution is continuous with Darwinian evolution through *supra-macromolecular cellular* evolution.
4. Biological cells operate as open thermodynamic systems and thus can "defeat" the second law of thermodynamics. Fundamentally, they can do it *only locally and temporarily*, with expenditures of external energies and evolution of heat. Hence, there is no "Schrödinger-like unease" between chemical thermodynamics and the living states of matter.
5. Cellular life can emerge and evolve only when *macromolecular crowding* becomes commensurate with screened electrostatic forces and with hydrogen bonding (including hydration and the related hydrophobic effect), over a distance of at about 1 nm.
6. Cell envelopes, metabolism, and genetics evolved *together with* the environment, as a complex cycling open thermodynamic system.
7. The statistical nature of the second law and of quantum mechanics, combined with the contingencies of cyclic environments, creates errors in the transmission of genetic information (molecular and supra-macromolecular errors in nucleoids and chromosomes), and thus, the progeny are not exact copies. This makes cellular *Darwinian evolution inevitable*—a fact or a law of nature.
8. The first persistent life came about when the nucleoid and the cell envelope unified in a *homeostatic* open thermodynamic chemical system, with increased dependence on chemical free energies of environmental nutrients. "Homeostatic" means the maintenance of repeatable spatiotemporal "architecture" during each cell cycle, as the cell responds to variable environmental conditions.
9. Prokaryotic "life" is a *cyclic state of matter* of being alive and having progeny. Viruses are not alive, because they do not operate as open thermodynamic systems.

These provisional observations appear more or less plausible. The physicochemical jigsaw puzzle shifts attention from particular prebiotic chemical reactions to supra-macromolecular pattern-forming processes driven by external cycling energies; it also shifts attention from the catalytic surfaces of specific minerals to the supra-macromolecular geochemistry of consolidating sediments that create "liquid" micro-spaces and "solid" permeable confinements. Some cyclic experiments are described in Section 6.6.5 from both bottom-up (chemical) and top-down (microbiological) standpoints. First, however, it is helpful to take a top-down view of today's prokaryotic cells *in vivo*, because we cannot avoid the assumption that "first" Archaean cells already had a similar degree of complexity 3.5 billion years ago.

6.6.4 WHY DO BACTERIAL CELLS APPEAR SO COMPLEX?

In the last 20 years, there has been a dramatic progress in understanding the structuring of prokaryotic cells during their cell cycle (Kyne and Crowley 2016); it involves spatiotemporal replication, localization, and segregation of the nucleoid, together with cytoplasmic proteins—a complex process "orchestrated" with the growth and division of the cell envelope. These advances have come from many directions, the most important being the following:

1. Fluorescence spectroscopic studies of diffusion using genetic fusions of the green fluorescent protein (GFP) with target proteins and nucleic acids. The "surprising" result has been the characterization of the cytoplasm as a crowded but "watery" medium, indicating fast diffusion of *small* metabolites, proteins, and nucleic acids (Verkman 2002); under osmotic stress, the diffusion patterns become significantly altered, particularly for larger biomacromolecules (Konopka et al. 2009; Mika and Poolman 2011).
2. The greater acceptance of (complex) effects of biomacromolecular crowding *in vivo* on the folding, stability, and association of biomacromolecules vs. conventional "dilute" biochemical protocols *in vitro* (Zimmerman and Minton 1993; Ellis 2001; Zhou et al. 2008; Sarkar et al. 2013; Boersma et al. 2015).
3. The discovery of "cytoskeleton" (eukaryotic-like) proteins in bacteria and their role in bacterial division processes (Lutkenhaus 2003; Michie and Löwe 2006).
4. The visualization of cellular ultrastructure by three-dimensional (3-D) cryotomography (Baumeister 2005; Nickell et al. 2006; Oikonomou and Jensen 2017) and by computer (and "artistic") modeling of bacterial cytoplasm (Elcock 2010; Goodsell 2010; Cossins et al. 2011; Yu et al. 2016).

These recent developments transformed our view of a prokaryotic cell from a "bag of enzymes" (Mathews 1993) to a highly crowded, yet well-regulated "architecture" of cytoplasmic biomacromolecules confined in a multilayered cell envelope that includes the bilayer plasma membrane (Baumeister 2005; Harold 2005; Silhavy et al. 2010).

This change of perspective has implications for directions in origins research. When the simplest life had been perceived as "...*a packet of proteins and nucleic acids, cofactors and so on...*" (Orgel 2004), it then seemed reasonable to concentrate on synthesizing the bag's components. However, bacterial cells are now known to be dynamically *nano-structured* in an energy-consuming spatiotemporal "architecture," which is substantially repeated during each cell cycle. Thus, it cannot be considered that bacterial cells are a mixture of molecules that can somehow "self-assemble" or that a simpler prebiotic mixture of similar molecules could have self-assembled and started a cell cycle on prebiotic Earth. Rather, life's emergence, point [B] in Figure 6.6.1, began with *dynamical phase-separation* of one or more localized and self-purifying processes from the totality of complex non-living cycling processes on prebiotic Earth.

In general, the spatiotemporal complexity of today's pro-karyotic cells can be described as follows:

1. *Multicomponent*, containing many kinds of small molecules, ionic salts, surfactant-like molecules, macromolecules, polyelectrolytes, and water.
2. *Phase-separated* from nutrient environments, that is, bounded by membranous layers, and phase-separated within, as a dynamic non-equilibrium sol-gel system.
3. *Crowded*, with a high total volume fraction of bio-macromolecules that create super-crowded micro-gels with vectorial electrolytic nano-channels that guide the diffusion of ions and low-molecular-weight metabolites.
4. *In disequilibrium*, both in chemical sense and in physical sense, that is, catalytically reacting (grow-ing) with inflows of water, ions, and nutrients from the environment and vice versa.
5. *In a re-emergent process*—an energy-consuming, thermodynamically open, "chemical engineering" process of imprecise cyclic replications and segrega-tions of the nucleoid into daughter cells.

The first four characteristics define any complex *non-living* chemical mixtures that occur naturally in planetary environments. Examples of prebiotic molecular complexity include the great variety of carbon compounds that have been identified so far in our solar system and observed in the distant universe (McDonald et al. 1991; Chyba and Sagan 1992; Dworkin et al. 2001; Bernstein 2006; Pizzarello 2006; Rhee et al. 2007; Carrasco et al. 2009; Ehrenfreund and Cami 2010) and the readily formed "tars" in non-enzymatic organic syntheses of prebiotic biomolecules and biopolymers (Miller 1953; Shapiro 1986; Schwartz 2007; Scherer et al. 2016). Another example of non-living complex matter is "dead organismal matter" in the process of reaching physicochemical equilibrium (Atkins 2011).

The fifth property—re-emergence—differentiates *com-plex living states* from complex chemical states, as described previously. This property is quintessentially biological, as re-emergence requires the maintenance of cellular identity during the cell cycle, that is, the chemical maintenance and continuation of the genome (nucleoid) and the physical main-tenance and continuation of the phenotype (the cell envelope).

Complex chemical mixtures, considered as non-equilibrium reactive open thermodynamic systems (e.g., permeable colloidal micro-spaces, chemical reactors, and biological cells), can either be approaching an equilibrium or be kinetically arrested away from equilibrium; or they can be in a steady state, chemically reacting but not evolving, their properties being independent of time; or they can be in a non-steady state, chemically reacting and evolving in a random, irreproducible fashion, contingent on the kinds and ranges of external inputs, that is, on the environment. Only when non-equilibrium complex molecular mixtures are continuously phase-separating and chemically reacting under *cyclic non-steady state conditions*, that is, repeatedly stoked with energy, their chemical evolution into living states becomes conceivable. These general points are next elaborated in greater detail, using textbook sources (Schaechter et al. 2005; Kim and Gadd 2008; Phillips et al. 2009) and the emerging advances in the understanding of dynamic cellular ultrastructure of bacterial cells.

6.6.4.1 MULTICOMPONENT AND IONIC COMPOSITION

Bacterial cells are composed of many kinds of chemi-cal compounds—ionic and hydrophobic in a wide range of molecular sizes, from nanometers to micrometers. The latter represents very long nucleic acids and cell wall biopolymers (Zimmerman and Murphy 2001). A great deal of biochemis-try has been concerned with their isolation and purification and with determinations of their chemical compositions, 3-D chemical structures, *in vitro* biochemical reactivity and re-assembly, and their cellular localizations and physiological functions *in vivo*.

In a bacterial cell, there are over 1000–2000 kinds of proteins, many kinds of RNAs, a single DNA polymer, and three to six kinds of phospholipids. There are also inorganic chemicals of non-biological origin that are important reactants or catalysts (widely available from the environment), chief being water and carbon dioxide and inorganic ions, for example, K^+, Na^+, Mg^{+2}, Ca^{+2}, $[HCO_3]^{-1}$, $[H_2PO_4]^{-1}$, $[HPO_4]^{-2}$, $[SO_4]^{-2}$, and Cl^{-1}. Physicochemical interactions between bacterial biomolecules and biomacromolecules are mediated by water and simple inorganic ions, and thus, cellular life depends critically on a buffered aqueous (multicomponent) electrolyte of relatively high ionic strength. The bacterial membrane is also ionic—it contains anionic lipids (Yeung et al. 2008), including highly charged cardiolipin (Romantsov et al. 2008). There are relatively few strictly uncharged biomolecules (and few, if any, biomacromolecules), except for "special cases" of proteins at their isoelectric points or zwitter-ionic lipids and similar compounds derived from choline. Uncharged "food" molecules, such as sugars, are transported via membrane proteins and then converted by kinases to anionic organic phosphates, in order to become available to cell's biochemistry. Finally, as viewed from the nutrient medium, the bacterial cell *in toto* is charged, as evidenced by cells' electrophoretic motion in an external electric field (van Loosdrecht et al. 1987; de Kerchove and Elimelech 2005).

6.6.4.2 MULTIPHASE SEPARATIONS

Phase separations of any kind imply classical physicochemical concepts of "high concentrations" of molecules, when their molecular size and shape and non-covalent molecular forces come into play, cf. the difference between the ideal gas law and van der Waals equation for real gases. The attractive forces responsible for the existence of liquid and solid phases, in general, are van der Waals forces and ion-dipole and similar Debye-type polar forces between small neutral molecules, as well as screened electrostatic forces between ions in solution, described by the Debye-Hückel theory.

However, the case of the multicomponent and ionic cytoplasm is much more complicated on account of the presence of biopolymers and their high "biomacromolecular crowding" *in vivo* (Ovadi and Srere 1991; Ellis 2001; Zhou et al. 2008). The designation "biomacromolecular crowding" reflects the high overall concentration of all biomacromolecular components. On a nanometer scale and microsecond (and longer) timescales, biomacromolecular surfaces can be considered as "smoothed-out," or averaged out, into four kinds of molecular "patches" that interact through attractive and repulsive non-covalent forces. These patches are hydrophilic (capable of hydrogen bonding), hydrophobic, and positively or negatively charged (Spitzer 2011; Wang et al. 2011; Laue 2012).

The two most important "cytological" attractive non-covalent forces are as follows:

1. Hydrogen bonding of water molecules with each other and with polar "patches" (ionic and hydrophilic) at folded biomacromolecular surfaces (Pauling et al. 1951; Eisenberg, 2003). The hydrogen-bonded network of water molecules brings about the hydrophobic effect—the "expulsion" of hydrophobic molecular surfaces from the network of hydrogen-bonded water molecules (Tanford 1980; Southall et al. 2002).
2. Screened and direct electrostatic attractions between cationic and anionic molecular "patches" of biomacromolecular surfaces (Schreiber and Fersht 1996; Halford 2009). The coulombic interactions are reflected in bimodal distributions of isoelectric points of many proteomes, with peaks in the alkaline pH region (negatively charged proteins) and in the acidic region (positively charged proteins), suggesting that electrostatically induced clustering (gelation) is common inside many cells (Kiraga et al. 2007; Spitzer and Poolman 2009). Similar electrostatic interactions account for interactions ("packaging") of the nucleoid with cationic proteins or polyamines inside a bacterial cell (Friedrich et al. 1988; Zimmerman 2006; Lee et al. 2015).

The effects of attractive forces are countered by repulsive non-covalent forces, notably by the excluded volume effect of biomacromolecular crowding (Zhou et al. 2008; Foffi et al. 2013) and by screened electrostatic repulsions between anionic molecular surfaces. At close separations below 1 nm, these screened repulsions can be very large (Spitzer 1984; Spitzer 2003), ensuring that the crowded cytoplasm does not collapse into an agglomerated, non-functional gel.

6.6.4.3 CELLULAR CROWDED COMPLEXITY

Generic textbook pictures of bacterial cells show four physicochemical phases (each being also multicomponent):

1. The hydrophobic cell envelope of a particular geometrical shape (coccoid, rod, etc.), the key functional part being a lipid bilayer studded with a high number of many kinds of hydrophobic membrane proteins. Nevertheless, other parts of cell envelopes may also be critical for survival (Silhavy et al. 2010). For example, the L-forms of bacteria, lacking a cell wall, would not generally thrive in planetary unconstrained environments; they require specialized environments for survival (Niemark 1986; Errington 2013).
2. The phase-separated nucleoid (Valkenburg and Woldringh 1984; Woldringh and Nanninga, 2006) portrayed like a bundle of "spaghetti," representing anionic DNA helices tied up by cationic proteins and amines into denser conformations.
3. The ribosomes, large phase-separated clusters of proteins and nucleic acids, are shown as tiny spheres scattered throughout the cytosol (Alberts et al. 2002) or attached to mRNA strands (polyribosomes) that are connected to DNA through RNA polymerase (Schaechter et al. 2005)
4. The "unstructured" cytoplasm—a "bag of enzymes."

At greater spatiotemporal resolutions, nucleoid double helices and proteins and their hetero-dimers and multimers (biomacromolecular complexes and "nano-machines") can be shown "randomly crowded," representing ~25% of the volume of the cytoplasm (Zimmerman and Trach 1991; Elcock 2010; Goodsell 2010). Simple physical models of such high crowding show that cytoplasmic biomacromolecules are essentially "touching" *in vivo* (Phillips et al. 2009; Spitzer 2011; Spitzer and Poolman 2013). They thus interact via repulsive and attractive non-covalent forces over short distance of about 1 nm, significantly reducing rotational diffusion (Wang et al. 2011), thereby ensuring functional dynamics of folding and association of biomacromolecular surfaces, as they are being synthesized.

Dynamic clustering of biomacromolecules *in vivo* has also been described from a top-down biochemical perspective as "quinary" structure (McConkey 1982; Monteith et al. 2015), as "metabolons," where reactants and products are shuttled between the catalytic sites within biomacromolecular clusters, without bulk diffusion in the cytoplasmic sol (Srere 1985; Ovadi and Srere 1991; Mathews 1993). Another view of the cytoplasm is that it is composed of functional modular "hyperstructures" (Hartwell et al. 1999; Norris et al. 2007) or various "omes" of system biology. For example, the FtsZ divisome (Rowlett and Margolin 2015) is synthesized and assembled before cell division takes places in *Escherichia coli*. The bacterial cytoplasm has been also described as a "sieve-like meshwork" of bacterial cytoskeleton and of other proteins (Mika et al. 2010), and as a glassy material liquefied by metabolism (Parry et al. 2014).

The bacterial crowding *in vivo* can significantly increase, up to ~50% volume fraction, in hyperosmotic media, when the cell loses water. The protein sensors and transporters within the cell envelope then become activated and import osmolytes (e.g., potassium ions and glycine betaine) that reduce the osmotic outflow of water and thus render plasmolysis less severe (Poolman et al. 2004; van den Bogaart et al. 2007; Wood 2011). Astonishingly, *E. coli* cells can re-adjust and resume growing, under such extreme "super-crowded" conditions, provided that the increase of crowding is performed in a stepwise sequential manner (Konopka et al. 2009). The stepwise process allows biomacromolecules enough time to re-structure their spatio-temporal architecture—to adapt to higher crowding conditions (Cayley et al. 1991). However, not all cells survive such severe mechanical stresses of viscoelastic "shrinking" brought about by a hyperosmotic shock.

Thus, under "normal" osmotic (environmental) conditions, non-covalent molecular forces organize a robust, amazingly well-orchestrated "ballet" of cytoplasmic molecular motions, which, remarkably, avoids any catastrophic (fatal) agglomerations of biomacromolecules. Nevertheless, when environmental conditions change over large ranges of variables, for example, of temperatures, the "orchestrated ballet" can partly (or altogether) collapse into irreversible folding (misfolding) and aggregations of proteins and nucleic acids (Ellis and Minton 2006; Boquist and Gröbner 2007; Pastore and Temussi 2011).

6.6.4.4 ELECTROLYTIC NATURE OF CELLULAR REGULATION

The earlier description of the physicochemical complexity of a bacterial cell—ionic and multicomponent, multiphase, and crowded and super-crowded—is made much more complicated by the fact that, during the cell cycle, there are about 1000–2000 concurrent and sequential biochemical reactions within the cell and between the cell and the environment. These biochemical reactions are well-synchronized and regulated to yield many physiological processes, such as the import of nutrients into the cell by the cell envelope and the replication and segregation of the nucleoid during the cell cycle, as well as many physiological responses to environmental stresses (Storz and Hengge 2011).

In order for such a complex cellular system to work reasonably reproducibly during the cell cycle (as it does), biochemical reactions have to be localized within the cell and within the cell envelope. Thus, biomacromolecules are localized as two-dimensional (2-D) "rafts" within the plasma membrane and cell envelope (López and Kolter 2010) and in "super-crowded" 3-D gels within the cytoplasm, as

hypothesized by the sol-gel model (Spitzer 2011; Spitzer and Poolman 2013). In the sol-gel model, the "super-crowded" biomacromolecular clusters (agglomerated micro-gels) are associated with (i) the cytoplasmic side of the cell envelope (which inter alia increases its mechanical stability), and (ii) the nucleoid in the middle, which extends all the way into the cell envelope. Similar model was used previously to explain very fast diffusion in mitochondria (Partikian et al. 1998). The functioning of this complex chemical system rests in the regulating nature of non-covalent molecular forces, brought into action by biochemical signaling reactions. Gel formation (biomacromolecular super-crowding) and the reverse processes of gel liquefaction are controlled by an increase or decrease of hydrophobic and screened electrostatic interactions. For example, methylations and de-phosphorylations increase non-covalent attractions (leading to gel formation), and the reverse reactions increase repulsions (leading to gel liquefaction).

In the super-crowded gelled regions, biomacromolecules are separated by narrow channels filled with percolating multi-ionic electrolyte solution that hydrates (stabilizes) biomacromolecular surfaces and thus maintains their biochemical functions. Such channels are theorized to have semi-conducting electrolytic properties, based on solutions of a system of differential Poisson-Boltzmann equations (Spitzer 1984; Spitzer 2003; Spitzer and Poolman 2005). The channels sort out and direct both inorganic ions and biochemical ions into different parts of the crowded cell according to their overall charge, for example, ATP^{-3}, ADP^{-2}, AMP^{-1}, $[H_2PO_4]^{-1}$, and $[HPO_4]^{-2}$. These channels can be considered as transient "microfluidic *electrolytic* integrated circuits," ultimately "programmed" by the nucleoid (genome) but contingently re-adjusting in response to environmental signals during the cell cycle—an aspect of cellular homeostasis, cf. jigsaw puzzle #7 (Figure 6.6.2b). There are many other signaling reactions, such as adenylations and acetylations, which, in addition to simple charge effects, modify the channel morphologies on the crucial 1-nm scale.

Biochemistry and phylogenetics suggest that the genetic code became "immutable" in the last universal common ancestors (LUCAs') populations. Because confinement is mandatory for a life to emerge, we assume that LUCAs also had semi-permeable "surfactant" bilayer, or layers, functionally similar to that of today's prokaryotes, that is, able to stabilize hydrophobic membrane proteins and the excrescences of the nucleoid and prevent uncontrolled diffusional access of environmental ions into the cytoplasm. The chief characteristic of cellular phospholipids, both bacterial and archaeal (Albers and Meyer 2011), is their two-chain hydrophobic tail that is known to favor flexible "flat" nano-phases of bilayers suitable to form micron-sized vesicles, whereas single-chain surfactants give rise to spherical micelles and, at high concentrations, to cylindrical nano-phases. It is noteworthy that *polar* tri-cyclic hydrocarbon fused rings and similar carbon compounds derived from polycyclic hydrocarbons could have been involved in the initial hydrophobic evolution of proto-cellular envelopes and hydrophobic membrane proteins, cf. Section 6.6.5.

In conclusion, today's bacteria exhibit an unprecedented physicochemical sophistication, which has enabled them to evolve in very different environments and exploit very different environmental energies. Remarkably, this sophistication is based on only a few "building blocks" of life: five nucleobases, phosphate esters based on two versions of ribose, 20 amino acids, and some phospholipids with two-tail hydrophobic chains. The ancient proto-prokaryotic cells exhibited such physicochemical sophistication already ~3.5 billion years ago, point [C] in Figures 6.6.1 and 6.6.2, when they began to emerge via crowded, confined, and self-purifying (cyclic) chemical processes represented by point [B] in Figures 6.6.1 and 6.6.2. These complex chemical processes eventually "*crystallized and annealed*"; Woese's *metaphorical* phrase of crystallization and annealing (Woese 1998) thus turns out to echo actual but complex physicochemical processes, which involved submicron "self-purification" (repeated *crystallizations*) within permeable microspaces under cyclic mechanical stress relaxations (*annealing*). The physicochemical nature of "becoming alive" thus becomes somewhat clearer.

6.6.5　EXPERIMENTAL PARADIGMS OF CYCLIC EVOLUTIONARY PROCESSES

The physicochemical jigsaw puzzle suggests an experimental paradigm, based on cyclic external energies that drive cyclic chemical processes within evolving compartments. Historical questions, such as whether the emergence of life was a rather sudden phenomenon, or whether it required long gradual processes or specialized geochemical locations, cannot be rigorously answered. It is thus unlikely that the chemical "bottom-up" part of the puzzle of life's emergence (Figure 6.6.2a) can be quantified and experimentally investigated to yield definitive conclusions; yet, some experiments are worth doing, to better understand the evolutionary potential of *non-living* cyclic chemical processes.

6.6.5.1　Bottom-Up Cyclic Chemistry

An underappreciated aspect of Stanley Miller approach (Miller 1953) is its open-ended character—a "see-what-happens" approach—to find out if any building blocks of life can be detected from electrical discharges in prebiotic atmospheres. The results were positive, showing the presence of amino acids, and they led to many variations of such experiments, including the re-analyses of old samples with new analytical methods (Bada and Lazcano 2003; Johnson et al. 2008). However, the *unconstrained* nature of Stanley Miller experiments has been subsequently relaxed in search for synthetic pathways toward all "building blocks" (amino acids, nucleobases, sugars, phosphate esters, etc.), transforming plausible natural processes of volcanic electrical discharges to less plausible multi-step strategies of synthetic organic chemistry (Shapiro 1986, 2007; Orgel 2004; Eschenmoser 2007; McCollom 2013; Patel et al. 2015).

6.6.5.1.1　Chemical Engineering Simulators of "Model Earths"

The original rationale of Stanley Miller experimentation can be re-established in designed chemical engineering simulators of "prebiotic Earths" (Deamer 2011), operating under cycling planetary conditions (Spitzer 2013b). By varying the initial compositional conditions, different process outcomes will materialize after a given period of operation, for example, after a week, a month, or a year, while implementing a specific observational and sampling protocols to characterize the overall process. This "take-the-bull-by-the-horns" approach could thus uncover the rates of chemical complexification of prebiotic habitable planets vs. initial compositions, that is, the prebiotic chemical evolution between [A] and [B] in Figure 6.6.1, where low initial chemical complexity is assumed (a higher chemical complexity of initial compositions, e.g., with polycyclic hydrocarbons, is also a plausible alternative). The results would illuminate the evolutionary potential of reactive dynamic pattern formation in *non-living*, phase-separating chemical systems, driven by cyclic external energies. As an aside, since 1953, chemical engineering processes underwent unprecedented computerization of hardware, such as load cells, pumps, and new sensors, which makes the design of chemical processes on "model Earths" much more feasible, though not simple.

6.6.5.1.2　Bottom-Up "Proof-of-Concept" Experiments

Before building an all-encompassing pilot plant simulator of a model prebiotic Earth (or any habitable planet), some preliminary "proof-of-concept" data could be obtained with simple experiments. The following questions and ideas may be useful in considering such preliminary experiments.

1. Did prebiotic Earth have only simple low-molecular-weight molecules at any point in its formation during the emergence of our solar planetary system? What would be the effects of adding polycyclic hydrocarbons?
2. How will planetary sciences select model initial compositions of the atmosphere, hydrosphere, and lithosphere and the specifics of cycling electromagnetic irradiation and seawater tides?
3. What surface-active compounds are likely to arise from the selected initial compositions? How will they create "micro-spaces," or sedimentary matrices?
4. What will be the role of poly-phosphates in stabilizing and dispersing inorganic and organic precipitates (in tidal sediments) in the nano- to micro-ranges of sizes? How will they evolve under cycling environmental conditions?
5. Did biochemical chirality (Ault 2004; Weissbuch and Lahav 2011) and the cytoplasmic excess of potassium (Mulkidjanian et al. 2012) become coupled in the cyclic chemical evolution of geochemical tidal sediments under crowding molecular conditions?

Based on these questions, at least two simple, "proof-of-concept" experiments can be designed, even though they are not directly related to the issues of life's origins (Spitzer 2013b).

The first one can address the question of the origin of homo-chirality and potassium ions in confined *non-living* cyclic chemical states. Because K^+ salts are generally less soluble than Na^+ salts and may crystallize first in chirally distinct crystals, as observed by Pasteur (Flack 2009), their cyclic dissolutions and crystallizations could enrich any microspaces with K^+ and create an enantiomeric excess of chiral organic anions at the same time. Thus, a mixture of potassium and sodium salts of racemic organic ions (e.g., glutamate) can be subjected to cyclic crystallizations and dissolutions in the presence of "micro-spaces," for example, pumice or artificial "sedimentary phases," while monitoring the concentration of K^+ and Na^+ ions and of the organic ions in both the "micro-space" phase and the supernatant solution. Could the surfaces of micro-spaces act as templates to enable or stimulate the creation of enantiomeric excess (Weissbuch and Lahav 2011)?

Second, the "goo" of tarry compounds from Stanley Miller experimentation could be cyclically fractionated and the fractions analyzed and evaluated for surface and interfacial activity and thus for the ability to form hydrophobic permeable micro-spaces or vesicles. These compounds may not represent classical surfactant structures of one or two long hydrophobic tails with polar heads but could be derived from polycyclic aromatic hydrocarbons, with some methyl and polar groups attached to fused carbon rings (Bernstein et al. 1999, 2002; Sandford et al. 2013). Such polycyclic molecular structures are similar to derivatives of rosin acids, some of which are efficient surfactants for emulsion polymerization (Blackley 1975). Paradoxically, rosin acids, such as abietic acid, are of biotic (plant) origin, though this is irrelevant from a functional cellular standpoint (discounting the possibility that polycyclic hydrocarbons could be bio-signatures of "cosmic plant life"). Thus, ancient proto-cell envelopes may not have had the traditional bilayer of today's cells but may have had a less "precise" multilayer morphology of hydrophobic nano-phases derived from polar polycyclic hydrocarbons.

6.6.5.2 Top-Down Cyclic Microbiology

The top-down "microbiological" paradigm (Figure 6.6.2b) assumes that ancient Archaean organisms had cellular biochemistry and molecular biology similar to today's prokaryotes, including the physicochemical complexity of confining cell envelopes. This paradigm is likely to yield more compelling results, compared with the "bottom-up" chemical approaches that have much deeper historical roots, which can be approached only indirectly, cf. the points [A]-[B]-[C, D] in Figure 6.6.1.

The top-down microbiological paradigm is based on Crick's suggestion to assemble living bacteria from their natural (separated) components (Crick 1966). Specifically, it is based on the question of whether bacterial *biotic* soups can be brought into a living state through temperature and hydration/dehydration cycles. Thus, the goal is to make sure that the

spatiotemporal architecture of a living physicochemical state is broken down and then recovered through cyclic evolutionary processes (Harold 2005; Davey 2011; Spitzer 2014). The first stage of this experimentation is the creation "dead biotic soups" (which, incidentally, could also illuminate the important medical issues of killing pathogenic bacteria).

6.6.5.2.1 Making "Biotic Soup"

Any bacterial population can be lysed, yielding a lysate for further characterization and experimentation. Importantly, the lysate has all the biomolecules in the right proportions, from which a "living state" ought to arise. The following questions could help with experimental designs:

1. *Growth conditions before killing.* Which dead bacterial populations are the likeliest to be "stoked" by temperature and hydration/dehydration cycles back into living states? Which nutritional medium to select or design?
2. *Methods of killing.* Which physical means of killing to choose? Starvation, a mild hypo-osmotic shock, various degrees of sonication, different rates of freezing at different water activities? Different rates of heating and boiling? And the combinations of such killing processes?
3. *Timing of killing.* At which point in the growth curve to "kill?" Mid-point exponential, early on in the population growth, with sufficient "food" leftover, or in the stationary stage, or some particular state of nutritional or environmental stress?
4. *How to characterize the biotic soup?* There will be ribosomes, plasmids, and nucleoids, all partly broken, with partly attached "broken" cell envelopes (Tremblay et al. 1969; Firshein 1989; Zimmerman and Murphy 2001); their kinds and concentrations will depend on particular selection 1., 2. and 3. above.
5. *Is the biotic soup really dead?* Important but not an easy experimental question (Davey 2011). When the cytoplasmic spatiotemporal architecture is only "partially" broken down, semi-living states of bacteria such as "being injured, sick, comatose" may arise and be amenable to recovery.

6.6.5.2.2 "Putting Humpty-Dumpty Together Again"

According to the notion that Earth's diurnal cycles and tidal hydration/dehydration cycles were the "designers" of living systems, such cycles may also perform "repeat" experiments with biotic soups in continuous evolutionary attempts to reconstruct a living system—"Putting Humpty-Dumpty together again" (Gierasch and Gershenson 2009). Both temperature and hydration-dehydration cycles *shape* supra-macromolecular colloidal structures needed for "life" through non-covalent attractive and repulsive interactions inherent in biomacromolecular crowding and *drive* some biochemical reactions more than others. Perhaps, the externally energized biotic soup "lives" when the temperature goes up and becomes dormant or dead when the temperature goes down, thus simulating the confined progenotic stages #4 and #5 of the jigsaw puzzle (Figure 6.6.2b).

Some questions to consider, regarding the dynamic cyclic reconstruction of living states, are as follows:

1. Could chemostat equipment and processes and PCR protocols be modified and adapted for recovering living states of bacterial soups? How could the "feeding" regimes of nutrients be synchronized with cooling/heating cycles?
2. Could experimental equipment in studying biofilms at high "crowding" be adapted for recovering living states?
3. What would be the starting volume fraction of all biomolecules and biomacromolecules?
4. What effects could be observed by increasing the ATP/ADP ratio?
5. How to determine biochemical conversions (the "growth of the biotic soup") and "define" operationally different kinds of living states? What would be the criteria to distinguish between the living and the dead; the comatose; the sick or injured; the dormant; the viable but non-culturable?

Any emergent living state may not readily materialize in the classical prokaryotic physical forms (coccoid, rod, etc.), because the formation of a *closed cell envelope with a lipid bilayer membrane* is expected to be the most challenging phenomenon to bring about and manipulate by cycling physicochemical gradients. The cycling experiments may have to be done in the "super-crowded" regime (Spitzer 2011; Spitzer and Poolman 2013), when "super-cells," with imperfect internal compartments and multiple proto-nucleoids, in effect "proto-biofilms," will initially start assembling and disassembling. Such unstable pre-Darwinian "barely living states" may evolve into LUCAs, that is, with fixed genetic code but not well-established heredity mediated by horizontal gene transfer (HGT) and maintained by cyclic energies.

Additional biotic soup experimentation can include the following: (i) Using the cycling biotic soup as a supporting medium for the replication of phages and of other viruses (inter alia demonstrating the "semi-living" status of a growing and cycling biotic soup). (ii) The mixing of different bacterial species in the initial biotic soup to genetically increase the chances and speed of emergence of living states and also to create "new" LUCAs, viruses, and bacterial species by "recombination" of existing genomes or their fragments. (iii) The development of characterization methods, for example, centrifugal fractionation (Schuster and Laue 1994; Harding and Rowe 2010; Rowe 2011).

6.6.6 BACTERIAL CELLS AS EVOLVING CHEMICAL MICRO-REACTORS

As the jigsaw puzzle suggests, prebiotic chemical evolution and organismal Darwinian evolution are continuous through *cellular evolution*; thus, molecular and supra-macromolecular aspects of cellular evolution are critical for understanding life's nature and its origins (Woese 1998; Koonin 2014).

Since both biological cells and chemical reactors are open thermodynamic systems, their operation and regulation can be usefully contrasted. Admittedly, a bacterial cell differs from an industrial reactor by obvious realities: (i) the walls of a reactor are fixed and are not part of chemical production as compared with the cell envelope; however, they are "permeable," as raw materials go in and the product gets out, and (ii) the solid-state computer with an appropriate software program that regulates and controls the reactor resides outside and is not part of the production either, as compared with the "viscoelastic nucleoid program" that is inside the cell and gets re-manufactured. In both cases, there are many sensors and regulators that monitor and control the process inside and outside. However, trying to understand bacterial cells as chemical reactors, or as computers (Bray 2009; Danchin 2010), can be taken only so far. Nevertheless, such analogies do illuminate the workings of a cell as a self-programmed chemical engineering process, which, as any such process, is prone to *stochastic and contingent errors*, malfunction, and even catastrophic upsets.

6.6.6.1 THE PROCESS ERRORS OF THE CELL CYCLE

The process errors of the bacterial cell cycle, when they occur during the replication and segregation of the nucleoid, account for Darwinian evolution; without errors, Darwinian evolution cannot operate. The errors arise during the template synthesis of complementary strands of DNA (molecular errors, such as point mutations); but they can also arise through *supra-macromolecular* mechanisms, when living cells fuse with environmental nucleic acids or with each other; such "colloidal errors" account for HGT. Regardless of the causes of the errors, they can be lethal and non-lethal and anything in between (Davey 2011). The lethal errors release biomolecules into the environment for potential re-use by other living cells (heterotrophy), and the non-lethal errors create Darwinian genetic variations on which natural selection can act (Barrick and Lenski 2013).

When the process errors in the cell cycle are *negligibly small*, one cellular micro-reactor will give rise to two copies with vanishingly small differences—only *ideal biological* cells are then produced, and Darwinian evolution cannot be observed. (In chemical engineering language, the production is within the product's specifications). Larger process errors (but non-fatal) make the difference between the two copies quite apparent—the bacterial cell will have evolved in the original Darwinian sense, yielding one or both new cells noticeably different but similar to the original cell. This means that the cellular DNA has changed, producing different phenotypic strains. (In chemical engineering language, the product is out of specification—it is a "different" product.) Very large errors bring about failures—the death of bacterial cells. Such errors come from large and sudden changes in the environment, when the process-control computer's capability becomes compromised—or in Darwinian language, the genome's computer-like performance has not yet reached the required evolutionary stage to ensure its own survival. (In chemical engineering language, the manufacturing process software has failed, as it was not designed to cope with such unexpected inputs.)

6.6.6.2 Is Cellular Evolution "Saltational"?

It is not inconceivable that large, normally fatal cellular errors may, though extraordinarily rarely, bring about very different but still functioning (perhaps poorly) cellular micro-reactors that can dynamically re-stabilize under cyclic environmental conditions and evolve into new kinds of cells, *for example,* by fusions of simple cells, incorporating internal lipid surfaces and multiple nucleoids. Eukaryotic cells did likely arise this way (Sagan 1967; Martin et al. 2015). We can thus distinguish three kinds of cellular Darwinian evolution, plus an ideal pro-karyotic cell cycle that can serve as a reference process, which does not evolve.

1. *Non-evolving ideal cell cycle* is characterized by: a) its operation and regulation (growth and division) are without errors, and b) the ideal cells are "infinitely" apart from each other, ensuring that they reproduce in a constant nutrient environment, with no interactions between the cells.
2. *Molecular micro-evolution* arises from errors in the template replication of DNA within any one cell cycle, when no nucleic acids pass through the cell envelope. This molecular mechanism corresponds approximately to the "gradualism" of Darwin's original conception, though even single-point mutations can have noticeable phenotypic effects.
3. *Molecular meso-evolution* arises from dehydrating-hydrating fusions between living cells and "dead" environmental DNA and RNA, including viruses, which evolved into natural bacterial transformation, transduction, and "sexual" mating. These supra-macromolecular mechanisms account for large changes in the genome, such as lateral or horizontal gene transfer (HGT).
4. *Molecular macro-evolution* arises from direct cell-to-cell fusions *in vivo*, for example, in biofilms, when at least one cell is alive and survives the fusion process. Such fusions give rise to complex proto-eukaryotic cells, involving multiple DNA nucleoids, different kinds of protein-RNA complexes ("ribosomes" and other cellular machinery), and new membranes.

The latter two kinds of cellular Darwinian evolutions are not related to errors in DNA template synthesis; they involve large-scale breaking, re-sealing, and re-organizations of cell envelopes, with concurrent melting, hybridization, and recombination of nucleic acids, driven by cycling temperature, water activity, and other gradients. Such large "errors" can be caused by physicochemical stresses (e.g., temperature and water activity) that microbial populations are exposed to in cycling and fluctuating environments. When such fusions survive and re-stabilize, they could be viewed as "saltational," as they come about by quite different mechanisms that generate new cellular structures, compared with molecular errors in the template syntheses of nucleic acids. Indeed, similar environmental fusions (breaking and sealing of cell envelopes in large biofilm formations) probably gave rise

to first simple multicellular organisms and thus accounted for some of the major transitions in evolution (Smith and Szathmáry 1999).

The above-mentioned terms, such as "macro-evolution," are similar to those used in classical (non-molecular) evolutionary theories (Sapp 2003); however, in this review, they have specific physicochemical meanings, related to a recently coined term "natural genetic engineering" (Shapiro 2011). The jigsaw puzzle thus reflects processes that are analogous to the methods of genetic engineering and the protocols of PCRs.

6.6.7 CONCLUSIONS AND PERSPECTIVES

A number of publications have recently promulgated a degree of disquiet about the refractory nature of origins research; they have begun to question its assumptions and its goals. A few examples illustrate these concerns:

1. *"The natural path from simple cosmic molecules to cells, from chemistry to biology, remains undiscovered."* (Harold 2014).
2. *...no convincing models explain how living cells formed from abiotic constituents...* (Stüeken et al. 2013).
3. *"Most chemists believe, as do I, that life emerged spontaneously from mixtures of molecules in the prebiotic Earth. How? I have no idea." "The cell is a bag with a Jell-O of reacting chemicals...a network of reactions, organized in space and time in ways we do not grasp"* (Whitesides 2007).
4. *"...questions concerning the biogenesis...have never been asked or addressed in a proper way"* (Luisi 2014).
5. *...we are not yet even at the beginning of the end of our quest...* (Sutherland 2017).

As the above opinions imply, origins research is undergoing a wide-ranging re-appraisal, indicating that some of the traditional assumptions may need a significant revision, even a revolution. The main advance in the origin-of-life research has been the rejection of an idea that complex molecules such as a protein or an RNA got synthesized by chance on prebiotic Earth (a near miracle). It is now recognized that some *evolutionary sequences* of events or milestones with plausible continuity between them ended up in life's emergence (Deamer and Fleischaker 1994; Lahav 1999; Fry 2000; Spitzer and Poolman 2009; Hud et al. 2013; Sutherland 2017; Spitzer 2017).

The physicochemical jigsaw puzzle represents evolutionary processes inherently continuous through daily "blind experiments" performed by cycling diurnal and tidal energies. The jigsaw puzzle identifies physicochemical requirements that have to be met for a life to emerge, in particular the unconditional requirements for compartmentalization and a large degree of macromolecular crowding (point [B] in Figure 6.6.1). The (numbered) jigsaw puzzle pieces, however, are only tentative evolutionary milestones to be worked out in greater detail. The very first early proto-life after the point [B] was chemically complex (diverse), with lower fidelity of DNA replication, while evolving toward today's "simplified" biochemistry of a few building blocks common to all life, the point [C]. The very first early life after the point [B] had a high "genetic temperature" (Woese 1998), or fast proto-species evolution and extinctions, when replication, transcription, and translation machineries evolved and stabilized (crystalized and annealed), together with the cell envelope, creating a homeostatic capability. However, the total chemical complexity on Earth started to increase very fast, represented by the sum of the cycling organismal matter *in vivo*, point [C], and dead organismal and inorganic matter, point [D], in Figure 6.6.1. This *evolutionary* chemical paradigm is different from the *constructive* (complexification only) paradigm, cf. Figure 6.6.1; the latter is an engineering paradigm, based on designs of orderly progression from simple molecules, which react to yield building blocks of today's biomacromolecules, which then polymerize and self-assemble into proto-cells that evolve into cellular life (e.g., Jortner 2006).

The bottom-up chemical part of the jigsaw puzzle, in the evolutionary path from [A] to [B] in Figure 6.6.1, can be studied via chemical simulators of prebiotic Earths and with some simple non-living cyclic systems to illustrate the evolutionary potential of chemical processes driven by cyclic energies. The microbiological top-down part of the jigsaw puzzle from [B] to [C, D] can be studied under the assumption that Archaean first organisms and today's bacteria were substantially similar. The jigsaw puzzle has thus been expanded to better describe the evolution of "becoming alive" all the way to "being alive and having progeny" in Figure 6.6.2b. These cellular phenomena can be investigated through microbiological experimentation with "biotic soups," using cyclic processes analogous to PCR and genetic engineering protocols, for example, cyclically fusing "environmental" DNA through the membrane into a prokaryotic cell, and similar fusion experiments with living cells.

As a caveat, it may turn out that the assumption of physi-cochemical, metabolic, and genetic similarity between today's bacteria and the first archaean organisms may not be a good one. The ancient organisms originated with the help of cyclic planetary energies, which, in effect, acted as a "scaffold" for a life to emerge, cf. the history of "chemical scaffolds" in the origin-of-life research (Fry 2000). Undoubtedly, cyclic environmental temperatures and changes in ionic strength aided the natural evolution of replication, hybridization, and recombinations of nucleic acids, as well as the natural "genetic engineering" of fusions of cells with each other and with environmental nucleic acids. When the nucleoid and the cell envelope jointly stabilized to create a *homeostatic* cell, the "scaffold" of cyclic environmental energies was partly replaced with nutrients' chemical free energies, ensuring life's persistence, with evolving metabolic and genetic cycles that could exploit many different "food" sources (geochemical sites). Which, if any, bacterial species can be taken as a 3.5-billion-year-old

living fossil, still significantly dependent on the "scaffold" of cyclic environmental gradients?

Lastly, the fundamental physics question raised by Schrödinger in 1944 can still be contemplated: "Does the statistical nature of the 2nd law and of quantum mechanics play any (major) role in the occurrence of cell cycle errors?" If so, such true random errors need be distinguished from other sources of replication errors, such as errors arising from cycling and variable environments. Experimentation with an ideal bacterial cell cycle could throw some light on this fundamental question.

Regardless of these historical concerns, the questions of life's emergence can be tackled experimentally in an evolutionary paradigm of continuous *cyclic chemical processes*, particularly in the "microbiological" part of the puzzle. Such processes have an inherent evolutionary nature, but their chemical nano-phase complexity and dependence on the "arrow of time" make their investigations challenging.

DEDICATION

This article is dedicated to the late Prof. Robert Shapiro. I met him only once in 2008; his observations on the predicaments of origins research were compelling.

REFERENCES

Alberts, B., D. Bray, K. Hopkin et al. 2010. *Essential Cell Biology*. New York: Garland Science.

Alberts, B., A. Johnson, J. Lewis et al. 2002. *Molecular Biology of the Cell*. New York: Garland Science.

Albers, S.-V., and B. J. Meyer. 2011. The archaeal cell envelope. *Nat. Rev. Microbiol.* 9: 414–426. doi:10.1038/nrmicro2576.

Atkins, P. 2011. *On Being*. Oxford, UK: Oxford University Press.

Atkins, P. 2007. *Four Laws that Drive the Universe*. Oxford, UK: Oxford University Press.

Atkins, P. 2003. *Galileo's Finger. The Ten Great Ideas of Science*. Oxford, UK: Oxford University Press.

Atkins, P. W. 1978. *Physical Chemistry*. San Francisco, CA: W. H. Freeman & Company.

Ault, A. 2004. The monosodium glutamate story: The commercial production of MSG and other amino acids. *J. Chem. Educ.* 81: 347–355.

Bada, J. L., and A. Lazcano. 2003. Prebiotic soup: Revisiting the Miller experiment. *Science* 300: 745–746.

Bak, P., and M. Paczuski. 1995. Complexity, contingency, and criticality. *Proc. Natl. Acad. Sci. U.S.A.* 92: 6689–6696.

Barrick, J. E., and R. E. Lenski. 2013. Genome dynamics during experimental evolution. *Nat. Rev. Genet.* 14: 827–839. doi:10.1038/nrg3564.

Baumeister, W. 2005. From proteomic inventory to architecture. *FEBS Lett.* 579: 933–937.

Bedau, M. A., and C. E. Cleland (Eds.). 2010. *The Nature of Life*. Cambridge, UK: Cambridge University Press.

Benner, S. A. 2010. Defining life. *Astrobiology* 10: 1021–1030.

Bernstein, M. 2006. Prebiotic materials from on and off the early Earth. *Philos. Trans. R. Soc. B* 361: 1689–1702.

Bernstein, M. P., S. A. Sandford, L. J. Allamandola, J. S. Gillette, S. J. Clemett, and, R. N. Zare. 1999. UV irradiation of polycyclic aromatic hydrocarbons in ices: Production of alcohols, quinones, and ethers. *Science* 283: 1135–1138.

Bernstein, M. P., J. E. Elsila, J. P. Dworkin, S. A. Sandford, L. J. Allamandola, and R. N. Zare. 2002. Side group addition to the polycyclic aromatic hydrocarbon coronene by ultraviolet photolysis in cosmic ice analog. *Astrophys. J.* 576: 1115–1120.

Blackley, D. C. 1975. *Emulsion Polymerization.* London, UK: Applied Science Publishers.

Boersma, A. J., I. S. Zuhorn, and B. Poolman. 2015. A sensor for quantification of macromolecular crowding in living cells. *Nat. Methods* 12: 227–229.

Boquist, M., and G. Gröbner. 2007. Misfolding of amyloidogenic proteins at membrane surfaces: The impact of macromolecular crowding. *J. Am. Chem. Soc.* 129: 14848–14849.

Brack, A. (Ed). 2000. *The Molecular Origins of Life: Assembling Pieces of the Puzzle.* Cambridge, UK: Cambridge University Press.

Bray, D. 2009. *Wetware: A Computer in Every Living Cell.* New Haven, CT: Yale University Press.

Cayley, S. B., A. Lewis, H. J. Guttman et al. 1991. Characterization of the cytoplasm of *Escherichia coli* K-12 as a function of external osmolarity. Implications for protein-DNA interactions *in vivo. J. Mol. Biol.* 222: 281–300.

Carrasco, N., I. Schmitz-Afonso, J. Y. Bonnet et al. 2009. Chemical characterization of Titan's tholins: Solubility, morphology and molecular structure revisited. *J. Phys. Chem. A* 113: 11195–11203.

Carroll, L. 1999. *Through the Looking-Glass and What Alice Found There.* Mineola, NY: Dover Publications.

Chyba, C., and C. Sagan. 1992. Endogenous production, exogenous delivery and impact-shock synthesis of organic molecules: An inventory for the origins of life. *Nature* 355: 125–132.

Cleaves, H. J. 2010. Hierarchical definitions in the origin of life. *Orig. Life Evol. Biosph.* 40: 489.

Cleland, C. E., and C. F. Chyba. 2002. Defining "life." *Origins Life Evol. Biosphere* 32: 387–393.

Cossins, B. P., M. P. Jacobson, and V. Guallar. 2011. A new view of the bacterial cytosol environment. *PLoS Comput. Biol.* 7(6): e1002066. doi:10.1371/journal.pcbi.1002066.

Crick, F. 1966. *Of Molecules and Men.* Seattle, WA: University of Washington Press.

Danchin, A. 2009. Bacteria as computers making computers. *FEMS Microbiol. Rev.* 33: 3–26.

Davey, H. M. 2011. Life, death and in-between: Meanings and methods in microbiology. *Appl. Environ. Microbiol.* 77: 5571–5576.

Dawkins, R. 1987. *The Blind Watchmaker.* London, UK: W.W. Norton & Company.

Deamer, D. 2011. *First Life.* Berkeley, CA: University of California Press.

Deamer, D. W., and G. R. Fleischaker, Eds. 1994. Origins of Life: *The Central Concepts.* Boston, MA: Jones & Bartlett Publishers.

de Kerchove, A. J., and M. Elimelech. 2005. Relevance of electrokinetic theory for "soft" particles to bacterial cells: Implications for bacterial adhesion. *Langmuir* 21: 6462–6472.

Doolittle, W. F. 1999. Phylogenetic classification and the universal tree. *Science* 284: 2124–2128.

Dworkin, J. P., D. W. Deamer, A. S. Sandford et al. 2001. Self-assembling amphiphilic molecules: Synthesis in simulated interstellar/precometary ices. *Proc. Natl. Acad. Sci. U.S.A.* 98: 815–819.

Ehrenfreund, P., and J. Cami. 2010. Cosmic carbon chemistry: From the interstellar medium to the early Earth. *Cold Spring Harb. Perspect. Biol.* doi:10.1101/cshperspect.a002097.

Eigen, M., and P. Schuster. 1979. *The Hypercycle.* Berlin, Germany: Springer Verlag.

Eisenberg, D. 2003. The discovery of α-helix and β-sheet, the principal structural features of proteins. *Proc. Natl. Acad. Sci. U.S.A.* 100: 11207–11210.

Elcock, A. H. 2010. Models of macromolecular crowding effects and the need for quantitative comparisons with experiment. *Curr. Opin. Struct. Biol.* 20: 1–11.

Ellis, R. J., and A. P. Minton. 2006. Protein aggregation in crowded environments. *Biol. Chem.* 387: 485–497.

Ellis, R. J. 2001. Macromolecular crowding-obvious but underappreciated. *Trends Biochem. Sci.* 26: 597–604.

Errington, J. 2013. L-form bacteria, cell walls and the origins of life. *Open Biol.* 3: 120143. doi:10.1098/rsob.120143.

Eschenmoser, A. 2007. The search for the chemistry of life's origin. *Tetrahedron* 63: 12821–12844.

Firshein, W. 1989. Role of the DNA/membrane complex in prokaryotic DNA replication. *Annu. Rev. Microbiol.* 43: 89–120.

Flack, H. D. 2009. Louis Pasteur's discovery of molecular chirality and spontaneous resolution in 1848, together with a complete review of his crystallographic and chemical work. *Acta Crystallogr. A* 65: 371–389. doi:10.1107/S0108767309024088.

Foffi, G., A. Pastore, F. Piazza, P. A. Temussi. 2013. Macromolecular crowding: Chemistry and physics meet biology (Ascona, Switzerland, 10–40 June, 2012). *Phys. Biol.* 10: 040301.

Friedrich, K., C. O. Gualerzi, M. Lammi, M. A. Losso, and C. L. Pon. 1988. Proteins from the prokaryotic nucleoid. Interaction of nucleic acids with the 15 kDa *Escherichia coli* histone-like protein H-NS. *FEBS Lett.* 229: 197–202.

Fry, I. 2000. *The Emergence of Life on Earth.* New Brunswick, NJ: Rutgers University Press.

Gierasch, L. M., and A. Gershenson. 2009. Post-reductionist protein science or putting Humpty Dumpty back together again. *Nat. Chem. Biol.* 5: 774–777.

Goodsell, D. S. (2010). *The Machinery of Life.* New York: Springer-Verlag.

Halford, S. E. 2009. An end to 40 years of mistakes in DNA-protein association kinetics? *Biochem. Soc. Trans.* 37: 343–348.

Harding, S. E., and A. J. Rowe. 2010. Insight into protein-protein interactions from analytical ultracentrifugation. *Biochem. Soc. Trans.* 38: 901–907.

Harold, F. M. 1986. *The Vital Force: A Study of Bioenergetics.* New York: W. H. Freeman & Company.

Harold, F. M. 2005. Molecules into cells: Specifying spatial architecture. *Microbiol. Mol. Biol. Rev.* 69: 544–564.

Harold, F. M. 2014. *In Search of Cell's History.* Chicago, IL: University of Chicago Press.

Hartwell, L. H., J. J. Hopfield, S. Leibler et al. 1999. From molecular to modular cell biology. *Nature* 402: C47–C52.

Hazen, R. M. 2005. *Genesis: The Scientific Quest for Life's Origins.* Washington, DC: Joseph Henry Press.

Higgs, P. G., and Lehman, N. 2015. The RNA World: Molecular cooperation at the origins of life. *Nat. Rev. Genet.* 16: 7–17.

Hinshelwood, C. 1946. *Chemical Kinetics of the Bacterial Cell.* London, UK: Oxford University Press.

Hud, N. V., B. J. Cafferty, R. Krishnamurthy, L. D. Williams. 2013. The origin of RNA and "my grandfather's axe." *Chem. Biol.* 20: 466–474.

Johnson, A. P., H. J. Cleaves II, J. P. Dworkin et al. 2008. The Miller volcanic spark discharge experiment. *Science* 322: 404. doi:10.1126/science.1161527.

Jortner, J. 2006. Conditions for the emergence of life on the early Earth: Summary and reflections. *Philos. Trans. R. Soc. B* 361: 1877–1891.

Kauffman, S. A. 1993. *The Origins of Order: Self-Organization and Selection in Evolution.* Oxford, UK: Oxford University Press.

Kim, B. H., and G. M. Gadd. 2008. *Bacterial Physiology and Metabolism*. Cambridge, UK: Cambridge University Press.

Kiraga, J., P. Mackiewicz, D. Mackiewicz et al. 2007. The relationship between the isoelectric point and length of proteins, taxonomy and ecology of organisms. *BMC Genomics* 8: 163. doi:10.1186/1471-2164-8-163.

Knoll, A. H. 2003. *Life on a Young Planet. The First Three Billion Years of Evolution on Earth*. Oxford, UK: Oxford University Press.

Konopka, M. C., K. A. Sochacki, B. Bratton et al. 2009. Cytoplasmic protein mobility in osmotically stressed *Escherichia coli*. *J. Bacteriol.* 191: 231–237.

Koonin, E. V. 2014. Carl Woese's vision of cellular evolution and the domains of life. *RNA Biol.* 11: 197–204.

Kyne, C., and P. B. Crowley. 2016. Grasping the nature of the cell interior: From physiological chemistry to chemical biology. *FEBS J.* 283: 3016–3028. doi:10.1111/febs.1374.

Lahav, N. 1999. *Biogenesis. Theories of Life's Origins*. Oxford, UK: Oxford University Press.

Lane, N. 2003. *The Vital Question. Energy, Evolution, and the Origins of Complex Life*. New York: W. W. Norton & Company.

Laue, T. 2012. Proximity energies: A framework for understanding concentrated solutions. *J. Mol. Recognit.* 25: 165–173.

Lazcano, A. 2008. Towards a definition of life: The impossible quest? *Space Sci. Rev.* 135: 5–10.

Lee, S. Y., C. J. Lim, P. Dröge et al. 2015. Regulation of bacterial DNA packaging in early stationary phase by competitive DNA binding of DPS and IHF. *Sci. Rep.* 5: 18146. doi:10.1038/srep18146.

López, D., and R. Kolter. 2010. Functional microdomains in bacterial membranes. *Genes Dev.* 24: 1893–1902.

Luisi, P. L. 1998. About various definitions of life. *Orig. Life Evol. Biosph.* 28: 613–622.

Luisi, P. L. 2003. Autopoiesis: A review and a reappraisal. *Naturwissenschaften* 90: 49–59.

Luisi, P. L. 2014. A new start from ground zero? *Orig. Life Evol. Biosph.* 44: 303–306. doi:10.1007/s11084-014-9386-1.

Lutkenhaus, J. 2003. Another cytoskeleton in the closet. *Cell* 115: 648–650.

Martin, W. F., S. Garg, and V. Zimorski. 2015. Endosymbiotic theories for eukaryote origin. *Philos. Trans. R. Soc. B* 370: 20140330. doi:10.1098/rstb.2014.0330.

Mathews, C. K. 1993. The cell—A bag of enzymes or a network of channels? *J. Bacteriol.* 175: 6377–6381.

McCollom, M. 2013. Miller-Urey and beyond: What have we learned about prebiotic organic synthesis reactions in the past 60 Years? *Annu. Rev. Earth Planet. Sci.* 41: 207–229.

McConkey, E. H. 1982. Molecular evolution, intracellular organization, and the quinary structure of proteins. *Proc. Natl. Acad. Sci. U.S.A.* 79: 3236–3240.

McDonald, G. D., B. N. Khare, W. R. Thompson et al. 1991. $CH_4/NH_3/H_2O$ spark tholin: Chemical analysis and interaction with Jovian aqueous clouds. *Icarus* 94: 354–367.

Mesler, B., and H. J. Cleaves II. 2016. *A Brief History of Creation. Science and the Search for the Origin of Life*. New York: W. W Norton & Company.

Mika, J. T., and B. Poolman. 2011. Macromolecule diffusion and confinement in prokaryotic cells. *Curr. Opin. Biotechnol.* 22: 117–126.

Mika J. T., G. van der Bogaart, L. Veenhof et al. 2010. Molecular sieving properties of the cytoplasm of *Escherichia coli* and consequences of osmotic stress. *Mol. Microbiol.* 77: 200–207.

Miller, S. L. 1953. A production of amino acids under possible primitive Earth conditions. *Science* 117: 528–529.

Mitchell, P. 1979. Compartmentation and communication in living systems. Ligand conduction: A general catalytic principle in chemical, osmotic, and chemiosmotic reaction systems. *Eur. J. Biochem.* 95: 1–20.

Michie, K. A., and J. Löwe. 2006. Dynamic filaments of the bacterial cytoskeleton. *Annu. Rev. Biochem.* 75: 467–492.

Monteith, W. B., Cohen, R. D., Smith, A. E. et al. 2015. Quinary structure modulates protein stability in cells. *Proc. Natl. Acad. Sci. U.S.A.* 112: 1739–1742.

Morowitz, H. J. 1992. *Beginnings of Cellular Life*. New Haven, CT: Yale University Press.

Morowitz, H. J., and Eric Smith, E. 2007. Energy flow and the organization of life. *Complexity* 13: 51–59.

Mulkidjanian, A. Y., A. Y. Bychkov, D. V. Dibrova et al. 2012. Open questions on the origin of life at anoxic geothermal fields. *Orig. Life Evol. Biosph.* 42: 507–516.

Newburgh, R., J. Peidle, and W. Rueckner. 2006. Einstein, Perrin, and the reality of atoms: 1905 revisited. *Am. J. Phys.* 74: 478–481.

Nicholls, D. G., and S. J. Ferguson. 2002. *Bioenergetics*, 3rd ed. London, UK: Academic Press.

Nickell, S., C. Kofler, A. P. Leis, and W. Baumeister. 2006. A visual approach to proteomics. *Nat. Rev. Mol. Cell Biol.* 7: 225–230. doi:10.1038/nrm1861.

Niemark, H. D. 1986. Origins and evolution of wall-less prokaryotes. In *The Bacterial L-forms*, S. Madoff (Ed.), pp. 21–42. New York: Marcel Dekker.

Norris, V., T. den Blaauwen, A. Cabin-Flaman et al. 2007. Functional taxonomy of bacterial hyperstructures. *Microbiol. Mol. Biol. Rev.* 71: 230–253.

Oikonomou, C. M., and G. J. Jensen. 2017. Cellular electron cryotomography: Toward structural biology in situ. *Annu. Rev. Biochem.* 86: 873–879.

Orgel, L. E. 2004. Prebiotic chemistry and the origin of the RNA World. *Crit. Rev. Biochem. Mol. Biol.* 39: 99–123.

Orgel, L. E. 2000. Self-organizing biochemical cycles. *Proc. Natl. Acad. Sci. U.S.A.* 97: 12503–12507.

Ovadi, J., and P. A. Srere. 1991. Macromolecular compartmentation and channeling. *Int. Rev. Cytol.* 192: 255–280.

Parry, B. R., I. V. Surovtsev, M. T. Cabeen et al. 2014. The bacterial cytoplasm has glass-like properties and is fluidized by metabolic activity. *Cell* 156: 183–194. doi:10.1016/j.cell.2013.11.028.

Partikian, A., B. Olveczky, R. Swaminathan et al. 1998. Rapid diffusion of green fluorescent protein in the mitochondrial matrix. *J. Cell Biol.* 140: 821–829.

Pascal, R., and A. Pross. 2015. Stability and its manifestation in the chemical and biological worlds. *Chem. Commun.* 51: 16160–16165.

Pastore, A., and P. A. Temussi. 2011. The two-faces of Janus: Functional interactions and protein aggregation. *Curr. Opin. Struct. Biol.* 22: 1–8.

Patel, B. H., C. Percivalle, D. J. Ritson et al. 2015. Common origins of RNA, protein and lipid precursors in a cyanosulfidic protometabolism. *Nat. Chem.* 7: 301–307. doi:10.1038/nchem.2202.

Pauling, L., R. B. Corey, and H. R. Branson. 1951. The structure of proteins: Two hydrogen-bonded helical configurations of the polypeptide chain. *Proc. Natl. Acad. Sci. U.S.A.* 37: 205–211.

Phillips, R., J. Kondev, and J. Theriot. 2009. *Physical Biology of the Cell*. New York: Garland Science.

Perutz, M. 1991. Physics and the riddle of life. In *Is Science Necessary? Essays on Science and Scientists*, M. Perutz (Ed.), pp. 242–259. Oxford, UK: Oxford University Press.

Pizzarello, S. 2006. The chemistry of life's origin: A carbonaceous meteorite perspective. *Acc. Chem. Res.* 39: 231–237.

Poolman, B., J. J. Spitzer, and M. Wood. 2004. Bacterial osmosensing: Roles of membrane structure and electrostatics in lipid-protein and protein-protein interactions. *Biophys. Biochim. Acta* 1666: 88–104.

Pross, A. 2005. On the chemical nature and origin of teleonomy. *Orig. Life Evol. Biosph.* 35: 383–394; doi:10.1007/s11084-005-2045-9.

Rhee, Y. M., T. J. Lee, M. S. Gudipati, L. J. Allamandola, and M. Head-Gordon. 2007. Charged polycyclic aromatic hydrocarbon clusters and the galactic extended red emission. *Proc. Natl. Acad. Sci. U.S.A.* 104: 5274–5278.

Robertson, M. P., and G. F. Joyce. 2012. The origins of the RNA world. *Cold Spring Harb. Perspect. Biol.* doi:10.1101/cshperspect.a003608.

Romantsov, T., L. Stalker, D. E. Culham et al. 2008. Cardiolipin controls the osmotic stress response and the subcellular location of transporter ProP in *Escherichia coli*. *J. Biol. Chem.* 283: 12314–12323.

Rowlett, V. W., and W. Margolin. 2015. The bacterial divisome: Ready for its close up. *Philos. Trans. R. Soc. B* 370: 2015.0028. doi:10.1098/rstb.2015.0028.

Rowe, A. J. 2011. Ultra-weak reversible protein-protein interactions. *Methods* 54: 157–166.

Sagan, L. 1967. On the origin of mitosing cells. *J. Theor. Biol.* 14: 255–274.

Sandford, S. A., M. P. Bernstein, and C. K. Materese. 2013. The infrared spectra of polycyclic aromatic hydrocarbons with excess peripheral H Atoms (Hn-PAHs) and their Relation to the 3.4 and 6.9 µm PAH Emission Features. *Astrophys. J. Suppl. Ser.* doi:10.1088/0067-0049/205/1/8.

Sapp, J. 2003. *Genesis. The Evolution of Biology*. Oxford, UK: Oxford University Press.

Sarkar, M., A. E. Smith, and G. J. Pielak. 2013. Impact of reconstituted cytosol on protein stability. *Proc. Natl. Acad. Sci. U.S.A.* 110: 19342–19347.

Schaechter, M., J. L. Ingraham, and F. C. Neidhardt. 2005. *Microbe*. Washington, DC: ASM Press.

Scherer, S., E. Wollrab, L. Codutti et al. 2016. Chemical analysis of a "Miller-type" complex prebiotic broth. *Orig. Life Evol. Biosph.* doi:10.1007/s11084-016-9528-8.

Schreiber, G., and A. R. Fersht. 1996. Rapid, electrostatically assisted association of proteins. *Nat. Struct. Biol.* 3: 427–431.

Schrödinger, E. 2012. *What is Life?* Cambridge, UK: Cambridge University Press.

Schuster, T. M., and T. M. Laue. 1994. *Modern Analytical Ultracentrifugation: Acquisition and Interpretation of Data for Biological and Synthetic Polymer Systems*. Boston, MA: Birkhäuser.

Schwartz, A. W. 2007. Intractable mixtures and the origin of life. *Chem. Biodivers.* 4: 656–664.

Shapiro, R. 1986. *Origins: A skeptic's Guide to the Creation of Life on Earth*. New York: Simon & Schuster.

Shapiro, R. 2007. A simpler origin of life. *Sci. Am.* 296(6): 46–53.

Shapiro, J. A. 2011. *Evolution. A View from the 21st Century*. Upper Saddle River, NJ: Financial Times Press Science.

Silhavy, T. J., D. Kahne, and S. Walker. 2010. The bacterial cell envelope. *Cold Spring Harb. Perspect. Biol.* 2: a000414. doi:10.1101/cshperspect.a000414.

Smith, J. M., and E. Szathmáry. 1999. *The Origins of Life*. Oxford, UK: Oxford University Press.

Solé, R. V., and B. Goodwin. 2000. *Signs of Life: How Complexity Pervades Biology*. New York: Basic Books.

Southall, N. T., K. A. Dill, and A. D. J. Haymet. 2002. A view of the hydrophobic effect. *J. Phys. Chem. B* 106: 521–533.

Spitzer, J. J. 1984. A re-interpretation of hydration forces. *Nature* 310: 396–397.

Spitzer, J. 2003. Maxwellian double layer forces: From infinity to contact. *Langmuir* 19: 7099–7111.

Spitzer, J. 2011. From water and ions to crowded biomacromolecules: *In vivo* structuring of a prokaryotic cell. *Microbiol. Mol. Biol. Rev.* 75: 491–506.

Spitzer, J. 2013a. Cycling physicochemical gradients as 'Evolutionary Drivers': From complex matter to complex living states. *bioRxiv*, 000786.

Spitzer, J. 2013b. Emergence of life from multicomponent mixtures of chemicals: The case for experiments with cycling physicochemical gradients. *Astrobiology* 13: 404–413.

Spitzer, J. 2014. The continuity of bacterial and physicochemical evolution: Theory and experiments. *Res. Microbiol.* 165: 457–461.

Spitzer, J. 2017. Emergence of life on Earth: A physicochemical jigsaw puzzle. *J. Mol. Evol.* 84: 1–7.

Spitzer, J., and B. Poolman. 2005. Electrochemical structure of the crowded cytoplasm. *Trends Biochem. Sci.* 30: 536–541.

Spitzer, J., and B. Poolman. 2009. The role of biomacromolecular crowding, ionic strength and physicochemical gradients in the complexities of life's emergence. *Microbiol. Mol. Biol. Rev.* 73: 371–388.

Spitzer, J., and B. Poolman. 2013. How crowded is the prokaryotic cytoplasm? *FEBS Lett.* 587: 2094–2098.

Spitzer, J., G. Pielak, and B. Poolman. 2015. Emergence of life: Physical chemistry changes the paradigm. *Biol. Direct* 10: 33. doi:101186/s13062-015-0060-y.

Srere, P. A. 1985. The metabolon. *Trends Biochem. Sci.* 10: 109–110.

Storz, G., and R. Hengge (Eds.). 2011. *Bacterial Stress Responses*, 2nd ed. Washington, DC: ASM Press.

Stüeken, E. E., R. E. Anderson, J. S. Bowman et al. 2013. Did life originate from a global chemical reactor? *Geobiology*. doi:10.1111/gbi.12025.

Sutherland, J. D. 2017. Studies on the origin of life—The end of the beginning. *Nat. Rev. Chem.* doi:10.1038/s41570-016-0012.

Szostak, J. W. 2012. Attempts to define life do not help to understand the origin of life. *J. Biomol. Struct. Dyn.* 29: 599–600.

Tanford, C. 1980. *The Hydrophobic Effect: The Formation of Micelles and Biological Membranes*. New York: Wiley-Interscience.

Tremblay, G. Y., M. J. Daniels, and M. Schaechter. 1969. Isolation of a cell membrane-DNA-nascent RNA complex from bacteria. *J. Mol. Biol.* 40: 65–74, IN1, 75–76.

Trifonov, E. N. 2012. Definition of life: Navigation through uncertainties. *J. Biomol. Struct. Dyn.* 29: 647–650.

Valkenburg, J. A. C., and C. L. Woldringh. 1984. Phase separation between nucleoid and cytoplasm in *Escherichia coli* as defined by immersive refractometry. *J. Bacteriol.* 160: 1151–1157.

van den Bogaart, G., N. Heermans, V. Krasnikov et al. 2007. Protein mobility and diffusive barriers in *Escherichia coli*: Consequences of osmotic stress. *Mol. Microbiol.* 64: 858–871.

van Loosdrecht, M. C. M., J. Lyklema, W. Norde et al. 1987. Electrophoretic mobility and hydrophobicity as a measure to predict the initial steps of bacterial adhesion. *Appl. Environ. Microbiol.* 53: 1898–1901.

Verkman, A. S. 2002. Solute and macromolecule diffusion in cellular aqueous compartments. *Trends Biochem. Sci.* 27: 27–33.

Wang, Q., A. Zhuravleva, and L. M. Gierasch. 2011. Exploring weak, transient protein-protein interactions in crowded *in vivo* environments by in-cell nuclear magnetic resonance spectroscopy. *Biochemistry* 50: 9225–9236.

Weissbuch, I., and M. Lahav. 2011. Crystalline architectures as templates of relevance to the origins of homochirality. *Chem. Rev.* 111: 3236–3267.

Whitesides, G. M. 2007. Revolutions in chemistry. *C&EN* 85(13): 12–17.

Whitesides, G. M., and R. F. Ismagilov. 1999. Complexity in chemistry. *Science* 284: 89–92.

Whitesides, G. M., and M. Boncheva. 2002. Beyond molecules: Self-assembly of mesoscopic and macroscopic components. *Proc. Natl. Acad. Sci. U.S.A.* 99: 4769–4774.

Woese, C. R. 1998. The universal ancestor. *Proc. Natl. Acad. Sci. U.S.A.* 95: 6854–6859.

Woese, C. R., and N. Goldenfeld. 2009. How the microbial world saved evolution from the Scylla of molecular biology and the Charybdis of the modern synthesis. *Microbiol. Mol. Biol. Rev.* 73: 14–21.

Woldringh, C. L., and N. Nanninga. 2006. Structural and physical aspects of bacterial chromosome segregation. *J. Struct. Biol.* 156: 273–283.

Wood, J. 2011. Osmotic stress. In *Bacterial Stress Responses*, G. Storz and R. Hengge (Eds.), pp. 133–156. Washington, DC: ASM Press.

Yeung, T., G. E. Gilbert, J. Shi et al. 2008. Membrane phosphatidylserine regulates surface charge and protein localization. *Science* 319: 210–213.

Yu, I., T. Mori, T. Ando et al. 2016. Biomolecular interactions modulate macromolecular structure and dynamics in atomistic model of a bacterial cytoplasm. *eLife* 5: e19274. doi:10.7554/eLife.19274.

Zhou, H.-X., G. Rivas, and A. P. Minton. 2008. Macromolecular crowding and confinement: Biochemical, biophysical and potential physiological consequences. *Annu. Rev. Biophys.* 37: 247–263.

Zimmerman, S. B. 2006. Shape and compaction of *Escherichia coli* nucleoids. *J. Struct. Biol.* 156: 255–261.

Zimmerman, S. B., and A. P. Minton. 1993. Macromolecular crowding: Biochemical, biophysical, and physiological consequences. *Annu. Rev. Biophys. Biomol. Struct.* 22: 27–65.

Zimmerman, S. B., and L. D. Murphy. 2001. Release of compact nucleoids with characteristic shapes from *Escherichia coli*. *J. Bacteriol.* 183: 5041–5049.

Zimmerman, S. B., and S. O. Trach. 1991. Estimation of macromolecular concentrations and excluded volume effects for the cytoplasm of *Escherichia coli*. *J. Mol. Biol.* 222: 599–620.

Zubay, G. L. 1998. *Biochemistry*. New York: McGraw-Hill Companies.

Section VII

Origin of Life

Early Compartmentalization—Coacervates and Protocells

7.1 Oparin's Coacervates

Vera M. Kolb

CONTENTS

7.1.1 INTRODUCTION: A BRIEF OVERVIEW OF OPARIN'S COACERVATES

In Chapter 1.1 of this handbook, a brief introduction to the importance of Oparin's work for astrobiology was provided. Most significantly, in 1924, he proposed a major hypothesis about the origin of life on Earth, which is now accepted as a key foundation of astrobiology (Oparin 1924, 1938, 1966). Oparin's hypothesis stated that the origin of life on Earth can be understood in terms of the laws of chemistry and physics, and thus, it does not need any explanations that involve supernatural causes. Oparin further proposed that life arose in the Earth's distant past by chemical reactions and physical processes under the specific conditions on the early Earth and over a long period of time. The significance of Oparin's work is covered throughout this handbook. This chapter focuses on Oparin's work on coacervates, which he proposed as a model for the primordial protocells.

Since his original proposal of coacervates, Oparin made experimental progress in preparing coacervates and studying their properties (Oparin 1938, 1966). He found out that coacervates are feasible as prebiotic compartmentalized systems and can serve as sites for the primitive metabolism. In addition, coacervates can undergo a primitive self-replication by splitting into the smaller units.

The coacervate model is that of a membraneless protocell, which, however, allows for the separation of the protocell content from the aqueous environment. Coacervates are formed in water and comprise either droplets or layers, in which the coacervate exists as a separate aqueous phase, which has a clear boundary from the rest of the aqueous medium. Such coacervates, according to Oparin, are capable of absorbing organic molecules from the aqueous environment. The absorbed molecules may react inside the coacervate to produce new chemicals. If coacervates absorb the appropriate catalysts from the environment, chemical reactions inside coacervates will be facilitated. As coacervates grow, due to the absorption of more chemicals from the environment, they become thermodynamically unstable and then split into the daughter cells. This process provides a primitive self-replication, which is driven by the well-understood physical forces, such as a thermodynamic instability of the system or its breakage upon a mechanical impact. In addition, the daughter cells acquire chemicals from the original cell, and thus, the chemical composition is preserved. Such a replicating system is not as reliable as a genetic system, which evolved later, but it represents a good prebiotic beginning. A faster growth of coacervates would result in a faster replication. Such coacervates would be selected. However, at some point, there will be a shortage of chemical food in the environment. Oparin proposed that coacervates would compete for the food. Coacervates that are able to develop chemical pathways to make their own food would be selected. Such coacervates would have a primitive metabolism. Thus, the Oparin's coacervate model involves natural selection at the chemical level and provides a feasible way for the chemical evolution to occur, in the absence of a more reliable genetic system.

7.1.2 THE OPARIN'S PROTOCELLS: DETAILS OF THE DEVELOPMENT OF HIS EXPERIMENTAL MODEL

Oparin's development of the coacervate model for protocells went through two stages. He first developed a general idea based on the science of the colloidal gels, and later, he introduced coacervates. This is described in the following two subsections, 7.1.2.1 and 7.1.2.2.

7.1.2.1 THE INITIAL IDEA: COLLOIDS AND GELS

In conceiving his protocell model, Oparin started with the knowledge that at that time was a "comparatively young science of colloid chemistry," in his own words (Oparin 1924). Oparin started from Graham's description of colloids from 1861 and developed it further. Graham's colloids are substances that usually give cloudy solutions and cannot pass through the vegetable or animal membranes. The examples that Graham gave include starch, proteins, gums, and mucus. Oparin linked these properties to the protoplasm, which, he stated, is mostly made up of "very large and complicated particles and therefore must give colloidal solutions" (Oparin 1924). Oparin then addressed precipitates of colloidal substances, which form as "clots or lumps of mucus or jelly" (Oparin 1924). Coagulates of gels have some structural features that resemble protoplasm. Oparin connected feeding of the living organisms to the ability of the colloidal coagula to extract various substances from the solution and absorb them. One example includes dyes. The absorbed substances penetrate the coagula, and some of them enter into the chemical reactions with them. Thus, the process in which various chemicals are absorbed into the coagula and react inside is analogous to the first stage of metabolism, namely the feeding.

Since starch and proteins tend to make colloidal solutions in water, Oparin thought that such substances would be good candidates for the primordial colloids. He proposed that carbohydrates and proteins were available on the early Earth. As it will be shown later, in Section 7.1.4, modern advances in prebiotic chemistry challenge this assumption, especially in regard to carbohydrates. Oparin stated that these colloidal solutions are not stable and that dissolved organic substances would have come out of the solution in the form of precipitates, coagula, or gels. He stated that when such separation occurred, the resulting organic body would become an "individual."

Oparin believed that the first piece of organic slime that was formed on the early Earth had many of the features that we now consider characteristic of life. Even if such slime could not metabolize fully, it could still nourish itself, by absorbing and assimilating substances from its environment, since such behavior is characteristic of organic gels (Oparin 1924).

Oparin introduced the concept of competition for nutrients between the bits of gel and the selection that resulted therefrom. He visualized bits of gel floating in a mixture of nutrients and absorbing them. Each bit of gel grew at the expense of the nutrients, but since the bits of gel were not identical in their chemical composition, they assimilated the material from the environment at different rates. The bits of gel that absorbed faster also grew faster, and their surface became larger. The larger pieces of gel were more likely to break up, owing to the mechanical reasons or surface tension. Such a breakage of the gel pieces gave new pieces that inherited the original gel structure. This was the basis of self-reproduction. As the cycle of growth and break up continued, changes that were acquired by the new gels were transmitted to their offspring. Oparin believed that this cycle provided a basis for selection of the gels with an improved absorption and assimilation of the nutrients.

As the nutrients became increasingly depleted, a battle for the nutrients took place. The bits of gel had the options of using the old ways of nutrition by acquiring organic substances from the medium or by "eating" (assimilating) other bits of gel, or they could develop a way of synthesizing organic nutrients from the inorganic precursors. Such synthetic ability would give them an edge in competition and would lead toward a more developed metabolism. Eventually, the original primitive gel bit forms were not competitive anymore and were wiped out by being eaten by the more advanced gel bit forms (Oparin 1924).

7.1.2.2 THE DEVELOPMENT OF THE IDEA: THE INTRODUCTION OF COACERVATES

In his later work, Oparin became much more specific in his model of the primordial protocells, and his gel bits became replaced by coacervates (Oparin 1938, 1966).

In his 1938 book, Oparin described work by de Jong from 1931 and 1932 on the coacervation phenomenon (Oparin 1938). In the aqueous solution of hydrophilic organic colloids, besides coagulation, another process occurs, which is termed coacervation. It includes separation into two layers, a fluid sediment, rich in the colloidal substance, termed the coacervate, and another liquid layer that is in equilibrium with it, termed the equilibrium liquid. Oparin described how, in many instances, the coacervate does not settle out immediately as a continuous fluid layer but, instead, remains in the form of microscopic droplets that are floating in the equilibrium liquid. These droplets may later fuse with each other to form a layer. Oparin contrasted the coacervates, which represent a fluid mass, with the coagulates, which do not. Most importantly, although in both the coacervate and the equilibrium liquid, the solvent is water, coacervates are sharply demarcated from the surrounding medium by a clearly discernible surface. This description of coacervates is still valid (e.g., Menger and Sykes 1998). Oparin clearly recognized the potential of coacervates to serve as compartmentalized primordial protocell systems.

In his 1938 book, Oparin described preparation of coacervates, in which he followed work by de Jong. Oparin gave a detailed preparation of complex coacervates from two or more colloidal solutions, which have opposite charges. The classical preparation is by mixing dilute solutions of gelatin and gum arabic and then adjusting the pH until coacervation occurs. This occurs at a pH below the isoelectric point of gelatin (pH 4.82). At such pH, gelatin will become positive, while the gum arabic will still be negatively charged. Coacervation is induced by such a change in pH, and droplets will separate from previously homogeneous solution. Oparin gave more examples of coacervates, such as those formed from gelatin and egg lecithin and from gelatin and protamine. Since coacervates have an inherent ability to absorb substances from the environment, such substances could react with the coacervate constituents. According to the experiments by de Jong, this would result in an increase in size and weight of the original coacervates. Oparin also described the influence of electrolytes, salts, temperature, and the pH on the formation of coacervates.

In his 1966 book, Oparin gave numerous other examples of complex coacervates that were made in his laboratory. These were prepared from complex biomolecules, and the pH was adjusted until a coacervation occurred (Oparin 1966). These will be discussed in Section 7.1.4.

7.1.3 OPARIN'S EXPERIMENTS: CHEMICAL REACTIONS INSIDE COACERVATES

Oparin went further in his studies of coacervates than just preparing them and observing their physical properties. In his more advanced work, he explored chemical reactions within the coacervates (Oparin 1966). This work was particularly well-summarized by Hanczyc and Fenchel (Fenchel 2002; Hanczyc 2009).

In an important experiment, a coacervate was prepared from gelatin and gum arabic so as to contain the phosphorylase enzyme. This was accomplished by exploiting the tendency of dissolved enzymes to become trapped inside coacervates. The phosphorylase enzyme catalyzes polymerization of glucose-1-phosphate into starch. When glucose-1-phosphate was added to the external medium of coacervate, it was first adsorbed and then incorporated into the coacervate. When this happened, the starch production occurred inside the coacervate. Phosphate, which is a by-product, diffused out of the coacervate. This important experiment demonstrated that a biologically relevant reaction can occur inside the coacervate. Further, while the enzyme phosphorylase and the starch molecule that was formed were trapped inside the coacervate, glucose-1-phosphate and phosphate could pass through the coacervate boundary, thus mimicking the cell membrane (Fenchel 2002; Hanczyc 2009).

In another experiment, a coacervate was prepared so as to contain two enzymes, phosphorylase and ß-amylase. Then, glucose-1-phosphate was added. As in the previously described experiment, starch formed inside the coacervate. However, since ß-amylase was present, it catalyzed the reaction of starch, in which maltose was released. Maltose diffused out of the coacervate into the external medium. This experiment modeled a two-step enzymatic pathway inside the coacervate (Hanczyc 2009).

Additional experiments were performed by Oparin to illustrate the ability of coacervates to act as chemical reactors for biologically important reactions and to mimic protocells (Oparin 1966, 1971). Such experiments include polyadenine synthesis, Nicotinamide Adenine Dinucleotide (NAD) redox reactions, and ascorbic acid oxidation (Hanczyc 2009). In one such experiment, the enzyme NADH (Nicotinamide Adenine

Dinucleotide H; the reduced form of NAD+) dehydrogenase was trapped inside the coacervate. Then, the redox dye methyl red and NADH were added to the external medium. When these substances diffused inside the coacervate, the NADH dehydrogenase catalyzed the reduction of methyl red, coupled to the oxidation of NADH to NAD. This experiment simulated a simple electron-transfer system (Fenchel 2002).

7.1.4 TYPES OF COMPONENTS OF OPARIN'S COACERVATES AND THEIR PREBIOTIC FAVORABILITY

First, details about the types of components of Oparin's coacervates are presented in Section 7.1.4.1. Then, prebiotic feasibility and favorability of these components are evaluated in Section 7.1.4.2.

7.1.4.1 CHEMICAL FEATURES OF THE COMPONENTS OF OPARIN'S COACERVATES

Most often, Oparin used the gelatin-gum arabic coacervate, but he also prepared numerous other types of coacervates in his laboratory (Oparin 1966). These are listed in Table 7.1.1.

In most of his experiments, Oparin followed a straightforward formula for the formation of complex coacervates, in which two biological macromolecules with a potential of

TABLE 7.1.1

Types of Coacervates Prepared in Oparin's Laboratory

Protein-Carbohydrate
Gelatin-gum arabic
Serum albumin-gum arabic
Histone-gum arabic
Salmon protamine-gum arabic
Sturgeon protamine-gum arabic
Clupein-gum arabic

Protein-Protein
Histone-serum albumin
Histone-gelatin
Clupein-gelatin

Protein-Nucleic acid
Clupein-RNA
Clupein-DNA
Histone-RNA
Histone-DNA

Protein-Carbohydrate-Nucleic Acid
Gelatin-gum arabic-DNA
Gelatin-gum arabic-RNA

Protein-Lipid
Gelatin-oleate

Multi-component Coacervate
Phosphorylase-histone-starch-gum arabic etc.

Source: Oparin, A. I., *Genesis and Evolutionary Development of Life* (published in Russian), Translated by E. Maass, English version published by Academic press in 1968, Academic Press, New York, 1966.

having opposite charges were mixed, and the pH was adjusted until the opposite charges were established.

Details about the types of Oparin's coacervates from Table 7.1.1 are given in the following paragraphs.

Gelatin-gum arabic coacervate consists of the protein and carbohydrate parts. Gelatin is a protein mixture formed by a partial hydrolysis of collagen extracted from animal skin, bones, cartilage, and ligaments (Hanczyc 2009). Gum arabic is an extract from the Acacia trees. It consists of a complex and variable mixture of arabinogalactan oligosaccharides, polysaccharides, and glycoproteins (Hanczyc 2009). Oparin also used serum albumin as another protein. In addition, he used histones, protamine, and clupein as biological proteins. These contain basic amino acids. Other constituents of his coacervates included nucleic acids RNA and DNA and the lipid component consisting of oleic acid.

7.1.4.2 Prebiotic Feasibility and Favorability of the Components of Oparin's Coacervates

What is in common to all Oparin's coacervates is that they are composed of complex biological macromolecules, which also have, in most instances, variable structures. Such complex structures were not feasible on the prebiotic Earth. This is the case with the oligo- and polysaccharides and other sugar-containing macromolecular structures, such as glycoproteins, RNA, and DNA. Complex carbohydrates that Oparin used cannot be considered as prebiotic materials, since there are no known simple polysaccharide syntheses under the simulated prebiotic conditions (Kolb et al. 2012; Kolb 2015, 2016). For the protein components that Oparin used, one may consider their simpler, prebiotically feasible versions. For example, instead of a protein, one could imagine a prebiotic proteinoid, which could be formed by condensation of prebiotic amino acids. Prebiotically feasible syntheses of such proteinoids have been described (Fox and Harada 1958; Fox 1964). However, capability of such proteinoids for making coacervates has not been evaluated. Oparin also used proteins containing basic amino acids. Such amino acids do not appear to be prebiotically favored, based on the report by McDonald and Storrie-Lombardi (2010). These authors concluded that lysine and arginine would not have been abundant on the prebiotic Earth, since most attempts to synthesize these amino acids under various prebiotic conditions were not successful, and they were not identified in the carbonaceous meteorites. The authors assumed that histidine, which is also classified as a basic amino acid since its imidazole side chain can act as a base, would behave mostly as neutral within proto-peptides under the plausible pHs of the prebiotic environments. Thus, the proteinoids, which would be based on the basic amino acids, do not appear to be prebiotically favored.

Most options from Table 7.1.1 include complex biomolecules whose prebiotic synthesis is not feasible or is not favored, because of the presence of carbohydrates or basic amino acids, respectively. Only one case, that of protein-lipid (Simoneit et al. 2007), could be potentially prebiotic in principle; however, the prebiotic versions of these components have not been tested for their coacervate-forming ability.

Considering this analysis, questions arise if coacervates can be made of molecules that are simpler than those that were chosen by Oparin and if such molecules could conceivably be synthesized under the simulated prebiotic conditions. A single compound, dioctyl sodium sulfosuccinate (AOT), an amphiphile, has been found to fulfill these requirements (Kolb et al. 2012; Kolb 2015, 2016), but there may be many others. For example, zwitterionic geminis, which are also single organic compounds, have been shown to make coacervates readily (Peresypkin and Menger 1999). However, their prebiotic feasibility has not been studied.

The AOT case is discussed in Section 7.1.5.

7.1.5 AN AOT-BASED PREBIOTICALLY FEASIBLE COACERVATE THAT CAN SERVE AS A CHEMICAL REACTOR

AOT, which is chemically dioctyl sodium sulfosuccinate, can make coacervates and was used in the study of the coacervate structure and properties (Menger and Sykes 1998). The chemical structure of AOT is shown in Figure 7.1.1.

Chemical features of AOT molecule comprise the branched alkyl chains, ester of succinic acid, and a sulfonate group. All these functional groups and their combinations have been found in various organic compounds on carbonaceous meteorites (Peltzer et al. 1984; Cronin and Pizzarello 1990; Cooper et al. 1992; Cronin and Chang 1993; Cronin et al. 1995; Cooper and Chang 1995; Cooper et al. 2001; Pizzarello and Huang 2002; Sephton 2002, 2005; Fuller and Huang 2003; Task Group 2007; Schmitt-Kopplin et al. 2010). Specifically, carboxylic acids are the most abundant free compounds in Murchison meteorite (Sephton 2005). Succinic acid is common on meteorites such as Murchison (Peltzer et al. 1984) and Tagish Lake (Pizzarello and Huang 2002). Organic sulfonates have been extracted from Murchison (Cooper and Chang 1995). Branched alkyl chains, in general, are more prevalent than the straight chains (Cronin and Pizzarello 1990). Organic alcohols are also found (Task Group 2007). Esters can be readily made by the solid-state processes on asteroids, which are the parent bodies of meteorites (Kolb 2012). Thus, AOT is a potentially prebiotic molecule (Kolb et al. 2012; Kolb 2015, 2016), although it has yet to be prepared in the laboratory under the simulated prebiotic conditions.

Preparation of an AOT coacervate in a form of two layers, the coacervate and the equilibrium liquid, was performed (Kolb et al. 2012). Figure 7.1.2 shows the appearance of such a coacervate.

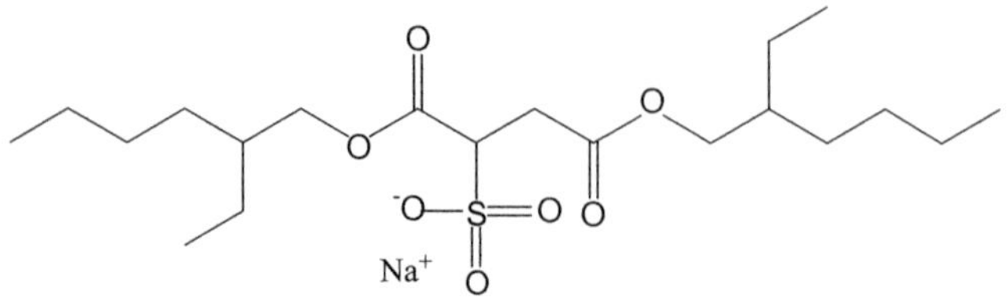

FIGURE 7.1.1 Chemical structure of AOT.

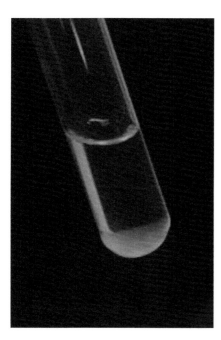

FIGURE 7.1.2 A photograph of the AOT coacervate showing the coacervate layer on the bottom and the equilibrium layer on the top.

Next, a chemical reaction was performed in the coacervate to mimic Oparin's experiments inside the coacervates. The reaction was chosen to fulfill the criteria of being prebiotically feasible, to be easy to perform, and to allow for an easy observation of the product. The Passerini reaction (Hooper and DeBoef 2009) was chosen (Kolb et al. 2012; Kolb 2015, 2016). Chemical equations for a typical Passerini reaction are shown in Figure 7.1.3.

The Passerini reaction is a multicomponent reaction in which water-insoluble starting materials react quickly and quantitatively in water to give a single product, which is also water-insoluble. The reaction is facilitated by the hydrophobic effects, in which the organic starting materials are driven toward each other to escape water. This brings them closer, which results in a fast and selective reaction. This represents one of many cases of the so-called "on-water" reactions (Lubineau et al. 1994; Pirrung and Sarma 2004; Narayan et al. 2005; Pirrung et al. 2008). Such reactions have only recently been studied for their prebiotic potential. (They are further covered in Section V of this handbook.) Starting materials for the Passerini reaction are all prebiotically feasible, based on the occurrence of such types of compounds in the meteorites. In addition, the product has a good potential to react further and contribute to the prebiotic chemical diversity. The Passerini reaction is quite fast and can be followed visually by observing the formation of a white solid product. When the mixture of the starting materials is introduced into the coacervate, a white product forms almost instantly. This is shown in Figure 7.1.4.

The Passerini reaction product was removed from the coacervate and identified as the correct product by its melting point and the Infra Red (IR) spectrum. The Passerini reaction was also run and observed and the product identified in the same way as in the Oparin's gelatin-gum arabic coacervate. This showed that the reaction is not dependent on the coacervate structure, at least not for these two examples.

In the conclusion of this section, the experiments done with AOT demonstrate that a prebiotically feasible coacervate can be found and that a prebiotically feasible reaction can be run in it (Kolb et al. 2012; Kolb 2015, 2016).

Benzaldehyde   *tert*-**Butylisocyanide**   **Benzoic acid**

Benzoic acid, *tert*-butylcarbamoyl-phenyl-methyl ester

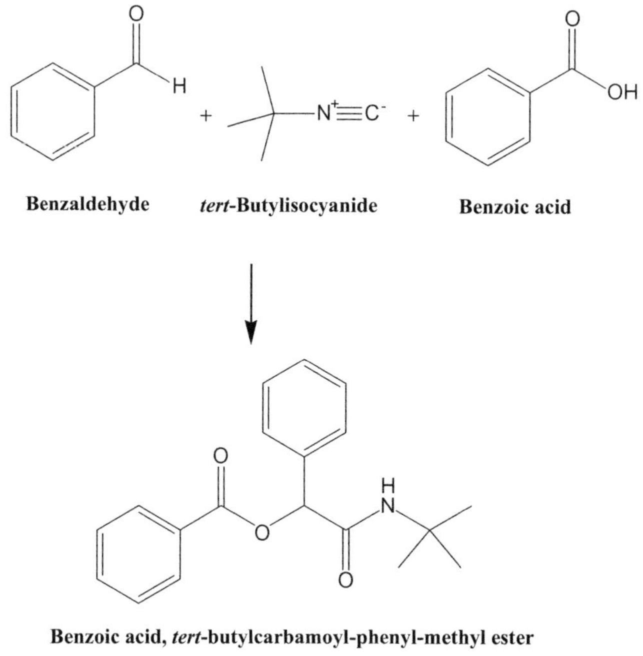

FIGURE 7.1.3 Chemical equations for the Passerini reaction.

FIGURE 7.1.4 A photograph of the AOT coacervate with the white solid product of the Passerini reaction in it.

7.1.6 OPARIN'S COACERVATE MODEL FOR PROTOCELLS: FROM THE ACCEPTANCE TO THE REJECTION AND BACK TO THE ACCEPTANCE

During the period in which Oparin proposed his coacervates and had done the experimental work with them, coacervates were considered a prebiotically plausible way to form protocells. This notion was supported by different types of experiments by Oparin and his co-workers and later by his followers (Oparin 1938, 1966, 1971; Evreinova et al. 1973, 1974, 1975, 1977; Gladilin et al. 1978; Oparin and Gladilin 1980). The types of such experiments were described in Section 7.1.3.

The proposal of coacervates as prebiotic protocells was revisited several times, notably by Walde et al. (1994). These authors briefly addressed Oparin's coacervates and then proposed that micelles and vesicles were better suited for self-reproduction than coacervates, since the former are, ostensibly, more stable. Walde et al. doubt the prebiotic importance of Oparin's coacervates:

"Oparin's reactions with coacervates represent an important reference in the field of the chemistry of life, but somehow their importance seems to be more historical than scientific. This is due to the fact that coacervates are generally regarded as dubious structures; in fact, they are devoid of thermodynamic stability (they generally precipitate in a short time and build a water-free layer...) and as such they are difficult to synthesize in terms of structure or of physicochemical properties" (Walde et al. 1994).

This claim that lack of the "water-free" layer of coacervates is not consistent with the accepted understanding of coacervates, by which a coacervate is in fact an aqueous layer. Further, coacervates are not devoid of thermodynamic stability. Still, the grim picture of "dubious" coacervates, as presented by Walde et al., as well as these authors' emphasis on micelles, may have inhibited further research on the prebiotic role of coacervates. It also appears that prebiotic chemists might have been unaware of the vigorous research on coacervates unrelated to the prebiotic objectives. Such research resulted in voluminous literature on coacervates in the fields of study such as colloids, polymers, physical chemistry, and pharmaceuticals (e.g., Singh and Burgess 1989; Burgess 1990; Burgess et al. 1991; Burgess and Singh 1993; Rabiskova et al. 1994; Menger and Sykes 1998; Menger et al. 2000; Menger 2011; Devi et al. 2017). Modern work on coacervates repeatedly demonstrated, by a variety of characterization methods, including the instrumental techniques, that coacervates are by no means "dubious" structures but are real and measurable, as shown in a recent review (Devi et al. 2017), among other sources. Also, coacervates can be quite stable. A statement that micelles are more stable than coacervates should not be generalized. In fact, molecules depart and reenter micelles at the microsecond timescale, and a mere dilution can destroy micelles instantly.

In a fascinating reversal of the micelles' fortunes, various authors have shown that the micelles can transform into coacervates under some conditions (Stuart et al. 1998; Wang et al. 1999, 2000; Dubin et al. 2008). For example, when simple salts are added to certain micelles above their critical micelle concentration, coacervates form in a stepwise process. At the beginning, micelles aggregate to form submicroscopic clusters, which then coalesce first into droplets and then into two separate layers, thus producing a coacervate. Therefore, to focus only on micelles (e.g., Walde et al. 1994; Luisi 1996), while disregarding coacervates, was becoming increasingly difficult to justify.

Recent work on coacervates as models for protocells has remedied this problem. Examples include work by Szostak and his co-workers (Jia et al. 2014) and Mann and his research group (Mann 2012; Li et al. 2014; Kumar et al. 2016; Martin et al. 2016; Qiao et al. 2017; Rodriguez-Arco et al. 2017). These recent works have shown that coacervates are excellent models for protocells and complex cell-like behavior. Coacervates have also been studied recently as models of membraneless organelles by Keating and her co-workers (Aumiller Jr. et al. 2016). Their work has also been quite fruitful. However, this modern work is focused on coacervates' ability to model protocells and membraneless organelles, without examining prebiotic feasibility of the coacervate components. Hopefully, the latter examination will be realized in the future.

7.1.7 SUMMARY

Oparin proposed coacervates as models for the primordial protocells. Coacervates absorb nutrients from the environment, a process that causes their growth. At some point, due to their thermodynamic instability or mechanical breakage, coacervates split into the daughter cells. This represents a primitive self-replication. Coacervates were initially selected based on the speed of their growth and self-replication. When food in the environment became scarce, coacervates that had the ability to synthesize their own food were selected. This provided a pathway to a primitive metabolism. Chemical features of the constituents of Oparin's coacervates were examined for their prebiotic feasibility and favorability. Modern developments in prebiotic chemistry do not lend much support to the latter for most constituents of Oparin's coacervates. Examples include complex carbohydrates and proteins that have basic amino acids. However, a simple amphiphile, AOT, which makes coacervates readily, is prebiotically feasible and has been proposed as a prebiotic substitute for the original components of the Oparin's coacervates. Further, the Passerini multicomponent reaction, a prebiotically feasible reaction, has been shown to occur inside the AOT coacervate, as well as in the Oparin's gelatin-gum arabic coacervate. This supports the Oparin's proposal that coacervates can act as primordial chemical reactors. Oparin's model of coacervates as protocells has gone in and out of favor over a period of time but has recently been revived by vigorous research demonstrating that coacervates are excellent models for protocells and also for the membraneless organelles. However, in these recent studies, chemical constituents of the coacervates are not prebiotically feasible.

DEDICATION

This chapter is dedicated to the memory of A. I. Oparin, whom I never met, but whose 1924 pamphlet on the origin of life influenced me to become a chemist when I was only 13 years old and an astrobiologist later in life. I felt inspired by the power of scientific explanations ever since I read his precious book, a small tattered paperback, which is now lost somewhere or perhaps was found by another interested 13-year-old.

REFERENCES

Aumiller, Jr., W. M., F. P. Cakmak, B. W. Davis, and C. D. Keating. 2016. RNA-based coacervates as a model for membraneless organelles: Formation, properties, and interfacial liposome assembly. *Langmuir* 32: 10042–10053.

Burgess, D. J. 1990. Practical analysis of complex coacervate systems. *J. Colloid. Interface Sci.* 140: 227–238.

Burgess, D. J., K. K. Kwok, and P. T. Megremis. 1991. Characterization of albumin-acacia complex coacervation. *J. Pharm. Pharmacol.* 43: 232–236.

Burgess, D. J., and O. N. Singh. 1993. Spontaneous formation of small sized albumin/acacia coacervate particles. *J. Pharm. Pharmacol.* 45: 586–591.

Cooper, G. W., and S. Chang. 1995. Isotopic measurements of organic sulfonates from the Murchison meteorite. *Lunar and Planetary Institute Meeting, LPS XXVI* 281.

Cooper, G. W., W. M. Onwo, and J. R. Cronin. 1992. Alkyl phosphonic acids and sulfonic acids in the Murchison meteorite. *Geochim. Cosmochim. Acta* 56: 4109–4115.

Cooper, G., N. Kimmich, W. Belisle, J. Sarinana, K. Brabham, and L. Garrel. 2001. Carbonaceous meteorites as a source of sugar-related organic compounds for the early Earth. *Nature* 414: 879–883.

Cronin, J. R., and S. Chang. 1993. Organic matter in meteorites: Molecular and isotopic analysis of the Murchison meteorite. In *The Chemistry of Life's Origins*, J. M. Greenberg, C. X. Mendoza-Gomez, and V. Pirronello (Eds.), pp. 209–258. Dordrecht, the Netherlands: Kluwer Academic Publ.

Cronin, J. R., G. W. Cooper, and S. Pizzarello. 1995. Characteristics and formation of amino acids and hydroxyl acids of the Murchison meteorite. *Adv. Space Res.* 15: 91–97.

Cronin, J. R., and S. Pizzarello. 1990. Aliphatic hydrocarbons of the Murchison meteorite. *Geochim. Cosmochim. Acta* 54: 2859–2868.

Devi, N., M. Sarmah, B. Khatun, and T. K. Maji. 2017. Encapsulation of active ingredients in polysaccharide-protein complex coacervates. *Adv. Colloid. Interface Sci.* 239: 136–145.

Dubin, P. L., Y. Li, and W. Jaeger. 2008. Mesophase separation in polyelectrolyte-mixed micelle coacervates. *Langmuir* 24: 4544–4549.

Evreinova, T. N., T. W. Mamontova, V. N. Karnauhov, S. B. Stephanov, and U. R. Hrust. 1974. Coacervate systems and origins of life. *Orig. Life* 5: 201–205.

Evreinova, T. N., E. V. Mel'nikova, A. G. Pogorelov, and B. L. Allakhverdov. 1975. Stabilization and coexistence of coacervate systems of different chemical composition. *Dokl. Akad. Nauk SSSR* 224: 239–241.

Evreinova, T. N., S. B. Stefanov, and T. V. Marmontova. 1973. Structure of coacervate drops in an electron microscope. *Dokl. Akad. Nauk SSSR* 208: 243–244.

Evreinova, T. N., E. V. Mel'nikova, B. L. Allakhverdov, and V. N. Karnaukhov. 1977. Coexistence of coacervate systems with different chemical composition. *Zhurnal Evolyutsionnoi Biokhimii i Fiziologii* 13: 330–335.

Fenchel, T. 2002. *Origin and Early Evolution of Life*, pp. 19–21. Oxford, UK: Oxford University Press.

Fox, S., and K. Harada. 1958. Thermal copolymerization of amino acids to a product resembling protein. *Science* 128 (3333): 1214.

Fox, S. W. 1964. Thermal polymerization of amino-acids and production of formed microparticles on lava. *Nature* 201: 336–337.

Fuller, M., and Y. Huang. 2003. Quantifying hydrogen-deuterium exchange of meteoritic dicarboxylic acids during aqueous extraction. *Meteorit. Planet. Sci.* 38: 357–363.

Gladilin, K. L., A. F. Orlovsky, D. B. Kirpotin, and A. I. Oparin. 1978. Coacervate drops as a model for precellular structures. *Origins of Life, Proceedings of ISSOL Meeting, 2nd*, Tokyo, Japan, pp. 357–362.

Hanczyc, M. M. 2009. The early history of protocells: The search for the recipe of life. In *Protocells: Bridging Nonliving and Living Matter*, S. Rasmussen, M. A. Bedau, L. Chen et al. (Eds.), pp. 3–17. Cambridge, MA: The MIT Press.

Hooper, M. H., and B. DeBoef. 2009. A Green multicomponent reaction for the organic chemistry laboratory. The aqueous Passerini reaction. *J. Chem. Ed.* 86: 1077–1079.

Jia, T. Z., C. Hentrich, and J. W. Szostak. 2014. Rapid RNA exchange in aqueous two-phase system and coacervate droplets. *Orig. Life Evol. Biosph.* 44: 1–12.

Kolb, V. M. 2012. On the applicability of solventless and solid state reactions to the meteoritic chemistry. *Int. J. Astrobiol.* 11: 43–50.

Kolb, V. M. 2015. Oparin's coacervates as an important milestone in chemical evolution. In *Instruments, Methods, and Missions for Astrobiology XVII*, R. B. Hoover, G. V. Levin, A. Yu. Rozanov, and N. C. Wickramasinghe (Eds.). *Proceedings of SPIE*, Vol. 9606. doi:10.1117/12.2180604.

Kolb, V. M. 2016. Origins of life: Chemical and philosophical approaches. *Evol. Biol.* 43: 506–515.

Kolb, V. M., M. Swanson, and F. M. Menger. 2012. Coacervates as prebiotic chemical reactors. In *Instruments, Methods, and Missions for Astrobiology XV*, R. B. Hoover, G. V. Levin, and A. Y. Rozanov (Eds.). *Proceedings of SPIE*, Vol. 8521, 85210E. doi:10.1117/12.928550.

Kumar, R. K., R. L. Harniman, A. J. Patil, and S. Mann. 2016. Self-transformation and structural reconfiguration in coacervate-based protocells. *Chem. Sci.* 7: 5879–5887.

Li, M., X. Huang, T. Y. D. Tang, and S. Mann. 2014. Synthetic cellularity based on non-lipid micro-compartments and protocell models. *Curr. Opin. Chem. Biol.* 22: 1–11.

Lubineau, A., J. Augé, and Y. Queneau. 1994. Water-promoted organic reactions. *Synthesis* 8: 741–760.

Luisi, P. L. 1996. Self-reproduction of micelles and vesicles: Models for the mechanisms of life from the perspective of compartmented chemistry. *Adv. Chem. Phys.* 92: 425–438.

Mann, S. 2012. Systems of creation: The emergence of life from nonliving matter. *Acct. Chem. Res.* 45: 2131–2141.

Martin, N., M. Li, and S. Mann. 2016. Selective uptake and refolding of globular proteins in coacervate microdroplets. *Langmuir* 32: 5881–5889.

Menger, F. M. 2011. Remembrances of self-assemblies past. *Langmuir* 27: 5176–5183.

Menger, F. M., A. V. Peresypkin, K. L. Caran, and R. P. Apkarian. 2000. A sponge morphology in an elementary coacervate. *Langmuir* 16: 9113–9116.

Menger, F. M., and B. M. Sykes. 1998. Anatomy of a coacervate. *Langmuir* 14: 4131–4137.

McDonald, G. D., and M. C. Storrie-Lombardi. 2010. Biochemical constraints in a protobiotic Earth devoid of basic amino acids: The "BAA (–) World."*Astrobiology* 10: 989–1000.

Narayan, S., J. Muldoon, M. G. Finn, V. V. Fokin, H. C. Kolb, and K. B. Sharpless. 2005. "On water": Unique reactivity of organic compounds in aqueous suspension. *Angew. Chem. Int. Ed.* 44: 3275–3279.

Oparin, A. I. 1924. *The Origin of Life* (published in Russian). Translated by A. Synge. English version published in 1994. In *Origins of life: The Central Concepts*, D. W. Deamer and G. P. Fleischaker (Eds.), pp. 31–71. Boston, MA: Jones and Bartlett.

Oparin, A. I. 1938. *Origin of Life*. Translated by S. Morgulis. Republication of the original publication by Macmillan Company by Dover Publications Inc. 2nd ed. 1965. New York: Dover.

Oparin, A. I. 1966. *Genesis and Evolutionary Development of Life* (published in Russian). Translated by E. Maass. English version published by Academic press in 1968. New York: Academic Press.

Oparin, A. I. 1971. Routes for the origin of the first forms of life. *Sub-Cell. Biochem.* 1: 75–81.

Oparin, A. I., and K. L. Gladilin. 1980. Evolution of self-assembly of probionts. *Biosystems* 12: 133–145.

Peltzer, E. T., J. L. Bada, G. Schesinger, and S. L. Miller. 1984. The chemical conditions on the parent body of the Murchison meteorite: Some conclusions based on amino, hydroxy and dicarboxylic acids. *Adv. Space Res.* 4: 69–74.

Peresypkin, A. V., and F. M. Menger. 1999. Zwitterionic geminis. Coacervate formation from a single organic compound. *Org. Lett.* 1: 1347–1350.

Pirrung, M. C., and K. D. Sarma. 2004. Multicomponent reactions are accelerated in water. *J. Amer. Chem. Soc.* 126: 444–445.

Pirrung, M. C., K. D. Sarma, and J. Wang. 2008. Hydrophobicity and mixing effects on select heterogeneous, water-accelerated synthetic reactions. *J. Org. Chem.* 73: 8723–8730.

Pizzarello, S., and Y. Huang. 2002. Molecular and isotopic analyses of Tagish Lake alkyl dicarbocylic acids. *Meteorit. Planet. Sci.* 37: 678–696.

Qiao, Y., R. Booth, and S. Mann. 2017. Predatory behavior in synthetic protocell communities. *Nat. Chem.* 9: 110–119.

Rabiskova, M., J. Song, F. O. Opawale, and D. J. Burgess. 1994. The influence of surface properties on uptake of oil into complex coacervate microcapsules. *J. Pharm. Pharmacol.* 46: 631–635.

Rodriguez-Arco, L., M. Li, and S. Mann. 2017. Phagocytosis-inspired behavior in synthetic protocell communities of compartmentalized colloidal objects. *Nat. Mater.* 16: 857–864.

Schmitt-Kopplin, P., Z. Gabelica, R. D. Gougeon et al. 2010. High molecular diversity of extraterrestrial organic matter in Murchison meteorite revealed 40 years after its fall. *Proc. Natl. Acad. Sci. U.S.A.* 107: 2763–2768.

Sephton, M. A. 2002. Organic compounds in carbonaceous meteorites. *Nat. Prod. Rep.* 19: 292–311.

Sephton, M. A. 2005. Organic matter in carbonaceous meteorites: Past, present and future research. *Philos. Trans. R. Soc. A* 363: 2729–2742.

Simoneit, B. R. T., A. I. Rushdi, and D. W. Deamer, 2007. Abiotic formation of acylglycerols under simulated hydrothermal conditions and self-assembly properties of such lipid products. *Adv. Space Res.* 40: 1649–1656.

Singh, O. N., and. D. J. Burgess. 1989. Characterization of albumin-alginic acid complex coacervation. *J. Pharm. Pharmacol.* 41: 670–673.

Stuart, M. A. C., N. A. M. Besseling, and R. G. Fokking. 1998. Formation of micelles with complex coacervate cores. *Langmuir* 14: 6846–6849.

Task Group on Organic Environment in the Solar System, National Research Council. 2007. Ch. 3 Meteorites. In *Exploring Organic Environments in the Solar System*, pp. 37–51. Washington, DC: The National Academies Press.

Walde, P., A. Goto, P.-A. Monnard, M. Wessicken, and P. L. Luisi. 1994. Oparin's reactions revisited: Enzymatic synthesis of poly(adenylic acid) in micelles and self-reproducing vesicles. *J. Amer. Chem. Soc.* 116: 7541–7547.

Wang, Y., K. Kimura, P. L. Dubin, and W. Jaeger. 2000. Polyelectrolyte-micelle coacervation: Effects of micelle surface charge density, polymer molecular weight, and polymer/surfactant ratio. *Macromolecules* 33: 3324–3331.

Wang, Y., K. Kimura, Q. Huang, P. L. Dubin, and W. Jaeger 1999. Effect of salt on polyelectrolyte-micelle coacervation. *Macromolecules* 32: 7128–7134.

Punam Dalai and Nita Sahai

CONTENTS

7.2.1   INTRODUCTION

The quest to understand the origins and evolution of life has led scientists to attempt the bottom-up assembly of an enclosed synthetic entity capable of performing both metabolism and self-replication functions (Figure 7.2.1).

The common features of all forms of life include a membrane enclosing the individual cell, self-replication, metabolism, and the use of homochiral molecules. The origin of the homochiral nature of life is a vast subject in its own right, so, in the present paper, we will focus on the emergence of the

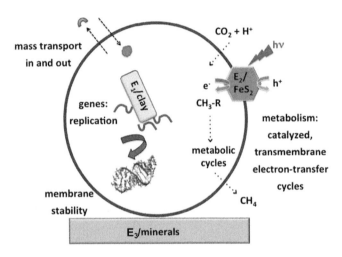

FIGURE 7.2.1   Schematic of model protocell with semi-permeable membrane and molecular machinery for replication and metabolism (Sahai et al. 2016). The protocell is shown at the mineral-water interface. Minerals (*e.g.*, montmorillonite clay, photocatalytic iron sulfide, and TiO_2) may have played the roles of prebiotic enzymes (E1–E3) in several functions. The role of minerals as reactants and catalysts in the OoL has been reviewed recently (Cleaves et al. 2012). Reproduced with permission from Consortium of Scientific Societies.

membrane, self-replication, and metabolism. The operative definition of life in the origins of life (OoL) field accepts the cell as the basic unit of life and the RNA, proteins, and membrane-forming molecules as the fundamental molecular building blocks of life (Kee and Monnard 2016). Simple enclosed bilayer membranes or vesicles composed of single-chain amphiphiles (SCAs) without encapsulated reactants were originally referred to as model protocells in the literature (Segré et al. 2001). Subsequently, the term was extended to include vesicle-peptide or vesicle-RNA intermediate states. In the present paper, the term "protocells" will be used to refer to simple lipid vesicles, to intermediate states, and to more complex entities. Indeed, Kee and Monnard (2016) have stated eloquently that, "one cannot speak of a protocell type, but rather a lineage of protocell systems that slowly evolved from simple self-assembled molecular systems towards pre-cellular entities capable of self-sustenance and self-replication."

In one school of thought in the OoL community, protocells are believed to have possessed membranes composed of SCAs rather than the phospholipids (PLs) typical of modern cells (Figure 7.2.2) (Deamer and Oró 1980; Oró et al. 1990; Monnard and Deamer 2002; Mansy et al. 2008; Albertsen et al. 2014; Dalai et al. 2016b; Fiore et al. 2017).

The prebiotic synthesis of PLs has been considered too complex (Hargreaves et al. 1977), whereas SCAs may have been present on early Earth because complex mixtures of organic compounds, including fatty acids (FAs), fatty alcohols, and polycyclic aromatic hydrocarbons (PAHs), have been discovered in carbonaceous chondrites that can form membrane like structures (Yuen and Kvenvolden 1973; Deamer and Pashley 1989) (Figure 7.2.3).

Single-chain amphiphiles have also been synthesized by Fischer–Tropsch type (FTT) reactions under simulated hydrothermal vent conditions (Nooner and Oró 1979; McCollom et al. 1999; Rushdi and Simoneit 2001; Foustoukos 2004;

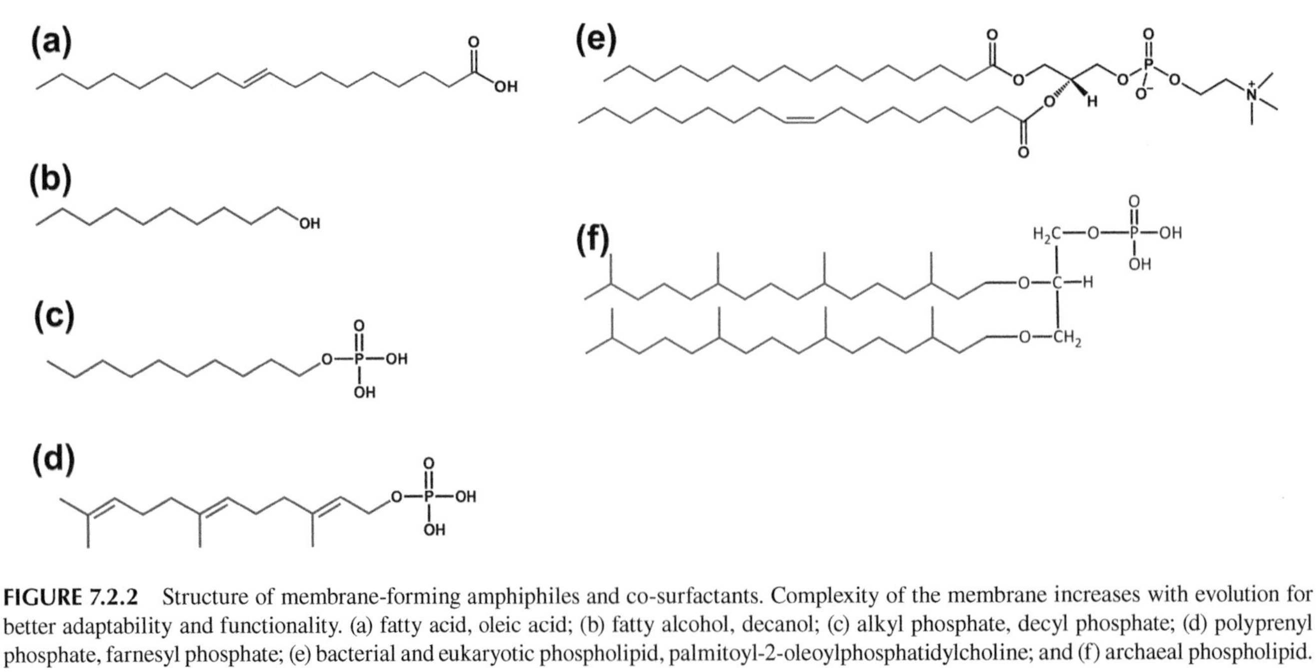

(a)

(b)

(c)

(d)

(e)

(f)

FIGURE 7.2.2   Structure of membrane-forming amphiphiles and co-surfactants. Complexity of the membrane increases with evolution for better adaptability and functionality. (a) fatty acid, oleic acid; (b) fatty alcohol, decanol; (c) alkyl phosphate, decyl phosphate; (d) polyprenyl phosphate, farnesyl phosphate; (e) bacterial and eukaryotic phospholipid, palmitoyl-2-oleoylphosphatidylcholine; and (f) archaeal phospholipid.

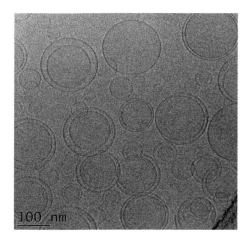

FIGURE 7.2.3 Cryo-transmission electron microscopy (TEM) image of synthetic FA vesicles resembling cell membranes in aqueous solution.

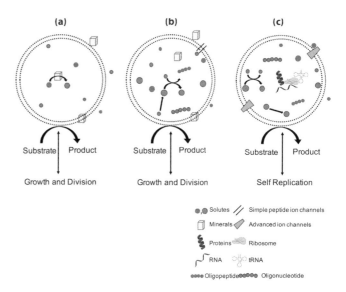

FIGURE 7.2.4 Schematic representation of model protocells of increasing complexity. (a) Simplified protocell that can only undergo metabolism; (b) protocell that can undergo metabolism and store genetic information; and (c) fully autonomous protocell with metabolism and replication.

McCollom 2013). Although the abiotic synthesis of PLs has been considered a major challenge in the OoL community, several studies have recently reported non-enzymatic synthesis of PLs and closely related compounds (Rao et al. 1982; Maheen et al. 2010; Patel et al. 2015; Devaraj 2017). The PL membranes into which self-reproducing catalysts are embedded are capable of further PL synthesis (Hardy et al. 2015). Thus, PLs and related compounds may have been present prebiotically, at least in trace amounts. Protocell membranes composed of amphiphilic peptides have also been proposed in the literature (Fox and Harada 1960; Fox 1980).

The survival, growth, and division of a protocell would have required mass and energy transfer across the membrane, presumably leading eventually to the emergence of advanced transmembrane transport machineries, metabolism, and self-replication. One may envisage several steps in this evolution of protocells (Figure 7.2.4).

First, SCAs produced on the early Earth by abiotic synthesis and/or exogenous delivery would have self-assembled into enclosed lipid-bound compartments (Figure 7.2.3). These vesicles or earliest protocells could encapsulate solute and colloids (molecules and mineral nanoparticles) from the environment. The solutes and colloids would be distributed both in the aqueous vesicle interior and in the hydrophobic space provided by the hydrocarbon tails of the membrane bilayer, depending on the relative hydrophilicity of the molecules. Subsequently, synergistic interactions between the encapsulated entities, membrane components, and extra-vesicular environment would eventually lead to the development of selective solute transport; autocatalytic networks (proteins and protoenzymes), resulting in energy transduction (proto-metabolism); and a rudimentary machinery (RNA and ribosomes) for self-replication. These processes would have been influenced by various environmental factors such as atmospheric and aqueous composition, temperature, and pressure.

In general, a requirement of any geochemical environment for the OoL is a disequilibrium system with sustained fluxes of mass and energy at least compared to the rates of organic

molecule synthesis and degradation. Many prebiotic syntheses require ultraviolet (UV) radiation, $^{*}OH$ free radicals or H_2O_2, lightning and wetting-drying cycles associated with solar energy, geothermal heat, and tidal action, which implies a sub-aerial or shallow sub-aqueous geochemical environment on Earth's surface. The prebiotic simple precursor compounds are envisioned to have been used by the protocell to build more complex polymers, eventually resulting in metabolism and self-replication machineries. Thus, a heterotrophic metabolism is suggested, often with the implicit assumption that the emergence of RNA and lipid membranes preceded metabolism, because some RNAs (e.g., ribozymes) can serve both genetic and catalytic roles. This is known as the "RNA world hypothesis" (Gilbert 1986; Joyce 2002; Bernhardt 2012). Other have argued against the RNA world and proposed that metabolism, which may be described as self-organizing autocatalytic biochemical cycles, emerged first. This is known as the "metabolism-first" or "protein world" hypothesis and generally implies a chemoautotrophic metabolism for protocells. Some compelling geochemical environments for a chemoautotrophic OoL include low-temperature, alkaline hydrothermal vents, also known as "white smokers," and high-temperature, acidic hydrothermal vents, also known as "black smokers" (Wächtershäuser 1988, 2000; Amend and Shock 2001). The large temperature, pH, and other chemical gradients between the hydrothermal vent fluid and ambient ocean water are proposed to provide the chemical and thermal energy for prebiotic organic molecule synthesis in the chemoautotrophic scenario. Despite tantalizing descriptions of the proposed steps involving mineral walls as the earliest "membranes" in the evolution of chemoautotrophic organisms at hydrothermal vents, few experimental studies exist, taking a bottom-up self-assembly approach to test these ideas; these studies provide rich opportunities for future investigations. In comparison, most experimental studies on protocell synthesis in the OoL field have focused on lipid-bound membranes in the heterotrophy perspective. In this chapter, therefore, we will focus primarily on a review of studies where organic molecules are believed to have formed protocell membranes.

We will first describe the thermodynamic requirements for the self-assembly of SCAs into aggregates with various morphologies (phases), including vesicles, followed by the effects of lipid chemistry and structure as well as environmental conditions such as solution composition (pH and dissolved ions or "salts"), temperature, and pressure on the stability of SCA vesicles as model protocells. We will then summarize studies of protocell growth and division and competition between different populations of protocells in the development of predator-prey relationships in a "protoecology." Examples of synergistic interactions between lipid membranes, solutes (inorganic ions, monomers, and short oligomers of nucleotides or amino acids), and mineral nanoparticles resulting in polymerization to longer oligomers, with the facility for folding and, hence, simple functions (*e.g.*, selective membrane transport and catalysis) will be presented. Finally, it will be proposed that increasing complexity leads to the co-emergence of ribosomes, transmembrane pumps, auto-catalytic networks (protometabolism), and self-replication. Before proceeding, we note that a large majority of experimental studies using lipid vesicles as model protocells have been conducted by proponents of the RNA-world hypothesis; far fewer experimental studies involving model protocells have examined the metabolism-first theory, utilizing, for example, catalytic peptides and transmembrane electron transfer reactions (TMETRs) to simulate protometabolism. This is reflected in the present review. As we will see below, a synergistic co-evolution of RNA, proteins, and lipids is a more likely scenario, given that a complex mixture of starting inorganic and organic compounds would have been present in any geochemical environment.

7.2.2 SELF-ASSEMBLING COMPARTMENTS

7.2.2.1 SINGLE-CHAIN AMPHIPHILES

Single-chain amphiphiles possess a polar headgroup and a single, hydrophobic FA tail. Under specific conditions of temperature and solvent, SCA molecules aggregate and self-assemble into various morphologies or phases (Chen and Walde 2010). The phases can transform into other phases when these conditions are changed. In aqueous solution, the polar headgroups are oriented toward the solution, whereas the tails prefer to associate with each other. The concentration at which SCAs aggregate is called the critical aggregate concentration (CAC) (Figure 7.2.5, Budin et al. 2012).

Below the CAC, the molecules exist as monomers. Above the CAC, if solution pH is much lower than the apparent pK_a of the headgroup, the aggregate forms a structureless clump called an oil droplet, in which the molecules are randomly arranged. At pHs >> apparent pK_a, the headgroups are deprotonated and form small (~10–50 nm), spherically structured aggregates

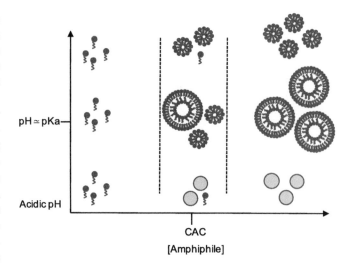

FIGURE 7.2.5 Schematic illustrating SCA phase transitions as a function of concentration and pH.

called micelles, which minimize headgroup-headgroup repulsion and maximize tail-tail interactions. A micelle has a single layer of amphiphilic molecules arranged in a sphere, with the headgroups oriented toward the aqueous solution and the hydrophobic tails in contact with each other. Around the apparent pK_a, the headgroups are partially protonated and can form large, spherical bilayer structures called vesicles. Vesicles may be unilamellar, which refers to a single bilayer, or multilamellar, which indicates that they possess multiple bilayers. The CAC is called the critical micelle concentration (CMC) or the critical vesicle concentration (CVC), depending on the phase regime of the system. The CAC is a thermodynamic property of a system.

For OoL protocell studies, vesicles are typically prepared by evaporating the chloroform solvent, which leaves behind a residue of a thin lipid film. The film is rehydrated with buffer solution to the desired concentration above the CVC, which causes the lipid to self-assemble into vesicles. The buffer pH is usually ~7–8.5 for FA self-assembly to yield vesicles. The suspension may be vortexed gently to break up lipid lumps and disperse the vesicles. Another way to prepare vesicles is to rehydrate the thin film with highly alkaline pH buffer (pH ~12), which results in the formation of micelles. This solution is then acidified to circum-neutral pH to yield vesicles. Both methods typically yield polydisperse (wide size distribution) vesicles that are predominantly in the 150- to 400-nm-size range, but a few are as large as 0.8–1.0 μm. The vesicles are also usually multilamellar. Extrusion through specific pore-size membrane produces unilamellar and more monodisperse vesicles.

If a specific solute such as a nucleotide, peptide, or fluorescent dye is to be encapsulated within the vesicles, it is added in the buffer during the dried-film rehydration step. The vesicle suspension is then passed through a separation (size exclusion) column, and fractions of vesicles that contain the solute are collected, as ascertained by Ultraviolet-Visible (UV-Vis) or fluorescent peaks for each fraction. The background solute is left behind on the column and elutes later.

7.2.2.2 STABILITY OF MODEL PROTOCELL MEMBRANES

It is assumed in the OoL literature that SCAs delivered exogenously by meteorites or endogenously synthesized at hydrothermal vents, if concentrated sufficiently to values above their CVCs, self-assembled to form vesicles in aqueous solution. Hydration-dehydration cycles simulating lagoons, saltmarshes, or tidal pools provide some geological environments where the dilution problem could be overcome by concentrating amphiphilic molecules, thus allowing protocell membrane self-assembly (Deamer and Barchfeld 1982). The stability of the protocell membrane would have depended on properties of the SCA as well as on environmental factors.

In OoL studies, stability is estimated in terms of the CVC and the integrity (or permeability) of the membrane to the solutes. While FA membranes demonstrate some behaviors such as growth and division, which are similar to modern PL membranes, FAs tend to have much higher CVCs (mM versus μM or nM). Also, SCA vesicles are much more permeable and sensitive to a wide range of environmental conditions than PL (especially phosphatidylcholine, which has a zwitterionic headgroup) membranes. It is expected that, as on modern Earth, the early Earth conditions would have spanned a wide range of dissolved ion concentrations, speciation, ionic strengths, pHs, and temperature conditions.

7.2.2.2.1 Effect of Acyl Chain on Critical Vesicle Concentration

Acyl chain length and the degree of unsaturation of the hydrocarbon chain are key factors in determining the CVC of a lipid at a given temperature. Short-chain FAs up to about C_{10}–C_{12} form bilayers at very high concentrations compared with longer-chain FAs with $\geq C_{18}$. For example, the CVC of decanoic acid (DA) is 20–40 mM compared with 20–200 μM for oleic acid (OA). For a given chain length, the CAC increases as the amount of unsaturation in the chain increases. The CVC of OA (C_{18}, 1 C=C) is 0.02–0.2 mM and that of the linolenic acid (LA, C_{18}, 3 C=C) is ~1 mM (Teo et al. 2011). Unsaturation increases the bilayer fluidity and mobility. Vesicle size also decreases significantly as the degree of unsaturation increases. The average sizes of OA and LA vesicles were ~200 nm and ~150 nm, respectively, after 5 days (Teo et al. 2011).

7.2.2.2.2 Effect of Co-surfactants on Critical Vesicle Concentration

Amphiphiles with tails longer than 10 carbon atoms have not been detected in meteorites. Although SCAs with tails up to C_{22} have been synthesized in simulated hydrothermal vent conditions, concentrating the lipid sufficiently to form vesicles may be difficult in an ocean environment. Hence, short-chain FA ($\leq C_{10}$) might be the appropriate choice for the earliest cell membranes. The presence of pure amphiphiles vesicles was probably unlikely in the natural environment, and mixed amphiphile vesicles are a more realistic approach for model protocells. The introduction of co-surfactants, such as fatty alcohols, fatty amines, and fatty glycerol monoesters, to FAs can significantly increase vesicle stability, as reflected

in a lower value of the CVC. Mixtures of FA of varying chain lengths also have lower CVC than the corresponding pure components. The CVC of pure DA is 20–40 mM, whereas the CVC is only to ~1 mM in the presence of 1-decanol (DOH) in 2:1 molar ratio of DA-DOH, thus allowing vesicles to form at a lower concentration (Maurer and Monnard 2009; Maurer et al. 2009; Sahai et al. 2017). A mixture of DA-OA in 95:5 molar ratio forms vesicles at 15 mM, which is significantly lower than that for DA alone (Budin et al. 2014).

The geological environments in which protocells would have self-assembled by wetting-drying cycles, which is required to sufficiently concentrate the lipids, would also have resulted in the intermittent concentration of ions dissolved in solution. Thus, high ionic concentrations would have been present even in sub-aerial aqueous environments. In such environments, as well as in salty oceans, the ability to form bilayers could have decreased due to the generally deleterious effects of salts, especially multivalent cations (valence >1), on lipid self-assembly. Happily, the stability of vesicles against higher ionic strength, pH, and extreme temperatures is also enhanced by co-surfactant addition, as summarized in Sections 7.2.2.2.3–7.2.2.2.5.

7.2.2.2.3 Salt Tolerance

Vesicles constitute a separate phase from water and stay in suspension. For charged headgroup SCAs, the vesicles remain suspended partly because of repulsion between vesicles. In the presence of dissolved ions, especially monovalent cations, charge screening allows vesicles to approach close enough, so that van der Waals interactions may take over and the vesicles may form amorphous aggregates and the lipid settles out of solution. Additionally, divalent cations may bind directly to negatively charged FA headgroups, thus causing charge neutralization and settling out of the lipids.

Modern seawater has mildly alkaline pH (~8.5) and an ionic strength of 0.5 M from the presence dissolved ions, mainly Na^+, Cl^-, Mg^{2+}, and SO_4^{2-}. In contrast, freshwater has a neutral pH and ionic strength of only about 0.1–1 mM, mainly due to Ca^{2+}, HCO_3^-, and soluble silica. The early oceans would have been enriched in dissolved Fe^{2+} from hydrothermal vents and Mg^{2+} from weathering of komatiite, basalt, and peridotite (Schoonen et al. 2004; Schoonen and Smirnov 2016). It is estimated that the early oceans were twice more saline than the present oceans (Knauth 2005) and were probably slightly acidic (pH 6) because of the equilibrium with high atmospheric level of carbon dioxide. Surface water would also have been acidic (Schoonen et al. 2004; Schoonen and Smirnov 2016). Evaporative environments on land would have concentrated all salts to even higher levels. The extensive presence of monovalent and divalent ions has also been detected directly on Mars, and the presence of a subsurface, briny ocean on Enceladus has been inferred indirectly.

Dissolved ions are required for various biological processes such as RNA polymerization, RNA folding, Na^+/H^+ transmembrane pumping involved in ATP generation, and co-factors of various enzymes. In the OoL context, magnesium is one of the most important divalent cations, as it is required at high concentrations (~75 mM) for both prebiotic, non-enzymatic, templated as well as prebiotic clay-catalyzed polymerization of RNA nucleotides (Ferris et al. 1996; Joshi et al. 2009; Joshi and Aldersley 2013; Adamala and Szostak 2013b). Recently, it has been demonstrated that Fe^{2+} can catalyze single-electron-transfer reactions by ribosomal RNA in anoxic atmosphere and at mildly acidic pH of 6.5 (Hsiao et al. 2013). Unfortunately, divalent ions have a deleterious effect on pure FA membranes, even at millimolar concentrations, owing to the interaction between deprotonated headgroups of FAs and the cation (Monnard et al. 2002; Chen et al. 2005). Hence, it is critical to understand the stability of protocell membranes in briny conditions.

Mixed SCA membranes are more robust than pure FA vesicles. The increased stability of mixed FA vesicles, consisting, for example, of DA-DOH or of DA-DN (decylamine), is due to hydrogen bonding provided by the co-surfactant to deprotonated carboxylate headgroup of FA (Monnard and Deamer 2002; Monnard et al. 2002). Moreover, addition of fatty alcohol co-surfactant decreases the overall negative charge on the bilayer, thus decreasing the electrostatic interaction between FA headgroup and cations. Decanol, when added to DA (DA-DOH 37:1 mM), decreases its sensitivity against monovalent cation (Na^+ and K^+) up to 400 mM; however, divalent cations (Mg^{2+} and Ca^{2+}) still cause the precipitation of such a system at ion concentrations of only 1–5 mM (Monnard et al. 2002). Similarly, the addition of DN to DA forms more salt-tolerant vesicles up to 100 mM $MgCl_2$ and $CaCl_2$ concentrations at acidic (pH 11) conditions (Namani and Deamer 2008), but the prebiotic plausibility of DN is not known. The increased stability of mixed DA-DN vesicles is because of the charge-charge repulsion between the cations and positively charged DN. Similarly, DA and glycerol monodecanoate (GMD) in 2:1 mixture was tolerant up to 100 mM Na^+ and 1 mM each of Mg^{2+} and Ca^{2+} (Monnard et al. 2002; Chen et al. 2005). Chen et al. (2005) observed that 4:1 mixtures of myristoleate (MA, C_{14}) and glycerol monomyristoleate (GMM) increased the vesicle stability against Mg^{2+} to 2–3 mM in contrast to MA alone (only 0.5 mM $MgCl_2$), as determined by dye leakage and turbidity assay.

Fatty acid (OA, MA-GMM [2:1] and DA-DOH-GMD [4:1:1]) vesicles could be protected from the disruptive effects of Mg^{2+} cation effects (up to 50 mM $MgCl_2$) in the presence of citrate (200 mM), where citrate chelated the Mg^{2+} (Adamala and Szostak 2013b). The high Mg^{2+} is necessary for non-enzymatic polymerization and replication of the encapsulated RNA. However, the prebiotic availability of such a high concentration of chelates is an open question. Indeed, the prebiotic availability of DN, MA, GMD, GMM, and OA is debatable; hence, one must view the model protocell studies as precisely that – a model. In a separate study, increased stabilization of DA vesicles was found in the presence of nucleobases and sugars in 300 mM NaCl (Black et al. 2013). It was proposed that the amine of the nucleobase or the hydroxyl groups of sugars form hydrogen bond with the carboxyl groups of the FAs, thus stabilizing the vesicles.

Dalai et al. (2017) have demonstrated that mixed FA-PL vesicles are immune to the presence of magnesium up to 40 mM $MgCl_2$. Oleic acid and palmitoyl-2-oleoylphosphatidylcholine (POPC) vesicles in various ratios were subject to increasing $MgCl_2$ concentrations. The Mg^{2+} concentration at which nearly all vesicles were disrupted, as estimated by fluorescence spectrophotometry, dynamic light scattering, and phase-contrast microscopy, was termed the "fatal" magnesium concentration ($[Mg^{2+}]_{fatal}$). The $[Mg^{2+}]_{fatal}$ was found to increase dramatically by an order of magnitude from ~5 mM for pure OA to ~15–30 mM for mixed OA-POPC vesicles (10:1 and 5:1, respectively). A $[Mg^{2+}]_{fatal}$ was not detectable for OA-POPC (3:1, 1:1) and pure POPC vesicles up to a magnesium concentration of 50 mM. The increased resistance to Mg^{2+} was due to selective binding of Mg^{2+} to OA and abstraction of OA from the mixed OA-POPC vesicle, thus relatively enriching the resultant vesicle in POPC, which is more tolerant to cations. Thus, Mg^{2+} acted as an environmental selection pressure in the evolution of mixed FA-PL membranes toward an increasingly PL-enriched composition, which is more similar to modern cell membranes. It was shown for the first time that the previously determined deleterious effect of divalent cations may actually have helped in membranes' evolution toward modern membranes in mixed-lipid systems, where the initially minor lipid component is more tolerant to cations.

7.2.2.2.4 pH Tolerance

Various extreme local pH conditions are known to be present on modern Earth. The crater lake of Kawah Ijen volcano, Indonesia, has a pH of <0.3 and is enriched in H^+, SO_4^{2-}, and Cl^-. Similarly, the monovalent ion-rich crater lakes of the East African Rift Valley are highly alkaline (pH 11-12). Submarine hydrothermal vents or black smokers emit fluids at high temperatures of ~250°C–400°C and very acidic pH values of ~2–3. The vent solutions, when mixed with cool, alkaline ocean water, create concentration, redox, and thermal gradients that can be an important site for organic synthesis reactions. In comparison, white smokers emit fluids at milder temperatures (60°C–90°C) and alkaline pH of ~9–11. All the above-mentioned sites representing extreme pH and temperature conditions are home for modern biodiversity.

Similarly, the pH of surface waters and the ocean on early Earth was probably mildly acidic (~6–6.5), because of equilibrium with a high partial pressure of atmospheric CO_2 gas and weathering of komatiite or tonalite rocks by surface waters. Highly acidic or highly alkaline solutions may have existed around the primitive analogs of hydrothermal vents. The presence of various monovalent and divalent soluble minerals, carbonates, silica, and clays on the Martian surface also suggests a variety of pH environments, and the inferred sub-surface saline ocean of Enceladus is believed to be highly alkaline (pH ~12).

Fatty acid membranes are sensitive to the pH of a solution. As discussed earlier, at concentrations above the CAC, FAs form an oily phase at acidic pH, whereas at alkaline pH, the fully deprotonated acids form micelles (Cistola et al. 1988).

Fatty acid vesicles are formed in the pH range of 7–9 (Figure 7.2.5; Cistola et al. 1988; Apel and Deamer 2005; Morigaki and Walde 2007). In general, vesicular structures are formed when the pH of the solution is near the apparent pK_a of the surfactant, which decreases the electrostatic repulsion between adjacent protonated and ionized carboxylates by hydrogen bonding (Apel et al. 2002). Thus, a change in the solution pH can induce phase transitions. Pure fatty alcohols do not form vesicles on their own in the normal pH range.

Adding fatty alcohol to FA increases the immunity of vesicles against alkaline pH 11, where micelles would normally exist (Apel et al. 2002). Hargreaves and Deamer (1978) first observed that mixed FA-alcohol (C_8–C_{18}) vesicles were stable under a wider range of pH than either pure lipid. This is because alcohol functional groups replace neutral carboxylic acid groups as proton donors to the deprotonated carboxylate of FA and alcohols do not deprotonate under alkaline conditions, thus providing stabilized hydrogen bonds at significantly higher pH conditions (Gebicki and Hicks 1976; Hargreaves and Deamer 1978; Apel et al. 2002; Monnard and Deamer 2002).

A major limitation of FA amphiphiles is that their self-assembly occurs in a narrow range of pH conditions around 7–9. Compared with the carboxylic acid group, a phosphoryl headgroup should have stability over a wider pH range, because phosphate is a multiprotic acid. Recently, phosphate amphiphiles (Figure 7.2.2) were synthesized, and it was shown that they form vesicles that are stable over a wide pH range (2–12) in the presence of co-surfactants such as DOH and DN (Albertsen et al. 2014). Polyprenyl phosphates (Figure 7.2.2) also showed pH-dependent vesicle formation over a pH range of 3–13 (Takajo et al. 2001; Gotoh et al. 2006; Streiff et al. 2007). The pH stability of polyprenyl phosphates also increases by mixing up to 50 mol% of polyprenyl alcohol (Streiff et al. 2007).

7.2.2.2.5 Temperature Tolerance

In addition to oil, vesicle, and micelle phases, which depend on pH at concentrations above the CAC, lipids also demonstrate temperature-dependent phase transitions. The fluidity and semi-permeability of the membrane in vesicles depends on a liquid-crystal structure; membrane becomes a less useful gel-like phase at lower temperatures. Generally, increasing temperatures increases membrane permeability and may also decompose organic molecules encapsulated within the vesicles.

Extant extremophiles have altered lipid structures to survive temperatures much different from the mesophiles, which thrive at 20°C–45°C. Many extremophiles are members of the archaeal domain of life. The hydrocarbon acyl chains of PLs and glycolipids are adapted to low-temperature conditions (−20°C to 10°C) in psychrophiles (Russell et al. 1995). The acyl chains in the lipids of these low-temperature-loving organisms demonstrate unsaturated and branched bonds, whereas the chain length decreases to maintain the membrane fluidity (Russell 1990). Additionally, there is increased incorporation of lipids with a lower gel-to-liquid

crystalline (e.g., 16:1Δ^9, palmitoleic acid) transition temperature (Russell 1984). Thermophiles contain a higher content of saturated straight- and branched-chain FAs, as well as membrane-spanning lipids, in their membranes to enhance the membrane stability at elevated temperatures (Chan et al. 1971).

It is widely accepted that the global Archaean Earth had average surface temperatures of ~50°C–85°C (Knauth 2005), most likely with local environments that reached freezing or temperatures >85°C at surface of hot springs or hydrothermal vents. Thermostability or change in permeability of OA vesicles as a function of temperature has been assessed by estimating leakage of encapsulated self-quenching fluorescent dyes (carboxyfluorescein or calcein) or adenine oligonucleotide (up to 10-mer) up to a temperature of 90°C. Vesicles of short-chain FA, such as DA, tend to have greater leakage at ~45°C in comparison with long-chain (<C_{14}) FA (Maurer et al. 2009). Single-chain amphiphiles with unsaturated bonds in their hydrocarbon chain are more susceptible to oxidation. Interestingly, lauric acid (C_{12}) vesicles are stable at higher temperatures but tend to precipitate at lower temperatures (30°C) (Monnard and Deamer 2002). In our laboratory, we have found that DA or DA-DOH (2:1) vesicles are highly leaky, even at room temperature (Sahai lab., unpublished results). In comparison, OA vesicles are relatively impermeable up to 90°C, but some leakage is detected at higher temperatures >90°C (Mansy and Szostak 2008). Leakage could be prevented by the addition of GMM as a co-surfactant. In experiments mimicking diurnal temperature cycles, vesicle containing encapsulated DNA was heated to 90°C and cooled to 20°C (for 1 minute at each temperature). The DNA strands inside the vesicle showed separation and reannealing, which are important properties for replication (Mansy and Szostak 2008). This result shows the potential for the amplification of information molecules inside the thermostable lipid compartment.

7.2.2.2.6 Pressure Tolerance

We are unaware of pressure-dependent stability date for SCA membranes. Phospholipid membranes exhibit a less sensitive response to changes in pressure. Pressure ~2 kbars in 1,2-dipalmitoyl-sn-glycero-3-phosphatidylcholine (DPPC) resulted in a change from liquid crystalline to gel phase and, finally, to an "interdigitated phase," accompanied by a 5% decrease in the bilayer volume in the latter two phases (Braganza and Worcester 1986). In the non-interdigitated phase, both ends of the hydrocarbon chains meet in the membrane midplane, thus forming a thick hydrophobic region. However, in the fully interdigitated membrane, the thickness of the membrane is greatly reduced and there is the loss of the midplane (Smith and Dea 2013). This change was a result of straightening of the acyl hydrocarbon chains (Stamatoff et al. 1978; Kato and Hayashi 1999). Such an effect would be potentially relevant in the hypothetical scenario where life evolved at oceanic hydrothermal vents after the earliest mineral-wall stage into the stage where the cells had started to produce PL membranes.

7.2.2.2.7 Effects of Minerals on Model Protocell Stability

Most extant prokaryotes live at the mineral/water interface to exploit the advantages offered by this microenvironment, such as inorganic nutrients for chemoautolithotrophs, reduced wave and current turbulence than in a free-standing column of water, and a surface for attachment to facilitate biofilm community living. It is, therefore, logical to propose that protocells may also have preferred to remain in contact with mineral surfaces. This raises the question of whether the self-assembly and permeability of SCA membranes are affected by mineral surface properties.

Sahai et al. (2017) examined the CVC and permeability of pure DA and of DA-DOH (2:1) vesicles at pH 7 and 8.5, respectively, in the presence of 10 minerals and 2 rocks representative of the early Earth. The minerals were chosen to represent a range of particle sizes, isoelectric points, and chemical compositions such as oxides, hydroxides, carbonates, sulfides, and aluminosilicates. It was found that negatively charged minerals, including silica, quartz, montmorillonite, anatase, rutile, and pyrite, did not affect the CVC of the lipids. Positively charged minerals such as γ-Al_2O_3, goethite, and siderite with surface coating of goethite increased the CVC three-fold at high mineral loadings (1 mg.mL^{-1} and higher) but not at lower loadings. It was also found that lipids adsorb to the negatively charged minerals and positively charged minerals, but adsorption on the latter was significantly greater. The positively charged minerals and adsorbed lipids at higher loadings tended to aggregate and settle out of solution, so more lipid had to be added to solution, thus apparently increasing the CVC. At lower mineral loadings, the lipid-mineral aggregates did not settle out appreciably. Thus, the apparent CVC can be affected by some minerals under specific conditions of high loading. Permeability was not affected by the presence of minerals, as determined by leakage of calcein fluorescent dye encapsulated within the vesicle, either at room temperature or up to 80°C.

In summary, increasing hydrocarbon chain length and the addition of co-surfactants with alcohol, amine, or phosphate headgroups enhance the stability of FA membranes toward extremes of pH, dissolved ion concentrations, and temperature.

7.2.2.3 OTHER COMPARTMENT-FORMING SYSTEMS

Besides SCAs, other molecules such as peptides have been explored as plausible membrane components. Amphiphilic peptides formed by heating a mixture of amino acids (aspartic acid, glutamic acid, and other 16 amino acids) at 160°C–180°C were shown to form membranous structures or microspheres that resembled protocells (Fox and Harada 1960). These structures were termed proteinoids, and it was suggested, though not shown, that they may have catalytic properties, which are required to develop a protometabolism (Fox 1980).

In another study, a proteolipid (protein-lipid complex) membrane supported onto mesoporous silica particles was synthesized and compared to lipid vesicles (Nordlund et al. 2009). These surface-templated mesoporous silica structures with proteolipid could maintain a proton electrochemical gradient across the membrane. This means that the membranes were proton-tight in the surrounding aqueous medium. These structures were developed for potential drug-delivery applications, but it is conceivable that similar architectures may have existed prebiotically.

Others have suggested that the earliest protocell boundaries were not composed of organic molecules at all but rather of mineral surfaces, specifically, pores in mineral chimneys at low-temperature hydrothermal vents (Russell and Hall 1997; Barge et al. 2014; Lane 2015; Saladino et al. 2016; Hansma 2017). The maintenance of a natural geochemical pH gradient across the pore wall is proposed to be a result of the flow of acidic ambient seawater and highly alkaline vent fluid. This is proposed to have provided the chemiosmotic energy for organic molecule synthesis, eventually including specialized ion channels and active pumps.

In a related but different idea, it has been proposed that "membraneless organelles" may have formed in the spaces between mica sheets (Hansma 2017). This mineral-bound space is suggested to have advantages over organic membranes, because they are not affected by the environmental factors as compared with fragile organic molecules. Hansma does not elaborate on what is meant by membraneless organelles, so we speculate here what these might have been. Normally, an organelle is a eukaryotic sub-cellular structure bounded by a lipid membrane that performs a specific function. The ribosome, a large RNA-protein macromolecular complex, is sometimes called a "membrane organelle" simply because it is large enough to be resolved by transmission electron microscopy (TEM) and is the fundamental machinery of the cell that provides a link between RNA and protein synthesis. The ribosome's structure and functions are highly conserved across all species of life, and its core is believed to be highly ancient, perhaps having evolved even at the protocell stage (Petrov et al. 2015). This idea is consistent with the co-evolution of functional RNA oligomers and peptides by synergistic catalysis, which has been proposed recently (Kaddour and Sahai 2014). It has been shown that Na-exchanged montmorillonite clay, when pretreated in a specific manner, is capable of catalyzing the polymerization of activated RNA mononucleotides in the presence of ~75 mM $MgCl_2$ and 0.1 M NaCl (Ferris et al. 1996; Joshi et al. 2009). Amino acid adsorption and oligomerization (up to 5–8 mers) in the presence of various clays was also studied (Hashizume, 2012; Dalai et al 2016a). Thus, combining the ideas of various workers, we speculate that, perhaps, a primitive precursor of the ribosome core evolved in the space between mica or clay sheets. On the other hand, the interlayer spacing in swelling clays is only ~1.2–1.8 nm, so the extent to which even a small, primitive ribosome could have been accommodated is limited. This is sheer speculation without any experimental or computational evidence but provides avenues for future research.

7.2.3 GROWTH AND DIVISION OF PROTOCELLS AND COMPETITION BETWEEN PROTOCELL POPULATIONS

One of the most important characteristics of a protocell is its ability to grow and subsequently divide to form daughter cells. Various factors affecting model protocell growth and division have been examined in the literature. Studies on the competition between different populations of model cells in primitive predator-prey type relationships, which may lead to some prebiotic evolutionary advantage for the predator protocells, are now beginning to be conducted.

7.2.3.1 MINERAL EFFECTS ON INITIAL RATE OF VESICLE FORMATION

Hanczyc et al. (2007) examined the potential effect of mineral surfaces on the rate of SCA vesicle formation by the acidification of an alkaline (pH ~12) micelle solution (see Section 7.2.2.1 on methods to prepare vesicles). It was found that the initial vesicle formation rates were enhanced in the presence of all the minerals studied and even a hydrophobic surface like polystyrene beads (Hanczyc et al. 2003, 2007). Furthermore, direct contact between the surface and lipid micelles or vesicles was apparently not required to obtain the surface-promotion effect. The authors did not identify any relationship with the chemical or physical properties of the surfaces. Furthermore, the mechanism was not established for the surface promotion effect.

Recently, Sahai et al. (2017) confirmed Hanczyc's findings, identified mineral-specific trends for initial rates of vesicle self-assembly from micelles and also for *de novo* vesicle formation by rehydration of a lipid thin film (see Section 7.2.2.1), and provided a model for the surface-promoting effect without direct contact. In detail, it was shown that the initial vesicle-formation rates correlated with isoelectric point and particle size of the minerals (Figure 7.2.6a).

The isoelectric point (IEP) is the pH at which the mineral surface has net zero charge. At the experimental pH (~7 and 8), some minerals are negatively charged while others are positively charged. Rapid lipid adsorption occurs on all mineral surfaces, as controlled by a combination of electrostatic, H-bonding, and van der Waals hydration forces. A greater amount of FA lipid adsorption was observed on positively

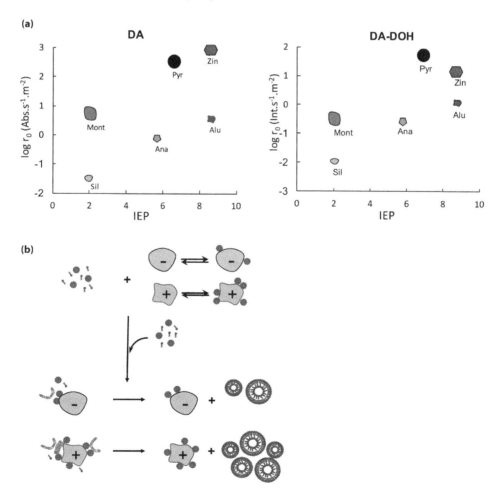

FIGURE 7.2.6 Mineral surface promoted vesicle formation. (a) Initial rate of DA and DA-DOH vesicle formation (r_0) normalized to the mineral surface area as a function of the IEP of the mineral. (b) Schematic of the proposed model for mineral enhancing effect on the rate of vesicle formation. (Modified from Sahai, N. et al., *Sci. Rep.*, 7, 43418, 2017.)

charged mineral than on negatively charged minerals due to electrostatic forces. Some FA adsorption occurred even on negatively charged minerals, despite electrostatic repulsion, because of H-bonding and van der Waals forces. The adsorbed lipid islands on the partially coated mineral were then proposed to serve as a template for more lipid to attach from solution and rapidly transform into vesicles (Figure 7.2.6b).

Hence, the initial rate of vesicle formation was greater on both negatively and positively charged minerals as compared with rates in the absence of any mineral.

The reason that catalysis works even without direct contact with the mineral surface is based on modified Derjaguin-Landau-Verwey-Overbeek (DLVO) theory for colloidal stability that includes the steric hydration force (Oleson and Sahai 2010; Israelachvili 2011). Depending on the mineral, two or more lipid bilayers (up to 12–15 nm away from the surface) may adsorb because of van der Waals interactions between the surface of the lipid bilayers and between the bilayers themselves. The physical size of the bilayers excludes solvent and counterions from the region immediately proximal to the mineral surface, thus effectively extending the electric double layer. Hence, the electrostatic effect of a mineral's surface charge is felt even up to ~12–15 nm away from the surface through two or three stacked lipid bilayers (Oleson and Sahai 2010). The surface-induced rate enhancement is an example of the "matrix effect," which was first identified by Luisi and co-workers (Blöchliger et al. 1998; Lonchin et al. 1999; Berclaz et al. 2001; Cheng and Luisi 2003; Thomas and Luisi 2005). In the studies involving minerals, the surfaces promote rapid initial formation of the lipid islands, which serve as the matrix or template for subsequent vesicle formation.

7.2.3.2 Size Effect

As described in Section 7.2.2.1, a polydisperse suspension of vesicles is formed when a dry lipid film is rehydrated. Protocells are also expected to have had a distribution of sizes. One may question whether medium-sized protocells (vesicles) grew at the expense of smaller vesicles by fusion or whether growth by uptake of monomers or micelles from the environment was the preferred mechanism. Conversely, did the largest protocells form during spontaneous self-assembly during film rehydration as a result of fusion and then subsequently divide to form smaller ones? What forces might have driven growth and division?

Interaction between vesicles of different size distribution was studied by Cheng and Luisi and co-workers (Blöchliger et al. 1998; Lonchin et al. 1999; Berclaz et al. 2001; Cheng and Luisi 2003; Thomas and Luisi 2005) and Szostak and coworkers (Chen and Szostak 2004b). It was found that pure FA or pure POPC vesicles of different sizes could coexist, without the smaller ones fusing with each other to form larger vesicles of a single size. The relative stability of smaller vesicles compared with larger ones was also probed using t-RNA-loaded POPC vesicles of different sizes. A small amount (10%) of positively charged surfactant (cetyltrimethylammonium bromide [CTAB]) was added to the vesicle suspension. The presence of CTAB was expected to cause vesicle aggregation. This was observed for the 160-nm-sized POPC vesicles, whereas the smaller vesicles (~80 nm) were stable against CTAB-induced aggregation (Thomas and Luisi 2005).

These findings were in accordance with the "kinetic trap" concept, which suggests that vesicles are not in thermodynamic equilibrium but rather are kinetically trapped systems (Yan et al. 2016). In a kinetically trapped system, small vesicles of different sizes occupy local energy minima and are prevented from fusing by large energy barriers between the various local minima (Figure 7.2.7a).

Therefore, small vesicles of different-size vesicles tend to remain unfused for a long time (Cheng and Luisi 2003; Yan et al. 2016). In contrast, the energy barriers between larger vesicles is small, so larger vesicles can divide to achieve a lower energy state more easily (Figure 7.2.7a) (Yan et al. 2016). Thus, FA vesicles grew when fed with FA micelles and divided easily by simple physical agitation (Figure 7.2.7b) (Chen and Szostak 2004b; Zhu and Szostak 2009).

The growth of preformed FA vesicles upon the addition of micelles occurred by two pathways. At low micelle-to-vesicle

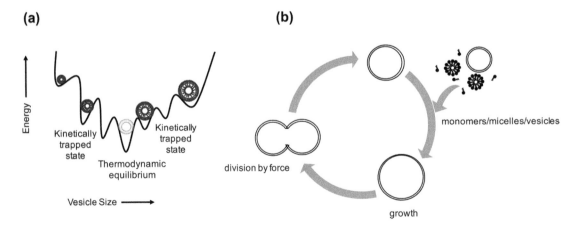

FIGURE 7.2.7 Growth and division of model protocells. (a) Schematic free-energy profile for vesicle fusion from smaller to larger equilibrium size or vesicle division from larger to smaller equilibrium size. (Modified from Yan, Y. et al., *Chem. Commun.*, 52, 11870–11884, 2016.) (b) Growth of vesicles by adding monomers, micelles or vesicles to preexisting vesicles and subsequent division induced by agitation.

ratio (<0.4), micelles were incorporated into the existing lipid bilayer (Chen and Szostak 2004b). However, when the micelle to vesicle ratio exceeded 0.4, micelles aggregated to form the "seeds" or templates for *de novo* vesicle formation. These vesicles could then grow and divide into daughter vesicles. The latter pathway was first reported by Luisi and coworkers, who termed it the "matrix effect." (Blöchliger et al. 1998; Lonchin et al. 1999; Berclaz et al. 2001; also see Section 7.2.3.3 on co-surfactant effect).

7.2.3.3 CO-SURFACTANT EFFECT

Phospholipid membranes are more robust than SCA membranes toward the deleterious effects of divalent cations and pH extremes and are also much less permeable. Hence, PL vesicles formed by small quantities of abiotically produced PLs may be considered to have an evolutionary advantage over less evolved SCA vesicles. The competition between OA and POPC to form vesicles when mixed together has been studied (Lonchin et al. 1999; Berclaz et al. 2001; Cheng and Luisi 2003; Fujikawa et al. 2005). Oleate vesicles were found to fuse to the POPC vesicles, eventually causing the disappearance of OA vesicles. This effect was observed regardless of the relative sizes of the OA and PL vesicles. Oleic acid micelles (small) were incorporated when added to a suspension of extruded (unidisperse size) POPC vesicles (large) (Lonchin et al. 1999; Berclaz et al. 2001). The resulting mixed vesicles were significantly smaller in size, suggesting that the mixed vesicles had presumably grown very large and then divided (Berclaz et al. 2001). When premixed MA-POPC micelles were added to large unilamellar MA vesicles, MA molecules were abstracted, resulting in a decrease in MA vesicle size (Fujikawa et al. 2005).

The above results showed that the size effect (*i.e.*, relative size of the FA vs. the PL phase) was not the factor controlling which phase grew. Rather, the FA phase was consistently at a disadvantage in the presence of a pure PL or mixed FA-PL phase. This effect can be explained based on the relative CVCs of FAs compared with that of PLs. The SCAs have CVCs that are three to six orders of magnitude larger than the CVCs of PLs, and mixed FA-PL systems with even a small relative PL content have substantially lower CVCs than pure FAs (see Section 7.2.2.2). Hence, it is the FA molecules that are incorporated into the PL or mixed FA-PL phase.

7.2.3.4 ENCAPSULATED SOLUTE EFFECT

It is generally believed that protocells containing a higher concentration of solutes would have had an advantage for growth over those containing a lower concentration, because this allows for a greater amount of products to be synthesized within the former population. But the question arises as to how different populations are created in the first place. When lipids self-assemble in a solution containing solutes, the concentration of solutes within the vesicles is equal to the bulk concentration outside the vesicles, and, moreover, all the vesicles are expected to have the same concentration.

In a fascinating study, Luisi et al. (2010) demonstrated that a solute, ferritin, when encapsulated in lipid vesicles, is not uniformly distributed. Rather, diverse vesicle populations having different amounts of encapsulated solute were identified. A large number of "empty" vesicles and a small population of ferritin-enriched vesicles were observed. Ferritin was chosen for encapsulation because it contains many electron-dense Fe^{3+} ions; this makes it suitable for visualization using cryo-TEM (Figure 7.2.8).

It was proposed that, if more ferritin molecules were randomly bound to the initially forming membrane, it might have slowed the rate of membrane closing, thus allowing even more solute to be encapsulated. Conversely, those initially forming membranes, which randomly did not bind ferritin molecules, closed quickly to form empty vesicles. These experiments demonstrated that solute encapsulation at high concentrations is possible even from dilute surroundings. If these solutes are nutrients- or information-carrying molecules, the concentration phenomenon could potentially allow the compartment to eventually develop functionality.

Protocell populations capable of exchanging nutrients and transfer of information may evolve as a result of Darwinian evolution (Szathmáry and Demeter 1987; Szostak et al. 2001). This was demonstrated for model protocells that encapsulate RNA or peptide oligomers (Chen et al. 2004; Adamala and Szostak 2013a). Fatty acid (OA or MA-GMM) vesicles with encapsulated RNA grew by incorporating membrane components from "relaxed" membranes that are devoid of nucleic acid. The growth of the vesicles was driven by the osmotic pressure exerted by the RNA inside the vesicle. Thus, the RNA-encapsulated vesicles showed growth advantage over empty vesicles. The advantage of solute-encapsulating vesicles over empty vesicles was also

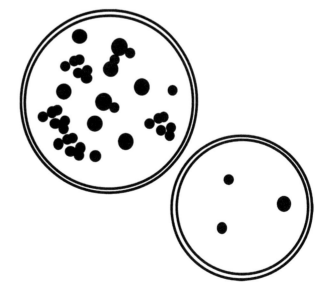

FIGURE 7.2.8 Schematic representation of solute-rich and relatively empty vesicles. The original work showed the cryo-TEM images of "superfilled" and empty unextruded OA-POPC vesicles with electron-dense ferritin molecules. (From Luisi, P. L. et al., *ChemBioChem*, 11, 1989–1992, 2010.)

demonstrated for a peptide solute. The OA vesicles encapsulating AcPheOEt, LeuNH$_2$, and Ser-His dipeptide catalyzed the intra-vesicular formation of a hydrophobic peptide, AcPheLeuNH$_2$ (Adamala and Szostak 2013a). Vesicles with the newly formed AcPheLeuNH$_2$ dipeptide, localized in the membrane bilayer, then grew at the expense of added FA micelles. In comparison, empty vesicles grew much less in size upon addition of FA micelles. Interestingly, AcPheLeuNH$_2$ peptide synthesis also occurred at a higher efficiency in vesicles than in bulk solution, potentially suggesting some catalytic property of the lipid monomer or the lipid membrane for peptide synthesis.

Homeostasis is a physiological process by which organisms maintain intra-cellular steady state such as constant pH and Na$^+$, K$^+$, and Ca^{2+} ion concentrations regardless of the extra-cellular environmental conditions. This is a critical feature of cells for survival in changing environmental conditions. Engelhart et al. (2016) have demonstrated a primitive type of homeostasis for ribozyme activity in MA-glycerol monomyristoleate (GMM)-dioleylphosphatidic acid (DOPA) vesicles. Ribozymes and short oligonucleotides (5–7 mers) were co-encapsulated at high concentrations in the vesicles. Enzymatic activity of the ribozyme was inhibited due to duplex formation between the ribozyme and complementary oligonucleotides. Subsequent addition of MA-GMM vesicles led to the growth of the MA-GMM-DOPA vesicles at the expense of the FA vesicles. The growth resulted in dilution of vesicle contents, thus inhibiting ribozyme and oligonucleotide complexation and regaining ribozyme activity. These results indicate that short oligonucleotides were capable of controlling ribozyme functionality preceding modern RNA regulatory machineries, under changing environmental conditions, in this case, lipid concentration.

Competition between two non-lipid populations consisting of coacervates and "proteinosomes" was also recently demonstrated (Qiao et al. 2017). Proteinosomes are membranes consisting of a closely packed monolayer of conjugated amphiphilic protein-polymer building blocks. The amphiphilic construct was composed of bovine serum albumin (BSA) and polymer [poly(N-isopropylacrylamide)], or PNIPAAM. The proteinosomes were loaded with sugar, DNA, or platinum nanoparticles as cargo. Coacervates were assembled from a mixture of cationic polymer [poly(diallyldimethylammonium chloride)] and anionic ATP and were loaded with protease. When mixed together, the proteases in the coacervates digested the protein boundary of the proteinosomes. Thus, the coacervates "fed" on the proteinosomes. Note, however, that in this study, the number of predators (coacervates) remained unchanged after consuming the prey (proteinosomes), thus differing from natural predator-prey relationships, where the predator population (number) grows by feeding on its prey. Also, the proteinosomes, coacervates, and their cargo are too complex to represent model protocells or contemporary cells, but some analogous systems may have existed.

7.2.4 MEMBRANE PERMEABILITY AND SOLUTE TRANSPORT

The transport of solutes across the plasma membrane, that is, of essential nutrients into the cell and of waste products out of the cell, is an essential feature for metabolism. Solutes are exchanged by both passive and active transport mechanisms in modern cells. In passive diffusion, molecules move down the concentration gradient, that is, from high to low concentration, to maintain the equilibrium between inside and outside the cell, without using energy. Passive transport occurs by both non-carrier- and carried-mediated diffusions. Water, ions, and gases such as CO$_2$ and O$_2$ are transported by simple non-carrier-mediated passive diffusion. Carrier-mediated passive transport requires channel proteins and carrier proteins for diffusion of ions and small organic molecules, for example, sugars. In contrast, active transport uses ATP as an energy source to pump molecules (e.g., Na$^+$/K$^+$) against the concentration gradient, from low to high concentration of the solute. Protons, amino acids, complex sugars, and lipids need specialized proteins, ion channels, and endo- and exocytosis mechanisms for active transportation of solutes. These specialized transport machineries for mass exchange likely evolved later in the evolutionary process. The earliest protocell membranes were likely composed of simple short-chain FAs, perhaps along with some small amount of prebiotically synthesized PLs. Short-tail FA (C ≤ 12) membranes are highly leaky. In the absence of complex transport machineries, nutrient (nucleotides, sugars, amino acids, ions, and dissolved gases) uptake into protocells and expulsion of waste products would have occurred through passive mechanisms, including simple diffusion, lipid flip-flop, defects in the bilayer, and transient pore mechanisms (Paula et al. 1996; Wilson and Pohorille 1996; Paula and Deamer 1999; Mansy et al. 2008). Regardless of the mechanism, the permeation rate across a membrane is dependent on the properties of both the membrane-forming lipid and the solute.

7.2.4.1 FACTORS CONTROLLING PERMEATION RATE

7.2.4.1.1 Lipid-Dependent Factors

The factors associated with the lipid molecule include the type of headgroup, hydrophobic chain length, and lipid packing density, which determine the membrane thickness, hydrophobicity, and fluidity (Paula and Deamer 1999; Mansy 2010). Phospholipids are composed of two acyl chains, which impart greater stability but slower dynamics than a single-chain FA in a membrane. Fatty acid membranes are orders of magnitude more permeable than PL membranes. For instance, the permeability coefficient of K$^+$ through OA membranes is 10^{-6} cm s^{-1} compared with 10^{-10}–10^{-12} cm s^{-1} for various phosphatidylcholine (C$_{14}$-C$_{24}$) membranes (Paula et al. 1996; Chen and Szostak 2004a). Amino acid permeability of PL membranes also depends on the acyl chain length, with shorter-chain lipids being more permeable, and is independent of the solution pH from pH 2–9 (Chakrabarti and Deamer 1992). Nucleotides are permeable across FA membranes in the presence of Mg^{2+} (Mansy et al. 2008).

7.2.4.1.2 Solute-Dependent Factors

Among solute properties, several factors such as solute charge, hydrophilicity, and isomeric structures affect permeability. Charged molecules are relatively impermeable to PL membranes because of the high energetic cost of desolvation, which is required to cross the hydrophobic membrane interior composed of the hydrocarbon lipid tails. This charge-dependent permeability could have been beneficial for accumulating biomolecules within the protocell compartment. Thus, phosphate in its neutral form at highly acidic pH of 1 has higher permeability (permeability coefficient, $P = 3 \times 10^{-11}$ cm s^{-1}) across a phosphatidylcholine membrane than in its monoanionic form at pH ~4 ($P = 5 \times 10^{-12}$ cm s^{-1}) (Chakrabarti and Deamer 1992). Lysine methyl ester, a derivative of lysine, which is neutral at higher pH values and positively charged at lower pH values, translocates faster ($P = 2.1 \times 10^{-2}$ cm s^{-1}) than does positively charged lysine ($P = 5.1 \times 10^{-12}$ cm s^{-1}) across PL membranes. Permeability of lysine methyl ester is dependent on the transmembrane pH gradient (Chakrabarti and Deamer 1992; Chakrabarti and Deamer 1994). Also, activated nucleotides can cross a FA vesicle membrane more rapidly than negatively charged natural nucleotides, as activated forms are less polar and have reduced negative charge (Mansy et al. 2008).

The relative hydrophilicity of a molecule also influences its permeation rate through PL membranes. Hydrophobic amino acids (e.g., phenylalanine and tryptophan) exhibited transmembrane permeability by two orders of magnitude greater than hydrophilic amino acids (e.g., glycine, serine, and lysine) (Naoi et al. 1977; Chakrabarti and Deamer 1992; Chakrabarti et al. 1992).

Isomeric structures of neutral sugars were also found to affect permeability across FA membranes (Sacerdote and Szostak 2005). Ribose was found to have much faster permeability than its diastereomers, arabinose and xylose. Thus, early cell membranes might have had a better access to ribose, which could have been used subsequently to synthesize nucleic acids inside the protocell. Using molecular dynamics simulations, Wei and Pohorille (2009) found that the free energy of transferring ribose in its β-pyranose form from water to the lipid bilayer was more favorable than the other aldopentoses. The difference in the free energy of transfer between ribose in its β-pyranose and its diastereomers was attributed to inter- and intramolecular interactions between consecutive exocyclic hydroxyl groups. However, their results showed that ribose crosses the bilayer in the β-pyranose form, and permeation of the biologically more relevant furan form was thermodynamically unfavorable. Thus, the simulation and experimental results are not entirely consistent.

Monnard and Deamer (2001) have demonstrated the permeation of ATP in the presence of a low concentration of magnesium (1.75 mM), through PL and mixed FA-PL membranes at the transition temperature of the lipid. While the vesicles are permeable to ATP monomer, it was found that dimers and large tRNA molecule showed neither uptake nor release (Monnard and Deamer 2001). They suggested that ATP passively diffused across bilayers according to the transient defect model (Mouritsen et al. 1995; Paula et al. 1996). Therefore, it may have been possible to supply ATP and other nutrients for polymerization inside the protocell in the absence of complex transport machineries. Surprisingly, increasing Mg^{2+} concentration inhibited ATP permeation in contrast to previous results (Petkau and Chelack 1972; Stillwell and Winter 1974), and this discrepancy was not further explained.

7.2.4.2 Mechanisms of Transport Across Fatty Acid Membranes

7.2.4.2.1 Lipid Flip-Flop

Flip-flop refers to the translocation of FA molecules from the outer leaflet to the inner leaflet of the bilayer membrane. Flip-flop dynamics are influenced by the chemical properties of lipid molecules. Flip-flop of a FA molecule is significantly faster (t$_{1/2}$ = milliseconds) than that of a PL molecule (t$_{1/2}$ = days) (McLean and Phillips 1981; Hamilton 2003). When a FA monomer in solution attaches to a vesicle, it first partitions into the outer leaflet and, subsequently, may flip into the inner leaflet of the membrane. Thus, a lipid molecule with a protonated headgroup from the outer membrane can be flipped into the inner membrane or vice versa. It was found that OA membranes cannot maintain a pH gradient in the presence of alkali-metal cations, and this phenomenon was attributed to rapid flip-flop dynamics (Chen and Szostak 2004a). However, the pH gradient could be maintained in the presence of a larger cation, arginine, for several hours, and this was attributed to arginine being non-permeating. The alkali metal ions followed the trend of decreasing permeability with increasing size through the OA membranes (Chen and Szostak 2004a). A high degree of flip-flop dynamics would have rendered protocells permeable to alkali cations and protons. Phospholipid vesicles have the advantage of maintaining a transmembrane pH gradient due to the slower flip-flop rate, as compared with FAs.

7.2.4.2.2 Transient Pores and Local Defects

The surface tension associated with a vesicle membrane may experience a temporary and local increase due to association with another surface or insertion of a peptide or other large molecule. The lipids rearrange in the bilayer to accommodate this change in surface tension, resulting in transient pores or local defects, to allow some solution to leak out, thus easing the increased tension. The lifetime of transient pores depends on the solution viscosity and lipid mobility. Studies examining the effects of membrane-spanning peptides and RNA oligomers are discussed in Section 7.2.7 on co-evolution of lipids, RNA, and proteins.

7.2.5 TRANSITION FROM SINGLE-CHAIN AMPHIPHILE TO PHOSPHOLIPID MEMBRANES

At some stage in the evolutionary process, simple protocell membranes composed of SCAs evolved into predominantly PL membranes. Budin and Szostak (2011) have shown that low levels of dioleoyl phosphatidic acid (DOPA), a PL, could drive the growth of mixed OA-DOPA vesicles, which was deemed an evolutionary step, because mixed OA-DOPA vesicles grew at the expense of pure OA vesicles. However, the resulting vesicles would have higher OA content than the starting vesicles, so they would still be subject to the disadvantages of FA membranes. Recently, the transition of mixed OA-POPC with low POPC content to OA-POPC with high POPC in the presence of Mg^{2+} as an environmental selection pressure was shown (Dalai et al. 2017). It was found that Mg^{2+} preferentially binds to and selectively abstracts OA from OA-POPC mixed lipid membranes, resulting in relatively more POPC in the vesicle. Thus, Mg^{2+} could change the composition of surviving vesicles toward PL enrichment. Additionally, vesicles with higher POPC content were shown to be more robust to the potential effects of divalent cations. These results were reconfirmed by Jin et al. (2018). Therefore, Mg^{2+} ions provided a clear selective pressure for the earliest steps in evolution toward modern cell membranes. The transition of FA vesicles to PL-enriched membranes would be limited by supply of prebiotically synthesized PL. These membranes would provide improved compartmentalization for peptide amphiphiles or replicating machineries to evolve protometabolism and replication cooperatively (discussed in Section 7.2.7).

Before the complex cell walls and biofilms of extant bacteria evolved possibly as armor against cytotoxicity of some minerals (Xu et al. 2012, 2013), the simple PL membranes of early (proto)cells would have come in contact with mineral surfaces. The stability of PL membranes at various mineral surfaces has been examined by studying model vesicles of zwitterionic DPPC and ditridecanoylphosphatidylcholine (DTPC), anionic dipalmitoylphosphatidylserine (DPPS), and cationic dipalmitoylethylphosphatidylcholine (DPEPC). The minerals examined were silica (amorphous silica and quartz), α-TiO$_2$ (rutile), α-Al$_2$O$_3$ (corundum) and mica. The PL-mineral interactions were studied at low and high ionic strengths (17 and 217 mM NaCl), in the presence and absence of dissolved Ca^{2+} ions (Oleson and Sahai 2008; Xu et al. 2009; Oleson et al. 2010, 2012). For the PLs with charged headgroups, adsorption was found to be driven primarily by electrostatic interactions with the mineral surface and was limited to a single adsorbed bilayer. The adsorption of DPPC was controlled in addition by van der Waals interactions, and the effect of solvent and counterion exclusion from the vicinity of the mineral surface ("hydration force"), which allowed multiple layers (up to three on alumina and up to two bilayers on silica) to be adsorbed at low ionic strength. Thus, the mineral was able to exert its electrostatic effect up to 12–15 nm away from the surface.

7.2.6 PROTOMETABOLISM

Most of the studies summarized previously assume the prebiotic availability of FAs, more difficult yet possible PL synthesis, and an RNA world. The existence of the RNA world has been questioned because non-enzymatic synthesis of polynucleotides, long enough to have catalytic functionality, is unlikely under plausible early Earth conditions. Also, RNA is easily hydrolyzed and would have been unstable (Shapiro 2006, 2007). Furthermore, almost all experiments involving nucleotide polymerization first require the monomers to be activated by the addition of a good leaving group. This process usually involves prebiotically implausible reactions, but in the lack of other, viable alternatives, it is implicitly agreed upon by workers in the field that the monomers were somehow activated (Kaddour and Sahai 2014). The "Polymer Transition" stage of the RNA world theory, in which proteins eventually took over the function of catalysis from ribozymes, has also been critiqued (Bowman et al. 2015).

Because of these difficulties associated with the RNA world hypothesis, it has been proposed that primitive metabolism, or "protometabolism," might be an answer to the first ordered chemistry capable of surviving. However, in contrast to the RNA world hypothesis, which is being actively examined experimentally, a very limited number of studies have examined protometabolism in model protocells. One of the problems is to define metabolism in a conceptually simple yet robust way that can be used to design experiments.

Metabolism includes the sum of all chemical reactions performed by an organism. These reactions include biosynthesis of the molecules required for replication and self-maintenance, a process known as anabolism, as well as release of energy from specific nutrient molecules in order to drive the biosynthetic pathways during the process of catabolism. The pathway for both synthesis and breakdown of a biomolecule is a complex network of reactions involving key intermediate molecules and is catalyzed by enzymes at almost every step. A key question in the OoL field is whether the earliest metabolism was chemoautotrophic or heteroautotrophic. In this section, we first provide a brief introduction to the common themes in modern metabolism, followed by a summary of different approaches in the OoL field to address the metabolism-first or protein-world scenario.

7.2.6.1 MODERN METABOLISM

The discussion in this section follows from an excellent introductory textbook by Tortora et al. (1995).

7.2.6.1.1 Shared Features of Metabolism

Modern metabolism in all forms of life share some key features. Anabolic reactions are often dehydration reactions and require energy (activation) to form new chemical bonds. Polymerization of monomers is a class of anabolic reactions. Catabolic reactions involve breakdown of molecules, often by hydrolysis, and release energy, *for example*, the breakdown of

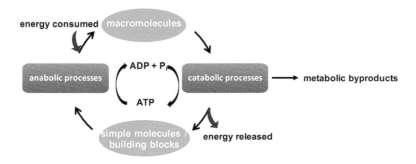

FIGURE 7.2.9 Schematic of modern metabolism showing the ATP-linked anabolic and catabolic processes. Reactions in both processes occur through a series of enzyme-mediated steps.

carbohydrates, in which CO_2 and H_2O are generated as waste products. The energy derived by catabolism is stored in the ATP molecule in the high energy P-O-P bond, wherein ADP combines with AMP to form ATP. Biosynthesis of molecules requires energy, which is provided by the breakdown of ATP back to ADP and AMP. Thus, ATP is a "concentrated" carrier of energy.

Metabolic reactions are usually electron-transfer (redox) reactions, which are often slow and need to be catalyzed. Specific enzymes are required to catalyze almost every intermediate compound in the complex anabolic or catabolic reaction networks (Figure 7.2.9).

Enzymes are assisted by cofactors, which are either organic compounds called coenzymes or metal ions. Nicotinamide adenine dinucleotide (NAD) and nicotinamide adenine dinucleotide phosphate (NADP) are the most important cofactors in redox reactions. They get reduced to NADH and NADPH, respectively. These electrons can be donated back to an oxidant in a subsequent step, thus converting the coenzymes back to their oxidized forms. Thus, NAD and NADP act as electron carriers or shuttles. They are utilized predominantly in anabolic and catabolic reactions, respectively. Inorganic ions also act as cofactors. For example, Mg^{2+} assists enzymes that catalyze the transfer of phosphate from ADP to another substrate or the transfer of a nucleotide triphosphate to a growing nucleic acid chain, with the release of the corresponding nucleotide diphosphate. Iron sulfide clusters (Fe_3S_4 and Fe_4S_4) are found at the active site of some enzymes such as ferredoxin, NADH dehydrogenase, and coenzyme Q-cytochrome c reductase. Glycolysis required five Mg^{2+}- and one Zn^{2+}-dependent enzymes (Belmonte and Mansy 2016).

7.2.6.1.2 ATP Synthesis (Phosphorylation) and Chemiosmosis

The generation of ATP is a key step in storing the chemical energy derived from catabolism in a form that can be used efficiently at a later stage for an anabolic reaction. The addition of a phosphate group to ADP or other substrate is called phosphorylation. Three mechanisms are utilized by all forms of life to generate ATP. In substrate-level phosphorylation, an existing phosphorylated compound containing a high-energy C-O-P bond transfers the energy to ADP, with the formation of ATP. In oxidative phosphorylation, electrons

are transferred from reduced organic or inorganic compounds (*e.g.*, glucose and CH_4) to electron carriers such as NAD in an electron chain to a terminal electron acceptor (*e.g.*, O_2 and CO_2). Energy is released in each electron transfer reaction of the electron transport chain. Some of this energy is used to convert ADP into ATP by a process known as chemiosmosis. The third mechanism for ATP generation is known as photophosphorylation. It is the mechanism in which organic compounds are synthesized from CO_2 as a carbon source during photosynthesis. Energy from UV light is transferred by light-trapping pigments such as chlorophyll and electron carriers (flavin mononucleotide (FMN), cytochromes, and ubiquinone or coenzyme Q) in an electron transport chain, finally to ATP. This process also involves chemiosmosis.

Chemiosmosis is a process in which the energy released by the movement of a solute (specifically, H^+) down a concentration gradient is used to synthesize ATP during oxidative phosphorylation or photophosphorylation. As electrons from NADH generated by photocatalysis (sunlight and chlorophyll) are passed down the electron chain, some of the energy is used to drive protons out of the PL membrane by using specific proteins called protons pumps. This builds up a high H^+ concentration outside the cells, which results in both a pH and an electrical charge gradient across the membrane. This electrochemical gradient is called the proton motive force. The H^+ then diffuse back across membrane, assisted by a special protein channel called ATP synthase. The down-gradient movement of H^+ releases energy, which drives ATP synthase to phosphorylate ADP to ATP inside the cell. In most prokaryotes, the relevant membrane is the plasma membrane, whereas it is the mitochondrial membrane in eukaryotes and the thylakoid membrane in cyanobacteria and eukaryotic chloroplasts. In summary, ATP generation requires (1) terminal electron donors such as pigment molecules and sunlight in photosynthesis, or glucose or reduced inorganic redox species such as H_2O, elemental S^0, Fe(II), Mn(II), H_2, NH_3, and CH_4; (2) electron transfer by electron carriers, such as NAD, NADP, and FMN; and (3) terminal electron acceptors. The electron acceptors may be O_2 in oxidative respiration, or oxidized inorganic species such as SO_4^{2-}, Fe(III), Mn(III), NO_3^-, CO_2, and even elemental S^0 in anaerobic respiration, or organic compounds in fermentation (Konhauser 2007). Aerobic and anaerobic respirations utilize substrate-level and oxidative phosphorylation to generate ATP for every glucose molecule metabolized. Aerobic respiration generates 36 (prokaryotic) or 38 (eukaryotic) ATP molecules per glucose molecule, and anaerobic respiration generates between 3 and 38 ATP molecules per glucose. Fermentation, which is also an anaerobic process, relies on substrate-level phosphorylation to generate two ATP molecules per glucose molecule.

7.2.6.1.3 Metabolic Classification

The metabolisms of extant organisms have been classified based on the electron source and the carbon source. Organisms that use light energy are called phototrophs, and those that use chemical energy from redox reactions are called chemotrophs. The organisms may be autotrophs, wherein the source of carbon is inorganic carbon (CO_2), or they may be heterotrophs, which use another organic compound (reduced carbon) to synthesize their biomolecules. Combinations of these modes yield different metabolic categories. Photoautotrophs use light as their energy source and CO_2 as the carbon source. Examples include photosynthetic purple and sulfur bacteria, cyanobacteria, algae, and plants. Photoheterotrophs, such as purple and green nonsulfur bacteria, use sunlight as their energy source and organic compounds for carbon source. The RNA world hypothesis for the OoL is generally consistent with a heterotrophic metabolism. Chemoautotrophs use electrons from inorganic compounds, and their carbon source is CO_2, for example, hydrogen, sulfur, iron, and nitrifying bacteria. Methanogens, acetogens, and methanotrophs are some examples of chemoautotrophic metabolisms that have been proposed to comprise the primordial metabolism (Martin and Russell 2007; Martin et al. 2008; Lane 2015). When both electrons and carbon sources are organic compounds, the organisms are called chemoheterotrophs, including many bacteria, all fungi, protozoa, and animals.

7.2.6.2 Conceptual Definition of Metabolism

Several conceptual definitions of metabolism have been attempted. The first metabolic pathways were described as involving monomers and small molecules in networks of an autocatalytic set (Segré et al. 2000; Shapiro 2006, 2007; Kauffman 2007). In an autocatalytic set, molecules involved in a reaction catalyze other reactions of the set. In Figure 7.2.10, under a constant supply of A and B reactant molecules, the initial products AB and BA can be formed. Once the starting products are formed, they can catalyze the synthesis of other molecules, *for example*, ABB from AB and B with BA as a catalyst. Thus, the reactions in the set become collectively autocatalytic (Segré et al. 2000; Shapiro 2006; Kauffman 2007).

Autocatalytic networks can develop inside a compartment selectively permeable to prebiotically formed molecules, a key process in the transition from chaos to an ordered form. These compartments could have been simple vesicles composed of amphiphilic organic molecules or mineral surfaces (Wächtershäuser 1988, 1990; Russell and Hall 1997; Lane 2015; Saladino et al. 2016; Hansma 2017).

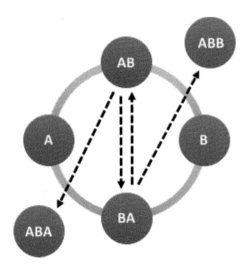

FIGURE 7.2.10 A simple autocatalytic set, with A and B as starting reactants. The dotted arrows represent catalysis. (Modified from Kauffman, S., *Orig. Life Evol. Biosph.*, 37, 315–322, 2007.)

Shapiro (2007) has defined five minimal requirements for small molecules to form a simple metabolism-based form of life, without the need for information-storing molecules such as RNA. A source of energy, such as a photocatalytic mineral and sunlight, is necessary, which drives an exothermic chemical reaction. A chemical reaction network must develop and become more complex to allow for adaptation and evolution. The rate of mass used in the reaction network to make complex organic molecules must be faster than the rate of degradation of these molecules in the environment, so that the system can grow and reproduce.

7.2.6.3 Core Metabolism

It has been proposed that a small reaction network pathway with connected internal sub-cycles, whose intermediates are capable of giving rise to all other biosynthetic molecules and is widely distributed among all species, must have formed a primitive core metabolism (Tortora et al. 2004). The tricarboxylic acid (TCA) or citric acid cycle fits this bill. The TCA is an amphibolic pathway (Figure 7.2.11).

It can function in a forward sense to produce intermediate molecules, used for amino acid and other biomolecule synthesis (anabolism). In reverse, the TCA can be used for ATP generation (catabolism). For the anaerobic conditions present on prebiotic Earth, the reverse citric acid cycle or reductive tricarboxylic acid cycle (rTCA) has been proposed to constitute the core universal metabolism (Hartman 1975; Corliss et al. 1980; Kandler and Stetter 1981; Wächtershäuser 1990; Smith and Morowitz 2016). Acetyl-coenzyme A (coA) and succinate are among the key molecules in this pathway. A small core group of metabolites or "sparse metabolism" consisting of acetate, pyruvate, oxaloacetate, succinate, and α-ketogluarate has been proposed as the "standard universal precursors" for all biosynthetic pathways (anabolism) (Smith and Morowitz 2016). This selection is believed to be

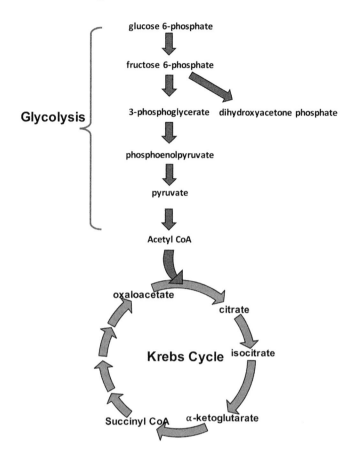

Glycolysis

glucose 6-phosphate

fructose 6-phosphate

3-phosphoglycerate   dihydroxyacetone phosphate

phosphoenolpyruvate

pyruvate

Acetyl CoA

oxaloacetate

citrate

Krebs Cycle isocitrate

Succinyl CoA   α-ketoglutarate

FIGURE 7.2.11 Schematic for an amphibolic pathway that involves both anabolism and catabolism.

based on chemical kinetic arguments. Smith and Morowitz (2016) recognize that while the rTCA is common in the eubacterial and eukaryotic domains, archaea possess the dicarboxylate-4-hydroxybutyrate cycle. This cycle shares the same reactions from acetyl-coA to succinate. Acetyl-coA is significant not only because it drives organic molecule synthesis but also because it reacts with phosphate to form acetyl phosphate, which may have been a precursor to ATP (Lane 2015).

A different primordial metabolism was proposed by other workers (Martin and Russell 2007; Nitschke and Russell 2013; Weiss et al. 2016). These authors proposed the Wood-Ljundahl (WL) pathway, which leads to acetyl-coA and is used by acetogens, methanogens, and methanotrophs. The WL pathway combines both catabolism and anabolism, *that is*, energy harvesting and carbon fixation, respectively, and requires H_2 as the terminal electron donor and CO_2 as the carbon source with iron sulfide clusters in protoenzymes.

As with most issues in the OoL field, there is also a different perspective on whether a core or sparse metabolism was a feature of protocells, at all. Virgo and Ikegami used computer algorithms of abstract model reactions to show that intersecting and branching autocatalytic reaction networks form easily, without the need for an initial catalytic molecule to be present. They argued that the earliest metabolisms were "maximal" and that the rTCA or other minimal metabolic systems were derived by later evolution (Virgo and Ikegami 2013).

7.2.6.4 Chemoautotrophic Protometabolism

7.2.6.4.1 Chemoautotrophy at Acidic, High-Temperature Hydrothermal Vents

Soon after the discovery of acidic hydrothermal vents, a chemoautotrophic origin of life on mineral surfaces was proposed. According to this model, the energy required to fix carbon is provided at high-temperature hydrothermal vent systems by iron-rich minerals such as pyrite (FeS_2), formed by the combination of pyrrhotite (FeS) and H_2S or HS^- (Corliss et al. 1980; Wächtershäuser 1988, 1990). These dark-colored sulfide minerals give such vents their nickname as "black smokers." Carbon monoxide or carbon dioxide was the proposed carbon source. The products of the reactions were proposed to remain fixed to the pyrite surface, where they subsequently developed a network of surface-based reactions in an early "surface metabolism." Following this hypothesis, aliphatic hydrocarbons, aromatic hydrocarbons, and carboxylic acids have been produced under simulated hydrothermal conditions by FTT reactions, in which a solid surface catalyzes the reduction of CO or CO_2 by H_2 (reviewed in McCollum et al. 1999; McCollom 2013; Dalai and Sahai 2016b). Amino acid oligomers (*e.g.*, alanine oligomers) were detected in experiments simulating hydrothermal systems from alanine monomer (Kawamura et al. 2005). Pyruvate has been synthesized from formic acid in the presence of pyrrhotite under hydrothermal vent conditions at 250°C and 50–20 MPa pressure (Cody et al. 2000). Reaction of pyruvate with H_2S, H_2, and NH_4^+ in the presence of various transition metal sulfide minerals, including pyrite, pyrrhotite, arsenopyrite, marcasite, and sphalerite, at 25°C–100°C, yielded a small number of molecules (pyruvate, lactate, thiolacetate, thiolacetate disulfide, thiolacetate persulfide, propionate, and alanine) rather than a large number of potential products (Novikov and Copley 2013). This result seems to be consistent with the concept of a small set of core metabolites or sparse metabolism. The carboxylic acids and their thiolated derivatives (thioacids and thioesters) produced in these types of reactions are high-energy molecules that are regarded as prebiotic analogs to acetyl-CoA, a key molecule in many biosynthetic pathways.

These are very interesting results, and the plausibility of such systems has been examined. In one study combining experiments with density functional theory modeling of the thermodynamics and kinetics of thioacetate and methylthioacetate hydrolysis, it was shown that these species could not have accumulated abiotically to concentrations sufficient for maintaining metabolic networks (Chandru et al. 2016). However, based on the kinetics of thioester exchange reaction, it was concluded elsewhere that theories invoking thioesters as prebiotic analogs of acetyl-coA can neither be accepted nor be rejected *a priori* (Bracher et al. 2011). The studies summarized here provide promising avenues for future research. An especially important step would be to show how these early metabolic molecules could be encapsulated and sustained within a membrane.

7.2.6.4.2 Chemoautotrophy at Alkaline, Low-Temperature Hydrothermal Vents

In the low-temperature hydrothermal vent theory, it is hypothesized that the earliest membranes were not organic, but rather, the walls of pores in the mineral chimneys precipitated at alkaline hydrothermal vents or white smoker vents (Russell and Hall 1997; Martin et al. 2008; Lane 2015; Figure 7.2.12).

The energy for organic molecule synthesis is proposed to have been provided by the geothermal and geochemical gradients at these vents. At such sites, highly alkaline (pH ~11–12) water is produced by hydrothermal alteration of seafloor peridotites that would come in contact with ambient seawater, which would have been mildly acidic (pH ~6–6.5) because of equilibrium with a CO_2-rich atmosphere. The flow of water with two very different pHs across the pores in the precipitated mineral chimneys would have produced an ancient geochemical equivalent of the chemiosmotic gradient, which is apparently a universal feature of all extant life-forms. This geochemical gradient would have provided the energy required for the synthesis of organic compound precursors from inorganic starting compounds such as dissolved CO_2, CH_4, and NH_3. Thus, a deep sub-aqueous, hydrothermal vent environment is proposed, eventually leading to the emergence of chemoautotrophic metabolism and replication.

The question of metabolism and encapsulation within a membrane has been approached creatively by proponents of a chemoautotrophic origin of life at low-temperature, alkaline hydrothermal vents, wherein the walls of pores in the mineral precipitates at the vent are proposed to have formed the first "mineral membranes" (Russell and Hall 1997; Russell and Martin 2004; Martin and Russell 2007; Martin et al. 2008; Lane 2015). This kind of vent produces highly alkaline fluids at temperatures of 60°C–90°C, and carbonate minerals

precipitated result in the so-called "white smokers." The process involves the reaction of seawater with the olivine and pyroxene in upper mantle peridotite and the formation of the secondary minerals, serpentine, magnetite, brucite, and carbonates, along with the release of H. The H_2 gas reacts with CO_2 in the seawater and reduces it to methane. Ferrous iron present in the ambient seawater would have formed iron sulfides. Russell and Hall (1997) first proposed that the walls of pores in the iron sulfide precipitates might have served as the first inorganic membranes. Under simulated white smoker hydrothermal conditions, "chemical gardens" were produced from the precipitation reaction in which the concentrations of the reactants (e.g., dissolved Fe^{2+} in the acidic primordial ocean and dissolved sulfide in the alkaline hydrothermal fluid) are far from equilibrium (Barge et al. 2014). The transition metal sulfide chemical gardens can generate electrochemical energy and act as a catalyst. Molecules produced in these conditions were proposed to be concentrated by the process of thermophoresis, wherein larger molecules will have lower kinetic energy than smaller ones and, with a thermal gradient, will accumulate in the cooler region. This was shown experimentally for various protonated states of phosphate ion (Duhr and Braun 2006; Reichl et al. 2014).

The mineral pore walls are proposed to have constituted a leaky membrane, which would have allowed H^+, OH^-, Na^+, Cl^-, and ions present in seawater to enter and leave the volume of space defined by the pore wherein the protocell is hosted. However, a proton gradient would have been maintained across the inside and outside of the pore space across the mineral wall, because it separates the acidic ocean water (pH 5–6) from the alkaline hydrothermal fluid (pH 9–11) (Lane 2015). The availability of energy in the form of the pH gradient was proposed to have played a key role in the development of metabolism based on acetyl-coA, which begins with the reaction of H_2 and CO_2, eventually leading to an intermediate stage involving an Na^+/H^+ antiporter and, finally, a free-living cell that could escape the vent environment. Acetogens and methanogens are proposed to have been the bacterial and archaeal ancestors. These are very interesting and provocative ideas that provide avenues for future experimental research.

7.2.6.5 HETEROTROPHIC PROTOMETABOLISM

Heterotrophs use external sources of organic molecules from their surroundings. These organisms are classified as chemoheterotrophs if the energy source is electrons from other organic compounds and as photoheterotrophs if the energy source is from pigment molecules activated by UV light. A large literature exists on the synthesis of lipids, nucleotides, and amino acids and their oligomers under Earth's surface conditions, without or with UV light as the energy source. Most conceptualizations of the RNA world require a heterotrophic protometabolism. For example, Mansy et al. (2008) have shown the permeability of charged and polar molecules such as activated nucleotides (AMP and ADP) and ribose across specific mixture of FA membranes (e.g., MA-GMM, DA-DOH-GMD, and OA). Eventually, the activated mononucleotides

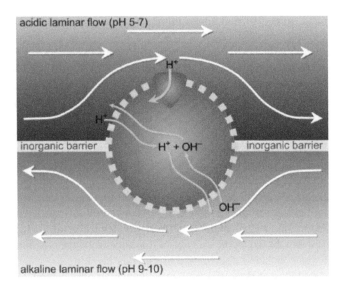

FIGURE 7.2.12 Hypothetical protocell with mineral pore walls, as membrane formed at a low-temperature alkaline hydrothermal vent. Ambient fluid flow of two different pHs establishes a chemiosmotic potential across the mineral wall. (From Sojo, V. et al., *PLoS Biol.*, 12, e1001926, 2014.)

can participate in oligomerization by copying the template encapsulated in the vesicle interior. Thus, it was possible for heterotrophic protocells to have utilized nutrients from their environment in the absence of any transport machinery. However, autotrophy was not completely ruled out, because it is also possible that metabolites produced inside the protocell may have leaked out and not have been available for use.

In an interesting new study, it was shown that iron-sulfur clusters can be formed simply by UV irradiation of ferrous ions and photolysis of organic thiols in solution. These Fe_2S_2 and Fe_4S_4 clusters are stabilized by cysteine-containing peptides (Bonfio et al. 2017). Thus, iron-sulfur bearing protoenzymes may have evolved on prebtioic Earth to support a heterotrophic protometabolism.

Some experimental attempts have been made to mimic a photoheterotrophic protometabolism by using lipid vesicles as model protocells. In general, an artificial photoheterotrophic energy-transducing system is composed of (1) an extra vesicular reductant/terminal electron donor (e.g., EtOH, EDTA, and ascorbic acid) analogous to water in natural photosynthesis, (2) a photosensitizer organic compound or photocatalytic mineral comparable to the calcium-manganese complex in chlorophyll (e.g., ruthenium bipyridinium, TiO_2, CdSe, and FeS_2), (3) a transmembrane electron shuttle incorporated into the bilayer (e.g., PAHs, which mimic the quinones in the photosynthesis electron transfer chain), and (4) an encapsulated oxidant or terminal electron acceptor analogous to O_2 (e.g., rhodium bipyridinium, $Rh(bpy)_3^{3+}$, or ferricyanide, $Fe(CN)_6^{3+}$, or NAD) (Figure 7.2.13).

Using PL (dihexadecyl phosphate [DHP]) vesicles, Fendler and coworkers demonstrated TMETRs in the presence of $Ru(bpy)_3^{3+}$ as photosensitizer and photocatalytic minerals or quantum dots (CdSe) (Tunuli and Fendler 1981; Horváth and Fender 1992) (Figure 7.2.13a). When irradiated by UV light, these mineral nanoparticles generate an electron-hole pair

and can pass the electrons via $Ru(bpy)_3^{3+}$ to various electron transfer reactions. A TMETR was also reported across an FA membrane by using only an intramembrane PAH and no photosensitizer (Figure 7.2.13b). Thus, the PAH appears to have played the role of both photosensitizer and transmembrane electron shuttle in an FA membrane (Cape et al. 2011). These authors showed that an anionic hydrophilic species such as ferricyanide could be retained for long periods inside the negatively charged DA membranes. The long retention time was ascribed to charge-charge repulsion between the anionic solutes and the carboxylate headgroups. Further, PAH probably increased the stability of the membrane (Groen et al. 2012). While transmembrane electron transport is a key step in the simulation of a photoheterotrophic metabolism, it is only a single reaction. The demonstration of a coupled reaction network inside a model protocell remains a grand challenge in the field.

In contrast to a TMETR, electron transfer reaction with the photocatalyst, oxidant, and reductant, all encapsulated within the volume of a PL membrane, has also been shown (Summers et al. 2009; Summers and Rodoni 2015) (Figure 7.2.13c). However, these were not transmembrane redox reactions.

7.2.6.6 PREBIOTIC PHOSPHORYLATION

Phosphorylation using the orthophosphate anion is generally thermodynamically unfavorable in aqueous medium and ambient temperatures, because it is a dehydration (condensation) reaction. Activated phosphorylated molecules such as ATP are the energy-rich molecules that conduct phosphorylation in modern metabolism. The importance of phosphorous bioenergetics and the requirement for compartmentalization have been highlighted (Kee and Monnard 2016). As we have seen previously, ATP is generated in extant life-forms by substrate-level phosphorylation, oxidation phosphorylation,

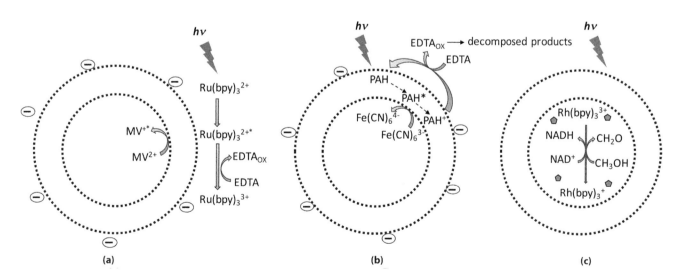

FIGURE 7.2.13 Schematic representation of model photoheterotrophic protometabolism across or within PL or FA vesicle membranes, as reported in the literature. Transmembrane electron transfer reactions across (a) a PL membrane with $Ru(bpy)_3^{3+}$ as a photosensitizer (Tunuli and Fendler 1981); (b) a FA membrane with PAH as a photosensitizer (Cape et al. 2011); and (c) redox reaction within PL vesicle without transmembrane electron transfer (Summers and Rodoni 2015). See text for details and description of abbreviations.

or photophosphorylation. The latter two mechanisms involve chemiosmosis and the phosphorylation of ADP to ATP, whereas substrate-level phosphorylation involves phosphorylated organic compound containing a high-energy C-O-P bond, transferring its phosphate group to ADP. Model prebiotic phosphorylation studies have been reviewed recently (Pasek and Kee 2011; Gull 2014). Several studies in the OoL field have attempted to mimic prebiotic substrate-level phosphorylation, where the high-energy bond or phosphate moiety is part of a phosphate mineral. Unfortunately, most common metal phosphate minerals such as hydroxyapatite are orthophosphates and do not possess high-energy bonds, and geologically common phosphate minerals are not very soluble. These two issues have been addressed in different ways in the OoL field.

Polyphosphates are oligomers of phosphate having high-energy P-O-P bonds similar to those in nucleoside di- and triphosphates. Inorganic pyrophosphate and trimetaphosphates have been synthesized in experiments, representing model volcanic environments, and pyrophosphate (P_2O_7) concentration of ~0.45 μM was measured in fumarole gases (Yamagata et al. 1991). Cyclic trimetaphosphate were also studied as potential phosphorylating organic molecules (Krishnamurthy et al. 1999). Trimetaphosphates and larger polyphosphates can phosphorylate adenosine, but pyrophosphate cannot (Schwartz and Ponnamperuma 1968). Diamidophosphate has been shown recently to have facile phosphorylation capability, but the molecule requires trimetaphosphate for its own synthesis (Gibard et al. 2017). Trimetaphosphates and higher polyphosphates are rare in ambient aqueous environments.

Another interesting proposal involves the mineral schreibersite [(Fe,Ni)P], which contains an active, reduced form of phosphorous (Pasek and Lauretta 2008). This mineral is found in meteorite impact sites and lightning strike sites. It was estimated that schreibersite might have contributed up to ~10% of the Earth's crustal phosphorus (Pasek and Lauretta 2008; Pasek et al. 2008; Bryant et al. 2013; Gull et al. 2014). Schreibersite oxidizes to form phosphite, hypophosphate, pyrophosphate, and orthophosphate, which, in turn, could have directed surface-mediated chemistry, *for example*, phosphorylation.

The low solubility of phosphates (as insoluble metal phosphates) has limited their use as phosphorylating agents (Gilbert 1986; Hud et al. 2013). However, eutectic solutions containing formamide increase the solubility of phosphates, and solutions containing urea promote condensation reactions because urea is hydrolyzed in the process. Costanzo et al. (2007) have demonstrated the phosphorylation of nucleosides by various phosphate minerals in liquid non-aqueous formamide environment. They have characterized phosphate minerals as inactive (*e.g.*, herderite Ca[BePO$_4$F]), low-level (*e.g.*, pseudomalachite $Cu^{2+}{}_5(PO_4)_2(OH)_4$), and active phosphorylating agents (*e.g.*, libethenite $Cu^{2+}{}_2(PO_4)(OH)$ and hydroxylapatite $Ca_5(PO_4)_3OH$)), based on their phosphorylating efficiency. Phosphorylating minerals act as phosphate donors to nucleosides, yielding 5'- and 3'-AMP in formamide conditions.

Urea:choline chloride (3:1) eutectic and urea-ammonium formate-water (1:2:4) eutectics or simple non-eutectic aqueous solutions have been employed successfully to phosphorylate biomolecules and to promote DNA and RNA secondary structure formation (Mamajanov et al. 2010; Albertsen et al. 2014; Gull et al. 2015; Burcar et al. 2016).

Although phosphorylation by monomeric phosphate anion is unfavorable at normal temperatures and phosphate concentrations, synthesis of phosphorylated biomolecules was achieved by using very high (1 M) phosphate concentration (Powner et al. 2009; Patel et al. 2015). Whether such a high concentration of phosphate could have been reached under plausible early Earth conditions is an open question.

7.2.7 CO-EVOLUTION OF PROTOMETABOLISM, REPLICATION AND MEMBRANES

The above-mentioned summary of studies involving model protocells in either the RNA world or the metabolism-first world shows that both hypotheses have several problems when actually tested experimentally. Thus, a growing number of workers are turning to the idea that the protocell membrane, RNA, and proteins may have co-evolved in a mutualistic manner. This makes sense from the point of view that natural environments are multicomponent systems. Charles Darwin first suggested that life may have spontaneously generated in a "warm little pond" in the presence of ammonia, phosphorus, and other inorganic salts with various energy sources (Peretó et al. 2009). Later, Oparin and Haldane independently developed the "primordial soup" hypothesis (Fry 2006; Kwok 2013). Further, simple organic molecules either formed abiotically on Earth and/or exogenously transferred to the Earth's surface could have accumulated in the primordial soup. These organic molecules adsorbed on the various minerals would have increased their local concentration. Thus, components needed for the OoL may have been present in the same local environment. Indeed, it is difficult to imagine how molecules produced under one specific set of solvent or pH or temperature conditions, say RNA nucleotide or polymer, would be transported and remain stable in another environment where some other essential molecule, such as a lipid, had been synthesized in order to provide the molecular building blocks of a protocell. Following this idea, simultaneous, one-pot abiotic synthesis of RNA mononucleotides, amino acids, and lipid precursors (*e.g.*, glycerol-1-phosphate) has been successfully demonstrated by reductive homologation (chain extension) of hydrogen cyanide driven by UV light (Patel et al. 2015), although a caveat is that a very high concentration of dissolved phosphate (up to 1 M) is required. Achieving such high concentrations under plausible early Earth environmental conditions is a challenge. Nonetheless, up to one order of magnitude less total dissolved phosphate (~80 mM) may be achievable under specific environmental conditions of temperature and specific atmospheric gas composition, including a particular carbon dioxide content after numerous wetting-drying cycles (Sahai and Schoonen unpublished data).

7.2.7.1 MEMBRANE-BOUND PEPTIDES AND NUCLEOTIDE OLIGOMERIC COMPLEXES

One example of co-evolution is offered by membrane-bound peptides. Membrane-spanning molecular complexes are proposed to have been the first "information-rich" molecules, which may have played important functions such as signal transduction and material transport across membranes by creating pores or defects (Pohorille et al. 2005; Bywater 2009; Wilson et al. 2014). These peptides would be more stabilized in lipid bilayer than in aqueous solution because of decreased interaction between water molecules and those N–H and C=O groups that are in the membrane, thus protecting the peptide from hydrolysis. More complex peptides with catalytic activity and channel-forming ability would have developed later in evolutionary history. The evolution of early membrane-bound peptides, which may have promoted transmembrane solute transport and energy transduction, could have eventually led to ATP synthases (Mulkidjanian et al. 2009).

The association of peptides with the protocell membrane may have affected protocellular processes such as enhancing RNA polymerization and replication. Positively charged arginine-rich peptides are well-known antimicrobial compounds to modern bacterial cells. These peptides associate with negatively charged cell membranes and hydrophobic amino acid portions of the peptide insert into the PL membrane leaflets (Bechara and Sagan 2013; Herce et al. 2014; Takeuchi and Futaki 2016). In a similar manner, basic and amphipathic/lipophilic peptides (with arginine, histidine, isoleucine, tryptophan, and phenylalanine residues) were found to associate with OA-POPC and pure POPC membranes (Kamat et al. 2015). In the process, the peptide was able to cross the PL bilayer and attract negatively charged RNA nucleotides encapsulated in the vesicle. This localized the RNA to the membrane. The extent of nucleotide localization was dependent on the number of positively charged and hydrophobic amino acid residues (Kamat et al. 2015). Increased cationic and hydrophobic characters in the peptide increased nucleotide localization. These results obtained in PL membranes were compared with FA (OA) membrane by using the same peptides. While some localization was observed, the extent was lesser in FA than in PL membranes. This is because of the electrostatic repulsion between the negatively charged RNA nucleotides in the vesicle and the OA in the inner leaflet of the membrane. Furthermore, strong electrostatic binding of the cationic peptide to the outer leaflet of the OA membrane compared with weaker interactions with the zwitterionic PL membrane would also have limited translocation of the peptide into the membrane. These experiments suggest a possible route by which RNA could have been localized at protocell membranes, thus increasing their concentration at the inner leaflet, ultimately favoring subsequent RNA polymerization or replication.

Yarus and coworkers have found that PL membrane-bound supramolecular (trimeric) RNA complexes increased vesicle permeability to Na^+ (Khvorova et al. 1999; Vlassov et al. 2001; Janas and Yarus 2003). Specific conformations of RNA were needed for the ability to bind and change permeability. The RNA molecules probably interacted with the positively charged choline moiety of the PL membranes and remained external to the vesicles. These experiments were conducted on pure PL membranes, which probably do not reflect the earliest protocells.

7.2.7.2 SYNERGISM AMONGST CATALYSTS AND CO-EVOLUTION OF THE MOLECULAR BUILDING BLOCKS OF LIFE

Non-enzymatic RNA polymerization promoted separately by minerals, lipids, and peptide catalysis has been reported (Ferris and Ertem 1992; Rajamani et al. 2008; Wieczorek et al. 2013). Various possible interactions between the reaction catalysts or promoters are possible when they are present together (Kaddour and Sahai 2014). For instance, in synergism, one catalyst helps the other catalyst and the combined catalytic effect in the system is greater than the effect by the addition of the individual catalysts. In antagonism, one catalyst inhibits or decreases the activity of the other. In an additive effect, each catalyst can operate without the influence by the other, and thus, the combined catalytic effect is similar to the effect when added together (Kaddour and Sahai 2014). Synergism among prebiotic oligonucleotides, oligopeptides, and lipid molecules may have initiated protometabolic network and the co-emergence of self-replication and protometabolism (Copley et al. 2007; Li et al. 2013).

7.2.7.3 THE RNA-PROTEIN WORLD AND THE RIBOSOMAL PEPTIDYL TRANSFERASE CENTER

One of the proposals within the RNA-world hypothesis is that RNA ribozymes were eventually replaced by protein enzymes; this is known as the polymer transition stage. This assumption has been challenged because it is deemed strange that all putative ribozymes except one, the ribosome, have been replaced by random Darwinian evolution (Fox 2010; Bowman et al. 2015; Petrov et al. 2015). Williams, Hud, and co-workers argue that it is more likely that the ribosome and, specifically, the peptidyl transferase center (PTC), which constitutes the catalytic core of the ribosome, predated other RNA ribozymes (Hud et al. 2013; Bowman et al. 2015). The PTC core is proposed to have grown by accretion, ultimately resulting in the ribosome as we know it.

The above-mentioned hypothesis is based on observations about the ribosome in extant organisms. The size of ribosomal RNA (rRNA) tends to increase with increasing complexity of organisms, and the large ribosomal subunit is significantly greater in eukaryotes than in prokaryotes. However, a common core, which is structurally very similar across organisms, is maintained (Melnikov et al. 2012; Petrov et al. 2015). In this common core, the regions of the PTC, the decoding center, and tRNA-binding sites have not changed during billions of years of evolution (Ben-Shem et al. 2010; Petrov et al. 2015).

Petrov et al. (2015) have presented a molecular-level model for the origin and evolution of the translation system. Their model explains the evolution of the large and small ribosomal subunits, tRNA, and messenger RNA (mRNA) by accretion and adding expansion segments. The ribosome is proposed to have survived by existing in symbiotic relationship between protein and nucleic acid, suggesting early cooperation between polymers (polynucleotide and polypeptide) of different functions. Similar RNA-protein mutualism has also been suggested by others (Li et al. 2013; Carter 2015).

7.2.8 FROM PROTOCELLULAR TO PROKARYOTIC MEMBRANES

The transition of protocell membranes, whether composed of SCAs or "mineral walls," to PL membranes was a key step in evolution. As discussed earlier, divalent cations may have acted as an environmental selection pressure in the transition of mixed FA-PL membranes, where small amounts of PLs were provided by abiotic synthesis, to predominantly PL membranes (Dalai et al. 2017). Many theories have been proposed for evolution and emergence of the three domains of life, namely, bacteria, archaea, and eukarya, with the widely accepted notion of the existence of the last universal common ancestor (LUCA) from which extant organisms have diverged (Woese and Fox 1977). These hypotheses are mainly based on the membrane composition of the LUCA.

Since bacteria and prokaryotes have PL membranes, it is widely accepted that the LUCA had a PL-based membrane. However, according to the phylogenetic tree of life, bacteria and archaea comprise the deepest branches of life and the archaeal cell membrane differs in both lipid composition and lipid biogenesis pathways (Koga and Morii 2007; Thomas and Rana 2007; Sojo et al. 2014). The bacterial and eukaryotic cell membranes are made up of FA chains linked to an sn-glycerol-3-phosphate (G3P) skeleton by ester bonds (Koga and Morii 2007; Thomas and Rana 2007). The archaeal cell membrane is generally composed of isoprenoid chains linked by ether bonds to an sn-glycerol-1-phosphate (G1P) backbone. The mevalonate or isoprenoid metabolic pathway operates in archaea, in which isopentenyl pyrophosphate and dimethylallyl pyrophosphate are produced, which are eventually used for isoprenoid synthesis involving prenyltransferases (Figure 7.2.14).

Biosynthesis of G3P and G1P is catalyzed by different enzymes that are not evolutionarily related (Koga et al. 1998). Further, the glycerol units of PLs in all archaea and bacteria have opposite stereochemistry. These differences in their membrane chemistry is considered as one of the great unsolved problems in biology (Peretó 2005).

According to Wächterhäuser (2003), the LUCA might have had an unstable membrane with both the G1P and G3P backbone, which slowly diverged into organisms with more stable homochiral membrane. Thus, the LUCA might have existed with both the G1PDH (G1P dehydrogenase) and G3PDH (G3P dehydrogenase). It has been proposed that environmental factors may have led to the divergence of archaea and bacteria with different hydrocarbon chains (Valentine 2007; Lombard et al. 2012; Koga 2014). Supporting this proposal is the observation that archaeal and bacterial membranes reflect some similarities in their biosynthetic pathways. The glycerol phosphate (GP) backbone in both domains is synthesized by a reduction of dihydroxyacetone phosphate (DHAP), using NADH (Jain et al. 2014). Further, the headgroup attaches through a cytidine-diphosphate intermediate in both archaea and bacteria (Koga 2014).

A different perspective is that the LUCA had a cell wall composed of mineral pore surfaces, and its survival depended on the natural pH gradient of seawater mixing with alkaline white smoker vent fluids (Lane 2015; Sections 7.2.2.3 and 7.2.6.4.2). With the eventual evolution of an Na+/H+ antiporter pump, bacteria and archaea were able to diverge and escape from an "attached" mode of existence at mineral surfaces (Sojo et al. 2014).

Finally, the eukaryotes presumably emerged as a consequence of endosymbiosis (Margulis 1970, 1981), in which a bacterial cell hosted an archaeal cell (Lake and Rivera 1994). However, a large number of variations of the endosymbiont theory have been proposed, which are outside the scope of the present chapter. The reader is referred to an excellent recent review on the topic (Martin et al. 2015).

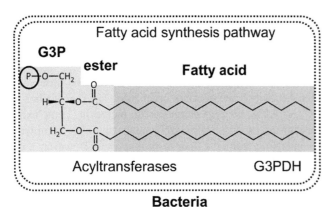

FIGURE 7.2.14 Comparison of bacterial and archaeal membrane phospholipid structures. (Modified from Jain, S. et al., *Front. Microbiol.*, 5, 1–16, 2014.)

7.2.9 SUMMARY

The membrane is the defining boundary of a cell, the basic unit of life. Compartmentalization offered many potential benefits to the emergence of protocells of increasing complexity, from simple amphiphilic vesicles or mineral pore walls to modern cell membranes. Model protocell membranes exhibit a wide spectrum of interesting properties similar to, yet distinct from, those of modern PL membranes. Concentration of solutes within the membrane, competition between different populations of cells, development of bioenergetics, and cooperative interactions between different molecules ultimately leading to more complex ones with simple functions have all been shown to be promoted by membranes. Many of these elegant experiments have been conducted in special solvents, solution compositions, or temperature. The co-evolution of the molecular building blocks of life, and the development of folding and function, in the presence of the many cations and anions in geochemical aqueous solutions is one of the challenges in the OoL field.

ACKNOWLEDGMENTS

The authors thank the present and former Sahai group members for interesting discussions. N.S. is grateful for financial support was provided by National Science Foundation-Earth Sciences (NSF-EAR) grant (EAR-1251479), the Simons Foundation Origins of Life Initiative award (290359), and "start-up" funds from the University of Akron.

REFERENCES

Adamala, K., and J. W. Szostak. 2013a. Competition between model protocells driven by an encapsulated catalyst. *Nat. Chem.* 5: 495–501.

Adamala, K., and J. W. Szostak. 2013b. Nonenzymatic template-directed RNA synthesis inside model protocells. *Science* 342: 1098–1100.

Albertsen, A. N., C. D. Duffy, J. D. Sutherland et al. 2014. Self-assembly of phosphate amphiphiles in mixtures of prebiotically plausible surfactants. *Astrobiology* 14: 462–472.

Amend, J. P., and E. L. Shock. 2001. Energetics of overall metabolic reactions of thermophilic and hyperthermophilic Archaea and bacteria. *FEMS Microbiol. Rev.* 25: 175–243.

Apel, C. L., and D. W. Deamer. 2005. The formation of glycerol monodecanoate by a dehydration/condensation reaction: Increasing the chemical complexity of amphiphiles on the early Earth. *Orig. Life Evol. Biosph.* 35: 323–332.

Apel, C. L., D. W. Deamer, and M. N. Mautner. 2002. Self-assembled vesicles of monocarboxylic acids and alcohols: Conditions for stability and for the encapsulation of biopolymers. *Biochim. Biophys. Acta Biomembranes* 1559: 1–9.

Barge, L. M., T. P. Kee, I. J. Doloboff et al. 2014. The fuel cell model of abiogenesis: A new approach to origin-of-life simulations. *Astrobiology* 14: 254–270.

Bechara, C., and S. Sagan. 2013. Cell-penetrating peptides: 20 years later, where do we stand? *FEBS Lett.* 587: 1693–1702.

Belmonte, L., and S. S. Mansy. 2016. Metal catalysts and the origin of life. *Elements.* doi:10.2113/gselements.12.6.413.

Ben-Shem, A., L. Jenner, G. Yusupova et al. 2010. Crystal structure of the eukaryotic ribosome. *Science* 330: 1203–1209.

Berclaz, N., E. Blöchliger, M. Müller et al. 2001. Matrix effect of vesicle formation as investigated by cryotransmission electron microscopy. *Phys. Chem. B* 105: 1065–1071.

Bernhardt, H. S. 2012. The RNA world hypothesis: The worst theory of the early evolution of life (except for all the others). *Biol. Direct* 7: 23.

Black, R. A., M. C. Blosser, B. L. Stottrup et al. 2013. Nucleobases bind to and stabilize aggregates of a prebiotic amphiphile, providing a viable mechanism for the emergence of protocells. *Proc. Natl. Acad. Sci. U.S.A.* 110: 13272–13276.

Blöchliger, E., M. Blocher, P. Walde et al. 1998. Matrix effect in the size distribution of fatty acid vesicles. *J. Phys. Chem. B* 102: 10383–10390.

Bonfio, C., L. Valer, S. Scintilla et al. 2017. UV-light-driven prebiotic synthesis of iron-sulfur clusters. *Nat. Chem.* doi:10.1038/nchem.2817.

Bowman, J. C., N. V. Hud, and L. D. Williams. 2015. The ribosome challenge to the RNA world. *J. Mol. Evol.* doi:10.1007/s00239-015-9669-9.

Bracher, P. J., P. W. Snyder, B. R. Bohall et al. 2011. The relative rates of thiol–thioester exchange and hydrolysis for alkyl and aryl thioalkanoates in water. *Orig. Life Evol. Biosph.* 41: 399–412.

Braganza, L. F., and D. L. Worcester. 1986. Hydrostatic pressure induces hydrocarbon chain interdigitation in single-component phospholipid bilayer. *Biochemistry* 25: 2591–2596.

Bryant, D. E., D. Greenfield, R. D. Walshaw et al. 2013. Hydrothermal modification of the Sikhote-Alin iron meteorite under low pH geothermal environments. A plausibly prebiotic route to activated phosphorus on the early Earth. *Geochim. Cosmochim. Acta* 109: 90–112.

Budin, I., A. Debnath, and J. W. Szostak. 2012. Concentration-driven growth of model protocell membranes. *J. Am. Chem. Soc.* 134: 20812–20819.

Budin, I., and J. W. Szostak. 2011. Physical effects underlying the transition from primitive to modern cell membranes. *Proc. Natl. Acad. Sci. U.S.A.* 108: 5249–5254.

Budin, I., N. Prwyes, N. Zhang et al. 2014. Chain-length heterogeneity allows for the assembly of fatty acid vesicles in dilute solutions. *Biophys. J.* 107: 1582–1590.

Burcar, B., M. Pasek, M. Gull et al. 2016. Darwins warm little pond: A one-pot reaction for prebiotic phosphorylation and the mobilization of phosphate from minerals in a urea-based solvent. *Angew. Chem. Int. Ed.* 55: 13249–13253.

Bywater, R. P. 2009. Membrane-spanning peptides and the origin of life. *J. Theor. Biol.* 261: 407–413.

Cape, J. L., P.-A. Monnard, and J. M. Boncella. 2011. Prebiotically relevant mixed fatty acid vesicles support anionic solute encapsulation and photochemically catalyzed trans-membrane charge transport. *Chem. Sci.* 2: 661–671.

Carter, C. W. 2015. What RNA world? Why a peptide/RNA partnership merits renewed experimental attention. *Life* 5: 294–320.

Chakrabarti, A. C., and D. W. Deamer. 1992. Permeability of lipid bilayers to amino acids and phosphate. *Biochim. Biophys. Acta* 1111: 171–177.

Chakrabarti, A. C., I. Clark-Lewis, P. R. Harrigan et al. 1992. Uptake of basic amino acids and peptides into liposomes in response to transmembrane pH gradients. *Biophys. J.* 61: 228–234.

Chakrabarti, A. C., and D. W. Deamer. 1994. Permeation of membranes by the neutral form of amino-acids and peptides—Relevance to the origin of peptide translocation. *J. Mol. Evol.* 39: 1–5.

Chan, M., R. H. Himes, and J. M. Akagi. 1971. Fatty acid composition of thermophilic, mesophilic, and psychrophilic clostridia. *J. Bacteriol.* 106: 876–881.

Chandru, K., A. Gilbert, C. Butch et al. 2016. The abiotic chemistry of thiolated acetate derivatives and the origin of life. *Sci. Rep.* 6: 29883. doi:10.1038/srep29883.

Chen, I. A., and J. W. Szostak. 2004a. Membrane growth can generate a transmembrane pH gradient in fatty acid vesicles. *Proc. Natl. Acad. Sci. U.S.A* 101: 7965–7970.

Chen, I. A., and J. W. Szostak. 2004b. A kinetic study of the growth of fatty acid vesicles. *Biophys. J.* 87: 988–998.

Chen, I. A., K. Salehi-Ashtiani, and J. W. Szostak. 2005. RNA catalysis in model protocell vesicles. *J. Am. Chem. Soc.* 127: 13213–13219.

Chen, I. A., R. W. Roberts, and J. W. Szostak. 2004. The emergence of competition between model protocells. *Science* 305: 1474–1476.

Chen, I., and P. Walde. 2010. From self-assembled vesicles to protocells. *Cold Spring Harb. Perspect. Biol.* 2: a002170.

Cheng, Z., and L. P. Luisi. 2003. Coexistence and mutual competition of vesicles with different size distributions. *J. Phys. Chem. B* 107: 10940–10945.

Cistola, D. P., J. A. Hamilton, D. Jackson et al. 1988. Ionization and phase behavior of fatty acids in water: Application of the Gibbs phase rule. *Biochemistry* 27: 1881–1888.

Cleaves, H. J., A. M. Scott, F. C. Hill et al. 2012. Mineral-organic interfacial processes: Potential roles in the origins of life. *Chem. Soc. Rev.* 41: 5502–5525.

Cody, G. D., N. Z. Boctor, T. R. Filley et al. 2000. Primordial carbonylated iron-sulfur compounds and the synthesis of pyruvate. *Science* 289: 1337–1340.

Copley, S. D., E. Smith, and H. J. Morowitz. 2007. The origin of the RNA world: Co-evolution of genes and metabolism. *Bioorg. Chem.* 35: 430–443.

Corliss, J. B., J. A. Baross, and S. E. Hoffman. 1980. Submarine hydrothermal systems: A probable site for the origin of life. School of Oceanography, Oregon State University Oceanography Spec. Pub. 80–88.

Costanzo, G., R. Saladino, C. Crestini et al. 2007. Nucleoside phosphorylation by phosphate minerals. *J. Biol. Chem.* 282: 16729–16735.

Dalai, P., H. L. Pleyer, H. Strasdeit et al. 2016a. The influence of mineral matrices on the thermal behavior of glycine. *Orig. Life Evol. Biosph.* —47: 427–452.

Dalai, P., H. Kaddour, and N. Sahai. 2016b. Incubating life: Prebiotic sources of organics for the origin of life. *Elements* 12: 401–406.

Dalai, P., P. Ustriyana, and N. Sahai. 2017. Aqueous magnesium as an environmental selection pressure in the evolution of phospholipid membranes on early Earth. *Geochim. Cosmochim. Acta* 223: 216–228 (revised).

Deamer, D. W., and G. L. Barchfeld. 1982. Encapsulation of macromolecules by lipid vesicles under simulated prebiotic conditions. *J. Mol. Evol.* 18: 203–206.

Deamer, D. W., and J. Oró. 1980. Role of lipids in prebiotic structures. *Biosystems* 12: 167–175.

Deamer, D. W., and R. M. Pashley. 1989. Amphiphilic components of the Murchison carbonaceous chondrite-surface-properties and membrane formation. *Orig. Life Evol. Biosph.* 19: 21–38.

Devaraj, N. K. 2017. In situ synthesis of phospholipid membranes. *J. Org. Chem.* 82: 5997–6005.

Duhr S., and D. Braun. 2006. Why molecules move along a temperature gradient. *Proc. Natl. Acad. Sci. U.S.A.* 103: 19678–19682.

Engelhart, A. E., K. P. Adamala, and J. W. Szostak. 2016. A simple physical mechanism enables homeostasis in primitive cells. *Nat. Chem.* 8: 448–453.

Ferris, J. P., A. R, Hill Jr., R. Liu et al. 1996. Synthesis of long prebiotic oligomers on mineral surfaces. *Nature* 381: 59–61.

Ferris, J. P., and G. Ertem. 1992. Oligomerization of ribonucleotides on montmorillonite: Reaction of the 5′-phosphorimidazolide of adenosine. *Science* 257: 1387–1389.

Fiore, M., O. Maniti, A. Girard-Egrot et al. 2017. Glass microsphere-supported giant vesicles as tools for observation of self-reproduction of lipid boundaries. *Angew. Chem. Int. Ed.* doi:10.1002/anie.201710708.

Foustoukos, D. I. 2004. Hydrocarbons in hydrothermal vent fluids: The role of chromium-bearing catalysts. *Science* 304: 1002–1005.

Fox, G. E. 2010. Origin and evolution of the ribosome. *Cold Spring Harb. Perspect. Biol.* 2: a003483.

Fox, S. W., and K. Harada. 1960. The thermal copolymerization of amino acids common to protein. *J. Am. Chem. Soc.* 82: 3745–3751.

Fox, S. W. 1980. Metabolic microspheres: Origins and evolution. *Naturwissenschaften* 67: 378–383.

Fry, I. 2006. The origins of research into the origins of life. *Endeavour* 30: 24–28.

Fujikawa, S. M., I. A. Chen, and J. W. Szostak. 2005. Shrink-wrap vesicles. *Langmuir* 21: 12124–12129.

Gebicki, J. M., and M. Hicks. 1976. Preparation and properties of vesicles enclosed by fatty acid membranes. *Chem. Phys. Lipids* 16: 142–160.

Gibard, G., S. Bhowmik, M. Karki et al. 2017. Phosphorylation, oligomerization and self-assembly in water under potential prebiotic conditions. *Nat. Chem.* doi:10.1038/nchem.2878.

Gilbert, W. 1986. Origin of life: The RNA world. *Nature* 319: 618.

Gotoh, M., A. Miki, H. Nagano et al. 2006. Membrane properties of branched polyprenyl phosphates, postulated as primitive membrane constituents. *Chem. Biodivers* 3: 434–455.

Groen, J., D. W. Deamer, A. Kros et al. 2012. Polycyclic Aromatic hydrocarbons as plausible prebiotic membrane components. *Orig. Life Evol. Biosph.* 42: 295–306.

Gull, M. 2014. Prebiotic phosphorylation reactions on the early Earth. *Challenges* 5: 193–212.

Gull, M., M. A. Mojica, F. M. Fernández et al. 2015. Nucleoside phosphorylation by the mineral schreibersite. *Sci. Rep.* 5: 17198. doi:10.1038/srep17198.

Hamilton, J. A. 2003. Fast flip-flop of cholesterol and fatty acids in membranes: Implications for membrane transport proteins. *Curr. Opin. Lipidol.* 14: 263–271.

Hanczyc, M. M., S. M. Fujikawa, and J. W. Szostak. 2003. Experimental models of primitive cellular compartments: Encapsulation, growth, and division. *Science* 302: 618–622.

Hanczyc, M. M., S. S. Mansy, and J. W. Szostak. 2007. Mineral surface directed membrane assembly. *Orig. Life Evol. Biosph.* 37: 67–82.

Hansma, H. G. 2017. Better than membranes at the Origin of Life? *Life* 7: 28.

Hardy, M. D., J. Yang, J. Selimkhanov et al. 2015. Self-reproducing catalyst drives repeated phospholipid synthesis and membrane growth. *Proc. Natl. Acad. Sci. U.S.A.* 112: 8187–8192.

Hargreaves, W. R., and D. W. Deamer. 1978. Liposomes from ionic, single-chain amphiphiles. *Biochemistry* 17: 3759–3768.

Hargreaves, W. R., S. J. Mulvihill, and D. W. Deamer. 1977. Synthesis of phospholipids and membranes in prebiotic conditions. *Nature* 266: 78–80.

Hartman, H. 1975. Speculations on the origin and evolution of metabolism. *J. Mol. Evol.* 4: 359–370.

Hashizume, H. 2012. Role of clay minerals in chemical evolution and the origins of life. In *Clay Minerals in Nature—Their Characterization, Modification and Application*, M. Valaskova and G. S. Martynkova (Eds.), pp. 191–208. Rijeka, Croatia: InTech.

Herce, H. D., A. E. Garcia, and M. C. Cardoso. 2014. Fundamental molecular mechanism for the cellular uptake of guanidinium-rich molecules. *J. Am. Chem. Soc.* 136: 17459–17467.

Horváth, O., and J. H. Fender. 1992. CdS-particle-mediated transmembrane photoelectron transfer in surfactant vesicles. *J. Phys. Chem.* 96: 9591–9594.

Hsiao, C., I.-C. Chou., C. D. Okafor et al. 2013. RNA with iron(II) as a cofactor catalyses electron transfer. *Nat. Chem.* 5: 525–528.

Hud, N. V., B. J. Cafferty, R. Krishnamurthy et al. 2013. The origin of RNA and "My grandfather's Axe." *Chem. Biol.* 20: 466–474.

Israelachvili, J. N. 2011. *Intermolecular and Surface Forces.* Burlington, MA: Academic Press.

Jain, S., A. Caforio, and A. J. M. Driessen. 2014. Biosynthesis of archaeal membrane ether lipids. *Front. Microbiol.* 5: 1–16.

Janas, T., and M. Yarus. 2003. Visualization of membrane RNAs. *RNA* 9: 1353–1361.

Jin, L., N. P. Kamat, S. Jena et al. 2018. Fatty acid/phospholipid blended membranes: A potential intermediate state in protocellular evolution. *Small* 14: 1704077.

Joshi, P. C., and M. F. Aldersley. 2013. Significance of mineral salts in prebiotic RNA synthesis catalyzed by montmorillonite. *J. Mol. Evol.* 76: 371–379.

Joshi, P. C., M. F. Aldersley, J. W. Delano et al. 2009. Mechanism of montmorillonite catalysis in the formation of RNA oligomers. *J. Am. Chem. Soc.* 131: 13369–13374.

Joyce, G. F. 2002. The antiquity of RNA-based evolution. *Nature* 418: 214–221.

Kaddour, H., and N. Sahai. 2014. Synergism and mutualism in non-enzymatic RNA polymerization. *Life* 4: 598–620.

Kamat, N. P., S. Tobé, I. T. Hill et al. 2015. Electrostatic localization of RNA to protocell membranes by cationic hydrophobic peptide. *Angew. Chem. Int. Ed.* 54: 11735–11739.

Kandler, O., and K. O. Stetter. 1981. Evidence for autotrophic CO_2 assimilation in *Sulfolobus brierleyi* via a reductive carboxylic acid pathway. *Zbl. Bakt. Hyg., I. Abt. Orig. C* 2: 111–121.

Kato, M., and R. Hayashi. 1999. Effects of high pressure on lipids and biomembranes for understanding high-pressure-induced biological phenomena. *Biosci. Biotechnol. Biochem.* 63: 1321–1328.

Kauffman, S. 2007. Question 1: Origin of life and the living state. *Orig. Life Evol. Biosph.* 37: 315–322.

Kawamura, K., T. Nishi, and T. Sakiyama. 2005. Consecutive elongation of alanine oligopeptides at the second time range under hydrothermal conditions using a microflow reactor system. *J. Am. Chem. Soc.* 127: 522–523.

Kee, T. P., and P.-A. Monnard. 2016. On the emergence of a proto-metabolism and the assembly of early protocells. *Elements* 12: 419–424. doi:10.2113/gselements.12.6.419.

Khvorova, A., Y.-G. Kwak, M. Tamkun et al. 1999. RNA's that bind and change the permeability of phospholipid membranes. *Proc. Natl. Acad. Sci. U.S.A.* 96: 10649–10654.

Knauth, L. P. 2005. Temperature and salinity history of Precambrian ocean: Implications for the course of microbial evolution. *Palaeogeogr. Palaeoclimatol. Palaeoecol.* 219: 53–69.

Koga, Y. 2014. From promiscuity to the lipid divide: On the evolution of distinct membranes in archaea and bacteria. *J. Mol. Evol.* 78: 234–242.

Koga, Y., and H. Morii. 2007. Biosynthesis of ether-type polar lipids in archaea and evolutionary considerations. *Microbiol. Mol. Biol. Rev.* 71: 97–120.

Koga, Y., T. Kyuragi, M. Nishihara et al. 1998. Did archaeal and bacterial cells arise independently from noncellular Precursors? A hypothesis stating that the advent of membrane phospholipid with enantiomeric glycerophosphate backbones caused the separation of the two lines of descent. *J. Mol. Evol.* 47: 631.

Konhauser, K. 2007. *Introduction to Geomicrobiology*, pp. 36–46. Oxford, UK: Blackwell Publishing.

Krishnamurthy, R., G. Arrhenius, and A. Eschenmoser. 1999. Formation of glycolaldehyde phosphate from glycolaldehyde in aqueous solution. *Orig. Life Evol. Biosph.* 29: 333–354.

Kwok, S. 2013. Where do we come from? In *Stardust*, pp. 1–9. Berlin, Germany: Springer.

Lake, J. A., and M. C. Rivera. 1994. Was the nucleus the first endosymbiont? *Proc. Natl. Acad. Sci. U.S.A.* 91: 2880–2881.

Lane, N. 2015. *The Vital Question: Energy, Evolution, and the Origins of Complex Life*, pp. 122–154. New York: W. W. Norton and Company.

Li, Q., A. Iqbal, M. Perc et al. 2013. Coevolution of quantum and classical strategies on evolving random networks. *PLoS One* 8: e68423.

Lombard, J., P. López-García, and D. Moreira. 2012. The early evolution of lipid membranes and the three domains of life. *Nat. Rev. Microbiol.* 10: 507–515.

Lonchin, S., P. L. Luisi, P. Walde et al. 1999. A matrix effect in mixed phospholipid/fatty acid vesicle formation. *J. Phys. Chem. B* 103: 10910–10916.

Luisi, P. L., M. Allegretti, T. P. de Souza et al. 2010. Spontaneous protein crowding in liposomes: A new vista for the origin of cellular metabolism. *ChemBioChem* 11: 1989–1992.

Maheen, G., G. Tian, Y. Wang et al. 2010. Resolving the enigma of prebiotic C-O-P bond formation: Prebiotic hydrothermal synthesis of important biological phosphate esters. *Heteroat. Chem.* 21: 161–167.

Mamajanov, I., A. E. Engelhart, H. D. Bean et al. 2010. DNA and RNA in anhydrous media: Duplex, triplex, and G-quadruplex secondary structures in a deep eutectic solvent. *Angew. Chem. Int. Ed.* 49: 6310–6314.

Mansy, S. S. 2010. Membrane transport in primitive cells. *Cold Spring Harb. Perspect. Biol.* 2: a002188.

Mansy, S. S., and J. W. Szostak. 2008. Thermostability of model protocell membranes. *Proc. Natl. Acad. Sci. U.S.A.* 105: 13351–13355.

Mansy, S. S., J. P. Schrum, M. Krishnamurthy et al. 2008. Template-directed synthesis of a genetic polymer in a model protocell. *Nature* 454: 122–125.

Margulis, L. 1970. *Origin of Eukaryotic Cells: Evidence and Research Implications for a Theory of the Origin and Evolution of Microbial, Plant and Animal Cells on the Precambrian Earth*, 349 pp. New Haven, CT: Yale University Press.

Margulis, L. 1981. *Symbiosis in Cell Evolution: Life and Its Environment on the Early Earth*. San Francisco, CA: W. H. Freeman.

Martin, W. F., S. Garg, and V. Zimorski. 2015. Endosymbiotic theories for eukaryote origin. *Phil. Trans. R. Soc. B* 370: 20140330.

Martin, W., J. Baross, D. Kelley et al. 2008. Hydrothermal vents and the origin of life. *Nat. Rev. Microbiol.* 6: 805–814.

Martin, W., and M. J. Russell. 2007. On the origin of biochemistry at an alkaline hydrothermal vent. *Phil. Trans. R. Soc. B* 362: 1887–1925.

Maurer, S. E., and P.-A. Monnard. 2009. Integration of primitive metabolic, genetic and structural protocell components under simulated early earth conditions. In *Astrobiology: Physical Origin, Biological Evolution and Spatial Distribution*, S. Hegedus and J. Csonka (Ed.), pp. 59–79. New York: Nova Science Publishers.

Maurer, S. E., D. W. Deamer, J. M. Boncella et al. 2009. Chemical evolution of amphiphiles: Glycerol monoacyl derivatives stabilize plausible prebiotic membranes. *Astrobiology* 9: 979–987.

McCollom, T. M. 2013. Miller-Urey and beyond: What have we learned about prebiotic organic synthesis reactions in the past 60 years? *Annu. Rev. Earth Planet. Sci.* 41: 207–229.

McCollom, T. M., G. Ritter, and B. R. T. Simoneit. 1999. Lipid synthesis under hydrothermal conditions by Fischer–Tropsch-Type reactions. *Orig. Life Evol. Biosph.* 29: 153–166.

McLean, L. R., and M. C. Phillips. 1981. Mechanism of cholesterol and phosphatidylcholine exchange or transfer between unilamellar vesicles. *Biochemistry* 20: 2893–2900.

Melnikov, S., A. Ben-Shem, N. G. de. Loubresse et al. 2012. One core, two shells: Bacterial and eukaryotic ribosomes. *Nat. Struct. Mol. Biol.* 196: 560–567.

Monnard, P.-A., and D. W. Deamer. 2002. Membrane self-assembly processes: Steps toward the first cellular life. *Anat. Rec.* 268: 196–207.

Monnard, P.-A., and D. W. Deamer. 2001. Nutrient uptake by protocells: A liposome model system. *Orig. Life Evol. Biosph.* 31: 147–155.

Monnard, P.-A., C. L. Apel, A. Kanavarioti et al. 2002. Influence of ionic inorganic solutes on self-assembly and polymerization processes related to early forms of life: Implications for a prebiotic aqueous medium. *Astrobiology* 2: 139–152.

Morigaki, K., and P. Walde. 2007. Fatty acid vesicles. *Curr. Opin. Colloid Interface Sci.* 12: 75–80.

Mouritsen, O. G., K. Jorgensen and T. Honger. 1995. Permeability of lipid bilayers near the phase transition. In *Permeability and Stability of Lipid Bilayers*, E. A. Disalvo and S. A. Simon (Eds.), pp. 137–160. Boca Raton, FL: CRC Press.

Mulkidjanian, A. Y., M. Y. Galperin, and E. V. Koonin. 2009. Co-evolution of primordial membranes and membrane proteins. *Trends Biochem. Sci.* 34: 206–215.

Namani, T., and D. W. Deamer. 2008. Stability of model membranes in extreme environments. *Orig. Life Evol. Biosph.* 38: 329–341.

Naoi, M., M. Naoi, T. Shimizu et al. 1977. Permeability of amino acids into liposomes. *Biochim. Biophysi. Acta* 471: 305–310.

Nitschke, W., and M. J. Russell. 2013. Beating the acetyl coenzyme A-pathway to the origin of life. *Phil. Trans. R. Soc. B* 368: 20120258.

Nooner, D. W., and J. Oró. 1979. Synthesis of fatty acids by a closed system Fischer–Tropsch process. *Adv. Chem.* 178: 159–171.

Nordlund, G., J. B. S. Ng, L. Bergström et al. 2009. A membrane-reconstituted multisubunit functional proton pump on mesoporous silica particles. *ACS Nano* 3: 2639–2646.

Novikov, Y., and S. D. Copley. 2013. Reactivity landscape of pyruvate under simulated hydrothermal vent conditions. *Proc. Natl. Acad. Sci. U.S.A.* 110: 13283–13288.

Oleson T. A., and N. Sahai. 2008. Oxide-dependent adsorption and self-assembly of dipalymitoylphosphatidylcholine, a cell-membrane phospholipid: Bulk adsorption isotherms. *Langmuir* 24: 4865–4873.

Oleson, T. A., and N. Sahai. 2010. Interaction energies between oxide surfaces and multiple phosphatidylcholine bilayers from extended-DLVO theory. *J. Colloid. Interf. Sci.* 352: 316–326.

Oleson T. A., N. Sahai N and J. A. Pedersen. 2010. Electrostatic effects on deposition of multiple phospholipid bilayers at oxide surfaces. *J. Colloid. Interf. Sci.* 352: 327–336.

Oleson T. A., D. J., Wesolowski, J. A. Dura et al. 2012. Neutron reflectvity study of phosphatidylcholine bilayers on sapphire (1 1 0). *J. Colloid. Interf. Sci.* 370: 192–200.

Oró, J., S. L. Miller, and A. Lazcano. 1990. The origin and early evolution of life on Earth. *Annu. Rev. Earth Planet. Sci.* 18: 317.

Pasek, M. A., and D. S. Lauretta. 2008. Extraterrestrial flux of potentially prebiotic C, N, and P to the early Earth. *Orig. Life Evol. Biosph.* 38: 5–21.

Pasek, M. A., T. P. Kee, D. E. Bryant et al. 2008. Production of potentially prebiotic condensed phosphates by phosphorus redox chemistry. *Angew. Chem. Int. Ed.* 47: 7918–7920.

Pasek, M. A., and T. P. Kee. 2011. On the origin of phosphorylated biomolecules. In *Origins of Life: The Primal Self-Organization*, R. Egel, D.-H. Lankenau, and A. Y. Mulkidjanian (Eds.), pp. 57–84. Berlin, Germany: Springer.

Patel, B. H., C. Percivalle, D. J. Ritson et al. 2015. Common origins of RNA, protein and lipid precursors in a cyanosulfidic proto-metabolism. *Nat. Chem.* 7: 301–307.

Paula, S., A. G. Volkov, A. N. VanHoek et al. 1996. Permeation of protons, potassium ions, and small polar molecules through phospholipid bilayers as a function of membrane thickness. *Biophys. J.* 70: 339–348.

Paula, S., and D. W. Deamer. 1999. Membrane permeability barriers to ionic and polar solutes. *Curr. Top. Membr.* 48: 77–95.

Peretó, J. 2005. Controversies on the origin of life. *Int. Microbiol.* 8: 23–31.

Peretó, J., J. L. Bada, and A. Lazcano. 2009. Charles Darwin and the origin of life. *Orig. Life Evol. Biosph.* 39: 395–406.

Petkau, A., and W. S., Chelack. 1972. Model lipid membrane permeability to ATP. *Can. J. Biochem.* 50: 615–619.

Petrov, A. S., B. Gulen, A. M. Norris et al. 2015. History of the ribosome and the origin of translation. *Proc. Natl. Acad. Sci. U.S.A.* 112: 15396–15401.

Pohorille, A., K. Schweighofer, and M. A. Wilson. 2005. The origin and early evolution of membrane channels. *Astrobiology* 5: 1–17.

Powner, M. W., B. Gerland, and J. D. Sutherland. 2009. Synthesis of activated pyrimidine ribonucleotides in prebiotically plausible conditions. *Nature* 459: 239–242.

Qiao, Y., M. Li, R. Booth et al. 2017. Predatory behaviour in synthetic protocell communities. *Nat. Chem.* 9: 110–119.

Rajamani, S., A. Vlassov, S. Benner et al. 2008. Lipid-assisted synthesis of RNA-like polymers from mononucleotides. *Orig. Life Evol. Biosph.* 38: 57–74.

Rao, M., J. Eichberg, and J. Oró. 1982. Synthesis of phosphatidylcholine under possible primitive earth conditions. *J. Mol. Evol.* 18: 196–202.

Reichl, M., M. Herzog, A. Götz, and D. Braun. 2014. Why charged molecules move across a temperature gradient: The role of electric Fields. *Phys. Rev. Lett.* 112: 198101.

Rushdi, A. I., and B. R. T. Simoneit. 2001. Lipid formation by aqueous Fischer–Tropsch-Type synthesis over a temperature range of 100 to 400°C. *Orig. Life Evol. Biosph.* 31: 103–118.

Russell, M. J., and A. J. Hall. 1997. The emergence of life from iron monosulphide bubbles at a submarine hydrothermal redox and pH front. *J. Geol. Soc. London* 154: 377–402.

Russell, M. J., and W. Martin. 2004. The rocky roots of the acetyl-CoA pathway. *Trends Biochem. Sci.* 29: 358–363.

Russell, N. J. 1984. Mechanisms of thermal adaptation in bacteria: Blueprints for survival. *Trends Biochem. Sci.* 9: 108–112.

Russell, N. J. 1990. Cold adaptation of microorganisms. *Phil. Trans. R. Soc. B* 326: 595–611.

Russell, N. J., R. I. Evans, P. F. Ter Steeg et al. 1995. Membranes as a target for stress adaptation. *Int. J. Food Microbiol.* 28: 255–261.

Sacerdote, M. G., and J. W. Szostak. 2005. Semipermeable lipid bilayers exhibit diastereoselectivity favoring ribose. *Proc. Natl. Acad. Sci. U.S.A.* 102: 6004–6008.

Sahai, N., H. Kaddour, and P. Dalai. 2016. The transition from geochemistry to biogeochemistry. *Elements* 12: 389–394.

Sahai, N., H. Kaddour, P. Dalai et al. 2017. Mineral surface chemistry and nanoparticle-aggregation control membrane self-assembly. *Sci. Rep.* 7: 43418. doi:10.1038/srep43418.

Saladino, R., G. Botta, B. Mattia Bizzarri et al. 2016. A global scale scenario for prebiotic chemistry: Silica-based self-assembled mineral structures and formamide. *Biochemistry* 55: 2806–2811.

Schoonen, M., and A. Smirnov. 2016. Staging life in an early warm 'seltzer' ocean. *Elements.* doi:10.2113/gselements.12.6.395.

Schoonen, M., A. Smirnov, and C. Cohn. 2004. A perspective on the role of minerals in prebiotic synthesis. *Ambio* 33: 539–551.

Schwartz, A., and C. Ponnamperuma. 1968. Phosphorylation on the primitive Earth: Phosphorylation of adenosine with linear polyphosphate salts in aqueous solution. *Nature* 218: 443.

Segré, D., D. Ben-Eli, and D. Lancet. 2000. Compositional genomes: Prebiotic information transfer in mutually catalytic noncovalent assemblies. *Proc. Natl. Acad. Sci. U.S.A.* 97: 4112–4117.

Segré, D., D. Ben-Eli, D. W. Deamer et al. 2001. The lipid world. *Orig. Life Evol. Biosph.* 31: 119–145.

Shapiro, R. 2006. Small molecule interactions were central to the origin of life. *Q. Rev. Biol.* 81: 105–126.

Shapiro, R. 2007. A simpler origin for life. *Sci. Am.* 296: 46–53.

Smith, E. A., and P. K. Dea. 2013. Differential scanning calorimetry studies of phospholipid membranes: The interdigitated gel phase. In *Applications of Calorimetry in a Wide Context— Differential Scanning Calorimetry, Isothermal Titration Calorimetry and Microcalorimetry*, A. A. Elkordy (Ed.), pp. 407–444. Rijeka, Croatia: InTech.

Smith, E., and H. J. Morowitz. 2016. *The Origin and Nature of Life on Earth: The Emergence of the Fourth Geosphere*, pp. 170–272. Cambridge, UK: Cambridge University Press.

Sojo, V., A. Pomiankowski, and N. Lane. 2014. A bioenergetic basis for membrane divergence in archaea and bacteria. *PLoS Biol.* 12: e1001926.

Stamatoff, J., D. Guillon, L. Powers et al. 1978. X-ray diffraction measurements of dipalmitoylphosphatidylcholine as a function of pressure. *Biochem. Biophys. Res. Commun.* 85: 724–728.

Stillwell, W., and H. C. Winter. 1974. The stimulation of diffusion of adenine nucleotides across bimolecular lipid membranes by divalent metal ions. *Biochem. Biophys. Res. Commun.* 56: 617–622.

Streiff, S., N. Ribeiro, Z. Wu et al. 2007. Primitive membrane from polyprenyl phosphates and polyprenyl alcohols. *Chem. Biol.* 14: 313–319.

Summers, D. P., and D. Rodoni. 2015. Vesicle encapsulation of a nonbiological photochemical system capable of reducing NAD+ to NADH. *Langmuir* 31: 10633–10637.

Summers, D. P., J. Noveron, and R. C. B. Basa. 2009. Energy transduction inside of amphiphilic vesicles: Encapsulation of photochemically active semiconducting particles. *Orig. Life Evol. Biosph.* 39: 127–140.

Szathmáry, E., and L. Demeter. 1987. Group selection of early replicators and the origin of life. *J. Theor. Biol.* 128: 463–486.

Szostak, J. W., D. P. Bartel, and P. L. Luisi. 2001. Synthesizing life. *Nature* 409: 387–390.

Takajo, S., H. Nagano, O. Dannenmuller et al. 2001. Membrane properties of sodium 2-and 6-(poly)prenyl-substituted polyprenyl phosphates. *New J. Chem.* 25: 917–929.

Takeuchi, T., and S. Futaki. 2016. Current understanding of direct translocation of arginine-rich cell-penetrating peptides and its internalization mechanisms. *Chem. Pharm. Bull.* 64: 1431–1437.

Teo, Y. Y., M. Misran, K. H. Low et al. 2011. Effect of unsaturation on the stability of C_{18} polyunsaturated fatty acids vesicles suspension in aqueous solution. *Bull. Korean Chem. Soc.* 32: 59–64.

Thomas, C. F., and P. L. Luisi. 2005. RNA selectively interacts with vesicles depending on their size. *J. Phys. Chem. B* 109: 14544–14550.

Thomas, J. A., and F. R. Rana. 2007. The influence of environmental conditions, lipid composition, and phase behavior on the origin of cell membranes. *Orig. Life Evol. Biosph.* 37: 267–285.

Tortora, G. J., B. R. Funke, C. L. Case et al. 1995. *Microbiology: An Introduction*, 5th ed. San Francisco, CA: Benjamin Cummings.

Tortora, G. J., B. R. Funke, C. L. Case et al. 2004. *Microbiology: An Introduction*, 12th ed. San Francisco, CA: Benjamin Cummings.

Tunuli, S. M., and J. H. Fendler. 1981. Aspects of artificial photosynthesis. Photosensitized electron transfer across bilayers, charge separation, and hydrogen production in anionic surfactant vesicles. *J. Am. Chem. Soc.* 103: 2507–2513.

Valentine, D. L. 2007. Adaptations to energy stress dictate the ecology and evolution of the Archaea. *Nat. Rev. Microbiol.* 5: 316–323.

Virgo, N., and T. Ikegami. 2013. Autocatalysis before enzymes: The emergence of prebiotic chain reactions. In *Advances in Artificial Life, ECAL 2013, Proceedings of the Twelfth European Conference on the Synthesis and Simulation of Living Systems*, P. Liò, O. Miglino, G. Nicosia, S. Nolfi, and M. Pavone (Eds.), pp. 240–247. Cambridge, MA: The MIT Press.

Vlassov, A., A. Khvorova, and M. Yarus. 2001. Binding and disruption of phospholipids bilayers by supramolecular RNA complexes. *Proc. Natl. Acad. Sci. U.S.A.* 98: 7706–7711.

Wächtershäuser, G. 2003. From pre-cells to Eukarya - A tale of two lipids. *Mol. Microbiol.* 47: 13–23.

Wächterhäuser, G. 2000. Origin of life. Life as we don't know it. *Science* 289: 1307–1308.

Wächtershäuser, G. 1988. Before enzymes and templates: Theory of surface metabolism. *Microbiol. Rev.* 52: 452–484.

Wächtershäuser, G. 1990. Evolution of the first metabolic cycles. *Proc. Natl. Acad. Sci. U.S.A.* 87: 200–204.

Wei, C. Y., and A. Pohorille. 2009. Permeation of membranes by ribose and its diastereomers. *J. Am. Chem. Soc.* 131: 10237–10245.

Weiss, M. C., F. L. Sousa, N. Mrnjavac et al. 2016. The physiology and habitat of the last universal common ancestor. *Nat. Microbiol.* 1: 16116.

Wieczorek, R., M. Dörr, A. Chotera et al. 2013. Formation of RNA phosphodiester bond by histidine-containing dipeptides. *ChemBioChem* 14: 217–223.

Wilson, M. A., and A. Pohorille. 1996. Mechanism of unassisted ion transport across membrane bilayers. *J. Am. Chem. Soc.* 118: 6580–6587.

Wilson, M. A., C. Wei, and A. Pohorille. 2014. Towards co-evolution of membrane proteins and metabolism. *Orig. Life Evol. Biosph.* 44: 357–361.

Woese, C. R., and G. E. Fox. 1977. Phylogenetic structure of the prokaryotic domain: The primary kingdoms. *Proc. Natl. Acad. Sci. U.S.A.* 74: 5088–5090.

Xu J., M. J. Stevens, T. A. Oleson et al. 2009. Role of oxide surface chemistry and phospholipid phase on adsorption and self-assembly: Isotherms and atomic force microscopy. *J. Phys. Chem. C* 113: 2187–2196.

Xu J., J. Campbell, N. Zhang et al. 2012. Evolution of bacterial biofilms as armor against mineral toxicity. *Astrobiology* 12: 785–798.

Xu J., N. Sahai, C. M. Eggleston et al. 2013. Reactive oxygen species at the oxide/water interface: Formation mechanisms and implications for prebiotic chemistry and the origin of life. *Earth Planet. Sci. Lett.* 363:156–167.

Yamagata, Y., H. Watanabe, M. Saitoh et al. 1991. Volcanic production of polyphosphates and its relevance to prebiotic evolution. *Nature* 352: 516–519.

Yan, Y., J. Huang, and B. Z. Tang. 2016. Kinetic trapping—A strategy for directing the self-assembly of unique functional nanostructures. *Chem. Commun.* 52: 11870–11884.

Yuen, G. U., and K. A. Kvenvolden. 1973. Monocarboxylic acids in Murray and Murchison carbonaceous meteorites. *Nature* 246: 301–303.

Zhu, T. F., and J. W. Szostak. 2009. Coupled growth and division of model protocell membranes. *J. Am. Chem. Soc.* 131: 5705–5713.

Section VIII

Origin of Life and Its Diversification. Universal Tree of Life. Early Primitive Life on Earth. Fossils of Ancient Microorganisms. Biomarkers and Detection of Life

8.1 The Progenote, Last Universal Common Ancestor, and the Root of the Cellular Tree of Life

Johann Peter Gogarten

CONTENTS

8.1.1 INTRODUCTION

8.1.1.1 THE TREE METAPHOR TO DESCRIBE EVOLUTION

Trees have a long history to depict genealogies of individuals (family trees) and interactions between living systems, including the biblical tree of life, the world tree in Mesoamerican cultures, and Yggdrasil in Norse mythology (Gogarten et al. 2008; Anonymous 2017). Mark Ragan in a well-illustrated article (Ragan 2009) describes trees and network diagrams that were used before Darwin to depict relationships between plants and animals. The first explicit description of evolution as a tree goes back to Jean-Baptiste Lamarck, although Peter Simon Pallas's description of the gradation between organisms came close to depicting evolution. Lamarck's tree-like diagram overlaid multiple chains of progress, each for a different lineage, into a single diagram (Lamarck 1809). Charles Darwin in his *Origin of Species* (Darwin 1859) provides a poetic description of the Tree of Life as an image of evolution, and the tree diagrams in his notebooks (Darwin 1836) illustrate the use of trees to depict shared ancestry. In his notebook, Darwin expressed concern about the tree metaphor, because in a botanical tree, the whole tree is alive, including the root, whereas in the tree-of-life image only the top layer is represented by living organisms (Darwin 1836; Olendzenski and Gogarten 2009), and his concern gained new prominence with the realization that molecular-based phylogenies usually only access data from extant or recently extinct species and that most organisms that ever existed on this planet belong to lineages that are now extinct (Zhaxybayeva and Gogarten 2004; Fournier et al. 2009; Fournier et al. 2015; Weigel 2017).

8.1.1.2 THE TREE IN LIGHT OF GENE TRANSFER

The exchange of genetic information between independent lines of descent and the fusion of independent lineages are deviations from a tree-like structure. The evolution of genomes is undoubtedly a highly reticulated network. While gene transfer is indeed rampant in most bacterial and archaeal lineages, and even appears to be frequent in eukaryotes (Soucy et al. 2015), most genes are transferred between closely related organisms (Andam and Gogarten 2011b). The transfer rate to organisms belonging to different orders and classes is several orders of magnitude lower than that of within genus transfers (Williams et al. 2012). Over short periods of time, most genes are passed on vertically (from mother to daughter cell). This majority signal over short periods of time can define the organismal lineage, even in the presence of rampant gene transfer. Garry Olsen used the metaphor of a rope to describe this. In a rope (the lineage), each fiber (the gene traveling through time) extends only over a very short distance. Nevertheless, the rope is a continuous reality. By analogy, the same can be said for the organismal lineage: Even if not a single gene travels through a lineage from beginning to end, the lineage can still be defined through the majority of genes being passed on over short time intervals (Zhaxybayeva and Gogarten 2004).

Instances of lineage fusion that occurred in case of the endosymbionts that evolved into mitochondria and plastids violate the tree paradigm, if one considers the host and the symbiont as equal contributors. However, these reticulations in the tree of live are rare (Martin and Herrmann 1998). In many or even all instances, the host cells genome dominates the gene content persisting in the symbiosis. In most instances, the phylogenetic signals resulting from shared ancestry and from biased gene transfer reinforce one another (Andam et al. 2010; Pace et al. 2012). Therefore, while the history of genomes is highly reticulated, hope persists that careful examination of the phylogenetic signal retained in molecular phylogenies may result in an improved inference of life's early history (Williams et al. 2011). The phylogenetic information provided through transferred genes that persist in the recipient lineage might improve the reconstruction of early cellular evolution (Huang and Gogarten 2006; Szollosi et al. 2012) and allows to correlate evolutionary events in different parts of the tree of life (Gogarten 1995). However, emphasis has to be placed on "careful," because the uncritical application of computational screening tools may lead to unwarranted conclusions (Gogarten and Deamer 2016).

8.1.1.3 Most Recent Common Ancestors, Roots, and Stem Groups

With the introduction of ribosomal RNA as a marker molecule, it became possible to place unicellular anucleate organisms onto a tree-like phylogeny (Woese and Fox 1977a). One immediate realization was that prokaryotes fall into two groups, now known as Bacteria and Archaea, previously labeled as Eubacteria and Archaebacteria. Willi Hennig developed a natural taxonomy, in which proper groups are defined by a common ancestor that is ancestor only to members of this group (Hennig 1966). Such a group is known as monophyletic. A group whose members all trace back to the same ancestor but this ancestor also gave rise to organisms that are not part of the group is known as paraphyletic, and is usually not considered as monophyletic. (However, see [Ashlock 1971], who defines monophyletic to include both para- and holophyletic groups.) In the now-traditional tree of life, the Archaea are the sister group to the eukaryotic nucleocytoplasm, and the ancestor of the bacteria and archaea is also the ancestor to the eukaryotes. Therefore, in a cladistics classification system, the prokaryotes are a paraphyletic group, and the name prokaryotes describes a grade (developmental stage) and not a proper group in a natural taxonomic system. However, naming groups based on properties has a long history and often makes intuitive sense, and many of these group labels are in common use (e.g., reptiles, whose most recent common ancestor (MRCA) is also ancestor to birds and mammals). Before the availability of molecular data, grades were often the only available classification scheme. For further discussion, see the exchange of opinions between Mayr (1998) and Woese (1998).

As an aside, a similar debate concerns the archaea: if the eukaryotic nucleocytoplasm emerged from within the archaeal domain ([Zaremba-Niedzwiedzka et al. 2017]; see [Da Cunha et al. 2017; Levasseur et al. 2017] for conflicting opinions), then either the archaea will need to be considered paraphyletic and/or eukaryotes and archaea need to be united into a single taxonomic group.

In considering a tree, or part of a tree, the MRCA of a group is the organism placed at the deepest (earliest) split between lineages leading to members of the group. In the field of evolutionary biology, this MRCA is often described as the root of the group. In case of the tree of life that includes all cellular organisms, the root of the tree of life refers to that organism that existed at the deepest split, and this organism is also known as the Last Universal Common Ancestor (LUCA), the MRCA of all cellular life, or the organismal cenancestor (Fitch and Upper 1987). This root of a phylogenetic tree is very different from a botanical tree, where the root is located at the base of the stem.

The stem group of a taxon is defined as the group of those extinct organisms that branch of the lineage leading the MRCA of that taxon. In case of the organismal LUCA, molecular evidence suggests that deeper branching lineages existed and contributed some genes through horizontal gene transfer to extant lineages (Fournier et al. 2009, 2015).

8.1.2 PLACING THE ROOT IN THE TREE OF CELLULAR LIFE

The tree of cells is embedded into network of gene trees. To find the root of a phylogenetic tree, different approaches are employed. The most common and widely accepted approach is to use an outgroup. In case of the tree of life, no organism can function as an outgroup, because in case of the tree of life, all organisms by definition are part of the ingroup. However, ancient gene duplications that occurred before LUCA can provide an outgroup to molecular phylogenies (Gogarten et al. 1989; Iwabe et al. 1989; Brown and Doolittle 1995; Gribaldo and Cammarano 1998). Dayhoff (1972) first had suggested the approach to use an ancient gene duplication to provide an outgroup. For most ancient duplicated genes, this approach places the root between bacteria on side of the root and the archaea and the eukaryotic nucleocytoplasm on the other side. Using the unrooted tree of the translation machinery as reference, the root is placed on the branch leading from the central trifurcation to the bacterial domain; therefore, this placement of the root is also known as placing the root on the bacterial branch. The placement of LUCA on the branch connecting bacteria and archaea was also inferred by using ancestral sequence reconstruction for ribosomal proteins and by identifying the place in the unrooted tree of ribosomal proteins that has an amino acid composition that likely reflects the signal from the assembly of the genetic code; that is, proteins that evolved before the genetic code evolved to include all of today's genetically encoded amino acids should not include the later-added amino acids in conserved positions (Fournier and Gogarten 2010). A few scientists place LUCA in different places in the tree of life, either within the bacteria (e.g., Cavalier-Smith 2002; Skophammer et al. 2007) or within the archaea (Di Giulio 2007; Kim and Caetano-Anollés 2011). Often, these rootings are based on singular characters, for example, split genes (Di Giulio 2007) and the outer membrane of the Gram-negative bacteria (Cavalier-Smith 2002) or on novel approaches to reconstruct evolution (Cejchan 2004; Skophammer et al. 2007; Swithers et al. 2011). While the placement of the root remains under debate, most scientists, including former critics of the rooting on the bacterial branch, agree that the root of the cellular tree of life is located between Bacteria and Archaea, and with the eukaryotic nucleocytoplasm grouping on the archaeal side of the tree (Puigbò et al. 2013; Forterre 2015).

8.1.3   THE PROGENOTE CONCEPT

Woese and Fox (1977b) defined the progenote as a hypothetical stage in evolution that existed before a strict coupling between geno- and phenotype had emerged, that is, before the typical features characteristic of today's prokaryotes had emerged: "Eucaryotes did arise from procaryotes, but only in the sense that the procaryotic is an organizational, not a phylogenetic distinction. In analogous fashion procaryotes arose from simpler entities. The latter are properly called progenotes, because they are still in the process of evolving the relationship between genotype and phenotype". The progenote concept is related to the Darwinian threshold (Woese 2002) or Darwinian transition (Goldenfeld et al. 2017). After living systems passed this threshold, natural selection acting on cells drove the optimization of transcription and translation of genetically encoded systems. Below this threshold, life is assumed to have been characterized by a high rate of gene exchange. A high gene transfer rate in communal entities also allows for smaller genomes (Lawrence 1999) and may help to avoid an error catastrophe (Eigen 1971; Biebricher and Eigen 2005). Kandler (1994) described the same concept under the term pre-cell populations.

8.1.4   INFERRING PROPERTIES OF THE LAST UNIVERSAL COMMON ANCESTOR

Confusion regarding the term progenote resulted from Woese and Fox's assumption that the most recent ancestor of the three domains was a progenote (Woese and Fox 1977). In their and Kandler's vision (Kandler 1994), all three cellular domains independently trace their ancestry back to the progenote phase of life (see Figure 8.1.1 panel A).

In contrast, the many properties shared by all cells, and the fact that all ancient duplicated genes, including the aminoacyl tRNA synthetases themselves (Wolf et al. 1999; Woese et al. 2000; Fournier et al. 2011), make use of the same 20 genetically encoded amino acids and the observation that among the ancient duplicated genes are two subunits of the ATP synthase that translate transmembrane gradients for ions (sodium or hydrogen ions) into chemical energy stored in the ATP molecule (Gogarten et al. 1989) suggest that the LUCA was a cellular entity not too dissimilar from a modern prokaryote that possessed ribosomes for mRNA-directed protein biosynthesis, aminoacyl tRNA synthetase proteins to charge tRNAs, and membranes that were used for chemiosmotic coupling (Gogarten and Taiz 1992) (see Figure 8.1.1 panel B).

As discussed in Section 8.1.1.2, gene transfer has not erased all information that can be used to trace the evolutionary history of cell, and in many instances, gene transfer has generated

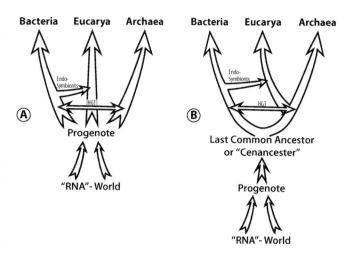

FIGURE 8.1.1   Two versions of the tree of life indicating the possible relations between the Last Universal Common Ancestor (LUCA) or cenancestor of all extant cellular life and the progenote phase of early evolution. The progenote describes a period in the evolution of cells before a strict coupling between geno- and phenotype had been established. During this phase, cells or pre-cells frequently exchanged genes with one another (see text for discussion). The sketch in panel A reflects the view popularized through the work of Carl Woese and Otto Kandler (Woese and Fox 1977b; Kandler 1994). In this version of the tree of life, the three domains of life evolved independently from the progenote. The sketch in panel B recognizes that the LUCA was a complex organism with an established coupling between geno- and phenotype (Gogarten and Taiz 1992; Delaye et al. 2005; Kim and Caetano-Anollés 2011; Goldman et al. 2013) and depicts the progenote as an earlier phase preceding the LUCA.

signals that allow to define groups and correlate evolutionary events in different parts of the tree. However, when trying to infer LUCA's properties from molecular phylogenies, wrong conclusions can be reached, when gene transfer events are mistaken for lines of vertical descent. In case of proteins with nearly universally distribution, such as amino acyl tRNA synthetase or the F/V/A-ATPases, careful analysis and comparison with other conserved proteins allow identification of branches in a molecular phylogeny that correspond to gene transfer event (Ibba et al. 1999; Wolf et al. 1999; Zhaxybayeva et al. 2005; Lapierre et al. 2006). However, this task is more challenging when genes are considered that do not have near-universal distribution. The automated detection of genes that were present in the LUCA faces two difficulties: either too many gene families are removed from consideration, because they show signs of horizontal gene transfer, or a gene transfer event that took place between archaea and bacteria is mistaken for the branch containing the LUCA (Gogarten and Deamer 2016). Weiss et al. (2016) used a computational pipeline to identify genes present in the LUCA. Many of the genes they identified as being present in the LUCA mistook a gene transfer event between the domains for the location of the LUCA. This is even true for genes that, according to more careful analyses, were present in the LUCA, such as the phenylalanine amino acyl tRNA synthetase (pheRS) subunits. The LUCA branch identified in Weiss et al. (2016) for these gene families corresponds to a known transfer of the archaeal version of this enzyme to Spirochetes (Andam and Gogarten 2011a), whereas most of the bacterial pheRS genes were not included in the analysis. At present, reliable inference of genes to be present in the LUCA is limited to genes with nearly universal distribution. Analysis of less conserved genes encoding functions in metabolism faces hurdles, such as the decay of phylogenetic information and difficulties to identify gene transfer events. However, progress in gene tree-species tree reconciliation (Boussau et al. 2013; Sjostrand et al. 2014; Bansal et al. 2015; Wen and Nakhleh 2017) may make further progress possible.

Another source for inferring properties of early life is the analysis of genes that diverged before LUCA. At present, these analyses are restricted to ATP synthase and aminoacyl tRNA synthetase subunits (Fournier and Gogarten 2007), and information gleaned from these data is restricted to the assembly of the genetic code and the way tRNAs were charged during the early expansion of the genetic code (Fournier et al. 2011). While sequence data for most gene families may not be sufficiently conserved for this type of analyses, consideration of conserved protein folds might provide alternative avenues to characterize pre-LUCA evolution (see Koonin et al. 2006).

REFERENCES

Andam, C. P., and J. P. Gogarten. 2011a. Biased gene transfer and its implications for the concept of lineage. *Biology Direct* 6: 47. doi:10.1186/1745-6150-6-47.

Andam, C. P., and J. P. Gogarten. 2011b. Biased gene transfer in microbial evolution. *Nature Reviews Microbiology* 9: 543–55. doi:10.1038/nrmicro2593.

Andam, C. P., D. Williams, and J. P. Gogarten. 2010. Natural taxonomy in light of horizontal gene transfer. *Biology and Philosophy* 25: 589–602. doi:10.1007/s10539-010-9212-8.

Anonymous. 2017. *Tree of Life*. https://en.wikipedia.org/wiki/Tree_of_life (Accessed July 27).

Ashlock, P. D. 1971. Monophyly and associated terms. *Systematic Zoology* 20: 63–69.

Bansal, M. S., Y.-C. Wu, E. J. Alm, and M. Kellis. 2015. Improved gene tree error correction in the presence of horizontal gene transfer. *Bioinformatics* 31: 1211–1218. doi:10.1093/bioinformatics/btu806.

Biebricher, C. K., and M. Eigen. 2005. The error threshold. *Virus Research* 107: 117–127.

Boussau, B., G. J. Szollosi, L. Duret, M. Gouy, E. Tannier, and V. Daubin. 2013. Genome-scale coestimation of species and gene trees. *Genome Research* 23: 323–330. doi:10.1101/gr.141978.112.

Brown, J. R., and W. F. Doolittle. 1995. Root of the universal tree of life based on ancient aminoacyl-tRNA synthetase gene duplications. *Proceedings of the National Academy of Sciences of the United States of America* 92: 2441–2445.

Cavalier-Smith, T. 2002. The neomuran origin of archaebacteria, the negibacterial root of the universal tree and bacterial megaclassification. *International Journal of Systematic and Evolutionary Microbiology* 52: 7–76.

Cejchan, P. A. 2004. LUCA, or just a conserved Archaeon? *Gene* 333: 47–50.

Da Cunha, V., M. Gaia, D. Gadelle, A. Nasir, and P. Forterre. 2017. Lokiarchaea are close relatives of *Euryarchaeota*, not bridging the gap between prokaryotes and eukaryotes. *PLoS Genetics* 13: e1006810. doi:10.1371/journal.pgen.1006810.

Darwin, C. 1836. *Charles Darwin's Notebooks, 1836-1844*, transcription published in 1987. Edited by P. H. Barrett. Ithaca, NY: Cornell University Press.

Darwin, C. 1859. *On the Origin of Species by Means of Natural Selection, or the Preservation of Favoured Races in the Struggle for Life*. London, UK: John Murray.

Dayhoff, M. O. 1972. *Atlas of Protein Sequence and Structure*, Vol. 5. Washington, DC: National Biomedical Research Foundation.

Delaye, L., A. Becerra, and A. Lazcano. 2005. The last common ancestor: what's in a name? *Origins of Life and Evolution of Biospheres* 35: 537–554.

Di Giulio, M. 2007. The tree of life might be rooted in the branch leading to Nanoarchaeota. *Gene* 401: 108–113.

Eigen, M. 1971. Selforganization of matter and the evolution of biological macromolecules. *Die Naturwissenschaften* 58: 465–523.

Fitch, W. M., and K. Upper. 1987. The phylogeny of tRNA sequences provides evidence for ambiguity reduction in the origin of the genetic code. *Cold Spring Harbor Symposia on Quantitative Biology* 52: 759–767.

Forterre, P. 2015. The universal tree of life: An update. *Frontiers in Microbiology* 6: 717. doi:10.3389/fmicb.2015.00717.

Fournier, G. P., and J. P. Gogarten. 2007. Signature of a primitive genetic code in ancient protein lineages. *Journal of Molecular Evolution* 65: 425–436. doi:10.1007/s00239-007-9024-x.

Fournier, G. P., and J. P. Gogarten. 2010. Rooting the ribosomal tree of life. *Molecular Biology and Evolution* 27: 1792–1801. doi:10.1093/molbev/msq057.

Fournier, G. P., C. P. Andam, and J. P. Gogarten. 2015. Ancient horizontal gene transfer and the last common ancestors. *BMC Evolutionary Biology* 15: 70. doi:10.1186/s12862-015-0350-0.

Fournier, G. P., C. P. Andam, E. J. Alm, and J. P. Gogarten. 2011. Molecular evolution of aminoacyl tRNA synthetase proteins in the early history of life. *Origins of Life and Evolution of Biospheres* 41: 621–632. doi:10.1007/s11084-011-9261-2.

Fournier, G. P., J. Huang, and J. P. Gogarten. 2009. Horizontal gene transfer from extinct and extant lineages: Biological innovation and the coral of life. *Philosophical Transactions of the Royal Society of London. Series B, Biological Sciences* 364: 2229–2239. doi:10.1098/rstb.2009.0033.

Gogarten, J. P. 1995. The early evolution of cellular life. *Trends in Ecology and Evolution* 10: 147–151.

Gogarten, J. P., and D. Deamer. 2016. Is LUCA a thermophilic progenote? *Nature Microbiology* 1. doi:10.1038/nmicrobiol.2016.229.

Gogarten, J. P., and L. Taiz. 1992. Evolution of proton pumping ATPases: Rooting the tree of life. *Photosynthesis Research* 33: 137–146. doi:10.1007/BF00039176.

Gogarten, J. P., G. P. Fournier, and O. Zhaxybayeva. 2008. Gene transfer and the reconstruction of life's early history from genomic data. *Space Science Reviews* 135: 115–131.

Gogarten, J. P., H. Kibak, P. Dittrich et al. 1989. Evolution of the vacuolar H+-ATPase: Implications for the origin of eukaryotes. *Proceedings of the National Academy of Sciences of the United States of America* 86: 6661–6665.

Goldenfeld, N., T. Biancalani, and F. Jafarpour. 2017. Universal biology and the statistical mechanics of early life. *Philosophical Transactions of the Royal Society A: Mathematical, Physical and Engineering Sciences* 375: 20160341. doi:10.1098/rsta.2016.0341.

Goldman, A. D., T. M. Bernhard, E. Dolzhenko, and L. F. Landweber. 2013. LUCApedia: a database for the study of ancient life. *Nucleic Acids Research* 41: D1079–D1082. doi:10.1093/nar/gks1217.

Gribaldo, S., and P. Cammarano. 1998. The root of the universal tree of life inferred from anciently duplicated genes encoding components of the protein-targeting machinery. *Journal of Molecular Evolution* 47: 508–516.

Hennig, W. 1966. *Phylogenetic Systematics*. Edited by R. Zangerl. Urbana, IL: University of Illinois Press.

Huang, J., and J. P. Gogarten. 2006. Ancient horizontal gene transfer can benefit phylogenetic reconstruction. *Trends in Genetics* 22: 361–366.

Ibba, M., A. W. Curnow, J. Bono, P. A. Rosa, C. R. Woese, and D. Soll. 1999. Archaeal aminoacyl-tRNA synthesis: Unique determinants of a universal genetic code? *Biological Bulletin* 196: 335–337.

Iwabe, N., K.-I. Kuma, M. Hasegawa, S. Osawa, T. M. Source, M. Hasegawat, S. Osawat, and T. Miyata. 1989. Evolutionary relationship of archaebacteria, eubacteria, and eukaryotes inferred from phylogenetic trees of duplicated genes. *Proceedings of the National Academy of Sciences of the United States of America Evolution* 86: 9355–9359. doi:10.1073/pnas.86.23.9355.

Kandler, O. 1994. The early diversification of life. *In Early Life on Earth*, S. Bengston (Ed.), pp. 152–509. New York: Columbia University Press.

Kim, K. M., and G. Caetano-Anollés. 2011. The proteomic complexity and rise of the primordial ancestor of diversified life. *BMC Evolutionary Biology* 11: 140. doi:10.1186/1471-2148-11-140.

Koonin, E. V., T. G. Senkevich, and V. V. Dolja. 2006. The ancient virus world and evolution of cells. *Biology Direct* 1: 29.

Lamarck, J.-B. 1809. *Philosophie Zoologique, Part II*, p. 463. Paris, France: Dentu, et l'Auteur.

Lapierre, P., R. Shial, and J. P. Gogarten. 2006. Distribution of F- and A/V-type ATPases in Thermus scotoductus and other closely related species. *Systematic and Applied Microbiology* 29: 15–23.

Lawrence, J. G. 1999. Gene transfer and minimal genome size. *In Size Limits of Very Small Microorganisms*, A. Knoll, M. J. Osborn, J. Baross, H. C. Berg, N. R. Pace, and M. Sogin (Eds.), pp. 32–38. Washington, DC: National Research Council.

Levasseur, A., V. Merhej, E. Baptiste, V. Sharma, P. Pontarotti, and D. Raoult. 2017. The rhizome of Lokiarchaeota illustrates the mosaicity of archaeal genomes. *Genome Biology and Evolution* 9: 2635–2639. doi:10.1093/gbe/evx208.

Martin, W., and R. G. Herrmann. 1998. Gene transfer from organelles to the nucleus: How much, what happens, and why? *Plant Physiology* 118: 9–17.

Mayr, E. 1998. Two empires or three? *Proceedings of the National Academy of Sciences of the United States of America* 95: 9720–9723.

Olendzenski, L., and J. P. Gogarten. 2009. Evolution of genes and organisms: The tree/web of life in light of horizontal gene transfer. *Annals of the New York Academy of Sciences* 1178: 137–145. doi:10.1111/j.1749-6632.2009.04998.x.

Pace, N. R., J. Sapp, and N. Goldenfeld. 2012. Phylogeny and beyond: Scientific, historical, and conceptual significance of the first tree of life. *Proceedings of the National Academy of Sciences of the United States of America* 109: 1011–1018. doi:10.1073/pnas.1109716109.

Puigbò, P., Y. I. Wolf, and E. V Koonin. 2013. Seeing the tree of life behind the phylogenetic forest. *BMC Biology* 11: 46. doi:10.1186/1741-7007-11-46.

Ragan, M. A. 2009. Trees and networks before and after Darwin. *Biology Direct* 4: 43. doi:10.1186/1745-6150-4-43.

Sjostrand, J., A. Tofigh, V. Daubin, L. Arvestad, B. Sennblad, and J. Lagergren. 2014. A Bayesian method for analyzing lateral gene transfer. *Systematic Biology* 63: 409–420. doi:10.1093/sysbio/syu007.

Skophammer, R. G., J. A. Servin, C. W. Herbold, and J. A. Lake. 2007. Evidence for a gram-positive, eubacterial root of the tree of life. *Molecular Biology and Evolution* 24: 1761–1768. doi: 10.1093/molbev/msm096.

Soucy, S. M., J. Huang, and J. P. Gogarten. 2015. Horizontal gene transfer: Building the web of life. *Nature Reviews. Genetics* 16: 472–482.

Swithers, K. S., G. P. Fournier, A. G. Green, J. P. Gogarten, and P. Lapierre. 2011. Reassessment of the lineage fusion hypothesis for the origin of double membrane bacteria. *PLoS One* 6. doi:10.1371/journal.pone.0023774.

Szollosi, G. J., B. Boussau, S. S. Abby, E. Tannier, and V. Daubin. 2012. Phylogenetic modeling of lateral gene transfer reconstructs the pattern and relative timing of speciations. *Proceedings of the National Academy of Sciences of the United States of America* 109: 17513–17518. doi:10.1073/pnas.1202997109.

Weigel, C. 2017. *Small Things Considered: To Not Get Lost in the Trees of Life (ToL)*. http://schaechter.asmblog.org/schaechter/2017/12/to-not-get-lost-in-the-trees-of-life-tol.html (Accessed December 19, 2017).

Weiss, M. C., F. L. Sousa, N. Mrnjavac, S. Neukirchen, M. Roettger, S. Nelson-Sathi, and W. F. Martin. 2016. The physiology and habitat of the last universal common ancestor. *Nature Microbiology* 1: 16116. doi:10.1038/nmicrobiol.2016.116.

Wen, D., and L. Nakhleh. 2017. Coestimating reticulate phylogenies and gene trees from multilocus sequence data. *Systematic Biology*. doi:10.1093/sysbio/syx085.

Williams, D., G. P. Fournier, P. Lapierre, K. S. Swithers, A. G. Green, C. P. Andam, and J. P. Gogarten. 2011. A rooted net of life. *Biology Direct* 6: 45. doi:10.1186/1745-6150-6-45.

Williams, D., J. P. Gogarten, and R. T. Papke. 2012. Quantifying homologous replacement of loci between haloarchaeal species. *Genome Biology and Evolution* 4: 1223–1244. doi:10.1093/gbe/evs098.

Woese, C. R. 1998. Default taxonomy: Ernst Mayr's view of the microbial world. *Proceedings of the National Academy of Sciences of the United States of America* 95: 11043–11046.

Woese, C. R. 2002. On the evolution of cells. *Proceedings of the National Academy of Sciences of the United States of America* 99: 8742–8747.

Woese, C. R., and G. E. Fox. 1977a. Phylogenetic structure of the prokaryotic domain: The primary kingdoms. *Proceedings of the National Academy of Sciences of the United States of America* 74: 5088–5090.

Woese, C. R., and G. E. Fox. 1977b. The concept of cellular evolution. *Journal of Molecular Evolution* 10: 1–6.

Woese, C. R., G. J. Olsen, M. Ibba, and D. Soll. 2000. Aminoacyl-tRNA synthetases, the genetic code, and the evolutionary process. *Microbiology and Molecular Biology Reviews* 64: 202–236.

Wolf, Y. I., L. Aravind, N. V. Grishin, and E. V. Koonin. 1999. Evolution of aminoacyl-tRNA synthetases – analysis of unique domain architectures and phylogenetic trees reveals a complex history of horizontal gene transfer events. *Genome Research* 9: 689–710.

Zaremba-Niedzwiedzka, K., E. F. Caceres, J. H. Saw, D. Bäckström, L. Juzokaite, E. Vancaester, K. W. Seitz, K. Anantharaman, et al. 2017. Asgard archaea illuminate the origin of eukaryotic cellular complexity. *Nature* 541: 353–358. doi:10.1038/nature21031.

Zhaxybayeva, O., and J. P. Gogarten. 2004. Cladogenesis, coalescence and the evolution of the three domains of life. *Trends in Genetics* 20: 182–187.

Zhaxybayeva, O., P. Lapierre, and J. P. Gogarten. 2005. Ancient gene duplications and the root(s) of the tree of life. *Protoplasma* 227: 53–64.

8.2 Horizontal Gene Transfer in Microbial Evolution

Johann Peter Gogarten and R. Thane Papke

CONTENTS

8.2.1 INTRODUCTION

Horizontal gene transfer (HGT) can be described as the transfer of genetic information between cells that are not in an ancestor-descendant relationship. In bacteria, this sharing of genetic information was first discovered in the 1940s (Tatum and Lederberg 1947). The comparison of molecular phylogenies with those of ribosomal RNA (rRNA) revealed that most gene families are not in complete agreement with the ribosomal tree of life (Hilario and Gogarten 1993; Woese et al. 2000). The best explanation for these conflicts is the HGT. While gene transfers between divergent lineages allow to correlate evolutionary events in different parts of the tree of life (Gogarten 1995), frequent and especially biased gene transfer also poses a problem for phylogenetic reconstruction (Gogarten et al. 2002). Advances in genome sequencing have revealed the HGT as the most important driving force in microbial evolution; however, gene transfer also occurs between eukaryotes and from bacteria and archaea to eukaryotes, often in the context of a close symbiotic association (see Soucy et al. 2015 for a recent review). In bacteria and archaea, transfer of small DNA fragments that integrate into the recipient genome through homologous recombination prevents genome-wide selective sweeps that would purge within-population diversity; the transfer of genes and operons can provide new capabilities to the recipients and allow them to move into new ecological niches. The HGT is also an innovative force in microbial evolution (see Swithers et al. 2012 for a recent review): it is the main process for gene family expansion in bacteria (Treangen and Rocha 2011), and it played a role in extending existing and assembling new metabolic pathways (Boucher et al. 2003; Fournier and Gogarten 2008; Khomyakova et al. 2011).

8.2.2 DEFINITIONS OF PAN- AND CORE GENOMES OF TAXONOMIC GENOMES (STRICT AND EXTENDED)

Only a decade ago, most biologists, including microbiologists, expected that genomes from members of the same species had about the same gene content. Species definitions in plants, animals, algae, and fungi are linked to type specimen, and this rule has been extended to microbial taxonomy: type strains are used in the description of bacterial and archaeal species; they are deposited in culture collections, and they and their genomes serve as important taxonomic and phylogenomic reference points (Kyrpides et al. 2014). However, in contrast to eukaryotes, the genome of an individual type specimen captures only a tiny fraction of the gene content within a population. The pan-genome of a taxon or group refers to the sum of all genes present in each individual (Tettelin et al. 2005; Lapierre and Gogarten 2009). Pan-genomes comprise the core genome, that is, the genes that are found in all members, and the accessory genome, that is, the genes that are present in only one or a few members of the group. Welch et al. (2002) provided the first illustration that genome content in bacteria changes rapidly. Comparing three *Escherichia coli* strains, they found the shared core to be less than 40% of the gene families present in all three genomes. More recently, the size of this core was further reduced to less than 10% of gene families present in 61 *Escherichia coli* genomes (Lukjancenko et al. 2010). When the number of gene families per genome encountered in a taxon is plotted as function of the number of genomes sampled, the resulting curve approaches a line that continues to rise. Pan-genomes that are characterized by these non-saturating rarefaction curves are considered open (Lapierre and Gogarten 2009; Lobkovsky et al. 2014; Puigbò et al. 2014). Obviously, pan-genomes cannot really have an infinite size, and rarefaction curves will saturate eventually. Baumdicker et al. (2012) provided a realistic estimation of the size of the pan-genome: taking population size and time since divergence into account, they estimated that the *Prochlorococcus* pan-genome contains about 58,000 genes—recently, this estimate was increased to 84,872 genes (Biller et al. 2015), whereas the individual *Prochlorococcus* genomes encode only about 2000 genes each.

The pan-genome concept was originally developed to describe the fluidity of prokaryotic genomes (Tettelin et al. 2005). Because HGT is more frequent between close relatives (Ravin 1963; Dykhuizen and Green 1991; Vulic et al. 1997; Gogarten et al. 2002; Andam and Gogarten 2011), the pan-genome also represents a set of genes that is more readily available via HGT to any member of the group. The pan-genome may then be thought of as a shared genetic resource of a population.

Large "open" pan-genomes are found for most bacterial and archaeal species. The most detailed study of population genomics was performed in *Prochlorococcus* (Kashtan et al. 2014; Biller et al. 2015), which showed that even in the presence of gene transfer within and between subpopulations, divergent subpopulations exist for long periods of time. These marine nanocyanobacteria provide about 50% of CO_2 fixation on this planet and constitute the largest know bacterial population (Lynch and Conery 2003; Baumdicker et al. 2012; Kashtan et al. 2014); the overall population is differentiated in high-light and low-light ecotypes, but within each of these ecotypes, many stable subpopulations exist (Kashtan et al. 2014). Gene flow between *Prochlorococcus* subpopulations, and between low-light-adapted *Prochlorococcus* and *Synechococcus* occurs frequently, transferring even DNA that encodes genes for ribosomal proteins and rRNA (Zhaxybayeva et al. 2009; Shapiro et al. 2012). For example, the internal transcribed spacer (ITS) region in the rRNA operon, while usually a good predictor and classifier for subpopulations, failed in two out of 96 instances compared with a strong coherent signal from whole genome sequences (Kashtan et al. 2014).

8.2.3 DESCRIPTION OF PROCESSES THAT LEAD TO GENES WITH LIMITED DISTRIBUTION

8.2.3.1 THE BLACK QUEEN HYPOTHESIS

The black queen hypothesis proposed by Morris et al. (2012) is built on the premise of "leaky" common good functions, that is, gene encoded processes that lead to products that benefit a population or community, not only the producer. This hypothesis suggests that leaky functions combined with selection for small genomes may lead to a situation in which these leaky functions are encoded in only a fraction of the genomes comprising the community. The name "Black Queen" originates from the card game "hearts," where two winning strategies exist, one is to acquire all hearts and the most valuable queen of spades, and the other strategy is to divest as many hearts and the queen of spades as possible. Applied to genomes, these strategies correspond to (A) maintaining a keystone genome that includes all necessary genes or (B) to lose as many genes encoding leaky functions as possible.

1. The cell can retain all genes encoding leaky functions (in the game of hearts, this strategy is known as "shooting the moon"). The cost is a large genome and, consequently, a lower growth rate and a decreasing frequency in the population. The advantage is that following a population bottleneck, all genes encoding leaky functions are available in the genome. If they exist, these members of a community may be thought of as analogous to keystone species.
2. The cell loses some or all of its leaky functions and increases its growth rate (in hearts, this represents the usual strategy of taking as few point cards as possible). The cell that no longer produces the leaky function relies on a common good provided by other cells. If a bottleneck occurs, a cell following this strategy is unlikely to survive on its own, provided the leaky product is not provided by other species. A possible outcome of all cells in a population following strategy B is that all members of a population cheat on some leaky functions.

Division of labor rather than cheating may be a more appropriate description, especially in those cases where a keystone genome is no longer present in the population and all individuals in the population are dependent on some common good produced by others (black queen hypothesis [Fullmer et al. 2015]). Widespread cheating can lead to the tragedy of the commons; Oliveira et al. (2014) found that stable cooperation through reciprocal exchange was unlikely to emerge in evolution under the models and parameters tested; however, experimental work by Morris et al. (2014) has shown that producers and consumers of the leaky product, in their case H_2O_2 detoxification, can enable the stable co-existence of two very similar organisms that use the same resources. Additionally, most bacterial and archaeal cells live in biofilms or small aggregates (Kolter and Greenberg 2006;

Stemmann and Boss 2012) and therefore are more likely to be close to cells with whom they share recent ancestry and are more likely to have the same genotype with respect to leaky functions. These neighborhood relations are expected to increase frequency-dependent selection on genes encoding these functions. For example, Drescher et al. (2014) showed that *Vibrio cholerae* can avoid the public goods dilemma by strengthening relationships between cells of the same genotype through creation of a thick biofilm, leading to larger benefits of the producers, when the overall concentration of the public good decreases.

Not all cheating leads to mutual collaboration. In addition to frequency-dependent selection, other mechanisms exist to slow the emergence of cheaters. In the production of common goods, for which lower levels of synthesis are detrimental for the whole population and that are under the control of quorum sensing, the emergence of cheaters that ignore the quorum sensing signal was found to be counter-selected through quorum sensing signals also controlling the synthesis of private goods (Dandekar et al. 2012), in this case adenosine catabolism.

8.2.3.2 THE RED QUEEN HYPOTHESIS—GENES ALTERING THE INTERACTIONS WITH VIRUSES AND OTHER PARASITES

Van Valen (1973) proposed the red queen hypothesis. It was proposed to describe the necessity for ongoing evolution in the arms race between biological species. Bacteria and archaea are under severe virus predation (Thurber 2009); in addition, members of the same population compete for limited resources. Therefore, selection pressure to evade predation and outcompete niche rivals causes a constant genetic arms race in cellular and viral populations, hence the analogy to the Red Queen from Lewis Carroll's *Through the Looking-Glass* (Carroll 1871; Carroll and Gardner 1993), who states that *it takes all the running you can do, to keep in the same place* (Van Valen 1973). The analysis of phage metagenomes and rank abundance curves indicated that phage predation follows the *kill the winner* strategy (Hoffmann et al. 2007), where successful strains in a population are targeted more frequently. The surprising stability of species composition, despite phage predation, suggests that cycling between different susceptible target cells is more frequent within a population than between populations from different species (Rodriguez-Brito et al. 2010). Consequently, within a population, host genes that are utilized by phage and virus to enter the cell are expected to turn over quickly, creating within-population diversity (Chaturongakul and Ounjai 2014), as will genes that are costly otherwise but provide anti-viral resistance (Hille and Charpentier 2016).

Chemicals and peptides that act as antibiotics are often produced by separate members of the same species and population, resulting in a fitness advantage of resistant bacteria. Conflict is more commonly observed between isolates from

different locations, and cooperation dominates between conspecifics (Cordero et al. 2012); however, this does not mean that conflict does not arise in populations: the observation by Cordero et al. of conflict between conspecifics being largely absent may be due to selection for antibiotic resistance having already successfully occurred within populations (*i.e.*, the non-resistant members of the population have died out).

8.2.3.3 SELFISH GENETIC ELEMENTS AND SELECTIVELY NEARLY NEUTRAL GENES RANDOMLY ACQUIRED

In most lineages, genomes constantly acquire and lose genes. Some genes provide a selective advantage to the recipient and will become fixed in the population. However, many of the transferred genes do not find permanent homes in recipient genomes (Lawrence and Ochman 1997; Mira et al. 2001). Among these genes are parasites (prophages) and selfish genetic elements. Most, but certainly not all (Lobkovsky et al. 2013), of the transferred genes are selectively neutral or nearly neutral to the recipient (Gogarten and Townsend 2005; Baumdicker et al. 2010; Haegeman and Weitz 2012). These genes do not persist in the genomes they "visit," and their loss also has little impact on fitness (Bolotin and Hershberg 2015, 2016).

8.2.3.4 NICHE-ADAPTING GENES AND WEAKLY SELECTED FUNCTIONS

In the case when a population is present in different neighboring ecological niches, genes that adapt their carrier to a particular niche, such as virulence genes and genes that encode metabolic pathways, may be present in a subpopulation only. In addition to spatial heterogeneity, many genes are used only temporarily under some circumstances; for example, a phosphatase that releases phosphate from extracellular macromolecules is under purifying selection only when inorganic phosphate is a limiting nutrient. Lawrence and Roth (1996) in developing the selfish operon theory describe these genes as encoding a weakly selected function.

8.2.3.5 COLLABORATION AND UNITS OF SELECTION

For many decades, especially since the formulation of the Gaia hypothesis by Lovelock and Margulis (Lovelock 1972; Lovelock and Margulis 2011), some considered microbial communities, and in the extreme all bacteria and archaea existing on Earth, as a single superorganism. This idea was spearheaded by Sorin Sonea (1988a, 1988b) and is based on the observation that bacteria and archaea can share genes between cells that are not in a parent—descendent relationship. Related is the question on units and levels of selection (gene, individual organism, population, and community) (Soucy et al. 2015). Dawkins (1976) introduced a gene-centered view of evolution, in which all genes are selfish, although most express their selfishness in collaboration with other genes to build an organism with increased fitness. Evolution can be studied and described by selection acting on individuals, resulting in changing gene frequencies within populations (Graur 2016). In the case of selfish genetic elements (non-cooperating genes, molecular parasites), at least initially, the benefit or cost to the individual is small, especially in case of self-splicing elements, and the element can spread in the population as a parasite, limited only by its transmissibility. Levels of selection are often intertwined, and selection at the gene level (e.g., spread of metal- or antibiotic-resistance genes or Ti plasmids that allow utilization of a particular food source) often also provides a strong benefit at the group level (e.g., avoiding severe bottlenecks, thereby maintaining within-population diversity); however, in most instances, it is doubtful that group-level selection was the driving force to share these genes (Olendzenski and Gogarten 2009; Naor et al. 2016). Life is a densely woven fabric: viruses, plasmids, and selfish genetic elements contribute to constructing the fibers, composition, and pattern of life's fabric.

8.2.4   MECHANISMS FOR GENE TRANSFER

Genes can be transferred horizontally by a variety of mechanisms. Traditionally, conjugation (DNA is transferred between cells through a specialized machinery), transduction (DNA is transferred through phages), and transformation (DNA is taken up from the environment into a cell) are distinguished. Related to these are gene transfer agents (GTAs) and cell fusion and, in eukaryotes, introgression and endosymbiotic (aka, intracellular) gene transfer. Some of these categories are not clearly separated, and, in many instances, rather different processes and machineries are lumped together into a single category. In case of multicellular eukaryotes, the fact that genes were transferred has been established in many instances, (e.g., Stewart et al. 2003; Graham et al. 2012). Mechanisms for these transfers have been proposed (e.g., the weak-link model, Huang 2013); however, the details for these transfers remain to be established.

8.2.4.1   CONJUGATION

Conjugation was first described by (Lederberg and Tatum 1946). Cells make contact through a pilus, and a newly synthesized strand of DNA is transferred through a pilus into the recipient cell. The dedicated conjugation machinery often is plasmid-encoded, and often, only the plasmid is transferred into the recipient. However, if the plasmid is integrated into the chromosome, or if the machinery is encoded on the chromosome (Derbyshire and Gray 2014), chromosomal DNA is transferred.

A single bacterial cell can possess multiple conjugation machineries, for example, *Agrobacterium tumefaciens* possesses a machinery that it uses to transfer T-DNA into plant cells (Joos et al. 1983) and another system for conjugation with other bacteria (Alt-Mörbe et al. 1996) that is under control of quorum sensing and following the successful transformation of a plant catalyzes the transfer of the Ti-plasmid between different *Agrobacterium* strains (White and Winans 2007).

8.2.4.2   CELL FUSION

Many archaeal cells often form intricate networks with connections between individual cells (Rosenshine et al. 1989; Stetter 2013). During mating in *Haloferax volcanii*, these connections lead to exchange of cytoplasm between the mating cells. Fused heterodiploid cells can be recovered, in which recombination between the two parental chromosomes has been detected (Naor et al. 2012).

8.2.4.3   TRANSDUCTION

Phage particles can pack host DNA in addition or instead of the phage DNA. In case of generalized transduction, a random piece of host DNA is packaged into the phage particle and delivered to a recipient cell; in specialized transduction, a prophage imprecisely excises itself from the host genome and delivers neighboring genes to the host. In addition, many phages include genes in their genome that do not play a function in phage replication. These genes have been termed morons, and in case of lysogenic phage, these genes can have a large impact on the host bacteria (Hendrix et al. 2000; Cumby et al. 2012). For example, the toxins in botulism, diphtheria, and cholera all are encoded as morons in prophage genomes (Brussow et al. 2004), reflecting the interrelation between selection on the phage, the bacterial symbiont, and the eukaryotic host.

8.2.4.4   GENE TRANSFER AGENTS

Gene transfer agents can be described as prophages that have lost the ability to recognize and preferentially package their own DNA. They were described in five different groups of bacteria and archaea. The best studied GTA is the one in *Rhodobacter capsulatus*. See Lang et al. (2017) for a recent review. The frequency of occurrence in some groups of Alphaproteobacteria argues that the GTAs are more than defective prophages; however, the mechanisms of GTA maintenance remains under debate (Lang et al. 2012; Omer et al. 2017). The GTA head is too small to package all of the GTA encoding genes, and the release of GTAs is accompanied by the lysis of the bacterium. The latter thus constitutes an ultimate fitness cost to the bacterium activating its GTA, and the former makes it impossible that a complete GTA gene cluster can be transferred to a new host.

8.2.4.5   TRANSFORMATION

Transformation is the uptake of exogenous DNA from the environment. This was first demonstrated by Fredrick Griffith (1928) in *Streptococcus pneumoniae*; the transforming principle was later shown to be DNA (Avery et al. 1944). Transformation has been described for both archaea and bacteria (Chimileski et al. 2014; Johnston et al. 2014). Some bacteria are naturally competent (Johnston et al. 2014), while others can be transformed after special treatment. Naturally competent bacteria often have an uptake bias in favor of DNA similar to their own, preferentially taking up DNA that contains motifs that bind to the uptake machinery (Mell and Redfield 2014). In other cases, DNA uptake is biased by being under the control of quorum sensing and occurring (by definition) only when there are many nearby conspecific donors that often were the victims of fratricide (Claverys and Håvarstein 2007; Borgeaud et al. 2015).

8.2.4.6 INTROGRESSION

In eukaryotes, hybridization between species, followed by repeated backcrosses to one of the parent species, allows for gene flow across species boundaries. This process is of concern in case of transgenic crops that grow in the neighborhood of their non-domesticated relatives (Stewart et al. 2003). Adaptive introgression also has occurred in human evolution through hybridization between archaic human lineages and modern humans (Racimo et al. 2016).

8.2.4.7 ENDOSYMBIOTIC OR INTRACELLULAR GENE TRANSFER

Gene transfer in eukaryotes is frequently between symbionts and their host. In case of the endosymbionts that evolved into mitochondria and plastids, only a few genes remain in the organellar genome; most genes are transferred to the nucleus, where they are transcribed and translated in the host cell's cytoplasm, and the encoded proteins are transported into the organelle. This transfer from endosymbiont to host has been described as intracellular or endosymbiotic gene transfer (Adams et al. 1999; Timmis et al. 2004). Many other transfers likely have happened from symbiont to host (see Soucy et al. 2015 for discussion); for example, many chlamydial genes found in Archaeplastida (plants and algae with primary plastids) were suggested to have contributed to integrating plastid and host (Huang and Gogarten 2007; Tyra et al. 2007), and in case of secondary plastids that evolved from endosymbiotic eukaryotic algae, gene transfer from many other sources occurred (Archibald et al. 2003); however, in many other cases, once the endosymbiont is lost from a lineage, only the transfer from a bacterium to a eukaryote remains detectable.

8.2.5 CONCLUSION

Gene transfer has turned the evolution of genomes into a network. The tree of cells is embedded in this network. Gene transfer is biased and can create patterns of apparent relationships between genomes that are undistinguishable from patterns due to shared ancestry. However, because a strong bias is toward close relatives, gene transfer tends to reinforce the patterns due to shared ancestry. Gene transfer allows for the spread of selfish genetic elements and the acquisition of neutral and slightly deleterious genes, but it also allows for the long-term persistence of weakly selected functions, turning the population pan-genome into shared genetic resource.

REFERENCES

Adams, K. L., K. Song, P. G. Roessler, J. M. Nugent, J. L. Doyle, J. J. Doyle, and J. D. Palmer. 1999. Intracellular gene transfer in action: Dual transcription and multiple silencings of nuclear and mitochondrial cox2 genes in legumes. *Proceedings of the National Academy of Sciences of the United States of America* 96: 13863–13868.

Alt-Mörbe, J., J. L. Stryker, C. Fuqua, P. L. Li, S. K. Farrand, and S. C. Winans. 1996. The conjugal transfer system of Agrobacterium tumefaciens octopine-type Ti plasmids is closely related to the transfer system of an IncP plasmid and distantly related to Ti plasmid vir genes. *Journal of Bacteriology* 178: 4248–4257.

Andam, C. P., and J. P. Gogarten. 2011. Biased gene transfer in microbial evolution. *Nature Reviews. Microbiology* 9: 543–555. doi:10.1038/nrmicro2593.

Archibald, J. M., M. B. Rogers, M. Toop, K.-I. Ishida, and P. J. Keeling. 2003. Lateral gene transfer and the evolution of plastid-targeted proteins in the secondary plastid-containing alga Bigelowiella natans. *Proceedings of the National Academy of Sciences of the United States of America* 100: 7678–7683. doi:10.1073/pnas.1230951100.

Avery, O. T., C. M. Macleod, and M. McCarty. 1944. Studies on the chemical nature of the substance inducing transformation of pneumococcal types: Induction of transformation by a deoxyribonucleic acid fraction isolated from pneumococcus type III. *The Journal of Experimental Medicine* 79: 137–158.

Baumdicker, F., W. R. Hess, and P. Pfaffelhuber. 2010. The diversity of a distributed genome in bacterial populations. *The Annals of Applied Probability* 20: 1567–1606.

Baumdicker, F., W. R. Hess, and P. Pfaffelhuber. 2012. The infinitely many genes model for the distributed genome of bacteria. *Genome Biology and Evolution* 4: 443–456. doi:10.1093/gbe/evs016.

Biller, S. J., P. M. Berube, D. Lindell, and S. W. Chisholm. 2015. Prochlorococcus: The structure and function of collective diversity. *Nature Reviews. Microbiology* 13: 13–27. doi:10.1038/nrmicro3378.

Bolotin, E., and R. Hershberg. 2015. Gene loss dominates as a source of genetic variation within clonal pathogenic bacterial species. *Genome Biology and Evolution* 7: 2173–2187. doi:10.1093/gbe/evv135.

Bolotin, E., and R. Hershberg. 2016. Bacterial intra-species gene loss occurs in a largely clocklike manner mostly within a pool of less conserved and constrained genes. *Scientific Reports* 6: 35168. doi:10.1038/srep35168.

Borgeaud, S., L. C. Metzger, T. Scrignari, and M. Blokesch. 2015. The type VI secretion system of *Vibrio cholerae* fosters horizontal gene transfer. *Science* 347: 63–67. doi:10.1126/science.1260064.

Boucher, Y., C. J. Douady, R. T. Papke, D. A. Walsh, M. E. Boudreau, C. L. Nesbo, R. J. Case, and W. F. Doolittle. 2003. Lateral gene transfer and the origins of prokaryotic groups. *Annual Review of Genetics* 37: 283–328.

Brussow, H., C. Canchaya, and W.-D. Hardt. 2004. Phages and the evolution of bacterial pathogens: From genomic rearrangements to lysogenic conversion. *Microbiology and Molecular Biology Reviews* 68: 560–602. doi:10.1128/MMBR.68.3.560-602.2004.

Carroll, L. 1871. *Through the Looking-Glass*, 1st ed. London, UK: Macmillan.

Carroll, L., and M. Gardner. 1993. *The Annotated Alice: Alice's Adventures in Wonderland & Through the Looking Glass*. New York City: Random House Value Publishing.

Chaturongakul, S., and P. Ounjai. 2014. Phage-host interplay: Examples from tailed phages and Gram-negative bacterial pathogens. *Frontiers in Microbiology* 5: 442. doi:10.3389/fmicb.2014.00442.

Chimileski, S., K. Dolas, A. Naor, U. Gophna, and R. T. Papke. 2014. Extracellular DNA metabolism in *Haloferax volcanii*. *Frontiers in Microbiology* 5: 57. doi:10.3389/fmicb.2014.00057.

Claverys, J.-P., and L. S. Håvarstein. 2007. Cannibalism and fratricide: Mechanisms and raisons d'être. *Nature Reviews Microbiology* 5: 219–229. doi:10.1038/nrmicro1613.

Cordero, O. X., H. Wildschutte, B. Kirkup, S. Proehl, L. Ngo, F. Hussain, F. Le Roux, T. Mincer et al. 2012. Ecological populations of bacteria act as socially cohesive units of antibiotic production and resistance. *Science* 337: 1228–1231. doi:10.1126/science.1219385.

Cumby, N., A. R. Davidson, and K. L. Maxwell. 2012. The moron comes of age. *Bacteriophage* 2: e23146. doi:10.4161/bact.23146.

Dandekar, A. A., S. Chugani, and E. P. Greenberg. 2012. Bacterial quorum sensing and metabolic incentives to cooperate. *Science* 338: 264–266. doi:10.1126/science.1227289.

Dawkins, R. 1976. *The Selfish Gene*. Oxford, UK: Oxford University Press.

Derbyshire, K. M., and T. A. Gray. 2014. Distributive conjugal transfer: New insights into horizontal gene transfer and genetic exchange in mycobacteria. *Microbiology Spectrum* 2. doi:10.1128/microbiolspec.MGM2-0022-2013.

Drescher, K., C. D. Nadell, H. A. Stone, N. S. Wingreen, and B. L. Bassler. 2014. Solutions to the public goods dilemma in bacterial biofilms. *Current Biology* 24: 50–55. doi:10.1016/j.cub.2013.10.030.

Dykhuizen, D. E., and L. Green. 1991. Recombination in *Escherichia coli* and the definition of biological species. *Journal of Bacteriology* 173: 7257–7268.

Fournier, G. P., and J. P. Gogarten. 2008. Evolution of acetoclastic methanogenesis in Methanosarcina via horizontal gene transfer from cellulolytic Clostridia. *Journal of Bacteriology* 190: 1124–1127. doi:10.1128/JB.01382-07.

Fullmer, M. S., S. M. Soucy, and J. P. Gogarten. 2015. The pan-genome as a shared genomic resource: Mutual cheating, cooperation and the black queen hypothesis. *Frontiers in Microbiology* 6: 728. doi:10.3389/fmicb.2015.00728.

Gogarten, J. P. 1995. The early evolution of cellular life. *Trends in Ecology and Evolution* 10: 147–151.

Gogarten, J. P., and J. P. Townsend. 2005. Horizontal gene transfer, genome innovation and evolution. *Nature Reviews Microbiology* 3: 679–687.

Gogarten, J. P., W. F. Doolittle, and J. G. Lawrence. 2002. Prokaryotic evolution in light of gene transfer. *Molecular Biology and Evolution* 19: 2226–2238.

Graham, L. A., J. Li, W. S. Davidson, and P. L. Davies. 2012. Smelt was the likely beneficiary of an antifreeze gene laterally transferred between fishes. *BMC Evolutionary Biology* 12: 190. doi:10.1186/1471-2148-12-190.

Graur, D. 2016. *Molecular and Genome Evolution*. Sunderland, MA: Sinauer Associates.

Griffith, F. 1928. The significance of pneumococcal types. *The Journal of Hygiene* 27: 113–159.

Haegeman, B., and J. S. Weitz. 2012. A neutral theory of genome evolution and the frequency distribution of genes. *BMC Genomics* 13: 196. doi:10.1186/1471-2164-13-196.

Hendrix, R. W., J. G. Lawrence, G. F. Hatfull, and S. Casjens. 2000. The origins and ongoing evolution of viruses. *Trends in Microbiology* 8: 504–508.

Hilario, E., and J. P. Gogarten. 1993. Horizontal transfer of ATPase genes – The tree of life becomes a net of life. *Biosystems* 31: 111–119.

Hille, F., and E. Charpentier. 2016. CRISPR-Cas: Biology, mechanisms and relevance. *Philosophical Transactions of the Royal Society of London. Series B, Biological Sciences* 371. doi:10.1098/rstb.2015.0496.

Hoffmann, K. H., B. Rodriguez-Brito, M. Breitbart, D. Bangor, F. Angly, B. Felts, J. Nulton, F. Rohwer et al. 2007. Power law rank-abundance models for marine phage communities. *FEMS Microbiology Letters* 273: 224–228. doi:10.1111/j.1574-6968.2007.00790.x.

Huang, J. 2013. Horizontal gene transfer in eukaryotes: The weak-link model. *BioEssays* 35: 868–875. doi:10.1002/bies.201300007.

Huang, J., and J. P. Gogarten. 2007. Did an ancient chlamydial endosymbiosis facilitate the establishment of primary plastids? *Genome Biology* 8: R99. doi:10.1186/gb-2007-8-6-r99.

Johnston, C., B. Martin, G. Fichant, P. Polard, and J.-P. Claverys. 2014. Bacterial transformation: Distribution, shared mechanisms and divergent control. *Nature Reviews Microbiology* 12: 181–196. doi:10.1038/nrmicro3199.

Joos, H., B. Timmerman, M. V Montagu, and J. Schell. 1983. Genetic analysis of transfer and stabilization of *Agrobacterium* DNA in plant cells. *The EMBO Journal* 2: 2151–2160.

Kashtan, N., S. E. Roggensack, S. Rodrigue, J. W. Thompson, S. J. Biller, A. Coe, H. Ding, P. Marttinen et al. 2014. Single-cell genomics reveals hundreds of coexisting subpopulations in wild Prochlorococcus. *Science* 344: 416–420. doi:10.1126/science.1248575.

Khomyakova, M., Ö. Bükmez, L. K. Thomas, T. J. Erb, and I. A. Berg. 2011. A methylaspartate cycle in haloarchaea. *Science* 331: 334–337. doi:10.1126/science.1196544.

Kolter, R., and E. P. Greenberg. 2006. Microbial sciences: The superficial life of microbes. *Nature* 441: 300–302. doi:10.1038/441300a.

Kyrpides, N. C., P. Hugenholtz, J. A. Eisen, T. Woyke, M. Göker, C. T. Parker, R. Amann, B. J. Beck et al. 2014. Genomic encyclopedia of bacteria and archaea: Sequencing a myriad of type strains. *PLoS Biology* 12: e1001920. doi:10.1371/journal.pbio.1001920.

Lang, A. S., O. Zhaxybayeva, and J. T. Beatty. 2012. Gene transfer agents: Phage-like elements of genetic exchange. *Nature Reviews Microbiology* 10: 472–482. doi:10.1038/nrmicro2802.

Lang, A. S., A. B. Westbye, and J. T. Beatty. 2017. The distribution, evolution, and roles of gene transfer agents in prokaryotic genetic exchange. *Annual Review of Virology* 4: 87–104. doi:10.1146/annurev-virology-101416-041624.

Lapierre, P., and J. P. Gogarten. 2009. Estimating the size of the bacterial pan-genome. *Trends in Genetics* 25: 107–110. doi:10.1016/j.tig.2008.12.004.

Lawrence, J. G., and H. Ochman. 1997. Amelioration of bacterial genomes: Rates of change and exchange. *Journal of Molecular Evolution* 44: 383–397.

Lawrence, J. G., and J. R. Roth. 1996. Selfish operons: Horizontal transfer may drive the evolution of gene clusters. *Genetics* 143: 1843–1860.

Lederberg, J., and E. L. Tatum. 1946. Gene recombination in *Escherichia coli*. *Nature* 158: 558.

Lobkovsky, A. E., Y. I. Wolf, and E. V. Koonin. 2013. Gene frequency distributions reject a neutral model of genome evolution. *Genome Biology and Evolution* 5: 233–242. doi:10.1093/gbe/evt002.

Lobkovsky, A. E., Y. I. Wolf, and E. V Koonin. 2014. Estimation of prokaryotic supergenome size and composition from gene frequency distributions. *BMC Genomics* 15: S14. doi:10.1186/1471-2164-15-S6-S14.

Lovelock, J. E. 1972. Gaia as seen through the atmosphere. *Atmospheric Environment* (1967) 6: 579–580. doi:10.1016/0004-6981(72)90076-5.

Lovelock, J. E., and L. Margulis. 2011. Atmospheric homeostasis by and for the biosphere: The gaia hypothesis. *Tellus A* 26. doi:10.3402/tellusa.v26i1-2.9731.

Lukjancenko, O., T. M. Wassenaar, and D. W. Ussery. 2010. Comparison of 61 sequenced *Escherichia coli* genomes. *Microbial Ecology* 60: 708–720. doi:10.1007/s00248-010-9717-3.

Lynch, M., and J. S. Conery. 2003. The origins of genome complexity. *Science* 302: 1401–1404.

Mell, J. C., and R. J. Redfield. 2014. Natural competence and the evolution of DNA uptake specificity. *Journal of Bacteriology* 196: 1471–1483. doi:10.1128/JB.01293-13.

Mira, A., H. Ochman, and N. A. Moran. 2001. Deletional bias and the evolution of bacterial genomes. *Trends in Genetics* 17: 589–596.

Morris, J. J., R. E. Lenski, and E. R. Zinser. 2012. The Black Queen Hypothesis: Evolution of dependencies through adaptive gene loss. *mBio* 3: e00036-12. doi:10.1128/mBio.00036-12.

Morris, J. J., S. E. Papoulis, and R. E. Lenski. 2014. Coexistence of evolving bacteria stabilized by a shared black queen function. *Evolution* 68: 2960–2971. doi:10.1111/evo.12485.

Naor, A., P. Lapierre, M. Mevarech, R. T. Papke, and U. Gophna. 2012. Low species barriers in halophilic archaea and the formation of recombinant hybrids. *Current Biology: CB* 22: 1444–1448. doi:10.1016/j.cub.2012.05.056.

Naor, A., N. Altman-Price, S. M. Soucy, A. G. Green, Y. Mitiagin, I. Turgeman-Grott, N. Davidovich, J. P. Gogarten et al. 2016. Impact of a homing intein on recombination frequency and organismal fitness. *Proceedings of the National Academy of Sciences of the United States of America* 113: E4654–E4661. doi:10.1073/pnas.1606416113.

Olendzenski, L., and J. P. Gogarten. 2009. Evolution of genes and organisms: The tree/web of life in light of horizontal gene transfer. *Annals of the New York Academy of Sciences* 1178: 137–145.

Oliveira, N. M., R. Niehus, and K. R. Foster. 2014. Evolutionary limits to cooperation in microbial communities. *Proceedings of the National Academy of Sciences of the United States of America* 111: 17941–17946. doi:10.1073/pnas.1412673111.

Omer, S., T. J. Harlow, and J. P. Gogarten. 2017. Does sequence conservation provide evidence for biological function? *Trends in Microbiology* 25. doi:10.1016/j.tim.2016.09.010.

Puigbò, P., A. E. Lobkovsky, D. M. Kristensen, Y. I. Wolf, and E. V Koonin. 2014. Genomes in turmoil: Quantification of genome dynamics in prokaryote supergenomes. *BMC Biology* 12: 66. doi:10.1186/s12915-014-0066-4.

Racimo, F., D. Marnetto, and E. Huerta-Sánchez. 2016. Signatures of archaic adaptive introgression in present-day human populations. *Molecular Biology and Evolution* 34: msw216. doi:10.1093/molbev/msw216.

Ravin, A. W. 1963. Experimental approaches to the study of bacterial phylogeny. *The American Naturalist* 97: 307–318. doi:10.2307/2458469.

Rodriguez-Brito, B., L. Li, L. Wegley, M. Furlan, F. Angly, M. Breitbart, J. Buchanan, C. Desnues et al. 2010. Viral and microbial community dynamics in four aquatic environments. *The ISME Journal* 4: 739–751. doi:10.1038/ismej.2010.1.

Rosenshine, I., R. Tchelet, and M. Mevarech. 1989. The mechanism of DNA transfer in the mating system of an archaebacterium. *Science* 245: 1387–1389.

Shapiro, B. J., J. Friedman, O. X. Cordero, S. P. Preheim, S. C. Timberlake, G. Szabó, M. F. Polz, and E. J. Alm. 2012. Population genomics of early events in the ecological differentiation of bacteria. *Science* 336: 48–51. doi:10.1126/science.1218198.

Sonea, S. 1988a. A bacterial way of life. *Nature* 331: 216.

Sonea, S. 1988b. The global organism: A new view of bacteria. *The Sciences* 28: 38–45.

Soucy, S. M., J. Huang, and J. P. Gogarten. 2015. Horizontal gene transfer: Building the web of life. *Nature Reviews. Genetics* 16. doi:10.1038/nrg3962.

Stemmann, L., and E. Boss. 2012. Plankton and particle size and packaging: From determining optical properties to driving the biological pump. *Annual Review of Marine Science* 4: 263–290. doi:10.1146/annurev-marine-120710-100853.

Stetter, K. O. 2013. A brief history of the discovery of hyperthermophilic life. *Biochemical Society Transactions* 41: 416–420. doi:10.1042/BST20120284.

Stewart, C. N., M. D. Halfhill, and S. I. Warwick. 2003. Transgene introgression from genetically modified crops to their wild relatives. *Nature Reviews Genetics* 4: 806–817. doi:10.1038/nrg1179.

Swithers, K. S., S. M. Soucy, and J. P. Gogarten. 2012. The role of reticulate evolution in creating innovation and complexity. *International Journal of Evolutionary Biology* 2012: 418964. doi:10.1155/2012/418964.

Tatum, E. L., and J. Lederberg. 1947. Gene recombination in the bacterium *Escherichia coli*. *Journal of Bacteriology* 53: 673–684.

Tettelin, H., V. Masignani, M. J. Cieslewicz, C. Donati, D. Medini, N. L. Ward, S. V. Angiuoli, J. Crabtree et al. 2005. Genome analysis of multiple pathogenic isolates of Streptococcus agalactiae: Implications for the microbial "pan-genome." *Proceedings of the National Academy of Sciences of the United States of America* 102: 13950–13955.

Thurber, R. V. 2009. Current insights into phage biodiversity and biogeography. *Current Opinion in Microbiology* 12: 582–587. doi:10.1016/j.mib.2009.08.008.

Timmis, J. N., M. A. Ayliffe, C. Y. Huang, and W. Martin. 2004. Endosymbiotic gene transfer: Organelle genomes forge eukaryotic chromosomes. *Nature Reviews Genetics* 5: 123–135. doi:10.1038/nrg1271.

Treangen, T. J., and E. P. C. Rocha. 2011. Horizontal transfer, not duplication, drives the expansion of protein families in prokaryotes. *PLoS Genetics* 7: e1001284. doi:10.1371/journal.pgen.1001284.

Tyra, H. M., M. Linka, A. P. Weber, and D. Bhattacharya. 2007. Host origin of plastid solute transporters in the first photosynthetic eukaryotes. *Genome Biology* 8: R212.

Van Valen, L. 1973. A new evolutionary law. *Evolutionary Theory* 1: 1–30.

Vulic, M., F. Dionisio, F. Taddei, and M. Radman. 1997. Molecular keys to speciation: DNA polymorphism and the control of genetic exchange in enterobacteria. *Proceedings of the National Academy of Sciences of the United States of America* 94: 9763–9767. doi:10.1073/pnas.94.18.9763.

Welch, R. A., V. Burland, G. Plunkett, P. Redford, P. Roesch, D. Rasko, E. L. Buckles, S. R. Liou et al. 2002. Extensive mosaic structure revealed by the complete genome sequence of uropathogenic *Escherichia coli*. *Proceedings of the National Academy of Sciences of the United States of America* 99: 17020–17024.

White, C. E., and S. C. Winans. 2007. Cell-cell communication in the plant pathogen Agrobacterium tumefaciens. *Philosophical Transactions of the Royal Society B: Biological Sciences* 362: 1135–1148. doi:10.1098/rstb.2007.2040.

Woese, C. R., G. J. Olsen, M. Ibba, and D. Soll. 2000. Aminoacyl-tRNA synthetases, the genetic code, and the evolutionary process. *Microbiology and Molecular Biology Reviews* 64: 202–236.

Zhaxybayeva, O., W. F. Doolittle, R. T. Papke, and J. P. Gogarten. 2009. Intertwined evolutionary histories of marine *Synechococcus* and *Prochlorococcus marinus*. *Genome Biology and Evolution* 2009: 325–339. doi:10.1093/gbe/evp032.

8.3 Viruses in the Origin of Life and Its Subsequent Diversification

Luis P. Villarreal and Frank Ryan

CONTENTS

8.3.1 BASIC CONCEPTS OF VIRUSES IN RELATION TO EVOLUTION

8.3.1.1 VIRUSES AND THE DEFINITION OF LIFE

It is strange that entities so small that they can only be visualized under the extreme magnification of the electron microscope may have played a key role in the origins of life and may have continued to play a key role in the evolution of biodiversity to the present day. We call these entities "viruses." But what are viruses? In reality, they are difficult to define. Most folks' opinions on viruses are somewhat negative. They associate them with disease, such as AIDS, smallpox, and polio. Life is usually defined as "cellular" in form and function. Viruses are non-cellular. Yet they have many properties of what we think of as organismal. They are "born" within the living cells of their hosts. They reproduce themselves, albeit with the assistance of host molecular chemistry. They can be killed by viricidal drugs and through a range of physical insults. The majority of viruses have genomes based on DNA, although a significant minority has genomes based on RNA. Only recently have we discovered that viruses outnumber all other organisms in the biosphere, including bacteria (Koonin and Dolja 2013). Indeed, the introduction of viral metagenomics has demonstrated that the dominant viruses in environmental communities are poorly represented in the traditional studies of cultured viruses in existing sequence databases (Rosario and Breitbart 2011). Viruses evolve through the same established mechanics of evolution as cellular life. They are commonly defined as "obligate parasites," but this definition is now seen to be too narrow to accommodate the wide range of interactive relationships that viruses share with their hosts. In recognizing viruses as organismal, Forterre and Prangishvili (2009) suggest that they represent capsid-encoding organisms, as opposed to the ribosomal-encoding organisms, which denote cellular life-forms. Others argue that, while useful, even this capsid-related definition may be too restrictive to cover the field of viruses (Koonin et al. 2015). To understand viruses, and this more comprehensive role of viruses in the origins of life and its subsequent diversity, we need to examine the fundamental processes of evolution.

8.3.1.2 The Basic Modalities of Evolution

When Darwin proposed his theory of natural selection to explain the origins and subsequent diversification of life, he knew that it depended on both a system of heredity—in his days referred to as "pedigree"—and some mechanism, or range of mechanisms, capable of altering heredity. Nature could only select for advantages in the struggle for existence between individuals, or populations, where it was presented with a range of variations between those individuals and populations. For Darwinian selection to operate, the advantageous variation had to be inherited by subsequent generations. Today, we know that somatic characters are determined by the particulate hereditary nature of genes, and those genes, together with the machinery that regulates them, are

intrinsic to the genetic and epigenetic inheritances that pass from parents to offspring. In sexually reproducing organisms, this involves a mixing of the heredity of the parents through sexual homologous recombination during the formation of the germ cell through meiosis. In Darwin's days, the particulate nature of genes was unknown. Instead, it was assumed that sexual mixing was somewhat akin to the blending of liquids and that the variation resulting from sexual mixing within a species was sufficient, over lengthy periods, to give rise to speciation. But by the early years of the twentieth century, with the growing understanding of genes and genetics, biologists realized that sexual recombination was not a credible source of the hereditary variation needed to give rise to speciation. Speciation demanded definable and measurable mechanisms capable of changing heredity—mechanisms sufficient to give rise to the dramatic differences that we now witness in the great diversity of genomes throughout biodiversity.

Evolutionary change does not begin with natural selection. It begins with genomic change arising in individuals, then spreading to family groups through asexual and sexual reproductions, and then being incorporated into the species gene pool through being selected as an advantage for survival by natural selection. Four distinct and definable mechanisms capable of changing heredity are listed in Table 8.3.1. These include mutation, epigenetic inheritance systems, genetic symbiosis, and hybridogenesis (MESH).

Mutation, in this classification, is defined as errors in copying DNA during meiosis and mitosis. We have gathered these mechanisms under the collective umbrella of "genomic creativity." We employ the term "genomic" and not merely "genetic" to include epigenetic inheritance systems, which do not necessarily imply change in DNA sequences, and we employ the term "creativity" to emphasize the fact that these mechanisms are creative in themselves. Mutation includes a variety of different patterns of change, including, for example, point mutations, where the error comprises the erroneous copying of a single nucleotide, and frameshift mutations,

---

TABLE 8.3.1

This Table Illustrates the Important Differences Between the Four Mechanisms that Create Hereditary Change as Part of the Evolutionary Process. Mutation Is Defined as Mistakes in Copying DNA During Meiosis and Mitosis. Epigenetic Change Implies Change in Epigenetic Inheritance Systems. Symbiogenesis Is Explained in the Text. Hybridogenesis Is Genetic Change Through Sexual Crossing Between Closely Related Species that Gives Rise to Reproductively Fertile Offspring

	"GENOMIC CREATIVITY" The Mechanisms of Hereditary Change in Evolution				
Operative Mechanism	Genetic Change	Nature of Genomic Change	Level at Which Selection Works	Pattern of Phylogeny	Amenable to Environmental Influences
Mutation	Yes	Usually random, cumulative	Mainly individual gene, organism	Linear, branching	No
Epigenetic	No	Change in gene expression	Epigenetic inheritance system	Linear, branching	Yes
Symbiogenesis (genetic)	Yes	Non-random, rapid	Holobiont	Reticulate	No
Hybridogenesis	Yes	Non-random, rapid	Hybrid genome	Reticulate	No

which derive from a deletion or insertion of nucleotides, which disrupts the subsequent triplet coding system and renders the protein non-functional.

Epigenetic inheritance systems are a less familiar, yet uniquely important source of hereditary variation, including methylation of C-G couplets in gene promoter areas, the histone code, RNA interference, and long- and short-chain non-coding RNAs. Hybridogenesis, which implies evolution through sexual crossing of dissimilar species, is also a powerful evolutionary force in its own right, with a single hybridization event capable of creating a major increase in genomic complexity in subsequent generations. Viruses may be involved in several of these mechanisms. For example, genetic symbiosis involving endogenous viruses has played a key role in the evolution of animal and plant genomes.

Symbiosis was first defined in the 1870s by de Bary as the "living together of dissimilar organisms." Today, this might be translated to significant interaction between different organisms, which are referred to as symbionts. When it gives rise to evolutionary change, this is termed "symbiogenesis." While sometimes mistakenly assumed to refer exclusively to mutualism, in fact, symbiosis embraces commensalism, where one or more organisms benefit from the association, without causing harm to their partners; parasitism, where the parasite benefits at the expense of the partner; as well as mutualism, where two or more partners benefit from the symbiotic relationship. Mutualisms often begin as parasitism and evolve to include stages somewhere between the two extremes. Mutualistic symbiosis involves life-supporting exchanges between partners. For example, mycorrhizal "root-fungus" interactions are vital to all terrestrial plants, where the plant benefits from the fungal absorption of water and minerals from the soil, meanwhile the fungus benefits from the energy-giving products of photosynthesis carried from leaves to roots. This is termed a "metabolic symbiosis." Some symbioses involve mutually supportive behavioral interactions. An example of this is seen in the feeder stations on the floor of the ocean, where large fish, such as sharks and groupers, benefit from having their skins and mouth cavities cleansed of parasites by shrimps and small fish. A third, and very important, level of symbiotic interaction is at the genetic level. At its simplest level, this may involve the transfer of a single pre-evolved gene, or a package of genes, from one evolutionary lineage to another. At its most powerful, it involves the union of whole disparate genomes, to create a novel "holobiontic genome," usually uniting the genome of a host with the whole genome of a microbial symbiont. This was the mechanism of origin of chloroplasts and mitochondria, which involved the union of microbial genomes with what is believed to have been a protist host some 2 billion or so years ago (McFadden 2001; Ryan 2009). It was also the mechanism of origin of the endogenous retroviruses (ERVs) that have repeatedly invaded the nuclear genomes of animals and mammals, in particular over the whole of their evolutionary time, with considerable implications for animal evolution (Ryan 2016). From these, and many other examples, viruses have engaged in a wide variety of symbioses with their hosts, including parasitisms, commensalisms, and mutualisms (Ryan 2007). For this reason, we suggest that it is no longer sufficient

to define viruses as "obligate parasites" but rather as "obligate symbionts." This widening of definition makes possible a more comprehensive enlightenment of the role of viruses in the origins and subsequent diversification of life.

8.3.1.3 Extrapolating to the Prebiotic Stage of Evolution

Darwin envisaged that natural selection must have operated from the very earliest stages of the evolution of life. Indeed, several of these basic mechanisms of evolution would readily extrapolate to the purported stage of prebiotic evolution that led to self-replicating chains of nucleotides. There has been much debate as to the nature of such prebiotic evolution, in particular the possibilities that might have led from basic chemistry to molecular evolution (Lazcano and Miller 1996; Cronin et al. 2017; Shirt-Ediss et al. 2017). The discovery that RNA can serve genetic, catalytic, structural, and regulatory roles led to the proposal of an RNA-based world as a key stage in prebiotic evolution (Rich 1962; Gilbert 1986). Proponents have outlined how, from the presumptive stage of competing RNA polynucleotides, the more complex stages of proteins, DNA, and protocellular life might have arisen (Alberts et al. 2002; Copley et al. 2007). In the half century or so, since this theory was first proposed, the idea has gained traction, provoking scientific debate and criticism (Robertson and Joyce 2012; Lehman 2015; Patel et al. 2015). For example, Bernhardt, who detailed the contradictions implicit in the argument and labeled it the "worst theory of the early evolution of life" was nevertheless forced to admit that it remains the best theory we can currently envisage (Bernhardt 2012). Mutation of RNA polynucleotide chains, through copying errors during replication, would be capable of giving rise to heritable change in the sequences of daughter chains, just as it does in biotic evolution today. Where two different pre-evolved chains coalesced to create a larger and more complex informational chain, this might be seen as the prebiotic equivalent of genetic symbiogenesis. If, as we can assume, natural selection then operated on the resultant mutants, or genetic holobionts, the more successful self-replicators would be more likely to dominate the local population. There are additional implications for the self-replicator stage of evolution that might be extrapolated from current theories of evolution and that are specific to viral patterns of evolution.

8.3.1.4 An Alternative Perspective of Evolution in the Form of Biological Information Systems

Philosophically inclined authors have offered an alternative to the traditional organismal perspective, proposing that evolutionary biology can be viewed as an epiphenomenon of integrated self-referential information management (Miller 2017). In extrapolating the biocommunicative approach to role of viruses in the tree of life, Villarreal and Witzany (2009) look at evolutionary genomics from the bioinformatic perspectives of biolinguistics and biosemiotics. Viruses share sequences with, and have a close affinity to, transposable elements (TEs), which are found throughout the genomes of living organisms. In examining the contribution of viruses and transposable elements to adaptation and evolution, Oliver and Green (2012) have developed a comprehensive and predictive hypothesis, the "TE-Thrust" hypothesis, which examines the holobiontic potential of such elements to generate genetic novelties. This includes an active mode using transposition—the mobility within and between genomes first mooted by Nobel Laureate Barbara McClintock (1956, 1984)—which includes the exaptation of TE sequences as promoters, exons, or genes. It also includes a passive mode, which when present in large homogeneous populations, can cause ectopic DNA recombination, resulting in genomic duplications, deletions, or rearrangements.

In developing the importance of the virosphere to the evolution of life and biodiversity, it is important to grasp that viruses not only transfer *genes* between themselves and their hosts. It is still much commoner for viruses to transfer *virus-derived information* (Villarreal 2014). Stable transfer of the entire informational content of a virus is a definition of viral persistence. Even transfer of partial (defective) viral information can lead to virus-host persistence at a population level. This stable colonization of host leading to viral persistence is in fact a form of genetic symbiosis.

8.3.1.5 An a Priori Role for "Aggressive Symbiosis"

Symbiosis, like nature, is assuredly not benign. When different organisms enter into a symbiotic evolutionary partnership, the partners arrive with pre-evolved abilities specific to each partner that, in the case of mutualism, contribute improved chances of survival to the other partners and thus to the survival of the symbiotic relationship. It seems somewhat counter-intuitive to consider that one of the properties that protoviruses, like their successor viruses, were capable of contributing is a lethal aggression. The fact that, again like their successor viruses, they inhabit the landscape of the genome of their host cells, or prototypical self-replicators, affords them a powerful means for such "aggressive symbiosis." We shall examine the concept of aggressive symbiosis in relation to the viruses of eukaryotes in a later section.

8.3.2 A VIROCENTRIC APPROACH TO THE ORIGINS OF LIFE

8.3.2.1 The Virus-First Model for the Origins of Life

The origin of viruses is essentially unknown, leading to a variety of theories for viral origins. Four main theories have been posited: the "virus first" theory proposes a primal origin in the prebiotic era of the Earth's evolution; the "reduction" hypothesis proposes a viral origin from a previous unicellular stage of evolution; the "escape" hypothesis proposes an origin from genetic material that escaped from the control of cellular life-forms to become parasitically self-driven; and a fourth theory proposes that viruses may have polyphyletic origins (Bremermann 1983;

Villarreal 2005; Forterre 2006; Koonin et al. 2006; Fisher 2010). While acknowledging that all four theories have their pros and cons, and also while accepting the possibility of polyphyletic origins for viruses as a whole, we favor a virus-first hypothesis for the origin of RNA viruses. A major criticism of the virus-first hypothesis has been the belief that viruses could not have evolved before there were cellular hosts for them to parasitize (Nasir et al. 2012). We suggest that this is based on a mistaken premise.

When the self-replicating stage of evolution is set up in microbiological laboratory settings, or as "digital organisms" settings in computer simulations, parasitic elements are seen to invade and then interact with the self-replicators (Bremermann 1983; Takeuchi and Hogewoeg 2008; Bansho et al. 2012; Zaman et al. 2014; Colizzi and Hogeweg 2016). We see such parasites of self-replicators as prototypes of RNA virus evolution. Those who promote the concept of viruses as originating from cellular ancestry would need to explain how several genes coding for key proteins involved in viral replication and morphogenesis are shared by RNA- and DNA-based viruses but are missing from cellular life-forms (Koonin et al. 2006). They would also need to explain how the vast majority of viral genes, and viral proteins, including the major capsid protein of icosahedral DNA- and RNA-based viruses, are found to have no cellular homologues (Prangishvili and Garrett 2004; Koonin et al. 2006).

The virus-first hypothesis is also supported by the strongly inverse relationship between genome size and mutation rates across all replication systems, suggesting that the earliest genomes were likely small and highly error prone—a situation that typifies RNA viruses (Holmes 2011). This prompts us to suggest that RNA viruses are likely to have evolved from ancestral forms that existed during the very earliest phase of protolife, that of self-replicating RNA-based molecules. In this, we are largely in agreement with Forterre (2006) and Koonin et al. (2006). We suggest that the virus-first model offers a logical construct for the evolution of viruses as well as their ever-present and continuing role in the evolution of their natural hosts.

We propose that the prebiotic stages of evolution must have operated in an environment of aggressively symbiotic protoviral genetic agents, the prebiotic equivalent of what, in the biotic era, we call the "virosphere" (Ryan 2009; Koonin and Dolja 2013; Villarreal 2014). Current opinion raises the possibility that life may have evolved in the vicinity of deep-sea hydrothermal vents. A search of such environments, at temperatures above 80°C, revealed a morphological diversity of virus-like particles greatly exceeding the number found in aquatic systems at lower temperatures. Viruses are potentially aggressive, a property that, conferred on a symbiotic partnership of virus and host, has powerful evolutionary potential (Ryan 1997). As aggressive symbionts, viruses are eminently capable of both killing and protecting their hosts. Such biological and evolutionary power, combined with the extraordinary genomic creativity of viruses, makes them the ultimate aggressive symbionts (Ryan 1997, 2009). The fact that self-replicating entities with properties suggestive of prototypical viruses so readily manifest in laboratory and computer simulations that resemble the presumptive RNA world and the fact they are such powerful players in the killing or survival sense, combined with their genomic creativity, suggest that virus-host interactivity is likely to have been a fundamental driving mechanism in the origins of life and also in its subsequent diversification. Viruses are adept at transferring genetic "information" between themselves and their hosts. Indeed, judging from metagenomic analysis, the transfer of genetic information is far commoner from virus to host rather than from host to virus (Koonin 2001). The only organisms with RNA-coded genomes today are RNA-based viruses, suggesting that we might derive helpful insight into the prototypical RNA world from the study of RNA-based viruses. One such clue is their potential to evolve as quasispecies in the form of genetically interrelated consortia.

8.3.2.2 The Concept of RNA-Based Quasispecies Behavior

In 1977, Eigen and Schuster introduced the concept of "quasispecies" as an explanation for the self-organization of clusters of prebiotic macromolecule chemicals, such as polynucleotides, in such a way as to conform to Darwinian natural selection (Eigen and Schuster 1977, 1978a, 1978b, 1979). This was subsequently extrapolated to the concept of "viral quasispecies theory," in which swarms of viruses, all closely related to one another through shared mutations, compete for existence in a highly mutagenic environment. While such a high mutation rate would normally be predicted to result in non-viable virions, quasispecies theory posits that, in the case of RNA viruses, it creates a "cloud of potentially beneficial mutations" at the population level, which affords the viral quasispecies an advantage of adapting to novel environments and existential challenges (Holland et al. 1992; Vignuzzi et al. 2006). The theory predicts that in such a scenario, quasispecies will compete with one another. The theory further predicts that a viral quasispecies operating at a low but evolutionarily neutral and highly connected (graphwise flat) region of the fitness landscape will outcompete a quasispecies located at a higher but narrower fitness peak, in which the surrounding mutants are unfit (Nimwegen et al. 1999; Wilke et al. 2001). Surprisingly, it has been observed experimentally that in the quasispecies situation, less fit viral mutants will even suppress more fit mutants (De la Torre and Holland 1990). This suggests that the evolving RNA virus population has a fitness associated with its quasispecies components rather than the conventional "master fittest" model. Quasispecies theory has been extrapolated to the evolutionary behavior of retroviruses, such as human immunodeficiency virus (HIV)-1, in infected patients. The virus-encoded reverse transcriptase enzyme has an error rate of about 10^{-4} to 10^{-3} per base, and thus, patients infected with HIV-1 are found to harbor a highly diverse viral population with many different mutant strains (Nowak 1992). We suggest that inherent to the evolutionary behavior of quasispecies is a collective function and group identity (Villarreal 2014). Indeed, we go further and suggest that a similar RNA-mediated group identity may have facilitated the origins of life during the presumptive RNA world stage.

We acknowledge that a virus-first emergence from the prototypical RNA world would the acquisition of "group identity" in the form of quasispecies consortia.

We have already suggested that, in the RNA world, the RNA-based self-replicators would inevitably be colonized by what would formerly have been termed RNA-based selfish elements; this we would now amend to aggressive RNA-based proto-symbionts behaving with quintessential virus-like behavior. These aggressive symbionts could function as a consortium, in the way clones of mutating retroviruses behave today, but to truly represent a concept of self, they would also need to have a second key development: the operation of what virologists call "an addiction module" that would equate to a prototypical equivalent of "self" (Villarreal, 2005, 2014). We shall deal with the addiction module strategy in more detail in relation to the prokaryotes later. Here, we argue that such key evolutionary features as group identity, addiction module, regulatory complexity, and virus-host ecology are fundamentally linked. In other words, we view the evolution of the whole as a key step in the establishment of life on Earth.

8.3.2.3 The Viral Role in the Evolution of Life

We propose that proto-viruses arose during the stage of prebiotic life. We now know that any genetic self-replicator, even a non-cellular prebiotic example, is susceptible to parasitism by other replicators adopting quintessentially virus-style behavior. This tendency for the "parasitic" replicators themselves to become further "parasitized" is a well-established phenomenon in virology (Villarreal 2005). Indeed, we can extend this understanding to the wider conceptualization of the symbiotic nature of viruses defined earlier. This virus-like invading sequence would potentially be capable not only of parasitic colonization type behavior but also of commensal and mutualistic interactions with the "host" self-replicator. This suggests that the earliest "proto-virus" may have come into being at the same time and quintessentially in a profound symbiotic interaction with the purported RNA world of self-replicators. Even at the pre-biotic stage of the RNA world, RNA-virus quasispecies patterns of behavior might offer further enlightenment on an important question in the evolution from chemical precursors to definable life.

8.3.2.4 The RNA "Gangen" Hypothesis

The capacity of a particular RNA virus quasispecies to outcompete and displace any former quasispecies was observed in the 1990s (Clarke et al. 1994). In fact, the quasispecies is fitter than any individual member because of complementation, sharing of gene products among its members, and interference with some members and non-members, and hence, including the preclusion of prior quasispecies. The importance of such an evolutionary pattern was confirmed when it was further observed that to attain disease-associated fitness, the RNA virus needed to generate a virus diversity that acted cooperatively (Domingo et al. 2012). From an understanding of such RNA-based quasispecies evolution, it was possible to construct a primary model of group identity, based on a complementary collective of stem-loop RNA molecules with group identity mediated by cooperative subfunctional agents that together provided both the replication and functional endonuclease features of an addiction module. This group identity, coupled with the distinguishing features of an addiction module, offers the possibility that it may have conferred on the evolving quasispecies the first primitive identification of "self." Thus, a society of subfunctional RNA agents offers a logical theory for the predecessor of the first RNA-based life-form. This is the proposed basis for the "RNA gangen hypothesis" (Villarreal 2014).

8.3.3 VIRUSES, PROKARYOTES, AND THE ECOLOGY

8.3.3.1 VIRUSES IN THE EVOLUTION OF PROKARYOTES

Prokaryotes are microscopic organisms that lack a nucleus and possess no internal organelles other than ribosomes. Their genomes comprise a confluent circle, which is attached to the bacterial cell wall at a single fixation point. Prokaryotes were the first known cellular life-forms to evolve on Earth, roughly 3.8 billion years ago. Today, their descendants comprise two of Woese's three domains of life, the Eubacteria and the Archaea. Woese painted a vivid scenario for the origins of cellular life when "genetic temperatures" were very high and the primitive cellular entities, which he called "progenotes," were very simple (Woese 1998). He assumed that the boundaries of "self" were less strictly defined in such early stages of organismal evolution, so that information processing systems were somewhat inaccurate and mutation and lateral genetic informational flow were elevated in relation to what we see today. Viruses are likely to have played an important role in this evolution of complexity, since viruses have an extraordinary capacity both to generate complexity in terms of neogenes and to transmit such novelty to their cellular hosts (Villarreal 2005). In the last decade, microbiologists have belatedly come to realize the vital role of the viruses of prokaryotes as "a major component of the biosphere" (Krupovic et al. 2011). This new understanding, which one eminent authority has labeled "the great virus comeback" (Forterre 2013), has come about through the expanding fields of evolutionary biology, genetics, genomics, metagenomics, and population dynamics, opening our vision to the fact that viruses are both essential agents within the tree of life (Villarreal and Witzany 2009) and key to understanding the complex dynamics of major ecologies. Today, we have come to realize that the non-cellular capsid-encoding world of viruses and the three domains of cellular life have been entwined in a complex labyrinth of formative evolutionary interactions since the dawn of evolution (Durzyńska and Goździcka-Jósefiak 2015).

The full range and taxonomy of prokaryotic viruses have undergone a major expansion, with consequent reclassification and renaming in recent years (Krupovic et al. 2016). Both the Eubacteria and the Archaea are "parasitized" by DNA- and RNA-based bacteriophage viruses, which follow two different cycles in how the parasitism—what we would term the aggressively symbiotic interaction—progresses. In the so-called "lytic cycle," the virus hijacks the bacterial genetic machinery, so the prokaryotic cell becomes a factory for the manufacture of daughter viruses, which are released into the ecology with the destruction (lysis) of the cell. In the "lysogenic" cycle, the virus inserts its DNA into the prokaryotic genome, where the viral insert is now termed a "prophage," and the phage DNA is reproduced in tandem with the prokaryotic host genome during the budding type of reproduction. Luria and Darnell, while acknowledging that bacteria and their viruses were likely to have shared a lengthy genetic inter-relationship, posited an origin of viruses after the origins of primitive cells and most likely deriving from mobile cellular genetic elements that were capable of transmissibility and self-replication from one cell to another (Luria and Darnell 1967). While it is likely that such creative evolutionary interactions between viruses and the prokaryotic domains have been ongoing and universal, there are difficulties with the Luria-Darnell theory of viral origins from mobile cellular genetic entities. Modern metagenomic analysis indicates that the direction of genetic transfer is far more commonly from virus to host (Koonin 2011). Forterre (2006), in an analysis of more than 250 cellular genomes from Archaea, Bacteria, and Eukarya, showed that most of the proteins detected in viral genomes have no cellular homologues. The most plausible explanation, deriving from this, is that viruses had a separate evolutionary origin from prokaryotic cellular life. The interesting recent discovery of a plasmid that transferred its genetic information to the host cell, an Antarctic haloarchaeon, was hailed by its authors as supporting the escape hypothesis for viral origins (Erdman et al. 2017). Perhaps, it does. But there is an alternative explanation, in which the evolution of all three cellular domains, together with their parasitizing viruses, was interactively complex, involving continuing and bi-directional genetic transfers that began at the very origins of life from protobiotic self-replicating chemicals and are continuing to the present day. We shall illustrate what we mean by this with some examples.

8.3.3.2 VIRAL PERSISTENCE AND AGGRESSIVE SYMBIOSIS

We have seen how viruses have an extraordinary ability to transfer genes, which might be regarded as genetic information, to their hosts (Koonin 2011). Stable transfer of the entire "information content" of the virus to the genome of the host could also be defined as "viral persistence" (Domingo et al. 1998). While viruses have, in the main, been considered from the perspective of disease, many viruses establish species-specific persistent, often unapparent, infections that are stable on an evolutionary timescale (Villarreal 2007, 2009a, 2009b). Such persistent infections can have large effects on relative reproductive fitness of competing host populations and, indeed, can promote host population survival. Viruses stably colonize their host genomes. In prokaryotes, virus killing and virus protection combine to become a truly creative and cooperative force. In effect, we are witnessing a combination of both symbiotic and competitive behaviors, which is increasingly seen as key to understanding how evolution works.

Such viral persistence leads to stable interactions at metabolic, behavioral, and genetic levels (Villarreal 2005; Villarreal and Ryan 2011). This virus-host dynamic is very ancient in its origins, and it is ongoing and inherently symbiotic. One remarkable example of such virus-host interactions is the above-mentioned "addiction module." P1 is a stable episomal DNA-based prophage of *Escherichia coli* frequently found in wild isolates, and it appears to be capable of interfering with infection of its host bacterium by other phages (Yarmolinsky 2004). First discovered by Yarmolinsky's group at the National Institutes of Health, it was found to be a persistent and lysogenic phage with a remarkable stability, which was mediated by an addiction module. This comprises a stable protein toxin and a less stable protein antitoxin that act in a curious but very effective coordination. Loss of phage presence in daughter generations of *E. coli* as a result of cell division does not lead to "curing" of the daughter cells of the "infection" but rather to the killing of the same cells by the remaining stable toxin. This dynamic phage-bacterium interaction is further complicated by involvement with the cell's own programmed death systems, which ensures maintenance of the P1 phage while also inhibiting its own lytic spread. The presence of P1 will also kill cells infected by other phages. Indeed, this system creates yet another primitive recognition of "self" in what might be seen as a forerunner of the acquisition of adaptive immunity (Villarreal 2005, 2009a, 2009b; Villarreal and Witzany 2009). Because a single virus type can be either lytic (harmful) or latent (protective) in a particular host, the virus itself has the capacity to promote a "virus mediated" addiction module by preventing the lysis of colonized host, as long as the latent virus is retained by the host, and this applies even to retention of a defective version of the virus. Thus, aggressive symbiosis emerges from stable virus colonization.

8.3.3.3 Viruses as Major Players in the Ecology

Over the last decade or so, we have come to realize that the virosphere, which comprises the junctional zones where viruses interact with their myriad hosts, spans all environments where life is to be found. Viruses are the most abundant biological entities in many different environments, exceeding the numbers of cellular life-forms, including prokaryotes, by one or two orders of magnitude (Koonin and Dolja 2013). Genetic diversity of viruses is commensurately enormous and might substantially exceed the genetic diversity of cellular organisms (Hambly and Suttle 2005; Rosario and Breitbart 2011; Koonin and Dolja 2013).

The oceans teem with prokaryotic viruses (Suttle 2007). Marine virologists have estimated that there are 10^{31} tailed bacteriophage viruses on Earth (Krupovic et al. 2011). Equally surprising is the discovery, through metagenomic analysis of the marine virosphere of four different oceanic regions, that most of the viral sequences extracted from these ecologies are unlike the known sequences in current databases. Global genetic diversity was very high, suggesting several hundred thousand of hitherto unrecognized species (Angly et al. 2006).

Such colossal numbers of phage viruses, with such high genetic diversity, make it likely that viruses play important roles in the planet's ecosystems, exerting a significant, perhaps dominant, force on the evolution of their bacterial and archaeal hosts. Thus, it is not altogether surprising that, besides being a major cause of disease and mortality in every life-form, viruses are also drivers of global geochemical cycles (Suttle 2005; Rosario and Breitbart 2011). Through prokaryotic lytic and lysogenic cycles, phage viruses impact the oceanic carbon, nitrogen, and phosphorus ecological cycles, having a major effect on the availability of nutrients in the marine ecosphere and terminating algal blooms (Fuhrman 1999; Wilhelm and Suttle 1999; Wommack and Colwell 2000). Identical bacteriophage sequences have also been found in a wide variety of different marine environments, suggesting that there is an extensive circulation of viral genes among distantly related host populations. It is unlikely that the presence and importance of viruses are confined to oceanic ecologies. Up to the present, we simply have not searched sufficiently for them elsewhere. But, with the expansion of metagenomics, this situation is changing.

8.3.3.4 A Viral Role in the Origins of Photosynthesis

The evolution of photosynthesis by cyanobacteria was a major step in the evolution of complex life. Cyanobacteria are one of the earliest branching groups of organisms on Earth. They are also the only known prokaryotes to carry out oxygenic photosynthesis—in other words, to capture the energy of sunlight for metabolic purposes—meanwhile producing oxygen to enter the ambient ecology as a by-product. The complexity of the photosynthesis machinery leaves no doubt that its origin and subsequent evolution involved multiple steps under constant selection pressure. This selective pressure came from two key factors: the need for energy to feed the cell's metabolic activities and the concomitant need to reduce the damaging effects of solar ultraviolet (UV) radiation, which was at more toxic levels than today because of the absence of an ozone shield (Mulkidjanian et al 2006). From the original photosynthetic cyanobacteria, the capacity for photosynthesis spread to other prokaryotes within the same taxonomic group and, further through a series of genetic symbiogenetic incorporations, to eukaryotic protists and beyond to the kingdom of the plants (Bonen and Doolittle 1975; Sagan 1967). Such repeated horizontal gene transfers suggest the involvement of viral transferases and ultimately phage viruses in what must have been a very complex and long-term process during the evolution of photosynthesis. The marine cyanobacterium *Prochlorococcus* is "infected by" three phage viruses from two viral families, *Mycoviridae* and *Podoviridae*. Analysis of the phage genomes reveals that the genes that encode the photosystem II core reaction center protein D1 and a high-light-inducible protein (HLIP), *hli*, which are central to oxygenic photosynthesis, are present in the viral genomes. The mycoviruses also contain additional *hli* genes, one of which encodes the second photosystem II core reaction center protein D2 (Lindell et al. 2004). The authors conclude that the bacteriophage viruses may be mediating the expansion of the *hli* gene family by transferring these genes back into their hosts after a period of evolution within the phage virus. The authors concluded that these gene transfers were likely to play a role in the fitness landscape of hosts and bacteriophage viruses in the surface oceans.

Phylogenetic examination of the genomes of the two key organelles, chloroplasts and mitochondria—themselves of microbial symbiogenetic origins in photosynthetic eukaryotes—has revealed historic bacteriophage symbiotic interaction. For example, mitochondrial RNA polymerase, DNA polymerase, and DNA primase are likely to have been contributed by T3/T7 bacteriophages, with authors suggesting that the prophage was present in the ancestral α-proteobacterium at the very origin of mitochondria (Filée and Forterre 2005). The same authors report another T3/T7 viral-like RNA polymerase in chloroplasts, indicating that strong selection pressure has favored replacement of some cellular proteins by viral alternatives during organelle evolution. This same group extended the examination more broadly to the role played by viruses in the evolution of their hosts, concluding that such is the level of lateral gene transfers from viruses to cells and non-orthologous gene replacements of cellular genes by viral ones that viruses should be considered major players in the evolution of cellular genomes (Filée et al. 2003).

8.3.4   VIRUSES IN THE EVOLUTION AND DIVERSIFICATION OF THE EUKARYOTES

8.3.4.1   VIRAL SYMBIOSIS IN THE ORIGINS AND DIVERSIFICATION OF THE EUKARYOTES

The explosive evolution of eukaryotic life, with its complexity and diversity, was accompanied by a parallel explosion of novel viral evolution, involving all three modalities of symbiotic interaction, namely parasitism, mutualism, and commensalism and including dynamic intermediate situations (Villarreal 2005; Ryan 2009; Witzany 2012). The eukaryotic viruses are much more varied than those of prokaryotes, being grouped into many different families, subfamilies, and genera, with further subdivisions based on the degree of antigenic similarity. Classification is imprecise beyond this point, since virus families cannot readily be placed into branching taxonomic trees; although modern genomic and protein analysis has discovered interesting relationships between some that may point to common ancestries in the remote past. Nomenclature is somewhat random, with some viruses named after the diseases they cause, for example, the pox (from "pocks" or skin ulcers) viruses, while others based on acronyms, such as the papovaviruses (from *pa*pilloma–*po*lyoma–*va*cuolating agent), and still others based on morphological features or even the places where they were first discovered (Collier and Oxford 1993). Today, viruses, and

virus-like "selfish genetic elements," are ubiquitous to life (Forterre 2013). The same author considers viruses to have polyphyletic, if also truly ancient, origins. We argue that all through the ages, from the self-replicators through 3 billion years exclusive to the prokaryotes and the evolution of the ensuing diversity of eukaryotes, the same deep interrelationship between viruses, and virus-linked genetic entities, and life has been essential and formative. This is amply illustrated when we consider the evolution of eukaryotic viruses.

Where, in the origin of viruses and the evolution from the stage of self-replicating chemicals to that of prokaryotic life, we inevitably encounter mysteries, the further evolution from prokaryotic viruses to the present-day world of eukaryotes has the benefit of potential genomic exploration of both stages and their viruses in the present day. In contrast to the prokaryotes, where the majority of viruses are double-stranded DNA-based bacteriophages, it is both curious and interesting that the majority of eukaryotic viruses are RNA-based (Koonin et al. 2015). Phylogenetic analysis reveals "tangible clues" to the origins of positive-strand RNA viruses of eukaryotes from double-stranded RNA-based bacteriophages or from positive-strand RNA viruses. Meanwhile, the same authors suggest that different families of double-stranded DNA viruses of eukaryotes may have originated from specific groups of bacteriophages. They conclude that the evolution of all classes of eukaryotic viruses is likely to have involved the amalgamation of various structural and replicative genetic modules deriving from different sources. Such an evolution would, in part, be facilitated through prolonged and extensive virus-host interactions, resulting in holobiontic genomic unions. Indeed, it is enlightening to consider some examples of such a pattern of evolution.

8.3.4.2   A POSSIBLE VIRAL ROLE IN THE ORIGIN OF THE EUKARYOTIC NUCLEUS

In 1967, Lynn Margulis, writing under her married name of Sagan, published a now-famous article, in which she proposed the "Serial Endosymbiotic Theory" for the origin of the eukaryotic cells, including their intrinsic organellar structures, mitochondria, plastids, and flagellae (Sagan 1967). This involved a key holobiontic genomic merger of a thermoplasma-like Archaean and Spirochaete-like eubacterium. This gave rise to an amitochondriate chimera capable of evolving to a primal eukaryotic protist, which generated the nucleus as a complex series of evolutionary steps deriving from a component of the "karyomastigont," an intracellular complex that ensured genetic continuity of the former bacterial symbionts (Margulis et al. 2000). Margulis's theory of endosymbiotic origins of mitochondria and plastids has withstood the test of time (López-García et al. 2017), though the same authors do not support the spirochaete origins of flagellae. The karyomastigont theory for the nucleus has also been challenged. In 2000, a viral origin was proposed for the eukaryotic replication proteins (Villarreal and DeFilippis 2000). The eukaryotic replicative DNA polymerases are dissimilar to those of eubacteria but bear similarities to the polymerases of large

DNA viruses that infect eukaryotes as well as the polymerases of T4 bacteriophages. In examining the DNA polymerase from phycodnavirus, which was chosen because it infects the primitive eukaryotes, microalgae, these authors found significant sequence similarities to the polymerases of eukaryotes and certain of their large DNA viruses. Subsequent reconstruction of a phylogenetic tree by these authors indicated that these algal viral DNA polymerases are near the root of the clade containing all eukaryotic DNA polymerase delta members and that this clade did not contain the polymerases of other DNA viruses.

In 2001, Bell put forward a hypothesis for a viral origin of the eukaryotic nucleus (Bell 2001), and the same author subsequently developed this to the "Viral Eukaryogenesis Hypothesis" (Bell 2009). A recent study supported this hypothesis with the report of the assembly of a nucleus-like structure during viral infection and replication by a *Pseudomonas chlororaphis* bacteriophage infecting a strain of *Pseudomonas* eubacterium, in which a phage-encoded tubulin-like cytoskeletal protein (PhuZ) formed a bipolar spindle that positioned the replicating viral DNA at the cell midpoint, apparently compartmentalizing and segregating phage and bacterial proteins according to function (Chaikeeratisak et al. 2017). Proteins involved in DNA replication and transcription were localized inside the compartment; meanwhile, proteins involved in translation and nucleotide synthesis were localized outside the compartment. In the words of the authors: "This subcellular compartmentalization of enzymes and DNA resembles the organization of eukaryotic cells containing a nucleus. In this, and the expanding modern scientific experience, viruses, formerly excluded from mainstream biological consideration, are now "taking centre stage in cellular evolution" (Claverie 2006).

This viral theory for the origin of the eukaryotic nucleus is challenged by López-García et al., above, who broadly support the Margulis's theory of nuclear origin from symbiogenetic fusion of archaea and bacteria, with the nucleus most likely deriving from an archaean of an Asgard or related lineage. Meanwhile, Witzany (2008) has proposed a viral origin for the telomeres and their enzymes, the telomerases, that play an important role in the physiological health of chromosomes.

8.3.4.3 Aggressive Plague Dynamics

Even in the case of the extreme interactions of virus and host that accompany an epidemic, or pandemic, plague virus and host can be seen to affect one another's evolution. Aggression *per se* may thus be seen to play an important role in the origins of some persisting evolutionary partnerships. Moreover, this capacity for aggression, initially directed at the subsequent partner species, may later be directed at a rival or prey of the partner species after the co-evolutionary association evolves. The virus-host dynamics in the myxomatosis epidemic of Australian and subsequently European rabbits is an illustrative example of this. The wild European rabbit, *Oryctolagus cuniculus*, was first introduced into Australia in 1859 as a source of food for British settlers. Lacking

natural predators, the rabbit population underwent an explosive expansion, leading to widespread destruction of agricultural grassland. In 1950, in a deliberate act of biological warfare, feral rabbits in southeastern Australia were infected with a myxoma pox virus of the genus *Leporipoxvirus* (Fenner and Kerr 1994), known to infect the South American rabbit, *Sylvilagus brasiliensis*.

The myxoma virus gives rise to a persistent infection in rabbits. After initial arrival, it never abandons its new host, generation after generation, co-evolving with—a virological term that is essentially the equivalent of symbiotic with—the host species. In its evolved relationship with the Brazilian rabbit, the virus caused little in the way of illness, but some strains of this virus were known to be exceedingly lethal to the European rabbit. High lethality strains were injected into a test group of rabbits, which were released into the wild. In this way, the innate aggression of a potentially "plague virus" was employed to control a "plague of rabbits."

Although no evolutionary experiment was planned, the methodology and outcome are now seen to provide a model with stark evolutionary implications. In retrospect, we might visualize the release of infected rabbits into the wild as creating a situation analogous to virgin contact between the two rabbit species in nature, with the Brazilian rabbit already symbiotic with its myxoma virus. Surprisingly, little happened after the first release of the infected rabbits. But the virus is spread in nature by biting insects. After proliferation of mosquitoes during a wet spring, the epidemic exploded. Within 3 months, 99.8% of the rabbits of southeast Australia, a land area the size of Western Europe, were exterminated. Had this been the result of a natural coming together of the two rabbit populations, it seems likely that viral aggression would have decimated the Brazilian rabbit's rival, thus allowing the Brazilian rabbit-plus-virus to dominate the ecology.

In this man-induced scenario, there was no evolutionary rival to take advantage of the cull. Instead, the aggressive symbiotic evolutionary dynamics of virus and host took a different course and a pattern we shall return to in relation to the dynamics of HIV-1 in humans. We must assume, in the rabbit example, that the myxoma virus culled the most susceptible genotypes of Australian rabbit population, selecting a minority genotype capable of surviving its lethal presence. Culling was followed by co-evolution between the virus and its new host, so that, within 7 years, the lethality of what was now an endemic infection was reduced to 25%. This same evolutionary symbiosis between rabbit and virus has continued to the present, when the rabbit population, still persistently infected with the myxoma virus, is now almost completely resistant to its former lethality. Indeed, the rabbit-plus-virus holobiont now occupies the ecology in large numbers.

8.3.4.4 A Successful Evolutionary Strategy of Aggressive Viral Symbiosis in Insects

Another exemplar of aggressive symbiosis as an efficient evolutionary strategy is that of the parasitoid wasps. These include approximately 25,000 species in symbiotic partnerships with approximately 20,000 species of polydnaviruses. In many instances, the wasps inject their eggs into the lepidopteran caterpillar prey. In some relationships, the virus inhabits the tissues of the wasp ovaries and the eggs are coated with virus as they emerge into the ovipositor. But in many other relationships, the viral genome has been integrated into the wasp genome (Whitfield 1990). These "unique symbiotic viruses," with a segmented genome of circular double-stranded DNA, replicate from integrated proviral DNA in the wasp ovary (Marti et al. 2003; Wyler and Lanzrein 2003) and are injected as fully infectious viruses along with the wasp egg into the host, where they are essential for the wasp larval survival. Genetic analysis of the wasp genomes reveals major change to incorporate the genomes of the symbiotic viruses. Analysis of the polydnaviruses has also revealed a complex reorganization of their genomes in an adaptive response to the holobiontic relationship (Espagne et al. 2004). The discovery of a conserved gene family in viruses of different wasp subfamilies led Provost et al. (2004) to propose a common phylogenetic origin of all such symbiotic partnerships. Estimations of the first establishment of the symbiosis, based on mitochondrial and nuclear sequencing, suggest a unique integration event 73.7 ± 10 million years ago (Belle et al. 2002; Whitfield 2002), illustrating powerful selection stability of the virus-wasp symbiosis over time.

It is important to grasp how such aggressive symbioses work. Selection is no longer working exclusively at rabbit or myxoma virus level, or at wasp or polydnavirus level, but also to a significant degree at the "holobiontic" level of rabbit-plus-virus and wasp-plus-virus. The symbiosis of parasitoid wasps with their viral partners is believed to be one of the most successful survival strategies among the insects.

8.3.4.5 Virulence versus Mutualism: Acute and Persistent Viral Strategies

Study of virulence in viral infections suggests that the level of virulence is predicated on the relationship between degree of virulence and evolutionary success in terms of infection rate and ultimately viral reproduction. In the case of some viruses, notoriously so in cases such as smallpox and pandemic influenza in humans, extreme virulence and high mortality would suggest that such virulence offers the most efficient route to viral reproduction. As we have seen with the Australian rabbit epidemic of myxomatosis, a similar high virulence and mortality characterized the virus-host interaction to begin with. But then, the virulence and mortality ameliorated as the virus-host relationship evolved to mutualism. It is noteworthy that both the symbiotic partnerships of the myxoma virus and rabbit and the parasitic wasp and its polydnavirus involve persistent infections—infections that never abandon the host once virus and host become associated.

Such persistent viral infections, extending through vast numbers of generations, allow the evolutionary time necessary for a mutualistic symbiosis to develop and consolidate. But even in the acute and highly virulent phase of a lethal virus pandemic, symbiotic interactions between virus and host are demonstrable. Thus, we find that even at the height of the AIDS pandemic, a multi-center study of the virus-host dynamics revealed a telling series of findings (Kiepiela et al. 2004). The rate of disease progression was strongly associated with the human leukocyte antigen HLA-B but not HLA-A allele expression. HLA-B is a gene coded within the major histocompatibility complex of the human genome—the part of the genome that determines "self" from foreign and governs the immune reaction to microbial invaders. The same authors concluded that substantially greater selection pressure was imposed on HIV-1 evolution by host HLA-B alleles than by HLA-A. They also concluded that HLA-B gene frequencies in the human population were those likely to be most influenced by HIV disease. In other words, at the height of the AIDS pandemic, virus and host were influencing the evolution of one another. This is quintessentially a symbiotic pattern of evolution.

8.3.4.6 Viruses of Plants

When, approximately half a billion years ago, the liverwort-like forerunners of plants first colonized dry land, the oceans already teemed with prokaryotic life, with its aggressively interactive virosphere (Wellman et al. 2003). Photosynthesis, with its consequent oxygenation of atmosphere, had long been established in the oceans. The colonization of the land masses with plants was equally prone to interaction with viruses as was the animal kingdom. Indeed, the first virus to be discovered was the *Tobacco mosaic virus* (TMV), which was identified as a "filterable virus" by the eminent Dutch microbiologist Beijerinck in 1898. Most of the historic work on plant viruses was directed at their potential for disease in food plants, which cause billions of dollars in loss of crop yields annually. From the taxonomic perspective, plant viruses have been grouped into 49 families and further subdivided into 73 genera: genomically, these include circular double-stranded DNA viruses; circular single-stranded DNA viruses; single-stranded RNA viruses; negative single-stranded RNA viruses; double-stranded RNA viruses; and unassigned viruses, satellites, and viroid.

Plants are sedentary. To transmit from one plant to another, plant viruses must make use of environmental factors, such as wind and moving water, and vectors such as biting insects, nematodes, plasmodiophorids, browsing animals, and seed and pollen dissemination—not to mention the potential for transfer of viruses through widespread human intrusion, including all manner of farming, into hitherto wild ecologies (Zaitlin and Palukaitis 2000; Gray and Banerjee 1999; Verchot-Lubicz 2003). As with animal viruses, most of the earlier studies of plant viruses have focused on viral pathogens, where the devastation of agriculture by emerging viral epidemics in cultivated crops appears to follow similar host-pathogen evolutions as seen in animals, whether resulting from familiar pathogens evolving new virulence in the same hosts or through the arrival of a totally new pathogen arising from intrusion into, or proximity to, a reservoir virus in a wild or natural ecology (Elena et al. 2011, 2014). Deeper evolutionary and ecological studies in plants and their viruses are largely still awaited. But we might reasonably infer that emerging plague viruses would cull the newly colonized host plant genotype, much as we have seen in animals, and, in the case of persisting viral infection, are likely to enter into long term co-evolutionary symbiosis with their hosts (Fraile and García-Arenal 2010).

8.3.4.7 EXAMPLES OF PLANT VIRUS EVOLUTION

Symbiotic partnerships sometimes include more than two partners. A well-known and ecologically important symbiosis involves viruses, bacteria, and plants intrinsic to the nitrogen cycle, in which soil-based rhizobial bacteria—also discovered by Beijerinck—provoke the development of root nodules in host plants, such as legumes. Here, the rhizobia fix atmospheric nitrogen into more complex organic nitrogenous compounds, which are then available for not only the host plants' metabolism but also uptake and digestion by browsing animals (Larenjo et al. 2014). Where the bacteria provide nitrogenous compounds to the plant, the plant also feeds the bacteria with carbon-rich organic compounds, constituting a metabolic symbiosis vital to the nitrogen cycle and thus biodiversity. However, many strains of soil-based rhizobia lack the fixation and nodulation genes necessary for this metabolic symbiosis. But they can gain this ability through a genetic symbiosis with a rhizobial strain known as *Mesorhizobium loti*, which carries the necessary genes within a 500-kb "symbiosis island" (Sullivan and Ronson 1998; Long 2001). Here, we discover a second twist to the story. The symbiosis island can only be transferred from a fixing to non-fixing strain of rhizobium through the contribution of a P4 bacteriophage integrase gene that is intrinsic to the symbiosis island. The integrase is the smoking gun that points to a preceding bacteriophage-rhizobial genetic symbiosis.

Virus colonization of hosts can also lessen the effects of environmental stress. For example, a tropical panic grass can survive the high soil temperatures of geothermal soils of the Yellowstone National Park through the triple symbiotic interplay between the plant, a fungus, and a virus within the fungus (Márquez et al. 2007). Plant viruses frequently make use of genetic recombination and re-assortment as driving forces in their evolution (Roossinck 1997). Plant RNA viruses also have highly error-prone replication mechanisms capable of giving rise to quasispecies fast-track evolution. As with animal genomes, a significant proportion of the genomes of higher plants consist of "transposable elements." A component of these transposable elements arises from genome colonization by pararetroviruses, which, contrary to what one might expect, are not RNA-based but double-stranded DNA viruses. But, like retroviruses, plant pararetroviruses are dependent on the enzyme reverse transcriptase for viral transcription in the host, albeit in a different series of genetic processes to those of retroviruses in animals. When a retrovirus or pararetrovirus inserts its genome into the germline of its host, it is known as an ERV or pararetrovirus. Endogenous pararetroviruses in plants may sometimes confer resistance to exogenous viruses—a familiar phenomenon in animals in relation to

ERVs (Staginnus et al. 2007). Plant pararetroviruses will also enter into systematic recombination with non-autonomous viral species, involving region-specific non-coding regulatory sequences, which gives rise to a symbiotic pattern of evolutionary adaptation (Chen et al. 2017).

Liu et al. (2011) have discovered widespread gene transfer from circular single-stranded DNA viruses to a broad range of eukaryotic genomes. For example, they found evidence that the replication initiation protein (Rep)-related sequences of geminiviruses, nanoviruses, and circoviruses have been frequently transferred to plants, fungi, animals, and protists. They also identified geminivirus-like transposable elements in the genomes of fungi and parvovirus-like transposable elements in the genomes of lower animals, providing evidence that eukaryotic transposons could derive from single-stranded DNA viruses. Fukuhara has reported a new distinct taxon of RNA-based double-stranded endornaviruses in plants such as bell pepper and rice, which have no obvious effects in terms of illness or phenotype of the host plants (Fukuhara 2015). They appear to be efficiently transmitted to the next generation via pollen and ova in a manner that suggests that they are genuine symbionts.

Chu et al. (2014) have found at least eight families of endogenous non-retroviral genes in the genomes of eudicot and monocot plants. Most were derived from double-stranded DNA viruses, but some appear to have been single-stranded DNA viruses and single-stranded RNA viruses. Most of these belonged to the families of *Partitiviridae, Chrysoviridae,* and *Geminiviridae*. While it is possible that these play some role in resistance to exogenous viral invasion, further study will be needed to examine their potential contribution to plant metabolism, physiology, and evolution. Roossinck et al. discovered a wide range of viruses in plants that similarly showed no obvious effects in terms of disease or phenotype (Roossinck 2015). We should recall that symbioses involving viruses include a very wide range of virus-host interactions, from outright parasitism to mutualism, but we should also remember that it includes the possibility of commensalism, in which a virus might take advantage of host physiology or genetics, without contributing any advantage or causing any disease.

Viral metagenomics is the study of viruses, without preset notions of disease, in environmental samples, using next-generation sequencing that produces very large data sets. Recently, the introduction of such metagenomic analysis to different environments has revealed a remarkable hidden world of viruses, involving domestic and wild plants and some interfaces between the two (Roossinck et al. 2015). In the words of Roossinck, who has helped to pioneer such metagenomic searches: "these... show that we have barely begun to appreciate the immense diversity of viruses that occur on our planet" (Stobbe and Roossinck 2014). Such metagenomic studies are revealing a large diversity and proliferation of viruses that appear to be present without obvious signs of plant pathology; these in turn are changing our perceptions of how the majority of viruses actually interact in a variety of symbiotic associations with their plant hosts (Roossinck 2012).

8.3.4.8 Giants and Minuscule Dependents: The Extremes of the Virosphere

When the Mimivirus, or *Acanthamoeba polyphaga mimivirus*, was first discovered inside an amoeba, it shocked the world of biology (Claverie 2006). Mimi had the typical genome of a virus, but it was much larger than usual. Indeed, it was so large that it was initially mistaken for a bacterium and was given the title of *Bradfordcoccus*, hence the origin of its common name, "*Mi*mmicking *Mi*crobe." It also proved to have a much more complex genome than most viruses, amassing more genes than some of the smallest bacteria. Mimi was the first of a growing plethora of giant viruses, or "macroviruses," to be discovered, including *Megavirus chilensis*, *Pandoravirus salinas*, and *Cafeteria roenbergensis*. Many of these infect amoebae, but *Cafeteria roenbergensis* virus infects a small bacterivorous marine flagellate that gave the virus its name. One astonishing pandoravirus was found in an amoeba living in the contact lens of a woman in Germany (Scheid et al. 2010). Within the virus, they further discovered a "virophage," labeled "Sputnik"—a virus parasitizing another virus. The discovery of these viral giants, which are now classed as the nucleocytoplasmic large DNA viruses, or the proposed order Megavirales, challenged many of the basic assumptions about viruses. In fact, detailed study suggests that they belong to three distinct groups that evolved from much smaller DNA virus ancestors that acquired many of their novel genes from genetic interchange with their eukaryotic hosts (Yutin et al. 2014).

At the opposite extreme, we find the family of miniscule viruses known as parvoviruses. These consist of three genera, two of which infect humans, causing gastroenteritis acquired from shellfish (Collier and Oxford 1993); meanwhile, other parvoviruses infect rodents, cats, dogs, and other animals. The parvoviruses are a mere 20 or so nanometers in diameter, and their genomes are so small that they can only replicate in the presence of another virus or helpful source of active DNA synthesis. They could be said to parasitize other viruses, and hence, they are also called dependoviruses. They are also the only DNA viruses with genomes consisting of single-stranded nucleic acid.

8.3.4.9 From Plagues to Symbionts: The Creative Retroviruses

In the early 1980s, one of the most dangerous plagues to emerge in the twentieth century arrived in America. Today, we know it as acquired immunodeficiency syndrome or AIDS. The causative virus is a retrovirus known as HIV-1. Retroviruses comprise a large and diverse family of RNA-based viruses, defined by common taxonomic denominators. HIV-1 is a member of the retrovirus subfamily known as lentiviruses, which also includes HIV-2 and the simian immunodeficiency virus, SIV-1. The first retrovirus to come to scientific attention was the avian sarcoma virus, discovered by Rous in 1911 during his research into the cause of avian sarcomas (Rous 1911). All retroviruses possess a viral envelope surrounding an icosahedral capsid,

FIGURE 8.3.1 This displays the quintessential structure of a retrovirus genome, including exogenous viruses, such as HIV-1, and endogenous viral loci within the human genome. It includes the multigene domains of *gag, pol,* and *env* as well as the single gene, protease, flanked at both ends by the genomic regulatory regions known as the 5′ and 3′ long terminal repeats, or LTRs.

which in turn encloses a helical nucleocapsid (Coffin 1990; Collier and Oxford 1993). The individual viruses, or virions, are small even for viruses at 80–130 nanometers. It is likely that retroviruses infect all the known vertebrate animals. In fact, we are coming to realize that their host range extends far beyond the vertebrates.

The retroviral genome consists of two identical molecules of single-stranded RNA, bound together noncovalently (Fields and Knipe 1990). All retroviruses, including exogenous and endogenous varieties, have a similar basic genomic structure, as seen in Figure 8.3.1. The genome comprises three genetic domains, each of which codes for multiple proteins.

For example, *gag* codes for matrix and core shell proteins; *pol* codes for enzymes such as reverse transcriptase, protease, ribonuclease and integrase; and *env* codes for the surface and transmembrane glycoproteins. The flanking long terminal repeats (LTRs) are regulatory regions that control the expression of the viral genes. But once inserted as ERVs into the host genome, these same LTRs are also capable of taking over the control of neighboring host genes. The key viral enzyme, reverse transcriptase, reverses the normal direction of genetic progression by converting the viral RNA of its genome to the homologous DNA, an initial step before the viral integrase incorporates the viral DNA homologue into the chromosomes of the infected host cell (Hindmarsh and Leis 1999).

In the parasitic cycle of retroviral infection, the host cell is a specific T4 lymphocyte, the T-helper cell, which carries a CD4 receptor on its cell surface. However, some retroviruses have the capability of inserting their genome in similar fashion into the chromosomes of the host germ cells. This has remarkable potential for future genetic symbiogenesis through the creation of what is now a "holobiontic genome"—a genome comprising the former host and retroviral evolutionary lineages brought into union through genetic symbiosis. A prevailing koala retroviral epidemic in eastern Australia is providing important information on the dynamics of epidemic retrovirus behavior and, in particular, the retroviral endogenization into host germlines (Tarlinton et al. 2006; Xu and Eiden 2015). The koala epidemic began in the northeastern Australia, more than a century ago, and is working its way southward over the eastern side of the country. All of the koalas in the north are now infected, as are a third of those in the south. The virus has killed millions of animals, largely through hematological malignancies such as leukemia and lymphoma. The virus is also rapidly endogenizing into the koala germline, with some animals already accumulating up to 46 proviral loci scattered throughout their chromosomes. Study of this extraordinary plague virus behavior is helping to throw light onto the evolution of animal genomes and, particularly, the human genome.

When, on February 12, 2001, two rival organizations announced simultaneously that they had completed the first comprehensive analysis of the human genome, there was astonishment at the fact that only some 1.5% of our DNA coded for proteins. This can be clearly seen in an updated pie chart of the DNA breakdown of the human genome shown in Figure 8.3.2.

How could such a small proportion of the genome code for what was presumed to be 80,000 to 100,000 proteins involved in human metabolism? This problem was subsequently solved by the discovery of splicing. The DNA coding sequences for genes involves a number of introns and exons, with only the exons contributing to coding. The introns play a complex role, in combination with other regulatory elements, that decides what particular cluster of introns actually codes for a protein. Thus, the protein-coding portion of the human genome, which is now thought to comprise roughly 24,000 genes, can potentially code for the presumed 80,000 to 100,000 proteins (Pontén et al. 2011). There was a second surprise with the initial decipherment of the genome, which is also illustrated in Figure 8.3.2. Some 9% of our human DNA is made up of sequences that are embedded endogenous retroviral loci. An additional 34% of the human DNA is made up of sequences known as LINEs and SINEs, which are closely related to retrovirus genomic sequences and are referred to as "retrotransposons." Meanwhile, half the total DNA, shown on the left of Figure 8.3.2, remains somewhat uncertain with regard to function, though much of this would appear to be involved in so-called non-coding RNA, which is involved in regulatory and epigenetic functioning. Contrary to the earlier belief that the viral and virus-related elements of the genome were mere "junk," making no contribution to phenotype (Orgel and Crick 1980), we now know that they have played a very important role in the evolution and function of the human genome.

8.3.4.10 THE SYMBIOTIC ROLE OF ENDOGENOUS RETROVIRUSES

If we examine the retroviral component of mammalian genomes, it consists of vast numbers of provirus genomic inserts scattered randomly throughout the chromosomes. Each insert is believed to represent an individual ancestor infected with an exogenous epidemic retrovirus (Ryan 2009). These viral genomic inserts into animal genomes are known as "endogenous retroviruses" (ERVs). Endogenous retroviruses are found in the genomes of every vertebrate on land. Indeed, they appear to have preceded the arrival of the first land vertebrates, being discovered in some varieties of farmed fish, where infection appears to cause systemic sickness and carcinogenesis. The increasing availability of fish genome sequences has allowed new insights into the diversity and host distribution of ERVs in fish, including Fugu, zebrafish, elephant shark, coelacanth, and lamprey. But the level of ERV sequences in fish genomes is generally much lower than that found in mammals, typically lower than 1% of total DNA (Naville and Volff 2016).

It is perhaps too early to speculate if fish ERVs might fulfill mutualistic functions with their hosts such as we see in mammals; however, a comprehensive recent study of 36 lineages of foamy-like retroviruses, in basal amphibians and fish, concluded that ray-finned fish foamy-like endogenous retroviruses (FLERVs) exhibited an overall co-speciation pattern with their hosts, but amphibian FLERVs might not (Aiewsakun and Katzourakis 2017). The same authors also reported several viral cross-class transmissions, involving lobe-finned fish, shark, and frog. They concluded that this major retroviral lineage and therefore retroviruses as a whole have an ancient marine origin at the time of, if not before, their jawed vertebrate hosts, more than 450 million years ago, in the Paleozoic Era.

Retroviruses suggestive of even more ancient origins have been found in the photosynthetic sea slug, *Elysia chorotica*, where the virus floods the tissues near to the end of the slug's life cycle (Pierce et al. 1999). The same viral manifestation, coinciding with the slug senescence or death, is seen in all *Elysia chlorotica* populations, regardless of location, including laboratory-kept specimens. Phylogenetic analysis of the *Elysia* retrovirus genome, and in particular its reverse transcriptase sequences, found sequence similarities with the reverse transcriptase of retrotransposons from the sea slug, *Aplysia californica*, and purple sea urchin, *Strongylocentrotus*

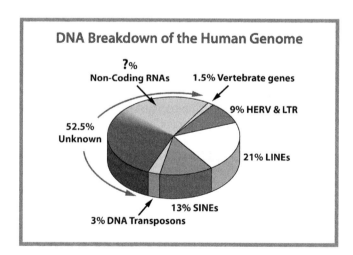

FIGURE 8.3.2 This shows an updated DNA breakdown of the human genome, showing the 1.5% that codes for proteins and the 9% comprising human endogenous retroviruses and their breakdown products, including isolated pairs of LTRs. It also shows the large sections labeled LINEs and SINEs, which are virus-related in their functions. The 3% labeled "DNA transposons" comprise relics of ancient DNA virus colonizations of the ancestral genome. The 52.5% labeled "unknown" contains sections of DNA that code for RNA molecules that are functional end products in themselves rather than messenger RNA, hence the term "non-coding" RNAs. Many of these are thought to contribute to genomic regulation in the form of long-noncoding RNAs and smaller regulatory RNAs of various sorts.

purpuratus. While the role of the retrovirus in the sea slug's life and death remains unknown (Pierce et al. 2016), the striking discovery of retroviruses in some of the earliest animals to evolve in the oceans suggests a very early origin of retroviruses; meanwhile, the very ubiquity of their presence in animals and their continuing evolutionary activity to the present day confirms their enormous importance to the evolution of the diversity of life on Earth.

When incorporated into the human genome, the retroviral inserts are referred to as "human endogenous retroviruses" (HERVs). Human endogenous retroviruses comprise between 30 and 50 families, depending on definition, and these families are further subdivided into more than 200 distinct groups and subgroups (Jurka 2000). Each of these is thought to represent an independent invasive lineage, implying that our primate ancestors have been subjected to a large series of independent HERV colonization, through exogenous retroviral epidemics. Although most of these colonizations took place more than 10 million years ago, a significant number occurred after the separation of the human lineage from that of chimpanzees. The families CERV 1 (PTERV1) and CERV 2 are unique to chimpanzees (Polaverapu et al. 2006), meanwhile at least 10 full-length HERV-K families are unique to humans (Medstrand and Mager 1998; Barbalescu et al 1999). When an ERV first enters the human genome, its loci will be suppressed by epigenetic mechanisms, such as methylation and histone modification. Over lengthy periods of time, many loci will be permanently suppressed by insertions or deletions—"indels"—or will be lost entirely from the genome during homologous sexual recombination. But epigenetic suppression is not permanent. Hence, during the lengthy period when epigenetic silencing predominates, there will opportunities for evolutionary interaction between the viral and host portions of what is now a holobiontic genome containing both mammalian and viral evolutionary lineages. Such interactions offer the potential for wide-scale genetic symbiogenesis.

Having long parasitized this same host as exogenous plague viruses, the now-endogenized viruses will have pre-evolved potentials for manipulation of host genetics, development and physiology. One important outcome would be the acquisition of new regulatory capability arising from the viral LTRs. Eukaryotic genomes are characterized more by regulatory complexity than by increase in gene numbers. Much of this complexity comes from virus-derived sequences, which, in addition to the influence of LTRs, also code for the production of small interacting regulatory RNAs, usually with stem-loop structures, which make an important contribution to the evolution and development of regulatory networks (Sundaram et al. 2014; Chuong et al. 2017).

Evolutionary biologists refer to the incorporation of viral elements into holobiontic evolution as "domestication" of viral genes and regulatory sequences. But it is important to grasp the limitations of such a term in these circumstances. The viral genes, and regulatory sequences, do not change to become similar to human genes and regulatory sequences. Indeed, their symbiotic contribution depends upon the fact that they retain those alien abilities evolved as part of their parasitic viral lineages, in order to contribute novel interactive potential to the now-holobiontic human-viral genome and organism. We are only beginning to realize the nature and extent of this holobiontic evolutionary potential for novel innovation at the level of genetics, and whole genomics, giving rise to innovative changes in reproduction, embryological development, immunology, and cellular physiology (Villarreal 2005; Ryan 2016). This symbiogenetic pattern of evolution has undoubtedly changed the course of animal and, in particular, mammalian and human evolution.

8.3.4.11 Endogenous Retroviruses in Reproduction

The first viral locus found to play a key role in human physiology was the ERVWE1 locus at chromosome 7 (7q21.2), from which the envelope gene, or *env*, was discovered to code for a viral protein, syncytin-1, which is important in human placentation (Mi et al. 2000; Mallet et al. 2004). Production of syncytin-1 is promoted by the viral locus 5'LTR, under the upstream regulatory control of a complex coordination of viral and vertebrate elements. The protein is strongly expressed in the placental trophoblasts, which fuse together to form a syncytium—a single confluent membrane studded with nuclei and with no junctional gaps between adjacent cell walls. In effect, syncytin has altered the fate of trophoblasts, converting them into syncytiotrophoblasts. This then forms the interface between the placenta and uterus, invading deep into the lining to create a very close contact between the maternal and fetal circulations, and with the syncytiotrophoblast, ensuring that all nutrients from the mother and waste from the fetus passes through cytoplasm rather than between cells. It also creates a better barrier, however fine, between the maternal and fetal circulations, helping to prevent immunological contact between maternal circulation and fetal tissues, which contain paternal antigens that would be treated as alien by the maternal immune system.

The syncytin-1 locus is common to the great apes, including gorillas, orangutans, chimpanzees, and humans and hence the ERV nomenclature rather than HERV. In an elegant study a few years later, a second fusiogenic endogenous retroviral protein, dubbed syncytin-2, was discovered. Expressed by the locus FRD on chromosome 6, syncytin-2 was also found to exhibit a powerful immunosuppressant function deeper to the syncytiotrophoblast layer of the placenta (Blaise et al. 2003). Today, we know of at least 12 different endogenous retroviral loci that play significant roles in human reproduction, with at least five involved in placentation, though the precise roles of some of the others remain to be determined (Villarreal and Ryan 2011). Mice were discovered to possess two different fusiogenic ERV syncytin-type genes, syncytin-A and syncytin-B, which play an important role in placentation. This led to the generation of knockout mice for syncytin-A and syncytin-B, whose embryonic placentas displayed major defects in cell-to-cell fusion, resulting in impaired embryo survival, confirming that syncytins are important for normal placental structure and function (Dupressoir et al. 2009).

Following the discovery of the human syncytins, Heidmann et al. screened a number of different mammalian clades for related endogenous placental retroviral *syncytin* genes to confirm a variety of syncytin variants associated with a wide range of clades that included Rodentia (*syncytin*-A and -B); Lagomorpha, which include rabbits (*syncytin-Orl*); Primates (*syncytin*-1 and -2); Carnivora (*syncytin-Car1*); Perissodactyls, which includes the horses; Chiroptera, which are bats; higher Ruminantia (*syncytin-Rum1*); Cetacea; Suina, which includes the pigs; Insectivora, which includes hedgehogs and shrews; Afrothera (*syncytin-Ten1*), which include elephants, aardvarks, and sea cows; and Xenarthra, which include anteaters, sloths, and armadillos (Cornelis et al. 2012, 2013, 214). Remarkably, the same group went on to demonstrate the presence of a syncytin in the South American opossum (*syncytin-Opo1*), a marsupial that undergoes a short-lived placentation (Cornelis et al. 2015). They further demonstrated a second non-fusogenic retroviral envelope gene that had been selectively conserved for more than 80 million years among all marsupials, including the South American opossum and the Australian tammar wallaby, and that possesses an immunosuppressive domain. This latter appeared to function much as synctytin-2 in great apes, to help suppress maternal rejection of paternal and thus potentially alien antigens in the fetus.

The discovery of two retroviruses in marsupials showing transient placentation would appear to be a key finding. Up to this point, there was uncertainty whether the viruses had played a key role in the origins of placentation or whether they arrived after the evolution of a more primitive placenta and helped to make it more efficient. In the words of these researchers: "The capture of a founding syncytin by an oviparous ancestor was pivotal for the emergence of placentation more than 150 million years ago (Cornelis et al. 2014)."

It is also possible that in this wide range of retroviruses involved in placentation, and more generally in reproduction, we may be witnessing competition between groups of retroviruses, with the newer arrivals in the process of selectively displacing more ancient rivals. While this is speculative, there is some hard evidence for competition involving the Jaagsiekte sheep retrovirus (JRSV) and sheep, in this case for a fierce battle of ego versus alter-ego for the territory of the viral host. The retrovirus donating the ruminant syncytin gene, which the Heidmann group labeled *syncytin-Rum1*, is not the only retrovirus involved in placentation in sheep and goats. Where the retrovirus carrying *syncytin-Rum1* is thought to have colonized the ruminant genome approximately 30 million years ago, a second retrovirus, the JSRV, is believed to have colonized the common ancestor of sheep and goats some 5 to 7 million years ago (Palmarini et al. 2004). The name, Jaagsiekte, is a translation from the Afrikaans for the "chasing illness," because it provokes breathlessness similar to that seen in sheep after they have been chased by a dog. The virus exists as two almost identical variants, the form endogenous to the sheep and goat genomes that contributes to placentation (Dunlap et al. 2006) and an epidemic exogenous retrovirus that causes a contagious lung cancer in sheep (Cousens et al. 2004). The two strains of the virus share 90% of their genetic sequences, so are likely to have a common origin. At some stage in history, perhaps between 400,000 and 1.3 million years ago, the JSRV endogenous virus evolved two

separate routes of blocking its "exogenous self" from gaining entry through the genital route. The exogenous virus then evolved a new route of infection through the lungs, where, unfortunately, its entire envelope gene protein has proved carcinogenic to sheep, particularly to lambs (Palmarini et al. 2004).

8.3.4.12 ENDOGENOUS RETROVIRUSES IN EMBRYONIC DEVELOPMENT

Developmental geneticists have long been aware of a strange phenomenon that takes place during the embryonic development of plants and animals. Methylation is one of the key epigenetic control regulators capable of switching off the expression of unwanted genes and other genetic sequences in the genomes of animals and plants. So, it was puzzling to discover that a major wave of genome-wide demethylation took place throughout a substantial portion of the genome of the cell during the earliest stage of embryonic development of plants and animals, including mammals (Eckardt 2006; Guo et al. 2014). This had the potential to activate normally suppressed endogenous retroviral entities and their related retrotransposons. In an experiment to explore this phenomenon, Spadafora exposed early murine embryos to the anti-reverse transcriptase drug, Nevirapine, to make the surprising discovery that it led to an irreversible arrest of development up to the four-cell stage (Spadafora 2008). Other researchers had already discovered that reverse transcriptase inhibition of early embryonic development resulted in substantial reprogramming of gene expression, involving both developmental and translational genes (Pittoggi et al. 2003). These curious findings suggested that ERVs, and/or LINE-type products, are involved in regulation and gene expression at the earliest stages of mammalian development.

The human endogenous retroviruses known as the HERV-Ks are the most recent viral lineages to colonize the primate lineage, including more than 10 groups that arrived into the human lineage after the divergence of humans and chimpanzees. These resulted in multiple HERV-K loci inserts with intact open reading frames scattered throughout the human chromosomes. Further investigation led to the discovery that DNA hypomethylation of the LTRs of the most recently acquired HERV-K integrations, accompanied by transactivation by OCT4 (a key definer of pluripotency during development), synergistically facilitates HERV-K reactivation and expression during this critical phase of human fetal development (Grow et al. 2015). Such retroviral expression appears to facilitate early embryogenesis, beginning with embryonic genome activation at the eight-cell stage, continuing through the emergence of epiblast cells in preimplantation blastocysts, and ceasing at the stage of embryonic stem cell derivation from blastocyst outgrowths.

There are roughly 1,000 loci of another endogenous virus group, the HERV-H, scattered through the human chromosomes. Some 231 of these are highly expressed in embryonic stem cells, where their LTRs, working in a coordinated way, act as a long noncoding RNA required for human embryonic stem cell identity. This causes the upregulation of *OCT4*, *SOX2*, and *NANOG*, which maintain stem cell pluripotency. When HERV-H expression is blocked, using RNAi, the stem cells lose their pluripotency and differentiate into cells that resemble fibroblasts (Lu et al. 2014).

This arena of research is still relatively new. We have already underlined the role of endogenous viral sequences in the development of regulatory networks. It seems likely that such networks will be found to contribute to the complex regulation of embryogenesis. Indeed, given that levels of colonization of the animals with ERVs, it seems likely that, given more time and exploration, many more such intimate and important interactions with host embryonic development will be discovered.

8.3.4.13 A Viral Contribution to Genomic Plasticity and Function

Recombination of viral loci on different chromosomes during meiosis has contributed to large-scale genetic deletions and duplications throughout our mammalian, hominid, and hominin evolution (Medstrand and Mager 1998; Hughes and Coffin 2001). Duplications of flanking endogenous retroviral loci appear to have contributed to the evolution of the extended major histocompatibility complex (Dawkins et al. 1999). A similar virally induced genomic plasticity may also have contributed to the surprising level of genetic variation, currently being observed between individual humans (Redon et al. 2006). A recent study of regulatory evolution of innate immunity showed that ERVs have shaped the evolution of a transcriptional network underlying the interferon response, which is a key element of innate immunity (Chuong et al. 2016). Meanwhile, ERVs have also contributed to many symbiotic interactions with host at post-developmental level, with implications for genetics, epigenetics, physiology, and health.

8.3.4.14 The Operation of Selection on the Virus-Host Holobiontic Genome

We have already explained that, in the holobiontic genome that arises from the symbiogenetic union of viral and host genomes, selection no longer operates solely at the level of viral or host genetic elements at individual level: for holobiontic survival, it must also operate at the level of the "holobiontic genome." This implies selection for former host or viral genes and other genetic sequences that enhance holobiontic survival and selection against former host or viral genes and other genetic products that threaten survival. More than half of the human-specific HERV-K LTRs actively promote host DNA transcription (Buzdin et al. 2006), suggesting that viral LTRs have replaced less efficient former host promoters. For example, the LTR of ERV-9 plays a key role in transcriptional control of the β-globin gene cluster in primates, including humans (Plant et al. 2001; Routledge and Proudfoot 2002). A range of virus-derived elements is also found in a large number of human protein-coding genes, where most are inserted into introns—the non-coding regions between the coding regions, known as exons—where they frequently effect gene function (Nekrutenko and Li 2001). Other LTRs have taken on the role of alternative promoters, or splice receptors, for example, in the control of the endothelin B receptor and apolipoprotein C-I genes and in the control of the human leptin receptor, which is involved in energy expenditure, production of sex hormones, and activation of hematopoietic cells (Kapitonov and Jurka 1999; Medstrand et al. 2001).

The observed viral roles fulfill the predictions of symbiogenetic evolution, as do the large repertoire of viral elements that code for a multiplicity of functions essential to the evolution and normal working of the genome. Even viral loci that have lost all of their protein-coding genetic elements, to be reduced to paired empty LTRs, are capable of playing a significant role in genomic regulation.

8.3.4.15 Viral Loci Expressing Viral Genes as Proteins in Human Tissues

A number of earlier studies have suggested that viral genes in the human genome might be expressed as proteins in the human tissue. We have powerful evidence for this in the viral role in placentation described previously. There is growing evidence that HERV genes may also be expressed as proteins in a variety of non-placental somatic cells, tissues, and organs. Most such studies have examined viral gene expression at messenger RNA (mRNA) level rather than the more difficult direct definition of actual protein expression within cells and tissues. However, expression of viral genes at RNA level may be an unreliable guide to protein expression because of differences in mRNA turnover, stability, and efficiency in translation. There is clearly a need to develop reliable techniques for detecting and measuring virally encoded proteins in cells, tissues, and organs. One such technique has been developed at The Rudbeck Laboratory, Uppsala University, and Uppsala University Hospital, Sweden, in an attempt to solve this problem. Focusing on the endogenous virus locus, known as ERV3, situated on chromosome 7q11.21, as a primary test system, they generated monospecific polyclonal antibodies capable of reliably detecting the expression of the ERV3-encoded Env proteins in human cells, tissues, and organs (Fei et al. 2014). This experiment found that ERV3-encoded Env proteins are expressed at substantive levels in placenta, testis, adrenal gland, corpus luteum, Fallopian tubes, sebaceous glands, astrocytes, bronchial epithelium, and the ducts of the salivary glands. Meaningful expression was also discovered in epithelial cells, fused macrophages, myocardium, and striated muscle. The authors concluded that the virus plays a significant role in human physiology and may also play a possible role in disease.

Preliminary results of a more comprehensive search for other endogenous viral gene expression as proteins in human cells, tissues, and organs look promising (personal communication). These results clearly suggest the need for more comprehensive investigation, along the lines of the former Human Genome and Human Proteome projects, to catalogue the total

HERV genome and HERV proteome expression in human reproduction, embryogenesis, and normal physiology.

8.3.5 CONCLUSION

While the origin of viruses as a whole remains uncertain, and may well be polphyletic, we have made a case for the likely origin of RNA-based viruses in the putative RNA world. It seems likely, at the very least, that DNA-based viruses existed at the very earliest stages of prokaryotic life, and it is overwhelmingly evident that DNA- and RNA-based viruses have made a major contribution to, and have been an essential component of, the subsequent evolution to biodiversity as we find it today. We humans, like all other cellular life-forms, inhabit a ubiquitous and intensively interactive virosphere that molds our environment and ecology and interacts with life in a quintessentially symbiological fashion at every level.

REFERENCES

Aiewsakun P. and A. Katzourakis. 2017. Marine origin of retroviruses in the early Palaeozoic Era. *Nat. Comm.* doi:10.1038/ncomms13954.

Alberts B., A. Johnson, J. Lewis et al. 2002. *The RNA World and the Origins of Life, from Molecular Biology of the Cell*, 4th ed. Garland Science, New York.

Angly F.E., B. Felts, M. Breitbart et al. 2006. The marine viromes of four oceanic regions. *PLoS Biol.* 4(11): 2121–2131.

Bansho Y., N. Ichihashi, Y. Kazuta et al. 2012. Importance of parasite RNA species repression for prolonged translation-coupled RNA self-Replication. *Chem. Biol.* 19: 478–487.

Barbalescu M., G. Turner, M.I. Seaman et al. 1999. Many human endogenous retrovirus K (HERV-K) proviruses are unique to humans. *Curr. Biol.* 9: 861–868.

Bell P.J.L. 2001. Viral eukaryogenesis: Was the ancestor of the nucleus a complex DNA virus? *J. Mol. Evol.* 53: 251–256.

Bell P.J.L. 2009. The viral eukaryogenesis hypothesis: A key role for viruses in the emergence of eukaryotes from a prokaryotic world environment. *Ann. N. Y. Acad. Sci.* 1178: 91–105.

Belle E., N.E. Beckage, J. Rousselet et al. 2002. Visualization of polydnavirus sequences in a parasitoid wasp chromosome. *J. Virol.* 76: 5793–5796.

Bernhardt H.S. 2012. The RNA hypothesis: The worst theory of the early evolution of life (except for all the others). *Biol. Direct* 7: 23. http//www.biology-direct.com/content/7/1/23.

Blaise S., N. de Parseval, L. Bénit et al. 2003. Genomewide screening for fusogenic human endogenous retrovirus envelopes identifies syncytin 2, a gene conserved on primate evolution. *Proc. Natl. Acad. Sci. U.S.A.* 100: 13013–13018.

Bonen L. and W.F. Doolittle. 1975. On the prokaryotic nature of the red algal chloroplasts. *Proc. Natl. Acad. Sci. U.S.A.* 72: 2310–2314.

Bremermann H.J. 1983. Parasites at the origin of life. *J. Math. Biol.* 16: 165–180.

Buzdin A., E. Kovalskaya-Alexandrova, E. Gogvadze et al. 2006. At least 50% of human-specific HERV-K (HML-2) long terminal repeats serve in vivo as active promoters for host nonrepetitive DNA transcription. *J. Virol.* 80: 10752–10762.

Chaikeeratisak V., K. Nguyen, K. Khanna et al. 2017. Assembly of a nucleus-like structure during viral replication in bacteria. *Science* 355: 194–197.

Chen S., H. Zheng and Y. Kishima. 2017. Genomic fossils reveal adaptation of non-autonomous pararetroviruses driven by concerted evolution of noncoding regulatory sequences. *PLoS Pathog.* doi:10.1371/journal.ppat.1006413.

Chu H., Y. Jo and W.K. Cho. 2014. Evolution of endogenous non-retroviral genes integrated into plant genomes. *Curr. Plant Biol.* 1: 55–59.

Chuong E.B., N.C. Elde and C. Feschotte. 2016. Regulatory evolution of innate immunity through co-option of endogenous retroviruses. *Science* 351: 1083–1087.

Chuong E.B., N.C. Elde and C. Feschotte. 2017. Regulatory activities of transposable elements: From conflicts to benefits. *Nat. Rev. Genet.* 18(2): 71–86.

Clarke D.K., E.A. Duarte, S.F. Elena et al. 1994. The red queen reigns in the kingdom of RNA viruses. *Proc. Natl. Acad. Sci. U.S.A.* 91:4821–4824.

Claverie J.-M. 2006. Viruses take center stage in cellular evolution. *Genome Biol.* 7: 110. doi:101186/gb-2006-7-6-110.

Coffin J.M. 1990. Retroviridae and their replication. In Fields B.N. and D.M. Knipe (Eds.), *Fields Virology*, Chapter 51. Raven Press, New York.

Colizzi E.S. and P. Hogeweg. 2016. Parasites sustain and enhance RNA-like replicators through spatial self-organisation. *PLoS Comput. Biol.* doi:10.1371/journal.pcbi.1004902.

Collier L. and J. Oxford. 1993. *Human Virology*. Oxford University Press, Oxford, UK.

Copley S.D., E. Smith and H.J. Morowitz. 2007. The origin of the RNA world: Co-evolution of genes and metabolism. *Bioorg. Chem.* 35(6): 430–443.

Cornelis G., O. Heidmann, S. Bernard-Stoecklin et al. 2012. Ancestral capture of syncytin-Car1, a fusogenic endogenous retroviral envelope gene involved in placentation and conserved in Carnivora. *Proc. Natl. Acad. Sci. U.S.A.* 109(7). www.pnas.org/cgi/doi/10.1073/pnas.1115346109.

Cornelis G., O. Heidmann, S.A. Degrelle et al. 2013. Captured retroviral envelope syncytin gene associated with unique placental structure of higher ruminants. *Proc. Natl. Acad. Sci. U.S.A.* www.pnas.org/cgi/doi/10.1073/pnas.1215787110.

Cornelis G., C. Vernochet, S. Malicorne et al. 2014. Retroviral envelope syncytin capture in an ancestrally diverged mammalian clade for placentation in the primitive Afrotherian tenrecs. *Proc. Natl. Acad. Sci. U.S.A.* www.pnas.org/cgi/doi/10.1073/pnas.1412268111.

Cornelis G., C. Vernochet, Q. Carradec et al. 2015. Retroviral envelope gene captures and syncytin exaptation for placentation in marsupials. *Proc. Natl. Acad. Sci. U.S.A.* www.pnas.org/cgi/doi/10.1073/pnas.1417000112.

Cousens C., J.V. Bishop, A.W. Philbey et al. 2004. Analysis of Integration sites of Jaagsiekte sheep retrovirus in ovine pulmonary adenocarcinoma. *J. Virol.* 78(16): 8506–8512.

Cronin L., A.C. Evans and D.A. Winkler (Eds.). 2017. From prebiotic chemistry to molecular evolution. This is a free access compendium of articles on the theme of prebiotic chemistry to molecular evolution. www.belstein–journals/bjoc/70.

Dawkins R., C. Leelayuwat, S. Gaudieri et al. 1999. Genomics of the major histocompatibility complex: Haplotypes, duplication, retroviruses and disease. *Immunol. Rev.* 167: 275–304.

De La Torre J.C. and J.J. Holland. 1990. RNA virus quasispecies populations can suppress vastly superior mutant progeny. *J. Virol.* 64(12): 6278–6281.

Domingo E., E. Baranowski, C.M. Ruiz-Jarabo et al. 1998. Quaisispecies structure and persistence of RNA viruses. *Emerg. Inf. Dis.* 4(4): 521–527.

Domingo E., J. Sheldon and C. Perales. 2012. Viral quasispecies evolution. *Microbiol. Mol. Biol. Rev.* 76: 159–216.

Dunlap K.A., M. Palmarini, M. Varela et al. 2006. Endogenous retroviruses regulate periimplantation placental growth and differentiation. *Proc. Natl. Acad. Sci. U.S.A.* 103(39): 14390–14395.

Dupressoir A., C. Vernochet, O. Bawa et al. 2009. Syncytin-A knockout mice demonstrate the critical role in placentation of fusogenic, endogenous retrovirus-derived, envelope gene. *Proc. Natl. Acad. Sci. U.S.A.* 106(2): 12127–12132.

Durzyńska J. and A. Goździcka-Jósefiak. 2015. Viruses and cells intertwined since the dawn of evolution. *Virol. J.* 12: 169. doi:10.1186/s12985-015-0400-7.

Eckardt N.A. 2006. Genetic and epigenetic regulation of embryogenesis. *Plant Cell.* 18: 781–784.

Eigen M. and P. Schuster. 1977. The hypercycle. A principle of natural self-organisation. Part A: Emergence of the hypercycle. *Naturwissenschaften* 64: 541–565.

Eigen M. and P. Schuster. 1978a. The hypercycle. A principle of natural self-organisation. Part B: The abstract hypercycle. *Naturwissenschaften* 65: 7–41.

Eigen M. and P. Schuster. 1978b. The hypercycle. A principle of natural self-organisation. Part C: The realistic hypercycle. *Naturwissenschaften* 65: 341–369.

Eigen M. and P. Schuster. 1979. *The Hypercycle: A Principle of Natural Self-Organization.* Springer, Berlin, Germany.

Elena S.F., A. Fraile and F. Garcia-Arenal. 2014. Evolution and emergence of plant viruses. *Adv. Virus Res.* doi:10.1016/B978-0-12-800098-4.00003-9.

Elena S.F., S. Bedhomme, P. Carrasco et al. 2011. The evolutionary genetics of emerging plant RNA viruses. *Mol. Plant-Microbe Interact.* 24(3): 287–293.

Erdmann S., B. Tschitschko, L. Zhong et al. 2017. A plasmid from an Antarctichaloarchaeon uses specialized membrane vesicles to disseminate and infect plasmid-free cells. *Nat. Microbiol.* doi:10.1038/s41564-017-0009-2.

Espagne E., C. Dupuy, E. Huguet et al. 2004. Genome sequence of a polydnavirus: Insights into symbiotic virus evolution. *Science* 306: 286–289.

Fei C., C. Atterby, P.-H. Edqvist et al. 2014. Detection of the human endogenous retrovirus ERV3 encoded Env-protein in human tissues using antibody-based proteomics. *J. R. Soc. Med.* 107(1): 22–29.

Fenner F. and P.J. Kerr. 1994. Chapter 13: Evolution of the pox viruses, including the coevolution of virus and host in myxomatosis. In *The Evolutionary Biology of Viruses.* Raven Press, New York.

Fields B.N. and D.M. Knipe. 1990. *Fields Virology*, 2nd ed. Raven Press, New York. Vol 1 - Part I: General Virology: 26.

Filée J. and P. Forterre. 2005. Viral proteins functioning in organelles: A cryptic origin? *Trends Microbiol.* 13(11): 510–513.

Filée J., P. Forterre and J. Laurent. 2003. The role played by viruses in the evolution of their hosts: A view based on informational protein phylogenies. *Res. Virol.* 154: 237–243.

Fisher S. 2010. Are RNA viruses vestiges of an RNA world? *J. Gen. Philos. Sci.* 41: 121–141.

Forterre P. 2006. The origin of viruses and their possible roles in major evolutionary transitions. *Virus Res.* 117: 5–16.

Forterre P. and D. Prangishvili. 2009. The great billion-year war between ribosome- and capsid-encoding organisms (cells and viruses) as the major source of evolutionary novelties. *Ann. N. Y. Acad. Sci.* 1178: 65–77.

Forterre P. 2013. The great virus comeback (translated from the French). *Biol. Aujourdhui* 207(3): 153–168.

Fraile A. and F. García-Arenal. 2010. The coevolution of plants and viruses: Resistance and pathogenicity. *Adv. Virus Res.* 76: 1–32.

Fuhrman J.A. 1999. Marine viruses and their biogeochemical and ecological effects. *Nature* 399: 541–548.

Fukuhara T. 2015. Unique symbiotic viruses in plants: Endornaviruses. *Uirusu* 65(2): 209–218.

Gilbert W. 1986. The RNA world. *Nature* 319: 618.

Gray S.M. and N. Banerjee. 1999. Mechanisms of arthropod transmission of plant and animal viruses. *Microbiol. Mol. Biol. Rev.* 63(1): 128–148.

Grow E.J., R.A. Flynn, S.L. Chavez et al. 2015. Intrinsic retroviral reactivation in human preimplantation embryos and pluripotent cells. *Nature* 522: 221–225.

Guo H., P. Zhu, L. Yan et al. 2014. The DNA methylation landscape of human early embryos. *Nature* 511: 606–610.

Hambly E. and C.A. Suttle. 2005. The virosphere, diversity, and genetic exchange within phage communities. *Curr. Opin. Microbiol.* 8: 444–450.

Hindmarsh P. and J. Leis. 1999. Retroviral DNA integration. *Microbiol. Mol. Biol. Rev.* 63(4): 836–843.

Holland J.J., J.C. De La Torre and D.A. Steinhauer. 1992. RNA virus populations as quasispecies. *Curr. Top. Microbiol. Immunol.* 176: 1–20.

Holmes E.C. 2011. What does virus evolution tell us about virus origins? *J. Virol.* 85(11): 5427–5251.

Hughes J.F. and J.M. Coffin. 2001. Evidence for genomic rearrangements mediated by human endogenous retroviruses during primate evolution. *Nat. Genet.* 29: 487–489.

Jurka J. 2000. Repbase update: A database and an electronic journal of repetitive elements. *Trends Genet.* 16 (9): 418–420.

Kapitonov V.V. and J. Jurka. 1999. The long terminal repeat of an endogenous retrovirus induces alternative splicing and encodes an additional carboxy-terminal sequence in the human leptin receptor. *J. Mol. Evol.* 48: 248–51.

Kiepiela P, A.J. Leslie, I. Honeyborne et al. 2004. Dominant influence of HLA-B in mediating the potential co-evolution between HIV and HLA. *Nature* 432: 769–774.

Koonin E.V. 2011. *The Logic of Chance: The Nature and Origin of Biological Evolution.* F.T. Press Science, Upper Saddle River, NJ.

Koonin E.V. and V.V. Dolja. 2013. A virocentric perspective on the evolution of life. *Curr. Opin. Virol.* 3(5): 546–557.

Koonin E.V., T.G. Senkevich and V.V. Dolja. 2006. The ancient virus world and the evolution of cells. *Biol. Direct.* doi:10.1186/1745-6150-1-29.

Koonin E.V., V.V. Dolja and M. Krupovic. 2015. Origins and evolution of viruses of eukaryotes: The ultimate modularity. *Virology* 479–480: 2–25.

Krupovic M., D. Prangishvili, R.W. Hendrix and D.H. Bamford. 2011. Genomics of bacterial and archaeal viruses: Dynamics within the prokaryotic virosphere. *Microbiol. Mol. Biol. Rev.* 75(4): 610–635.

Krupovic M., B.E. Dutilh, E.M. Adriaenssens et al. 2016. Taxonomy of prokaryotic viruses: Update from the ICTV bacterial and archaeal viruses subcommittee. *Arch. Virol.* 161: 1095–1099.

Larenjo M., A. Alexandre and S. Oliveira. 2014. Legume growth-promoting rhizobia: An overview on the *Mesorhizobium* genus. *Microbiol. Res.* 169: 2–17.

Lazcano A. and S.L. Miller. 1996. The origin and early evolution of life: Prebiotic chemistry and the pre-RNA world and time. *Cell* 85: 793–798.

Lehman N. 2015. The RNA World: 4,000,000,050 years old. *Life* 5: 1583–1586.

Lindell D., M.B. Sullivan, Z.I. Johnson et al. 2004. Transfer of photosynthesis genes to and from Prochlorococcus viruses. *Proc. Natl. Acad. Sci. U.S.A.* 101(30): 11013–11018.

Liu H., Y. Fu, B. Li et al. 2011. Widespread horizontal gene transfer from circular single-stranded DNA viruses to eukaryotic genomes. *BMC Evol. Biol.* 11: 276. http://www.biomedcentral.com/1471-2148/11/276.

Long S.R. 2001. Genes and signals in the rhizobium-legume symbiosis. *Plant Physiol.* 125: 69–72.

López-García P., L. Eme and D. Moreira. 2017. Symbiosis in eukaryotic evolution. *J. Theor. Biol.* 434: 20–33.

Lu X., F. Sachs, L. Ramsay et al. 2014. The retrovirus HERVH is a long noncoding RNA required for human embryonic stem cell identity. *Nat. Struct. Mol. Biol.* 21(4): 423–425.

Luria S.E. and J.E. Darnell. 1967. *General Virology.* John Wiley & Sons, New York.

Mallet F., O. Bouton, S. Prudhomme et al. 2004. The endogenous retroviral locus ERVWE1 is a bona fide gene involved in hominoid placental physiology. *Proc. Natl. Acad. Sci. U.S.A.* 101: 1731–1736.

Margulis L., M.F. Dolan and R. Guerrero. 2000. The chimeric eukaryote: Origin of the nucleus from the karyomastigont in amitochondriate protists. *Proc. Natl. Acad. Sci. U.S.A.* 97(13): 6954–6959.

Márquez L.M., R.S. Redman, R.J. Rodriguez and M.J. Roossinck. 2007. A virus in a fungus in a plant: Three-way symbiosis required for thermal tolerance. *Science* 315: 513–515.

Marti D., C. Grossniklaus-Bürgin, S. Wyder et al. 2003. Ovary development and polydnavirus morphogenesis in the parasitic wasp *Chelonus inanitus*. I. Ovary morphogenesis, amplification of viral DNA and ecdysteroids titres. *J. Gen. Virol.* 84: 1141–1150.

McClintock B. 1956. Controlling elements and the gene. *Cold Spring Harb. Symp. Quant. Biol.* 21: 197–216.

McClintock B. 1984. The significance of responses of the genome to challenge. *Science* 226: 792–801.

McFadden G.I. 2001. Chloroplast origin and integration. *Plant Physiol.* 125: 50–53.

Medstrand P. and D.L. Mager. 1998. Human-specific integrations of the HERV-K endogenous retrovirus family. *J. Virol.* 72: 9782–9787.

Medstrand P, J.R. Landry and D.L. Mager. 2001. Long terminal repeats are used as alternative promoters for the endothelin B receptor and apolipoprotein C-I genes in humans. *J. Biol. Chem.* 276: 1896–903.

Mi S., X. Lee and X. Li et al. 2000. Syncytin is a captive retroviral envelope protein involved in human placental morphogenesis. *Nature* 403: 785–789.

Miller W.B. 2017. Biological information systems: Evolution as cognition-based information management. *Prog. Biophys. Mol. Biol.* doi:10.1016/j.pbiomolbio.2017.11.005.

Mulkidjanian A.Y., E.V. Koonin, K.S. Makarova et al. 2006. The cyanobacterial genome core and the origin of photosynthesis. *Proc. Natl. Acad. Sci. U.S.A.* 103(35): 13126–13131.

Nasir A., K. M. Kim and G. Caetano-Annolés. 2012. Viral evolution. Primordial cellular origins and late adaptation to parasitism. *Mob. Genet. Elem.* 2(5): 247–252.

Naville M. and J.-N. Volff. 2016. Endogenous retroviruses in fish genomes: From relics of past infections to evolutionary innovations? *Front. Microbiol.* 7. doi:10.3389/fmicb.2016.01197.

Nekrutenko A. and W.-H. Li. 2001. Transposable elements are found in a large number of human protein-coding genes. *Trends Genet.* 17: 619–21.

Nimwegen E.V., J.P. Crutchfield and M. Huynan. 1999. Neutral evolution of mutational robustness. *Proc. Natl. Acad. Sci. U.S.A.* 96: 9716–9720.

Nowak M.A. 1992. What is a quasispecies? *TREE* 7: 118–121.

Oliver K.R. and W.K. Green. 2012. Transposable elements and viruses as factors in adaptation and evolution: An expansion and strengthening of the TE-thrust hypothesis. *Ecol. Evol.* doi:10.1002/ece3.400.

Orgel L.E. and F.H.C. Crick. 1980. Selfish DNA: The ultimate parasite. *Nature* 284: 604–607.

Palmarini M., M. Mura and T.E. Spencer 2004. Endogenous betaretroviruses of sheep: Teaching new lessons in retroviral interference and adaptation. *J. Gen. Virol.* 85: 1–13.

Patel B.H., C. Percivalle, D.J. Ritson et al. 2015. Common origins of RNA, protein and lipid precursors in a cyanosulfidic protometabolism. *Nat. Chem.* 7: 301–307.

Pierce S.K., T.K. Maugel, M.E. Rumpho et al. 1999. Annual viral expression in a sea slug population: Life cycle control and symbiotic chloroplast maintenance. *Biol. Bull.* 197: 1–6.

Pierce S.K., P. Mahadevan, S.E. Massey et al. 2016. A preliminary molecular and phylogenetic analysis of the genome of a novel endogenous retrovirus in the sea slug *Elysia chlorotica*. *Biol. Bull.* 231: 236–244.

Pittoggi C., I. Sciamanna, E. Mattei et al. 2003. Role of endogenous reverse transcriptase in murine early embryo development. *Mol. Reprod. Dev.* 66: 225–236.

Plant K.E., S.J. Routledge and N.J. Proudfoot. 2001. Intergenic transcription in the human beta-globin gene cluster. *Mol. Cell. Biol.* 21: 6507–6514.

Polaverapu N., N.J. Bowen and J.F. McDonald. 2006. Identification, characterization and comparative genomics of chimpanzee endogenous retroviruses. *Genome Biol.* 7. doi:10.1186/gb-2006-7-6-r51.

Pontén F., J.M. Schwenk, A. Asplund and P.H. Edqvist. 2011. The Human Protein Atlas as a proteomic resource for biomarker discovery. *J. Intern. Med.* 270: 428–446.

Prangishvili D. and R.A. Garrett. 2004. Exceptionally diverse morphotypes and genomes of crenarcheal hyperthermophilic viruses. *Biochem. Soc. Trans.* 32(2): 204–208.

Provost B., P. Varricchio, E. Arana et al. 2004. Bracoviruses contain a large multigene family coding for protein tyrosine phosphatases. *J. Virol.* 130: 90–103.

Redon R., S. Ishikawa, K.R. Fitch et al. 2006. Global variation in copy number in the human genome. *Nature* 444: 444–454.

Rich A. 1962. On the problems of evolution and biochemical information transfer. In *Horizons in Biochemistry*, Kasha M. and B. Pullman (Eds.). Academic Press, New York, pp. 103–106.

Robertson M.P. and G.F. Joyce 2012. The origins of the RNA world. *Cold Spring Harb. Perspect. Biol.* 4(5): a003608. doi:10.1101/cshperspect.a003608.

Roossinck M.J. 1997. Mechanisms of plant virus evolution. *Ann. Rev. Phytopathol.* 35: 191–209.

Roossinck M.J. 2012. Plant virus metagenomics: Biodiversity and ecology. *Annu. Rev. Genet.* 46: 357–367.

Roossinck M.J. 2015. Plants, viruses and the environment: Ecology and mutualism. *Virology* 479–480: 271–277.

Roossinck M.J., D.P. Martin and P. Roumagnac. 2015. Plant virus metagenomics: Advances in virus discovery. *Phytopathology* 105: 716–727.

Rosario K. and M. Breitbart. 2011. Exploring the viral world through metagenomics. *Curr. Opin. Virol.* 1(1): 289–297.

Rous R. 1911. Transmission of a malignant new growth by means of a cell-free filtrate. *JAMA* 56: 198.

Routledge S.J. and N.J. Proudfoot. 2002. Definition of transcriptional promoters in the human beta globin locus control region. *J. Mol. Biol.* 323: 601–611.

Ryan F. 1997. *Virus X*. Little Brown Publishers, Boston, MA.

Ryan F. 2009. *Virolution*. Harpercollins Publishers, London, UK.

Ryan F. 2007. Viruses as symbionts. *Symbiosis* 44: 11–21.

Ryan F.P. 2016. Viral symbiosis and the holobiontic nature of the human genome. *APMIS* 124: 11–19.

Sagan L. 1967. On the origin of mitosing cells. *J. Theor. Biol.* 14: 255–274.

Scheid P., B. Haröder and R Michel. 2010. Investigations of an extraordinary endosymbiont in *Acanthamoeba* sp.: Development and replication. *Parasitol. Res.* 106(6): 1371–1377.

Shirt-Ediss B., S. Murillo-Sánchez and K. Ruiz-Mirazo 2017. Framing major prebiotic transitions as stages of protocell development: Three challenges for origins-of-life research. *Beilstein J. Org. Chem.* 13: 1388–1395.

Spadafora C. 2008. A reverse transcriptase-dependent mechanism plays central roles in fundamental biological processes. *Syst. Biol. Reprod. Med.* 54: 11–21.

Staginnus C., W. Gregor, M.F. Mette et al. 2007. Endogenous pararetroviral sequences in tomato (*Solanum lycopersicum*) and related species. *BMC Plant Biol.* 7: 24. doi:10.1186/1471-2229-7-24.

Stobbe A.H. and M.J. Roossinck. 2014. Plant virus metagenomics: What we know and why we need to know more. *Front. Plant Sci.* 5: 150. doi:10.3389/fpls.2014.00150.

Sullivan J.T. and C.W. Ronson. 1998. Evolution of rhizobia by acquisition of a 500-kb symbiosis island that integrates into a phe-tRNA gene. *Proc. Nat. Acad. Sci. U.S.A.* 95: 5145–5149.

Sundaram V., Y. Cheng, Z. Ma et al. 2014. Widespread contribution of transposable elements to the innovation of gene regulatory networks. *Genome Res.* 24: 1963–1976.

Suttle C.A. 2007. Marine viruses – major players in the global ecosystem. *Nat. Rev. Microbiol.* 5: 801–812.

Suttle C.A. 2005. Viruses in the sea. *Nature* 437: 356–361.

Pontén F., J.M. Schwenk, A. Asplund and P.H. Edqvist. 2011. The Human Protein Atlas as a proteomic resource for biomarker discovery. *J. Int. Med.* 270: 428–446.

Takeuchi N. and P, Hogeweg. 2008. Evolution of complexity in RNA-like replicators systems. *Biol. Direct.* 3: 11. doi:10.1186/1745-6150-3-11.

Tarlinton R.E., J. Meers and P.R. Young. 2006. Retroviral invasion of the koala genome. *Nature* 442: 79–81.

Verchot-Lubicz J. 2003. Soilborne viruses: Advances in virus movement, virus induced gene silencing and engineered resistance. *Physiol. Mol. Plant Pathol.* 62: 55–63.

Vignuzzi M., J.K. Stone, J.J. Arnold et al. 2006. Quasispecies diversity determines pathogenesis through cooperative interactions in a viral population. *Nature* 439: 344–348.

Villarreal L.P. 2007. Virus-host symbiosis mediated by persistence. *Symbiosis* 44: 1–9.

Villarreal L.P. 2005. *Viruses and the Evolution of Life*. ASM Press, Washington, DC.

Villarreal L.P. and G. Witzany. 2009a. Viruses are essential agents within the roots and stem of the tree of life. *J. Theoret. Biol.* doi:10.1016/j.jtbi.2009.10.014.

Villarreal L.P. 2009b. Persistence pays: How viruses promote host group survival. *Curr. Opin. Microbiol.* 12: 1–6.

Villarreal L.P. 2014. Force for ancient and recent life: Viral and stem-loop RNA consortia promote life. *Ann. N. Y. Acad. Sci.* 1341: 25–34.

Villarreal L.P. and V.R. DeFilippis. 2000. A Hypothesis for DNA viruses as the Origin of Eukaryotic Replication Proteins. *J. Virol.* 74(15): 7079–7084.

Villarreal L.P. and F. Ryan. 2011. Viruses in host evolution: General principles and future extrapolations. *Curr. Top. Virol.* 9: 79–90.

Villarreal L.P. and G. Witzany. 2009. Viruses are essential agents within the roots and stem of the tree of life. *J. Theor. Biol.* doi:10.1016/j.jtbi.2009.10.014.

Wellman C.H., P.L. Osterloff and U. Mohiuddin. 2003. Fragments of the earliest land plants. *Nature* 425: 282–285.

Whitfield J.B. 1990. Parasitoids, polydnaviruses and endosymbiosis. *Parasitology* 6: 381–384.

Whitfield J.B. 2002. Estimating the age of the polydnavirus/braconid wasp symbiosis. *Proc. Natl. Acad. Sci. U.S.A.* 99(11): 7508–7513.

Wilhelm S.W. and C.A. Suttle. 1999. Viruses and nutrient cycles in the sea. *Bioscience* 49: 781–788.

Wilke C.O., J.L. Wang, C. Ofria et al. 2001. Evolution of digital organisms at high mutation rates leads to survival of the flattest. *Nature* 412: 331–333.

Witzany G. 2008. The viral origins of telomeres and telomerases and their important role in eukaryogenesis and genome maintenance. *Biosemiotics.* doi:10.1007/s12304-008-9018-0.

Witzany G. 2012. *Viruses: Essential Agents of Life*. Springer, Dordrecht, the Netherlands.

Woese C. 1998. The universal ancestor. *Proc. Natl. Acad. Sci. U.S.A.* 95(12): 6854–6859.

Wommack K.E. and R.R. Colwell. 2000. Viroplankton: Viruses in aquatic ecosystems. *Microbiol. Mol. Biol. Rev.* 64: 69–114.

Wyler T. and B. Lanzrein. 2003. Ovary development and polydnavirus morphogenesis in the parasitic wasp Chelonus inanitus. II. Ultrastructural analysis of calyx cell development, virion formation and release. *J. Gen. Virol.* 84(5): 1151–1163.

Xu W. and M.V. Eiden. 2015. Koala Retroviruses: Evolution and disease dynamics. *Ann. Rev. Virol.* 2: 119–134.

Yarmolinsky M.B. 2004. Bacteriophage in retrospect and in prospect. *J. Bacteriol.* 186: 7025–7028.

Yutin N., Y.I. Wolf and V. Koonin. 2014. Origin of giant viruses from smaller DNA viruses not from a fourth domain of cellular life. *Virology* 466–467: 38–52.

Zaman L., J.R. Meyer, S. Devangam, et al. 2014. Coevolution drives the emergence of complex traits and promotes evolvability. *PLoS Biol.* 12(12): e1002023. doi:10.1371/journal.pbio.1002023.

Zaitlin M. and P. Palukaitis. 2000. Advances in understanding plant viruses and virus diseases. *Ann. Rev. Phytopath.* 38: 117–143.

8.4 Carl R. Woese and the Journey toward a Universal Tree of Life

Bruce W. Fouke, Killivalavan Solai, Shelby L. Jones,
Elizabeth M. Smith, Kyle W. Fouke, Kaitlyn E. Fouke,
Claudia C. Lutz, Mayandi Sivaguru, and Glenn A. Fried

CONTENTS

8.4.1 INTRODUCTION

> If I have seen further than others, it is because I was looking in the right direction.
>
> **Carl R. Woese**
> *2010 (personal communication)*

All of life is composed of "stardust" that came into existence with the advent of the Big Bang event billions of years ago. These celestial raw materials, relentlessly transformed and reshaped, provided molecules that are now the building blocks for the breathtaking biodiversity inhabiting our planet (Lane 2015). Yet, even this awe-inspiring display of life is but a small sampling, a modern snapshot, of the trillions upon trillions of organisms that have flourished and gone extinct during geological time on Earth and likely other planets. A more in-depth understanding of the origin and evolution of life on Earth provides an essential context with which to probe the primary mission objectives of NASA Astrobiology (Hays 2015; Domagal-Goldman et al. 2016), which include the following: How does life begin and evolve? Does life exist elsewhere in the universe? What is the future of life on Earth and beyond?

Concepts of human genealogy, established by the ancient Egyptians, Chinese, Greeks, Arabs, Romans, and all other peoples of the world, have long been the centerpiece for determining social structure, wealth distribution, royalty bloodlines, and governance. Therefore, an invaluable starting point from which to consider evolution is the idea of a family tree—an idea that is familiar and relevant to all people. Anyone can sketch a few branches of their own family and know what they mean: your grandparents gave birth to your parents, who gave birth to you. Everyone on every branch in your tree is descended from, and connected to, everyone else. Similar diagrams of lineage and descent are pervasive in the biological sciences (Barton et al. 2007). In this case, however, the objective is to understand evolutionary relationships and gain insight regarding the shape and structure of the ultimate family tree that relates all living entities through geological time. Reproducible science that is predictive in nature, the very definition of good science, requires a robust tree of life applicable to the past, present, and future of all of life on Earth.

Remarkably, we now actually *do* know much more about the tree of life, based on the revolutionary work that begun nearly 60 years ago by Carl R. Woese, who was a professor in the Department of Microbiology at the University of Illinois Urbana-Champaign (Figure 8.4.1). His research literally established a new structure for the tree of life for our planet, as well as hypothesized the sequence of events surrounding its origin (Sapp 2009). The long and winding road of Woese's discoveries about the tree of life reached a pivotal point when classical Darwinian evolutionists, rooted in macro-level observations, were challenged by unexpected new discoveries that were driven by the advent of new micro-scale recombinant deoxyribonucleic acid (DNA) technologies. The progression of Woese's work is also a story of how the life and physical sciences themselves were required to change and adapt as new discoveries were made. This article presents the progression of scientific inquiry that led Woese to establish a three-domain universal tree of life

FIGURE 8.4.1 Photograph of Carl R. Woese at the light table on which he painstakingly analyzed molecular patterns of microbial ribosomes that were radioactively exposed on x-ray film. These tedious and methodical measurements led to the identification the 16S rRNA gene sequence, the fundamental yardstick for measuring evolutionary relatedness. (Courtesy of Jason Lindsay, College of Liberal Arts & Sciences, University of Illinois Urbana-Champaign, Urbana, IL.)

(Bacteria, Archaea, and Eucarya; Figure 8.4.2). The result is that we now have fundamentally new integrated biological approach to better understanding the environmental sustainability, energy exploration, and production, as well as make bold new advances in medicine and space exploration.

8.4.2 REVOLUTION OF THE TREE OF LIFE

Science has long endeavored to establish a tree of life that would encompass all single-celled and multicellular life, human and otherwise (Table 8.4.1). This effort in Western science began in 1674, when Antonie van Leeuwenhoek developed the first microscope (Gest 2004). A textile merchant and self-taught scientist in Delft, The Netherlands, Leeuwenhoek is considered the father of modern-day microbiology. His revolutionary research tool was more like a powerful magnifying lens than an actual microscope, composed of a small brass plate, a hand-pulled and ground spherical glass lens, and set screws that positioned a needle point on which a microscopic sample was held. Despite its simplicity, this revolutionary device allowed Leeuwenhoek to make the first observation of microscopic life in lake water, organisms that he called "cavorting wee beasties" and published in the *Proceedings of the Royal Academy of Science*. In recognition of this fundamental contribution, one of the most prestigious awards in microbiology to this day is the Leeuwenhoek Medal—an honor bestowed on Woese in 1992.

The next major advancement toward establishing a tree of Life took place in 1735, when Carl Linnaeus, a Swedish scientist (also known as Carl von Linné), established a systematic approach (*taxonomy*) for describing and classifying living organisms in a publication he entitled *Systema Naturae* (Huneman 2007). The endeavor required many years of

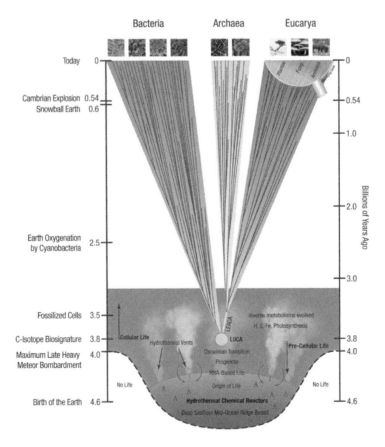

FIGURE 8.4.2 Universal tree of life, integrating the deep seafloor origin of life, geological time, and key Earth events with the three-domain (Bacteria, Archaea, and Eucarya) structure of life.

TABLE 8.4.1
Timeline of the Development of the Tree of Life (ToL) Model

Year	Scientist's Name	Tool	Historical Importance
1674	Antonie van Leeuwenhoek	First, simple microscope (no stains)	Known as the father of microbiology, he used his microscope invention to make the first discovery of microbes that were living in lake water.
1735	Carl Linnaeus	Field and lab observations	Created a taxonomic system for classifying organisms. Developed a three-kingdom model, *Animalia*, *Vegetabile*, and *Lapideum* (the inanimate world). Developed the binomial nomenclature of *genus* and *species*.
1859	Charles Darwin	Field and lab observations	Darwin, inspired by the ideas of Alfred Wallace, further developed and described patterns of biodiversity. This included proposing the process of natural selection, yet it did not include microorganisms.
1866	Ernst Haeckel	Field and lab human observations	Established the first full ToL that included microbes (*Monera*) at the base of a tree that had three branches: *Plantae*, *Animalia*, and *Protista*. Also proposed that evolutionary history is recorded in developmental stages of individuals. (*Ontogeny recapitulates phylogeny*)
1938	Édouard Chatton	Sophisticated microscope with stains	Established two evolutionary "empires": the single-celled Prokaryotes without a nucleus, which were thought to directly evolve into the single- and multi-celled Eukaryotes with a nucleus.
1959	Whittaker	Field and lab observations; biochemistry	Established a five-kingdom tree of life model, rooted by *Monera*, which branched into *Protista*, *Animalia*, *Plantae*, and *Fungi*.
1977 and 1990	Carl Woese	Ribosomal RNA molecular sequencing	Identified analysis of the ribosome (*molecular phylogeny*) as a sensitive tool for measuring the evolutionary relatedness of all living organisms. Results indicated a tree of life with three domains: *Bacteria*, *Archaea*, and *Eucarya*.

Source: Baumgartner, L.K., and Pace, N.R., Current taxonomy in classroom instruction, *The Science Teacher*, pp. 46–51, 2007; Barton, N.H. et al., *Evolution*, Cold Spring Harbor Laboratory Press, Cold Spring Harbor, NY, 833 p, 2007; Sapp, J., *The New Foundations of Evolution*, Oxford University Press, Oxford, UK, 425 p, 2009; Pace, N.R. et al., *Proc. Natl. Acad. Sci. U.S.A.*, 109, 1011–1018, 2012.

revision and development, spanning from inception of the idea during his education at the University of Leiden, The Netherlands, through his appointment as a professor at Uppsala University, Sweden. During this time, *Systema Naturae* grew from a thin pamphlet into a multi-volume treatise. Linnaeus's taxonomy organized all living things into just two groups or kingdoms, Animalia and Vegetabile. The third kingdom was that of Lapideum, which he conceived of as the inanimate world of rocks, minerals, and fossils. Linnaeus's taxonomy was based on the shared traits of plants and animals, rather than the type of environment in which the organism lived, which was the widely accepted context for taxonomy in the scientific community at that time. Linnaeus originally based his taxonomy on reproductive organs, later expanding it to incorporate all morphological characteristics of an organism. Importantly, Linnaeus dramatically simplified and systematized the naming of organisms into just two words—genus and species.

The middle of the nineteenth century saw explosive developments in the quest to understand the history of life. The groundbreaking works of Alfred Russel Wallace and Charles Darwin were the first to introduce evolutionary theory and the concepts of natural selection and fitness (Barton et al. 2007). Darwin's 1859 publication of *Origin of Species* was the first to link the branching patterns of heredity with evolutionary process and recommended that taxonomic classifications be based on these evolutionary relationships. However, *Origin of Species* focused exclusively on animals and plants and did not include single-celled microorganisms. It was not until an 1866 publication by Ernst Haeckel, a professor at the University of Jena, Germany, included microbes for the first time into the tree of life (Richards 2008). Haeckel strongly supported evolutionary theory and is well known for his masterful illustrations of life-forms, as well as his hypothesis that "phylogeny recapitulates ontogeny"—the observation that an organism's evolutionary history is often expressed in its early developmental stages. Haeckel constructed the first comprehensive tree of life in his publication *Generelle Morphologie der Organismen*, in which he described three great branches he called Plantae, Animalia, and Protista. Importantly, Haeckel included unicellular microorganisms, which he called Monera, at the base of the tree as part of the Protista kingdom.

From its first invention by Robert Hooke of the Royal Society of London in 1665, through its ever-increasing technical sophistication into the early twentieth century, the compound microscope facilitated advancement in the observation of cells (Gest 2004). This allowed Édouard Chatton, in 1938, to propose that all of life can be organized into two great "Empires"—*Eukaryotes*, plants and animals with cells that have a nucleus containing genetic material, and *Prokaryotes,* whose microbial cells have no nucleus (Woese et al. 1990). This work further fueled Haeckel's notion of a linear tree of life in which more "complex" and larger multicellular eukaryotic life evolved directly along a single trunk from the small less complex single-celled prokaryotic life (Figure 8.4.3). Classification and ecological thinking were then fundamentally recast in the middle of the twentieth century by Robert H. Whittaker, who completed his graduate education at the University of Illinois Urbana-Champaign before eventually becoming a professor in the Department of Biology at Cornell University. Initially published in 1959 and later refined and expanded in 1969, Whittaker's work established a five-kingdom structure for the tree of life that included the Animalia, Plantae, Fungi, Protista, and Monera (Figure 8.4.4). This followed the primary tree structure of Haeckel and his rooted separation of Monera, while also reflecting Whittaker's inclination that "functional" kingdoms should be based on community ecosystem biochemistry and ecology (Barton et al. 2007).

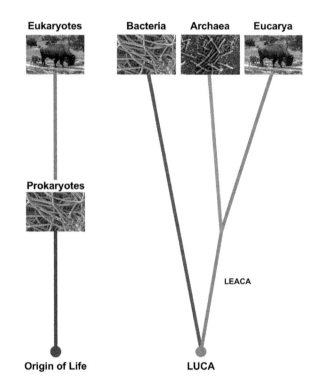

FIGURE 8.4.3   Sketches illustrating the distinct differences between the Chatton Prokaryote-Eukaryote tree of life (left), as opposed to the Woese three-domain universal tree of life (right).

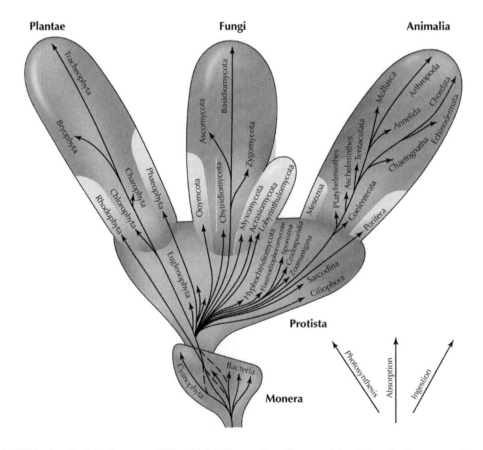

FIGURE 8.4.4   The Whittaker five-kingdom tree of life, which indicates direct lineage of the Animalia, Plantae, and Fungi from the Monera and Protista. (Modified from Barton, N.H. et al., *Evolution*, Cold Spring Harbor Laboratory Press, Cold Spring Harbor, NY, 833 p, 2007.)

8.4.3 ESTABLISHMENT OF MOLECULAR PHYLOGENY

Prior to Woese's work, the scientific community's view was predicated upon Chatton's proposal that there were just two types of life on Earth—Eukaryotes and Prokaryotes (Figure 8.4.3). At this time, microbes were sorted into one or the other of these grouping based on two criteria—their biochemical (metabolic) capabilities when grown in the laboratory (called *isolates* or *pure cultures*) and their shapes and structures when viewed under the microscope. While culturing provides invaluable information on how microorganisms function, it is estimated that less than 1% of the microbes on Earth can be cultured in the laboratory (Pace et al. 2012). Furthermore, even if cultivated, the full microbial community with which a species interacts in nature can only rarely be reproduced in the laboratory. Furthermore, microbes are generally tiny and indistinguishable spheres, rods, or filaments with similar internal physical structure. Therefore, even when the biochemical and morphological attributes of microbes were determined with microscopic analysis and culturing, they still did not provide a crisp and unambiguous measure of evolutionary relatedness. This is akin to using traits like hair color and height in people to determine their relatedness—they tell you something but not enough to be a reliable indicator of family ties.

Woese therefore began his scientific quest with the basic idea that, in order to create a universal tree of life, some kind of objective measure of evolutionary relatedness needed to be made that was guaranteed to be found in the cells of every single living creature. Born in Syracuse, New York, Woese originally studied mathematics and physics at Amherst College and then received a Ph.D. in biophysics, studying viruses under the direction of Professor Earnest Pollard at Yale University in 1953. Woese then worked in the General Electric Research Laboratory and was given an unprecedented instant-tenure faculty position in microbiology at University of Illinois Urbana-Champaign, driven in large part by Professor Sol Spiegelman. As a result of his background, Woese was a relative outsider to classical biology and had no preconceived agenda. He did not set out to overturn the foundations of modern classification, but instead simply wanted taxonomy to better reflect evolutionary relatedness (Sapp 2009).

The word phylogeny is derived from the Greek roots "phylon" meaning stem and "genesis" meaning origin. Molecular phylogenetics therefore seeks to establish evolutionary relatedness by measuring and comparing the chemical composition of key biomolecules within cells (Pace et al. 2012). The *genetic code* is the set of three-letter combinations, in which each letter represents a nitrogen-containing nucleotide base pair (A-T or adenine-thymine, G-C or guanine-cytosine), and each triplet corresponds to a specific amino acid building block. Woese believed that a universal tree of life was possible because the genetic code is universally consistent across all lifeforms on Earth (Woese 1965). Humans have the same genetic code as bison, which have the same genetic code as bears, moose, eagles, rattle snakes, beetles, lodge pole pines, and all other living organisms that have been encountered. In addition, the genetic code is extremely good at minimizing errors

(Butler et al. 2009). Furthermore, if a permanent alteration is made in the nucleotide sequence of an organism's genome (a *mutation*), due to either damage or errors, the resulting amino acids are still close to the ones that should have originally been made. While the genetic code has evolved toward an optimal state, is close to being optimal, and itself has evolved somewhat, its robust stability and consistency through most of life's history permit development of an accurate and comprehensive molecular phylogeny (Woese 1965).

In 1964, just a few years after becoming a professor in microbiology at University of Illinois Urbana-Champaign, Woese began an ambitious program to construct a molecular phylogeny for all of life. This was while a major focus in the rest of the scientific world was on furthering the work of American biologist James Watson and English physicist Francis Crick on the double-helical structure of DNA and the genetic code (Sapp 2009). However, from early on, Woese clearly understood that this level of structure was not sufficient to understand biological functions. Woese went so far as to write a personal note to Crick about these ideas in 1969, but Crick responded that his goals were likely impossible to reach, that his proposed work would therefore likely offer limited results, and that it would be hard for him to receive federal funding for this work. This would be a harbinger of precisely how the rest of the scientific community would react to Woese's avant-garde work in future years.

Woese was undeterred. He continued his work by considering what cells do for a living. Cells must reproduce to make future generations, they must metabolize by extracting energy from the environment, and they must learn how to do these things from the information stored and passed along in their genetic composition. Conceptualized as the *Dogma of Molecular Biology* by Crick, the key observation is that a cell ultimately functions and survives by making proteins, which they do by implementing information represented in the chemical structure of DNA. To accomplish this, the cell first reads (*transcribes*) the DNA into messenger ribonucleic acid (mRNA), an information-carrying strand, which is then processed by a complex organic molecule (a *molecular machine*) called the ribosome—the cell's protein factory. Ribosomes interpret the genetic code echoed from DNA to mRNA and string together a corresponding sequence of amino acids (a process called *translation*) to form the proteins that are essential to life. So, Woese's reasoning was that every cell must have a ribosome in order to produce proteins and survive. Therefore, to understand the evolutionary history of the cell, Woese targeted the chemical composition (*sequence*) of the various parts of the ribosome, which was most likely to be present in every living organism on Earth. He summarized these early ideas in his 1967 book *The Genetic Code: The Molecular Basis for Genetic Expression*. Woese came to appreciate that, in order to fully derive and understand biological function (especially of microorganisms), one needed to know the structure, as well as the motions and vibrations of the molecular components of the ribosome (Sapp 2009).

The ribosome is an incredibly complicated molecular machine that has somewhat-variable amounts of protein and RNA. Just as

Darwin sought to measure what he called essential evolutionary characteristics, Woese similarly chose to measure the essential chemical composition of the ribosomal RNA (rRNA) gene sequence. The idea of molecular evolution was that the age of genes could be read like a clock, a concept first proposed by Francis Crick in 1958 and later by Linus Pauling and Emil Zuckerkandl in 1965 (Sapp 2009). The information stored in molecules of nucleic acids will, as a result of random mutations, consistently change through time, like the steady beat of a ticking clock. Study of the pattern of these changes in information molecules, whether they be amino acids or nucleic acids, is therefore like studying a permanently running evolutionary timepiece. With this in mind, Woese believed that the rRNA gene sequence molecule was the ultimate universal chronometer with which to construct the universal tree of life (Woese et al. 1990). To establish this evolutionary clock, Woese labored in relative isolation from the larger scientific community, relying on the help of just a select few graduate students, postdoctoral researchers, and colleagues. This is extremely rare today, as large laboratory groups and their collaborative networks are now conducting the majority of molecular studies.

In the early 1960s, a brand new, manually time-consuming, and, in fact, hazardous radioactive tagging technique called *molecular sequencing* became available (Sapp 2009). Originally developed by Fred Sanger for protein analysis, this technique was crucial to Crick's development of ideas about how DNA encodes for proteins and later earned Sanger the Nobel Prize in Chemistry in 1958. These cutting-edge techniques were learned and brought to the Woese's laboratory as the result of a collaboration between a graduate student mentored by Woese and a postdoctoral fellow who had formerly worked in the Sanger laboratories. Using the then state-of-the-art technique, they found that the information-packed small submit (SSU) of the rRNA was enough to determine common ancestry. The more similar the nucleotide sequences of specific regions of the chromosome in two organisms, the more closely related they are on the tree of life; the sequences found within the SSU transcript were particularly informative. However, at the time, it took months to collect a single SSU rRNA fingerprint. When sorted according to particle size settling time during centrifugation, the SSU rRNA is called the 16S rRNA gene sequence. The S unit (10^{-13} seconds) is named after the Swedish chemist Theodor Svedberg, who was awarded a 1926 Nobel Prize for developing the techniques of quantitative centrifugation of macromolecules. The 16S rRNA gene sequence is approximately 1542-nucleotide base pairs in total length, yet Sanger sequencing measured only short 13–15 nucleic acid base-pair-long reads. Woese pushed ahead to collect and analyze these short molecular sequences from as many different living organisms as possible, after tagging them with radioactive phosphate (^{32}P) and separating them using two-dimensional gel electrophoresis. The resulting gels were exposed and measured on large sheets of photographic x-ray film and analyzed on IBM computer punch cards. Woese labored for thousands of hours in solitude, hunched over a light box in a darkened laboratory, to read the x-ray films.

By looking at these small differences in these 16S rRNA gene sequences by using statistical analyses, Woese could track the most likely evolutionary pathway that might explain their differences and similarities. Woese was eventually joined by George Fox, his postdoctoral research associate, who collaborated closely with the laboratory of Illinois Professor of Microbiology Ralph Wolfe. Wolfe's laboratory specialized in culturing and biochemically analyzing microorganisms that produce natural gas or methane in environments with no oxygen (*anaerobic methanogens*). A sample, collected in spring of 1976 on the University of Illinois Veterinary Medicine farms, was taken from a surgically implanted opening (*fistula*) into the complex four-chambered stomach (*rumen*) of a cow. It was from this rumen sample that the first 16S rRNA gene sequences of methanogens were extracted and sequenced and were used to identify a whole new branch of life—the Archaea. However, Woese and Fox needed more evidence to prove that the Archaea existed. Although classified until that time as Bacteria, methanogens lack a porous outer cell wall layer composed of peptidoglycan. Upon review of the literature, Woese and his colleagues found that a few other select prokaryotes also lacked peptidoglycan, had similar metabolisms and fats (lipids) to methanogens, and thrived in extreme environments created by salt (*halophiles* in evaporating wetlands) and heat (*thermophiles* in Yellowstone hot springs). Woese and Fox acquired samples of these organisms and soon accumulated enough data to prove their similarity in 16S rRNA gene sequence composition. Thus, the Urkaryotes, Bacteria and Archaeabacteria (the new third branch of life) were established and called "urkingdoms" (a taxonomic ranking above that of Whittaker's kingdoms; Woese and Fox 1977).

These surprising results, which contextualized the five Whittaker kingdoms and departed from the eukaryote-prokaryote dichotomy of the early twentieth century, were first published in the *Proceedings of the National Academies of Science* and reported on the front page of the New York Times on Thursday November 3, 1977. The world now knew for the first time of Woese and his universal tree of life. Yet ironically, these scientific results were much more widely embraced abroad than they were by the scientific community in the United States. The world's first scientific conference on Archaebacteria was held in Munich, Germany, in 1981. Later, in 1990, Woese and his colleagues Otto Kandler and Mark Wheelis formally renamed the urkingdoms as the three domains of life—Archaea, Bacteria, and Eucarya (Figure 8.4.2). Studies of taxonomy, phylogeny, and evolution were now unleashed to vigorously conduct experimentation in the laboratory and in the field and objectively measure evolutionary relatedness, and thus, the direction and applicability of scientific discovery would never be the same again. Today, the debate about appropriate targets to explore relatedness among cellular life continues and can be generally reduced to information systems (such as the ribosome or other protein synthesis machinery) or physiological structures (such as subunits of the ATP synthase complex; Williams and Embley 2014). In this article, the conceptual model of the tree of life follows the informational genes of the small subunit of the ribosome (16S rRNA gene sequences) as the organizing principle for universally determining evolutionary relatedness (Figure 8.4.2).

8.4.4 EMERGENCE OF LIFE

The next step is to consider how geological time encapsulates and frames the progressive events comprising the universal tree of life (Figure 8.4.2). A practical understanding requires acknowledgment that there are various geological "clocks" that exist in nature. These tools measure the age of the Earth by taking advantage of the decay of radioactive isotopes and cooling, changing magnetic fields in lava flows that emerge from mid-ocean ridges, layering of sedimentary rocks and soil and ice deposits, the analysis of fossils trapped within these layers, tree rings, chemical analysis of layered skeletons, and the monitoring of impact craters on the moon and other planets within our solar system (Gradstein et al. 2012). All of these approaches, taken together, make it possible to accurately quantify the age of the Earth and the universe. Within this are multiple frames of reference that link geological and biological time, ranging from billionths of a second to billions of years.

Understanding the emergence of life begins with discerning its most ancient forms. Even if all of the DNA, RNA, protein, and other molecular components of a living cell are extracted and then mixed back together, humankind remains incapable of bringing this naturally perfect cocktail of ingredients back to life (Sapp 2009; Lane 2015). Scientific experiments, famously such as those of Stanley Miller and Harold Urey at the University of Chicago in 1952, have produced amino acids as well as a host of other organic molecules. These tests added electrical energy, reactive mineral surfaces, and other ingredients to prebiotic aqueous chemical mixtures under conditions that were thought at the time (but no longer) to mimic those found on the ancient Earth. This explains why amino acids, the building blocks of proteins, are commonly found on the surface of meteors. It also resoundingly proves that simply having the right water chemistry and an input of energy is not sufficient to generate life.

Two of the most ancient "fossil" entities on modern-day planet Earth are H_2O molecules and the genetic code. Much progress has been made recently in deciphering the physical and chemical conditions in which the earliest life may have been jump-started (Lane 2015). The primordial Earth was a terrifyingly violent place of ongoing meteor impact strikes and resulting extreme environmental conditions (Knoll et al. 2012). Although still intensely debated, the window of opportunity for life to gain a foothold is thought to have spanned from the latest Hadean into the early Archean, approximately 4.3–3.8 billion years before present (BYA). This is also the time frame in which the Earth's magnetic field had become adequately developed, so as to provide a protective shield from the arrival of deadly galactic cosmic rays and solar particles. The older age limit is based on work suggesting that the Earth and Moon, as well as our entire solar system, were hit with a barrage of massive asteroids called the Late Heavy Bombardment, which peaked at about 4.0 BYA and declined until approximately 3 BYA (Abramov and Mojzsis 2009).

The earliest fossil evidence for life on Earth occurs no later than about 3.8 BYA in the form of chemofossils. Carbon-isotope changes preserved in ancient sedimentary rocks (metamorphosed shales) are consistent with those created by the metabolic activity of modern photosynthetic cyanobacteria (Mojzsis et al. 1996; Bell 2015). Iron ore deposits in sedimentary rock, known as banded iron formations (BIFs), are also believed to be indicative of microbial metabolic activity. These BIFs are thought to have originally been formed by marine microbes capable of anoxygenic photosynthesis, which later evolved to generate oxygenic photosynthesis carried out by aquatic cyanobacteria. The Earth's oceans and atmosphere were then charged with oxygen derived from cyanobacterial photosynthesis by about 2.5 BYA to create the worldwide Great Oxygenation Event (GOE; Lyons et al. 2014).

The first physical fossils on Earth can be found in the form of microbial cells trapped in silica crystals (Knoll 2003). Cyanobacteria and other microbes also formed *stromatolites*, many of which are finely layered sedimentary rock deposits formed when small sedimentary grains are embedded in the sticky mucous (also called extracellular polymeric substances [EPS]) of microbial mats over time. This combination of mineral-entombed cells, carbon-isotope fractionation, extensive iron deposition, and stromatolites provides strong evidence of the presence and metabolic activity of ancient Bacteria, Archaea, and Eucarya. The evolutionary chronometer of the genetic code indicates all of life evolved from a single common ancestor called the last universal common ancestor (LUCA; Weiss et al. 2016). Following the LUCA, two forms of cellular life emerged: thermophilic bacteria and another unique organism called the Last Eucarya and Archaea Common Ancestor (LEACA) that eventually evolved into Archaea and Eucarya (Pace et al. 2012).

The Earth's oceans and atmosphere were charged with oxygen derived from cyanobacterial photosynthesis by about 2.5 BYA to create the worldwide GOE (Knoll et al. 2012). At approximately 600 million years ago (MYA), ice covered the world's oceans, including the equatorial zones, to create a "Snowball Earth." A rise in greenhouse gases, primarily derived from volcanic activity, then brought the planet back to warmth. At only 542 MYA, the Cambrian Explosion took place, in which large multicellular Eucarya (metazoans) with external skeletons (invertebrates) suddenly evolved on the seafloor of oceans around the world. The fossil record generally presented in science classrooms begins with this invertebrate Cambrian Explosion and then goes on to include the first and last occurrences (radiations and extinctions) of macroscopic plants and vertebrate animals. This recent approximately half-a-billion-year stretch of geological time is the physical realm of the fossil record that has been, and continues to be, extensively studied to establish early prototypes of the tree of life (most without microorganisms).

In contrast to the shallow, warm, land-based ponds of "Primordial Soup" envisioned by Darwin, it is much more likely that life began in the more stable environments of deep seafloor crustal spreading centers, which would have been sheltered from many of the direct effects of meteor bombardments. In this sheltered deep ocean realm, microbial life would have been able to originate and evolve even during the tapering of meteor bombardment (Russell et al. 2010, 2013). Immense volumes of seawater move down and through fractures in the newly formed basaltic rock that makes up the oceanic crust at the core of the spreading ridges. These water-rock interactions create highly acidic (pH 2–3) iron-sulfide black smoker vents that are widely distributed along spreading centers in the modern ocean and reach temperatures of 350°C—too hot to sustain life. However, cooler off-ridge hydrothermal vents also occur, such as the present-day Lost City hydrothermal field west of the mid-Atlantic Ridge at about 900 m water depth. In these locations, vent water is cooler and much more alkaline (70°C–90°C, pH 9–11) after circulating through and chemically reacting with ultramafic igneous crustal rocks (low in silicon dioxide and high in manganese and iron) and their overlying thin layer of marine sediments. This produces huge volumes of diagenetically altered rock deposits called serpentine. In the process, seawater is reduced by the huge supply of iron available in the crust, releasing sulfur and the type of H_2-CO_2 redox couples thought to be essential for the origin of life.

At these Lost City vents, limestone chimneys have formed over the last 100,000 years that tower 60 m above the surrounding seafloor. They are built of aragonite and calcite crystals, as well as other minerals formed during serpentinization that include double-layer hydroxides (green rust), which rapidly encrust filamentous microbial mats (Russell et al. 2010, 2013). The hot chemically charged water flows up and out of the vent to immediately react with cold deep seawater, which also drives the precipitation of iron and nickel. This creates an environment of extreme heat loss (*thermal gradients*) and rapid chemical reactions that are rapidly changing and strongly unbalanced (far from *equilibrium*). It is under these conditions of dynamic, indeed frantic, carbon and energy exchange where equilibrium is trying to be restored and where life is believed to have begun (Figure 8.4.2). The energy release and exchange in these settings is something like the whirling dervish performances of Cirque du Soleil. Here, however, instead of trapeze flyers, tightrope walkers, and contortionists, metals serve to catalyze an array of basic chemical reactions that in turn form a variety of organic biomolecules.

These coupled environments of sub-seafloor serpentinization and alkaline vent travertine precipitation are the natural chemical reactor vessels in which the biomolecules of life were first synthesized (Russell et al. 2010, 2013). In the process, small crystalline mineral chambers with microscopic channels (*ram jets*) formed in the serpentine and travertine mineral deposits, where biomolecules came together from the rapidly outflowing vent water to create free-floating RNA-like molecules. This earliest *RNA-based life* created a massive collective of genetic and biochemical material that was openly free flowing and without boundaries. Although nearly impossible for us humans to comprehend, there was no "individuality," because there were no cellular entities and no structures to separate and distinguish distinct living things. Woese, as well as Crick and Leslie Orgel, founded this idea that nucleic acids played a central role in the origin of life, which buoyed the notion of a RNA-based life precursor to cellular life (Figure 8.4.2)—a world in which nucleic acids served as both chemical catalysts and information transmitters. This suggests that one of the most enduring and realistic definitions of life is simply "the ability to evolve." As a result, instead of "survival of the fittest" that requires individuality, success is better defined by "survival by those who are fit the best" to the environmental conditions at hand. The origin of life was therefore all about the open and complete sharing of genetic information and the strategies for survival that resulted from the ability to evolve. Woese's work was the first to imply that there was life, and evolution, long before there were genes and before there were cells enclosed with an outer membrane (Sapp 2009).

Woese surmised that the next evolutionary step of life occurred when membranes incrementally began to close around and capture RNA and other molecules within cells. This created primitive cell-like entities, which Woese named the *progenote*, that were radically different from modern-day living cells (Figure 8.4.2). It is thought that this progression from original deep seafloor RNA synthesis through the progenote, Darwinian Transition, and the LUCA all took place in an exceedingly short time span around approximately 4.0 BYA (Woese 1998a; Woese et al. 1990; Pace et al. 2012; Lane 2015). Among other things, the progenote did not have either a fully formed cell wall membrane or the capacity to translate RNA into proteins. Thus, the linkages between genetic composition and outward cell structure and metabolism that are essential to modern living were only just in the process of being established at this time. Only late-stage versions of the progenote proto-cells would have evolved the complex capability of translation. In addition, the progenote necessarily evolved quickly, leading Woese and Fox and later workers to reason that this was driven in large part by horizontal gene transfer—the ability to move genetic material laterally between living cells in a single generation, rather than only vertically between separate generations of cells. Since then, horizontal gene transfer has been found to operate within and between all forms of life. A profound evolutionary event then took place that Woese called the *Darwinian Transition*, when the progenote evolved into modern-type cells. The resulting cells had a fully encapsulating and well-developed cell wall, as well as all of the complicated translational molecular machinery required to synthesize proteins. The LUCA was therefore the anchor point for the genetic code and the entire tree of life, but it was not the start of life (Figure 8.4.2).

The universal tree of life (Figure 8.4.2) illustrates several key concepts that are of critical importance to understanding the structure and evolutionary history of life on earth. These were surprising and fully unexpected, given that the phylogenetic relationships are only fully revealed on the molecular level. The first is that all three domains of life (Bacteria, Archaea, and Eucarya) evolved nearly simultaneously very early in geological time at approximately 4.0–3.8 BYA window (Figure 8.4.2). Another is that the single-celled Archaea are more closely phylogenetically related to the multicellular Eucarya than they are to their fellow single-celled Bacteria. However, Eucaryotic cells are generally 10–20 times larger than those of Bacteria and Archaea, and Eukarya have genetic material organized within a nucleus, and both bacterial and archaeal cells generally have the same shape and appearance. Only the Archaea are capable of methanogenesis, but they are also metabolically diverse. Archaea are often thought of as being able to live in environments of extreme temperature, acidity, and salinity (*extremophiles*). However, this is not always the case; they are now commonly detected in common low-temperature natural and man-made environments. In addition, archaeal cell walls do not contain the peptidoglycan found in bacterial cell walls. Furthermore, Archaea have different membrane lipid bonding and fatty acids from both Bacteria and Eucarya.

8.4.5 EVOLUTIONARY WORLDS COLLIDE

Woese was confident that his findings on the evolution of the ribosome would revolutionize the very foundations on which modern biology is built and practiced (Sapp 2009). However, the process of sharing these results with the world got off to a rocky start. This was in part because Woese had labored in anonymity for so many years and did not frequently attend conferences; thus, he was unknown to most of the scientific community. Furthermore, comparison of oligonucleotide catalogs is not as straightforward as the direct comparison of DNA and RNA sequences that can be done today. These obstacles led to a general lack of understanding and skepticism in the scientific world about the validity of Woese's approaches. In addition, and not due to anything Woese had done, the New York Times made several errors in the reporting of its 1977 front-page story—thus many scientists viewed this as a failed attempt at publicity. And finally, molecular phylogeny showed that Whittaker's five-kingdom configuration (Figure 8.4.4), on which so much of biology had been anchored, was in reality a series of branches on the top portions of the universal tree of life (Figure 8.4.2). This was not, and in some sectors still is not, well received by many practicing biologists.

The *status quo* confronting Woese is perhaps best exemplified by a remarkable and protracted discourse he had with Ernst Mayr—one of the foremost classical Darwinian evolutionists of the twentieth century. These communications were published in 1998 in the *Proceedings of the National Academy of Sciences*. The controversy centered on Mayr's belief that the great morphological diversity of the eukaryotic world would certainly overwhelm any molecular and biochemical diversity present in the microbial world. In other words, how could differences in the chemical composition of a single gene possibly compare to the immense morphological diversity of animals and plants? Thus, Mayr's argument viewed molecular

phylogeny as a competing type of taxonomy and thus represented a challenge to long-held views on the assumed primitive simplicity of the microbial world. Classical evolutionists like Mayr saw phylogenetic taxonomy as something distinct and separate from the study of evolutionary processes, which was generally thought to be the bailiwick of geneticists (Sapp 2009).

Mayr, emulating Darwin, shifted the question of evolution to the origin of species and away from the big universal tree of life (Mayr 1998). Darwin had argued that in order to have a deep evolutionary phylogeny, one had to focus exclusively on the essential characteristics of an organism. Darwin's world was one of plants and animals; the accessible way to classify these living things at the time was by choosing easily observable morphological traits. Darwin focused on embryonic characters, because those characters would evolve gradually and might in fact be stable and highly conserved in evolutionary time. The problem was that this approach was not applicable for microbes. Microbes didn't have embryos, or great morphologies, or any obvious trait that was useful for Darwinian classification.

Woese therefore viewed things differently, seeing the debate as being over methodology—Darwinian morphologists use direct human observation, while molecular phylogeneticists use unseen genes and biochemistry detected only with sensitive instruments. Mayr and the classical evolutionists wanted to maintain the prokaryotes as a single "monophyletic" group because of their visible characteristics. Woese pointed out that this meant that they did not fully understand that the 16S rRNA gene sequence had shown that the Archaea are as phylogenetically distinct from Bacteria as they are from Eucarya (Woese 1998b; Sapp 2009). Furthermore, a significant amount of microbiology research in the previous two decades had focused on one representative microbe, *Escherichia coli*, which could easily be cultured and manipulated in the laboratory. The mutual exclusion and opposed differences (*dichotomy*) set up by Mayr between the eukaryotes and prokaryotes reinforced the notion that knowing everything about *E. coli* was sufficient to understand everything about prokaryotes in general, including the Archaea. Woese's molecular phylogeny, combined with the detailed biochemical analyses of archaeal metahanogens, proved that this was fully incorrect. Yet, the scientific community resisted, as evidenced when Woese and Gary Olsen at Illinois submitted a proposal to sequence an archaean but were turned down on the grounds that a sequence of *E. coli* bacteria already existed and one didn't need to sequence any more of these things. If you've seen one microorganism, you've seen them all.

Further strife later arose within the field of molecular phylogeny itself, over the potential effects of horizontal gene transfer (Woese 2004). Some believed that extensive horizontal gene transfer would obliterate generational lines of descent (*lineages*), thus believing there was no fundamental core structural phylogeny that the microbial world had followed. Conversely, Woese and his colleagues saw horizontal gene transfer as a powerful integrative force that rapidly shared and integrated genetic components (Pace et al. 2012). However, they did not envision it as being capable of blurring the lineages that were so robustly shown by the universal tree of life. In doing so, they made a distinction between the more ancient informational genes that involved transcription and translation and the more recent operational genes that governed metabolism and were more prone to horizontal gene transfer.

The scientific persistence and dogged personal commitment of Carl R. Woese eventually succeeded. As his publications and supporting data became more and more known and widely accepted in the late 1990s, the first generations of high-throughput DNA sequencing technologies also became available. Complete archaeal genome sequences immediately substantiated them as a unique third domain of life. From this point forward, Woese's accomplishments were recognized by the most prestigious of international science organizations, including the Macarthur Fellowship, membership in the US National Academy of Sciences and the Royal Netherlands Academy of Arts and Sciences, the Leeuwenhoek Medal, the US National Medal of Science, and the Royal Swedish Academy of Sciences Crafoord Award in Biosciences. Carl R. Woese fundamentally changed the way in which we perceive life on planet Earth, and he established an irreplaceable molecular insight that is now at the forefront of scientific inquiry in all sectors of modern society.

8.4.6 MOVING FORWARD

As the technologies required for high-throughput sequencing become more rapid and provide ever-deeper sequence coverage and the cost of sequencing continues to precipitously drop in price, the pace at which new hypotheses are proposed regarding the structure of the universal tree of life will continue to accelerate. An important focus of several recent deep sequencing efforts has been to better understand the appearance of the LUCA and the origination of the Eucarya. As would be expected given the history of evolutionary biology itself, these studies have been surprising, inspiring, highly controversial, and hotly debated. As a case in point, genomic DNA from a suite of water samples collected at a mid-ocean ridge vent called Loki's Castle, located on the deep-sea floor between Norway, Greenland, and Svalbard, have been heavily sequenced (Spang et al. 2015; Zaremba-Niedzwiedzka et al. 2017). This work has resulted in the identification of a group of Archaea called the Asgard Group with genetic structures and predicted membrane remodeling capabilities that are closely related to those of Eucarya, although it is critical to note that none of these microorganisms are yet in culture. Metagenomic analyses of one of these groups, called Lokiarchaea, suggest that it was directly evolved from Eucarya at about 1 Bya later in earth history than previously thought. If correct, this would imply a two-domain structure for the tree of life (called the eocyte or merger model; Raymann et al. 2015), wherein the Eucarya are a branch of the Archaea rather than representing a separate domain.

The multicellular Eucarya, which include animals, plants, and fungi, are extremely complicated in their own right, and their origins have long been disputed. This is equally true for the algae, yeast, and amoeba, which are single-celled versions of the Eucarya (Barton et al. 2007). The discovery of the Lokiarchaea, in addition to variations in evolutionary relationships suggested by other marker genes, has led some to believe that the Eucarya evolved from the merger of Archaea and a Bacteria via endosymbiotic fusion (Martin et al. 2015). However, these studies are controversial, because they suggest they transfer of genes between cells and across domains (horizontal gene transfer [HGT]), resulting in overprinting and resetting of a significant portion of the LUCA's original genomic composition. They also assume that viruses originated from the LUCA (Forterre 2014). Martin et al. identified only 355 genes, which are not enough to drive the biosynthesis of nucleotide and amino acids. Ribosomal proteins that underwent HGT were also omitted. Regardless, their interpretation is that the original suite of genes derived from the LUCA results in a two-domain tree of life.

Hug et al. (2016) generated an extensive new suite of genome sequences from more than 1,000 uncultivated organisms and merged these with previously published sequences. These were then aligned and concatenated with a suite of 16 ribosomal proteins from each identified organism, creating what they argue is a higher-resolution tree of life than that derived from the single 16S rRNA gene sequence. Results substantiate the dominance of bacterial diversification and identify many new evolutionary radiation events that could eventually impact the basic Woese three-domain tree of life model. However, Hug et al. do not use this data set to try to decipher the evolutionary placement of the Eucarya. Woese established a three-domain universal tree of life model (Figure 8.4.2) based on changes in the ribosome, which indicates that the deeply rooted Archaea split from the Eucarya at ~3.5 Bya (Figure 8.4.2; Woese et al. 1990). Many other subsequent studies have followed suit, stating that there is more certainty in ribosomal DNA than in concatenated genes. In addition, reevaluation of the Lokiarchaea 16S rRNA gene sequences places them into a group of environmental Crenarchaeota. Furthermore, Lokiarchaea utilize a unique membrane lipid that is different from the lipids used by Eucarya and Bacteria. Conversely, citing protein phylogeny and incorporation of RNA polymerase phylogenies, other researchers have recently concluded that Lokiarchaea and the other members of the Asgard Superphylum are a firmly rooted branch within the Archaea domain (Da Cunha et al. 2017; Zaremba-Niedzwiedzka et al. 2017).

These types of invaluable scientific discussions and controversies will continue to take place as more and more organisms are found and analyzed (both on Earth and potentially other planets). The reoccurring question will therefore be how to keep the scientific evaluations of the tree of life moving forward with each new discovery (Forterre 2016). As in the case of the Lokiarchaea, the first step toward full analysis and resulting tree of life reevaluation will be to isolate the microorganisms in question. To date, the Asgard Group has only been detected using molecular techniques. A full determination of evolutionary relatedness will require having Lokiarchaea and the others in pure culture in the laboratory. Simply put, our ideas of evolution must continually evolve with the data being collected, but those data need to include a full determination of biogeochemistry, metabolism, ecology, and closure of complete genomes, accompanied by completion of the full suite of DNA-RNA-protein *Omics* analyses. The tree of life is not yet complete and will be a work in progress for decades to come. However, the Woese three-domain universal tree of life remains reliable, because rRNA genes are some of the most stable and conserved. This provides a consistently defendable and reproducible framework, against which new interpretations and hypotheses can be rigorously tested.

8.4.7 SUMMARY

The three-domain universal tree of life established by Carl R. Woese fundamentally reset the stage of evolutionary theory and has forever changed the face of modern biology (Woese 2004). His new vision of cells that are capable of extensive collective sharing of genetic materials, and their universal possession of partially conserved genes that can serve as a measuring stick of evolutionary relatedness, resulted in the identification of the Archaea and a fundamentally new understanding of the origin and early evolution of life. The NASA was the sole federal funding agency that made this research possible. The resulting universal tree of life is now the fundamental roadmap used to probe the most fundamental of challenges facing society regarding the environment, energy, medicine, and the search for life throughout the cosmos.

ACKNOWLEDGMENTS

Financial support for this review was provided from the National Aeronautics and Space Administration (NASA) through the NASA Astrobiology Institute Education and Public Outreach initiative under Cooperative Agreement No. NNA13AA91A issued by the Science Mission Directorate. We also thank Claudia Lutz and Isaac Cann for their invaluable editorial evaluations. Conclusions in this study are those of the authors, and they do not necessarily reflect those of the funding agency.

REFERENCES

Abramov, O., and Mojzsis, S., 2009. Microbial habitability of the Hadean Earth during the late heavy bombardment. *Nature* 459: 7245.

Barton, N.H., Briggs, D.E.G., Eisen, J.A., Goldstein, D.B., and Patel, N.H., 2007. *Evolution.* Cold Spring Harbor Laboratory Press, Cold Spring Harbor, NY, 833 p.

Baumgartner, L.K., and Pace, N.R., 2007. Current taxonomy in classroom instruction. *The Science Teacher*, pp. 46–51.

Bell, E.A., 2015. Potentially biogenic carbon preserved in a 4.1 billion-year-old zircon. *Proceedings of the National Academy of Sciences of the United States of America* 112(47): 14518–14521.

Butler, T., Goldenfeld, N., Mathew, D., and Luthey-Schulten, Z., 2009. Extreme genetic code optimality from a molecular dynamics calculation of amino acid polar requirement. *Physical Review E* 79: 060901.

Da Cunha, V., Gala, M., Gadelle, D., Nasir, A., and Forterre, P., 2017. Lokiarchaea are close relatives of Euryarchaeota, not bridging the gap between prokaryotes and eukaryotes. *PLoS Genetics* 13(6): e1006810.

Domagal-Goldman, S.D. et al., 2016. The astrobiology primer v2.0. *Astrobiology* 16(8): 561–653.

Forterre, P., 2014. Cellular domains and viral lineages. *Trends in Microbiology*, 22(10): 554–558.

Forterre, P., 2016. The universal tree of life: An update. *Frontiers in Microbiology* 6: 717.

Gest, H., 2004. The discovery of microorganisms by Robert Hooke and Antoni Van Leeuwenhoek, Fellows of the Royal Society. *Notes and Records of the Royal Society of London* 58(2): 187–201.

Gradstein, F.M., Ogg, J.G., Schmitz, M.D., and Ogg, G.M., 2012. *The Geologic Time Scale 2012.* Elsevier B.V., Amsterdam, the Netherlands, 1176 p.

Hays, L., 2015. *NASA Astrobiology Strategy.* NASA Headquarters, Washington, DC, 236 p., NASA/SP-2015-3710.

Hug, L.A. et al., 2016. A new view of the tree of life. *Nature Microbiology* 1: 16048.

Huneman, P. 2007. Understanding purpose: Kant and the philosophy of biology. University of Rochester Press, Rochester, NY, 191 p.

Jardine, N., Secord, J.A., and Spary, E.C., 2000. *Cultures of Natural History.* Cambridge University Press, Cambridge, UK, 470 p.

Knoll, A., 2003. *Life on a Young Planet: The First Three Billion Years of Evolution on Earth.* Princeton University Press, Princeton, NJ, 277 p.

Knoll, A.H., Canfield, D.E., and Konhauser, K., 2012. *Fundamentals of Geobiology.* John Wiley & Sons, Hoboken, NJ, 443 p.

Lane, N., 2015. *The Vital Question: Energy, Evolution and the Origins of Complex Life.* W.W. Norton & Co., New York, 368 p.

Lyons, T.W., Reinhard, C.T., and Planavsky, N.J., 2014. The rise of oxygen in Earth's early ocean and atmosphere. *Nature Review* 506: 307–315.

Martin, W.F., Garg, S., and Zimorski, V., 2015. Endosymbiotic theories for eukaryote origin. *Philosophical Transactions of the Royal Society of London B Biological Sciences* 370(1678): 20140330.

Mayr, E., 1998. Two empires or three? *Proceedings of the National Academy of Sciences of the United States of America* 95: 9720–9723.

Mojzsis, S.J., Arrhenius, G., McKeegan, K.D., Harrison, T.M., Nutman, A.P., and Friend, C.R.L., 1996. Evidence for life on Earth before 3,800 million years ago. *Nature* 384: 55–59.

Pace, N.R., Sapp, J., and Goldenfeld, N., 2012. Phylogeny and beyond: Scientific, historical and conceptual significance of the first tree of life. *Proceedings of the National Academy of Sciences of the United States of America* 109: 1011–1018.

Raymann, K., Brochier-Armanet, C., and Gribaldo, S., 2015. The two-domain tree of life is linked to a new root for the Archaea. *Proceedings of the National Academies of Science of the United States of America* 112(21): 6670–6675.

Richards, R.J., 2008. *The Tragic Sense of Life: Ernst Haeckel and the Struggle over Evolutionary Thought.* University of Chicago Press, Chicago, IL, 545 p.

Russell, M.J., Hall, A.J., and Martin, W., 2010. Serpentinization and its contribution to the energy for the emergence of life. *Geobiology* 8: 355–371.

Russell, M., Nitschke, W., and Branscomb, E., 2013. The inevitable journey to being. *Philosophical Transactions of the Royal Society B* 368: 20120254.

Sapp, J., 2009. *The New Foundations of Evolution*. Oxford University Press, Oxford, UK, 425 p.

Spang, A., Saw, J.H., Jorgensen, S.L., Zaremba-Niedzwiedzka, K., Martijn, J., Lind, A.E., van Eijk, R., Schleper, C., Guy, L., and Ettema, T.J.G., 2015. Complex archae bridge the gap between prokaryotes and eukaryotes. *Nature* 521(7551): 173–179.

Weiss, M.C., Sousa, F.L., Mrnjavac, N., Neukirchen, S., Roetther, M., Nelson-Sathi, S., and Martin, W.F., 2016. The physiology and habitat of the last universal common ancestor. *Nature Microbiology* 1: 16116.

Williams, T.A., and Embley, T.M., 2014. Archaeal "dark matter" and the origin of eukaryotes. *Genome Biology and Evolution* 6(3): 474–481.

Woese, C.R., and Fox, G.E., 1977. Phylogenetic structure of the prokaryotic domain: The primary kingdoms. *Proceedings of the National Academy of Sciences* 74: 5088–5090.

Woese, C.R., Kandler, O., and Wheelis, M.L., 1990. Towards a natural system of organisms: Proposal for the domains Archaea, Bacteria, and Eucarya. *Proceedings of the National Academy of Sciences of the United States of America* 87: 4576–4579.

Woese, C.R., 1965. On the evolution of the genetic code. *Proceedings of the National Academy of Sciences of the United States of America* 54: 1546.

Woese, C.R., 1998a. The universal ancestor. *Proceedings of the National Academy of Sciences of the United States of America* 95: 6854–6859.

Woese, C.R., 1998b. Default taxonomy: Ernst Mayr's view of the microbial world. *Proceedings of the National Academy of Sciences of the United States of America* 95: 11043–11046.

Woese, C.R., 2004. A new biology for a new century. *Microbiology and Molecular Biology Review* 68(2): 173–186.

Zaremba-Niedzwiedzka, K. et al., 2017. Asgard archaea illuminate the origin of eukaryotic cellular complexity. *Nature* 541: 352.

8.5 Fossils of Ancient Microorganisms

Kenichiro Sugitani

CONTENTS

8.5.1 INTRODUCTION

Fossils of microorganisms that can be identified under the microscope are called microfossils. Microfossils provide invaluable information about the evolution of life, stratigraphic correlation, and reconstruction of paleoenvironments. Except for well-known Ediacaran macrofossils and a few other examples, fossil records identified from the Precambrian era older than 0.54 Ga are all microfossils. Study of Precambrian microfossils started with the discovery of morphologically diverse microfossils from the 1.9 Ga sedimentary rocks of the Gunflint Formation (Barghoorn and Tyler 1965; Cloud 1965), and since then, numerous efforts have been made to search for older microfossils and other biosignatures, such as microbial sedimentary structures (stromatolites), molecular biomarkers, bio-mineralization, and isotopic signatures. In particular, cellularly preserved microfossils that provide us with direct images of ancient life have been fascinating research targets. However, there has been great debate over the biogenicity of Archean fossil-like microstructures, including how to discriminate genuine microfossils from pseudo-microfossils. This issue is still heavily debated, and new discoveries of Archean biosignatures, including doubtful ones, are appearing in succession. Therefore, it is worth reviewing Archean microfossil records, based on previously published informative reviews (Schopf and Walter 1983; Altermann and Kazmierczak 2003; Schopf 2006; Schopf et al. 2007; Wacey 2009, 2012). In this chapter, I provide a detailed review of representative microfossils and possible microfossils from the early (Meso- and Paleo-) Archean successions in the Isua Supracrustal Belt, Greenland; Nuvvuagittuq Belt, Canada; Kaapvaal Craton, South Africa; and Pilbara Craton, Western Australia, with reference to the early Earth's surface environments, criteria for biogenicity of Archean microbe-like structures, and some examples of pseudofossils. Detailed discussions are given for problematic microfossils, characterized by their large eukaryotic size and lenticular shape, that are incomparable with extant prokaryotes. Finally, frameworks and hints for further studies are presented.

8.5.2 SURFACE ENVIRONMENT OF THE EARLY EARTH

Although the Archean surface environment has not yet been fully revealed and controversies are continuing, it was closely related to metabolisms and habitats of the contemporaneous organisms. Therefore, it is worth mentioning this topic briefly before describing microfossils.

8.5.2.1 ACTIVE VOLCANISMS

Early Archean surface environments were distinctly different to those from the modern and even the Proterozoic environments, with the presence of extensively active volcanism, as indicated by rock records, dominated by mafic to ultramafic volcanic rocks (Lowe 1999; Van Kranendonk et al. 2006). Ultramafic lava known as komatiite is indicative of a much higher temperature of the upper mantle than today (Berry et al. 2008). Greater active volcanism in the early Archean than later periods means that hydrothermal activities were more widespread and more intense in the early Archean period. Intense hydrothermal activities likely resulted in alteration of rocks, represented by silicification and carbonization, and provided nutrients essential to metabolisms of primitive life.

8.5.2.2 HIGHER TEMPERATURE

Owing to active volcanism and low rate of organic carbon burial, assumed from presumed much lower biomass, the atmospheric concentration of carbon dioxide in the Archean was much higher than today. This, together with possible higher concentrations of other greenhouse gases such as methane, compensated the lower solar luminosity and might have even kept the Earth's surface temperature higher than today. Isotopic studies of oxygen and silicon of Archean cherts (quartz) suggest that the temperatures of seawater at that time were between 55°C and 85°C (Knauth and Lowe 2003; Robert and Chaussidon 2006). Such estimate had some uncertainties due to unknown seawater oxygen isotopic compositions. More recently, however, Tartèse and co-workers (2017) demonstrated that the seawater oxygen isotopic composition did not change through time. Combined with the chert oxygen isotopes, the authors indicated that the temperature of ca. 3.5 Ga ocean bottom-water was ~50°C–60°C higher than today. Although some researchers argued a temperate climate for the Archean (Hren et al. 2009; Blake et al. 2010), it seems more likely that the Archean seawater temperature was significantly higher than today.

8.5.2.3 SEAWATER COMPOSITIONS

It is widely accepted that the Archean seawater was enriched in silica and ferrous iron. Silica and iron enrichment is evidenced by extensive deposition of cherts and banded iron formations in the Archean (Siever 1992; Sugitani 1992; Maliva et al. 2005). Silica enrichment in the Archean ocean was attributed to the absence of organisms requiring silica as a nutrient, as well as intense chemical weathering on land due to high concentrations of atmospheric CO_2 and a higher flux of dissolved silica from hydrothermal vents. It is also suggested that there was ferrous iron enrichment, at least in the bottom water of stratified oceans, which can be attributed to higher hydrothermal activities and a presumed anoxia. Shibuya and co-workers (2010), on the other hand, argued that the sub-seafloor Archean hydrothermal system produced highly alkaline, SiO_2-rich, Fe-poor fluids, contrastive to most of modern seafloor hydrothermal system along mid-ocean ridges. Concentrations of other redox-sensitive components such as sulfate could have varied significantly between the surface and bottom waters and even much more locally. However, as a whole, sulfate concentrations in the Archean ocean were likely much lower than in the present ocean (Habicht et al. 2002; Schröder et al. 2008).

8.5.2.4 ASTEROID IMPACTS

Approximately 4.1–3.8 Ga ago, extensively large numbers of asteroids collided with the Earth (the Late Heavy Bombardment [LHB]). Sedimentary records in the Paleoarchean, suggest that asteroid impacts had still occurred after the LHB more frequently than late periods (Byerly et al. 2002; Lowe et al. 2003, 2014; Glikson et al. 2016). Extensive investigations in the Barberton greenstone belt revealed that eight large (at least 20–70 km across) asteroid or meteorite impacts had occurred ~3.47–3.23 Ga (Lowe et al. 2014). It is suggested that such impacts triggered the modern-type plate tectonics, also causing tsunamis, heating of the atmosphere, and boiling and even evaporation of ocean surface water (Lowe and Byerly 2015). Such impact events plausibly influenced early evolution of life; however, details are still unknown (Lowe et al. 2003).

8.5.3 ROCKS PRESERVING FOSSILIZED MICROBES

Ancient biosignatures have been identified from a wide range of rocks, including those of relatively low metamorphic grade such as cherts and carbonates, siliciclastic rocks such as shales and sandstones, volcaniclastic rocks, and even volcanic rocks. Chert composed of microcrystalline quartz is formed by various processes, including chemical precipitation of silica, silicification of non-siliceous sediments and igneous rocks, and accumulation of siliceous tests such as those of diatoms and radiolaria. In the Archean, when silica-secreting organisms were absent, cherts were formed by chemical precipitation or replacement of non-siliceous rocks/sediments (silicification) (Duchač and Hanor 1987; Sugitani et al. 1998) (Figure 8.5.1).

Chemically precipitated chert has long been regarded as the most promising rock type for hosting ancient cellularly preserved

FIGURE 8.5.1 Marble Bar Chert in the Pilbara Craton, Western Australia, representing Archean cherts formed by chemical precipitation of silica (reddish and white band) and silicification of fine-grained clastic sediments (dark-gray band and vein). See, for example, Sugitani (1992)

microfossils. Indeed, Archean microfossils have been identified mostly in chert, as described later. Other rock types are also known to host biosignatures. For example, stromatolites, which are shallow-water sedimentary structures characterized by a complicated architecture of laminae and morphology, are perhaps the most reliable non-organic biosignatures. These forms were highly abundant in the Precambrian. Walter (1976) has provided the most comprehensive work describing these structures and their potential significance. Stromatolites are preserved almost exclusively as carbonates (Walter 1976; Allwood et al. 2006), whereas microbially induced sedimentary structures are identified mainly in siliciclastic sediments (Noffke 2010). Interestingly, volcanic rocks may preserve ancient biosignatures as traces of endolithic microbes (McLoughlin et al. 2007); however, this topic is out of the focus of this chapter. Rocks of high metamorphic grade and/or subjected to intense alteration such as metasomatism could preserve biosignatures although it should be reminded that possible biosignatures in such problematic rocks have been subjected to severe criticisms and skepticisms (Bridgwater et al. 1981; Fedo et al. 2006).

8.5.4 IDENTIFICATION OF MICROBE-LIKE STRUCTURES AS MICROFOSSILS

In this section, criteria for discriminating microfossils from non-microfossils are discussed. Controversies over Archean microfossil-like structures, represented by the *Isuasphaera* controversy (see Section 8.5.6.1), have resulted in the development of sophisticated approaches to microfossil analyses. The more stringent the criteria applied and satisfied, the more confidently the biogenicity can be claimed. However, too rigid criteria, suggested, for example, by Brasier and co-workers (2002, 2005, 2006), are undesirable. The authors claimed that Archean microfossil-like structures needed to be considered as non-biological, until the non-biological origin of the structures is refuted. This "null hypothesis" approach, however, is not constructive and should not be applied to Archean paleobiology, because it is virtually impossible to test "all possible alternatives" for biological origins of microfossil-like objects. Indeed, this approach is feared to be "playing one's own work." Also, the null-hypothesis approach requires substantial analyses of structures under consideration by using the most updated analytical techniques, available at only limited institutions. It is therefore believed that criteria for biogenicity should be basically tested by conventional methodologies, and it seems realistic to employ a series of prefixes such as "pseudo-," "dubio-," "possible-," "probable-," and "genuine-" in order to express the reliability of the biological origin of the object under consideration. In the following sections, the criteria integrated from those previous studies (Schopf and Walter 1983; Buick 1990; Brasier et al. 2002, 2005, 2006; Hofmann 2004; Sugitani et al. 2007; Wacey 2009) are shown, along with some new concepts.

8.5.4.1 GEOLOGICAL CONTEXT

Microstructures should occur in sedimentary rocks, preferably of low metamorphic grade from well-known early Archean

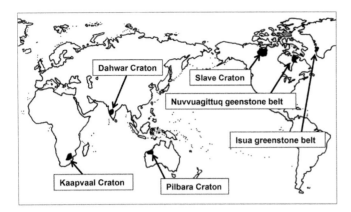

FIGURE 8.5.2 Cratons and greenstone belts containing early Archean sedimentary rocks.

terranes (Figure 8.5.2). Direct determination of the age of the host rock is desirable. This, however, can be substituted by establishing the age of the unit from which rock specimens containing the microbe-like structures were collected. It is essential to demonstrate that the host rock is definitely a part of a geographically extended Archean sedimentary succession, because duricrusts that form through modern cementation of sand and gravel by silica (silcrete) and carbonate (calcrete) could be mistaken as ancient sediments.

8.5.4.2 Indigenousness and Syngenecity

Microstructures should be demonstrated to be indigenous to the host rocks. This is particularly important when microstructures are detected as objects separated by acid maceration of host rocks (extraction of organic matter by HF-HCl digestion at room temperature), during which pollens and other modern organic materials could contaminate the sample. In other words, if structures under consideration are identified within petrographic thin sections, their indigenousness is basically guaranteed. Syngenecity means that the structures are embedded within the primary mineral phase but not within significantly later-formed pores, fractures, and veins (Figure 8.5.3a). Oehler and Cady (2014) describe the importance of syngenecity in detail, along with new techniques for assessing both biogenicity and syngenecity.

8.5.4.3 Biological Context: Size and Its Range

The size of unicellular microorganisms varies greatly, from the minimum value of ~200 nm for *Mycoplasma* up to the maximum of ~20 cm for coenocytic *Xenophyophore*. In the strict sense, therefore, size criterion merely gives the minimum value. In general, microorganisms that reproduce simply by binary fission tend to have a narrow size distribution, which is expected to have a Divisional Dispersion Index (DDI) smaller than 5; such a population should also fit to a Gaussian, normal distribution (Schopf 1976). However, a population with a wide size range does not always mean that the structure is abiogenic. Microorganisms that reproduce by multiple fissions, by asymmetric divisions, and by producing spores could result in populations characterized by a wide range of sizes. Also, it should not be overlooked that fossil populations can be composed of plural taxa, characterized by different size ranges.

8.5.4.4 Biological Context: Shape

The shape of unicellular microorganisms is very diverse; however, most prokaryotic cells are coccoid-, rod-, or filament-shaped. Abiogenic processes also can produce various morphologies, as represented by silica-witherite biomorphs (García-Ruiz et al. 2004). Thus, similarities to extant microorganisms should not be taken as weighted evidence for biogenicity. Rather than specific shapes, incompleteness is a key feature for assessing biogenicity. Cell envelopes (also called plasma membranes, cell walls, and cell capsules) are composed of polymeric materials and thus have plasticity. This means that cell shape could be deformed to various degrees. Therefore, shapes of microorganisms, particularly fossilized ones, tend to vary from complete to incomplete (i.e., being deformed) to various degrees even within a single colony. Thus, such a feature is a key part of identifying microstructures as microfossils.

Cells, if not all, display various elaborate morphologies and textures, such as an organelle, flagellum, and/or appendages. Thus, elaboration is a key criterion for biogenicity. Although delicate structures such as organelles and flagellum are difficult to preserve, protrusions and ornaments could be preserved, as evidenced by some organic-walled Proterozoic microfossils (Figure 8.5.3b) (Javaux et al. 2003; Butterfield 2005; Moczydłowska et al. 2011). Whatever their origins and functions are, the more elaborate the structures, the more likely would be their biogenicity. Morphological elaboration, assumingly related to reproduction, is strong evidence for the biogenicity of the fossil-like structure.

8.5.4.5 Biological Context: Occurrence

Microstructures should be abundant, hopefully occurring as clusters (colonies) (Figure 8.5.3c). Unicellular microorganisms can reproduce very rapidly if environmental conditions such as temperature, pH, and concentrations of nutrients are suitable. Therefore, abundant occurrence is advantageous in assessing biogenicity, because biogenicity can be tested statistically by using the size distribution and the DDI, along with various taphonomic features, as described in the sections above and below. Additionally, a close spatial relationship of microstructures to carbonaceous laminae, which most likely represent degraded biofilms, can be presented as strong evidence for biogenicity (Figure 8.5.3d).

8.5.4.6 BIOLOGICAL CONTEXT: TAPHONOMY

Polymerized compounds that comprise cell envelopes would be degraded to smaller compounds after death of cells by hydrolysis and by heterotrophic metabolisms. Such degradation tends to progress heterogeneously, as exemplified by the different preservation status of microfossils in the same bed (Figure 8.5.3e).

8.5.4.7 BIOLOGICAL CONTEXT: CHEMICAL AND ISOTOPIC COMPOSITIONS

Carbonaceous composition is a key geochemical criterion for the biogenicity of microbe-like structures. During degradation and thermal maturation, hydrogen, oxygen, sulfur, and nitrogen, which comprise organic materials making up cell components, are eventually released, resulting in

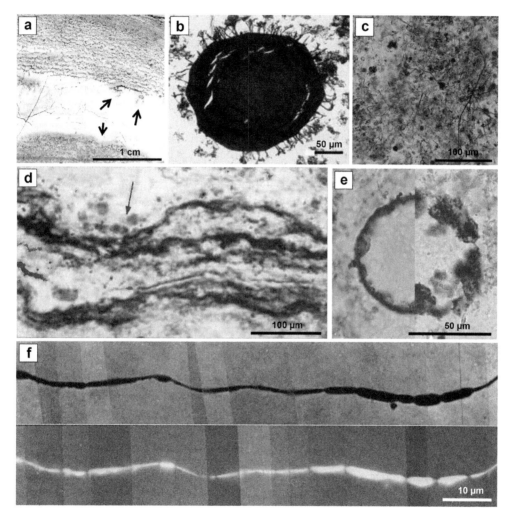

FIGURE 8.5.3 (a) Photomicrograph of vein composed of megaquartz and microquartz parallel to local lamination that locally displays coliform textures (arrows). Though probably formed during very early diagenesis, such a vein can be mistaken as a syndepositional layer. (b) Process-bearing acritarch, *Shuiyousphaeridium macroreticulatum*, from the Ruyang Group, northern China. (Javaux, E.J. et al., *Geobiology*, 2, 121–132, 2004. With permission from Blackwell publishing Ltd.) (c) Dense populations of filamentous and spheroidal microfossils comprising a Paleoproterozoic microbial colony (1.9 Ga Gunflint Chert). (d) Association of spheroidal microfossils (arrow) with irregularly wrinkled carbonaceous laminae (biofilms), from the middle Proterozoic Balbirini Dolomite (McArthur Group) of Australia. (From Oehler, D.Z., *Alcheringa*, 2, 269–309, 1978; Courtesy of D. Oehler.) (e) Composite photo of contrastively preserved walls of spheroidal microfossils from the 3.0 Ga Farrel Quartzite in the Pilbara Craton, Western Australia. The right side shows a degraded and partially granular wall, whereas the left side shows hyaline with regular dimples. (f) Filamentous bacterial microfossil replaced by pyrite, from the middle Proterozoic Barney Creek Formation (McArthur Group) of Australia. (From Oehler, J.H., *Alcheringa*, 1, 315–349, 1977.) The upper and lower photographs represent images in transmitted and reflected light, respectively. (Courtesy of D. Oehler.)

residues composed dominantly of carbon (Oberlin 1984). While an association of C, S, and N, established at the sub-micron scale with NanoSIMS, with the carbonaceous structures could be strong evidence for their biogenicity (Oehler et al. 2006, 2009, 2010), it must be cautioned that organic matter could be mobilized and redistributed in rock matrix (Marshall et al. 2012; Lepot et al. 2013), potentially producing structures that could mimic cells (Brasier et al. 2002). Though not direct evidence, detection of hydrocarbon biomarkers from host rocks could support biogenicity of microbe-like structures. It is again noteworthy that syngenicity of biomarker molecules in Archean rocks has raised controversies and requires strictly controlled protocols from sampling through to analyses (Oehler and Cady 2014). Another note is that carbonaceous matter comprising microfossils could be replaced by other materials such as hematite and sulfide (Figure 8.5.3f) (Oehler 1977; Rasmussen 2000; Sugitani et al. 2007). Therefore, non-carbonaceous compositions of microbe-like structures do not readily discard their biogenicity.

Light (< −20 per mil Pee Dee Belemnite (PDB)) carbon isotopic compositions ($\delta^{13}C_{PDB}$) have long been taken as a key feature in claiming the biogenicity of putative microfossils. However, this should not be taken as diagnostic, because isotopic fractionation and resultant formation of organic carbon with very light isotopic values could be produced through inorganic processes (McCollum and Seewald 2006). Yet, this is still an important feature for assessing biogenicity, particularly if carbon isotopic signatures associated with fossil-like structures are varied across micron scale and are texture-specific (Lepot et al. 2013; Williford et al. 2016; Morag et al. 2016). Carbon isotopic values of extant organisms are so diverse and depend on various factors such as the availability of inorganic carbon. Therefore, it is generally difficult to specify metabolisms based on isotopic values alone.

8.5.5 PSEUDOFOSSILS

Prior to describing representative early Archean microfossils and possible microfossils, some examples of pseudofossils in Archean successions are described. To date, some well-described Archean pseudo-microfossils include pyrite trails in cherts (Knoll and Barghoorn 1974), microtubes in volcanic ashes (Lepot et al. 2011), filamentous structures composed of stacked phyllosilicates in chert (Brasier et al. 2015), volcanic vesicles (Wacey et al. 2018a, 2018b) and filamentous structures mimicked by intragranular cracks or fractures in chert (Marshall et al. 2011; Bower et al. 2016). Here, three spheroids and one filamentous structure mimicking microfossils, identified in the author's collections, are described. All specimens are from the 3.4 Ga Strelley Pool Formation in the Pilbara Craton, Western Australia.

8.5.5.1 SILICA SPHERULITE COATED WITH MEMBRANOUS CARBONACEOUS MATTER

Some Archean cherts are thought to originate from chemically precipitated silica gel. In such cherts, micron-scale spherulitic structures (chalcedonic quartz spherulites) can form by nucleation on the surface of organic matter, as demonstrated previously (Oehler and Schopf 1971; Oehler 1976). This structure can be diagnostic for a sedimentary origin of chert, which is preferred for the assessment of biogenicity. On the other hand, this structure, which has a relatively clear carbonaceous rim, mimics fossilized spheroid cells (Buick 1990; Fig. 3). Figure 8.5.4a shows hemispheric silica bodies coated by membranous (not particulate) carbonaceous matter. Although an abiogenic origin is evident from their clearly hemispheric shape and occurrence in a vein, the presence of such objects should be reminded when considering the biogenicity of spheroids with carbonaceous wall.

8.5.5.2 SPHEROID MOLDS OF DISSOLVED SULFIDE GRAINS

Sulfide, represented by pyrite (FeS_2), is a common accessory mineral contained in Archean sedimentary rocks, particularly carbonaceous black chert. It occurs as spheroid grains, equant cubic crystals, and their aggregates. Sulfide could be oxidized and dissolved to leave molds, as shown in Figure 8.5.4b and c. The molds of spheroid grains are outlined by very fine black particles, possibly carbonaceous matter, mimicking fossilized spheroid cells.

8.5.5.3 CLUSTERED FILAMENTOUS OPAQUE MATERIAL

Possibly abiogenic filamentous structures are also found (Figure 8.5.4d). Filaments are highly sinuous and comprise a cluster 50 μm across. Widths of most filaments are constant, around 1 μm, but tapering filaments are also present. These filaments are composed of opaque materials, obscuring any inner structures. Thus, hollowness and septation, key features in assessing the biogenicity of filamentous structures, cannot be identified. Their abiogenic origin is suggested from the presence of dispersed similar opaque structures of different width and lengths, without any fragmented features.

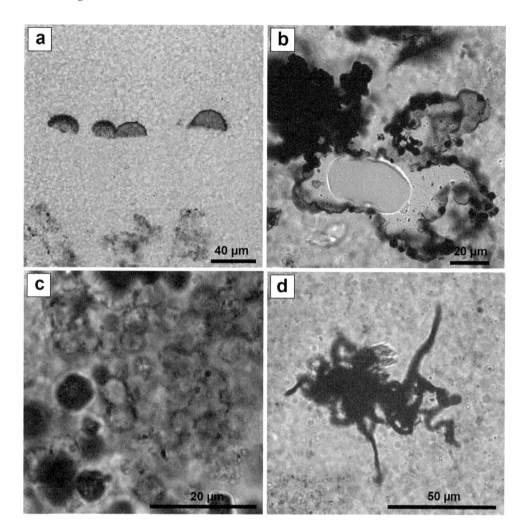

FIGURE 8.5.4 Pseudofossils from the 3.4 Ga Strelley Pool Formation in the Pilbara Craton, Western Australia. (a) Organic-walled silica hemispheres in clear chert vein. (b) Irregular spheroidal cavity formed by dissolution of sulfide cluster, walled by a thin possible organic film. (c) Spheroids probably formed by dissolution of sulfide grains (opaque areas). Remained materials are possible organic matter. (d) Cluster of sinuous filamentous opaque minerals (probably sulfide), like "Caput Medusae."

8.5.6 EXAMPLES OF ARCHEAN MICROFOSSILS, POSSIBLE MICROFOSSILS, AND QUESTIONABLE MICROFOSSILS

8.5.6.1 Isua Supracrustal Belt, Greenland

Pflug (1978) described cell-like structures in ca. 3.8 Ga. rocks collected from the Isua Supracrustal Belt, Greenland. The ellipsoidal structures generally range from 10 to 40 μm in diameter and have hollow interiors partly filled with organic matter or pyrite (Figure 8.5.5a and b). The author claimed that the structures were biogenic and named them *Isuasphaera*. The similarity of *Isuasphaera* to yeasts was suggested. Subsequently, Pflug and Jaeschke-Boyer (1979)

performed more detailed structural and chemical analyses, including Raman microprobe analysis, where the carbonaceous compositions of walls of *Isuasphaera* specimens were demonstrated. These studies have since been subjected to criticism. Bridgwater and co-authors (1981) denied the biogenicity of *Isuasphaera* because of the context of intense geochemical alterations and high-grade (up to amphibolite-facies) metamorphism of host rocks, which were unlikely to preserve fragile microorganisms. Also, these authors suggested that limonite-stained fluid inclusions had been misinterpreted as microfossils. Roedder (1981), on the other hand, suggested that the microbe-like objects were neither limonite-stained fluid inclusions nor microfossils and reinterpreted them as limonite-stained cavities. It seems uncertain whether

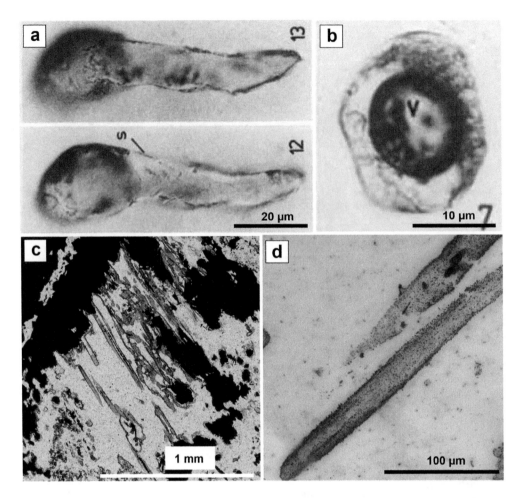

FIGURE 8.5.5 (a) Two images of *Isuasphaera isua* at different focus depths. Thin, elongated wall indicated by S is interpreted as sheath. (b) Another type of *Isuasphaera isua* with central void (v) interpreted as gas vacuole. (a and b: With kind permission from Springer Science+Business Media: *Naturwissenschaften*, Yeast-like microfossils detected in oldest sediments of the earth, 65, 1978, 611–615, Pflug, H.D.) (c and d) Hematitic filaments and tubes in early Archean (>3.8 Ga) ferruginous sedimentary rocks from the Nuvvuagittuq greenstone belt, Canada. (Courtesy of M.S. Dodd.)

these authors examined exactly the same objects (holotypes), and thus, the controversies might not have been to the point. More recently, Appel and co-workers (2003) emphasized that the host meta-chert was subjected to extreme stretching deformation, making it unlikely that syn-depositional spherical objects were preserved, and suggested that the microbe-like objects reported by Pflug (1978) probably resulted from pre-Quaternary weathering.

8.5.6.2 THE NUVVUAGITTUQ GREENSTONE BELT, CANADA

Dodd and co-workers (2017) reported filaments and tubes from at least 3,770 million and potentially 4,280 million years old rocks from the Nuvvuagittuq Belt, Canada. The structures are composed dominantly of hematite and occur in seafloor-hydrothermal-vent-related precipitates (Figure 8.5.5c and d). The authors emphasized that the structures were similar in morphology and associated materials to modern and younger equivalents (Little et al. 1999; Hein et al. 2008; Li et al. 2012). Carbon isotopic values of carbonate and carbonaceous matter are also consistent with their biogenic origin. The authors claimed that

the structures were oxidized biomass that inhabited a very ancient submarine hydrothermal environment, based on the null hypothesis. As stated earlier, the author (K.S.) does not subscribe to a "null-hypothesis approach." Furthermore, their morphological similarity to younger and modern filamentous microfossils provides only limited evidence for biogenicity, because of their morphological simplicity. At this stage, the hematitic structures cannot be readily accepted as of a biological origin.

8.5.6.3 KAAPVAAL CRATON, SOUTH AFRICA

The Kaapvaal Craton is one of the major Archean cratons in the southern part of the African Shield. Microfossils and other biosignatures such as stromatolites and microbially induced sedimentary structures have been reported from the Barberton greenstone belt and Late-Archean sedimentary units. The Barberton greenstone belt consists of the Swaziland Supergroup, which is composed of the 3.55–3.27 Ga Onverwacht, the 3.26–3.23 Ga Fig Tree, and the 3.22–3.10 Ga Moodies Groups, from all of which microfossils and possible microfossils have been reported.

8.5.6.3.1 The Onverwacht Group

The Onverwacht Group is dominated by komatiitic and basaltic volcanic and volcaniclastic rocks, with minor amounts of dacitic volcaniclastic rocks and cherts (Lowe 1999). Microfossils and other biosignatures have been reported mostly from cherts enriched in carbonaceous matter. Fossil-like microstructures from the Onverwacht Group were reported as early as the 1960s and 1970s (Engel et al. 1968; Nagy and Nagy 1969; Brooks et al. 1973; Muir and Hall 1974). Reported structures were morphologically diverse, including spheroids and filaments, among others. However, these earlier findings were later questioned (Schopf 1976; Schopf and Walter 1983). On the other hand, carbonaceous filaments, spheroids, and spindle-lens-shaped structures reported by other investigators (Knoll and Barghoorn 1977; Walsh 1992; Walsh and Lowe 1985; Westall et al. 2001) have been widely accepted as possible to genuine microfossils. In addition, Kremer and Kaźmierczak (2017) recently described new microfossil assemblages. Selected specimens are described in this section.

Walsh (1992) described large granular spheroids and ellipsoids around 20 µm (up to a maximum of 70 µm) and spindles around 30 µm (up to a maximum of 140 µm) in size, in addition to filaments similar to those described previously by Walsh and Lowe (1985) (Figure 8.5.6a and b). Their sizes are much larger than many prokaryotic cells, and they often have thick wall. Based on the assumed frequent large asteroid impact events in

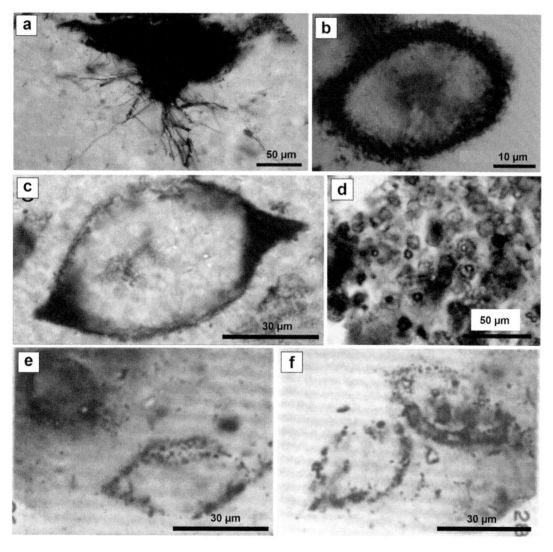

FIGURE 8.5.6 Microfossils and possible microfossils from the Onverwacht Group (a–d) and the Fig Tree Group (e, f), South Africa. (a) Filamentous microfossils from the Kromberg Formation. (From Walsh, M.M., and Lowe, D.R., *Nature*, 314, 530–532, 1985.) (b) Granular-walled large spheroid from the Kromberg Formation. (From Walsh, M.M., *Precam. Res.*, 54, 271–293, 1992.) (a and b: Courtesy of M.M. Walsh.) (c) Lenticular microfossils from the Kromberg Formation discovered independently by M.M. Walsh and K. Sugitani, respectively. (From Oehler, D.Z. et al., *Precam. Res.*, 296, 112–119, 2017.) (d) Cluster of cell-like bodies, which are interpreted as remains of variously degraded colonies of cyanobacteria-like microbes. (Reprinted from *Precam. Res.*, 295, Kremer, B., and Kaźmierczak, J., Cellularly preserved microbial fossils from ~3.4 Ga deposits of South Africa: A testimony of early appearance of oxygenic life? 117–129, Copyright 2017, with permission from Elsevier.) (e and f) Possibly poorly preserved lenticular microfossils, equivalent to that shown in (c). (Reprinted from *Rev. Palaeobot. Palynol.*, 5, Pflug, H.D., Structured organic remains from the Fig Tree Series (Precambrian) of the Barberton Mountain Land (South Africa), 9–29, Copyright 1967, with permission from Elsevier.)

the early Archean (e.g., Lowe et al. 2014), Walsh (1992) suggested the possibility that these large granular fossil-like structures represent envelopes containing endospores tolerant against the harsh environments of the early Earth. The described "spindles," which are more likely lenticular in shape, were recently re-identified, and their biogenicity was confirmed by individual isotopic analyses (Oehler et al. 2017) (Figure 8.5.6c). Structures similar in appearance have also been reported from the 3.4 Ga Strelley Pool Formation and the 3.0 Ga Farrel Quartzite in the Pilbara Craton, Western Australia, as described later (Sugitani et al. 2007, 2010). Kremer and Kaźmierczak (2017) described large (~150–700 μm in length) clusters composed of small carbonaceous cell-like bodies in massive and weakly laminated black cherts with $\delta^{13}C_{PDB}$ values of ~−25 per mil for bulk carbonaceous matter (Figure 8.5.6d).

These cell-like bodies range in size from 3 to 12 μm, show various taphonomic features, and are associated with Al-K-Mg-Fe silicate material, known also from both modern and fossil microbial mats (Douglas 2005). The authors give an in-depth biological interpretation of their benthic-planktonic life cycle and cyanobacterial affinity, such as *Microcystis*.

8.5.6.3.2 The Fig Tree Group

The Fig Tree Group consists of terrigenous clastic rocks, with dacitic to rhyodacitic volcaniclastic and volcanic rocks. The sedimentary successions are composed of graywackes, slates, shales, chert, jasper, and banded iron formation. Possible microfossils were recorded from cherts.

Several reports of the Fig Tree microbe-like structures were published in the 1960s (Barghoorn and Schopf 1966; Pflug 1967; Schopf and Barghoorn 1967). Here, possible microfossils reported by Schopf and Barghoorn (1967) and Pflug (1967) are described.

Schopf and Barghoorn (1967) examined thin sections of carbonaceous black cherts and identified spheroidal dark-colored organic bodies ($n = 28$) that range from 15.6 to 23.3 μm in diameter. Their sizes are significantly larger than the matrix micro-quartz. The spheroids, with walls up to 1-μm thick, were interpreted as to often exhibit a reticulate surface texture, have an irregular mass of organic matter inside, and display various taphonomic stages; however, resolutions of the photographs are not satisfactory.

Pflug (1967) examined cherts and shales of the Fig Tree Group, using both chemical and optical methods. Various morphological types of carbonaceous structures were identified, including opaque, egg-shaped bodies, ellipsoidal bodies characterized by a somewhat transparent wall, disc- or lens-shaped bodies, globular bodies, and filamentous structures. Globular-type objects ranged from 5 to 70 μm across, with the majority larger than 20 μm. Pflug (1967) described that these often occur as colonies and exhibit differentiations such as pores and internal structures. As resolutions of the photographs are not satisfactory, some of the interpretations cannot be verified. It is, however, noteworthy that the described globular objects include morphologies comparable to lenticular microfossils (Figure 8.5.6e and f) described from the older Onverwacht Group (Walsh 1992; Oehler et al. 2017) and the ca. 3.4 Ga Strelley Pool Formation and the ca. 3.0 Ga Farrel Quartzite of the Pilbara Supergroup, Western Australia (Sugitani et al. 2007, 2010; Oehler et al. 2017).

8.5.6.3.3 The Moodies Group

The Moodies Group consists dominantly of alluvial to shallow-marine sandstone with subordinate conglomerate, shale, siltstone, iron formation, and volcanic rocks and provides one of the oldest records of a tide (Heubeck and Lowe 1994; Eriksson and Simpson 2000; Heubeck et al. 2013). Microfossils were discovered from shale, siltstone and chert lenses in sandstone; two reports of microfossils are known to date (Javaux et al. 2010; Homann et al. 2016).

Javaux and co-workers (2010) reported spheroidal microfossils from the lowermost Clutha Formation (Figure 8.5.7a). These occur in gray shales and siltstones that were deposited in shallow-water environments, above the wave base. This finding is significant for the following reasons. First, the host

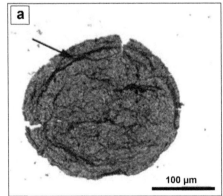 

FIGURE 8.5.7 Microfossils from the Moodies Group, South Africa. (a) Organic-walled carbonaceous spheroidal microfossil extracted from siliciclastic rock (shale/siltstone). The arrow shows concentric fold. (With kind permission from Springer Science+Business Media: *Nature*, Organic-walled microfossils in 3.2-billion-year-old shallow-marine siliciclastic deposits, 463, 2010, 934–938, Javaux, E.J. et al.) (b) Meshwork of filamentous molds embedded in chert. Image obtained by secondary electron microscope. (Homann et al. 2016; Courtesy of M. Homann.)

clastic rocks have long been regarded to have low potential to preserve ancient microfossils. Second, the microfossils are large, up to 300 µm in diameter. Third, the microfossils have a recalcitrant organic wall and can be extracted by conventional acid (HF-HCl) maceration. Homann et al. (2016) identified abundant cylindrical and hollow filamentous molds from 0.3 to 0.5 µm in diameter in lens-shaped chert, with kerogenous laminae, in siliciclastic rock (Figure 8.5.7b). The filaments are bent in various directions and are regularly segmented. Carbon isotopic values of associated kerogenous laminae are significantly light ($\delta^{13}C_{PDB}$ = −26.5 per mil, on average). The authors interpreted the lens-shaped chert as early-silicified cavities formed beneath microbial mats and the filamentous molds as cavity-dwelling microbes. It is suggested that the microbes protected themselves from the supposed intense ultraviolet radiation on the early Earth (Catling and Claire 2005).

8.5.6.4 THE PILBARA CRATON, WESTERN AUSTRALIA

The Pilbara Craton is composed of the East Pilbara, Regal, Karratha, Sholl, and Kurrana Terranes and the De Grey Supergroup (Van Kranendonk et al. 2002, 2006; Hickman and Van Kranendonk 2008; Hickman 2012). Most of the microfossils and possible microfossils have been reported from the Pilbara Supergroup, including the ca. 3.52–3.42 Ga Warrawoona Group, the ca. 3.4 Ga Strelley Pool Formation, the ca. 3.24 Ga Sulfur Springs Group, and the ca. 3.0 Ga Gorge Creek Group of the De Grey Supergroup. In addition, the ca. 3.2 Ga Dixon Island Formation in the Western Pilbara is known to contain possible microfossils.

8.5.6.4.1 The Warrawoona Group

The Warrawoona Group is dominated by mafic to ultramafic volcanic rocks. Subordinate components include felsic volcanic and sedimentary rocks such as chert, evaporite, carbonates, siliciclastic rocks, and volcaniclastic rocks (Van Kranendonk et al. 2006). Microfossils and possible microfossils have been reported from four different rock units, including the Dresser Formation (~3,490 Ma), the Mount Ada Basalt (~3,470 Ma), the Apex Basalt (~3,460 Ma), and the Panorama Formation (~3,450 Ma); these are described in the subsequent paragraphs.

The Dresser Formation is composed dominantly of vari-colored bedded cherts, with minor carbonates and barite. Komatiitic basalt and chert veins also occur (Van Kranendonk et al. 2008). The depositional environment is somewhat controversial. Buick and Dunlop (1990) suggested deposition in a closed to semi-closed shallow coastal basin, whereas Van Kranendonk and co-workers claimed an active volcanic caldera setting (Van Kranendonk 2006; Van Kranendonk et al. 2008). Noffke and co-workers (2013) also suggested a shallow to subaerial environment, including sabkha. Others described microfossils and possible microfossils from bedded chert and chert veins (Dunlop et al. 1978; Ueno et al. 2001a, 2001b; Glikson et al. 2008).

Morphologically diverse hollow carbonaceous spheroids 1.2–12 µm in diameter were obtained by acid maceration of carbonaceous cherts (Dunlop et al. 1978). The spheroids are solitary, paired, and rarely in chains. Some show splitting and rupturing (Figure 8.5.8a and b).

Biogenicity of the Dresser spheroids was argued based on composition, variations in morphology, and DDI. Additionally,

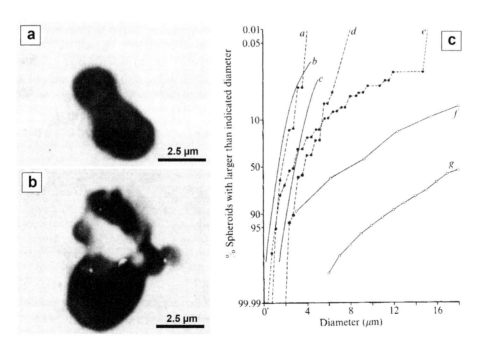

FIGURE 8.5.8 Microfossils from the Dresser Formation of the Warrawoona Group in the Pilbara Craton. (a and b) Carbonaceous spheroids extracted by acid maceration, showing paired morphology and ruptured split. (c) Probability plots of carbonaceous spheroids from the Dresser Formation and extant microalgae: a, gray-black spheroids; b, *Chlorella* sp; c, *Gloeocapsa* sp; d, spheroids with splits; e, North Pole spheroids (total); f, North Pole abiogenic spheroids; and g, synthetic quartz spheres. (With kind permission from Springer Science+Business Media: *Nature*, A new microfossil assemblage from the Archaean of Western Australia, 274, 1978, 676–678, Dunlop, J.S.R. et al.)

Dunlop and co-workers (1978) employed a size probability plot (Knoll and Barghoorn 1977) to claim their biogenicity. The results show that two distinct morphological types, separately plotted, comprise steeply inclined curves that indicate a narrow size distribution, comparative to the extant algae *Chlorella* and *Gloeocapsa* (Figure 8.5.8c). However, these structures are now generally considered to be viscous bitumen droplets or mineralic non-biological spheroids (Awramik et al. 1983; Schopf and Walter 1983; Buick 1990; Wacey 2009).

Ueno et al. (2001a, 2001b) described carbonaceous filamentous structures of several types from hydrothermal chert veins and bedded cherts of this formation. The structures include solitary spiral filaments, radiating clusters of threads, and tubular filaments ~10-μm thick (Figure 8.5.9a). Analyses of carbon isotopic values of these filaments by using secondary-ion mass spectrometry (SIMS) give significantly light values ($\delta^{13}C < -30$ per mil). It is suggested that the values can be interpreted as products of the Calvin cycle or reductive acetyl-coenzyme A (CoA) pathway. Supposed habitats for these microfossils also suggest that they represent chemoautotrophs. Evidence for microbial methanogenesis in this formation was also obtained from analyses of fluid inclusions (Ueno et al. 2006). Acid maceration of bedded black chert also suggested the presence of different types of microbes in the Dresser Formation (Glikson et al. 2008). Several types of carbonaceous objects, including tiny cell-like bodies (less than 3 μm to submicron in diameter), were obtained; however, their biogenicity has not yet been fully demonstrated owing to the small size. It should be finally noted that organic matter extracted from the Dresser chert has some biogenic traits (long-chain aliphatic hydrocarbons characterized by an odd-over-even carbon number dominance) (Derenne et al. 2008).

In bedded chert collected from the Mount Ada Basalt in the North Pole area, Awramik and co-workers (1983) identified carbonaceous spheroids and filaments and suggested their prokaryotic affinity. One of the described filaments possesses biological elaboration and taphonomy. It is relatively thick (~5 μm), acutely folded, and regularly septate (Figure 8.5.9b), consistent with biogenicity. Buick questioned this finding, in the context of sample locality and syngenicity (Buick 1984), leading to a published debate (Awramik et al. 1988; Buick 1988). However, Grey and co-workers (2010) resolved uncertainty about the sample locality; the cherts hosting the structures were indeed in the Mount Ada Basalt.

The Apex Basalt, together with the Panorama Formation, comprises the Salgash Subgroup. This volcanic unit is composed dominantly of basalt, komatiitic basalt, and serpentinized peridotite, with local dolerite sills. Felsic volcaniclastic rocks and chert occur as minor components. Microfossil-like structures were reported from the chert collected from the Chinaman Creek locality (Schopf and Packer 1987; Schopf 1993). The structures are carbonaceous and filamentous and were classified into 11 taxa (Schopf 1993). The authors suggested the possibility of their cyanobacterial (*Oscillatoria*) affinity, based on morphological similarity, and inferred inhabitation in shallow-water environment. However, the original rock specimen was subsequently revealed to be from a hydrothermal chert vein within the Apex Basalt. Furthermore, it was shown that the distribution of carbonaceous matter was at least partially controlled by crystal growth. The formation process of the host rock appears to be not suitable for life, and microbe-like morphologies could have been produced by abiological processes (Brasier et al. 2002, 2005, 2006; Van Kranendonk and Pirajno 2004; Van Kranendonk 2006; Sforna et al. 2014). Pinti and co-authors (2009) also demonstrated that the chert in the Apex Basalt was repeatedly hydrothermally altered. Marshall and co-workers (2012) revealed that carbonaceous matter in the Apex chert had multiple generations, clearly associated with hydrothermal processes, indicating that indigenousness of carbonaceous matter cannot always be guaranteed. Marshall and co-workers (2011) also examined segmented structures, supposedly similar to those described by Schopf (1993), but in a different Apex chert sample, and showed that they were fractures composed of a series of quartz and hematite crystals. Schopf and Kudryavtsev (2011) pointed out the many differences between the mineral veins of Marshall and co-authors (2011) and the carbonaceous structures analyzed by Schopf and colleagues in numerous publications. However, some Apex microfossil-like structures such as the *Archaeoscillatoriopsis* and *Primaevifilum* paratypes (Schopf 1993) were also revealed to be composed of vermiform phyllosilicate grains and adsorbed carbonaceous matter (Brasier et al. 2015). Bower and co-workers (2016) demonstrated that *Eoleptonema apex* (Schopf 1993) was a sheet-shaped pseudofossil composed of an intragranular crack infilled with carbon. These studies may lead readers to wonder if all of the Apex structures described by Schopf and Packer (1987) and later by Schopf (1993) are, in fact, pseudofossils. However, the analyses were not always performed on the holotype specimens (Schopf and Kudryavtsev 2011). To date, the only holotype whose biogenicity was directly rejected is *Eoleptonema apex* (Bower et al. 2016). The *Primaevifilum amoenum* are indeed cylindrical and "hollow" and are composed of a uniseriate sequence of individual carbonaceous compartments, most likely biogenic in origin (Figure 8.5.9c and d) (Schopf and Kudryavtsev 2009). This has recently been demonstrated by analyses of carbon isotopic compositions of individual microstructures (Schopf et al., 2018).

Also, it should not be overlooked that organic matter extracted from the Apex chert has some biogenic traits (De Gregorio et al. 2009). Thus, the author (K.S.) suggests that it is possible that some other Apex microstructures could be biogenic. The author's observation of the genuine microfossils from the Pilbara Craton, whose biogenicity has been established by multidisciplinary studies, as discussed later, indicates that microfossils composed originally of carbonaceous matter can be replaced by other materials such as hematite (Sugitani et al. 2007) or pyrite (Figure 8.5.3f). Thus, non-carbonaceous compositions of microbe-like structures in the Apex chert demonstrated by some critics do not readily exclude their biogenicity.

From the Kitty's Gap, 50 km northeast of Marble Bar in the Pilbara Craton, fossil-like microstructures were reported from silicified volcaniclastic sediments (Westall et al. 2006, 2011). The volcanic-sedimentary succession there is thought to be

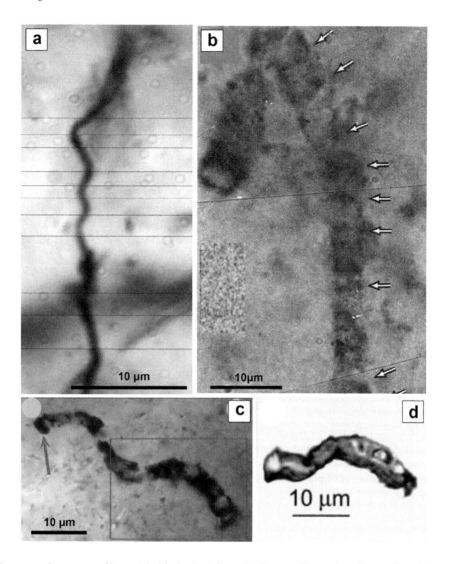

FIGURE 8.5.9 (a) Sinuous carbonaceous filament in black chert from the Dresser Formation. (From Ueno, Y. et al., *Int. Geol. Rev.*, 43, 196–212, 2001; Courtesy of Y. Ueno.) (b) Acute-folded, relative thick septate filament from chert of the Mount Ada Basalt in the North Pole area, Pilbara Craton. The arrows show septation. (Reprinted from *Precam. Res.*, 20, Awramik, S.M. et al., Filamentous fossil bacteria from the Archean of Western Australia, 357–374, Copyright 1983, with permission from Elsevier.) (c) Septate filament (*Primaevifilum amoenum*) from the Apex chert in the Pilbara Craton. The arrow points to the rounded terminus. (d) Three-dimensional Raman image indicating cylindrical morphology of the rectangular portion of (c). (Reprinted from *Precam. Res.*, 158, Schopf, J.W. et al., Evidence of Archean life: Stromatolites and microfossils, 141–155, Copyright 2007, with permission from Elsevier.)

correlative to the Panorama Formation of the Salgash Subgroup (Van Kranendonk et al. 2002). Chert at the Kitty's Gap is cut by numerous chert veins and originates from silicification of the volcaniclastic sediments. The depositional setting was closely associated with hydrothermal activities. Westall and co-workers (2006, 2011) identified microstructures on the vapor hydrofluoric acid (HF)-etched surface. Carbonaceous microstructures include filaments less than 0.3 μm in width, coccoids 0.4–0.8 μm in diameter, and rods ~1 μm in length. They occasionally comprise colony-like clusters associated with extracellular polymeric substance (EPS). Morphology (shape and size), composition, and occurrence appear to be all consistent with their biogenicity; however, some doubts can be casted on their syngenicity based on the absence of corresponding structures in petrographic thin sections (Wacey 2009).

8.5.6.4.2 The Strelley Pool Formation

The Strelley Pool Formation is an independent formation that does not belong to the higher stratigraphic unit. It has been identified widely in the East Pilbara Terrane, and its depositional area is assumed to have been at least 30,000 km² (Hickman 2008). This formation is typically 8–11 m thick, but locally, it can be up to 100 m thick. It is composed of various types of rock, including siliciclastic sedimentary rocks, carbonates (mainly bedded and stromatolitic dolomite), cherts, and volcaniclastic rocks. The environments of deposition were various but mainly shallow-water environment, such as a rocky coastal shoreline and beach environment; shallow-water marine carbonate platform; possible sabhka; and tidal flat and alluvial fan, all with or without hydrothermal inputs (Allwood et al. 2006, 2007, 2010; Hickman 2008;

Sugitani et al. 2015b). Schopf and Packer (1987) and Schopf (2006) first described microbe-like structures in carbonaceous chert from the Strelley Pool Formation. The sheathed structures are up to over 50 μm across and are composed of multiple spheroids. Such morphological elaboration appears to be consistent with the biogenicity.

Promising microbe-like structures were later reported by Sugitani and co-authors (2010), who described morphologically diverse carbonaceous microstructures in carbonaceous cherts of sedimentary origin from the Strelley Pool Formation. These were discovered from three remote localities in the Panorama, Warralong, and Goldsworthy greenstone belts. The lithostratigraphies at these localities are different to each other, meaning different depositional settings. Nevertheless, a similar morphological assemblage of microstructures, including spheroids, lenses, filaments, and films, occurs. This preliminary report for the Strelley Pool microfossils was followed by a discovery of similar structures from a new locality in the Panorama greenstone belt, which was 8 km away from the original locality (Sugitani et al. 2013). Microfossil-bearing black cherts in the Panorama greenstone belt and the Warralong greenstone belt were deposited in a shallow marine environment (intertidal to sub-tidal zone), and those in the Goldsworthy greenstone belt were possibly deposited in a terrestrial hydrothermal field near a coastal zone (Sugitani et al. 2010, 2013, 2015b).

Spheroids are mostly less than 15 μm in diameter, but rarely, large ones, >50 μm in diameter, also occur. Spheroids with intermediate sizes are also present. Small and intermediate spheroids often comprise colony-like clusters (Figure 8.5.10a and b), whereas very large ones occur solitarily (Figure 8.5.10c and d). Colonies of intermediate-sized spheroids are not common, but where they do occur, these exhibit unique habits. One example is a colony composed of several spheroids of nearly the same size (~20 μm across), with a mutual compress contact; it is further characterized by being embedded within a foam-like envelope (Figure 8.5.10e). Other interesting specimens include an elongated colony composed of spheroids of different sizes, with a budding-like attachment of smaller hemispheroids (Figure 8.5.10f).

Lenses are mostly 20–60 μm in length along the major dimension, and some can be up to 100 μm. The structures are composed of a central body and surrounding sheet-like appendage, which is called a flange (Figure 8.5.11a) (Sugitani et al. 2007). These lens-shaped structures can be extracted by acid maceration. Observation of extracted specimens, in addition to those in petrographic thin sections, revealed their minor morphological variations such as ellipticity of the polar view, transparency of the central body, and flange width and texture (Sugitani et al. 2010, 2013, 2015a) (Figure 8.5.11b–h), possible reflection of speciation (Sugitani et al. 2018). Lenses often comprise colony-like clusters of various shapes and pairs or chains (Figure 8.5.12a–c).

Biogenicity of these spheroids and lenses has been demonstrated by their carbonaceous composition; size distribution; taphonomic features; occurrences suggestive of vital activities such as reproduction, texture-specific, and significant negative carbon isotopic value; the presence of nitrogen- and oxygen-rich

molecules in lenticular structures and alveolar or hollow internal texture of the central body (Figure 8.5.12d–f) (Sugitani et al. 2010, 2013, 2015a; Lepot et al. 2013; Alleon et al. 2018).

Possible biogenic filaments are rare and occur solitarily, without any spatial relationships (Sugitani et al. 2013). They are relatively thick, up to 20 μm in width, and up to 200 μm in length. The structures are hollow, bent, and have notched edges (Figure 8.5.13a). They were likely reworked from the original site of formation. The other type of filament has been discovered from Anchor Ridge in the Panorama greenstone belt, from which lenticular microfossils were originally described (Sugitani et al. 2010). The thin (~2 μm wide), sinuous, and hollow filaments are identified within facies. showing a carbonaceous cobweb structure (Figure 8.5.13b), which is inter-layered with distinct facies (characterized by fine lamination of carbonaceous matter and enrichment of sulfides) containing spheroidal and lenticular microfossils. Recently, Schopf interpreted that the cobweb structures are composed largely of degraded filaments and silicified sulfate nodules (Schopf et al. 2017). Combined with sulfur isotopic data and resemblance to younger equivalents (Schopf et al. 2015), the authors argued that the filaments represent sulfur bacteria.

In addition to the previously described chert-hosted microfossils, possible biogenic carbonaceous spheroids, ellipsoids, and tubular filaments were reported from the basal sandstone of this formation in the East Strelley greenstone belt (Figure 8.5.13c and d) (Wacey et al. 2011). The structures were described to show various taphonomic features and occur as colony-like chains and clusters, with reproduction-like habits. Carbon isotopic values of wall-comprising carbonaceous matter, associated with nitrogen, range from −33‰ to −46‰. Possible affinity to sulfur-metabolizing bacteria was suggested (Wacey et al. 2011, 2014a).

8.5.6.4.3 The Sulfur Springs Group

The Sulfur Springs Group is composed mainly of felsic to mafic-ultramafic volcanics and volcaniclastic rocks, with minor amounts of sedimentary rocks. The Kangaroo Caves Formation of this group, from which microbe-like structures were described, consists mainly of sedimentary rocks, including for example, breccia, banded iron formation, layered chert, and shale, associated with widely spaced black chert hydrothermal veins and massive sulfide deposits. The massive sulfide deposits have possibly been deposited on a deep (>1000 m) seafloor, from high (~300°C)-temperature hydrothermal solutions (Vearncombe et al. 1995). Filamentous and spheroidal possible microfossils have been reported (Rasmussen 2000; Duck et al. 2007).

These possible fossil-like microstructures include thin pyritic filaments that occur in colloform-textured microcrystalline quartz masses in massive sulfide deposits (Rasmussen 2000). The sinuous filaments, up to 300 μm long, have a uniform length-wise diameter of 0.7–0.9 μm. They also exhibit an intertwined habit and were oriented differently in different portions, suggesting behavioral variations in different microenvironments (Figure 8.5.13e). Rasmussen interpreted the filaments as probable microfossils replaced by pyrite and suggested their affinity to hyperthermophilic chemoautotrophs (Rasmussen 2000). Later, these filaments were revealed to have internal carbon and nitrogen patches and nano-pores (Wacey et al. 2014b) (Figure 8.5.13f). Such additional evidence appears to provide convincing evidence for their biogenicity; however, Wacey and co-workers (2014b) posed some doubts, referring to their morphological simplicity.

Acid maceration of sedimentary rock overlying the massive sulfide deposit also yielded other types of possible microfossils (Duck et al. 2007). Obtained materials were bundles of filamentous, tubular carbonaceous structures and associated carbonaceous nanospheres, characterized by significantly light bulk carbon isotopic values ($\delta^{13}C_{PDB}$ = −26.8‰ to −34.0‰).

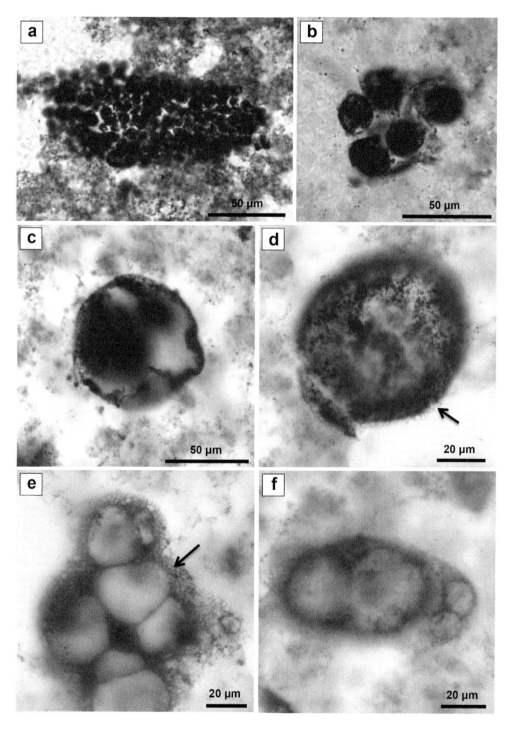

FIGURE 8.5.10 Spheroidal microfossils from the 3.4 Ga Strelley Pool Formation in the Pilbara Craton, Western Australia. (a) Massive spheroidal colony composed of small spheroids. (b) Colony composed of several spheroids of intermediate size (~15 μm), associated with film-like object. (c) Large, quartz-filled spheroid with irregularly wrinkled wall. (d) Large spheroid with granular wall and short fluff (arrow). (e) Colony composed of large spheroids, with compressed mutual contact and foam-like envelope (arrow). (f) Colony composed of spheroids of different sizes, enveloped by fluffy material.

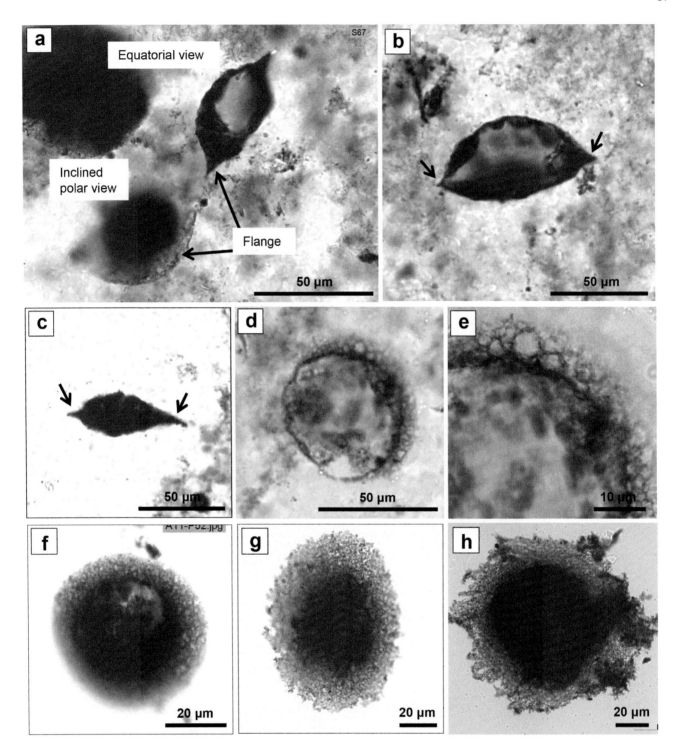

FIGURE 8.5.11 Lenticular microfossils from the 3.4 Ga Strelley Pool Formation in the Pilbara Craton, Western Australia. (a) Two adjacent specimens of lenticular microfossils set in different directions in petrographic thin section, allowing the interpretation of the three-dimensional morphology of these lenticular microfossils. (b) Equatorial view of lenticular microfossil with hollow (quartz-filled) interior, characterized by short flange (arrows). (c) Equatorial view of lenticular microfossil with asymmetric lengths of flange (arrows). (d) Polar view of lenticular microfossil with translucent wall. (e) Enlarged image of the specimen in (d), showing clear reticulation on its flange. (f–h) Polar views of lenticular microfossils extracted from host cherts by acid maceration. Note the variations of flange textures (f, reticulation; g, granular; and h, fluffy and oblateness). (From Sugitani, K. et al., *Geobiology*, 13, 507–521, 2015.)

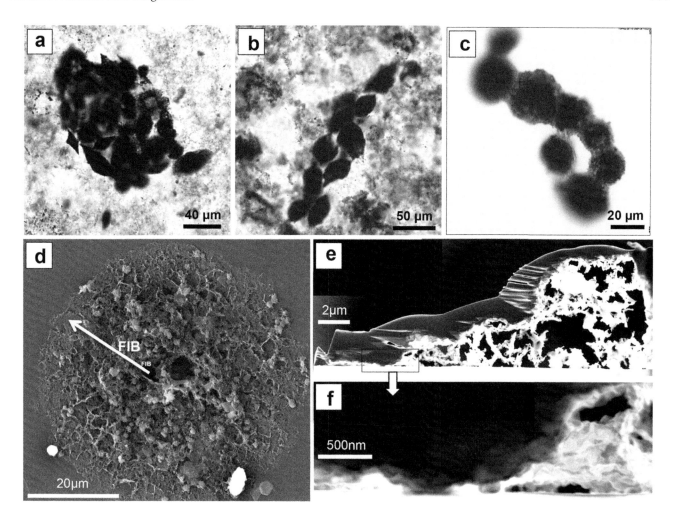

FIGURE 8.5.12 Lenticular microfossils from the 3.4 Ga Strelley Pool Formation in the Pilbara Craton, Western Australia. (From Sugitani, K. et al., *Geobiology*, 13, 507–521, 2015.) (a) Tightly packed and randomly oriented colony. (b) Linearly aligned colony. (c) Chain obtained by acid maceration. (d) Scanning electron microscope (SEM) image of extracted lenticular microfossil. The arrow indicates the Focused Ion Beam (FIB) sectioning used to sub-sample for transmission electron microscope (TEM) analysis. (e and f) TEM images of the section shown by arrow in (d).

These structures are similar to both modern and ancient microbes documented from seafloor hydrothermal systems, but the biogenicity has not yet been fully demonstrated.

8.5.6.4.4 The Dixon Island Formation

The Dixon Island Formation in the Regal Terrane of the West Pilbara comprises sedimentary succession of rhyolitic tuff, black shale, and varicolored cherts. The succession overlies basalts showing pillow structures. Basalts and overlying sediments were extensively veined by carbonaceous black cherts, suggesting that the deposition of this formation was closely related to hydrothermal activity. The paleodepth was assumed to have been ~500–2000 m (Kiyokawa et al. 2006, 2014). The Dixon Island Formation represents a Mesoarchean seafloor hydrothermal system. Identified microbe-like structures include iron-rich spheroids in stromatolite-like ferruginous beds and carbonaceous structures of five morphological types; spiral-shaped filaments, rod-shaped filaments, cell-shaped filaments, dendritically stalked filaments, and spheroids in carbonaceous black chert. Among them, hollow filaments of

50–100 μm in length and 10 μm in width, with length-wise uniformity, are likely biogenic.

8.5.6.4.5 The Gorge Creek Group

The Gorge Creek Group, contrastive to the Warrawoona Group described earlier, is dominated by sedimentary rocks, which include chert, jasper, and banded iron formation, as well as terrigenous clastic rocks such as shale, siltstone, sandstone, and conglomerate. To date, microfossils have been reported from the Farrel Quartzite (Van Kranendonk et al. 2006), which is the lowest unit of the Gorge Creek Group. The Farrel Quartzite in the Goldsworthy greenstone belt is composed dominantly of very coarse- to coarse-grained sandstone, including quartzite and minor conglomerate. Mafic to ultramafic volcaniclastic beds, evaporite beds, fine-grained clastic rock, and layers of black chert comprise minor components. This sedimentary unit varies from several to 80 m in thickness along strike (Sugitani et al. 2003, 2007) and is conformably overlain by a thick (>100 m) unit of chert and banded iron formation, which is correlative to the Cleaverville Formation.

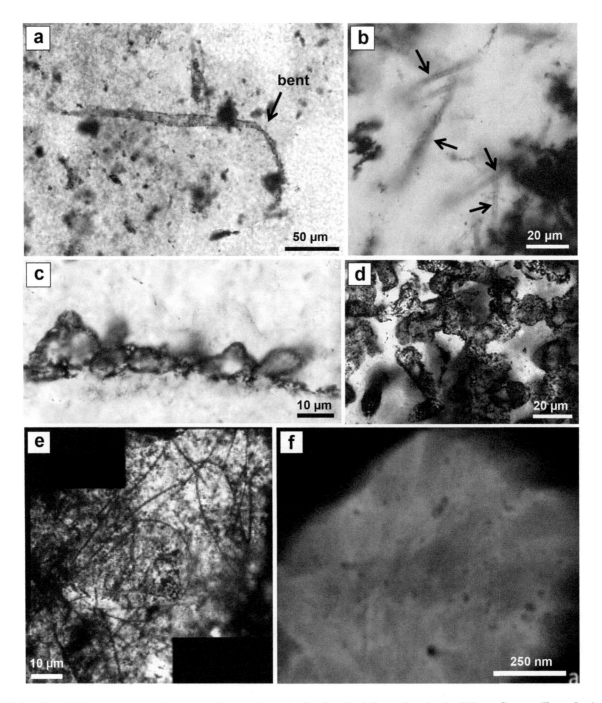

FIGURE 8.5.13 (a) Bent, tubular carbonaceous filament from the Strelley Pool Formation, in the Pilbara Craton. (From Sugitani, K. et al., *Precam. Res.*, 226, 59–74, 2013.) (b) Narrow tubular sinuous filamentous microfossils (arrows) from cobweb-like structured facies in carbonaceous black chert of the Strelley Pool Formation. (Schopf, J.W. et al., *Precam. Res.*, 299, 309–318, 2017.) (c) Spheroids and (d) tubular filaments in silica-filled pore spaces in the basal sandstone. (From Wacey, D. et al., *Nat. Geosci.*, 4, 698–702, 2011; Courtesy of D. Wacey.) (e) Variously oriented curved filaments composed of pyrite from the Sulfur Springs Group. (With kind permission from Springer Science+Business Media: *Nature*, Filamentous microfossils in a 3,235-million-year-old volcanogenic massive sulphide deposit, 405, 2000, 676–679, Rasmussen, B.) (f) TEM image of a section of the pyrite filament in (a) showing porous interior (Wacey et al. 2014b, with permission from Elsevier.)

Microfossils were discovered from black cherts at the uppermost portion of the Farrel Quartzite (Sugitani et al. 2007). The microfossil-bearing black cherts and associated evaporite beds can be traced for ca. 7 km along strike.

The microfossil-bearing black cherts exhibit micron-scale spherulitic structures and locally cross-lamination, identified by the distribution of fine carbonaceous particles, carbonaceous clots, and microfossils. As described earlier, the spherulitic structure in cherts indicates that they originated from a silica gel, but not the silicification of non-siliceous sediments. Cross-lamination also indicates deposition under the influence of fluid flows. Aqueous deposition of the microfossil-bearing cherts is also consistent with features of rare-earth elements and yttrium, which are best interpreted as mixing of seawater and low-temperature hydrothermal fluids, and/or as terrestrial run-off (Sugahara et al. 2010). On the other hand, Retallack and co-authors (2016) argued that the black chert originated from a paleosol and the described microfossils could be assigned to modern soil-inhabiting microbes. The authors, however, overlooked the previously described diagnostic features, and their paleosol model is unfounded (Sugitani et al. 2017).

The cellularly preserved microfossil assemblage identified in this unit is basically similar to that of the Strelley Pool Formation; spheroids and lenses were identified. Filamentous structures of equivocal origin are also present. Most spheroids are less than 15 μm in diameter and often comprise colony-like clusters, which are rarely associated with film-like objects (Figure 8.5.14a–c). Large spheroids, from 40 μm up to 80 μm in diameter, are also present (Figure 8.5.14d–h) and occasionally comprise colonies together with smaller spheroids (Figure 8.5.14i). The walls of spheroids exhibit various taphonomic features, from granular to hyaline and from smooth to irregular wrinkles. In contrast, fluff and regular wrinkles identified in some specimens are likely primary features (Figures 8.5.3e and 8.5.14e). The Farrel Quartzite lenticular structures are similar to those from the Strelley Pool Formation; however, minor morphological and textural variations are more conspicuous in the Farrel Quartzite lenses. Flange is texturally various (translucent, hyaline, reticulated, and striated) (Sugitani et al. 2009a) (Figure 8.5.15a–c). Also,

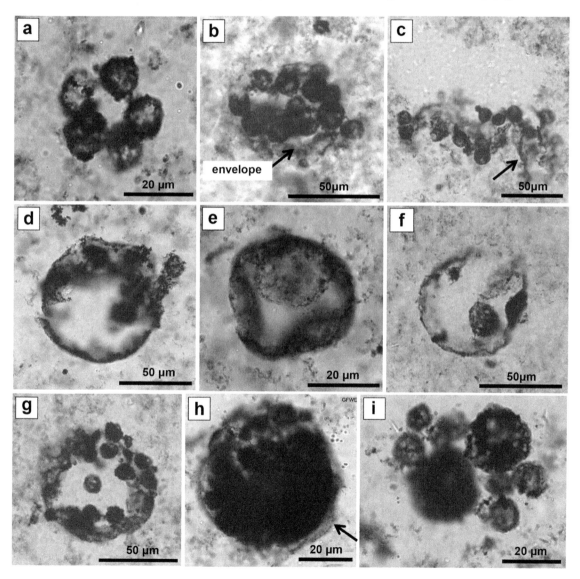

FIGURE 8.5.14 Spheroidal microfossils from the Farrel Quartzite of the Gorge Creek Group in the Pilbara Craton, Western Australia. (Sugitani, K. et al., *Precam. Res.*, 158, 228–262, 2007; Sugitani, K. et al., *Precam. Res.*, 173, 50–59, 2009.) (a) Spherical colony composed of regularly arranged small spheroids (~10 μm). (b) Colony composed of small spheroids with thin envelope (arrow). (c) Small spheroids attached with hyaline film (arrow). (d) Large spheroid with smooth but partially broken wall. (e) Large spheroid with folded wall. (f) Large spheroid showing broken habit, with a smaller spheroid inside. (g) Large spheroid with wrinkled wall, associated with small spheroids inside and outside the wall. (h) Possibly partially double-walled (arrow) spheroid containing small spheroids inside. (i) Colony composed of spheroids of various sizes.

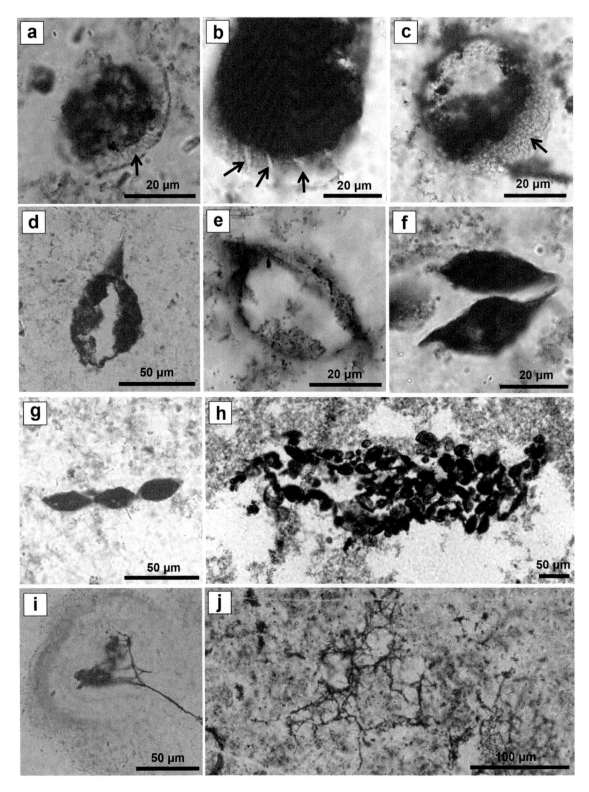

FIGURE 8.5.15 Lenticular microfossils and filaments from the Farrel Quartzite. (Sugitani, K. et al., *Precam. Res.*, 158, 228–262, 2007; Sugitani, K. et al., *Astrobiology*, 9, 603–615, 2009.) (a) Inclined polar view of lenticular microfossil with a hyaline translucent flange (arrow). (b) Inclined polar view of lenticular microfossil characterized by striated (arrows) flange. (c) Inclined polar view of lenticular microfossil with hollow (quartz-filled) central body and reticulated flange (arrow). (d) Equatorial view of lenticular microfossil with one-sided flange. (e) Three-dimensional (3D)-reconstructed image of lenticular microfossil showing asymmetric relationship between flange and central body. (f) Obliquely paired lenticular microfossils. (g) Linearly aligned triplet of lenticular microfossils. (h) Large colony composed of lenticular and spheroidal microfossils. (i) Carbonaceous filaments in colloform-textured portion in vein, such as that shown in Figure 8.5.3a. (j) Entangled, thin carbonaceous filaments observed in specific portion of the fossil-bearing chert bed.

the width is asymmetric in some specimens (Figure 8.5.15d) (Sugitani et al. 2007), while the shape of the central body is both symmetric and asymmetric (Sugitani et al. 2009a) (Figure 8.5.15e). Like the spheroids, lenses often comprise colonies, although they can occur solitarily or in pairs or chains (Figure 8.5.15f–h).

Biogenicity of the Farrel Quartzite spheroids and lenses was fully demonstrated (Sugitani et al. 2007, 2009a, 2009b; Grey and Sugitani 2009; Oehler et al. 2009, 2010; Schopf et al. 2010; House et al. 2013; Delarue et al., 2018). Their host chert is primary, deposited in an aqueous environment, which is suitable for life. The structures are composed dominantly of carbon, associated with minor amounts of N and S, which are representative organic elements. Carbon isotopic values of individual specimens are adequately negative, within the range of extant organisms, and are different from those of dispersed carbonaceous clots in the matrix. Nitrogen isotopic ratios are highly variable and morpho-specific. Walls comprising the structures are assumed to have been flexible and breakable, like cell membranes and walls. The structures are prolific: up to 100 or more, in a petrographic thin section 2.5 × 3.4 cm wide and 30 µm thick. Their size distributions are relatively narrow, and they often comprise pairs, chains, and colony-like clusters, likely results of reproduction. Most of the structures, if not all, are composed of membranous organic matter and can be extracted by acid maceration. To date, no such abiogenic models have been presented that could produce such complex organic-walled microstructures and their assemblages.

Filamentous microstructures from the Farrel Quartzite include solid-looking threads less than 1 µm in diameter and a hollow tube-like structure ca. 20 µm in diameter, of which only one specimen has been identified. Thus, the thread-like structures are described here. These exhibit two distinct occurrences: one occurs in colloform-textured portions composed of microcrystalline quartz in cavity fills and veins (Figure 8.5.15i), while the other consists of a dense population of threads in a specific layer that is a few millimeters thick, where the other types of microstructures (spheroids and lenses) do not occur (Figure8.5.15j). Threads range widely from a few tens of microns to several hundred microns in length, and their length-wise thicknesses appear to be uniform. The thin filamentous microstructures are often bent and entangled. It is equivocal whether these thread-like structures, especially those in cavity fills and veins, represent fossilized microbes. Threads occurring in a dense population are more likely microfossils, but the possibility that these threads are fossilized fibrillar extracellular polymeric substances cannot be excluded.

8.5.7 IMPLICATIONS FOR HABITATS OF ARCHEAN MICROBES

As described in this chapter, the depositional environments of rocks hosting early Archean microfossils are quite diverse, including shallow to subaerial evaporitic basins, coastal hydrothermal fields, intertidal to supratidal zones, and relatively deep ocean floor with intense hydrothermal activities.

The occurrence of non-cellularly preserved biosignatures, such as stromatolites and microbially induced sedimentary structures, is mostly restricted to sediments deposited from shallow platform to supratidal and even terrestrial settings, such as a sabkha (e.g., Allwood et al. 2006; Wacey 2010; Noffke et al. 2013; Heubeck 2009). On the whole, records of habitats for Archean organisms appear to be dominantly from shallow-water environments. This, however, does not imply that deep-sea and open-ocean regions at that time were not important habitats for microbes. Sedimentary successions of deep-sea and open-ocean facies in the Precambrian are represented by banded iron formations, which may have been formed by iron-oxidizing bacteria (Kappler et al. 2005; Posth et al. 2013). It should also be reminded that morphologically preserved biosignatures may represent only a part of the Archean biosphere.

8.5.8 WHAT DO THE ARCHEAN "LARGE" MICROFOSSILS TELL?

It is now fully demonstrated that early Archean microfossil assemblages are quite diverse and include significantly large microfossils: spheroids or lenses larger than 20 µm are common and can be up to 300 µm across. Such a large size may hint of a eukaryotic affinity. In this section, the validity of this interpretation and strategies to elucidate biological affinities of large Archean microfossils are discussed. The timing of the emergence of eukaryotes is one of the most important milestones in the evolution of early life and is one of the unresolved great problems in astrobiology.

8.5.8.1 The Oldest Records of Eukaryotes

In general, microbes in the Archean are considered to be prokaryotic, which is partly based on that the metabolism of eukaryotes is primarily aerobic and that the Earth's atmosphere began to be oxygenated from around 2.45 Ga (the Great Oxidation Event [GOE]). Although the oldest record of eukaryotic fossils has not yet been settled, widely accepted and reliable eukaryotic microfossils can be found in the Paleoproterozoic, such as large organic-walled spheroidal microfossils that have regular geometric ornaments and/or protrusions (Figure 8.5.3b) (Javaux et al. 2003; Knoll et al. 2006). However, Butterfield (2014) suggested that the stem group of eukaryotes could extend back to the early Archean. Fossils and possible fossils of assumed eukaryotic affinity, which are older than 2.0 Ga, have been reported. Some of these structures include 2.1 Ga megascopic coiled filaments (*Grypania spiralis*) (Han and Runnegar 1992), 2.1 Ga pyrite-replaced megascopic structures interpreted to be colonial organisms with a coordinated growth pattern (El Albani et al. 2010), and ~2.8–2.7 Ga organic-walled tubular microstructures interpreted to be equivalent to modern siphonous green or yellow-green microalgae (Kaźmierczak et al. 2016). However, a eukaryotic affinity of these older materials has not yet been widely accepted.

8.5.8.2 Criteria for Eukaryotic Fossils and Application to Large Archean Microfossils

Knoll and co-workers (2006) listed three criteria for fossilized eukaryotic cells: (1) having an acid-resistant organic wall (envelope), (2) having micron-scale ornaments and processes (Figures 8.5.3b and 8.5.16a, b), and (3) having significantly large (>20 μm) cell size. Microfossils under consideration meeting all three criteria are likely eukaryotic. These criteria are based on the following reasons. Most prokaryotic cells are smaller than 10 μm, whereas those of eukaryotes are larger than 10 μm. Such limited cell size of prokaryotes is considered to be a consequence of intracellular metabolism, regulated by the diffusion of nutrients, without active transportation by a cytoskeleton. Micron-scale ornaments and processes were also interpreted to be formed only under regulation by a cytoskeleton (Knoll et al. 2006). While many eukaryotic algae can produce acid-resistance recalcitrant walls, prokaryotes cannot, except for some types of cyanobacteria. Furthermore, eukaryotic cell walls are multilayered, whereas prokaryotic cell walls are not (Javaux et al. 2004) (Figure 8.5.16c).

Spheroidal and lenticular microfossils discovered from the Kaapvaal Craton, South Africa, and the Pilbara Craton, Western Australia, are large (>20 μm and up to 300 μm) and have acid-resistant organic walls. Regularly arranged micron-scale ornamentation cannot be identified, but textures such as reticulation, striations, and regular wrinkles observed in some specimens could be related to a cytoskeleton-regulated texture; however, the primary origin of these features has yet to be fully established. Flanges, sheet-like appendages that surround the central body, are a far simpler feature compared with the protrusions of Proterozoic eukaryotic microfossils, but again, their formation was likely controlled by the cytoskeleton. Although TEM analyses revealed that envelopes

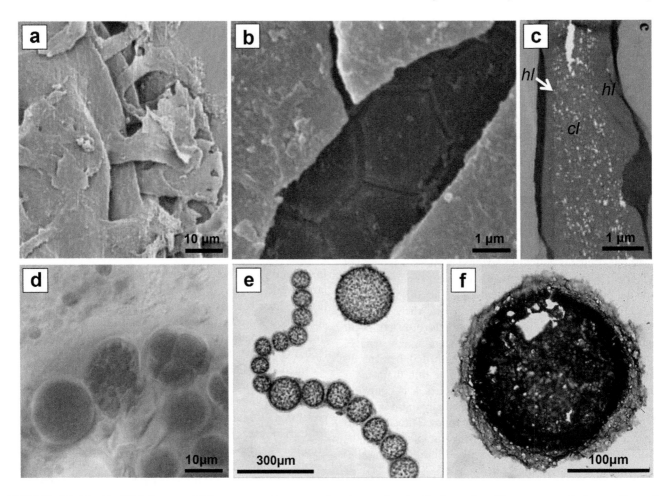

FIGURE 8.5.16 (a) SEM image of Mesoproterozoic process-bearing acritarch, *Syuiyousphaeridum macroreticulatum*, from the Ruyang Group, northern China. Note that the processes are flaring, furcating processes. (b) SEM image of the outer and inner walls of *Syuiyousphaeridum macroreticulatum*. (c) TEM image of a multilayered wall ultrastructure of two homogenous layers (*hl*) that sandwich a thick porous central layer (*Cl*). (a–c: Javaux, E.J. et al., *Geobiology*, 2, 121–132, 2004. With permission from Blackwell Publishing Ltd.) (d) Cyanobacteria containing numerous baeocytes (the center), likely related to genus *Chroococcopsis*. (Courtesy of Sergei Shalygin.) (e) Large chain-forming sulfur bacteria *Thiomargarita namibiensis*. (Reprinted from *Syst. Appl. Microbiol.*, 34, Salman, V. et al., A single-cell sequencing approach to the classification of large, vacuolated sulfur bacteria, 243–259, Copyright 2011, with permission from Elsevier.) (f) Lenticular microfossil (*Pterospermopsimorpha pileiformis*) extracted by acid maceration, from the lower Mesoproterozoic Kaltasy Formation, East European Platform. (Reprinted from *Precam. Res.*, 278, Sergeev, V.N. et al., Microfossils from the lower Mesoproterozoic Kaltasy Formation, East Europian Platform, 87–107, Copyright 2016, with permission from Elsevier.)

of some acid-extracted specimens of lenticular and spheroidal microfossils were single-layered (Sugitani et al. 2015b; Javaux et al. 2010), the eukaryotic affinity of large Archean microfossils, particularly flanged lenticular ones, may be worth considering. However, the reality does not allow such a straightforward interpretation, as briefly reviewed in the next section.

8.5.8.3 BIG BACTERIA AND BACTERIAL CYTOSKELETON

Large cell size is not a monopoly of eukaryotic cells. Large (>20 μm) prokaryotic spheroidal cells are known for cyanobacteria, sulfur bacteria, and gut-symbiont heterotrophic bacteria (*Epulopiscium fishelsoni*). Spheroidal cyanobacteria such as *Dermocarpella* and *Stanieria* (Waterbury and Stanier 1978; Angert 2005; Schirrmeister et al. 2016) can be 20 μm or more in diameter (Figure 8.5.16d). However, such a large size is measured for mother cells containing multiple baeocytes inside; these cells reproduce by multiple fission (Angert 2005). Large spheroidal (and ellipsoidal) cells are also found for some genera of sulfur bacteria such as *Thiomargarita, Achromatium*, and *Thiovulum* (Schulz et al. 1999; Schulz and Jørgensen 2001). *Thiomargarita namibiensis*, for example, can reach 750 μm in diameter and can form chains (Figure 8.5.16e); however, it should be noted that their cytoplasm exists only peripherally and their cells are volumetrically dominated by a vacuole, and in the case of *Achromatium*, cells are filled with many large calcite inclusions (Schulz et al. 1999; Schulz and Jørgensen 2001). Vacuoles and inclusions in such sulfur bacteria reduce the volume of active cytoplasm, allowing metabolisms dependent on molecular diffusion. Additionally, a newly identified taxon of sulfur bacteria (*Ca. Thiomargarita nelsonii*) includes specimens that form a colony of cells surrounded by a "rigid" envelop up to ~90 μm in diameter (Salman et al. 2011, 2013).

The cytoskeleton has long been specific to eukaryotic cells, but in the last decade, homologs of elements of the eukaryotic cytoskeletal have been identified in prokaryotes (Shih and Rothfield 2006; Celler et al. 2013), and their involvement in cell growth, cell morphogenesis, cell division, DNA partitioning, and cell motility were implicated. Furthermore, new elements specific to prokaryotes have been identified. Such facts caution us against making an employment of criteria for eukaryotic affinity related to cytoskeletons described earlier; the functions of bacterial cytoskeltons have not yet been fully revealed. Obviously, these facts should be considered in discussing biological affinities of large early Archean microfossils.

8.5.8.4 FRAMEWORK FOR CONSIDERING BIOLOGICAL AFFINITY OF LARGE MICROFOSSILS

Considering the inevitable incompleteness involved in the criteria for classifying eukaryotic cells in previous sections, it is difficult at this stage to assign large Archean microfossils specifically to eukaryotes or prokaryotes. Their biological affinities are equivocal and could be cyanobacteria, sulfur bacteria or eukaryotes, and potentially other extinct lineages. Nevertheless, the following points would be helpful for further examination of large Archean microfossils.

8.5.8.4.1 Early Evolution of Oxygenic Photosynthesis

It is evident that shallow to subaerial habitats had already been explored by microbes by 3.0 Ga. But, what are the implications of this? The most straightforward speculation is that microbes with the ability to photosynthesize had already evolved and diversified, as suggested by, for example, Tice and Lowe (2004, 2006a, 2006b). The timing of the evolution of oxygen-producing photoautotrophs is controversial, but it could be back before the Great Oxidation Event (GOE) and even to 3.0 Ga and earlier (e.g., Rosing 1999; Nisbet and Sleep 2001; Rosing and Frei 2004; Buick 2008; Crowe et al. 2013; Lyons et al., 2014; Homann et al. 2015; Schirrmeister et al. 2016; Delarue et al. 2018). The evolution of oxygen-producing photoautotrophs earlier than 3.0 Ga is not necessarily in conflict with the widely accepted scenario that atmospheric oxygen began to increase around 2.45 Ga (the GOE). Oxygen produced by photosynthesis in aquatic environments was not necessarily released to the atmosphere but could have been consumed by heterotrophs and by reductants such as Fe^{2+}. Also, for example, in a closed to semi-closed basin, where reductants could have been consumed, local enrichment in O_2 could be possible (Oxygen Oasis; e.g., Olson et al. 2013). Even in the "fully oxygenated" modern Earth, redox conditions are highly variable. As exemplified by thermally stratified water masses such as in dam reservoirs, oxygenic conditions can be closely associated with anoxygenic conditions (Bellanger et al. 2004). Plausibly, local oxygenic environments could have developed on the early Earth, which might have provided opportunities for the early evolution of eukaryotes.

8.5.8.4.2 Constraints from Life Cycle and Lifestyle

It is essential to demonstrate the stage of the life cycle at which a large microfossil is preserved. If a large microfossil represents a juvenile stage, but not aged cells with vacuoles or baeocytes, the cell size could give some constraints on the biological affinity. Sugitani and co-workers (2015a, 2018), for example, demonstrated that lenticular microfossils of the Strelley Pool Formation represent vegetative cells that reproduce simply by binary fission. This is inferred from the presence of a series of paired specimens corresponding to various stages of binary fission, from fused to point-to-point attachment, and also from that colonial clusters of lenticular microfossils do not contain other morphological types. Sugitani and co-authors (2015a) also estimated the cell volume of lenticular microfossils to be 7000 μm³, assuming that they have an ellipsoidal shape with half axes of 15, 15, and 7.5 μm. This volume corresponds to a spheroid with a ~23 μm diameter. Sugitani and co-workers (2009b) also suggested that some of the large spheroidal microfossils from the 3.0 Ga Farrel Quartzite might represent mother cells containing multiple daughter cells (baeocyte) inside, such as some cyanobacteria, and others might represent vesicles containing possible endospore (Figure 8.5.14f–h). It was implied that some of the lenticular microfossils from the Farrel Quartzite reproduced by multiple fissions (Sugitani 2012). Consequently, it should not be overlooked that the significance of the size of microfossils changes, depending on the stage of life cycle preserved. Also, House and co-workers (2013) and Oehler and co-workers (2017) suggested that Meso- and Paleoarchean lenticular microfossils were planktonic, based on their stable carbon isotopic compositions (determined at the micron scale by SIMS) as well as their characteristic shape with wing-like appendages (flanges) that are thought to be advantageous, perhaps, for floating. Such an assumed life cycle could also place some constraints on the physiology of the microfossils.

8.5.8.4.3   Constraints from Ecological Niches

Assumed ecological niches could place some constraint on the biological affinity of large Archean microfossils. Javaux and co-workers (2010), for example, considered the biological affinity of organic-walled large spheroidal microfossils from the Moodies Group of South Africa, comparing these to extant large prokaryotes, the symbiont *Epulopiscium* sp., myxobacteria sporangioles, *Thiomargarita namibiensis*, and some cyanobacteria. These candidates, except cyanobacteria, were discarded, as their ecological niches are nutrient-rich gut, soil, and sulfur-rich environments, distinct from that expected for the Moodies microbes. Furthermore, they have no ability to produce a recalcitrant envelope. Javaux and co-workers (2010) suggested the possible affinity of the Moodies large spheroidal microfossils to cyanobacteria, as cyanobacteria could produce large cysts and envelopes, and the Moodies microbes were thought to have inhabited the photic zone.

8.5.8.4.4   Harsh Surface Environments of the Archean

Surface environments of the Archean Earth are presumably quite different to those of the Phanerozoic and even the Proterozoic periods. Thus, these environments are discussed in the section, related to large Archean microfossils. Walsh (1992), for example, implied large spheroids and ellipsoids discovered from the Kaapvaal Craton to be spores that had a great advantage on Archean volcanic platforms, where microbes were subjected to violent volcanisms, and would have grown within limited periods of volcanic quiescence. Additionally, the formation of spores might have been indispensable for microbes to survive through repeated asteroid impacts during the early Archean (Byerly et al. 2002; Lowe et al. 2003, 2014; Lowe and Byerly 2015). Oehler and co-workers (2017) also emphasized that Paleo- and Mesoarchean lenticular microfossils with robust walls, as shown by their thick and acid-resistant nature, had advantages for surviving on a young planet experiencing strong ultraviolet (UV) radiation (Catling and Claire 2005), as well as sudden environmental changes such as heating and even evaporation of ocean surface water, crustal fracturing, and extremely high energy waves (tsunamis) caused by large asteroid impacts (Lowe et al. 2003; Lowe 2013; Lowe and Byerly 2015).

8.5.8.4.5   Morphological Comparison with Younger Equivalents

Comparative study with morphologically similar microfossils from younger successions and even extant microbes would be an effective approach to elucidate the biological affinity and phylogeny; however, morphological similarity should not be considered diagnostically. Sugitani and co-workers (2007, 2009a, 2009b, 2015a), for example, have suggested that lenticular microfossils are morphologically similar to mature phycoma with a flange-like ala of an extant prasinophyte alga, *Pterosperma* (Parke et al. 1978; Tappan 1980). Strikingly well-preserved microfossils equivalent to phycoma of *Pterosperma* were reported from the Lower Devonian Rhynie chert (Kustatscher et al. 2014). Though biological affinities are not well known, spheroidal acritarchs with flange are also present in Meso- to Neoproterozoic successions (Samuelsson 1997; Samuelsson et al. 1999; Vorob'eva et al. 2015; Sergeev et al. 2016) (Figure 8.5.16f).

8.5.9   LESSONS FROM PREVIOUS STUDIES AND CONTROVERSIES ON ARCHEAN MICROFOSSILS

As exemplified by controversies on the biogenicity of some described microstructures from cherts in the Apex Basalt, Mount Ada Basalt, and the Isua Supracrustal Belt, studies of Archean microbe-like structures are frequently subjected to rigorous criticisms. Criticism of Archean microfossils has arisen mainly from the geologic histories of host rocks being unsuitable for preserving traces of life. Controversies are complicated when analyses are unable to be made on exactly the same material (as in the case of the Apex microstructures and possibly *Isuasphere*), and they can also be complicated when the three-dimensional structure of an object under study is not correctly assessed. These previous studies and controversies provide some lessons for Archean micropaleontology and the search for life on other planets:

1. Geological context is viewed as important, more than necessary. To prove the biogenicity of microbe-like objects in rocks collected from a region that has been subjected to severe deformation, metasomatism and metamorphism require tremendous lines of evidence.
2. Analyses to help determine the biogenicity of fossil-like objects should be made on exactly identical materials. Otherwise, irrelevant arguments and confusions may occur in scientific society. It should also be reminded that micro-scale objects are heterogeneously distributed in rocks, and thus, even hand specimens collected from, for example, a 30-cm-thick unit, do not always contain exactly the same objects. Also, the presence of morphologically similar well-defined pseudofossils should not be taken directly as counter-evidence for the biogenicity of the structures in question; however, it should

also not be overlooked. Such logic may mislead scientific societies and induce barren controversies (see Section 8.5.6.4.1).

3. Criteria for biogenicity need to be applied appropriately. For example, DDI and narrow size distribution are well-defined criteria. However, these criteria require a prerequisite that microbes reproduced simply by binary fission. As described in this chapter, Archean microbes had already established various reproductive styles and complex life cycles, including multiple fissions and endospore formation. Even if a negative result is obtained from these criteria, the biogenicity of microbe-like structures under consideration should not be readily discarded.

4. Systematic criticism and skepticism, which can be applied to one's own materials as well as others', are indispensable for studies of Archean microbe-like structures that are less-well-preserved compared with younger microfossils. The requirement of excessive analyses to confirm the biogenicity based on the "null-hypothesis approach," which in many cases can be performed only at limited institutions, also restricts sound progress of Archean paleobiology and astrobiology.

8.5.10 SUMMARY

This chapter describes representative early (Paleo- and Meso-) Archean microfossils and possible microfossils, referring in particular to surface environments of the ancient Earth, rock types potentially preserving microfossils, and various criteria for identifying Archean microbe-like structures as microfossils. To date, the oldest possible microfossils are recently reported hematite tubes and filaments from at least 3,770-million and potentially 4,280-million-year-old rocks in the Nuvvuagittuq Belt, Canada; however, their biogenicity is awaiting confirmation through further studies. In either case, it is evident from the cellularly preserved microfossils and possible microfossils that until 3.0 Ga, microbes of various morphologies and various metabolisms had already emerged and inhabited a range of environments, from deep-sea hydrothermal vent systems to shallow water and even terrestrial hydrothermal fields. Inhabitation in shallow to sub-aerial habitats provides indirect evidence for the early evolution of photosynthesis. Anoxygenic photosynthesis could have initiated as early as 3.4 Ga and oxygenic photosynthesis as early as 3.0 Ga. Thus, recent studies have provided evidence for the diversification of microbes during the first billion years since the emergence of life around 4.0 Ga and, at the same time, have revealed the presence of large, organic-walled lenticular microfossils in the Archean that appear to be out of place in an evolutionary context. However, the eukaryotic features of these lenticular microfossils should not be taken as reason for questioning their biogenicity. Discoveries of large prokaryotes with complex shapes have been accumulating. Also, there is no guarantee that evolution progressed linearly, especially in the Archean. Thus, the possibility cannot entirely

be excluded that the peculiar, large, and lenticular microbes represent extinct taxon, without phylogenetic relationship to any extant organisms. Despite the difficulties in studying Archean microfossils, in comparison with Proterozoic ones, it is worth to challenge exploring biosignatures in such deep time on Earth. As a result of controversies surrounding newly discovered materials, as well as revisions of previously reported materials, sophisticated methodologies with which we can identify cellularly preserved microfossils have been developed. These new technologies could also be applied to the search for life on other planets. Finally, the possibility cannot be eliminated that large microfossils similar to those described in detail here may be discovered in future Mars Missions, considering "…., compared to early Earth, early Mars might have had a greater supply of biologically useable energy and was perhaps, by implication, a better place for the origin of life" (Kirschvink et al. 2001).

ACKNOWLEDGMENT

I deeply acknowledge Dr. Vera Kolb for her invitation to this significant contribution and for editorial handling and Dr. Dorothy Z. Oehler for her critical and constructive comments to the earlier draft. Special thanks to Erica Barlow, who did detailed English editing and helped in finalizing the manuscript.

REFERENCES

Altermann, W. and Kazmierczak, J., 2003. Archean microfossils: A reappraisal of early life on Earth. *Res. Microbiol.* 154: 611–617.

Alleon, J., Bernard, S., Le Guillou, C., Beyssac, O., Sugitani, K., and Robert, F., 2018. Chemical nature of the 3.4 Ga Strelley Pool microfossils. *Geochem. Persp. Lett.* 7: 37–42.

Allwood, A.C., Kamber, B.S., Walter, M.R., Burch, I.W., and Kanik, I., 2010. Trace elements record depositional history of an Early Archean stromatolitic carbonate platform. *Chem. Geol.* 270: 148–163.

Allwood, A.C., Walter, M.R., Kamber, B.S., Marshall, C.P., and Burch, I.W., 2006. Stromatolite reef from the Early Archaean era of Australia. *Nature* 441: 714–718.

Allwood, A.C., Burch, I., and Walter, M.R., 2007. Stratigraphy and facies of the 3.43 Ga Strelley Pool Chert in the southwestern North Pole Dome, Pilbara Craton, Western Australia. *Geol. Surv. West. Aust.*, Record 2007/11.

Angert, E.R., 2005. Alternatives to binary fission in bacteria. *Nature Rev. Microbiol.* 3: 214–224.

Appel, P.W.U., Moorbath, S., and Myers, J.S., 2003. Isuasphaera isua (Pflug) revisited. *Precam. Res.* 126: 309–312.

Awramik, S.M., Schopf, J.W., and Walter, M.R., 1983. Filamentous fossil bacteria from the Archean of Western Australia. *Precam. Res.* 20: 357–374.

Awramik, S.M., Schopf, J.W., and Walter, M.R., 1988. Carbonaceous filaments from North Pole, Western Australia: Are they fossil bacteria in Archaean stromatolites? A discussion. *Precam. Res.* 39: 303–309.

Barghoorn, E.S. and Tyler, S.A., 1965. Microorganisms from the Gunflint Chert. *Science* 147: 563–575.

Barghoorn, E.S. and Schopf, J.W., 1966. Microorganisms three billion years old from the Precambrian of South Africa. *Science* 152: 758–763.

Bellanger, B., Huon, S., Steinmann, P., Chabaux, F., Velasquez, F., Vallès, V., Arn, K., Clauer, N., and Mariotti, A., 2004. Oxic-anoxic conditions in the water column of a tropical freshwater reservoir (Peña-Larga dam, NW Venezuela). *Appl. Geochem.* 19: 1295–1314.

Berry, A.J., Danyushevsky, L.V., O'Neil, H. St C., Newville, M., and Sutton, S.R., 2008. Oxidation state of iron in komatiitic melt inclusions indicates hot Archaean mantle. *Nature* 45: 960–963.

Blake, R.E., Chang, S.J., and Lapland, A., 2010. Phosphate oxygen isotopic evidence for a temperate and biologically active Archaean ocean. *Nature* 464: 1029–1032.

Bower, D.M., Steele, A., Fries, M.D., Green, O.R., and Lindsay, J.F., 2016. Raman imaging spectroscopy of a putative microfossil from the ~3.46 Ga Apex chert: Insights from quartz grain orientation. *Astrobiology* 16: 169–180.

Brasier, M.D., Antcliffe, J., Saunders, M., and Wacey, D., 2015. Changing the picture of Earth's earliest fossils (3.5-1.9 Ga) with new approaches and new discoveries. *Proc. Nat. Acad. Sci. USA* 112: 4859–4864.

Brasier, M.D., Green, O.R., Jephcoat, A.P., Kleppe, A.K., Van Kranendonk, M.J., Lindsay, J.F., Steele, A., and Grassineau, N.V., 2002. Questioning the evidence for Earth's oldest fossils. *Nature* 416: 76–81.

Brasier, M.D., Green, O.R., Lindsay, J.F., McLoughlin, N., Steele, A., and Stoakes, C., 2005. Critical testing of Earth's oldest putative fossil assemblage from the ~3.5 Ga Apex chert, Chinaman Creek, Western Australia. *Precam. Res.* 140: 55–102.

Brasier, M., McLoughlin, N., Green, O., and Wacey, D., 2006. A fresh look at the fossil evidence for early Archaean cellular life. *Phil. Trans. Roy. Soc.* B361: 887–902.

Bridgwater, D., Allaart, J.H., Schopf, J.W., Klein, C., Walter, M.R., Barghoorn, E.S., Strother, P., Knoll, A.H., and Gorman, B. E., 1981. Microfossil-like objects from the Archaean of Greenland: A cautionary note. *Nature* 289: 51–53.

Brooks, J., Muir, M.D., and Shaw, G., 1973. Chemistry and morphology of Precambrian microorganisms. *Nature* 244: 215–217.

Buick, R., 1984. Carbonaceous filaments from North Pole, Western Australia: Are they fossil bacteria in Archaean stromatolites? *Precam. Res.* 24: 157–172.

Buick, R., 1988. Carbonaceous filaments from North Pole, Western Australia: Are they fossil bacteria in Archaean stromatolites? A reply. *Precam. Res.* 39: 311–317.

Buick, R., 1990. Microfossil recognition in Archean rocks: An appraisal of spheroids and filaments from a 3500 m.y. old chert-barite unit at North Pole, Western Australia. *Palaios* 5: 441–459.

Buick, R., 2008. When did oxygenic photosynthesis evolve? *Phil. Trans. Roy. Soc.* B363: 2731–2743.

Buick, R. and Dunlop, J.S.R., 1990. Evaporitic sediments of Early Archaean age from the Warrawoona Group, North Pole, Western Australia. *Sedimentology* 37: 247–277.

Butterfield, N.J., 2014. Early evolution of the Eukaryota. *Palaeontology* 58: 5–17.

Butterfield, N.J., 2005. Probable Proterozoic fungi. *Paleobiology* 31: 165–182.

Byerly, G.R., Lowe, D.R., Wooden, J.L., and Xie, X., 2002. An Archean impact layer from the Pilbara and Kaapvaal Cratons. *Science* 297: 1325–1327.

Catling, D.C. and Clairem M.W., 2005. How Earth's atmosphere evolved to an oxic state: A status report. *Earth Planet. Sci. Lett.* 237: 1–20.

Celler, K., Koning, R.I., Koster, A.J., and van Wezel, G.P., 2013. Multidimensional view of the bacterial cytoskeleton. *J. Bacteriol.* 195: 1627–1636.

Cloud Jr., P.E., 1965. Significance of the Gunflint (Precambrian) microflora. *Science* 148: 27–35.

Crowe, S.A., Døssing, L.N., Beukes, N.J., Bau, M., Kruger, S.J., Frei, R., and Canfield, D.E., 2013. Atmospheric oxygenation three billion years ago. *Nature* 501: 535–538.

De Gregorio, B.T., Sharp, T.G., Flynn, G.J. Wirick, S., and Hervig, R.L., 2009. Biogenic origin for Earth's oldest putative microfossils. *Geology* 37: 631–634.

Delarue, F., Robert, F., Sugitani, K., Tartèse, R., Duhamel, R., and Derenne, S., 2018. Nitrogen isotope signatures of microfossils suggest aerobic metabolism 3.0 Gyr ago. *Geochem. Persp. Lett.* 7: 32–36.

Derenne, S., Robert, F., Skrzypczak-Bonduelle, A., Gourier, D., Binet, L., and Rouzaud, J.-N., 2008. Molecular evidence for life in the 3.5 billion year old Warrawoona chert. *Earth Planet. Sci. Lett.* 272: 476–480.

Douglas, S., 2005. Mineralogical footprints of microbial life. *Am. J. Sci.* 305: 503–525.

Dodd, M.S., Papineau, D., Grenne, T., Slack, J.F., Rittner, M., Pirajno, F., O'Neil, J., and Little, C.T.S. 2017. Evidence for early life in Earth's oldest hydrothermal vent precipitates. *Nature* 543: 60–64.

Duchač, K.C. and Hanor, J.S., 1987. Origin and timing of the metasomatic silicification and of an early Archean komatiite sequence, Barberton Mountain Land, South Africa. *Precam. Res.* 37: 125–146.

Duck, L.J., Glikson, M., Golding, S.D., and Webb, R.E., 2007. Microbial remains and other carbonaceous forms from the 3.24 Ga Sulphur Springs black smoker deposit, Western Australia. *Precam. Res.* 154: 205–220.

Dunlop, J.S.R., Milne, V.A., Groves, D.I., and Muir, M.D., 1978. A new microfossil assemblage from the Archaean of Western Australia. *Nature* 274: 676–678.

El Albani, A., Bengston, S., Canfield, D.E. et al., 2010. Large colonial organisms with coordinated growth in oxygenated environments 2.1 Gyr ago. *Nature* 466: 100–104.

Engel, A.E.J., Nagy, B., Nagy, L.A., Engel, C.G., Kremp, G.O.W., and Drew, C.M., 1968. Alga-like forms in Onverwacht Series, South Africa: Oldest recognized lifelike forms on Earth. *Science* 161: 1005–1008.

Eriksson, K.A. and Simpson, E.L., 2000. Quantifying the oldest tidal record: The 3.2 Ga Moodies Group, Barberton Greenstone Belt, South Africa. *Geology* 28: 831–834.

Fedo, C.M., Whitehouse, M.J., and Kamber, B.S., 2006. Geological constraints on detecting the earliest life on Earth: A perspective from the Ealry Archaean (older than 3.7 Gyr) of southwest Greenland. *Phil. Trans. Roy. Soc. Lond.* B361: 851–867.

García-Ruiz, J.M., Carnerup, A., Christy, A.G., Welham, N.J., and Hyde, S.T., 2004. Morphology: An ambiguous indicator of biogenicity. *Astrobiology* 2: 353–369.

Glikson, M., Duck, L.J., Golding, S.D., Hofmann, A., Bolhar, R., Webb, R., Baiano, J.C.F., and Sly, L.I., 2008. Micorbial remains in some earliest Earth rocks: Comparison with a potential modern analogue. *Precam. Res.* 164: 187–200.

Glikson, A., Hickman, A., Evans, N.J., Kirkland, C.L., Park, J.-W., Rapp, R., and Romano, S., 2016. A new ~3.46 Ga asteroid impact ejecta unit at Marble Bar, Pilbara Craton, Western Australia: A petrological, microprobe and laser ablation ICPMS study. *Precam. Res.* 279: 103–122.

Grey, K., Roberts, F.I., Freeman, M.J., Hickman, A.H., Van Kranendonk, M.J., and Bevan, A.W.R., 2010. Management plan for State Geoheritage Reserves. *Geol. Sur. West. Aust.* Record 2010/13, 23p.

Grey, K. and Sugitani, K., 2009. Palynology of Archean microfossils (c. 3.0 Ga) from the Mount Grant area, Pilbara Craton, Western Australia: Further evidence of biogenicity. *Precam. Res.* 173: 60–69.

Habicht, K.S., Gade, M., Thamdrup, B., Berg, P., and Canfield, D.E., 2002. Calibration of sulfate levels in the Archean ocean. *Science* 298: 2372–2374.

Han, T.M. and Runnegar, B., 1992. Megascopic eukaryotic algae from the 2.1-billion-year-old Negaunee Iron-Formation, Michigan. *Science* 257: 232–235.

Heubeck, C., 2009. An early ecosystem of Archean tidal microbial mats (Moodies Group, South Africa, ca. 3.2 Ga). *Geology* 37: 931–934.

Heubeck, C. and Lowe, D.R., 1994. Depositional and tectonic setting of the Archean Moodies Group. Barberton greenstone belt, South Africa. *Precam. Res.* 68: 257–290.

Hein, J.R., Clague, D.A., Koski, R.A., Embley, R.W., and Duham, R.E., 2008. Metalliferous sediment and a silica-hematite deposit within the Blanco Fracture Zone, Northeast Pacific. *Mar. Geores. Geotechnol.* 26: 317–339.

Heubeck, C., Engelhardt, J., Byerly, G.R., Zeh, A., Sell, B., Luber, T., and Lowe, D.R., 2013. Timing of deposition and deformation of the Moodies Group (Barberton Greenstone Belt, South Africa): Very-high-resolution of Archaean surface processes. *Precam. Res.* 231: 236–262.

Hickman, A.H., 2008. Regional review of the 3426-3350 Ma Strelley Pool Formation, Pilbara Craton, Western Australia. *Geol. Surv. West. Aust.* Record 2008/15.

Hickman, A.H., 2012. Review of the Pilbara Craton and Fortescue Basin, Western Australia: Crustal evolution providing environments for early life. *Island Arc* 21: 1–31.

Hickman, A.H. and Van Kranendonk, M.J., 2008. Archean crustal evolution and mineralization of the northern Pilbara Craton – A field guide. *Geol. Surv. West. Aust.* Record 2008/13.

Hofmann, H.J., 2004. Archean microfossils and abiomorphs. *Astrobiology* 4: 135–136.

Homann, M., Heubeck, C., Airo, A., and Tice, M.M., 2015. Morphological adaptations of 3.22 Ga-old tufted microbial mats to Archean coastal habitats (Moodies Group, Barberton Greenstone Belt, South Africa). *Precam. Res.* 266: 47–64.

Homann, M., Heubeck, C., Bontognali, T.R.R., Bouvier, A.-S., Baumgartner, L.P., and Airo, A. (2016). Evidence for cavity-dwelling microbial life in 3.22 Ga tidal deposits. *Geology* 44: 51–54.

House, C.H., Oehler, D.Z., Sugitani, K., and Mimura, K., 2013. Carbon isotopic analyses of ca. 3.0 Ga microstructures imply planktonic autotrophs inhabited Earth's early oceans. *Geology* 41: 651–654.

Hren, M.T., Tice, M.M., and Chamberlain, C.P., 2009. Oxygen and hydrogen isotope evidence for a temperate climate 3.42 billion years ago. *Nature* 462: 205–208.

Javaux, E.J., Marshall, C.P., and Bekker, A., 2010. Organic-walled microfossils in 3.2-billion-year-old shallow-marine siliciclastic deposits. *Nature* 463: 934–938.

Javaux, E.J., Knoll A.H., and Walter M.R., 2003. Recognizing and interpreting the fossils of early eukaryotes. *Orig. Life Evol. Biosphere* 33: 75–94.

Javaux, E.J., Knoll, A.H., and Walter, M.R., 2004. TEM evidence for eukaryotic diversity in mid-Proterozoic oceans. *Geobiology* 2: 121–132.

Kappler, A., Pasquero, C., Konhauser, K.O., and Newman, D.K., 2005. Deposition of banded iron formations by anoxygenic phototrophic Fe(II)-oxidizing bacteria. *Geology* 33: 865–868.

Kaźmierczak, J., Kremer, B., Altermann, W., and Franchi, I., 2016. Tubular microfossils from ~2.8 to ~2.7 Ga –old lacustrine deposits of South Africa: A sign for early origin of eukaryotes? *Precam. Res.* 286: 180–194.

Kirschvink, J.L., and Weiss, B.P., 2001. Mars, Panspermia, and the origin of life: Where did it all begin? *Palaeontol. Electr.* 4: 8–15.

Kiyokawa, S., Ito, T., Ikehara, M., and Kitajima, F., 2006. Middle Archean volcano-hydrothermal sequence: Bacterial microfossil-bearing 3.2 Ga Dixon Island Formation, coastal Pilbara terrane, Australia. *Geol. Soc. Am. Bull.* 118: 3–22.

Kiyokawa, S., Koge, S., Ito, T., and Ikehara, M., 2014. An ocean-floor carbonaceous sedimentary sequence in the 3.2-Ga Dixon Island Formation, coastal Pilbara terrane, Western Australia. *Precam. Res.* 255: 124–143.

Knauth, L.P. and Lowe, D.R., 2003. High Archean climatic temperature inferred from oxygen isotope geochemistry of cherts in the 3.5 Ga Swaziland Supergroup, South Africa. *Geol. Soc. Am. Bull.* 115: 566–580.

Knoll, A.H., 2014. Paleobiological perspectives on early eukaryotic evolution. *Cold Spring Harb. Perspec. Biol.* doi:10.1101/cshperspect.a016121.

Knoll, A.H. and Barghoorn, E.S., 1974. Ambient pyrite in Precambrian chert: New evidence and a theory. *Proc. Nat. Acad. Sci.* USA 71: 2329–2331.

Knoll, A.H. and Barghoorn, E.S., 1977. Archean microfossils showing cell division from the Swaziland System of South Africa. *Science* 198: 396–398.

Knoll, A.H., Javaux, E.J., Hewitt, D., and Cohen, P., 2006. Eukaryotic organisms in Proterozoic oceans. *Phil. Trans Roy. Soc.* B361: 1023–1038.

Kremer, B. and Kaźmierczak, J., 2017. Cellularly preserved microbial fossils from ~3.4 Ga deposits of South Africa: A testimony of early appearance of oxygenic life? *Precam. Res* 295: 117–129.

Kustatscher, E., Dotzler, N., Taylor, T.N., and Krings, M., 2014. Microfossils with suggested affinities to the Pyramimonadales (Pyramimonadophyceae Chlorophyta) from the Lower Devonian Rhynie chert. *Acta Palaeobot.* 54: 163–171.

Lepot, K., Williford, K.H., Ushikubo, T., Sugitani, K., Mimura, K., Spicuzza, M.J., and Valley, J.W., 2013. Texture-specific isotopic compositions in 3.4 Gyr old organic matter support selective preservation in cell-like structures. *Geochim. Cosmochim. Acta* 112: 66–86.

Lepot, K., Benzerara, K., and Philippot, P., 2011. Biogenic versus metamorphic origins of diverse microtubes in 2.7 Gyr old volcanic ashes: Multi-scale investigations. *Earth Planet. Sci. Lett.* 312: 37–47.

Li, J., Zhou, H., Peng, X., Wu, Z., Chen, S., and Fang, J., 2012. Microbial diversity and biomineralization in low-temperature hydrothermal iron-silica-rich precipitates of the Lau Basin hydrothermal field. *FEMS Microbiol. Ecol.* 81: 205–216.

Little, C.T.S., Herrington, R.J., Haymon, R.M., and Danelian, T., 1999. Early Jurassic hydrothermal vent community from the Franciscan Complex, San Rafael Mountains, *California*. *Geology* 27: 167–170.

Lowe, D.R., 2013. Crustal fracturing and chert dike formation triggered by large meteorite impacts, ca. 3.260 Ga, Barberton greenstone belt, South Africa. *Geol. Soc. Am. Bull.* 125: 894–912.

Lowe, D.R., 1999. Geologic evolution of the Barberton greenstone belt and vicinity. *Geol. Soc. Am. Spec. Pap.* 329: 287–312.

Lowe, D.R., Byerly, G.R., Kyte, F.T., Shukolyukov, A., Asaro, F., and Krull, A., 2003. Spherule beds 3.47-3.24 billion years old in the Barberton greenstone belt, South Africa: A record of large meteorite impacts and their influence on early crustal and biological evolution. *Astrobiology* 3: 7–48.

Lowe, D.R. and Byerly, G.R., 2015. Geologic record of partial ocean evaporation triggered by giant asteroid impacts, 3.29-3.23 billion years ago. *Geology* 43: 535–538.

Lowe, D.R., Byerly, G.R., and Kyte, F.T., 2014. Recently discovered 3.42-3.23 Ga impact layers, Barberton Belt, South Africa: 3.8 Ga detrital zircons, Archean impact history, and tectonic implications. *Geology* 42: 747–750.

Lyons, T.W., Reinhard, C.T., and Planavsky, N.J., 2014. The rise of oxygen in Earth's early ocean and atmosphere. *Nature* 506: 307–315.

Maliva, R.G., Knoll, A.H., and Simonson, B.M., 2005. Secular change in the Precambrian silica cycle: Insights from chert petrology. *Geol. Soc. Am. Bull.* 117: 835–845.

Marshall, C.P., Emry, J.R., and Marshall, A.O., 2011. Haematite pseudomicrofossils present in the 3.5-billion-year-old Apex Chert. *Nat. Geosci.* 4: 240–243.

Marshall, A.O., Emry, J.R., and Marshall, C.P., 2012. Multiple generations of carbon in the Apex Chert and implications for preservation of microfossils. *Astrobiology* 12: 160–166.

McCollum, T.M. and Seewald, J.S., 2006.Carbon isotope composition of organic compounds produced by abiotic synthesis under hydrothermal conditions. *Earth Planet. Sci. Lett.* 243: 74–84.

McLoughlin, N., Brasier, M.D., Wacey, D., Green, O.R., and Perry, R.S., 2007. On biogenicity criteria for endolithic microborings on early Earth and beyond. *Astrobiology* 7: 10–26.

Morag, N., Williford, K.H., Kitajima, K., Philippot, P., Van Kranendonk, M.J., Lepot, K., Thomazo, C., and Valley, J.W., 2016. Microstructure-specific carbon isotopic signatures of organic matter from ~3.5 Ga cherts of the Pilbara Craton support a biologic origin. *Precam. Res.* 275: 429–449.

Moczydłowska M., Landing, E., Zang, W., and Palacios, T., 2011. Proterozoic phytoplankton and timing of Chlorophyte algae origins. *Paleontology* 54: 721–733.

Muir, M.D. and Hall, D.O., 1974. Diverse microfossils in Precambrian Onverwacht group rocks of South Africa. *Nature* 252: 376–378.

Nagy, B. and Nagy, L.A., 1969. Early Pre-Cambrian Onverwacht microstructures: Possibly the oldest fossil on Earth? *Nature* 223: 1226–1229.

Nisbet, E.G. and Sleep, N.H., 2001. The habitat and nature of early life. *Nature* 409: 1083–1091.

Noffke, N., 2010. *Geobiology: Microbial Mats in Sandy Deposits from the Archean era to Today.* Berlin, Germany: Springer.

Noffke, N., Christian, D., Wacey, D., and Hazen, R.M., 2013. Microbially induced sedimentary structures recording an ancient ecosystem in the ca. 3.48 billion-year-old Dresser Formation, Pilbara, Western Australia. *Astrobiology* 13: 1103–1124.

Oberlin, A., 1984. Carbonization and graphitization. *Carbon* 22: 521–541.

Oehler, J.H., 1977. Microflora of the H.Y.C. pyritic shale member of the Barney Creek Formation (McArthur Group), middle Proterozoic of northern Australia. *Alcheringa* 1: 315–349.

Oehler, J.H., 1976. Hydrothermal crystallization of silica gel. *Geol. Soc. Am. Bull.* 87: 1143–1152.

Oehler, D.Z., 1978. Microflora of the middle Proterozoic Balbirini Dolomite (McArthur Group) of Australia. *Alcheringa* 2: 269–309.

Oehler, D.Z. and Cady, S.L., 2014. Biogenicity and syngeneity of organic matter in ancient sedimentary rocks: Recent advances in the search for evidence of past life. *Challenges* 5: 260–283.

Oehler, J.H. and Schopf, J.W., 1971. Artificial microfossils: Experimental studies of permineralization of blue-green algae in silica. *Science* 174: 1229–1231.

Oehler, D.Z., Robert, F., Mostefaoui, S., Meibom, A., Selo, M., and McKay, D.S., 2006. Chemical mapping of Proterozoic organic matter at submicron spatial resolution. *Astrobiology* 6: 838–850.

Oehler, D.Z., Robert, F., Walter, M.R., Sugitani, K., Allwood, A., Meibom, A., Mostefaoui, S., Selo, M., Thomen, A., and Gibson, E.K., 2009. NanoSIMS: Insights to biogenicity and syngeneity of Archaean carbonaceous structures. *Precam. Res.* 173: 70–78.

Oehler, D.Z., Robert, F., Walter, M.R., Sugitani, K., Meibom, A., Mostefaoui, S. and Gibson, E.K., 2010. Diversity in the Archaean biosphere: New insights from NanoSIMS. *Astrobiology* 10: 413–424.

Oehler, D.Z., Walsh, M.M., Sugitani, K., Liu, M.-C., and House, C.H., 2017. Large and robust lenticular microorganisms on the young Earth. *Precam. Res.* 296: 112–119.

Olson, S.L., Kump, L.R., and Kasting, J, F., 2013. Quantifying the areal extent and dissolved concentrations of Archean oxygen oasis. *Chem. Geol.* 362: 35–43.

Parke, M., Boalch, G.T., Jowett, R., and Harbour, D.S., 1978. The Genus Pterosperma (Prasinophyceae): Species with a single equatorial ala. *J. Mar. Biol. Assoc. UK* 58: 239–276.

Pflug, H.D., 1967. Structured organic remains from the Fig Tree Series (Precambrian) of the Barberton Mountain Land (South Africa). *Rev. Palaeobot. Palynol.* 5: 9–29.

Pflug, H.D., 1978. Yeast-like microfossils detected in oldest sediments of the earth. *Naturwissenschaften* 65: 611–615.

Pflug, H.D. and Jaeschke-Boyer, H., 1979. Combined structural and chemical analysis of 3,800-Myr-old microfossils. *Nature* 280: 483–486.

Pinti, D.L., Mineau, R., and Clement, V., 2009. Hydrothermal alteration and microfossil artefacts of the 3,465-million-year-old Apex chert. *Nat. Geosci.* 2: 640–643.

Posth, N.R., Konhauser, K.O., and Kappler, A., 2013. Microbiological processes in banded iron formation deposition. *Sedimentology* 60: 1733–1754.

Rasmussen, B., 2000. Filamentous microfossils in a 3,235-million-year-old volcanogenic massive sulphide deposit. *Nature* 405: 676–679.

Retallack, G.J., Krinsley, D.H., Fischer, R., Razink, J.J., and Langworthy, K.A., 2016. Archean coastal-plain paleosols and life on land. *Gondwana Res.* 40: 1–20.

Rosing, M.T., 1999. ^{13}C-depleted carbon microparticles in >3700-Ma sea-floor sedimentary rocks from West Greenland. *Science* 283: 674–676.

Robert., F. and Chaussidon, M.A., 2006. A palaeotemperature curve for the Precambrian oceans based on silicon isotopes in cherts. *Nature* 443: 969–972.

Roedder, E. 1981. Are the 3,800-Myr-old Isua objects microfossils, limonite-stained fluid inclusions, or neither? *Nature* 293: 459–462.

Rosing, M.T. and Frei, R., 2004. U-rich Archaean sea-floor sediments from Greenland – indications of > 3700 Ma oxygenic photosynthesis. *Earth Planet. Sci. Lett.* 217: 237–244.

Salman, V., Amann, R., Girnth, A.-C., Polerecky, L., Bailey, J.V., Høgslund, S., Jessen, G., Pantoja, S., and Schult-Vogt, H.N., 2011. A single-cell sequencing approach to the classification of large, vacuolated sulfur bacteria. *Syst. Appl. Microbiol.* 34: 243–259.

Salman, V., Bailey, J.V., and Teske, A., 2013. Phylogenetic and morphologic complexity of giant sulphur bacteria. *Antonie van Leeuwenhoek* 104: 169–189.

Samuelsson, J., 1997. Biostratigraphy and palaeobiology of Early Neoproterozoic strata of the Kola Peninsula, Northwest Russia. *Norsk Geologisk Tidsskrift* 77: 165–192.

Samuelsson, J., Dawes, P.R., and Vidal, G., 1999. Organic-walled microfossils from the Proterozoic Thule Supergroup, Northwest Greenland. *Precam. Res.* 96: 1–23.

Schirrmeister, B.E., Sanchez-Baracaldo, P., and Wacey, D., 2016. Cyanobacterial evolution during the Precambrian. *International Journal of Astrobiology* 15: 187–204.

Schopf, J.W., 1976. Are the oldest 'fossils', fossils? *Orig. Life* 7: 19–36.

Schopf, J.W., 1993. Microfossils of the Early Archean Apex Chert: New evidence of the antiquity of life. *Science* 260: 640–646.

Schopf, J.W., 2006. Fossil evidence of Archaean life. *Phil. Trans. Roy. Soc. Lond.* B361: 869–885.

Schopf, J.W. and Barghoorn, E.S., 1967. Algal-like fossils from the Early Precambrian of South Africa. *Science* 156: 508–512.

Schopf, J.W. and Kudryavtsev, A.B., 2009. Confocal laser scanning microscopy and Raman imagery of ancient microscopic fossils. *Precam. Res.* 173: 39–49.

Schopf, J.W. and Kudryavtsev, A.B., 2011. Biogenicity of Apex Chert microstructures. *Nature Geosci.* 4: 346–347.

Schopf, J.W., Kitajima, K., Spicuzza, M.J., Kudryavtsev, A.B., and Valley J.W., 2018. SIMS analyses of the oldest known assemblage of microfossils document their taxon-correlated carbon isotope compositions. *Proc. Nat. Acad. Sci. USA* 115: 53–58.

Schopf, J.W., Kudryavtsev, A.B., Walter, M.R., Van Kranendonk, M.J., Williford, K.H., Kozdon, R., Valley, J.W., Gallardo, V.A., Espinoza, C., and Flannery, D.T., 2015. Sulfur-cycling fossil bacteria from the 1.8-Ga Duck Creek Formation provide promising evidence of evolution's null hypothesis. *Proc. Nat. Acad. Sci. USA* 112: 2087–2092.

Schopf, J.W., Kudryavtsev, A.B., Czaja, A.D., Tripathi, A.B., 2007. Evidence of Archean life: Stromatolites and microfossils. *Precam. Res.* 158, 141–155.

Schopf, J.W. and Packer, B.M., 1987. Early Archean (3.3-billion to 3.5-billion-year-old) microfossils from Warrawoona Group, Australia. *Science* 237: 70–73.

Schopf, J.W. and Walter, M.R., 1983. Archean microfossils: New evidence of ancient microbes. In *Earth's Earliest Biosphere, Its Origin and Evolution*, J.W. Schopf (Ed.), pp. 214–239. Princeton, NJ: Princeton University Press.

Schopf, J.W., Kudryavtsev, A.B., Osterhout, J.T., Williford, K.H., Kitajima, K., Valley, J.W., and Sugitani, K., 2017. An anaerobic ~3400 Ma shallow-water microbial consortium: Presumptive evidence of Earth's Paleoarchean anoxic atmosphere. *Precam. Res.* 299: 309–318.

Schopf, J.W., Kudryavtsev, A.B., Sugitani, K., and Walter, M.R., 2010. Precambrian microbe-like pseudofossils: A promising solution to the problem. *Precam. Res.* 179: 191–205.

Schröder, S., Bekker, A., Beukes, N.J., Strauss, H., and van Niekerk, H.S., 2008. Rise in seawater sulphate concentration associated with the Paleoproterozoic positive carbon isotope excursion: Evidence from sulphate evaporites in the ~2.2-2.1 Gyr shallow-marine Lucknow Formation, South Africa. *Terra Nova* 20: 108–117.

Schulz, H.N., Brinkhoff, T., Ferdelman, T.G., Mariné, M.H., Teske, A., and Jorgensen, B.B., 1999. Dense populations of a giant sulfur bacterium in Namibian shelf sediments. *Science* 284: 493–495.

Schulz, H.N. and Jørgensen, B.B., 2001. Big Bacteria. *Ann. Rev. Microbiol.* 55: 105-137.

Sergeev, V.N., Knoll, A.H., Vorob'eva, N.G., and Sergeeva, N.D., 2016. Microfossils from the lower Mesoproterozoic Kaltasy Formation, East Europian Platform. *Precam. Res.* 278: 87–107.

Sforna, M.C., van Zuilen, M.A., and Philippot, P., 2014. Structural characterization by Raman hyperspectral mapping of organic carbon in the 3.46 billion-year-old Apex chert, Western Australia. *Geochim. Cosmochim. Acta* 124: 18–33.

Shih, Y.-L. and Rothfield, L., 2006. The bacterial cytoskeleton. *Microbiol. Molec. Biol. Rev.* 70: 729–754.

Shibuya, T., Komiya, T., Nakamura, K., Takai, K., and Maruyama, S., 2010. Highly alkaline, high-temperature hydrothermal fluids in the early Archean ocean. *Precam. Res.* 182: 230–238.

Siever, R., 1992. The silica cycle in the Precambrian. *Geochim. Cosmochim. Acta* 56: 3265–3272.

Sugahara, H., Sugitani, K., Mimura, K., Yamashita, F., and Yamamoto, K., 2010. A systematic rare-earth elements and yttrium study of Archean cherts at the Mount Goldsworthy greenstone belt in the Pilbara Craton: Implications for the origin of microfossil-bearing black cherts. *Precam. Res.* 177: 73–87.

Sugitani, K., 1992. Geochemical characteristics of Archean cherts and other sedimentary rocks in the Pilbara Block, Western Australia: Evidence for Archean seawater enriched in hydrothermally-derived iron and silica. *Precam. Res.* 57: 21–47.

Sugitani, K., 2012. Life cycle and taxonomy of Archean flanged microfossils from the Pilbara Craton, Western Australia. *34th International Geological Congress (IGC): AUSTRALIA 2012*, 17.3, #257.

Sugitani, K., Grey, K., Allwood, A.C., Nagaoka, T., Mimura, K., Minami, M., Marshall, C.P., Van Kranendonk, M.J., and Walter, M.R., 2007. Diverse microstructures from Archaean chert from the Mount Goldsworthy – Mount Grant area, Pilbara Craton, Western Australia: Microfossils, dubiomicrofossils, or pseudofossils? *Precam. Res.* 158: 228–262.

Sugitani, K., Grey, K., Nagaoka, T., and Mimura, K., 2009a. Three-dimensional morphological and textural complexity of Archean putative microfossils form the northeastern Pilbara Craton: Indications of biogenicity of large (>15µm) spheroidal and spindle-like structures. *Astrobiology* 9: 603–615.

Sugitani, K., Grey, K., Nagaoka, T., Mimura, K., and Walter, M.R., 2009b. Taxonomy and biogenicity of Archaean spheroidal microfossils (ca. 3.0 Ga) from the Mount Goldsworthy-Mount Grant area in the northeastern Pilbara Craton, Western Australia. *Precam. Res.* 173: 50–59.

Sugitani, K., Kohama, T., Mimura, K., Takeuchi, M., Senda, R., and Morimoto, H., 2018. Speciation of Paleoarchean life demonstrated by analysis of the morphological variation of lenticular microfossils from the Pilbara Craton, Australia. *Astrobiology* 18: 1057–1070.

Sugitani, K., Lepot, K., Nagaoka, T., Mimura, K., Van Kranendonk, M., Oehler, D.Z., and Walter, M.R., 2010. Biogenicity of morphologically diverse carbonaceous microstructures from the ca. 3400 Ma Strelley Pool Formation, in the Pilbara Craton, Western Australia. *Astrobiology* 10: 899–920.

Sugitani, K., Mimura, K., Nagaoka, T., Lepot, K., and Takeuchi, M., 2013. Microfossil assemblage from the 3400Ma Strelley Pool Formation in the Pilbara Craton, Western Australia: Results from a new locality. *Precam. Res.* 226: 59–74.

Sugitani, K., Mimura, K., Suzuki, K., Nagamine, K., and Sugisaki, R., 2003. Stratigraphy and sedimentary petrology of an Archean volcanic-sedimentary succession at Mt. Goldsworthy in the Pilbara Block, Western Australia: Implications of evaporite (nahcolite) and barite deposition. *Precam. Res.* 120: 55–79.

Sugitani, K., Mimura, K., Takeuchi, M., Lepot, K., Ito, S., and Javaux, E.J., 2015a. Early evolution of large micro-organisms with cytological complexity revealed by microanalyses of 3.4 Ga organic-walled microfossils. *Geobiology* 13: 507–521.

Sugitani, K., Mimura, K., Takeuchi, M., Yamaguchi, T., Suzuki, K., Senda, R., Asahara, Y., Wallis, S., and Van Kranendonk, M.J., 2015b. A Paleoarchean coastal hydrothermal field inhabited by diverse microbial communities: The Strelley Pool Formation, Pilbara Craton, Western Australia. *Geobiology* 13: 522–545.

Sugitani, K., Van Kranendonk, M.J., Oehler, D.Z., House, C.H., and Walter, M.R., 2017. Comment: Archean coastal-plain paleo-sols and life on land. *Gondwana Res.* 44: 265–269.

Sugitani, K., Yamamoto, K., Adachi, M., Kawabe, I., and Sugisaki, R., 1998. Archean cherts derived from chemical, biogenic and clastic sedimentation in a shallow restricted basin: Examples from the Gorge Creek Group in the Pilbara Block. *Sedimentology* 45: 1045–1062.

Tappan, H., 1980. *The Paleobiology of Plant Protists*. San Francisco, CA: W.H. Freeman and Co.

Tartèse, R., Chaussidon, M., Gurenko, A., Delarue, F., and Robert, F., 2017. Warm Archean oceans reconstructed from oxygen iso-tope composition of early-life remnants, *Geochem. Perspec. Lett.* 3: 55–65,

Tice, M.M. and Lowe, D.R., 2004. Photosynthetic microbial mats in the 3,416-Myr-old-ocean. *Nature* 431: 549–552.

Tice, M.M. and Lowe, D.R., 2006a. The origin of carbonaceous matter in pre-3.0 Ga greenstone terraines: A review and new evidence from the 3.42 Ga Buck Reef Chert. *Earth-Sci. Rev.* 76: 259–300.

Tice, M.M. and Lowe, D.R., 2006b. Hydrogen-based carbon fixation in the earliest known photosynthetic organisms. *Geology* 34: 37–40.

Ueno, Y., Isozaki, Y., Yurimoto, H. and Maruyama, S., 2001a. Carbon isotopic signatures of individual Archean microfossils (?) from Western Australia. *Int. Geol. Rev.* 43: 196–212.

Ueno, Y., Maruyama, S., Isozaki, Y. and Yurimoto, H., 2001b. Early Archean (ca. 3.5 Ga) microfossils and ^{13}C-depleted carbo-naceous matter in the North Pole area, Western Australia: Field occurrence and geochemistry. In *Geochemistry and the Origin of Life*, S. Nakashima et al. (Eds.), pp. 203–236.Tokyo, Japan: Universal Academy Press.

Ueno, Y., Yamada, K., Yoshioka, N., Maruyama, S. and Isozaki, Y., 2006. Evidence from fluid inclusions for microbial methano-genesis in the early Archaean era. *Nature* 440: 516–519.

Van Kranendonk, M.J., 2006. Volcanic degassing, hydrothermal cir-culation and the flourishing of early life on Earth: A review of the evidence from c. 3490-3240 Ma rocks of the Pilbara Supergroup, Pilbara Craton, Western Australia. *Earth-Sci. Rev.* 74: 197–240.

Van Kranendonk, M.J., Hickman, A.H., Smithies, R.H., Nelson, D.N., and Pike, G., 2002. Geology and tectonic evolution of the Archaean North Pilbara terrain, Pilbara Craton, Western Australia. *Econ. Geol.* 97: 695–732.

Van Kranendonk, M.J., Hickman, A.H., Smithies, R.H., Williams, I.R., Bagas, L. and Farrell, T.R., 2006. Revised lithostratig-raphy of Archean supracrustal and intrusive rocks in the northern Pilbara Craton, Western Australia. *West. Aust. Geol. Sur.*Record 2006/15.

Van Kranendonk, M.J., Philippot, P., Lepot, K., Bodorkos, S., and Pirajno, F., 2008. Geological setting of Earth's oldest fossils in the ca. 3.5 Ga Dresser Formation, Pilbara Craton, Western Australia. *Precam. Res.* 167: 93–124.

Van Kranendonk, M.J. and Pirajno, F., 2004. Geological setting and geochemistry of metabasalts and alteration zones associated with hydrothermal chert ± barite deposits in the ca. 3.45 Ga Warrawoona Group, Pilbara Craton, Australia. *Geochem. Explor. Environ. Anal.* 4: 253–278.

Vearncombe, S., Barely, M.E., Groves, D.I., McNaughton, N.J., Mikuchki, E.J., and Vearncombe, J.R., 1995. 3.26 Ga black smoker type mineralization in the Strelley Pool Belt, Pilbara Craton, Western Australia. *J. Geol. Soc. Lond.* 152: 587–590.

Vorob'eva, N.G., Sergeeve, V.N., and Petrov, P.Y., 2015. Koutikan Formation assemblage: A diverse organic-walled microbiota in the Mesoproterozoic Anabar succession, northern Siberia. *Precam. Res.* 256: 201–222.

Wacey, D., 2009. *Early Life on Earth: A practical guide*. Heidelberg, Germany: Springer.

Wacey, D., 2010. Stromatolites in the ~3400 Ma Strelley Pool Formation, Western Australia: Examining biogenicity from the macro- to the nano-scale. *Astrobiology* 10: 381–395.

Wacey, D., 2012. Earliest evidence for life on Earth: An Australian perspective. *Aust. J. Earth Sci.* 59: 153–166.

Wacey, D., Kilburn, M.R., Saunders, M., Cliff, J., and Brasier, M.D., 2011. Microfossils of sulphur-metabolizing cells in 3.4-billion-year-old rocks of Western Australia. *Nat. Geosci.* 4: 698–702.

Wacey, D., McLoughlin, N., Whitehouse, M.J., and Kilburn, M.R., 2014a. Two coexisting sulfur metabolisms in a ca. 3400 Ma sandstone. *Geology* 38: 1115–1118.

Wacey, D., Noffke, N., Saunders, M., Guagliardo, P., and Pyle, D.M., 2018a. Volcanogenic pseudo-fossils from the ~3.48 Ga Dresser Formation, Pilbara, Western Australia. *Astrobiology* 18: 539–555.

Wacey, D., Saunders, M., and Kong, C., 2018b. Remarkably preserved tephra from the 3430 Ma Strelley Pool Formation, Western Australia: Implications for the interpretation of Precambrian microfossils. *Earth Planet. Sci. Lett.* 487: 33–43.

Wacey, D., Saunders, M., Cliff, J., Kilburn, M.R., Kong, C., Barley, M.E., and Brasier, M.D., 2014b. Geochemistry and nano-structure of a putative ~3240 million-year-old black smoker biota, Sulphur Springs Group, Western Australia. *Precam. Res.* 249: 1–12.

Walsh, M.M., 1992. Microfossils and possible microfossils from early Archean Onverwacht Group, Barberton Mountain Land, South Africa. *Precam. Res.* 54: 271–293.

Walsh, M.M. and Lowe, D.R., 1985. Filamentous microfossils from the 3,500-Myr-old Onverwacht Group, Barberton Mountain Land, South Africa. *Nature* 314: 530–532.

Walter, M.R., ed. 1976. *Stromatolite*. Amsterdam, the Netherlands: Elsevier, 789 pp.

Waterbury, J.B. and Stanier, R.Y., 1978. Patterns of growth and development in pleurocapsalean cyanobacteria. *Microbiol. Rev.* 42: 2–44.

Westall, F., de Vries, S.T., Nijman, W. et al., 2006. The 3.446 Ga "Kitty's Gap Chert", an early Archean microbial ecosystem. *Geol. Soc. Am Bul. Spec. Pap.* 405: 105–131.

Westall, F., de Wit, M.J., Dann, J., van der Gaast, S., de Ronde, C.E.J., and Gerneke, D., 2001. Early Archean fossil bacteria and biofilms in hydrothermally-influenced sediments from the Barberton greenstone belt, South Africa. *Precam. Res.* 106: 93–116.

Westall, F., Foucher, F., Cavalazzi, B. et al., 2011. Volcaniclastic habitats for early life on Earth and Mars: A case study from ~3.5 Ga-old rocks from the Pilbara, Australia. *Planet. Space Sci.* 59: 468–479.

Williford, K.H., Ushikubo, T., Lepot, K., Kitajima, K., Hallmann, C., Spicuzza, M.J., Kozdon, R., Eigenbrode, J.L., Summons, R.E., and Valley, J.W., 2016. Carbon and sulfur isotopic signatures of ancient life and environment at the microbial scale: Neoarchean shales and carbonates. *Geobiology* 14: 105–128.

8.6 Biomarkers and Their Raman Spectral Signatures

An Analytical Challenge in Astrobiology

Howell G.M. Edwards

CONTENTS

8.6.1 INTRODUCTION

The prime target for two forthcoming space missions, the European Space Agency (ESA)/Roscosmos' ExoMars 2020 and the National Aeronautics and Space Administration (NASA's) Mars 2020, is to acquire indisputable evidence for life signatures on another planet (Mars) and to assess whether these have arisen from extant or extinct sources. However, this simple statement generates two very important and directly relevant philosophical questions: firstly, how do we define life in the first place, and, secondly, how would we recognize its presence from chemical signatures, especially from the degraded chemical residues remaining from extinct organisms in the planetary geological record, by using remote analytical instrumentation? In this exercise, we should recognize the possibility that any extraterrestrial organism identified in future space exploration could have originated on Earth and may have been transported to our planetary neighbors in the solar system either by our own intervention, reinforcing the need for the most stringent planetary protection protocols for our spacecraft and landers, or through natural *panspermia* processes, which could include the deposition of chemical building bricks through delivery by meteorites, comets, and asteroids. The precise definition of "life" itself is actually rather elusive, and many initial attempts to do so have been eventually deemed to be not very satisfactory (Cleland and Chyba 2000; Shapiro and Schulze-Makuch 2009; Bedau 2010; Benner 2010; Tirard et al. 2010). After much debate, the NASA definition of life simply as "a self-sustaining system capable of Darwinian evolution"

incorporates a molecular genesis with replicative procedures and avoids several pitfalls of alternative definitions, which are based solely upon the perceived ability of the biological system to reproduce. Whilst the attempts to define life have been addressed comprehensively, both philosophically and scientifically, the scientific basis of the environmental requirements necessary to sustain emergent life have been summarized: the presence of an energy source, however derived; the presence of liquid water; and the presence of elements carbon, hydrogen, nitrogen, oxygen, phosphorus, and sulfur (the so-called CHNOPS elements) in readily assimilable forms are all seen to be primary and necessary candidates for the successful establishment and subsequent evolution of life. However, this statement masks several factors that can conspire to restrict or defeat the establishment of the ongoing generation of life, such as the presence of extremes of temperature; desiccation; damaging insolation by low wavelength; high-energy radiation in the ultraviolet region of the electromagnetic spectrum; and environmental exposure to highly reactive chemical toxins, which can destroy bioorganic molecules, such as peroxide ions, hydroxyl radicals, and perchlorates, in addition to the compromising coordination effects of heavy metal ions.

To counter the potentially serious life-destroying scenarios of these adverse environmental conditions, terrestrial organisms such as cyanobacteria, endolithic, chasmolithic, and epilithic lichens have developed sophisticated biochemical strategies, which basically involve the harnessing of their geological substrates in niche environments where life can not only exist but multiply and evolve. The production of unique suites of chemical protectants designed to combat a range of external stresses in hostile environments has been recognized as being critically important for the survival of cyanobacterial species in stressed terrestrial conditions (Cockell and Knowland 1999; Holder et al. 2000; Wynn-Williams and Edwards 2000a, b, 2002; Edwards 2010). This factor is actually the primary objective and the key to the analytical chemical search for life experiments that have been proposed for the first phase specialized and dedicated planetary rover Raman spectroscopic instrumentation that will be described here for the ExoMars 2020 mission (Rull et al. 2017; Vago et al. 2017). Hence, the analytical chemical perspective for a remote "search for life" astrobiological mission centres upon the ability to detect and recognize in a niche geological matrix the characteristic spectral biosignatures from these protective biomolecules, henceforth known as *biomarkers*. At this stage, we should also attempt to differentiate between scientific missions that search for evidence of life-sustaining water—however necessary that would be perceived in its astrobiological role—and those that are truly searching for extant life signals or for the residual surviving biomarkers where life has now become extinct. Where life has ceased to exist in a particularly hostile environment, the analytical detection of these surviving biosignatures is paramount, and this could be achieved primarily through the observation of signals from the preserved biomolecules and secondarily from the modifications made to the geological and mineral matrix by the biological colonies as part of their survival strategies (Wynn-Williams et al. 1999; Wynn-Williams and Edwards 2000a, b, 2002; Edwards 2010).

8.6.2 THE ROLE OF RAMAN SPECTROSCOPY IN LIFE DETECTION

Since the first reports in the literature of the successful Raman spectroscopic analyses of the biological colonization of toxic mineral pigments in Renaissance frescoes (Edwards et al. 1991a, 1991b) and that of an endolithic colonization of Beacon Sandstone from the Antarctic Peninsula (Edwards et al. 1997a, 1997b), the analytical chemical interrogation of extremophilic systems situated within a variety of geological substrates has resulted in novel information being provided about the nature of the synthetic protective chemicals and the strategies being employed for survival of the colonies in terrestrial stressed environments by using this technique. At first, the identification of the chemical complexes formed as a result of the reaction of epilithic lichen metabolic waste products upon calcareous substrates in the form of oxalates was a primary objective of the Raman spectroscopic analysis (Edwards et al. 1994, 1998, 2003a, 2003b, 2003c, 2004; Russ et al. 1995; Holder et al. 2000; Edwards 2004). It was not until much later that the characterization of the organic metabolic and protective by-products in epiliths and cryptoendoliths was accomplished through the adoption of laser excitation, using a range of wavelengths, and comparison with the spectra derived from laboratory extracted materials (Russell et al. 1998; Edwards et al. 2000, 2003d, 2003e; Jorge-Villar et al. 2003; Jorge-Villar and Edwards 2005), which had been characterised using a combination of

structural analytical techniques. As noted previously, the successful survival of extremophilic colonies in stressed terrestrial environments is dependent upon their adaptation to the prevailing conditions of high or low temperatures; extreme desiccation; high-energy ultraviolet insolation; high or low barometric pressures; extremes of pH, in the range from <1 to >12; and the presence of toxic chemicals and ions such as mercury (II), antimony (III), lead (II), barium (II), arsenic (III), and copper (II), which are found in many minerals and rocks. It was also apparent that the production of protective chemicals by extremophiles as a response to environmental stresses was supported by their ability to adapt their geological matrices and substrates, often resulting in modified biogeological signatures, which have been shown to remain in the geological record long after the extinction of the biological colonies.

The detection of the presence of life in terrestrial extreme scenarios such as oceanic vents, meteorite impact craters; glacial icefields; sabkhas; salterns; and desiccated hot and cold deserts such as the Antarctic, Atacama, Negev, and Mojave deserts, have revealed key molecular entities such as carotenoids, scytonemin, phycocyanins, and metabolic products of colonies, which are considered indubitably to be biosignatures. The key point here is that these biochemicals have been synthesized by biological organisms for their protection and survival in conditions of extreme desiccation, pressure, temperature ranges, chemical toxicity, and radiation insolation, so, indeed, they can be considered as "signs of life" for a such a forensic analytical investigation. What now needs to be considered for the transition to remote planetary studies are the following questions:

- The environmental conditions that potential life forms are going to experience on our neighbouring planets and their satellites are expected to be significantly more severe than we can find on Earth. What is the implication for the survival of biological colonies on Mars, for example?
- In the Darwinian scheme of *adaptation for survival*, would the survivors have created new molecules that represent a range of novel biosignatures, which we may not recognize from terrestrial studies undertaken hitherto?
- If life did develop extraterrestrially, would the worsening of environmental conditions result in its immediate extinction or would the adaptation for surface survival, given time, have taken a new course, such as the colonization of subsurface strata to escape surface radiation, coupled with the requisite geological host matrix modifications—and how could we then access these sites remotely?
- If life was initiated and was then extinguished, would there be a biosignature left in the planetary geological record, which could be interrogated successfully by remote investigative instrumentation?
- What would be the effect of exposure of the potential biomarkers created by now-extinct organisms to the toxic chemicals and radiation that could degrade them molecularly? And then, would such degraded

entities be recognized as descendants of their true original biomarker molecules? For example, the chemical degradation of biomolecular complex-ring systems such as polyaromatic hydrocarbons and heterocyclic porphyrins to simple fused-ring aromatic hydrocarbons and purines, and eventually even to elemental carbon, might result in chemically stressed environments.

- What type of remote analytical instrumentation would be best for the securing of unambiguous evidence of molecular biomarkers in the geological record that would be recognized as being indicative of extinct or extant life on the planetary surface or subsurface?

It is in this context that we should now explore the advantages that Raman spectroscopy can offer to the interrogation of specimens of biogeological matrices in search for life experiments. Basically, these can be summarized as follows:

- Raman spectroscopy involves the focussing of a monochromatic laser beam on a sample in the ultraviolet, visible, and near-infrared regions of the electromagnetic spectrum between about 300-nm and 100-nm wavelength. Whereas most of the radiation scattered is in the form of unchanged frequency (Rayleigh scattering), the resultant Raman spectrum arises from the much weaker intensity Stokes and anti-Stokes scattered radiations, which occur with changed frequencies, shifted between about 0 cm^{-1} and 3500 cm^{-1} from the incident or Rayleigh frequency, which form a characteristic Raman spectrum comprising vibrational bands occurring in this wavenumber region. The most important analytical information arising from these frequency-shifted bands is that each chemical species (molecule or molecular ion) has a distinctive and unique Raman series of Raman spectral bands, which can be effectively utilized to identify it from a "molecular spectroscopic fingerprint." The Raman bands arise from molecular moieties or entities found within the specimen, such as organic amide, carbonyl, aromatic rings, and aliphatic carbon-carbon unsaturation, and this applies equally well to inorganic ions found in minerals, such as sulphate, carbonate, nitrate, and metal oxide vibrations (Long 2002).
- The Raman spectrum between 0 cm^{-1} and 3500 cm^{-1} contains complementary structural information arising from mixtures of both inorganic and organic components in the same specimen, without having to effect chemical or mechanical methods of separation—making it ideal and quite specific for the interrogation of biogeological specimens that have been described earlier.
- The Raman spectroscopic examination of a specimen can be effected microscopically through the marriage of one of the oldest scientific pieces of

equipment, the microscope, and one of the newest, the laser, operating over a range of monochromatic wavelengths. In this way, depending upon the wavelength of the incident laser beam, the "footprint" at the sample can be of the order of only several microns up to about 100 microns; the high values of the incident power density (irradiance) possible for the focussed laser at the specimen should be realised for the smallest footprints, being greater than 1 MW cm^{-2}, which could be extremely damaging for sensitive materials. The practising Raman spectroscopist must therefore evaluate the desirability of using the highest level of incident laser irradiance to produce the strongest observable Raman effect, against the likely possibility of inflicting thermal damage upon the specimen. Normally, therefore, one errs on the side of safety and uses well-tried laser power levels and microscope magnifications from existing laboratory work with similar samples, and this has proved to be a trustworthy *modus operandi* for examination of biogeological specimens, even though mineral samples are found to be quite robust to higher irradiance levels of laser illumination.

- The Raman spectral data obtained from the laser illumination of samples generally comes from the surface regions, unless particular transparency to the incoming laser radiation in mineral crystals is noted, when subsurface sampling can be made to the depth of a few centimetres or more. In this way, the Raman spectra of biological colonies of cyanobacteria that exist some 5 cm below the surface of clear selenite crystals have been reported *in situ* and without fracture of the surrounding crystal matrix (Edwards et al. 2005).

8.6.3 WHY RAMAN SPECTROSCOPY?

It was appreciated at the outset that a major advantage of Raman spectroscopy as an analytical technique for the study of extremophilic colonization and survival was its ability to interrogate a system microscopically, across a horizontal or vertical transect, without any physical separation or chemical and mechanical pretreatments such as coating, grinding, and polishing (Long 2002). Hence, information about the interaction between the biological and geological components is accessible naturally, without extraction or modification of the biogeological system. This was appreciated in the analysis of works of art that had been subjected to lichen deterioration and required conservation (Edwards and Chalmers 2005).

The recognition of key definitive Raman spectral signatures in the same spectrum from the organic and inorganic components is a desirable outcome for the projected deployment of a miniaturized spectrometer unit as part of the life detection instrumentation suite on the planetary rover vehicle on the ESA/Roscosmos ExoMars mission scheduled for 2020 (Rull et al. 2017; Vago et al. 2017). In this mission, the Raman spectrometer will be a first-pass probe of selected samples from the Martian surface and subsurface (down to a depth of 2 metres below the surface) for the detection of extremophilic spectral signatures from extinct or extant colonies in niche environments. The assimilation of Raman spectral data and signatures into a database relating to extremophiles from terrestrial Mars analogue sites is therefore of vital importance, and this has been undertaken in recent years and is still ongoing. Similarly, Raman spectroscopic analyses of other extreme terrestrial sites that have relevance to the remote exploration of other planets and their satellites in our solar system, such as glaciers, deep-sea smokers, thermal vents, hot geysers, and snowfields, have been reported. These extreme terrestrial sites have an ambience with planetary icy moons such as Europa, Titan, Io, and Enceladus, which are also noteworthy in this respect.

The purpose of the present chapter is to survey and critically examine comprehensively the existing wealth of Raman spectral data and their resultant interpretation for terrestrial extremophiles; from this analysis, the following deductions can then be made:

- The commonality of Raman spectral signatures between various sites and extreme environments (Jorge-Villar and Edwards 2005).
- The factors that affect the recording of Raman spectral signatures from the extremophilic colonization of terrestrial sites (Edwards et al. 2003a, 2003c).
- The definition of Raman data for evidence of extinct or extant life in the geological record.

8.6.4 THE ANALYTICAL CHALLENGE FOR REMOTE SPECTROSCOPY

The analytical challenge facing Martian search for life extraterrestrial experiments involving remote spectroscopic instrumentation can be summarized as follows:

1. Can the spectral biosignatures that we recognize and have assimilated into a definitive database from terrestrial Mars analogue studies be assumed to be chemically identical? It has already been explained that the terrestrial biosignatures derive from suites of biochemical that have been synthesized by extremophilic organisms and bacterial colonies in response to external environmental conditions and stimuli. It is generally accepted that the so-called terrestrial Mars analogue sites approximate to only those conditions that can be expected on Mars itself—it can be stated categorically that nowhere on Earth do the hostile conditions pertaining on Mars match: the high insolation of ultraviolet and gamma radiations, extremely oxidising chemical toxicity in the form of perchlorates and free radicals, deficiency of nutrients, very low temperatures (about −150°C) and daily temperature range of around 170°C, and, of course, water/ice and water vapor deficiencies. Given

this apparent new range of external influences, we should expect chemical adaptation to have perhaps sought novel directions to enhance survival, with consequent new biosignatures that are not present in terrestrial extremophilic organisms. Having said this, it has been reasonably proposed that it would be impossible to expect organic molecules to survive on Mars because of the low-wavelength ionising radiation insolation and the presence of chemical species such as hydroxyl radicals, hydrogen peroxide, and perchlorates. Despite this, the recent detection of organic molecules, such as chlorobenzenes, in the Sheepbed Mudstone at Gale Crater by the NASA rover, albeit in trace concentrations, gives hope for their survival (Freissinet et al. 2015).

2. Extant *versus* extinct life: can Raman spectroscopy differentiate between the molecular signatures in living systems and those signatures from once-living systems that have not survived in their hostile environment? This argument can be reduced to a simple, rather philosophical consideration: the successful Raman spectral detection of molecules in a search for life experiment is dependent on the molecular composition of the target, which is the same for living and non-living systems. Of course, in living systems, the possible temporal degradation of molecular biomarkers is balanced by their continuous replacement, whereas the environmental destruction of the same biomarkers in an abiotic system will eventually result in their disappearance. The term "biomarkers" should not therefore be assumed to have a biotic significance—these molecules are not "living," and therefore, their detection should not be assumed to be an unambiguous indicator of extant life.

3. A major problem facing the remote spectroscopic detection of biomarkers in a Martian niche environment is the survivability of these from long-dead colonies: this can be rephrased as: how long can reactive chemical species such as carotenoids, scytonemin, phycocyanins, sugars, and chlorophylls survive the conditions of extreme temperatures, exposure to ionising radiation, and chemical toxicity on Mars, even in protected niche environments? This will be considered in the next section.

8.6.5 THE GREAT ENIGMA: HOW WILL WE RECOGNIZE THE PRESENCE OF EXTANT OR EXTINCT LIFE USING REMOTE ANALYTICAL INSTRUMENTATION?

In our discourse, we have considered the necessity of essential adaptation of a geological niche in otherwise-hostile environments for the stabilization and survival strategy of a living organism in early planetary scenarios. For extinct life, which once comprised perhaps cyanobacterial colonies, we are faced with two problems in recognition: one of these is the detection of unambiguous chemical signatures, which were produced during their existence, and the second is the survival of these chemical entities in recognisable quantities and forms after the death of the host colonies. We have considered the former earlier in this article, but the latter is an unknown parameter, which itself creates a further question: what is the reality of survival of the intact chemical biosignatures over periods of geological time? Current plans for search for life missions to the planet Mars, for example, represented by the NASA's Mars 2020 and ESA/Roscosmos' ExoMars 2020 missions, target geological areas and rocks from the earliest period in Martian history, known as the Noachian, extending from about 4,600 million years ago to approximately 3,500 million years ago and perhaps extending into the Hesperian period to about 3000 million years ago, when it is believed that Mars had a climate not too dissimilar from that of Earth and had significant liquid water on its surface in the form of Martian oceans. What should be appreciated is that rocks from the latest period in Martian geological history, the Amazonian, from 1800 million years ago to the present, are not considered conducive to the preservation of life signatures, because of the intense bombardment of the Martian surface by meteorites and volcanic activity and the loss of the Martian atmosphere caused by a collision with other large bodies in the solar system in the later Hesperian period. A possible location for spectral signatures for residual extant or extinct life, therefore, would be the subsurface interrogation of potential niche environments (Ellery et al. 2003, 2004). The ExoMars mission is specifically targeting this aspect with an onboard drill that will be capable of accessing and retrieving specimens from 2 metres below the Martian surface. Biosignatures in the subsurface environment, protected from the chemical toxicity present on the surface of Mars, arising from the intense ionising radiation, will be additionally preserved from deteriorative low-wavelength radiation effects owing to iron (III) oxide on the surface (Clark 1998).

We are thus faced with a very real problem in the use of remote analytical instrumentation, geared up to the detection of extraterrestrial biosignatures from extinct life, in that we need to consider how the pristine biochemical entities can be preserved in their geological environment and also how they degrade (Westall et al. 2015). Thus far, we have considered only the chemical identification of life signatures: the other possibility is the observation of residual fossilized shapes and structures in the geological record, which can be indicative of once-extant life forms, the geomorphological signature. The presence of fossilised bacterial signatures in the terrestrial geological record, along with associated minerals, such as pyrites, haematite, pyrolusite, and magnetite, gives credence to their recognition extraterrestrially by light microscopy: the first exciting release of the discovery of morphological shapes characteristic of fossilized bacteria in Martian geology came for the Martian meteorite found in 1984 in the Allen Hills in Antarctica, ALH 84001, dating from about 18 million years ago (McKay et al. 1996). After much controversial debate, this apparently conclusive evidence for life on Mars was questioned and was filed as only a possibility, as the seemingly definitive structures attributed to fossilized cellular nanobacteria some 20–100 nm or so in

length could possibly have been produced by the methodology of production of the thin sample of meteorite taken for petrological examination and its subsequent polishing process. ALH 84001, an Shergotty-Nahkla-Chassigny (SNC) class meteorite of mass 1.93 kg, was believed to have been ejected into space from the *Valles Marineris* region of Mars by meteorite impact, was captured by Earth's gravity, and was found in Antarctica on December 27, 1984. The recent announcement in *Nature* (Dodd et al. 2017) of the morphological evidence for terrestrial fossilized bacterial structures dating back to 4220 million years ago at Nuvvuagittuq in the Canadian Arctic has revived this interest. It has also moved the concept of a start-up of life on Earth some 500 million years earlier than was once thought and now to within some 400 million years of our own planet's creation. However, this still begs the question as to whether or not biological signatures would have survived this long: in this context, a specimen from the ancient Australian site at Pilbara showed evidence for chemical biosignatures, using laboratory Raman spectroscopic equipment, in 3600-million-year-old geological deposits (Pullan et al. 2008). The spectral bands for carotenoids and porphyrin key biomarkers were observed together from a small 10-micron sample footprint encased in the ancient stromatolites. Other regions of the same sample similarly gave signatures of carotenoids, scytonemin, and chlorophyll in various admixtures. A possible explanation of the appearance of these biomarkers in such an ancient specimen was that they could have resulted from later cyanobacterial colonization of the stromatolites, which, of course, challenges any idea that these biosignatures could be commensurate with the geological age of the sediments. It is clear that what is now needed is a combined analytical study of a geomorphological signature to examine the presence of chemical biosignatures, which could match the visual record; this could establish the reality of assessment of survival of such biosignatures over an extended geological timeframe appropriate to the planned survey of the Martian geological record, and similarly, their detection would inform the attribution of a biological origin to what otherwise may be doubtful structures, as in the case of ALH 84001. As Garcia-Ruiz (1999) has correctly stated:

Morphology alone cannot be used unambiguously as a tool for primitive life detection.... the interpretation of morphology is notoriously subjective.

8.6.6 KEY BIOMARKERS AND THEIR RAMAN SPECTROSCOPIC SIGNATURES

The Raman spectra of the key biomarkers and minerals relevant to Mars can be broadly classified into the following wavenumber ranges: 3200–3450 cm^{-1}, containing the characteristic stretching modes of OH and NH groups; 2800–3150 cm^{-1}, containing the characteristic CH stretching modes of aromatic and aliphatic organic compounds; 1500–1650 cm^{-1}, containing the characteristic C=C stretching modes of aromatics, alkenes, and carotenoids; 1300–1400 cm^{-1}, containing the characteristic CH$_2$ and NH bending modes; and 1050–1200 cm^{-1}, containing the characteristic C-C stretching modes. In the lower-wavenumber range, 100–1100 cm⁻¹, we would expect to see the vibrational modes of heavy-metal oxides, such as Fe_2O_3, and of the inorganic molecular ions sulphate, silicate, phosphate, and carbonate. It would be facile to suggest, however, that it is possible to separate organic signatures from inorganic minerals, since the Raman spectra of both contain features that occur in several of the ranges specified earlier, and the situation is exacerbated even more with biogeologically modified material; for example, metal oxalates, formed from the metabolic decomposition products of organisms in calcareous substrates, have bands near 1500, 900, and 500 cm⁻¹ and carbonates have bands near 1100, 700, and 300 cm⁻¹.

Also, there is no precise, unchanging, and defined value for any particular vibrational band in a biomarker entity, irrespective of the molecular species, since molecular structural, environmental, and crystal field effects conspire to change the wavenumbers of an identifying group significantly. Hence, calcium carbonate, $CaCO_3$, has a C-O stretching band at 1086 cm⁻¹, which in dolomite, $CaMg(CO_3)_2$, occurs at 1094 cm⁻¹ and in magnesite, $MgCO_3$, occurs at 1115 cm⁻¹. In the potential list of organic biomarkers, carotenoids provide a key group of target molecules, and here the strongest signature, namely that of the C=C stretching vibration, can have the spectral value between 1535 cm⁻¹ and 1508 cm⁻¹, dependent upon the length of conjugation in the alkene chain. Also, it is quite rational to make a molecular structural diagnosis on the basis of several cognitive

Raman bands, rather than one alone, so the selection of several wavenumber ranges for the characterization of molecular entities and functional groups is a normal procedure in Raman spectroscopy.

From comprehensive Raman spectroscopic studies, which have been reported in the scientific literature for terrestrial extremophilic systems and their colonization of geological strata in some of the most extreme environments on Earth (Jorge-Villar and Edwards 2005), it is possible to construct a working database of characteristic signatures from which it should be possible to identify analogues on Mars from remote Raman spectral data. The terrestrial data comprise laboratory experiments and, where appropriate, *in situ* experiments carried out in field expeditions to, amongst others, the Antarctic, Canadian Arctic, Svalbard, Spitsbergen, Atacama Desert, Negev Desert, Dead Sea, Mojave Desert, Death Valley, Pilbara North Pole Dome, Barberton, Kilauea, and the Dolomites (Wynn-Williams and Edwards 2000a, b; Vitek et al. 2012). The results are summarized in Table 8.6.1 (taken from Edwards et al. 2014), where the key Raman bands of some of the most important molecules are listed, with their characteristic features suitable for identification in mixtures highlighted in bold type. Naturally, this list cannot be totally inclusive, and several molecules have not been listed, but the molecules given in Table 8.6.1 represent a good cross-section for the construction of a comprehensive Raman database for biomarker detection in hostile environments.

TABLE 8.6.1

Raman Bands of Biomarkers and Associated Geomarkers in Extremophile Examples and Their Chemical Formulae

Name	Formula													
Calcite	$CaCO_3$		**1086**	**712**	**282**	156								
Aragonite	$CaCO_3$		**1086**	**704**	**208**	154								
Dolomite	$CaMg(CO_3)_2$		**1098**	**725**	**300**	177								
Magnesite	$MgCO_3$		**1094**	**738**	**330**	**213**	119							
Hydromagnesite	$Mg_5(CO_3)_4(OH)_2\ 4H_2O$		**1119**	**728**	326	**232**	202	184	147					
Gypsum	$CaSO_4\ 2H_2O$	1133	**1007**		669	628	**492**	413						
Anhydrite	$CaSO_4$		**1015**	674	628	500	416							
Quartz	SiO_2		1081	1064	808	796	696	500	542	**463**	354	263	**206**	128
Haematite	Fe_2O_3				**610**	500	**411**	**293**	245	**226**				
Limonite	$FeO(OH)nH_2O$			693	**555**	481	**393**	**299**	**203**					
Apatite	$Ca_5(PO_4)_3(F,Cl,OH)$			1034	**963**	**586**	**428**							
Weddellite	$Ca(C_2O_4)2H_2O$	1630	**1475**	1411	**910**	869	597	**506**	**188**					
Whewellite	$Ca(C_2O_4)H_2O$	1629	**1490**	**1463**	1396	942	**896**	865	596	521	**504**	223	207	**185** 141
Chlorophyll	$C_{55}H_{72}O_5N_4Mg$			1438	1387	**1326**	1287	1067	1048	**988**	**916**	**744**	**517**	351
c-Phycocyanin	$C_{36}H_{38}O_6N_4$	1655	**1638**	1582	1463	**1369**	1338	**1272**	1241	1109	1054	**815**	**665**	499
Beta-carotene	$C_{40}H_{56}$		**1515**	**1155**	1006									
Rhizocarpic acid	$C_{26}H_{23}O_6$	**1665**	1610	**1595**	**1518**	**1496**	1477	1347	1303	1002	944	902	768	448
Scytonemin	$C_{36}H_{20}N_2O_4$	1605	**1590**	**1549**	1444	**1323**	1283	1245	**1172**	1163	984	752	675	574 270
Calycin	$C_{18}H_{10}O_5$	1653	**1635**	**1611**	**1595**	**1380**	1344	1240	1155	1034	**960**	878	498	484
Paretin	$C_{16}H_{12}O_5$	**1671**	1631	1613	**1153**	1387	1370	**1277**	1255	**926**	571	519	467	**458**
Usnic acid	$C_{18}H_{16}O_7$	**1694**	1627	**1607**	**1322**	**1289**	1192	1119	**992**	959	846	602	540	
Emodin	$C_{15}H_{10}O_5$	**1659**	1607	1577	1557	**1298**	**1281**	942	**565**	**467**				
Atranorin	$C_{19}H_{18}O_8$	**1666**	**1658**	1632	**1303**	**1294**	1266	**588**						
Pulvinic dilactone	$C_{18}H_{10}O_4$	**1672**	**1603**	1455	**1405**	1311	**981**	**504**						

Source: Edwards, H.G.M. et al., *Philos. Trans. Roy. Soc. A*, 372, 20141093, 2014.

8.6.6.1 SELECTED BIOMARKERS AND THEIR RAMAN SPECTRA

From the biomarker molecules listed in Table 8.6.1 (Edwards et al. 2014), we shall select here just two for a more detailed discussion in this paper, namely, carotenoids and scytonemin. Then, a similar discussion will follow on carbon, the end-of-line result of degradative processes, which may have been operating upon these biomarkers that have been exposed to chemical change over many thousands or millions of years.

8.6.6.1.1 Carotenoids

In certain areas of the Earth's surface, the colored carotene pigments can be seen as evidence of an active biological colonisation: an example of this phenomenon is seen in Figure 8.6.1, which shows the pink colour of carotenoids in the halophilic cyanobacterial colonization of Don Juan Pond in Antarctica—the very high ionic calcareous salt content, comprising 44% by weight of salts, mainly calcium chloride and sodium chloride,

FIGURE 8.6.1 Don Juan Pond in the McMurdo Dry Valleys, Victoria Land, Antarctica, showing a pink bloom from halophilic cyanobacterial colonization. This is a hypersaline lake and is recognized as the most salt-laden liquid water on earth, containing 44% salinity and comprising almost 3.7 mol kg^{-1} and 0.5 mol kg^{-1} of CaCl$_2$ and NaCl, respectively, at a temperature of $-51°C$. (Courtesy of Dr. David Wynn-Williams, British Antarctic Survey, Cambridge, UK.)

of this location maintains one of the lowest temperatures on our planet; the water here is still in its liquid phase at approximately $-50°C$, yet the biology is clearly thriving!

Several biomolecules have been identified as potential biomarkers for the detection of signs of extant or extinct life in geological niches in hostile terrestrial environments, and carotenoids are recognized as a prime marker in this context: currently, some 500 members of the carotenoid family are known terrestrially, each possessing a conjugated chain of C=C and C–C linkages with pendant methyl groups and aliphatic or aromatic end units, which can be cyclic, as in the case of beta-carotene, or can comprise complex groups based on sugars, such as spiroxanthin. Their Raman spectroscopic detection is based upon key molecular signatures centred on the conjugated C=C moieties, which give rise to two characteristic Raman vibrations due to C=C stretching at about 1520 cm^{-1} and C–C stretching at about 1150 cm^{-1}, both of strong intensity, and a weaker feature at about 1005 cm^{-1}, which arises from a C=C–H bending mode. The presence of these bands in a biogeological system is well established as a diagnostic confirmation of the presence of carotenoids even in complex specimens, which can exhibit other bands due to accessory light-harvesting pigment, chlorophyll, and low-wavelength radiation protectants such as scytonemin (Edwards et al. 2014).

The importance of the Raman spectroscopic diagnostic analysis of biomarkers such as carotenoids in a variety of geological, biogeological, and biological systems on ExoMars 2020 rests upon their unambiguous identification and attribution to chemical species, and this depends upon several factors, including the Raman cross-section or molecular scattering coefficient, which results in a relatively high-band intensity from even small amounts of relevant material in the specimen; the placement of the characteristic bands in regions that are not subject to potential interference from other molecular species present, so minimising the risk of their escaping detection; and the spectroscopic interpretation of their molecular origins upon which the reliable band assignments are based (Marshall and Marshall 2010). All of these points are generally applicable to carotenoids, and indeed, their detectability is singularly and specifically enhanced by several orders of magnitude by using laser excitation in the green region of the spectrum through the operation of a resonance Raman effect, which is a critical factor in the 532-nm laser excitation used on ExoMars 2020 and an appropriate advantage for the detection of carotenoids through the resonance Raman enhancement of the scattered Raman signals. Figure 8.6.2 shows the Raman spectra of two distinct carotenes from an extremophilic cyanobacterial colony inside a basaltic lava vacuole (Jorge-Villar and Edwards 2010).

These spectra have been obtained with green laser excitation and are an example of the operation of resonance Raman scattering enhancement upon carotenoids. In this resonance experiment, no chlorophyll was detected, but in contrast, using near-infrared 785-nm excitation in an out-of-resonance condition, the presence of these carotenoids was recorded

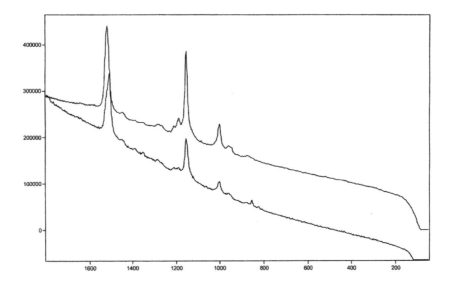

FIGURE 8.6.2   Raman spectra of two distinct carotenoids, beta-carotene and astaxanthin, from cyanobacterial colonization inside a vacuole in basalt found in Svalbard, Arctic Norway.

less strongly but along with other biomarkers such as chlorophyll and scytonemin. The detection of the three main characteristic bands of a generic carotenoid will be undoubtedly critical for the successful deployment of the Raman instruments on Mars; however, it is equally important scientifically to be able to match the signals received remotely from Mars to the molecular structure of the carotenoid concerned. This is not such a routine task, as several factors need to be considered before an unequivocal conclusion about the species of carotenoid can be deduced from the wavenumber positions of the characteristic bands, since these are dependent upon the molecular environment. Hence, one cannot assign precisely the characteristic wavenumbers of the Raman bands in beta-carotene, for example, based upon a standard specimen, since these are changed by its admixture with other molecules—so, the C=C stretching band of beta-carotene cannot be assigned a single wavenumber, but this can lie within the range 1511–1523 cm^{-1}. A further problem arises when two or more carotenoids are present in the same target specimen—both C=C stretching wavenumbers can lie within this narrow range and may not be fully resolved under the operational spectrometer conditions. In laboratory instrumentation, where the spectral resolution may be 1 cm^{-1} or smaller, this may not be a problem, but most miniaturized Raman spectrometers have defined spectral resolutions, which are cited as high as 10 cm^{-1}, and this clearly could potentially compromise the identification of the individual molecular species. A possible marker as to the presence of more than one carotenoid in a system under study would be given by the relative intensities of the three characteristic C=C, C–C, and CCH features described earlier: for a range of carotenoids, the wavenumber positions of these features lie approximately in the narrow regions 1532–1500, 1160–1152, and 1008–1004 cm^{-1}, respectively, and the relative intensities are approximately 1.5:1:0.2. However, depending upon the spectral resolution and relative proportions of two or more carotenoids in admixture in a system, these relative intensities are changed and could also

involve peak wavenumbers sensibly different from the analogous values observed in each individual component (Jehlicka et al. 2014).

8.6.6.1.2 Scytonemin

Scytonemin is a sunscreen pigment secreted as part of the outer protective sheath of cyanobacteria; it absorbs ultraviolet B (UVB) (280–320 nm) and ultraviolet C (UVC) (190–280 nm) radiations, is resistant to degradation, and has been adopted as a distinct and selectively exclusive biomarker for cyanobacteria (Edwards et al. 2000, 2003a; Varnali et al. 2009; Varnali and Edwards 2014a, b). Scytonemin is the parent of a growing number of potential substituted biomarker derivatives (Varnali and Edwards 2010, 2013, 2014a, b), based on this unique dimeric structure, including an interesting and novel range of iron complexes (Varnali and Edwards 2010, 2013). Cyanobacterial organisms within Arctic and Antarctic rocks illustrate the capacity of life to survive under intensely cold and dry conditions, and an example of a cyanobacterial halotrophic colonization of *Nostoc* and *Gloeocapsa,* some 5 mm inside a selenite (gypsum) crystal from breccia deposits at the rim of the Haughton meteorite crater in Devon Island, Arctic Canada, was analysed using confocal Raman microscopy and 785-nm excitation (Edwards et al. 2005). The Raman spectrum of scytonemin is shown in Figure 8.6.3.

8.6.6.1.3 Carbon

The degradation of biomolecules in hostile geological environments eventually produces carbon, whose signatures in the Raman spectrum have been well described. The idea that it is possible to differentiate between biotic and abiotic carbon formation from the shape of the so-called D and G Raman bands, characteristic of sp^3 and sp^2 hybridized carbons, effectively represented by structures typical of diamond and graphite, respectively, has had a long and rather controversial history in the literature, which is still raging today (Kudryavtsev et al. 2001; Brasier et al. 2002; Pasteris and Wopenka 2002, 2003;

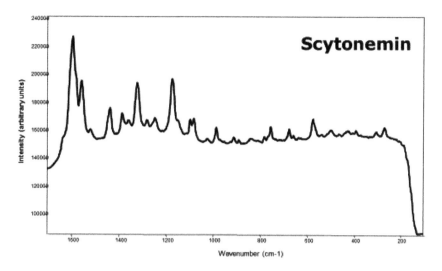

FIGURE 8.6.3 Raman spectrum of cyanobacterial colony inside a selenite crystal from breccia at edge of Haughton Crater, Devon Island, Canadian Arctic, with the characteristic bands of scytonemin.

Schopf et al. 2002; Schopf and Kudryatsev 2005); a summary of the extensive literature on this subject is provided by Marshall et al. (2010) and Marshall and Marshall (2010, 2014), who have defined a methodology for rigorously discriminating between these different carbon sources. However, if we consider the requirements of such fine spectroscopic work that has to be undertaken in the field by using portable miniaturized instrumentation, with all the sacrifices that have had to be made from laboratory versions, additionally performed under extremely hostile conditions that are mirrored nowhere else terrestrially, then one must pose the question: is it possible with current instrumentation to be able to differentiate between abiotic and biotic carbon, especially with portable devices? The result then is that the observation of characteristic carbon signatures does not constitute the presence of extinct life. So, in our definition, carbon cannot be a true biomarker. The same analogy applies to other biochemicals, which have been synthesised abiotically in the laboratory, such as amino acids, sugars, porphyrins, and proteins. This is the reason that these molecules are not presented as true biomarkers in Table 8.6.1.

REFERENCES

Bedau, M.A. 2010. An Aristotelean account of minimal chemical life. *Astrobiology* 10: 1011–1020.

Benner, S.A. 2010. Defining life. *Astrobiology* 10: 1021–1030.

Brasier, M.D., O.R. Green, A.P. Jephcoat, A.T. Kleppe, M.J. Van Kranendonk, J.F. Lindsay, A. Steele, and N.V. Grassineau. 2002. Questioning the evidence for Earth's oldest fossils. *Nature* 416: 76–81.

Clark, B.C. 1998. Surviving the limits to life at the surface of Mars. *Journal of Geophysical Research: Planets* 103: 28545–28556.

Cleland, C.E., and C.F. Chyba. 2002. Defining life. *Origins of Life and Evolution of the Biosphere* 32: 387–393.

Cockell, C.S., and J.R. Knowland. 1999. Ultraviolet screening compounds. *Biological Reviews* 74: 311–345.

Dodd, M.S., D. Papineau, T. Carenne, J.F. Slack, M. Rittner, F. Piragrio, J. O'Neil, and C.T.S. Little. 2017. Evidence for early Life in Earth's oldest hydrothermal vent precipitates. *Nature* 543: 60–64.

Edwards, H.G.M. 2004. Raman spectroscopic protocol for the molecular recognition of key biomarkers in astrobiological exploration. *Origins of Life and Evolution of the Biosphere* 34: 3–11.

Edwards, H.G.M. 2010. Raman spectroscopic approach to analytical astrobiology: The detection of key geological and biomolecular markers in the search for life. *Philosophical Transactions of the Royal Society A* 368: 3059–3066.

Edwards, H.G.M., and J.M. Chalmers (Eds.). 2005. *Raman Spectroscopy in Archaeology and Art History*. Cambridge, UK: Royal Society of Chemistry.

Edwards, H.G.M., K.A.E. Edwards, D.W. Farwell, I.R. Lewis, and M.R.D. Seaward. 1994. An approach to stone and fresco lichen biodeterioration through FT-Raman microscopic investigation of thallus-substratum encrustations. *Journal of Raman Spectroscopy* 25: 99–103.

Edwards, H.G.M., D.W. Farwell, and M.R.D. Seaward. 1991a. Raman spectra of oxalates in lichen encrustations on Renaissance frescoes. *Spectrochimica Acta* 47A: 1531–1539.

Edwards, H.G.M., D.W. Farwell, M.R.D. Seaward, and C. Giacobini. 1991b. Preliminary Raman microscopic analyses of a lichen encrustation involved in the biodeterioration of Renaissance frescoes in central Italy. *International Biodeterioration* 27: 1–9.

Edwards, H.G.M., F. Garcia-Pichel, E.M. Newton, and D.D. Wynn-Williams. 2000. Vibrational Raman spectroscopic study of scytonemin, the UV-protective cyanobacterial pigment. *Spectrochimica Acta, Part A* 56: 193–200.

Edwards, H.G.M., J.M. Holder, N.C. Russell, and D.D. Wynn-Williams. 1997a. Biodeterioration of rocks by lichens in hostile Antarctic environments studied by FT-Raman spectroscopy. In *Spectroscopy of Biological Molecules: Modern Trends*, P. Carmona, R. Navarro, and A. Hernanz (Eds.), pp. 509–510. Dordrecht, the Netherlands: Kluwer Academic.

Edwards, H.G.M., J.M. Holder, and D.D. Wynn-Williams. 1998. Comparative FT-Raman spectroscopy of *Xanthoria* epilithic lichen-substratum systems from temperate and Antarctic habitats. *Soil Biology Biochemistry* 30: 1947–1953.

Edwards, H.G.M., I.B. Hutchinson, R. Ingley, and J. Jehlicka. 2014. Biomarkers and their Raman spectroscopic signatures: A spectral challenge for analytical astrobiology. *Philosophical Transactions of the Royal Society A* 372: 20141093.

Edwards, H.G.M., S.E Jorge-Villar, J.A Parnell, C.S. Cockell, and P. Lee. 2005. Raman spectroscopic studies of cyanobacterial gypsum halotrophs and their relevance for sulfate deposits on Mars. *Analyst* 130: 917–923.

Edwards, H.G.M., E.M. Newton, D.L. Dickensheets, and D.D. Wynn-Williams. 2003a. Raman spectroscopic detection of biomolecular markers from Antarctic materials: evaluation for putative Martian habitats. *Spectrochimica Acta, Part A* 59: 2277–2290.

Edwards, H.G.M., E.M. Newton, and D.D. Wynn-Williams. 2003b. Molecular structural Studies of lichen substances II: Atranorin, gyrophoric acid, fumarprotocetraric acid, rhizocarpic acid, calycin, pulvinic dilactone and usnic acid. *Journal of Molecular Structure* 651–653: 27–37.

Edwards, H.G.M., E.M. Newton, D.D. Wynn-Williams, D. Dickensheets, C. Schoen, and C. Crowder. 2003c. Laser wavelength selection for Raman spectroscopy of microbial pigments *in situ* in Antarctic desert ecosystem analogues of former habitats on Mars. *International Journal of Astrobiology* 1: 333–348.

Edwards, H.G.M., E.M. Newton, D.D. Wynn-Williams, and R.I. Lewis-Smith. 2003d. Nondestructive analysis of pigments and other organic compounds in lichens using Fourier-Transform Raman spectroscopy: A study of Antarctic epilithic lichens. *Spectrochimica Acta, Part A* 59: 2301–2309.

Edwards, H.G.M., N.C. Russell, and D.D. Wynn-Williams. 1997b. FT-Raman spectroscopic and scanning electron microscopic study of cryptoendolithic lichens from Antarctica. *Journal of Raman Spectroscopy* 28: 685–690.

Edwards, H.G.M., D.D. Wynn-Williams, S.J. Little, L.F.C. de Oliveira, C.S. Cockell, and J.C. Ellis-Evans. 2004. Stratified response to environmental stress in a polar lichen characterised with FT-Raman microscopic analysis. *Spectrochimica Acta, Part A* 60: 2029–2033.

Edwards, H.G.M., D.D. Wynn-Williams, E.M. Newton, and S.J. Coombes. 2003e. Molecular structural studies of lichen substances I: Parietin and emodin. *Journal of Molecular Structure* 648: 49–59.

Ellery, A., J.A. Parnell, A. Steele et al. 2003. Astrobiological instrumentation for Mars—the only Way is down! *International Journal of Astrobiology* 1: 365–380.

Ellery, A., D.D. Wynn-Williams, J. Parnell, H.G.M. Edwards, and D. Dickensheets. 2004. The role of Raman spectroscopy as an astrobiological tool in the exploration of Mars. *Journal of Raman Spectroscopy* 35: 441–457.

Freissinet, C., D.P. Glavin, P.R. Mahaffy et al. 2015. Organic molecules in the Sheepbed mudstone, Gale Crater, Mars. *Journal of Geophysical Research: Planets* 120: 495–514.

Garcia-Ruiz, J.M. 1999. Morphological behavior of inorganic precipitation systems. *SPIE Proceedings, 3755, Instruments, Methods and Missions for Astrobiology II 74*, December 30, 1999, R.B. Hoover (Ed.). doi:10.1117/12.375088.

Holder, J.M., D.D. Wynn-Williams, F. Rull Perez, and H.G.M. Edwards. 2000. Raman spectroscopy of pigments and oxalates *in situ* within epilithic lichens: *Acarospora* from the Antarctic and Mediterranean. *New Phytology* 145: 271–280.

Jehlicka, J., H.G.M. Edwards, K. Osterrothova, J. Novotna, L. Nedbalova, J. Kopecky, I. RNemec, and A. Oren. 2014. Potential and limits of Raman spectroscopy for carotenoid determination in microorganisms: Implications for astrobiology. *Philosophical Transactions of the Royal Society A* 372: 20140199.

Jorge-Villar, S.E., and H.G.M. Edwards. 2005. Raman spectroscopy in astrobiology. *Analytical & Bioanalytical Chemistry* 384: 100–113.

Jorge-Villar, S.E., and H.G.M. Edwards. 2010. Raman spectroscopy of volcanic lavas and inclusions of relevance to astrobiological exploration. *Philosophical Transactions of the Royal Society A* 368: 3127–3135. doi:10.1098/rsta.2010.0102.

Jorge-Villar, S.E., H.G.M. Edwards, D.D. Wynn-Williams, and M.R. Worland. 2003. FT-Raman spectroscopic analysis of an Antarctic endolith. *International Journal of Astrobiology* 1: 349–355.

Kudryavtsev, A.B., J.W. Schopf, D.G. Agresti, and T.J. Wdowiak. 2001. *In situ* laser-Raman imagery of precambrian microscopic fossils. *Proceedings of the National Academy of Sciiences of the United States of America* 98: 823–826.

Long, D.A. 2002. *The Raman Effect*. Chichester, UK: John Wiley and Sons.

Marshall, C.P., H.G.M. Edwards, and J. Jehlicka. 2010. Understanding the application of Raman spectroscopy to the detection of traces of life. *Astrobiology* 10: 229–243.

Marshall, C.P., and A. Olcott Marshall. 2010. The potential of Raman spectroscopy for the analysis of diagenetically transformed carotenoids. *Philosophical Transactions of the Royal Society A* 368: 3137–3144.

Marshall, C.P., and A. Olcott Marshall. 2014. Raman spectroscopy as a screening tool for ancient life detection on Mars. *Philosophical Transactions of the Royal Society A* 372: 20140195. doi:10.10198/rsta.2014.0195.

McKay, D.S., E.K. Gibson Jr., K.L. Thomas-Keptra, H. Vali, C. S Romanek, S.J. Clemett, X.D.F. Chillier, C.R. Maechling, and R.N. Zare.1996. Search for past life on Mars: Possible relict biogenic activity in Martian meteorite ALH 84001. *Science* 273(5572): 924–930.

Pasteris, J.D., and B. Wopenka. 2002. Images of the Earth's earliest fossils? *Nature* 420: 476–477.

Pasteris, J.D., and B. Wopenka. 2003. Necessary, but not sufficient: Raman identification of disordered carbon as a signature of ancient life. *Astrobiology* 3: 727–738.

Pullan, D., B.A. Hofmann, F. Westall et al. 2008. Identification of morphological biosignatures in Martian analogue field specimens using *in situ* planetary instrumentation. *Astrobiology* 8: 119–156.

Rull, F., S. Maurice, I. Hutchinson et al. 2017. The Raman laser spectrometer (RLS) for the EXOMARS rover mission to Mars. *Astrobiology* 17(6/7): 627–654.

Russ, J., R.L. Palma, D.H. Loyd, D.W. Farwell, and H.G.M. Edwards.1995. Analysis of the rock accretions in the Lower Pecos Region of Southwest Texas. *Geoarchaeology* 10: 43–63.

Russell, N.C., H.G.M. Edwards, and D.D. Wynn-Williams. 1998. FT-Raman Spectroscopic Analysis of endolithic microbial communities from Beacon sandstone in Victoria Land, Antarctica. *Antarctic Science* 10: 63–74.

Schopf, J.W., and A.B. Kudryavtsev. 2005. Three-dimensional Raman imagery of precambrian microsopic organisms. *Geobiology* 3: 1–12.

Schopf, J.W., A.B. Kudryavtsev, D.G. Agresti, T.J. Wdowiak, and A.D. Czaja. 2002. Laser-Raman imagery of Earth's earliest fossils. *Nature* 416: 73–76.

Shapiro, R., and D. Schulze-Makuch. 2009. The search for alien life in our Solar System: Strategies and priorities. *Astrobiology* 9: 1–9.

Tirard, S., M. Morange, and A. Lazcano. 2010. The definition of life: A brief history of elusive scientific endeavour. *Astrobiology* 10: 1003–1009.

Vago, J.L., F. Westall, A. Coates et al. 2017. Habitability on early Mars and the search for biosignatures with the ExoMars rover. *Astrobiology* 17(6/7): 471–510.

Varnali, T., and H.G.M. Edwards. 2010a. *Ab initio* calculations of scytonemin derivatives of relevance to extremophile characterisation by Raman spectroscopy. *Philosophical Transactions of the Royal Society A* 368: 3193–3204.

Varnali, T., and H.G.M. Edwards. 2010b. Iron-scytonemin complexes: DFT calculations on new UV protectants for terrestrial cyanobacteria and astrobiological implications. *Astrobiology* 10: 711–716.

Varnali, T., and H.G.M. Edwards. 2013a. A potential new biosignature of life in iron-rich extreme environments: An iron (III) complex of scytonemin and proposal for its identification using Raman spectroscopy. *Planetary and Space Sciences* 82–83: 128–133.

Varnali, T., and H.G.M. Edwards. 2013b. Theoretical study of novel complexed structures for methoxy derivatives of scytonemin: Potential biomarkers in iron-rich stressed environments. *Astrobiology* 9: 861–869.

Varnali, T., and H.G.M. Edwards. 2014a. Raman spectroscopic identification of scytonemin and its derivatives as key biomarkers in stressed environments. *Philosophical Transactions of the Royal Society A* 372: 20140197.doi:10.1098.rsta.2014.0197.

Varnali, T., and H.G.M. Edwards. 2014b. Scytonin, a novel cyanobacterial photoprotective pigment: Calculations of Raman spectroscopic biosignatures. *Journal of Molecular Modelling* 20: 1–8.

Varnali, T., H.G.M. Edwards, and M.D. Hargreaves. 2009. Scytonemin: Molecular structural studies of a key extremophilic biomarker for astrobiology. *International Journal of Astrobiology* 8: 133–140.

Vitek, P, J. Jehlicka, H.G.M. Edwards, I.B. Hutchinson, C. Ascaso, and J. Wierzchos. 2012. The miniaturized Raman system and detection of traces of life in halite from the Atacama Desert: Some considerations for the search for life signatures on Mars. *Astrobiology* 12: 1–5.

Westall, F., F. Foucher, N. Bost et al. 2015. Biosignatures on Mars: What, where and how? Implications for the search for Martian life. *Astrobiology* 15: 1–32.

Wynn-Williams, D.D., and H.G.M. Edwards. 2000a. Antarctic eco-systems as models for extra-terrestrial surface habitats. *Planetary and Space Sciences* 48: 1065–1075.

Wynn-Williams, D.D., and H.G.M. Edwards. 2000b. Proximal analysis of regolith habitats and protective biomolecules *in situ* by laser Raman spectroscopy: Overview of terrestrial Antarctic habitats and Mars analogs. *Icarus* 144: 486–503.

Wynn-Williams, D.D., and H.G.M. Edwards. 2002. Environmental UV-radiation: Biological strategies for protection and avoidance. In *Astrobiology: The Quest for the Origins of Life*, G. Horneck and C. Baumstarck-Khan (Eds.), pp. 245–260. Berlin, Germany: Springer-Verlag.

Wynn-Williams, D.D., H.G.M. Edwards, and F. Garcia-Pichel. 1999. Functional biomolecules of Antarctic stromatolitic and endolithic cyanobacterial communities. *European Journal of Phycology* 34: 381–391.

Fossilization of Bacteria and Implications for the Search for Early Life on Earth and Astrobiology Missions to Mars

Frances Westall and Keyron Hickman-Lewis

CONTENTS

8.7.1   INTRODUCTION

Microorganisms (herein the term "bacteria" *sensu lato* is used generically for microorganisms, including viruses) are the most common organisms on the surface of the Earth and the only organisms to live at depth in the subsurface. They are also the organisms that can survive the most extreme environmental conditions known today, including (to a certain extent) the space environment. Despite their generally simple morphologies, small size, and lack of hard parts, bacteria can nevertheless be themselves fossilized or leave a variety of traces in the rock record that are capable of withstanding geological processing, thereby providing direct evidence for life as far back in time as the Paleoarchean (~3.7–3.3 billion years ago, Ga). Generally, these traces are preserved in microcrystalline silica (chert), which has the ability to contemporaneously encapsulate organic material. However, the range of environments that could preserve biosignatures, particularly in deep time, include a full traverse of stratiform bedded cherts and siliciclastic settings, shallow-water sandstones and carbonates, volcanogenic edifices and deposits such as pillow basalts, hydrothermal vent deposits and chemical sediments, and terrestrial riverine and lacustrine environments (Nisbet, 2000; Brasier et al. 2011; Westall 2016). The often-very-great ages of these traces mean that they have survived the many geological complications of Earth's history, enduring physical, hydraulic, tectonic, and metamorphic events that, at times, could have completely obliterated them. Despite their great age, a small number of Paleoarchean terranes have survived relatively unscathed. For studies of early life, these include notably the East Pilbara terrane of Western Australia and the Barberton greenstone belt of South Africa, and more are appearing, as global warming removes protective glacial shields in certain locations, such as Greenland (indeed, there is a peculiar benefit in global warming for geologists), but this chapter will focus mostly on well-preserved and definitive morphological biosignatures.

Archean (~4.0–2.5 Ga) terranes host all the known types of biosignatures, be they morphological fossils of cells, colonies, biofilms, or other biological constructions, such as microbial mats and stromatolites; the degraded chemical traces of organic components of the original microbial communities (biomarkers); or even geochemical metabolic by-products, such as biominerals, records of specific isotopic fractionation, and trace or rare-earth element enrichments (Westall and Cavalazzi 2011; Mustard et al. 2013). These biosignatures document a time period when the Earth was, in itself, an extreme environment: without oxygen, with high ultraviolet (UV) radiation flux, warmer oceans, slightly acidic seas, highly active in terms of volcanism and hydrothermalism, and subject to frequent impacts of extra-terrestrial bolides (Hofmann 2011; Arndt and Nisbet 2012; Westall et al. 2018). Life on the early Earth was therefore what we would today consider extremophilic. This concept should, perhaps, be taken with a pinch of salt, since life, as we know it, could not have appeared on the modern Earth because of the presence of oxygen and, because these anaero-bic microorganisms are most likely among the most frequent life forms in the universe, it is we, modern-day terrans, who are the true extremophiles!

Notwithstanding, the very fact that early life on Earth was anaerobic, extremophilic, and relatively simple means that it provides ideal analogues for understanding the potential dis-tribution of life on other planets and satellites, such as Mars (Westall et al. 2015a). Early life can inform us about what we might find on the Red Planet, how it might be preserved, how to search for it, and how to confirm its biogenic origin. In determining biogenicity, Precambrian paleontology provides the perfect philosophical analogue and has set a benchmark of stringency against which potential extraterrestrial biosig-natures will need to be tested. Furthermore, we will not only be searching for fossilized traces of life on Mars. Tantalizing trace amounts of methane in the Martian atmosphere, observed from the Earth (Mumma et al. 2009), from Martian orbit (Formisano et al. 2004), and *in situ* by the Mars Science Laboratory *Curiosity* rover (Webster et al. 2015), may have been, and may still be, produced by Martian microorganisms; however, of course, there are also abiotic crustal processes that could account for this methane. Thus, extant life could have found a habitable niche and refuge in the upper crust of Mars and could, under particular circumstances, be brought back up to the surface. Therefore, missions to Mars need to search for both fossil and extant traces of life, the biosigna-tures of which are similar but not identical and for which the means of detection and analysis are different.

8.7.2 THE FOSSILIZATION OF BACTERIA

Living microorganisms are composed of three major organic components: a replication molecule, for example, DNA; mol-ecules for doing work in the cell and allowing it to conduct its essential functions, for example, proteins; and molecules that provide a confined and controlled environment for metabolic and replication activity, such as lipids forming a semi-permeable membrane. The identification of unknown living organisms is largely based on a battery of techniques, including lab on a chip, various -omics approaches (genomics, proteomics, or metabolomics), and polymerase chain reaction (PCR) to study small segments of DNA, all of which allow clas-sification of organisms according to their individual molecular

signatures. Direct cell cultivation is a further method by which microbial biosignatures can be understood; however, to date, less than 2% of all microorganisms have been successfully cultivated (Wade 2002).

For the fossil record, this same suite of approaches can-not be used normally. DNA, for instance, is estimated to have a maximum longevity of 0.4–1.5 Ma under even excep-tional preservational conditions while retaining sufficient signatures for analysis by sequencing technology (Willerslev et al. 2004). As a result, the range of techniques with which ancient life is studied differs greatly from (micro)-biological approaches. For many decades, direct observa-tion was the only method by which traces of early life could be studied. Limited macroscopic to microscopic observa-tions of cells, colonies, biofilms, microbial mats, and other morphological manifestations of microbes, including micro-bial constructs, such as stromatolites, organo-sedimentary structures, and microbially induced sedimentary structures (MISS; *cf.* Noffke 2010), nonetheless, yielded an impres-sive array of identified structures, the validities of some of which have survived even recent criticism (e.g., Knoll and Barghoorn 1977; Walsh 1992). However, advancements in our understanding of biogeochemistry and the application of techniques capable of estimating Earth-Life interac-tions in ancient rocks have significantly widened the range of possible biological indicators. We are thus able to delin-eate biological indicators, both living and fossil, into three classes: morphological, organic, and metabolic. The latter includes the precipitation of minerals as a side effect of changes in the immediate environment of the microbes by their metabolic activity, together with the dissolution of minerals; the leaching of minerals to obtain necessary nutrients or elements; and the concomitant patterns of enrichments and depletion that result from these processes and that are probably biological. Manifestations of micro-bial presence are therefore manifold (Summons et al. 2011; Westall and Cavalazzi 2011). To some degree, all of these features can be preserved in fossilized form in the rock record, with, necessarily, a certain quota of alteration after natural degradation and transformation into fossil status, as well as post-diagenetic and post-taphonomic alterations by geological processes and the effects of geological time.

8.7.2.1 THE PROCESS OF FOSSILIZATION

The series of processes affecting microorganisms upon death and their potential preservation in the rock record is termed *taphonomy*. Upon cell death, the release of enzymes breaks down the cell wall, allowing the internal contents to escape into the environment. In general, most dead cellular material is immediately degraded and taken up by heterotrophic organ-isms that oxidize the organic molecules as a source of energy. It is estimated that less than 10% of cellular organic matter is preserved in the rock record. Thus, in order for organisms to be preserved at the cellular level, they generally need to be (1) rapidly encapsulated or replaced by a mineral that replicates cell shape (permineralization) or (2) rapidly entombed by anoxic sediments, permafrost, or ice, which essentially "mummifies" the cells. In both cases, stabilization of the cell wall by the chelation of metals, such as Fe (Ferris et al. 1988; Orange et al. 2011), can prolong the "life time" of the cell walls, thus allowing more time for the formal fossilization process. Most frequently, a third means of preservation occurs, whereby (3) their degraded organic components become trapped within a mineral matrix.

Finally, in order to pass into the rock record and be preserved for significantly long geological timescales, the mineralized or mummified microbial matter or their degraded components need to be lithified into sediments. With increasing age, there is more likelihood of a particular geological formation being destroyed by tectonic, mass wasting, or weathering processes. Indeed, there remain only a few enclaves of rocks dating from the early Earth, and none of these are older than 4 Ga. The preserved early Earth rocks dating from the Early Archean (subdivided into the Eoarchean and Paleoarchean) stretch the geological record to about 3.9 Ga. All have been subjected to varying degrees of metamorphism. The oldest (Eoarchean) sedimentary formations occur in Canada and Greenland. These rocks have witnessed up to amphibolite or granulite facies metamorphism and are therefore significantly altered. Understanding their protoliths and the geochemical signatures preserved within is challenging. Nevertheless, carbon in the form of graphite exhibiting isotopic signatures consistent with those produced by microbial fractionation occurs in a number of these highly metamorphosed formations (e.g., Rosing 1999; Tashiro et al. 2017). Once fossilized, biosignatures do not automatically enter into the rock record to be preserved in deep time. For example, even though microbial mats can be calcified, calcification can also be reversed depending on diurnal cycles (Dupraz et al. 2009). Indeed, modern stromatolites form largely from sediment trapping within their sticky extracellular polymeric substance (EPS), as opposed to calcification.

Also, among these ancient metamorphosed terranes are two that are significantly better preserved, having generally undergone no greater than lower greenschist metamorphic grade. These two terranes are the Kaapvaal craton in South Africa and the East Pilbara craton in Australia, both dating back to ~3.5 Ga. Sedimentary successions within greenstone belts in these cratons contain a large variety of well-preserved traces of life, including those fossilized by mineral encapsulation and mummification, together with more generic, degraded organic matter. The fossilized microbial matter therein is well preserved because of the very early (almost immediate) encapsulation by silica, which, by virtue of widespread hydrothermal activity, was present at elevated concentrations in the already-Si-rich seas. The early Earth was hydrothermally very active, and microbial materials provided a suitable substrate for mineral precipitation. Thus, the cells, colonies, biofilms, mats, and generic degraded organic matter were rapidly fixed in a mineral matrix. This pervasive silicification affected not only the layers of sediments but also the underlying volcanic rocks (Hofmann 2011), thus further stabilizing the sediment pile. Finally, the terranes onto which the

sediments were deposited were themselves tectonically stabilized by an underlying keel (Van Kranendonk et al. 2015).

In the case of Mars, on the other hand, the small size of the planet (approximately half the diameter of the Earth) vastly reduces the potential temperatures and pressures encountered in typical burial metamorphism on Earth. As a result, Martian sediments of Noachian age (>3.8 Ga), that is, almost the time equivalent of the Early Archean, are far less metamorphosed than Archaean sediments. An estimated metamorphic grade of prehnite-pumpellyite (the lowest grade of burial metamorphism) is common for surface rocks on Mars and is largely due to heat from hydrothermal activity generated by impacts (McSween et al. 2015). Noachian biosignatures, especially organic biosignatures, are thus likely to be little altered on Mars.

8.7.2.2 Mineral Encapsulation

During encapsulation by a mineral—a process defined as biologically influenced fossilization (Li et al. 2013)—ions in the immediate environment of the cell or other microbial products, such as extracellular polymeric substances that co-occur with the degraded products of dead organisms, are passively fixed (chelated) to functional groups, such as carboxyls, hydroxyls, and phosphoryls, in the organic material. Fixation can also occur by means of cation bridging, whereby the cation, for instance, Fe, acts as a link between the organic functional group and the fossilizing mineral. Initial fixation occurs immediately, within the first 24 hours of exposure to the fossilizing medium (Orange et al. 2009). After initial templating by the organic substrate, polymerization of the fossilizing mineral results in complete encapsulation of the organic substrate, cell, or EPS (Figure 8.7.1a and b). Interestingly, EPS is more readily "fossilizable" than most individual cells because of the greater abundance of organic molecules with functional groups (Westall et al. 2000). Cyanobacterial sheaths and Gram-positive bacteria are also rich in functional groups and, therefore, relatively more fossilizable than bacteria with a different cell wall construction and composition, such as Gram-negative bacteria and Archaea. For example, it has been shown that fossilization of the Gram-positive *Bacillus laterosporus* (with a peptidoglycan-rich outer cell envelope) by silica produced a thick, robust crust around the organism, compared with a thin, delicate crust around the Gram-negative *Pseudomonas fluorescens, P. vesicularis,* and *P. acidovorans,* suggesting that microorganisms with a peptidoglycan-rich cell wall were more likely to enter into the rock record than more delicately encrusted forms (Westall 1997).

Even among microorganisms with similar cell wall characteristics, there are differing reactions to mineral encapsulation. The thermophilic Archaea *Pyrococcus abyssi* has been fossilized by silica, with delicate reproduction of its cell wall structure (Figure 8.7.1a), whereas another thermophile from the Archaea, *Methanocaldococcus jannaschii,* lysed almost immediately in the presence of silica (Orange et al. 2009). No cellular remains of the latter were preserved, but the EPS with which it was associated (that also included the degraded

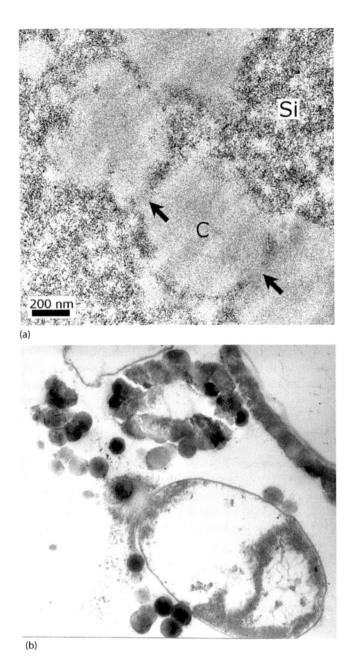

FIGURE 8.7.1 Examples of mineral-encapsulated microbial fossils. (a) *Pyrococcus abyssi*, a member of the domain Archaea, after 4 months of experimental fossilization in silica (Si), showing a chain of cells (C) whose cell walls have been silicified (and are still visible at the boundaries of juxtaposed cells (arrows). (Courtesy of F. Orange, 2008.) (b) Four months of experimental silicification of a cocktail of marine microorganisms from the domain bacteria showing the differential aspect of mineral encapsulation. The upper cell exhibits robust silica encapsulation in the form of spherical silica crystallites, while still conserving some degraded organic components inside. The lower cell has resisted encapsulation; almost all of its wall is clear of the silica spheroids; and the degraded remnants of its cytoplasm are preserved within the cell. Field of view = 3 µm. (Courtesy of F. Westall, 1993.)

breakdown products of the lysed organisms) was faithfully encrusted by silica. Interestingly, it was shown that exposure of the *M. jannaschii* cells to metal cations prior to exposure to the fossilizing solution conferred sufficient stability of the cell wall as to permit the fossilization of this organism, with Fe^{3+} being more effective than Ca^{2+}, Pb^{2+}, Zn^{2+}, or Cu^{2+} (Orange et al. 2011).

A further artificial fossilization study—in the case of a Mars-analogue microorganism, the *Yersinia intermedia*

MASE-LG-1 strain, isolated from the basalt-hosted, cold, acidic Lake Graenavatn in Iceland—produced interesting results of relevance to fossilization in astrobiology (Gaboyer et al. 2017). *Yersinia intermedia* MASE-LG-1, a Proteobacterium, was stressed under a combination of Mars-like environmental conditions, namely desiccation, UV radiation, and exposure to perchlorates, and both stressed and unstressed cells were fossilized using silica or the calcium sulfate mineral gypsum. Both minerals have relevance for Mars since silica is a typical product of hydrothermal fluids and was the agent of fossilization on the early Earth under conditions similar to those of the Noachian Mars, whereas sulfates are a common mineral phase on Mars from Hesperian time onward. In the experiment, prior environmental stress made no difference to the fossilization potential of *Yersinia,* but there was a difference in the rate and manner with which the two minerals fossilized the microorganism. While silica rapidly fixed to the cell wall and embedded cells in a thick, mineral precipitate, fossilization by gypsum occurred much more slowly. On the other hand, it appears that gypsum precipitation occurred both inside and outside the cells, while silica remained only on the outside of the cell. This is in contrast to earlier observations documenting the permeation of microbial cells by silica, often with silica precipitating around collapsed cytoplasm within the cells (Westall et al. 1995).

Fossilization or mineral encapsulation of microbial communities in the natural environment is a common phenomenon and can be considered a natural stress. Microorganisms can change the expression of their genome during mineralization, for example, silicification, to control membrane exchange (Iwai et al. 2010). Moreover, in artificial fossilization experiments, there are always some cells that appear to survive for significant periods of time. Approximately 10% of the cells could still be alive after several months of exposure to a silica solution (Gaboyer et al. 2017); however, this of course depends upon the rapidity of mineralization. There are thus strategies that microorganisms use to adapt to survive stressful situations. While the majority of cells do not survive beyond initial exposure to the agent of mineralization, some do.

8.7.2.3 THE ORGANIC COMPONENT

What happens to the organic matter of microorganisms that have been encapsulated by a mineral? In an aerobic environment, the organics are oxidized, leaving a mineral cast, but in an anaerobic environment, such as on the early Earth or Mars (and many euxinic environments on Earth today), the organic component would be preserved but undergo gradual degradation over time. Investigations of the first stages of degradation show a breakdown a rapid change in the composition of the organic components during fossilization. Changes in the composition and quantity of amino acids, monosaccharides, and fatty acids of silicified *M. jannaschii* (that lyses rapidly on exposure to silica; Orange et al. 2009) were monitored over a 1-year period (Figure 8.7.2) (Orange et al. 2012). Although lysed, the cell components of *M. jannaschii* formed part of the silicified EPS. After 1 year, of these components, the amino and fatty acids were relatively well preserved, while the monosaccharides broke down into smaller fragments, in the process, thus increasing the amino acid fraction. In this way, organic matter either directly associated with a fossilized cell or colonies/biofilms of cells can be preserved in degraded form, together with the degraded breakdown products of lysed cells trapped in a mineral cement.

The alteration and maturation of biogenic organic matter with time have been well-studied in the oil industry. Going

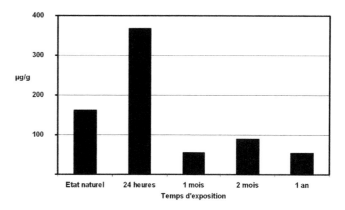

FIGURE 8.7.2 Change in the quantity of sugars present in a fossilized mixture of *Methanocaldococcus jannaschii* over the period of 1 year of fossilization with silica (the X-axis shows variation with time: temps d'exposition—exposure time [to silica], Etat naturel—natural state, 24 heures—24 hours, 1 mois—1 month, 2 mois—2 months, 1 an—1 year). The initial increase is due to the fact that *M. jannaschii* rapidly lyses upon exposure to silica, with the degraded components joining the cellular extracellular polymeric substances (EPS). Continued fossilization leads to the breakdown of sugars with time.

through various breakdown stages related to thermal maturation and the effects of burial pressure or higher pressures related to tectonic transport into the crust and mantle, biogenic organic matter becomes increasing mature. The structure and composition of the molecules change with the loss of H, O, N, S, and functional groups, until they become refractory, aromatic molecules, generically termed kerogen. Finally, under high degrees of metamorphism, kerogen becomes graphite or even diamond, as surface sediments containing organic matter are transported into the mantle by plate tectonics (e.g., Nemchin et al. 2008). Despite the degree of degradation, it is still sometimes possible to distinguish a specific biomolecule precursor in ancient kerogen (which, in this case, is termed a biomarker). The oldest confirmed eukaryotic biomarker comes from 1.6-Ga-old Barney Creek sediments in Australia (Summons et al. 1988). Biomarker study is notoriously difficult because of the problem of contamination, not just during sampling and handling but also during transport of younger kerogen through porous horizons to older strata. This has been suggested to be the case with purported biomarkers, 2-methylhopanoids, documented in 2.7 Ga rocks (Brocks et al. 1999; Summons et al. 1999), that have been shown to originate from biomarker contaminants in mature hydrocarbons (French et al. 2015). Indeed, it is possible that no known Archean strata is within the window of thermal maturity for syngenetic biomarker preservation (French et al. 2015).

Although organic molecules can be preserved in deep time, they occur as highly degraded fragments of the former biological macromolecules. Nevertheless, certain biological characteristics can still be identified, such as isomeric, molecular weight, and isotopic fractionations of carbon, and used for determining the biogenic origin of the remnant carbon (Summons et al. 2011; Vago et al. 2017).

8.7.2.4 "Mummified" Microorganisms

Some of the most spectacular microfossils of, especially, the Proterozoic era are acritarchs. These are organic-walled cells preserved without mineral encrustation, simply by entombment in fine-grained anaerobic sediments. It is clear that the kinds of cells that could be thus preserved need to have a robust outer cell envelope, and, in fact, acritarchs are believed to be the remains of protists, eukaryotic organisms (Knoll et al. 2006). They were deposited as a component of fine-grained sediments in anaerobic sedimentary basins. Of course, with time and geological processing, the organic components will degrade to become refractory molecules, as described earlier. While being relatively common in the Proterozoic Era, small acritarchs of as-yet-unknown affinity have also been described from 3.2 Ga sediments in the Barberton greenstone belt (Javaux et al. 2010, Figure 8.7.3), which we will revisit later in this chapter.

8.7.2.5 Fossilized Communities

Most of the artificial fossilization experiments described earlier concern single strains of microbes; however, in the natural environment, a wide range of microbes coexist in communities, either free floating in the water column or attached to a substrate. Experimental silicification of a natural cocktail of marine microorganisms has documented that they exhibit a wide range of responses to fossilization that corroborate some of the experiments described earlier: certain spirochetes resisted fossilization for many months, some microorganisms exhibited delicate silica crusts, while others were thickly encrusted (Westall et al. 1995). The differential responses of microorganisms to fossilization is important, because it tells us that what we find in the rock record is only a fraction of the original communities. What is preserved in the rock record, whether it be mineral casts alone or morphologically preserved organisms complete with their degraded and preserved organic components, can be considered a bulk signature in terms of analysis in a rock.

Another aspect of the fossilization of microbial communities is that the composition of the communities evolves with time. Taking a modern phototrophic microbial mat as an example, while the primary producers are phototrophs (primarily oxygenic but also anoxygenic photosynthesizers), the mat is host to a cohort of different species that occupy microscopic ecological niches and whose niche can change on a diurnal timescale, from day to night (Bolhuis et al. 2014). Phototrophic mats can be fossilized, and stromatolites and microbial mats are the striking evidence of this. However, cellular microfossils or the traces of their former presence (palimpsest microstructures) are seldom preserved, while the mats are often characterized as layers of trapped detrital particles and bio-precipitated minerals. In rare cases, organic matter associated with the mats (or biofilms) can be investigated, but this organic matter must be appraised as a degraded bulk sample of all the microbial components of the original mat or biofilm—phototrophs and heterotrophs—and cannot be related back to one specific type of microorganism.

8.7.2.6 Fossilized Viruses

It has been estimated that there are far more viral particles on Earth than bacterial cells, and viruses are considered to be as ancient as bacteria (Moelling 2017). Since bacteria are naturally fossilized in the environment, it is not inconceivable that viruses would also be fossilized in a similar fashion. Indeed, in the natural environment, the iron mineralization of viruses has been observed in the acid waters of Rio Tinto, Spain (Kyle et al. 2008). One of the first fossilization experiments on viruses showed precipitation of silica on the bacteriophage T4 after a few days in simulated hot spring conditions (Laidler and Stedman 2010). Furthermore, experimental silicification of viruses from extremophilic Archaea *Sulfolobus islandicus* rod-shaped virus 2 (SIRV2), *Thermococcus prieurii* virus 1 (TPV1), and *Pyrococcus abyssi* virus 1 (PAV1) over a period of several months showed that silica precipitation on DNA proteins and some cell envelopes occurred in a species-dependent manner (Orange et al. 2011); however, the resultant fossils did not reliably reproduce any particularly identifiable morphology. Moreover, the silicified parts were extremely small. It is therefore likely that fossil viruses would be very difficult to identify, even though they may be as common as fossil bacteria in the rock record.

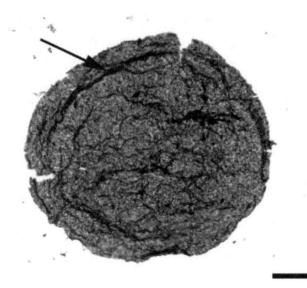

FIGURE 8.7.3 "Mummified" microbial cell from fine-grained sediments in the Moodies Group, Barberton greenstone belt. The cell wall was sufficiently robust to resist mechanical breakdown in these sediments. Scale bar = 50 μm. (Reproduced from Javaux, E.J. et al., *Nature*, 463, 934–938, 2010. With permission.)

8.7.2.7 Other Biosignatures

A priori it is not expected that processes as ephemeral as microbial metabolism could be preserved. However, the effects of microbial metabolic activities on their

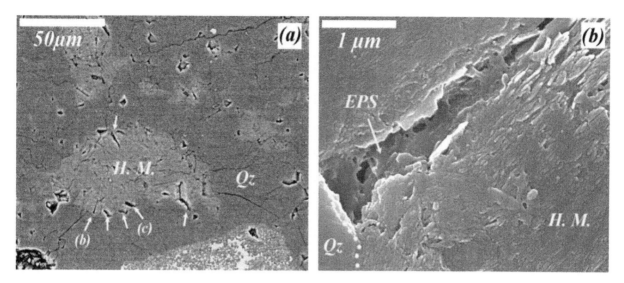

FIGURE 8.7.4 Fossilized evidence of microbial corrosion. (a) Silicified, volcanic particle (light gray) altered to hydromuscovite (H.M.), embedded within a lithifying silica cement (Qz). The arrows point to tunnels within the volcanic particle (backscattered scanning electron micrograph). (b) Close-up view of a corrosion tunnel tapering from the outer edge of the grain inward. Note the infilling of silicified amorphous material interpreted as extracellular polymeric substances (EPS). Qz: quartz; H.M.: hydromuscovite. (Reprinted from Foucher, F. et al., *Icarus*, 207, 616–630, 2010. With permission.)

immediate surroundings can be encapsulated into the rock record. For example, enzymes emanating from chemolithotrophs can corrode the underlying substrate to liberate elements essential for life (Figure 8.7.4). Microbial corrosion of metals is of particular interest to ships and the oil industry, costing industry large sums. Microorganisms using oxidation or reduction of sulfur play a large role in these processes (Iverson 1987). Microorganisms are also implicated in the weathering of silicate minerals on the surface and in the upper crust. The mechanisms of mineral corrosion include physical disaggregation and the effects of inorganic and organic acids, such as oxalate, as well as siderophores and proteins (Barker et al. 1997).

Minerals can be precipitated either directly by microorganisms or indirectly because of the presence of biological organic matter. Magnetite (Fe_3O_4), for example, can be indirectly precipitated extracellularly by biologically mediated mineralization (Bazylinski and Moskowitz 1997), while greigite (Fe_3S_4) can be formed through biologically controlled mineralization, that is, specifically synthesized and deposited in a particular location in the cell. Another example of relevance to astrobiology and, in particular, biosignatures, is the precipitation of carbonate. Biologically controlled precipitation of carbonate is known from hard-shelled mollusks, but biologically (in this case, microbially) mediated carbonate precipitation can lead to the encapsulation of microbial cells or the precipitation of calcium carbonate crystals within microbial EPS. Precambrian stromatolites were largely formed through the calcification of the EPS associated with the mat layers formed by phototrophic microorganisms (e.g., Pentecost 1990; Hofmann et al. 1999).

8.7.3 IMPLICATIONS FOR THE SEARCH FOR FOSSIL TRACES OF LIFE ON THE EARLY EARTH

8.7.3.1 THE EARLY MICROBIAL ENVIRONMENT

Microorganisms are vital components of many surface and subsurface environments, and the study of their traces in deep time needs, accordingly, to take into account the paleoenvironmental and geological characteristics of the biome that they inhabited. Indeed, understanding environmental context is an essential part of interpretation of fossilized traces of life. Consequently, study of early life requires an understanding of the regional and local environments prevalent on the early Earth. This has been described in greater detail in other works (Nisbet and Sleep 2001; Arndt and Nisbet 2012; Pearce et al. 2018; Westall et al. 2018), but it is useful to briefly review the salient characteristics here. The early Earth was a truly alien world: an anoxic planet characterized by high heat flow from the mantle—the result of some combination of the latent heat of accretion, the decay of now-extinct radioactive nucleides, and the unique (proto)tectonic system—that resulted in vigorous volcanic activity and pervasive hydrothermal circulation within the crust (Westall et al. 2018). pH values in the early oceans were likely variable: while the ocean would have been largely acidic as a result of its interaction with a volcanically derived CO_2 atmosphere, smaller-scale alkaline conditions occurred in sediment pore spaces due to alteration of (ultra) mafic minerals and in locations where alkaline hydrothermal fluids effused at the surface. In some oceanographic regimes, it is likewise possible that the alteration of mafic strata and sediments would have allowed the precipitation of veneers of carbonate sediments, as espoused in the "soda lake" model

(Kempe and Degens 1985). As noted earlier, only a very small portion of the sedimentary record of the early Earth has been preserved, and this represents almost exclusively shallow water basins and the platformal surrounds of sub-aerial volcanic systems. The CO_2-rich atmosphere and lack of ozone imply that UV radiation flux to the surface of the Earth was high, up to 1000 W/m^2 (DNA weighted, cf. Cockell and Raven 2004) in the worst-case scenario and ~54 W/m^2 in more clement cases, compared with 1 W/m^2 at present. Thus, at a global scale, the early Earth can be considered an extreme environment, and, necessarily, its microbial inhabitants were extremophiles. Infinitely small amounts of oxygen were produced by abiotic processes, this oxygen being poison to strict anaerobes.

While the earlier brief description may suggest that the early Earth was a homogeneously hostile world, that description pertains to the global environment. At the microbial scale, that is, in the sediments and on the sediment/rock surfaces, around hydrothermal vents and springs, and, possibly, even in the water column, there seems to have lived a wide variety of thermophiles, acidophiles, alkaliphiles, halophiles, and other extremophilic and extremotolerant microorganisms. Although only a small fraction of the microbial community in any habitat is preserved in the rock record, and despite the many effects of degradation during geological processing, the record of early life is nonetheless surprisingly rich. This is partly due to the lack of reworking of sediments by bioturbation prior to the Cambrian Explosion of animal life (Brasier 2009). More significantly, however, this biotic richness is due to the fact that most well-preserved Archean sediments were very rapidly silicified, silica being one of the most stable minerals at the surface of the Earth. Once silicified, it is difficult to chemically alter the lithified rock; for this reason, cherts provide high-fidelity archives of geochemistry that can endure over billions of years (van Zuilen et al. 2007; Ledevin et al. 2014; Gourcerol et al. 2016). Silica is certainly the *nec plus ultra* mineral for microbial fossilization. There is, nonetheless, a slight shortcoming in silicification and that is the dilution factor. The sometimes-significant dilution of the organic and elemental components of biosignatures mean that, while well-preserved, quite sophisticated methods and instrumentation are needed for their characterization.

Although a wide variety of minerals are known to encapsulate microbial matter, including ice, halite, sulfates, carbonates, phyllosilicates, silica, hematite, and phosphates (Summons et al. 2011), certain minerals are evidently metastable under changing conditions and alter over geological time. Ice is an excellent preserver of organic matter—as documented by the famous mummified fossil Ötzi, preserved in an alpine glacier (Müller et al. 2003), or the mammoths preserved in Siberian permafrost—but, unless on a cold or icy planet or satellite, such as Mars, Enceladus, or Europa, ice is not long-lasting. Salt has been used through historical (and possibly prehistorical) time to preserve foodstuffs, but halite is metastable and hygroscopic. It is subject to contamination, as demonstrated by the debate surrounding the discovery of bacteria in 250-Ma-old salt deposits (Vreeland et al. 2000),

and now is considered to be the result of contamination (Graur and Pupko 2001). Notwithstanding, a recent review concluded that halite can be stable on geological timescales (hundreds of My) and could, with reasonable fidelity, conserve microorganisms (Jaakkola et al. 2016). In terms of deep geological time and the search for traces of life from the early Earth, halite as such is not preserved, but halide crystals coating a fossilized microbial biofilm, both impregnated and pseudomorphed by hydrothermal silica, have been described in 3.33 Ga sediments from the Barberton greenstone belt in South Africa (Westall at al. 2006a, 2011a). With respect to sulfates, it has been claimed that DNA fragments were recovered from fossilized cyanobacteria in Miocene-age gypsum (Panieri et al. 2010), although gene sequencing from ancient strata is fraught with contamination issues. Irrespective of whether the identified DNA fragments are syngenetic with the gypsum, the morphological remains of the cyanobacteria are certainly preserved and authentic. As with halite, gypsum, as such, has not survived into deep time; however, silica-pseudomorphed gypsum has been noted in ancient rocks, for example, associated with ancient biofilms (Westall et al. 2006a).

8.7.3.2 Case Studies from Early Life on Earth

In what follows, we present a small selection of some cases of exceptional preservation of microbial biosignatures from the early Earth, emphasizing the positive and negative aspects of their processes of fossilization on the eventual biosignature, and how the record of natural fossilization gives credence to the experimental approaches and findings of fossilization studies outlined earlier. The interested reader could also seek more thorough overviews of the full range of biosignatures from the ancient Earth (e.g., Wacey 2009; Westall and Cavalazzi 2011) and their relationships with the environment of early life (Nisbet and Sleep 2001; Westall 2016; Pearce et al. 2018).

Carbonates: Carbonates are common preservers of microbial material, going back into deep time. Although also often pseudomorphed by silica, carbonate in the form of dolomite occurs associated with phototrophic microbial structures in Proterozoic and Early Archean rocks (Lowe 1980; Beukes and Lowe 1989; Hofmann et al. 1999; Allwood et al. 2006). The most iconic Archean stromatolites are certainly those of the Strelley Pool Formation of the East Pilbara terrane (Figure 8.7.5), from which well-described, morphologically varied stromatolites of almost universally accepted biological significance (Hofmann et al. 1999; Van Kranendonk et al. 2003; Allwood et al. 2006, 2007; Wacey 2010) are known from at least two major localities in the East Strelley greenstone belt (the "Trendall locality" and "McPhee Creek"—the latter formerly "Strelley West"—as reported in the literature) and from the Panorama greenstone belt. REE+Y signals from well-preserved stromatoidal carbonates (Van Kranendonk

FIGURE 8.7.5 Cross-section view of silicified stromatolites from the Trendall locality of the 3.43 Ga Strelley Pool Chert, Western Australia. Fieldwork hammer (~30 cm) for scale. (Courtesy of F. Westall, 2000.)

et al. 2003), together with the multiple morphologies of stromatolite-like structures present in a single carbonate unit (Allwood et al. 2006, 2007; Wacey 2010), are highly persuasive of a biogenic origin for these structures and of a plausible geochemical environment for the development of a stromatolitic ecosystem. In addition to biologically driven carbonate (now dolomite) sedimentation during stromatolite growth, chemical precipitation certainly played a significant role in the preservation of these stromatolites, given the widespread evaporite crystal pseudomorphs in intimate association with stromatolitic laminae, which may have formed mineral crusts, aiding the preservation of the laminae (Allwood et al. 2007). The inclusion of allogenic (detrital) and authigenic mineral phases within laminations also suggests that the microbial biostabilization helped to preserve the original organo-sedimentary structure (Allwood et al. 2007). This "trapping and binding" behavior is a widely acknowledged mechanism by which otherwise-fragile organo-sedimentary constructs may leave visible traces in the fossil record (Noffke et al. 2006, 2013; Noffke 2010; Hickman-Lewis et al. 2016, 2018). The fossilization of stromatolites of biogenic origin in carbonates of the Strelley Pool Formation can be conclusively summed up as a multi-mechanism procedure, involving biomediated chemical precipitation and mechanical sedimentation in an environment where the propensity for the precipitation of carbonate and associated evaporite phases was enhanced. For this reason, comparison with younger forms (e.g., *Conophyton*, *Jacutophyton*, and *Thyssagetes*, dependent upon morphology) in better-preserved Proterozoic and Phanerozoic sediments has enabled and justified a deeper understanding of the development and fossilization of organo-sedimentary structures such as these.

Silica: As noted earlier, silica presents an unprecedented agent of preservation when considering biosignatures from deep time. The elevated Si concentrations of ocean water throughout the Precambrian (Siever 1992; Maliva et al. 2005) enhanced the predisposition both to siliceous chemical sedimentation and to the incorporation of silica within sediments to chertify them rapidly, no doubt, in large part, due to the considerably elevated influx of Si-rich hydrothermal waters into Precambrian seas (Westall et al. 2018). Even the carbonate-preserved organo-sedimentary structures of the East Strelley greenstone belt were preserved due to synsedimentary or later silicification that pseudomorphed the carbonate mineralogy (Allwood et al. 2007; Wacey 2010). In deep time, however, chertified sediments are the typical repository for biosignatures. The following three examples from the Paleoarchean showcase the superb preservation that can be afforded by silica and explain the key mechanistic aspects ensuring this preservation:

First, within the 3.22 Ga Moodies Group, a diverse ecosystem has been reported that likely occupied multiple trophic levels. In now-silica-filled cavities beneath kerogenous microbial mat laminae, well-preserved chains of moulds with cell-like dimensions, morphologies, and habits have been reported (Homann et al. 2016; Figure 8.7.6). Transverse cross-sections of these features show tubular morphologies encapsulated by silica and are testament to the preservational potential of silica at even the micron scale. Second, in the 3.33 Ga Kromberg Formation, an exceptionally preserved microbial biofilm coated with a thin layer of silica is inferred to have been preserved by Si-rich fluids emanating from a proximal hydrothermal system (Westall et al. 2011a). Synchrotron Micro X-ray adsorption near edge studies (µXANES) mapping and spot analyses demonstrated concentrations of sulfur in kerogenous material, within which the S-bearing molecule thiophene was identified. Together with preserved nanoscale domains of aragonite crystallites that are syngenetic with the formation of the biofilm, it was argued that such a suite of high-resolution geochemical data—faithfully preserved over long geological timescales—supports an originally diverse consortium of mat-building organisms, including sulfate-reducing bacteria. Third, silicified microclastic cherts from the 3.46 Ga stratiform "Apex chert" preserve low-relief, wrinkle-like structures that, when imaged by high-resolution three-dimensional X-ray tomographic techniques, depict preserved filament-like structures on the micron scale (Hickman-Lewis et al. 2017). These structures are directly comparable to the filament-like constructs, indicating the former presence of bacteria in both modern and fossil microbial mats (e.g., Noffke 2010), and were

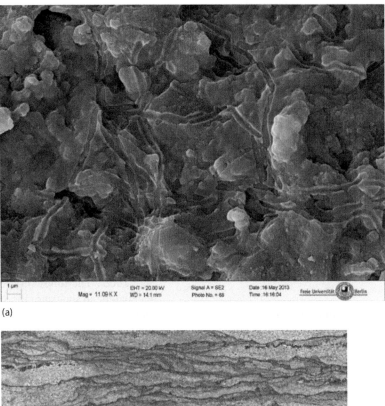

(a)

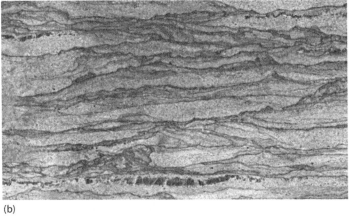

(b)

FIGURE 8.7.6 Silicified microscopic and macroscopic biosignatures from the 3.22 Ga Moodies Group clastic sediments, Barberton green-stone belt, South Africa. (a) Moulds of chains of rod-shaped of microorganisms in the silica infill of a subaerial vadose cavity in the Moodies Group sediments. Here, silica has encapsulated the organisms whose organic remains were subsequently completely degraded. (b) Slab of rock cut transversally to the sediment bedding planes, in which the microbial mat-stabilized sedimentary surfaces are evidenced by the dark (organic-rich), wrinkly layering. Field of view = ~25 cm. (Reproduced from Homann, M. et al., *Geology*, 44, 51–54, 2016. With permission.)

therefore presented as a type of MISS within lami-nae that pass many other necessary morphological biogenicity criteria, again demonstrating the preser-vation potentials of silica at the finest scales.

The cavities in which the Moodies Group microfos-sils occur resemble gas-filled, fenestral hollows in tidal environments, which, forming through some combination of tidally driven hydraulic pumping and biologically driven gas accumulation (Homann et al. 2016), were certainly dynamic environments on short timescales. That vestiges of their micro-bial inhabitants are conserved is therefore a clear indication of the rapidity of the silicification in these coelobiontic microbiomes. Microbial remains from the Kromberg Formation, exhibiting both the volume preservation of a biofilm and syngenetic

metastable mineral phases (e.g., aragonite) within, support that silicification was initiated rapidly while the biofilm was still alive and metabolizing (Westall et al. 2011a). The ancient MISS-like structures in the stratiform "Apex chert" being preserved in volume and including turgid, three-dimensional fabric ele-ments (Hickman-Lewis et al. 2017) strongly sug-gest that the silicification of this putative microbial system occurred penecontemporaneously with its growth and that particularly rapid silicification can indeed circumvent the morphological altera-tion caused to biosignatures by compression dur-ing diagenetic processes. These three examples of exceptional biosignature preservation demonstrate the fine-scale fidelity of preservation afforded thereby ensuring rapid entry of morphological and geochemical biosignatures into the fossil record, and consequently the status of microcrystalline silica as an exceptional mineralization medium for the most ancient rock record.

This siliceous entombment is similarly beautifully exhibited by the microfossil communities of the Kitty's Gap Chert. Coccoidal microfossils occurring in monolayer coatings surrounding volcanic particles (Figure 8.7.7), together with microbial corrosion features (Figure 8.7.4), have been well-documented in terms of their morphological and geochemical characteristics (Westall et al. 2006b, 2011b; Foucher et al. 2010). The preservation of these coccoidal microfossils is such that they have been captured in multiple life modes: cell division, cell lysis, and at presumably multiple growth stages (Westall et al. 2011b). Further high-resolution scanning electron microscopy (SEM) imaging of corroded volcanic clasts has shown tunnels approximately 0.5–1 μm in diameter and up to ~20 μm in length that taper inward into their host clasts and are filled with a silicified amorphous material, interpreted as extracellular polymeric substance (Foucher et al. 2010). Fossils such as these, preserved from oligotrophic conditions, provide apposite analogue organisms for consideration in the search for life on other celestial bodies, where nutrient-rich conditions may not have prevailed. We will return to this idea later in the chapter.

Phosphate: A further means of high-fidelity preservation of organic fossils is cryptocrystalline phosphate,

FIGURE 8.7.7 Variable preservation of a colony of silicified microbial cells from the 3.446 Ga Kitty's Gap Chert, Western Australia. The large cell in the center of the image was undergoing cell division at the time of fossilization, while those above it were already lysed and partially degraded at the moment of fossilization. The crystals surrounding and embedding the cells are now quartz but would originally have been amorphous, hydrothermal silica gel. (Courtesy of F. Westall, 2002.)

which is, as with chert, a means of preserving fossils in three dimensions. Within the scope of the current study—the harmonization between our understanding of the fossilization processes on the early Earth of relevance to potentially fossilized organisms on Mars—there is regrettably little judicial sense to discussing phosphatization. Phosphate-preserved microfossils on Earth date back no further than around 1 Ga (the lacustrine Torridon biota of Scotland), by which time any planetary (and likely microbial) similarity between Earth and Mars had necessarily ceased to exist. The preserved organisms are eukaryotic and parts of a broad and diverse ecosystem, the likes of which would not be predicted for Mars. Furthermore, P-rich deposits on Mars are often linked more to magmatic origins than to sedimentary origins and therefore not with environments clement to life. The interested reader may wish to read further into phosphate fossilization on Earth, for which a wealth of literature exists (e.g., Wilby and Briggs 1997; Strother et al. 2011; Strother and Wellman 2016).

"Mummified" fossils: Apart from the fine-scale preservation afforded by silica and phosphate, one instance of interpreted "mummified" biogenic features with morphological preservation at the nano-scale has been recorded from the Archean in the form of acritarchs in the Moodies Group (Javaux et al. 2010; Figure 8.7.3). These acritarchs are preserved in layered, laminated, silicified shales, and siltstones, which entombed and compressed them parallel to bedding planes. The highly wrinkled appearance of the preserved cell wall-like structures has been attributed to taphonomic degradation and collapse of the carbonaceous material through loss of turgor pressure (Javaux et al. 2010).

Sulfides: It is worthwhile concluding this section by examining two further, but less well-recognized, preservational media from the ancient rock record. In traverses of environmental settings conducive to the preservation of microbial biosignatures, deep-sea, black smoker settings and pillow-basalt hosted fluid-rock systems have frequently been noted as possible biosignature-rich deposits (Brasier et al. 2011; Westall and Cavalazzi 2011) but are hindered by their rare occurrence in the rock record and the simplicity and ambiguity of the biosignatures preserved therein.

Only one convincing example of biosignatures in a black smoker-type deposit is reported from the Archean rock record: the Sulphur Springs volcanogenic massive sulfide ore deposit (Rasmussen 2000). Although ore-forming fluids in this locality may have reached 300°C (beyond the known thermotolerant capacities of life), lower temperatures would have characterized the distal locations from which the putative microfossils were found. Pyritic filaments of up to 300 μm length and 0.5–2 μm constant diameter occur in early silica generations; are of unbranched morphology, intertwine and curve; and are densely distributed in the rock. These characters are consistent with a biological origin. A fossilization model was proposed: filaments were engulfed by hydrothermal silica from proximal vent systems, after which increasing fluid temperatures due to a nearby intruding granitic body led to the replacement of the silica by pyrite (Rasmussen 2000). Unfortunately, this preservational model is somewhat incomplete in terms of geochemistry, and, to our knowledge, no experimental replication of the proposed model has been conducted. Furthermore, the abiogenic null hypothesis that these features are instead abiotic mineral growths has not been disproven. Despite later recollection of material from Sulphur Springs and advocacy for the biologically promising character of these putative microfossils (Wacey 2009; Brasier et al. 2011), no further data has been presented, and thus, the fossilization and preservational processes of bacteria in black smoker-type systems remain understudied.

Microbial corrosion: Finally, in a suite of studies, microbial trace fossils have been proposed from the rims of pillow basalts in the Archean rock records of South Africa and Western Australia (Furnes et al. 2004, 2007; Banerjee et al. 2006, 2010). These microtubular structures, of up to 10 μm in diameter and 200 μm in length, bear resemblance to modern microbial etching features in seafloor basalts, for example, in terms of their irregular growth patterns, carbon isotope fractionations, and petrological contexts. The ancient examples are preserved by titanite, which is often a metamorphic phase; however, enrichments in carbon, nitrogen, and phosphorus have also been reported in the microtubules themselves. The titanite preservation has been ascribed to water cycling through cavities, enhancing the growth and excretion processes of the chasmolithic inhabitants while simultaneously replacing and sealing the tubules (Furnes et al. 2007). Although modern etching structures in basalt and hyaloclastite rims are doubtless of biological origin, a set of geochemical-geochronological studies has cast doubt on the syngenicity of the ancient examples (e.g., Grosch and McLoughlin 2014, 2015). In the absence of detailed experimental work to understand the preservation of these kinds of biosignatures, these structures remain enigmatic, and their relevance to the preservation of biotic processes on the early Earth remains unconfirmed.

In contrast, two recent studies have documented biocorrosion in the ancient volcanic sediments. The Kitty's Gap Chert example mentioned earlier reveals tunnels running from the rims of clasts inward and infilled with what is interpreted to be silicified EPS, and this is a sample within which silicified microbial colonies have been observed on the surfaces of the volcanic particles (Foucher et al. 2010). Another recent study of tubular structures in volcanic glass showed the co-occurrence of organic material and titanium (Wacey et al., 2017), i.e. the kind of signature needed in the Furnes et al. (2007) study.

Implications of early biosignature study for the search for life on Mars: In this section, we have attempted to outline, through several relevant examples, the links between processes known from artificial fossilization experiments and the representation of these reactions in the ancient fossil record. Despite great ages of more than 3 Ga in some cases, the appearance and nature of preservation of some exceptional biosignatures corroborate experimental procedures. This correlative understanding is important for understanding the links between organisms, their environments, and their diagenesis and taphonomy. In turn, the recognition of these processes is of fundamental importance to assessing the biogenicity of potentially controversial microfossils and microbial constructs. The relevance of the understanding of early life to astrobiology is axiomatic: Archaean life records the early, and perhaps initial, stages of evolution following the emergence of life on Earth and preserves a record of biotas and their paleoenvironments, which needs to be interpreted through use of multiple appropriate techniques.

Before advancing to the discussion of the search for life on Mars, one final consideration should be raised in terms of biosignature detection. Although we have herein recapitulated a series of widely accepted, well-described, and comprehensively understood biosignatures for early life, these represent "exceptions," not the "rule." Biosignatures from the Archean provide both a paleontological analogue for potential Martian life and a philosophical analogue, which incorporates aspects of both epistemology and semiotics, and thus, the scrutiny that we apply to Archean biosignatures should provide a baseline for the scrutiny applied to putative Martian biosignatures. Many Early Archean biosignatures have been proposed and analyzed on weak geological and geochemical foundations, and yet, many more have been misinterpreted, despite having been subjected to the efforts of the finest terrestrial laboratories, by virtue of their enigmatic, ambiguous, and obfuscated geological-geochemical signals. The assessment of biosignatures *in situ* on Mars will present new, and even greater, challenges. The lessons from the Allan Hills 84001 meteorite (McKay et al. 1996), in which Martian bacterial fossils were proposed and subsequently disproven, underline the challenges involved in assessing the biosignatures, even in terrestrial laboratories, with the most sophisticated instrumentation. Some researchers have already started to postulate comparisons between sedimentary fabrics on Earth and Mars—strata resembling microbially induced sedimentary structures in the Gillespie Lake Member (Noffke 2015) and digitate structures resembling biogenic hot spring deposits in Columbia Hills (Ruff and Farmer 2016)—and thus the lessons of caution, consideration, multidisciplinary consensus, correlative multi-scalar approaches, and conscious, prudent circumspection in biosignature interpretation have never been more important.

8.7.4 ASTROBIOLOGY MISSIONS TO MARS

8.7.4.1 Habitability on Mars

Water on Mars: Understanding early life on Earth, including the environmental parameters of its habitat, the nature of early life, its preservation, and, finally, the methods used to detect it, is an essential aid to the search for life on Mars. Unlike Earth, Mars has not been permanently habitable, and there is considerable question as to the degree of habitability on the planet throughout its geological history. There is evidence for a significant amount of water on the surface of the planet during its early history (Figure 8.7.8) but not a global ocean, while the later history of Mars is characterized by an overall gradual decrease in habitability, punctuated by sporadic and localized peaks in aqueous activity during the Hesperian (e.g., the catastrophic outflow channels), before reaching essentially dry desert conditions from the Late Hesperian through the Amazonian (Carr and Head 2010). The habitability of Mars has thus changed drastically through geological time, however. Mars is half the size of the Earth and, on a global scale, would have lost its habitable conditions at the surface at around 3.8 Ga owing to the demise of the internal dynamo that created a radiation shield protecting the atmosphere and volatiles from erosion by the solar wind and specific solar events (Jakosky et al. 2015). The degradation in habitable conditions on Mars was further punctuated by events with more local effects, such as the large-scale volcanism associated with the growth of the Tharsis Massif, impacts and catastrophic outflow events, and potential tsunami.

At the present day, much water is believed to reside on Mars in a thick subsurface cryosphere, beneath which liquid water may be trapped. Breaching of

FIGURE 8.7.8   Artist's depiction of early Mars, with water at the surface. (Courtesy of Ittiz, Own work, https://commons.wikimedia.org/w/index.php?curid=7861829.)

the cryosphere, for example, by impacts, could lead to the catastrophic outflow events that characterized the Hesperian (Carr and Head 2010). There have also been many theories concerning the existence of major ocean in the Northern Plains (e.g., Clifford and Parker 2001), as well as large bodies of water accumulating in topographic lows, such as the giant Hellas impact basin; however, debate concerns the timing of their existence. During the Noachian, when atmospheric pressure was higher, the surface water inventory was likely to have been higher (Figure 8.7.8), whereas later, during the Hesperian, catastrophic outflows were perhaps more likely. Even after these events, as the planet was becoming increasingly dry, Amazonian Mars saw occasional liquid water at the surface. Even today, runnels formed by possibly briny water appear on south-facing slopes at the equator (recurring slope lineae, McEwen et al. 2015) and layers of water ice mixed with CO_2 ice occur at the poles (Carr and Head 2010; Orosei et al. 2018). Unlike the Earth, Mars undergoes large changes in obliquity because it lacks the stabilizing influence of a large moon. Thus, the higher latitudes can be periodically warmed by sunlight, and frozen subsurface water can be released.

The evidence for water on Mars is found not only in the abundant geomorphological features, for example, rivers, run-off channels, catastrophic channels, lakes, deltas feeding into lakes, and possibly even oceans and seas, but also in abundant mineralogical evidence. A recent review (Bishop et al. 2018) summarized the evidence for surface aqueous alteration of mostly volcanic lithologies, as well as subsurface alteration by hydrothermal solutions. Distinct clay facies suggest, for instance, that the thick and widespread sequences of Fe/Mg clays overlain by Al species with Si in a number of locations on Mars, for example, Mawrth Vallis, were the result of an event or events, such as an impact, that led to increased water vapor in the atmosphere and slight, regional warming culminating in intense chemical weathering. The *in situ* measurements by the Mars Exploration Rovers (MER) and Curiosity rover have also detected evidence for hydrous minerals (Grotzinger et al. 2005, 2014). In addition, orbital radar mapping of the planet has documented ice (in the form of hydrogen detection) (Plaut et al. 2000), whereas groundwater ice was detected by the Phoenix lander near the northern ice cap.

The history of Mars is therefore vastly different to that of the permanently habitable Earth. Nevertheless, although there are significant differences in the global-scale habitability of the two planets throughout geological time to the present day, on a microbial scale, the environment of early Mars (pre-Noachian, Noachian, and early Hesperian, i.e., >~3.6 Ga) can be considered to have been somewhat similar to that of the early Earth (Westall et al. 2013, 2015a, b).

The presence of liquid water at various times during the history of Mars, even up to the present day, means that it could have been potentially habitable, if the presence of water was co-located with a source of carbon and a source of energy. Even the small and ephemeral recurring slope lineae are considered to be potentially habitable for the purposes of planetary protection (Rettberg et al. 2016). This variation in habitability in terms of spatial and temporal scales has been termed "punctuated habitability" (Westall et al. 2015b).

An important consideration to keep in mind with respect to the concept of habitability is that its definition changes depending upon whether one is considering the emergence of life, the development of already-established and viable cells, or situations in which or life can simply survive (but not flourish), that is, either in a state of dormancy or life when growing very slowly. In addition, another valid concept for Mars is that of uninhabited habitats (Cockell et al. 2012), which refer to environments that hosted the requirements for life (carbon, water, energy, and other essential elements) but remained uninhabited because viable microorganisms were unable to reach that habitat. For example, the recurring slope lineae on Mars today are most likely uninhabited habitats.

Habitability and the emergence of life: With respect to the appearance or origin of life, we have no idea of the length of time necessary to go from abiotic ingredients to the first living species on Earth. Moreover, we also do not know exactly where the first cells could have emerged. The various scenarios proposed, from hydrothermal vents, pumice rafts, coastal volcanic plash pools, and hydrothermally flushed sediments to subaerial springs, have been recently reviewed, with the conclusion that the interface between hot to warm hydrothermal fluids and reactive volcanic minerals and rock fragments would have been a favorable location for prebiotic chemical reactions, leading to the emergence of the first cells (Westall et al. 2018). Importantly, the role of mineral surfaces in prebiotic chemistry has become increasingly appreciated (e.g., Hazen and Sverjensky 2010; Dass et al. 2016; Westall et al. 2018). Mineral surfaces can concentrate organic molecules and contribute to their structural organization and configuration as well as to their confinement (Dass et al. 2018). It is thought that the prebiotic chemical reactions that culminated in the first living organisms must have occurred over short timescales, otherwise the reactions may reverse and the increasing complexification of the molecules may become inviable. The definition of "short timescales" in that context has yet to be extensively discussed; on the early, largely oceanic Earth, upon which the ingredients for life were, compared with Mars, unlimited, the question could be considered academic. Various estimates have been proposed, from tens or hundreds of thousands of years to even a several million years (Westall et al. 2018). On Mars, however, this question is of critical importance because the emergence of life depends upon whether a particular environment hosted the ingredients of life for sufficient time to allow prebiotic chemistry to work. It has been suggested that an impact crater about 100 km in diameter could have contained water and could have been host to hydrothermal activity for up to a few million years, long enough for life to emergence, because the other essential ingredients, that is, carbon molecules, the life essential elements (H, N, O, P, S, and transition elements), and reactive mineral surfaces would have been available (Westall et al. 2015a, b). It is probable that the major source of carbon molecules would have been exogenous, likely carbonaceous meteorites and micrometeorites, together with a minor endogenous source from the crust and possibly atmosphere. The kinds of volcanic materials on early Mars, namely iron and magnesium-rich mafic and ultramafic basalts (*sensu lato*), were similar to those on the early Earth and were sources of bioelements as well as of energy from redox reactions at their surfaces. Thus, life could have emerged on Mars, possibly in multiple, isolated locations and at different points in time, as long as the ingredients of life (organic molecules, water, energy sources, hydrothermal activity, other nutrients, and physico-chemical gradients) were co-located for the length of time required for prebiotic chemistry to achieve protocellular complexity. An established cellular community can then rapidly colonize niches that are ephemerally habitable for a brief period of time (hours to week) or restricted in small spaces (centimeters to kilometers).

Punctuated habitability and its consequences: It is becoming increasingly clear that, even though early Mars had water, it was likely either generally cold and wet or cold and dry (Bishop et al. 2018). The more clement conditions of the early Earth did not exist on the Red Planet. The emergence of life on Mars might therefore be likened to a start-and-stop process. Under these conditions, it is unlikely that life could ever have evolved into complex organisms with sophisticated metabolic strategies, and putative Martian life would likely not have evolved beyond "primitive" chemosynthesis, that is, organisms obtaining their energy from the oxidation of either carbon or inorganic sources. Comparison with the early Earth shows us that the highest microbial biomass occurs in the vicinity of nutrient-rich hydrothermal sources, while oligotrophic environments away from these sources are host only to chemolithoautotrophs, whose biomass is correspondingly feeble (Westall et al. 2011b, 2015a, 2015b). In the latter case, the more limited supply of nutrients (organic or inorganic substrates that could be oxidized as an energy source) meant that organisms were required to obtain energy from sources that were not exclusively hydrothermal, for example, redox reactions at the surfaces of

volcanic detrital particles constituting the sediments. The limited potential for biomass development and the distribution of biosignatures has important notable implications for the promise of life on Mars.

Phototrophs, on the other hand, obtain their energy from a more or less inexhaustible source, sunlight, and are therefore able to reproduce more rapidly, produce cells that are larger, and, thus, provide more varied and more readily identifiable biosignatures. The only limitations to the development of phototrophs are access to sunlight and substrates. It is interesting to note that, on the early Earth, even though there was no ozone layer and the calculated flux of UV radiation (DNA-weighted) was between 54 and 1000 times greater than that reaching the surface today, phototrophs were common at least by ~3.5 Ga, when the fossil record is sufficiently well-preserved to identify them (Buick et al. 1981; Hofmann et al. 1999; Westall et al. 2006a, 2011a; Noffke et al. 2013; Hickman-Lewis et al. 2018).

Restricted habitability and cell development: What would happen to organisms when the spatially restricted habitable environment on Mars becomes uninhabitable? This is where our understanding of the limits of terrestrial life and the strategies that microbes adopt to flourish or survive are important (McKay 2014). Table 8.7.1 provides a list of the currently known physico-chemical limits of life. This shows that Martian microbes could have survived for a period of time in an environment that was drying out, where the activity of water was gradually becoming restrictive, and probably where saline conditions were increasing, together with exposure to harmful UV radiation. The ways in which microorganisms cope with extreme environments are both energy- and nutrient-intensive. For example, *Deinococcus radiodurans*, an organism able to

---

TABLE 8.7.1

The Limits of Life

Parameter	Limit	Note
Lowest temperature	−18°C	Limited by liquid water associated with thin films along ice crystal boundaries or saline solutions
Highest temperature	122°C	Limited by protein stability
Maximum pressure	1100 atm	
pH	0–12.5	
Salinity	Saturated NaCl	Depends on the salt and the temperature
Water activity	0.65	
UV radiation	≥1000 J m^{-2}	*Deinococcus radiodurans*

Source: McKay, C.P., *Proc. Natl. Acad. Sci.*, 111, 12628–12633, 2014.

survive high doses of radiation and concomitant des-iccation, has rapid cell metabolism, multiple DNA-repair mechanisms, and a turnover rate of one cell division every 30 minutes.

The habitable environment of early Mars was much more variable than that of the early Earth and included desertification and adaptation to more saline environments, as well as to the presence of chlorides. Today, Mars is considered to be a desert, but the study of life in modern extreme environments on Earth allows us to consider the possibility that viable cells on Mars could survive and maybe still flourish in the subsurface (Boston et al. 1992) and for short periods of time throughout the history of Mars in isolated, ephemeral habitable niches at the surface (Rettberg et al. 2016). If life did not emerge indepen-dently on Mars, it is still possible that extremophile terrestrial organisms could briefly colonize ephem-eral habitats or the subsurface, hence the concern with planetary protection (Rettberg et al. 2016).

Dormant or slow-growing cells occupy only the amount of space required by the cell or colony; however, it is not known how long dormant or slow-growing cells can remain viable. Cells have been reportedly resuscitated from 250 My salt deposits (Vreeland et al. 2000), although this observation has been considered to be due to contamination (Graur and Pupko 2001). Studies of 600-Ka-old permafrost showed that the DNA in the cells contained therein could still be replicated, but this was not achieved in older permafrost (Johnson et al. 2007). These studies suggest that low-level gene repair in a very slowly metabolizing viable cell is better for cell via-bility than dormancy. The genomes of spores, for example, degrade with time because of spontaneous chemical reactions, such as hydrolysis and oxida-tion, that finally become fatal, preventing the cell from germinating (Johnson et al. 2007). Another example of slow-growing cells of relevance to plan-etary habitability are those living in the Earth's crust or deep in sediments. Deep-sea sediments host viable cells down to depths of nearly 2 km; in this oligotrophic environment, apart from a low-nutrient source, another major problem is water availability, and cells have a metabolic turnover of about one cell division every 1000 years (Ciobanu et al. 2014)! The fractured crust is a slightly different environ-ment, in which water infiltrating deep fractures has entrained microbes over long periods of time to depths of between 3 km and 4 km (Onstott 2016). A fracture system, sometimes as small as 1 cm, is an oasis, and, although relatively nutrient-rich, the microbial biomass in the subsurface is lower than that at the surface. Autotrophy and methane cycling are the main means of support of these deep sub-surface microorganisms. Cell turnover times range from one to hundreds of years.

8.7.4.2 Astrobiology on Mars

In terms of missions and observations, the earlier discussion of variable, punctuated habitability and its consequences on the development of life makes is clear that a potentially habit-able body does not necessarily need to be globally habitable, and much of the potential active biomass may be either hidden from view, for example, in the subsurface, or not sufficiently "evolved" to produce the manifest biosignatures of photo-trophic organisms, for instance, biolaminations, microbially induced sedimentary structures, and stromatolites, that are potentially easier to identify than subtle chemotrophic expres-sions. Indeed, if life is extinct at the surface of a planet, as is probably the case for Mars, the search for its remains will be akin to looking for a needle in a haystack, as a result of punc-tuated habitability. One pivotally important point to keep in mind, known to scientists but perhaps more difficult for the general public and decision-makers, is that *not* finding traces of life is not a showstopper. Finding traces of past life on Mars will require almost serendipitous landing in the right location, and there are limitations to the high-resolution coverage that can be achieved from orbit. Finding traces of extant life on the planet will require either deep drilling or some other serendip-itous event that brings viable cells from the subsurface to the surface—and, at present, planetary protection regulations pre-clude missions to these ephemeral environments (e.g., recur-rent slope lineae) (Rettberg et al. 2016). Lack of observation of the atmospheric indicators of life on a planet or exoplanet may mean only that, if inhabited, the life forms had not developed oxygenic photosynthesis or an equivalent metabolism.

> *The Viking mission:* The first astrobiology mission to Mars, landing in 1976, comprised two landers, Viking 1 in Chryse Planitia and Viking 2 in Utopia Planitia (https://mars.nasa.gov/programmissions/missions/past/viking/). These two landers (Figure 8.7.9) carried a series of experiments to search for extant life on Mars. The gas chromatograph-mass spectrom-eter (GC-MS) determined the chemical components that evolved during the heating of untreated Martian regolith. The number of organic molecules found was insignificant (Biemann et al. 1977; Biemann 2007), a result that had profound significance for the inter-pretation of other experiments, such as the labeled release experiment, in which Martian regolith doped with ^{14}C-labeled water was analyzed for potential CO_2 production related to microbial metabolic activ-ity (Levin and Straat 1976). Although an initial gas release was observed, it was not reproduced. From this observation, together with the indication of no organics in the regolith, given by the GC-MS result, the results appeared to be inconclusive (Levin and Straat 1976). The intervening years have, however, seen an increase in controversy over the labeled release experiments, with their proponents suggest-ing that the experiments did, indeed, find traces of extant Martian organisms (Bianciardi et al. 2012);

FIGURE 8.7.9   Carl Sagan with a model of a Viking lander. Carl Sagan was a strong supporter of the search for life on Mars. (Courtesy of http://solarsystem.nasa.gov/multimedia/display.cfm?IM_ID=244.)

however, this view is not common consensus. The gas exchange experiment used a GC to test for consumption or release of gases by microbial metabolism from doped Martian regolith. Although CO_2 was evolved, its release is believed to have been due to oxidation of organics present in the nutrient used to dope the regolith by oxidized minerals in the samples, such as Fe_2O_3; thus, this experiment also produced negative results (Oyama and Berdahl 1977). The pyrolytic release experiment was designed to test for evidence of phototrophy, that is, microbial metabolism fueled by sunlight (Horowitz et al. 1976). Subsequent studies have shown that there are oxidizing components in the Martian regolith that could have interacted with the ingredients used during the Viking experiments and during the heating of the regolith. These include *in situ* detection of perchlorates in the Phoenix mission (Hecht et al. 2009), as well as experimental studies that showed that the radiolysis of perchlorates in the Martian regolith could also reproduce many of the Viking results (Quinn et al. 2013). The problem of perchlorates has been well-illustrated by the on-going Mars Science Laboratory mission and the analysis of organics in Gale Crater by the Sample Analysis at Mars (SAM) instrument (e.g., Freissinet et al. 2015).

The Mars Exploration Rover mission: Two rovers landed on Mars in 2003 to search for signs of water: Opportunity in Meridiani Planum and Spirit in Gusev Crater. A spectral signature suggesting the presence of hematite was the stimulus for the Meridiani Planum location; however, in fact, Opportunity with an instrumental payload, including a camera and thermal emission, Mössbauer, and alpha particle X-ray spectrometers (together with other instruments) found nodules of hematite in layered rocks, suggestive of the presence of water (Squyres et al. 2006). Spirit landed in what was believed to have been a water-filled crater. Although Spirit stopped working a number of years ago (at the time of writing, Opportunity is still going strong), it provided evidence for limited aqueous alteration (thin films of water) of a basically basaltic lithology in the plains area (Arvidson et al. 2004; McSween et al. 2004); however, its investigation of the rocks in the Columbia Hills area showed increased evidence of water alteration and the formation of clay minerals, as well as sulfates and jarosite (Ming et al. 2006). The conclusion of the Spirit rover operation is thus that Gusev Crater may have once been filled by a lake, but as conditions became drier, more evaporitic lithologies became dominant.

One curious discovery of the Spirit rover concerned high silica deposits at Home Plate, widely accepted to be hydrothermal in origin (Ruff and Farmer 2016). These authors hypothesize that these deposits could represent traces of past Martian life, based on comparison with terrestrial hot spring analogues.

The Phoenix mission: The Phoenix mission comprised a lander that landed close to the northern polar ice cap of Mars. Its objectives were to study the history of water in the polar region and to search for evidence of a habitable zone and assess the biological potential of the ice-soil boundary. Water ice and water vapor were detected, as were calcium carbonate, aqueous minerals, and salts, all of which were likely deposited in the presence of water (Smith et al. 2010). Apart from this first direct detection of water (ice) on Mars, the most significant observation from Phoenix was the surprising significant presence of perchlorate in the soil (Hecht et al. 2009). This fundamental discovery has influenced the interpretation of previous and subsequent search for organic molecules on Mars.

Mars Science Laboratory (MSL) and the Curiosity rover: This mission has provided abundant evidence for the presence of habitable conditions in Gale Crater (Grotzinger et al. 2014) in the form of sediments deposited in fluvio-lacustrine conditions and detailed evidence of aqueous alteration of volcanic detrital deposits. The goals of this mission are to determine the role of water, to study Martian climate through changes in the deposits infilling Gale Crater, and to better understand the geology of Mars. Additional observations and measurements (e.g., environmental) will help to plan for a future manned mission to Mars.

Gale Crater was chosen because of the diversity of environments and sediments infilling the crater and also because of its potential habitability (Golombek at al. 2012). Well-preserved sedimentary structures, together with what we know of its mineralogy, are allowing the most detailed analysis of habitability and environmental changes in any one location on Mars, thanks to the well-equipped instrumental payload, which is capable of making analyses of the elemental, mineralogical, and organic compositions of sediments drilled from a depth of about 10 cm into the rock (Grotzinger et al. 2014).

The mission also has three specific goals of relevance to astrobiology involving (i) study of the nature and inventory of organic carbon compounds; (ii) investigation of the chemical building blocks of life (carbon, hydrogen, nitrogen, oxygen, phosphorus, and sulfur); and (iii) identification of potential biosignatures (Summons et al. 2011). Since the Viking mission, MSL is the first to actively search for the potential for life and signs of life. With respect to organic molecules, as noted earlier, the presence of perchlorates in the Martian surface has hindered the analysis of organics but the SAM team, with their GC-MS instrument, have valiantly revealed that Martian organics in the form of chlorinated compounds, for example, chlorobenzene and dichloroalkanes, occur in the Sheepbed Member of Yellowknife Bay (Freissinet et al. 2015), while Eigenbrod et al. (2018) report a range of aromatic and aliphatic molecules, including the S-containing

molecule thiophene in the Mojave and Confidence Hills samples. Evidence for the elemental building blocks of life has also been given (Grotzinger et al. 2014). Importantly, the observations of the Curiosity rover have underlined the influence of groundwater alteration and the precipitation of impure sulfate- and silica-rich deposits as veins or horizons, followed by further alteration of the primary alteration products to produce pure sulfate veins (Schwenzer et al. 2016).

Future astrobiology missions to Mars: There are two upcoming missions whose primary goal is to find traces of life on Mars. The European-Russian ExoMars 2020 mission aims at finding traces of primarily past life (without excluding the possibility of extant life) within its geological context (Vago et al. 2017), an exploratory strategy well-demonstrated by the Curiosity rover. Following the reasoning outlined earlier regarding the punctuated habitability of Mars and the likelihood that Martian life is chemotrophic rather than phototrophic (*cf.* Westall et al. 2015a) and concerned by the modeled destruction of especially labile organics by radiation in the surface of Mars (Kminek and Bada 2006), the ExoMars rover is equipped with a 2-meter drill in order to obtain sediments and organics from the subsurface, that is, those that have been protected from surface radiation (Figure 8.7.10). While there is always the possibility of detecting morphological evidence of microbial life (for instance, clotted-textured, chemotrophic biofilms in the vicinity of hydrothermal effluent), it is more likely that chemical biosignatures could be detected by a cousin of the SAM instrument, Mars Organic Molecule Analayser (MOMA), which incorporates laser desorption with mass spectroscopy, as well as GC-MS. As with SAM, the chemical signatures sought include isomeric selectivity (enantiomeric excess and diastereoisomeric and structural isomer preference), molecular weight characteristics (uneven distribution patterns of C number, concentration and $\delta^{13}C$ of structurally related compounds, repeated constitutional subunits, and systematic isotopic ordering at molecular and group level), and isotopic fractionations of carbon (Vago et al. 2017). Despite the enhanced potential for the preservation of organic molecules deeper in the subsurface, MOMA will still be confronted with the problem of perchlorates and their effects on GC-MS analysis of organic matter, as with the SAM instrument on Curiosity. The detailed detective work undertaken by the SAM team will feed into analysis by MOMA. In addition, laser-desorption-mass spectrometry (LD-MS) will aim to obviate this problem, since the laser will release entrapped molecules, especially more refractory ones, without heating samples in the way that those analyzed by the GC-MS instrument are.

ExoMars 2020 will land in one of two broadly similar locations on the edge of Chryse Planitia: Mawrth

FIGURE 8.7.10   Artist's impression of the ExoMars 2020 rover, with its landing platform in the background. Note the drill (horizontal black cylinder) on the front of the rover. The drill will be able to take a sample from a depth of 2 meters. (Courtesy of ESA, Paris, France.)

Valley or Oxia Planum. Both have been characterized, by orbital mapping, to have diverse clay mineralogy at the surface, the supposed evidence of changing environmental conditions at the end of the Noachian (Bishop et al. 2018). Clay mineralogy is of interest for what it tells us about past habitability, as well as for the potential of clays to trap organic molecules. If these locations were habitable and inhabited, the organic remains of past Martian life could be entrapped within the clayey lithologies (Vago et al. 2017).

The NASA Mars 2020 mission will be the first step in returning samples from Mars for study in terrestrial laboratories. It has sophisticated instrumentation for mapping the elemental and mineralogical compositions of cleaned rock surfaces with the instruments Planetary Instrument for X-ray Lithochemistry (PIXL) and Scanning Habitable Environments with Raman and Luminescence for Organics and Chemicals (SHERLOC). Therein, the combination of macro- and micro-scale mineralogical and elemental contexts, with determination of the presence of organics, will aid in the choice of samples for collection (at 5 cm) and storage. The return mission is planned with collection at a later date by a fetch rover and eventual return to Earth.

8.7.5   CONCLUSION

Investigations of traces of life on early Earth and on Mars have many points in common. On a microbial scale, the habitable environments were similar, although they became increasingly desert-like and uninhabitable on the post-Noachian Mars. In both cases, the life forms were (or are expected to be) prokaryotes: on Earth, both chemotrophs and phototrophs, whereas on Mars, likely only chemotrophs owing of the lack of long-term, clement habitable environments. Understanding the preservation of signatures of life through experimentation and observation of microbial fossils throughout the terrestrial geological record shows that microbes could be preserved on Mars in a multitude of fashions but that the probability of finding chemotrophic biosignatures would be higher in the vicinity of Si-rich hydrothermal activity than in distal oligotrophic environments. Past missions to Mars have underlined its habitability in terms of the past presence of water, nutrients, life essential elements, and energy sources. They have also highlighted the difficulty in analyzing organic molecules in Martian materials because of the oxidizing conditions at the surface of the planet and the presence of perchlorates in the Martian regolith. Nevertheless, the ongoing MSL mission and the upcoming ExoMars 2020 and Mars 2020 missions will build on the information available to date. ExoMars 2020—with its 2-meter drill—may be able to detect and analyze organics protected from the harsh surface conditions. Its results could feed into the Mars 2020 mission, providing added information to aid sampling for future collection and return to Earth.

We live in exciting times for astrobiology on Mars, and the prospect of sample return missions heralds the advent of increased technological advances, with the eventual goal of studying precious material brought back from the Red Planet.

REFERENCES

Allwood, A.C., Walter, M.R., Burch, I.W., Kamber, B.S., 2007. 3.43 billion-year-old stromatolite reef from the Pilbara Craton of Western Australia: Ecosystem-scale insights to early life on Earth. *Precambrian Research* 158, 198–227.

Allwood, A.C., Walter, M.R., Kamber, B.S., Marshall, C.P., Burch, I.W., 2006. Stromatolite reef from early Archaean era of Australia. *Nature* 441, 714–718.

Arndt, N.T., Nisbet, E.G., 2012. Processes on the young Earth and the habitats of early life. *Annual Review of Earth and Planetary Sciences* 40, 521–549.

Arvidson, R.E., Anderson, R.V., Bartlett, P. et al., 2004. Localization and physical properties experiments conducted by Spirit at Gusev Crater. *Science* 305, 821–824.

Banerjee, N.R., Furnes, H., Meuhlenbachs, K., Staudigel, H., de Wit, M.J., 2006. Preservation of ca. 3.4–3.5 Ga microbial biomarkers in pillow lavas and hyaloclastites from the Barberton Greenstone Belt, South Africa. *Earth and Planetary Science Letters* 241, 707–722.

Banerjee, N.R., Izawa, M.R.M., Sapers, H.M., Whitehouse, M.J., 2010. Geochemical biosignatures preserved in microbially altered basaltic glass. *Surface and Interface Analysis* 43, 452–457.

Barker, W.W., Welch, S.S., Banfield, J.F., 1997. Biochemical weathering of silicate minerals. In: Banfield, J.F. and Nealson, K.H. (Eds.), *Geomicrobiology: Interactions Between Microbes and Minerals*, Reviews in Mineralogy, Vol. 35. Mineralogical Society of America, Washington, DC, pp. 391–428.

Bazylinski, D.A., Moskowitz, B.M., 1997. Microbial biomineralization of magnetic iron minerals: Microbiology, magnetism and environmental significance. In: Banfield, J.F. and Nealson, K.H. (Eds.), *Geomicrobiology: Interactions Between Microbes and Minerals*, Reviews in Mineralogy, Vol. 35. Mineralogical Society of America, Washington, DC, pp. 181–223.

Beukes, N.J., Lowe, D.R., 1989. Environmental control on diverse stromatolite morphologies in the 3000 Myr Pongola Supergroup, South Africa. *Sedimentology* 36, 383–397.

Bianciardi, G., Miller, J.D., Straat, P.A., Levin, G.V., 2012. Complexity analysis of the Viking labeled release experiments. *International Journal of Aeronautics and Space Sciences* 13, 14–26.

Biemann, K., 2007. On the ability of the Viking gas chromatograph–mass spectrometer to detect organic matter. *Proceedings of the National Academy of Sciences* 104, 10310–10313.

Biemann, K., Oro, J., Toulmin, P. et al., 1977. The search for organic substances and inorganic volatile Compounds in the surface of Mars. *Journal of Geophysical Research* 82(28), 4641–4658.

Bishop, J., Fairén, A., Michalski, J.R. et al., 2018. Surface clay formation during short-term warmer and wetter conditions on a largely cold ancient Mars. *Nature Astronomy* 2, 206–213.

Bolhuis, H., Cretoiu, M.S., Stal, L.J., 2014. Molecular ecology of microbial mats. *FEMS Microbiology Ecology* 90(2), 335–350.

Boston, P.J., Ivanov, M.V., McKay, C.P., 1992. On the possibility of chemosynthetic ecosystems in subsurface habitats on Mars. *Icarus* 95, 300–308.

Brasier, M.D., 2009. *Darwin's Lost World: The Hidden History of Animal Life*. Oxford University Press, Oxford, UK, 288 pp.

Brasier, M.D., Wacey, D., McLoughlin, N., 2011. Taphonomy in temporally unique settings: An environmental traverse in search of the earliest life on Earth. In P.A. Allison and D.J. Bottjer (Eds.), *Taphonomy: Process and Bias through Time*, Topics in Geobiology, Vol. 32. Springer, New York, p. 487.

Brocks, J.J., Logan. G.A., Buick, R., Summons, R.E., 1999. Archean molecular fossils and the early rise of eukaryotes. *Science* 285(5430), 1033–1036.

Buick, R., Dunlop, J., Groves, D., 1981. Stromatolite recognition in ancient rocks: An appraisal of irregularly laminated structures in an Early Archean chert-barite unit from North Pole, Western Australia. *Alcheringa* 5, 161–181.

Carr, M.H., Head, J.W., 2010. Geologic history of Mars. *Earth and Planetary Science Letters* 294, 185–203.

Ciobanu, M.C., Burgaud, G., Dufresne, A. et al., 2014. Microorganisms persist at record depths in the subseafloor of the Canterbury Basin. *The ISME Journal* 8, 1370–1380.

Clifford, S.M., Parker, T.J., 2001. The evolution of the Martian hydrosphere: Implications for the fate of a primordial ocean and the current state of the northern plains. *Icarus* 154, 40–79.

Cockell, C.S., Balme, M., Bridges, J.C., Davila, A., Schwenzer, S.P., 2012. Uninhabited habitats on Mars. *Icarus* 217(1), 184–193.

Cockell, C.S, Raven, J.A., 2004. Zones of photosynthetic potential on Mars and the early Earth. *Icarus* 169, 300–310.

Dass, A.V., Hickman-Lewis, K., Brack, A., Kee, T.P., Westall, F., 2016. Stochastic prebiotic chemistry within realistic geological systems. *ChemistrySelect* 1, 4906–4926.

Dass, A.V., Jaber, M., Brack, A., Kee, T.P., Georgelin, T., Westall, F., 2018. Potential role of inorganic confined environments in prebiotic phosphorylation. *Life* 8(1), E7.

Dupraz, C., Reid, R.P., Braissant, O., Decho, A.W., Norman, R.S., Visscher, P.T., 2009. Processes of carbonate precipitation in modern microbial mats. *Earth Science Review* 96, 141–162.

Eigenbrod, J.L., Summons, R.E., Steele, A., et al., 2018. Organic matter preserved in 3-billion-year-old mudstones at Gale crater, Mars. *Science*, 360, 1096–1101.

Ferris, F.G., Fyfe, W.S., Beveridge, T.J., 1988. Metallic ion binding by Bacillus subtilis: Implications for the fossilization of microorganisms. *Geology* 16, 149–152.

Formisano, V., Atreya, S., Encrenaz, T. et al., 2004. Detection of methane in the atmosphere of Mars. *Science* 306, 1758–1761.

Foucher, F., Westall, F., Brandstätter, F. et al., 2010. Testing the survival of microfossils in artificial martian sedimentary meteorites during entry into Earth's atmosphere: The STONE 6 experiment. *Icarus* 207, 616–630.

Freissinet, C., Glavin, D.P., Mahaffy, P.R. et al., 2015. Organic molecules in the Sheepbed Mudstone, Gale Crater, Mars. *Journal of Geophysical Research Planets* 120, 495–514.

French, K.L., Hallmann, C., Hope, J.M. et al., 2015. Reappraisal of hydrocarbon biomarkers in Archean rocks. *PNAS* 112, 5915–5920.

Furnes, H., Banerjee, N.R., Muehlenbachs, K., Staudigel, H., de Wit, M.J., 2004. Early life recorded in Archean pillow lavas. *Science* 304, 578–581.

Furnes H., Banerjee, N.R., Staudigel, H. et al., 2007. Bioalteration textures in recent to mesoarchean pillow lavas: A petrographic signature of subsurface life in oceanic igneous rocks. *Precambrian Research* 158, 156–176.

Gaboyer, F., Le Milbeau, C., Bohmeier, M. et al., 2017. Mineralization and preservation of an extremotolerant bacterium isolated from an early Mars analog environment. *Nature Scientific Reports* 7, 8775.

Golombek, M., Grant, J., Kipp, D. et al., 2012. Selection of the Mars Science Laboratory landing site. *Space Science Reviews* 170, 641–737.

Gourcerol, B., Thurston, P.C., Kontak, D.J., Biczok, J., 2016. Depositional setting of Algoma-type banded iron formation. *Precambrian Research* 281, 47–79.

Graur D., Pupko T., 2001. The Permian bacterium that isn't. *Molecular Biology and Evolution* 18:1143–1146.

Grosch, E.G., McLoughlin, N., 2014. Reassessing the biogenicity of Earth's oldest trace fossil with implications for biosignatures in the search for early life. *Proceedings of the National Academy of Sciences of the United States of America* 111, 8380–8385.

Grosch, E.G., McLoughlin, N., 2015. Questioning the biogenicity of titanite mineral trace fossils in Archean pillow lavas. *Proceedings of the National Academy of Sciences of the United States of America* 112, E3090–E3091.

Grotzinger, J.P., Arvidson, R.E., Bell, J.F. et al., 2005. Stratigraphy and sedimentology of a dry to wet eolian depositional system, Burns formation, Meridiani Planum, Mars. *Earth and Planetary Science Letters* 240, 11–72.

Grotzinger, J.P., Sumner, D.Y., Kah, L.C. et al., 2014. A habitable fluvio-lacustrine environment at Yellowknife Bay, Gale Crater, Mars. *Science* 343, 1242777.

Hazen, R.M., Sverjensky, D.A., 2010. Mineral surfaces, geochemical complexities, and the origins of life. *Cold Spring Harbor Perspectives in Biology* 2, a002162.

Hecht, M.H., Kounaves, S.P., Quinn, R.C. et al., 2009. Detection of perchlorate and the soluble chemistry of martian soil at the Phoenix lander site. *Science* 325(5936), 64–67.

Hickman-Lewis, K., Cavalazzi, B., Foucher, F., Westall, F., 2018. Most ancient evidence for life in the Barberton greenstone belt: Microbial mats and biofabrics of the ~3.47 Ga Middle Marker horizon, *Precambrian Research* 312, 45–67.

Hickman-Lewis, K., Garwood, R.J., Brasier, M.D. et al., 2016. Carbonaceous microstructures of the 3.46 Ga stratiform 'Apex chert', Chinaman Creek locality, Pilbara, Western Australia. *Precambrian Research* 278, 161–178.

Hickman-Lewis, K., Garwood, R.J., Withers, P.J., Wacey, D., 2017. X-ray microtomography as a tool for investigating the petrological context of Precambrian cellular remains. In *Earth System Evolution and Early Life: A Celebration of the Work of Martin Brasier*. Geological Society, London, UK, Special Publication 448.

Hofmann, A., 2011. Archaean hydrothermal systems in the Barberton greenstone belt and their significance as a habitat for early life. In Golding, S.D., Glikson, M. (Eds), *Earliest Life on Earth: Habitats, Environments and Methods of Detection*. Springer, Dordrecht, the Netherlands, pp. 51–78.

Hofmann, H.J., Grey, K., Hickman, A.H., Thorpe, R.I., 1999. Origin of 3.45 Ga coniform stromatolites in Warrawoona Group, Western Australia. *GSA Bulletin* 111, 1256–1262.

Homann, M., Heubeck, C., Bontognali, T.R.R., Bouvier, A.-S., Baumgartner, L.P., Airo, A., 2016. Evidence for cavity-dwelling microbial life in 3.22 Ga tidal deposits. *Geology* 44, 51–54.

Horowitz, N., Hobby, G.L., Hubbard, J.S., 1976. The Viking carbon assimilation experiments - Interim report. *Science* 194, 1321–1322.

Iverson, I.P., 1987. Microbial corrosion of metals. *Advances in Applied Microbiology* 32, 1–36.

Iwai, S., Doi, K., Fujino, Y., Nakazono, T., Fukuda, K., Motomura, Y., Ogata, S., 2010. Silica deposition and phenotypic changes to Thermus thermophilus cultivated in the presence of supersaturated silica. *The ISME Journal* 4, 809–816.

Jaakkola, S.T., Ravantti, J.J., Oksanen, H.M., Bamford, D.H., 2016. Buried alive: Microbes from ancient halite. *Trends in Microbiology* 24, 148–160.

Jakosky, B., Grebowsky, J.M., Luhmann, J.G., Brain, D.A., 2015. Initial results from the MAVEN mission to Mars. *Geophysical Research Letters* 42, 8791–8802.

Javaux, E.J., Marshall, C.P., Bekker, A., 2010. Organic-walled microfossils in 3.2-billion-year-old shallow-marine siliciclastic deposits. *Nature* 463, 934–938.

Johnson, S.S., Hebsgaard, M.B., Christensen, T.R. et al., 2007. Ancient bacteria show evidence of DNA repair. *Proceedings of the National Academy of Sciences of the United States of America* 104, 14401–14405.

Kempe, S., Degens, E.T., 1985. An early soda ocean? *Chemical Geology* 53, 95–108.

Kminek, G., Bada, J.L., 2006. The effect of ionizing radiation on the preservation of amino acids on Mars. *Earth and Planetary Science Letters* 245, 1–5.

Knoll, A.H., Barghoorn, E.S., 1977. Archean microfossils showing cell division from the Swaziland System of South Africa. *Science* 189, 396–398.

Knoll, A.H., Javaux, E.J., Hewitt, D., Cohen, P., 2006. Eukaryotic organisms in Proterozoic oceans. *Philosophical Transactions of the Royal Society of London B: Biological Sciences* 361(1470), 1023–1038.

Kyle, J.E., Pedersen, K., Ferris, F.G., 2008. Virus mineralization at low pH in the Rio Tinto, Spain. *Geomicrobiology Journal* 25, 338–345.

Laidler, J.R., Stedman, K.M., 2010. Virus silicification under simulated hot spring conditions. *Astrobiology* 10, 569–576.

Ledevin, M., Arndt, N., Simionovici, A., Jaillard, E., Ulrich, M., 2014. Silica precipitation triggered by clastic sedimentation in the Archean: New petrographic evidence from cherts of the Kromberg type section, South Africa. *Precambrian Research* 255, 316–334.

Levin, G.V., Straat, P.A., 1976. Viking labeled release biology experiment: Interim results. *Science* 194, 1322–1329.

Li, J., Benzerara, K., Bernard, S., Beyssac, O. 2013. The link between biomineralization and fossilization of bacteria: Insights from field and experimental studies. *Chemical Geology* 359, 49–69.

Lowe, D.R., 1980. Stromatolites 3,400-Myr old from the Archean of Western Australia. *Nature* 284, 441–443.

Maliva, R.G., Knoll, A.H., Simonson, B.M., 2005. Secular change in the Precambrian silica cycle: Insights from chert petrology. *Geological Society of America Bulletin* 117, 835–845.

McEwen, A.S., Dundas, C.M., Mattson, S.S. et al., 2015. Recurring slope lineae in equatorial regions of Mars. *Nature Geoscience* 7, 53–58.

McKay, C.P., 2014. Requirements and limits for life in the context of exoplanets. *Proceedings of the National Academy of Sciences of the United Stated of America* 111, 12628–12633.

McKay, D.S., Gibson Jr., E.K., Thomas-Keprta, K.L. et al., 1996. Search for past life on Mars: Possible relic biogenic activity in martian meteorite ALH 84001. *Science* 273, 924–930.

McSween, H.Y., Arvidson, R.E., Bell, J.F. et al., 2004. Basaltic rocks analyzed by the Spirit Rover in Gusev Crater. *Science* 305(5685), 842–845.

McSween, H.Y., Labotka, T.C., Viviano-Beck, C.E., 2015. Metamorphism in the Martian crust. *Meteoritics* 50, 590–603.

Ming, D.W., Mittlefehldt, D.W., Morris, R.V. et al., 2006. Geochemical and mineralogical indicators for aqueous processes in the Columbia Hills of Gusev crater, Mars. *Journal of Geophysical Research* 111, E02S12.

Moelling, K., 2017. *Viruses: More Friends than Foes*. World Scientific Publishing, Singapore, 420 pp.

Müller, W., Fricke, H., Halliday, A.N., McCulloch, M.T., Wartho, J.-A., 2003. Origin and Migration of the Alpine Iceman. *Science* 302(5646), 862–866.

Mumma, M.J., Villanueva, G.L., Novak, R.E. et al., 2009. Strong release of methane on Mars in northern summer 2003. *Science* 323, 1041–1045.

Mustard, J.F., Adler, M., Allwood, A.C. et al., 2013. Report of the Mars 2020 Science Definition Team, July 1, 2013, Mars Exploration Program Analysis Group (MEPAG). http://mepag.jpl.nasa.gov/reports/MEP/Mars_2020_SDT_Report_Final.pdf.

Nemchin, A.A. Whitehouse, M.J., Menneken, M., Geisler, T., Pidgeon, R.T., Wilde, S.A., 2008. A light carbon reservoir recorded in zircon-hosted diamond from the Jack Hills. *Nature* 454, 92–95.

Nisbet, E.G., 2000. The realms of Archaean life. *Nature* 405, 625–626.

Nisbet, E.G., Sleep, N.H., 2001. The habitat and nature of early life. *Nature* 409, 1083–1091.

Noffke, N., 2010. *Geobiology: Microbial Mats in Sandy Deposits from the Archaean Era to Today.* Springer-Verlag, Berlin, Germany, p. 194.

Noffke, N., 2015. Ancient sedimentary structures in the <3.7 Ga Gillespie Lake Member, Mars, that resemble macroscopic morphology, spatial associations, and temporal succession in terrestrial microbialites. *Astrobiology* 15, 169–192.

Noffke, N., Christian, D., Wacey, D., Hazen, R.M., 2013. Microbially induced sedimentary structures recording an ancient ecosystem in the ca. 3.48 billion-year-old Dresser Formation, Pilbara, Western Australia. *Astrobiology* 13, 1103–1124.

Noffke, N., Eriksson, K.A., Hazen, R.M., Simpson, E.L., 2006. A new window into Archean life: Microbial mats in Earth's oldest siliciclastic tidal deposits (3.2 Ga Moodies Group, South Africa). *Geology* 34, 253.

Onstott, T.C., 2016. *Deep Life: The Hunt for the Hidden Biology of Earth, Mars, and beyond.* Princeton University Press, Princeton, NJ, 512 pp.

Orange, F., Chabin, A., Gorlas, A., Lucas-Staat, S., Geslin, C., Le Romancer, M., Prangishvili, D., Forterre, P., Westall, F., 2011. Experimental fossilisation of viruses from extremophilic Archaea. *Biogeosciences* 8, 1465–1475.

Orange, F., Disnar, J.-P., Gautret, P. et al., 2012. Preservation and evolution of organic matter during experimental fossilisation of the hyperthermophilic archaea Methanocaldococcus jannaschii. *Origins of Life and Evolution of Biosphere* 42, 587–609.

Orange, F., Westall, F., Disnar, J.-R. et al., 2009. Experimental silicification of the extremophilic Archaea Pyrococcus abyssi and Methanocaldococcus jannaschii: Applications in the search for evidence of life in early Earth and extraterrestrial rocks. *Geobiology* 7, 403–418.

Orosei, R., Lauro, S.E., Pettinelli, E., et al. 2018. Radar evidence for subglacial liquid water on Mars. *Science Reports.* doi:10.1126/Scienceaar7268(2018).

Oyama, V.I., Berdahl, B.J., 1977. The Viking gas exchange experiment results from Chryse and Utopia surface samples. *Journal of Geophysical Research* 82, 4669–4676.

Panieri, G., Lugli, S., Manzi, V., Roveri, M., Schrieber, B.C., Palinska, K.A., 2010. Ribosomal RNA gene fragments from fossilized cyanobacteria identified in primary gypsum from the late Miocene, Italy. *Geobiology* 8, 101–111.

Pearce, B.K.D., Tupper, A.S., Pudritz, R.E., Higgs, P.G., 2018. Constraining the time interval for the origin of life on Earth. *Astrobiology* 18, 343–364.

Pentecost, A. 1990. Calcification processes in algae and cyanobacteria. In: Riding, R. (Ed.), *Calcareous Algae and Stromatolites.* Springer, Berlin, Germany, pp. 1–20.

Plaut, J.J., Picardi, G., Safaeinili, A. et al., 2000. Subsurface radar sounding of the south polar layered deposits of Mars. *Science* 316(5821), 92–95.

Quinn, R.C., Martucci, H.F.H., Miller, S.R., Bryson, C.E., Grunthaner, F.J., Grunthaner, P.J., 2013. Perchlorate radiolysis on Mars and the origin of Martian soil reactivity. *Astrobiology* 13, 515–520.

Rasmussen, B., 2000. Filamentous microfossils in a 3,235-million-year-old volcanogenic massive sulphide deposit. *Nature* 405, 676–679.

Rettberg, P., Anesio, A.M., Baker, V.R. et al., 2016. Planetary protection and Mars special regions–a suggestion for updating the definition. *Astrobiology* 16, 119–125.

Rosing, M.T., 1999. ^{13}C-depleted carbon microparticles in >3700-Ma sea-floor sedimentary rocks from West Greenland. *Science* 283(5402), 674–676.

Ruff, S.W., Farmer J.D., 2016. Silica deposits on Mars with features resembling hot spring biosignatures at El Tatio in Chile. *Nature Comunications* 7, 13554.

Schwenzer, S.P., Bridges, J.C., Wiens, R.C. et al., 2016. Fluids during diagenesis and sulfate vein formation in sediments at Gale crater, Mars. *Meteoritics and Planetary Science* 51, 2175–2202.

Siever, R., 1992. The silica cycle in the Precambrian. *Geochimica et Cosmochimica Acta* 56, 3265–3272.

Smith, P., Tamppari, L.K., Arvidson, R. et al., 2010. H$_2$O at the Phoenix landing site. *Science* 325, 58–61.

Squyres, S.W., Arvidson, R.E., Blaney, D.L. et al., 2006. Rocks of the Columbia Hills. *Journal of Geophysical Research Planets* 111, E02S11.

Strother, P.K., Battison, L., Brasier, M.D., Wellman, C.H., 2011. Earth's earliest non-marine eukaryotes. *Nature* 473, 505–509.

Strother, P.K., Wellman, C.H., 2016. Palaeoecology of a billion-year-old non-marine cyanobacterium from the Torridon Group and Nonesuch Formation. *Palaeontology* 59, 89–108.

Summons, R.E., Amend, J.P., Bish, D. et al., 2011. Preservation of Martian organic and environmental records: Final report of the Mars biosignature working group. *Astrobiology* 11, 157–181.

Summons, R.E., Jahnke, L.L., Hope, J.M., Logan, G.A., 1999. 2-Methylhopanoids as biomarkers for cyanobacterial oxygenic photosynthesis. *Nature* 400, 554–557.

Summons, R.E., Powell T.G., Boreham C.J., 1988. Petroleum geology and geochemistry of the Middle Proterozoic McArthur Basin, northern Australia: III. Composition of extractable hydrocarbons. *Geochimica et Cosmochimica Acta* 52, 1747–1763.

Tashiro, T., Ishida, A., Hori, M. et al., 2017. Early trace of life from 3.95 Ga sedimentary rocks in Labrador, Canada. *Nature* 549, 526–518.

Vago, J.L., Westall, F., Pasteur Instrument Teams et al., 2017. Habitability of early Mars and the search for biosignatures with the ExoMars rover. *Astrobiology* 17, 471–510.

Van Kranendonk, M.J., Smithies, R.H., Griffin, W.L. et al., 2015. Making it thick: A volcanic plateau model for Paleoarchean continental lithosphere of the Pilbara and Kaapvaal cratons. In: Roberts, N.M.W., Van Kranendonk, M., Parman, S., Shirey, S., Clift, P.D. (Eds.), *Continent Formation through Time.* Geological Society, London, UK, Special Publications 389, pp. 83–112.

Van Kranendonk, M.J., Webb, G.E., Kamber, B.S., 2003. Geological and trace element evidence for a marine sedimentary environment of deposition and biogenicity of 3.45 Ga stromatolitic carbonates in the Pilbara Craton, and support for a reducing Archaean ocean. *Geobiology* 1, 91–108.

van Zuilen M.A., Chaussidon M., Rollion-Bard C., Marty B., 2007. Carbonaceous cherts of the Barberton greenstone belt, South Africa: Isotopic, chemical and structural characteristics of individual microstructures. *Geochimica et Cosmochimica Acta* 71, 655–669.

Vreeland R.H., Rosenzweig W.D., Powers D.W., 2000. Isolation of a 250 million-year-old halotolerant bacterium from a primary salt crystal. *Nature* 407, 897–900.

Wade, W., 2002. Unculturable bacteria—the uncharacterized organisms that cause oral infections. *Journal of the Royal Society of Medicine* 95(2), 81–83.

Wacey, D., 2009. Early life on earth, a practical guide. In: Landman, N.H., Harries, P.J. (Eds.), *Topics in Geobiology*, Vol. 31. Springer, Heidelberg, Germany.

Wacey, D., 2010. Stromatolites in the ~3400 Ma Strelley Pool Formation, Western Australia: Examining biogenicity from the macro- to the nano-scale. *Astrobiology* 10, 381–395.

Wacey, D., Battison, L., Garwood, R.J., Hickman-Lewis, K., Brasier, M.D., 2016. High-resolution techniques for the study of the morphology and chemistry of Proterozoic microfossils. In Brasier, A.T., McIlroy, D., McLoughlin, N. (Eds.), *Earth System Evolution and Early Life: A Celebration of the Work of Martin Brasier*. Geological Society, London, UK, Special Publication 448, pp. 81–104.

Wacey, D., Fisk, M., Saunders, M., Eiloart, K., Kong, C., 2017. Critical testing of potential cellular structures within microtubes in 145 Ma volcanic glass from the Argo Abyssal Plain. *Chemical Geology* 466, 575–587.

Walsh, M.M., 1992. Microfossils and possible microfossils from the Early Archean Onverwacht Group, Barberton Mountain Land, South Africa. *Precambrian Research* 54, 271–293.

Webster, C.R., Mahaffy, P.R., Atreya, S.K. et al., 2013. Mars methane detection and variability at Gale crater. *Science* 347(6220), 415–417.

Westall, F., 1997. The influence of cell wall composition on the fossilization of bacteria and the implications for the search for early life forms. In: Cosmovici, C., Bowyer, S., Werthimer, D. (Eds.), *Astronomical and Biochemical Origins and the Search for Life in the Universe*. Editori Compositrici, Bologna, Italy, pp. 491–504.

Westall, F., 2016. Microbial palaeontology and the origin of life: A personal approach. *Bollettino della Società Paleontologica Italiana* 55, 85–103.

Westall, F., Boni, L., Guerzoni, E., 1995. The experimental silicification of microorganisms. *Paleontology* 38, 495–528.

Westall, F., Campbell, K.A., Bréhéret, J.G. et al., 2015a. Archean (3.33Ga) microbe-sediment systems were diverse and flourished in a hydrothermal context. *Geology* 43, 615–618.

Westall, F., Cavalazzi, B., 2011. Biosignatures in rocks. In Thiel, V. (Ed.), *Encyclopedia of Geobiology*. Springer, Berlin, Germany, pp. 189–201.

Westall, F., Cavalazzi, B., Lemelle, L. et al., 2011a. Implications of in situ calcification for photosynthesis in a ~3.3 Ga-old microbial biofilm from the Barberton Greenstone Belt, South Africa. *Earth and Planetary Science Letters* 310, 468–479.

Westall, F., de Ronde, C.E.J., Southam, G. et al., 2006a. Implications of a 3.472–3.333 Gyr-old subaerial microbial mat from the Barberton greenstone belt, South Africa for the UV environmental conditions on the early Earth. *Philosophical Transactions of the Royal Society B: Biological Sciences* 361, 1857–1875.

Westall, F., de Vries, S.T., Nijman, W. et al., 2006. The 3.466 Ga "Kitty's Gap Chert", an early Archean microbial ecosystem. *Geological Society of America Special Paper* 405, 105–131.

Westall, F., Foucher, F., Bost, N. et al., 2015b. Biosignatures on Mars: What, where, and how? Implications for the search for Martian life. *Astrobiology* 15, 998–1029.

Westall, F., Foucher, F., Cavalazzi, B. et al., 2011b. Volcaniclastic habitats for early life on Earth and Mars: A case study from ~3.5 Ga-old rocks from the Pilbara, Australia. *Planetary and Space Science* 59, 1093–1106.

Westall, F., Hickman-Lewis, K., Hinman, N. et al., 2018. A hydrothermal-sedimentary context for the origin for life. *Astrobiology* 18, 259–293.

Westall, F., Loizeau, D., Foucher, F. et al., 2013. Habitability on Mars from a microbial point of view. *Astrobiology* 13, 887–897.

Westall, F., Steele, A., Toporski, J., Walsh, M., Allen, C., Guidry, S., McKay, D., Gibson, E., Chafetz, H., 2000. Polymeric substances and biofilms as biomarkers in terrestrial materials: Implications for extraterrestrial samples. *Journal of Geophysical Research* 105, 24511–24527.

Wilby, P.R., Briggs, D.E.G., 1997. Taxonomic trends in the resolution of detail preserved in fossil phosphatized soft tissues. *Geobios* 30, 493–502.

Willerslev, E., Hansen, A.J., Rønn, R. et al., 2004. Long-term persistence of bacterial DNA. *Current Biology* 14, R9–R10.

Section IX

Life under Extreme Conditions—Microbes in Space

9.1 Extremophiles and Their Natural Niches on Earth

Aharon Oren

CONTENTS

9.1.1 INTRODUCTION

There is hardly any place on our planet, however extreme, that is not inhabited by at least some types of living organisms adapted to the environment. Thus, the physico-chemical boundaries for life on Earth are very broad. One can almost state that any place where a minimal amount of liquid water is available can support life (Mazur 1980; Madigan 2000; Xu and Glansdorff 2007; Stan-Lotter 2012). We find microbial life in very cold and very hot environments, even at temperatures exceeding 100°C in undersea springs, where the boiling point of water is elevated because of the hydrostatic pressure. We know microorganisms adapted to life in concentrated acids and in highly alkaline lakes. Life is possible in salt-saturated brines, under high levels of ionizing radiation and under pressure in the deepest parts of the ocean. Microorganisms even live in deserts, hot as well as cold, thriving on very small amounts of water that become available occasionally. The lowest water activity that still allows active growth of microorganisms appears to be around 0.63–0.64 (Stevenson et al. 2015), but many types can survive prolonged periods of desiccation until conditions again become suitable for growth.

There are also many types of microorganisms that we can call "polyextremophiles": organisms adapted to more than one environmental extreme. Thus, we know haloalkaliphiles that grow only at salt concentrations approaching saturation and at pH >8.5–9; there are thermoacidophiles that thrive only in hot acid solutions, thermophiles that prefer the hydrostatic pressure of the deep sea, and even haloalkaliphilic thermophilic anaerobes that inhabit the sediments of some shallow soda lakes in tropical area (Capece et al. 2013; Seckbach et al. 2013).

Extremophiles are found in all three domains of life: Archaea, Bacteria, and Eukarya. The archaeal domain is particularly rich in organisms adapted to life at the extremes: high salt, high temperature, hot and acidic, hypersaline and alkaline, and more. The special structure of the archaeal lipid membrane, with ether bonds linking the glycerol moieties with the hydrophobic carbon chains, and even the possibility realized in some Archaea of a monolayer membrane, in which the glycerol moieties at both sides of the membrane are bridged by covalent bonds only, make the Archaea especially suitable for life in the hottest environments. But also among the Bacteria and the eukaryotic microorganisms, we find halophiles, thermophiles, acidophiles, alkaliphiles, and other extremophilic types.

This chapter provides an overview of the kind of extremophilic microorganisms found on our planet and of the environments they inhabit. Further information can be found in many monographs and review articles (e.g., Kushner 1978; Da Costa et al. 1989; Horikoshi and Grant 1991, 1998; Kristjansson and Hreggvidsson 1995; Madigan 2000; Wharton 2002; Rainey and Oren 2006a, 2006b; Pikuta et al. 2007; Horikoshi 2010). Further literature on specific types of extremophiles is referred to in the subsequent sections.

9.1.2 HIGH-TEMPERATURE ENVIRONMENTS AND THERMOPHILIC MICROORGANISMS

There is no lack of hot environments on planet Earth. The best studied ones are associated with volcanic activity: hot springs are abundant, for example, in the western USA (e.g., in Yellowstone National Park), New Zealand, Iceland, Japan, and Italy. Many hot springs are also highly acidic due to the oxidation of reduced sulfur compounds. Even higher temperatures prevail in the deep-sea hydrothermal vents, where the in situ temperature often exceeds 100°C and the ambient pressure prevents the water from boiling (Stetter 1996). The anoxic, sulfide-, and mineral-loaded waters emitted by the so-called "black smoker" chimneys in marine hydrothermal vents can reach temperatures of ~350°C. Surface soils in hot deserts may heat up to temperatures as high as 70°C during daytime. Similar temperatures were also measured in compost piles and silage, heated as a result of the microbial activity of fermentative bacteria. Microbial life at high temperature may also abound deep within the crust of the Earth (Gold 1992).

All these environments have become rich sources of isolates of thermophilic (generally defined as organisms with an optimum growth temperature >45°C) and hyperthermophilic microorganisms (with optimum growth temperatures >80°C) of many kinds. Some of these have found interesting biotechnological applications (Burgess et al. 2007).

Since the pioneering studies by Tom Brock of the hot springs of Yellowstone National Park (Brock 1978), our knowledge of the diversity of life at high temperatures has greatly increased (Baross and Deming 1995). No eukaryote appears to be able to complete its life cycle above 60°C (Clarke 2014). No protists are known that live above 56°C, and the upper temperature limit for the growth of algae and fungi is about 60°C and 62°C, respectively. Photosynthesis by prokaryotes, oxygenic (by cyanobacteria) as well as anoxygenic, can proceed up to 70°C–73°C. Some chemoorganotrophic and chemolithotrophic Bacteria can grow at temperatures up to 95°C–100°C. At the highest temperatures enabling life (the highest value recorded today being 122°C), only members of the domain Archaea survive.

A detailed discussion of the mechanisms enabling thermophiles and hyperthermophiles, not only to survive at the temperature extremes but also to actively grow, in many cases even with very short doubling times, is beyond the scope of this chapter. The mechanisms by which proteins and various biological cofactors and organic intermediates are stabilized at extreme temperatures are only now becoming clear (Adams 1993; Wiegel and Adams 1998; Madigan and Oren 1999). The differences in amino acid composition of thermophilic and non-thermophilic proteins are often slight: subtle changes in the amino acid sequence of a protein can have dramatic effects on its thermotolerance. But there is a general trend that thermophilic enzymes have increased ionic bonding between basic and acidic amino acids and also have a highly hydrophobic interior that prevents unfolding. In addition, solutes such as di-inositol phosphate, diglycerol phosphate, and mannosyl-glycerate may help to stabilize proteins against denaturation. Some aerobic and anaerobic extreme thermophiles can grow within a span of more than 40°C. Such organisms may have two sets of key enzymes, whose synthesis is regulated by temperature, enabling them to growth in both the mesophilic and thermophilic ranges (Wiegel 1990). Archaeal lipids, with ether bonds connecting the glycerol moiety, are far more stable than the bacterial and eukaryotic lipids with ester bonds that are easily hydrolyzed at high temperatures. And the possibility

of rigid monolayer membranes spanning the entire width of the cell membrane with covalent bond adds even more stability at temperatures at which conventional membranes cannot function (van de Vossenberg et al. 1998; Daniel and Cowan 2000). At high temperatures, the stability of the DNA double helix is increased by elevated salt concentrations, polyamines, cationic proteins, and supercoiling, rather than by manipulation of C-G ratios. RNA stability is achieved both by formation of secondary structure and by covalent modifications (Daniel and Cowan 2000). In vivo, the half-lives of both RNA and DNA of thermophilic organisms are usually longer than the values estimated in vitro. Different mechanisms may be involved in the stabilization of RNAs and DNAs, both based on the intrinsic chemical structures of the nucleic acids, on their interactions with other biomolecules, and on enzymes for detecting and repairing the DNA damage or for renewing damaged RNA molecules (Grosjean and Oshima 2007).

Table 9.1.1 lists representative examples of thermophilic and hyperthermophilic microorganisms and their properties. Thermophiles can be found in different phylogenetic lineages and physiological groups. In addition to the cultivated minority, there is a large uncultivated majority yet to be characterized. This became clear already since the first pioneering 16S rRNA-based studies in hot springs: for example, a study of archaeal 16S rRNA gene sequences retrieved from the Obsidian Pool hot spring in Yellowstone National Park environment yielded many lineages for which we do not have cultivated representatives even today. These include the proposed Korarchaeota phylum (Barns et al. 1994; Hugenholtz et al. 1998). In their natural habitats, hyperthermophiles form complex food webs, and their metabolic potential includes various types of aerobic and anaerobic respiration, chemoautotrophy, and different modes of fermentation (Huber et al. 2000a; Reysenbach et al. 2001). Even a symbiotic or parasitic lifestyle is possible at extremely high temperatures, as displayed by *Nanoarchaeum equitans*, a very small (about 0.4-μm diameter) organism that represents a separate lineage within the Archaea and lives at 90°C on another extreme thermophile, *Ignicoccus* (Crenarchaeota) (Huber et al. 2002). This non-exhaustive overview of the known types of thermophiles and hyperthermophiles does not include organisms inhabiting hot and acidic environments, hot and alkaline environments, and hot environment under high hydrostatic pressure. Such microorganisms are discussed in later sections in this chapter.

As stated earlier, the upper temperature limit of bacterial thermophiles is much below that of the most thermophilic Archaea. Examples of thermophilic Bacteria are members of the *Aquificales* and the *Thermotogales*, as well as the genus *Thermus*. Isolates of *Thermus aquaticus* growing at 70°C–75°C were retrieved from thermal springs in Yellowstone National Park and in California. *Thermus* was also found in man-made thermal habitats such as domestic water-heating systems. Members of the genus tolerate temperatures up to 79°C (Brock and Freeze 1969). Also, other phyla of Bacteria contain thermophiles. An example is *Thermaerobacter marianensis*

TABLE 9.1.1
Representative Examples of Thermophilic Microorganisms, Their Taxonomic Affiliation, Habitat, and Properties. The Table Includes Thermoacidophiles, Thermophilic Piezophiles, Halothermoalkaliphilic Species, and Radiation-Resistant Thermophiles Listed Also in Other Tables in This Chapter

Name	Affiliation	Isolated from	Properties of the Environment	Limits of Existence	Other Properties	References
	Archaea					
Pyrodictium abyssi	Crenarchaeota	Sea bottom near Mexico, Iceland	Shallow hydrothermal vents	80°C–110°C (opt. 97°C)	Ferments carbohydrates and proteins; hydrogen stimulates growth; reduces elemental sulfur	Pley et al. (1991)
Pyrodictium brockii, Pyrodictium occultum	Crenarchaeota	Vulcano, Italy	Submarine solfataric field	Up to 110°C (opt. 105°C)	Autotrophic on hydrogen and elemental sulfur	Stetter et al. (1983, 1986)
Pyrolobus fumarii	Crenarchaeota	Hydrothermal vent, Mid-Atlantic Ridge, 3,650 m depth	Hydrothermally heated black smoker wall	90°C–113°C (opt. 106°C), pH 4.0–6.5 (opt. 5.5), 1%–4% salt. Survives 1-hour autoclaving at 121°C. Grows at 25 MPa	Facultative aerobic chemolithoautotroph oxidizing hydrogen with nitrate, thiosulfate, or low concentrations of oxygen as electron acceptors	Blöchl et al. (1997)
Staphylothermus marinus	Crenarchaeota	Vulcano, Italy	Geothermally heated marine sediment	Opt. 95°C–92°C, max. 98°C	Anaerobic heterotrophic, fermentative	Fiala et al. (1986)
Sulfolobus acidocaldarius	Crenarchaeota	Yellowstone National Park, USA and acidic hot springs elsewhere	Acidic hot springs	55°C–85°C (opt. 70°C–75°C), pH 0.9–5.8 (opt. 2–3)	Aerobic, heterotrophic on simple organic compounds; facultative autotroph on elemental sulfur	Doemel and Brock (1970); Brock et al. (1972)
Archaeoglobus profundus	Euryarchaeota	Hydrothermal system, Guaymas Basin, Mexico	Black smoker hydrothermal vent and surrounding sediments	Up to 90°C	Anaerobic, oxidizing hydrogen with sulfate or thiosulfate as electron acceptors; mixotrophic, also requiring organic carbon	Burggraf et al. (1990)
Ferroplasma acidiphilum	Euryarchaeota	Bioleaching pilot plant, Russia	Leaching plant for gold-containing arsenopyrite/pyrite ore	pH 1.3–2.2 (opt. 1.7); 15°C–45°C	Aerobic, obligately autotrophic	Golyshina et al. (2000); Golyshina and Timmis (2005)
Methanocaldococcus infernus	Euryarchaeota	Mid-Atlantic Ridge, 3,000 m depth	Hydrothermal vent	55°C–91°C (opt. 85°C)	Methanogenic on hydrogen/carbon dioxide; reduces elemental sulfur	Jeanthon et al. (1998)
Methanocaldococcus jannaschii	Euryarchaeota	East Pacific Rise, 2,600 m depth	Hydrothermal vent—white smoker chimney	Opt. 85°C	Chemolithoautotroph methanogenic on hydrogen/carbon dioxide	Jones et al. (1983)
Methanopyrus kandleri	Euryarchaeota	Guaymas Basin, Gulf of California; Kolbeinsey Ridge, Iceland	Hydrothermal vents	84°C–110°C (opt. 98°C)	Chemolithoautotroph methanogenic on hydrogen/carbon dioxide	Kurr et al. (1991)

(Continued)

TABLE 9.1.1 (Continued)

Representative Examples of Thermophilic Microorganisms, Their Taxonomic Affiliation, Habitat, and Properties. The Table Includes Thermoacidophiles, Thermophilic Piezophiles, Halothermoalkaliphilic Species, and Radiation-Resistant Thermophiles Listed Also in Other Tables in This Chapter

Name	Affiliation	Isolated from	Properties of the Environment	Limits of Existence	Other Properties	References
Methanopyrus kandleri strain 116	*Euryarchaeota*	Central Indian Ridge	Black smoker fluid, hydrothermal field	Max. 116°C at 0.4 MPa; 122°C at 20 MPa	Chemolithoautotroph methanogenic on hydrogen/carbon dioxide	Takai et al. (2008)
Methanotorris formicicus	*Euryarchaeota*	Central Indian Ridge	Black smoker hydrothermal chimney	55°C–83°C (opt. 75°C)	Methanogenic on hydrogen/carbon dioxide or formate	Takai et al. (2004a)
Palaeococcus ferrophilus	*Euryarchaeota*	Deep Sea, Japan	Hydrothermal vent	60°C–88°C (opt. 83°C); 0.1–60 MPa (opt. 30 MPa)	Anaerobic chemoorganotroph oxidizing proteinaceous compounds with elemental sulfur or ferrous iron as electron acceptor	Takai et al. (2000)
Picrophilus oshimae	*Euryarchaeota*	Northern Japan	Acidic soil (pH <0.5) heated to 55°C by solfataric gases	pH opt. 0.7, growth also at pH 0; 45°C–65°C (opt. 60°C)	Heterotrophic, aerobic	Schleper et al. (1995a, 1996); Futterer et al. (2004)
Picrophilus torridus	*Euryarchaeota*	Northern Japan	Acidic soil (pH <0.5) heated to 55°C by solfataric gases	pH opt. 0.7, growth also at pH 0; 45°C–65°C (opt. 60°C)	Heterotrophic, aerobic	Schleper et al. (1995a, 1996); Futterer et al. (2004)
Pyrococcus abyssi	*Euryarchaeota*	Hydrothermal vent, SW Pacific, depth 2,000 m	Hydrothermal vent fluid	67°C–102°C, opt. 96°C at 0.1 MPa. Upper temperature 105°C at 20 MPa	Anaerobic heterotrophic, fermentative	Erauso et al. (1993)
Pyrococcus furiosus	*Euryarchaeota*	Vulcano, Italy	Geothermally heated marine sediment	70°C–103°C (opt. 100°C)	Anaerobic heterotrophic, fermentative	Fiala and Stetter (1986)
Pyrococcus horikoshii	*Euryarchaeota*	Okinawa Trough, NE Pacific Ocean, 1,395 m depth	Hydrothermal vent	Opt. 98°C (max. 102°C). Prolonged survival at 105°C	Grows on proteins and amino acids; growth is stimulated by elemental sulfur	González et al. (1998)
Thermococcus barophilus	*Euryarchaeota*	Mid-Atlantic Ridge, 3,550 m depth	Hydrothermal vent	75°C–100°C; opt. 85°C. Obligate barophilic at >95°C	Anaerobic, grows on complex organic compounds; elemental sulfur enhances growth	Marteinsson et al. (1999)
Thermococcus gammatolerans *Thermococcus marinus* *Thermococcus radiotolerans*	*Euryarchaeota*	Guaymas Basin, Mid-Atlantic Ridge	Hydrothermal chimneys	55°C–95°C (opt. 88°C), survive 20–30 kGy of radiation	Anaerobic, oxidizes proteinaceous compounds with elemental sulfur as electron acceptor	Jolivet et al. (2003a, 2004)
Thermococcus fumicolans	*Euryarchaeota*	North Fiji Basin	Deep-sea hydrothermal vent	Opt. 85°C, pH 8.5, 2%–4% salt.	Anaerobic, respiring proteins with elemental sulfur as electron acceptor	Godfroy et al. (1996)

(Continued)

TABLE 9.1.1 (Continued)

Representative Examples of Thermophilic Microorganisms, Their Taxonomic Affiliation, Habitat, and Properties. The Table Includes Thermoacidophiles, Thermophilic Piezophiles, Halothermoalkaliphilic Species, and Radiation-Resistant Thermophiles Listed Also in Other Tables in This Chapter

Name	Affiliation	Isolated from	Properties of the Environment	Limits of Existence	Other Properties	References
Thermococcus hydrothermalis	*Euryarchaeota*	East Pacific Rise	Deep-sea hydrothermal vent	Opt. 85°C, pH 6, 2%–4% salt.	Anaerobic, respiring amino acids with elemental sulfur as electron acceptor	Godfroy et al. (1996)
Thermococcus peptonophilus	*Euryarchaeota*	Western Pacific Ocean	Deep-sea hydrothermal vent	60°C–100°C (opt. 85°C–90°C). Opt. 30–45 MPa. Opt. temp. shifts from 85°C at 30 MPa to 90°C–95°C at 45 MPa	Proteins, amino acids; elemental sulfur stimulates	González et al. (1995); Canganella et al. (1997)
Thermococcus profundus	*Euryarchaeota*	Okinawa Trough, NE Pacific Ocean, 1,395 m depth	Hydrothermal vent	50°C–90°C (opt. 80°C)	Grows on proteins and amino acids; elemental sulfur is used as the electron acceptor	Kobayashi et al. (1994)
	Bacteria					
Thermus aquaticus	*Deinococcus-Thermus*	Yellowstone National Park, USA	Freshwater hot spring	Up to 79°C (opt. 70°C, min. 40°C); opt. pH 7.5–7.8	Heterotrophic, aerobic	Brock and Freeze (1969)
Truepera radiovictrix	*Deinococcus-Thermus*	Azores Islands	Hot-spring runoff	Opt. 50°C, pH 7.5–9.5. survives 5.0 kGy	Aerobic, heterotrophic	Albuquerque et al. (2005)
Anaerobranca horikoshii	*Firmicutes*	Yellowstone National Park, USA	Soil from geothermal pools	pH 6.9–10.3 (opt. 8.5); 34°C–66°C (opt. 57°C)	Anaerobic proteolytic, fermentative	Engle et al. (1995)
Natranaerobius thermophilus	*Firmicutes*	Wadi Natrun, Egypt	Hypersaline soda lake	35°C–65°C (opt. 53°C), pH 8.3–10.6 (opt. 9.5), 3.1–4.9 M Na$^+$ (opt. 3.3 M)	Anaerobic, fermentative	Mesbah et al. (2007b); Mesbah and Wiegel (2011)
Natranaerobius trueperi	*Firmicutes*	Wadi Natrun, Egypt	Hypersaline soda lake	26°C–55°C (opt. 52°C), pH 7.8–11.0 (opt. 9.9); 4.1–5.4 M Na$^+$ (opt. 3.9 M)	Anaerobic, fermentative	Mesbah and Wiegel (2009)
Natronovirga wadinatrunensis	*Firmicutes*	Wadi Natrun, Egypt	Hypersaline soda lake	26°C–56°C (opt. 51°C), pH 8.5–11.1 (opt. 9.9); 3.1–5.3 M Na$^+$	Anaerobic, fermentative	Mesbah and Wiegel (2009)
Thermaerobacter marianensis	*Firmicutes*	Mariana Trench	Depth of 10,897 m	50°C–80°C, opt. 75°C–76°C)	Aerobic, heterotrophic	Takai et al. (1999)
Thermobrachium celere	*Firmicutes*	New Zealand, Italy, USA, Argentina, Germany	Geothermally and anthropogenically heated environments	pH 5.4–9.5 (opt. 8.2); 43°C–75°C (opt. 66°C)	Anaerobic proteolytic, fermentative	Engle et al. (1996)
Marinitoga piezophila	*Thermotogae*	East Pacific Rise	Hydrothermal chimney (26 MPa)	Opt. 40 MPa, 45°C–70°C (opt. 65°C).	Anaerobic, chemoorganotrophic, uses carbohydrates and proteins, with elemental sulfur as electron acceptor	Alain et al. (2002)

(*Firmicutes*), which grows optimally at 74°C–76°C and up to 80°C as a strict aerobe that grows on yeast extract, peptone, cellulose, starch, chitin, or casein (Takai et al. 1999).

Hyperthermophiles are found in both the Crenarchaeota and the Euryarchaeota phyla. Some hyperthermophilic Archaea obtain their energy by fermentation of organic compounds. Examples are the marine isolates *Pyrococcus furiosus*, which grows on starch, maltose, peptone, and complex organic substrates (Fiala and Stetter 1986); *Pyrococcus abyssi*, fermenting peptides or mixtures of amino (Erauso et al. 1993); and *Pyrodictium abyssi*, a species of disk-shaped cells that appear entrapped within a mycelium-like network of fibers, grows at 110°C, and ferments carbohydrate and proteins (Pley et al. 1991). *Pyrodictium* species can also grow chemoautotrophically on hydrogen as energy source, with elemental sulfur as the electron acceptor (Stetter et al. 1983).

Many more anaerobic hyperthermophilic Archaea depend on elemental sulfur or are stimulated by it (Stetter et al. 1986; Kletzin et al. 2004). Examples are *Thermococcus* spp., which grow optimally at 80°C–85°C on substrates such as tryptone, peptone, and amino acid mixtures and either require elemental sulfur (*T. profundus*) (Kobayashi et al. 1994) or are greatly stimulated by it (*T. fumicolans*, *T. hydrothermalis*, and *T. peptonophilus*) (González et al. 1995; Godfroy et al. 1996, 1997). *Staphylothermus marinus* is another elemental sulfur-dependent archaeon that ferments yeast extract in the presence of sulfur to acetate, isovalerate, CO_2, and H_2S (Fiala et al. 1986). Sulfur also enhances growth of *Pyrococcus horikoshii* (González et al. 1998). Some aerobic thermophilic Archaea, such as *Acidianus* and *Sulfolobus*, depend on elemental sulfur as the electron donor for energy generation.

There are also thermophilic Archaea that use sulfate as the electron acceptor for anaerobic respiration. Species of the genus *Archaeoglobus* have been isolated from the walls of smokers' chimneys and from sediments or marine hydrothermal systems. They can grow heterotrophically on a range of organic electron donors, while reducing sulfate or thiosulfate chemolithoautotrophically on hydrogen, carbon dioxide, and thiosulfate, or as mixotrophs that require molecular hydrogen in addition to an organic carbon source such as acetate (Stetter 1988; Burggraf et al. 1990). Measurements of dissimilatory sulfate reduction above 100°C in the hot deep-sea sediments near the hydrothermal vents of the Guaymas Basin in the Gulf of California showed the process to occur at temperatures up to 110°C, with an optimum rate at 103°C–106°C (Jørgensen et al. 1992). Even arsenate and selenate can serve as the electron acceptors for certain hyperthermophilic Archaea. An isolate affiliated with the *Thermoproteales* was isolated from a hot spring near Naples, Italy, that grew chemolithoautotrophically, with carbon dioxide as carbon source; hydrogen as the electron donor; and arsenate, selenate, thiosulfate, or elemental sulfur as the electron acceptor. Arsenate was reduced to arsenite; during growth on selenate, elemental selenium was formed. Anaerobic chemoorganotrophic growth was also possible with hydrogen as the energy source (Huber et al. 2000b).

Some of the most thermophilic Archaea are methanogens. A hyperthermophilic subsurface lithoautotrophic microbial ecosystem dominated by methanogens was found beneath a deep-sea hydrothermal field in the Central Indian Ridge. Both cultivation and cultivation-independent analyses suggested that members of the *Methanococcales* dominate the methanogenic community (Takai et al. 2004b). All the isolated hyperthermophilic methanogens grow on hydrogen and carbon dioxide for methane generation, and some have very short generation times. These include *Methanotorris formicicus*, isolated from a black smoker hydrothermal vent and able to use formate as an alternative energy source (Takai et al. 2004a); *Methanococcus infernus*, isolated from a deep-sea hydrothermal vent in the Mid-Atlantic Ridge (optimum growth at 85°C) (Jeanthon et al. 1998); *Methanocaldococcus jannaschii*, isolated from a "white smoker" chimney on the East Pacific Rise (doubling time 26 min at 85°C) (Jones et al. 1983); and *Methanopyrus kandleri*, which grows up to 110°C, with an optimum at 98°C (Kurr et al. 1991).

Over the years, the upper temperature known to support life has increased from the boiling point of water to the current value of 122°C. For some years, *Pyrolobus fumarii* was considered the most thermotolerant organism. This crenarchaeote, isolated from the wall of a deep-sea black smoker, does not grow below 90°C, grows up to 113°C (optimum 106°C), and survives autoclaving at 121°C (Blöchl et al. 1997; Stetter 1999). An unnamed and still incompletely characterized formate-oxidizing and iron-reducing archaeon closely related to *Pyrodictium* and *Pyrobaculum*, and known as "Strain 121," reportedly grows even at 121°C (Kashefi and Lovley 2003; Cowan 2004). Today, the record for growth at high temperatures is held by the chemolithotrophic methanogen *Methanopyrus kandleri* strain 116, isolated from a deep-sea hydrothermal vent. When incubated at high pressure (20 MPa), it grows up to 122°C (Takai et al. 2008).

A short time after the discovery of the deep-sea hot vents and the life associated with them, a report was published, claiming that "black smoker" bacteria may be capable of chemolithotrophic growth under in situ vent pressure of ~27 MPa at temperatures of at least 250°C (Baross and Deming 1983). This observation is today considered an artifact. Based on the limited thermal stability of metabolites in aqueous solution, the theoretical maximum temperature for life was recently estimated at 150°C–180°C (Bains et al. 2015).

9.1.3 COLD ENVIRONMENTS AND PSYCHROPHILIC MICROORGANISMS

Permanently cold environments are abundant on planet Earth. Most of the deep waters of the oceans have a constant temperature of 1°C–3°C. The Arctic and the Antarctic are permanently cold, and parts of northern countries such as Siberia-Russia and Canada are permanently frozen ("permafrost"). All these environments support life of cold-adapted, psychrophilic, and psychrotolerant microorganisms (Deming 2002, 2007; Miller and Whyte 2011; Goordial et al. 2013). Cryoenvironments also include glaciers, ice caps, sea ice, and cold lakes and ponds. The search for life elsewhere in the universe and the environmental changes on Earth as a result of global warming have greatly stimulated research on the microbiology of cold ecosystems (Deming 2002).

At low temperatures, metabolism is slow. But the primary constraint to life at low temperatures is the availability of liquid water. When the cytoplasm becomes frozen, the cell may remain viable, but metabolic activity is no longer possible. The freezing point of the intracellular water thus determines the lowest temperature for growth. Accumulation of organic and inorganic solutes that act as "antifreeze" agents (antifreeze proteins and certain sugars) can, to some extent, lower the freezing temperature of the intracellular water. Because the presence of liquid water in cryoenvironments is often due to the freezing-point depression properties of solutes, many cold-adapted microorganisms also tolerate osmotic stress caused by high salinity (Cowan et al. 2007).

Table 9.1.2 summarizes the properties of a few well-characterized psychrophilic bacteria. True psychrophiles are generally defined as such organisms that grow optimally below 15°C and that still can grow below 0°C. Organisms that show growth below 0°C but have their optimum above 15°C–20°C are called psychrotolerant or psychrotrophs (Cowan et al. 2007). The ability of psychrophiles and psychrotrophs to grow at low temperatures depends on adaptive changes in their cellular proteins and lipids (Russell 1990; D'Amico et al. 2006). Psychrophilic enzymes often have a high content of α-helix and little β-sheet secondary structure, enabling a higher flexibility at low temperatures. Cold-active enzymes also tend to have a high polar and a low hydrophobic amino acid content and lower numbers of weak hydrogen and ionic bonds. Unsaturated fatty acids may abound in the cell membrane. Thus, *Polaromonas vacuolata* may contain up to 74%–79% of 16:1 ω7c, the highest level of this fatty acid reported in any species. In addition it contains 75–95 of 18:1 (Irgens et al.

1996). A high content of polyunsaturated fatty acid is often found in psychrophiles, including members of the Bacteria domain, where occurrence of polyunsaturated fatty acids is very uncommon. Thus, *Psychroflexus torquis* (*Bacteroidetes*) isolated from Antarctic sea ice possesses the unusual ability to synthesize the polyunsaturated fatty acids eicosapentaenoic acid (20:5 ω-3) and arachidonic acid (20:4 ω-6) (Bowman et al. 1998).

The lowest growth temperature recoded for the heterotrophic gas vacuolate *Psychromonas ingrahamii* (*Gammaproteobacteria*), isolated from sea ice of Alaska, is −12°C, with a generation time of 240 h (Breezee et al. 2004). Another organism adapted to low temperatures is *Polaromonas vacuolata* isolated from Antarctic marine waters: it grows optimally at 4°C, with a minimum at 0°C and a maximum at 12°C (Irgens et al. 1996). Other species able to grow at low temperatures are *Psychrobacter cryohalentis* and *Psychrobacter arcticus* isolated from Siberian permafrost: they grow at temperatures from −10°C to +3°C and salinities up to 1.7 M (Bakermans et al. 2006). *Planococcus halocryophilus*, isolated from permafrost, even grows slowly at −15°C, the lowest growth temperature documented for any bacterium (Mykytczuk et al. 2013).

Microbial activity may be possible at even lower temperatures. During respiration measurements in cultures of the Antarctic glacial isolates *Sporosarcina* sp. B5 and *Chryseobacterium* sp. V3519-10 within ice at temperatures from −4°C to −33°C, initial high metabolic rates were followed by lower rates, when the cells entered a non-reproductive dormant metabolic state, but metabolism was sustained by viable cells, as quantified via culturability, reduction of 5-cyano-2,3-ditoyl tetrazolium chloride (CTC), and LIVE/DEAD staining. ^{14}C-labeled acetate was actively respired by these organisms in polycrystalline ice at −5° in the presence

TABLE 9.1.2

Representative Examples of Psychrophilic Microorganisms, Their Taxonomic Affiliation, Habitat, and Properties. The Table Includes a Piezophilic Psychrophilic Species Listed Also in Table 9.1.6

Name	Affiliation	Isolated from	Properties of the Environment	Limits of Existence	Other Properties	References
	Bacteria					
Psychroflexus torquis	*Bacteroidetes*	Antarctica	Sea ice	0°C to <20°C (opt. 10°C–15°C)	Aerobic, heterotrophic	Bowman et al. (1998)
Colwellia psycherythraea	*Proteobacteria*	Atlantic Ocean, Puerto Rico Trench	Seawater depth of 7,410 m	At 2°C growth at 37–103 MPa (opt. 75 MPa); best growth at 10°C and 94 MPa	Aerobic, heterotrophic	Deming et al. (1988)
Psychrobacter arcticus, *Psychrobacter cryohalentis*	*Proteobacteria*	Siberia	Permafrost	−10°C to +30°C, 0–1.7 M NaCl	Aerobic, heterotrophic	Bakermans et al. (2006)
Psychromonas ingrahamii	*Proteobacteria*	Alaska	Sea ice	−12°C to +10°C; 0.2–2.1 M NaCl needed; weak growth up to 3.4 M NaCl	Aerobic, heterotrophic	Breezee et al. (2004)

of abundant nutrients, in spite of the low temperatures and the physical ice matrix. *Sporosarcina* even increased in cell numbers and biomass under these conditions (Bakermans and Skidmore 2011a, 2011b). Measurements of [³H]-leucine incorporation into macromolecules by the marine psychrophilic bacterium *Colwellia psychrerythraea* showed incorporation into protein down to −20°C. Presence of exopolysaccharides enhanced the rates (Junge et al. 2006).

Cultivation-independent characterization of the microbial communities within permanently cold marine sediments by means of 16S rRNA gene clone libraries showed a predominance of sequences (43.4%); this can be attributed to sulfate reducers (*Desulfotalea* and others). Additional groups frequently encountered were *Desulfuromonas palmitatis* and groups affiliated with the myxobacteria and with *Bdellovibrio*. Many *Gammaproteobacteria* clones retrieved were related to symbiotic or free-living sulfur oxidizers (Ravenschlag et al. 1999).

Much research on microbial communities at low temperatures was performed in Antarctica (Vincent 1988). Based on ATP measurements, between 3×10^6 and 2×10^9 live bacterial cells per gram soil were calculated to be present in Antarctic soils at temperatures between −0.5°C and +3.8°C (Cowan et al. 2002). Bacterial populations of 200–5,000 cells mL⁻¹ were present in melting Antarctic snow sampled in January 1999 and 2000. 16S rRNA genes affiliated with *Deinococcus* were recovered from this environment. Low but significant rates of bacterial DNA and protein syntheses were measured in the snow at temperatures of −12°C to −17°C (Carpenter et al. 2000). Liquid pockets of concentrated brine within Antarctic sea ice may support active microbial communities, but also the ice itself harbors a diverse community (Thomas and Dieckmann 2002). A cultivation-independent, 16S rRNA gene-based survey of the communities present in Antarctic sea ice showed dominance of *Alphaproteobacteria*, *Gammaproteobacteria*, and *Bacteroidetes*, with other groups also present, showing that the ice contains niches conducive to the proliferation of a diverse array of psychrophilic bacterial species (Bowman et al. 1997).

The perennial ice that covers the Antarctic lakes in the McMurdo Dry Valleys is an interesting habitat for microbial communities that develop within liquid water inclusions in response to solar heating of internal aeolian-derived sediments. Within these complex microbial consortia embedded in the lake ice cover, photosynthesis, nitrogen fixation, and decomposition of organic matter occur (Priscu et al. 1998).

Small communities of prokaryotes were found even within ice at a depth of 3,590 m below the Vostok research station in Antarctica. Between 2.8×10^3 cells and 3.6×10^4 cells per milliliter were detected, mainly affiliated with the *Proteobacteria* and the *Actinobacteria*, but no activity could be detected based on incorporation of organic substrates or bicarbonate (Priscu et al. 1999). Ice below ~3,500 m consists of refrozen water from Lake Vostok, accreted to the bottom of the glacial ice. In the accretion ice, the carbon and energy sources in the veins within the ice can maintain significant numbers of actively metabolizing cells (Price 2000).

Attempts to assess microbial activities by using CTC as a probe for O_2-based respiration were also made in Arctic wintertime sea-ice cores (temperature gradient −2°C to −20°C and salinity of the brine inclusions 38–209 ppt). The CTC-active bacteria (0.5%–4% of the total) and cells detectable by rRNA probes (18%–86% of the total count) were found in all samples, including the coldest ones (−20°C). The percentage of active bacteria (probably affiliated with the *Bacteroidetes*) associated with particles increased with decreasing temperature (Junge et al. 2004).

A highly intriguing Antarctic ecosystem is Blood Falls, McMurdo Dry Valleys, a subglacial outflow from the Taylor Glacier. The falls have a high content of iron and sulfate, likely due to interactions of the subglacial brine with the iron-rich bedrock. The bacterial 16S rRNA gene clone library was dominated by a phylotype affiliated with *Thiomicrospira arctica*, a psychrophilic marine autotrophic sulfur oxidizer. Different representatives of the *Betaproteobacteria*, *Deltaproteobacteria*, and *Gammaproteobacteria* and the *Bacteroidetes* were also found. The brine below the Taylor Glacier hosts an anoxic microbial ecosystem with a limited supply of organic carbon, and based on an active sulfur cycle, with Fe(III) serving as the terminal electron acceptor (Mikucki and Priscu 2007; Mikucki et al. 2009).

The terrestrial permafrost environment and its microorganisms can be considered as a model for conditions for life on other cold planets. The deep cold biosphere is the most stable environment for microorganisms, and possible fluctuations should be explained by geological events only. Cold-adapted microorganisms may survive within permafrost at subzero temperatures over geological time and resume their activities upon thawing. Studies of microbial communities in permafrost sediments of different lithology and age suggest that the temperature and the length of exposure define the ratio between cells that can rapidly resume activity and deep-resting "viable but non-culturable" cells (Vorobtova et al. 1997). Microorganisms may survive within a film of unfrozen water enveloping soil particles that protects them from freezing (Gilichinsky 1997, 2002). Brine lenses called cryopegs are found in Siberian permafrost, and these were formed and isolated from ancient marine sediment layers of the Arctic Ocean ~100,000–120,000 years ago. Owing to its high salinity (170–300 g/L), the water remains liquid at the in situ temperature of −10°C, but even at −15°C, uptake of [¹⁴C]-glucose could be measured. Different types of psychrophilic prokaryotes, including aerobes, fermentative anaerobes, sulfate reducers, acetogens, and methanogens, could be retrieved from these cryopegs (Gilichinsky et al. 2003).

Metabolic activity within Siberian permafrost samples during incubation at different temperatures between +5°C and −20°C was measured by incorporation of ¹⁴C-labeled acetate into lipids. Minimum estimated doubling times ranged from 1 day (at 5°C) to 20 days (at −10°C) to ca. 160 days (at −20°C) (Rivkina et al. 2000). Metabolic activity and growth were tested in a number of bacterial isolates from Siberian permafrost as a function of temperature. Endospore-forming Gram-positive isolates proved unable to grow or metabolize below 0°C. Other Gram-positive isolates showed metabolic activity at lower temperatures but did not show any growth at −10°C. However, one Gram-negative isolate grew at −10°C, with a doubling time of 39 days. When the temperature decreased below 4°C, more energy was generally used for cell maintenance than for growth (Bakermans et al. 2003).

9.1.4 MICROORGANISMS IN EXTREMELY ACIDIC ENVIRONMENTS

Environments with pH values below 2–3 are not very abundant on our planet (Schleper et al. 1995b). Many of these are associated with volcanic activity: hot sulfur springs, mud pots, etc. found in Italy; Iceland; New Zealand; Yellowstone National Park, USA; and a few other sites. The low pH of such solfataras is often due, at least in part, to thermophilic chemoautotrophic microorganisms that oxidize elemental sulfur and other reduced sulfur compounds to sulfuric acid. At moderate temperatures, microbial activity is generally responsible for the formation of low-pH environments. Bacterial fermentations such as the lactic acid fermentation can locally lower the pH to very acidic values, but most often, the low pH is generated by microbial oxidation of reduced sulfur compounds such as sulfide, elemental sulfur, and pyrite by chemoautotrophic microorganisms that produce sulfuric acid. Many of these organisms are highly acid-tolerant or even obligately acidophilic. Formation of "acid mine drainage" in mining areas where sulfur-containing ores are brought to the surface. Here, microbially mediated oxidation of pyrite often results in the formation of highly acidic wastewaters with a high content of iron and other metal ions (Johnson and Hallberg 2003). The very low pH (~2) and the high concentrations of heavy metals in the waters of the 100-km-long Tinto River (Huelva, southwestern Spain) are not directly associated with recent mining activity, but they originate in the same way as modern acid mine drainage: by the activity of chemolithotrophs obtaining their energy from complex sulfides present in the Iberian Pyrite Belt (González-Toril et al. 2003). Such activity of sulfur-oxidizing autotrophic prokaryotes is not always disadvantageous: acidification resulting from the activity of chemoautotrophic sulfur-oxidizing bacteria is exploited for the commercial recovery of copper and other metals from sulfur-containing ores.

Acidophilic microorganisms are very diverse, physiologically as well as phylogenetically; they occur in all three domains of life (Johnson 1998, 2007; Hallberg and Johnson 2001; Johnson and Aguilera 2016) (Table 9.1.3). The unicellular green alga *Dunaliella acidophila*, isolated from highly acidic waters and soils, is one of the most acidophilic organisms known, as it grows between pH 0 and 3, with an optimum at pH 1 (Pick 1999). The intriguing thermophilic acidophilic alga *Cyanidium* and related organisms found in acidic hot springs are discussed later in this section. There are extremely acidophilic representatives both in the *Euryarchaeota* branch and in the *Crenarchaeota* branch of the Archaea, and extreme acidophiles are also found in different bacterial phyla (*Firmicutes*, *Actinobacteria*, *Proteobacteria*, *Nitrospirae*, and *Aquificae*).

Many extreme acidophiles are unable to grow or even survive at neutral pH. *Acidithiobacillus thiooxidans* (*Proteobacteria*),

which can grow down to pH 0.5, does not grow above pH 4–6 (Temple and Colmer 1951; Kelly and Wood 2000). The most extreme case is the thermophilic (optimum temperature ~60°C) genus *Picrophilus* (*Euryarchaeota*), with a growth optimum at pH 0.5. Its members can even grow at pH −0.06, but the cells lyse above pH 4 (Schleper et al. 1995a).

All known acidophilic microorganisms maintain their intracellular pH value close to neutral, and their intracellular enzymatic machinery is not adapted to low pH. The cells must therefore maintain a proton concentration gradient of four or even five orders of magnitude, using powerful proton pumps in the cytoplasmic membrane (Schleper et al. 1995b).

The microbial diversity in acid mine drainage and other low-pH environments was explored using both cultivation-dependent and cultivation-independent methods. A search for eukaryotic microorganisms, fungi as well as protists, in warm (30°C–50°C) and extremely acidic (pH 0.8–1.38) and metal-rich (Fe, Zn, As, Cu) mine drainage from the Richmond Mine (Iron Mountain, California), based on 18S rRNA gene sequences, showed the presence of a lineage of Rhodophyta (red algae) and amoeboid organisms (Vahlkampfiidae and Heterolobosea), as well as fungal groups associated with the Dothideomycetes and the Eurotiomycetes (Baker et al. 2004). The prokaryotic communities within slime biofilms that had developed on exposed pyrite ore within the same mine were dominated by *Leptospirillum* spp. (*L. ferriphilum* and smaller numbers of *L. ferrodiazotrophum*) and *Ferroplasma acidarmanus*, *Sulfobacillus*, and *Acidimicrobium-/Ferrimicrobium*-related species (Johnson 2007). This abandoned mine has become a model system to test metagenomics techniques in a relatively simple, low-diversity ecosystem, enabling reconstruction of near-complete genomes of a *Leptospirillum* sp. and a *Ferroplasma* sp. (Tyson et al. 2004). *Ferroplasma acidarmanus*, a relative of *Ferroplasma acidiphilum*, is an archaeal iron-oxidizing extreme acidophile capable of growth at pH 0, found in slime streamers, and attached to pyrite surfaces in this mine (Edwards et al. 2000). *Ferroplasma* spp. are facultative anaerobes, growing optimally at pH 1.0–1.7 and coupling chemoorganotrophic growth on organic substrates to the reduction of ferric iron (Dopson et al. 2004; Golyshina and Timmis 2005). Cultivation-independent studies in the Tinto River in Spain yielded 16S rRNA sequences very similar to those in acid mine drainage and in abandoned mine sites elsewhere in the world: *Leptospirillum* spp., *Acidithiobacillus ferrooxidans*, *Acidiphilium* spp., *Ferrimicrobium acidiphilum*, *Ferroplasma acidiphilum*, and *Thermoplasma acidophilum*. Archaea made up only a minor fraction of the community (González-Toril et al. 2003; Johnson 2012).

Macroscopically visible growth as streamers in acidic, metal-rich waters in an abandoned copper mine and in a chalybeate spa in North Wales was found to consist of 80%–90% of *Betaproteobacteria*, as based on fluorescence in situ hybridization studies. A single species dominated in the copper mine streamers; in the spa water, this organism was accompanied by a novel acidophilic autotrophic iron oxidizer, closely related to the neutrophilic *Gallionella ferruginea* (Hallberg et al. 2006).

TABLE 9.1.3

Representative Examples of Acidophilic Microorganisms, Their Taxonomic Affiliation, Habitat, and Properties. The Table Includes Thermoacidophilic and Haloacidophilic Species Listed Also in Other Tables in This Chapter

Name	Affiliation	Isolated from	Properties of the Environment	Limits of Existence	Other Properties	References
	Archaea					
Sulfolobus acidocaldarius	*Crenarchaeota*	Yellowstone National Park, USA and acidic hot springs elsewhere	Acidic hot springs	55°C–85°C (opt. 70°C–75°C); pH 0.9–5.8 (opt. 2–3)	Aerobic, heterotrophic on simple organic compounds; facultative autotroph on elemental sulfur	Doemel and Brock (1970); Brock et al. (1972)
Sulfurisphaera ohwakuensis	*Crenarchaeota*	Japan	Acidic hot spring	pH 1.0–5.0 (opt. 2.0); 63°C–92°C (opt. 84°C)	Facultatively anaerobic, heterotrophic	Kurosawa et al. (1998)
"*Ferroplasma acidarmanus*"	*Euryarchaeota*	Iron mountain, California	Water from a pyrite mine	pH opt. 1.0–1.7, growth down to pH 0; opt. 35°C–42°C	Chemoheterotroph or chemomixotroph, facultative anaerobic, reducing ferric ions	Edwards et al. (2000); Dopson et al. (2004)
Ferroplasma acidiphilum	*Euryarchaeota*	Bioleaching pilot plant, Russia	Leaching plant for gold-containing arsenopyrite/ pyrite ore	pH 1.3–2.2 (opt. 1.7); 15°C–45°C	Aerobic, obligately autotrophic	Golyshina et al. (2000)
Halarchaeum acidiphilum	*Euryarchaeota*	Japan	Commercial solar salt	pH 4.0–4.6 (opt. 4.4–4.5); 18–30 NaCl (opt. 21%–24%)	Aerobic, heterotrophic	Minegishi et al. (2008, 2010)
Picrophilus oshimae	*Euryarchaeota*	Northern Japan	Acidic soil (pH <0.5) heated to 55°C by solfataric gases	pH opt. 0.7, growth also at pH 0; 45°C–65°C (opt. 60°C)	Heterotrophic, aerobic	Schleper et al. (1995a, 1996); Futterer et al. (2004)
Picrophilus torridus	*Euryarchaeota*	Northern Japan	Acidic soil (pH <0.5) heated to 55°C by solfataric gases	pH opt. 0.7, growth also at pH 0; 45°C–65°C (opt. 60°C)	Heterotrophic, aerobic	Schleper et al. (1995a, 1996); Futterer et al. (2004)
	Bacteria					
Leptospirillum ferrooxidans	*Nitrospirae*	Armenia	Copper mine	pH 1.5–4.0 (opt. 2.5–3.0)	Aerobic; ferrous ions as electron donor for chemolithoautotrophic growth	Hippe (2000)
Acidithiobacillus ferrooxidans	*Proteobacteria*	West Virginia and Pennsylvania	Coal mine drainage	pH 1.4–6.0 (opt. 2.5–5.8)	Aerobic; ferrous ions or reduced sulfur compounds as electron donor for chemolithoautotrophic growth	Temple and Colmer (1951)
	Eukarya					
Dunaliella acidophila	*Chlorophyceae*	Czech Republic	Acidic water near a sulfur spring	pH 0.3–3.0; NaCl up to 0.5 M	Photoautotrophic	Pick (1999)
Cyanidium caldarium	*Rhodophyta*	Yellowstone National Park, USA; Japan, New Zealand	Acidic hot springs in volcanic areas	pH 0.2–4.0; 40°C–56°C	Photoautotrophic	Doemel and Brock (1970); Seckbach (1994)

Life at low pH and high temperature is also diverse (Hedlund et al. 2016). Volcanic activity results in the formation of hot springs, and these are often acidic, especially when reduced-sulfur compounds are present. Hot springs can be inhabited by algae of the species *Cyanidium caldarium* and related organisms of the genera *Galdieria* and *Cyanidioschyzon* (Rhodophyta), growing optimally at pH 2–3 and tolerating values as low as pH 0.2 (Seckbach 1994, 1999; Toplin et al. 2008). Field studies in the algae's natural habitats, $^{14}CO_2$ incorporation measurements of natural populations, and cultivation experiments have shown an upper temperature limit of 55°C–60°C for *C. caldarium* (Doemel and Brock 1970).

Early studies of the acidic hot springs of Yellowstone National Park, USA, led to the discovery of the first thermoacidiophilic archaeon, *Sulfolobus acidocaldarius* (*Crenarchaeota*) (Brock et al. 1972). This facultative autotroph can obtain energy from the oxidation of sulfur or from a variety of simple organic compounds. It grows between pH 0.9 and 5.8, with an optimum at pH 2–3 and a temperature optimum of 70°C–75°C. *Sulfolobus* and related organisms have been isolated from many natural acidic thermal habitats worldwide, terrestrial as well as aquatic (e.g., Brierley and Brierley 1973). A relative of *Sulfolobus* that tolerates even higher temperatures is *Sulfurisphaera ohwakuensis*, isolated in Japan; it grows up to 92°C, with a minimum pH of 1 (Kurosawa et al. 1998).

Thermoacidophiles with an even lower pH optimum and minimum are found in the genus *Picrophilus* (*Euryarchaeota*). *Picrophilus torridus* and *P. oshimae*, isolated from solfataric locations in Japan, are heterotrophs that grow even below pH 0 at up to 65°C (Schleper et al. 1995a, 1996; Futterer et al. 2004).

A culture-independent study of rocks in a Yellowstone National Park hot spring demonstrated the presence of endolithic microorganisms in rock pore water at pH 1.0. These included organisms phylogenetically affiliated with the genus *Mycobacterium* (Actinobacteria) (Walker et al. 2005).

Deep-sea hydrothermal vents that emit hot water of low pH (<4.5) are inhabited by Archaea of the Deep-sea Hydrothermal Vent Euryarchaeota 2 (DHVE2) group, obligate thermoacidophilic sulfur- or iron-reducing heterotrophs capable of growing from pH 3.3 to 5.8 at 55°C–75°C (Reysenbach et al. 2006).

9.1.5 ALKALINE ENVIRONMENTS AND ALKALIPHILIC MICROORGANISMS

Permanently alkaline lakes ("soda lakes"), whose high pH is caused by geological-geochemical processes, are found on all continents. Some of the best investigated ones are Mono Lake (California), Lake Magadi (Kenya) and other East African soda lakes, the Wadi Natrun lakes in Egypt, and a number of soda lakes in China and Tibet. Many of these alkaline lakes also contain high salt concentrations. Thus, the organisms inhabiting them must be adapted both to high pH and to high salinity (Grant and Tindall 1986; Grant and Jones 2000; Jones and Grant 2000). The Wadi Natrun lakes have become an excellent model for the study of the ecology, physiology, and taxonomy of aerobic and anaerobic haloalkaliphilic microorganisms (Oren 2013). Many soils are alkaline, and many

isolates of alkaliphilic bacteria (here defined as organisms that grow optimally above pH 8 and usually between pH 9 and 10) have been isolated from soils, where transient alkaline conditions may be generated as a result of biological activity (Grant et al. 1990). Alkaline conditions are also commonly encountered in the surface layers of lakes and ponds, where, during daytime, CO_2 becomes depleted as a result of photosynthesis by planktonic algae and cyanobacteria.

A great diversity of alkaliphilic microorganisms is known, both phylogenetically and physiologically. The biogeochemical cycles of carbon, nitrogen, sulfur, and other elements can function at high pH, under both aerobic and anaerobic conditions (Horikoshi 1999; Banciu and Sorokin 2013). Thus, a diverse anaerobic community in the sediments of Lake Magadi (Kenya) and other soda lakes with pH ~10 was found to perform all functions known for neutral-pH environments (Zhilina and Zavarzin 1994; Zavarzin and Zhilina 2000). The eutrophic alkaline desert lakes of Wadi Natrun in Egypt (pH ~11, >300 g/L salt) are inhabited by cyanobacteria, anoxygenic phototrophs, haloalkaliphilic Archaea, sulfate reducers, and many other types of microorganisms (Imhoff et al. 1979; Mesbah et al. 2007a; Oren 2013), including an interesting group of anaerobic fermentative organisms, that, in addition to being adapted to life at high pH and high salinity, have markedly thermophilic properties (Mesbah and Wiegel 2008). The properties of these organisms are further discussed in the last paragraph of this section. Searles Lake (California), a salt-saturated lake at pH 9.8, even supports a full biogeochemical cycle of arsenic, including chemoautotrophic oxidation of As(III) to As(V) in the oxic layer and As(V) respiration in the anaerobic layers (Oremland et al. 2005). Haloalkaliphilic bacteria performing such transformation have been isolated: *Alkalilimnicola ehrlichii* (*Gammaproteobacteria*) from Mono Lake oxidizes arsenite, with nitrate or oxygen as the electron acceptor. It can also use hydrogen, sulfide, or thiosulfate as an electron donor (Hoeft et al. 2007). *Desulfohalophilus alkaliarsenatis* (*Deltaproteobacteria*) from Searles Lake can respire both sulfate and arsenate (Switzer Blum et al. 2012).

Alkaliphilic microorganisms maintain their intracellular pH close to neutrality. As energy transformation is commonly based on the outward pumping of protons, the alkaliphiles must thus maintain an inverted pH gradient. Membrane-linked bioenergetic processes are then based on an exceptionally large membrane potential. Still, protons are used as the coupling ions between the respiratory chain and ATP synthase in haloalkaliphilic Archaea (Falb et al. 2005) and in most alkaliphilic Bacteria. But many cases have been reported, especially in haloalkaliphilic anaerobes, where energy transformation processes are linked with the transmembrane transport of sodium ions rather than protons. Thus, Na^+-dependent ATPases, Na^+-driven flagellar movement, and Na^+-driven solute transport systems have been documented. Accordingly, many alkaliphiles are strictly dependent on sodium ions for growth (Banciu and Sorokin 2013).

Table 9.1.4 shows the properties of representative alkaliphilic Archaea and Bacteria available in culture. Hypersaline alkaline lakes such as Lake Magadi and the Wadi Natrun lakes are colored pink-red by dense communities of haloalkaliphilic

TABLE 9.1.4

Representative Examples of Alkaliphilic Microorganisms, Their Taxonomic Affiliation, Habitat, and Properties. The Table Includes Thermophilic Alkaliphiles, Haloalkaliphilic, and Halothermoalkaliphilic Species Listed Also in Other Tables in This Chapter

Name	Affiliation	Isolated from	Properties of the Environment	Limits of Existence	Other Properties	References
	Archaea					
Halorubrum vacuolatum	*Euryarchaeota*	Lake Magadi, Kenya	Hypersaline alkaline lake	pH 8.5–10.5 (opt. 9.5); NaCl needed 2.5–5.2 M (opt. 3.5 M)	Aerobic, heterotrophic	Mwatha and Grant (1993)
Methanosalsum zhilinae	*Euryarchaeota*	Wadi Natrun, Egypt	Hypersaline alkaline lake sediment	pH 8–10 (opt. 8.7–9.5); NaCl needed 0.2–2.1 M (opt. 0.4–0.7 M)	Anaerobic, methanogenic on methanol or methylated amines	Mathrani et al. (1988)
Natronococcus occultus	*Euryarchaeota*	Lake Magadi, Kenya	Hypersaline alkaline lake	Opt. pH 9.5–10.0; NaCl needed 1.4–5.2 M (opt. 3.5–3.6 M)	Aerobic, heterotrophic	Tindall et al. (1984)
Natronomonas pharaonis	*Euryarchaeota*	Wadi Natrun, Egypt	Hypersaline alkaline lake	pH 7–10 (opt. pH 8.5–9.5); 2–5.2 M NaCl (opt. 3.5 M)	Aerobic, heterotrophic	Soliman and Trüper (1982); Falb et al. (2005)
	Bacteria					
Anaerobranca horikoshii	*Firmicutes*	Yellowstone National Park, USA	Soil from geothermal pools	pH 6.9–10.3 (opt. 8.5); 34°C–66°C (opt. 57°C)	Anaerobic proteolytic, fermentative	Engle et al. (1995)
Natranaerobius thermophilus	*Firmicutes*	Wadi Natrun, Egypt	Hypersaline soda lake	35°C–65°C (opt. 53°C), pH 8.3–10.6 (opt. 9.5); 3.1–4.9 M Na$^+$ (opt. 3.3 M)	Anaerobic, fermentative	Mesbah et al. (2007); Mesbah and Wiegel (2011)
Natranaerobius trueperi	*Firmicutes*	Wadi Natrun, Egypt	Hypersaline soda lake	26°C–55°C (opt. 52°C); pH 7.8–11.0 (opt. 9.9); 4.1–5.4 M Na$^+$ (opt. 3.9 M)	Anaerobic, fermentative	Mesbah and Wiegel (2009)
Natronovirga wadinatrunensis	*Firmicutes*	Wadi Natrun, Egypt	Hypersaline soda lake	26°C–56°C (opt. 51°C); pH 8.5–11.1 (opt. 9.9); 3.1–5.3 M Na$^+$	Anaerobic, fermentative	Mesbah and Wiegel (2009)
Thermobrachium celere	*Firmicutes*	New Zealand, Italy, USA, Argentina, Germany	Geothermally and anthropogenically heated environments	pH 5.4–9.5 (opt. 8.2); 43°C–75°C (opt. 66°C)	Anaerobic proteolytic, fermentative	Engle et al. (1996)
Tindallia magadii	*Firmicutes*	Lake Magadi, Kenya	Hypersaline soda lake	pH 7.5–10.5 (opt. 8.5)	Anaerobic, fermentative	Kevbrin et al. (1998)
Alkalilimnicola ehrlichii	*Proteobacteria*	Mono Lake, California	Hypersaline soda lake	pH 7.3–10.0 (opt. 9.3); salinity 15–190 g/L (opt. 30 g/L)	Facultatively aerobic (oxygen or nitrate as electron acceptor), chemoautotrophic (arsenite, hydrogen, sulfide, or thiosulfate as electron donor), or heterotrophic	Hoeft et al. (2007)
"*Desulfohalophilus alkaliarsenatis*"	*Proteobacteria*	Searles Lake, California, USA	Hypersaline alkaline lake sediment	55–330 g/L (opt. 200 g/L); pH 7.8–9.7 (opt. 9.3)	Anaerobic, respiring sulfate or arsenate	Switzer Blum et al. (2012)

Archaea of the class *Halobacteria* (Oren 2002a). Examples are *Natronomonas pharaonis* isolated from the Wadi Natrun lakes (Soliman and Trüper 1982) and the genera *Natronobacterium* and *Natronococcus* from the East African soda lakes (Tindall et al. 1984; Mwatha and Grant 1993). Alkaliphilic halophilic methanogenic Archaea have also been isolated: *Methanosalsum zhilinae* from the Wadi Natrun lakes produces methane from methanol, methylated amines, and dimethylsulfide (Mathrani et al. 1988).

In the domain Bacteria, we also find alkalithermophiles and even haloalkalithermophiles. The group of anaerobic fermentative thermophilic (optimum 57°C–66°C) alkaliphiles (optimum pH 8.2–8.5 and growing up to pH 9.5–10.3) of *Thermobrachium* and *Anaerobranca* is especially interesting (Wiegel 1998; Kevbrin et al. 2004; Wiegel and Kevbrin 2004). *Thermobrachium celere* (*Firmicutes*) is a proteolytic, obligately anaerobic, moderately alkaliphilic (optimum pH 8.2, growing up to pH 9.5) fermentative bacterium that under optimal conditions has a doubling time of only 10 minutes. It was isolated from geothermally and anthropogenically heated environments (Engle et al. 1996). *Anaerobranca horikoshii* from Yellowstone National Park has similar properties (Engle et al. 1995).

It may be argued that the anaerobic fermentative halothermoalkaliphiles live at the physicochemical boundary for life (Mesbah and Wiegel 2008, 2012; Bowers et al. 2009). The sediments of the Wadi Natrun lakes have yielded isolates such as *Natronovirga wadinatrunensis*, *Natranaerobius thermophilus*, and *Natranaerobius trueperi* (*Firmicutes*; pH optima 9.5–9.9; temperature optima 51°C–53°C; and Na$^+$ requirement and tolerance 3.1–5.3 M) (Mesbah et al. 2007b; Mesbah and Wiegel 2009). Analysis of the genome of *N. thermophilus* revealed that the organism may combine two mechanisms for adaptation to high salinity: accumulation of potassium ions together with accumulation of organic osmotic solutes (Mesbah and Wiegel 2011). Cultivation-independent characterization of the microbial community in the water and sediments of the Wadi Natrun lakes shows that they harbor a unique and novel prokaryotic diversity, different from that described from other alkaline, athalassohaline lakes (Mesbah et al. 2007a).

9.1.6 HYPERSALINE ENVIRONMENTS AND HALOPHILIC MICROORGANISMS

Hypersaline environments include salt lakes such as Great Salt Lake (Utah, USA), the Dead Sea on the border between Israel and Jordan, and many other natural hypersaline lakes worldwide. There are also man-made hypersaline lakes: saltern evaporation and crystallizer ponds for the production of solar salt from seawater by evaporation. In addition, there are saline and hypersaline soils. In many of these environments, the ionic composition reflects that of seawater from which the brines were derived ("thalassohaline environments"), while in other cases ("athalassohaline brines"), the chemical composition of the salts is greatly different from that of seawater. Examples are the Dead Sea, dominated by divalent cations (Mg and Ca) rather than by monovalent cations (Na and K),

with very low sulfate concentrations and a relatively low pH (~6), and soda lakes such as Mono Lake (California, USA) and the East African lakes such as Lake Magadi in Kenya, with high pH (9–10), high carbonate/bicarbonate concentrations, and very low concentrations of divalent cation concentrations. All these environments are inhabited by diverse communities of high-salt-adapted microorganisms. Even rock salt deposited many millions of years ago may be inhabited by microorganisms; some of them may even have survived from the time the salt was formed by evaporation of seawater (Oren 2002a, 2016).

Microorganisms living at high salt concentrations need to maintain their cytoplasm at least osmotically equivalent to their surroundings, in order not to lose water. Maintenance of a turgor pressure even requires a higher osmotic pressure within the cells. Different halophiles have developed two fundamentally different strategies to allow life at the extremely high osmotic pressures exerted by salt-saturated brines. One is based on the accumulation of salts inside the cells at concentrations no less than those in the outside medium. The second strategy involves exclusion of salts from the cytoplasm to a large extent and synthesis or accumulation of simple organic solutes ("osmotic solutes" or "compatible solutes") to balance the osmotic pressure of the medium. Thus, the highly halotolerant unicellular green alga *Dunaliella* produces glycerol for osmotic stabilization. Many halophilic and halotolerant representatives of the Bacteria and some halophilic Archaea (e.g., the halophilic methanogens) produce solutes such ectoine, glycine betaine, simple sugars, certain amino acids, and amino acid derivatives. Use of the "low-salt, high-organic-solutes-in" strategy does not require a far-going adaptation of the intracellular osmotic machinery to the presence of high salt concentrations (Galinski 1995; Ventosa et al. 1998; Oren 2000). A different option, realized by the aerobic halophilic Archaea of the class *Halobacteria* and by a limited number of members of the Bacteria such as *Salinibacter* (*Bacteroidetes*) (Antón et al. 2002; Oren et al. 2002) and the obligatory anaerobic *Halanaerobiales* (*Firmicutes*) (Oren et al. 1997), is to accumulate molar concentrations of KCl for osmotic balance. This "salt-in" strategy of osmotic adaptation is possible only when the intracellular enzymatic machinery is fully adapted to function in the presence of high salt concentrations. Thus, many proteins of extreme halophiles such as *Halobacterium* and *Salinibacter* even require molar concentrations of salt for activity and stability. They generally show a very high excess of acidic amino acids (glutamate and aspartate) over basic amino acids (lysine and arginine) and show additional molecular adaptations to stabilize the structure of their enzymes and other proteins and to optimize their activity at high salt concentrations. At low salt concentrations, such enzymes denature and lose their activity (Lanyi 1974; Madigan and Oren 1999; Mevarech et al. 2000; Oren 2000, 2002a, 2002b).

The functional diversity of halophilic/halotolerant organisms that grow at salt concentrations >100 g/L is great: they include oxygenic and anoxygenic phototrophs, aerobic heterotrophs, fermenters, denitrifiers, sulfate reducers, chemoautotrophic sulfur oxidizers, and methanogens. But not all types of metabolism known from low-salt environments are encountered at salt concentrations approaching NaCl saturation (~300 g/L). Examples of metabolic types missing at the highest salinities are methanogens that produce methane from acetate or from hydrogen/carbon dioxide, dissimilatory sulfate reduction with oxidation of acetate, and autotrophic nitrification. This is probably due to the high energy cost of the adaptation to life at high salt: the upper salinity limit at which each dissimilatory process takes place is correlated with the amount of energy generated and the energetic cost of osmotic adaptation (Oren 1999, 2002b, 2011).

Phylogenetically, we find halophilic and highly halotolerant microorganisms in all three domains of life. Highly halotolerant unicellular green algae of the genus *Dunaliella* are the main or sole primary producers in salt lakes in which the total salt concentration exceeds 250–300 g/L. We also know halophilic fungi (e.g., *Hortaea werneckii* and *Wallemia ichthyophaga*) and halophilic protozoa of different types. Representative types of halophilic prokaryotes are listed in Table 9.1.5.

Representatives of the archaeal class *Halobacteria* (Euryarchaeota) are the extreme halophiles par excellence: red-pink pigmented organisms that characteristically color the brines

TABLE 9.1.5

Representative Examples of Halophilic Microorganisms, Their Taxonomic Affiliation, Habitat, and Properties. The Table Includes Haloalkaliphiles, Halothermoalkaliphilic Species, and a Haloacidophile Listed Also in Other Tables in This Chapter

Name	Affiliation	Isolated from	Properties of the Environment	Limits of Existence	Other Properties	References
		Archaea				
Halarchaeum acidiphilum	*Euryarchaeota*	Commercial solar salt	Marine salt production facilities	pH 4.0–4.6 (opt. 4.4–4.5); 3.1–5.1 M NaCl (opt. 3.6–4.1 M)	Aerobic, heterotrophic	Minegishi et al. (2008, 2010)
Halobacterium salinarum	*Euryarchaeota*	Different locations worldwide	Salt lakes, salterns, salted proteinaceous products	3–5.2 M NaCl (opt. 3.5–4.5 M)	Aerobic heterotrophic; potential for fermentative growth and photoheterotrophy	Oren (2002a); Gruber et al. (2004)
Haloquadratum walsbyi	*Euryarchaeota*	Spain, Australia	Solar salterns	2.4–6.2 M NaCl (opt. >3.1 M)	Aerobic heterotrophic, possibly also photoheterotrophic	Burns et al. (2007)
Halorubrum lacusprofundi	*Euryarchaeota*	Deep Lake, Antarctica	Cold hypersaline lake	1.5–5.1 M NaCl (opt. 2.6–3.4 M). Opt. growth at 31°C–37°C, slow growth at 4°C	Aerobic, heterotrophic	Franzmann et al. (1988)
Halorubrum vacuolatum	*Euryarchaeota*	Lake Magadi, Kenya	Hypersaline alkaline lake	pH 8.5–10.5 (opt. 9.5); 2.5–5.2 M NaCl needed (opt. 3.5 M)	Aerobic, heterotrophic	Mwatha and Grant (1993)
Methanosalsum zhilinae	*Euryarchaeota*	Wadi Natrun, Egypt	Hypersaline alkaline lake sediment	pH 8–10 (opt. 8.7–9.5); 0.2–2.1 M NaCl needed (opt. 0.4–0.7 M)	Anaerobic, methanogenic on methanol or methylated amines	Mathrani et al. (1988)
Natronococcus occultus	*Euryarchaeota*	Lake Magadi, Kenya	Hypersaline alkaline lake	Opt. pH 9.5–10.0; 1.4–5.2 M NaCl needed (opt. 3.5–3.6 M)	Aerobic, heterotrophic	Tindall et al. (1984)
Natronomonas pharaonis	*Euryarchaeota*	Wadi Natrun, Egypt	Hypersaline alkaline lake	pH 7–10 (opt. pH 8.5–9.5); 2–5.2 M NaCl needed (opt. 3.5 M)	Aerobic, heterotrophic	Soliman and Trüper (1982); Falb et al. (2005)
		Bacteria				
Salinibacter ruber	*Bacteroidetes*	Mallorca, Spain	Solar salterns	2.6–5.1 M NaCl (opt. 4.3–5.1 M)	Aerobic heterotrophic; potential for photoheterotrophy	Antón et al. (2002)
Natranaerobius thermophilus	*Firmicutes*	Wadi Natrun, Egypt	Hypersaline soda lake	35°C–65°C (opt. 53°C); pH 8.3–10.6 (opt. 9.5); 3.1–4.9 M Na$^+$ (opt. 3.3 M)	Anaerobic, fermentative	Mesbah et al. (2007); Mesbah and Wiegel (2011)

(Continued)

TABLE 9.1.5 (Continued)

Representative Examples of Halophilic Microorganisms, Their Taxonomic Affiliation, Habitat, and Properties. The Table Includes Haloalkaliphiles, Halothermoalkaliphilic Species, and a Haloacidophile Listed Also in Other Tables in This Chapter

Name	Affiliation	Isolated from	Properties of the Environment	Limits of Existence	Other Properties	References
Natranaerobius trueperi	*Firmicutes*	Wadi Natrun, Egypt	Hypersaline soda lake	26°C–55°C (opt. 52°C); pH 7.8–11.0 (opt. 9.9); 4.1–5.4 M Na$^+$ (opt. 3.9 M	Anaerobic, fermentative	Mesbah and Wiegel (2009)
Natronovirga wadinatrunensis	*Firmicutes*	Wadi Natrun, Egypt	Hypersaline soda lake	26°C–56°C (opt. 51°C); pH 8.5–11.1 (opt. 9.9); 3.1–5.3 M Na$^+$	Anaerobic, fermentative	Mesbah and Wiegel (2009)
"*Desulfohalophilus alkaliarsenatis*"	*Proteobacteria*	Searles Lake, California, USA	Hypersaline alkaline lake sediment	55–330 g/L total salts (opt. 200 g/L); pH 7.8–9.7 (opt. 9.3)	Anaerobic, respiring sulfate or arsenate	Switzer Blum et al. (2012)
Alkalilimnicola ehrlichii	*Proteobacteria*	Mono Lake, California	Hypersaline soda lake	pH 7.3–10.0 (opt. 9.3); salinity 15–190 g/L; total salts (opt. 30 g/L)	Facultatively aerobic (oxygen or nitrate as electron acceptors), chemoautotrophic (arsenite, hydrogen, sulfide, or thiosulfate as electron donors), or heterotrophic	Hoeft et al. (2007)

of saltern crystallizer ponds, the north arm of Great Salt Lake, and other hypersaline lakes red due to their presence in large numbers. Among the neutrophilic members of the group, the best known ones include *Halobacterium salinarum* (Gruber et al. 2004) and the unusual flat, square-shaped *Haloquadratum walsbyi*, first isolated only in 2004 but now known as one of the dominant components of many hypersaline brines (Bolhuis et al. 2006; Burns et al. 2007). These organisms typically require at least 200–250 g/L salt for growth, and they even grow at salt saturation. The interesting *Salinibacter ruber* (*Bacteroidetes*) shares many properties with the *Halobacteria* group, including its "high-salt-in" mode of osmotic adaptation, in spite of its disparate phylogenetic position (Antón et al. 2002). Table 9.1.5 also lists several alkaliphilic members of the *Halobacteria* (*Natronomonas pharaonis*, *Natronococcus occultus*, and *Halorubrum vacuolatum*) isolated from the alkaline hypersaline lakes of the Wadi Natrun in Egypt and Lake Magadi in Kenya. These were discussed further in the previous section on alkaliphilic microorganisms. Growth of such extreme halophiles at low pH is rarely observed. A rare exception is *Halarchaeum acidiphilum*, a non-pigmented member of the *Halobacteria*, isolated from a sample of commercial solar salt that did not have any acidic properties. It grows at pH 4.0–6.0 (optimum at pH 4.4–4.5) and 15°C–45°C (Minegishi et al. 2008, 2010). Alkaliphilic halophiles are also found among the Bacteria. An example is *Alkalilimnicola ehrlichii* from Mono Lake (Hoeft et al. 2007).

As shallow hypersaline basins in tropical areas can reach quite high temperatures during daytime, there is also a niche for thermophilic halophiles. Some aerobic halophilic Archaea

(*Haloarcula quadrata, Haloferax elongans, Haloferax mediterranei,* and *Natronolimnobius aegyptiacus*) have temperature optima between 51°C and 55°C (Bowers and Wiegel 2011). *Halothermothrix orenii*, a halophilic, thermophilic (optimum 60°C), fermentative, strictly anaerobic bacterium, was isolated from the sediment of a Tunisian salt lake (Cayol et al. 1994). The interesting haloanaerobic alkalithermophiles discovered in the sediments of the Wadi Natrun (Egypt) soda lakes (*Natronovirga* and *Natranaerobius* species) were discussed in the previous section on life at high pH.

On the Antarctic continent, there are a number of unfrozen hypersaline lakes that harbor interesting communities of cold-adapted halophiles. Deep Lake, Antarctica, was the source of isolation of *Halorubrum lacusprofundi* (Franzmann et al. 1988). Metagenomic studies showed that *Hrr. lacusprofundi* shares its ecosystem with three additional types of halophilic Archaea of different genera. These organisms have extensively exchanged genes and have even shared long contiguous pieces of DNA (DeMaere et al. 2013; Williams et al. 2014).

Different types of athalassohaline brine are also inhabited by diverse communities of microorganisms (Oren 2013). While Na and K ions stabilize proteins, Ca and Mg ions at high concentrations disrupt biological structures, based on the "Hofmeister series" of stabilizing ("kosmotropic ions") and destabilizing ("chaotropic") ions (Hofmeister 1888). The waters of the Dead Sea, with >2 M Mg, >0.5 M Ca, and <1.5 M Na, have now become too toxic to support extensive life. In 2007, small but diverse communities of *Halobacteria* could still be detected using a cultivation-independent approach

(Bodaker et al. 2010). Since then, the continuing drying out of the lake led to a further increase in the concentrations of Mg and Ca and a further decrease in the ratio of kosmotropic vs. chaotropic cations. It is well possible that the brines of the lake have, by now, become devoid of life. However, no recent studies have been devoted to the exploration of the current status of the lake's biota. Some fungi, including *Hortaea werneckii*, *Eurotium amstelodami*, *Eurotium chevalieri,* and *Wallemia ichthyophaga*, are extremely tolerant to high concentrations of $MgCl_2$ (up to 2.1 M) or $CaCl_2$ (up to 2.0 M), without compensating kosmotropic salts, but no obligate "chaophilic" fungi have yet been discovered (Zajc et al. 2014).

The brines of Discovery Basin, a deep anoxic basin on the bottom of the Mediterranean Sea (depth ~3,585 m) are almost saturated with $MgCl_2$, present at a concentration of ~5 M. Examination of the occurrence of active microbial life along the gradient from normal seawater to the $MgCl_2$ brines on the bottom of the basin suggested that, in the absence of kosmotropic solutes, the upper $MgCl_2$ concentration for life is about 2.3 M (van der Wielen et al. 2005; Hallsworth et al. 2007).

9.1.7 HOT AND COLD DESERTS AND THEIR MICROBIAL COMMUNITIES

Life depends on water. Therefore, areas with low water availability are harsh habitats for life. This is true for hot deserts as well as for very cold desert environments such as the Dry Valleys in Antarctica. Because of the relevance of such environments as model systems to evaluate the possibilities for life in space, relatively much research has been devoted to the presence and activities of microorganisms in the driest deserts on Earth.

Mars-like soils in the most arid region of the Atacama Desert in Chile contain very little organic material, and in most samples, the numbers of colony-forming heterotrophic bacteria were below the detection limits of dilution plating. Decomposition of organic material in these soils is probably due mainly to non-biological processes (Navarro-González et al. 2003). Solar ultraviolet-B (UVB) radiation kills even the most resistant microorganisms within a few hours of exposure to the conditions of the Atacama. However, endospores of *Bacillus subtilis* (*Firmicutes*) and conidia of *Aspergillus niger* (Ascomycota) incubated in the dark in the Atacama Desert for up to 15 months survived to a significant extent (Dose et al. 2001).

Microorganisms can survive in deserts, hot as well as cold ones, as endolithic communities below the surface of sandstone and other porous rocks. There, they are relatively protected from extreme desiccation and from excess light. The organisms rapidly switch their metabolic activities on and off in response to environmental changes. In hot desert rocks, the endolithic community consists entirely of cyanobacteria (mainly *Chroococcidiopsis*) and other prokaryotes. Growth is slow, but the communities can survive for very long times. Production of pigments,

exopolysaccharides, and osmoprotectants may also reduce the level of stress (Friedmann 1980; Omelon 2016). Characterization of hypolithic communities of *Chroococcidiopsis* and associated heterotrophs colonizing translucent stones in Atacama revealed that each stone supported a number of unique 16S rRNA gene-defined genotypes. In the most arid zone of the Atacama Desert, such hypolithic cyanobacteria are rarely found (Warren-Rhodes et al. 2006).

The Dry Valleys of Eastern Antarctica are the coldest and driest deserts on Earth (Cowan et al. 2007). Yet, a narrow subsurface zone of certain rock types is colonized by microorganisms (Friedmann 1980; Cowan et al. 2010). Lichens, growing between the crystals of porous rocks, are the dominant compound of this community (Friedmann 1982); the endolithic communities of the cold deserts of Antarctica thus differ from those of the hot deserts by dominance of eukaryotes rather than prokaryotes (Friedmann 1980). A large number of black, mostly meristematic fungi were isolated from these lichen-dominated communities. Most isolates were affiliated with the Dothideomycetidae, genera *Friedmanniomyces* and *Cryomyces*. All had thick melanized cell walls and could produce exopolysaccharides (Selbmann et al. 2005). A small-subunit rRNA gene clone library prepared from a cryptoendolithic community from the McMurdo Dry Valleys yielded 51 phylotypes: 46 bacterial and 5 of eukaryotes; no Archaea were detected. In the lichen-dominated community, three rRNA sequences of types known to be associated with lichens, attributed to a fungus, a green alga, and a chloroplast, accounted for over 70% of the clones. Other sequences belonged to a member of the *Alphaproteobacteria* that may be capable of aerobic anoxygenic photosynthesis and to a distant relative of *Deinococcus* (de la Torre et al. 2003). Using GeoChip-based functional gene arrays, autotrophic, heterotrophic, and diazotrophic life strategies were identified in the microbial communities. The rocky substrates supported greater functional diversity in stress-response pathways than in the surrounding soils (Chan et al. 2013). But also, the soils contain unexpectedly high levels of microbial biomass, as estimated by using bioluminescent ATP assays. Estimates were as high as 3×10^6–4×10^8 cells per gram of surface soil in Antarctic desert sites (Cowan et al. 2002). This community is phylogenetically diverse (Cary et al. 2010).

9.1.8 MICROBIAL LIFE AT HIGH HYDROSTATIC PRESSURE

The mean depth of the oceans is about 4 km, equivalent to a pressure of 400 atmospheres or 40 MPa, and the deepest parts of the oceans are more than 10-km deep. Microorganisms living in such environments must withstand pressures of >100 MPa. The deep sea is inhabited not only by microorganisms: a variety of marine animals is adapted to life at high pressures. As most of the deep sea is also cold (typically 1°C–3°C), microorganisms living there must possess both psychrophilic and piezophilic/barophilic) or piezotolerant/barotolerant properties (Yayanos 1995, 2000; Horikoshi 1998). Deep-sea bacteria adapted to life at high pressure and low temperature generally grow slowly, so that the overall microbial activity in the deep-sea environment is low (Jannasch and Taylor 1984; Jannasch and Wirsen 1984).

Because of the complex technical demands of performing experiments at hundreds of atmospheres of pressure, without decompression that kills obligately piezophilic microorganisms, our understanding of the special adaptations of such organisms is limited (Kato et al. 2004).

Microorganisms isolated from low-pressure environments generally do not grow well when exposed to high hydrostatic pressure. In *Escherichia coli*, DNA-binding proteins often display pressure-sensitive binding properties. Translation, involving nucleic acid-protein interactions, is another pressure-sensitive process (Bartlett et al. 2007). Bacteria from the ocean generally grow fastest at pressures close to those present at the depth from which they were isolated (Yayanos et al. 1982), showing a high degree of adaptation to the environment. Thus, a spirillum isolated from a depth of 5,700 m grew optimally, with generation times of 4–13 h, at about 50 MPa and 2°C–4°C. When grown at the same temperatures and at atmospheric pressure, the generation time was 3–4 days (Yayanos et al. 1979). A bacterium isolated from an amphipod retrieved (with decompression) from a depth of 10,476 m in the Mariana Trench did not grow at 38 MPa and showed optimal growth at 69 MPa (generation time 25 h at 2°C). At 103.5 MPa, close to the in situ pressure, the generation time at 2°C was ~33 h (Yayanos et al. 1981). Obligate piezophiles unable to grow at atmospheric pressure also exist: isolates related to *Shewanella* and *Moritella* (*Gammaproteobacteria*), retrieved from the Mariana Trench, the Japan Trench, and the Philippine Trench, require 70–80 MPa pressure for optimal growth and do not grow below 50 MPa (Kato and Bartlett 1997; Kato et al. 1998). Isolate MT41 from the Mariana Trench does not grow below 50 MPa, has its optimum at 70 MPa, and tolerates at least 100 MPa (Yayanos et al. 1981; Yayanos 2000). Different isolates of barophilic and barotolerant bacteria were also obtained from deep-sea sediment samples (Kato et al. 1995).

Molecular studies of barophilic members of the genus *Shewanella* enabled the identification of several genes whose expression is pressure-regulated (Kato and Bartlett 1997). Further molecular genetic studies using *Photobacterium profundum* led to the identification of genes important for pressure sensing or pressure adaptation, including genes required for fatty acid unsaturation, the membrane protein genes *toxR* and *rseC*, and the DNA recombination gene *recD*. Genes for the production of omega-3 polyunsaturated fatty acids are frequently found in deep-sea bacteria (Bartlett 1999).

The deep-sea hot vents are hotspots of life on the sea bottom. As discussed in Section 9.1.2, such hot vents have been a treasure trove for the discovery of interesting thermophilic

Archaea and Bacteria. As such hot vents are typically found at depths of several kilometers below the sea surface, the organisms inhabiting them must also be barophilic or at least barotolerant. The properties of some of these pressure-tolerant thermophiles are listed in Table 9.1.6. Temperature tolerance and pressure requirement are often interrelated: the higher the pressure, the higher the temperature minimum and maximum for growth (Pledger et al. 1994). A well-known case is *Methanopyrus kandleri* strain 116, a methanogenic archaeon isolated from a hydrothermal vent at the Central Indian Ridge (depth 2.5 km), reported to grow at 122°C, the highest growth temperature reported thus far, but only when incubated at high pressure (20 MPa). At 0.4 MPa, no growth was obtained above 116°C (Takai et al. 2008).

Other examples of such barotolerant thermophiles are as follows:

- *Pyrolobus fumarii*, isolated a depth of 3,650 m from a hydrothermally heated black smoker wall at the Mid-Atlantic Ridge. It can grow up to 113°C and up to at least 25 MPa pressure (Blöchl et al. 1997).
- *Pyrococcus abyssi*, isolated from 2-km-deep hydrothermal vents in the North Fiji basin. The upper growth temperature (~102°C) is extended by at least 3°C when cells are cultivated under the in situ hydrostatic pressures of 20 MPa (Erauso et al. 1993).
- *Marinitoga piezophila*, isolated from a 2,630-m deep hydrothermal chimney sample collected from the East-Pacific Rise. Its growth is enhanced by hydrostatic pressure; the optimal pressure for growth is 40 MPa, that is, higher than the 26 MPa pressure at the sampling site (Alain et al. 2002).
- *Thermococcus peptonophilus*, isolated from a hydrothermal vent in the western Pacific Ocean at a depth of 1,380 m. Its optimal growth temperature shifted from 85°C at 30 MPa to 90°C–95°C at 45 MPa. Cell viability following starvation was enhanced at 30 and 45 MPa as compared with atmospheric pressure (Canganella et al. 1997).
- *Thermococcus barophilus*, isolated from a hydrothermal vent site on the Mid-Atlantic Ridge (3,550-m depth). It grew at 48°C–95°C under atmospheric pressure, but for growth above 95°C, a pressure of 15.0–17.5 MPa was required. The optimal temperature for growth was 85°C at both high (40 MPa) and low (0.3 MPa) pressures. At 85°C, the growth rate was twice as high at the in situ hydrostatic pressure as compared with that at low pressure (Marteinsson et al. 1999).
- *Palaeococcus ferrophilus*, isolated from a deep-sea hydrothermal vent chimney in the Ogasawara-Bonin Arc, Japan, growing optimally at 83°C and 30 MPa (Takai et al. 2000).

TABLE 9.1.6

Representative Examples of Piezophilic Microorganisms, Their Taxonomic Affiliation, Habitat, and Properties. The Table Includes Thermophilic and Psychrophilic Species Listed Also in Other Tables in This Chapter

Name	Affiliation	Isolated from	Properties of the Environment	Limits of Existence	Other Properties	References
	Archaea					
Methanopyrus kandleri strain 116	*Euryarchaeota*	Central Indian Ridge	Black smoker fluid, 2,450 m depth, hydrothermal field	Max. 116°C at 0.4 MPa; 122°C at 20 MPa	Chemolithoautotroph methanogenic on hydrogen/carbon dioxide	Takai et al. (2008)
Palaeococcus ferrophilus	*Euryarchaeota*	Deep-Sea, Japan	Hydrothermal vent, 1.338 m depth	60°C–88°C (opt. 83°C); 0.1–60 MPa (opt. 30 MPa)	Anaerobic chemoorganotroph oxidizing proteinaceous compounds with elemental sulfur or ferrous iron as electron acceptor	Takai et al. (2000)
"*Pyrococcus abyssi*"	*Euryarchaeota*	Hydrothermal vent, SW Pacific	Hydrothermal vent fluid, 2,000 m depth	67°C–102°C, opt. 96°C at 0.1 MPa. Upper temp. 105°C at 20 MPa	Anaerobic heterotrophic, fermentative	Erauso et al. (1993)
Thermococcus peptonophilus	*Euryarchaeota*	Western Pacific Ocean	Deep-sea hydrothermal vent	60°C–100°C (opt. 85°C–90°C). Opt. 30–45 MPa. Opt. temp. shifts from 85°C at 30 MPa to 90°C–95°C at 45 MPa	Grows on proteins and amino acids; growth is stimulated by elemental sulfur	González et al. (1995); Canganella et al. (1997)
	Bacteria					
Colwellia psychrerythraea	*Proteobacteria*	Atlantic Ocean, Puerto Rico Trench	Seawater, 7,410 m depth	At 2°C, growth at 37–103 MPa (opt. 75 MPa); best growth at 10°C and 94 MPa	Aerobic, heterotrophic	Deming et al. (1988)
Marinitoga piezophila	*Thermotogae*	East Pacific Rise	Hydrothermal chimney, 2,630 m depth	Opt. 40 MPa; 45°C–70°C, opt. 65°C	Anaerobic, chemoorganotrophic, uses carbohydrates and proteins with elemental sulfur as electron acceptor	Alain et al. (2002)

9.1.9 LIFE AT HIGH LEVELS OF IONIZING RADIATION

Another extreme environmental condition to which some microorganisms show a surprisingly high level of resistance is ionizing radiation, including radioactivity. The study of the mechanisms enabling microorganisms to tolerate high radiation levels is highly relevant for astrobiology. Table 9.1.7 presents some of the best known radiation-resistant microorganisms from the bacterial and the archaeal domains.

The genus best known for its extreme radiation tolerance is *Deinococcus*, which forms a deep phylogenetic lineage with the thermophilic *Thermus* and relatives. The first member of the genus, *Deinococcus radiodurans*, is still unsurpassed with respect to its tolerance toward high radiation levels: it survives exposure to >5 kGy of gamma radiation and to 1500 J m^{-2} of UV radiation. For comparison, exposure to 5 Gy is lethal for humans. Strains of *D. radiodurans* were originally isolated from canned meat that had been irradiated (Anderson et al. 1956). But the species has also been found in radioactive waste. *Deinococcus* strains have been isolated from desert soil (Rainey et al. 2005) and from many other environments. 16S rRNA gene sequences of *Deinococcus* were also found in South Pole snow (Carpenter et al. 2000). One of the secrets of *Deinococcus* is its extraordinary ability to repair damaged DNA, not only single-strand breaks but also double-strand breaks (Battista 1997; Rothschild 1999; Cox and Battista 2005). It also has special ways to package its DNA and to protect its proteins against damage (Daly 2009).

Studies on the desiccation resistance of ionizing-radiation-resistant organisms suggest that the ability of microorganisms

TABLE 9.1.7

Representative Examples of Radiation-resistant Microorganisms, Their Taxonomic Affiliation, Habitat and Properties. The Table Includes Thermophilic Species Also Listed in Other Tables in This Chapter

Name	Affiliation	Isolated from	Properties of the Environment	Limits of Existence	Other Properties	References
	Archaea					
Thermococcus gammatolerans, *Thermococcus marinus*, *Thermococcus radiotolerans*	*Euryarchaeota*	Guaymas Basin, Mid-Atlantic Ridge	Hydrothermal chimneys	55°C–95°C (opt. 88°C), survive 20–30 kGy radiation	Anaerobic, oxidizes proteinaceous compounds with elemental sulfur as electron acceptor	Jolivet et al. (2003a, 2004)
	Bacteria					
Rubrobacter radiotolerans	*Actinobacteria*	Japan	Hot spring	10% survival at ~12 kGy gamma radiation	Aerobic, heterotrophic	Ferreira et al. (1999)
Deinococcus radiodurans	*Deinococcus-Thermus*	USA	Ground pork and beef	Tolerates >5 kGy gamma radiation	Aerobic, heterotrophic	Brooks and Murray (1981)
Truepera radiovictrix	*Deinococcus-Thermus*	Azores Islands	Hot spring runoff	Opt. 50°C; pH 7.5–9.5; survives 5.0 kGy	Aerobic, heterotrophic	Albuquerque et al. (2005)

to repair their DNA might be a response to DNA damage caused by prolonged desiccation rather than ionizing radiation (Mattimore and Battista 1996). This is supported by the higher proportions of ionizing-radiation-resistant bacteria found in arid soils as compared with soils from less arid regions (Rainey et al. 2005). However, cells of *D. radiodurans* incubated in the dark in the Atacama Desert for up to 15 months poorly survived, because they were inactivated at relative humidity between 40% and 80%, which typically occurs during desert nights (Dose et al. 2001).

Other radiation-resistant bacteria include *Rubrobacter* spp. (*Actinobacteria*) (Suzuki et al. 1988; Ferreira et al. 1999). The desiccation-tolerant cyanobacterium *Chroococcidiopsis*, found in cold and hot deserts, is also highly resistant to ionizing radiation and possesses efficient repair mechanisms for damaged DNA (Billi et al. 2000). Following exposure to simulated space and Martian conditions, *Chroococcidiopsis* spp. from hot deserts in Chile survived better than a strain from Antarctica (Billi et al. 2011).

The ability to tolerate high radiation levels is also found in a number of thermophilic microorganisms. One of these is *Truepera radiovictrix*, a relative of *Deinococcus*, isolated from hot spring runoff on the Azores Islands. It grows optimally at 50°C, and 60% of the cells survived 5.0 kGy of radiation (Albuquerque et al. 2005). Resistance against gamma radiation resistance was also found in a number of archaeal thermophiles, including *Pyrococcus furiosus* (DiRuggiero et al. 1997) and *Pyrococcus abyssi* (Jolivet et al. 2003b). Following a dose of 2.5 kGy, the *P. furiosus* 2-Mb chromosome was fragmented into pieces, ranging from 500 kb to shorter than 30 kb, but was fully restored upon incubation at 95°C (DiRuggiero et al. 1997). Other radiation-resistant thermophilic Archaea are *Thermococcus gammatolerans* from hydrothermal chimney samples at the Guaymas Basin (Jolivet et al. 2003a) and

T. marinus and *T. radiotolerans*, also retrieved from deep-sea hydrothermal vents (Jolivet et al. 2004).

9.1.10 FINAL COMMENTS

The microorganisms on Earth can colonize at nearly every place on the planet, thanks not only to the fact that they can utilize almost any energy source available but also to the fact that the microbial world can adapt to extremes of temperature (hot as well as cold), pH (from pH <0 to pH >10–11), salt concentration and presence of chaotropic ions, hydrostatic pressure, radiation, and often combinations of one or more of these potentially stressful environmental factors.

As shown in the tables presented in the different sections, most extremophilic microorganisms listed are Archaea (Cavicchioli 2002). Above 90°C–100°C, the only life forms known to develop are Archaea. They probably are the only ones that can maintain the selective permeability of their membranes because of the special structure of their membrane lipids. But members of the Bacteria can also grow at the highest salt concentrations; they can thrive at high and low pH and at low temperatures, and they are superior to the Archaea in their tolerance of high radiation levels. Even some simple eukaryotes (algae and fungi) can grow at extremes of salinity, drought, and pH.

Where there is liquid water on Earth, virtually no matter what the physical and chemical conditions are, there is life. There are only few exceptions. When in deep-sea hydrothermal vents, the extremely high temperature of the water is incompatible with the stability of proteins, nucleic acids, lipid membranes, and other biomolecules, or when the concentration of chaotropic ions in the absence of stabilizing kosmotropic ions is too high for stability of biological structures, life cannot be sustained, in spite of the presence of water in the liquid state.

Our understanding of the possibilities and limitations of life "as we know it" on Earth and of the properties of extremophilic microorganisms of all kinds must be the basis for any search for similar life elsewhere in space (Nealson 1997; Rothschild and Mancinelli 2001).

ACKNOWLEDGMENTS

The author's current studies of life at high salt concentrations are supported by grants no. 343/13 and 2221/15 from the Israel Science Foundation.

REFERENCES

Adams, M. W. W. 1993. Enzymes and proteins from organisms that grow near and above 100°C. *Annu. Rev. Microbiol.* 47: 627–658.

Alain, K., V. T. Marteinsson, M. L. Miroshnichenko, E. A. Bonch-Osmolovskaya, D. Prieur, and J.-L. Birrien. 2002. *Marinitoga piezophila* sp. nov., a rod-shaped thermo-piezophilic bacterium isolated under high hydrostatic pressure from a deep-sea hydrothermal vent. *Int. J. Syst. Evol. Microbiol.* 52: 1331–1339.

Albuquerque, L., C. Simões, M. F. Nobre et al. 2005. *Truepera radiovictrix*, gen. nov., sp. nov., a new radiation resistant species and the proposal of *Trueperaceae* fam. nov. *FEMS Microbiol. Lett.* 247: 161–169.

Anderson, A. W., H. C. Nordan, R. F. Cain et al. 1956. Studies on radio-resistant micrococcus. I. Isolation, morphology, cultural characteristics, and resistance to gamma radiation. *Food Technol.* 10: 575–577.

Antón, J., A. Oren, S. Benlloch, F. Rodríguez-Valera, R. Amann, and R. Rosselló-Mora. 2002. *Salinibacter ruber* gen. nov., sp. nov., a novel extreme halophilic member of the Bacteria from saltern crystallizer ponds. *Int. J. Syst. Evol. Microbiol.* 52: 485–491.

Bains, W., Y. Xiao, and C. Yu. 2015. Prediction of the maximum temperature for life based on the stability of metabolites to decomposition in water. *Life* 5: 1054–1100.

Baker, B. J., M. A. Lutz, S. C. Dawson, P. L. Bond, and J. F. Banfield. 2004. Metabolically active eukaryotic communities in extremely acidic mine drainage. *Appl. Environ. Microbiol.* 70: 6264–6271.

Bakermans, C., H. L. Ayala-del-Rio, M. A. Ponder et al. 2006. *Psychrobacter cryohalentis* sp. nov. and *Psychrobacter arcticus* sp. nov., isolated from Siberian permafrost. *Int. J. Syst. Evol. Microbiol.* 56: 1285–1291.

Bakermans, C., and M. L. Skidmore. 2011a. Microbial respiration in ice at subzero temperatures (−4 to −33°C). *Environ. Microbiol. Rep.* 4: 774–782.

Bakermans, C., and M. L. Skidmore. 2011b. Microbial metabolism in ice and brine at -5°C. *Environ. Microbiol.* 13: 2269–2278.

Bakermans, C., A. I. Tsapin, V. Souza-Egipsy, D. A. Gilichinsky, and J. H. Nealson. 2003. Reproduction and metabolism at −10°C of bacteria isolated from Siberian permafrost. *Environ. Microbiol.* 5: 321–326.

Banciu, H., and D. Y. Sorokin. 2013. Adaptation in haloalkaliphiles and natronophilic bacteria. In *Polyextremophiles: Life Under Multiple Forms of Stress*, J. Seckbach, A. Oren, and H. Stan-Lotter (Eds.), pp. 123–178. Dordrecht, the Netherlands: Springer.

Barns, S. M., R. E. Fundyga, M. W. Jeffries, and N. R. Pace. 1994. Remarkable archaeal diversity detected in a Yellowstone National Park hot spring environment. *Proc. Natl. Acad. Sci. USA* 91: 1609–1613.

Baross, J. A., and J. W. Deming. 1983. Growth of "black smoker" bacteria at temperatures of at least 250°C. *Nature* 303: 423–426.

Baross, J. A., and J. W. Deming. 1995. Growth at high temperatures: Isolation and taxonomy, physiology, and ecology. In *The Microbiology of Deep-Sea Hydrothermal Vents*, D. M. Karl (Ed.), pp. 169–217. Boca Raton, FL: CRC Press.

Bartlett, D. H. 1999. Microbial adaptations to the psychrosphere/piezosphere. *J. Mol. Microbiol. Biotechnol.* 1: 93–100.

Bartlett, D. H., F. M. Lauro, and E. A. Eloe. 2007. Microbial adaptation to high pressure. In *Physiology and Biochemistry of Extremophiles*, C. Gerday, and N. Glansdorff (Eds.), pp. 333–348. Washington, DC: ASM Press.

Battista, J. R. 1997. Against all odds: The survival strategies of *Deinococcus radiodurans*. *Annu. Rev. Microbiol.* 51: 203–224.

Billi, D., E. I. Friedmann, K. G. Hofer, M. G. Caiola, and R. Ocampo-Friedmann. 2000. Ionizing-radiation resistance in the desiccation-tolerant cyanobacterium *Chroococcidiopsis*. *Appl. Environ. Microbiol.* 66: 1489–1492.

Billi, D., E. Viaggiu, C. S. Cockell, E. Rabbow, G. Horneck, and S. Onofri. 2011. Damage escape and repair in dried *Chroococcidiopsis* spp. from hot and cold deserts exposed to simulated space and Martian conditions. *Astrobiology* 11: 65–73.

Blöchl, E., R. Rachel, S. Burggraf, D. Hafenbradl, H. W. Jannasch, and K. O. Stetter. 1997. *Pyrolobus fumarii*, gen. and sp. nov., represents a novel group of archaea, extending the upper temperature limit for life to 113°C. *Extremophiles* 1: 14–21.

Bodaker, I., I. Sharon, M. T. Suzuki et al. 2010. The dying Dead Sea: Comparative community genomics in an increasingly extreme environment. *ISME J.* 4: 399–407.

Bolhuis, H., P. Palm, A. Wende et al. 2006. The genome of the square archaeon *Haloquadratum walsbyi*: Life at the limits of water activity. *BMC Genomics* 7: 169.

Bowers, K. J., N. M. Mesbah, and J. Wiegel. 2009. Biodiversity of polyextremophilic Bacteria: Does combining the extremes of high salt, alkaline pH and elevated temperature approach a physicochemical boundary for life? *Saline Syst.* 5: 9–15.

Bowers, K. J., and Wiegel, J. 2011. Temperature and pH optima of extremely halophilic archaea: A mini-review. *Extremophiles* 15: 119–128.

Bowman, J. P., S. A. McCammon, M. V. Brown, D. S. Nichols, and T. A. McMeekin. 1997. Diversity and association of psychrophilic bacteria in Antarctic sea ice. *Appl. Environ. Microbiol.* 63: 3068–3078.

Bowman, J. P., S. A. McCammon, T. Lewis et al. 1998. *Psychroflexus torquis* gen. nov., sp. nov., a psychrophilic species from Antarctic sea ice, and reclassification of *Flavobacterium gondwanense* (Dobson et al. 1993) as *Psychroflexus gondwanense* gen. nov., comb. nov. *Microbiology UK* 144: 1601–1609.

Breezee, J., J. Cady, and J. T. Staley. 2004. Subfreezing growth of the sea ice bacterium "*Psychromonas ingrahamii*". *Microb. Ecol.* 47: 300–304.

Brierley, C. L., and Brierley, J. A. 1973. A chemoautotrophic and thermophilic microorganism isolated from an acid hot spring. *Can. J. Microbiol.* 19: 183–188.

Brock, T. D. 1978. *Thermophilic Microorganisms and Life at High Temperatures*. New York: Springer-Verlag.

Brock, T. D., K. M. Brock, R. T. Belly, and R. L. Weiss. 1972. *Sulfolobus*: A new genus of sulfur-oxidizing bacteria living at low pH and high temperature. *Arch. Mikrobiol.* 84: 54–68.

Brock, T. D., and H. Freeze. 1969. *Thermus aquaticus* gen. n. and sp. n., a non-sporulating extreme thermophile. *J. Bacteriol.* 98: 289–297.

Brooks, B. W., and R. G. E. Murray. 1981. Nomenclature for "Micrococcus radiodurans" and other radiation-resistant cocci: *Deinococcaceae* fam. nov. and *Deinococcus* gen. nov., including five species. *Int. J. Syst. Bacteriol.* 31: 353–360.

Burgess, E. A., I. D. Wagner, and J. Wiegel. 2007. Thermal environments and biodiversity. In *Physiology and Biochemistry of Extremophiles*, C. Gerday, and N. Glansdorff (Eds.), pp. 13–29. Washington, DC: ASM Press.

Burggraf, S., H. W. Jannasch, B. Nicolaus, and K. O. Stetter. 1990. *Archaeoglobus profundus*, sp. nov., represents a new species within the sulfate-reducing archaebacteria. *Syst. Appl. Microbiol.* 13: 24–28.

Burns, D. G., P. H. Janssen, T. Itoh et al. 2007. *Haloquadratum walsbyi* gen. nov., sp. nov., the square haloarchaeon of Walsby isolated from salterns in Australia and Spain. *Int. J. Syst. Evol. Microbiol.* 57: 387–392.

Canganella, F., J. M. Gonzalez, M. Yanagibayashi, C. Kato, and K. Horikoshi. 1997. Pressure and temperature effects on growth and viability of the hyperthermophilic archaeon *Thermococcus peptonophilus. Arch. Microbiol.* 168: 1–7.

Capece, M. C., E. Clark, J. K. Saleh et al. 2013. Polyextremophiles and the constraints for terrestrial habitability. In *Polyextremophiles: Life Under Multiple Forms of Stress*, J. Seckbach, A. Oren, and H. Stan-Lotter (Eds.), pp. 7–59. Dordrecht, the Netherlands: Springer.

Carpenter, E. J., S. Lin, and D. G. Capone. 2000. Bacterial activity in South Pole snow. *Appl. Environ. Microbiol.* 66: 4514–4517.

Cary, S. C., I. R. McDonald, J. E. Barrett, and D. A. Cowan. 2010. On the rocks: The microbiology of Antarctic dry valleys. *Nat. Rev. Microbiol.* 8: 129–138.

Cavicchioli, R. 2002. Extremophiles and the search for extraterrestrial life. *Astrobiology* 2: 281–292.

Cayol, J.-L., B. Ollivier, B. K. C. Patel, G. Prensier, J. Guézennec, and J.-L. Garcia. 1994. Isolation and characterization of *Halothermothrix orenii* gen. nov., sp. nov., a halophilic, thermophilic, fermentative, strictly anaerobic bacterium. *Int. J. Syst. Evol. Microbiol.* 44: 534–540.

Chan, Y., J. D. Van Nostrand, J. Zhou et al. 2013. Functional ecology of an Antarctic Dry Valley. *Proc. Natl. Acad. Sci. USA* 110: 8990–8995.

Clarke, A. 2014. The thermal limits to life on Earth. *Int. J. Astrobiol.* 13: 141–154.

Cowan, D. A. 2004. The upper temperature for life – where do we draw the line? *Trends Microbiol.* 12: 58–60.

Cowan, D. A., A. Casanueva, and W. Stafford. 2007. Ecology and biodiversity of cold-adapted microorganisms. In *Physiology and Biochemistry of Extremophiles*, C. Gerday, and N. Glansdorff (Eds.), pp. 119–132. Washington, DC: ASM Press.

Cowan, D. A., N. Khan, S. B. Pointing, and C. Cary. 2010. Diverse hypolithic refuge communities in the McMurdo Dry Valleys. *Antarct. Sci.* 22: 714–720.

Cowan, D. A., N. J. Russell, A. Mamais, and D. M. Sheppard. 2002. Antarctic Dry Valley mineral soils contain unexpectedly high levels of microbial biomass. *Extremophiles* 6: 431–436.

Cox, M. M., and J. R. Battista. 2005. *Deinococcus radiodurans* – The consummate survivor. *Nat. Rev. Microbiol.* 3: 882–892.

Da Costa, M. S., J. C. Duarte, and R. A. D. Williams (Eds.). 1989. *Microbiology of Extreme Environments and Its Potential for Biotechnology*. London, UK: Elsevier Applied Science.

D'Amico, S., T. Collins, J. C. Marx, G. Feller, and C. Gerday. 2006. Psychrophilic microorganisms: Challenges for life. *EMBO Rep.* 7: 385–389.

Daly, M. J. 2009. A new perspective on radiation resistance based on *Deinococcus radiodurans. Nat. Rev. Microbiol.* 7: 237–245.

Daniel, R. M., and D. A. Cowan. 2000. Biomolecular stability and life at high temperatures. *Cell. Mol. Life Sci.* 57: 250–264.

de la Torre, J. R., B. M. Goebel, E. I. Friedmann, and N. R. Pace. 2003. Microbial diversity of cryptoendolithic communities from the McMurdo Dry Valleys, Antarctica. *Appl. Environ. Microbiol.* 69: 3858–3867.

DeMaere, M. Z., T. J. Williams, M. A. Allen et al. 2013. High level of intergenera gene exchange shapes the evolution of haloarchaea in an isolated Antarctic lake. *Proc. Natl. Acad. Sci. USA* 110: 16939–16944.

Deming, J. W. 2002. Psychrophiles and polar regions. *Curr. Opin. Microbiol.* 5: 301–309.

Deming, J. W. 2007. Life in ice formations at very cold temperatures. In *Physiology and Biochemistry of Extremophiles*, C. Gerday, and N. Glansdorff (Eds.), pp. 133–144. Washington, DC: ASM Press.

Deming, J. W., L. K. Somers, W. L. Straube, D. G. Schwartz, and M. T. Macdonell. 1988. Isolation of an obligately barophilic bacterium and description of a new genus, *Colwellia* gen. nov. *Syst. Appl. Microbiol.* 10: 152–160.

DiRuggiero, J., N. Santangelo, Z. Nackerdien, J. Ravel, and F. T. Robb. 1997. Repair of extensive ionizing-radiation DNA damage at 95°C in the hyperthermophilic archaeon *Pyrococcus furiosus. J. Bacteriol.* 179: 4643–4645.

Doemel, W. N., and T. D. Brock. 1970. The upper temperature limit of *Cyanidium caldarium. Arch. Mikrobiol.* 72: 326–332.

Dopson, M., C. Baker-Austin, A. Hind, J. P. Bowman, and P. L. Bond. 2004. Characterization of *Ferroplasma* isolates and *Ferroplasma acidarmanus* sp. nov., extreme acidophiles from acid mine drainage and industrial bioleaching environments. *Appl. Environ. Microbiol.* 70: 2079–2088.

Dose, K., A. Bieger-Dose, B. Ernst et al. 2001. Survival of microorganisms under the extreme conditions of the Atacama Desert. *Orig. Life Evol. Biosph.* 31: 287–303.

Edwards, K. J., P. L. Bond, T. M. Gihring, and J. F. Banfield. 2000. An archaeal iron-oxidizing extreme acidophile important in acid mine drainage. *Science* 287: 1796–1799.

Engle, M., Y. Li, F. Rainey et al. 1996. *Thermobrachium celere* gen. nov., sp. nov., a rapidly growing thermophilic, alkalitolerant, and proteolytic obligate anaerobe. *Int. J. Syst. Bacteriol.* 46: 1025–1033.

Engle, M., Y. Li, C. R. Woese, and J. Wiegel. 1995. Isolation and characterization of a novel alkalitolerant thermophile, *Anaerobranca horikoshii* gen. nov., sp. nov. *Int. J. Syst. Bacteriol.* 45: 454–461.

Erauso, G., A.-L. Reysenbach, A. Godfroy et al. 1993. *Pyrococcus abyssi* sp. nov., a new hyperthermophilic archaeon isolated from a deep-sea hydrothermal vent. *Arch. Microbiol.* 160: 338–349.

Falb, M., F. Pfeiffer, P. Palm et al. 2005. Living with two extremes: Conclusions from the genome sequence of *Natronomonas pharaonis. Genome Res.* 15: 1336–1343.

Ferreira, A. C., M. F. Nobre, E. D. Moore et al. 1999. Characterization and radiation resistance of new isolates of *Rubrobacter radiotolerans* and *Rubrobacter xylanophilus. Extremophiles* 3: 235–238.

Fiala, G., and K. O. Stetter. 1986. *Pyrococcus furiosus* sp. nov. represents a novel genus of marine heterotrophic archaebacteria growing optimally at 100°C. *Arch. Microbiol.* 145: 56–61.

Fiala, G., K. O. Stetter, H. W. Jannasch, T. A. Langworthy, and J. Madon. 1986. *Staphylothermus marinus* sp. nov. represents a novel genus of extremely thermophilic submarine heterotrophic archaebacteria growing up to 98°C. *Syst. Appl. Microbiol.* 8: 106–113.

Franzmann, P. D., E. Stackebrandt, K. Sanderson et al. 1988. *Halobacterium lacusprofundi* sp. nov., a halophilic bacterium isolated from Deep Lake, Antarctica. *Syst. Appl. Microbiol.* 11: 20–27.

Friedmann, E. I. 1980. Endolithic microbial life in hot and cold deserts. *Origins Life Evol. B.* 10: 223–235.

Friedmann, E. I. 1982. Endolithic microorganisms in the Antarctic cold desert. *Science* 215: 1045–1053.

Futterer, O., A. Angelov, H. Liesegang et al. 2004. Genome sequence of *Picrophilus torridus* and its implications for life around pH 0. *Proc. Natl. Acad. Sci. USA* 101: 9091–9096.

Galinski, E. A. 1995. Osmoadaptation in bacteria. *Adv. Microb. Physiol.* 37: 272–328.

Gilichinsky, D. 2002. Permafrost as a microbial habitat. In *Encyclopedia of Environmental Microbiology*, G. Bitton (Ed.), pp. 932–956. New York: Wiley.

Gilichinsky, D., E. Rivkina, V. Shcherbakova, K. Laurinavichius, and J. M. Tiedje. 2003. Supercooled water brines within permafrost – An unknown ecological niche for microorganisms: A model for astrobiology. *Astrobiology* 3: 331–341.

Gilichinsky, D. A. 1997. Permafrost as a microbial habitat: Extreme for the Earth, favorable in space. In *Proceedings of SPIE 3111, Instruments, Methods, and Missions for the Investigation of Extraterrestrial Microorganisms*, p. 472. Optical Science, Engineering and Instrumentation '97, San Diego, CA, doi:10.1117/12.278803.

Godfroy, A., F. Lesongeur, G. Raguénès et al. 1997. *Thermococcus hydrothermalis* sp. nov., a new hyperthermophilic archaeon isolated from a deep-sea hydrothermal vent. *Int. J. Syst. Bacteriol.* 47: 622–626.

Godfroy, A., J. R. Meunier, J. Guezennec et al. 1996. *Thermococcus fumicolans* sp. nov., a new hyperthermophilic archaeon isolated from a deep-sea hydrothermal vent in the North Fiji Basin. *Int. J. Syst. Bacteriol.* 46: 1113–1119.

Gold, T. 1992. The deep, hot biosphere. *Proc. Natl. Acad. Sci. USA* 89: 6045–6049.

Golyshina, O. V., T. A. Pivovarova, G. I. Karavaiko et al. 2000. *Ferroplasma acidiphilum* gen. nov., sp. nov., an acidophilic, autotrophic, ferrous-iron-oxidizing, cell-wall-lacking, mesophilic member of the *Ferroplasmaceae* fam. nov., comprising a distinct lineage of the Archaea. *Int. J. Syst. Evol. Microbiol.* 50: 997–1006.

Golyshina, O. V., and Timmis, K. N. 2005. *Ferroplasma* and relatives, recently discovered cell wall-lacking archaea making a living in extremely acid, heavy metal-rich environments. *Environ. Microbiol.* 7: 1277–1288.

González, J. M., C. Kato, and K. Horikoshi. 1995. *Thermococcus peptonophilus* sp. nov., a fast-growing, extremely thermophilic archaebacterium isolated from deep-sea hydrothermal vents. *Arch. Microbiol.* 164: 159–164.

González, J. M., Y. Masuchi, F. T. Robb et al. 1998. *Pyrococcus horikoshii* sp. nov., a hyperthermophilic archaeon isolated from a hydrothermal vent at the Okinawa Trough. *Extremophiles* 2: 123–130.

González-Toril, E., E. Llobet-Brossa, E. O. Casamayor, R. Amann, and R. Amils. 2003. Microbial ecology of an extreme acidic environment, the Tinto River. *Appl. Environ. Microbiol.* 69: 4853–4865.

Goordial, J., G. Lamarche-Gagnon, C.-Y. Lay, and L. Whyte. 2013. Life out in the cold: Life in cryoenvironments. In *Polyextremophiles: Life Under Multiple Forms of Stress*, J. Seckbach, A. Oren, and H. Stan-Lotter (Eds.), pp. 337–363. Dordrecht, the Netherlands: Springer.

Grant, W. D., and B. E. Jones. 2000. Alkaline environments. In *Encyclopedia of Microbiology*, 2nd ed., J. Lederberg (Ed.), pp. 126–133. London, UK: Academic Press.

Grant, W. D., W. E. Mwatha, and B. E. Jones. 1990. Alkaliphiles: Ecology, diversity and applications. *FEMS Microbiol. Rev.* 75: 255–270.

Grant, W. D., and B. J. Tindall. 1986. The alkaline saline environment. In *Microbes in Extreme Environments*, R. A. Herbert, and G. A. Codd (Eds.), pp. 25–54. London, UK: Academic Press.

Grosjean, H., and T. Oshima. 2007. How nucleic acids cope with high temperature. In *Physiology and Biochemistry of Extremophiles*, C. Gerday, and N. Glansdorff (Eds.), pp. 39–56. Washington, DC: ASM Press.

Gruber, C., A. Legat, M. Pfaffenhuemer et al. 2004. *Halobacterium noricense* sp. nov., an archaeal isolate from a bore core of an alpine Permian salt deposit, classification of *Halobacterium* sp. NRC-1 as a strain of *H. salinarum* and emended description of *H. salinarum*. *Extremophiles* 8: 431–439.

Hallberg, K. B., K. Coupland, S. Kimura, and D. B. Johnson. 2006. Macroscopic streamer growths in acidic, metal-rich mine waters in North Wales consist of novel and remarkably simple bacterial communities. *Appl. Environ. Microbiol.* 72: 2022–2030.

Hallberg, K. B., and D. B. Johnson. 2001. Biodiversity of acidophilic prokaryotes. *Adv. Appl. Microbiol.* 49: 37–84.

Hallsworth, J. E., M. M. Yakimov, P. N. Golyshin et al. 2007. Limits of life in $MgCl_2$-containing environments: Chaotropicity defines the window. *Environ. Microbiol.* 9: 801–813.

Hedlund, B. P., S. C. Thomas, J. A. Dodsworth, and C. L. Zhang. 2016. Life in high-temperature environments. The microbiology of extremely acidic environments. In *Manual of Environmental Microbiology*, 4th ed., M. V. Yates, C. H. Nakatu, R. V. Miller, and S. D. Pillai (Eds.), Chapter 4.3.4. Washington, DC: ASM Press.

Hippe, H. 2000. *Leptospirillum* gen. nov. (ex Markosyan 1972), nom. rev., including *Leptospirillum ferrooxidans* sp. nov. (ex Markosyan 1972), nom. rev. and *Leptospirillum thermoferrooxidans* sp. nov. (Golovacheva *et al.* 1992). *Int. J. Syst. Evol. Microbiol.* 50: 501–503.

Hoeft, S. E., J. Switzer Blum, J. F. Stolz et al. 2007. *Alkalilimnicola ehrlichii* sp. nov. a novel arsenite-oxidizing, haloalkaliphilic gammaproteobacterium capable of chemoautotrophic or heterotrophic growth with nitrate or oxygen as the electron acceptor. *Int. J. Syst. Evol. Microbiol.* 57: 504–512.

Hofmeister, F. 1888. Zur Lehre von der Wirkung der Salze. Zweite Mittheilung. *Arch. Exp. Pathol. Pharmakol.* 24: 247–260.

Horikoshi, K. 1998. Barophiles: Deep-sea microorganisms adapted to an extreme environment. *Curr. Opin. Microbiol.* 1: 291–295.

Horikoshi, K. (Ed.) 1999. *Alkaliphiles*. Reading, UK: Harwood Academic.

Horikoshi, K. (Ed.) 2010. *Extremophiles Handbook*. Tokyo, Japan: Springer-Verlag.

Horikoshi, K., and W. D. Grant (Eds.). 1991. *Superbugs. Microorganisms in Extreme Environments*. Tokyo, Japan: Japan Scientific Societies Press.

Horikoshi, K., and W. D. Grant (Eds.) 1998. *Extremophiles, Microbial Life in Extreme Environments*. New York: Wiley-Liss Publishers.

Huber, H., M. J. Hohn, R. Rachel, T. Fuchs, V. C. Wimmer, and K. O. Stetter. 2002. A new phylum of Archaea represented by a nanosized hyperthermophilic symbiont. *Nature* 417: 63–67.

Huber, R., H. Huber, and K. O. Stetter. 2000a. Towards the ecology of hyperthermophiles: Biotopes, new isolation strategies and novel metabolic properties. *FEMS Microbiol. Rev.* 24: 615–623.

Huber R., M. Sacher, A. Vollman, H. Huber, and D. Rose. 2000b. Respiration on arsenate and selenate by hyperthermophilic Archaea. *Syst. Appl. Microbiol.* 23: 305–314.

Hugenholtz, P., C. Pitulle, K. L. Hershberger, and N. R. Pace. 1998. Novel division level bacterial diversity in a Yellowstone hot spring. *J. Bacteriol.* 180: 366–376.

Imhoff, J. F., H. G. Sahl, G. S. H. Soliman, and H. G. Trüper. 1979. The Wadi Natrun: Chemical composition and microbial mass developments in alkaline brines of a eutrophic desert lakes. *Geomicrobiol. J.* 1: 219–234.

Irgens, R. L., J. J. Gosink, and J. T. Staley. 1996. *Polaromonas vacuolata* gen. nov., sp. nov., a psychrophilic, marine, gas vacuolate bacterium from Antarctica. *Int. J. Syst. Bacteriol.* 46: 822–826.

Jannasch, H. W., and C. D. Taylor. 1984. Deep-sea microbiology. *Annu. Rev. Microbiol.* 38: 487–515.

Jannasch, H. W., and C. O. Wirsen. 1984. Variability of pressure adaptation in deep-sea bacteria. *Arch. Microbiol.* 139: 281–288.

Jeanthon C., S. L'Haridon, A. L. Reysenbach et al. 1998. *Methanococcus infernus* sp. nov., a novel extremely thermophilic lithotrophic methanogen isolated from a deep-sea hydrothermal vent. *Int. J. Syst. Bacteriol.* 48: 913–919.

Johnson, D. B. 1998. Biodiversity and ecology of acidophilic microorganisms. *FEMS Microbiol. Ecol.* 27: 303–317.

Johnson, D. B. 2007. Physiology and ecology of acidophilic microorganisms. In *Physiology and Biochemistry of Extremophiles*, C. Gerday, and N. Glansdorff (Eds.), pp. 257–270. Washington, DC: ASM Press.

Johnson, D. B. 2012. Geomicrobiology of extremely acidic subsurface environments. *FEMS Microbiol. Ecol.* 81: 2–12.

Johnson, D. B., and A. Aguilera. 2016. The microbiology of extremely acidic environments. In *Manual of Environmental Microbiology*, 4th ed., M. V. Yates, C. H. Nakatu, R. V. Miller, and S. D. Pillai (Eds.), Chapter 4.3.1. Washington, DC: ASM Press.

Johnson, D. B., and K. B. Hallberg. 2003. The microbiology of acidic mine waters. *Res. Microbiol.* 154: 466–473.

Jolivet, E., E. Corre, S. L'Haridon, P. Forterre, and D. Prieur. 2004. *Thermococcus marinus* sp. nov., and *Thermococcus radiotolerans* sp. nov., two hyperthermophilic archaea from deep-sea hydrothermal vents that resist ionizing radiation. *Extremophiles* 8: 219–227.

Jolivet, E., S. L'Haridon, E. Corre, P. Forterre, and D. Prieur. 2003a. *Thermococcus gammatolerans* sp. nov., a hyperthermophilic archaeon from a deep-sea hydrothermal vent that resists ionizing radiation. *Int. J. Syst. Evol. Microbiol.* 53: 847–851.

Jolivet, E., F. Matsunaga, Y. Ishino, P. Forterre, D. Prieur, and H. Myllykallio. 2003b. Physiological responses of the hyperthermophilic archaeon "Pyrococcus abyssi" to DNA damage caused by ionizing radiation. *J. Bacteriol.* 185: 3958–3961.

Jones, B. E., and W. D. Grant. 2000. Microbial diversity and ecology of soda lakes. In *Journey to Diverse Microbial Worlds*, J. Seckbach (Ed.), pp. 177–190. Dordrecht, the Netherlands: Kluwer Academic Publishers.

Jones, W. J., J. A. Leigh, F. Mayer, C. R. Woese, and R. S. Wolfe. 1983. *Methanococcus jannaschii* sp. nov., an extremely thermophilic methanogen from a submarine hydrothermal vent. *Arch. Microbiol.* 136: 254–261.

Jørgensen, B. B., M. F. Isaksen, and H. W. Jannasch. 1992. Bacterial sulfate reduction above 100°C in deep-sea hydrothermal vent sediments. *Science* 258: 1756–1757.

Junge, K., H. Eicken, and J. W. Deming. 2004. Bacterial activity at -2 to -20°C in Arctic wintertime sea ice. *Appl. Environ. Microbiol.* 70: 550–557.

Junge, K., H. Eicken, B. D. Swanson, and J. W. Deming. 2006. Bacterial incorporation of leucine into protein down to -20°C with evidence for potential activity in subeutectic saline ice formations. *Cryobiology* 52: 2163–2181.

Kashefi, K., and D. R. Lovley. 2003. Extending the upper temperature limit for life. *Science* 301: 934.

Kato, C., and D. H. Bartlett. 1997. The molecular biology of barophilic bacteria. *Extremophiles* 1: 111–116.

Kato, C., L. Li, Y. Nogi, Y. Nakamura, J. Tamaoka, and K. Horikoshi. 1998. Extremely barophilic bacteria isolated from the Mariana Trench, Challenger Deep, at a depth of 11,000 meters. *Appl. Environ. Microbiol.* 64: 1510–1513.

Kato, C., T. Sato, and K. Horikoshi. 1995. Isolation and properties of barophilic and barotolerant bacteria from deep-sea mud samples. *Biodivers. Conserv.* 4: 1–9.

Kato, C., T. Sato, Y. Nogi, and K. Nakasone. 2004. Piezophiles: High pressure-adapted marine bacteria. *Mar. Biotechnol.* 6: S195–S201.

Kelly D. P., and A. P. Wood. 2000. Reclassification of some species of *Thiobacillus* to the newly designated genera *Acidithiobacillus* gen. nov., *Halothiobacillus* gen. nov. and *Thermithiobacillus* gen. nov. *Int. J. Syst. Evol. Microbiol.* 50: 511–516.

Kevbrin, V. V., C. S. Romanek, and J. Wiegel. 2004. Alkalithermophiles: A double challenge from extreme environments. In *Cellular Origins, Life in Extreme Habitats and Astrobiology*, J. Seckbach (Ed.), pp. 395–412. Dordrecht, the Netherlands: Kluwer Academic Publishers.

Kevbrin, V. V., T. N. Zhilina, F. A. Rainey, and G. A. Zavarzin. 1998. *Tindallia magadii* gen. nov., sp. nov.: An alkaliphilic anaerobic ammonifier from soda lake deposits. *Curr. Microbiol.* 37: 94–100.

Kletzin, A., T. Urich, F. Müller, T. M. Bandeiras, and C. M. Gomes. 2004. Dissimilatory oxidation and reduction of elemental sulfur in thermophilic Archaea. *J. Bioenerg. Biomembr.* 36: 77–91.

Kobayashi, T., Y. Kwak, T. Akiba, T. Kudo, and K. Horikoshi. 1994. *Thermococcus profundus* sp. nov., a new hyperthermophilic archaeon isolated from a deep-sea hydrothermal vent. *Syst. Appl. Microbiol.* 17: 232–236.

Kristjansson, J. K., and G. O. Hreggvidsson. 1995. Ecology and habitats of extremophiles. *World J. Microbiol. Biotechnol.* 11: 17–25.

Kurosawa N., Y. H. Itoh, T. Iwai et al. 1998. *Sulfurisphaera ohwakuensis* gen. nov., sp. nov., a novel extremely thermophilic acidophile of the order *Sulfolobales*. *Int. J. Syst. Bacteriol.* 49: 451–456.

Kurr, M., R. Huber, H. König et al. 1991. *Methanopyrus kandleri* gen. and sp. nov. represents a novel group of hyperthermophilic methanogens, growing at 110°C. *Arch. Microbiol.* 156: 239–247.

Kushner, D. J. (Ed.) 1978. *Microbial Life in Extreme Environments*. London, UK: Academic Press.

Lanyi, J. K. 1974. Salt-dependent properties of proteins from extremely halophilic bacteria. *Bacteriol. Rev.* 38: 272–290.

Madigan, M. T. 2000. Bacterial habitats in extreme environments. In *Journey to Diverse Microbial Worlds: Adaptation to Exotic Environments*, J. Seckbach (Ed.), pp. 61–72. Dordrecht, the Netherlands: Kluwer Academic Publishers.

Madigan, M. T., and A. Oren. 1999. Thermophilic and halophilic extremophiles. *Curr. Opin. Microbiol.* 2: 265–269.

Marteinsson, V. T., J. L. Birrien, A.-L. Reysenbach et al. 1999. *Thermococcus barophilus* sp. nov., a new barophilic and hyperthermophilic archaeon isolated under high hydrostatic pressure from a deep-sea hydrothermal vent. *Int. J. Syst. Bacteriol.* 49: 351–359.

Mathrani, I. M., D. R. Boone, R. A. Mah, G. E. Fox, and P. P. Lau. 1988. *Methanohalophilus zhilinae* sp. nov., an alkaliphilic, halophilic, methylotrophic methanogen. *Int. J. Syst. Bacteriol.* 38: 139–142.

Mattimore, V., and J. R. Battista. 1996. Radioresistance of *Deinococcus radiodurans*: Functions necessary to survive ionizing radiation are also necessary for desiccation. *J. Bacteriol.* 178: 633–637.

Mazur, P. 1980. Limits to life at low temperatures and at reduced water contents and water activities. *Orig. Life* 10: 137–159.

Mesbah, N. M., S. H. Abou-El-Ela, and J. Wiegel. 2007a. Novel and unexpected prokaryotic diversity in water and sediments of the alkaline, hypersaline lakes of the Wadi An Natrun, Egypt. *Microb. Ecol.* 54: 598–617.

Mesbah, N. M., D. B. Hedrick, A. D. Peacock, M. Rohde, and J. Wiegel. 2007b. *Natranaerobius thermophilus* gen. nov., sp. nov., a halophilic, alkalithermophilic bacterium from soda lakes of the Wadi An Natrun, Egypt, and proposal of *Natranaerobiaceae* fam. nov. and *Natranaerobiales* ord. nov. *Int. J. Syst. Evol. Microbiol.* 57: 2507–2512.

Mesbah, N. M., and J. Wiegel. 2008. Life at extreme limits: The anaerobic halophilic alkalithermophiles. *Ann. N.Y. Acad. Sci.* 1125: 44–57.

Mesbah, N. M., and J. Wiegel. 2009. *Natronovirga wadinatrunensis* gen. nov., sp. nov. and *Natranaerobius trueperi* sp. nov., halophilic, alkalithermophilic microorganisms from soda lakes of the Wadi An Natrun, Egypt. *Int. J. Syst. Evol. Microbiol.* 59: 2042–2048.

Mesbah, N. M., and J. Wiegel. 2011. Halophiles exposed concomitantly to multiple stressors: Adaptive mechanisms of halophilic alkalithermophiles. In *Halophiles and Hypersaline Environments*, A. Ventosa, A. Oren, and Y. Mah (Eds.), pp. 249–273. Berlin, Germany: Springer.

Mesbah, N. M., and J. Wiegel. 2012. Life under multiple extreme conditions: Diversity and physiology of the halophilic alkalithermophiles. *Appl. Environ. Microbiol.* 78: 4074–4082.

Mevarech, M., F. Frolow, and L. M. Gloss. 2000. Halophilic enzymes: Proteins with a grain of salt. *Biophys. Chem.* 86: 155–164.

Mikucki, J. A., A. Pearson, D. T. Johnston et al. 2009. A contemporary microbially maintained subglacial ferrous "ocean". *Science* 324: 397–400.

Mikucki, J. A., and J. C. Priscu. 2007. Bacterial diversity associated with Blood Falls, a subglacial outflow from the Taylor Glacier, Antarctica. *Appl. Environ. Microbiol.* 73: 4029–4039.

Miller, R. V., and L. G. Whyte (Eds.). 2011. *Polar Microbiology: Life in the Deep Freeze*. Washington, DC: ASM Press.

Minegishi, H., A. Echigo, S. Nagaoka, M. Kamekura, and R. Usami. 2010. *Halarchaeum acidiphilum* gen. nov., sp. nov., a moderately acidophilic haloarchaeon isolated from commercial solar salt. *Int. J. Syst. Evol. Microbiol.* 60: 2513–2516.

Minegishi, H., T. Mizuki, A. Echigo, T. Fukushima, M. Kamekura, and R. Usami. 2008. Acidophilic haloarchaeal strains are isolated from various solar salts. *Saline Syst.* 4: 16.

Mwatha, W. E., and W. D. Grant. 1993. *Natronobacterium vacuolata* sp. nov., a haloalkaliphilic archaeon isolated from Lake Magadi, Kenya. *Int. J. Syst. Bacteriol.* 43: 401–404.

Mykytczuk, N. C., S. J. Foote, C. R. Omelon, G. Southam, C. W. Greer, and L. G. Whyte. 2013. Bacterial growth at -15°C; molecular insights from the permafrost bacterium *Planococcus halocryophilus* Or1. *ISME J.* 7: 1211–1226.

Navarro-González, R., F. A. Rainey, P. Molina et al. 2003. Mars-like soils in the Atacama Desert, Chile, and the dry limit of microbial life. *Science* 302: 1018–1021.

Nealson, K. H. 1997. The limits of life on Earth and searching for life on Mars. *J. Geophys. Res.* 102: 23675–23686.

Omelon, C. R. 2016. Endolithic microorganisms and their habitats. In *Their World: A Diversity of Microbial Environments*, Advances in Environmental Microbiology, 1st ed., C. J. Hurst (Ed.), pp. 171–201. Cham, Switzerland: Springer.

Oremland, R. S., T. R. Kulp, J. Switzer Blum et al. 2005. A microbial arsenic cycle in a salt-saturated, extreme environment. *Science* 308: 1305–1308.

Oren, A. 1999. Bioenergetic aspects of halophilism. *Microbiol. Mol. Biol. Rev.* 63: 334–348.

Oren, A. 2000. Life at high salt concentrations. In *The Prokaryotes: An Evolving Electronic Resource for the Microbiological Community*, 3rd ed., M. Dworkin, S. Falkow, E. Rosenberg, K.-H. Schleifer, and E. Stackebrandt (Eds.). New York: Springer-Verlag.

Oren, A. 2002a. *Halophilic Microorganisms and Their Environments*. Dordrecht, the Netherlands: Kluwer Scientific Publishers.

Oren, A. 2002b. Diversity of halophilic microorganisms: Environments, phylogeny, physiology, and applications. *Indust. Microbiol. Biotechnol.* 28: 56–63.

Oren, A. 2011. Thermodynamic limits to microbial life at high salt concentrations. *Environ. Microbiol.* 13: 1908–1923.

Oren, A. 2013. Life in magnesium- and calcium-rich hypersaline environments: Salt-stress by chaotropic ions. In *Polyextremophiles: Life Under Multiple Forms of Stress*, J. Seckbach, A. Oren, and H. Stan-Lotter (Eds.), pp. 217–232. Dordrecht, the Netherlands: Springer.

Oren, A. 2016. Life in high-salinity environments. In *Manual of Environmental Microbiology*, 4th ed., M. V. Yates, C. H. Nakatu, R. V. Miller, and S. D. Pillai (Eds.), Chapter 4.3.2. Washington, DC: ASM Press.

Oren, A., M. Heldal, and S. Norland. 1997. X-ray microanalysis of intracellular ions in the anaerobic halophilic eubacterium *Haloanaerobium praevalens*. *Can. J. Microbiol.* 43: 588–592.

Oren, A., M. Heldal, S. Norland, and E. A. Galinski. 2002. Intracellular ion and organic solute concentrations of the extremely halophilic bacterium *Salinibacter ruber*. *Extremophiles* 6: 491–498.

Pick, U. 1999. *Dunaliella acidophila* – A most extreme acidophilic alga. In *Enigmatic Microorganisms and Life in Extreme Environmental Habitats*, J. Seckbach (Ed.), pp. 465–478. Dordrecht, the Netherlands: Kluwer Academic Publishers.

Pikuta, E. V., R. B. Hoover, and J. Tang. 2007. Microbial extremophiles at the limits of life. *Crit. Rev. Microbiol.* 33: 183–209.

Pledger, R. J., B. C. Crump, and J. A. Baross. 1994. A barophilic response by two hyperthermophilic, hydrothermal vent Archaea: An upward shift in the optimal temperature and acceleration of growth rate at supra-optimal temperatures by elevated pressure. *FEMS Microbiol. Ecol.* 24: 233–242.

Pley, U., J. Schipka, A. Gambacorta et al. 1991. *Pyrodictium abyssi* sp. nov. represents a novel heterotrophic marine archaeal hyperthermophile growing at 110°C. *Syst. Appl. Microbiol.* 14: 245–253.

Price, P. B. 2000. A habitat for psychrophiles in deep Antarctic ice. *Proc. Natl. Acad. Sci. USA* 97: 1247–1251.

Priscu, J. C., E. E. Adams, W. B. Lyons et al. 1999. Geomicrobiology of subglacial ice above Lake Vostok, Antarctica. *Science* 286: 2141–2144.

Priscu, J. C., C. H. Fritsen, E. E. Adams et al. 1998. Perennial Antarctic lake ice: An oasis for life in a polar desert. *Science* 280: 2095–2098.

Rainey, F. A., and A. Oren. 2006a. Extremophile microorganisms and the methods to handle them. In *Extremophiles—Methods in Microbiology*, Vol. 35, F. A. Rainey, and A. Oren (Eds.), pp. 1–25. Amsterdam, the Netherlands: Elsevier.

Rainey, F. A., and A. Oren (Eds.) 2006b. *Extremophiles—Methods in Microbiology*, Vol. 35. Amsterdam, the Netherlands: Elsevier.

Rainey, F. A., K. Ray, M. Ferreira et al. 2005. Extensive diversity of ionizing radiation-resistant bacteria recovered from a Sonoran Desert soil and the description of 9 new species of the genus *Deinococcus* from a single soil sample. *Appl. Environ. Microbiol.* 71: 5225–5235.

Ravenschlag, K., K. Sahm, J. Pernthaler, and R. Amann. 1999. High bacterial diversity in permanently cold marine sediments. *Appl. Environ. Microbiol.* 65: 3982–3989.

Reysenbach, A.-L., Y. Liu, A. B. Banta et al. 2006. A ubiquitous thermoacidophilic archaeon from deep-sea hydrothermal vents. *Nature* 442: 444–447.

Reysenbach, A.-L., M. Voytek, and R. Mancinelli (Eds.). 2001. *Thermophiles: Biodiversity, Ecology, and Evolution*. Dordrecht, the Netherlands: Kluwer Academic Publishers.

Rivkina, E. M., E. I. Friedmann, C. P. McKay, and D. A. Gilichinsky. 2000. Metabolic activity of permafrost bacteria below the freezing point. *Appl. Environ. Microbiol.* 66: 3230–3233.

Rothschild, L. J. 1999. Microbes and radiation. In *Enigmatic Microorganisms and Life in Extreme Environmental Habitats*, J. Seckbach (Ed.), pp. 549–562. Dordrecht, the Netherlands: Kluwer Academic Publishers.

Rothschild, L. J., and R. L. Mancinelli. 2001. Life in extreme environments. *Nature* 409: 1092–1101.

Russell, N. J. 1990. Cold adaptation of microorganisms. *Phil. Trans. R. Soc. London B* 326: 595–608; Discussion: 608–611.

Schleper, C., G. Pühler, I. Holz et al. 1995a. *Picrophilus* gen. nov., fam. nov.: A novel aerobic, heterotrophic, thermoacidophilic genus and family comprising archaea capable of growth around pH 0. *J. Bacteriol.* 177: 7050–7059.

Schleper, C., G. Pühler, H.-P. Klenk, and W. Zillig. 1996. *Picrophilus oshimae* and *Picrophilus torridus* fam. nov., gen. nov., sp. nov., two species of hyperacidophilic, thermophilic, heterotrophic, aerobic archaea. *Int. J. Syst. Bacteriol.* 46: 814–816.

Schleper, C., G. Pühler, B. Kühlmorgen, and W. Zillig. 1995b. Life at extremely low pH. *Nature* 375: 741–742.

Seckbach, J. (Ed.). 1994. *Evolutionary pathways and enigmatic algae:* Cyanidium caldarium *(Rhodophyta) and related cells.* Dordrecht, the Netherlands: Kluwer Academic Publishers.

Seckbach, J. 1999. The Cyanidiophyceae: Hot spring acidophilic algae. In *Enigmatic Microorganisms and Life in Extreme Environmental Habitats*, J. Seckbach (Ed.), pp. 425–435. Dordrecht, the Netherlands: Kluwer Academic Publishers.

Seckbach, J., A. Oren, and H. Stan-Lotter. 2013. *Polyextremophiles: Life Under Multiple Forms of Stress.* Dordrecht, the Netherlands: Springer.

Selbmann, L., G. S. de Hoog, A. Mazzagalia, E. I. Friedmann, and S. Onofri. 2005. Fungi at the edge of life: Cryptoendolithic black fungi from Antarctic desert. *Stud. Mycol.* 51: 1–32.

Soliman, G. S. H., and H. G. Trüper. 1982. *Halobacterium pharaonis* sp. nov., a new extremely haloalkaliphilic archaebacterium with low magnesium requirement. *Zentralbl. Bakteriol. Hyg. Abt. I Orig.* C3: 318–329.

Stan-Lotter, H. 2012. Physico-chemical boundaries of life. In *Adaptation of Microbial Life to Environmental Extremes*, H. Stan-Lotter, and S. Fendrihan (Eds.), pp. 1–19. Vienna, Austria: Springer.

Stetter, K. O. 1988. *Archaeoglobus fulgidus* gen. nov., sp. nov. a new taxon of extremely thermophilic *Archaebacteria*. *Syst. Appl. Microbiol.* 10: 172–173.

Stetter, K. O. 1996. Hyperthermophilic prokaryotes. *FEMS Microbiol. Rev.* 18: 149–158.

Stetter, K. O. 1999. Extremophiles and their adaptation to hot environments. *FEBS Lett.* 452: 22–25.

Stetter, K. O., H. König, and E. Stackebrandt. 1983. *Pyrodictium* gen. nov., a new genus of submarine disc-shaped sulfur reducing archaebacteria growing optimally at 105°C. Syst. Appl. Microbiol. 4: 535–551.

Stetter, K. O., A. Segrer, W. Zillig et al. 1986. Extremely thermophilic sulfur-metabolizing archaebacteria. *Syst. Appl. Microbiol.* 7: 393–397.

Stevenson, A., J. A. Cray, J. P. Williams et al. 2015. Is there a common water-activity limit for the three domains of life? *ISME J.* 9: 1333–1351.

Suzuki, K.-I., M. D. Collins, E. Iijima, and K. Komagata. 1988. Chemotaxonomic characterization of a radiotolerant bacterium, *Arthrobacter radiotolerans*: Description of *Rubrobacter radiotolerans* gen. nov., comb. nov. *FEMS Microbiol. Lett.* 52: 33–40.

Switzer Blum, J., T. R. Kulp, S. Han et al. 2012. *Desulfohalophilus alkaliarsenatis* gen. nov., sp. nov., an extremely halophilic sulfate- and arsenate-respiring bacterium from Searles Lake, California. *Extremophiles* 16: 727–742.

Takai, K., A. Inoue, and K. Horikoshi. 1999. *Thermaerobacter marianensis* gen. nov., sp. nov., an anaerobic extremely thermophilic marine bacterium from the 11,000 m deep Mariana Trench. *Int. J. Syst. Bacteriol.* 49: 619–628.

Takai, K., K. Nakamura, T. Toki et al. 2008. Cell proliferation at 122°C and isotopically heavy CH_4 production by a hyperthermophilic methanogen under high-pressure cultivation. *Proc. Natl. Acad. Sci. USA* 105: 10949–10954.

Takai, K., K. H. Nealson, and K. Horikoshi. 2004a. *Methanotorris formicicus* sp. nov., a novel extremely thermophilic, methane-producing archaeon isolated from a black smoker chimney in the Central Indian Ridge. *Int. J. Syst. Evol. Microbiol.* 54: 1095–1100.

Takai, K., A. Sugai, T. Itoh, and K. Horikoshi. 2000. *Palaeococcus ferrophilus* gen. nov., sp. nov., a barophilic, hyperthermophilic archaeon from a deep-sea hydrothermal vent chimney. *Int. J. Syst. Evol. Microbiol.* 50: 489–500.

Takai, K., G. Toshitaka, U. Tsunogai et al. 2004b. Geochemical and microbiological evidence for a hydrogen-based, hyperthermophilic subsurface lithoautotrophic microbial ecosystem (HyperSLiME) beneath an active deep-sea hydrothermal field. *Extremophiles* 8: 269–282.

Temple, K. L., and Colmer, A. R. 1951. The autotrophic oxidation of iron by a new bacterium, *Thiobacillus ferrooxidans*. *J. Bacteriol.* 62: 605–611.

Thomas, D. N., and G. S. Dieckmann. 2002. Antarctic sea ice – A habitat for extremophiles. *Science* 295: 641–644.

Tindall, B. J., H. N. M. Ross, and W. D. Grant. 1984. *Natronobacterium* gen. nov. and *Natronococcus* gen. nov., two new genera of haloalkaliphilic archaebacteria. *Syst. Appl. Microbiol.* 5: 41–57.

Toplin, J. A., T. B. Norris, C. R. Lehr, T. R. McDermott, and R. W. Castenholz. 2008. Biogeographic and phylogenetic diversity of thermoacidophilic Cyanidiales in Yellowstone National Park, Japan and New Zealand. *Appl. Environ. Microbiol.* 74: 2822–2344.

Tyson, G. W., J. Chapman, P. Hugenholtz et al. 2004. Community structure and metabolism through reconstruction of microbial genomes from the environment. *Nature* 428: 37–43.

van der Wielen, P. W. J. J., H. Bolhuis, S. Borin et al. 2005. The enigma of prokaryotic like in deep hypersaline anoxic basins. *Science* 307: 121–123.

van de Vossenberg, J. L. C. M., A. J. M. Driessen, and W. N. Konings. 1998. The essence of being extremophilic: The role of the unique archaeal membrane lipids. *Extremophiles* 2: 163–170.

Ventosa, A., J. J. Nieto, and A. Oren. 1998. Biology of aerobic moderately halophilic bacteria. *Microbiol. Mol. Biol. Rev.* 62: 504–544.

Vincent, W. F. 1988. *Microbial Ecosystems of Antarctica.* Cambridge, UK: Cambridge University Press.

Vorobtova, E., V. Soina, M. Gorlenko et al. 1997. The deep cold biosphere: Facts and hypothesis. *FEMS Microbiol. Rev.* 20: 277–290.

Walker, J. J., J. R. Spear, and N. R. Pace. 2005. Geobiology of a microbial endolithic community in the Yellowstone geothermal environment. *Nature* 434: 1011–1014.

Warren-Rhodes, K. A., K. L. Rhodes, S. B. Pointing et al. 2006. Hypolithic cyanobacteria, dry limit of photosynthesis and microbial ecology in the hyperarid Atacama Desert. *Microb. Ecol.* 52: 389–398.

Wharton, D. A. 2002. *Life at the Limits: Organisms in Extreme Environments.* Cambridge, UK: Cambridge University Press.

Wiegel, J. 1990. Temperature spans for growth: A hypothesis and discussion. *FEMS Microbiol. Rev.* 75: 155–169.

Wiegel, J. 1998. Anaerobic alkalithermophiles, a novel group of extremophiles. *Extremophiles* 2: 257–267.

Wiegel, J., and M. W. W. Adams (Eds.). 1998. *Thermophiles: The Keys to Molecular Evolution and the Origin of Life?* London, UK: Taylor & Francis Group.

Wiegel, J., and V. Kevbrin. 2004. Diversity of aerobic and anaerobic alkalithermophiles. *Biochem. Soc. Trans.* 32: 193–198.

Williams, T. J., M. A. Allen, M. A. DeMaere et al. 2014. Microbial ecology of an Antarctic hypersaline lake: Genomic assessment of ecophysiology among dominant haloarchaea. *ISME J.* 8: 1645–1658.

Xu, Y., and N. Glansdorff. 2007. Lessons from extremophiles: Early evolution and border conditions of life. In *Physiology and Biochemistry of Extremophiles*, C. Gerday, and N. Glansdorff (Eds.), pp. 409–421. Washington, DC: ASM Press.

Yayanos, A. A. 1995. Microbiology to 10,500 meters in the deep sea. *Annu. Rev. Microbiol.* 49: 777–805.

Yayanos, A. A. 2000. Deep-sea bacteria. In *Journey to Diverse Microbial Worlds*, J. Seckbach (Ed.), pp. 164–174. Dordrecht, the Netherlands: Kluwer Academic Publishers.

Yayanos, A. A., A. S. Dietz, and R. Van Boxtel. 1979. Isolation of a deep-sea barophilic bacterium and some of its growth characteristics. *Science* 205: 808–810.

Yayanos, A. A., A. S. Dietz, and R. Van Boxtel. 1981. Obligately barophilic bacterium from the Mariana Trench. *Proc. Natl. Acad. Sci. USA* 78: 5212–5215.

Yayanos, A. A., A. S. Dietz, and R. Van Boxtel. 1982. Dependence of reproduction rate on pressure as a hallmark of deep-sea bacteria. *Appl. Environ. Microbiol.* 44: 1356–1361.

Zajc, J., S. Džeroski, D. Kocev et al. 2014. Chaophilic or chaotolerant fungi: A new category of extremophiles? *Front. Microbiol.* 5: 708.

Zavarzin, G. A., and T. N. Zhilina. 2000. Anaerobic chemotrophic alkaliphiles. In *Journey to Diverse Microbial Worlds*, J. Seckbach (Ed.), pp. 191–208. Dordrecht, the Netherlands: Kluwer Academic Publishers.

Zhilina, T. N., and G. A. Zavarzin. 1994. Alkaliphilic anaerobic community at pH 10. *Curr. Microbiol.* 29: 109–112.

9.2 Microbes in Space

Kasthuri Venkateswaran

CONTENTS

9.2.1 INTRODUCTION

The National Research Council (NRC) Committee's Decadal Survey on Biological and Physical Sciences in Space reported that "microbial species that are uncommon, or that have significantly increased or decreased in number, can be studied in a 'microbial observatory (MO)' on the International Space Station (ISS)" (NRC 2011), and numerous NASA-sponsored studies were performed in response to the NRC recommendations. Furthermore, the NRC's decadal survey committee recommended that, for MO studies, NASA should: "(a) Capitalize on the technological maturity, low cost, and speed of genomic analyses and the rapid generation time of microbes to monitor the evolution of microbial genomic changes in response to the selective pressures present in the spaceflight environment; (b) Study changes in microbial populations from the skin and feces of the astronauts, plant and plant growth media, and environmental samples taken from surfaces and the atmosphere of the ISS; and (c) Establish an experimental program targeted at understanding the influence of the spaceflight environment on defined microbial populations." There is a significant knowledge gap in understanding how the spaceflight environment affects microbial population dynamics; however, this also represents an opportunity to study how microbial populations evolve and how these populations affect crew health, as well as the associated engineering risks of closed environments for long-term space habitation.

To reduce this knowledge gap, this study has sampled several defined ISS environmental surfaces for three consecutive flights, collected air samples, and characterized microbiome by utilizing traditional cultivation procedures as well as state-of-the-art molecular technologies such as amplicon-based targeted sequencing for microbiome and shotgun metagenome sequencing. In addition, "omics" of several microbial species (bacteria and fungi) that could be cultured were characterized for their novelty, antimicrobial resistance (AMR), and virulence profiles. Furthermore, select microbial species were also exposed outside the space station to characterize their survival capabilities in response to outer-space radiation conditions.

The ISS is a closed system inhabited by microorganisms originating from life support systems, cargo contaminants, and humans. Microorganisms associated with the ISS are exposed to selective pressures from the extreme space environment. In this chapter, attempts were made to discuss microbial (microbiome) and functional (metagenome) diversity of various elements of ISS environments (air and surface). Microbial population estimations using culture-dependent and culture-independent analyses were documented. In addition, descriptions of novel microbial species, their antimicrobial, and virulence characteristics were reported.

9.2.2 SIGNIFICANCE OF THE STUDY

The ISS is a built environment that has been specially designed to meet its physical characteristics and habitation requirements. The microgravity, exposure to space radiation, and increased carbon dioxide concentrations are conditions that are unique to the ISS. Furthermore, a large proportion of the space within the ISS environment is occupied by astronauts. Learning the makeup of the microbial community within this closed system will improve safety and maintenance procedures. The approach seeks to characterize the viable microbiome of the built environment of the ISS when molecular techniques are used and then determine if the built environments of cleanrooms on Earth designed for space exploration are a suitable simulation of the ISS environment.

The microbiome of environmental surfaces on the ISS was characterized so that its relationship to crew and hardware maintenance can be analyzed. The results of these studies will help future habitation scientists ascertain the habitability of the ISS environment over time. In addition, the results validate the usefulness of measuring viable cell diversity and population size at a variety of sites in order to identify the sites that would require more rigorous cleaning regiment. Finally, the results can be used as a basis for comparisons with future human habitation sites and to facilitate the development of enhancements to the ISS that would minimize negative effects on the health of its inhabitants.

The ISS Microbial Observatory (ISS-MO) experimental flight projects used both advanced molecular microbial community analyses and traditional culture-based methods to generate a microbial census of the ISS environment. Substantial insights into how spaceflight changes the population of both beneficial and potentially harmful microbes were documented from the extensive microbial census generated by the omics methodologies. The omics of the spaceflight samples were placed in the NASA GeneLab system, a one-stop data repository, which will enable future space biologists to datamine and design suitable spaceflight experiments.

9.2.3 OVERVIEW OF THE "MICROBES IN SPACE"

In the past decade, NASA invested resources to develop capabilities in measuring microbial survival and proliferation in space, as well as in estimating the composition of microorganisms in closed system such as the ISS. In order to comprehensively understand the microbiome of the ISS environmental surfaces, microbial tracking studies were undertaken from simulated systems on the ground and the ISS. In addition, efforts were taken to maximize the use of both the interior (cabin) and the exterior (for mounting science instruments) of the ISS platform for elucidating microbial survival under space conditions. Such microbiological characterizations were intended to explore not only problematic microorganisms but also beneficial microbes for the production of novel bioactive compounds under microgravity. In this regard, known microbial strains were safely taken to the ISS to study the effect of microgravity, and samples were brought from the ISS to characterize them. In addition, environmental samples collected were explored for the microbial diversity of the closed system by using state-of-the-art molecular techniques. Figure 9.2.1 shows some of the approaches taken by NASA and executed at Jet Propulsion Laboratory (JPL; Pasadena, CA) with reference to space microbiology investigations.

9.2.4 MICROBIOME OF SIMULATED HUMAN HABITAT SYSTEM AND CREW RESUPPLY SERVICE VEHICLE (GROUND CONTROL)

A safe, enclosed habitat is needed for potential future human missions to the Moon or Mars, as well as for continued human presence in the ISS. However, long-term confinement results in a lowered immune response, leading to a potential threat to the inhabitants of a closed system from potential microbial contamination. It is critical that the design of safer habitats for crew members is informed by the lessons gathered from characterizing analogous habitats. A key question to address is how the presence of humans affects the accumulation of microorganisms in an enclosed habitat. Isolated by high-efficiency particulate arrestance (HEPA) filtration, the inflatable lunar/Mars analog habitat (ILMAH) is a closed system that was used to simulate ISS conditions and potential human habitats on other planets, except for the exchange of air between outdoor and indoor environments. Although it was mainly commissioned to assess the physiological, psychological, and immunological effects of living in isolation, it was also available for microbiological studies.

9.2.4.1 BACTERIOME OF INFLATABLE LUNAR/MARS ANALOG HABITAT

Both molecular technologies and traditional microbiological techniques were used to catalog microbial succession within the ILMAH. Surface samples were gathered at various locations over the 30-day human occupation of the ILMAH to capture a full range of viable and potential opportunistic pathogenic bacterial population. The cultivable, viable, and metabolically active microbial populations were estimated using traditional cultivation, propidium monoazide

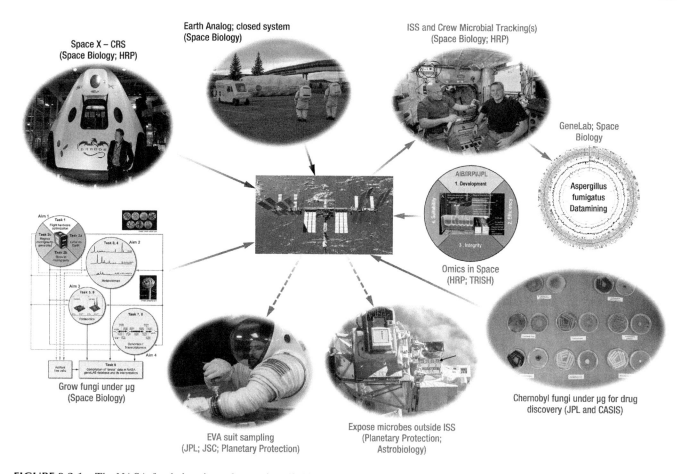

FIGURE 9.2.1 The NASA-funded projects that are interlinking International Space Station as a test bed to measure microbial community of a closed system, to utilize this unique structure to discover novel microbial byproducts, as well as to exploit the distinctive facility in characterizing microbial life in extraterrestrial conditions.

(PMA)–quantitative polymerase chain reaction (qPCR), and adenosine triphosphate (ATP) assays, respectively. Targeted amplicon sequencing was performed to reveal the microbial dynamics and community profiles of samples collected at different sites and times within the ILMAH.

Statistical analyses showed that the length of human occupation strongly affected the bacterial community profiles. The microbial diversity of the samples taken before human occupation (Day 0) was very different from those taken at three later times. Members of Proteobacteria (esp. *Oxalobacteraceae* and *Caulobacteraceae*) and Firmicutes (especially *Bacillaceae*) were most abundant before human occupation (Day 0), whereas other members of Firmicutes (*Clostridiales*) and Actinobacteria (especially *Corynebacteriaceae*) were abundant during the 30-day occupation. Samples treated with PMA (a DNA-intercalating dye for selective detection of viable microbial population) had a significantly lower microbial diversity compared with samples that were not treated with PMA.

Statistical analyses of the samples showed that the community structure changed significantly over time, particularly when comparing the bacteriomes existing before human occupation of the habitat (Day 0 sampling) with those after human occupation (Day 13, Day 20, and Day 30 samplings). Actinobacteria (mainly *Corynebacteriaceae*) and Firmicutes (mainly *Clostridiales Incertae Sedis* XI and

Staphylococcaceae) were shown to increase with the length of human occupation. These results indicate that, within a closed habitat, there is a strong link between human presence and succession of microbial diversity. Therefore, we must develop methods and countermeasure tools that can effectively maintain a safe and healthy environment within a closed habitat for future human missions to the Moon or Mars, or beyond.

9.2.4.2 Mycobiome of Inflatable Lunar/Mars Analog Habitat

Fungi can not only be hazardous to inhabitants but also deteriorate the habitat. Consequently, it is important to characterize and understand the possible changes and succession of fungal species. The results of studying these mycobiome changes in the presence of humans will facilitate the development of countermeasures that can safeguard the health of crew members living in a closed habitat.

Both state-of-the-art molecular technologies and traditional microbiological techniques were used to characterize the succession of fungi. Surface samples were gathered at various locations over the 30-day human occupation of the ILMAH to capture both the total and viable fungal populations of common environmental and opportunistic pathogenic species. The community structure and fluctuation of the mycobiome were measured at various times and locations within the ILMAH by using the internal transcribed spacer region-based iTag Illumina sequencing. Statistical analysis of the samples showed that the viable fungal community structure increased in diversity and decreased in fungal burden over the length of the human occupation. The fungal profiles of the samples collected at Day 20 were distinct from those of samples collected both before and after Day 20. Analysis indicates that the population of viable fungal families such as *Davidiellaceae*, *Teratosphaeriaceae*, *Pleosporales*, and *Pleosporaceae* increased over the course of the human occupation. On a genus level, *Epiccocum*, *Alternaria*, *Pleosporales*, *Davidiella*, and *Cryptococcus* showed increased abundance over the occupation time. The results of the study show that the length of human presence within the closed habitat affects the overall fungal diversity; therefore, it can be concluded that the closed habitat has to be properly maintained to ensure the health of crew members and to preserve the habitat from deterioration.

9.2.4.3 MICROBIOME OF CRS VEHICLE

Several crew resupply service (CRS) vehicles developed by commercial companies were analyzed for their microbiological characteristics. The microbial burden of the CRS vehicles' surfaces ranged from no counts to 10^4 CFU/m^2, with an average count being 10^3 CFU/m^2. The bacterial and fungal counts were similar, and only 3 out of 15 locations sampled yielded any counts. When bacterial population was detected in a given location, fungal presence was also noticed. Similarly, most of the locations were devoid of any cultivable microbial population. When amplicon-based Illumina sequencing as well as metagenome sequencing approaches were conducted, only 2 out of 15 samples yielded any results.

The cultivable microbial diversities associated with the inside surface of the CRS vehicles included *Acidovorax* sp., *Arthrobacter siccitolerans*, *Bacillus methylotrophicus*, *Brevibacterium frigoritolerans*, *Compostimonas suwonensis*, *Flavobacterium oceanosedimentum*, *Hymenobacter* sp., *Massilia* sp., *Massilia haematophila*, *Methylobacterium tardum*, *Microbacterium natoriense*, *Microbacterium paraoxydans*, and *Sphingomonas asaccharolytica*. Likewise, cultivable fungal diversity of the outside surface of the CRS vehicles included *Alternaria brassicae*, *Aureobasidium pullulans*, *Cryptococcus luteolus*, *Cryptococcus rajasthanensis*, *Diozegia catarinonii*, *Laurilia taxodii*, *Pithomyces chartarum*, *Rhodotorula mucilaginosa*, and *Talaromyces variabilis*.

9.2.5 MICROBIAL CHARACTERIZATION OF ISS ATMOSPHERE OBSERVED VIA HEPA FILTER PARTICULATES

9.2.5.1 MICROBIAL POPULATIONS

Although there is an ever-growing number of ways to detect microorganisms in a given sample by using molecular techniques, a quick and robust method to assess the differential viability of the microbial cells, as a function of phylogenetic lineage, has yet to be developed. A PMA treatment, coupled with downstream qPCR and pyrosequencing analyses, was used to capture the frequency, diversity, and distribution of viable microorganisms observed via HEPA filter particles and other debris collected from the crew quarters of the ISS. The cultured bacterial counts were found to be higher in the ISS samples than the cultured fungal counts. Rapid molecular analyses were used to provide an estimate of viable population, and the results showed a five-fold increase in bacterial (qPCR-PMA assay) burden when compared with the cultured bacterial population. Bacterial diversity varied with the methods used to identify the stains. The ribosomal nucleic acid (NA)-based technique used to identify the cultivable microbes resulted in only four to eight bacterial species in the ISS samples, but the amplicon-based sequencing used to estimate the viable bacterial diversity detected by the PMA-pyrosequencing method resulted in 12 to 23 bacterial taxa, with the majority being members of actinobacterial genera (*Propionibacterium* and *Corynebacterium*) and *Staphylococcus*. Sample fractions that were not treated with PMA (inclusive of both live and dead cells) resulted in greater diversity than samples that were treated with PMA, yielding 94 to 118 bacterial taxa and 41 fungal taxa. Although the deep sequencing capability of the molecular analysis expanded the knowledge of microbial diversity, the cultivation assay is also important because some of the spore-forming microorganisms were detected only by using the culture-based method, and some of these microorganism were novel (Venkateswaran et al. 2014).

9.2.5.2 COMPARATIVE MICROBIOME PROFILES OF ISS-HEPA AND EARTH-BASED CLEANROOM-HEPA FILTERS

Both traditional cultivation and state-of-the-art PMA-qPCR assays were used to estimate viable microbial populations with the samples collected from the HEPA filters of the ISS and two cleanrooms at the JPL. The amplicon-targeted (16S rRNA gene) Illumina sequencing provided information on microbial diversity and allowed for a comparison of ISS and Earth cleanroom microbiomes. Results of statistical analyses indicated that members of the phyla Actinobacteria, Firmicutes, and Proteobacteria were dominant in the samples examined, but their abundance varied. Actinobacteria were dominant in the ISS samples, whereas Proteobacteria were least common. However, Proteobacteria were dominant in the JPL cleanroom samples. The viable bacterial populations seen by PMA treatment were greatly decreased, but the PMA treatment did not appear to affect the viable bacterial diversity associated with each sampling site. The results of this study strongly suggest that specific microorganisms associated with human skin make a substantial contribution to the microbiome on the ISS, but it is not the case for the cleanrooms at the JPL. For example, *Corynebacterium* and *Propionibacterium* (Actinobacteria) species are dominant on the ISS in terms of viable and total bacterial community composition, but *Staphylococcus* (Firmicutes) species are not (Checinska et al. 2015).

9.2.5.3 Comparative Metagenome Profiles of ISS-HEPA and Earth-Based Cleanroom-HEPA Filters

Previous studies have reported on the microbial composition of particulates from the ISS HEPA filters (Checinska et al. 2015), but the functional genomics have been characterized only recently (Be et al. 2017). The functional genomics are important because this approach will help to identify constituents that may potentially affect human health and operational mission success. This study analyzes the metagenome profiles at both species- and gene-level resolutions. The ISS-HEPA filter and dust samples were analyzed and compared with samples collected in an Earth-based cleanroom. In addition, dominant, virulent, and novel microorganisms were characterized using metagenome mining. The whole-genome sequences (WGS) of select cultivable strains isolated from these samples were extracted from the metagenome and compared. Results showed that *Corynebacterium ihumii* GD7 dominated the ISS samples at the species level and that ISS samples had lower overall microbial diversity than the cleanroom samples. When the study looked for microbial genes that affect human health, such as antimicrobial resistance and virulence genes, it was found that the ISS samples had a larger number of relevant gene categories when compared with the cleanroom samples. Strain-level cross-sample comparisons were made for *Corynebacterium*, *Bacillus*, and *Aspergillus* to look for possible distinctions in the dominant strain between samples. Species-level analyses of the ISS and cleanroom samples showed distinct differences, meaning that the cleanroom population is not necessarily representative of the space habitation environment found on the ISS (Be et al. 2017). The observation and study of the overall population of viable microorganisms and the functional diversity that is innate to the unique environment of an enclosed habitat are critical to the evaluation of the conditions needed to ensure the long-term health of its human occupants.

9.2.5.4 Round Spore-Forming Bacteria Isolated from Spacecraft Assembly and ISS Environments

Bacterial spores are extremely recalcitrant and notoriously difficult to eradicate from contaminated surfaces. The spore's resistance derives from (1) existing in a metabolically dormant state, and (2) a series of protective structures that encase the interior-most compartment, the core, which houses the spore chromosome. The spore has a very different structure than that of a growing cell, with a number of unique features and constituents found in no other cell type (Figure 9.2.2). Organized as a series of concentric shells, starting from the outside and moving inward, spore layers include the coat, outer membrane, cortex, germ cell wall, inner membrane, and core. In some species, an additional layer, referred to as the exosporium, envelops the entire spore (Figure 9.2.2-3 and 9.2.2-4) (Driks 2002b; La Duc et al. 2004; Lai et al. 2003; Probst et al. 2009; Redmond et al. 2004; Vaishampayan et al. 2008; Venkateswaran et al. 2003; Waller et al. 2004).

The exosporium is composed largely of proteins and glycoproteins (Daubenspeck et al. 2004; Lai et al. 2003; Redmond et al. 2004; Todd et al. 2003), and while the function of these proteins and of the exosporium is unclear, preliminary studies suggest a role in protecting the spore from ultraviolet (UV) and other environmental stresses (La Duc et al. 2004; Newcombe et al. 2005; Venkateswaran et al. 2003).

In addition, several novel bacteria have been isolated from spacecraft-associated surfaces, whose spores are often observed to have what appears to be an "extraneous layer (EL; Figure 9.2.2)" (Probst et al. 2009; Venkateswaran et al. 2003). In the case of *Bacillus nealsonii* and *Bacillus horneckiae*, scanning electron microscopy reveals a fibrous material that appears to connect spores together (Figure 9.2.2-5 and 9.2.2-7). However, the exact structure and molecular composition of these matrices are not known. Since the EL appears to be an abundant material that encases the spores within it, it is likely that, at a minimum, the EL also contributes to spore resistance by acting as a passive shield.

While the spore coat and exosporium/extraneous layer (E/EL) are known, or likely, to be important to spore rigidity and resistance, their specific roles are quite dissimilar. The spore coat is well studied in *Bacillus subtilis* (a species that lacks the exosporium) (Driks 1999, 2002a, 2002b; Kuwana et al. 2002; Lai et al. 2003) and has been shown to be important in resistance to chemicals and exogenous lytic enzymes that can degrade the spore cortex. *Bacillus pumilus* SAFR-032 exhibits greatly elevated resistance to UV radiation in comparison with *B. subtilis* (Driks 1999; Nicholson et al. 2000; Setlow et al. 2000), and both lack any semblance of an E/EL. This type of resistance is almost certainly not due solely to the spore core (Driks and Setlow 2000). It is highly likely that *B. pumilus* SAFR-032 encodes coat proteins that have specialized roles in resistance and are absent in *B. subtilis*. The roles of the E/EL in resistance remain unclear and are the focus of significant research. Nonetheless, as already argued, it is highly likely that these structures provide, at a minimum, passive resistance to the spore. The spore core is the analogue of the growing cell's protoplast, for it is here that most spore enzymes, ribosomes, and NAs are housed. Several features of the core protect essential molecules during spore dormancy. First, core water content is low, a major factor in resistance to wet heat (Gerhardt and Marquis 1989). Second, abundant dipicolinic acid in the core significantly enhances survival in the presence of wet heat and/or UV radiation (Slieman and Nicholson 2001). Third, small, acid-soluble proteins bind to and stabilize the spore DNA (Driks 2002a; Setlow 1992, 1995). Since the core composition of a spore does not appear to vary greatly among species, while coat and E/EL composition does (Driks 1999; Driks and Setlow 2000; Giorno et al. 2007), the most likely reason that some species (such as *B. pumilis* SAFR-032 and *Bacillus odysseyi* ATCC PTA-4993[T]) survive planetary protection-relevant simulated space conditions is the protective effect of the coat, the E/EL, or both. The coat and exosporium are composed predominantly made of protein (Driks 2002a; Kornberg et al. 1968; Strange and Dark 1956; Sylvestre et al. 2002). Therefore, the E/EL team will focus its analysis on the protein components of the coat and the E/EL.

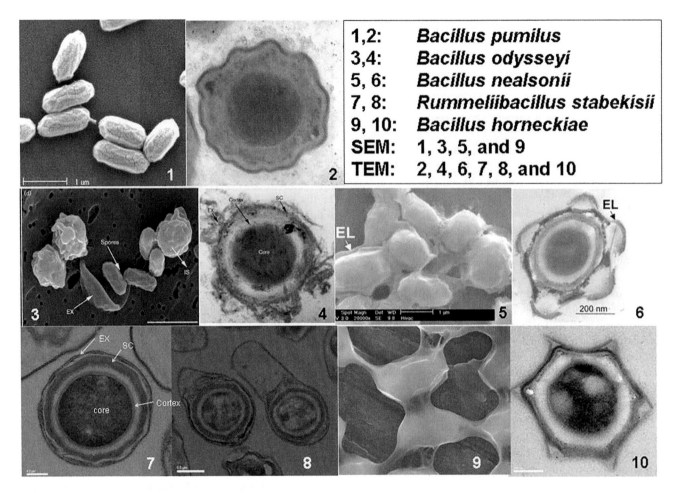

FIGURE 9.2.2 Electron micrographs of exosporium/extraneous layer (E/EL)-bearing spores. 1, 2: *Bacillus pumilus* (no-E/EL); 3: *Bacillus odysseyi* (EX+); 4: *Rummeliibacillus stabekisii* (EX+); 5, 6: *Bacillus nealsonii* (EL+); 7, 8: *Bacillus horneckiae* (EL+). IS: intact spores, SC: spore coat, EX: exosporium (3 and 4), and EL: extraneous layer (5–8).

Among the novel spore-forming bacteria associated with spacecraft assembly and ISS environmental surfaces described, the highly resistant to Mars UV conditions *B. pumilus* SAFR-032 strain was isolated from the JPL spacecraft assembly facility (Newcombe et al. 2005). Similarly, other round spore-forming bacterial species were isolated from surfaces of the Mars Odyssey Orbiter (*B. odysseyi* ATCC PTA-4993[T]; Figure 9.2.2-2 and 9.2.2-3), from Phoenix spacecraft (*Rummeliibacillus stabekisii* NRRL B-51320[T] [Figure 9.2.2-4] [Vaishampayan et al. 2008] and *B. horneckiae* 1P10SC[T] [Figure 9.2.2-7 and 9.2.2-8] [Probst et al. 2009]), and from the natural particulate fall-out of the JPL spacecraft assembly facility (*B. nealsonii* ATCC BAA-519[T]; Figure 9.2.2-5 and 9.2.2-6) (La Duc et al. 2004; Venkateswaran et al. 2003).

Similar to spacecraft assembly facilities, ISS environmental samples also yielded novel round spore-forming bacterial species (Figure 9.2.3) described as *Solibacillus kalamii*. The 16S rRNA gene sequence analyses of this novel species showed highest similarity to several *Solibacillus* species, but DNA–DNA hybridization analysis revealed that *S. kalamii* had DNA-relatedness values in the range of 41%–47%. Based on the phylogenetic analysis, strain ISSFR-015[T]

belongs to the genus *Solibacillus*. The polyphasic taxonomic data, including low DNA–DNA hybridization values, and the chemotaxonomic analysis confirmed that strain ISSFR-015[T] represents a novel species, for which the name *Solibacillus kalamii* sp. nov. is given (Checinska Sielaff et al. 2017). Furthermore, the whole genomes of several of these round spore-forming bacteria were reported (Gioia et al. 2007; Seuylemezian et al. 2017, 2018).

9.2.5.5 A NOVEL CLADE OF *BACILLUS CEREUS SENSU LATO* GROUP BELONGS TO *BACILLUS ANTHRACIS* ISOLATED FROM ISS ENVIRONMENTS

The microbial characterization of the ISS samples isolated 11 *Bacillus* strains (two from the Kibo Japanese experimental module, four from the U.S. segment, and five from the Russian module), and the whole genomes of these strains were sequenced (Venkateswaran et al. 2017a). A comparative analysis of the 16S rRNA gene sequence of these strains showed the highest similarity (>99%) to the *Bacillus anthracis–B. cereus–B. thuringiensis* group. The conventional fatty acid composition, polar lipid profile, isoprenoid quinones type, and

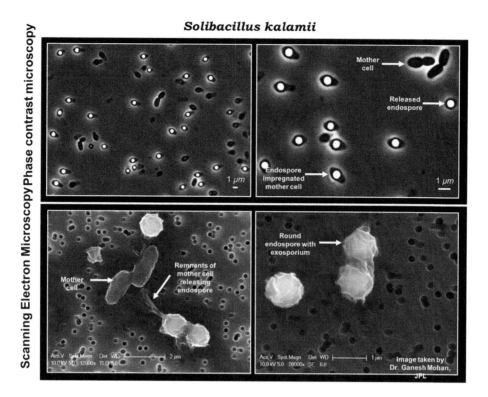

FIGURE 9.2.3 The scanning electron microscopy and phase-contrast microscopy revealed the presence of endospores, round endospores with exosporium, and mother cells of *Solibacillus kalamii*.

peptidoglycan type of the cell wall were similar to those in the *B. cereus* sensu lato group. The multi-locus sequence typing placed these strains far from any described members of the *B. cereus* sensu lato group. On the basis of lacking genes encoding protective antigen (*pag;* pXO1 marker), capsular antigen (*cap;* pXO2 marker), motile cells, and lysing sheep blood (enterotoxin), the ISS isolates were determined as not *B. anthracis*. The digital DNA–DNA hybridization analyses showed that ISS isolates are closer to *B. anthracis* (86%–88%) than to *B. cereus* (44%–47%) and *B. thuringiensis* (44%–46%). However, all isolates contained the *plcR* ancestral "C" allele and lacked significant hits to pXO1 and pXO2 plasmids and toxins specific to *B. anthracis*. Similarly, the genome analysis of these ISS isolates showed no *B. thuringiensis* biomarker, *cry* protein. The DNA G + C content of the strain ISSFR-003ᵀ was 35.4 mol%. Based on phylogenetic analysis, these strains belong to the genus *Bacillus*. However, the collective phenotypic traits and genomic evidence were the reasons to exclude the ISS isolates from *B. anthracis*. Nevertheless, multilocus sequence typing and whole-genome SNP analyses placed these isolates in a clade that is distinct from previously described members of the *B. cereus sensu lato* group but closely related to *B. anthracis* (Venkateswaran et al. 2017b). The spore-forming *B. cereus sensu lato* group consists of pathogenic (*B. anthracis*), food poisoning (*B. cereus*), and biotechnologically useful (*B. thuringiensis*) microorganisms; their presence in a closed system such as the ISS might be a concern for the health of crew members. The pathogenicity and virulence properties of these potential pathogens should be characterized in detail in

order to develop suitable countermeasures for long-term future missions and to provide a better understanding of microorganisms associated with space missions.

9.2.5.6 OMICS CHARACTERIZATION OF VIRULENT ISS *ASPERGILLUS FUMIGATUS* STRAINS

In addition to the bacterial characterization, examining the traits and diversity of fungal isolates would increase the understanding of how fungi may adapt to microgravity environments and how this adaptation may affect their interactions with humans in a closed habitat. Two isolates, ISSFT-021 and IF1SW-F4, of *Aspergillus fumigatus* collected from the ISS were characterized, and the results were compared with the experimentally established clinical isolates Af293 and CEA10 (Knox et al. 2016). The WGS of ISSFT-021 and IF1SW-F4 showed 54,960 and 52,129 single-nucleotide polymorphisms, respectively, compared with Af293, which is consistent with observed genetic heterogeneity among sequenced *A. fumigatus* isolates from diverse clinical and environmental sources. In order to determine if the strain collected from the ISS adapted to its unique environment, the in vitro growth characteristics, secondary metabolite (SM) production, and susceptibility to chemical stresses were compared with those of the clinical strain, but the comparison did not show any exceptional differences between the ISS and clinical strains. However, a virulence assessment in a neutrophil-deficient larval zebrafish model of invasive aspergillosis revealed that both ISSFT-021 and IF1SW-F4 were significantly more lethal than Af293 and CEA10.

The proteomic profiles of the *A. fumigatus* strains were characterized to understand differentially expressed proteins that could be related to the enhanced virulence. Differentially expressed proteins in ISS-isolated strains ISSFT-021 and IF1SW-F4 and clinical isolates Af293 and CEA10 were studied once the total protein was extracted from each strain. Extracted proteins were labeled using tandem mass tag (TMT) and analyzed via LC-MS/MS, and then, a spectra analysis was performed using the Proteome Discoverer with the *A. fumigatus* Af293 database. The abundance ratios of all identified proteins were normalized to the Af293 clinical isolate, which enabled the detection of 553, 464, and 626 upregulated and 314, 289, and 317 downregulated proteins in CEA10, ISSFT-021, and IF1SW-F4 strains, respectively (Blachowicz et al. submitted). To reveal the differentially expressed proteins in space strains only, the Af293 standardized proteins of ISSFT-021 and IF1SW-F4 were further normalized to CEA10, which facilitated the detection of 60 upregulated and 32 downregulated proteins in both space strains simultaneously (FC > |2|). Analysis of the upregulated proteins showed that 14 proteins were associated with carbohydrate metabolic processes, eight with stress responses, five with secondary metabolism and toxins biosynthesis, and two with pathogenesis, whereas five, three, one, and zero proteins were downregulated in these categories, respectively. Only the space strains showed an increased abundance of proteins associated with cellular amino acid metabolic process [6], lipid cellular homeostasis [3], metabolic processes [2], pathogenesis [2], and translation [2]. A search of the FungiDB showed that significantly over-represented upregulated biological processes included secondary metabolic processes (40% of all upregulated proteins), carbohydrate metabolic processes (23%), and response to chemical (~15%), whereas significantly over-represented downregulated processes included carbohydrate metabolic processes (15%), response to heat (6%), and mRNA metabolic processes (6%).

The metabolomic profiles were characterized to understand the production of SMs associated with ISS-isolated *A. fumigatus* strains and were compared with clinical isolates (Knox et al. 2016). An inspection of the SM profiles of ISSFT-021, IF1SW-F4, CEA10, and Af293 showed distinctive chemical signatures for each strain under the condition tested. When SM profiles were compared with those of Af293, an increase in fumigaclavine A production was observed in IF1SW-F4 (p = 0.0001) but not in ISSFT-021, whereas a significant decrease in fumigaclavine C production was noticed in both strains. Fumiquinazoline production increased in ISSFT-021 but not in IF1SW-F4 compared with the two controls. Pyripyropene A production increased in both ISSFT-021 and IF1SW-F4. Observed SM yields for CEA10 were lower than that for any other strain, with a decrease in production of all but two compounds (pyripyropene A and fumagillin) observed. Nevertheless, as production of SM in media does not necessarily replicate the pattern of production of SMs during infections, the potential in vivo SM profiles of the ISS strains remain unclear.

9.2.6 MICROBIAL CHARACTERIZATION OF ISS ENVIRONMENTAL SURFACES (CHECINSKA ET AL. 2018 UNPUBLISHED)

9.2.6.1 MICROBIAL BURDEN ESTIMATION AND CULTIVABLE MICROBIAL DIVERSITY

Cultivable bacterial population. Both traditional culture counts and molecular methods were used to analyze bacteria and fungi samples collected from eight different locations of ISS surfaces. The first two flight sampling sessions (1 and 2) resulted in similar bacterial counts that ranged from 10^4 to 10^9 CFU/m^2. However, the third flight sampling session resulted in at least 2-logs greater bacterial populations. Three phyla (Actinobacteria, Firmicutes and Proteobacteria) were identified from the 134 bacterial isolates characterized for their phylogenetic affiliations. *Staphylococcus*, *Pantoea*, and *Bacillus* were the most dominant genera at the genus level (Hendrickson et al. 2017).

Cultivable fungal population. The average fungal burden was 10^6 to 10^7 CFU/m^2. No difference was noticed in the density of the fungal population of flight samples 1 and 3, but the density of the fungal population of flight sample 2 was at least 1 log less than samples 1 and 3. Fungal population ranged from 10^5 to 10^7 CFU/m^2, and no statistically significant spatial variation in their abundances was noticed. Five different genera were observed among the 64 fungal strains that were identified using molecular phylogeny targeting internal transcribed spacer (ITS) sequences. *Rhodotorula mucilaginosa* and *Penicillium chrysogenum* were dominant among the fungal species. *Rhodotorula mucilaginosa* was found in all location samples from flights 1 and 2, except for location 2 from flight 1. However, *R. mucilaginosa* and *P. chrysogenum* were not found in any samples from flight 3. Furthermore, because it has been reported to be a causative agent for human pathogenicity, it should be noted that *Cryptococcus* (yeast) was also isolated (Furman-Kuklinska et al. 2009; Rimek et al. 2004). Opportunistic human pathogen (*Aspergillus fumigatus*) and plant pathogen (*Fusarium oxysporum*) were also isolated in low numbers. *Aspergillus fumigatus* is a saprophytic, filamentous fungus that is commonly found both outdoors and indoors, and it is able to adapt to various environmental conditions and form airborne conidia that are the inoculum for a variety of diseases in immunocompromised hosts (Hendrickson et al. 2017).

Viable microbial population as measured by PMA-qPCR. The total (dead and alive) and viable bacterial and fungal gene copy numbers were measured using PMA-treated (viable) and non-PMA-treated samples. Flight 1 PMA-treated samples (viable) compared with non-treated samples (total) resulted in an average of an approximately 50% reduction in bacterial burden, with flight 2 samples resulting in an approximately 75% reduction, but fungal numbers did not change drastically. Both PMA-untreated samples (total) and PMA-treated samples (viable) demonstrated the same trends in microbial load at the various sampling locations. The PMA-qPCR assay was used to estimate the total viable bacterial burden, but there were no statistically significant differences in microbial load from the three flight samples. Flight samples 1 and 2 resulted in similar copy numbers of bacteria and fungi, but flight 3 samples resulted in higher viable bacterial (~60%) and lower fungal (~2-logs less) abundance. It was found that approximately 46% of the viable population were cultivated, when a comparison was made between the cultivable bacterial density and the viable bacterial population detected by the PMA-qPCR (Hendrickson et al. 2017).

9.2.6.2 Opportunistic Pathogens of ISS Microbial Population and Their Antibiotic Resistance

Seven (*Staphylococcus aureus*, *S. haemolyticus*, *S. hominis*, *Acinetobacter pittii*, *Enterobacter* sp., *Klebsiella quasipneumoniae*, and *Pantoea conspicua*) of the 30 different bacterial species cultivated are classified as biosafety level 2 (BSL-2) bacteria. The cupola area samples from the ISS resulted in the isolation of multiple strains of *A. pitti*, with one strain tested using various antimicrobial agents and showing resistance to cefazolin, cefoxitin, oxacillin, penicillin, and rifampicin. The space toilet samples from the ISS resulted in several *Enterobacter* isolates, which were found to be resistant to multiple drugs. The *Enterobacter* species belongs to the coliform group of bacteria; these are reported to be pathogenic, causing opportunistic infections in immunocompromised hosts, and are commonly associated with urinary and respiratory tract infections (Mezzatesta et al. 2012). The cupola, space toilet, and exercise platform samples from the ISS resulted in *Klebsiella* isolates. *Klebsiella quasipneumoniae* is a blood-borne opportunistic pathogen (Brisse et al. 2014). When tested with various antibiotics, *K. quasipneumoniae* strain was found to be resistance to all of them, except for tobramycin. *Pantoea conspicua* was only found in one sampling location on the ISS during two different flights, but it is the second most prevalent species. When tested with various antibiotics, *P. conspicua* was found to be resistant to erythromycin, oxacillin, penicillin, and rifampin. *Pantoea conspicua* was originally isolated from human blood (Brady et al. 2010).

Staphylococcus isolates were the most prevalent cultivable bacteria found from the ISS surface samples, with *Staphylococcus aureus* being the most abundant in all ISS surface samples. Although this species is a common human commensal (Skawinska et al. 2009), it can lead to various types of minor skin infections, bacteremia, or scalded skin syndrome,

particularly for those who are immunocompromised (Archer 1998). The Vitek 2 system (BioMerieux, France) was used for additional testing of the *S. aureus* isolates, and it was found that they were not resistant to methicillin (i.e., not Methicillin-resistant *Staphylococcus aureus* [MRSA] strains). However, most of the *S. aureus* isolates were resistant to penicillin, and some of them were resistant to erythromycin, gentamycin, and tobramycin. A few isolates became resistant to rifampicin during the study. *Staphylococcus haemolyticus* and *S. hominis* belong to coagulase-negative staphylococci (Barros et al. 2012; Mendoza-Olazarán et al. 2013). One *S. hominis* strain was tested and found to be resistant to penicillin and erythromycin, but the *S. haemolyticus* strain was found not to be resistant to these antibiotics. It has been reported that all three *Staphylococcus* species become methicillin-resistant by acquiring the Staphylococcal Cassette Chromosome *mec* (SCC*mec*) (Katayama et al. 2000), but the methicillin-resistant phenotype was not observed in strains that were isolated from samples collected on ISS surfaces (Urbaniak et al. 2018).

The isolation of *A. fumigatus* is noteworthy among the fungi, because it can cause pathologies ranging from allergic asthma to invasive aspergillosis. Several pathogenic characteristics of the ISS isolates were compared with those of two experimentally established clinical isolates (Nierman et al. 2005; Rizzetto et al. 2013). Virulence assessment in a larval zebrafish model of invasive aspergillosis showed that both strains from the ISS were significantly more lethal than the two clinical isolates (Knox et al. 2016). The omics of strains from the ISS may be able to expose the molecular mechanisms behind the increased virulence. If it turns out that the increased virulence is a result of exposure to microgravity, then NASA should consider creating countermeasures that would protect astronaut health, because it has been reported that the immune systems of astronauts are compromised under microgravity (Crucian et al. 2014). *Cryptococcus laurentii* was previously considered to be a saprophyte and non-pathogenic to humans, but infections caused by cryptococci have been increasingly recognized. It seems to be an important pathogen in cases where immunity has been diminished (Furman-Kuklinska et al. 2009). Furthermore, a case of meningitis caused by *Cryptococcus adeliensis* has been reported in a patient who had acute myeloid leukemia and was undergoing allogeneic peripheral blood stem cell transplantation (Rimek et al. 2004).

9.2.6.3 ISS Microbiome Profiles as Seen via Targeted Amplicon Sequencing

Up until now, traditional culture techniques have been used for obligatory microbial monitoring and for observational studies of spacecraft and space stations. However, this study also makes use of molecular techniques to analyze microbial communities sampled from various defined ISS locations over three flight missions (Checinska et al. 2018 in submission). The amplicon sequencing was able to detect more bacterial phyla than the culture-based analyses was able to detect, but the number of identified fungal phyla was similar, regardless of the technique used. In general, amplicon sequencing showed temporal distribution across the three flights but not spatial distribution for the bacteriome, whereas the mycobiome did not show any difference.

The dominant organisms found from the samples of the ISS environmental surfaces were those that are associated with humans, with some of them considered to be opportunistic pathogens on Earth. It is unknown if these organisms could cause disease in astronauts on the ISS, since that would depend on the health of the astronaut and how these organisms function while in the unique environment of the ISS. Nevertheless, the detection of these potential disease-causing organisms emphasizes the value of further genomic and transcriptomic studies to analyze how these ISS microbes function in the space environment and how they may affect the health of astronauts. Because no differences were detected in the composition and richness of the community between PMA-treated (intact cells) and non-PMA-treated (both dead and alive microbes) samples, it implies that the DNA analyzed from these possible opportunistic pathogens found on the ISS is intact/viable and these are not dead organisms. It should be noted that approximately 46% of intact/viable bacteria and 40% of intact/viable fungi could be cultured with the culture media used during this study. These percentages are high when compared with the only 1%–10% of intact/viable microorganisms that can be cultured using samples collected from spacecraft assembly cleanrooms on Earth (Ghosh et al. 2010). One possible reason could be that the ISS is not deprived of nutrients (unlike spacecraft assembly cleanrooms), hermetically sealed, and relatively young (~25 years) in terms of human (maximum six astronauts at a given time) and cargo transport (~4–6 times per year), which are the only methods by which these organisms are transferred.

Many of the organisms identified from the ISS samples form biofilms, such as members of the genus *Acinetobacter* (*Moraxellaceae*), *Sphingomonas*, *Bacillus* (*Bacillaceae*), *Burkholderiales*, *Corynebacterium* (*Corynebacteriaceae*), and *Klebsiella* (*Enterobacteriaceae*) (Biteen et al. 2016), as well as the fungi *Penicillium, Aspergillus, Cryptococcus*, and *Rhodotorula* (Nunes et al. 2013). These biofilms are known to promote resistance to antibiotics and therefore could be problematic for the health of astronauts if they become infected. Biofilm formation on the ISS itself could also affect the stability of the infrastructure by causing mechanical blockages, reducing heat transfer efficiency, and inducing microbial-influenced corrosion (Beale et al. 2014). Some of the microorganisms identified from the ISS samples that have been associated with microbial-induced corrosion on Earth include *Methylobacterium, Sphingomonas, Bacillus, Penicillium*, and *Aspergillus* (Dai et al. 2016; Pavissich et al. 2010; Rajasekar and Ting 2010; Yang et al. 1998), but what function they play in corrosion aboard the ISS has not been determined. Understanding the potential ability of these microorganisms to form biofilms and the amount of actual biofilm formation on a spacecraft during extended space missions is crucial to

ensuring the structural stability of the crew vehicle when regular indoor maintenance cannot be performed easily.

Members of the family *Staphylococcaceae* and *Enterobacteriaceae* were the most dominant organisms found on samples of the ISS surfaces from the U.S. module, which is similar to the results published for samples from the Japanese module of the ISS (Ichijo et al. 2016); these organisms were found in almost every sample, regardless of the method of detection. Both of these organisms are associated with humans, with *Staphylococcaceae* being commonly found on the skin and in the nasal passage and *Enterobacteriaceae* being commonly linked with the gastrointestinal tract. These two taxa are also commonly found in fitness centers (Mukherjee et al. 2014), office buildings (Rintala et al. 2008), and hospitals (Lax et al. 2017), implying that the microbiome of the ISS is similar to that of other built environments on Earth in that it is influenced by the presence of humans (Mayer et al. 2016). Additional studies examining how long these organisms can survive on the surfaces of the ISS and how easily they can be transferred from one person to the next in space can help to develop countermeasures to reduce the spread of infections among astronauts during simultaneous or even separate flight missions.

Methylobacteriaceae/Methylobacterium was also dominant across samples from various surfaces on the ISS and has been found in NASA spacecraft assembly cleanrooms (Moissl et al. 2007), hospital intensive care units (ICUs) (Poza et al. 2012), and the MARS500 habitat (Schwendner et al. 2017). It is a robust organism that can withstand harsh conditions, such as ionizing radiation and strong cleaning detergents. *Moraxellaceae*, another abundant organism found on samples from the ISS, is also resilient in harsh conditions, and it has been found in higher relative abundances in spacecraft assembly cleanrooms (Bashir et al. 2016), areas of the home that are exposed to many chemicals (i.e., washing machine) (Savage et al. 2016), and deep-sea sediment of inactive hydrothermal vents (Zhang et al. 2016).

9.2.6.4 ISS Metagenome Profiles to Elucidate Functional Diversity

Analysis of metagenomic characteristics of ISS samples targeted both total (dead and live cells) and viable (after PMA treatment) microorganisms. The metagenomic analysis indicates that the AMR footprint mirrors a nosocomial environment. Three pathways that govern the ISS antimicrobial resistance scenario are beta-lactam resistance, cationic antimicrobial peptide (CAMP) resistance, and vancomycin resistance. The study indicates a gradual increase in the AMR profile from the first to the third flight sampling. In-depth metabolic pathway analysis was attempted to examine the parts played by the multiple genes involved in the AMR functions. The shotgun metagenome sequence analysis demonstrated that beta-lactam resistance was prevalent due to a penicillin-binding protein and beta-lactamase enzyme, whereas a two-component system, *omp*R histidine kinase, was key to CAMP resistance. Vancomycin resistance was primarily due to the presence of alanine racemase and support by other factors. Microbial genes were annotated using a curated Kyoto Encyclopedia of Genes and Genomes (KEGG) database. Read-based analysis indicates that the phage, prophage, transposable element, and plasmid gradually increase from the first to the third flight sampling. The findings of the metagenomics study will help NASA understand the functional characteristics associated with the ISS environment, thereby enabling the development of suitable countermeasures to eradicate problematic microorganisms and ensure the health of the crew.

9.2.6.5 FUTURE STUDIES: IN SITU MEASUREMENT OF "OMICS IN SPACE"

The omics in space (OIS) project that is in progress will develop instrumentations for extracting NAs and sequencing methodologies for inflight detection and measurement of several biomolecules related to physiological and immunological effects related to spaceflight (e.g., aging, crew health, elevated antimicrobial resistance, and virulence).

Since Sanger's founding discovery of dideoxynucleotide sequencing, researchers have successfully developed automated instruments to rapidly sequence the DNA of biological materials. The generation of large data sets was problematic throughout the 1990s, but next-generation sequencing (NGS) platforms emerged in 2005 to solve this issue and to lower cost barriers. The interplay and optimization of chemistry, engineering, software, and molecular biology with NGS development enabled the examination of previously unresolvable biological questions. The ability to generate genome-scale data sets, thoroughly reviewed elsewhere (Mardis 2013), is transforming the nature of biological inquiries and investigations. The NGS biomolecule sequencer (BSeq), using the MinION platform (Ashton et al. 2015; Jain et al. 2015) and PCR instruments (SmartCyler and RAZOR), was adopted by NASA last year; it has successfully sequenced libraries that had been prepared on Earth and shipped to the ISS (Castro-Wallace et al. 2016; McIntyre et al. 2016). A limitation in this study was that samples had to be prepared on Earth before sequencing could be performed on the ISS, as there are no technologies onboard the ISS to process samples, extract biomolecules, and prepare libraries.

The Translational Research Institute for Space Health (TRISH) administered by NASA selected the OIS project to address these limitations by developing an automated Sample-Processing Instrumentation (SPI) for NA extraction that is streamlined, minimizing crew time, reducing the amount of contamination between samples, and producing consistent results. Factors such as the quantity and physical characteristics of the NAs and the desired application(s) will be addressed by preparing high-quality sequencing libraries (Head et al. 2014). The OIS project will generate microbiome data (targeted for bacteria, fungi, and viruses) to address microbial composition, metagenome (crew and environments) for functional characteristics, and miRNA for immunological

function. Moreover, the OIS project will examine epigenetic data to better understand radiation-induced damage, cellular disruption, and aging of astronauts.

The mission of the TRISH is to lead a national effort to apply innovative terrestrial research to human risk-mitigation strategies for long-duration space exploration missions. The OIS project will target biomolecules that may be responsible for physiological and immunological conditions that affect humans during spaceflight missions. The OIS project is a consortium of investigators from various NASA centers, industries, and academia to solve microgravity-related sample-processing issues. This is a 4-year TRISH Core Research Project (TCore), consisting of an interdisciplinary team of omics researchers, experts in biomolecule assay development, and clinicians. Three science principal investigators (PIs) will lead the team and ensure that everyone works in a coordinated fashion.

Advances in molecular biology, including the MinION system recently demonstrated on the ISS through the BSeq project, will enable inflight detection of biomolecules of physiological and psychological stressors (Castro-Wallace et al. 2016). Molecular technologies can be used to identify candidate biomolecules and biomarkers for a variety of medical conditions and could be used to investigate/explore spaceflight-associated physiological changes or to measure the efficacy of countermeasures. The objective of the proposed OIS project is to develop omics capabilities suitable for use in the spaceflight environment. The OIS project will include ground-based state-of-the art gold standard methods and measurements of OIS to reveal changes that are relevant to the mission.

The OIS project will generate vast amount of data by utilizing molecular methods in space to provide TRISH with an opportunity to understand the omics changes under microgravity conditions and microbial communities of both the crew and the habitat. The state-of-the-art molecular methods will allow researchers to not only describe the changes in microbial communities of the ISS but also distinguish whether these biological signatures would negatively affect crew health and engineering systems. Focused and targeted molecular approaches are likely to reveal a subset of novel and medically important microbes posing particular threats to crew health. Archiving of omics data and their availability for extensive datamining may feed into functional genomics activities. It is expected that a range of metadata will allow NASA to correlate microbial community information with temporal or environmental changes.

The development of an all-encompassing, integrated OIS database will enable various phylogenetic- and pathogenic-based strategies to screen for, and identify, specific subsets of microorganisms (e.g., dominating viral and microbial pathogens, as well as those that bear resistance traits relevant to antibiotics). This data set will (1) create a capability for NASA to compare fluctuating microbial communities to "baseline" standards; (2) enable more accurate assessments of crew health associated with mission planning; (3) allow evidence-based development of future bio-load management policies, particularly for long-duration missions; and (4) capitalize on parallel research from non-NASA institutions efforts (Hoisington et al. 2014; Ma et al. 2014).

9.2.7 UTILIZATION OF ISS AND INFLUENCE OF MICROGRAVITY IN THE PRODUCTION OF NOVEL DRUG DISCOVERY

Fungi are a great source of natural products that are of use to humankind in the pharmaceutical industry, agriculture, and beyond. Antibiotics (penicillin and cephalosporins), immunosuppressants (cyclosporine A), statins (lovastatin), and citric acid are only a few examples of molecules used by humans. The advance in genome sequencing and its increased availability revealed that the number of known fungal SMs is significantly lower than the predicted gene clusters. These orphan clusters are either dormant under laboratory conditions or expressed at extremely low levels. Artificial induction of silent clusters, such as co-culturing with bacteria or fungi, gene deletion, and heterologous expression in different host, is a subject of study for many scientific groups around the world.

The Natural Products under Microgravity (NPμG) project seeks to investigate whether unique conditions of the ISS, such as microgravity and enhanced radiation, would trigger changes in the molecular, biochemical, and genetic suit of selected filamentous fungi, resulting in the production of novel SM. This approach would lead to *novel drug discovery that potentially benefits astronauts' health*. In the ISS-NPμG project, fungal strains isolated from the Chernobyl Nuclear Power Plant (ChNPP) accident were also screened for the secretion of natural products that could be beneficial for biomedical and agricultural applications. Particular radiation-tolerant microorganisms were selected, because they were known to produce valuable natural products, their genomic sequences contain SM pathways, or they display radiotropism. Seven fungi selected for the flight experiment were isolated after the explosion of Reactor 4 of the ChNPP in 1986, and one was isolated from the ISS. It has been hypothesized that growing fungi under potentially stressful microgravity conditions would induce the production of novel SMs, because orphan clusters would be activated.

Total genomic DNA was extracted from each strain by using a PowerSoil® DNA Isolation Kit (MoBio), following the manufacturer's instructions. The whole-genome sequences of these eight fungal strains were obtained by shotgun sequencing performed on an Illumina HiSeq2500 platform with a paired-end module. The draft genomes and their preliminary genomic characterizations were already published (Singh et al. 2017). The single-nucleotide polymorphism analyses and comparative genomics of the space-grown and ground-grown (control) fungal strains are in progress.

The SM analysis of the ISS and ground-grown strains was carried out immediately after sample retrieval from space. Three plugs taken in triplicate from each strain were extracted with methanol and a methanol:dichloromethane 1:1 mixture (Sanchez and Wang 2012). The extracted samples were analyzed using liquid chromatography-mass spectrometry (LC/MS) (Sanchez and Wang 2012). The resulting spectra revealed significant differences in SM production in *Cladosporium sphaerospermum* IMV 00045 (radiotrophic) and *Aureobasidium pullulans* IMV 00882 that were space-exposed as compared with the ground control. The SMs produced by each strain are being characterized by large-scale extraction, high-pressure liquid chromatography (HPLC) purification, and nuclear magnetic resonance (NMR) spectroscopy. The compound produced by *A. pullulans* IMV 00882 was identified as BK223-C, and two compounds produced by *Cladosporium cladosporioides* IMV 00236 (radiotrophic) were identified as viriditoxin and its derivative (Liu et al. 2016). These compounds may be novel, and further characterizations are in progress.

9.2.8 EXPLOITING OUTSIDE PLATFORM OF ISS IN MEASURING SURVIVAL OF MICROORGANISMS UNDER SPACE CONDITIONS

To avoid forward contamination from terrestrial microorganisms and uphold the scientific integrity of future missions to find life on other planets, it is crucial to characterize and attempt to eradicate these microorganisms from exploratory spacecraft and landing vehicles. Microorganisms that hitchhike on a spacecraft can end up contaminating other celestial bodies, making it difficult for scientists to determine whether a life-form existed on another planet or was introduced by the arrival of explorers. Consequently, it is crucial to the success of future exploration to know what types of terrestrial microorganisms can survive on a spacecraft or landing vehicle.

Presently, spacecraft landing on planets where life might exist (e.g., Mars) must meet requirements for a maximum allowable level of microbial life, or bioburden (Benardini et al. 2014). The determination of what constitutes acceptable levels was based on studies of how various life-forms survive exposure to the unique conditions of the space environment. If it is possible to reduce the numbers to acceptable levels (a proxy for cleanliness), then the assumption is that the life-forms will not survive under harsh space conditions. However, the latest research has revealed that some microbes are tougher than expected and that others may use various protective mechanisms to survive exposure to the space environment. Three recently published peer-reviewed scientific articles used the ISS to examine the risks of interplanetary exchange of organisms and are given in the subsequent paragraphs and discussed in detail.

In the first study (Vaishampayan et al. 2012), *B. pumilus* SAFR-032 spores were flown to the ISS and exposed to a variety of space conditions via the European Technology Exposure Facility (EuTEF). After 18 months of exposure in the EXPOSE facility of the European Space Agency (ESA) on EuTEF under dark space conditions, SAFR-032 spores showed a survival rate of 10%–40%. For comparison, when these spores were kept aboard the ISS under dark, simulated Martian atmospheric conditions, they showed a survival rate of 85%–100%. In contrast, when SAFR-032 spores were exposed to UV (>110 nm) radiation for the same length of time and under the same conditions as used on the EXPOSE facility, there was a 7-log reduction in viability. A parallel experiment was conducted on Earth with identical samples under simulated space conditions. Spores exposed to ground simulations showed less of a reduction in viability when compared with the "real space" exposed spores (~ 3-log reduction in viability for "UV-Mars," and ~ 4-log reduction in viability for "UV-Space"). A comparative proteomics analysis indicated that a higher concentration of proteins that provide resistant traits (superoxide dismutase) was found in space-exposed spores when compared with the ground control spores. In addition, the first-generation cells and spores derived from space-exposed samples showed higher resistance to UVC when compared with the ground control spores. The data generated are important for calculating the probability and mechanisms of microbial survival in space conditions and for assessing microbial contaminants as risks for forward contamination from terrestrial microorganisms and the validity of in situ life detection.

In the second study (Horneck et al. 2012), the spores of *B. pumilus* SAFR-032 and *B. subtilis* 168, another spore-forming bacterium, were dried on pieces of spacecraft-quality aluminum and exposed to space vacuum, cosmic and extraterrestrial solar radiations, and temperature fluctuations on EuTEF for 18 months. These spores were also simultaneously exposed to a simulated Martian atmosphere at the ISS by using EuTEF. Most of the organisms exposed to solar UV radiation in space and in the Mars spectrum were killed, but when UV rays were filtered out and samples were kept in the dark, approximately 50% or more of those spores exposed to other space- and Mars-like conditions survived. These results show that it is likely that spores could survive the spaceflight to Mars if they are shielded from solar irradiation, which may be possible in a tiny pocket of the spacecraft surface or underneath a layer of other spores.

The third experiment (Onofri et al. 2012) placed rock-colonizing cellular organisms in the EuTEF facility for 1.5 years to test a theory of how organisms might move from one planet to another, known as lithopanspermia. This theory postulates that rocks ejected from a planet by impact with another object (e.g., a meteor) can carry organisms on their surface through space and then land on another planet, thereby transferring these organisms from one planet to the next. In this experiment, researchers selected organisms that have adapted to the extreme environments of their natural habitats on Earth; they found that some are also able to survive in the even more hostile environment of outer space. The process of lithopanspermia would require thousands or even millions of years to occur, much longer than the experiment's 18 months, but the results demonstrate the resilience of these organisms in space and suggest that rocks traveling through space could indeed carry life between planets.

Future space exploration missions can apply the results of these experiments to develop methods to reduce the risk of forward contamination, which will prevent the misidentification of a hitchhiking organism as a native of planet being explored.

9.2.9 CONCLUSION

The results of the ISS Environmental Omics project (basic science) should be leveraged to enhance the health and well-being of astronauts living in an enclosed habitat. To put it simply, the ISS Environmental Omics research seeks to translate the results of fundamental research into medical practice (pathogen detection) and meaningful health outcomes (countermeasure development). The omics data sets were placed in the NASA GeneLab bioinformatics environment, which consists of a database, computational tools, and improved methods, and are already made available to the scientific research community to encourage innovation (https://genelab.nasa.gov/). Appropriate countermeasures should be continuously developed to maintain the stability of the closed system and to ensure the health of the crew. NASA's approach will assist spacefaring nations in recognizing the functionality of microbiomes associated with a close habitat, thereby facilitating the development of appropriate countermeasures. The AMR information could result in mitigation strategies to ensure the health of astronauts during long-duration space missions, when a return trip to Earth for treatment is not feasible or possible.

The results of the microbial monitoring with omics technologies give NASA the ability to precisely assess the spectrum of microorganisms associated with the closed habitat and to maintain crew health. In addition to providing overall microbial profiles, the molecular omics approach can identify the microbial taxa that pose particular threats to the health of the crew and spacecraft systems. The extensive microbial census produced by the omics methodologies allowed researchers to gain substantial insights into spaceflight-induced changes in the populations of beneficial and/or potentially harmful microbes.

ACKNOWLEDGMENTS

Part of the research described in this publication was carried out at the JPL, California Institute of Technology, under a contract with NASA. The author would like to thank astronauts Captain Terry Virts and J. Williams for collecting samples aboard the ISS and the Implementation Team at NASA Ames Research Center for coordinating this effort. Contributions from the Microbial Tracking team are acknowledged. The team includes Aleksandra Checinska Sielaff, Nitin Singh, Ganesh B. M. Mohan, Camilla Urbaniak, Adriana Blachowicz, Kenneth Frey, Jonathan E. Allen, Satish Mehta, Nicholas H. Bergman, Fathi Karouia, David J. Smith, Duane L. Pierson, Jay Perry, Nancy Keller, George E. Fox, Crystal Jaing, Clay Wang, and Tamas Torok. © 2018 California Institute of Technology. Government sponsorship acknowledged.

FUNDING

These projects were funded by several NASA funding: 2012 Space Biology NNH12ZTT001N grant nos. 19-12829-26 and 19-12829-27 under Task Order NNN13D111T award to KV, which also funded post-doctoral fellowship for ACS, NS, GMM, and AB; NNX15AB49G for CW and KV; NNX17AE50G for NK and KV; 80NSSC18K0113 (previously NNX15AJ29G) for CJ, KV, SM, and DJS; and NASA Space Biology program sponsored NASA Postdoctoral Program (NPP) post-doc for CU (2016 to present).

COMPETING INTERESTS

None.

ETHICS APPROVAL AND CONSENT TO PARTICIPATE

Not applicable.

DISCLAIMER

The author declares that all these materials were published elsewhere as peer-reviewed journal articles, and scientific contents are compiled as a book chapter.

REFERENCES

Archer GL (1998) *Staphylococcus aureus*: A well-armed pathogen. *Clin Infect Dis* 26:1179–1181.

Ashton PM, Nair S, Dallman T, Rubino S, Rabsch W, Mwaigwisya S (2015) MinION nanopore sequencing identifies the position and structure of a bacterial antibiotic resistance island. *Nat Biotechnol* 33. doi:10.1038/nbt.3103.

Barros EM, Ceotto H, Bastos MCF, dos Santos KRN, Giambiagi-deMarval M (2012) *Staphylococcus haemolyticus* as an important hospital pathogen and carrier of methicillin resistance genes. *J Clin Microbiol* 50(1):166–168. doi:10.1128/JCM.05563-11.

Bashir M, Ahmed M, Weinmaier T, Ciobanu D, Ivanova N, Pieber TR, Vaishampayan PA (2016) Functional metagenomics of spacecraft assembly cleanrooms: Presence of virulence factors associated with human pathogens. *Front Microbiol* 7:1321. doi:10.3389/fmicb.2016.01321.

Be NA, Avila-Herrera A, Allen JE, Singh N, Checinska Sielaff A, Jaing C, Venkateswaran K (2017) Whole metagenome profiles of particulates collected from the International Space Station. *Microbiome* 5(1):81. doi:10.1186/s40168-017-0292-4.

Beale DJ, Morrison PD, Key C, Palombo EA (2014) Metabolic profiling of biofilm bacteria known to cause microbial influenced corrosion. *Water Sci Technol* 69(1):1–8. doi:10.2166/wst.2013.425.

Benardini JN, 3rd, La Duc MT, Ballou D, Koukol R (2014) Implementing planetary protection on the atlas v fairing and ground systems used to launch the Mars Science Laboratory. *Astrobiology* 14(1):33–41. doi:10.1089/ast.2013.1011.

Biteen JS, Blainey PC, Cardon ZG, Chun M, Church GM, Dorrestein PC, Fraser SE et al. (2016) Tools for the Microbiome: Nano and Beyond. *ACS Nano* 10(1):6–37. doi:10.1021/acsnano.5b07826.

Brady CL, Cleenwerck I, Venter SN, Engelbeen K, De Vos P, Coutinho TA (2010) Emended description of the genus *Pantoea*, description of four species from human clinical samples, *Pantoea septica* sp. nov., *Pantoea eucrina* sp. nov., *Pantoea brenneri* sp. nov. and *Pantoea conspicua* sp. nov., and transfer of *Pectobacterium cypripedii* (Hori 1911) Brenner et al. 1973 emend. Hauben et al. 1998 to the genus as *Pantoea cypripedii* comb. nov. *Int J Syst Evol Microbiol* 60(Pt 10):2430–2440. doi:10.1099/ijs.0.017301-0.

Brisse S, Passet V, Grimont PA (2014) Description of *Klebsiella quasipneumoniae* sp. nov., isolated from human infections, with two subspecies, *Klebsiella quasipneumoniae* subsp. *quasipneumoniae* subsp. nov. and *Klebsiella quasipneumoniae* subsp. *similipneumoniae* subsp. nov., and demonstration that *Klebsiella singaporensis* is a junior heterotypic synonym of *Klebsiella variicola*. *Int J Syst Evol Microbiol* 64(Pt 9):3146–3152. doi:10.1099/ijs.0.062737-0.

Castro-Wallace SL, Chiu CY, John KK, Stahl SE, Rubins KH, McIntyre ABR, Dworkin JP et al. (2016) Nanopore DNA sequencing and genome assembly on the International Space Station. *bioRxiv*. doi:10.1101/077651.

Checinska A, Probst AJ, Vaishampayan P, White JR, Kumar D, Stepanov VG, Fox GE et al. (2015) Microbiomes of the dust particles collected from the International Space Station and Spacecraft Assembly Facilities. *Microbiome* 3(1). doi:10.1186/s40168-015-0116-3.

Checinska Sielaff A, Kumar RM, Pal D, Mayilraj S, Venkateswaran K (2017) *Solibacillus kalamii* sp. nov., isolated from a high-efficiency particulate arrestance filter system used in the International Space Station. *Int J Syst Evol Microbiol* 67(4):896–901. doi:10.1099/ijsem.0.001706.

Crucian BE, Zwart SR, Mehta S, Uchakin P, Quiriarte HD, Pierson D, Sams CF, Smith SM (2014) Plasma cytokine concentrations indicate that in vivo hormonal regulation of immunity is altered during long-duration spaceflight. *J Interferon Cytokine Res* 34(10):778–786. doi:10.1089/jir.2013.0129.

Dai X, Wang H, Ju L-K, Cheng G, Cong H, Newby B-mZ (2016) Corrosion of aluminum alloy 2024 caused by Aspergillus niger. *Int Biodeterior Biodegrad* 115:1–10. doi:10.1016/j.ibiod.2016.07.009.

Daubenspeck JM, Zeng H, Chen P, Dong S, Steichen CT, Krishna NR, Pritchard DG, Turnbough CL (2004) Novel oligosaccharide chains of the collagen-like region of BclA, the major glycoprotein of the *Bacillus anthracis* exosporium. *J Biol Chem* 279:30945–30953.

Driks A (1999) *Bacillus subtilis* spore coat. *Microbiol Mol Biol Rev* 63(1):1–20.

Driks A (2002a) Maximum shields: The assembly and function of the bacterial spore coat. *Trends Microbiol* 10(6):251–254.

Driks A (2002b) Overview: Development in bacteria: Spore formation in *Bacillus subtilis*. *Cell Mol Life Sci* 59(3):389–391.

Driks A, Setlow P (2000) Morphogenesis and properties of the bacterial spore. In: Brun YV, Shimkets LJ (Eds), *Prokaryotic Development*. American Society for Microbiology, Washington, DC, pp. 191–218.

Furman-Kuklinska K, Naumnik B, Mysliwiec M (2009) Fungaemia due to *Cryptococcus laurentii* as a complication of immunosuppressive therapy—A case report. *Adv Med Sci* 54(1):116–119. doi:10.2478/v10039-009-0014-7.

Gerhardt P, Marquis RE (1989) Spore thermoresistance mechanisms. In: Smith I, Slepecky RA, Setlow P (Eds), *Regulation of Prokaryotic Development*. American Society for Microbiology, Washington, DC, pp. 43–63.

Ghosh S, Osman S, Vaishampayan P, Venkateswaran K (2010) Recurrent isolation of extremotolerant bacteria from the clean room where Phoenix spacecraft components were assembled. *Astrobiology* 10(3):325–335. doi:10.1089/ast.2009.0396.

Gioia J, Yerrapragada S, Qin X, Jiang H, Igboeli OC, Muzny D, Dugan-Rocha S et al. (2007) Paradoxical DNA repair and peroxide resistance gene conservation in Bacillus pumilus SAFR-032. *PLoS One* 2(9):e928. doi:10.1371/journal.pone.0000928.

Giorno R, Bozue J, Cote C, Wenzel T, Moody K-S, Ryan M, Wang R et al. (2007) Morphogenesis of the *Bacillus anthracis* spore coat. *J Bacteriol* 189:691–705.

Head SR, Komori HK, LaMere SA, Whisenant T, Van Nieuwerburgh F, Salomon DR, Ordoukhanian P (2014) Library construction for next-generation sequencing: Overviews and challenges. *Biotechniques* 56(2):61-passim. doi:10.2144/000114133.

Hendrickson R, Lundgren P, Malli-Mohan GB, Urbaniak C, Benardini JN, Venkateswaran K (2017) Comprehensive measurement of microbial burden in nutrient-deprived cleanrooms. In: *47th International Conference on Environmental Systems; ICES-2017-177*, Charleston, SC, ICES.

Hoisington A, Maestre JP, King MD, Siegel JA, Kinney KA (2014) The impact of sampler selection on characterizing the indoor microbiome. *Build Environ*. doi:10.1016/j.buildenv.2014.04.021.

Horneck G, Moeller R, Cadet J, Douki T, Mancinelli RL, Nicholson WL, Panitz C et al. (2012) Resistance of bacterial endospores to outer space for planetary protection purposes—experiment PROTECT of the EXPOSE-E mission. *Astrobiology* 12(5):445–456. doi:10.1089/ast.2011.0737.

Ichijo T, Yamaguchi N, Tanigaki F, Shirakawa M, Nasu M (2016) Four-year bacterial monitoring in the International Space Station-Japanese Experiment Module "Kibo" with culture-independent approach. *NPJ Microgravity* 2:16007. doi:10.1038/npjmgrav.2016.7.

Jain M, Fiddes IT, Miga KH, Olsen HE, Paten B, Akeson M (2015) Improved data analysis for the MinION nanopore sequencer. *Nat Methods* 12. doi:10.1038/nmeth.3290.

Katayama Y, Ito T, Hiramatsu K (2000) A new class of genetic element, *Staphylococcus* cassette chromosome *mec*, encodes methicillin resistance in *Staphylococcus aureus*. *Antimicrob Agents Chemother* 44(6):1549–1555.

Knox BP, Blachowicz A, Palmer JM, Romsdahl J, Huttenlocher A, Wang CCC, Keller NP, Venkateswaran K (2016) Characterization of *Aspergillus fumigatus* isolates from air and surfaces of the International Space Station. *mSphere* 1(5). doi:10.1128/mSphere.00227-16.

Kornberg A, Spudich JA, Nelson DL, Deutscher M (1968) Origin of proteins in sporulation. *Ann Rev Bioch* 37:51–78.

Kuwana R, Kasahara Y, Fujibayashi M, Takamatsu H, Ogasawara N, Watabe K (2002) Proteomics characterization of novel spore proteins of *Bacillus subtilis*. *Microbiology* 148:3971–3982.

La Duc MT, Satomi M, Venkateswaran K (2004) *Bacillus odysseyi* sp. nov., a round-spore-forming bacillus isolated from the Mars Odyssey spacecraft. *Int J Syst Evol Microbiol* 54(Pt 1):195–201.

Lai EM, Phadke ND, Kachman MT, Giorno R, Vazquez S, Vazquez JA, Maddock JR, Driks A (2003) Proteomic analysis of the spore coats of *Bacillus subtilis* and *Bacillus anthracis*. *J Bacteriol* 185:1443–1454.

Lax S, Sangwan N, Smith D, Larsen P, Handley KM, Richardson M, Guyton K et al. (2017) Bacterial colonization and succession in a newly opened hospital. *Science Transl Med* 9(391). doi:10.1126/scitranslmed.aah6500.

Liu Y, Kurtán T, Yun Wang C, Han Lin W, Orfali R, Müller WEG, Daletos G, Proksch P (2016) Cladosporinone, a new viriditoxin derivative from the hypersaline lake derived fungus *Cladosporium cladosporioides*. *J Antibiot* 69:702. doi:10.1038/ja.2016.11.

Ma Y, Madupu R, Karaoz U, Nossa CW, Yang L, Yooseph S, Yachimski PS, Brodie EL, Nelson KE, Pei Z (2014) Human papillomavirus community in healthy persons, defined by metagenomics analysis of human microbiome project shotgun sequencing data sets. *J Virol* 88(9):4786–4797. doi:10.1128/jvi.00093-14.

Mardis ER (2013) Next-generation sequencing platforms. *Annu Rev Anal Chem (Palo Alto Calif)* 6:287–303. doi:10.1146/annurev-anchem-062012-092628.

Mayer T, Blachowicz A, Probst AJ, Vaishampayan P, Checinska A, Swarmer T, de Leon P, Venkateswaran K (2016) Microbial succession in an inflated lunar/Mars analog habitat during a 30-day human occupation. *Microbiome* 4(1):1–17. doi:10.1186/s40168-016-0167-0.

McIntyre AB, Rizzardi L, Yu AM, Alexander N, Rosen GL, Botkin DJ, Stahl SE et al. (2016) Nanopore sequencing in microgravity. *npj Microgravity* 2:16035. doi:doi:10.1038/npjmgrav.2016.35.

Mendoza-Olazarán S, Morfin-Otero R, Rodríguez-Noriega E, Llaca-Díaz J, Flores-Treviño S, González-González GM, Villarreal-Treviño L, Garza-González E (2013) Microbiological and molecular characterization of *Staphylococcus hominis* isolates from blood. *PLOS One* 8(4):e61161.

Mezzatesta ML, Gona F, Stefani S (2012) *Enterobacter cloacae* complex: Clinical impact and emerging antibiotic resistance. *Future Microbiol* 7(7):887–902. doi:10.2217/fmb.12.61.

Moissl C, Osman S, La Duc MT, Dekas A, Brodie E, DeSantis T, Venkateswaran K (2007) Molecular bacterial community analysis of clean rooms where spacecraft are assembled. *FEMS Microbiol Ecol* 61(3):509–521. doi:10.1111/j.1574-6941.2007.00360.x.

Mukherjee N, Dowd SE, Wise A, Kedia S, Vohra V, Banerjee P (2014) Diversity of bacterial communities of fitness center surfaces in a U.S. metropolitan area. *Int J Environ Res Public Health* 11(12):12544–12561. doi:10.3390/ijerph111212544.

Newcombe DA, Schuerger AC, Benardini JN, Dickinson D, Tanner R, Venkateswaran K (2005) Survival of spacecraft-associated microorganisms under simulated martian UV irradiation. *Appl Environ Microbiol* 71(12):8147–8156.

Nicholson WL, Munakata N, Horneck G, Melosh HJ, Setlow P (2000) Resistance of *Bacillus* endospores to extreme terrestrial and extraterrestrial environments. *Microbiol Mol Biol Rev* 64(3):548–572.

Nierman WC, Pain A, Anderson MJ, Wortman JR, Kim HS, Arroyo J, Berriman M et al. (2005) Genomic sequence of the pathogenic and allergenic filamentous fungus Aspergillus fumigatus. *Nature* 438(7071):1151–1156. doi:10.1038/nature04332.

NRC (2011) *Committee for the Decadal Survey on Biological Physical Sciences in Space: Recapturing a Future for Space Exploration: Life and Physical Sciences Research for a New Era*. The National Academies Press, Washington, DC.

Nunes JM, Bizerra FC, Ferreira RCe, Colombo AL (2013) Molecular identification, antifungal susceptibility profile, and biofilm formation of clinical and environmental *Rhodotorula* species isolates. *Antimicrob Agents Chemother* 57(1):382–389. doi:10.1128/aac.01647-12.

Onofri S, de la Torre R, de Vera J, Ott S, Zucconi L, Selbmann L, Scalzi G, Venkateswaran K, Rabbow E, Horneck G (2012) Hitchhikers between planets? – Rock-colonising organisms survive 1.5 years in outer space. *Astrobiology* 12(5):508–516.

Pavissich JP, Vargas IT, Gonzalez B, Pasten PA, Pizarro GE (2010) Culture dependent and independent analyses of bacterial communities involved in copper plumbing corrosion. *J Appl Microbiol* 109(3):771–782. doi:10.1111/j.1365-2672.2010.04704.x.

Poza M, Gayoso C, Gomez MJ, Rumbo-Feal S, Tomas M, Aranda J, Fernandez A, Bou G (2012) Exploring bacterial diversity in hospital environments by GS-FLX Titanium pyrosequencing. *PloS one* 7(8):e44105. doi:10.1371/journal.pone.0044105.

Probst A, Vaishampayan P, Ghosh S, Osman S, Krishnamurthi S, Mayilraj S, Venkateswaran K (2009) *Bacillus horneckiae* sp. nov., isolated from a clean room where the Phoenix spacecraft was assembled. *Int J Syst Evol Microbiol* 60(Pt 5):1031–1037. doi:10.1099/ijs.0.008979-0.

Rajasekar A, Ting Y-P (2010) Microbial corrosion of aluminum 2024 aeronautical alloy by hydrocarbon degrading bacteria *Bacillus cereus* ACE4 and *Serratia marcescens* ACE2. *Ind Eng Chem Res* 49(13):6054–6061. doi:10.1021/ie100078u.

Redmond C, Baillie LW, Hibbs S, Moir AJ, Moir A (2004) Identification of proteins in the exosporium of *Bacillus anthracis*. *Microbiol Immunol* 150:355–363.

Rimek D, Haase G, Lück A, Casper J, Podbielski A (2004) First report of a case of meningitis caused by *Cryptococcus adeliensis* in a patient with acute myeloid leukemia. *J Clin Microbiol* 42(1):481–483. doi:10.1128/JCM.42.1.481-483.2004.

Rintala H, Pitkaranta M, Toivola M, Paulin L, Nevalainen A (2008) Diversity and seasonal dynamics of bacterial community in indoor environment. *BMC Microbiol* 8:56. doi:10.1186/1471-2180-8-56.

Rizzetto L, Giovannini G, Bromley M, Bowyer P, Romani L, Cavalieri D (2013) Strain dependent variation of immune responses to *A. fumigatus*: Definition of pathogenic species. *PLoS One* 8(2):e56651. doi:10.1371/journal.pone.0056651.

Sanchez JF, Wang CC (2012) The chemical identification and analysis of *Aspergillus nidulans* secondary metabolites. *Methods Mol Biol* 944:97–109. doi:10.1007/978-1-62703-122-6_6.

Savage AM, Hills J, Driscoll K, Fergus DJ, Grunden AM, Dunn RR (2016) Microbial diversity of extreme habitats in human homes. *PeerJ* 4:e2376. doi:10.7717/peerj.2376.

Schwendner P, Mahnert A, Koskinen K, Moissl-Eichinger C, Barczyk S, Wirth R, Berg G, Rettberg P (2017) Preparing for the crewed Mars journey: Microbiota dynamics in the confined Mars500 habitat during simulated Mars flight and landing. *Microbiome* 5(1):129. doi:10.1186/s40168-017-0345-8.

Setlow B, McGinnis KA, Ragkousi K, Setlow P (2000) Effects of major spore-specific DNA binding proteins on *Bacillus subtilis* sporulation and spore properties. *J Bacteriol* 182(24):6906–6912.

Setlow P (1992) I will survive: Protecting and repairing spore DNA. *J Bacteriol* 174:2737–2241.

Setlow P (1995) Mechanisms for the prevention of damage to DNA in spores of *Bacillus* species. *Annu Rev Microbiol* 49:29–54.

Seuylemezian A, Singh NK, Vaishampayan P, Venkateswaran K (2017) Draft genome sequence of *Solibacillus kalamii*, isolated from an air filter aboard the International Space Station. *Genome Announc* 5(35):e00696-17. doi:10.1128/genomeA.00696-17.

Seuylemezian A, Vaishampayan P, Cooper K, Venkateswaran K (2018) Draft genome sequences of *Acinetobacter* and *Bacillus* strains isolated from spacecraft-associated surfaces. *Genome Announc* 6(6). doi:10.1128/genomeA.01554-17.

Singh NK, Blachowicz A, Romsdahl J, Wang C, Torok T, Venkateswaran K (2017) Draft genome sequences of several fungal strains selected for exposure to microgravity at the International Space Station. *Genome Announc* 5(15):e01602-16. doi:10.1128/genomeA.01602-16.

Skawinska O, Kuhn G, Balmelli C, Francioli P, Giddey M, Perreten V, Riesen A, Zysset F, Blanc DS, Moreillon P (2009) Genetic diversity and ecological success of *Staphylococcus aureus* strains colonizing humans. *Appl Environ Microbiol* 75(1):175–183.

Slieman TA, Nicholson WL (2001) Role of dipicolinic acid in survival of *Bacillus subtilis* spores exposed to artificial and solar UV radiation. *Appl Environ Microbiol* 67(3):1274–1279.

Strange RE, Dark FA (1956) The composition of the spore coats of *Bacillus megatherium*, *B. subtilis* and *B. cereus*. *Biochemical J* 62:459–465.

Sylvestre P, Couture-Tosi E, Mock M (2002) A collagen-like surface glycoprotein is a structural component of the *Bacillus anthracis* exosporium. *Mol Microbiol* 45:169–178.

Todd SJ, Moir AJ, Johnson MJ, Moir A (2003) Genes of *Bacillus cereus* and *Bacillus anthracis* encoding proteins of the exosporium. *J Bacteriol* 185:3373–3378.

Urbaniak C, Sielaff AC, Frey KG, Allen JE, Singh N, Jaing C, Wheeler K, Venkateswaran K (2018) Detection of antimicrobial resistance genes associated with the International Space Station environmental surfaces. *Nat Sci Rep* 8(1):814. doi:10.1038/s41598-017-18506-4.

Vaishampayan P, Miyashita M, Ohnishi A, Satomi M, Rooney A, Duc MTL, Venkateswaran K (2008) Description of *Rummeliibacillus stabekisii* gen. nov., sp. nov. and reclassification of *Bacillus pycnus* Nakamura et al. 2002 as *Rummeliibacillus pycnus* comb. nov. *Intern J Syst Evol Microbiol* 59(Pt 5):1094–1099. doi:10.1099/ijs.0.006098-0.

Vaishampayan P, Rabbow E, Horneck G, Venkateswaran K (2012) Survival of *Bacillus pumilus* spores for a prolonged period of time in real space conditions. *Astrobiology* 12(5):487–497.

Venkateswaran K, Checinska Sielaff A, Ratnayake S, Pope RK, Blank TE, Stepanov VG, Fox GE et al. (2017a) Draft genome sequences from a novel clade of *Bacillus cereus* Sensu Lato strains, isolated from the International Space Station. *Genome Announc* 5(32). doi:10.1128/genomeA.00680-17.

Venkateswaran K, Kempf M, Chen F, Satomi M, Nicholson W, Kern R (2003) *Bacillus nealsonii* sp. nov., isolated from a spacecraft-assembly facility, whose spores are gamma-radiation resistant. *Int J Syst Evol Microbiol* 53(Pt 1):165–172.

Venkateswaran K, Singh NK, Checinska Sielaff A, Pope RK, Bergman NH, van Tongeren SP, Patel NB et al. (2017b) Non-toxin-producing *Bacillus cereus* strains belonging to the *B. anthracis* clade isolated from the International Space Station. *mSystems* 2(3). doi:10.1128/mSystems.00021-17.

Venkateswaran K, Vaishampayan P, Cisneros J, Pierson DL, Rogers SO, Perry J (2014) International Space Station environmental microbiome—Microbial inventories of ISS filter debris. *Appl Microbiol Biotechnol* 98(14):6453–6466. doi:10.1007/s00253-014-5650-6.

Waller LN, Fox N, Fox KF, Fox A, Price RL (2004) Ruthenium red staining for ultrastructural visualization of a glycoprotein layer surrounding the spore of *Bacillus anthracis* and *Bacillus subtilis*. *J Microbiol Methods* 58(58):23–30.

Yang SS, Lin JY, Lin YT (1998) Microbiologically induced corrosion of aluminum alloys in fuel-oil/aqueous system. *J Microbiol Immunol Infect* 31(3):151–164.

Zhang L, Kang M, Xu J, Xu J, Shuai Y, Zhou X, Yang Z, Ma K (2016) Bacterial and archaeal communities in the deep-sea sediments of inactive hydrothermal vents in the Southwest India Ridge. *Sci Rep* 6:25982. doi:10.1038/srep25982.

9.3 Virus Evolution and Ecology
Role of Viruses in Adaptation of Life to Extreme Environments

Marilyn J. Roossinck

CONTENTS

9.3.1 INTRODUCTION

Adaptation to extreme environments can be a rapid process, especially after a cataclysmic event such as a volcanic eruption, a wildfire, or a severe drought due to climatic disruptions. In most cases, organisms are unable to adapt, and die. However, in many extreme environments we find organisms thriving, even organisms that don't normally live under extreme conditions. How does this adaptation occur? Rapid adaptation cannot be accommodated by the slow changes of Darwinian evolution, but microbes can provide new genetic material to facilitate rapid changes. For example, plants were able to thrive on land only through the help of fungal endophytes, and fungi are found in plants that have adapted to extreme environments such as geothermal, high salinity, and drought-ravaged soils (Rodriguez and Redman 2008). Another source of novel genetic information is viruses. Although viruses are often thought of as strictly pathogenic, there have been a number of studies in recent years that demonstrate the beneficial nature of many viruses (Roossinck 2011, 2015, Virgin 2014, Roossinck and Bazán 2017). Viruses are probably the most neglected entities in the field of astrobiology (Berliner et al. 2018).

Unlike all other extant life, viruses utilize either RNA or DNA as their basic genomic material, and their genomes can be single-stranded (ss) or double-stranded (ds). Archaea are the only organisms in which all of the known viruses have DNA genomes. Although there has been some indication of an RNA virus from a metagenomic study, it has not been verified (Deimer and Stedman 2012).

Viruses of bacteria and archaea are intimately involved in the ecology of their hosts (Gustavsen et al. 2014); however, their roles in adaptation to extreme environments are less clear. Plants use viruses to confer tolerance to heat, cold, and drought, and bacteria and archaea that thrive in extreme environments are also often infected with potentially beneficial viruses. In addition, viruses are responsible for horizontal gene transfer among many organisms across kingdoms; this has led to further rapid adaptation (see Chapter 8.3).

9.3.2 PLANT ADAPTATIONS

Vascular plants have a narrow range of temperature tolerance and generally cannot grow in soil above 45°C (Brock 1985). However, plants can be found around the geothermal areas of Yellowstone National Park at temperatures well above 50°C (Stout and Al-Niemi 2002), as well as other volcanic areas in the world. In Costa Rica, plants can be found thriving in volcanic soils at temperatures of 65°C (unpublished). These plants are colonized by endophytic fungi that are required for thermotolerance (Redman et al. 2002). However, in the plants in Yellowstone, the fungus does not act alone. It is, in turn, infected by a small RNA virus that is also required: when fungi are cured of the virus, they no longer confer thermal tolerance (Márquez et al. 2007). Closely related fungal endophytes from nearby plants growing in normal soils are not infected by the virus. In experimental studies, the fungal-virus combination could confer heat tolerance to both monocots and dicots, indicating that the players in this interaction are ancient (Márquez et al. 2007).

Plant can also adapt to other extreme environments, such as hypersaline soils, extreme cold, and extreme drought. Examples are found in all of these conditions that include fungal endophytes (Rodriguez et al. 2008), but the viral status of these fungi has not been carefully studied. However, plant viruses also can play a major role in drought tolerance. In experimental infections, several plant viruses were found to confer drought tolerance to infected plants (Xu et al. 2008). These viruses are generally pathogenic under normal conditions but convert to mutualists under drought stress (Bao and Roossinck 2013). Field studies have confirmed the potential role of viruses in drought tolerance (Davis et al. 2015). One mechanism for drought tolerance imparted by *Cucumber mosaic virus* (CMV) involves RNA silencing (Westwood et al. 2013), although other factors may also be involved; a variety of metabolites were elevated in virus-infected plants experiencing drought stress (Xu et al. 2008).

Viruses may also be involved in plant tolerance to cold. *Cucumber mosaic virus* conferred cold tolerance to red beet plants in an experiment designed to mimic freezing nighttime temperatures at the beginning or end of a normal growing season (Xu et al. 2008).

9.3.3 ANIMALS IN EXTREME ENVIRONMENTS

Very little work has been done on the potential role of viruses in the adaptation of animals in extreme environments. In general, viruses found in polar animals have been assumed to be pathogens. Recently, Smeele et al. used high-throughput sequencing to screen a variety of vertebrate life in the extreme environment of the Antarctic seas, including seals, penguins, and a few seabirds, as well as summarized what has been found to date through serology and early sequencing methods. Numerous viruses were identified, all within the families that one would expect from what is known about wildlife in tropical and temperate zones. There is no evidence either for or against a role of viruses in adaptation of macroanimal life to extreme environments (Smeele et al. 2018).

9.3.4 MICROBES IN THE EXTREME

The last decade has seen an explosion in virus discovery, largely due to advances in sequence technologies. Most of these studies have been metagenomic in nature, where an environmental sample is enriched for viruses or virus sequences and massively sequenced (see volume 239 of the journal *Virus Research*, a themed issue on deep sequencing in virology [Berkhout et al. 2017]). The early studies were done in the sea (Suttle 2005), but many studies have been conducted in extreme environments, including deep-sea vents, hypersaline regions, extreme drought conditions, and polar or arctic seas and lakes. Discovery leads to a deeper understanding of what is there, and we now know that viruses are the most diverse and widely distributed entities on earth. However, it is still challenging to link discovery to function, and in many cases, it is not clear who the viral host is. A majority of virus-related sequences are still novel, meaning they do not have any similarities to

anything known, and this complicates understanding. In this section, I review the studies of prokaryotic viruses in extreme environments and detail a few that give clues as to what the viruses might be doing in the process of adaptation.

Viruses of prokaryotes can be lytic, lysogenic, or chronic. In lytic infections, the virus replicates to high levels until the cell becomes full and bursts. In lysogenic cycles, the viral genome is integrated into the host and replicated along with the host. Lysogenic viruses often protect their host from superinfection by related viruses. A virus can convert from lysogenic to lytic due to a number of external inducers. Chronic infections have not been found in bacteria but are common archaea. The virus does not integrate into the host genome but remains in the host cell and replicates at a low level. The lytic cycle is often credited with controlling host populations, while the lysogenic cycle is often associated with immunity. The chronic cycle is not well understood, but if the virus is providing essential functions to the host, a chronic cycle would be likely to facilitate that.

The microbes in extreme environments are dominated by archaea, but there are numerous bacterial species as well. A review of archaeal viruses published in 2014 described only 117 archaeal viruses that had been characterized (Dellas et al. 2014). By now, there are more, but the number is still very limited. Viruses have only been identified in a small subset of the diversity of archaea. This most likely reflects a lack of looking rather than a lack of viruses (Snyder et al. 2015). Little is known about virus-host interactions, beyond population regulation, and not much is known about how they impact their hosts or are involved in adaptation to extreme environments (Krupovic et al. 2018).

9.3.4.1 NOVELTY OF ARCHAEAL VIRUSES

The morphology of archaeal viruses is extremely diverse. There are archaeal viruses that are structurally related to bacterial virus, but they are genetically unrelated, and there are also many with novel morphology (Figure 9.3.1).

Those similar to bacterial viruses have similar life cycles, but the novel viruses are very different (Pietilä et al. 2014). This could be because they evolved before the division of the three domains of life and hence are fossils of the viral world (Prangishvili 2015). Even though very few viruses of archaea have been characterized, they are pushing the boundaries of our understanding of virus structures and genomes. For example, *Aeropyrum* coil-shaped virus, from the hyperthermophilic archaeon *Aeropyrum pernix*, has a completely novel structure of a hollow cylindrical coiling fiber. Moreover, it has the largest known ssDNA genome of nearly 25,000 nucleotides (Mochizuki et al. 2012). The hyperthermophilic archaeon *Halorubrum lacusprofundi* releases membrane-surrounded vesicles from the cell that contain a plasmid genome that can then infect plasmid-free cells; this makes it more similar to a virus than to a plasmid (Erdmann et al. 2017), bridging the gap between different mobile elements. Viruses and related entities in archaea have been dubbed as the mobilome (Lossourarn et al. 2015). The need for further discovery is

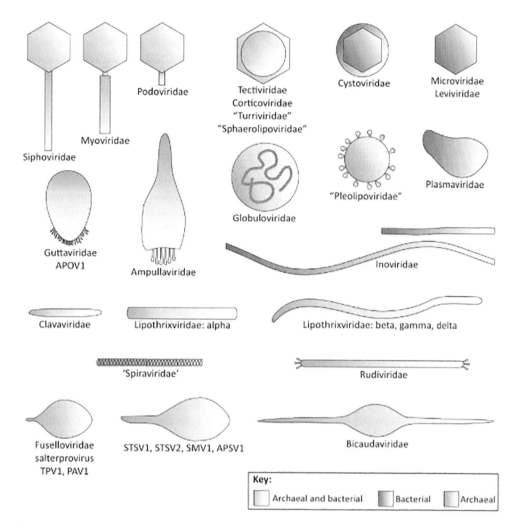

FIGURE 9.3.1 The viruses of archaea have dramatically increased our knowledge of virus structure. The shapes in blue are shared between bacteria and archaea but are genetically distinct. (From Pietilä, M.K. et al., *Trends Microbiol.*, 22, 324–344, 2014.)

exemplified by a resent analysis of the acidic hot springs of Yellowstone National Park, which found over 100 new virus groups in this single environment (Bolduc et al. 2015).

9.3.4.2 Tools for Deeper Understanding

Metagenomics have dramatically increased our knowledge base for viruses, but they are not without drawbacks. There are no universal genes for viruses, so all sequencing must be done with random priming that does not bias the results. This is especially important in extreme environments, where so many of the viruses have no similarity to known viruses. For example, a metagenomic analysis of microbes in a high-sulfur saline lake that could mimic conditions on Mars found many bacteria and archaea, but there was no evidence of viruses. However, it is not clear if viruses were really missing, as the lack of detection was based on looking for similarities with known viruses (Pontefract et al. 2017).

Evidence of past infections by viruses are revealed by the presence of Clustered Regularly Interspaced Short Palindromic Repeats, or CRISPRs, the adaptive immune system in bacteria and archaea (Horvath and Barrangou 2010). When prokaryotes are infected by viruses, they generate short sequences, identical to the viral sequences, that are incorporated into CRISPR genes. Upon reinfection with the same or a related virus, CRISPRs target the foreign genomes for degradation. CRISPRs have been found in about 40% of sequenced bacterial genomes and about 90% of archaeal genomes. Metagenomic studies show a significantly higher level of CRISPRs in thermophiles (bacteria and archaea) than in their counterparts in mesophilic organisms (Cowan et al. 2015).

Researchers often use the virus:prokaryote ratio (VPR) to assess the ecology of viruses in metagenomic studies. However, these relationships vary across environments and in different studies. In extreme environments, VPRs have not been studied very much, but in saline environments, this number varies from 1 to 100, and in thermal environments, it varies from 1 to 20 (Parikka et al. 2016). This could be due to the instability of virus particles in hot environments. Most infections in thermophilic archaea are lysogenic or chronic. Haloviruses are more often lytic. The VPR may be a reflection of these relationships but needs to be interpreted with caution, as the environment may dramatically affect virus stability too (Parikka et al. 2016).

9.3.4.3 HYPERTHERMIC AND HYPERSALINE ENVIRONMENTS

Hypersaline environments are populated with archaea, bacteria, and eukaryotic microbes. Nearly 100 viruses have been identified in the archaeal halophyte family of Halobacteriaceae, whereas only about 10 are known to infect bacterial halophytes. Unlike thermal vent viruses, the archaeal viruses in halophytes are mostly of *Caudovirales* type, that is, they have head-tail morphology. These viruses also have very broad host ranges. Eukaryotic halophytic virus in the Mimivirus group have also been identified (Atanasova et al. 2015). In a different approach, metagenomic studies of halophytes used fosmid clones of *Haloquadratum walsbyi* genomes; these studies show the viral infection history in the CRISPR genes. They then sequenced viruses from same environment that had matching sequences and recovered 42 nearly complete viral genomes, most with some similarity to *Caudovirales*. No *Fuselloviridae* were found, although these are the most common viruses of thermophilic archaea (Garcia-Heredia et al. 2011). Using a combination of viruses in the environment and their counterpart signatures in host CRISPRs clearly identifies the host-virus relationships.

Looking at a gradient of salinity from brackish to near saturation in sites in West Africa, bacterial and archaeal abundance increased with higher salt concentrations, as did the viruses, but while a majority of viruses at the brackish level were lytic, at the highest levels of salinity, most viruses were lysogenic (Bettarel et al. 2011).

9.3.4.4 POLAR ENVIRONMENTS

Polar seas and lakes have high levels of microbes and viruses. The VPR levels were measured in an Antarctic lake at different times of year, when the lake was completely frozen over or experienced surface thaws. The VPR ranged from about two to eight and was higher at 4.5-meter depths than at 1-meter depths. In addition, the ratio was highest in mid-January, when the lake had the least amount of ice (Rochera et al. 2017). RNA viruses sampled over 4 years in an Antarctic lake showed very little change in the population structure of one virus, but three other related viruses showed much greater diversity. However, when isolated from cyanobacterial mats rather than water, the diversity was very low (López-Bueno et al. 2015). Studies of polar freshwater lakes in Norway found a dominance of DNA viruses with small circular DNA genomes, not related to viruses from most other environments, but some had similarity to Antarctic viruses. In a few cases, they found the same virus (90% similarity) in Arctic and Antarctic environments (deCárcer et al. 2015). This is surprising and not easily explained, but it implies that transfer has occurred fairly recently between these environments.

In a saline permafrost site in Alaska, a cryopeg from 7 meters deep was analyzed for bacteria and viruses. Compared with the surrounding ice, the bacteria counts were very high. The viruses were mostly lysogenic and had less diversity than those in the surrounding surface area. The viruses were related to marine phage, supporting a marine origin for the cryopeg (Colangelo-Lillis et al. 2016).

9.3.4.5 DESERT ENVIRONMENTS

Under an extreme drought and heat environment, several bacteriophages were identified, but most of these were lysogenic, only becoming detectable after incubation of the host bacteria and induction in the lab. This differs from a cold desert environment, where phage particles are observable in soil samples (Prestel et al. 2013), perhaps because the phage may be very labile in the heat, similar to what is seen in deep vents. Metagenomic analyses of four ponds within the Sahara Desert found that virus diversity was lowest in the site most impacted by human activity (Fancello et al. 2013). In the barren soils in the cold deserts of Antarctica that were sampled for DNA viruses, tailed phages were abundant, and diversity varied based on altitude and soil chemistry. Phycodnaviruses and mimiviruses (viruses of algae and amoeba, respectively) were also found in the sites (Adriaenssens et al. 2017). Phage populations in the dry valleys of Antarctica may be important for bacterial population regulation (Wei et al. 2015).

9.3.4.6 VIRUS ROLES IN ADAPTATION

In some extreme environments, such as deep-sea vents and soda lakes, the metabolism is dominated by sulfur oxidizers. The soda lakes of East Africa represent an extreme environment with high salinity, pH, and temperature. Soda lakes host numerous bacteria and archaeal species, including nitrifiers and sulfur oxidizers (Grant and Jones 2016). Sulfur-oxidizing bacteria are also abundant in marine hydrothermal vents and oxygen minimum zones. In a recent study of DNA viruses infecting sulfur-oxidizing bacteria, 15 out of 18 had genes involved in sulfur metabolism, indicating that they provide essential functions to their host bacteria in adapting to these extreme environments (Anantharaman et al. 2014).

The Namib Desert in southwestern Africa is home to hypoliths, microbial communities that inhabit the underside of translucent rocks. In this hyperarid environment, cyanobacteria are prevalent, and the viruses were dominated by dsDNA viruses, with a majority of these in the *Siphoviridae*, but these differ significantly from the cyanophage found in marine environments. Auxiliary metabolic genes that are thought to assist the host in metabolism in photosynthesis are common in marine cyanophage (Lindell et al. 2005, Avrani et al. 2012). In the hypolith environment, auxiliary genes for nucleotide metabolism and for an unknown function (*phoH*) were common in the viruses, indicating that they are involved in adaptation to the extreme environment. *PhoH* may be involved in phosphate uptake, but this is unproven (Adriaenssens et al. 2015).

9.3.4.7 Experimental Strategies

Very few studies have included any experimental work to look at virus adaptation to extreme environments. However, Singhal et al. adapted RNA bacteriophage phi6 to high-temperature growth by either gradually increasing temperatures or rapidly increasing temperatures. Gradual increase was more successful. Mutants in end populations were not all thermo-adapted; however, some had simply increased their replication rates (Singhal et al. 2017).

In experimental studies with an icosahedral virus with an internal membrane, *Haloarcula californiae* icosahedral virus 1, the virus could experimentally infect its archaeal halophyte host under low-salt conditions as well as conditions mimicking the environment from where it was isolated (Demina et al. 2016).

9.3.5 CONCLUSIONS

The critical role of viruses in the evolution of life is widely accepted (see Chapter 8.3), and we might expect that evidence for this would be common in extreme environments. However, so little work has been done in this area that it is difficult to make a strong case. In plants, the most detailed studies have unequivocally demonstrated the roles of viruses in adaptation to heat, drought, and cold stress. Virtually, nothing has been done in the area of animal adaptation. In microbes, most of the work in extreme environments is in the era of discovery. Tools are being developed to try to take metagenomics to a level of ecological understanding (Simmonds et al. 2017, Dolja and Koonin 2018), but these have not been applied to any extent in extreme environments. However, the discovery itself lays an important groundwork for further understanding, and it seems almost certain that microbial viruses are also playing a fundamental role in the adaptations of their hosts.

REFERENCES

Adriaenssens, E. M., R. Kramer, M. W. VanGoethem, T. P. Makhalanyane, I. Hogg, and D. A. Cowan. 2017. Environmental drivers of viral community composition in Antarctic soils identified by viromics. *Microbiome* 5:83.

Adriaenssens, E. M., L. VanZyl, P. DeMaayer, E. Rubagotti, E. Rybicki, M. Tuffin, and D. A. Cowan. 2015. Metagenomic analysis of the viral community in Namib Desert hypoliths. *Environ. Microbiol.* 17(2):480–495.

Anantharaman, K., M. B. Duhaime, J. A. Breier, K. A. Wendt, B. M. Toner, and G. J. Dick. 2014. Sulfur oxidation genes in diverse deep-sea viruses. *Science* 344:757–760.

Atanasova, N. S., H. M. Oksanen, and D. H. Bamford. 2015. Haloviruses of archaea, bacteria, and eukaryotes. *Curr. Opin. Microbiol.* 25:40–48.

Avrani, S., D. A. Schwartz, and D. Lindell. 2012. Virus-host swinging party in the oceans. *Mob. Genet. Elem.* 2(2):88–85.

Bao, X., and M. J. Roossinck. 2013. A life history view of mutualistic viral symbioses: Quantity or quality for cooperation? *Curr. Opin. Microbiol.* 16:514–518.

Berkhout, B., N. Beerenwindel, and E. Domingo (Eds.). 2017. Deep sequencing in virology. *Virus Res.* 239:1–180.

Berliner, A. J., T. Mochizuki, and K. M. Stedman. 2018. Astrovirology: Viruses at large in the universe. *Astrobiology* 18(2):207–223.

Bettarel, Y., T. Bouvier, C. Bouvier, C. Carré, A. Desnues, I. Domaizon, S. Jacquet, A. Robin, and T. Sime-Ngando. 2011. Ecological traits of planktonic viruses and prokaryotes along a full-salinity gradient. *FEMS Microbial. Ecol.* 76:360–372.

Bolduc, B., J. F. Firth, A. Mazurie, and M. J. Young. 2015. Viral assemblage composition in Yellowstong acidic hot springs assessed by network analysis. *ISME J.* 9:2162–2177.

Brock, T. D. 1985. Life at high temperatures. *Science* 230(4722):132–138.

Colangelo-Lillis, J., H. Eicken, S.D. Carpenter, and J.W. Deming. 2016. Evidence or marine origin and microbial-viral habitability of sub-zero hypersaline aqueous includsions within permafroms near Barrow, Alaska. *FEMS Microbial. Ecol.* 92:fiw053.

Cowan, D. A., J.-B. Ramond, T. P. Makhalanyane, and P. DeMaayer. 2015. Metgenomics of extreme environments. *Curr. Opin. Microbiol.* 25:97–102.

Davis, T. S., N. A. Bosque Pérez, N. E. Foote, T. Magney, and S. D. Eigenbrode. 2015. Environmentally dependent host-pathogen and vector-pathogen interactions in the *Barley yellow dwarf virus* pathosystem. *J. Appl. Ecol.* 52:1392–1401.

deCárcer, D. A., A. López-Bueno, D. A. Pearce, and A. Alcamí. 2015. Biodiversity and distribution of polar freshwater DNA viruses. *Sci. Adv.* 1:e1400127.

Deimer, G. S., and K. M. Stedman. 2012. A novel virus genome discovered in an extreme environment suggests recombination between unrelated groups of RNA and DNA viruses. *Biol. Direct* 7:14.

Dellas, N., J. C. Snyder, B. Bolduc, and M. J. Young. 2014. Archael viruses: diversity, replication, and structure. *Ann. Rev. Virol.* 1:399–426. doi:10.1146/annurev-virology-031413-085357.

Demina, T. A., M. K. Peitilä, J. Svirskaite, J. J. Ravantti, N. S. Atanasova, D. H. Bamford, and H. M. Oksanen. 2016. Archaeal *Haloarcula californiae* icosahedral virus 1 highlights conserved elements in ocsahedram membrane-containing DNA viruses from extreme environments."*mBio* 7(4):e00699-16.

Dolja, V. V., and E. V. Koonin. 2018. Metagenomics reshapes the concepts of RNA virus evolution by reveraling extensive horizontal virus transfer. *Virus Res.* 244:36–52.

Erdmann, S., B. Tschitschko, L. Zhong, M. Raftery, and R. Cavicchioli. 2017. A plasmid from an Antarctic haloarchaeon uses specialized membrane vesicles to disseminate and infect plasmid-free cells. *Nat. Microbiol.* 2:1446–1455.

Fancello, L., S. Trape, C. Robert, M. Boyer, N. Popgeorgiev, D. Raoult, and C. Desnues. 2013. Viruses in the desert: A metagenomic survey of viral communities in four perennial ponds of the Mauritanian Sahara. *ISME J.* 7:359–369.

Garcia-Heredia, I., A.-B. Martin-Cuadrado, F. J. M. Mojica, F. Santos, A. Mira, J. Antón, and F. Rodriguez-Valera. 2011. Reconstructing viral genomes from the environment using fosmid clones: The case of haloviruses. *PLOS ONE* 7(3):e0033802.

Grant, W. D., and B. E. Jones. 2016. Bacteria, Archaea and Viruses of Soda Lakes. In *Soda Lakes of East Africa*, M. Schagerl (Ed.), pp. 97–148. Switzerland: Springer.

Gustavsen, J. A., D. M. Winget, X. Tian, and C. A. Suttle. 2014. High temporal and spatial diversity in marine RNA viruses implies that they have an important role in mortality and structureing plankton communities. *Front. Microbiol.* 5:702.

Horvath, P., and R. Barrangou. 2010. CRISPR/Cas, the immune system of bacteria and archaea. *Science* 327:167–170.

Krupovic, M., V. Cvirkaite-Krupovic, J. Iranzo, D. Prangishvili, and E. V. Koonin. 2018. Virsues of archaea: Structural, functional, environmental and evolutionary genomics. *Virus Res.* 244:181–193.

Lindell, D., J. D. Jaffe, Z. I. Johnson, G. M. Church, and S. W. Chisholm. 2005. Photosynthesis genes in marine viruses yield proteins during host infection. *Nature* 438:86–89.

López-Bueno, A., A. Rastrojo, R. Peiró, M. Arenas, and A. Alcamí. 2015. Ecological connectivity shapes quasispecies structure of RNA viruses in an Antarctic lake. *Mol. Ecol.* 24:4812–4825.

Lossourarn, J., S. Dupont, A. Gorlas, C. Mercier, N. Bienvenu, E. Marguet, P. Forterre, and C. Geslin. 2015. An abyssal molilome: Viruses, plasmids and vesicles from deep-sea hydrothermal vents. *Res. Microbiol.* 166:742–752.

Márquez, L. M., R. S. Redman, R. J. Rodriguez, and M. J. Roossinck. 2007. A virus in a fungus in a plant—three way symbiosis required for thermal tolerance. *Science* 315:513–515.

Mochizuki, T., M. Krupovic, G. Pehau-Arnaudet, Y. Sako, P. Forterre, and D. Prangishvili. 2012. Archael virus with exceptional virion architecture and the largest single-stranded DNA genome. *Proc. Natl. Acad. Sci.* 109(33):13386–13391.

Parikka, K. J., M. LeRomancer, N. Wauters, and S. Jacquet. 2016. Deciphering the virus-to-prokaryote ratio (VPR)" insights into virus-host relationships in a variety of ecosystems. *Biol. Rev.* 92:1081–1100.

Pietilä, M. K., T. A. Demina, N. S. Antanasova, H. M. Oksanen, and D. H. Bamford. 2014. Archaeal viruses and bacteriophages: Comparisons and contrasts. *Trends Microbiol.* 22(6):324–344.

Pontefract, A., T. F. Zhu, V. K. Walker, H. Hepburn, C. Lui, M. T. Zuber, G. Ruvkun, and C. E. Carr. 2017. Microbial diversity in a hyperaline sulfate lake: A terrestrial analog of ancient mars. *Front. Microbiol.* 8:1819.

Prangishvili, D. 2015. Archaeal viruses: Living fossils of the ancient virosphere? *Ann.NY Acad. Sci.* 1341:35–40.

Prestel, E., C. Regeard, S. Salamitou, J. Neveu, and M. S. DuBow. 2013. The bacteria and bacteriophages from a Mesquite Flats site of the Death Valley Desert. *Antonie van Leeuwenhoek* 103:1329–1341.

Redman, R. S., K. B. Sheehan, R. G. Stout, R. J. Rodriguez, and J. M. Henson. 2002. Thermotholerance generated by plant/fungal symbiosis. *Science* 298:1581.

Rochera, C., A. Quesada, M. Toro, E. Rico, and A. Camacho. 2017. Plankton assembly in an ultra-oligotrophic Antarctic lake over the summer transition from the ice-cover to the ice-free period: A size spectra approach. *Polar Sci.* 11:72–82.

Rodriguez, R. J., J. Henson, E. VanVolkenburgh, M. Hoy, L. Wright, F. Beckwith, Y.-O. Kim, and R. S. Redman. 2008. Stress tolerance in plants via habitat-adapted symbiosis. *ISME J.* 2:404–416.

Rodriguez, R., and R. Redman. 2008. More than 400 million years of evolution and some plants still can't make it on their own: plant stress tolerance via fungal symbiosis. *J. Exp. Bot.* 59(5):1109–1114.

Roossinck, M. J. 2011. The good viruses: Viral mutualistic symbioses. *Nat. Rev.Microbiol.* 9(2):99–108.

Roossinck, M. J. 2015. Move over, bacteria! Viruses make their mark as mutualistic microbial symbionts. *J. Virol.* 89(3):1–3.

Roossinck, M. J., and E. R. Bazán. 2017. Symbiosis: Viruses as intimate partners. *Annu. Rev. Virol.* 4:123–139.

Simmonds, P., M. J. Adams, M. Breitbart et al. 2017. Virus taxonomy in the age of metagenomics. *Nat. Rev. Microbiol.* 15:161–168.

Singhal, S., C. M. Leon Guerrero, S. G. Whang, E. M. McClure, H. G. Busch, and B. Kerr. 2017. Adaptations of an RNA virus to increasing thermal stress. *PLOS ONE* 12(12):e0189602.

Smeele, Z. E., D. G. Ainley, and A. Varsani. 2018. Viruses associated with Antarctic wildlife: from serology based detection to identification of genomes using high throughpur sequencing. *Virus Res.* 243:91–105.

Snyder, J. C., B. Bolduc, and M. J. Young. 2015. 40 years of archaeal virology: Expanding viral diversity. *Virology* 479–480:369–378.

Stout, R. G., and T. S. Al-Niemi. 2002. Heat-tolerant flowering plants of active geothermal areas in Yellowstone National Park. *Ann. Bot.* 90:259–267.

Suttle, C. A. 2005. Viruses in the sea. *Nature* 437:356–361.

Virgin, H. W. 2014. The virome in mammalian physiology and disease. *Cell* 157:142–150.

Wei, S. T. S., C. M. Higgins, E. M. Andriaenssens, D. A. Cowan, and S. B. Pointing. 2015. Genetic signatures indicate widespread antibiotic resistance and phage infection in microbial communities of the McMurdo Dry Valleys, East Antarctica. *Polar Biol.* 38:919–925.

Westwood, J. H., L. McCann, M. Naish et al. 2013. A viral RNA silencing suppressor interferes with abscisic acid-mediated signalling and induces drought tolerance in *Arabidopsis thaliana. Mol. Plant Pathol.* 14(2):158–170.

Xu, P., F. Chen, J. P. Mannas, T. Feldman, L. W. Sumner, and M. J. Roossinck. 2008. Virus infection improves drought tolerance. *New Phytol.* 180:911–921.

Section X

Habitability
Characteristics of Habitable Planets

10.1 The Evolution of Habitability
Characteristics of Habitable Planets

Charles H. Lineweaver, Aditya Chopra, and Sarah R. N. McIntyre

CONTENTS

Habitability is a commonly used word. Its usage is usually vague...

(Cockell et al. 2016)

The number one goal of NASA's astrobiology roadmap is to "understand the nature and distribution of habitable environments in the universe" (Des Marais et al. 2008, NASA 2011). The search for habitable planets simultaneously addresses two distinct questions: Which planets are habitable for extraterrestrial life of any kind? Which planets could we inhabit? These are existential questions because our survival may depend on our ability (1) to find and understand aliens (Stevenson and Large 2017) and (2) to find a second habitable Earth in order to become a multi-planet species (Musk 2017) (Figure 10.1.1).

10.1.1 EARLIER VIEWS OF HABITABLE PLANETS

Our notions of habitability can be traced back to pre-scientific analogies:

to consider the Earth the only populated world in infinite space is as absurd as to assert that in an entire field sown with millet only one grain will grow.

(Metrodorus of Chios, pupil of Democritus, fifth century BC)

Ancient fanciful speculation (e.g., Lucian of Samosata second century AD) developed into modern speculation (e.g., Nicolas of Cusa 1440, Bruno 1584, Kepler 1634, de Fontenelle 1686, Huygens 1698, see also Duhem 1987), like Metrodorus

FIGURE 10.1.1 Martians circa 1908. "There are certain features in which they are likely to resemble us. And as likely as not they will be covered with feathers or fur. It is no less reasonable to suppose instead of a hand, a group of tentacles or proboscis like organs." (Drawing by William R. Leigh, "The Things That Live on Mars" by H.G. Wells, *Cosmopolitan Magazine*, March 1908.)

years earlier, William Herschel (the discoverer of Uranus) believed in an infinity of inhabited planets:

> Since stars appear to be suns, and suns, according to the common opinion, are bodies that serve to enlighten, warm, and sustain a system of planets, we may have an idea of numberless globes that serve for the habitation of living creatures.

(Kawaler and Veverka 1981, Herschel 1795)

According to Clerke (1893), Herschel also thought the Sun might be inhabited, since it was a "cool, dark solid globe clothed in luxuriant vegetation and richly stored with inhabitants protected by a heavy cloud-canopy from the intolerable flare of the upper luminous region." The idea of the Sun being habitable is currently out of favor. For more on earlier ideas of habitable worlds, see Wallace (1904), Brooke (1977), Dick (1984), Crowe (1999), and Sullivan III and Carney (2007).

10.1.2 THE PHYSICS OF HABITABILITY

10.1.2.1 CIRCUMSTELLAR HABITABLE ZONES, ATMOSPHERES, AND WATER

Wet rocky Earth-like planets in the circumstellar habitable zones (CHZs) of their host stars are the focus of much current research (Kasting and Catling 2003, Gaidos et al. 2005,

Nisbet et al. 2007, Zahnle et al. 2007, Lammer et al. 2009, Kopparapu et al. 2013, Seager 2013, Cockell et al. 2016, Kaltenegger 2017). Figure 10.1.2 shows the accretion of the Earth and estimates of the CHZ. The wide orange band labeled "circumstellar habitable zone" (CHZ) has a positive slope because the luminosity of the Sun is increasing. Compared to Hart's (1979) CHZ (blue line), the width of the CHZ has increased considerably due to estimates of the thermally stabilizing effect of the carbonate-silicate cycle (Walker et al. 1981), but also see Section 10.1.4.

Since all life on Earth needs liquid water during some part of its life cycle, and the surface of the Earth is covered with it, the presence of liquid water on a planet's surface is taken as a necessary (but not sufficient) condition for life (Domagal-Goldman et al. 2016). Liquid water can exist only within the relatively narrow range of pressures and temperatures, indicated in Figure 10.1.3. If we are interested in water-based life on the surface of a planet, then the surface of that planet, like the surface of the Earth, must lie within the blue region.

10.1.2.2 INSOLATION, ESCAPE VELOCITY, AND ATMOSPHERES

The concept of a CHZ (e.g., Kasting et al. 1993) is primarily useful because it is based on two observables from an exoplanet detection: the distance from the host star and the luminosity of the host star. These tell us whether a planet is in,

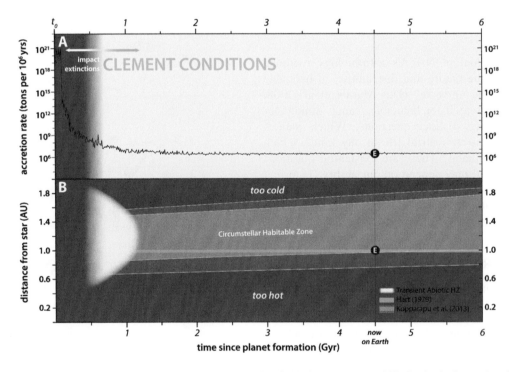

FIGURE 10.1.2 The accretion of the Earth (panel A) and circumstellar habitable zones (panel B). In A, the heavy bombardment during the first ~0.5 billion years includes the Moon-forming impact and many other impacts capable of sterilizing life (Maher and Stevenson 1988, Sleep et al. 1989, 2001, Davies and Lineweaver 2005, Cockell et al. 2012). The first few hundred million years of the formation of any rocky planet may be so impactful that the emergence of life is frustrated (hence the initial red color, which shades into the yellow clement conditions on the right side of panel A). Panel B shows several versions of circumstellar habitable zones (CHZs). The orange bands show two modern estimates of a CHZ from Kopparapu et al. (2013). These CHZs are more than 10 times wider than the narrow continuously habitable zone (thin horizontal blue band) computed by Hart (1979) before Walker et al. (1981). The Earth is plotted as a blue dot at 4.5 Gyr. (Modified from Chopra, A., and Lineweaver, C.H., *Astrobiology*, 16, 7–22, 2016.)

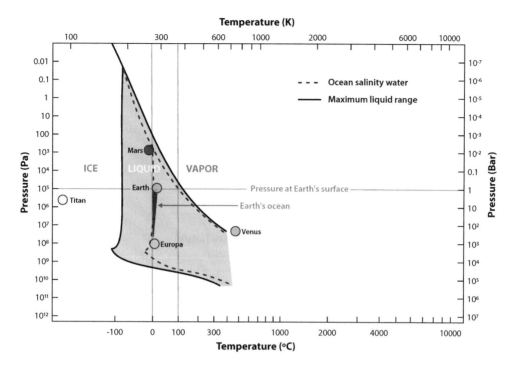

FIGURE 10.1.3 The phase diagram of H_2O. The dotted contours circumscribe ocean salinity water, while the solid contours circumscribe the maximum range of liquid water. The thin vertical sliver of dark blue represents the Earth's oceans. The surfaces of Mars, Venus, and Titan are plotted. The surface of Mars has a pressure just below the triple point pressure of ocean salinity water but well within the maximum liquid range. The subsurface water in Europa is plotted. (Modified from Jones, E.G., and Lineweaver, C.H., *Astrobiology*, 10, 349–361, 2010.)

or close to, the CHZ. These two parameters can be usefully combined into a single parameter: insolation (the amount of light falling on a planet). Although the CHZ depends only on insolation, the presence of a liquid at the surface of a rocky body depends on that rocky body having an atmosphere with a surface pressure higher than the triple point pressure of the substance of interest.

The presence of an atmosphere and the contents of that atmosphere depend pre-dominantly on insolation and surface gravity. For example, the Moon has the same insolation as the Earth, and so, the Moon is in the CHZ, but it is not large enough to hold onto an atmosphere. The Moon, Mercury, Ganymede, Ceres, and Vesta do not appear in Figure 10.1.3, because these bodies are not massive enough to hold onto an atmosphere with a surface pressure larger than the triple point of water. The presence or absence of an atmosphere is an important pre-requisite for surface water. Since the presence of an atmosphere depends on a planet's escape velocity, to take the next step in understanding surface water and habitability, we need to plot escape velocity versus insolation. In a groundbreaking paper, Zahnle and Casting (2017) made such a plot (Figure 10.1.4).

In the lower left of Figure 10.1.4, Sedna's insolation is so low that its mean surface temperature is 12 K. It is thought to have a neon atmosphere. All the other gases are condensed out as ices on its surface. At their triple point pressures, neon condenses at 25 K, nitrogen at 77 K, methane at 91 K, carbon dioxide at 217 K, and water at 273 K. Thus, as we move diagonally up the brown band to higher surface gravities, higher

insolations, and, on average, higher surface temperatures, the more volatile ices begin to seasonally sublimate—Pluto, Triton, and Titan have nitrogen and methane atmospheres. As we move further up diagonally, CO_2 and then H_2O also begin to contribute to atmospheric gases. Further up, liquid water becomes so abundant that we are in the realm of water worlds or ocean planets (e.g., Léger 2004). In the upper right, escape velocities are so high that hydrogen and helium atmospheres are maintained through planet formation. The initial abundance of H and He is ~100 times more abundant than any rock-forming material, so these planets will be gaseous hot Jupiters and hot Saturns rather than rocks with atmospheres—and therefore are of less interest to habitable-planet hunters.

If a body has a high eccentricity, like Sedna (e ~0.8), then its insolation is variable. This can produce a "see saw" atmosphere that freezes out around aphelion to produce an icy surface with no atmosphere. Around perihelion, the ice revaporizes and reinstates the atmosphere. Eris's eccentricity of e~0.4 produces a nitrogen seesaw atmosphere. Triton's low eccentricity gives it a stable insolation and therefore a stable nitrogen atmosphere. Large bodies with high escape velocities and low insolations (lower right in Figure 10.1.4) will have atmospheres that dominate to such an extent that they don't have an identifiable rocky surface (e.g., Jupiter, Saturn, Uranus, and Neptune).

Detection of a planet by both radial velocity and transit photometry yields an estimate of both mass and radius, from which density can be estimated. Densities reveal whether a planet is rocky (like Mercury, Venus, Earth, and Mars) or

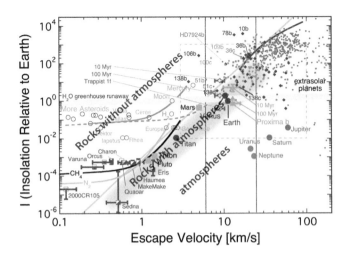

FIGURE 10.1.4 Insolation and escape velocities. The diagonal turquoise line seems to be the empirical boundary between rocks with and without atmospheres. Rocks in the upper left have no atmospheres—their escape velocities are too low and their insolations are too high to keep atmospheres. Rocks to the lower right have escape velocities high enough and insolations low enough to keep atmospheres. Rocks in the far lower right have so much atmosphere that we call them gaseous planets (i.e., Jupiter and Saturn). The brown region indicates rocky planets or moons with atmospheres. Starting from the right, the closer a rock is to the turquoise line, the more tenuous is its atmosphere. For example, Titan (red dot) has a relatively thick atmosphere, while closer to the turquoise boundary, Mars, Triton, Pluto, and Ganymede have almost negligible atmospheres. The horizontal yellow band corresponds to the insolation assumed to characterize the CHZ. Notice that the Moon is in the CHZ but is also in the "rocks without atmospheres" region. The small blue dots in the upper right are mostly gaseous hot Saturns and hot Jupiters with insolation ~3 orders of magnitude larger than the isolation of Saturn and Jupiter. The next figure shows a blowup of the region within the dashed rectangle. (Modified from Figure 1 of Zahnle, K.J., and Catling, D.C. *Astrophys. J.*, 843, 122, 2017.)

gaseous (like Jupiter and Saturn) or something else. Most planets with masses less than ~2.5 $M_\oplus$ or with radii less than ~1.4 $R_\oplus$ are rocky (Rogers 2015, Kaltenegger 2017, but see Fulton et al. 2017). If they are to the right of the turquoise line in Figure 10.1.4, they will have atmospheres. If they are in the CHZ insolation region (yellow), these become the most interesting potentially habitable Earth-like planets.

For larger planets, somewhere in the mass range $2.5\,M_\oplus < M < 4\,M_\oplus$ or in the radius range $1.4\,R_\oplus < R < 1.6\,R_\oplus$, surface gravity becomes large enough to hold H_2O and H_2, producing lower densities. Thus, we do not expect planets with radii more than ~2 $R_\oplus$ to be rocky. The escape velocity v'_{esc} of a planet with a radius R' and density ρ' is:

$$v'_{esc} = v_{esc\,\oplus} \left(\frac{R'}{R_\oplus}\right)\sqrt{\left(\frac{\rho'}{\rho_\oplus}\right)} \qquad (10.1.1)$$

Thus, for the largest rocky planets with twice the radius of the Earth and about the same density, we have:

$$v'_{esc} \approx 2\,v_{esc\,\oplus} \approx 22\,\text{km/sec} \qquad (10.1.2)$$

This value is indicated with a vertical black line in Figures 10.1.4 and 10.1.5. At the low-mass end, to be in the

yellow CHZ insolation and to be to the right of the turquoise line, a planet needs an escape velocity of half Earth's or ~6 km/sec. At the same density as the Earth, this corresponds to a planet with half the Earth's radius. This 6 km/sec escape velocity is also indicated with a vertical black line in Figures 10.1.4 and 10.1.5. The notional "habitable" green region is between these limits and centered on the CHZ insolation band. Both Mars and Venus are to the right of the turquoise boundary and have atmospheres, but their insolation is just below and just above the yellow CHZ, respectively. If Mars had been more massive, it would have had a larger escape velocity and a thicker atmosphere and would presumably have been habitable longer for water-based life at its surface. If Venus had been less massive (lower escape velocity), it would have had a thinner atmosphere that presumably would be less susceptible to a runaway greenhouse and would probably have been more habitable. These are the considerations that notionally tilt the green habitable region (for water-based life) compared with the horizontal yellow insolation levels of the CHZ. Proxima Centauri b and TRAPPIST-1f are currently within this notional habitable zone, but they may have lost all their initial water because of the long and luminous pre-main sequence evolutionary paths of their M-star hosts (Ribas et al. 2016, Barnes et al. 2018, Grimm et al. 2018).

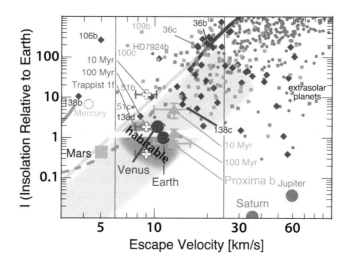

FIGURE 10.1.5 Blow up of the region around the Earth, from the previous figure. The vertical black line on the right indicates the surface gravity of a roughly Earth-composition planet with a radius approximately twice the Earth's radius. The vertical black line on the left is surface gravity of a planet just enough to hold onto an atmosphere in the CHZ. This corresponds to a rocky planet with approximately half the Earth's radius. See text below for discussion of the time dependence of the three points for Proxima Centauri b and TRAPPIST-1f. (Modified from Figure 1 of Zahnle, K.J., and Catling, D.C. *Astrophys. J.*, 843, 122, 2017.)

10.1.2.3 THE M-STAR HABITABILITY PROBLEM (EARLY DEVOLATILIZATION)

Low-mass M-stars are the most numerous stars in the universe, and therefore, they may host the largest number of habitable planets. TRAPPIST-1 and Proxima Centauri are M dwarfs with 9% and 12% of the mass of the Sun, respectively. Pre-main sequence Hayashi tracks tell us that during the first few hundred million years, the luminosity of these stars decreased by ~2 orders of magnitude (Ramirez and Kaltenegger 2015). Thus, when one plots the insolation as a function of time of the two planets (TRAPPIST-1f and Proxima Centauri b), which are currently in or near the CHZ of these two stars, their insolation decreases by the factors of 38 and 7, respectively (compare the three points for each planet in Figures 10.1.4 and 10.1.5). Accordingly, if a rocky planet is currently in the CHZ of an M dwarf, during the first half a billion years of its existence, it experienced an insolation (and an extreme UV flux) probably large enough to devolatilize its surface, removing most or all its water (Luger and Barnes 2015, Zahnle and Catling 2017). This effect is larger for lower-mass stars, as can be seen in Figure 10.1.5 by the larger spread in insolation between the three TRAPPIST-1f points compared with the three Proxima Centauri b points.

The CHZ insolation and atmosphere-securing escape velocities plotted in Figure 10.1.4 are important factors affecting habitability. A planet's position along the brown-shaded region of Figure 10.1.4 can help us make educated guesses about atmospheric compositions based on the condensation temperatures of the half dozen likely candidates for atmospheric gases (Ne, N_2, CH_4, CO_2, H_2O, and H_2).

The detailed volatile content of the atmosphere controls both albedo and greenhouse warming and thus strongly influences the habitability of a rocky planet (Catling and Kasting 2017). But volatile content is volatile. The strength, rapidity, and universality of abiotic positive feedbacks and the rapid evolution of the atmosphere, probably within the first billion years, as happened for Venus and Mars (Jones et al. 2011), can lead to temperatures too hot or cold for life (through runaway greenhouse or runway glaciation) and loss of liquid water (through atmospheric escape of hydrogen), which can preclude long-term planetary habitability (Chopra and Lineweaver 2016).

The variability in the initial inventories of ices and volatiles (e.g., H_2O) of CHZ planets is probably quite large, because it is the result of a small number of impacts with the largest volatile-rich planetesimals. This variability is unobservable in exoplanets but may be large enough to strongly modify the assumed effects of a given insolation. A good example of this is the 750°C surface temperature of Venus, dominated not by its insolation or effective temperature but by its greenhouse gas composition and a runaway greenhouse effect. The detailed atmospheric compositions and greenhouse gas fractions of rocky exoplanets in the CHZ of their host stars are not observables and will remain guesswork for the next decade or so.

A variety of other astrophysical and geological factors may play important roles in modifying habitability, including volcanism (Ramirez and Kaltenegger 2017), plate tectonics (Valencia et al. 2007; Noack and Breuer 2014; Noack et al. 2014; Foley 2015), bulk chemistry (Jellinek and Jackson 2015), albedo (Joshi and Haberle 2012), magnetic fields (Christensen 2010; Stevenson 2010; Tarduno et al. 2015), eccentricity (Dressing et al. 2010; Linsenmeier et al. 2015), tidal heating (Driscoll and Barnes 2015; Barr et al. 2018), tidal locking (Yang et al. 2013; Barnes 2017), impact events (Cockell et al. 2012; Lupu et al. 2014), and stellar type (Fritz et al. 2014; Ramirez and Kaltenegger 2015).

10.1.3 USING THE EARTH AS A HABITABILITY CALIBRATOR

The radius of the Earth is about 6000 km. The thickness of the biosphere (Δx) is about 6 km (Figure 10.1.6). Thus, only a tiny fraction of the Earth is habitable:

$$\frac{\text{Thickness of biosphere } (\Delta x)}{\text{Radius of Earth } (R)} \approx \frac{6 \text{ km}}{6000 \text{ km}} \approx \frac{1}{1000} \quad (10.1.3)$$

or in terms of fractional volume:

$$\frac{\text{Volume}_{\text{biosphere}}}{\text{Volume}_{\text{Earth}}} \approx \frac{4\pi R^2 \Delta x}{\frac{4}{3}\pi R^3} \approx \frac{3\Delta x}{R} \approx \frac{3}{1000} \quad (10.1.4)$$

Thus, an overwhelming majority of the volume of the Earth is uninhabitable, and not even all the liquid water on Earth is habitable (Jones and Lineweaver 2010). Even the ~6-km-thin bioshell of the Earth is not equally habitable. The density of life (as a proxy for habitability) on the surface of the Earth, from polar caps and deserts to rain forests, varies by many orders of magnitude (Figure 10.1.7, Lineweaver and Chopra 2012a).

Earthlife (Feinberg and Shapiro 1980) is NOT everywhere. The entire volume of the Earth beneath a depth of ~5 km is a temperature desert, with temperatures above the 122°C maximum temperature of life (Takai et al. 2008). Also, the polar caps with temperatures below −20°C are cold-temperature deserts. Lack of water is the bottleneck for life in the traditional deserts such as the Atacama and Sahara. But there may also be places on Earth where a paucity of carbon or phosphorus or sulfur precludes the existence of life (Harrison et al. 2013). These chemical deserts have not been well-explored.

The Earth thus contains environments over the entire spectrum, from completely uninhabitable to very habitable (and richly inhabited). So, the Earth is both uninhabitable and habitable, depending on location. Similarly, nuances in understanding habitability can be made if we discourage the common zero-dimensional dichotomy, habitable or uninhabitable, and talk rather about the extent of habitability. Habitability is at least a one-dimensional continuous variable (Figure 10.1.8). Discussion of planets more habitable than the Earth ("superhabitability") makes a useful contribution to this adjustment (Heller and Armstrong 2014). In this view, the CHZ has no sharp boundaries (sometimes characterized by the inappropriate precision of three significant figures). To remove the unrealistically close connection between the words "habitable zone planet" and "planet with life," the phrase "circumstellar habitable zone" should be replaced by "circumstellar temperate zone," as in Gillon et al. (2017).

One large clue that we do not understand about the habitability of the Earth is that the vast majority of life-forms on Earth cannot be cultivated. We do not know what are their requirements to stay alive. Apparently, isolation from other organisms and diets of agar-agar in petri dishes are not habitable environments for most organisms. It is for this reason that metagenomics (as opposed to the genomics of cultured organisms) has been able to find entirely new phyla in the tree of life (Hug et al. 2016).

Culturing problems are issues of micro-habitability. One can also discuss habitable zones at other size scales (Nisbet et al. 2007). For example, one can identify the galactic

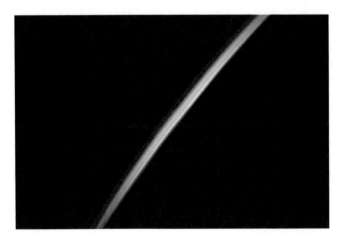

FIGURE 10.1.6 Photograph of the Earth's atmosphere at sunrise on February 9, 2016, taken by astronaut Scott Kelly from the International Space Station.

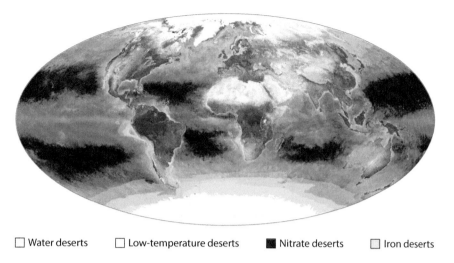

FIGURE 10.1.7 Four types of deserts where life in Earth's bioshell is relatively rare. If we take biomass density as a proxy for the habitability of the surface of the Earth, then the habitability of the surface varies by many orders of magnitude. (From Lineweaver, C.H., and Chopra, A., *Annu. Rev. Earth Planet. Sci.*, 40, 597–623, 2012.)

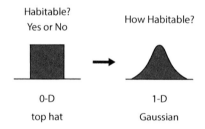

FIGURE 10.1.8 A conceptual transition is needed from a zero-dimensional model of a circumstellar habitable zone to a more nuanced Gaussian distribution, which can parametrize the extent of habitability.

habitable zone—regions of the galaxy most conducive to life (e.g., Lineweaver et al. 2004) or, more speculatively, discuss the habitable island universes in the multiverse—universes in the multiverse, that are most conducive to life (Ćirković 2012, Lineweaver 2014).

10.1.4 THE BIOLOGY OF HABITABILITY

In our discussion of CHZs, insolation, surface gravity, and atmospheres, we have only considered the physics and chemistry of habitability. However, if it is the case that only inhabited planets are habitable—if life plays a major role in modifying its environment to make it habitable—then we have so far ignored what may be the dominant parameter controlling the habitability of a planet: the life on it.

10.1.4.1 THE CASE FOR A GAIAN BOTTLENECK

The pre-requisites and ingredients for life seem to be abundantly available in the universe. However, we have yet to find any evidence for extraterrestrial life. A common explanation

for this is a low probability for the emergence of life (an emergence bottleneck), notionally due to the intricacies of the molecular recipe. An alternative Gaian bottleneck explanation argues that if life emerges on a planet, it only rarely evolves quickly enough to regulate greenhouse gases and albedo, thereby maintaining surface temperatures compatible with liquid water and habitability (Chopra and Lineweaver 2016). Such a Gaian bottleneck suggests that (1) extinction is the cosmic default for most life that has ever emerged on the surfaces of wet rocky planets in the universe and (2) rocky planets need to be inhabited to remain habitable (Figure 10.1.9).

The most important data needed to constrain, validate, or invalidate the Gaian Bottleneck model will probably come from estimates of the strength of the abiotic negative feedback of the carbonate-silicate cycle in the first billion years of Earth's history, when the area of continental crust (and therefore the amount of sub-aerial silicate weathering) was probably negligible. If negative feedback from silicate weathering is to create a stable CHZ during the first billion years, several conditions need to be fulfilled. Sub-aerial or sub-aqueous silicate weathering needs to be strong enough, to make its negative feedback dominate the positive feedbacks from a runaway greenhouse or a runaway ice-albedo glaciation.

Biotic regulation could provide the necessary level of negative feedback (lower panel of Figure 10.1.10). However, the emergence of metabolisms and ecosystems that could regulate planetary-scale greenhouse gases or albedo may be a rare and quirky result of evolution—like vertebrate heads, mammalian hair, bird feathers, or human language. Dependence on the evolution of such quirks may present a Gaian bottleneck to the persistence of life on inherently unstable planets. In the Gaian bottleneck model, the maintenance of planetary habitability is a property more associated with a quirky and unusually rapid evolution of biological regulation of surface volatiles than with the luminosity and distance to the host star.

Emergence Bottleneck

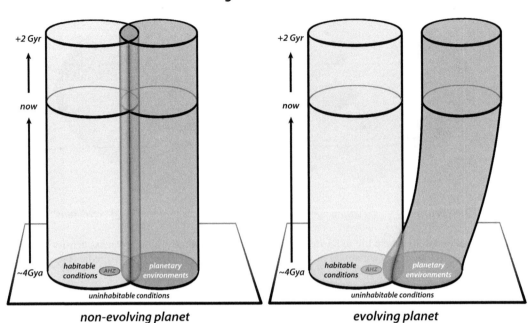

No Emergence Bottleneck

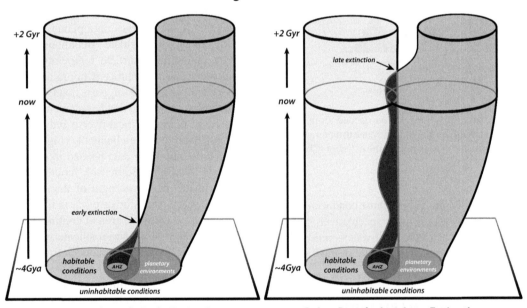

FIGURE 10.1.9 Emergence bottleneck (top panel) versus no emergence bottleneck (bottom panel). The yellow tubes represent habitable conditions that we have assumed to be unchanging through time. The orangish spot at the base of the yellow tubes, labeled "AHZ" (abiogenesis habitable zone), represents a more restrictive set of conditions necessary and sufficient for life to get started from non-life. The blue tubes represent the environments on a planet as a function of time (~4 billion years ago at the base, and ~2 billion years into the future at the top). The environments on emergence bottleneck planets (blue tubes in the top panel) do not overlap with the AHZ, so life is unable to get started. Even if such planets have habitable conditions, they are uninhabited by life that evolved in situ. No emergence bottleneck planets (bottom panel) are planets in which life can get started, since their environments initially overlap with the AHZ. The Gaian bottleneck is the idea that even if the emergence of life is a common occurrence (as in the two blue tubes on the bottom panel), the evolution of biotic regulation of greenhouse gases and/or albedo (needed to maintain habitable conditions and life) is a rare and random adaptation (blue tube on the bottom right). The most common default on wet rocky planets would be early extinction (blue tube on the bottom left). On such planets, life emerges but never acquires globally adaptive properties that permit biotic regulation of an inherently unstable surface environment. This is an extension to the planetary level of the idea that most mutations are maladaptive. (From Doolittle, W.F., *J. Theor. Biol.*, 434, 11–19, 2017.)

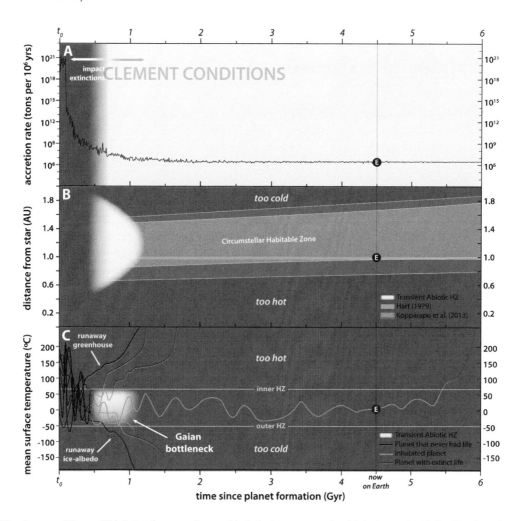

FIGURE 10.1.10   Same as Figure 10.1.2, but here, we have added the bottom panel, which shows the Gaian Bottleneck model, in which, unlike the Earth, most planets go either runaway greenhouse and get too hot or runaway ice-albedo and get too cold, to have liquid water at their surfaces. The yellow region in panel C between 0.5 Gyr and 1 Gyr is where life could have gotten started—after the early heavy bombardment but before the runaway greenhouse or runaway ice-albedo.

10.1.4.2   THE ABIOGENESIS HABITABLE ZONE: CONDITIONS REQUIRED TO START LIFE

We are unsure how life got started (Oparin 1968, Lahav 1999, Peretó 2005, Sutherland 2017), but there are several plausible non-mutually-exclusive candidates: metabolism first, also known as the garbage bag world (e.g., Dyson 1999); compartmentalization first (Deamer 1997, Segre and Lancet 2000); and information first, also known as the RNA world (Gilbert 1986, Joyce 2012).

The abiogenesis habitable zone (AHZ) could depend on impact events of a heavy bombardment, hydrothermal activities (enhanced on younger, more tectonically active planets), lightning, or large tides producing cycling between hydration and dehydration (Figure 10.1.11) (Brandes et al. 1998). To make more specific educated guesses about the requirements to get Earthlife started, we need to find out more about the metabolism and environment of the last universal common ancestor (LUCA). Possible scenarios of proto-biological molecular evolution before the LUCA have been proposed (Mann 2013).

The problem of how and where Earthlife got started can be approached through genomics (e.g., Weiss et al. 2016) or by trying to identify promising potential AHZ environments. The two most popular scenarios for the origin of Earthlife are alkaline hydrothermal vents in the deep sea (e.g., Martin and Russell 2007, Smith and Morowitz 2016) and freshwater hydrothermal systems on land (Djokic et al. 2017, Damer et al. 2017). Hydrothermal vents (acidic or alkaline) seem to harbor a promising variety of prebiotic chemistries (Holm and Andersson 2000; Orgel 2004; Padgett 2012; Sojo et al. 2016). See also Joyce and Orgel (1993), Wächtershäuser (2000, 2008), de Duve (2000), Burmeister (2000), Cleaves (2015), and McDonald (2015). Since our ideas about the Earth's AHZ are quite speculative, our ideas about the origin of life elsewhere and a universal AHZ are even more so.

10.1.4.3   THE EVOLUTION OF HABITABILITY

Bananas can't survive in a desert. Cacti can't survive in a rain forest. In our discussion of habitability, it seems we need to specify the kind of life we are talking about. If we're curious about whether a planet is habitable or not (or how habitable it is), we need to know what life-form we are talking about. We need to ask the question: Habitable for whom?

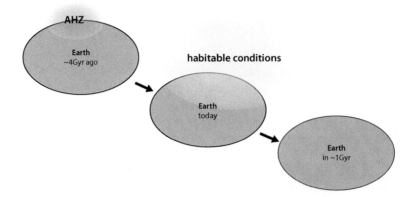

FIGURE 10.1.11   The abiogenesis habitable zone (AHZ) and habitable conditions. In Figure 10.1.9, we assumed that the AHZ was a subset of the yellow tubes representing habitable conditions. Here, we drop that assumption. The three ovals represent the evolution of the environments on Earth. Four billion years ago, there were environments on Earth in which Earthlife got started. Despite the Earth evolving away from the AHZ, habitable conditions evolved and were maintained. A billion years from now, the Earth will no longer have habitable conditions, probably because of the loss of hydrogen and therefore the loss of the ability to maintain surface water.

The concept of habitability is inextricably coupled to the concept of life (but see Cockell 2011). Cockell et al. (2016) suggested to define habitability as "the ability to support the activity of at least one known organism." However, for microbiologists, who can culture only a tiny fraction of known bacteria, this definition is operationally difficult. The number of known organisms is growing, but our ability to culture, isolate, and characterize them is severely limited (Rinke et al. 2013). This being the case, how can we know whether a given environment is habitable, even for some of the most common organisms on Earth?

For astrobiologists, primarily interested in the habitability of wet rocky Earth-like planets, the limitations of Cockell et al.'s (2016) definition are severe. If we are interested only in the question: "Could any known organism on Earth, live on that planet?," then we are left with the obvious problem that organisms do not live alone. Individual organisms are obligately embedded in ecological webs made of other mutually dependent life-forms. An obvious example is the dependence of aerobic organisms (e.g., us) on the availability of molecular oxygen produced by photosynthesizers.

The presence of liquid water at the surface of a planet could be an abiotic property set by temperature and pressure (as is usually assumed), or it could be a biotic feature regulated by a biosphere (Harding and Margulis 2010). The effect of physical parameters such as insolation and escape velocity can be strongly modified by biotic regulation of albedo and greenhouse gas content of the atmosphere. In other words, the existence of feedback between life and its environment—and even the distinction between life and environment—is problematic. The biotic regulation of the carbon-silicate cycle is an example (Walker et al. 1981, Schwartzman and Volk 1989). If life plays a strong role in creating a habitable planet, then it is possible that to be habitable, a planet needs to be inhabited (Kasting 2010, Goldblatt 2016, Hall 2016, however see also Cockell 2011).

If life and environment are in a tight feedback, then we should not conceive of them as being separate. We cannot compute the width of a CHZ with any useful precision without taking biotic regulation into consideration. Circumstellar habitable zone computations seek to define the region within which a planet should be capable of harboring life…

> … yet whether the planet is inhabited will determine whether the climate may be habitable at any given distance from the star. This matters because future life detection missions may use habitable zone boundaries in mission design.
>
> **(Goldblatt 2016)**

It is generally agreed that the very early Earth was uninhabitable. A few hundred million years later, life got started, and since we are here today, the Earth necessarily became habitable. It is also generally agreed that in 1 or 2 billion years from now, the Earth will have lost its water and be uninhabitable again. Obviously, habitability changes and evolves under the influence of an evolving planet and the evolving life on it.

10.1.4.4   DEFINITIONS OF LIFE AND DEFINITIONS OF HABITABILITY ARE NOT HELPFUL

It seems obvious to many people that if we want to find life—if we want to develop an instrument to identify life—and if we want to find an unambiguous biosignature that can be remotely detected (Lovelock 1965, Meadows and Seager 2010, Seager et al. 2016, Krissanen-Totten et al. 2018), then we need to be as specific as possible about how life differs from non-life. This suggests that we need a definition of life.

Biology textbooks are full of definitions of life. Most definitions have been sculpted around the flora, fauna, and fungi we know on Earth. They usually mention growth, self-regulation, self-reproduction, and chemical complexity (Lineweaver and Chopra 2012b). Much thought has gone into the development of an operational definition of life suitable for application elsewhere in the universe (Sagan 1970, Joyce 1994, Chyba and McDonald 1995, Gould 1995, Nealson and Conrad 1999, Cleland and Chyba 2002, 2007, Koshland 2002, Lineweaver 2006, Benner 2010, Kolb 2010, Bains 2014). None is universally accepted (Popa 2015).

The title of an influential article by Jack Szostak (2012) is "Attempts to Define Life Do Not Help to Understand the Origin of Life," in which he argues that if one wants to understand the origin of Earthlife, it is not useful to define it. After all, one is trying to understand what is probably a relatively complicated and drawn-out transition from non-life to what we now call life.

> if the emergence of life is seen as the stepwise (but not necessarily slow) evolutionary transition between the non-living and the living, then it may be meaningless to draw a strict line between them… as argued long ago by Immanuel Kant, precise definitions may be achievable in mathematics and philosophy, but empirical concepts (as is the case for what life is) can only be made explicit through descriptions that depend on the historical context (Fry 2000).
>
> **(Tirard et al. 2010)**

Like democracy and eyeballs, life itself has a history and, therefore, according to Nietzsche, cannot be defined:

> Only that which has no history can be defined
>
> **(Nietzsche 1887)**

An all-encompassing goal for astrobiology is to understand cosmic evolution—cosmic history—and since it is history, precise definitions are not the way to go.

This conundrum is very similar to the problem of defining *Homo sapiens*. "There never was an Australopithecus mother who gave birth to a Homo child." (Dawkins 2017). As long as we were unaware of *Homo erectus*, Neanderthals, Denisovans, Australopithecines, and the large number of ancestors and extinct cousins that connect us to our common ancestor with chimps, we could naively congratulate ourselves for having workable definitions of *Homo sapiens* that depend for their success on our ignorance of these transitional forms. But *Homo sapiens,* like all life-forms (and, we postulate like life itself), has a history. The more we find out about the history of the evolution of life and what might be called transitional forms, the more meaningless it becomes to draw a strict line between life and non-life (Szostak 2012).

> The whole system of labeling species with discontinuous names is geared to a time slice, such as the present, in which ancestors have been conveniently expunged from our awareness. If by some miracle every ancestor were preserved as a fossil, discontinuous naming would be impossible.
>
> **(Dawkins 2017)**

We can apply Dawkins' reasoning about species more generically to life:

> The whole system of labeling things life or non-life is geared to a time slice, such as the present, in which transitional forms are no longer present or are ignored. If by some miracle every ancestor, and every transitional form were preserved as a fossil, discontinuous naming of some things "life" and other things "non-life" would be impossible.

These same caveats about defining life apply to attempts to define habitability. Since the concept of "habitability" depends on the concept of "life," it may be meaningless to draw a strict line between habitable and non-habitable. If there are no minimal properties required for a system to be considered alive, it follows that there are no minimal properties required for an environment to be considered habitable.

Habitability becomes more precise if we are talking about habitable for us. But then, it becomes less precise because we modify our environments. In a spacesuit, the surface of the Moon is habitable. Without one, it is uninhabitable. Humans are not alone in this respect. Life has a long tradition of modifying its environment, from the formation of the first lipid bilayers to termite nests and beaver dams (Odling-Smee et al. 2013). In trying to understand life and habitability, studying the important feedback between life and its environment and their coevolution will bear more fruit than insisting that what we are looking for has a precise definition (Chopra and Lineweaver 2018).

10.1.5 OUTSTANDING ISSUES

Studies of the habitability of a planet and the characteristics of habitable planets are problematic because of our ignorance about the following issues:

1. What is life? If we don't understand what life elsewhere is, we cannot understand what habitability elsewhere is. How can we talk about habitability without having a reasonable idea of what life is?
2. What are the conditions under which life can originate? Does it make sense to talk about habitability on a planet on which life could not originate?
3. What are the conditions that maintain life? Since we have very little idea about the limits of life elsewhere, we have very little idea about what could maintain life elsewhere.
4. Negative feedbacks are necessary to prevent Earth-like planets from becoming runaway greenhouses and runaway glaciations and thus uninhabitable. Are these feedbacks abiotic, biotic, or both? Are there different negative feedbacks over the habitable lifetime of a planet?
5. It is not easy to detect something if you don't know what you are looking for. As we learn more about exoplanet atmospheres, the ambiguity between biotic and abiotic chemical disequilibrium may not go away.

REFERENCES

Bains, W. 2014. What do we think life is? A simple illustration and its consequences. *International Journal of Astrobiology*, 13(2), 101–111.

Barnes, R. 2017. Tidal locking of habitable exoplanets. *Celestial Mechanics and Dynamical Astronomy*, 129(4), 509–536.

Barnes, R., R. Deitrick, R. Luger et al. 2018. The habitability of Proxima Centauri b I: Evolutionary scenarios. *Astrobiology*, 18(2), 133–189.

Barr, A.C., Dobos, V., Kiss, L.L. 2018. Interior structures and tidal heating in the TRAPPIST-1 planets. *Astronomy & Astrophysics*, 613, A37.

Benner, S.A. 2010. Defining life. *Astrobiology*, 10(10), 1021–1030.

Brandes, J.A., Boctor, N.Z., Cody, G.D. et al. 1998. Abiotic nitrogen reduction in the early Earth. *Nature*, 395, 365–367.

Brooke, J.H. 1977. Natural Theology and the plurality of worlds: Observations on the Brewster-Whewell debate. *Annals of Science*, 34, 3.

Bruno, G. 1584. *The Ash Wednesday Supper*. The Hague, the Netherlands: Mouton & Co.

Burmeister, J. 2000. Self-replication and autocatalysis. In *The Molecular Origins of Life: Assembling Pieces of the Puzzle*, A. Brack (Ed.), pp. 295–311. Cambridge, UK: Cambridge University Press.

Catling, D.C., and Kasting, J.F. 2017. *Atmospheric Evolution on Inhabited and Lifeless Worlds*. Cambridge, UK: Cambridge University Press.

Chopra, A., and Lineweaver, C.H. 2016. The case for a Gaian bottleneck: The biology of habitability. *Astrobiology*, 16(1), 7–22.

Chopra, A., and Lineweaver, C.H. 2018. The cosmic evolution of biochemistry. In *Habitability of the Universe Before Earth*, R. Gordon and A. Sharov (Eds.), Volume 1 in the series: Astrobiology: Exploring Life on Earth and Beyond, series editors: P.H. Rampelotto, J. Seckbach, and R. Gordon. pp. 75–87. Amsterdam, the Netherlands: Elsevier.

Christensen, U.R. 2010. Dynamo scaling laws and applications to the planets. *Space Science Reviews*, 152(1–4), 565–590.

Chyba, C.F., and McDonald, G.D. 1995. The origin of life in the solar system: Current issues. *Annual Review of Earth and Planetary Sciences*, 23(1), 215–249.

Ćirković, M.M. 2012. *The Astrobiological Landscape: Philosophical Foundations of the Study of Cosmic Life*. Cambridge, UK: Cambridge University Press.

Cleaves, H.J. 2015. Prebiotic synthesis of biochemical compounds: An overview. In *Astrobiology: An Evolutionary Approach*, V.M. Kolb (Ed.), pp. 83–117. Boca Raton, FL: CRC Press/Taylor & Francis.

Cleland, C.E., and Chyba, C.F. 2002. Defining life. *Origins of Life and Evolution of Biospheres*, 35, 333–343.

Cleland, C.E., and Chyba, C.F. 2007. Does 'life' have a definition? In *Planets and Life: The Emerging Science of Astrobiology*, W.T. Sullivan III and J.A. Baross (Eds.), pp. 119–131. Cambridge, UK: Cambridge University Press.

Clerke, A. 1893. *History of Astronomy During the Nineteenth Century*. London, UK: Adam and Charles Black.

Cockell, C.S. 2011. Vacant habitat in the Universe. *Trends in Ecology & Evolution*, 26, 73–80.

Cockell, C.S., Bush, T., Bryce, C. et al. 2016. Habitability: A review. *Astrobiology*, 16(1), 89–117.

Cockell, C.S., Voytek, M.A., Gronstal, A.L. et al. 2012. Impact disruption and recovery of the deep subsurface biosphere. *Astrobiology*, 12, 231–246.

Crowe, M.J. 1999. *The Extraterrestrial Life Debate, 1750-1900*. Mineola, NY: Dover.

Damer, B.F., Deamer, D.W., Van Kranendonk, M.J., and Djokic, T. 2017. An origin of life in terrestrial fresh water hydrothermal pools [abstract]. In *Astrobiology Science Conference*, April 24–28, Mesa, AZ. Abstract 3220.

Davies, P.C.W., and Lineweaver, C.H. 2005. Finding a second sample of life on earth. *Astrobiology*, 5(2), 154–163.

Dawkins, R.D. 2017. The dead hand of Plato. In *Science in the Soul*, G. Somerscales (Ed.), pp. 287–296. London, UK: Bantam Press.

Deamer, D. 1997. The first living systems: A bioenergetics perspective. *Microbiology and Molecular Biology Review*, 61, 239–261.

De Duve, C. 2000. Clues from present-day biology: The thioester world. In *The Molecular Origins of Life: Assembling Pieces of the Puzzle*, A. Brack (Ed.), pp. 219–236. Cambridge, UK: Cambridge University Press.

de Fontenelle, B.B. 1686. *Conversations on the Plurality of Worlds*. Berkeley, CA: University of California Press.

Des Marais, D.J., Nuth III, J.A., Allamandola, L.J. et al. 2008. The NASA astrobiology roadmap. *Astrobiology*, 8, 715–730.

Dick, S.J. 1984. *Plurality of Worlds: The Origins of the Extraterrestrial Debate from Democritus to Kant*. Cambridge, UK: Cambridge University Press.

Djokic, T., Van Kranendonk, M.J., Campbell, K.A. et al. 2017. Earliest signs of life on land preserved in ca. 3.5 hot spring deposits. *Nature Communications*, 8, 15263.

Domagal-Goldman, S.D., Wright, K.E. Adamala, K. et al. 2016. The astrobiology primer v2.0. *Astrobiology*, 16, 561–653.

Doolittle, W.F. 2017. Darwinizing Gaia. *Journal of Theoretical Biology*, 434, 11–19.

Dressing, C.D., Spiegel, D.S., Scharf, C.A., Menou, K., and Raymond, S.N. 2010. Habitable climates: The influence of eccentricity. *The Astrophysical Journal*, 721(2), 1295.

Driscoll, P.E., and Barnes, R. 2015. Tidal heating of Earth-like exoplanets around M stars: Thermal, magnetic, and orbital evolutions. *Astrobiology*, 15(9), 739–760.

Duhem, P. 1987. *Medieval Cosmology: Theories of Infinity, Place, Time, Void, and the Plurality of Worlds*. Translated and edited by R. Ariew. Chicago, IL: The University of Chicago Press.

Dyson, F. 1999. *Origins of Life*, 2nd ed. Cambridge, UK: Cambridge University Press.

Feinberg, G., and Shapiro, R. 1980. *Life Beyond Earth: The Intelligent Earthlings Guide to Life in the Universe*. New York: William Morrow.

Foley, B.J. 2015. The role of plate tectonic–climate coupling and exposed land area in the development of habitable climates on rocky planets. *The Astrophysical Journal*, 812(1), 36.

Fritz, J., Bitsch, B., Kührt, E. et al. 2014. Earth-like habitats in planetary systems. *Planetary and Space Science*, 98, 254–267.

Fry, I. 2000. *The Emergence of Life on Earth: A Historical and Scientific Overview*. New Brunswick, NJ: Rutgers University Press.

Fulton B.J., Petigura E.A., Howard A.W. et al. 2017. The California-Kepler Survey. III. A Gap in the Radius Distribution of Small Planets. *The Astronomical Journal*, 154, 109.

Gaidos, E., Deschenes, B., Dundon, L., Fagan, K., Menviel-Hessler, L., Moskovitz, N., and Workman, M. 2005. Beyond the principle of plentitude: A review of terrestrial planet habitability. *Astrobiology*, 5(2), 100–126.

Gilbert, W. 1986. The RNA world. *Nature*, 319, 618.

Gillon, M., Triaud, A.H., Demory, B.O., Jehin, E., Agol, E., Deck, K.M., and Bolmont, E. 2017. Seven temperate terrestrial planets around the nearby ultracool dwarf star TRAPPIST-1. *Nature*, 542(7642), 456–460.

Goldblatt, C. 2016. The inhabitable paradox: How habitability and inhabitancy are inseparable. arXiv:1603:00950v1.

Gould, S.J. 1995. What is life? As a problem in history. In *What is Life? The Next Fifty Years*, M.P. Murphy and L.A.J. O Neill (Eds.), pp. 25–39. Cambridge, UK: Cambridge University Press.

Grimm, S.L., Demory, B.-O., Gillon, M. et al. 2018. The nature of the TRAPPIST-1 exoplanets. *Astronomy & Astrophysics*. 613, A68.

Hall, S. 2016. Which came first on Earth – Habitability or Life? *Scientific American*. https://www.scientificamerican.com/article/which-came-first-on-earth-habitability-or-life/.

Harding, S., and Margulis, L. 2010. Water Gaia: 3.5 thousand million years of wetness on planet Earth. In *Gaia in Turmoil: Climate Change, Biodepletion, and Earth Ethics in an Age of Crisis*, E. Crist and H.B. Rinker (Eds.), pp. 41–60. Cambridge, MA: MIT Press.

Harrison, J.P. Gheeraert, N., Tsigelnitskiy, D., and Cockell, C.S. 2013. The limits for life under multiple extremes. *Trends in Microbiology*, 21, 204–212.

Hart, M.H. 1979. Habitable zones about main sequence stars. *Icarus*, 37, 351–357.

Heller, R., and Armstrong, J. 2014. Superhabitable worlds. *Astrobiology*, 14, 50–66.

Herschel, W. 1795. III. On the nature and construction of the sun and fixed stars. *Philosophical Transactions of the Royal Society of London*, 85, 46–72.

Holm, N.G., and E.M. Andersson. 2000. Hydrothermal systems. In *The Molecular Origins of Life: Assembling Pieces of the Puzzle*, A. Brack (Ed.), pp. 86–99. Cambridge, UK: Cambridge University Press.

Hug, L.A., Baker, B.J., Anantharaman, K. et al. 2016. A new view of the tree of life. *Nature Microbiology*, 1, 16048.

Huygens, C. 1698. The Celestial Worlds Discovered: Or Conjectures Concerning the Inhabitants, Plants and Productions of the Worlds in the Planets. London, printed for Timothy Childe.

Jellinek, A.M., and Jackson, M.G. 2015. Connections between the bulk composition, geodynamics and habitability of Earth. *Nature Geoscience*, 8(8), 587–593.

Jones, E.G., and Lineweaver, C.H. 2010. To what extent does terrestrial life "follow the water."*Astrobiology*, 10(3), 349–361.

Joshi, M.M., and Haberle, R.M. 2012. Suppression of the water ice and snow albedo feedback on planets orbiting red dwarf stars and the subsequent widening of the habitable zone. *Astrobiology*, 12(1), 3–8.

Jones, E.G., Lineweaver, C.H., and Clarke, J.D. 2011. An extensive phase space for the potential Martian biosphere. *Astrobiology*, 11(10), 1017–1033.

Joyce, G.F. 1994. Foreword. In *Origins of Life: The Central Concepts*, D.W. Deamer and G. Fleischaker (Eds.). Boston, MA: Jones and Bartlett.

Joyce, G.F. 2012. Bit by bit: The Darwinian basis of life. *PLoS Biology*, 10(5), e1001323. doi:10.1371/journal.pbio.1001323.

Joyce, G.F., and Orgel, L.E. 1993. Prospects for understanding of RNA world. In *The RNA World: The Nature of Modern RNA Suggests a Prebiotic RNA World*, R.F. Gesteland and J.F. Atkins (Eds.). Cold Spring Harbor, NY: Cold Spring Harbor University Press.

Kasting, J. 2010. *How to Find a Habitable Planet*. Princeton, NJ: Princeton University Press.

Kasting, J.F., and Catling, D.C. 2003. Evolution of a habitable planet. *Annual Review of Astronomy and Astrophysics*, 41(1), 429–463.

Kasting, J.F., Whitmire, D.P., and Reynolds, R.T. 1993. Habitable zones around main sequence stars. *Icarus*, 101(1), 108–128.

Kaltenegger, L. 2017. How to characterize habitable worlds and signs of life. *Annual Review of Astronomy and Astrophysics*, 55, 4333–4485.

Kawaler, S., and Veverka, J. 1981. The habitable sun: One of William Herschel's stranger ideas. *Journal of the Royal Astronomical Society of Canada*, 75(1), 46–55.

Kepler, J. 1634. *Somnium*, Sagani Silesiorum, Francofurti.

Kolb, V.M. 2010. On the applicability of dialetheism and philosophy of identity to the definition of life. *International Journal of Astrobiology*, 9, 131–136.

Kopparapu, R.K., Ramirez, R., Kasting, J.F. et al. 2013. Habitable zones around main-sequence stars: New estimates. *Astrophysical Journal*, 765(2), 131.

Koshland, D.E. 2002. The seven pillars of life. *Science*, 295, 2215–2216.

Krissanen-Totten, J., Olson, S., and Catling, D.C. 2018. Disequilibrium biosignatures over Earth history and implications for detecting exoplanet life. *Science Advances*, 4, 5747.

Lahav, N. 1999. *Biogenesis: Theories of Life's Origin*. Oxford, UK: Oxford University Press.

Lammer, H., Bredehöft, J.H., Coustenis, A., Khodachenko, M.L., Kaltenegger, L., Grasset, O., and Wahlund, J.E. 2009. What makes a planet habitable? *The Astronomy and Astrophysics Review*, 17(2), 181–249.

Léger, A. 2004. A new family of Planets? "Ocean Planets." *Icarus*, 169(2), 499–504.

Lineweaver, C.H. 2006. We have not detected extraterrestrials, or have we? In *Life as We Know It. Cellular Origins and Life in Extreme Habitats and Astrobiology*, vol. 10, J. Seckbach (Ed.), pp. 445–457. Dordrecht, the Netherlands: Springer.

Lineweaver, C.H. 2014. Review of Ćirković 2012. *Origins of Life and Evolution of Biospheres*. doi:10.1007/s11084-014-9369-2.

Lineweaver, C.H., and Chopra, A. 2012a. The habitability of our Earth and other Earths: Astrophysical, geochemical, geophysical, and biological limits on planet habitability. *Annual Review of Earth and Planetary Sciences*, 40, 597–623.

Lineweaver, C.H., and Chopra, A. 2012b. What can life on earth tell us about life in the universe? In *Genesis—In The Beginning: Precursors of Life Chemical Models and Early Biological Evolution*, J. Seckbach (Ed.), pp. 799–815. Dordrecht, the Netherlands: Springer.

Lineweaver, C.H., Fenner, Y., and Gibson, B.K. 2004. The galactic habitable zone and the age distribution of complex life in the Milky Way. *Science*, 303(5654), 59–62.

Linsenmeier, M., Pascale, S., and Lucarini, V. 2015. Climate of Earth-like planets with high obliquity and eccentric orbits: Implications for habitability conditions. *Planetary and Space Science*, 105, 43–59.

Lovelock, J.E. 1965. A physical basis for life detection experiments. *Nature*, 207, 568–570.

Lucian of Samosata. 2nd Century AD. *True History (or True Stories)*.

Luger, R., and Barnes, R. 2015. Extreme water loss and abiotic O2 buildup on planets throughout the habitable zone of M dwarfs. *Astrobiology*, 15, 119–143.

Lupu, R.E., Zahnle, K., Marley, M.S., Schaefer, L., Fegley, B., Morley, C., Cahoy, K., Freedman, R., and Fortney, J.J. 2014. The atmospheres of Earthlike planets after giant impact events. *The Astrophysical Journal*, 784(1), 27.

Maher, K.A., and Stevenson, D.J. 1988. Impact frustration of the origin of life. *Nature*, 331, 612–614.

Mann, S. 2013. The origins of life: Old problems, new chemistries. *Angewandte Chemie International Edition*, 52, 155–162.

Martin, W., and Russell, M.J. 2007. On the origin of biochemistry at an alkaline hydrothermal vent. *Philosophical Transactions of the Royal Society B*, 362, 1887–1925.

McDonald, G.D. 2015. Biochemical pathways as evidence for prebiotic syntheses. In *Astrobiology: An Evolutionary Approach*, V.M. Kolb (Ed.), pp. 119–147. Boca Raton, FL: CRC Press/ Taylor & Francis.

Meadows, V., and Seager, S. 2010. Terrestrial planet atmospheres and biosignatures. In *Exoplanets*, S. Seager (Ed.), pp. 441–470. Tucson, AZ: University of Arizona Press.

Musk, E. 2017. Making humans a multi-planetary species. *New Space*, 5(2), 46–61.

NASA. 2011. Astrobiology: Roadmap. Archived from the original on January 17. https://web.archive.org/web/20110117011137/ http://astrobiology.arc.nasa.gov/roadmap/g1.html.

Nealson, K.H., and Conrad, P.G. 1999. Life: Past, present and future. *Philosophical Transactions of the Royal Society of London B*, 354, 1923–1939.

Nicolas of Cusa. 1440. *On Learned Ignorance*. Minneapolis, MN: The Arthur J. Banning Press.

Nietzsche, F. 1887. *On the Genealogy of Morality*. Edited by K. Ansell-Pearson, translated by C. Diethe, Cambridge Texts in the History of Political Thought. Cambridge, UK: Cambridge University Press.

Nisbet, E., Zahnle, K., Gerasimov, M.V. et al. 2007. Creating habitable zones, at all scales, from planets to mud micro-habitats, on Earth and on Mars. *Space Science Reviews*, 129(1–3), 79–121.

Noack, L., and Breuer, D. 2014. Plate tectonics on rocky exoplanets: Influence of initial conditions and mantle rheology. *Planetary and Space Science*, 98, 41–49.

Noack, L., Godolt, M., von Paris, P., Plesa, A.C., Stracke, B., Breuer, D., and Rauer, H. 2014. Can the interior structure influence the habitability of a rocky planet? *Planetary and Space Science*, 98, 14–29.

Odling-Smee, J., Erwin, D., Palkovacs, E.P., Feldman, M.W., and Laland, K.N. 2013. Niche construction theory: A practical guide for ecologists. *Quarterly Review of Biology*, 88, 3–28.

Oparin, A.I. 1968. Genesis and evolutionary development of life. Translated by E. Maass. New York: Academic Press.

Orgel, L.E. 2004. Prebiotic chemistry and the origin of the RNA world. *Critical Reviews in Biochemistry and Molecular Biology*, 9, 99–123.

Padgett, J.F. 2012. Autocatalysis in chemistry and the origin of life. In *The Emergence and Organizations and Markets*, J.F. Padgett and W.W. Powell (Eds.), pp. 33–69. Princeton, NJ: Princeton University Press.

Peretó, J. 2005. Controversies on the origin of life. *International Microbiology*, 8, 23–31.

Popa, R. 2015. Elusive definition of life: A survey of main ideas. In *Astrobiology: An Evolutionary Approach*, V.M. Kolb (Ed.), pp. 325–348. Boca Raton, FL: CRC Press/Taylor & Francis.

Ramirez, R.M., and Kaltenegger, L. 2015. The habitable zones of pre-main-sequence stars. *The Astrophysical Journal Letters*, 797(2), L25.

Ramirez, R.M., and Kaltenegger, L. 2017. A volcanic hydrogen habitable zone. *The Astrophysical Journal Letters*, 837(1), L4.

Ribas, I., Bolmont, E., Selsis, R. et al. 2016. The habitability of Proxima Centraui b. I Irradiation, rotation and volatile inventory from formation to the present. *Astronomy & Astrophysics*, 596, A111.

Rinke, C., Schwientek, P., Sczyrba, A. et al. 2013. Insights into the phylogeny and coding potential of microbial dark matter. *Nature*, 499, 431–437.

Rogers, L. 2015. Most 1.6 Earth-radius planets are not rocky. *Astrophysical Journal*, 801, 41.

Sagan, C. 1970. Life. In *Encyclopaedia Britannica*. Reprinted in *Encyclopaedia Britannica*, 1998, Vol. 22, 964–981.

Schwartzman, D., and Volk, T. 1989. Biotic enhancement of weathering and the habitability of the Earth. *Nature*, 340, 457–460.

Seager, S. 2013. Exoplanet habitability. *Science*, 340, 577–581.

Seager, S., Bains, W., and Petkowski, J.J. 2016. Toward a list of molecules as potential biosignature gases for the search for life on exoplanets and applications to terrestrial biochemistry. *Astrobiology*, 16(6), 465–485.

Segre, D., and Lancet, D. 2000. Composing life. *EMBO Reports*, 1(3), 217–222.

Sleep, N.H., Zahnle, K.J., Kasting, J.F., and Morowitz, H.J. 1989. Annihilatiosn of ecosystems by large asteroid impacts on the early Earth. *Nature*, 342, 139–142.

Sleep, N.H., Zahnle, K.J., and Neuhoff, P.S. 2001. Initiation of clement surface conditions on the early Earth. *PNAS*, 98, 3666–3672.

Smith, E., and Morowitz, H. 2016. *The Origin and Nature of Life on Earth: The Emergence of the Fourth Geosphere*. Cambridge, UK: Cambridge University Press.

Sojo, V., Herschy, B., Whicher, A., Camprubi, E., and Lane, N. 2016. The origin of life in alkaline hydrothermal vents. *Astrobiology*, 16, 181–197.

Stevenson, D.J. 2010. Planetary magnetic fields: Achievements and prospects. *Space Science Reviews*, 152(1–4), 651–664.

Stevenson, D.S., and Large, S. 2017. Evolutionary exobiology: Towards the qualitative assessment of biological potential on exoplanets. *International Journal of Astrobiology*. doi:10.1017/S1473550417000349.

Sullivan III, W.T., and Carney, D. 2007. History of astrobiological ideas. In *Planets and Life: The Emerging Science of Astrobiology*, W.T. Sullivan III, and J.A. Baross (Eds.), pp. 9–45. Cambridge, UK: Cambridge University Press.

Sutherland, J.D. 2017. Studies on the origin of life – the end of the beginning. *Nature Reviews Chemistry*, 1, 0012.

Szostak, J.W. 2012, Attempts to define life do not help to understand the origin of life. *Journal of Biomolecular Structure and Dynamics*, 29(4), 599–600.

Takai, K., Nakamura, K., Toki, T. et al. 2008. Cell proliferation at 122°C and isotopically heavy CH_4 production by a hyperthermophilic methanogen under high-pressure cultivation. *Proceedings of the National Academy of Sciences of the United States of America*, 105, 10949–10954.

Tarduno, J.A., Cottrell, R.D., Davis, W.J., Nimmo, F., and Bono, R.K. 2015. A Hadean to Paleoarchean geodynamo recorded by single zircon crystals. *Science*, 349(6247), 521–524.

Tirard, S., Morange, M., and Lazcano, A. 2010. The definition of life: A brief history of an elusive scientific endeavor. *Astrobiology*, 10(10), 1003–1009.

Valencia, D., O'Connell, R.J., and Sasselov, D.D. 2007. Inevitability of plate tectonics on super-Earths. *The Astrophysical Journal Letters*, 670(1), L45.

Wächtershäuser, G. 2000. Origin of life in an iron-sulfur world. In *The Molecular Origins of Life: Assembling Pieces of the Puzzle*, A. Brack (Ed.), pp. 206–218. Cambridge, UK: Cambridge University Press.

Wächtershäuser, G. 2008. On the chemistry and evolution of the pioneer organism. In *Origin of Life: Chemical Approach*, P. Herdewijn and M.V. Kisakürek (Eds.), pp. 61–79. Zürich, Switzerland: VHCA.

Walker, J.C., Hays, P.B., and Kasting, J.F. 1981. A negative feedback mechanism for the long-term stabilization of Earth's surface temperature. *Journal of Geophysical Research: Oceans*, 86(C10), 9776–9782.

Wallace, A.R. 1904. *Man's Place in the Universe*, p. 6. London, UK: Chapman & Hall.

Weiss, M.C., Sousa, F.L., Mrnjavac, N., Neukirchen, S., Roettger, M., Nelson-Sathi, S., and Martin, W.F. 2016. The physiology and habitat of the last universal common ancestor. *Nature Microbiology*, 1, 16116.

Yang, J., Cowan, N.B., and Abbot, D.S. 2013. Stabilizing cloud feedback dramatically expands the habitable zone of tidally locked planets. *The Astrophysical Journal Letters*, 771(2), L45.

Zahnle, K., Arndt, N., Cockell, C., Halliday, A., Nisbet, E., Selsis, F., and Sleep, N.H. 2007. Emergence of a habitable planet. *Space Science Reviews*, 129(1–3), 35–78.

Zahnle, K.J., and Catling, D.C. 2017. The cosmic shoreline: The evidence that escape determines which planets have atmospheres, and what this may mean for Proxima Centauri b. *Astrophysical Journal*, 843, 122.

Section XI

Intelligent Life in Space

History, Philosophy, and SETI
(Search for Extraterrestrial Intelligence)

11.1 Mind in Universe
On the Origin, Evolution, and Distribution of Intelligent Life in Space

David Dunér

CONTENTS

11.1.1 INTRODUCTION

It is an exhilarating thought to stand on a clear night, gazing at the countless numbers of stars above and wondering if there could be someone out there who—in that very same moment—wonders if we exist. To think about the existence of life out there—be it unicellular organisms or perhaps even strange, more complex creatures—is indeed astonishing. The possibility that these extraterrestrial beings might also think, feel, dream, intend, strive, and wonder about their own existence would be even more mind-boggling. And moreover, if we could exchange thoughts with each other about our existence and about what it means to live, feel, and think, that would be the most overwhelming experience of the human mind.

Even supposing there is nobody out there who wonders whether we exist or not, we know at least that we think that we exist. Because we think about our own existence and our own place in the universe, one could say that the universe is aware of its own existence and that the universe thinks. Through the self-conscious human being, the universe can be considered self-conscious and able to reflect upon itself. It seems that the universe is not only bio-friendly (Davies 2007) but also *cogito-friendly*. The universe is not only living; it contains not just self-reproductive entities, but it also has apparently led to self-reproductive organisms that are able to reflect upon the universe they live in, upon their place in that world, and upon their own thoughts and existence.

This chapter concerns exactly this: the thinking universe or intelligence in space. It concerns the presence of intelligence and its possible origin, evolution, and distribution. We know at least that we think, but are there other creatures out there that think, and would it be possible to understand them? I raise three main questions about mind in the universe in this chapter. The first concerns *terrestrial intelligence*: What happens to the human mind when it tries to understand extraterrestrial phenomena and environments? The second concerns *extraterrestrial intelligence*: Are there intelligent living beings out there, and if so, what makes them possible? What are the necessary or sufficient conditions for intelligence to evolve? The third question concerns *terrestrial-extraterrestrial interaction*, or in other words: If there are other intelligent beings out there, could we understand them, and they understand us, and could it be possible to mutually exchange thoughts between the two of us? The following text could be seen as an attempt not only to reach outward to the stars and to the minds that we hope to find out there, it also reaches inward into the uncharted depths of the human mind. This chapter concerns the human mind in front of the unknown as well as the evolution of an unknown potential mind in the unknown space—and finally, the mutual understanding of minds in the universe.

11.1.1.1 ASTROCOGNITION

The scientific search for life in space, called astrobiology, tries to find the answers to the questions about the origin, evolution, and distribution of life in the universe (Bertka 2009). How did intelligent life emerge, how did it evolve, and does it exist elsewhere in the universe? Astrobiology is, according to the 1996 NASA Strategic Plan's definition, "the study of the living Universe" (NASA 1998; Chyba and Hand 2005). In this chapter, I go a step further and ask not only for the conditions for life but also for the conditions for life with awareness or self-awareness, that is, *intelligent life*, about the origin, evolution, and distribution of intelligent life in the universe. This scientific endeavor could be called *astrocognition*, which is the study of the origin, evolution, and distribution of intelligent life in the universe. What necessary or sufficient conditions must be in place in order for more complex intelligence to emerge and evolve? And does it exist elsewhere in the universe?

The multidisciplinary field of astrocognition—first proposed in 2009 (Dunér 2011a)—could be generally defined as "the study of the origin, evolution and distribution of cognition in the Universe," or simply "the study of the thinking Universe." Astrocognition concerns the cognitive processes in space, the origin and evolution of cognitive abilities, and the distribution of cognition in extraterrestrial environments (Dunér 2011a; Osvath 2013). "Cognition" could, in this context, be understood as the ability of processing sensory inputs for action in the environment. If we are discussing the existence of extraterrestrial intelligence in space, then we should take into account the research within cognitive science and affiliated research areas, in order to find answers to these questions: What is needed for higher cognitive skills to evolve? What physical, biological, societal, cultural, and other environmental factors shape cognition? What cognitive abilities are needed for a living organism to be able to manipulate its environment or, in other words, to develop technology? Like the astrobiology field, of which it is a part, astrocognition requires a multidisciplinary approach that brings together cognitive science, philosophy of mind, animal cognition, semiotics, linguistics, anthropology, cultural studies, history of science, computer science, neuroscience, evolutionary theory, physics, astronomy, space technology, and other fields of research.

Astrocognition could give us further theoretical, scientific knowledge of how we encounter the unknown and how the human mind interacts with space and the environment around us. It could also give us a better understanding of the evolution and prerequisites needed for cognition to emerge. These achievements will be valuable, even though we might never find intelligent life in outer space. From this, we would learn more about human cognition and how it has been developed here on Earth. The perhaps most important outcome is that possible encounters with other minds here on Earth or on extrasolar foreign planets in deep space among billions of stars and galaxies can give a more universal answer to the question: "What is thinking?"

11.1.2 TERRESTRIAL INTELLIGENCE

The human desire for exploration and human encounters with the unknown are a fundamental part of the cultural history of humankind, from the first stumbling steps on the African plains to the recent explorations of our globalized and urbanized world. From the very dawn of the hominids to the days of the modern human being, this mutable terrestrial being has expanded in ever-widening circles of spatial consciousness, in an endeavor to climb over mountains to the next valley, cross vast oceans, and fly through the air. The next small step for man, or giant leap for humankind, that of going far beyond the atmosphere and gravity of the Earth to the unknown extrasolar space, is decisive, but that too is part of the long history of humankind.

This preference for the unknown and for new things and new experiences—"neophilia"—is probably one of the explanations for human expansion and human endeavors such as art and science. This enabled us to gain more knowledge of the world around us, but it also led to severe cognitive challenges, which in its turn led to new thoughts and ways of thinking. The exploration of extraterrestrial environments and the possible contact with other forms of life and civilizations will most likely change our thinking, conceptual frameworks, and belief systems. In the following, I will discuss the first main question concerning mind in the universe: the challenges of terrestrial intelligence. What cognitive challenges are we likely going to face when we encounter the unknown (Dunér 2011a; Dunér and Sonesson 2016a)? In other words, it concerns the challenges our Earth-bound perceptual, cognitive, and psychological capacities face in a space context (Pálsson 2009).

In the course of everyday events and encounters, the human mind has been enabled, through an evolutionary process, to understand, interact, deal with, and adapt to the environment of this particular planet and to the minds of other human beings. Thus, the human brain is well adapted to the biological, ecological, and physical characteristics of our planet, as well as the cultural, social, and cognitive characteristics of the tellurian species *Homo sapiens*. But how would the human mind, which is entirely a product of a terrestrial environment, function in extraterrestrial environments and understand extraterrestrial minds?

The success of our search for intelligent life beyond our solar system will depend on, first, obviously, that there really exists intelligent life to be discovered, and second, that we have the technology to discover it. But that is not enough. The discovery depends on—and this is the most critical—the capability of the human brain and the organization and efficiency of that systematic search for knowledge that we call science, which is a product of the socio-cultural history of our species (Dunér 2016a). Our endeavor depends on human cognition and our ability to understand and interpret what we observe in our surrounding world.

Even though intelligent life might not exist out there, it is we human beings, with our brains, bodies, and cultures, who are searching for it. The scientific endeavors of astrobiology and interstellar communication are dependent on the human mind and human culture (Dunér 2011a, 2013a; Dunér et al. 2013). Astrobiologists have brains, for sure; they are using cognitive tools that are a result of the bio-cultural coevolution of human cognitive abilities. Certain cognitive processes are at work when astrobiologists encounter unknown things, when they interpret their observational data, when they gather and classify it, and when they make conclusions. This does not go on in subjective isolation. Astrobiologists live in a culture, in a certain time in history, and in a specific research environment and collaborate with other thinking beings (Vakoch and Harrison 2011; Billings 2012; Dick 2012; Race et al. 2012; Vakoch 2013; Traphagan 2015a, 2016).

Cognitive science can give clues to how we understand and think about the universe and reveal new perspectives on human encounters with the unknown. In short, cognitive science studies how the external world as well as our inner worlds are represented and how we use cognitive tools for our thinking, such as language, image schemas, mental maps, metaphors, and categories. Cognitive science also studies how we use and interpret signs, objects, drawings, images, etc., to enhance communication. It is about perception, attention, memory, learning, consciousness, reasoning, and other abilities that we include in what is called "thinking."

Cutting-edge thought in cognitive science portrays the mind as embodied, extended, distributed, and situated. According to the theory of the embodied mind (Varela et al. 1991; Lakoff and Johnson 1999; Krois 2007; Thompson 2007), the mind is not something independent and detached from the body. There is no "brain in a vat," to borrow a turn of phrase from Hilary Putnam's famous thought experiment (Putnam 1981). Instead, we think with the body. Therefore, bodies of other kinds and evolutionary backgrounds, like those that might exist on other planets in the universe, will have other minds and ways of thinking. The brain not only needs the body but also the surrounding world in order to function efficiently. The environment has an active role in driving cognitive processes. According to the extended-mind hypothesis of Andy Clark and David Chalmers (Clark and Chalmers 1998; Clark 2008), the mind leaks in various substantive ways into the environment. The boundary between self and non-self, and between self and world, is one that is never fixed but is constantly being re-negotiated. In a similar way, distributed cognition (Hutchins 1995) stresses the non-localizability of much of cognition: that we are using our environment and tools for enhancing thinking and that we place our ideas and memories in things, such as books and computers. Mind is not isolated; it is situated, in space and time, in a specific physical and social environment.

The search for extraterrestrial intelligence is not just about possible intelligent life-forms out there but also about ourselves and our ability to perceive and interpret our surrounding world, in phenomenological terms, our lifeworld (Dunér and Sonesson 2016b). Space is "the ultimate mirror we hold up to ourselves," as Parthemore (2013, p. 83) puts it. In confronting the depths of the cosmos, we confront the unseen depths in ourselves. In other words, astrobiology and astrocognition challenge our everyday conception of ourselves as human thinking beings in the universe. In the following, I will discuss what is "intelligence" and highlight some cognitive functions or capacities that are in work when a terrestrial mind confronts the unknown.

11.1.2.1 Intelligence

What is intelligence? When we search the skies for extraterrestrial intelligence, we should at least have some ideas of what kind of phenomena we are looking for (Chick 2014; Marino 2015b; Ruse 2015; Schneider 2015). Within the Search for Extraterrestrial Intelligence (SETI) community, this question has been rather overlooked. The Drake equation (Drake 2011; Vakoch and Dowd 2015; Dunér 2017), and much of the SETI research, has instead an operative, pragmatic understanding of "intelligence" and just looks for a civilization able to transmit electromagnetic waves. "Intelligence" is, in that sense, "the ability to transmit electromagnetic waves." This is of course not what we mean when we recognize something as "intelligent" in ordinary life. Nor is it sufficient to help us understand the distribution of intelligence in space. In order to detect content-rich and meaningful signals from an extraterrestrial civilization, it is not sufficient to rest on the belief that the "intelligent" being is able to construct advanced devices; a creature that constructs meaningful messages should also be able and willing to communicate, that is, to share experiences through a medium.

A recent attempt to define intelligence in connection to extraterrestrial life and evolution of intelligence is more elaborated. Ted Peters defines intelligence in terms of seven traits: interiority, intentionality, communication, adaptation, problem solving, self-reflection, and judgment (Peters 2017). Even microbes exhibit the first four traits; humans, along with some other animals, exhibit all of them. Where there's life, there's intelligence, so to speak, according to Peters. Intelligence seems to be a matter of degree rather than of kind. However, to this list of traits, I would add a most critical one: intersubjectivity, which I will explain in more detail in the following. Intersubjectivity, the ability to understand other minds, is an important trait in order to explain intelligence and how an intelligent creature can evolve complex communication, civilization, and technology (Dunér 2014).

The definition of intelligence has commonly been connected to problem solving (Sternberg 1985, 2002). As such, intelligence is understood as the ability to solve problems, to make rational choices, to reason logically, and to handle the constraints and limitations of time, space, and materials. However, it is not sufficient to describe intelligence as a mere problem solving in order to explain the development of advanced technology. As a broader concept, cognition includes not only the abilities that we call rational, logical, or intelligent. An important part of what it is to be intelligent is to have emotional skills, a capacity to emotionally appraise the relevant environment with attraction, disgust, etc., and to respond to socializing, bounding, coupling, etc., in a social group. By this, I mean that intelligence is an efficient strategy to cope with, not only the physical but also—which is important—the social environment.

There are two features (among many other possible ones) of an "intelligent" life-form that are particularly important. First, it can imagine things not existing, that is, things, events, etc., not present in time or space, right in front of the thinking subject. Second, an intelligent being is also able to engage in intersubjective interactions, understand other minds, and imagine and envision what they will do, what they feel, and how they reason. To be intelligent is to have intersubjective skills, to be able to understand and make interferences about other minds. If the extraterrestrial being that we encounter is lacking these two abilities, it would probably not have complex communication and advanced technology, and we would not be able to communicate with it. Without the ability of imaging non-existing things, it would not be able to construct abstract concepts, anticipate future needs, and invent new technology. Without intersubjective skills, it would not be able to collaborate for achieving shared goals.

"Intelligence" could be explained—rather than a mere problem solving—as a kind of cognitive flexibility, an ability to adjust to changes in the physical and socio-cultural environment. It is an evolved mental flexibility required to survive and reproduce within a specific environment. This includes the capability of representing activities and being able to make inner models of reality and other minds. "Extraterrestrial intelligence" is actually a rather misleading and narrow concept, referring, for the most part, to the problem solving and rational reasoning skills and not to those mental abilities, for example, intersubjectivity, that are indispensable for a life-form to have civilization, culture, and technology. Thus, instead of searching for extraterrestrial intelligence, we should search for extraterrestrial cognitive flexibility. We should search for a certain cognitive flexibility that is a result of a more or less unique adaptation to the specific habitable environment in which it has its origin.

Within in the human mind, which is a product of a biocultural coevolution of cognitive abilities of our terrestrial species, there are a number of cognitive tools by which we can understand and handle the world around us. Next, I will highlight three cognitive functions that are particularly striking in the human search for life in the universe, when astrobiologists try to interpret and understand encountered unknown phenomena: *perception*, how we interpret what our senses convey and how the information we get from the surrounding world is processed in our minds; *conceptualization*, how we form concepts of life, how we define and categorize things, and the relation between our concepts and our knowledge of the world; and finally, *analogy*, how we see similarities between things and, with inductive, analogical reasoning, go from what we know to what we do not know.

11.1.2.2 Perception

Where we are in time and space is essential for cognition. What the senses convey have to be interpreted through means of specific cognitive processes. The cognitive agent is never just a passive recipient of images and information from the surrounding world. Instead, the brain actively searches out patterns in what is conveyed to it through the senses and interprets them through a process that is determined by both biological and cultural factors. Striking examples of this *epistemic perception* are the maps of Venus and Mars from the seventeenth century and onward that delineated the surface of the planets (Dunér 2013c). We see what we expect to find. In 1877, at the Brera Observatory in Milan, Giovanni Schiaparelli recorded a detailed network of canals on Mars. This finding was confirmed by the American astronomer Percival Lowell, who detected hundreds of Martian canals that he interpreted as an artificial irrigation system (Lowell 1895). Quite soon, it became clear that this was a mistake. However, it was neither the telescopes nor the eyes of the observers that caused the fallacy. It was the interpretation of the sensory inputs that led them wrong. The Martian canals existed only in the mind of the observers.

In the optical observations of distant worlds, preconceived understanding often shapes the interpretation of what the observers see. Through their senses, the observers receive impressions from outer space, and they collect and collate information by using their sight. The interpretation of what is observed is based on a preconceived understanding, concepts, and prior knowledge. Observations are not separate from theory. In contrast to the tendency to place excessive trust in "objective" observations, many philosophers of science, such as N.R. Hanson, Thomas Kuhn, and Michael Polanyi, have emphasized that observations are theory-laden and that there is no sharp dichotomy between observation and theory in scientific research (Crowe 1986). We need theories to understand what we see, and our preconceptions and expectations lead us in one or another direction, sometimes the wrong course. It is not uncommon to see what we expect to find.

Perception is not a neutral, objective, and realistic recording of reality. This conceptual or epistemic perception implies an identification of what is seen and takes place by applying our concepts to visual perceptions, or in other words, concepts affect what we see, and, should we lack any concept of a specific phenomenon, then it will be difficult to distinguish it among all our impressions. The world distorts our concepts, and the concepts distort our world.

11.1.2.3 Conceptualization

To have concepts is to be able to categorize things, to see a thing as something different from something else, but also to see similarities between things (Lakoff 1990; Taylor 2003). All living creatures seem to categorize the environment in terms of edible versus inedible, benign versus harmful, and so forth. Categorization becomes more complex in human cognition. The human mind tends to categorize and see hierarchies and similarities between things, such as species and genera in taxonomy (Berlin 1992; Dunér 2013b) or the classification of stars and other astronomical objects. We need categories and invent new ones in order to handle the world around us; otherwise, we would fall into a chaotic abyss of unsorted impressions.

Constructing concepts in order to be able to think and talk about the new phenomena encountered is a major task for astrobiological research. Astrobiologists use a wide range of concepts, such as biosignature, habitability, habitable zones (see, e.g., Kane and Gelino 2012), Earth analogues, exoplanets (Perryman 2012), and other concepts and terms inherited from already well-established scientific disciplines. These concepts together form that multidisciplinary field of research we call astrobiology. The most debated and discussed conceptual issue in astrobiology is the concept of life (Losch 2017). If we are searching for non-terrestrial life, we ought to know what it is we are looking for and what characteristics it might have. If life is a recipe, what are the essential ingredients and which are optional? Should these criteria pertain to metabolism, entropy, genes, reproduction, or something else? So far, the debate has intuitively employed an Aristotelian conception of definition (Aristotle, *Posterior Analytics*, 2.3.90b30–31), in which a "definition" is a limited list of characteristics that are both necessary and sufficient for something to be of the type of object it is and from which all the characteristics of the object originate. In our daily lives, however, we make relatively little use of Aristotelian-type definitions and depend much more on prototypes (Rosch 1975, 1978). Dogs, cats, and horses may seem to be more typical representatives for "life" than arsenic microbes. The debate on the definition of life could certainly benefit from the insights of contemporary philosophy and cognitive science about human categorization. The problem of defining life has to do with how the human mind categorizes things.

Science—and this is certainly true for astrobiology as well—concerns concepts. We need names and abstract concepts in order to be able to talk and reason about objects, structures, processes, etc., that we gather from our senses, through observations and experiments. When we are using these terms, they need to be reasonably well defined and some sort of consent needs to be established, so we can agree on what we talk about. We need definitions. One might think that these concepts already exist out there—the task is just to discover them. However, the input we get from the surrounding world has to be processed by our brains and depends on the cognitive abilities we possess. The scientific concepts we use are not just dependent on the particular characteristics of human cognition that is a product of a biological evolution of the human brain, it is also a product of a specific cultural evolution here on Earth and many generations of natural philosophers and scientists in the history of human, terrestrial science (Dick 1982; Crowe 1986, 2008; Dick 1996; Dunér 2012, 2016b; Dunér et al. 2012, 2013, 2016).

Future discoveries in astrobiology will most likely challenge our categorizations and definitions, that is to say, our preconception of what the world is and not is. So, we should be prepared to re-categorize and redefine our concepts. Future exobiological systematics and taxonomy will face problems concerning categorization, identification, and description. The taxonomy of future extraterrestrial fauna and flora will be a product of the human mind. It is our human way of understanding the world, by using certain categories and concepts, that makes some problems for us, when we are dealing with phenomena, far beyond the well-known of our daily lives.

11.1.2.4 ANALOGY

Through the history of astrobiology, we find a certain common form of argumentation: the analogy, the thought moving from what we know to what we do not know (Dunér 2013c). An analogical argument could be explained as a search for similarities, that is, a way of selecting features in the source domain that are to be mapped onto the target domain and of transferring relevant properties from the source to the target. If x has the properties $P_1, P_2, P_3, P_4 \ldots P_n$, and there is a y that has P_1, P_2, and P_3, we may conclude that it also has P_4. If we know there is an x that has these qualities and we discover a y that also has some of these qualities, then we conclude that all y also have the quality that we are seeking, P_4, or formulated in first-order logic: $\exists x(P_1x \wedge P_2x \wedge P_3x \wedge P_4x \ldots P_nx) \wedge \exists y(P_1y \wedge P_2y \wedge P_3y) \Rightarrow \forall y(P_4y)$.

The challenge is then to select the correct and relevant salient features from an infinite number of possible ones in the source domain, which will then be transferred to and mapped onto the target domain. In the *Sidereus Nuncius* from 1610, Galileo showed, based on his telescopic observations and analogical reasoning, that the Moon had mountains and therefore had the same solid, opaque, and rugged nature as the Earth (Galilei 1610; Spranzi 2004). If the Moon and Earth are similar in many respects, why should the former lack life, if the latter has it? In some sense, astrobiology as a whole is one single, great analogy. Starting from the one particular type of life we happen to know something about, namely life on Earth, we proceed to search for life on other planets. We predominantly look for life as we know it: something needing oxygen, liquid water, being based mainly on carbon, inhabiting a planet of a certain magnitude, atmosphere, chemistry, temperature, etc.; something that revolves within the habitable zone, at a certain distance, period, eccentricity, inclination, etc., of a solar-type star (a G2 main sequence star of 4.5 Gyr of age), which in turn has to be of a certain size and temperature, and so on.

There are a number of problems with such an analogical argument: first of all, do we know all the necessary or sufficient conditions for life? Another problem is that we restrict ourselves to one particular sort of life, that is, the one we happen to know—life as we know it—and might overlook other forms of "weird life"—life as we do not know (Davies et al. 2009). The third problem lies in the unpredictability of evolution that there are some stochastic events involved in the evolutionary process. The question is: If all ingredients of life are in place, will that inevitably, by necessity lead to the emergence of life? Is life a natural manifestation of matter (cf. de Duve 1995).

Logically speaking, analogical arguments are invalid. However, in providing us with some point of departure, they still might hold some heuristic benefits in the search for life. For example, studies of analogue sites on Earth might give us valuable knowledge that could support and guide us toward future discoveries of life in space (Martins et al. 2017). What we are actually looking for is something that reminds us of ourselves, something similar to us. Life, however, might be very different from what we imagine. The history of science is actually a history of surprises. The world we are living in turned out to be very different from what we first thought: richer, more complicated, more advanced, more peculiar, and more astonishing than what we could have dreamt of. This will also be true for astrobiology. Future discoveries in astrobiology will take us completely by surprise.

11.1.2.5 Summary

From these cognitive theories concerning the human mind, we can conclude that encounters with the unknown outer space will change our spatial consciousness. Furthermore, it will change our thinking, conceptions, categories, belief systems, culture, and meanings of things. What we have come to believe so far through science and human cogitation will face anomalies. The old concepts, categories, systems, and beliefs will most likely fall short when we try to understand these new unfamiliar things. Our thinking, science, and belief systems will then have to be revised, which will lead to adjustments, adaptations, and compromises. A task for astrocognition is to search for the limits of our bio-culturally evolved, earthly brains to try to find out what we can know and what we are likely to encounter in the future. As thinking beings, we are earthbound and historically constrained. Our intellect does not transcend space. Instead, our cognition is situated in space.

One day, we might encounter another living planet. The travelogues and descriptions of these new worlds will inform us about ourselves and our place in the universe and how we interpret and understand the "reality" around us. An independent world outside us might exist, but we will never be able to reach it without filtering it through our minds. To conclude, if we find extraterrestrial life, we can be sure that this will change our thinking about how we perceive the world and our place in the living universe. It will change our science and culture forever.

11.1.3 EXTRATERRESTRIAL INTELLIGENCE

Perhaps, mused the philosopher in Bernard de Fontenelle's *Entretiens sur la pluralité des mondes* (Fontenelle 1686), there are astronomers on Jupiter, and maybe, we cause them to engage in scientific quarrels, so that some Jovian philosophers must defend themselves when they put forward the ludicrous opinion that we exist. Their telescopes are directed toward us, as ours are toward them, "that mutual curiosity, with which the inhabitants of these Planets consider each other, and demand the one of the other, *What world is that? What people inhabit it?*" (Fontenelle 1701, p. 93, 1767, p. 207)

The search for extraterrestrial intelligence includes hypotheses and imaginations of foreign worlds and of the kind of creatures that might inhabit these worlds (Traphagan 2015b). The question is: What are we supposing when we human beings are searching for "extraterrestrial intelligence"? What are we searching for? Basically, there are four characteristics of possible extraterrestrial life-forms that we have in mind and that are presumed in our search. We assume that "they" are "intelligent," "social," and "communicative" and have "technology" (Dunér 2017). To some extent, we actually search for ourselves, for an intelligent life-form similar to us, and for something to which we can relate and establish a mutual understanding. We assume that they are social beings willing to and intelligent enough to be able to invent advanced technology for interstellar communication aimed for transferring information that would be possible to be decoded by the receiver. However, SETI research has mostly focused on the detection methods, especially radio and optical searches (Schuch 2011), rather than seriously addressing the questions of who and what extraterrestrial intelligence could be, which in its turn would direct us to where and how we should search for it (Cabrol 2016). The questions here are: How has intelligent life evolved? What are sociability, culture, and advanced technology? Why do we have these abilities? Why should we expect that the extraterrestrials have them?

11.1.3.1 Evolution

Astrocognition starts from a fundamental, basic premise, *the evolutionary astrocognitive premise*: Cognition in the universe develops through evolutionary processes of adaption to a specific but changing environment and the challenges it presents. The search for life in the universe has, to a great extent, highlighted how strongly the evolution of life and environment are intertwined (Golding and Glikson 2011; Schulze-Makuch et al. 2015; Cabrol 2016) and that the coevolution of life and environment determines the uniqueness of an extraterrestrial life-form (Watson 1999; Irwin and Schulze-Makuch 2001; Kooijman 2004; Dietrich et al. 2006). The same could be said of cognition and environment. We should not only search for environmental habitability, but "environmental cogitability."

Under certain conditions, which we are only beginning to understand, the environmental pressures force the cognitive agent to evolve toward more complex and flexible cognition. On Earth, we find that intelligence seems—in the same way as vision, aerial locomotion and other abilities—to have emerged several times, apparently independently, in the course of evolution and in separate evolutionary lines, that is, convergent or parallel evolution (Seed et al. 2009; Flores Martinez 2014). One might then argue that analogous planetary biospheres would lead to, through natural selection, the emergence of similarly adapted life-forms. Similar intelligent behavior could partly be an adaption to analogous biospheres. A certain biological convergence might be expected, but cultural convergence much less so, in respect to the arbitrary nature of symbolic thinking and communication.

Due to changing environmental pressures on other planets, sometimes similar and sometimes utterly different those that formed life on Earth, we cannot presume that extraterrestrial life would follow the same path as terrestrial life. Why a life-form evolves advanced cognition needs to be explained (Crowe 2015; Marino 2015a). One could bear in mind that complex cognitive skills are not necessary features of a living creature in order for it to sustain life. Advanced cognitive processing is costly and energy-consuming and not necessarily the most efficient way of securing continued survival of the genes of an organism. However, it is evident that cognitive flexibility has evolutionary benefits and is a good strategy for the adaptation to a changing environment. Intelligence or cognitive flexibility has thus emerged through a bio-cultural coevolution of the embodied mind, due to its benefits for survival, orientation, and adaptation to a variable environment in a Darwinian struggle for existence. Intelligence is thus the ability to respond to changes in the environment with flexibility and success, and a plasticity of learning, that is, to be able to learn from experience, of its own or others.

The more intelligent or cognitively flexible species on Earth, such as primates, dolphins, and corvids, share some important peculiarities. First, they are social animals and have a high degree of social complexity. Second, they are comparatively good at adapting to different environments and diets. If we could better understand the processes behind the rapid brain evolution that began a few million years ago on Earth—the encephalization in the Phanerozoic (Bogonovich 2011; Carter 2012), that is, the increased brain size relative to body size, over time—then we could use this knowledge to formulate astrocognitive theories on the evolution of intelligence in space (Osvath 2013).

Cognition has largely evolved as an adaptation to certain problems that the ancestors of a particular organism faced during the evolution of the species (see, e.g., Gärdenfors 2006). That is, the cognitive processor of the organism is adapted to, first, the physical and biological environment of their celestial body in order to understand and interpret, interact and deal with, and orientate itself in the particular physical and biological environment, in relation to its specific conditions, such as planetary orbit, gravitation, light conditions, atmosphere, radiation, temperature, chemistry, geology, ecology, and biota. Second, the cognition of an organism is also adapted to the minds and culture of its conspecifics, in order for it to understand and interact with other individuals, and to understand emotions, thoughts, and motives, etc., in a psychological and sociological interplay that forms that particular exoculture.

This bio-cultural coevolution of cognition is what can explain the emergence of advanced cognitive skills. Central here is the view of the mind as embodied, situated, enactive, and distributed, in other words, the coevolution of cognition and environment. Mind is thus a product of the interaction between the body and the surrounding environment, both the physical and cultural environments. And the very specific coevolution of cognition and environment, and the stochastic

events it is subdued to, forms the intelligent species of a particular exoplanet. This random nature of bio-cultural coevolution would most likely result in very distinct and unique forms of cognition in outer space. The stochastic events in the history of a species will give each intelligent life-form its unique fingerprint. The extraterrestrial intelligence and its thought processes that we might encounter one day would be completely alien to us. Extraterrestrial minds, like terrestrial minds, have adapted to their specific environment and the specific social interactions between the minds of their species. SETI research is, to a large extent, an endeavor to understand how intelligent life interacts with its environment and communicates information about its perspective on the surrounding world.

11.1.3.2 SOCIABILITY

The evolution of complex life on Earth is dependent on certain major evolutionary transitions, where a group of individuals that could replicate independently began to cooperate and formed a new, more complex form of life. Complexity is, in this context, commonly defined in terms of functional parts that take on more tasks and have more functional interactions (Smith and Szathmáry 1995; Corning and Szathmáry 2015; Levin 2017). One of the transitions on Earth was when individual insects began to collaborate for a common goal, in an alignment of interests, in a eusocial society. As Levin et al. (2017, p. 5) have stated, "Complexity requires different parts or units working together towards a common goal or purpose." In order to unite for a shared goal, the group needs to decrease conflict between the individuals. A society of societies aligns interests and eliminates conflicts; here, different social colonies collaborate, with different specialized tasks, and finally become completely dependent on each other.

The evolution of cognition is not fully elucidated, but it seems that the orientation in the social environment has played a particularly important role in the encephalization. According to the social-brain hypothesis (Dunbar 1996, 1998), there is a correlation between the size of an animal's social group and the size of its brain, leading to the conclusion that social behavior drives encephalization. The benefits of living in social groups, according to Dunbar, are that they provide defense against predators and enhance the defense of resources (Dunbar 2001; Chick 2014). A huge part of the brainpower will thus be used for handling social relations and mutual interactions within a group.

A congregation of a number of individuals, what we call a "social group" or "society," has evolutionary benefits for the individuals as well as for the entire group. The efficiency in how social organisms manage to collaborate, or use their sociability skills, increases their chances for survival. In complex social systems, "individuals frequently interact in many different contexts with many different individuals, and often repeatedly interact with many of the same individuals over time" (Freeberg et al. 2012a, p. 1785, 2012b). The complex social structure of the group is probably a very important drive for the emergence of intelligence, brain size, and communicative complexity. The brain has increased in capacity in order to tackle different kinds of social relations. Complex social worlds are like selective environments, driving species toward increased cognitive processing ability, which in its turn leads to higher social complexity, and when social complexity increases, it gives rise to a greater selection pressure on individuals for cognitive skills; that feedback produces even more social complexity, and so on (Bogonovich 2011).

In short, intelligent species are social species. Sociability and the social context enhance the adaption to the physical environment and make the individuals less vulnerable to a hostile environment. These social skills cannot subsist without a cognitive capacity to understand, feel, and share experiences of other minds. Beings, which we would recognize as intelligent in space, would be aware of themselves and of other minds and be able to share experiences, actions, information, and mental content. In other words, an extraterrestrial being technologically capable of transmitting and receiving interstellar messages would have intersubjective skills. In contemporary cognitive science, intersubjectivity has become a key concept for understanding not only empathy and altruism but also intelligence, sociability, and communication (Gärdenfors 2008a, 2008b; Gillespie 2009; Hrdy 2009; Gillespie and Cornish 2010; Tylén et al. 2010; Fusaroli et al. 2012; Gentilucci et al. 2012). In its most general definition, intersubjectivity can be explained as the *"sharing of experiences* about objects and events" (Brinck 2008, p. 116). To be more precise, Zlatev et al. (2008, p. 1) describe intersubjectivity as "the sharing of experiential content (e.g., feelings, perceptions, thoughts, and linguistic meanings) among a plurality of subjects." Intersubjectivity is an indispensable requisite for the evolution of intelligence, sociability, communication, and advanced technology (Dunér 2014). An extraterrestrial life-form that we could recognize as "intelligent" is, in a way or another, social and able to handle complex social relations and would have intersubjective skills for interpretation, envision, and sharing the behavior and mind of others.

11.1.3.3 Technology

Within the SETI community, advanced technology has often been understood heuristically as "technology for interstellar communication" (Raulin Cerceau 2015; Shostak 2015; Frank and Sullivan 2016). Behind the SETI conception of technology is the tendency of viewing advanced technology as a sort of applied science, as a product of the rational, inventive mind. This is, of course, a very narrow definition of technology that does not fully explain what technology is, why we have it, and where it comes from. In a more general sense, I would rather describe technology as ways of manipulating the surrounding world, using objects in the environment outside the body, in order to strengthen the genetically given capacities, such as body strength, perception, and cognition. Technology gives the intelligent organism the capability to manipulate the environment, in order to make it easier to live in it and to adapt

the environment to fit the organism, instead of adapting the organism to the environment.

To answer the question of how advanced technology might evolve, we have to turn to studies in the evolution of cognition—how hominids began using and manipulating their environment—and to studies in the history of technology—how the cultural evolution of *Homo sapiens sapiens* led to higher technology (Donald 1991; Tomasello 1999; Steels 2004; Richerson and Boyd 2005; Dennett 2009; Rospars 2013). Characteristic of human social interaction is the ability to learn from others, that is, culture, the existence of intraspecies group differences in behavioral patterns and repertoires, which has evolved through transmission of learned behavior and knowledge that are not biologically encoded, or in other words, the ability to transfer information from generation to generation that does not use the genetic code for the transfer but is learned, taught, and transferred by a multitude of communicative and cultural devices and artifacts, such as language, signs, pictures, sounds, and objects (Tomasello 1999; Sinha 2009, p. 292). Accordingly, culture presupposes enduring joint beliefs or common knowledge.

Technology, resting on specific cognitive abilities, is, to a large extent, a social phenomenon, a product of the cultural evolution. Without social interaction, joint beliefs, intersubjectivity, and information transfer—in one word, without culture—science and technology would not have arisen on this planet. The rise of civilization involved closeness, interaction of many individuals, and exchanges of ideas, products, and experiences that paved the way for a technological society. Culture, the ability to learn from others within a society of high social and communicational complexity, is what made advanced technology, as well as science, possible.

Thus, technology for interstellar communication is not just a form of advanced tool making that has its necessary conditions in physical or biological factors (cf. Casti 1989; Chick 2014). I would also maintain that there are three sociocognitive capacities that characterize advanced complex technology and that are crucial for its development. First, a sustainable *complex social system*, with a regulated system for collaboration, such as ethics; second, *complex communication*, for collaboration and abstract conceptualization; and third, a high degree of *distributed cognition*. All these three capacities require intersubjective skills.

Cooperation within a complex social system is one of the reasons behind the emergence of higher technology. Cooperation, in turn, requires some fundamental cognitive and communicative functions. Cooperation regarding a striving for some detached non-present goal requires advanced coordination of the inner worlds of the individuals, that is, intersubjective skills. In order to achieve advanced technology within a civilization, the individuals have to cooperate in joint activities, where they are sharing goals and attentions. By coordinating their roles when working toward a specific goal, they achieve a joint intention. To this, we can add that they must be able to engage in prospective planning to anticipate the future, that is, have the capacity to represent future needs, a prospective thinking, or "mental time travel" (Suddendorf and Corballis 1997; Roberts 2007).

A complex communicative system is needed in order to handle the social complexity, to facilitate collaboration, and to transfer information between the individuals, which are indispensable requisites for the innovation, development, and maintenance of complex technology. The communicative system must enable the users to construct abstract concepts and symbols and to generalize and discuss things and events not existent, that have ceased to exist, and have not yet come into being.

Important for the emergence of advanced technology is also the organisms' degree of distributed cognition, that is, the ability to use external objects and/or other minds to enhance thinking (Hutchins 1995). As pointed earlier, the mind is extended to the environment, and the environment takes an active role in driving the cognitive processes (Clark and Chalmers 1998; Clark 2008). The ability to construct external cognitive artifacts is significant in human cognition (Norman 1991; Norman 1993; Malafouris and Renfrew 2010; Malafouris 2012). These organism-independent artifacts, in Donald's terms, exograms, compensate for the limitations of the biological memory and could be regarded as an externalization and materialization of memory (Donald 1991, 2008, 2010). The distributed cognition strengthens the organism's inborn sensory equipment and gives it a set of devices for thinking. An advanced technological civilization cannot rest on the cognitive flexibility of a few individuals but needs as many as possible sharing their knowledge, cooperating, and completing different specialized tasks.

A civilization with advanced technology is a complex social system, which entails a high degree of communicative complexity and high degree of cognitive flexibility. Such a socially complex extraterrestrial civilization would have many individuals rather than a few, a high rather than low density, many different member roles rather than a few roles, and many directional relations rather than a few. Many individuals entail greater collective brainpower. A high density entails more frequent and faster interactions between individuals. Many different member roles entail a distributed and specialized cognitive processing. And finally, complex societies have greater diversity of directional relations. These four characteristics of social complexity will enhance the emergence of advanced technology.

To conclude, if there were intelligent beings able to communicate with advanced technology, they would probably have a complex social system, complex communication, and a high degree of distributed cognition. A fundamental requisite for social and communicative complexity is intersubjectivity. A technologically advanced civilization could then be described as a social and communicative organization that facilitates cooperation and regulates and prevents conflicts in order to achieve specific joint intentions for the manipulation of its environment as an adaption of itself to the prevailing physical conditions or as a way of adapting its environment for its own purposes.

11.1.3.4 Summary

Intelligence, regarded as a cognitive flexibility that enhances the chance of survival of a Darwinian creature, has evolved due to an adaptation to a changing environment, both the physical and social environments, through a bio-cultural coevolution. One of the most needed abilities of a cognitively flexible creature is intersubjectivity, the ability to understand other minds. The capability of intersubjectivity decides the degree of a society's social and communicative complexities, which in turn are fundamental requisites for the emergence of advanced technology.

Then comes the question of the distribution of intelligence in space. We might confer that there are many habitable worlds in outer space that show cognitability. But the issue is connected to our chances to detect them, and this is a question of simultaneity. In the Drake equation, factor L, the length of time for which such civilizations release detectable signals into space (Chick 2015; Dunér 2015, 2017), has to do with survival, sustainability, and societal organization. It is about how an extraterrestrial intelligent life-form can create a sustainable society in equilibrium with the physical and biological environment, so that it can survive for a time period long enough to significantly increase the probability of detection. Another way of formulating L is as the number of Earth years that a cognitively flexible extraterrestrial life-form with advanced capability to manipulate its environment can manage to maintain a social organization that enables it to voluntary transmit electromagnetic radiation that we can detect. L is often regarded as the most difficult factor to estimate in the Drake equation (Shostak 2009; Denning 2011; Dominik and Zarnecki 2011; Penny 2011; Dunér 2015). We know only one global advanced technological civilization—our own—and we have not yet seen the end of it. The difficulty has to do not only with the fact that we lack empirical data about other technological civilizations, but also with the fact that we are dealing with very complex questions about how self-conscious beings organize their social structure.

The lifetime of an advanced technological civilization seems to be connected to how cognitively flexible creatures are able to control the power of their technology, find equilibrium with their environmental resources, and organize their society in order to prevent it from breaking down. In order to deal with destructive behavior within a society, an advanced technological civilization must have advanced intersubjective skills to understand other subjects; it must have a high degree of communicative complexity to sustain and strengthen the intersubjective interactions between its members, including long experience of communicating with a diversity of groups and species; and finally, it must have arrived at some sort of reliable regulation system for behavior, or what can be called "ethics." L is, in other words, about how a congregation of cognitively flexible Darwinian organisms successfully manages the discrepancy between its capacity of manipulating its environment and its regulatory system for collaboration: L is a measure of this ratio.

11.1.4 TERRESTRIAL-EXTRATERRESTRIAL INTERACTION

11.1.4.1 COMMUNICATION

Communication can be regarded as a sharing of mental states, and the expression as information about a mental state (Østergaard 2012; cf. Gärdenfors 2013). Communication is a way of transferring mental content from mind to mind, where the encoding by the producer of a message is followed by a decoding by the recipient (Saint-Gelais 2014). As such, communication involves interpretation, where the interpreter endows meaning to the message. Communication is not about "things out there"; it is about our *conceptualization* of the things "out there" or rather our phenomenological experiences of the Lifeworld. According to the "social complexity hypothesis" for communication, "groups with complex social systems require more complex communicative systems to regulate interactions and relations among group members" (Freeberg et al. 2012a, p. 1785). Social complexity, in other words, leads to communicative complexity. Elsewhere, I have discussed the cognitive foundations of interstellar communication (Dunér 2011b) and maintained that communication is based on cognitive abilities embodied in an organism that has developed through an evolutionary and socio-cultural process by interacting with its specific environment (see also Arbib 2013; Holmer 2013). One of the most crucial cognitive abilities for language acquisition is, yet again, intersubjectivity. Communication presupposes shared knowledge or perhaps better, shared experiences.

If we extend communicational interactions beyond Earth, the question naturally arises: How could communication be possible between intelligent beings of different environments that differ physically, biologically, and culturally and have developed through separate evolutionary lines? This is the *cognitive-semiotic problem of interstellar communication* (Dunér 2011b). The circumstances in which we will have no kinship—that we will not likely share similar bodies or, even less, cultures—or even similar physical environments will have far-reaching consequences for how we will be able to construct and interpret messages from distant civilizations. The usual strategy to overcome the problem of interstellar communication has been to try to construct a message that is a universal symbolic information transfer, which is independent of context, time, and human nature (Sagan 1973; Vakoch 2011). This can be called *the universal-transcendental interstellar message objective*. However, this strategy and its requirements are not reconciled with what we presently know about cognition, communication, and evolution. Particularly, it presupposes the universality it is aimed at, and thereby ignores the facticity of evolution and symbolization as embodied and situated. It ignores the context that the living organisms, with their cognition and communication, are planet-bound and constrained by certain physical conditions. It leaves out time and history—the evolution, the phylogenetic, ontogenetic, and cultural-historical development—in which the organisms are evolving. Finally, it ignores the nature of the communicators—that they have bodies and brains evolved

in interaction with their environment. Building on the basic observation in cognitive science that our cognitive and communicational skills are embodied, situated in and adapted to our terrestrial environment, we cannot exclude the context, the situation, space, time, and human nature if we would like to construct comprehensible interstellar messages.

In order to solve the interstellar communication problem, we need the insights of cognitive science, evolutionary theory, semiotics, hermeneutics, and history. It could be argued that the problem of interstellar communication is not just a problem within natural science but a humanistic problem in its true sense, a human problem. It is we humans who will send and receive and code and decode the messages. The communicative problem also has to do with history, understood on the most basic level as the interaction of organisms with their environment over time. Cognition and communication are results of time and of history, both evolutionary history and socio-cultural history. Communication is thereby not something pre-given, but it rather evolves in interplay with the environment, in dialogue of agent with agent. In our case, this process took millions of years (Donald 1991; Christiansen and Kirby 1997; Deacon 1997; Donald 2001; Tomasello 2008).

Human communication, whether it consists of lingual, symbolic, or bodily expressions, is dependent on the inner workings of our brains and how humans interact with their physical, biological, and cultural environments. Accordingly, communication is a bio-cultural hybrid, a changing product of the genetic-cultural coevolution. Communication is therefore a situated practice. It is constrained by its surroundings and is adapted to specific circumstances. This means that we cannot exclude the situation where the message is performed and the physical, biological, and socio-cultural contexts of the communicators. We are planet-bound creatures.

Intelligence, as has been explained, could be seen as evolved mental gymnastics, which is required for a particular organism to survive and reproduce within its specific environment. This includes the capability of representing activities and being able to make inner models of reality. By using symbols, an intelligent creature could engage in abstract thinking detached from the environment. If the extraterrestrials are intelligent, then they probably have some kind of symbolization abilities and abstract thinking detached from the environment, with which they can reason about things that do not exist and things that are not right in front of them, facing their senses, in a specific moment in time. A very effective tool for symbolizing thought is our communicational devices. John Taylor describes language as a set of resources that are available to the language user for the symbolization of thought and the communication of those symbolizations (Taylor 2002). It facilitates thinking about the things and events that are not immediately in front of us, engaging with our senses, as well as about those things and events that, seemingly, could never exist outside of fiction. It frees us ever further from the here-and-now and lets us better contemplate the might-have-been. It allows us to share ideas and mental states. Yet, it rests on the cognitive abilities that are a result of bio-cultural coevolution here on Earth.

The cognition of an extraterrestrial life-form is likewise adapted to a specific environment, and the way this cognitively skilled organism expresses and communicates its thoughts, perspectives, and interpretations of its environment is a result of its specific bio-cultural coevolution. How we, and the aliens, transfer meaning in different ways, I would say, is the result of our dissimilar evolutions, bodily and cognitive constructions, and socio-cultural histories. The construction and interpretation of the symbols are dependent upon how our brains work, what our bodies are like, how we interact with our environment, how our sensations are processed, and the history of our culture. So, we can conclude that communication is based on cognitive abilities embodied in the organism that has developed through an evolutionary and socio-cultural process in interaction with its specific environment. In other words, our communication is adapted to an earthly environment and for communication with our conspecifics. Our communicational and symbolic skills have evolved through an evolutionary and cultural-historical process here on Earth and are thereby constrained by our human bodies, terrestrial environment, and the socio-cultural characteristics of our species. So, our human communication is, in fact, maladapted to interstellar communication. This understanding of human cognition is crucial for future interstellar communication and should be taken into account in order to be able to transfer messages to other minds in the universe.

11.1.4.2 SEMIOTICS

Inasmuch as interstellar communication is thought to be an exercise in coding and decoding signs, then the relevance of semiotics should be obvious (Vakoch 1998). In particular, cognitive semiotics studies the meaning-making structured by the use of different sign vehicles and the properties of meaningful interactions with the surrounding environment, both with the physical and social environments (Sonesson 2007, 2009; Zlatev 2012, 2015; Dunér and Sonesson 2016b). A key concept here is the bio-cultural coevolution, or in other words, meaning-making is a result of not only a natural, biological-genetic evolution, but also a cultural evolution, as well as the incessant interaction between the mind and its environment (Richerson and Boyd 2005). The interstellar communication problem is very much a semiotic problem: how meaning can be transferred and interpreted. The first problem that arises in a situation of interstellar communication is realizing that it really is a message at all (Sonesson 2013). Some regularity and order and finding a repetition in the pattern are not enough. We have to understand that someone has an intention with it that we should understand as a message. Next comes the problem of deciphering what the message means.

Some kind of vehicle for transferring the mental content is needed, that is, signs. The study of meaning-making, semiotics, concerns *signs*, which can be said to be something that we interpret as having meaning. According to Charles Sanders Peirce, one of the main figures in semiotics, "A sign, or *representamen*, is something that stands to somebody for something in some respect or capacity" (Peirce 1932, p. 135; Saint-Gelais 2014). The sign, as *expression*, stands for something, its *object*. The sign does not include its meaning; rather, the meaning is attributed through elaboration of an *interpreter*. So, for something to be meaningful, an interpreter is needed, a human being (or other meaning-making creatures), who endows the sign a meaning. A physical phenomenon is meaningless so far as there is no one to recognize it as meaningful. For example, phenomena that we call biosignatures become meaningful phenomena when we interpret them as containing a meaning by making a connection between the expression and the object, in other words, between the "biosignature" and "the living organism" (Dunér 2019). With signs, we make sense of the world, approach it, get access to it, and differentiate things from each other. Correspondingly, the biosignatures we detect are one way among many other ways of making sense of the data we receive from outer space. And these biosignatures exist so far as we find them meaningful. In that perspective, the biosignatures are not solely "out there"; instead, they are, to a great extent, in our minds or rather in the interaction between our minds and the outer world. It is in our meaning-making practices that the "biosignatures" become biosignatures. The astrobiologist searching for biosignatures is a sort of semiotician, an astrobio-semiotician, trying to establish connections between expressions and objects in the universe. Semiotics of biosignatures concerns qualities and categories, as well as the search for rules and regularities within such a nomothetic science as astrobiology concerned with generalities. More generally, semiotics of biosignatures concerns the meaning-making processes of astrobiology.

The *semiosis* is the sign process, how the signs operate in the production of meaning. In that sense, when the astrobiologist is interpreting biosignatures, he or she is involved in a meaning-producing semiosis. This semiosis is triadic: it contains expression, object, and interpreter—which in our case respond to "biosignature," "life," and "astrobiologist," respectively. Depending on how the interpreter makes or interprets the connection between the expression and the object, we have basically three types of sign relations, following Peirce, *icon*, *index*, and *symbol*. The meaning of the relation between expression and content that the interpreter experiences is based on similarity (iconicity); proximity (indexicality); or habits, rules, or conventions (symbolicity). An icon is a sign relation based on similarity, where the expression shares some of the object's properties. An index is when the expression has some contiguity with the object. And finally, a symbol is when the connection between expression and object is just a mere convention.

The problem of interstellar communication lies not so much in the physical or technological constraints, even though they very much challenge our scientific and technological skills, but in the cognitive and semiotic problems that interstellar message decoding provokes (Vakoch 1998; Dunér 2011b, 2014; Sonesson 2013). It might be suggested that icons and indices would be easier to decode, depending on their respective similarity and contiguity. However, their decoding rests on our biological sensory constraints and perceptual interpretations and our ability to recognize the connection between the expression and its object. The interpretation of indices requires empirical knowledge of the recurrent connection between the sign and what it refers to. The perceptual world consists of a profuse amount of potential indexicality, even though we do not yet recognize these indices as signs with meaning.

In order to reach more complex communication, a social organism not just needs attention, imitation, mimetic skills, and iconic and indexical signs, but it also needs to be able to use symbolic signs. If the extraterrestrials are intelligent, they probably have some kind of symbolization abilities and abstract thinking detached from the environment, with which they can reason about things not existent, non-present, senseless, timeless, and abstract. In other words, to reach a higher degree of communicative complexity, they need signs where the expression is separated from the content, that is, symbols. Icons and indices are signs that have some *non-arbitrary* similarity or contiguity with the signified, in contrast to the symbols' completely *arbitrary* relation. For example, the word "life" has no causal link to what it stands for, nor does it resemble what it signifies. There are no intrinsic relationships between the expression and the content whatsoever. It is the interpreters (the ones that construct the message and the ones that decode them, respectively) that join the expression and the content together and establish the connection between them. And the matching between the transmitters' and the receivers' interpretation of the symbols is by no means self-evident. We may figure out the reference of the signal but will probably have severe problems in understanding extraterrestrial symbols. It is the message's expression rather than its content that becomes the difficulty for the interpreter. In symbols, there is a gap between the sign and the meaning. Nothing in the physical appearance of the sign gives any clue to its object; they are instead linked by an arbitrary correlation.

The problem with symbols is that they are conventional, or arbitrary, as Ferdinand de Saussure called them (Saussure 1916). They are detached representations and, as such, dependent on cultural and social interactions that create some specific regularities that have their origin in more or less stochastic habits, conventions, etc., of the species (Sonesson and Dunér 2016). The sign (the expression) and the signified (the content) have no intrinsic connection. It is not impossible to imagine that the aliens would have certain knowledge about their environment that in its content is similar to our own knowledge of mathematics, physics, or chemistry. However, their expression of it would most likely be very different from ours. This basic semiotic distinction between expression and content is neglected in most human attempts at interstellar messaging. For all its ingenuity, Freudenthal's *Lingua Cosmica* is probably an effort in vain (Freudenthal 1960).

11.1.4.3 SUMMARY

Our communication and symbolization have evolved through an evolutionary and cultural-historical process here on Earth and are thereby constrained by our human bodies, terrestrial environment, and the socio-cultural characteristics of our species. And likewise, a potential information transfer containing a symbolic message from an alien civilization would be constrained by the bio-cultural coevolution of the extraterrestrial intelligence that coded it. The semiosis of sign use, its triadic nature of expression, object, and interpreter, exposes the meaning-making processes behind biosignatures and interstellar communication.

11.1.5 CONCLUSION

In this chapter, I have discussed the question of mind in space, the thinking universe. We know for a fact that there are beings in the universe who think—the human species. But are there also other beings out there who cogitate about their own existence and perhaps even about our existence? And would it be possible to share and transfer mental content between various thinking beings in the universe? In order to answer these questions, I have suggested astrocognition as a new research field within astrobiology. Astrocognition explores the origin, evolution, and distribution of intelligent life in the universe, based on findings from cognitive science, linguistics, semiotics, history, philosophy, evolutionary theory, and allied sciences. This research will hopefully give us a better understanding of how the human mind interacts with unknown environments and phenomena, about the origin and evolution of intelligence, and what kind of environments intelligence needs in order to emerge on a planet and to what extent it might be distributed in outer space.

In the first section, concerning terrestrial intelligence, I discussed the cognitive phenomena involved in human exploration of the unknown, when humans search for extraterrestrial life and intelligence. First, I stated that intelligence is something more than a kind of logical, rational problem solving. Instead, I explained it as a cognitive flexibility, an ability to adjust to changes in the physical and socio-cultural environments, which include intersubjective skills. I highlighted three cognitive functions that are particularly prominent in the history of astrobiology, which says something about human cognition facing the unknown, that is, perception, conceptualization, and analogy. The sensory input is interpreted through our preconceived concepts, theories, and prior knowledge; that our concepts, categories, and definitions are challenged; and that we commonly engage in an analogical reasoning from what we know to what we do not know.

The next section, about extraterrestrial intelligence, dealt with the evolution of intelligence in outer space and the extent to which sociability and technology affect it. The starting point is that cognition evolves through an evolutionary process as an adaption to a changing physical and social environment. This bio-cultural coevolution of cognition explains the emergence of advanced cognitive skills. One of the most important drives for intelligence seems to be the organism's need to orientate itself in the social environment and to handle the increasing social complexity of the group. An indispensable requisite for the evolution of intelligence, sociability, communication, and advanced technology is intersubjectivity, the ability to understand other minds. Technology is not a mere applied science; rather, it could be defined as ways of manipulating the surrounding world by using objects in the environment in order to strengthen the genetically given capacities and to adapt the environment to fit the organism, instead of the other way around. Resting on specific cognitive abilities, technology is, to a large extent, a social phenomenon, a product of the cultural evolution, that is, through the transmission of learned behavior and knowledge that are not biologically encoded. An intelligent being that has managed to develop advanced technology would likely have a complex social system, complex communication, and a high degree of distributed cognition.

In the final section, about terrestrial-extraterrestrial interaction, I discussed the question of whether two distinct creatures that share neither biology, environment, and evolution nor culture and history could be able to understand each other. The sharing of mental content, communication, is a product of a particular bio-cultural coevolution of a species, adapted to a specific physical and social environment. Cognitive semiotics, the study of meaning-making, is a key to understand the semiosis involved in astrobiology and astrocognition, such as biosignatures and interstellar communication. The problem of interstellar communication is, to a great extent, a semiotic problem, concerning how meaning is transferred and interpreted. Symbols, which are indispensable for abstract, complex reasoning, are particularly challenging due to their detached, arbitrary nature. To conclude, this chapter aimed to lay the ground for an emerging research field studying the origin, evolution, and distribution of intelligence in the universe. It concerned the thinking universe, what it means to think, why we think, and how we could understand other thinking beings, resting on the belief that the greatest discovery in a human's life is the encounter with the thoughts of another human being. Humankind's encounter with the thoughts of other thinking beings from another world would surely be the greatest discovery in the history of the human mind.

REFERENCES

Arbib, M. A. 2013. Evolving an extraterrestrial intelligence and its language-readiness. In *The History and Philosophy of Astrobiology: Perspectives on the Human Mind and Extraterrestrial Life*, D. Dunér, J. Parthemore, E. Persson, and G. Holmberg (Eds.), pp. 139–156. Newcastle upon Tyne, UK: Cambridge Scholars Publishing.

Aristotle. 1996. *Posterior Analytics*; *Topica*. H. Tredennick, and E. S. Forster (Eds.). London, UK: Heinemann.

Berlin, B. 1992. *Ethnobiological Classification: Principles of Categorization of Plants and Animals in Traditional Societies*. Princeton, NJ: Princeton University Press.

Bertka, C. M. (Ed.). 2009. *Exploring the Origin, Extent, and Future of Life, Philosophical, Ethical and Theological Perspectives*. Cambridge, UK: Cambridge University Press.

Billings, L. 2012. Astrobiology in culture: The search for extraterrestrial life as 'Science.' *Astrobiology* 12(10):966–975.

Bogonovich, M. 2011. Intelligence's likelihood and evolutionary time frame. *International Journal of Astrobiology* 10:113–122.

Brinck, I. 2008. The role of intersubjectivity in the development of intentional communication. In *The Shared Mind: Perspectives on Intersubjectivity*, J. Zlatev et al. (Eds.), pp. 115–140. Amsterdam, the Netherlands: Benjamins.

Cabrol, N. A. 2016. Alien mindscapes: A perspective on the search for extraterrestrial intelligence. *Astrobiology* 16(9):1–16.

Carter, B. 2012. Hominid evolution: Genetics versus memetics. *International Journal of Astrobiology* 11:3–13.

Casti, J. L. 1989. *Paradigms Lost: Tackling the Unanswered Mysteries of Modern Science*. New York: Avon Books.

Chick, G. 2014. Biocultural prerequisites for the development of interstellar communication. In *Archaeology, Anthropology, and Interstellar Communication*, D. A. Vakoch (Ed.), pp. 203–226. Washington, DC: NASA.

Chick, G. 2015. Length of time such civilizations release detectable signals into space, L, 1961 to the present. In *The Drake Equation: Estimating the Prevalence of Extraterrestrial Life through the Ages*, D. A. Vakoch, and M. F. Dowd (Eds.), pp. 270–297. Cambridge, UK: Cambridge University Press.

Christiansen, M. H., and S. Kirby (Eds.). 1997. *Language Evolution*. Oxford, UK: Oxford University Press.

Chyba, C. F., and K. P. Hand. 2005. Astrobiology: The study of the living universe. *Annual Review of Astronomy and Astrophysics* 43:31–74.

Clark, A. 2008. *Supersizing the Mind: Embodiment, Action, and Cognitive Extension*. Oxford, UK: Oxford University Press.

Clark, A., and D. Chalmers. 1998. The extended mind. *Analysis* 58(1):7–19.

Corning, P. A., and E. Szathmáry. 2015. Synergistic selection: A Darwinian frame for the evolution of complexity. *Journal of Theoretical Biology* 371:45–58.

Crowe, M. J. 1986. *The Extraterrestrial Life Debate 1750–1900: The Idea of a Plurality of Worlds from Kant to Lowell*. Cambridge, UK: Cambridge University Press.

Crowe, M. J. 2008. *The Extraterrestrial Life Debate, Antiquity to 1915: A Source Book*. Notre Dame, IN: University of Notre Dame.

Crowe, M. J. 2015. Fraction of life-bearing planets on which intelligent life emerges, f_i, pre-1961. In *The Drake Equation: Estimating the Prevalence of Extraterrestrial Life through the Ages*, D. A. Vakoch, and M. F. Dowd (Eds.), pp. 163–180. Cambridge, UK: Cambridge University Press.

Davies, P. 2007. *Cosmic Jackpot: Why Our Universe is Just Right for Life*. Boston, MA: Houghton Mifflin.

Davies, P. C. W., S. A. Benner, C. E. Cleland, C. H. Lineweaver, C. P. McKay, and F. Wolfe-Simon, F. 2009. Signatures of a shadow biosphere. *Astrobiology* 9(2):241–249.

Deacon, T. 1997. *The Symbolic Species: The Co-evolution of Language and the Brain*. New York: Norton.

de Duve, C. 1995. *Vital Dust: Life as a Cosmic Imperative*. New York: Basic Books.

Dennett, D. C. 2009. The evolution of culture. In *Cosmos and Culture: Cultural Evolution in a Cosmic Context*, S. J. Dick, and M. L. Lupisella (Eds.), pp. 125–143. Washington, DC: NASA History Series.

Denning, K. 2011. 'L' on earth. In *Civilizations Beyond Earth: Extraterrestrial Life and Society*, D. A. Vakoch, and A. A. Harrison (Eds.), pp. 74–86. New York: Berghahn Books.

Dick, S. J. 1982. *Plurality of Worlds: The Origins of the Extraterrestrial Life Debate from Democritus to Kant*. Cambridge, UK: Cambridge University Press.

Dick, S. J. 1996. *The Biological Universe: The Twentieth-Century Extraterrestrial Life Debate and the Limits of Science*. Cambridge, UK: Cambridge University Press.

Dick, S. J. 2012. Critical issues in the history, philosophy, and sociology of astrobiology. *Astrobiology* 12(10):906–927.

Dietrich, L. E. P., M. Michael, and D. K. Newman. 2006. The coevolution of life and earth. *Current Biology* 16:pR395–pR400.

Dominik, M., and J. C. Zarnecki. 2011. The detection of extra-terrestrial life and the consequences for science and society. *Philosophical Transactions of the Royal Society A: Mathematical, Physical and Engineering Sciences* 369(1936):499–507.

Donald, M. 1991. *Origins of the Modern Mind: Three Stages in the Evolution of Culture and Cognition.* Cambridge, MA: Harvard University Press.

Donald, M. W. 2001. *A Mind So Rare: The Evolution of Human Consciousness.* New York: Norton.

Donald, M. W. 2008. A view from cognitive science. In *Was ist der Mensch?* D. Ganten, V. Gerhardt, J. C. Heilinger, and J. Nida-Rümelin (Eds.), pp. 45–49. Berlin, Germany: de Gruyter.

Donald, M. W. 2010. The exographic revolution: Neuropsychological sequelae. In *The Cognitive Life of Things: Recasting the Boundaries of the Mind*, L. Malafouris, and C. Renfrew (Eds.), pp. 71–79. Cambridge, MA: McDonald Institute for Archaeological Research.

Drake, F. 2011. The search for extra-terrestrial intelligence. *Philosophical Transactions of the Royal Society A: Mathematical, Physical and Engineering Sciences* 369(1936):633–643.

Dunbar, R. 1996. *Grooming, Gossip and the Evolution of Language.* Cambridge, MA: Harvard University Press.

Dunbar, R. 1998. The social brain hypothesis. *Evolutionary Anthropology* 6(5):178–190.

Dunbar, R. 2001. Brains on two legs: Group size and the evolution of intelligence. In *Tree of Origin: What Primate Behavior Can Tell Us about Human Social Evolution*, F. B. M. de Waal (Ed.), pp. 173–191. Cambridge, MA: Harvard University Press.

Dunér, D. 2011a. Astrocognition: Prolegomena to a future cognitive history of exploration. In *Humans in Outer Space – Interdisciplinary Perspectives*, U. Landfester, N. L. Remuss, K. U. Schrogl, and J. C. Worms (Eds.), pp. 117–140. Vienna, Austria: Springer.

Dunér, D. 2011b. Cognitive foundations of interstellar communication. In *Communication with Extraterrestrial Intelligence*, D. A. Vakoch (Ed.), pp. 449–467. Albany, NY: State University of New York Press.

Dunér, D. 2012. Introduction: The history and philosophy of astrobiology. *Astrobiology* 12(10):901–905.

Dunér, D. 2013a. Extraterrestrial life and the human mind. In *The History and Philosophy of Astrobiology: Perspectives on Extraterrestrial Life and the Human Mind*, D. Dunér, J. Parthemore, E. Persson, and G. Holmberg (Eds.), pp. 7–31. Newcastle upon Tyne, UK: Cambridge Scholars Publishing.

Dunér, D. 2013b. The language of cosmos: The cosmopolitan endeavour of universal languages. In *Sweden in the Eighteenth-Century World: Provincial Cosmopolitans*, G. Rydén (Ed.), pp. 41–65. Farnham, UK: Ashgate.

Dunér, D. 2013c. Venusians: The planet Venus in the 18th-century extraterrestrial life debate. *The Journal of Astronomical Data*, 19(1): 145–167.

Dunér, D. 2014. Interstellar intersubjectivity: The significance of shared cognition for communication, empathy, and altruism in space. In *Extraterrestrial Altruism: Evolution and Ethics in the Cosmos*, D. A. Vakoch (Ed.), pp. 139–165. Dordrecht, the Netherlands: Springer.

Dunér, D. 2015. Length of time such civilizations release detectable signals into space, L, pre-1961. In *The Drake Equation: Estimating the Prevalence of Extraterrestrial Life through the Ages*, D. A. Vakoch, and M. F. Dowd (Eds.), pp. 241–269. Cambridge, UK: Cambridge University Press.

Dunér, D. 2016a. Science: The structure of scientific evolutions. In *Human Lifeworlds: The Cognitive Semiotics of Cultural Evolution*, D. Dunér, and G. Sonesson (Eds.), pp. 229–266. Pieterlen, Switzerland: Peter Lang.

Dunér, D. 2016b. Swedenborg and the plurality of worlds: Astrotheology in the eighteenth century. *Zygon: Journal of Religion and Science* 51(2):450–479.

Dunér, D. 2017. On the plausibility of intelligent life on other worlds: A cognitive-semiotic assessment of $f_i \cdot f_c \cdot L$. *Environmental Humanities* 9(2):433–453.

Dunér, D. 2019. The history and philosophy of biosignatures. In *Biosignatures for Astrobiology*, F. Westall and B. Cavalazzi (Eds.). Cham, Switzerland: Springer.

Dunér, D., C. Malaterre, and W. Geppert. 2016. The history and philosophy of the origin of life. *International Journal of Astrobiology* 15(4):1–2.

Dunér, D., J. Parthemore, E. Persson, and G. Holmberg (Eds.). 2013. *The History and Philosophy of Astrobiology: Perspectives on Extraterrestrial Life and the Human Mind.* Newcastle upon Tyne, UK: Cambridge Scholars Publishing.

Dunér, D., E. Persson, and G. Holmberg (Eds.). 2012. The history and philosophy of astrobiology. *Astrobiology* 12(10): 901–1016.

Dunér, D., and G. Sonesson, 2016a. Encounters: The discovery of the unknown. In *Human Lifeworlds: The Cognitive Semiotics of Cultural Evolution*, D. Dunér, and G. Sonesson (Eds.), pp. 267–300. Pieterlen, Switzerland: Peter Lang.

Dunér, D., and G. Sonesson, Ed. 2016b. *Human Lifeworlds: The Cognitive Semiotics of Cultural Evolution.* Pieterlen, Switzerland: Peter Lang.

Flores Martinez, C. L. 2014. SETI in the light of cosmic convergent evolution. *Acta Astronautica* 104(1):341–349.

Fontenelle, B. L. B. d. *Entretiens sur la pluralité des mondes.* Amsterdam, the Netherlands: Mortier; new ed. Amsterdam, the Netherlands: Pierre Mortier, 1701; trans., *Conversations on the Plurality of Worlds.* London, UK: Thomas Caslon, 1767.

Frank, A., and W. T. Sullivan III. 2016. A new empirical constraint on the prevalence of technological species in the universe. *Astrobiology* 16(5): 359–362.

Freeberg, T. M., R. I. M. Dunbar, and T. J. Ord. 2012a. Social complexity as a proximate and ultimate factor in communicative complexity. *Philosophical Transactions of Royal Society. Series B: Biological Sciences* 367:1785–1801.

Freeberg, T. M., T. J. Ord, and R. I. M. Dunbar. 2012b. The social network and communicative complexity: Preface to theme issue. *Philosophical Transactions of Royal Society. Series B: Biological Sciences* 367:1782–1784.

Freudenthal, H. 1960. *Lincos: Design of a Language for Cosmic Intercourse.* Amsterdam, the Netherlands: North-Holland.

Fusaroli, R., P. Demuru, and A. M. Borghi. 2012. The intersubjectivity of embodiment. *Journal of Cognitive Semiotics* 4(1):1–5.

Galilei, G. 1610. Siderevs nvncivs magna, longeqve admirabilia spectacula pandens, … Venice, Italy: Thomas Baglionus; trans. W. R. Shea, *Galileo's Sidereus Nuncius, or, A Sidereal Message.* Sagamore Beach, MA: Science History Publications, 2009.

Gärdenfors, P. 2006. *How Homo Became Sapiens: On the Evolution of Thinking.* Oxford, UK: Oxford University Press.

Gärdenfors, P. 2008a. Evolutionary and developmental aspects of intersubjectivity. In *Consciousness Transitions – Phylogenetic, Ontogenetic and Physiological Aspects*, H. Liljenström, and P. Århem (Eds.), pp. 281–385. Amsterdam, the Netherlands: Elsevier.

Gärdenfors, P. 2008b. The role of intersubjectivity in animal and human cooperation. *Biological Theory* 3(1):1–12.

Gärdenfors, P. 2013. The evolution of semantics: Sharing conceptual domains. In *The Evolutionary Emergence of Language: Evidence and Inference*, R. Botha (Ed.). Oxford, UK: Oxford University Press.

Gentilucci, M., C. Gianelli, G. C. Campione, and F. Ferri. 2012. Intersubjectivity and embodied communication systems. *Journal of Cognitive Semiotics* 4(1):124–137.

Gillespie, A. 2009. The intersubjective nature of symbols. In *Symbolic Transformations: The Mind in Movement Through Culture and Society: Cultural Dynamics of Social Representation*, B. Wagoner (Ed.), pp. 23–37. London, UK: Routledge.

Gillespie, A., and F. Cornish. 2010. Intersubjectivity: Towards a dialogical analysis. *Journal for the Theory of Social Behaviour* 40(1):19–46.

Golding, S. D., and M. Glikson. 2011. *Earliest Life on Earth: Habitats, Environments and Methods of Detection*. Dordrecht, the Netherlands: Springer.

Holmer, A. 2013. Greetings earthlings!: On possible features of exolanguage. In *The History and Philosophy of Astrobiology: Perspectives on the Human Mind and Extraterrestrial Life*, D. Dunér, J. Parthemore, E. Persson, and G. Holmberg (Eds.), pp. 157–183. Newcastle upon Tyne, UK: Cambridge Scholars Publishing.

Hrdy, S. B. 2009. *Mothers and Others: The Evolutionary Origins of Mutual Understanding*. Cambridge, MA: Belknap Press of Harvard University Press.

Hutchins, E. 1995. How a cockpit remembers its speeds. *Cognitive Science* 19:265–288.

Irwin, N. I., and D. Schulze-Makuch. 2001. Assessing the plausibility of life on other worlds. *Astrobiology* 1(2):143–160.

Kane, S. R., and D. M. Gelino. 2012. The habitable zone and extreme planetary orbits. *Astrobiology* 12(10):940–945.

Kooijman, S. A. L. M. 2004. On the co-evolution of life and its environment. In *Scientists Debate Gaia: The Next Century*, S. H. Schneider, J. R. Miller, E. Crist, and P. J. Boston (Eds.), pp. 343–351. Cambridge, MA: MIT Press.

Krois, J. et al. (Eds.). 2007. *Embodiment in Cognition and Culture*. Amsterdam, the Netherlands: Benjamins.

Lakoff, G. 1990. *Women, Fire, and Dangerous Things: What Categories Reveal about the Mind*. Chicago, IL: University of Chicago Press.

Lakoff, G., and M. Johnson. 1999. *Philosophy in the Flesh: The Embodied Mind and Its Challenge to Western Thought*. New York: Basic Books.

Levin, S. R., T. W. Scott, H. S. Cooper, and S. A. West. 2017. Darwin's aliens. *International Journal of Astrobiology*. doi:10.1017/S1473550417000362.

Losch, A. (Ed.). 2017. *What is Life?: On Earth and Beyond*. Cambridge, UK: Cambridge University Press.

Lowell, P. 1895. *Mars*. Boston, MA: Houghton, Mifflin.

Malafouris, L. 2012. Linear B as distributed cognition: Excavating a mind not limited by the skin. In *Excavating the Mind: Cross-Sections through Culture, Cognition and Materiality*, N. Johannsen, M. Jessen, and H. Juel Jensen (Eds.), pp. 69–84. Aarhus, Denmark: Aarhus University Press.

Malafouris, L, and C. Renfrew (Eds.). 2010. *The Cognitive Life of Things: Recasting the Boundaries of the Mind*. Cambridge, UK: The McDonald Institute Monographs.

Marino, L. 2015a. Fraction of life-bearing planets on which intelligent life emerges, f_i, 1961 to the present. In *The Drake Equation: Estimating the Prevalence of Extraterrestrial Life through the Ages*, D. A. Vakoch, and M. F. Dowd (Eds.), pp. 181–204. Cambridge, UK: Cambridge University Press.

Marino, L. 2015b. The landscape of intelligence. In *The Impact of Discovering Life Beyond Earth*, S. J. Dick (Ed.), pp. 95–112. Cambridge, UK: Cambridge University Press.

Martins, Z. et al. 2017. Earth as a tool for astrobiology – A European perspective. *Space Science Reviews* 209:43–81.

NASA. 1998. *NASA Strategic Plan 1998: With 1999 Interim Adjustments*. http://www.hq.nasa.gov/office/codez/plans/NSP99.pdf.

Norman, D. A. 1991. Cognitive artifacts. In *Designing Interaction: Psychology at the Human-Computer Interface*, J. M. Carroll (Ed.), pp. 17–38. Cambridge, UK: Cambridge University Press.

Norman, D. A. 1993. Cognition in the head and in the world. *Cognitive Science* 17:1–6.

Østergaard, S. 2012. Imitation, mirror neurons and material culture. In *Excavating the Mind: Cross-Sections through Culture, Cognition and Materiality*, N. Johannsen, M. Jessen, and H. Juel Jensen (Eds.), pp. 25–38. Aarhus, Denmark: Aarhus University Press.

Osvath, M. 2013. Astrocognition: A cognitive zoology approach to potential universal principles of intelligence. In *The History and Philosophy of Astrobiology: Perspectives on the Human Mind and Extraterrestrial Life*, D. Dunér, J. Parthemore, E. Persson, and G. Holmberg (Eds.), pp. 49–65. Newcastle upon Tyne, UK: Cambridge Scholars Publishing.

Pálsson, G. 2009. Celestial bodies: Lucy in the sky. In *Humans in Outer Space – Interdisciplinary Odysseys*, L. Codignola, and K. U. Schrogl (Eds.), pp. 69–81. Vienna, Austria: Springer.

Parthemore, J. 2013. The 'final frontier' as metaphor for mind: Opportunities to re-conceptualize what it means to be human. In *The History and Philosophy of Astrobiology: Perspectives on the Human Mind and Extraterrestrial Life*, D. Dunér, J. Parthemore, E. Persson, and G. Holmberg (Eds.), pp. 139–156. Newcastle upon Tyne, UK: Cambridge Scholars Publishing.

Peirce, C. S. 1932. *Collected Papers 2: Elements of Logic*. Cambridge, MA: Belknap Press of Harvard University Press.

Penny, A. 2011. The lifetime of scientific civilizations and the genetic evolution of the brain. In *Civilisations Beyond Earth: Extraterrestrial Life and Society*, D. A. Vakoch, and A. A. Harrison (Eds.), pp. 60–73. New York: Berghahn Books.

Perryman, M. 2012. The history of exoplanet detection. *Astrobiology* 12(10):928–939.

Peters, T. 2017. Where there's life there's intelligence. In *What is Life?: On Earth and Beyond*, A. Losch (Ed.), pp. 236–259. Cambridge, UK: Cambridge University Press.

Putnam, H. 1981. *Reason, Truth, and History*. Cambridge, UK: Cambridge University Press.

Race, M. et al. 2012. Astrobiology and society: Building an interdisciplinary research community. *Astrobiology* 12(10):958–965.

Raulin Cerceau, F. 2015. Fraction of civilizations that develop a technology that releases detectable signs of their existence into space, f_c, pre-1961. In *The Drake Equation: Estimating the Prevalence of Extraterrestrial Life through the Ages*, D. A. Vakoch, and M. F. Dowd (Eds.), pp. 205–226. Cambridge, UK: Cambridge University Press.

Richerson, P. J., and R. Boyd. 2005. *Not By Genes Alone: How Culture Transformed Human Evolution*. Chicago, IL: University of Chicago Press.

Roberts, W. A. 2007. Mental time travel: Animals anticipate the future. *Current Biology* 17(11):R418–R420.

Rosch, E. 1975. Cognitive representations of semantic categories. *Journal of Experimental Psychology: General* 104:192–233.

Rosch, E. 1978. Principles of categorization. In *Cognition and Categorization*, E. Rosch, and B. B. Lloyd (Eds.), pp. 27–48. Hillsdale, NJ: Erlbaum.

Rospars, J. P. 2013. Trends in the evolution of life, brains and intelligence. *International Journal of Astrobiology* 12(3):186–207.

Ruse, M. 2015. 'Klaatu barada nikto' – or, do they really think like us? In *The Impact of Discovering Life Beyond Earth*, S. J. Dick (Ed.), pp. 175–188. Cambridge, UK: Cambridge University Press.

Sagan, C. (Ed.). 1973. *Communication with Extraterrestrial Intelligence.* Cambridge, MA: MIT Press.

Saint-Gelais, R. 2014. Beyond linear B: The metasemiotic challenge of communication with extraterrestrial intelligence. In *Archaeology, Anthropology, and Interstellar communication*, D. A. Vakoch (Ed.), pp. 78–93. Washington, DC: NASA.

Saussure, F. d. 1916. *Cours de linguistique générale.* Paris, France: Payot.

Schneider, S. 2015. Alien minds. In *The Impact of Discovering Life Beyond Earth*, S. J. Dick (Ed.), pp. 189–206. Cambridge, UK: Cambridge University Press.

Schuch, H. P. 2011. Project Ozma: The birth of observational SETI. In *Searching for Extraterrestrial Intelligence: SETI Past, Present, and Future*, H. P. Schuch (Ed.), pp. 13–18. Berlin, Germany: Springer.

Schulze-Makuch D., L. N. Irwin, and A. G. Fairén. 2015. Extraterrestrial life: What are we looking for? In *Astrobiology: An Evolutionary Approach*, V. M. Kolb (Ed.), pp. 399–412. Boca Raton, FL: CRC Press.

Seed, A. M., N. J. Emery, and N. S. Clayton. 2009. Intelligence in corvids and apes: A case of convergent evolution? *Ethology* 115:401–420.

Shostak, S. 2009. The value of 'L' and the cosmic bottleneck. In *Cosmos and Culture: Cultural Evolution in a Cosmic Context*, S. J. Dick, and M. L. Lupisella (Eds.), pp. 399–414. Washington, DC: NASA History Series.

Shostak, S. 2015. Fraction of civilizations that develop a technology that releases detectable signs of their existence into space, f_c, 1961 to the present. In *The Drake Equation: Estimating the Prevalence of Extraterrestrial Life Through the Ages*, D. A. Vakoch, and M. F. Dowd (Eds.), pp. 227–240. Cambridge, UK: Cambridge University Press.

Sinha, C. 2009. Language as a biocultural niche and social institution. In *New Directions in Cognitive Linguistics*, V. Evans, and S. Pourcel (Eds.), pp. 289–309. Amsterdam, the Netherlands: Benjamins.

Smith, J. M., and E. Szathmáry. 1995. The major evolutionary transitions. *Nature* 374(6519):227–232.

Sonesson, G. 2007. From the meaning of embodiment to the embodiment of meaning. In *Body, Language and Mind. Vol. 1: Embodiment*, T. Zimke, J. Zlatev, and R. Frank (Eds.), pp. 85–128. Berlin, Germany: Mouton.

Sonesson, G. 2009. The view from Husserl's lectern: Considerations on the role of phenomenology in cognitive semiotics. *Cybernetics and Human Knowing* 16(3–4):107–148.

Sonesson, G. 2013. Preparations for discussing constructivism with a Martian (the second coming). In *The History and Philosophy of Astrobiology: Perspectives on the Human Mind and Extraterrestrial life*, D. Dunér, J. Parthemore, E. Persson, G. Holmberg (Eds.), pp. 189–204. Newcastle upon Tyne, UK: Cambridge Scholars Publishing.

Sonesson, G., and D. Dunér. 2016. The cognitive semiotics of cultural evolution. In *Human Lifeworlds: The Cognitive Semiotics of Cultural Evolution*, D. Dunér, and G. Sonesson (Eds.), pp. 7–21. Pieterlen, Switzerland: Peter Lang.

Spranzi, M. 2004. Galileo and the mountains of the moon: Analogical reasoning, models and metaphors in scientific discovery. *Journal of Cognition and Culture* 4(3):451–483.

Steels, L. 2004. Social and cultural learning in the evolution of human communication. In *Evolution of Communication Systems*, D. K. Oller, and U. Griebel (Eds.), pp. 69–90. Cambridge, MA: MIT Press.

Sternberg, R. J. 1985. *Beyond IQ: A Triarchic Theory of Intelligence.* Cambridge, MA: Cambridge University Press.

Sternberg, R. J. 2002. The search for criteria: Why study the evolution of intelligence. In *The Evolution of Intelligence*, R. J. Sternberg, and J. C. Kaufman (Eds.), pp. 1–7. Mahwah, NJ: Lawrence Erlbaum Associates Publishers.

Suddendorf, T., and M. C. Corballis. 1997. Mental time travel and the evolution of human mind. *Genetic, Social and General Psychology Monographs* 123(2):133–167.

Taylor, J. R. 2002. *Cognitive Grammar.* Oxford, UK: Oxford University Press.

Taylor, J. R. 2003. *Linguistic Categorization.* Oxford, UK: Oxford University Press.

Thompson, E. 2007. *Mind in Life: Biology, Phenomenology and the Sciences of Mind.* Cambridge, MA: Harvard University Press.

Tomasello, M. 1999. *The Cultural Origins of Human Cognition.* Cambridge, MA: Harvard University Press.

Tomasello, M. 2008. *Origins of Human Communication.* Cambridge, MA: MIT Press.

Traphagan, J. 2015a. Equating culture, civilization, and moral development in imagining extraterrestrial intelligence: Anthropocentric assumptions? In *The Impact of Discovering Life Beyond Earth*, S. J. Dick (Ed.), pp. 127–142. Cambridge, UK: Cambridge University Press.

Traphagan, J. 2015b. *Extraterrestrial Intelligence and Human Imagination: SETI at the Intersection of Science, Religion, and Culture.* Cham, Switzerland: Springer.

Traphagan, J. 2016. *Science, Culture and the Search for Life on Other Worlds.* Cham, Switzerland: Springer.

Tylén, K., E. Weed, M. Wallentin, A. Roepstoorf, and C. D. Frith. 2010. Language as a tool for interacting minds. *Mind and Language* 25:3–29.

Vakoch, D. A. 1998. Constructing messages to extraterrestrials: An exosemiotic perspective. *Acta Astronautica* 42(10–12):697–704.

Vakoch, D. A. (Ed.). 2011. *Communication with Extraterrestrial Intelligence.* Albany, NY: State University of New York Press.

Vakoch, D. A. (Ed.). 2013. *Astrobiology, History, and Society: Life Beyond Earth and the Impact of Discovery.* Berlin, Germany: Springer.

Vakoch, D. A., and M. F. Dowd, Eds. 2015. *The Drake Equation: Estimating the Prevalence of Extraterrestrial Life Through the Ages.* Cambridge, UK: Cambridge University Press.

Vakoch, D. A., and A. H. Harrison (Eds.). 2011. *Civilization Beyond Earth: Extraterrestrial Life and Society.* New York: Berghahn Books.

Varela, F. J., E. Thompson, and E. Rosch. 1991. *The Embodied Mind: Cognitive Science and Human Experience.* Cambridge, MA: MIT Press.

Watson, A. J. 1999. Coevolution of the earth's environment and life: Goldilocks, Gaia and the anthropic principle. *Geological Society Special Publication* 150:75–88.

Zlatev, J. 2012. Cognitive semiotics: An emerging field for the transdisciplinary study of meaning. *Public Journal of Semiotics* 4(1):2–24.

Zlatev, J. 2015. Cognitive semiotics. In *International Handbook of Semiotics*, P. Trifonas (Ed.), pp. 1043–1068. Berlin, Germany: Springer.

Zlatev, J., T. P. Racine, C. Sinha, and E. Itkonen (Eds.). 2008. *The Shared Mind: Perspectives on Intersubjectivity.* Amsterdam, the Netherlands: Benjamins.

11.2 Where Are They? Implications of the Drake Equation and the Fermi Paradox

Nikos Prantzos

CONTENTS

11.2.1 INTRODUCTION

Most of the effort in establishing the chances for the existence of life elsewhere in the universe has been put on evaluating the various astronomical and biological factors that may lead to the emergence of life on the surface of terrestrial-type exoplanets around other stars in the Milky Way. Twenty years after the discovery of the first exoplanet, we have now started to have sufficient data to allow us to think that such objects are relatively common in the Galaxy (Petigura et al. 2018). However, we do not yet have any evidence about the existence of other lifeforms, even at the microscopic level.

Historically, the main argument for the existence of life forms and intelligent beings beyond Earth has been formulated by Metrodorus, disciple of Epicurus, in the third century BC: *To consider the Earth as the only populated world in infinite space is as absurd as to assert that in an entire field sown with millet only one grain will grow.* The idea of space being infinite, with an infinite number of atoms populating it and composing its various objects, was a key ingredient of the atomistic philosophy of Leucippus, Democritus, and Epicurus (Furley 1987).

The argument is also invoked today by proponents of extraterrestrial intelligence (ETI), essentially unaltered; however, the concept of infinity is not used any more (because it is difficult to handle and it may lead to paradoxes, e.g., in an infinite universe, everything—including ourselves—could exist in an infinite number of copies). Our Galaxy contains about 100 billion stars, a number considered by some—mostly astronomers—to be large enough as to make Metrodorus' argument applicable to the Milky Way. Others, however—especially evolutionary

biologists—are not impressed by that number and remain skeptical concerning ETI (Simpson 1964, Mayr 1985).

In the second half of the twentieth century, the debate on ETI was largely shaped by the Drake equation and the Fermi paradox. The former was proposed in 1961 by American astronomer Frank Drake and became ever since the key quantitative tool to evaluate the probabilities for radio-communication with extraterrestrial intelligence (CETI), with the early evaluations being overly optimistic. The latter was formulated a decade earlier by Italian physicist Enrico Fermi, but it remained virtually unknown until 1975, when it was independently re-discovered twice (Hart 1975, Viewing 1975). In a concise form, it opposes a healthy skepticism to the optimistic views on ETI: (*if there are many of them*), *where are they?*

In this chapter, I discuss the physical meaning of the Drake equation and analyze it in a way revealing the implications of its last term, namely the lifetime of technologically evolved civilizations, the importance of which is often overlooked. The analysis allows us to define a region in the parameter space of the Drake equation, where the so-called "great silence" (Brin 1983) may be easily understood. The adopted framework is also applied to the analysis of the Fermi paradox, under the assumption of a simplified scheme for the colonization of the Galaxy. From the joint analysis of the Drake equation and the Fermi paradox, it appears that for sufficiently long-lived civilizations, colonization of the Galaxy appears to be the only reasonable option to gain knowledge about other life-forms. The analysis allows one to define a corresponding region in the parameter space of the Drake equation where the Fermi paradox holds.

11.2.2 INTRODUCING THE DRAKE EQUATION

Scientific study of the search for ETI has a short history, going back about half a century ago. In an article published in *Nature*, physicists Giuseppe Cocconi and Philip Morisson (1959) suggested that microwaves (high-frequency radio waves) are the best vector for interstellar communication. They can penetrate the terrestrial atmosphere and can pass through the dust and gas clouds of the Milky Way disk. In contrast, visible photons, our traditional window upon the universe, are absorbed by these clouds, limiting communication through optical telescopes to relatively short distances (of a few thousand light-years). In addition, radio telescopes can survey the skies 24 hours a day, even in broad daylight and under thick cloud cover. Another advantage of microwaves is that they carry little energy, implying that less energy would be expended using them for communication. Last but not least, our Galaxy emits little in the microwave region of the electromagnetic spectrum, compared with other radio frequencies; thus, the background noise would be reduced. While recognizing the difficulty of the enterprise, Cocconi and Morisson concluded: "Few will deny the profound importance, practical and philosophical, which the detection of interstellar communications would have. We therefore feel that a discriminating search for signals deserves a considerable effort. The probability of success is difficult to estimate; but if we never search the chance of success is zero."

These considerations opened up the modern era in the plurality of worlds debate. The first to apply these ideas was Frank Drake, at the Green Bank National Astronomy Observatory in the USA (Drake 1960). He set up the first systematic search for extraterrestrial signals, called *Project Ozma*. The project was named after the queen of the imaginary land of Oz, a distant and inaccessible place in the story by Frank Baum. In 1960, the Green Bank telescope spent 4 months looking for radio signals from two nearby stars, ε Eridani and τ Ceti—both at about 12 light-years away—with no success (Drake 1961). The next year, Drake hosted a meeting on the Search for Extraterrestrial Intelligence (SETI) with 10 invitees, including Philip Morisson and Carl Sagan, and then 27. In preparing the agenda of the meeting, Drake established his famous equation, trying to evaluate the number of detectable civilizations in our Galaxy.

11.2.3 AN ANALYSIS OF THE DRAKE EQUATION

In its original formulation, the Drake equation reads

$$N = R_* \, f_p \, n_e \, f_l \, f_i \, f_T \, L \qquad (11.2.1)$$

where R_* is the rate of star formation in the Galaxy (i.e., number of stars formed per unit time), f_p is the fraction of stars with planetary systems, n_e is the average number of planets around each star, f_l is the fraction of planets where life developed, f_i is the fraction of planets where intelligent life developed, and f_T is the fraction of planets with technological civilizations. Obviously, N and L are intimately connected:

if N is the number of radio-communicating civilizations—as in the original formulation by Drake—then L is *the average duration of the radio-communication phase* of such civilizations (and not their total lifetime, as sometimes incorrectly stated). On the other hand, if N is meant to be the number of technological or space-faring civilizations, then L represents the duration of the corresponding phase.

The Drake formula obviously corresponds to the equilibrium solution of an equation similar to the well-known equation of radioactivity for the decay rate D of a number N of radioactive nuclei: $D = dN/dt = -N/L$, where L is the lifetime of those nuclei. In the steady state, where the production rate P is equal to the decay rate D, one has $N = P \, L$. In a similar vein, the product of all the terms of the Drake formula, except L, can be interpreted as the production rate P of radio-communicating (or technological or space-faring) civilizations in the Galaxy.

It should be noticed that time does not appear explicitly in the terms of the Drake equation. This may be problematic, since we are interested in the present-day number of civilizations $N(t_0)$, while the stars harboring such civilizations were formed at time $t_0 - T$, where T is probably a substantial fraction of the age of the Galaxy ($T{\sim}4.5$ Gy in the case of our civilization), that is, at a time where the rate of star formation $R(t_0 - T)$ was, perhaps, very different from the present-day star formation rate $R(t_0)$. However, in the case of the Milky Way, which is an Sbc-type spiral galaxy, there is evidence for a quiescent evolution at quasi-constant pace over billions of years (Kennicutt and Evans 2012). In that case, the average star formation rate $<R>$ over the Galaxy age A~10 Gy can be considered as a fairly good approximation of $R(t)$ at any time t, and, consequently, the equilibrium solution approximates well the real situation. Notice that the solution $N < 1$ is acceptable in that case, meaning that the interval Δt between the occurrence of two such civilizations in the Galaxy is larger than their average duration L; in other terms, there is just one civilization in the Galaxy emerging at time t and lasting for a duration L, but the next one emerges at $t + \Delta t$, where $\Delta t > L$. We shall discuss quantitatively that case in Section 11.2.5. On the other hand, if $R_*(t)$ varies widely in time, for example, in elliptical galaxies—which formed practically all their stars in the first couple of Gy—then the equilibrium solution does not apply, and instead of the Drake formula, one should use simply $N(t) = N_* \, exp(-(t - T)/L)$, where N_* is the total number of stars in that galaxy, formed in the first couple of Gy.

In the 50 years since the formulation of the Drake equation, a fairly popular game consisted of attributing plausible numerical values to its various terms to estimate N. Unsurprisingly, estimates of different authors varied by many orders of magnitude: from a few million (Cameron 1963, Sagan 1963, Shklovskii and Sagan 1966), down to $N < 100$ (von Hoerner 1962, Jones 1981). It is significant to notice that typical values in the early years were substantially larger than those found in later times. In some cases, additional terms were added to the original ones to account for new astronomical factors (e.g., Ksanfomality 2004) or for intermediate steps in the development of a civilization, and even statistical treatments were considered, to account for dispersion in the values of the relevant parameters (e.g., Maccone 2010b, Glade et al. 2012).

Despite the apparent sophistication of such efforts, the game is rather meaningless, because, until a few years ago, only $R_*(t_0)$ could be estimated from observations. Only in the past few years, statistics to estimate the second and third terms became available. The next four terms will remain unknown until we have a good theory on the emergence of life, intelligence, and technology or even better until such phenomena are detected beyond Earth (but then, very few will care about the Drake equation anymore, since the goal will have been achieved). However, despite its inability to help calculate N—which can have, in principle, any value between 1 and, say, 10^8—the Drake equation has been extremely useful, as it provided a framework allowing us to formulate our knowledge/thoughts/educated guesses about a very complex phenomenon such as the development of life and intelligence in the astrophysical setting of the Milky Way.

11.2.4 REFORMULATING THE DRAKE EQUATION

In a recent study (Prantzos 2013), a different approach was adopted: instead of introducing additional terms in Equation 11.2.1, its seven terms were condensed to only three. The aim was twofold: (1) to illustrate quantitatively some implications of the number N for SETI and CETI, and (2) to use exactly the same framework for a quantitative assessment of the Fermi paradox (see Section 11.2.7).

The Drake equation is rewritten as

$$N = R_{ASTRO}\, f_{BIOTEC}\, L \tag{11.2.2}$$

where $R_{ASTRO} = R\, f_P\, n_e$ represents the production rate of habitable planets (determined through astrophysics) and $f_{BIOTEC} = f_l\, f_i\, f_T$ represents the product of all chemical, biological, and sociological factors leading to the development of a technological civilization. Obviously, $f_{BIOTEC} \leq 1$; its maximum possible value $f_{BIOTEC} = 1$ requires $f_l = f_i = f_T = 1$ (a rather implausibly optimistic combination), but there is no constrain on its lower value.

The astrophysical factor R_{ASTRO} is expected to be reasonably constrained in the foreseeable future. Indeed, its first term, R_*, is already constrained by observations in the Milky Way to be ~4 stars/year (Chomiuk and Povich 2011 give ~1.9 $M_\odot$/year for the present-day star formation rate, and there are ~2 stars per $M_\odot$ in a normal stellar initial mass function, like the one of Kroupa 2002). However, its average past value was probably higher by a factor of 2, and we shall adopt the value of 4 $M_\odot$/year, which corresponds to an average star production rate of $R^* = <R>$~8 stars/year; this average star formation (SF) rate reproduces well the stellar mass of ~4 × 10^{10} $M_\odot$ or the ~10^{11} stars of the Milky Way if assumed to hold for the age of the Galaxy A~10 Gy.

We shall assume that only 10% of those stars are appropriate for harboring habitable planets, because their mass has to be smaller than 1.1 $M_\odot$, that is, they have to be sufficiently long-lived (with main sequence lifetimes larger than 4.5 Gy as to give enough time for the development of intelligence and technology) and larger than 0.7 $M_\odot$, to possess circumstellar habitable zones outside the tidally locked region (Selsis 2007). This leaves aside the most numerous class of stars, namely the low-mass red dwarfs: their intense and time-varying activity and their small circumstellar habitable zone largely balance the effect of their larger number with respect to solar-type stars. In any case, considering them would increase the planet numbers by a factor of 10–20, but this would change the conclusions little, given the extremely large uncertainties of the other factors of the Drake equation, as illustrated in a recent analysis (Wandel 2017).

A recent analysis of the Kepler satellite data (Petigura et al. 2018) finds that the statistics currently available on extra-solar planets around solar-type stars point to ~20% of the surveyed stars possessing super-Earths (planets in the ~2–10 $M_\oplus$ range) with orbital periods P = 10–100 days. This fraction may be considered to describe the product $f_p\, n_e$ in the Drake equation, corresponding to *stars with continuously habitable Earth-like planets* (i.e., orbiting their star continuously within the circumstellar habitable zone). This is, admittedly, a fairly optimistic estimate, its only merit being that it imposes a plausible upper limit on the fraction of such solar-type stars. Combined with the aforementioned formation rate of 0.7–1.1 $M_\odot$ stars, it leads to R_{ASTRO} ~0.1 habitable planet per year. We shall adopt this rounded value here, and we shall investigate the space of the remaining parameters f_{BIOTEC} and L, which are totally unknown at present.

In Figure 11.2.1, the re-formulated Drake equation appears in a graphical form, after fixing $R_{ASTRO} = 0.1$/y and putting L and f_{BIOTEC} in the ordinate and abscissa, respectively. In this log-log diagram, a given value of the product N of Equation 11.2.2 is represented by a straight quasi-diagonal line. For instance, $N = 1$ is obtained for both $f_{BIOTEC} = 0.1$ (high probability of appearance of a communicating civilization) and $L = 10^2$ year (short lifetime in that phase) or for $f_{BIOTEC} = 10^{-8}$ (small probability) and $L = 10^9$ year (extremely long lifetime). In both cases, $N = 1$ civilization exists permanently in the Galaxy, but in the former case, it is replaced by the next one every century, whereas in the latter, every Gy. It is interesting to notice the typical distances (appearing also in Figure 11.2.1) between such civilizations, assuming that they are uniformly distributed in the Milky Way, as discussed in Section 11.2.6: even for a million civilizations, typical distances are of several hundred light-years, making contact—either by radio signals or space probes—difficult. This is the topic of Sections 11.2.6 and 11.2.8.

Notice that these numbers assume that all civilizations have similar values of L, that is, the dispersion ΔL in L is much smaller than L itself. This need not be the case. Statistical treatments, considering $\Delta L \sim L$, and canonical distributions have been recently applied (Maccone 2010b, Glade et al. 2012). However, in any case, the unknown mean value of L plays a more important role than the equally unknown form of its distribution (see also Barlow 2013).

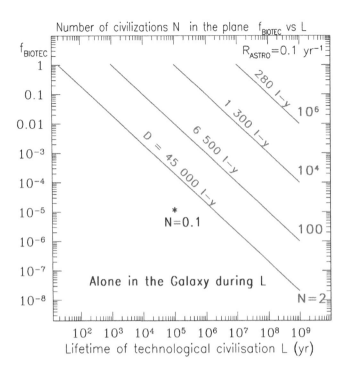

FIGURE 11.2.1 Number of civilizations *N* in the plane *f* versus *L*, assuming R_{ASTRO} = 0.1/year. A given number *N* is represented by a straight diagonal line. Typical distances *D* (in light-years) between civilizations for each *N*, obtained from the discussion of Section 11.2.6 (Figure 11.2.4), are indicated on the corresponding lines. The lower left region corresponds to *N* < 1. The potential meaning of that region (represented by the point *N* = 0.1) is illustrated in Figure 11.2.2 and discussed in Section 11.2.5.

11.2.5 LONELY HEARTS IN THE MILKY WAY?

The region below the line *N* = 1 is not necessarily void. In principle, it corresponds to values of *N* < 1, but such values may have a physical, albeit probabilistic, meaning, as illustrated in Figure 11.2.2: they may represent the fraction *f* of the time span *T* (between the appearance of two successive civilizations) that is occupied by the typical lifetime *L*: *f* = *L*/*T*. In Figure 11.2.2, this is illustrated by assuming arbitrarily that *L* = 10^5 year and *T* = 10^6 year. Ten civilizations appear within 10 million years, with an individual lifetime of 100,000 year: their summed lifetime is 1 My, that is, they exist for 1/10 = 0.1 of that time span. For an external observer, the probability of finding a technological civilization anytime in the Galaxy is then 0.1.

The case illustrated in Figure 11.2.2 is rather depressive: a technological civilization may emerge in the Galaxy and live as long as a hundred-thousand years (10 times more than the total human history and a thousand times longer than our technological era). Still, during its whole lifetime, the civilization would be alone in the Milky Way: a gap 10 times longer would separate it from both its predecessor and its successor. None of those civilizations would ever know of the existence of the others, especially if they remain confined to their planets (or within their solar systems).

Our own civilization may be such a lonely heart in the Milky Way. The implications of that situation are obvious: only a systematic research by unmanned probes or manned spaceships could provide evidence for or against the existence of other, perhaps now extinct, civilizations. Before discussing the connection of this conclusion with the Fermi paradox in Section 11.2.7, we shall consider another aspect of the Drake equation: the constraints it puts on some aspects of Galactic civilizations seeking communication, namely their probability of emergence and their lifetime.

11.2.6 CONSTRAINTS FROM CETI

The original intention of Frank Drake was to evaluate the chances for radio communication with ETI. However, his formula was mainly (or, rather, exclusively) used to evaluate the number of co-existing ETI in the Galaxy today, which is not exactly the same thing. The original task requires to take one more step and connect somehow the lifetime of the civilizations (last term in the formula) to the typical distances between such civilizations, to account for the finite speed of electromagnetic signals. As it turns out, those distances are related not only to the dimensions of the Galaxy but also to number *N* and, implicitly, to lifetime *L*. These relations then impose constraints on *N* and its components, f_{BIOTEC} and *L*.

To evaluate typical distances between Galactic civilizations, it may be assumed that, to a first approximation, the Galactic disk is described by a cylinder of radius R_G = 10 kpc and height *h* = 1 kpc, where the *N* civilizations of the Drake formula are distributed uniformly (Figure 11.2.3). A better approximation would be to consider the exponential profile of the stellar disk, but the conclusions depend little on such assumptions and much more on the unknown factors of the Drake formula.

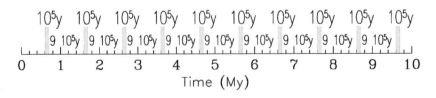

FIGURE 11.2.2 If *N* < 1, civilizations cannot coexist in the Galaxy. Still, the Drake equation has a meaning, albeit in a statistical sense only. Here is an example with *N* = 0.1, which may correspond to log(*f*) = −5 log(*L*) = 5 in Figure 11.2.1. A given technological civilization lives for *L* = 10^5 year and becomes extinct, leaving the Galaxy void from intelligent life for a period of Δ*t* = 9 *L* = 900,000 year, until the appearance of the next civilization. In that case, the probability of finding a civilization in the Galaxy during any period >>*L* is 0.1.

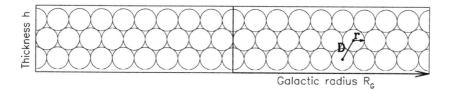

FIGURE 11.2.3   Filling the Galaxy with N civilizations located at average distances D from each other.

By equating the volume of the Galactic cylinder $V = \pi R_G^2 h$ with the sum of N volumes of spheres of average radius r occupied by each civilization, one obtains the average distance (D) between two civilizations as $D = 2\, r = 2\, (3\, V/4\, \pi\, N)^{1/3}$ for the case where $D < h$. In the case of a small number of civilizations (say $N < 1000$), it turns out that $D > h,$ and a more appropriate expression is then $D = 2\, r = 2\, R_G/\sqrt{N}$.

The values of the average distances D as a function of the number of civilizations N appear in Figure 11.2.4, where the astronomical factor of the Drake formula is fixed to $R_{ASTRO} = 0.1/y$. One can appreciate the magnitude of the distances between civilizations, which plays an important role in the discussion (and understanding) of the great silence and the Fermi paradox. For less than a thousand civilizations, typical distances are larger than 3000 light-years, while for a million civilizations, they are of the order of 350 light-years (see also Figure 11.2.1). Notice that N is the number of civilizations coexisting in the Galaxy during their lifetime L.

For each number N, there is a minimum value L_{MIN}, corresponding to $f_{BIOTEC} = 1$ in Equation 11.2.2 for the adopted value of $R_{ASTRO} = 0.1/$year. Obviously, communication between neighboring civilizations requires their duration L to be larger than twice the travel time D/c (where c is the light speed) of radio waves. An inspection of Figure 11.2.4 shows that if there are less than a few hundred co-existing civilizations in the Galaxy, their radio-emission phase has to last longer than $\sim 10^4$ years to allow them to establish radio communication.

The analysis of the previous paragraph holds for the maximum possible value of the biotechnological factor $f_{BIOTEC} = 1$,

which has a corresponding minimal value of L for a given N. In order to explore the whole range of possible values for f_{BIOTEC} and L, the diagram presented already in Figure 11.2.1 was used (Prantzos 2013), as shown in Figure 11.2.5: for the adopted value of R_{ASTRO}, a given value of N (and thus a corresponding value of D) represents a line in the f_{BIOTEC} versus L diagram, with $N = 2$ being a lower limit for communication.

The time required for such a communication, involving an exchange of radio signals, is $T = 2D/c$. Using the average distances D derived earlier, the condition for radio communication ($T < L$) is found to bound the f_{BIOTEC} versus L diagram from the left, according to

$$N > 48\, R_G^2 h /\left(Lc\right)^3 \text{ for } D < h, \text{ and } N > 16\, R_G^2 /\left(Lc\right)^2 \text{ for } D > h.$$

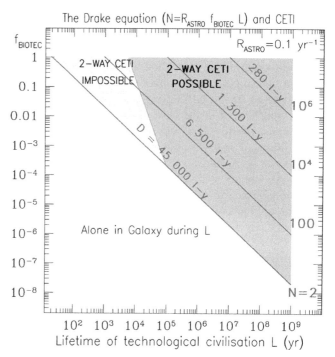

FIGURE 11.2.5   The parameter space of the bio-sociological factor $f_{BIOTEC} = f_L f_I f_C$ versus the average lifetime L of a technological civilization. Calculations are done by assuming a value $R_{ASTRO} = R_* f_P n_e = 0.1$ habitable planet per year for the astronomical factor (see text) in the Drake equation. This leads to values for the number N of technological civilizations indicated by the parallel diagonal lines. $N = 2$ is the minimum required for two-way communication during L. The shaded region allows for two-way communication between two civilizations during L; that is, it satisfies the condition $L > 2D/c$, where D is the average distance between civilizations (see Figure 11.2.4) and c the velocity of light. (Adapted from Prantzos, N., *Int. J. Astrobiol.*, 12, 246–253, 2013. With permission.)

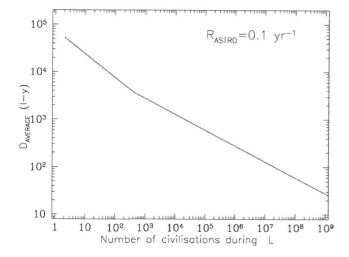

FIGURE 11.2.4   Average distances D (continuous curve) between civilizations as a function of their total number N in the Galaxy, assuming that they are uniformly distributed in the Milky Way disk.

In those expressions, distances are expressed in light-years and durations are expressed in years, thus allowing one to omit the light speed of radio-signal propagation ($c = 1$ light-year per year). By replacing N through the Drake formula (Equation 11.2.2), the derived constraints can also be expressed as a relationship between the factors L and f_{BIOTEC} and the dimensions of the Galaxy $f_{BIOTEC} L^4 > 48 R_G^2 h / R_{ASTRO}$ for $D < h$ and $f_{BIOTEC} L^3 > 16 R_G^2 / R_{ASTRO}$ for $D > h$, respectively.

An inspection of Figure 11.2.5 shows that, for the adopted value of $R_{ASTRO} = 0.1$ habitable planets per year (an upper limit), communications are possible only for civilizations spending at least 10,000 years in the radio-communication phase. If $f_{BIOTEC} = 1$, there are about 1000 such civilizations in the Galaxy, and our chances of eavesdropping the communications of our closest neighbors are not negligible. But if $f_{BIOTEC} = 10^{-3}$, there are just a few civilizations in the Galaxy, and chances of eavesdropping appear insignificant. Of course, the larger the duration of the radio-communication phase, the larger the N (for the same value of f_{BIOTEC}), and chances improve considerably. However, it should be noticed that there is a large region of the considered log-log parameter space (for all $L < 10,000$ year, irrespectively of f_{BIOTEC}) for which communications are impossible or there is no second radio-emitting civilization in the Galaxy during L.

One may conclude then that it is improbable that we communicate with (or eavesdrop) civilizations of a level either similar to ours ($L \sim 100$ year of radio emission) or even radio-emitting for a few thousands of years. Only civilizations emitting for much longer timescales have chances to be detected by our SETI programs. On the other hand, it may well be that in the case of $f_{BIOTEC} < 10^{-2}$, we may spend thousands of years—and in the case of $f_{BIOTEC} < 10^{-5}$, even millions of years—searching for radio signals with no effect, since we would be alone in the Galaxy during that period. As already stressed, those numbers correspond to the optimistic case of $R_{ASTRO} = 0.1$ habitable planet per year. It seems obvious that such considerations will affect the strategy of any extraterrestrial civilizations searching their siblings.

At this point, it should be noticed that it is a rather futile exercise to try to imagine the kind of communications that technological civilizations older than a few centuries might have. In particular, the problem of understanding alien messages has been given serious attention by soviet astronomers in the 1960s and 1970s—as properly emphasized in the monograph by Sheridan (2009)—but it appears to be virtually absent from the recent literature on the subject.

11.2.7 EARLY HISTORY AND ANALYSIS OF THE FERMI PARADOX

The origin of this famous paradox is documented by Eric Jones in his 1985 Los Alamos report (Jones 1985): it appears that Fermi formulated his question: Where is everybody? during a lunchtime conversation at Los Alamos in the summer of 1950 with colleagues Emil Konopinsky, Herbert York, and Edward Teller. Its early history is related by Stephen Webb in his book *Where is everybody?* perhaps the most complete investigation today of the solutions proposed to solve the paradox (Webb 2014). Webb provides a detailed account of the early ideas on the paradox, including its earlier discovery by Russian father of astronautics K. Tsiolkovski in 1933, its independent re-discovery by M. Hart in 1975 and its identification as a paradox by D. Viewing in 1975. In recognition of those early contributions to the debate, Webb calls it the Tsiolkovski–Fermi–Hart–Viewing paradox. We shall keep the shorter term *Fermi paradox* here for convenience.

The basic idea of the "paradox" was already formulated more than three centuries ago, albeit in an inconclusive way. In 1686, French novelist Bernard le Bovier de Fontenelle (who later became secretary of the French Academy of Sciences) published his best-selling book *Entretiens sur la Pluralité des Mondes* (*Conversations on the Plurality of Worlds*). It is often considered as the first popular science book, and it is written in the form of dialogues between the author and a charming and ingenuous marchioness. The marchioness counters the author's assertion that "intelligent beings exist in other worlds, for instance the Moon" by the retort: "If this were the case, the Moon's inhabitants would already have come to us before now." Fontenelle can only argue that the time to master space travel is probably too long and that the Moon's inhabitants "at this time are maybe exercising themselves; when they shall be more able and more experienced, we may see them... after all, the Europeans did not arrive in America till nearly at the end of six thousand years" (in Fontenelle's days, the universe was thought to be 6000 years old, based on biblical accounts). Note that Fontenelle's time argument is completely symmetric between Earth's and Moon's inhabitants; that is, they both have at most 6000 years to master space travel, but there is no hint as to whether one of the two civilizations is more advanced than the other; thus, the absence of people from the Moon on Earth could not really be considered as a surprise, since earthlings had not visited the Moon either. The modern argument by Fermi explicitly assumes, by virtue of the Copernican principle, that some of the extraterrestrial civilizations are (considerably) older than ours, and therefore, they have had enough time to spread into the Galaxy and reach our planet, while our own civilization is unable to do so at present.

For more than 10 years after the now famous lunchtime discussion at Los Alamos, there appears to be no written trace of Fermi's question (at least to my knowledge). The earliest trace I am aware of is a footnote in a paper published by American astronomer Carl Sagan: This possibility has been seriously raised before, for example, by Enrico Fermi, in a now-rather-well-known dinner-table discussion at Los Alamos during the Second World War, when he introduced the problem with the words *Where are they?* (Sagan 1963). Sagan provides no information on his sources, making it difficult to know where he got such erroneous information about a dinner-table discussion during the Second World War. A few years later, the phrase *Where are they?*, attributed to Enrico Fermi but without any comment, appeared a few years later in the book of I. Shklovskii and C. Sagan *Intelligent Life in the Universe* (Shklovskii and Sagan 1966). Obviously, Sagan realized that the question might have some profound implications for the CETI endeavor, but he could not or did not wish to brood over them.

Any paradox is based on at least one invalid assumption. The logical statement of Fermi's paradox may run as follows (Prantzos 2000):

A. Our civilization is not the only technological civilization in the Galaxy.
B. Our civilization is in every way "average" or typical. In particular, it is not the first to have appeared in the Galaxy, it is not the most technologically advanced, and it is not the only one seeking to explore the cosmos and communicate with other civilizations.
C. Interstellar travel, although impossible for us today, is not too difficult for civilizations slightly more advanced than ours. Some extraterrestrial civilizations have mastered this kind of travel and undertaken a galactic colonization program, either with or without self-replicating robots (Tipler 1981).
D. Galactic colonization is a relatively fast undertaking and could be achieved in a relatively small fraction of Galaxy's age (less than a few hundred million years).

If hypotheses A to D are valid, one clearly deduces that "they should be here." Supporters of ETI reject at least one of the assumptions C and D, and some even go so far as to deny B, in order to save their key hypothesis A. In contrast, their opponents uphold the plausibility of C and D, whilst completely rejecting B. By virtue of the Copernican principle, hypothesis A should then be rejected as well. Of course, if the Copernican principle is inapplicable in case B, hypothesis A can still be saved (but at what a price!).

Most of the arguments in the debate on the Fermi paradox are of sociological nature (see Webb 2014 for a thorough, albeit not exhaustive, census) and concern assumptions B and D. Here, I will assume that assumptions A to C are valid, and I will explore the consequences of D, namely Galactic colonization by a technologically developed civilization.

The issue of the timescale for Galactic colonization in the context of the Fermi paradox has been quantitatively explored by various techniques using population dynamics (Hart 1975, Viewing 1975, Newman and Sagan 1981), percolation analysis (Landis 1998), or Monte Carlo methods (e.g., Jones 1981, Bjoerk 2011). Depending on the underlying assumptions, quite different results are found on the timescales of Galactic colonization. In most cases, timescales are found to be much smaller than the age of the Galaxy (e.g., Jones 1981), thus substantiating the Fermi paradox, while in a few cases, much larger values are found (e.g., Bjoerk 2011). In fact, it has been argued (Wiley 2011) that none of the assumptions underlying the aforementioned models is strong enough to render the conclusions robust.

11.2.8 THE FERMI PARADOX IN TERMS OF THE DRAKE FORMULA

The Fermi paradox may be illustrated in a fairly simple way (Prantzos 2013), by casting it into the same form as the Drake equation. It is assumed that once a civilization masters the techniques of interstellar travel, it starts a thorough colonization/exploration of its neighborhood for a period L. Colonization proceeds in a directed way, that is, it concerns only stars harboring nearby habitable planets, which are detected before the launching of the spaceships. According to the discussion in Section 11.2.2, an upper limit to the total number of such interesting stars in the Galaxy (i.e., by assuming $R_{ASTRO} = 0.1$ per year) is $n_I \sim 10^9$ stars, one hundredth of the total number of Galactic stars; their average density is then $\rho \sim 3 \times 10^{-5}$ stars per (l-year)^3 and their average distances $D \sim 30$ l-year (from Figure 11.2.4).

The colonization front expands outward at an average velocity $v = \beta c$, where β is a fraction of light speed c. Ships are sent to new stars not from the mother planet but from the colonized planets in the wave front, and they are launched after a time interval t_{PREP} following colony foundation. This gives enough time to the colonizers to prepare the next colonizing mission after their installation on the planet. Notice that v is the *effective velocity of the colonization front* and not the velocity of the interstellar ships; the latter is given by $v_{SHIP} = D/(D/v - t_{PREP})$ and has to be larger than v. A quantitatively study of the above scheme (Prantzos 2013) finds that full exploration of the Galaxy by a single civilization and its offsprings would take approximately $\Delta T = 4 \times 10^5$, 4×10^6, and 4×10^7 years for values of $\beta = 0.1$, 0.01, and 0.001, respectively (Prantzos 2013). And even if some of the colonies within the expanding spheres die out, the spheres need not remain partially hollow for a long time, since (some of) the active colonies could send new ships back and revive the dying colonies.

The model adopted here for the expansion of a civilization in the Galaxy corresponds to the so-called *Coral* model, describing how corals grow in the sea (Bennett and Shostak 2007), and has been used (Maccone 2010a) in an attempt to provide a statistical description of Galactic colonization. In both cases, a factor $k = 1/2$ is introduced in front of the fraction defining the ship speed to account for the zig-zag motion of the colonizing ships in space (sometimes colonizing back some stellar systems previously left behind or re-activating dying colonies). Introducing this factor should slightly change the numerical results presented here. Alternatively, the results should remain the same, under the assumption of a speed twice as large as the values originally adopted.

The analysis of the previous paragraphs concerns the colonization of the Galaxy in the case of a single colonizing civilization, starting from a single place in the Galaxy (a place close to the Galactic center was assumed here, but the results would not be very different if any other place was chosen). In Figure 11.2.6, the analysis is extended to the case of N independent civilizations—as given by the Drake formula—in a similar form as the one adopted for the analysis of CETI in

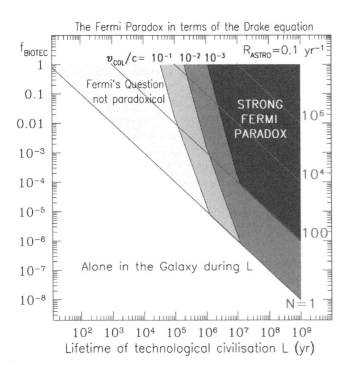

The Fermi Paradox in terms of the Drake equation

FIGURE 11.2.6 The Fermi paradox presented in the f_{BIOTEC} versus L plane, in terms of the factors appearing in the Drake formula. In the shaded region, space-faring civilizations may collectively explore/colonize the whole Galaxy within their average duration L, assuming that the colonization front expands with average velocity $v/c = 0.1$, 0.01, and 0.001, respectively (thick blue lines). The Fermi paradox applies in the whole shaded region. The region of the "Strong Fermi Paradox" is discussed in the text. (Adapted from Prantzos, N., *Int. J. Astrobiol.*, 12, 246–253, 2013. With permission.)

Figure 11.2.5. It is checked whether the Galactic volume can be filled within time L by $N = R_{ASTRO} f_{BIOTEC} L$ spheres of radius $r = \beta cL$ for $\beta = 0.1$, 0.01, and 0.001, respectively. If it can be filled, then every corner of the Galaxy should be visited by at least one of the N civilizations within its space-faring phase, and the Fermi paradox holds for the corresponding values of f_{BIOTEC} and L. Notice that here, the lifetime L corresponds to a single civilization and includes all its offspring colonies; in other terms, the colonies do not count as different civilizations, since they do not originate in an independent way.

Figure 11.2.6 shows that, depending on β and L, full colonization of the Galaxy by the N civilizations may or may not be possible during L. For instance, in the ultra-optimistic case of $f_{BIOTEC} = 1$, a thousand civilizations would need only a few tens of thousands of years to collectively colonize the Galaxy at $v/c = 0.1$, while a single civilization would need a hundred times longer. For values of f_{BIOTEC} and L outside the shaded region, the Fermi paradox is no more a paradox: civilizations are too rare or too short-lived to fully colonize the Galaxy within the duration L of their space-exploring phase. Within the shaded region, the Fermi paradox holds, since such civilizations can, in principle, colonize the galaxy, and they should have found us already.

In the case of a small number of civilizations, any one of the sociological ideas put forward to explain the Fermi paradox may be valid: indeed, some civilizations may never master space travel or never wish to colonize or to perturb other, less mature, ones, or they may abandon their colonization effort shortly after they started it, etc. (see Webb 2014). However, such arguments appear hardly plausible in the case of a large number of independent civilizations (Prantzos 2000). It shall be assumed here, somewhat arbitrarily, that for $N > 100$, such sociological arguments can hold for some but not for the majority of civilizations. In an equally arbitrary way, it will be assumed that an expansion velocity of $v/c = 0.001$ for the colonization front is much more reasonable (less demanding) than $v/c = 0.1$. Those assumptions define a region in the f_{BIOTEC} versus L plane (bounded from below by $N = 100$ and from the left by $v/c = 0.001$), where the Fermi paradox is arguably on a solid basis (strong Fermi paradox), whereas the situation is less clear for the remaining shaded region (weak Fermi paradox).

It has been argued (Ostriker and Turner 1986) that, even if advanced technological civilizations are common, they are unlikely to fully occupy the Galaxy, because, at some point of their expansion, their mutual interactions could reduce the pace of colonization, leaving some portions of the Galaxy unoccupied for periods of the order of L. They base their arguments on a mathematical analysis, drawing from the ideas of theoretical ecology, and they suggest that the Earth may be found in such an unoccupied region, thus providing an(other) explanation of the Fermi paradox. Somewhat counterintuitively, they claim that the complexity of the system describing population expansion allows one to predict its behavior with a reasonably high degree of confidence. Unfortunately, at the current stage of our knowledge, it is impossible to know whether Galactic colonization is better described by simple models (such as the Coral model adopted here) or by more sophisticated ones, such as theirs.

11.2.9 SUMMARY

For several decades, the Drake formula played an important role in the search for extraterrestrial life, providing a framework to formulate our current understanding about a very complex phenomenon such as the development of life and intelligence in the astrophysical setting of the Milky Way. However, in evaluating the number of technological civilizations N, emphasis was put mostly on the astrophysical and biotechnological factors describing the emergence of a civilization (R_{ASTRO} and f_{BIOTEC}, respectively) and much less on the lifetime L, which is related to its decline. In contrast, there has been no attempt to quantitatively evaluate the conditions where the Fermi paradox is indeed a paradox (although it is reported that Fermi had done so), while the role of factor L clearly appears to be important in that case.

The Drake formula and the Fermi paradox can be analyzed in the same framework, using a diagram of f_{BIOTEC} versus L (for a given value of R_{ASTRO}) in Figures 11.2.5 and 11.2.6. This type of analysis (first proposed in Prantzos 2013) leads—under some assumptions—to quantitative conclusions, emphasizing the role of the civilization lifetime L. The shaded regions in those figures—allowing CETI and full colonization of the Galaxy, respectively—are of similar size; that is, they correspond to approximately the same range of values of f_{BIOTEC} and L. This occurs despite the fact that communication velocities are at light speed c, while maximum velocities of colonization fronts are $v = 0.1\ c$ (in the most optimistic case). The reason is that CETI involves at least two signals exchanged over distances $D = 2r$, that is, a total distance of $4r$ crossed at light velocity c, which takes just 40% of the time required by the colonization front to cross radius r at velocity $v = 0.1\ c$. This difference by a factor of ~2 between the effective velocities of the communication carriers (electromagnetic signals vs. starships) is barely visible in the shaded regions of Figures 11.2.5 and 11.2.6. Of course, $v = 0.1\ c$ is a huge velocity for the colonization front; velocities hundreds of times smaller appear much more reasonable.

It should be noticed that, in this discussion, our own civilization is treated as a test particle rather than as a typical one. The lifetime L (~100 year) of our terrestrial technological civilization is far smaller than the typical values required either for CETI or for Galaxy colonization.

Although sending and receiving radio signals is certainly a much easier enterprise than launching starships, the latter should not be too difficult for a $>10^4$-year-old technological civilization (minimum age to have reasonable chances for CETI according to Figure 11.2.5). Assuming that such civilizations wish to learn about other life-forms (intelligent or not), one may ask what kind of action they might adopt: just content themselves with a program of radio-signal emission/reception or undertake a serious effort of interstellar colonization? The former, although less demanding, might provide no results for several more 10^4 year (and even for millions of years, in case there are just a few such civilizations in the Galaxy at that period). The latter, even at the slow pace of $0.001\ c$, would bring within their grasp, in that same period, thousands of stars, harboring perhaps interesting life-forms or even civilizations as advanced as the Egyptians or the Greeks, yet unable to radio-communicate. Obviously, the benefits would be far greater in the latter case, pushing most (if not all) technological civilizations older than a few 10^4 year to undertake a serious program of interstellar exploration. And even if some of them abandon the program for various reasons, the effort of their neighbors could compensate for that.

In summary, it appears that although radio communications constitute a natural means for SETI for civilizations younger than a few thousand years, older civilizations should rather develop extensive programs of interstellar colonization, because this is the only way to achieve undisputable evidence (either for or against the existence of ETI) within their lifetime L. In those conditions, the Fermi paradox appears all the more paradoxical: if, as the SETI proponents claim, there are literally thousands of such advanced civilizations wishing to establish contact, where are they?

Today, the *Plurality of worlds*, as the field was known in the antiquity, is as controversial as ever. Arguments on both sides ("It is unlikely that we are alone, in view of the Copernican principle and of such a large number of stars in the Galaxy" and "If there are so many of them, where are they?") are of statistical kind. They are consequently of little importance, for

statistics cannot be based on the single case provided by life on Earth. Detection of life signatures on another planet would be a powerful reason to undertake interstellar travel, at first by sending unmanned probes. Detection of some extraterrestrial civilization would undoubtedly be one of the major landmarks in the history of mankind. On the other hand, non-detection of ETI signals, even after millennia of research, would never prove that there were no extraterrestrial civilizations. But it would be a reason to prepare ourselves for a life of cosmic solitude.

REFERENCES

Barlow, M. 2013. Galactic exploration by directed self-replicating probes, and its implications for the Fermi paradox. *International Journal of Astrobiology* 12, 63–68.

Bennett, J., and Shostak S. 2007. *Life in the Universe*, 2nd ed. San Francisco, CA: Pearson - Addison Wesley.

Bjoerk, R. 2011. Exploring the Galaxy using space probes. arXiv:astro-ph/0701238.

Brin, G. D. 1983. The great silence—The controversy concerning extraterrestrial intelligent life. *Quarterly Journal Royal Astronomical Society* 24, 283–309.

Cameron, A. C. W. 1963. Communication with extraterrestrial intelligence on other worlds. *Sky and Telescope* 26, 258–261.

Chomiuk, L., and Povich, M. S. 2011. Toward a unification of star formation rate determinations in the Milky way and other galaxies. *The Astronomical Journal* 142, 197.

Cocconi, G., and Morrison, P. 1959. Searching for interstellar communications. *Nature* 184, 844–846.

Drake, F. 1960. How can we detect radio-transmissions from distant planetary systems? *Sky and Telescope* 19, 140–143.

Drake, F. 1961. Project Ozma. *Physics Today* 14, 40–42.

Furley, D. J. 1987. *The Greek Cosmologists Vol 1: The Formation of the Atomic Theory and its Earliest Critics*. Cambridge, UK: Cambridge University Press.

Glade, N. Ballet, P., and Bastien, O. 2012. A stochastic process approach of the Drake equation parameters. *International Journal of Astrobiology* 11, 103–108.

Hart, M. 1975. Explanation for the absence of extraterrestrials on Earth. *Quarterly Journal of the Royal Astronomical Society* 16, 128.

Jones, E. 1981. Interstellar migration and settlement. *Icarus* 46, 328–336.

Jones, E. M. 1985. Where is everybody? an account of Fermi's question (Los Alamos Report).

Kennicutt Jr., R. C., and Evans II, N. J., 2012. Star formation in the Milky way and nearby galaxies. *Annual Review of Astronomy and Astrophysics* 50, 531–608.

Kroupa, P. 2002. The Initial mass function of stars: Evidence for uniformity in variable systems. *Science* 295, 82–91.

Ksanfomality, L. V. 2004. The Drake equation may need new factors based on peculiarities of planets of Sun-like stars. *Proceedings of IAU Symposium #202*, A. Penny (Ed.), pp. 458–464.

Landis, G. A. 1998. The Fermi paradox: An approach based on percolation theory. *Journal of the British Interplanetary Society* 51, 163–166.

Maccone, C. 2010a. The statistical Fermi paradox. *Journal of the British Interplanetary Society* 63, 222–239.

Maccone, C. 2010b. The statistical Drake equation. *Acta Astronautica* 67, 1366–1383.

Mayr, E. 1985. The probability of extraterrestrial intelligent life. In *Extraterrestrials: Science and Alien Intelligence*, E. Regis (Ed.). Cambridge, UK: Cambridge University Press, pp. 23–30.

Newman, W. I., and Sagan, C 1981. Galactic civilizations - population dynamics and interstellar diffusion. *Icarus* 46, 293–327.

Ostriker, J., and Turner, E. 1986. The inclusion of interaction terms into population dynamics equations of interstellar colonization. *Journal of the British Interplanetary Society* 39, 141.

Petigura E. et al. 2018. The California-Kepler survey. IV. Metal-rich stars host a greater diversity of planets. *The Astronomical Journal* 155, 89.

Prantzos, N. 2000. *Our Cosmic Future - Humanity's Fate in the Universe*. Cambridge, UK: Cambridge University Press.

Prantzos, N. 2013. A joint analysis of the Drake equation and the Fermi paradox. *International Journal of Astrobiology* 12, 246–253.

Sagan C. 1963. Direct contact among galactic civilizations by relativistic interstellar spaceflight. *Planetary and Space Science* 11, 485–499.

Selsis, F. 2007. *Habitability: The Point of View of an Astronomer. Lectures in Astrobiology*. Springer-Verlag, Berlin, Germany, pp. 199–223.

Sheridan, M. 2009. SETI's scope (how the search for extraterrestrial intelligence became disconnected from new ideas about extraterrestrials). PhD Thesis, Drew University.

Shklovskii I., and Sagan C. 1966. *Intelligent Life in the Universe*. New York: Random House.

Simpson, G. G. 1964. The non-prevalence of humanoids. *Science* 143, 769–775.

Tipler, F. 1981. Extraterrestrial intelligent beings do not exist. *Quarterly Journal Royal Astronomical Society* 21, 267–281.

Viewing, D. 1975. Directly interacting extra-terrestrial technological communities. *British Interplanetary Society Journal* 28, 735–744.

Von Hoerner, S. 1962. The general limits of space travel. *Science* 137, 18–23.

Wandel, A. 2017. How far are extraterrestrial life and intelligence after Kepler? *Acta Astronautica* 137, 498–503.

Webb, S. 2014. *Where is Everybody? Seventy-Five Solutions to the Fermi Paradox*. Cham, Switzerland: Springer.

Wiley, K. 2011. The Fermi paradox, self-replicating probes, and the interstellar transportation bandwidth. arXiv:1111.6131v1.

11.3 SETI
Its Goals and Accomplishments

Eric J. Korpela

CONTENTS

11.3.1 INTRODUCTION

The search for extraterrestrial intelligence, or SETI, has existed as a discipline since Cocconi and Morrison (1959) published a rationale for a search for narrow-band radio transmission from nearby stars.[1] The following year Drake (1961) performed initial observations of two nearby stars, Tau Ceti and Epsilon Eridani, using a 26-meter parabolic dish antenna tuned to 1.42 GHz. No extraterrestrial signals were detected. In the nearly six decades following, SETI endeavors have expanded in number of targets, sky coverage, and spectral coverage.

11.3.2 BASIC MOTIVATION: THE DRAKE EQUATION

The basic motivation for searching for extraterrestrial intelligence (ETI) often begins with an estimate of the number of civilizations in the galaxy with whom we could communicate. The Drake equation is used to provide that estimate, using quantities that could, in principle, be determined. The standard Drake equation is given as:

$$N = R_* f_p n_e f_l f_i f_c L$$

where:

R_* is the average galactic star formation rate

f_p is the fraction of stars that have planets

n_e is the number of planets per star with planets that could support life

f_l is the fraction of planets that could support life that actually develops life

f_i is the fraction of planets that develops life where intelligence arises

f_c is the fraction of planets where intelligence arises and a communicating civilization arises

L is the average lifetime of a communicating civilization

In principle, we can simply determine those quantities and arrive at an estimate of the number of civilizations we could communicate with now. In practice, it is much more difficult than that. These values have significant uncertainty. For example, while the current value of the galactic star formation rate is well known, it is not the appropriate value to use for R_*. The appropriate value would be a weighted average of the galactic star formation rate, with the weighting based upon the likelihood that communicating intelligence would exist now on a planet around a star of a given age. This is, of course, unknown as the timescales for the origin of life and for biological evolution are unknown. It is entirely possible, and maybe even likely, that Earth is atypical in the time it took to reach those milestones (Korpela 2004). While some of the parameters are getting more certain (e.g., f_p and n_e), others have likelihood distributions that are more accurately expressed as roughly uniform between some limits in logarithmic space (f_l, f_i, f_c, and L). The values are uncertain enough that the value of N is relatively unconstrained, with little reason to prefer $N \sim 750,000$ over $N \sim 5 \times 10^{-7}$ (although the prior number is beginning to be approached by some SETI surveys for some definitions of "communicating").

In essence, N is unconstrained. While some may find this to be discouraging, it is clear that the parameter space is very wide and that the most efficient way to obtain knowledge of the number of communicating civilizations is a direct search for communicating civilizations, hence SETI.

11.3.3 SETI, A DEFINITION

Before proceeding further, it is necessary to define SETI beyond the literal meaning. Intelligence is not something that can be detected directly. In nearby organisms (e.g., another human), it is identified by observing behavior that is presumed to require thought. Such behavior might be a complex set of operations requiring long-term planning. At interstellar distances, such behavior is more difficult to observe. We typically find ourselves limited to artifacts. Such artifacts could be signals encoded in electromagnetic or other radiation. They could be physical objects launched by an ETI. They could even be an object or set of objects so large that they effect the appearance of planets, stars, or even galaxies in their vicinity. At this point, I will attempt to dispense with terms such as civilization and species, as such terms presume, to some extent, that ETI will be like us: organized groups of independent biological organisms. While it is often the default assumption that is used by astrobiology researchers, the universe may surprise us. The most common type of "civilization" might consist of a single electronic intelligence. In cases where I use the term civilization, it will be associated with citations that use the term.

11.3.3.1 DISTINCTION FROM METI

It is important to distinguish SETI from messaging extraterrestrial intelligence (METI). Also known as active SETI, METI involves beaming transmissions toward nearby stars thought likely to permit the evolution of life. SETI is a purely observational science, and many SETI proponents do not encourage directed messaging prior to a SETI detection (Azua-Bustos et al. 2015). The justification for this position stems from several bases. The primary concerns are that we have no way to predict the motivations or behaviors of an ETI apart from comparison to our own troubled history. Any ETI we do encounter is likely to be far more technologically advanced than our own. SETI is still in its infancy and has not even completed a systematic study of nearby stars across a wide range of electromagnetic frequencies; therefore, many SETI scientists feel that more study is necessary before directed transmissions are sent. There is also strong consensus among those opposed to directed transmissions that a small group of scientists should not take actions that could have significant consequences for future generations without at least getting explicit consent from intergovernmental groups such as the United Nations.

In response, proponents of METI often suggest that we have been sending signals for a significant time. As of this publication, omnidirectional radio signals from Earth have passed 0.008% of the stars in the galaxy, which implies we could have been detected through past radio emissions presuming $N \gtrsim 12,500$. However, omnidirectional signals are much weaker than the directional signals proposed for METI. A detection of the proverbial "I Love Lucy" rerun at this extreme range would require ~ 700 km^2 of collecting area at the current extreme distances of these transmissions. For smaller values of N, the probability that our radio emissions have been detected is much smaller.

Radio emissions are not the only potentially detectable signatures of intelligent life on Earth.

Industrial activity has released anthropogenic chemicals such as chlorofluorocarbons (CFCs) into the atmosphere in quantities that might be detectable in the spectrum of the Earth; however, this signature is of more recent genesis than radio signals. Large terrestrial cities have thermal infrared (IR) and visible light signatures which may be detectable at interstellar distances given a large enough telescope.

11.3.4 DETECTABLE ASPECTS OF INTELLIGENCE

Let us consider the aspects of intelligence that can be detected. As we've mentioned, direct observation of behavior at interstellar distances is unlikely, and, therefore, any evidence of intelligence is likely to be transmitted. Transmitted, in this case, does not necessarily indicate electromagnetic radiation. A physical object can also be transmitted, albeit at a lower speed.

11.3.4.1 Energetics of Directed Transmissions, Beacons, and Leakage

Transmissions are generally divided into three types. The first, directed transmissions, are transmitted toward one or more chosen targets. This conserves energy at the transmitter by concentrating the transmission in a small solid angle $\Omega_A = \int \Phi(\Omega) d(\Omega)$, where $\Phi(\Omega)$ is the emission pattern, oriented toward the target. The second, a beacon, is an omnidirectional transmission covering 4π steradians. It's easy to see that, for a given transmitted power P, the average power of the signal in the beam of a directed transmission is a factor of $\frac{4\pi}{\Omega_A}$ stronger. This gain (Γ) of a diffraction limited telescope can be approximated as:

$$\Gamma = \frac{4\pi}{\Omega_A} = \frac{4\pi A_e}{\lambda^2}$$

where A_e is the effective area, defined such that $A_e \Omega_A \equiv \lambda^2$.[2] For a perfectly efficient, diffraction-limited telescope, the effective area is approximately the geometric area. For a non-diffraction-limited telescope, telescopes with multiple apertures, or those with less-than-perfect optical efficiency, $A\Omega_A \neq \lambda^2$, a calculated value of Ω_A and efficiency must be used to determine gain. Because of the λ^{-2} dependence of gain, in general, the gain of optical telescopes of even moderate sizes is huge compared with that of single-dish radio telescopes.

Because of the energetics involved, it is often presumed that an ETI will send directed transmissions toward promising targets rather than utilizing omnidirectional beacons. This may be especially true if the transmitting intelligence has observational capabilities that allow them to detect biosignatures and weak technosignatures.

The final type of transmission, leakage, represents transmission of information as a by-product of activity of an intelligent being or beings, as opposed to intentional transmissions. Leakage transmissions can include omnidirectional signals used for local communication such as radio or television broadcasts. It should be noted, however, that current terrestrial telescopes would be unable to detect such transmissions at distances much larger than 1 parsec. More promising types of leakage transmissions would be directional signals, satellite communications, interplanetary communications within a planetary system, and radars. While these directed transmissions are expected to be stronger than the omnidirectional leakage transmissions, they are also expected to be intermittent, with duty cycles inversely proportional to the transmitter gain. That makes repeated detection of high-gain leakage transmissions unlikely, unless there is a geometrically favorable arrangement of transmitter, receiver, and the distant observer. For example, Siemion et al. (2014) suggest searches for signals be conducted for when exoplanets are in conjunction, as viewed from Earth. Large telescopes are still required for this type of search, as the extraterrestrials would

presumably use only as much signal power as is necessary for their short-range communication.

11.3.4.2 Physical Artifacts versus Radiated Transmissions

It has been suggested that it is energetically favorable to send physical artifacts rather than radiated emissions. Physical artifacts, if they can be decelerated at their destination, have the benefit of persisting, potentially for billions of years. No such method of decelerating radiated emissions exists, as far as our current technology is concerned. In addition, storage densities in physical structures can be large; 100 Gb per gram is easily achievable with current technologies. Transmission across interstellar distances can be achieved with non-relativistic velocities, resulting in low costs. The drawback is primarily higher latency between sending and receiving a reply due to slower carrier velocity. Because of the lower velocity and the persistence, it is common to conceive of physical artifact transmission as a one-way information transport, commonly carrying a near-complete encyclopedia of an ETI's knowledge. Others have conceived of a biological library capable of seeding life-forms on planets in the destination system. The typical behavior of such a physical message from an ETI upon reaching the destination system is difficult to predict. It could use stellar gravity to redirect its trajectory toward another star system. During such a flyby it might attract attention by directing electromagnetic signals toward planetary bodies. Alternatively, it could achieve a stable orbit within the destination system. If the purpose of the artifact is communication with other intelligences, it could sit idle for a long period of time, emitting a beacon only at specified intervals or when intelligent activity is detected. In either case, such a message might be difficult to detect for a civilization of limited technology.

People on Earth have deliberately transmitted four passive artifacts for the purpose of communication with extraterrestrials, although none is directed toward any specific star system. Pioneers 10 and 11 each contain a gold-anodized plaque etched with a description of the location of the solar system relative to galactic pulsars, a schematic of the solar system showing Earth as the origin, and scale images of male and female humans and the Pioneer spacecraft (Sagan et al. 1972).

Voyagers 1 and 2 each contain a gold-plated aluminum disk engraved in the style of an audio record. A stylus for playing the recording is provided. On one side is engraved a schematic description of the decoding method and the position of the solar system relative to pulsars. The record contains both audio and image data. There are greetings in multiple languages, music from many cultures, and sounds of nature and civilization. The images include diagrams explaining mathematical and scientific concepts, pictures of solar system objects, images of human anatomy and function, images of Earth landscapes, and images of culture and technology

(Sagan et al. 1978). Although not deliberate messages, to date, seven additional artifacts have been sent on escape trajectories from the solar system, where they may eventually confuse extraterrestrials as to their purpose. These include the New Horizons spacecraft, the third propulsion stages that propelled New Horizons, Pioneer 10, and both Voyagers, and two small de-spin weights from New Horizons. All are on escape trajectories from the solar system.

Our ability to detect equivalent artifacts is quite limited. An object as small as Voyager 2 would need to pass very close to (less than a few times the lunar distance [LD] from) Earth to be detected, unless it was an active source of optical or radio emission. The first detected object known to originate outside the solar system, A/2017 U1 ('Oumuamua) had a size of $230 \times 35 \times 35$ meters and was detected at a distance of 0.22 AU (85 LD) (Jewitt et al. 2017).

11.3.4.3 Sensitivity Constraints for Radiated Signals

For a noise-limited system, the minimum detectable signal flux is given by the rather obvious relation

$$F_{min} = n_{thresh}\sigma_F$$

where σ_F is the relative error in measurements due to noise within the instrument and internal and external backgrounds and n_{thresh} is chosen to limit the false alarm rate to a manageable level. The functional form of σ_F depends strongly on the instrumentation and its noise processes and the distribution of noise events. In general, it can include thermal noise processes, atmospheric and/or stellar backgrounds, and photon-counting statistics. The statistical form of these noise processes will also vary from instrument to instrument. Whereas a 5 σ threshold may be adequate for a Gaussian noise distribution, a 25 σ threshold might be necessary for an exponential distribution.

For future discussions of sensitivity in this article, we shall concentrate on integrated flux rather than spectral flux density, under the assumption that we are able to integrate across the entire spectral distribution of a signal.

11.3.4.4 Detection of Extraterrestrial Intelligence

To declare a detection of an extraterrestrial signal or artifact, it's important that the evidence points to an extraterrestrial intelligent origin and that natural origin be ruled out as much as possible. An extraterrestrial radiated signal must be repeatedly detected using multiple instruments, if possible at multiple telescopes. Single observations of short-duration events, such as the "Wow! signal" (Ehman 2011), should not be considered a detection until they are detected repeatedly. It's also important that a radiated signal be localized to a single source in the sky. The terrestrial environment is filled with sources of transient anthropogenic electromagnetic signals that appear to come from essentially random locations on the sky.

For any signal or artifact, intelligent extraterrestrial origin should be the final hypothesis rather than the first. When 'Oumuamua was detected, the media rapidly propagated the idea that it was an extraterrestrial spaceship, an idea that was even spurred on by some scientists. Yet, even a simple analysis would indicate that comets and asteroids that have been ejected from stellar systems are likely to outnumber extraterrestrial spaceships by factors of a billion. The media attention prompted SETI scientists to observe 'Oumuamua. Unsurprisingly, they detected no evidence of signals.

11.3.5 WHERE TO LOOK

11.3.5.1 A Choice of Carrier: Photons, Neutrinos, and Gravitational Waves

As humans, one of our most important sensory inputs, the eye, is based upon processing inputs of electromagnetic radiation (photons). As human technology has progressed, the use of the eye as an input for communications has increased. Some of our most important technological advances have been devices that turn information into visual stimuli that can be processed by the brain (i.e., writing) and devices that turn information into electromagnetic radiation that can be transmitted over long distances for later sensory processing (e.g., radio and television). Therefore, it's natural to conceive of interstellar data transmission using similar devices.

The benefit of using photons as a carrier is that photons interact electromagnetically with nearly all forms of normal matter. That makes it relatively easy to build devices that collect and measure photons and to build devices that deflect or guide photons into those devices. The drawback of using photons as a carrier is that photons interact electromagnetically with nearly all forms of normal matter. Atmospheres are opaque to many frequencies that would otherwise be useful for transmission. Interstellar matter can also block clear transmission because of opacity, dispersion, scintillation, refraction, and high natural backgrounds.

Because photon detectors are ubiquitous in astronomy, they will comprise the majority of the discussions in this chapter. We'll divert here into a short discussion of non-electromagnetic options for interstellar communications.

11.3.5.2 Neutrinos and Gravitational Waves

Photons are not the only type of radiation that can travel interstellar distances and be detected; they are simply the easiest to create and detect. Other options include neutrinos. Neutrinos interact with normal matter through the weak nuclear force. As is implied in the name, the weak nuclear force is a weaker interaction than the electromagnetic interaction. That makes it difficult to build devices that create directed neutrinos (a particle accelerator) and very difficult to build devices that detect neutrinos (e.g., Super-Kamiokande) (Abe et al. 2011). It is virtually impossible to build devices that redirect or focus neutrinos (none known). Neutrino-detection efficiencies of these devices are of order 10^{-16} at the low energies typical of solar neutrinos. At higher energies, neutrino detection can be significantly higher ($\sim 10^{-5}$). Silagadze (2008) postulates that a 200 TeV muon collider operating 6 parsecs from the Earth would be detectable with an existing neutrino detector, provided it created neutrinos at a rate of $10^{14}\,\text{s}^{-1}$ and was pointed within 50 miliarcsecond (mas) of the Earth. For random orientations, the probability of such an alignment is $P \sim 1.5 \times 10^{-14}$, which would imply purposeful targeting, should such a detection ever be made. Lacki (2015) postulates that extremely advanced civilizations might build particle accelerators to probe physical processes at the Plank scale. Such accelerators might emit $10^{24}\,\text{eV}$ neutrinos. Such a neutrino could interact in the upper atmosphere and be detected by existing cosmic ray detectors. No cosmic rays of such high energy have yet been detected. An additional possibility would be the use of gravitational waves as the information carrier. The interaction between gravitational waves and matter is very weak, and therefore, such waves are typically detected by measuring the distortion of spacetime between two points by using a laser interferometer many kilometers in size. The natural phenomena detected with gravitational wave detectors include merging black holes visible at hundreds of Mpc distance (Abbott et al. 2016) and merging neutron stars visible at tens of Mpc distance (Abbott et al. 2017).[3] Although they are detectable with current technology, gravitational waves that can be detected at interstellar distances are difficult to generate. An interstellar gravitational wave transmitter might consist of a pair of closely orbiting neutron stars with some way of feeding energy into the orbits, in order to modulate the orbital period, and therefore the gravitational wave frequency. Such a transmitter is well beyond our current technology. As gravitational waves are additive in a manner similar to electromagnetic waves, it should be possible to build a directional transmitter array from synchronized orbiting masses. This does not remove the need for stellar mass neutron stars or black holes in the transmitter, as there are no known stable states of matter of lower mass that would provide the necessary densities to generate significant gravitational radiation. Gravitational wave generators that do not use mass to generate the waves require speculative extensions to physics and, for now, will remain science fiction.

11.3.5.3 Target Choices

Much consideration goes into target selection for SETI observations, with much of the preference going to nearby solar-type stars. There may be some validity to such a choice, as the only known technological civilization lives on a planet around such a star. Turnbull and Tarter (2003a,b) developed "HabCat: a Catalog of Nearby Habitable Stellar Systems" as a potential target list for SETI on the Allen Telescope Array (ATA). They selected for main sequence stars with low stellar activity and variability, stellar age greater than 3 Gyr, and luminosities less than 2.5 L. Because their choices depended upon *Hipparcos* parallaxes, the volume sampled in the catalog is strongly limited by stellar luminosity. Low-luminosity

K and M stars found in the catalog are very nearby. Most of the catalog consists of brighter late F and early G stars, thus their catalog is heavily biased toward solar-type stars.

For small values of Drake *N*, being limited to nearby stars is not necessarily appropriate. Other suggestions have been put forward. Some have suggested that faint cool M-dwarfs may have benefits for development of life (Tarter et al. 2007). They are abundant, representing more than 75% of the stars in the galaxy. They are long-lived, with lifetimes in the trillions of years for the lowest stellar masses. High stellar activity early in the life of the star, resulting in high ultraviolet (UV) and X-ray fluxes, could be a catalyst for generation of complex organic molecules necessary for the development of life. A potential disadvantage of these stars is the narrow width of the habitable zone, which may make it less likely that planets will be found in the habitable zone. The proximity of the habitable zone to the stars results in tidally locked planetary rotation for any planets in the habitable zone. This may result in conditions that make it difficult for life to thrive. The stellar activity of these stars early in their lifetimes is likely to be sufficient to strip atmospheres, which may deplete the planets of volatile elements (H, O, and N) necessary for terrestrial-type life.

It is possible that advanced life either has altered its environment to the point where we can't detect its star (see 11.3.9) or has no need to live near stars and planets, only visiting them to collect raw materials. In that case, concentrating solely on stars may be a mistake.

The Breakthrough Listen Green Bank Telescope survey (Isaacson et al. 2017) has chosen a more diverse sample of locations, although still concentrating on stars. Their target list includes all known stars within 5 pc, the 100 nearest giants, and main sequence stars between 5 and 50 pc with spectral types from A-M, with 100 chosen per unit 0.1 magnitude of B-V color, unless fewer were available. To this, they added 139 external galaxies.

Prior to Breakthrough Listen, most radio SETI observing by the Berkeley SETI team, through its projects *SERENDIP* (Search for Extraterrestrial Radio Emissions from Nearby Developed Intelligent Populations) and *SETI@home* (Korpela et al. 2011a; Bowyer et al. 2016), has been conducted in a commensal or piggyback mode, generally directed where other observers using the telescope have scheduled observations. At Arecibo, this has resulted in participating in both pointed observations of specific targets and surveys conducting observations of large regions of the sky.

11.3.6 RADIO AND MICROWAVE SETI

The earliest forms of serious SETI searched for radio-frequency transmissions by using instrumentation attached to radio telescopes. Because radio technology was one of the first methods of potential interstellar communications developed by humans, it seems a likely mechanism for extraterrestrials interested in communicating with developing species. In addition to being simple to deploy, it's quite easy to use radio to create signals that are easily distinguished from natural sources of radiation.

Natural sources of radiation are typically broadband, with signal bandwidths of tens of kilohertz to gigahertz. Since no known natural source of extremely narrow band (<1 Hz) emissions exists, if such a signal is found, it can be presumed to be artificial. Narrow bandwidth also economizes transmission power. The energy required to outshine background sources is narrower. Another way of outshining background sources is that, rather than putting energy into a narrow band, the transmitter could put lots of power into a pulse of short duration. This is less likely to be distinguishable from natural sources, however, as pulsars and fast radio bursts do give out short powerful bursts.

11.3.6.1 Basic Concepts: Detectors, Bandwidths, and Channels

In its simplest form, a radio telescope consists of an antenna, a receiver system, and back-end instrumentation. The purpose of the antenna is to collect the radio waves, which are added in-phase and directed toward the receiver. The receiver system amplifies the weak RF signals and converts them into higher-amplitude electrical signals. Typically, a receiver processes two polarizations, either crossed linear polarizations or left and right circular polarization. These electrical signals may be converted into signals of lower frequency by mixing them with a sinusoidal oscillator. This prevents the need that the back-end instrumentation be able to process the RF signals directly at the Nyquist rate.[4] A wide variety of back-end instrumentation is typically available. These back ends preserve differing amounts of the information that is received from the front end.

Direct digitization and recording of the down converted signal can preserve nearly all of the information present and, in principle, can reproduce the receiver voltages versus time to high precision. The drawback is that large amounts of data must be stored, typically four bytes (two per polarization) per unit time, with time being the inverse of the Nyquist rate. Therefore, a 100 MHz bandwidth will typically require 400 MB/s of recording capacity. Such recording techniques are used with very long baseline interferometry (VLBI), where preserving phase and timing relations is extremely important. It's also used in some SETI experiments. Breakthrough Listen (Enriquez et al. 2017) utilizes this method for its initial data capture, but processes the data quickly to a format that is thousands of times smaller. *SETI@home* (Korpela et al. 2001) utilizes this technique for recording data, but uses a recording technique that only requires 4 or 8 bits per complex sample.

The basic instrumentation for radio SETI is a spectrometer. Typically, these spectrometers generate spectra where the relationship between channel width and integration time is $\Delta v_{\mathrm{chan}} \Delta t = 1$. There are few known natural processes that generate signals with that relationship; therefore, if one is seen, it could be a first indication of an artificial signal.

11.3.6.2 Backgrounds and Opacities

There are some issues that limit the sensitivity of radio observation by increasing the noise within a measurement. The first is quantization noise, which is dependent upon the number of bits used to sample a signal. The easiest visualization is to imagine a sine wave sampled at 1 bit per sample. In essence, the sine wave is converted into a square wave by the sampling, which moves power out of the primary frequency and into the Fourier components of a square wave. For single-bit digitization, this results in a signal power loss of 36%. For 2-bit, the loss is 19% and drops rapidly to less than 1% for quantizations of 5 bits and higher (Vleck and Middleton 1966; Thompson et al. 2007).

Other limits are set by the noise processes present in interstellar and intergalactic spaces. At low frequencies, the noise processes are dominated by the emissions from energetic particles interacting with the galactic magnetic field. At higher frequencies, molecular rotational and vibrational transactions provide the dominant noise source. At frequencies between about 500 MHz and 20 GHz, there is a region known as the terrestrial microwave window. In this region, both the atmosphere and the galactic interstellar medium (ISM) are relatively transparent. Natural noise in this region is dominated by the 3K cosmic background radiation. However, thermal noise processes in telescope systems are often significantly larger than this, with receiver contributions of 20 to 40K. The sum of all the thermal noise sources is referred to as system temperature, T_{sys}. Figure 11.3.1 shows the microwave window in schematic form.

A portion of the microwave window known as the "water hole" has been the subject of significant interest among SETI scientists. It is bounded at the low-frequency end by the 1.42 GHz magnetic dipole transition of Hı and at the high-frequency end by the 1.67 GHz transition of the OH radical. The Hı transition is used by scientists to map the structure of the galaxy. The OH transition is similarly useful in studying molecular gas in the galaxy. It is presumed that ETI would utilize these transitions as well, and therefore, we would expect that a transmission on these frequencies is more likely to be

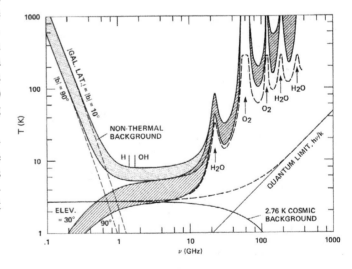

FIGURE 11.3.1 The terrestrial microwave window. (From Morrison, P. et al., *The Search for Extraterrestrial Intelligence, SETI*, NASA Special Publication, 419, 1977.)

seen by extraterrestrial scientists. In addition, the combination of H and OH results in water, which is an important compound for life, at least on Earth.

Sensitivity for detection of an extraterrestrial signal is dominated by the noise processes discussed previously. We define system equivalent flux density for a single dish as:

$$S_{sys} = \frac{2k_B T_{sys}}{A_e}$$

and the minimum detectable flux for a narrow signal as:

$$F_{min} = n_{thresh}\sigma_F = n_{thresh}S_{sys}\sqrt{\frac{\Delta v_{ch}}{n_{pol}\Delta t}} \qquad (11.3.1)$$

where:

Δv_{ch} is the spectral channel bandwidth.

Δt is the observation time.

n_{pol} is the number of polarizations measured.

A 100-meter telescope with a 30 K system temperature would have S_{sys} ~10 Jansky in 3 Hz channels, with 30-second dual polarization observations, and a 20 σ threshold would have F_{min} ~ 5 × 10^{-25}W m^{-2}. This is, coincidentally, about the spectral resolution and per-spectrum integration used by the Breakthrough Listen project for its high-resolution data product.

11.3.6.3 Radio Frequency Interference

Perhaps an even larger problem for detecting ETI is that technosignatures surround us. Our radio telescopes are embedded in the middle of a technological civilization that uses radio frequencies, both narrow band and wide band, for communication and for object detection and ranging. A significant part of SETI science involves developing techniques to distinguish between local interference and potential extraterrestrial signals. This involves many techniques. On-source/off-source observation is one technique in which either a separate telescope or receiver is used simultaneously to gather data at a location away from a suspected ETI source or the telescope is moved between off source and on source positions. Any signal detected in both observations can be rejected as interference. Some of these techniques are discussed by Korpela et al. (2011b).

Signal properties can also be used to identify interference. The rotating Earth imparts a Doppler drift to a signal. If this drift is not seen, a signal can be presumed to be local. Satellites, on the other hand, pass overhead at a range of velocities and accelerations in a wide range of possible drift rates, which makes them more difficult to reject. An ETI would need to be aware of the exact distance to the Earth, the rotation rate of the Earth, and the location of the telescope on the Earth, in order to correct for this effect. The ISM is partially ionized, which results in a frequency-dependent delay in broadband pulses. If this delay is not

seen, a pulsed signal can be presumed to be local. We will discuss these effects further in the next sections. As interferometric arrays of radio telescopes are being used more frequently for SETI, new capabilities are being developed. Beamformers can produce nearly arbitrary beam shapes that allow null regions to be placed over radio frequency interference (RFI) sources (e.g., satellites) in the field of view. If multiple beamformers are available, both on target and off target observations can be conducted simultaneously. If a signal is seen from a target, a null can be placed over it to verify the signal disappears. See (Harp 2013) for further discussion of these methods.

11.3.6.4 The Doppler Drift Problem

There are some issues that limit the sensitivity of radio observation to narrow-band signals. One primary concern is the Doppler drift caused by accelerations of the transmitter or receiver. For a receiver located on the surface of the Earth, the dominant acceleration of the receiver is imparted by the rotation of the Earth. This acceleration is centripetal. The highest drift would occur for a receiver located at the equator, observing a target overhead. This acceleration of 3.4 cm s^{-2} ~ $10^{-10}c$ s^{-1} may not seem significant, and for most astronomical observations, it is not. But when dealing with channel widths near 1 Hz, the drifts become problematic. At L-band (1.4 GHz), this corresponds to a frequency drift in any received signal of 0.16 Hz s^{-1}. If uncorrected, this could cause a reduction in sensitivity by limiting the effective exposure time to no more than $\Delta t \leq \frac{\Delta v_{ch}}{\dot{v}}$, or equivalently increasing channel bandwidth to $\Delta v_{ch} \geq \dot{v}\Delta t$. Keeping in mind that the relation $\Delta v_{chan}\Delta t \geq 1$ still holds, the minimum channel width for an uncorrected L-band spectrometer is Δv_{ch} ~ 0.4 Hz and Δt ~ 2.5 sec. Because of this, most SETI spectrometers have operated with channel bandwidths between 0.5 and 3 Hz. By Equation 11.3.1, this means that they are operating at sub-optimal sensitivity. It is possible to perform this correction coherently on the incoming signal to regain full sensitivity and allow for narrower channel widths. Unfortunately, because this process was not implemented with SETI in mind, at most observatories, the frequency steps implemented for the local oscillator are often too large (≥10 Hz) to be useful.

There's a worse Doppler drift problem, the drift of the transmitter frequency due to its own motions. An extraterrestrial may take care of this problem for us. If the transmissions are directed specifically at Earth, they can be corrected by the sender for the motions of the transmitter. Omnidirectional beacons could be constructed from multiple antennas, each corrected for velocity changes along its direction of transmission (Korpela 2011). It's also possible that uncorrected transmitters are what an ETI may decide to build. Since the parameters of motion of the transmitter are unknown, SETI projects generally perform a blind search to likely Doppler drift rates.

There are two ways to perform this search. The most common method is to use an incoherent dechirp process, essentially shifting spectra by the drift rate to find drifting signals. This can be implemented as a tree-search algorithm, which reduces the complexity to $O(n\log n)$ from the $O(n^2)$ of the naive implementation (Siemion et al. 2013). Because of the quantization of frequency and time, there is a sensitivity loss associated with this method, as the drift results in energy bleeding into adjacent frequency bins during the spectral integration. This results in an effective bin width of $\Delta v_{eff} \sim \Delta v_{ch}$ and a corresponding loss in sensitivity per Equation 11.3.1; in practice, this search looks through hundreds of potential drift rates.

The option is to use a coherent dechirp method on the time-domain (baseband) data. This method involves the complex multiplication of the data by a chirp function, effectively translating the data into an accelerating frame. Because this is done in the time domain, no loss of frequency resolution results, and the process can be performed for arbitrarily small differences in drift rate. For *SETI@home*, this is done for drift rate differentials as small as 10^{-3} Hz s^{-1} over the range -100 Hz s^{-1} to $+100$ Hz s^{-1}, which enables *SETI@home* to utilize 0.075 Hz spectral bin while still preserving sensitivity over the full Doppler drift range.

It is important to note that the behavior of a drifting signal is expected to vary over time. A signal not corrected at the source would have velocity components due to planetary rotation, due to orbital motions of the planet, and due to gravitational influences of other planets or satellites. These are likely to add up to 10 s to 100 s of km/s, which translates to long-term frequency shifts of 10 s or 100 s of kHz. This can add to the difficulty of detecting a signal.

11.3.6.5 The Dispersion Problem

An equivalent problem to the Doppler drift problem exists when trying to detect pulsed radio emissions. The ISM is partially or fully ionized. A broadband radio pulse traveling through the ISM will interact with free electrons and will be delayed. The amount of this delay is dependent on frequency, such that:

$$\Delta t = 8.3 \, \mu s \, \frac{(v - v_o)[\text{MHz}]}{v_o^3[\text{GHz}]} DM[\text{pc cm}^{-3}]$$

where v_o is a reference frequency and DM is the integrated electron density along the line of sight, given in units of pc cm^{-3}. So, a pulse coming from a distance of 10 pc through an electron density of 0.1 cm^{-3} at a frequency of 1 GHz would have a dispersion of about 8.3 μs MHz^{-1}. Again, if the dispersion measure to a source is unknown, a tree search can be performed over a wide range of dispersions. This search can be performed incoherently, with the accompanying loss of sensitivity and a loss of potential time resolution, or coherently, requiring significantly more computation. Nearly every such SETI search is performed incoherently with time resolutions greater than 64 μs and spectral resolutions of megahertz. The only coherent search of note is the *Astropulse* search, which performed a search for pulses as short as 400 ns with the 305-meter Arecibo telescope (Von Korff et al. 2013).

11.3.7 OPTICAL AND INFRARED SETI

11.3.7.1 Photon-Counting Instruments

Because the photon energies in the IR and optical are high enough to result in an electron being liberated via the photoelectric effect, most optical and IR SETI experiments use low-noise photon-counting detectors, such as charge coupled devices (CCDs), avalanche photodiodes (APDs), or photomultiplier tubes.

For an observation of a source with a photon flux integrated over the detection bandpass of F_{source} photons/cm^2 with a photon-counting detector of N_{total} counts with non-zero sky and/or stellar background (N_{sky}), detector background (N_{det}), and read noise (N_{read}), the measured flux can be expressed as:

$$F_{source} = \frac{1}{A_{eff}t}\left(N_{total} - N_{sky} - N_{det} - N_{read}\right)$$

In the case where only shot noise is a significant error contribution and backgrounds levels can be determined with high precision:

$$\sigma_F = \sqrt{\frac{F_{source}}{A_{eff}t} + \Omega_r \frac{dF_b}{d\Omega}\frac{1}{A_{eff}t} + \frac{F_{stellar}}{A_{eff}t} + \frac{R_{det}}{A_{eff}^2 t} + \frac{N_{read}}{A_{eff}^2 t^2}}$$

where:

A_{eff} is the effective area of the telescope.
Ω_r is the solid angle of the integrated region.
$\frac{dF_b}{d\Omega}$ is the background surface brightness.
R_{det} is background count rate coincident with the signal.
$F_{stellar}$ is coincident stellar flux.

Fluxes here are measured in units of ph m^{-2} s^{-1}. Obvious takeaways are that (1) in general, the F_{min} will go as $(A_{eff} t)^{-\frac{1}{2}}$, and (2) angular resolution is important for reducing background, as are low-noise detectors.

In the case of a diffraction limited instrument, $A_{eff} = \pi R^2 \varepsilon$, where R is the aperture radius and ε is the instrument's integrated efficiency, including optical and detector efficiency. When detector and readout noise are also low, σ_F becomes:

$$\sigma_F = \frac{1}{\sqrt{\pi \varepsilon t R}}\sqrt{F_{source} + F_{stellar} + \frac{\pi}{2}\left(\frac{\lambda}{R}\right)^2 \frac{dF_b}{d\Omega}} \quad (11.3.2)$$

The basic types of detection systems are unchanged from the radio case. We can observe for a spectral signature: lots of power put into a single frequency (or a small number of frequencies), or for a transient signature: a detectable change in flux on short timescales.

11.3.7.2 The Problems: The Atmosphere, Dust, and Backgrounds

Some of the difficulties involved in optical and IR SETI are the opacities of the media through which the radiation propagates. The Earth's atmosphere presents difficulties with turbulence, which results in scintillation (twinkling) and blurring of stellar images. The total effect is known as "seeing" and is expressed as the angular blur in arcseconds that it adds to an image beyond what would be expected if the telescope were diffraction-limited. This can be combated with adaptive optics. However, at this time, those techniques are restricted to larger telescopes, which are not often made available for SETI observations.

The atmosphere is not entirely transparent at optical and IR wavelengths. Figure 11.3.2 shows the atmospheric transmittance from sea level in the optical and IR ranges. Significant spectral regions of high opacity are present. To avoid that opacity, it is necessary to get above the atmosphere. Some opacity can be avoided by locating observatories at high altitude. Others require balloon- or space-based observatories. Again, getting use of such observatories for SETI observations is problematic.

The opacity of the ISM is also a significant concern at X-ray, UV, and optical wavelengths. At photon energies above 13.6 eV, the opacity tends to be dominated by ionization of interstellar atoms, which leads to very high opacities near their ionization edges (Cruddace et al. 1974). Below 13.6 eV, the opacity tends to be dominated by interstellar dust. Dust opacities are highest in the UV and decrease rapidly in the IR, as wavelengths become larger than the size of the dust grains.

Given Equation 11.3.2, a 2-meter diameter telescope with 85% optical efficiency and a 75% efficiency noiseless detector would have:

$$\frac{F_{source}}{\sigma_F} \sim 140\sqrt{t[\sec]F_{source}}$$

in the absence of background, as expected for Poisson statistics. For ground-based applications, sky backgrounds (~21.8 m at V band, or equivalently 150 ph s^{-1}m^{-2} integrated across the band, is considered a dark sky), combined with

seeing (~0.5–3″) and/or aperture size (arcseconds to arcminutes) and spectral resolution (fractions of an Å to 100s of nm), are often the dominant constraints. The sky background in a 3″ circular aperture alone would result in a background rate of ~2100 s^{-1} in our hypothetical 2-meter telescope. Observations coincident with a star compound this rate, with even a 15th magnitude star providing a background count rate of 1.6 × 10^5 s^{-1}. In this case, σ_F is dominated by the $F_{stellar}$ term, and the signal-to-noise ratio is:

$$\frac{F_{source}}{\sigma_F} = \frac{F_{source}A_{eff}t}{\sqrt{F_{stellar}A_{eff}t}}$$

which, for our example, evaluates as $0.006F_{stellar}\sqrt{t[s]}$ and translates to the obvious statements that more intense signals and longer exposures increase signal to noise, and higher backgrounds decrease it. The less obvious conclusion is that emissions need to be intense (and, of course, variable) to be noticeable in a single broadband detector.

The means for combating these backgrounds on both the sending and receiving ends are similar to those used in radio SETI. If we are looking for bright laser lines, we would use a spectrograph to create a spectrum, and we would look for bright narrow lines in the spectrum, either within the stellar spectrum or adjacent to it. The V band has a spectral width of 880 Å (88 nm), so a spectral resolution of 0.1 Å essentially reduces the stellar background rate per bin by a factor of 8800.[5] This technique is explored in more detail by Tellis and Marcy (2017).

11.3.7.3 PULSE AND COINCIDENCE AND ANTICOINCIDENCE DETECTORS

There are significant reasons to expect that pulsed lasers might be used for extraterrestrial communications. The

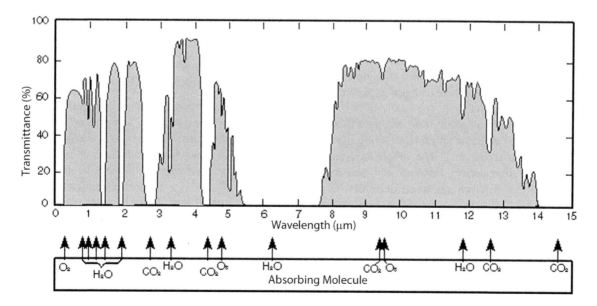

FIGURE 11.3.2 Optical and IR transmissions of the Earth's atmosphere versus wavelength.

largest existing terrestrial pulse laser, the National Ignition Facility, produces a pulse of 1.3 petawatt 1.053 μm IR light lasting a few nanoseconds, with total pulse energy of about 4 MJ. A similar laser attached to an astronomical telescope could be used to send a pulse of light that would outshine the Sun at interstellar distances. Even a moderate, 1″ focus would illuminate a region 10 AU in diameter at a distance of 10 pc with a pulse consisting of 12 photons per square meter (Stolz 2012). A pulse of that intensity could be detectable, even with current optical and IR detector technologies.

If we are looking for short-duration pulses, we look for bursts of photons on short timescales among the relatively constant rate backgrounds. Our hypothetical system, with its 1.6×10^5 s^{-1} count rate, results in a photon detection on average, every 6.25 μs. If we integrate for 100 ns, the probability of detecting photons is governed by Poisson statistics. The likelihood of detecting n or more photons in an exposure is $P(n, Rt)$, where $P()$ is the incomplete gamma function, R is the photon rate, and t is the integration time. Of course, in real photon-counting systems, some dead time is associated with each detection, which alters the statistics. Assuming zero dead time for our hypothetical system, the likelihood of detecting 10 photons in 100 ns is of order 3^{-25}, which would be one false alarm every 10 billion years. That would seem to be acceptable for detecting short-duration bright pulses from pulsed laser systems. Unfortunately, there are many instrumental, artificial, and natural sources of pulses that can be mistaken for extraterrestrial signals. These include cosmic rays hitting the detector, radiation from cosmic ray showers in the atmosphere, strobes, power line glitches, detector noise processes such as "after-pulsing," and even RFI affecting the electronics. The false alarm rates of single detector systems are often dominated by these effects.

There are many ways of combating these effects. One way of combating cosmic rays is to surround the photodetector with an anticoincidence shield, often constructed of a plastic that scintillates and fluoresces when a cosmic ray passes through it. One or more photomultipliers detect cosmic ray signatures and veto any simultaneous photon events from the main detector. It's often less costly to build a coincidence detection system of two or three detectors to detect coincidence in the emission from the source being observed.

In a three-detector coincidence system, the light from the source passes through two beam splitters, dividing the light between the three photodetectors. The detection circuitry detects and counts coincidences between any two detectors and coincidences between all three detectors. Since the detectors are not colinear, it's unlikely that a cosmic ray could pass through all three. There is the potential of being triggered by light from the sky (cosmic ray Cerenkov radiation, strobes) that is not extraterrestrial. In the low rate ($Rt \ll 1$) limit, the probability of a false alarm in a coincidence system, given random photon arrival times, is simply $P(1, Rt/N)^N$, where N is the number of detectors. So, for our hypothetical telescope with triple detector and 100 ns

coincidence time, the false alarm probability per interval is 1.5×10^{-7} or about 1 per second. To adjust the false alarm rate, one could reduce the coincidence time, choose a fainter source, or increase the number of detectors. A 1-ns coincidence time would result in a false alarm rate of one every two hours. Such a system used in the optical is described in more detail by Wright et al. (2001). A similar IR system is described by Maire et al. (2016). The availability of arrays of photon-counting detectors such as APDs could allow the development of systems that use anticoincidence between different pixels in the detector array to reject any signal that impacts more than one location on the detector simultaneously.

11.3.8 THE MISSING BANDS: SPACE-BASED SETI

One of the biggest missing links in the electromagnetic spectrum for SETI are the bands to which the Earth's atmosphere is opaque. From the gamma rays to the near UV, no radiation penetrates the atmosphere. Similarly, in the IR, atmospheric molecules create broad absorption bands, through which no radiation penetrates. We can speculate that an ETI that wishes to communicate with other intelligences that have advanced to a spacefaring state might transmit on these frequencies.

While it is cost-prohibitive to place a telescope in space for the purpose of SETI, evidence of ETI might be hiding in the data of existing space telescopes. Many surveys in the UV and X-ray have sources that have not been identified with a known counterpart. For example, a combined analysis of the Extreme Ultraviolet Explorer (*EUVE*) extreme ultraviolet (EUV) and Röntgensatellit (*ROSAT*) soft X-ray surveys (Lampton et al. 1997) found 61 sources for which no obvious counterpart was identified. Some of these were later determined to be neutron stars, white dwarfs, or flare stars. It's unlikely that any are ETI sending X-ray messages, but we offer it as an example of the data sets that are available. Corbet (2016) provides a discussion of other types of X-ray SETI, using other existing X-ray facilities and data sets.

11.3.9 PHYSICAL ARTIFACT DETECTION AT INTERSTELLAR DISTANCE

Dyson (1960) suggested that an extraterrestrial civilization might build a giant structure to capture and utilize most or all of the visible emissions of a star, radiating only waste heat. A Dyson sphere would be detectable as a stellar-luminosity ~300K featureless blackbody source if the structure completely encases the star, or as a star with a significant IR excess in the case of a partial sphere. If civilizations capable of such feats of engineering were common, we could likely detect them in IR surveys. Searches of the IR sky sensitive to both complete (Carrigan 2009) and partial (Conroy and Werthimer 2003, unpublished[6]; Globus et al. 2003, unpublished[7]; Jugaku and Nishimura 2004). Dyson spheres have been conducted by using the Infrared Astronomical Satellite (*IRAS*) data set. While these have detected a few interesting candidates, none have been shown to indicate intelligent origin. Two high-sensitivity searches for Dyson spheres are ongoing, one searching for IR signatures of Dyson spheres in the *WISE* survey (Wright et al. 2014a,b; Griffith et al. 2015) and another examining *Kepler* light curves for unusual transit eclipses that could indicate a partial Dyson sphere (Marcy et al. 2014).

A civilization advanced enough to trap most of the available output of their host star ($\sim 4 \times 10^{33}$ ergs^{-1}) meets the definition of Kardashev Type II on the Kardashev (1964) scale of technological advancement of a civilization. A Type I civilization uses energy on the scale of current terrestrial civilization ($\sim 4 \times 10^{19}$ ergs^{-1}). Given the possibility of extinction, it seems likely that Kardashev Type II civilizations are much rarer than those capable of sending small spacecraft into orbit and that a continuum of possible engineering stages should exist. We might consider a civilization with significant ability to modify its planetary environment but not capable of building astronomical unit (AU) scale structures as Type I.5.

Carrigan (2012) recently summarized several other potential archaeological signatures for civilizations at various levels on the Kardashev scale, concluding that apart from electromagnetic communications, these are generally undetectable with current human technologies. Arnold (2005) suggests that an extraterrestrial civilization could deliberately create planetary-scale structures with unique shapes or periodicities that could be detected when the structure transits the star and could be distinguished from a planetary transit with current technology. Korpela et al. (2015) explored the possibility that even smaller structures could be detectable. They considered

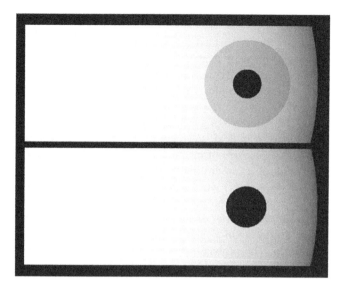

FIGURE 11.3.3 Top: simulated image of a planet surrounded with orbiting mirrors transiting in front of an M5 star. Bottom: same as above, but for an isolated planet large enough to produce a transit of the same depth.

the case where an ETI was using orbiting mirrors to illuminate the dark side of a tidally locked planet (see Figures 11.3.3 and 11.3.4). The differences in light curves between the case of a larger planet versus a smaller planet with orbiting mirrors could not be detected with existing instrument but would be detectible in some cases with the *James Webb Space Telescope* (*JWST*).

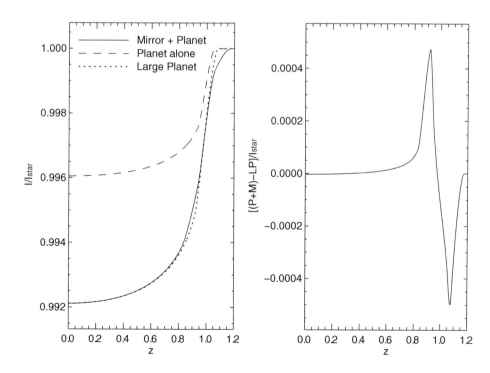

FIGURE 11.3.4 Left: transit light curves for a planet orbiting an M5 star (dashed line: *P*), the same planet surrounded by orbiting mirrors extending to three planetary radii (solid line: *P* + *M*), and a planet without mirrors that is large enough to produce a light curve with the same depth of transit (dotted line: large planet [LP]). Right: difference between transit light curve with mirrors (*P* + *M*) and the one for the solitary LP, relative to the stellar intensity.

11.3.10 THE FERMI PARADOX: A FINAL WORD?

The lack of any confirmed defections of ETI by SETI experiments is often described as a failure of SETI to produce results, rather than success in reducing the parameter space in which an ETI could remain hidden. Often this is mentioned in relation to the so-called "Fermi paradox," which is neither a paradox nor a concept that originated with Fermi (Gray 2016). A typical statement of this "paradox" is: Since a civilization capable of interstellar travel could colonize the galaxy within a few million years and since the galaxy is 13 billion years old, why is there no evidence of a galaxy-wide civilization? The possible reasons for this have been discussed widely. Any of the Drake parameters could lead to $N \gg 1$, resulting in a galaxy where Earth is the first or only civilization. The probability of tool-using intelligence arising could be low. L could be small, resulting in civilizations that don't last long enough for interstellar colonization. The end result of technological evolution could be machine intelligence that is not driven by biological desires to reproduce, expand, colonize, or even communicate. In the end, this is all speculation. The definition of success in science is not "Did I find what I was looking for?" but merely "Do I understand more than I did yesterday?"

ENDNOTES

1 Earlier attempts to detect extraterrestrial transmissions originating within the solar system did occur; however, they were performed on radio frequencies that do not penetrate the Earth's ionosphere. [NewYorkTimes 1920]
2 This should be distinguished from the standard radio astronomy definition of gain, $G = \frac{A_e}{2k}$, usually given in units of K/Jy to provide a frequency-independent gain in units more appropriate for radio astronomers.
3 Mpc = 10^6 parsec (pc) = 3.1×10^{22} m.
4 The Nyquist sampling rate is equal to twice the sampled bandwidth for real valued data and equal to the bandwidth for complex valued data.
5 A spectrometer will also have a non-unity efficiency, typically of order 10% for an Echelle spectrograph, which will further reduce both the background and source rates.
6 From http://seti.berkeley.edu/IR Excess Search retrieved 9 July, 2015.
7 Mentioned in Carrigan (2009).

BIBLIOGRAPHY

Azua-Bustos, A., Dyson, G., Korpela, E. J. et al. 2015. *Regarding Messaging to Extraterrestrial Intelligence (METI)/Active Searches for Extraterrestrial Intelligence* http://setiathome.berkeley.edu/meti statement final.html.

Abbott, B. P., Abbott, R., Abbott, T. D. et al. 2016. Observation of gravitational waves from a binary black hole merger. *Phys. Rev. Lett.*, 116(6):061102. doi:10.1103/PhysRevLett.116.061102.

Abbott, B. P., Abbott, R., Abbott, T. D., et al. 2017. GW170817: Observation of gravitational waves from a binary neutron star inspiral. *Phys. Rev. Lett.*, 119(16):161101. doi:10.1103/PhysRevLett.119.161101.

Abe, K., Abgrall, N., Ajima, Y., et al. 2011. Indication of electron neutrino appearance from an accelerator-produced off-axis muon neutrino beam. *Phys. Rev. Lett.*, 107:041801. doi:10.1103/PhysRevLett.107.041801.

Arnold, L. F. A. 2005. Transit light-curve signatures of artificial objects. *Astrophys. J.*, 627:534–539. doi:10.1086/430437.

Bowyer, S., Lampton, M., Korpela, E., et al. 2016. The SERENDIP III 70 cm search for extraterrestrial intelligence. *ArXiv e-prints* arXiv:1607.00440.

Carrigan, R. A. 2012. Is interstellar archeology possible? *Acta Astronaut.*, 78:121–126. doi:10.1016/j.actaastro.2011.12.002.

Carrigan, Jr., R. A. 2009. IRAS-based whole-sky upper limit on Dyson spheres. *Astrophys. J.*, 698:2075–2086. doi:10.1088/0004-637X/698/2/2075.

Cocconi, G. and Morrison, P. 1959. Searching for interstellar communications. *Nature*, 184:844–846. doi:10.1038/184844a0.

Corbet, R. H. D. 2016. SETI at X-ray energies–Parasitic searches from astrophysical observations. *ArXiv e-prints* arXiv:1609.00330.

Cruddace, R., Paresce, F., Bowyer, S., and Lampton, M. 1974. On the opacity of the interstellar medium to ultrasoft X-rays and extreme-ultraviolet radiation. *Astrophys. J.*, 187:497–504. doi:10.1086/152659.

Drake, F. D. 1961. Project Ozma. *Phys. Today*, 14:40–46. doi:10.1063/1.3057500.

Dyson, F. J. 1960. Search for artificial stellar sources of infrared radiation. *Science*, 131:1667–1668. doi:10.1126/science.131.3414.1667.

Ehman, J. R. 2011. "Wow!" – A Tantalizing Candidate. In Shuch, H.P.(Ed.),*Searching for Extraterrestrial Intelligence: SETI Past, Present, and Future (The Frontiers Collection)*, pp. 47. Springer, Heidelberg, Germany. doi:10.1007/978-3-642-13196-7_4.

Enriquez, J. E., Siemion, A., Foster, G. et al. 2017. The breakthrough listen search for intelligent life: 1.1–1.9 GHz observations of 692 nearby stars. *Astrophys. J.*, 849:104. doi:10.3847/1538-4357/aa8d1b.

Gray, R. 2016. The so-called Fermi paradox is misleading, flawed, and harmful. *Astrobiology*, 16:741–743. doi:10.1089/ast.2016.0823.rcm.

Griffith, R. L., Wright, J. T., Maldonado, J., Povich, M. S., Sigurdsson, S., and Mullan, B. 2015. The G infrared search for extraterrestrial civilizations with large energy supplies. III. The reddest extended sources in WISE. *Astrophys. J. Suppl.*, 217:25. doi:10.1088/0067-0049/217/2/25.

Harp, G. R. 2013. Using multiple beams to distinguish radio frequency interference from SETI signals. *ArXiv e-prints* arXiv:1309.3826.

Isaacson, H., Siemion, A. P. V., Marcy, G. W., et al. 2017. The breakthrough listen search for intelligent life: Target selection of nearby stars and galaxies. *Publ. Astron. Soc. Pac.*, 129(5):054501. doi:10.1088/1538-3873/aa5800.

Jewitt, D., Luu, J., Rajagopal, J., et al. 2017. Interstellar interloper 1I/2017 U1: Observations from the NOT and WIYN telescopes. *Astrophys. J.*, 850:L36. doi:10.3847/2041-8213/aa9b2f.

Jugaku, J. and Nishimura, S. 2004. A search for Dyson spheres around late-type stars in the solar neighborhood. In Norris, R. and Stootman, F. (Eds.), *Bioastronomy 2002: Life Among the Stars*, volume 213 of *IAU Symposium*, pp. 437. ads:2004IAUS..213..437J.

Kardashev, N. S. 1964. Transmission of information by extraterrestrial civilizations. *Soviet Astron.*, 8:217 http://adsabs.harvard.edu/abs/1964SvA.....8..217K.

Korpela, E., Werthimer, D., Anderson, D., Cobb, J., and Leboisky, M. 2001. SETI@home-massively distributed computing for SETI. *Comput. Sci. Eng.*, 3(1):78–83. doi:10.1109/5992.895191.

Korpela, E. J. 2004. Statistics of one: What Earth can and can't tell us about life in the universe. In *Bioastronomy 2004: Habitable Worlds*, volume 4 of *Astrobiology*, pp. 266. viXra:1108.0003.

Korpela, E. J. 2011. Distributed processing of SETI data. In Shuch, H. P., Editor, *Searching for Extraterrestrial Intelligence: SETI Past, Present, and Future (The Frontiers Collection)*, pp. 183–200. Springer, Heidelberg, Germany. doi:10.1007/978-3-642-13196-7_11.

Korpela, E. J. 2012. SETI@home, BOINC and volunteer distributed computing. *Ann. Rev. Earth Planet. Sci.*, 40(1):69–87. doi:10.1146/annurev-earth-040809-152348.

Korpela, E. J., Anderson, D. P., Bankay, R., et al. 2011a. Status of the UC-Berkeley SETI efforts. In Hoover, R. B., Davies, P. C. W., Levin, G. V., and Rozanov, A. Y. (Eds.), *Instruments, Methods, and Missions for Astrobiology XIV*, volume 8152 of *Proc. SPIE*, pp. 815212 (8 p) doi:10.1117/12.894066.

Korpela, E. J., Cobb, J., Lebofsky, M. et al. 2011b. Candidate identification and interference removal in SETI@home. In Vakoch, D. A., Editor, *Communication with Extraterrestrial Intelligence (CETI)*, pp. 37–44. SUNY Press, Albany, NY, arXiv:1109.1595.

Korpela, E. J., Sallmen, S. M., and Greene, D. L. 2015. Indications of technology in planetary transit light curves–Dark-side illumination. *Astrophys. J.*, 809:139. doi:10.1088/0004-637X/809/2/139.

Lacki, B. C. 2015. SETI at planck energy: When particle physicists become cosmic engineers. *ArXiv e-prints* arXiv:1503.01509.

Lampton, M., Lieu, R., Schmitt, J. H. M. M. et al. 1997. An all-sky catalog of faint extreme ultraviolet sources. *Astrophys. J. Suppl.*, 108:545–557. doi:10.1086/312965.

Maire, J., Wright, S. A., Dorval, P., et al. 2016. A near-infrared SETI experiment: commissioning, data analysis, and performance results. In *Ground-based and Airborne Instrumentation for Astronomy VI*, volume 9908 of *Proc. SPIE*, pp. 990810. doi:10.1117/12.2232861.

Marcy, G. W., Isaacson, H., Howard, A. W. et al. 2014. Masses, radii, and orbits of small *Kepler* planets: The transition from gaseous to rocky planets. *Astrophys. J. Suppl.*, 210:20. doi:10.1088/0067-0049/210/2/20.

Morrison, P., Billingham, J., and Wolfe, J. 1977. *The Search for Extraterrestrial Intelligence, SETI*. NASA Special Publication, 419 ads:1977NASSP.419.....M. Washington, DC.

Sagan, C., Drake, F. D., Druyan, A., Ferris, T., Lomberg, J., and Salzman Sagan, L. 1978. *Murmurs of Earth: The Voyager Interstellar Record*. Random House, NY. ads:1978mevi.book.....S.

Sagan, C., Salzman Sagan, L., and Drake, F. 1972. A message from Earth. *Science*, 175:881–884. doi:10.1126/science.175.4024.881.

Siemion, A. P. V., Benford, J., Cheng-Jin, J., et al. 2014. Searching for extraterrestrial intelligence with the square kilometer array. In *Astronomy and Astrophsics with the Square Kilometer Array*, volume 116 of *Proc. Sci.* arXiv:1412.4867.

Siemion, A. P. V., Demorest, P., Korpela, E., et al. 2013. A 1.1 to 1.9 GHz SETI survey of the *Kepler* field: I. A search for narrow-band emission from select targets. *Astrophys. J.*, 767:94. doi:10.1088/0004-637X/767/1/94.

Silagadze, Z. K. 2008. SETI and muon collider. *Acta. Phys. Pol. B*, 39:2943. arXiv:0803.0409.

Stolz, C. J. 2012. The National Ignition Facility: The path to a carbon-free energy future. *Philos. Trans. R. Soc. London, Ser. A*, 370:4115–4129. doi:10.1098/rsta.2011.0260.

Tarter, J. C., Backus, P. R., Mancinelli, R. L. et al. 2007. A reappraisal of the habitability of planets around M dwarf stars. *Astrobiology*, 7:30–65. doi:10.1089/ast.2006.0124.

Tellis, N. K. and Marcy, G. W. 2017. A search for laser emission with megawatt thresholds from 5600 FGKM stars. *Astron. J.*, 153:251. doi:10.3847/1538-3881/aa6d12.

Thompson, A. R., Emerson, D. T., and Schwab, F. R. 2007. Convenient formulas for quantization efficiency. *Radio Sci.*, 42:RS3022. doi:10.1029/2006RS003585.

Turnbull, M. C. and Tarter, J. C. 2003a. Target selection for SETI. I. A catalog of nearby habitable stellar systems. *Astrophys. J. Suppl.*, 145:181–198. doi:10.1086/345779.

Turnbull, M. C. and Tarter, J. C. 2003b. Target selection for SETI. II. *Tycho* 2 dwarfs, old open clusters, and the nearest 100 stars. *Astrophys. J. Suppl.*, 149:423–436. doi:10.1086/379320.

Vleck, J. H. V. and Middleton, D. 1966. The spectrum of clipped noise. *Proc. IEEE*, 54(1):2–19. doi:10.1109/PROC.1966.4567.

Von Korff, J., Demorest, P., Heien, E., et al. 2013. Astropulse: A search for microsecond transient radio sgnals using distributed computing. I. Methodology. *Astrophys. J.*, 767:40. doi:10.1088/0004-637X/767/1/40.

Wright, J. T., Griffith, R. L., Sigurdsson, S., Povich, M. S., and Mullan, B. 2014a. The Ĝ infrared search for extraterrestrial civilizations with large energy supplies. II. Framework, Strategy, and First Result. *Astrophys. J.*, 792:27. doi:10.1088/0004-637X/792/1/27.

Wright, J. T., Mullan, B., Sigurdsson, S., and Povich, M. S. 2014b. The G infrared search for extraterrestrial civilizations with large energy supplies. I. Background and justification. *Astrophys. J.*, 792:26. doi:10.1088/0004-637X/792/1/26.

Wright, S. A., Drake, F., Stone, R. P., Treffers, D., and Werthimer, D. 2001. Improved optical SETI detector. In Kingsley, S. A. and Bhathal, R., Editors, *The Search for Extraterrestrial Intelligence (SETI) in the Optical Spectrum III*, volume 4273 of *Proc. SPIE*, pp. 173–177. doi:10.1117/12.435376.

11.4 Humanistic Implications of Discovering Life Beyond Earth

Steven J. Dick

CONTENTS

11.4.1 INTRODUCTION

The remarkable discoveries in astrobiology in the last few decades, including extremophile life on Earth, ocean worlds such as Europa and Enceladus, and thousands of planets beyond our solar system, as well as ongoing programs searching for extraterrestrial intelligence, all raise an urgent question: what is the impact of discovering life beyond Earth on humanity? In this chapter, we discuss possible approaches to answering this question, as well as critical issues and potential impacts under a variety of discovery scenarios. The humanistic implications of astrobiology fall under the broad heading of astrobiology and society, a suite of issues raised already in NASA's first Astrobiology Roadmap (NASA 1998), elaborated in subsequent versions (NASA 2003; Des Marais et al. 2008), still present in its current Astrobiology Strategy document (Hays 2015), and recognized as well in other international documents that guide research in astrobiology (Horneck et al. 2006). In a more general sense, these humanistic questions are a subset of an important suite of issues in science and society. In the same way that social, ethical, and legal aspects are being studied for frontier areas of science and technology such as the Human Genome Project, nanotechnology, and artificial intelligence, the impact of the discovery of life beyond Earth deserves serious and systematic attention. Indeed, the World Economic Forum (2013) has declared the discovery of life beyond Earth as one of five X factors—emerging concerns for planet Earth of possible future importance, but with unknown consequences. This research must be a combined effort encompassing the sciences, social sciences, and humanities and should become an integrated part of the astrobiological endeavor.

11.4.2 APPROACHES

At first thought, the problem of determining astrobiological impacts would seem to be intractable, and in one important sense, it is. We certainly cannot predict what the impacts will be, any more than we can predict with certainty other outcomes that involve human behavior. But this is no excuse: as in other areas such as economics, technology, and science, what we *can* do is provide guidelines for impact under certain scenarios by using systematic methods. In this section, we look at three approaches: history, discovery, and analogy, with the idea that these might be used to inform impact scenarios.

11.4.2.1 HISTORY

Numerous times in the last four centuries of telescopic astronomy, Earthlings believed that extraterrestrial life had been detected. Galileo's 1610 landmark telescopic lunar observations had barely been published when Kepler enthusiastically conjectured that one particularly circular crater must be an artificial construction of lunar inhabitants. Two centuries later, the famous "Moon Hoax" of 1835 placed lunarians on the Moon, supposedly based on the latest telescopic observations of John Herschel. Sixty years further on, even as H. G. Wells was penning his *War of the Worlds*, the astronomer Percival Lowell argued that there were canals on Mars, built by a dying civilization managing its water resources. Though largely discredited by his death in 1916, the idea had not dimmed so much that when Orson Welles broadcast his radio version of *War of the Worlds* on Halloween eve, 1938, a considerable reaction ensued as some Americans believed that a real Martian invasion was underway. Thirty years later, when strange pulses were detected from the heavens (soon dubbed "pulsars"), astronomers briefly but seriously considered the "Little Green Men" hypothesis. Finally, in 1996, NASA announced that it had evidence of fossil life from Mars in the form of the now famous and infamous Mars rock dubbed ALH84001.

Each one of these episodes holds lessons (Dick 2018a), but perhaps, most relevant is the last and most recent: the reaction to the claimed discovery of nanofossils in the Mars rock ALH84001, discovered in the Antarctic in 1984. That episode reveals the public, media, and scientific reaction to even the most minimal claimed discovery: not a microbe or complex life or intelligence, but nanofossils so small that they could only be seen by a scanning electron microscope. From the time the rock was recognized as Martian in origin, almost 3 years of intense study had taken place before NASA scientists dared to make the announcement about potential fossil life in the summer of 1996 (McKay et al. 1996; Dick and Strick 2004). For fear of leaks, during that time, scientists at Johnson Space Center, where the work was being done, were circumspect about discussing their results even within NASA. By April 1996, they had enough confidence to submit an article to the prestigious journal *Science*, at which time NASA Headquarters began to play a role. The claim was closely held at the Headquarters' level until mid-July, when peer review was complete and publication was scheduled for mid-August. On July 30, the NASA Administrator informed the White House, and NASA began receiving news inquiries on August 1. Although NASA had planned for the announcement to coincide with the publication of the *Science* article in mid-August, these news inquiries forced NASA to move the announcement forward to August 7. The timing of the announcement was also influenced by political events, since the Republican National Convention was scheduled to take place at about the time the announcement was originally scheduled.

In these days of social media, it is highly unlikely that such a discovery could be kept secret as long as the Mars rock claims were kept under wraps. And it is certain that political calculations will be factored into any decision to announce, or not announce. Already at the press conference, the claims were subject to skepticism from scientists. Within 5 weeks, Congressional hearings were held on life on Mars; within 4 months, symposia on the implications were being held, including a high-level meeting involving the vice president of the United States, the National Academy of Sciences, the NASA Administrator, and scholars from a wide array of disciplines. On the scale of months and years, the more technical and scholarly examination of the evidence and its implications was carried out. The consensus today is that the putative nanofossils are not biogenic in origin, though some still hold to the original claim.

In short, the ALH84001 episode constitutes a real-life scenario of what might happen under even a minimal discovery scenario of fossilized life: reluctance about confirmation and announcement, attempts at secrecy whether because of uncertainty or competition, a media frenzy that would now be greatly accelerated by social media, scientific excitement and skepticism played out in the media, conferences and journals, the involvement of the government and high-level scientific institutions, and a general public clamoring for answers to one of the great questions of science—one with profound implications for humanity at the individual and global levels. While it is true the Mars rock announcement and subsequent events were relevant to a specific time, culture, and set of circumstances, in broad outline, these events are likely to be mimicked by any discovery of extraterrestrial life. The reaction of government institutions, the media, and the public will occur side by side with the reaction among scientists, who will subject the discovery to their exacting standards. Moreover, these events will likely take place over an extended period of time, a characteristic also reinforced by the nature of scientific discovery discussed in the next section. The point is that we can learn lessons from history: the study of this and other episodes of reaction to putative discoveries can inform what we might expect in the future, even if in the end it must be in the form of guidelines, not predictions.

11.4.2.2 Discovery

The concept of discovery is important for at least two reasons when discussing the impact of astrobiology. First, the very nature of scientific discovery as a process provides insights into how the discovery of life beyond Earth and its consequences will play out; and second, the societal impact will very much depend on the discovery scenario. We therefore need to consider the nature of discovery and parse under what circumstances the discoveries may take place. The general question "What is the societal impact of discovering extraterrestrial life?" makes no sense unless we are talking about a specific discovery scenario and understand the nature of scientific discovery.

Contrary to common expectations, history shows that scientific discovery is far from the Eureka moment it is often thought to be. Rather, discovery is an extended process involving the efforts of many individuals across multiple stages of detection and interpretation, before true understanding finally dawns, perhaps years or decades later. "Discovery is a process and must take time," scientist-philosopher Thomas Kuhn wrote already in 1962 in his famous *Structure of Scientific Revolutions*. In the half-century since Kuhn's pronouncement (Kuhn 1962a, 1962b), scholars have repeatedly documented the truth of discovery's extended and complex nature. It has been subjected to historical, psychological, and sociological analyses, and philosophers have parsed its conceptual meanings as well (Dick 2013). They have also insisted that discovery is not only a technical but also a social process, involving real people with real emotions that have real effects on the final results—and their impacts on society.

The extended nature of discovery is evident, whether looking at the discovery of elementary particles such as the Higgs boson in physics, the elements in chemistry, or new classes of objects in astronomy, in short, in any area of science. To take only one example, in astronomy, Galileo is commonly credited with discovering the rings of Saturn. But when Galileo turned his telescope to Saturn on July 30, 1610, he was totally perplexed, thinking he had discovered some strange sort of satellite (van Helden 1974). It was only in March 1655 that Christiaan Huygens turned his large telescope toward Saturn and interpreted what Galileo had seen as handles or "anses" (we would say "rings") that changed appearance depending on the angle of observation from Earth. Lacking knowledge of dynamics, Huygens in turn had no idea how such rings could physically exist and persist around Saturn. That understanding came only in the mid-nineteenth century, when James Clerk Maxwell provided a basic dynamical explanation. It would be most accurate to say that the full discovery of Saturn's rings took centuries and that new discoveries about the rings continue to be made with spacecraft such as Cassini. The same can be said of comets, exoplanets, supernovae, spiral galaxies, or any of the other 82 classes of astronomical objects that have been identified (Dick 2013). Extended discovery applies as well to phenomena such as the expansion of the universe, the discovery of radio waves, and the three-degree cosmic background radiation. The discovery of extraterrestrial life will be no exception.

Moreover, the history of discovery indicates that there is very often a pre-discovery phase, during which the true nature of an object, signal, or phenomenon goes unrecognized or unreported or during which theory only indicates that the phenomenon *should* exist. This is true not only for pre-telescopic astronomy and pre-microscopic biology, but also after these technologies have reached advanced stages. We are perhaps not surprised that Galileo did not recognize his observations of the apparent protuberances of Saturn as rings or that the intriguing nebulous objects in the sky were not known to be external galaxies until the twentieth century. But the same is true today, as objects are not immediately recognized for what they truly are, or are theorized long before they are found (think black holes). The phenomenon of life beyond Earth is still in a pre-discovery phase. Whether anything has been observed that turns out to be an indication of life, or life itself, remains to be seen. The pre-discovery phase is bookended on the other side of discovery by a post-discovery phase, including short-term reaction and long-term impact. As seen in the Mars rock episode described previously, in this final phase, the media will play a crucial role, with all the complex interactions with scientists, government, and the general public.

The discovery of extraterrestrial life, whether microbial in our solar system or intelligent in our galaxy, will therefore be an extended process, with the general structure displayed in Figure 11.4.1 As in other astronomical discoveries, discovery will likely be preceded by a pre-discovery phase, in which the microbe, biosignature, or electromagnetic signal is at first

The Anatomy of Discovery: An Extended Process

Discovery

Detection	Interpretation	Understanding

Technological, Conceptual and Social Roles at Each Stage

Pre-Discovery	Post-Discovery
• Theory	• Issues of credit & reward
• Casual or Accidental observations	• How do discoveries end?
• Classification of Phenomena (Harvard spectral types)	• Classification of "The Thing Itself" (MK spectral types)

FIGURE 11.4.1 The extended structure of discovery, showing its three stages of detection, interpretation, and basic understanding, as well as pre-discovery and post-discovery phases. Discovery also has a microstructure consisting of conceptual, technical, and social roles.

not recognized as extraterrestrial life. The discovery will then go through phases of detection and interpretation before understanding is achieved. These phases could be long or short, depending on the evidence, and might result in success or failure. How society reacts to the discovery will be affected by its extended nature, as evidence is presented, interpreted, and understood. This basic structure of discovery is essential to any understanding of its impact.

Just as essential is the insight that the reaction to the discovery of alien life in the post-discovery phase will be very much scenario-dependent. Table 11.4.1 summarizes possible discovery scenarios, showing direct and indirect encounters in terrestrial and extraterrestrial environments. In other words, we could encounter life *directly* here on Earth or in space, and we could find life *indirectly* here on Earth or in space. Indirect

TABLE 11.4.1

Discovery Scenarios for Contact with Extraterrestrial Life

	Terrestrial	Extraterrestrial
Direct	**Encounter type 1**	**Encounter type 2**
	Accidental contamination by sample or astronaut return	Human space exploration
	Panspermia or interplanetary matter transfer (ALH 84001)	
	Alien space exploration	
	UFOs	
Indirect	**Encounter type 4**	**Encounter type 3**
	Shadow alien biosphere	Robotic space exploration (Viking, Europa probes, etc)
	Unknown alien microbes	Biosignatures
	Earth orbit or vicinity	SETI—radio and Electromagnetic (EM) Spectrum
	Artifact on Earth or vicinity	Artifact in space

extraterrestrial contact—labeled encounter type 3 in Table 11.4.1—is where most of the action is found today, in the form of spacecraft observations, the detection of biosignatures on exoplanets, and search for extraterrestrial intelligence (SETI) programs. Direct encounters of types 1 (terrestrial) and 2 (extraterrestrial) are the most unlikely to occur in the near future, because they imply alien spaceflight to Earth or human spaceflight beyond the solar system. Unidentified flying objects (UFOs) as extraterrestrial spaceships remain a logical possibility, dependent on evidence, which is not yet very strong. A type 3 indirect encounter is the most likely, or at least the most anticipated, scenario, since remote detection of microbes beyond Earth by robotic space exploration is one of the main goals of NASA's astrobiology program, the search for biosignatures is an active research goal, and several SETI programs are in progress. A type 4 encounter raises interesting possibilities not often discussed, including the discovery of an alien artifact on Earth or its vicinity. In its static form, Table 11.4.1 lays out scenarios, but in its dynamic form, proceeding clockwise from encounter types 1 to 4, it arguably depicts impacts from their strongest to their weakest.

Each of these scenarios have triggered a large number of science fiction novels, beginning substantially with H. G. Wells' *War of the Worlds* more than a century ago, accelerating in the second half of the twentieth century with the rise of the Space Age and continuing unabated today, in prototype novels and movies such as Arthur C. Clarke's *2001: A Space Odyssey*, Carl Sagan's *Contact*, and Stanislaw Lem's *Solaris*. The best of science fiction literature lays out thoughtful scenarios about discovery and possible impacts, and this should not be forgotten as a source of ideas. In the more sober analysis of impacts, not only science but also numerous disciplines beyond science must become involved. Much more could be said about the nature of discovery and discovery scenarios for extraterrestrial life. Suffice to say here that it will be a large and daunting research project. But the essential point is that the reaction will be extended and will very much depend on the scenario.

11.4.2.3 ANALOGY

Analogy represents a third approach to studying the impact of discovering life beyond Earth. Far from its reputation as a "fuzzy way of thinking," many scholars now see analogy as an essential and omnipresent mode of reasoning (Hesse 1966; Hofstadter 2001; Bartha 2010; Hofstadter and Sander 2013). Analogy has been defined as "a structural or functional similarity between two domains of knowledge," more precisely "a mapping of knowledge from one domain (the base) into another (the target) such that a system of relations that holds among the base objects also holds among the target objects" (Gentner and Jeziorski 1993). At its best, analogy is a cognitive mechanism that allows us to learn and solve problems by drawing a map from the known to the unknown. In a broader sense, analogical reasoning is helpful in discovering, developing, explaining, or evaluating scientific theories. As philosopher Vidal (2014) summarizes its potential, "such reasoning enables us to propose new hypotheses, and thus discover new

phenomena. These new hypotheses trigger us to develop new experiments and theories." All researchers on analogy caution that analogy is not a proof but a heuristic, a method that can serve and has served as an aid to learning and discovery. They also caution that the relation between any two systems needs to be examined in detail before an analogy or disanalogy can be tested. Scholars have also offered principles of analogical reasoning meant to avoid its many pitfalls; these principles include the structural consistency of what is being compared, the depth of any relation, avoidance of mixed analogies, and the realization that analogy is not causation (Gentner and Jeziorski 1993).

Four analogies, corresponding to different discovery scenarios and timescales, serve to illustrate the usefulness of this approach: the microbe analogy, the culture contact analogy, the transmission/translation analogy, and the worldview analogy. Because the discovery of microbes beyond Earth in any form potentially represents a revolution in biology, analogical candidates might include revolutionary ideas such as Darwinian evolution by natural selection, the role of DNA in genetics, and the discovery of extremophile microorganisms. But perhaps, no biological discovery closer approximates the potential fallout than the discovery of terrestrial microbes themselves, representing a new world of life on Earth, the microcosmos (Margulis and Sagan 1986). While Robert Hooke's and Antony van Leeuwenhoek's discovery of microbes in the second half of the seventeenth century excited considerable interest among both scientists and the public, they led nowhere during the lifetime of Hooke or Leeuwenhoek and for several generations thereafter (Gest 2004). Once microscopes had been considerably improved, however, the impact was swift and profound, spinning off the field of bacteriology. Today, the discovery of the microcosmos has greatly enlarged our view of life, carrying what had been a macro-science into the domain of the very small. Three centuries after Hooke and Leeuwenhoek, microbes once again radically changed biologist's view of life with the discovery of extremophiles, microorganisms functioning under conditions far beyond what had been considered the limits of life on Earth. These discoveries in turn led to something even more surprising—a complete reclassification of life (Sapp 2009). Surely, the discovery of life beyond Earth would do the same. A detailed look at this analogy and others may provide a more robust guide to the potential reaction to the discovery of microbes beyond Earth, taking into account the many cultural differences that now exist.

A very different scenario—the discovery of extraterrestrial intelligence—raises a very different set of analogies involving culture contacts. Again, one must distinguish between direct and indirect contacts. History abounds with direct contacts between cultures, usually with unhappy effects. The Western mind naturally goes to analogies such as the disastrous European contact with the Aztecs and Incas as part of the sixteenth-century Age of Discovery (Elliott 2006). But outside the tradition of Western expansionism, the less well-known great voyages of the fifteenth-century Ming China treasure fleets 50 years before Columbus, led by Admiral Zheng He, offer a very different scenario.

In contrast to the case of Cortes and the Aztecs, Chinese culture contacts resulted in new embassies in the Near East, deification (to this day) of Zheng He and his cult in South-East Asia, and the publication of works enlarging Chinese knowledge of the oceans and landmasses visited (Dreyer 2007). The point is that there are other models of culture contact than the destructive ones usually cited, including Jesuit models of culture contact that were not disastrous. Hoerder (2002) documents hundreds of culture contacts over the last thousand years; the sixteenth-century European exploits cover only a few of the book's 800 pages. Moreover, the contacts were arguably an equal mix of good, bad, and indifferent. Today, culture contacts under the name of globalism are also such a mix, as were the culture contacts in the first era of globalization dating from the fifteenth to the twelfth century BC in the Aegean, Egypt, and Near Eastern civilizations (Cline 2004). While it is understandable for Westerners to focus on our own recent history from the Age of Discovery, to use the Aztecs and Incas as the sole or chief analogies for contact with extraterrestrial intelligence is to ignore the bulk of history. While we cannot draw definitive conclusions even based on a small sample of culture contacts, an analysis by historians and anthropologists of characteristics common to all culture contacts is a respectable research program that could pay significant dividends. Beyond physical culture clashes, these contacts illustrate the difficulties among cultures of communication and even grasping certain concepts such as the "soul" and "immortality" (Kuznicki 2011).

A third analogy for the discovery of life beyond Earth is the "decipherment/translation" analogy (Dick 1995; Finney and Bentley 2014). If, as is widely assumed, first contact with intelligence beyond Earth turns out to be indirect, in the form of an electromagnetic signal that is deciphered with significant information transmitted, the flow of information between civilizations across time is a more appropriate analog. Particularly compelling is the transmission of Greek knowledge to the Latin West by way of the Arabs in the twelfth and thirteenth centuries, only one example of what Arnold Toynbee (1957) has called in the terrestrial context "encounters between civilizations in time." Others have pointed out that a better analogy might be decipherment of the Mayan glyphs, since we will not know the alien language (Finney and Bentley 2014). The two analogies might profitably be combined into a more robust decipherment/translation analogy, since both are likely to be phases in the same process.

Even if an extraterrestrial message is not deciphered, and perhaps even in the case that "only" microbial life is discovered constituting a second genesis, a change in worldview would likely gradually take place. In the long term, the discovery of microbial or intelligent life beyond Earth might be analogous to grand changes in scientific worldviews, exemplified in the Copernican worldview originated in the sixteenth century, the Darwinian worldview of the nineteenth century, or the Shapley-Hubble worldview of the galactic universe in the twentieth century. The gradual construction of worldviews, and their influence on our thinking, is a deep philosophical problem requiring more research (Vidal 2007, 2014). The impact of scientific worldviews has been studied extensively (Smith 1982; Blumenberg 1987; Bowler 1989), and the rich literature on worldviews cannot help but illuminate the problem of the impact of discovering extraterrestrial life.

Despite the obvious need for caution and the lack of predictive value, analogies can indeed serve as solid guidelines to cosmic encounters with alien life. In sorting good analogies from bad, the guiding principle should be what I call the "Goldilocks principle of Analogy": *Analogy must not be so general as to be meaningless, nor so specific as to be misleading. The middle "Goldilocks" ground is where analogies may serve as useful guideposts.* On the one hand, it does little good to argue that both science and religion are searching for our place in the universe, when one addresses the natural world and the other invokes the supernatural—differences so great as to swamp any comparison whatsoever. And it does little good to argue that because there is life on Earth, there is life out there—the very thing we seek to demonstrate. On the other hand, it is hopelessly naïve to expect that contact with extraterrestrial intelligence will change our worldviews in ways precisely mirroring past discoveries, culture contacts, decipherment efforts, or revolutions in thought, leading us to reiterate that under no circumstances will analogy predict the future.

11.4.3 CRITICAL ISSUES

In order to address seriously the impact of astrobiology, we must not assume that life beyond Earth will be similar to ours, whether microbial or intelligent. Nor when it comes to SETI strategy can we assume that human knowledge is universal, even in the realm of mathematics and science. Envisioning impact requires us to act locally but think universally. In this section, we address some of these issues to indicate how important it is to take these problems into account if we are going to discover life and deal with its impact in a realistic way.

11.4.3.1 Transcending Anthropocentrism

Transcending anthropocentrism is easily recognized as an important issue. But how do we move beyond our own preconceptions and "get out of our heads" to study concepts in astrobiology that are perhaps entirely divorced from our everyday experience on Earth? The question is far from academic and indeed becomes very practical when searches for extraterrestrial life and intelligence require a specific idea of the object of the search. The 100-million-dollar Viking Mars lander biology package could not have been built without some specific strategy and technology to search for a preconceived type of life (in this case life that metabolizes). Search for extraterrestrial intelligence (SETI) and messaging extraterrestrial intelligence (METI) programs assume a certain mode of intelligence that involves technology, usually technology that will reveal itself in the electromagnetic spectrum. High technology in turn assumes a culture and civilization sophisticated enough to produce it, which assumes in turn that aliens have a social organization in some way similar to ours.

And communication with such intelligence makes assumptions about technology, language, metaphysics, universals, and the nature of knowledge. Our conception of life and intelligence, culture and civilization, and technology and communication are exemplars of foundational concepts that need to be critically examined in an extraterrestrial context.

Among the most basic concepts is life itself. Attempts to define life have a long and checkered history, with some concluding that such a definition is impossible in the absence of a general theory of living systems (Cleland and Chyba 2007) and others arguing that this is definitional pessimism (Smith 2016). These conclusions are important but theoretical. A more practical way forward is to begin with life on Earth and speculate how life beyond Earth might differ in specific respects, especially in its biochemical scaffolding, genetics, solvents, and metabolism. One widely accepted definition of life is a "self-sustaining chemical system capable of Darwinian evolution," where Darwinian evolution at the molecular level refers to descent with modification by natural selection in a replicating genetic system such as RNA or DNA on Earth (Joyce 1994). The "scaffolding" for this replication system on Earth is based on carbon, hydrogen, oxygen, and nitrogen (CHON), with carbon as the anchor and water as the solvent. But what if the environmentally available biochemical scaffolding is silicon or ammonia? What if the genetic system involves proteins instead of DNA? And what if the solvent is nitrogen or methane? All of these possibilities have been considered, with conclusions that are still controversial (Bains 2004; Ward and Benner 2007).

Moreover, with the decades-ago discovery of microbes thriving under extreme conditions of temperature, pressure, acidity, radioactivity, and other parameters, our conception of the environment in which life might emerge has significantly broadened. New views of the diversity of metabolism, such as chemosynthetic ecosystems like hydrothermal vents deep in the ocean, have broadened our conceptions of life even more.

These questions regarding the origin and evolution of life are multiplied when it comes to the evolution and nature of intelligence. Our usual concept of intelligence in the context of SETI is dominated by the Drake paradigm, whose central icon is the Drake equation, including the crucial term representing the fraction of planets on which intelligence arises (Vakoch and Dowd 2015). Because the equation goes on to winnow this down to the number of technologically communicative civilizations, we normally think of intelligence as an entity that can conceive and build the requisite technology for such communication. This, however, is a very restricted concept of intelligence, even if a practical one in terms of our ability to detect it. In short, at the outset, we need to distinguish intelligence from technologically communicative intelligence.

The difficulties with the slippery concept of intelligence are immediately revealed by posing the simple question "when did intelligence arise on Earth?" Even if posed to biological or paleontological experts, the answers would wildly diverge, because the story of the evolution of intelligence on Earth depends entirely on how broadly one defines intelligence. Is it hominid intelligence, mammalian intelligence, or some earlier

ancestor that is the first "real" intelligence? The only reasonable answer is that there was no one day, year, or even period in the history of life where intelligence was lacking and then suddenly it appeared. Because it is not a binary choice, it follows that there was no one life-form that suddenly possessed the quality we call intelligence. It was part of the evolutionary process, just as physical morphology was part of that process; indeed, the two are inextricably intertwined (Marino 2015).

Such an expansive concept of intelligence opens the way to much broader investigations. While most research has focused on the evolution of intelligence in hominids on Earth over the last 5 to 7 million years, for astrobiology, the evolution of intelligence in a more general sense is of primary importance. Here, broader definitions of intelligence apply, such as "the ability to respond flexibly and successfully to one's environment and to learn from experience," or, put another way, "the ability to respond with an appropriate behavior in a given context" (Bogonovich 2011; Rospars 2013). Such broad definitions clearly apply to a wide range of animals. Accordingly, research has been undertaken, especially on dolphins and other mammals, and has even extended to cephalopods such as octopuses. Issues such as dolphin communication, brain encephalization, convergent brain evolution, and progressive trends in the history of life have all been studied in the context of astrobiology. By some definitions, there is a good deal of consensus that the higher primates and some cetaceans such as dolphins and whales share some of our reasoning abilities.

Thus, the landscapes of life and intelligence may extend far beyond our normal anthropocentric views of terrestrial life. Table 11.4.2 summarizes some general principles for life, namely biochemistry, morphology, and cognitive capacity, all of which may operate through evolution according to physical conditions much broader than we now realize, just as the extremophiles on Earth have expanded our view of the possibilities of life. In addition to taking into account the scaffolding, genetic material, and solvents available at any given location, for more complex life, we need to consider body patterns in terms of size, shape, and mass such as

---

TABLE 11.4.2
Principles for Possible Types of Life Beyond Earth

Biochemistry/Genetics/Solvents	Morphology
Carbon//DNA/Water	Body patterns
Carbon/RNA/Water	Size
Carbon/Protein/Water	Shape
Silicon/?/Nitrogen	Mass
Ammonia/?/Methane	
Space-Time Filters	**Cognitive Capacity**
Geometry	Sensory apparatus
Mathematics	Consciousness
Sense of time	Cognition
	Biological intelligence
	Postbiological intelligence
	Collective "Hive" or "Swarm" intelligence

originated in the Cambrian explosion, all closely associated with issues of chance, necessity, and evolutionary convergence. For intelligence, we need to consider scientific issues about what constitutes intelligence, as well as age-old philosophical issues of mind, body, and consciousness, and again issues of convergence.

Arguably, we should think even further out of the box of intelligence by emphasizing that the universe may be postbiological, given the timescales involved (Dick 2003; Schneider 2015). The key realization here is that cultural evolution is not often taken seriously enough in imagining what extraterrestrial intelligence might be like. Beyond extrapolations from Earth (Moravec 1988; Kurzweil 2005), there are evolutionary reasons for believing the universe may already have transitioned from biologicals to postbiologicals, possibly in the form of artificial intelligence. I have formalized this concept in what I call the Intelligence Principle: *The maintenance, improvement and perpetuation of knowledge and intelligence is the central driving force of cultural evolution, and to the extent intelligence can be improved, it will be improved.* Failure to do so may cause cultural evolution to cease to exist in the presence of competing forces (Dick 2003). Table 11.4.2 also lists another form of intelligence sometimes associated with social insects, a collective or distributed form often known as hive, or swarm intelligence. Finally, Table 11.4.2 lists "space-time filters" as one of the principles for cognitive life. We discuss this in the next section on the universality of knowledge.

In the same way, our concepts of life and intelligence should be scrutinized and our ideas of culture, civilization, technology and communication should be subjected to critical analysis, with an eye toward what is universal and what is contingent. Are there cultural universals? (Ashkenazi 2017; Traphagan 2015). Does technology converge? (Basalla 2006). Would aliens communicate in the same way as we do? (Traphagan 2011, 2015; Vakoch 2011). Tackling these questions is a serious task for historians, anthropologists, and the social sciences, and these issues have sometimes already been addressed in fields such as the philosophy of technology (Kelly 2010) and the philosophy of the social sciences (Rosenberg 2016). This is not an academic exercise but an effort that will increase our chances of discovering life and evaluating its impact.

11.4.3.2 THE UNIVERSALITY OF HUMAN KNOWLEDGE

The contemplation of extraterrestrial intelligence almost immediately raises an even broader set of questions about the validity and universality of human knowledge itself. The crucial question is how much of our knowledge—especially our seemingly "objective" natural science and mathematics—is contingent on the history of evolution on planet Earth and how much is universal? Even more problematic is: would our approaches to the social sciences and humanities—the "human sciences"—be recognizable to aliens? These questions inform two practical matters: how much of our knowledge would actually be useful in assessing the alien in the event of a discovery? And how might our knowledge be universalized if we find life beyond Earth, in the sense of placing our possibly parochial ideas in a much broader and perhaps more objective context?

In asking whether human knowledge, in general, is universal, we can draw on research from at least three fields: philosophy of knowledge, cognitive science, and evolutionary biology. The philosophy of knowledge, also known as epistemology in one of its many aspects, normally proceeds with a discussion of the relative merits of the schools of rationalism and empiricism, realism and idealism, and their variations and elaborations by philosophers over time. Most relevant here is the eighteenth-century German philosopher Immanuel Kant, who posited a category of knowledge he termed synthetic *a priori* judgments (Kuehn 2002). He claimed that these judgments, which include mathematics, geometry, and God, depend on our cognitive apparatus, so that these claims are about reality "only as it is experienced by beings such as we are... these claims cannot be claims about the world as it is independent of our conceptual apparatus." In other words, in Kant's view, our experiential knowledge must first pass through a set of a priori "forms" or filters. Space and time (Table 11.4.2) are forms of our "sensibilities," without which we cannot view the world of our experience. They are objective for us but not for all beings. While few philosophers today would accept the details of Kant's epistemology, almost all would endorse his conclusion that our knowledge of the external world is compromised by the human filters through which our perceptions must pass. In short, just as in the sixteenth century Copernicus placed the Sun in the center of the world rather than the Earth, already, in the eighteenth century, Kant placed the human mind at the center of reality only *as we perceive it.* So, the human mind may not be the center of reality, as perceived by extraterrestrials.

Cognitive science brings to bear on the problem of universal knowledge disciplines such as neuroscience, animal cognition, and artificial intelligence in an attempt to understand the nature of the human mind. The idea of embodied cognition (Lakoff 2003) proposes that cognition is evolutionarily related to the body it inhabits, a claim that seems to be validated by some studies in animal cognition (Griffin 2001). Such embodied cognition in turn relates directly to knowledge in the epistemological sense of how we know things. When extended to extraterrestrials, David Dunér has coined the term "astrocognition" to study the origin and evolution of cognitive abilities in extraterrestrial environments, as well as a variety of other issues related to space (Dunér 2011, 2013). In the context of the evolution of cognition, the basic premise is that cognitive abilities are adapted to "the physical and biological environment of their celestial body in order to understand and interpret, interact and deal with, and orientate itself in the particular physical and biological environment, in relation to its specific conditions." Embodied cognition and its extension to astrocognition are also in keeping with the similar idea of "situated cognition." Thus, an argument can be made that the search for intelligence beyond Earth is meaningless, unless we understand the underlying cognitive basis of intelligence.

If the body shapes mind, it also shapes thought, and extraterrestrials will have knowledge of the external world very different from us—just as Kant suspected on different grounds. Whatever term is used, whether embodied mind theory, situated cognition, or astrocognition, this is a field wide open for future development based on real data.

Finally, evolutionary biology, along with everything else it encompasses, sheds light on the evolution of the mind. Whereas embodied mind theory states that cognition depends on the body in which it is embedded, evolutionary epistemology now claims that since the body is a product of evolution and cognition is related to the body, cognition will also depend on the environment in which it evolves. In the words of philosopher of biology Michael Ruse (1998), it means "taking Darwin seriously," not only for biology but also for all of its by-products. In this view, all the tools of evolutionary biology not only can but must be applied to cognition, and epistemology itself is a product of evolution. Epistemology without evolution is seen as bankrupt. The evolution of cognitive processes during the long history of life on Earth can now be studied. Just as some see evolutionary biology as the starting point for an entire suite of cultural evolution, possibilities such as evolutionary psychology—the commonsense but controversial idea that our psychology and behavior have evolved over time according to natural selection—so also for epistemology.

The implications of evolutionary epistemology for astrobiology are clear: as an evolutionary product of its environment, not only the mind but also its ways of knowing will be different, depending on the long history of life under any given planetary conditions. Whether Kant or embodied cognition or the extension of evolutionary biology, the conclusion is the same for extraterrestrials: if all minds are embodied and Darwinian factors are at work in epistemology, then it is at least possible that our knowledge is not their knowledge, our science is not their science, and our mathematics is not their mathematics.

Science and mathematics are good tests for our ideas of the universality of human knowledge, since most people would consider them the most objective and universal aspects of all our knowledge. But even their objectivity has been called into question. The philosopher Rescher (1985), for example, has emphasized the potential diversity of science in a universal context: the *mathematics* used for alien science might be very different; the *orientation* of alien science might be different in the sense that any particular extraterrestrial intelligence might concentrate on social science rather than natural science, or their natural science might be very different; and the alien *conceptualization* of science might be different, embracing radically different cognitive points of view, just as our conceptual apparatus (think relativity and quantum mechanics) is radically different from a century ago.

While we can easily see why an Earth-bound and locally invented system like theology might not be the same among extraterrestrials as among Earthlings—diverse as theologies are even here on Earth—the non-universality of science and mathematics is a more difficult pill to swallow. But the philosophy of knowledge, cognitive science, and evolutionary biology all suggest the same conclusion: the different cognitive structures bound to exist among extraterrestrials leave open the real possibility that human knowledge, including even science and mathematics, might not be the same as theirs. And these differences might be enough to affect SETI programs in terms of both discovery and communications—perhaps another solution of the Fermi paradox of why we have not yet found extraterrestrials, and perhaps cannot, unless we find a way to transcend our anthropocentrism.

11.4.3.3 Envisioning Impact

Another critical issue is how to frame the societal impact of discovering of life beyond Earth, who should be involved in this framing, and how to make it systematic rather than sporadic. The question of impact has received increasing attention in the last two decades, since John Billingham, the head of NASA's SETI program at the time, convened a series of workshops on "The Cultural Aspects of SETI" (CASETI) on the eve of the inauguration of NASA SETI observations in 1992 (Billingham et al. 1999). For its time, this was an admirable effort, and the following decades saw additional sporadic efforts, both individual and in group settings. During the formulation and initiation of the first Astrobiology Roadmap in 1998 (Des Marais et al. 2008), calls were made for the study of cultural impacts of astrobiology (Dick 2000a), and in 1999, NASA Ames Research Center organized a workshop on the societal implications of astrobiology (Harrison and Connell 2001). Other organizations, including the John Templeton Foundation and the Foundation for the Future, organized meetings on the subject at about the same time (Dick 2000b; Harrison and Dick 2000; Tough 2000). Interest has increased in the last decade, notably with the American Association for the Advancement of Science (AAAS) series of workshops sponsored by its program on Dialogue on Science, Ethics, and Religion (Bertka 2009); several meetings at the Royal Society of London (Dominik and Zarnecki 2011); and a series of sessions at the American Anthropological Association (Vakoch 2009). While most of the attention has focused on the impact of the discovery of extraterrestrial intelligence, the AAAS volume also addresses the quite different scenario of the impact of discovery of microbial life. Race et al. (2012) has taken the lead in marshalling the astrobiology, social sciences, and humanities communities to address these issues in the context of the Astrobiology Roadmap, with the support of the NASA Astrobiology Institute. NASA has also advanced the study of the humanistic aspects of astrobiology by establishing the Baruch S. Blumberg NASA/Library of Congress Chair in Astrobiology, which held an international symposium on the impact of astrobiology at the Library of Congress in 2014 (Dick 2015). Individual efforts have also concentrated on different aspects of the problem (Davies 1995; Harrison 1997; Achenbach 1999; Harrison and Dick 2000; Michaud 2007; Impey et al. 2013; Vakoch 2013; Dick 2015, 2018b; Race 2015), including a comparison to the impact of other scientific endeavors such as biotechnology (Race 2007).

The results of these studies have been to demonstrate the serious impact that the discovery of extraterrestrial life could have on society, especially in the case of extraterrestrial intelligence. There is no consensus among these disparate studies. What is needed is a general framework within which these studies might be incorporated. One such framework was suggested almost a half-century ago in the much broader context of the impact of science and technology (Ladrière 1977), though it has received little attention. In the most basic sense, just as we have distinguished pre-discovery, discovery, and post-discovery stages in the discovery process, the Belgian philosopher and scientist Jean Ladrière demonstrated that we should distinguish pre-impact, impact, and post-impact phases, making explicit the simple but undeniable fact that any impact scenario consists of a past, present, and future. Before impact, a certain set of assumptions, ideas, and values (we would also say worldviews) is in place. These are then in flux for an extended period, before the assimilation of new assumptions, ideas, and values is completed. During this period of flux, reinforcing and competing worldviews are also in play. In Ladrière's rather unwieldy terminology, the process of impact involves a "destructuration" phase, during which the old order is torn down, and a "restructuration" phase, during which the new order is built up. We refer to these simply as destructuring and restructuring stages, each of which will have political, economic, and cultural components (Table 11.4.3).

During the destructuring phase, a new discovery of large-enough magnitude affects all three components of society (political, economic, and cultural) and either challenges or confirms them. This is followed by growing acceptance of the new information and its implications for the future, which feeds into society's values and meanings. Finally, a new idea of the future is established, a future that may be rich in possibility and transformable by human activity or, alternatively, threatening and out of our control. In short, a critical mass of novelty is reached, bursting with potential that begs to be

reintegrated. This triggers the restructuring phase, in which there is a reintegration and reunification of culture and its subsystems based on the new meanings and values. Some aspects of culture may accept the new system of values and meanings, while others reject it. This is followed by criticism and resolution of conflicting views. And finally, in the post-impact, there is restoration of a unity of vision—to the extent that such unity can ever exist in a multicultural world.

The Jesuit astronomer Stoeger (1996) has proposed that Ladrière's ideas may be useful as a way of understanding astronomy's impact on culture. This seemingly simple schema of destructuring and restructuring, which may apply to large and small impacts over the long and short terms, also provides a potentially powerful framework within which we can discuss the impact of discovering life beyond Earth under whatever scenario. It resonates with the rise and fall of worldviews that we will discuss in the next chapter, including stages in worldview development ranging from presentation based on observation to elaboration, opposition, exploration of implications outside the field, general acceptance, and final confirmation. It also parallels in some ways Thomas Kuhn's paradigms, though the latter are now so laden with baggage that it is perhaps best to avoid the complications of that comparison. Finally, the destructuring and restructuring process is arguably inherent at a very practical level in the long-term cycles of the global world order (Kissinger 2014). It is obviously an idealized view, bound to be complicated by the messiness of human behavior, not to mention the problem of overlapping discoveries.

11.4.4 SOCIETAL IMPACTS OF DISCOVERING EXTRATERRESTRIAL INTELLIGENCE

In this section, we explore possible impacts on humanity of finding life beyond Earth, using the concept of worldviews as our framework (Vidal 2014). For our purposes here, we use worldview in the sense of the German *Weltanschauung*, defined as how we see the world from a variety of perspectives and carrying with it the implication that worldviews have impact on our daily lives. Each of us sees the world in a different way; our individual worldviews are a mix of political, cultural, religious, philosophical, and scientific components, among others. Given this framework, our task of exploring the impact of discovering life beyond Earth then becomes an exploration of its impact on current worldviews and the generation of new worldviews. The three worldview levels in Figure 11.4.2 thus provide the organizing principle for this section: the cosmological, the theological, and the cultural.

11.4.4.1 COSMOLOGICAL WORLDVIEWS: THE RISE OF THE BIOLOGICAL UNIVERSE

Cosmological worldviews have always affected the human mind. The ordered geocentric universe in Dante's *Divine Comedy,* with its moral lessons played out in a cosmological worldview of nested spheres, with the Earth at the center and

TABLE 11.4.3

The Anatomy of Impact

Pre-Impact	**Established assumptions, ideas, and values**	
Impact	**Destructuring phase (Disruption of cultural elements)**	Political economic cultural proportional to magnitude of discovery
	• Challenge or confirm values	
	• Growing acceptance	
	• New vision of future	
	Restructuring phase (Rebuilding cultural elements)	
	• Acceptance or rejection of new elements	
	• Criticism and move toward resolution	
	• Unity of vision restored	
Post-impact	Acceptance of new assumptions, ideas, and values	Over decades and centuries

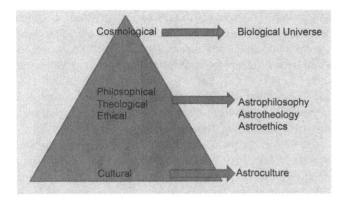

FIGURE 11.4.2 A hierarchy of worldviews, in which cosmological worldviews trickle down and profoundly affect various levels of human culture. The discovery of life beyond Earth may result in a transition from our current physical cosmology to one based on a biological universe. Philosophical, religious, and ethical worldviews may be generalized and transformed into astrophilosophy, astrotheology, and astroethics. And the transformation of human culture to astroculture may be accelerated beyond the initial thrust of the Space Age.

the sphere of fixed stars at the periphery, provided the backdrop for medieval life. The heliocentric universe upended that framework, prompting John Donne to say that "'Tis all in peeces, all cohaerence gone." We are now at a turning point similar to that of the sixteenth and seventeenth centuries. Over the last four centuries since Copernicus, the idea of a purely physical universe has gradually given way to the idea of a universe filled with life, what I have termed "the biological universe" (Dick 1996). Those championing this idea hold that planetary systems are common; that wherever conditions are favorable, life will originate and evolve, and in its strongest form; and that this evolution will likely culminate in intelligence. In its most robust form, the biological universe makes claims about the large-scale nature of the universe, especially that life is one of its basic properties. It is therefore more than an idea, more than another theory or hypothesis. It is sufficiently fundamental and comprehensive to qualify as a worldview of its own, one that has implications for all humanity. Moreover, because it is testable, it is a scientific worldview, perhaps best described as a cosmology. And because it combines the biological with the physical, it may accurately be termed "the biophysical cosmology" (Dick 1989).

The potential impact of this biological universe is best approached by distinguishing several possibilities. As many scientists have argued, it is entirely possible that the universe is full of microbes but lacks complex life or intelligence. Such a *microbial biological universe* would be a fundamental discovery in the history of science. If found to be of independent origin, the discovery of such microbes would open the possibility of a universal biology in the same way we now believe we have a universal physics. It is tempting to say that the impact of a microbial biological universe would be primarily scientific. But the reaction in 1996, when nanofossils were claimed in the Mars rock ALH84001, clearly indicates that the impact will be much broader than that, at least over the

short term. If microbial life is common, it may indicate that, given enough time as we have had on Earth, intelligence may develop. But the discovery of microbial life in itself would be the greatest discovery in the history of science.

An *intelligent biological universe* is one where intelligent life is a common feature. We can argue about the meaning of "common," or in its reverse "rare Earth," formulation, about the meaning of "rare" (Ward and Brownlee 2000). But in a universe so large that even if one in a million planets have intelligence, millions of planets with intelligence would still occur even in our own galaxy, the semantic argument over the definition of common or rare does seem rather pointless. More urgent is a discussion of what it means to be sentient, intelligent, or superintelligent (Bostrom 2014); what interaction with such entities implies in terms of ethical standards; and, beyond that, what issues of law and diplomacy might be expanded into space law and metalaw (Haley 1963).

The *anthropic biological universe* holds that life and mind are intimately connected with the physical universe (Barrow and Tipler 1986). The physical and biological universes that are obviously already connected in the sense biology cannot exist without a prior physical substrate. But scientists and philosophers alike are increasingly aware of possible deep connections between the physical universe of planets, stars, and galaxies and the biological universe of life, mind, and intelligence. The more we learn about the universe, the more we find that it is even more mysterious than we thought—that the universe is not only subject to mathematical laws, but also finely tuned for life. In its deepest structure and most fundamental properties such as the physical constants, the universe seems tailor-made for life. Or, to put it another way, the universe is "biofriendly," as witnessed by our existence and the possible existence of other life beyond Earth (Davies 2007).

The microbial, intelligent, and anthropic versions of the biological universe, and the postbiologial universe as well, each represents different worldviews with different implications for humanity. They can also be viewed in terms of human destiny, usually seen in theological terms, but also in more secular terms. In a microbial biological universe, aside from its scientific value, we may be dealing primarily with the instrumental rather than intrinsic value of life. Instrumental value in this case implies that humans might be mainly interested in the utility of extraterrestrial microbes for medicines and other purposes, just as has been the case with newly discovered extremophile organisms on Earth. That would be a subject of major ethical and practical discussion. In such a microbial biological universe, human destiny would be to explore the universe as the Lords of creation—at least until sentience or intelligence was discovered near or far.

In an intelligent biological universe, by contrast, human destiny is to interact with extraterrestrials, whether in struggle or to join the galactic club. This is the stuff of science fiction, but in case of an actual discovery, it would very quickly become a practical problem. As we have emphasized earlier, the outcome would very much depend on the scenario. In an anthropic biological universe, the questions turn more to the philosophical. The obvious question is why is the universe so biofriendly? Could it be just a coincidence? For most people who have thought deeply about the problem, this seems too good to be true. Could it be that a supernatural intelligence fine-tuned the universe, the God hypothesis? Many would say so, but it depends on whether the supernatural is part of your metaphysics. Could a highly evolved *natural* intelligence have fine-tuned the universe? This possibility has received some attention, but is difficult to prove. Or perhaps, a "multiverse" exists, an ensemble of universes, and we happen to live in a universe that is suitable for life. Whatever version of the anthropic universe we except, and whatever is the explanation for it, the anthropic biological universe is a cosmological worldview, as new and expansive as any in the past. Finally, the postbiological universe (Dick 2003) has barely been discussed in the terrestrial context, much less the extraterrestrial.

If after many centuries, it turns out that we find none of these versions of the biological or postbiological universe, then we will have determined that we live in a physical universe in which humans are unique or nearly unique. In that case, human destiny will be to explore this universe and perhaps to spread humanity throughout its vast spaces. We need not invoke or repeat the sins of manifest destiny in doing so. If we are indeed alone, we need not worry about harming the "natives." But there will be many questions about exploiting and harming the space environment for human purposes.

In summary, the impact of actually discovering life—as opposed to conjecturing and theorizing about it based on preconceptions—will be to strengthen and consolidate the new worldview called the biological universe, or its cultural extension the postbiological universe, all as a part of the epic of cosmic evolution. Whether it is microbial, biological, anthropic, or postbiologial, or whether some mix of these or something yet to be discovered, remains to be seen. But in any case, with the discovery of alien life, our cosmological worldviews will have been transformed in proportion to the complexity of life discovered and in proportion to the degree to which the universe is anthropic or biocentric.

11.4.4.2 Theological Worldviews: From Theology to Astrotheology

Below the cosmological level in our framework of Figure 11.4.2 looms a perplexing variety of theological worldviews, a dominating feature of human cultures. Whatever the other effects on cultural elements at a level below the theological, the impact of the discovery of life beyond Earth on these theologies is likely to be pervasive, foundational, and personal. Such impacts will vary depending on the nature of the discovery and the particular religion involved. Because theological worldviews are so deeply held by such a large part of the population, changes in worldviews at this level—whether those worldviews are objectively true or not—are sure to have profound effects. Conversely, however, they also have the potential to help Earthlings cope with the discovery in whatever form it takes.

The effect of what we have called the intelligent biological universe on theologies, particularly Christian theology, has long been the subject of sporadic debate since Copernicus made the Earth a planet and the planets potential Earths (Dick 1996). In subsequent centuries, three choices were logically open to Christians, who pondered the question of other worlds: they could reject other worlds, reject Christianity, or attempt to reconcile the two. Historically, all three of these possibilities came to pass in the eighteenth and nineteenth centuries (Crowe 1986). In the twentieth century, the relation of alien life to theology—most often Christian theology—has also been discussed, as events have unfolded in the SETI and as new discoveries or reputed discoveries have been made in NASA's astrobiology program. Debate has centered around Christian doctrines such as incarnation and redemption, Christian practices such as baptism (Consolmagno and Mueller 2014; Consolmagno 2015), and the nature of God. The subject has been debated in books (O'Meara 2012; Wilkinson 2013; Weintraub 2014; Peters et al. 2018), articles (Bertka 2013; Peters 2011, 2013a, 2013b), and meetings (Bertka 2009; Dick 2000b), with no consensus. Some argue for an expansion of Christian doctrine and practice, while others conclude that the discovery of life beyond Earth would be fatal to the one-to-one relationship between God and humanity inherent in most Western religions.

Unlike Christianity, no substantial historical tradition exists in Judaism or Islam, the other two "Adamist" religions analogous to the Christian tradition on this subject. In his "Jewish exotheology," Rabbi Norman Lamm (1978) did point to some historical precedents and express his viewpoint that extraterrestrials would not be problematic, much less fatal, for the Jewish religion. In the Islamic tradition, second only to Christianity, with its 1.5 billion adherents, scholars have pointed out that some particulars of Islam would prove problematic in their geocentric and anthropocentric aspects: the importance of Mecca for prayer and travel; praying five times per day on worlds with potentially vastly different day lengths; and, in general, following the teachings of Muhammad, who was born and lived on Earth. Thus, while Muslims do not have to deal with incarnation, they still have some of the same problems brought about by the anthropocentric nature of their religion.

Beyond the Adamist traditions, there has been even less discussion, but one ambitious overview touching on more than a dozen religions (Weintraub 2014) indicates that Eastern religions might have less trouble adapting than Western religions. Weintraub finds that the least impact on these would be to Hinduism (900 million members) and Buddhism (400 million members), which do not posit a single godhead with a relationship to humanity. Each of these religions might absorb, and even welcome, the discovery of extraterrestrial life. Indeed, some religions such as Mormonism already have extraterrestrials as part of their doctrine, an artifact of the time of their founding. Belying the seemingly simple question of the effect of astrobiology on theology, these findings point to one main conclusion, in some ways obvious from the outset: each religion will have to deal with the discovery of extraterrestrial life in its own way. There will not be one effect, but many.

It is also possible that the discovery of life beyond Earth will result in new theologies. One of these, known as cosmotheology (Dick 2000c, 2018b), is in its most general sense a theology that takes into account what we know about the universe based on science. It is therefore a naturalistic cosmotheology, but it is not coextensive with scientism, because it does not imply that science is the only way to understand the world. Cosmotheology has six bedrock principles: (1) humanity is in no way physically central in the universe; (2) humanity is not central biologically, mentally, or morally in the universe; (3) we must take into account the probability that humanity is near the bottom in the great chain of beings in the universe; (4) cosmotheology must be open to radically new conceptions of God, not necessarily the God of the ancient near East or the God of the human imagination, but a natural God grounded in cosmic evolution; (5) cosmotheology must have a moral dimension, extended to embrace all species in the universe—a reverence and respect for life in any form; and (6) cosmotheology can and should transform our ideas about human destiny. With regard to the latter, although human destiny has most often been couched in divine terms, as in Reinhold Niebuhr's *The Nature and Destiny of Man* (1941) or Pierre Lecomte du Nouy's best-selling *Human Destiny* (1947), or, indeed, as in the entire Christian theology, it need not be linked to the supernatural. Rather, it can be linked to the process and endpoint of cosmic evolution. Taken as a whole, this version of astrotheology can also be seen in the tradition of religious naturalism, but specifically formulated to take into account the realities of cosmic evolution and the many possibilities inherent in life in the universe.

All of these streams of thought have resulted in what is now a considerable scholarly discussion of the subject, known variously as exotheology (Peters 1995), astrotheology (Peters 2014; Peters et al. 2018), and cosmotheology (Dick 2000c, 2018b). Such terminological diversity is not unusual in the early stages of a new intellectual endeavor.

11.4.4.3 Cultural Worldviews: From Culture to Astroculture

At the broad base of the pyramid of worldviews of Figure 11.4.2, culture is being transformed into astroculture at an ever-accelerating pace, in its broadest sense an increasing awareness of our place in the universe. "Astroculture" is a relatively new umbrella concept used to describe the array of images, events, and media reactions that "ascribe meaning to outer space while stirring both the individual and the collective imagination" (Geppert 2012). The concept may be expanded to argue that, while different perspectives on space may exist in different cultures, humanity as a whole is increasingly creating and immersed in an overarching astroculture that transcends national boundaries, a kind of *global astroculture*. While this transformation is much broader than the possibility of life beyond Earth, the idea of alien life has, for some time, been a major component of culture (at least Western culture) and is sure to accelerate if a discovery of life beyond Earth is made.

Over the last 50 years, the idea that Carl Sagan embodied in the phrase "the cosmic connection" (Sagan 1973, 2000) has become more and more a part of our daily lives and will even more in the future, as our cosmic consciousness increases.

The cosmic connection of which Sagan spoke in 1973 not only embraced cosmic evolution, life beyond Earth, and the exploration of space in a hundred different forms, it also emphasized the most intimate connection of all: that we are made of stardust. The cosmic connection has only increased over the last several decades, as we have viewed spectacular imagery from the Hubble Space Telescope and other observatories, as well as absorbed the results of other spacecraft that have revealed our place in space and time. Within this grand context of cosmic evolution, the search for life in the universe finds its place, becoming in the process a dominant part of astroculture.

The discovery of life in any form will also impact broader aspects of culture. Given the impact that even the possibility of alien life has had on science fiction, the effect on literature in general will likely be enormous if extraterrestrial life is discovered. In the same way as literature was affected in the wake of the Copernican theory with John Donne and other writers, surely, the biological universe in any of its forms will do the same in ever more expressive form. The discipline of history will be affected, as our attempts at "universal history" in the tradition of Spengler, Toynbee, and Fukuyama will become only subsets of a much larger and truly universal history. Beyond theological-philosophical worldviews, philosophy will also change, as venerable questions such as the nature of objective knowledge are finally answered, or at least advanced or expanded. In short, the transformation of culture to global astroculture will witness the slow dismantling of one worldview and the reconstruction of a new one, in accordance with the Ladrierean framework discussed previously. In the process, new values will also emerge, feeding into established ethical systems and even giving rise to new ones.

11.4.5 ASTROETHICS: INTERACTING WITH ALIEN LIFE

The Space Age has required the contemplation and development of ethical standards for a variety of issues, including terraforming of planets, resource utilization, near-Earth asteroid threats, space exploration, and planetary protection. The discovery of extraterrestrial life would certainly raise these concerns to a new level, giving rise to a new discipline, sometimes termed astroethics (Impey et al. 2013; Peters 2013a, 2013b). Does Mars belong to the Martians, even if the Martians are only microbes? What do we say in response to an alien message, and who speaks for Earth? How do we treat an alien in a "close encounter of the third kind?" In short, whether we discover alien microbes or advanced alien life, we will immediately be faced with the problem of how to interact.

Before we can act in any situation that involves life, it is first important to assess the moral status of the organisms involved (Cleland and Wilson 2015). This is no easy task, since we are ambiguous even about relations with animals on Earth. But over the last few centuries, a good deal of thought has been given to the subject of moral status of Earth organisms and the idea of intrinsic value on which it is often based. This has resulted in an entire spectrum of frameworks or theories for moral status, an ethical landscape that has been presented with broad brushstrokes by the Swedish philosopher Persson (2012). He distinguishes four ethical theories of moral status, in order of increasing inclusivity: *anthropocentrism*, in which only humans have moral status; *sentientism*, in which all and only sentient beings have moral status; *biocentrism* in which all and only living beings have moral status; and *ecocentrism*, in which all living beings, species, and ecosystems, and perhaps even nonliving matter, have moral status. Which of these frameworks we accept will determine how we treat alien life.

Contemplating encounters with alien life tremendously expands our ethical horizons. In the case of intelligent aliens, it also encompasses not just the problem of how we might treat them but also how aliens might act or react. In other words, it is not just a question of our ethics. What about their ethics? Is there any basis for inferring whether alien intelligence might be good or bad? Might there be such a thing as a universal ethics, perhaps in the form of the Golden Rule, as Rolston (2014) suggests, or simply a reverence for life, as Albert Schweitzer famously taught? Or is Star Trek's "Prime Directive" of non-intervention a naive one-way street, a recipe for our own extinction? In short, to expand one of the best-known statements of Martin Luther King to an extraterrestrial context, does the arc of the moral universe indeed bend toward justice? Or perhaps, more precisely, does the moral arc of the universe bend toward justice? This is the domain of altruism, which surprisingly has also been the subject of considerable thought in an extraterrestrial context, though with no decisive conclusions (Vakoch 2014b).

As with theories of moral status, the answers to these questions about altruism will inform our actions in real-world contacts with alien life under different scenarios. By contemplating these issues, and certainly by putting them into practice in the event of the discovery of life beyond Earth, we will not only address an important "wild card" problem in our near or far future but also transform our thinking by moving from an anthropocentric ethic toward a "cosmocentric" ethic, one that establishes the universe and all of its life as a priority in a value system, rather than just humans or even terrestrial life in general (Lupisella and Logsdon 1997; Lupisella 2009, 2016, in press). This is compatible with certain brands of astrotheology such as cosmotheology, which includes as its fifth principle a moral dimension, extended to include all species in the universe—a reverence and respect for life that we find difficult enough to foster on Earth. For both our Earthly and celestial concerns, a cosmocentric ethic will surely be a shift in worldview as radical as any of those we discussed in the last chapter.

The need for a basic cosmocentric ethic is already growing in proportion to the growth of our cosmic consciousness and the development of astroculture, as discussed previously. At one level, the novel ethical challenges that arise as we explore the extraterrestrial environment argue for a cosmos-wide ethical view. At another level, we need such an ethic because the cosmos potentially grounds our ethics in some objective way, or at least serves as a shared frame of reference with other intelligent beings, or even because the cosmos and

life are intimately entwined. The answer turns on how constrained we consider our terrestrial ethics, but a cosmic perspective is surely in order as we expand our views of the space environment, including (and especially) life. Such a view will not happen overnight, but perhaps, humanity will increase its awareness in stages, as we encounter the universe in increasingly intimate ways that will become a basic part of what it means to be human, or post-human.

11.4.6 THE ROAD AHEAD: PREPARING FOR DISCOVERY

Two practical questions remain: how do we prepare for the discovery of life, and what should we do if we find it? These questions have been sporadically raised at the highest levels of government, usually in the context of astrobiology programs at NASA, but also at other space agencies such as the European Space Agency. Arguably, preparing for astrobiological discovery and its aftermath is no less urgent than other science policy issues with the potential to impact all of society.

The questions here are legion, and potentially Earth-shaking. Who should take the lead in preparing for discovery? What do we do if life is actually discovered, microbial or intelligent, near or far? Should national governments be in charge or international political and scientific institutions, scientists and social scientists, ethicists and theologians, or some mix thereof? How do we prevent contamination of potential microbes on Mars, Europa, Enceladus, or other habitable sites in the solar system, and (more perhaps more urgently from most Earthlings' point of view), how do we protect our planet from back-contamination in the event of the discovery of microbial life? If a message is received as a result of a successful SETI program, should we answer? If so, who speaks for Earth? Should we initiate messages as part of a METI program? If so, what should we say, and who, if anyone, should control what is said? These questions are only the leading edge of the many decisions that will have to be made once alien life is actually discovered. And each discovery scenario will have its own unique problems and solutions.

While we can debate how likely any extraterrestrial life discovery scenario might be, in the last few decades, developments in astrobiology have made the discovery of life beyond Earth more and more feasible. Clearly, the time is ripe for serious discussion of strategies and policies in the event of discovery. Any analysis needs to keep in mind the two primary scenarios we distinguished earlier: the discovery of microbial life and intelligent life, each with its multiple modes of near and remote discovery and direct or indirect impacts. While policies do not always survive contact with the actual events, it is indisputably prudent to contemplate them in advance, especially when the implications for Earth are so all encompassing as in the discovery of alien life. We need not only to "look forward," but also to develop policies to achieve the goal of being fully proactive—and to have the best chance of having a positive impact on society when and if alien life is discovered under a variety of scenarios. Scientists and scholars in the humanities and social sciences can and do play a major role in formulating policy (Michaud 2007; Race 2010, 2015; Lupisella 2015). Because there are many other stakeholders in society, they may not be the final arbiters of policy, much less the major players in implementing it. This highlights the importance of not only of continuing to develop strategies and policies, but also of having the political mandate to implement them when the time comes.

REFERENCES

Achenbach, J. 1999. *Captured by Aliens: The Search for Life and Truth in a Very Large Universe*. New York: Simon & Schuster.

Ashkenazi, M. 2017. *What We Know About Extraterrestrial Intelligence: Foundations of Xenology*. Cham, Switzerland: Springer.

Bains, W. 2004. Many chemistries could be used to build living systems, *Astrobiology*, 4: 137–167.

Barrow, J. D. and F. J. Tipler. 1986. *The Anthropic Cosmological Principle*. Oxford, UK: Oxford University Press.

Bartha, P. 2010. *By Parallel Reasoning: The Construction and Evaluation of Analogical Arguments*. New York: Oxford University Press.

Basalla, G. 2006. *Civilized Life in the Universe: Scientists on Intelligent Extraterrestrials*. Oxford, UK: Oxford University Press.

Bertka, C. Ed. 2009. *Exploring the Origin, Extent, and Future of Life: Philosophical, Ethical and Theological Perspectives*. Cambridge, UK: Cambridge University Press.

Bertka, C. M. 2013. Christianity's response to the discovery of extraterrestrial intelligent life: Insights from science and religion and the sociology of religion, in In *Astrobiology, History, and Society*. Berlin, Germany: Springer, pp. 329–340.

Billingham, J., R. Heyns, D. Milne et al., Eds. 1999. *Social Implications of the Detection of an Extraterrestrial Civilization*. Mountain View, CA: SETI Press.

Blumenberg, H. 1987. *The Genesis of Copernican World*, trans. R. M. Wallace. Cambridge, MA: MIT Press.

Bogonovich, M. 2011. Intelligence's likelihood and evolutionary time frame, *International Journal of Astrobiology*, 10: 113–122.

Bostrom, N. 2014. *Superintelligence: Paths, Dangers, Strategies*. Oxford, UK: Oxford University Press.

Bowler, P. 1989. *Evolution: The History of an Idea*. Berkeley, CA: University of California Press.

Cleland, C. E. and Christopher Chyba. 2007. Does 'life' have a definition? in Sullivan and Baross (2007), pp. 119–131.

Cleland, C. E. and E. M. Wilson 2015. The moral subject of astrobiology, in Dick (2015), pp. 207–221

Cline, E. H. 2014. *1177 B.C.: The Year Civilization Collapsed*. Princeton, NJ: Princeton University Press.

Consolmagno, G. 2015. Would you baptize an extraterrestrial? in Dick (2015), 233–243.

Consolmagno, G. and P. Mueller. 2014. *Would You Baptize an Extraterrestrial?* New York: Random House.

Crowe, M. J. 1986. *The Extraterrestrial Life Debate, 1750–1900: The idea of a Plurality of Worlds from Kant to Lowell*. Cambridge, UK: Cambridge University Press.

Davies, P. 1995. *Are We Alone? Philosophical Implications of the Discovery of Extraterrestrial Life*. New York: Basic Books.

Davies, P. 2007. Cosmic jackpot: Why our universe is just right for life. Boston, MA: Houghton Mifflin.

Des Marais, D., Joseph A. Nuth III, L. Allamandola et al. 2008. The NASA astrobiology roadmap, *Astrobiology*, 8: 715–730.

Dick, S. J. 1989. The concept of extraterrestrial intelligence–An emerging cosmology, *Planetary Report*, 9: 13–17.

Dick, S. J. 1995. Consequences of success in SETI: Lessons from the history of science, in Shostak, *Progress in the Search for Extraterrestrial Life*, pp. 521–532.

Dick, S. J. 1996. *The Biological Universe: The Twentieth Century Extraterrestrial Life Debate and the Limits of Science*. Cambridge, UK: Cambridge University Press.

Dick, S. J. 2000a. Cultural aspects of astrobiology: A preliminary reconnaissance at the turn of the millennium, in Lemarchand and Meech (2000). San Francisco, CA: Astronomical Society of the Pacific, pp. 649–659.

Dick, S. J. 2000b. *Many Worlds: The New Universe, Extraterrestrial Life and the Theological Implications*. Philadelphia, PA: Templeton Press.

Dick, S. J. 2000c. Cosmotheology: Theological implications of the new universe, in Dick (2000b), *Many Worlds: The New Universe, Extraterrestrial Life and the Theological Implications*, pp. 191–210.

Dick, S. J. 2003. Cultural evolution, the postbiological universe and SETI, *International Journal of Astrobiology*, 2: 65–74; reprinted as Bringing culture to cosmos: The postbiological universe, in Dick and Lupisella, Eds. (2009), pp. 463–488.

Dick, S. J. 2013. *Discovery and Classification in Astronomy: Controversy and Consensus*. Cambridge, UK: Cambridge University Press.

Dick, S. J. Ed. 2015. *The Impact of Discovering Life Beyond Earth*. Cambridge, UK: Cambridge University Press.

Dick, S. J. 2018a. *Astrobiology, Discovery, and Society*. Cambridge, UK: Cambridge University Press.

Dick, S. J. 2018b. Toward a constructive naturalistic cosmotheology, in Peters (2018).

Dick, S. J. and Lupisella, M. L. Eds. 2009. *Cosmos & Culture: Cultural Evolution in a Cosmic Context*. Washington, DC: NASA, online at http://history.nasa.gov/SP-4802.pdf.

Dick, S. J. and J. E. Strick. 2004. *The Living Universe: NASA and the Development of Astrobiology*. New Brunswick, NJ: Rutgers University Press.

Dominik, M. and J. C. Zarnecki. 2011. The detection of extraterrestrial life and the consequences for science and society, *Philosophical Transactions of the Royal Society A*, 369, 1936.

Dreyer, E. L. 2007. *Zheng He: China and the Oceans in the Early Ming Dynasty, 1405–1433*. New York: Pearson Longman.

Dunér, D. 2011. Astrocognition, in *Humans in Outer Space – Interdisciplinary Perspectives*, U. Landfester et al. Eds. Wien, Austria: Springer Verlag, pp. 117–140.

Dunér, D. 2013. Extraterrestrial life and the human mind, in Dunér et al. (2013), pp. 1–25.

Dunér, D., Joel Parthemore, E. Persson, and G. Holmberg, Eds. 2013. *The History and Philosophy of Astrobiology: Perspectives on Extraterrestrial Life and the Human Mind*. Newcastle-upon-Tyne, UK: Cambridge Scholars Publishing.

Elliott, J. H. 2006. *Empires of the Atlantic World: Britain and Spain in America, 1492–1830*. New Haven, CT: Yale University Press.

Finney, B. and J. Bentley. 2014. A tale of two analogues: Learning at a distance from the ancient Greeks and Maya and the problem of deciphering extraterrestrial radio transmissions, in Vakoch (2014b), pp. 65–78.

Gentner, D. and M. Jeziorski. 1993. The shift from metaphor to analogy in western science, in A. Ortony (Ed.), *Metaphor and Thought*. Cambridge: Cambridge University Press, pp. 447–480, online at http://www.psych.northwestern.edu/psych/people/faculty/gentner/newpdfpapers/GentnerJeziorski93.pdf.

Geppert, A. C. T. Ed. 2012. *Imagining Outer Space: European Astroculture in the Twentieth Century*. New York: Palgrave Macmillan.

Gest, H. 2004. The discovery of microorganisms by Robert Hooke and Antoni van Leeuwenhoek, fellows of the Royal Society, *Notes and Records of the Royal Society of London*, 58(2): 187–201.

Griffin, D. R. 2001. *Animal Minds: Beyond Cognition to Consciousness*. Chicago, IL: University of Chicago Press.

Haley, A. G. 1963. *Space Law and Government*. New York: Appleton-Century-Crofts.

Harrison, A. A. 1997. *After Contact: The Human Response to Extraterrestrial Life*. New York: Plenum.

Harrison, A. A. and S. Dick. 2000. Contact: Long-term implications for humanity, in *Tough (2000)*, pp. 7–31.

Harrison, A. A. and K. Connell. 2001. *Workshop on the Societal Implications of Astrobiology*. Moffett Field, CA: NASA Ames Research Center. Online at http://astrobiology.arc.nasa.gov/workshops/societal/.

Hays, L. Ed. 2015. *Astrobiology Strategy* (NASA), available at https://nai.nasa.gov/media/medialibrary/2015/10/NASA_Astrobiology_Strategy_2015_151008.pdf.

Hesse, M. B. 1966. *Models and Analogies in Science*. Notre Dame, IN: University of Notre Dame Press, pp. 1–9.

Hoerder, D. 2002. *Cultures in Contact: World Migrations in the Second Millennium*. Durham, NC: Duke University Press.

Hofstadter, D. 2001. Analogy as the core of cognition, in D. Gentner, K. J. Holyoak, and B. N. Kokinov (Eds.), *The Analogical Mind: Perspectives from Cognitive Science*. Cambridge, MA: The MIT Press, pp. 499–538.

Hofstadter, D. and E. Sander. 2013. *Surfaces and Essences: Analogy as the Fuel and Fire of Thinking*. New York: Basic Books.

Horneck, G. et al. 2016. AstRoMap European astrobiology roadmap, *Astrobiology*, 16, 201–243.

Impey, C., A. Spitz, and W. Stoeger, Eds. 2013. *Encountering Life in the Universe: Ethical Foundations and Social Implications of Astrobiology*. Tucson, AZ: University of Arizona Press.

Joyce, G. G. 1994. Foreword, in D. W. and G. Fleischaker Eds., *Origins of Life: The Central Concepts*. Boston, MA: Jones & Bartlett Learning.

Kelly, K. 2010. *What Technology Wants*. New York: Viking.

Kissinger, H. 2014. *World Order*. New York: Penguin Books.

Kuehn, M. 2002. *Kant: A Biography*. Cambridge, UK: Cambridge University Press.

Kuhn, T. S. 1962a. *The Structure of Scientific Revolutions*. Chicago, IL: University of Chicago Press, expanded edition, 1970, p. 55.

Kuhn, T. S. 1962b. The historical structure of scientific discovery, *Science*, 136, 760–764, reprinted in Kuhn, *The Essential Tension: Selected Studies in Scientific Tradition and Change*. Chicago, IL: University of Chicago Press, 1977, pp. 165–177.

Kurzweil, R. 2005. *The Singularity is Near: When Humans Transcend Biology*. New York: Viking.

Kuznicki, J. T. 2011. The Inscrutable Names of God, in (Vakoch, 2011), 202–213.

Ladrière, J. 1977. *The Challenge Presented to Cultures by Science and Technology*. Paris, France: UNESCO.

Lakoff, G. 2003. How the body shapes thought: Thinking with an all-too-human brain, in A. Sanford (Ed.), *The Nature and Limits of Human Understanding*. London, UK: T & T Clark, pp. 49–74.

Lamm, N. 1978. The religious implication of extraterrestrial life, in A. Cannell and C. Domb (Eds.), *Challenge: Torah Views on Science and Its Problems*. New York, pp. 354–398.

Lemarchand, G. A. and K. J. Meech Eds. 2000. *Bioastronomy'99: A New Era in Bioastronomy*. San Francisco, CA: Astronomical Society of the Pacific.

Lupisella, M. 2009. Cosmocultural Evolution: The Coevolution of Culture and Cosmos and the Creation of Cosmic Value, In S. J. Dick and M. L. Lupisella (Eds.) *Cosmos and Culture: Cultural Evolution in a Cosmic Context*, NASA, pp. 321–356.

Lupisella, M. 2015. *Life, Intelligence, and the Pursuit of Value*, in Dick (2015), pp. 159–174.

Lupisella, M. 2016. Cosmological theories of value: Relationalism and connectedness as foundations for cosmic creativity, in J. S. J. Schwartz and T. Milligan (Eds.), *The Ethics of Space Exploration*. Basel, Switzerland: Springer, 75–91.

Lupisella, M. In press. *Cosmological Theories of Value: Exploring Meaning in Cosmic Evolution*. Springer.

Lupisella, M. and John Logsdon. 1997. Do we need a cosmocentric ethic? *Paper IAA-97-IAA.9.2.09*. Turin, Italy: IAC.

McKay, D. S., E. K. Gibson Jr., K. L. Thomas-Keprta et al. 1996. Search for past life on mars: Possible relic biogenic activity in Martian meteorite ALH84001, *Science*, 273(5277): 924–930. (New Series.)

Margulis, L. and D. Sagan 1986. *Microcosmos: Four Billion Years of Microbial Evolution*. Berkeley, CA: University of California Press.

Marino, L. 2015. The landscape of intelligence, in Dick (2015), pp. 94–112.

Meech, K. J., J. V. Keane, M. Mumma et al. Eds. 2009. *Bioastronomy 2007: Molecules, Microbes and Extraterrestrial Life*. San Francisco, CA: Astronomical Society of the Pacific.

Michaud, M. A. G. 2007. *Contact with Alien Civilizations: Our Hopes and Fears about Encountering Extraterrestrials*. New York: Copernicus.

Moravec, H. 1988. *Mind Children: The Future of Robot and human Intelligence*. Cambridge, MA: Harvard University Press.

NASA. 1998. NASA astrobiology roadmap. https://nai.nasa.gov/media/roadmap/1998/

NASA. 2003. NASA astrobiology roadmap. https://nai.nasa.gov/media/roadmap/2003/

O'Meara, T. F. 2012. *Vast Universe: Extraterrestrial Life and Christian Revelation*. Collegeville, MN: Liturgical Press.

Persson, E. 2012. The moral status of extraterrestrial life, *Astrobiology*, 12, 976–984.

Peters, T. 1995. Exo-theology: Speculations on extraterrestrial life. In J. R. Lewis (Ed.), *The Gods Have Landed: New Religions from Other Worlds*. Albany, NY: State University of New York Press, pp. 187–206.

Peters, T. 2011. The implications of the discovery of extra-terrestrial life for religion, *Philos. Trans. R. Soc., A* 369: 644–655.

Peters, T. 2013a. Would discovery of ETI provoke a religious crisis? In Vakoch (2013), pp. 341–355.

Peters, T. 2013b. Astroethics: Engaging extraterrestrial intelligent life-forms, in Impey et al., Eds., *Encountering Life in the Universe: Ethical Foundations and Social Implications of Astrobiology*. pp. 200–221.

Peters, T. 2014. Astrotheology: A constructive proposal, *Zygon*, 49, 443–457.

Peters, T., M. Hewlett, J. Moritz, and R. J. Russell, Eds. 2018. *Astrotheology: Science and Theology Meet Extraterrestrial Life*. Eugene, OR: Cascades Books.

Race, M. S. 2007. Societal and ethical concerns, in Sullivan and Baross (2007), 483–497.

Race, M. S. 2010. The implications of discovering extraterrestrial life: Different searches, different issues, in Bertka (2010), pp. 205–219.

Race, M. S. 2015. Preparing for the discovery of extraterrestrial life: Are we ready? in Dick (2015), pp. 262–285.

Race, M. S., K. Denning, C. Bertka et al. 2012. Astrobiology and society: Building an interdisciplinary research community, *Astrobiology*, 12(10): 958–965.

Rescher, Ns. 1985. Extraterrestrial science, in E. Regis Jr. (Ed.), *Extraterrestrials: Science and Alien Intelligence.* Cambridge, UK: Cambridge University Press, pp. 83–116.

Rolston III, H. 2014. Terrestrial and extraterrestrial altruism, in Vakoch (2014b), pp. 211–222.

Rosenberg, A. 2016. *Philosophy of Social Science.* Boulder, CO: Westview Press.

Rospars, J. P. 2013. Trends in the evolution of life, brains and intelligence, *International Journal of Astrobiology*, 12, 186–207.

Ruse, M. 1998. *Taking Darwin Seriously.* Amherst, NY: Prometheus Books.

Sagan, C. 1973. *The Cosmic Connection.* New York: Doubleday & Company.

Sagan, C. 2000. *Carl Sagan's Cosmic Connection: An Extraterrestrial Perspective.* Cambridge, UK: Cambridge University Press.

Sapp, J. 2009. *The New Foundations of Evolution on the Tree of Life.* Oxford, UK: Oxford University Press.

Schneider, S. 2015. Alien minds, in Dick (2015), pp. 189–206.

Shostak, S. 1995. *Progress in the Search for Extraterrestrial Life.* San Francisco, CA: ASP.

Smith, K. 2016. Life is hard: Countering definitional pessimism concerning the definition of life. *International Journal of Astrobiology*, 15(4): 277–289.

Smith, R. 1982. *The Expanding Universe: Astronomy's "Great Debate."* Cambridge, UK: Cambridge University Press.

Stoeger, W. R. 1996. Astronomy's integrating impact on culture: A Ladrierean hypothesis, *Leonardo: Journal of the International Society for the Arts, Sciences and Technology*, 29(2): 151–154.

Sullivan, W. T. III and J. A. Baross Eds. 2007. *Planets and Life: The Emerging Science of Astrobiology.* Cambridge, UK: Cambridge University Press.

Tough, A. Ed. 2000. *When SETI Succeeds: The Impact of High-Information Contact.* Bellevue, WA: Foundation for the Future, online at http://www.futurefoundation.org/documents/hum_pro_wrk1.pdf.

Toynbee, A. 1957. *A Study of History*, Abridgement by D. C. Sovervell, vol. 2. London, UK: Oxford University Press.

Traphagan, J. 2011. Culture, meaning and interstellar message construction, in Vakoch (2011), pp. 469–485.

Traphagan, J. 2015. *Extraterrestrial Intelligence and Human Imagination: SETI at the Intersection of Science, Religion, and Culture.* Heidelberg, Germany: Springer.

Vakoch, D. 2009. Anthropological contributions to the search for extraterrestrial intelligence, in Meech et al. (2009), pp. 421–427.

Vakoch, D. Ed. 2011. *Communication with Extraterrestrial Intelligence.* Albany, NY: SUNY Press.

Vakoch, D. Ed. 2013. *Astrobiology, History and Society: Life Beyond Earth and the Impact of Discovery.* Berlin, Germany: Springer-Verlag, pp. 329–340.

Vakoch, D. Ed. 2014a. *Archaeology, Anthropology and Interstellar Communication.* Washington, DC: NASA.

Vakoch, D. 2014b. *Extraterrestrial Altruism: Evolution and Ethics in the Cosmos.* Heidelberg, Germany: Springer.

Vakoch, D. A. and M. Dowd Eds. 2015. *The Drake Equation: Estimating the Prevalence of Extraterrestrial Life through the Ages.* Cambridge, UK: Cambridge University Press.

Van Helden, A. 1974. Saturn and his Anses, *Journal for the History of Astronomy*, 5, 155–121.

Vidal, C. 2007. An enduring philosophical agenda: Worldview construction as a philosophical method, online at http://cogprints.org/6048.

Vidal, C. 2014. *The Beginning and the End: The Meaning of Life in a Cosmological Perspective.* New York: Springer.

Ward, P. D. and D. Brownlee. 2000. *Rare Earth: Why Complex Life is Uncommon in the Universe.* New York: Springer-Verlag.

Ward, P. D. and S. A. Benner. 2007. Alien biochemistries, in Sullivan and Baross, pp. 537–544.

Weintraub, D. 2014. *Religions and Extraterrestrial Life: How Will We Deal with It?* Heidelberg, Germany: Springer.

Wilkinson, D. 2013. *Science, Religion, and the Search for Extraterrestrial Intelligence.* Oxford: Oxford University Press.

World Economic Forum. 2013. *Global Risks 2013*, L. Howard, Ed. Geneva, Switzerland: World Economic Forum, 2013), at http://reports.weforum.org/global-risks-2013/section-five/x-factors/#hide/img-5.

Section XII

Exoplanets, Exploration of Solar System, the Search for Extraterrestrial Life in Our Solar System, and Planetary Protection

12.1 Exoplanets

Methods for Their Detection and Their Habitability Potential

Ken Rice

CONTENTS

12.1.1 INTRODUCTION

Extrasolar planets, or exoplanets, are planetary-mass bodies in orbit around stars other than the Sun. To date, we have confirmed detections of more than 3700 exoplanets, ranging from objects with masses more than 10 times that of Jupiter to ones with masses, and radii, less than that of the Earth.

Discovering what are very faint objects around very bright host stars is, of course, challenging. As an illustration, Figure 12.1.1 shows the 2M1207 system. It consists of a planetary-mass object (red object in lower-left region) in orbit around what is known as a brown dwarf (see, e.g., Chauvin et al. 2005). A brown dwarf is an object that does not become massive enough to ignite nuclear fusion in its core and, therefore, does not become a star.

In this case (Biller and Close 2007), the brown dwarf host has a mass 25 times that of Jupiter and a luminosity 500 times less than that of the Sun. The planetary companion has a mass of about 4 Jupiter masses and, because it is still young, is considerably brighter than Jupiter. The separation between the planet and its host is also more than 10 times greater than the separation between Jupiter and the Sun.

Now imagine trying to detect a planet that is fainter than that shown in Figure 12.1.1 and more than 10 times closer to a host that is 500 times brighter. It is clearly hugely challenging, and although it is sometimes possible to directly image an exoplanet around a Sun-like star, most of the exoplanets detected to date have been detected using indirect, rather than direct, methods.

In the rest of this chapter, we will focus on the various methods that have been used to detect exoplanets and also

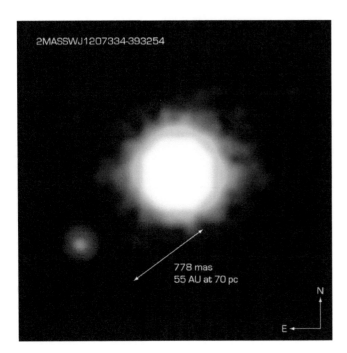

2MASSWJ1207334-393254

778 mas
55 AU at 70 pc

N

E

FIGURE 12.1.1 Image showing the 2MASS J12073346-3932539 system, which includes a planetary-mass body (red object in the lower-left region of the figure) and a brown dwarf host. (Courtesy of European Southern Observatory [ESO].)

what we can learn about these planets from these different methods. We'll focus, initially, on methods that indirectly detect exoplanets but will also explore direct detection, which is becoming more and more successful. We will also discuss what we now know about the properties and characteristics of exoplanets and what we may expect to discover in the coming years and will examine the prospects for detecting planets that may potentially be habitable.

12.1.2 DETECTION METHODS

12.1.2.1 PULSAR TIMING

Even though we often talk in terms of planets orbiting their parent stars, in reality, the star and planets all orbit the common center of mass of the system. In most planetary systems, the star has, by far, the most mass, and so, the center of mass will be located near, or sometimes inside, the star. The motion of the star around the center of mass is a consequence of the gravitational influence of the planets on their parent star. Therefore, one way to detect exoplanets is to observe stars to see if they are moving in a way consistent with them having planetary companions.

The first known exoplanets were detected in this way but were not actually detected around a Sun-like star; they were detected around a pulsar. A pulsar is a rapidly spinning compact object, typically the remnant core of a dead, high-mass star, a star that originally had a mass at least eight times that of the Sun.

Pulsars are highly magnetized and can emit a beam of electromagnetic radiation. Because the radiation beam is not aligned with the spin axis, the beam can sweep past the Earth, producing what we observe as pulses, hence the name.

If the pulsar is an isolated object, then the interval between these pulses will be very regular. If, however, it is the host to a planet, or planets, then because it will also be in orbit around the center of mass of the system, it will sometimes be slightly further from the Earth than at other times. Consequently, there will be variations in the pulse timings that can then be used to infer the presence of a planetary companion.

The first such detection was announced in 1992 (Wolszczan and Frail 1992) and was a planetary system composed of three planets orbiting a 1.4 solar-mass pulsar. The innermost planet has a mass only slightly larger than that of the Earth's moon, while the other two have masses a few times that of the Earth. They orbit the pulsar with orbital periods of 25, 66, and 98 days.

In this case, however, the pulsar likely formed from the merger of two white dwarfs, the remnant cores of stars with masses similar to that of the Sun. Therefore, rather than the planets being primordial, they probably formed from a disc of material generated during the merger of these two white dwarfs (Podsaidlowski 1993).

So, not only was this detection not of planets around a star like the Sun, these planets were also probably never in orbit around a main-sequence star, having formed well after the death of the stars that later became their pulsar host. Such planets also turn out to be rare, with very few others having been found. They are, however, still the first planets to be found outside our solar system and demonstrate how they can be detected via their influence on their parent object.

12.1.2.2 ASTROMETRY

Given that, in a planetary system, both the planets and the host star orbit the common center of mass, one way we might expect to be able to detect planets would be to actually observe the motion of the star around the center of mass of the system. Doing so would require the measuring the angular motion of a planetary host star by using high-precision astrometry.

However, the angular motion of a planetary host star is typically very small. For example, the Sun executes an orbit about the center of mass (due to the gravitational influence of the planets in the solar system), with a radius of about 0.005 AU. When viewed from a distance of 10 parsecs (1 parsec being 206,265 times the distance from the Sun to the Earth), the Sun would appear to have an angular motion of about 0.001 arcsecs, or 2.8×10^{-7} degrees. We have not yet been able to make such measurements. However, European Space Agency (ESA's) *Gaia* mission, which launched in 2013, will soon have collected enough data that it will be possible to detect planets via astrometric measurements of the motion of their host stars.

12.1.2.3 RADIAL VELOCITY/DOPPLER WOBBLE

The first exoplanet found in orbit around a Sun-like star was announced in Mayor and Queloz (1995). The method used to detect this planet is known as the radial velocity, or Doppler wobble, method. Stars have a spectrum with absorption lines that depend on the various atomic, and molecular, species in their atmospheres.

If a star is moving relative to us, then the wavelengths of these spectral lines shift in a way that depends on the relative line-of-sight velocity of the star. This is known as the Doppler effect. If the star is moving away from us, the lines will shift to longer wavelengths (red-shifted). If the star is moving toward us, they will shift to shorter wavelengths (blue-shifted).

However, if a star has a companion—such as a planet—it will orbit the common center of mass of the system. This means that at some times, it will be coming toward us, while at other times, it will be moving away. As illustrated in Figure 12.1.2, the spectral lines will, therefore, be red-shifted when the star is moving away from us and blue-shifted when it is moving toward us.

This effect can then be used to determine the radial (or line-of-sight) velocity of the star. If measurements are taken at different times, then one can determine how this radial velocity varies with time and can plot a radial velocity curve, as shown in Figure 12.1.3 (taken from Rowan et al. 2016).

Figure 12.1.3 shows the radial velocity of the Sun-like star HD32963 in meters per second (m/s), plotted against time, in days. It shows an amplitude of just over 10 m/s, with a period of 2372 days (6.5 years). In this case, the observations were taken over a period of more than 6.5 years, but the data has been phase-folded, so as to represent it as a single orbital period.

Figure 12.1.3 clearly shows that the star is executing some kind of orbit and indicates that there must be a companion. The period of the orbit can be determined directly from the radial velocity curve. The radial velocity measurements are also estimated using the spectrum of the star itself. Hence, we can also use the stellar spectra to determine what type of star it is and its mass. From the mass of the star, and the

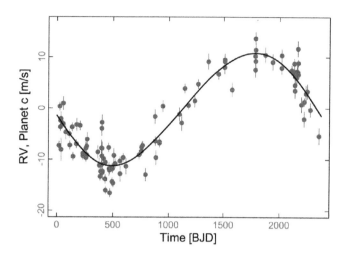

FIGURE 12.1.3 Radial velocity curve for the star HD32963 showing a radial velocity variation, with an amplitude of just over 10 m/s and a period of 6.5 years. (From Rowan, D., *Astrophys. J.*, 817: 104, 2016.)

period of the orbit, we can use Kepler's laws to infer the orbital radius of the planet, typically called the semimajor axis.

We can also determine the mass of the companion by using conservation of linear momentum. The radial velocity measurements give the line-of-sight velocity of the star. The companion's orbital radius and orbital period can be used to determine its velocity. Combining this with the mass of the star can be used to estimate the companion's mass.

As already noted, Figure 12.1.3 shows a system with an orbital period of 6.5 years. The central star has a mass of 1.03 solar masses, which indicates that the companion must have orbital radius (semimajor axis) of 3.4 AU, where 1 AU is an astronomical unit, the average distance from the Sun to the Earth. This means that this planet orbits its parent star in between where Mars and Jupiter orbit the Sun. The amplitude of the radial velocity curve indicates that the mass of the companion is around 0.7 Jupiter masses.

There is, however, one caveat. The radial velocity method only allows us to determine the radial, or line-of-sight, velocity of the star. We do not know the actual inclination of the orbit. If it is inclined relative to our line of sight, then the radial velocity we measure will be smaller than the actual orbital velocity of the star. The mass we estimate for the companion will therefore be smaller than its actual mass, and so, this estimate is typically taken to be a lower limit to the mass of the companion. However, it doesn't depend very strongly on inclination. Hence, the estimated mass will often be reasonably close to the actual mass.

One final thing that can be determined using radial velocity measurements is the eccentricity of the orbit. The eccentricity indicates how circular, or non-circular, the orbit is. An eccentricity close to 0 would indicate an almost circular orbit, while an eccentricity close to 1 would indicate an extremely eccentric orbit.

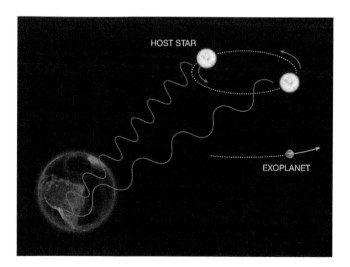

FIGURE 12.1.2 Illustration showing how the spectral lines of a star with a planetary companion will shift toward the blue side when the star is moving toward us and then toward the red side when it is moving away. (Courtesy of ESO.)

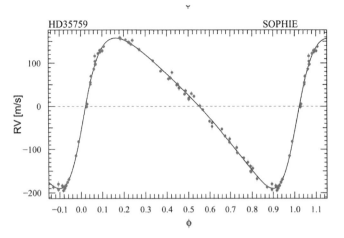

FIGURE 12.1.4 Radial velocity curve for the star HD35759. The non-symmetric nature of the light curve indicates that the orbit is eccentric, rather than circular. In this case, $e = 0.389$. (From Hebrard, G. et al., *Astron. Astrophys.*, 588, A145, 2016.)

The radial velocity curve shown in Figure 12.1.3 is quite symmetric and indicates that this system has a reasonably circular orbit ($e = 0.07$). Figure 12.1.4, however, shows the radial velocity curve for HD35759 (Hebrard et al. 2016) and indicates that the star is moving faster at some points in its orbit than in others. In this case, the orbit of the planet is quite eccentric ($e = 0.389$).

12.1.2.3.1 "Hot" Jupiters—the Case of 51 Pegasi B

The system illustrated in Figure 12.1.3 was announced in 2015 and comprises a roughly Sun-like star, with a roughly Jupiter-like planet that has an orbital period of 6.5 years—a bit smaller than the orbital period of Jupiter around the Sun (11 years). To a certain extent, it is a Jupiter analogue.

However, the first exoplanet detected via the radial velocity method around a Sun-like star was actually announced in Mayor and Queloz (1995) and was found around the star 51 Pegasus (hence the name, 51 Pegasi b). What was somewhat astonishing about this planet is that, although the mass estimate suggested a mass similar to that of Jupiter, the orbital period was only 4.23 days. In other words, this Jupiter-like planet is orbiting its parent star closer than Mercury is to the Sun.

Part of this is simply that it is easier to find planets that are closer to their parent stars than planets, of a similar mass, that are further out. This is partly a consequence of having to wait for at least an orbital period to be confident of a detection; orbital periods increase with orbital distance, and so, it takes longer to detect a planet on a wider orbit. Also, close-in planets induce larger radial velocities on their parent stars than planets that are further out and, hence, are easier to detect.

However, despite this selection effect, a reasonable number of such close-in planets—now known as "hot" Jupiters—have been found, with about 1% of all solar-like stars hosting such companions (Wang et al. 2015). Their origin is still not completely clear, but we currently think that they probably formed further from their star and then migrated through the planet-forming disc, or were scattered via dynamical interactions with other bodies, onto the orbits they now occupy.

12.1.2.3.2 Some Relevant Numbers

A Jupiter-mass planet orbiting a Sun-like star at 5 AU will induce a stellar velocity of 13 m/s. The most accurate radial velocity spectrometers today, such as the High Accuracy Radial velocity Planet Searcher North (HARPS-N) on the 3.6-m Telescopio Nazionale Galileo (TNG) on La Palma, can accurately measure stellar radial velocities down to about 1 m/s (Cosentino et al. 2014).

This means that we can quite easily detect Jupiter analogues, the only complication being that such planets take about 11 years to complete an orbit. Planets with similar masses, but with closer-in orbits, induce even larger radial velocities and have shorter orbital periods. Such planets are, therefore, now reasonably easy to detect.

If we consider Earth-mass planets, then current instruments can only detect radial velocity signatures from Earth-mass planets if they orbit very close to their parent stars. For example, to induce a radial velocity of 1 m/s on a Sun-like star will require an Earth-mass planet orbiting 100 times closer to its star than the Earth is to the Sun. To detect a true Earth-analogue (1 Earth mass orbiting a Sun-like star, with an orbital period of 1 year) will require measuring a radial velocity signature of only 10 cm/s. This is currently not possible, and so, confirming a true Earth analogue will have to wait for future instruments and methods. This will require to be able to detect such a small radial velocity signature and will also require to be able to extract this signal from data that will include stellar noise of comparable amplitude.

12.1.2.4 The Transit Method

The radial velocity method (discussed in Section 12.1.2.2) can be used to determine a number of properties of a planetary system, in particular, orbital period, orbital radius, planetary mass, and orbital eccentricity. Another method for detecting, and characterizing, exoplanets is the transit method. This involves observing a large number of stars, with the goal of measuring small dips in brightness caused by something passing between us and the star being observed.

Figure 12.1.5 illustrates the basics of the transit method. Most stars have a reasonably constant brightness. Hence, if there is a small dip in brightness, this may indicate that something has passed between the star and us. If this dip repeats periodically, then it can indicate that something is in orbit around the star, potentially a planetary-mass companion.

One might expect that it is pretty straightforward to detect planetary companions by using the transit method. It turns out, however, that there are various complications. As mentioned in Section 12.1.1, a brown dwarf is an object that is more massive than something regarded as a planet but isn't massive enough to become a star. Its radius, however, is very similar to Jupiter's radius. Therefore, a transit of a brown dwarf can look similar to that of a massive planet. Similarly, a grazing eclipse by a stellar companion can block as much

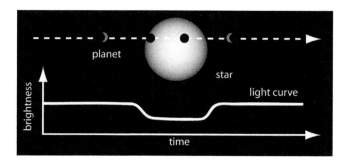

FIGURE 12.1.5 Illustration showing the transit method for detecting exoplanets. If a planet passes in front of its parent star, it will block some of the star's light, which we will measure as a small dip in brightness. (Courtesy of NASA AMES, Mountain View, CA.)

light as a full planetary transit. Very distant binary stars that happen to almost align with the star being observed can also produce apparent periodic dips of the target star.

However, it is possible to eliminate these false positives, and the transit method has become the most successful of the exoplanet detection methods, accounting for more than 2500 of the 3700 known exoplanets.

12.1.2.4.1 What Can the Transit Method Tell Us?

As with the radial velocity method, the transit method can tell us the orbital period of the companion. If we know the type of star, then we can also estimate its mass and radius. Using Kepler's laws, the orbital period can be used to estimate the orbital distance. The amount of light blocked also tells us the ratio of the cross-sectional area of the planet to the cross-sectional area of the star. From this, given the radius of the star, we can determine the radius of the planet.

Figure 12.1.6 shows a transit lightcurve for the HD209458 system (Brown et al. 2001), the first planet detected transiting its parent star (Charbonneau et al. 2000). It was generated by combining lightcurves from four different

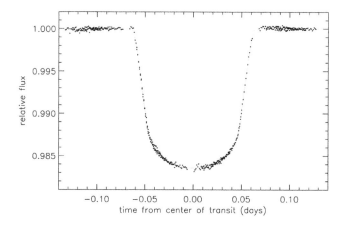

FIGURE 12.1.6 Transit lightcurve for HD209458. The transiting planet (HD209458b) has an orbital period of 3.52 days, and, when it transits, it blocks about 1.7% of the starlight. Given the radius of the star (1.146 Solar radii), this indicates that the planet has a radius of 1.347 Jupiter radii. (From Brown, T.M. et al., *Astrophys. J.*, 552, 699–709, 2001.)

transits. In this case, the planet (HD209458b) is found to have an orbital period of 3.52 days. The transit of the planet across the face of its parent star takes just under 3 hours and blocks about 1.7% of the starlight. Given that the star has a radius 1.146 times that of the Sun means that the companion has a radius 1.347 times that of Jupiter.

As mentioned earlier, a radius similar to that of Jupiter does not mean that the companion is of planetary mass; brown dwarfs can have similar radii. However, HD209458b was actually first detected via the radial velocity method, which means that there was already a mass estimate.

Normally, this would be a minimum mass, since the inclination of the orbit would be unknown. However, this planet also transits its host star, which means that the planet orbits in a plane that is very close to our line of sight. The mass estimate is therefore going to be very close to the actual mass of the companion. In this case, the radial velocity measurements indicate a mass of about 0.7 Jupiter masses.

Therefore, from the transit observations, we have an object with a radius slightly bigger than that of Jupiter, while the radial velocity measurements indicate a mass slightly below that of Jupiter. Clearly, this is what we would normally call a gas giant: a planet like Jupiter that is primarily composed of a hydrogen-rich, gaseous envelope/atmosphere. As we will discuss later in this chapter, combining different detection methods can provide additional information about the systems being observed.

12.1.2.4.2 Some Relevant Numbers

A Jupiter-like planet has a radius about 10 times less than that of a Sun-like star. The transit of a Jupiter-like planet will therefore block about 1% of the starlight from a Sun-like star (see, e.g., Figure 12.1.6). Detecting such signals is quite possible, even from the ground. In fact, one of the most successful transit projects is the Wide-Angle Search for Planets (WASP—Street et al. 2003), which uses camera lenses. This allows the WASP team to survey a reasonably large patch of the sky containing many stars. Just over 150 exoplanets have been detected by the WASP team, mostly exoplanets with radii similar to that of Jupiter.

An Earth-like planet has a radius about 10 times less than that of Jupiter and would therefore only block 0.01% of the starlight from a Sun-like star. This isn't really possible from the ground but is possible from space. NASA's Kepler satellite has now detected more than 1500 exoplanets via the transit method, in excess of 200 of which have sizes similar to that of the Earth.

Figure 12.1.7 shows the known planets by size (as of January 2018) and goes from planets slightly larger than Jupiter to those smaller than the Earth (data taken from http://exoplanets.eu). What it indicates is that we now have quite a large number of planets similar to, and even smaller than, the Earth. What is also interesting is that Figure 12.1.7 indicates that the most common type of exoplanet is probably one with a size between that of the Earth and Neptune (super-Earths and sub-Neptunes), a type of planet that we do not find in our solar system.

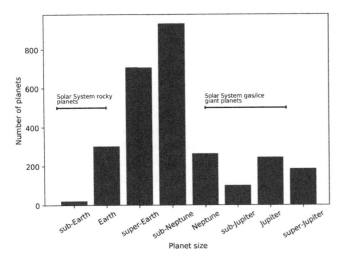

FIGURE 12.1.7 Bar chart showing the number of exoplanets of various sizes, from ones smaller than the Earth (sub-Earth) to those larger than Jupiter (super-Jupiter). What this indicates is that we have now found quite a large number with sizes similar to, or even smaller than, the Earth. Also, the most common exoplanet appears to have a size between that of the Earth and Neptune (super-Earths and sub-Neptunes). There are no such planets in our solar system. (Data from http://exoplanet.eu.)

One final thing to consider here is that a planet will only transit if it happens to pass between us (the observer) and its parent star. The probability of a transit depends on the radius of the star and on the distance of the planet from the star; the further a planet is from a star, the less likely it is to transit. Detecting transits therefore requires observing many stars. The first phase of NASA's Kepler mission, for example, observed more than 100,000 stars (Borucki et al. 2006) and detected just over 2500 planets.

This does not, however, mean that only 2500 of Kepler's target stars host planets. Given that the transit probably decreases with increasing planet orbital radius, it is very unlikely that a planet with an orbital radius greater than about 1 AU will be detected. Similarly, a large number of closer-in planets may simply not transit, because their orbit means that they will not pass between their parent star and us. As we will discuss in a later section, this can be used to infer the planet occurrence rate.

12.1.2.5 GRAVITATIONAL MICROLENSING

The final indirect detection method that we will briefly discuss is based on Einstein's Theory of General Relativity. Gravity is the force between two objects that depends on their masses and on the distance between them. The original formulation, presented by Isaac Newton, essentially suggested that this force acted instantaneously. This violates Einstein's Theory of Special Relativity, which suggests that nothing can travel faster than the speed of light.

Einstein's Theory of General Relativity proposed that objects with mass act to distort spacetime (which is really just the fusion of the three spatial dimensions with the time dimension) and that gravity is simply a manifestation of this

curved spacetime. Essentially, massive objects in the universe will tend to "fall" toward other massive objects because of this curvature of spacetime.

Similarly, light will also be affected by this curvature of spacetime. Light follows a path that takes the least time and, as a consequence of the curvature of spacetime, will appear to be bent if it passes near an object that has mass. We can then use this to detect planets around stars.

This is illustrated in the diagram in Figure 12.1.8. If a distant star happens to pass behind a nearer star, when viewed from the Earth, the nearer star can act like a lens, focusing some of the more distant star's light onto the Earth. In principle, this will produce two images of the more distant (source) star. However, telescopes today do not have the resolution to actually observe this. What actually happens is that as the source star moves behind the lens star, more and more of its light will be focused onto the Earth and it will appear to get much brighter. As it passes out from behind the lens star, it will then get fainter and fainter, until it returns to its original brightness. This can take many days, potentially even a few months.

If, however, the lens star happens to have a planet and this planet happens to be in the right place, it can provide an additional magnification that will last for a relatively short amount of time, typically a few hours to a day.

For example, Figure 12.1.9 shows the lightcurve for the OGLE-2005-BLG-390 microlensing event, which occurred in 2005 (Beaulieu et al. 2006). The event lasted for about 50 days and was due to the light from the distant star being mangnified by a closer star (as illustrated in Figure 12.1.8). In this case, the lens star does have a planet, which causes the additional magnification—lasting for approximately 1 day—about 10 days after July 31, 2005. This event was analyzed, and it indicated that the planet has a mass of about 5.5 Earth masses and an orbital radius of about 2.6 AU (i.e., about 2.6 times further from its parent star than the Earth is from the Sun).

What makes this an interesting method is that it is able to detect planets with masses similar to that of the Earth. However, because such alignments are very rare, we typically search in the direction of the center of our galaxy (the Milky Way), because there are lots of stars in that direction and therefore

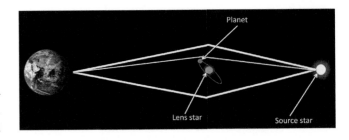

FIGURE 12.1.8 Diagram illustrating how we can use gravitational microlensing to detect planets around other stars. If two stars line up, when viewed from the Earth, the nearer star (lens star) can focus the light from the further star (source star), making it appear to get brighter. If the lens star also hosts a planet, the planet can provide an additional magnification that can then be used to infer its presence and some of its properties.

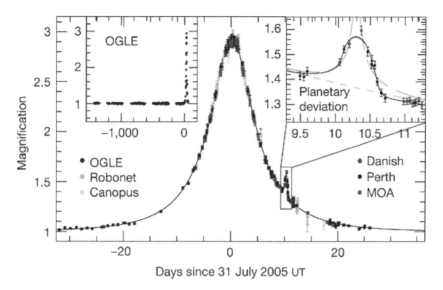

Days since 31 July 2005 UT

FIGURE 12.1.9 Lightcurve for the OGLE-2005-BLG-390 microlensing event. This shows the overall magnification event, lasting about 50 days, and the additional magnification, about 10 days after July 31, 2005, which is indicative of the lens star hosting a planet with a mass of about 5.5 Earth masses and orbiting at about 2.6 AU. (From Beaulieu, J.-P. et al., *Nature* 439: 437–440, 2006.)

such chance alignments become more likely. A consequence of this, however, is that such events will typically show a planetary signature only if the planet lies a few astronomical units from its host star (i.e., if the orbital radii is greater than that of the Earth around the Sun). Hence, this method can potentially find Earth-mass planets that are cool.

A downside of this method, though, is that such an event will at most happen once for a particular system, and these stars are faint, so there is not really any possibility of doing follow-up observations. So, we can get estimates for the planet's mass and orbital radius but cannot really characterize the system any further. However, they do provide an additional sample of exoplanets and also probe a region of parameter space (cool, low-mass planets) that cannot currently be probed by other methods.

12.1.2.6 DIRECT IMAGING

The previous sections have discussed indirect methods for detecting exoplanets. Here, we will discuss their direct detection. An advantage of this is that rather than inferring the planets' properties from their influence on their parent stars, we can infer them through direct observations. In particular, we can make direct observations that will help us understand the properties of their atmospheres. As we will discuss later, we can also infer some atmospheric properties by using spectroscopic observations of transiting exoplanets. However, the direct observation is likely to play a key role in the characterization of exoplanet atmospheres.

Directly imaging exoplanets involves surveying nearby young (<200 million years old) stars to look for faint companions. Typically, observations are taken at different times to establish if the star and its potential companion are moving together and also to establish if the companion appears to be orbiting the star.

Typically, when making observations with a telescope, the telescope is rotated so as to fix the orientation of what is being observed in the resulting image. However, when trying to directly detect planets, rotation is often turned off. If the star is placed in the center of the image, any other sources in the image will appear to move as the telescope tracks the target star. However, the noise, which comes from the structure, and optics, of the telescope itself, will remain fixed. The star, and this noise, can then be removed, enhancing any other real objects in the image. These can then be analyzed to see if they are planets in orbit around the target star. This process is known as angular differential imaging (ADI).

The HR8799 planetary system (shown in Figure 12.1.10) was one of the first to be directly detected (Marois et al. 2008). It was found to initially have three and then four (Marois et al. 2010) planetary companions with masses of between 5 and 7 Jupiter masses. The closest orbits at about 14.5 AU, while the furthest orbits at about 68 AU. The central star has a mass about 1.5 times that of the Sun and is probably about 30 million years old.

This quite nicely illustrates the constraints associated with directly imaging extrasolar planets. Currently, it can only really detect relatively massive planets (masses above that of Jupiter) on wide orbits (orbital radii greater than about 10 AU) and that are still quite young. This is because we typically observe in the infrared, where we detect thermal emission from the planet itself, rather than reflected light from the parent star. Planets with masses below that of Jupiter are simply not bright enough to be detected. Also, these planets tend to cool as they age and hence become fainter. Therefore, we can only detect them when they are relatively young (younger than a few hundred million years). Also, even though we can remove much of the bright central star from the resulting images, it is still difficult to find planetary companions that are closer than about 10 AU.

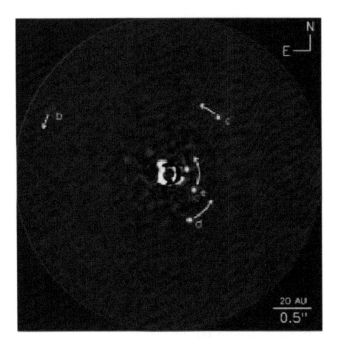

FIGURE 12.1.10 Image showing the HR8799 system. It has four planets (HR8799b, c, d, and e). The closest (e) orbits at about 14.5 AU, while the furthest orbits at about 68 AU. The planets' masses are estimated to be between 5 and 7 Jupiter masses. (From NRC-HIA, C. Marois, and Keck Observatory.)

12.1.2.6.1 Some Relevant Numbers

The direct search for companions to stellar bodies has already uncovered a reasonable number of objects. If we restrict it to those with planetary mass (typically less than about 13 Jupiter masses) and in orbit around a star (rather than having a brown dwarf host), the numbers are quite modest. We now know of about 10 such objects. As mentioned, they typically have masses above a few Jupiter masses and orbit at radii beyond about 10 AU.

The above, however, is mostly a selection effect; we simply don't have the ability to detect lower-mass planets on orbits that are closer to their parent star. Given that directly imaging a planet allows us to potentially characterize its atmospheric properties, we would like to be able to do so for smaller planets on closer-in orbits. This would allow us to directly probe their potential habitability. This, however, will have to wait for another generation of telescopes and instruments, some of which, such as 30-m class, ground-based telescopes, will become available within the next decade.

12.1.3 WHERE HAVE WE EXPLORED?

We have now covered the main exoplanet detection methods. In most cases, exoplanets are detected via indirect methods (astrometry, radial velocity, transits, timing, and microlensing), but we are starting to be able to directly image some planetary-mass objects. Each method, however, typically works in some circumstances but not in others.

Figure 12.1.11 illustrates the different detection methods and the regions in which they can detect exoplanets (Dominik 2010). The y-axis represents the planet's mass, while the x-axis is the planet's temperature, increasing as it gets closer to its host star. Orange shows the region accessible by the transit technique. It is possible to detect planets with small radii (and, hence, low masses). However, the probability of a transit occurring decreases with increasing orbital radius, and, therefore, it becomes increasing unlikely that we will observe a transit as the planet's orbital radius increases. Therefore, the transit method might be able to detect small planets, but it can currently only really do so for planets with orbital radii inside about 1 AU.

The yellow shows the region accessible by the radial velocity/Doppler wobble method. It can detect relatively low-mass planets but not quite as low as is possible with the transit method. However, the influence of the planet on the star decreases as the planet's orbital radius increases. Therefore, the mass of a planet that can be detected by the Doppler wobble method increases as the planet's orbital radius increases.

The light blue shows the region in which gravitational microlensing can detect exoplanets. It can detect quite-low-mass planets (down to masses similar to that of the Earth). However, because of the geometry of these microlensing events, it typically finds exoplanets with orbital radii greater than that of the Earth (cool low-mass planets). Finally, purple shows where direct imaging can detect exoplanets. As discussed earlier, this method preferentially finds massive planets on wide orbits.

The red circle illustrates the region around a Sun-like star in which we would find something analogous to the Earth, a rocky planet on which liquid water could exist on its surface. It is, however, a region of parameter space that we currently

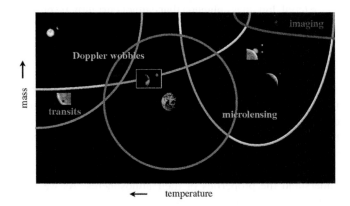

FIGURE 12.1.11 Diagram illustrating the different regions of planet mass-orbital radius space accessible to the various exoplanet detection methods. In the diagram, orbital-radius is expressed as Temperature, since the closer a planet orbits to its host star, the higher its temperature. The region of parameter space that is still not accessible is exoplanets with Earth-like masses orbiting a Sun-like star at an orbital radii similar to that of the Earth. (From Dominik, M., *Gen. Relativ. Grav.*, 42: 2075–2100, 2010.)

can't easily explore. We can find planets more massive than the Earth with similar orbital radii, and we can find planets of a similar mass, either closer to their parent star or further from their parent star. Finding a true Earth analogue will, however, probably have to wait for the next generation of telescopes and instruments. On the other hand, as we will discuss later, it is already possible to find (using radial velocity and transit measurements) planets around stars less massive than the Sun that orbit in a zone where liquid water could exist on their surface.

Figure 12.1.12 shows the actual exoplanet data and also illustrates what is shown in Figure 12.1.11. It shows planet mass (in Jupiter masses) against orbital separation (in AU). The red circles are those exoplanets detected via the transit method. They can go down to quite low masses but show few detections beyond 1 AU. The blue circles are the radial velocity exoplanets, which show some exoplanets with low masses at small separations but only shows the more massive planets at larger separations. The green triangles are the microlensing exoplanets, which tend to be found beyond 1 AU, while the yellow squares are the massive, wide-orbit planets detected via direct imaging. As also illustrated in Figure 12.1.11, a region that we cannot yet explore is Earth-mass planets ($M = 0.0032$ Jupiter masses) at about the same distance as the Earth is from the Sun (~1 AU).

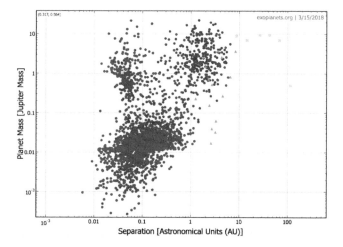

FIGURE 12.1.12 Planet mass (in Jupiter masses) against orbital separation (in astronomical units, or AU) for all known exoplanets. The different colors are for the different detection methods: red circles (transit), blue circles (radial velocity), green triangles (microlensing), and yellow squares (direct imaging). This illustrates, as discussed in the text, what type of planets each method can uncover and also illustrates that we are not yet able to detect true Earth analogues—a ~0.003 Jupiter mass planet at ~1 AU around a Sun-like star. (From NASA Exoplanet Archive. This research has made use of the NASA Exoplanet Archive, which is operated by the California Institute of Technology, under contract with the National Aeronautics and Space Administration under the Exoplanet Exploration Program. With permission.)

12.1.4 EXOPLANET CHARACTERISTICS

As discussed previously, the different detection methods provide different information about the planets that they detect. For example, both the transit and radial velocity methods can determine the planet's orbital period (and, hence, orbital radius), but the former also determines the planet's physical radius, while the latter can be used to estimate the planet's mass. If you combine the two, you can then get an estimate for the planet's density and, hence, internal composition.

There are now many exoplanets for which we have both mass and radius estimates. Initially, these were mostly planets that had masses and radii consistent with them being gas giants, like Jupiter or Saturn. In some cases, however, these turned out to have radii somewhat bigger than expected (Bodenheimer et al. 2001), with some having radii almost twice that of Jupiter (Hebb et al. 2009). In the absence of any additional energy sources, gas giant planets more than a few billion years old should not have radii more than 1.2 times that of Jupiter (Fortney et al. 2007).

However, most of these inflated gas giant exoplanets are very close to their parent stars and, hence, are heavily irradiated. Being so strongly irradiated might either reduce the rate at which these planets cool and, hence, shrink, or actually deposit energy in the planetary interior, inflating their radii (Lopez and Fortney 2016).

With the advent of NASA's Kepler mission, we now also know of numerous planets with radii very similar to that of the Earth. In some cases, we have been able, using ground-based radial velocity measurements, to estimate their masses. Figure 12.1.13 shows a mass radius plot for all exoplanets with masses below 20 Earth masses and with a mass precision of better than 20% (Lopez-Morales et al. 2016).

The curves in Figure 12.1.13 show different possible internal compositions, from a fully iron planet (100% Fe) to one that would be almost entirely water (100% H_2O). It also shows Earth and Venus (between 35% and 40% iron) for reference. There are now a number of known exoplanets with compositions similar to that of Earth and Venus. However, the colors indicate the planet's likely surface temperature, based on the incident stellar flux (relative to that of the Earth). All of those with Earth-like internal compositions orbit very close to their parent stars and therefore will have surface temperatures much higher than that of the Earth. For example, K-78b (or, Kepler-78b) orbits a star similar to the Sun, has an Earth-like internal composition, but has an orbital period of only 8.5 hours (Pepe et al. 2013). It, therefore, probably has a surface temperature in excess of 2000 K.

Figure 12.1.13 also shows that there are a number of super-Earths that have substantial amounts of water. However, if water makes up more than 1% of the planet's mass, then the pressure at the bottom of a water layer will be high enough to form high-pressure ice polymorphs (Levi et al. 2014), which means that there will be no direct contact between the liquid ocean and the planetary interior. This will probably have a

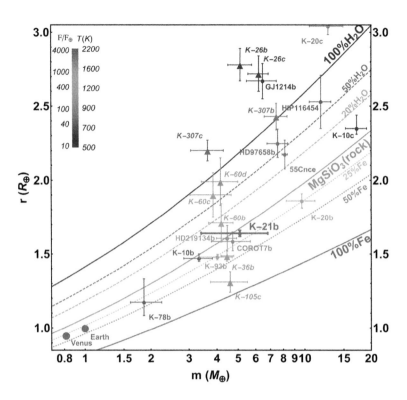

FIGURE 12.1.13 Mass-radius diagram for all exoplanets with masses less than 20 Earth masses and with a mass precision of better than 20%. The curves show different possible internal compositions. Earth and Venus are shown for reference. (From Lopez-Morales, M., et al., *Astron. J.*, 152, 204, 2016.)

substantial impact on the chemical composition of the liquid ocean, which will almost certainly impact the potential habitability of such planets (Kite and Ford 2018).

There also appears to be an approximate gap in the exoplanet radius distribution. Planets tend to have radii below about 1.5 Earth radii, or above 2 Earth radii; there are currently very few with radii between 1.5 and 2 Earth radii (Fulton et al. 2017). Those below 1.5 Earth radii tend to be rocky, while those above 2 Earth radii are either water-rich or have substantial volatile envelopes (Rogers 2015). If we think that habitability requires a predominantly rocky planet, then that would suggest that such planets will preferentially have radii less than 1.5 Earth radii.

12.1.5 EXOPLANET OCCURRENCE RATES

Now that we have a large sample of known exoplanets, we can start to estimate actual occurrence rates, the mean number of planets, within some parameter range, per star. For example, we expect about 10% of solar-like stars to host gas giant planets orbiting within 3 AU (Cumming et al. 2008). The occurrence rate for close-in giant planets ("hot" Jupiters) is considerably smaller, with only about 1% of solar-like stars hosting such planets (Wright et al. 2012). The occurrence rate of very wide orbit giant planets is also low (Biller et al. 2013), potentially also as low as 1% for those with masses between 5 and 13 Jupiter masses (Bowler and Nielsen 2018).

For smaller planets, the occurrence rate tends to be higher, with about 15% of solar-like stars hosting planets with masses

between 3 and 10 Earth masses and with orbital periods less than 50 days (Howard et al. 2010). Overall, we might expect about 50% of solar-like stars to host a planet with an orbital period of less than 85 days (Fressin et al. 2013).

It's also possible to estimate the occurrence rate of Earth-size planets in a zone where they have the potential to be habitable (i.e., where liquid water could exist on their surface). The result does depend on assumptions about where they might be habitable, but Petigura et al. (2013) suggest that at least 6% of Sun-like stars host a planet with a size comparable to that of the Earth, and that orbits in a zone where liquid water could exist on its surface.

However, we do have to be slightly careful. This is partly based on extrapolating from the known population of small exoplanets. Many of these are very close to their parent stars, are heavily irradiated, and, hence, may initially have had substantial volatile envelopes that they have now lost (Fulton et al. 2017). Therefore, these may not be part of a population of rocky planets that extends to regions where they may be habitable (Lopez and Rice 2018). Consequently, estimates that do not take this into account may suggest a higher occurrence rate than is actually the case.

12.1.6 THE HABITABLE ZONE

As already indicated previously, the habitable zone is typically taken to be the region, around a star, in which a planet could have liquid water on its surface and, hence, could potentially support life. We do have to be slightly careful when

FIGURE 12.1.14   Figure showing the habitable zone (based on the analysis by Kopparapu et al. 2013) around different types of stars. It also shows the solar system planets and a sample of known exoplanets that happen to lie in their stars' habitable zones. (From Kasting, J.F. et al., *Proc. Natl. Acad. Sci. USA*, 111, 12641–12646, 2014.)

using this, because being in the habitable zone doesn't mean that a planet will support life, and being outside it does not necessarily mean that it cannot support life.

The exact region in which a planet could have liquid water on its surface depends on a number of factors, including the type of star, the atmospheric composition of the planet, and how much of the incoming stellar flux it reflects back into space.

A simple estimate might suggest that it extends—around a Sun-like star—from about 0.5 AU to about 2 AU (i.e., from half the distance between the Sun and the Earth to twice the distance between the Sun and the Earth). However, this would imply that both Venus and Mars (in our solar system) lie within the habitable zone, yet neither is regarded as currently habitable. A more accurate estimate might suggest that it is somewhat narrower (Kasting et al. 1993; Kopparapu et al. 2013). However, if a planet is highly reflective, or has a weak greenhouse effect, it could sustain liquid water on its surface even when quite close to its parent star (Zsom et al. 2013). Similarly, a planet with a low reflectivity and/or a strong greenhouse effect could sustain liquid water on its surface even if quite far from its parent star, in extreme cases potentially out to ~10 AU around a Sun-like star (Pierrehumbert and Gaidos 2011). This would, however, require a hydrogen-rich atmosphere. For a planet on which carbon dioxide drives the greenhouse effect, the outer edge is probably well within 2 AU (Kopparapu et al. 2013).

Figure 12.1.14, from Kasting et al. (2014), and based on the work of Kopparapu et al. (2013), illustrates the habitable zone for different types of stars. It also shows the solar system planets and a selection of known extrasolar planets that happen to lie within their star's habitable zone.

What the Figure shows is that around a Sun-like star, the zone probably extends from just inside where Earth is in the solar system to just inside the orbit of Mars. Stars more massive than the Sun are brighter, and hence, the habitable zone is at larger orbital radii. Conversely, stars less massive than the Sun are fainter, and so, the habitable zone moves to smaller orbital radii. This has implications that we will discuss in more detail later.

12.1.6.1   KNOWN POTENTIALLY HABITABLE EXOPLANETS

As mentioned previously, it is not possible to precisely define the habitable zone, and, hence, it is not possible to determine exactly how many known exoplanets lie within such a zone. However, we could be optimistic and assume that the habitable zone is quite broad (~0.75 AU to ~1.84 AU) and that the range of planets that could sustain liquid water on their surface is also broad (planet mass below about 5 Earth masses and/or planet radius less than 2.5 Earth radii). Given this, there are already about 40 known exoplanets that could have liquid water on their surface.

There are, however, indications that most planets with radii larger than 1.5 Earth radii are not rocky (Rogers 2015; Fulton et al. 2017). Similarly, a more conservative estimate of the habitable zone is probably also preferred (Kopparapu et al. 2013). Consequently, a more reasonable estimate would suggest about 13 known exoplanets in the habitable zones of their parent stars, most of which are around stars less massive than the Sun, known as M- or K-dwarfs.

12.1.6.2   POTENTIALLY HABITABLE PLANETS AROUND LOW-MASS STARS

As mentioned in Section 12.1.6.1, a conservative estimate of the number of known exoplanets in a potentially habitable zone indicates that most orbit stars less massive than the Sun. This, however, is probably mostly a selection effect. As shown in Figure 12.1.14, the habitable zone around a star less massive than the Sun is closer to the star than it is around a Sun-like star. This means that it is easier to find planets in the habitable zone around lower-mass stars than it is around higher-mass stars. They are more likely to transit, and they induce a larger radial velocity on their parent star.

The next major exoplanet mission is NASA's Transiting Exoplanet Survey Satellite (TESS), due to launch later in 2018. It will survey 200,000 stars and is expected to find around 1700 transiting exoplanets, of which just over 500 will have radii less than twice that of the Earth (Sullivan et al. 2015). Potentially 50 of these planets will lie in their star's habitable zone, and most of these planets' parent stars will be M- or K-dwarfs, stars with masses less than that of the Sun. So, TESS will probably enhance the number of known exoplanets in the habitable zone of these lower-mass stars (Ballard 2018).

However, since the habitable zone around stars less massive than the Sun is closer to the star than around a Sun-like star, planets in the habitable zone around M- and K-dwarfs are likely to be tidally locked (Barnes 2017). This means that they will have a side that always faces their parent star, just like the same side of the Moon always faces the Earth.

How such a configuration will influence the atmospheric properties of a rocky planet in the habitable zone around an M-dwarf is unclear. It was originally thought (Dole 1964) that tidal locking would lead to the night side being so cold that it would freeze out and potentially collapse the atmosphere. However, recent work has suggested that atmospheric transport (Joshi et al. 1997), cloud feedbacks (Yang et al. 2013), and/or ocean heat transport (Hu and Yang 2014) could keep the night side warm enough to prevent such a collapse of the atmosphere.

Hence, such planets may indeed be habitable and may well be good direct imaging targets that would allow us to look for signatures of life in their atmospheric features (Rugheimer et al. 2015).

12.1.6.2.1 Proxima Centauri B

In 2016, it was announced that Proxima Centauri, the Sun's closest stellar neighbor, hosts a potentially rocky planet orbiting in the region where liquid water could exist on its surface (Anglada-Escudé et al. 2016). The planet has a mass of around 1.3 Earth masses, but orbits a star that has a mass just over one-tenth that of the Sun. The planet also orbits this star at a distance of only 0.05 AU—20 times to closer to its parent star than the Earth is to the Sun.

There are some indications that atmospheric conditions of Proxima Centauri b might be reasonably insensitive to changes in stellar flux (Boutle et al. 2017), which would indicate quite a large range of parameters over which it might have conditions suitable for the presence of liquid water on its surface. On the other hand, relative to the Earth, Proxima Centauri b is heavily irradiated by extreme ultraviolet (XUV) and x-rays. High-energy photons can cause photo-dissociation of water, the hydrogen from which can then escape if it becomes sufficiently energetic. In fact, XUV heating of the upper atmosphere can also lead to the loss of other volatile elements. Hence, a heavily irradiated planet could lose the atmospheric elements necessary for life. The outcome will,

however, depend on its actual evolutionary history (Meadows et al. 2018), and there are indications that Proxima Centauri b could still have conditions suitable for liquid water to exist on its surface, despite being exposed to far more XUV and x-ray irradiation than the Earth (Ribas et al. 2016).

That Proxima Centauri b is both orbiting the Sun's closest stellar neighbor and in its parent star's habitable zone makes it a fascinating target for observations. Directly observing Proxima Centauri b will, however, probably have to wait for the next generation of ground-based telescope, such as the 39-m European Extremely Large Telescope (Turbet et al. 2016). However, very-high-contrast imaging, together with high-resolution spectroscopy, could be used to observe variations in reflected starlight as the planet orbits its parent star (Snellen et al. 2015), which may provide information about the orbital inclination and the planet's albedo (reflectivity).

Also, given that a tidally locked planet will probably have a dayside to night side temperature gradient, infrared observations with the James Webb Space Telescope (JWST—due for launch in 2020) could be used to detect thermal phase variations (Kreidberg and Loeb 2016). In fact, these thermal phase variations will depend on whether or not the planet has an atmosphere that can redistribute heat from the dayside to the night side. Hence, these observations could give an indication of whether or not Proxima Centauri b actually has an atmosphere.

12.1.6.3 Potentially Habitable Planets Around Sun-Like Stars

As mentioned previously, most of the known potentially habitable exoplanets are around stars less massive than the Sun. Also, we expect that most of the potentially habitable exoplanets that we will detect in the next few years will also mostly be around stars less massive than the Sun.

However, as mentioned in Section 12.1.6.2, this is mostly a selection effect; it is easier to find rocky planets in the habitable zones of M- and K-dwarfs than it is to find rocky planets in the habitable zone of a Sun-like star (also known as a G-dwarf). It seems likely that at least a few percent of Sun-like stars host rocky planets in a region where liquid water could exist on their surface (Petigura et al. 2013).

Detecting such planets will, however, probably have to wait for the ESA's PLATO mission, which is due to launch in 2026. PLATO will survey a large number of bright nearby stars, with the goal of detecting planetary transits. In particular, it will survey enough stars that it is likely to recover a sample of Earth analogues, planets with masses and radii similar to that of the Earth and orbiting a Sun-like star at about 1 AU (Rauer et al. 2014). The targets will also be bright enough that we may be able use other observational methods to more accurately determine masses and, consequently, their internal composition.

12.1.6.4 Determining Habitability

As already mentioned, simply finding rocky planets in the habitable zones of other stars does not mean that they are indeed habitable. This will require other observations to try and identify biosignatures. I won't discuss possible biosignatures, as that is covered elsewhere in this volume. However, it is clear that detecting biosignatures will require characterizing the atmospheres of these potentially habitable exoplanets.

We can, however, already say something about exoplanet atmospheres. For close-in exoplanets, we can use transit spectroscopy (Seager 2008) and secondary eclipse spectroscopy (Wilkins et al. 2014). Both these look for changes in the stellar spectrum as a planet move in front of, or behind, its parent star. Such observations have indicated the presence of sodium (Snellen et al. 2008), water vapor (Birkby et al. 2013), and clouds (Evans et al. 2013) in the atmospheres of "hot" Jupiters.

We can even sometimes extend this to lower-mass planets. There are observations suggesting the presence of clouds in the atmosphere of a super-Earth (Kreidberg et al. 2014), and observations indicating that some of the planets in the TRAPPIST-1 system do not have cloud-free, hydrogen-dominated atmospheres. This suggests that these planets are most likely rocky (de Wit et al. 2018). TRAPPIST-1 is a seven-planet system orbiting a very-low-mass star and in which a number of the planets lie within the star's habitable zone (Gillon et al. 2017). With JWST, it may even be possible to use transit spectroscopy to detect ozone in the atmospheres of some of the TRAPPIST-1 planets (Barstow and Irwin 2016).

As already mentioned in relation to Proxima Centauri b, we can also use phase-curve variations to infer something about an exoplanet's atmosphere. As a close-in planet orbits its host star, the side facing toward us will go from being the hot, irradiated dayside to the much colder night side. This can produce a detectable thermal phase variation. The form of this phase curve depends on the presence of an atmosphere and on its properties and so can be used to infer—or rule out—if a close-in exoplanet has an atmosphere (Seager and Deming 2009).

The recently selected ESA's Ariel mission, due to launch in the late 2020s, will use transit, eclipse, and phase-curve spectroscopy to characterize the atmospheres of hundreds of exoplanets. This will include gas giants (Jupiter-like), Neptune-like, super-Earths, and Earth-sized planets around a range of host-star types (Tinetti et al. 2017). It will primarily focus on hot, or warm, exoplanets, with a particular goal of understanding how atmospheres form and how the chemical composition of these exoplanet atmospheres relates to the properties of the host star.

Since planets move much faster than their parent stars, their spectral lines will also be shifted with respect to those from the star. We can therefore use high-resolution spectroscopy to try to identify the planet's spectral lines and hence learn something about its atmospheric composition (Snellen et al. 2013, 2015). Doing so for rocky planets may, however, require us to wait for the European Extremely Large Telescope (E-ELT), due to start operations in the mid-2020s.

Ultimately, however, we would probably like to directly image planets, so as to determine their spectra, and to infer something about their potential habitability. This, however, remains extremely challenging, and the ability to do so depends on both the contrast ratio between the planet and its parent star and their angular separation. Currently, we can directly observe young, giant planets on wide orbits, both because they are quite bright (planet-to-star contrast ratios of about 10^{-4}) and because the angular separation between the planet and its host star can be quite large. There are already indications, for example, that clouds are present in the atmospheres of some of the directly imaged giant planets (Bonnefoy et al. 2016).

Imagers on 30-m class telescopes (due to beginning of the operating in the next decade) will probably be able to directly image young Jupiter analogues (i.e., Jupiter-mass planets on orbits a few astronomical units from their parent star) and, potentially, rocky planets in the habitable zones of very-low-mass stars (Turbet et al. 2016). However, to directly observe a true Earth analogue (Earth mass/size planet at about 1 AU around a Sun-like star) may require large space telescopes that are currently only being planned, such as NASA's Luvoir mission, which has yet to be formally selected and would only launch in the mid-2030s, at the earliest.

12.1.7   CONCLUSIONS

We now have a very large sample of known exoplanets in orbit around stars other than the Sun. These range from planets larger than Jupiter to ones smaller than the Earth. The various detection methods allow us to probe a wide range of parameter space, in both planet mass and orbital radius, but the one region that we are not yet able to probe is Earth-sized planets orbiting at about 1 AU around a Sun-like star. This region will become accessible through future space-based exoplanet missions.

However, we do now have a small sample of planets orbiting within their stars' habitable zone, defined as being the region in which liquid water could exist on the planet's surface. Most of these planets are, however, orbiting stars less massive than the Sun. The habitability of such planets is still unclear, but we will start uncovering atmospheric characteristics of such planets with the next generation of both ground-based and space-based instruments.

Determining if an exoplanet is habitable, or not, may well turn out to be even beyond the next generation of astrophysical instruments. However, we will almost certainly, in the near future, have a much better understanding of exoplanet atmospheres, what will be required in order to further improve this understanding, and what would be required to actually detect potential biosignatures.

REFERENCES

Anglada-Escudé, G., P.J. Amado, J. Barnes et al. 2016. A terrestrial planet candidate in a temperate orbit around Proxima Centauri. *Nature* 536: 437–440.

Ballard, S. 2018. Predicted number, multiplicity, and orbital dynamics of TESS M dwarf exoplanets. *The Astrophysical Journal*. Submitted.

Barnes, R. 2017. Tidal locking of habitable exoplanets. *Celestial Mechanics and Dynamical Astronomy* 129: 509–536.

Barstow, J.K., and P.G.J. Irwin. 2016. Habitable worlds with JWST: Transit spectroscopy of the TRAPPIST-1 system? *Monthly Notices of the Royal Astronomical Society* 461: L92–L96.

Beaulieu, J.-P., D.P. Bennett, P. Fouqué, A. Williams et al. 2006. Discovery of a cool planet of 5.5 Earth masses through gravitational microlensing. *Nature* 439: 437–440.

Biller, B.A., M.C. Liu, Z. Wahhaj et al. 2013. The Gemini/NICI planet-finding campaign: The frequency of planets around young moving group stars. *The Astrophysical Journal* 777(2): 160.

Biller, B.A., and L.M. Close. 2007. A direct distance and luminosity determination for a self-luminous giant exoplanet: The trigonometric parallax to 2MASSW J1207334-393254Ab. *The Astrophysical Journal* 669: L41–L44.

Birkby, J.L., R.J. de Kok, M. Brogi et al. 2013. Detection of water absorption in the day side atmosphere of HD189733b using ground-based high-resolution spectroscopy at 3.2 μm. *Monthly Notices of the Royal Astronomical Society* 436: L35–L39.

Bodenheimer, P., D.N.C. Lin, and R.A. Mardling. 2001. On the tidal inflation of short-period extrasolar planets. *The Astrophysical Journal* 548: 466–472.

Bonnefoy, M., A. Zurlo, J.L. Baudino et al. 2016. First light of the VLT planet finder SPHERE. IV. Physical and chemical properties of the planets around HR8799. *Astronomy & Astrophysics* 587: A58.

Borucki, W.J., D. Koch, G. Basri, T. Brown et al. 2006. The Kepler mission: A transit-photometry mission to discover terrestrial planets. In: *Planetary Systems and Planets in Systems*. Eds. S. Udry, W. Benz, and R. von Steiger. ISSI Scientific Reports: ESA/ISSI. pp. 207–220.

Boutle, I.A., N.J. Mayne, B. Drummond et al. 2017. Exploring the climate of Proxima B with the Met Office Unified Model. *Astronomy & Astrophysics* 601: A120.

Bowler, B.P., and E.L. Nielsen. 2018. Occurrence rates from direct imaging surveys. In press.

Brown, T.M., D. Charbonneau, R.L. Gilliland, R.W. Noyes, and A. Burrows. 2001. Hubble space telescope time-series photometry of the transiting planet of HD209458. *The Astrophysical Journal* 552: 699–709.

Charbonneau, D., T.M. Brown, D.W. Latham, and M. Mayor. 2000. Detection of planetary transits across a Sun-like star. *The Astrophysical Journal* 529: L45–L48.

Chauvin, G., A.-M. Lagrange, C. Dumas et al. 2005. Giant planet companion to 2MASSW J1207334-393254. *Astronomy and Astrophysics* 438: L25–L28.

Cosentino, R., C. Lovis, F. Pepe et al. 2014. HARPS-N @ TNG, two year harvesting data: Performances and results. *Proceedings of the SPIE* 9147. id. 91478C.

Cumming, A., R.P. Butler, G.W. Marcy et al. 2008. The Keck planet search: Detectability and the minimum mass and orbital period distribution of extrasolar planets. *Publications of the Astronomical Society of the Pacific* 120: 531–554.

de Wit, J., H.R. Wakeford, N.K. Lewis et al. 2018. Atmospheric reconnaissance of the habitable-zone Earth-sized planets orbiting TRAPPIST-1. *Nature Astronomy* 2: 214–219.

Dole, S.H. 1964. *Habitable Planets for Man*. Blaisdell Publishing Company, New York.

Dominik, M. 2010. Studying planet populations by gravitational microlensing. *Gen. Relativ. Grav.* 42: 2075–2100.

Evans, T.M., F. Pont, D.K. Sing et al. 2013. The deep blue color of HD 189733b: Albedo measurements with Hubble Space Telescope/Space Telescope Imaging Spectrograph at visible wavelengths. *The Astrophysical Journal Letters* 772: L16.

Fortney, J.J., M.S. Marley, and J.W. Barnes. 2007. Planetary radii across five orders of magnitude in mass and stellar insolation: Application to transits. *The Astrophysical Journal* 659: 1661–1672.

Fressin, F., G. Torres, D. Charbonneau et al. 2013. The false positive rate of Kepler and the occurrence of planets. *The Astrophysical Journal* 766: 81.

Fulton, B.J., E.A. Petigura, A.W. Howard et al. 2017. The California-Kepler Survey. III. A gap in the radius distribution of small planets. *The Astronomical Journal* 154: 109.

Gillon, M., A.H.M.J. Triaud, B.-O. Demory, et al. 2017. Seven temperate terrestrial planets around the nearby ultracool dwarf star TRAPPIST-1. *Nature* 542: 456–460.

Hebb, L., A. Collier-Cameron, B. Loeillet, D. Pollacco et al. 2009. WASP-12b: The hottest transiting extrasolar planet yet discovered. *The Astrophysical Journal* 693: 1920–1928.

Hebrard, G., L. Arnold, T. Forveille et al. 2016. The SOPHIE search for northern extrasolar planets. X. Detection and characterization of giant planets by the dozen. *Astronomy & Astrophysics* 588: A145.

Howard, A.W., J.A. Johnson, G.W. Marcy et al. 2010. The occurrence and mass distribution of close-in super-Earths, Neptunes, and Jupiters. *Science* 330: 653.

Hu, Y., and J. Yang. 2014. Role of ocean heat transport in climates of tidally locked exoplanets around M dwarf stars. *Proceedings of the National Academy of Science of the United States of America* 111: 629–634.

Joshi, M.M., R.M. Haberle, and R.T. Reynolds. 1997. Simulations of the atmospheres of synchronously rotating terrestrial planets orbiting M dwarfs: Conditions for atmospheric collapse and the implications for habitability. *Icarus* 129: 450–465.

Kasting, J.F., D.P. Whitmire, R.T. Reynolds. 1993. Habitable Zones around Main Sequence Stars. *Icarus* 101: 108–128.

Kasting, J.F., R. Kopparapu, R.M. Ramirez, and C.E. Harman. 2014. Remote life-detection criteria, habitable zone boundaries, and the frequency of Earth-like planets around M and late K stars. *Proceedings of the National Academy of Sciences of the United States of America* 111: 12641–12646.

Kite, E.S., and E.B. Ford. 2018. Habitability of exoplanet water-worlds. Submitted.

Kopparapu, R.K., R. Ramirez, J.F. Kasting et al. 2013. Habitable zones around main-sequence stars: New estimates. *The Astrophysical Journal* 765: 131.

Kreidberg, L., and A. Loeb. 2016. Prospects for characterizing the atmosphere of Proxima Centauri b. *The Astrophysical Journal Letters* 832: L12.

Kreidberg, L., J. Bean, J. Désert et al. 2014. Clouds in the atmosphere of the super-Earth exoplanet GJ1214b. *Nature* 505: 69–72.

Levi, A., D. Sasselov, and M. Podolak. 2014. Structure and dynamics of cold water super-Earths: The case of occluded CH_4 and its outgassing. *The Astrophysical Journal* 792: 125.

Lopez, E.D., K. Rice. 2018. How formation time-scales affect the period dependence of the transition between rocky super-Earths and gaseous sub-Neptunes and implications for etaEarth. *Monthly Notices of the Royal Astronomical Society* 479: 5303–5311.

Lopez, E., and J.J. Fortney. 2016. Re-inflated warm Jupiters around red giants. *The Astrophysical Journal* 818: 4.

Lopez-Morales, M., R.D. Haywood, J.L. Coughlin, L. Zeng et al. 2016. Kepler-21b: A rocky planet around a V = 8.25 magnitude star. *The Astronomical Journal* 152: 204.

Marois, C., B. Zuckerman, Q. Konopacky et al. 2010. Images of a fourth planet orbiting HR 8799. *Nature* 468: 1080–1083.

Marois, C., B. Macintosh, T. Barman, B. Zuckerman et al. 2008. Direct imaging of multiple planets orbiting the star HR8799. *Science* 322: 1348.

Mayor, M., and D. Queloz. 1995. A Jupiter-mass companion to a solar-type star. *Nature* 378: 355–359.

Meadows, V.S., G.N. Arney, E.W. Schwieterman et al. 2018, The habitability of Proxima Centauri b: Environmental states and observational discriminants. *Astrobiology* 18: 133–189.

Pepe, F., A. Collier Cameron, D.W. Latham, E. Molinari et al. 2013. An Earth-sized planet with an Earth-like density. *Nature* 503: 377–380.

Petigura, E.A., A.W. Howard, and G.W. Marcy. 2013. Prevalence of Earth-size planets orbiting Sun-like stars. *Proceedings of the National Academy of Sciences of the United States of America* 110: 19273–19278.

Pierrehumbert, R.T., and E. Gaidos. 2011. Hydrogen greenhouse planets beyond the habitable zone. *The Astrophysical Journal* 734: L13.

Podsaidlowski, P. 1993. Planet formation scenarios. In: *Planets around Pulsars; Proceedings of the Conference.* California Institute of Technology, Pasadena, CA, April 30–May 1, 1992. pp. 149–165.

Rauer, H., C. Catala, C. Aerts et al. 2014. The PLATO 2.0 mission. *Experimental Astronomy* 38: 249–330.

Ribas, I., E. Bolmont, F. Selsis et al. 2016. The habitability of Proxima Centauri b. I. Irradiation, rotation and volatile inventory from formation to the present. *Astronomy & Astrophysics* 596: A111.

Rogers, L. 2015. Most 1.6 Earth-radius planets are not rocky. *The Astrophysical Journal* 801: 41.

Rowan, D., S. Meschiari, G. Laughlin et al. 2016. The Lick-Carnegie Exoplanet Survey: HD 32963—A new Jupiter analog orbiting a Sun-like star. *The Astrophysical Journal* 817: 104.

Rugheimer, S., L. Kaltenegger, A. Segura et al. 2015. Effect of UV radiation on the spectral fingerprints of Earth-like planets orbiting M stars. *The Astrophysical Journal* 809: 57.

Seager, S. 2008. Exoplanet transit spectroscopy and photometry. *Space Science Reviews* 135: 345–354.

Seager, S., and D. Deming. 2009. On the method to infer an atmosphere on a tidally locked super Earth exoplanet and upper limits to GJ 876d. *The Astrophysical Journal* 703: 1884–1889.

Snellen, I.A.G., R.J. de Kok, R. le Poole, M. Brogi, J. Birkby et al. 2013. Finding extraterrestrial life using ground-based high-dispersion spectroscopy. *The Astrophysical Journal* 764: 182.

Snellen, I., R. de Kok, J.L. Birkby et al. 2015. Combining high-dispersion spectroscopy with high contrast imaging: Probing rocky planets around our nearest neighbors. *Astronomy & Astrophysics* 576: A59.

Snellen, I.A.G., S. Albrecht. E.J.W. de Mooij, and R.S. Le Poole. 2008. Ground-based detection of sodium in the transmission spectrum of exoplanet HD209458b. *Astronomy & Astrophysics* 487: 357–362.

Street, R.A., D.L. Pollaco, A. Fitzsimmons et al. 2003. SuperWASP: Wide angle search for planets. *Scientific Frontiers in Research on Extrasolar Planets, ASP Conference Series.* Vol. 294. Eds. D. Deming and S. Seager, Astronomical Society of the Pacific, San Francisco, CA. pp. 405–408.

Sullivan, P.W., J.N. Winn, Z.K. Berta-Thompson et al. 2015. The transiting exoplanet survey satellite: Simulations of planet detections and astrophysical false positives. *The Astrophysical Journal* 809: 77.

Tinetti, G., P. Drossart, P. Eccleston et al. 2017. The science of ARIEL. *European Planetary Science Congress 2017.* Riga, Latvia, id. EPSC2017-713.

Turbet, M., J. Leconte, F. Selsis et al. 2016. The habitability of Proxima Centauri b. II. Possible climates and observability. *Astronomy & Astrophysics* 596: A112.

Wang, J., D.A. Fischer, E.P. Horch, and X. Huang. 2015. On the occurrence rate of hot Jupiters in different stellar environments. *The Astrophysical Journal* 799: 229.

Wilkins, A.N., D. Deming, N. Madhusudhan et al. 2014. The emergent 1.1–1.7 μm spectrum of the exoplanet CoRoT-2b as measured using the Hubble Space Telescope. *The Astrophysical Journal* 783: 113.

Wolszczan, A., and D.A. Frail. 1992. A planetary system around the millisecond pulsar PSR1257+12. *Nature* 355: 145–147.

Wright, J.T., G.W. Marcy, A.W. Howard et al. 2012. The frequency of hot Jupiters orbiting nearby Solar-type stars. *The Astrophysical Journal* 753: 160.

Yang, J., N.B. Cowan, and D.S. Abbot. 2013. Stabilizing cloud feedback dramatically expands the habitable zone of tidally locked planets. *The Astrophysical Journal* 771: L45.

Zsom, A., S. Seager, J. de Wit, and V. Stamenković. 2013. Toward the minimum inner edge distance of the habitable zone. *The Astrophysical Journal* 778: 109.

12.2 Solar System Exploration
Small Bodies and Their Chemical and Physical Conditions

Hikaru Yabuta

CONTENTS

12.2.1 INTRODUCTION

Primitive small bodies, such as asteroids, meteorites, and comets, are the remnants of planetesimals, which did not grow large enough to become planets, and thus, they are thought to preserve the precursor materials in the early solar system 4.5 Gyr ago. These small bodies are originally derived from accretion of interstellar dusts that are composed of an amorphous silicate, refractory organic material, and ice (Greenberg and Li 1997). Afterwards, interaction of organics, water, and minerals in protoplanetary disk and within planetesimals due to a variety of locations and chemical and physical processes (e.g., thermal, photochemical, and aqueous processes) produced compositional and geological diversity of small bodies. Thus, the investigation of these small bodies enables us to understand the history of the solar system formation. As the traditional theory of solar system formation is in the process of revising by the recent hypotheses, such as Nice model (Gomes et al. 2005) and Grand Tack model (Walsh et al. 2012), it is becoming increasingly important to substantiate the distributions and chemical evolution of organic molecules and water in the solar system for determining origins of life and planetary habitability.

To date, our understanding of chemical and physical processes in small bodies has been significantly improved by the extensive chemical analyses of organic materials and minerals in carbonaceous chondritic meteorites, interplanetary dust particles (IDPs), and Antarctic micrometeorites (AMMs) (see review in Alexander et al. 2017; Glavin et al. 2018; Yabuta et al. 2018). However, there have remained uncertainties in the source and geological information for the chemical evolution, which are not clearly recorded in the fallen extraterrestrial materials. In order to link the cosmochemical features of meteorites, IDPs, and AMMs with geology of their original parent bodies, small bodies' sample return/landing explorations that integrate the outcomes from observation and sample analyses are essential. Another advantage of small bodies exploration is that one can obtain contamination-free extraterrestrial samples.

This chapter reviews (1) the achievements from the past small bodies' sample return/landing exploration missions (*Stardust*, *Hayabusa*, and *Rosetta*), (2) the scientific goals of the ongoing asteroid sample return missions (e.g., *Hayabusa2* and the Origins, Spectral Interpretation, Resource Identification, Security-Regolith Explorer [*OSIRIS-REx*]), and (3) the scientific strategies of future missions (e.g., *DESTINY+*). Integrated understanding of the missions that explore the small bodies in different evolution stages will be able to establish solar system science covering from interstellar medium to planetesimals.

12.2.2 *STARDUST* COMET DUST SAMPLE RETURN MISSION

Stardust is the first sample return mission of a comet, developed by the National Aeronautics and Space Administration (NASA). The *Stardust* spacecraft was launched in February 1999, collected the dust particles from comet 81P/Wild 2 in January 2004, and returned the samples to the Earth in January 2006. Comet 81P/Wild 2 is a Jupiter-family comet that was brought into the inner solar system due to perturbations by Jupiter in September 1974. The comet is a 5-km-size oblate body from which jets of gas and dusts were observed, and it contained craters, mesas, and cliffs that were probably formed before its injection into the inner solar system (Figure 12.2.1a) (Brownlee et al. 2004).

Dust particles released from comet 81P/Wild 2 were collected by the spacecraft during its flyby at 6.1 km/s. In order to reduce possible alteration of cometary dust components upon high-velocity impact of dusts on the spacecraft, ultra-low-density silica aerogels (0.03 g/cm³) were used as a capture material. The silica aerogel capture cells (2 × 4 × 3 cm) were mounted into the tennis-racket-shaped dust collector with aluminum grids in the sample return capsule.

The impacted comet dusts on the dust collector produced impact craters on the aluminum grids and impact tracks (Figure 12.2.1b) with various sizes and shapes in the silica aerogels; this is reflected by diversity in size and density of the comet dust particles (Horz et al. 2006; Tsuchiyama et al. 2008). In comparison with hypervelocity impact experiments, it was suggested that carrot-type impact track is derived from a refractory (e.g., metal-rich) grain, and a bulbous impact track is derived from a volatile-rich particle. The dusts themselves were split into a number of smaller particles from the impact tracks. For the laboratory analyses, a thin section of silica aerogel containing the impact tracks, called *keystone*, was cut by needles mounted on a micro-manipulator, and the comet dust particles were carefully extracted at NASA curation facility.

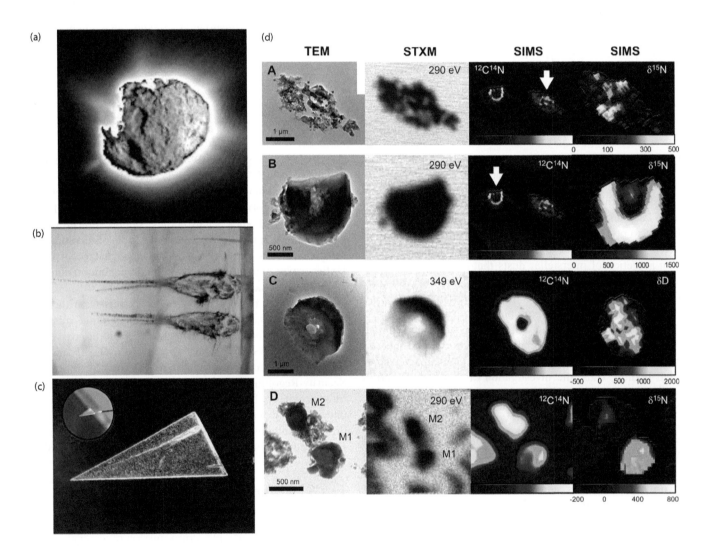

FIGURE 12.2.1 (a) Image of comet 81P/Wild 2 taken by the navigation camera during the close approach phase of Stardust's flyby of the comet. Credit: NASA. (b) Image of the impact tracks formed by two comet particles penetrating the silica aerogel mounted in Stardust spacecraft's comet dust collector. Credit: NASA. (c) Image of a wedge-shaped aerogel slice containing comet dust particles, called a keystone. A specialized silicon pickle fork is then used to remove the keystone from the remaining aerogel for further analysis. Credit: NASA. (d) Transmission electron microscopy (TEM) (left-most column) and scanning transmission x-ray microscope (STXM) (middle-left column) images of and secondary ion mass spectrometry (SIMS) maps of CN⁻ (middle-right column) and δ¹⁵N or δD (right-most column) of extraterrestrial organic materials. (A) Organic section in comet Wild 2 dust particles. (B) Organic nanoglobule section in comet Wild 2 dust particles. (C) Organic nanoglobule in comet Wild 2 dust particles. (D) Two organic globules (M1 and M2) found in the insoluble organic residue from Murchison meteorite (De Gregorio et al. 2011).

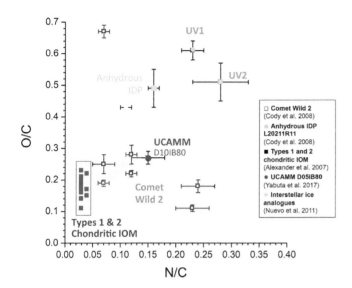

FIGURE 12.2.2 N/C versus O/C ratios of organics in the comet Wild 2 dust particles (□, Cody et al., 2008), the anhydrous IDP L20211R11 (▨, Cody et al., 2008), types types 1 and 2 chondritic IOM (■, Alexander et al., 2007), the UCAMM D05IB80 (●, this study), and the UV irradiation products from interstellar analogues (●, Nuevo et al., 2011) (UV1 $H_2O:CH_3OH:CO:NH_3 = 100:50:1:1$, UV2 $H_2O:CH_3OH:CO:NH_3:C_3H_8 = 100:50:1:1:10$). The ratios were estimated from the fitting of C-, N-, and O-XANES spectra.

Organic materials in some of the comet 81P/Wild 2 dust particles are poor in aromatic carbon, while they are enriched in aliphatic carbon and nitrogen- and/or oxygen-bearing functional groups, such as nitriles, amides, carboxyls, and alcohols (Sandford et al. 2006; Cody et al. 2008). These elemental and molecular compositions are distinct from insoluble organic matter (IOM) from primitive carbonaceous chondrites (Figure 12.2.2). Other comet 81P/Wild 2 dust particles show similar organic functional group chemistry to those of chondritic IOM, and some of the organic nanoglobules (Figure 12.2.1d) found in *Stardust* samples are highly aromatic (De Gregorio et al. 2011). These diversities in organic molecular compositions suggest that Wild 2 was not equilibrated and contains organic materials not found in meteorites or their asteroid parent bodies. On the other hand, hydrogen isotopic compositions of the organics in the 81P/Wild 2 dust particles are in a similar range of those in IDPs and carbonaceous chondrites (McKeegan et al. 2006); however, the highest value of hydrogen isotopic composition ($\delta D = 2200‰$) seen to date is an order of magnitude lower than that of other extraterrestrial materials. Nitrogen isotopic compositions of the organic material in the 81P/Wild 2 dust particles are terrestrial or slightly enriched ($\delta_{15}N = -55‰ \sim +70‰$), except for a highly ${}^{15}N$-enriched organic nanoglobule ($+136‰ \pm 15‰$) (De Gregorio et al. 2011). Thus, it is suggested that a major fraction of the refractory organics in the comet is of nebular origin.

Polycyclic aromatic hydrocarbons (PAHs) ranging from one ring to six rings are seen in both comet 81P/Wild 2 dust particles and carbonaceous chondrites (Sandford et al. 2006). Methylamine, ethylamine, and glycine detected from the comet-exposed aerogel capture media are thought to be the products from energetically processed icy grains containing NH_3, CH_4, and C_2H_6 (Glavin et al. 2008). Detection of only this one amino acid from comet 81P/Wild 2 is a great contrast to the detection of >70 kinds of amino acids from carbonaceous chondrites; however, this is probably largely due to the fact that *Stardust* samples are so small that the amino acids' abundances are below current detection limits. Another possible reason is that formation of amino acids did not occur efficiently in a cometary icy body, since the many reactions that form amino acids require liquid water (e.g., Strecker synthesis), although some amino acids can be made directly by ice irradiation.

Mineralogy of comet 81P/Wild 2 dust particles (e.g., silicates, glass with embedded metals and sulfides [GEMS]-like objects, Fe-Ni sulfides, and Fe-Ni metals) was mostly similar to that of anhydrous IDPs, while their compositions were diverse, indicating a wide range of formation condition in the solar system (Zolensky et al. 2006). Finding of crystalline silicates, high-temperature minerals, together with amorphous grains in the cometary dusts (Zolensky et al. 2006; Keller et al. 2006), indicated that large-scale mixing occurred in the early solar system (e.g., Brownlee et al. 2006). Discovery of chondrule-like objects from the Wild 2 dust particles further demonstrated the transportation of material from the inner to the outer solar system (Nakamura et al. 2008). According to the densities ($0.80-5.96$ g cm^{-3}) of the cometary dust particles estimated by synchrotron X-ray-based three-dimensional structure analyses of the impact tracks, crystalline particles account for approximately 5 vol% (20 wt%) of the Wild 2 dust particles, which is consistent with the values of chondrules and calcium-aluminum inclusions (CAIs) (Iida et al. 2010).

Although little evidence of parent body aqueous alteration was reported from Wild 2 comet dust particles, several records of weak aqueous alteration, such as the presence of carbonate (Mikouchi et al. 2007), sulfides (e.g., cubanite grain [$CuFe_2S_3$], a pyrrhotite [$(Fe, Ni)_{1-x}S$]/pentlandite [$(Fe, Ni)_9S_8$] assemblage, and a pyrrhotite/sphalerite [$(Fe, Zn)S$] assemblage) (Berger et al. 2011), cosmic symplectite (Nguyen et al. 2017), and magnetite (Hicks et al. 2017), have been reported.

12.2.3 *HAYABUSA* S-TYPE ASTEROID SAMPLE RETURN MISSION

Hayabusa is the Japan Aerospace Exploration Agency (JAXA)'s first mission to collect asteroid samples and return them to Earth for the laboratory analyses. The mission explored the near-Earth S-type (stony) asteroid 25143 Itokawa. The spacecraft was launched in May 2003 and arrived at Itokawa in September 2005 (Fujiwara et al. 2006). Itokawa is a 500-m-size asteroid. Its density was low (1.9 g/cm³), and its porosity was estimated 40%. This asteroid is composed of two parts, the head and the body, and it has boulder-rich regions and smooth regions (Figure 12.2.3a–c). Based on the shape, boulders, low density, and high porosity, Itokawa was regarded as a rubble-pile asteroid, which was formed by re-cumulation of the fragments resulted from a collisional disruption of its parent body (Fujiwara et al. 2006). Near-infrared (IR) spectra of Itokawa were similar to those of an ordinary LL5 chondrite (Abe et al. 2006). However, the slight spectral difference between Itokawa

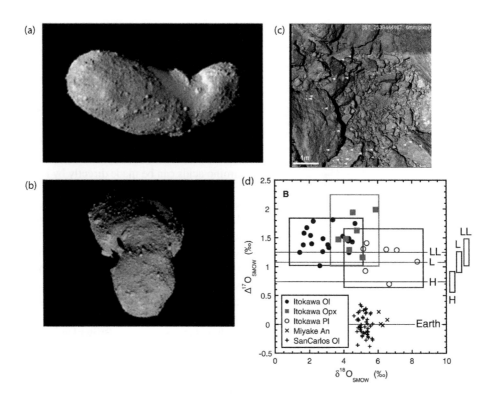

FIGURE 12.2.3   (a) Image of Itokawa taken from about 8 km away from the asteroid. Credit: JAXA. (b) Image of Itokawa taken from 4.4 km away from the asteroid. Credit: JAXA. (c) Close-up image of Itokawa on the east side at ranges below 2 km and down to 63 m. Large gravels generally overlie smaller particles. (Miyamoto et al. 2007). (d) Oxygen isotopic compositions of Itokawa minerals compared with those of a forsterite crystal from San Carlos, Arizona, USA, and an anorthite crystal from Miyake-Jima, Japan. Isotope variation defined by 2σ for each mineral is shown by a rectangle with a color of the corresponding symbol. Ol, olivine; Opx, orthopyroxene; Pl, plagioclase; An, anorthite; Miyake, Miyake-Jima. A mass fractionation line of the average O isotopic composition of LL chondrite group is shown as a reference. Variations (2σ) of whole-rock $\Delta^{17}O_{SMOW}$ values for H, L, and LL chondrite groups are shown to the right of the graph (Yurimoto et al. 2011).

and an LL5 chondrite is due to space-weathering processes on the surface of asteroid Itokawa (Hiroi et al. 2006).

The spacecraft faced several engineering troubles during its navigation and touchdown on the asteroid, which postponed the return date. Despite those difficulties, the spacecraft managed to return the sample canister to Earth in June 2010. At the Institute of Space and Astronautical Science (ISAS) curation facility at JAXA, 1534 particles of the asteroid Itokawa were identified (Nakamura et al. 2011), and the 1-year preliminary examination of the collected Itokawa particles was implemented.

For the first time, the direct link between an asteroid (an S-type asteroid) and meteorites (ordinary LL5-6 chondrites) was proved by comparison of the oxygen-isotopic compositions (Figure 12.2.3d) (Yurimoto et al. 2011), mineralogy (Nakamura et al. 2011; Noguchi et al. 2011), petrology (Tsuchiyama et al. 2011), and chemistry (Ebihara et al. 2011) between the Itokawa particles and meteorites. Mineralogical investigations indicated that the highly equilibrated Itokawa particles, which are mainly composed of olivine, low-Ca pyroxene, and plagioclase, experienced intense thermal metamorphism approximately at 800°C and cooled slowly to 600°C on its parent body larger than 20 km (Nakamura et al. 2011; Tsuchiyama et al. 2011). Those sample analyses also revealed the detailed processes of the asteroid surface, which were not seen in meteorites.

The size and three-dimensional shape distributions of the Itokawa particles were similar to those of fragments generated in laboratory impact experiments but they were not similar to lunar regolith particles. This result suggested that the asteroid surface experienced repetitive impact processes and abrasion of grains (Tsuchiyama et al. 2011). Fe-rich nanoparticles were observed in the rims of the surfaces of Itokawa particles, and they were regarded as space-weathering products based on the similarities with those of lunar soils (Noguchi et al. 2011). Additional observation of the surface morphology indicated that the space-weathered rim with blisters was formed mainly by solar wind irradiation (Figure 12.2.4) (Matsumoto et al. 2015). The helium, neon, and argon isotopic compositions of the Itokawa particles were close to the solar wind compositions (Nagao et al. 2011). Various profiles of noble gas release are reflected by repeated solar wind implantation, in addition to a preferential loss of helium due to friction among the particles or sputtering by other solar wind particles (Nagao et al. 2011). According to the cosmic exposure ages estimated by neon isotopic compositions (Nagao et al. 2011) and the crater retention age estimated by a new crater-size scaling law (Tatsumi and Sugita 2018) (Figure 12.2.5), global resurfacing of Itokawa in the main belt occurred 3–33 million years after the complete breakup of its parent body with collisions.

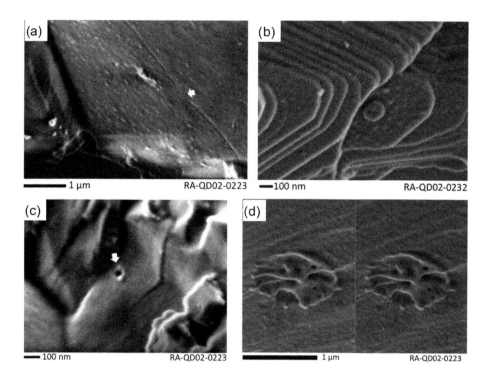

FIGURE 12.2.4 (a) Secondary electron (SE) Field Emission Scanning Electron Microscopy (FE-SEM) image of fine spotted structures (blisters) on a space-weathered rim on the wavy and stepped surface of an Itokawa particle (RA-QD02-0223), (b) SE FE-SEM image showing blisters on a concentric stepped surface of low-Ca pyroxene on an Itokawa particle (RA-QD02-0232), (c) SE FE-SEM image showing a pit or a crater-like structure of about 50 nm in diameter (indicated with an arrow) on a low-Ca pyroxene surface of an Itokawa particle (RA-QD02-0223), and (d) SE FE-SEM stereo image of another crater-like structure some 1 μm in diameter on an opposite surface of the same particle, indicating particle motion due to the regolith activity (Matsumoto et al. 2015).

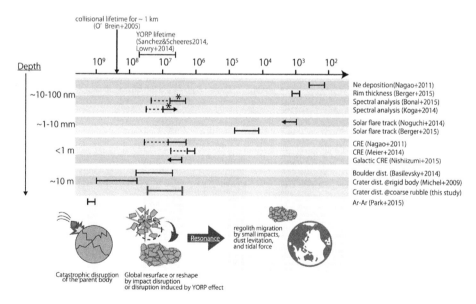

FIGURE 12.2.5 Comparison among Itokawa surface ages estimated from analyses of returned samples and the remote-sensing data. The solid lines indicate original literature values based on near-earth orbits. Dashed lines indicate ages recalculated with a reduced solar wind flux in the main belt. The age estimates with asterisks may be macroscopic surface ages (Tatsumi and Sugita, 2018).

The preliminary examinations of carbonaceous materials/ organic compounds for the Itokawa particles were also conducted; however, the carbon contents in the particles were so low that they were below the detection limits of the analytical instruments (Naraoka et al. 2012). Aside from the Itokawa particles, 58 unknown carbonaceous particles were collected from a sample catcher (Yada et al. 2014). In order to determine their origins, the coordinated analytical investigations for the several carbonaceous particles have been carried out (Ito et al. 2014; Yabuta et al. 2014; Uesugi et al. 2014; Kitajima et al. 2015; Naraoka et al. 2015). As a result, it was revealed that those carbonaceous particles were lack of any robust isotopic or mineralogical evidence of extraterrestrial origin and that they were very likely terrestrial contamination derived from the degradation products of polymer materials used in the spacecraft and/or biological origin. The result that indigenous organic compounds from Itokawa have not been identified is reasonable, considering that the parent body of Itokawa experienced 600°C–800°C, which could have thermally decomposed or dehydrated if there were organics and/or water.

12.2.4 *ROSETTA* COMET RENDEZVOUS MISSION

The European Space Agency (ESA's) *Rosetta* mission is the first mission to rendezvous with Jupiter-family comet 67P/Churyumov–Gerasimenko (Figure 12.2.6a and b) and land on its surface by using a small lander (Philae). The spacecraft was launched in March 2004, arrived at and released Philae to comet 67P/Churyumov–Gerasimenko in 2014, and carried out observation for 14 months till December 2015. Remote-sensing observation and in situ analyses of organic molecules and volatiles were made using the Visible, Infrared, and Thermal Imaging Spectrometer (VIRTIS), the Cometary Secondary Ion Mass Analyser (COSIMA), and the Rosetta Orbiter Spectrometer for Ion and Neutral Analysis (ROSINA) instruments on the *Rosetta* mothership and the Cometary Sampling and Composition (COSAC) experiment and Ptolemy gas analyzer equipped on the Philae lander. Altogether, the Rosetta mass spectrometer detected about 60 volatile species in the coma of the comet. This includes the approximately two dozen parent species that had been seen from the ground observation at ultraviolet (UV), IR, and radio wavelengths.

The VIRTIS showed that water is largely absent from the surface of 67P/Churyumov–Gerasimenko, except for small spots of water ice (Figure 12.2.6c) (Auger et al. 2015), indicating that the surface is dehydrated by solar heating (Capaccioni

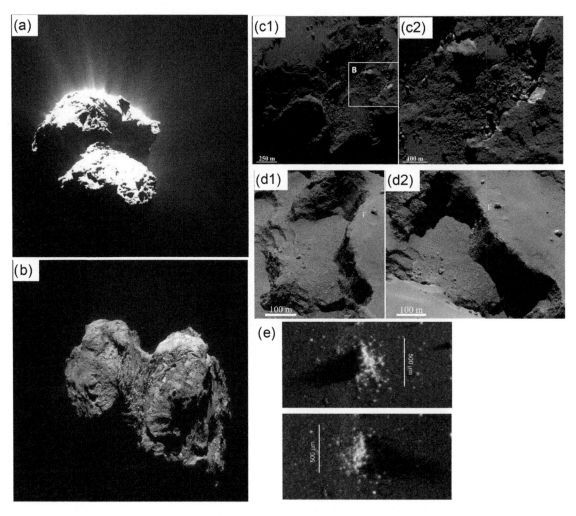

FIGURE 12.2.6 (a) Image of Comet 67P/Churyumov–Gerasimenko taken from a distance of about 170 km. Credit: ESA/Rosetta/MPS for OSIRIS Team MPS/UPD/LAM/IAA/SSO/INTA/UPM/DASP/IDA. (b) Image of Comet 67P/Churyumov–Gerasimenko taken from a distance of 67.6 km. Credit: ESA/Rosetta/NavCam. (c) Color-composite images of possible water-ice patches associated with a scarp east of the region that has been illuminated daily for several months on the surface of comet 67P/ Churyumov–Gerasimenko (Auger et al. 2015). (d) The two views of the cliff edge of comet 67P/ Churyumov–Gerasimenko taken on different dates. The feature likely leads to the slumping of material. Image was taken from a distance of 30 km (Vincent et al. 2015). (e) One of the dust particles (500-μm size) that was collected by the collecting target of COSIMA (Schulz et al. 2015).

et al. 2015; Quirico et al. 2016). For the first time, dark refractory organic materials, which may be associated with opaque minerals (e.g., Fe–Ni metals and FeS), were discovered on the cometary surface. Its reflectance spectra with low albedo features are distinct from those of IOM in carbonaceous chondrites but are close to the aliphatic- and carboxyl-rich organic materials produced by UV irradiation of cometary ice analogs (Quirico et al. 2016).

The COSIMA detected 27,000 dust particles of comet 67P/Churyumov–Gerasimenko and acquired mass spectra of 200 particles of them (Figure 12.2.6e) (Fray et al. 2016). They revealed that the Na/Mg ratio was as high as those of Leonids and Perseid meteor showers, but they were much higher than those of carbonaceous chondrites (Figure 12.2.7a) (Fray et al. 2016). Thus, it was concluded that comet 67P/Churyumov–Gerasimenko dusts were similar to IDPs and AMMs but they were not similar to meteorites. Carbon-bearing ions were detected by both positive and negative ion modes, but unlike other known organic molecule series, the detected ions were only less than m/z (mass) 50, indicating that the organic material in the 67P/Churyumov–Gerasimenko is refractory, hydrogen-poor aromatic structure (Fray et al. 2016). The result appears to be in line with those obtained by VIRTIS. According to the comparison of the mass spectra between 67P/Churyumov–Gerasimenko and carbonaceous chondrites, the H/C ratios

of 67P/Churyumov–Gerasimenko are higher than those of carbonaceous chondrites (Figure 12.2.7b). In addition, C/Si ratios of comet 67P/Churyumov–Gerasimenko dust particles estimated from the Time-of-Flight Secondary Ion Mass Spectrometry (ToF-SIMS) data were similar to that of comet 1P/Halley, while the values were much higher than those of chondritic-porous IDPs and carbonaceous CI chondrites (Figure 12.2.7e) (Bardyn et al. 2017). This difference suggests that the comet records the initial compositions of precursor molecules in the outer region of protoplanetary disk, before the nebula processing and/or parent body alteration.

The COSAC mass spectrometer analyzed volatile and organic compounds in sniffing mode after Philae's touchdown on the comet 67P/Churyumov–Gerasimenko surface. Of the 16 kinds of molecules tentatively identified (Goesmann et al. 2015), 12 are the expected molecules that have been observed by ground observations: H_2O, CH_4, HCN, CO, methylamine (CH_3NH_2), acetonitrile (CH_3CN), isocyanic acid (HNCO), etc. Newly observed molecules include methyl isocyanate (CH_3NCO), acetone ($CH_3(CO)CH_3$), propionaldehyde (C_2H_5CHO), and acetamide ($CH_3(CO)NH_2$). Sulfur-bearing molecules were not detected. Unexpectedly, typical cometary volatiles, NH_3, CO_2, and formaldehyde (HCHO), were not detected, possibly because the measured area (nucleus surface) was depleted in volatile components. In contrast, CO_2 was

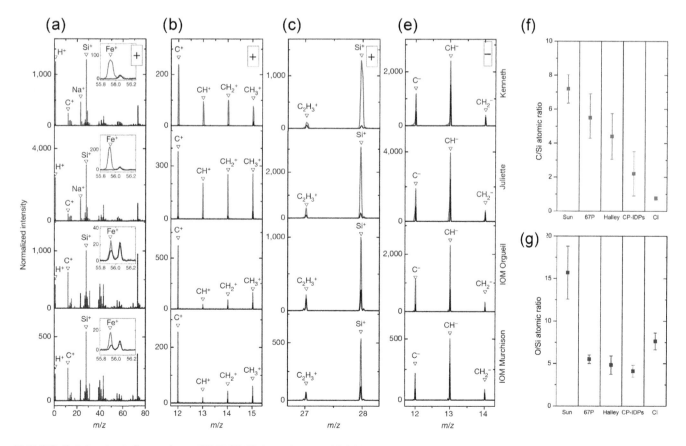

FIGURE 12.2.7 (a–d) Comparison of ToF-SIMS data of comet 67P/ Churyumov–Gerasimenko dust particles (Kenneth and Juliette) and IOM from Orgueil and Murchison carbonaceous chondritic meteorites (Fray et al. 2016). (f, g) Comparison of (f) carbon and (g) oxygen elemental ratios relative to silicon in the protosolar disc (Sun), comet 67P/ Churyumov–Gerasimenko dust particles (based on the data from COSIMA), comet Halley dust particles, chondritic-porous IDPs, and carbonaceous CI chondrites (Bardyn et al. 2017).

detected in abundance by Ptolemy gas analyzer, which measured coma gases. Formation of the identified nitrogen- and oxygen-bearing molecules can be explained by UV irradiation and/or radiolysis of cometary ice.

In addition, the ROSINA detected abundant CO_2 and ethane relative to water from the coma of 67P/Churyumov–Gerasimenko (Le Roy et al. 2015). Also of note is the detection of abundant molecular oxygen at the 4% level. Because protoplanetary disk models suggest that O_2 can only form in the gas phase at mid layers and because of the strong correlation of O_2 with H_2O, it has been suggested that the O_2 is preserved from the interstellar medium (Bieler et al. 2015).

It has been expected that the sources of volatile inventory to the terrestrial planets were recorded in the hydrogen isotopic compositions of small bodies. The ROSINA revealed that the D/H ratio of water from 67P/Churyumov–Gerasimenko [$(5.3 \pm 0.7) \times 10^{-4}$] was approximately three times the terrestrial value (Altwegg et al. 2015). If comet 67P/Churyumov–Gerasimenko is representative of all comets, then it is unlikely that comets are the major source of Earth's water. However, given the large variation in D/H seen in comets so far, no firm conclusion can yet be drawn. An asteroidal origin for terrestrial water is more likely, based on the similarity in D/H between carbonaceous chondrites and terrestrial ocean water (Alexander et al. 2012).

12.2.5 *HAYABUSA2* C-TYPE ASTEROID SAMPLE RETURN MISSION

Hayabusa2 is a C-type (carbonaceous) asteroid sample return mission of JAXA to explore the near-Earth C-type asteroid 162173 Ryugu (1999 JU3). The scientific goal of *Hayabusa2* is to investigate organic molecules and water are considered as origin of Earth life and ocean, as well as the major building blocks of the solar system.

The *Hayabusa2* spacecraft was launched on December 3, 2014, and arrived at the asteroid Ryugu on June 27, 2018. During its 18-month stay, remote-sensing observations will be carried out with the on-board instruments. *Hayabusa2* is planned to collect asteroid samples from up to three sites. It is planed that one of the three locations to collect samples will be around the artificial crater created by the small carry-on impactor (SCI) (Saiki et al. 2016), a new component of the *Hayabusa2*, which enables collection of internal materials (i.e., impact ejecta) of Ryugu. The collected samples will be returned to Earth in December 2020.

The sample catcher and container of *Hayabusa2* basically follow those used in *Hayabusa*; a 5-g Ta projectile will be shot at 300 m/s upon touchdown, and the ejecta will be transferred into a sample catcher through an extendable sampler horn under microgravity (Tachibana et al. 2014). Several improvements have been added for collecting volatiles and organics, such as aluminum metal-seal system and gas-sampling interface (Okazaki et al. 2016). The sampling recovery is estimated at a level of hundreds of milligrams.

Ryugu has been thought to be similar to the CM chemical group of carbonaceous chondrite, according to the grand-based observation that a 0.7-μm absorption feature in a reflectance spectrum, which is derived from iron-bearing phyllosilicates (e.g., serpentine), was detected (Vilas 2008). On the other hand, the other ground-based observations did not detect a clear feature of 0.7 μm (Lazzaro et al. 2013; Moskovitz et al. 2013). Thus, hydrous minerals may be distributed locally and/or during a limited time period (Kameda et al. 2015). Their reflectance spectra can also be explained by a combination of the heated Murchison meteorite at different temperatures (Hiroi et al. 1993), and thus the surface of the asteroid Ryugu may have experienced heterogeneous heating, for example, impact, space weathering, and solar radiation heating.

Hayabusa2 mission will perform three types of multi-scale observations: remote-sensing instruments, rover, and laboratory sample analyses. Remote sensing will be carried out by the four onboard instruments: an optical navigation camera (ONC), a near-infrared spectrometer (NIRS3), a thermal infrared imager (TIR), and a light detection and ranging (LIDAR) (Table 12.2.1). These instruments will perform kilometer to meter scale of geological and topographical survey almost all over the Ryugu and provide the global map data of the physical properties and chemical compositions of the asteroid surface, which will be used for the landing site selection. A hopping rover developed by Deutsches Zentrum für Luft- und Raumfahrt (DLR) and Centre national d'études spatiales (CNES), Mobile Asteroid Surface Scout (MASCOT), is composed of four instruments, a camera (Cam), a near-infrared microscope (MicrOmega), a MASCOT Radiometer (MARA), and a magnetometer (MaG), and plays roles of both observation and sample analyses, with centimeter to micrometer scale, helping to bridge between the two different scales. Laboratory sample analyses will conduct the high-precision and high-sensitivity quantitative measurements of elemental, isotopic, mineralogical, and molecular compositions of the returned asteroid samples with micrometer to nanometer scale, for determination of origin, evolution, and chronology of the solar system.

12.2.6 *OSIRIS-REx* B-TYPE ASTEROID SAMPLE RETURN MISSION

Origins, Spectral Interpretation, Resource Identification, Security, Regolith Explorer (OSIRIS-REx) is NASA's sample return mission to target a B-type carbonaceous asteroid (101955) Bennu (500-m size) (Lauretta 2016). OSIRIS-REx and Hayabusa2 share the same scientific goal. Bennu has a relatively featureless spectrum that was similar to that of a carbonaceous CM1 chondrite (Clark et al. 2011). The OSIRIS-REx spacecraft was launched in September 2016 and will arrive at Bennu in 2018. Observation of Bennu will be carried out by the OSIRIS-REx camera suite (OCAMS), OSIRIS-REx laser altimeter (OLA), the visible and infrared spectrometer (OVIRS), and thermal emission spectrometer (OTES). More than 60 g of regolith will be sampled by blowing a N_2 gas jet from a sampler head of the articulated positioning arm (touch-and-go sample acquisition mechanism [TAGSAM]) to the regolith and by collecting them into the collection chamber. The asteroid sample will be returned to Earth in 2023.

TABLE 12.2.1
Remote-Sensing Instruments and Data Products

	ONC-T	TIR	NIRS3	LIDAR
Wavelengths	390, 480, 550, 700, 860, 950 nm	8–12 μm	1.8–3.2 μm	
View angle	$5.7° \times 5.7°$	H16 × V12	$0.11° \times 0.11°$	1 mrad = 0.057
Pixels	1024 × 1024	320 × 240	1	1
Spatial resolution	1 m/pix (@10 km) 1 cm/pix (@100m)	20 m	38 m	20 m
Physical properties of asteroid surface	Shape modelHapke roughness-parameters	Maps of (1) Thermal inertia (2) Grain size (3) Maximum temperature		Distance Roughness
Hydrous minerals	Reflectance spectra and 1- and 7-band image data of: (1) 0.7-μm absorption depth		NIR reflectance spectra and spectral feature maps of: (1) 3-μm band depth (2) 3-μm band center (3) Near infrared albedo	
Organic carboncontents	(2) Albedo at 390 and 550 nm			
Secondary processes	(3) Spectral slope (480-860 nm) (4) Spectral slope in UV		(4) Spectral slope (1.8 μm to 2.6 μm)	
Safety and sampling recovery	(5) Boulder sizefrequency	Same as above (1)-(3)		Same as above

12.2.7 FUTURE SMALL-BODY MISSION: *DESTINY+*

DESTINY+(Demonstration and Experiment of Space Technology for INterplanetary voYage Phaethon fLyby dUSt science) is JAXA's mission to flyby of Geminids parent (3200) Phaethon and in situ dust analyses; this mission is in a pre-project phase selected as JAXA's Epsilon class small program and it is planned to launch in 2022. *DESTINY+* aims to understand (i) the physical and chemical properties and origins of interstellar dusts and IDPs and (ii) the dust ejection mechanism of the asteroid Phaethon for elucidating the nature and origin of cosmic dusts brought to the early Earth in the context of exogenous delivery of life's building blocks (Arai et al. 2018).

Phaethon has been known as an active asteroid from which recurrent dust ejection is observed at its perihelion (0.14 au) (e.g., Jewitt and Li, 2010; Arai et al. 2018). The unique feature of the small body will be an important clue to unveil the genetic link between primitive asteroids and comets. The reflectance spectra of Phaethon are similar to those of B-type asteroids (e.g., Licandro et al. 2007). The spectra are partially similar to the spectral features of heated carbonaceous CI/CM chondrites (Licandro et al. 2007), carbonaceous CK chondrites (Clark et al. 2010), and carbonaceous chondrites with larger grain sizes (e.g., Cloutis et al. 2012). The surface geology of Phaethon will be observed by panchromatic telescopic camera (TCAP) and visual (VIS) - near infrared (NIR) multiband camera (MCAP) (Arai et al. 2018).

The *in situ* dust analyzer (*DESTINY+* Dust Analyzer, DDA) is a time-of-flight mass spectrometer developed by Stuttgart University, Germany, with heritage of Cassini Cosmic Dust Analyzer (CDA) onboard Cassini (e.g. Srama, 2014). The great advantage of DDA is to obtain the flux, masses, velocities, orbits, and compositions for the individual dust particles at one time, which will enable the identification of interstellar dusts, cometary IDPs and asteroidal IDPs during the cruise phase. That is to say, *DESTINY+* dust science will have an opportunity to comprehensively decipher the chemical evolution from interstellar medium to solar nebula and planetesimals in the early Solar System.

12.2.8 CONCLUSION

Small bodies sample return/on-site analysis missions will surely continue to be regarded as one of the significant astrobiological approaches after Rosetta, Hayabusa2, and OSIRIS-REx. In future, it is expected to plan to collect not only primitive asteroids but also the small bodies in various stages of evolution, ranging from more primitive icy small bodies (comets) to differentiated asteroids, for better understanding of how diverse the chemical and physical processes in the early solar system occurred. In other words, explorations of *both* volatile suppliers (e.g., primitive asteroids and comets) and building blocks of the Earth system will be required for answering the question where we came from. It is therefore important to expand chemical analyses strategies, so that we are able to handle any type of returned samples. It is worth improving next-generation analytical techniques targeting a large sample that retains geological information. Destructive techniques (e.g., combustion and pyrolysis) and wet chemistry (e.g., for soluble organic molecules and trace elements) were not positively applied in the past missions, but they are powerful techniques to obtain the bulk compositions of returned samples, which provide insights on the formation of target bodies.

Therefore, future technological development in close connection with extraterrestrial material research, as well as synergy among the development of sample collection/preservation techniques, curation facility, and laboratory analysis of returned samples, will be extremely important. Collecting larger amount of samples, underground samples of the target bodies, and/or various states of materials (e.g., ice, liquid, and gas) will be of scientific value. In particular, it is necessary that researchers from observation, sample analyses, and engineering work together to share the common picture of multi-scale small-body science.

REFERENCES

Alexander, C'M. O. D., Bowden, R., Fogel, M. L., Howard, K. T., Herd, C. D. and Nittler, L. R. 2012. The provenances of asteroids, and their contributions to the volatile inventories of the terrestrial planets. *Science* 337: 721–723.

Alexander, C'M. O. D., Cody, G. D., De Gregorio, B. T., Nittler, L. R. and Stroud, R. M. 2017. The nature, origin and modification of insoluble organic matter in chondrites, the major source of Earth's C and N. *Chemie der Erde* 77: 227–256.

Altwegg, K., Balsiger, H., Bar-Nun, A. et al. 2015. 67P/Churyumov-Gerasimenko, a Jupiter family comet with a high D/H ratio. *Science* 347: doi:10.1126/science.1261952

Auger, A.-T., Groussin, O., Jorda, L. et al. 2015. Geomorphology of the Imhotep region on comet 67P/Churyumov-Gerasimenko from OSIRIS observations. *Astronomy & Astrophysics* 583: A35. doi:10.1051/0004-6361/201525947

Abe, M., Takagi, Y., Kitazato, K. et al. 2006. Near-infrared spectral results of asteroid Itokawa from the Hayabusa spacecraft. *Science* 312: 1334–1338.

Altwegg, K., Balsiger, H., Bar-Nun, A. et al. 2015. 67P/Churyumov–Gerasimenko, a Jupiter family comet with a high D/H ratio. *Science* 347: doi:10.1126/science.1261952

Arai, T., Kobayashi, M., Ishibashi, K. et al. 2018. DESTINY+ mission: Flyby of Geminids parent asteroid (3200) Phaethon and in-situ analyses of dust accreting on the Earth. *49th Lunar and Planetary Science Conference*, Abstract #2570.

Bardyn, A., Baklouti, D., Cottin, H. et al. 2017. Carbon-rich dust in comet 67P/Churyumov-Gerasimenko measured by COSIMA/Rosetta. *Monthly Notices of the Royal Astronomical Society* 469: S712–S722.

Berger, E. L., Zega, T. J., Keller, L. P. and Lauretta, D. S. 2011. Evidence for aqueous activity on comet 81P/Wild 2 from sulfide mineral assemblages in Stardust samples and CI chondrites. *Geochimica et Cosmochimica Acta* 75: 3501–3513.

Bieler, A., Altwegg, K., Balsiger, H. et al. 2015. Abundant molecular oxygen in the coma of comet 67P/Churyumov-Gerasimenko. *Nature* 526: 678–683.

Brownlee, D. E., Horz, F., Newburn, R. L. et al. 2004. Surface of young Jupiter family comet 81P/Wild 2: view from the Stardust Spacecraft. *Science* 304: 1764–1769.

Brownlee, D. E., Tsou, P., Aléon, J. et al. 2006. Comet 81P/Wild 2 under a microscope. *Science* 314: 1711–1716.

Capaccioni, F., Coradini, A., Filacchione, G. et al. 2015. The organic-rich surface of comet 67P/Churyumov-Gerasimenko as seen by VIRTIS/Rosetta. *Science* 347: doi:10.1126/science.aaa0628

Clark, B. E., Ziffer, J., Nesvorny, D. et al. 2010. Spectroscopy of B-type asteroids: Subgroups and meteorite analogs. *Journal of Geophysical Research* 115: E06005.

Clark, B. E., Binzel, R. P., Howell, E. S. et al. 2011. Asteroid (101955) 1999 RQ36: Spectroscopy from 0.4 to 2.4 mu m and meteorite analogs. *Icarus* 216: 462–475.

Cloutis, E. A., Hudon, P., Hiroi, T., Gaffey, M. J. 2012. Spectral reflectance properties of carbonaceous chondrites: 7. CK chondrites. *Icarus* 221: 911–924. doi:10.1016/j.icarus.2012.09.017.

Cody, G. D., Ade, H., Alexander, C. M. O'D. et al. 2008. Quantitative organic and light-element analysis of comet 81P/Wild 2 particles using C-, N-, and O-XANES. *Meteoritics & Planetary Science* 43: 353–365.

De Gregorio, B. T., Stroud, R. M., Cody, G. D., Nittler, L. R., Kilcoyne, A. L. D. and Wirick, S. 2011. Correlated microanalysis of cometary organic grains returned by Stardust. *Meteoritics and Planetary Science* 46: 1376–1396.

Ebihara, M., Sekimoto, S., Shirai, N. et al. 2011. Neutron Activation Analysis of a Particle Returned from Asteroid Itokawa. *Science* 333: 1119–1121.

Fray, N., Bardyn, A., Cottin, H. et al. 2016. High-molecular-weight organic matter in the particles of comet 67P/Churyumov-Gerasimenko. *Nature* 538: 72–74.

Fujiwara, A., Kawaguchi, J., Yeomans, D. K. et al. 2006. The rubble-pile asteroid Itokawa as observed by Hayabusa. *Science* 312: 1330–1334.

Glavin, D. P., Dworkin, J. P. and Sandford, S. A. 2008. Detection of cometary amines in samples returned by Stardust. *Meteoritics and Planetary Science* 43: 399–413.

Glavin, D. P., Alexander, C. M. O'D., Aponte, J. C., Dworkin, J. P., Elsila, J. E. and Yabuta, H. 2018. The origin and evolution of organic matter in carbonaceous chondrites and links to their parent bodies. In: *Primitive Meteorites and Asteroids: Physical, Chemical, and Spectroscopic Observations Paving the Way to Exploration* (Ed. Abreu N.) Elsevier, pp. 205–271.

Goesmann, F., Rosenbauer, H., Bredehöft, J. H. et al. 2015. Organic compounds on comet 67P/Churyumov-Gerasimenko revealed by COSAC mass spectrometry. *Science* 349: doi:10.1126/science.aab0689.

Gomes, R., Levison, H. F., Tsiganis, K., Morbidelli, A. 2005. Origin of the cataclysmic Late Heavy Bombardment period of the terrestrial planets. *Nature* 435: 466–469.

Greenberg J. M., Li A. 1997. Silicate core-organic refractory mantle particles as interstellar dust and as aggregated in comets and stellar disks. *Advances in Space Research* 19: 981–990.

Hicks, L. J., MacArthur, J. L., Bridges, J. C. et al. 2017. Magnetite in comet Wild 2: Evidence for parent body aqueous alteration. *Meteoritics & Planetary Science* 52: 2075–2096.

Hiroi T., Pieters C. M., Zolensky M. E. and Lipschutz M. E. 1993. Evidence of thermal metamorphism on the C, G, B, and F asteroids. *Science* 261: 1016–1018.

Hiroi, T., Abe, M., Kitazato, K. et al. 2006. Developing space weathering on the asteroid 25143 Itokawa. *Nature* 443: 56–58.

Hörz, F., Bastien, R., Borg, J. et al. 2006. Impact features on Stardust: implications for comet 81P/Wild 2 dust. *Science* 314: 1716–1719.

Iida, Y., Tsuchiyama, A., Kadono, T. et al. 2010. Three-dimensional shapes and Fe contents of Stardust impact tracks: A track formation model and estimation of comet Wild 2 coma dust particle densities. *Meteoritics & Planetary Science* 45: 1302–1319.

Ito, M., Uesugi, M., Naraoka, H. et al. 2014. H, C, and N isotopic compositions of Hayabusa category 3 organic samples. *Earth, Planets and Space* 66: 91. doi:10.1186/1880-5981-66-91

Jewitt, D., Li, J. 2010. Activity in Geminid parent (3200) Phaethon. *The Astronomical Journal* 140: 1519–1527. doi:10.1088/0004-6256/140/5/1519.

Kameda, S., Suzuki, H., Cho, Y. et al. 2015. Detectability of hydrous minerals using ONC-T camera onboard the Hayabusa2 spacecraft. *Advances in Space Research* 56: 1519–1524.

Keller, L. P., Bajt, S., Baratta, G. A. et al. 2006. Infrared spectroscopy of comet 81P/Wild 2 samples returned by stardust. *Science* 314: 1728–1731.

Kitajima, F., Uesugi, M., Karouji, Y. et al. 2015. A micro-Raman and infrared study of several Hayabusa category 3 (organic) particles. *Earth, Planets and Space* 67: 20. doi.10.1186/s40623-015-0182-6

Lauretta, D. S., Balram-Knutson, S. S., Beshore, E. et al. 2017. OSIRIS-REx: Sample return from Asteroid (101955) Bennu. *Space Science Reviews*. doi:10.1007/s11214-017-0405-1

Lazzaro D., Barucci M. A., Perna D., Jasmim F. L., Yoshikawa M. and Carvano J. M. F. 2013. Rotational spectra of (162173) 1999 JU3, the target of the Hayabusa2 mission. *Astronomy & Astrophysics* 549. doi:10.1051/0004-6361/201220629

Le Roy, L., Altwegg, K., Balsiger, H. et al. 2015. Inventory of the volatiles on comet 67P/Churyumov-Gerasimenko from Rosetta/ROSINA. *Astronomy & Astrophysics* 583: A1. doi:10.1051/0004-6361/201526450

Licandro, J., Campins, H., Mothé-Diniz, T., Pinilla-Alonso, N., de León, J. 2007. The nature of comet-asteroid transition object (3200) Phaethon. *Astronomy & Astrophysics* 461: 751–757. doi:10.1051/0004-6361:20065833.

Matsumoto, T., Tsuchiyama, A., Uesugi, K. et al. 2016. Nanomorphology of Itokawa regolith particles: Application to space-weathering processes affecting the Itokawa asteroid. *Geochimica et Cosmochimica Acta* 187: 195–217.

McKeegan, K. D., Aléon, J., Bradley, J. et al. 2006. Isotopic compositions of cometary matter returned by Stardust. *Science* 314: 1724–1728.

Mikouchi, T., Tachikawa, O., Hagiya, K. et al. 2007. Mineralogy and crystallography of comet 81P/Wild 2 particles. *Lunar Planet. Sci.* XXXVIII. Abstract#1946.

Miyamoto, H., Yano, H., Scheeres, D. J. et al. 2007. Regolith migration and sorting on asteoid Itokawa. *Sciecexpress*. doi:10.1126/science.1134390

Moskovitz, N. A., Abe, S., Pan, K. S. et al. 2013. Rotational characterization of Hayabusa II target Asteroid (162173) 1999 JU3. *Icarus* 224: 24–31.

Nagao, K., Okazaki, R., Nakamura, T. et al. 2011. Irradiation History of Itokawa Regolith Material Deduced from Noble Gases in the Hayabusa Samples. *Science* 333: 1128–1131.

Nakamura, T., Noguchi T, Tsuchiyama A. et al. 2008. Chondrulelike objects in short-period comet 81P/Wild 2. *Science* 321: 1664–1667.

Nakamura T., Noguchi T., Tanaka M. et al. 2011. Mineralogy and Thermal History of Itokawa Surface Particles Recovered by Hayabusa Mission. *Meteoritics & Planetary Science* 46: A172–A172.

Naraoka, H., Mita, H., Hamase, K. et al. 2012. Preliminary organic compound analysis of microparticles returned from Asteroid 25143 Itokawa by the Hayabusa mission. *Geochemical Journal* 46: 61–72.

Naraoka, H., Aoki, D., Fukushima, K. et al. 2015. ToF-SIMS analysis of carbonaceous particles in the sample catcher of the Hayabusa spacecraft. *Earth, Planets Space* 67: 67. doi:10.1186/s40623-015-0224-0

Nguyen, A. N., Berger, E. L., Nakamura-Messenger, K., Messenger, S. and Keller, L. P. 2017. Coordinated mineralogical and isotopic analyses of a cosmic symplectite discovered in a comet 81P/Wild 2 sample. *Meteoritics & Planetary Science* 52: 2004–2016.

Noguchi, T., Nakamura, T., Kimura, M. et al. 2011. Incipient space weathering observed on the surface of Itokawa dust particles. *Science* 333: 1121–1125.

Okazaki, R., Sawada, H., Yamanouchi, S. et al. 2017. Hayabusa2 sample catcher and container: Metal-seal system for vacuum encapsulation of returned samples with volatiles and organic compounds recovered from C-type asteroid Ryugu. *Space Science Reviews* 208: 107–124.

Quirico, E., Moroz, L.V., Schmitt, B. et al. 2016. Refractory and semi-volatile organics at the surface of comet 67P/Churyumov Gerasimenko: Insights from the VIRTIS/Rosetta imaging spectrometer. *Icarus* 272: 32–47.

Saiki, T. Imamura, H., Arakawa, M. et al. 2017. The Small Carry-on Impactor (SCI) and the Hayabusa2 impact experiment. *Space Science Reviews* 208: 165–186.

Sandford, S. A., Aleon, J., Alexander, C.M.O.D. et al. 2006. Organics captured from comet 81P/Wild 2 by the Stardust spacecraft. *Science* 314: 1720–1724.

Schulz, R., Hilchenbach, M., Langevin, Y. et al. 2015. Comet 67P/Churyumov-Gerasimenko sheds dust coat accumulated over the past four years. *Nature* 518: 216–218.

Srama, R., Ahrens, T. J., Altobelli, N. et al. 2004. The Cassini cosmic dust analyzer. *Space Science Reviews* 114: 465–518.

Tachibana, S., Abe, M., Arakawa, M. et al. 2014. Hayabusa2: Scientific importance of samples returned from C-type near-Earth asteroid (162173) 1999 JU(3). *Geochemical Journal* 48: 571–587.

Tatsumi, E. and Sugita, S. 2018. Cratering efficiency on coarse-grain targets: Implications for the dynamical evolution of asteroid 25143 Itokawa. *Icarus* 300: 227–248.

Tsuchiyama, A., Nakamura, T., Okazaki, T. et al. 2009. Three-dimensional structures and elemental distributions of Stardust impact tracks using synchrotron microtomography and X-ray fluorescent analysis. *Meteoritics & Planetary Science* 44: 1203–1224.

Tsuchiyama, A., Uesugi, M., Matsushima, T. et al. 2011. Three-dimensional structure of Hayabusa samples: Origin and evolution of Itokawa regolith. *Science* 333: 1125–1128.

Uesugi, M., Naraoka, H., Ito, M. et al. 2014. Sequential analysis of carbonaceous materials in Hayabusa-returned samples for the determination of their origin. *Earth, Planets and Space* 66: 102. doi:10.1186/1880-5981-66-102

Vilas, F. 2008. Spectral characteristics of Hayabusa 2 near-Earth asteroid targets 162173 1999 JU3 and 2001 QC34. *The Astronomical Journal* 135: 1101–1105.

Walsh, K. J., Morbidelli, A., Raymond, S. N., O'Brien, D. P., Mandell, A. M. 2012. Populating the asteroid belt from two parent source regions due to the migration of giant planets—"The Grand Tack". *Meteoritics & Planetary Science* 47: 1–7.

Yabuta, H., Uesugi, M., Naraoka, H. et al. 2014. X-ray absorption near edge structure spectroscopic study of Hayabusa category 3 carbonaceous particles. *Earth, Planets Space* 66: 156. doi:10.1186/s40623-014-0156-0

Yabuta, H., Sandford, S. A. and Meech, K. J. 2018. Cometary organic molecules and volatiles. *Elements* 14: 101–106.

Yada, T., Fujimura, A., Abe, M. et al. 2014. Hayabusa-returned sample curation in the planetary material sample curation facility of JAXA. *Meteoritics & Planetary Science* 49: 135–153.

Yurimoto, H., Abe, K., Abe, M. et al. 2011. Oxygen isotopic compositions of asteroidal materials returned from Itokawa by the Hayabusa mission. *Science* 333: 1116–1119.

Zolensky, M. E., Zega, T. J., Yano, H. 2006. Mineralogy and petrology of comet 81P/wild 2 nucleus samples. *Science* 314: 1735–1739.

12.3 Solar System Exploration
Icy Moons and Their Habitability

Steven D. Vance

CONTENTS

12.3.1 INTRODUCTION

Earth's ocean is unique in the solar system. No other world orbiting our Sun has a surface covered in liquid water. By contrast, subsurface oceans appear to be relatively common in the outer solar system. Three giant oceans have been confirmed in moons of Jupiter: Europa, Ganymede, and Callisto. Saturn hosts two, and possibly four, worlds with subsurface oceans: Enceladus and Titan. For discussion of Saturn's moons Mimas and Dione, we direct the reader to work by Beuthe (2016). Beyond Saturn, Neptune's moon Triton likely contains an ocean, as does the Kuiper-Belt object Pluto. These worlds contain the first prerequisite for life as it exists on Earth. From the handful of additional available constraints obtained by spacecraft missions, some of these worlds also appear to provide the chemical conditions needed to sustain life. Planned spacecraft missions will begin to look for signs of habitability and extant life in some of these icy ocean worlds, and missions to follow in the ensuing decades may have the chance to sample living organisms directly. This chapter focuses on the astrobiological context of icy ocean worlds. For an excellent review of their physical workings and details on different means for detecting subsurface oceans, the reader is directed to the paper by Nimmo and Pappalardo (2016) (Figures 12.3.1 and 12.3.2).

12.3.2 EVIDENCE FOR OCEANS AND PROSPECTS FOR HABITABILITY

12.3.2.1 EUROPA

A present-day ocean has been inferred from the magnetic field induced within the near subsurface by Jupiter's changing field, as observed during multiple flybys of the *Galileo* spacecraft (Kivelson et al., 2000). This measurement reveals only that a global electrically conductive layer exists near the surface, consistent with salty water. The overall conductance (the product of conductivity and thickness) is not well constrained. Thus, based on the induced magnetic field alone, Europa's ocean may be highly saline and only a few kilometers thick or more dilute than Earth's ocean and hundreds of kilometers thick. The latter seems more likely, based on geophysical measurements of Europa's bulk structure.

Galileo and *Voyager* spacecraft measurements determined that Europa has a large bulk density and small gravitational moment of inertia (Table 12.3.1).

From these, it can be inferred that Europa is strongly differentiated, likely hosting a metallic core underneath a rocky mantle. The H_2O covering Europa is 80- to 170-km thick (Anderson et al., 1998). The corresponding pressure at the seafloor is as high as ~200 MPa, not much higher than in the

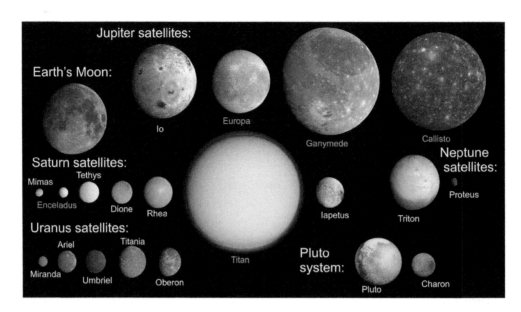

FIGURE 12.3.1 Some prominent worlds of the outer solar system, with Earth's Moon for comparison. Nine of them show evidence for extant subsurface oceans. (Courtesy of NASA/JPL/SSI/JHUAPL/SwRI data, processed by Emily Lakdwalla, Ted Stryk, Gordan Ugarkovic, and Jason Perry.)

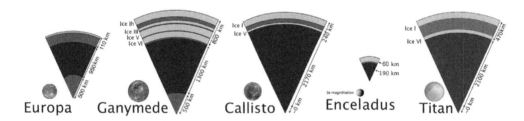

FIGURE 12.3.2 Basic interior structures of known icy ocean worlds based on observed bulk densities and gravitational moments of inertia. Layer thicknesses are not provided with precisions commensurate with known radii (Table 12.3.1), because the interpreted thicknesses depend on the assumed material properties. (Modified from Vance, S. D. et al., *J. Geophys. Res. Planets*, 123, 180–205, 2018; see that reference for more detailed exploration of interior structure and composition, as they relate to the habitability of icy ocean worlds.)

TABLE 12.3.1
Internal Structure Properties of Confirmed Ocean Worlds, Obtained from Robotic Missions of Exploration

	Radius (km)	Density (kg m⁻³)	Moment of Inertia
Europa[a]	1565.0 ± 8.0	2989 ± 46	0.346 ± 0.005
Ganymede[a]	2631 ± 1.7	1942.0 ± 4.8	0.3115 ± 0.0028
Callisto[a]	2410.3 ± 1.5	1834.4 ± 3.4	0.3549 ± 0.0042
Enceladus[b,c]	252.1 ± 0.2	1609 ± 5	0.335
Titan[c]	2574.73 ± 0.09	1879.8 ± 0.004	0.3438 ± 0.0005

Source: [a]Schubert, G. et al., Interior composition, structure and dynamics of the Galilean satellites, In F. Bagenal et al. (Eds.), *Jupiter: The Planet, Satellites and Magnetosphere*, pp. 281–306, Cambridge University Press, Cambridge, UK, 2004; [b]Thomas, P., *Icarus*, 208, 395–401, 2010; Iess, L. et al., *Science*, 344, 78–80, 2014; [c]Jacobson, R. et al. *Astronom. J.*, 132, 2520–2526, 2006; Iess, L. et al., *Science*, 337, 457–459, 2012.

deepest parts of Earth's ocean. Given the expected long history of tidal heating of Europa, it seems likely that the interior has fully differentiated to create a metallic core (Hussmann and Spohn, 2004).

Europa's geology suggests a high heat flux and thus a thick ocean. The icy surface appears to be geologically active; *Galileo* imaging may have even captured evidence of active resurfacing (Schmidt et al., 2011). The ice is perhaps <10-km thick. Tectonic activity has remade the entire icy surface, layering it with linear fractures and broader banded features. Based on the incomplete imaging coverage, the driver for tectonic resurfacing of Europa's ice may be plate tectonics and subduction analogous to the dynamics of Earth's mantle (Greenberg et al., 1998; Kattenhorn and Prockter, 2014).

Extrusion of brines in chaotic terrains (Collins and Nimmo, 2009; Schmidt et al., 2011; Kattenhorn and Prockter, 2014) may create conditions analogous to geyser formation and eruption on Earth. This is exciting from a standpoint of mission design, because it may be possible to sample materials from the ocean, without landing on the surface. To date, three separate *Hubble* Space Telescope observations have shown evidence for eruptive emissions, water jetting hundreds of kilometers above Europa's surface (Roth et al., 2014; Sparks et al., 2016, 2017). These intermittent detections suggest, by contrast with Saturn's moon Enceladus, that eruptions from Europa are infrequent. Alternatively, the emissions may not escape Europa's higher gravity and so may not be as easily detected with available instrumentation. Modeling suggests that dense particles and gas emissions from Europa rarely exceed the 100 km in altitude (Southworth et al., 2015).

12.3.2.2 GANYMEDE

Ganymede is the largest icy ocean world and is altogether different from Europa and from Earth. Ganymede's low density means that it contains hundreds of kilometers of water, like the other large satellites Callisto and Titan. However, Ganymede is more differentiated than the other large satellites, with a low moment of inertia (Table 12.3.1), corresponding to an H_2O thickness of up to 900 km.

Based on models of the interior structure that account for the diverse properties of water and ice, and that account for the freezing point suppression of various salts that may be dissolved in the ocean, Ganymede's seafloor is covered with hundreds of kilometers of high-pressure ice (III, V, VI; Vance et al., 2014, 2018), unless the ocean is very warm and saline. The ice Ih covering Ganymede is at least 30-km thick and possibly thicker than 150 km if Ganymede cooled efficiently during its long history (Hussmann and Spohn, 2004; Bland et al., 2009; Kimura et al., 2009). Its young surface (Collins et al., 2013) and evidence for active resurfacing (Hammond and Barr, 2014) also support the notion of a warm interior.

Ganymede has an endogenous dipolar magnetic field (Schubert and Soderlund, 2011). This unique feature among icy ocean worlds is consistent with a hot interior with a molten iron core that creates a dynamo similar to Earth's (e.g., Bland et al., 2008; Kimura et al., 2009). The implied high heat flux is perhaps more consistent with the thinner-ice scenario. *Hubble* tracking of the oscillating auroral ovals (Saur et al., 2015) provides further evidence for Ganymede's ocean, placing minimum bounds on the salinity of roughly 1 g kg^{-1} (as estimated from Hand and Chyba, 2007, Figure 12.3.1).

12.3.2.3 CALLISTO

Callisto may be nearly frozen. Its icy surface appears to have been geologically inactive for the past 4 Gyr; it is heavily cratered and darkened by dusty materials that were not completely characterized by Voyager or Galileo (Moore et al., 2004). Despite its apparent inactivity, *Galileo* flybys revealed an induced magnetic response to Jupiter's field (Khurana et al., 1998; Zimmer et al., 2000) and thus evidence for a present-day ocean. The ice covering probably undergoes solid-state convection under the upper 100 km or more of warm ice below, making Callisto the only known example of solid-state convection under a fully stagnant lid (McKinnon,

2006). If Callisto's ocean is nearly frozen, its remaining fluids should approach a eutectic composition, the composition with the lowest freezing point. Suppression of the freezing point may mean that high-pressure ices are mostly absent in Callisto. Densification of the fluids means that where such ices occur, they may be buoyantly unstable.

12.3.2.4 ENCELADUS

Though tiny by comparison with the other known ocean worlds, Enceladus is cryovolcanically active (e.g., Spencer and Nimmo, 2013). Its south polar plumes dissipate >10 GW of heat, which can only be explained by tidal flexing of the interior (Choblet et al., 2017) Possible mechanisms driving the plumes include flowing liquids within the ice (Roberts and Nimmo, 2008; Kite and Rubin, 2016), boiling liquids reminiscent of geysers or volcanoes on Earth (Postberg et al., 2009), or some combination thereof (Ingersoll and Nakajima, 2016; Nakajima and Ingersoll, 2016). The eruptive output is modulated with Enceladus's orbit (33-hour period; Hedman et al., 2013); fluxes are highest when the conduits at the south pole are under tension from tidal forces.

The ocean's pH is constrained from direct sampling of gas and dust from the plumes, based on the abundance of CO_2 (Glein et al., 2015). Silica nanograins found in ice particles from the plumes (Hsu et al., 2015) are best explained as forming in the rocky interior from hydrothermal rock-water interactions in alkaline solution above 90°C. A fully water-permeable interior would facilitate such water-rock interactions and has been predicted on the basis of thermal fracturing (Vance et al., 2007, 2016a) and gravity models (Vance et al., 2018; Waite et al., 2017). Taken at face value, thermal fracturing should enable water-rock alteration to produce the high hydrogen and methane concentrations, observed in the Enceladus plumes (Waite et al., 2009). The high inferred ocean pH (>11), consistent with active serpentinization (Glein et al., 2015), could imply that the ocean is relatively young—perhaps owing to recent migration into tidal resonance with Dione (Cuk et al., 2016)—or that serpentinization has proceeded very slowly.

12.3.2.5 TITAN

Saturn's largest moon (comparable in size to planet Mercury) has a hydrocarbon-rich atmosphere (1.4 atm N_2 at its surface; Teanby et al., 2012; Hörst, 2017). The surface covering of solid and liquid hydrocarbons at 92 K generates hydrological features familiar on Earth: channels carved by rains (Turtle et al., 2011; Burr et al., 2012) and rivers flowing into hydrocarbon seas that grow and shrink with Saturn's decade-long seasons (Aharonson et al., 2009; Hayes et al., 2011). Titan's Earthlike atmosphere and hydrological cycle invite questions of "weird life," whether alternative biological frameworks might have evolved to support life in non-aqueous liquid solutions (Schulze-Makuch and Grinspoon, 2005; Schulze-Makuch et al., 2011; Stevenson et al., 2015).

Titan's internal heat has typically been assumed radiogenic, but it appears that some amount of tidal flexing occurs.

A time-variable component to Titan's gravity field was revealed by *Cassini* radio science measurements (Iess et al., 2012). The changing tides are consistent with deformation on timescales of days. Any resulting tidal heating requires an explanation for why Titan's high orbital eccentricity has not been damped out. Conveniently, the same scenario for explaining present-day serpentinization on Enceladus—recent orbital disruption of Saturn's satellites—might explain the high eccentricity of tidally forced Enceladus (Cuk et al., 2016). A recent onset of tidal heating might also explain the outgassing of the atmosphere, which should have escaped if it formed 4 Gya (Yung et al., 1984).

In addition to the organic hydrosphere on its surface, Titan appears to have a global liquid-water ocean under its icy lithosphere. The high tidal potential Love number k_2 requires the icy lithosphere to be thinner than about 100 km and requires the underlying ~400-km layer to be 1,200–1,400 kg m^{-3}. The dense lower layer is best explained by a very salty ocean (Mitri et al., 2014). The density structure of the watery upper part of Titan is difficult to reconcile with the high gravitational moment of inertia (Table 12.3.1), because it implies a low-density interior (<2,500 kg m^{-3}), inconsistent with a purely rocky core resulting from differentiation; a high-density ocean also implies that ice VI may be the only high-pressure phase in Titan's ocean (Vance et al., 2018). Additional clues to the nature of Titan's subsurface water ocean come from the putative Schumann resonance observed by the *Huygens* lander deployed by *Cassini*. This low-frequency electromagnetic oscillation between Saturn's magnetic field, Titan's atmosphere, and an electrically conducting subsurface liquid layer provides further evidence for an ocean beneath an ice shell 55- to 85-km thick (Béghin et al., 2012).

How much exchange occurs between Titan's organic surface and aqueous interior is unknown (e.g., Fortes, 2000; Fortes et al., 2007; Grindrod et al., 2008). Surface geology, to date, has provided few clues. The majority of landforms can be interpreted as exogenic features modified only by fluvial and aeolian processes (Moore and Pappalardo, 2011). Surface hydrocarbons interact with the near subsurface (Hayes et al., 2008) and may reduce the shear strength of surface materials to enable contractional tectonism (Liu et al., 2016b). Recent analyses of the fuller set of *Cassini* radar-imaging data have revealed potential tectonic features (Liu et al., 2016a), but these also do not require complete overturning of the lithosphere. Outgassing from the interior may have occurred episodically as Titan cooled and its ices thickened and methane-enriched clathrates migrated upward through the ice by solid-state convection (Tobie et al., 2006). However, solid-state convection may or may not occur in Titan's frozen lithosphere (Mitri and Showman, 2008, also see the discussion of convection in ice Ih later in this chapter). Even in the absence of ongoing convection, resurfacing of the lithosphere has been proposed to occur through periodic eruptions of ammonia-water blebs formed from 10-km cracks at the base of the lithosphere (Mitri et al., 2008). If such eruptions occur, mass balance requires displacement of materials toward the ocean, but these need not include lighter hydrocarbons and organics from the surface.

Because of the many open questions pertaining to Titan's surface and the nature of its lithosphere, it is difficult to evaluate the composition and habitability of its ocean. Methane, ethane, and ammonia clathrates may have a role in regulating ocean chemistry and atmospheric composition (Choukroun et al., 2010; Choukroun and Grasset, 2010). To a lesser extent, organic materials figure into the composition and dynamics of other icy ocean worlds in the solar system.

12.3.2.6 Triton

Neptune's moon Triton is undergoing extreme orbital degradation. Its orbit is retrograde opposite the direction of Neptune's orbit around the sun, highly inclined, and has near-zero eccentricity. These imply an energetic orbital-tidal evolution (Chyba et al., 1989), potentially beginning with the moon's capture from a binary pair (Agnor and Hamilton, 2006). Triton's surface has arcuate fractures reminiscent of cycloids on Europa (e.g., Kattenhorn and Hurford, 2009). Plumes of nitrogen gas and frozen particles may be linked with fractures (Collins and Nimmo, 2009) and thus to an active ice shell. Although Triton's geology has been attributed to exogenic processes (e.g., changes in solar insulation), arguments have been made that favor a warm interior, perhaps including a liquid water ocean, caused by obliquity tides (Nimmo and Spencer, 2015). Stresses from the gradual shrinking of Triton's orbit toward Neptune may drive ongoing activity (Prockter et al., 2005; Correia, 2009), such as hypothesized for Phobos, the tiny moon of Mars (Hurford et al., 2016).

12.3.2.7 Pluto

The *New Horizons* mission revealed Pluto to be an extremely cold ocean world. Its surface, like Triton's, is covered in nitrogen. Solid nitrogen in the Sputnik Planitia region appears to be undergoing mobile-lid convection (McKinnon et al., 2016). The low thermal conductivity of nitrogen and the presence of anti-freeze materials such as ammonia and methanol may explain the persistence of a liquid water ocean under Sputnik Planitia (Keane et al., 2016; Nimmo et al., 2016)

12.3.3 HABITABILITY OF ICY OCEAN WORLDS, BY ANALOGY WITH EARTH

The composition (salinity, pH, and redox state) of an icy ocean world determines the types of metabolism that will be energetically favored and the environmental stresses that organisms must endure (Marion et al., 2003). The ocean's pH and associated salinity result from the geochemical evolution of the world in question; a low-pH ocean would be dominated by sulfate anions, whereas a neutral or basic ocean would be dominated by chlorides (Zolotov, 2008; Zolotov and Kargel, 2009; Figure 12.3.3).

The first chemical models of icy moons predicted that they would have eutectic ocean compositions dominated by sulfates (Lewis, 1971; Kargel, 1991; Kargel et al., 2000).

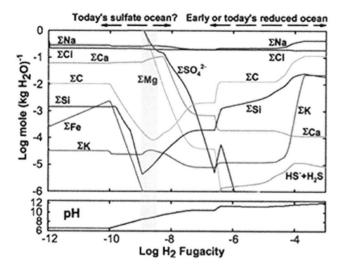

FIGURE 12.3.3 Europa's modeled ocean salinity and pH versus redox state. The vertical bar notes the transition from a chloride-dominated to a sulfate-dominated ocean. (Modified from Zolotov, M., Oceanic composition on Europa: Constraints from mineral solubilities, in *Lunar and Planetary Institute Science Conference Abstracts*, vol. 39, p. 2349, 2008.)

However, a reducing and chloride-dominated ocean might be expected instead, if oxidants were not delivered to the ocean (Hand et al., 2007; Zolotov and Kargel, 2009). The high water content of ocean worlds would seem to imply more dilute compositions (Zolotov and Shock, 2001; McKinnon and Zolensky, 2003; Figure 12.3.3). On the other hand, greater depths of water-rock interaction may cancel this effect, leading to effective water-rock ratios more similar to Earth's.

Ocean composition measurements cannot be based on surface imaging or atmospheric sampling alone, due to unknown fractionation within the ice. On Europa, chlorides inferred from surface infrared reflectance spectra of Europa's surface (Brown and Hand, 2013; Fischer et al., 2015; Ligier et al., 2016; Trumbo et al., 2017) are not a conclusive proxy for its ocean salinity, because chlorides fractionate relative to sulfates in ice shelves (Cragin et al., 1986) and sea ice (Gjessing et al., 1993; Maus et al., 2011). Thus, additional constraints from geophysical measurements are needed (Vance et al., 2018).

Cassini provided the first semi-direct measurement of the salinity and redox state of an icy ocean world. Sodium detected in ice particles from Enceladus captured in Saturn's E-ring is consistent with an NaCl-rich ocean of unknown salinity.

Particle (dust) and gas measurements from the Enceladus plumes indicate a reduced ocean with a pH in the range of 9–11(Glein et al. 2015; Waite et al. 2017). This provides a first indication that Enceladus is currently habitable: methanogenesis $(CO_2(aq)+4H_2(aq)\rightarrow CH_4(aq)+2H_2O)$ has a chemical affinity of 50–120 ± 10 kJ per mol CH_4 at 1 bar pressure and 273 K temperature at pH 9–11 for the hydrogen/water mixing ratio of 0.004–0.02 measured in the plume. That is, methanogenesis generates excess Gibbs energy under these conditions and could thus be a reaction used by life.

12.3.3.1 REDOX BALANCE

Earth's biogeochemical cycles rely on the mantle to provide reduced materials (e.g., Hayes and Waldbauer, 2006). This is balanced by oxidants produced mainly by photosynthesis (e.g., Catling and Claire, 2005). On icy ocean worlds, tides and irradiation may facilitate similar reservoirs of redox energy. Prior published work on Europa illustrates this (Hand et al., 2007; Vance et al., 2007, 2016a). Except in the case of Titan, where the atmosphere is an additional chemical interface, the ice covering ocean worlds in the solar system is the direct interface to space. Photochemistry and impacts may be key sources of oxidants for the ocean, with fluxes depending on both the rate of production at the surface and the efficiency of delivery to liquid regions in the interior. The production and movement of liquids within the ices may thus be essential to the possible presence of life.

12.3.3.1.1 Oxidants Produced by Surface Irradiation

Ultraviolet light and energetic particles oxidize the icy surface or atmosphere by dissociating water, causing the loss of lighter hydrogen to space. Among icy ocean worlds, oxidant production is predicted to be largest at Europa because of its active geology and intense surface irradiation (Cooper et al., 2001; Johnson, 2004; Hand et al., 2007). The radiation belts of the gas giants capture charged particles from the solar wind. {Europa, Ganymede, Callisto} receive averages of {125, 6, 0.6} mW m^{-2} (Cooper et al., 2001; Johnson, 2004). Enceladus and Triton receive less flux than Ganymede (Cooper et al., 2009).

An oxidant's delivery to the ocean depends on its generation at the surface and the efficiency of transport through the ice. The delivery of ionizable materials is expected to occur by impact gardening of micrometeorites and by geodynamic churning (e.g., Kattenhorn and Prockter, 2014).

12.3.3.1.2 Oxidant Fluxes to Europa's Ocean

Irradiation generates oxygen-rich materials $(O_2, H_2O_2, SO_4, SO_2, CO_2, ...)$ on Europa to depths greater than 10 cm (Hand et al., 2007). If the delivery time is less than the surface age of Europa (<100 Myr), this yields a delivery rate for O_2 in the range of 10^9 mol yr^{-1}. However, Greenberg (2010) and Pasek and Greenberg (2012) suggest that geological overturn will oxidize the entire ice after 2 Gyr if the rate of irradiation is constant through time. This will lead to a higher oxidant delivery rate in the current era, up to 2×10^{11} mol yr^{-1}. Vance et al. (2016a) assume that the relevant geophysical overturn time is 30 Myr, based on findings suggesting near-surface liquids (Schmidt et al., 2011) and liquids deeper within the ice (Kalousová et al., 2014) and possible plate tectonic-like behavior (Kattenhorn and Prockter, 2014). The upper limit of 3×10^{11} mole O_2 yr^{-1} (Greenberg, 2010) is within two orders of magnitude of Earth's net photosynthetic oxygen flux of 2×10^{13} mole O_2 yr^{-1} (Catling and Claire, 2005).

The oxidant flux into a 100-km ocean, integrated over 1 Gyr, is comparable to that in Earth's ocean (0.1–0.2 mmoles O_2/L Schmidtko et al., 2017) and well above the minimum oxygen concentrations found in some regions of Earth's ocean

(Childress and Seibel, 1998; Hand et al., 2007). If the ice delivers oxidants on timescales of less than a few Myr and there are no buffers to neutralize them in the ocean, the concentration of oxidants may exceed that of Earth's ocean by factors of 10. However, the complementary fluxes of reductants generated by water-rock alteration in Europa's interior may match or exceed the fluxes of oxidants or may have done so earlier in Europa's history.

The delivery of oxidants from the surface to the ocean depends most likely on the mechanics of solid state convection in ice Ih. A high Rayleigh number is expected for ice shells exceeding 10-km thickness. Thus, solid-state convection should occur (e.g., Mitri et al., 2008; Kalousová et al., 2014, 2016). The rheology of ices is somewhat well-known (Durham et al. 2001), but grain size dependence and intragranular effects complicate predictions of onset and vigor of convection. Non-linear rheology has been a focus for studies of convection in icy ocean worlds (e.g., Barr and Pappalardo, 2005). The characteristic ice grain size determines the viscosity of the ice and thus the activation, vigor, and longevity of convection (e.g., Barr and McKinnon, 2007). In addition, the viscosity may increase as tidal and other stresses in the ice increase the mean grain size, so-called dynamic recrystallization. The influences of large-scale and intragranular impurities on ice dynamics (Mitri and Showman, 2008; Mitri et al., 2008; Choukroun and Sotin, 2012; Choukroun et al., 2013; McCarthy and Cooper, 2016) may limit the growth of grains and thus facilitate transport of materials through the ice.

12.3.3.1.3 Reductant Fluxes to Europa's Ocean

The fluxes of reducing material from the silicate interiors of ocean worlds, in the solar system and beyond, may be maintained by low- (~100°C) or high-temperature (>350°C) water-rock alteration. High-temperature hydrothermal fluids on Earth have a global mass flux $Q_{ht} \approx 3 \times 10^{13}$ kg yr^{-1} and produce $F_{H2} \approx 0.06$ Tmol H$_2$ yr^{-1} (Sleep and Bird, 2007). Earth's total flux of hydrothermal fluids is $Q_{hydr} = 10^{15}$ to >10^{17} kg yr^{-1} (Johnson and Pruis, 2003; Nielsen et al., 2006).

The production of high-temperature hydrothermal H$_2$ can be estimated based on global fluid fluxes, as a function of H_{mantle}: $F_{H2,ht} = n_{H2}Q_{ht}$, with $H_{ht} = 0.1H_{hydr} = 0.05H_{mantle}$. High-temperature mass flux is:

$$Q_{ht} = \frac{H_{ht}}{c_w \Delta T} \quad (12.3.1)$$

where $\Delta T = 350$°C and $c_w = 4$ kJ kg^{-1} is the heat capacity of water at 350°C and pressures around 100 MPa (Wagner and Pruss, 2002). A representative value for the hydrogen content of high-temperature hydrothermal fluids on Earth is $n_{H2} = 2$–6 mmol H$_2$ kg^{-1} (Holland, 2002).

The heat produced in Europa's mantle (Lowell and DuBose 2005; H_{mantle}) sets an upper bound for high-temperature hydrothermal reductant flux of ~0.03 Tmol H$_2$ yr^{-1}, comparable to Earth's flux. Figure 12.3.4 shows $F_{H2,ht}$, as a function of mantle tidal heating, up to $H_{mantle} = 3.5$ TeraWatts (TW). Europa's maximum global hydrothermal fluid mass flux has been

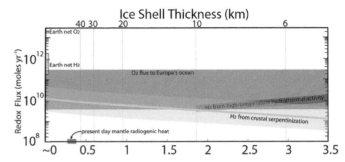

FIGURE 12.3.4 Present-day fluxes of H$_2$ and O$_2$ to Europa's ocean versus tidal heating in the rocky interior. Top axis: equilibrium thickness of a thermally conductive ice shell heated from below. (From Vance, S. D. et al., *Geophys. Res. Lett.*, 43, 4871–4879, 2016. With permission.)

computed as (Lowell and DuBose, 2005) $Q_{hydr} = 10^7$ kg s^{-1}, or ~3 × 10^{14} kg yr^{-1}. They assume $H_{hydr} = 0.5$ and $H_{mantle} = 5$ TW, with H_{mantle} accounting for the majority of Europa's tidal heating. This extreme is unlikely today (Tobie et al., 2003; Sotin et al., 2009) but probably occurred early in Europa's history (Pasek and Greenberg, 2012).

The flux of hydrogen from low-temperature serpentinization has been computed from thermal fracturing, assuming an Earth-like oxidation state of the rocky interior. Europa's predicted present-day hydrogen production (Figure 12.3.4; Vance et al., 2016a) is nearly an order of magnitude larger than Earth's on a tidal heating in Europa's mantle (TW) (Table 12.3.2).

A big-picture result of studies to date, coupling to studies of life's origin on Earth, is the possibility that alkaline serpentinizing systems fostered an independent origin of life on Europa and other icy ocean worlds and that vigorous ecosystems might arise at the ice-ocean interface, where redox gradients are strongest due to the delivery of oxidant-rich fluids from the overlying ice (Russell et al., 2014, 2017).

12.3.3.2 Redox Fluxes in Larger Icy Ocean Worlds

In ocean worlds with very thick ice shells, high fluxes of chemical energy may be difficult to sustain, thus limiting the production and transport of oxidants from the surface to the underlying ocean. This is illustrated most profoundly for Callisto (Section 3.2.3) but also pertains to the thick ice covering of Ganymede and Titan. Despite the expected slow overturning rates relative to Europa, the extent of oxidant production and delivery to the oceans of these worlds deserves further investigation.

Reductant fluxes may also be limited in the larger ocean worlds due to the high overburden pressures at the hundreds of kilometers of depth, where rocks occurring in those worlds may limit the extent of water-rock interaction (Vance et al., 2007; Byrne et al., 2018). However, the low density of Titan's interior (Vance et al., 2018) would seem to require extensive mineral hydration. Clearly, this is a paradox that needs to be addressed. Recent work (Journaux et al., 2013, 2017; Kalousová et al., 2018) suggests that high pressure ices are less of a barrier to the transmission of materials into the

TABLE 12.3.2

Heat and Hydrogen fluxes in Europa, Enceladus, and Earth, based on Vance et al. (2007), including present-day fracture depth (z), and rate (ż). R_P and M_P are the radius and mass of the planetary body, respectively. d is the ocean depth. T_o and T_z are the average temperatures at the surfaces of the bodies, and at the depth of the fracture front (z). H_{rad} is the computed heat production from radioactive decay of U, Th, and K. H_{serp} is the heat from serpentinization of material exposed by new fracturing (at rate ż). Hydrogen fluxes from serpentinization (FH_2) are listed as a ratio to Earth's in terms of relative aerial fluxes and as global flux in *Tmol* per year. The predicted fracture depth for a present-day 100-km-deep-ocean on Europa is 24 km. Europa produces up to 50 Gmoles of H_2 per year from water-rock reactions with newly exposed material. This is 4× the calculated flux for Earth

	R_P	M_P	d	T_o	T_z	P_z	z	$\dot{z}$	Hrad	$\frac{H_{Serp}}{H_{rad}}$	FH_2	
	(km)	(×10²⁰ kg)	(km)	(°C)	(°C)	(MPa)	(km)	$\frac{\mu m}{yr}$	$\frac{mw}{m^2}$	(%)	$\frac{F}{F_{Earth}}$	$\frac{Tmol}{yr}$
Earth	6371	59742.00	4	0	114	237	6	0.3	60.0	0.00	1	0.2
Europa	1565	487.00	100	0	71	248	24	1.0	9.0	0.07	4	0.05
Enceladus	252	0.73	80	−20	0	45	172	—	0.6	7.0	12	0.003

overlying ocean, providing room to speculate that chemical gradients across thick ice layers and oceans may create conditions suitable for life.

12.3.4 DYNAMICS OF OCEANS, ICES, ROCK, AND NEEDED FUTURE WORK, INCLUDING LABORATORY STUDIES

It seems likely that ice-covered oceans will be strongly convecting. Because they are heated from below and cooled from above, and because of their great depths and the low viscosity of water, they have a high predicted Rayleigh number ($Ra \gg 1000$) (e.g., Vance and Brown, 2005; Vance and Goodman, 2009; Travis et al., 2012; Travis and Schubert, 2015). In fact, recent progress in fluid dynamics suggests that turbulent convection may be the rule, organizing the heat transport to be primarily equatorial (Soderlund et al., 2014). A caveat to this prior work is that it did not account for dissipation due to flow along seafloor and sub-ice topography. Such dissipation could be a significant source of tidal heat but is currently not well understood, in large part due to the uncertainty in the topography of the ice (Tyler, 2008, 2014; Chen et al., 2014). Though the vigor of turbulent convection awaits further studies and confirmation by future missions, it is important to consider, because strong turbulence could erase Taylor columns at moderate latitudes that might otherwise permit a more or less direct connection between seafloor hydrothermal systems and the underside of the ice (Goodman et al., 2004). Pole-to-equator variations in the thickness of the ice, known to occur on Titan and Enceladus (McKinnon, 2015; Corlies et al., 2017), and possible for Europa (Nimmo et al., 2007), could give rise to melting where the ice is thick and refreezing where it is thin. This redistribution of ice also implies the dilution of the ocean's salinity where ice melts and concentration

where it forms. The result of this lateral flux of ice and salt could be an additional driver for overturning circulation, analogous to thermohaline circulation in Earth's oceans (Zhu et al., 2017).

An organizing theme of the study of icy ocean worlds is the occurrence of conditions familiar on Earth at atypical pressures and temperatures (Figure 12.3.5). As discussed previously, the very high pressures and temperatures of the larger worlds may impede the occurrence of water-rock interactions. Thermodynamic conditions of pressure, temperature, and composition are primary drivers for exotic properties of icy ocean worlds. Experiments and theory continue to reveal new phenomena that await confirmation or refutation by future missions. This work has an excellent synergy with studies of Earth's carbon cycle and water-rock interactions in subduction zones and at even greater depths.

Geochemical modeling has benefited in recent years from extrapolation of water's dielectric properties beyond 0.5 GPa (Sverjensky et al., 2014a, 2014b). Studies of high-pressure aqueous systems are revealing a change in the organization of water around dissolved ions at $P > 0.4$ GPa, with a pseudo-phase transition to the so-called "high-pressure water," facilitating the increased association of ions. This implies a fundamental shift in the shape of the Gibbs energy surface. On this basis, chemical reactions predicted by extrapolation from lower pressures should be viewed critically. Also, the resulting decrease in electrical conductivity with pressure is directly relevant to planned efforts to measure the induced magnetic field response of Ganymede's deep ocean (Vance et al., 2018).

New laboratory measurements have been obtained in recent years, mitigating the need for extrapolation of chemical properties from standard pressures and temperatures (e.g., Bollengier et al., 2013; Mantegazzi et al., 2013; Vance and Brown, 2013; Bezacier et al., 2014). This work has produced large data sets,

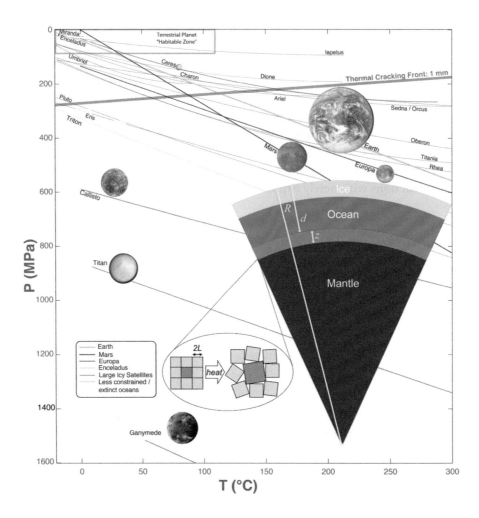

FIGURE 12.3.5   Pressures and temperature profiles for known and putative ocean worlds used in the fracture model. The inset schematic, (Modified from Neveu, M. et al. *J. Geophys. Res. Planet.*, 120, 123–154, 2015.), illustrates the mechanism for cracking due to thermal expansion anisotropy. (From Vance, S. D. et al., *Geophys. Res. Lett.*, *43*, 4871–4879, 2016. With permission.)

mainly comprising surfaces of sound speed in pressure (P) and temperature (T) at various compositions. Equations of state in Gibbs energy and chemical potential are derived from these surfaces by integration in pressure and temperature. The application of modern geophysical inverse theory, aided by modern computational capabilities, has proven to be a key tool in such efforts (Brown, 2018). The use of standard spline fits, which serve as local basis functions, has enabled a shift away from the global damped polynomial approach (e.g., Lemmon and Jacobsen, 2005) that requires refitting when new data sets are added to the fits.

Laboratory studies are also yielding surprising findings regarding the possible retention of volatile materials in the rocky interiors of icy ocean worlds. For example, diamond anvil cell studies of the CO_2-H_2O system have revealed the stability of crystalline carbonic acid at higher pressures than previously predicted (Abramson et al., 2017a, 2017b). A possible implication of this work is the more efficient retention of carbon in Earth's crust and in the rocky interiors of other ocean worlds.

Clearly, there is much work yet to do in laboratories and offices on Earth.

12.3.5   FUTURE PROSPECTS AND MISSIONS

As a new decade of planetary exploration approaches, missions are being planned and formulated to investigate the habitability of ocean worlds and possibly to look directly for signs of life.

Europa will be visited in the mid to late 2020s by NASA's planned Europa Clipper mission (Pappalardo et al., 2016), which will be able to search for and characterize plumes. It will also map nearly the entire surface at better than 500 m/pixel spatial resolution, at wavelengths spanning the ultraviolet to thermal infrared, and peer through the ice with an ice-penetrating radar at radio frequencies of 9 MHz high frequency (HF) and 60 MHz very high frequency (VHF). These investigations will thus address the mystery of how Europa's ice works, the composition of its ocean, and its potential habitability.

Whereas the Europa Clipper concept focuses on the context of Europa's habitability, a studied Europa Lander (Hand et al., 2017) would directly sample a young region on Europa's surface, with the goal of searching for signs of life. It would achieve this by making use of modern mass spectrometric techniques, Raman spectroscopy, and high-resolution imaging to map the context of organics on the surface and seismology to improve our understanding of the regional ice and deeper interior.

The JUpiter Icy satellite Explorer (JUICE) is planned to orbit Ganymede in the late 2020s, with a complement of instruments similar to those on the Europa Clipper (Grasset et al., 2013). At present, the JUICE mission plan calls for also for executing 2 science flybys of neighboring Europa and 12 flybys of Callisto, during the approach to Ganymede orbit.

A NASA mission dedicated to exploring Titan is under concept and technology development ("Phase A"[1]) as a New Frontiers class mission (Turtle et al., 2017). The Dragonfly lander would leverage Titan's dense atmosphere and low gravity, using an eight-rotor copter system to fly to multiple landing sites, studying and sampling different landforms in Titan's equatorial region. Such a mission could reveal Titan's compositional evolution and habitability by studying its hydrocarbon dunes (Hand et al., 2018). It could also probe Titan's radial interior structure by using seismology.

Missions to Uranus and Neptune were prioritized for the current decade. Completing the detailed reconnaissance of the large planets after the fashion of *Galileo* and *Cassini* remains a high priority (Hofstadter et al., 2017). Flagship-class missions to Uranus and Neptune could map the remaining potential ocean worlds, constrain their sizes and densities, and use the tilted magnetic fields of their respective parent ice giants to search for induced magnetic responses indicative of subsurface oceans.

ACKNOWLEDGMENTS

This work was performed at the Jet Propulsion Laboratory, California Institute of Technology, under a contract with NASA, and was supported by the Icy Worlds node of NASA's Astrobiology Institute (13-13NAI7 2-0024). Copyright 2018. All rights reserved.

ENDNOTE

[1] https://nodis3.gsfc.nasa.gov/displayDir.cfm?t=NPR&c=7120&s=5E.

REFERENCES

Abramson, E. H., Bollengier, O., and Brown, J. M. (2017a). The water-carbon dioxide miscibility surface to 450°C and 7 GPa. *American Journal of Science*, 317(9), 967–989. doi:10.2475/09.2017.01.

Abramson, E. H., Bollengier, O., and Brown, J. M. (2017b). Water-carbon dioxide solid phase equilibria at pressures above 4 GPa. *Scientific Reports*, 7(1). doi:10.1038/s41598-017-00915-0.

Agnor, C., and Hamilton, D. (2006). Neptune's capture of its moon Triton in a binary-planet gravitational encounter. *Nature*, 441(7090), 192–194.

Aharonson, O., Hayes, A. G., Lunine, J. I., Lorenz, R. D., Allison, M. D., and Elachi, C. (2009). An asymmetric distribution of lakes on titan as a possible consequence of orbital forcing. *Nature Geoscience*, 2(12), 851–854. doi:10.1038/ngeo698.

Anderson, J., Schubert, G., Jacobson, R., Lau, E., Moore, W., and Sjogren, W. (1998). Europa's differentiated internal structure: Inferences from four Galileo encounters. *Science*, 281(5385), 2019–2022.

Barr, A., and Pappalardo, R. (2005). Onset of convection in the icy Galilean satellites: Influence of rheology. *Journal of Geophysical Research*, 110. doi:10.1029/2004JE002371.

Barr, A. C., and McKinnon, W. B. (2007). Convection in ice I shells and mantles with self-consistent grain size. *Journal of Geophysical Research*, 112, E02012. doi:10.1029/2006JE002781.

Béghin, C., Randriamboarison, O., Hamelin, M., Karkoschka, E., Sotin, C., Whitten, R. C., Berthelier, J. J., Grard, R., and Simões, F. (2012). Analytic theory of Titan's Schumann resonance: Constraints on ionospheric conductivity and buried water ocean. *Icarus*, 218(2), 1028–1042.

Beuthe, M. (2016). Crustal control of dissipative ocean tides in Enceladus and other icy moons. *Icarus*, 280, 278–299. doi:10.1016/j.icarus.2016.08.009.

Bezacier, L., Le Menn, E., Grasset, O., Bollengier, O., Oancea, A., Mezouar, M., and Tobie, G. (2014). Experimental investigation of methane hydrates dissociation up to 5 GPa: Implications for Titan's interior. *Physics of the Earth and Planetary Interiors*, 229, 144–152.

Bland, M., Showman, A., and Tobie, G. (2008). The production of Ganymede's magnetic field. *Icarus*, 198(2), 384–399.

Bland, M., Showman, A., and Tobie, G. (2009). The orbital–thermal evolution and global expansion of Ganymede. *Icarus*, 200(1), 207–221.

Bollengier, O., Choukroun, M., Grasset, O., Le Menn, E., Bellino, G., Morizet, Y., Bezacier, L., Oancea, A., Taffin, C., and Tobie, G. (2013). Phase equilibria in the H₂O–CO₂ system between 250–330 K and 0–1.7 GPa: Stability of the CO₂ hydrates and H₂O-ice VI at CO₂ saturation. *Geochimica et Cosmochimica Acta*, 119, 322–339.

Brown, J. M. (2018). Local basis function representations of thermodynamic surfaces: Water at high pressure and temperature as an example. *Fluid Phase Equilibria*, 463, 18–31. doi:10.1016/j.fluid.2018.02.001.

Brown, M., and Hand, K. (2013). Salts and radiation products on the surface of Europa. *The Astronomical Journal*, 145(4), 110.

Burr, D. M., Taylor Perron, J., Lamb, M. P., Irwin, R. P., Collins, G. C., Howard, A. D., Sklar et al. (2012). Fluvial features on Titan: Insights from morphology and modeling. *Geological Society of America Bulletin*, 125(3–4), 299–321. doi:10.1130/B30612.1.

Byrne, P. K., Regensburger, P. V., Klimczak, C., Bohnenstiehl, D. R., Hauck, S. A., II, Dombard, A. J., and Hemingway, D. J. (2018). The geology of the rocky bodies inside Enceladus, Europa, Titan, and Ganymede. In *49th Lunar and Planetary Science Conference*, (p. Abstract #2905). Houston: Lunar and Planetary Institute. URL http://www.lpi.usra.edu/meetings/lpsc2018/pdf/2905.pdf.

Catling, D. C., and Claire, M. W. (2005). How Earth's atmosphere evolved to an oxic state: a status report. *Earth and Planetary Science Letters*, 237(1), 1–20.

Chen, E., Nimmo, F., and Glatzmaier, G. (2014). Tidal heating in icy satellite oceans. *Icarus*, 229, 11–30.

Childress, J. J., and Seibel, B. A. (1998). Life at stable low oxygen levels: Adaptations of animals to oceanic oxygen minimum layers. *Journal of Experimental Biology*, 201(8), 1223–1232.

Choblet, G., Tobie, G., Sotin, C., Běhounková, M., Cadek, O., Postberg, F., and Souček, O. (2017). Powering prolonged hydrothermal activity inside enceladus. *Nature Astronomy*, 1(12), 841–847. doi:10.1038/s41550-017-0289-8.

Choukroun, M., and Grasset, O. (2010). Thermodynamic data and modeling of the water and ammonia-water phase diagrams up to 2.2 GPa for planetary geophysics. *The Journal of Chemical Physics*, 133, 144502.

Choukroun, M., Grasset, O., Tobie, G., and Sotin, C. (2010). Stability of methane clathrate hydrates under pressure: Influence on outgassing processes of methane on Titan. *Icarus*, 205(2), 581–593.

Choukroun, M., Kieffer, S. W., Lu, X., and Tobie, G. (2013). Clathrate hydrates: Implications for exchange processes in the outer solar system. In S. M. Gudipati and J. C. Castillo-Rogez (Eds.), *The Science of Solar System Ices* (pp. 409–454). Springer, New York.

Choukroun, M., and Sotin, C. (2012). Is Titan's shape caused by its meteorology and carbon cycle? *Geophysical Research Letters*, *39*(4), L04201.

Chyba, C. F., Jankowski, D. G., and Nicholson, P. D. (1989). Tidal evolution in the Neptune-Triton system. *Astronomy and Astrophysics*, *219*, L23–L26.

Collins, G., and Nimmo, F. (2009). *Chaotic Terrain on Europa* (p. 259). University of Arizona Press, Tucson, AZ.

Collins, G. C., Patterson, G. W., Head, J. W., Pappalardo, R. T., Prockter, L. M., Lucchitta, B. K., and Kay, J. P. (2013). Global geologic map of Ganymede. *USGS Scientific Investigations Map*, 3237.

Cooper, J., Cooper, P., Sittler, E., Sturner, S., and Rymer, A. (2009). Old faithful model for radiolytic gas-driven cryovolcanism at Enceladus. *Planetary and Space Science*, *57*(13), 1607–1620.

Cooper, J. F., Johnson, R. E., Mauk, B. H., Garrett, H. B., and Gehrels, N. (2001). Energetic ion and electron irradiation of the icy Galilean satellites. *Icarus*, *149*(1), 133–159.

Corlies, P., Hayes, A. G., Birch, S. P. D., Lorenz, R., Stiles, B. W., Kirk, R., Poggiali, V., Zebker, H., and Iess, L. (2017). Titan's topography and shape at the end of the Cassini mission. *Geophysical Research Letters*, *44*(23), 11,754–11,761. doi:10.1002/2017GL075518.

Correia, A. C. M. (2009). Secular evolution of a satellite by tidal effect: Application to triton. *The Astrophysical Journal*, *704*(1), L1–L4. doi:10.1088/0004-637X/704/1/L1.

Cragin, J. H., Gow, A. J., and Kovacs, A. (1986). Chemical fractionation of brine in the McMurdo Ice Shelf, Antarctica. *Journal of Glaciology*, *32*(112), 307–313.

Cuk, M., Dones, L., and Nesvorný, D. (2016). Dynamical evidence for a late formation of Saturn's moons. *The Astrophysical Journal*, *820*(2), 97. doi:10.3847/0004-637X/820/2/97.

Durham, W., Kirby, S., and Stern, L. (2001). Rheology of water ice—Applications to satellites of the outer planets. *Annual Review of Earth and Planetary Sciences*, *29*, 295–330.

Fischer, P. D., Brown, M. E., and Hand, K. P. (2015). Spatially resolved spectroscopy of Europa: The distinct spectrum of large-scale chaos. *The Astronomical Journal*, *150*(5), 164.

Fortes, A., Grindrod, P., Trickett, S., and Vocadlo, L. (2007). Ammonium sulfate on Titan: Possible origin and role in cryovolcanism. *Icarus*, *188*, 139–153.

Fortes, A. D. (2000). Exobiological implications of a possible ammonia–water ocean inside Titan. *Icarus*, *146*(2), 444–452.

Gjessing, Y., Hanssen-Bauer, I., Fujii, Y., Kameda, T., Kamiyama, K., and Kawamura, T. (1993). Chemical fractionation in sea ice and glacier ice. *Bulletin of Glacier Research*, *11*, 1–8.

Glein, C. R., Baross, J. A., and Waite, J. H. (2015). The pH of Enceladus' ocean. *Geochimica et Cosmochimica Acta*, *162*, 202–219.

Goodman, J. C., Collins, G. C., Marshall, J., and Pierrehumbert, R. T. (2004). Hydrothermal plume dynamics on Europa: Implications for chaos formation. *Journal of Geophysical Research-Planets*, *109*(E3), E03008, doi:10.1029/2003JE002073.

Grasset, O., Dougherty, M., Coustenis, A., Bunce, E., Erd, C., Titov, D., Blanc, M. et al. (2013). JUpiter ICy moons Explorer (JUICE): An {ESA} mission to orbit Ganymede and to characterise the Jupiter system. *Planetary and Space Science*, *78*, 1–21. http://www.sciencedirect.com/science/article/pii/S0032063312003777.

Greenberg, R. (2010). Transport rates of radiolytic substances into Europa's ocean: Implications for the potential origin and maintenance of life. *Astrobiology*, *10*(3), 275.

Greenberg, R., Geissler, P., Hoppa, G., Tufts, B. R., Durda, D. D., Pappalardo, R., Head, J. W., Greeley, R., Sullivan, R., and Carr, M. H. (1998). Tectonic processes on Europa: Tidal stresses, mechanical response, and visible features. *Icarus*, *135*(1), 64–78.

Grindrod, P., Fortes, A., Nimmo, F., Feltham, D., Brodholt, J., and Vočadlo, L. (2008). The long-term stability of a possible aqueous ammonium sulfate ocean inside Titan. *Icarus*, *197*(1), 137–151.

Hammond, N. P., and Barr, A. C. (2014). Formation of Ganymede's grooved terrain by convection-driven resurfacing. *Icarus*, *227*, 206–209.

Hand, K., Carlson, R., and Chyba, C. (2007). Energy, chemical disequilibrium, and geological constraints on Europa. *Astrobiology*, *7*(6), 1006–1022.

Hand, K., and Chyba, C. (2007). Empirical constraints on the salinity of the Europan ocean and implications for a thin ice shell. *Icarus*, *189*(2), 424–438.

Hand, K. P., Murray, A. E., Garvin, J. B., Brinckerhoff, W. B., Christner, B. C., Edgett, K. S., Ehlmann, B. L. et al. (2017). *Report of the Europa Lander Science Definition Team*. Technical report, Jet Propulsion Laboratory, California Institute of Technology.

Hand, K. P., Turtle, E. P., Barnes, J. W., Lorenz, R. D., MacKenzie, S. M., Cable, M. L., Neish, C. D. et al. (2018). Dragonfly and the exploration of Titan's astrobiological potential. *LPI Contributions* (p. 2430).

Hayes, A., Aharonson, O., Callahan, P., Elachi, C., Gim, Y., Kirk, R., Lewis, K. et al. (2008). Hydrocarbon lakes on Titan: Distribution and interaction with a porous regolith. *Geophysical Research Letters*, *35*(9). doi:10.1029/2008GL033409.

Hayes, A., Aharonson, O., Lunine, J., Kirk, R., Zebker, H., Wye, L., Lorenz, R. et al. (2011). Transient surface liquid in Titan's polar regions from Cassini. *Icarus*, *211*(1), 655–671.

Hayes, J. M., and Waldbauer, J. R. (2006). The carbon cycle and associated redox processes through time. *Philosophical Transactions of the Royal Society B: Biological Sciences*, *361*(1470), 931–950.

Hedman, M., Gosmeyer, C., Nicholson, P., Sotin, C., Brown, R., Clark, R., Baines, K., Buratti, B., and Showalter, M. (2013). An observed correlation between plume activity and tidal stresses on Enceladus. *Nature*, *500*, 182–184.

Hofstadter, M., Simon, A., Reh, K., Elliott, J., Niebur, C., and Colangeli, L. (2017). Ice giants pre-decadal study final report. *JPL D-100520*.

Holland, H. D. (2002). Volcanic gases, black smokers, and the great oxidation event. *Geochimica et Cosmochimica Acta*, *66*(21), 3811–3826. http://www.sciencedirect.com/science/article/B6V66-4729T04-8/2/977249fad18330da291be5985d41.

Hörst, S. M. (2017). Titan's atmosphere and climate. *Journal of Geophysical Research: Planets*, *122*(3), 432–482. doi:10.1002/2016JE005240.

Hsu, H. W., Postberg, F., Sekine, Y., Shibuya, T., Kempf, S., Horányi, M., Juhász, A. et al. (2015). Ongoing hydrothermal activities within Enceladus. *Nature*, *519*(7542), 207–210.

Hurford, T., Asphaug, E., Spitale, J., Hemingway, D., Rhoden, A., Henning, W., Bills, B., Kattenhorn, S., and Walker, M. (2016). Tidal disruption of Phobos as the cause of surface fractures. *Journal of Geophysical Research: Planets, 121*, 1054–1065.

Hussmann, H., and Spohn, T. (2004). Thermal–orbital evolution of Io and Europa. *Icarus*, *171*(2), 391–410.

Iess, L., Jacobson, R., Ducci, M., Stevenson, D., Lunine, J., Armstrong, J., Asmar, S., Racioppa, P., Rappaport, N., and Tortora, P. (2012). The tides of Titan. *Science*, *337*(6093), 457–459.

Iess, L., Stevenson, D. J., Parisi, M., Hemingway, D., Jacobson, R. A., Lunine, J. I., Nimmo et al. (2014). The gravity field and interior structure of Enceladus. *Science, 344*(6179), 78–80.

Ingersoll, A. P., and Nakajima, M. (2016). Controlled boiling on Enceladus. 2. Model of the liquid-filled cracks. *Icarus, 272*, 319–326.

Jacobson, R., Antreasian, P., Bordi, J., Criddle, K., Ionasescu, R., Jones, J., Mackenzie, R. et al. (2006). The gravity field of the saturnian system from satellite observations and spacecraft tracking data. *The Astronomical Journal, 132*(6), 2520–2526.

Johnson, H. P., and Pruis, M. J. (2003). Fluxes of fluid and heat from the oceanic crustal reservoir. *Earth and Planetary Science Letters, 216*(4), 565–574.

Johnson, R. (2004). The magnetospheric plasma-driven evolution of satellite atmospheres. *The Astrophysical Journal Letters, 609*, L99–L102.

Journaux, B., Daniel, I., Caracas, R., Montagnac, G., and Cardon, H. (2013). Influence of NaCl on ice VI and ice VII melting curves up to 6 GPa, implications for large icy moons. *Icarus, 226*(1), 355–363. doi:10.1016/j.icarus.2013.05.039.

Journaux, B., Daniel, I., Petitgirard, S., Cardon, H., Perrillat, J. P., Caracas, R., and Mezouar, M. (2017). Salt partitioning between water and high-pressure ices. Implication for the dynamics and habitability of icy moons and water-rich planetary bodies. *Earth and Planetary Science Letters, 463*, 36–47. doi:10.1016/j.epsl.2017.01.017.

Kalousová, K., Sotin, C., Choblet, G., Tobie, G., and Grasset, O. (2018). Two-phase convection in Ganymede's high-pressure ice layer—Implications for its geological evolution. *Icarus, 299*, 133–147. doi:10.1016/j.icarus.2017.07.018.

Kalousová, K., Souček, O., Tobie, G., Choblet, G., and Cadek, O. (2014). Ice melting and downward transport of meltwater by two-phase flow in Europa's ice shell. *Journal of Geophysical Research: Planets, 119*. doi:10.1002/2013JE004563.

Kalousová, K., Souček, O., Tobie, G., Choblet, G., and Cadek, O. (2016). Water generation and transport below Europa's strike-slip faults. *Journal of Geophysical Research: Planets, 121*, 2444–2462.

Kargel, J. (1991). Brine volcanism and the interior structures of asteroids and icy satellites. *Icarus, 94*, 368–390.

Kargel, J. S., Kaye, J. Z., Head, J. W., III, Marion, G. M., Sassen, R., Crowley, J. K., Ballesteros, O. P., Grant, S. A., and Hogenboom, D. L. (2000). Europa's crust and ocean: Origin, composition, and the prospects for life. *Icarus, 148*(1), 226–265.

Kattenhorn, S., and Hurford, T. (2009). Tectonics of Europa. In R. T. Pappalardo, W. B. McKinnon, K. Khurana (Eds.), *Europa* (pp. 199–236). University of Arizona Press, Tucson, AZ.

Kattenhorn, S. A., and Prockter, L. M. (2014). Evidence for subduction in the ice shell of Europa. *Nature Geoscience, 7*(10), 762–767.

Keane, J. T., Matsuyama, I., Kamata, S., and Steckloff, J. K. (2016). Reorientation and faulting of Pluto due to volatile loading within Sputnik Planitia. *Nature, 540*(7631), 90–93.

Khurana, K., Kivelson, M., Stevenson, D., Schubert, G., Russell, C., Walker, R., and Polanskey, C. (1998). Induced magnetic fields as evidence for subsurface oceans in Europa and Callisto. *Nature, 395*(6704), 777–780.

Kimura, J., Nakagawa, T., and Kurita, K. (2009). Size and compositional constraints of Ganymede's metallic core for driving an active dynamo. *Icarus, 202*(1), 216–224.

Kite, E. S., and Rubin, A. M. (2016). Sustained eruptions on Enceladus explained by turbulent dissipation in tiger stripes. *Proceedings of the National Academy of Sciences, 113*(15), 3972–3975.

Kivelson, M., Khurana, K., Russell, C., Volwerk, M., Walker, R., and Zimmer, C. (2000). Galileo magnetometer measurements: A stronger case for a subsurface ocean at Europa. *Science, 289*, 1340–1343.

Lemmon, E. W., and Jacobsen, R. T. (2005). A new functional form and new fitting techniques for equations of state with application to pentafluoroethane (HFC-125). *Journal of Physical and Chemical Reference Data, 34*(1), 69–108.

Lewis, J. (1971). Satellites of the outer planets: Their physical and chemical nature. *Icarus, 15*(2), 174–185.

Ligier, N., Poulet, F., Carter, J., Brunetto, R., and Gourgeot, F. (2016). Vlt/sinfoni observations of Europa: New insights into the surface composition. *The Astronomical Journal, 151*(6), 163.

Liu, Z. Y. C., Radebaugh, J., Harris, R. A., Christiansen, E. H., Neish, C. D., Kirk, R. L., Lorenz, R. D. et al. (2016a). The tectonics of titan: Global structural mapping from Cassini radar. *Icarus, 270*, 14–29.

Liu, Z. Y. C., Radebaugh, J., Harris, R. A., Christiansen, E. H., and Rupper, S. (2016b). Role of fluids in the tectonic evolution of titan. *Icarus, 270*, 2–13.

Lowell, R. P., and DuBose, M. (2005). Hydrothermal systems on Europa. *Geophysical Research Letters, 32*(5), L05202. doi:10.1029/2005GL022375.

Mantegazzi, D., Sanchez-Valle, C., and Driesner, T. (2013). Thermodynamic properties of aqueous NaCl solutions to 1073 K and 4.5 GPa, and implications for dehydration reactions in subducting slabs. *Geochimica et Cosmochimica Acta, 121*, 263–290. doi:10.1016/j.gca.2013.07.015.

Marion, G. M., Fritsen, C. H., Eicken, H., and Payne, M. C. (2003). The search for life on Europa: Limiting environmental factors, potential habitats, and Earth analogues. *Astrobiology, 3*(4), 785–811.

Maus, S., Müller, S., Büttner, J., Brütsch, S., Huthwelker, T., Schwikowski, M., Enzmann, F., and Vähätalo, A. (2011). Ion fractionation in young sea ice from Kongsfjorden, Svalbard. *Annals of Glaciology, 52*(57), 301–310.

McCarthy, C., and Cooper, R. F. (2016). Tidal dissipation in creeping ice and the thermal evolution of Europa. *Earth and Planetary Science Letters, 443*, 185–194.

McKinnon, W. (2006). On convection in ice I shells of outer solar system bodies, with detailed application to Callisto. *Icarus, 183*(2), 435–450.

McKinnon, W. B. (2015). Effect of Enceladus's rapid synchronous spin on interpretation of Cassini gravity. *Geophysical Research Letters, 42*(7), 2137–2143.

McKinnon, W. B., Nimmo, F., Wong, T., Schenk, P. M., White, O. L., Roberts, J., Moore, J. et al. (2016). Convection in a volatile nitrogen-ice-rich layer drives Pluto's geological vigour. *Nature, 534*(7605), 82–85.

McKinnon, W. B., and Zolensky, M. E. (2003). Sulfate content of Europa's ocean and shell: Evolutionary considerations and some geological and astrobiological implications. *Astrobiology, 3*(4), 879–897.

Mitri, G., Meriggiola, R., Hayes, A., Lefevre, A., Tobie, G., Genova, A., Lunine, J. I., and Zebker, H. (2014). Shape, topography, gravity anomalies and tidal deformation of Titan. *Icarus, 236*, 169–177.

Mitri, G., and Showman, A. (2008). Thermal convection in ice-I shells of Titan and Enceladus. *Icarus, 193*, 387–396.

Mitri, G., Showman, A., Lunine, J., and Lopes, R. (2008). Resurfacing of Titan by ammonia–water cryomagma. *Icarus, 196*, 216–224.

Moore, J., Chapman, C., Bierhaus, E., Greeley, R., Chuang, F., Klemaszewski, J., Clark, R. et al. (2004). Callisto. In F. Bagenal et al. (Eds.), *Jupiter. The Planet, Satellites and Magnetosphere*, vol. 1, pp. 397–426. Cambridge University Press, Cambridge, UK.

Moore, J., and Pappalardo, R. (2011). Titan: An exogenic world? *Icarus, 212*, 790–806.

Nakajima, M., and Ingersoll, A. P. (2016). Controlled boiling on Enceladus. 1. Model of the vapor-driven jets. *Icarus, 272*, 309–318.

Neveu, M., Desch, S. J., and Castillo-Rogez, J. C. (2015). Core cracking and hydrothermal circulation can profoundly affect Ceres' geophysical evolution. *Journal of Geophysical Research: Planets, 120*(2), 123–154. doi:10.1002/2014JE004714.

Nielsen, S. G., Rehkämper, M., Teagle, D. A., Butterfield, D. A., Alt, J. C., and Halliday, A. N. (2006). Hydrothermal fluid fluxes calculated from the isotopic mass balance of thallium in the ocean crust. *Earth and Planetary Science Letters, 251*(1), 120–133.

Nimmo, F., Hamilton, D., McKinnon, W., Schenk, P., Binzel, R., Bierson, C., Beyer, R. et al. (2016). Reorientation of Sputnik Planitia implies a subsurface ocean on Pluto. *Nature, 540*, 94–96.

Nimmo, F., and Pappalardo, R. (2016). Ocean worlds in the outer solar system. *Journal of Geophysical Research: Planets, 121*(8), 1378–1399.

Nimmo, F., and Spencer, J. (2015). Powering Triton's recent geological activity by obliquity tides: Implications for Pluto geology. *Icarus, 246*, 2–10.

Nimmo, F., Thomas, P., Pappalardo, R., and Moore, W. (2007). The global shape of Europa: Constraints on lateral shell thickness variations. *Icarus, 191*(1), 183–192.

Pappalardo, R., Prockter, L., Senske, D., Klima, R., Fenton Vance, S., and Craft, K. (2016). Science objectives and capabilities of the NASA Europa Mission. In *Lunar and Planetary Science Conference*, vol. 47, p. 3058.

Pasek, M. A., and Greenberg, R. (2012). Acidification of Europa's subsurface ocean as a consequence of oxidant delivery. *Astrobiology, 12*(2), 151–159.

Postberg, F., Kempf, S., Schmidt, J., Brilliantov, N., Beinsen, A., Abel, B., Buck, U., and Srama, R. (2009). Sodium salts in E-ring ice grains from an ocean below the surface of Enceladus. *Nature, 459*(7250), 1098–1101.

Prockter, L., Nimmo, F., and Pappalardo, R. (2005). A shear heating origin for ridges on Triton. *Geophysical Research Letters, 32*(14), L14202.

Roberts, J., and Nimmo, F. (2008). Tidal heating and the long-term stability of a subsurface ocean on Enceladus. *Icarus, 194*, 675–689.

Roth, L., Saur, J., Retherford, K. D., Strobel, D. F., Feldman, P. D., McGrath, M. A., and Nimmo, F. (2014). Transient water vapor at Europa's south pole. *Science, 343*(6167), 171–174.

Russell, M. J., Barge, L. M., Bhartia, R., Bocanegra, D., Bracher, P. J., Branscomb, E., Kidd, R. et al. (2014). The drive to life on wet and icy worlds. *Astrobiology, 14*(4), 308–343.

Russell, M. J., Murray, A. E., and Hand, K. P. (2017). The possible emergence of life and differentiation of a shallow biosphere on irradiated icy worlds: The example of Europa. *Astrobiology, 17*(12), 1265–1273. doi:10.1089/ast.2016.1600.

Saur, J., Duling, S., Roth, L., Jia, X., Strobel, D. F., Feldman, P. D., Christensen, U. R. et al. (2015). The search for a subsurface ocean in Ganymede with Hubble Space Telescope observations of its auroral ovals. *Journal of Geophysical Research: Space Physics, 120*, 1715–1737.

Schmidt, B., Blankenship, D., Patterson, G., and Schenk, P. (2011). Active formation of chaos terrain over shallow subsurface water on Europa. *Nature, 479*(7374), 502–505.

Schmidtko, S., Stramma, L., and Visbeck, M. (2017). Decline in global oceanic oxygen content during the past five decades. *Nature, 542*(7641), 335–339. doi:10.1038/nature21399.

Schubert, G., Anderson, J., Spohn, T., and McKinnon, W. (2004). Interior composition, structure and dynamics of the Galilean satellites. In F. Bagenal et al. (Eds.), *Jupiter: The Planet, Satellites and Magnetosphere*, pp. 281–306. Cambridge University Press, Cambridge, UK.

Schubert, G., and Soderlund, K. (2011). Planetary magnetic fields: Observations and models. *Physics of the Earth and Planetary Interiors, 187*(3), 92–108.

Schulze-Makuch, D., and Grinspoon, D. H. (2005). Biologically enhanced energy and carbon cycling on titan? *Astrobiology, 5*(4), 560–567.

Schulze-Makuch, D., Haque, S., de Sousa Antonio, M. R., Ali, D., Hosein, R., Song, Y. C., Yang, J. et al. (2011). Microbial life in a liquid asphalt desert. *Astrobiology, 11*(3), 241–258.

Sleep, N., and Bird, D. (2007). Niches of the pre-photosynthetic biosphere and geologic preservation of Earth's earliest ecology. *Geobiology, 5*(2), 101–117.

Soderlund, K., Schmidt, B., Wicht, J., and Blankenship, D. (2014). Ocean-driven heating of Europa's icy shell at low latitudes. *Nature Geoscience, 7*(1), 16–19.

Sotin, C., Tobie, G., Wahr, J., and McKinnon, W. (2009). Chap. Tides and tidal heating on Europa. In R. T. Pappalardo, W. B. McKinnon, and K. K. Khurana (Eds.), *Europa*, pp. 85–117. University of Arizona Press, Tucson, AZ.

Southworth, B., Kempf, S., and Schmidt, J. (2015). Modeling Europa's dust plumes. *Geophysical Research Letters, 42*(24). doi:10.1002/2015GL066502.

Sparks, W., Hand, K., McGrath, M., Bergeron, E., Cracraft, M., and Deustua, S. (2016). Probing for evidence of plumes on Europa with HST/STIS. *The Astrophysical Journal, 829*(2), 121.

Sparks, W. B., Schmidt, B. E., McGrath, M. A., Hand, K. P., Spencer, J., Cracraft, M., and Deustua, S. E. (2017). Active cryovolcanism on Europa? *The Astrophysical Journal Letters, 839*(2), L18.

Spencer, J. R., and Nimmo, F. (2013). Enceladus: An active ice world in the Saturn system. *Annual Review of Earth and Planetary Sciences, 41*(1), 693.

Stevenson, J., Lunine, J., and Clancy, P. (2015). Membrane alternatives in worlds without oxygen: Creation of an azotosome. *Science advances, 1*(1), e1400067.

Sverjensky, D. A., Harrison, B., and Azzolini, D. (2014a). Water in the deep Earth: The dielectric constant and the solubilities of quartz and corundum to 60 kb and 1200 C. *Geochimica et Cosmochimica Acta, 129*, 125–145.

Sverjensky, D. A., Stagno, V., and Huang, F. (2014b). Important role for organic carbon in subduction-zone fluids in the deep carbon cycle. *Nature Geoscience, 7*(12), 909–913.

Teanby, N. A., Irwin, P. G. J., Nixon, C. A., de Kok, R., Vinatier, S., Coustenis, A., Sefton-Nash, E., Calcutt, S. B., and Flasar, F. M. (2012). Active upperatmosphere chemistry and dynamics from polar circulation reversal on titan. *Nature, 491*(7426), 732–735. doi:10.1038/nature11611.

Thomas, P. (2010). Sizes, shapes, and derived properties of the saturnian satellites after the Cassini nominal mission. *Icarus, 208*(1), 395–401.

Tobie, G., Choblet, G., and Sotin, C. (2003). Tidally heated convection: Constraints on Europa's ice shell thickness. *Journal of Geophysical ResearchPlanets, 108*(E11), 5124.

Tobie, G., Lunine, J., and Sotin, C. (2006). Episodic outgassing as the origin of atmospheric methane on Titan. *Nature, 440*(7080), 61–4.

Travis, B., Palguta, J., and Schubert, G. (2012). A whole-moon thermal history model of Europa: Impact of hydrothermal circulation and salt transport. *Icarus*, *218*, 1006–1019.

Travis, B., and Schubert, G. (2015). Keeping Enceladus warm. *Icarus*, *250*, 32–42.

Trumbo, S. K., Brown, M. E., Fischer, P. D., and Hand, K. P. (2017). A new spectral feature on the trailing hemisphere of Europa at 3.78 μm. *The Astronomical Journal*, *153*(6), 250.

Turtle, E., Barnes, J., Trainer, M., Lorenz, R., MacKenzie, S., and Hibbard, K. (2017). Exploring titan's prebiotic organic chemistry and habitability. *LPI Contributions*, p. 1958.

Turtle, E., Perry, J., Hayes, A., Lorenz, R., Barnes, J., McEwen, A., West, R. et al. (2011). Rapid and extensive surface changes near Titan's equator: Evidence of April showers. *Science*, *331*(6023), 1414.

Tyler, R. (2014). Comparative estimates of the heat generated by ocean tides on icy satellites in the outer solar system. *Icarus*, *243*, 358–385. http://www.sciencedirect.com/science/article/pii/S0019103514004539.

Tyler, R. H. (2008). Strong ocean tidal flow and heating on moons of the outer planets. *Nature*, *456*(7223), 770–772.

Vance, S., Bouffard, M., Choukroun, M., and Sotin, C. (2014). Ganymede's internal structure including thermodynamics of magnesium sulfate oceans in contact with ice. *Planetary and Space Science*, *96*, 62–70.

Vance, S., and Brown, J. (2005). Layering and double-diffusion style convection in Europa's ocean. *Icarus*, *177*(2), 506–514.

Vance, S., and Brown, J. M. (2013). Thermodynamic properties of aqueous $MgSO_4$ to 800 MPa at temperatures from −20 to 100°C and concentrations to 2.5 mol kg^{-1} from sound speeds, with applications to icy world oceans. *Geochimica et Cosmochimica Acta*, *110*, 176–189. doi:10.1016/j.gca.2013.01.040.

Vance, S., and Goodman, J. (2009). Chap. Oceanography of an IceCovered Moon. In R. T. Pappalardo, W. B. McKinnon, and K. K. Khurana (Eds.), *Europa*, pp. 459–482. Arizona University Press, Tucson, AZ.

Vance, S., Hand, K., and Pappalardo, R. (2016a). Geophysical controls of chemical disequilibria in Europa. *Geophysical Research Letters*. doi:10.1002/2016GL068547.

Vance, S., Harnmeijer, J., Kimura, J., Hussmann, H., deMartin, B., and Brown, J. M. (2007). Hydrothermal systems in small ocean planets. *Astrobiology*, *7*(6), 987–1005.

Vance, S. D., Hand, K. P., and Pappalardo, R. T. (2016b). Geophysical controls of chemical disequilibria in Europa. *Geophysical ResearchLetters*,*43*(10),4871–4879.doi:10.1002/2016GL068547.

Vance, S. D., Panning, M. P., Stähler, S., Cammarano, F., Bills, B. G., Tobie, G., Kamata, S. et al. (2018). Geophysical investigations of habitability in ice-covered ocean worlds. *Journal of Geophysical Research: Planets, 123*, 180–205. doi:10.1002/2017JE005341.

Wagner, W., and Pruss, A. (2002). The IAPWS formulation 1995 for the thermodynamic properties of ordinary water substance for general and scientific use. *Journal of Physical and Chemical Reference Data*, *31*(2), 387–535.

Waite, J. H., Glein, C. R., Perryman, R. S., Teolis, B. D., Magee, B. A., Miller, G., Grimes, J. et al. (2017). Cassini finds molecular hydrogen in the Enceladus plume: Evidence for hydrothermal processes. *Science*, *356*(6334), 155–159. doi:10.1126/science.aai8703.

Waite, J. H., Lewis, W. S., Magee, B. A., Lunine, J. I., McKinnon, W. B., Glein, C. R., Mousis, O. et al. (2009). Liquid water on Enceladus from observations of ammonia and ^{40}Ar in the plume. *Nature*, *460*(7254), 487–490. doi:10.1038/nature08153.

Yung, Y. L., Allen, M., and Pinto, J. P. (1984). Photochemistry of the atmosphere of Titan – Comparison between model and observations. *The Astrophysical Journal Supplement Series*, *55*, 465. doi:10.1086/190963.

Zhu, P., Manucharyan, G. E., Thompson, A. F., Goodman, J. C., and Vance, S. D. (2017). The influence of meridional ice transport on Europa's ocean stratification and heat content. *Geophysical Research Letters*, *44*(12), 5969–5977. doi:10.1002/2017GL072996.

Zimmer, C., Khurana, K. K., and Kivelson, M. G. (2000). Subsurface oceans on Europa and Callisto: Constraints from Galileo magnetometer observations. *Icarus*, *147*(2), 329–347.

Zolotov, M. (2008). Oceanic composition on Europa: Constraints from mineral solubilities. In *Lunar and Planetary Institute Science Conference Abstracts*, vol. 39, p. 2349.

Zolotov, M., and Shock, E. (2001). Composition and stability of salts on the surface of Europa and their oceanic origin. *Journal of Geophysical Research*, *106*(E12), 32815–32828.

Zolotov, M. Y., and Kargel, J. (2009). On the chemical composition of Europa's icy shell, ocean, and underlying rocks. In R. T. Pappalardo, W. B. McKinnon, and K. Khurana (Eds.), *Europa*, pp. 431–458. University of Arizona Press, Tucson, AZ.

12.4 Searching for Extraterrestrial Life in Our Solar System

Benton C. Clark

CONTENTS

12.4.1   INTRODUCTION

Beyond humankind's relentless quest for learning about various life forms that populate our planet, including our fellow human beings, we also seem to have a fundamental desire to know where else there may be life in the universe. Are we alone?

An extension of this desire, from the viewpoint of scientists, is to learn how another form of life might be constituted. Will it be tantalizingly similar to our own biochemistry? Or, will it be so radically different that we could not have imagined its existence? These two outcomes span the range of possibilities. In either case, however, the results will have far-reaching implications.

The search for life elsewhere is still in its infancy. Indeed, when it was finally discovered that the surface of Venus is so hot as to be inhospitable to life, attention turned to Mars. Flyby missions and the Mariner 9 orbiter at Mars (see https://www.nasa.gov/missions) dispelled the ideas of canals and the waxing and waning of crop patterns, unceremoniously demolishing

earlier claims by astronomers of evidence of intelligent life based on their telescopic observations from Earth. Five years later, the Viking landers touched down on opposite sides of the red planet, obtaining nearly identical results in analyzing Martian soils (Klein 1978; Horowitz 1986; https://mars.nasa.gov/programmissions/missions/past/viking/)

Now, four decades later, the life detection results are still in dispute because of the challenges of making *in situ* measurements of an unknown surface within the practical constraints on instrumentation designed for space flight.

Because liquid water is considered essential to life as we know it, subsequent missions to Mars (see https://www.nasa.gov/topics/journeytomars/index.html) have focused on exploring the geologic and atmospheric history of the planet, with an emphasis from the viewpoint of when and where was the planet habitable. More recently, the evidence has been mounting that some of the larger satellites of the giant planets, as well as large asteroids such as Ceres, currently or in the past, had oceans beneath icy surfaces. Enceladus, and now possibly

Europa, harbor plumes of ejected water that are accessible to space exploration missions.

It now seems time to again begin the earnest search for extant life within our solar system and to face the challenges that must be overcome to reach for the ultimate discovery that resonates so strongly with the science community and the public at large.

12.4.2 CRITERIA FOR LIFE

Living things are intrinsically different from the inanimate objects that populate our universe. Life is a phenomenon that is, in many ways, unlike all other phenomena. Yet, the ancients had difficulty understanding the differences. For example, "gods" were attributed to phenomena such as fire, volcanism, and even the ocean itself. Only through painstaking scientific research were these phenomena demystified and explained by the natural laws deduced from the study of physics, chemistry, biology, and mathematics. The criteria for determining whether each given observation or test is definitive for life, or simply circumstantial, must be carefully examined and justified.

12.4.2.1 WHAT ARE WE TRYING TO FIND?

Even though various forms of life are often clearly identifiable by their appearance or their actions, the reality is that these characteristics alone are generally insufficient. Camouflage hides some—for example, fish and lizards that look like rocks; see Figure 12.4.1.

Yet, other non-living entities can have most of the attributes of higher forms of life (e.g., animals and other beings with intelligence). The painting forming the basis of Figure 12.4.2 was created during advocacy of a 1979 or 1981 follow-on mission after Viking-1 and -2, employing the third flight vehicle but replacing footpads with tracks originally developed as

FIGURE 12.4.1 Viking Lander-1 touched down on a surprising abundance of rocks of all shapes and sizes, up to boulder-sized. (Courtesy of NASA, Washington, DC.)

candidates for the Apollo lunar buggy. This would have created a roving vehicle that may have discovered salts or evidence of thermal activity. In the context of discovering life, would an alien civilization mistake a robot for a living being?

Far more consequential is that there is an enormous and essential portion of the biosphere that consists of microbes, invisible to our eyesight because of the limited resolution of the human eyeball as well as the fact that the size of most bacteria is comparable to or less than the wavelengths of visible light. Viruses are generally smaller and can typically be "seen" only with the aid of the electron microscope.

Furthermore, the natural geologic world creates objects that can be confused with organisms. Spherical concretions form in aqueous sediments. Tektites, geodes, and other ellipsoidal objects form without the aid of biological activity. Manganese oxides crystallize in dendritic patterns reminiscent of biological forms such as plant root systems or colonies of filamentous microorganisms.

Yet, the biological science community seems to agree to disagree on what should be the precise definition of what constitutes a living organism. Numerous attempts at a "definition of life" have been put forth (Kolb, V. M. 2018, Chapter 2.1 this handbook), yet there is no universal agreement on a single formulation.

The so-called NASA definition is "Life is a self-sustained chemical system capable of undergoing Darwinian evolution" (Joyce 1994). Although this is an admirable synopsis of the characteristics of life as we know it, it has some quite limited aspects and is also very difficult to put into practice in a first exploration for life on an alien world. For example, the term "self-sustained" can be interpreted as independence from everything else, including other organisms. Yet, parasites and predators are forms of life that are everywhere and, in many cases, would be easier to detect than their victims (e.g., mice vs. earthworms and seeds). We also now know that more than 99% of all microbial species cannot be successfully cultivated in the laboratory, presumably because of mutual interdependence on other microbes. Finally, we also know that organisms in our biosphere depend on the primary productivity of photosynthesizers, powered by sunlight, and that all organisms have requirements for nutrients in various forms (dissolved key elements, atmospheric gases, etc.). The term "self-sustained" might possibly be interpreted as the ability to utilize energy from the environment to accomplish evolution.

The NASA definition goes on to specify a "chemical system" as being necessary. This is also perhaps too restrictive. We already know that biological systems are not strictly chemical but also often involve electrical properties (e.g., photosynthesis electron transport, membrane interactions with ions, and mitochondrial proton pump) as well as sensor systems and mechanical processes. It could even be envisioned that some forms of life may involve more electrical processes than chemical ones.

The term "Darwinian evolution" in the NASA definition is meant to point out the central feature of life compared with non-life, that is, it can evolve to future versions that are different, and sometimes more capable, than the existing version of

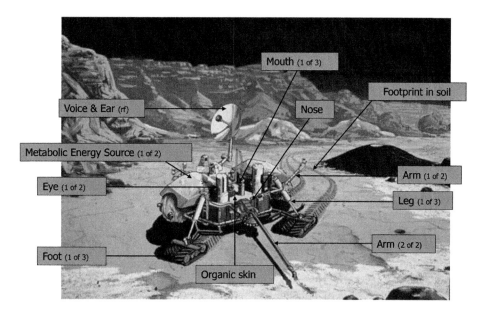

FIGURE 12.4.2 Earthling roaming Mars?

Artist's painting of proposed Viking-3 rover mission, showing anthropomorphic similarities to animal characteristics, including purposeful movement, internal energy source, and ingestion of soil. The "skin" is a resilient paint applied to bounce off abrasive wind-blown sand; the "nose" is the inlet to the GCMS; the "mouth" is the inlet to the VLBI; and the energy sources, producing heat, are the Radioisotope Thermoelectric Generators (RTGs), which power the "alien." (Courtesy of Martin Marietta.)

the organism. Clearly, evolution is the hallmark of life-forms, as compared with objects such as rocks or phenomena such as lightning. These latter examples can in fact change over time (rocks can weather and a lightning bolt changes course during the development of the strike), but they do so along known pathways that follow the laws of physics and chemistry. On the other hand, the specification of "Darwinian" evolution pays homage to the conclusion by Darwin and Wallace that evolution in our biosphere is driven by random chance and natural selection of successful progeny. There is, however, no fundamental impediment to Lamarckian evolution, whereby changes that occur during the lifetime of a single organism in adapting to its environment can be immediately passed on to future generations. Indeed, proving whether a given instance of evolution is either Darwinian or Lamarckian can be extremely difficult. Clearly, the evolution of human civilization is intended to be more Lamarckian than Darwinian (although sometimes difficult to discern).

In Chapter 2.2 of this handbook (Clark 2018), I offer a more generalized and universal definition of life. It also includes evolution as the central essence of a living system but explicitly calls for energy transduction, which is something that can be tested for in a realistic experiment. It also points out that not all living organisms are reproductive (e.g., worker bees, mules, and sterile human beings).

Perhaps, the most important qualification is that organisms can become dormant until environmental conditions change for the better. Bacterial spores, as well as seeds, are ubiquitous. They are so passive, however, as to be non-detectable, except perhaps by their intriguing morphology. Yet, they can be alive, although the only way to find out, in general, is to provide them with a "suitable environment," which coaxes them

out of dormancy into activities of growth and reproduction. This feature of so many organisms, especially microbes, can be of extreme importance in designing investigations to test for life. What constitutes a suitable environment can be somewhat pre-judged but only verified by actual experimentation.

The conclusion from defining life is that it is not at all trivial to prove that something is truly alive. First, it must be observed to be reproductive. This may require experimentation with different environments, including not only the physical conditions (temperature, pressure, etc.) but also the provision of nutrients. Next, it must be shown to be capable of evolving. This is yet more challenging because of the slow course of evolution, generation to generation, and the need to identify phenotypical traits that can be tracked. Testing a putative organism to prove that it satisfies a definition of life can be difficult, especially in the context of a space mission with limited resources and sometimes limited duration. Clearly, we need to consider other approaches that can be used to "screen" for the possible presence of living entities to choose which samples to study more exhaustively.

12.4.2.2 Pragmatic Approaches

Obvious signs of life include evidences of intelligence, such as the canals and croplands imagined by early astronomers using the limited observational techniques available at that time. If, when Viking touched down on the surface of Mars for the first time, the images had shown shrub brush, cacti, or a freeway overpass bridge in the distance, both the public and the science community would have quickly concluded that there was convincing evidence of life. Instead, the images were of rocks and soils, just like some deserts on Earth.

This was, in fact, not surprising to most scientists, who expected Martian life to be limited to microbial organisms in the soil. The life-detection payload selected for the Viking lander missions was designed to test Martian soils in a multiplicity of ways. Originally, there were four separate biology experiments with incubated samples and a gas chromatograph-mass spectrometer (GCMS) to test for organics in the soils.

Years before the Viking landings, Nobel Prize winner Joshua Lederberg opined "A negative assay for organic materials would preclude biology" (Space Studies Board [SSB] 1965, p. 129). This opinion was shared by many of the members of the biology team, such that when the GCMS team reported detecting no organics in Martian soils above about one part per million at either landing site (Biemann et al. 1977), there was pessimism that living (or dead) organisms could be in the samples. The Viking pronouncement of no organics in soils was later challenged (Navarro-Gonzalez and McKay 2011), and subsequent analysis of Martian sedimentary rocks and soils by the Sample Analysis at Mars (SAM) instrument on the Curiosity rover have detected chlorinated (Freissinet et al. 2015) and sulfonated (Eigenbrode et al. 2018) organics in Martian soil, albeit at extremely low levels.

The Viking Lander Biology Instrument (VLBI) originally included the "Wolf Trap" experiment (Space Studies Board [SSB]) 1965, p. 497), which would have incubated Martian soil in water to measure pH and use optical scattering to seek evidence for growth and reproduction of microbes, a common experiment in beginning microbiology courses. Because of the increased costs and complexity encountered during development of the VLBI instrument, this one experiment was "descoped" (i.e., eliminated) from the mission. The three remaining experiments focused on attempts to detect metabolic activity of various types but could not readily detect reproduction and certainly not evolution of the putative Martian microbes. These three experiments are shown schematically in Figure 12.4.3.

Up to four incubation chambers were provided for each experiment. Because water would boil at these temperatures under the low ambient Martian atmospheric pressure, an inert cover gas was added after delivery of soil to raise the pressures in the incubations chambers.

One of the experiments measured the composition of gases in the headspace of its incubation chamber, with the aid of a gas chromatograph. This "Gas Exchange" (GEx) experiment provided the soil with a generous supply of nutrients: an aqueous solution containing virtually all the amino acids used by biology on Earth (in both D and L forms), as well as nucleosides, vitamins, several simple metabolites, and numerous key ions and trace elements as salts. When the soil was allowed to be humidified, before introduction of the liquid "chicken soup," it was found to release oxygen (Klein 1978), which indicated that there was a strongly active chemical oxidant in the soil. Once wetted, there was no release of any metabolic gases (or uptake of Martian gases or the added CO_2). These results were taken as a lack of metabolic activity in the soil samples.

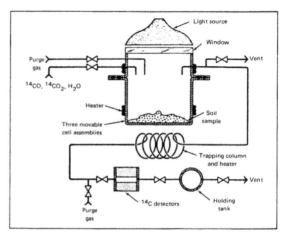

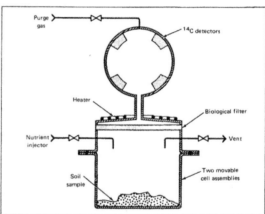

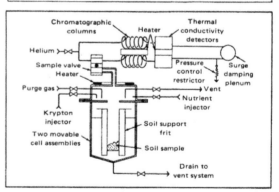

FIGURE 12.4.3 The Viking experiments were elegantly simple and seemingly straightforward. Top, Pyrolytic Release (PR) experiment; middle, Labeled Release (LR) experiment; and bottom, Gas Exchange (GEx) experiment. (Courtesy of NASA SP-334, p. 58.)

Two of the experiments utilized radiolabeled ^{14}C compounds and looked for transformation from one molecular form to another. In the case of the "Pyrolytic Release" (PR) experiment, soil was incubated with labeled CO_2 and CO, and then, after removal of these gases, the soil was pyrolyzed to release any carbon that might have been metabolically fixed (incorporated into organic molecules). Oxidation of the thermally volatilized organics was achieved with the aid of heated CuO. This procedure could be performed in the dark or with a solar simulator to stimulate any photosynthetic capability of the sample. Except for one particular experiment considered anomalous, the incorporated gas was very small or negligible.

The "Labeled Release" (LR) experiment (Levin and Straat 1976, 2016) added to its aliquot of soil a small amount of aqueous nutrient containing five different organic compounds and measured the evolution of gas containing the ^{14}C. A positive result was obtained. As hoped, there was a sharp release of gas, as the putative organisms metabolized one or more of the nutrients, and perhaps multiplied, with an eventual decrease in the *rate* of gas production because nutrient was consumed, as portrayed in Figure 12.4.4.

When a fresh soil sample was heated to high temperature (160°C) to sterilize it, this response did not occur, indicating the putative organisms had been killed or otherwise inactivated. This seemed to be convincing evidence of life. However, additional experiments on Mars showed that when fresh nutrient was added after a first response, there was a complete lack of evolution of new gas. The expectations were additional metabolic activity and creation of gas (upper points of Figure 12.4.4). If there was, in addition to growth, actual multiplication of organisms (reproduction), the rate of activity would have increased even more. In fact, there was no positive response, which is a negative indicator for life and has been pointed out a few times before (Horowitz 1986; Schuerger and Clark 2007). Instead, the total gas actually decreased somewhat, and there were smaller but quite noticeable fluctuations in the head gas concentration on a diurnal basis, which are indications that the solution created was dissolving a portion of the gas. These results led to two camps—those who favored an interpretation as metabolic activity of organisms and those who sought geochemical explanations. As summarized in Table 12.4.1, it seemed possible to explain the various results either way.

Overall, the results of the Viking biology experiments were viewed as being negative for life (Klein 1978 1999; Horowitz 1986), and there began an extended series of experiments by others to emulate the LR results by adding various oxidizing agents to soils.

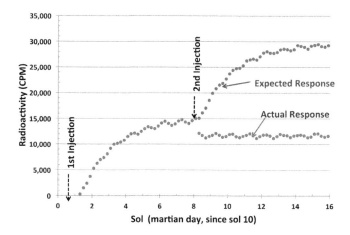

FIGURE 12.4.4 Diagram of actual and expected results from the Labeled Release biology experiment. See text and Table 12.4.1 for additional explanation. (Basis: Figure 6 of Levin, G.V. and Straat, P.A., *Science*, 194, 1322–1329, 1976.)

TABLE 12.4.1
Explanations Offered for LR Results

- First injection produces strong signal
  Life: just as expected
  Non-life: one or more oxidants in soil
- Second injection
  Life: liquid re-absorbs released gas, and organisms have died
  Non-life: soil oxidant was consumed (destroyed by first injection)
- Diurnal oscillation of amount of gas in headspace
  Life: circadian rhythm
  Non-life: temperature-dependent uptake in soil (e.g., carbonate)
- Heat sterilization treatments
  Life: aha! Proof positive of signs of life!
  Non-life: oxidant is heat labile, or release of H_2O from soil destroys it

12.4.2.3 Future Approaches

The NASA "Ladder of Life Detection" is provided on its astrobiology website: https://astrobiology.nasa.gov/research/life-detection/ladder/

The purposes of this very useful chart have been to capture and summarize current scientific thinking on approaches to life detection and to stimulate new thought on potential techniques and their advantages, disadvantages, and potential ambiguities in their results. As a result, the chart is subject to continuous update. Most assessments within the chart are adjectival, rather than any attempt at numerical evaluations. Although absolute metrics might be desirable, they would undoubtedly have large error bars, such that synthesizing overall numerical scores would almost assuredly have very large error bars (propagation of errors), undermining confidence in assessing the relative merit of each of the approaches. A variety of criteria, including practicality, reliability, cost, precision, accuracy, sensitivity, and uniqueness in interpretation, render judgments questionable, at best. However, attempting objective judgments on *relative strengths* can be fruitful in setting a basis for the selection of one approach over another. Different missions may take different approaches to searching for life. It must be remembered that "life as we *don't* know it" mandates a broadly based line of attack to the problem. If Martian life is chemically Earth-like, whether because our biochemistry is the predominant option that nature has available or because of transport between the two planets (panspermia), then the tests can be much more focused. Since we have no overt evidence for discarding either of these cases, there are grounds for both strategies, even on the same mission.

The "ladder rungs" are keyed to the various properties entered in the first column of each row (or "rung"). At the highest level is the goal of demonstrating evolution, which, as we discussed earlier, would be extraordinarily challenging if the life-form had not yet even been detected, isolated, or cultivated in the laboratory. An indirect method of detecting evolution would be, however, a discovery of two different types of organisms that differ from one another in significant ways, such as their response to stimuli, robustness to environmental stressors, and/or morphological differences in appearance in structure or colonization. If these observations imply the presence of two or more different species, then evolution may be inferred.

The second rung is "Growth and Reproduction." These are two separate properties, with reproduction being most significant. However, a spherical organism needs to increase only by 26% in diameter (sometimes difficult to measure this accurately) to double its volume and hence be ready for division. Even more challenging is when the progeny cell is born much smaller in size and then grows to maturity. For micron-sized organisms, and below, morphology can be evaluated well only with short-wavelength microscopy, such as ultraviolet (UV), x-ray, and electron microscopes. Without cultivation in the laboratory, the number of organisms that occur naturally in soils is relatively small, with mean free distance of 100 cell diameters between them if dispersed uniformly. This means there must be extensive searches to locate individual cells, which is either a major data load of images sent back to Earth or else invokes the use of smart algorithms or artificial intelligence on site to choose the images of most interest to send back to Earth.

The third rung of the ladder is "Metabolism." As discussed previously, this was the primary approach of the Viking experiments, which have the advantage that neither morphology nor even quantity of organisms is critically important. Furthermore, it can target extant, active, living organisms—the Holy Grail of astrobiology. A potential pitfall is the possibility of non-biologically mediated reactions that can mimic likely biochemical ones. This is, of course, the root cause of the lingering controversy over whether the Viking LR measurement results were indicative of organisms or rather of unexpected chemical ingredients in the global soils.

This rung can benefit from the use of isotopes as tracers. Viking employed radioactively labeled (^{14}C beta emitter) organic substrates to trace chemical reactions. Modern approaches may be able to utilize rare but non-radioactive isotopes such as deuterium, ^{13}C, ^{15}N, ^{17}O, ^{36}S, and/or ^{58}Fe to trace chemical changes. This may require purification of sought-after products and will also benefit very much from high-resolution mass spectrometry, possibly combined with very advanced tuned-laser techniques (e.g., polycyclic aromatic hydrocarbon [PAH] detection). A dilemma is which soluble substrates to provide (e.g., saccharides, amino acids, amphiphiles, and energy-rich molecules such as ATP or thioesters) and which bioactive gases to test (e.g., H_2, H_2S, CO, CO_2, CH_4, NH_3, and NO_x).

A sub-rung of the ladder's "Metabolism" category would be discovery of the juxtaposition of redox pairs of reactants, the logic being that if life did not exist, the reactants could remain chemically stable due to kinetic inhibition, whereas life takes advantage of such occurrences and catalyzes their reaction to obtain free energy in a controlled and useful way. However, there are numerous examples of ecologies in which such seeming anomalies exist, such as the abundance of organic matter over vast surfaces of planet Earth while surrounded by atmospheric oxygen gas, without reacting unless a "combustion temperature" is provided (e.g., lightning). Although there is a major portion of the biosphere, the respiring organisms,

which does, in fact, attempt to eliminate this disequilibrium, the action of the oxygenic photosynthesizing organisms predominates. Paradoxically, the existence of such unexploited redox couples is often invoked by astrobiologists as evidence of "habitability" and hence the likelihood of finding life. Finally, the interactions of the natural non-living world should also have eliminated many non-equilibrium situations, such as the joint presence of water, CO_2, and Ca- and Mg-containing minerals, which should be thermodynamically converted to various states of hydrated carbonates. Yet, carbonates on Mars are relatively rare, and it is the widespread and powerful biological activity on Earth that is invoked as a key enabler for these and many other kinetically inhibited mineral conversions. Consequentially, the existence of unexpected redox couples or ingredients can be interpreted in multiple ways and requires a full understanding of the current as well as past environments at the location of interest.

Below these three major attributes of life (evolution, reproduction, and metabolism), the ladder labels the subsequent rungs as "Suspicious Biomaterials." Following the NASA definition of life, this means the presence of biochemicals that are of paramount importance for life as we know it (DNA, RNA, amino acids, proteins, and lipids). Specific tests for individual molecules or classes of molecules can be readily devised, although not all are straightforward to implement for spaceflight.

Some of these tests are not uniquely positive for either extant or extinct life. For example, a whole range of amino acids and other organic molecules of biochemical importance has been found to occur naturally in carbonaceous chondrite meteorites and some in comets and interstellar space (Kwok 2016). Polycyclic aromatic hydrocarbons are abundant in biological systems, yet also widespread in the universe.

Hopanes are often invoked as important biological tracers, because they are so resistant to degradation. However, hopanoids are not universal among life-forms, being at very low abundance in some plants and low or absent in one entire branch of the three main branches of life, the archaea microorganisms (Brassell et al. 1981; Belin et al. 2018).

Lipids are present in carbonaceous chondrites, but the pattern (even vs. odd molecular weight) is different in biological systems compared with natural non-biological processes such as Fischer-Tropsch synthesis of organic molecules from CO and H_2 (Georgiou and Deamer 2014). Life on Earth utilizes only a restricted set of amino acids compared with the three-fold or more larger collection of amino acids found in carbonaceous meteorites.

In a planet's atmosphere, gases that are unstable with respect to chemical reaction with the environment, or the action of UV light, must have a source to replenish them. Methane can be biogenic, or it can be a by-product of the serpentinization reaction of water with ultramafic minerals.

Another pattern diagnostic of terrestrial life is chirality, with unequal enantiomeric pairs of molecules. It is expected that other biological systems will exhibit chirality in certain molecular forms. This was a test that could not have been accomplished by the Viking GCMS, because the necessary pre-processing of samples was not incorporated. However, the SAM instrument on the Curiosity rover has this capability, as does the MOMA instrument on the ExoMars rover (Goetz et al. 2016).

Aside from organic compounds, there are certain elements, especially transition metals, that are incorporated into protein-based enzymes and were probably involved early in the origin and evolution of life because of their strong catalytic properties for certain reactions. These elements include Fe, Mn, Ni, V, Cu, Zn, and several others, especially when complexed with S. Phosphorus also plays key roles in several aspects of the known biology (nucleic acids, phospholipids, ATP, and others). It is only rarely, however, that an enrichment of such elements remains as a tracer of biological activity, because their relatively low concentration in biological systems tends to mirror their natural geochemical occurrences.

Aside from advanced instruments for gas and liquid chromatography, as well as very-high-resolution mass spectrometry, various so-called "agnostic approaches" are also under consideration. In this more generalist approach, Johnson et al. (2018) advocate techniques that rely on assessing chemical hierarchy by conducting binding experiments on candidate organic assemblages and sequencing of the resulting complex polymers. This is one step beyond looking for patterns in one molecular class, such as amino acids and lipids, and, in principle, might provide a pathway to determining if information storage is occurring, the basis of genetics and evolution.

12.4.3 HABITATS

In evaluations of habitability, various locales in the solar system have different degrees of likelihood as incubators of life. These evaluations have tended to be highly terrestrial-centric. For example, our biosphere includes various forms of life capable of existing over extremely broad temperature ranges but primarily restricted to the range at which H_2O is in the liquid state. Perhaps, a non-terrestrial life-form may have achieved the capability to live off deeply frozen ice, by sufficient harvesting of solar or some other form of energy. Nonetheless, because mission priorities must optimize the chances of successful outcomes, it has become prudent to adopt guidelines based on terrestrial experience.

12.4.3.1 BACKGROUND AND HISTORY

When the Viking mission was first formulated and approved, there was very little knowledge of where would be the most promising landing sites for finding life on Mars. On the one hand, it was believed that peak temperatures more compatible with life were restricted to the lower latitudes. At the same time, the fact that water ice was apparent in the polar regions was a factor favoring higher latitudes. As a result, the two nuclear-powered Vikings landed at 22° N and 47° N. However, the primary driving factor in each case was the perceived safety of the chosen landing site (low surface slopes, minimum rock hazards, etc.). The original landing site for Viking-1 was revealed, via Viking Orbiter imagery, to have geomorphological features indicating past flooding. This imagery was not of

sufficiently high resolution to detect the presence or absence of boulders, but the presumption was that the terrain might be too rugged for a safe touchdown. Although the pre-selected site was apparently where water had flowed heavily in the past, the plans were aborted and the landing target moved to where there was no such apparent activity in the past. Decades later, the Mars Pathfinder mission landed successfully with its airbags at the original prime landing site.

The Viking mantra was "If life is anywhere, life is everywhere," which was based on the fact that microbial activity is found in soils virtually everywhere on Earth, including hot and cold deserts (Antarctica). Now that we know the soil of Mars is global in nature (Clark et al. 1982; Gellert and Clark 2015), and therefore that the two Vikings sampled the same type of material, we can speculate that there must be different environments that might be more favorable to life. The Viking robot arm did manage to successfully push away a rock and sample the soil that had been protected underneath that rock, which was a valiant attempt at a second environment (Moore et al. 1987). The concern with the global soil unit, of course, is that the periodic global dust storms levitate soil and cause it to be fully exposed to sterilization by the EUV, which reaches down to the surface of Mars because of the lack of the equivalent of Earth's stratospheric ozone layer that absorbs the more energetic photons. A total of four different samples were taken for the VLBI instruments on each lander, some of which included testing of soils taken from shallow and deeper depths below the surface (Moore et al. 1987).

Viking also attempted, unsuccessfully, to break open rocks for sampling, although little was known at the time about endolithic communities in Antarctica and other deserts. The salt-enriched duricrust fragments were also not tested.

Ocean worlds are quite different. Originally, there was one putative ocean world, Europa, and that was unproven. However, the surface lineaments and clear evidence of an icy crust were intriguing clues. To sample the ocean itself, it was thought to be necessary to find cracks that opened periodically. Now, it appears that some, if not most, ocean worlds have breaches through their outer ice shells, which allow liquid to spew forth, creating plumes of material that reach high altitudes. This is certainly the case for Enceladus, and apparently for Europa as well. Such plumes are somewhat difficult to discover without thorough searches, so it remains unanswered at this time how many such worlds may be ejecting oceanic material into space. Because these bodies have liquid water, which is almost surely in contact with a core containing minerals and carbonaceous matter, they are considered habitable.

12.4.3.2 WHERE TO SEARCH: MARS

The attitudes for where to land on Mars have changed dramatically since Viking. The "life everywhere" mantra has, itself, died. Orbital and rover missions at Mars since Viking, as well as the Viking Orbiter imagery, have discovered a vast diversity of hydrologic landforms on Mars. Remote sensing of mineralogy by infrared spectroscopy has likewise discovered

FIGURE 12.4.5 Mud-cracked material crisscrossed by veins of calcium sulfate salt, termed "Old Soaker," discovered in Gale Crater, Mars. (Courtesy of NASA Curiosity Rover.)

variegated deposits of minerals such as clays and salts, which can only be produced by long-term aqueous alteration of igneous rocks. An abundance of areas that have had substantial quantities of water, as rivers or lakes or perhaps even oceans, is now apparent. Layered sediments containing clays, salts, ferric oxides or interbedded with several constituents, have been discovered. In Gale crater, the Curiosity rover has discovered strong evidence not only of a great lake, but also of inflow fans and evidence for waxing and waning of water levels, complete with areas populated with mud cracks, Figure 12.4.5.

Higher latitudes are now looked upon with greater interest because of the proven presence of near-surface ice, discovered by the Phoenix lander at 68° N. Not until more than a decade after the Viking landings was it realized that Mars undergoes major cycles, whereby its obliquity (axis tilt) systematically changes and, at some point, induces much warmer climates at higher latitudes (Ward 1992). For this reason, Mars, which is now cold was once, and will be again, warm enough to melt the ice that pervades the soil.

The relentless bombardment by cosmic rays, unimpeded by a geomagnetic field and raining down through the thin atmosphere of Mars, has resulted in destruction or conversion of environmental organic compounds originally present from meteoritic in-fall, abiotic synthesis, and possibly life. For this reason, there is strong interest in drilling into the regolith to some depth where this ionizing radiation has been attenuated. The European Space Agency (ESA's) ExoMars rover slated for launch in 2020 will have a drill capable of penetrating to a depth of up to 2 meters.

One version of locations of special interest is the feature termed recurrent slope lineae (RSL). The RSLs are gullies, located on steep slopes, which have shown one or more change during observation by orbital imagery. Changes seem to occur mainly in local springtime, implicating the possibility that ice or salt-rich brines are involved. They actually may be less interesting than this, simply reflecting downslope movement of relatively dry soil, triggered by temperature changes. However, because of the possible involvement of

liquid water, they have been tentatively identified as special regions (SP) where terrestrial (and Martian) organisms might flourish. Exploring such locations in the search for life is of great interest but very challenging for a variety of reasons.

Ultimately, it will be desirable to sample material at such a depth that the natural geothermal gradient inside Mars reaches the temperature of, or above, 273 K, such that H_2O ice will melt (or shallower, if freezing-point depressing brines are involved). Although it would be quite challenging for a robotic space mission to drill so deep, ~2 or 3 km, there might be areas that provide access to material from such depths, such as crater ejecta. Assuming Mars has a cryosphere with warm ice or liquid water at these depths, this would be an ideal place for organisms to have survived, as surface conditions worsened due to today's less hospitable surface environment, with its cold, oxidizing atmosphere, lack of liquid water, and UV radiation. Water-rock serpentinization reactions involving ultramafic minerals can result in the production of hydrogen or methane, which would be important metabolites for a subsurface biota.

Other potential habitats that have been suggested include organisms adapted for caves, as well as biota that might live in cracks (chasmoliths), inside rocks (endoliths), inside moist salt deposits (halophiles), or on pro-solar slopes at high latitudes, where surface ice is nearby.

12.4.3.3 Where to Search: Ocean Worlds

It is now clear that there are many ocean worlds, and some, such as Enceladus and apparently Europa, have active plumes, where the water can expel and expand into the vacuum of space. The Cassini mission has already successfully flown through the plumes of Enceladus and detected not only the water (ice grains) but also silicates, salt (NaCl), and organics (Postberg et al. 2011, 2017). This was one of the most spectacular discoveries of the Cassini mission to Saturn and its moons, and the instruments used to make these discoveries were not at all optimized for this opportunity.

NASA's "Europa Clipper" mission is currently in its design phase. In addition to accomplishing multiple close flybys, it may have the opportunity to fly through plumes. A Europa Lander has also been under intensive study (Hand et al. 2016). The ESA's JUpiter ICy moons Explorer (JUICE) mission is already under development and will make detailed remote-sensing observations of three of Jupiter's large satellites, which are all thought to be ocean worlds: Ganymede, Callisto, and Europa.

Future missions could consider drilling down into the icy crusts of these worlds, although the best opportunity for accessing unfrozen water (brine?) would probably be to find a location where the crust has natural access, via cracks where plumes are emitted or where a recent impact has breached the surface. The most ambitious of far-future missions would be special submarines designed to descend the ocean depths to study stratification and ultimately to reach the presumed rocky cores where the source of organics and other nutrients may exist.

12.4.3.4 WHERE TO SEARCH: ICY WORLDS

Some planetary objects may contain abundant ice in their interiors, although not currently liquid. In addition to abundant subsurface ice, the dwarf planet Ceres has clays and salts, clearly implicating the past presence of liquid water (Prettyman et al. 2017). Comets are rich in organic compounds, as well as ices. Although the general model of comet formation does not allow for temperatures warm enough for liquid water to have existed, there are some models that invoke radioactive heating, which might have caused some liquid phase at depth early in their history. There are even some minerals found in comet particles (e.g., cubanite, magnetite, and pentlandite) that may be indicative of aqueous alteration (Brownlee 2014), the presumption being that they were produced on other bodies and were incorporated later. Nonetheless, comets as well as carbonaceous chondrite meteorites and their asteroidal parent bodies are of considerable interest in the origin of life, because they may have provided the seed organics from which life arose.

12.4.4 METABOLISM AND GROWTH/REPRODUCTION

The Viking PR and LR life detection experiments focused on anabolism and catabolism of simple carbon compounds, while the GEx experiment was seeking evidence of metabolism that involved uptake and/or release of simple gaseous products.

12.4.4.1 BEYOND VIKING

Many more experiments directed toward metabolism were possible and were incorporated in a proposed experiment for subsequent flight, prior to even the landing of the Vikings (Radmer and Kok 1971). This experiment incorporated a mass spectrometer and a large number of incubation chambers with a variety of metabolites to probe for various metabolic activities, including the LR experiment as well as nitrogen fixation and various tests of nitrate, sulfate, sulfide, and phosphate metabolism. Although many current approaches often include mass spectrometry, their focus is typically on biomarkers rather than real-time active processing of organics and other biochemicals to search for active metabolism.

12.4.4.2 PROVIDE A SUITABLE ENVIRONMENT

One major difficulty in testing for biological activity lies in finding the conditions under which it will be active, as opposed to dormant. The diversity in metabolic repertoires that we encounter in various species on Earth may not be prevalent elsewhere, nor the ability to flourish in as many different environments. In consideration of the cold Martian environment, Viking used low incubation temperatures (typically +6°C to +26°C)—above freezing, but well below the traditional 25°C to 37°C temperature optima that encompass a wide variety of terrestrial organisms. Also, because the degree of wetting needed was uncertain, the aqueous solutions were added

incrementally (1:5 water/soil ratio for the LR experiment) or through a porous barrier (GEx humidification, then 2:1 water/soil ratio) to gradually wet the soil and permit moisture gradients to occur (i.e., don't "drown" the putative organisms). The PR experiments were conducted alternatively under both wet and dry conditions (Brown et al. 1978).

Other considerations for incubation media are such factors as pH and salt concentrations. The ionic strength of the solution and the nature of the ions may be important (Na^+, Mg^{++}, Cl^-, SO_4^-, etc.) for eliciting metabolic activity.

As with many organisms on Earth, there could be a lag phase before activity responds. Reproduction could be slow or even absent. Metabolic activity as an indicator has the advantage that it could be detected well before reproduction itself would occur, especially if the organisms had been dormant.

We now know, of course, that a very large fraction of microbes that inhabit natural environments have never been successfully cultivated under laboratory conditions. However, if they are metabolically active, then their products would contribute to the overall response.

12.4.5 MEASUREMENT TECHNIQUES AND INSTRUMENTS

Analyses of Martian materials for evidence of life can be conducted in numerous ways. To the extent that there are obvious patterns that cannot be explained by geological processes, an orbiting spacecraft could discover evidence of life. Notwithstanding the absence of the purported canals, the so-called "face on Mars" was claimed to be evidence akin to the Egyptian pyramids but was eventually revealed to be jumbled terrain, which, with the aid of shadowing and poorer resolution imagery, appeared somewhat face-like. Orbital observations of atmospheric composition, such as the early claimed detections of methane, can be used as evidence and will be a key objective in future studies of exoplanets. However, investigations done at ground level, and sub-surface, are expected to provide the most convincing evidence for the presence or absence of reliable clues to life.

12.4.5.1 REAL-TIME IN SITU ANALYSIS

Molecular analysis remains one of the key approaches for future studies of Mars in the search for life. Virtually, all landed science missions have included mass spectrometers to probe Martian samples for organic compounds, with the exception of the M2020 rover. Over the decades, mass spectrometers built for spaceflight have improved in both resolution and sensitivity. In addition, pre-analysis chemical processing of samples has been incorporated, with the ExoMars rover MOMA experiment incorporating three different types of derivatization agents to render targeted classes of molecules, such as amino acids, more volatile, so that they can be analyzed (Goetz et al. 2016). MOMA also incorporates laser desorption and ionization to replace the more conventional pyrolysis to release organics. This technique avoids the problem of oxidation of organics during analysis by the perchlorate (and perhaps other oxidants) within Martian global soils. The Europa Clipper mission also includes mass spectrometers of advanced design.

Element analyzers using x-ray fluorescence (XRF) have flown on nearly all landed missions, although the ExoMars rover will rely on Raman spectroscopy to identify minerals. On the M2020 rover, the PIXL instrument will employ micro-beam XRF analysis to search for minor and trace mineralizations and possible evidence of biological activity via, for example, vanadium or manganese hot spots.

High-resolution imagers have been incorporated post-Viking. Although their resolution is typically a few tens of microns per pixel, far too crude to identify typical microbial cell forms, they are very useful to image microbe-formed structures such as two-dimensional (2-D) microbial mats or three-dimensional (3-D) stromatolites. Their main use, of course, has been to substitute for the hand lens that no field geologist would be without.

12.4.5.2 Sample Return

In many respects, the ideal approach to searching for life is to simply bring back to Earth selected samples for more detailed analysis. Not unlike the early explorers, notably including Charles Darwin, the collection of specimens in the field for return to well-stocked laboratories with highly experienced scientists and technicians is eminently to be desired. This allows the full panoply of measurement resources, including the crews who operate the state-of-the-art analytical systems in unique, specialized laboratories. Some instruments simply are difficult to miniaturize. The classical example, to cite an extreme absurdity, is the synchrotron (which occupies many acres for the primary accelerator). Perhaps surprisingly, attempts at miniaturizing a scanning electron microscope (SEM) for spaceflight have so far been unsuccessful, perhaps partially because its greatest value is in analyzing thin sections of materials. Even the petrologist's standard sample approach of making thin sections for analysis, using a variety of types of microscopes, element mapping systems, and diffractometers, has not yet been developed for flight. A dilemma with such equipment is that there is a seemingly infinite number of thin sections that could be prepared and analyzed, invoking petabytes of data just to understand one geologic location's complex history.

Finding microfossils is similarly challenging. Some previous studies (SSB (Space Studies Board) 1977) have gone so far as to recommend only sample return missions for continuing the search for life in Martian samples.

12.4.6 CHALLENGES OF FLIGHT MISSIONS

As is broadly well known by participants in space missions, there are many challenges to the engineering development of the spacecraft and rockets that make such activities possible. Even more challenging, often, are the scientific instruments and sample processing devices because of the strong motivation to continue to incorporate state-of-the-art advances in

capabilities while, at the same time, needing to stay within the boundaries that constrain spaceflight.

12.4.6.1 Instrument Challenges

The most significant challenge for instrumentation has traditionally been and continues to be the miniaturization of laboratory techniques, which sometimes is measured in meters, rather than centimeters in size, and is tens or hundreds of kilograms, as opposed to a few to twenty or so maximum kilograms. Indeed, mass continues to be a salient metric because costing models are generally proportionally anchored to this benchmark parameter. Volume is also generally related to mass because overall density seldom exceeds that of water, in spite of higher-density components, because of intrinsic void spaces in the packaging (e.g., between optical elements, spacing of detectors from sources, and voids between components of sample acquisition and handling systems).

Other major constraints include power consumption, generally just 10 or a few tens of watts available, once the mission is beyond Earth orbit and especially if more remote than Mars from the sun. Typically, however, if the instrument does not need to be operated continuously, then the critical parameter is electrical energy (W-hr) per measurement sequence, rather than power consumption, because the orbiters, landers, or rovers have sizeable batteries, which can provide higher power wattage than the average it must achieve per day. For example, rovers and landers typically go into low-power modes ("nap," "sleep," and "deep sleep") whenever possible to achieve the low average power that must be less than what the solar arrays or nuclear power sources provide, so that the battery can be recharged. An instrument with very high power requirements will therefore not be allowed to operate as often as desired, compared with one that is so low in power that it may be allowed to operate virtually continuously.

Scientific investigations must be robust to the rigors of spaceflight, which include not only the ultrahigh vacuum of space but also wide temperature ranges for instrument operation and even wider ranges for instrument survivability. For optically based systems, there may be requirements to survive unplanned direct impingement by sunlight. Such systems must also contend with "stray light," whereby sunrays undergo reflection or diffuse scattering from multiple locations on the spacecraft to reach the sensor and cause undesired background or faux signals.

Other issues are data quantity, especially when averaged over the course of the mission, and also the permissible data latency. If large amounts of data must be returned immediately, it will drive the telecommunication system more than if that data can be spread out over a period of days or weeks (sometimes, months, if from very deep space). With modern flash memory storage, the quantity of data stored onboard is no longer such a limited resource as it was in the earlier history of space exploration. For enormous data sets, such as image cubes that involve both high spatial resolution mapping and high-resolution spectroscopy for each image element (pixel), an approach that is starting to gain more serious interest is onboard processing to select and return more refined results to Earth, without all the raw data. Radar imaging is another example where onboard processing is needed. Although imaging missions originally strove to return all data with no compression, so that radiometric accuracy would never be sacrificed, modern data compression algorithms produce images that are scientifically useful with as low as two bits per pixel average. For operations purposes (e.g., planning of driving routes for a rover), ever-greater losses in fidelity are permissible. The engineering cameras on the M2020 Mars rover will be of higher resolution and in color; this invokes data issues, because the telecommunications pipeline from Mars to Earth still relies on short relay periods per sol, limited to when one of the Mars orbiters happens to pass overhead within range of the surface asset.

A much greater limitation, due to the orbiter-relay based architecture, is the latency of data. Thus, new results from a rover on Mars typically occur only once or twice per day, and the data may therefore be more than one sol old before the ground can react to the results of driving. The longer-term solution, which could dramatically speed up exploration, will be when aerostationary satellites (aka geostationary satellites around Earth) can monitor a Mars ground asset with a continuous line of sight. A few multiple synchronous satellites will be needed, however, to cover all longitudes of Mars, or else, new missions would be constrained by the pre-existing positioning of one single satellite.

Cost is a very major concern for instrumentation, although with high-enough importance, the cost of an instrument that is central to the topmost mission science objectives can be much higher than that of the instruments making ancillary measurements. A traditional metric is that science instruments cost approximately one order of magnitude more, per kilogram, than the host spacecraft itself. Some science instruments cost much more.

The more an instrument costs, the greater the concern for its reliability. In general, although the spacecraft may have two redundant copies of every "black box" function, so that if one box fails the other can be switched on, that is almost never the case with science instruments. Both a spacecraft's subsystems and its scientific payload are operated for many hundreds of hours before launch to ensure that "infant mortality" failure modes or defects are found, if they exist. Likewise, all the software modes of operation are fully checked out. Due to a number of spacecraft failures during the faster-better-cheaper (FBC) era, spacecraft are also operated before flight in off-nominal conditions as part of "stress-testing" to establish robustness and margins against failures.

Achieving an instrument configuration that is functional yet can be accommodated by the location and viewing limitations intrinsic to the design of the spacecraft is termed the "packaging" issue. If the instrument requires low-temperature cooling of its sensors via use of a thermal radiator, that radiator will need a relatively unobstructed or clear view of deep space. These are the issues that can only be solved by close interaction between instrument designers and spacecraft designers.

The VLBI life detection instrument on Viking is a classic example of constraints-driven packaging, which resulted in complexity and high cost. As an anecdote, I was once delivering terrestrial samples potentially simulating Martian soils, based on his measurements of the soil composition, to George Hobby, a member of the PR investigation team. The meeting place was a particular biology laboratory at the California Institute of Technology. Reaching the laboratory, and looking through the window panes in the door, the experiment did not appear to be hosted in that lab. However, upon entering, it was seen that the experimental setup was in one small corner, spread out horizontally over a single lab bench. The PR experiment was actually relatively simple compared to wet chemistry analysis or multi-step chemical synthesis lab setups. However, in the VLBI flight instrument, the PR experiment had to be reconfigured into a small, three-dimensional version, with convoluted tubing around the incubation chamber, solar light source, heaters, valves, etc., as evident in Figure 12.4.6.

For space experiments, the valves themselves are a major challenge, since they must be greatly miniaturized yet resistant to leaks and backflow, while being highly reliable and ultra-clean. Ironically, although the Viking landers would be operating in Martian sunlight and on the very cold Martian surface, it was found necessary to incorporate a solar simulator (xenon lamp) into the packaging and also thermoelectric coolers for the incubation chambers. The latter were necessary because the dissipative heat from nearby electronics and the thermally controlled lander compartment would raise temperatures above those desired (the interior of the Viking thermal compartment, where the VLBI and other equipment were located, was maintained near "room temperature" by waste heat from the nuclear power sources—the radioisotope thermoelectric generators).

Requirements for cleanliness and minimization of contamination by organic molecules, by living organisms, and also by dead organisms (the "dead bug bodies" problem) are typically severe compared with the already-existing standards of clean room operations in assembling hardware and spacecraft. Assays of biologically active contaminants are routine for Mars missions, which are intended to reach the surface, requiring monitoring assays for organisms as well as other types of contamination. For life detection sampling and instrumentation, such cleanliness levels are daunting (Viking-equivalent, after heat-soak "sterilization") but achievable if clearly identified and pursued.

The bioload assessment comes under the cognizance of the Planetary Protection Office (PPO) of each space agency. The organic cleanliness is specified by the science team for the mission and implemented by the spacecraft provider. Fairén et al. (2017) have argued for a relaxation of requirements for bioburden levels, because ever since Viking, the cost of implementation has seemed to preclude missions for searching for active life. The Viking landers went through a dry heat "sterilization" procedure, which required that they be sealed in a bioshield, and operated properly after being subjected to 112°C temperatures for 26 hours in a special oven, Figure 12.4.7.

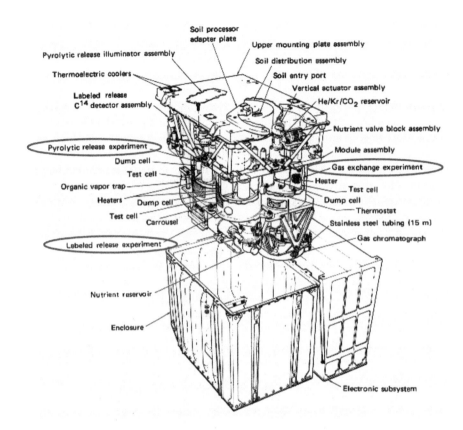

FIGURE 12.4.6   Complexity of the Viking Lander Biology Instrument, and its three main experiments. A major challenge was packaging all the required functions within the limited volume available and its cubical configuration. (From Reproduction of figure 3 from Brown, F.S. et al., *Rev. Sci. Instrum.*, 49, 139, 1978, with the permission of AIP Publishing.)

FIGURE 12.4.7   Viking lander inside bioshield, being moved into oven for dry heat "sterilization" prior to mating with the rocket for launch. (Courtesy of NASA, Washington, DC.)

A major problem is finding organizations with experience in designing, building, and testing high-performance science instruments for flying in space, with little or no previous heritage for that instrument. Often, this unique instrument expertise is to be found in laboratories within a few limited universities, NASA laboratories, and even fewer industrial organizations. As a result of the relatively minimal competition, the principal investigator (P.I.) who invents/conceives the investigation and associated instrumentation is at the mercy of the engineering organization with which she or he is teamed. Thus, it is important that the investigation P.I. retain an appropriate level of decision authority to safeguard scientific performance, within the bounds of the constraints of the sponsoring space agency (typically, mainly instrument cost but also that the instrument be accommodated at reasonable cost, complexity, and reliability by the spacecraft implementer). It is also important that the P.I. plan on less, typically much less, time available for on-going science activities because of the need to devote significant efforts for maintaining the forward progress in instrument development and learning the protocols and exigencies of mission operations. In some cases, the P.I. may lose significant portions of career time for a duration of 10 or more years. This dedication is in the face of the risk of spacecraft or instrument failure, which is real, although reliability has steadily improved in deep space exploration (with the exception of the period during which the FBC cost caps caused loss of several missions). In the end, since most missions are groundbreaking and filled with new discoveries, the dedication required may be compensated by being one of the most rewarding portions of their overall careers.

12.4.6.2 ENVIRONMENT UNCERTAINTIES

One major drawback to the Viking search for life was that although much was already known about the Martian environment at the surface (freezing surface temperatures, low pressure, low humidity, and extreme UV flux) before landing, there were other characteristics neither known, nor envisioned. For example, the extremely fine nature of Martian dust (4-μm characteristic size, a factor of 10 below the mean size of lunar soils) was not fully appreciated, and the life detection techniques were tested against only a very limited number of soil analogs. The relatively high concentrations of sulfates, chlorates, and perchlorates were not expected—the XRF spectrometer (XRFS) instrument had been calibrated for more than a dozen common geochemical elements, but not at all for S or Cl. In particular, it was not *a priori* expected that there would be salts or relatively strong oxidants in the soil. It is now appreciated that oxidants such as perchlorate presumably react with the small quantities of organics in Martian soil at the pyrolysis temperatures used to volatilize organics (and also release some inorganic molecules). Perchlorate, especially when converted into other products by long-term cosmic ray damage, may have provided the oxidant(s) that produced the reactions in the LR experiments on Viking (Quinn et al. 2013; Georgiou et al. 2017).

With the high sulfate content of Martian soil, it would have been logical to add hydrogen (H_2) gas to incubation chambers.

Numerous microorganisms on Earth can metabolize it, especially using sulfate as the electron acceptor, and hydrogen (H_2) may have been an important greenhouse gas, which allowed early Mars to be warmer and wetter (Wordsworth and Pierrehumbert 2013; Wordsworth et al. 2017). Although the GC column in the GEx experiment could analyze for H_2 gas, the only bioactive gas purposely supplied to the incubation chamber was CO_2. Ironically, just adjacent to the VLBI instrument was the GCMS instrument, which contained a relatively large tank of hydrogen gas at high pressure, utilized as the carrier gas for its gas chromatograph portion of the experiment. There was no possibility of tapping that gas to test for sulfate reduction or methanogenesis, for example, because the two instruments were built by separate organizations and had no cross-coupling of capabilities. Likewise, since Mars is a very sulfur-rich planetary surface, it is regrettable that there was no attempt to introduce H_2S into the incubation chambers, especially since this volcanogenic gas may have provided the earliest electron donor for non-oxygenic photosynthesis on Earth, and similarly on Mars. Many other constituents of the Martian soil, some still unknown (chlorites? peroxysulfates? other oxidants?), may be only fully understood when samples are brought back to Earth for much more extensive analysis.

Although much remains to be learned, and Viking had only limited tests, we already have tantalizing evidence for the suitability for an origin of life on Mars. First and foremost, Mars was at one time eminently habitable, as several missions have proven. This is true because of the availability of not only liquid H_2O, and the clement temperatures that go with it, but also the entire panoply of carbon, hydrogen, nitrogen, oxygen, phosphorus, and sulfur (CHNOPS) essential elements (available in the atmosphere, global soil, or both). Most importantly, as pointed out previously, in addition to CO_2, there may have been abundant H_2 gas (and perhaps CH_4) in Mars' early history, because these are the greenhouse gases now considered as the most likely candidates for facilitating the warm climate in spite of the faint early Sun. These gases are, of course, of enormous interest as participants in early metabolic architectures for microorganisms. With a highly reducing atmosphere, other important gases such as H_2S and NH_3 may have also been available.

Life should have arisen on Mars, according to current knowledge. At least, it should have had a probability of occurrence competitive with the Earth. Several different theories of the chemical and physical conditions that could facilitate abiotic chemical evolution are providing promising avenues for plausible routes to life. One issue is what might be the initial endowment of organic matter. Contributions of exogenous organics are often attributed to carbonaceous asteroids, meteorites, and comets. Both Mars and Earth may have accreted a late veneer relatively rich in these materials. With Mars closer to the asteroid belt and outer solar system, it has had the opportunity to accrete organics as well or even better than Earth.

According to the Sutherland/Powner scenario (Sutherland 2017), a cyanosulfidic set of chemical pathways can account for all of the most fundamental constituents of living systems.

Mars is clearly S-rich compared with Earth and may at one time have been equally endowed with (H, C, and N) ingredients. These reactions are catalytically favored if copper ion is available as a catalyst. On Mars, examples of Cu enrichments have now been reported by both the APXS instrument analyses (Berger et al. 2017) and ChemCam instrument analyses (Goetz et al. 2018) in Gale Crater.

Other studies of chemical pathways invoke the need for borate availability to stabilize RNA (Benner et al. 2012). Boron is rare in igneous minerals, but boron enrichments in the lacustrine environment at Gale Crater have now been also discovered on Mars (Gasda et al. 2017).

Hydrothermal activity has been often invoked as a triggering precursor to the origin of life. Multiple causes of such activity are on Mars, including volcanism, intrusive magma chambers, and/or the thermal energy deposited by the abundant impacts that formed the widespread craters. Hydrothermal occurrences on land are favored by some researchers as abodes for the origin of life (Deamer and Georgiou 2015; Damer 2016). At Home Plate in Gusev Crater, an apparent hydrothermal edifice, the Mars Exploration Rover mission (MER) Spirit rover, has discovered high silica deposits analogous to the microbe-rich hot springs at El Tatio (Ruff and Farmer 2016).

Mars has apparently been eminently suited as a host for the early origin of life. Whether Earth or Mars was the original source of life in our solar system remains an open question.

12.4.6.3 Sample Return Challenges

Not unlike typical scientific field work, a sample return mission itself is a major undertaking, especially since it needs to be conducted at a distance of some tens of millions of kilometers. For spaceflight, the cost of a roundtrip mission to the ocean worlds or to the surface of Mars is somewhere between three and ten times more expensive than a one-way mission and also is less likely to succeed because there are more activities that can go wrong and result in a lack of total mission success (samples safely in labs on Earth). Thus, reliability can command exceptional attention, which translates into additional cost for the overall program. Whether on Earth or in space, it is also always the case that there are far more samples of interest in the field location than can be brought back. Thus, there is a significant amount of triage that must be undertaken, yet without the portable and powerful analytical equipment (relatively speaking) and versatile sampling tools that might be taken into the field on Earth.

All sample return missions suffer from this shortfall in potentially finding and returning the "Rosetta stone" sample. Yet, they must try. At the same time, it is incumbent to also collect sufficient samples that are representative of the broad context of the site being studied. This is compounded by the kaleidoscopic variety of sedimentary samples that we now know populate the Martian surface.

For spacecraft missions, it has been learned over and over again that it is the mechanical moving parts (mechanisms) that are the weakest link—the most likely items to fail. Thus, rovers have had problems with wheels and sampling tools such as drills, and other spacecraft have had problems with gimbal mechanisms that position antennae to communicate back to Earth or reaction wheels, which orient the spacecraft for targeting their cameras. Thus, it is to be expected that one of the greatest challenges to success is simply the collection, storage, and transfer of samples. A simple rock hammer to collect chips of larger rocks is not available on rovers. It is difficult to drill into small rocks. Drilling deeper than 1 meter invokes special challenges in design and reliability.

Landing on a planetary object can be challenging, and Mars is about the most difficult place to land in the solar system because its thin atmosphere invokes the necessity of a heatshield and affects the trajectory during descent but is not sufficiently dense to slow down the lander by parachutes alone to a safe touchdown velocity. Terminal descent propulsion is therefore mandatory, after extraction of the vehicle from its enclosing aeroshell. On Earth and Titan, a parachute is all that is needed for safe landing, for example, of sample return capsules or astronauts. On the moon or other airless bodies such as asteroids, cometary nuclei, and most planetary satellites, the descent and landing can be done straightforwardly and accurately with a propulsion system alone. Landing is simpler but sampling is not, which is why the OSIRIS-REx mission will employ touch-and-go techniques and a rapid sample acquisition mechanism.

Operations to collect samples from a small body are also problematic, because the low gravity and the specter of a weak regolith might not allow successful anchoring, complicating any attempts at sampling to depth by drilling or excavation.

Collecting samples by flying through a plume of material is simpler, but the flyby velocities can severely damage or destroy any included biota or their biomarkers. As a result, additional propulsion or special time-consuming maneuvers may need to be made to reduce flythrough velocities to less than 2 or 3 km/s, in an effort to avoid compromising the integrity of the constituents of the plume if they are organic and/or alive.

Contamination control is another major issue, especially when searching for life, which may be sparse in abundance and perhaps delicate. When taking multiple samples at a complex site, cross-contamination can be a concern. For example, there may be multiple sample containers, but the drill bits may be limited and hence can cause some material from a previous sample to contaminate subsequent samples.

Molecular cleanliness of the spacecraft, with particular attention to the chain of sampling mechanisms, is of paramount importance. Organic-based lubricants must be avoided or fully contained. Many materials that would normally be used, such as plastics and epoxies, must be screened or not used at all. Many elements and their isotopes are also important for scientific analysis, which imposes stringent constraints on the permissible level of metallic particulates.

Of great concern is contamination by terrestrial organisms, including dormant or dead ones. This is also non-trivial and requires special cleaning as well as constant monitoring of cleanliness to ensure that the controls put into place are effective. Clean room operations must include education of the workers as well as extensive use of outer garments, face masks, taped gloves, tucked booties, etc., to achieve the specified goals.

These requirements for cleanliness are generally set by the project, under the guidance of the science team. For biological contamination, there is an additional layer of responsibility, however, which is the PPO, whose primary aim is to ensure that the total bioburden, which is transported via the spacecraft to the target location (Mars, or ocean world), does not inadvertently transfer too many viable terrestrial organisms so as to potentially contaminate the target for all future time. Stringent requirements and protocols are set by the PPO, which are part of the launch readiness approval process.

Once samples are acquired in clean containers, their integrity must be maintained. Seals must not fail. Temperatures should be controlled. For example, in missions that collect samples for organic analysis, a maximum temperature of $+50°C$ is typically set (Stardust and OSIRIS-REx), although sample return from a comet's surface may require even lower temperatures. The Mars 2020 rover mission (M2020) has set a requirement for a maximum temperature of $+60°C$ for the samples stored on the Martian surface. Although this allows preservation of many hydrated sedimentary minerals, it is well above the temperatures that were found to alter the response of the LR experiment on Viking, simulating sterilization (Levin and Straat 2016). Hence, it would behoove that mission to protect some soil samples against temperature excursions to this level, such as by shallow burial, until retrieval for return to Earth. Otherwise, any metabolic-like activity by putative organisms or labile inorganic oxidants could be destroyed and result in an inability to convincingly explain the results of the Viking analyses.

The PPO is also charged with responsibility of imposing the requirement on the Flight Project organization that the returned samples, which may contain biological activity, be carefully controlled so as to not present a threat to the Earth's biosphere. This hazard could be more than just issues of human health, because a Martian microbe might be unusually hardy and able to propagate in cold regions on Earth, out of control, like many species from formerly isolated locations on Earth, which cause damage of various kinds once transported elsewhere ("invasive species" problem). This issue of "back contamination" can be extremely serious, because if insufficient precautions are taken to ensure the reliability of containment of samples in returning to Earth, the approval of the final phases of the mission may be in jeopardy.

Once on Earth, any samples suspected of harboring alien life-forms will be kept in special containment facilities, similar to the Biosafety Level-4 facilities used to study Ebola and other dangerous biological agents. Such facilities will be an even greater technical challenge for extraterrestrial samples, because they must not only prevent release of sample (restriction of outward flow) but also must prevent organics and organisms from the laboratories and researchers themselves from contaminating the samples (restriction of inward flow). Traditional techniques for contamination control, such as maintaining pressure differentials, will need to be augmented. This also greatly restricts the number of scientists

and, to some extent, the number of analytical techniques that can be brought to bear in studying the retrieved samples. It has often been proposed that some sub-portions of the samples be subjected to sterilizing processes (heat or radiation) and then released to the larger science community. This can become an issue if there is no agreement on what a satisfactory sterilizing protocol should be, taking into account the fact that the putative life-forms may not react the same as our DNA/protein-based organisms.

12.4.7 SOCIETAL ISSUES

With the vast public interest, if life is found on another world, will come questions and concerns. Is the discovery valid? What is the full extent of it? Could it be dangerous to humans, to the environment, or to our economic underpinnings?

These are questions that will be raised if life is found and especially if it will be brought to our home planet. Although future public reactions are almost impossible to predict, it will be prudent to consider the possibility and ramifications of major public alarm, possibly fueled by some group of scientists, however small, with professional qualifications but espousing dire concerns.

12.4.7.1 CONFIRMATION OF LIFE

When asked whether there is life on Mars, scientists can only point to the continually increasing evidence for habitability, especially in ancient times, on Mars. It is always difficult, sometimes impossible, to prove a negative—that is, in this case, to prove that life never has existed on Mars or even that it cannot exist today. Likewise, as some investigators have found, it can be very difficult to prove that life exists, in spite of positive results based on preconceived notions about the planet. This has been true for several cases, ranging from the telescopic maps of apparent activities of intelligent life, to the Viking LR results, and to the detailed studies of putative microfossils in meteorite ALH84001 from Mars.

For scientists, it will require a preponderance of evidence of different sorts, most likely including morphological evidence of complexity, with cellular characteristics, as well as chemical evidence such as organic molecule patterns and isotopic anomalies. These criteria only assert evidence for past life. To affirm extant life-forms, it will be necessary to exhibit active metabolic functions of growth and reproduction, and presumably, eventually, evolution.

To the general public, microbial life will not be nearly as exciting for them as it will be for the biosciences community. However, life-forms of any type will nonetheless create global interest, because it will verify the hypothesis that life is to be found elsewhere in the universe, perhaps not even rarely. If it is learned that life-forms on Mars are so similar to terrestrial life that interplanetary panspermia (Clark 1985, 2001; Horneck and Bueckerl985; Melosh 1988) is a candidate explanation, and *not* forward contamination, there will arise the equally intriguing question: on which planet did it first arise? Are we all Martians?

12.4.7.2 Future Fallout

As much as it may be difficult to prove the existence of an alien life-form, it is even more challenging to prove the negative—that is, that no life of any kind exists in any retrieved samples. By extrapolation, it is likewise difficult to prove that an alien organism is wholly benign in all respects.

The quantities of returned samples will be very limited, with dozens of scientific analytical techniques waiting in the wings for their chance to investigate all properties of the precious cargo. Furthermore, there will purposely be a variety of sample types, ranging from rocks and soils to consolidated sediments, including, hopefully, some organic-rich samples. All must be considered dangerous, invoking a battery of tests to be performed multiple times because of the multiple sample types. Until such samples can be shown to be benign with respect to potential hazards to our civilization, the sending of astronauts to the red planet may be delayed.

In spite of all the impediments to a search for life on Mars, and elsewhere in the solar system, the value of the search should not be ignored. Space exploration occurs because of the endorsement of the general public and the tax revenues they provide for funding these projects. Technological feats, such as landing rovers and driving them around the planet, are inspirational. Although news media often report that "scientists drove the rover to…", the reality is that it is engineers who actually plan and "drive" the detailed routes for reaching the locations the scientists request. Likewise, it takes a consortium of engineers with a variety of talents to design the spacecraft, rocket, rover, and even the science instruments. Again, unbeknownst to the general public, it is seldom the engineers who actually construct the designs they develop but rather highly trained and skilled technicians who carefully follow the detailed plans and instructions for building, assembling and testing the hardware and software that will conduct the next mission.

The results of Viking, which were widely regarded as negative, set Mars exploration back for the next two decades. When I asked Carl Sagan if the life issue was dead and we needed to find new rationales for exploring the red planet, he responded that the prospect for life was so important that it would continue to be the underpinnings of Mars exploration. He was correct. In recent decades, the emphasis has focused on discovering if Mars was truly habitable at one time, with the new mantra of "Follow the Water." This goal has now been proven dramatically, with the discovery of bedded salt-rich sediments by MER rover Opportunity, of pure salt layers by MER rover Spirit, as well as of montmorillonite-clay-like compositions at both MER exploration sites. The Mars Science Laboratory mission (MSL) rover Curiosity has found yet more evidence of aqueous activity, as have numerous orbiter observations by Mars Express and Mars Reconnaissance Orbiter. An emphasis on geological formations and geochemistry has become central, while the question of evidence for extinct or extant life has receded into the background.

However, maintaining the past level of support for further exploration will not be advanced if scientists must attempt to convey to the public the subtleties of "eolian versus lacustrine cross-bedding" or "instances of juxtaposed metabolic redox couples." As advanced in the introduction, now is the time for rejuvenation of the search for evidences of life on Mars and, if it cannot be found, to seek to understand why not. Likewise, it is time to begin seriously the study of the ocean worlds to the same end.

ACKNOWLEDGMENTS

It is with special gratitude that I acknowledge the encouragement, advice, and editorial assistance of V. M. Kolb, without which this article would not have been possible.

REFERENCES

The web sites cited in this paper are from NASA and are regularly maintained and updated. They were accessed on 4/10/2018.

https://www.nasa.gov/missions

https://mars.nasa.gov/programmissions/missions/past/viking/

https://www.nasa.gov/topics/journeytomars/index.html

https://astrobiology.nasa.gov/research/life-detection/ladder/

Belin, B. J., N. Busset, E. Giraud, A. Molinaro, A. Silipo, and D. K. Newman. 2018. Hopanoid lipids: from membranes to plant–bacteria interactions. *Nat. Rev. Microbiol.* doi:10.1038/nrmicro.2017.173.

Benner, S. A., H. J. Kim, and M. A. Carrigan. 2012. Asphalt, water, and the prebiotic synthesis of ribose, ribonucleosides, and RNA. *Acct. Chem. Res.* 45: 2025–2034.

Berger, J. A., M. E. Schmidt, R. Gellert et al. 2017. Zinc and germanium in the sedimentary rocks of Gale Crater on Mars indicate hydrothermal enrichment followed by diagenetic fractionation. *J. Geophysical Res.: Planets*, 122(8): 1747–1772. doi:10.1002/2017JE005290.

Biemann, K., J. Oro, P. Toulmin III et al. 1977. The search for organic substances and inorganic volatile compounds in the surface of Mars. *J. Geophys. Res.*, 82(28): 4641–4658. doi:10.1029/JS082i028p04641.

Brassell, S. C., A. M. K. Wardroper, I. D. Thomson, J. R. Maxwell, and G. Eglinton. 1981. Specific acyclic isoprenoids as biological markers of methanogenic bacteria in marine sediments. *Nature* 290: 693–696. doi:10.1038/290693a0.

Brown, F. S., H. E. Adelson, M. C. Chapman et al. 1978. The biology instrument for the Viking Mars mission. *Rev. Sci. Instrum.* 49(2): 139–182.

Brownlee, D. 2014. The stardust mission: Analyzing samples from the edge of the solar system. *Annu. Rev. Earth Planet. Sci.* 42: 179–205. doi:10.1146/annurev-earth-050212-12420.

Clark, B. C. 2001. Planetary interchange of bioactive material: Probability factors and implications. *Orig. Life Evol. Biosph.* 31: 185–197.

Clark, B. C. 1985. Barriers to the interchange of biologically active material between Earth and Mars. Abstract/Poster. *Orig. Life Evol. Biosph.* 16: 410.

Clark, B. C., A. K. Baird, R. J. Weldon et al. 1982. Chemical composition of Martian fines. *J. Geophys. Res.* 87: 10059–10067.

Damer, B. 2016. A field trip to the Archaean in search of Darwin's warm little pond. *Life* 6: 21 doi:10.3390/life6020021.

Deamer, D. W., and C. D. Georgiou. 2015. Hydrothermal conditions and the origin of cellular life. *Astrobiology* 15: 1091–1095. doi:10.1089/ast.2015.1338.

Eigenbrode, J. L., R.E. Summons, A. Steele et al. 2018. Organic matter preserved in 3-billion-year-old mudstones at Gale crater, Mars. *Science* 360: 1096–1101. doi:10.1126/science.aas9185.

Fairén, A. G., V. Parro, D. Schulze-Makuch, and L. Whyte. 2017. Searching for life on Mars before it is too late. *Astrobiology* 17(10). doi:10.1089/ast.2017.1703.

Freissinet, C., D. P. Glavin, P. R. Mahaffy et al. 2015. Organic molecules in the Sheepbed Mudstone, Gale Crater, Mars. *J. Geophys. Res. Planets* 120: 495–514. doi:10.1002/2014JE004737.

Gasda, P. J., E. B. Haldeman, R. C. Wiens et al. 2017. In situ detection of boron by ChemCam on Mars. *Geophys. Res. Lett.* 44: 8739–8748. doi:10.1002/2017GL074480.

Gellert, R. and B. C. Clark. 2015. In situ compositional measurements of rocks and soils with the alpha particle X-ray spectrometer on NASA's Mars rovers. *Elements* 11(1): 39–44. doi:10.2113/gselements.11.1.39.

Georgiou, C. D., D. Zisimopoulos, E. Kalaitzopoulou, and R. C. Quinn. 2017. Radiation-driven formation of reactive oxygen species in oxychlorine-containing Mars surface analogues. *Astrobiology* 17: 319–336. doi:10.1089/ast.2016.1539.

Georgiou, C. D., and D. W. Deamer. 2014. Lipids as universal biomarkers of extraterrestrial life. 2014. *Astrobiology* 14: 541–549. doi:10.1089/ast.2013.1134.

Goetz, W., W. B. Brinckerhoff, R. Arevalo et al. 2016. MOMA: The challenge to search for organics and biosignatures on Mars. *Int. J. Astrobiol.* 15: 239–250. doi:10.1017/S1473550416000227.

Goetz, W., V. Payre, R. C. Wiens et al. 2018. Detection of copper by the ChemCam instrument onboard the Curiosity rover in Gale Crater, Mars. *Extended Abstract 2679*. LPSC.

Hand, K.P., A. E. Murray, and J. B. Garvin. 2016. Europa lander study 2016 report. *JPL D-97667*.

Horneck, G., and H. Buecker. 1985. Can microorganisms withstand the multistep trial of interplanetary transfer? *Orig. Life Evol. Biosph.* 31: 414.

Horowitz, N. H. 1986. *To Utopia and Back: The Search for Life in the Solar System*. New York: W. H. Freeman.

Johnson, S. S., E. V. Anslyn, H. V. Graham, P. R. Mahaffy, and A. D. Ellington. 2018. Fingerprinting non-Terran biosignatures. *Astrobiology*. Published Online: 8 March 2018 doi:10.1089/ast.2017.1712.

Joyce, G. F. 1994. Forward, in origins of life: The central concepts, Eds. D. W. Deamer and G. R. Fleischaker, xi. Boston, MA: Jones & Bartlett.

Klein, H. P. 1999. Did Viking discover life on Mars. *Orig. Life Evol. Biosph.* 29: 625–631.

Klein, H. P. 1978. The Viking biological experiments on Mars. *Icarus* 34: 666–674.

Kwok, S. 2016. Complex organics in space from Solar System to distant galaxies. *Astron. Astrophys. Rev.* 24: 8. doi:10.1007/s00159-016-0093-y.

Levin, G. V., and P. A. Straat. 1976. Viking labeled release biology experiment: Interim results. *Science* 194: 1322–1329. doi:10.1126/science.194.4271.1322.

Levin, G. V., and P. A. Straat. 2016. The case for extant life on Mars and its possible detection by the Viking labeled release experiment. *Astrobiology* 16: 798–810. doi:10.1089/ast.2015.1464.

Melosh, H. J. 1988. The rocky road to panspermia. *Nature* 332: 687–688.

Moore, H. J., R. E. Hutton, G. D. Clow, and C. R. Spitzer. 1987. Physical properties of the surface materials at the Viking landing sites on Mars. *U.S. Geological Survey Professional Paper 1389*.

Navarro-Gonzalez, R., and C. P. McKay. 2011. Reply to comment by Biemann and Bada on reanalysis of the Viking results suggests perchlorate and organics at midlatitudes on Mars. *J. Geophys. Res.* 116: E12002. doi:10.1029/20IIJE003854.

Postberg, F., J. Schmidt, J. Hillier, S. Kempf and R. Srama. 2011. A salt-water reservoir as the source of a compositionally stratified plume on Enceladus. *Nature* 474: 620–622.

Postberg, F., N. Khawaja, S. Kempf et al. 2017. Complex organic macromolecular compounds in ice grains from Enceladus. *48th LPSC, Abstract no. 1964*.

Prettyman, T. H, N. Yamashita, M. J. Toplis et al. 2017. Extensive water ice within Ceres' aqueously altered regolith: Evidence from nuclear spectroscopy. *Science* 355: 55–59. doi:10.1126/science.aah6765.

Quinn, R. C., H. F. H. Martucci, S. R. Miller, C. E. Bryson, F. J. Grunthaner, and P. J. Grunthaner. 2013. Perchlorate radiolysis on Mars and the origin of martian soil reactivity. *Astrobiology* 13: 515–520. doi:10.1089/ast.2013.0999.

Radmer. R., and B. Kok. 1971. A unified procedure for the detection of life on Mars. *Science* 174: 233–239.

Ruff, S.W. and J. D. Farmer. 2016. Silica deposits on Mars with features resembling hot spring biosignatures at El Tatio in Chile. *Nat. Commun.* 7, Article number 13554. doi:10.1038/ncomms13554.

Schuerger, A. C. and B. C. Clark. 2007. Viking biology experiments: Lessons learned and the role of ecology in future Mars life-detection experiment. In *Strategies of Life Detection* (Eds.) O. Botta et al. pp. 233–243. doi:10.1007/978-0-387-77516-6_16.

SSB (Space Studies Board). 1965. Biology and the exploration of Mars. *SSB Report 1296* (October, 1965).

SSB (Space Studies Board). 1977. *Post-Viking Biological Investigations of Mars*. Washington DC: Committee on Planetary Biology and Chemical Evolution, National Academy of Sciences.

Sutherland, J. D. Studies on the origin of life–The end of the beginning. 2017. *Nat. Rev. Chem.* 1: 1–7.

Ward, W. R. 1992. Long-term orbital and spin dynamics of Mars. In *Mars* (Eds.) H. H. Kieffer et al. pp. 298–320, Tucson, AZ: University of Arizona Press.

Wordsworth, R., and R. Pierrehumbert. 2013. Hydrogen-nitrogen Greenhouse warming in Earth's early atmosphere. *Science* 339: 64–67. doi:10.1126/science.1225759.

Wordsworth, R., Y. Kalugina, S. Lokshtanov et al. 2017. Transient reducing greenhouse warming on early Mars. *Geophysical. Res. Lett.* 44: 665–671. doi:10.1002/2016GL071766.

12.5 Planetary Protection

Catharine A. Conley

CONTENTS

12.5.1   INTRODUCTION

12.5.1.1   OVERVIEW OF PLANETARY PROTECTION CONSIDERATIONS

Previous sections of this book have reviewed in great detail the historical and ongoing efforts to understand the origin and evolution of life on Earth, as well as the potential for life to exist on other planets. One of the critical questions in searching for life elsewhere is to ensure that contamination from Earth does not interfere with detection of extraterrestrial signals—this could lead to interpretation of results that are either "false-positive" (e.g., reports of fossil Mars life in the meteorite ALH84001 that was subsequently shown to be abiotic; see Steele et al. 2010) or "false-negative" (e.g., failure to identify indications of Mars organic compounds in data from NASA's Viking missions to Mars; see Glavin et al., 2013; Freissinet et al., 2015). The potential for Earth contamination to interfere with scientific investigations of other planetary objects and the need for international standards to reduce this risk were recognized internationally around the launch of Sputnik (e.g., Committee on Contamination by Extraterrestrial Exploration [CETEX], 1958, 1959). Very rapidly thereafter, guidelines and practices were put in place to prevent contamination; these are collectively known today as "planetary protection," as described in detail by Meltzer (2011). Planetary protection policy and guidelines are founded in the best available scientific knowledge and maintained by the International Council for Science to provide consensus standards, for use by all countries and space exploration activities, as part of complying with United Nations (UN) treaty obligations. The policy and guidelines are regularly updated to reflect new scientific discoveries, following a process that continues to this day.

12.5.1.2 Relationship to Astrobiology and Planetary Exploration

All missions coming in close proximity to other planets have a potential to introduce Earth contamination, which could reduce confidence in scientific conclusions relating to extraterrestrial life detection. In addition, the introduction of Earth organisms capable of surviving for long periods in a dormant state could put in jeopardy future human goals such as settlement or terraforming, which are entirely outside the scope of near-term scientific missions or other exploration activities. Control of Earth contamination on planetary spacecraft is a technological challenge that has been surmounted on multiple occasions, most notably during NASA's Viking program in the 1970s (Daspit et al., 1988) and the European Space Agency (ESA's) ExoMars missions ongoing today. "Forward contamination" is the term used to describe these processes, which have risks that can be determined and quantified—at least to the extent that we understand the capabilities of Earth organisms and non-living contamination to be present on spacecraft, and can persist after introduction to planetary environments.

In contrast, potential risks resulting from the introduction of extraterrestrial organisms into the environment of the Earth, termed "backward contamination," are entirely unknown and currently unquantifiable, because, to date, we have no evidence regarding the characteristics of possible extraterrestrial life and thus no basis to assess pathogenicity or ecological consequences. From a policy and regulatory standpoint, these uncertainties about the actual risks of backward contamination are compounded by challenges associated with detecting extraterrestrial life, if it is present. The two Viking and the Mars Science Laboratory (MSL) missions to Mars carried sensitive instruments to detect metabolic activity (Viking) or organic compounds (Viking and MSL) on Mars (e.g., Glavin et al., 2013; Freissinet et al., 2015). Each of these missions returned data indicating levels of Earth contamination that exceeded detection limits for Mars organics, and these data were interpreted as non-detections of Mars life—however, in the case of the Viking Life Detection Package, one of the principal investigators still publishes papers disputing this conclusion (e.g., DiGregorio et al., 1997).

In 1964, a decade prior to the Viking missions, the US Space Science Board (SSB) was asked to evaluate backward contamination issues for NASA's Apollo program and future Mars missions. It noted that "negative findings could provide a sense of security which might well be false" (Space Science Board [SSB], 1964). The possibility that extraterrestrial organisms could be hazardous, either to the environment of the Earth or to humans directly, makes detecting them at very high sensitivity a primary concern for ensuring the safety of the Earth.

12.5.1.3 Interfaces with Wider Society

On Earth, humans have a long history of transporting biological organisms from one location to another, which has often caused major disease outbreaks and/or ecosystem disruption, as reviewed in Mann (2011) and many others. It is not knowable, until we have an example of extraterrestrial life, whether that life could become pathogenic or disrupt the ecosystems that humans rely on. In contrast, we do know that Earth organisms inadvertently transported on spacecraft could interfere with future objectives of human settlers, as they have done on Earth in the past. Avoiding the accidental transport of Earth organisms, before decisions are made to introduce them, is a long-term concern for planetary protection, because one single release of a self-replicating entity into a habitable environment can engender a persistent population and thus cause permanent contamination.

For the reasons mentioned above, it is essential to address planetary protection at the highest levels of global societal decision-making, which for 60 years has been done via the UN and the International Council for Science. In addition, to ensure that one bad actor does not cause permanent damage to everyone, individual nations have obligations under the 1967 Outer Space Treaty to take appropriate measures to ensure that all exploration of other planets carried out under their auspices, whether governmental or private/commercial, follows the same guidelines. Space exploration is the first effort in human history for which we, as a global society, have recognized the potential risks of contamination before it happened, and planetary protection is the first time humanity as a species has taken responsible steps to prevent contamination of pristine environments even before our very first efforts to explore them.

12.5.2 PLANETARY PROTECTION: DEFINITION AND SCOPE

Planetary protection covers explicitly the search for extraterrestrial life and also the potential for Earth life to interfere with future human objectives: the focus of planetary protection is exclusively on biological contamination and does not address other kinds of contamination such as radiation or physical detritus. Due to the very high level of concern for protecting the Earth on which we all live, relative to the more limited concern about contaminating other planets, the guidelines and policy for planetary protection are divided, conceptually, on the basis of whether spacecraft are only traveling outward to other planetary bodies and therefore could cause forward contamination, or whether there is an expectation that hardware will return back to Earth, possibly carrying extraterrestrial material, which could release backward contamination.

12.5.2.1 FORWARD CONTAMINATION CONCERNS

Spacecraft traveling to another planetary body (moon, asteroid, comet, etc.) have the potential to cause forward contamination by depositing organic material and/or organisms from Earth onto the target object. Forward contamination is of concern for planetary protection only to the extent that contaminants could interfere with scientific or other human objectives at the target object. The vast majority of objects in the solar system are known to support conditions inhospitable to all Earth life (e.g., no atmosphere, too irradiated, and too dry), and therefore, spacecrafts going to them are not controlled to prevent introduction of Earth organisms. To the extent that a non-habitable object is of scientific interest for understanding the origin and evolution of life in the solar system, which involves studying whatever organic compounds could be present, a mission may be required to provide a list of materials present on the spacecraft and to report locations where spacecraft hardware is left at end of mission. Missions going to non-habitable targets are required to provide only a straightforward set of documentation about hardware composition, spacecraft operations, and final disposition.

In contrast, when a planetary body has the potential to provide a habitat for Earth organisms, spacecraft traveling to them are required to avoid introducing Earth organisms into habitats: this involves applying very strict decontamination procedures to hardware that could introduce Earth organisms and also (or instead) avoiding contact of contaminated hardware with potentially habitable environments. Currently, three solar system objects are of concern for contamination by Earth life: the planet Mars, and the moons Europa around Jupiter and Enceladus around Saturn. Other planetary bodies may be added to this list as additional potential habitats are identified, and requirements to prevent contamination of them should be put in place. Missions going to habitable targets are required, as part of as part of mission formulation, to submit and receive approval for detailed plans that describe proposed decontamination procedures; during hardware assembly and launch operations, they will undergo regular inspections and extensive reviews to confirm compliance prior to launch, and after launch, they continue to ensure contamination avoidance during spacecraft operations, as well as to provide information about relevant research findings and report final hardware disposition.

12.5.2.2 BACKWARD CONTAMINATION CONCERNS

When spacecraft hardware is being brought back to Earth after contact with another planetary body, it is of primary concern to ensure the safety of the Earth's biosphere and everything that lives in it. Again, the focus is on biological contamination, so the same conceptual distinction is made relating to

habitability and the potential for extraterrestrial life—defined as biochemistries that could function in the surface physical environment (temperature, pressure, etc.) of Earth—with very different levels of concern around preventing release of material from "potentially habitable" versus "non-habitable" target objects. Missions returning from all non-habitable target objects, including the Earth's Moon and near-Earth asteroids, follow planetary protection guidelines to provide documentation of the target's non-habitability during mission planning phases and receive no further planetary protection requirements once appropriate review is completed.

As with forward contamination, restrictions are only imposed on spacecraft returning from objects that might host extraterrestrial life, which is *by definition* considered biohazardous until tests demonstrate otherwise—currently, this includes the three "potentially habitable" objects: Mars, Europa, and Enceladus. If other solar system objects are found, in future, to host physical conditions that could support biochemistries also potentially active on Earth, then these other objects would be added to the short list requiring stringent precautions for sample return. Missions sending hardware intended for return from potentially habitable targets are required stringently to limit contamination from Earth that could interfere with detecting extraterrestrial life, as well as all requirements appropriate to the particular outbound mission—these missions are also required to maintain an archive and detailed record of potential Earth contaminants and the provenance of samples collected, in a format that can be provided to a pre-return "Earth Safety" review process that will be carried out at international level (e.g., Haltigin et al. 2018; Kminek et al., 2017).

Appropriate documentation on the return status of all missions also needs to be carried through whatever local/national approval processes apply to a re-entry event into the Earth's atmosphere.

12.5.2.3 INTERNATIONAL, NATIONAL, AND AGENCY-LEVEL OBLIGATIONS AND RESPONSIBILITIES

The current set of accepted international consensus guidelines on how to implement planetary protection is held by a permanent committee of the International Council for Science, the Committee on Space Research (COSPAR) (Kminek et al., 2017). The 1967 Outer Space Treaty, more accurately the "United Nations Treaty on Principles Governing the Activities of States in the Exploration and Use of Outer Space, including the Moon and Other Celestial Bodies," is the major international agreement governing how treaty signatories ("States Parties to the Treaty") go about exploring and using outer space (United Nations [UN], 1966). This treaty has been signed by all countries involved or interested in space exploration, numbering over 100 and including North Korea (United Nations [UN], 2018).

The United Nations Committee on the Peaceful Uses of Outer Space (UN-COPUOS) is the committee of the United Nations in which discussion of matters related to the treaty takes place. The UN-COPUOS, in its 60th meeting report to the UN General Assembly in 2017, *noted the long-standing role of COSPAR in maintaining the planetary protection policy as a reference standard for spacefaring nations and in guiding compliance with article IX of the Outer Space Treaty* (United Nations Committee on the Peaceful Uses of Outer Space [UN-COPUOS], 2017). This recognition reiterates the role that COSPAR has played over the 60 years since its creation in 1957 (described in Section 12.5.3.1 and reviewed in Meltzer, 2011).

States parties to the Outer Space Treaty, by signing and ratifying the document, agree to abide by all 17 treaty articles, including Article IX which reads, in part:

> *States Parties to the Treaty shall pursue studies of outer space, including the moon and other celestial bodies, and conduct exploration of them so as to avoid their harmful contamination and also adverse changes in the environment of the Earth resulting from the introduction of extraterrestrial matter and, where necessary, shall adopt appropriate measures for this purpose.*

The COSPAR planetary protection policy addresses *harmful contamination* in the context of forward contamination and *adverse changes in the environment of the earth resulting from the introduction of extraterrestrial matter* in the context of backward contamination. As noted previously, other forms of contamination or environmental damage are not addressed by the COSPAR planetary protection policy or guidelines.

Other articles of the Outer Space Treaty impose additional obligations on States Parties to the treaty that are relevant to planetary protection. Article VI requires that:

> *States Parties to the Treaty shall bear international responsibility for national activities in outer space, including the moon and other celestial bodies, whether such activities are carried on by governmental agencies or by non-governmental entities.*

Further, States Parties must assure that:

> *The activities of non-governmental entities in outer space, including the moon and other celestial bodies, shall require authorization and continuing supervision by the appropriate State Party to the Treaty.*

In general, each State Party to the Outer Space Treaty has assigned the responsibility to ensure compliance with international planetary protection guidelines to their national space agency, which, in some cases, also requires cooperation with other national agencies that regulate activities within the Earth's atmosphere. In the case of the ESA, which is a regional organization with multiple national members, the responsibility of each State Party to ensure compliance with planetary protection policy and guidelines, for ESA missions in which they participate, has been transferred to ESA. Individual States Parties are also responsible for developing internal processes to provide the required *authorization and continuing supervision* of the actions of their non-governmental entities.

12.5.3 A BRIEF HISTORY OF PLANETARY PROTECTION

12.5.3.1 EARLY CONCERNS

Fictional accounts of interplanetary travel causing biological contamination, including ecological damage and pathogenicity (e.g., Wells, 1898), reflect historical experiences from European colonialism and the Columbian Expansion. For related historical reasons, "planetary quarantine" was the term used instead of "planetary protection" prior to the mid-1980s, but the practices and precedents are identical. Planetary protection as a practice began after World War II, with concerns expressed by scientists involved in organizing the 1957–1958 International Geophysical Year. In preparatory discussions, the international scientific community recognized that advances in rocketry would soon permit artificial satellites to be launched from Earth to the Earth's Moon and other planets. Concerns about potential biological contamination were first raised at the 7th International Astronautical Congress held in Rome, Italy, in September 1956 (Phillips, 1974; Meltzer, 2011). Both the United States (US) and the Soviet Union (USSR) announced plans to launch Earth-orbiting satellites for scientific research purposes, with the USSR launching two Sputnik satellites in 1957 and the US launching the Explorer and Vanguard satellites in 1958, as reviewed in Doyle and Skoog (2012).

Starting in 1958, with the formation of UN-COPUOS and COSPAR, as well as multiple national space agencies, extensive discussion of how to prevent biological contamination by planetary spacecraft took place in the international community, including Europe, the US, and the USSR, facilitated by interactions associated with the International Geophysical Year (e.g., Committee on Contamination by Extraterrestrial Exploration [CETEX], 1958, 1959). Within the US, the National Academy of Sciences (NAS) convened a working group known as the West Coast Committee on Extraterrestrial Exploration (WESTEX), chaired by Melvin Calvin and including Joshua Lederberg and the graduate student Carl Sagan, that supported the NAS' newly formed Space Science Board as well as COSPAR. Discussions held during WESTEX meetings, as documented in their final report and appendices (Space Science Board [SSB], 1959), addressed all the concerns of planetary protection that inform policy today, emphasizing both their historical basis and global scope.

Statements made in 1959 by this committee, on the topic of transporting extraterrestrial materials to Earth, are surprisingly pertinent in providing clarity to current debates:

> *We know of many unhappy examples of biological competition from the introduction of new organisms into fresh niches—e.g., many insect pests in the US; rabbits and prickly pear in Australia, smallpox into the New World, and syphilis into Europe. Even the relatively limited damage of these incidents should not be duplicated as a byproduct of space research.* (WESTEX Report, pg. 17)

Finally, it may be remarked that the task of evaluating the potential hazard of a planetary biota will be multiplied if this has to be isolated from organisms inadvertently transferred from Earth. (WESTEX Report, pg. 18)

12.5.3.2 EVOLUTION OF PLANETARY PROTECTION POLICY AND GUIDELINES

During the late 1950s and early 1960s, frequent discussions were held in the international arena that informed the development of guidelines on how to respond to policy-level concerns about planetary contamination and potential consequences for scientific research and other human endeavors (reviewed in Phillips, 1974; Meltzer, 2011). The need for a risk-based approach that included the careful sterilization of spacecraft hardware was recognized very early, and methods for accomplishing this were proposed and evaluated (e.g., Space Science Board [SSB], 1959, 1964). Early missions implemented an approach based on a "probability of contamination" model, with information gained from each mission, leading to refinements in policy and guidelines over time.

In the 1980s, COSPAR accepted a conceptual shift in the formulation of planetary protection policy, moving from an explicit risk-based probabilistic approach to the "by exception" approach used today (COSPAR internal decision memo No. 7/84, accepting the proposals in DeVincenzi et al., 1983). This shift responded to accumulated scientific data, showing that most solar system objects were not contaminable by Earth life and, therefore, by inference, were also unlikely to host extraterrestrial life that could be biohazardous to the Earth. Four categories of possible outbound missions were described, as in Table 12.5.1, determined by the level of interest in the target object for understanding the origin and evolution of life and also the mission operations to be performed.

Missions to objects not of concern for understanding the origin and evolution of life are assigned Planetary Protection Category I, with no further documentation or other requirements. Missions to objects that do not provide natural habitats for Earth life but that could retain organic and prebiotic compounds of scientific interest are assigned Planetary Protection Category II, and limited documentation of mission operations is required—this includes the vast majority solar system objects. Missions to solar system objects that could provide habitats for Earth life—as noted previously and currently include Mars, Europa, and Enceladus—are assigned Planetary Protection Category III if hardware is not intended to contact the target object (flyby and orbiter missions) and assigned Planetary Protection Category IV if hardware is planned to contact the target (probe and lander missions). Because of the considerable interest in Mars as a target of human exploration beyond purely scientific investigations, and the more extensive information available about the planet, Planetary Protection Category IV landed missions to Mars are further divided on the basis of landing site and mission objectives, as described in the "future missions" section later.

In addition, a fifth category of "Earth Return" missions was established, recognizing explicitly the much higher priority placed on protecting the biosphere of the Earth than elsewhere, as noted in Table 12.5.1. This Earth Return category, Planetary Protection Category V, is divided into "Restricted" and "Unrestricted" Earth Return, determined by evaluating scientific data supporting the hypothesis that samples to be brought to Earth contain no extraterrestrial life, which is by definition considered biohazardous. For most solar system objects, this is accomplished by responding to the six questions listed in Table 12.5.2, about conditions on those objects and the natural influx of material to Earth.

When data are inconclusive or support the presence of possible habitats, as is true for Mars, Europa, and Enceladus, then a designation of Planetary Protection Category V "Restricted Earth Return" is given. Planetary Protection Category V "Restricted Earth Return" applies to all missions involved in a sample return effort—this is to ensure that information needed to support the pre-return Earth safety analysis and post-return biohazard test protocol is captured and retained by early outbound missions as well as the final return leg. Each mission carrying hardware intended for possible future return to Earth, including missions that emplace hardware possibly to be retrieved by future mission activities, also receives requirements appropriate to that particular orbiter or lander mission.

TABLE 12.5.1
Planetary Protection Mission Categories

	Planet Priority	Mission Type	Mission Category
A	Not of direct interest for understanding the process of chemical evolution. No protection of such planets is warranted.	Any	I
B	Of significant interest relative to the process of chemical evolution, but only a remote chance that contamination by spacecraft could jeopardize future exploration. Documentation is required.	Any	II
C	Of significant interest relative to the process of chemical evolution and/or the origin of life or for which scientific opinion provides a significant chance of contamination that could jeopardize a future biological experiment. Substantial documentation and mitigation are required.	Flyby, Orbiter	III
		Lander, Probe	IV
All	Any solar system body	Earth Return *Restricted or Unrestricted*	V

TABLE 12.5.2
Six Questions for Restricted Earth Return

1. Does the preponderance of scientific evidence indicate that there was never liquid water in or on the target body?
2. Does the preponderance of scientific evidence indicate that metabolically useful energy sources were never present?
3. Does the preponderance of scientific evidence indicate that there was never sufficient organic matter (or CO_2 or carbonates and an appropriate source of reducing equivalents) in or on the target body to support life?
4. Does the preponderance of scientific evidence indicate that subsequent to the disappearance of liquid water, the target body has been subjected to extreme temperatures (i.e., >160°C)?
5. Does the preponderance of scientific evidence indicate that there is or was sufficient radiation for biological sterilization of terrestrial life-forms?
6. Does the preponderance of scientific evidence indicate that there has been a natural influx to Earth, for example, via meteorites, of material equivalent to a sample returned from the target body?

12.5.3.3 PLANETARY PROTECTION POLICY APPLIED TO EARLY PROGRAMS

12.5.3.3.1 Soviet Decontamination Efforts

Soviet scientists were involved, from the earliest international discussions, in raising concerns relevant to planetary protection, and it was reported to the international community that early Soviet lunar missions did comply with the nascent guidelines being developed by COSPAR. These announcements had significant influence on decisions made within the US to ensure that early NASA missions would be decontaminated, as documented in a memo dated September 14, 1959, from the executive director of the SSB, Hugh Odishaw, to the first NASA Administrator T. Keith Glennan (Space Science Board [SSB], 1959, pg. 84), and also in the final report of WESTEX, which states:

We applaud the respect for these considerations on the part of the USSR in the light of Academician Topchiev's announcement that Lunik-II has been decontaminated. (Space Science Board [SSB], 1959, pg. 13)

Throughout the 1960s, it was reported in public that Soviet spacecraft sent to Mars and Venus had been "sterilized," but there was significant uncertainty in the West as to what this actually meant, as reviewed in Meltzer (2011). Questions were raised, in the international community, about the benefit of implementing stringent sterilization protocols on some spacecraft, if other spacecraft had already delivered Earth organisms to the same target—as if a single contamination event would render completely useless all subsequent efforts to limit additional contamination. This is a little like someone asking "Why should we keep brushing our teeth, after we've eaten our first candy-bar?"

12.5.3.3.2 Ranger

A number of NASA's early missions were managed by the Jet Propulsion Lab (JPL) in Pasadena, California, which, along with all other NASA facilities, was instructed to ensure appropriate sterilization of planetary spacecraft, in October 1959, following the recommendations from the SSB mentioned previously (see Hall, 1977 for details). Prior to this, JPL had established a program to develop methods for spacecraft sterilization, which presented a paper at the 10th International Astronautical Congress, containing the statement:

Sterilizing space probes is an engineering nuisance, however, the same ordeal has confronted surgical crews for quite some time. In both instances, anticipation of the task is necessary. (Davies and Comuntzis, 1960, included in Space Science Board [SSB], 1959).

Paradoxically, at the same time, JPL also began designing a multi-purpose planetary spacecraft bus, called "Vega," that did not include any sterilization-tolerance requirements in the design constraints.

This design was adopted for use in the Ranger program in 1959, by which time it had already undergone considerable preliminary testing and refinement (Hall, 1977). In 1960, JPL staff proposed a sterilization protocol that involved subjecting components, subsystems, and the assembled spacecraft system to 125°C before shipment to the launch site, with a final ethylene oxide gas treatment applied at the launch site to eliminate organisms that might have recontaminated spacecraft surfaces during transport and launch preparations. Despite the plan to apply a system-level heat treatment, design constraints for the Ranger program, including those provided to subcontractors and instrument contributors, did not include heat tolerance among the requirements, and very little testing was done, during development of Ranger spacecraft hardware, to evaluate the tolerances of spacecraft components and subsystems to heat treatment.

Despite the October 1959 memo requiring spacecraft sterilization, and associated 1960 protocol, only a few of the components selected for the Ranger spacecraft, and none of the early engineering or flight models, were tested for compatibility with heat sterilization treatment. The first hardware to undergo a heat sterilization protocol was the flight model of the first lander spacecraft, Ranger 3. Multiple materials' incompatibilities and failed components were identified during subsequent testing, which required extensive re-work. At the time, JPL reported:

Although no failures are directly traceable to heat damage, it is felt that heat sterilization does shorten the expected life of electronic components. (Hall, 1977, pg. 124)

Today, military specifications for high-reliability hardware require an operational high-temperature burn-in phase, to eliminate defective components at risk for early failure (MIL-STD-810G, 2008).

After the launch of Ranger 3, which included a prelaunch ethylene oxide gas treatment (tolerance to which was also not mentioned in design constraints), the spacecraft "performed flawlessly" (Hall, 1977, pg. 147), but incorrect commands transmitted from ground control caused the spacecraft to lose contact, without accomplishing any mission objectives other than impact on the Moon. Despite this successful performance, the reaction from Ranger project managers was to attribute the failure to heat sterilization (Hall, 1977, pg. 125):

> Although lacking firm evidence that this requirement caused the equipment failures, JPL now requested and received more waivers from NASA Headquarters on heat sterilizing certain crucial components.

Rangers 4 and 5 underwent only partial heat sterilization treatments, though they were subjected to pre-launch ethylene oxide gas because this was not considered a risk to spacecraft hardware—yet both missions were unsuccessful, with failures in spacecraft bus control systems despite electronic components having been exempted from heat treatment.

In 1962, after the failure of Ranger 5, the program was extensively reorganized, with new management and a much-strengthened quality-assurance program, and the program received approval to cease all sterilization treatments. In 1963, a few months prior to the planned launch of Ranger 6, it was discovered that a particular type of diode, used by the hundreds throughout each of the Ranger spacecraft, was often defective and susceptible to shorting in microgravity, *with potentially disastrous consequences* (Hall, 1977, pg 197). Potentially defective components were replaced extensively in Rangers 6 through 9, with much better quality control on the replacements—even so, the cameras on the Ranger 6 spacecraft did not function, though the rest of the mission was accomplished successfully. Rangers 7, 8, and 9 were considered to be fully successful and returned imagery that was of considerable interested to both the scientific community and the general public, as well as useful to the Apollo program for landing site selection.

The Ranger program was instrumental in establishing processes in space mission formulation that balance scientific and engineering concerns, as well as quality-control programs and interfaces for project management, that have subsequently become widely implemented on NASA's robotic missions. The influence of the Ranger program on the development (or lack thereof) of standard approaches for spacecraft sterilization is not so well recognized but is still considerable.

12.5.3.3.3 Apollo

The Apollo program remains the most complex effort in space exploration attempted prior to 2018, involving astronauts landing on another planetary body and returning to Earth with samples, as well as both samples and astronauts being subject to isolating containment and analytical and biohazard testing after return to Earth. Apollo was, applying current planetary protection policy, a Planetary Protection Category V Restricted Earth Return campaign, including all the additional health and safety concerns associated with human spaceflight. The history of the Apollo program has

been covered extensively elsewhere (e.g., Launius, 1994), so only two aspects of the Apollo program are mentioned here, as being of particular relevance to planetary protection.

In the words of US President Kennedy, the Apollo program was established for "landing a man on the moon and returning him safely to earth"—more for purposes of political and technological positioning than for scientific investigation (Launius, 1994). Even before the inception of the Apollo program, it was understood that the return of astronauts and extraterrestrial materials back to Earth had the potential to introduce extraterrestrial biohazards, as reviewed previously and by Meltzer (2011), and the US Government recognized that these risks needed to be controlled. In 1963, after questions were asked about backward contamination in Congress (Meltzer 2011), President Kennedy signed National Security Action Memo 235, on *Large-Scale Scientific or Technological Experiments with Possible Adverse Environmental Effects* (JFK Library, 1993), which established a process that required presidential approval prior to conducting any such experiments, in consultation with the NAS and other relevant federal agencies (e.g., the State Department). In particular:

> Experiments which by their nature could result in domestic or foreign allegations that they might have such effects will be included in this category even though the sponsoring agency feels confident that such allegations would in fact prove to be unfounded.

and

> international scientific bodies or intergovernmental organizations may be consulted in the case of those experiments that might have adverse environmental effects beyond the U.S.

This memo, declassified in 1993, was applicable both to US nuclear testing activities and to the Apollo program.

Following public expressions of concern about backward contamination, NASA consulted with the US Public Health Service, which assigned a liaison officer to support the development of a "quarantine" program. This resulted in the formation of an "Interagency Committee on Back Contamination" (ICBC), chartered to provide oversight of both astronaut quarantine and curation of lunar materials. The committee included the three US regulatory agencies covering public health, agriculture, and the environment, as well as two additional "interested agencies": the NAS and NASA. A formal interagency agreement was established that required high-level interagency consultation prior to acting on any decision that was not *in accordance with the unanimous recommendation of the agencies represented on the Interagency Committee on Back Contamination*. NASA is not a regulatory agency, so this structure ensured that the regulatory agencies would exert effective oversight, even though over half the individual members of the ICBC were NASA staff (Radley and Rosen, 1969).

The interagency coordination framework that was established during the Apollo program, which was founded on the best-available scientific advice and included the regulatory agencies responsible for ensuring the health and safety of humans, animals, and agricultural activities, and the wider environment, was sound. In practice, the ICBC did perform oversight of the Apollo program's sample return activities and issued determinations regarding astronaut quarantine and biohazard testing of lunar materials, including termination of the quarantine program after Apollo 14. However, it was very fortunate that lunar samples, in fact, are not biohazardous to astronauts or the Earth, because the implementation of quarantine measures during the Apollo program would not have been adequate to prevent release. Reluctance on the part of those responsible for implementing the program was motivated by disputes over jurisdiction and authority, cost concerns, and a rather widespread perception within the space exploration community that precautions were unnecessary, as reviewed in Meltzer (2011). Many very valuable lessons can be learned from the Apollo program, both on practices that were surprisingly foresighted, and should be replicated, and on aspects of sample return and post-return operations that would benefit from improvement.

One rather famous example, which would be good not to repeat, involves biological analyses performed on the Surveyor 3 camera, which were claimed to show survival of Earth organisms after traveling round trip to the Moon (Rummel et al., 2011). The Surveyor robotic mission landed on the Moon in 1967 and was not subject to decontamination procedures for planetary protection. The Apollo 12 mission landed near the Surveyor 3 site in 1969, and astronauts collected hardware from the Surveyor 3 lander spacecraft, including a camera, for return to Earth. The Apollo 12 astronauts and other lunar samples that had been collected were subject to quarantine and containment procedures; in contrast, the camera from Surveyor 3 was placed in a laminar flow hood and subjected to biological sampling. The organism *Streptococcus mitis*, commonly found in the human respiratory tract and rapidly killed by desiccation, was the only organism isolated from the camera and was found in only one sample collected very late in the sampling period. The sampling process was filmed, and these films document that the "sterile technique" practices used by the technicians collecting samples would not, today, be considered adequate to maintain sterile culture conditions. In addition, a photographer leaned into the hood and took close-up still images, just prior to biological sampling, of the location on the Surveyor 3 camera, from which the *S. mitis* organism was collected. The appearance of *S. mitis* in cultures from the Surveyor 3 camera prompted a review, during which the Surveyor 3 spacecraft engineering model was found possibly to have been contaminated by *S. mitis*. Following this observation, it was concluded that the organism collected after return must have survived the round-trip travel between the Earth and the Moon, including years of exposure on the lunar surface.

Several logical fallacies, of concern to planetary protection, can be identified in this procession of events, which would be better avoided in the future. First, given that lunar quarantine procedures were supposed to be in place, how is it that the Surveyor 3 camera was sampled in what was effectively a shirt-sleeve environment, with a photographer present in street clothes rather than within containment? Second, the conclusion that a desiccation-sensitive organism must have survived several years of exposure to hard vacuum should have required some additional supporting data, beyond the merely circumstantial—the principle "Extraordinary claims require extraordinary evidence" does apply. Finally, the collection of unsterilized hardware by astronauts, who are subject to quarantine, adds the risk that they could be exposed to pathogenic Earth organisms from the collected hardware, possibly invalidating the purpose of a quarantine altogether. *Streptococcus mitis*, though mostly harmless, is a facultative human pathogen that can cause infective endocarditis—which was not, in the 1960s, an easy disease to diagnose. The trajectory of the Apollo program, and the future of human spaceflight, might have been quite different if, during quarantine, the Apollo 12 astronauts had come down with fever, bruising, exhaustion, stroke, and possibly heart or kidney failure.

12.5.3.3.4 Viking

If one early robotic exploration program stands out from all others for effective planetary protection compliance, that is the Viking program, which in the 1970s sent NASA's first lander missions to Mars (Daspit et al., 1988). The Viking program was managed by NASA Langley Spaceflight Center in Virginia, with prime contractor support from the Martin Marietta Corporation, and the lander spacecraft carried scientific instruments contributed by several academic institutions as well as other US government facilities. Each of the two Viking landers was transported to Mars by a Viking orbiter spacecraft, which carried replicate sets of scientific experiments. The two Viking landers returned the first meteorological, physical, and seismological measurements, as well as performed both metabolism- and chemistry-based life detection experiments, from two locations on the surface of Mars (Ezell and Ezell, 1984).

The Viking landers carried a suite of instruments designed to detect chemical constituents of possible Martian organisms, as well as metabolic indicators of possible Mars life. Both of these instrument payloads received cleanliness requirements that were more stringent than the rest of the Viking lander spacecraft (Daspit et al., 1988), to protect the integrity of the scientific results obtained from them. The chemistry-based experiment utilized a gas chromatograph-mass spectrometer (GC-MS) instrument to measure gases that evolved as Martian regolith was heated, results from which were interpreted at the time to indicate the presence of cleaning fluids used before launch from Earth. However, in 2017, it was reported that low-abundance peaks from the Viking GC-MS data indicate the presence of compounds, in the Viking samples, that were definitively identified by the Curiosity rover's Sample Analysis at Mars instrument as being Martian in origin (e.g., Glavin et al., 2013; Freissinet et al., 2015). The metabolic experiments included several culture cells that provided conditions predicted to support growth of Martian organisms, as well as the ability to heat samples of Mars regolith to temperatures expected to inactivate any organisms, as discussed in

DiGregorio et al. (1997). The results from this suite of experiments, known as the "Life Detection Package," definitely detected heat-labile chemical reactivity but were inconclusive as to whether this was consistent with biological metabolism. At the time, because the GS-MS instrument data were interpreted to indicate the absence of organic material in the regolith samples, the Viking results were interpreted as a failure to detecting Mars life, although this interpretation has subsequently been questioned (e.g., Navarro-Gonzalez et al., 2006).

The Viking program undertook the most stringent implementation of planetary protection requirements ever attempted, involving component or subsystem-level heat treatment to reduce microbial populations present inside the items or materials; careful cleaning of hardware surfaces to reduce the levels of heat-resistant microbes to fewer than 300 per square meter of spacecraft surface; packaging of the entire assembled lander and heatshield into a "bioshield" that was overpressured through launch, to prevent recontamination after heat microbial reduction; and finally, shortly prior to launch, a full-system heat microbial reduction treatment of the lander spacecraft inside the bioshield for over 40 hours at 112°C, which was demonstrated to reduce the levels of viable microbial contamination on lander surfaces by four orders of magnitude. The key to the successful implementation of this effort was the complete acceptance of planetary protection requirements on the part of project staff, including the Project Manager James Martin and the Project Scientist Gerald Soffen, both at NASA Langley, with strong support from program staff at NASA Headquarters (Daspit et al., 1988).

From the very beginning of the program, planetary protection requirements were fully captured in the design constraints for the landed hardware and integrated with controls on approved parts and materials, as well as testing, assembly, and operational procedures, that were followed by almost all hardware contributors. Viking Program staff at Langley were well-educated about planetary protection, having consulted with members of WESTEX, and studied heat-sterilization of hardware since 1964 (Ezell and Ezell, 1984) as well as attended meetings of the planetary protection advisory committee for several years prior to the start of the program in 1969 (Daspit et al., 1988). Although questions were raised about the need for heat microbial reduction prior to the start of the Viking program, once that was confirmed, on the basis of life-detection science contamination concerns as well as international policy considerations, planetary protection was managed as just another element required for project success.

From the start, it was recognized that electronic components and other spacecraft materials were not necessarily pre-qualified to tolerate temperatures planned for the final system-level heat microbial reduction, so the majority of parts and materials was acquired and tested in bulk. Issues related to heat treatments appeared on the project manager's "Top 10 Concerns" lists early during the program (Ezell and Ezell, 1984), but all of these were retired, by appropriate qualification, substitution, or process modification, by mid-1972, prior to the assembly of spacecraft hardware. During hardware assembly, an extensive model testing program was carried out; it included evaluation of heat tolerance, starting with smaller and then larger subsystems and continuing with full lander system qualification models. No significant issues were found after both of the flight systems underwent the full-system heat microbial reduction, validating the general assessment by Viking managers that the testing and qualification program significantly reduced risk and increased the reliability of the spacecraft (Daspit et al., 1988).

It is certainly true that the Viking requirement for heat tolerance required extensive component testing and replacement of susceptible parts and materials in heritage hardware, which might appear to increase program cost. However, multiple Viking managers observe that the bulk purchases and extensive testing program also had cost benefits, in reducing the need to evaluate components case by case and then re-work subsystems after they failed. For new hardware subsystems, requirements for heat tolerance could be addressed during the design phases and if done well should not have increased cost significantly, beyond the difference in component prices (Daspit et al., 1988). When these design constraints were not addressed adequately, both cost and schedule slips were the result, as demonstrated for the camera system originally contracted to the company TRW (Ezell and Ezell, 1984).

In the case of the GC-MS instrument, that was contributed by JPL, heat-tolerance requirements were not addressed effectively during the design phases, and a number of problems with this instrument were reported. The GC-MS instrument on the Viking 2 lander, which was located at higher latitude and therefore experienced greater thermal cycling, failed during the extended mission phase, with one engineer on that team attributing the problem to weaknesses induced by the pre-launch full-system heat microbial reduction. However, other Viking managers noted that the GC-MS team was reluctant to accept the heat-tolerance requirements, to a greater extent than other hardware providers (see Daspit et al., 1988). It is possible that this reluctance on the part of some JPL engineers to embrace heat sterilization is influenced by corporate memory of the Ranger program discussed previously.

It is sometimes claimed that planetary protection required 10% of the total Viking program budget, with an implication that this is an unreasonable cost for a mission that both protects the target of exploration and performs experiments searching for extraterrestrial life (e.g., Fairén et al. 2018). However, as described previously, it is not actually possible to pull out a subset of the Viking expenditures and label it as "planetary protection," because implementation was so completely integrated into overall project management. It is more relevant, for students of astrobiology, to understand that the total Viking program budget was about $1 billion, in 1975 dollars, with the Life Detection Package costing $59 million and the GC-MS costing $41 million (Ezell and Ezell, 1984). These actual costs demonstrate that the two instruments on Viking, addressing research of the greatest interest to astrobiology, together cost 20% of the total program budget—including all the effort to establish how to accomplish full-system heat microbial reduction and other associated cleaning processes. More recent estimates of the cost to retrofit a modern Mars rover for full-system heat microbial reduction, done independently by ESA and NASA, suggest that this cost has remained at the equivalent of one major science instrument (e.g., Rummel and Conley, 2018). Whether the equivalent of one science instrument is a reasonable cost to answer key research questions in astrobiology is an important advocacy question for the astrobiology community to address.

It is a consistent conclusion among people involved in the Viking program that the requirements for full-system heat microbial reduction provided significant benefits to the reliability and success of the missions and was not as difficult to implement as initially anticipated. The most important lesson to learn from historical missions, for planetary protection in the future, was summarized very well by the Viking Project Manager, James Martin (quoted in Daspit et al., 1988):

There are many young people at JPL and elsewhere working on these problems, and they are so enthused by the technologies they're working on that they don't realize the extent of the impact PP and Contamination control requirements can have or how that impact can enlarge as a problem the longer it is neglected. The PP requirements should be expressed in a more forthright and regulatory sense, so that people aren't allowed to forget about it until five or ten years from now when they suddenly discover that 'you really can't make a widget that you can heat.

12.5.4 APPLICATION TO MISSIONS IN PREPARATION

12.5.4.1 OUTER PLANETS MISSIONS

12.5.4.1.1 Probability of Contamination

It has been suggested that the greatest volume of environments providing conditions suitable for the growth of microbial life from Earth could be in the outer solar system, in the form of subsurface liquid water within the moons of the giant planets, and also any water that might remain liquid in the interiors of Kuiper Belt objects. Such habitable environments are of concern for planetary protection, to the extent that the probability of introducing a single viable Earth organism into a liquid water environment must be held lower than 1×10^{-4} per mission, which applies to both lander and orbiter/flyby missions and must take into account spacecraft and operational reliability. Fortunately for mission planners, it is thought to be extremely difficult to access potential liquid water volumes within the vast majority of objects in the outer solar system: a recent analysis of Ganymede, which is the object currently considered to have the most-accessible water after Europa and Enceladus, suggested that viable Earth organisms deposited on the surface would have a probability much lower than 1×10^{-4} of reaching subsurface liquid water (Grasset et al., 2013).

For this reason, missions that would not encounter Europa or Enceladus receive no *a priori* restrictions on hardware cleanliness, for planetary protection; rather, these missions are required to ensure, by careful operation of the spacecraft, that any liquid water bodies discovered during the mission are not contacted by spacecraft hardware, even after the mission ends. This requirement was applied, in the past, to the missions that discovered potential habitats within Europa and Enceladus: the Galileo spacecraft was de-orbited into Jupiter,

and the Cassini spacecraft was de-orbited into Saturn, in order to protect the newly discovered watery moons of those planets. The Juno mission to Jupiter, which will not encounter Europa during the prime mission, plans to dispose the spacecraft into Jupiter—but they also performed calculations to show that, should the deorbit maneuver fail and the inactive Juno spacecraft impact Europa at some point in the future, the impact energy would be so high as to incinerate all remaining viable organisms, in a sufficiently high fraction of the total impact cases in Monte Carlo simulations that the $<1 \times 10^{-4}$ probability requirement was met (Bernard et al., 2013).

12.5.4.1.2 Cleanliness and Life Detection

The detection of extraterrestrial life is of relevance to planetary protection because extraterrestrial organisms are by definition considered biohazards: this is based on long experience from Earth that introduced species can become invasive or pathogenic. For this reason, outbound-only missions to outer solar system objects that carry life detection investigations do not receive additional cleanliness requirements from planetary protection on the basis of instrument capabilities, beyond the 1×10^{-4} probability of contaminating liquid water and potential habitats.

In contrast, additional organic and biological cleanliness requirements will be imposed on sample return missions from potentially habitable environments in the outer solar system (currently Enceladus or Europa), to ensure adequate levels of confidence in the results of the post-return biohazard test protocol. During the Apollo program, sample collection hardware was carefully cleaned and samples were (in principle) quarantined until a biohazard test protocol had been completed. Subsequent to Apollo, a notional "Draft Test Protocol for Detecting Biohazards" has been established (Rummel et al., 2002), during the course of multiple rounds of studies preparing for missions that would collect samples and return from Mars. This draft test protocol requires additional refinement (e.g., Kminek et al., 2014), to establish requirements on the level of statistical confidence needed to make a determination that extraterrestrial samples are "safe for release"—which is a risk assessment that needs to be made at a global societal level, not solely by space scientists or space agency staff (e.g., Haltigin et al., 2018). Confidence that the test protocol is not generating false-negative results would, of course, be reduced in proportion to the amount of Earth contamination present in collected samples, and it depends on the capabilities of instruments used to make measurements.

The development of biohazard test protocols and associated contamination requirements is ongoing.

12.5.4.2 MARS MISSIONS

12.5.4.2.1 Planetary Protection Requirements for Mars: It's Complicated

The planet Mars has been explored more extensively than any planet in the solar system other than Earth; likewise, Mars is also the focus of a wider diversity of human interests. From the earliest discussions of Mars exploration, the potential for future human colonization/settlement and the possibility of terraforming Mars were taken seriously, and the potential for biological contamination from Earth to interfere with these long-term goals was recognized (e.g., Space Science Board [SSB], 1959). In response to the rapidly increasing scientific information about Mars and recognizing the diversity of short-term and long-term goals in Mars exploration, planetary protection requirements for Mars have undergone more discussion and refinement than those for any other target of exploration.

Spacecraft and hardware not intended to impact Mars are allowed to meet planetary protection requirements by avoiding impact onto Mars. Launched hardware items that were not maintained in a controlled clean environment—e.g., the upper stages of launch vehicles—are required to avoid impact onto Mars, for a period of 50 years after launch, at a probability of 1×10^{-4}. It is preferred that spacecraft intended to orbit or flyby Mars should also avoid Mars impact for a period of 50 years, though with a relaxed probabilistic constraint—this involves no greater cleanliness than assembly in controlled environments typical for spaceflight hardware assembly. If, however, impact avoidance is not feasible for orbiter spacecraft, due, for example, to aerobraking, then orbiter missions may demonstrate that they deliver no more organisms to Mars than are permitted on landed hardware that is expected to break open on impact.

Current planetary protection requirements for the cleanliness of hardware landing on Mars depend both on where the hardware is intended to land and on the purpose of the hardware. The Viking program imposed additional cleanliness requirements on some hardware, to ensure the integrity of the life detection experiments by removing contamination from Earth that could interfere with the detection of signals from Mars. No indications were observed of Earth organisms being present, so the terminal heat microbial reduction and additional cleaning of the Life Detection Package did accomplish the goal of eliminating Earth life. However, despite the program's best efforts, signals interpreted as Earth contaminants were the most abundant compounds measured by the Viking GC-MS instruments. At the time, it was assumed that these Earth contaminants were the only compounds present and that no organics from Mars had been detected (e.g., Space Science Board [SSB], 1977). Because the Viking landers were not collecting Mars samples for return to Earth, this interpretation did not put in jeopardy the safety of the Earth, which is of paramount importance to planetary protection.

12.5.4.2.2 "Average" Mars

After the Viking program demonstrated that most of the Martian surface is cold and dry, providing very limited resources that could support Earth life, cleanliness requirements for Mars landers were relaxed, to eliminate the four-order-of-magnitude heat-reduction step and protection from recontamination implemented on Viking (DeVincenzi et al., 1996). This basic level of cleanliness is required for all Mars missions, and missions receive a designation of Planetary Protection Category IVa if they do not carry life detection instruments or plan operations to access more-habitable locations on Mars.

Before the terminal full-system heat reduction, the Viking landers had been cleaned to 300 heat-resistant organisms per square meter of spacecraft surface over about 1000 square meters of area and also carried approximately 200,000 heat-resistant organisms in the interior of the spacecraft. The number of heat-susceptible organisms was not measured as part of the requirement, because any heat treatment that killed resistant organisms would eliminate susceptible organisms at a much higher rate. These relaxed requirements of less than 300 heat-resistant organisms per square meter, less than 300,000 heat-resistant organisms over all surfaces exposed to the environment of Mars, and less than 500,000 heat-resistant organisms in total (this includes organisms inside hardware that could break open on impact, such as heat shields, backshells, and parachutes), are necessary to meet Planetary Protection Category IVa.

12.5.4.2.3 Special Regions on Mars

Following orbital observations in the 2000s that confirmed the presence of active gully systems and water ice in the near-subsurface of Mars, the Viking requirement for four-log microbial reduction with recontamination prevention was reinstated for missions targeting locations on Mars where environmental conditions have the potential to provide, at least transiently, temperatures and available water that could support growth of Earth life. So-called Mars Special Regions are, as of 2017, defined as locations where temperatures reach above −28°C and "water activity" (1/relative humidity) reaches above 0.5 (equivalent to 50% relative humidity). These regions include features on Mars where the presence of such conditions is in question (Kminek et al., 2017). Each project planning a landed mission to Mars is required to do an analysis of their proposed landing sites, to establish whether Special Regions might be present: only if Special Regions are not within a landing ellipse that includes 3 standard deviations of targeting error after parachute opening (3-sigma) would a project be designated Planetary Protection Category IVa. Missions that do target areas with potential Special Regions inside the 3-sigma landing ellipse are designated Planetary Protection Category IVc and are required to use microbial reduction processes and recontamination prevention to achieve Viking-equivalent cleanliness of <0.03 viable Earth organisms per square meter of exposed spacecraft surface. Missions intended to access Special Regions outside the landing ellipse are required to clean at least the subsystems used for such access to Viking-equivalent levels and protect them from recontamination by other spacecraft hardware.

It is relevant to note that the definition of Special Regions, originally adopted in the mid-2000s, does not follow historical precedent in setting conservative limits on parameters of concern to planetary protection, with room for subsequent relaxation: in the 2010s, the temperature limit for Special Regions had to be reduced from −25°C to −28°C, after additional data on the capabilities of Earth organisms were obtained.

12.5.4.2.4 Mars Life Detection

The highest concern for planetary protection is to avoid "harmful contamination" of the environment of the Earth, which includes imposing appropriate constraints on future round-trip missions, robotic and human. In this context, false-negative results from in situ experiments to detect Mars life are of considerable concern, because future requirements would be set based on an incorrect assessment of potential risk. One example of the consequences of false-negative results is that Mars exploration was put on hold for 20 years because "Viking had shown there were no organics" (Space Science Board [SSB], 1977)—yet recent data from the Sample Analysis at Mars (SAM) instrument informed a re-evaluation of Viking data, which identified Martian organic compounds not recognized previously (e.g., Glavin et al., 2013; Freissinet et al., 2015).

Missions to Mars that carry instruments capable of detecting Mars *life forms, precursors, and remnants*, in the language of the COSPAR policy (Kminek et al., 2017) automatically also are capable of detecting biological and organic contamination from Earth and receive a separate designation of Planetary Protection Category IVb. To address the greater policy-level concern about Mars life (by definition considered biohazardous) being brought to Earth, life detection missions are required, in addition to landing site constraints, to ensure that hardware subsystems with a potential to contaminate the life detection experiments meet the equivalent of the Viking lander overall requirements: cleaned to 300 resistant organisms per square meter of surface area, protected from recontamination, and reduced by four orders of magnitude. As noted previously and in histories of the Viking Program, this level of cleanliness is less than what was required for the Viking GC-MS and Life Detection Package, which was set based on the scientific objectives of those instrument (Daspit et al., 1988). Currently, rather than setting specific numerical limits on non-viable Earth contamination, planetary protection policy specifies that such requirements be set *driven by the nature and sensitivity of the particular life detection experiments* (Kminek et al., 2017), with the expectation that appropriate limits would be established by the instrument teams and project management during payload selection and accommodation, and subsequently monitored for compliance, along with all other planetary protection requirements. This is an effective approach for *in situ* experiments, but when samples are returned to Earth, additional considerations pertain.

12.5.4.2.5 Mars Sample Return

Mars is of astrobiological interest as a potential habitat for extraterrestrial life and therefore has the potential to host life that is, by definition, considered potentially biohazardous to the Earth. For this reason, missions planned to collect samples from Mars and bring them to Earth receive the designation "Planetary Protection Category V Restricted Earth Return," which means that spacecraft hardware must be cleaned to levels that will ensure that Earth contamination is not introduced into Mars samples in quantities that could invalidate the biohazard test protocol that will be performed after return to Earth (e.g., Haltigin et al., 2018). This is in addition to meeting all requirements appropriate for mission operations at Mars, as well as ensuring that documentation of potential contamination is collected and provided for pre-return Earth Safety analyses

and reviews. During the return phase of the mission, samples are required to be contained, so as to ensure a probability of less than 1×10^{-6} that a particle greater than 10 nanometers in size is released into the environment of the Earth, including particles adhering to spacecraft surfaces, as discussed in Haltigin et al. (2018). In the special case of sample return from one of the moons of Mars, additional calculations would be needed to assess the amount of material from Mars present in the collected samples, due to recent impact ejection events, such as the 60-km Mojave impact crater (Chappaz et al., 2013).

After Mars samples have landed on Earth, it is required to contain them at levels equivalent to the strictest biosafety level (BSL-4/P-4, used for, e.g., ebola virus) and also protect them from Earth contamination while early analyses and the biohazard test protocol are performed (Rummel et al., 2002). This is required to ensure that the risk to the Earth from retaining Mars samples is acceptably low. For Restricted Earth Return missions, "acceptably low" is a regulatory determination that has not yet been fully developed—currently, there is a containment requirement but not a biohazard test confidence requirement. Containment procedures are required to ensure that the risk of releasing extraterrestrial material into the Earth's environment is kept under one in a million, which was recommended based on comparison with other risks that human societies accept (European Science Foundation [ESF], 2012).

12.5.4.3 Human Missions

Human missions to other planetary targets almost invariably involve a return trip to Earth—as such, they would receive requirements equivalent to Planetary Protection Category V. Current planetary protection policy includes guidelines for human missions that focus predominantly on forward contamination. Protection of the Earth must be ensured, but specific practices to accomplish this have not been established—the Apollo program provides useful precedent in some areas and cautionary experience in others.

As with robotic missions, planetary protection concerns for human missions are divided conceptually by "habitable" vs. "non-habitable" targets: when humans explore non-habitable targets, neither contamination of the astronauts or the Earth nor contamination of the target by Earth life is relevant. Mars is the only feasible target for human exploration that is considered as potentially hosting native extraterrestrial life and is potentially "habitable" for Earth microbes; thus, human missions to Mars have received the most policy-level consideration (e.g., Space Studies Board [SSB], 2002; Conley and Rummel, 2010; Kminek et al., 2017; Haltigin et al., 2018).

It is recognized that human missions are unlikely to be fully contained; thus, provisions must be made to address contamination of astronauts by Mars material, as well as contamination of Mars by Earth microbes. By analogy, with the requirements for Special Regions, approaches have been proposed for early human missions that involve allowing a greater level of contamination in areas where humans travel, while protecting Special Regions to currently required levels. The degree of separation needed would be based on the technical capabilities of the hardware and human support systems, which currently are not well-established. As is routine for planetary protection, the collection of additional scientific information may lead to relaxation of requirements, once it is demonstrated that the initial level of stringency is not needed.

Two additional factors pertain to the human exploration of Mars, which are not currently covered in planetary protection policy—yet these issues will need to be addressed at a policy level, before such events occur that require a response from the global society. The first question is the extent to which the environment of Mars might be toxic to astronauts or Earth microbes, in the absence of a Mars biota—this needs to be understood in order to assess the potential for false-positive indications of biohazards in samples from Mars. Measurements made on Mars suggest that the dust contains ~1 weight-percent of oxy-chlorine compounds, which are known to be toxic to humans and some Earth microbes. An understanding of the potential toxicity of Mars dust may be important to assess overall risk of human missions to Mars and to provide information both to the astronauts performing missions and to the societies that pay for them.

The second question, which is explicitly excluded from planetary protection policy but still needs to be addressed, is the extent to which Earth microbes could change during spaceflight, resulting in increased biohazard after they return to Earth. This issue was raised by the NASA Advisory Council but has not yet received significant attention, despite experiments from the International Space Station that seem to indicate increased virulence in *Salmonella* grown in microgravity (Sarker et al. 2010).

Societal perception and acceptance of these risks will be greatly facilitated by providing accurate and complete information, starting as early as possible in the process of mission development.

12.5.5 OPEN ISSUES IN INTERNATIONAL POLICY

12.5.5.1 Relationship of International Policy to National Obligations

The 1967 Outer Space Treaty, by its title, provides for COSPAR, also apply more broadly than just scientific exploration. Whether the COSPAR guidelines represent legally binding "customary law" has not yet been tested in court, although their recognition in 2017 by UN-COPUOS as a *long-standing... reference standard... in guiding compliance with article IX of the Outer Space Treaty* could strengthen that case. All the major international agreements on space exploration have elements that support planetary protection practices (Achilleas and Crapart, 2003), with the state responsible for space objects also being responsible for mitigating *harmful contamination* caused by them. In addition, the international framework of environmental law, although focused on terrestrial activities, includes language that could reasonably be extended to backward contamination brought to Earth by sample return missions (Achilleas and Crapart, 2003).

In addition, the 1992 Convention on Biological Diversity, Article 3, states the principle that states have:

> *the responsibility to ensure that activities within their jurisdiction or control do not cause damage to the environment of other States or of areas beyond the limits of national jurisdiction.*

Whether such areas extend beyond the environment of the Earth has again not yet been tested in court, but it has been suggested that a plausible case could be made (Achilleas and Crapart, 2003).

Supporting these international obligations, some countries also have national laws that address environmental protection aspects of space activities. For example, the Russian Federation, in Decree No. 5663-1, states the principle of *provision of safety in space activity, including protection of the environment;* and prohibits actions *to create harmful contamination of outer space which leads to unfavourable changes of the environment,* (UNOOSA, retrieved 2018). Although the US has not yet passed legislation addressing planetary protection, several sections of the Code of Federal Regulations describe the applicability of the National Environmental Protection Act to Earth Return missions. In addition, the 1975 Presidential Directive that superseded the Apollo-era NSAM-235, PD/NSC-25 on *Scientific or Technological Experiments with Possible Large-scale Adverse Environmental Effects and Launch of Nuclear Systems into Space* specifies a consultation process that would need to be followed prior to implementing a Category V Restricted Earth Return Mission.

12.5.5.2 Application of Planetary Protection Guidelines to Commercial Missions

It is only recently that non-governmental entities have developed capabilities to engage in activities of concern for planetary protection, and the mechanisms for ensuring *authorization and continuing supervision* (UN, 1966) of private space exploration are still under development. Within the US, the commercial launch approval process overseen by the Federal Aviation Administration requires consultation with the State Department and NASA, and planetary protection compliance for commercial launches has been addressed. Other countries may have more straightforward internal structures for establishing legal and/or regulatory frameworks for commercial space exploration.

With the increasing interest and activity in private space exploration, and considering the potential for global negative consequences due to the release of a hazardous extraterrestrial entity, formal processes, both nationally and internationally, will undoubtedly be needed. Although planetary protection is the aspect of commercial space exploration with the broadest potential consequences, the establishment of an appropriate regulatory framework for commercial space activities is an issue that extends far beyond planetary protection. For example: if one commercial entity identifies a valuable target and another entity succeeds in reaching the same target more rapidly, how will rights and responsibilities for use of that target be determined?

12.5.5.3 International Concerns Around Restricted Earth Return

Over the past decade, the iMARS Working Group, chartered by the International Mars Exploration Working Group, has explored requirements for Mars Sample Return and made considerable progress on developing an architecture for handling Mars samples brought to Earth (Haltigin et al., 2018). The major concern for backward contamination is to prevent release of extraterrestrial organisms over the entire time for which extraterrestrial material is stored on Earth, thus the precautions taken involve two steps (e.g., Rummel et al., 2002; Kminek et al., 2014). First, the hardware returning to Earth must be carefully designed to reduce the risk of accidental breach to an acceptably low level, until the hardware is placed in a properly designed containment facility. In addition, the samples brought to Earth must undergo careful testing for possible extraterrestrial life, which is, by definition, considered biohazardous. This "Biohazard Test Protocol" must be performed early during the post-return sample analysis period, to minimize the risk of accidental release due to failures in containment. To ensure that the test protocol can detect extraterrestrial biology with high sensitivity, hardware used to collect extraterrestrial samples must be carefully cleaned of Earth contamination, and information retained, starting with the earliest mission phases, about the potential for and types of contamination that could be introduced. This set of requirements was recommended during the US Apollo program, to minimize the potential for "false negative" results due to the masking low levels of extraterrestrial biosignatures by higher levels of Earth contamination (e.g., Space Science Board [SSB] 1964).

The most significant long-lasting consequence of space exploration for human society and the environment of the Earth would be backward contamination, the release of a novel extraterrestrial entity that has adverse effects—this was recognized as a potential problem in the 1950s and is explicitly prohibited by Article IX of the Outer Space Treaty (UN, 1966). It has been recommended that COSPAR guidelines specify a probability of no more than "one in a million" that potentially viable extraterrestrial material be released into the Earth's environment, a limit that was proposed after a survey of risk/benefit evaluations that are considered acceptable for other human activities (European Science Foundation (ESF), 2012; Haltigin et al., 2018). As mentioned previously, many spacefaring nations have national laws on environmental protection, as well as the existing international agreements—but the specific applicability of national and/or international regulatory frameworks to Restricted Earth Return efforts remains to be elaborated.

12.5.6 SOCIETAL CONSIDERATIONS

12.5.6.1 Ambiguity in a "Detection" of Extraterrestrial Life

Planetary exploration is currently done almost entirely by using taxpayer funding—even companies that have planetary missions in preparation, such as Moon Express, receive considerable government funding. Responsible use of these funds requires that the benefits of investment in planetary exploration outweigh the risk of loss, including loss due to lack of investment elsewhere. This is a societal decision, not a scientific or engineering one, so societal perception of exploration activities and scientific conclusions is critical. After the Viking missions reported a non-detection of Mars life, no missions were sent to Mars for several decades, despite continued interest on the part of spaceflight engineers and Mars scientists (Space Science Board [SSB], 1977), because funds were allocated elsewhere. If it had been recognized at the time that the Viking mass spectrometers did, in fact, detect Mars organic compounds, albeit at low levels (Glavin et al., 2013; Freissinet et al., 2015), the trajectory of Mars exploration would likely have been quite different.

The need to ensure accuracy in the conclusions of a life/biohazard detection protocol is of much greater importance when the samples being analyzed have been brought back to Earth. There has recently been some confusion regarding contamination control requirements supporting a Biohazard Test Protocol, resulting from conflation of concerns associated with "false positive" and "false negative" results (e.g., Space Studies Board [SSB] 2018). From the standpoint of interpreting scientific data, this results from ambiguity about which "null hypothesis" is being tested (Kminek et al., 2014). Astrobiologists and scientists interested in detecting extraterrestrial life will err on the side of not announcing a finding of "life" until they have high confidence in the detection: they are testing the null hypothesis "there is **no** life in these samples." If Earth contamination is detected but there are not signals clearly indicative of extraterrestrial biosignatures, this would be interpreted as a "non-detection" of extraterrestrial life. This approach is consistent with careful scientific discovery, and it also ensures that no harm is done to scientists' or space agencies' reputations, from holding over-enthusiastic press conferences.

12.5.6.2 The Potential for Unidentified Biohazards

In contrast, concerns for Earth Safety require avoiding the release of potential biohazards into the environment of the Earth, and therefore, they need to test the contrary null hypothesis: to ensure safety of the Earth if samples are released, it is necessary to disprove the hypothesis "there is **extraterrestrial** life in these samples" (e.g., Kminek et al., 2014; Haltigin et al., 2018). When testing this null hypothesis, the detection of any Earth contamination would indicate the presence of biosignatures that could also be associated with extraterrestrial biohazards, and very careful further testing would be required to evaluate this possibility at acceptable levels of statistical significance. Samples can certainly be analyzed in containment, while biohazard testing is ongoing—in fact, many of the measurements made will inform both scientific and regulatory communities—but the risk that a breach in containment could release a potential biohazard increases the longer the question remains unanswered. This is a societal question, as much as a scientific one: the public will be affected, if a breach occurs, so they will also want to know how safe the samples are. The space agencies bringing extraterrestrial samples to Earth have an obligation and responsibility to provide accurate answers.

12.5.6.3 QUARANTINE AND SETTLEMENT

The concern about detection of potential biohazards does not stop after the first Mars sample return mission, but the challenge of differentiating Mars life from Earth contamination will become progressively more difficult. Robotic exploration can be accomplished while maintaining stringent levels of cleanliness; however, once human missions to Mars are initiated, the concerns increase in ways that are outside the purview of planetary protection. All organisms known to be hazardous to humans are from Earth: this includes overt pathogens (e.g., *Clostridium tetani*) and environmental hazards (e.g., *Stachybotrys* fungi). Despite cleaning protocols, pathogenic organisms have been identified on the International Space Station (e.g., Lang et al., 2017), though fortunately, they have not yet caused illness in astronauts. Societal concerns could change rapidly, if an astronaut were to become ill from a spaceborne pathogen—as human spaceflight becomes more common, policies to control transport of Earth organisms will need to be developed.

Once humans attempt to settle elsewhere, those environments will facilitate independent evolution, and the possibility of exchanging biohazardous organisms that are originally from Earth will need to be addressed. Current and historical quarantine procedures may not be adequate, when people, with their microbes, want to go back and forth. If an infectious disease or hazardous organism has never been present in an environment, what rules should be imposed to control introduction? How should they be different for independently replicating (e.g., most bacteria and archaea) vs. obligate-parasite (e.g., viruses and some eukaryotic parasites) entities? What about quiescent or zoonotic infections that have no vector, for example, a human with trichinosis or malaria? Should the rules be different for infections that could spread, given an appropriate environment, such as *Giardia* and *Clostridium difficile*? All of these issues should be considered, at least at a policy level, before the actual circumstances develop.

12.5.7 SUMMARY AND CONCLUSIONS

This chapter provides a brief summary of planetary protection policy and implementation, including the current context and historical aspects of its development as well as outlining applicability to select future missions. The issues that planetary protection addresses range from the highly technical, including astrobiological and other scientific questions, as well as engineering implementation, to the purely societal and legal, including concerns over invasive species and negotiation of international regulations. As with most technical and academic endeavors, the wider context in which researchers carry out their studies may have an enormous impact on the research that can be performed. It is important for all

educated citizens but particularly for researchers involved in esoteric and expensive taxpayer-funded work, to participate in the social and cultural conversations that surround them.

REFERENCES

Achilleas, P. and L. Crapart. 2003. *Legal Issues on Planetary Protection and Astrobiology.* Contract: IDEST (Institute of Space and Telecommunications Law), University Paris Sud – 11/European Space Agency.

Bernard, D.E., R.D. Abelson, J.R. Johannesen, T. Lam, W.J. McAlpine, and L.E. Newlin. 2013. Europa planetary protection for Juno Jupiter Orbiter. *Advances in Space Research,* 52(3), 547–568. doi:10.1016/j.asr.2013.03.015.

Chappaz, L., H.J. Melosh, M. Vaquero, and K.C. Howell. 2013. Transfer of impact ejecta material from the surface of Mars to Phobos and Deimos. *Astrobiology,* 13(10), 963–980. doi:10.1089/ast.2012.0942.

Committee on Contamination by Extraterrestrial Exploration (CETEX). 1958. News of science. *Science,* 128(3329), 887–891. doi:10.1126/science.128.3329.887.

Committee on Contamination by Extraterrestrial Exploration (CETEX). 1959. Contamination by extraterrestrial exploration. *Nature,* 183, 925.

Conley, C.A. and J.D. Rummel. 2010. Planetary protection for human exploration of Mars. *Acta Astronautica,* 66(5–6), 792–797, doi:10.1016/j.actaastro.2009.08.015.

Daspit, L., J. Stern, and J. Martin. 1988. *Lessons Learned From the Viking Planetary Quarantine and Contamination Control Experience.* Contract NASW-4355 for NASA Headquarters, Washington, DC. Available at https://planetaryprotection. nasa.gov/documents/.

Davies R.W. and M.G. Comuntzis. 1960. The Sterilization of Space Vehicles to Prevent Extraterrestrial Biological Contamination. In: Hecht F. (eds.) Xth International Astronautical Congress London 1959 / X. Internationaler Astronautischer Kongress / Xe Congrès International d'Astronautique. Springer, Berlin, Germany.

DeVincenzi, D.L., P.D. Stabekis, and J.B. Barengoltz. 1983. A proposed new policy for planetary protection. *Advances in Space Research,* 3(8), 13–21.

DeVincenzi, D.L., P.D. Stabekis, and J.B. Barengoltz. 1996. Refinement of planetary protection policy for Mars missions. *Advances in Space Research,* 18(1–2), 311–316. doi:10.1016/0273-1177(95)00821-U.

DiGregorio, B.E., L. Gilbert, and S. Patricia. 1997. *Mars: The Living Planet.* Frog, Ltd. and North Atlantic Books, Berkeley, CA.

Doyle, S.E. and A.I. Skoog. 2012. *The International Geophysical Year.* Initiating International Scientific Space Cooperation International Astronautical Federation, Paris, France.

European Science Foundation (ESF). 2012. *Mars Sample Return Backward Contamination–Strategic Advice and Requirements.* European Space Sciences Committee. Available at http://www. essc.esf.org/list-of-publications/.

Ezell, E.C. and L.N. Ezell. 1984. *On Mars: Exploration of the Red Planet, 1958–1978.* Document SP-4212, National Aeronautics and Space Administration, Washington, DC.

Fairén, A.G., V. Parro, D. Schulze-Makuch, and L. Whyte. 2018. Is searching for Martian life a priority for the Mars Community? *Astrobiology,* 18(2), 101–107. doi:10.1089/ast.2017.1772.

Freissinet, C., D.P. Glavin, P.R. Mahaffy, K.E. Miller, J.L. Eigenbrode, R.E. Summons, A.E. Brunner et al. 2015. Organic molecules in the Sheepbed Mudstone, Gale Crater, Mars. *Journal of Geophysical Research: Planets,* 120, 495–514. doi:10.1002/2014JE004737.

Glavin, D.P., C. Freissinet, K.E. Miller, J.L. Eigenbrode, A.E. Brunner, A. Buch, B. Sutter et al. 2013. Evidence for perchlorates and the origin of chlorinated hydrocarbons detected by SAM at the Rocknest aeolian deposit in Gale Crater. *Journal of Geophysical Research: Planets*, 118, 1955–1973. doi:10.1002/jgre.20144.

Grasset, O., E. Bunce, A. Coustenis, M.K. Dougherty, C. Erd, H. Hussmann, R. Jaumann, and O. Prieto-Ballesteros. 2013. Review of exchange processes on Ganymede in view of its planetary protection categorization. *Astrobiology*, 13, 991–1004. doi:10.1089/ast.2013.1013.

Hall, R.C. 1977. *Lunar Impact: A History of Project Ranger*. NASA SP-4210, NASA History Office, National Aeronautics and Space Administration, Washington, DC.

Haltigin, T., C. Lange, R. Mugnolo, and C. Smith (co-chairs), H. Amundsen, P. Bousquet, C. Conley et al. 2018. iMARS Phase 2–A draft mission architecture and science management plan for the return of samples from Mars. *Astrobiology*, 18(S1).

JFK Library. 1993. NASM-235, downloaded from https://www.jfklibrary.org/Asset-Viewer/1DrKlFNAHUaRi22CwONwAw.aspx.

Kminek, G., C. Conley, C.C. Allen, D. Bartlett, D. Beaty, L. Benning, R. Bhartia et al. 2014. Report of the workshop for life detection in samples from Mars. *Life Sciences in Space Research*, 2, 1–5. doi:10.1016/j.lssr.2014.05.001.

Kminek, G., C. Conley, V. Hipkin, and H. Yano. 2017. COSPAR's planetary protection policy. *Space Research Today*, 200, 12–25.

Lang, J.M., D.A. Coil, R.Y. Neches, W.E. Brown, D. Cavalier, M. Severance, J.T. Hampton-Marcell, J.A. Gilbert, and J.A. Eisen. 2017. A microbial survey of the International Space Station (ISS). *PeerJ Bioinformatics and Genomics Section*, 5, e4029. doi:10.7717/peerj.4029.

Launius, R.D. 1994. *Apollo: A Retrospective Analysis. Monographs in Aerospace History*. Number 3, National Aeronautics and Space Administration, Washington, DC.

Mann, C.C. 2011. *1493: Uncovering the New World Columbus Created*. Knopf, New York.

Meltzer, M. 2011. *When Biospheres Collide: A History of NASA's Planetary Protection Program*. NASA SP-2011-4243 National Aeronautics and Space Administration, Washington, DC.

MIL-STD-810G. 2008. *Department of Defense Test Method Standard: Environmental Engineering Considerations and Laboratory Tests*. Available at http://everyspec.com/MIL-STD/MIL-STD-0800-0899/MIL-STD-810G_12306/.

Navarro-Gonzalez, R., K.F. Navarro, J. de la Rosa, E. Iniguez, P. Molina, L.D. Miranda, P. Morales et al. 2006. The limitations on organic detection in Mars-like soils by thermal volatilization–gas chromatography–MS and their implications for the Viking results. *Proceedings of the National Academy of Sciences of the United States*, 103(44), 16089–16094. doi:10.1073/pnas.0604210103.

Phillips, C.R. 1974. *The Planetary Quarantine Program Origins and Achievements, 1956–1973*. Document SP-4902, National Aeronautics and Space Administration, Washington, DC.

Radley, R. and S. Rosen. 1969. Lunar receiving laboratory. *Syracuse/NASA Program Regulations in Space, NASA CR-109164*. 6224-WP-7, Syracuse University, Syracuse, NY.

Rummel, J.D. and C.A. Conley. 2018. Inadvertently finding earth contamination on Mars should not be a priority for anyone. *Astrobiology*, 18(2), 108–115. doi:10.1089/ast.2017.1785.

Rummel, J.D., J.H. Allton, and D. Morrison. 2011. *A Microbe on the Moon? Surveyor III and Lessons Learned for Future Sample Return Missions*. The Importance of Solar System Sample Return Missions to the Future of Planetary Science, Lunar and Planetary Science Institute, Houston, TX, presentation abstract 5023.

Rummel, J., M. Race, D.L. DeVincenzi, P.J. Schad, P. Stabekis, M. Viso, and S.E. Acevedo. 2002. *A Draft Test Protocol for Detecting Possible Biohazards in Martian Samples Returned to Earth*. National Aeronautics and Space Administration, Washington, DC.

Sarker, S., C.M. Ott, J. Barrila, and C.A. Nickerson. 2010. Discovery of spaceflight-related virulence mechanisms in salmonella and other microbial pathogens: Novel approaches to commercial vaccine development. *Gravitational and Space Research*, 23(2).

Steele, A., M.D. Fries, H.E.F. Amundsen, B.O. Mysen, M.L. Fogel, M. Schweizer, and N.Z. Boctor. 2010. Comprehensive imaging and Raman spectroscopy of carbonate globules from Martian meteorite ALH 84001 and a terrestrial analogue from Svalbard. *Meteoritics and Planetary Science*, 42(9), 1549–1566.

Space Science Board (SSB). 1959. *Summary Report of WESTEX*. National Academy of Sciences, National Research Council.

Space Science Board (SSB). 1964. *Conference on Potential Hazards of Back Contamination from the Planets*. National Academy of Sciences, National Research Council.

Space Science Board (SSB). 1977. *Post-Viking Biological Investigations of Mars*. Committee on Planetary Biology and Chemical Evolution, Space Science Board, Assembly of Mathematical and Physical Sciences, National Research Council, Washington, DC.

Space Studies Board (SSB). 2002. *Safe on Mars Precursor Measurements Necessary to Support Human Operations on the Martian Surface*. National Academies Press, Washington, DC. Available at https://www.nap.edu/catalog/10360/safe-on-mars-precursor-measurements-necessary-to-support-human-operations.

Space Studies Board (2018). *Review and Assessment of Planetary Protection Policy Development Processes*. National Academies Press, Washington, DC.

United Nations (UN). 1966. *United Nations Treaty on Principles Governing the Activities of States in the Exploration and Use of Outer Space, including the Moon and Other Celestial Bodies*. Available at http://www.unoosa.org/oosa/en/ourwork/spacelaw/treaties/outerspacetreaty.html.

United Nations (UN). 2018. *Status of International Agreements Relating to Activities in Outer Space as at 1 January 2018*. Available at http://www.unoosa.org/oosa/en/ourwork/spacelaw/treaties/status/index.html.

United Nations Committee on the Peaceful Uses of Outer Space (UN-COPUOS). 2017. *Report of the Committee on the Peaceful Uses of Outer Space, 60th Session*. Available at http://www.unoosa.org/oosa/oosadoc/data/documents/2017/a/a7220_0.html.

United Nations Office of Outer Space Affairs 2018. Translation of Russian Federation Decree Number 5663-1. Available at http://www.unoosa.org/oosa/en/ourwork/spacelaw/nationalspacelaw/russian_federation/decree_5663-1_E.html.

Wells, H.G. 1898. *War of the Worlds*. W. Heinemann, London, UK.

Index

Note: Page numbers in italic and bold refer to figures and tables respectively.

Printed and bound by CPI Group (UK) Ltd, Croydon, CR0 4YY

24/10/2024

01778292-0018